The Entire Electromagnetic Spectrum

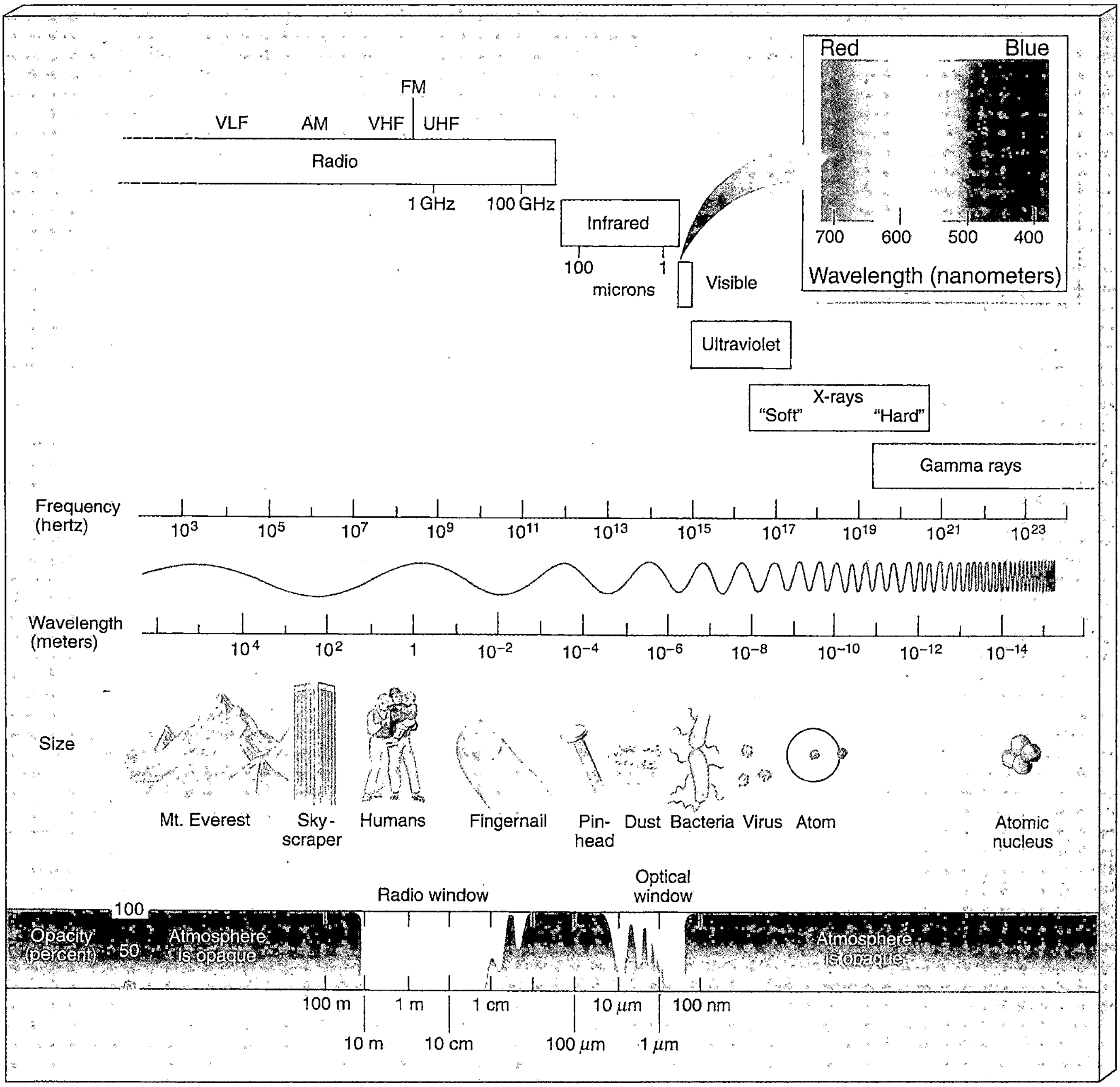

	Red	Blue

Wavelength (nanometers)
700   600   500   400

FM

VLF   AM   VHF │ UHF

Radio

1 GHz   100 GHz

Infrared

100   1
microns   **Visible**

Ultraviolet

X-rays
"Soft"   "Hard"

Gamma rays

Frequency (hertz)
10^3 10^5 10^7 10^9 10^{11} 10^{13} 10^{15} 10^{17} 10^{19} 10^{21} 10^{23}

Wavelength (meters)
10^4 10^2 1 10^{-2} 10^{-4} 10^{-6} 10^{-8} 10^{-10} 10^{-12} 10^{-14}

Size

Mt. Everest   Sky-scraper   Humans   Fingernail   Pin-head   Dust   Bacteria   Virus   Atom   Atomic nucleus

Radio window

Optical window

Opacity (percent)   100   50

Atmosphere is opaque   Atmosphere is opaque

100 m   1 m   1 cm   10 μm   100 nm

10 m   10 cm   100 μm   1 μm

ASTRONOMY TODAY

ABOUT THE AUTHORS

ERIC CHAISSON

Eric holds a doctorate in Astrophysics from Harvard University, where he spent ten years on the faculty of Arts and Sciences. For five years, Eric was a Senior Scientist and Director of Educational Programs at the Space Telescope Science Institute and Adjunct Professor of Physics at Johns Hopkins University. He then joined Tufts University, where he is now Professor of Physics, Professor of Education, and Director of the Wright Center for Innovative Science Education. He has written nine books on astronomy, which have received such literary awards as the Phi Beta Kappa Prize, two American Institute of Physics Awards, and Harvard's Smith-Weld Prize for Literary Merit. He has published more than 100 scientific papers in professional journals, and has also received Harvard's Bok Prize for original contributions to astrophysics.

STEVE MCMILLAN

Steve holds a bachelor's and master's degree in Mathematics from Cambridge University and a doctorate in Astronomy from Harvard University. He held post-doctoral positions at the University of Illinois and Northwestern University, where he continued his research in theoretical astrophysics, star clusters, and numerical modeling. Steve is currently Distinguished Professor of Physics at Drexel University and a frequent visiting researcher at Princeton's Institute for Advanced Study and the University of Tokyo. He has published over 40 scientific papers in professional journals.

ASTRONOMY TODAY

Fourth Edition

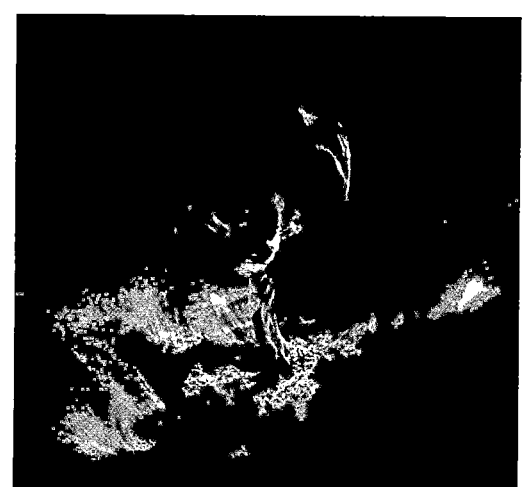

Eric Chaisson
TUFTS UNIVERSITY

Steve McMillan
DREXEL UNIVERSITY

Prentice Hall
Upper Saddle River, New Jersey 07458

Library of Congress Cataloging-in-Publication Data

Chaisson, Eric.
    Astronomy Today/Eric Chaisson, Steve McMillan.—4th ed.
      p. cm
    Includes index.
    ISBN 0-13-091542-4
    1. Astronomy.   I. McMillan, S. (Stephen).   II. Title.
    QB43.3.C48 2002
    520—dc21   2001037479

Executive Editor: Alison Reeves
Assistant Editor: Christian Botting
Marketing Manager: Erik Fahlgren
Development Editor: Donald Gecewicz
Production Editor: Joanne Hakim
Vice President of Production
  and Manufacturing: David W. Riccardi
Executive Managing Editor: Kathleen Schiaparelli
Assistant Managing Editor, Media Science: Alison Lorber
Manager of Formatting: Jim Sullivan
Electronic Production Specialist/
  Electronic Page Makeup: Joanne Del Ben
Director, Creative Services: Paul Belfanti
Creative Director: Carole Anson
Interior Designer: Carole Anson
Cover Designer: Bruce Kenselaar
Editor in Chief, Development: Ray Mullaney
Art Studio: Artworks
  Senior Manager: Patty Burns
  Production Manager: Ronda Whitson

Manager, Production Technologies: Matt Haas
Project Coordinator: Connie Long
Illustrators: Mark Landis, Jay McElroy, Kathryn Anderson
Manufacturing Manager: Trudy Pisciotti
Assistant Manufacturing Manager: Michael Bell
Art Manager: Grace Hazeldine
Art Editor: Adam Velthaus
Editorial Assistants: Eileen Harrity and Eileen Nee
Copy Editor: Jocelyn Phillips
Photo Editor: Beth Boyd
Photo Coordinator: Anthony Arabia
Photo Researcher: Christine Pulo
Media Production Support: Elizabeth Wright
CD Project Managers: Gretel Seham, Nicole Bush
CD Production Manager: William Johnson
CD Production: Elizabeth Gschwind, Lawrence LaRaia
CD Proofreader: Sheree Van Vreede
Companion Website Project Manager: Eileen McDonough
CD Testing: Nancy Segalini, Giselli Redondo

 © 2002, 1999, 1996, 1993 by Prentice-Hall, Inc.
Upper Saddle River, NJ 07458

All rights reserved. No part of this book may be
reproduced, in any form or by any means,
without permission in writing from the publisher.

Printed in the United States of America

10 9 8 7 6 5 4 3 2 1

COLLEGE ISBN 0-13-091542-4

SCHOOL ISBN 0-13-094334-7

Pearson Education LTD., *London*
Pearson Education Australia PTY, Limited, *Sydney*
Pearson Education Singapore, Pte. Ltd.
Pearson Education North Asia Ltd., *Hong Kong*
Pearson Education Canada, Ltd., *Toronto*
Pearson Educación de Mexico, S.A. de C.V.
Pearson Education—*Japan, Tokyo*
Pearson Education Malaysia, Pte. Ltd.
Pearson Education, *Upper Saddle River, New Jersey*

BRIEF CONTENTS

PART 1 ASTRONOMY AND THE UNIVERSE

1   Charting the Heavens: The Foundations of Astronomy   3
2   The Copernican Revolution: The Birth of Modern Science   33
3   Radiation: Information from the Cosmos   61
4   Spectroscopy: The Inner Workings of Atoms   85
5   Telescopes: The Tools of Astronomy   107

PART 2 OUR PLANETARY SYSTEM

6   The Solar System: An Introduction to Comparative Planetology   143
7   Earth: Our Home in Space   165
8   The Moon and Mercury: Scorched and Battered Worlds   195
9   Venus: Earth's Sister Planet   227
10   Mars: A Near Miss for Life?   249
11   Jupiter: Giant of the Solar System   275
12   Saturn: Spectacular Rings and Mysterious Moons   301
13   Uranus, Neptune, and Pluto: The Outer Worlds of the Solar System   327
14   Solar System Debris: Keys to Our Origin   355
15   The Formation of Planetary Systems: The Birth of Our World   383

PART 3 STARS AND STELLAR EVOLUTION

16   The Sun: Our Parent Star   405
17   Measuring the Stars: Giants, Dwarfs, and the Main Sequence   437
18   The Interstellar Medium: Gas and Dust Among the Stars   467
19   Star Formation: A Traumatic Birth   489
20   Stellar Evolution: The Life of a Star   515
21   Stellar Explosions: Novae, Supernovae, and the Formation of the Heavy Elements   543
22   Neutron Stars and Black Holes: Strange States of Matter   567

PART 4 GALAXIES AND COSMOLOGY

23   The Milky Way Galaxy: A Grand Design   599
24   Normal Galaxies: The Large-Scale Structure of the Universe   629
25   Active Galaxies and Quasars: Limits of the Observable Universe   661
26   Cosmology: The Big Bang and the Fate of the Universe   691
27   The Early Universe: Toward the Beginning of Time   715
28   Life in the Universe: Are We Alone?   739

CONTENTS

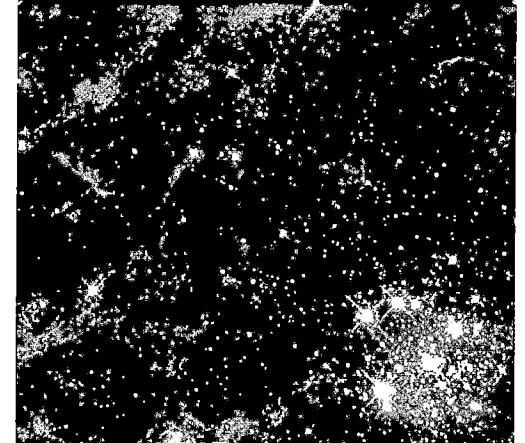

Preface xv

PART ONE
ASTRONOMY AND THE UNIVERSE

1   Charting the Heavens

The Foundations of Astronomy   3

1.1   Our Place in Space   4
1.2   The Obvious View   6
1.3   Earth's Orbital Motion   9
 • **Animation** The Earth's Seasons
1.4   The Motion of the Moon   15
1.5   The Measurement of Distance   24
       Chapter Review   28
 ■ **More Precisely 1-1** Angular Measure   11
 ▣ **More Precisely 1-2** Celestial
                          Coordinates   14
 ▣ **More Precisely 1-3** Astronomical
                          Timekeeping   22
 ▣ **More Precisely 1-4** Measuring Distances
                          with Geometry   26
 ■ **Discovery 1-1** Sizing Up Planet Earth   27

2   The Copernican Revolution

The Birth of Modern Science   33

2.1   Ancient Astronomy   34
2.2   The Geocentric Universe   36
2.3   The Heliocentric Model of the Solar
       System   39
2.4   The Birth of Modern Astronomy   41
2.5   The Laws of Planetary Motion   43
2.6   The Dimensions of the Solar System   48
2.7   Newton's Laws   49

       Chapter Review   56
 ▣ **Discovery 2-1** The Foundations of the
       Copernican Revolution   40
 ■ **Discovery 2-2** The Scientific Method   44
 ▣ **More Precisely 2-1** Some Properties of
                          Planetary Orbits   46
 ▣ **More Precisely 2-2** The Moon Is
                          Falling!   50
 ▣ **More Precisely 2-3** Weighing the Sun   56

3   Radiation

Information from the Cosmos   61

3.1   Information from the Skies   62
3.2   Waves in What?   65
3.3   The Electromagnetic Spectrum   70
3.4   The Distribution of Radiation   72
 ◦ **Animation** The Planck Spectrum
3.5   The Doppler Effect   77
       Chapter Review   79
 ■ **Discovery 3-1** The Wave Nature
                     of Radiation   66
 ■ **More Precisely 3-1** The Kelvin
                          Temperature Scale   73
 ■ **More Precisely 3-2** More About the
                          Radiation Laws   74

4   Spectroscopy

The Inner Workings of Atoms   85

4.1   Spectral Lines   86
 ◦ **Animation** Classical Hydrogen Atom I
 • **Animation** Classical Hydrogen Atom II
4.2   The Formation of Spectral Lines   91
4.3   Molecules   96
4.4   Spectral-Line Analysis   97

Chapter Review   103
- More Precisely 4-1 The Energy Levels
  of the Hydrogen Atom   98

5   Telescopes

The Tools of Astronomy   107

5.1   Optical Telescopes   108
5.2   Telescope Size   114
5.3   High-Resolution Astronomy   118
5.4   Radio Astronomy   124
5.5   Interferometry   127
5.6   Space-Based Astronomy   130
5.7   Full-Spectrum Coverage   136
- Animation Light and Data Path
  Chapter Review   138
- Discovery 5-1 The *Hubble Space
  Telescope*   122
- Discovery 5-2 The *Chandra* X-Ray
  Observatory   134

PART TWO
OUR PLANETARY SYSTEM

6   The Solar System

An Introduction to Comparative
  Planetology   143

6.1   An Inventory of the Solar System   144
6.2   Planetary Propterties   146
- Animation Astronomical Ruler
6.3   The Overall Layout of the Solar
  System   148

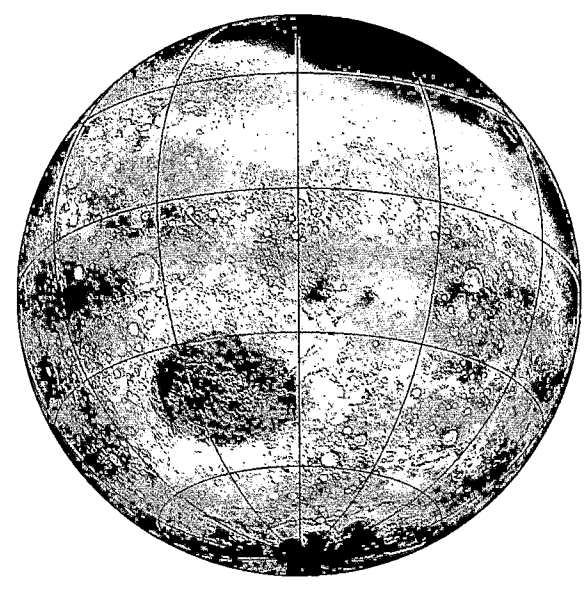

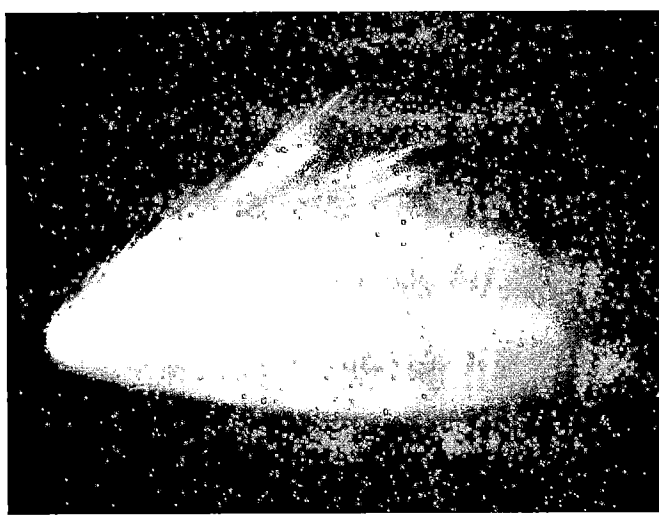

6.4   Terrestrial and Jovian Planets   150
- Animation Terrestrial Planets Part I
- Animation Jovian Planets Part I
6.5   Interplanetary Debris   151
6.6   Spacecraft Exploration of the Solar
  System   153
  Chapter Review   159
- More Precisely 6-1 Computing Planetary
  Properties   147
- Discovery 6-1 The Titus–Bode
  "Law"   149
- Discovery 6-2 Gravitational
  "Slingshots"   154

7   Earth

Our Home in Space   165

7.1   Overall Structure of Planet Earth   166
7.2   Earth's Atmosphere   167
7.3   Earth's Interior   171
7.4   Surface Activity   175
7.5   Earth's Magnetosphere   183
7.6   The Tides   187
  Chapter Review   190
- More Precisely 7-1 Why Is the Sky
  Blue?   169
- Discovery 7-1 Earth's "Rapidly" Spinning
  Core   174
- More Precisely 7-2 Radioactive
  Dating   180
- More Precisely 7-3 Tidal Forces   188

8 The Moon and Mercury

Scorched and Battered Worlds 195

8.1 Orbital Properties 196
8.2 Physical Properties 198
8.3 Surface Features on the Moon
    and Mercury 199
8.4 Rotation Rates 202
8.5 Lunar Cratering and Surface
    Composition 206
8.6 The Surface of Mercury 214
8.7 Interiors 215
• **Animation** Terrestrial Planets Part II
8.8 The Origin of the Moon 217
8.9 Evolutionary History of the Moon
    and Mercury 218
    Chapter Review 222
■ **Discovery 8-1** Lunar Exploration 204
■ **More Precisely 8-1** Why Air Sticks
                        Around 208
■ **Discovery 8-2** Exploring the Moon on
                        a Shoestring 220

9 Venus

Earth's Sister Planet 227

9.1 Orbital Properties 228
9.2 Physical Properties 229
• **Animation** The Phases of Venus
• **Animation** The Rotation of Venus
9.3 Long-Distance Observations of Venus 230
9.4 The Surface of Venus 232
9.5 The Atmosphere of Venus 240
9.6 Venus's Magnetic Field and Internal
    Structure 243
    Chapter Review 244

10 Mars

A Near Miss for Life? 249

10.1 Orbital Properties 250
10.2 Physical Properties 252
10.3 Long-Distance Observations of Mars 252
10.4 The Surface of Mars 253
• **Animation** The Rotation of Mars
• **Animation** Mars North Pole
10.5 The Martian Atmosphere 265
10.6 Martian Internal Structure 268

10.7 The Moons of Mars 269
    Chapter Review 270
■ **Discovery 10-1** Martian Canals? 259
■ **Discovery 10-2** Life on Mars? 266

11 Jupiter

Giant of the Solar System 275

11.1 Orbital and Physical Properties 276
11.2 The Atmosphere of Jupiter 278
• **Animation** Galileo Mission to Jupiter
11.3 Internal Structure 283
• **Animation** Jovian Planets Part II
11.4 Jupiter's Magnetosphere 285
11.5 The Moons of Jupiter 287
• **Animation** Galileo Flyby of Io
• **Animation** Io Cutaway
• **Animation** Jupiter's Moon Europa
• **Animation** Jupiter's Moon Ganymede
11.6 Jupiter's Ring 296
    Chapter Review 296
■ **Discovery 11-1** Almost a Star? 284

12 Saturn

Spectacular Rings and Mysterious Moons 301

12.1 Orbital and Physical Properties 302
12.2 Saturn's Atmosphere 302
• **Animation** Saturn Storm
12.3 Saturn's Interior and Magnetosphere 307
12.4 Saturn's Spectacular Ring System 309
12.5 The Moons of Saturn 315
    Chapter Review 322
■ **Discovery 12-1** New Moons or Ring
                        Debris at Saturn? 317

13 Uranus, Neptune, and Pluto

The Outer Worlds of the Solar System   327

13.1   The Discovery of Uranus   328
13.2   The Discovery of Neptune   329
13.3   Physical Properties of Uranus
          and Neptune   329
13.4   The Atmospheres of Uranus
          and Neptune   332
13.5   Magnetospheres and Internal Structure   335
13.6   The Moon Systems of Uranus
          and Neptune   337
13.7   The Rings of the Outermost Jovian
          Planets   342
13.8   The Discovery of Pluto   345
   • **Animation** Orbits of Neptune and Pluto
   • **Animation** Historical Observations of Pluto
   • **Animation** Mutual Eclipses of Charon
          and Pluto
13.9   Physical Properties of Pluto   346
13.10 The Origin of Pluto   349
          Chapter Review   350
   ■ **Discovery 13-1** Surface Detail on Planet
          Pluto   349

14 Solar System Debris

Keys to Our Origin   355

14.1   Asteroids   356
14.2   Comets   362
   • **Animation** Anatomy of a Comet I
   • **Animation** Anatomy of a Comet II
14.3   Meteoroids   372
   • **Animation** Asteroid Comet Breakup

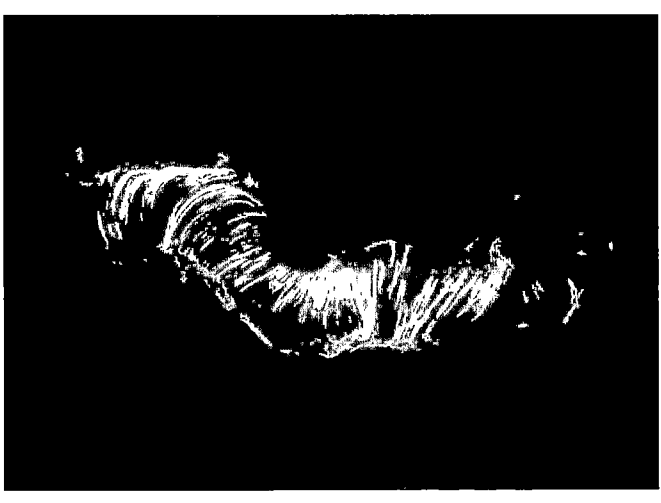

   • **Animation** Rotating Comet Hale–Bopp
          Nucleus
          Chapter Review   378
   ■ **Discovery 14-1** What Killed the
          Dinosaurs?   368
   ■ **Discovery 14-2** Comets Hyakutake
          and Hale-Bopp   374

15 The Formation of Planetary Systems

The Birth of Our World   383

15.1   Modeling the Origin of Our Solar
          System   384
15.2   The Condensation Theory   385
   ◦ **Animation** Beta Pictoris Warp
   ◦ **Animation** Solar System Formation
15.3   The Differentiation of the Solar
          System   392
15.4   The Role of Catastrophes   395
15.5   Planets Beyond the Solar System   396
          Chapter Review   401
   ▣ **More Precisely 15-1** The Concept
          of Angular Momentum   390
   ■ **Discovery 15-1** The Kuiper Belt and the
          Search for a Tenth Planet   400

PART THREE
STARS AND STELLAR EVOLUTION

16 The Sun

Our Parent Star   405

16.1   Physical Properties of the Sun   406
16.2   The Solar Interior   409
16.3   The Solar Atmosphere   413
16.4   The Active Sun   418
   • **Animation** Solar Flare Event
   ◦ **Animation** Coronal Mass Ejection
16.5   The Heart of the Sun   425
16.6   Observations of Solar Neutrinos   429
          Chapter Review   432
   ■ **Discovery 16-1** *SOHO:* Eavesdropping
          on the Sun   410
   ■ **More Precisely 16-1** Fundamental
          Forces   427
   ▣ **Discovery 16-2** Solar–Terrestrial
          Relations   431

17 Measuring the Stars

Giants, Dwarfs, and the Main Sequence   437

17.1 The Distances to the Stars   438
17.2 Stellar Motion   439
17.3 Luminosity and Apparent Brightness   441
  • **Animation** The Inverse Square Law
17.4 Stellar Temperatures   445
17.5 Stellar Sizes   449
17.6 The Hertzsprung–Russell Diagram   450
17.7 Extending the Cosmic Distance Scale   454
17.8 Stellar Masses   457
   Chapter Review   461
  ■ **Discovery 17-1** The *Hipparcos*
                       Mission   442
  ■ **More Precisely 17-1** More on the
                       Magnitude Scale   446
  ■ **More Precisely 17-2** Estimating Stellar
                       Radii   452

18 The Interstellar Medium

Gas and Dust Among the Stars   467

18.1 Interstellar Matter   468
18.2 Emission Nebulae   473
  • **Video** Gaseous Pillars of Star Birth
  ◦ **Animation** M16 Eagle Nebula
  ◦ **Video** Orion Nebula Mosaic
18.3 Dark Dust Clouds   478
18.4 21-Centimeter Radiation   481
18.5 Interstellar Molecules   482
   Chapter Review   485
  ■ **Discovery 18-1** Ultraviolet Astronomy
     and the "Local Bubble"   470

19 Star Formation

A Traumatic Birth   489

19.1 Star-Forming Regions   490
19.2 The Formation of Stars Like the Sun   492
  • **Animation** Evolution of A 1-Solar-Mass
     Star
19.3 Stars of Other Masses   497
19.4 Observations of Cloud Fragments
     and Protostars   499
  • **Animation** Bi-Polar Outflow

19.5 Shock Waves and Star Formation   504
19.6 Star Clusters   505
   Chapter Review   510
  ■ **Discovery 19-1** Observations of Brown
                       Dwarfs   498
  ■ **Discovery 19-2** Eta Carinae   509

20 Stellar Evolution

The Life of a Star   515

20.1 Leaving the Main Sequence   516
20.2 The Evolution of a Sunlike Star   517
  ◦ **Animation** Death of the Sun Part I
  ◦ **Animation** Death of the Sun Part II
20.3 The Death of a Low-Mass Star   523
  • **Video** Formation of Helix Nebula
  • **Animation** Bi-Polar Planetary Nebula
  ◦ **Animation** H–R Diagram Tracks Stellar
                   Evolution
20.4 Evolution of Stars More Massive
     than the Sun   529
20.5 Observing Stellar Evolution in Star
     Clusters   532
20.6 The Evolution of Binary-Star Systems   535
   Chapter Review   538
  ■ **More Precisely 20-1** The CNO
                       Cycle   518
  ■ **Discovery 20-1** Mass Loss from Giant
                       Stars   524
  ■ **Discovery 20-2** Learning Astronomy
                       from History   530

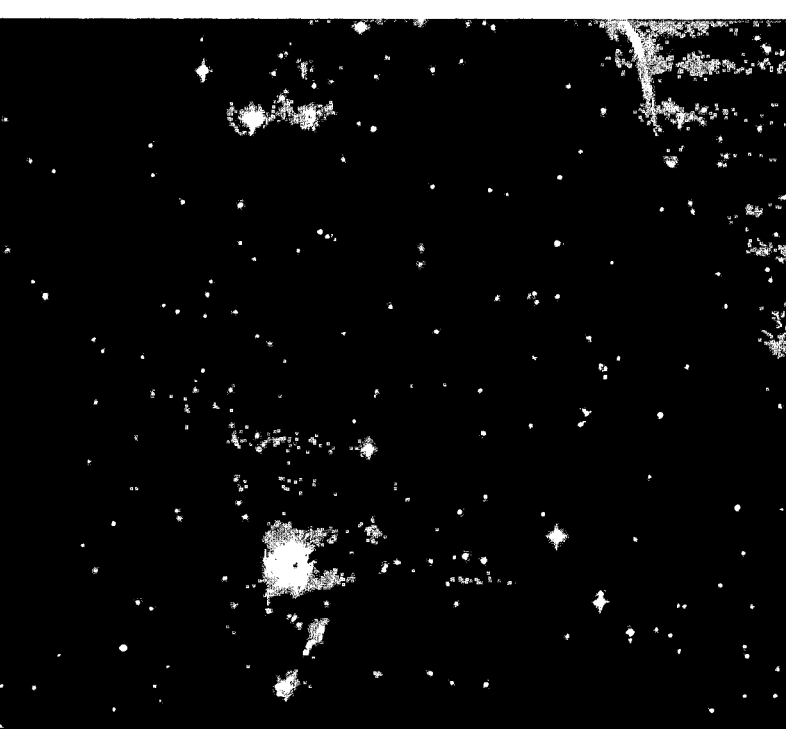

21 Stellar Explosions

Novae, Supernovae, and the Formation of the Heavy Elements 543

21.1 Life after Death for White Dwarfs 544
21.2 The End of a High-Mass Star 545
 ° **Animation** Recurrent Nova
21.3 Supernova Explosions 548
21.4 The Formation of the Elements 552
 ° **Animation** Supernova Explosion
 ° **Animation** Composition and Structure of the Ring of the Supernova 1987A
 • **Animation** Shockwaves Hit the Ring around Supernova 1987A
21.5 The Cycle of Stellar Evolution 561
 Chapter Review 562
 ▪ **Discovery 21-1** Supernova 1987A 554
 ▪ **Discovery 21-2** The Crab Nebula in Motion 558

22 Neutron Stars and Black Holes

Strange States of Matter 567

22.1 Neutron Stars 568
22.2 Pulsars 569
22.3 Neutron-Star Binaries 572
22.4 Gamma-Ray Bursts 575
22.5 Black Holes 580
22.6 Black Holes and Curved Space 582
22.7 Space Travel Near Black Holes 585

22.8 Observational Evidence for Black Holes 589
 • **Animation** Black Hole Geometry
 Chapter Review 594
 ▪ **Discovery 22-1** Gravity Waves 578
 ▪ **More Precisely 22-1** Einstein's Theories of Relativity 582
 ▪ **More Precisely 22-2** Tests of General Relativity 586

PART FOUR
GALAXIES AND COSMOLOGY

23 The Milky Way Galaxy

A Grand Design 599

23.1 Our Parent Galaxy 600
23.2 Measuring the Milky Way 600
 ° **Animation** Cepheid Star in Distant Galaxy
23.3 The Large-Scale Structure of Our Galaxy 607
23.4 The Formation of the Milky Way 610
23.5 Galactic Spiral Arms 612
23.6 The Mass of the Milky Way Galaxy 616
23.7 The Galactic Center 620
 Chapter Review 625
 ▪ **Discovery 23-1** Early Computers 606
 ▪ **Discovery 23-1** Density Waves 614
 ▪ **Discovery 23-1** Cosmic Rays 624

24 Normal Galaxies

The Large-Scale Structure of the Universe 629

24.1 Hubble's Galaxy Classification 630
24.2 The Distribution of Galaxies in Space 637
24.3 Galaxy Masses 642
24.4 Galaxy Formation and Evolution 645
 ° **Animation** Cluster Merger
 ° **Animation** Hubble Deep Field Zoom Sequence
 • **Animation** Starburst Galaxy
24.5 Hubble's Law 651
 ° **Animation** Collision of Two Spiral Galaxies
 Chapter Review 655
 ▪ **Discovery 24-1** Colliding Galaxies 646

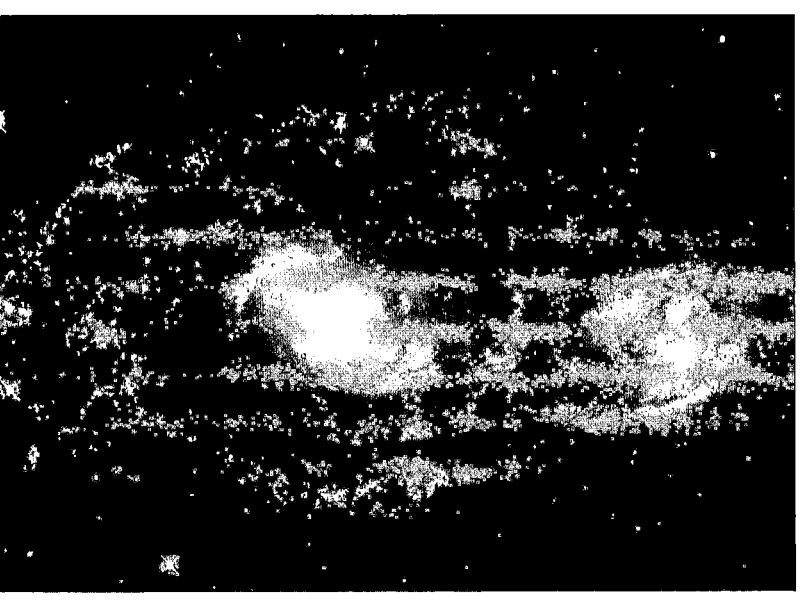

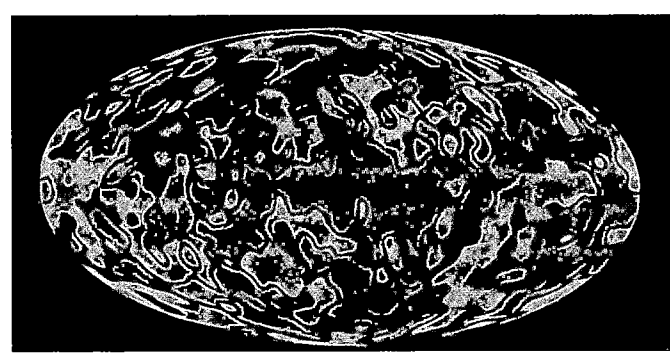

25 Active Galaxies and Quasars

Limits of the Observable Universe 661

25.1 Beyond the Local Realm 662
25.2 Properties of Active Galaxies 663
 ◦ **Animation** Cosmic Jets
25.3 Quasi-Stellar Objects 669
 • **Animation** Birth of a Quasar
25.4 The Central Engine of an Active
 Galaxy 673
 ◦ **Animation** Supermassive Black Hole
 ◦ **Animation** Active Galaxy
 ◦ **Animation** Black Hole in Galaxy M87
25.5 Quasars as Cosmic Probes 678
 • **Animation** Gravitational Lensing
 • **Animation** Dark Matter
25.6 Active Galaxy Evolution 683
 Chapter Review 687
 ■ **More Precisely 25-1** Faster-Than-Light
 Velocities? 670
 ■ **More Precisely 25-2** Relativistic Redshifts
 and Look-Back Time 680
 ■ **Discovery 25-1** BL Lac Objects 686

26 Cosmology

The Big Bang and the Fate
 of the Universe 691

26.1 The Universe on the Largest Scales 692
 ◦ **Animation** Cosmic Structure
26.2 The Expanding Universe 693
 • **Animation** Big Bang
26.3 The Fate of the Universe 700
26.4 Will the Universe Expand Forever? 702
26.5 The Geometry of Space 707
26.6 The Cosmic Microwave Background 708

 Chapter Review 711
 ■ **Discovery 26-1** Stunning Views of Deep
 Space 696
 ◘ **Discovery 26-2** The Cosmological
 Constant 704
 ▣ **More Precisely 26-1** Curved Space 710

27 The Early Universe

Toward the Beginning of Time 715

27.1 Back to the Big Bang 716
27.2 The Evolution of the Universe 718
27.3 The Formation of Nuclei and Atoms 721
27.4 The Inflationary Universe 726
27.5 The Formation of Structure in the
 Universe 730
 Chapter Review 735
 ■ **More Precisely 27-1** More on
 Fundamental Forces 722

28 Life in the Universe

Are We Alone? 739

28.1 Cosmic Evolution 740
28.2 Life in the Solar System 745
28.3 Intelligent Life in the Galaxy 747
28.4 The Search for Extraterrestrial
 Intelligence 751
 Chapter Review 754
 ■ **Discovery 28-1** The Virus 741

APPENDICES A-1
Appendix 1: Scientific Notation A-1
Appendix 2: Astronomical Measurement A-2
Appendix 3: Tables A-3
Appendix 4: Using SkyChart III A-9

Glossary G-1
Answers to Concept Check Questions AK-1
Answers to Self-Test Questions AK-5
Index I-1
Star Charts S-1

PREFACE

A stronomy continues to enjoy a golden age of exploration and discovery. Fueled by new technologies and novel theoretical insights, the study of the cosmos has never been more exciting. We are pleased to have the opportunity to present a representative sample of the known facts, evolving ideas, and frontier discoveries in astronomy today.

This book is written for students who have taken no previous college science courses and who will likely not major in physics or astronomy. The text is suitable for both one-semester and two-semester courses. We present a broad view of astronomy, straightforwardly descriptive and without complex mathematics. The absence of sophisticated mathematics, however, in no way prevents discussion of important concepts. Rather, we rely on qualitative reasoning as well as analogies with objects and phenomena familiar to the student to explain the complexities of the subject without oversimplification. We have tried to impart the enthusiasm that we feel about astronomy, and to awaken students to the marvelous universe around us.

In teaching astronomy to nonscientists, as in writing this book, we are not seeking to convert students to careers in astronomy or even science in general. Instead, we strive to reach the wider audience of students who are majoring in many other worthwhile fields. We want to encourage these students to become scientifically literate members of modern society—to appreciate new developments in the world of science, to understand what scientists do for a living and its importance, to make informed judgments regarding national initiatives in science and the public funding of scientific projects, and to vote intelligently in our democratic, increasingly technological world.

We are very gratified that the first three editions of this text have been so well received by many in the astronomy education community. In using those earlier texts, many of you—teachers and students alike—have sent us helpful feedback and constructive criticisms. From these, we have learned to better communicate both the fundamentals and the excitement of astronomy. Many improvements inspired by your comments have been incorporated into this new edition.

Organization and Approach

Our overall organization follows the popular and effective "Earth-out" progression. We have found that most students, especially those with little scientific background, are much more comfortable studying the (relatively familiar) solar system before tackling stars and galaxies. Thus, Earth is the first object we discuss in detail. With Earth and the Moon as our initial planetary models, we move through the solar system, drawing on comparative planetology to provide an understanding of the many varied worlds we encounter. We conclude our coverage of the solar system with a discussion of its formation, a line of investigation that leads directly into a study of our Sun.

With the Sun as our model star, we then broaden the scope of our discussion to include stars in general—their properties, their evolutionary histories, and their varied fates. This journey naturally leads us to coverage of the Milky Way Galaxy, which in turn serves as an introduction to our treatment of other galaxies. Finally, we reach the subject of cosmology and the large-scale structure and dynamics of the universe as a whole. Throughout, we strive to emphasize the dynamic nature of the cosmos—virtually every major topic, from planets to quasars, includes a discussion of how those objects formed and how they evolve.

We continue to place much of the needed physics in the early chapters—an approach derived from years of experience teaching thousands of students. Additional physical principles are developed as needed later, both in the text narrative and in the boxed *More Precisely* features (described below). We feel strongly that this is the most economical and efficient means of presentation. However, we acknowledge that not all instructors feel the same way. Accordingly, we have made the treatment of physics, as well as the more quantitative discussions, as modular as possible, so that these topics can be deferred to later stages of an astronomy course if desired. In addition, we have included as much modern astronomy as possible in the introductory chapters. These chapters are likely to engage students only if they are made to realize how simple physical principles provide the keys to our understanding of a vast and otherwise incomprehensible universe.

New and Revised Material

The text has been extensively updated in content since the third edition. Most chapters have been significantly changed, and several have seen major reorganization. Among the many changes are:

* New boxes in Chapter 1 on astronomical timekeeping and distance measurement.
* Expanded discussion in Chapter 2 of Newton's discovery of the law of gravity.
* Updated material in Chapter 5 on adaptive optics, Subaru, Gemini, the VLT, and infrared and optical interferometry; new material on the *Chandra* mission; updates (and a conclusion) to the *CGRO* story.

- New material in Chapter 6 on measuring planetary properties; updates on the *Galileo, Cassini,* and *Mars Global Surveyor* missions.

- Expanded material in Chapter 6 on *Clementine* and *Lunar Prospector,* with updates on their important findings, including the possibility of ice at the lunar poles.

- Greatly expanded coverage in Chapter 10 of *Mars Global Surveyor* and the many scientific results that have come from it; the possibility of (past or present) liquid water on Mars; an update on the Martian Meteorite controversy.

- Updates in Chapter 11 on the *Galileo/GEM* mission, including the latest results on the possible existence of a liquid water ocean below Europa's icy surface; discussion of the magnetic fields of the Galilean moons.

- Coverage of the many new moons of Jupiter, Saturn, and Uranus.

- Expanded coverage of Pluto and the Kuiper belt in Chapters 13–15.

- Updates in Chapter 14 on asteroid numbers and the properties of near-Earth objects; coverage of the *NEAR* mission and its exploration of Eros.

- Substantially updated coverage of solar system formation in Chapter 15, including disk instabilities, planetary migration, and their implications for extrasolar planetary systems.

- New section in Chapter 15 on extrasolar planets, with updated material on the latest observations.

- Incorporation of results from the *Yohkoh, SOHO,* and *TRACE* missions into Chapter 16.

- The latest experimental results in the search for the missing solar neutrinos (Chapter 16).

- Use of *Hipparcos* data in Chapter 17 and throughout the text; new H–R diagram based on *Hipparcos* measurements; discussion of future astrometry missions and their implications.

- Updated information in Chapter 17 on the numbers and mass distribution of stars in our Galaxy.

- Extensive revision of the material on stellar mass determination in Chapter 17.

- Expanded discussion in Chapter 18 of the Local Bubble.

- Updated information in Chapter 19 on brown dwarfs; new material on jets and outflows in star formation.

- New coverage in Chapter 20 of the end-states of stellar and binary evolution; discussion of blue stragglers; more examples of familiar stars in specific evolutionary stages.

- New section and latest results on gamma-ray bursts in Chapter 22; discussion of intermediate-mass and supermassive black holes.

- Latest results in Chapter 23 on Sgr A* and the Galaxy's central black hole.

- Expanded and substantially revised coverage in Chapter 24 of galaxy collisions, hierarchical merging, and galaxy evolution; updated discussion of the measurement of *Hubble's* constant.

- Streamlined discussion in Chapter 25 of active galaxies and quasars; revised discussion of active galaxy evolution.

- New material in Chapter 25 on quasar absorption lines and the Lyman-alpha forest; expanded discussion of gravitational lensing, including the construction of dark-matter maps from lensing of background galaxies.

- Extensive rewriting of Chapter 26 to include recent observations of cosmic acceleration and discussion of "dark energy;" revised discussions of the cosmological constant and the "age controversy."

- New material in Chapter 26 on *HDF-S* and the *Chandra Deep Field.*

- Consistent distances and times in Chapters 25–27, assuming a flat universe with dark matter and dark energy.

- Expanded discussion in Chapter 27 of inflation, dark energy, and structure formation; results from the Boomerang experiment suggesting a flat universe.

- Updated coverage of Europa, Mars, interstellar organic molecules, and extrasolar planets in Chapter 28.

The Illustration Program

Visualization plays an important role in both the teaching and the practice of astronomy, and we continue to place strong emphasis on this aspect of our book. We have tried to combine aesthetic beauty with scientific accuracy in the artist's conceptions that adorn the text, and we have sought to present the best and latest imagery of a wide range of cosmic objects. Each illustration has been carefully crafted to enhance student learning; each is pedagogically sound and tied tightly to the nearby discussion of important scientific facts and ideas. For this edition, the illustration program has been extensively revised and updated, resulting in more than 100 figures that show the latest imagery and the results learned from them.

Full Spectrum Coverage and Spectrum Icons

Increasingly, astronomers are exploiting the full range of the electromagnetic spectrum to gather information about the cosmos. Throughout this book, images taken at radio, infrared, ultraviolet, X-ray, or gamma-ray wavelengths are used to supplement visible-light images. As it is sometimes difficult (even for a professional) to tell at a glance which images are visible-light photographs and which are false-color images created with other wavelengths, each photo in the text is provided with an icon that identifies the wavelength of electromagnetic radiation used to capture the image and reinforces the connection between wavelength and radiation properties.

Compound Art ▷

It is rare that a single image, be it a photograph or an artist's conception, can capture all aspects of a complex subject. Wherever possible, multiple-part figures are used in an attempt to convey the greatest amount of information in the most vivid way:

- Visible images are often presented along with their counterparts captured at other wavelengths.

- Interpretive line drawings are often superimposed on or juxtaposed with real astronomical photographs, helping students to really "see" what the photographs reveal.

- Breakouts—often multiple ones—are used to zoom in from wide-field shots to closeups so that detailed images can be understood in their larger context.

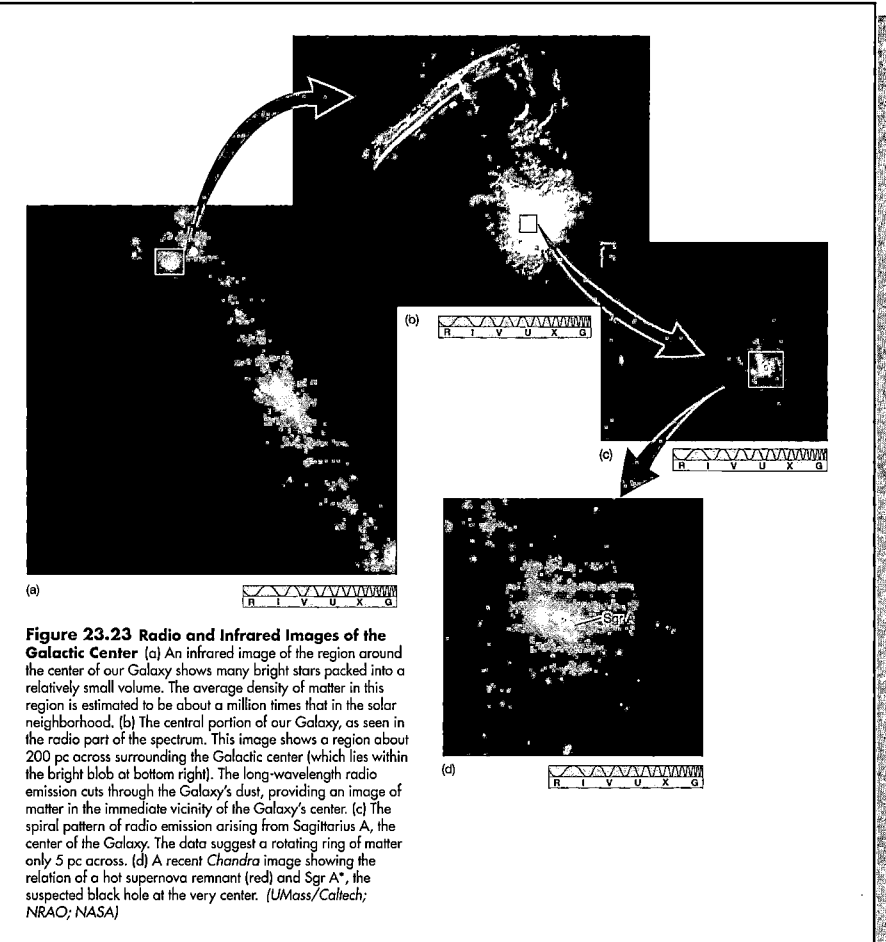

Figure 23.23 Radio and Infrared Images of the Galactic Center (a) An infrared image of the region around the center of our Galaxy shows many bright stars packed into a relatively small volume. The average density of matter in this region is estimated to be about a million times that in the solar neighborhood. (b) The central portion of our Galaxy, as seen in the radio part of the spectrum. This image shows a region about 200 pc across surrounding the Galactic center (which lies within the bright blob at bottom right). The long-wavelength radio emission cuts through the Galaxy's dust, providing an image of matter in the immediate vicinity of the Galaxy's center. (c) The spiral pattern of radio emission arising from Sagittarius A, the center of the Galaxy. The data suggest a rotating ring of matter only 5 pc across. (d) A recent *Chandra* image showing the relation of a hot supernova remnant (red) and Sgr A*, the suspected black hole at the very center. *(UMass/Caltech; NRAO; NASA)*

▲ Explanatory Captions

Students often review a chapter by "looking at the pictures." For this reason, the captions in this book are often a bit longer and more detailed than those in other texts.

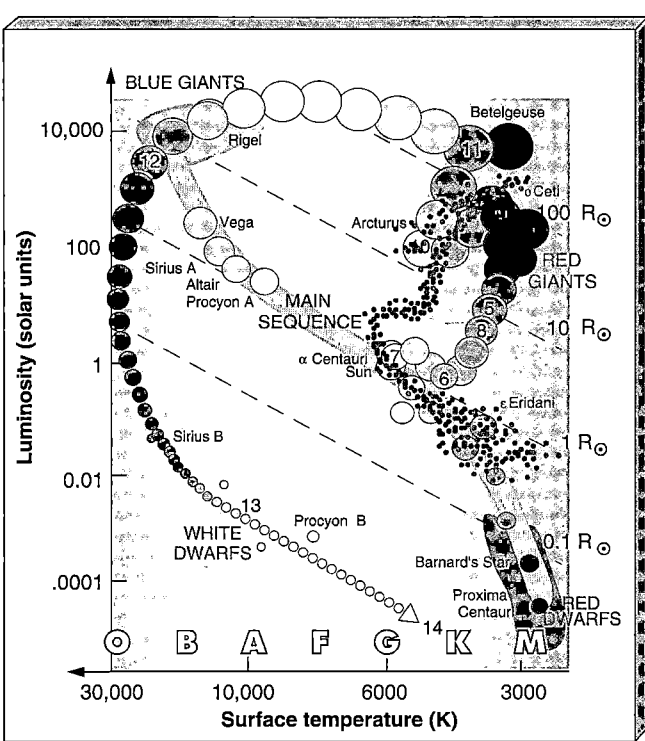

◁ H–R Diagrams and the Cosmic Distance Ladder

All of the H–R diagrams, assembled and drawn by Lola Judith Chaisson, are presented in a uniform format, using real data wherever possible. The goal is to make it easy for students to compare theoretical and observational results presented across several different chapters. The cosmic distance ladder is a theme that spans the text, and we use an evolving set of standard figures to illustrate how distance measurement techniques fit into and ultimately drive our understanding of the cosmos.

Acetate Overlays

Two unique sets of transparent acetate overlays dramatically illustrate two key pedagogical elements of the book. The H–R diagram overlays demonstrate to students how astronomers organize information about the stars and track their evolutionary histories. The cosmic distance scale overlays summarize, in simplified form, the main methods used by astronomers to chart their way among increasing scales in the universe.

Other Pedagogical Features

As with many other parts of our textbook, instructors have helped guide us toward what is most helpful for effective student learning. With their assistance, we have revised both our in-chapter and end-of-chapter pedagogical apparatus to increase its utility to students.

Learning Goals. Studies indicate that beginning students often have trouble prioritizing textual material. For this reason, a few (typically five or six) well-defined Learning Goals are provided at the start of each chapter. These help students to structure their reading of the chapter and then test their mastery of key facts and concepts. The Learning Goals are numbered and cross-referenced to key sections in the body of each chapter. This in-text highlighting of the most important aspects of the chapter also helps students to review. They are organized and phrased in such a way as to make them objectively testable, affording students a means of gauging their own progress.

Concept Checks. New to this edition, we have ▶ incorporated into each chapter a number of "Concept Checks"—key questions that require the reader to reconsider some of the material just presented or attempt to place it into a broader context. Answers to the Concept Check questions are provided at the end of the book.

Cross-Links. In astronomy, as in many scientific disci- ▶ plines, almost every topic seems to have some bearing on almost every other. In particular, the connection between the specifically astronomical material and the physical principles set forth early in the text is crucial. Practically everything in Chapters 6–28 of this text rests on the foundation laid in the first five chapters. For example, it is important that students, when they encounter the discussion of high-redshift objects in Chapter 25, recall not only what they just learned about Hubble's law in Chapter 24 but also refresh their memories, if necessary, about the inverse-square law (Chapter 17), stellar spectra (Chapter 4), and the Doppler shift (Chapter 3). Similarly, the discussions of the mass of binary-star components (Chapter 17) and of galactic rotation (Chapter 23) both depend on the discussion of Kepler's and Newton's laws in Chapter 2. Throughout, the discussion of new astronomical objects relies heavily on comparison with topics introduced earlier.

It is essential to remind students of these links so they can recall the principles on which later discussions rest and, if necessary, review them. To reinforce these connections, "cross links" have been inserted throughout the text—symbols that mark key intellectual bridges between material in different chapters. The links are denoted by the symbol ∞ and, together with a section reference

Key Terms. Like all subjects, astronomy has its own special vocabulary. To aid student learning, the most important astronomical terms are boldfaced at their first appearance in the text. Each boldfaced key term is also incorporated in the appropriate chapter summary, together with the page number where it was defined. In addition, a full alphabetical glossary, defining each key term and locating its first use in the text, appears at the end of the book.

Planetary Data Boxes. Providing concise summaries of planetary properties within the body of each chapter, the data are repeated in the Appendix for easy reference. Similar data boxes are provided for both the Moon and the Sun. All planetary data are drawn from the database maintained by the Solar System Dynamics Group at the Jet Propulsion Laboratory.

☑ Concept Check
- Explain, in terms of Newton's laws of motion and gravity, why planets orbit the Sun.

27.5 The Formation of Structure in the Universe

Just as stars form from inhomogeneities in interstellar clouds, galaxies, galaxy clusters, and larger structures are believed to have grown from small density fluctuations in the matter of the expanding universe. ∞ (Sec. 19.1) Given the conditions in the universe during the atomic and galactic epochs (Table 27.1), cosmologists calculate that regions of higher-than-average density that contained more than about a million times the mass of the Sun would have begun to contract. There was thus a natural tendency for million-solar-mass "pregalactic" objects to form. In Chapter 24 we saw a little of how these pregalactic fragments might have interacted and merged to form galaxies. ∞ (Sec. 24.4) Here, we concern ourselves mostly with the formation of structure on much larger scales.

(a hyperlink on the accompanying CD-ROM), signal to students that the topic under discussion is related in some significant way to ideas developed earlier, and direct them to material that they might wish to review before proceeding.

Discovery Boxes. Exploring a ▶
wide variety of interesting supplementary topics, these features have
been expanded and renamed from
the "Interludes" of previous editions
to better reflect their goal of providing the reader with insight into
how scientific knowledge evolves.

DISCOVERY 5-2

The *Chandra* X-Ray Observatory ⟨Animation⟩

In the summer of 1999, the Space Shuttle *Columbia* carried into space the largest and most sophisticated X ray telescope ever built. Known simply as *Chandra* after the late Indian-American astrophysicist Subramanyan Chandrasekhar, the *Chandra X-ray Observatory (CXO)* is the third of NASA's "Great Observatories" to be deployed in space (the other two being *Hubble* and *Compton*). But unlike the others that hug Earth in low orbits only hundreds of kilometers in altitude, *Chandra* was boosted by onboard rockets to a much higher, elliptical orbit; its farthest point from Earth, 140,000 km, reaches almost one-third of the way to the Moon. The telescope's orbital period is 64 hours. Such an orbit takes *Chandra* well above most of the interfering van Allen belts girdling Earth and allows more efficient observations of the cosmos.

The first figure below, at left, shows the *Chandra* spacecraft in its final stages of construction in 1998. The technicians give perspective to this bus-sized craft which contains computers, antennae, and recorders to transmit and receive information between *Chandra* and ground stations, as well as finely polished, barrel-shaped mirrors to guide the X rays to a precise focus (see Figure 5.32). The front end of the telescope is at the bottom of the photograph and the scientific instruments at the top. Those instruments include a high-resolution camera to take images of X rays and a spectrometer to record the energy of those

X rays. Data collected by *Chandra's* instruments are relayed from orbit to scientists at the mission control center in Cambridge, Massachusetts.

The second figure below shows an example of *Chandra's* scientific capabilities; many more will be found throughout this textbook. The object shown is a supernova remnant—a debris field of scattered, glowing gases that were once part of a massive star. This "object," known as Cas A, is all that remains of a star in the constellation Cassiopeia. Its explosion was observed about 320 years ago. For scale, the debris is spread across some 10 light-years; Cas A itself is roughly 10,000 light-years from Earth.

The image shows what *Chandra* saw when it targeted Cas A for its first image taken in X rays. The region, barely visible in the optical part of the spectrum, is awash in brilliantly glowing X rays—large amounts of energy emitted at wavelengths to which the human eye is insensitive. (The X rays detected by *Chandra* in orbit are converted to radio waves that are then sent to ground stations, after which they are converted into visible images by astronomers for display and analysis.) This X-ray image shows that the gas in the wisps of ejected stellar material has temperatures of nearly 50 million kelvins, and the bright white point at the very center of the debris might well be a black hole—all of which will be studied in more detail in Chapters 21 and 22.

(NASA)

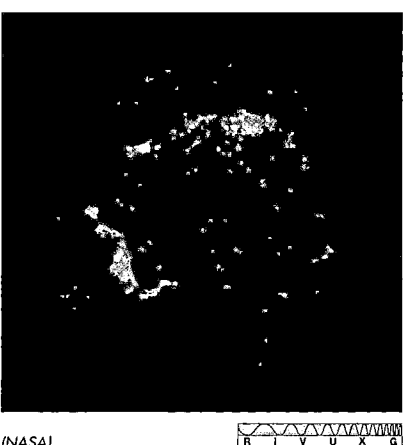
(NASA)

MORE PRECISELY 2-1

Some Properties of Planetary Orbits

Two numbers—semi-major axis and eccentricity—are all that are needed to describe the size and shape of a planet's orbital path. From them we can derive many other useful quantities. Two of the most important are the planet's *perihelion* (its point of closest approach to the Sun) and its *aphelion* (greatest distance from the Sun). From the definitions presented in the text, it follows that if the planet's orbit has semi-major axis a and eccentricity e, its perihelion is at a distance $a(1-e)$ from the Sun, while its aphelion is at $a(1+e)$. These points and distances are illustrated in the accompanying figure.

EXAMPLE: A (hypothetical) planet with a semi-major axis of 400 million km and an eccentricity of 0.5 (the eccentricity of the ellipse shown in the diagram) would range between $400 \times (1-0.5) = 200$ million km and 400×3 $(1+0.5) = 600$ million km from the Sun over the course of one complete orbit. With $e = 0.9$, the range would be 40–760 million km, and so on.

No planet has an orbital eccentricity as large as 0.5—the planet with the most eccentric orbit is Pluto, with $e = 0.249$ (see Table 2.1). However, many meteoroids, and all

comets (see Chapter 14) have eccentricities considerably greater than this. In fact, most comets visible from Earth have eccentricities very close to $e = 1$. Their highly elongated orbits approach within a few astronomical units of the Sun at perihelion, yet these tiny frozen worlds spend most of their time far beyond the orbit of Pluto.

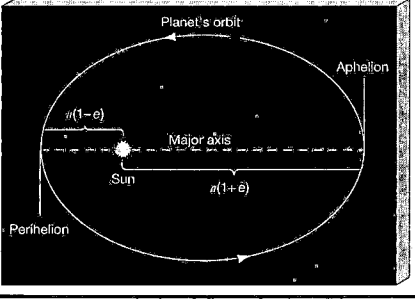

◀ **More Precisely Boxes.** These provide quantitative treatments of subjects discussed qualitatively in the text, or explore the physics of astronomical processes in greater detail. Removing these more challenging topics from the main flow of the narrative and placing them within a separate modular element of the chapter design (so that they can be covered in class, assigned as supplementary material, or simply left as optional reading for those students who find them of interest) affords instructors greater flexibility in setting the level of their coverage.

Chapter Summaries. The chapter summaries, a primary review tool for the student, have been revised and streamlined for the fourth edition. All key terms introduced in each chapter are listed again, in context and in boldface, along with page references to the text discussion.

Questions, Problems, and Projects.
Other elements of the end-of-chapter material have also seen substantial reorganization and expansion:

*   Each chapter incorporates 30 Self-Test questions, equally divided between "True/False" and "Fill-in-the-Blank" formats, designed to allow students to assess their understanding of the chapter material. Answers to *all* questions appear at the end of the book.

*   Each chapter also has 20 Review and Discussion questions (increased from 15 in the previous edition), which may be used for in-class review or for assignments. As with the Self-Test questions, the material needed to answer Review questions may be found within the chapter. The Discussion questions explore particular topics more deeply, often asking for opinions, not just facts. As

with all discussions, these questions usually have no single "correct" answer.

*   ▼ The end-of-chapter **Problems** have been expanded to contain 15 questions, spanning a somewhat broader range of difficulty than those in the previous edition. The level of difficulty is indicated on each problem. Answers to the Problems, particularly for the more sophisticated ones, are not necessarily contained verbatim within the chapter, but all the information required to solve the problems is in the text. Solutions for the odd-numbered problems are given at the end of the book. New for this edition, we have put algorithmic versions of the problems in a Practice Problems module of the text's Companion Website: The values in each problem change each time the student uses a chapter's problem set, thereby giving students unlimited practice in solving numeric problems.

*   Each chapter includes a few (2–4) Projects meant to get the student out of the classroom and looking at the sky, although some entail research in libraries or other extracurricular activities.

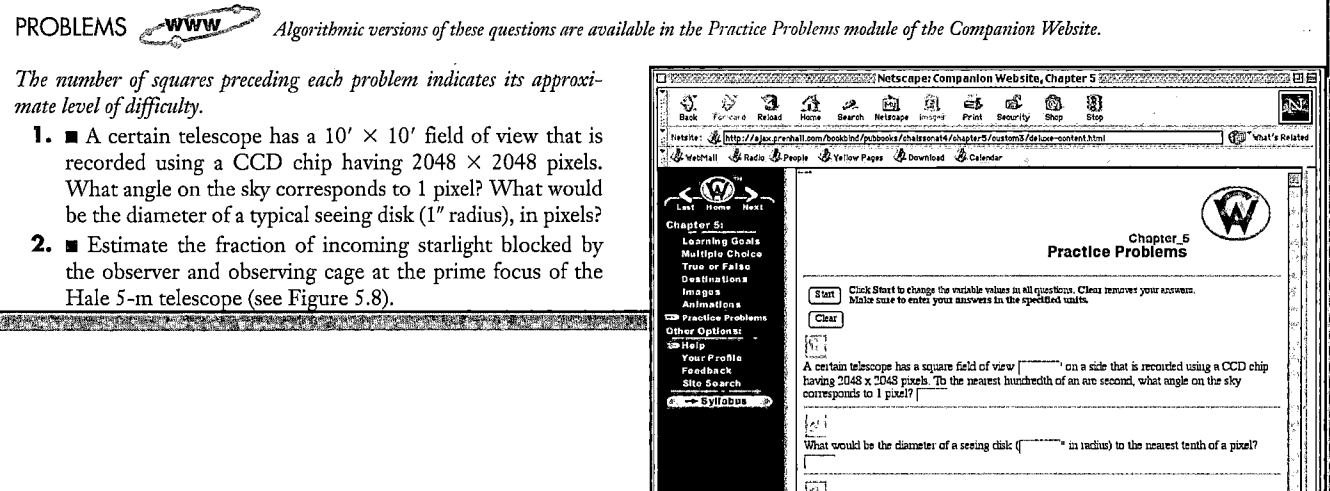

New to this edition are **Collaborative Exercises** ▶ provided by Tim Slater at Montana State University. These interactive collaborative learning group activities are designed to encourage students to work together on reasoning tasks. They are designed specifically for use in the classroom, allow for multiple correct solution pathways, and give students some degree of choice in how to approach solving a complex, multifaceted problem. Most tasks require 7–12 minutes to complete.

COLLABORATIVE EXERCISES

1. **Radio Interferometer.** Determine the maximum size interferometer your group could build if you placed 2 m radio telescopes where each group member lives.
2. **Observing New Stars in the Orion Region.** Your group has been assigned to observe the region of the sky around Orion to look for bright new stars hidden in molecular

RESEARCHING ON THE WEB  *To complete the following exercises, go to the online Destinations module for Chapter 5 on the Companion Website for Astronomy Today 4/e.*

1. Access the "Buying a First-Time Telescope" page and determine which telescopes can be purchased for about $1000.
2. Access the "Introduction to Light Pollution" page and determine the five most common contributors to light pollution.

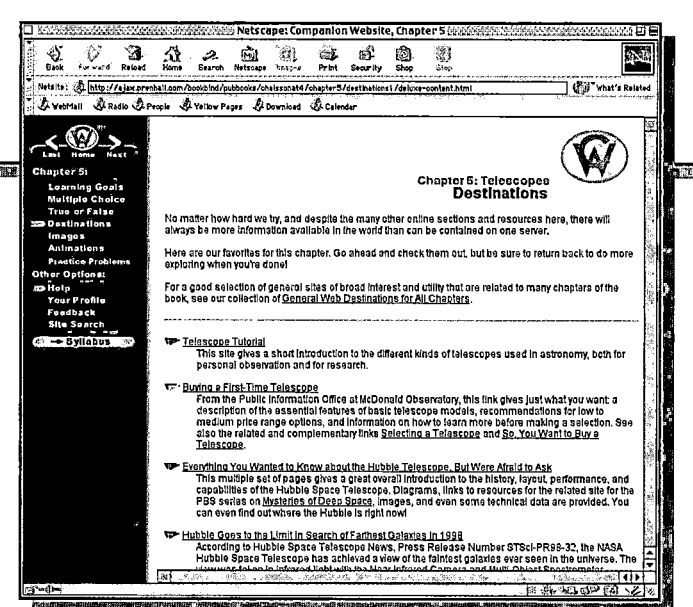

▲ Also new to this edition and provided by Tim Slater are **"Researching on the Web"** exercises. Using stable and maintained Internet resources hyperlinked from the Destinations module on the text's Companion Website, students are assigned analysis tasks using current and real-time Internet data sources. Quite different from the single correct answer tasks often used, many of these engaging tasks yield different but correct results for students accessing data resources on different days.

SKYCHART III PROJECTS  *The SkyChart III Student Version planetarium program on which these exercises are based is included as a separately executable program on the CD in the back of this text.*

1. ■ Change your view to a place in the Southern Hemisphere. Identify the brightest stars and most prominent constellations in the southern sky. Can you identify a southern pole star, the equivalent to the North Star, Polaris?
2. ■ Change your viewing location to Buenos Aires on January 1, 2001 at 12:00 a.m. Print a chart in which your view is

scope and adjust the parameters for field width to match those of your telescope. If you wish to drive a telescope with the software, it will be necessary first to deselect *Emulate real telescope connection* and to align the telescope according to normal procedures for that telescope. Either with a real telescope or with emulation, locate an object of interest on

▲ Finally, each chapter now concludes with **SkyChart III Projects** by Erik Bodegom and Sean Goe of Portland State University and Duane Ingram of Rock Valley College. These exercises are based on the SkyChart III Student Version planetarium software, which is included as a separately executable program on the e-book CD in the back of the text. Square bullets preceding each exercise indicate its approximate level of difficulty. Appendix 4 provides general instructions for the software, and specific menu commands are included with each exercise. Instructor's notes for all the exercises are included in a password-protected part of the text's Companion Website.

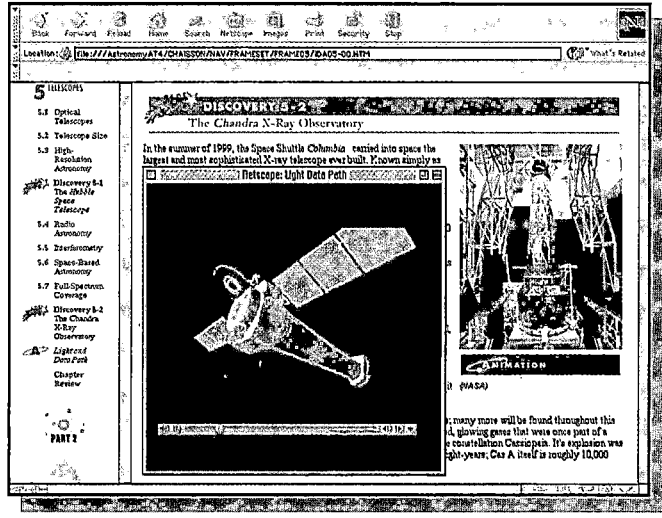

e-Book CD and Companion Website

◀ **e-Book CD.** Each copy of the text comes with a free e-book CD. This is a browser-based version of the text with extensive hyperlinks (over 3000 throughout the text), 61 videos integrated with relevant text discussions, and links to our Companion Website, which is organized by text chapter and updated monthly. We have added 12 new, exclusive animations from the Wright Center at Tufts University and redigitized all the videos and animations for larger size and higher resolution, and now they all have narrations. A script to facilitate use of the e-Book CD under Unix is available at:

ftp://ftp.prenhall.com/pub/esm/physics.s-085/chaisson/bg

SkyChart III Student Version. Also included on the e-Book CD as a separately executable program is a student version of the SkyChart III planetarium software by Southern Stars Systems. This planetarium program accurately simulates and displays the sky as it currently appears, as well as thousands of years in the past or future. The sky can be viewed from any place on Earth, or from any object in the solar system or beyond. The program includes a database of 300,000 fully customizable objects based on NASA's SKY2000 Master Star Catalog, showing all stars up to about magnitude 10.

Companion Website. Our Companion Website at http://www.prenhall.com/chaisson organizes Destinations (links to related websites), additional Images and Animations, Multiple Choice, True/False, and Labeling exercises, and algorithmic versions of the text's end-of-chapter Problems on a chapter-by-chapter basis. The Destinations, Images, and Animations modules are updated monthly to keep links current and provide information on significant new discoveries. All of the exercises and problems are interactive, meaning that the student answers the questions on-line and then receives immediate scoring and feedback, including text section references for any areas the student needs to study further if they answered a given question incorrectly.

Additional Supplementary Material

This edition of *Astronomy Today* is accompanied by an outstanding set of instructional aids.

Comets Published annually at the beginning of each academic year and available free to adopters, *Comets* is a unique kit that includes a collection of slides, videos, and *New York Times* articles on events and discoveries that have occurred since the publication of the prior year's *Comets* kit. The slide kit contains 28 new slides from NASA, JPL, STScI, GSFC, *HST* Comet LINEAR Investigation Team, APL, JPL, ESA, *Hubble* Heritage Team, IPAC, European Southern Observatory, SDSS/Astrophysical Research Consortium, and the U.S. Department of Defense. Custom animations prepared by the Wright Center for Science Visualization and many other videos of new discoveries and animations from various sources, including NASA, STScI, APL/NRL, ESA, Stanford Lockheed Institute for Space Research, and JPL are provided in both CD and DVD formats. The collection of *New York Times* articles, called "Themes of the Times," is published twice yearly and is available free in quantity for your students using either Chaisson/McMillan text. A newsletter provides a cross reference between all the materials in the *Comets* kit and corresponding chapters of both Chaisson/McMillan texts, as well as annotations describing the subject and source of each slide and video in the kit. ISBN 0-13-093801-7

Instructor's Resource Manual By Leo Connolly (California State University at San Bernardino). This manual provides an overview of each chapter, pedagogical tips, useful analogies, suggestions for classroom demonstrations, writing questions, answers to the end-of-chapter Review and Discussion questions and Problems. New features include an expanded introduction with an overview of how to utilize the IRM, an index of demonstrations, applications of the writing questions, sample assignments, discussion of common student misconceptions, teaching notes for the collaborative exercises, and a list of selected readings for each chapter. ISBN 0-13-093796-7

Media Portfolio CD-ROM By Suzanne Willis (Northern Illinois University). This flexible, easy-to-use tool contains a wealth of photographs, line art, animations, and videos to use in class lectures. Instructors can easily search, access, and organize the materials according to their lecture outlines and add their own visuals and lecture notes. The hybrid CD contains all of the line art and photographs from *Astronomy Today 4e*, as well as the animations and videos that are on the e-Book CD in the back of the student text. In addition, the Image Viewer incorporates slides from the current and past editions of *Comets*. ISBN 0-13-093791-6

Acetates and Slides A set of approximately 260 images from the text—almost double the number offered for the third edition—are available as a package of color acetates or 35mm slides, and are available free to qualified adopters. ISBN 0-13-093780-0 (Slide set) ISBN 0-13-093803-3 (Transparency pack)

Test Item File An extensive file of over 2600 test questions, newly compiled and revised for the fourth edition by Wayne Wooten (Pensacola Junior College), is offered free upon adoption. Available in both printed and electronic formats (Macintosh or Windows format). The fourth edition Test Item File has been thoroughly revised, including many new Multiple Choice and Essay questions for added conceptual emphasis. Overall, approximately 600 new questions have been added. ISBN 0-13-093798-3

Prentice Hall Custom Test Available for Macintosh and Windows, Prentice Hall Custom Test allows educators to create and tailor the exam to their own needs. With the On-line Testing option, exams can also be administered on-line, and data can then be automatically transferred for evaluation. A comprehensive desk reference guide is included, along with on-line assistance. ISBN 0-13-093797-5 (Macintosh) ISBN 0-13-093799-1 (Windows)

Science on the Internet By Andrew Stull and Harry Nickla. A guide to general science resources on the Inter-

net. Everything you need to know to get yourself on-line and browsing the World Wide Web!
ISBN 0-13-028253-7

Acknowledgments

Throughout the many drafts that have led to this edition, we have relied on the critical analysis of many colleagues.

Their suggestions ranged from the macroscopic issue of the book's overall organization to the minutiae of the technical accuracy of each and every sentence. We have also benefited from much good advice and feedback from many users of the third edition of the text and from our shorter book, *Astronomy: A Beginner's Guide to the Universe*. To these many helpful colleagues, we offer our sincerest thanks.

Stephen G. Alexander
Miami University of Ohio

Robert H. Allen
University of Wisconsin, La Crosse

Timothy C. Beers
University of Evansville

William J. Boardman
Birmingham Southern College

Donald J. Bord
University of Michigan, Dearborn

Elizabeth P. Bozyan
University of Rhode Island

Anne Cowley
Arizona State University

Bruce Cragin
Richland College

Ed Coppola
Community College of Southern Nevada

David Curott
University of North Alabama

Norman Derby
Bennington College

John Dykla
Loyola University, Chicago

Kimberly Engle
Drexel University

Martin Goodson
Delta College

David G. Griffiths
Oregon State University

Donald Gudehus
Georgia State University

Marilynn Harper
Delaware County Community College

Clint D. Harper
Moorpark College

Susan Hartley
University of Minnesota, Duluth

Joseph Heafner
Catawaba Valley Community College

Fred Hickok
Catonsville Community College

Darren L. Hitt
Loyola College, Maryland

F. Duane Ingram
Rock Valley College

Steven D. Kawaler
Iowa State University

William Keel
University of Alabama

Marvin Kemple
Indiana University-Purdue University, Indianapolis

Mario Klairc
Midlands Technical College

Kristine Larsen
Central Connecticut State University

Andrew R. Lazarewicz
Boston College

Robert J. Leacock
University of Florida

Larry A. Lebofsky
University of Arizona

M. A. Lohdi
Texas Tech University

Michael C. LoPresto
Henry Ford Community College

Phillip Lu
Western Connecticut State University

Fred Marschak
Santa Barbara College

Chris Mihos
Case Western Reserve University

Milan Mijic
California State University, Los Angeles

Richard Nolthenius
Cabrillo College

Edward Oberhofer
University of North Carolina, Charlotte

Andrew P. Odell
Northern Arizona University

Gregory W. Ojakangas
University of Minnesota, Duluth

Ronald Olowin
Saint Mary's College of California

Robert S. Patterson
Southwest Missouri State University

Cynthia W. Peterson
University of Connecticut

Andreas Quirrenback
University of California, San Diego

Richard Rand
University of New Mexico

James A. Roberts
University of North Texas

Gerald Royce
Mary Washington College

Malcolm P. Savedoff
University of Rochester

John C. Schneider
Catonsville Community College

Harry L. Shipman
University of Delaware

C. G. Pete Shugart
Memphis State University

Stephen J. Shulik
Clarion University

Tim Slater
Montana State University

Don Sparks
Los Angeles Pierce College

Maurice Stewart
Williamette University

Jack W. Sulentic
University of Alabama

Andrew Sustich
Arkansas State University

Stephen R. Walton
California State University, Northridge

Peter A. Wehinger
University of Arizona

Louis Winkler
Pennsylvania State University

Robert L. Zimmerman
University of Oregon

The publishing team at Prentice Hall has assisted us at every step along the way in creating this text. Much of the credit for getting the project completed on time goes to our Executive Editor, Alison Reeves, who has successfully navigated us through the many twists, turns, and "absolute final deadlines" of the publishing world. We have been fortunate to work once again with Production Editor Joanne Hakim, who has done another truly excellent job of tying together the threads of this very complex project. Special thanks go also to Development Editor Don Gecewicz, who has skillfully helped us update and improve the manuscript. Many thanks to Eileen Harrity, Editorial Assistant, for all of her help trafficking manuscripts, reviews, page proofs, and accuracy-checker comments; and to Christian Botting for arranging and managing a complete set of revised supplements. Alison Lorber, Elizabeth Wright, Gretel Seham, Nicole Bush, and William Johnson all worked diligently and tirelessly on the e-book CD and Companion Website so that they could be available the same time as the text, which is no small feat.

Our special thanks go to all who helped us create materials both in the book itself and the supplements package: Duane Ingram (Rock Valley College), and Eric Bodegom and Sean Goe (Portland State University) for creating the SkyChart III Projects; Tim Slater (Montana State University) for providing the Collaborative Exercises and Researching on the Web exercises; Alan Sill (Texas Tech University) and Dale Etheridge (Community College of Southern Nevada) for updating and revising the Multiple Choice and True/False modules of the Companion Website; David McKenzie (Montana State University) for monthly updates on an ongoing basis of the Destinations, Images, and Animations modules of the Companion Website; Leo Connolly (California State University, San Bernardino) for providing the Labeling questions for the Companion Website, as well as for hyperlink cross references for the e-book CD and revising and expanding the Instructor's Manual; Wayne Wooten (Pensacola Junior College) for updating and expanding the Test Item File; Suzanne Willis (Northern Illinois University) for her help in preparing the Image Viewer CD; Dan Schiller (Prentice Hall) for collecting and assembling important articles for the "Themes of the Times" supplement from *The New York Times*; and to Ray Villard (STScI) for again providing the new slides, videos, and animations for the annual *Comets* update kit.

ASTRONOMY TODAY

1 CHARTING THE HEAVENS

The Foundations of Astronomy

LEARNING GOALS

Studying this chapter will enable you to:

1　Explain the concept of the celestial sphere and the conventions of angular measurement that enable us to locate objects in the sky.

2　Describe how the Sun, the Moon, and the stars appear to change their positions from night to night and from month to month.

3　Account for these apparent motions in terms of the actual motions of Earth and the Moon.

4　Show how the relative motions of Earth, the Sun, and the Moon lead to eclipses.

5　Explain the simple geometric reasoning that allows astronomers to measure the distances and sizes of faraway objects.

 Visit http://www.prenhall.com/chaisson for additional annotated images, animations, and links to related sites for this chapter.

One of the most awe-inspiring sights in the universe is that of a spiral galaxy. Here, this face-on view of the galaxy NGC 1232, located in the southern constellation Eridanus, shows its spread of stars, gas, and dust over a vast expanse some 150,000 light-years across. Despite its distance of about 100 million light-years, modern telescopes allow study of its intricate details—including the older, reddish stars in the central regions and the younger, bluish stars in its spiral arms. *(ESO)*

The Big Picture: Galaxies are candidates for the grandest, most beautiful structures in the universe. They are colossal collections of typically a hundred billion stars, held together by gravity within one loose structure. There are approximately 40 billion galaxies in the observable universe. One of the great unsolved mysteries in science is the fact that astronomers do not fully understand how galaxies originated.

Nature offers no greater splendor than the starry sky on a clear, dark night. Silent and jeweled with the constellations of ancient myth and legend, the night sky has inspired wonder throughout the ages—a wonder that leads our imaginations far from the confines of Earth and the pace of the present day, and out into the distant reaches of space and cosmic time itself. Astronomy, born in response to that wonder, is built on two of the most basic traits of human nature: the need to explore and the need to understand. Through the interplay of curiosity, discovery, and analysis—the keys to exploration and understanding—people have sought answers to questions about the universe since the earliest times. Astronomy is the oldest of all the sciences, yet never has it been more exciting than it is today.

1.1 Our Place in Space

Of all the scientific insights attained to date, one stands out boldly: Earth is neither central nor special. We inhabit no unique place in the universe. Astronomical research, especially within the past few decades, strongly suggests that we live on what seems to be an ordinary rocky *planet* called Earth, one of nine known planets orbiting an average *star* called the Sun, a star near the edge of a huge collection of stars called the Milky Way Galaxy, which is one *galaxy* among countless billions of others spread throughout the observable universe. To begin to get a feel for the relationships among these very different objects, consult Figures 1.1 through 1.4 and put them in perspective by studying Figure 1.5.

We are connected to the most distant realms of space and time not only by our imaginations but also through a common cosmic heritage. Most of the chemical elements in our bodies were created billions of years ago in the hot centers of long-vanished stars. Their fuel supply spent, these giant stars died in huge explosions, scattering the elements created deep within their cores far and wide. Eventually, this matter collected into clouds of gas that slowly collapsed to give birth to new generations of stars. In this way, the Sun and its family of planets formed nearly five billion years ago. Everything on Earth embodies atoms from other parts of the universe and from a past far more remote than the beginning of human evolution. Elsewhere, other beings—perhaps with intelligence much greater than our own—may at this very moment be gazing in wonder at their own night sky. Our own Sun may be nothing more than an insignificant point of light to them, if it is visible at all. Yet if such beings exist, they must share our cosmic origin.

Simply put, the **universe** is the totality of all space, time, matter, and energy. **Astronomy** is the study of the universe. It is a subject unlike any other, for it requires us to profoundly change our view of the cosmos and to consider matter on scales totally unfamiliar from everyday experience. Look again at the galaxy in Figure 1.3. This galaxy, whose catalog name is NGC 4414, is a swarm of about a hundred billion stars—more stars than the number of people who have ever lived on Earth. The entire assemblage is spread across a vast expanse of space 100,000 **light-years** in diameter. A light-year is the *distance* traveled by light in a year, at a speed of about 300,000 kilometers per second. Multiplying, we find that a light-year is equal to 300,000 kilometers/second × 86,400 seconds/day × 365 days, or about 10 trillion kilometers (roughly six trillion miles). Typical galactic systems are truly "astronomical" in size. For comparison, Earth's roughly 13,000 km diameter is less than 1/20 of a light-*second*.

The light-year is a convenient unit introduced by astronomers to help them describe immense distances. We will encounter many such "custom" units in our studies. As discussed in more detail in Appendix 1, astronomers frequently augment the standard SI (Système Internationale) metric system with new units tailored to the particular problem at hand.

A thousand (1000), a million (1,000,000), a billion (1,000,000,000), and even a trillion (1,000,000,000,000)—these words occur regularly in everyday speech. But let's take a moment to understand the magnitude of these numbers and appreciate the differences among them. One thousand is easy enough to understand: At the rate of one number per second, you could count to a thousand in 1000 seconds—about 16 minutes. However, if you wanted to count to a million, you would need more than two weeks of counting at the rate of one number per second, 16 hours per day (allowing eight hours per day for sleep). To count from one to a billion at the same rate of one number per second and 16 hours per day would take nearly 50 years—the better part of an entire human lifetime.

In this book we consider spatial domains spanning not just billions of kilometers but billions of light-years, objects containing not just trillions of atoms but trillions of stars, time intervals of not just billions of seconds or hours but billions of years. You will need to become familiar with—and comfortable with—such enormous numbers. A good way to begin is learning to recognize just how much larger than a thousand is a million, and how much larger still is a billion. Appendix 1 explains the convenient method used by scientists for writing and manipulating very large and very small numbers. If you are unfamiliar with this method, please read Appendix 1 carefully—the *scientific notation* described there will be used consistently throughout our text, beginning in Chapter 2.

15,000 kilometers

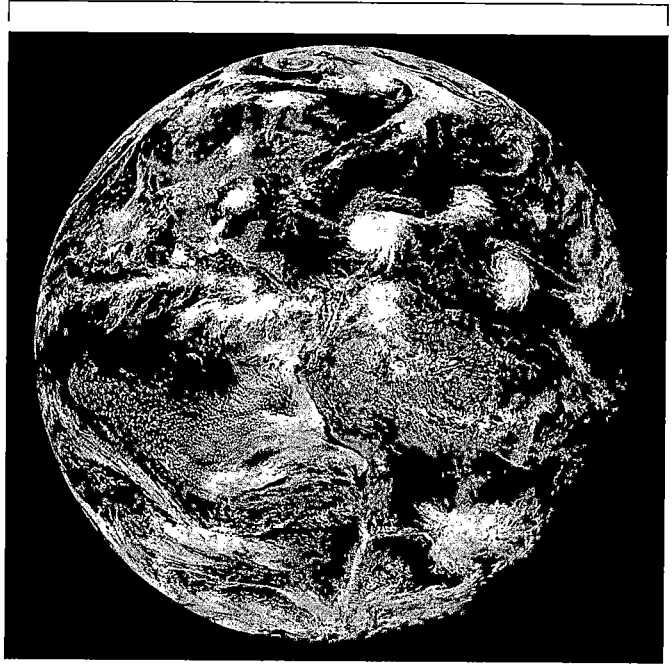

R   I   V   U   X   G

Figure 1.1 Earth Earth is a planet, a mostly solid object, though it has some liquid in its oceans and core, and gas in its atmosphere. In this view, you can clearly see the North and South American continents. *(NASA)*

1,500,000 kilometers

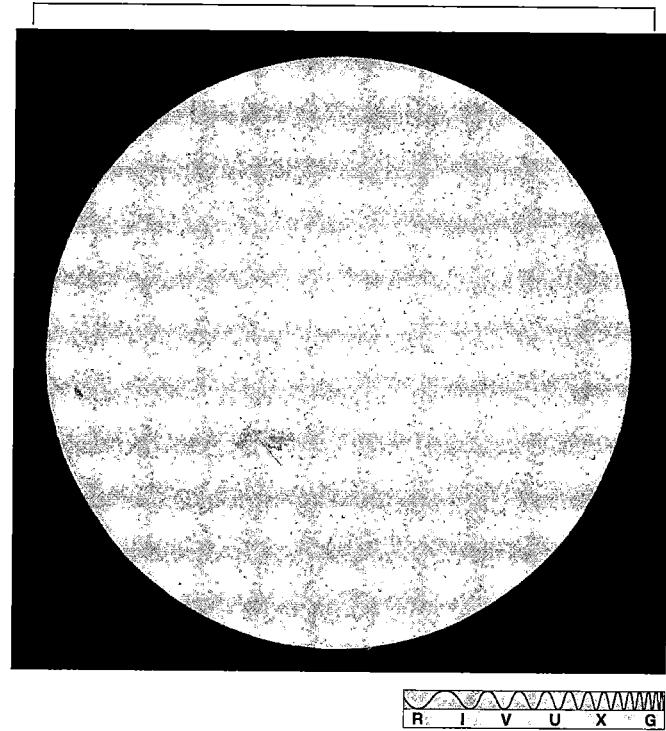

R   I   V   U   X   G

Figure 1.2 The Sun The Sun is a star, a very hot ball of gas. Much bigger than Earth, the Sun is held together by its own gravity. *(AURA)*

About 1000 quadrillion kilometers, or 100,000 light-years

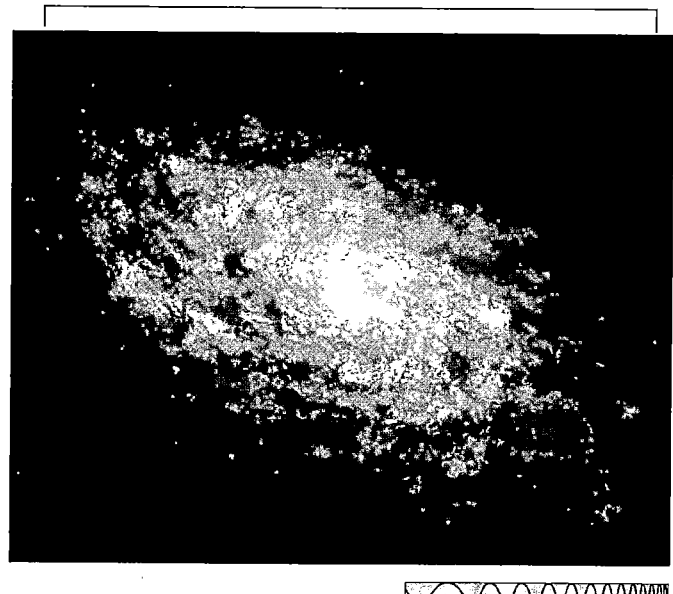

R   I   V   U   X   G

Figure 1.3 Galaxy A typical galaxy is a collection of a hundred billion stars, each separated by vast regions of nearly empty space. This galaxy is the 4,414th entry in the *New General Catalog of Astronomical Objects*—NGC 4414 for short. Our Sun is a rather undistinguished star near the edge of another such galaxy, called the Milky Way. *(NASA)*

About 1,000,000 light-years

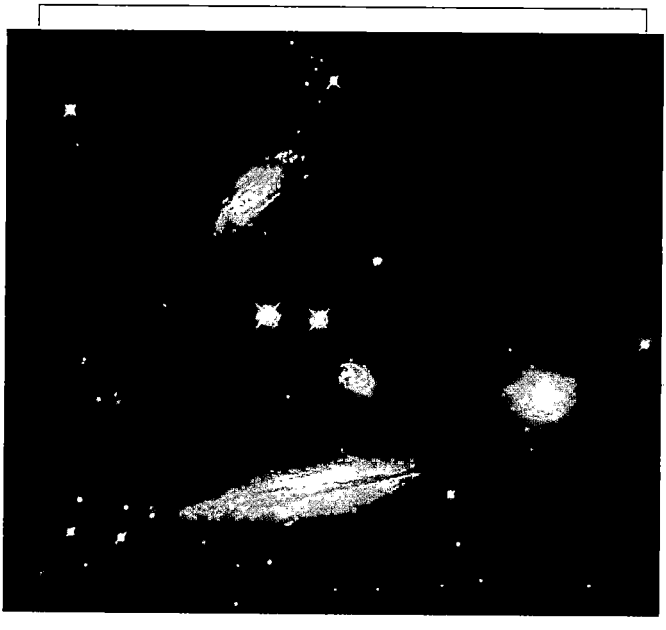

R   I   V   U   X   G

Figure 1.4 Galaxy Cluster This photograph shows a typical cluster of galaxies, roughly a million light-years from Earth. Each galaxy contains hundreds of billions of stars, probably planets, and possibly living creatures. *(NASA)*

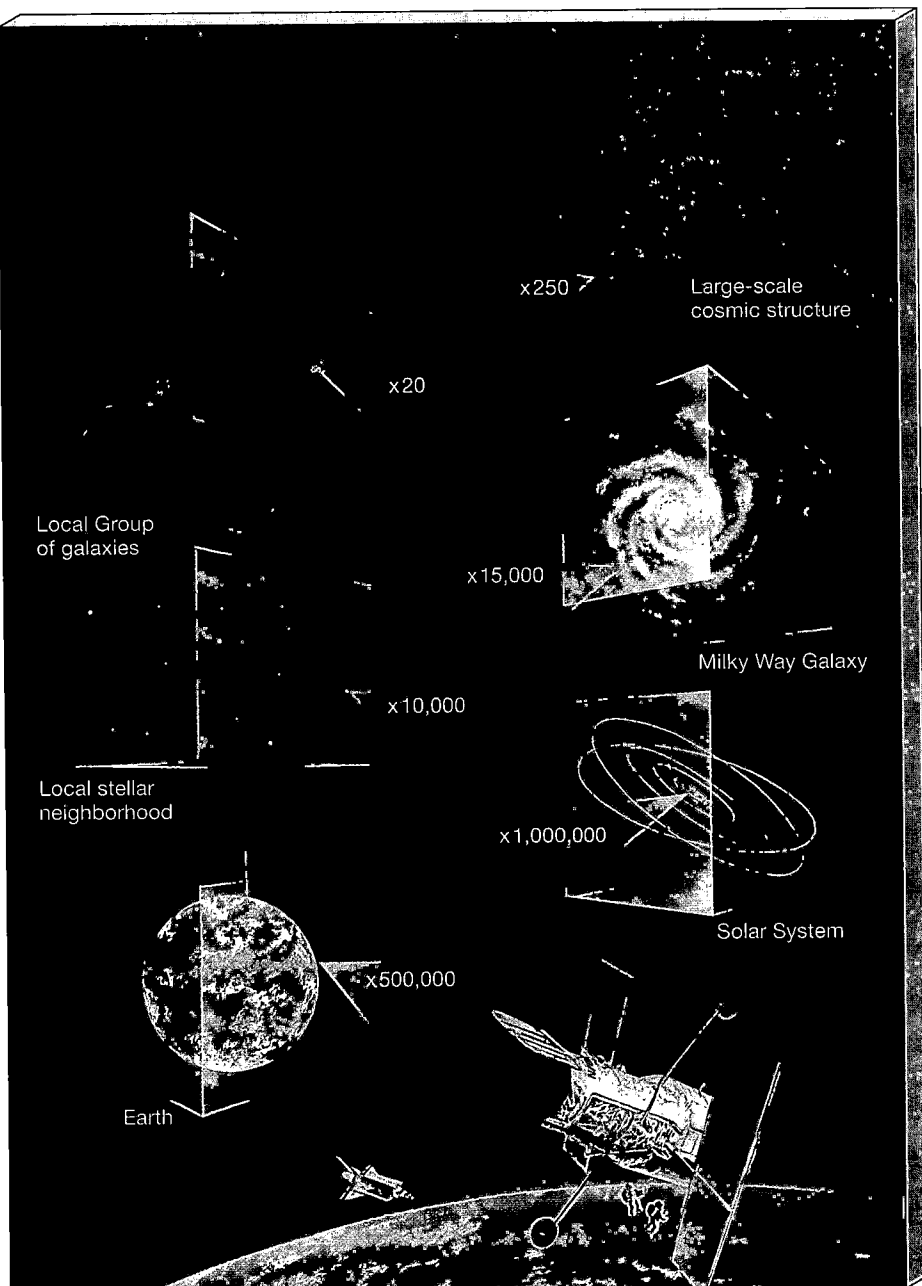

Figure 1.5 Size and Scale This artist's conception puts each of the previous figures in perspective. The bottom of this figure shows spacecraft (and astronauts) in Earth orbit, a view that widens progressively in each of the next five cubes drawn from bottom to top—Earth, the planetary system, the local neighborhood of stars, the Milky Way Galaxy, and the closest cluster of galaxies. The top image depicts the distribution of galaxies in the universe on scales of hundreds of millions of light-years. The numbers indicate approximately the increase in scale between successive images. *(D. Berry)*

1.2 The "Obvious" View

How have we come to know the universe around us? How do we know the proper perspective sketched in Figure 1.5? Let's start with some very basic observations. Our study of the universe, the modern science of astronomy, begins simply by looking at the night sky.

CONSTELLATIONS IN THE SKY

Between sunset and sunrise on a clear night we can see about 3000 points of light. If we include the view from the opposite side of Earth, nearly 6000 stars are visible to the unaided eye. A natural human tendency is to see patterns and relationships between objects even when no true connection exists, and people long ago connected the brightest stars into configurations called **constellations**, which ancient astronomers named after mythological beings, heroes, and animals—whatever was important to them. Figure 1.6 shows a constellation especially prominent in the nighttime sky from October through March: the hunter named Orion. Orion was a mythical Greek hero famed, among other things, for his amorous pursuit of the Pleiades, the seven daughters of the giant Atlas. According to Greek mythology, to protect the Pleiades from Orion, the gods placed them among the stars, where Orion nightly stalks them across the sky. Many constellations have similarly fabulous connections with ancient lore.

Perhaps not surprisingly, the patterns have a strong cultural bias—the astronomers of ancient China saw mythical figures different from those seen by the ancient Greeks, the Babylonians, and the people of other cultures, even though they were all looking at the same stars in the night sky. Interestingly, different cultures often made the same basic *groupings* of stars, despite widely varying interpretations of what they saw. For example, the group of seven stars usually known in North America as "the Dipper" is known as "the Wagon" or "the Plough" in western Europe. The ancient Greeks regarded these same stars as the tail of "the Great Bear," the Egyptians saw them as the leg of an ox, the Siberians as a stag, and some Native Americans as a funeral procession.

Early astronomers had very practical reasons for studying the sky. Some constellations served as navigational guides. The star Polaris (part of the Little Dipper) indicates north, and the near-constancy of its location in the sky, from hour to hour and night to night, has aided travelers for centuries. Other constellations served as primitive calendars to predict planting and harvesting seasons. For example, many cultures knew that the appearance of certain stars on the horizon just before daybreak signaled the beginning of spring and the end of winter.

In many societies, people came to believe that there were other benefits in being able to trace the regularly changing positions of heavenly bodies. The relative positions of stars and planets at a person's birth were carefully studied by *astrologers*, who used the data to make predictions

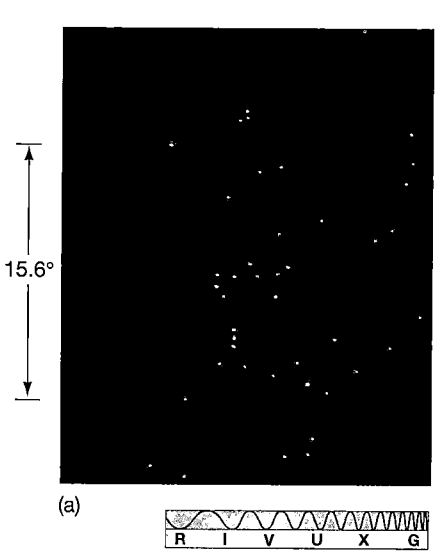

15.6°

(a)

R I V U X G

(b)

Figure 1.6 Constellation Orion (a) A photograph of the group of bright stars that make up the constellation Orion. (b) The stars are connected to show the pattern visualized by the Greeks: the outline of a hunter. You can easily find this constellation in the winter sky by identifying the line of three bright stars in the hunter's "belt." (c) The true three-dimensional relationships among the brightest stars (the Greek letters indicate brightness—see *More Precisely 1-2*). The distances were determined by the *Hipparcos* satellite in the early 1990s (Chapter 17). *(S. Westphal)*

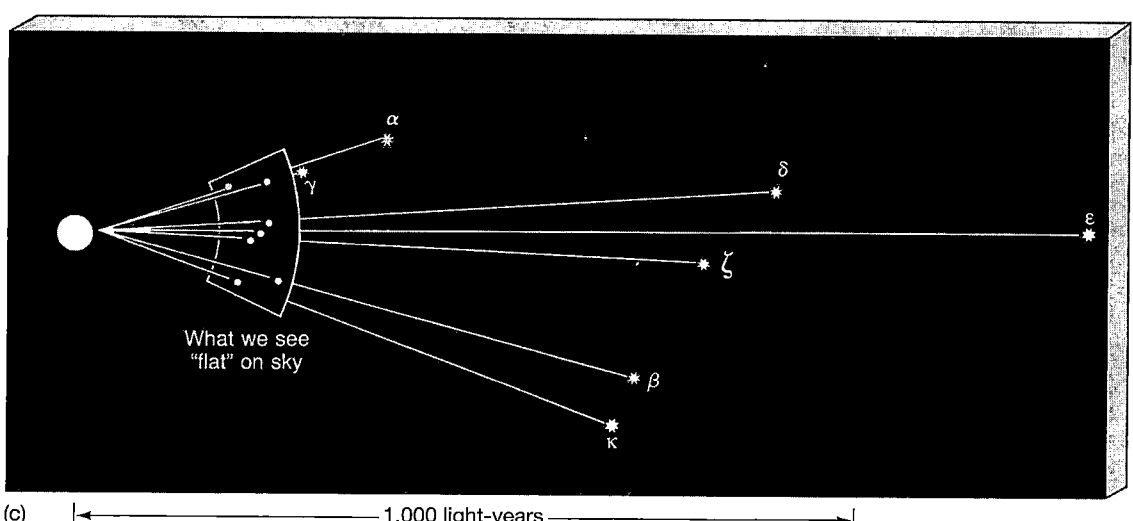

What we see "flat" on sky

(c) |◄———————— 1,000 light-years ————————►|

about that person's destiny. Thus, in a sense, astronomy and astrology arose from the same basic impulse—the desire to "see" into the future—and, indeed, for a long time they were indistinguishable from one another. Today most people recognize that astrology is nothing more than an amusing diversion. (Although millions still study their horoscope in the newspaper every morning!) Nevertheless, the ancient astrological terminology—the names of the constellations and many terms used to describe the locations and motions of the planets—is still used throughout the astronomical world.

Generally speaking, as illustrated in Figure 1.6(c) for the case of Orion, the stars that make up any particular constellation are not actually close to one another in space, even by astronomical standards. They merely are bright enough to observe with the naked eye and happen to lie in roughly the same direction in the sky as seen from Earth. Still, the constellations provide a convenient means for astronomers to specify large areas of the sky, much as geologists use continents, or politicians use voting precincts to identify certain localities on planet Earth. In all, there are 88 constellations, most of them visible from North America at some time during the year. Figure 1.7 shows how the conventionally defined constellations cover a portion of the sky in the vicinity of Orion.

THE CELESTIAL SPHERE

⬛ Over the course of a night, the constellations seem to move smoothly across the sky from east to west, but ancient skywatchers were well aware that the *relative* locations of

stars remain unchanged as this nightly march takes place. It was natural for those observers to conclude that the stars must be firmly attached to a **celestial sphere** surrounding Earth—a canopy of stars resembling an astronomical painting on a heavenly ceiling. Figure 1.8 shows how early astronomers pictured the stars as moving with this celestial sphere as it turned around a fixed, unmoving Earth. Figure 1.9 shows how all stars appear to move in circles around a point very close to the star Polaris (better known as the Pole Star or North Star). To the ancients, this point represented the axis around which the entire celestial sphere turned.

From our modern standpoint, the apparent motion of the stars is the result of the spin, or **rotation**, not of the celestial sphere but of Earth. Polaris indicates the direction—due north—in which Earth's rotation axis points. Even though we now know that the celestial sphere is an incorrect description of the heavens, we still use the idea as a convenient fiction that helps us visualize the positions of stars in the sky. The points where Earth's axis intersects the celestial sphere are called the **celestial poles**. In the Northern Hemisphere, the north celestial pole lies directly above Earth's North Pole. The extension of Earth's axis in the opposite direction defines the south celestial pole, directly above Earth's South Pole. Midway between the north and south celestial poles lies the **celestial equator**, representing the intersection of Earth's equatorial plane with the celestial sphere. These parts of the celestial sphere are marked on Figure 1.8.

When discussing the locations of stars "on the sky," astronomers naturally talk in terms of *angular* positions and separations. *More Precisely 1-1* presents some basic

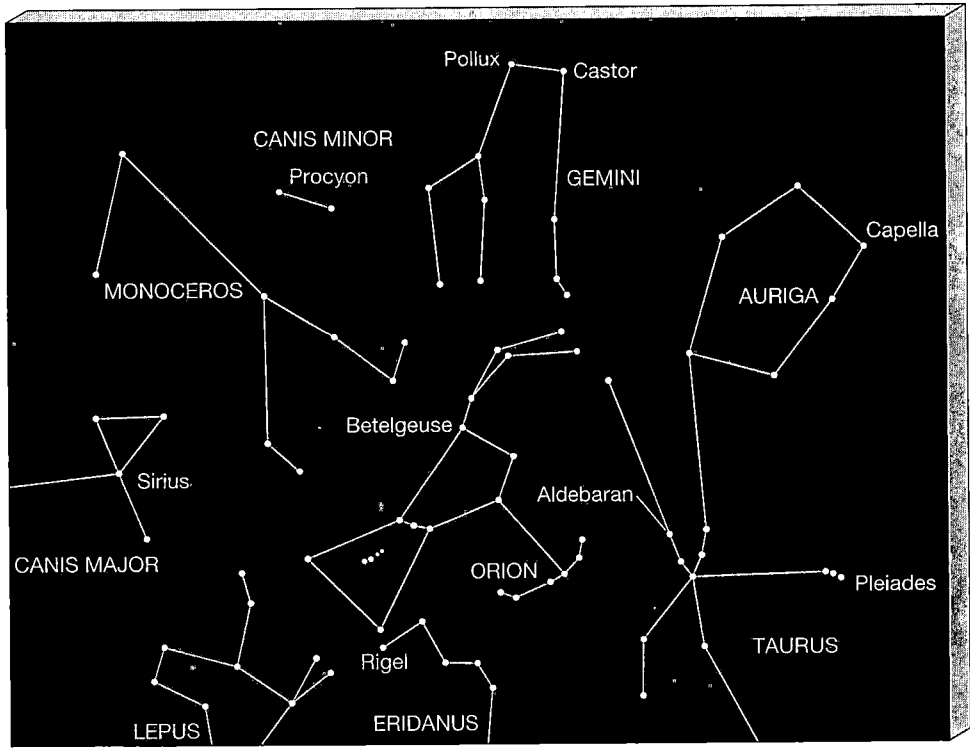

Figure 1.7 Constellations Near Orion The region of the sky conventionally associated with the constellation Orion, together with some neighboring constellations (labeled in all capital letters). Some prominent stars are also labeled in lowercase letters. The 88 constellations span the entire sky, so that every astronomical object lies in precisely one of them.

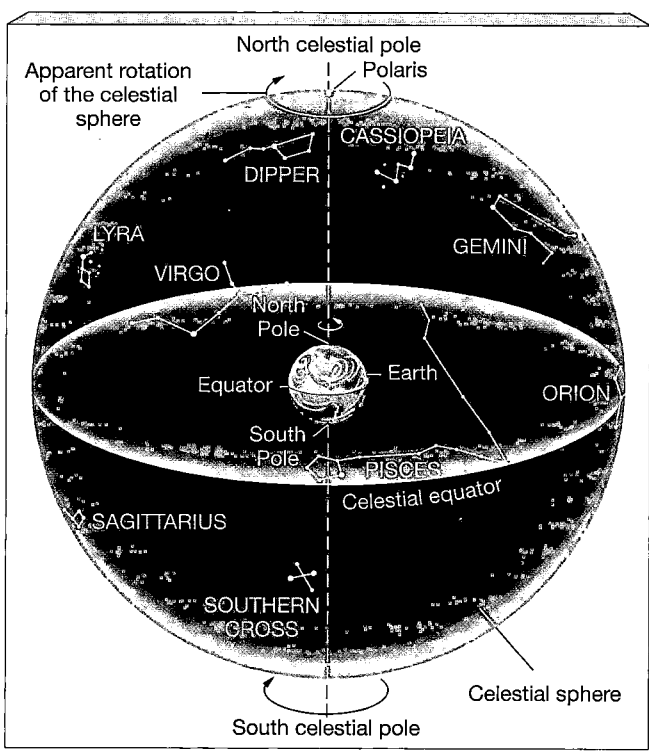

Figure 1.8 Celestial Sphere Planet Earth sits fixed at the hub of the celestial sphere, which contains all the stars. This is one of the simplest possible models of the universe, but it doesn't agree with all the facts that astronomers know about the universe.

information on angular measure. *More Precisely 1-2* discusses in more detail the system of coordinates used to specify stellar positions.

☑️ Concept Check

■ Why do astronomers find it useful to retain the fiction of the celestial sphere when describing the sky? What vital piece of information about stars is lost when we talk about their locations "on" the sky?

1.3 Earth's Orbital Motion

DAY-TO-DAY CHANGES

2 We measure time by the Sun. Because the rhythm of day and night is central to our lives, it is not surprising that the period of time from one noon to the next, the 24-hour solar day, is our basic social time unit. The daily progress of the Sun and the other stars across the sky is known as *diurnal motion*. As we have just seen, it is a consequence of Earth's rotation. But the stars' positions in the sky do *not* repeat themselves exactly from one night to the next. Each night, the whole celestial sphere appears to be shifted a little relative to the horizon, compared with the night before. The easiest way to confirm this difference is by noticing the stars that are visible just after sunset or just before dawn. You will find that they are in slightly different locations from those of the previous night. Because of this shift, a day measured by the stars—called a **sidereal day** after the Latin word *sidus*, meaning "star"—differs in length from a solar day. Evidently, there is more to the apparent motion of the heavens than simple rotation.

The reason for the difference between a solar day and a sidereal day is sketched in Figure 1.10. It is a result of the fact that Earth moves in two ways simultaneously: It rotates on its central axis while at the same time **revolving** around the Sun. Each time Earth rotates once on its axis, it also moves a small distance along its orbit about the Sun. Earth therefore has to rotate through slightly more than 360° (360 degrees—see *More Precisely 1-1*) for the Sun to return to the same apparent location in the sky. Thus, the interval of time between noon one day and noon the next (a solar day) is slightly greater than one true rotation period (one sidereal day). Our planet takes 365 days to orbit the Sun, so the additional angle is 360°/365 = 0.986°. Because Earth, rotating at a rate of 15° per hour, takes about 3.9 minutes to rotate through this angle, the solar day is 3.9 minutes longer than the sidereal day (that is, one sidereal day is roughly 23^h56^m long).

Figure 1.9 Northern Sky A time-lapse photograph of the northern sky. Each trail is the path of a single star across the night sky. The duration of the exposure is about five hours. (How can you tell that this is so?) The center of the concentric circles is near the North Star, Polaris, whose short, bright arc is prominently visible. *(AURA)*

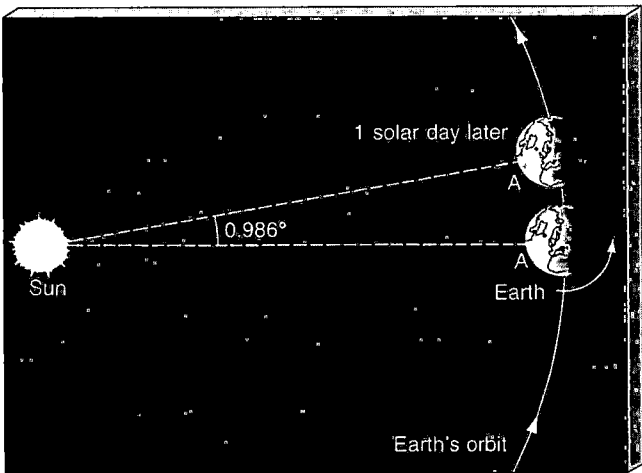

Figure 1.10 Solar and Sidereal Days The difference between a solar and a sidereal day can be easily explained once we understand that Earth revolves around the Sun at the same time as it rotates on its axis. A solar day is the time from one noon to the next. In that time, Earth also moves a little in its solar orbit. Because Earth completes one circuit (360°) around the Sun in one year (365 days), it moves through nearly 1° in one day. Thus, between noon at point A on one day and noon at the same point the next day, Earth actually rotates through about 361°. Consequently, the solar day exceeds the sidereal day (360° rotation) by about four minutes. Note that the diagram is not drawn to scale, so the true 1° angle is in reality much smaller than shown here.

For more on astronomical timekeeping, and for a working definition of the *mean solar day*, the time measured by our clocks, consult *More Precisely 1-3*.

SEASONAL CHANGES

Figure 1.11(a) illustrates the major stars visible from most locations in the United States on clear summer evenings. The brightest stars—Vega, Deneb, and Altair—form a conspicuous triangle high above the constellations Sagittarius and Capricornus, which are low on the southern horizon. In the winter sky, however, these stars are replaced as shown in Figure 1.11(b) by several well-known constellations that include Orion, Leo, and Gemini. In the constellation Canis Major lies Sirius (the Dog Star), the brightest star in the sky. Year after year, the same stars and constellations return, each in its proper season. Every winter evening, Orion is high overhead; every summer, it is gone. (For more detailed maps of the sky at different seasons, consult the star charts at the end of the book.)

These regular seasonal changes occur because of Earth's revolution around the Sun: Earth's darkened hemisphere faces in a slightly different direction in space each evening. The change in direction is only about 1° per night (Figure 1.10)—too small to be easily noticed with the naked eye from one evening to the next, but clearly noticeable over the course of weeks and months, as illustrated in Figure 1.12. After six months, Earth has reached the opposite side of its orbit, and we face an entirely different

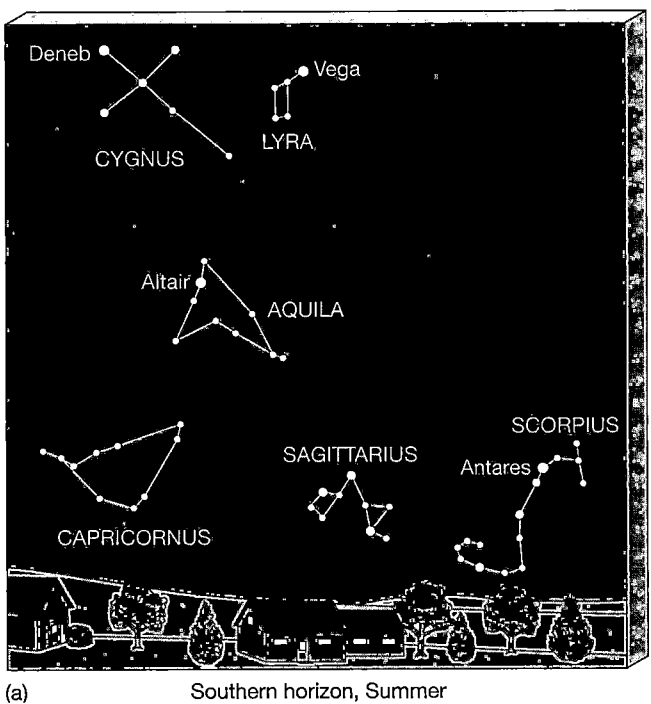

(a)　　　Southern horizon, Summer

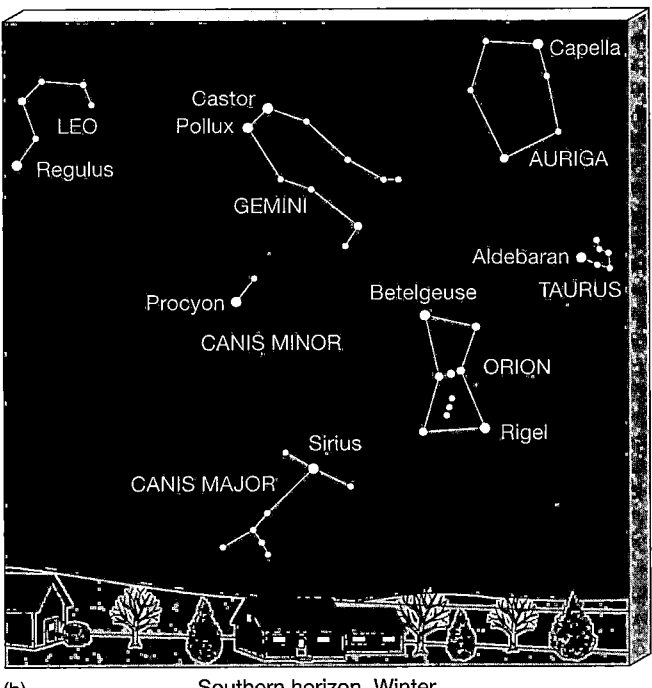

(b)　　　Southern horizon, Winter

Figure 1.11 Typical Night Sky (a) A typical summer sky above the United States. Some prominent stars (labeled in lowercase letters) and constellations (labeled in all capital letters) are shown. (b) A typical winter sky above the United States.

group of stars and constellations at night. Because of this motion, the Sun appears (to an observer on Earth) to move relative to the background stars over the course of a year. This apparent motion of the Sun on the sky traces out a path on the celestial sphere known as the **ecliptic**.

The 12 constellations through which the Sun passes as it moves along the ecliptic—that is, the constellations we would see looking in the direction of the Sun, if they weren't overwhelmed by the Sun's light—had special significance for astrologers of old. These constellations are collectively known as the **zodiac**.

As illustrated in Figure 1.13(a), the ecliptic forms a great circle on the celestial sphere, inclined at an angle of 23.5° to the celestial equator. In reality, as illustrated in Figure 1.13(b), the plane of the ecliptic is *the plane of Earth's orbit around the Sun*. Its tilt is a consequence of the *inclination* of our planet's rotation axis to its orbital plane.

The point on the ecliptic where the Sun is at its northernmost point above the celestial equator is known as the **summer solstice** (from the Latin words *sol*, meaning "sun," and *stare*, "to stand"). As shown in Figure 1.13, it represents the point in Earth's orbit where our planet's North Pole points closest to the Sun. This occurs on or near June 21—the exact date varies slightly from year to year because the actual length of a year is not a whole number of days. As Earth rotates, points north of the equator spend the greatest fraction of their time in sunlight on that date, so the summer solstice corresponds to the longest day of the year in the Northern Hemisphere and the shortest day in the Southern Hemisphere.

Six months later, the Sun is at its southernmost point, or the **winter solstice** (December 21)—the shortest day in the Northern Hemisphere and the longest in the Southern Hemisphere. These two effects—the height of the Sun

MORE PRECISELY 1-1

Angular Measure

Size and scale are often specified by measuring lengths and angles. The concept of length measurement is fairly intuitive to most of us. The concept of *angular measurement* may be less familiar, but it too can become second nature if you remember a few simple facts.

- A full circle contains 360 *degrees* (360°). Thus, the half-circle that stretches from horizon to horizon, passing directly overhead and spanning the portion of the sky visible to one person at any one time, contains 180°.

- Each 1° increment can be further subdivided into fractions of a degree, called *arc minutes*. There are 60 arc minutes (written 60′) in one degree. (The term "arc" is used to distinguish this angular unit from the unit of time.) Both the Sun and the Moon project an angular size of 30 arc minutes on the sky. Your little finger, held at arm's length, does about the same, covering about a 40′ slice of the 180° horizon-to-horizon arc.

- An arc minute can be divided into 60 *arc seconds* (60″). Put another way, an arc minute is 1/60 of a degree, and an arc second is 1/60 × 1/60 = 1/3600 of a degree. An arc second is an extremely small unit of angular measure—it is the angular size of a centimeter-sized object (a dime, say) at a distance of about two kilometers (a little over a mile).

The accompanying figure illustrates this subdivision of the circle into progressively smaller units.

Don't be confused by the units used to measure angles. Arc minutes and arc seconds have nothing to do with the measurement of time, and degrees have nothing to do with temperature. Degrees, arc minutes, and arc seconds are simply ways to measure the size and position of objects in the universe.

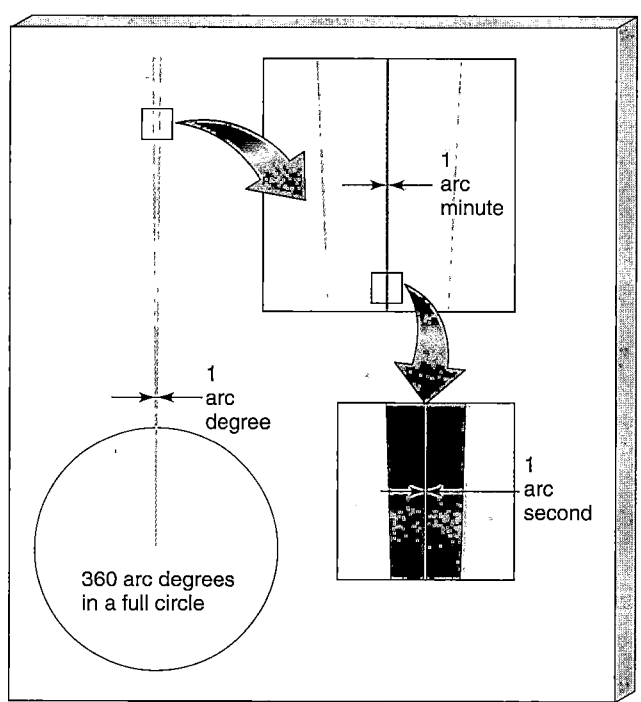

The angular size of an object depends both on its actual size and on its distance from us. For example, the Moon, at its present distance from Earth, has an angular diameter of 0.5°, or 30′. If the Moon were twice as far away, it would appear half as big—15′ across—even though its actual size would be the same. Thus, *angular size by itself is not enough to determine the actual diameter of an object—the distance must also be known*. We return to this topic in more detail in *More Precisely 1-3*.

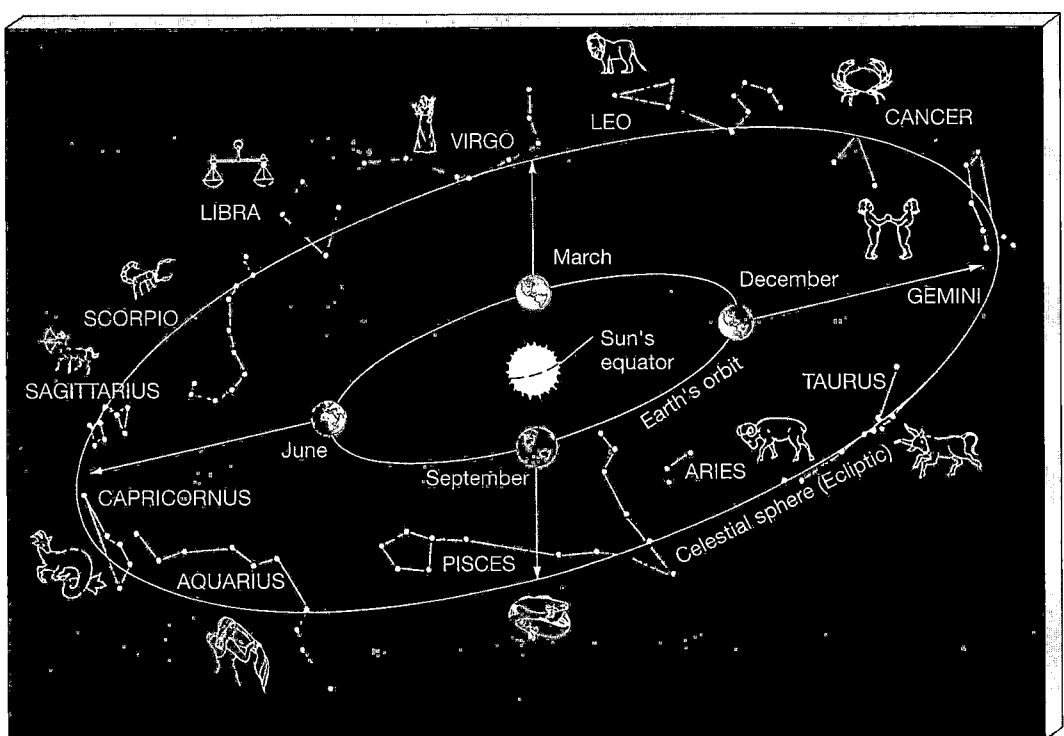

Figure 1.12 The Zodiac The view of the night sky changes as Earth moves in its orbit about the Sun. As drawn here, the night side of Earth faces a different set of constellations at different times of the year. The twelve constellations named here comprise the astrological zodiac.

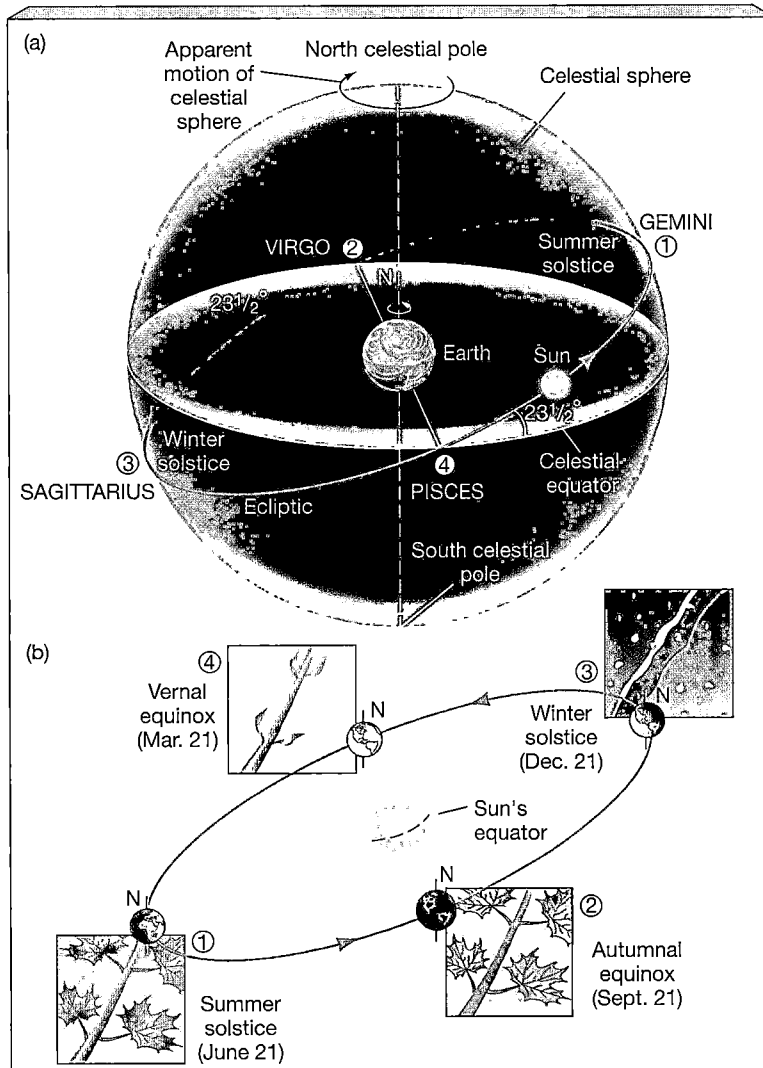

Figure 1.13 Seasons (a) The apparent path of the Sun on the celestial sphere and (b) its actual relation to Earth's rotation and revolution. The seasons result from the changing height of the Sun above the horizon. At the summer solstice (the points marked 1), the Sun is highest in the sky, as seen from the Northern Hemisphere, and the days are longest. In the "celestial sphere" figure (part a), the Sun is at its northernmost point on its path around the ecliptic; in reality (b), the summer solstice corresponds to the point on Earth's orbit where our planet's North Pole points most nearly toward the Sun. The reverse is true at the winter solstice (point 3). At the vernal and autumnal equinoxes, day and night are of equal length. These are the times when, as seen from Earth (a), the Sun crosses the celestial equator. They correspond to the points in Earth's orbit when our planet's axis is perpendicular to the line joining Earth and Sun (b).

above the horizon and the length of the day— combine to account for the **seasons** we experience. In summer in the Northern Hemisphere, the Sun is high in the sky and the days are long, so temperatures are generally much higher than in winter, when the Sun is low and the days are short.

The two points where the ecliptic intersects the celestial equator are known as **equinoxes**. On those dates, day and night are of equal duration. (The word *equinox* derives from the Latin for "equal night.") In the fall (in the Northern Hemisphere), as the Sun crosses from the Northern into the Southern Hemisphere, we have the **autumnal equinox** (on September 21). The **vernal equinox** occurs in northern spring, on or near March 21, as the Sun crosses the celestial equator moving north. Because of its association with the end of winter and the start of a new growing season, the vernal equinox was particularly important to early astronomers and astrologers. It also plays an important role in human timekeeping. The interval of time from one vernal equinox to the next—365.242 mean solar days (see *More Precisely 1-3*)—is known as one **tropical year**.

LONG-TERM CHANGES

3 Earth has many motions—it spins on its axis, it travels around the Sun, and it moves with the Sun through our Galaxy. We have just seen how some of these motions can account for the changing nighttime sky and the changing seasons. In fact, the situation is even more complicated. Like a spinning top that rotates rapidly on its own axis while that axis slowly revolves about the vertical, Earth's axis changes its *direction* over the course of time (although the angle between the axis and a line perpendicular to the plane of the ecliptic always remains close to 23.5°). Illustrated in Figure 1.14, this change is called **precession**. It is caused by torques (twisting forces) on Earth due to the gravitational pulls of the Moon and the Sun, which affect our planet in much the same way as the torque due to Earth's own gravity affects the top. During a complete cycle of precession—about 26,000 years—Earth's axis traces out a cone.

The time required for Earth to complete exactly one orbit around the Sun, relative to the stars, is called a **sidereal year**. One sidereal year is 365.256 mean solar days long—about 20 minutes longer than a tropical year. The reason for this slight difference is Earth's precession. Recall that the vernal equinox occurs when Earth's rotation axis is perpendicular to the line joining Earth and Sun, and the Sun is crossing the celestial equator moving from south to north. In the absence of precession, this would occur exactly once per sidereal orbit, and the tropical and sidereal years would be identical. However, because of the slow precessional shift in the orientation of Earth's rotation axis, the instant when the axis is next perpendicular to the line from Earth to the Sun occurs slightly *sooner* than we would otherwise expect. Consequently, the vernal

equinox drifts slowly around the zodiac over the course of the precession cycle.

The tropical year is the year that our calendars measure. If our timekeeping were tied to the sidereal year, the seasons would slowly march around the calendar as Earth precessed—13,000 years from now, summer in the Northern Hemisphere would be at its height in late February! By using the tropical year, we ensure that July and August will always be (northern) summer months. However, in 13,000 years' time, Orion will be a summer constellation.

☑ Concept Check

■ In astronomical terms, what are "summer" and "winter," and why do we see different constellations during those seasons?

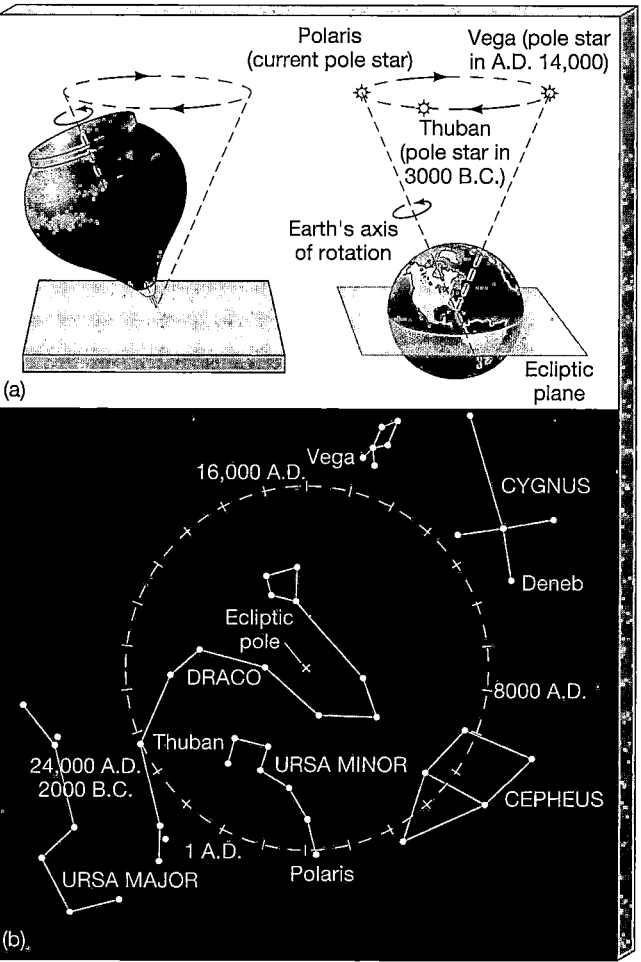

Figure 1.14 Precession (a) Earth's axis currently points nearly toward the star Polaris. About 12,000 years from now— nearly halfway through one cycle of precession—Earth's axis will point toward a star called Vega, which will then be the "North Star." Five thousand years ago, the North Star was a star named Thuban in the constellation Draco. (b) The dashed circle shows the precessional path of the north celestial pole among some prominent Northern stars. Tick marks indicate intervals of 1000 years.

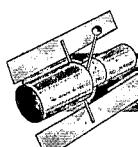

MORE PRECISELY 1-2

Celestial Coordinates

The simplest method of locating stars in the sky is to specify their constellation and then rank the stars in it in order of brightness. The brightest star is denoted by the Greek letter α (alpha), the second brightest by β (beta), and so on (see Figure 1.6). Thus, the two brightest stars in the constellation Orion—Betelgeuse and Rigel—are also known as α Orionis and β Orionis, respectively. (More recently, precise observations show that Rigel is actually brighter than Betelgeuse, but the names are now permanent.) Similarly, Sirius, the brightest star in the sky (see Appendix 3), which lies in the constellation Canis Major (the Great Dog), is denoted α Canis Majoris (or α CMa for short); the (present) Pole Star, in Ursa Minor (the Little Bear), is also known as α Ursae Minoris (or α UMi), and so on. Because there are many more stars in any given constellation than there are letters in the Greek alphabet, this method is of limited utility. However, for naked-eye astronomy, where only bright stars are involved, it is quite satisfactory.

For more precise measurements, astronomers find it helpful to lay down a system of *celestial coordinates* on the sky. If we think of the stars as being attached to the celestial sphere centered on Earth, then the familiar system of latitude and longitude on Earth's surface extends naturally to cover the sky. The celestial analogs of latitude and longitude on Earth's surface are called *declination* and *right ascension*, respectively. The accompanying figure illustrates the meanings of right ascension and declination on the celestial sphere and compares them with longitude and latitude on Earth.

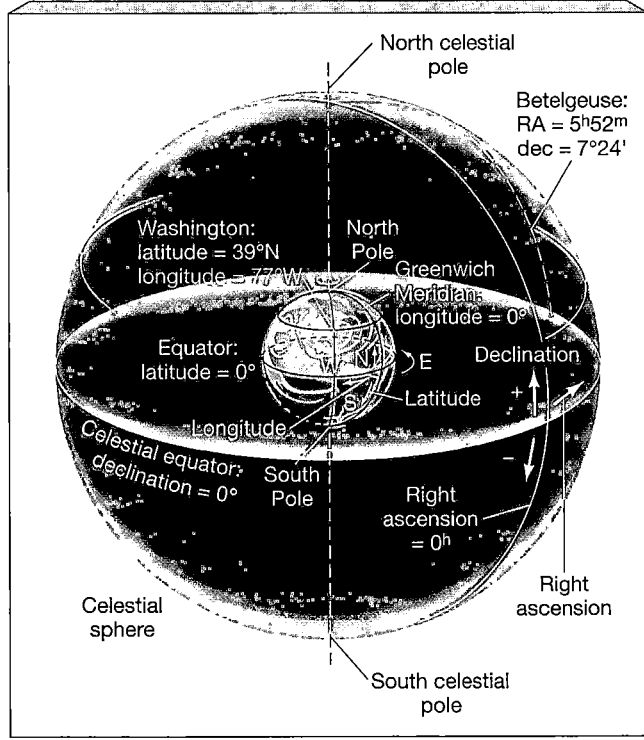

- Declination (dec) is measured in degrees (°) north or south of the celestial equator, just as latitude is measured in degrees north or south of Earth's equator. (See *More Precisely 1-1* for a discussion of angular measure.) Thus, the celestial equator is at a declination of 0°, the north celestial pole is at +90°, and the south celestial pole is at −90° (the minus sign here just means "south of the celestial equator").

- Right ascension (RA) is measured in units called *hours*, *minutes*, and *seconds*, and it increases in the eastward direction. The angular units used to measure right ascension are constructed to parallel the units of time, in order to assist astronomical observation. The two sets of units are connected by the rotation of Earth (or of the celestial sphere). In 24 hours, Earth rotates once on its axis, or through 360°. Thus, in a time period of one hour, Earth rotates through $360°/24 = 15°$, or 1^h. In one minute of time, Earth rotates through an angle of $1^m = 15°/60 = 0.25°$, or 15 arc minutes (15'). In one second of time, Earth rotates through an angle of $1^s = 15'/60 = 15$ arc seconds (15''). As with longitude, the choice of zero right ascension (the celestial equivalent of the Greenwich Meridian) is quite arbitrary—it is conventionally taken to be the position of the Sun in the sky *at the instant of the vernal equinox.*

Right ascension and declination specify locations on the sky in much the same way as longitude and latitude allow us to locate a point on Earth's surface. For example, to find Washington on Earth, look 77° west of the Greenwich Meridian (the line on Earth's surface with a longitude of zero) and 39° north of the equator. Similarly, to locate the star Betelgeuse on the celestial sphere, look $5^h52^m0^s$ east of the vernal equinox (the line on the sky with a right ascension of zero) and 7°24' north of the celestial equator. The star Rigel, also mentioned earlier, lies at $5^h13^m36^s$ (RA), −8°13' (dec). Thus, we have a quantitative alternative to the use of constellations in specifying the positions of stars in the sky. Just as latitude and longitude are tied to Earth, right ascension and declination are fixed on the celestial sphere. Although the stars appear to move across the sky because of Earth's rotation, their celestial coordinates remain *constant* over the course of a night.

Actually, because right ascension is tied to the position of the vernal equinox, the celestial coordinates of any given star slowly drift due to Earth's *precession* (see Section 1.4). Since one cycle of precession takes 26,000 years to complete, this represents a shift of a little over 0.1'' on a night-to-night basis—a small angle, but one that must be taken into account in high-precision astronomical measurements. Rather than deal with slowly changing coordinates for every object in the sky, astronomers conventionally correct their observations to the location of the vernal equinox at some standard epoch (such as, 1 January 1950 or 1 January 2000).

1.4 The Motion of the Moon

2 **3** The Moon is our nearest neighbor in space. Apart from the Sun, it is the brightest object in the sky. Like the Sun, the Moon appears to move relative to the background stars. Unlike the Sun, however, the Moon really does revolve around Earth. It crosses the sky at a rate of about 12° per day, which means it moves an angular distance equal to its own diameter—30 arc minutes—in about an hour.

LUNAR PHASES

The Moon's appearance undergoes a regular cycle of changes, or **phases**, taking a little more than 29 days to complete. (The word *month* is derived from the word *moon*.) Figure 1.15 illustrates the appearance of the Moon at different times in this monthly cycle. Starting from the *new Moon*, which is all but invisible in the sky, the Moon appears to *wax* (or grow) a little each night and is visible as a growing *crescent* (panel 1 of Figure 1.15). One week after new Moon, half of the lunar disk can be seen (panel 2). This phase is known as a *quarter Moon*. During the next week, the Moon continues to wax, passing through the *gibbous* phase (panel 3) until, two weeks after new Moon, the *full Moon* (panel 4) is visible. During the next two weeks, the Moon *wanes* (or shrinks), passing in turn through the gibbous, quarter, and crescent phases (panels 5–7), eventually becoming new again.

The Moon doesn't actually change its size and shape from night to night, of course. Its full circular disk is present at all times. Why then don't we always see a full Moon? The answer to this question lies in the fact that, unlike the Sun and the other stars, the Moon emits no light of its own. Instead, it shines by reflected sunlight. As illustrated in Figure 1.15, half of the Moon's surface is illuminated by the Sun at any instant. However, not all of the Moon's sunlit face can be seen because of the Moon's position with respect to Earth and the Sun. When the Moon is full, we see the entire "daylit" face because the Sun and the Moon are in opposite directions from Earth in the sky. In the case of a new Moon, the Moon and the Sun are in almost the same part of the sky, and the sunlit side of the Moon is oriented away from us. At new Moon, the Sun must be almost behind the Moon, from our perspective.

As it revolves around Earth, the Moon's position in the sky changes with respect to the stars. In one **sidereal month** (27.3 days), the Moon completes one revolution and returns to its starting point on the celestial sphere, having traced out a great circle in the sky. The time required for the Moon to complete a full cycle of phases, one **synodic month**, is a little longer—about 29.5 days. The synodic month is a little longer than the sidereal month for the same reason that a solar day is slightly longer than a sidereal day: Because of Earth's motion around the Sun, the Moon must complete slightly more than one full revolution to return to the same phase in its orbit (Figure 1.16).

ECLIPSES

4 From time to time—but only at new or full Moon—the Sun and the Moon line up precisely as seen from Earth, and we observe the spectacular phenomenon known as an **eclipse**. When the Sun and the Moon are in exactly *opposite* directions, as seen from Earth, Earth's shadow sweeps across the Moon, temporarily blocking the Sun's light and darkening the Moon in a **lunar eclipse**, as illustrated in Figure 1.17. From Earth, we see the curved edge of Earth's shadow begin to cut across the face of the full Moon and slowly eat its way into the lunar disk. Usually, the alignment of the Sun, Earth, and Moon is imperfect, so the shadow never completely covers the Moon. Such an occurrence is known as a **partial lunar eclipse**. Occasionally, however, the entire lunar surface is obscured in a **total lunar eclipse**, such as that shown in the inset of Figure 1.17. Total lunar eclipses last only as long as is needed for the Moon to pass through Earth's shadow—no more than about 100 minutes. During that time, the Moon often acquires an eerie, deep red coloration—the result of a small amount of sunlight reddened by Earth's atmosphere (for the same reason that sunsets appear red—see *More Precisely 7-1*) and refracted (bent) onto the lunar surface, preventing the shadow from being completely black.

When the Moon and the Sun are in exactly the *same* direction, as seen from Earth, an even more awe-inspiring event occurs. The Moon passes directly in front of the Sun, briefly turning day into night in a **solar eclipse.** In a *total solar eclipse*, when the alignment is perfect, planets and some stars become visible in the daytime as the Sun's light is reduced to nearly nothing. We can also see the Sun's ghostly outer atmosphere, or *corona* (Figure 1.18a, p. 18).* In a *partial solar eclipse*, the Moon's path is slightly "off center," and only a portion of the Sun's face is covered. In either case, the sight of the Sun apparently being swallowed up by the black disk of the Moon is disconcerting even today. It must surely have inspired fear in early observers. Small wonder that the ability to predict such events was a highly prized skill.

Unlike a lunar eclipse, which is simultaneously visible from all locations on Earth's night side, a total solar eclipse can be seen from only a small portion of Earth's daytime side. The Moon's shadow on Earth's surface is about 7000 kilometers wide—roughly twice the diameter

---

*Actually, although a total solar eclipse is undeniably a spectacular occurrence, the visibility of the corona is probably the most important astronomical aspect of such an event today. It enables us to study this otherwise hard-to-see part of our Sun (see Chapter 16).

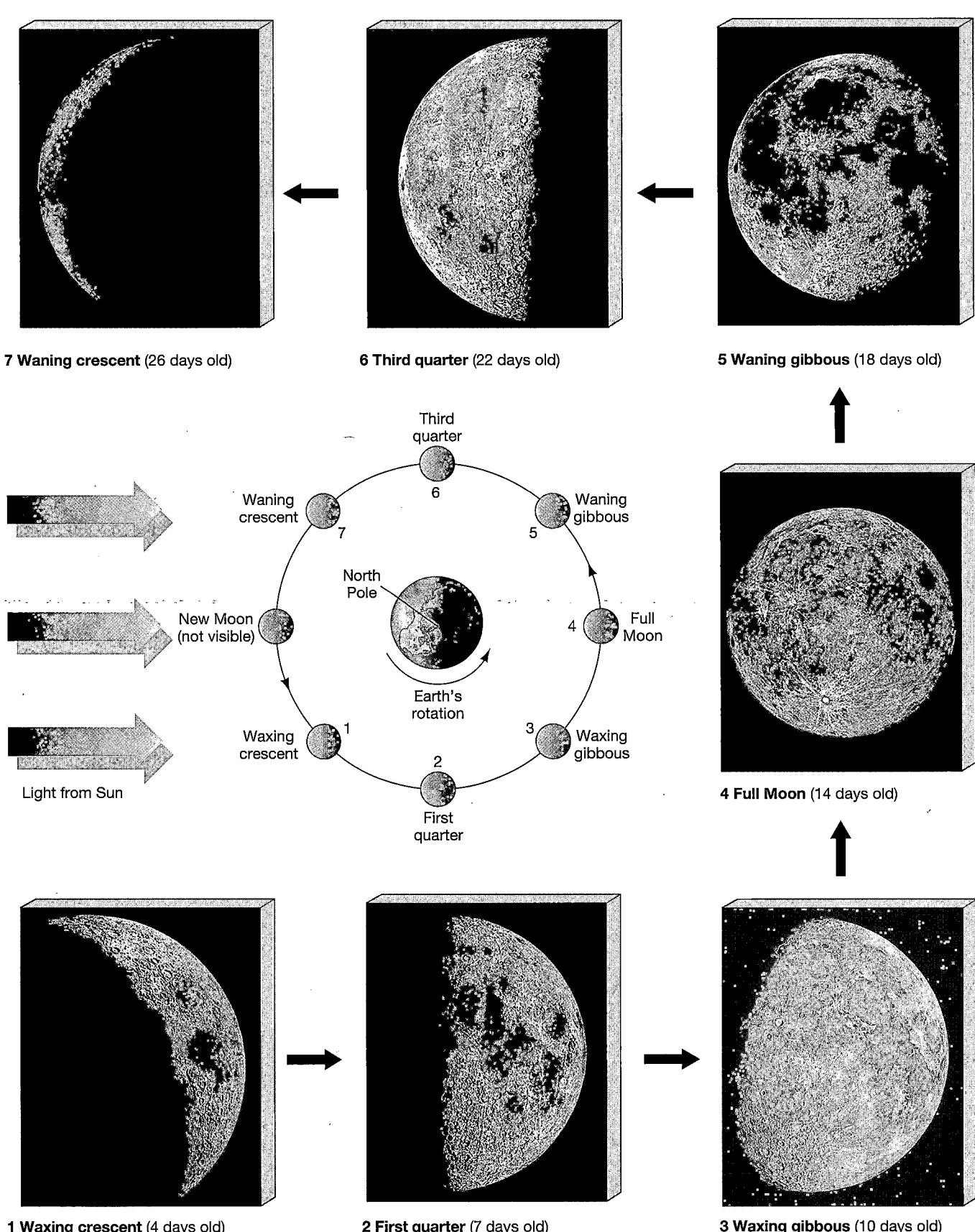

7 Waning crescent (26 days old)

6 Third quarter (22 days old)

5 Waning gibbous (18 days old)

Third
quarter

Waning
crescent

Waning
gibbous

North
Pole

New Moon
(not visible)

Full
Moon

Earth's
rotation

Waxing
crescent

Waxing
gibbous

Light from Sun

First
quarter

4 Full Moon (14 days old)

1 Waxing crescent (4 days old)

2 First quarter (7 days old)

3 Waxing gibbous (10 days old)

Figure 1.15 Lunar Phases Because the Moon orbits Earth, the visible fraction of the sunlit face differs from night to night. The complete cycle of lunar phases takes 29.5 days to complete. *(UC/Lick Observatory)*

R   I   V   U   X   G

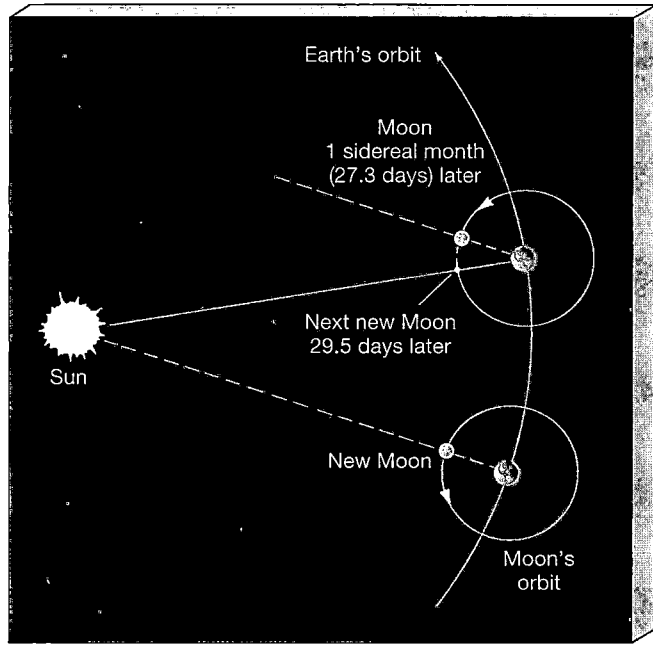

Figure 1.16 Sidereal Month The difference between a synodic and a sidereal month stems from the motion of Earth relative to the Sun. Because Earth orbits the Sun in 365 days, in the 29.5 days from one new Moon to the next (one synodic month), Earth moves through an angle of approximately 29°. Thus the Moon must revolve more than 360° between new Moons. The sidereal month, which is the time taken for the Moon to revolve through exactly 360°, relative to the stars, is about two days shorter.

of the Moon (Figure 1.19). Outside of that shadow, no eclipse is seen. However, within the central region of the shadow, called the **umbra**, the eclipse is total. Within the shadow but outside the umbra, in the **penumbra**, the eclipse is partial, with less and less of the Sun obscured the farther one travels from the shadow's center. The connections between the umbra, the penumbra, and the relative locations of Earth, Sun, and Moon are illustrated in Figure 1.20. One of the reasons that total solar eclipses are rare is that although the penumbra is some 7000 kilometers across, the umbra is always very small. Even under the most favorable circumstances, its diameter never exceeds 270 kilometers. Because the shadow sweeps across Earth's surface at over 1700 kilometers per hour, the duration of a total eclipse at any given point can never exceed 7.5 minutes.

The Moon's orbit around Earth is not exactly circular. Thus, the Moon may be far enough from Earth at the moment of an eclipse that its disk fails to cover the disk of the Sun completely, even though their centers coincide. In that case, there is no region of totality—the umbra never reaches Earth at all, and a thin ring of sunlight can still be seen surrounding the Moon. Such an occurrence, called an **annular eclipse**, is depicted in Figures 1.18(b) and 1.20. Roughly half of all solar eclipses are annular.

Why isn't there a solar eclipse at every new Moon and a lunar eclipse at every full Moon? The answer is that the Moon's orbit is slightly inclined to the ecliptic (at an angle of 5.2°), so the chance that a new (or full) Moon will occur just as the Moon happens to cross the ecliptic plane (so Earth, Moon, and Sun are perfectly aligned) is quite low. Figure 1.21 on p. 20 illustrates some possible configurations of the three bodies. If the Moon happens to lie above or below the plane of the ecliptic when new (or full), a

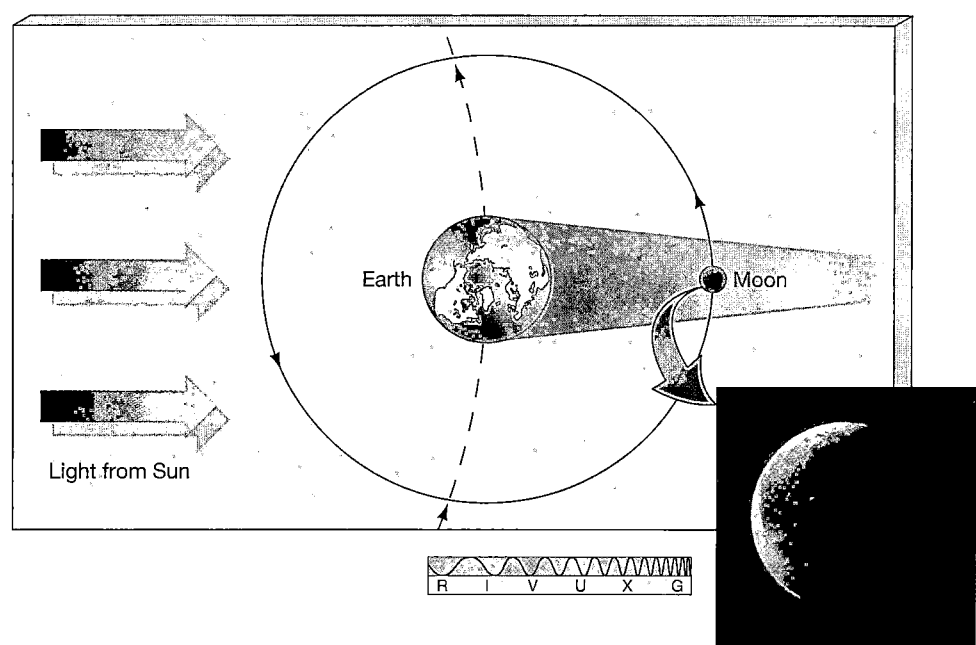

Figure 1.17 Lunar Eclipse A lunar eclipse occurs when the Moon passes through Earth's shadow. At these times we see a darkened, copper-colored Moon, as shown in the inset photograph. The red coloration is caused by sunlight being reddened and deflected by Earth's atmosphere onto the Moon's surface. An observer on the Moon would see Earth surrounded by a bright but narrow ring of orange sunlight. Note that this figure is not drawn to scale, and only Earth's umbra (see text and Figure 1.19) is shown. *(Inset: G. Schneider)*

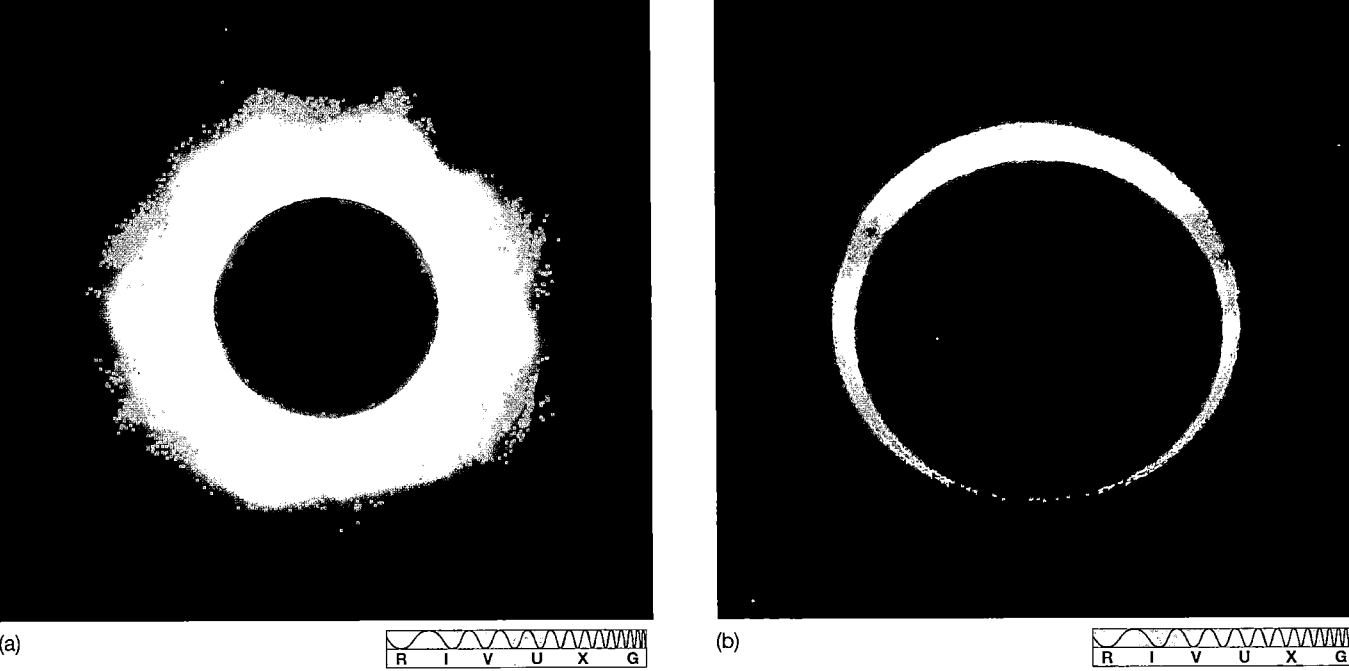

(a)

(b)

R I V U X G

R I V U X G

Figure 1.18 Solar Eclipse (a) During a total solar eclipse the Sun's corona becomes visible as an irregularly shaped halo surrounding the blotted-out disk of the Sun. This was the August 1999 eclipse, as seen from the banks of the Danube River near Sofia, Bulgaria.
(b) During an annular eclipse, the Moon fails to completely hide the Sun, so a thin ring of light remains. No corona is seen in this case because even the small amount of the Sun still visible completely overwhelms the corona's faint glow. This was the December 1973 eclipse, as seen from Algiers. (The gray fuzzy areas at top left and right are clouds in Earth's atmosphere.)
(Bencho Angelov; G. Schneider)

solar (or lunar) eclipse cannot occur. Such a configuration is termed *unfavorable* for producing an eclipse. In a *favorable* configuration, the Moon is new or full just as it crosses the ecliptic plane, and eclipses are seen. Unfavorable configurations are much more common than favorable ones, so eclipses are relatively rare events.

As indicated on Figure 1.21(b), the two points on the Moon's orbit where it crosses the ecliptic plane are known as the *nodes* of the orbit. The line joining them, which is also

the line of intersection of Earth's and the Moon's orbital planes, is known as the *line of nodes*. When the line of nodes is not directed toward the Sun conditions are unfavorable for eclipses. However, when the line of nodes briefly lies along Earth–Sun line, eclipses are possible. These two periods, known as *eclipse seasons*, are the only times at which an eclipse can occur. Notice that there is no guarantee that an eclipse *will* occur. For a solar eclipse, we must have a new Moon during an eclipse season. Similarly, a lunar eclipse

Figure 1.19 Lunar Shadow Photograph taken by an Earth-orbiting weather satellite of the Moon's shadow projected onto Earth's surface (off the coast of Baja California) during the total solar eclipse of July 11, 1991. *(Inset: NOAA)*

can occur only at full Moon during an eclipse season. Figure 1.22 shows the location and duration of all total and annular eclipses of the Sun between 2000 and 2020. Because we know the orbits of Earth and the Moon to great accuracy, we can predict eclipses far into the future.

The solar eclipses that we do see highlight a remarkable cosmic coincidence. Although the Sun is many times farther away from Earth than is the Moon, it is also much larger. In fact, the ratio of distances is almost exactly the same as the ratio of sizes, so the Sun and the Moon both

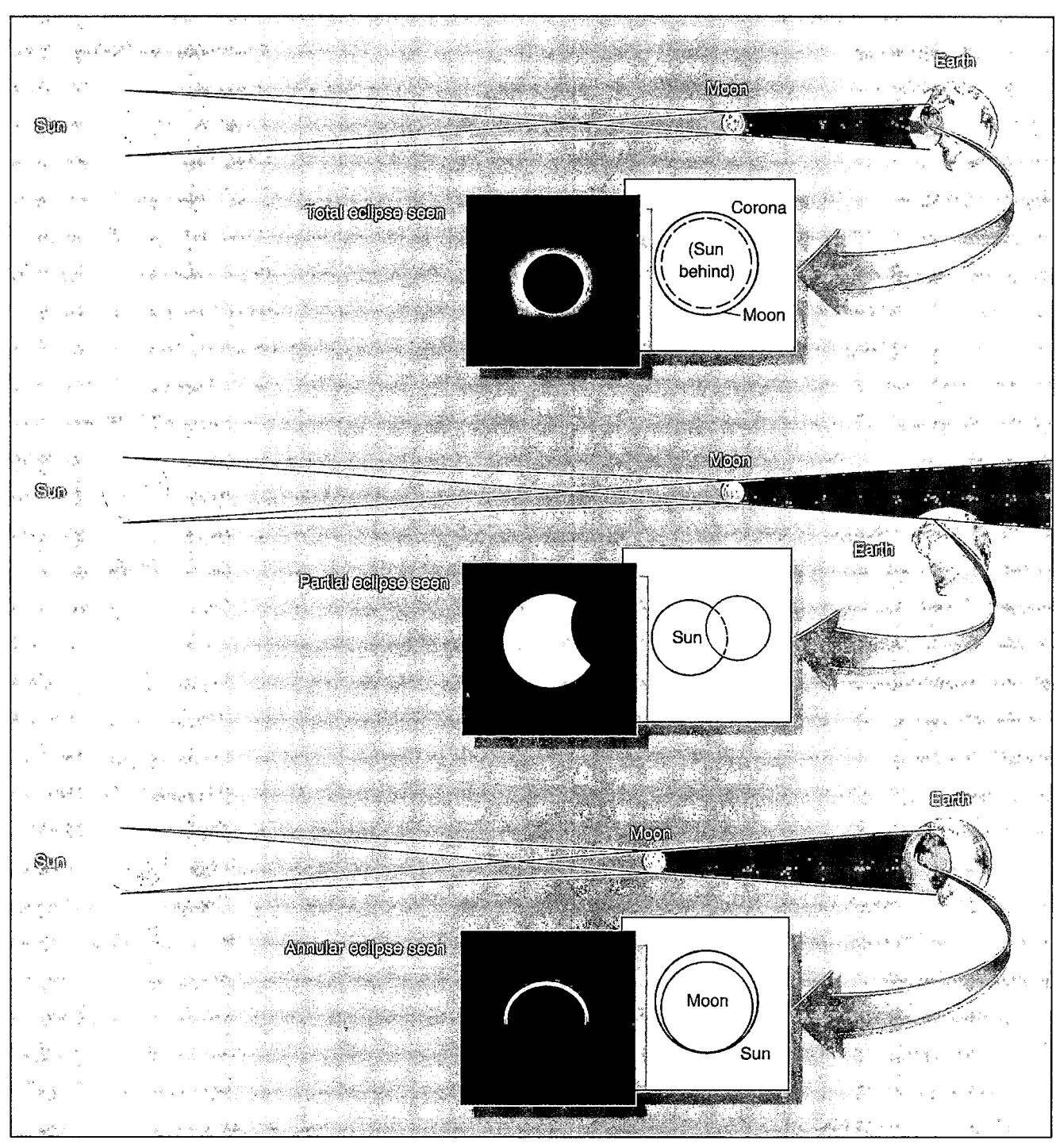

Figure 1.20 Solar Eclipse Types The Moon's shadow on Earth during a solar eclipse consists of the umbra, where the eclipse is total, and the penumbra, where the Sun is only partially obscured. If the Moon is too far from Earth at the moment of the eclipse, there is no region of totality; instead, an annular eclipse is seen. *(Inset: G. Schneider)*

have roughly the *same* angular diameter—about half a degree seen from Earth. Thus, the Moon covers the face of the Sun almost exactly. If the Moon were larger, we would never see annular eclipses, and total eclipses would be much more common. If the Moon were a little smaller, we would see only annular eclipses.

The gravitational tug of the Sun causes the Moon's orbital orientation, and hence the direction of the line of

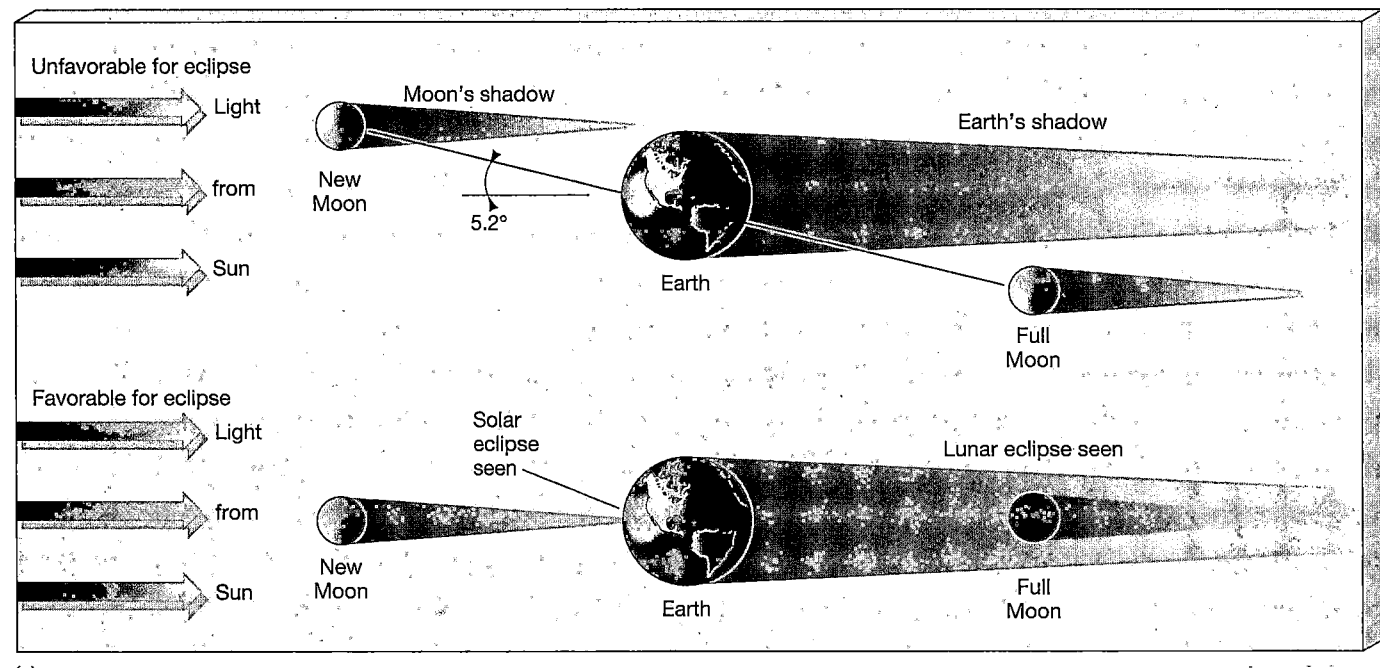

(a)

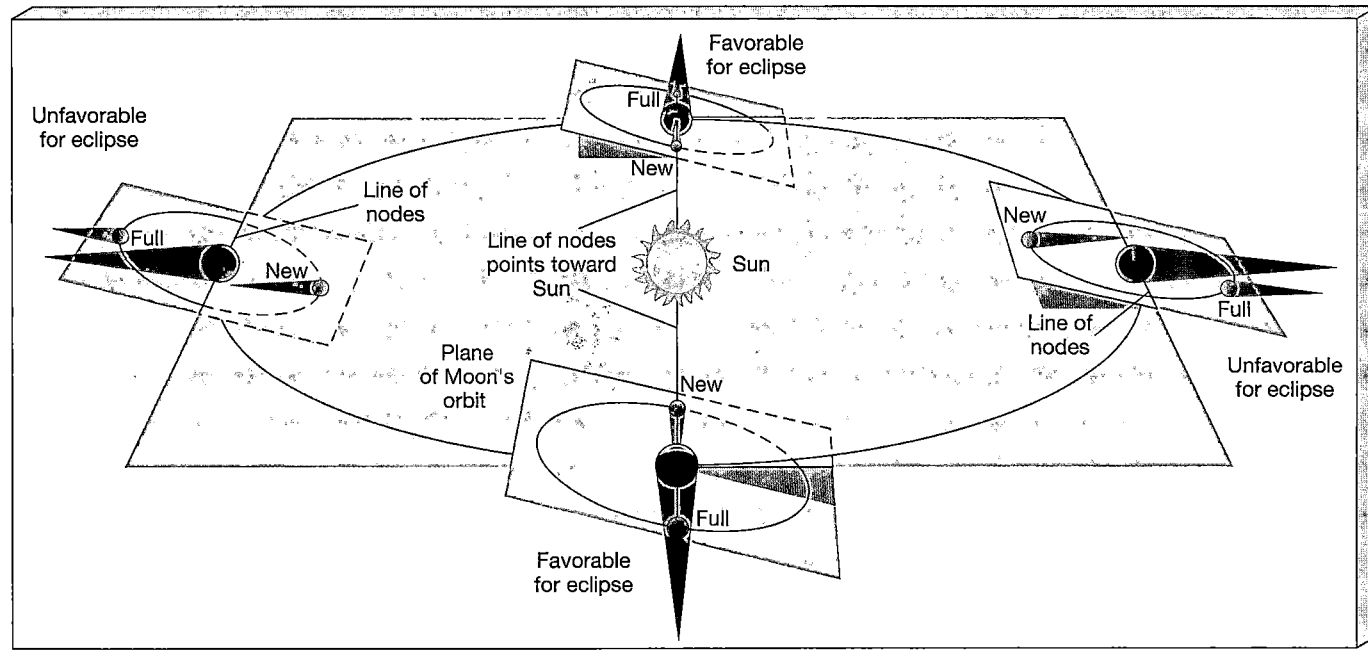

(b)

Figure 1.21 Eclipse Geometry (a) An eclipse occurs when Earth, Moon, and Sun are precisely aligned. If the Moon's orbital plane lies in exactly the plane of the ecliptic, this alignment would occur once a month. However, the Moon's orbit is inclined at about 5° to the ecliptic, so not all configurations are favorable for producing an eclipse. (b) For an eclipse to occur, the line of intersection of the two planes must lie along the Earth–Sun line. Thus, eclipses can occur only at specific times of the year. Only the umbra of each shadow is shown, for clarity (see Figure 1.19).

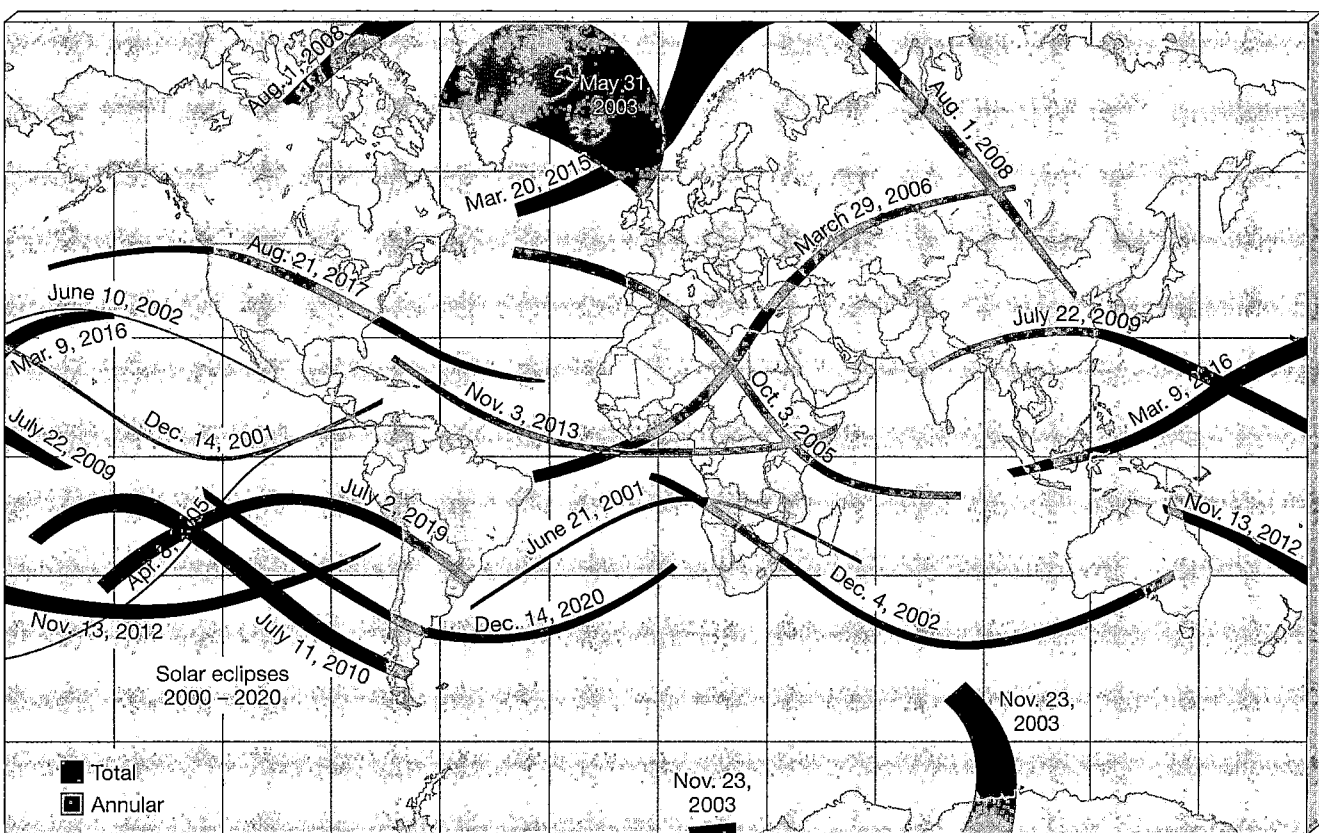

Figure 1.22 Eclipse Tracks Regions of Earth that will see total or annular solar eclipses between the years 2000 and 2020. Each track represents the path of the Moon's umbra across Earth's surface during an eclipse. The width of the track depends upon latitude on Earth and upon the distance from Earth to the Moon during the eclipse. High-latitude tracks are broader because sunlight strikes Earth's surface at an oblique angle near the poles (and also because of the projection of the map). The closer the Moon is to Earth during a total eclipse, the wider the umbra (see Figure 1.19).

nodes, to change slowly with time. As a result, the time interval between one orbital configuration with the line of nodes pointing at the Sun and the next (with the Moon crossing the ecliptic in the same sense in each case) is not exactly 1 year, but instead is 346.6 days—one *eclipse year*. Thus the eclipse seasons gradually progress backward through the calendar, occurring about 19 days earlier each year. This phenomenon is known as the *regression of the line of nodes*.

In 1991 the eclipse seasons were in January and July; on July 11 a total eclipse actually occurred, visible in Hawaii, Mexico, and parts of Central and South America. Three years later, in 1994, the eclipse seasons were in May and October; on May 10 an annular eclipse was visible across much of the continental United States. By 1999, the eclipse seasons had drifted into February and August, and on August 11 much of Europe and southern Asia was treated to the last total eclipse of the millennium.

The combination of the eclipse year and the Moon's synodic period leads to an interesting long-term cycle in solar (and lunar) eclipses. A simple calculation shows that 19 eclipse years is almost exactly 223 lunar months. Thus, every 6585 solar days (actually 18 years, 11.3 days) the "same" eclipse recurs. This repetition is called the *Saros cycle*. Well known to ancient astronomers, it undoubtedly was the key to their "mystical" ability to predict eclipses!

The regression of the line of nodes of the Moon's orbit causes Earth's rotation axis to wobble slightly, changing the angle between Earth's axis and the ecliptic by plus or minus nine arc seconds every 18.6 years. This additional motion, which is superimposed on Earth's precession, is known as *nutation*.

✓ Concept Check

■ What types of solar eclipses would you expect to see if Earth's distance from the Sun were to double? What if the distance became half its present value?

MORE PRECISELY 1-3

Astronomical Timekeeping

The motions of Earth and the Moon define the basic time units—day, month, and year—by which we measure our lives. Today the SI unit of time, the second, is defined by ultra-high-precision atomic clocks maintained in the National Bureau of Standards in Washington, D.C., and in other sites around the world. However, the time they measure is still based on the time determined by astronomical events. Let's take a moment to consider in a little more detail how days, months, and years are defined and measured.

At any location on Earth, a *solar day* may be simply defined as the time between one noon and the next. Here, "noon" means the instant when the Sun crosses the *meridian*—an imaginary line on the celestial sphere through the north and south celestial poles, passing directly overhead at the given location. This is the time that a sundial would measure. Unfortunately, this most direct measure of time has two serious drawbacks—the length of the solar day varies throughout the year, and the definition of noon varies from place to place on Earth.

Recall that the solar day is the result of a "competition" between Earth's rotation and its revolution around the Sun—Earth's revolution means that our planet must rotate through a little more than 360° between one noon and the next (see Figure 1.10). However, while Earth's rotation rate is virtually constant, the rate of revolution—or

more specifically, the rate at which the Sun traverses the celestial sphere as it moves along the ecliptic—is not, for two reasons (illustrated in the exaggerated diagrams below).

First, Earth's orbit is not exactly circular (see Chapter 2), and our orbital speed is not constant—Earth moves more rapidly than average when closer to the Sun, more slowly when farther away—so the speed at which the Sun moves along the ecliptic varies with time. Second, because the ecliptic is inclined to the celestial equator, the *eastward* component of the Sun's motion on the celestial sphere depends on the time of year (and note that this would be the case even if Earth's orbit were circular and the first point did not apply). At the equinoxes, the Sun's path is inclined to the equator, and only the Sun's eastward progress across the sky contributes to part of the motion. At the solstices, however, the motion is entirely eastward.

The combination of these effects means that the solar day varies by roughly half a minute over the course of the year—not a large variation, but unacceptable for astronomical and many other purposes. The solution is to define a *mean solar day*, originally cast in terms of a fictitious *mean Sun* that moves around the celestial equator at constant speed, but which in effect is just the average solar day over an entire year. This is the day our clocks (atomic

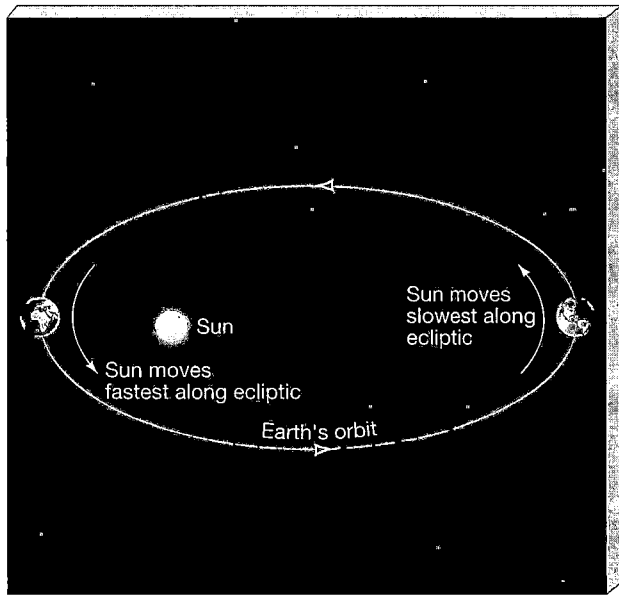

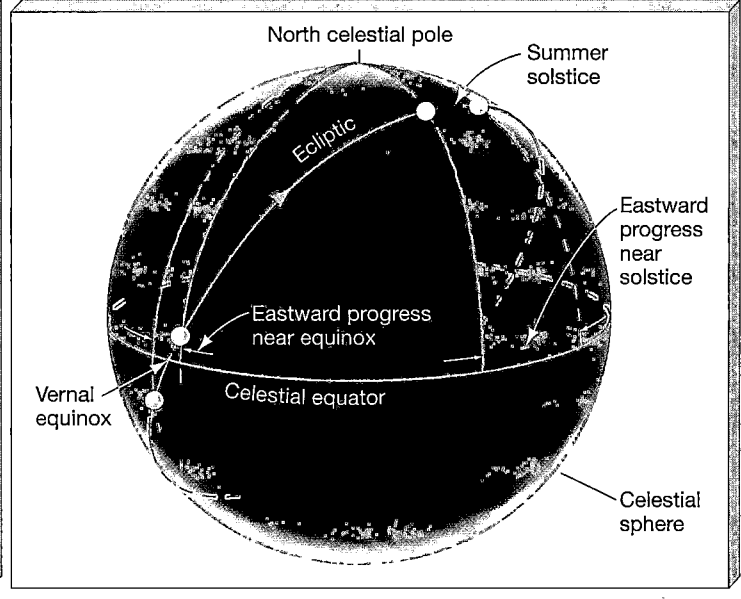

or otherwise) measure, and it is by definition constant. One second is $1/24 \times 1/60 \times 1/60 = 1/86{,}400$ mean solar days.

Now all of our clocks tick at a constant rate, but we are not quite out of the woods in our search for a standard of time. The above definition is still *local*—observers at different longitudes see noon at different times, so even though their clocks keep pace, they all tell different times. In 1883, driven by the need for uniform and consistent times in long-distance travel and communications (the railroads and the telegraph), the continental U.S. was divided into four standard *time zones*. Within each zone, everyone adopted the mean solar time corresponding to a specific meridian within the zone. Since 1884, *standard time* has been used around the world. The global standard time zones are shown below.

By convention, the reference meridian is taken to be 0° longitude (the Greenwich meridian). In the United States, Eastern, Central, Mountain, and Pacific time zones keep the mean solar time of longitudes 75° W, 90° W, 105° W, and 120° W, respectively. Hawaii and Alaska keep the time of longitude 150° W. In some circumstances, it is even more convenient to adopt a single time zone: *universal time* (formerly known as *Greenwich mean time*) is simply the mean solar time at the Greenwich meridian.

Having defined the day, let's now turn our attention to how those days fit into the year—in other words, to the construction of a *calendar*. The "year" in question is the *tropical year*, tied to the changing seasons. It is convention-

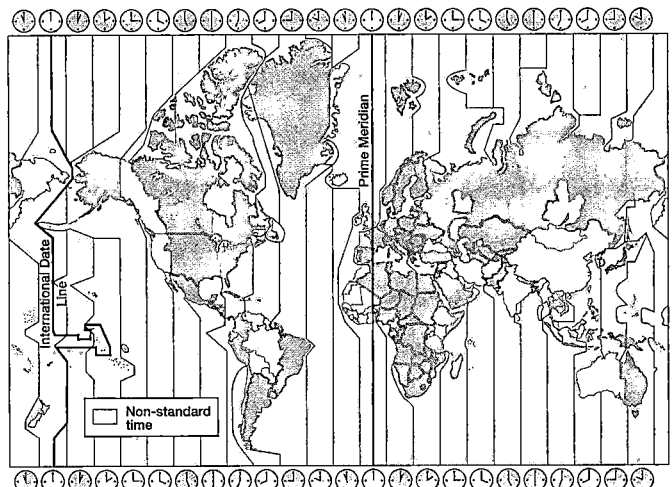

Non-standard time

ally divided into *months*, which were originally defined by the phases of the Moon. The basic problem with all this is that a synodic (lunar) month does not contain a whole number of mean solar days, and the tropical year does not contain a whole number of months or days. Many ancient calendars were lunar, and the variable number of days in modern months may be traced back to attempts to approximate the 29.5-day synodic month by alternating 29- and 30-day periods. The modern (Western) calendar retains months as convenient subdivisions, but they have no particular lunar significance.

One tropical year is 365.2422 mean solar days long, and hence cannot be represented by a whole number of calendar days. The solution is to vary the number of days in the year, much as early calendars varied the number of days in the month, to ensure the correct *average* length of the year. In 46 B.C. the Roman emperor Julius Caesar decreed that the calendar would include an extra day every fourth year—a *leap year*—ensuring that the average year would be 365.25 days long. This *Julian calendar* was a great improvement over earlier lunar calendars, which had 354 ($= 12 \times 29.5$) days in the year, necessitating the inclusion of an extra month every three years!

Still, the Julian year was not exactly equal to the tropical year and, over time, the calendar drifted relative to the seasons. By A.D. 1582, the difference amounted to 10 days, and Pope Gregory XIII instituted another reform, first skipping the extra 10 days (resetting the vernal equinox back to March 21), then changing the rule for leap years to omit the extra day on years that were multiples of 100, but retain it in multiples of 400. The effect of this is that the average year became 365.2425 mean solar days long—good to one day in 3300 years. The idea of "losing" ten days at the behest of the Pope was unacceptable to many countries, with the result that the *Gregorian calendar* was not fully accepted for several centuries. Britain and the American colonies finally adopted it in 1752. Russia abandoned the Julian calendar only in 1917, after the Bolshevik revolution, at which time they had to skip 13 days to come into agreement with the rest of the world.

The most recent modern correction to the Gregorian calendar has been to declare that the years 4000, 8000, etc., will *not* be leap years, improving its accuracy to one day in 20,000 years—good enough for most of us to make it to work on time!

1.5 The Measurement of Distance

5 We have seen a little of how astronomers track and record the positions of the stars on the sky. But knowing the direction to an object is only part of the information needed to locate it in space. Before we can make a systematic study of the heavens, we must find a way of measuring *distances*, too. One distance-measurement method, called **triangulation**, is based on the principles of Euclidean geometry and finds widespread application today in both terrestrial and astronomical settings. Surveyors use these age-old geometric ideas to measure indirectly the distance to faraway objects. It forms the foundation of the family of distance-measurement techniques making up the **cosmic distance scale**.

Imagine trying to measure the distance to a tree on the other side of a river. The most direct method is to lay a tape across the river, but that's not the simplest way. A smart surveyor would make the measurement by visualizing an *imaginary* triangle (hence *triangulation*), sighting the tree on the far side of the river from two positions on the near side, as illustrated in Figure 1.23. The simplest possible triangle is a right triangle, in which one of the angles is exactly 90°, so it is usually convenient to set up one observation position directly opposite the object, as at point A. The surveyor then moves to another observation position at point B, noting the distance covered between points A and B. This distance is called the **baseline** of the imaginary triangle. Finally, the surveyor, standing at point B, sights toward the tree and notes the angle at point B between this sightline and the baseline. Knowing the value of one side (AB) and two angles (the right angle, at point A and the angle at point B) of the right triangle, the surveyor geometrically constructs the remaining sides and angles and establishes the distance from A to the tree.

To use triangulation to measure distances, a surveyor must be familiar with *trigonometry*, the mathematics of geometrical angles and distances. However, even if we knew no trigonometry at all, we could still solve the problem by graphical means, as shown in Figure 1.24. Suppose that we pace off the baseline AB, measuring it to be 450 meters, and measure the angle between the baseline and the line from B to the tree to be 52°, as illustrated in the figure. We can transfer the problem to paper by letting one box on our graph represent 25 meters on the ground. Drawing the line AB on paper, completing the other two sides of the triangle, at angles of 90° (at A) and 52° (at B), we measure the distance on paper from A to the tree to be 23 boxes—that is, 575 meters. We have solved the real problem by *modeling* it on paper. The point to remember here is this: Nothing more complex than basic geometry is needed to infer the distance, the size, and even the shape of an object too far away or inaccessible for direct measurement.

Obviously, for a fixed baseline, the triangle becomes longer and narrower as the tree's distance from A increases. Narrow triangles cause problems because it becomes hard to measure the angles at A and B with sufficient accuracy. The measurements can be made easier by "fattening"

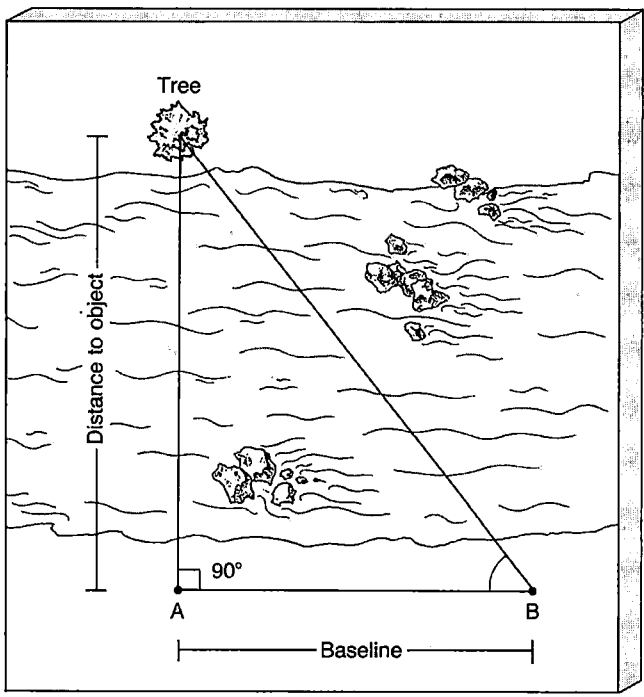

Figure 1.23 Triangulation Surveyors often use simple geometry and trigonometry to estimate the distance to a faraway object. By measuring the angles at A and B and the length of the baseline, the distance can be calculated without the need for direct measurement.

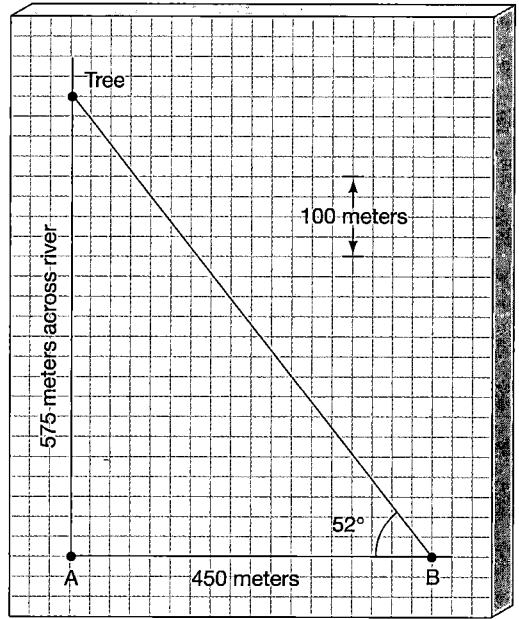

Figure 1.24 Geometric Scaling We don't even need trigonometry to estimate distances indirectly. Scaled estimates, like this one on a piece of paper, often suffice.

the triangle—that is, by lengthening the baseline—but there are limits on how long a baseline we can choose in astronomy. For example, consider an imaginary triangle extending from Earth to a nearby object in space, perhaps a neighboring planet. The triangle is now extremely long and narrow, even for a relatively nearby object (by cosmic standards). Figure 1.25(a) illustrates a case in which the longest baseline possible on Earth—Earth's diameter, measured from point A to point B—is used. In principle, two observers could sight the planet from opposite sides of Earth, measuring the triangle's angles at A and B. However, in practice it is easier to measure the third angle of the imaginary triangle. Here's how.

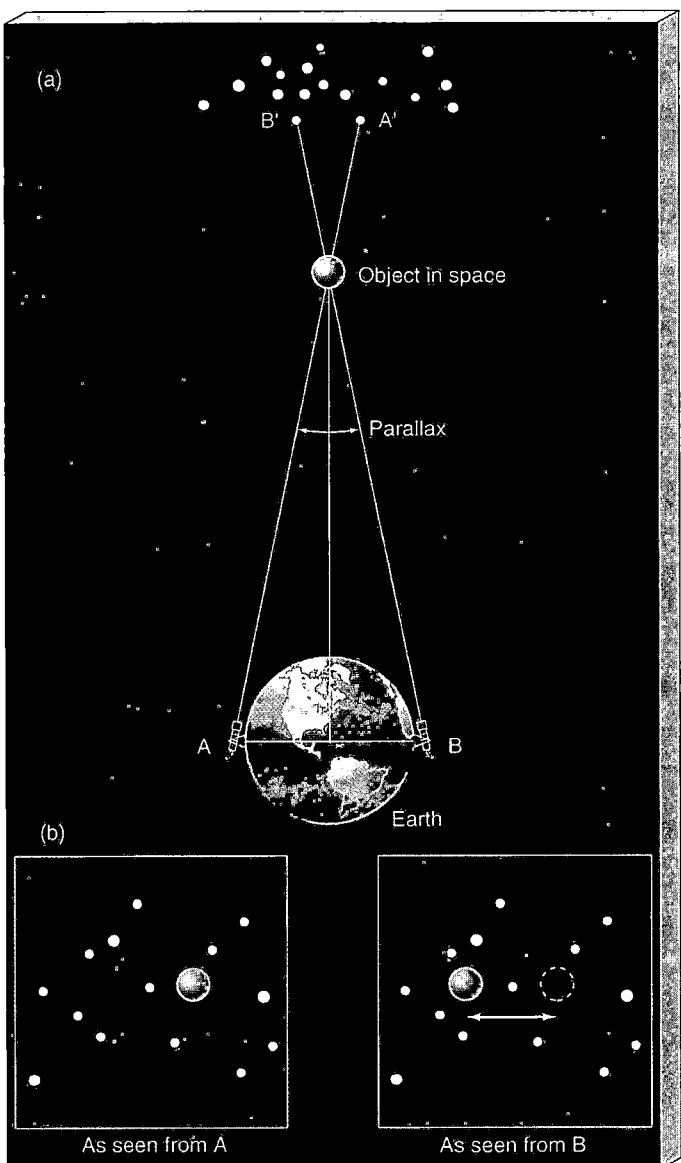

Figure 1.25 Parallax (a) This imaginary triangle extends from Earth to a nearby object in space (such as a planet). The group of stars at the top represents a background field of very distant stars. (b) Hypothetical photographs of the same star field showing the nearby object's apparent displacement, or shift, relative to the distant, undisplaced stars.

The observers sight toward the planet, taking note of its position *relative to some distant stars* seen on the plane of the sky. The observer at point A sees the planet at apparent location A′ relative to those stars, as indicated in Figure 1.25(a). The observer at B sees the planet at point B′. If each observer takes a photograph of the appropriate region of the sky, the planet will appear at slightly different places in the two images. The planet's photographic image is slightly displaced, or shifted, relative to the field of distant background stars, as shown in Figure 1.25(b). The background stars themselves appear undisplaced because of their much greater distance from the observer. This apparent displacement of a foreground object relative to the background as the observer's location changes is known as **parallax**. The size of the shift in Figure 1.25(b), measured as an angle on the celestial sphere, is the third, small angle in Figure 1.25(a).

In astronomical contexts, the parallax is usually very small. For example, the parallax of a point on the Moon, viewed using a baseline equal to Earth's diameter, is about 2°; the parallax of the planet Venus at closest approach (45 million km), is just 1′.

The closer an object is to the observer, the larger the parallax. Figure 1.26 illustrates how you can see this for

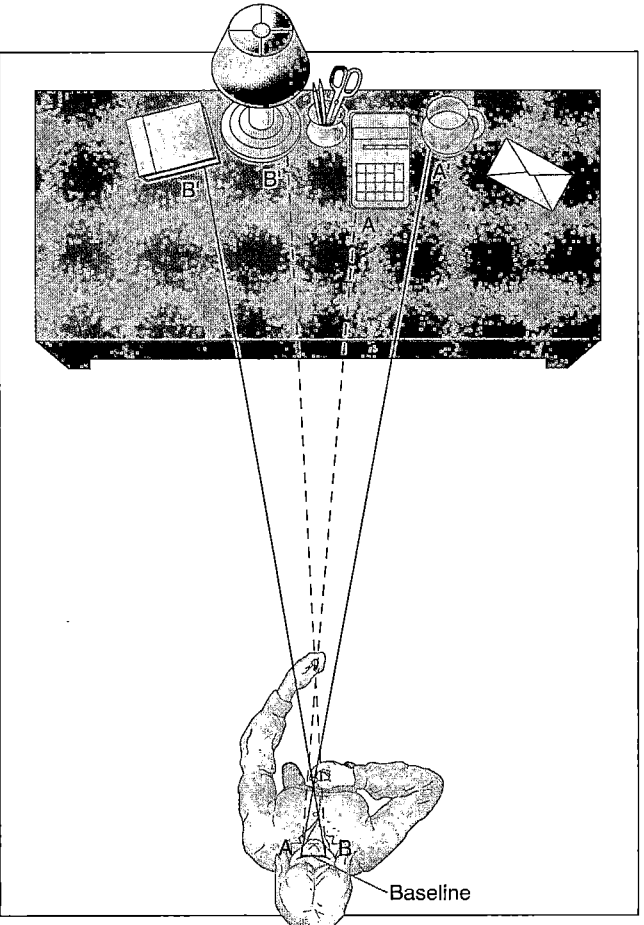

Figure 1.26 Parallax Geometry Parallax is inversely proportional to an object's distance. An object near your nose has a much larger parallax than an object held at arm's length.

yourself. Hold a pencil vertically in front of your nose and concentrate on some far-off object—a distant wall, perhaps. Close one eye, then open it while closing the other. You should see a large shift of the apparent position of the pencil projected onto the distant wall—a large parallax. In this example, one eye corresponds to point A, the other eye to point B, the distance between your eyeballs to the baseline, the pencil to the planet, and the distant wall to a remote field of stars. Now hold the pencil at arm's length,

corresponding to a more distant object (but still not as far away as the even more distant stars). The apparent shift of the pencil will be less. By moving the pencil farther away, we are narrowing the triangle and decreasing the parallax (making accurate measurement more difficult). If you were to paste the pencil to the wall, corresponding to the case where the object of interest is as far away as the background star field, blinking would produce no apparent shift of the pencil at all.

MORE PRECISELY 1-4

Measuring Distances with Geometry

We can convert baselines and parallaxes into distances using arguments made by the Greek geometer Euclid. The figure below represents Figure 1.25(a), but we have changed the scale and added the circle centered on the target planet and passing through our baseline on Earth.

To compute the planet's distance, we note that the ratio of the baseline AB to the circumference of the large circle shown in the figure must be equal to the ratio of the parallax to one full revolution, 360°. Since the radius of the large circle is 2π times the distance to the planet (where π—the Greek letter "pi"—is approximately equal to 3.142), it follows that

$$\frac{\text{baseline}}{2\pi \times \text{distance}} = \frac{\text{parallax}}{360°},$$

from which we find

$$\text{distance} = \text{baseline} \times \frac{(360°/2\pi)}{\text{parallax}}.$$

The angle $360°/2\pi \approx 57.3°$ in the above equation is usually called 1 *radian*.

EXAMPLE: Two observers 1000 km apart looking at the Moon might measure (using the photographic technique described in the text) a parallax of 9.0 arc minutes—that is, 0.15°. It then follows that the distance to the Moon is 1000 km × (57.3/0.15) ≈ 380,000 km. (More accurate

measurements, based on laser ranging using equipment left on the lunar surface by *Apollo* astronauts, yield a mean distance of 384,000 km.)

Knowing the distance to an object, we can then determine many other properties. For example, by measuring the object's *angular diameter*, we can compute its size. The figure below illustrates the geometry involved.

Notice that this is basically the same diagram as the previous one, except that now the angle (the angular diameter) and distance are known, instead of the angle (the parallax) and baseline. The same reasoning as before then allows us to calculate the diameter:

$$\frac{\text{diameter}}{2\pi \times \text{distance}} = \frac{\text{angular diameter}}{360°},$$

so

$$\text{diameter} = \text{distance} \times \frac{\text{angular diameter}}{57.3°}.$$

EXAMPLE: The Moon's angular diameter is measured to be about 31 arc minutes—a little over half a degree. From the preceding discussion, it follows that the Moon's actual diameter is 380,000 km × (0.52°/57.3°) ≈ 3450 km. A more precise measurement gives 3476 km.

Study the above reasoning carefully. Simple measurements such as these form the basis for almost every statement made in this book about size and scale in the universe.

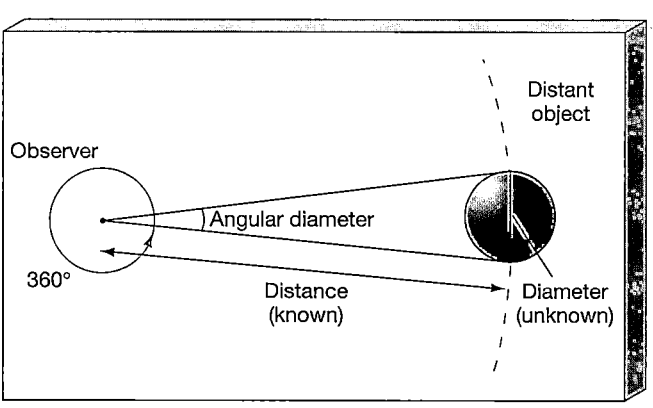

The amount of parallax is thus inversely proportional to an object's distance. Small parallax implies large distance, large parallax implies small distance. Knowing the amount of parallax (as an angle) and the length of the baseline, we can easily derive the distance through triangulation. *More Precisely 1-4* explores the connection between angular measure and distance in more detail, showing how we can use elementary geometry to determine both the distances and the dimensions of faraway objects.

Surveyors of the land routinely use such simple geometric techniques to map out planet Earth (see *Discovery 1-1* for an early example). As surveyors of the sky, astronomers use the same basic principles to chart the universe.

Concept Check

■ Why is elementary geometry essential for measuring distances in astronomy?

Sizing Up Planet Earth

In about 200 B.C. a Greek philosopher named Eratosthenes (276–194 B.C.) used simple geometric reasoning to calculate the size of our planet. He knew that at noon on the first day of summer, observers in the city of Syene (now called Aswan), in Egypt, saw the Sun pass directly overhead. This was evident from the fact that vertical objects cast no shadows and sunlight reached to the very bottoms of deep wells, as shown in the insets in the accompanying figure. However, at noon of the same day in Alexandria, a city 5000 *stadia* to the north, the Sun was seen to be displaced slightly from the vertical. (The *stadium* was a Greek unit of length, roughly equal to 0.16 km—the modern town of Aswan lies about 780 km south of Alexandria.) Using the simple technique of measuring the length of the shadow of a vertical

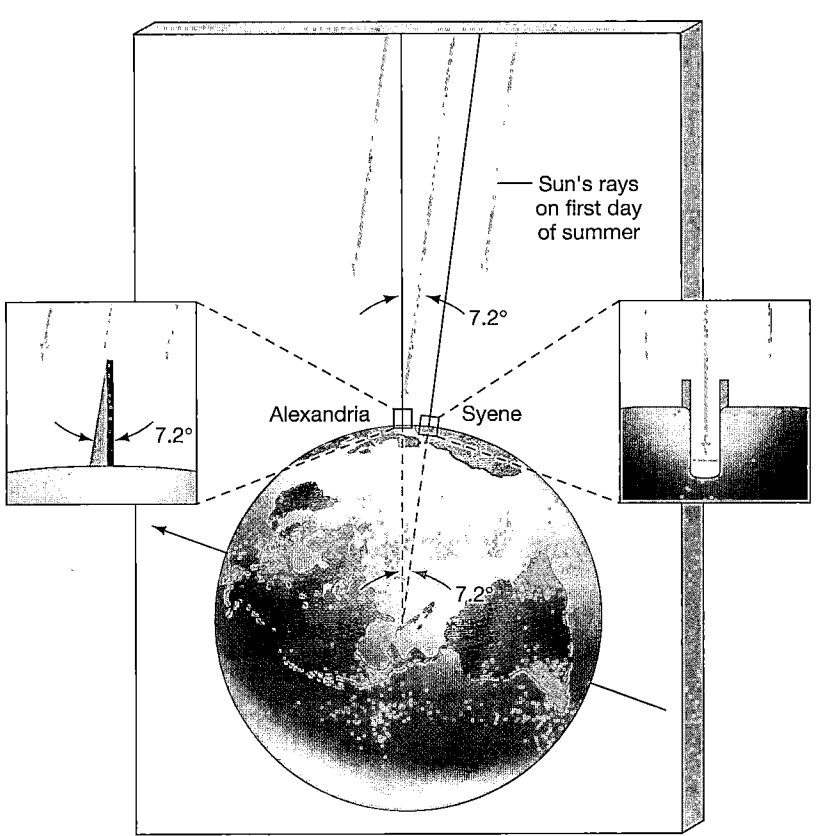

— Sun's rays on first day of summer

7.2°

Alexandria   Syene

7.2°

7.2°

stick and applying elementary trigonometry, Eratosthenes determined the angular displacement of the Sun from the vertical at Alexandria to be 7.2°.

What could have caused this discrepancy between the two measurements? It was not the result of measurement error—the same results were obtained every time the observations were repeated. Instead, as illustrated in the figure, the explanation is simply that Earth's surface is not flat, it is *curved*. Our planet is a sphere. Eratosthenes was not the first person to realize that Earth is spherical—the philosopher Aristotle had done that over 100 years earlier (see *Discovery 2-2*)—but he was apparently the first to build on this knowledge, combining geometry with direct measurement to infer the size of our planet. Here's how he did it:

Rays of light reaching Earth from a very distant object, such as the Sun, travel almost parallel to one another. Consequently, as shown in the figure, the angle measured at Alexandria between the Sun's rays and the vertical (that is, the line joining Alexandria to the center of Earth) is equal to the angle between Syene and Alexandria, as seen from Earth's center. (For the sake of clarity, this angle has been exaggerated in the drawing.) As discussed in *More Precisely 1-3*, the size of this angle in turn is proportional to the fraction of Earth's circumference that lies between Syene and Alexandria:

$$\frac{7.2°}{360°} = \frac{5000 \text{ stadia}}{\text{Earth's circumference}}.$$

Earth's circumference is therefore 50 × 5000, or 250,000 stadia, or about 40,000 km. Earth's radius is therefore 250,000/2π stadia, or 6366 km. The correct values for Earth's circumference and radius, now measured accurately by orbiting spacecraft, are 40,070 km and 6378 km, respectively.

Eratosthenes' reasoning was a remarkable accomplishment. More than 20 centuries ago, he estimated the circumference of Earth to within one percent accuracy, using only simple geometry. A person making measurements on only a small portion of Earth's surface was able to compute the size of the entire planet on the basis of observation and pure logic—an early triumph of scientific reasoning.

Chapter Review

SUMMARY

The **universe** (p. 4) is the totality of all space, time, matter, and energy. **Astronomy** (p. 4) is the study of the universe. A widely used unit of distance in astronomy is the **light-year** (p. 4), the distance traveled by a beam of light in one year.

Early observers grouped the thousands of stars visible to the naked eye into patterns called **constellations** (p. 6). These patterns have no physical significance, although they are a useful means of labeling regions of the sky. The nightly motion of the stars across the sky is the result of Earth's **rotation** (p. 8) on its axis. Early astronomers, however, imagined that the stars were attached to a vast **celestial sphere** (p. 8) centered on Earth and that the motions of the heavens were caused by the rotation of the celestial sphere about a fixed Earth. The points where Earth's rotation axis intersects the celestial sphere are called the north and south **celestial poles** (p. 8). The line where Earth's equatorial plane cuts the celestial sphere is the **celestial equator** (p. 8).

The time from one noon to the next is called a **solar day** (p. 9). The time between successive risings of any given star is one **sidereal day** (p. 9). Because Earth **revolves** (p. 9) around the Sun, the solar day is a few minutes longer than the sidereal day. Because Earth orbits the Sun, we see different constellations at different times of the year, and the Sun appears to move relative to the stars. The Sun's yearly path around the celestial sphere, or equivalently, the plane of Earth's orbit around the Sun, is called the **ecliptic** (p. 11). The constellations lying along the ecliptic are collectively called the **zodiac** (p. 11). Because Earth's axis is inclined to the ecliptic plane, we experience **seasons** (p. 13), depending on which hemisphere (Northern or Southern) happens to be "tipped" toward the Sun. At the **summer solstice** (p. 11), the Sun is highest in the sky, and the length of the day is greatest. At the **winter solstice** (p. 11), the Sun is lowest, and the day is shortest. At the **vernal** (p. 13) and **autumnal equinoxes** (p. 13), Earth's rotation axis is perpendicular to the line joining Earth to the Sun, so day and night are of equal length. The interval of time from one vernal equinox to the next is one **tropical year** (p. 13).

The Moon emits no light of its own. It shines by reflected sunlight. As the Moon orbits Earth, we see **lunar phases** (p. 15) as the amount of the Moon's sunlit face visible to us varies. At full Moon, we can see the entire illuminated side. At quarter Moon,

only half the sunlit side can be seen. At new Moon, the sunlit face points away from us, and the Moon is nearly invisible from Earth. The time between successive full Moons is one **synodic month** (p. 15). The amount of time for the Moon to return to the same position in the sky, relative to the stars, is one **sidereal month** (p. 15). Because of Earth's motion around the Sun, the synodic month is about two days longer than the sidereal month.

The time required for Earth to complete one orbit around the Sun, relative to the stars, is one **sidereal year** (p. 13). In addition to its rotation about its axis and its revolution around the Sun, Earth has many other motions. One of the most important of these is **precession** (p. 13), the slow "wobble" of Earth's axis due to the influence of the Moon. As a result, the sidereal year is slightly longer than the tropical year, and the particular constellations that happen to be visible during any given season change over the course of thousands of years.

A **lunar eclipse** (p. 15) occurs when the Moon enters Earth's shadow. The eclipse may be **total** (p. 15), if the entire Moon is (temporarily) darkened, or **partial** (p. 15), if only a portion of the Moon's surface is affected. A **solar eclipse** (p. 15) occurs when the Moon passes between Earth and the Sun, so that a small part of Earth's surface is plunged into shadow. For observers in the **umbra** (p. 17), the entire Sun is obscured, and the solar eclipse is total. In the **penumbra** (p. 17), a partial solar eclipse is seen. If the Moon happens to be too far from Earth for its disk to completely hide the Sun, an **annular eclipse** (p. 17) occurs. Because the Moon's orbit around Earth is slightly inclined with respect to the ecliptic, solar and lunar eclipses do not occur every month, but only a few times per year.

Surveyors on Earth use **triangulation** (p. 24) to determine the distances to faraway objects. Astronomers use the same technique to measure the distances to planets and stars. The **cosmic distance scale** (p. 24) is the family of distance-measurement techniques by which astronomers chart the universe. **Parallax** (p. 25) is the apparent motion of a foreground object relative to a distant background as the observer's position changes. The larger the **baseline** (p. 24), the distance between the two observation points, the greater the parallax. The same basic geometric reasoning is used to determine the sizes of objects whose distances are known.

SELF-TEST: TRUE OR FALSE?

_____ **1.** The light-year is a measure of distance.

_____ **2.** The number 2×10^9 is equal to two billion.

_____ **3.** The stars in a constellation are physically close to one another.

_____ **4.** The star Polaris always lies precisely at the north celestial pole.

_____ **5.** Constellations are no longer used by astronomers.

_____ **6.** The solar day is longer than the sidereal day.

_____ **7.** The constellations lying along the ecliptic are collectively referred to as the zodiac.

_____ **8.** The seasons are caused by the precession of Earth's axis.

_____ **9.** The vernal equinox marks the beginning of spring.

_____ **10.** The new phase of the Moon cannot be seen because it always occurs during the daytime.

_____ **11.** A lunar eclipse can occur only during the full phase.

_____ **12.** Solar eclipses are possible during any phase of the Moon.

_____ **13.** An annular eclipse is a type of eclipse that occurs every year.

_____ **14.** Eclipses can occur only during winter and summer months.

_____ **15.** The parallax of an object is inversely proportional to its distance.

SELF-TEST: FILL IN THE BLANK

1. A _____ is a glowing ball of gas held together by gravity.
2. A _____ is a collection of hundreds of billions of stars.
3. Rotation is the term used to describe the motion of a body around some _____.
4. To explain the daily and yearly motions of the heavens, ancient astronomers imagined that the Sun, Moon, stars, and planets were attached to a rotating _____.
5. The solar day is measured relative to the Sun; the sidereal day is measured relative to the _____.
6. The apparent path of the Sun across the sky is known as the _____.
7. On December 21, known as the _____, the Sun is at its _____ point on the celestial sphere.
8. Declination measures the position of an object north or south of the _____.

9. An arc second is _____ (give the fraction) of an arc minute.
10. When the Sun, Earth, and Moon are positioned to form a right angle at Earth, the Moon is seen in the _____ phase.
11. A _____ eclipse can be seen by about half of Earth at once.
12. As seen from Earth, the Sun and the Moon have roughly the same _____.
13. The distance to an object can be determined, for a known baseline, by measuring its _____.
14. The size of an object can be determined, if we know its distance, by measuring its _____.
15. The radius of _____ was first measured by Eratosthenes in 200 B.C.

REVIEW AND DISCUSSION

1. Compare the size of Earth with that of the Sun, the Milky Way Galaxy, and the entire universe.
2. What does an astronomer mean by "the universe?"
3. How big is a light-year?
4. What is a constellation?
5. Why are constellations useful for mapping the sky?
6. Why does the Sun rise in the east and set in the west each day? Does the Moon also rise in the east and set in the west? Why? Do stars do the same? Why?
7. How and why does a day measured by the Sun differ from a day measured by the stars?
8. How many times in your life have you orbited the Sun?
9. Why do we see different stars at different times of the year?
10. Why are there seasons on Earth?
11. What is an equinox?
12. What is precession, and what causes it?

13. If one complete hemisphere of the Moon is always lit by the sun, why do we see different phases of the Moon?
14. What causes a lunar eclipse? A solar eclipse?
15. Why aren't there lunar and solar eclipses every month?
16. Do you think an observer on another planet might see eclipses? Why or why not?
17. What is parallax? Give an everyday example.
18. Why is it necessary to have a long baseline when using triangulation to measure the distances to objects in space?
19. What two pieces of information are needed to determine the diameter of a faraway object?
20. If you traveled to the outermost planet in our solar system, do you think the constellations would appear to change their shapes? What would happen if you traveled to the next-nearest star? If you traveled to the center of our Galaxy, could you still see the familiar constellations found in Earth's night sky?

PROBLEMS  *Algorithmic versions of these questions are available in the Practice Problems module of the Companion Website.*

The number of squares preceding each problem indicates its approximate level of difficulty.

1. ■ In one second, light leaving Los Angeles reaches approximately as far as (a) San Francisco, about 500 km; (b) London, roughly 10,000 km; (c) the Moon, 384,000 km; (d) Venus, 45,000,000 km from Earth at closest approach; or (e) the nearest star, about three light-years from Earth. Which is correct?
2. ■ (a) Write the following numbers in scientific notation (see Appendix 1 if you are unfamiliar with this notation): 1000; 0.000001; 1001; 1,000,000,000,000,000; 123,000; 0.000456. (b) Write the following numbers in "normal" numerical form: 3.16×10^7; 2.998×10^5; 6.67×10^{-11}; 2×10^0. (c) Calculate: $(2 \times 10^3) + 10^{-2}$; $(1.99 \times 10^{30})/(5.98 \times 10^{24})$; $(3.16 \times 10^7) \times (2.998 \times 10^5)$.

3. ■■ How, and by roughly how much, would the length of the solar day change if Earth's rotation were suddenly to reverse direction?
4. ■ The vernal equinox is now just entering the constellation Aquarius. In what constellation will it lie in the year A.D. 10,000?
5. ■■ What would be the length of the synodic month if the Moon's sidereal orbital period were (a) one week (seven solar days); (b) one (sidereal) year?
6. ■ Through how many degrees, arc minutes, or arc seconds does the Moon move in (a) one hour of time; (b) one minute; (c) one second? How long does it take for the Moon to move a distance equal to its own diameter?
7. ■■ Given the data presented in the text, estimate the speed (in km/s) at which the Moon moves in its orbit around Earth.

8. ■ A surveyor wishes to measure the distance between two points on either side of a river, as illustrated in Figure 1.23. She measures the distance AB to be 250 m and the angle at B to be 30°. What is the distance between the two points?

9. ■ At what distance is an object if its parallax, as measured from either end of a 1000-km baseline, is (a) 1°; (b) 1′; (c) 1″?

10. ■ Given that the angular size of Venus is 55″ when the planet is 45,000,000 km from Earth, calculate Venus's diameter (in kilometers).

11. ■ Calculate the parallax, using Earth's diameter as a baseline, of the Sun's nearest neighbor, Proxima Centauri, which lies 4.3 light-years from Earth.

12. ■ Estimate the angular diameter of your thumb, held at arm's length.

13. ■ The Moon lies roughly 384,000 km from Earth and the Sun lies 150,000,000 km away. If both have the same angular size, as seen from Earth, how many times larger than the Moon is the Sun?

14. ■ Given that the distance from Earth to the Sun is 150,000,000 km, through what distance does Earth move in (a) a second, (b) an hour, (c) a day?

15. ■ What angle would Eratosthenes have measured (see *Discovery 1-1*) had Earth been flat?

COLLABORATIVE EXERCISES

1. Chasing Solar Eclipses. Consider the text figure showing solar eclipse paths over a world map. As a group, write a description of which eclipse your group would most like to observe together, where and when you would go to observe it, and fully explain why you selected the date and site you did.

2. Measuring Diameters from the Surface. Eratosthenes used simple geometric reasoning to calculate Earth's size using shadows. As a group, create a sketch and an accompanying written description showing exactly how his measurements would lead to a different result using one of Jupiter's moons.

3. Parallax Measurements. If the angular width of your thumb at arm's length is about 1/2 of a degree, determine the angular size of four different objects in the room selected by your group members. Provide a sketch with an organized data table.

4. Astrophotographs from Distant Planets. Consider the semi-circular star trails shown in the time-lapse photograph (Fig. 1.9) of the northern sky. What was the exposure time used for the photograph? How long would you need to take a similar picture from a different planet of your group's choosing? Fully explain your answer.

RESEARCHING ON THE WEB

To complete the following exercises, go to the online Destinations module for Chapter 1 on the Companion Website for Astronomy Today 4/e.

1. Access the "Coordinate Conversions" page and determine the J2000 Right Ascension and Declination Coordinates for the following targets: Orion nebula, Pleiades, Andromeda galaxy, and Jupiter.

2. Access the "Moon Phase" page and determine what will be the illuminated fraction and age in days of the Moon on your next two birthdays. Include a sketch of the Moon's appearance.

3. Access the "Solar Eclipse Path Predictions" page and determine the maximum duration of totality for the next total

solar eclipse and write a rationale for where the best location to observe the eclipse would be.

4. Access the "List of Constellation Facts and Figures" page and describe the Messier objects that are located in your constellation most closely associated with your horoscope birth-sign.

PROJECTS

1. Go to a country location on a clear dark night. Imagine patterns among the stars, and name the patterns yourself. Note (or better yet, draw) the locations of these stars with respect to trees or buildings in the foreground. Do this every week or so for a couple of months, and be sure to look at the same time every night. What happens?

2. Find the star Polaris, also known as the North Star, in the evening sky. Identify any separate pattern of stars in the

same general vicinity of the sky. Wait several hours, at least until after midnight, and then locate Polaris again. Has Polaris moved? What has happened to the nearby pattern of stars? Why?

3. Hold your little finger out at arm's length. Can you cover the disk of the Moon? The Moon projects an angular size of 30′ (half a degree); your finger should more than cover it. How can you apply this fact in making sky measurements?

SKYCHART III PROJECTS

The SkyChart III Student Version planetarium program on which these exercises are based is included as a separately executable program on the CD in the back of this text.

1. ▣ Familiarity with the night sky starts with learning how to recognize significant constellations. SkyChart III can be very helpful in this endeavor. Set the *COMPUTATION/Location* for your area and set a convenient date and time for you to observe the stars. Do not attempt initially to learn all of the constellations, but start with the prominent ones currently visible. Concentrate your attention on the constellations depicted in the star charts on pages 7–10 of your text.

Make a careful sketch of the night sky that faithfully represents the size, shape and relative locations of the constellations. Prepare the sketch so you can carry it with you to the field when you view the stars. In your sketch, include the circumpolar constellations and asterisms that are always visible in the Northern Hemisphere. While the hand sketch is important in helping you appreciate the location and size of the various star groupings; it is also helpful to have a printed copy of the screen. To print out a copy of the screen, select *FILE/Print Setup* and select *Landscape* under *Orientation*. When *FILE/Print* is selected, accept the recommendation to not print colors and to print black lines on a white background.

The Big Dipper (an asterism) is a good place to start when attempting to find your way around the stars. Orient the display with north up, and carefully note the orientation of Polaris with respect to the pointer stars of the Big Dipper. Then note how you can project off the Big Dipper in the opposite direction to find Leo. You will find also that there are other useful projections that lead from the Big Dipper to objects of interest. When the Moon is visible, it will be in a different position each evening, providing a reference point that can be useful. A dark sky is not necessary to view constellations. In fact, the constellations typically are composed of the brightest stars, actually making it easier to see them from areas with moderate light pollution. In a dark sky, so many stars are visible that making out even familiar constellations can sometimes present a challenge.

2. ▣ Set up and print observation charts for a viewing session. Use the *COMPUTATION* menu to choose the *Location* and *Date & Time*. If it is not preset, you will need to enter the *Longitude*, *Latitude*, and *Altitude* of your location. Identify on your charts the brightest stars and most recognizable constellations. Identify galaxies, nebulae, and clusters.

3. ▣ Figure 1.12 illustrates the zodiac. Set up charts identifying which constellation of the zodiac marks the beginning of each season. Draw the *Ecliptic: VIEW/Coordinates/Ecliptic; VIEW/Center Planet/Sun.* Use the *ANIMATION* menu, with an appropriate *Time Step*, to observe the changing of the Zodiac. Keep track of time in the bottom left corner.

4. ▣ The constellations as we see them are not only unique to our culture but also unique to our point of reference. Use the *COMPUTATION* menu to *View From* the star Sirius, a star in the neighboring constellation of Canis Major. Can you explain the distortions seen in both Orion and Canis Major?

5. ▣ Simulate the total solar eclipse that will cross the central United States in 2017. Select *COMPUTATION/Location* and configure SkyChart III for St. Louis at longitude −90 30 00 and latitude +38 45 00. Under *COMPUTATION/Date & Time* set the *Local date and time* to 2017/08/21 and 11:30:00 with *Time zone* to −6.0. Select *VIEW/Center Planet/Sun* to center on the Sun. The Sun will now remain in the center of the screen as time and zoom are changed. Zoom in with *Pg Up* on the keyboard until you have a field of view of approximately 5°. You should see the Moon poised near the Sun. Animate the scene with one-minute time steps to observe motion of the Moon with respect to the Sun during this total solar eclipse.

6. ▣▣ Observe the August 21, 2017 solar eclipse as seen from a vantage point on the Sun. Configure SkyChart III by opening *COMPUTATION/Location* and selecting *View from Object*. Click on *Select* and *Search* for Sun. Turn off the horizon mask with *DRAW/Horizon Mask*. Center on Earth with *VIEW/Center/Planet/Earth*. Since you are viewing the Moon from the distance of the Sun, it will be necessary to use a more powerful telescope than is necessary to view the Moon from Earth. Zoom in until the field is approximately 1/60°. With time set for approximately 11:30 a.m., the moon will be seen poised between Earth and the Sun. Animate the scene with time steps of one minute and watch the Moon advance across the surface of Earth. It should be obvious why the eclipse is visible only in certain locations on Earth at any one time.

7. ▣▣ Simulate the precession of Earth's axis with SkyChart III by centering on Polaris and using animation to step through time in 100-year increments. Under the pull-down menu *DRAW* select *Stars, Constellations, Chart Legend, Object Labels,* and *Grid Lines.* Lock Polaris in the center of the screen by selecting *VIEW/Center Object,* typing in Polaris and clicking on *Find*. When Polaris is located, click on *Select.* Choose *VIEW/180° Field.* Select *ANIMATION/100 Years* and start *Animation Forward.* You can also run the animation backward to observe how difficult it might have been for navigation in the Dark Ages when there was no distinct North Star. When was Vega our North Star and when will it be again? Turn on *DRAW/Mouse Coordinates* to measure the angular separation of objects. Determine how close Polaris is to being true north today by pointing the cursor to true north and while holding the left button down, move the cursor over to Polaris. The angular separation is provided in a set of numbers in the upper left corner of the screen. How close was Polaris to being the North Star when Columbus made his famous voyage in 1492?

 *In addition to the Practice Problems and Destinations modules, the Companion Website at http://www.prenhall.com/chaisson provides for each chapter an additional true-false, multiple choice, and labeling quiz, as well as additional annotated images, animations, and links to related Websites.*

2 THE COPERNICAN REVOLUTION

The Birth of Modern Science

LEARNING GOALS

Studying this chapter will enable you to:

1 Relate how some ancient civilizations attempted to explain the heavens in terms of Earth-centered models of the universe.

2 Summarize the role of Renaissance science in the history of astronomy.

3 Explain how the observed motions of the planets led to our modern view of a Sun-centered solar system.

4 Sketch the major contributions of Galileo and Kepler to the development of our understanding of the solar system.

5 State Kepler's laws of planetary motion.

6 Explain how Kepler's laws enable us to construct a scale model of the solar system, and explain the technique used to determine the actual size of the planetary orbits.

7 State Newton's laws of motion and universal gravitation and explain how they account for Kepler's laws.

8 Explain how the law of gravitation enables us to measure the masses of astronomical bodies.

 Visit http://www.prenhall.com/chaisson for additional annotated images, animation, and links to related sites for this chapter.

This magnificent true-color image shows a group of stars surrounded by glowing nebulosity, about 150,000 light-years away. Known officially as Hodge 301, the cluster of stars at lower right is part of the Tarantula Nebula, a "spider's nest" of massive, brilliant stars that are about to explode. In fact, the red and green wispiness marks the debris of previously detonated stars, called supernovae. (STScI)

The Big Picture: Astronomers today unquestionably know a great deal more about the subject of astronomy than did astronomers of long ago. The reason is that we have better equipment to aid our eyes and other senses, enabling us to perceive the cosmos in ways that the ancients could have only imagined. This one image alone would surely have astounded them—for its superb color, its faint nebulosity, and its fine detail.

Living in the Space Age, we have become accustomed to the modern view of our place in the universe. Images of our planet taken from space leave little doubt that Earth is round, and no one seriously questions the idea that we orbit the Sun. Yet there was a time, not so long ago, when some of our ancestors maintained that Earth was flat and lay at the center of all things. Our view of the universe—and of ourselves—has undergone a radical transformation since those early days. Earth has become a planet like many others, and humankind has been torn from its throne at the center of the cosmos and relegated to a rather unremarkable position on the periphery of the Milky Way Galaxy. But we have been amply compensated for our loss of prominence—we have gained a wealth of scientific knowledge in the process. The story of how all this came about is the story of the rise of the scientific method and the genesis of modern astronomy.

2.1 Ancient Astronomy

1 Many ancient cultures took a keen interest in the changing nighttime sky. The records and artifacts that have survived until the present make that abundantly clear. But unlike today, the major driving force behind the development of astronomy in those early societies was probably neither scientific nor religious. Instead, it was decidedly practical and down to earth. Seafarers needed to navigate their vessels, and farmers had to know when to plant their crops. In a real sense, then, human survival depended on knowledge of the heavens. The ability to predict the arrival of the seasons, as well as other astronomical events, was undoubtedly a highly prized, perhaps jealously guarded, skill.

In Chapter 1 we saw that the human brain's ability to perceive patterns in the stars led to the "invention" of constellations as a convenient means of labeling regions of the celestial sphere. ∞ (Sec. 1.2) The realization that these patterns returned to the night sky at the same time each year met the need for a practical means of tracking the seasons. Widely separated cultures all over the world built large and elaborate structures to serve, at least in part, as primitive calendars. In some cases, the keepers of the secrets of the sky enshrined their knowledge in myth and ritual, and these astronomical sites were often also used for religious ceremonies.

Perhaps the best known such site is *Stonehenge*, located on Salisbury Plain in England, and shown in Figure 2.1. This ancient stone circle, which today is one of the most popular tourist attractions in Britain, dates from the Stone Age. Researchers believe it was an early astronomical observatory of sorts—not in the modern sense of the term (a place for making new observations and discoveries) but

Figure 2.1 Stonehenge This remarkable site in the south of England was probably constructed as a primitive calendar and almanac. The fact that the largest stones were carried to the site from many miles away attests to the importance of this structure to its Stone Age builders. The inset shows sunrise at Stonehenge on the summer solstice. As seen from the center of the stone circle, the Sun rose directly over the "heel stone" on the longest day of the year. *(English Heritage)*

rather a kind of three-dimensional calendar or almanac, enabling its builders and their descendants to identify important dates by means of specific celestial events. Its construction apparently spanned a period of about 17 centuries, beginning around 2800 B.C. Additions and modifications continued to about 1100 B.C., indicating its ongoing importance to the Stone Age and later Bronze Age people who built, maintained, and used Stonehenge. The largest stones shown in Figure 2.1 weigh up to 50 tons and were transported from quarries many miles away.

Many of the stones are aligned so that they point toward important astronomical events. For example, the line joining the center of the inner circle to the so-called heel stone, set some distance from the rest of the structure, points in the direction of the rising Sun on the summer solstice. Other alignments are related to the rising and setting of the Sun and the Moon at various other times of the year. The accurate alignments (within a degree or so) of the stones of Stonehenge were first noted in the eighteenth century, but it is only relatively recently—in the second half of the twentieth century, in fact—that the scientific community has credited Stone Age technology with the ability to carry out such a precise feat of engineering. While some of Stonehenge's purposes remain uncertain and controversial, the site's function as an astronomical almanac seems well established. Although Stonehenge is the most impressive and the best preserved, other stone circles, found all over Europe, are believed to have performed similar functions.

Many other cultures are now known to have been capable of similarly precise accomplishments. The Big Horn Medicine Wheel in Wyoming (Figure 2.2a) is similar to Stonehenge in design—and, presumably, intent—although it is somewhat simpler in execution. The Medicine Wheel's alignments with the rising and setting Sun and with some bright stars indicate that its builders—the Plains Indians—had much more than a passing familiarity with the changing nighttime sky. Figure 2.2(b) shows the Caracol temple, built by the Mayans around A.D. 1000 on Mexico's Yucatan peninsula. This temple is much more sophisticated than Stonehenge, but it probably played a similar role as an astronomical observatory. Its many windows are accurately aligned with astronomical events, such as sunrise and sunset at the solstices and equinoxes and the risings and settings of the planet Venus. Astronomy was of more than mere academic interest to the Mayans, however. Caracol was also the site of countless human sacrifices, carried out when Venus appeared in the morning or evening sky.

The ancient Chinese also observed the heavens. Their astrology attached particular importance to "omens" such as comets and "guest stars"—stars that appeared suddenly in the sky and then slowly faded away—and they kept careful and extensive records of such events. Twentieth-century astronomers still turn to the Chinese records to obtain observational data recorded during the Dark Ages (roughly from the fifth to the tenth century A.D.), when turmoil in Europe largely halted the progress of Western science. Perhaps the best-known guest star was one that appeared in A.D. 1054 and was visible in the daytime sky for many months. We now know that the event was actually a *supernova*: the explosion of a giant star, which scattered most of its mass into space (see Chapter 21). It left behind

(a)

(b)

Figure 2.2 Observatories in the Americas (a) The Big Horn Medicine Wheel, in Wyoming, was built by the Plains Indians. Its spokes and other features are aligned with risings and settings of the Sun and other stars. (b) Caracol temple in Mexico. The many windows of this Mayan construct are aligned with astronomical events, indicating that at least part of Caracol's function was to keep track of the seasons and the heavens. *(G. Gerster/Comstock)*

a remnant that is still detectable today, nine centuries later. The Chinese data are a prime source of historical information for supernova research.

A vital link between the astronomy of ancient Greece and that of medieval Europe was provided by astronomers in the Islamic world (Figure 2.3). For six centuries, from the depths of the Dark Ages to the beginning of the Renaissance, Islamic astronomy flourished and grew, preserving and augmenting the knowledge of the Greeks. The Arab influence on modern astronomy is subtle but quite pervasive. Many of the mathematical techniques involved in trigonometry were developed by Muslim astronomers in response to practical problems, such as determining the precise dates of holy days or the direction of Mecca from any given location on Earth. Astronomical terms like "zenith" and "azimuth" and the names of many stars, such as Rigel, Betelgeuse, and Vega, all bear witness to this extended period of Muslim scholarship.

Astronomy is not the property of any one culture, civilization, or era. The same ideas, the same tools, and even the same misconceptions have been invented and reinvented by human societies all over the world, in response to the same basic driving forces. Astronomy came into being because people believed that there was a practical benefit in being able to predict the positions of the stars, but its roots go much deeper than that. The need to understand where we came from, and how we fit into the cosmos, is an integral part of human nature.

Figure 2.3 Arab Astronomers at Work During the Dark Ages, much scientific information was preserved and new discoveries were made by astronomers in the Arab world, as depicted in this illustration from a medieval manuscript. *(The Granger Collection)*

2.2 The Geocentric Universe

The Greeks of antiquity, and undoubtedly civilizations before them, built models of the universe. The study of the workings of the universe on the largest scales is called *cosmology*. Today, cosmology entails looking at the universe on scales so large that even entire galaxies can be regarded as mere points of light scattered throughout space. To the Greeks, however, the universe was basically the *solar system*—namely, the Sun, Earth, Moon, and the planets known at that time. The stars beyond were surely part of the universe, but they were considered to be fixed, unchanging beacons on the celestial sphere. The Greeks did not consider the Sun, the Moon, and the planets to be part of this mammoth celestial dome, however. Those objects had patterns of behavior that set them apart.

Greek astronomers observed that over the course of a night, the stars slid smoothly across the sky. Over the course of a month, the Moon moved smoothly and steadily along its path on the sky relative to the stars, passing through its familiar cycle of phases. Over the course of a year, the Sun progressed along the ecliptic at an almost constant rate, varying little in brightness from day to day. In short, the behavior of both Sun and Moon seemed fairly simple and orderly. But ancient astronomers were also aware of five other bodies in the sky—the planets Mercury, Venus, Mars, Jupiter, and Saturn—whose behavior was not so easy to grasp. Their motions ultimately led to the downfall of an entire theory of the solar system and to a fundamental change in humankind's view of the universe.

Planets do not behave in as regular and predictable a fashion as the Sun, Moon, and stars. They vary in brightness, and they don't maintain a fixed position in the sky. Unlike the Sun and Moon, the planets seem to wander around the celestial sphere—indeed, the word planet derives from the Greek word *planetes*, meaning "wanderer." Planets never stray far from the ecliptic and generally traverse the celestial sphere from west to east, like the Sun. However, they seem to speed up and slow down during their journeys, and at times they even appear to loop back and forth relative to the stars, as shown in Figure 2.4. In other words, there are periods when a planet's eastward motion (relative to the stars) stops, and the planet appears to move westward in the sky for a month or two before reversing direction again and continuing on its eastward journey. Motion in the eastward sense is usually referred to as *direct*, or *prograde*, motion; the backward (westward) loops are known as **retrograde motion**.

Like the Moon, the planets produce no light of their own; instead, they shine by reflected sunlight. Ancient astronomers correctly reasoned that the apparent brightness of a planet in the night sky is related to its distance from Earth—planets appear brightest when closest to us. However, the planets Mars, Jupiter, and Saturn are always brightest during the retrograde portions of their orbits.

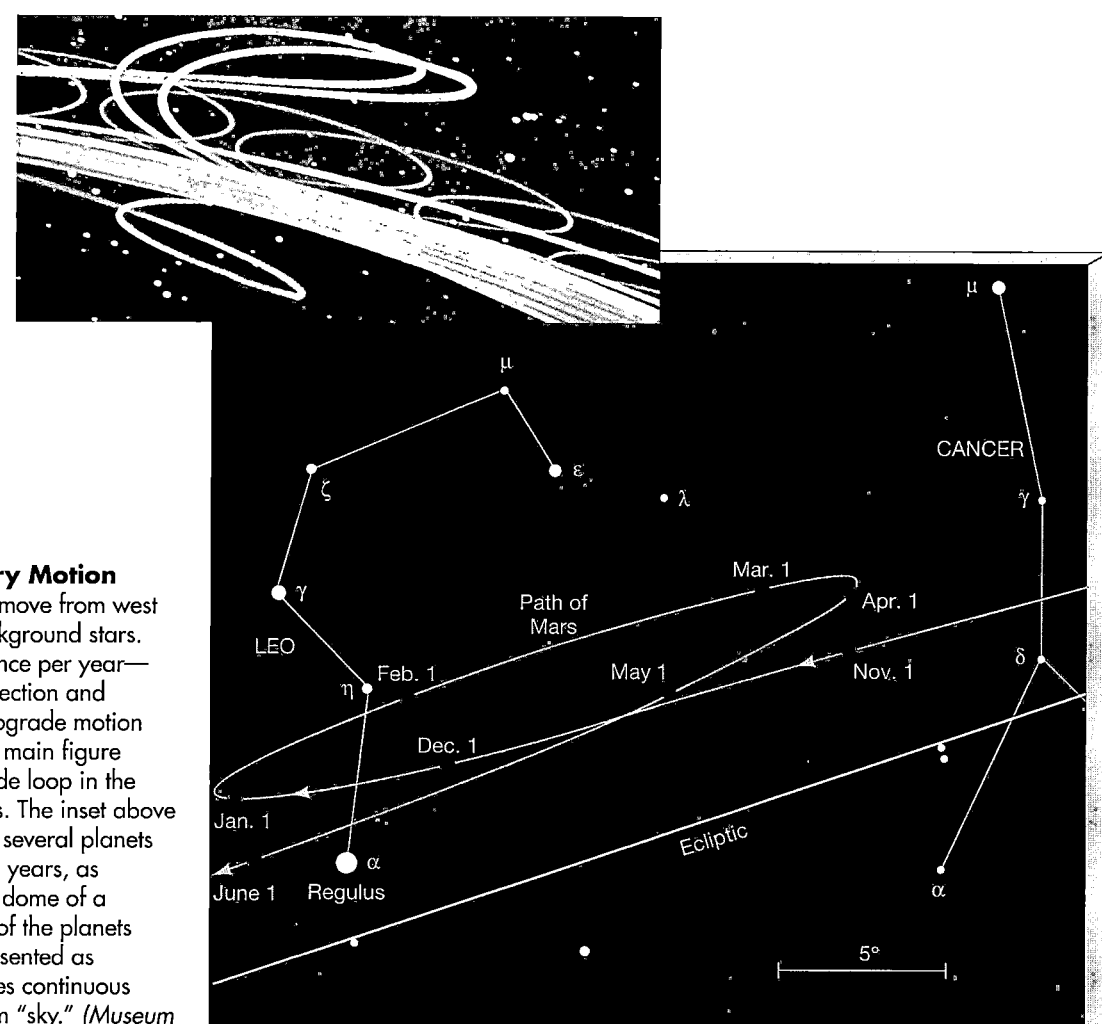

Figure 2.4 Planetary Motion
Most of the time, planets move from west to east relative to the background stars. Occasionally—roughly once per year—however, they change direction and temporarily undergo retrograde motion before looping back. The main figure shows an actual retrograde loop in the motion of the planet Mars. The inset above depicts the movements of several planets over the course of several years, as reproduced on the inside dome of a planetarium. The motion of the planets relative to the stars (represented as unmoving points) produces continuous streaks on the planetarium "sky." (Museum of Science, Boston)

The challenge facing astronomers was to explain the observed motions of the planets and to relate those motions to the variations in planetary brightness.

The earliest models of the solar system followed the teachings of the Greek philosopher Aristotle (384–322 B.C.) and were **geocentric**, meaning that Earth lay at the center of the universe and all other bodies moved around it. (Figures 1.7 and 1.10a illustrate the basic geocentric view.) ⟳ (Sec. 1.2) These models employed what Aristotle, and Plato before him, had taught was the perfect form: the circle. The simplest possible description—uniform motion around a circle with Earth at its center—provided a fairly good approximation to the orbits of the Sun and the Moon, but it could not account for the observed variations in planetary brightness or their retrograde motion. A more complex model was needed to describe the planets.

In the first step toward this new model, each planet was taken to move uniformly around a small circle, called an **epicycle**, whose *center* moved uniformly around Earth on a second and larger circle, known as the **deferent** (Figure 2.5). The motion was now composed of two separate circular orbits, creating the possibility that, at some rate circular orbits, creating the possibility that, at some

times, the planet's apparent motion could be retrograde. Also, the distance from the planet to Earth would vary, accounting for changes in brightness. By tinkering with the relative sizes of epicycle and deferent, with the planet's speed on the epicycle, and with the epicycle's speed along the deferent, early astronomers were able to bring this "epicyclic" motion into fairly good agreement with the observed paths of the planets in the sky. Moreover, this model had good predictive power, at least to the accuracy of observations at the time.

However, as the number and the quality of observations increased, it became clear that the simple epicyclic model was not perfect. Small corrections had to be introduced to bring it into line with new observations. The center of the deferents had to be shifted slightly from Earth's center, and the motion of the epicycles had to be imagined uniform with respect to yet another point in space, not Earth. Around A.D. 140, a Greek astronomer named Ptolemy constructed perhaps the best geocentric model of all time. Illustrated in simplified form in Figure 2.6, it explained remarkably well the observed paths of the five planets then known, as well as the paths of the Sun and the

Moon. However, to achieve its explanatory and predictive power, the full **Ptolemaic model** required a series of no fewer than 80 distinct circles. To account for the paths of the Sun, Moon, and all nine planets (and their moons) that

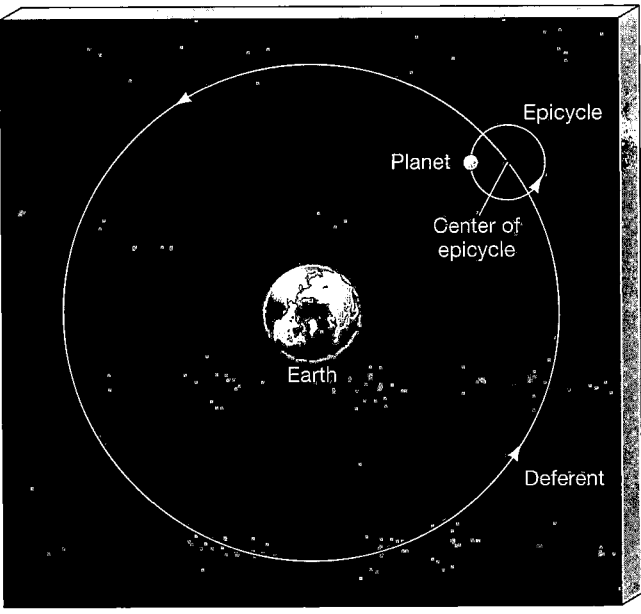

Figure 2.5 Geocentric Model In the geocentric model of the solar system, the observed motions of the planets made it impossible to assume that they moved on simple circular paths around Earth. Instead, each planet was thought to follow a small circular orbit (the epicycle) about an imaginary point that itself traveled in a large, circular orbit (the deferent) about Earth.

we know today would require a vastly more complex set. Nevertheless, Ptolemy's text on the topic, *Syntaxis* (better known today by its Arabic name *Almagest*—"the greatest"), provided the intellectual framework for all discussion of the universe for well over a thousand years.

Today, our scientific training leads us to seek simplicity, because simplicity in the physical sciences has so often proved to be an indicator of truth. We would regard the intricacy of a model as complicated as the Ptolemaic system as a clear sign of a fundamentally flawed theory. With the benefit of hindsight, we now recognize that the major error lay in the assumption of a geocentric universe. This was compounded by the insistence on uniform circular motion, whose basis was largely philosophical, rather than scientific, in nature.

Actually, history records that some ancient Greek astronomers reasoned differently about the motions of heavenly bodies. Foremost among them was Aristarchus of Samos (310–230 B.C.), who proposed that all the planets, including Earth, revolve around the Sun and, furthermore, that Earth rotates on its axis once each day. This, he argued, would create an *apparent* motion of the sky—a simple idea that is familiar to anyone who has ridden on a merry-go-round and watched the landscape appear to move past in the opposite direction. However, Aristarchus's description of the heavens, though essentially correct, did not gain widespread acceptance during his lifetime. Aristotle's influence was too strong, his followers too numerous, his writings too comprehensive. The geocentric model went largely unchallenged until the sixteenth century A.D.

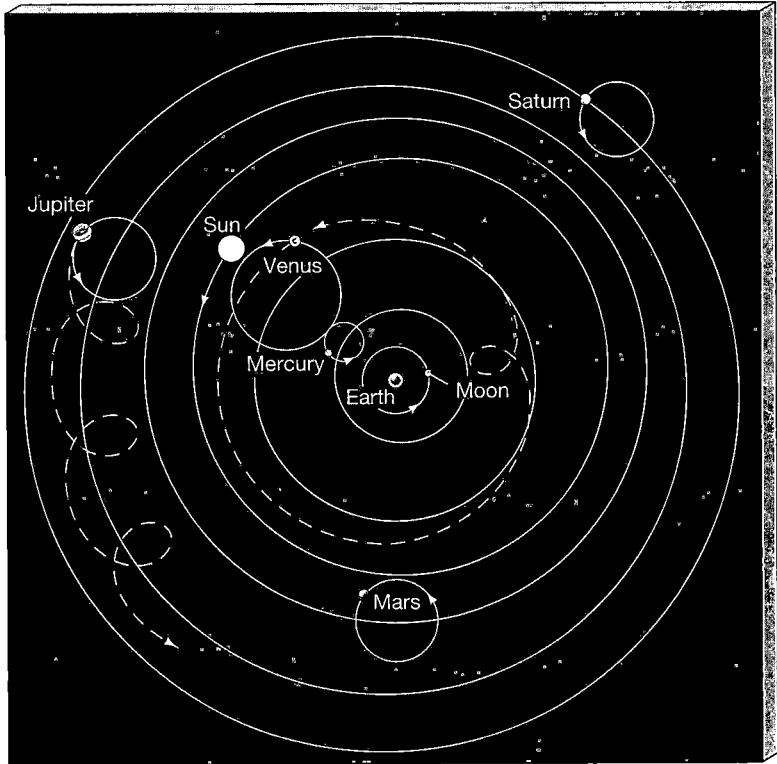

Figure 2.6 Ptolemy's Model The basic features, drawn roughly to scale, of the geocentric model of the inner solar system that enjoyed widespread popularity prior to the Renaissance. The planets' deferents were considered to move on spheres lying within the celestial sphere that held the stars. The celestial sphere carried all interior spheres around with it, but the planetary (and solar) spheres had additional motions of their own, causing the Sun and planets to move relative to the stars. To avoid confusion, partial paths (dashed) of only two planets, Venus and Jupiter, are drawn here.

The Aristotelian school did present some simple and (at the time) compelling arguments in favor of their views. First, of course, Earth doesn't *feel* as if it's moving. And if it were, wouldn't there be a strong wind as we move at high speed around the Sun? Then again, considering that the vantage point from which we view the stars changes over the course of a year, why don't we see stellar parallax? Nowadays we might be inclined to dismiss the first two points as merely naïve, but the third is a valid argument and the reasoning is essentially sound. We now know that there *is* stellar parallax as Earth orbits the Sun. However, because the stars are so distant, it amounts to less than 1″, even for the closest stars. Early astronomers simply would not have noticed it. We will encounter many other instances in astronomy where correct reasoning has led to the wrong conclusions because it was based on inadequate data.

2.3 The Heliocentric Model of the Solar System

2.3 The Ptolemaic picture of the universe survived, more or less intact, for almost 13 centuries, until a sixteenth-century Polish cleric, Nicholas Copernicus (Figure 2.7), rediscovered Aristarchus's **heliocentric** (Sun-centered) model and showed how, in its harmony and organization, it provided a more natural explanation of the observed facts than did the tangled geocentric cosmology. Copernicus asserted that Earth spins on its axis and, like the other planets, orbits the Sun. Only the Moon, he said, orbits Earth. Not only does this model explain the observed daily and seasonal changes in the heavens, as we saw in Chapter 1, but it also naturally accounts for planetary retrograde motion and brightness variations. The critical realization that Earth is not at the center of the universe is now known as the **Copernican revolution**. The seven crucial statements that form its foundation are summarized in *Discovery 2-1*.

Figure 2.8 shows how the Copernican view explains both the varying brightness of a planet (in this case, Mars) and its observed looping motions. If we suppose that Earth moves faster than Mars, then every so often Earth "overtakes" that planet. Mars will then appear to move backward in the sky, in much the same way as a car we overtake on the highway seems to slip backward relative to us. Notice that in the Copernican picture the planet's looping motions are only apparent. In the Ptolemaic view, they were real.

Copernicus's major motivation for introducing the heliocentric model was simplicity. Even so, he was still influenced by Greek thinking and clung to the idea of circles to model the planets' motions. To bring his theory into agreement with observations, he was forced to retain the idea of epicyclic motion, although with the deferent centered on the Sun rather than on Earth, and with smaller epicycles than in the Ptolemaic picture. Thus, he retained unnecessary complexity and actually gained little in accu-

racy over the geocentric model. The heliocentric model did rectify some small discrepancies and inconsistencies in the Ptolemaic system, but for Copernicus, the primary attraction of heliocentricity was its simplicity, its being "more pleasing to the mind." His theory was more something he *felt* than he could *prove*. To the present day, scientists still are guided by simplicity, symmetry, and beauty in modeling all aspects of the universe.

Despite the support of some observational data, neither his fellow scholars nor the general public easily accepted Copernicus's model. For the learned, heliocentricity went against the grain of much previous thinking and violated many of the religious teachings of the time, largely because it relegated Earth to a noncentral and undistinguished place within the solar system and the universe. And Copernicus's work had little impact on the general populace of his time, at least in part because it was published in Latin (the standard language of academic discourse at the time), which most people could not read. Only long after Copernicus's death, when others—notably Galileo Galilei—popularized his ideas, did the Roman Catholic Church take them seriously enough to bother banning them. Copernicus's writings on the heliocentric universe were placed on the *Index of Prohibited Books* in 1616, 73 years after they were first published. They remained there until the end of the eighteenth century.

✅ Concept Check

■ How do the geocentric and heliocentric models of the solar system differ in their explanations of planetary retrograde motion?

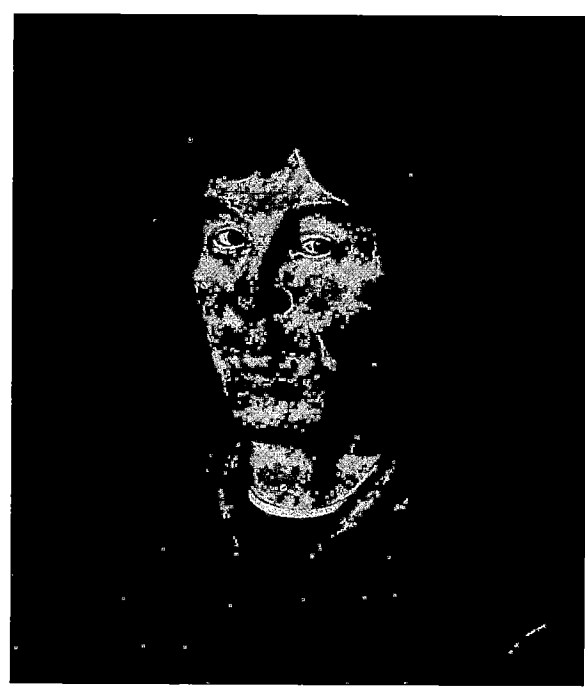

Figure 2.7 Nicholas Copernicus (1473–1543).
(E. Lessing/Art Resource, NY)

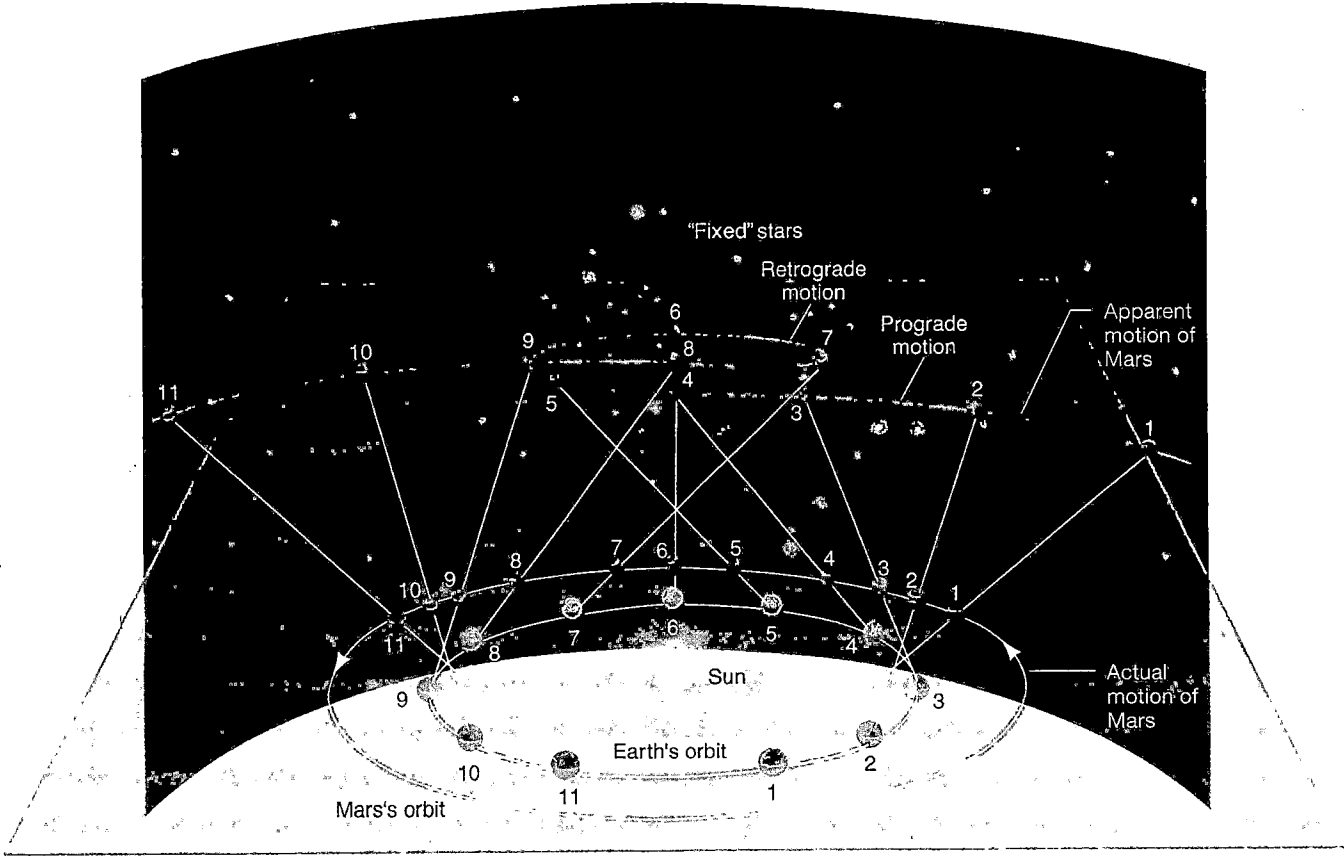

Figure 2.8 Retrograde Motion The Copernican model of the solar system explains both the varying brightnesses of the planets and the phenomenon of retrograde motion. Here, for example, when Earth and Mars are relatively close to one another in their respective orbits (as at position 6), Mars seems brighter. When they are farther apart (as at position 1), Mars seems dimmer. Also, because the line of sight from Earth to Mars changes as the two planets orbit the Sun, Mars appears to loop back and forth in retrograde motion. The line of sight changes because Earth, on the inside track, moves faster in its orbit than does Mars. The white curves are the actual planetary orbits. The apparent motion of Mars, as seen from Earth, is shown as the red curve.

DISCOVERY 2-1

The Foundations of the Copernican Revolution

The following seven points are essentially Copernicus's own words. The italicized material is additional explanation.

1. The celestial spheres do not have just one common center. *Specifically, Earth is not at the center of everything.*

2. The center of Earth is not the center of the universe but is instead only the center of gravity and of the lunar orbit.

3. All the spheres revolve around the Sun. *By spheres, Copernicus meant the planets.*

4. The ratio of Earth's distance from the Sun to the height of the firmament is so much smaller than the ratio of Earth's radius to the distance to the Sun that the distance to the Sun is imperceptible when compared with the height of the firmament. *By firmament, Copernicus meant the distant stars. The point he was making is that the stars are very much farther away than the Sun.*

5. The motions appearing in the firmament are not its motions but those of Earth. Earth performs a daily rotation around its fixed poles while the firmament remains immobile as the highest heaven. *Because the stars are so far away, any apparent motion we see in them is the result of Earth's rotation.*

6. The motions of the Sun are not its motions but the motion of Earth. *Similarly, the Sun's apparent daily and yearly motion are actually due to the various motions of Earth.*

7. What appears to us as retrograde and forward motion of the planets is not their own but that of Earth. *The heliocentric picture provides a natural explanation for retrograde planetary motion, again as a consequence of Earth's motion.*

2.4 The Birth of Modern Astronomy

In the century following the death of Copernicus and the publication of his theory of the solar system, two scientists—Galileo Galilei and Johannes Kepler—made indelible imprints on the study of astronomy. Contemporaries, they were aware of each other's work and corresponded from time to time about their theories. Each achieved fame for his discoveries and made great strides in popularizing the Copernican viewpoint, yet in their approaches to astronomy they were as different as night and day.

GALILEO'S HISTORIC OBSERVATIONS

2.4 Galileo Galilei (Figure 2.9) was an Italian mathematician and philosopher. By his willingness to perform experiments to test his ideas—a rather radical approach in those days (see *Discovery 2-2*)—and by embracing the brand-new technology of the telescope, he revolutionized the way in which science was done, so much so that he is now widely regarded as the father of experimental science.

The telescope was invented in Holland in the early seventeenth century. Hearing of the invention (but without having seen one), Galileo built a telescope for himself in 1609 and aimed it at the sky. What he saw conflicted greatly with the philosophy of Aristotle and provided much new data to support the ideas of Copernicus.*

In fact, Galileo had already abandoned Aristotle in favor of Copernicus, although he had not published these beliefs at the time he began his telescopic observations.

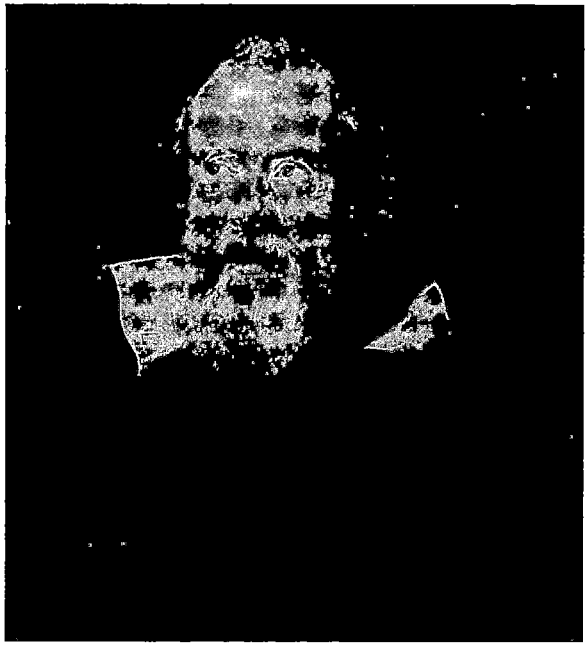

Figure 2.9 Galileo Galilei (1564–1642). *(Art Resource, NY)*

Using his telescope, Galileo discovered that the Moon had mountains, valleys, and craters—terrain in many ways reminiscent of that on Earth. Looking at the Sun (something that should *never* be done directly, and which may have eventually blinded Galileo), he found imperfections—dark blemishes now known as *sunspots*. These observations ran directly counter to the orthodox wisdom of the day. By noting the changing appearance of these sunspots from day to day, Galileo inferred that the Sun *rotates*, approximately once per month, around an axis roughly perpendicular to the ecliptic plane.

Galileo also saw four small points of light, invisible to the naked eye, orbiting the planet Jupiter and realized that they were moons. To Galileo, the fact that another planet had moons provided the strongest support for the Copernican model. Clearly, Earth was not the center of all things. He also found that Venus showed a complete cycle of phases, like those of our Moon (Figure 2.10), a finding that could be explained only by the planet's motion around the Sun. These observations were more strong evidence that Earth is not the center of all things and that at least one planet orbited the Sun.

In 1610, Galileo published a book called *Sidereus Nuncius (The Starry Messenger)*, detailing his observational findings and his controversial conclusions supporting the Copernican theory. In reporting and interpreting the wondrous observations made with his new telescope, Galileo was directly challenging both the scientific orthodoxy and the religious dogma of his day. He was (literally) playing with fire—he must certainly have been aware that only a few years earlier, in 1600, the astronomer Giordano Bruno had been burned at the stake in Rome, in part for his heretical teaching that Earth orbited the Sun. However, by all accounts, Galileo delighted in publicly ridiculing and irritating his Aristotelian colleagues. In 1616 his ideas were judged heretical, Copernicus's works were banned by the Roman Catholic Church, and Galileo was instructed to abandon his astronomical pursuits.

But Galileo would not desist. In 1632 he raised the stakes by publishing *Dialogue Concerning the Two Chief World Systems*, which compared the Ptolemaic and Copernican models. The book presented a discussion among three people: one of them a dull-witted Aristotelian, whose views time and again were roundly defeated by the arguments of one of his two companions, an articulate proponent of the heliocentric system. To make the book accessible to a wide popular audience, Galileo wrote it in Italian rather than Latin. These actions brought Galileo into direct conflict with the authority of the Church. Eventually, the Inquisition forced him, under threat of torture, to retract his claim that Earth orbits the Sun, and he was placed under house arrest in 1633. He remained imprisoned for the rest of his life. Not until 1992 did the Church publicly forgive Galileo's "crimes." But the damage to the orthodox view of the universe was done, and the Copernican genie was out of the bottle once and for all.

Figure 2.10 Venus Phases (a) The phases of Venus, rendered at different points in the planet's orbit. If Venus orbits the Sun and is closer to the Sun than is Earth, as Copernicus maintained, then Venus should display phases, much as our Moon does. As shown here, when directly between Earth and the Sun, Venus's unlit side faces us, and the planet is invisible to us. As Venus moves in its orbit (at a faster speed than Earth moves in its orbit), progressively more of its illuminated face is visible from Earth. Note also the connection between orbital phase and the apparent size of the planet. Venus seems much larger in its crescent phase than when it is full because it is much closer to us during its crescent phase. (The insets at bottom left and right are actual photographs of Venus at two of its crescent phases.) (b) The Ptolemaic model (see also Figure 2.6) is unable to account for these observations. In particular, the full phase of the planet cannot be explained. Seen from Earth, Venus reaches only a "fat crescent" phase, then begins to wane as it nears the Sun.

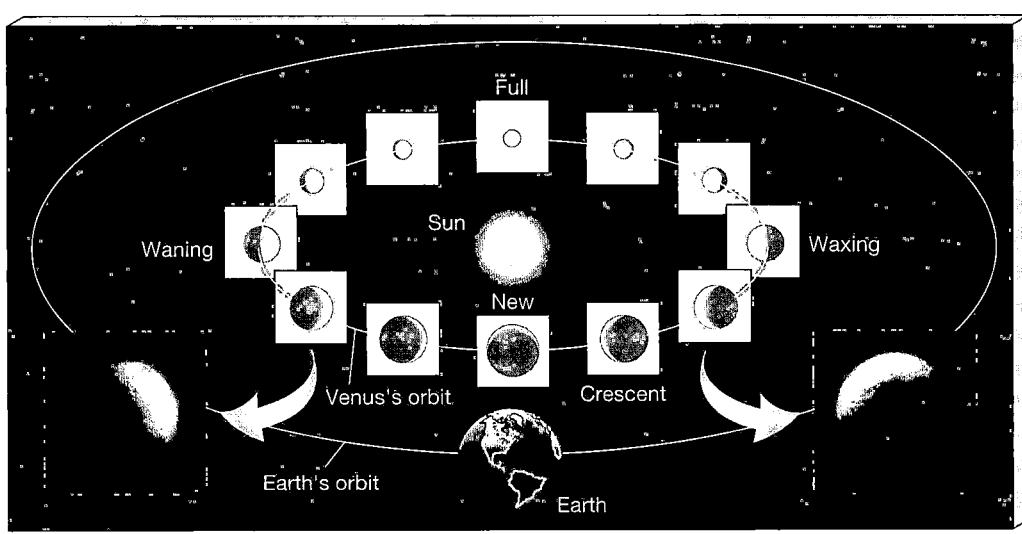

(a)

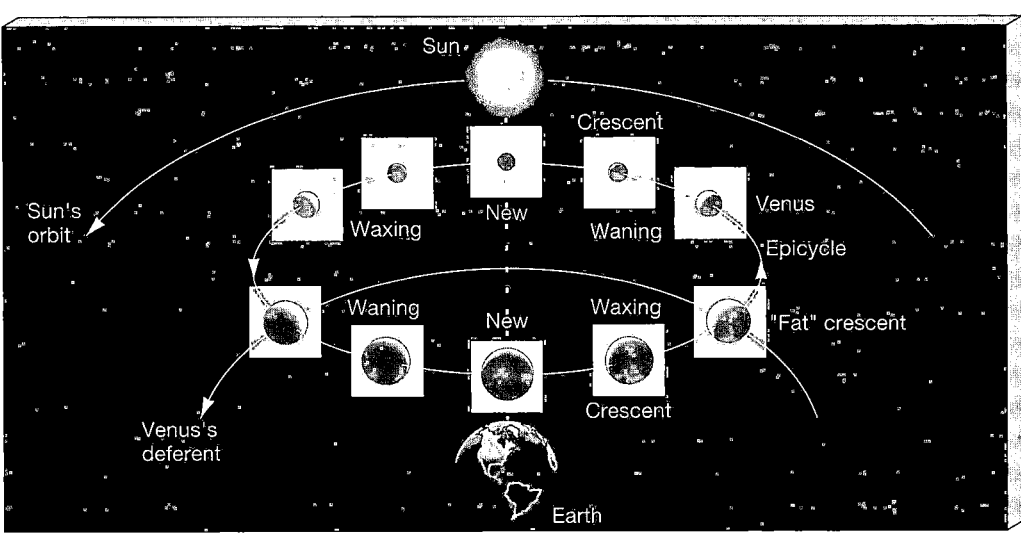

(b)

THE ASCENDANCY OF THE COPERNICAN SYSTEM

Although Renaissance scholars were correct, they could not *prove* that our planetary system is centered on the Sun, or even that Earth moves through space. Direct evidence for this was obtained only in the early eighteenth century, when astronomers discovered the *aberration of starlight*—a slight (20″) shift in the observed direction to a star, caused by Earth's motion perpendicular to the line of sight. Additional proof came in the mid-nineteenth century, with the first unambiguous measurement of stellar parallax. Further verification of the heliocentricity of the solar system came gradually, with innumerable observational tests that culminated with the expeditions of our unmanned space probes of the 1960s, 1970s, and 1980s. The development and eventual acceptance of the heliocentric model were milestones in human thinking. This removal of Earth from any position of great cosmic significance is generally known, even today, as the *Copernican principle*.

The Copernican revolution is a good example of how the scientific method, though affected at any given time by the subjective whims, human biases, and even sheer luck of researchers, does ultimately lead to a definite degree of objectivity. Over time, many groups of scientists checking, confirming, and refining experimental tests can neutralize the subjective attitudes of individuals. Usually one generation of scientists can bring sufficient objectivity to bear on a problem, though some especially revolutionary concepts are so swamped by tradition, religion, and politics that more time is necessary. In the case of heliocentricity, objective confirmation was not obtained until about three centuries after Copernicus published his work and more than 2000 years after Aristarchus had proposed the concept. Nonetheless, objectivity *did in fact* eventually prevail, and our knowledge of the universe has expanded immeasurably as a result.

☑ **Concept Check**

◘ In what ways did Galileo's observations of Venus and Jupiter conflict with the prevailing view at the time?

2.5 The Laws of Planetary Motion

At about the same time as Galileo was becoming famous—or notorious—for his pioneering telescopic observations and outspoken promotion of the Copernican system, Johannes Kepler (Figure 2.11), a German mathematician and astronomer, was developing the laws of planetary motion that now bear his name. Galileo was in many ways the first "modern" observer. He used emerging technology, in the form of the telescope, to achieve new insights into the universe. In contrast, Kepler was a pure theorist. His groundbreaking work that so clarified our knowledge of planetary motion was based almost entirely on the observations of others, principally an extensive collection of data compiled by Tycho Brahe (1546–1601), Kepler's employer and, arguably one of the greatest observational astronomers that has ever lived.

BRAHE'S COMPLEX DATA

2 4 Tycho, as he is often called, was both an eccentric aristocrat and a skillful observer. Born in Denmark, he was educated at some of the best universities in Europe, where he studied astrology, alchemy, and medicine. Most of his observations, which predated the invention of the telescope by several decades, were made at his own observatory, named *Uraniborg*, in Denmark (Figure 2.12). There, using instruments of his own design, Tycho maintained meticulous and accurate records of the stars, planets, and other noteworthy celestial events (including a comet and a supernova, the appearance of which helped convince Tycho that the Aristotelean view of the universe could not be correct).

In 1597, having fallen out of favor with the Danish court, Tycho moved to Prague as Imperial Mathematician of the Holy Roman Empire. Prague happens to be fairly close to Graz, in Austria, where Kepler lived and worked. Kepler joined Tycho in Prague in 1600 and was put to work trying to find a theory that could explain Brahe's planetary data. When Tycho died a year later, Kepler inherited not only Brahe's position but also his most priceless possession: the accumulated observations of the planets, spanning several decades. Tycho's observations, though made with the naked eye, were nevertheless of very high quality. In most cases, his measured positions of stars and planets were accurate to within about 1'. Kepler set to work seeking a unifying principle to explain in detail the motions of the planets, without the need for epicycles. The effort was to occupy much of the remaining 29 years of his life.

Kepler had already accepted the heliocentric picture of the solar system. His goal was to find a simple and

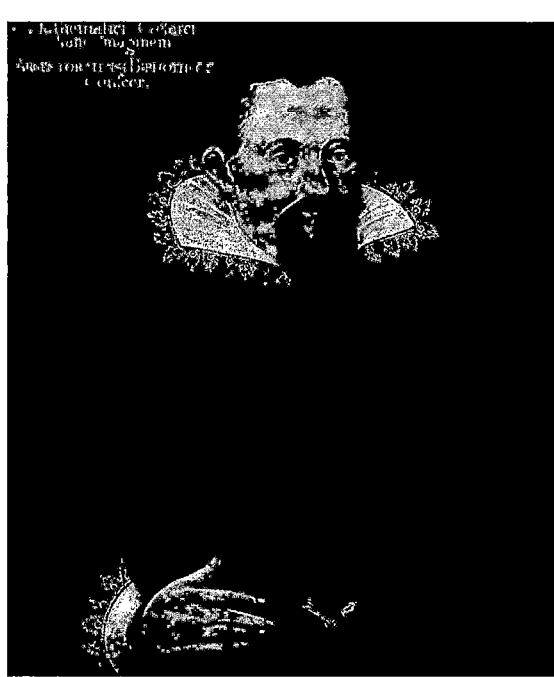

Figure 2.11 Johannes Kepler (1571–1630). (E. Lessing/Art Resource, NY)

Figure 2.12 Tycho Brahe in his observatory *Uraniborg,* on the island of Hveen in Denmark. Brahe's observations of the positions of stars and planets on the sky were the most accurate and complete set of naked-eye measurements ever made. *(The Royal Ontario Museum, © ROM)*

elegant description of the solar system, within the Copernican framework, that fit Tycho's complex mass of detailed observations. In the end, he found it necessary to abandon Copernicus's original simple idea of circular planetary orbits. However, even greater simplicity emerged as a result. After long years of studying Brahe's planetary data and many false starts and blind alleys, Kepler developed the laws of planetary motion that now bear his name.

Kepler determined the shape of each planet's orbit by triangulation—not from different points on Earth, but from different points on Earth's orbit, using observations made at many different times of the year. ↪ (Sec. 1.5) By using a portion of Earth's orbit as a baseline, Kepler was able to measure the relative sizes of the other planetary orbits. Noting where the planets were on successive nights, he found the speeds at which the planets move. We do not know how many geometric shapes Kepler tried for the orbits before he hit upon the correct one. His difficult task was made even more complex because he had to determine Earth's own orbit, too. Nevertheless, he eventually succeeded in summarizing the motions of all the known planets, including Earth, in just three laws, the **laws of planetary motion**.

KEPLER'S SIMPLE LAWS

5 *Kepler's first law* has to do with the *shapes* of the planetary orbits:

> 1. The orbital paths of the planets are elliptical (*not* circular), with the Sun at one focus.

An **ellipse** is simply a flattened circle. Figure 2.13 illustrates a means of constructing an ellipse using a piece of string and two thumbtacks. Each point at which the string is pinned is called a **focus** (plural: *foci*) of the ellipse. The long axis of the ellipse, containing the two foci, is known as the *major axis*. Half the length of this long axis is referred to as the **semi-major axis**; it is a measure of the ellipse's size. The **eccentricity** of the ellipse is the ratio of the distance between the foci to the length of the major axis.

DISCOVERY 2-2

The Scientific Method

The earliest known models of the universe were based largely on imagination and mythology, and made little attempt to explain the workings of the heavens in terms of known earthly experience. However, history shows that some early scientists did come to realize the importance of careful observation and testing to the formulation of their theories. The success of their approach changed, slowly but surely, the way science was done and opened the door to a fuller understanding of nature.

As knowledge from all sources was sought and embraced for its own sake, the influence of logic and reasoned argument grew and the power of myth diminished. People began to inquire more critically about themselves and the universe. They realized that thinking about nature was no longer sufficient; looking at it was also necessary. Experiments and observations became a central part of the process of inquiry. To be effective, a *theory*—the framework of ideas and assumptions used to explain some set of observations and make predictions about the real world—must be continually tested. If experiments and observations favor it, a theory can be further developed and refined, but if they do not, it must be rejected, no matter how appealing it originally seemed.

The process is illustrated (very) schematically in the accompanying figure. This new approach to investigation, combining thinking and doing—that is, theory and experiment—is often known as the *scientific method*. It separates science from pseudoscience, fact from fiction.

Notice that there is no "end point" to the process depicted in the figure. A theory can be invalidated by a single wrong prediction, but no amount of observation or experimentation can ever prove it correct. Theories simply become more and more widely accepted as their predictions are repeatedly confirmed.

The notion that theories must be tested and may be proven wrong sometimes leads people to dismiss their importance. We have all heard the expression, "Of course, it's only a theory," used to undermine an idea that someone finds unacceptable. Don't be fooled by this abuse of the concept! Gravity (Section 2.7) is only a theory, but calculations based on it have guided human spacecraft throughout the solar system. Electromagnetism (Chapter 3) and quantum mechanics (Chapter 4) are theories too, yet they form the foundation for most of twentieth (and twenty-first) century technology.

In astronomy we are rarely afforded the luxury of performing experiments to test our theories, so observation becomes vitally important. The birth of modern science is usually associated with the Renaissance (and indeed, scientific inquiry can legitimately be said to have blossomed since the time of the Copernican revolution). However, one of the first documented uses of the scientific method

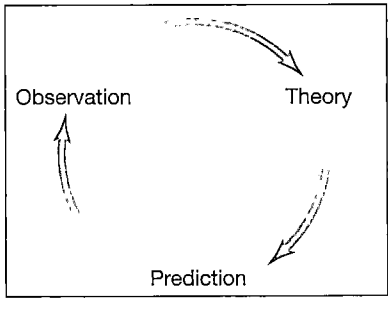

Note that, while the Sun resides at one focus, the other focus is empty and has no particular physical significance.

The length of the semi-major axis and the eccentricity are all we need to describe the size and shape of a planet's orbital path (see *More Precisely 2-1*). A circle is a special kind of ellipse in which the two foci happen to coincide, so the eccentricity is zero. The semi-major axis of a circle is simply its radius. In fact, no planet's elliptical orbit is nearly as elongated as the one shown in Figure 2.13. With two exceptions (the paths of Mercury and Pluto), planetary orbits in our solar system have such small eccentricities that our eyes would have trouble distinguishing them from true circles. Only because the orbits are so nearly circular were the Ptolemaic and Copernican models able to come as close as they did to describing reality.

Kepler's substitution of elliptical for circular orbits was no small advance. It amounted to abandoning an aesthetic bias—the Aristotelian belief in the perfection of the circle—that had governed astronomy since Greek antiquity. Even Galileo Galilei, not known for his conservatism in

scholarly matters, clung to the idea of circular motion and never accepted that the planets move on elliptical paths.

Kepler's second law, illustrated in Figure 2.14, addresses the speed at which a planet traverses different parts of its orbit:

> II. An imaginary line connecting the Sun to any planet sweeps out equal areas of the ellipse in equal intervals of time.

While orbiting the Sun, a planet traces the arcs labeled A, B, and C in Figure 2.14 in equal times. Notice, however, that the distance traveled by the planet along arc C is greater than the distance traveled along arc A or arc B. Because the time is the same and the distance is different, the speed must vary. When a planet is close to the Sun, as in sector C, it moves much faster than when farther away, as in sector A.

By taking into account the relative speeds and positions of the planets in their elliptical orbits about the

---

in an astronomical context was performed by Aristotle (384–322 B.C.) nearly 25 centuries ago.

Aristotle noticed that, during a lunar eclipse, when Earth is positioned between the Sun and the Moon, it casts a curved shadow onto the Moon's surface. The accompanying figure shows a series of photographs taken during a recent lunar eclipse. Earth's shadow is indeed slightly curved. This is what Aristotle must have seen and recorded so long ago.

Because the observed shadow seemed always to be an arc of the same circle, Aristotle concluded that Earth, the cause of the shadow, must be a sphere. On the basis of this *hypothesis*—this possible explanation of the observed facts—he then went on to predict that any and all future lunar eclipses would show that Earth's shadow was curved, regardless of the orientation of our planet. That prediction has been tested every time a lunar eclipse has occurred. It has yet to be proved wrong. Aristotle was not the first person to argue that Earth is round, but he was apparently the first to offer evidence of it using the lunar-eclipse method.

The reasoning procedure Aristotle used forms the basis of all scientific inquiry today. Observation, theory,

and testing are the cornerstones of the scientific method, a technique whose power will be demonstrated again and again throughout our text. Scientists throughout the world today use an approach that relies heavily on testing ideas. They gather data, form a working hypothesis that explains the data, then proceed to test the implications of the hypothesis using experiment and observation. Eventually, one or more "well-tested" hypotheses may be elevated to the stature of physical laws and come to form the basis of a theory of even broader applicability. The new predictions of the theory will in turn be tested, as scientific knowledge continues to grow.

Experiment and observation are integral parts of the process of scientific inquiry. Untestable theories, or theories unsupported by experimental evidence, rarely gain any measure of acceptance in scientific circles. Used properly over a period of time, this rational, methodical approach enables us to arrive at conclusions that are mostly free of the personal bias and human values of any one scientist. The scientific method is designed to yield an objective view of the universe we inhabit.

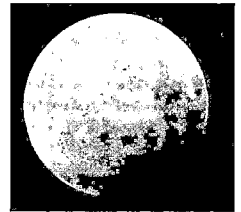 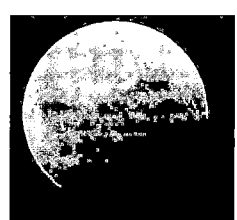 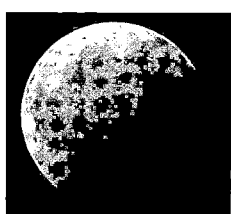 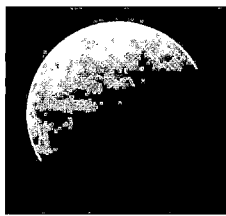 

(G. Schneider)

R I V U X G

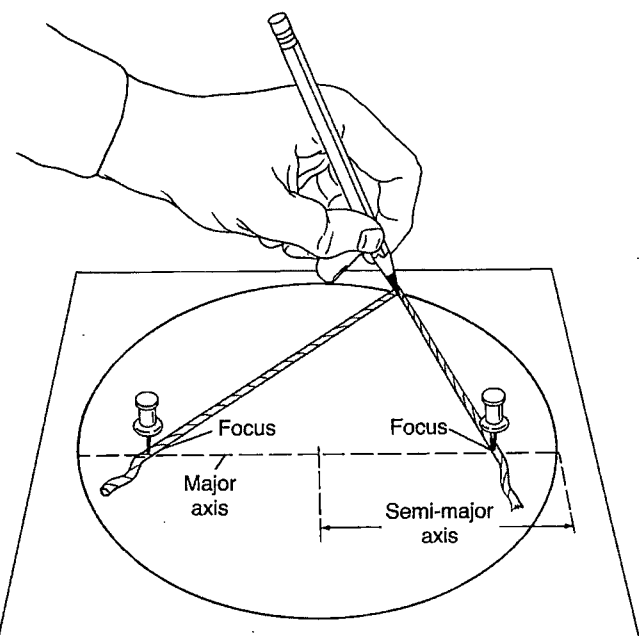

Figure 2.13 Ellipse An ellipse can be drawn with the aid of a string, a pencil, and two thumbtacks. The wider the separation of the foci, the more elongated, or eccentric, is the ellipse. In the special case where the two foci are at the same place, the drawn curve is a circle.

Sun, Kepler's first two laws explained the variations in planetary brightness and some observed peculiar nonuniform motions that could not be accommodated within the assumption of circular motion, even with the inclusion of epicycles. Gone at last were the circles within circles that rolled across the sky. Kepler's modification of the Copernican theory to allow the possibility of elliptical orbits both greatly simplified the model of the solar system and at the same time provided much greater predictive accuracy than had previously been possible. Note, too, that these laws are not restricted to planets. They apply to *any* orbiting object. Spy satellites, for example, move very rapidly as they swoop close to Earth's surface not because they are propelled with powerful on-board rockets but because their highly eccentric orbits are governed by Kepler's laws.

Kepler published his first two laws in 1609, stating that he had proved them only for the orbit of Mars. Ten years later, he extended them to all the then-known planets (Mercury, Venus, Earth, Mars, Jupiter, and Saturn) and added a third law relating the size of a planet's orbit to its sidereal orbital **period**—the time needed for the planet to complete one circuit around the Sun. *Kepler's third law* states that

III. The square of a planet's orbital period is proportional to the cube of its semi-major axis.

This law becomes particularly simple when we choose the (Earth sidereal) year as our unit of time and the *astronomi-*

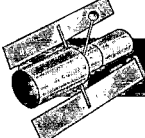

MORE PRECISELY 2-1

Some Properties of Planetary Orbits

Two numbers—semi-major axis and eccentricity—are all that are needed to describe the size and shape of a planet's orbital path. From them we can derive many other useful quantities. Two of the most important are the planet's *perihelion* (its point of closest approach to the Sun) and its *aphelion* (greatest distance from the Sun). From the definitions presented in the text, it follows that if the planet's orbit has semi-major axis a and eccentricity e, its perihelion is at a distance $a(1 - e)$ from the Sun, while its aphelion is at $a(1 + e)$. These points and distances are illustrated in the accompanying figure.

EXAMPLE: A (hypothetical) planet with a semi-major axis of 400 million km and an eccentricity of 0.5 (the eccentricity of the ellipse shown in the diagram) would range between $400 \times (1 - 0.5) = 200$ million km and 400×3 $(1 + 0.5) = 600$ million km from the Sun over the course of one complete orbit. With $e = 0.9$, the range would be 40–760 million km, and so on.

No planet has an orbital eccentricity as large as 0.5—the planet with the most eccentric orbit is Pluto, with $e = 0.249$ (see Table 2.1). However, many meteoroids, and all

comets (see Chapter 14) have eccentricities considerably greater than this. In fact, most comets visible from Earth have eccentricities very close to $e = 1$. Their highly elongated orbits approach within a few astronomical units of the Sun at perihelion, yet these tiny frozen worlds spend most of their time far beyond the orbit of Pluto.

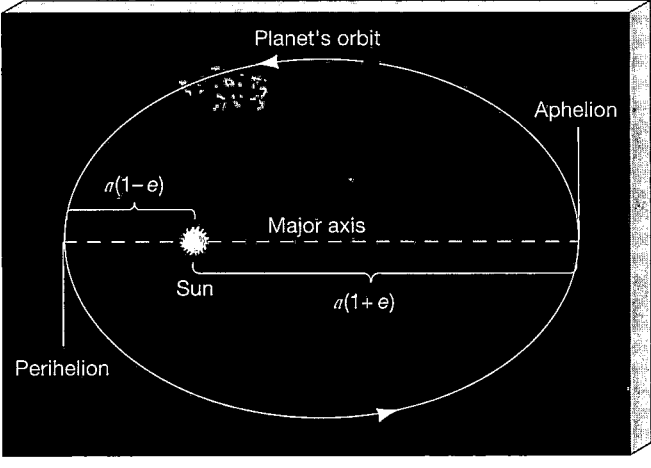

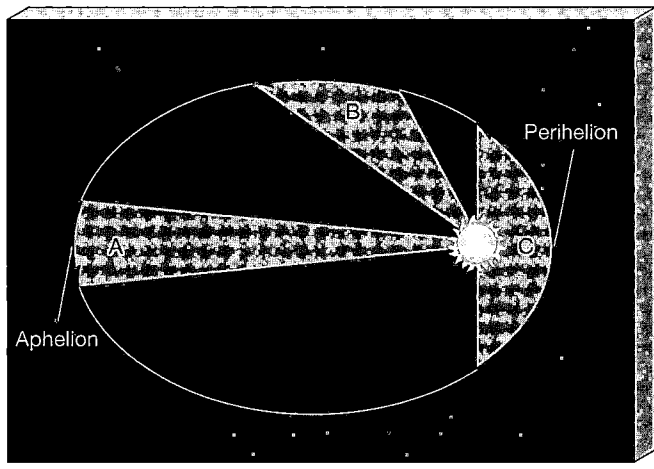

Figure 2.14 Kepler's Second Law Equal areas are swept out in equal intervals of time. The three shaded areas (A, B, and C) are equal. Note that an object would travel the length of each of the three red arrows in the same amount of time. Therefore, planets move faster when closer to the Sun.

cal unit as our unit of length. One **astronomical unit** (A.U.) is the semi-major axis of Earth's orbit around the Sun—essentially the average distance between Earth and the Sun. Like the light-year, the astronomical unit is custom-made for the vast distances encountered in astronomy. Using these units for time and distance, we can write Kepler's third law for any planet as

$$P^2 \text{ (in Earth years)} = a^3 \text{ (in astronomical units),}$$

where P is the planet's sidereal orbital period, and a is the length of its semi-major axis. The law implies that a planet's "year" P increases more rapidly than does the size of its orbit a. For example, Earth, with an orbital semi-major axis of 1 A.U., has an orbital period of one Earth year. The planet Venus, orbiting at a distance of roughly 0.7 A.U., takes only 0.6 Earth years—about 225 days—to complete

one circuit. By contrast, Saturn, almost 10 A.U. from the Sun, takes considerably more than 10 Earth years—in fact, nearly 30 years—to orbit the Sun just once.

Table 2.1 presents basic data describing the orbits of the nine planets now known. Renaissance astronomers knew these properties for the innermost six planets and used them to construct the currently accepted heliocentric model of the solar system. The second column presents each planet's orbital semi-major axis, measured in astronomical units; the third column gives the orbital period, in Earth years. The fourth column lists the planets' orbital eccentricities. For purposes of verifying Kepler's third law, the fifth column lists the ratio P^2/a^3. As we have just seen, the third law implies that this number should always equal one in the units used in the table.

The main points to be grasped from Table 2.1 are these: (1) With the exception of Mercury and Pluto, the planets' orbits are nearly circular (that is, their eccentricities are close to zero); and (2) the farther a planet is from the Sun, the greater is its orbital period, in agreement with Kepler's third law to within the accuracy of the numbers in the table. (The small but significant deviations of P^2/a^3 from one in the cases of Uranus and Neptune are caused by the gravitational attraction between those two planets; see Chapter 13.) Most important, note that Kepler's laws are obeyed by *all* the known planets, *not just by the six on which he based his conclusions.*

The laws developed by Kepler were far more than mere fits to existing data. They also made definite, testable predictions about the future locations of the planets. Those predictions have been borne out to high accuracy every time they have been tested by observation—the hallmark of any scientific theory.

Concept Check

▣ Do Kepler's laws apply to comets orbiting the Sun? Do they apply to the moons of Jupiter?

TABLE 2.1 Some Solar System Dimensions

PLANET	ORBITAL SEMI-MAJOR AXIS, a (astronomical units)	ORBITAL PERIOD, P (Earth years)	ORBITAL Eccentricity, e	P^2/a^3
Mercury	0.387	0.241	0.206	1.002
Venus	0.723	0.615	0.007	1.001
Earth	1.000	1.000	0.017	1.000
Mars	1.524	1.881	0.093	1.000
Jupiter	5.203	11.86	0.048	0.999
Saturn	9.537	29.42	0.054	0.998
Uranus	19.19	83.75	0.047	0.993
Neptune	30.07	163.7	0.009	0.986
Pluto	39.48	248.0	0.249	0.999

2.6 The Dimensions of the Solar System

Kepler's laws allow us to construct a scale model of the solar system, with the correct shapes and *relative* sizes of all the planetary orbits, but they do not tell us the *actual* size of any orbit. We can express the distance to each planet only in terms of the distance from Earth to the Sun. Why is this? Because Kepler's triangulation measurements all used a portion of Earth's orbit as a baseline, distances could be expressed only relative to the size of that orbit, which was itself undetermined. Thus our model of the solar system would be like a road map of the United States showing the *relative* positions of cities and towns but lacking the all-important scale marker indicating distances in kilometers or miles. For example, we would know that Kansas City is about three times more distant from New York than it is from Chicago, but we would not know the actual mileage between any two points on the map.

If we could somehow determine the value of the astronomical unit—in kilometers, say—we would be able to add the vital scale marker to our map of the solar system and compute the exact distances between the Sun and each of the planets. We might propose using triangulation to measure the distance from Earth to the Sun directly. However, we would find it impossible to measure the Sun's parallax using Earth's diameter as a baseline. The Sun is too bright, too big, and too fuzzy for us to distinguish any apparent displacement relative to a field of distant stars. To measure the Sun's distance from Earth, we must resort to some other method.

Before the middle of the twentieth century, the most accurate measurements of the astronomical unit were made using triangulation on the planets Mercury and Venus during their rare *transits* of the Sun—that is, during the brief periods when those planets passed directly between the Sun and Earth (as shown for the case of Mercury in Figure 2.15). Because the time at which a transit occurs can be determined with great precision, astronomers can use this information to make accurate measurements of a planet's position in the sky. They can then use simple geometry to compute the distance to the planet by combining observations made from different locations on Earth, as discussed earlier in Chapter 1. ⟳ (Sec. 1.5) For example, the parallax of Venus at closest approach to Earth, as seen from two diametrically opposite points on Earth (separated by about 13,000 km), is about 1 arc minute (1/60°)—at the limit of naked-eye capabilities but easily measurable telescopically. Using the second formula presented in *More Precisely 1-4*, we find that this parallax represents a distance of 13,000 km × 57.3°/(1/60°), or approximately 45,000,000 km.

Knowing the distance to Venus, we can compute the magnitude of the astronomical unit. Figure 2.16 is an idealized diagram of the Sun–Earth–Venus orbital geometry. The planetary orbits are drawn as circles here, but in reality they are slight ellipses. This is a subtle difference, and

we can correct for it using detailed knowledge of orbital motions. Assuming for the sake of simplicity that the orbits are perfect circles, we see from the figure that the distance from Earth to Venus at closest approach is approximately 0.3 A.U. Knowing that 0.3 A.U. is 45,000,000 km makes determining 1 A.U. straightforward—the answer is 45,000,000/0.3, or 150,000,000 km.

The modern method for deriving the absolute scale (that is, the scale expressed in kilometers, rather than just relative to Earth's orbit) of the solar system uses radar rather than triangulation. The word **radar** is an acronym for **ra**dio **d**etection **a**nd **r**anging. In this technique, radio waves are transmitted toward an astronomical body, such as a planet. (We cannot use radar ranging to measure the distance to the Sun directly because radio signals are absorbed at the solar surface and not reflected to Earth.) The returning echo indicates the body's direction and range, or distance, in absolute terms—that is, in kilometers rather than in astronomical units. Multiplying the 300-second round-trip travel time of the radar signal (the time elapsed between transmission of the signal and reception of the echo) by the speed of light (300,000 km/s, which is also the speed of radio waves), we obtain twice the distance to the target planet.

Venus, whose orbit periodically brings it closest to Earth, is the most common target for radar ranging. The round-trip travel time (for example, at closest approach, as indicated by the wavy lines on Figure 2.16) can be measured with high precision—in fact, well enough to determine the

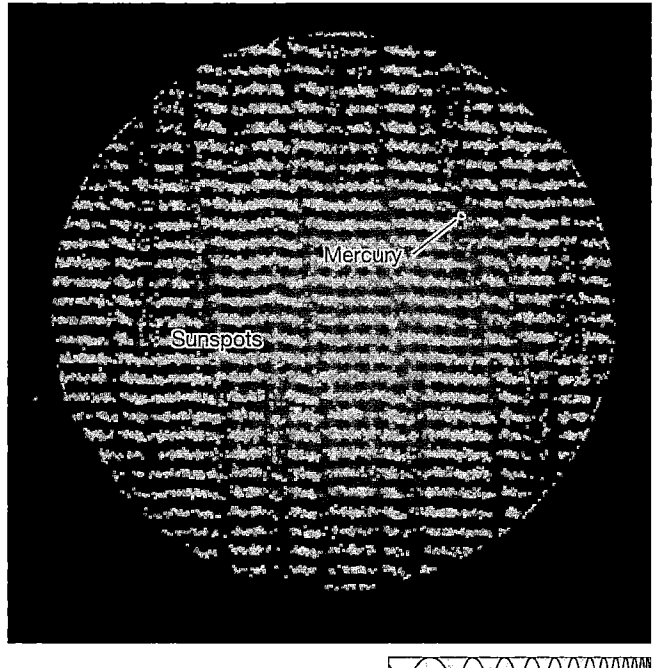

Figure 2.15 Solar Transit The transit of Mercury across the face of the Sun. Such transits happen only about once per decade, because Mercury's orbit does not quite coincide with the plane of the ecliptic. Transits of Venus are even rarer, occurring only about twice per century. *(AURA)*

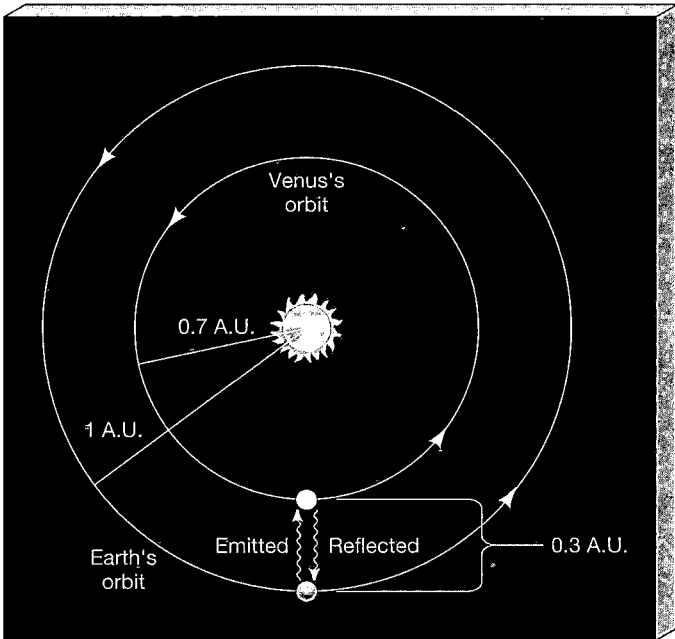

Figure 2.16 Astronomical Unit Simplified geometry of the orbits of Earth and Venus as they move around the Sun. The wavy lines represent the paths along which radar signals might be transmitted toward Venus and received back at Earth at the particular moment (chosen for simplicity) when Venus is at its minimum distance from Earth. Because the radius of Earth's orbit is 1 A.U. and that of Venus is about 0.7 A.U., we know that this distance is 0.3 A.U. Thus, radar measurements allow us to determine the astronomical unit in kilometers.

planet's distance to an accuracy of about 1 km. In this way, the astronomical unit is now known to be 149,597,870 km. We will use the rounded-off value of 1.5×10^8 km in this text.

Having determined the value of the astronomical unit, we can reexpress the sizes of the other planetary orbits in terms of more familiar units, such as miles or kilometers. The entire scale of the solar system can then be calibrated to high precision.

☑ Concept Check

■ Why don't Kepler's laws tell us the value of the astronomical unit?

2.7 Newton's Laws

⁷ Kepler's three laws, which so simplified the solar system, were discovered *empirically*. In other words, they resulted solely from the analysis of observational data and were not derived from any theory or mathematical model. Indeed, Kepler did not have any appreciation for the physics underlying his laws. Nor did Copernicus understand the basic reasons *why* his heliocentric model of the solar system worked. Even Galileo, often called the father

of modern physics, failed to understand why the planets orbit the Sun (although Galileo's work laid vital groundwork for Newton's theories—see *More Precisely 2-2*).

What prevents the planets from flying off into space or from falling into the Sun? What causes them to revolve about the Sun, apparently endlessly? To be sure, the motions of the planets obey Kepler's three laws, but only by considering something more fundamental than those laws can we really understand these motions. The heliocentric system was secured when, in the seventeenth century, the British mathematician Isaac Newton (Figure 2.17) developed a deeper understanding of the way *all* objects move and interact with one another as they do.

THE LAWS OF MOTION

Isaac Newton was born in Lincolnshire, England, on Christmas Day in 1642, the year Galileo died. Newton studied at Trinity College of Cambridge University, but when the bubonic plague reached Cambridge in 1665, he returned to the relative safety of his home for two years. During that time he made probably the most famous of his discoveries, the law of gravity (although it is but one of the many major scientific advances for which Newton was responsible). However, either because he regarded the theory as incomplete or possibly because he was afraid that he would be attacked or plagiarized by his colleagues, he did not tell anyone of his monumental achievement for almost 20 years. It was not until 1684, when Newton was discussing with Edmund Halley (of Halley's comet fame) the leading astronomical problem of the day—*Why* do the planets move according to Kepler's laws?—that he astounded his companion by

Figure 2.17 Isaac Newton (1642–1727). *(The Granger Collection)*

MORE PRECISELY 2-2

The Moon Is Falling!

The story of Isaac Newton seeing an apple fall to the ground and "discovering" gravity is well known, in one form or another, to most high-school students. However, the real importance of Newton's observation was his realization that, by observing falling bodies on Earth and elsewhere, he could *quantify* the properties of the gravitational force and deduce the mathematical form of his law of gravitation.

Galileo Galilei had demonstrated some years earlier, by the simple experiment of dropping different objects from a great height (the top of the Tower of Pisa, according to lore) and noting that they hit the ground at the same time (at least, to the extent that air resistance was unimportant), that gravity causes the *same* acceleration in all bodies, regardless of mass. Since acceleration is proportional to force divided by mass (Newton's second law), this meant that the gravitational force on one body due to another had to be directly proportional to the first body's mass. Applying the same reasoning to the other body and using Newton's third law, it follows that the force must also be proportional to the mass of the second body, hence the two "mass" terms in the law of gravity. (The experimental finding that the gravitational force is precisely proportional to mass is now known as the *equivalence principle*. It forms an essential part of the modern theory of gravity; see *More Precisely 22-1*).

What about the "inverse-square" part of Newton's law? At Earth's surface, the acceleration due to gravity is approximately 9.80 m/s². It is denoted by the letter *g*. Where else other than on Earth could a falling body be seen? As illustrated in the accompanying figures, Newton realized that he could tell how gravity varies with distance by studying another object influenced by our planet's gravity—the Moon. Here's how he did it.

Let's assume for the sake of simplicity that the Moon's orbit around Earth is circular. As shown in the second figure, even though the Moon's orbital *speed* is constant, its *velocity* (red arrows) is not—the direction of the Moon's motion is steadily changing. In other words, the Moon is *accelerating*, constantly falling toward Earth. In fact, the acceleration of any body moving with speed *v* in a circular orbit of radius *r* may be shown to be

$$a = \frac{v^2}{r},$$

always directed toward the center of the circle (toward Earth, in the case of the Moon). This acceleration is sometimes called *centripetal* ("center-seeking") acceleration. You probably already have an intuitive experience for this equation—just think of the acceleration you feel as you take a tight corner (small *r*) at high speed (large *v*) in your car.

Knowing the Moon's distance *r* = 384,000 km (measured by triangulation), and the Moon's sidereal orbit period *P* = 27.3 days, Newton computed the Moon's orbital speed *v* = 2π*r*/*P* = 1.02 km/s, and hence determined its acceleration to be *a* = 0.00272 m/s², or 0.000278 *g*. ⚏ *(More Precisely 1-4)* Thus, Newton found, the Moon, lying 60 times farther from Earth's center than the apple falling from the tree in his garden (taking Earth's radius to be 6400 km), experiences an acceleration 3600, or 60², times smaller. In other words, *the acceleration due to gravity is inversely proportional to the square of the distance*. Isaac Newton's application of simple geometric reasoning and some very basic laws of motion resulted in a breakthrough that would revolutionize astronomers' view of the solar system, and pave the way for humanity's exploration of the universe.

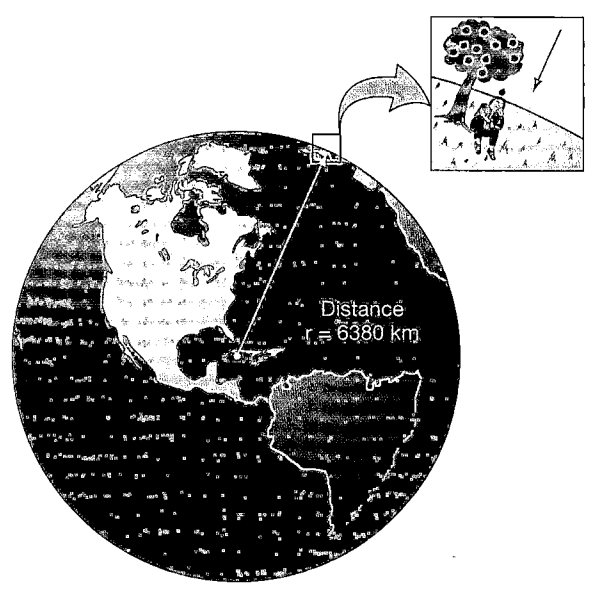

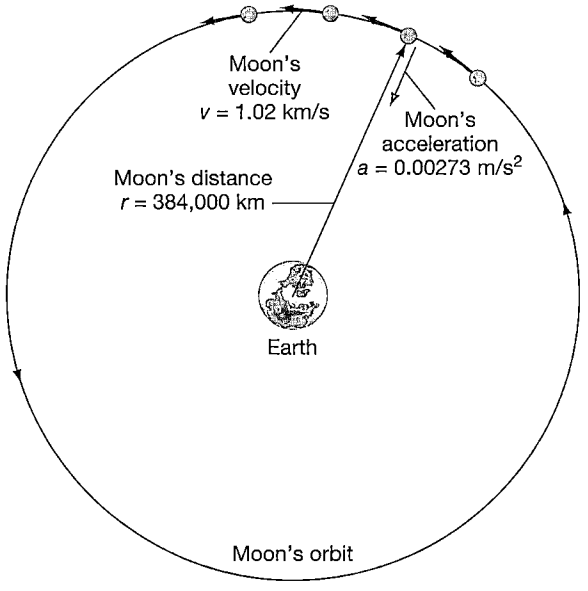

revealing that he had solved the problem in its entirety nearly two decades before!

Prompted by Halley, Newton published his theories in perhaps the most influential physics book ever written: *Philosophiae Naturalis Principia Mathematica* (*The Mathematical Principles of Natural Philosophy*—what we would today call "science"), usually known simply as Newton's *Principia*. The ideas expressed in that work form the basis for what is now known as **Newtonian mechanics**. Three basic laws of motion, the law of gravity, and a little calculus (which Newton also developed) are sufficient to explain and quantify virtually all the complex dynamic behavior we see on Earth and throughout the universe.

Figure 2.18 illustrates *Newton's first law of motion*:

I. Every body continues in a state of rest or in a state of uniform motion in a straight line unless it is compelled to change that state by a force acting on it.

The first law simply states that a moving object will move forever in a straight line unless some external **force**—a push or a pull—changes its direction of motion. For example, the object might glance off a brick wall or be hit with a baseball bat; in either case, a force changes the original motion of the object. The tendency of an object to keep moving at the same speed and in the same direction unless acted upon by a force is known as **inertia**. A familiar measure of an object's inertia is its **mass**—loosely speaking, the total amount of matter it contains. The greater an object's mass, the more inertia it has, and the greater is the force needed to change its state of motion.

Newton's first law implies that it requires no force to maintain motion in a straight line with *constant velocity*—that is, motion with constant speed and constant direction in space. This contrasts sharply with the view of Aristotle, who maintained (incorrectly) that the natural state of an object was to be *at rest*—most probably an opinion based on Aristotle's observations of the effect of friction. In our discussion we will neglect friction—the force that slows balls rolling

along the ground, blocks sliding across tabletops, and baseballs moving through the air. In any case, it is not an issue for the planets because there is no appreciable friction in outer space. The fallacy in Aristotle's argument was first realized and exposed by Galileo, who conceived of the notion of inertia long before Newton formalized it into a law.

The rate of change of the velocity of an object—speeding up, slowing down, or simply changing direction—is called its **acceleration**, and is the subject of *Newton's second law*, which states that the acceleration of an object is directly proportional to the applied force and inversely proportional to its mass:

II. When a force F acts on a body of mass m, it produces in it an acceleration a equal to the force divided by the mass. Thus, $a = F/m$, or $F = ma$.

Thus, the greater the force acting on the object, or the smaller the mass of the object, the greater its acceleration. If two objects are pulled with the same force, the more massive one will accelerate less; if two identical objects are pulled with different forces, the one experiencing the greater force will accelerate more. In honor of Newton, the SI unit of force is named after him. By definition, 1 newton (N) is the force required to cause a mass of 1 kilogram to accelerate at a rate of one meter per second every second (1 m/s²). One newton is approximately 0.22 pounds.

Finally, *Newton's third law* simply tells us that forces cannot occur in isolation:

III. To every action there is an equal and opposite reaction.

In other words, if body A exerts a force on body B, then body B necessarily exerts a force on body A that is equal in magnitude, but oppositely directed.

Only in extreme circumstances—when velocities approach the speed of light—do Newton's laws break down, and this fact was not realized until the twentieth century, when Albert Einstein's theories of relativity once again

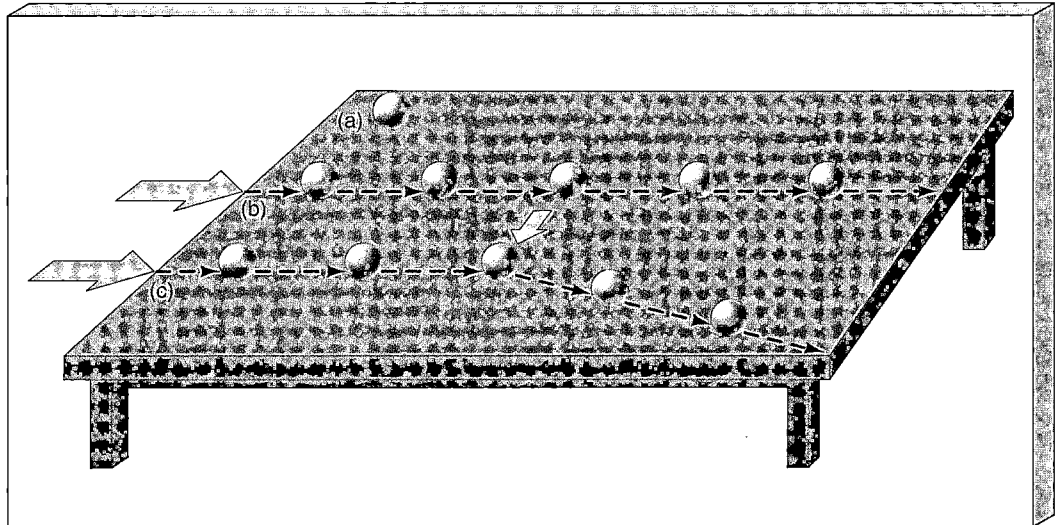

Figure 2.18
Newton's First Law An object at rest will remain at rest (a) until some force acts on it (b). It will then remain in that state of uniform motion until another force acts on it. The arrow in (c) shows a second force acting at a direction different from the first, which causes the object to change its direction of motion.

revolutionized our view of the universe (see Chapter 22). Most of the time, however, Newtonian mechanics provides an excellent description of the motion of planets, stars, and galaxies through the cosmos.

GRAVITY

Forces may act *instantaneously* or *continuously*. The force from a baseball bat that hits a home run can be thought of as being instantaneous. A good example of a continuous force is the one that prevents the baseball from zooming off into space—**gravity**, the phenomenon that started Newton on the path to the discovery of his laws. Newton hypothesized that any object having mass always exerts an attractive *gravitational force* on all other massive objects. The more massive an object, the stronger its gravitational pull.

Consider a baseball thrown upward from Earth's surface, as illustrated in Figure 2.19. In accordance with Newton's first law, the downward force of Earth's gravity continuously modifies the baseball's velocity, slowing the initial upward motion and eventually causing the ball to fall back to the ground. Of course, the baseball, having some mass of its own, also exerts a gravitational pull on Earth. By Newton's third law, this force is equal and opposite to the *weight* of the ball (the force with which Earth attracts it). But, by Newton's second law, Earth has a much greater effect on the light baseball than the baseball has on the much more massive Earth. The ball and Earth feel the same gravitational force, but Earth's *acceleration* is much smaller.

Now consider the trajectory of the same baseball batted from the surface of the Moon. The pull of gravity is about one-sixth as great on the Moon as on Earth, so the baseball's velocity changes more slowly—a typical home run in a ballpark on Earth would travel nearly half a mile on the Moon. The Moon, less massive than Earth, has less gravitational influence on the baseball. The magnitude of the gravitational force, then, depends on the *masses* of the attracting bodies. In fact, the force is *directly proportional* to the product of the two masses.

Studying the motions of the planets reveals a second aspect of the gravitational force. At locations equidistant from the Sun's center, the gravitational force has the same strength, and it is always directed toward the Sun. Furthermore, detailed calculation of the planets' accelerations as they orbit the Sun reveals that the strength of the Sun's gravitational pull decreases in proportion to the *square* of the distance from the Sun. (See also *More Precisely 2-2* for the line of reasoning that originally led Newton to this conclusion.) The force of gravity is said to obey an **inverse-square law.** As shown in Figure 2.20(b), inverse-square forces decrease rapidly with distance from their source. For example, tripling the distance makes the force $3^2 = 9$ times weaker, while multiplying the distance by five results in a force that is $5^2 = 25$ times weaker. Despite this rapid decrease, the force never quite reaches zero. The gravitational pull of an object having some mass can never be completely extinguished.

We can combine the preceding statements about mass and distance to form a law of gravity that dictates the way in which *all* material objects attract one another:

> Every particle of matter in the universe attracts every other particle with a force that is directly proportional to the product of the masses of the particles and inversely proportional to the square of the distance between them.

As a proportionality, Newton's law of gravity is

$$\text{gravitational force} \propto \frac{\text{mass of object 1} \times \text{mass of object 2}}{\text{distance}^2}$$

(The symbol $\propto$ here means "is proportional to.") The rule for computing the force F between two bodies of masses m_1 and m_2, separated by distance r, is usually written more compactly as

$$F = \frac{Gm_1m_2}{r^2}$$

The quantity G is known as the *gravitational constant*, or often simply as Newton's constant. It is one of the fundamental constants of the universe. Its value has been measured in extremely delicate laboratory experiments as 6.67×10^{-11} newton meter^2/kilogram^2 (N m^2/kg^2).

To Newton, gravity was a force that acted at a distance, with no obvious way in which it was actually transmitted from place to place. Newton was not satisfied with this explanation, but he had none better. To appreciate the modern view of gravity, consider any piece of matter having some

Figure 2.19 Gravity A ball thrown up from the surface of a massive object such as a planet is pulled continuously by the gravity of that planet (and, conversely, the gravity of the ball continuously pulls the planet).

mass—it could be smaller than an atom or larger than a galaxy. Extending outward from this object in all directions is a **gravitational field** produced by the matter. We now regard such a field as a property of space itself—a property that determines the influence of one massive object on another. All other matter "feels" the field as a gravitational force.

PLANETARY MOTION

⑧ The mutual gravitational attraction between the Sun and the planets, as expressed by Newton's law of gravity, is responsible for the observed planetary orbits. As depicted

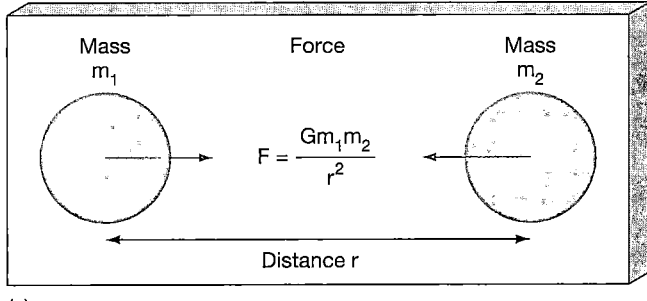

(a)

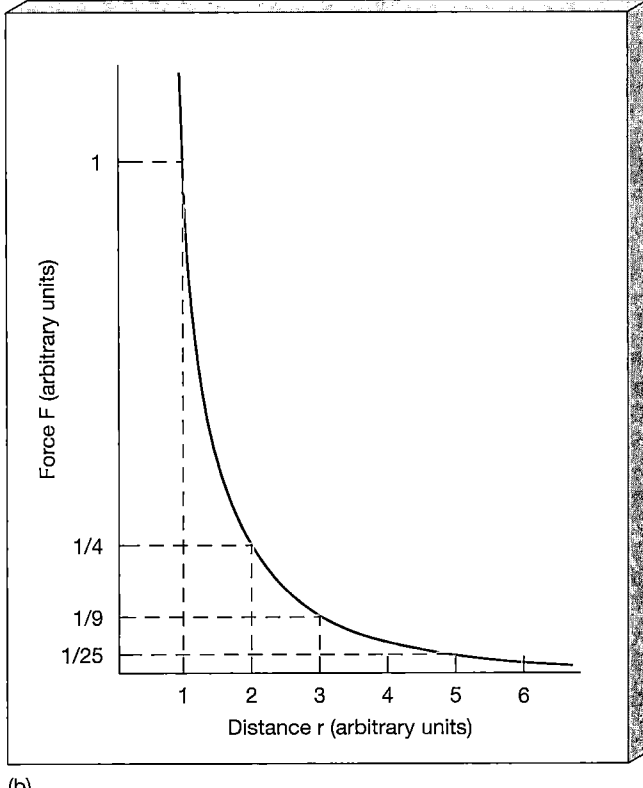

(b)

in Figure 2.21, this gravitational force continuously pulls each planet toward the Sun, deflecting its forward motion into a curved orbital path. Because the Sun is much more massive than any of the planets, it dominates the interaction. We might say that the Sun "controls" the planets, not the other way around.

The Sun–planet interaction sketched here is analogous to what occurs when you whirl a rock at the end of a string above your head. The Sun's gravitational field is your hand and the string, and the planet is the rock at the end of that string. The tension in the string provides the force necessary for the rock to move in a circular path. If you were suddenly to release the string—which would be like eliminating the Sun's gravity—the rock would fly away along a tangent to the circle, in accordance with Newton's first law.

In the solar system, at this very moment, Earth is moving under the combined influence of these two effects: the competition between gravity and inertia. The net result is a stable orbit, despite our continuous rapid motion through space. (In fact, Earth orbits the Sun at a speed of about 30 km/s, or about 70,000 mph. Verify this for yourself by calculating how fast Earth must move to complete a circle of radius 1 A.U.—and hence of circumference 2π A.U., or 940 million km—in 1 year, or 3.2×10^7 seconds. The answer is 9.4×10^8 km/3.2×10^7 s, or 29.4 km/s.) *More Precisely 2-3* describes how astronomers can use Newtonian mechanics and the law of gravity to quantify planetary motion and measure the masses of Earth, the Sun, and many other astronomical objects by studying the orbits of objects near them.

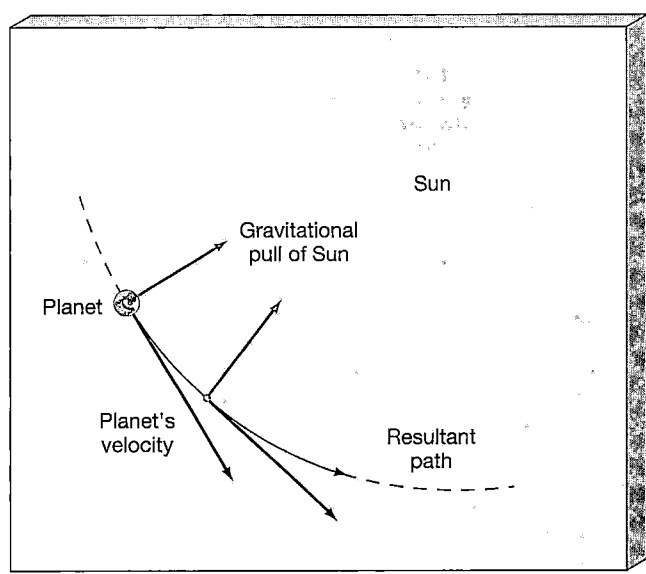

Figure 2.20 Gravitational Force (a) The gravitational force between two bodies is proportional to the mass of each and is inversely proportional to the square of the distance between them. (b) Inverse-square forces rapidly weaken with distance from their source. The strength of the gravitational force decreases with the square of the distance from the Sun, but never quite reaches zero, no matter how far away from the Sun.

Figure 2.21 Solar Gravity The Sun's inward pull of gravity on a planet competes with the planet's tendency to continue moving in a straight line. These two effects combine, causing the planet to move smoothly along an intermediate path, which continually "falls around" the Sun. This unending tug-of-war between the Sun's gravity and the planet's inertia results in a stable orbit.

KEPLER'S LAWS RECONSIDERED

Newton's laws of motion and the law of universal gravitation provided a theoretical explanation for Kepler's empirical laws of planetary motion. Kepler's three laws follow directly from Newtonian mechanics, as solutions to the equations describing the motion of a body moving in response to an inverse-square force. However, just as Kepler modified the Copernican model by introducing ellipses rather than circles, so too did Newton make corrections to Kepler's first and third laws. It turns out that a planet does not orbit the exact center of the Sun. Instead, both the planet and the Sun orbit their common **center of mass**. Because the Sun and the planet feel equal and opposite gravitational forces (by Newton's third law), the Sun must also move (by Newton's first law), driven by the gravitational influence of the planet. The Sun is so much more massive than any planet that the center of mass of the planet–Sun system is very close to the center of the Sun, which is why Kepler's laws are so accurate. Thus, Kepler's first law becomes

> I. The orbit of a planet around the Sun is an ellipse, with the *center of mass of the planet–Sun system* at one focus.

As shown in Figure 2.22, however, the center of mass for two objects of comparable mass does not lie within either object. For identical masses (Figure 2.22a), the orbits are identical ellipses, with a common focus located midway between the two objects. For unequal masses (as in Figure 2.22b), the elliptical orbits still share a focus and both have the same eccentricity, but the more massive object moves more slowly and on a tighter orbit. (Note that Kepler's second law, as stated earlier, continues to apply without modification to each orbit separately, but the *rates* at which the two orbits sweep out area are different.) In the extreme case of a planet orbiting the much more massive Sun (see Figure 2.22c), the path traced out by the Sun's center lies entirely within the Sun itself.

The change to Kepler's third law is also small in the case of a planet orbiting the Sun, but very important in other circumstances, such as the orbital motion of two stars that are gravitationally bound to each other. Following through the mathematics of Newton's theory, we find that the true relationship between the semi-major axis a (measured in astronomical units) of the planet's orbit relative to the Sun and its orbital period P (in Earth years) is

$$P^2 \text{ (in Earth years)} = \frac{a^3 \text{ (in astronomical units)}}{M_{\text{total}} \text{ (in solar units)}},$$

where M_{total} is the *combined* mass of the two objects. Notice that Newton's restatement of Kepler's third law preserves the proportionality between P^2 and a^3, but now the proportionality includes M_{total}, so it is *not* quite the same for all the planets. The Sun's mass is so great, however, that the differences in M_{total} among the various combinations of the Sun and the other planets are almost unno-

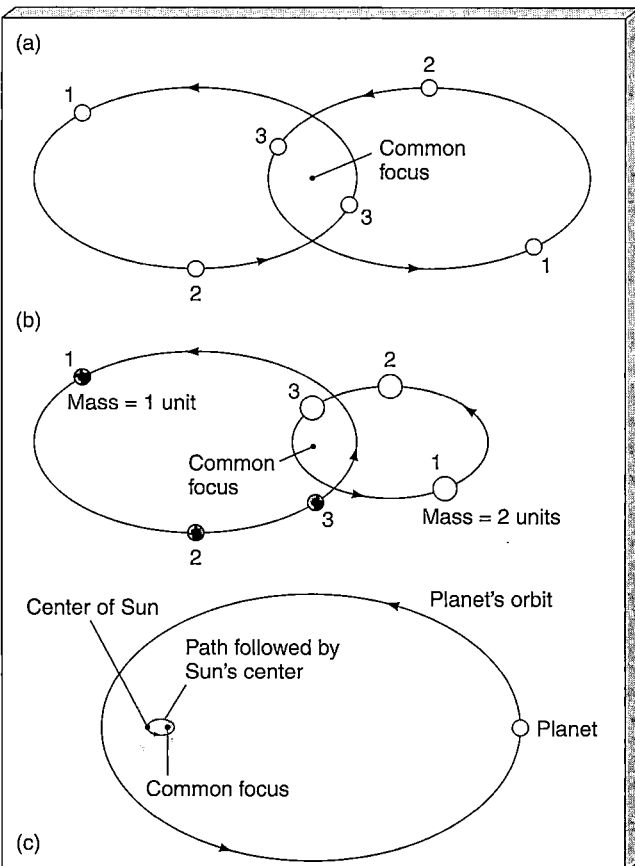

Figure 2.22 Orbits (a) The orbits of two bodies (stars, for example) with equal masses, under the influence of their mutual gravity, are identical ellipses with a common focus. That focus is not at the center of either star but instead is located at the center of mass of the pair, midway between them. The positions of the two bodies at three different times are indicated by the pairs of numbers. (Notice that a line joining the bodies always passes through the common focus.) (b) The orbits of two bodies, one of which is twice as massive as the other. Again, the elliptical orbits have a common focus, and the two ellipses have the same eccentricity. However, in accordance with Newton's laws of motion, the more massive body moves more slowly, and in a smaller orbit, staying closer to the center of mass (at the common focus). In this particular case, the larger ellipse is twice the size of the smaller one. (c) In this extreme case of a hypothetical planet orbiting the Sun, the common focus of the two orbits lies inside the Sun.

ticeable, so Kepler's third law, as originally stated, is a very good approximation. This modified form of Kepler's third law is true in all circumstances, inside or outside the solar system.

ESCAPING FOREVER

The law of gravity that describes the orbits of planets around the Sun applies equally well to natural moons and artificial satellites orbiting any planet. All our Earth-orbiting, human-made satellites move along paths governed by a combination of the inward pull of Earth's gravity and the forward motion gained during the rocket launch. If the rocket initially imparts enough speed to the satellite, it can go into orbit. Satellites not given enough speed at launch (such as intercontinental ballistic missiles, ICBMs) fail to achieve orbit and fall back to Earth (see Figure 2.23). (Technically, ICBMs actually do orbit Earth's attracting center, but their orbits intersect Earth's surface.)

Some space vehicles, such as the robot probes that visit other planets, attain enough speed to escape our planet's gravitational field and move away from Earth forever. This speed, known as the **escape speed**, is about 41 percent greater (actually, $\sqrt{2} = 1.414 \ldots$ times greater) than the speed of a circular orbit at any given radius.* At less than the escape speed, the old adage "What goes up must come down" (or at least stay in orbit) still applies. At more than the escape speed, however, a spacecraft will leave Earth for good. Planets, stars, galaxies—all gravitating bodies—have escape speeds. No matter how massive the body, gravity decreases with distance. As a result, the escape speed diminishes with increasing separation. The farther we go from Earth (or any gravitating body), the easier it becomes to escape.

*In terms of the formula presented in More Precisely 2-3, the escape speed is given by $v_{escape} = \sqrt{\dfrac{2GM}{r}}$.

The speed of a satellite in a circular orbit just above Earth's atmosphere is 7.9 km/s (roughly 18,000 mph). The satellite would have to travel at 11.2 km/s (about 25,000 mph) to escape from Earth altogether. If an object exceeds the escape speed, its motion is said to be **unbound**, and the orbit is no longer an ellipse. In fact, the path of the spacecraft relative to Earth is a related geometric figure called a *hyperbola*. If we simply change the word *ellipse* to *hyperbola*, the modified version of Kepler's first law still applies, as does Kepler's second law. (Kepler's third law does not extend to unbound orbits because it doesn't make sense to talk about a period in those cases.)

Newton's laws explain the paths of objects moving at any point in space near any gravitating body. These laws provide a firm physical and mathematical foundation for Copernicus's heliocentric model of the solar system and for Kepler's laws of planetary motions. But they do much more than that. Newtonian gravitation governs not only the planets, moons, and satellites in their elliptical orbits but also the stars and galaxies in their motion throughout our universe—as well as apples falling to the ground.

☑ Concept Check

◼ Explain, in terms of Newton's laws of motion and gravity, why planets orbit the Sun.

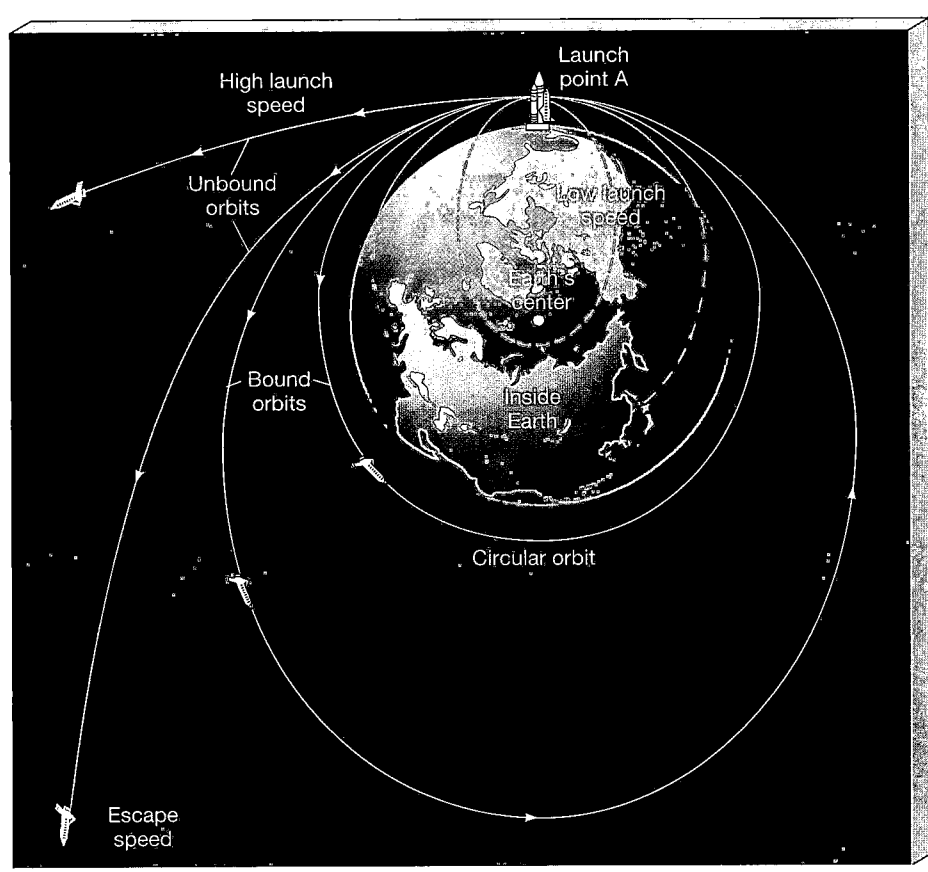

Figure 2.23 Escape Speed The effect of launch speed on the trajectory of a satellite. With too low a speed at point A, the satellite will simply fall back to Earth. Given enough speed, however, the satellite will go into orbit—it "falls around Earth." As the initial speed at point A is increased, the orbit will become more and more elongated. When the initial speed exceeds the escape speed, the satellite will become unbound from Earth and will escape along a hyperbolic trajectory.

We can use Newtonian mechanics to calculate some useful formulae relating the properties of planetary orbits to the mass of the Sun. Again for simplicity, let's assume that the orbits are circular (not a bad approximation in most cases, and Newton's laws easily extend to cover the more general case of eccentric orbits). Consider a planet of mass m moving at speed v in an orbit of radius r around the Sun, of mass M. The planet's acceleration (see More Precisely 2-2) is

$$a = \frac{v^2}{r},$$

so, by Newton's second law, the force required to keep it in orbit is

$$F = ma = \frac{mv^2}{r}.$$

Comparing this with the gravitational force due to the Sun, it follows that

$$\frac{mv^2}{r} = \frac{GmM}{r^2},$$

so the circular orbit speed is

$$v = \sqrt{\frac{GM}{r}}.$$

Dividing this speed into the circumference of the orbit $(2\pi r)$, we obtain a form of Kepler's third law (equivalent to the formula presented in the text):

$$P = 2\pi\sqrt{\frac{r^3}{GM}}.$$

where $P = 2\pi r/v$ is the orbital period.

Now we turn the problem around. Because we have measured G in the laboratory on Earth and because we know the length of a year and the size of the astronomical unit, we can use Newtonian mechanics to *weigh* the Sun. Rearranging the above equation to read

$$M = \frac{rv^2}{G}$$

and substituting the known values of $v = 30$ km/s, $r = 1$ A.U. $= 1.5 \times 10^{11}$ m, and $G = 6.7 \times 10^{-11}$ N m^2/kg^2, we calculate the mass of the Sun to be 2.0×10^{30} kg—an enormous mass by terrestrial standards. Similarly, knowing the distance to the Moon and the length of the (sidereal) month, we can measure the mass of Earth to be 6.0×10^{24} kg.

In fact, this is how basically *all* masses are measured in astronomy. Because we can't just go out and attach a scale to an astronomical object when we need to know its mass, we must look for its gravitational influence on something else. This principle applies to planets, stars, galaxies, and even clusters of galaxies—very different objects but all subject to the same physical laws.

Chapter Review

SUMMARY

Geocentric (p. 37) models of the universe were based on the assumption that the Sun, the Moon, and the planets all orbit Earth. The most successful and long-lived of these was the Ptolemaic model (p. 38). Unlike the Sun and the Moon, planets sometimes appear to temporarily reverse their direction of motion (from night to night) relative to the stars and then resume their normal "forward" course. This phenomenon is called retrograde motion (p. 36).To account for retrograde motion within the geocentric picture, it was necessary to suppose that planets moved on small circles called epicycles (p. 37), whose centers orbited Earth on larger circles called deferents (p. 37). The heliocentric (p. 39) view of the solar system holds that Earth, like all the planets, orbits the Sun. This model accounts for retrograde motion and the observed size and brightness variations of the planets in a much more straightforward way than the geocentric model. The widespread realization during the Renaissance that the solar system is Sun centered, and not Earth centered, is known as the

Copernican revolution (p. 39), in honor of Nicholas Copernicus, who laid the foundations of the modern heliocentric model.

Galileo Galilei is often regarded as the father of experimental science. His telescopic observations of the Moon, the Sun, Venus, and Jupiter played a crucial role in supporting and strengthening the Copernican picture of the solar system. Johannes Kepler improved on Copernicus's model with his three laws of planetary motion (p. 44): (1) Planetary orbits are ellipses (p. 44), with the Sun at one focus (p. 44). (2) A planet moves faster as its orbit takes it closer to the Sun. (3) The semi-major axis (p. 44) of the orbit is related in a simple way to the planet's orbit period (p. 46). Most planets move on orbits whose eccentricities (p. 44) are quite small, so their paths differ only slightly from perfect circles. The distance from Earth to the Sun is called the astronomical unit (p. 47). Nowadays, the astronomical unit is determined by bouncing radar (p. 48) signals off the planet Venus and measuring the time the signal takes to return.

Isaac Newton succeeded in explaining Kepler's laws in terms of a few general physical principles, now known as **Newtonian mechanics** (p. 51). The tendency of a body to keep moving at constant velocity is called **inertia** (p. 51). The greater the body's **mass** (p. 51), the greater its inertia. To change the velocity, a **force** (p. 51) must be applied. The rate of change of velocity, called **acceleration** (p. 51), is equal to the applied force divided by the body's mass. To explain planetary orbits, Newton postulated that **gravity** (p. 52) attracts the planets to the Sun. Every object with any mass is surrounded by a **gravitational field** (p. 53), whose

strength decreases with distance according to an **inverse-square law** (p. 52). This field determines the gravitational force exerted by the object on any other body in the universe. Newton's laws imply that a planet does not orbit the precise center of the Sun, but instead that both the planet and the Sun orbit the common **center of mass** (p. 54) of the two bodies. For an object to escape from the gravitational pull of another, its speed must exceed the **escape speed** (p. 55) of the second body. In this case, the motion is said to be **unbound** (p. 55), and the orbital path is no longer an ellipse, although it is still described by Newton's laws.

SELF-TEST: TRUE OR FALSE?

1. Aristotle was first to propose that all planets revolve around the Sun.
2. The teachings of Aristotle remained unchallenged until the eighteenth century A.D.
3. Ptolemy was responsible for a geocentric model that was successful at predicting the positions of the planets, Moon, and the Sun.
4. The heliocentric model of the universe holds that Earth is at the center, and everything else moves around it.
5. Kepler's discoveries regarding the orbital motion of the planets were based mainly on observations made by Copernicus.
6. The Sun's location in a planet's orbit is at the center.
7. The semi-major axis of an orbit is half the major axis.
8. A circle has an eccentricity of zero.
9. The astronomical unit is a distance equal to the semi-major axis of Earth's orbit around the Sun.
10. The speed of a planet orbiting the Sun is independent of the planet's position in its orbit.
11. Kepler's laws work for only the six planets known in his time.
12. Kepler never knew the true distances between the planets and the Sun, only their relative distances.
13. Galileo's observations of the sky were made with the naked eye.
14. Using his laws of motion and gravity, Newton was able to prove Kepler's laws.
15. You throw a baseball to someone; before the ball is caught, it is temporarily in orbit around Earth's center.

SELF-TEST: FILL IN THE BLANK

1. Stonehenge was used as a _____ by people in the Stone Age.
2. Accurate records of comets and "guest" stars were kept over many centuries by _____ astrologers.
3. The astronomical knowledge of ancient Greece was kept alive and augmented by _____ astronomers.
4. The apparent "backward" (westward motion) of the planets Mars, Jupiter, or Saturn in the sky relative to the stars is known as _____ motion.
5. Observation, theory, and testing are the cornerstones of the _____.
6. The heliocentric model was reinvented by _____.
7. Central to the heliocentric model is the assertion that the observed motions of the planets and the Sun are the result of _____ motion around the Sun.
8. Kepler's laws were based on observational data obtained by _____.
9. Kepler discovered that the shape of an orbit is an _____, not a _____, as had previously been believed.
10. Kepler's third law relates the _____ of the orbital period to the _____ of a planet's semi-major axis.
11. Galileo discovered _____ orbiting Jupiter, the _____ of Venus, and the Sun's rotation from observations of _____.
12. The modern method of measuring the astronomical unit uses _____ measurements of a planet or asteroid.
13. Newton's first law states that a moving object will continue to move in a straight line with constant speed unless acted upon by a _____.
14. Newton's law of gravity states that the gravitational force between two objects depends on the _____ of their masses and inversely on the _____ of their separation.
15. Newton discovered that, in Kepler's third law, the orbital period depends on the semi-major axis and on the sum of the _____ of the two objects involved.

REVIEW AND DISCUSSION

1. What contributions to modern astronomy were made by Chinese and Islamic astronomers during the Dark Ages of medieval Europe?
2. Briefly describe the geocentric model of the universe.
3. The benefit of our current knowledge lets us see flaws in the Ptolemaic model of the universe. What is its basic flaw?
4. What was the great contribution of Copernicus to our knowledge of the solar system? What was still a flaw in the Copernican model?
5. What is a theory? Can a theory ever be proved to be true?
6. When were Copernicus's ideas finally accepted?
7. What is the Copernican principle?

8. What discoveries of Galileo helped confirm the views of Copernicus, and how?

9. Briefly describe Kepler's three laws of planetary motion.

10. How did Tycho Brahe contribute to Kepler's laws?

11. If radio waves cannot be reflected from the Sun, how can radar be used to find the distance from Earth to the Sun?

12. How did astronomers determine the scale of the solar system prior to the invention of radar?

13. What does it mean to say that Kepler's laws are empirical?

14. What are Newton's laws of motion and gravity?

15. List the two modifications made by Newton to Kepler's laws.

16. Why do we say that a baseball falls toward Earth, and not Earth toward the baseball?

17. Why would a baseball go higher if it were thrown up from the surface of the Moon than if it were thrown with the same velocity from the surface of Earth?

18. In what sense is the Moon falling toward Earth?

19. What is the meaning of the term escape speed?

20. What would happen to Earth if the Sun's gravity were suddenly "turned off"?

PROBLEMS  *Algorithmic versions of these questions are available in the Practice Problems module of the Companion Website.*

The number of squares preceding each problem indicates its approximate level of difficulty.

1. ■ Tycho Brahe's observations of the stars and planets were accurate to about one arc minute (1'). To what distance does this angle correspond at the distance of (a) the Moon; (b) the Sun; and (c) Saturn (at closest approach)?

2. ■■ To an observer on Earth, through what angle will Mars appear to move relative to the stars over the course of 24 hours, when the two planets are at closest approach? Assume for simplicity that Earth and Mars move on circular orbits of radii 1.0 A.U. and 1.5 A.U., respectively, in exactly the same plane. Will the apparent motion be prograde or retrograde?

3. ▢ Using the data in Table 2.1, show that Pluto is closer to the Sun at perihelion (the point of closest approach to the Sun in its orbit) than Neptune is at any point in its orbit.

4. ■■ An asteroid has a perihelion distance of 2.0 A.U. and an aphelion distance of 4.0 A.U. Calculate its orbital semi-major axis, eccentricity, and period.

5. ■■ A spacecraft has an orbit that just grazes Earth's orbit at aphelion and just grazes Venus's orbit at perihelion. Assuming that Earth and Venus are in the right places at the right times, how long will the spacecraft take to travel from Earth to Venus?

6. ■■ Halley's comet has a perihelion distance of 0.6 A.U. and an orbital period of 76 years. What is its aphelion distance from the Sun?

7. ■■ What is the maximum possible parallax of Mercury during a solar transit, as seen from either end of a 3000-km baseline on Earth?

8. ■ How long would a radar signal take to complete a round trip between Earth and Mars when the two planets are 0.7 A.U. apart?

9. ■ Jupiter's moon Callisto orbits the planet at a distance of 1.88 million km. Callisto's orbital period about Jupiter is 16.7 days. What is the mass of Jupiter? [Assume that Callisto's mass is negligible compared with that of Jupiter, and use the modified version of Kepler's third law (Section 2.7).]

10. ■■ The Sun moves in a roughly circular orbit around the center of the Milky Way Galaxy, at a distance of 26,000 light-years. The orbit speed is approximately 220 km/s. Calculate the Sun's orbital period and centripetal acceleration, and use these numbers to estimate the mass of our galaxy.

11. ■■ At what distance from the Sun would a planet's orbital period be one million years? What would be the orbital period at a distance of one light-year?

12. ■ The acceleration due to gravity at Earth's surface is 9.80 m/s². What is the gravitational acceleration at altitudes of (a) 100 km; (b) 1000 km; (c) 10,000 km? Take Earth's radius to be 6400 km.

13. ■▢ What would be the speed of a spacecraft moving in a circular orbit at each of the three altitudes listed in the previous problem? In each case, how does the centripetal acceleration (*More Precisely 2-2*) compare with the gravitational acceleration?

14. ■ Use Newton's law of gravity to calculate the force of gravity between you and Earth. Convert your answer, which will be in newtons, to pounds using the conversion 4.45 N equals one pound. What do you normally call this force?

15. ■ The Moon's mass is 7.4 × 10²² kg and its radius is 1700 km. What is the speed of a spacecraft moving in a circular orbit just above the lunar surface? What is the escape speed from the Moon?

COLLABORATIVE EXERCISES

1. Tourist Attraction or Sacred Ground. Your group has been asked to arbitrate a dispute between a tour bus company and a nearby Native American tribe. The dispute surrounds an ancient medicine wheel recently discovered by a team of university archeologists. Using sketches as neces-

sary, compose a legal brief that describes what a medicine wheel is designed to do astronomically and summarize the opposing positions of the two groups.

2. Galileo's Observations. Your group should select what it believes to be Galileo's single most important astronomical observation, why it was most important, and explain what he observed using sketches.

3. **Binary Star Observing Proposal.** Write a one-page proposal for "telescope-time" to determine the total mass of stars in a binary star system. Explain what a binary star system is and describe exactly which measurement each member in your group will need to make in order to completely determine the star system's total mass.

4. **Galileo's Dialogue.** Galileo's *Dialogue Concerning the Two World Chief Systems* described fictional conversations between three people. Create a short play using this style describing Kepler's Laws of Planetary Motion using each person in your group.

RESEARCHING ON THE WEB  *To complete the following exercises, go to the online Destinations module for Chapter 2 on the Companion Website for* Astronomy Today 4/e.

1. Access the "Kepler's Laws Java Orbit Simulator" and, while holding a piece of paper over the screen, trace the orbits and show that this particular simulator is or is not a valid depiction of Kepler's Second Law by determining the area swept out at perigee and apogee.

2. Access the "Keplerian Elements Tutorial" and describe the difference between a satellite's orbital inclination and its argument of perigee.

3. Access the "Laws of Science Abused by Science Fiction" page and list the five most commonly misused physics laws in science fiction.

PROJECTS

1. Look in an almanac for the date of opposition of one or all of these bright planets: Mars, Jupiter, and Saturn. At opposition, these planets are at their closest points to Earth and are at their largest and brightest in the night sky. Observe these planets. How long before opposition does each planet's retrograde motion begin? How long afterward does it end?

2. Draw an ellipse. (See Figure 2.13, p. 46.) You'll need two pins, a piece of string, and a pencil. Tie the string in a loop and place it around the pins. Place the pencil inside the loop and run it around the inside of the string, holding the loop taut. The two pins will be at the foci of the ellipse. What is the eccentricity of the ellipse you have drawn?

3. Use a small telescope to replicate Galileo's observations of Jupiter's four largest moons. Note the moons' brightnesses and their locations with respect to Jupiter. If you watch over a period of several nights, draw what you see; you'll notice that these moons change their positions as they orbit the giant planet. Check the charts given monthly in *Astronomy* or *Sky & Telescope* magazines to identify each moon you see.

SKYCHART III PROJECTS  *The SkyChart III Student Version planetarium program on which these exercises are based is included as a separately executable program on the CD in the back of this text.*

1. ▣ Observe the apparent motion of Jupiter against the background stars by centering the display on Jupiter with *VIEW/Center Planet/Jupiter*. Deselect *DRAW/Horizon Mask, Constellations,* and *Grid Lines.* With a *180° Field of View* set *ANIMATION/Trail For/Jupiter.* Set *DRAW/Sky Background/Black* and set the time to approximately midnight. Animate the image with time steps of one week beginning April 1, 2001, until approximately September 1, 2004. If the animation runs too fast on your computer, use time steps of one day. During this time, count how many times the motion of Jupiter stops and reverses itself. You will observe that after moving in retrograde motion for a while, Jupiter will again stop and reverse its motion, illustrating the epicycle the ancients imagined.

2. ▣▣ The Mesopotamians were well aware of the cycles of the heavens. Venus (the Star of Ishtar) was symbolized by an eight-pointed star. It turns out there is a reason for the number eight. Venus experiences a cycle of approximately eight years. Venus is located at the same location with respect to the "fixed" stars every eight years. In this time, it has switched back and forth between morning and evening star five times. Using an appropriate *Time Step* and using the *Trails* feature in the *ANIMATION* menu, show that this period is correct and that Venus switches back and forth five times. Center on the vernal equinox (R.A. 00 00 and Dec. 00 00, *Equatorial Coordinates*).

 *In addition to the Practice Problems and Destinations modules, the Companion Website at http://www.prenhall.com/chaisson provides for each chapter an additional true-false, multiple choice, and labeling quiz, as well as additional annotated images, animations, and links to related Websites.*

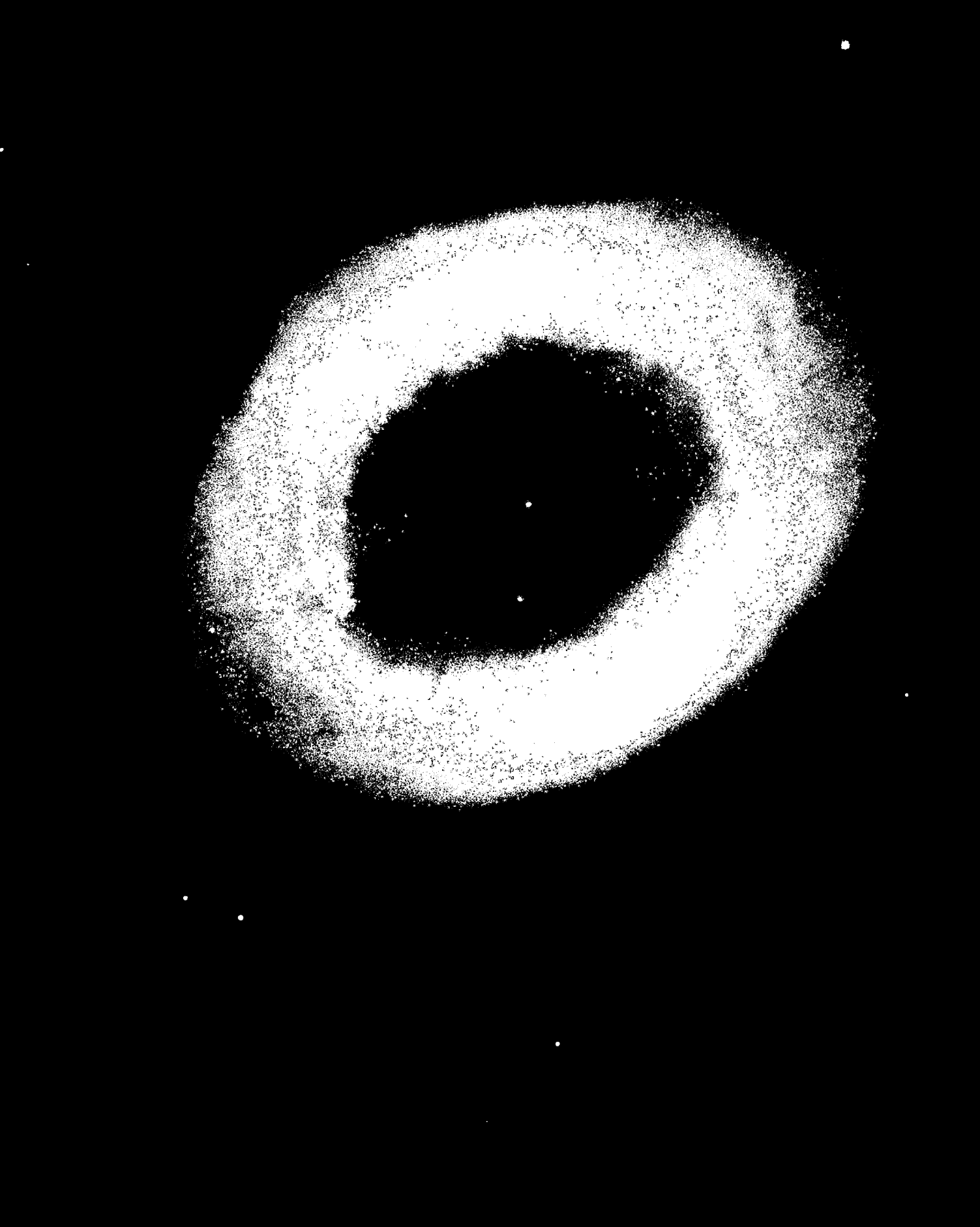

3 RADIATION

Information from the Cosmos

LEARNING GOALS

Studying this chapter will enable you to:

1 Discuss the nature of electromagnetic radiation, and tell how that radiation transfers energy and information through interstellar space.

2 Describe the major regions of the electromagnetic spectrum and explain how the properties of Earth's atmosphere affect our ability to make astronomical observations at different wavelengths.

3 Explain what is meant by the term "blackbody radiation" and describe the basic properties of such radiation.

4 Tell how we can determine the temperature of an object by observing the radiation that it emits.

5 Show how the relative motion of a source of radiation and its observer can change the perceived wavelength of the radiation, and explain the importance of this phenomenon to astronomy.

Visit http://www.prenhall.com/chaisson for additional annotated images, animation, and links to related sites for this chapter.

The Ring Nebula in the constellation Lyra is one of the most magnificent objects in our galaxy. Seen here glowing in the light of its own fading radiation, the nebula is actually the expanding atmosphere of a nearly dead star. The dying dwarf star can be seen at the center of the ring of hydrogen-rich gas, shown here in true color. Owing to the nebula's distance of some 5000 light-years, its apparent size is just 1/100th that of the full Moon. It is too dim to see with the naked eye. (STScI)

The Big Picture: Stars evolve from birth to maturity to death, much like living things, but on vastly longer timescales. Our own star, the Sun, is about mid-way through its evolution. In another 5 billion years, the Sun will swell late in life to resemble the object shown here. By that time, humanity will be long gone from Earth—either voluntarily, as our descendants move out into the wider Universe, or involuntarily, having perished in the atmosphere of our dying parent star. (Astronomical Society of the Pacific)

Astronomical objects are more than just things of beauty in the night sky. Planets, stars, and galaxies are of vital significance if we are to fully understand our place in the big picture—the "grand design" of the universe. Each object is a source of information about the material aspects of our universe—its state of motion, its temperature, its chemical composition, even its past history. When we look at the stars, the light we see actually began its journey to Earth decades, centuries—even millennia—ago. The faint rays from the most distant galaxies have taken billions of years to reach us. The stars and galaxies in the night sky show us the far away and the long ago. In this chapter we begin our study of how astronomers extract information from the light emitted by astronomical objects. These basic concepts of radiation are central to modern astronomy.

3.1 Information from the Skies

Figure 3.1 shows a galaxy in the constellation Andromeda. On a dark, clear night, far from cities or other sources of light, the Andromeda Galaxy, as it is generally called, can be seen with the naked eye as a faint, fuzzy patch on the sky, comparable in diameter to the full Moon. Yet the fact that it is visible from Earth belies this galaxy's enormous distance from us. It lies roughly 2.5 million *light-years* away. An object at such a distance is truly inaccessible in any realistic human sense. Even if a space probe could miraculously travel at the speed of light, it would need 2.5 million years to reach this galaxy and 2.5 million more to return with its findings. Considering that civilization has existed on Earth for fewer than 10,000 years, and its prospects for the next 10,000 are far from certain, even this unattainable technological feat would not provide us with a practical means of exploring other galaxies. Even the farthest reaches of our own galaxy, "only" a few tens of thousands of light-years distant, are effectively off-limits to visitors from Earth, at least for the foreseeable future.

LIGHT AND RADIATION

Given the impossibility of traveling to such remote parts of the universe, how do astronomers know anything about objects far from Earth? How do we obtain detailed information about any planet, star, or galaxy too distant for a personal visit or any kind of controlled experiment? The answer is that we use the laws of physics, as we know them here on Earth, to interpret the **electromagnetic radiation** emitted by these objects. *Radiation* is any way in which energy is transmitted through space from one point to another without the need for any physical connection between those two locations. The term *electromagnetic* just means that the energy is carried in the form of rapidly fluctuating *electric* and *magnetic* fields (to be discussed in more detail later in Section 3.2). Virtually all we know about the universe beyond Earth's atmosphere has been gleaned from painstaking analysis of electromagnetic radiation received from afar.

Visible light is the particular type of electromagnetic radiation to which our human eyes happen to be sensitive. As light enters our eye, the cornea and lens focus it onto

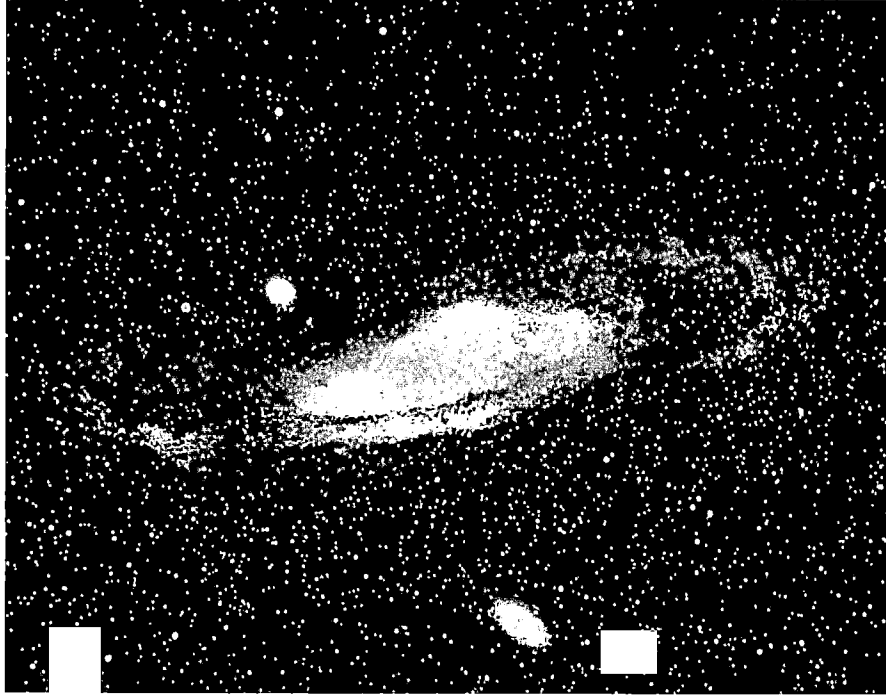

Figure 3.1 Andromeda The pancake-shaped Andromeda Galaxy lies about 2.5 million light-years away, according to the most recent distance measurements. It contains a few hundred billion stars. *(T. Hallas)*

the retina, whereupon small chemical reactions triggered by the incoming energy send electrical impulses to the brain, producing the sensation of sight. But there is also *invisible* electromagnetic radiation, which goes completely undetected by our eyes. **Radio, infrared,** and **ultraviolet** waves, as well as **X rays** and **gamma rays,** all fall into this category. Recognize that, despite the different names, the words *light, rays, radiation,* and *waves* really refer to the same thing. The names are just historical accidents, reflecting the fact that it took many years for scientists to realize that these apparently very different types of radiation are in reality one and the same physical phenomenon. Throughout this text, we will use the general terms "light" and "electromagnetic radiation" more or less interchangeably.

WAVE MOTION

Despite the early confusion still reflected in current terminology, scientists now know that all types of electromagnetic radiation travel through space in the form of *waves.* To understand the behavior of light, then, we must know a little about wave motion.

Simply stated, a **wave** is a way in which energy is transferred from place to place without physical movement of material from one location to another. In wave motion, the energy is carried by a disturbance of some sort. This *disturbance,* whatever its nature, occurs in a distinctive repeating pattern. Ripples on the surface of a pond, sound waves in air, and electromagnetic waves in space, despite their many obvious differences, all share this basic defining property.

Imagine a twig floating in a pond. A pebble thrown into the pond at some distance from the twig disturbs the surface of the water, setting it into up-and-down motion. This disturbance will propagate outward from the point of impact in the form of waves. When the waves reach the twig, some of the pebble's energy will be imparted to it, causing the twig to bob up and down. In this way, both energy and *information*—the fact that the pebble entered the water—are transferred from the place where the pebble landed to the location of the twig. We could tell that a pebble (or, at least, some object) had entered the water just by observing the twig. With a little additional physics, we could even estimate the pebble's energy.

A wave is not a physical object. No water traveled from the point of impact of the pebble to the twig—at any location on the surface, the water surface simply moved up and down as the wave passed. What, then, *did* move across the surface of the pond? As illustrated in Figure 3.2, the

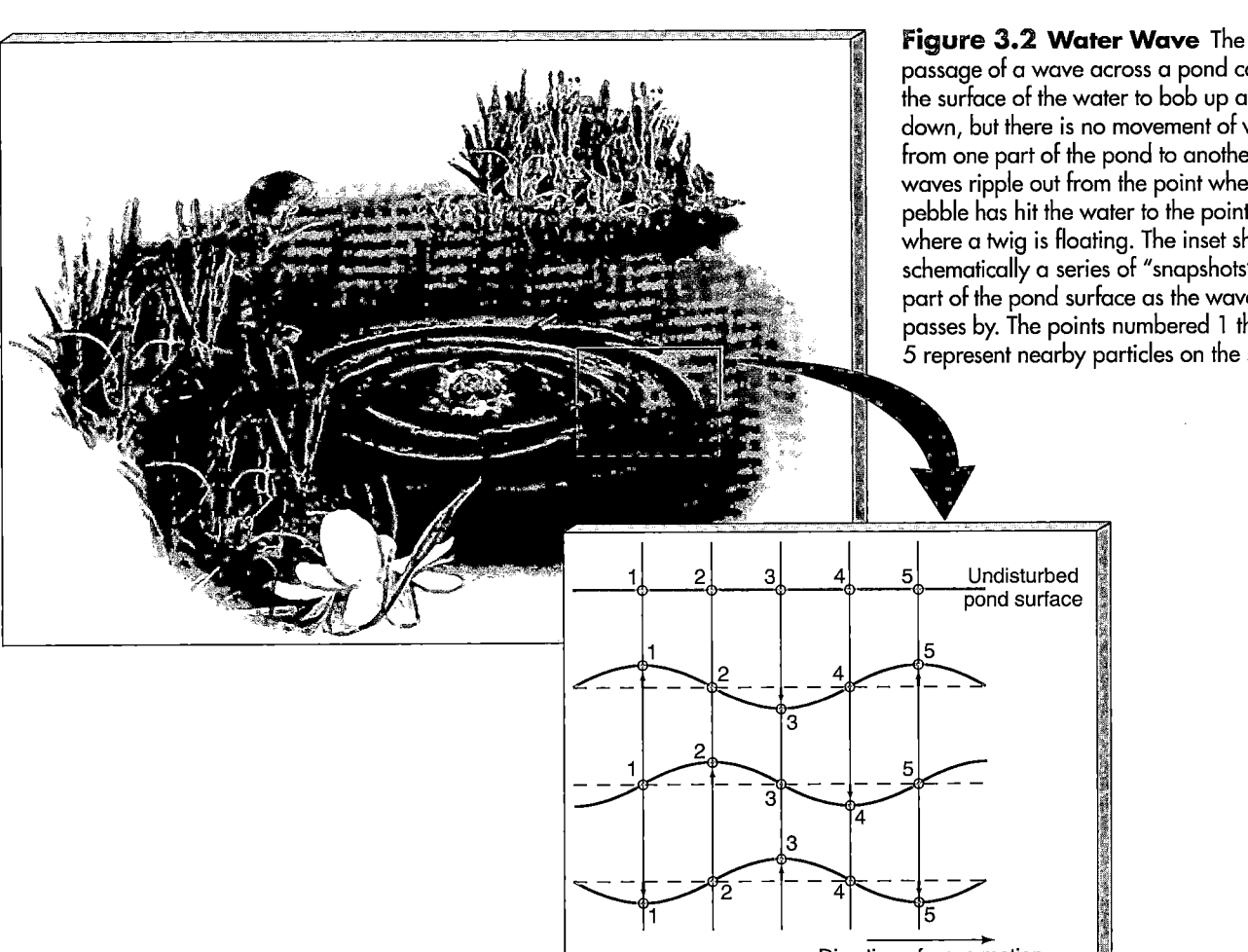

Figure 3.2 Water Wave The passage of a wave across a pond causes the surface of the water to bob up and down, but there is no movement of water from one part of the pond to another. Here waves ripple out from the point where a pebble has hit the water to the point where a twig is floating. The inset shows schematically a series of "snapshots" of part of the pond surface as the wave passes by. The points numbered 1 through 5 represent nearby particles on the surface.

answer is that the wave was the *pattern* of up-and-down motion. This pattern was transmitted from one point to the next as the disturbance moved across the water.

Figure 3.3 shows how wave properties are quantified and illustrates some standard terminology. The *wave period* is the number of seconds needed for the wave to repeat itself at some point in space. The **wavelength** is the number of meters needed for the wave to repeat itself at a given moment in time. It can be measured as the distance between two adjacent wave *crests*, two adjacent wave *troughs*, or any other two similar points on adjacent wave cycles (for example, the points marked X in the figure). The maximum departure of the wave from the undisturbed state—still air, say, or a flat pond surface—is called its **amplitude**.

The number of wave crests passing any given point per unit time is called the wave's **frequency**. If a wave of a given wavelength moves at high speed, then many crests pass per second and the frequency is high. Conversely, if the same wave moves slowly, then its frequency will be low. The frequency of a wave is just one divided by the wave's period:

$$\text{wave frequency} = \frac{1}{\text{wave period}}.$$

Frequency is expressed in units of inverse time (cycles per second), called hertz (Hz) in honor of the nineteenth-century German scientist Heinrich Hertz, who studied the properties of radio waves. Thus a wave with a period of 5 s has a frequency of $(1/5)$ cycles/s $= 0.2$ Hz, meaning that one wave crest passes a given point in space every five seconds.

A wave moves a distance equal to one wavelength in one wave period. The product of wavelength and frequency therefore equals the *wave velocity:*

$$\text{wavelength} \times \text{frequency} = \text{velocity}.$$

Thus, if the wave in our earlier example had a wavelength of 0.5 m, its velocity is $(0.5 \text{ m}) \times (0.2 \text{ Hz}) = 0.1$ m/s. Wavelength and wave frequency are *inversely* related—doubling one halves the other.

Before about 1800, scientists were divided in their opinions about the nature of light. Some believed that light was a wave phenomenon, while others maintained that light was in reality a stream of particles that moved in straight lines. Given the experimental apparatus available at the time, neither camp could find conclusive evidence to disprove the other theory. *Discovery 3-1* discusses some more wave properties of importance to modern astronomers, and describes how their detection in experiments using visible light early in the nineteenth century finally tilted the balance of scientific opinion in favor of the wave theory.

THE COMPONENTS OF VISIBLE LIGHT

White light is a mixture of colors, which we conventionally divide into six major hues—red, orange, yellow, green, blue, and violet. As shown in Figure 3.4, we can separate a beam of white light into a rainbow of these basic colors—called a *spectrum* (plural, *spectra*)—by passing it through a prism. This experiment was first reported by Isaac Newton over 300 years ago. In principle, the original beam of white light could be recovered by passing the spectrum through a second prism to recombine the colored beams.

What determines the color of a beam of light? The answer is its frequency (or equivalently, its wavelength). We see different colors because our eyes react differently to electromagnetic waves of different frequencies. A prism splits a beam of light up into separate colors because light rays of different frequencies are bent, or *refracted*, slightly differently as they pass through the prism—red light the least, violet light the most. Red light has a frequency of roughly 4.3×10^{14} Hz, corresponding to a wavelength of about 7.0×10^{-7} m. Violet light, at the other end of the visible range, has nearly double the frequency—7.5×10^{14} Hz—and (since the speed of light is the same in either case) just over half the wavelength—4.0×10^{-7} m. The other colors we see have frequencies and wavelengths intermediate between these two extremes, spanning the entire *visible spectrum* shown in Figure 3.4. Radiation outside this range is invisible to human eyes.

Scientists often use a unit called the *nanometer* (nm) when describing the wavelength of light (see Appendix 2). There are 10^9 nanometers in one meter. An older unit called the *angstrom* ($1 \text{ Å} = 10^{-10}$ m $= 0.1$ nm) is also widely used. (The unit is named after the nineteenth-century Swedish

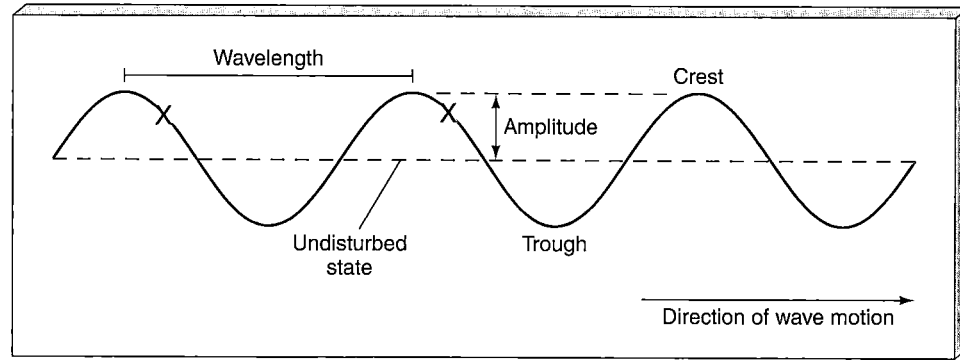

Figure 3.3 Wave Properties
Representation of a typical wave, showing its direction of motion, wavelength, and amplitude. In one wave period, the entire pattern shown here moves one wavelength to the right.

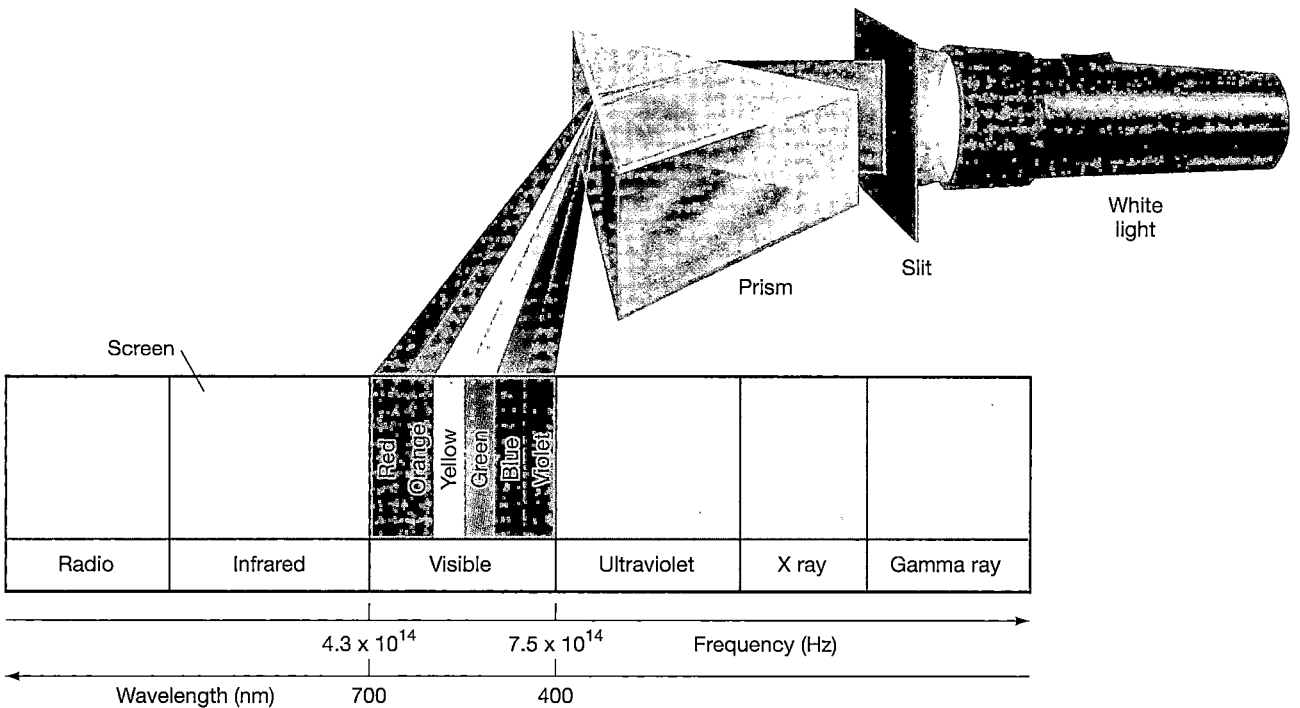

Figure 3.4 Visible Spectrum While passing through a prism, white light splits into its component colors, spanning red to violet in the visible part of the electromagnetic spectrum. The slit narrows the beam of radiation. The image on the screen is just a series of different-colored images of the slit. Human eyes are insensitive to radiation of wavelength shorter than 400 nm or longer than 700 nm, but radiation outside the visible range is easily detected by other means.

physicist Anders Ångstrom—pronounced "ong.strem.") However, in SI units, the nanometer is preferred. Thus, the visible spectrum covers the wavelength range from 400 nm to 700 nm (4000 Å to 7000 Å). The radiation to which our eyes are most sensitive has a wavelength near the middle of this range, at about 550 nm (5500 Å), in the yellow-green region of the spectrum. It is no coincidence that this wavelength falls within the range of wavelengths at which the Sun emits most of its electromagnetic energy—our eyes have evolved to take greatest advantage of the available light.

3.2 Waves in What?

Ⅱ Waves of radiation differ fundamentally from water waves, sound waves, or any other waves that travel through a material medium. Radiation needs *no* such medium. When light radiation travels from a distant galaxy, or from any other cosmic object, it moves through the virtual vacuum of space. Sound waves, by contrast, cannot do this; if we were to remove all the air from a room, conversation would be impossible. Communication by flashlight or radio, however, would be entirely feasible.

The ability of light to travel through empty space was once a great mystery. The idea that light, or any other kind of radiation, could move as a wave through nothing at all seemed to violate common sense, yet it is now a cornerstone of modern physics.

INTERACTIONS BETWEEN CHARGED PARTICLES

To understand more about the nature of light, consider for a moment an *electrically charged* particle, such as an **electron** or a **proton**. Like mass, electrical charge is a fundamental property of matter. Electrons and protons are elementary particles—"building blocks" of atoms and all matter—that carry the basic unit of charge. Electrons are said to carry a *negative* charge, whereas protons carry an equal and opposite *positive* charge.

Just as a massive object exerts a gravitational force on any other massive body (as we saw in Chapter 2), an electrically charged particle exerts an *electrical* force on every other charged particle in the universe. ∞ (Sec. 2.7) Buildup of electrical charge (a net imbalance of positive over negative, or vice versa) is what causes "static cling" on your clothes when you take them out of a hot clothes dryer, or the shock you sometimes feel when you touch a metal door frame on a particularly dry day.

Unlike the gravitational force, which is always attractive, electrical forces can be either attractive or repulsive. As illustrated in Figure 3.5(a), particles with *like* charges

The Wave Nature of Radiation

Until the early nineteenth century, debate raged in scientific circles regarding the true nature of light. On the one hand, the particle, or *corpuscular*, theory, first expounded in detail by Isaac Newton, held that light consisted of tiny particles moving in straight lines at the speed of light. Different colors were presumed to correspond to different particles. On the other hand, the *wave* theory, championed by the seventeenth-century Dutch astronomer Christian Huygens, viewed light as a wave phenomenon, in which color was determined by frequency, or wavelength. During the first few decades of the nineteenth century, growing experimental evidence that light displayed three key wave properties—*diffraction*, *interference*, and *polarization*—argued strongly in favor of the wave theory.

Diffraction is the deflection, or "bending," of a wave as it passes a corner or moves through a narrow gap. As depicted in the figure below, a sharp-edged hole in a barrier seems at first glance to produce a sharp shadow, as we might expect if radiation were composed of rays or particles moving in perfectly straight lines. Closer inspection, however, reveals that the shadow actually has a "fuzzy" edge, as shown in this photograph at right of the diffraction pattern pro-

duced by a small circular opening. We are not normally aware of such effects in everyday life because diffraction is generally very small for visible light. For any wave, the amount of diffraction is proportional to the ratio of the wavelength to the width of the gap. The longer the wave-

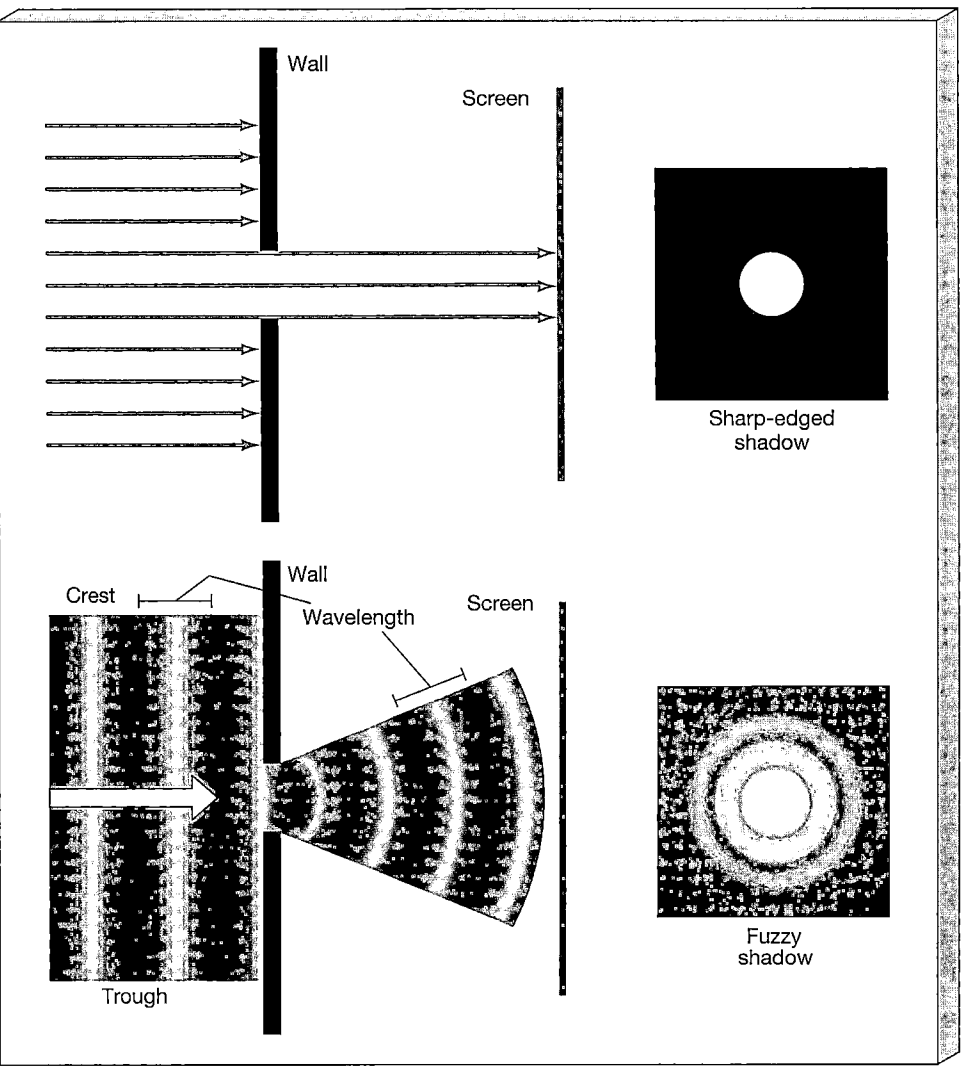

(that is, both negative or both positive—for example, two electrons or two protons) repel one another. Particles with *unlike* charges (that is, having opposite signs—an electron and a proton, say) attract.

How is the electrical force transmitted through space? Extending outward in all directions from any charged particle is an **electric field**, which determines the electrical force exerted by the particle on all other charged particles

in the universe (Figure 3.5b). The strength of the electric field, like the strength of the gravitational field, decreases with increasing distance from the charge according to an inverse-square law. By means of the electric field, the particle's presence is "felt" by all other charged particles, near and far.

Now suppose our particle begins to vibrate, perhaps because it becomes heated or collides with some other par-

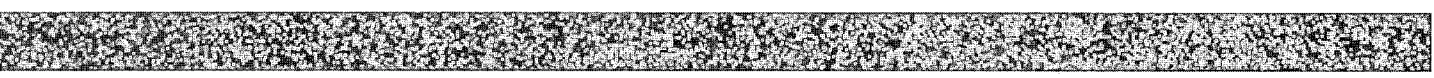

length and/or the smaller the gap, the greater the angle through which the wave is diffracted. Thus, visible light, with its extremely short wavelengths, shows perceptible diffraction only when passing through very narrow openings. (The effect is much more noticeable for sound waves, however—no one thinks twice about our ability to hear people even when they are around a corner and out of our line of sight.)

Interference is the ability of two or more waves to reinforce or cancel each other. The second figure shows two identical waves moving through the same region of space. In the first part, the waves are positioned so that their crests and troughs exactly coincide. The net effect is that the two wave motions reinforce each other, resulting in a wave of greater amplitude. This is known as *constructive interference*. In the second part of the figure, the two waves exactly cancel, so no net motion remains. This is *destructive interference*. As with diffraction, interference between waves of visible light is not noticeable in everyday experience; however, today it is easily measured in the laboratory (as shown in the diagram on the right).

Finally, the phenomenon known as *polarization* of light is also readily understood in terms of the description of electromagnetic waves presented in the text. Normally, light waves are randomly oriented—the electric field in Figure 3.7 may vibrate in any direction perpendicular to the direction of wave motion—and we say the radiation is unpolarized. Most natural objects emit unpolarized radiation. Under some circumstances, however, the electric fields can become aligned—all vibrating in the same plane as the radiation moves through space, and the radiation is said to be polarized. On Earth we can produce polarized light by passing unpolarized light through a Polaroid filter, which has specially aligned elongated molecules that allow the passage of only those waves having electric fields oriented in some specific direction. Reflected light is often polarized, which is why sunglasses constructed with suitably oriented Polaroid filters can be effective in blocking glare.

Diffraction and interference play critical roles in many areas of observational astronomy, including telescope design (Chapter 5). The polarization of starlight provides astronomers with an important technique for probing the properties of interstellar gas (Chapter 18). All three phe-

nomena are predicted by the wave theory of light. The particle theory did not predict them; in fact it predicted that they should *not* occur. Until the early 1800s, the technology was inadequate to resolve the issue. However, by 1830 experimenters had reported unequivocal measurement of each, convincing most scientists that the wave theory was the proper description of electromagnetic radiation. It would be almost a century before the particle description of radiation would resurface, but in radically different form, as we will see in Chapter 4.

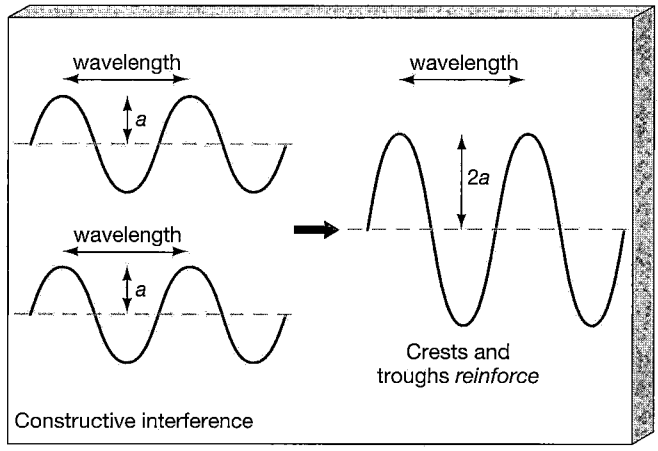

Constructive interference

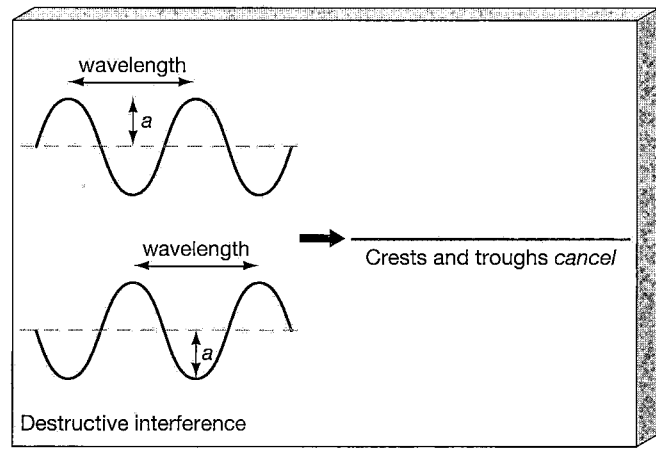

Destructive interference

ticle. Its changing position causes its associated electric field to change, and this changing field in turn causes the electrical force exerted on other charges to vary (Figure 3.5c). If we measure the change in the force on these other charges, we learn about our original particle. Thus, *information about the particle's state of motion is transmitted through space via a changing electric field.* This *disturbance* in the particle's electric field travels through space as a wave.

ELECTROMAGNETIC WAVES

The laws of physics tell us that a **magnetic field** must accompany every changing electric field. Magnetic fields govern the influence of *magnetized* objects on one another, much as electric fields govern interactions between charged particles. The fact that a compass needle always points to magnetic north is the result of the interaction

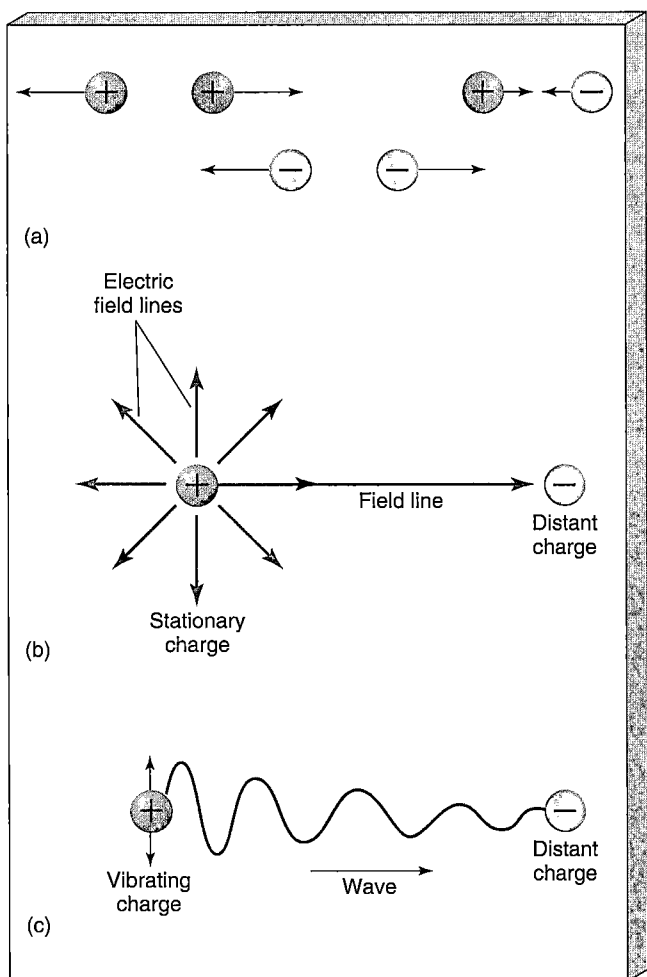

(a)

Electric
field lines

Field line

Distant
charge

Stationary
charge

(b)

Vibrating
charge

Wave

Distant
charge

(c)

Figure 3.5 Charged Particles (a) Particles carrying like electrical charges repel one another, whereas particles carrying unlike charges attract. (b) A charged particle is surrounded by an electric field, which determines the particle's influence on other charged particles. We represent the field by a series of field lines. (c) If a charged particle begins to vibrate back and forth, its electric field changes. The resulting disturbance travels through space as a wave.

between the magnetized needle and Earth's magnetic field (Figure 3.6). Magnetic fields also exert forces on *moving* electric charges (that is, electric currents)—electric meters and motors rely on this basic fact. Conversely, moving charges *create* magnetic fields (electromagnets are a familiar example). In short, electric and magnetic fields are inextricably linked to one another: a change in either one necessarily creates the other.

Thus, as illustrated in Figure 3.7, the disturbance produced by our moving charge actually consists of vibrating electric *and* magnetic fields, always oriented perpendicular to one another and moving together through space. These fields do not exist as independent entities; rather, they are different aspects of a single physical phenomenon: **electromagnetism.** Together, they constitute an electromagnetic wave that carries energy and information from one part of the universe to another.

Now consider a real cosmic object—a star, say. When some of its charged contents move around, their electric fields change, and we can detect that change. The resulting electromagnetic ripples travel outward in waves, requiring

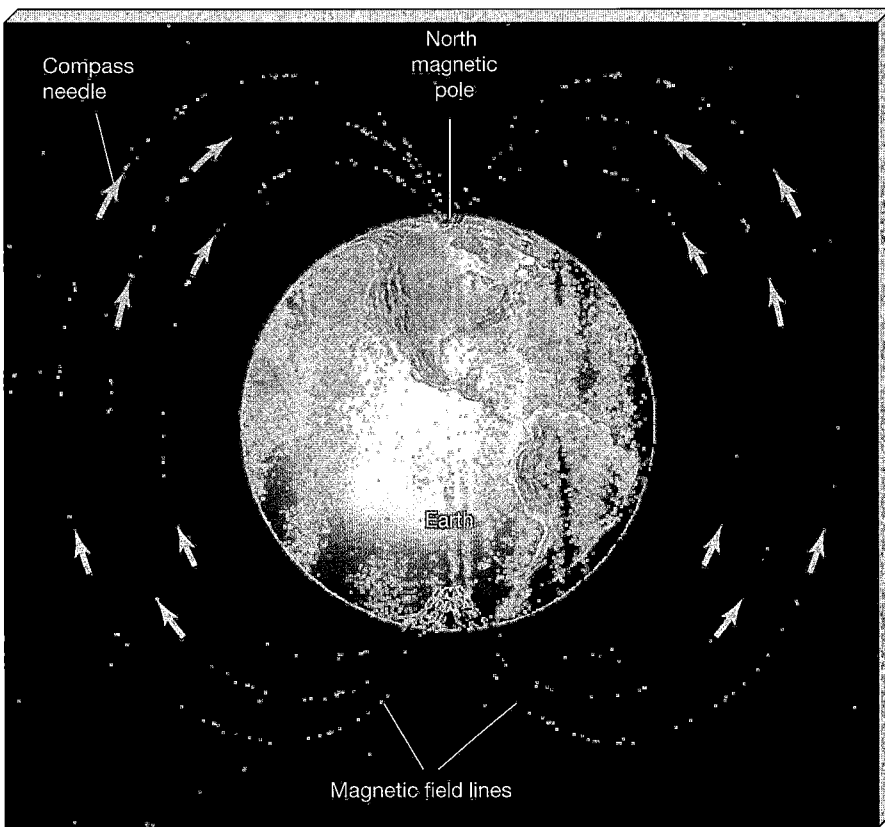

Compass
needle

North
magnetic
pole

Earth

Magnetic field lines

Figure 3.6 Magnetism Earth's magnetic field interacts with a magnetic compass needle, causing the needle to become aligned with the field—that is, to point toward Earth's north (magnetic) pole. The north magnetic pole lies at latitude 80° N, longitude 107° W, some 1140 km from the geographic North Pole.

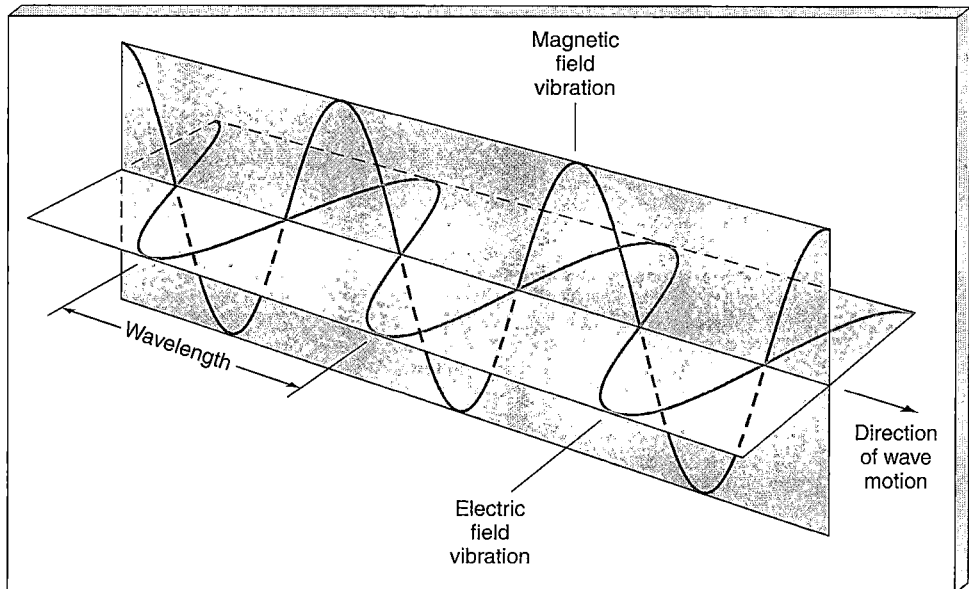

Figure 3.7 Electromagnetic Wave Electric and magnetic fields vibrate perpendicular to each other. Together they form an electromagnetic wave that moves through space at the speed of light in the direction perpendicular to both the electric and the magnetic fields comprising it.

no material medium in which to travel. Small charged particles, either in our eyes or in our experimental equipment, eventually respond to the electromagnetic field changes by vibrating in tune with the received radiation. This response is how we detect the radiation—and how we see. Figure 3.8 shows a more familiar example of information being transferred by electromagnetic radiation. A television transmitter causes electric charges to oscillate up and down a metal rod near the tower's top, thereby generating electromagnetic radiation. This radiation can be detected by rooftop antennas. In the metal rods of the receiving antenna, electric charges respond by vibrating in time with the transmitted wave frequency. The information carried by the pattern of vibrations is then converted into sound and pictures by your TV set.

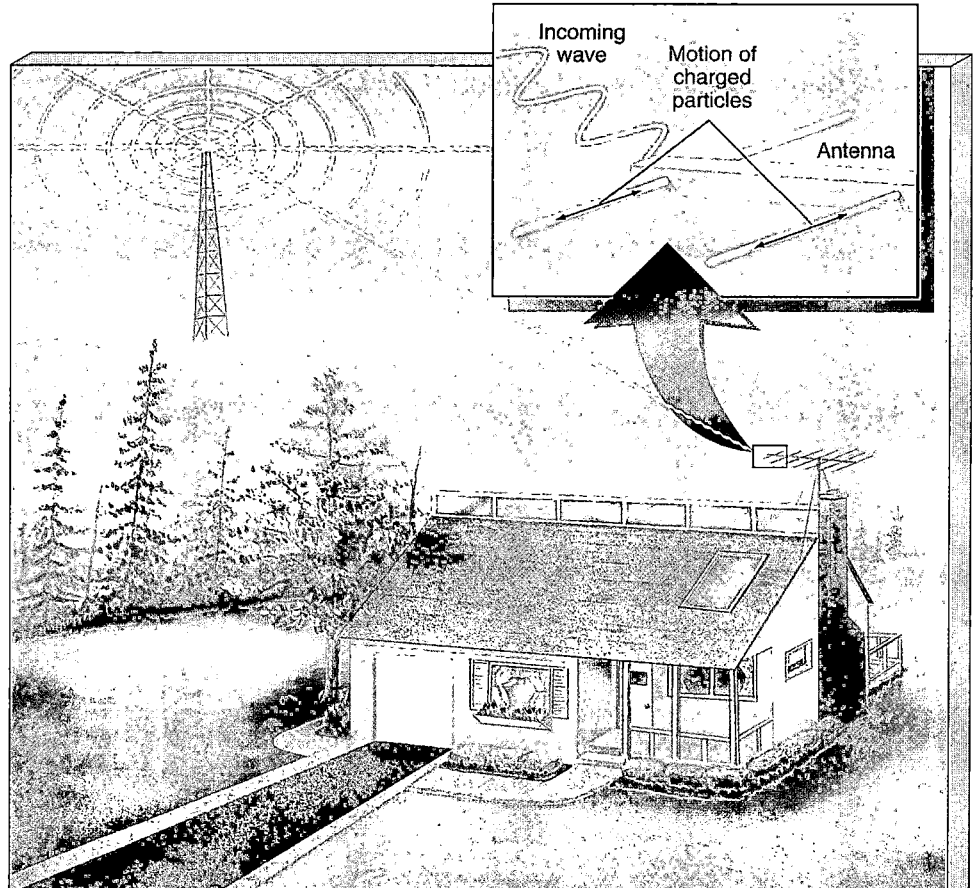

Figure 3.8 Television Signal Charged particles in an ordinary household television antenna vibrate in response to electromagnetic radiation broadcast by a distant transmitter. The radiation is produced when electric charges are made to oscillate in the transmitter's emitting antenna. The vibrations in the receiving antenna "echo" the oscillations in the transmitter, allowing the original information to be retrieved.

How *quickly* does one charge feel the change in the electromagnetic field when another, begins to move? This is an important question, because it is equivalent to asking how fast an electromagnetic wave travels. Does it propagate at some measurable speed, or is it instantaneous? Both theory and experiment tell us that all electromagnetic waves move at a very specific speed—the **speed of light** (always denoted by the letter c). Its exact value is 299,792.458 km/s in a vacuum (and somewhat less in material substances such as air or water). We will round this value off to $c = 3.00 \times 10^5$ km/s. This is an extremely high speed. In the time needed to snap your fingers (about a tenth of a second) light can travel three-quarters of the way around our planet! If the currently known laws of physics are correct, then the speed of light is the fastest speed possible (see *More Precisely 22-1*).

The speed of light is very large, but it is still *finite*. That is, light does not travel instantaneously from place to place. This fact has some interesting consequences for our study of distant objects. It takes time—often lots of time—for light to travel through space. The light we see from the nearest large galaxy—the Andromeda Galaxy, shown in Figure 3.1—left that object about 2.5 million years ago, around the time our first human ancestors appeared on planet Earth. We can know nothing about this galaxy as it exists today. For all we know, it may no longer even exist! Only our descendants, 2.5 million years into the future, will know if it exists now. So as we study objects in the cosmos, remember that the light we see left those objects long ago. We can never observe the universe as it is—only as it was.

☑ Concept Check

▣ What is light? List some similarities and differences between light waves and waves on water or in air.

3.3 The Electromagnetic Spectrum

Figure 3.9 plots the entire range of electromagnetic radiation, illustrating the relationships among the different "types" of electromagnetic radiation listed earlier. We see that the only characteristic distinguishing one from another is wavelength, or frequency. To the low-frequency, long-wavelength side of visible light lie *radio* and *infrared* radiation. Radio frequencies include radar, microwave radiation, and the familiar AM, FM, and TV bands. We perceive infrared radiation as heat. At higher frequencies (shorter wavelengths) are the domains of *ultraviolet, X-ray,* and *gamma-ray* radiation. Ultraviolet radiation, lying just beyond the violet end of the visible spectrum, is responsible for suntans and sunburns. X rays are perhaps best known for their ability to penetrate human tissue and reveal the state of our insides without resorting to surgery. Gamma rays are the shortest-wavelength radiation. They are often associated with radioactivity and are invariably damaging to living cells they encounter.

All these spectral regions, including the visible spectrum, collectively make up the **electromagnetic spectrum**. Remember that, despite their greatly differing wavelengths and the different roles they play in everyday life on Earth, all are basically the same phenomenon, and all move at the same speed—the speed of light, c.

Figure 3.9 is worth studying carefully, as it contains a great deal of information. Note that wave frequency (in hertz) increases from left to right, and wavelength (in meters) increases from right to left. Scientists often disagree on the "correct" way to display wavelengths and frequencies in diagrams of this type. When picturing wavelengths and frequencies, this book consistently adheres to the convention that frequency increases toward the *right*.

Notice also that the wavelength and frequency scales in Figure 3.9 do not increase by equal increments of 10. Instead, successive values marked on the horizontal axis differ by *factors of 10*—each is 10 times greater than its neighbor. This type of scale, called a *logarithmic* scale, is often used in science to condense a large range of some quantity into a manageable size. Had we used a linear scale for the wavelength range shown in Figure 3.9, the figure would have been many light-years long! Throughout the text we will often find it convenient to use a logarithmic scale to compress a wide range of some quantity onto a single, easy-to-view plot.

Figure 3.9 shows that wavelengths extend from the size of mountains for radio radiation to the size of an atomic nucleus for gamma-ray radiation. The box at the upper right emphasizes how small the visible portion of the electromagnetic spectrum is. Most objects in the universe emit large amounts of invisible radiation. Indeed, many of them emit only a tiny fraction of their total energy in the visible range. A wealth of extra knowledge can be gained by studying the invisible regions of the electromagnetic spectrum. To remind you of this important fact and to identify the region of the electromagnetic spectrum in which a particular observation was made, we have attached a spectrum icon—an idealized version of the wavelength scale in Figure 3.9—to every astronomical image presented in this text.

Only a small fraction of the radiation produced by astronomical objects actually reaches Earth's surface because of the *opacity* of our planet's atmosphere. Opacity is the extent to which radiation is blocked by the material through which it is passing—in this case, air. The more opaque an object is, the less radiation gets through it: Opacity is just the opposite of transparency. Earth's atmospheric opacity is plotted along the wavelength and frequency scales at the bottom of Figure 3.9. The extent of shading is proportional to the opacity. Where the shading is greatest, no radiation can get in or out. Where there is no shading at all, the atmosphere is almost completely transparent.

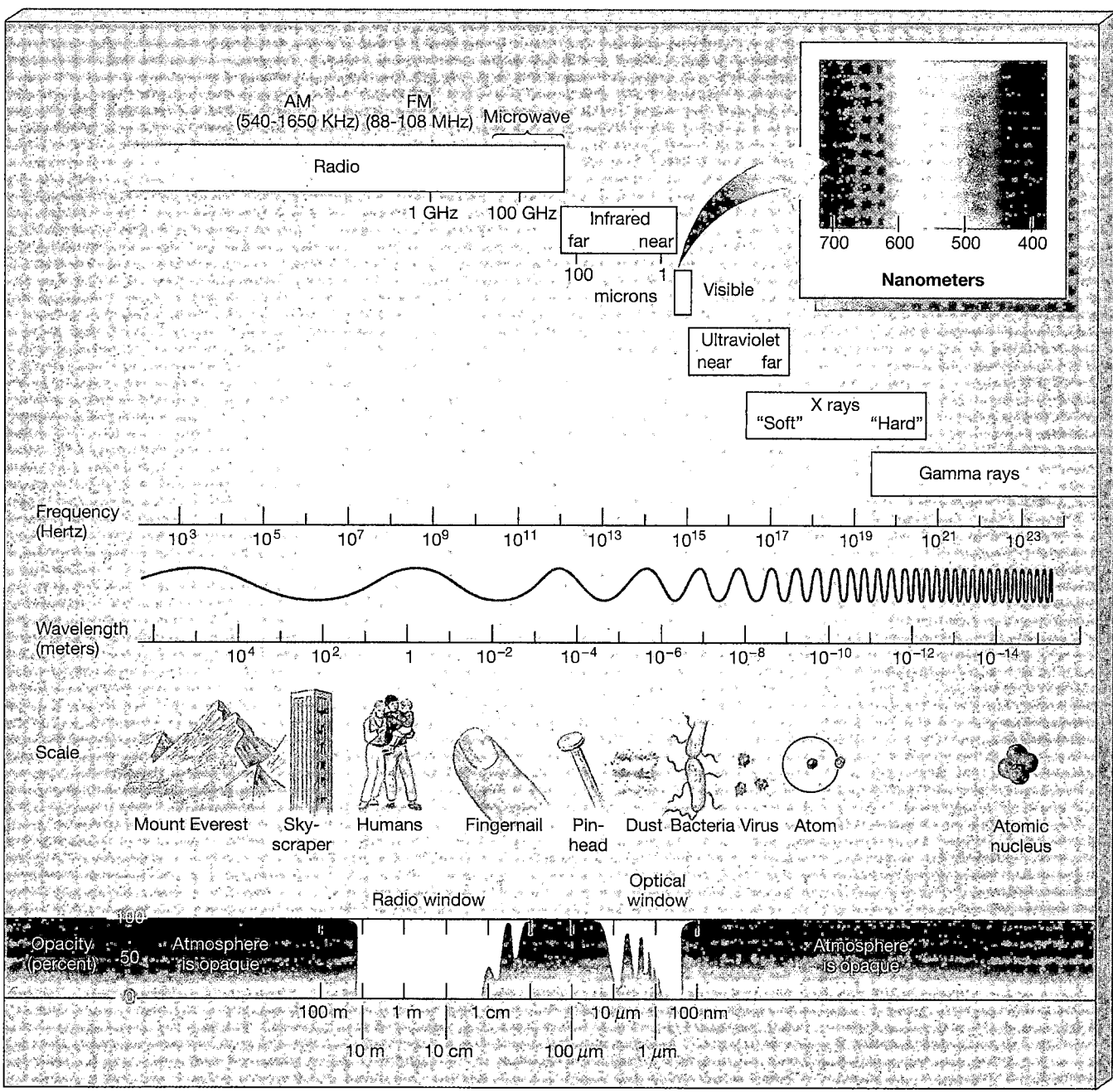

Figure 3.9 Electromagnetic Spectrum The entire electromagnetic spectrum, running from long-wavelength, low-frequency radio waves, to short-wavelength, high-frequency gamma rays.

What causes opacity to vary along the spectrum? Certain atmospheric gases absorb radiation very efficiently at some wavelengths. For example, water vapor (H_2O) and oxygen (O_2) absorb radio waves having wavelengths less than about a centimeter, while water vapor and carbon dioxide (CO_2) are strong absorbers of infrared radiation. Ultraviolet, X-ray, and gamma-ray radiation are completely blocked by the *ozone layer* (O_3) high in Earth's atmosphere (see Section 7.3). A passing but unpredictable source of atmospheric opacity in the visible part of the spectrum is the blockage of light by atmospheric clouds.

In addition, the interaction between the Sun's ultraviolet radiation and the upper atmosphere produces a thin, electrically conducting layer at an altitude of about 100 km. The *ionosphere*, as this layer is known, reflects long-wavelength radio waves (wavelengths greater than about 10 m) as well as a mirror reflects visible light. In this way, extraterrestrial waves are kept out, and terrestrial waves—such as those produced by AM radio stations—are kept in. (That's why it is possible to transmit some radio frequencies beyond the horizon—the broadcast waves bounce off the ionosphere.)

The effect of atmospheric opacity is that there are only a few *spectral windows*, at well-defined locations in the electromagnetic spectrum, where Earth's atmosphere is transparent. In much of the radio and in the visible portions of the spectrum, the opacity is low and we can study the universe at those wavelengths from ground level. In parts of the infrared range, the atmosphere is partially transparent, so we can make certain infrared observations from the ground. Moving to the tops of mountains, above as much of the atmosphere as possible, improves observations. In the rest of the spectrum, however, the atmosphere is opaque. Ultraviolet, X-ray, and gamma-ray observations can be made only from above the atmosphere, from orbiting satellites.

☑ Concept Check

■ In what sense are radio waves, visible light, and X rays one and the same phenomenon?

3.4 The Distribution of Radiation

③ *All* macroscopic objects—fires, ice cubes, people, stars—emit radiation at all times, regardless of their size, shape, or chemical composition. They radiate mainly because the microscopic charged particles they are made up of are in constantly varying random motion, and whenever charges change their state of motion, electromagnetic radiation is emitted. The **temperature** of an object is a direct measure of the amount of microscopic motion within it (see *More Precisely 3-1*). The hotter the object—that is, the higher its temperature—the faster its constituent particles move and the more energy they radiate.

THE BLACKBODY SPECTRUM

Intensity is a term often used to specify the amount or strength of radiation at any point in space. Like frequency and wavelength, intensity is a basic property of radiation. No natural object emits all its radiation at just one frequency. Instead, the energy is generally spread out over a range of frequencies. By studying how the intensity of this radiation is distributed across the electromagnetic spectrum, we can learn much about the object's properties.

Figure 3.10 illustrates schematically the distribution of radiation emitted by any object. The curve peaks at a single, well-defined frequency and falls off to lesser values above and below that frequency. Note that the curve is not shaped like a symmetrical bell that declines evenly on either side of the peak. The intensity falls off more slowly from the peak to lower frequencies than it does on the high-frequency side. This overall shape is characteristic of the radiation emitted by *any* object, regardless of its size, shape, composition, or temperature.

The curve drawn in Figure 3.10 is the radiation-distribution curve for a mathematical idealization known as a *blackbody*—an object that absorbs all radiation falling on it. In a steady state, a blackbody must reemit the same amount of energy it absorbs. The **blackbody curve** shown in the figure describes the distribution of that reemitted radiation. (The curve is also known as the *Planck curve*, after Max Planck, whose mathematical analysis of such thermal emission in 1900 played a key role in modern physics.) No real object absorbs and radiates as a perfect blackbody. However, in many cases, the blackbody curve is a good approximation to reality, and the properties of blackbodies provide important insights into the behavior of real objects.

THE RADIATION LAWS

④ The blackbody curve shifts toward higher frequencies (shorter wavelengths) and greater intensities as an object's temperature increases. Even so, the *shape* of the curve remains the same. This shifting of radiation's peak frequency with temperature is familiar to us all: Very hot glowing objects, such as toaster filaments or stars, emit visible light. Cooler objects, such as warm rocks or household radiators, produce invisible radiation—warm to the touch but not glowing hot to the eye. These latter objects emit most of their radiation in the lower-frequency infrared part of the electromagnetic spectrum (Figure 3.9).

Imagine a piece of metal placed in a hot furnace. At first, the metal becomes warm, although its visual appearance doesn't change. As it heats up, it begins to glow dull red, then orange, brilliant yellow, and finally white. How do we explain this? As illustrated in Figure 3.11, when the metal is at room temperature (300 K—see *More Precisely 3-1* for a discussion of the Kelvin temperature scale), it emits only invisible infrared radiation. As the metal becomes hotter, the peak of its blackbody curve shifts toward higher frequencies. At 1000 K, for instance, most of the

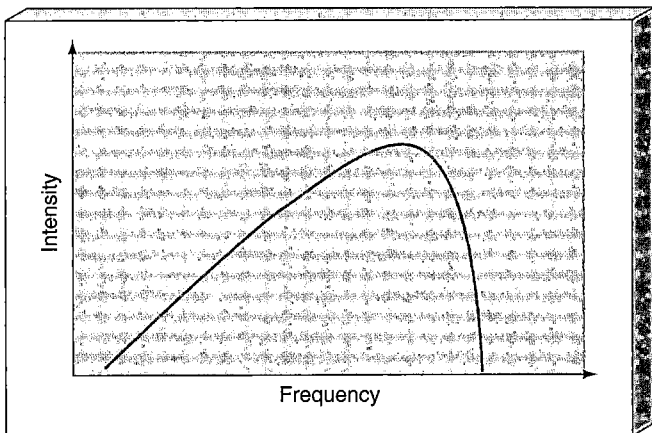

Figure 3.10 Ideal Blackbody Curve The blackbody, or Planck, curve represents the distribution of the intensity of radiation emitted by any object.

The Kelvin Temperature Scale

The atoms and molecules that make up any piece of matter are in constant random motion. This motion represents a form of energy known as *thermal energy*—or, more commonly, *heat*. The quantity we call *temperature* is a direct measure of this internal motion: The higher an object's temperature, the faster, on average, the random motion of its constituent particles. The temperature of a piece of matter specifies the average thermal energy of the particles it contains.

Our familiar Fahrenheit temperature scale, like the archaic English system in which length is measured in feet and weight in pounds, is of somewhat dubious value. In fact, the "degree Fahrenheit" is now a peculiarity of American society. Most of the world uses the Celsius scale of temperature measurement (also called the centigrade scale). In the Celsius system, water freezes at 0 degrees (0°C) and boils at 100 degrees (100°C), as illustrated in the accompanying figure.

There are, of course, temperatures below the freezing point of water. Although we know of no matter anywhere in the universe that is actually this cold, temperatures can in theory reach as low as $-273.15°C$. This is the temperature at which atomic and molecular motion all but ceases. It is convenient to construct a temperature scale based on this lowest possible temperature, or *absolute zero*. Scientists commonly use such a scale, called the *Kelvin scale* in honor of the nineteenth-century British physicist Lord Kelvin. Since it takes absolute zero as its starting point, the Kelvin scale differs from the Celsius scale by 273.15°. In this book, we round off the decimal places and simply use

$$\text{kelvins} = \text{degrees Celsius} + 273.$$

Thus,

- All thermal motion ceases at 0 kelvins (0 K).
- Water freezes at 273 kelvins (273 K).
- Water boils at 373 kelvins (373 K).

Note that the unit is "kelvins," or "K," *not* "degrees kelvin" or "°K." (Occasionally, the term "degrees absolute" is used instead.)

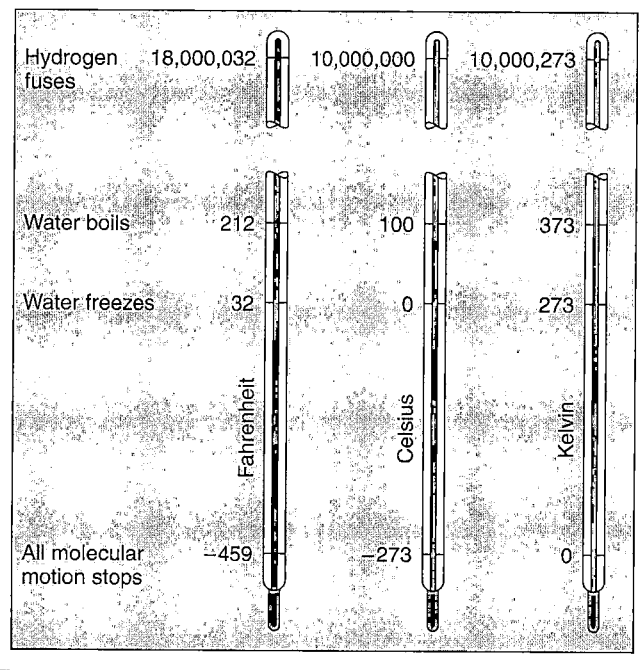

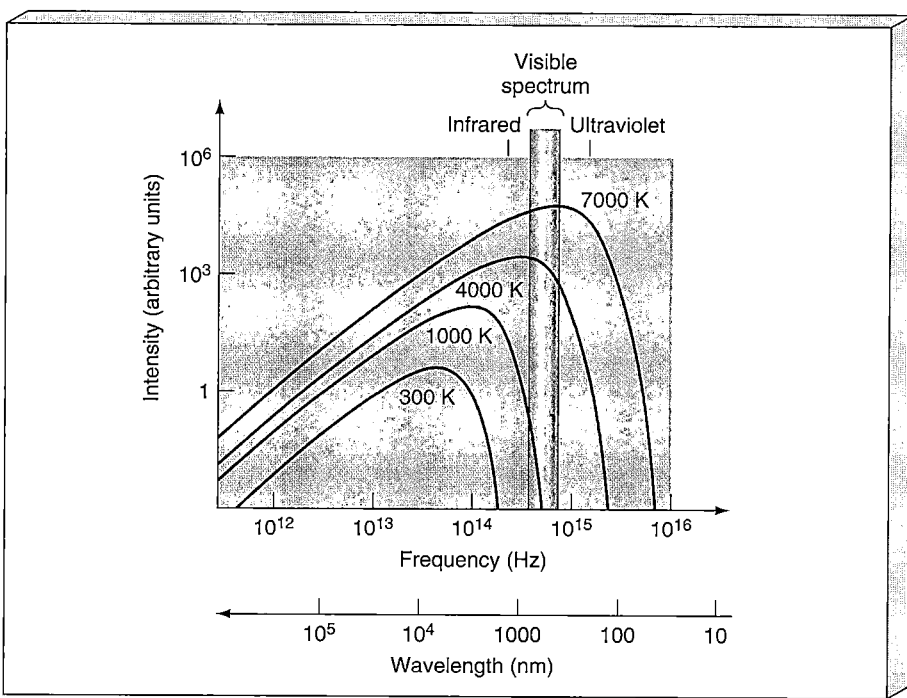

Figure 3.11 Blackbody Curves
As an object is heated, the radiation it emits peaks at higher and higher frequencies. Shown here are curves corresponding to temperatures of 300 K (room temperature), 1000 K (beginning to glow dull red), 4000 K (red hot), and 7000 K (white hot).

Animation

emitted radiation is still infrared, but now there is also a small amount of visible (dull red) radiation being emitted (note in Figure 3.11 that the high-frequency portion of the 1000 K curve just overlaps the visible region of the graph).

As the temperature continues to rise, the peak of the metal's blackbody curve moves through the visible spectrum, from red (the 4000 K curve) through yellow. The metal eventually becomes white hot because, when its blackbody curve peaks in the blue or violet part of the spectrum (the 7000 K curve), the low-frequency tail of the curve extends through the entire visible spectrum (to the left in Figure 3.11), meaning that substantial amounts of green, yellow, orange, and red light are also emitted. Together, all these colors combine to produce white.

From studies of the precise form of the blackbody curve we obtain a very simple connection between the wavelength at which most radiation is emitted and the ab-

solute temperature (that is, the temperature measured in kelvins) of the emitting object:

$$\text{wavelength of peak emission} \propto \frac{1}{\text{temperature}}.$$

This relationship, called **Wien's law**, is discussed in more detail in *More Precisely 3-2*.

Simply put, Wien's law tells us that the hotter the object, the bluer its radiation. For example (see Figure 3.13), an object with a temperature of 6000 K emits most of its energy in the visible part of the spectrum, with a peak wavelength of 480 nm. At 600 K, the object's emission would peak at a wavelength of 4800 nm, well into the infrared portion of the spectrum. At a temperature of 60,000 K, the peak would move all the way through the visible spectrum to a wavelength of 48 nm, in the ultraviolet range.

MORE PRECISELY 3-2

More About the Radiation Laws

As mentioned in Section 3.4, Wien's law relates the temperature T of an object to the wavelength λ_{max} at which it emits the most radiation. (The Greek letter λ—lambda—is conventionally used to denote wavelength.) Mathematically, if we measure T in kelvins and λ_{max} in centimeters, we find that

$$\lambda_{max} = \frac{0.29 \text{ cm}}{T}.$$

We could also convert Wien's law into an equivalent statement about frequency f, using the relation $f = c/\lambda$, but the law is most commonly stated in terms of wavelength and is probably easier to remember that way.

EXAMPLE: For a blackbody with the same temperature as the surface of the Sun, approximately 6000 K, the wavelength of maximum intensity is $(0.29/6000)$ cm, or 480 nm, corresponding to the yellow-green part of the visible spectrum. A cooler star with a temperature of 3000 K has a peak wavelength of $(0.29/3000)$ cm ≈ 970 nm, just longward of the red end of the visible spectrum, in the near infrared. The blackbody curve of a star with a temperature of 12,000 K peaks at 242 nm, in the near ultraviolet, and so on.

Stellar temperatures are typically measured in thousands of kelvins, so it is helpful to rewrite the above equation in more "stellar" units:

$$\lambda_{max} = \frac{2900 \text{ nm}}{T \text{ (thousands of kelvins)}}.$$

We can also give Stefan's law a more precise mathematical formulation. With T measured in kelvins, the total amount of energy emitted per square meter of its surface per second (a quantity known as the *energy flux*, F) is given by

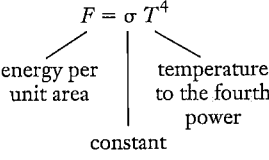

$$F = \sigma T^4$$

energy per unit area | constant | temperature to the fourth power

The constant σ (the Greek letter sigma) is known as the *Stefan-Boltzmann constant*, or often just Stefan's constant, after Josef Stefan, the Austrian scientist who formulated the equation.

The SI unit of energy is the *joule* (J). Probably more familiar is the closely related unit called the *watt* (W), which measures power—the *rate* at which energy is emitted or expended by an object. One watt is the emission of one joule per second. For example, a 100-W lightbulb emits energy (mostly in the form of infrared and visible light) at a rate of 100 J/s. In these units, the Stefan-Boltzmann constant has the value $\sigma = 5.67 \times 10^{-8}$ W/m$^2 \cdot$K^4.

EXAMPLE: Notice just how rapidly the energy flux increases with increasing temperature. A piece of metal in a furnace, when at a temperature of 3000 K, radiates energy at a rate of about 5.67×10^{-8} W/m$^2 \cdot$K$^4 \times (0.01$ m$)^2 \times (3000$ K$)^4 = 460$ W for every square centimeter of its surface area. Doubling its temperature to 6000 K, the surface temperature of the Sun (so that it becomes yellow-hot, by Wien's law) increases the energy emitted by a factor of 16 (four "doublings"), to 7.3 *kilo*watts (7,300 W) per square centimeter.

Notice also that the law relates to energy emitted *per unit area*. The flame of a blowtorch is considerably hotter than a bonfire, but the bonfire emits far more energy *in total* because it is much larger.

Figure 3.12 The Sun at Many Wavelengths Three images of the Sun, made using (a) radio waves, (b) infrared radiation, and (c) visible light, are shown here in false color. By studying the similarities and differences among these views of the same object, acquired on the same day, astronomers can find important clues to its structure, composition, and surface activity. Although most sunlight is emitted in the form of infrared and visible radiation, a wealth of information about our parent star can be obtained by studying it in other regions of the electromagnetic spectrum. *(NRAO; AURA)*

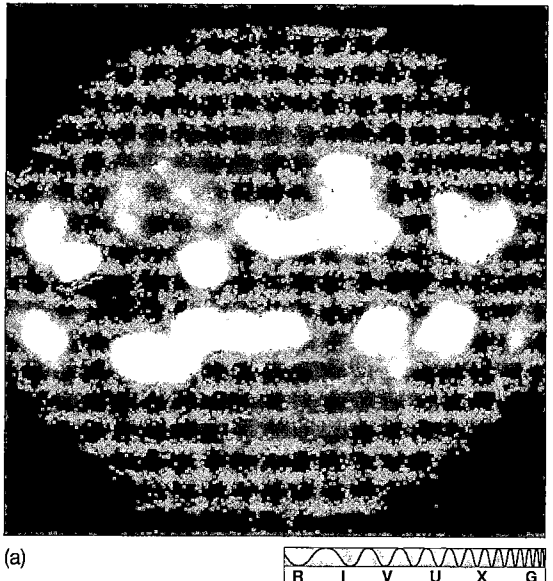

(a)

It is also a matter of everyday experience that, as the temperature of an object increases, the *total* amount of energy it radiates (summed over all frequencies) increases rapidly. For example, the heat given off by an electric heater increases very sharply as it warms up and begins to emit visible light. Careful experimentation leads to the conclusion that the total amount of energy radiated per unit time is actually proportional to the fourth power of the object's temperature:

$$\text{total energy emission} \propto \text{temperature}^4.$$

This relation is called **Stefan's law**. This is discussed further in *More Precisely 3-2*. From the form of Stefan's law we can see that the energy emitted by a body rises dramatically as its temperature increases. Doubling the temperature causes the total energy radiated to increase by a factor of $2^4 = 16$; tripling the temperature increases the emission by $3^4 = 81$, and so on.

ASTRONOMICAL APPLICATIONS

No known natural terrestrial objects reach temperatures high enough to emit very-high-frequency radiation. Only human-made thermonuclear explosions are hot enough for their spectra to peak in the X-ray or gamma-ray range. (Most human inventions that produce short-wavelength, high-frequency radiation, such as X-ray machines, are designed to emit only a specific range of wavelengths and do not operate at high temperatures. They are said to produce a *nonthermal* spectrum of radiation.) Many extraterrestrial objects, however, do emit copious quantities of ultraviolet, X-ray, and even gamma-ray radiation. Figure 3.12 shows a familiar object—our Sun—as it appears when viewed using radiation from different parts of the electromagnetic spectrum.

Astronomers often use blackbody curves as thermometers to determine the temperatures of distant objects. For example, study of the solar spectrum makes it possible to measure the temperature of the Sun's surface. Observations of the radiation from the Sun at many frequencies yield a curve shaped somewhat like that shown in Figure 3.10. The Sun's curve peaks in the visible part of the electromagnetic spectrum; the Sun also emits a lot of infrared and a little ultraviolet radiation. Using Wien's law, we find that the temperature of the Sun's surface is approximately 6000 K. (A more precise measurement, applying

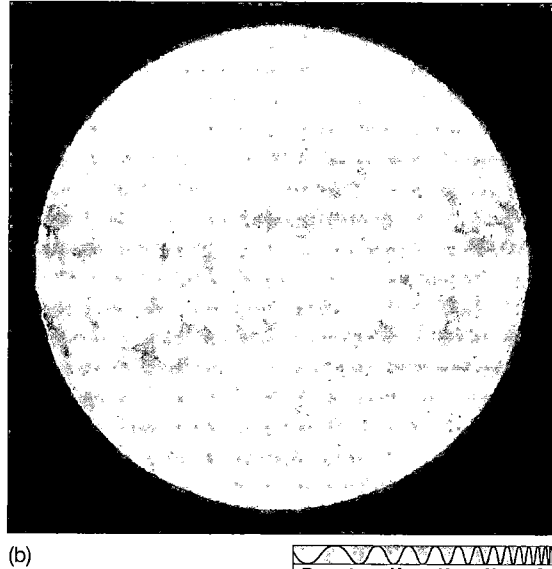

(b)

(c)

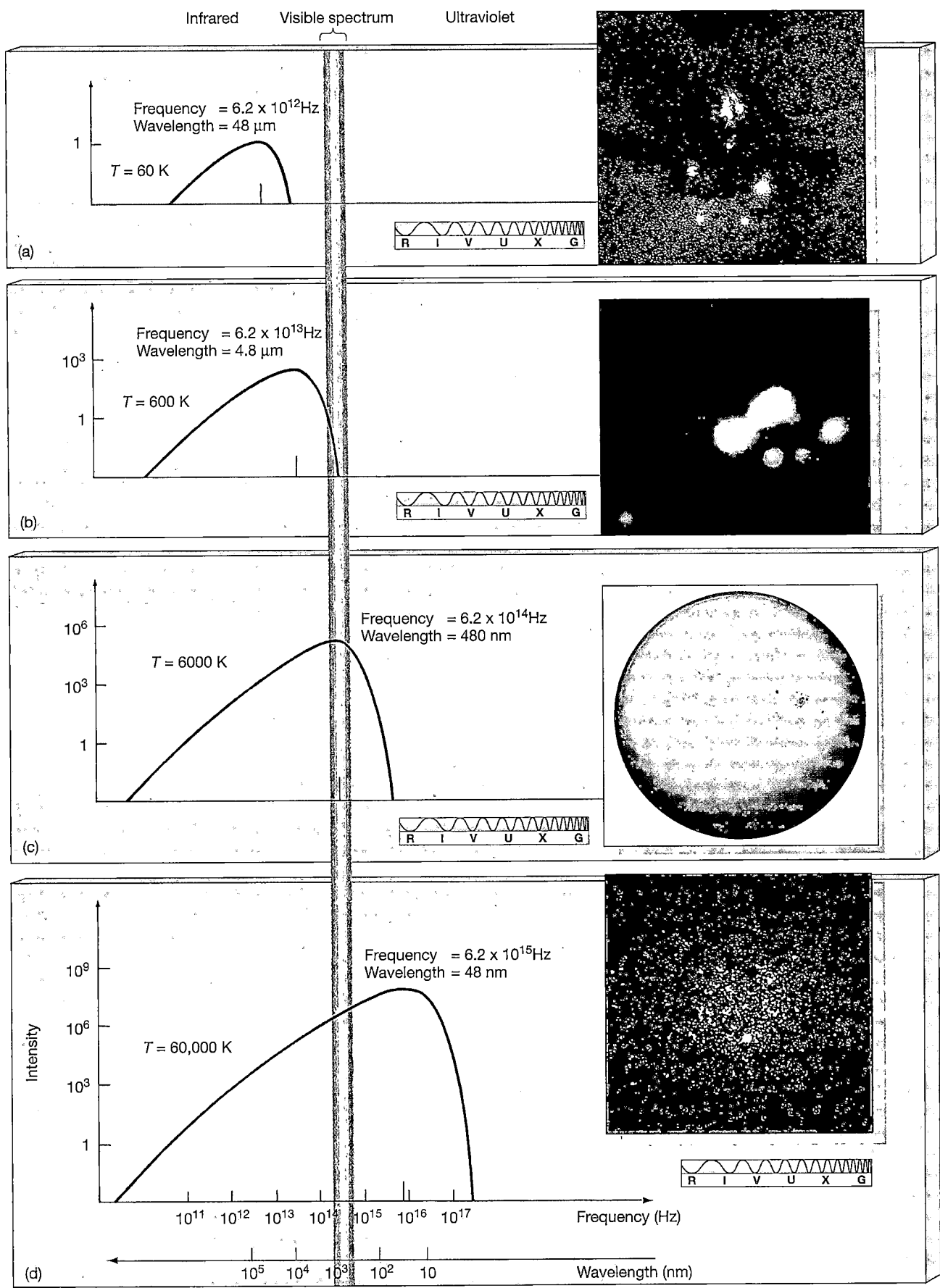

Infrared   Visible spectrum   Ultraviolet

(a)
Frequency   = 6.2 x 10^{12}Hz
Wavelength = 48 µm

$T = 60$ K

1

R I V U X G

(b)
Frequency   = 6.2 x 10^{13}Hz
Wavelength = 4.8 µm

$T = 600$ K

10^3

1

R I V U X G

(c)
Frequency   = 6.2 x 10^{14}Hz
Wavelength = 480 nm

$T = 6000$ K

10^6

10^3

1

R I V U X G

(d)
Frequency   = 6.2 x 10^{15}Hz
Wavelength = 48 nm

$T = 60,000$ K

10^9

10^6

10^3

1

Intensity

10^{11} 10^{12} 10^{13} 10^{14} 10^{15} 10^{16} 10^{17} Frequency (Hz)

10^5 10^4 10^3 10^2 10   Wavelength (nm)

R I V U X G

Figure 3.13 Astronomical Thermometer Comparison of blackbody curves for four cosmic objects. The frequencies and wavelengths corresponding to peak emission are marked. (a) A cool, invisible galactic gas cloud called Rho Ophiuchi. At a temperature of 60 K, it emits mostly low-frequency radio radiation. (b) A dim, young star (shown red in the inset photograph) near the center of the Orion Nebula. The star's atmosphere, at 600 K, radiates primarily in the infrared. (c) The Sun's surface, at approximately 6000 K, is brightest in the visible region of the electromagnetic spectrum. (d) Some very bright stars in a cluster called Omega Centauri, as observed by a telescope aboard a space shuttle. At a temperature of 60,000 K, these stars radiate strongly in the ultraviolet. *(Harvard College Observatory; J. Moran; AURA; NASA)*

Wien's law to the blackbody curve that best fits the solar spectrum, yields a temperature of 5800 K.)

Other cosmic objects have surfaces very much cooler or hotter than the Sun's, emitting most of their radiation in invisible parts of the spectrum (Figure 3.13). For example, the relatively cool surface of a very young star may measure 600 K and emit mostly infrared radiation. Cooler still is the interstellar gas cloud from which the star formed; at a temperature of 60 K, such a cloud emits mainly long-wavelength radiation in the radio and infrared parts of the spectrum. The brightest stars, by contrast, have surface temperatures as high as 60,000 K and hence emit mostly ultraviolet radiation.

Concept Check

▪ Describe, in terms of the radiation laws, how and why the appearance of an incandescent lightbulb changes as you turn a dimmer switch to increase its brightness from "off" to "maximum."

3.5 The Doppler Effect

▣ Imagine a rocket ship launched from Earth with enough fuel to allow it to accelerate to speeds approaching that of light. As the ship's speed increased, a remarkable thing would happen (Figure 3.14). Passengers would notice that the light from the star system toward which they were traveling seemed to be getting *bluer*. In fact, *all* stars in front of the ship would appear bluer than normal, and the greater the ship's speed, the greater the color change would be. Furthermore, stars behind the vessel would seem *redder* than normal, while stars to either side would be unchanged in appearance. As the spacecraft slowed down and came to rest relative to Earth, all stars would resume their usual appearance. The travelers would have to conclude that the stars had changed their colors not because of any real change in their physical properties but because of the spacecraft's own *motion*.

This phenomenon is not restricted to electromagnetic radiation and fast-moving spacecraft. Waiting at a railroad crossing for an express train to pass, most of us have had the experience of hearing the pitch of a train whistle change from high shrill (high frequency, short wavelength) to low blare (low frequency, long wavelength) as the train approaches and then recedes. This motion-induced change in the observed frequency of a wave is known as the **Doppler effect**, in honor of Christian Doppler, the nineteenth-century Austrian physicist who first explained it in 1842. Applied to cosmic sources of electromagnetic radiation, it has become one of the most important measurement techniques in all of modern astronomy. Here's how it works:

Imagine a wave moving from the place where it is created toward an observer who is not moving with respect to the wave source, as shown in Figure 3.15(a). By noting the distances between successive wave crests, the observer can determine the wavelength of the emitted wave. Now suppose that the wave source is moving. As illustrated in Figure 3.15(b), because the source moves between the times of emission of one wave crest and the next, successive wave crests in the direction of motion of the source will be seen to be *closer together* than normal, whereas crests behind the source will be more widely spaced. An observer in front of the source will therefore measure a *shorter* wavelength than normal, while one behind will see a *longer* wavelength. (The numbers indicate successive wave crests emitted by the source and the location of the source at the instant each wave crest was emitted.)

The greater the relative speed of source and the observer, the greater the observed shift. If the other velocities involved are not too large compared to the wave speed—less than a few percent, say—we can write down a particularly simple formula for what the observer sees. In terms of the net velocity of *recession* between source and observer, the apparent wavelength and frequency (measured by the observer) are related to the true quantities (emitted by the source) as follows:

$$\frac{\text{apparent wavelength}}{\text{true wavelength}} = \frac{\text{true frequency}}{\text{apparent frequency}}$$

$$= 1 + \frac{\text{recession velocity}}{\text{wave speed}}.$$

A positive recession velocity means that the source and the observer are moving apart; a negative value means that they are approaching. The wave speed is the speed of light c in the case of electromagnetic radiation. For most of this text, the assumption that the recession velocity is small compared to the speed of light will be a good one. Only when we discuss the properties of black holes (Chapter 22) and the structure of the universe on the largest scales (Chapters 25 and 26) will we have to reconsider this formula.

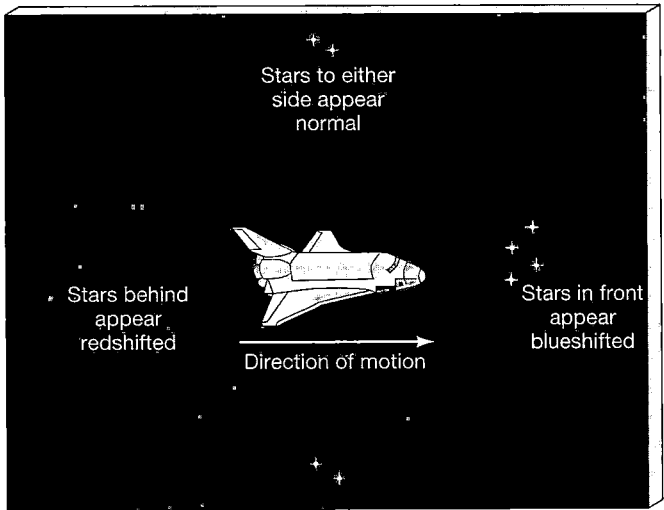

Figure 3.14 High-Speed Observers Observers in a fast-moving spacecraft will see the stars ahead of them seem bluer than normal, while those behind are reddened. The stars have not changed their properties—the color changes are the result of the observers' motion relative to the stars.

Note that in Figure 3.15 the *source* is shown in motion (as in our train analogy), whereas in our earlier spaceship example (Figure 3.14) the *observers* were in motion. For electromagnetic radiation, the result is the same in either case—only the *relative* motion of source and observer matters. Note also that only motion along the line joining source and observer—known as *radial* motion—appears in the above equation. Motion *transverse* (perpendicular) to the line of sight has no significant effect.*

A wave measured by an observer situated in front of a moving source is said to be *blueshifted*, because blue light has a shorter wavelength than red light. Similarly, an observer situated behind the source will measure a longer-than-normal wavelength—the radiation is said to be *redshifted*. This terminology is used even for invisible radiation, for which "red" and "blue" have no meaning. Any

*In fact, Einstein's theory of relativity (see Chapter 22) implies that when the transverse velocity is comparable to the speed of light, a wavelength change, called the transverse Doppler shift, does occur. For most terrestrial and astronomical applications, however, this shift is negligibly small, and we will ignore it here.

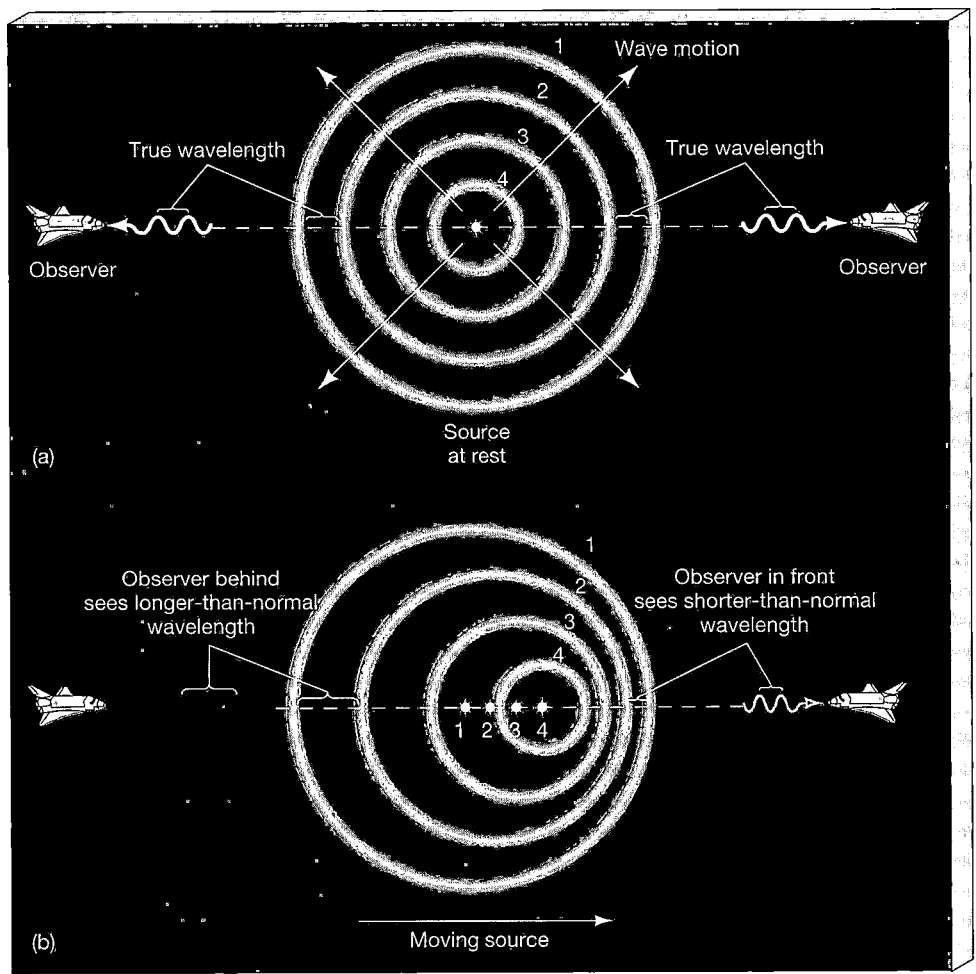

(a)

(b)

Figure 3.15 Doppler Effect (a) Wave motion from a source toward an observer at rest with respect to the source. The four numbered circles represent successive wave crests emitted by the source. At the instant shown, the fifth wave crest is just about to be emitted. As seen by the observer, the source is not moving, so the wave crests are just concentric spheres (shown here as circles). (b) Waves from a moving source tend to "pile up" in the direction of motion and be "stretched out" on the other side. (The numbered points indicate the location of the source at the instant each wave crest was emitted.) As a result, an observer situated in front of the source measures a shorter-than-normal wavelength—a blueshift—while an observer behind the source sees a redshift. In this diagram the source is shown in motion. However, the same general statements hold whenever there is any relative motion between source and observer.

shift toward shorter wavelengths is called a blueshift, and any shift toward longer wavelengths is called a redshift. For example, ultraviolet radiation might be blueshifted into the X ray part of the spectrum or redshifted into the visible; infrared radiation could be redshifted into the microwave range, and so on.

Because c is so large—300,000 km/s—the Doppler effect is extremely small for everyday terrestrial velocities. For example, consider a source receding from the observer at Earth's orbital speed of 30 km/s, a velocity much greater than any encountered in day-to-day life. A beam of blue light would be shifted by only 30 km/s/300,000 km/s = 0.01 percent, from 400 nm to 400.04 nm—a very small change indeed, and one that the human eye cannot distinguish. (It is easily detectable with modern instruments, though.)

The importance of the Doppler effect to astronomers is that it allows them to determine the speed of any cosmic object along the line of sight simply by determining the extent to which its light is redshifted or blueshifted. Suppose that the beam of blue light just mentioned is observed to have a wavelength of 399 nm instead of the 400 nm with which it was emitted. (Let's defer to the next chapter the question of *how* an observer might know the wavelength of the emitted light.) Using the above equation, the observer could calculate the source's radial velocity to be $399/400 - 1 = -0.0025$ times the speed of light. In other

words, the source is *approaching* the observer at a speed of 0.0025 c, or 750 km/s. The basic reasoning is simple but very powerful. The motions of nearby stars and distant galaxies—even the expansion of the universe itself—have all been measured in this way.

Motorists stopped for speeding on the highway have experienced another, much more down-to-earth, application. Police radar measures speed by means of the Doppler effect, as do the radar guns used to clock the velocity of a pitcher's fastball or a tennis player's serve. Notice, incidentally, that the Doppler effect depends only on the relative motion of source and observer; it does not depend on *distance* in any way.

In practice, it is hard to measure the Doppler shift of an entire blackbody curve, simply because it is spread over many wavelengths, making small shifts hard to determine with any accuracy. However, if the radiation were more narrowly defined and took up just a narrow "sliver" of the spectrum, then precise measurements of Doppler effect *could* be made. We will see in the next chapter that in many circumstances this is precisely what does happen, making the Doppler effect one of the observational astronomer's most powerful tools.

☑ Concept Check

◘ How can the Doppler effect be used in determining the mass of a distant star?

Chapter Review

Visible light (p. 62) is a particular type of **electromagnetic radiation** (p. 62) and travels through space in the form of a **wave** (p. 63). A wave is characterized by its period, the length of time taken for one complete cycle; its **wavelength** (p. 64), the distance between successive wave crests; and its **amplitude** (p. 64), which measures the size of the disturbance associated with the wave. A wave's **frequency** (p. 64) is simply one divided by the wave period—it counts the number of wave crests that pass a given point in one second. Diffraction, interference, and polarization are properties of radiation that mark it as a wave phenomenon.

Electrons (p. 65) and **protons** (p. 65) are elementary particles that carry equal and opposite electrical charges. Any electrically charged object is surrounded by an **electric field** (p. 66) that determines the force it exerts on other charged objects. Like gravitational fields, electric fields decrease as the square of the distance from their source. When a charged particle moves, information about that motion is transmitted throughout the universe by the particle's changing electric field. The information

travels in the form of a wave at the **speed of light** (p. 70). Both electric and **magnetic fields** (p. 67) are involved, so the phenomenon is known as **electromagnetism** (p. 68).

A beam of white light is bent, or refracted, as it passes through a prism. Different frequencies of light within the beam are refracted by different amounts, so the beam is split up into its component colors—the visible spectrum. The color of visible light is simply a measure of its wavelength—red light has a longer wavelength than blue light. The entire **electromagnetic spectrum** (p. 70) consists of (in order of increasing frequency) **radio waves, infrared radiation,** visible light, **ultraviolet radiation, X rays,** and **gamma rays** (p. 63). The opacity of Earth's atmosphere—the extent to which it absorbs radiation—varies greatly with wavelength. Only radio waves, some infrared wavelengths, and visible light can penetrate the atmosphere and reach the ground from space.

The **temperature** (p. 72) of an object is a measure of the speed with which its constituent particles move. The intensity of radiation of different frequencies emitted by a hot object has a characteristic distribution, called a **blackbody curve** (p. 72), that depends only on the temperature of the object. **Wien's law** (p. 74) tells us that the wavelength at which the object radiates most energy is inversely proportional to its temperature. **Stefan's law** (p. 75) states that the total amount of energy radiated is proportional to the fourth power of the temperature.

Our perception of the wavelength of a beam of light can be altered by our velocity relative to the source. This motion-induced change in the observed frequency of a wave is called the **Doppler effect** (p. 77). Any net motion away from the source causes a redshift —a shift to lower frequencies—in the received beam. Motion toward the source causes a blueshift. The extent of the shift is directly proportional to the observer's radial velocity relative to the source.

SELF-TEST: TRUE OR FALSE?

_____ **1.** Light, radio, ultraviolet, and gamma rays are all forms of electromagnetic radiation.

_____ **2.** Sound is a familiar type of electromagnetic wave.

_____ **3.** The amount of diffraction increases with increasing wavelength.

_____ **4.** Interference occurs when one wave is brighter than another; the fainter wave cannot be observed.

_____ **5.** Electromagnetic waves cannot travel through a perfect vacuum.

_____ **6.** Electromagnetic waves all travel at the same speed, the speed of light.

_____ **7.** Visible light makes up the greatest part of the entire electromagnetic spectrum.

_____ **8.** Ultraviolet light has the shortest wavelength of any electromagnetic wave.

_____ **9.** A blackbody emits all its radiation at one wavelength or frequency.

_____ **10.** A perfect blackbody emits exactly as much radiation as it absorbs from outside.

_____ **11.** The shape of a blackbody curve depends on the temperature of the body.

_____ **12.** The frequency at which the blackbody curve peaks increases with temperature.

_____ **13.** Objects moving away from an observer are redshifted because they actually turn red.

_____ **14.** The Doppler effect occurs for all types of wave motion.

_____ **15.** An object emitting radiation, moving transverse to the line of sight, produces no appreciable Doppler effect.

SELF-TEST: FILL IN THE BLANK

1. The speed of a radio wave is _____ km/s.

2. The _____ of a wave is the distance between any two adjacent wave crests.

3. The _____ of a wave is measured in units of hertz (Hz).

4. _____ is the ability of a wave to "bend around corners."

5. When a charged particle moves, information about this motion is transmitted through space by means of its changing _____ and _____ fields.

6. The visible spectrum ranges from _____ to _____ in wavelength.

7. Light with a wavelength of 700 nm is perceived to be _____ in color.

8. Earth's atmosphere has low opacity for three forms of electromagnetic radiation. They are _____, _____, and _____.

9. The peak of an object's emitted radiation occurs at a frequency or wavelength determined by the object's _____.

10. The lowest possible temperature is _____ K.

11. Water freezes at _____ K.

12. Because the Sun emits its peak amount of radiation at about 480 nm, its temperature must be about _____ K.

13. Two identical objects have temperatures of 1000 K and 1200 K, respectively. It is observed that one of the objects emits twice as much radiation as the other. Which one is it? _____.

14. If an astronomical object is observed to emit X rays, it is reasonable to assume its temperature is very _____.

15. When an observer and an object emitting radiation move toward each other, the observer sees the radiation shifted to _____ wavelengths.

REVIEW AND DISCUSSION

1. What is a wave?

2. Define the following wave properties: period, wavelength, amplitude, frequency.

3. What is the relationship between wavelength, wave frequency, and wave velocity?

4. What is diffraction, and how does it relate to the behavior of light as a wave?

5. What's so special about c?

6. Name the colors that combine to make white light. What is it about the various colors that causes us to perceive them differently?

7. What effect does a positive charge have on a nearby negatively charged particle?

8. Compare and contrast the gravitational force with the electric force.

9. Describe the way in which light leaves a star, travels through the vacuum of space, and finally is seen by someone on Earth.

10. Why is light referred to as an electromagnetic wave?

11. What do radio waves, infrared radiation, visible light, ultraviolet radiation, X rays, and gamma rays have in common? How do they differ?

12. In what regions of the electromagnetic spectrum is the atmosphere transparent enough to allow observations from the ground?

13. What is a blackbody? What are the main characteristics of the radiation it emits?

14. What does Wien's law reveal about stars in the sky?

15. What does Stefan's law tell us about the radiation emitted by a blackbody?

16. In terms of its blackbody curve, describe what happens as a red-hot glowing coal cools.

17. What is the Doppler effect, and how does it alter the way in which we perceive radiation?

18. How do astronomers use the Doppler effect to determine the velocities of astronomical objects?

19. A source of radiation and an observer are traveling through space at precisely the same velocity, as seen by a second observer. Would you expect the first observer to measure a Doppler shift in the light received from the source?

20. If Earth were completely blanketed with clouds and we couldn't see the sky, could we learn about the realm beyond the clouds? What forms of radiation might be received?

PROBLEMS  *Algorithmic versions of these questions are available in the Practice Problems module of the Companion Website.*

The number of squares preceding each problem indicates its approximate level of difficulty.

1. ■ A sound wave moving through water has a frequency of 256 Hz and a wavelength of 5.77 m. What is the speed of sound in water?

2. ■ What is the wavelength of a 100-MHz ("FM 100") radio signal?

3. ■ What would be the frequency of an electromagnetic wave having a wavelength equal to Earth's diameter? In what part of the electromagnetic spectrum would such a wave lie?

4. ▨ Estimate the frequency of an electromagnetic wave having a wavelength equal to the size of the period at the end of this sentence. In what part of the electromagnetic spectrum would such a wave lie?

5. ▨ What would be the wavelength of an electromagnetic wave having a frequency equal to the clock speed of an 800 MHz personal computer? In what part of the electromagnetic spectrum would such a wave lie?

6. ▨ The blackbody emission spectrum of object A peaks in the ultraviolet region of the electromagnetic spectrum, at a wavelength of 200 nm. That of object B peaks in the red region, at 650 nm. Which object is hotter and, according to Wien's law, how many times hotter is it? According to Stefan's law, how many times more energy per unit area does the hotter body radiate per second?

7. ■ Normal human body temperature is about 37°C. What is this temperature in kelvins? What is the peak wavelength emitted by a person with this temperature? In what part of the spectrum does this lie?

8. ■■ Estimate the total amount of energy you radiate to your surroundings.

9. ■ The Sun has a temperature of 5800 K, and its blackbody emission peaks at a wavelength of approximately 500 nm. At what wavelength does a protostar with a temperature of 1000 K radiate most strongly?

10. ▨ Two otherwise identical bodies have temperatures of 300 K and 1500 K, respectively. Which one radiates more energy, and by what factor does its emission exceed the emission of the other body?

11. ■■ According to the Stefan-Boltzmann law, how much energy is radiated into space per unit time by each square meter of the Sun's surface (see *More Precisely 3-2*)? If the Sun's radius is 696,000 km, what is the total power output of the Sun?

12. ■ Radiation from the nearby star Alpha Centauri is observed to be reduced in wavelength (after correction for Earth's orbital motion) by a factor of 0.999933. What is the radial velocity of Alpha Centauri relative to the Sun?

13. ▨ At what radial velocity, and in what direction, would a spacecraft have to be moving for a radio station on Earth transmitting at 100 MHz to be picked up by a radio tuned to 99.9 MHz?

14. ■▨ A space traveler is approaching the Sun at a speed of 100 km/s and is observing a 700-nm red laser beam coming from Earth. If his trajectory lies in the same plane as Earth's orbit, what will be the minimum and maximum wavelengths he observes as Earth orbits the Sun?

15. ■■■ Imagine you are observing a spacecraft moving in a circular orbit of radius 100,000 km around a distant planet. You happen to be located in the plane of the spacecraft's orbit. You find that the spacecraft's radio signal varies periodically in wavelength between 2.99964 m and 3.00036 m. Assuming that the radio is broadcasting normally, at a constant wavelength, what is the mass of the planet?

COLLABORATIVE EXERCISES

1. **Radio Station Wavelength.** Determine the wavelength from your group's favorite radio station and calculate how long it takes for the radio waves to arrive at your current location from the radio station.

2. **Doppler Shift.** Calculate what the new wavelength for a 2 kHz automobile horn would be for a car moving first toward you and then away from you at a speed agreed upon by your group.

RESEARCHING ON THE WEB  *To complete the following exercises, go to the online Destinations module for Chapter 3 on the Companion Website for Astronomy Today 4/e.*

1. Access the "Emission Mechanisms" page and determine which major astronomical objects can be observed at each of the various wavelengths of the electromagnetic spectrum.

2. Access the "Temperature Conversion Calculator" to calculate the approximate current outside temperature, inside room temperature, and average human body temperature in both Celsius and Kelvin temperature scales.

3. Access "NASA's SkyView Virtual Observatory" and obtain images of the Andromeda Galaxy from each observatory dataset and determine in which dataset the galaxy is most easily visible.

PROJECTS

1. Locate the constellation Orion. Its two brightest stars are Betelgeuse and Rigel. Which of these is the hotter star? Which is cooler? How can you tell? Which of the other stars scattered across the night sky are hot, and which are cool?

2. Stand near (but not too near!) a train track or busy highway and wait for a train or traffic to pass by. Can you notice the Doppler effect in the pitch of the engine noise or whistle blowing? How does the sound frequency depend on (a) speed and (b) the train's motion toward or away from you?

SKYCHART III PROJECTS  *The SkyChart III Student Version planetarium program on which these exercises are based is included as a separately executable program on the CD in the back of this text.*

1. ■ For your location, determine if and when the constellation Orion is above the horizon today. Print a chart that clearly shows Orion. When does it rise? When does it set?

2. ■ The two most prominent stars in Orion are Betelgeuse and Rigel. Locate them. Betelgeuse is a cool red giant and appears red to the naked eye. Rigel is a hot blue star. The quantitative way of measuring the color or color index is by taking the blue magnitude minus the visible magnitude. A star's color index is a number that indicates the star's temperature; the larger the index, the cooler the star. The magnitude scale is explained in Chapter 17. The important and counterintuitive fact to know is that the larger the number is, the fainter the star is. When a star has a negative magnitude, it will appear brighter than a star with a positive magnitude. Blue and visible magnitudes are measured using filters that pass light in the violet/blue region and in the green/yellow region respectively. Obtain the color index of these two stars from the *Object Information* box in SkyChart. Explain the cause of the difference in the color index between the two stars.

 *In addition to the Practice Problems and Destinations modules, the Companion Website at http://www.prenhall.com/chaisson provides for each chapter an additional true-false, multiple choice, and labeling quiz, as well as additional annotated images, animations, and links to related Websites.*

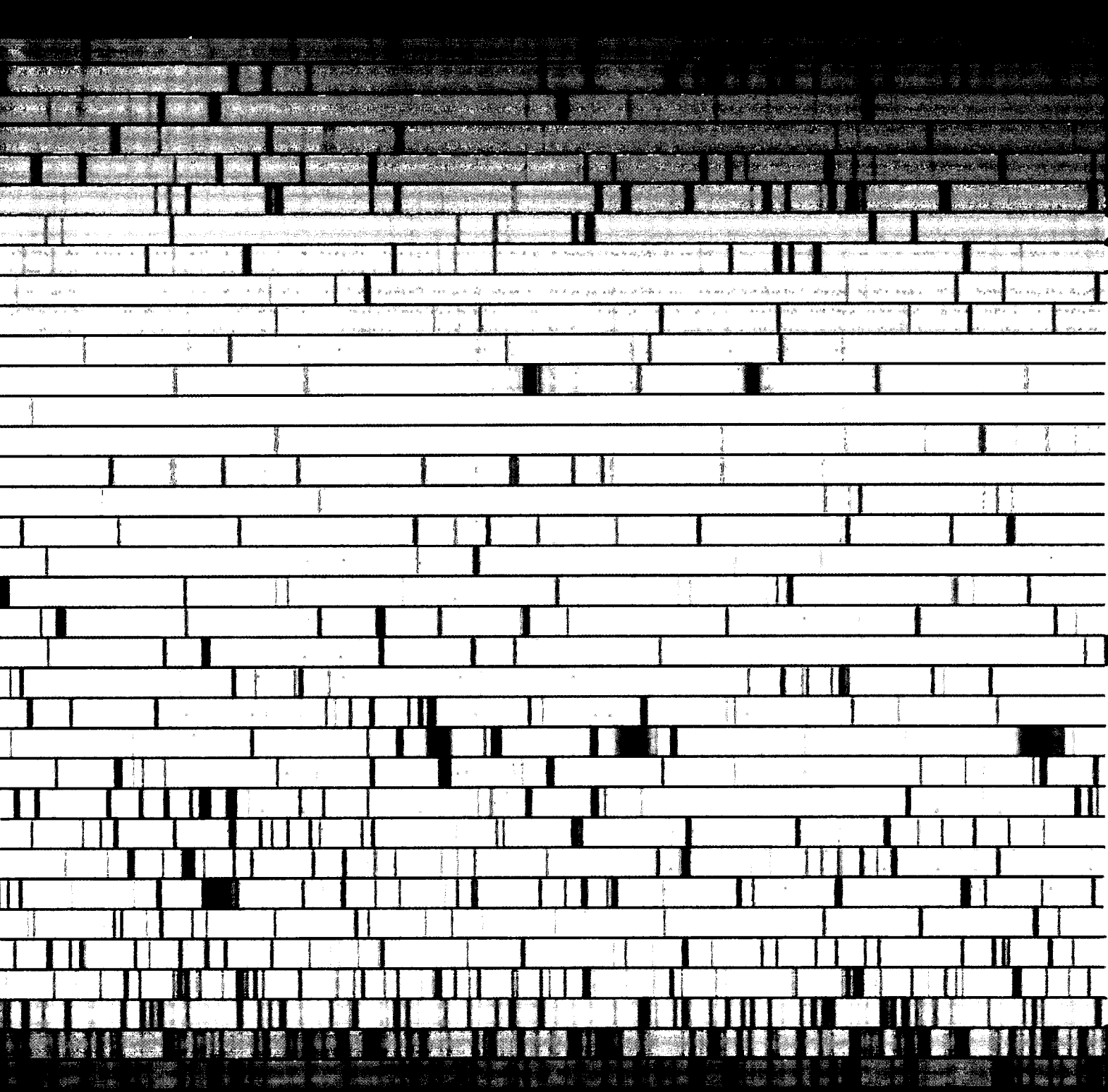

4 SPECTROSCOPY

The Inner Workings of Atoms

LEARNING GOALS

Studying this chapter will enable you to:

1. Describe the characteristics of continuous, emission, and absorption spectra and the conditions under which each is produced.

2. Explain the relation between emission and absorption lines and what we can learn from these lines.

3. Specify the basic components of the atom and describe our modern conception of its structure.

4. Discuss the observations that led scientists to conclude that light has particle as well as wave properties.

5. Explain how electron transitions within atoms produce unique emission and absorption features in the spectra of those atoms.

6. Describe the general features of spectra produced by molecules.

7. List and explain the kinds of information that can be obtained by analyzing the spectra of astronomical objects.

 Visit http://www.prenhall.com/chaisson for additional annotated images, animations, and links to related sites for this chapter.

This illustration shows the complete visible spectrum of the giant star Arcturus. By passing the star's light through a prism, the light is spread out into its component colors, displayed here wrapped around row after row from red at top left to blue at bottom right. Each of the 50 rows covers 8 nanometers in wavelength, for complete coverage of the visible spectrum from 300 to 700 microns. The many dark lines are caused by light from the hot star being absorbed by specific atoms and ions in the star's cooler atmosphere. *(NOAO)*
The Big Picture: Spectroscopy is a powerful observational technique enabling scientists to infer the nature of matter by the way it emits or absorbs radiation. Not only can spectroscopy reveal the chemical composition of distant stars and yield knowledge of how they shine, it can also provide a wealth of information about the birth, evolution, and death of myriad objects in the Universe.

The wave description of radiation allowed nineteenth-century astronomers to begin to decipher the information reaching Earth from the cosmos in the form of visible and invisible light. However, early in the twentieth century, it became clear that the wave theory of electromagnetic phenomena was incomplete—some aspects of light simply could not be explained purely in wave terms. When radiation interacts with matter on atomic scales, it does so not as a continuous wave but in a jerky, discontinuous way—in fact, as a particle. With this discovery, scientists quickly realized that atoms, too, must behave in a discontinuous way, and the stage was set for a scientific revolution that has affected virtually every area of modern life. In astronomy, the observational and theoretical techniques that enable researchers to determine the nature of distant atoms by the way they emit and absorb radiation are now the indispensable foundation of modern astrophysics.

4.1 Spectral Lines

In Chapter 3 we saw something of how astronomers can analyze electromagnetic radiation received from space to obtain information about distant objects. A vital step in this process is the formation of a *spectrum*—splitting the incoming radiation into its component wavelengths. But in reality, *no* cosmic object emits a perfect blackbody spectrum like those discussed earlier. ∞ (Sec. 3.4) All spectra deviate from this idealized form—some by only a little, others by a lot. Far from invalidating our earlier studies, however, these deviations contain a wealth of detailed information about physical conditions in the source of the radiation. Because spectra are so important, let's examine in more detail how astronomers obtain and interpret them.

Radiation can be analyzed with an instrument known as a **spectroscope**. In its most basic form, this device consists of an opaque barrier with a slit in it (to define a beam of light), a prism (to split the beam into its component colors), and an eyepiece or screen (to allow the user to view the resulting spectrum). Figure 4.1 shows such an arrangement. The research instruments called *spectrographs*, or *spectrometers*, used by professional astronomers are rather more complex, consisting of a telescope (to capture the radiation), a dispersing device (to spread it out into a spectrum), and a detector (to record the result). Despite their greater sophistication, however, their basic operation is conceptually similar to the simple spectroscope shown in the figure.

In many large instruments the prism is replaced by a device called a *diffraction grating*, consisting of a sheet of transparent material with many closely spaced parallel lines ruled on it. The spacing between the lines is typically a few microns (10^{-6} m), comparable to the wavelength of visible light. The spaces act as many tiny openings, and light is diffracted as it passes through the grating (or is reflected from it, depending on the design of the device). ∞ (*Discovery 3-1*) Because different wavelengths of elec-

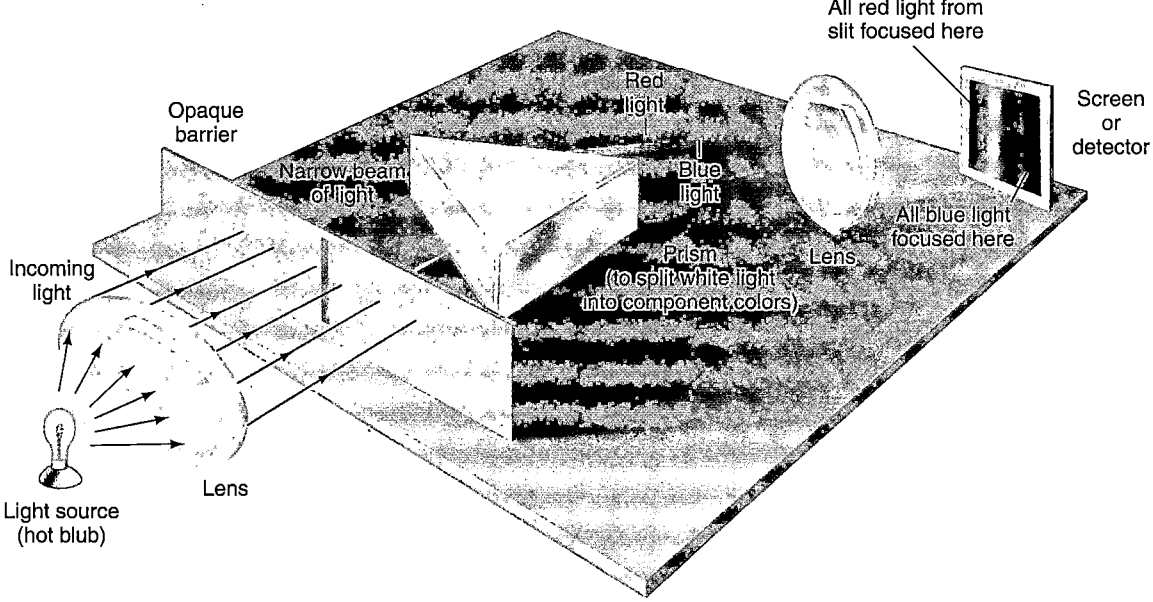

Figure 4.1 Spectroscope Diagram of a simple spectroscope. A small slit in the mask on the left allows a narrow beam of light to pass. The light passes through a prism and is split up into its component colors. The resulting spectrum can be viewed through an eyepiece or simply projected onto a screen.

tromagnetic radiation are diffracted by different amounts on encountering the grating, the effect is to split a beam of light into its component colors. You are probably more familiar with diffraction gratings than you think—the "rainbow" of colors seen in light reflected from a compact disk is the result of precisely this process.

EMISSION LINES

The spectra we encountered in Chapter 3 are examples of **continuous spectra**. A lightbulb, for example, emits radiation of all wavelengths (mostly in the visible range), with an intensity distribution that is well described by the blackbody curve corresponding to the bulb's temperature. ∞ (Sec. 3.4) Viewed through a spectroscope, the spectrum of the light from the bulb would show the familiar rainbow of colors, from red to violet, without interruption, as presented in Figure 4.2(a).

Not all spectra are continuous, however. For instance, if we took a glass jar containing pure hydrogen gas and passed an electrical discharge through it (a little like a lightning bolt arcing through Earth's atmosphere), the gas would begin to glow—that is, it would emit radiation. If we were to examine that radiation with our spectroscope, we would find that its spectrum consists of only a few bright lines on an otherwise dark background, quite unlike the continuous spectrum described for the incandescent lightbulb. Figure 4.2(b) shows this schematically. A more detailed rendering of the spectrum of hydrogen appears in the top panel of Figure 4.3. The light produced by the hydrogen in this experiment does *not* consist of all possible colors but instead includes only a few narrow, well-defined **emission lines**—narrow "slices" of the continuous spectrum. The black background represents all the wavelengths *not* emitted by hydrogen.

After some experimentation we would also find that although we could alter the *intensity* of the lines—for example, by changing the amount of hydrogen in the jar or the strength of the electrical discharge—we could not alter their *color* (in other words, their frequency or wavelength). The pattern of spectral emission lines is a property of the element hydrogen. Whenever we perform this experiment, the same characteristic colors result.

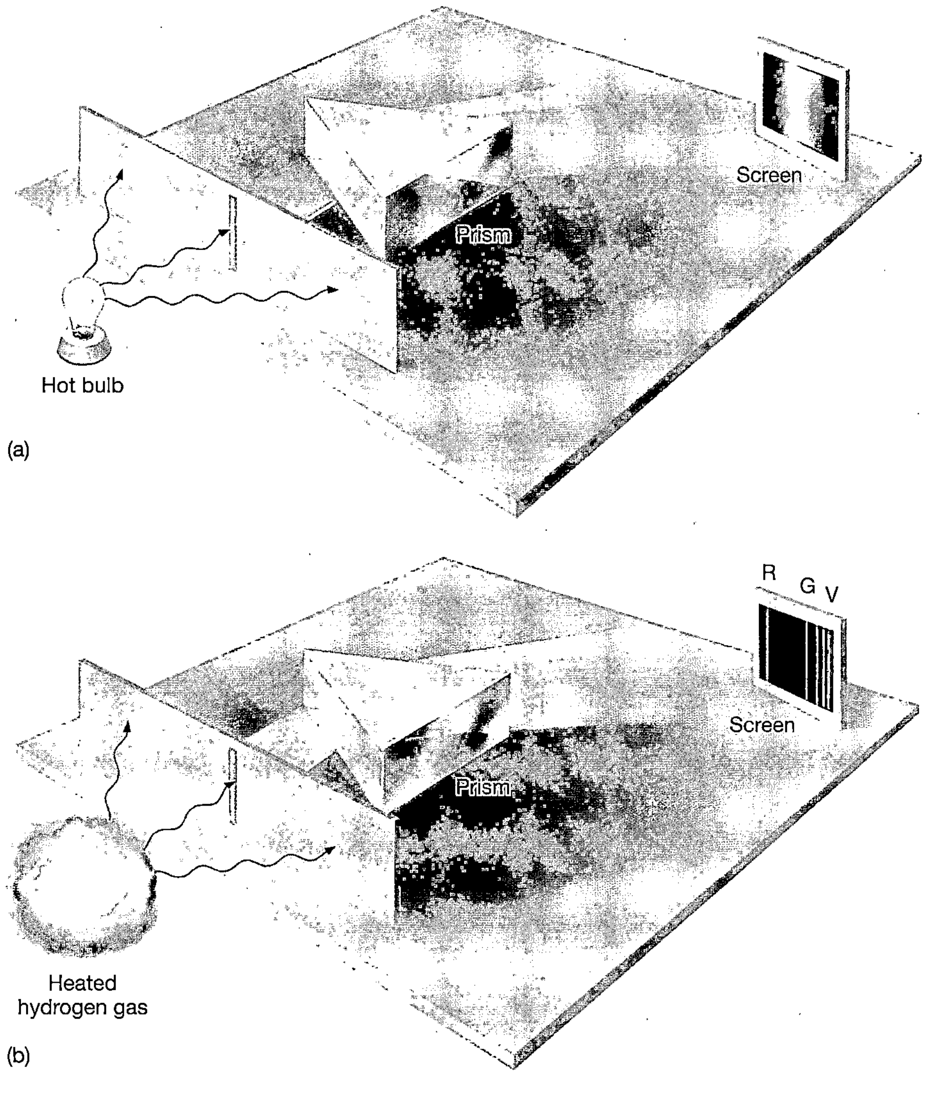

(a)

Figure 4.2 Continuous and Emission Spectra
When passed through a slit and split up by a prism, light from a source of continuous radiation (a) gives rise to the familiar rainbow of colors. By contrast, the light from excited hydrogen gas (b) consists of a series of distinct bright spectral lines called emission lines. (The focusing lenses have been omitted for clarity.)

(b)

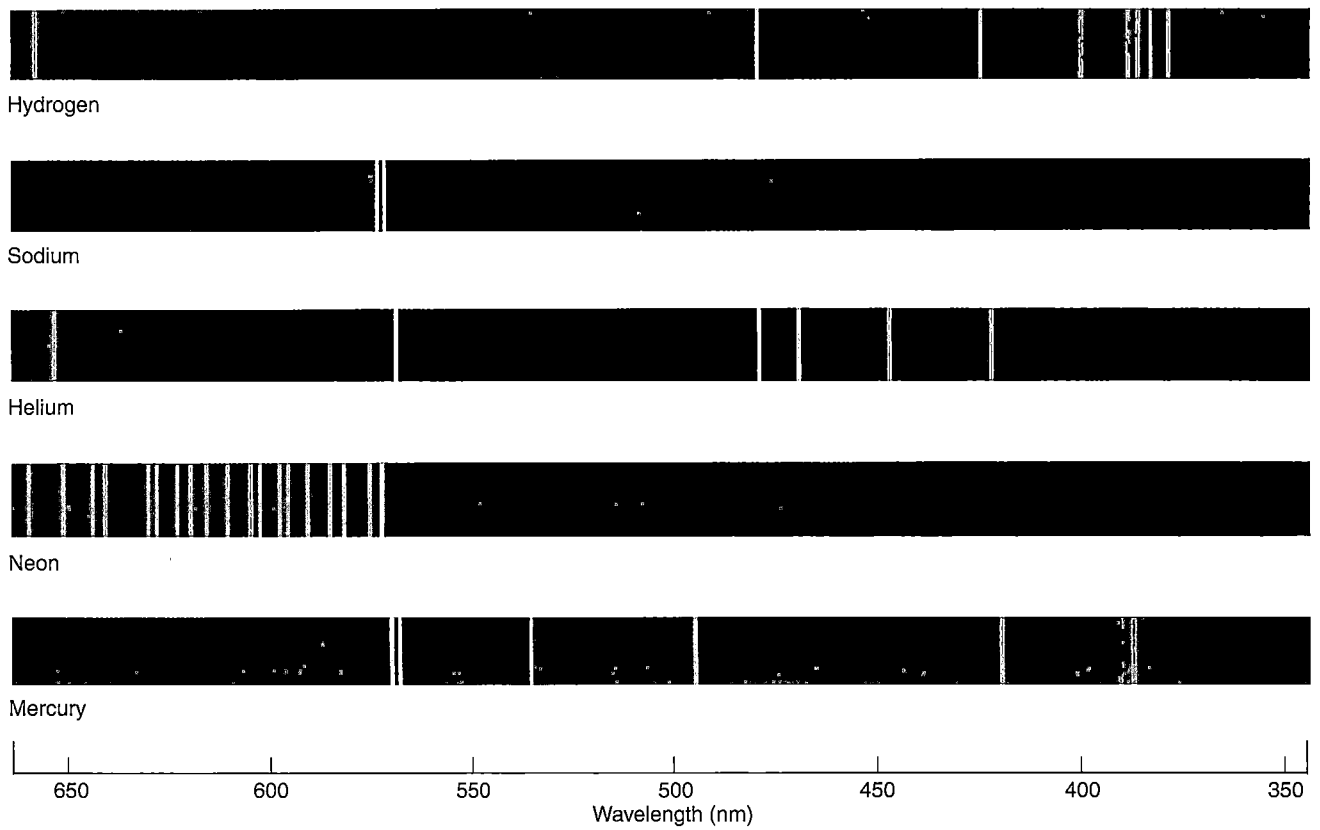

Figure 4.3 Elemental Emission The emission spectra of some well-known elements. In accordance with the convention adopted throughout this text, frequency increases to the right. *(Wabash Instrument Corp.)*

By the early nineteenth century scientists had carried out similar experiments on many different gases. By vaporizing solids and liquids in a flame, they extended their inquiries to include materials that are not normally found in the gaseous state. Sometimes the pattern of lines was fairly simple, sometimes it was very complex, but it was always *unique* to that element. Even though the origin of the lines was not understood, researchers quickly realized that the lines provided a one-of-a-kind "fingerprint" of the substance under investigation. They could detect the presence of a particular atom or molecule (a group of atoms held together by chemical bonds—see Sec. 4.4) solely through the study of the light it emitted. Scientists have accumulated extensive catalogs of the specific wavelengths at which many different hot gases emit radiation. The particular pattern of the light emitted by a gas of a given chemical composition is known as its **emission spectrum**. Examples of the emission spectra of some common substances are shown in Figure 4.3.

ABSORPTION LINES

When sunlight is split by a prism, at first glance it appears to produce a continuous spectrum. However, closer scrutiny with a spectroscope shows that the solar spectrum is in-terrupted vertically by a large number of narrow dark lines, as shown in Figure 4.4. We now know that many of these lines represent wavelengths of light that have been removed (absorbed) by gases present either in the outer layers of the Sun or in Earth's atmosphere. These gaps in the spectrum are called **absorption lines**.

The English astronomer William Wollaston first noticed the solar absorption lines in 1802. They were studied in greater detail about 10 years later by the German physicist Joseph von Fraunhofer, who measured and cataloged over 600 of them. They are now referred to collectively as *Fraunhofer lines*. Although the Sun is by far the easiest star to study, and so has the most extensive set of observed absorption lines, similar lines are known to exist in the spectra of all stars.

At around the same time as the solar absorption lines were discovered, scientists found that absorption lines could also be produced in the laboratory by passing a beam of light from a source that produces a continuous spectrum through a cool gas, as shown in Figure 4.5. They quickly observed an intriguing connection between emission and absorption lines: The absorption lines associated with a given gas occur at precisely the *same* wavelengths as the emission lines produced when the gas is heated.

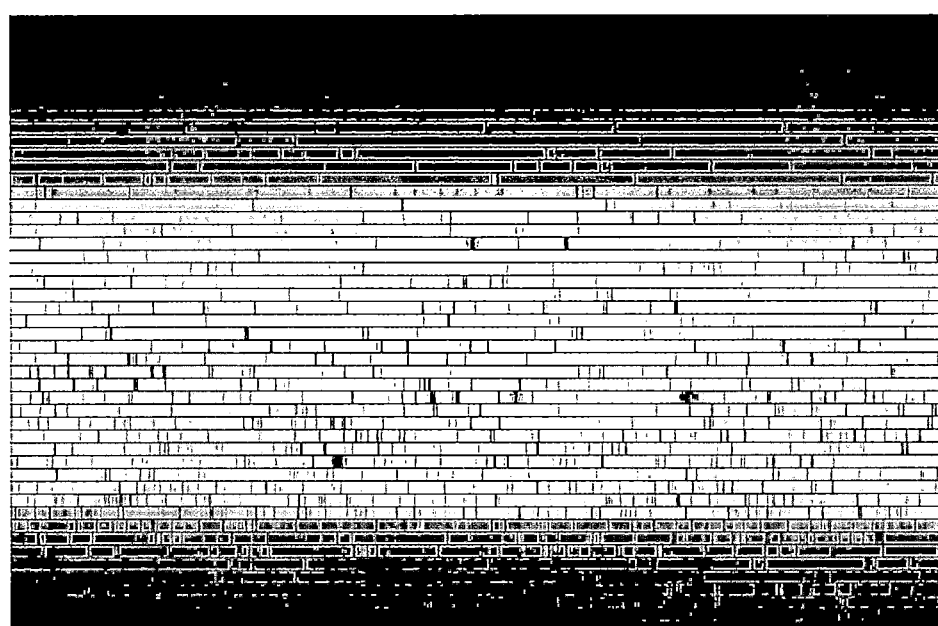

Figure 4.4 Solar Spectrum This visible spectrum of the Sun shows hundreds of dark absorption lines superimposed on a bright continuous spectrum. Here, the scale extends from long wavelengths (red) at the upper left to short wavelengths (blue) at the lower right. *(AURA)*

As an example, consider the element sodium, whose emission spectrum appears in Figure 4.3. When heated to high temperatures, a sample of sodium vapor emits visible light strongly at just two wavelengths—589.9 nm and 589.6 nm—lying in the yellow part of the spectrum. When a continuous spectrum is passed through some relatively cool sodium vapor, two sharp, dark absorption lines appear at precisely the same wavelengths. The emission and absorption spectra of sodium are compared in Figure 4.6, clearly showing the relation between emission and absorption features.

KIRCHHOFF'S LAWS

2 The analysis of the ways in which matter emits and absorbs radiation is called **spectroscopy**. One early spectroscopist, the German physicist Gustav Kirchhoff, summarized the observed relationships among the three types of spectra—continuous, emission line, and absorp-

tion line—in 1859. He formulated three spectroscopic rules, now known as **Kirchhoff's laws**, governing the formation of spectra:

1. A luminous solid or liquid, or a sufficiently dense gas, emits light of all wavelengths and so produces a *continuous spectrum* of radiation.

2. A low-density hot gas emits light whose spectrum consists of a series of bright *emission lines*. These lines are characteristic of the chemical composition of the gas.

3. A cool thin gas absorbs certain wavelengths from a continuous spectrum, leaving dark *absorption lines* in their place superimposed on the continuous spectrum. Once again, these lines are characteristic of the composition of the intervening gas—they occur at precisely the same wavelengths as the emission lines produced by that gas at higher temperatures.

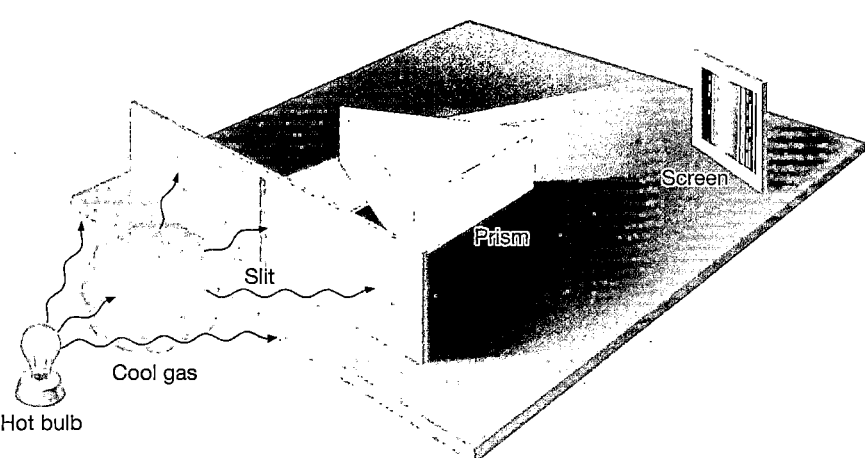

Figure 4.5 Absorption Spectrum When cool gas is placed between a source of continuous radiation, such as a hot lightbulb, and a detector, the resulting spectrum consists of a continuous spectrum crossed by a series of dark absorption lines. These lines are formed when the intervening gas absorbs certain wavelengths (colors) from the original beam. The absorption lines appear at precisely the same wavelengths as the emission lines that would be produced if the gas were heated to high temperatures.

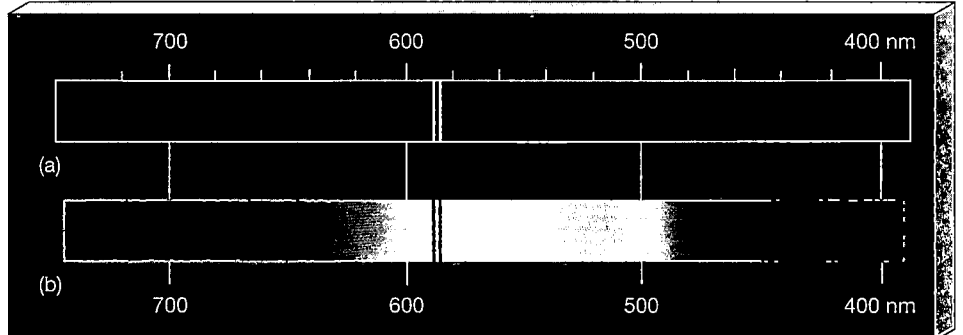

Figure 4.6 Sodium Spectrum
(a) The characteristic emission lines of sodium. The two bright lines in the center appear in the yellow part of the spectrum. (b) The absorption spectrum of sodium. The two dark lines appear at exactly the same wavelengths as the bright lines in the sodium emission spectrum.

Figure 4.7 illustrates Kirchhoff's laws and the relationship between absorption and emission lines. When viewed directly, the light source, a hot solid (the filament of the bulb), has a continuous (blackbody) spectrum. When the light source is viewed through a cloud of cool hydrogen gas, a series of dark absorption lines appear, superimposed on the spectrum at wavelengths characteristic of hydrogen. The lines appear because the light at those wavelengths is absorbed by the hydrogen. As we will see later in this chapter, the absorbed energy is subsequently reradiated into space, but in all directions, not just the original direction of the beam. Consequently, when the cloud is viewed from the side against an otherwise dark background, a series of faint emission lines is seen. These

lines contain the energy lost by the forward beam. If the gas was heated to incandescence, it would produce stronger emission lines at precisely the same wavelengths.

ASTRONOMICAL APPLICATIONS

By the late nineteenth century, spectroscopists had developed a formidable arsenal of techniques for interpreting the radiation received from space. Once astronomers knew that spectral lines were indicators of chemical composition, they set about identifying the observed lines in the solar spectrum. Almost all the lines in light from extraterrestrial sources could be attributed to known elements.

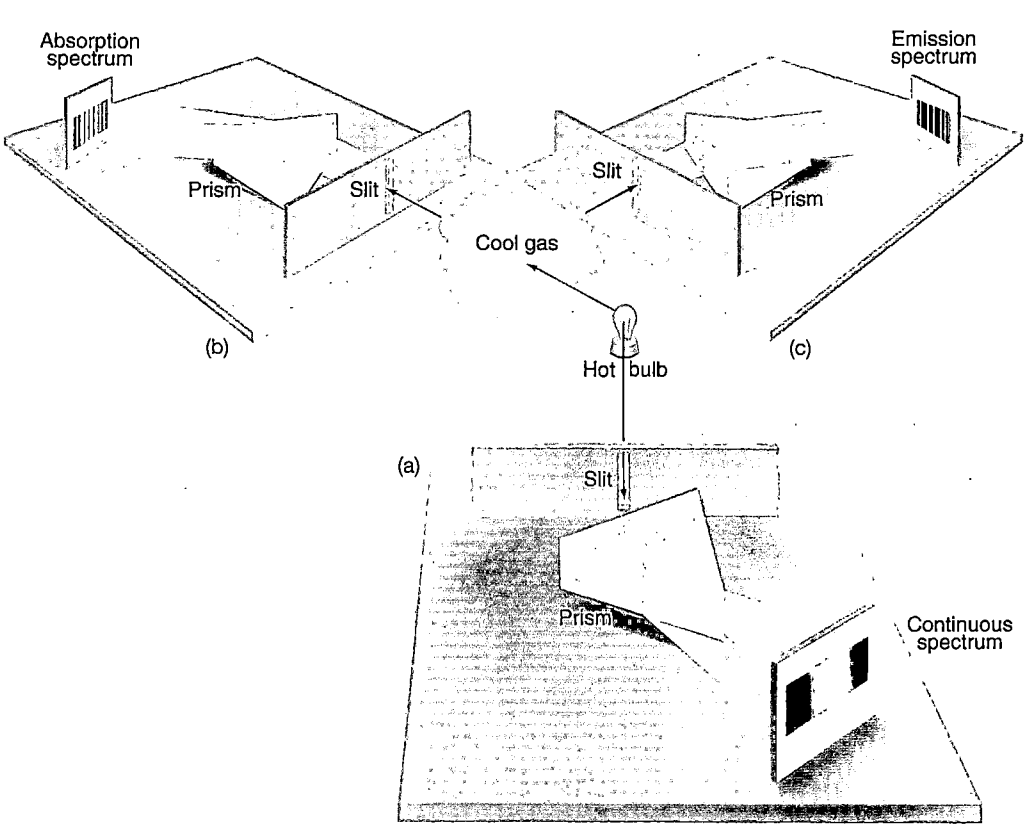

Figure 4.7 Kirchhoff's Laws A source of continuous radiation, here represented by a lightbulb, is used to illustrate Kirchhoff's laws of spectroscopy. (a) The unimpeded beam shows the familiar continuous spectrum of colors. (b) When the source is viewed through a cloud of hydrogen gas, a series of dark hydrogen absorption lines appears in the continuous spectrum. These lines are formed when the gas absorbs some of the bulb's radiation and reemits it in random directions. Because most of the reemitted radiation does not go through the slit, the effect is to remove the absorbed radiation from the light that reaches the screen at left. (c) When the gas is viewed from the side, a fainter hydrogen emission spectrum is seen, consisting of reemitted radiation. The absorption lines in (b) and the emission lines in (c) have the same wavelengths.

For example, many of the Fraunhofer lines in sunlight are associated with the element iron, a fact first recognized by Kirchhoff and coworker Robert Bunsen (of Bunsen burner fame) in 1859. However, some unfamiliar lines also appeared in the solar spectrum. In 1868, astronomers realized that those lines must correspond to a previously unknown element. It was given the name helium, after the Greek word *helios*, meaning "Sun." Not until 1895, almost three decades after its detection in sunlight, was helium discovered on Earth. (A laboratory spectrum of helium is included in Figure 4.3.)

Yet for all the information that nineteenth-century astronomers could extract from observations of stellar spectra, they still lacked a theory explaining how the spectra themselves arose. Despite their sophisticated spectroscopic equipment, they knew scarcely any more about the physics of stars than did Galileo or Newton. To understand how spectroscopy can be used to extract detailed information about astronomical objects from the light they emit, we must delve more deeply into the processes that produce line spectra.

☑ Concept Check

■ What are absorption and emission lines, and what do they tell us about the properties of the gas producing them?

4.2 The Formation of Spectral Lines

By the start of the twentieth century, physicists had accumulated substantial evidence that light sometimes behaves in a manner that cannot be explained by the wave theory. As we have just seen, the production of absorption and emission lines involves only certain very specific frequencies or wavelengths of light. This would not be expected if light behaved like a continuous wave and matter always obeyed the laws of Newtonian mechanics. Other experiments conducted around the same time strengthened the conclusion that the notion of radiation as a wave was incomplete. It became clear that when light interacts with matter on very small scales, it does so not in a continuous way but in a discontinuous, "stepwise" manner. The challenge was to find an explanation for this unexpected behavior. The eventual solution revolutionized our view of nature and now forms the foundation for all of physics and astronomy—indeed for virtually all modern science.

ATOMIC STRUCTURE

3 To explain the formation of emission and absorption lines, we must understand not just the nature of light but also the structure of **atoms**—the microscopic building

blocks from which all matter is constructed. Let's start with the simplest atom of all—hydrogen. A hydrogen atom consists of an electron with a negative electrical charge, orbiting a proton carrying a positive charge. The proton forms the central **nucleus** (plural: nuclei) of the atom. The hydrogen atom as a whole is electrically neutral. The equal and opposite charges of the proton and the orbiting electron produce an electrical attraction that binds them together within the atom.

How does this picture of the hydrogen atom relate to the characteristic emission and absorption lines associated with hydrogen gas? If an atom emits some energy in the form of radiation, that energy has to come from somewhere within the atom. Similarly, if energy is absorbed, it must cause some internal change. It is reasonable (and correct) to suppose that the energy emitted or absorbed by the atom is associated with changes in the motion of the orbiting electron.

The first theory of the atom to provide an explanation of hydrogen's observed spectral lines was set forth by the Danish physicist Niels Bohr in 1912. Now known simply as the *Bohr model* of the atom, its essential features are as follows. First, there is a state of lowest energy—the **ground state**—which represents the "normal" condition of the electron as it orbits the nucleus. Second, there is a maximum energy that the electron can have and still be part of the atom. Once the electron acquires more than that maximum energy, it is no longer bound to the nucleus, and the atom is said to be **ionized**; an atom missing one or more of its electrons is called an **ion**. Third, and most important (and also least intuitive), between those two energy levels the electron can exist only in certain sharply defined energy states, often referred to as *orbitals*.

This description of the atom contrasts sharply with the predictions of Newtonian mechanics, which would permit orbits with *any* energy, not just at certain specific values. In the atomic realm such discontinuous behavior is the norm. In the jargon of the field, the orbital energies are said to be **quantized**. The rules of *quantum mechanics*, the branch of physics governing the behavior of atoms and subatomic particles, are far removed from everyday experience.

In Bohr's model, each electron orbital was pictured as having a specific radius, much like a planetary orbit in the solar system, as shown in Figure 4.8. However, the modern view is not so simple. Although each orbital *does* have a precise energy, the electron is now envisioned as being smeared out in an "electron cloud" surrounding the nucleus, as illustrated in Figure 4.9. We cannot tell "where" the electron is—we can only speak of the *probability* of finding it in a certain location within the cloud. It is common to speak of the average distance from the cloud to the nucleus as the "radius" of the electron's orbit. When a hydrogen atom is in its ground state, the radius of the orbit is about 0.05 nm (0.5 Å). As the orbital energy increases, the radius increases, too. For the sake of

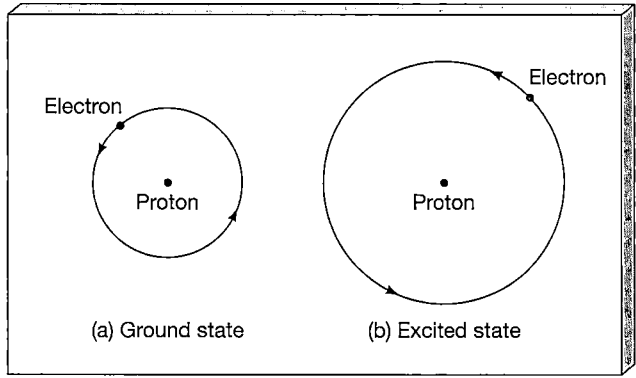

Figure 4.8 Bohr Atom An early-twentieth-century conception of the hydrogen atom pictured its electron orbiting the central proton in a well-defined orbit, rather like a planet orbiting the Sun. Two electron orbits of different energies are shown. (a) The ground state. (b) An excited state.

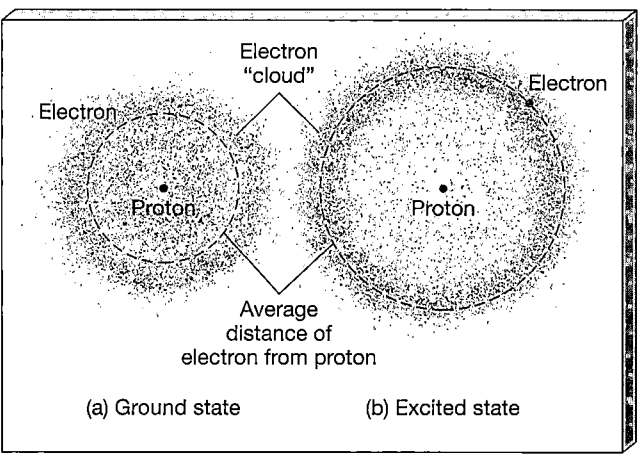

Figure 4.9 Modern Atom The modern view of the hydrogen atom sees the electron as a "cloud" surrounding the nucleus. The same two energy states are shown as in Figure 4.8.

clarity in the diagrams that follow, we will represent electron orbitals as solid lines, but always bear in mind that Figure 4.9 is a more accurate depiction of reality.

Atoms do not always remain in their ground state. An atom is said to be in an **excited state** when an electron occupies an orbital at a greater than normal distance from its parent nucleus. An atom in such an excited state has a greater than normal amount of energy. The excited state with the lowest energy (that is, the one closest in energy to the ground state) is called the *first excited state*, that with the second-lowest energy is the *second excited state*, and so on. An atom can become excited in one of two ways: by absorbing some energy from a source of electromagnetic radiation or by colliding with some other particle—another atom, for example. However, the electron cannot stay in a higher orbital forever; the ground state is the only level where it can remain indefinitely. After about 10^{-8} s, an excited atom returns to its ground state.

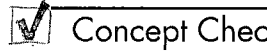 **Concept Check**

■ In what ways do electron orbits in an atom differ from planetary orbits around the Sun?

RADIATION AS PARTICLES

4 Because electrons may exist only in orbitals having specific energies, atoms can absorb only specific amounts of energy as their electrons are boosted into excited states. Likewise, they can emit only specific amounts of energy as their electrons fall back to lower energy states. Thus, the amount of light energy absorbed or emitted in these processes *must correspond precisely to the energy difference between two orbitals*. The atom's quantized energy levels require that light must be absorbed and emitted in the form of dis-

tinct "packets" of electromagnetic radiation, each carrying a specific amount of energy. We call these packets **photons**. A photon is, in effect, a "particle" of electromagnetic radiation.

The idea that light sometimes behaves not as a continuous wave but as a stream of particles was proposed by Albert Einstein in 1905 to explain a number of experimental results then puzzling physicists. Further, Einstein was able to quantify the relationship between the two aspects of light's double nature. He found that the energy carried by a photon had to be proportional to the *frequency* of the radiation:

photon energy $\propto$ radiation frequency.

Thus, for example, a "deep red" photon having a frequency of 4×10^{14} Hz (or a wavelength of approximately 750 nm) has half the energy of a violet photon of frequency of 8×10^{14} Hz (wavelength = 375 nm), and 500 times the energy of an 8×10^{11} Hz (wavelength = 375 μm) microwave photon.

The constant of proportionality in the above relation is now known as *Planck's constant*, in honor of the German physicist Max Planck, who determined its numerical value. It is always denoted by the symbol h, and the equation relating photon energy E to radiation frequency f is usually written

$$E = h f.$$

Like the gravitational constant G and the speed of light c, Planck's constant is one of the fundamental physical constants of the universe.

In SI units, the value of Planck's constant is a very small number: $h = 6.63 \times 10^{-34}$ joule seconds (J·s). Consequently, the energy of a single photon is tiny. Even a very-high-frequency gamma ray (the most energetic type of electromagnetic radiation) with a frequency of 10^{22} Hz has an energy of just $(6.63 \times 10^{-34}) \times 10^{22} \approx 7 \times 10^{-12}$ J—

about the same energy carried by a flying gnat. Nevertheless, this energy is more than enough to damage a living cell. The basic reason that gamma rays are so much more dangerous to life than visible light is that each gamma-ray photon typically carries millions, if not billions, of times more energy than a photon of visible radiation.

The equivalence between photon energy and photon frequency, or wavelength, completes the connection between atomic structure and atomic spectra. Atoms absorb and emit radiation at characteristic wavelengths determined by their own particular internal structure. Because this structure is *unique* to each element, the colors of the absorbed and emitted photons—that is, the spectral lines we observe—are characteristic of that element *and only that element*. The spectrum we see is thus a unique identifier of the atom involved.

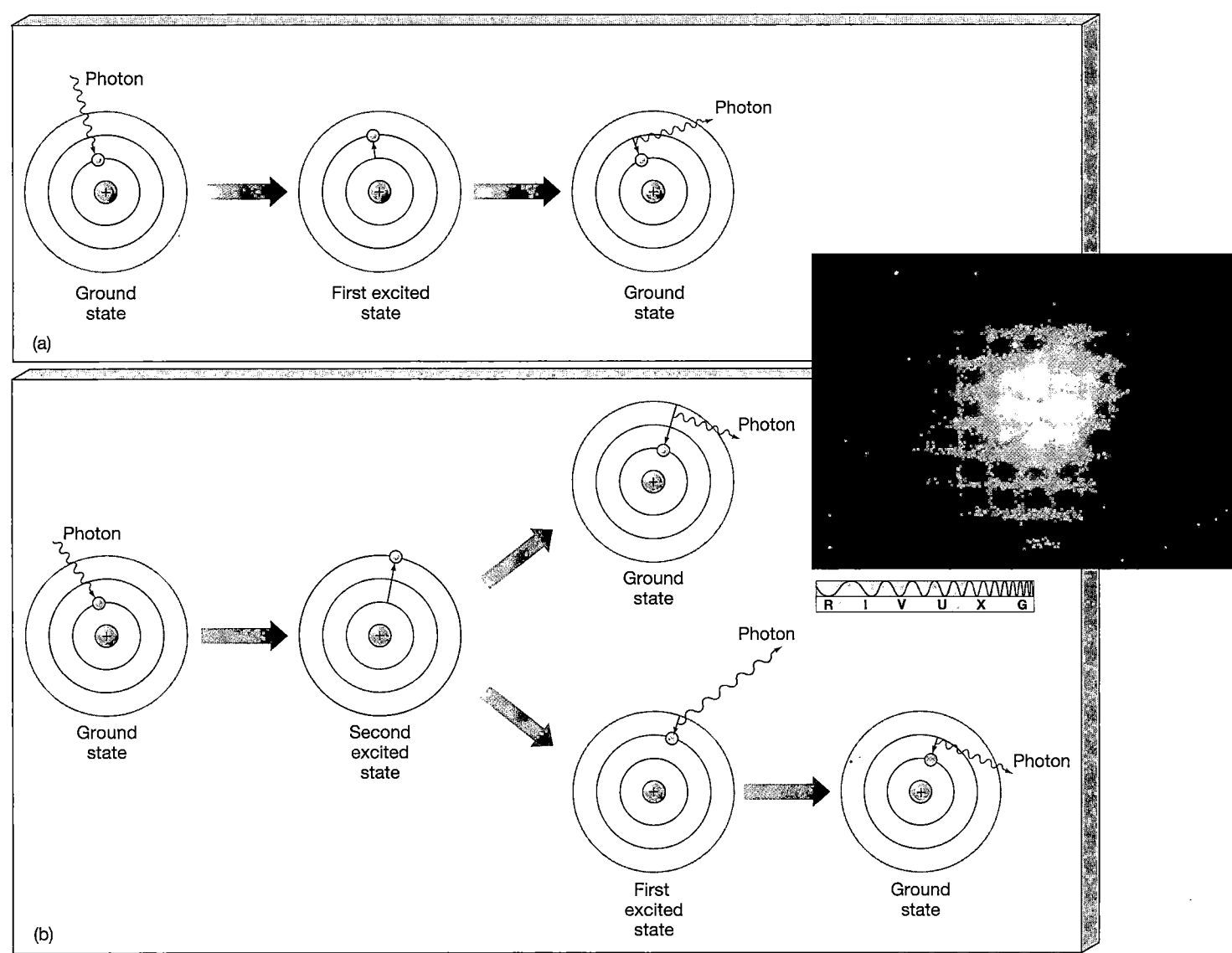

Figure 4.10 Atomic Excitation (a) Diagram of a photon being absorbed by a hydrogen atom (left), causing the momentary excitation of that atom (center) into its first excited state. After about 10^{-8} s, the atom returns to its ground state, accompanied by the emission of a photon of the same energy as the original photon (right). (b) Absorption of a higher-energy photon may also boost the atom into a higher excited state, from which there may be several possible paths back to the ground state. (Remember that the sharp lines used for the orbitals here and in similar figures that follow are intended merely as a schematic representation of the electron energy levels and are not meant to be taken literally. In actuality, electron orbitals are "clouds," as shown in Figure 4.9.) As ultraviolet photons from a hot star pass through surrounding hydrogen gas, many are absorbed by the gas, boosting its atoms into excited states. Electrons in the second excited state can fall to the first excited state on their way back to the ground state (the lower path in part b). This transition produces radiation in the visible region of the spectrum—the 656.3-nm red glow that is characteristic of excited hydrogen gas. The object shown in the inset, designated N81, is an emission nebula: an interstellar cloud consisting largely of hydrogen gas excited by extremely hot stars (as seen in white at the center). *(Inset: NASA)*

Many people are confused by the idea that light can behave in two such different ways. To be truthful, modern physicists don't yet fully understand *why* nature displays this wave–particle duality. Nevertheless, there is irrefutable experimental evidence for both of these aspects of radiation. Environmental conditions ultimately determine which description—wave or stream of particles—better fits the behavior of electromagnetic radiation. As a general rule of thumb, in the macroscopic realm of everyday experience, radiation is more usefully described as a wave, whereas in the microscopic domain of atoms it is best characterized as a series of particles.

THE SPECTRUM OF HYDROGEN

Figure 4.10 illustrates schematically the absorption and emission of photons by a hydrogen atom. Figure 4.10(a) shows the atom absorbing a photon and making a transition from the ground state to the first excited state, then emitting a photon of precisely the same energy and dropping back to the ground state. The energy difference between the two states corresponds to an ultraviolet photon, of wavelength 121.6 nm (1216 Å).

Absorption may also boost an electron into an excited state higher than the first excited state. Figure 4.10(b) depicts the absorption of a more energetic (higher-frequency, shorter-wavelength) ultraviolet photon, this one having a wavelength of 102.6 nm (1026 Å). Absorption of this photon causes the atom to jump to the *second* excited state. As before, the atom returns rapidly to the ground state, but this time it can do so in one of two possible ways:

1. It can proceed directly back to the ground state, in the process emitting an ultraviolet photon identical to the one that excited the atom in the first place.

2. Alternatively, the electron can *cascade* down one orbital at a time. If this occurs, the atom will emit *two* photons: one with an energy equal to the difference between the second and first excited states, and the other with an energy equal to the difference between the first excited state and the ground state.

Either possibility can occur, with roughly equal probability. The second step of the cascade process (2) produces a 121.6-nm ultraviolet photon, just as in Figure 4.10(a). However, the first transition—the one from the second to the first excited state—produces a photon with a wavelength of 656.3 nm (6563 Å), which is in the visible part of the spectrum. This photon is seen as red light. An individual atom—if one could be isolated—would emit a momentary red flash. This is the origin of the red line in the hydrogen spectrum shown in Figure 4.3.

The inset in Figure 4.10 shows an astronomical object whose red coloration is the result of precisely this process. As ultraviolet photons from a young hot star pass through the surrounding cool hydrogen gas out of which the star recently formed, some photons are absorbed by the gas, boosting its atoms into excited states or ionizing them completely. The 656.3-nm red glow characteristic of excited hydrogen gas results as the atoms cascade back to their ground states. This process is called *fluorescence*.

Absorption of additional energy can boost the electron to even higher orbitals within the atom. As the excited electron cascades back down to the ground state, the atom may emit many photons, each with a different energy and hence a different wavelength, and the resulting spectrum shows many spectral lines. In a sample of heated hydrogen gas, at any instant atomic collisions ensure that atoms are found in many different excited states. The complete emission spectrum therefore consists of wavelengths corresponding to all possible transitions between those states and states of lower energy.

In the case of hydrogen, all transitions ending at the ground state produce ultraviolet photons. However, downward transitions ending at the *first* excited state give rise to spectral lines in or near the visible portion of the electromagnetic spectrum (Figure 4.3). The energy levels and spectrum of hydrogen are discussed in more detail in *More Precisely 4-1*.

KIRCHHOFF'S LAWS EXPLAINED

5 Let's reconsider our earlier discussion of emission and absorption lines in terms of the model just presented. In Figure 4.7 a beam of continuous radiation shines through a cloud of hydrogen gas. The beam contains photons of all energies, but most of them cannot interact with the gas—the gas can absorb only those photons having just the right energy to cause a change in an electron's orbit from one state to another. All other photons in the beam—with energies that cannot produce a transition—do not interact with the gas at all but pass through it unhindered. Photons having the right energies are absorbed, excite the gas, and are removed from the beam. This is the cause of the dark absorption lines in the spectrum of Figure 4.7(b). These lines are direct indicators of the energy differences between orbitals in the atoms making up the gas.

The excited gas atoms return rapidly to their original states, each emitting one or more photons in the process. We might think, then, that although some photons from the beam are absorbed by the gas, they are quickly replaced by reemitted photons, with the result that we could never observe the effects of absorption. In fact, this is not the case, for two reasons. First, while the photons not absorbed by the gas continue on directly to the detector, the reemitted photons can leave in *any* direction. Most of the reemitted photons leave at angles that do not take them through the slit and on to the detector, and so they are effectively lost from the original beam. Second, as we have just seen, electrons may cascade back to the ground state, emitting several lower-energy photons instead of a single photon equal in energy to the one originally absorbed.

The net result of these processes is that some of the original energy is channeled into photons of many different colors, moving in many different directions. A second detector looking at the cloud from the side would record the reemitted energy as an emission spectrum, as in Figure 4.7(c). (A spectrum of the object shown in the inset of Figure 4.10, called an *emission nebula*, would show the same thing.) Like the absorption spectrum, the emission spectrum is characteristic of the gas, not of the original beam.

Absorption and emission spectra are created by the same atomic processes. They correspond to the same atomic transitions. They contain the same information about the composition of the gas cloud. In the laboratory we can move our detector and measure both. In astronomy we cannot change our vantage point, so the type of spectrum we see depends on our chance location with respect to both the source and the intervening gas cloud.

MORE COMPLEX SPECTRA

All hydrogen atoms have basically the same structure—a single electron orbiting a single proton—but, of course, there are many other kinds of atoms, each kind having a unique internal structure. The number of protons in the nucleus of an atom determines the **element** that it represents. Just as all hydrogen atoms have a single proton, all oxygen atoms have eight protons, all iron atoms have 26 protons, and so on.

The next simplest element after hydrogen is helium. The central nucleus of the most common form of helium is made up of two protons and two **neutrons** (another kind of elementary particle having a mass slightly larger than that of a proton but having no electrical charge). Two electrons orbit this nucleus. As with hydrogen and all other atoms, the "normal" condition for helium is to be electrically neutral, with the negative charge of the orbiting electrons exactly canceling the positive charge of the nucleus (Figure 4.11a).

More complex atoms contain more protons (and neutrons) in the nucleus and have correspondingly more orbiting electrons. For example, an atom of carbon, shown in Figure 4.11(b), consists of six electrons orbiting a nucleus containing six protons and six neutrons. As we progress to heavier and heavier elements, the number of orbiting electrons increases, and the number of possible electron transitions rises rapidly. The result is that very complicated spectra can be produced. The complexity of atomic spectra generally reflects the complexity of the atoms themselves. A good example is the element iron, which contributes nearly 800 of the Fraunhofer absorption lines seen in the solar spectrum (Figure 4.4).

Atoms of a single element such as iron can yield many lines for two main reasons. First, the 26 electrons of a normal iron atom can make an enormous number of different transitions among available energy levels. Second, many iron atoms are *ionized*, with some of their 26 electrons stripped away. The removal of electrons alters an atom's electromagnetic structure, and the energy levels of ionized iron are quite different from those of neutral iron. Each new level of ionization introduces a whole new set of spectral lines. Besides iron, many other elements, also in different stages of excitation and ionization, absorb photons at visible wavelengths. When we observe the entire Sun, all these atoms and ions absorb simultaneously, yielding the rich spectrum we see.

The power of spectroscopy is most apparent when many different gases are mixed together because it enables us to study one kind of atom or ion to the exclusion of all others simply by focusing on specific wavelengths of radiation. By identifying the superimposed absorption and emission spectra of many different atoms, we can determine the cloud's composition (and much more—see Section 4.4). Figure 4.12 shows an actual spectrum observed coming from a cosmic object. As in Figure 4.10, the characteristic red glow of this *emission nebula* comes

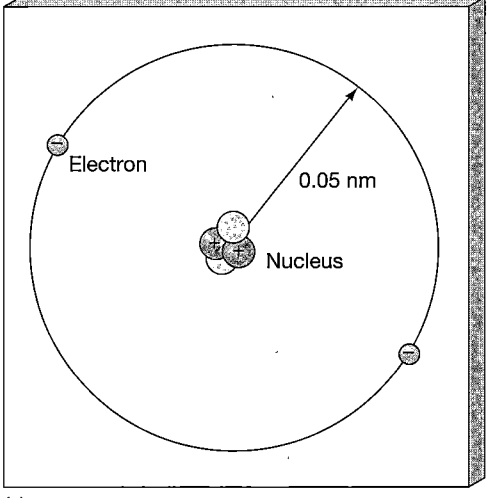
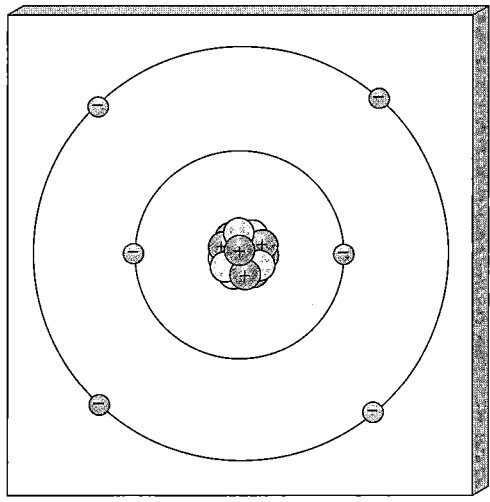

(a) (b)

Figure 4.11 Helium and Carbon (a) A helium atom in its normal ground state. Two electrons occupy the lowest-energy orbital around a nucleus containing two protons and two neutrons. (b) A carbon atom in its normal ground state. Six electrons orbit a six-proton, six-neutron nucleus, two in an inner orbital, with the other four at a greater distance from the center.

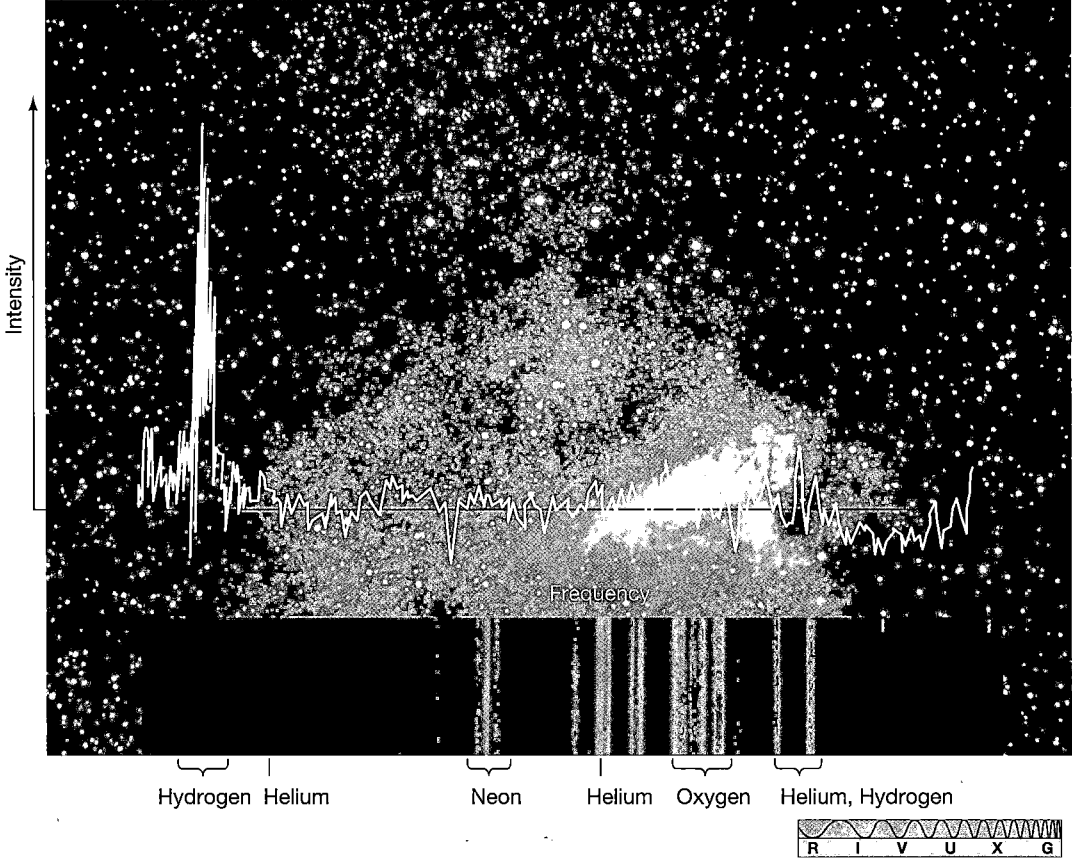

Figure 4.12 Emission Nebula The visible spectrum of the hot gases in a nearby star-forming region known as the Omega nebula (M17). Shining by the light of several very hot stars, the nebula produces a complex spectrum of bright and dark lines (bottom), also shown here as an intensity trace from red to blue (center). (ESO)

from the Hα transition in hydrogen, the nebula's main constituent.

Spectral lines occur throughout the entire electromagnetic spectrum. Usually, electron transitions among the lowest orbitals of the lightest elements, such as hydrogen and helium, produce visible and ultraviolet spectral lines. Transitions among very highly excited states of hydrogen and other elements can produce spectral lines in the infrared and radio parts of the electromagnetic spectrum. Conditions on Earth make it all but impossible to detect these radio and infrared features in the laboratory, but they are routinely observed coming from space. Electron transitions among lower energy levels in heavier, more complex elements produce X-ray spectral lines. These lines have been observed in the laboratory; some have also been observed in stars and other cosmic objects.

Concept Check

■ How does the structure of an atom determine the atom's emission and absorption spectra?

4.3 Molecules

⑤ A **molecule** is a tightly bound group of atoms held together by interactions among their orbiting electrons—interactions that we call *chemical bonds*. Much like atoms, molecules can exist only in certain well-defined energy states, and again like atoms, molecules produce emission or absorption spectral lines when they make a transition from one state to another. Because molecules are more complex than individual atoms, the rules of molecular physics are also more complex. Nevertheless, as with atomic spectral lines, painstaking experimental work over many decades has determined the precise frequencies (or wavelengths) at which millions of molecules emit and absorb radiation.

In addition to the lines resulting from electron transitions, molecular lines result from two other kinds of change not possible in atoms: molecules can *rotate*, and they can *vibrate*. Figure 4.13 illustrates these basic molecular motions. Molecules rotate and vibrate in specific ways. Just as with atomic states, only certain spins and vibrations are allowed by the rules of molecular physics. When a molecule *changes* its rotational or vibrational

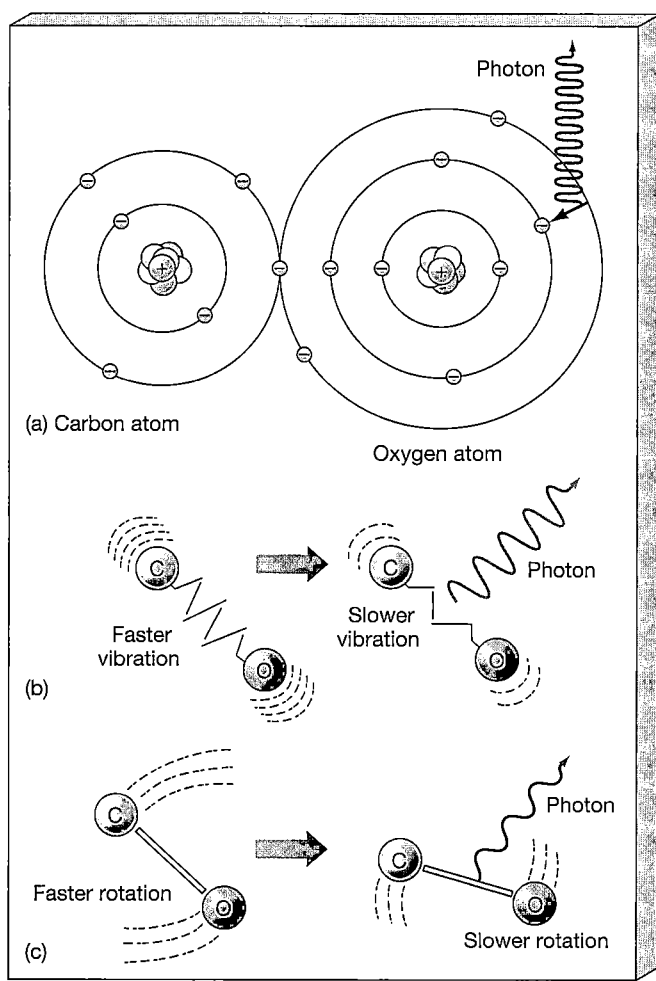

(a) Carbon atom

Oxygen atom

Photon

Faster
vibration

Slower
vibration

Photon

(b)

Faster rotation

Photon

(c)

Slower rotation

Figure 4.13 Molecular Emission Molecules can change in three ways while emitting or absorbing electromagnetic radiation. Sketched here is the molecule carbon monoxide (CO) experiencing (a) a change in electron arrangement in which an electron in the outermost orbital of the oxygen atom drops to a lower-energy state, (b) a change in vibrational state, and (c) a change in rotational state.

■ Changes in molecular *vibration* produce infrared spectral features.

■ Changes in molecular *rotation* produce spectral lines in the radio part of the electromagnetic spectrum (smallest energy changes).

Molecular lines usually bear little resemblance to the spectral lines associated with their component atoms. For example, Figure 4.14(a) shows the emission spectrum of the simplest molecule known—molecular hydrogen. Notice how different it is from the spectrum of atomic hydrogen shown in part (b) of the figure.

✓ Concept Check

■ What kinds of internal changes within a molecule can cause radiation to be emitted or absorbed?

state, a photon is emitted or absorbed. Spectral lines characteristic of the specific kind of molecule result. These lines are molecular fingerprints, just like their atomic counterparts, enabling researchers to identify and study one kind of molecule to the exclusion of all others. As a rule of thumb, we can say that

■ *Electron transitions* within molecules produce visible and ultraviolet spectral-line features (largest energy changes).

4.4 Spectral-Line Analysis

7 Astronomers apply the laws of spectroscopy in analyzing radiation from beyond Earth. A nearby star or a distant galaxy takes the place of the lightbulb in our previous examples. An interstellar cloud or a stellar (or even planetary) atmosphere plays the role of the intervening cool gas; and a spectrograph attached to a telescope replaces our simple prism and detector. We began our study of electromagnetic radiation by stating that virtually all we know about planets, stars, and galaxies is gleaned from studies of the light we receive from them, and we have presented some of the ways in which that knowledge is obtained. Here we describe a

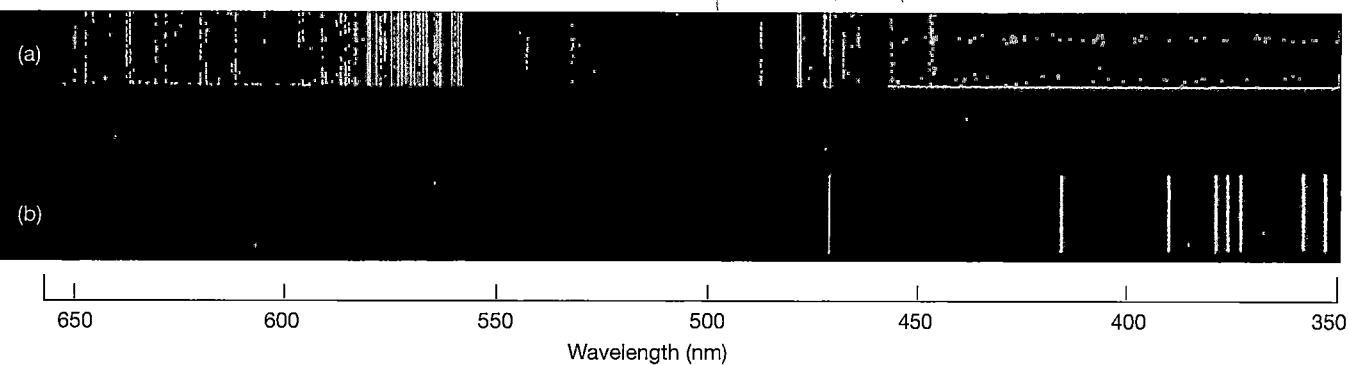

(a)

(b)

650   600   550   500   450   400   350

Wavelength (nm)

Figure 4.14 Hydrogen Spectra (a) The spectrum of molecular hydrogen. Notice how it differs from the spectrum of the simpler atomic hydrogen (b). *(Bausch & Lomb Inc.)*

MORE PRECISELY 4-1

The Energy Levels of the Hydrogen Atom

By observing the emission spectrum of hydrogen and using the connection between photon energy and color first suggested by Einstein (Sec. 4.2), Niels Bohr determined early in the twentieth century what the energy differences between the various energy levels must be. Using that information, he was then able to infer the actual energies of the excited states of hydrogen.

A unit of energy often used in atomic physics is the *electron volt* (eV). Its name derives from the amount of energy gained by an electron when it moves through an electric potential of one volt. For our purposes, however, it is just a convenient quantity of energy, numerically equal to 1.60×10^{-19} J (joule)—roughly half the energy carried by a single photon of red light. The minimum amount of energy needed to ionize hydrogen from its ground state is 13.6 eV. Bohr numbered the energy levels of hydrogen starting at the ground state, with level 1 the ground state, level 2 the first excited state, and so on. He found that by assigning zero energy to the ground state, the energy of any state (the nth, say) could be written as follows:

$$E_n = 13.6 \left(1 - \frac{1}{n^2} \right) \text{ eV.}$$

Thus, the ground state has energy $E_1 = 0$ (by our definition), the first excited state has energy $E_2 = 13.6 \times (1 - \frac{1}{4}) = 10.2$ eV, the second excited state has energy $E_3 = 13.6 \times (1 - \frac{1}{9})$ = 12.1 eV, and so on. Notice that there are infinitely many excited states between the ground state and the energy at which the atom is ionized, crowding closer and closer together as n becomes large and E_n approaches 13.6 eV.

EXAMPLE: Using Bohr's formula for the energy of each electron orbital, we can reverse his reasoning and calculate the energy associated with a transition between any two given states. To boost an electron from the second state to the third, an atom must be supplied with $E_3 - E_2 = 1.89$ eV of energy, or 3.03×10^{-19} J. Using the formula $E = hf$ presented in the text, we find that this corresponds to a photon with a frequency of 4.57×10^{14} Hz, having a wavelength of 656 nm and lying in the red portion of the spectrum. (A more precise calculation gives the value 656.3 nm

reported in the text.) Similarly, the jump from level 3 to level 4 requires $E_4 - E_3 = 13.6 \times (\frac{1}{3^2} - \frac{1}{4^2}) = 13.6 \times (\frac{1}{9} - \frac{1}{16}) = 0.66$ eV of energy, corresponding to an infrared photon with a wavelength of 1880 nm, and so on. A handy conversion between photon energies E in electron volts and wavelengths λ in nanometers is

$$E(\text{eV}) = \frac{1240}{\lambda \text{ (nm)}}.$$

The accompanying diagram summarizes the structure of the hydrogen atom. The various energy levels are depicted as a series of circles of increasing radius, representing increasing energy. The electronic transitions between these levels (indicated by arrows) are conventionally grouped into families, named after their discoverers, that define the terminology used to identify specific spectral lines. (Note that the spacings of the energy levels are not drawn to scale here, to provide room for all labels on the diagram. In reality, the circles should become more and more closely spaced as we move outward.)

Transitions starting from or ending at the ground state (level 1) form the *Lyman series*. The first is *Lyman alpha* (Lyα), corresponding to the transition between the first excited state (level 2) and the ground state. As we have seen, the energy difference is 10.21 eV, and the Lyα photon has a wavelength of 121.6 nm (1216 Å). The Lyβ (beta) transition, between level 3 (the second excited state) and the ground state, corresponds to an energy change of 12.10 eV and a photon of wavelength 102.6 nm (1026 Å). Lyγ (gamma) corresponds to a jump from level 4 to level 1, and so on. The accompanying table shows how we can calculate the energies, frequencies, and wavelengths of the photons in the Lyman series using the formulae given previously. All Lyman-series energies lie in the ultraviolet region of the spectrum.

The next series of lines, the *Balmer series*, involves transitions down to (or up from) level 2, the first excited state. All the Balmer series lines lie in or close to the visible portion of the electromagnetic spectrum. Because they form the most easily observable part of the hydrogen spectrum and were the first to be discovered, these lines are often re-

few of the ways in which the properties of emitters and absorbers can be determined by careful analysis of radiation received on (or near) Earth. We will encounter other important examples as our study of the cosmos unfolds.

A SPECTROSCOPIC THERMOMETER

Stars are very hot, especially deep down in their cores, where the temperature is measured in millions of

kelvins. Because of the heat, atoms are fully ionized. Electrons travel freely through the gas, unbound to any nucleus, and the spectrum of radiation is continuous. However, at the relatively cool stellar surface some atoms retain a few, or even most, of their orbital electrons. As discussed previously, by matching the spectral lines we see with the laboratory spectra of known atoms, ions, and molecules, we can determine a star's chemical composition.

ferred to simply as the "Hydrogen" series and denoted by the letter H. As with the Lyman series, the individual transitions are labeled with Greek letters. An Hα photon (level 3 to level 2) has a wavelength of 656.3 nm and is red, Hβ (level 4 to level 2) has a wavelength of 486.1 nm (green), Hγ (level 5 to level 2) has a wavelength of 434.1 nm (blue), and so on. The most energetic Balmer series photons have energies that place them just beyond the blue end of the visible spectrum, in the near ultraviolet.

The classification continues with the *Paschen series* (transitions down to or up from the second excited state), the *Brackett series* (third excited state), and the *Pfund series* (fourth excited state). Beyond that point, infinitely many other families exist, moving farther and farther into the infrared and radio regions of the spectrum, but they are not referred to by any special names. A few of the transitions making up the Lyman and Balmer (Hydrogen) series are marked on the figure. Astronomically, these are most important sequences.

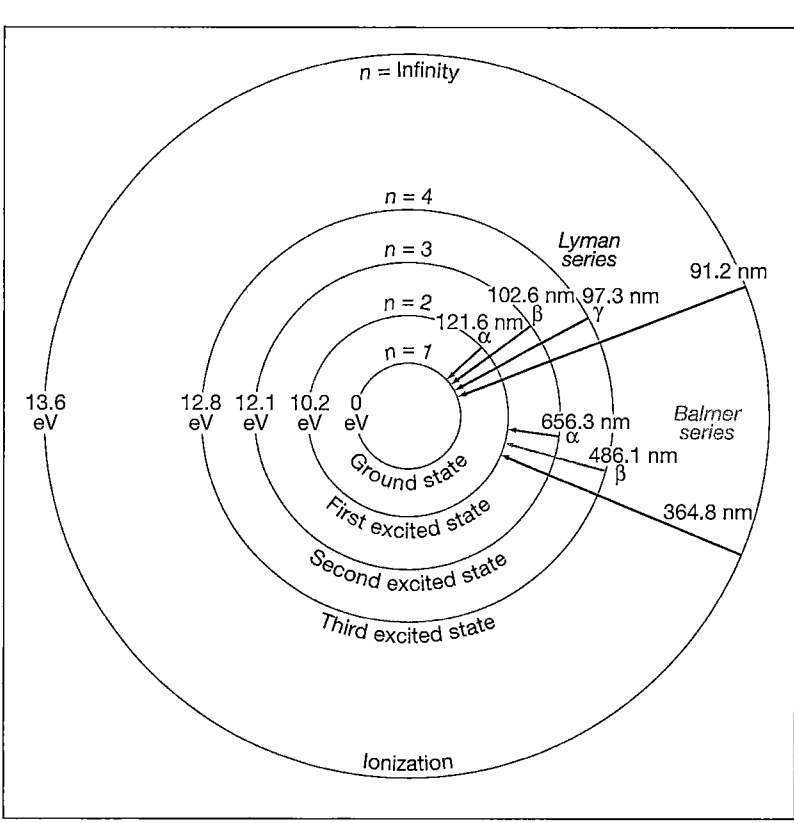

TRANSITION	ENERGY (eV)	FREQUENCY (10^{15} Hz)	WAVELENGTH (nm)
Lyα (2↔1)	$13.6 \times (1 - \frac{1}{4}) = 10.2$	$\dfrac{10.2 \times 1 \text{ eV}}{b} = 2.46$	$\dfrac{1240}{10.2} = 122$
Lyβ (3↔1)	$13.6 \times (1 - \frac{1}{9}) = 12.1$	2.92	103
Lyγ (4↔1)	$13.6 \times (1 - \frac{1}{16}) = 12.8$	3.08	97.3
Lyδ (5↔1)	$13.6 \times (1 - \frac{1}{25}) = 13.1$	3.15	95.0
⋮			
Ionization	$13.6 \times (1 - 0) = 13.6$	3.28	91.2

The strength of a spectral line (brightness or darkness, depending on whether the line is seen in emission or absorption) depends in part on the *number* of atoms giving rise to the line—the more atoms there are to emit or absorb photons of the appropriate frequency, the stronger the line. But line strength also depends on the *temperature* of the gas containing the atoms, because temperature determines how many atoms at any instant are in the right orbital to undergo any particular transition.

Consider the absorption of radiation by hydrogen atoms in an interstellar gas cloud or in the outer atmosphere of a star. If all the hydrogen were in its ground state—as it would be if the temperature were relatively low—then the only transitions that could occur would be the Lyman series (see *More Precisely 4-1*), resulting in absorption lines in the ultraviolet portion of the spectrum. Thus, astronomers would observe *no* visible hydrogen absorption lines (for example, the Balmer series) in the spectrum of this object, not because there was no hydrogen but because there would be no hydrogen atoms in the first excited state (as required to produce visible absorption features).

The spectrum of our own Sun is a case in point. Because the temperature of the Sun's atmosphere is a relatively cool 5800 K (as we saw in Chapter 3), few hydrogen atoms have electrons in any excited state. ∞ (Sec. 3.4) Hence, in the Sun, visible hydrogen lines are quite weak—that is, of low intensity compared with the same lines in many other stars—even though hydrogen is by far the most abundant element there.

As the temperature rises, atoms move faster and faster. More and more energy becomes available in the form of collisions, and more and more electrons are boosted into an excited state. At any instant, then, some atoms are temporarily in an excited state and so are capable of absorbing at visible or longer wavelengths. As the number of atoms in the first excited state increases, lines in the Balmer series become more and more evident in the spectrum. Eventually, a temperature is reached at which *most* of the atoms are in the first excited state, simply because of their frequent energetic collisions with other atoms in the gas. At this point the Balmer lines are at their strongest (and the Lyman lines are much weaker).

At even higher temperatures, most atoms are kicked beyond the first excited state into higher-energy orbitals, and new series of absorption lines are seen, while the strength of the Balmer series declines again. Eventually, the temperature becomes so high that most hydrogen is ionized, and no spectral lines are seen at all.

Spectroscopists have developed mathematical formulae that relate the number of emitted or absorbed photons to the energy levels of the atoms involved and the temperature of the gas. Once an object's spectrum is measured, astronomers can interpret it by matching the observed intensities of the spectral lines with those predicted by the formulas. In this way, astronomers can refine their measurements of both the composition *and* the temperature of the gas producing the lines. These temperature measurements are generally much more accurate than crude estimates based on the radiation laws and the assumption of blackbody emission. ∞ (Sec. 3.4)

THE DOPPLER EFFECT AND LINE BROADENING

The spectra of many atoms and ions are well known from laboratory measurements. Often, however, a familiar pattern of lines appears, but the lines are displaced from their usual locations. In other words, a set of spectral lines may be recognized as belonging to a particular element, except that they are all offset—blueshifted or redshifted—by the same amount from their normal wavelengths. Such shifts are produced by the Doppler effect. ∞ (Sec. 3.5) They thus allow astronomers to measure how fast the source of the radiation is moving along the line of sight from the observer (its *radial velocity*). For example, suppose the 486.1 nm Hβ line of

hydrogen in the spectrum of a distant galaxy is received on Earth at a wavelength of 485 nm—blueshifted to a shorter wavelength. We can compute the galaxy's line-of-sight velocity relative to Earth by using the Doppler equation presented in Section 3.5. We find that

$$\frac{\text{apparent wavelength (485 nm)}}{\text{true wavelength (486.1 nm)}} = 1 + \frac{\text{recession velocity}}{\text{speed of light, } c},$$

so the recession velocity is $(485/486.1 - 1) \times c = -680$ km/s. In other words, the galaxy is *approaching* us at a speed of 680 km/s.

The structure of the lines themselves reveals still more information. At first glance the emission lines shown earlier may seem uniformly bright, but more careful study shows that this is in fact not the case. As illustrated in Figure 4.15, the line brightness is greatest at the center and falls off toward either side. We stressed earlier that photons are emitted and absorbed at very precise energies, or frequencies. Why then aren't spectral lines extremely narrow, occurring only at specific wavelengths? This *line broadening* is not the result of some inadequacy of our experimental apparatus. It is caused by the *environment* in which the emission or absorption occurs. For definiteness, Figure 4.15 and subsequent figures refer to emission lines, but the ideas apply equally well to absorption features.

Several physical mechanisms can broaden spectral lines. The most important again involve the Doppler effect. Imagine a hot gas cloud containing individual atoms in random thermal motion in every possible direction, as illustrated in Figure 4.16(a). If an atom happens to be moving away from us as it emits a photon, that photon is redshifted by the Doppler effect—we do not record it at the precise wavelength predicted by atomic physics, but rather at a slightly

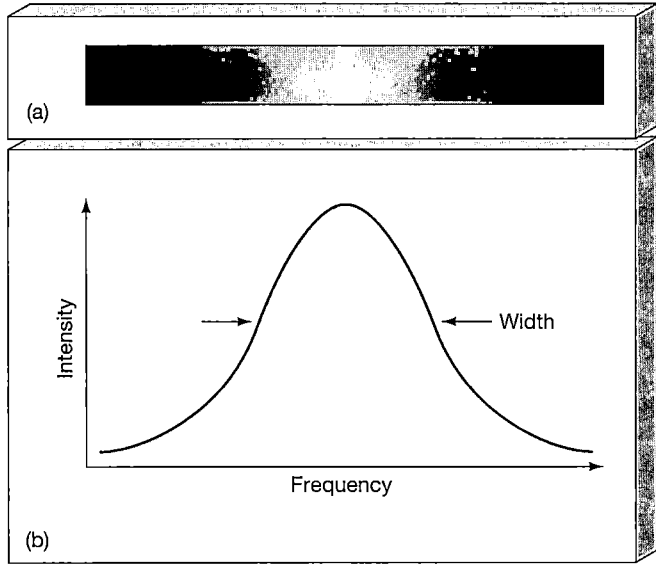

Figure 4.15 Line Profile By tracing the changing brightness across a typical emission line (a) and expanding the scale, we obtain a graph of its intensity versus wavelength (b).

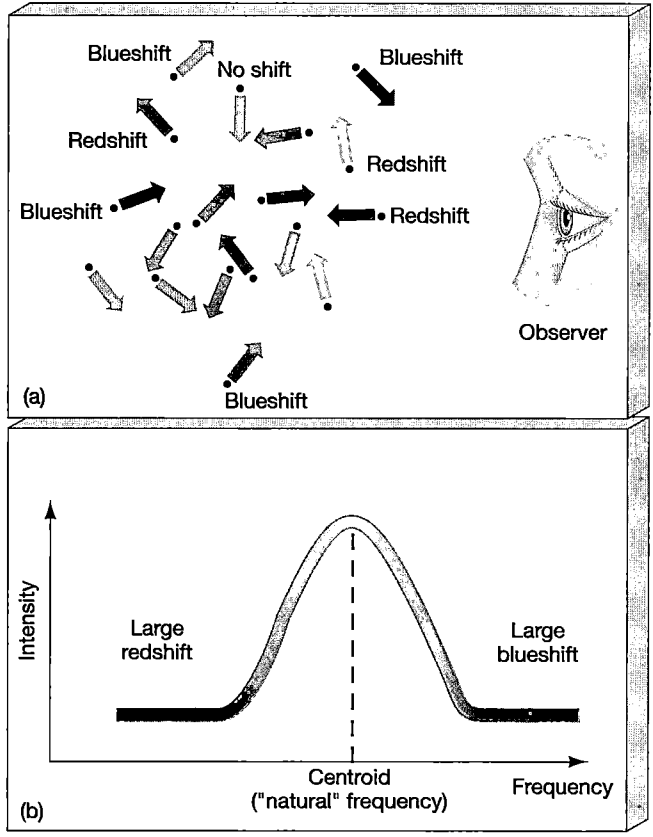

(a)

(b)

Figure 4.16 Thermal Broadening Atoms moving randomly (a) produce broadened spectral lines (b) as their individual redshifted and blueshifted emission lines merge in our detector. The hotter the gas, the greater the degree of thermal broadening.

longer wavelength. The extent of this redshift is proportional to the atom's instantaneous velocity away from the detector. Similarly, if the atom is moving toward us at the instant of emission, its light is blueshifted. In short, because of thermal motion within the gas, emission and absorption lines are observed at frequencies slightly different from those we would expect if all atoms in the cloud were motionless.

Most atoms in a typical cloud have small thermal velocities, so in most cases the line is Doppler-shifted just a little. Only a few atoms have large shifts. As a result, the center of a spectral line is much more pronounced than its "wings," producing a bell-shaped spectral feature like that shown in Figure 4.16(b). Thus, even if all atoms emitted and absorbed photons at only one precise wavelength, the effect of their thermal motion would be to smear the line out over a range of wavelengths. The hotter the gas, the larger the spread of Doppler motions and the greater the width of the line. ⊂⊃ (*More Precisely 3-1*) By measuring a line's width, astronomers can estimate the temperature of the gas producing it.

Rotation produces a similar effect. Consider an astronomical object (a planet, a star, or even an entire galaxy) oriented so that we see it spinning, as sketched in Figure 4.17. Photons emitted from the side spinning toward us are blueshifted by the Doppler effect. Photons emitted

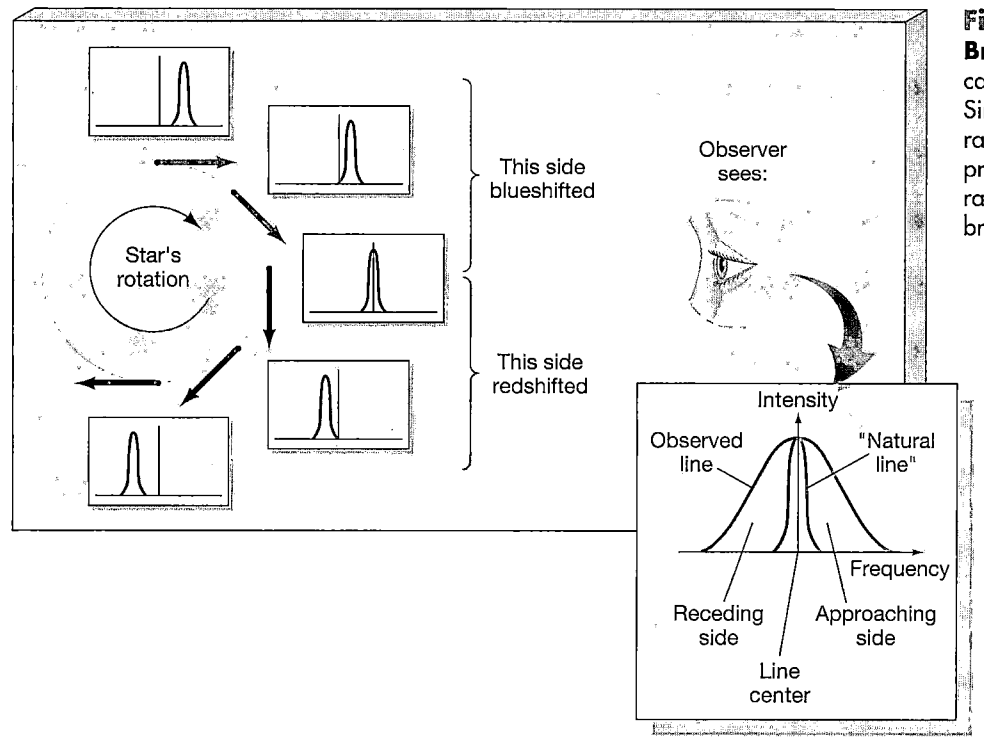

Figure 4.17 Rotational Broadening The rotation of a star can cause spectral line broadening. Since most stars are unresolved, light rays from all parts of the star merge to produce broadened lines. The more rapid the rotation, the greater the broadening.

from the side spinning away from us are redshifted. Very often, the object under study is so small or far away that our equipment cannot distinguish, or *resolve*, different parts from one another—all the emitted light is blended together in our detector, producing a net broadening of the observed spectral lines. (If we could resolve the object, of course, we would be able to distinguish the redshifted and blueshifted sides and measure the rotation rate directly.) Note that this broadening has *nothing* to do with the temperature of the gas producing the lines, and is generally superimposed on the thermal broadening just discussed. The faster the object spins, the more rotational broadening we see.

Yet another mechanism that can cause line broadening is gas *turbulence*, which exists when the gas in an interstellar cloud is not at rest or flowing smoothly but instead is seething and churning in eddies and vortices of many sizes. Motion of this type also causes Doppler shifts of spectral lines, but lines from different parts of the cloud are shifted more or less randomly. Just as in the case of rotation, if our equipment is unable to resolve the cloud, a net broadening of its observed spectral lines results. The faster the internal motion, the greater the broadening observed. Remember that, again, the internal motion has nothing to do with the temperature of the gas.

Other broadening mechanisms do not depend on the Doppler effect at all. For example, if electrons are moving between orbitals while their parent atom is colliding with another atom, the energy of the emitted or absorbed photons changes slightly, "blurring" the spectral lines. This mechanism occurs most often in dense gases, where collisions are most frequent. It is usually referred to as *collisional broadening*. The amount of broadening increases as the density of the emitting or absorbing gas rises.

Finally, *magnetic fields* can also broaden spectral lines, via a process called the *Zeeman effect*. The electrons and nuclei within atoms behave as tiny spinning magnets, and the basic emission and absorption rules of atomic physics change slightly whenever atoms are immersed in a magnetic field, as is the case in many stars to greater or lesser extents. The result is a slight "splitting" of a spectral line, which blurs into an overall line broadening. Generally, the stronger the magnetic field, the more pronounced the spectral-line broadening.

Additional analysis is usually required to determine the precise cause of line broadening. For example, if we

TABLE 4.1 Spectral Information Derived from Starlight

OBSERVED SPECTRAL CHARACTERISTIC	INFORMATION PROVIDED
Peak frequency or wavelength (continuous spectra only)	Temperature (Wien's law)
Lines present	Composition, temperature
Line intensities	Composition, temperature
Line width	Temperature, turbulence, rotation speed, density, magnetic field
Doppler shift	Line-of-sight velocity

know the temperature of the emitting gas by other means (perhaps by comparing intensities of different spectral lines, as discussed earlier), then we can calculate how much of the broadening is due to thermal motion. In addition, it is often possible to distinguish between the various broadening mechanisms described here by studying the detailed *shapes* of the lines.

Given sufficiently sensitive equipment, there is almost no end to the wealth of data that can be obtained from starlight. Table 4.1 lists some basic measurable properties of an incoming beam of radiation, and indicates what sort of information can be obtained from them. It is important to realize, however, that deciphering the extent to which each of the factors just described influences a spectrum can be a very difficult task. Typically, the spectra of many elements are superimposed on one another, and several competing physical effects are occurring simultaneously, each modifying the spectrum in its own way. The challenge facing astronomers is to unravel the extent to which each mechanism contributes to spectral-line profiles and so obtain meaningful information about the source of the lines. In the next chapter we will discuss some of the means by which astronomers obtain the data they need in their quest to understand the cosmos.

Concept Check

■ Why is it so important for astronomers to analyze spectral lines in detail?

Chapter Review

SUMMARY

A **spectroscope** (p. 86) is a device for splitting a beam of radiation into its component frequencies and delivering them to a screen or detector for detailed study. Many hot objects emit a **continuous spectrum** (p. 87) of radiation, containing light of all wavelengths. A hot gas may instead produce an **emission spectrum** (p. 88), consisting only of a few well-defined **emission lines** (p. 87) of specific frequencies, or colors. Passing a continuous beam of radiation through cool gas will produce **absorption lines** (p. 88) at precisely the same frequencies as are present in the gas's emission spectrum. **Kirchhoff's laws** (p. 89) describe the relationships among these different types of spectra. The emission and absorption lines produced by each element are unique—they provide a "fingerprint" of that element. The study of the spectral lines produced by different substances is called **spectroscopy** (p. 89). Spectroscopic studies of the Fraunhofer lines in the solar spectrum yield detailed information on the Sun's composition.

Atoms (p. 91) are made up of negatively charged electrons orbiting a positively charged heavy **nucleus** (p. 91) consisting of positively charged protons and electrically neutral **neutrons** (p. 95). In normal circumstances the number of orbiting electrons equals the number of protons in the nucleus, and the atom as a whole is electrically neutral. The number of protons in the nucleus determines the type of **element** (p. 95) that the atom represents. The Bohr model of the atom was an early attempt to explain how atoms can produce emission and absorption line spectra. An atom has a minimum-energy **ground state** (p. 91), representing its "normal" condition. If an orbiting electron is given enough energy, it can escape from the atom, which is then said to be **ionized** (p. 91). Between these two states, the electron can exist only in certain well-defined **excited states** (p. 92) each with a very specific energy. The electron's energy is said to be **quantized** (p. 91). In the modern view the electron is envisaged as being spread out in a "cloud" around the nucleus, but still with a sharply defined energy.

Electromagnetic radiation exhibits both wave and particle properties. Particles of radiation are called **photons** (p. 92). The energy of a photon depends on its color—it is directly proportional to the photon's frequency. As electrons move between energy levels within an atom, the difference in the energy between the states is emitted or absorbed in the form of photons. Because the energy levels have definite energies, the photons also have definite energies, and hence colors, which are characteristic of the type of atom involved.

Molecules (p. 96) are groups of two or more atoms bound together by electromagnetic forces. Like atoms, molecules exist in energy states that obey rules similar to those governing the internal structure of atoms. Again like atoms, when molecules make transitions between energy states, they emit or absorb a characteristic spectrum of radiation that identifies them uniquely.

Astronomers apply the laws of spectroscopy in analyzing radiation from beyond Earth. Several physical mechanisms can broaden spectral lines. The most important is the Doppler effect, which occurs because stars are hot and their atoms are in motion, or because the object being studied is rotating or in turbulent motion.

SELF-TEST: TRUE OR FALSE?

_____ **1.** Emission spectra are characterized by narrow bright lines of different colors.

_____ **2.** Imagine an emission spectrum produced by a container of hydrogen gas. Changing the amount of hydrogen in the container will change the colors of the lines in the spectrum.

_____ **3.** In the previous question, changing the gas in the container from hydrogen to helium will change the colors of the lines occurring in the spectrum.

_____ **4.** An absorption spectrum appears as a continuous spectrum interrupted by a series of dark lines.

_____ **5.** The wavelengths of the emission lines produced by an element are different from the wavelengths of the absorption lines produced by the same element.

_____ **6.** Gustav Kirchhoff is credited with the discovery of blackbody radiation.

_____ **7.** The density of the hot gas producing an emission spectrum must be very high.

_____ **8.** The energy of a photon is inversely proportional to the wavelength of the radiation.

_____ **9.** The ground state of an atom is that in which the electron is in its lowest energy level (orbital).

_____ **10.** An electron can have any energy within an atom so long as it is above the ground state energy.

_____ **11.** An atom can remain in an excited state indefinitely.

_____ **12.** Emission and absorption lines correspond to the specific energy differences between orbitals in an atom.

_____ **13.** The number of electrons in an atom or ion determines the identity of the element it represents.

_____ **14.** More than one element or molecule can have the same emission or absorption spectrum.

_____ **15.** Spectral lines of hydrogen are relatively weak in the Sun because the Sun contains relatively little hydrogen.

SELF-TEST: FILL IN THE BLANK

1. A _____ is a glass wedge that disperses light into a spectrum.
2. Blackbody radiation is an example of a _____ spectrum.
3. Fraunhofer discovered absorption lines in the _____.
4. A continuous spectrum can be produced by a luminous solid, liquid, or _____ gas.
5. An absorption spectrum is produced when a _____ gas lies in front of a source of continuous radiation.
6. Light behaves both as a wave and as a _____.
7. The experiment that demonstrated the _____ caused Einstein to realize that light does not always behave like a wave.
8. Protons carry a _____ charge; electrons carry a _____ charge.
9. When one or more electrons are stripped from the atom, the atom is said to be _____.

10. An electron moves to a higher energy level in an atom after it _____ a photon of a specific energy.
11. When an electron moves to a lower energy level in an atom, it _____ a photon of a specific energy.
12. The "specific energy" of the photon referred to in the two previous questions is exactly equal to the energy _____ between the energy levels involved.
13. Changes in molecular vibrational states generally produce spectral features in the _____ part of the electromagnetic spectrum.
14. Changes in molecular rotational states generally produce spectral features in the _____ part of the electromagnetic spectrum.
15. High temperatures, rotation, and magnetic fields all tend to _____ spectral lines.

REVIEW AND DISCUSSION

1. What is spectroscopy? Explain how astronomers might use spectroscopy to determine the composition and temperature of a star.
2. Describe the basic components of a simple spectroscope.
3. What is a continuous spectrum? An absorption spectrum?
4. Why are gamma rays generally harmful to life forms, but radio waves generally harmless?
5. What is a photon?
6. In the particle description of light, what is color?
7. In what ways does the Bohr model of atomic structure differ from the modern view?
8. Give a brief description of a hydrogen atom.
9. What does it mean to say that a physical quantity is quantized?
10. What is the normal condition for atoms? What is an excited atom? What are orbitals?
11. Why do excited atoms absorb and reemit radiation at characteristic frequencies?
12. How are absorption and emission lines produced in a stellar spectrum? What information might absorption lines in the

spectrum of a star reveal about a cloud of cool gas lying between us and the star?
13. According to Kirchhoff's laws, what are the necessary conditions for a continuous spectrum to be produced?
14. Explain how a beam of light passing through a diffuse cloud may give rise to both absorption and emission spectra.
15. Why is the Hα absorption line of hydrogen in the Sun relatively weak, even though the Sun has abundant hydrogen?
16. How do molecules produce spectral lines unrelated to the movement of electrons between energy levels?
17. How does the intensity of a spectral line yield information about the source of the line?
18. How can the Doppler effect cause broadening of a spectral line?
19. Describe what happens to a spectral line from a star as the star's rotation rate increases.
20. List three properties of a star that can be determined from observations of its spectrum.

PROBLEMS  *Algorithmic versions of these questions are available in the Practice Problems module of the Companion Website.*

The number of squares preceding each problem indicates its approximate level of difficulty.

1. ■ What is the energy (in electron volts—see *More Precisely 4-1*) of a 450-nm blue photon? A 200-nm ultraviolet photon?
2. ■ What is the energy (in electron volts) of a 100 GHz (1 gigahertz = 10^9 Hz) microwave photon?
3. ■ What is the wavelength of a 2-eV red photon? Repeat your calculation for an 0.1-eV infrared photon and a 5000-eV (5 keV) X ray.

4. ■ How many times more energy has an 1-nm gamma ray than a 10-MHz radio photon?
5. ■ How many times longer in wavelength is a 100-MHz radio photon than a 100-eV X ray?
6. ■ Calculate the energy change in the transition responsible for the left-hand yellow sodium line shown in Figure 4.6.
7. ■■ Calculate the wavelength and frequency of the radiation emitted by the electronic transition from the tenth to the ninth excited state of hydrogen. In what part of the electro-

magnetic spectrum does this radiation lie? Repeat the question for transitions from the 100th to the 99th excited state, and from the 1000th to the 999th excited state.

8. ■■ How many different photons (that is, photons of different frequencies) can be emitted as a hydrogen atom in the third excited state falls back, directly or indirectly, to the ground state? What are their wavelengths?

9. ■■ List all the spectral lines of hydrogen that lie in the visible range (taken to run from 400 to 700 nm in wavelength).

10. ▣ A distant galaxy is receding from Earth with a radial velocity of 3000 km/s. At what wavelength would its Lyα line be received by a detector above Earth's atmosphere?

11. ■ The Hα line of a certain star is received on Earth at a wavelength of 656 nm. What is the star's radial velocity with respect to Earth?

12. ■■ In a demonstration of the photoelectric effect, suppose that a minimum energy of 5×10^{-19} J (3.1 eV) is required

to dislodge an electron from the metal surface. What is the minimum frequency (and longest wavelength) of radiation for which the detector registers a response?

13. ■■ At a temperature of 5800 K, hydrogen atoms in the solar atmosphere have typical random speeds of about 12 km/s. Assuming that the broadening is simply the result of atoms moving toward us or away from us at this random speed, estimate the thermal width (in nanometers) of the 656.3-nm solar Hα line.

14. ▣■ Turbulent motion in a radio-emitting gas cloud broadens a 1.2-GHz (1 gigahertz = 10^9 Hz) radio line to a width of 0.5 MHz. Estimate the average speed (see Problem 13) at which the gas within the cloud is moving.

15. ▣■▣ Estimate how fast, in revolutions per day, the Sun would have to rotate in order for rotational broadening to be comparable to thermal broadening in determining line widths (see problem 13; the radius of the Sun is roughly 700,000 km).

RESEARCHING ON THE WEB

To complete the following exercises, go to the online Destinations module for Chapter 4 on the Companion Website for Astronomy Today 4/e.

1. Access the "Java-Based Display of Spectra of Gas Discharges" page and make a sketch comparing the spectra of hydrogen, helium, and two additional spectra of your choice.

2. Access the "Atomic Physics" page and determine which spectral lines emit in the ultraviolet and which limes emit in the infrared if Balmer lines emit in the optical.

3. Access the "Light Tour" page and determine the wavelength region for light with wavelengths of 1, 100, 1000, 10,000, 100,000, and 1,000,000 angstroms.

PROJECTS

1. Find a spectrum of the Sun that also has a wavelength scale alongside. Figure 16.8 is a good example; however, you may want to enlarge it on a copying machine. Select various absorption lines and determine their wavelengths by interpolation. Now, try to identify the element that produced these lines. Use a reference of lines such as found in Moore's *A Multiplet Table of Astrophysical Interest*. Other references may be found in the astronomy, chemistry, or physics sections of your library. Work with the darkest lines before trying the fainter lines.

2. Use a handheld spectroscope, available through Learning Technologies Inc. While in the shade, point the spectroscope at a white cloud or white piece of paper that is in direct sunlight. Look for the absorption lines in the Sun's spectrum. Note their wavelength from the scale inside the spectroscope. Compare your list with the Fraunhofer lines given in many physics, astronomy, or chemistry reference books.

 *In addition to the Practice Problems and Destinations modules, the Companion Website at http://www.prenhall.com/chaisson provides for each chapter an additional true-false, multiple choice, and labeling quiz, as well as additional annotated images, animations, and links to related Websites.*

5 TELESCOPES

The Tools of Astronomy

LEARNING GOALS

Studying this chapter will enable you to:

1 Sketch and describe the basic designs of the major types of optical telescopes used by astronomers.

2 Explain why very large telescopes are needed for most astronomical study, and specify the particular advantages of reflecting telescopes for astronomical use.

3 Describe how Earth's atmosphere affects astronomical observations, and discuss some of the current efforts to improve ground-based astronomy.

4 Discuss the advantages and disadvantages of radio astronomy compared to optical observations.

5 Explain how interferometry can enhance the usefulness of astronomical observations.

6 List the other types of nonvisible radiation currently being exploited for astronomical observation and summarize the advantages, limitations, and chief uses of each.

7 Tell why it is important to make astronomical observations in different regions of the electromagnetic spectrum.

 Visit http://www.prenhall.com/chaisson for additional annotated images, animations, and links to related sites for this chapter.

This composite photograph shows two of the premier optical telescopes available to astronomers today. At the top, overflying Earth is HST—the *Hubble Space Telescope*—in orbit some 600 kilometers above the surface of our planet; operating well above Earth's atmosphere, it has superb angular resolution. At the bottom, in the Atacama Desert high in the Chilean Andes is the new VLT—the *Very Large Telescope*—which is actually an array of four large telescopes working together to view the universe with equally superb resolution. *(ESO)*

The Big Picture: Telescopes are time machines and astronomers, in a sense, are historians. Telescopes aid our eyes and senses, enabling us not only to see to much greater distances than are accessible to the naked eye alone, but also to perceive cosmic objects that emit radiation at wavelengths to which our human senses are completely blind. Without telescopes, we would know only a tiny fraction of the information presented in this text.

At its heart, astronomy is an observational science. More often than not, observations of cosmic phenomena precede any clear theoretical understanding of their nature. As a result, our detecting instruments—our telescopes—have evolved to observe as broad a range of wavelengths as possible. Until the middle of the twentieth century, telescopes were limited to visible light. Since then, technological advances have expanded our view of the universe to all regions of the electromagnetic spectrum. Some telescopes are sited on Earth, whereas others must be placed in space, and design considerations vary widely from one part of the spectrum to another. Whatever the details of their construction, however, telescopes are devices whose basic purpose is to collect electromagnetic radiation and deliver it to a detector for detailed study.

5.1 Optical Telescopes

In essence, a **telescope** is a "light bucket" whose primary function is to capture as many photons as possible from a given region of the sky and concentrate them into a focused beam for analysis. An *optical* telescope is one designed specifically to collect the wavelengths that are visible to the human eye. Optical telescopes have a long history, reaching back to the days of Galileo in the early seventeenth century. They are probably also the best-known type of telescope, so it is fitting that we begin our study of astronomical hardware with these devices. However, when we turn our attention to telescopes designed to capture and analyze radiation in other, *invisible*, regions of the electromagnetic spectrum, we will find that many of these instruments share the same basic design.

Modern astronomical telescopes have evolved a long way from Galileo's simple apparatus. Their development over the years has seen a steady increase in *size* for one simple, but very important, reason: Large telescopes can gather and focus more radiation than can their smaller counterparts, allowing astronomers to study fainter objects and to obtain more detailed information about bright ones. This fact has played a central role in determining the design of contemporary instruments.

REFLECTING AND REFRACTING TELESCOPES

Optical telescopes fall into two basic categories—*reflectors* and *refractors*. Figure 5.1 shows how a **reflecting telescope** uses a curved mirror to gather and concentrate a beam of light. The mirror, usually called the *primary mirror* because telescopes often contain more than one mirror, is constructed so that all light rays arriving parallel to its axis (the imaginary line through the center of and perpendicular to the mirror), regardless of their distance from that axis, are reflected to pass through a single point, called the *focus*. The distance between the primary mirror and the focus is the *focal length*. In astronomical contexts, the focus of the primary mirror is referred to as the **prime focus.**

A **refracting telescope** uses a lens to focus the incoming light. **Refraction** is when a beam of light bends as it passes from one transparent medium (for example, air) into another (such as glass). For example, consider how a pencil half immersed in a glass of water looks bent. The pencil is straight, of course, but the light by which we see it is bent—refracted—as that light leaves the water and enters the air. When that light then enters our eyes, we perceive the pencil as being bent. Figure 5.2(a) illustrates the process and shows how a prism can be used to change the direction of a beam of light. As illustrated in Figure 5.2(b), we can think of a lens as a series of prisms combined in such a way that all light rays striking the lens parallel to the axis are refracted to pass through the focus.

Astronomical telescopes are often used to make **images** of their field of view. Figure 5.3 illustrates how this is accomplished, in this case by the mirror in a reflecting telescope. Light from a distant object (in this case, a comet) reaches us as parallel, or very nearly parallel, rays. Any ray of light entering the instrument parallel to the telescope's axis strikes the mirror and is reflected through the prime focus. Light coming from a slightly different direction—inclined slightly to the axis—is focused to a slightly different point. In this way, an image is formed

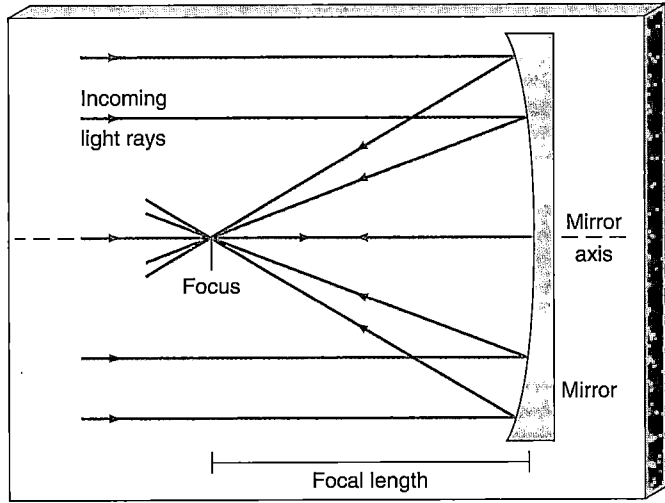

Figure 5.1 Reflecting Mirror A curved mirror can be used to focus to a single point all rays of light arriving parallel to the mirror axis. Light rays traveling along the axis are reflected back along the axis, as indicated by the arrowheads pointing in both directions. Off-axis rays are reflected through greater and greater angles the farther they are from the axis, so that they all pass through the same point—the focus.

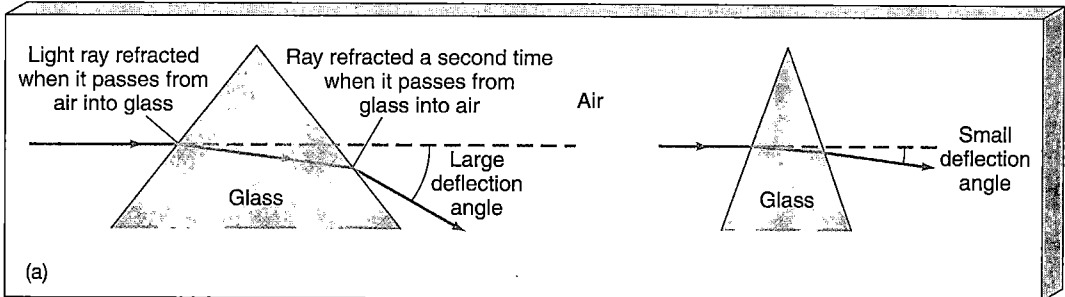

Figure 5.2 Refracting Lens (a) Refraction by a prism changes the direction of a light ray by an amount that depends on the angle between the faces of the prism. (b) A lens can be thought of as a series of prisms. A light ray traveling along the axis of a lens is unrefracted as it passes through the lens. Parallel rays arriving at progressively greater distances from the axis are refracted by increasing amounts, in such a way that all are focused to a single point.

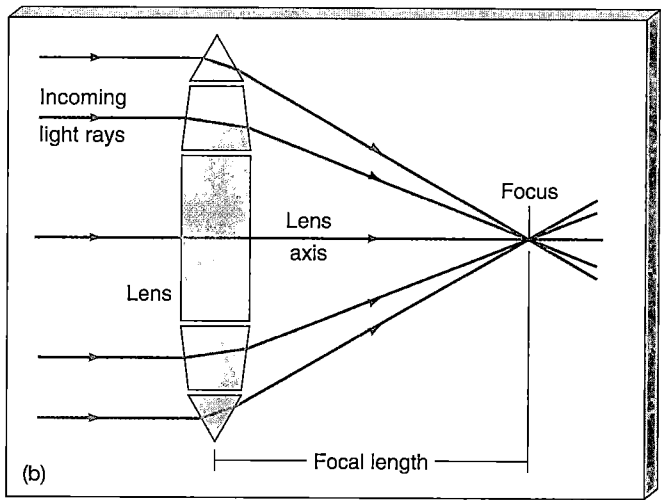

near the prime focus. Each point on the image corresponds to a different point in the field of view.

The prime-focus images produced by large telescopes are actually quite small—the image of the entire field of view may be as little as 1 cm across. Often, the image is magnified with a lens known as an *eyepiece* before being observed by eye or, more likely, recorded as a photograph or digital image. The angular diameter of the magnified image is much greater than the telescope's field of view,

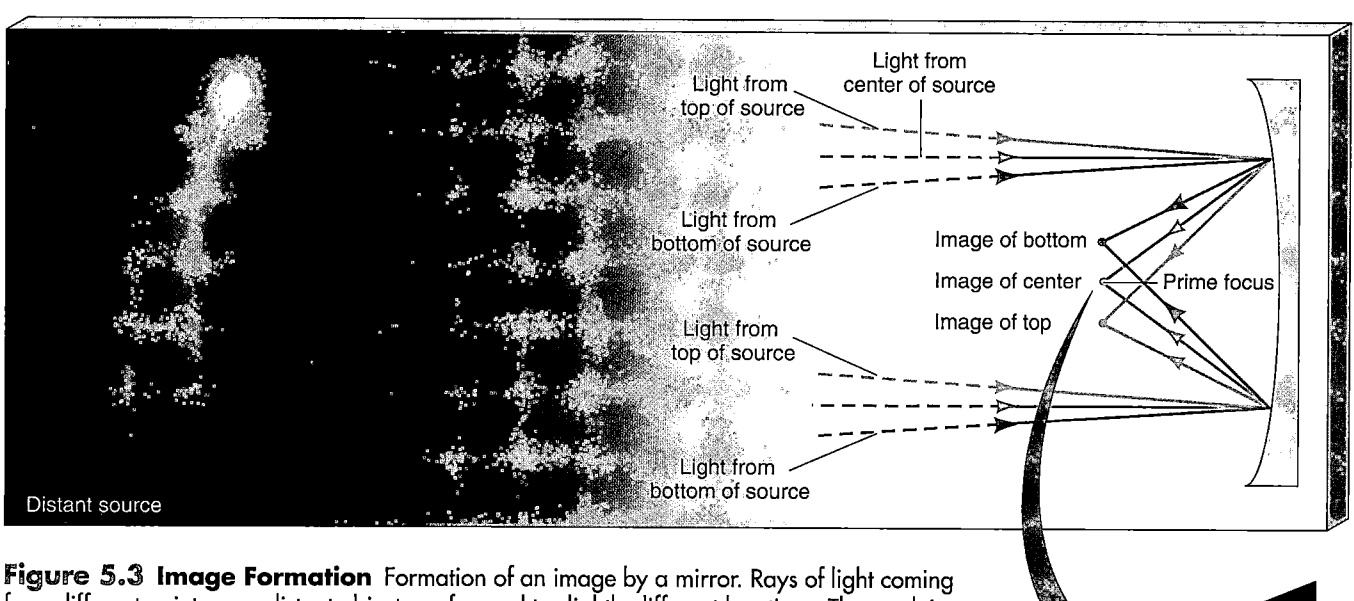

Figure 5.3 Image Formation Formation of an image by a mirror. Rays of light coming from different points on a distant object are focused to slightly different locations. The result is that an image of the object is formed around the prime focus. Notice that the image is inverted (that is, upside down).

allowing much more detail to be discerned. Figure 5.4(a) shows the basic design of a simple reflecting telescope, illustrating how a small secondary mirror and eyepiece are used to view the image. Figure 5.4(b) shows how a refracting telescope accomplishes the same function.

The two telescope designs shown in Figure 5.4 achieve the same result—light from a distant object is captured and focused to form an image. On the face of it, then, it might appear that there is little to choose between the two in deciding which type to buy or build. However, as telescope size has steadily increased over the years (for reasons to be discussed in Section 5.2), a number of important factors have tended to favor reflecting instruments over refractors:

1. The fact that light must pass through the lens is a major disadvantage of refracting telescopes. Just as a prism disperses white light into its component colors, the lens in a refracting telescope tends to focus red and blue light differently. This deficiency is known as *chromatic aberration*. Figure 5.5 shows how chromatic aberration occurs and indicates how it affects the image of a star. Careful design and choice of materials can largely correct chromatic aberration, but it is very difficult to eliminate entirely. This problem obviously does not occur with mirrors.

2. As light passes through the lens, some of it is absorbed by the glass. This absorption is a relatively minor problem for visible radiation, but it can be severe for infrared and ultraviolet observations because glass blocks most of the radiation in those regions of the electromagnetic spectrum. Again, this problem does not affect mirrors.

3. A large lens can be quite heavy. Because it can be supported only around its edge (so as not to block the incoming radiation), the lens tends to deform under its own weight. A mirror does not have this drawback because it can be supported over its entire back surface.

4. A lens has two surfaces that must be accurately machined and polished—which can be very difficult—but a mirror has only one.

For these reasons, *all* large modern telescopes use mirrors as their primary light gatherers. Figure 5.6 shows the world's largest refractor, installed in 1897 at the Yerkes Observatory in Wisconsin and still in use today. It has a lens diameter of just over 1 m (40 inches). By contrast, some new reflecting telescopes have mirror diameters in the 10-m range, and larger instruments are on the way.

Most of the considerations listed above for optical instruments are equally applicable to observations made in the infrared part of the spectrum. Indeed, many large ground-based optical facilities are also used extensively for infrared work. (Recall from Figure 3.9 that, while Earth's atmosphere blocks most infrared information, there are several fairly broad spectral windows through which ground-based infrared observations can be made.)

TELESCOPE DESIGN

Figure 5.7 shows some basic reflecting telescope designs. Radiation from a star enters the instrument, passes down the main tube, strikes the primary mirror, and is reflected back toward the prime focus, near the top of the tube.

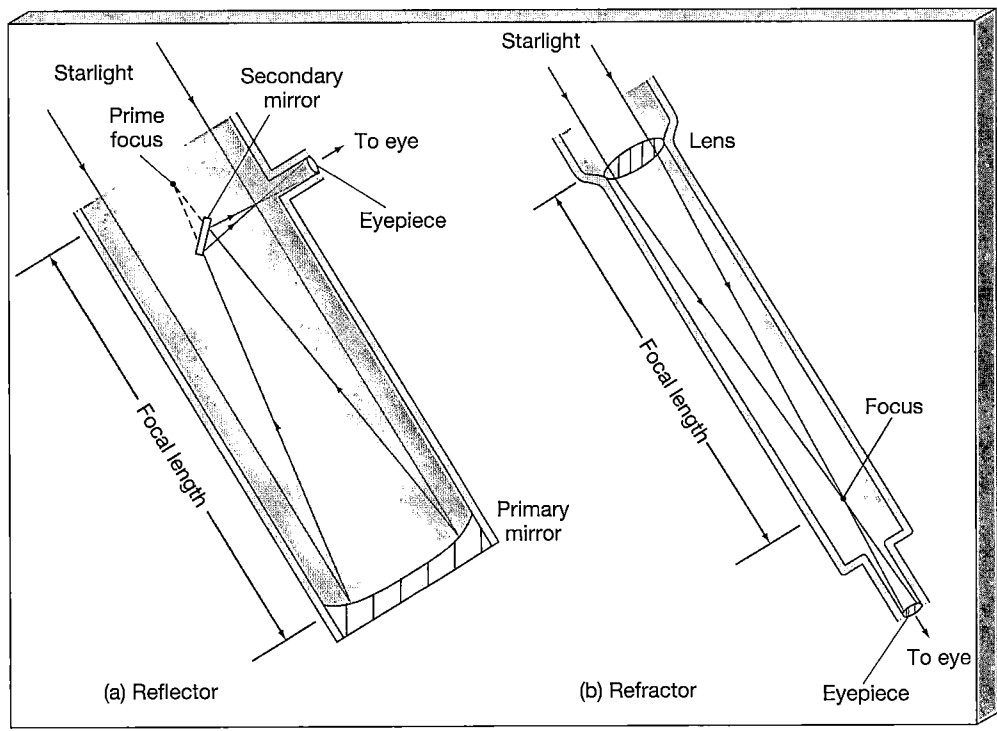

Figure 5.4 Reflectors and Refractors Comparison of (a) reflecting and (b) refracting telescope systems. Both types are used to gather and focus cosmic radiation—to be observed by human eyes or recorded on photographs or in computers. In both cases the image formed at the focus is viewed with a small magnifying lens called an eyepiece.

Figure 5.5 Chromatic Aberration A prism bends blue light more than it bends red light, so the blue component of light passing through a lens is focused slightly closer to the lens than the red component. As a result, the image of an object acquires a colored "halo," no matter where we place our detector.

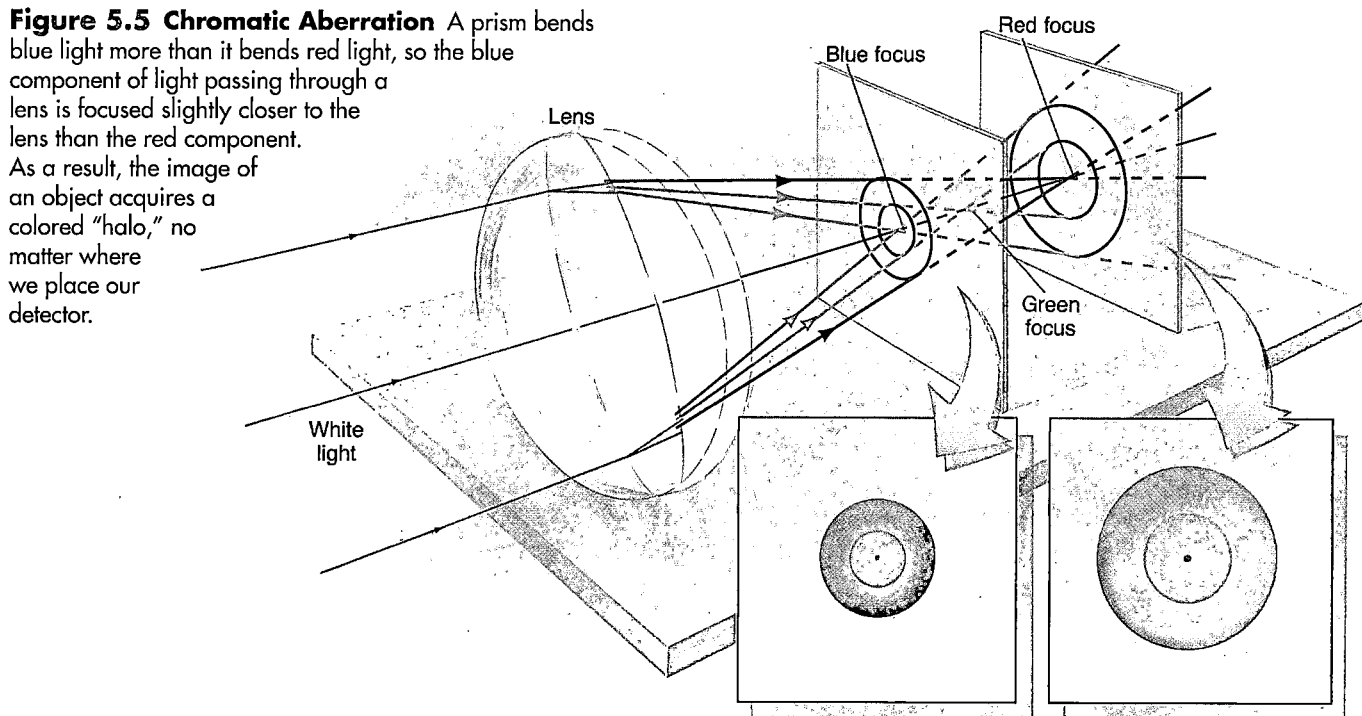

Sometimes astronomers place their recording instruments at the prime focus. However, it can be very inconvenient, or even impossible, to suspend bulky pieces of equipment there. More often, the light is intercepted on its path to the focus by a *secondary mirror* and redirected to a more convenient location, as in Figure 5.7(b) through (d).

In a **Newtonian telescope** (named after Sir Isaac Newton, who invented this particular design), the light is intercepted before it reaches the prime focus and is deflected by 90°, usually to an eyepiece at the side of the instrument. This is a particularly popular design for smaller reflecting telescopes, such as those used by amateur astronomers.

Alternatively, astronomers may choose to work on a rear platform where they can use equipment, such as a spectroscope, that is too heavy to hoist to the prime focus. In this case, light reflected by the primary mirror toward the prime focus is intercepted by a smaller secondary mirror, which reflects it back down through a small hole at the center of the primary mirror. This arrangement is known as a **Cassegrain telescope** (after Guillaume Cassegrain, a French lensmaker). The point behind the primary mirror where the light from the star finally converges is called the *Cassegrain focus*.

A more complex observational configuration requires starlight to be reflected by several mirrors. As in the Cassegrain design, light is first reflected by the primary mirror toward the prime focus and reflected back down the tube by a secondary mirror. A third, much smaller, mirror then reflects the light into an environmentally controlled laboratory. Known as the *coudé* room (from the French word for "bent"), this laboratory is separate from

Figure 5.6 Largest Refracting Telescope Photograph of the Yerkes Observatory's 40-inch-diameter refractor. *(Yerkes Observatory)*

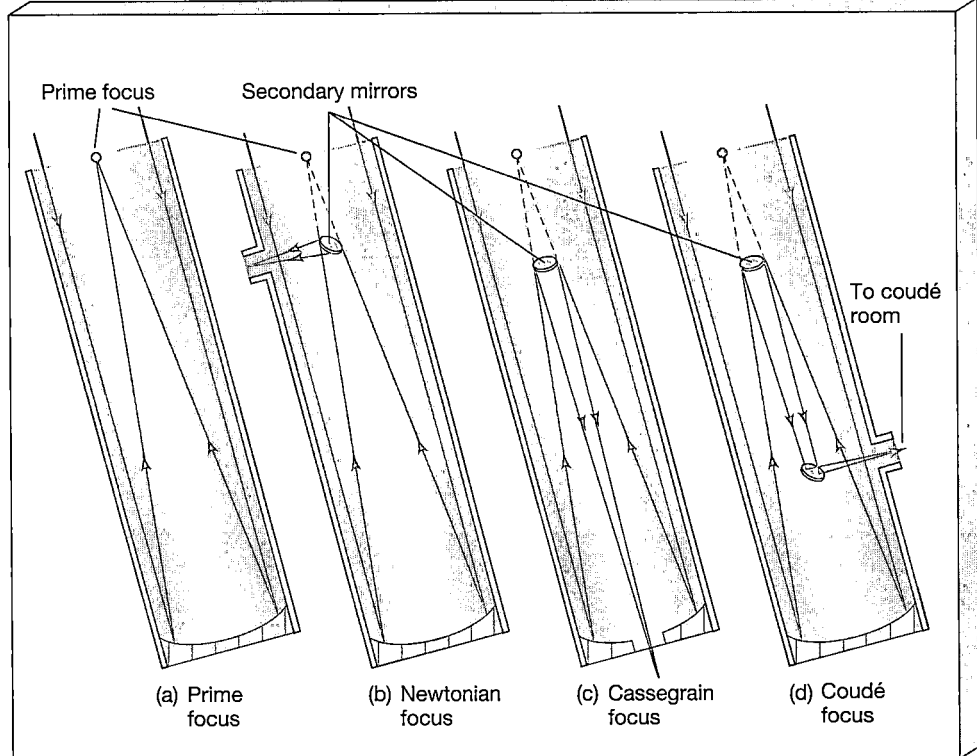

Prime focus Secondary mirrors

To coudé room

(a) Prime focus (b) Newtonian focus (c) Cassegrain focus (d) Coudé focus

Figure 5.7 Reflecting Telescopes Four reflecting telescope designs: (a) prime focus, (b) Newtonian focus, (c) Cassegrain focus, and (d) coudé focus. Each uses a primary mirror at the bottom of the telescope to capture radiation, which is then directed along different paths for analysis. Notice that the secondary mirrors shown in (c) and (d) are actually slightly diverging, so that they move the focus outside the telescope.

the telescope itself, enabling astronomers to use very heavy and finely tuned equipment that could not possibly be lifted to either the prime focus or the Cassegrain focus. The light path to the coudé room lies along the axis of the telescope's mount—that is, the axis around which the telescope rotates as it tracks objects across the sky—so that the light path does not change as the telescope moves.

To illustrate some of these points, Figure 5.8 depicts the Hale 5-m (200-inch) diameter optical telescope on California's Mount Palomar. As the size of the person drawn in the observer's cage at the prime focus indicates, this is indeed a very large telescope. In fact, for almost three decades after its dedication in 1948, the Hale telescope was the largest in the world. It was at or near the forefront of astronomical research for much of the latter half of the twentieth century. Observations were made at the prime, the Cassegrain, or the coudé focus, depending on the needs of the user.

IMAGING AND NONIMAGING DETECTORS

Large reflectors are good at forming images of narrow fields of view, where all the light that strikes the mirror surface moves almost parallel to the axis of the instrument. However, if the light enters at an appreciable angle, it cannot be accurately focused, degrading the overall quality of the image. The effect (called *coma*) worsens as we move farther from the center of the field of view. Eventually, the image quality is reduced to the point where it is no longer

usable. The distance from the center to where the image becomes unacceptable defines the useful field of view of the telescope—typically, only a few arc minutes for large instruments.

A design that overcomes this problem is the *Schmidt telescope*, named after its inventor, Bernhard Schmidt, who built the first such instrument in the 1930s. The telescope uses a correcting lens, which sharpens the final image of the entire field of view. Consequently, a Schmidt telescope is well suited to producing wide-angle photographs, covering several degrees of the sky. The design of the Schmidt telescope results in a curved image that is not suitable for viewing with an eyepiece, so the image is recorded on a specially shaped piece of photographic film. For this reason, the instrument is often called a *Schmidt camera* (Figure 5.9).

When a photographic plate is placed at the focus to record an image of the field of view, the telescope is acting in effect as a high-powered camera. However, this is by no means the only light-sensitive device that can be placed at the focus to analyze the radiation received from space. When very accurate and rapid measurements of light intensity are required, a device known as a **photometer** is used. A photometer measures the total amount of light received in all or part of the image. When only part of the image is under study, the region of interest is selected simply by masking out the rest of the field of view. Using a photometer often means "throwing away" spatial detail, but in return more infor-

Figure 5.8 Palomar Telescope (a) Artist's illustration of the 200-inch-diameter Hale optical telescope on Mount Palomar in California. (b) A photograph of the telescope. (c) Astronomer Edwin Hubble in the observer's cage at the Hale prime focus. *(California Institute of Technology)*

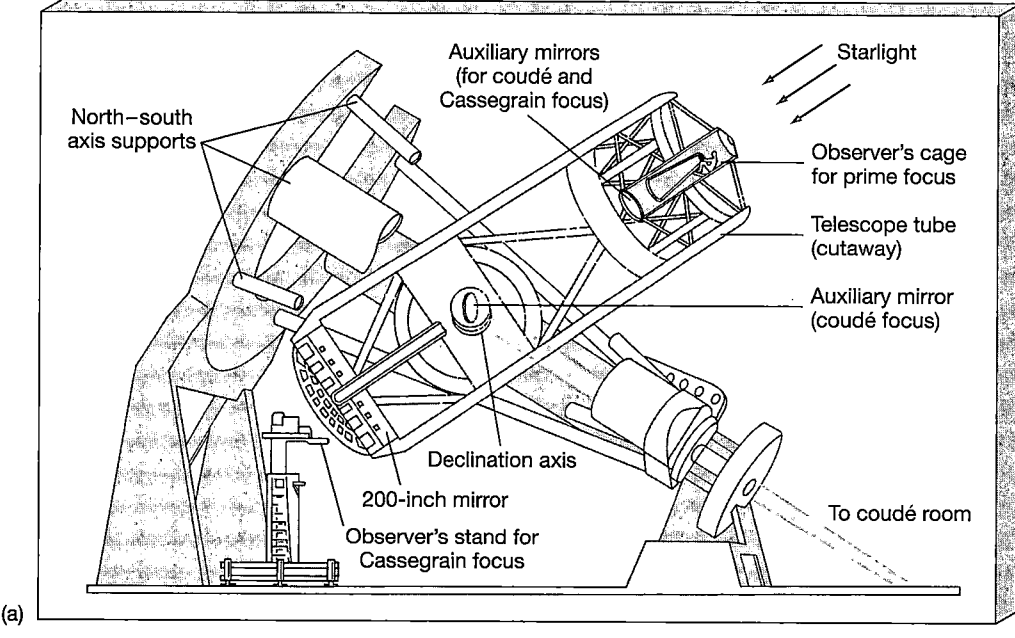

Auxiliary mirrors (for coudé and Cassegrain focus)

Starlight

North–south axis supports

Observer's cage for prime focus

Telescope tube (cutaway)

Auxiliary mirror (coudé focus)

Declination axis

200-inch mirror

To coudé room

Observer's stand for Cassegrain focus

(a)

(b)

(c)

mation is obtained about the intensity and time variability of a particular source, such as a pulsating star or a supernova explosion.

Often, astronomers want to study the *spectrum* of the incoming light. Large **spectrometers** work in tandem with optical telescopes. Light collected by the primary mirror may be redirected to the coudé room, defined by a narrow slit, dispersed (split into its component colors) using a prism or a diffraction grating, then sent on to a detector—a process not so different in concept from the operation of the simple spectroscope described in Chapter 4. ∞ (Sec. 4.1) The spectrum can be studied in real time (that is, as it happens) or stored on a photographic plate (or, more commonly nowadays, on a computer disk) for later analysis.

✅ Concept Check

■ Why do all modern telescopes use mirrors to gather and focus light?

Figure 5.9 Schmidt Telescope The Schmidt camera of the European Southern Observatory in Chile. *(ESO)*

5.2 Telescope Size

2 Astronomers generally prefer large telescopes over small ones, for two main reasons. The first has to do with the amount of light a telescope can gather—its *light-gathering power*. The second is related to the amount of detail that can be seen—the telescope's *resolving power*.

LIGHT-GATHERING POWER

One important reason for using a larger telescope is simply that it has a greater **collecting area**, which is the total area of a telescope capable of gathering radiation. The larger the telescope's reflecting mirror (or refracting lens), the more light it collects, and the easier it is to measure and study an object's radiative properties. Astronomers spend much of their time observing very distant—and hence very *faint*—cosmic sources. In order to make detailed observations of such objects, very large telescopes are essential. Figure 5.10 illustrates the effect of increasing telescope size by comparing images of the Andromeda Galaxy taken with two different instruments. A large collecting area is particularly important for spectroscopic work, as the received radiation in that case must be split into its component wavelengths for further analysis.

The observed brightness of an astronomical object is directly proportional to the area of our telescope's mirror and therefore to the *square* of the mirror diameter. Thus, a 5-m telescope will produce an image 25 times as bright as a 1-m instrument because a 5-m mirror has $5^2 = 25$ times the collecting area of a 1-m mirror. We can also think of this relationship in terms of the length of *time* required for a telescope to collect enough energy to create a recognizable image on a photographic plate. Our 5-m telescope will produce an image 25 times faster than the 1-m device because it gathers energy at a rate 25 times greater. Put another

way, a 1-hour exposure with a 1-m telescope is roughly equivalent to a 2.4-minute exposure with a 5-m instrument.

Until the 1980s the conventional wisdom was that telescopes with mirrors larger than five or six meters in diameter were simply too expensive and impractical to build. The problems involved in casting, cooling, and polishing a huge block of quartz or glass to very fine tolerances (typically less than the width of a human hair) were just too great. However, new high-tech manufacturing techniques, coupled with radically new mirror designs, make the construction of telescopes in the 8- to 12-m range almost a routine matter. Experts can now make large mirrors much lighter for their size than had previously been believed feasible and can combine many smaller mirrors into the equivalent of a much larger single-mirror telescope. Several large-diameter instruments now exist, and many more are planned.

Currently, the largest operating optical telescopes are the twin Keck instruments atop Mauna Kea in Hawaii (see Figure 5.11), administered jointly by the California Institute of Technology and the University of California. Each telescope combines 36 hexagonal 1.8-m mirrors into the equivalent collecting area of a single 10-m reflector. The first Keck telescope became fully operational in 1992; the second was completed in 1996. The high altitude and large size of these devices makes them particularly well suited for detailed spectroscopic studies of very faint objects, in both the optical and infrared parts of the spectrum. (Mauna Kea's 4-km altitude minimizes atmospheric absorption of infrared radiation, making this site one of the finest locations on Earth for infrared astronomy.)

Numerous other large telescopes can be seen in Figure 5.11. Some are designed exclusively for infrared work; others, like Keck, operate in both the optical and the infrared. To the right of the Keck domes is the 8.3-m Subaru (the Japanese name for the Pleiades) telescope, operated by the National Astronomical Observatory of Japan. It saw "first light" in 1999. In the distance is the 8.1-m Gemini

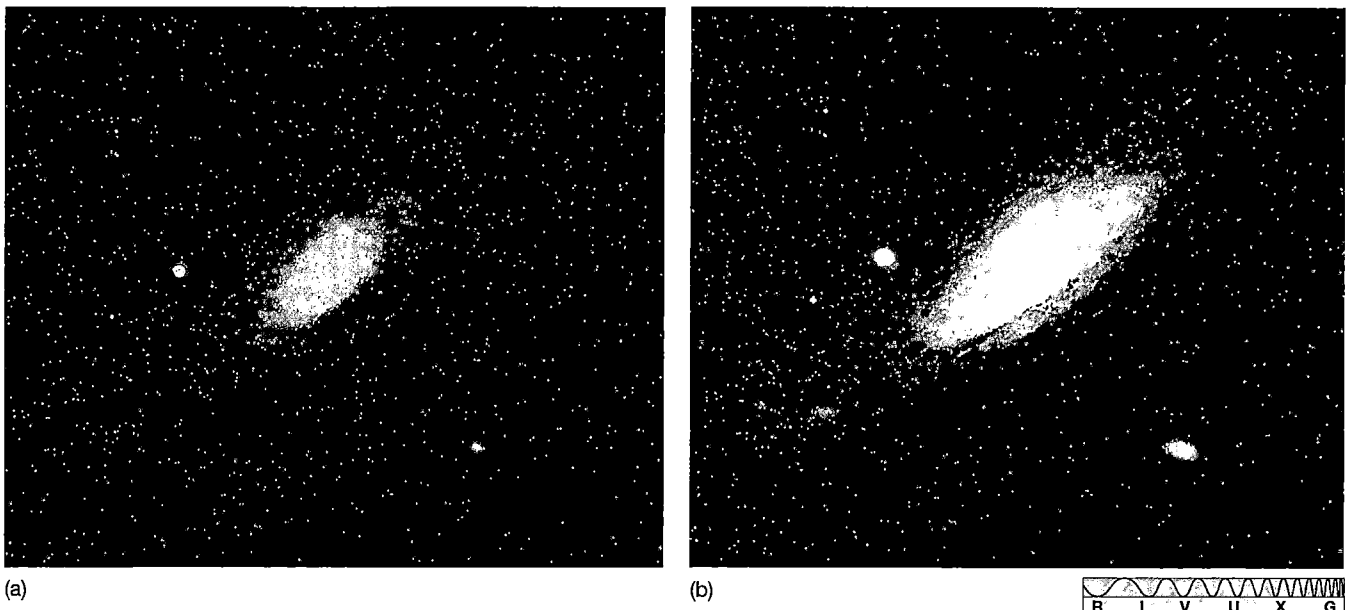

(a)

(b)

| R | I | V | U | X | G |

Figure 5.10 Sensitivity Effect of increasing telescope size on an image of the Andromeda Galaxy. Both photographs had the same exposure time; Image (b) was taken with a telescope twice the size of that used to make (a). Fainter detail can be seen as the diameter of the telescope mirror increases because larger telescopes are able to collect more photons per unit time. *(AURA)*

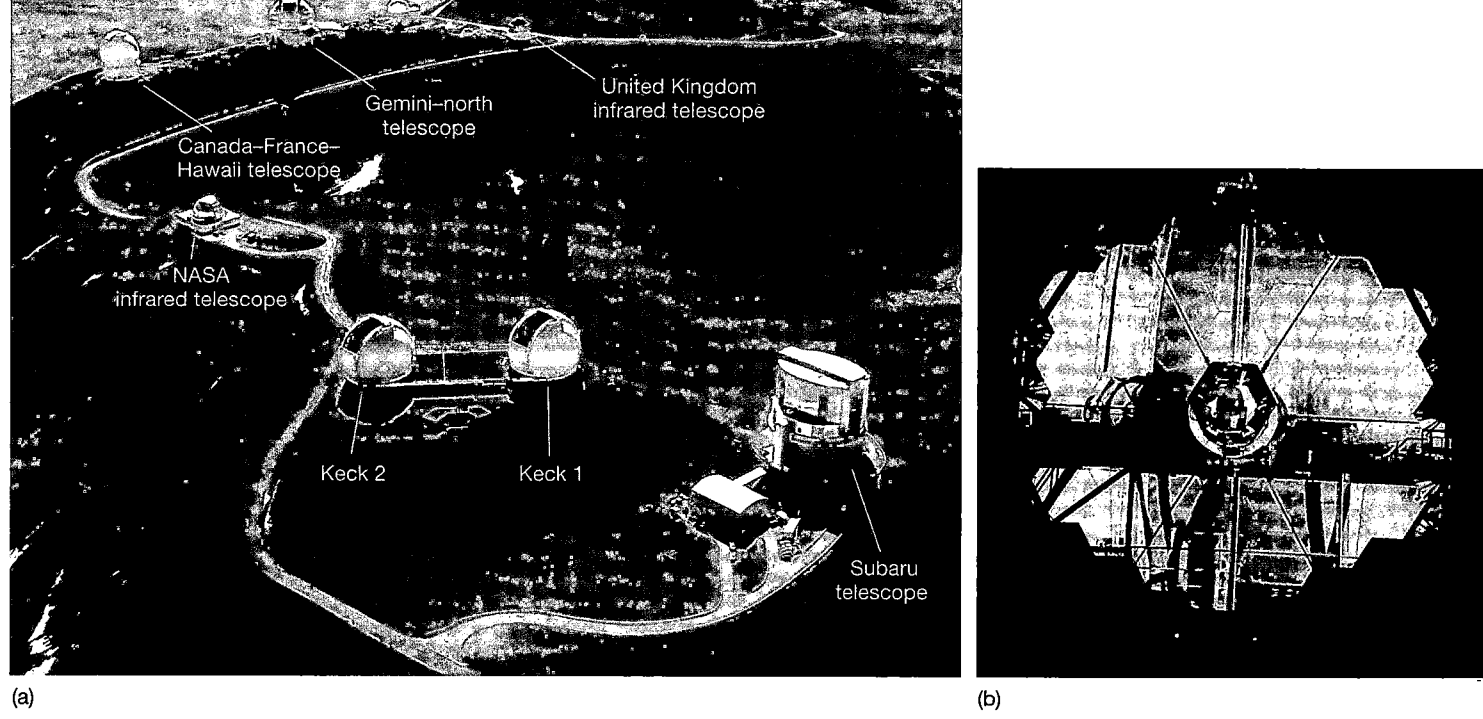

(a)

(b)

Figure 5.11 Mauna Kea Observatory (a) The world's highest ground-based observatory, at Mauna Kea, Hawaii, is perched atop an extinct volcano more than 4 km above sea level. Among the domes visible in the picture are those that house the Canada-France-Hawaii 3.6-m telescope, the 8.1-m Gemini North instrument, the 2.2-m telescope of the University of Hawaii, Britain's 3.8-m infrared facility, and the twin 10-m Keck telescopes. To the right of the twin Kecks is the Japanese 8.3-m Subaru telescope. Several of the largest telescopes are marked. The thin air at this 4-km-altitude site guarantees less atmospheric absorption of incoming radiation and hence a clearer view than at sea level, but the air is so thin that astronomers must occasionally wear oxygen masks while working. (b) The 10-m mirror in the first Keck telescope. Note the technician in orange coveralls at center. *(R. Wainscoat; R. Underwood/W. M. Keck Observatory)*

North instrument, also completed in 1999 by a consortium of seven nations including the U.S. Its twin, Gemini South, in the Chilean Andes, should become operational in 2001. In terms of total available collecting area, the largest telescope currently available is the European Southern Observatory's optical-infrared Very Large Telescope (VLT), located at Cerro Paranal, in Chile (Figure 5.12). It consists of four separate 8.2-m mirrors that can function as a single instrument. The third mirror was completed in 2000, the fourth in 2001.

RESOLVING POWER

A second advantage of large telescopes is their finer **angular resolution**. In general, *resolution* refers to the ability of any device, such as a camera or telescope, to form distinct, separate images of objects lying close together in the field of view. The finer the resolution, the better we can distinguish the objects and the more detail we can see. In astronomy, where we are always concerned with angular measurement, "close together" means "separated by a small angle on the sky," so angular resolution is the factor that determines our ability to see fine structure. Figure 5.13 illustrates how the appearance of two objects—stars, say—might change as the angular resolution of our telescope varies. Figure 5.14 illustrates the result of increasing resolving power with views of the Andromeda Galaxy at several different resolutions.

What limits a telescope's resolution? One important factor is *diffraction*, the tendency of light, and all other waves for that matter, to bend around corners. ⊙ (*Discov-*

ery 3-1) Because of diffraction, when a parallel beam of light enters a telescope, the rays spread out slightly, making it impossible to focus the beam to a sharp point, even with a perfectly constructed mirror. Diffraction introduces a certain "fuzziness," or loss of resolution, into the optical system. The degree of fuzziness—the minimum angular separation that can be distinguished—determines the angular resolution of the telescope. The amount of diffraction is proportional to the wavelength of the radiation divided by the diameter of the telescope mirror. As a result we can write, in convenient units,

$$\text{angular resolution (arc sec)} = 0.25 \, \frac{\text{wavelength } (\mu m)}{\text{mirror diameter (m)}},$$

where $1 \, \mu m$ (1 micron) $= 10^{-6}$ m (see Appendix 2). Thus, for a given telescope size, the amount of diffraction increases in proportion to the wavelength used. Observations in the infrared or radio range are often limited by its effects. For light of any given wavelength, large telescopes produce less diffraction than small ones.

For a given telescopic size, the amount of diffraction increases in proportion to the wavelength used. Observations in the infrared or radio range are often limited by its effects. For example, according to the formula above, in an otherwise perfect observing environment, the best possible angular resolution of blue light (with a wavelength of 400 nm) that can be obtained using a 1-m telescope is about $0.25'' \times (0.4/1) = 0.1''$. This quantity is known as the *diffraction-limited* resolution of the telescope. But if we were to use our 1-m telescope to make observations in the

Figure 5.12 VLT Observatory Located at the Paranal Observatory in Atacama, Chile, the European Southern Observatory's Very Large Telescope (VLT) is the world's largest optical telescope. It comprises four 8.2-m reflecting telescopes, which can be used in tandem to create the effective area of a single 16-m mirror. *(ESO)*

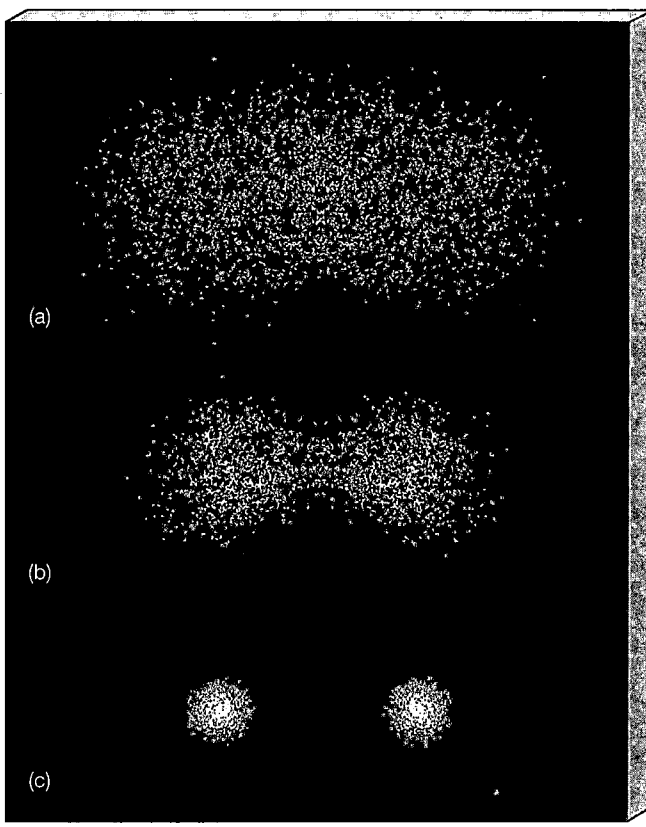

Figure 5.13 Resolving Power Two comparably bright light sources become progressively clearer when viewed at finer and finer angular resolution. When the angular resolution is much poorer than the separation of the objects, as at the top, the objects appear as a single fuzzy "blob." As the resolution improves, the two sources become discernible as separate objects.

near infrared, at a wavelength of 10 μm (10,000 nm), the best resolution we could obtain would be only 2.5″. A 1-m radio telescope operating at a wavelength of 1 cm would have an angular resolution of just under 1°.

For light of any given wavelength, large telescopes produce less diffraction than small ones. A 5-m telescope observing in blue light would have a diffraction-limited resolution five times finer than the 1-m telescope just discussed—about 0.02″. A 0.1-m (10-cm) telescope would have a diffraction limit of 1″ and so on. For comparison, the angular resolution of the human eye in the middle of the visual range is about 0.5′.

☑ Concept Check

■ Give two reasons why astronomers need to build very large telescopes.

Figure 5.14 Resolution Detail becomes clearer in the Andromeda Galaxy as the angular resolution is improved some 600 times, from (a) 10′, to (b) 1′, (c) 5″, and (d) 1″. *(AURA)*

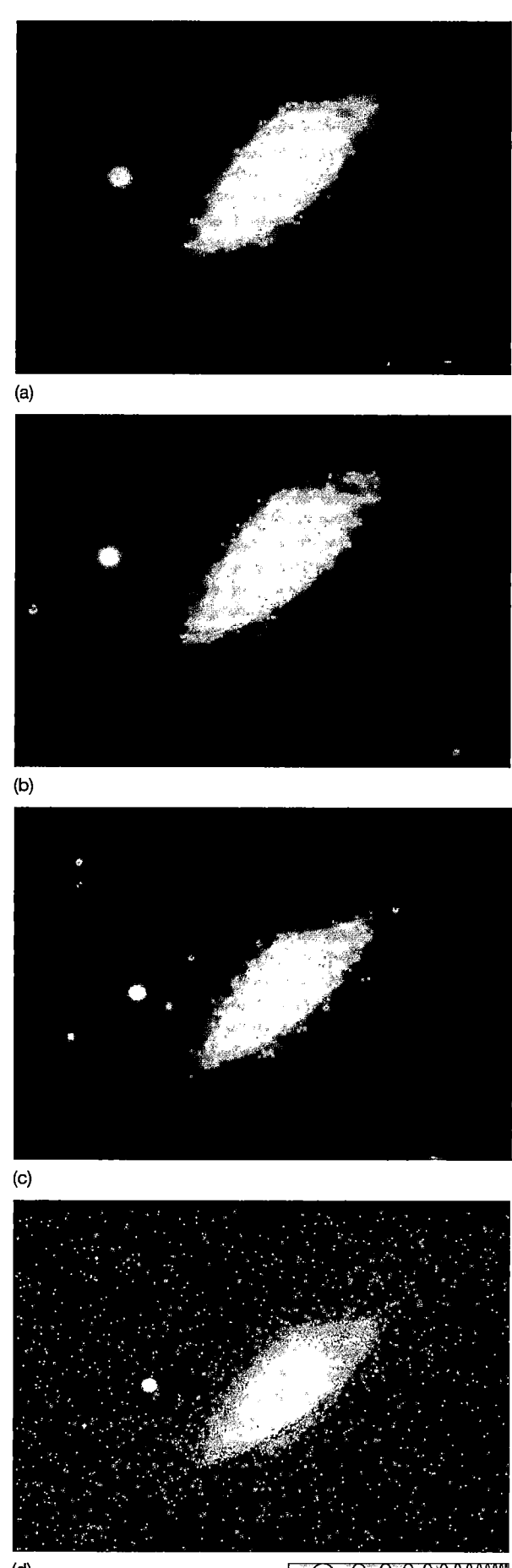

(a)

(b)

(c)

(d)

R   I   V   U   X   G

5.3 High-Resolution Astronomy

Even large telescopes have limitations. For example, according to the discussion in the preceding section, the 5-m Hale telescope should have an angular resolution in blue light of around 0.02″. In practice, however, it cannot do better than about 1″. In fact, apart from instruments using special techniques developed to examine some particularly bright stars, no ground-based optical telescope built before 1990 can resolve astronomical objects to much better than 1″. The reason is Earth's turbulent atmosphere, which blurs the image even before the light reaches our instruments. In recent years, great strides have been made in overcoming this obstacle. Telescopes have been placed above the atmosphere, and computers are playing an increasingly important role in both telescope operation and image processing.

ATMOSPHERIC BLURRING

As we observe a star, atmospheric turbulence produces continual small changes in the optical properties of the air between the star and our telescope (or eye). The light from the star is refracted slightly, and the stellar image dances around on the detector (or on our retina). This continual deflection is the cause of the well-known "twinkling" of stars. It occurs for the same reason that objects appear to shimmer when viewed across a hot roadway on a summer day.

On a good night at the best observing sites, the maximum amount of deflection produced by the atmosphere is slightly less than 1″. Consider taking a photograph of a star. After a few minutes of exposure time (long enough for the intervening atmosphere to have undergone many small random changes), the image of the star has been smeared out over a roughly circular region an arc second or so in diameter. Astronomers use the term **seeing** to describe the effects of atmospheric turbulence. The circle over which a star's light (or the light from any other astronomical source) is spread is called the **seeing disk**. Figure 5.15 illustrates the formation of the seeing disk for a small telescope.*

Atmospheric turbulence has less effect on light of longer wavelengths—ground-based astronomers generally

*In fact, for a large instrument—more than about 1 m in diameter—the situation is more complicated, because rays striking different parts of the mirror have actually passed through different turbulent atmospheric regions. The end result is still a seeing disk, however.

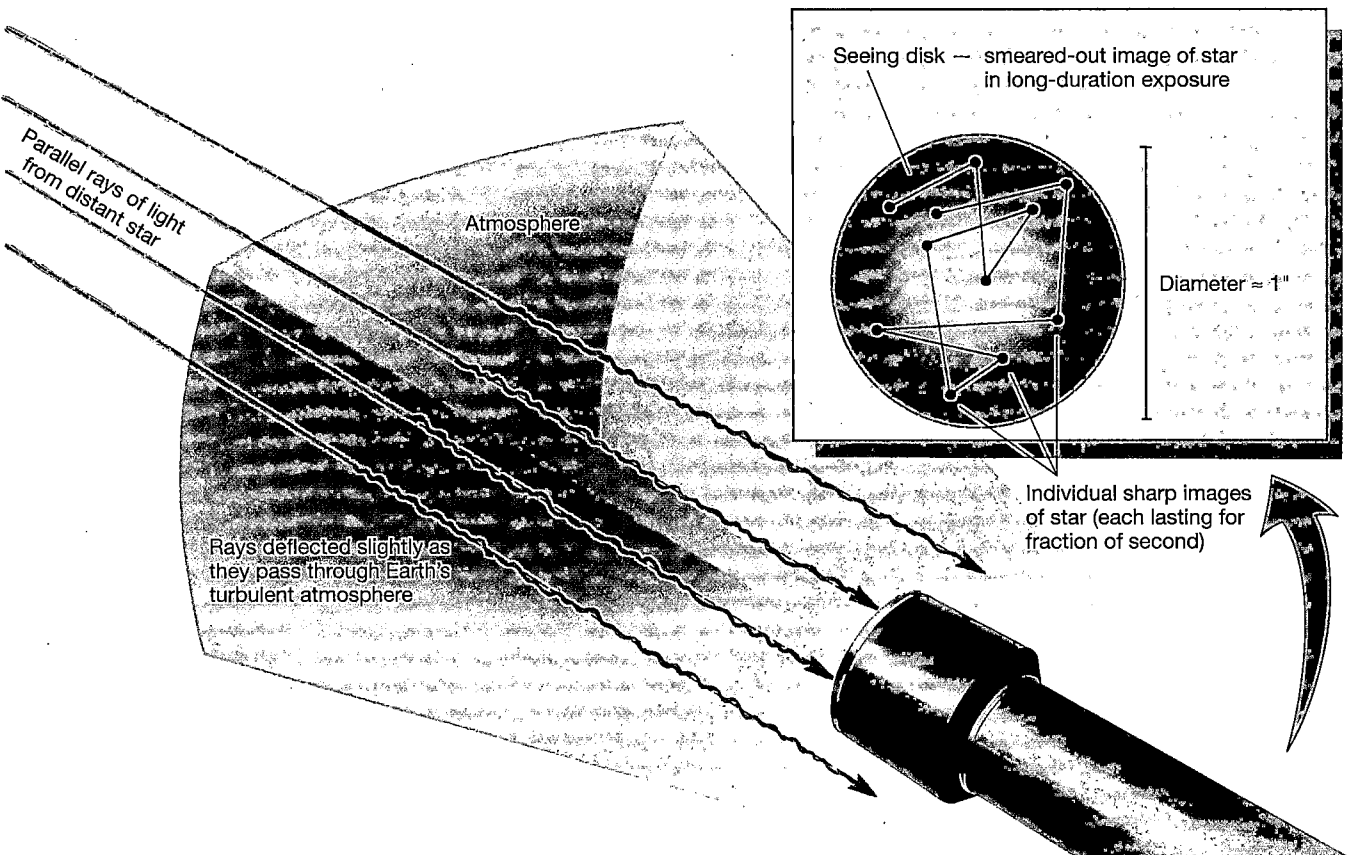

Figure 5.15 Atmospheric Turbulence Photons from a distant star strike the detector in a telescope at slightly different locations because of turbulence in Earth's atmosphere. Over time, the individual photons cover a roughly circular region on the detector, and even the pointlike image of a star is recorded as a small disk, called the seeing disk.

(a)

Figure 5.16 European Southern Observatory (a) Located in the Andes Mountains of Chile, the European Southern Observatory at La Silla is run by a consortium of European nations. Numerous domes house optical telescopes of different sizes, each with varied support equipment, making this one of the most versatile observatories south of the equator. (b) The largest telescope at La Silla—the New Technology Telescope, a 3.5-m state-of-the-art active optics device. *(ESO)*

"see" better in the infrared. However, counteracting this improvement in image quality is the fact that the atmosphere is wholly or partially opaque over much of the infrared range. ∞ (Sec. 3.3) For these reasons, to achieve the best possible observing conditions, telescopes are sited on mountaintops (to get above as much of the atmosphere as possible) in regions of the world where the atmosphere is known to be fairly stable and relatively free of dust, moisture, and light pollution from cities.

In the continental United States, these sites tend to be in the desert Southwest. The U.S. National Observatory for optical astronomy in the Northern Hemisphere, completed in 1973, is located high on Kitt Peak near Tucson, Arizona. The site was chosen because of its many dry, clear nights. Seeing of less than 1″ from such a location is regarded as good, and seeing of a few arc seconds is tolerable for many purposes. Even better conditions are found on Mauna Kea, Hawaii (Figure 5.11), and at numerous sites in the Andes Mountains of Chile (Figures 5.12 and 5.16), which is why many large telescopes have recently been constructed at these exceptionally clear locations.

An optical telescope placed in orbit about Earth or on the Moon could obviously overcome the limitations imposed by the atmosphere on ground-based instruments. Without atmospheric blurring, extremely fine resolution—close to the diffraction limit—can be achieved, subject only to the engineering restrictions of building or placing large structures in space. The *Hubble Space Telescope* (*HST*; named for one of America's most notable astronomers, Edwin Hubble) was launched into Earth orbit by NASA's space shuttle *Discovery* in 1990 (see *Discovery 5-1*). This telescope has a 2.4-m mirror, with a (blue-light) diffraction

(b)

limit of only 0.05″, giving astronomers a view of the universe as much as 20 times sharper than that normally available from even much larger ground-based instruments.

IMAGE PROCESSING

Computers play a vital role in observational astronomy. Most large telescopes today are controlled either by computers or by operators who rely heavily on computer assistance, and images and data are recorded in a form that can be easily read and manipulated by computer programs.

It is becoming rare for photographic equipment to be used as the primary means of data acquisition at large observatories. Instead, electronic detectors known as **charge-coupled devices**, or **CCDs**, are in widespread use. Their output goes directly to a computer. A CCD (Figure 5.17) consists of a wafer of silicon divided into a two-dimensional array of many tiny picture elements, known as **pixels**. When light strikes a pixel, an electric charge builds up on the device. The amount of charge is directly proportional to the number of photons striking each pixel—in other words, to the intensity of the light at that point. The charge buildup is monitored electronically, and a two-dimensional image is obtained. A CCD is typically a few square centimeters in area and may contain several million pixels, generally arranged on a square grid. As the technology improves, both the areas of CCDs and the number of pixels they contain continue to increase. Incidentally, the technology is not limited to astronomy—many home video cameras contain CCD chips similar in basic design to those in use at the great astronomical observatories of the world.

CCDs have two important advantages over photographic plates, which were the staple of astronomers for over a century. First, CCDs are much more *efficient* than photographic plates, recording as many as 90 percent of the photons striking them, compared with less than five percent for photographic methods. This means that a CCD image can show objects 10 to 20 times fainter than a photograph made using the same telescope and the same exposure time. Alternatively, a CCD can record the same level of detail in less than a tenth of the time required by photographic techniques, or record that detail with a much smaller telescope. Second, CCDs produce a faithful representation of an image in a digital format that can be placed directly on magnetic tape or disk, or even sent across a computer network to an observer's home institution.

Computers are also widely used to reduce *background noise* in astronomical images. Noise is anything that corrupts the integrity of a message, such as static on an AM radio or "snow" on a television screen. The noise corrupting telescopic images has many causes. In part, it results from faint, unresolved sources in the telescope's field of view and from light scattered into the line of sight by Earth's atmosphere. It can also be caused by electronic "hiss" within the detector. Whatever the origin of noise, its characteristics can be determined (for example, by observing a part of the sky where there are no known sources of radiation) and its effects partially removed with the aid of high-speed computers, allowing astronomers to see features that would otherwise remain hidden.

Using computer processing, astronomers can also compensate for known instrumental defects and even correct some effects of bad seeing. In addition, the computer can often carry out many of the relatively simple but tedious and time-consuming chores that must be performed before an image (or spectrum) reaches its final "clean" form. Figure 5.18 illustrates how computerized image-processing techniques were used to correct for known instrumental problems in the *Hubble Space Telescope*, allowing much of the planned resolution of the telescope to be recovered even before its repair in 1993.

NEW TELESCOPE DESIGN

The latest techniques for producing ultrasharp images take these ideas of computer control and image processing several stages further. By analyzing the image formed by a telescope *while the light is still being collected*, it is now possible to adjust the telescope from moment to moment to avoid or compensate for the effects of mirror distortion, temperature changes in the dome, and even the effects of atmospheric turbulence. By these means, some recently constructed telescopes have achieved resolutions very close to their theoretical (diffraction) limits.

Even under conditions of perfect seeing, most telescopes would not achieve diffraction-limited resolution. The temperature of the mirror or in the dome may fluctuate slightly during the many minutes or even hours required for the image to be exposed, and the precise shape of the mirror may change slightly as the telescope tracks a source across the sky. The effect of these changes is that

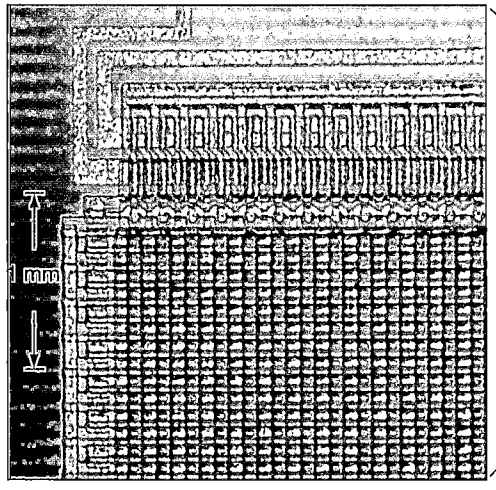

(a)

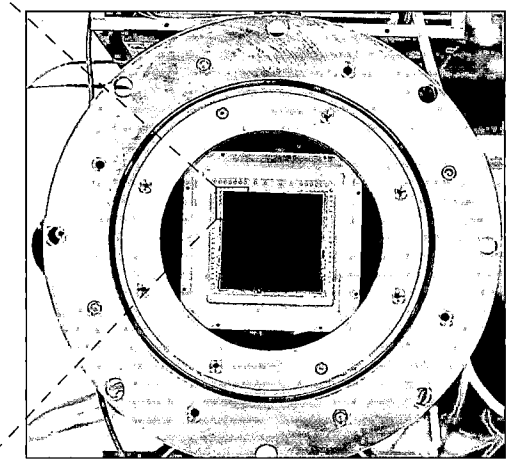

(b)

Figure 5.17 CCD Chip A charge-coupled device consists of hundreds of thousands, or even millions, of tiny light-sensitive cells, or pixels, usually arranged in a square array. Light striking a pixel causes an electrical charge to build up on it. By electronically reading out the charge on each pixel, a computer can reconstruct the pattern of light—the image—falling on the chip. (a) Detail of a CCD array. (b) A CCD chip mounted for use at the focus of a telescope. *(AURA; R. Wainscoat/Peter Arnold)*

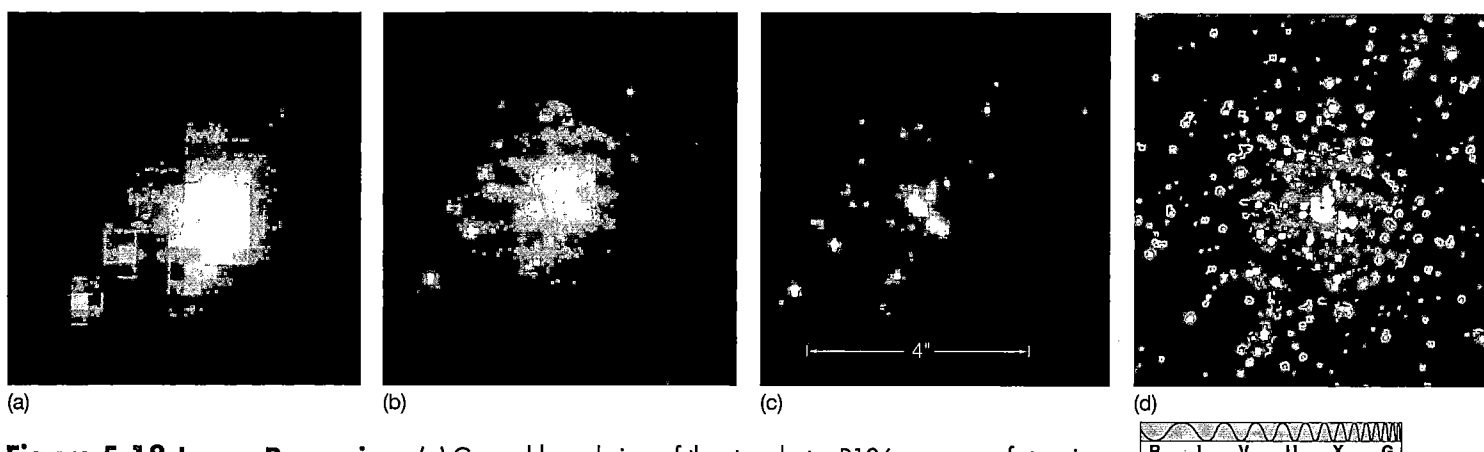

(a)   (b)   (c)   (d)

Figure 5.18 Image Processing (a) Ground-based view of the star cluster R136, a group of stars in the Large Magellanic Cloud (a nearby galaxy). (b) The "raw" image of this same region as seen by the *Hubble Space Telescope* in 1990, before the repair mission. (c) The same image after computer processing that partly compensated for imperfections in the mirror. (d) The same region as seen by the repaired *HST* in 1994, here observed at a somewhat bluer wavelength. *(AURA/NASA)*

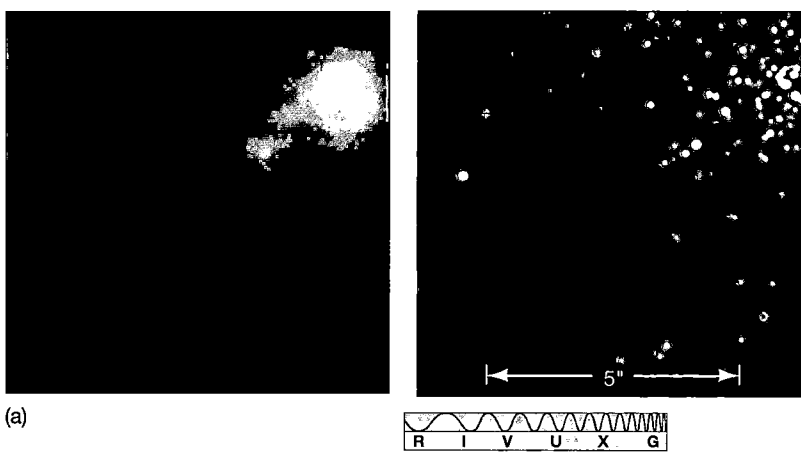

(a)

Figure 5.19 Active Optics (a) These false-color infrared photographs of part of the star cluster R136—the same object shown in Figure 5.18—contrast the resolution obtained without the active optics system (left image) with that achievable when the active optics system is in use (right image). These were both taken with the NTT telescope shown in Figure 5.16(b). (b) A hexagonal mirror segment destined for one of the Keck telescopes undergoes shaping and polishing. The unusually thin glass is backed by push-pull pistons that can adjust the precise configuration of the segment during observations to attain improved resolution. *(ESO/Caltech)*

(b)

the mirror's focus may shift from minute to minute, blurring the eventual image in much the same way as atmospheric turbulence creates a seeing disk (Figure 5.15). At the best observing sites, the seeing is often so good that these tiny effects may be the main cause of image blurring. The collection of techniques aimed at controlling these environmental and mechanical fluctuations is known as **active optics.**

The first telescope designed to incorporate active optics was the New Technology Telescope (NTT), constructed in 1989 at the European Southern Observatory in Chile (NTT is the most prominent instrument visible in Figure 5.16). This 3.5-m instrument, employing the latest in real-time telescope controls, achieved a resolution of about 0.5″ by making minute modifications to the tilt of its mirror as its temperature and orientation changed, thus maintaining the best possible focus at all times. Figure 5.19(a) shows how active optics can improve image resolution. Active optics techniques now include improved dome design to control airflow, precise control of the mirror temperature, and the use of actuators (pistons) behind the mirror to maintain its precise shape at all times (Figure 5.19b). All of the large telescopes described earlier include active optics systems, improving their resolution to a few tenths of an arc second. NTT itself was upgraded in 1997 and has reported resolution as sharp as 0.15″ under the best conditions.

With active optics systems in place, Earth's atmosphere once again becomes the main agent limiting a telescope's resolution. Remarkably, even this problem can

DISCOVERY 5-1

The *Hubble Space Telescope*

The *Hubble Space Telescope (HST)* is the largest, most complex, most sensitive observatory ever deployed in space. At over $4.5 billion (including the cost of three missions to service and refurbish the system), it is also the most expensive scientific instrument ever constructed. Built jointly by NASA and the European Space Agency, *HST* was designed to allow astronomers to probe the universe with at least 10 times finer resolution and with some 30 times greater sensitivity to light than existing Earth-based devices. *HST* is operated remotely from the ground; there are no astronauts aboard the telescope, which orbits Earth about once every 95 minutes at an altitude of about 600 km (380 miles).

The telescope's overall dimensions approximate those of a city bus or railroad tank car—13 m (43 feet) long, 12 m (39 feet) across with solar arrays extended, and 11,000 kg (12.5 tons when weighed on Earth). At the heart of *HST* is a 2.4-m-diameter mirror designed to capture optical, ultraviolet, and infrared radiation before it reaches Earth's murky atmosphere. The accompanying figure shows the telescope being lifted out of the cargo bay of the space shuttle *Discovery* in the spring of 1990.

The telescope reflects light from its large mirror back to a smaller, 0.3-m, secondary mirror, which sends the light through a small hole in the main mirror and into the aft bay of the spacecraft. There, any of five major scientific instruments wait to analyze the incoming radiation. Most of these instruments are about the size of a telephone booth. They are designed to be maintained by NASA astronauts, and indeed, most of the telescope's instruments have been upgraded or replaced since *HST* was launched.

Soon after launch, astronomers discovered that the telescope's primary mirror had been polished to the wrong shape. The mirror is too flat by 2 μm, about 1/50 the width of a human hair, making it impossible to focus light as well as expected. This optical flaw (known as *spherical aberration*) meant that *HST* was not as sensitive as designed, although it could still see many objects in the universe with unprecedented resolution. In 1993 astronauts aboard the space shuttle *Endeavour* visited *HST* and succeeded in repairing some of its ailing equipment (see the accompany-

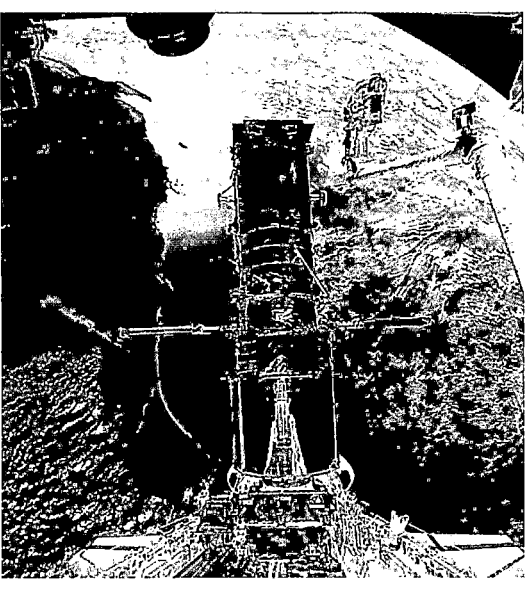

(NASA)

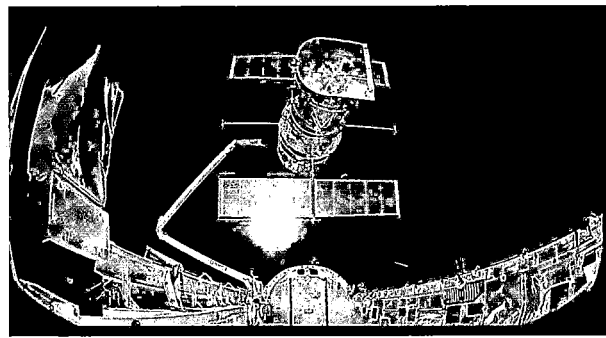

(NASA)

now be addressed, using an approach known as **adaptive optics**. This technique actually deforms the shape of the mirror's surface, under computer control, while the image is being exposed, in order to undo the effects of atmospheric turbulence. Adaptive optics presents formidable theoretical and technological problems, but the rewards are so great that they have been the subject of intense research since the 1980s. Declassified SDI ("Star Wars") technology in the 1990s provided an enormous boost to this effort. In the experimental military system shown in Figure 5.20(a), lasers probe the atmosphere above the telescope, returning information about the air's swirling motion to a computer that modifies the mirror thousands of times per second to compensate for poor seeing. Astronomical adaptive optics systems generally do not use lasers to gauge atmospheric conditions. Instead, they monitor standard stars in the field of view, constantly adjusting the mirror's shape to preserve those stars' appearance.

Figure 5.20(b) compares the results of a pair of observations of a nearby double star called Castor. Spectroscopic observations long ago revealed the double nature of this object, which appears to be a binary-star system. The image on the left shows the star system as seen through an ordinary, moderate-sized telescope—an oblong blur combining the light from the two stars, spread over several arc

ing figure). They replaced *Hubble*'s gyroscopes to help the telescope point more accurately, installed sturdier versions of the solar panels that power the telescope's electronics, and—most importantly—inserted an intricate set of small mirrors (each about the size of a coin) to compensate for the faulty primary mirror. *Hubble*'s resolution is now close to the original design specifications, and the telescope has regained much of its lost sensitivity. Additional service missions were performed in 1997 and 1999 to upgrade instruments and repair damaged systems.

A good example of *Hubble*'s scientific capabilities today can be seen by comparing the two images of the spiral galaxy M100, shown here. On the left is perhaps the best

ground-based photograph of this beautiful galaxy, showing rich detail and color in its spiral arms. On the right, to the same scale and orientation, is an *HST* image showing improvement in both resolution and sensitivity. (The chevron-shaped field of view is caused by the corrective optics inserted into the telescope; an additional trade-off is that *Hubble*'s field of view is smaller than those of ground-based telescopes.) The inset shows *Hubble*'s exquisite resolution of small fields of view. Many spectacular examples of its remarkable data appear throughout this book.

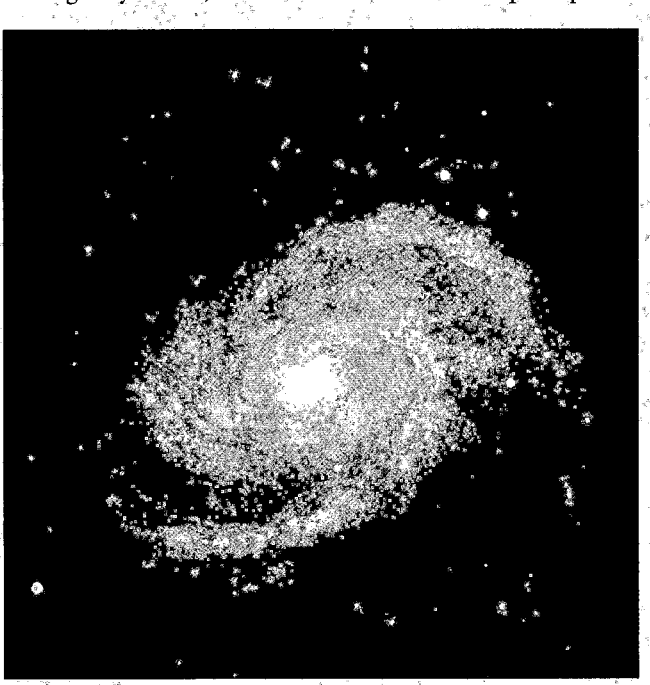

(D. Malin/Anglo-Australian Telescope)

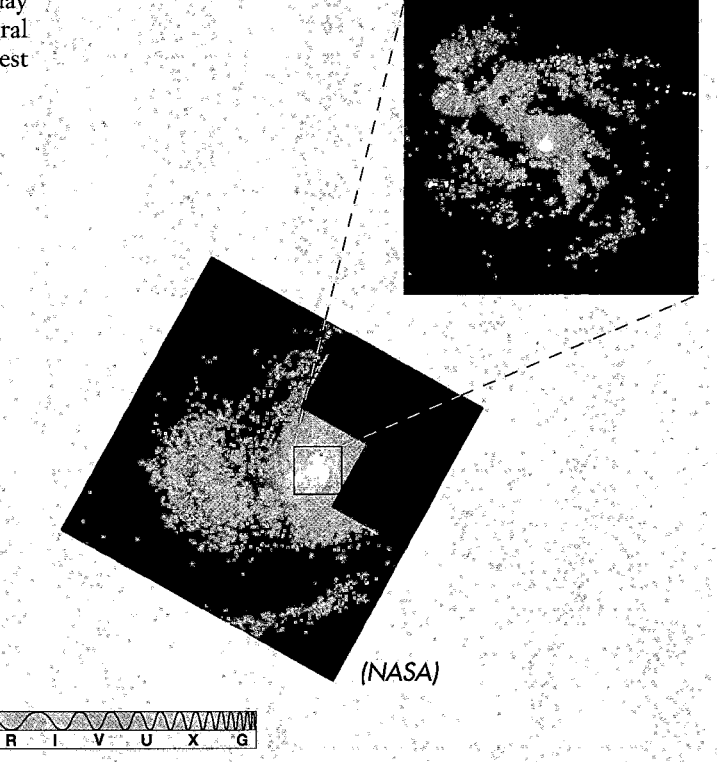

(NASA)

seconds. On the right, with the adaptive optics system turned on, one star is clearly distinguishable from the other. The two stars are separated by less than an arc second; the resolution, with adaptive optics turned on, is about 0.1".

Adaptive corrections are somewhat easier to apply in the infrared part of the spectrum, and infrared adaptive optics systems already exist in many large telescopes. Gemini and Subaru have reported adaptive optics resolutions of 0.09" and 0.07", respectively, in the near infrared—not yet at the diffraction limit, but already better than the resolution of *HST* at the same wavelengths. Both Keck and the VLT incorporate adaptive-optics instrumentation that

will ultimately be capable of producing diffraction-limited images at near-infrared wavelengths. Visible-light adaptive optics has been demonstrated experimentally, and some telescopes may incorporate the technology within the next few years. Remarkably, it may soon be possible to have the "best of both worlds," achieving with large ground-based optical telescopes the kind of resolution presently attainable only from space.

☑ Concept Check

▣ What steps do optical astronomers take to overcome the obscuring and blurring effects of Earth's atmosphere?

(a)

5.4 Radio Astronomy

In addition to the visible radiation that penetrates Earth's atmosphere on a clear day, radio radiation also reaches the ground. Indeed, as shown in Figure 3.9, the radio window in the electromagnetic spectrum is much wider than the optical window. ∞ (Sec. 3.3) Because the atmosphere is no hindrance to long-wavelength radiation, radio astronomers have built many ground-based **radio telescopes** capable of detecting cosmic radio waves. These devices have all been constructed since the 1950s—radio astronomy is a much younger subject than optical astronomy.

The field originated with the work of Karl Jansky at Bell Labs in 1931, but only after the technological push of World War II did it grow into a distinct branch of astronomy. Jansky was engaged in a study of shortwave-radio interference when he discovered a faint static "hiss" that had no apparent terrestrial source. He noticed that the strength of the hiss varied in time and that its peak occurred about four minutes earlier each day. He soon realized that the peaks were coming exactly one *sidereal day* apart and correctly inferred that the hiss was not of terrestrial origin but came from a definite direction in space. That direction is now known to correspond to the center of our Galaxy. It took over a decade, and the realization by astronomers that interstellar gas could actually be observed at radio wavelengths, for the full importance of his work to be appreciated, but today Jansky is regarded as the father of radio astronomy.

ESSENTIALS OF RADIO TELESCOPES

Figure 5.21 shows the world's largest steerable radio telescope, the large 105-m- (340-foot-) diameter telescope lo-

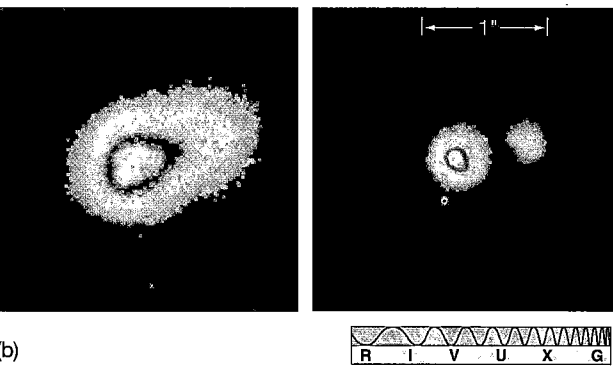

(b)

R I V U X G

Figure 5.20 Adaptive Optics (a) Until the early 1990s, the Starfire Optical Range at Kirtland Air Force Base in Albuquerque, New Mexico, was one of the U.S. Air Force's most closely guarded secrets. Here, laser beams probe the atmosphere above the system's 1.5-m telescope, allowing minute computer-controlled changes to be made to the mirror surface thousands of times each second. (b) The improvement in seeing produced by such systems can be dramatic, as can be seen in these images acquired at another military observatory atop Mount Haleakala in Maui, Hawaii, employing similar technology. The uncorrected image (left) of the double star Castor is a blur spread over several arc seconds, with little hint of its binary nature. With adaptive compensation applied (right), the resolution is improved to a mere 0.1" and the two components are clearly resolved. *(R. Ressmeyer; MIT Lincoln Laboratory)*

cated at the National Radio Astronomy Observatory in West Virginia. Although much larger than reflecting optical telescopes, most radio telescopes are built in basically the same way. They have a large, horseshoe-shaped mount supporting a huge curved metal dish that serves as the collecting area. The dish captures cosmic radio waves and reflects them to the focus, where a receiver detects the signals and channels them to a computer. Conceptually, the operation of a radio telescope is similar to the operation of an optical reflector with the detecting instruments placed at the prime focus (Figure 5.7a). However, unlike optical instruments, which can detect all visible wavelengths simultaneously, radio detectors normally register only a narrow band of wavelengths at any one time. To observe radiation at another radio frequency, we must retune the equipment, much as we tune a television set to a different channel.

Radio telescopes must be built large partly because cosmic radio sources are extremely faint. In fact, the total amount of radio energy received by Earth's entire surface is less than a trillionth of a watt. Compare this with the roughly 10 *million* watts our planet's surface receives in the form of infrared and visible light from any of the bright stars visible in the night sky. To capture enough radio energy to allow detailed measurements to be made, a large collecting area is essential.

Because of diffraction, the angular resolution of radio telescopes is generally quite poor compared with that of their optical counterparts. Typical wavelengths of radio waves are about a million times longer than those of visible light, and these longer wavelengths impose a corresponding crudeness in angular resolution. (Recall from Section 5.2 that the longer the wavelength, the greater the amount of diffraction.) Even the enormous sizes of radio dishes only partly offset this effect. The large radio telescope shown in Figure 5.21 can achieve resolution of about 1' when receiving radio waves having wavelengths of around 3 cm. However, it was designed to operate most efficiently (that is, it is most sensitive to radio signals) at wavelengths closer to 1 cm, where the resolution is approximately 20". The best angular resolution obtainable with a single radio telescope is about 10" (for the largest instruments operating at millimeter wavelengths)—at least 100 times coarser than the capabilities of some large optical systems.

Radio telescopes can be built so much larger than their optical counterparts because their reflecting surface need not be as smooth as is necessary for shorter-wavelength light waves. Provided that surface irregularities (dents, bumps, and the like) are much smaller than the wavelength of the waves to be detected, the surface will reflect them without distortion. Because the wavelength of visible radiation is short (less than 10^{-6} m), very smooth mirrors are needed to reflect the waves properly, and it is difficult to construct very large mirrors to such exacting tolerances. However, even rough metal surfaces can accurately focus 1-cm waves, and radio waves of wavelength a meter or more can be reflected and focused

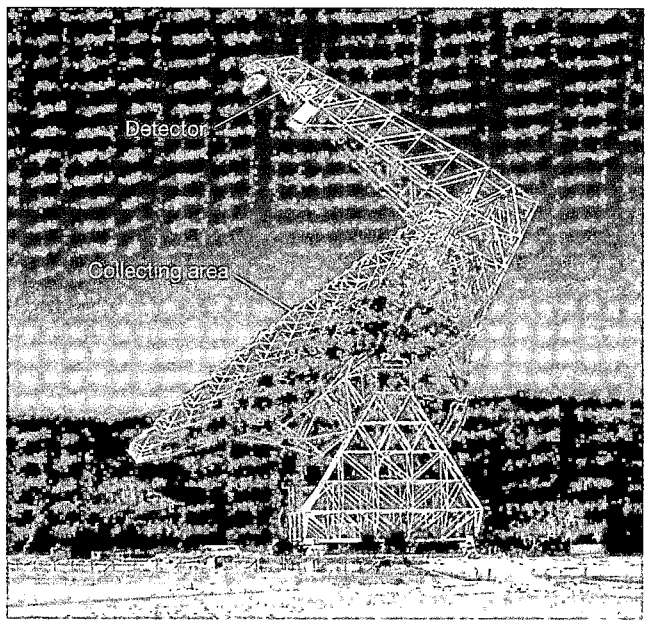

Figure 5.21 Radio Telescope This is the world's largest fully steerable radio telescope, the 105-m-diameter device at the National Radio Astronomy Observatory in Green Bank, West Virginia. It is 150 meters tall—taller than the Statue of Liberty and nearly as tall as the Washington Monument. *(NRAO)*

perfectly well by surfaces having irregularities even as large as your fist.

Figure 5.22 shows the world's largest and most sensitive radio telescope, located in Arecibo, Puerto Rico. Approximately 300 m (1000 feet) in diameter, the surface of the Arecibo telescope spans nearly 20 acres. Constructed in 1963 in a natural depression in the hillside, the dish was originally surfaced with chicken wire, which was lightweight and cheap. Although fairly rough, the chicken wire was adequate for proper reflection because the openings between adjacent strands of wire were much smaller than the long-wavelength radio waves to be detected.

The entire Arecibo dish was resurfaced in 1974 with thin metal panels, and upgraded in 1997 so that it can now be used to study shorter-wavelength radio radiation. Since the 1997 upgrade, the panels can be adjusted to maintain a precise spherical shape to an accuracy of about 3 mm over the entire surface. At a frequency of 5 GHz (corresponding to a wavelength of 6 cm—the shortest wavelength that can be studied given the properties of the dish surface), the telescope's angular resolution is about 1'. The huge size of the dish creates one distinct disadvantage, however. The Arecibo telescope cannot be pointed very well to follow cosmic objects across the sky. The detectors can move roughly 10° on either side of the focus, restricting the telescope's observations to those objects that happen to pass within about 20° of overhead as Earth rotates.

Arecibo is an example of a rough-surfaced telescope capable of detecting long-wavelength radio radiation. At the other extreme, Figure 5.23 shows the 36-m-diameter Haystack dish in northeastern Massachusetts. It is constructed of polished aluminum and maintains a parabolic curve to an accuracy of about a millimeter all the way across its solid surface. It can reflect and accurately focus radio radiation with wavelengths as short as a few millimeters. The telescope is contained within a protective shell, or radome, that protects the surface from the harsh New England weather. It acts much like the protective dome of an optical telescope, except that there is no slit through which the telescope "sees." Incoming cosmic radio signals pass virtually unimpeded through the radome's fiberglass construction.

THE VALUE OF RADIO ASTRONOMY

Despite the inherent disadvantage of relatively poor angular resolution, radio astronomy enjoys many advantages. Radio telescopes can observe 24 hours a day. Darkness is not needed for receiving radio signals because the Sun is a relatively weak source of radio energy, so its emission does not swamp radio signals arriving at Earth from elsewhere in the sky. In addition, radio observations can often be made through cloudy skies, and radio telescopes can detect the longest-wavelength radio waves even during rain or snowstorms. Poor weather causes few problems because the wavelength of most radio waves is much larger than the typical size of atmospheric raindrops or snowflakes. Optical

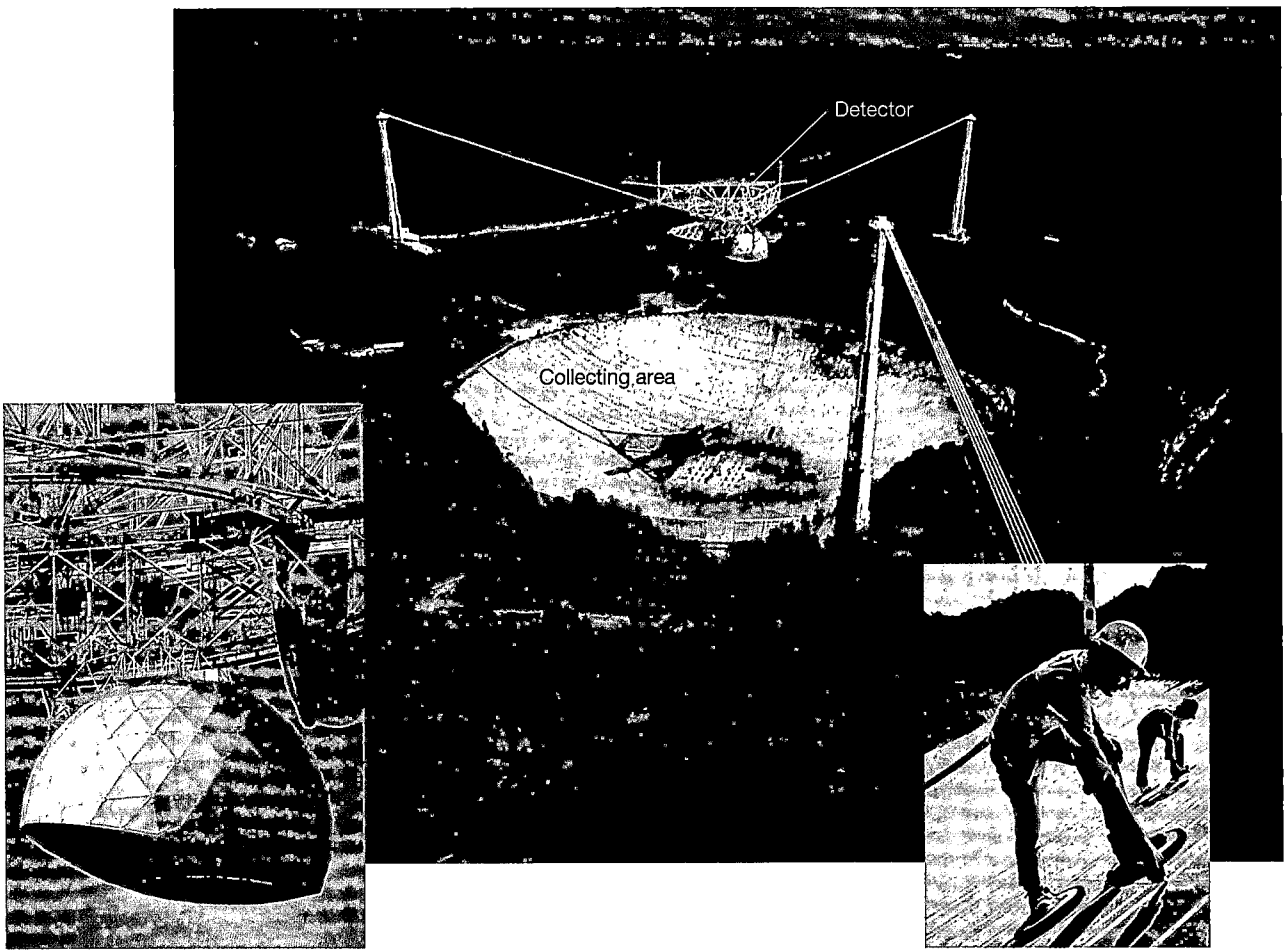

Figure 5.22 Arecibo Observatory An aerial photograph of the 300-m-diameter dish at the National Astronomy and Ionospheric Center near Arecibo, Puerto Rico. The receivers that detect the focused radiation are suspended nearly 150 m (about 45 stories) above the center of the dish. The left insert shows a close-up of the radio receivers hanging high above the dish. The right insert shows technicians adjusting the dish surface. *(D. Parker; T. Acevedo/NAIC; Cornell University)*

astronomy cannot be done under these conditions because the wavelength of visible light is smaller than a raindrop, a snowflake, or even a minute water droplet in a cloud.

However, perhaps the greatest value of radio astronomy (and, in fact, of all invisible astronomies) is that it opens up a whole new window on the universe. There are two main reasons for this. First, just as objects that are bright in the visible part of the spectrum (the Sun, for example) are not necessarily strong radio emitters, many of the strongest radio sources in the universe emit little or no visible light. Second, visible light may be strongly absorbed by interstellar dust along the line of sight to a source. Radio waves, on the other hand, are generally unaffected by intervening matter. Many parts of the universe cannot be seen at all by optical means but are easily detectable at longer wavelengths. The center of the Milky Way Galaxy is a prime example of such a totally invisible region—our knowledge of the Galactic center is based al-

most entirely on radio and infrared observations. Thus, these observations not only afford us the opportunity to study the same objects at different wavelengths, but also allow us to see whole new classes of objects that would otherwise be completely unknown.

Figure 5.24 shows an optical photograph of the Orion Nebula (a huge cloud of interstellar gas) taken with the 4-m telescope on Kitt Peak. Superimposed on the optical image is a radio map of the same region, obtained by scanning a radio telescope back and forth across the nebula and taking many measurements of radio intensity. The map is drawn as a series of contour lines connecting locations of equal radio brightness, similar to pressure contours drawn by meteorologists on weather maps or height contours drawn by cartographers on topographic maps. The inner contours represent stronger radio signals, the outside contours weaker signals.

The radio map in Figure 5.24 has many similarities to the visible-light image of the nebula. For instance, the radio emission is strongest near the center of the optical image and declines toward the nebular edge. But there are also subtle differences between the radio and optical images. The two differ mainly toward the upper left of the main cloud, where visible light seems to be absent, despite the existence of radio waves. How can radio waves be detected from locations not showing any light emission? The answer is that this particular nebular region is known to be especially dusty in its top left quadrant. The dust obscures the short-wavelength visible light but not the long-wavelength radio radiation. Our radio map allows us to see the true extent of this cosmic source.

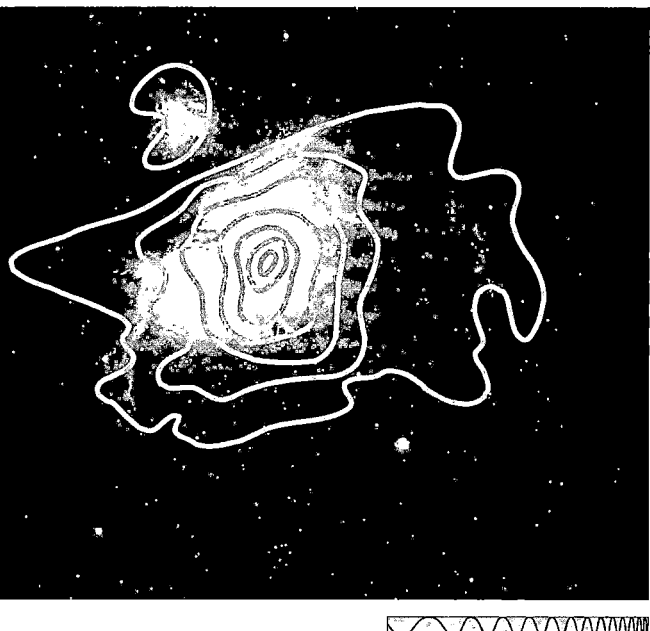

Figure 5.24 Orion Nebula in Radio and Visible The Orion Nebula is a star-forming region about 1500 light-years from Earth. (The nebula is located in the constellation Orion and can be seen in Figure 1.6.) The bright regions in this photograph are stars and clouds of glowing gas. The dark regions are not empty, but their visible emission is obscured by interstellar matter. Superimposed on the optical image is a radio contour map (white lines) of the same region. Each curve of the contour map represents a different intensity of radio emission. The resolution of the optical image is about 1″; that of the radio map is 1′. *(background photo: AURA)*

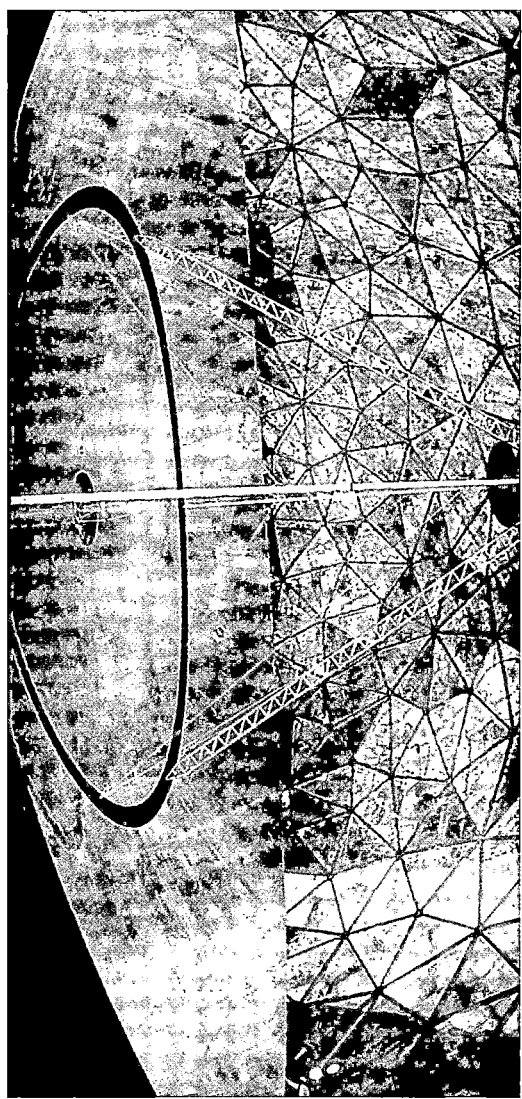

Figure 5.23 Haystack Observatory Photograph of the Haystack dish, inside its protective radome. For scale, note the engineer standing at the bottom. Also note the dull shine on the telescope surface, indicating its smooth construction. Haystack is a poor optical mirror but a superb radio telescope. Accordingly, it can be used to reflect and accurately focus radiation having short radio wavelengths, even as small as a fraction of a centimeter. *(MIT)*

Concept Check

■ In what ways does radio astronomy complement optical observations?

5.5 Interferometry

The main disadvantage of radio astronomy compared with optical work is its relatively poor angular resolution. However, in some circumstances, radio astronomers can overcome this problem by using a technique known as **interferometry**. This technique makes it possible to produce radio images of angular resolution higher than can be achieved with even the best optical telescopes, on Earth or in space.

In interferometry, two or more radio telescopes are used in tandem to observe the *same* object at the *same* wavelength and at the *same* time. The combined instruments together make up an **interferometer**. Figure 5.25 shows a large interferometer—many separate radio telescopes working together as a team. By means of electronic cables or radio links, the signals received by each antenna in the array making up the interferometer are sent to a central computer that combines and stores the data. The technique works by analyzing how the waves interfere with each other when added together. ∞ (*Discovery 3-1*) If the detected radio waves are in step, they combine constructively to produce a strong signal. If the signals are not in step, they destructively interfere and cancel each other. As the antennas track their target, a pattern of peaks and troughs emerges. After extensive computer processing, this pattern translates into a high-resolution image of the target object.

An interferometer is in essence a substitute for a single huge antenna. As far as resolving power is concerned, the effective diameter of an interferometer is the distance between its outermost dishes. In other words, two small dishes can act as opposite ends of an imaginary but huge single radio telescope, dramatically improving the angular resolution. For example, resolution of a few arc seconds can be achieved at typical radio wavelengths (such as 10 cm), either by using a single radio telescope 5 km in diameter (which is impossible to build) or by using two or more much smaller dishes separated by 5 km and connected electronically. The larger the distance separating the telescopes—the longer the *baseline* of the interferometer—the better the resolution attainable. Large interferometers like the instrument shown in Figure 5.25 now routinely attain radio resolution comparable to that of optical images. Figure 5.26 compares an interferometric radio map of a nearby galaxy with a photograph of that same galaxy made using a large optical telescope. The

radio clarity is superb—much better than the radio contour map of Figure 5.24.

Astronomers have created radio interferometers spanning very great distances, first across North America and later between continents. A typical very-long-baseline interferometry experiment (usually known by the acronym VLBI) might use radio telescopes in North America, Europe, Australia, and Russia to achieve angular resolution on the order of 0.001″. It seems that even Earth's diameter is no limit. Radio astronomers have successfully used an antenna in orbit, together with several antennas on the ground, to construct an even longer baseline and achieve still better resolution. Proposals exist to place interferometers entirely in Earth orbit, and even on the Moon.

Although the technique was originally developed by radio astronomers, interferometry is no longer restricted to the radio domain. Radio interferometry became feasible when electronic equipment and computers achieved speeds great enough to combine and analyze radio signals from separate radio detectors without loss of data. As the technology has improved, it has become possible to apply the same methods to higher-frequency radiation. Millimeter-wavelength interferometry has already become an established and important observational technique, and both the Keck telescopes and the VLT are expected to be used for infrared interferometry within the next few years.

Optical interferometry is currently the subject of intensive research. In 1997 a group of astronomers in Cambridge, England, succeeded in combining the light from three small optical telescopes to produce a single, re-

(a)   (b)

Figure 5.25 VLA Interferometer This large interferometer is made up of 27 separate dishes spread along a Y-shaped pattern about 30 km across on the Plain of San Augustin near Socorro, New Mexico. The most sensitive radio device in the world, it is called the Very Large Array, or VLA for short. (b) A close-up view from ground level of some of the VLA antennas. Notice that the dishes are mounted on railroad tracks so that they can be repositioned easily. *(NRAO)*

markably clear, image. Each telescope of the Cambridge Optical Aperture Synthesis Telescope (COAST) was only 0.4 m in diameter, but with the mirrors positioned 6 m apart, the resulting resolution was a stunning 0.01″—bet-

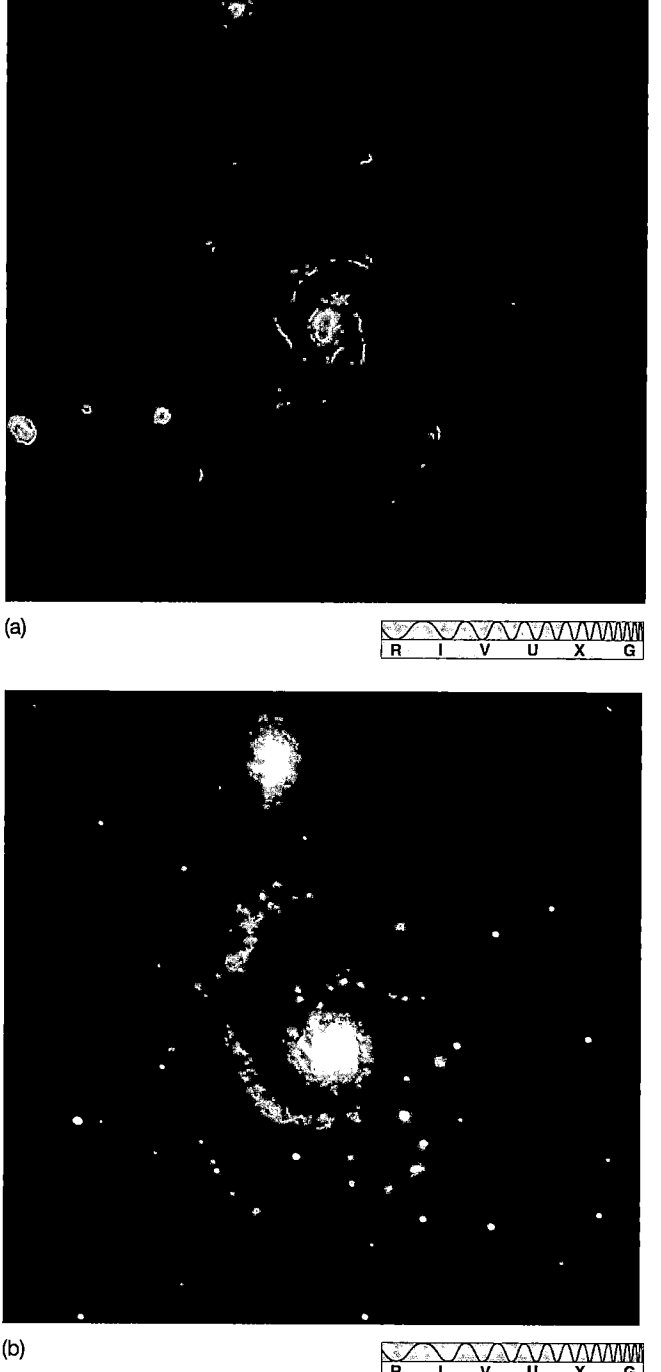

(a)

(b)

Figure 5.26 Radio–Optical Comparison (a) VLA radio "photograph" (or radiograph) of the spiral galaxy M51, observed at radio frequencies with an angular resolution of a few arc seconds; (a) shows nearly as much detail as (b), an actual (light) photograph of that same galaxy made with the 4-m Kitt Peak optical telescope. *(NRAO/AURA)*

ter by far than the resolution of the best adaptive-optics systems on the ground, or of *HST* operating above Earth's atmosphere. This enabled astronomers to "split" the binary star Capella, whose two member stars are separated by only 0.05″ and is therefore normally seen from the ground only as a slightly oblong blur. With the COAST array, the two stars are cleanly and individually separated. Figure 5.27 shows some of the telescopes of this interferometer and some sample results. NASA plans to place an optical interferometer—the *Space Interferometry Mission*, or *SIM*—in Earth orbit by 2010.

✅ **Concept Check**

- What is the main reason for the poor angular resolution of radio telescopes? How do radio astronomers overcome this problem?

(a)

0.05″

(b)

Figure 5.27 Interferometer Array (a) Some of the telescopes of the COAST interferometer can be seen here. (b) This extremely high-resolution image of the double-star Capella was made in the near infrared at 830 nm. *(MRAO)*

5.6 Space-Based Astronomy

Optical and radio astronomy are the oldest branches of astronomy, but since the 1970s there has been a virtual explosion of observational techniques covering the rest of the electromagnetic spectrum. Today, all portions of the spectrum are studied, from radio waves to gamma rays, to maximize the amount of information available about astronomical objects. As noted earlier, the types of astronomical objects that can be observed differ quite markedly from one wavelength range to another. Full-spectrum coverage is essential not only to see things more clearly but even to see some things at all. Because of the transmission characteristics of Earth's atmosphere, astronomers must study most wavelengths other than optical and radio from space. The rise of these "other astronomies" has therefore been closely tied to the development of the space program.

INFRARED ASTRONOMY

Infrared studies are a very important component of modern observational astronomy. Generally, **infrared telescopes** resemble optical telescopes, but their detectors are designed to be sensitive to longer-wavelength radiation. Indeed, as we have seen, many ground-based "optical" telescopes are also used for infrared work, and some of the most useful infrared observing is done from the ground (for example, from Mauna Kea—see Figure 5.11), even though the radiation is somewhat diminished in intensity by our atmosphere.

As with radio observations, the longer wavelength of infrared radiation often enables us to perceive objects partially hidden from optical view. As a terrestrial example of the penetrating properties of infrared radiation, Figure 5.28(a) shows a dusty and hazy region in California, hardly viewable optically, but easily seen using infrared radiation. Figure 5.28(b) shows a similar comparison for an astro-

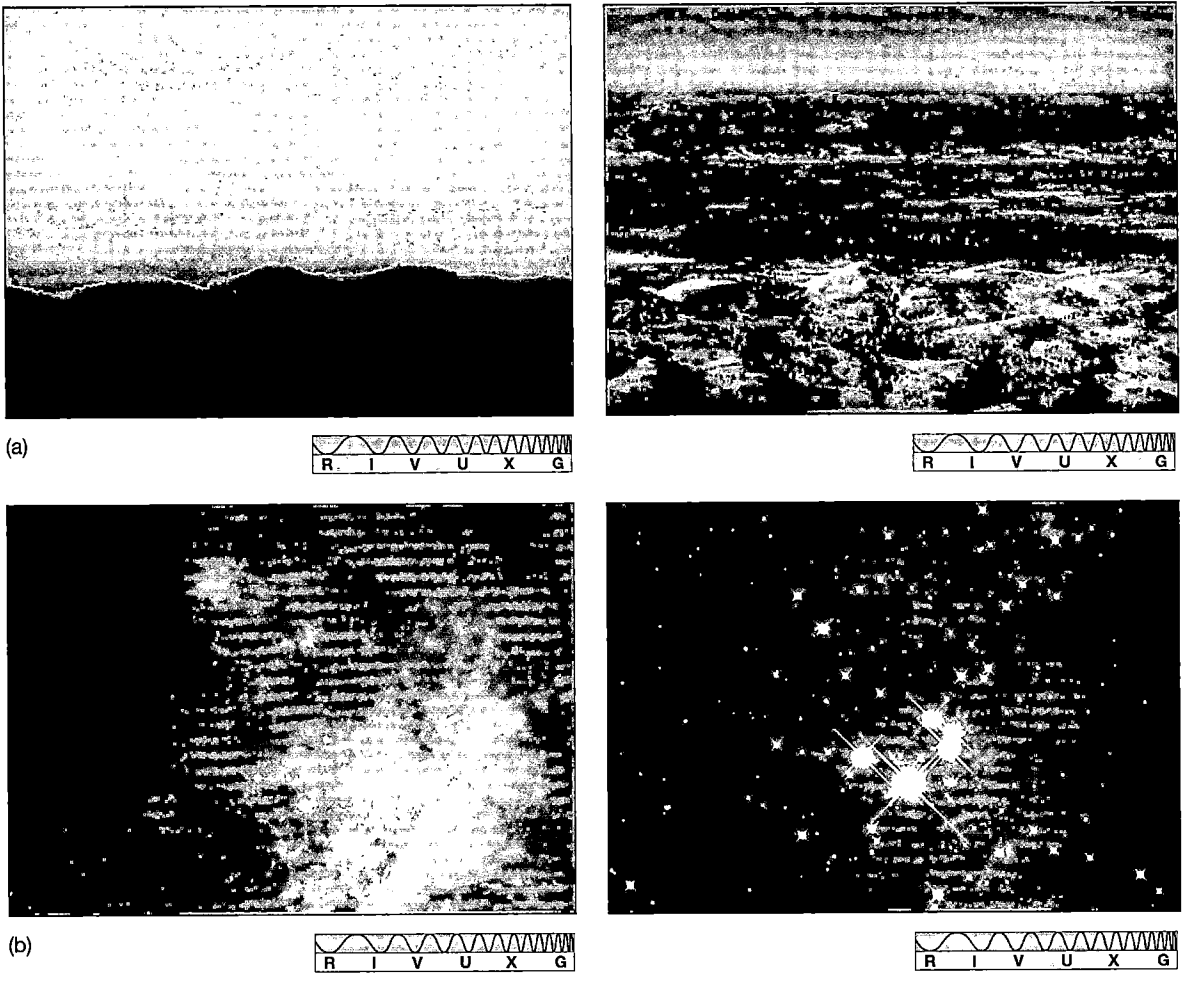

Figure 5.28 Smog Revealed (a) An optical photograph (left) taken near San Jose, California, and an infrared photo (right) of the same area taken at the same time. Longer-wavelength infrared radiation can penetrate smog much better than short-wavelength visible light. (b) Two views of an especially dusty part of our Galaxy, in optical (left) and infrared (right) light. This is the central region of the Orion Nebula, seen much more clearly in the infrared band. *(Harvard Observatory; NASA)*

nomical object—the dusty regions of the Orion Nebula, where much visible light is hidden behind interstellar clouds, but which is clearly distinguishable in the infrared.

Astronomers can make still better infrared observations if they can place their instruments above most or all of Earth's atmosphere. Improvements in balloon-, aircraft-, rocket-, and satellite-based telescope technologies have made infrared research a very powerful tool with which to study the universe (see Figure 5.29). However, as might be expected, the infrared telescopes that can be carried above the atmosphere are considerably smaller than the massive instruments found in ground-based observatories. A groundbreaking facility in this part of the spectrum was the *Infrared Astronomy Satellite* (*IRAS*), shown in Figure 5.29(b). Launched into Earth orbit in 1983 but now inoperative, this British/Dutch/U.S. satellite housed a 0.6-m mirror with an angular resolution as fine as 30″. (As usual, the resolution depended on the precise wavelength observed.) Its sensitivity was greatest for radiation in the 10-μm to 100-μm range.

During its 10-month lifetime (and long afterward—the data archives are still heavily used even today), *IRAS* contributed greatly to our knowledge of the clouds of Galactic matter destined to become stars and, possibly, planets. These regions of interstellar gas are composed of warm gas that cannot be seen with optical telescopes or adequately studied with radio telescopes. Because much of the material between the stars has a temperature between a few tens and a few hundreds of kelvins, Wien's law tells us that the infrared domain is the natural portion of the electromagnetic spectrum in which to study it. ∞ (Sec. 3.4) Throughout the text we will encounter many findings made by this satellite about comets, stars, galaxies, and the scattered dust and rocky debris found among the stars.

Figure 5.30(a) shows an *IRAS* image of the Orion region. The image is represented in *false color*, a technique commonly used for displaying images taken in nonvisible light. The colors do not represent the actual wavelength of the radiation emitted, but instead some other property of the source, in this case temperature, descending from white to red to black. The whiter regions thus denote higher temperatures and hence greater strength of infrared radiation. Figure 5.30(b) shows the same region photographed in (true color) visible light. At about 1′ angular resolution, the fine details of the Orion nebula evident in the visible portion of the earlier image (Figure 5.24) cannot be perceived. Nonetheless, astronomers can extract much useful information about this object and others like it from infrared observations. Star-forming clouds of warm dust and gas, and extensive groups of bright young stars, completely obscured at visible wavelengths, are seen.

Unfortunately, by Wien's law, telescopes themselves also radiate strongly in the infrared unless they are cooled to nearly absolute zero. ∞ (Sec. 3.4) The end of *IRAS's* mission came not because of any equipment malfunction or unexpected mishap but simply because its supply of liquid helium coolant ran out. *IRAS's* own thermal emission then overwhelmed the radiation it was built to detect.

Only a few long-duration infrared satellites have followed the *IRAS* mission. In November 1995 the European Space Agency launched the 0.6-m *Infrared Space Observatory* (*ISO*). The mission ended in May 1998. During its lifetime, *ISO's* imaging and spectroscopic observations refined and extended the pioneering work of *IRAS*. Today, the instrument package aboard the *Hubble Space Telescope* includes a near-infrared camera and spectroscope. NASA plans to deploy the 0.85-m *Space Infrared Telescope Facility* (*SIRTF*) in 2002. Like *ISO*, *SIRTF's* detectors will be designed to operate at wavelengths in the 3-μm to 200-μm range.

(a)   (b)

Figure 5.29 Infrared Telescopes (a) A gondola containing a 1-m infrared telescope (lower left) is readied for its balloon-borne ascent to an altitude of about 30 km, where it will capture infrared radiation that cannot penetrate Earth's atmosphere. (b) An artist's conception of the *Infrared Astronomy Satellite* (*IRAS*), placed in orbit in 1983. This 0.6-m telescope surveyed the infrared sky at wavelengths ranging from 10–100 μm. During its 10 months of operation it greatly increased astronomers' understanding of many different aspects of the universe—from the formation of stars and planets to the evolution of galaxies. *(SAO; NASA)*

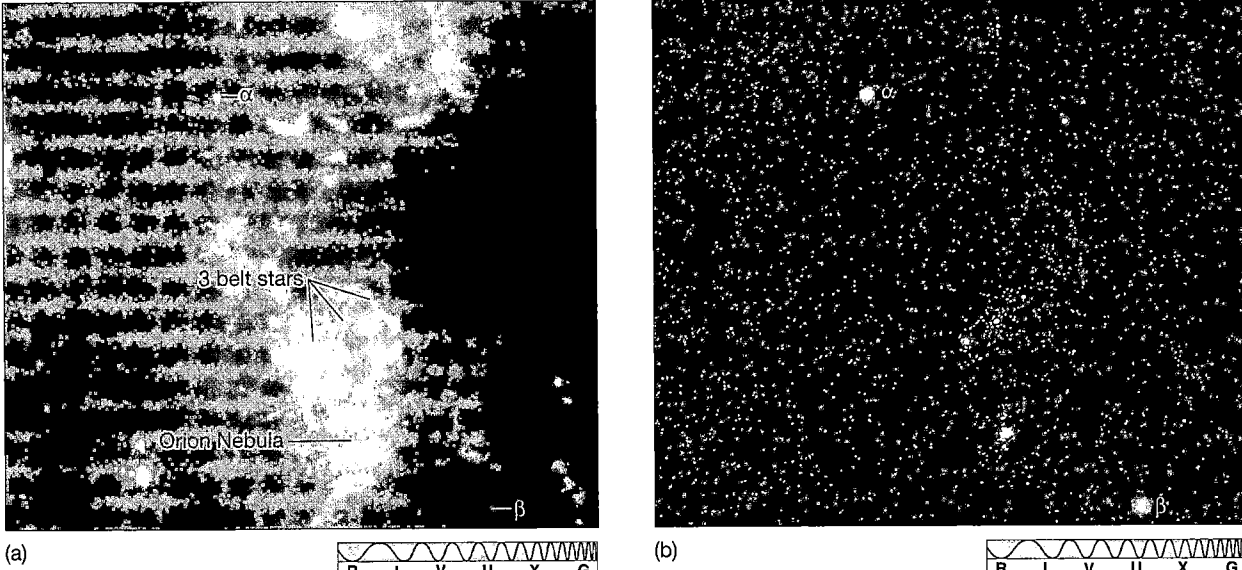

Figure 5.30 Infrared Image (a) This infrared image of the Orion Nebula and its surrounding environment was made by the *Infrared Astronomy Satellite*. In this false-color image, colors denote different temperatures, descending from white to red to black. The whiter regions denote greater intensity of infrared radiation. (b) The same region photographed in visible light. The labels α and β refer, respectively, to Betelgeuse and Rigel, the two brightest stars in the constellation. Note how the red star Betelgeuse is easily seen in the infrared (part a), while the blue star Rigel is very faint. *(NASA; J. Sanford)*

ULTRAVIOLET ASTRONOMY

On the short-wavelength side of the visible spectrum lies the ultraviolet domain. This region of the spectrum, extending in wavelength from 400 nm (blue light) down to a few nanometers ("soft" X rays), has only recently begun to be explored. Because Earth's atmosphere is partially opaque to radiation below 400 nm and is totally opaque below about 300 nm (in part because of the ozone layer), astronomers cannot conduct any useful ultraviolet observations from the ground, not even from the highest mountaintop. Rockets, balloons, or satellites are therefore essential to any **ultraviolet telescope**—a device designed to capture and analyze this high-frequency radiation.

One of the most successful ultraviolet space missions was the *International Ultraviolet Explorer* (*IUE*), placed in Earth orbit in 1978 and shut down for budgetary reasons in late 1996 (see *Discovery 18-1*). Like all ultraviolet telescopes, its basic appearance and construction were quite similar to optical and infrared devices. Several hundred astronomers from all over the world used *IUE*'s near-ultraviolet spectroscopes to explore a variety of phenomena in planets, stars, and galaxies. In subsequent chapters we will learn what this relatively new window on the universe has shown us about the activity and even the violence that seems to pervade the cosmos. Today, the *Far Ultraviolet Spectrographic Explorer* (*FUSE*) satellite,

launched in 1999, is extending *IUE*'s work into the "far" ultraviolet (around 100 nm) regime. In addition, the *Hubble Space Telescope* (*Discovery 5-1*), best known as an optical telescope, is also a superb imaging and spectroscopic ultraviolet instrument.

Figure 5.31(a) shows an image of a supernova remnant—the remains of a violent stellar explosion that occurred some 150,000 years ago—obtained by the *Extreme Ultraviolet Explorer* (*EUVE*) satellite. Launched in 1992 and terminated (again for budgetary reasons) in 2000, *EUVE* operated at the short-wavelength (1–50 nm) end of the ultraviolet range, making it sensitive to phenomena involving high temperatures (hundreds of thousands to millions of kelvins) or other energetic events. ∞ (Sec. 3.4) *EUVE* mapped out our local cosmic neighborhood as it presents itself in the far ultraviolet, and radically changed astronomers' conception of interstellar space in the vicinity of the Sun.

An alternative means of placing astronomical payloads into (temporary) Earth orbit is provided by NASA's space shuttles. In December 1990 and again in March 1995, a shuttle carried aloft the *Astro* package of three ultraviolet telescopes. An *Astro* image of a nearby galaxy is shown in Figure 5.31(b). Astronomical shuttle missions offer a potentially very flexible way for astronomers to get instruments into space, without the long lead times and great expense of permanent satellite missions like *HST*.

HIGH-ENERGY ASTRONOMY

High-energy astronomy studies the universe as it presents itself to us in X rays and gamma rays—the types of radiation whose photons have the highest frequencies and hence the greatest energies. How do we detect radiation of such short wavelengths? First, it must be captured high above Earth's atmosphere because none of it reaches the ground. Second, its detection requires the use of equipment fundamentally different in design from that used to capture the relatively low energy radiation discussed up to this point.

The difference in the design of **high-energy telescopes** comes about because X rays and gamma rays cannot be reflected easily by any kind of surface. Rather, these rays tend to pass straight through, or be absorbed by, any material they strike. When X rays barely graze a surface, however, they can be reflected from it in a way that yields an image, although the mirror design is fairly complex (see Figure 5.32). For gamma rays, no such method of producing an image has yet been devised. Present-day gamma-ray telescopes simply point in a specified direction and count photons received.

In addition, detection methods using photographic plates or CCD devices do not work well for hard X rays and gamma rays. Instead, individual photons are counted by electronic detectors on board an orbiting device, and the results are then transmitted to the ground for further processing and analysis. Furthermore, the number of photons in the universe seems to be inversely related to frequency. Trillions of visible (starlight) photons reach the detector of an optical telescope on Earth each second, but hours or even days are sometimes needed for a single gamma-ray photon to be recorded. Not only are these photons hard to focus and measure, they are also very scarce.

The *Einstein Observatory*, launched by NASA in 1978, was the first X-ray telescope capable of forming an image of its field of view. During its two-year lifetime, this spacecraft drove major advances in our understanding of high-energy phenomena throughout the universe. Its observational database is still heavily used. More recently, the German *ROSAT* (short for Röntgen Satellite, after Wilhelm Röntgen, the discoverer of X rays) was launched in 1991. During its seven-year lifetime, it generated a wealth of high-quality observational data (Figure 5.33). It was turned off in 1999, a few months after its electronics were irreversibly damaged when the telescope was accidentally pointed too close to the Sun.

In July 1999, NASA launched the *Advanced X-Ray Astrophysics Facility* (*AXAF*, renamed *Chandra* after launch, in honor of the Indian astrophysicist Subramanyan Chandrasekhar). With greater sensitivity, a wider field of view, and better resolution than either *Einstein* or *ROSAT*, *Chandra* is providing high-energy astronomers with new levels of observational detail. *Discovery 5-2* discusses the *Chandra* mission in more detail. The European *X-ray Multi-Mirror* satellite (now known as *XMM-Newton*) was launched in December 1999. *XMM* is more sensitive than *Chandra* (that is, it can detect fainter X-ray sources), but it has significantly poorer angular resolution (5″ compared to 0.5″ for *Chandra*), making the two missions complementary to one another.

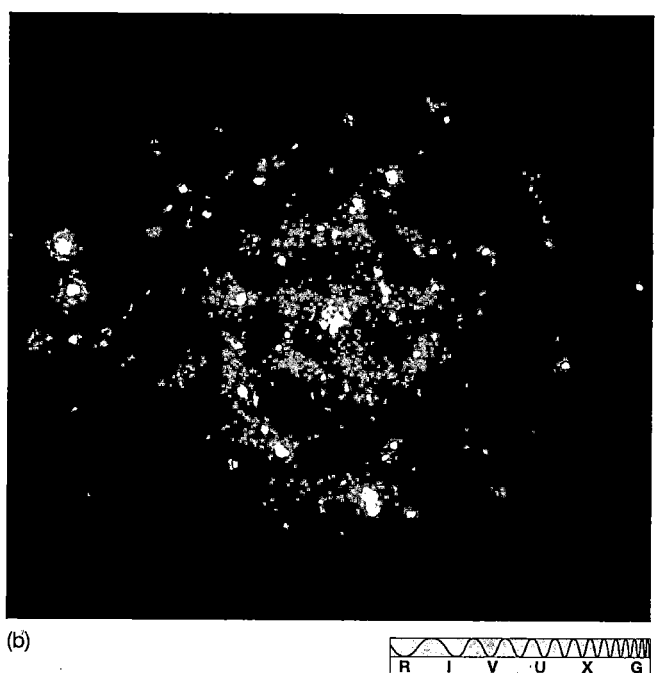

(a) [spectrum bar: R I V U X G]
(b) [spectrum bar: R I V U X G]

Figure 5.31 Ultraviolet Images (a) A camera on board the *Extreme Ultraviolet Explorer* satellite captured this image of the Cygnus Loop supernova remnant, the result of a massive star blowing itself virtually to smithereens. The release of energy was prodigious, and the afterglow lingered for millennia. The glowing field of debris shown here within the telescope's circular field of view lies some 2700 light-years from Earth. Based on the velocity of the outflowing debris, astronomers estimate that the explosion itself must have occurred about 150,000 years ago. (b) This false-color image of the spiral galaxy M74 was made by an ultraviolet telescope aboard the *Astro* payload carried by a space shuttle in 1995. *(NASA)*

DISCOVERY 5-2

The *Chandra* X-Ray Observatory

In the summer of 1999, the Space Shuttle *Columbia* carried into space the largest and most sophisticated X ray telescope ever built. Known simply as *Chandra* after the late Indian-American astrophysicist Subramanyan Chandrasekhar, the *Chandra X-ray Observatory* (*CXO*) is the third of NASA's "Great Observatories" to be deployed in space (the other two being *Hubble* and *Compton*). But unlike the others that hug Earth in low orbits only hundreds of kilometers in altitude, *Chandra* was boosted by onboard rockets to a much higher, elliptical orbit; its farthest point from Earth, 140,000 km, reaches almost one-third of the way to the Moon. The telescope's orbital period is 64 hours. Such an orbit takes *Chandra* well above most of the interfering van Allen belts girdling Earth and allows more efficient observations of the cosmos.

The first figure below, at left, shows the *Chandra* spacecraft in its final stages of construction in 1998. The technicians give perspective to this bus-sized craft which contains computers, antennae, and recorders to transmit and receive information between *Chandra* and ground stations, as well as finely polished, barrel-shaped mirrors to guide the X rays to a precise focus (see Figure 5.32). The front end of the telescope is at the bottom of the photograph and the scientific instruments at the top. Those instruments include a high-resolution camera to take images of X rays and a spectrometer to record the energy of those

X rays. Data collected by *Chandra*'s instruments are relayed from orbit to scientists at the mission control center in Cambridge, Massachusetts.

The second figure below shows an example of *Chandra*'s scientific capabilities; many more will be found throughout this textbook. The object shown is a supernova remnant—a debris field of scattered, glowing gases that were once part of a massive star. This "object," known as Cas A, is all that remains of a star in the constellation Cassiopeia. Its explosion was observed about 320 years ago. For scale, the debris is spread across some 10 light-years; Cas A itself is roughly 10,000 light-years from Earth.

The image shows what *Chandra* saw when it targeted Cas A for its first image taken in X rays. The region, barely visible in the optical part of the spectrum, is awash in brilliantly glowing X rays—large amounts of energy emitted at wavelengths to which the human eye is insensitive. (The X rays detected by *Chandra* in orbit are converted to radio waves that are then sent to ground stations, after which they are converted into visible images by astronomers for display and analysis.) This X-ray image shows that the gas in the wisps of ejected stellar material has temperatures of nearly 50 million kelvins, and the bright white point at the very center of the debris might well be a black hole—all of which will be studied in more detail in Chapters 21 and 22.

(NASA)

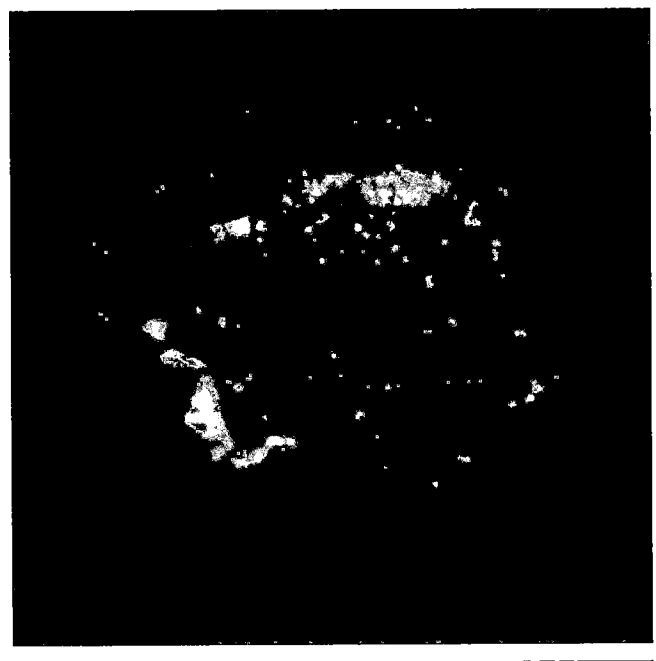
(NASA)

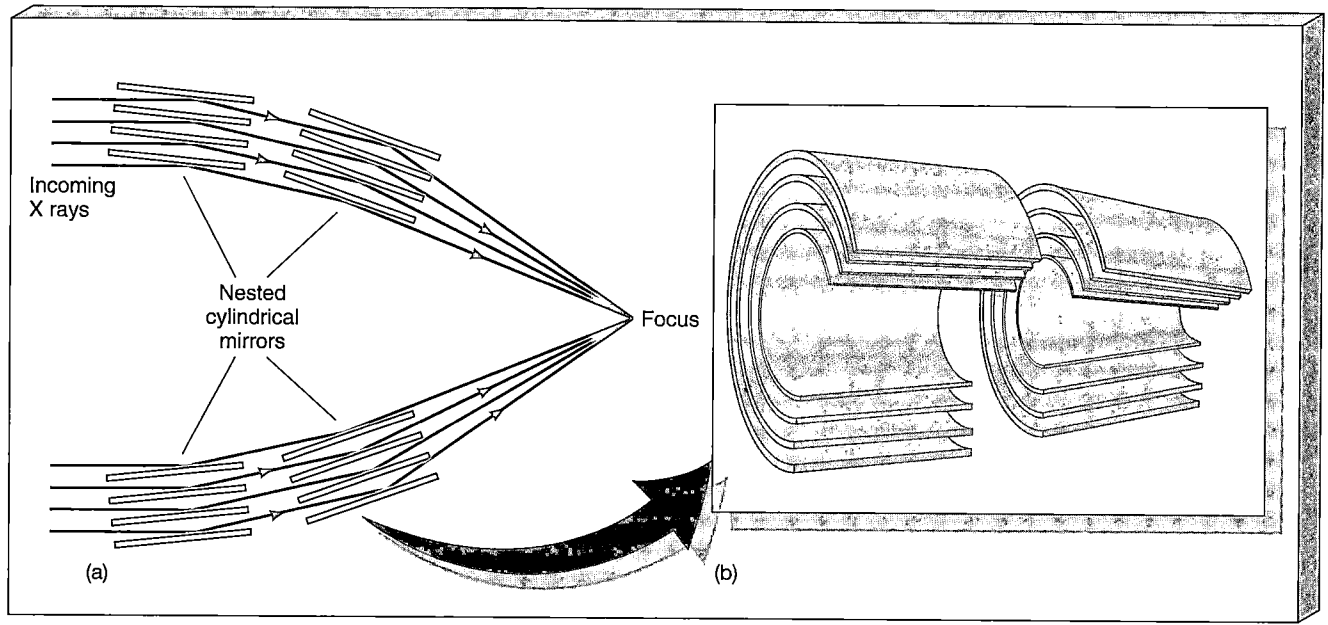

Figure 5.32 X-Ray Telescope The arrangement of mirrors in an X-ray telescope allows X rays to be reflected at grazing angles and focused to form an image.

Gamma-ray astronomy is the youngest entrant into the observational arena. As just mentioned, imaging gamma-ray telescopes do not exist, so only fairly coarse (1° resolution) observations can be made. Nevertheless, even at this resolution, there is much to be learned. Cosmic gamma rays were originally detected in the 1960s by the U.S. *Vela* series of satellites, whose primary mission was to monitor illegal nuclear detonations on Earth. Since then, several X-ray telescopes have also been equipped with gamma-ray detectors.

By far the most advanced instrument was the *Compton Gamma-Ray Observatory* (*CGRO*), launched by the space shuttle in 1991. At 17 tons, it was the largest astronomical payload ever launched by NASA at that time. *CGRO* scanned the sky and studied individual objects in much greater detail than had previously been attempted. Figure 5.34 shows the satellite in low Earth orbit, along with a false-color gamma-ray image of a highly energetic outburst in the nucleus of a distant galaxy. The mission ended on June 4, 2000, when, following a failure of one of the satellite's three gyroscopes, NASA opted for a controlled reentry and dropped *CGRO* into the Pacific Ocean.

☑ Concept Check

■ List some scientific benefits of placing telescopes in space. What are the drawbacks of space-based astronomy?

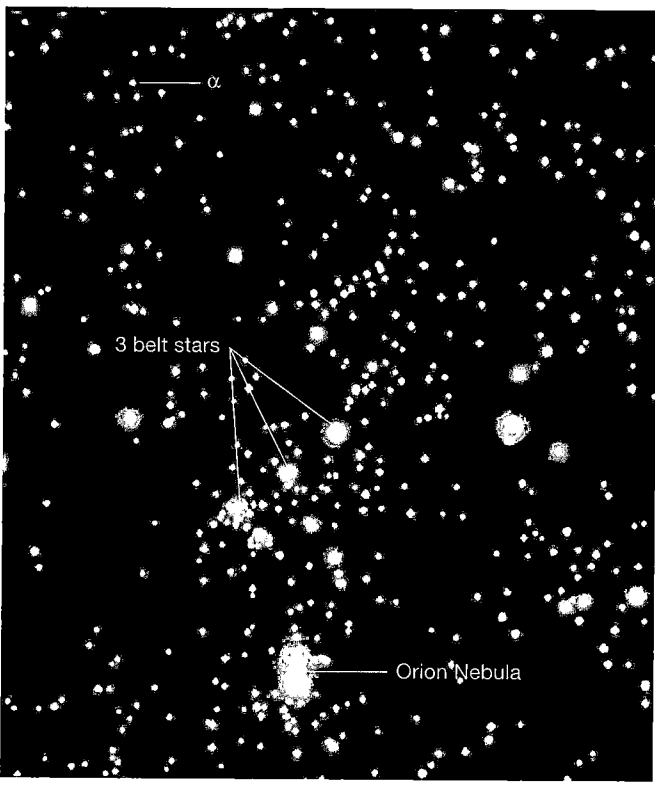

Figure 5.33 X-Ray Image An X-ray image of the Orion region, taken by the *ROSAT* X-ray satellite. (Compare with Figures 1.6, 5.24, and 5.30.) Objects that are most intensely emitting X-rays are colored blue, including the three stars of Orion's belt and the glowing nebula below them at bottom center in the photograph. Note that the star Betelgeuse (marked as α) now appears relatively faint; compare Figure 5.28. *(European Space Agency)*

(a)

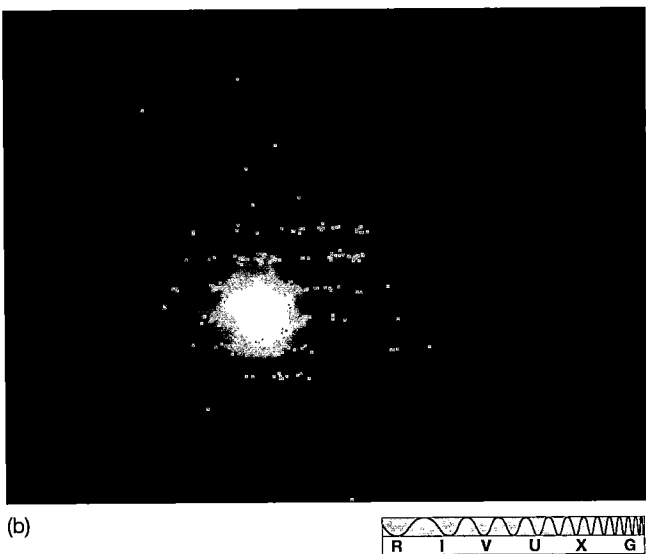

(b)

| R | I | V | U | X | G |

Figure 5.34 Gamma-Ray Observatory (a) This photograph of the 17-ton *Compton Gamma-Ray Observatory* (*CGRO*; named after an American gamma-ray pioneer) was taken by an astronaut during the satellite's deployment from the space shuttle *Atlantis* over the Pacific Coast of the United States. (b) A typical false-color gamma-ray image—this one showing a violent event in the distant galaxy 3C279, also known as a "gamma-ray blazar." *(NASA)*

5.7 Full-Spectrum Coverage

Table 5.1 lists the basic regions of the electromagnetic spectrum and describes objects typically studied in each frequency range. Bear in mind that the list is far from exhaustive and that many astronomical objects are now routinely observed at many different electromagnetic wavelengths. As we proceed through the text, we will discuss more fully the wealth of information that high-precision astronomical instruments can provide us.

It is reasonable to suppose that the future holds many further improvements in both the quality and the availability of astronomical data and that many new discoveries will be made. The current and proposed pace of technological progress presents us with the following exciting

prospect: Early in the twenty-first century, if all goes according to plan, it will be possible, for the first time ever, to make *simultaneous* high-quality measurements of any astronomical object at *all* wavelengths, from radio to gamma ray. The consequences of this development for our understanding of the workings of the universe may be little short of revolutionary.

As a preview of the sort of comparison that full-spectrum coverage allows, Figure 5.35 shows a series of images of our own Milky Way Galaxy. They were made by several different instruments, at wavelengths ranging from radio to gamma ray, over a period of about five years. By comparing the features visible in each, we immediately see how multi-wavelength observations can complement one another, greatly extending our perception of the dynamic universe around us.

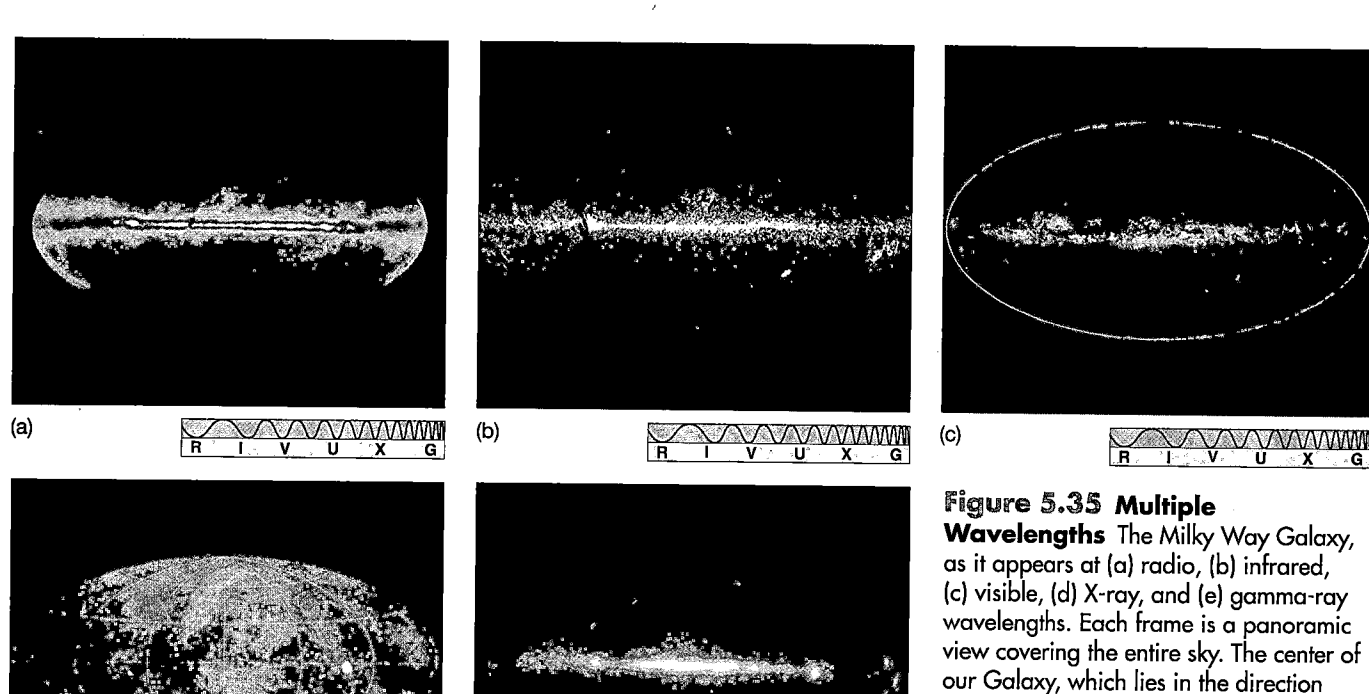

(a) R I V U X G
(b) R I V U X G
(c) R I V U X G
(d) R I V U X G
(e) R I V U X G

Figure 5.35 Multiple Wavelengths The Milky Way Galaxy, as it appears at (a) radio, (b) infrared, (c) visible, (d) X-ray, and (e) gamma-ray wavelengths. Each frame is a panoramic view covering the entire sky. The center of our Galaxy, which lies in the direction of the constellation Sagittarius, is at the center of each map. *(NRAO; NASA; Lund Observatory; K. Dennerl/and W. Voges; NASA)*

TABLE 5.1 Astronomy at Many Wavelengths

WAVELENGTH/FREQUENCY RANGE	GENERAL CONSIDERATIONS	COMMON APPLICATIONS (CHAPTER REFERENCE)
Radio	Can penetrate dusty regions of interstellar space Earth's atmosphere largely transparent to these wavelengths Can be detected in the daytime as well as at night High resolution at long wavelengths requires very large telescopes or interferometers	Radar studies of planets (2,9) Planetary magnetic fields (11) Interstellar gas clouds and molecules (18) Center of the Milky Way Galaxy (23) Galactic structure (23, 24) Active galaxies (25) Cosmic background radiation (27)
Infrared	Can penetrate dusty regions of interstellar space Earth's atmosphere only partially transparent to IR radiation, so some observations must be made from space	Star formation (19,20) Cool stars (20) Center of the Milky Way Galaxy (23) Active galaxies (25) Large-scale structure in the universe (24, 27)
Visible	Earth's atmosphere transparent to visible light	Planets (7–14) Stars and stellar evolution (17, 20, 21) Galactic structure (23, 24) Galaxies (24) Active galaxies (25) Large-scale structure in the universe (24, 27)
Ultraviolet	Earth's atmosphere is opaque to UV radiation, so observations must be made from space	Interstellar medium (19) Hot stars (21)
X ray	Earth's atmosphere opaque to X rays, so observations must be made from space Special mirror configurations needed to form images	Stellar atmospheres (16) Neutron stars and black holes (22) Hot gas in galaxy clusters (24) Active galactic nuclei (25)
Gamma ray	Earth's atmosphere opaque to gamma rays, so observations must be made from space Cannot form images	Neutron stars (22) Active galactic nuclei (25)

Chapter Review

SUMMARY

A **telescope** (p. 108) is a device designed to collect as much light as possible from some distant source and deliver it to a detector for detailed study. **Reflecting telescopes** (p. 108) use a mirror to concentrate and focus the light. **Refracting telescopes** (p. 108) use a lens; **refraction** (p. 108) is the bending of light as it passes from one medium to another. All astronomical telescopes larger than about 1 m in diameter use mirrors in their design. The **prime focus** (p. 108) of a telescope is the point where the incoming beam is focused and where analysis instruments may be placed. The **Newtonian** (p. 111) and **Cassegrain telescope** (p. 111) designs employ secondary mirrors to avoid placing heavy equipment at the prime focus. The light collected by a telescope may be processed in a number of ways. It can be made to form an **image** (p. 108), a **photometer** (p. 112) may be used to make detailed measurements of the energy received, or a **spectrometer** (p. 113) may study its spectrum.

The light-gathering power of a telescope depends on its **collecting area** (p. 114), which is proportional to the square of the mirror diameter. To study the faintest sources of radiation, astronomers must use large telescopes. Another important aspect of a telescope is its **angular resolution** (p. 116), the ability to distinguish between light sources lying close together on the sky. One limitation on the resolution of a telescope is diffraction, which makes it impossible to focus a beam perfectly. The amount of diffraction is proportional to the wavelength of the radiation under study and inversely proportional to the size of the mirror. Thus, at any given wavelength, larger telescopes suffer least from the effects of diffraction. The resolution of most ground-based optical telescopes is actually limited by **seeing** (p. 118)—the blurring effect of Earth's turbulent atmosphere, which smears the pointlike images of stars out into **seeing disks** (p. 118) a few arc seconds in diameter. Radio and space-based telescopes do not suffer from atmospheric effects, so their resolution is determined by the effects of diffraction.

Most modern telescopes now use **charge-coupled devices**, or **CCDs** (p. 119), instead of photographic plates to collect their data. The field of view is divided into an array of millions of **pixels** (p. 120) that accumulate an electric charge when light strikes them. CCDs are many times more sensitive than photo-graphic plates, and the resultant data are easily saved directly on disk or tape for later image processing.

Using **active optics** (p. 121), in which a telescope's environment and focus are carefully monitored and controlled, and **adaptive optics** (p. 122), in which the blurring effects of atmospheric turbulence are corrected for in real time, astronomers can now come close to diffraction-limited resolution in some ground-based instruments.

Radio telescopes (p. 124) are conceptually similar in construction to optical reflectors. However, radio telescopes are generally much larger than optical instruments, for two reasons. First, the amount of radio radiation reaching Earth from space is tiny compared with optical wavelengths, so a large collecting area is essential. Second, the long wavelengths of radio waves mean that diffraction severely limits the resolution unless large instruments are used.

In order to increase the effective area of a telescope, and hence improve its resolution, several separate instruments may be combined into a device called an **interferometer** (p. 127). Using **interferometry** (p. 127), radio telescopes can produce images sharper than those from the best optical telescopes. Infrared interferometers are under construction, and optical interferometric systems are under active development.

Infrared telescopes (p. 130) and **ultraviolet telescopes** (p. 132) are generally similar in design to optical systems. Infrared studies in some parts of the infrared range can be carried out using large ground-based systems. Ultraviolet astronomy must be carried out from space. **High-energy telescopes** (p. 133) study the X-ray and gamma-ray regions of the electromagnetic spectrum. X-ray telescopes can form images of their field of view, although the mirror design is more complex than for lower-energy instruments. Gamma-ray telescopes simply point in a certain direction and count photons received. Because the atmosphere is opaque at these short wavelengths, both types of telescopes must be placed in space.

Radio and other nonoptical telescopes are essential to studies of the universe because they allow astronomers to probe regions of space that are completely opaque to visible light and to study the many objects that emit little or no optical radiation at all.

SELF-TEST: TRUE OR FALSE?

_____ **1.** The primary purpose of any telescope is to produce an enormously magnified image of the field of view.

_____ **2.** A refracting telescope cannot form an image of its field of view.

_____ **3.** A Newtonian telescope has no secondary mirror.

_____ **4.** A Cassegrain telescope has a hole in the middle of the primary mirror to allow light reflected from its secondary mirror to reach a focus behind the primary mirror.

_____ **5.** The term "seeing" is used to describe how faint an object can be detected by a telescope.

_____ **6.** The primary advantage to using the _Hubble Space Telescope_ is the increased amount of "nighttime" available to it.

_____ **7.** One of the primary advantages of CCDs over photograph plates is their high efficiency in detecting light.

_____ **8.** The _Hubble Space Telescope_ can observe objects in the optical, infrared, and ultraviolet parts of the spectrum.

_____ **9.** The Keck telescopes contain the largest single mirrors ever produced.

_____ **10.** Radio telescopes are large, in part to improve their angular resolution, which is poor because of the long wavelengths at which they observe.

_____ **11.** Radio telescopes are large, in part because the sources of radio radiation they observe are very faint.

_____ **12.** Radio telescopes have to have surfaces as smooth as those in optical telescope mirrors.

_____ **13.** Infrared astronomy must be done from space.

_____ **14.** Because the ozone layer absorbs ultraviolet light, astronomers must make observations in the ultraviolet from the highest mountaintops.

_____ **15.** Gamma-ray telescopes employ the same basic design as optical instruments.

SELF-TEST: FILL IN THE BLANK

1. A telescope that uses a lens to focus light is called a _____ telescope.

2. A telescope that uses a mirror to focus light is called a _____ telescope.

3. All large modern telescopes are of the _____ type.

4. The light-gathering power of a telescope is determined by the _____ of its mirror or lens.

5. The angular resolution of a telescope is limited by the _____ of the telescope and the _____ of the radiation being observed.

6. The angular resolution of ground-based optical telescopes is more seriously limited by Earth's _____ than by diffraction.

7. The best optical telescopes on Earth can see angular detail down to about _____ arc second.

8. CCDs produce images in _____ form that can be easily transmitted, stored, and later processed by computers.

9. Active optics and adaptive optics are both being used to improve the _____ of ground-based optical telescopes.

10. All radio telescopes are of the _____ design.

11. An _____ is two or more telescopes used in tandem to observe the same object, in order to improve angular resolution.

12. _____, _____, and _____ astronomy can be done only from above Earth's atmosphere.

13. An object with a temperature of 300 K would be best observed with an _____ telescope.

14. The mirrors in x-ray telescopes are different in design from those in optical instruments because X rays tend to be _____, rather than reflected, by solid surfaces.

15. Gamma-ray telescopes are unable to form _____ of their fields of view.

REVIEW AND DISCUSSION

1. Cite two reasons why astronomers are continually building larger and larger telescopes.

2. List three advantages of reflecting telescopes over refracting telescopes.

3. What and where are the largest optical telescopes in use today?

4. How does Earth's atmosphere affect what is seen through an optical telescope?

5. What advantages does the _Hubble Space Telescope_ have over ground-based telescopes? List some disadvantages.

6. What are the advantages of a CCD over a photographic plate?

7. What is image processing?

8. Is the resolution of a 2-m optical telescope on Earth's surface more limited by atmospheric turbulence or by the effects of diffraction?

9. How do astronomers use active optics to improve telescope resolution?

10. How do astronomers use adaptive optics to improve telescope resolution?

11. Why do radio telescopes have to be very large?

12. What kind of astronomical objects can we best study with radio techniques?

13. What is interferometry, and what problem in radio astronomy does it address?

14. Is interferometry limited to radio astronomy?

15. Compare the highest resolution attainable with optical telescopes with the highest resolution attainable with radio telescopes (including interferometers).

16. Why do infrared satellites have to be cooled?

17. Are there any ground-based ultraviolet observatories?

18. In what ways do the mirrors in x-ray telescopes differ from those found in optical instruments?

19. What are the main advantages of studying objects at many different wavelengths of radiation?

20. Our eyes can see light with an angular resolution of 1′. Suppose our eyes detected only infrared radiation, with 1° angular resolution. Would we be able to make our way around on Earth's surface? To read? To sculpt? To create technology?

PROBLEMS  *Algorithmic versions of these questions are available in the Practice Problems module of the Companion Website.*

The number of squares preceding each problem indicates its approximate level of difficulty.

1. ▪ A certain telescope has a 10′ × 10′ field of view that is recorded using a CCD chip having 2048 × 2048 pixels. What angle on the sky corresponds to 1 pixel? What would be the diameter of a typical seeing disk (1″ radius), in pixels?

2. ▫ Estimate the fraction of incoming starlight blocked by the observer and observing cage at the prime focus of the Hale 5-m telescope (see Figure 5.8).

3. ▪ A 2-m telescope can collect a given amount of light in one hour. Under the same observing conditions, how much time would be required for a 6-m telescope to perform the same task? A 12-m telescope?

4. ▫ A certain space-based telescope can achieve (diffraction-limited) angular resolution of 0.05″ for red light (wavelength 700 nm). What would its resolution be (a) in the infrared, at 3.5 μm, and (b) in the ultraviolet, at 140 nm?

5. ▪ Based on the numbers given in the text, estimate the angular resolution of the Gemini North telescope (a) in red light (700 nm) and (b) in the near infrared (2 μm).

6. ▫▫ Two identical stars are moving in a circular orbit around one another, with an orbital separation of 2 A.U. ∞ (Sec. 2.6) The system lies 200 light-years from Earth. If we happen to view the orbit face-on, how large a telescope would we need to resolve the stars, assuming diffraction-limited optics at a wavelength of 2 μm?

7. ▫▫ What is the greatest distance at which *HST*, operating in blue light (400 nm), could resolve the stars in the previous question?

8. ▪ The photographic equipment on a telescope is replaced by a CCD. If the photographic plate records five percent of the light reaching it, while the CCD records 90 percent, how much time will the new system take to collect as much information as the old detector recorded in a 1-hr exposure?

9. ▪ The Moon lies about 380,000 km away. To what distances do the angular resolutions of the Hale telescope (1″), *HST* (0.05″), and a radio interferometer (0.001″) correspond at that distance?

10. ▫ The Andromeda Galaxy lies about 2.5 million light-years away. To what distances do the angular resolutions of the Hale telescope (1″), *HST* (0.05″), and a radio interferometer (0.001″) correspond at that distance?

11. ▪ Based on collecting areas, how much more sensitive would you expect the Arecibo telescope (Figure 5.22) to be, compared with the large telescope in Figure 5.21?

12. ▪ What would be the equivalent single-mirror diameter of a telescope constructed from two separate 10-m mirrors? Four separate 8-m mirrors?

13. ▫ Estimate the angular resolutions of (a) a radio interferometer with a 5000-km baseline, operating at a frequency of 5 GHz, and (b) an infrared interferometer with a baseline of 50 m, operating at a wavelength of 1 μm.

14. ▫▫ An instrument aboard an orbiting gamma ray telescope detects a burst of energy coming from some distant source. The *Chandra* x-ray telescope observes the same event. Both detectors record the location of the event on the sky, to within the accuracy of their respective measurements. You are about to make a follow-up observation using a large optical telescope with a 5′ field of view. Which position—the x-ray or the gamma-ray measurement—should you use, and why?

15. ▫▪ The resolution of the *Chandra* telescope is about 1″ for photons of energy 1 keV. ∞ (*More Precisely 4-1*) The effective collecting diameter of the mirror assembly is 1.2 m. Based on these numbers, is *Chandra* diffraction-limited?

COLLABORATIVE EXERCISES

1. Radio Interferometer. Determine the maximum size interferometer your group could build if you placed 2 m radio telescopes where each group member lives.

2. Observing New Stars in the Orion Region. Your group has been assigned to observe the region of the sky around Orion to look for bright new stars hidden in molecular clouds. Explain which telescope described in the text would be your group's first choice.

3. Multiple Wavelengths. As a group, decide which image of the Milky Way Galaxy from Figure 5.35—Multiple Wavelengths provides the most interesting information and explain your reasoning.

RESEARCHING ON THE WEB  *To complete the following exercises, go to the online Destinations module for Chapter 5 on the Companion Website for Astronomy Today 4/e.*

1. Access the "Buying a First-Time Telescope" page and determine which telescopes can be purchased for about $1000.

2. Access the "Introduction to Light Pollution" page and determine the five most common contributors to light pollution.

3. Access the "What is Adaptive Optics?" page and list the two limitations adaptive optics is designed to overcome and two observatories that are currently using adaptive optics.

PROJECTS

1. Here's how to take some easy pictures of the night sky. You will need a location with a clear, dark sky, a 35-mm camera with a standard 50-mm lens, tripod, and cable release, a watch with a seconds display visible in the dark, and a roll of high-speed color film. Set your camera to the "bulb" setting for the exposure and attach the cable release so you can take a long exposure. Set the focus on infinity. Point the camera to a favored constellation, seen through your viewfinder, and take a 20–30-second exposure. Don't touch any part of the camera or hold on to the cable release during the exposure; minimize all vibration. Keep a log of your shots. When finished, have the film developed in the standard way.

2. For some variation, vary your exposure times, use different films, take hours-long exposures for star trails, use different lenses such as wide-angle or telephoto, place the camera piggyback on a telescope that is tracking, and take exposures that are a few minutes long. Experiment and have fun!

SKYCHART III PROJECTS

The SkyChart III Student Version planetarium program on which these exercises are based is included as a separately executable program on the CD in the back of this text.

1. ■ Change your view to a place in the Southern Hemisphere. Identify the brightest stars and most prominent constellations in the southern sky. Can you identify a southern pole star, the equivalent to the North Star, Polaris?

2. ■ Change your viewing location to Buenos Aires on January 1, 2001 at 12:00 a.m. Print a chart in which your view is centered on the northern horizon. Now change to Miami at the same time and print a chart centered on the southern horizon. Place these charts side by side and focus your attention on the constellation Ophiuchus. How does Ophiuchus appear differently to two observers viewing from Miami and Buenos Aires? Can you explain the difference?

3. ■ Change your location to view from the Keck telescope in Mauna Kea, Hawaii (19 50 N, −155 28 E, 4123 m), and print out one or more observation charts. Now view from the Melipal telescope in Cerro Paranal, Chile (−24 38 N; −70 24 E, 2635m), and print out matching charts. What major objects can be seen with the Keck telescope that cannot be seen from Melipal? What major objects can be seen from the Melipal telescope but never from Keck?

4. ■■ Set the telescope control in SkyChart III to simulate viewing through a telescope with the following. Open *TELESCOPE/Display* and deselect *Camera field width*. Set *Finder field width* to 3 00 00 and *Eyepiece field width* to 1 00 00 to illustrate the fields seen through the finder scope and the eyepiece of a typical amateur telescope. Open *TELE-SCOPE/Connect*, select *Emulate real telescope connection*, and click on *OK*. Whenever you click *TELESCOPE/Connect/ OK*, the *Telescope Control* window will open.

    If you have access to a computerized telescope supported by the software, use the computer to drive the telescope and adjust the parameters for field width to match those of your telescope. If you wish to drive a telescope with the software, it will be necessary first to deselect *Emulate real telescope connection* and to align the telescope according to normal procedures for that telescope. Either with a real telescope or with emulation, locate an object of interest on the screen, such as Mars or Jupiter, and click on the object. This will open the *Object Info* window and prime the system so that clicking on the *Go To* button in the *Telescope Control* box will drive the telescope to the selected celestial object. Click on *Center* in the *Object Info* box and zoom in until you see the two concentric circles centered on the selected object. The resulting screen illustrates how the field of view in the telescope is related to the surrounding scene.

5. ■■ With SkyChart III configured for simulation of a telescope (see previous project), center on Polaris with *VIEW/Center Object* and type in Polaris. Click on Polaris to open the *Object Info* box and click on *Go To* in the *Telescope Control* box. Alternatively, if Polaris is visible on the screen, simply point the mouse at Polaris. Click on Polaris and click on *Center* in the *Object Info* box. Zoom in to a *1° Field* to see the field of view in a typical amateur telescope. You should see that Polaris separates into a pair of binary stars. Measure the separation of Polaris and its binary companion. In the field, if you have a question as to whether a star you have selected as Polaris really is Polaris, look at it through a telescope to see if it is a binary. It is not so obvious here, but the companion star is actually much fainter (magnitude +8.19) than Polaris (magnitude +1.89), so it may require a careful study to see it in a telescope.

 *In addition to the Practice Problems and Destinations modules, the Companion Website at http://www.prenhall.com/chaisson provides for each chapter an additional true-false, multiple choice, and labeling quiz, as well as additional annotated images, animations, and links to related Websites.*

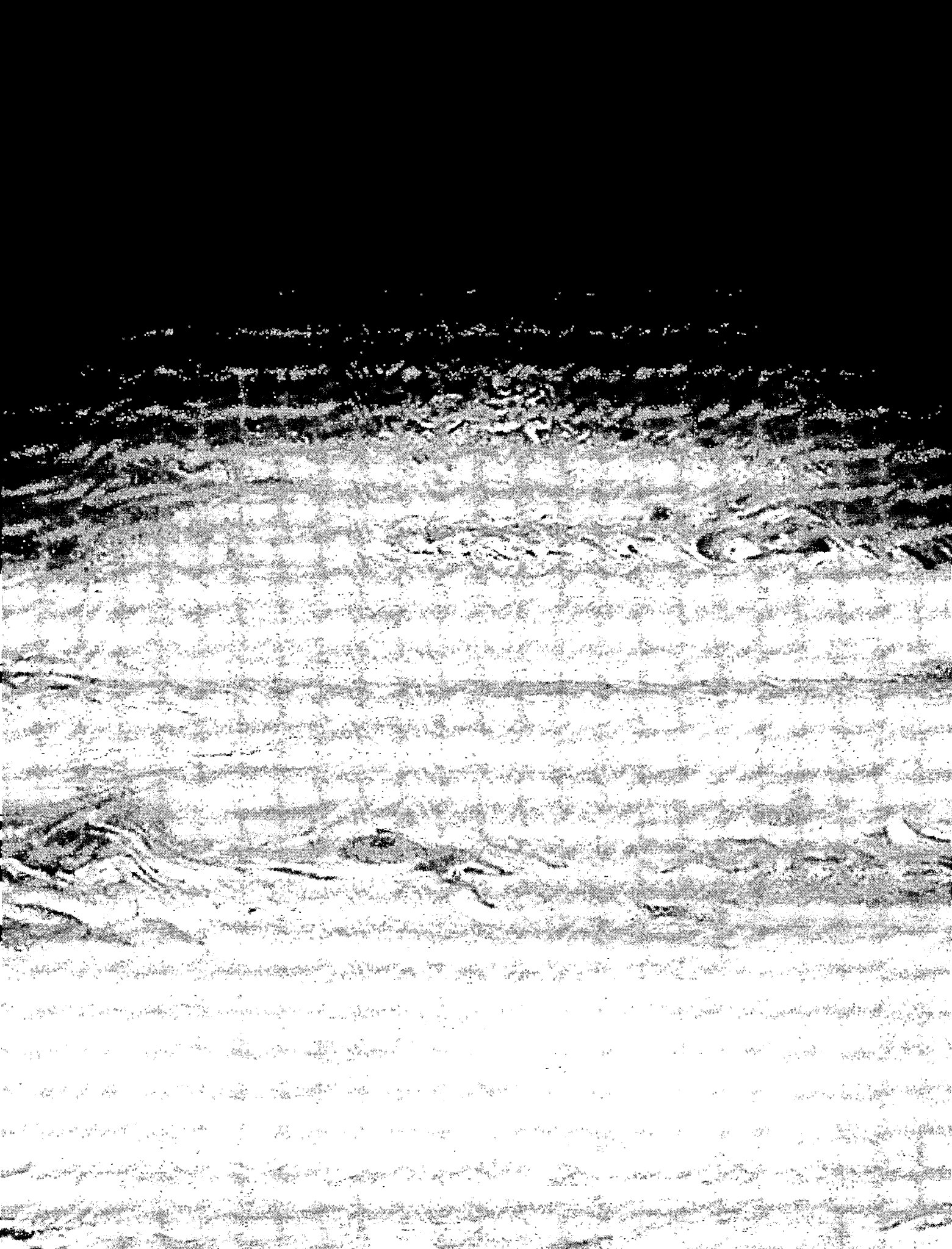

THE SOLAR SYSTEM

An Introduction to Comparative Planetology

LEARNING GOALS

Studying this chapter will enable you to:

1   Discuss the importance of comparative planetology to solar system studies.

2   Describe the overall scale and structure of the solar system.

3   Summarize the basic differences between the terrestrial and the jovian planets.

4   Identify and describe the major nonplanetary components of the solar system.

5   Describe some of the spacecraft missions that have contributed significantly to our knowledge of the solar system.

 Visit http://www.prenhall.com/chaisson for additional annotated images, animations, and links to related sites for this chapter.

To many observers, Jupiter is the most fascinating of all the planets in our solar system. Up close in a true-color image, as seen here while the *Cassini* spacecraft glided past in late 2000, we see the north polar region of the planet. Unlike its equatorial regions that show alternating light and dark belts, the high-latitude areas seen here appear mottled. The intricate structures represent clouds of different heights, thicknesses, and chemical compositions. *(JPL)*

The Big Picture: The engineering feats of the modern space age allow us to probe many of the diverse residents of the solar system in great detail. The premier planetary mission now underway, called *Cassini-Huygens*, left Earth in late 1997 and will arrive at Saturn in mid-2004. On its way, while bypassing Jupiter, it took the above image and radioed it back to Earth. The robot spacecraft is a joint venture of NASA and the European Space Agency, and is programmed to conduct 27 experiments when it reaches the ringed planet.

In less than a single generation, we have learned more about the solar system—the Sun and everything that orbits it—than in all the centuries that went before. By studying the planets, their moons, and the countless fragments of material that orbit in interplanetary space, astronomers have gained a richer outlook on our own home in space. Instruments aboard unmanned robots have taken close-up photographs of the planets and their moons and in some cases have made on-site measurements. The discoveries of the past few decades have revolutionized our understanding not only of our present cosmic neighborhood but also of its history, for our solar system is filled with clues to its own origin and evolution.

6.1 An Inventory of the Solar System

The Greeks and other astronomers of old were aware of the Moon, the stars, and five planets—Mercury, Venus, Mars, Jupiter, and Saturn—in the night sky. ∞ (Sec. 2.2) They also knew of two other types of heavenly objects that were clearly neither stars nor planets. *Comets* appear as long, wispy strands of light in the night sky that remain visible for periods of up to several weeks, then slowly fade from view. *Meteors*, or "shooting stars," are sudden bright streaks of light that flash across the sky, usually vanishing less than a second after they first appear. These transient phenomena must have been familiar to ancient astronomers, but their role in the "big picture" of the solar system was not understood until much later.

Human knowledge of the basic content of the solar system remained largely unchanged from ancient times until the early seventeenth century, when the invention of the telescope made more detailed observations possible. Galileo Galilei was the first to capitalize on this new technology (his simple telescope is shown in Figure 6.1). Galileo's discovery of the phases of Venus and of four moons orbiting Jupiter early in the seventeenth century helped change forever humankind's vision of the universe. ∞ (Sec. 2.4) As technological advances continued, knowledge of the solar system improved rapidly. Astronomers began discovering objects invisible to the unaided human eye. By the end of the nineteenth century, astronomers had found Saturn's rings (1659), the planets Uranus (1758) and Neptune (1846), many planetary moons, and the first *asteroids*—"minor planets" orbiting the Sun, mostly in a broad band (called the *asteroid belt*) lying between Mars and Jupiter. Ceres, the largest asteroid and the first to be sighted, was discovered in 1801. A large telescope of mid-nineteenth-century vintage is shown in Figure 6.2.

The twentieth century brought continued improvements in optical telescopes. One more planet (Pluto) was discovered, along with three more planetary ring systems,

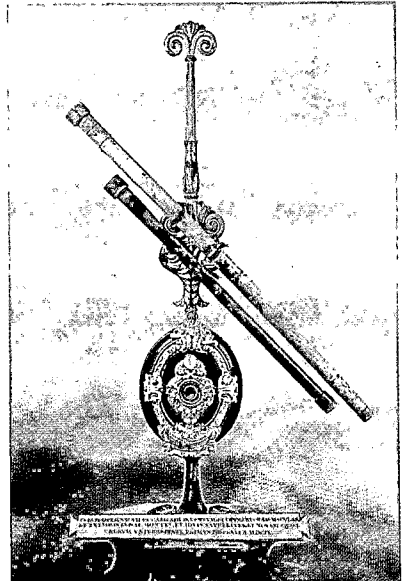

Figure 6.1 Early Telescope The telescope with which Galileo made his first observations was simple, but its influence on astronomy was immeasurable. *(Muzeo della Scienza, Florence/Scala/Art Resource, NY)*

Figure 6.2 Nineteenth-Century Telescope By the mid-nineteenth century, telescopes had improved enormously in both size and quality. Shown here is the instrument built and used by Irish nobleman and amateur astronomer the Earl of Rosse. *(Birr Scientific & Heritage Foundation)*

Figure 6.3 Lunar Exploration An *Apollo* astronaut does some lunar geology—prospecting near a huge boulder in the Mare Serenitatis during the final manned mission to the Moon in 1972. *(NASA)*

dozens of moons, and thousands of asteroids. The century also saw the rise of both nonoptical astronomy—especially radio and infrared—and spacecraft exploration, each of which has made vitally important contributions to the field of planetary science. Astronauts have carried out experiments on the Moon (see Figure 6.3), and numerous unmanned probes have left Earth and traveled to all but one of the other planets. Figure 6.4 shows the *Sojourner* robot roaming the Martian surface in 1997.

As currently explored, our **solar system** is known to contain one star (the Sun), nine planets, 91 moons (at last count) orbiting those planets, six asteroids larger than 300 km in diameter, tens of thousands of smaller (but well-

studied) asteroids, myriad comets a few kilometers in diameter, and countless *meteoroids* less than 100 m across. The list will undoubtedly grow as we continue to explore our cosmic neighborhood.

As we proceed through the solar system in the next few chapters, we will seek to understand how each planet compares with our own and what each contributes to our knowledge of the solar system as a whole. We will use the powerful perspective of **comparative planetology**—comparing and contrasting the properties of the diverse worlds we encounter—to understand better the conditions under which planets form and evolve. Our goal is to develop (in Chapter 15) a comprehensive theory of the origin and evolution of our planetary system that explains all, or at least most, of its observed properties. We will seek to answer basic questions such as: Why did planet X evolve in one way, while planet Y turned out completely differently? And why are the planets' orbits so orderly, when their individual properties are not? In addressing these issues, we will find many similarities and common features among planets. However, each planet will also present new questions and afford unique insights into the ways planets work.

As we unravel the origin of our solar system, we may also hope to learn something about planetary systems beyond our own. Since the mid-1990s, astronomers have detected dozens of "extrasolar" planets orbiting other stars, and many new planets are discovered each year (see Chapter 15). These observations provide critical tests of modern theories of planet formation. Before the discovery of extrasolar planets, those theories had necessarily been based on observations of our own solar system. Now astronomers have a whole new set of "proving grounds" in which to compare theory with reality.

Curiously, the present data suggest that many of the newly discovered systems have properties rather different

Figure 6.4 Martian Lander The Mars rover *Sojourner*, seen at left using a device called an Alpha Proton X-Ray Spectrometer to determine the chemical composition of a large Martian rock, nicknamed Yogi by mission scientists. For scale, the rock is about 1 m tall, and lies some 10 m from *Sojourner's* mother ship, *Pathfinder* (foreground). *(NASA)*

from our own, adding fuel to the long-standing debate among astronomers on the prevalence of planets like Earth, and the possible existence of life as we know it elsewhere in the universe. However, it will be some time before astronomers can make definitive statements about the existence (or otherwise) of planetary systems like our own.

✓ Concept Check

■ In what ways might observations of extrasolar planets help us understand our own solar system?

6.2 Planetary Properties

Table 6.1 lists some basic orbital and physical properties of the nine planets, with a few other solar-system objects (the Sun, the Moon, an asteroid, and a comet) included for comparison. Most of these quantities can be determined using methods described in Chapters 1 and 2. We present here some of the simpler techniques for making these measurements. Note that the Sun, having more than 1000 times the mass of the next most massive object (the planet Jupiter), is clearly the dominant member of the solar system. In fact, the Sun contains about 99.9 percent of all solar system material. The planets—including our own—are insignificant in comparison. *More Precisely 6-1* makes these methods a little more concrete by applying them to observations of the planet Jupiter.

■ The *distance* of each planet from the Sun is known from Kepler's laws once the scale of the solar system is set by radar-ranging on Venus. ⊂⊃ (Sec. 2.6)

■ A planet's (sidereal) *orbital period* can be measured from repeated observations of its location on the sky, so long as Earth's own motion around the Sun is properly taken into account.

■ A planet's *radius* is found by measuring its angular size and then applying elementary geometry. ⊂⊃ (Sec. 1.5)

■ The *masses* of planets with moons may be calculated by application of Newton's laws of motion and gravity, just by observing the moons' orbits around the planets. (*More Precisely 2-3*) The *sizes* of those orbits, like the sizes of the planets themselves, are determined by geometry.

■ The masses of Mercury and Venus (as well as those of our own Moon and the asteroid Ceres) are a little harder to determine accurately because these bodies have no natural satellites of their own. Nevertheless, it is possible to measure their masses by careful observations of their gravitational influence on other planets or nearby bodies. Mercury and Venus produce small but measurable effects on each other's orbits, as well as that of Earth. The Moon also causes small "wobbles" in Earth's motion as the two bodies orbit their common center of mass.

■ These techniques for determining mass were available to astronomers well over a century ago. Today the masses of most of the objects in Table 6.1 have been very accurately measured through their gravitational interaction with artificial satellites and space probes launched from

---

TABLE 6.1 Properties of Some Solar System Objects

OBJECT	ORBITAL SEMIMAJOR AXIS (A.U.)	ORBITAL PERIOD (Earth years)	MASS (Earth masses)	RADIUS (Earth radii)	NUMBER OF KNOWN SATELLITES	ROTATION PERIOD* (days)	AVERAGE DENSITY (kg/m³)	(g/cm³)
Mercury	0.39	0.24	0.055	0.38	0	59	5400	5.4
Venus	0.72	0.62	0.82	0.95	0	−243	5200	5.2
Earth	1.0	1.0	1.0	1.0	1	1.0	5500	5.5
Moon	—	—	0.012	0.27	—	27.3	3300	3.3
Mars	1.52	1.9	0.11	0.53	2	1.0	3900	3.9
Ceres (asteroid)	2.8	4.7	0.00015	0.073	0	0.38	2700	2.7
Jupiter	5.2	11.9	318	11.2	28	0.41	1300	1.3
Saturn	9.5	29.4	95	9.5	30	0.44	700	0.7
Uranus	19.2	84	15	4.0	21	−0.72	1300	1.3
Neptune	30.1	164	17	3.9	8	0.67	1600	1.6
Pluto	39.5	248	0.002	0.2	1	−6.4	2100	2.1
Comet Hale-Bopp	180	2400	1.0 × 10⁻⁹	0.004	—	0.47	100	0.1
Sun	—	—	332,000	109	—	25.8	1400	1.4

A negative rotation period indicates retrograde (backward) rotation relative to the sense in which all planets orbit the Sun.

Computing Planetary Properties

Let's look a little more closely at some of the methods used to determine the physical properties of a planet. The accompanying figure shows the planet Jupiter and Europa, one of its inner moons. The orbits of both Jupiter and Earth have been measured very precisely so, at any instant, the distance between the two planets is accurately known. However, since Jupiter's orbital semimajor axis is 5.2 A.U. and that of Earth is 1 A.U., for definiteness in what follows we will simply assume a distance of 4.2 A.U., or about 630,000,000 km. ∞ (Sec. 2.6) This corresponds roughly to the point of closest approach between the two planets.

Jupiter's angular diameter at this distance is 46.8 arc seconds. From *More Precisely 1-4*, it follows that the planet's actual diameter is

$$\text{diameter (2R)} = \text{distance} \times \frac{\text{angular diameter}}{57.3^\circ}$$

$$= (4.2 \times 150,000 \text{ km}) \times \frac{(46.8/3600)^\circ}{57.3^\circ}$$

$$= 143,000 \text{ km},$$

so Jupiter's radius is R = 71,500 km.

Jupiter's moon Europa is observed to orbit the planet with an orbital period of $P = 3.55$ days. The orbit is circular, with an angular radius of 3.66 arc minutes, as seen from Earth. Converting this to a distance in kilometers as above, we find that the radius of Europa's orbit around Jupiter is $r = 671,000$ km. The moon's orbital speed is $V = 2\pi r/P = 13.7$ km/s. Applying Newton's Laws, we then determine Jupiter's mass to be ∞ (*More Precisely 2-3*)

$$M = \frac{r V^2}{G} = 1.90 \times 10^{27} \text{ kg}.$$

Finally, dividing Jupiter's mass by the volume of a sphere of radius R, $\frac{4}{3}\pi R^3 = 1.53 \times 10^{24}$ m³, we obtain a

density of 1240 kg. This number differs from the figure listed in Table 6.1 because Jupiter is in reality not perfectly spherical. It is *flattened* somewhat at the poles, so our expression for the volume is not quite correct. When we take the flattening into account, the actual volume is a little lower than the number given here, and the density correspondingly higher.

We have determined several important physical properties of Jupiter by combining observations made from Earth with our knowledge of simple geometry and Newton's laws of motion. In fact, Jupiter is perhaps the easiest planet to study in this way as it is big, relatively close, and has several easy-to-see satellites. Nevertheless, these fundamental techniques are applicable throughout the solar system—indeed, throughout the entire universe. Prior to the Space Age, they formed the basis for virtually all astronomical measurements of planetary properties.

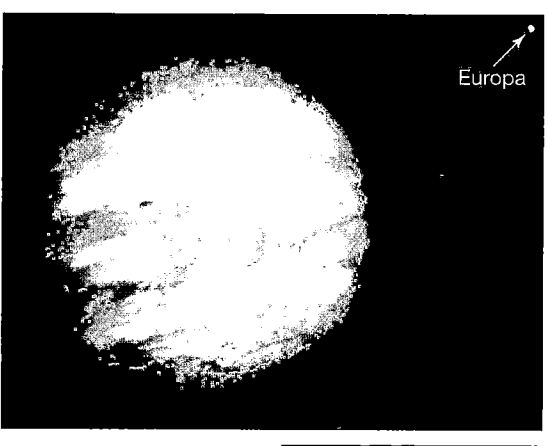

(NASA)

Earth. Only in the case of Ceres is the mass still poorly known, mainly because that asteroid's gravity is so weak.

A planet's *rotation period* may in principle be determined simply by watching surface features appear and disappear again as the planet rotates. However, for most planets this is difficult to do, as their surfaces are hard to see or may even be nonexistent. Mercury's surface features are hard to distinguish, the surface of Venus is completely obscured by clouds, and Jupiter, Saturn, Uranus, and Neptune have no solid surfaces at all—their atmospheres simply thicken, and eventually become liquid, as we descend deeper and deeper below the visible clouds. We will describe the methods used to measure these planets' rotation periods in later chapters.

The final columns in Table 6.1 list the *average density* of each object. **Density** is a measure of the "compactness" of matter. Average density is computed by divid-

ing an object's mass (in kilograms, say) by its volume (in cubic meters, for example). For example, we can easily compute Earth's average density. Earth's mass, as determined from observations of the Moon's orbit, is approximately 6.0×10^{24} kg. ∞ (*More Precisely 2-3*) Earth's radius R is roughly 6400 km, so its volume is $\frac{4}{3}\pi R^3 \approx 1.1 \times 10^{12}$ km³, or 1.1×10^{21} m³. ∞ (*Discovery 1-1*) Dividing Earth's mass by its volume, we obtain an average density of approximately 5500 kg/m³. *On average*, then, there are about 5500 kilograms of Earth matter in every cubic meter of Earth volume. For comparison, the density of ordinary water is 1000 kg/m³, rocks on Earth's surface have densities in the range 2000–3000 kg/m³, and iron has a density of some 8000 kg/m³. Earth's atmosphere (at sea level) has a density of only a few kilograms per cubic meter. Because many working astronomers are more familiar with the CGS (centimeter-gram-second)

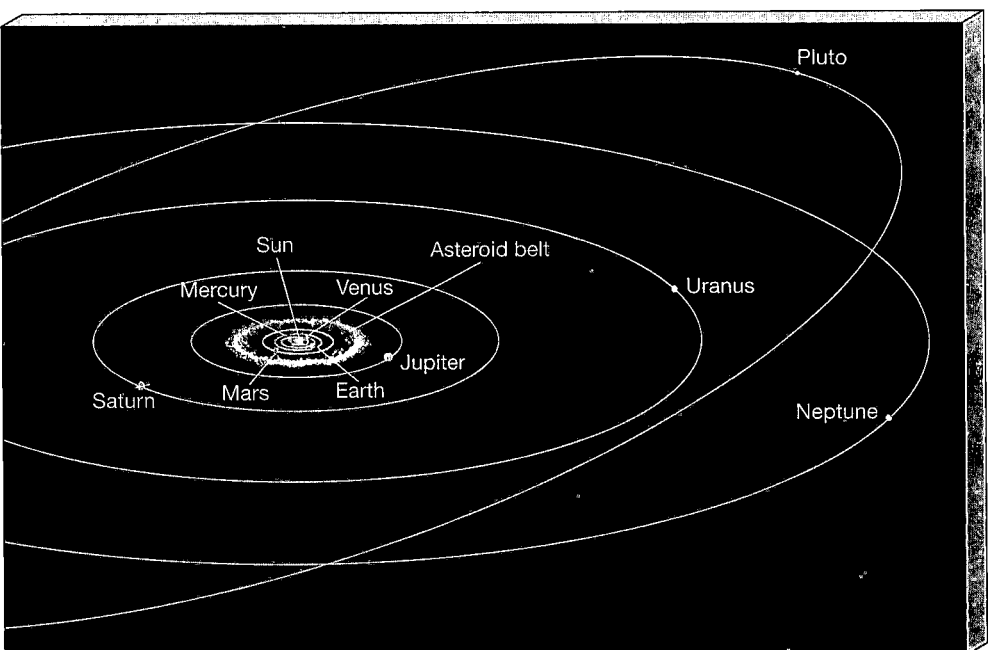

Figure 6.5 Solar System
Might future space voyagers travel far enough from Earth to gain this perspective on our solar system? Except for Mercury and Pluto, the orbits of the planets lie nearly in the same plane. As we move out from the Sun, the distance between the orbits of the planets increases. The entire solar system spans nearly 80 A.U.

unit of density (grams per cubic centimeter, g/cm^3, where 1 kg/m^3 = 1000 g/cm^3), Table 6.1 lists density in both SI and CGS units.

Concept Check

■ How do astronomers go about determining the bulk properties (masses, radii, and densities) of distant planets?

6.3 The Overall Layout of the Solar System

🔍 By earthly standards, the solar system is immense. The distance from the Sun to Pluto is about 40 A.U., almost a million times Earth's radius and roughly 15,000 times the distance from Earth to the Moon. Yet, despite the solar system's vast extent, the planets all lie very close to the Sun, astronomically speaking. Even the diameter of Pluto's orbit is less than 1/1000 of a light-year, whereas the next nearest star is several light-years distant.

The planet closest to the Sun is Mercury. Moving outward, we encounter in turn Venus, Earth, Mars, Jupiter, Saturn, Uranus, Neptune, and Pluto. In Chapter 2 we saw the basic properties of the planets' orbits. Their paths are all ellipses, with the Sun at (or very near) one focus. ∞ (Sec. 2.5) Most planetary orbits have low eccentricities. The exceptions are the innermost and the outermost worlds, Mercury and Pluto. Accordingly, we can reasonably think of most planets' orbits as circles centered on the Sun. The orbits of the major bodies in the solar system are illustrated in Figure 6.5.

All the planets orbit the Sun in the same sense—counterclockwise as seen from above Earth's North Pole—and in nearly the same plane as Earth (that is, the ecliptic plane). Mercury and Pluto deviate somewhat from this rule. Their orbital planes lie at 7° and 17° to the ecliptic, respectively. Still, we can think of the solar system as being quite flat. Its "thickness" perpendicular to the plane of the eclip-

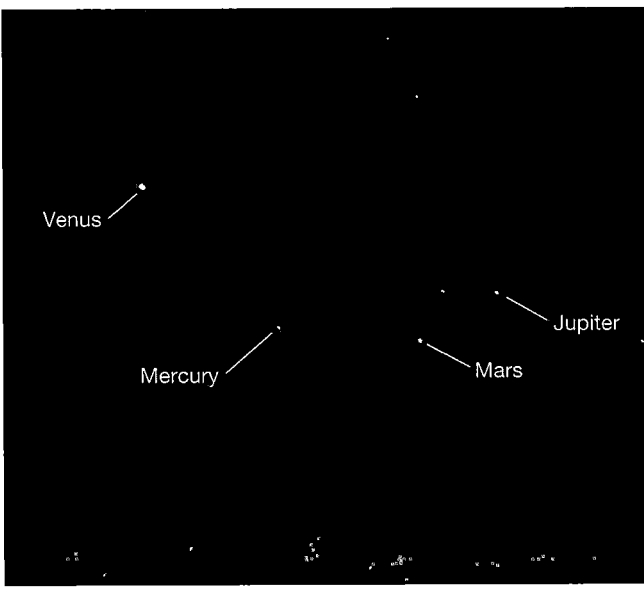

Figure 6.6 Planets During Eclipse Taken from Hawaii during the July 1991 eclipse of the Sun, this single photograph shows four planets—Mercury, Venus, Mars, and Jupiter. Because they all orbit in nearly the same plane, it is possible for them all to appear (by chance) in the same region of the sky, as seen from Earth. *(S. Westphal)*

DISCOVERY 6-1

The Titius–Bode "Law"

A close look at the list of semimajor axes of the known planets (see Table 6.1) reveals considerable regularity in their orbits. The *spacing* of the orbits increases more or less geometrically as we move out from the Sun: At any point in the list, the distance to the next planet out is about twice that to the next planet in.

Is there some underlying structure to the solar system? In 1766, in search of an answer to that question, the German astronomer Johann Titius came up with a simple formula that "predicted" quite well the orbits of the then-known planets, Mercury through Saturn. Johann Bode, a better-known astronomer of the day, later popularized this relationship among the planets' orbits, so that it now is usually known as the Titius–Bode law, or even just Bode's law. We must emphasize that it is not a law at all, in the scientific sense, but rather just a rule for determining the approximate orbital semimajor axes of the planets. Nevertheless, the "law" part of the name has stuck.

The rule for determining the planets' orbits is as follows. Start with 0.4, the distance (in A.U.) from the Sun to Mercury. Then add to it, successively, 0.3 (to arrive at Venus, with an orbital semimajor axis of 0.7 A.U.), 0.3 (to reach Earth, at 1.0 A.U.), 0.6 (Mars: 1.6 A.U.), 1.2 (?: 2.8 A.U.), 2.4 (Jupiter: 5.2 A.U.), and 4.8 (Saturn: 10.0 A.U.) to arrive at the (approximate) orbital distances of the known planets, plus one extra "planet" between Mars and Jupiter. The relation between successive orbits is easily seen—after Venus, the added term (that is, the interplanetary spacing) simply doubles at each step. If we number the planets starting at Mercury (with $n = 1$), then a simple formula for calculating the semimajor axis a_n of the nth orbit is

$$a_n = 0.4 + 0.3 \times 2^{n-2}$$

valid for $n = 2$ (Venus), 3 (Earth), 4 (Mars), ... (the formula fails for Mercury, for which we would have to take $n = -\infty$).

The test of any theory is its predictive power. According to the Titius–Bode rule, the next three planets beyond Saturn (corresponding to $n = 8$, 9, and 10) should lie at 19.6, 38.8, and 77.2 A.U. When Uranus was discovered in 1781, it fell close to the prediction, at 19.2 A.U. In 1801,

the discovery of the asteroid Ceres at 2.8 A.U., in the $n = 5$ "slot" seemed to support the notion that the rule contained some deep insight into our planetary system. Speculation was rife that the next planet would also lie where the "law" decreed it should. Sadly, the rule fails for Neptune, which lies only 30 A.U. from the Sun. However, if we ignore Neptune, Pluto is in just about the right place!

Few people today take the rule sufficiently seriously to bother to look for a scenario in which Neptune started out in the "right" orbit and somehow got moved farther in toward the Sun, perhaps leaving Pluto behind. Still, the regularity of the "law" was appealing to astronomers of the eighteenth and nineteenth centuries, as it suggested some fundamental harmony in the structure of the solar system, and, to some extent, that appeal remains today.

The Titius–Bode rule is not the result of any known planetary interaction, nor is it even very accurate. It may simply express the fact that planetary orbits tend to become more widely spaced as we move outward from the Sun, and the fit to some particular orbits may be mere coincidence. In addition, there is no evidence of similar rules for extrasolar planetary systems, although it must be admitted that current data are very limited, as they include only the most massive planets. Nevertheless, some astronomers still suspect that the Titius–Bode rule may be telling us *something* about the formation of our solar system—they just don't know quite *what*.

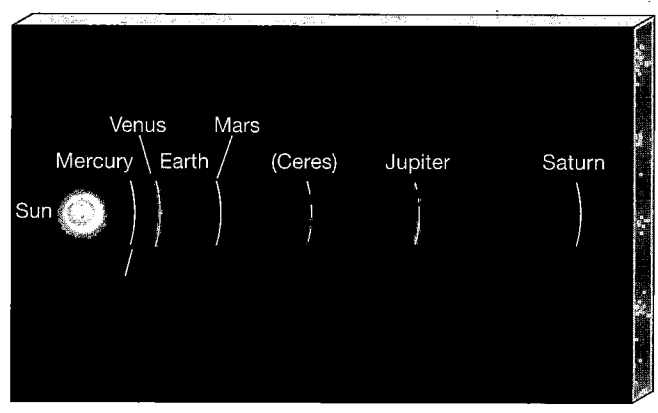

tic is less than $\frac{1}{50}$ the diameter of Pluto's orbit. If we were to view the planets' orbits from a vantage point in the ecliptic plane about 50 A.U. from the Sun, only Pluto's orbit would be noticeably tilted. Figure 6.6 is a photograph of planets Mercury, Venus, Mars, and Jupiter taken during the July 1991 solar eclipse. These four planets can (occasionally) be found in the same region of the sky in large part because their orbits lie nearly in the same plane in space.

The planetary orbits are not evenly spaced: the orbits get farther and farther apart as we move farther out from

the Sun. Nevertheless, there is a certain regularity in their spacing. In the eighteenth century a fairly simple rule, now known as the Titius–Bode law (*Discovery 6-1*), seemed to "predict" the radii of the planetary orbits remarkably well. Even the asteroid belt between Mars and Jupiter appeared to have a place in the scheme, which excited great interest among astronomers and numerologists alike. There is apparently no simple explanation for this empirical "law." Today, it is regarded more as a curiosity than as a fundamental property of the solar system, although many theorists still feel that it may hold some deep, but as yet undiscovered, insight into the way the solar system formed.

Concept Check

■ In what sense is the solar system "flat"?

6.4 Terrestrial and Jovian Planets

On large scales, the solar system presents us with a sense of orderly motion. The planets move nearly in a plane, on almost concentric (and nearly circular) elliptical paths, in the same direction around the Sun, at steadily increasing orbital intervals. However, the individual properties of the planets themselves are much less regular.

Figure 6.7 compares the major planets with one another and with the Sun. A clear distinction can be drawn between the inner and the outer members of our planetary system based on densities and other physical properties. The inner planets—Mercury, Venus, Earth, and Mars—are small, dense, and *rocky* in composition. The outer worlds—Jupiter, Saturn, Uranus, and Neptune (but not Pluto)—are large, of low density, and *gaseous*.

Because the physical and chemical properties of Mercury, Venus, and Mars are somewhat similar to Earth's, the four innermost planets are called the **terrestrial planets**. (The word *terrestrial* derives from the Latin word *terra*, meaning "land" or "earth.") The larger outer planets—Jupiter, Saturn, Uranus, and Neptune—are all similar to one another chemically and physically (and very different from the terrestrial worlds). They are labeled the **jovian** planets, after Jupiter, the largest member of the group. (The word *jovian* comes from *Jove*, another name for the Roman god Jupiter.) The jovian worlds are all much larger than the terrestrial planets, and quite different from them in both composition and structure.

The four terrestrial planets all lie within about 1.5 A.U. of the Sun. All are small and of relatively low mass, and all have generally rocky composition and solid surfaces. Beyond that, however, the similarities end:

■ All four terrestrial planets have atmospheres, but the atmospheres are about as dissimilar as we could imagine, ranging from a near-vacuum on Mercury to a hot, dense inferno on Venus.

■ Earth alone has oxygen in its atmosphere and liquid water on its surface.

■ Surface conditions on the four planets are quite distinct from one another, ranging from barren, heavily cratered terrain on Mercury to widespread volcanic activity on Venus.

■ Earth and Mars spin at roughly the same rate—one rotation every 24 (Earth) hours—but Mercury and Venus both take months to rotate just once, and Venus rotates in the opposite sense from the others.

■ Earth and Mars have moons, but Mercury and Venus do not.

■ Earth and Mercury have measurable magnetic fields, of very different strengths, whereas Venus and Mars have none.

Finding the common threads in the evolution of these four diverse worlds is no simple task! Comparative planetology will be our indispensable guide as we proceed through the coming chapters.

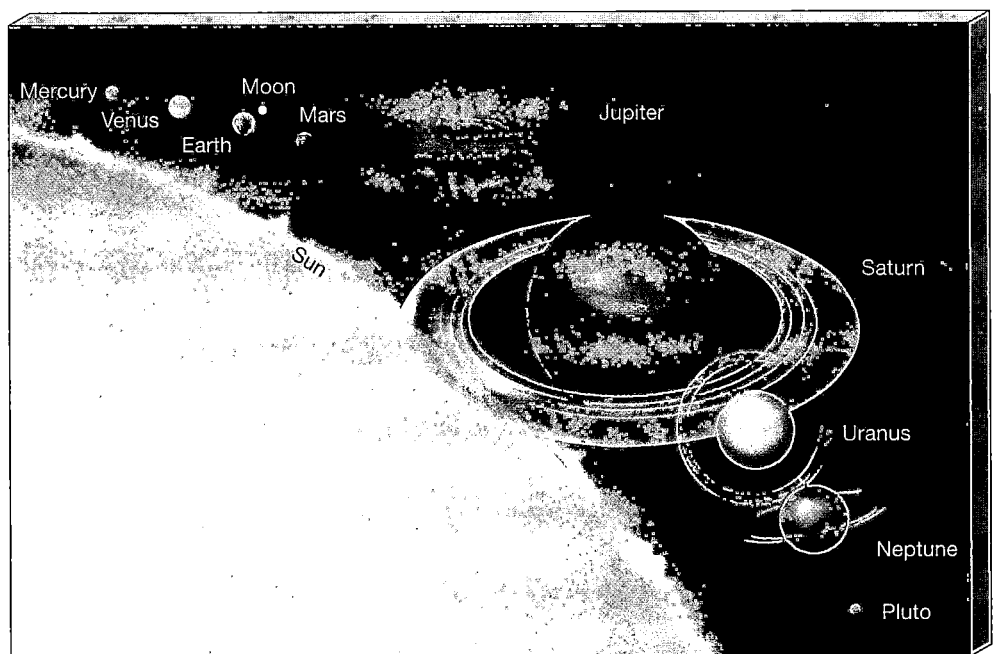

Figure 6.7 Sun and Planets Diagram, drawn to scale, of the relative sizes of the planets and our Sun. Notice how much larger the jovian planets are than Earth and the other terrestrials, and how much larger still is the Sun.

Animation   *Animation*

Comparing the average densities of the terrestrial planets allows us to say something about their overall *compositions*. However, before making the comparison, we must first take into account how the weight of overlying layers compresses the interiors of the planets to different extents. When we do this, we find that the *uncompressed densities* of the terrestrial worlds—the densities they would have in the absence of any compression due to their own gravity—decrease as we move outward from the Sun: 5300, 4400, 4400, and 3800 kg/m^3 for Mercury, Venus, Earth, and Mars, respectively. The amount of compression is greatest for the most massive planets, Earth and Venus, and much less for Mercury and Mars. Based in part on these figures, planetary scientists conclude that Earth and Venus are quite similar in overall composition. Mercury's higher density implies that it contains a higher proportion of some dense material—most likely nickel or iron. The lower density of Mars probably means that it is deficient in that same material.

Yet for all their differences, the terrestrial worlds still seem very similar when compared with the jovian planets. Perhaps the simplest way to express the major differences between the terrestrial and jovian worlds is to say that the jovian planets are everything the terrestrial planets are not. Table 6.2 compares and contrasts some key properties of these two planetary classes.

The terrestrial worlds lie close together, near the Sun; the jovian worlds are widely spaced through the outer solar system. The terrestrial worlds are small, dense, and rocky; the jovian worlds are large and gaseous, being made up predominantly of hydrogen and helium (the lightest elements), which are rare on the inner planets. The terrestrial worlds have solid surfaces; the jovian worlds have none (their dense atmospheres thicken with depth, eventually merging with their liquid interiors). The terrestrial worlds have weak magnetic fields, if any; the jovian worlds all have strong magnetic fields. The terrestrial worlds have only three moons among them; the jovian worlds have many moons each, no two of them alike and none of them like our own. Furthermore, all the jovian planets have *rings*, a feature unknown on the terrestrial planets. Finally, all four jovian worlds are thought to contain large, dense "terrestrial" cores some 10 to 15 times the mass of Earth. These cores account for an increasing fraction of each planet's total mass as we move outward from the Sun.

Beyond the outermost jovian planet, Neptune, lies one more small world, frozen and mysterious. Pluto doesn't fit well into either planetary category. Indeed, there is ongoing debate among planetary scientists as to whether it should be classified as a planet at all. In both mass and composition, it has much more in common with the icy jovian moons than with any terrestrial or jovian planet. Many astronomers suspect that it may in fact be the largest member of a newly recognized class of solar system objects that reside beyond the jovian worlds (see Chapter 14). In 1999 the International Astronomical

TABLE 6.2 Comparison of the Terrestrial and Jovian Planets

TERRESTRIAL PLANETS	JOVIAN PLANETS
close to the Sun	far from the Sun
closely spaced orbits	widely spaced orbits
small masses	large masses
small radii	large radii
predominantly rocky	predominantly gaseous
solid surface	no solid surface
high density	low density
slower rotation	faster rotation
weak magnetic fields	strong magnetic fields
few moons	many moons
no rings	many rings

Union, which oversees the rules for classifications in astronomy, decided that Pluto should, for now at least, still be called a planet. However, that status may well change as the makeup of the outer solar system becomes better understood.

☑ Concept Check

▫ Why do astronomers draw such a clear distinction between the inner and the outer planets?

6.5 Interplanetary Debris

In the vast space among the nine known major planets move countless chunks of rock and ice, all orbiting the Sun, many on highly eccentric paths. This final component of the solar system is the collection of *interplanetary matter*—cosmic "debris" ranging in size from the relatively large asteroids, through the smaller comets and even smaller meteoroids, down to the smallest grains of interplanetary dust that litter our cosmic environment.

The dust arises when larger bodies collide and break apart into smaller pieces that, in turn, collide again and are slowly ground into microscopic fragments, which eventually settle into the Sun or are swept away by the *solar wind*, a stream of energetic charged particles that continually flows outward from the Sun and pervades the entire solar system. The dust is quite difficult to detect in visible light, but infrared studies reveal that interplanetary space contains surprisingly large amounts of it. Our solar system is an extremely good vacuum by terrestrial standards, but positively dirty by the standards of interstellar or intergalactic space.

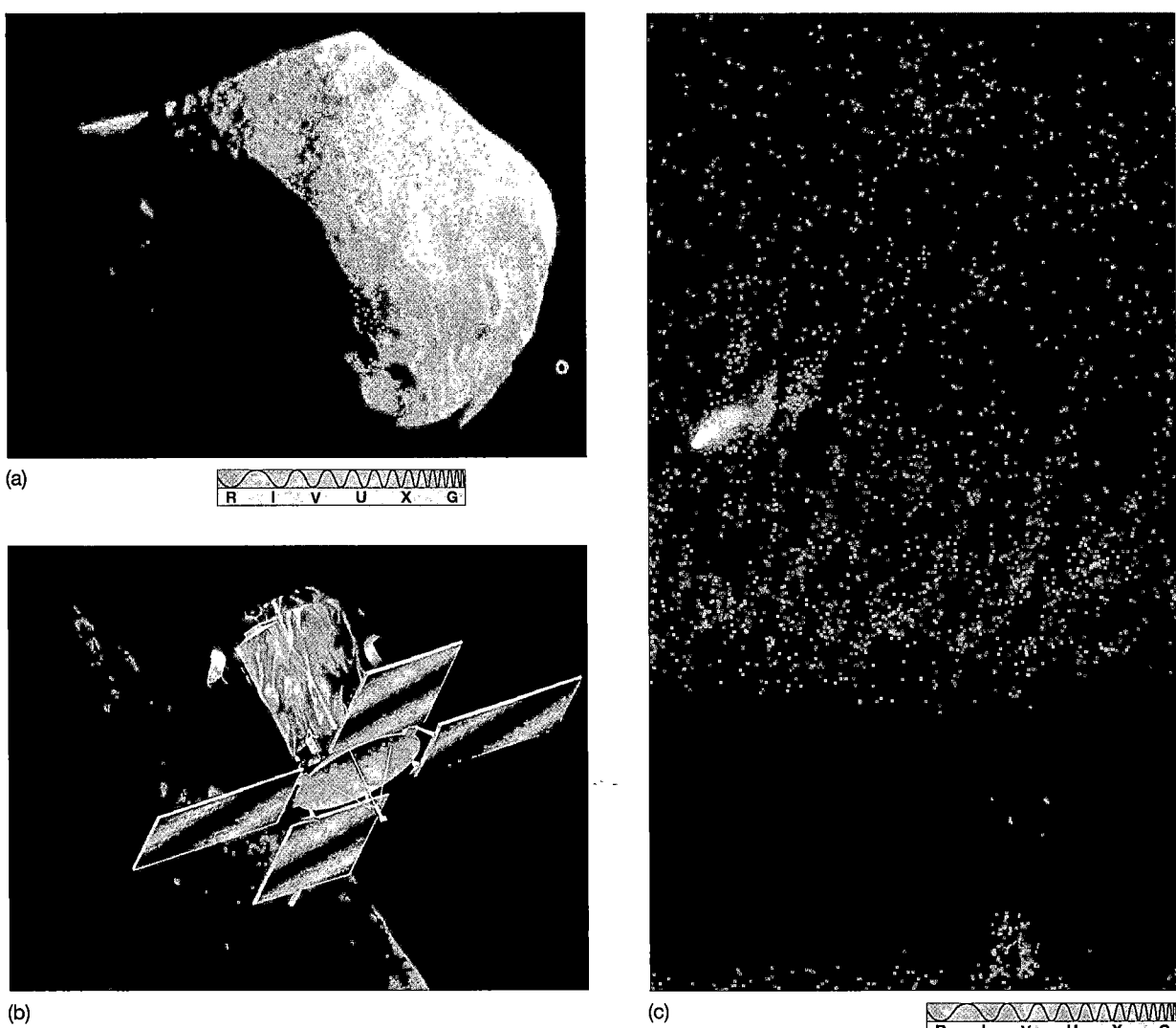

(a)

(b)

(c)

Figure 6.8 Asteroid and Comet (a) Asteroids, like meteoroids, are generally composed of rocky material. This asteroid, Mathilde, is about 52 km long; it was photographed by the *NEAR* (*Near Earth Asteroid Rendezvous*) spacecraft, shown in (b) as an artist's conception. (c) Comet Hale-Bopp, seen as it approached the Sun in 1997. Most comets are composed largely of ice, and so tend to be relatively fragile. Its vaporized gas and dust form the tail, here extending away from the Sun for nearly a quarter of the way across the sky. *(NASA; J. Lodriguss)*

Asteroids (Figure 6.8a) and meteoroids are generally rocky in composition, somewhat like the outer layers of the terrestrial planets. The distinction between the two is simply a matter of size—anything larger than 100 m in diameter (corresponding to a mass of about 10,000 tons) is conventionally termed an asteroid; anything smaller is a meteoroid. Their total mass is much less than that of Earth's Moon, so these objects play no important role in the present-day workings of the planets or their moons. Yet they are of crucial importance to our studies, for they are the keys to answering some very fundamental questions about our planetary environment and what the solar system was like soon after its birth. Many of these bodies

are made of material that has evolved hardly at all since the early days of the solar system. (In addition, they often conveniently deliver themselves right to our doorstep, in the form of meteorites, allowing us to study them in detail without having to fetch them from space.)

Comets are quite distinct from the other small bodies in the solar system. They are generally icy rather than rocky in composition (although they do contain some rocky material) and typically have diameters in the 1–10 km range. They are quite similar in chemical makeup to some of the icy moons of the outer planets. Even more so than the asteroids and meteoroids, comets represent truly ancient material—the vast majority have probably not changed in any significant way since their formation long ago along with the rest of the solar system (see Chapter 15). Comets striking Earth's atmosphere do not reach the surface intact, so we do not have actual samples of cometary material. However, they do vaporize and emit radiation as their highly elongated orbits take them near the Sun (see Figure 6.8c). Astronomers can determine a comet's makeup by spectroscopic study of the radiation it gives off as it is destroyed. ⊂⊃ (Sec. 4.2)

✔ Concept Check

■ Why are astronomers so interested in interplanetary matter?

6.6 Spacecraft Exploration of the Solar System

📚 Since the 1960s dozens of unmanned space missions have traveled throughout the solar system. All the planets but Pluto have been visited and probed at close range, and spacecraft have visited numerous comets and asteroids.

The first landing on an asteroid (by the *NEAR* spacecraft—see Figure 6.8b) occurred in February 2001. The impact of these missions on our understanding of our planetary system has been nothing short of revolutionary. In the next few chapters we will see many examples of the marvelous images radioed back to Earth. Here, we focus on just a few of these remarkable technological achievements.

THE *MARINER 10* FLYBYS OF MERCURY

In 1974, the U.S. spacecraft *Mariner 10* came within 10,000 km of the surface of Mercury, sending back high-resolution images of the planet. These photographs, which showed surface features as small as 150 m across, dramatically increased our knowledge of the planet. For the first time, we saw Mercury as a heavily cratered world, in many ways reminiscent of our own Moon.

Mariner 10 was launched from Earth in November 1973 and was placed in an eccentric 176-day orbit about the Sun, aided by a gravitational assist (see *Discovery 6-2*) from the planet Venus (Figure 6.9). In that orbit, *Mariner 10*'s nearest point to the Sun (perihelion) is close to Mercury's path, and its farthest point away (aphelion) lies between the orbits of Venus and Earth (*More Precisely 2-1*). The 176-day period is exactly two Mercury years, so the

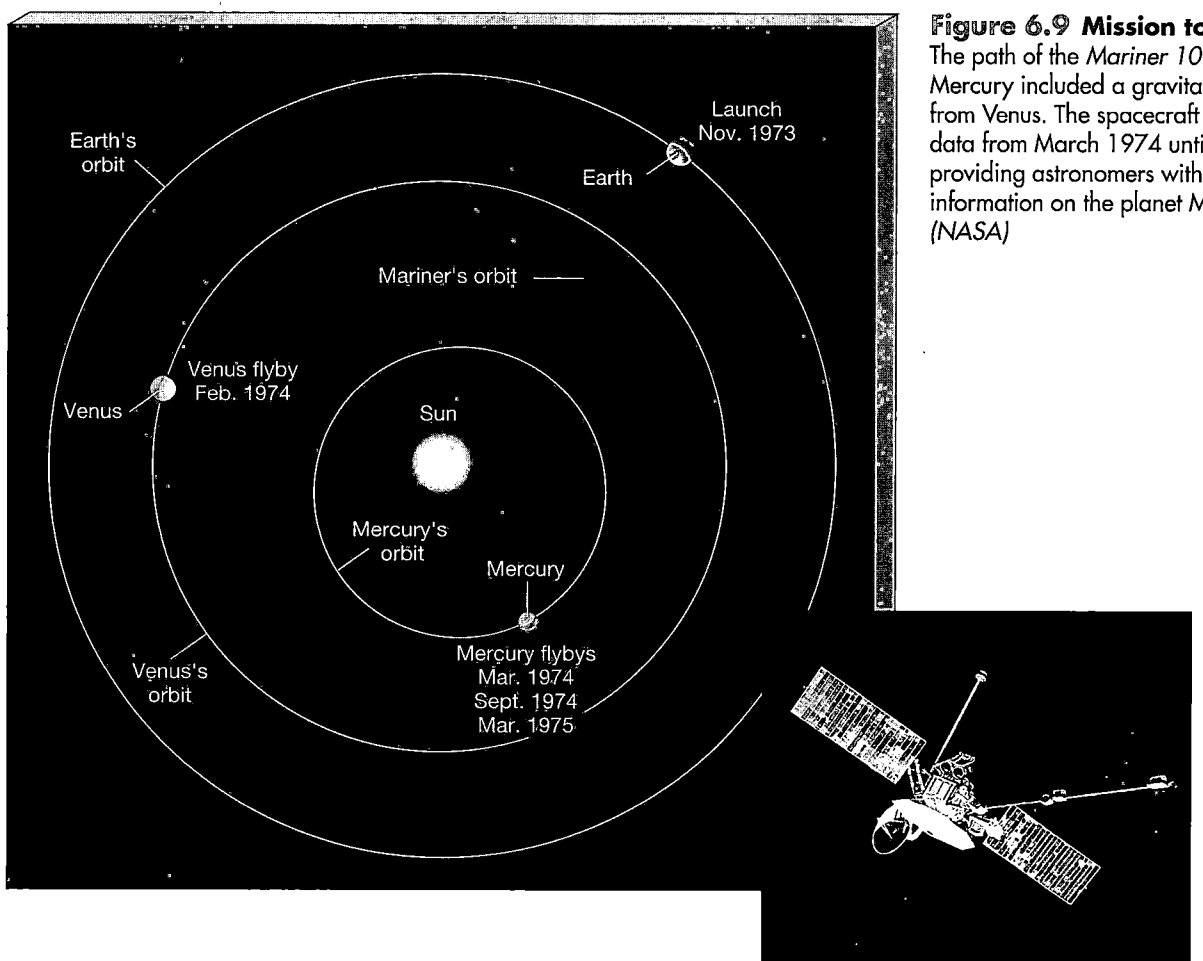

Figure 6.9 Mission to Mercury
The path of the *Mariner 10* probe to Mercury included a gravitational boost from Venus. The spacecraft (inset) returned data from March 1974 until March 1975, providing astronomers with a wealth of information on the planet Mercury. *(NASA)*

DISCOVERY 6-2

Gravitational "Slingshots"

Celestial mechanics is the study of the motions of gravitationally interacting objects, such as planets and stars, applying Newton's laws of motion to understand the intricate movements of astronomical bodies. ∞ (Sec. 2.7) Computerized celestial mechanics lets astronomers calculate planetary orbits to high precision, taking the planets' small gravitational influences on one another into account. Even before the computer age, the discovery of one of the outermost planets, Neptune, came about almost entirely through studies of the distortions of Uranus's orbit that were caused by Neptune's gravity.

Celestial mechanics is also an essential tool for scientists and engineers who wish to navigate manned and unmanned spacecraft throughout the solar system. Robot probes can now be sent on stunningly accurate trajectories, expressed in the trade with such slang phrases as "sinking a corner shot on a billion-kilometer pool table." Near-flawless rocket launches, aided by occasional midcourse changes in flight paths, now enable interplanetary navigators to steer remotely controlled spacecraft through an imaginary "window" of space just a few kilometers wide and a billion kilometers away.

However, sending a spacecraft to another planet requires a lot of energy—often more than can be conveniently provided by a rocket launched from Earth or safely transported in a shuttle for launch from orbit. Faced with these limitations, mission scientists often use their knowledge of celestial mechanics to carry out "slingshot" maneuvers, which can boost an interplanetary probe into a more energetic orbit and also aid navigation toward the target, all at no additional cost!

The accompanying figure illustrates a gravitational slingshot, or *gravity assist*, in action. A spacecraft approaches a planet, passes close by, then escapes along a new trajectory. Obviously, the spacecraft's *direction* of motion is

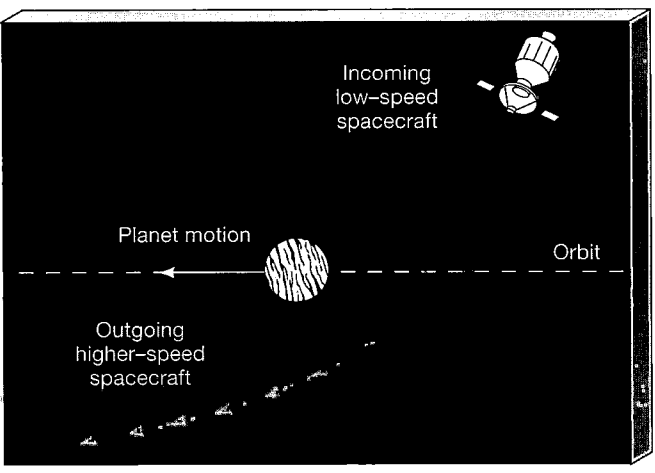

changed by the encounter. Less obvious, the spacecraft's *speed* is also altered as the planet's gravity propels the spacecraft in the direction of the planet's motion. By careful choice of incoming trajectory, the craft can speed up (by passing "behind" the planet, as shown) *or* slow down (by passing in front), by as much as twice the planet's orbital speed. Of course, there is no "free lunch"—the spacecraft gains energy from, or loses it to, the planet's motion, causing its orbit to change ever so slightly. However, since planets are so much more massive than spacecraft, the effect on the planet is tiny.

Such a slingshot maneuver has been used many times in missions to both the inner and the outer planets, as illustrated in the second figure, which shows the trajectories of the *Voyager* spacecraft through the outer solar system. The gravitational pulls of these giant worlds whipped the craft around at each visitation, enabling flight controllers to get considerable extra "mileage" out of the probes. *Voy-*

spacecraft revisits Mercury roughly every six months. However, only on the first three encounters—in March 1974, September 1974, and March 1975—did the spacecraft return data. After that, the craft's supply of maneuvering fuel was exhausted. In total, over 4000 photographs, covering about 45 percent of the planet's surface, were radioed back to Earth during the mission's active lifetime. The remaining 55 percent of Mercury is still unexplored.

No new missions have been sent to Mercury since *Mariner 10*. NASA plans a return in 2009, when the *Messenger* probe will be placed in orbit around the planet to map its entire surface at much higher resolution than was possible with *Mariner*. Both the European and the Japanese Space Agencies also have plans to place spacecraft in orbit around Mercury at roughly the same time.

EXPLORATION OF VENUS

In all, some 20 spacecraft have visited Venus since the 1970s, far more than have spied on any other planet in the solar system. The Soviet space program took the lead in exploring Venus's atmosphere and surface, while American spacecraft have performed extensive radar mapping of the planet from orbit. The American *Mariner 2* and *Mariner 5* missions passed within 35,000 km of the planet in 1962 and 1967, and *Mariner 10* grazed Venus at a distance of 6000 km en route to its rendevous with Mercury. During roughly the same period, the Soviet *Venera* (derived from the Russian word for Venus) program got under way, and the Soviet *Venera 4* through *Venera 12* probes parachuted into the planet's atmosphere between 1967 and 1978.

ager 1 is now high above the plane of the solar system, having been deflected up and out of the ecliptic plane following its encounter with Saturn. *Voyager 2* continued on for a "Grand Tour" of the four jovian planets. It is now outside the orbit of Pluto.

More recently, the *Galileo* mission to Jupiter, which was launched in 1989 and arrived at its target in 1995, received *three* gravitational assists en route—one from Venus and two from Earth. Once in the Jupiter system, *Galileo*

used the gravity of Jupiter and its moons to propel it through a complex series of maneuvers designed to bring it close to all the major moons as well as to the planet itself (see Figure 6.15). Every encounter with a moon had a slingshot effect—sometimes accelerating and sometimes slowing the probe, but each time moving it onto a different orbit—and every one of these effects was carefully calculated long before *Galileo* ever left Earth.

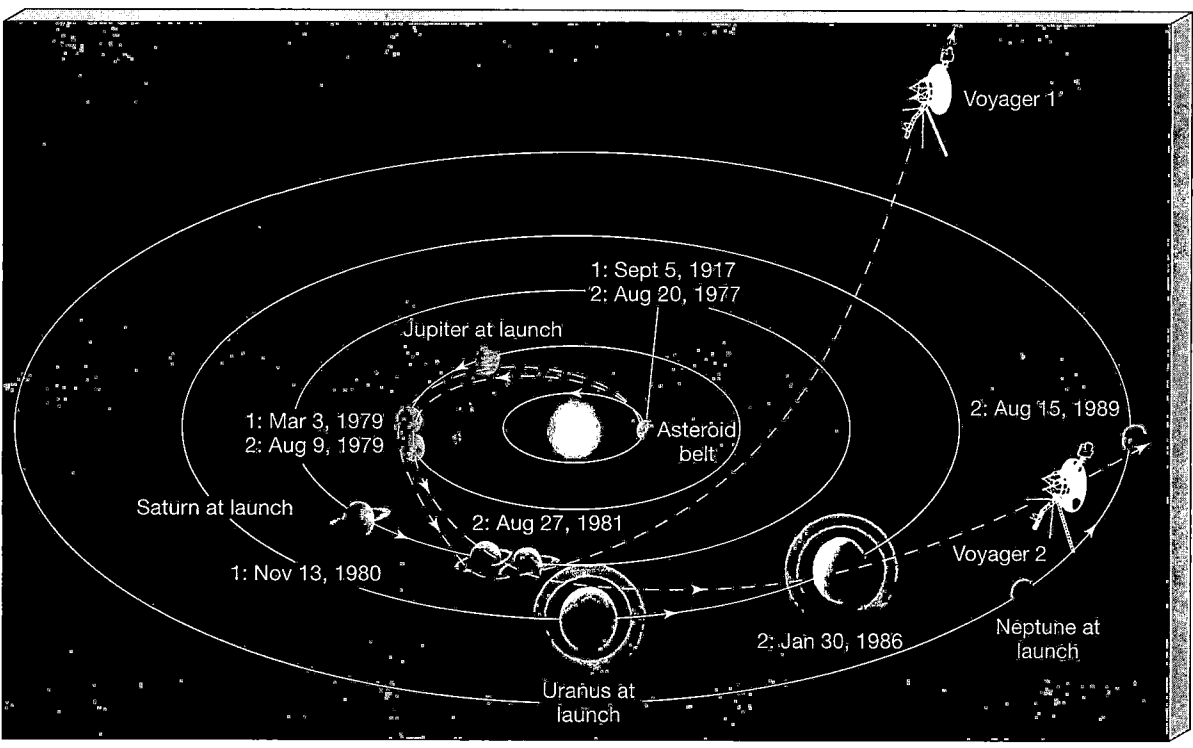

The early *Venera* probes were destroyed by enormous atmospheric pressures before reaching the surface, but in 1970, *Venera 7* (Figure 6.10) became the first spacecraft to soft-land on the planet. During the 23 minutes it survived on the surface, it radioed back information on atmospheric pressure and temperature. Since then, a number of *Venera* landers have transmitted photographs of the surface back to Earth and have analyzed the atmosphere and the soil. None of them survived for more than an hour in the planet's hot, dense atmosphere. The data they sent back make up the entirety of our direct knowledge of Venus's surface. In 1983 the *Venera 15* and *Venera 16* orbiters sent back detailed radar maps (at about 2-km resolution) of large portions of Venus's northern hemisphere.

The U.S. *Pioneer Venus* mission in 1978 placed an orbiter at an altitude of some 150 km above Venus's surface

and dispatched a "multiprobe" consisting of five separate instrument packages into the planet's atmosphere. During its hour-long descent to the surface, the probe returned information on the variation of density, temperature, and chemical composition with altitude in the atmosphere. The orbiter's radar produced images of most of the planet's surface.

The most recent U.S. mission was the *Magellan* probe (shown in Figure 6.11), which entered orbit around Venus in August 1990. It completed its first 243-day mapping cycle (the time required for Venus to rotate once beneath the probe's orbit) in May 1991, its radar imaging system achieving spatial resolution at least 10 times better than the best data previously obtained. It could distinguish objects as small as 120 m across and measure vertical distances to within less than 50 m. Between 1991 and 1994

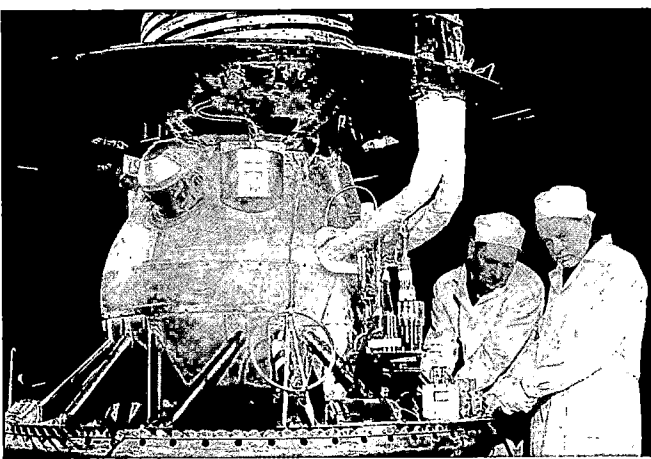

Figure 6.10 Venus Lander One of the Soviet *Venera* landers that reached the surface of Venus. The design was essentially similar for all the surface missions. Note the heavily armored construction, necessary to withstand the harsh conditions of high temperature and crushing atmospheric pressure on the planet's surface. *(Sovfoto/Eastfoto)*

the probe mapped 98 percent of the surface of Venus with unprecedented clarity and made detailed measurements of the planet's gravity, rendering all previous data virtually obsolete. The mission ended in October 1994 with a (planned) plunge into the planet's dense atmosphere, sending back one final stream of high-quality data. Many theories of the processes shaping the planet's surface have had to be radically altered or abandoned completely because of *Magellan*'s data. We will see many examples of *Magellan*'s spectacular imagery and learn more about the mission's scientific findings in Chapter 9.

EXPLORATION OF MARS

Both NASA and the Soviet (now Russian) Space Agency have Mars exploration programs that began in the 1960s. However, the Soviet effort was plagued by a string of technical problems, along with a liberal measure of plain bad luck. As a result, almost all the detailed planetary data we have on Mars has come from unmanned U.S. probes.

The first spacecraft to reach the Red Planet was *Mariner 4*, which flew by Mars in July 1965. The images sent back by the craft showed large numbers of craters caused by impacts of meteoroids with the planet's surface, but nothing of the Earthlike terrain some scientists had expected to find. Flybys in 1969 by *Mariner 6* and *Mariner 7* confirmed these findings, leading to the conclusion that Mars was a geologically dead planet having a heavily cratered, old surface. Studies of Mars received an enormous boost with the arrival in November 1971 of the *Mariner 9* orbiter. The craft mapped the entire Martian surface at a resolution of about 1 km, and it rapidly became clear that here was a world far more complex than the dead

planet imagined only a year or two previously. *Mariner 9*'s maps revealed vast plains, volcanoes, drainage channels, and canyons. All these features were completely unexpected, given the data provided by the earlier missions. These new findings paved the way for the next step—actual landings on the planet's surface.

The two U.S. *Viking* spacecraft arrived at Mars in mid-1976. *Viking 1* and *Viking 2* each consisted of two parts. An orbiter mapped the surface at a resolution of about 100 m (about the same as the resolution achieved by *Magellan* on Venus), and a lander (see Figure 6.12) descended to the surface and performed a wide array of geological and biological experiments. *Viking 1* touched down on Mars on July 20, 1976. *Viking 2* arrived in September of the same year. By any standards, the *Viking* mission was a complete success; the orbiters and landers returned a wealth of long-term data on the Martian surface and atmosphere. *Viking 2* stopped transmitting data in April 1980. *Viking 1* continued to operate until November 1982.

The relative positions of Mars and Earth in their respective orbits mean that it is most favorable to launch a spacecraft from Earth roughly every 26 months. It then takes six to nine months for the craft to arrive at Mars. In August 1993 the first U.S. probe since *Viking*—*Mars Observer*, which was designed to radio back detailed images of the planet's surface and provide data on the Martian atmosphere, gravity, and magnetic field—exploded just before entering Mars' orbit. A replacement for *Mars Observer*, called *Mars Global Surveyor*, was launched from Earth in 1996, arriving at Mars in late 1997. This space-

Figure 6.11 *Magellan* Orbiter The U.S. *Magellan* spacecraft is launched from the space shuttle *Atlantis* in May 1989 on a mission to explore the planet Venus. *(NASA)*

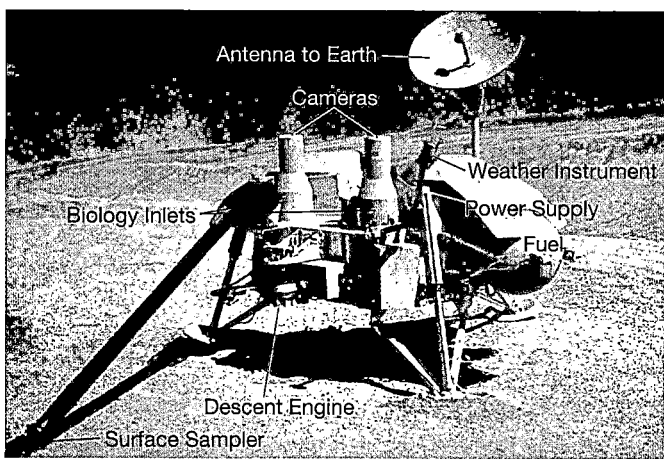

Figure 6.12 Viking Lander A *Viking* lander, here being tested in the Mojave Desert prior to launch. *(NASA)*

craft is currently orbiting the planet, scanning it with cameras and other sensors, and charting the Martian landscape. The mission is scheduled to end in 2002.

Mars Global Surveyor was followed (and in fact overtaken) by *Mars Pathfinder*, which arrived at Mars in late June 1997. On July 4 *Pathfinder* parachuted an instrument package to the Martian surface. Near the ground, the parachute fell away and huge airbags deployed, enabling the robot to bounce softly to a safe landing. Side panels opened, and out came a small six-wheeled minirover, called *Sojourner* (Figure 6.13). During the unexpectedly long lifetime of its mission (it lasted almost three months instead of the anticipated one), the lander performed measurements of the Martian atmosphere and atmospheric dust while *Sojourner* roamed the Martian countryside at a rate of a few meters per day, carrying out chemical analyses of the soil and rocks within about 50 m of the parent craft. In addition, over 16,000 images of the region were returned to Earth.

NASA's next two Mars exploration missions met with failure. In September 1999 *Mars Climate Orbiter* missed its proper orbit insertion point and instead burned up in the

Martian atmosphere, apparently as a result of navigational commands being sent to the spacecraft's onboard computer in English, rather than in metric, units. In December of the same year, the sister mission *Mars Polar Lander* failed to reestablish contact with Earth after entering the Martian atmosphere en route to a planned soft landing on the planet's south polar icecap. The reason is still unknown.

Despite these recent setbacks, the U.S. Mars program has now moved into high gear, with an ambitious program of missions spanning the first decade of the twenty-first century. A new orbiter, called *Mars Odyssey* is scheduled to reach Mars in October 2001. The year 2003 promises to be a particularly busy one in Mars exploration, as NASA will place two *Sojourner*-like landers on the planet's surface, while the European Space Agency will launch its *Mars Express* mission, consisting of an orbiter and a third lander. In 2007, if all goes according to plan, a NASA landing craft will return the first ever samples of Martian surface material to Earth (or at least Earth orbit—the issues of possible contamination of the sample by Earth bacteria, or vice versa, are still being debated).

NASA also has plans for manned missions to Mars. However, the enormous expense (and danger) of such an undertaking, coupled with the belief of many astronomers that unmanned missions are economically and scientifically preferable to manned missions, make the future of these projects uncertain at best.

MISSIONS TO THE OUTER PLANETS

Two pairs of U.S. spacecraft launched in the 1970s—the *Pioneer* and *Voyager* missions—revolutionized our knowledge of Jupiter and the jovian planets. *Pioneer 10* and *Pioneer 11* were launched in March 1972 and April 1973, respectively, arriving at Jupiter in December 1973 and December 1974.

The *Pioneer* spacecraft took many photographs and made numerous scientific discoveries. Their orbital trajectories also allowed them to observe the polar regions of Jupiter in much greater detail than later missions would achieve. In addition to their many scientific accomplishments, the *Pioneer* craft also played an important role as

Figure 6.13 Sojourner on Mars The Mars rover *Sojourner*, part of the *Mars Pathfinder* mission, shown here shortly after reaching the Martian surface on July 4, 1997. The parachute and the airbags used in the landing surround the vehicle. *(NASA)*

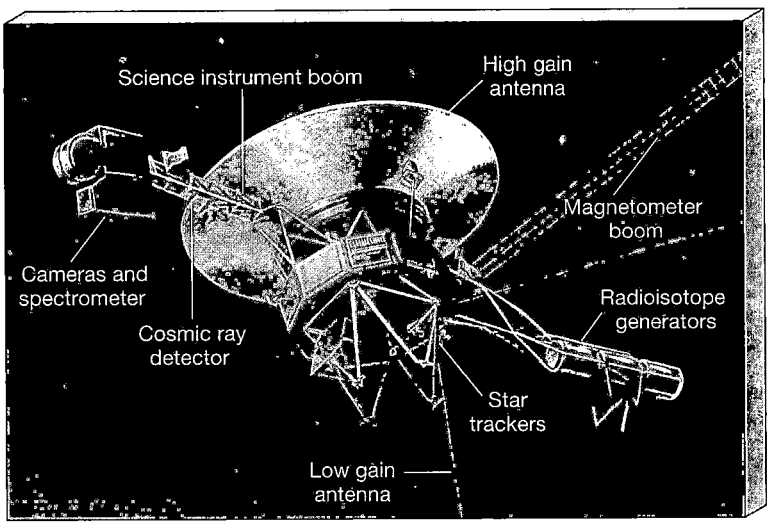

Figure 6.14 Voyager The *Voyager* spacecraft. *Voyager 1* and *Voyager 2* (shown here) were identical. *(NASA)*

"scouts" for the later *Voyager* missions. The *Pioneer* series demonstrated that spacecraft could travel the long route from Earth to Jupiter without colliding with debris in the solar system. They also discovered—and survived—the perils of Jupiter's extensive radiation belts (somewhat like Earth's Van Allen belts, but on a much larger scale). In addition, *Pioneer 11* used Jupiter's gravity to propel it along the same trajectory to Saturn that the *Voyager* controllers planned for *Voyager 2*'s visit to Saturn's rings.

The two *Voyager* spacecraft (see Figure 6.14) left Earth in 1977, reaching Jupiter in March (*Voyager 1*) and July (*Voyager 2*) of 1979 to study the planet and its major satel-

lites in detail. Each craft carried sophisticated equipment to study the planet's magnetic field, as well as radio, visible-light, and infrared sensors to analyze its reflected and emitted radiation. Both *Voyager 1* and *Voyager 2* used Jupiter's gravity to send them on to Saturn. *Voyager 1* was programmed to visit Titan, Saturn's largest moon, and so did not come close enough to the planet to receive a gravity-assisted boost to Uranus. However, *Voyager 2* went on to visit both Uranus and Neptune in a spectacularly successful "Grand Tour" of the outer planets. The data returned by the two craft are still being analyzed today. Like *Pioneer 11*, the two *Voyager* craft are now headed out of the solar system, still sending data as they race toward interstellar space. The figure in *Discovery 6-2* shows the past and present trajectories of the *Voyager* spacecraft.

The most recent mission to Jupiter is the U.S. *Galileo* probe, launched by NASA in 1989. It arrived at its target in 1995 after a rather roundabout route involving a gravity assist from Venus and two from Earth itself. The mission consisted of an orbiter and an atmospheric probe. The probe descended into Jupiter's atmosphere in December 1995, slowed by a heat shield and a parachute, making measurements and chemical analyses as it went. The orbiter executed a complex series of gravity-assisted maneuvers through Jupiter's moon system, returning to some moons already studied by *Voyager* and visiting others for the first time (see Figure 6.15). Some of *Galileo*'s main findings are described in Chapter 11. The mission was scheduled to end in December 1997, but was so successful that NASA extended its lifetime for four more years to obtain even more detailed data on Jupiter's inner moons. As of mid 2001, the spacecraft was still operational. NASA's next planned visit to Jupiter is the 2003 *Europa Orbiter*

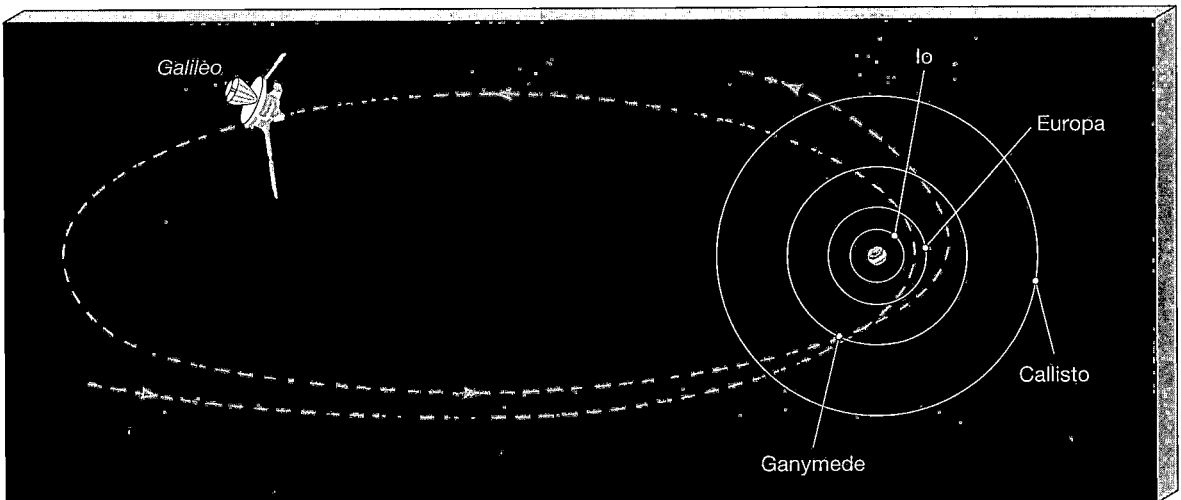

Figure 6.15 Galileo at Jupiter This is a highly simplified diagram showing *Galileo* swinging around some of the interior moons of Jupiter. Each time it reconnoitered the moons, the probe would enter a slightly different orbit, allowing it to see more detail during hundreds of close encounters throughout the mission.

mission, dedicated to a detailed study of Europa, one of the planet's inner moons.

In October 1997 NASA launched the *Cassini* mission to Saturn. The launch (from Cape Canaveral) sparked controversy because of fears that the craft's plutonium power source might contaminate parts of our planet following an accident either during launch or during a subsequent gravity assist from Earth in 1999—one of four needed for the

craft to reach Saturn's distant orbit (Figure 6.16). When the craft reaches its destination in 2004, it will dispatch a probe built by the European Space Agency into the atmosphere of Titan, Saturn's largest moon, and orbit among the planet's moons for four years, much like *Galileo* at Jupiter. If experience with *Galileo* is any guide, *Cassini* will likely resolve many outstanding questions about the Saturn system, but it is sure to pose many new ones too.

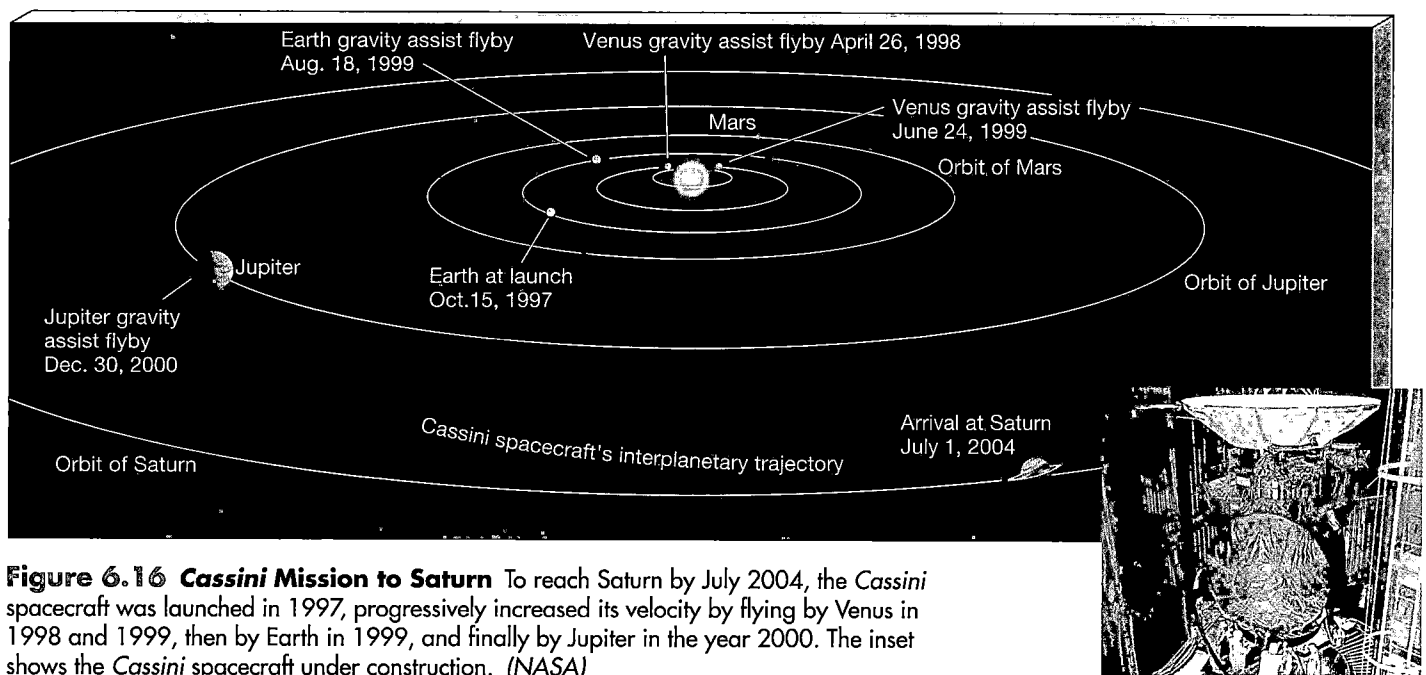

Figure 6.16 *Cassini* **Mission to Saturn** To reach Saturn by July 2004, the *Cassini* spacecraft was launched in 1997, progressively increased its velocity by flying by Venus in 1998 and 1999, then by Earth in 1999, and finally by Jupiter in the year 2000. The inset shows the *Cassini* spacecraft under construction. *(NASA)*

Chapter Review

SUMMARY

The **solar system** (p. 145) consists of the Sun and everything that orbits it, including the nine major planets, the moons that orbit them, and the many small bodies found in interplanetary space. The science of **comparative planetology** (p. 145) compares and contrasts the properties of the diverse bodies found in the solar system and elsewhere to understand better the conditions under which planets form and develop. The asteroids, or "minor planets," are small bodies, none of them larger than Earth's Moon, most of which orbit in a broad band called the asteroid belt between the orbits of Mars and Jupiter. Comets are chunks of ice found mostly in the outer solar system. Their importance to planetary astronomy lies in the fact that they are thought to be "leftover" material from the formation of the solar

system and therefore contain clues to the very earliest stages of its development.

The major planets orbit the Sun in the same sense—counterclockwise as viewed from above Earth's North Pole—on roughly circular orbits that lie close to the ecliptic plane. The orbits of the innermost planet, Mercury, and the outermost, Pluto, are the most eccentric and have the greatest orbital inclination. The spacing between planetary orbits increases as we move outward from the Sun.

Density (p. 147) is a convenient measure of the compactness of any object. The average density of a planet is obtained by dividing the planet's total mass by its volume. The innermost four planets in the solar system have average densities comparable to

Earth's and are generally rocky in composition. The outermost planets have much lower densities than the terrestrial worlds and, with the exception of Pluto, are made up mostly of gaseous or liquid hydrogen and helium.

Planetary scientists divide the eight large planets (excluding Pluto) in the solar system on the basis of their densities and composition, into the rocky **terrestrial planets** (p. 150)—Mercury, Venus, Earth, and Mars—which lie closest to the Sun, and the gaseous **jovian planets** (p. 150)—Jupiter, Saturn, Uranus, and

Neptune, which lie at greater distances. Compared with the terrestrial worlds, the jovian planets are larger and more massive, rotate more rapidly, and have stronger magnetic fields. In addition, the jovian planets all have ring systems and many moons orbiting them. All the major planets, with the exception of Pluto, have been visited by unmanned space probes. Spacecraft have landed on Venus and Mars. In many cases, the spacecrafts' trajectories have included "gravitational assists" from one or more planets to reach their destinations.

SELF-TEST: TRUE OR FALSE?

_____ **1.** Moons are small bodies that orbit the Sun.

_____ **2.** Most planets orbit the Sun in nearly the same plane as Earth.

_____ **3.** The Titius–Bode law, giving the spacing of planetary orbits, is not a true law at all and is not very accurate.

_____ **4.** The largest planets also have the largest densities.

_____ **5.** The total mass of all the planets is about half the mass of the Sun.

_____ **6.** A planet with a density of 5000 kg/m^3 most likely has a gaseous composition.

_____ **7.** The other terrestrial planets all rotate more rapidly than Earth.

_____ **8.** The jovian planets all rotate more rapidly than Earth.

_____ **9.** The solar wind occurs on Earth during days of exceptional solar heating.

_____ **10.** _Mars Polar Lander_ took samples of the planet's polar ice caps.

_____ **11.** All planets have moons.

_____ **12.** The landing of the Soviet _Venera_ 7 on the planet Venus was notable because it was the first time any spacecraft had ever soft-landed on another planet.

_____ **13.** _Mars Pathfinder_ is the last U.S. spacecraft scheduled to visit Mars.

_____ **14.** Both _Voyager_ missions visited all four jovian planets.

_____ **15.** The _Cassini_ mission was designed to explore the atmosphere and moon system of Jupiter.

SELF-TEST: FILL IN THE BLANK

1. The major bodies orbiting the Sun are known as _____.

2. The _____ are bodies that orbit the Sun between the orbits of Mars and Jupiter.

3. The two planets with the highest eccentricities and orbital tilts are _____ and _____.

4. The inner planets Mercury, Venus, Earth, and Mars are known as the _____ planets.

5. The outer planets Jupiter, Saturn, Uranus, and Neptune are known as the _____ planets.

6. The largest and most massive planet in the solar system is _____.

7. Asteroids and meteoroids have a _____ composition, in contrast with comets, which have an _____ composition.

8. Asteroids are somewhat similar in composition to the _____ planets.

9. Comets have compositions similar to the _____ moons of the _____ planets.

10. The _Mariner 10_ spacecraft was sent by _____ (country) to photograph the planet _____.

11. The U.S _Magellan_ probe mapped the entire surface of Venus using _____.

12. The _Viking 1_ and _Viking 2_ missions sent orbiters and landers to the planet _____.

13. Jupiter was first visited by the U.S. spacecraft _____.

14. The U.S. spacecraft _____ is the only probe to have visited each of the giant outer planets.

15. Spacecraft visiting one planet often use a _____ assist to visit another planet.

REVIEW AND DISCUSSION

1. Name and describe all the different types of objects found in the solar system. Give one distinguishing characteristic of each. Include a mention of interplanetary space.

2. What is the order of the planets, from the closest to the farthest from the Sun?

3. What is comparative planetology? Why is it useful? What is its ultimate goal?

4. What are extrasolar planets?

5. Compare and contrast Kepler's laws with the Titius–Bode "law" (_Discovery 6-1_). Why are Kepler's laws considered to be true natural laws, whereas the Titius–Bode "law" is not?

6. Why is it necessary to know the distance to a planet in order to determine its mass?

7. List some ways in which the solar system is an "orderly" place.

8. What are some "disorderly" characteristics of the solar system?

9. Which are the terrestrial planets? Why are they given this name?

10. Which are the jovian planets? Why are they given this name?

11. Name three important differences between the terrestrial planets and the jovian planets.

12. What is uncompressed density, and what does it tell us about the terrestrial planets?

13. Compare the properties of Pluto given in Table 6.1 with the properties of the terrestrial and jovian planets in Table 6.2. What do you conclude regarding the classification of Pluto as either a terrestrial or jovian planet?

14. Why are asteroids and meteoroids important to planetary scientists?

15. Comets generally vaporize upon striking Earth's atmosphere. How then do we know their composition?

16. Why has our knowledge of the solar system increased greatly in recent years?

17. How and why do scientists use gravity assists to propel spacecraft through the solar system?

18. Which planets have been visited by spacecraft from Earth? On which ones have spacecraft actually landed?

19. Why do you think *Galileo* and *Cassini* took such circuitous routes to Jupiter and Saturn, while *Pioneer* and *Voyager* did not?

20. How do you think NASA's new policy of building less complex, smaller, and cheaper spacecraft—with shorter times between design and launch—will affect future exploration of the outer planets? Will missions like *Galileo* and *Cassini* be possible in future?

PROBLEMS  *Algorithmic versions of these questions are available in the Practice Problems module of the Companion Website.*

The number of squares preceding each problem indicates its approximate level of difficulty.

1. ■ Choose two of the objects listed in Table 6.1 and calculate their densities. Compare your results with those given in the table. Assume that all objects are spherical.

2. ■ Use Newton's law of gravity to compute your weight (a) on Earth, (b) on Mars, (c) on the asteroid Ceres, and (d) on Jupiter (neglecting temporarily the absence of a solid surface on this planet!). ⊂⊃ (Sec. 2.6)

3. ■ Only Mercury, Mars, and Pluto have orbits that deviate significantly from circles. Calculate the perihelion and aphelion distances from the Sun of these planets. (*More Precisely 2-1*)

4. ■■■ At closest approach, the planet Neptune lies roughly 29.1 A.U. from Earth. At that distance, its angular diameter is 2.3″. Its moon Triton moves in a circular orbit with an angular diameter of 33.6″ and a period of 5.9 days. Use these data to compute the radius, mass, and density of Neptune, and compare with the figures given in Table 6.1.

5. ■■■ Use the data given in Table 6.1 to calculate the angular diameter of Saturn when it lies 9 A.U. from Earth. Saturn's moon Titan is observed to orbit 3.1′ from the planet. What is Titan's orbital period?

6. ■ According to the Titius–Bode "law," (*Discovery 6-1*) where should the ninth, tenth, and eleventh planets in the solar system lie?

7. ■ Suppose the average mass of each of the 7000 asteroids in the solar system is about 10^{17} kg. Compare the total mass of all asteroids with the mass of Earth.

8. ■ Assuming a roughly spherical shape and a density of 3000 kg/m³, estimate the diameter of an asteroid having the average mass given in the previous question.

9. ■■ A *short-period comet* is conventionally defined as a comet having an orbital period of less than 200 years. What is the maximum possible aphelion distance for a short-period comet with a perihelion of 0.5 A.U.? Where does this place the comet relative to the outer planets?

10. ■ How many times has *Mariner 10* now orbited the Sun?

11. ■■ What is the aphelion distance of *Mariner 10*'s orbit?

12. ■■ The asteroid Icarus has a perihelion distance of 0.2 A.U. and an orbital eccentricity of 0.7. What is its aphelion distance from the Sun?

13. ■■■ A spacecraft has an orbit that just grazes Earth's orbit at perihelion and that of Mars at aphelion. What is its orbital eccentricity and semimajor axis? How long does it take to go from Earth to Mars? (This is the so-called *minimum energy orbit* for a craft leaving Earth and reaching Mars. Assume circular planetary orbits for simplicity.)

14. ■■■ Earth and Mars were at closest approach in March 1997. *Mars Pathfinder* was launched on December 4, 1996, and arrived at Mars in late June 1997. Sketch the orbits of the two planets and the trajectory of the spacecraft. Be sure to indicate on your diagram the location of Mars at launch and of Earth when *Pathfinder* reached Mars. (For simplicity, neglect the eccentricity of Mars's orbit in your sketch.)

15. ■■ How long would it take for a radio signal to complete the round-trip between Earth and Saturn? Assume that Saturn is at its closest point to Earth. How far would a spacecraft orbiting the planet in a circular orbit of radius 100,000 km travel in that time? Do you think mission control could maneuver the spacecraft in real time—that is, control all its functions directly from Earth?

COLLABORATIVE EXERCISES

1. **Planetary Density.** Calculate the average density of any planet of your group's choosing, besides Earth, using the planetary radius and planetary mass as provided in the appendix and compare it with Earth's average density.

2. **Classification of Pluto.** Tables 6.1 and 6.2 list properties of the eight largest planets. As a group, decide if Pluto should be classified as a planet or not and explain your reasoning.

3. **Solar System Exploration.** As a group, decide which of the space missions described in the text produced the most interesting results about the solar system and explain your reasoning.

RESEARCHING ON THE WEB

To complete the following exercises, go to the online Destinations module for Chapter 6 on the Companion Website for Astronomy Today 4/e.

1. Access the "Earth Viewer" page and request a color composite showing clouds, land and sea temperatures, and ice. Note the data and UTC time and determine which continents currently are primarily in daylight, which are primarily in darkness, and which continent has the most amount of cloud cover.

2. Access the "Current Events and Research" pages from The Electronic Volcano and list the three locations of the three most recent eruptions.

PROJECTS

1. You can begin to visualize the ecliptic—the plane of the planets' orbits—just by noticing the path of the Sun throughout the day and of the full Moon in the course of a single night. It helps if you watch from one spot, such as your backyard or a rooftop. It's also good to have a general notion of direction. (West is where the Sun sets!) You will see that the movements of the Sun and Moon are confined to a narrow pathway across our sky. The planets also travel along this path. The motion of the Sun, Moon, and planets is a two-dimensional reflection of the three-dimensional plane of our solar system.

2. Once you get a feeling for the whereabouts of the ecliptic, try locating the North Star. Knowing the direction to ce-

lestial north makes it easier to imagine the motion of the planets in the plane of the solar system. Don't worry about being too precise. Just get a sense of the ecliptic as a kind of merry-go-round of planets—that we on Earth also ride!

3. Go to your library and find what planetary missions are in progress or are planned, other than those described in the text. Apart from the United States, what other countries have space agencies actively engaged in planetary exploration? What nonplanetary missions are in progress or planned?

SKYCHART III PROJECTS  *The SkyChart III Student Version planetarium program on which these exercises are based is included as a separately executable program on the CD in the back of this text.*

1. ■ Observe the apparent motion of Venus and Mercury by centering the display on the Sun with *VIEW/Center Planet/Sun.* Deselect *DRAW/Horizon Mask, Constellations,* and *Grid Lines.* With a 180° *Field of View* set *ANIMATION/Trail For/Venus* then also select a trail for Mercury. Set *DRAW/Sky Background/Black.* Animate the image with time steps of one day. In interpreting the simulation, remember that you are viewing the events from Earth. If you wish to look at the motion of the inner planets from a more distant vantage point, set location to Jupiter. Then turn off other trails and turn on a trail for Earth. With the animation, you will see the path Earth follows in its orbit around the Sun.

2. ■ Set SkyChart III to display the planets, stars and constellations. Set *DRAW/Sky Background/Color* and select *DRAW/Horizon Mask.* Set time for a convenient viewing time, and using *VIEW/Center Planet,* select the planet you wish to view. Adjust the date and time to bring the planet of choice into view when the Sun is below the horizon. Using this technique, determine when each of the naked-eye planets is visible, and record the constellation in which each will appear when you can observe.

3. ■ View the solar system from the Sun. From the *COMPUTATION* menu, choose *Location/View from Object/Select Sun* at an offset distance of 60 A.U., R.A. 18 00 00, and Dec. 66 00 00.0. Zoom in all the way to the Sun and center on the Sun. Show coordinate grids. Observe the Sun over a time span of a day. What do you observe? Zoom out so you see the inner planets and turn on the trails (*Time Step* of three days). Show a different number of points for each of the planets.

4. ■ Use the *ANIMATION* menu and the Trail option to record the physical positions of outer planets. To see the fainter planets, increase the limiting magnitude for planets to 18, in *Draw/Symbols & Grids.* On the *COMPUTATION* menu, select *Precision,* and choose *Use VFP.* Print a chart displaying one full cycle for the outer planets.

5. ■■ Following the previous exercise, measure the eccentricity of Pluto's orbit by measuring the aphelion and perihelion distance. See *More Precisely 2-1.*

 *In addition to the Practice Problems and Destinations modules, the Companion Website at http://www.prenhall.com/chaisson provides for each chapter an additional true-false, multiple choice, and labeling quiz, as well as additional annotated images, animations, and links to related Websites.*

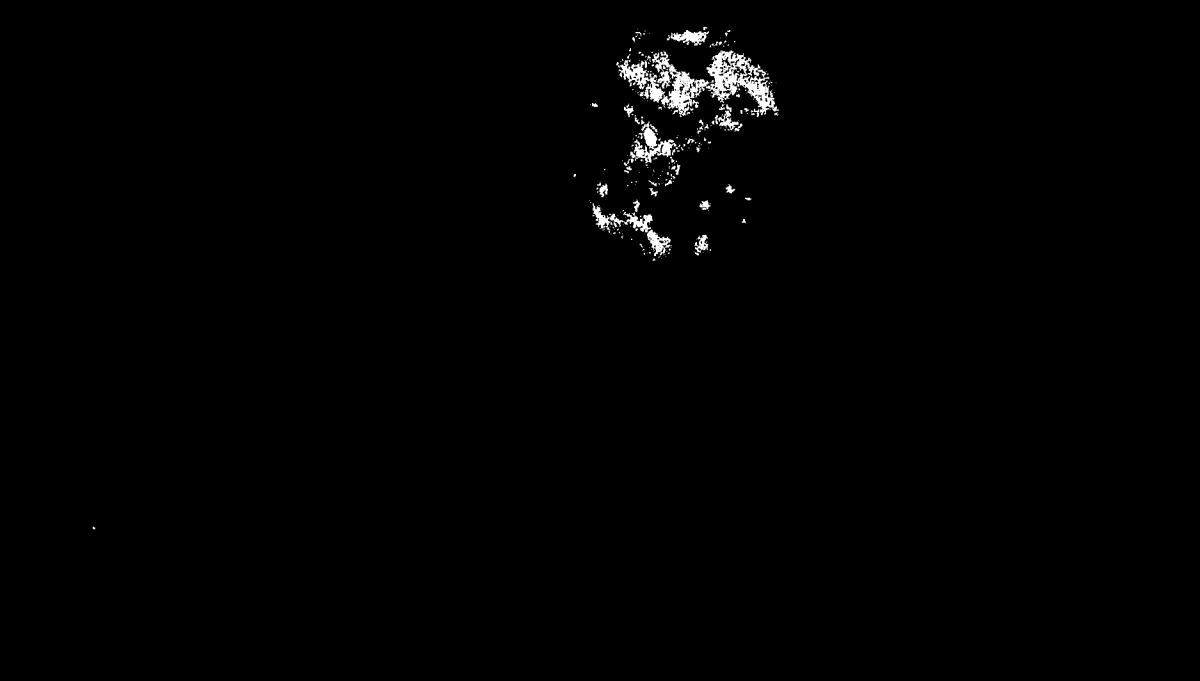
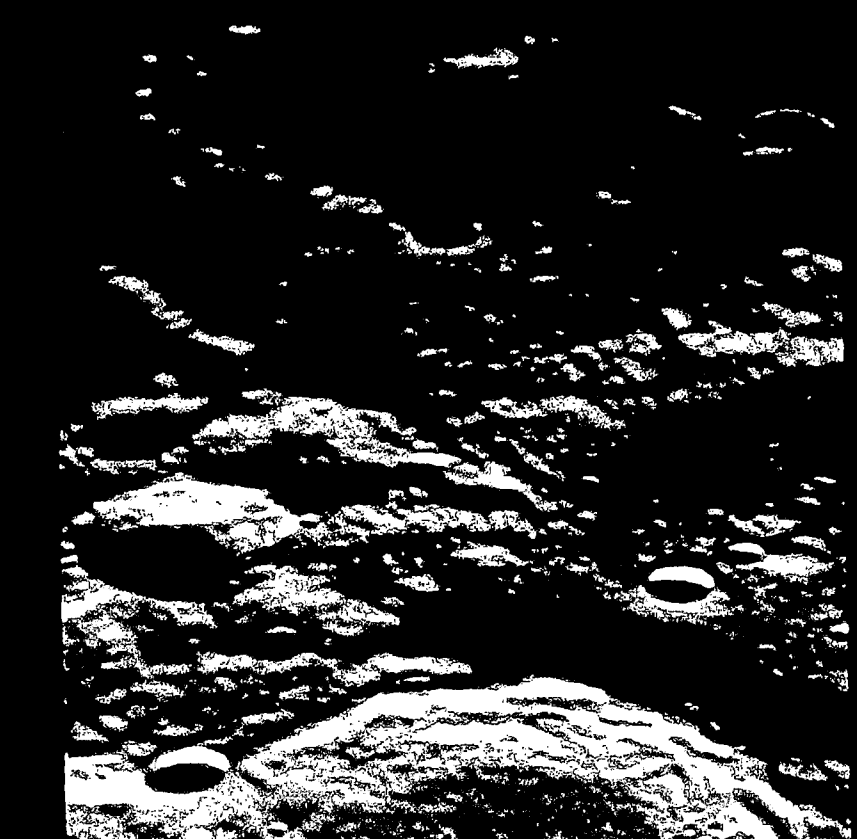

7 EARTH

Our Home in Space

LEARNING GOALS

Studying this chapter will enable you to:

1. Summarize the overall physical properties of planet Earth.

2. Explain how Earth's atmosphere helps to heat us as well as protect us.

3. Outline our current model of Earth's interior structure and describe some of the experimental techniques used to establish this model.

4. Summarize the evidence for the phenomenon of "continental drift" and discuss the physical processes that drive it.

5. Discuss the nature and origin of Earth's magnetosphere.

6. Describe how both the Moon and the Sun influence Earth's oceans.

 Visit http://www.prenhall.com/chaisson for additional annotated images, animations, and links to related sites for this chapter.

Earth's place in space was captured brilliantly in this image radioed back by the robot spacecraft *Clementine* while orbiting the Moon in the mid-1990s. The crater in the foreground is called Plaskett, spanning about 100 kilometers near the Moon's north pole. In the background, the African continent can be clearly seen, nearly cloud-free. *(U.S. Dept. of Defense)*

The Big Picture: Photographs like this one, showing Earth hovering in space like a "small blue marble," help grant an appreciation for our place in space. Akin to the phenomenon of sunrise from our vantage point on Earth, this image neatly captures lunar "Earthrise"—quite a different perspective for residents of Earth. Only during the space age—less than 50 years old—has humankind begun to leave its parent planet and venture into the nearby cosmos. Don't just look at this photo; mentally place yourself on that fragile planet.

Earth is the best-studied terrestrial planet. From the matter of our world sprang life, intelligence, culture, and all the technology we now use to explore the cosmos. We ourselves are "Earthstuff" as much as are rocks, trees, and air. Now, as humanity begins to explore the solar system, we can draw on our knowledge of Earth to aid our understanding of the other planets. By cataloging Earth's properties and attempting to explain them, we set the stage for our comparative study of the solar system. Every piece of information we can glean about the structure and history of our own world plays a potentially vital role in helping us understand the planetary system in which we live. If we are to appreciate the universe, we must first come to know our own planet. Our study of astronomy begins at home.

7.1 Overall Structure of Planet Earth

⬚ The Earth Data box on p. 174 lists in detail some of Earth's physical and orbital properties. These data are determined using techniques conceptually similar to those presented in Chapter 6. ⬚ (Sec. 6.2) Throughout the body of this text, we will use rounded-off numbers whenever possible, taking our planet's mass and radius to be 6.0×10^{24} kg and 6400 km, respectively.

Earth's average density is around 5500 kg/m³. This simple measurement allows us to make a very important deduction about the interior of our planet. The water that makes up much of Earth's surface has a density of 1000 kg/m³, and the rock beneath us on the continents, as well as on the seafloor, has a density in the range 2000–4000 kg/m³. We can immedi-

ately conclude that, because the surface layers have densities much less than the average, much denser material must lie deeper, under the surface. Hence we should expect that much of Earth's interior is made up of very dense matter, far more compact than the densest continental rocks on the surface.

Based on measurements made in many different ways—using aircraft in the atmosphere, satellites in orbit, gauges on the land, submarines in the ocean, and drilling gear below the rocky crust—scientists have built up the following overall picture of our planet. As indicated in Figure 7.1, our planet may be divided into six main regions. In Earth's interior, a thick **mantle** surrounds a smaller, two-part **core**. At the surface we have a relatively thin **crust**, comprising the solid continents and the seafloor, and the **hydrosphere**, which contains the liquid oceans and accounts for some 70 percent of our planet's total surface area. An **atmosphere** of air lies just above the surface. At much greater altitudes, a

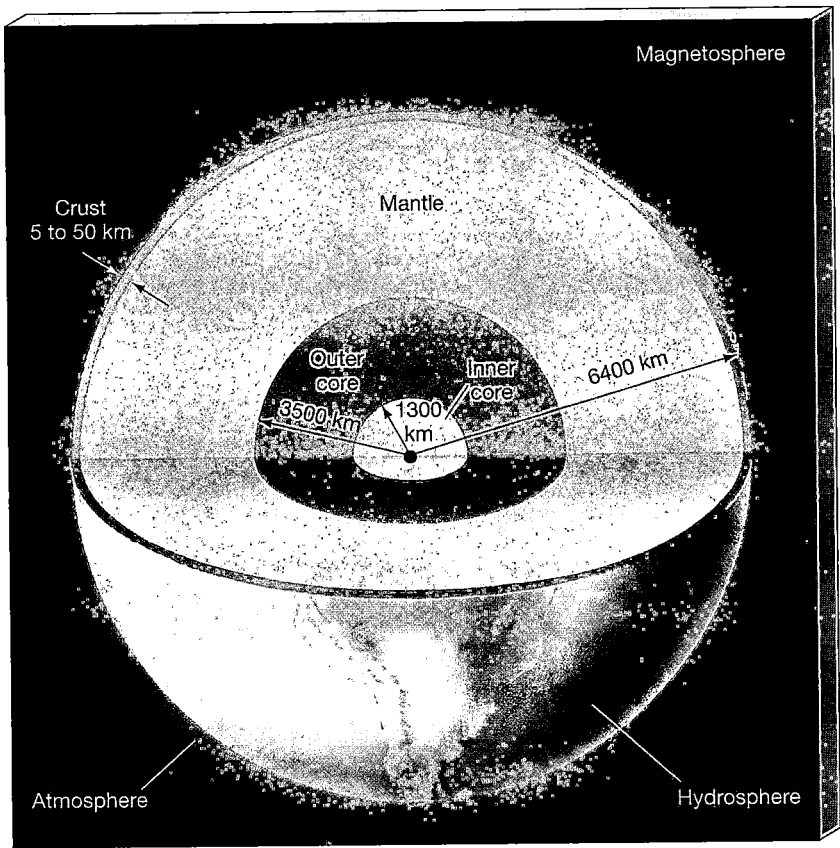

Figure 7.1 Earth The main regions of planet Earth. At Earth's center lies our planet's inner core, about 2600 km in diameter. Surrounding the inner core is an outer core, some 7000 km across. Most of the rest of Earth's 13,000-km diameter is taken up by the mantle, which is topped by a thin crust only a few tens of kilometers thick. The liquid portions of Earth's surface make up the hydrosphere. Above the hydrosphere and solid crust lies the atmosphere, most of it within 50 km of the surface. Earth's outermost region is the magnetosphere, extending thousands of kilometers out into space.

zone of charged particles trapped by our planet's magnetic field forms Earth's **magnetosphere**. Virtually all our planet's mass is contained within the surface and interior. The gaseous atmosphere and the magnetosphere contribute hardly anything—less than 0.1 percent—to the total.

7.2 Earth's Atmosphere

From a human perspective, probably the most important aspect of Earth's atmosphere is that we can breathe it. Air is a mixture of gases, the most common of which are nitrogen (78 percent by volume), oxygen (21 percent), argon (0.9 percent), and carbon dioxide (0.03 percent). Water vapor is a variable constituent, making up anywhere from 0.1–3 percent, depending on location and climatic conditions. The presence of a large amount of oxygen makes our atmosphere unique in the solar system, and the presence of even trace amounts of water and carbon dioxide plays a vital role in the workings of our planet.

ATMOSPHERIC STRUCTURE

Figure 7.2 shows a cross section of our planet's atmosphere. Compared with Earth's overall dimensions, the extent of the atmosphere is not great. Half of it lies within 5 km of the surface, and all but 1 percent is found below 30 km.

The portion of the atmosphere below about 12 km is called the *troposphere*. Above it, extending up to an altitude of 40 to 50 km, lies the *stratosphere*. Between 50 and 80 km from the surface lies the *mesosphere*. Above about 80 km, in the *ionosphere*, the atmosphere is kept partly ionized by solar ultraviolet radiation. These various atmospheric regions are distinguished from one another by the temperature gradient (decreasing or increasing with altitude) in each.

Atmospheric density decreases steadily with increasing altitude, and as the right-hand vertical axis in Figure 7.2 shows, so does pressure. Climbing even a modest mountain—4 or 5 km high, say—clearly demonstrates the thinning of the air in the troposphere. Climbers must wear oxygen masks when scaling the tallest peaks on Earth.

The troposphere is the region of Earth's (or any other planet's) atmosphere where *convection* occurs, driven by the heat of Earth's warm surface. **Convection** is the constant upwelling of warm air and the concurrent downward flow of cooler air to take its place, a process that physically transfers heat from a lower (hotter) to a higher (cooler) level. In Figure 7.3, part of Earth's surface is heated by the Sun. The air immediately above the warmed surface is heated, expands a little, and becomes less dense. As a result, the hot air becomes buoyant and starts to rise. At higher altitudes, the opposite effect occurs: The air gradually cools, grows denser, and sinks back to the ground. Cool air at the surface rushes in to replace the hot buoyant air. In this way, a circulation pattern is established. These

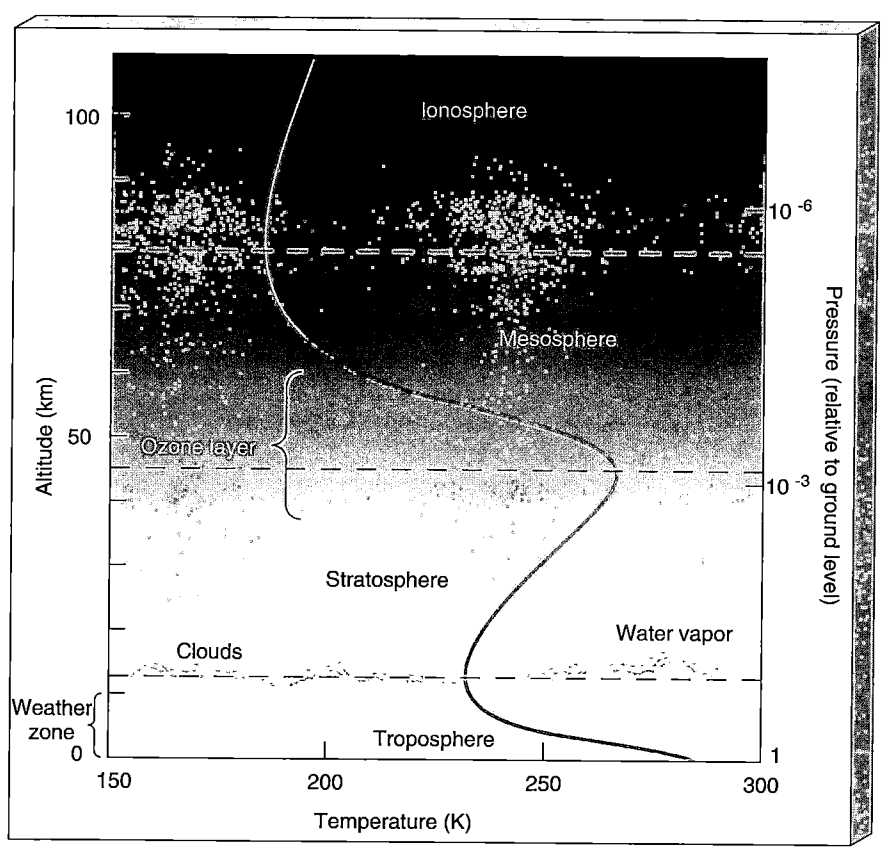

Figure 7.2 Earth's Atmosphere
Diagram of Earth's atmosphere, showing the changes of temperature and pressure from the surface to the bottom of the ionosphere. Note that the various regions are defined by the temperature *gradient*—increasing or decreasing with altitude—in each.

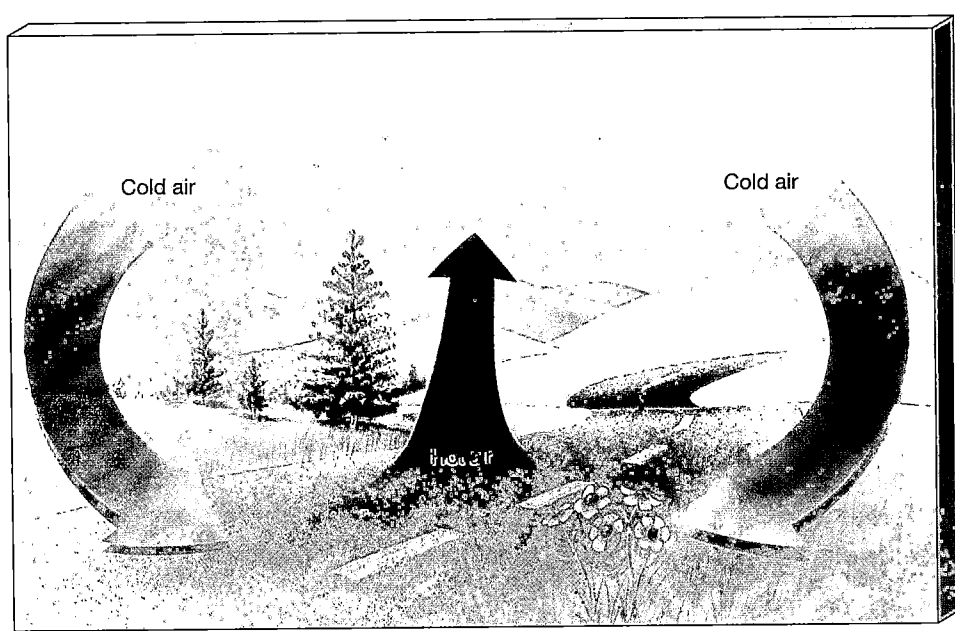

Figure 7.3 Convection
Convection occurs whenever cool fluid overlies warm fluid. The resulting circulation currents are familiar to us as the winds in Earth's atmosphere caused by the solar-heated ground. Hot air rises, cools, and falls repeatedly. Eventually, steady circulation patterns with rising and falling currents are established and maintained, provided that the source of heat (the Sun in the case of the atmosphere) remains intact.

convection cells of rising and falling air not only contribute to atmospheric heating but are also responsible for surface winds. This constant churning motion is responsible for all the weather we experience.

Atmospheric convection can also create clear-air turbulence—the bumpiness we sometimes experience on aircraft flights. Ascending and descending parcels of air, especially below fluffy clouds (themselves the result of convective processes), can cause a choppy ride. For this reason, passenger aircraft tend to fly above most of the turbulence, at the top of the troposphere or in the lower stratosphere, where the atmosphere is stable and the air is calm.

Straddling the boundary between the stratosphere and the mesosphere is the **ozone layer** where, at an altitude of around 50 km, incoming solar ultraviolet radiation is absorbed by atmospheric ozone and nitrogen. (Ozone is a form of oxygen [O_3] consisting of three oxygen atoms combined into a single molecule. Ultraviolet radiation breaks ozone down, forming molecular oxygen [O_2] again.) The ozone layer is one of the insulating spheres that serve to shield life on Earth from the harsh realities of outer space. Not so long ago, scientists judged space to be hostile to advanced life forms because of what is missing out there—breathable air and a warm environment. Now, most scientists regard outer space as harsh because of what is *present* out there—fierce radiation and energetic particles, both of which are injurious to human health. Without the protection of the ozone layer, advanced life (at least on Earth's surface) would be at best unlikely and at worst impossible.

Above about 100 km, in the ionosphere, the atmosphere is significantly ionized by the high-energy portion of the Sun's radiation spectrum, which breaks down molecules into atoms and atoms into ions. The degree of ionization increases with altitude. The presence of many free electrons makes this region of the upper atmosphere a good conductor of electricity, and this conductivity renders the ionosphere highly reflective to certain radio wavelengths. ⚙ (Sec. 3.3) The reason that AM radio stations can be heard well beyond the horizon is that their signals bounce off the ionosphere before reaching the receiver. FM signals cannot be received from stations over the horizon, however, because the ionosphere is transparent to the somewhat shorter wavelengths of radio waves in the FM band.

SURFACE HEATING

2 Much of the Sun's radiation manages to penetrate Earth's atmosphere, eventually reaching the ground. Most of this energy takes the form of visible radiation—ordinary sunlight. Essentially all of the solar radiation not absorbed by or reflected from clouds in the upper atmosphere is absorbed by Earth's surface. The result is that our planet's surface and most objects on it heat up considerably during the day. Earth cannot absorb this solar energy indefinitely, however. If it did, the surface would soon become hot enough to melt, and life on our planet would not exist.

As it heats up, Earth's surface reradiates much of the absorbed energy. This reemitted radiation follows the usual blackbody curve discussed in Chapter 3. ⚙ (Sec. 3.4) As the surface temperature rises, the amount of energy radiated increases rapidly, according to Stefan's law. Eventually, Earth radiates as much energy back into space as it receives from the Sun, and a stable balance is struck. In the absence of any complicating effects, this balance would be achieved at an average surface temperature of about 250 K ($-23°C$). At that temperature, Wien's law tells us that most of the reemitted energy is in the form of infrared (heat) radiation.

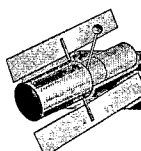

MORE PRECISELY 7-1

Why Is the Sky Blue?

Is the sky blue because it reflects the color of the ocean, or is the ocean blue because it reflects the color of the surrounding sky? The answer is the latter, and the reason has to do with the way that light is *scattered* by air molecules and minute dust particles. By scattering we mean the process by which radiation is absorbed and then reradiated by the material through which it passes.

As sunlight passes through our atmosphere, it is scattered by gas molecules in the air. The British physicist Lord Rayleigh first investigated this phenomenon about a century ago, and today it bears his name—it is known as Rayleigh scattering. The process turns out to be very sensitive to the wavelength of the light involved.

Rayleigh found that blue light is much more easily scattered than red light, in essence because the wavelength of blue light (400 nm) is closer to the size of air molecules than the wavelength of red light (700 nm). He went on to prove that the amount of scattering is actually inversely proportional to the *fourth* power of the wavelength. Dust particles also preferentially scatter blue light by an amount inversely proportional to the wavelength.

EXAMPLE: Let's compare the relative scattering of blue (400 nm) and red (700 nm) light by atmospheric molecules and dust. For Rayleigh scattering, blue light is scattered $(700/400)^4 \approx 9.4$ times more efficiently than red light. That is, blue photons are almost 10 times more likely to be scattered out of a beam of sunlight (taken out of the forward beam and redirected to the side) than are red photons. For scattering by dust, the corresponding factor is $(700/400) = 1.75$—not as big a differential, but still enough to have a large effect when the air happens to be particularly dirty.

Consequently, with the Sun at a reasonably high elevation, the blue component of incoming sunlight will scatter much more than any other color component. Thus, some blue light is removed from the line of sight between us and the Sun and may scatter many times in the atmosphere before eventually entering our eyes, as shown in the accompanying figure. Red or yellow light is scattered relatively little and arrives at our eyes predominantly along the line of sight to the Sun. The net effect is that the Sun is "reddened" slightly, because of the removal of blue light, while the sky away from the Sun appears blue. In outer space, where there is no atmosphere, there is no Rayleigh scattering of sunlight, and the sky is black (although, as we will see in Chapter 18, light from distant stars is reddened in precisely the same way as it passes through clouds of interstellar gas and dust).

At dawn or dusk, with the Sun near the horizon, sunlight must pass through much more atmosphere before reaching our eyes—so much so, in fact, that the blue component of the Sun's light is almost entirely scattered out of the line of sight, and even the red component is diminished in intensity. Accordingly, the Sun itself appears orange—a combination of its normal yellow color and a reddishness caused by the subtraction of virtually all of the blue end of the spectrum—and dimmer than at noon. At the end of a particularly dusty day, when weather conditions or human activities during the daytime hours have raised excess particles into the air, short-wavelength Rayleigh scattering can be so heavy that the Sun appears brilliantly red. Reddening is often especially evident when we look at the westerly "sinking" summer Sun over the ocean, where seawater molecules have evaporated into the air, or during the weeks and months after an active volcano has released huge quantities of gas and dust particles into the air—as was the case in North America when the Philippine volcano Mount Pinatubo erupted in 1991.

(NCAR)

Orbital semi-major axis	1.00 A.U.
	149.6 million km
Orbital eccentricity	0.017
Perihelion	0.98 A.U.
	147.1 million km
Aphelion	1.02 A.U.
	152.1 million km
Mean orbital speed	29.79 km/s
Sidereal orbital period	1.000038 tropical years
Orbital inclination to the ecliptic	0.01°
Mass	5.976×10^{24} kg
Equatorial radius	6378 km
Mean density	5520 kg/m³
Surface gravity*	9.80 m/s²
Escape speed	11.2 km/s
Sidereal rotation period	0.9973 solar days
Axial tilt[†]	23.45°
Magnetic axis tilt relative to rotation axis	11.5°
Mean surface temperature	290 K
Number of moons	1

*Acceleration due to gravity at Earth's surface.

[†]Angle between Earth's rotation axis and perpendicular to Earth's orbital plane (the plane of the ecliptic).

But there are complications. Infrared radiation is partially blocked by Earth's atmosphere. The primary reason for this is the presence of molecules of water vapor and carbon dioxide, which absorb very efficiently in the infrared portion of the spectrum. Even though these two gases are only trace constituents of our atmosphere, they manage to absorb a large fraction of all the infrared radiation emitted from the surface. Consequently, only some of that radiation escapes back into space. The remainder is trapped within our atmosphere, causing the temperature to increase.

This partial trapping of solar radiation is known as the **greenhouse effect**. The name comes from the fact that a very similar process operates in a greenhouse. Sunlight passes relatively unhindered through glass panes, but much of the infrared radiation reemitted by the plants is blocked by the glass and cannot get out. Consequently, the interior of the greenhouse heats up, and flowers, fruits, and vegetables can grow even on cold wintry days.* The radiative processes that determine the temperature of Earth's atmosphere are illustrated in Figure 7.4. Earth's greenhouse effect makes our planet almost 40 K hotter than would otherwise be the case.

The magnitude of the greenhouse effect is very sensitive to the concentration of so-called *greenhouse gases* (that is, gases that absorb infrared radiation efficiently) in the atmosphere. Carbon dioxide and water vapor are the most important of these, although other atmospheric gases (such as methane) also contribute. The amount of carbon dioxide in Earth's atmosphere is increasing, largely as a result of the burning of fossil fuels (principally oil and coal) in the industrialized world. Carbon dioxide levels have increased by over 20 percent in the last century, and they are continuing to rise at a present rate of 4 percent per decade. In Chapter 9 we will see how a runaway increase in carbon dioxide levels in the atmosphere of the planet Venus has radically altered conditions on its surface, causing its temperature to rise to over 700 K. Although no one is predicting that Earth's temperature will ever reach that of Venus, many scientists now believe that this increase, if left unchecked, may result in global temperature increases of several kelvins over the next half century—enough to cause dramatic, and possibly catastrophic, changes in our planet's climate.

ORIGIN OF EARTH'S ATMOSPHERE

Why is our atmosphere made up of its present constituents? Why is it not composed entirely of nitrogen, say, or of carbon dioxide, like the atmospheres of Venus and

---

* Although this process does contribute to warming the interior of a greenhouse, it is not the most important effect. A greenhouse works mainly because its glass panes prevent convection from carrying heat up and away from the interior. Nevertheless, the name "greenhouse effect" to describe the heating effect due to Earth's atmosphere has stuck.

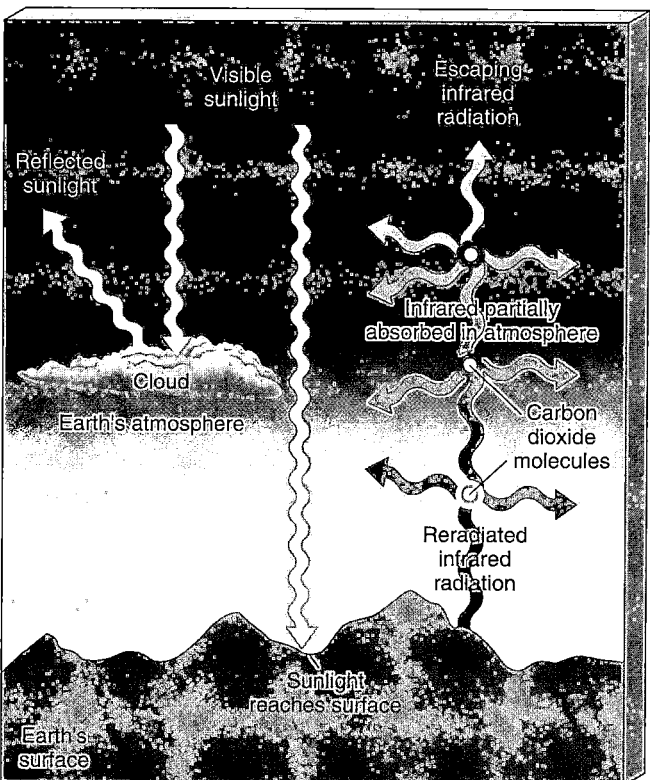

Figure 7.4 Greenhouse Effect Sunlight that is not reflected by clouds reaches Earth's surface, warming it up. Infrared radiation reradiated from the surface is partially absorbed by water vapor and carbon dioxide in the atmosphere, causing the overall surface temperature to rise.

Mars? The origin and development of Earth's atmosphere was a fairly complex and lengthy process.

When Earth first formed, any *primary atmosphere* it might have had would have consisted of the gases most common in the early solar system. These were light gases, such as hydrogen, helium, methane, ammonia, and water vapor—a far cry from the atmosphere we enjoy today. Almost all this light material, and especially any hydrogen or helium, escaped into space during the first half-billion or so years after Earth was formed. (For more information on how planets retain or lose their atmospheres, consult *More Precisely 8-1.*)

Subsequently, Earth developed a *secondary atmosphere*, which was *outgassed* from the planet's interior as a result of volcanic activity. Volcanic gases are rich in water vapor, methane, carbon dioxide, sulfur dioxide, and compounds containing nitrogen (such as nitrogen gas, ammonia, and nitric oxide). Solar ultraviolet radiation decomposed the lighter, hydrogen-rich gases, allowing the hydrogen to escape, and liberated much of the nitrogen from its bonds with other elements. As Earth's surface temperature fell, the water vapor condensed and oceans formed. Much of the carbon dioxide and sulfur dioxide became dissolved in the oceans or combined with surface rocks. Oxygen is such

a reactive gas that any free oxygen that appeared at early times was removed as quickly as it formed. An atmosphere consisting largely of nitrogen slowly appeared.

The final major development in the story of our planet's atmosphere is known so far to have occurred only on Earth. *Life* appeared in the oceans more than 3.5 billion years ago, and organisms eventually began to produce atmospheric oxygen. The ozone layer formed, shielding the surface from the Sun's harmful radiation. Eventually, life spread to the land and flourished. The fact that oxygen is a major constituent of the present-day atmosphere is a direct consequence of the evolution of life on Earth.

✓ Concept Check

■ Why is the greenhouse effect important for life on Earth?

7.3 Earth's Interior

Although we reside on Earth, we cannot easily probe our planet's interior. Drilling gear can penetrate rock only so far before breaking. No substance used for drilling—even diamond, the hardest known material—can withstand the pressure below a depth of about 10 km. That's rather shallow compared with Earth's 6400-km radius. Fortunately, geologists have developed other techniques that indirectly probe the deep recesses of our planet.

SEISMIC WAVES

A sudden dislocation of rocky material near Earth's surface—an *earthquake*—causes the entire planet to vibrate a little. Earth literally rings like a bell. These vibrations are not random, however. They are systematic waves, called **seismic waves** (after the Greek word for "earthquake"), that move outward from the site of the quake. Like all waves, they carry information. This information can be detected and recorded using sensitive equipment—a *seismograph*—designed to monitor Earth tremors.

Decades of earthquake research have demonstrated the existence of many kinds of seismic waves. Two are of particular importance to the study of Earth's internal structure. First to arrive at a monitoring site after a distant earthquake are the primary waves, or *P-waves*. These are *pressure* waves, a little like ordinary sound waves in air, that alternately expand and compress the material medium through which they move. Seismic P-waves usually travel at speeds ranging from 5–6 km/s and can travel through both liquids and solids. Some time later (the actual delay depends on the distance from the earthquake site), secondary waves, or *S-waves*, arrive. These are *shear* waves. Unlike P-waves, which vibrate the material through which they pass back and forth along the direction of travel of the wave, S-waves cause side-to-side motion, more like waves in a guitar string. The two types of waves are illustrated in Figure 7.5. S-waves normally travel through Earth's interior at 3–4 km/s; they cannot travel through liquid, which absorbs them.

The speeds of both P- and S-waves depend on the density of the matter through which the waves are traveling. Consequently, if we can measure the time taken for the waves to move from the site of an earthquake to one or more monitoring stations on Earth's surface, we can determine the density of matter in the interior. Figure 7.6 illustrates some P- and S-wave paths away from the site of an earthquake. Seismographs located around the world measure the times of arrival as well as the strengths of the seismic waves. Both observations contain much useful information—both about the earthquake itself and about Earth's interior through which the waves pass. Notice that the waves do not travel in straight lines through the planet. Because the wave velocity varies with depth, the waves bend as they move through the interior.

A particularly important result emerged after numerous quakes were monitored several decades ago: Seismic stations on the side of Earth opposite a quake never detect

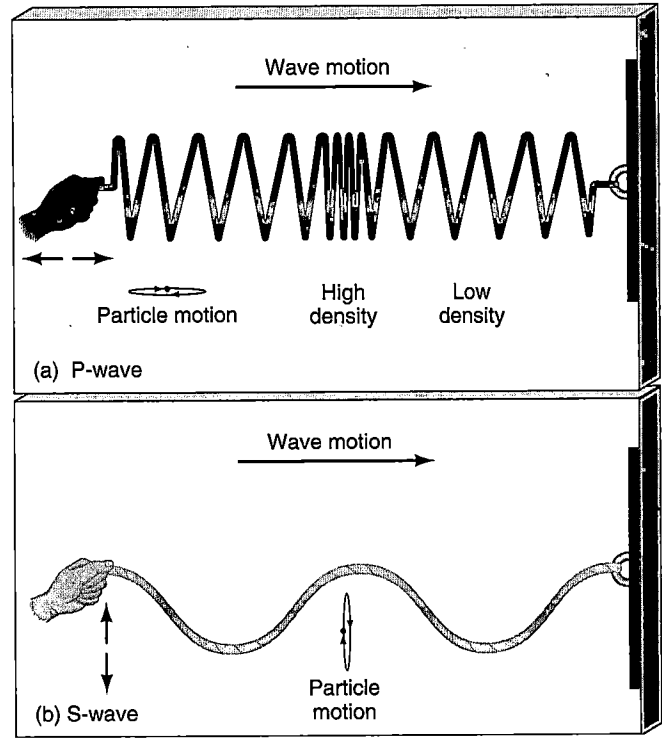

Figure 7.5 P- and S-waves (a) A pressure (P-) wave traveling through Earth's interior causes material to vibrate in a direction parallel to the direction of motion of the wave. Material is alternately compressed and expanded. (b) A shear (S-) wave produces motion perpendicular to the direction in which the wave travels, pushing material from side to side. Also shown is the motion of one typical particle. In case (a) the particle oscillates forward and backward about its initial position. In (b) the particle moves from side to side.

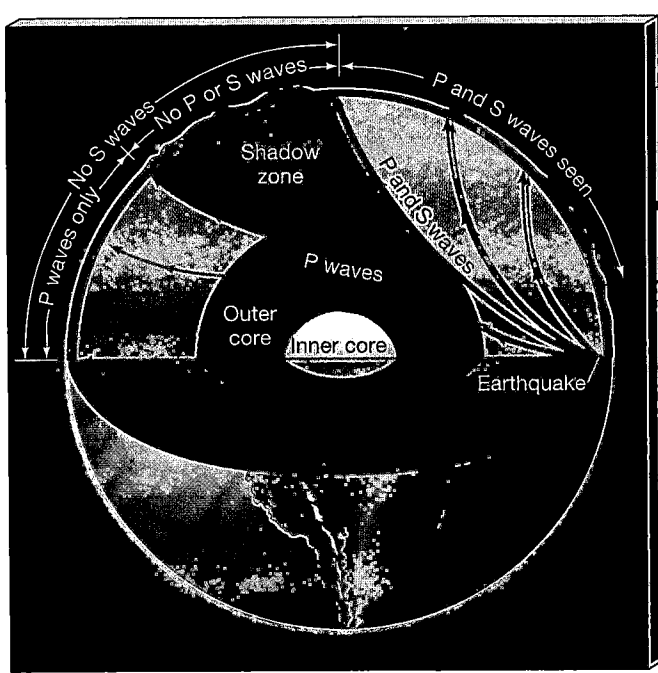

Figure 7.6 Seismic Waves Earthquakes generate pressure (P, or primary) and shear (S, or secondary) waves that can be detected at seismographic stations around the world. S-waves are not detected by stations "shadowed" by the liquid core of Earth. P-waves do reach the side of Earth opposite the earthquake, but their interaction with Earth's core produces another shadow zone, where no P-waves are seen.

S-waves—these waves are blocked by material within Earth's interior. Further, while P-waves always arrive at stations diametrically opposite the quake, there are parts of Earth's surface that they cannot reach (see Figure 7.6). Most geologists believe that S-waves are absorbed by a liquid core at Earth's center and that P-waves are "refracted" at the core boundary, much as light is refracted by a lens. The result is the S- and P-wave "shadow zones" we observe. The fact that every earthquake exhibits these shadow zones is the best evidence that the core of our planet is hot enough to be in the liquid state. *Discovery 7-1* presents some more seismic data on the structure of Earth's core.

The radius of the core, as determined from seismic data, is about 3500 km. In fact, very faint P-waves *are* observed in the P-wave shadow zone indicated in Figure 7.6. These are believed to be reflected off the surface of a solid **inner core**, of radius 1300 km, lying at the center of the liquid **outer core**.

MODELING EARTH'S INTERIOR

Because earthquakes occur often and at widespread places across the globe, geologists have accumulated a large amount of data about shadow zones and seismic-wave properties. They have used these data, along with direct knowledge of surface rocks, to build mathematical models of Earth's interior.

Figure 7.7 presents a model that most scientists accept. Earth's outer core is surrounded by a thick mantle and topped with a thin crust. The mantle is about 3000 km thick and accounts for the bulk (80 percent) of our planet's volume. The crust has an average thickness of only 15 km—a little less (around 8 km) under the oceans and somewhat more (20–50 km) under the continents. The average density of crust material is around 3000 kg/m³. Density and temperature both increase with depth. Specifically, from Earth's surface to its very center, the density increases from roughly 3000 kg/m³ to a little more than 12,000 kg/m³, while the temperature rises from just under 300 K to well over 5000 K. Much of the mantle has

a density midway between the densities of the core and crust: about 5000 kg/m³.

The high central density suggests to geologists that the inner parts of Earth must be rich in nickel and iron. Under the heavy pressure of the overlying layers, these metals (whose densities under surface conditions are around 8000 kg/m³) can be compressed to the high densities predicted by the model. The sharp density increase at the mantle–core boundary results from the difference in composition between the two regions. The mantle is composed of dense but *rocky* material, compounds of silicon and oxygen. The core consists primarily of even denser *metallic* elements. There is no similar jump in density or temperature at the inner core boundary—the material there simply changes from the liquid to the solid state.

The model suggests that the core must be a mixture of nickel, iron, and some other lighter element, possibly sulfur. Without direct observations, it is difficult to be absolutely certain of the light component's identity. All geologists agree that much of the core must be liquid. The existence of the shadow zone demands this (and, as we will see, our current explanation of Earth's magnetic field relies on it). However, despite the high temperature, the pressure near the center—about 4 million times the atmospheric pressure at Earth's surface—is high enough to force the material there into the solid state.

Because geologists have been unable to drill deeper than about 10 km, no experiment has yet recovered a sample of Earth's mantle. However, we are not entirely ignorant of the mantle's properties. In a *volcano*, hot lava upwells from below the crust, bringing a little of the mantle to us and providing some inkling of Earth's interior. The chemical makeup and physical state of newly emerged lava are generally consistent with predictions based on the model sketched in Figure 7.7.

The composition of the upper mantle is probably quite similar to the iron-magnesium-silicate mixtures known as *basalt*. You may have seen some dark gray basaltic rocks scattered across Earth's surface, especially near volcanoes. Basalt is formed as mantle material upwells from Earth's interior as lava, then cools and solidifies. With a density between 3000 kg/m³ and 3300 kg/m³, basalt contrasts with the lighter *granite* (density 2700–3000 kg/m³) that constitutes much of the rest of Earth's crust. Granite is richer than basalt in the light elements silicon and aluminum, which explains why the surface continents do not sink into the interior. Their low-density composition lets the crust "float" atop the denser matter of the mantle and core below.

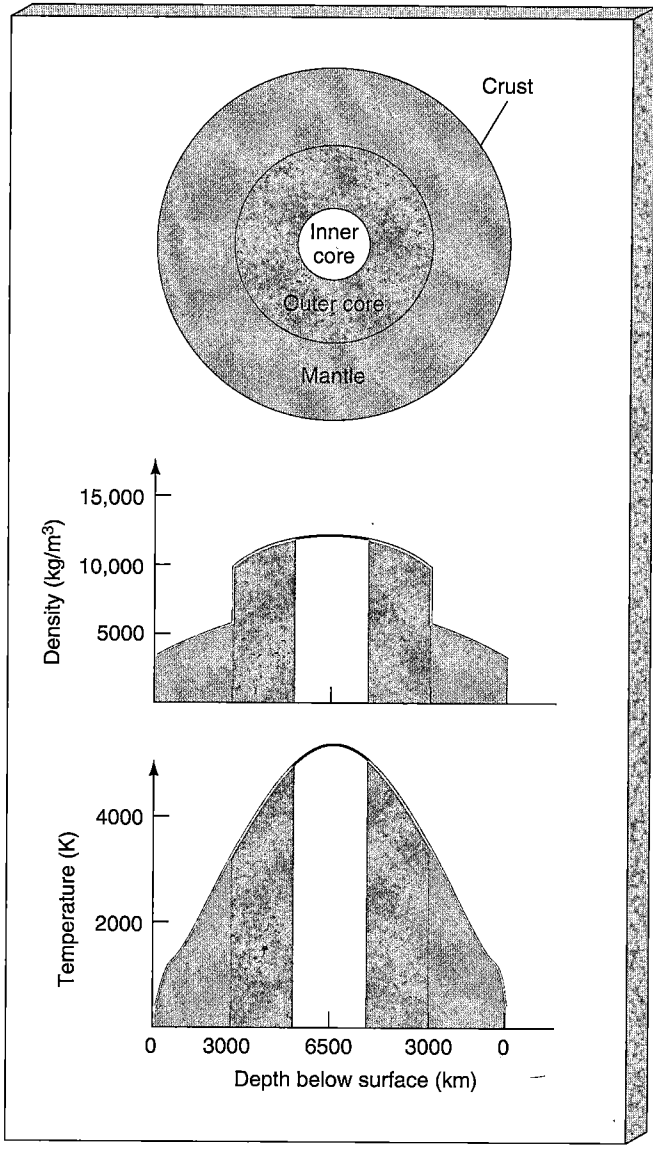

Figure 7.7 Earth's Interior Computer models of Earth's interior imply that the density and temperature vary considerably through the mantle and the core. Note the sharp density discontinuity between Earth's core and mantle.

DIFFERENTIATION

Earth, then, is not a homogeneous ball of rock. Instead, it has a layered structure, with a low-density crust at the surface, intermediate-density material in the mantle, and a high-density core. Such variation in density and composition is known as **differentiation**.

DISCOVERY 7-1

Earth's "Rapidly" Spinning Core

Seismic waves racing out from the site of earthquakes allow geologists to map the interior makeup of planet Earth (Section 7.3). Recent analyses of old earthquake data suggest that Earth's inner core is spinning slightly *faster* than the rest of our planet. The whole planet rotates in the same direction (west to east), but the inner core takes about $\frac{2}{3}$ of a second less to complete its daily rotation than the bulk of the matter surrounding it—which means that it gains a quarter turn each century.

Geologists arrived at this rather surprising result by carefully timing seismic P-waves as they moved through Earth's interior. Actually, the scientists used historical data acquired over the past few decades, mainly from 38 earthquakes that occurred in the south Atlantic Ocean and whose seismic vibrations had to pierce the inner core on their way to monitoring stations in Alaska. As discussed in the text, only P-waves can make it through the liquid outer core, to be detected on the opposite side of Earth.

For many years, geologists have known that P-waves travel through the inner core about 3–4 percent faster along north–south pathways than along those close to the plane of the equator. This difference probably results from the orientation of the iron atoms in the solid inner core; the atoms there must be aligned, as in a crystal lattice. Where the atomic alignment parallels the motion of the P-waves, those P-waves travel just a little faster—a little like going "with the grain" rather than against it.

The new finding is that the route of the fastest seismic waves through Earth's deep interior is gradually shifting eastward, implying that the inner core has some motion slightly different from the rest of the planet. In the accompanying figure, the red region is the solid inner core, the orange region the liquid outer core. The dashed lines passing through the core depict the paths of the fastest seismic waves from 1900 to 1996. The data indicate that the axis of the "fast track" for the P-waves is shifting by about 1° per year relative to the crust above. That axis now intersects Earth's surface in the Arctic Ocean northeast of Siberia; some 30 years ago it was 33° farther west. Gradual as it may be, this shift amounts to about 20 km at the edge of

the inner core—around a million times faster than the typical rates (centimeters per year) at which the continents creep across Earth's surface.

These discoveries may help us understand how the inner core is continuing to grow by condensing from the molten outer core as Earth cools, and how these two regions interact to produce Earth's magnetic field. As noted in Section 7.5, the dynamo theory of magnetism requires rotating, electrically conducting matter within Earth's interior. Detailed modeling had also predicted that a rapidly moving jet stream of partly molten matter—a "mushy zone"—is probably established at the base of the outer core, causing the inner core to "superrotate." Now it seems that this differential rotation pattern has been found.

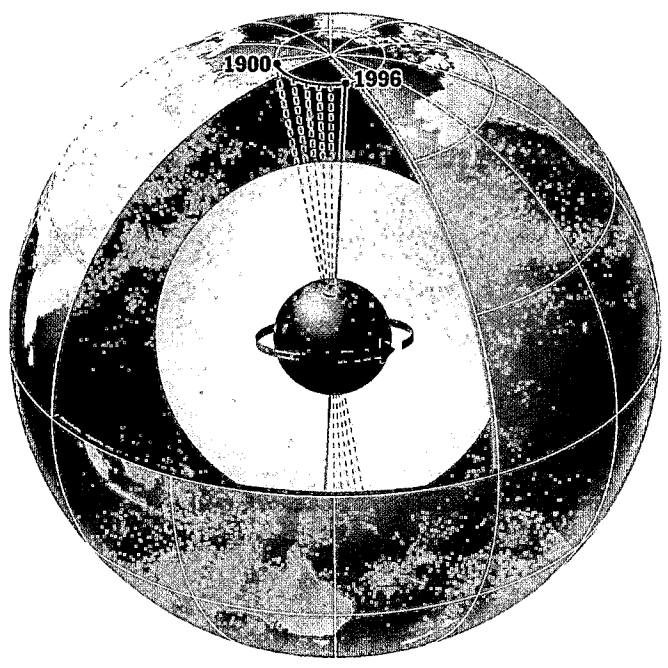

(D. Rohr/Columbia University)

Why isn't our planet just one big, rocky ball of uniform density? The answer appears to be that much of Earth was *molten* at some time in the past. As a result, the higher-density matter sank to the core and the lower-density material was displaced toward the surface. A remnant of this ancient heating exists today: Earth's central temperature is nearly equal to the surface temperature of the Sun. What processes were responsible for heating the entire planet to this extent? To answer this question, we must try to visualize the past.

We will see in Chapter 15 that when Earth formed 4.6 billion years ago, it did so by capturing material from its

surroundings, growing in mass as it swept up "preplanetary" chunks of matter in its vicinity. As the young planet grew, its gravitational field strengthened and the speed with which newly captured matter struck its surface increased. This process generated a lot of heat—so much, in fact, that Earth may already have been partially or wholly molten by the time it reached its present size. As Earth began to differentiate and heavy material sank to the center, even more gravitational energy was released, and the interior temperature must have increased still further.

Later, Earth continued to be bombarded with debris left over from the formation process. At its peak some 4 billion years ago, this secondary bombardment was probably intense enough to keep the surface molten, but only down to a depth of a few tens of kilometers. Erosion by wind and water has long since removed all trace of this early period from the surface of Earth, but the Moon still bears visible scars of the onslaught.

A second important process for heating Earth soon after its formation was **radioactivity**—the release of energy by certain rare heavy elements, such as uranium, thorium, and plutonium (see *More Precisely* 7-2). These elements release energy and heat their surroundings as their complex heavy nuclei decay (break up) into simpler lighter ones. While the energy produced by the decay of a single radioactive atom is tiny, Earth contained a lot of radioactive atoms, and a lot of time was available. Rock is such a poor conductor of heat that the energy would have taken a very long time to reach the surface and leak away into space, so the heat built up in the interior, adding to the energy left there by Earth's formation. Provided that enough radioactive elements were originally spread throughout the primitive Earth, rather like raisins in a cake, the entire planet—from crust to core—could have melted and remained molten for about a billion years. That's a long time by human standards, but not so long in the cosmic scheme of things. Measurements of the ages of some surface rocks indicate that Earth's crust finally began to solidify roughly 700 million years after it originally formed. Radioactive heating did not stop at this point, of course. It continued even after Earth's surface cooled and solidified. But radioactive decay works in only one direction, always producing lighter elements from heavier ones.

Once gone, the heavy and rare radioactive elements cannot be replenished.

The early source of heat diminished with time, allowing the planet to cool over the past 4 billion years. In so doing, Earth cooled from the outside in, much like a hot potato, since regions closest to the surface could most easily unload their excess heat into space. In this way, the surface developed a solid crust, and the differentiated interior attained the layered structure now implied by seismic studies.

☑ Concept Check

- How would our knowledge of Earth's interior be changed if our planet were geologically inactive, like the Moon?

7.4 Surface Activity

Earth is geologically alive today. Its interior seethes and its surface constantly changes. Figure 7.8 shows two indicators of surface geological activity: a volcano, where molten rock and hot ash upwell through fissures or cracks in the surface, and (the aftermath of) an earthquake, which occurs when the crust suddenly dislodges under great pressure. Catastrophic volcanoes and earthquakes are relatively rare events these days, but geological studies of rocks, lava, and other surface features imply that surface activity must have been more frequent, and probably more violent, long ago.

Figure 7.8 Geological Activity (a) An active volcano on Mount Kilauea in Hawaii. Kilauea seems to be a virtually ongoing eruption. Other eruptions, such as that of Mount St. Helens in Washington State on May 18, 1980, are catastrophic events that can release more energy than the detonation of a thousand nuclear bombs. (b) The aftermath of an earthquake that claimed more than 5000 lives and caused billions of dollars' worth of damage in Kobe, Japan, in January 1995. (P. Chesley/Tony Stone Images; H. Yamaguchi/Sygma)

(a)

(b)

CONTINENTAL DRIFT

🔷 Many traces of past geological events are scattered across our globe. Erosion by wind and water has wiped away much of the evidence for ancient activity, but modern exploration has documented the sites of most of the recent activity, such as earthquakes and volcanic eruptions. Figure 7.9 is a map of the currently active areas of our planet. The red dots represent sites of volcanism or earthquakes. Nearly all these sites have experienced surface activity within this century, some of them suffering much damage and the loss of many lives.

The intriguing aspect of Figure 7.9 is that the active sites are not spread evenly across our planet. Instead, they trace well-defined lines of activity, where crustal rocks dislodge (as in earthquakes) or mantle material upwells (as in volcanoes). In the mid-1960s, it became clear that these lines are really the outlines of gigantic "plates," or slabs of Earth's surface.* Most startling of all, the plates are slowly moving—literally drifting around the surface of our planet. These plate motions have created the surface mountains, oceanic trenches, and other large-scale

features across the face of planet Earth. In fact, plate motions have shaped the continents themselves. The process is popularly known as "continental drift." The technical term for the study of plate movement and its causes is **plate tectonics.** The major plates of the world are marked on Figure 7.9.

The plates are not simply slowing to a stop after some ancient initial movements. Rather, they are still drifting today, although at an extremely slow rate. Typically, the speeds of the plates amount to only a few centimeters per year—about the same rate as your fingernails grow. This is well within the measuring capabilities of modern equipment. Curiously, one of the best ways of monitoring plate motion on a global scale is by making accurate observations of very distant astronomical objects. Quasars (see Chapter 25), lying many hundreds of millions of light-years from Earth, will never show any measurable apparent motion on the sky stemming from their own motion in space. Thus any apparent change in their position (after correction for Earth's motion, of course) can be interpreted as arising from the motion of the telescope—or of the continental plate on which it is located. On smaller scales, laser-ranging and other techniques now routinely track the relative motion of plates in many populated areas, such as California. During the course of Earth history, each plate has had plenty of time to move large distances, even at its sluggish pace.

Not all volcanoes are associated with plate boundaries. The Hawaiian Islands, located near the center of the Pacific plate, are a case in point. They are associated with a "hot spot" in Earth's upper mantle that melts the crust above it. Over millions of years the motion of the Pacific plate across the underlying hot spot has resulted in a chain of volcanic islands.

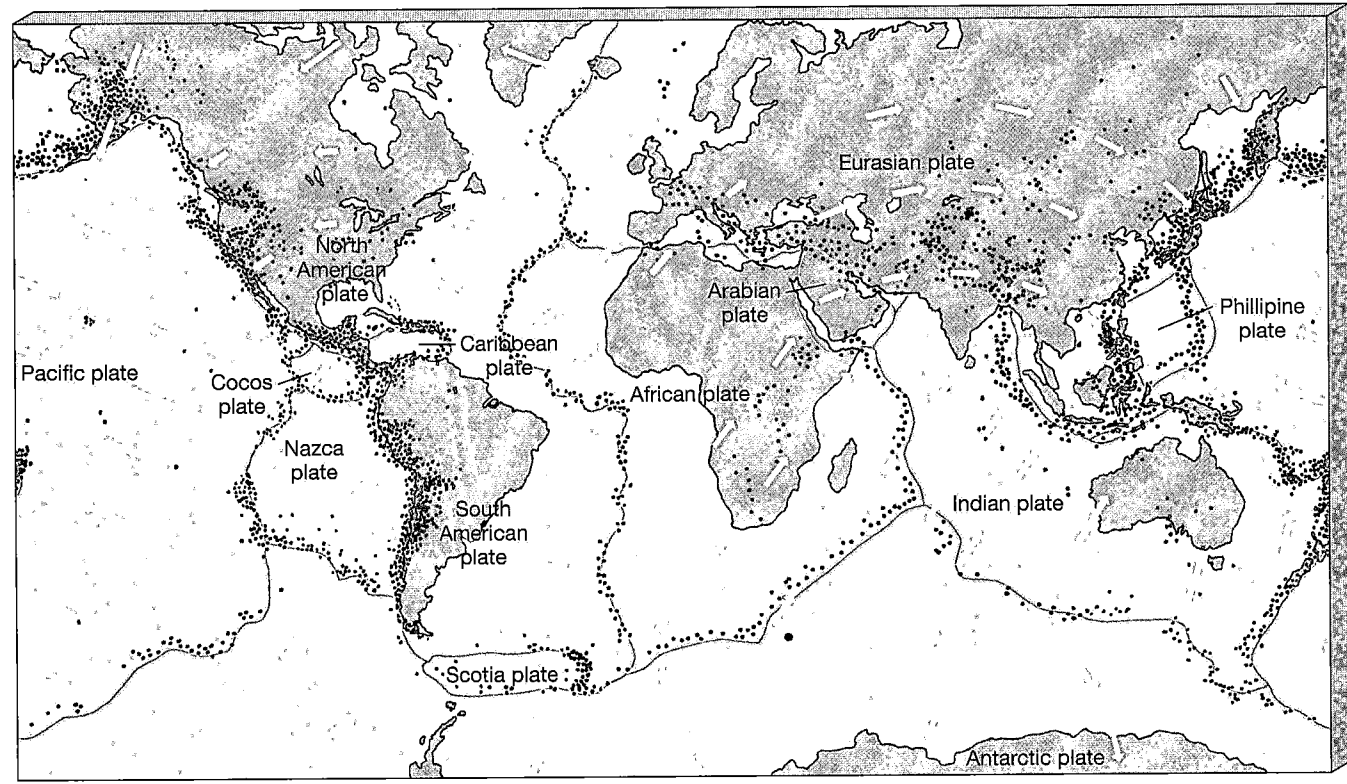

Figure 7.9 Global Plates Red dots represent active sites where major volcanoes or earthquakes have occurred in the twentieth century. Taken together, the sites outline vast "plates" that drift around on the surface of our planet. The white arrows show the general directions of the plate motions.

For example, a drift rate of only 2 cm per year can cause two continents (for example, Europe and North America) to separate by some 4000 km over the course of 200 million years. That may be a long time by human standards, but it represents only about 5 percent of the age of Earth.

A common misconception is that the plates are the continents themselves. Some plates are indeed made mostly of continental landmasses, but other plates are made of a continent plus a large part of an ocean. For example, the Indian plate includes all of India, much of the Indian Ocean, and all of Australia and its surrounding south seas (see Figure 7.9). Still other plates are mostly ocean. The seafloor itself is the slowly drifting plate, and the oceanic water merely fills in the depressions between continents. The southeastern portion of the Pacific Ocean, called the Nazca plate, contains no landmass at all. For the most part, the continents are just passengers riding on much larger plates.

Taken together, the plates make up Earth's **lithosphere**, which contains both the crust and a small part of the upper mantle. The lithosphere is the portion of Earth that undergoes tectonic activity. The semisolid part of the mantle over which the lithosphere slides is known as the **asthenosphere**. The relationships between these regions of Earth are shown in Figure 7.10.

EFFECTS OF PLATE MOTION

As the plates drift around, we might expect collisions to be routine. Indeed, plates do collide. But unlike two au-tomobiles that collide and then stop, the surface plates are driven by enormous forces. They do not stop easily. Instead, they just keep crunching into one another. Figure 7.11(a) shows a collision currently occurring between two continental landmasses: The subcontinent of India, on the prow of the northward-moving Indian plate, is crashing into the landmass of Asia, located on the Eurasian plate (see Figure 7.9). The resulting folds of rocky crust create mountains—in this case the snow-covered Himalayan mountain range at upper right. A peak like Mount Everest (Figure 7.11b) represents a portion of Earth's crust that has been lifted over 8800 m by the slow but inexorable force produced when one plate plows into another.

Not all colliding plates produce mountain ranges. At other collision locations, called *subduction zones,* one plate slides under the other, ultimately to be destroyed as it sinks into the mantle. Subduction zones are responsible for most of the deep trenches in the world's oceans.

Nor do all plates experience head-on collisions. As noted by the arrows of Figure 7.9, many plates slide or shear past one another. A good example is the most famous active region in North America—the San Andreas Fault in California (Figure 7.12). The site of much earthquake activity, this fault marks the boundary where the Pacific and North American plates are rubbing past each other. The motion of these two plates, like that of moving parts in a poorly oiled machine, is neither steady nor smooth. The sudden jerks that occur when they do move

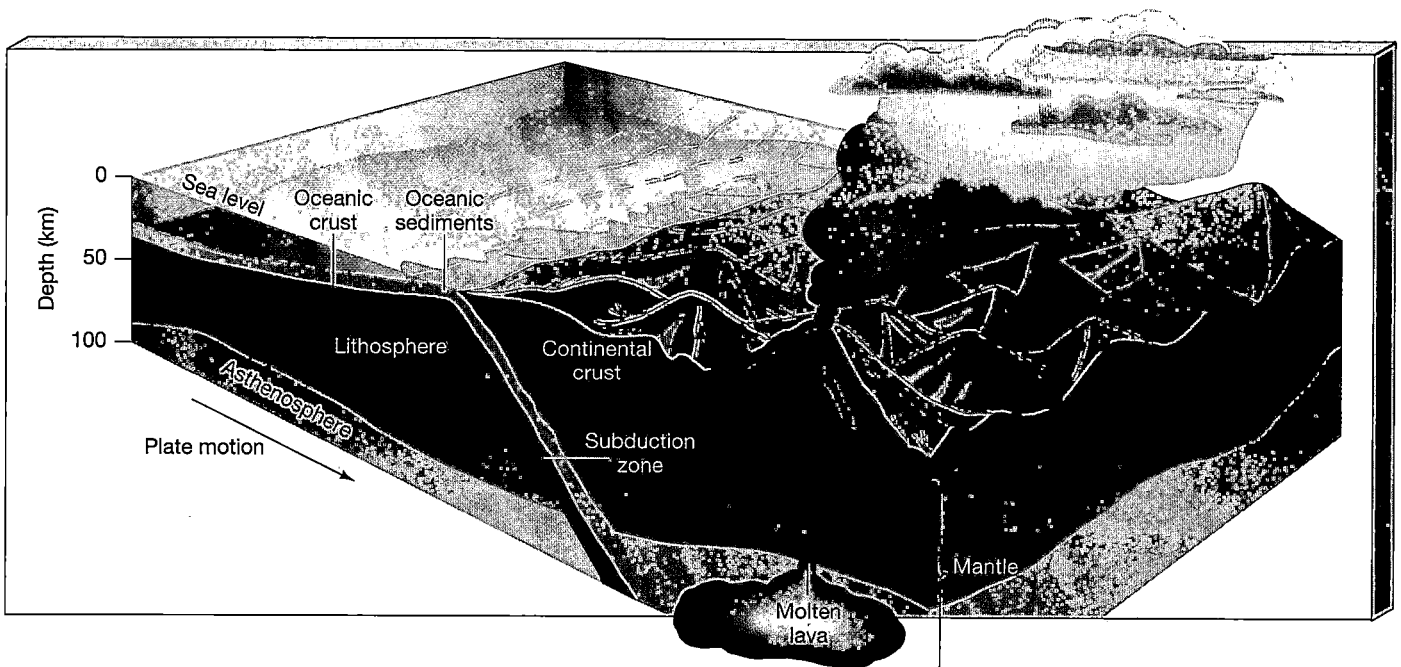

Figure 7.10 Earth's Upper Mantle The outer layers of Earth's interior. The rocky lithosphere comprises both the crust and part of Earth's upper mantle. It is typically between 50 and 100 km thick. Below it lies the asthenosphere, a relatively soft part of the mantle over which the lithosphere slips.

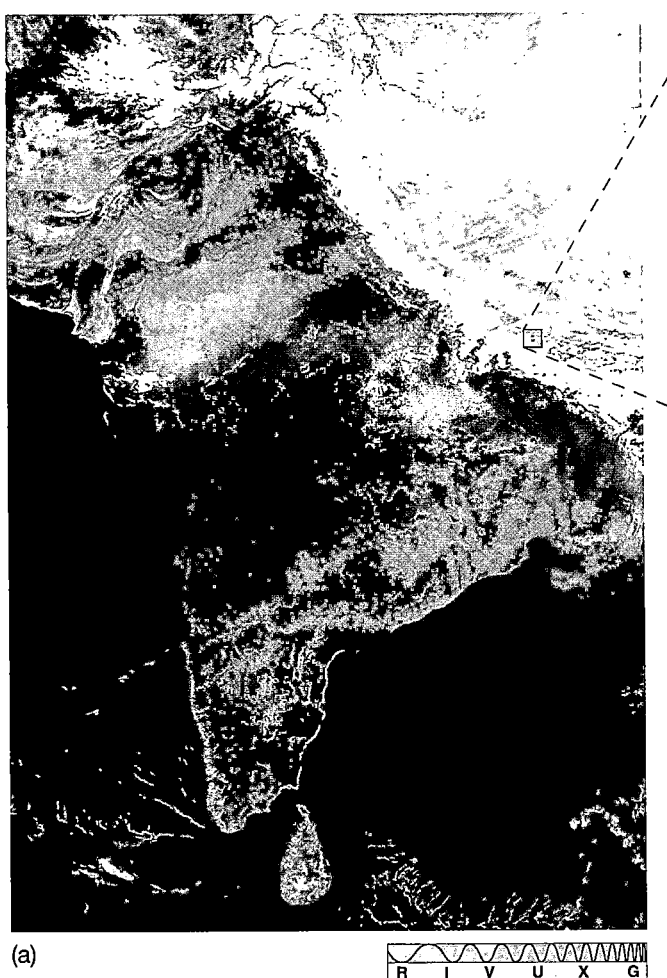

(a)
R   I   V   U   X   G

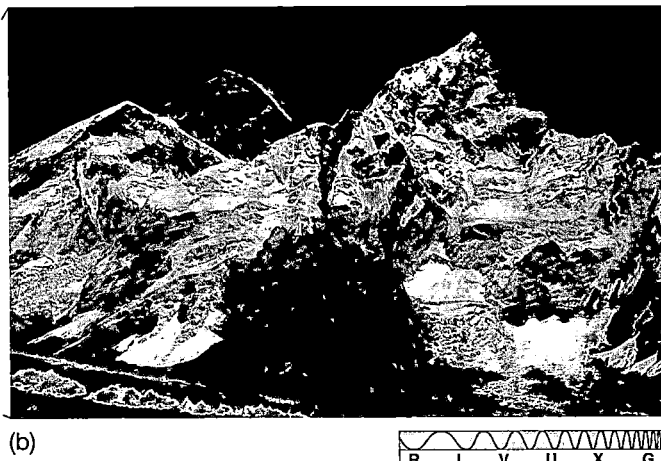

(b)
R   I   V   U   X   G

Figure 7.11 Himalayas Mountain building results largely from plate collisions. (a) The subcontinent of India, imaged here in infrared light from orbit, lies at the northernmost tip of the Indian plate. As this plate drifts northward, the Indian landmass collides with Asia, on the Eurasian plate. The impact causes Earth's crust to buckle and fold, thrusting up the Himalayan mountain range (covered with snow at the upper right). (b) The results of this process can be seen in this view of Mount Everest (the dark peak in the background).

against each other are often strong enough to cause major earthquakes.

At still other locations, the plates are moving apart. As they recede, new mantle material wells up between them, forming *midocean ridges*. Notice in Figure 7.9 the major boundary separating the North and South American plates from the Eurasian and African plates, marked by the thin strip down the middle of the Atlantic Ocean. Discovered after World War II by oceanographic ships studying the geography of the seafloor, this giant fault is called the Mid-Atlantic Ridge. It extends, like a seam on a giant baseball, all the way from Scandinavia in the North Atlantic to the latitude of Cape Horn at the southern tip of South America. The entire ridge is a region of seismic and volcanic activity, but the only major part of it that rises above sea level is the island of Iceland.

Robot submarines have retrieved samples of the ocean floor at a variety of locations on either side of the Mid-Atlantic Ridge, and the ages of the samples have been measured using radioactive dating techniques. As depicted in Figure 7.13, the ocean floor closest to the ridge is relatively young, whereas material farther away, on either side, is older—exactly as we would expect if hot molten matter is upwelling and solidifying as the plates on either side drift

apart. The Atlantic Ocean has apparently been growing in this way for the past 200 million years, the oldest age found on any part of the Atlantic seafloor.

Other studies of the Mid-Atlantic Ridge have yielded important information about Earth's magnetic field. As hot mantle material (carrying traces of iron) emerges from cracks in the oceanic ridges and solidifies, it becomes slightly magnetized, retaining an imprint of Earth's magnetic field *at the time it cooled*. Thus, the ocean floor has preserved within it a record of Earth's magnetism during past times, rather like a tape recording. Figure 7.14 is a diagram of a small portion of the Atlantic Ocean floor near Iceland. Earth's current magnetism is oriented in the familiar north–south fashion, and when samples of ocean floor close to the ridge are examined, the iron deposits are oriented just as expected—north–south. This is the "young" basalt that upwelled and cooled fairly recently. However, samples retrieved farther from the ridge, corresponding to older material that upwelled long ago, are often magnetized with the *opposite* orientation. As we move away from the ridge, the imprinted magnetic field flips back and forth, more or less regularly and symmetrically on either side of the ridge.

Scientists believe that these different magnetic orientations were caused by reversals in Earth's magnetic field

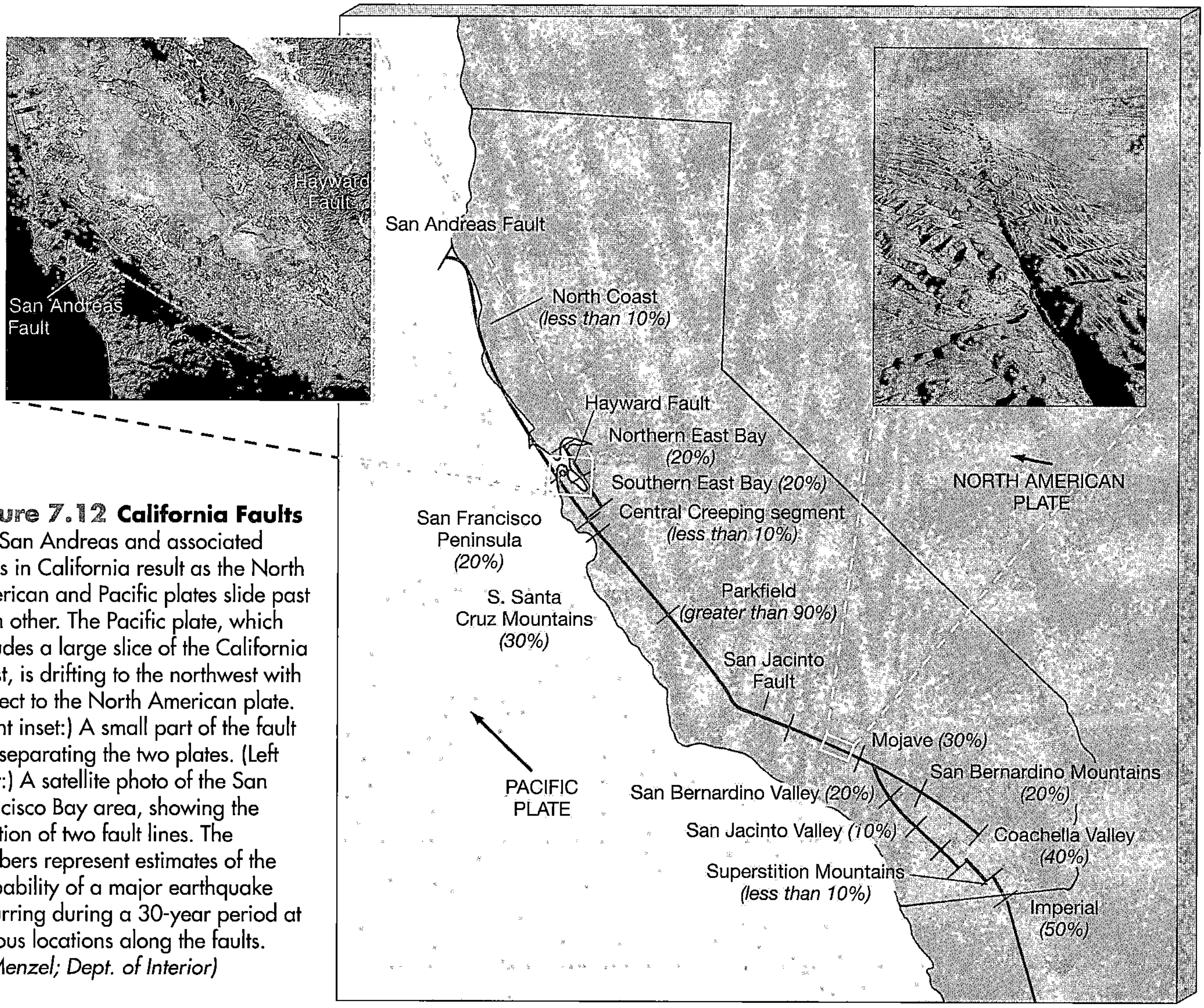

Figure 7.12 California Faults
The San Andreas and associated faults in California result as the North American and Pacific plates slide past each other. The Pacific plate, which includes a large slice of the California coast, is drifting to the northwest with respect to the North American plate. (Right inset:) A small part of the fault line separating the two plates. (Left inset:) A satellite photo of the San Francisco Bay area, showing the location of two fault lines. The numbers represent estimates of the probability of a major earthquake occurring during a 30-year period at various locations along the faults.
(P. Menzel; Dept. of Interior)

San Andreas Fault

North Coast
(less than 10%)

Hayward Fault

Northern East Bay
(20%)

Southern East Bay (20%)

San Francisco
Peninsula
(20%)

Central Creeping segment
(less than 10%)

NORTH AMERICAN
PLATE

S. Santa
Cruz Mountains
(30%)

Parkfield
(greater than 90%)

San Jacinto
Fault

Mojave (30%)

San Bernardino Mountains
(20%)

PACIFIC
PLATE

San Bernardino Valley (20%)

San Jacinto Valley (10%)

Coachella Valley
(40%)

Superstition Mountains
(less than 10%)

Imperial
(50%)

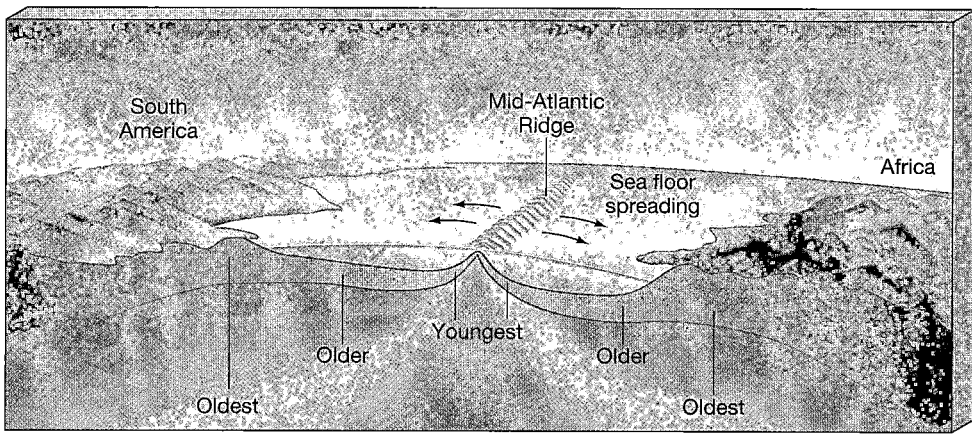

South
America

Mid-Atlantic
Ridge

Sea floor
spreading

Africa

Youngest

Older

Older

Oldest

Oldest

Figure 7.13 Seafloor Spreading Samples of ocean floor retrieved by oceanographic vessels are youngest close to the Mid-Atlantic Ridge and progressively older farther away.

that occurred as the plates drifted away from the central ridge. Working backward, we can use the fossil magnetic field to infer the past positions of the plates, as well as the orientation of Earth's magnetism. In addition to providing

strong support for the idea of seafloor spreading, these measurements, when taken in conjunction with the data on seafloor age, allow us to time our planet's magnetic reversals. On average, Earth's magnetic field reverses itself

MORE PRECISELY 7-2

Radioactive Dating

In Chapter 4 we saw that atoms are made up of electrons and nuclei, and that nuclei are composed of protons and neutrons. ⊙ (Sec. 4.2) The number of protons in a nucleus determines which element it represents. However, the number of neutrons can vary. In fact, most elements can exist in several *isotopic* forms, all containing the same number of protons but different numbers of neutrons in their nuclei. The particular nuclei we have encountered so far—the most common forms of hydrogen, helium, carbon, and iron—are all *stable*. For example, left alone, a carbon-12 nucleus, consisting of six protons and six neutrons, will remain unchanged forever. It will not break up into smaller pieces, nor will it turn into anything else.

Not all nuclei are stable, however. Many nuclei—such as carbon-14 (containing 6 protons and 8 neutrons), thorium-232 (90 protons, 142 neutrons), uranium-235 (92 protons, 143 neutrons), uranium-238 (92 protons, 146 neutrons), and plutonium-241 (94 protons, 147 neutrons)—are inherently *unstable*. Left alone, they will eventually break up into lighter "daughter" nuclei, in the process emitting some elementary particles and releasing some energy. The change happens spontaneously, without any external influence. This instability is known as *radioactivity*. The energy released by the disintegration of the radioactive elements just listed is the basis for nuclear fission reactors (and atomic bombs).

Unstable heavy nuclei achieve greater stability by disintegrating into lighter nuclei, but they do not do so immediately. Each type of "parent" nucleus takes a characteristic amount of time to decay. The *half-life* is the name given to the time required for half of a sample of parent nuclei to disintegrate. Notice that this is really a statement of probability. We cannot say *which* nuclei will decay in any given half-life interval, only that half of them are expected to do so. If a given sample of material has half-life T, then we can write down a simple expression for the amount of material remaining after time t:

$$\text{fraction of material remaining} = (1/2)^{t/T}$$

Thus, if we start with a billion radioactive nuclei embedded in a sample of rock, a half-billion nuclei will remain after one half-life, a quarter-billion after two half-lives, and so on.

Every radioactive isotope has its own half-life, and most of them are now well known from studies conducted since the 1950s. For example, the half-life of uranium-235 is 713 million years, and that of uranium-238 is 4.5 billion years. Some radioactive elements decay much more rapidly, others much more slowly, but these two types of uranium are particularly important to geologists because their half-lives are comparable to the age of the solar system. The accompanying figure illustrates the half-lives and decay reactions for four unstable heavy nuclei.

The decay of unstable radioactive nuclei into more stable *daughter* nuclei provides us with a useful tool for measuring the ages of any rocks we can get our hands on. The first step is to measure the amount of stable nuclei of a given kind (for example, lead-206, which results from the decay of uranium-238). This amount is then compared with the amount of remaining unstable parent nuclei (in this case, uranium-238) from which the daughter nuclei descended. Knowing the rate (or half-life) at which the disintegration occurs, the age of the rock then follows directly. If half of the parent nuclei of some element have decayed, so that the number of daughter nuclei equals the number of parents, the age of the rock must be equal to the half-life of the radioactive nucleus studied. Similarly, if only a quarter of the parent nuclei remain (three times as many daughters as parents), the rock's age is twice the half-life of that element, and so on.

In practice, ages can be determined by these means to within an accuracy of a few percent. The most ancient rocks on Earth are dated at 3.9 billion years old. These rare specimens have been found in Greenland and Labrador.

EXAMPLE: Suppose that careful analysis of a sample of rock reveals that, for every nucleus of uranium-238 remaining in the sample, there are 0.41 lead-206 nuclei. If we assume that there was no lead-206 initially present, and hence that every lead-206 nucleus is the decay product of a

---

roughly every half-million years. Current theory suggests that such reversals are part of the way in which all planetary magnetic fields are generated. As we will see in Chapter 16, a similar phenomenon (with a reversal time of approximately 11 years) is also observed on the Sun.

WHAT DRIVES THE PLATES?

What process is responsible for the enormous forces that drag plates apart in some locations and ram them together in others? The answer is probably convection—the same phys-

ical process we encountered earlier in our study of the atmosphere. Figure 7.15 is a cross-sectional diagram of the top few hundred kilometers of our planet's interior. It depicts roughly the region in and around a midocean ridge. There the ocean floor is covered with a layer of sediment—dirt, sand, and dead sea organisms that have fallen through the seawater for millions of years. Below the sediment lies about 10 km of granite, the low-density rock that makes up the crust. Deeper still lies the upper mantle, whose temperature increases with depth. Below the base of the lithosphere, at a depth of perhaps 50 km, the temperature is sufficiently high

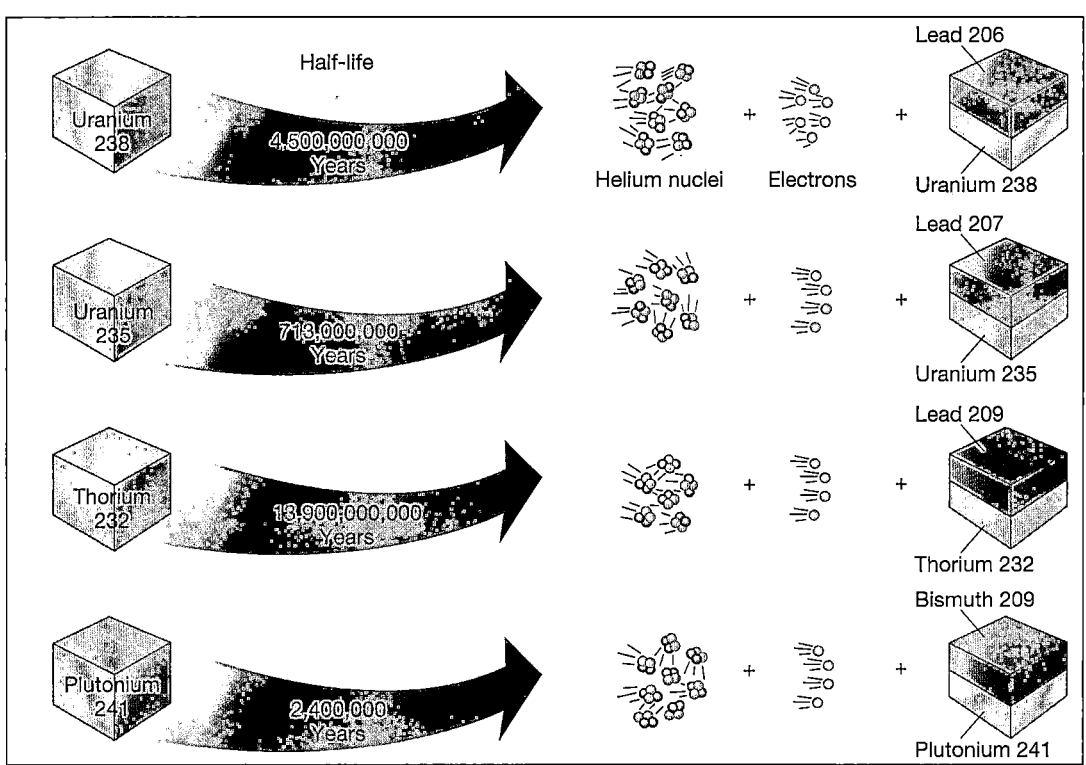

nucleus of uranium-238, we can easily calculate the fraction of uranium-238 nuclei remaining. The answer is

fraction of uranium-238 $=$

$$\frac{1}{1 + 0.41} = 0.71 \approx \sqrt{\frac{1}{2}} = (1/2)^{0.5}.$$

From the previous equation, it follows that the elapsed time must be 0.5 times the half-life of uranium-238, or 2.25 billion years. If we were to repeat the analysis with uranium-235 and lead-207, we would expect to find consistent results, within the measurement errors: 2.25 billion years is $2250/713 = 3.2$ uranium-235 half-lives, so only $1/2^{3.2} \approx 11$ percent of any uranium-235 should re-

main—daughter lead-207 nuclei should outnumber parent uranium-235 nuclei by more than eight to one.

The radioactive-dating technique rests on the assumption that the rock has remained *solid* while the radioactive decays have been going on. If the rock melts, there is no particular reason to expect the daughter nuclei to remain in the same locations their parents had occupied, and the whole method fails. Thus, radioactive dating indicates the time that has elapsed since the last time the rock in question solidified. Hence the 3.9-billion-year value represents only a portion—a lower limit—of the true age of our planet. It does not measure the duration of Earth's molten existence.

that the mantle is soft enough to very slowly flow, although it is not molten. This region is the asthenosphere.

This is a perfect setting for convection—warm matter underlying cool matter. The warm mantle rock rises, just as hot air rises in our atmosphere. Sometimes, the rock squeezes up through cracks in the granite crust. Every so often, such a fissure may open in the midst of a continental landmass, producing a volcano such as Mount St. Helens or possibly a geyser like those at Yellowstone National Park. However, most such cracks are on the ocean floor. The Mid-Atlantic Ridge is a prime example.

Not all the rising warm rock in the upper mantle can squeeze through cracks and fissures. Some warm rock cools and falls back down to lower levels. In this way, large circulation patterns become established within the upper mantle, as depicted in Figure 7.15. Riding atop these convection patterns are the plates. The circulation is extraordinarily sluggish. Semisolid rock takes millions of years to complete one convection cycle. Although the details are far from certain and remain controversial, many researchers believe that the large-scale circulation patterns near plate boundaries drive the motion of the plates.

Figure 7.14 Magnetic Reversals Samples of basalt retrieved from the ocean floor often show Earth's magnetism to have been oriented oppositely from the current north–south magnetic field. This simplified diagram shows the ages of some of the regions in the vicinity of the Mid-Atlantic Ridge (see Figure 7.9), together with the direction of the fossil magnetic field. The colored areas have the current orientation; they are separated by regions of reversed magnetic polarity.

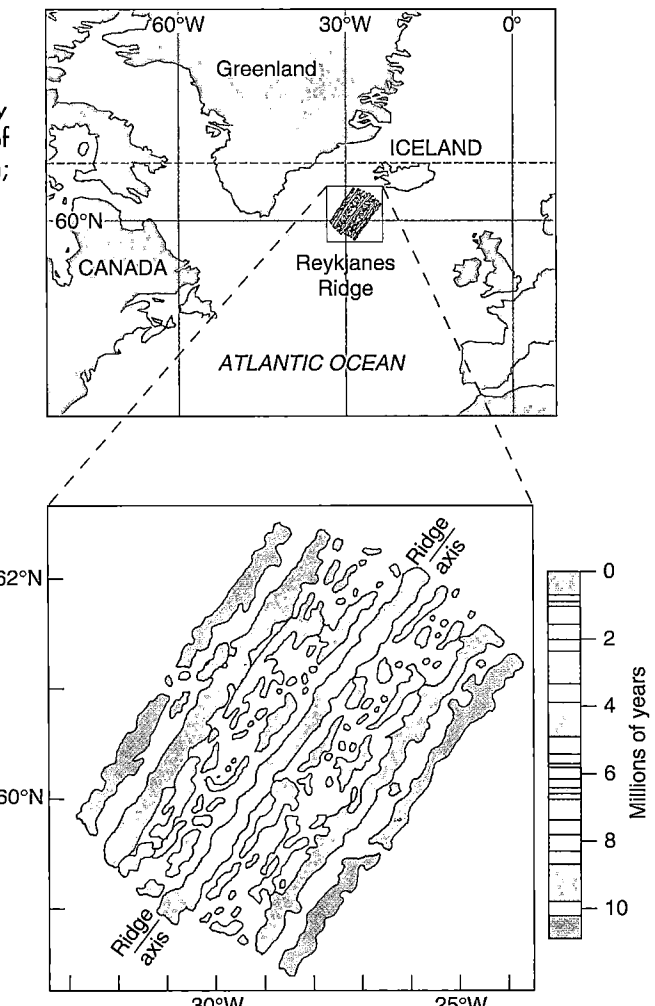

PAST CONTINENTAL DRIFT

Figure 7.16 illustrates how all the continents nearly fit together like pieces of a puzzle. Note especially how the Brazilian coast meshes with the Ivory Coast of Africa. In fact, most of the continental landmasses in the Southern Hemisphere fit together remarkably well. Following the arrows on Figure 7.9 backward, we can see that the fits are roughly consistent with the present motions of the plates involved. The fits would be even better if we were to consider the entire continental shelves (the continental borders, which are underwater) instead of just the portions that happen to stick up above sea level.

Geologists now believe that sometime in the past a single gargantuan landmass dominated our planet. This ancestral supercontinent, known as Pangaea (meaning "all lands"), is shown in Figure 7.16(a). The rest of the planet was presumably covered with water. The present locations of the continents, along with measurements of their current drift rates, suggest that Pangaea was the major land feature on Earth approximately 200 million years ago. Dinosaurs, which were then the dominant form of life, could have sauntered from Russia to Texas via Boston without getting their feet wet. Indeed, it was the existence of similar-looking fossils on opposite sides of the Atlantic Ocean,

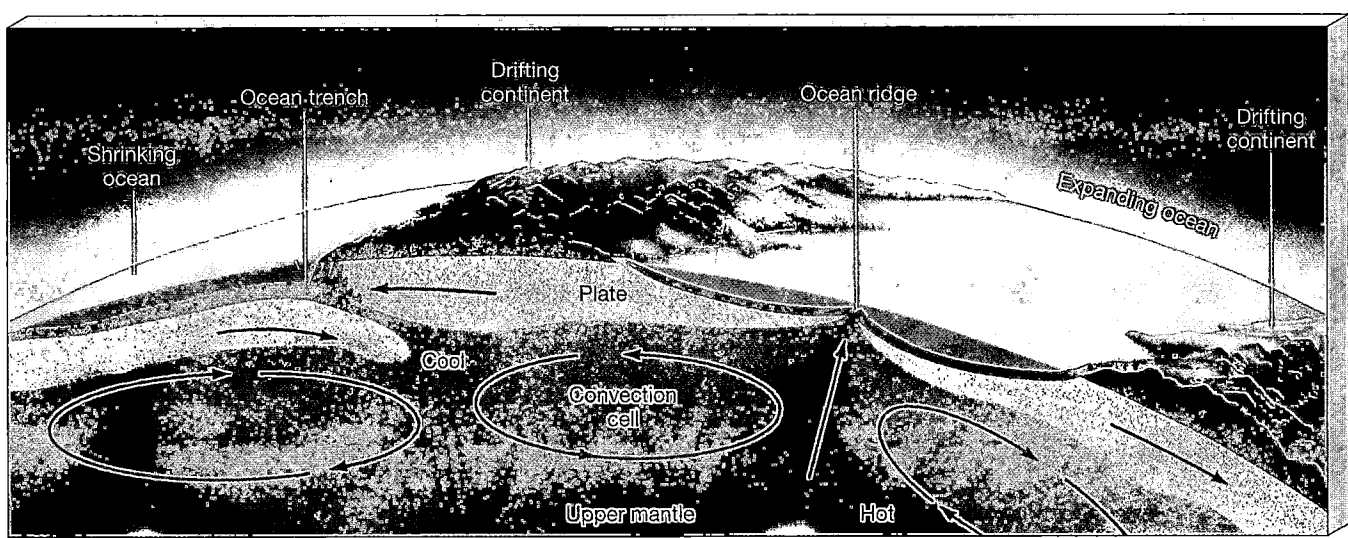

Figure 7.15 Plate Drift The motion of Earth's tectonic plates is probably caused by convection—in this case, giant circulation patterns in the upper mantle that drag the plates across the surface.

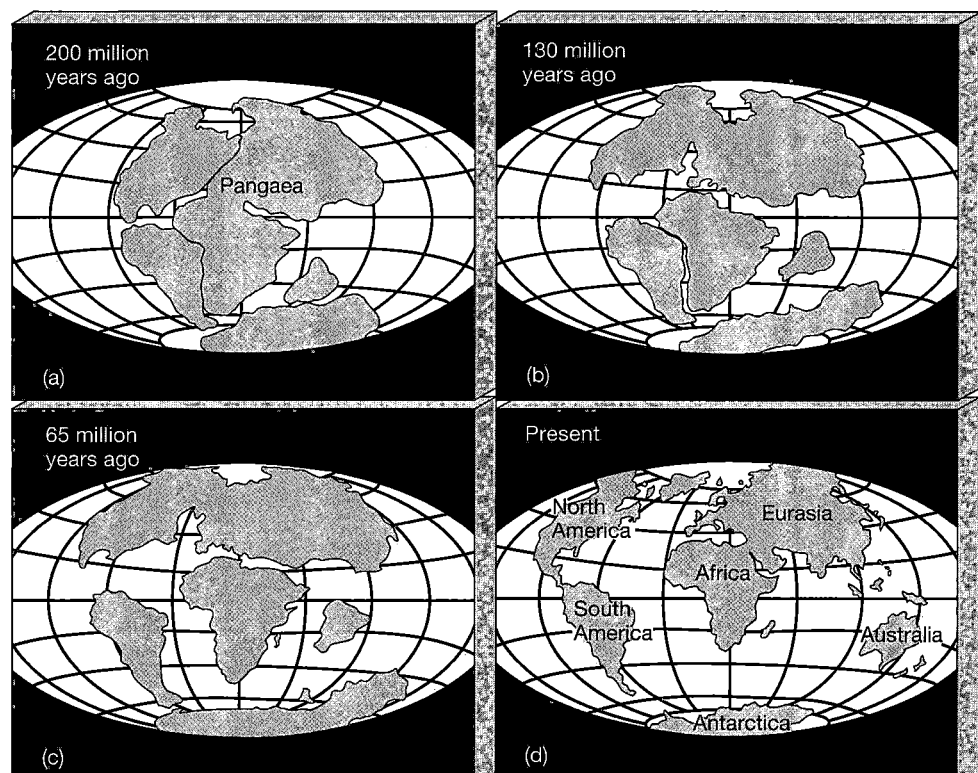

(a) 200 million years ago — Pangaea

(b) 130 million years ago

(c) 65 million years ago

(d) Present — North America, Eurasia, Africa, South America, Australia, Antarctica

Figure 7.16 Pangaea Given the current estimated drift rates and directions of the plates, we can trace their movements back into the past. About 200 million years ago, they would have been at the approximate positions shown in (a). The continents' current positions are shown in (d).

together with the striking geographical fit of the continents, that first led scientists to suggest the idea of continental drift early in the twentieth century. However, it was some time—in fact, almost half a century—before the scientific community began to take this revolutionary theory seriously. The other frames in Figure 7.16 show how Pangaea split apart, its separate pieces drifting across Earth's surface, eventually becoming the familiar continents we know today.

Of course, there is nothing particularly special about a time 200 million years in the past. We do not suppose that Pangaea remained intact for some 4 billion years after the crust first formed, only to break up so suddenly and so (relatively) recently. It is much more likely that Pangaea itself came into existence after an earlier period during which other plates, carrying widely separated continental masses, were driven together by tectonic forces, merging their landmasses to form a single supercontinent. It is quite possible that there has been a long series of "Pangaeas" stretching back in time over much of Earth's history, as tectonic forces have continually formed, destroyed, and re-formed our planet's land masses. There will probably be many more.

Concept Check

□ Describe the causes and some consequences of plate tectonics on Earth.

7.5 Earth's Magnetosphere

Discovered by artificial satellites launched in the late 1950s, Earth's magnetosphere extends far above the atmosphere. Simply put, the magnetosphere is the region around a planet that is influenced by that planet's magnetic field. Sketched in Figure 7.17, Earth's magnetic field is similar in overall structure to the field of a gigantic bar magnet and completely surrounds our planet. The magnetic field lines, which indicate the strength and direction of the field at any point in space, run from south to north, as indicated by the white arrowheads in the figure.

The north and south *magnetic poles*, where the magnetic field lines intersect Earth's surface vertically, are roughly aligned with Earth's spin axis. Neither pole is fixed relative to our planet, however—both drift at a rate of some 10 km per year—nor are the poles symmetrically placed. At present, Earth's magnetic north pole lies in northern Canada, at a latitude of about 80°N, almost due north of the center of North America; the magnetic south pole lies at a latitude of about 60°S, just off the coast of Antarctica south of Adelaide, Australia.

Earth's magnetosphere contains two doughnut-shaped zones of high-energy charged particles, one located about 3000 km and the other 20,000 km above Earth's surface. These zones are named the **Van Allen belts**, after the American physicist whose instruments on board one of the first satellites initially detected them. We call them

Figure 7.17
Magnetic Field Earth's magnetic field resembles that of an enormous bar magnet situated inside our planet. The arrows on the field lines indicate the direction in which a compass needle would point. Note that the bar is slightly misaligned with Earth's rotation axis.

"belts" because they are most pronounced near Earth's equator and because they completely surround the planet. Figure 7.18 shows how these invisible regions envelop Earth except near the North and South Poles.

The particles that make up the Van Allen belts originate in the solar wind. Traveling through space, neutral particles and electromagnetic radiation are unaffected by Earth's magnetism, but electrically charged particles are strongly influenced. As illustrated in Figure 7.19, a magnetic field exerts a force on a moving charged particle, causing the particle to spiral around the magnetic field lines. In this way, charged particles—mainly electrons and protons—from the solar wind can become trapped by Earth's magnetism. Earth's magnetic field exerts electromagnetic control over these particles, herding them into the Van Allen belts. The outer belt contains mostly electrons; the much heavier protons accumulate in the inner belt.

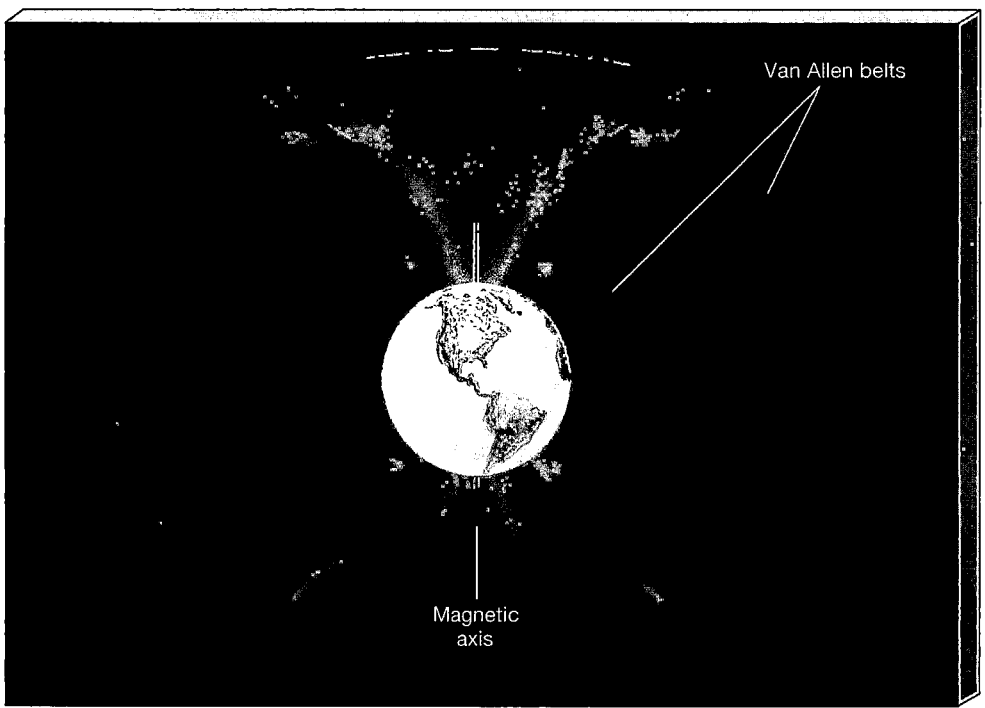

Van Allen belts

Magnetic
axis

Figure 7.18 Van Allen Belts High above Earth's atmosphere, the magnetosphere (lightly shaded blue area) contains at least two doughnut-shaped regions (heavily shaded violet areas) of magnetically trapped charged particles. These are the Van Allen belts.

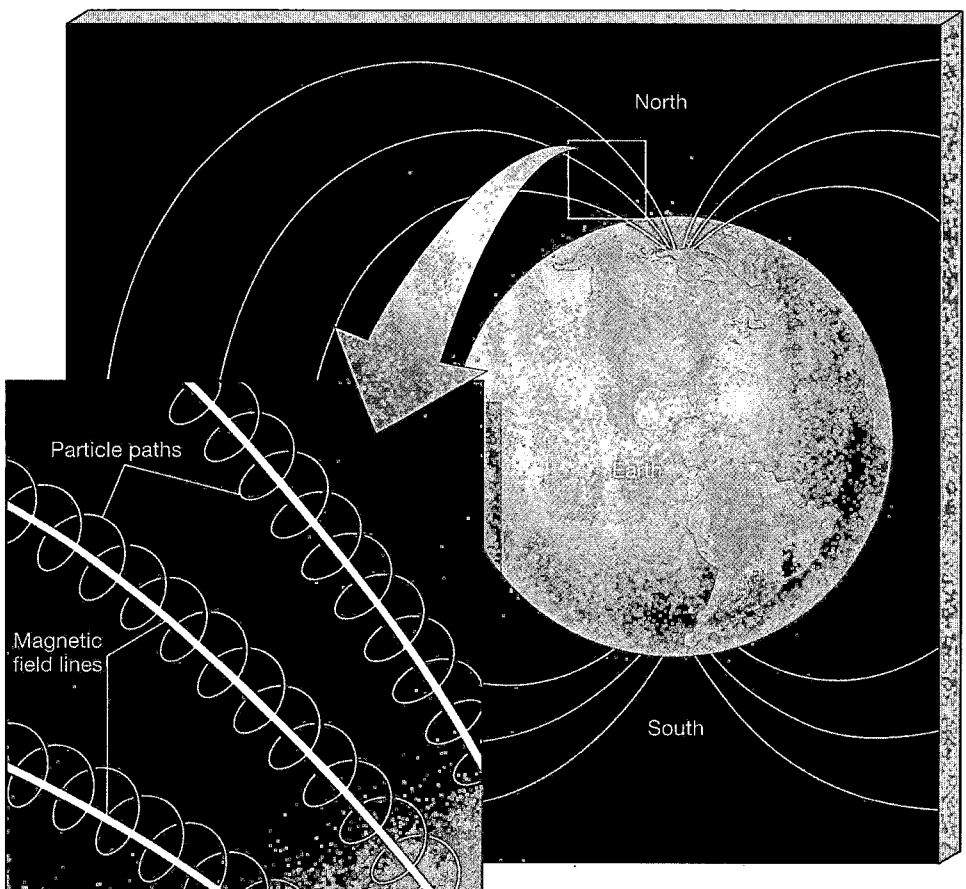

Figure 7.19 Magnetic Field Effect A charged particle in a magnetic field spirals around the field lines. The convergence of the field lines near Earth's magnetic poles causes the particles to be reflected back toward the other pole. Thus, charged particles tend to become "trapped" by Earth's magnetic field.

We could never survive unprotected in the Van Allen belts. Unlike the lower atmosphere, on which humans and other life forms rely for warmth and protection, much of the magnetosphere is subject to intense bombardment by large numbers of high-velocity, and potentially very harmful, charged particles. Colliding violently with an unprotected human body, these particles would deposit large amounts of energy wherever they made contact, causing severe damage to living organisms. Without sufficient shielding on the *Apollo* spacecraft, for example, the astronauts might not have survived the passage through the magnetosphere on their journey to the Moon.

Particles from the Van Allen belts often escape from the magnetosphere near Earth's north and south magnetic poles, where the field lines intersect the atmosphere. Their collisions with air molecules create a spectacular light show called an **aurora** (Figure 7.20). This colorful display results when atmospheric molecules, excited upon collision with the charged particles, fall back to their ground states and emit visible light. Many different colors are produced because each type of atom or molecule can take one of several possible paths as it returns to its ground state. ∞ (Sec. 4.2) Aurorae are most brilliant at high latitudes, especially inside the Arctic and Antarctic circles. In the north, the spectacle is called the *aurora borealis*, or *Northern Lights*. In the south, it is called the *aurora australis*, or *Southern Lights*.

Occasionally, particularly after a storm on the Sun (see Chapter 16), the Van Allen belts can become distorted by the solar wind and overloaded with many more particles than normal, allowing some particles to escape prematurely and at lower latitudes. For example, in North America, the aurora borealis is normally seen with any regularity only in northern Canada and Alaska. However, at times of greatest solar activity, the display has occasionally been seen as far south as the southern United States.

In reality, Earth's magnetosphere is not nearly as symmetrical as depicted in Figure 7.17. Satellites have mapped its true shape. As shown in Figure 7.21, the entire region of trapped particles is quite distorted, forming a teardrop-shaped cavity. On the sunlit (daytime) side of Earth, the magnetosphere is compressed by the flow of high-energy particles in the solar wind. The boundary between the magnetosphere and this flow is known as the *magnetopause*. It is found at about 10 Earth radii from our planet. On the side opposite the Sun, the field lines are extended away from Earth, with a long tail often reaching beyond the orbit of the Moon.

What is the origin of the magnetosphere and the Van Allen belts within it? Earth's magnetism is not really the result of a huge bar magnet lying within our planet. In fact, geophysicists believe that Earth's magnetic field is not a "permanent" part of our planet at all. Instead, it is thought to be

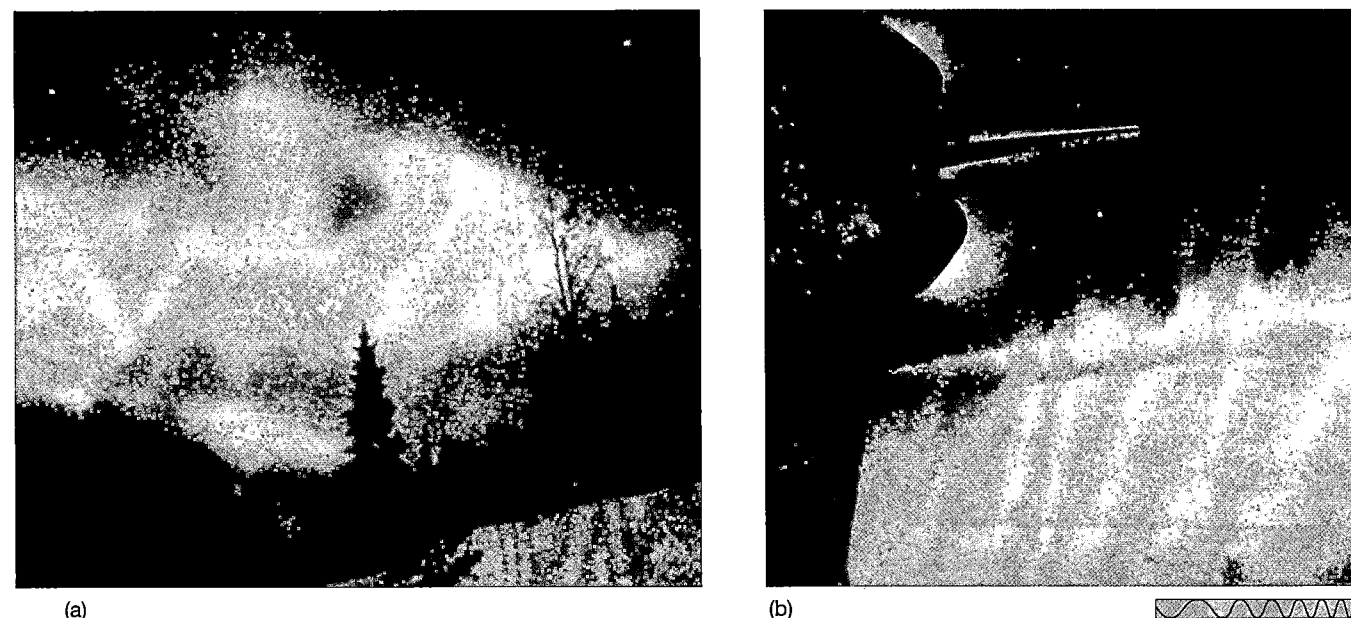

(a)

(b)

Figure 7.20 Aurorae (a) A colorful aurora rapidly flashes across the sky like huge wind-blown curtains glowing in the dark. The aurora is created by the emission of light radiation after magnetospheric particles collide with atmospheric molecules. The colors are produced as excited ions, atoms, and molecules recombine and cascade back to their ground states. (b) The aurora high above Earth, as photographed from a space shuttle (visible at left). *(NSF; NASA)*

continuously generated within the outer core and exists only because Earth is rotating. As in the dynamos that run industrial machines, Earth's magnetism is produced by the spinning, electrically conducting, liquid metal core deep within our planet. The theory that explains planetary (and other) magnetic fields in terms of rotating, conducting material flowing in the planet's interior is known as *dynamo theory.*

Both rapid rotation *and* a conducting liquid core are needed for this mechanism to work. This connection between internal structure and magnetism is very important for studies of the other planets in the solar system: We can tell a lot about a planet's interior simply by measuring its magnetic field.

Earth's magnetic field plays an important role in controlling many of the potentially destructive charged parti-

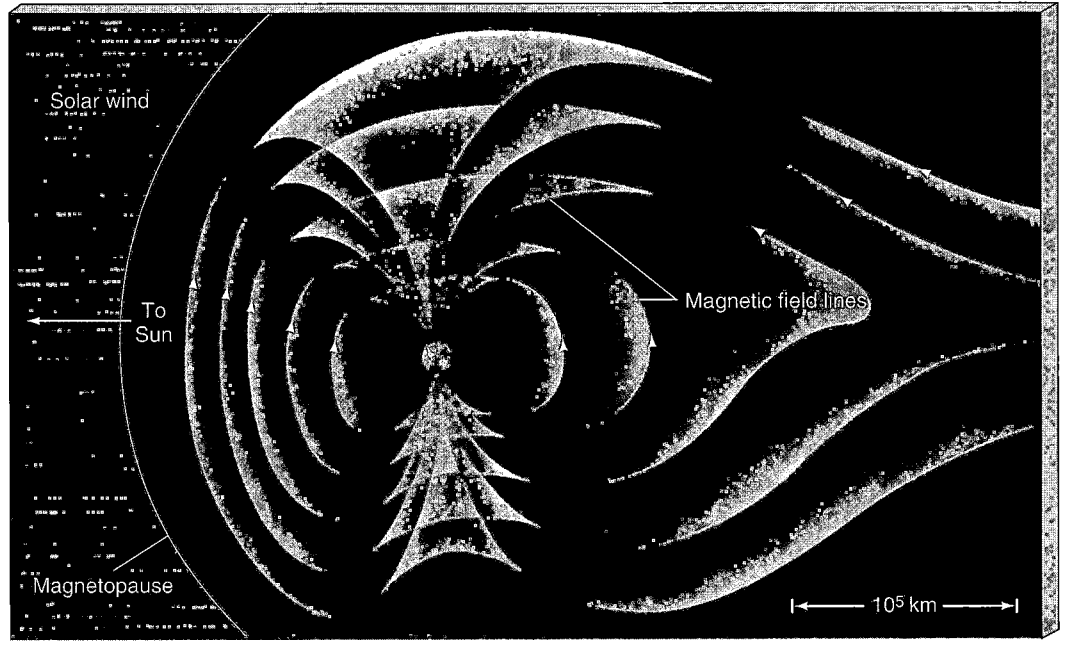

Figure 7.21 Earth's Magnetosphere Earth's real magnetosphere is actually greatly distorted by the solar wind, with a long tail extending from the nighttime side of Earth well into space.

cles that venture near our planet. Without the magnetosphere, Earth's atmosphere—and perhaps the surface, too—would be bombarded by harmful particles, possibly damaging many forms of life on our planet. Some researchers have even suggested that had the magnetosphere not existed in the first place, life might never have arisen at all on our planet.

✅ Concept Check

◘ What does the existence of a planetary magnetic field tell us about a planet's interior?

7.6 The Tides

Earth is unique among the planets in that it has large quantities of liquid water on its surface. Approximately three-quarters of Earth's surface is covered by water, to an average depth of about 3.6 km. Only 2 percent of the water is contained within lakes, rivers, clouds, and glaciers. The remaining 98 percent is in the oceans.

A familiar hydrospheric phenomenon is the daily fluctuation in ocean level known as the **tides**. At most coastal locations on Earth, there are two low tides and two high tides each day. The "height" of the tides—the magnitude of the variation in sea level—can range from a few centimeters to many meters, depending on the location on Earth and time of year. The height of a typical tide on the open ocean is about a meter, but if this tide is funneled into a narrow opening such as the mouth of a river, it can become much higher. For example, at the Bay of Fundy, on the U.S.–Canada border between Maine and New Branswick, the high tide can reach nearly 20 m (approximately 60 feet, or the height of a six-story building) above the low-tide level. An enormous amount of energy is contained in the daily motion of the oceans. This energy is constantly eroding and reshaping our planet's coastlines. In some locations, it has been harnessed as a source of electrical power for human activities.

GRAVITATIONAL DEFORMATION

✍ What causes the tides? A clue comes from the observation that they exhibit daily, monthly, and yearly cycles. In fact, the tides are a direct result of the gravitational influence of the Moon and the Sun on Earth. We have already seen how gravity keeps Earth and Moon in orbit about each other, and both in orbit around the Sun. ∞ (Sec. 2.7) For simplicity, let us first consider just the interaction between Earth and the Moon.

Recall that the strength of the gravitational force depends on the distance separating any two objects. Thus, the Moon's gravitational attraction is greater on the side of Earth

that faces the Moon than on the opposite side, some 12,800 km (Earth's diameter) farther away. This difference in the gravitational force is small—only about 3 percent—but it produces a noticeable effect, a **tidal bulge**. As illustrated in Figure 7.22, Earth becomes slightly elongated, with the long axis of the distortion pointing toward the Moon.

Earth's oceans undergo the greatest deformation because liquid can most easily move around on our planet's surface. (A bulge *is* actually raised in the solid material of Earth, but it is about a hundred times smaller than the oceanic bulge.) Thus, the ocean becomes a little deeper in some places (along the line joining Earth to the Moon) and shallower in others (perpendicular to this line). The daily tides we experience result as Earth rotates beneath this deformation.

The variation of the Moon's gravity across Earth is an example of a *differential force*, or **tidal force**. The *average* gravitational interaction between two bodies determines their orbit around each other. However, the *tidal* force, superimposed on that average, tends to deform the bodies themselves. As discussed in more detail in *More Precisely 7-3*,

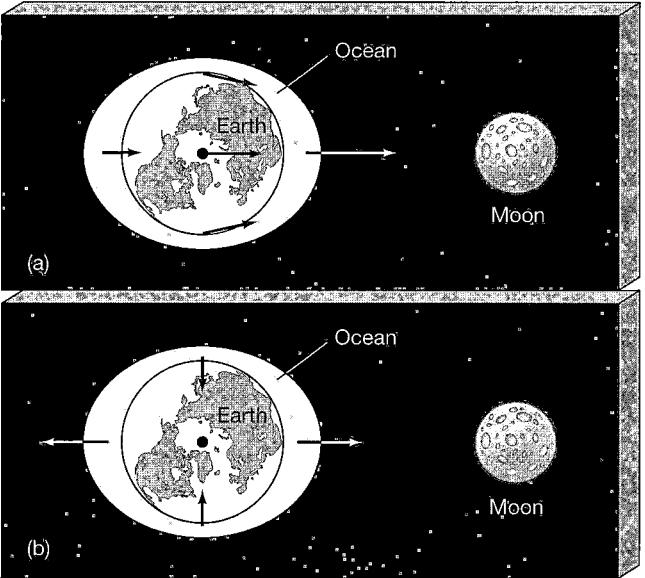

Figure 7.22 Lunar Tides This exaggerated illustration shows how the Moon induces tides on both the near and the far sides of Earth. The lengths of the straight arrows indicate the relative strengths of the Moon's gravitational pull on various parts of Earth. (a) The lunar gravitational forces acting on several different locations on and in Earth. The force is greatest on the side nearest the Moon and smallest on the opposite side. (b) The difference between the lunar forces experienced at those same locations and the force acting on Earth's center. The arrows represent the force with which the Moon tends to pull matter away from, or squeeze it toward, the center of our planet. Material on the side of Earth nearest the Moon tends to be pulled away from the center, and material on the far side is "left behind," and a bulge is formed. High and low tides result, twice per day, as Earth rotates beneath the bulges in the oceans.

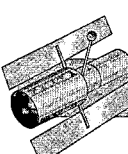

MORE PRECISELY 7-3

Tidal Forces

Consider the gravitational force exerted on Earth by the Moon. As discussed in the text, the *tidal* force is the name we give to the *variation* of the Moon's gravitational pull from one part of Earth to another. Specifically, as illustrated in the figure below (see also Figure 7.22), it is the *difference* between the actual force exerted on a chunk of Earth matter and the force exerted on the same chunk if it were placed at Earth's center.

For example, the Moon's gravitational pull on a 70-kg human located on Earth's surface with the Moon directly overhead (i.e., on the right in the figure) is

$$\frac{G M \times 70 \text{ kg}}{(r - R)^2} = 2.44 \times 10^{-3} \text{ newton,}$$

where the Moon's mass M is taken to be 7.3×10^{22} kg, the distance r from Earth to the Moon is 380,000 km, and Earth's radius R is 6400 km. (Recall that the newton is the SI unit of force.) The force on the same person located at Earth's center would be $G M \times 70 \text{ kg}/R^2 = 2.36 \times 10^{-3}$ newton. The Moon's tidal force is the small difference between these two forces.

With the help of a little algebra, we can write down an equation for the tidal force exerted by the Moon (of mass M and distance r) on a body (of mass m) placed at various locations on Earth's surface (assuming that Earth's

radius R is much less than r). When the body is placed along the line joining Earth to the Moon, the tidal force is directed *away* from Earth's center (that is, vertically up), and has magnitude

$$\text{tidal force} = \frac{2 \, G M m R}{r^3}.$$

(Verify for yourself that, for $m = 70$ kg, this tidal force is indeed equal to the difference between the two forces just calculated.) When the body is placed on a line through Earth's center perpendicular to the Earth–Moon line, the force has a magnitude one-half of the above value, and is directed *toward* Earth's center (that is, vertically down). Tidal forces tend to stretch bodies along the line joining them and compress them perpendicular to that line. The result is the characteristic tidal bulge shown in Figure 7.22.

Note that, unlike the familiar inverse-square law of gravity, the tidal force falls off as the inverse *cube* of the distance between the bodies involved. This rapid decline with increasing distance means that an object has to be very close, or very massive, in order to have a significant tidal effect on another.

EXAMPLE: Let's calculate the tidal forces exerted by first the Moon and then the Sun on a 1-kg mass placed on Earth's surface directly between the centers of Earth and the other body (i.e., at the peak of the bulge). For comparison, Earth's gravitational pull on the body (the body's *weight*) is $G \times M_{\text{Earth}} \times 1 \text{ kg} / R^2_{\text{Earth}} \approx 9.8$ newton (using the numbers in the Earth Data box). The above equation tells us that the tidal force due to the Moon is $2G \times M_{\text{Moon}} \times R_{\text{Earth}}/r^3_{\text{Moon}} \approx 1.1 \times 10^{-6}$ newton—tiny compared to Earth's pull, but nevertheless important, as we have seen. For the Sun, the result is $2G \times M_{\text{Sun}} \times R_{\text{Earth}}/r^3_{\text{Sun}} \approx 5.1 \times 10^{-7}$ newton. Even though the Sun is 375 times farther away from Earth than is the Moon, the Sun's mass is so much greater (by a factor of 27 million) that its tidal influence is still significant—about half that of the Moon.

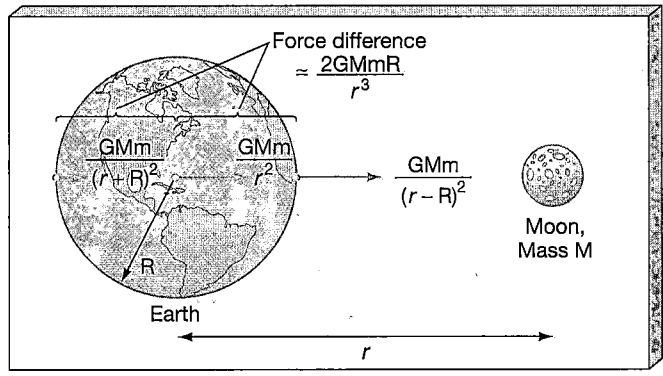

the tidal influence of one body on another diminishes very rapidly with increasing distance—in fact, as the inverse *cube* of the separation. For example, if the distance from Earth to the Moon were to double, the tides resulting from the Moon's gravity would decrease by a factor of eight. We will see many situations in this book where tidal forces are critically important in understanding astronomical phenomena. Notice that we still use the word *tidal* in these other contexts, even though we are not discussing oceanic tides and possibly not even planets at all.

Notice in Figure 7.22 that the side of Earth opposite the Moon also experiences a tidal bulge. The different gravitational pulls—greatest on that part of Earth closest to the Moon, weaker at Earth's center, and weakest of all on Earth's opposite side—cause average tides on opposite sides of our planet to be approximately equal in height. On the side nearer the Moon, the ocean water is pulled slightly toward the Moon. On the opposite side, the ocean water is left behind as Earth is pulled closer to the Moon. Thus, high tide occurs *twice*, not once, each day at any given location.

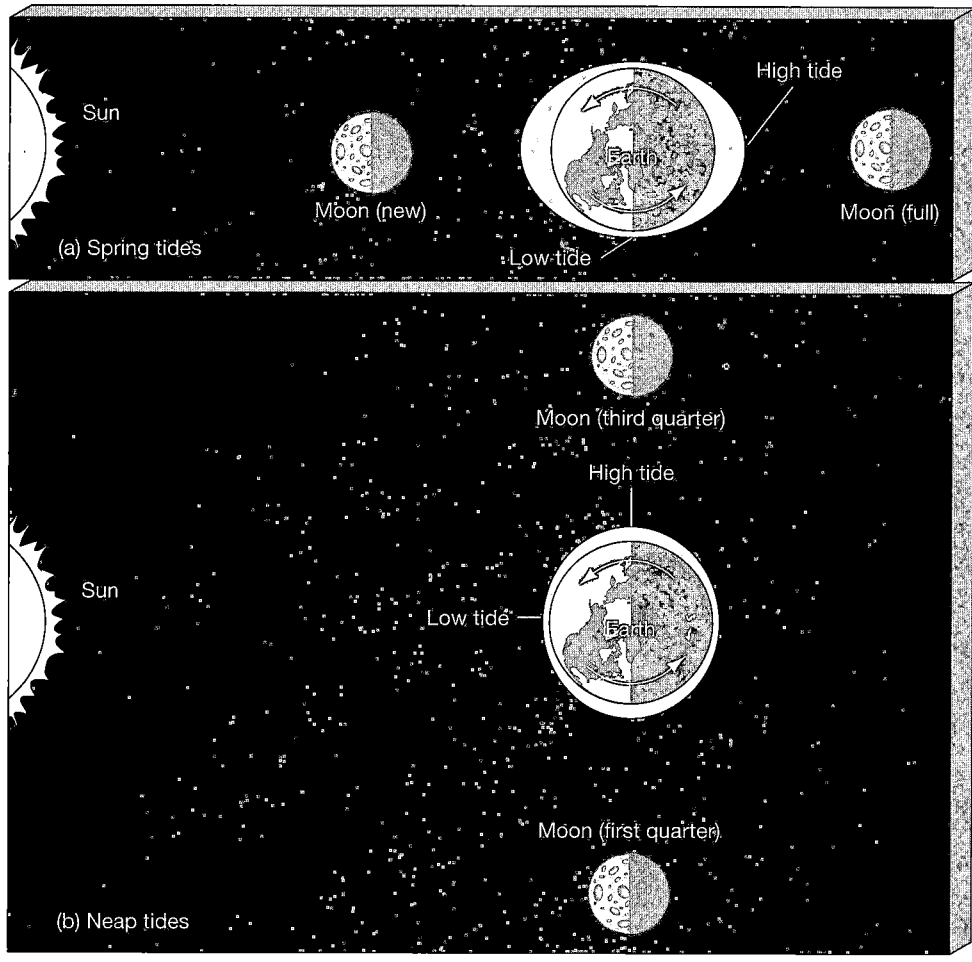

Figure 7.23 Solar and Lunar Tides The combined effects of the Sun and the Moon produce variations in the high and low tides. (a) When the Moon is either full or new, Earth, Moon, and Sun are approximately aligned, and the tidal bulges raised in Earth's oceans by the Moon and the Sun reinforce one another. (b) At first- or third-quarter Moon, the tidal effects of the Moon and the Sun partially cancel, and the tides are smallest. The Moon's tidal effect is greater than that of the Sun, so the net bulge points toward the Moon.

Both the Moon and the Sun exert tidal forces on our planet. Thus, instead of one tidal bulge, there are actually two—one pointing toward the Moon, the other toward the Sun—and the interaction between them accounts for the changes in the height of the tides over the course of a month or a year. When Earth, the Moon, and the Sun are roughly lined up (Figure 7.23a), the gravitational effects reinforce one another, and so the highest tides are generally found at times of new and full moons. These tides are known as *spring tides*. When the Earth–Moon line is perpendicular to the Earth–Sun line (at the first and third quarters; Figure 7.23b), the daily tides are smallest. These are termed *neap tides*.

EFFECT OF TIDES ON EARTH'S ROTATION

Earth rotates once on its axis (relative to the stars) in 23^h 56^m—one sidereal day. However, we know from fossil measurements that Earth's rotation is gradually slowing down, causing the length of the day to increase by about 1.5 milliseconds (ms) every century—not much on the scale of a human lifetime, but over millions of years, this steady slowing of Earth's spin adds up. At this rate, half a

billion years ago, the day was just over 22 hours long and the year contained 397 days.

A number of natural biological clocks lead us to the conclusion that Earth's spin rate is decreasing. For example, each day a growth mark is deposited on a certain type of coral in the reefs off the Bahamas. These growth marks are similar to the annual rings found in tree trunks, except that in the case of coral, the marks are made daily, in response to the day–night cycle of solar illumination. However, they also show yearly variations as the coral's growth responds to Earth's seasonal changes, allowing us to perceive annual cycles. Coral growing today shows 365 marks per year, but ancient coral shows many more growth deposits per year. Fossilized reefs that are five hundred million years old contain coral with nearly 400 deposits per year of growth.

Why is Earth's spin slowing? The main reason is the tidal effect of the Moon. In reality, the tidal bulge raised in Earth by the Moon does *not* point directly at the Moon, as was shown in Figure 7.22. Instead, because of the effects of friction, both between the crust and the oceans and within Earth itself, Earth's rotation tends to drag the tidal bulge around with it, causing the bulge to be displaced by a small angle from the Earth–Moon line, in the same direction as

Earth's spin (Figure 7.24). The net effect of the Moon's gravitational pull on this slightly offset bulge is to *reduce* our planet's rotation rate. At the same time, the Moon is spiraling slowly away from Earth, increasing its average distance from our planet by about 4 cm per year.

This process will continue until Earth rotates on its axis at exactly the same rate as the Moon orbits Earth. At that time the Moon will always be above the same point on Earth and will no longer lag behind the bulge it raises. Earth's rotation period will be 47 of our present days, and the distance to the Moon will be 550,000 km (about 43 percent greater than at present). However, this will take a very long time—many billions of years—to occur.

☑ Concept Check

◘ In what ways do tidal forces differ from the familiar inverse-square force of gravity?

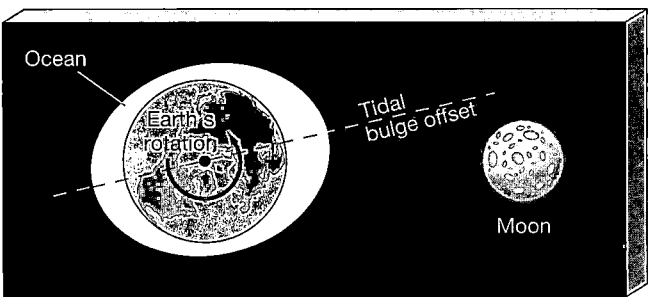

Figure 7.24 Tidal Bulge The tidal bulge raised in Earth by the Moon does not point directly at the Moon. Instead, because of the effects of friction, the bulge points slightly "ahead" of the Moon, in the direction of Earth's rotation. (The magnitude of the effect is greatly exaggerated in this diagram.) Because the Moon's gravitational pull on the near-side part of the bulge is greater than the pull on the far side, the overall effect is to decrease Earth's rotation rate.

Chapter Review

SUMMARY

The six main regions of Earth are (from inside to outside) a central metallic **core** (p. 166), which is surrounded by a thick rocky **mantle** (p. 166) and topped with a thin **crust** (p. 166). The liquid oceans on our planet's surface make up the **hydrosphere** (p. 166). Above the surface is the **atmosphere** (p. 166), which is composed primarily of nitrogen and oxygen. Higher still lies the **magnetosphere** (p. 167), where charged particles from the Sun are trapped by Earth's magnetic field.

Earth's atmosphere, composed primarily of nitrogen (78 percent), oxygen (21 percent), argon (0.9 percent), and carbon dioxide (0.03 percent), thins rapidly with altitude. **Convection** (p. 167) is the process by which heat is moved from one place to another by the upwelling or downwelling of a fluid, such as air or water. Convection occurs in the troposphere, the lowest region of Earth's atmosphere. It is the cause of surface winds and weather. Incoming solar ultraviolet radiation is absorbed in the **ozone layer** (p. 168). The ionosphere is kept ionized by high-energy radiation and particles from the Sun.

The **greenhouse effect** (p. 170) is the absorption and trapping by atmospheric gases (primarily carbon dioxide and water vapor) of infrared radiation emitted by Earth's surface. Incoming visible light from the Sun is not significantly absorbed by these gases. By making it more difficult for Earth to radiate its energy back into space, the greenhouse effect makes our planet's surface some 40 K warmer than would otherwise be the case. The air we breathe is not Earth's original atmosphere. It originated in material that was outgassed from our planet's interior by volcanoes and was then altered by solar radiation and, finally, by the emergence of life.

We study Earth's interior by observing how **seismic waves** (p. 171), produced by earthquakes just below Earth's surface, travel through the mantle. We can also study the upper mantle by analyzing the material brought to the surface when a volcano erupts. Seismic studies and mathematical modeling indicate that Earth's iron core consists of a solid **inner core** (p. 172) surrounded by a liquid **outer core** (p. 172). Earth's center is extremely hot—about the same temperature as the surface of the Sun. The density at Earth's center is much greater than the density of surface rocks.

The process by which heavy material sinks to the center of a planet while lighter material rises to the surface is called **differentiation** (p. 173). The differentiation of Earth implies that our planet must have been at least partially molten in the past. One way in which this could have occurred is by the heat released during Earth's formation and subsequent bombardment by material from interplanetary space. Another possibility is the energy released by the decay of **radioactive** (p. 175) elements present in the material from which Earth formed.

Earth's surface is made up of about a dozen enormous slabs, or plates. The slow movement of these plates across the surface is called continental drift, or **plate tectonics** (p. 176). Earthquakes, volcanism, and mountain building are associated with plate boundaries, where plates may collide, move apart, or rub against one another. The motion of the plates is thought to be driven by convection in Earth's mantle. The rocky upper layer of Earth that makes up the plates is the **lithosphere** (p. 177). The semisolid region in the upper mantle over which the plates slide is called the **asthenosphere** (p. 177). Evidence for past plate motion can be found in the geographical fit of continents, in the fossil record, and in the ages and magnetism of surface rocks.

Earth's magnetic field extends far beyond the surface of our planet. Charged particles from the solar wind are trapped by Earth's magnetic field lines to form the **Van Allen belts** (p. 183) that surround our planet. When particles from the Van Allen belts hit Earth's atmosphere, they heat and ionize the atoms there, causing them to glow in an **aurora** (p. 185). According to dynamo theory, planetary magnetic fields are produced by the motion of rapidly rotating, electrically conducting fluid (such as molten iron) in the planet's core.

The daily **tides** (p. 187) in Earth's oceans are caused by the gravitational effect of the Moon and the Sun, which raise **tidal**

bulges (p. 187) in the hydrosphere. The tidal effect of the Moon is almost twice that of the Sun. The size of the tides depends on the orientations of the Sun and the Moon relative to Earth. A differential gravitational force is always called a **tidal force** (p. 187), even when no oceans or even planets are involved. The tidal interaction between Earth and the Moon is causing Earth's spin to slow.

SELF-TEST: TRUE OR FALSE?

_____ **1.** Earth's average density is less than the density of water.

_____ **2.** The ozone layer is the warmest part of Earth's atmosphere.

_____ **3.** Earth's atmosphere is composed primarily of oxygen.

_____ **4.** Most of Earth's atmosphere lies within 30 km of the surface.

_____ **5.** A continued rise in the level of greenhouse gases in Earth's atmosphere will cause our planet's temperature to increase.

_____ **6.** P-waves can travel through both liquid and solid material; S-waves travel only through solid material.

_____ **7.** Geologists obtain most of their information about Earth's mantle by drilling deep into our planet's interior.

_____ **8.** Earth's core temperature is comparable to the Sun's surface temperature.

_____ **9.** Samples of Earth's core are available from volcanoes.

_____ **10.** Motion of the crustal plates is driven by convection in Earth's upper mantle.

_____ **11.** When plates collide, they simply come to rest and fuse together.

_____ **12.** Earth's magnetic field is the result of our planet's large, permanently magnetized iron core.

_____ **13.** Earth's magnetosphere extends about 5000 km above our planet's surface.

_____ **14.** There is one high tide and one low tide per day at any given coastal location on Earth.

_____ **15.** Because of the tides, Earth's rotation rate is slowing down.

SELF-TEST: FILL IN THE BLANK

1. Earth's surface rocks are _____ (more/less/equally) dense than our planet's average density.

2. Of Earth's crust, mantle, outer core, and inner core, which layer is the thinnest? _____

3. Seventy-eight percent of Earth's atmosphere is _____; 21 percent is _____.

4. The troposphere is where the process of _____ occurs.

5. Sunlight is absorbed by Earth's surface and is reemitted in the form of _____ radiation.

6. The primary greenhouse gases in Earth's atmosphere are _____ and _____.

7. Earth's secondary atmosphere was outgassed by _____.

8. Oxygen in Earth's atmosphere is the result of the appearance of _____.

9. Observations of S- and P-waves have confirmed that Earth's inner core is _____, and the outer core is _____.

10. Crustal rocks are made up primarily of low-density _____; the upper mantle is composed of slightly higher-density _____.

11. For differentiation to have occurred, Earth's interior must, at some time in the past, have been largely _____.

12. Continental drift, volcanism, earthquakes, faults, and mountain building can all be explained by the process known as _____.

13. When trapped electrons and protons in the magnetosphere collide with the upper atmosphere, they produce an _____.

14. Earth is unique among the planets in that it has _____ on its surface.

15. Tides are due to the _____ in the gravitational forces due to the Moon and the Sun from one side of Earth to the other.

REVIEW AND DISCUSSION

1. By comparison with Earth's average density, what do the densities of the water and rocks in Earth's crust tell us about Earth's interior?

2. What is Rayleigh scattering? What is its most noticeable effect for us on Earth?

3. How do geologists use earthquakes to obtain information about Earth's interior?

4. Compare and contrast P-waves and S-waves, and explain how they are useful to geologists.

5. What is the greenhouse effect, and what effect does it have on Earth's surface temperature?

6. Give two reasons why geologists believe that part of Earth's core is liquid.

7. What clue does Earth's differentiation provide to our planet's history?

8. What is convection? What effect does it have on (a) Earth's atmosphere? (b) Earth's interior?

9. How does radioactivity allow us to estimate Earth's age?

10. How did radioactive decay heat Earth early in its history? When did this heating end?

11. What process is responsible for the surface mountains, oceanic trenches, and other large-scale features on Earth's surface?

12. Discuss how distant quasars, lying hundreds of millions of light-years from Earth, are used to monitor the motion of Earth's tectonic plates.

13. What conditions are needed to create a dynamo in Earth's interior? What effect does this dynamo have?

14. Give a brief description of Earth's magnetosphere, and tell how it was discovered.

15. How does Earth's magnetosphere protect us from the harsh realities of interplanetary space?

16. How do we know that Earth's magnetic field has undergone reversals in the past? How do you think Earth's magnetic field reversals might have affected the evolution of life on our planet?

17. Explain how the Moon produces tides in Earth's oceans.

18. If the Moon had oceans like Earth's, what would the tidal effect be like there? How many high and low tides would there be during a "day"? How would the variations in height compare with those on Earth?

19. If Earth had no moon, do you think we would know anything about tidal forces?

20. Is the greenhouse effect operating in Earth's atmosphere helpful or harmful? Give examples. What are the consequences of an enhanced greenhouse effect?

PROBLEMS  *Algorithmic versions of these questions are available in the Practice Problems module of the Companion Website.*

The number of squares preceding each problem indicates its approximate level of difficulty.

1. ■ Verify that Earth's orbital perihelion and aphelion, mean orbital speed, surface gravity, and escape speed are correct as listed in the Earth Data box on p. 170.

2. ■ What would Earth's surface gravity and escape speed be if the entire planet had a density equal to that of the crust, say, 3000 kg/m³?

3. ■ Approximating Earth's atmosphere as a layer of gas 7.5 km thick, with uniform density 1.3 kg/m³, calculate its total mass. Compare this with Earth's mass.

4. ■■■ As discussed in the text, without the greenhouse effect, Earth's average surface temperature would be about 250 K. With the greenhouse effect, it is some 40 K higher. Use this information and Stefan's law to calculate the fraction of infrared radiation leaving Earth's surface that is absorbed by greenhouse gases in the atmosphere. ∞ (Sec. 3.3)

5. ■■ Most of Earth's ice is found in Antarctica, where permanent ice caps cover approximately 0.5 percent of Earth's total surface area and are 3 km thick, on average. Earth's oceans cover roughly 71 percent of our planet, to an average depth of 3.6 km. Assuming that water and ice have roughly the same density, estimate by how much sea level would rise if global warming were to cause the Antarctic ice caps to melt.

6. ■ Following an earthquake, how long would it take a P-wave, moving in a straight line with a speed of 5 km/s, to reach Earth's opposite side?

7. ■ Based on the data presented in the text, estimate the fractions of Earth's volume represented by (a) the inner core, (b) the outer core, (c) the mantle, and (d) the crust.

8. ■ At 3 cm/yr, how long would it take a typical plate to traverse the present width of the Atlantic Ocean, about 6000 km?

9. ■ In a certain sample of rock it is found that 25 percent of uranium-238 nuclei have decayed into lead-206. Estimate the age of the rock sample.

10. ■■ A second sample of rock is found to contain three times as many lead-207 nuclei as uranium-235 nuclei. What ratio of uranium-238 to lead-206 nuclei would you expect?

11. ■■■ Astronauts in orbit are weightless because they are falling freely in Earth's gravitational field, but they are still subject to tidal forces. Use the formula given in *More Precisely 7-3* to calculate the relative acceleration due to tidal forces of two masses in low Earth orbit, placed 1 m apart along a line extending radially outward from Earth's center. Compare this acceleration with the acceleration due to gravity at Earth's surface.

12. ■■ You are standing on Earth's surface, and the Moon is on the horizon. By what fraction is your weight changed due to the Moon's tidal gravitational force?

13. ■■ You are standing on Earth's surface, and the full Moon is directly overhead. By what fraction is your weight decreased due to the combination of the Sun's and the Moon's tidal gravitational forces?

14. ■■■ The planet Jupiter exerts a strong tidal force on its innermost moon, Io. Using the formula given in *More Precisely 7-3*, compare Jupiter's tidal force on Io with the gravitational force on Io's surface due to the moon's own gravity. Io orbits at a distance of 422,000 km. Its mass and radius are 9.0×10^{22} kg and 3600 km, respectively. Jupiter's mass is 1.9×10^{27} kg.

15. ■■ Compare the magnitude of the tidal gravitational force on Earth due to Jupiter with that due to the Moon. Assume an Earth–Jupiter distance of 4.2 A.U. Based on your answer, do you think that the tidal stresses caused by a "cosmic convergence"—a chance alignment of the four jovian planets, so that they all appear from Earth to be in exactly the same direction in the sky—would have any noticeable affect on our planet?

COLLABORATIVE EXERCISES

1. **Earth's Interior Scale Model.** Using a ruler and self-stick or taped-on labels, create a scale model of Earth on the shortest member of your group. Use the group member's height in inches divided by Earth's diameter (12,800 km) as the scale factor along with Figure 7.1. For example, if the selected group member is 65 inches tall then the 50 km maximum depth of Earth's crust is (65 inches/12,800 km) × (50 km) = 0.25 inches from the top of the head and 0.25 inches from the bottom of the feet.

2. **Global Warming.** Reach an agreement with your group about how the U.S. government should become involved in reducing the amount of greenhouse gases released from international industry that might result in an enhanced greenhouse effect called *global warming* and create a short press release.

RESEARCHING ON THE WEB  *To complete the following exercises, go to the online Destinations module for Chapter 7 on the Companion Website for* Astronomy Today *4/e.*

1. Access the Current Weather section of the "Daily Planet" page and determine the current surface dew point temperature at your present location.

2. Access the "Earth Observation Movie Studio" page and determine the approximate latitudes on Earth currently experiencing the most and least rainfall.

3. Access the "Aurora" page and use the latest three-day forecast to determine tomorrow's predicted geomagnetic conditions.

PROJECTS

1. Go to a sporting goods store and get a tide table; many stores near the ocean provide them free. Choose a month and plot the height of one high and one low tide versus the day of the month. Now mark the dates when the primary phases of the Moon occur. How well does the phase of the Moon predict the tides?

2. Go to a library and read about global warming. How much carbon dioxide is produced each year by human activities? How does this compare with the total amount of carbon dioxide in Earth's atmosphere? What natural processes tend to reduce the level of atmospheric carbon dioxide? Do all scientists agree that global warming is an inevitable consequence of carbon dioxide production? What political initiatives are currently under way to address the problem?

SKYCHART III PROJECTS  *The SkyChart III Student Version planetarium program on which these exercises are based is included as a separately executable program on the CD in the back of this text.*

1. ◼ Set the date and time in SkyChart III to just before the vernal equinox, such as 6:00 AM on March 20, 2002. Enable grid lines and center on the Sun. Zoom to a 5° field of view. Go to *DRAW/Symbols & Grids* and select *Draw Grid Lines* with *Equator* and *Ecliptic* enabled. *DRAW/Horizon Mask*, *Constellations* should be deselected and *DRAW/Sky Background/Black* should be selected. With animation set for one-hour steps, use F6 to step forward and observe the Sun approaching the intersection of the equator and the ecliptic. The time at which the Sun crosses this intersection is the vernal equinox.

2. ◼ With SkyChart III configured as described for viewing the vernal equinox, set field of view to 180° and time steps to one day. Using F6 to step through time. Note how the Sun travels along the ecliptic and reaches its most northerly excursion on June 21, 2002. The Sun arrives at a right ascension, R.A., of 6 h when it is at its most northerly excursion. This is summer solstice. Continue to advance time

with animation on until the Sun arrives at December 21, 2002. Note that the Sun arrives at its most southerly excursion at R.A. of 18 h. This is the winter solstice.

3. ◼ On a human time scale, the stars seem fixed in place. Any observed motion is due mostly to the motion of Earth. Earth's spin makes it appear that the stars in the Northern Hemisphere orbit Polaris as in Figure 1.8. Center your view on Polaris, and using the animation tools, reproduce Figure 1.8(b).

4. ◼◼ Using a setup similar to the foregoing exercise, you can measure the sidereal day by trailing one star and determining the time it takes that star to complete one full orbit around Polaris. Turn on the trail for Yildun, a neighboring star in Ursa Minor. Before starting the animation, record the exact local time. How many hours does it take to return exactly to the beginning? You might have to backtrack (*Animate Back*) with a smaller time step to determine it exactly. How close is your answer to the actual sidereal day? What is the main source of error?

 *In addition to the Practice Problems and Destinations modules, the Companion Website at http://www.prenhall.com/chaisson provides for each chapter an additional true-false, multiple choice, and labeling quiz, as well as additional annotated images, animations, and links to related Websites.*

8 THE MOON AND MERCURY

Scorched and Battered Worlds

LEARNING GOALS

Studying this chapter will enable you to:

1 Specify the general characteristics of the Moon and Mercury, and compare them with those of Earth.

2 Describe the surface features of the Moon and Mercury, and recount how they were formed by dynamic events early in their history.

3 Explain how the Moon's rotation is influenced by its orbit around Earth, and Mercury's by its orbit around the Sun.

4 Explain how observations of cratering can be used to estimate the age of a body's surface.

5 Compare the Moon's interior structure with that of Mercury.

6 Summarize the various theories for the formation of the Moon, and indicate which is presently considered most likely.

7 Discuss how astronomers have pieced together the story of the Moon's evolution, and compare its evolutionary history with that of Mercury.

 Visit http://www.prenhall.com/chaisson for additional annotated images, animations, and links to related sites for this chapter.

The magnificent desolation of the lunar surface is disturbed in this photograph by the cart at the center and the bootprints in the foreground—stark testimony to a visit by men and their machines. In this redigitized image of an old movie frame taken by an *Apollo-17* astronaut, we see the Taurus-Littrow Hills behind the "lunar buggy"—a carlike contraption brought to the Moon to allow the astronauts to venture far from their Lunar Excursion Module in which they landed after a four-day trip from Earth. *(NASA)*

The Big Picture: America's manned exploration of the Moon was arguably the greatest engineering accomplishment of the twentieth century, indeed one of the greatest of all time. Nine missions were launched to the Moon, a dozen astronauts were landed, and all returned safely to Earth. The *Apollo* program ended in 1972 as quickly as it had begun ten years before—largely because of political posturing at the height of the Cold War. There are currently no plans to return to our nearest neighbor in space.

The Moon is Earth's only natural satellite. Mercury, the smallest terrestrial world, is the planet closest to the Sun. Despite their different environments, however, these two bodies have many similarities—indeed, at first glance, you might even mistake one for the other. Both have heavily cratered, ancient surfaces, littered with boulders and pulverized dust. With no atmosphere to moderate variations in solar heating, each experiences wild temperature swings from day to night. Both are geologically dead. In short, the Moon and Mercury differ greatly from Earth, but it is precisely those differences that make these desolate worlds so interesting to planetary scientists. Why is the Moon so unlike our own planet, despite its nearness to us, and why does planet Mercury apparently have so much more in common with Earth's Moon than with Earth itself? In this chapter we explore the properties of these two worlds as we begin our comparative study of the planets and moons that make up our solar system.

8.1 Orbital Properties

We begin our study of the Moon and Mercury by examining their orbits. Knowledge of these will, in turn, aid us in determining and explaining the other properties of these worlds. Detailed orbital and physical data are presented in the Moon Data box (p. 202) and the Mercury Data box (p. 219).

THE MOON

Parallax methods, described in Chapter 1, can provide us with quite accurate measurements of the distance to the Moon, using Earth's diameter as a baseline. ∞ (Sec. 1.5) Radar and laser ranging yield more accurate distances. The Moon is much closer than any of the planets, and the radar echo bounced off the Moon's surface is strong. A radio telescope receives the echo after about a 2.56-second wait. Dividing this time by two (to account for the round-trip taken by the signal) and multiplying it by the speed of light (300,000 km/s) gives us a mean distance of 384,000 km. (The actual distance at any specific time depends on the Moon's location in its slightly elliptical orbit around Earth.)

Current laser-ranging technology, using reflectors placed on the lunar surface by *Apollo* astronauts (see *Discovery 8-1*) to reflect laser beams fired from Earth, allows astronomers to measure the round-trip time with submicrosecond accuracy. Repeated measurements have allowed astronomers to determine the Moon's orbit to within a few centimeters. This precision is necessary for programming unmanned spacecraft to land successfully on the lunar surface.

MERCURY

Viewed from Earth, Mercury never strays far from the Sun: The planet's 0.4 A.U. orbital semimajor axis means that its angular distance from the Sun never exceeds 28°. Consequently, the planet is visible to the naked eye only when the Sun's light is blotted out—just before dawn or just after sunset (or, much less frequently, during a total solar eclipse), and it is not possible to follow Mercury through a full cycle of phases. In fact, although Mercury was well-known to ancient astronomers, they originally believed that this companion to the Sun was two different

objects, and the connection between the planet's morning and evening appearances took some time to establish. However, later Greek astronomers were certainly aware that the "two planets" were really different alignments of a single body. Figure 8.1, a photograph taken just after sunset, shows Mercury above the western horizon, along with three other planets and the Moon.

Because Earth rotates at a rate of 15° per hour, Mercury is visible for at most 2 hours on any given night, even under the most favorable circumstances. For most observers at

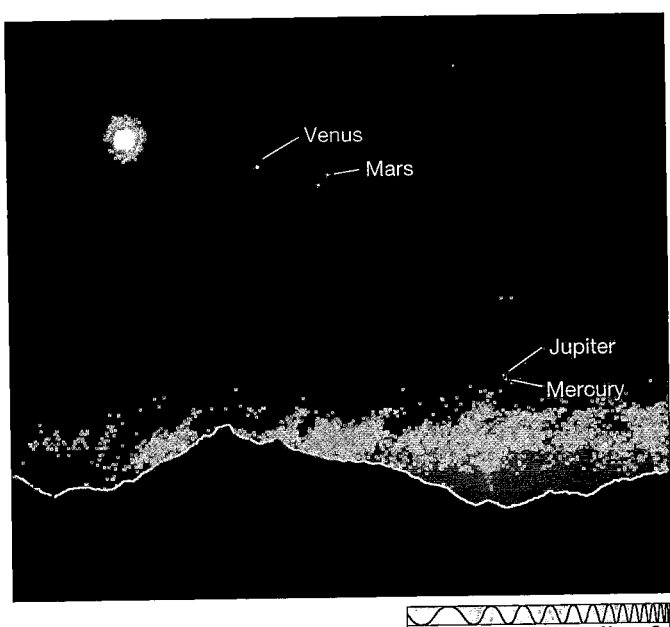

Figure 8.1 Evening Sky Four planets, together with the Moon, are visible in this photograph taken shortly after sunset. To the right of the Moon (top left) is the brightest planet, Venus. A little farther to the right is Mars, with the star Regulus just below and to its left. At the lower right, at the edge of the Sun's glare, are Jupiter and Mercury. (The moon appears round rather than crescent-shaped because the "dark" portion of its disk is indirectly illuminated by sunlight reflected from Earth. This "earthshine," relatively faint to the naked eye, is exaggerated in the overexposed photographic image.) *(J. Sanford/Science Photo Library/Photo Researchers, Inc.)*

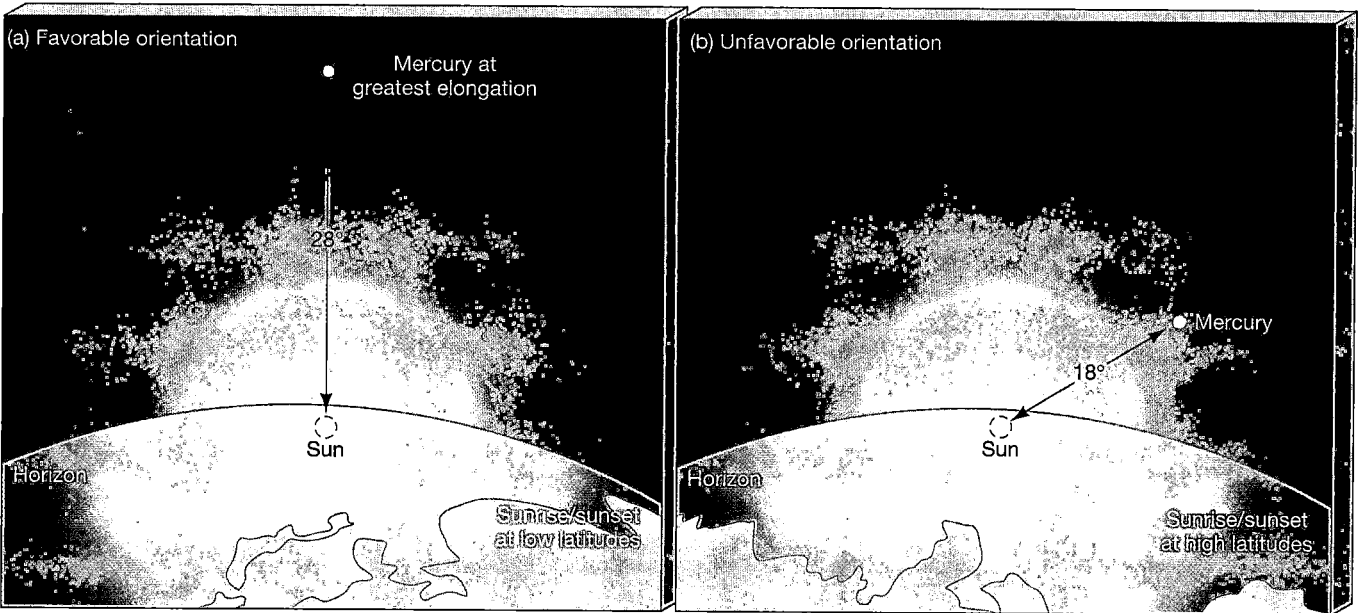

Figure 8.2 Visibility of Mercury Favorable and unfavorable orientations of Mercury's orbit result from different Earth orientations and observer locations. (a) At greatest elongation, Mercury lies some 28° from the Sun. The planet is most easily visible when it also lies high above the horizon. (b) At the most unfavorable orientations, Mercury is close to both the Sun and the horizon. Mercury is shown here at perihelion, 18° from the Sun.

most times of the year, Mercury is considerably less than 28° above the horizon, so it is generally visible for a much shorter period (see Figure 8.2). Nowadays, large telescopes can filter out the Sun's glare and observe Mercury even during the daytime, when the planet is higher in the sky and atmospheric effects are reduced. (The amount of air that the light from the planet has to traverse before reaching our telescope decreases as the height above the horizon increases.) In fact, some of the best views of Mercury have been ob-

tained in this way. The naked-eye or amateur astronomer is generally limited to nighttime observations, however.

In all cases, it becomes progressively more difficult to view Mercury the closer (in the sky) its orbit takes it to the Sun. The best images of the planet therefore show a "half Mercury," close to its maximum angular separation from the Sun, or *maximum elongation*, as illustrated in Figure 8.3. (A planet's elongation is just its angular distance from the Sun, as seen from Earth.)

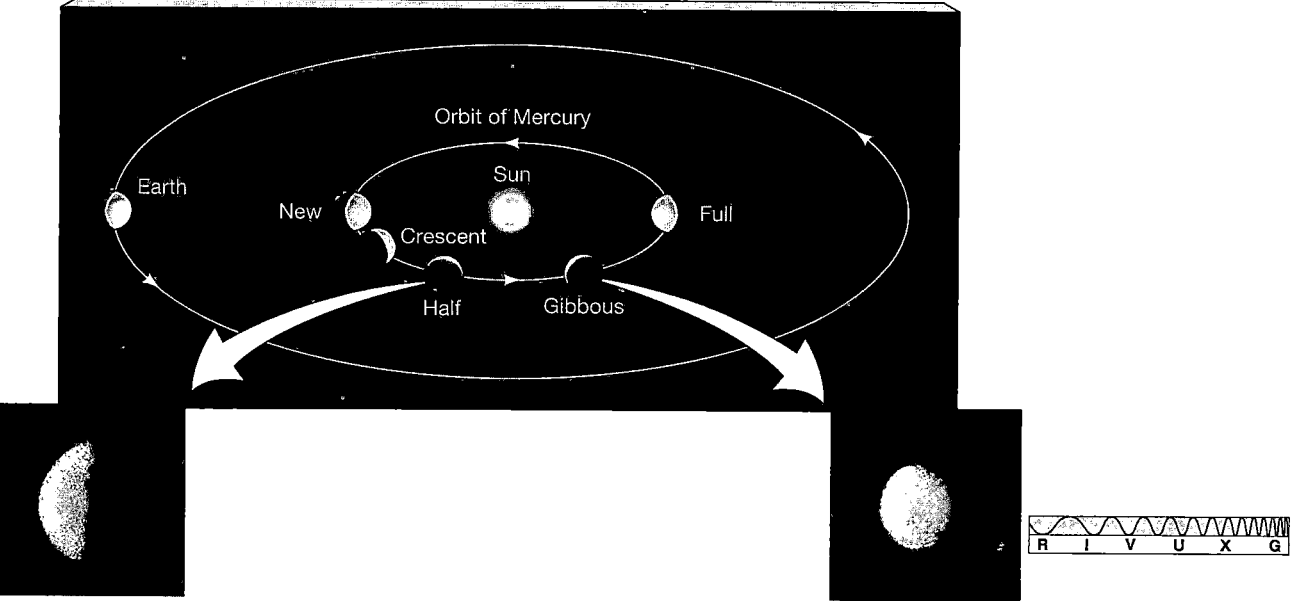

Figure 8.3 Mercury Phases Some views of Mercury at different points along its orbit. The best images of the planet are taken when it is at its maximum elongation (greatest apparent distance from the Sun) and show a "half Mercury." (Compare with Figure 2.14a.) *(R. Beebe)*

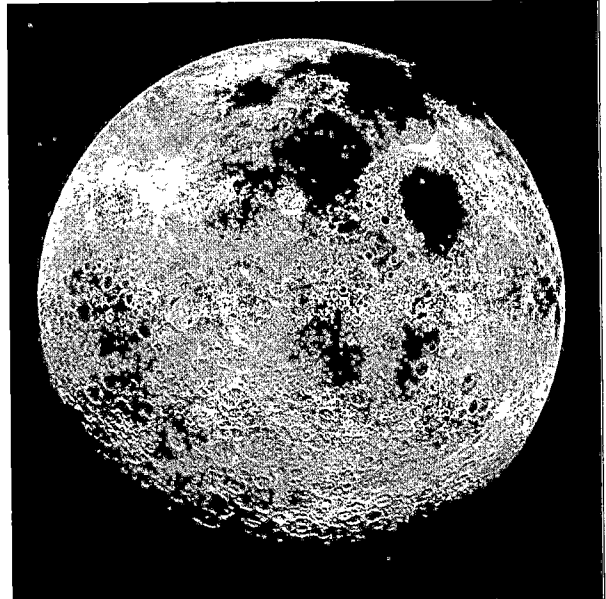

MOON DATA

Orbital semimajor axis*	384,000 km
Orbital eccentricity	0.055
Perigee†	363,000 km
Apogee‡	406,000 km
Mean orbital speed	1.02 km/s
Sidereal orbital period	27.3 solar days
Synodic orbital period	29.5 solar days
Orbital inclination to the ecliptic	5.2°
Greatest angular diameter, as seen from Earth	32.9'
Mass	7.35 × 10²² kg 0.012 (Earth = 1)
Equatorial radius	1738 km 0.27 (Earth = 1)
Mean density	3340 kg/m³ 0.61 (Earth = 1)
Surface gravity	1.62 m/s² 0.17 (Earth = 1)
Escape speed	2.38 km/s
Sidereal rotation period	27.3 solar days
Axial tilt	6.7°
Surface magnetic field	(no detectable global field)
Surface temperature	100–400 K

*Orbit around Earth.
†Closest approach to Earth.
‡Greatest distance from Earth.

8.2 Physical Properties

❶ From Earth, the Moon's angular diameter is about 0.5°. Knowing that and the distance to the Moon, we can easily calculate our satellite's true size, as discussed in Chapter 1. ∞ (Sec. 1.5) The Moon's radius is about 1700 km, roughly one-fourth that of Earth. More precise measurements yield a lunar radius of 1738 km. We can determine Mercury's radius by similar reasoning. At its closest approach to Earth, at a distance of about 0.52 A.U., Mercury's angular diameter is measured to be 13″ (arc seconds), implying a radius of about 2450 km, or 0.38 Earth radii. More accurate measurements by unmanned space probes yield a result of 2440 km.

As mentioned in Chapter 6, even before the Space Age, the masses of both the Moon and Mercury were already quite well known from studies of their effects on Earth's orbit. ∞ (Sec. 6.2) The mass of the Moon is 7.3 × 10²² kg, approximately 1/80 the mass of Earth. The mass of Mercury is 3.3 × 10²³ kg—about 0.055 Earth masses.

The Moon's average density of 3300 kg/m³ contrasts with the average Earth value of about 5500 kg/m³, suggesting that the Moon contains fewer heavy elements (such as iron) than Earth. However, despite its many other similarities to the Moon, Mercury's mean density is 5400 kg/m³, only slightly less than that of Earth. Assuming that surface rocks on Mercury are of similar density to surface rocks on Earth and the Moon, we are led to the conclusion that the interior of Mercury must contain a lot of high-density material, most probably iron. In fact, since Mercury is considerably less massive than Earth, its interior is squeezed less by the weight of overlying material, so Mercury's iron core must actually contain a much larger fraction of the planet's mass than in our own planet. ∞ (Sec. 6.2) Because the Moon and Mercury are so much less massive than Earth, their gravitational fields are also weaker. The force of gravity on the lunar surface is only about one-sixth that on Earth; Mercury's surface gravity is a little stronger—about 0.4 times Earth's. Thus, an astronaut weighing 180 lb on Earth would weigh a mere 30 lb on the Moon and 72 lb on Mercury. Those bulky spacesuits used by the *Apollo* astronauts on the Moon were not nearly as heavy as they appeared!

Astronomers have never observed any appreciable atmosphere on the Moon or Mercury, either spectroscopically from Earth or during close approaches by spacecraft. This is a direct consequence of these bodies' weak gravitational fields, as discussed in *More Precisely 8-1*. Simply put, a massive object has a better chance of retaining an atmosphere, because the more massive an object, the larger the speed needed for atoms or molecules to escape. The Moon's escape speed is only 2.4 km/s, compared with 11.2 km/s for Earth; Mercury's escape speed is 4.2 km/s. Any primary atmospheres these worlds had initially, or secondary atmospheres that appeared later, are gone forever. ∞ (Sec. 7.2) During its flybys of Mercury in 1974 and 1975, the U.S. space probe *Mariner 10* found traces of what was at first thought to be an atmosphere on Mercury. ∞ (Sec. 6.6) However, this gas is now known to be temporarily trapped hydrogen and helium stolen from the solar wind. Mercury captures this gas and holds it for just a few weeks. Both the Moon and Mercury have extremely tenuous (less than a trillionth the density of Earth's atmosphere) envelopes of sodium and potassium. Scientists believe that these atoms are torn out of the surface rocks following impacts with high-energy particles in the solar wind; they do not constitute a true atmosphere in any sense. Thus, neither the Moon nor Mercury has any protection against the harsh environment of interplanetary space. This fact is crucial in understanding their surface evolution and present-day appearance.

Lacking the moderating influence of an atmosphere, both the Moon and Mercury experience wide variations in surface temperature. Noontime temperatures at the Moon's equator can reach 400 K, well above the boiling point of water. Because of its proximity to the Sun, Mercury's daytime temperature is even higher—radio observations of the planet's thermal emission indicate that it can reach 700 K. ∞ (Sec. 3.4) But at night or in the shade, temperatures on both worlds fall to about 100 K, well below water's freezing point. Mercury's 600 K temperature range is the largest of any planet or moon in the solar system.

☑ Concept Check

■ Why do the Moon and Mercury have no significant atmospheres, unlike Earth?

8.3 Surface Features on the Moon and Mercury

LUNAR TERRAIN

🔍 The first observers to point their telescopes at the Moon, most notable among them Galileo Galilei, noted large dark areas, resembling (they thought) Earth's oceans. They also saw light-colored areas resembling the continents. Both types of regions are clear in Figure 8.4, a mosaic of the full Moon. The light and dark surface features are also evident to the naked eye, creating the face of the familiar "man in the moon."

Today we know that the dark areas are not oceans but extensive flat areas that resulted from lava flows during a much earlier period of the Moon's evolution. Nevertheless, they are still called **maria**, a Latin word meaning "seas" (singular: *mare*). There are 14 maria, all roughly circular. The largest of them (Mare Imbrium) is about 1100 km in diameter. The lighter areas, originally dubbed *terrae*, from the Latin word for "land," are now known to be elevated several kilometers above the maria. Accordingly, they are usually called the lunar **highlands**.

Figure 8.4 Full Moon, Near Side A photographic mosaic of the full Moon, north pole at the top. Because the Moon emits no visible radiation of its own, we can see it only by the reflected light of the Sun. Some prominent maria are labeled. *(UC/Lick Observatory)*

The smallest lunar features we can distinguish with the naked eye are about 200 km across. Telescopic observations further resolve the surface into numerous bowl-shaped depressions, or **craters** (after the Greek word for "bowl"). Most craters apparently formed eons ago primarily as the result of meteoritic impact. Craters are particularly clear in Figures 8.5(a) and (b) near the *terminator* (the line that separates day from night on the surface), where the Sun is low in the sky and casts long shadows that enable us to distinguish quite small surface details.

Due to the blurring effects of our atmosphere, the smallest lunar objects that telescopes on Earth's surface can resolve are about 1 km across (see Figure 8.5c). Much more detailed photographs have been taken by orbiting spacecraft and, of course, by visiting astronauts. Figure 8.6 is a view of some lunar craters taken from an orbiting spacecraft, showing features as small as 500 m across. Craters are found everywhere on the Moon's surface, although they are much more prevalent in the highlands. They come in all sizes—the largest are hundreds of kilometers in diameter; the smallest are microscopic.

Based on studies of lunar rock brought back to Earth by *Apollo* astronauts and unmanned Soviet landers, geologists have identified important differences in both *composition* and *age* between the highlands and the maria. The

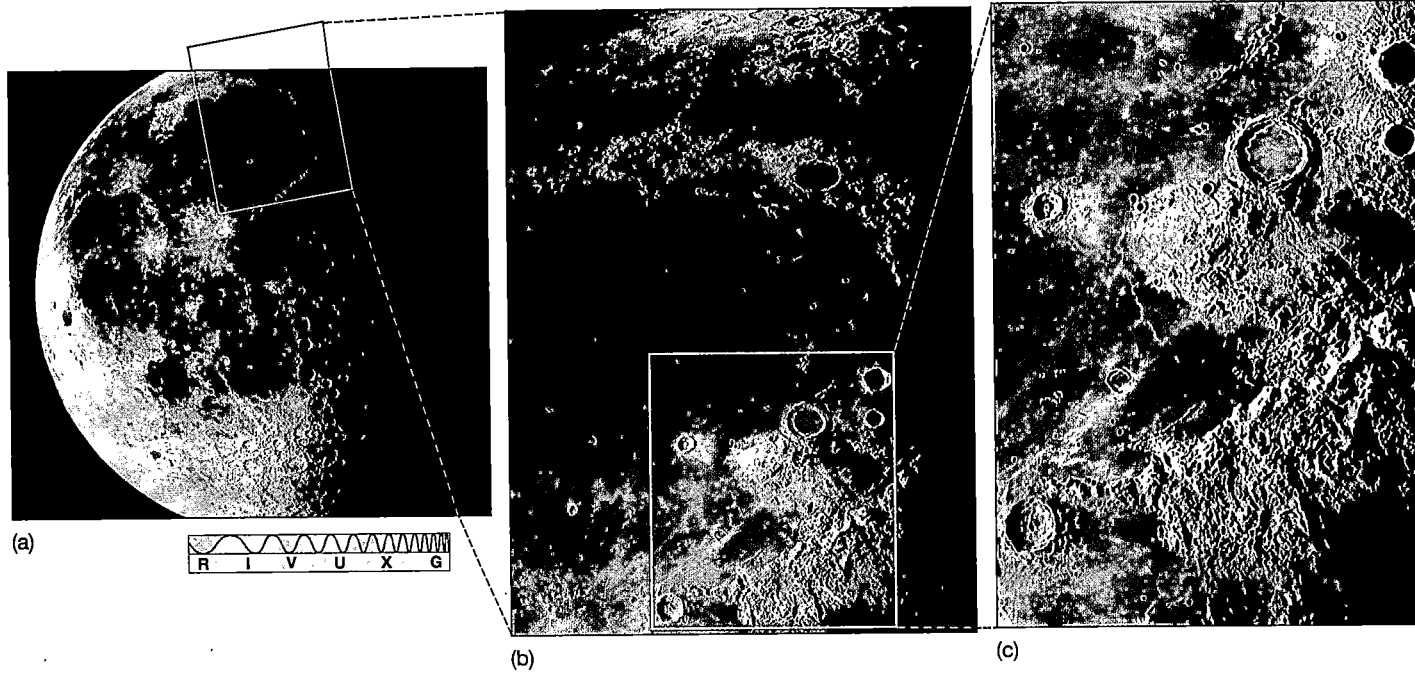

(a)

(b)

(c)

Figure 8.5 Moon, Close Up (a) The Moon near third quarter. Surface features are much more visible near the *terminator*, the line separating light from dark, where sunlight strikes at a sharp angle and shadows highlight the topography. The light-colored crater just below the lower left corner of the breakout box is Copernicus. (b) Magnified view of a region near the terminator, as seen from Earth through a large telescope. The central dark area is Mare Imbrium, ringed at bottom right by the Apennine mountains. (c) An enlargement of a portion of (b). The crater at the bottom left is Eratosthenes; Archimedes is at top center. The smallest craters visible here have diameters of about 2 km. *(UC/Lick Observatory; Caltech)*

highlands are made largely of rocks rich in aluminum, making them lighter in color and lower in density (2900 kg/m³). The maria's basaltic matter contains more iron, giving it a darker color and greater density (3300 kg/m³). Loosely speaking, the highlands represent the Moon's crust, while the maria are made of mantle material. Maria

rock is quite similar to terrestrial basalt, and geologists believe that it arose on the Moon much as basalt did on Earth, through the upwelling of molten material through the crust. Radioactive dating indicates ages of 4 to 4.4 billion years for highland rocks, and from 3.2 to 3.9 billion years for those from the maria.

Figure 8.6 Moon from *Apollo* The Moon, as seen from the *Apollo 8* orbiter during the first human circumnavigation of the Moon in 1968. Craters ranging in size from 50 km to 500 m (also the width of the long fault lines) can be seen. *(NASA)*

All the Moon's significant surface features have names. The 14 maria bear fanciful Latin names—Mare Imbrium ("Sea of Showers"), Mare Nubium ("Sea of Clouds"), Mare Nectaris ("Sea of Nectar"), and so on. Most mountain ranges in the highlands bear the names of terrestrial mountain ranges—the Alps, the Carpathians, the Apennines, the Pyrenees, and so on. Most of the craters are named after great scientists or philosophers, such as Plato, Aristotle, Eratosthenes, and Copernicus.

Because the Moon rotates once on its axis in exactly the same time it takes to complete one orbit around Earth, the Moon has a "near" side, which is always visible from Earth, and a "far" side, which never is (see Section 8.4). To the surprise of most astronomers, when the far side of the Moon was mapped, first by Soviet and later by U.S. spacecraft (see *Discovery 8-1*), no major maria were found there. The lunar far side (Figure 8.7) is composed almost entirely of highlands. This fact has great bearing on our theory of how the Moon's surface terrain came into being, for it implies that the processes involved could *not* have been entirely internal in nature. Earth's presence must somehow have played a role.

☑ Concept Check

■ Describe three important ways in which the lunar maria differ from the highlands.

THE SURFACE OF MERCURY

2 Mercury is difficult to observe from Earth because of its closeness to the Sun. Even with a fairly large telescope, we see it only as a slightly pinkish disk. Figure 8.8 is one of the few photographs of Mercury taken from Earth that shows any evidence of surface markings. Astronomers could only speculate about the faint, dark markings in the days before *Mariner 10*'s arrival. We now know that these markings are much like those seen when gazing casually at Earth's Moon. The largest ground-based telescopes can resolve surface features on Mercury about as well as we can perceive features on the Moon with our unaided eyes.

In 1974, *Mariner 10* approached within 10,000 km of the surface of Mercury, sending back high-resolution images of the planet. ∞ (Sec. 6.6) These photographs, which showed surface features as small as 150 m across, revolutionized our knowledge of the planet. Figures 8.9 (a) and (b) show views of Mercury taken by *Mariner 10* from a distance of about 200,000 km.

As discussed in Chapter 6, *Mariner 10* did not go into orbit around Mercury, but instead was placed in a somewhat eccentric orbit around the Sun that brought it close to the planet every 176 days—exactly two Mercury years. ∞ (Sec. 6.6) However, the peculiar combination of Mercury's orbital period and rotation rate (discussed in more detail in Section 8.4) meant that *Mariner* saw the *same* face of the planet at each approach. As a result, less than half of the planet's surface has been mapped. Together, the two mosaics in Figure 8.9 cover the known surface of Mercury. No similar photographs exist

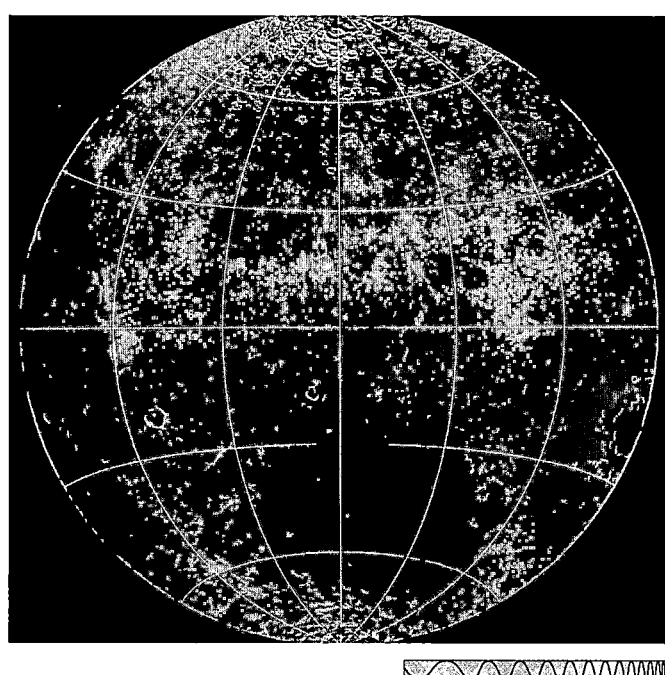

Figure 8.7 Full Moon, Far Side The far side of the Moon, as seen by the *Clementine* military spacecraft. The large, dark region at center bottom outlines the south pole–Aitken Basin, the largest and deepest impact feature known in the solar system. This image shows only one or two small maria on the far side. *(Dept. of Defense)*

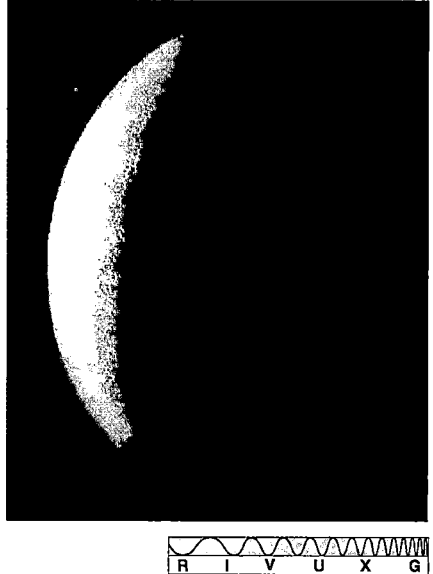

Figure 8.8 Mercury Photograph of Mercury taken from Earth with one of the largest ground-based optical telescopes. Only a few surface features are discernible. *(Palomar Observatory/Caltech)*

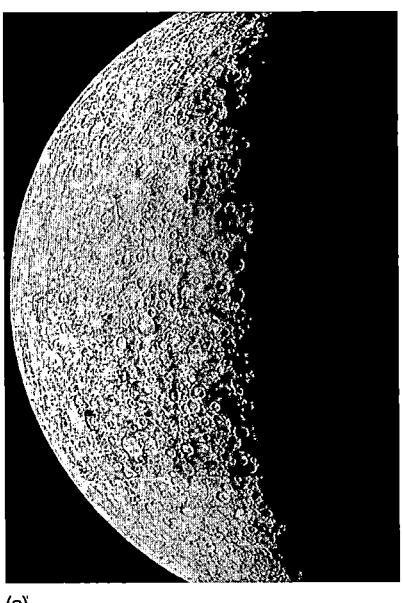

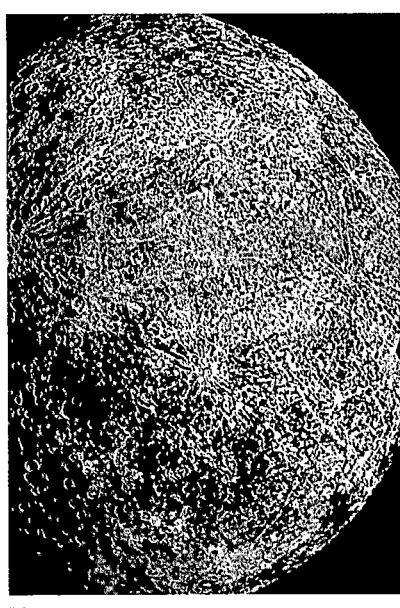

(a)
(b)

Figure 8.9 Mercury, Up Close
(a) Mercury is imaged here as a mosaic of photographs taken by the *Mariner 10* spacecraft in the mid-1970s during its approach to the planet. At the time, the spacecraft was some 200,000 km away. (b) *Mariner 10's* view of Mercury as it sped away from the planet after each encounter. Again, the spacecraft was about 200,000 km away when the photographs making up this mosaic were taken. *(NASA)*

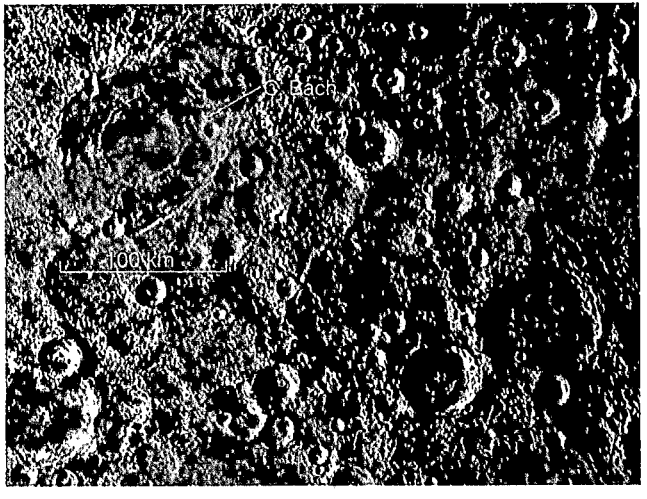

Figure 8.10 Mercury, Very Close Another photograph of Mercury by *Mariner 10*, this time from about 20,000 km above the planet's surface. The double-ringed crater at the upper left, named C. Bach, is about 100 km across; it exemplifies many of the large craters on Mercury, which tend to have double, rather than single, rings. The reason is not yet understood. *(NASA)*

of the hemisphere that happened to be in shadow during each encounter.

Figure 8.10 shows a higher-resolution photograph of the planet from a distance of 20,000 km. The similarities to the Moon are striking. We see no sign of clouds, rivers, dust storms, or other aspects of weather. Much of the cratered surface bears a strong resemblance to the Moon's highlands. Mercury, however, shows few extensive lava flow regions akin to the lunar maria.

8.4 Rotation Rates

THE ROTATION OF THE MOON

3 As mentioned above, the Moon's rotation period is precisely equal to its period of revolution about Earth—27.3 days—so the Moon keeps the same side facing Earth at all times (see Figure 8.11). To an astronaut standing on the Moon's near-side surface, Earth would appear almost stationary in the sky (although its daily rotation would be clearly evident). This condition, in which the spin of one body is precisely equal to (or *synchronized* with) its revolution around another body, is known as a **synchronous orbit**. The fact that the Moon is in a synchronous orbit around Earth is no accident. It is an inevitable consequence of the gravitational interaction between those two bodies.

Just as the Moon raises tides on Earth, Earth also produces a tidal bulge in the Moon. Indeed, because Earth is so much more massive, the tidal force on the Moon is about 20 times greater than that on Earth, and the Moon's tidal bulge is correspondingly larger. In Chapter 7 we saw how lunar tidal forces are causing Earth's spin to slow and how, as a result, Earth will eventually rotate on its axis at the same rate as the Moon revolves around Earth. ∞ (Sec. 7.6) Earth's rotation will not become synchronous with the Earth–Moon orbital period for hundreds of billions of years. In the case of the Moon, however, the process has already gone to completion. The Moon's much larger tidal deformation caused it to evolve into a synchronous orbit long ago, and the Moon is said to have become *tidally locked* to Earth. Most of the moons in the solar system are similarly locked by the tidal fields of their parent planets.

Actually, the size of the lunar bulge is too great to be produced by Earth's present-day tidal influence. The ex-

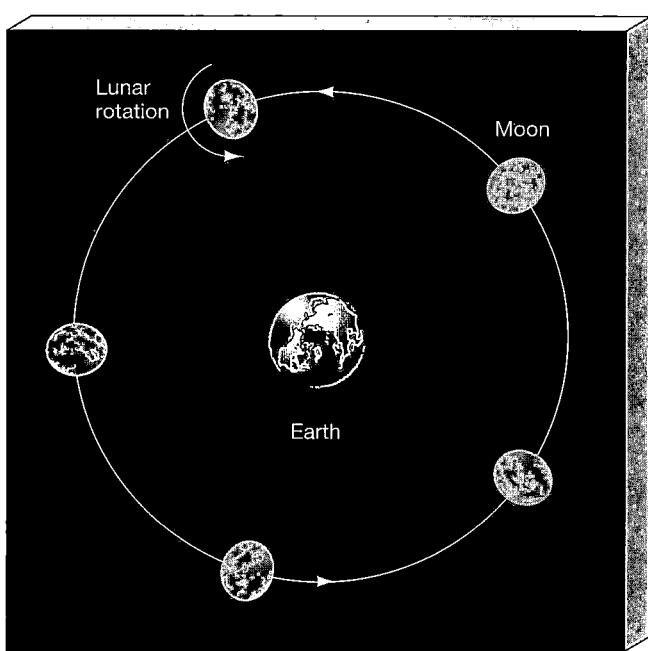

Figure 8.11 Lunar Elongation The Moon is slightly elongated in shape, with its long axis perpetually pointing toward Earth. (The elongation is highly exaggerated in this diagram.)

planation seems to be that, long ago, the distance from Earth to the Moon may have been as little as two-thirds of its current value, or about 250,000 km. Earth's tidal force on the Moon would then have been more than three times greater than it is today and could have accounted for the Moon's elongated shape. The resulting distortion could have "set" when the Moon solidified, thus surviving to the present day, while at the same time accelerating the synchronization of the Moon's orbit.

MEASUREMENT OF MERCURY'S SPIN

③ In principle, the ability to discern surface features on Mercury should allow us to measure its rotation rate simply by watching the motion of a particular region around the planet. In the mid-nineteenth century, an Italian astronomer named Giovanni Schiaparelli did just that. He concluded that Mercury always keeps one side facing the Sun, much as our Moon perpetually presents only one face to Earth. The explanation suggested for this synchronous rotation was the same as for the Moon—the tidal bulge raised in Mercury by the Sun had modified the planet's rotation rate until the bulge always pointed directly at the Sun. Although the surface features could not be seen clearly, the combination of Schiaparelli's observations and a plausible physical explanation was enough to convince most astronomers, and the belief that Mercury rotates synchronously with its revolution about the Sun (that is, once every 88 Earth days) persisted for almost half a century.

In 1965, astronomers making radar observations of Mercury from the Arecibo radio telescope in Puerto Rico (see Figure 5.22) discovered that this long-held view was in error. The technique they used is illustrated in Figure 8.12, which shows a radar signal reflecting from the surface of a hypothetical planet. Let's imagine, for the purpose of this discussion, that the pulse of outgoing radiation is of a single frequency. The returning pulse bounced off the planet is very much weaker than the outgoing signal. Beyond this change, the reflected signal can be modified in two important ways. First, the signal as a whole may be redshifted or blueshifted as a consequence of the Doppler effect, depending on the overall radial velocity of the planet with respect to Earth. ∞ (Sec. 3.5) We will assume for simplicity that this velocity is zero, so that, on average, the frequency of the reflected signal is the same as the outgoing beam. Second, if the planet is rotating, the radiation reflected from the side of the planet moving toward us returns at a slightly higher frequency than the radiation reflected from the receding side. (Think of the two hemispheres as being separate sources of radiation and moving at slightly different velocities, one toward us and one away.) The effect is very similar to the rotational line broadening discussed in Chapter 4 (see Figure 4.18), except that in this case the radiation we are measuring was not emitted by the planet but only reflected from its surface. ∞ (Sec. 4.4) What we see in the reflected signal is a spread of frequencies on either side of the original frequency. By measuring the extent of that spread, we can determine the planet's rotational speed.

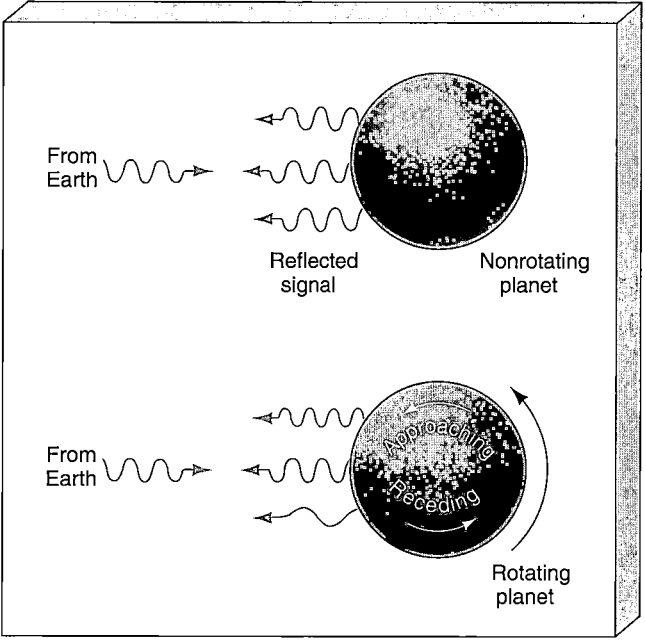

Figure 8.12 Planetary Radar A radar beam reflected from a rotating planet yields information about both the planet's overall motion and its rotation rate. Compare Figure 4.17.

DISCOVERY 8-1

Lunar Exploration

The Space Age began in earnest on October 4, 1957, with the launch of the Soviet satellite *Sputnik 1*. Thirteen months later, on January 4, 1959, the Soviet *Luna 1*, the first human-made craft to escape Earth's gravity, passed the Moon. *Luna 2* crash-landed on the surface in September of that year, and *Luna 3* returned the first pictures of the far side a month later. The long-running *Luna* series established a clear Soviet lead in the early "space race" and returned volumes of detailed information about the Moon's surface. Several of the *Luna* missions landed and returned surface material to Earth.

The U.S. lunar exploration program got off to a rocky start. The first six attempts in the *Ranger* series, between 1961 and 1964, failed to accomplish their objective of just hitting the Moon. The last three were successful, however. *Ranger 7* collided with the lunar surface (as intended) on June 28, 1964. Five U.S. *Lunar Orbiter* spacecraft, launched in 1966 and 1967, were successfully placed in orbit around the Moon, and they relayed back to Earth high-resolution images of much of the lunar surface. Between 1966 and 1968, seven *Surveyor* missions soft-landed on the Moon and performed detailed analyses of the surface.

Many of these unmanned U.S. missions were performed in support of the manned *Apollo* program. On May 25, 1961, at a time when the U.S. space program was in great disarray, President John F. Kennedy declared that the United States would "send a man to the Moon and return him safely to Earth" before the end of the decade, and the *Apollo* program was born. On July 20, 1969, less than 12 years after *Sputnik* and only 8 years after the statement of the program's goal, *Apollo 11* commander Neil Armstrong became the first human to set foot on the Moon, in Mare Tranquilitatis (Sea of Tranquility). Three and a half years later, on December 14, 1972, scientist-astronaut Harrison Schmitt, of *Apollo 17*, was the last.

The astronauts who traveled in pairs to the lunar surface in each lunar lander (shown in the photograph below) performed numerous geological and other scientific studies on the surface. The later landers brought with them a "lunar rover"—a small golf cart–sized vehicle that greatly expanded the area the astronauts could cover. Probably the most important single aspect of the *Apollo* program was the collection of samples of surface rock from various locations on the Moon. In all, some 382 kg of material was returned to Earth. Chemical analysis and radioactive dating of these samples revolutionized our understanding of the Moon's surface history—no amount of Earth-based observations could have achieved the same results.

Each *Apollo* lander also left behind a nuclear-powered package of scientific instruments called *ALSEP* (*Apollo Lunar Surface Experiments Package*; shown opposite) to

(NASA)

In this way, the Arecibo researchers found that the rotation period of Mercury is not 88 days, as had previously been believed, but 59 days, exactly two-thirds of the planet's orbital period. Because there are exactly three rotations for every two revolutions, we say that there is a 3:2 *spin–orbit resonance* in Mercury's motion. In this context, the term **resonance** just means that two characteristic times—here Mercury's day and year—are related to each other in a simple way. An even simpler example of a spin–orbit resonance is the Moon's orbit around Earth. In that case, the rotation is synchronous with the revolution, and the resonance is said to be 1:1.

Figure 8.13 illustrates some implications of Mercury's curious rotation for a hypothetical inhabitant of the planet. Mercury's solar day—the time from noon to noon, say—is two Mercury years long! The Sun stays "up" in the black

Mercury sky for almost three Earth months at a time, after which follows nearly three Earth months of darkness. At any given point in its orbit, Mercury presents the same face to the Sun not every time it revolves but *every other* time.

Figure 8.13 Mercury's Rotation Mercury's orbital and rotational motions combine to produce a day that is two Mercury years long. The arrow represents an observer standing on the surface of the planet. At day 0, it is noon for our observer, and the Sun is directly overhead. By the time Mercury has completed one full orbit around the Sun and moved from day 0 to day 88, it has rotated on its axis exactly 1.5 times, so that it is now midnight at the observer's location. After another complete orbit, it is noon once again. The eccentricity of Mercury's orbit is not shown in this simplified diagram.

monitor the solar wind, measure heat flow in the Moon's interior, and, perhaps most important, record lunar seismic activity. With several *ALSEP*s on the surface, scientists could determine the location of "moonquakes" by triangulation and map the Moon's inner structure, obtaining information critical to our understanding of the Moon's evolution.

By any standards, the *Apollo* program was a spectacular success. It represents a towering achievement of the human race. The project goals were met on schedule and within budget, and our knowledge of the Moon, Earth, and the solar system increased enormously. But the "Age of *Apollo*" was short-lived. Public interest quickly waned. Over half a billion people breathlessly watched on television as Neil Armstrong set foot on the Moon, yet barely three years later, when the program was abruptly canceled for largely political (rather than scientific, technological, or economic) reasons, the landings had become so routine that they no longer excited the interest of the American public. Unmanned space science moved away from the Moon and toward the other planets, and the manned space program foundered. Perhaps one of the most amazing—and saddest—aspects of the *Apollo* program is that today, nearly three decades later, *no* nation on Earth (including the United States) has the desire, the capability, or the money to repeat the feat.

In 1994, the small U.S. military satellite *Clementine* was placed in lunar orbit, and performed a detailed survey of the lunar surface. In January 1998, NASA returned to the Moon for the first time in a quarter century with the launch of *Lunar Prospector*, another small satellite on a one-year mission to study the Moon's structure and origins. As discussed in more detail in *Discovery 8-2*, both missions were very successful, and have amply demonstrated the wealth of information that can be obtained by low-budget spacecraft. Following the spectacular end of the *Lunar*

Prospector mission in 1999 (see Section 8.5), there are currently no active probes in lunar orbit.

Plans do exist to establish permanent human colonies on the Moon, either for commercial ventures, such as mining, or for scientific research. Proposals have also been made to site large optical, radio, and other telescopes on the lunar surface. Such instruments, which could be constructed larger than Earth-based devices, would enjoy perfect seeing and no light pollution. None of these projects is scheduled to become reality in the near future, although the discovery of water on the lunar surface alleviates at least one major logistical problem associated with such an undertaking. After a brief encounter with humankind, the Moon is once again a lifeless, unchanging world.

(NASA)

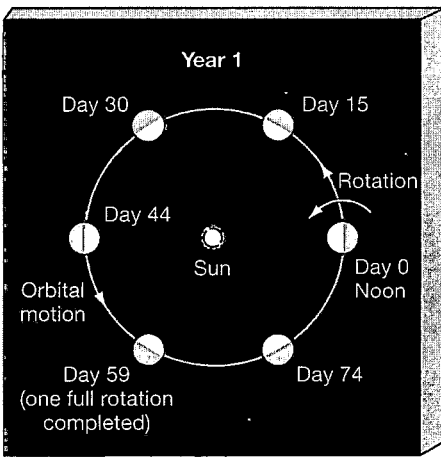

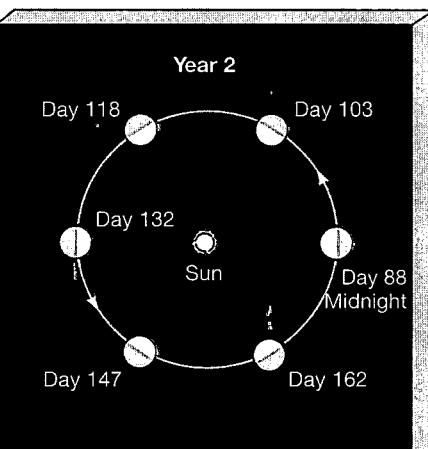

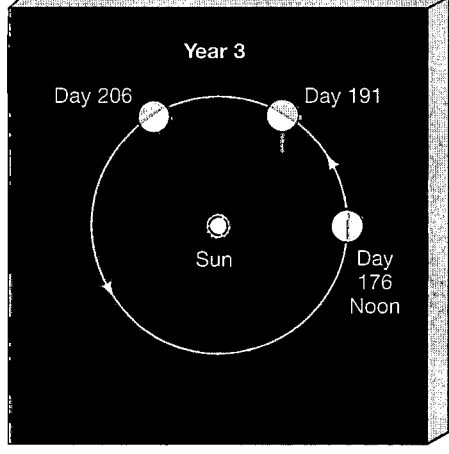

EXPLANATION OF MERCURY'S ROTATION

Mercury's 3:2 spin–orbit resonance did not occur by chance. What mechanism establishes and maintains it? In the case of the Moon orbiting Earth, the 1:1 resonance is explained as the result of tidal forces. In essence, the lunar rotation period, which probably started off much shorter than its present value, has lengthened so that the tidal bulge created by Earth is fixed relative to the body of the Moon. Tidal forces (this time due to the Sun) are also responsible for Mercury's 3:2 resonance, but in a much more subtle way.

Mercury cannot settle into a 1:1 resonance because its orbit around the Sun is quite eccentric. By Kepler's second law, Mercury's orbital speed is greatest at perihelion (closest approach to the Sun) and least at aphelion (greatest distance from the Sun). ⚼ (Sec. 2.3) A moment's thought shows that because of these variations in the planet's orbital speed, there is no way that the planet (rotating at a constant rate) can remain in a synchronous orbit. If its rotation were synchronous near perihelion, it would be too rapid at aphelion, while synchronism at aphelion would be too slow at perihelion.

Tidal forces always act to try to synchronize the rotation rate with the instantaneous orbital speed, but such synchronization cannot be maintained over Mercury's entire orbit. What happens? The answer is found when we realize that tidal effects diminish very rapidly with increasing distance. The tidal forces acting on Mercury at perihelion are much greater than those at aphelion, and perihelion "won" the struggle to determine the rotation rate. In the 3:2 resonance, Mercury's orbital and rotational motion are almost exactly synchronous *at perihelion*, so that particular rotation rate was naturally "picked out" by the Sun's tidal influence on the planet. Notice that even though Mercury rotates through only 180° between one perihelion and the next (see Figure 8.13), the appearance of the tidal bulge is the *same* each time around.

The motion of Mercury is one of the simplest nonsynchronous resonances known in the solar system. Astronomers now believe that these intricate dynamic interactions are responsible for much of the fine detail observed in the motion of the solar system. Examples of resonances can be found in the orbits of many of the planets, their moons, their rings, and in the asteroid belt.

The Sun's tidal influence also causes Mercury's rotation axis to be exactly perpendicular to its orbit plane. As a result, and because of Mercury's eccentric orbit and the spin–orbit resonance, some points on the surface get much hotter than others. In particular, the two (diametrically opposite) points on the equator where the Sun is directly overhead at perihelion get hottest of all. They are called the *hot longitudes*. The peak temperature of 700 K mentioned earlier occurs at noon at those two locations. At the *warm longitudes*, where the Sun is directly overhead at aphelion, the peak temperature is about 150 K cooler—a mere 550 K.

By contrast, the Sun is always on the horizon as seen from the planet's poles, so temperatures there never reach the sizzling levels of the equatorial regions. Earth-based radar studies carried out during the 1990s suggest that Mercury's polar temperatures may be as low as 125 K and that, despite the planet's scorched equator, the poles may be covered with extensive sheets of water ice. (See Section 8.5 for similar findings regarding the Moon.)

✓| Concept Check

■ How has gravity influenced the rotation rates of the Moon and Mercury?

8.5 Lunar Cratering and Surface Composition

On Earth, the combined actions of wind and water erode our planet's surface and reshape its appearance almost daily. Coupled with the never-ending motion of Earth's surface plates, the result is that most of the ancient history of our planet's surface is lost to us. The Moon, in contrast, has no air, no water, no plate tectonics, and no ongoing volcanic or seismic activity. Consequently, features dating back almost to its formation are still visible today.

METEORITIC IMPACTS

🔁 The primary agent of change on the lunar surface is interplanetary debris, in the form of *meteoroids*. This material, much of it rocky or metallic in composition, is strewn throughout the solar system, orbiting the Sun in interplanetary space, perhaps for billions of years, until it happens to collide with some planet or moon. ⚼ (Sec. 6.5) On Earth, most meteoroids burn up in the atmosphere, producing the streaks of light known as *meteors*, or "shooting stars." But the Moon, without an atmosphere, has no protection against this onslaught. Large and small meteoroids zoom in and collide with the surface, sometimes producing huge craters. Over billions of years, these collisions have scarred, cratered, and sculpted the lunar landscape. Craters are still being formed today—even as you read this—all across the surface of the Moon.

Meteoroids generally strike the Moon at speeds of several kilometers per second. At these speeds, even a small piece of matter carries an enormous amount of energy—for example, a 1-kg object hitting the Moon's surface at 10 km/s releases as much energy as the detonation of 10 kg of TNT. As illustrated in Figure 8.14, impact of a meteoroid with the surface causes sudden and tremendous pressures to build up, heating the normally brittle rock and deforming the ground like heated plastic. The ensuing explosion pushes previously flat layers of rock up and out, forming a crater.

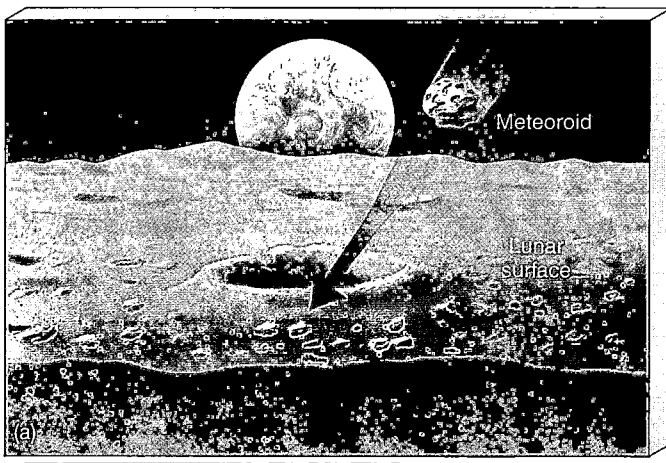

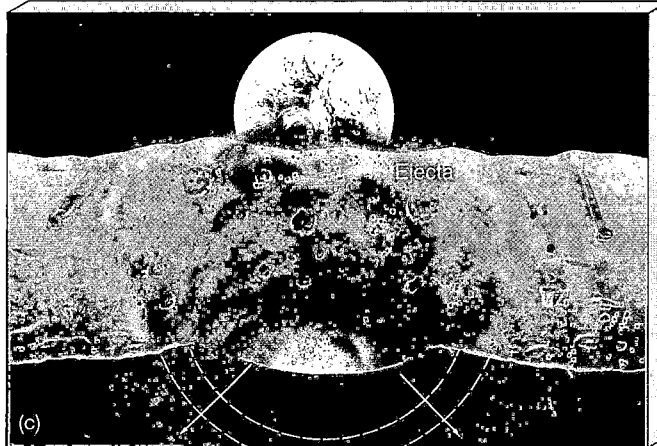

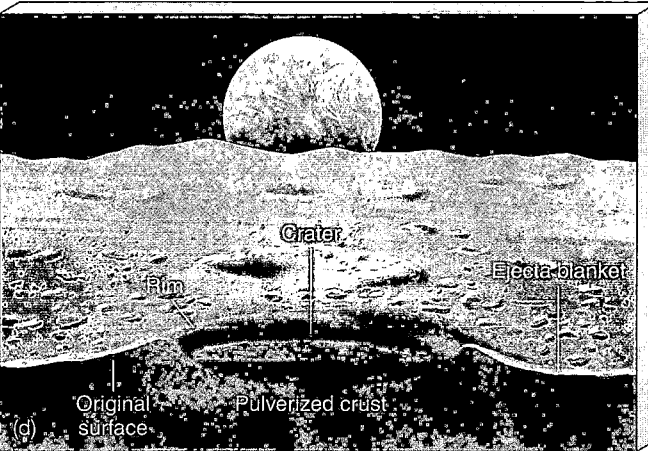

Figure 8.14 Meteoroid Impact Several stages in the formation of a crater by meteoritic impact. (a) A meteoroid strikes the surface, releasing a large amount of energy. (b, c) The resulting explosion ejects material from the impact site and sends shock waves through the underlying surface. (d) Eventually, a characteristic crater surrounded by a blanket of ejected material results.

The diameter of the eventual crater is typically 10 times that of the incoming meteoroid; the crater depth is about twice the meteoroid's diameter. Thus, our 1-kg meteoroid, measuring perhaps 10 cm across, would produce a crater about 1 m in diameter and 20 cm deep. Shock waves from the impact pulverize the lunar surface to a depth many times that of the crater itself. Many of the rock samples brought back by the *Apollo* astronauts show patterns of repeated shattering and melting—direct evidence of the violent shock waves and high temperatures produced in meteoritic impacts. The material thrown out by the explosion surrounds the crater in a layer called an *ejecta blanket*. The ejected debris ranges in size from fine dust to large boulders. Figure 8.15(a) shows the result of one particularly large meteoritic impact on the Moon. As shown in Figure 8.15(b), the larger pieces of ejecta may themselves form secondary craters.

In addition to the bombardment by meteoroids with masses of a gram or more, a steady "rain" of *micrometeoroids* (debris with masses ranging from a few micrograms up to about 1 gram) also eats away at the structure of the lunar surface. Some examples can be seen in Figure 8.16, a photomicrograph of some glassy "beads" brought back to Earth by *Apollo* astronauts. The beads themselves were formed during the explosion following a meteoroid impact, when surface rock was melted, ejected, and rapidly cooled. However, several of them also display fresh miniature craters caused by micrometeoroids that struck the beads after they had cooled and solidified.

In fact, the *rate* of cratering decreases rapidly with crater size—fresh large craters are scarce, but small craters are very common. The reason for this is simple: There just aren't very many large chunks of debris in interplanetary space, so their collisions with the Moon are rare. At present average rates, one new 10-km (diameter) lunar crater is formed roughly every 10 million years, a new meter-sized crater is created about once a month, and centimeter-sized craters are formed every few minutes.

CRATERING HISTORY OF THE MOON

Astronomers can use the known ages of Moon rocks to estimate the rate of cratering in the past. As we have seen, the heavily cratered highlands are older than the less-cratered maria, but the difference in cratering is not simply a matter of exposure time. Astronomers now believe that the Moon, and presumably the entire inner solar system, experienced a sudden sharp drop in meteoritic bombardment

MORE PRECISELY 8-1
Why Air Sticks Around

Why do some planets and moons have atmospheres, while others do not, and what determines the composition of the atmosphere if one exists? Why does a layer of air, made up mostly of nitrogen and oxygen, lie just above Earth's surface? After all, experience shows that most gas naturally expands to fill all the volume available. Perfume in a room, fumes from a poorly running engine, and steam from a tea kettle all rapidly disperse until we can hardly sense them. Why doesn't our planet's atmosphere similarly disperse by floating away into space?

The answer is that *gravity* holds it down. Earth's gravitational field exerts a pull on all the atoms and molecules in our atmosphere, preventing them from escaping. However, gravity is not the only influence acting, for if it were, all of Earth's air would have fallen to the surface long ago. *Heat* competes with gravity to keep the atmosphere buoyant. Let's explore this competition between gravity and heat in a little more detail.

All gas molecules are in constant random motion. The temperature of any gas is a direct measure of this motion—the hotter the gas, the faster the molecules are moving. ∞ (*More Precisely 3-1*) The Sun continuously supplies heat to our planet's atmosphere, and the resulting rapid movement of heated molecules produces *pressure*, which tends to oppose the force of gravity, preventing our atmosphere from collapsing under its own weight.

An important measure of the strength of a body's gravity is its *escape speed*—the speed needed for any object to escape forever from its surface. ∞ (Sec. 2.7) This speed increases with increased mass or decreased radius of the parent body (often a moon or a planet). In convenient units, it can be expressed as

escape speed (in km/s)

$$= 11.2 \sqrt{\frac{\text{mass of body (in Earth masses)}}{\text{radius of body (in Earth radii)}}}.$$

(see the Earth Data box on p. 170). Thus, if the *mass* of the parent body is quadrupled, the escape speed doubles. If the parent body's *radius* quadruples, then the escape speed is halved.

To determine whether a planet will retain an atmosphere, we must compare the planet's escape speed with the *molecular speed*, which is the average speed of the gas particles making up the atmosphere. This speed actually depends not only on the temperature of the gas but also on the mass of the individual molecules—the hotter the gas or the smaller the molecular mass, the higher the average speed of the molecules:

average molecular speed (in km/s)

$$= 0.157 \sqrt{\frac{\text{gas temperature (K)}}{\text{molecular mass (hydrogen atom mass)}}}.$$

Thus, increasing the absolute temperature of a sample of gas by a factor of four— for example, from 100 K to 400 K—doubles the average speed of its constituent molecules. And, at a given temperature, molecules of hydrogen (H_2: molecular mass = 2) in air move, on average, four times faster than molecules of oxygen (O_2: molecular mass = 32), which are 16 times heavier.

EXAMPLE: For nitrogen (N_2: molecular mass = 28) and oxygen (O_2: molecular mass = 32) in Earth's atmosphere, where the temperature near the surface is nearly 300 K, the above formula yields average molecular speeds of

nitrogen: speed $= 0.157$ km/s $\times \sqrt{\dfrac{300}{28}} = 0.51$ km/s

oxygen: speed $= 0.157$ km/s $\times \sqrt{\dfrac{300}{32}} = 0.48$ km/s

These speeds are far smaller than the 11.2 km/s needed for a molecule to escape into space. As a result, Earth is able to retain its nitrogen-oxygen atmosphere. On the whole, our

---

about 3.9 billion years ago. The highlands solidified and received most of their craters before that time, whereas the maria solidified afterward. The rate of cratering has been declining slowly ever since.*

The great basins that comprise the maria are thought to have been created during the final stages of the heavy meteoritic bombardment just described, between about 4.1 and 3.9 billion years ago. Subsequent volcanic activity

*Recent detailed studies of lunar rock samples returned by the Apollo missions now suggest that the slow decline may have ended about 400–500 million years ago, when the cratering rate increased by about a factor of 4, back to the levels of about 3 billion years earlier. The cause of the increase, if real, is unknown, but the fact that it can be inferred at all illustrates the importance of the Apollo samples.

filled the craters with lava, creating the formations we see today. In a sense, then, the maria *are* oceans—ancient seas of molten lava, now solidified.

Not all these great craters became flooded with lava, however. One of the youngest craters is the Orientale Basin (Figure 8.15a), which formed about 3.9 billion years ago. It did not undergo much subsequent volcanism, and we can recognize its structure as an impact crater rather than as another mare. Similar "unflooded" basins are seen on the lunar far side (Figure 8.7).

Apart from meteorites found on Earth, the Moon is the only solar-system object for which we have accurate age measurements. However, studies of lunar cratering provide astronomers with an important alternative means

planet's gravity simply has more influence than the heat of our atmosphere.

In reality the situation is a little more complicated than a simple comparison of speeds. Atmospheric molecules can gain or lose speed by bumping into one another or by colliding with objects near the ground. Thus, although we can characterize a gas by its average molecular speed, the molecules do not *all* move at the same speed, as illustrated in the accompanying figure. A tiny fraction of the molecules in any gas have speeds much greater than average—one molecule in two million has a speed more than three times the average, and one in 10^{16} exceeds the average by more than a factor of five. This means that at any instant, *some* molecules are moving fast enough to escape, even when the average molecular speed is much less than the escape speed. The result is that all planetary atmospheres slowly leak away into space.

Don't be alarmed—the leakage is usually very gradual! As a rule of thumb, if the escape speed from a planet ex-

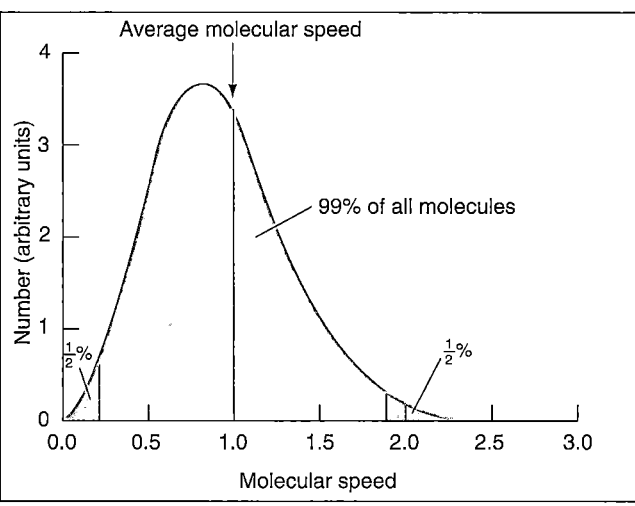

ceeds the average speed of a given type of molecule by a factor of six or more, then molecules of that type will not have escaped from the planet's atmosphere in significant quantities in the 4.6 billion years since the solar system formed. Conversely, if the escape speed is less than six times the average speed of molecules of a given type, then most of them will have escaped by now, and we should not expect to find them in the atmosphere.

For air on Earth, the mean molecular speeds of oxygen and nitrogen that were just computed are comfortably below one-sixth of the escape speed. However, if the Moon originally had an Earthlike atmosphere, that lunar atmosphere would have been heated by the Sun to much the same temperature as Earth's air today, so the average molecular speed would have been about 0.5 km/s. Because the Moon's escape speed is only 2.4 km/s—less than six times the average molecular speed—any original lunar atmosphere long ago dispersed into interplanetary space. Mercury's escape speed is 4.2 km/s. However, its peak surface temperature is around 700 K, corresponding to an average molecular speed for nitrogen or oxygen of about 0.8 km/s, more than one-sixth of the escape speed, so there has been ample time for those gases to escape.

EXAMPLE: We can also use these arguments to understand some aspects of atmospheric *composition*. Hydrogen molecules (H: molecular mass = 2) move, on average, at about 1.9 km/s in Earth's atmosphere at sea level, so they have had time to escape since our planet formed (6 × 1.9 km/s = 11.4 km/s, which is greater than Earth's 11.2 km/s escape speed). Consequently, we find very little hydrogen in Earth's atmosphere today. However, on the planet Jupiter, with a lower temperature (about 100 K), the speed of hydrogen molecules is correspondingly lower—about 1.1 km/s. At the same time, Jupiter's escape speed is 60 km/s, over five times higher than on Earth. For those reasons, Jupiter has retained its hydrogen—in fact, hydrogen is the dominant ingredient of Jupiter's atmosphere.

of estimating ages in the solar system. By counting craters on a planet, moon, or asteroid and using the Moon to calibrate the numbers, an approximate age for the surface can be obtained. In fact, this is how most of the ages presented in the next few chapters are determined. Note that, as with radioactive dating, this technique only measures the time since the surface in question last solidified—all cratering is erased and the clock is reset if the rock melts. ⟨⟩ (*More Precisely 7-2*)

LUNAR DUST

Meteoroid collisions with the Moon are the main cause of the layer of pulverized ejecta—also called lunar dust, or

regolith (meaning "fine rocky layer")—that covers the lunar landscape to an average depth of about 20 m. This microscopic dust has a typical particle size of about 0.01 mm. In consistency, it is rather like talcum powder or ready-mix dry mortar. Figure 8.17 shows an *Apollo* astronaut's bootprint in the regolith. The regolith is thinnest on the maria (10 m) and thickest on the highlands (over 100 m deep in places).

The constant barrage from space results in a slow but steady erosion of the lunar surface. The soft edges of the craters visible in the foreground of Figure 8.18 are the result of this process. In the absence of erosion, those features would still be as jagged and angular today as they were just after they formed. Instead, the steady buildup of

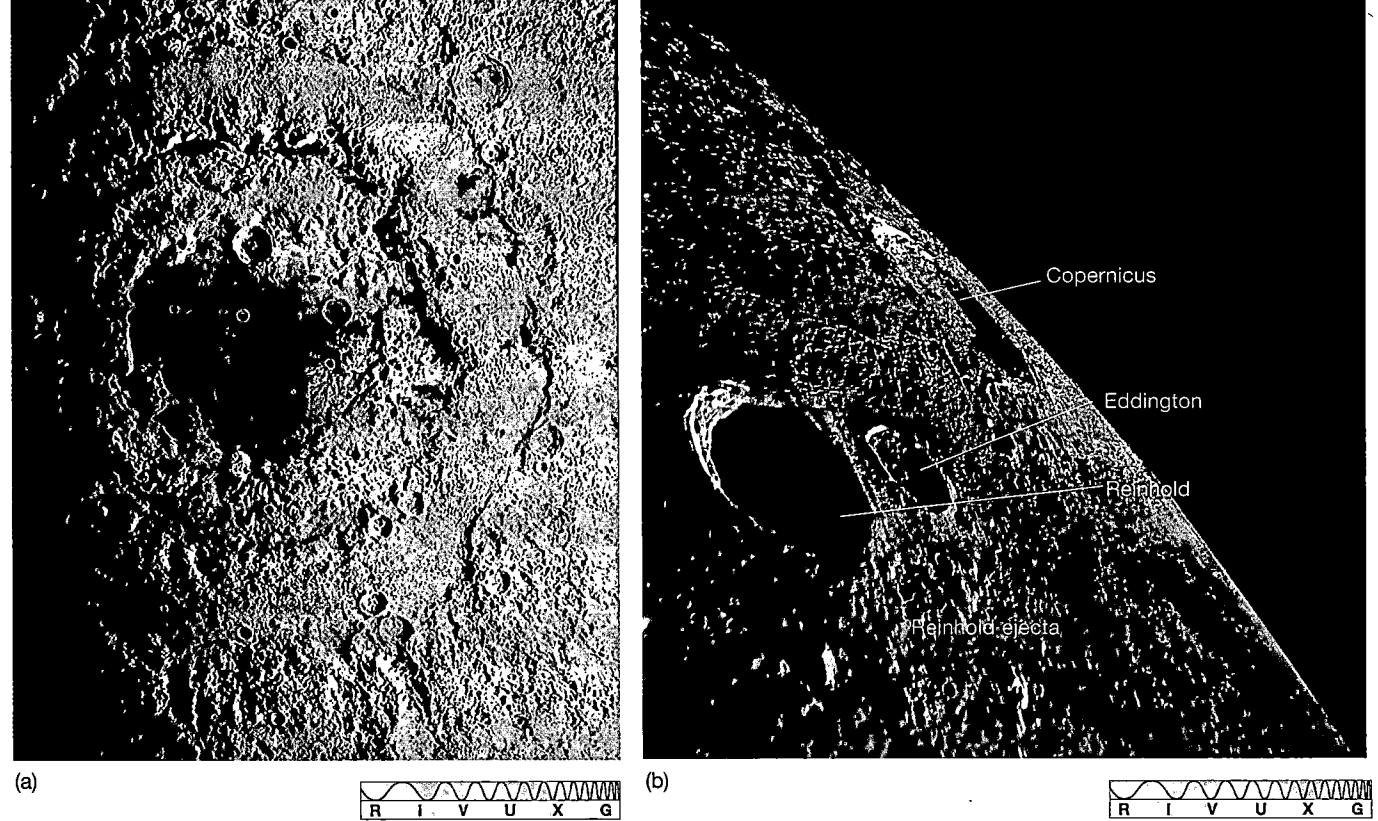

(a)

(b)

Figure 8.15 Large Lunar Craters (a) A large lunar crater, called the Orientale Basin. The impact that produced this crater thrust up much surrounding matter, which can be seen as concentric rings of cliffs called the Cordillera Mountains. The outermost ring is nearly 1000 km in diameter. (b) Two smaller craters called Reinhold and Eddington sit amid the secondary cratering resulting from the impact that created the 90-km-wide Copernicus crater (near the horizon) about a billion years ago. The ejecta blanket from crater Reinhold, 40 km across, and in the foreground, can be seen clearly. View is looking northeast from the lunar module during the *Apollo 12* mission. *(NASA)*

dust due to innumerable impacts has smoothed their outlines, and will probably erase them completely in about 100 million years.

Given the known dependence of cratering rate on crater size, planetary scientists can calculate how many small craters they would expect to find, given the numbers of large craters actually observed. When they do this, they find a shortage of craters less than about 20 m deep. These "missing" craters have been filled in by erosion over the lifetime of the Moon. This gives us a very rough estimate of the average erosion rate—about 5 m per billion years, or roughly 1/10,000 the rate on Earth.

The current lunar erosion rate is very low because meteoritic bombardment on the Moon is a much less effective erosive agent than are wind and water on Earth. For comparison, the Barringer Meteor Crater (Figure 8.19) in the Arizona desert, one of the largest meteoroid craters on Earth, is only 50,000 years old but has already undergone noticeable erosion. It will probably disappear completely in a mere million years, quite a short time geologically. If a crater that size had formed on the Moon even 4 billion

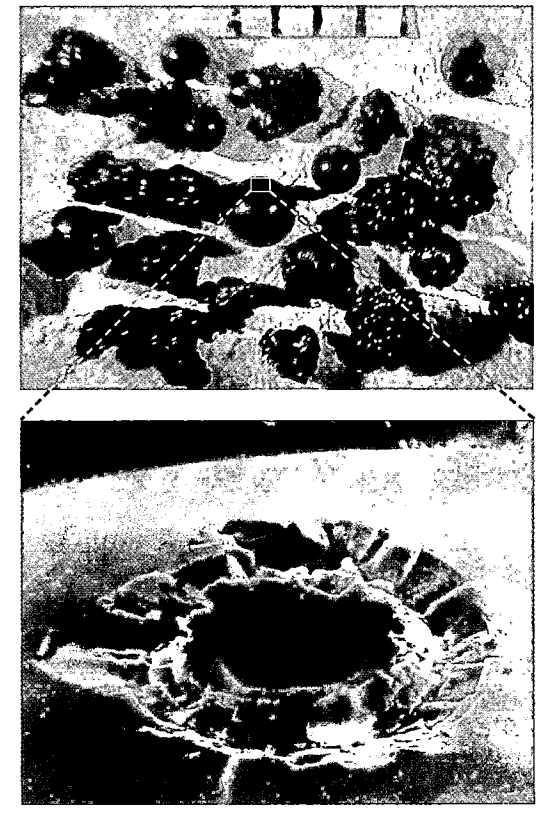

Figure 8.16 Microcraters Craters of all sizes litter the lunar landscape. Some shown here, embedded in glassy beads retrieved by *Apollo* astronauts, measure only 0.01 mm across. (The scale at the top is in millimeters.) *(NASA)*

Section 8.5 Lunar Cratering and Surface Composition 211

Figure 8.17 Regolith Photograph of an *Apollo* astronaut's bootprint in the lunar dust. The astronaut's weight has compacted the regolith to a depth of a few centimeters. *(NASA)*

years ago, it would still be plainly visible today. Even the shallow bootprint shown in Figure 8.17 is likely to remain intact for several million years.

LUNAR ICE?

In contrast to Earth's soil, the lunar regolith contains no organic matter like that produced by biological organisms. No life whatsoever exists on the Moon. Nor were any fossils found in *Apollo* samples. Lunar rocks are barren of life and apparently always have been. NASA was so confident of this that the astronauts were not even quarantined on their return from the last few *Apollo* landings. Furthermore, all the lunar samples returned by the U.S. and Soviet Moon programs were bone dry—they didn't even contain minerals having water molecules locked within their crystal structure. Terrestrial rocks, by contrast, are almost always one or two percent water. The main reason for this lack of water is the high (up to 400 K) daytime temperatures found over most of the lunar surface.

Some regions of the Moon *are* thought to contain water, however, in the form of ice. As early as the 1960s, some scientists had considered the theoretical possibility that ice might be found near the lunar poles. Since the Sun never rises more than a few degrees above the horizon, as seen from the Moon's polar regions, temperatures on the permanently shaded floors of craters near the poles never exceed about 100 K. Consequently, those scientists theorized, water ice there could have remained permanently frozen since the very early days of the solar system, never melting or vaporizing and hence never escaping into space.

In November 1996, mission controllers of the *Clementine* spacecraft (see *Discovery 8-2*) reported that radar echoes captured by *Clementine* from an old, deep crater near the lunar south pole suggested deposits of water ice at a depth of a few meters. In early March, 1998, NASA an-

Figure 8.18 Lunar Surface The lunar surface is not entirely changeless. Despite the complete lack of wind and water on the airless Moon, the surface has still eroded a little under the constant "rain" of impacting meteoroids, especially micrometeoroids. Note the soft edges of the craters visible in the foreground of this image. In the absence of erosion, these features would be as jagged and angular today as they were when they formed. (The twin tracks were made by the *Apollo* lunar rover.) *(NASA)*

nounced that sensitive equipment on board the *Lunar Prospector* mission had confirmed *Clementine*'s findings, and in fact had detected large amounts of water ice—possibly totaling trillions of tons—at both lunar poles. At first it appeared that the ice was mainly in the form of tiny crystals mixed with the lunar regolith, spread over many tens of thousands of square kilometers of deeply shadowed crater floors. However, subsequent analysis of the data suggests that much of the ice may exist in the form of smaller but more concentrated "lakes" of nearly pure material lying perhaps half a meter below the surface.

The *Lunar Prospector* discovery of lunar ice was indirect; the instruments on board the spacecraft actually detected the presence of hydrogen (H), whose existence was taken as evidence of water (H_2O). In an attempt to gain more direct information about lunar ice, NASA scientists decided to end the *Lunar Prospector* mission in a very spectacular way. As the spacecraft neared the end of its lifetime,

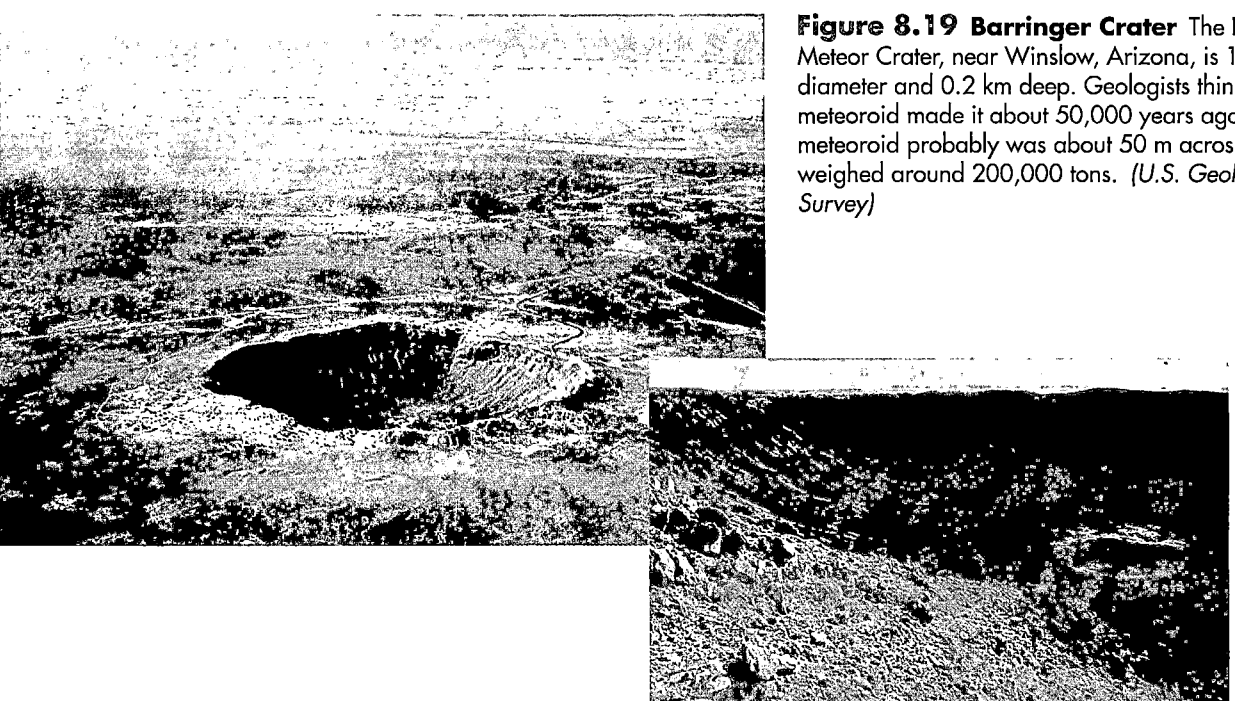

Figure 8.19 Barringer Crater The Barringer Meteor Crater, near Winslow, Arizona, is 1.2 km in diameter and 0.2 km deep. Geologists think a large meteoroid made it about 50,000 years ago. The meteoroid probably was about 50 m across and weighed around 200,000 tons. (U.S. Geological Survey)

it was directed to crash into one of the deep craters in which the ice was suspected to hide. Figure 8.20 shows the intended impact site and the trajectory to be taken by the satellite as it approached the surface. The hope was that the *Hubble Space Telescope* and ground-based telescopes on Earth might detect spectroscopic signatures of water vapor released by the impact. No water vapor was seen, although mission planners had stressed in advance that there were so many uncertainties involved in this effort that the probability of success was low—less than 10 percent. Thus, no conclusion can be drawn from the fact that water was not directly observed. Lunar ice remains a strong possibility, but its existence has not yet been definitively proven.

Assuming it does exist, where did all this ice come from? Most likely it was brought to the lunar surface by meteoroids and comets (we will see in Chapter 15 that this is the likely origin of Earth's water too). Any ice that survived the impact would have been scattered across the surface. Over most of the Moon, that ice would have rapidly vaporized and escaped, but in the deep basins near the poles it survived and built up over time. Whatever its origin, the polar ice may be a crucial component of any serious attempt at human colonization of the Moon: The anticipated cost of transporting a kilogram of water from Earth to the Moon is between $2,000 and $20,000, prompting one *Clementine* scientist to describe the lunar ice deposits as "possibly the most valuable piece of real estate in the solar system."

LUNAR VOLCANISM

Only a few decades ago, debate raged in scientific circles about the origin of lunar craters, with most scientists of the opinion that the craters were the result of volcanic activity.

We now know that almost all lunar craters are actually meteoritic in origin. However, a few apparently are not. Figure 8.21 shows an intriguing alignment of several craters in a *crater-chain* pattern so straight that it is very unlikely to have been produced by the random collision of meteoroids with the surface. Instead, the crater chain probably marks the location of a subsurface fault—a place where cracking or shearing of the surface once allowed molten matter to well up from below. As the lava cooled, it formed a solid "dome" above each fissure. Subsequently, the underlying lava receded and the centers of the domes collapsed, forming the craters we see today. Similar features have been observed on Venus by the orbiting *Magellan* probe (see Chapter 9).

Many other examples of lunar volcanism are known, both in telescopic observations from Earth and in the close-up photographs taken during the *Apollo* missions. Figure 8.22 shows a volcanic **rille**, a ditch where molten lava once flowed. There is good evidence for surface volcanism early in the Moon's history, and volcanism explains the presence of the lava that formed the maria. However, whatever volcanic activity once existed on the Moon ended long ago. The measured ages for rock samples returned from the Moon are all greater than 3 billion years. (Recall from *More Precisely 7-2* that the radioactivity clock starts "ticking" when the rock solidifies.) Apparently, the maria solidified over 3 billion years ago and the Moon has been dormant ever since.

☑ Concept Check

■ How has meteoritic bombardment affected the surface of the Moon?

(a)

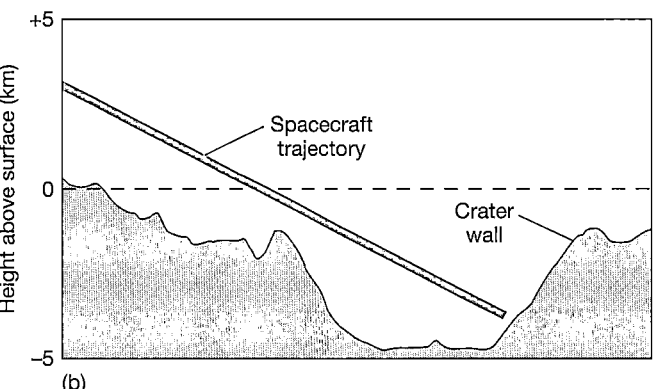

Figure 8.20 Prospector Impact (a) The intended impact site for the *Lunar Prospector* spacecraft was a deep crater close to the Moon's south pole. The purpose of the impact was to release water vapor for spectroscopic study by telescopes on or near Earth. (b) The trajectory of the spacecraft was designed to have it impact near the crater floor. *(Dept. of Defense)*

(b)

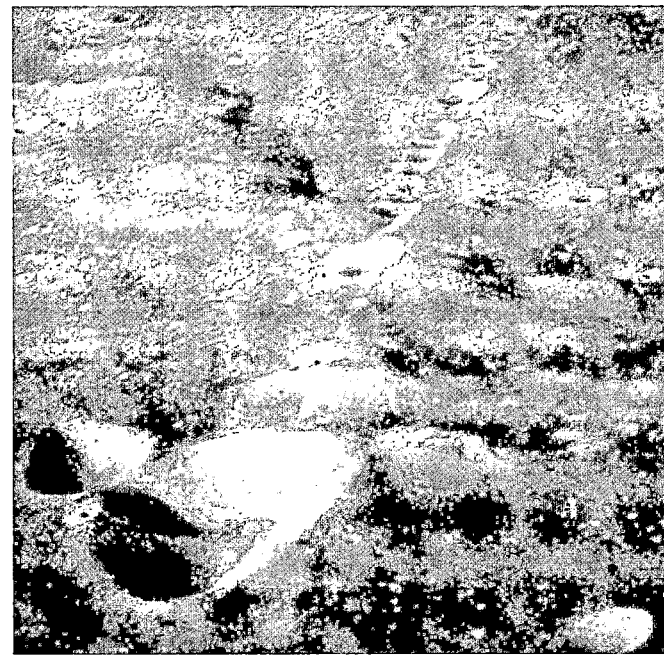

Figure 8.21 Crater Chain This "chain" of well-ordered craters was photographed by an *Apollo 14* astronaut. The largest crater, called Davy, is located on the western edge of Mare Nubium. The entire field of view measures about 100 km across. *(NASA)*

Figure 8.22 Lunar Volcanism A volcanic rille, photographed from the *Apollo 15* spacecraft orbiting the Moon, can be seen clearly here (bottom and center) winding its way through one of the maria. Called Hadley Rille, this system of valleys runs along the base of the Apennine Mountains (lower right) at the edge of the Mare Imbrium (to the left). Autolycus, the large crater closest to the center, spans 40 km. The shadow-sided, most prominent peak at the lower right, Mount Hadley, rises almost 5 km high. *(NASA)*

8.6 The Surface of Mercury

2 Like craters on the Moon, almost all craters on Mercury are the result of meteoritic bombardment. However, Mercury's craters are less densely packed than their lunar counterparts, and there are extensive **intercrater plains**. The crater walls are generally not as high as those on the Moon, and the ejected material appears to have landed closer to the impact site, exactly as we would expect on the basis of Mercury's stronger surface gravity.

One likely explanation for Mercury's relative lack of craters is that the older craters were filled in by volcanic activity, in much the same way as the Moon's maria filled in older craters as they formed. However, the intercrater plains do not look much like maria—they are much lighter in color and not as flat. Still, most geologists believe that volcanism did occur in Mercury's past, obscuring the old craters. The details of how Mercury's landscape came to look the way it does remain unexplained. The apparent absence of rilles or other obvious features associated with very-large-scale lava flows, along with the light color of the lava-flooded regions, suggest that Mercury's volcanic past was different from the Moon's.

Mercury has at least one type of surface feature not found on the Moon. Figure 8.23 shows a **scarp**, or cliff, on the surface that does not appear to be the result of volcanic or other familiar geological activity. The scarp cuts across several craters, which indicates that whatever produced it occurred *after* most of the meteoritic bombardment was over. Mercury shows no evidence for crustal motions like plate tectonics on Earth. The scarps, of which several are known from the *Mariner* images, probably formed when the planet's interior cooled and shrank long ago, much as wrinkles form on the skin of an old shrunken apple. Based on the amount of cratering observed in the surrounding terrain (as discussed in the previous section), astronomers estimate that the scarps probably formed about 4 billion years ago. Figure 8.24 shows what may have been the last great geological event in the history of Mercury—an immense bull's-eye crater called the Caloris Basin, formed eons ago by the impact of a large asteroid. (The basin is so called because it lies in Mercury's "hot longitudes"—see Section 8.3—close to the planet's equator; *calor* is the Latin word for "heat.") Because of the orientation of the planet during *Mariner 10*'s flybys, only half of the basin was visible. The center of the crater is off the left-hand side of the photograph. Compare this basin with the Orientale Basin on the Moon (Figure 8.15a). The impact-crater structures are quite similar, but even here there is a mystery: The patterns visible on the Caloris floor are unlike any seen on the Moon. Their origin, like the composition of the floor itself, is unknown. So large was the impact that created the Caloris Basin that it apparently sent strong seismic waves reverberating throughout the entire planet.

R I V U X G

Figure 8.23 Mercury's Surface Discovery Scarp on Mercury's surface. This appears to be a compressional feature that formed when the planet's crust cooled and contracted early in its history, causing a crease in the surface. This scarp, running diagonally across the center of the frame, is several hundred kilometers long and up to 3 km high in places. (Another scarp can be seen in the top left of the figure, close to the horizon.) *(NASA)*

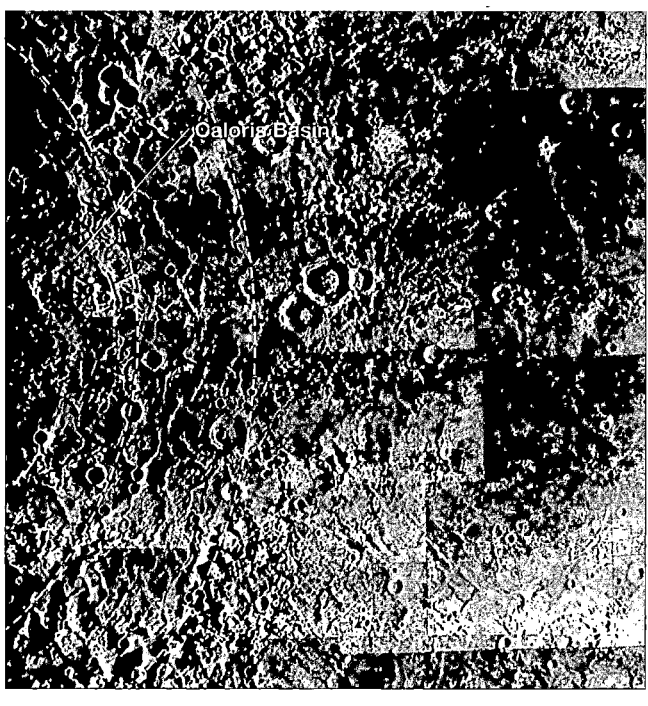

R I V U X G

Figure 8.24 Mercury's Basin Mercury's most prominent geological feature—the Caloris Basin—measures about 1400 km across and is ringed by concentric mountain ranges that reach more than 3 km high in places. This huge circular basin, only half of which can be seen (at the left) in this *Mariner 10* photo, is similar in size to the Moon's Mare Imbrium and spans more than half of Mercury's radius. *(NASA)*

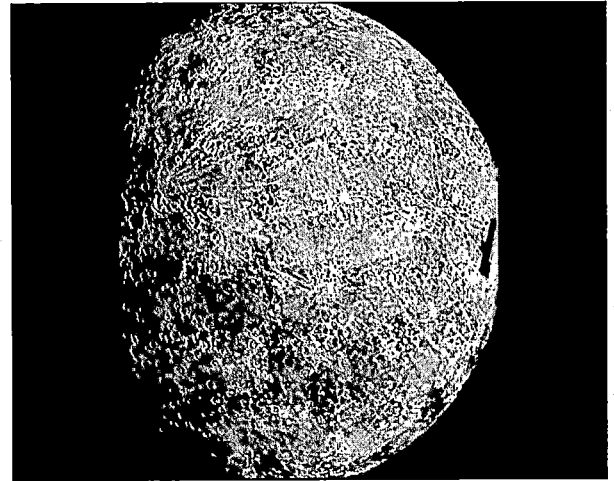

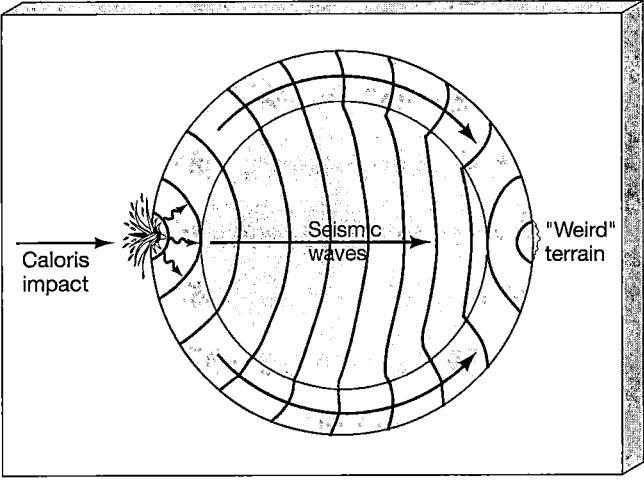

Figure 8.25 Weird Terrain The refocusing of seismic waves after the Caloris Basin impact may have created the weird terrain on the opposite side of the planet. *(NASA)*

Orbital semimajor axis	0.39 A.U.
	57.9 million km
Orbital eccentricity	0.206
Perihelion	0.31 A.U.
	46 million km
Aphelion	0.47 A.U.
	69.8 million km
Mean orbital speed	47.9 km/s
Sidereal orbital period	88.0 solar days
	0.241 tropical years
Synodic orbital period*	115.9 solar days
Orbital inclination to the ecliptic	7.00°
Greatest angular diameter, as seen from Earth	13″
Mass	3.30×10^{23} kg
	0.055 (Earth = 1)
Equatorial radius	2440 km
	0.38 (Earth = 1)
Mean density	5430 kg/m³
	0.98 (Earth = 1)
Surface gravity	3.70 m/s²
	0.38 (Earth = 1)
Escape speed	4.2 km/s
Sidereal rotation period	58.6 solar days
Axial tilt	0.0°
Surface magnetic field	0.011 (Earth = 1)
Magnetic axis tilt relative to rotation axis	< 10°
Mean surface temperature	100–700 K
Number of moons	0

The planet's apparent orbital period, taking Earth's own motion into account; specifically, the mean time between one closest approach to Earth and the next.

On the opposite side of Mercury from Caloris there is a region of oddly rippled and wavy surface features, often referred to as *weird* (or *jumbled*) terrain. Scientists believe that this terrain was produced when seismic waves from the Caloris impact traveled around the planet and converged on the diametrically opposite point, causing large-scale disruption of the surface there, as illustrated in Figure 8.25.

☑ Concept Check

■ How do scarps on Mercury differ from geological faults on Earth?

8.7 Interiors

THE MOON

5 The Moon's average density, about 3300 kg/m³, is similar to the measured density of lunar surface rock, virtually eliminating any chance that the Moon has a large, massive, and very dense nickel–iron core like that of Earth. In fact, the low density implies that the entire Moon is actually deficient in iron and other heavy metals compared to our planet.

There is no evidence for any large-scale lunar magnetic field. *Lunar Prospector* detected some very weak surface magnetic fields—less than 1/1000 of Earth's field—apparently associated with some large impact basins, but these are not thought to be related to conditions in the lunar core. As we saw in Chapter 7, researchers believe that planetary magnetism requires a rapidly rotating liquid metal core, like Earth's. ∞ (Sec. 7.5) Thus, the absence of a lunar magnetic field could be a consequence of the Moon's slow rotation, the absence of a liquid core, or both.

Data from the gravity experiment aboard *Lunar Prospector*, combined with measurements made by the probe's magnetometers as the Moon passed through Earth's magnetic "tail" (see Figure 7.21) imply that the Moon may have a small iron core perhaps 300 km in radius. Near the center, the temperature may be as low as 1500 K, too cool to melt rock. However, seismic data collected by sensitive equipment left on the surface by *Apollo* astronauts (see *Discovery 8-1*) suggest that the inner parts of the core may be at least partially molten, implying a somewhat higher temperature. Our knowledge of the Moon's deep interior is still quite limited.

Outside the core, the lunar interior seems to be of almost uniform density, although it is chemically differentiated—that is, the chemical properties change from core to surface. As depicted schematically in Figure 8.26, the core is surrounded by a roughly 400-km-thick inner mantle of semisolid rock having properties similar to Earth's asthenosphere. Above these regions lies an outer mantle of solid rock, some 900–950 km thick, topped by a 60–150-km crust (considerably thicker than that of Earth). Together these layers constitute the Moon's lithosphere. The crust material, which forms the lunar highlands, is lighter than the mantle, which is similar in chemical composition to the lunar maria.

The crust on the lunar far side is *thicker* than that on the side facing Earth. If we assume that lava takes the line of least resistance in getting to the surface, then we can readily understand why the far side of the Moon has no large maria—volcanic activity did not occur on the far side simply because the crust was too thick to allow it to occur

there. But *why* is the far-side crust thicker? The answer is probably related to Earth's gravitational pull. Just as heavier material tends to sink to the center of Earth, the denser lunar mantle tended to sink below the lighter crust in Earth's gravitational field. The effect of this was that the crust and the mantle became slightly off center with respect to each other. The mantle was pulled a little closer to Earth, while the crust moved slightly away. Thus, the crust became thinner on the near side and thicker on the far side.

MERCURY

5 Mercury's magnetic field, discovered by *Mariner 10*, is about 1/100 that of Earth. Actually, the discovery that Mercury has any magnetic field at all came as a surprise to planetary scientists. Having detected no magnetic field in the Moon (and, in fact, none in Venus or Mars, either), they had expected Mercury to have no measurable magnetism. Mercury certainly does not rotate rapidly, and it may lack a liquid metal core, yet a magnetic field undeniably surrounds it. Although weak, the field is strong enough to deflect the solar wind and create a small magnetosphere around the planet.

Scientists have no clear understanding of the origin of Mercury's magnetic field. If it is produced by ongoing dynamo action, as in Earth, then Mercury's core must be at least partially molten. Yet the absence of any recent surface geological activity suggests that the outer layers are solid to a considerable depth, as on the Moon. It is difficult to reconcile these two considerations in a single theoretical model of Mercury's interior. If the field is being generated dynamically, Mercury's slow rotation may at

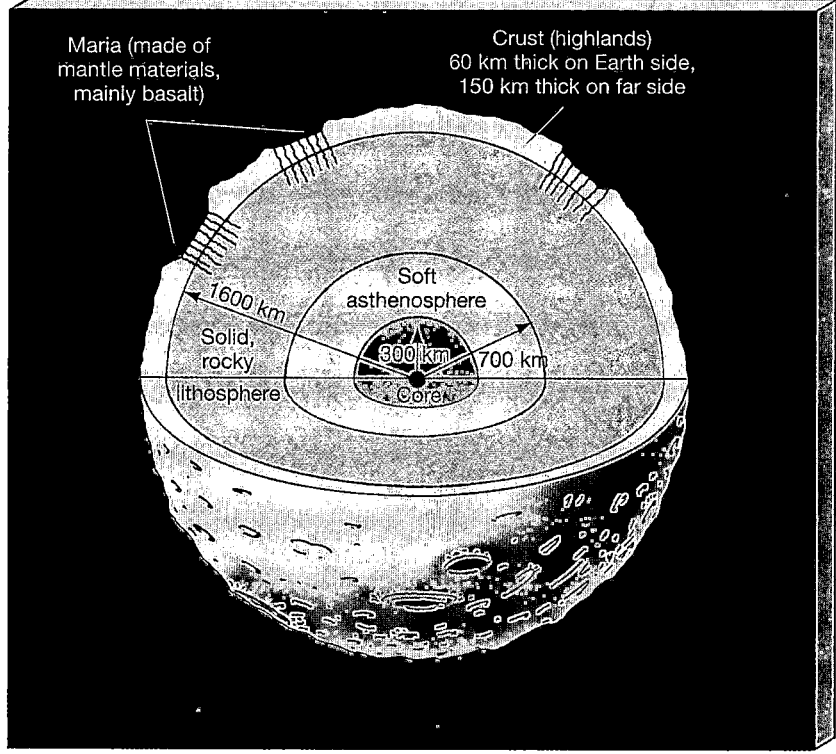

Maria (made of mantle materials, mainly basalt)

Crust (highlands) 60 km thick on Earth side, 150 km thick on far side

Soft asthenosphere

1600 km

Solid, rocky lithosphere

300 km 700 km

Core

Figure 8.26 Lunar Interior Cross-sectional diagram of the Moon. Unlike Earth's rocky lithosphere, the Moon's is very thick—about 1000 km. Below the lithosphere is the inner mantle, or lunar asthenosphere, similar in properties to that of Earth. At the center lies the core, which may be partly molten.

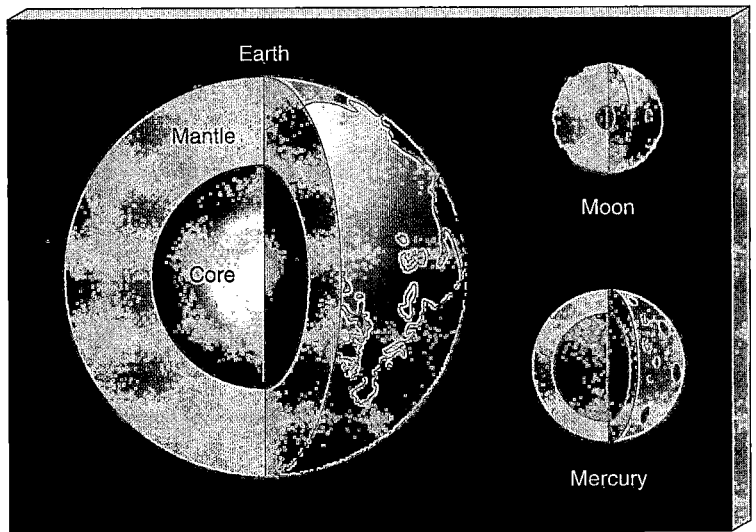

least account for the field's weakness. Alternatively, Mercury's current weak magnetism may simply be the remnant of an extinct dynamo—the planet's iron core may have solidified long ago but still bears a permanent magnetic imprint of the past. The models are inconclusive on this issue, and no spacecraft is scheduled to revisit Mercury until at least 2009.

Mercury's magnetic field and large average density together imply that the planet is differentiated. Even without the luxury of seismographs on the surface, we can infer that most of its interior must be dominated by a large, heavy, iron-rich core with a radius of perhaps 1800 km. Whether that core is solid or liquid remains to be determined. Probably a less dense lunarlike mantle lies above this core, to a depth of about 500 to 600 km. Thus, about 40 percent of the volume of Mercury, or 60 percent of its mass, is contained in its iron core. The ratio of core volume to total planet volume is greater for Mercury than for any other object in the solar system. Figure 8.27 illustrates the relative sizes and internal structures of Earth, the Moon, and Mercury.

✓ Concept Check

◻ Why would we not expect strong magnetic fields on the Moon or Mercury?

8.8 The Origin of the Moon

Ⓖ Over the years, many theories have been advanced to account for the origin of the Moon. However, both the similarities *and* the differences between the Moon and Earth conspire to confound many promising attempts to explain the Moon's existence.

One theory (the *sister*, or *coformation*, theory) suggests that the Moon formed as a separate object near Earth in much the same way as our own planet formed—the "blob"

of material that eventually coalesced into Earth gave rise to the Moon at about the same time. The two objects thus formed as a double-planet system, each revolving about the common center of mass. Although once favored by many astronomers, this idea suffers from a major flaw: The Moon differs in both density and composition from Earth, making it hard to understand how both could have originated from the same preplanetary material.

A second theory (the *capture* theory) maintains that the Moon formed far from Earth and was later captured by it. In this way, the density and composition of the two objects need not be similar, for the Moon presumably materialized in a quite different region of the early solar system. The objection to this theory is that the Moon's capture would be an extraordinarily difficult event; it might even be an impossible one. Why? Because the mass of our Moon is so large relative to that of Earth. It is not that our Moon is the largest natural satellite in the solar system, but it is unusually large compared with its parent planet. Mathematical modeling suggests that it is quite implausible that Earth and the Moon could have interacted in just the right way for the Moon to have been captured during a close encounter sometime in the past. Furthermore, although there are indeed significant composition differences between our world and its companion, there are also many similarities—particularly between the mantles of the two bodies—that make it unlikely that they formed entirely independently of one another.

A third, older, theory (the *daughter*, or fission, theory) speculates that the Moon originated out of Earth itself. The Pacific Ocean basin has often been mentioned as the place from which protolunar matter may have been torn—the result, perhaps, of the rapid spin of a young, molten Earth. Indeed, there are some chemical similarities between the matter in the Moon's outer mantle and that in Earth's Pacific basin. However, this theory offers no solution to the fundamental mystery of how Earth could possibly have been spinning so fast that it ejected an object as large as our Moon. Also, computer simulations indicate that the ejection of the Moon into a stable orbit simply would not have occurred. As a result, the daughter theory, in this form at least, is no longer taken seriously.

Today, many astronomers favor a hybrid of the capture and daughter themes. This idea—often called the *impact* theory—postulates a collision by a large, Mars-sized object with a youthful and molten Earth. Such collisions may have been quite frequent in the early solar system (see Chapter 15). The

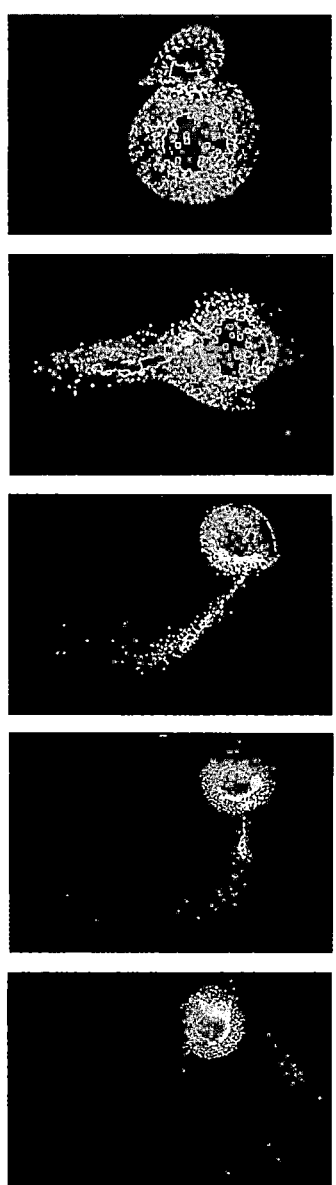

Figure 8.28 Moon Formation This sequence shows a simulated collision between Earth and an object the size of Mars. The sequence proceeds top to bottom and zooms out dramatically. The arrow in the final frame shows the newly formed Moon. *(W. Benz and A. Cameron)*

collision presumed by the impact theory would have been more a glancing blow than a direct impact. The matter dislodged from our planet then reassembled to form the Moon.

Computer simulations of such a catastrophic event show that most of the bits and pieces of splattered Earth could have coalesced into a stable orbit. Figure 8.28 shows some of the stages of one such calculation. If Earth had already formed an iron core by the time the collision occurred, then the Moon would indeed have ended up with a composition similar to that of Earth's mantle. During the collision, any iron core in the impacting object itself would have been left behind in Earth, eventually to become part of Earth's core. Thus both the Moon's overall similarity to that of Earth's mantle and its lack of a dense central core are naturally explained. Over the past decade, planetary scientists have come to realize that collisions such as this probably played important roles in the formation of all the terrestrial planets.

☑ Concept Check

■ How does the currently favored theory of the Moon's origin account for the Moon's depletion in heavy material compared to Earth and the similarity in composition between the lunar crust and that of Earth?

8.9 Evolutionary History of the Moon and Mercury

THE MOON

⛏ Given all the data, can we construct a reasonably consistent history of the Moon? The answer seems to be yes. Many specifics are still debated, but a consensus exists. Refer to Figure 8.29 while studying the following details.

The Moon formed about 4.6 billion years ago (see Chapter 15). The approximate age of the oldest rocks discovered in the lunar highlands is 4.4 billion years, so we know that at least part of the crust must already have solidified by that time and survived to the present. At formation, the Moon was already depleted in heavy metals compared with Earth.

During the earliest phases of the Moon's existence—roughly the first half billion years or so—meteoritic bombardment must have been frequent enough to heat and remelt most of the *surface* layers of the Moon, perhaps to a depth of 400 km in places. The early solar system was surely populated with lots of interplanetary matter, much of it in the form of boulder-sized fragments that were capable of generating large amounts of energy on collision with planets and their moons. But the intense heat derived from such collisions could not have penetrated very far into the lunar interior. Rock simply does not conduct heat well.

This situation resembles the surface melting we suspect occurred on Earth from meteoritic impacts during the first billion years or so. But the Moon is much less massive than Earth, and did not contain enough radioactive elements to heat it much further. Radioactivity probably heated the Moon a little, but not sufficiently to transform it from a warm, semisolid object to a completely liquid one. The chemical differentiation now inferred in the Moon's interior must have occurred during this period. If the Moon has a small iron core, that core also formed at this time.

About 3.9 billion years ago, around the time that Earth's crust solidified, the heaviest phase of the meteoritic bombardment ceased. The Moon was left with a solid crust, which would ultimately become the highlands, dented with numerous large basins, soon to flood with lava and become the maria (Figure 8.29a). Between 3.9 and 3.2 billion years ago, lunar volcanism filled the maria with the basaltic material we see today. The age of the youngest maria—3.2 billion years—indicates the time when the volcanic activity subsided. The maria are the sites of the last extensive lava flows on the Moon, over 3 billion years ago. Their smoothness, compared with the older, more rugged highlands, disguises their great age.

Small objects cool more rapidly than large ones because their interior is closer to the surface, on average. Being so small, the Moon rapidly lost its internal heat to space. As a consequence, it cooled much faster than Earth. As the Moon cooled, the volcanic activity ended as the thickness of the solid surface layer increased. With the exception of a few meters of surface erosion from eons of meteoritic bombardment (Figure 8.29c), the lunar landscape has remained more or less structurally frozen for the past 3 billion years. The Moon is dead now, and it has been dead for a long time.

MERCURY

7 Like the Moon, Mercury seems to have been a geologically dead world for much of the past 4 billion years. On both the Moon and Mercury, the lack of ongoing geological activity is a consequence of a thick solid mantle that prevents volcanism or tectonic motion. Because of the *Apollo* program, the Moon's early history is much better understood than Mercury's, which remains somewhat speculative. Indeed, what we do know about Mercury's history is gleaned mostly through comparison with the Moon.

When Mercury formed some 4.6 billion years ago, it was already depleted of lighter, rocky material. We will see later that this was largely a consequence of its location in the hot inner regions of the early solar system, although it is possible a collision stripped away some of its light mantle. During the next half-billion years, it melted and differentiated, like the other terrestrial worlds. It suffered the same intense meteoritic bombardment as the Moon. Being more massive than the Moon, Mercury cooled more slowly, so its crust was thinner and volcanic activity more common at early times. More craters were erased, resulting in the intercrater plains found by *Mariner 10*.

As the planet's large iron core formed and then cooled, the planet began to shrink, compressing the crust. This compression produced the scarps seen on Mercury's surface and may have prematurely terminated volcanic activity by squeezing shut the cracks and fissures on the surface. Thus Mercury did not experience the extensive volcanic outflows that formed the lunar maria. Despite its larger mass and greater internal temperature, Mercury has probably been geologically inactive for even longer than the Moon.

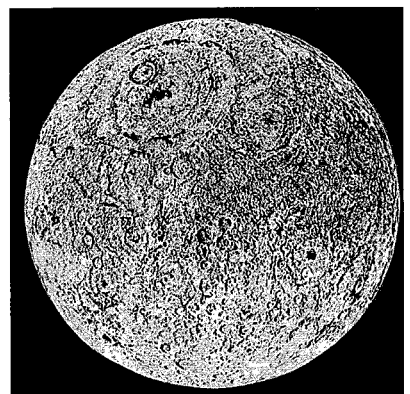  

(a) 4 billion years ago   (b) 3 billion years ago   (c) Today

Figure 8.29 Lunar Evolution Paintings of the Moon (a) about 4 billion years ago, after much of the meteoritic bombardment had subsided and the surface had somewhat solidified; (b) about 3 billion years ago, after molten lava had made its way up through surface fissures to fill the low-lying impact basins and create the smooth maria; and (c) today, with much of the originally smooth maria now heavily pitted with craters formed at various times within the past 3 billion years. *(U.S. Geological Survey)*

DISCOVERY 8-2

Exploring the Moon on a Shoestring

During the 1990s, the watchwords for unmanned exploration of space have been "smaller, faster, cheaper." Unlike previous generations of space missions, the emphasis now is on creating small-scale, lightweight systems that can be designed and built rapidly and cheaply, affording mission planners much greater flexibility, both in designing follow-up missions and in quickly changing mission parameters as circumstances warrant. Two lunar exploration satellites—*Clementine* and *Lunar Prospector*—have demonstrated the power of this approach. Both have returned high-quality science for total costs of about $70 million each, a small fraction of the cost of most other planetary spacecraft, which normally carry price tags of hundreds of millions, sometimes even billions, of dollars.

The *Clementine* satellite was sent to the Moon in 1994 by the U.S. Defense Department, largely to test some new sensing devices developed for the ballistic missile defense program. This was the first lunar mission by any nation since the crew of *Apollo 17* left the Moon in 1972. *Clementine* originally was a code word for a military-classified space project known as the Deep-Space Program Science Experiment. The vehicle and its onboard suite of instruments were designed to test the feasibility of miniaturizing a complex spacecraft, its engineering subsystems, and its sophisticated sensors for use in deep space. The spacecraft's total mass was less than 150 kg. *Clementine's* technical design was a product of the Strategic Defense Initiative—the Star Wars program—now known as the Ballistic Missile Defense Organization. Its target—the Moon—was of no interest to the military, other than being a convenient, known object in the cosmic neighborhood.

The lunar portion of the mission was a spectacular success, although a follow-on mission to map an asteroid had to be canceled after a computer malfunction caused an onboard thruster to fire until it had used up all its fuel, leaving the spacecraft spinning out of control. *Clementine* made the first digital global map of the Moon, at very high resolution. In two months of operation, its sensors took over 2.5 million images with a clarity at least 10 times better than NASA's most sophisticated planetary camera in the 1990s—the one aboard the *Galileo* mission to Jupiter—

and *Clementine's* highest-resolution camera imaged the Moon with up to 100 times better resolution than *Galileo* can achieve. *Clementine* was able to obtain global coverage of the Moon at visible, ultraviolet, and infrared wavelengths, and also was able to sense the Moon in 11 different spectral bands. In addition, the small craft carried lidar devices (the visible equivalent of radar), able to pulse the lunar surface and listen for echoes. In all, more than 50 advanced lightweight technologies were demonstrated on this powerful dwarf spacecraft.

Two examples of *Clementine's* data are shown below. The first figure shows a mosaic of about 1500 images centered on the Moon's south pole. The bottom half is part of the near side of the Moon, as seen from Earth; the top half is the far side that we never see from home. For scale, the double-ringed crater at upper left, called Schroedinger, has an outer diameter of 320 km. (Note that the circular image is rather misleading—only a portion of the Moon is shown here.) The dark region at the pole is an old, permanently shadowed depression. The second illustration is an altimetry map showing the mountains and valleys in nearly the same region. The colors indicate the ups and downs in the terrain, and reveal that the Moon is a good deal rougher than had previously been thought. The big, dark feature at center is named Aitken Basin. Its rim (yellow, outlined with black dashed circle) spans some 2000 km; the basin itself (mostly purple) averages 10 km deep. This huge depression is the largest impact basin known in the solar system.

NASA's *Lunar Prospector* was launched in January 1998. Designed and built in just 22 months and weighing 295 kg fully fueled, the spacecraft orbited the Moon for 18 months, probing the lunar surface and interior with an array of onboard instruments: A *gamma-ray spectrometer* mapped the abundances of certain elements on the Moon's surface; two sensitive *magnetometers* probed the Moon's extremely weak magnetic field; a *neutron spectrometer* searched for water ice by detecting the element hydrogen on the Moon's surface; an *alpha-particle spectrometer* searched for particles emitted by radioactive gases leaking out of the lunar interior; and a *Doppler gravity experiment*

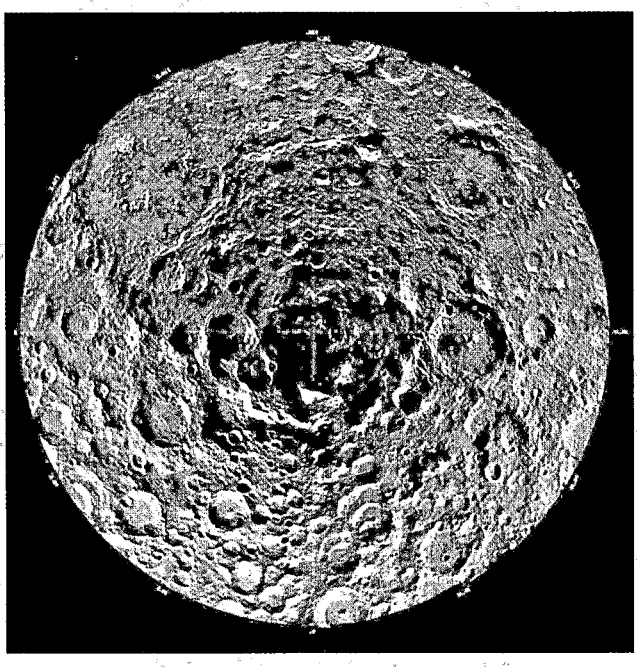

made detailed measurements of the Moon's gravitational field. *Lunar Prospector* did not carry a camera. The spacecraft's instrument package was designed in part to complement the imaging and radar capabilities of *Clementine*.

By far the most publicized aspect of the *Lunar Prospector* mission was its search for water ice at the lunar poles. Radar observations made by *Clementine* of the deep depression at the Moon's south pole (at the center of the first image above) had suggested deposits of water ice at a depth of a few meters. *Lunar Prospector's* neutron spectrometer confirmed the presence of large amounts of hydrogen (and presumably therefore water) there although, as discussed in more detail in the text, a follow-up attempt to detect water directly was unsuccessful.

Between them, *Clementine* and *Lunar Prospector* made detailed maps of the Moon's surface, charted the chemical composition of the lunar crust, and measured the Moon's magnetic and gravitational fields, all at unprecedented resolution. In doing so, they have allowed scientists to construct much more accurate models of the Moon's interior and to probe its past history with greater precision. But beyond the scientific results, important as they are, these spacecraft may have done something even more far-reaching. By demonstrating that major scientific findings can come from low-cost, fast-turnaround missions, they may have changed forever the way planetary scientists explore the solar system.

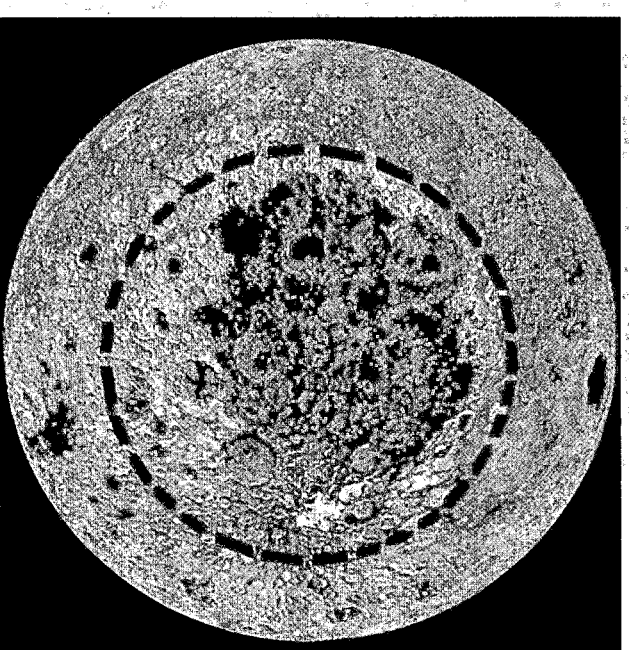

(Dept. of Defense)

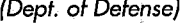
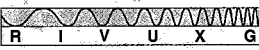

Chapter Review

SUMMARY

Both the Moon and Mercury are airless, virtually unchanging worlds that experience extremes in temperature. Mercury has no permanent atmosphere, although it does have a thin envelope of gas temporarily trapped from the solar wind. The main surface features on the Moon are the dark **maria** (p. 199) and the lighter-colored **highlands** (p. 199). Highland rocks are less dense than rocks from the maria, and are believed to represent the Moon's crust. Maria rocks are thought to have originated in the lunar mantle. The surfaces of both the Moon and Mercury are covered with **craters** (p. 199) of all sizes, caused by impacting meteoroids. Meteoritic impacts are the main source of erosion on the surfaces of both worlds. The lunar highlands are older than the maria and are much more heavily cratered. The rate at which craters are formed decreases rapidly with increasing crater size.

The high day-side temperatures and cold night-side temperatures on the Moon and Mercury result from the absence of significant heat conduction or atmospheric blanketing on the planet. Sunlight strikes the polar regions of both the Moon and Mercury at such an oblique angle that temperatures there are very low, with the result that both bodies may have significant amounts of water ice near the poles.

The tidal interaction between Earth and the Moon is responsible for the Moon's **synchronous orbit** (p. 202), in which the same side of the Moon always faces our planet. The large lunar equatorial bulge probably indicates that the Moon once rotated more rapidly and orbited closer to Earth. Mercury's rotation rate is strongly influenced by the tidal effect of the Sun. Because of Mercury's eccentric orbit, the planet rotates not synchronously but exactly three times for every two orbits around the Sun. The condition in which a body's rotation rate is simply related to its orbit period around some other body is known as a spin–orbit **resonance** (p. 204).

The Moon's surface consists of both rocky and dusty material. Lunar dust, called regolith, is made mostly of pulverized lunar rock, mixed with a small amount of material from impacting meteorites. Evidence for past volcanic activity on the Moon is found in the form of solidified lava channels called **rilles** (p. 212). Mercury's surface features bear a striking similarity to those of the Moon. The planet is heavily cratered, much like the lunar highlands. Among the differences between Mercury and the Moon are Mercury's lack of lunarlike maria, its extensive **intercrater plains** (p. 214), and the great cracks, or **scarps** (p. 214), in its crust. The plains were caused by extensive lava flows early in Mercury's history. The scarps were apparently formed when the planet's core cooled and shrank, causing the surface to crack. Mercury's evolutionary path was similar to that of the Moon for half a billion years after they both formed. Mercury's volcanic period probably ended before that of the Moon.

The absence of a lunar atmosphere and any present-day lunar volcanic activity are both consequences of the Moon's small size. Lunar gravity is too weak to retain any gases, and lunar volcanism was stifled by the Moon's cooling mantle shortly after extensive lava flows formed the maria more than 3 billion years ago. The crust on the far side of the Moon is substantially thicker than the crust on the near side. As a result, there are almost no maria on the lunar far side. Mercury has a large impact crater called the Caloris Basin, whose diameter is comparable to the radius of the planet. The impact that formed it apparently sent violent shock waves around the entire planet, buckling the crust on the opposite side.

The Moon's average density is not much greater than that of its surface rocks, probably because the Moon cooled more rapidly than the larger Earth and solidified sooner, so there was less time for differentiation to occur, although the Moon probably has a small iron-rich core. The lunar crust is too thick and the mantle too cool for plate tectonics to occur. Mercury's average density is considerably greater—similar to that of Earth—implying that Mercury contains a large high-density core, probably composed primarily of iron. The Moon has no measurable large-scale magnetic field, a consequence of its slow rotation and lack of a molten metallic core. Mercury's weak magnetic field seems to have been "frozen in" long ago when the planet's iron core solidified.

The most likely explanation for the formation of the Moon is that the newly formed Earth was struck by a large (Mars-sized) object. Part of the impacting body remained behind as part of our planet. The rest ended up in orbit as the Moon.

SELF-TEST: TRUE OR FALSE?

_____ **1.** Laser-ranging can determine the distance to the Moon to an accuracy of a few centimeters.

_____ **2.** Mercury and Earth have about the same average density.

_____ **3.** The Moon rotates on its axis about once per week.

_____ **4.** Mercury has a very small orbital eccentricity.

_____ **5.** Neither Mercury nor the Moon has any significant atmosphere.

_____ **6.** Both the Moon and Mercury have nighttime low temperatures of 300 K, well below the freezing point of water.

_____ **7.** Telescopes on Earth could see the astronauts and the lunar landers on the Moon during the _Apollo_ missions.

_____ **8.** Mercury has few lava-flow regions like the lunar maria.

_____ **9.** Mercury's solar day is actually longer than its solar year.

_____ **10.** Large craters are formed frequently today on the surface of the Moon.

_____ **11.** Some volcanic activity continues today on the surface of the Moon.

_____ **12.** Mercury's craters are more densely packed than are craters on the Moon.

_____ **13.** Scarps are found only on Mercury, not on the Moon.

_____ **14.** The Moon has an extensive iron core.

_____ **15.** Unlike the Moon, Mercury is differentiated.

SELF-TEST: FILL IN THE BLANK

1. The most accurate method for determining the distance to the Moon is by _____.
2. Mercury can be seen only just before _____ or just after _____.
3. The radius of the Moon is about _____ Earth's radius; the radius of Mercury is about _____ that of Earth. (Give your answers as simple fractions, not decimals.)
4. Because the Moon's average density is so much lower than Earth's average density, we can infer that the Moon must contain less _____.
5. Mercury's iron core contains a _____ fraction of the planet's total mass than does Earth's core.
6. Mercury's daytime temperature is higher than the Moon's because it is _____.
7. The _____ on the Moon are dark, flat, roughly circular regions hundreds of kilometers in diameter.

8. Craters on the Moon and Mercury are primarily due to _____.
9. Mercury's rotation rate was first measured using _____.
10. Although Mercury's daytime temperatures are always very hot, it may still be possible for it to have sheets of water ice at its _____.
11. The crater produced by the impact of a meteoroid on the Moon has a typical diameter about _____ times the diameter of the meteoroid.
12. The lunar crust is significantly _____ than Earth's crust.
13. The lunar maria's dark, dense rock originally was part of the lunar _____.
14. Mercury's _____ is about 1/100 that of Earth and was originally thought not to exist at all.
15. The most likely scenario for the formation of the Moon is a collision between Earth and a _____.

REVIEW AND DISCUSSION

1. How is the distance to the Moon most accurately measured?
2. Why is Mercury seldom seen with the naked eye?
3. Why did early astronomers think that Mercury was two separate planets?
4. Employ the concept of escape speed to explain why the Moon and Mercury have no significant atmospheres.
5. In what sense are the lunar maria "seas"?
6. Why is the surface of Mercury often compared with that of the Moon? List two similarities and two differences between the surfaces of Mercury and the Moon.
7. What does it mean to say that the Moon is in a synchronous orbit around Earth? How did the Moon come to be in such an orbit?
8. What does it mean to say that Mercury has a 3:2 spin–orbit resonance? Why didn't Mercury settle into a 1:1 spin–orbit resonance with the Sun as the Moon did with Earth?
9. What is a scarp? How are scarps thought to have formed? Why do scientists believe that the scarps formed after most meteoritic bombardment ended?
10. What is the primary source of erosion on the Moon? Why is the average rate of lunar erosion so much less than on Earth?
11. What evidence do we have for ice on the Moon?
12. Name two pieces of evidence indicating that the lunar highlands are older than the maria.

13. In contrast with Earth, the Moon and Mercury undergo extremes in temperature. Why?
14. How is Mercury's evolutionary history like that of the Moon? How is it different?
15. Describe the theory of the Moon's origin favored by many astronomers.
16. Because the Moon always keeps one face toward Earth, an observer on the moon's near side would see Earth appear almost stationary in the lunar sky. Still, Earth would change its appearance as the Moon orbited Earth. How would Earth's appearance change?
17. The best place to aim a telescope or binoculars on the Moon is along the terminator line, the line between the Moon's light and dark hemispheres. Why? If you were standing on the lunar terminator, where would the Sun be in your sky? What time of day would it be if you were standing on Earth's terminator line?
18. Where on the Moon would be the best place from which to make astronomical observations? What would be this location's advantage over locations on Earth?
19. Explain why Mercury is never seen overhead at midnight in Earth's sky.
20. How is the varying thickness of the lunar crust related to the presence or absence of maria on the Moon?

PROBLEMS  *Algorithmic versions of these questions are available in the Practice Problems module of the Companion Website.*

The number of squares preceding each problem indicates its approximate level of difficulty.

1. ■ How long does a radar signal take to travel from Earth to Mercury and back when Mercury is at its closest point to Earth?
2. ■ The Moon's mass is 1/80 that of Earth, and the lunar radius is 1/4 Earth's radius. Based on these figures, calculate the total weight on the Moon of a 100-kg astronaut with a 50-kg spacesuit and backpack, relative to his weight on Earth.
3. ■ What would be the same astronaut's weight on Mercury?
4. ■ Based on the data presented in the Moon Data box (p. 202), verify the values given for the Moon's perigee (minimum distance from Earth) and apogee (maximum dis-

tance from Earth), and estimate the Moon's minimum and maximum angular diameter, as seen from Earth. Compare these values with the angular diameter of the Sun (of actual diameter 1.4 million km), as seen from a distance of 1 A.U.

5. ■ What is the angular diameter of the Sun, as seen from Mercury, at perihelion? At aphelion?

6. ■ The *Hubble Space Telescope* has a resolution of about 0.05″. What is the size of the smallest feature it can distinguish on the surface of the Moon (distance = 380,000 km)? On Mercury, at closest approach to Earth?

7. ■ What was the orbital period of the *Apollo 11* command module, orbiting 10 km above the lunar surface?

8. ■■ Compare the gravitational tidal acceleration of the Sun on Mercury (at perihelion, solar mass = 2×10^{30} kg) with the tidal effect of Earth on the Moon (at perigee) ∞ (Sec. 7.6).

9. ■■■ Mercury's average orbital speed around the Sun is 47.9 km/s. Use Kepler's second law to calculate Mercury's speed (a) at perihelion and (b) at aphelion ∞ (Sec. 2.4). Convert these speeds to angular speeds (in degrees per day), and compare them with Mercury's 6.1°/day rotation rate.

10. ■■ Calculate the lengths of a sidereal and a solar day on Mercury if the planet were in a 4:3 spin–orbit resonance instead of the 3:2 resonance actually observed.

11. ■■ Assume that a planet will have lost its initial atmosphere by the present time if the average molecular speed exceeds 1/6 of the escape speed (see *More Precisely 8-1*). What would Mercury's mass have to be in order for it to still have a nitrogen (molecular weight 28) atmosphere?

12. ■■ With the same assumptions as the previous question, estimate the minimum molecular mass that might still be found in Mercury's atmosphere.

13. ■■ Using the rate given in the text for the formation of 10-km craters on the Moon, estimate how long would be needed for the entire Moon to be covered with new craters of this size. How much higher must the cratering rate have been in the past to cover the entire lunar surface with such craters in the 4.6 billion years since the Moon formed?

14. ■■ Repeat the previous question, for meter-sized craters.

15. ■ Using the data given in the text, calculate how long erosion would take to obliterate (a) the bootprint in Figure 8.17, (b) the Barringer Meteor Crater in Figure 8.19, (c) lunar crater Reinhold in Figure 8.15(b).

COLLABORATIVE EXERCISES

1. **New Views of Mercury.** The craters on the Moon are named after great scientists and philosophers. As a group, propose new names for the 10 largest craters found on Mercury when its "other" side is imaged by the Mercury *MESSENGER* mission in 2007 and explain your reasoning.

2. **Outpost on the Moon.** The anticipated cost of transporting a gallon of water from Earth to the Moon is $15,000.

Estimate the cost of taking a single-day's supply of water for your group to the Moon by determining how much water each of the group members use in a single day.

3. **Lunar Mineral Rights.** As a group, decide who owns the rights to mine mineral resources from the Moon and explain your reasoning.

RESEARCHING ON·THE WEB  *To complete the following exercises, go to the online Destinations module for Chapter 8 on the Companion Website for* Astronomy Today 4/e.

1. Access the "Top 10 Scientific Discoveries from *Apollo*" page and describe the three most interesting scientific discoveries from the *Apollo* missions to the Moon.

2. Access the "Virtual Reality Moon Phase Pictures" page and determine the phase of the Moon on the day you were born.

3. Access the "Mercury Fact Sheet" page from NASA and determine the maximum and minimum distances Mercury is from Earth.

PROJECTS

1. Observe the Moon during an entire cycle of phases. When does the Moon rise, set, and appear highest in the sky at each major phase? What is the interval of time between each phase?

2. If you have binoculars, turn them on the Moon when it appears at twilight and when it appears high in the sky. Draw pictures of what you see. What differences do you notice in your two drawings? What color is the Moon seen near the horizon? What color is the Moon seen high in the sky? Why is there a difference?

3. Watch the Moon over a period of hours on a night when you can see one or more bright stars near it. Estimate how many Moon diameters it moves per hour, relative to the stars. Knowing the Moon is about 0.5° in diameter, how many degrees per hour does it move? What is your estimate of its orbital period?

4. Try to spot Mercury in morning or evening twilight. (Hint: As seen from the Northern Hemisphere, the best evening apparitions of the planet take place in the spring, and the best morning apparitions take place in the fall.)

SKYCHART III PROJECTS

The SkyChart III Student Version planetarium program on which these exercises are based is included as a separately executable program on the CD in the back of this text.

1. ▣ With SkyChart III configured for normal viewing, center on the Moon at midnight. Set *Draw/Horizon Mask* deselect. Use a field of view of 3° 45 minutes and time step of one day. Use F6 to step through a month and observe the changes in the phases of the Moon. Repeat until you have a firm grasp on the relationship between phase and orientation with respect to the Sun.

Set *DRAW/Mouse Coordinates* and determine the position of the Moon. Advance time one day and again determine the position of the Moon. Note that it should advance 360°/27.2 days = 13.2° per day, or approximately 0.5° per hour against the background stars, while the stars advance approximately one degree per day. On some clear evening with a Moon, make careful observation of the position of the Moon and then again four hours later. You should be able to see the progress the Moon has made against the star background just as you do with SkyChart III.

2. ▣ Start viewing the Moon just before the full lunar eclipse of May 15, 2003. If you advance the time slowly, you will notice a dimming of the Moon. What are the magnitude and the illumination of the Moon at that time? Change your point of view to be directly above the ecliptic plane at 60 A.U. from the Sun. Verify that at the time of the lunar eclipse you can draw a line from the Moon, through the center of Earth to the Sun.

3. ▣ When one observes the full Moon rising, it appears to be huge. Later in the evening, the same full Moon will appear to be smaller. This is an illusion. From your location, plot the size of the Moon for one month. Use a time interval of one hour (Select *ANIMATION/Set Time Step* and enter: "0 01:00:00"). Observe the size at the Moon's rise and at midnight. At midnight, the Moon will be larger because it is closer to where you are standing (unless you are on one of the poles). Can you explain why?

4. ▣ Configure SkyChart III with a 90° field of view and center on Mercury. Set *ANIMATE/Trail For/Mercury* with *Record apparent positions against fixed stars*. Animate with one-day steps and observe the motion of Mercury against the background stars. Note how the orbit passes above and below the plane of the ecliptic. Allow the animation to proceed for at least a year to observe the varied patterns of the motion. Explain the motion by adding the effects of observing from a moving platform, Earth, to the regular elliptical orbit of Mercury.

5. ▣▣ Determine the day of the next new Moon, first quarter Moon, full Moon, and last quarter Moon. You can export the relevant data using: *File/Export/Ephemeris as Text* and then analyze it using a spreadsheet.

 *In addition to the Practice Problems and Destinations modules, the Companion Website at http://www.prenhall.com/chaisson provides for each chapter an additional true-false, multiple choice, and labeling quiz, as well as additional annotated images, animations, and links to related Websites*

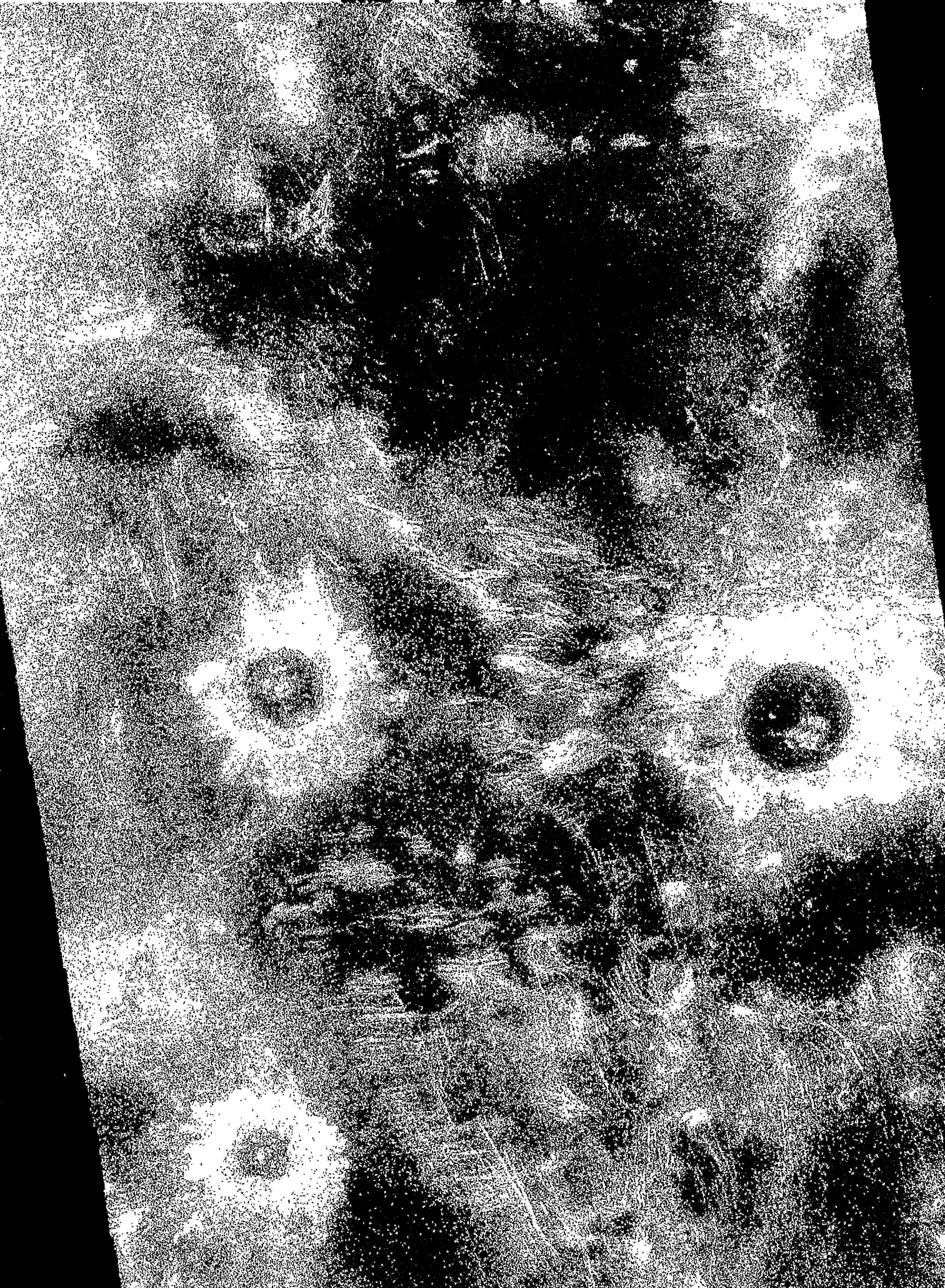

9 VENUS

Earth's Sister Planet

LEARNING GOALS

Studying this chapter will enable you to:

1   Summarize Venus's general orbital and physical properties.

2   Explain why Venus is hard to observe from Earth and how we have obtained more detailed knowledge of the planet.

3   Compare the surface features and geology of Venus with those of Earth and the Moon.

4   Describe the characteristics of Venus's atmosphere and contrast it with that of Earth.

5   Explain why the greenhouse effect has produced conditions on Venus very different from those on Earth.

6   Describe Venus's magnetic field and internal structure.

 Visit http://www.prenhall.com/chaisson for additional images, animations, and links to related sites for this chapter.

Venus is completely enshrouded in clouds, so its surface cannot be seen in visible light from above the atmosphere. This false-color radar image of a highly fractured plain was made by the *Magellan* spacecraft that mapped the planet in the 1990s; it measures about 400 km wide. Seen here is a unique triple crater, possibly the result of a single meteoroid that broke up in the dense Venusian atmosphere before striking the surface. The impacts created radar-bright blankets of ejecta around each crater, colored here in white-yellow; the blue areas are probably smooth, low-lying lava beds, the red area an uplifted plateau. (JPL)

The Big Picture: Often called Earth's sister planet, Venus is nothing like Earth when it comes to surface temperature; at 730 K, the temperature on Venus is high enough to melt lead. We now know that Venus's climate, like Earth's, has varied over time—largely the result of geologic activity and atmospheric change. What we do not know very well is why Venus became so very much hotter than Earth—or if Earth could someday heat up similarly.

Venus seems almost a carbon copy of our own world. The two planets are similar in size, density, and chemical composition. They orbit at comparable distances from the Sun. At formation, they must have been almost indistinguishable from one another. Yet they are now about as different as two terrestrial planets can be. Whereas Earth is a vibrant world, teeming with life, Venus is an uninhabitable inferno, with a dense, hot atmosphere of carbon dioxide, lacking any trace of oxygen or water. Somewhere along their respective evolutionary paths, Venus and Earth diverged, and diverged radically. How did this occur? What were the factors leading to Venus's present condition? Why are Venus's surface, atmosphere, and interior so different from Earth's? In answering these questions, we will discover that a planet's environment, as well as its composition, can play a critical role in determining its future.

9.1 Orbital Properties

⏹ Venus is the second planet from the Sun. Its orbit lies within Earth's, so Venus, like Mercury, is always found fairly close to the Sun in the sky—Venus is never seen more than 47° from the Sun. Given Earth's rotation rate of 15° per hour, this means that Venus is visible above the horizon for at most three hours before the Sun rises or after it sets. Because we can see Venus from Earth only just before sunrise or just after sunset, the planet is often called the "morning star" or the "evening star," depending on where it happens to be in its orbit. Figure 9.1 shows Venus in the western sky just after sunset. The Venus Data box on p. 232 lists some of the planet's orbital and physical properties.

Venus is the third brightest object in the entire sky (after the Sun and the Moon). It appears more than 10 times brighter than the brightest star, Sirius. You can see Venus even in the daytime, if you know just where to look. On a moonless night away from city lights, Venus casts a faint shadow. The planet's brightness stems from the fact that Venus is highly reflective. Nearly 70 percent of the sunlight reaching Venus is reflected back into space (compare this with roughly 10 percent in the case of Mercury and the Moon). Most of the sunlight is reflected from clouds high in the planet's atmosphere.

We might expect Venus to appear brightest when it is "full"—that is, when we can see the entire sunlit side. However, because Venus orbits between Earth and the Sun, Venus is full when it is at its greatest distance from us, 1.7 A.U. away on the other side of the Sun (an alignment known as *superior conjunction*, where the term "conjunction" simply indicates that two objects are close together on the sky), as illustrated in Figure 9.2. When Venus is closest to us, the planet is at the new phase, lying between Earth and the Sun (*inferior conjunction*), and we again can't see it, because now the sunlit side faces away from us—only a thin ring of sunlight, caused by refraction in Venus's atmosphere, surrounds the planet. As Venus moves away from inferior conjunction, more and more of it becomes visible, but its distance from us also continues to increase. Venus's maximum brightness, as seen from Earth, actually occurs about 36 days before or after closest approach to our planet. At that time, Venus is about 39° from the Sun and 0.47 A.U. from Earth, and we see it as a rather fat crescent.

9.2 Physical Properties

RADIUS, MASS, AND DENSITY

We can determine Venus's radius from simple geometry, just as we did for Mercury and the Moon. ⌐ᴗ (Sec. 8.2) At closest approach, when Venus is only 0.28 A.U. from us, its angular diameter is 64″. From this observation, we can determine its radius to be about 6000 km. More accurate measurements from spacecraft give a value of 6052 km, or 0.95 Earth radii.

Like Mercury, Venus has no moon. Before the Space Age, astronomers calculated its mass by indirect means—through studies of its small gravitational effect on the orbits of the other planets, especially Earth. Now that

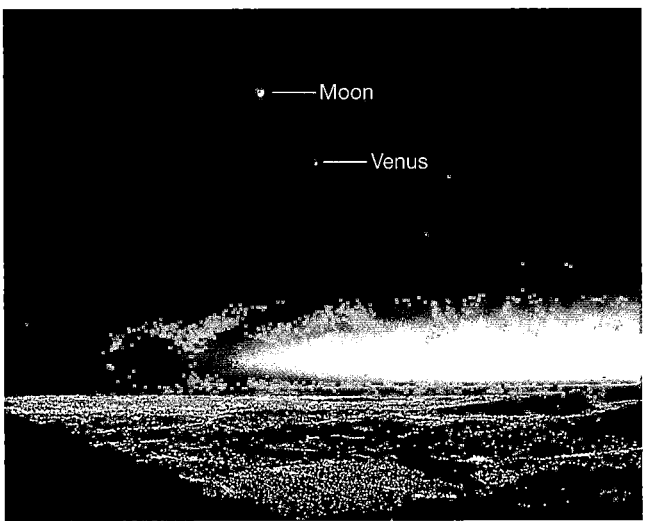

Figure 9.1 Venus at Sunset The Moon and Venus in the western sky just after sunset. Venus clearly outshines even the brightest stars in the sky. (J. Schad/Photo Researchers, Inc.)

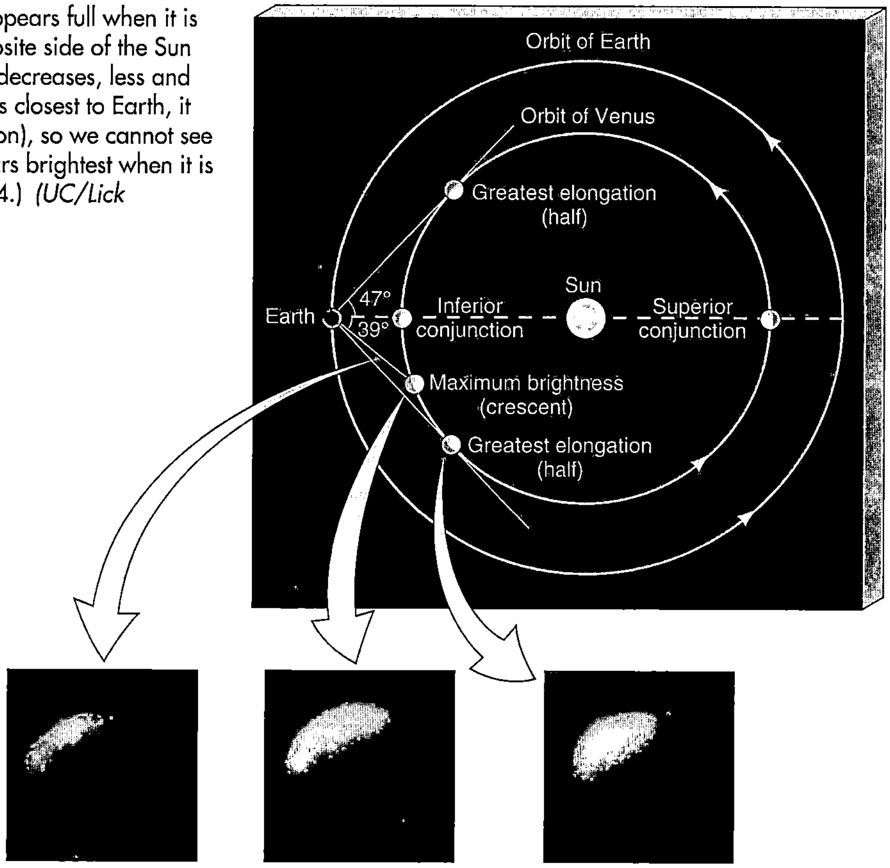

Figure 9.2 Venus Brightness Venus appears full when it is at its greatest distance from Earth, on the opposite side of the Sun from us (superior conjunction). As its distance decreases, less and less of its sunlit side becomes visible. When it is closest to Earth, it lies between us and the Sun (inferior conjunction), so we cannot see the sunlit side of the planet at all. Venus appears brightest when it is about 39° from the Sun. (Compare Figure 2.14.) *(UC/Lick Observatory)*

spacecraft have orbited the planet, we know its mass very accurately: 4.9×10^{24} kg, or 0.82 the mass of Earth.

From its mass and radius we find that Venus's average density is 5200 kg/m³. As far as these bulk properties are concerned, then, Venus seems very similar to Earth. If the planet's overall composition were similar to Earth's, we could then reasonably conclude that the planet's internal structure and evolution were basically Earthlike. We will review what evidence there is on this subject later in this chapter.

ROTATION RATE

2 The same clouds whose reflectivity makes Venus so easy to see in the night sky also make it impossible for us to discern any surface features on the planet, at least in visible light. As a result, until the advent of suitable radar techniques in the 1960s, astronomers did not know the rotation period of Venus. Even when viewed through a large optical telescope, the planet's cloud cover shows few features (see Figure 9.5), and attempts to determine Venus's rotation by observing the cloud deck were frustrated by the rapidly changing nature of the clouds themselves. Some astronomers argued for a 25-day period, while others favored a 24-hour cycle. Controversy raged until, to the surprise of all, radar observers announced that the Doppler broadening of their returned echoes implied a sluggish 243-day rotation period. ⊙ (Sec.

8.3) Furthermore, Venus's spin was found to be *retrograde*—that is, in a sense opposite that of Earth and most other solar system objects, and opposite that of Venus's orbital motion.

Planetary astronomers conventionally define "north" and "south" for individual members of the solar system by the condition that planets *always* rotate from west to east. With this definition, Venus's retrograde spin means that the planet's north pole lies *below* the ecliptic plane, unlike any of the other terrestrial worlds. Venus's axial tilt—the angle between its equatorial and orbital planes—is 177.4° (compared with 23.5° in the case of Earth). However, astronomical images of solar-system objects conventionally place objects lying above the ecliptic plane at the top of the frame. Thus, with the preceding definition of north and south, all the images of Venus shown in this chapter have the *south* pole at the top.

Figure 9.3 illustrates Venus's retrograde rotation and compares it with the rotation of its neighbors, Mercury, Earth, and Mars. Because of the planet's slow retrograde rotation, its solar day (noon to noon) is quite different from its sidereal rotation period of 243 Earth days (the time for one "true" rotation relative to the stars). ⊙ (Sec. 1.3) In fact, as illustrated in Figure 9.4, one Venus day is a little more than half a Venus year (225 Earth days).

Why is Venus rotating "backward," and why so slowly? At present, the best explanation planetary scientists can offer is that early in Venus's evolution, the planet was struck by a large body, much like the one that may have hit

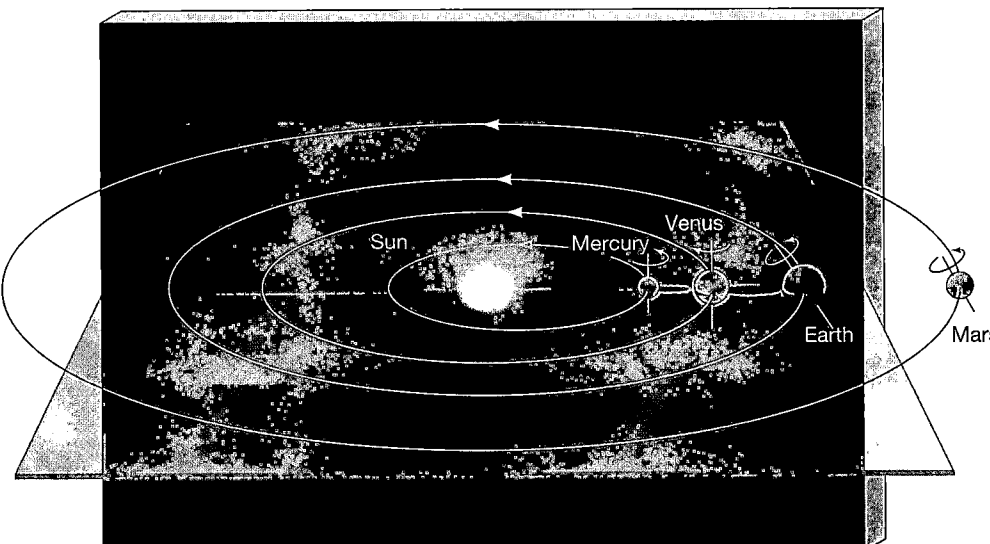

Figure 9.3 Terrestrial Planets' Spins The inner four planets of the solar system—Mercury, Venus, Earth, and Mars—display widely differing rotational properties. Although all orbit the Sun in the same sense and in nearly the same plane, Mercury's rotation is slow and prograde (in the same sense as the orbital motion), the rotations of Earth and Mars are fast and prograde, while Venus's is slow and retrograde. Venus rotates clockwise as seen from above the plane of the ecliptic in the direction of the north celestial pole.

Earth and formed the Moon, and that impact was sufficient to reduce the planet's spin almost to zero. ∞ (Sec. 8.8) Whatever its cause, the planet's rotation poses practical problems for Earthbound observers. It turns out that Venus rotates almost exactly five times between one closest approach to Earth and the next. As a result, *Venus always presents nearly the same face to Earth at closest approach*. This means that observations of the planet's surface cover one side—the one facing us at closest approach—much more thoroughly than the other side, which we can see only when the planet is close to its maximum distance from Earth.

This nearly perfect 5:1 resonance between Venus's rotation and orbital motion is reminiscent of the Moon's

synchronous orbit around Earth, and Mercury's 3:2 spin–orbit resonance with the Sun. ∞ (Sec. 8.3) However, no known interaction between Earth and Venus can account for this odd state of affairs. Earth's tidal effect on Venus is tiny, and is much less than the Sun's tidal effect in any case.

Furthermore, the key word in the above sentence is *near*. A resonance, if it existed, would require that the number of rotations per relative orbit be *exactly* five. The discrepancy amounts to less than 3 hours in 584 days (Venus's synodic period relative to Earth—see p. 232), but it appears to be real, and if that is so, then no resonance exists. Astronomers hate to appeal to coincidence to explain their observations, but the case of Venus's rotation appears to be just that. For now we are simply compelled to accept this strange coincidence without explanation.

Concept Check

◘ What is peculiar about Venus's rotation, and why does Venus rotate this way?

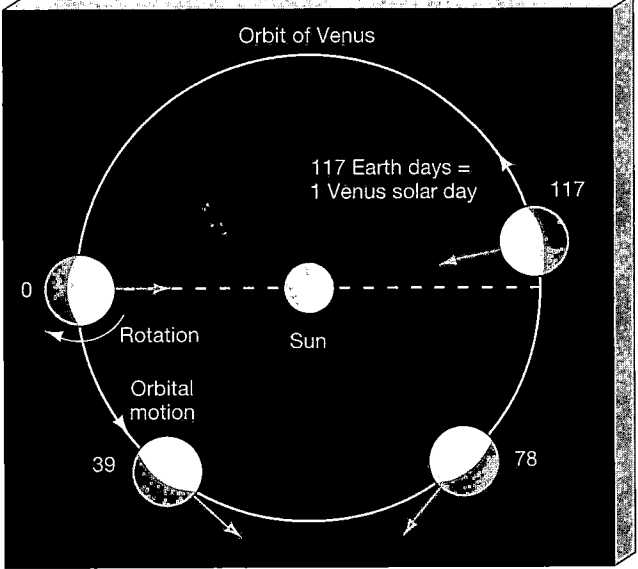

Figure 9.4 Venus Solar Day Venus's orbit and retrograde rotation combine to produce a Venus solar day equal to 117 Earth days, or slightly more than half a Venus year. The numbers in the figure mark time in Earth days.

9.3 Long-Distance Observations of Venus

⬡ Because Venus, of all the other planets, most nearly matches Earth in size, mass, and density, and because its orbit is closest to us, it is often called Earth's sister planet. But unlike Earth, Venus has a dense atmosphere and thick clouds that are opaque to visible radiation, making its surface completely invisible from the outside at optical wavelengths. Figure 9.5 shows one of the best photographs of Venus taken with a large telescope on Earth. The planet resembles a white-yellow disk and shows occasional hints of cloud circulation.

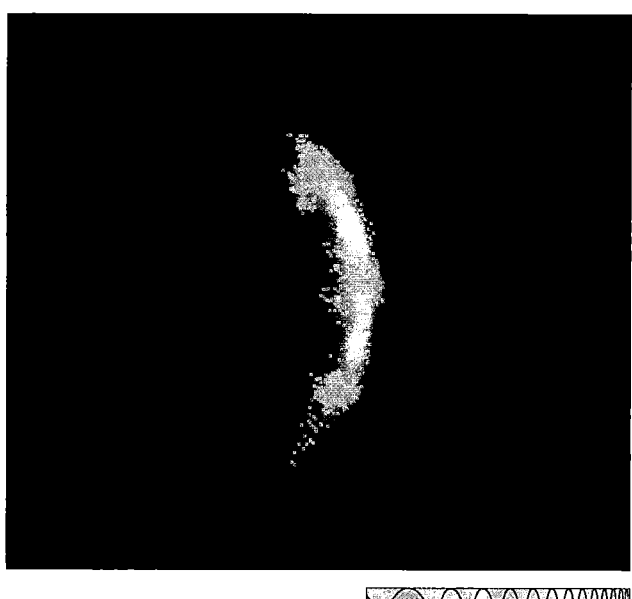

Figure 9.5 Venus This photograph, taken from Earth, shows Venus with its creamy yellow mask of clouds. *(AURA)*

measure its thermal energy emission. Unlike visible light, radio waves easily penetrate the cloud layer, and they gave the first indication of conditions on or near the surface. The radiation emitted by the planet has a blackbody spectrum characteristic of a temperature near 730 K! ∞ (Sec. 3.4) Almost overnight, the popular conception of Venus changed from that of a lush tropical jungle to an arid, uninhabitable desert.

Radar observations of the surface of Venus are routinely carried out from Earth using the Arecibo radio telescope. ∞ (Sec. 5.4) With careful signal processing, this instrument can achieve a resolution of a few kilometers, but it can adequately cover only a small fraction (roughly 25 percent) of the planet. The telescope's view of Venus is limited by the planet's peculiar near-resonance just described, and also because radar reflections from regions near the "edge" of the planet are hard to obtain. However, the Arecibo data can usefully be combined with information from probes orbiting Venus to build up a detailed picture of the planet's surface. Only with the arrival of the *Magellan* probe were more accurate data obtained.

Atmospheric patterns are much more evident when examined with equipment capable of detecting ultraviolet radiation. Some of Venus's atmospheric constituents absorb this high-frequency radiation, greatly increasing the cloud contrast. Figure 9.6 is an ultraviolet image taken in 1979 by the U.S. *Pioneer Venus* spacecraft at a distance of 200,000 km from the planet. The large, fast-moving cloud patterns resemble Earth's high-altitude jet stream more than the great cyclonic whirls characteristic of Earth's low-altitude clouds. The upper deck of clouds on Venus rotates around the planet in just 4 days—much faster than the planet itself.

Early spectroscopic studies of sunlight reflected from Venus's clouds revealed the presence of large amounts of carbon dioxide but provided little evidence for any other atmospheric gases. Until the 1950s, astronomers generally believed that observational difficulties alone prevented them from seeing other atmospheric components. The hope lingered that Venus's clouds were actually predominantly water vapor, like those on Earth, and that below the cloud cover Venus might be a habitable planet similar to our own. Indeed, in the 1930s scientists had measured the temperature of the atmosphere spectroscopically at about 240 K, not much different from our own upper atmosphere. ∞ (Sec. 4.4) Calculations of the surface temperature—taking into account the cloud cover and Venus's proximity to the Sun, and assuming an atmosphere much like our own—suggested that Venus should have a surface temperature only 10 or 20 degrees higher than Earth's.

These hopes for an Earth-like Venus were dashed in 1956, when radio observations of the planet were used to

Figure 9.6 Venus, Up Close Venus as photographed by the *Pioneer* spacecraft's cameras some 200,000 km from the planet. This image was made by capturing solar ultraviolet radiation reflected from the planet's clouds, which are probably composed mostly of droplets of sulfuric acid (the highly corrosive acid used in car batteries). *(NASA)*

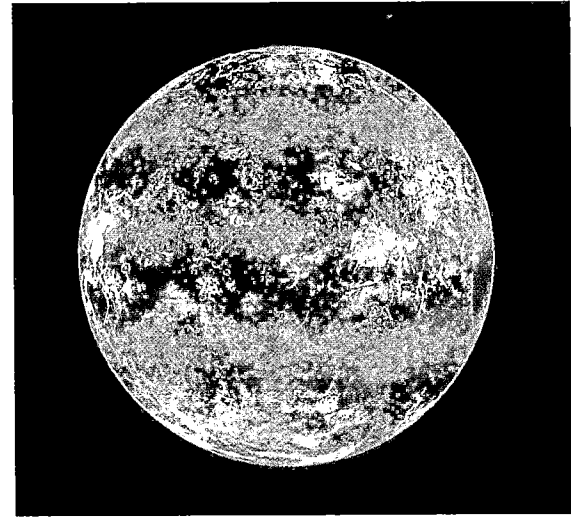

Orbital semimajor axis	0.72 A.U.
	108.2 million km
Orbital eccentricity	0.007
Perihelion	0.72 A.U.
	107.5 million km
Aphelion	0.73 A.U.
	108.9 million km
Mean orbital speed	35.0 km/s
Sidereal orbital period	224.7 solar days
	0.615 tropical years
Synodic orbital period	583.9 solar days
Orbital inclination to the ecliptic	3.39°
Greatest angular diameter, as seen from Earth	64″
Mass	4.87×10^{24} kg
	0.82 (Earth = 1)
Equatorial radius	6052 km
	0.95 (Earth = 1)
Mean density	5240 kg/m³
	0.95 (Earth = 1)
Surface gravity	8.87 m/s²
	0.91 (Earth = 1)
Escape speed	10.4 km/s
Sidereal rotation period	−243.0 solar days*
Axial tilt	177.4°
Surface magnetic field	< 0.001 (Earth = 1)
Magnetic axis tilt relative to rotation axis	—
Mean surface temperature	730 K
Number of moons	0

A negative sign denotes retrograde rotation.

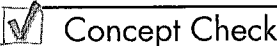

☑ **Concept Check**

■ Why did early studies of Venus lead astronomers to such an inaccurate picture of the planet's surface conditions?

9.4 The Surface of Venus

③ Although its clouds are thick and the terrain below them totally shrouded, we are by no means ignorant of Venus's surface. Detailed radar observations have been made both from Earth and from the *Venera, Pioneer Venus* and *Magellan* spacecraft. ∞ (Sec. 6.6) Analysis of the radar echoes yields a map of the planet's surface. Except for the last two figures, all the views of Venus in this section are "radargraphs" (as opposed to photographs) created in this way. As Figure 9.7(a) illustrates, the early maps of Venus suffered from poor resolution; however, more recent probes, especially *Magellan*, have provided much sharper views. As in all the *Magellan* images, the light areas in Figure 9.7(b) represent regions where the surface is rough and efficiently scatters *Magellan*'s sideways-looking radar beam back to the detector. Smooth areas tend to reflect the beam off into space instead and so appear dark.

LARGE-SCALE TOPOGRAPHY

Figure 9.8(a) shows basically the same *Pioneer Venus* data of Venus as Figure 9.7(a), except that it has been flattened out into a more conventional map. The altitude of the surface relative to the average radius of the planet is indicated by the use of color, with white representing the highest elevations, blue the lowest. (Note that the blue has nothing to do with oceans, and white does not indicate snow-capped mountains!) Some of the planet's main features are labeled in Figure 9.8(c). Figure 9.8(b) shows a map of Earth to the same scale and at the same spatial resolution.

The surface of Venus appears to be relatively smooth, resembling rolling plains with modest highlands and lowlands. Two continent-sized features, called Ishtar Terra and Aphrodite Terra (named after the Babylonian and Greek counterparts, respectively, of Venus, the Roman goddess of love), adorn the landscape, and these contain mountains comparable in height to those on Earth. The elevated "continents" occupy only 8 percent of Venus's total surface area. For comparison, continents on Earth make up about 25 percent of the surface. The remainder of Venus's surface is classified as lowlands (27 percent) or rolling plains (65 percent), although there is probably little geological difference between the two terrains.

Note that, while Earth's tectonic plate boundaries are plainly evident in Figure 9.8(b), no similar features can be seen in Figure 9.8(a). ∞ (Sec. 7.5) There appears to be no large-scale plate tectonics on Venus. Ishtar Terra ("Land of Ishtar") lies in the southern high latitudes (at the *tops* of Figures 9.7a and 9.8a—recall our earlier discussion of

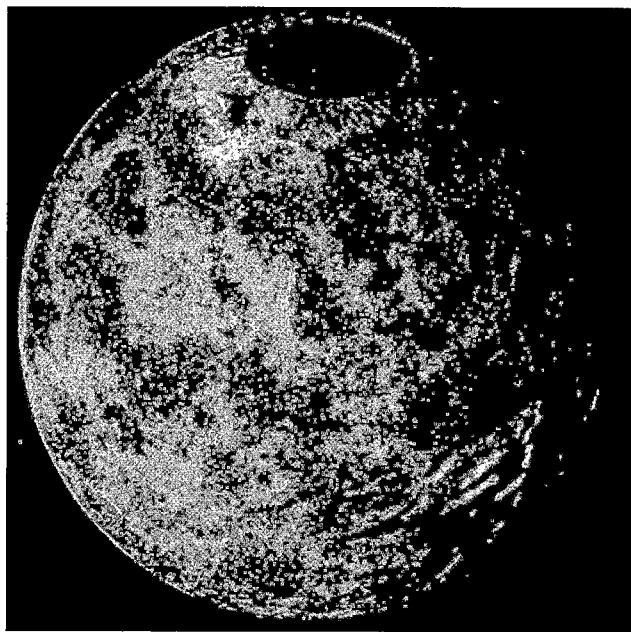

(a)

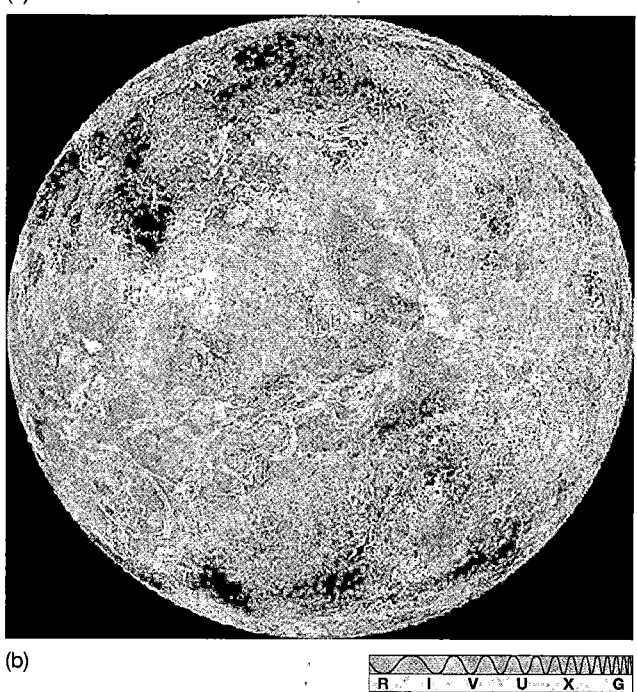

(b)

Figure 9.7 Venus Mosaics (a) This image of the surface of Venus was made by a radar transmitter and receiver on board the *Pioneer* spacecraft, which is still in orbit about the planet but now inoperative. The two continent-sized landmasses are named Ishtar Terra (upper left) and Aphrodite (lower right). Colors represent altitude: blue is lowest, red highest. The spatial resolution is about 25 km. (b) A planetwide mosaic of *Magellan* images, colored in roughly the same way as part (a). Aphrodite Terra is at the center of this image. *(NASA)*

Venus's retrograde rotation). The projection used in Figure 9.8 makes Ishtar Terra appear larger than it really is—it is actually about the same size as Australia. This landmass is dominated by a great plateau known as Lakshmi Planum (Figure 9.9), which is some 1500 km across at its widest point. This plain is ringed by mountain ranges, including the Maxwell Montes range, which contains the highest peak in the planet, rising some 14 km above the level of Venus's deepest surface depressions. Again for comparison, the highest point on Earth (the summit of Mount Everest) lies about 20 km above the deepest section of Earth's ocean floor (Challenger Deep, at the bottom of the Marianas Trench on the eastern edge of the Philippines plate).

Figure 9.9(a) shows a large-scale *Venera* image of Lakshmi Planum, at a resolution of about 2 km. The "wrinkles" are actually chains of mountains, hundreds of kilometers long and tens of kilometers apart. The red area immediately to the right of the plain is Maxwell Montes. On the western (right-hand) slope of the Maxwell range lies a great crater, called Cleopatra, about 100 km across. Figure 9.9(b) shows a *Magellan* image of Cleopatra. Cleopatra was originally thought to be volcanic in origin. However, close-up views of the crater's structure have led planetary scientists to conclude that it is meteoritic in origin, although some volcanic activity was associated with its formation. Notice the dark (smooth) lava flow emerging from within the inner ring and cutting across the outer rim at the upper right.

It is now conventional to name features on Venus after famous women—Aphrodite, Ishtar, Cleopatra, and so on. However, the early nonfemale names (for example, Maxwell Montes, named after the Scottish physicist James Clerk Maxwell) predating this convention have stuck, and they are unlikely to change. Venus's other continent-sized formation, Aphrodite Terra, is located on the planet's equator and is comparable in size to Africa. Before *Magellan*'s arrival, some researchers had speculated that Aphrodite Terra might have been the site of something akin to seafloor spreading on Earth—a region where two lithospheric plates moved apart and molten rock rose to the surface in the gap between them, forming an extended ridge. With the low-resolution data then available, the issue could not be settled at the time. The *Magellan* images now seem to rule out even this small-scale tectonic activity, and the Aphrodite region shows no signs of spreading. Figure 9.10(a) shows a portion of Aphrodite Terra called Ovda Regio. The crust appears buckled and fractured, with ridges running in two distinct directions across the image, suggesting that large compressive forces are distorting the crust. There seem to have been repeated periods when extensive lava flows occurred. The dark regions are probably solidified lava flows. Some narrow lava channels, akin to rilles on the Moon, also appear. ∞ (Sec. 8.5) Such lava channels appear to be quite common on Venus. Unlike lunar rilles, however, they can be extremely long—hundreds or even thousands of kilometers (Figure 9.10b). These lava "rivers" often have lava "deltas" at their mouths, where they deposited their contents into the surrounding plains.

Figure 9.11 shows a series of angular cracks in the crust, thought to have formed when lava welled up from a deep fissure, flooded the surrounding area, and then retreated below the planet's surface. As the molten lava withdrew, the thin new crust of solidified material collapsed under its own weight, forming the cracks we now see. Even taking into account the temperature and composition differences between Venus's crust and Earth's,

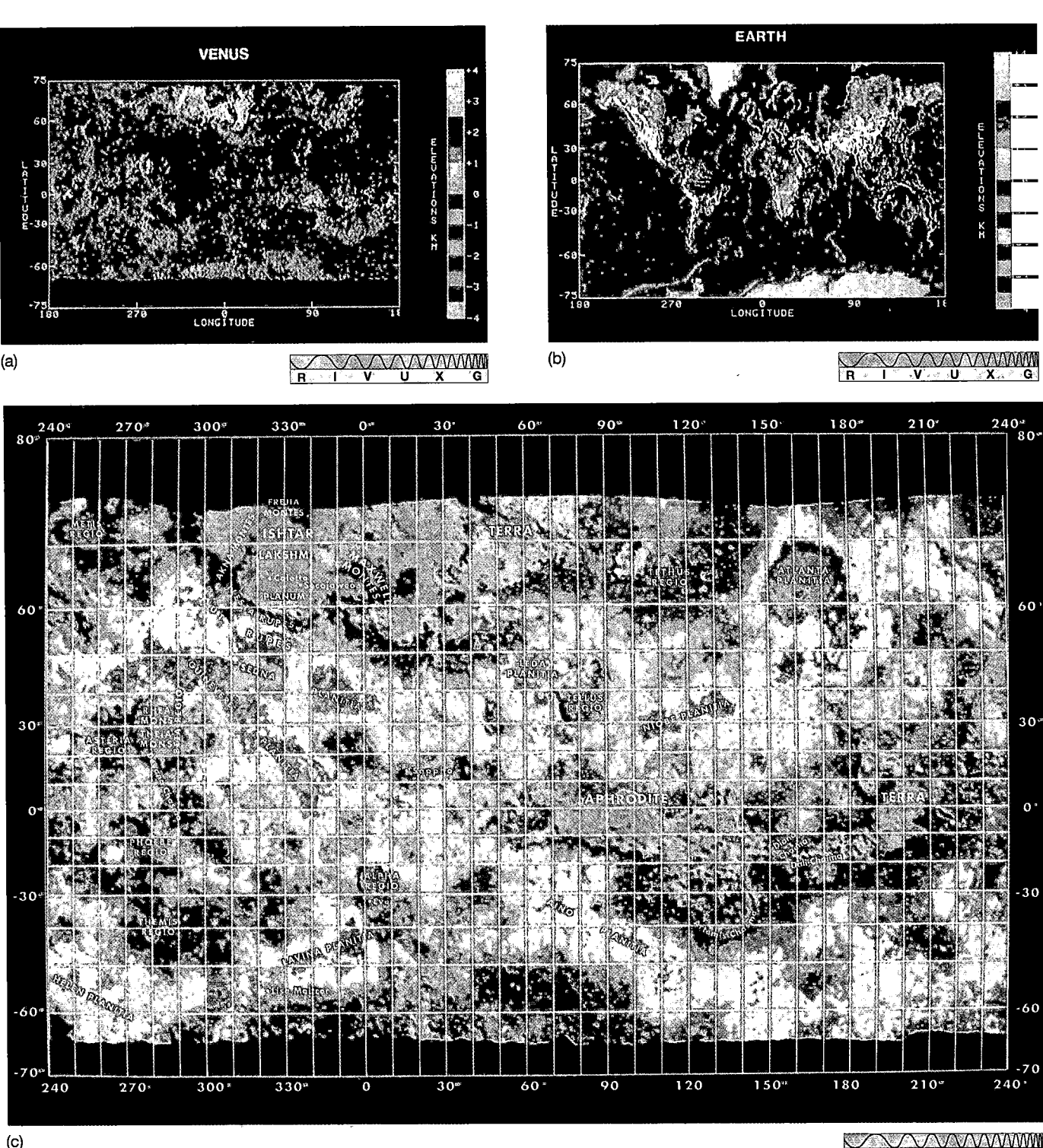

Figure 9.8 Venus Maps (a) Radar map of the surface of Venus, based on *Pioneer Venus* data. Color represents elevation according to the scale at the right. (b) A similar map of Earth, at the same spatial resolution. (c) Another version of (a), with major surface features labeled. Compare with Figure 9.7, and notice how the projection exaggerates the size of surface features near the poles. *(NASA)*

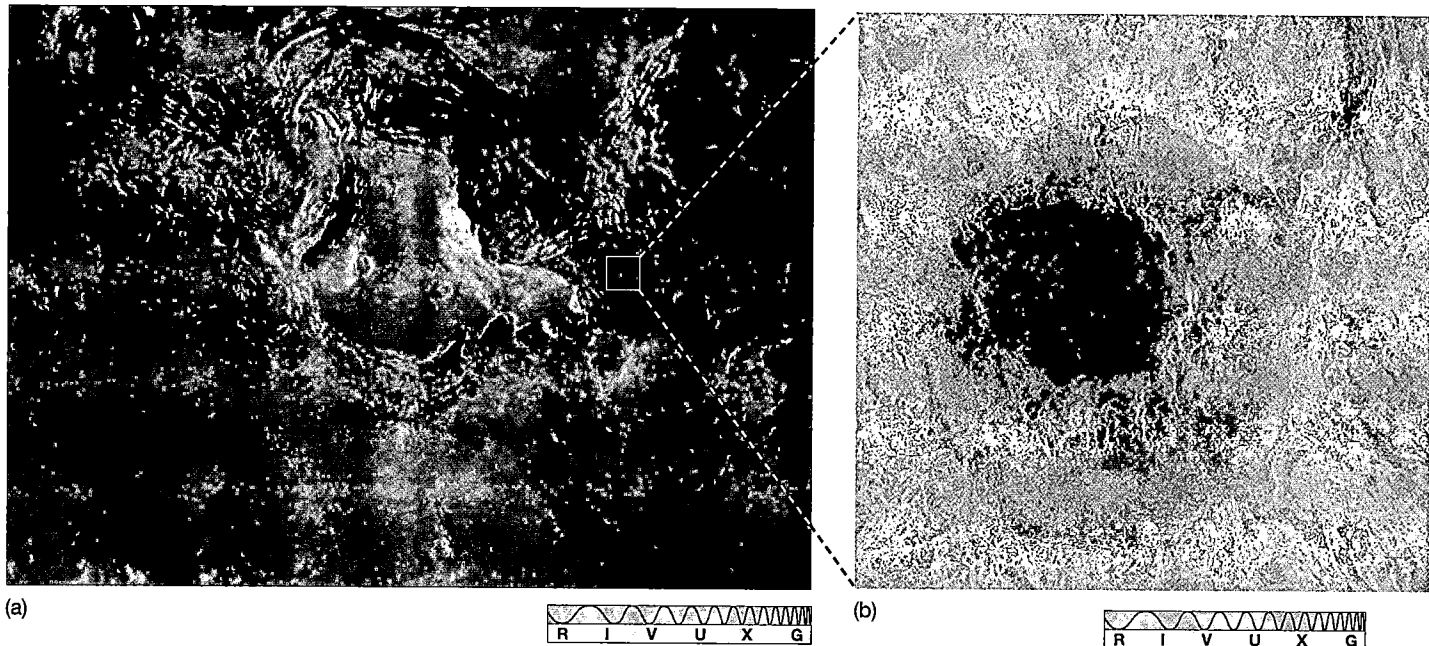

(a)

(b)

Figure 9.9 Ishtar Terra (a) A *Venera* orbiter image of a plateau in Ishtar Terra known as Lakshmi Planum. The Maxwell Montes mountain range (red) lies on the western margin of the plain, near the right-hand edge of the image. A meteor crater named Cleopatra is visible on the western slope of the Maxwell range. Note the two larger craters in the center of the plain itself. (b) A *Magellan* image of Cleopatra showing a double-ringed structure that identifies it to geologists as an impact crater. *(NASA)*

(a)

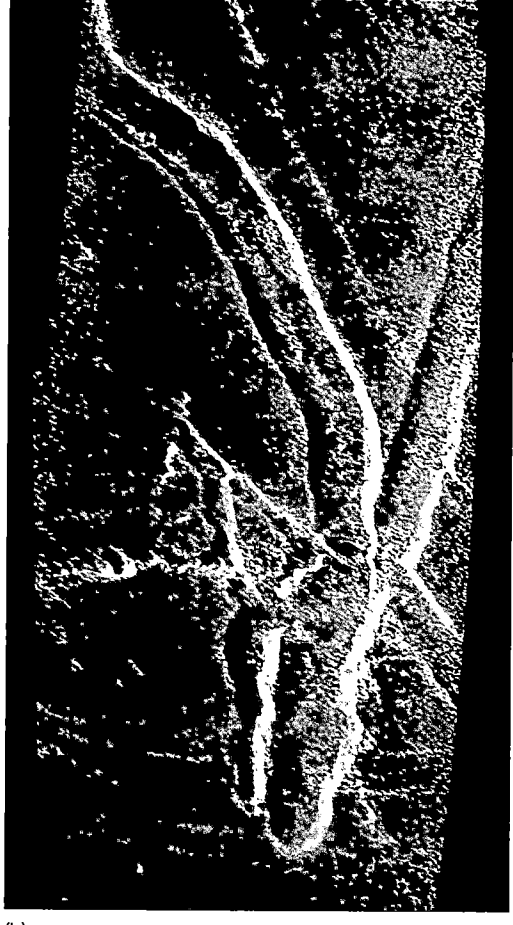

Figure 9.10 Aphrodite Terra (a) A *Magellan* image of Ovda Regio, part of Aphrodite Terra. The intersecting ridges indicate repeated compression and buckling of the surface. The dark areas represent regions that have been flooded by lava upwelling from cracks like those shown in Figure 9.11. (b) This lava channel in Venus's south polar region, known as Lada Terra, extends for nearly 200 km. *(NASA)*

(b)

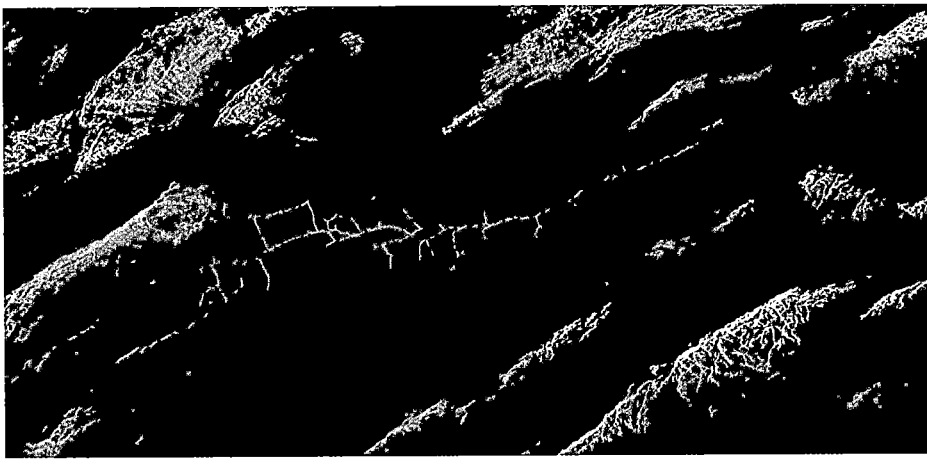

Figure 9.11 Lava Flows These cracks in Venus's surface, detected by *Magellan* in another part of Aphrodite Terra, have allowed lava to reach the surface and flood the surrounding terrain. The dark regions are smooth lava flows. The network of fissures visible here is about 50 km long. *(NASA)*

this terrain is not at all what we would expect at a spreading site similar to the Mid-Atlantic Ridge. ∞ (Sec. 7.5) Although there is no evidence for plate tectonics on Venus, it is likely that the stresses in the crust that led to the large mountain ranges were caused by convective motion within Venus's mantle, the same basic process that drives Earth's plates. ∞ (Sec. 7.5) Lakshmi Planum, for example, is probably the result of a "plume" of upwelling mantle material that raised and buckled the planet's surface.

VOLCANISM AND CRATERING

Many areas of Venus have extensive volcanic features. Figure 9.12 shows a series of seven pancake-shaped lava

domes, each about 25 km across. They probably formed when lava oozed out of the surface, formed the dome, and then withdrew, leaving the crust to crack and subside. Lava domes such as these are found in numerous locations on Venus.

Most volcanoes on the planet are of the type known as **shield volcanoes**. Two large shield volcanoes, called Sif Mons and Gula Mons, are shown in Figure 9.13. Shield volcanoes, such as the Hawaiian Islands on Earth, are not associated with plate boundaries. Instead, they form when lava wells up through a "hot spot" in the crust, and are built up over long periods of time by successive eruptions and lava flows. A characteristic of shield volcanoes is the formation of a *caldera*, or crater, at the summit when the underlying lava withdraws and the surface collapses. The

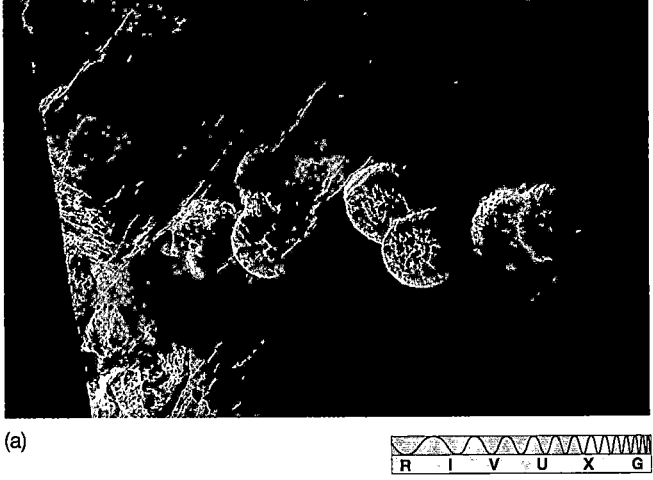

(a)

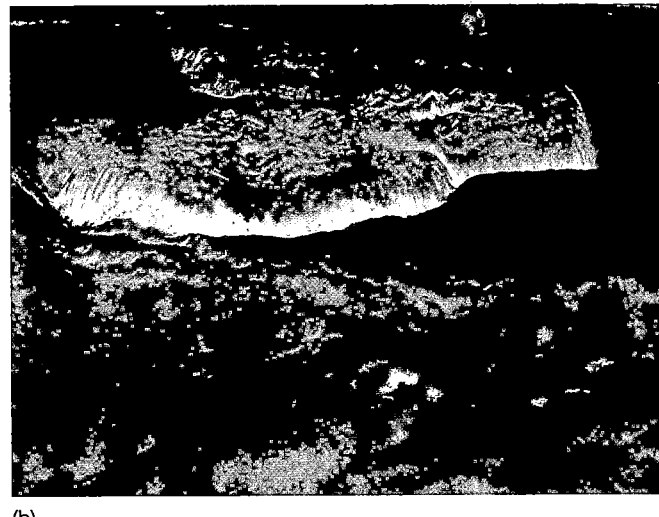

(b)

Figure 9.12 Lava Dome (a) These dome-shaped structures resulted when viscous molten rock bulged out of the ground and then retreated, leaving behind a thin solid crust that subsequently cracked and subsided. *Magellan* has found features like this in several locations on Venus. (b) A three-dimensional representation of four of the domes. The computer view is looking toward the right from near the center of the image in part (a). Color is based on data returned by Soviet *Venera* landers. *(NASA)*

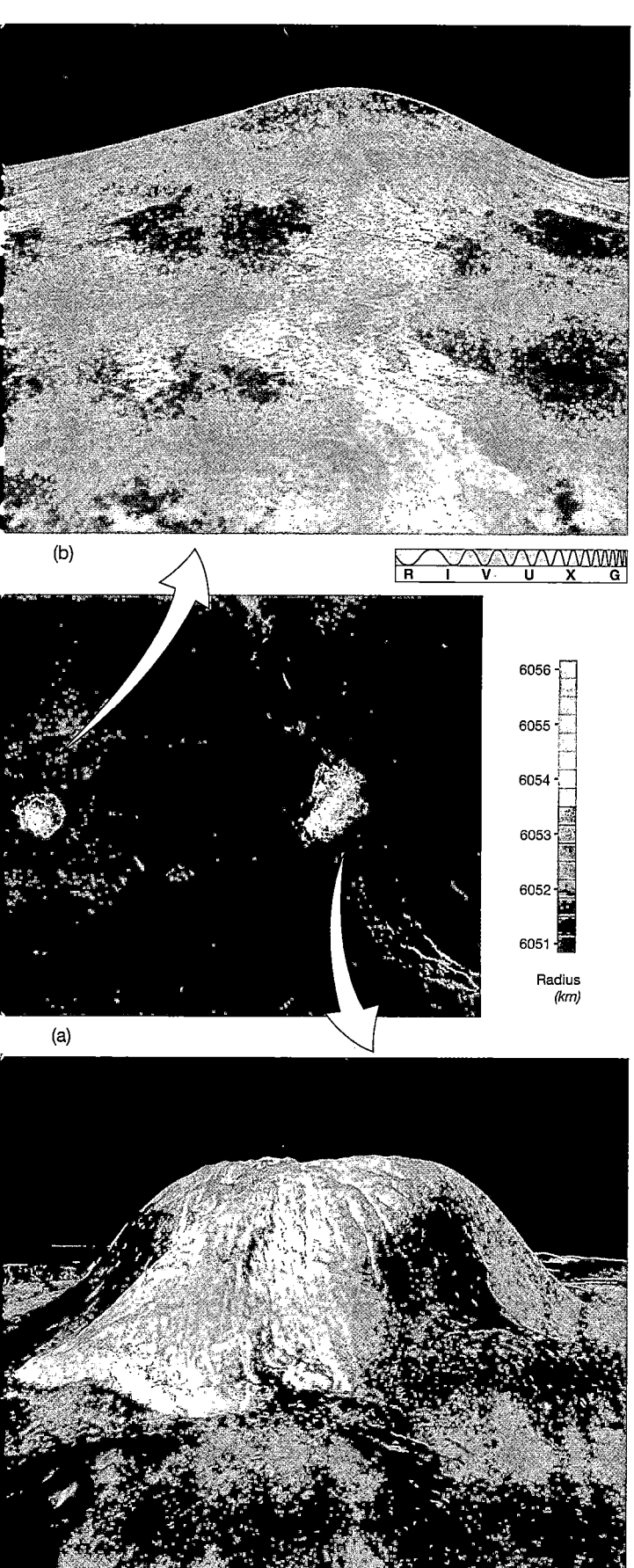

(b)

(a)

(c)

Radius
(km)

Figure 9.13 Venus Volcanism (a) Two larger volcanoes, known as Sif Mons (left) and Gula Mons, appear in this *Magellan* image. Color indicates height above a nominal planetary radius of 6052 km, ranging from purple (−1 km, the level of the surrounding plain) to orange (corresponding to an altitude of about 4 km). The two volcanic calderas at the summits are about 100 km across. (b) A computer-generated view of Sif Mons, as seen from ground level. (c) Gula Mons, as seen from ground level. In (b) and (c), the colors are based on data returned from Soviet landers, and the vertical scales have been greatly exaggerated (by about a factor of 40); Venus is actually a remarkably flat place. *(NASA)*

distribution of volcanoes over the surface of Venus appears random—quite different from the distribution on Earth, where volcanic activity clearly traces out plate boundaries (see Figure 7.9)—strongly supporting the idea that plate tectonics is absent on Venus.

The largest volcanic structures on Venus are huge, roughly circular regions known as **coronae**. A large corona, called Aine, can be seen in Figure 9.14, another large-scale mosaic of *Magellan* images. Coronae are unique to Venus. They appear to have been caused by upwelling mantle material, perhaps similar to the uplift that resulted in Lakshmi Planum but on a somewhat smaller scale. They generally have volcanoes both in and around them, and closer inspection of the rims usually shows evidence for extensive lava flows into the plains below.

There is overwhelming evidence for past surface activity on Venus. Has this activity now stopped, or is it still going on? Two pieces of indirect evidence suggest that volcanism continues today. First, the level of sulfur dioxide above Venus's clouds shows large and fairly frequent fluctuations. It is quite possible that these variations result from volcanic eruptions on the surface. If so, volcanism may be the primary cause of Venus's thick cloud cover. Second, both the *Pioneer Venus* and the *Venera* orbiters observed bursts of radio energy from Aphrodite and other regions of the planet's surface. These bursts are similar to those produced by lightning discharges that often occur in the plumes of erupting volcanoes on Earth, again suggesting ongoing activity. However, while these pieces of evidence are quite persuasive, they are still only circumstantial. No "smoking gun" (or erupting volcano) has yet been seen, so the case for active volcanism is not yet complete.

Not all the craters on Venus are volcanic in origin. Some, like Cleopatra (Figure 9.9b), were formed by meteoritic impact. Large impact craters on Venus are generally circular, but those less than about 15 km in diameter can be quite asymmetric in appearance. Figure 9.15(a) shows a *Magellan* image of a relatively small meteoritic impact crater, about 10 km across, in Venus's southern hemisphere. Geologists believe the light-colored region to be

Figure 9.14 Venus Corona This corona, called Aine, lies in the plains south of Aphrodite Terra; it is about 300 km across. It is probably the result of mantle material that upwelled, causing the surface to bulge outward. Note the pancake-shaped lava domes at top, the many fractures in the crust around the corona, and the large impact craters with their surrounding white (rough) ejecta blankets that stud the region. *(NASA)*

R I V U X G

the ejecta blanket—material ejected from the crater following the impact. The odd shape may be the result of a large meteoroid's breaking up just before impact into pieces that hit the surface near one another. This seems to be a fairly common fate for medium-sized bodies (1 km or so in diameter) that plow through Venus's dense atmosphere. Figure 9.15(b) shows the largest known impact feature on Venus, the 280-km diameter crater Mead. Its double-ringed structure is in many ways similar to the Moon's Mare Orientale (Figure 8.15). Numerous impact craters (identifiable by their ejecta blankets) can also be discerned in Figure 9.14.

Venus's atmosphere is sufficiently thick that small meteoroids do not reach the ground, so there are no impact craters smaller than about 3 km across. Atmospheric effects probably also account for the observed scarcity of impact craters less than 25 km in diameter. Overall, the rate of formation of large-diameter craters on Venus's surface seems to be only about one-tenth that in the lunar maria. Applying the same crater-age estimates to Venus as we do to Earth and the Moon suggests that much of the surface of Venus is quite young—less than a billion years old. Some planetary scientists have suggested that some areas, such as the region shown in Figure 9.13 are even younger—perhaps as little as 200 or 300 million years. Although erosion by the planet's atmosphere may play some part in obliterating surface features, the main agent is volcanism, which appears to have

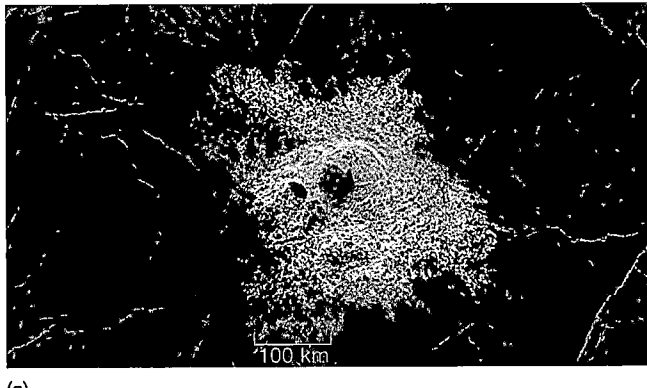

(a)

Figure 9.15 Impact Cratering on Venus (a) A *Magellan* image of an impact crater in Venus's southern hemisphere. The peculiar kidney shape seems to be the result of a meteoroid that fragmented just prior to impact. The dark regions in the crater may be pools of solidified lava. (b) Venus's largest crater, named Mead after anthropologist Margaret Mead, is about 280 km across. Bright (rough) regions clearly show its double-ringed structure. *(NASA)*

R I V U X G

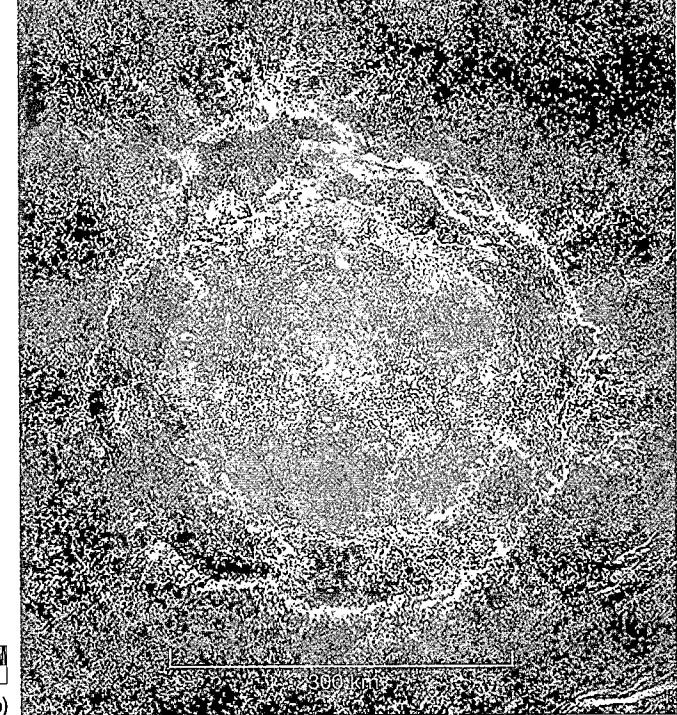

(b)

"resurfaced" much of the planet about 500 hundred million years ago.

DATA FROM THE SOVIET LANDERS

The 1975 soft landings of the Soviet *Venera 9* and *Venera 10* spacecraft directly established that Venus's surface is very dry and dusty. Figure 9.16(a) shows one of the first photographs of the surface of Venus radioed back to Earth. Each craft lasted only about an hour before overheating, their electronic circuitry literally melting in this planetary oven. Typical rocks in the photo measure about 50 cm by 20 cm across—a little like flagstones on Earth. Sharp-edged and slablike, they show little evidence of erosion. Apparently they are quite young rocks, again supporting the idea of ongoing surface activity of some kind.

Later *Venera* missions took more detailed photographs, as shown in Figure 9.16(b). The presence of small rocks and finer material indicates the effects of erosive processes. These later missions also performed simple chemical analyses of the surface of Venus. The samples studied by *Venera 13* and *Venera 14* were predominantly basaltic in nature, again implying a volcanic past. However, not all the rocks were found to be volcanic. The *Venera 17* and *Venera 18* landers also found surface material resembling terrestrial granite, probably (as on Earth) part of the planet's ancient crust.

☑ Concept Check

■ Are volcanoes on Venus mainly associated with the movement of tectonic plates, as on Earth?

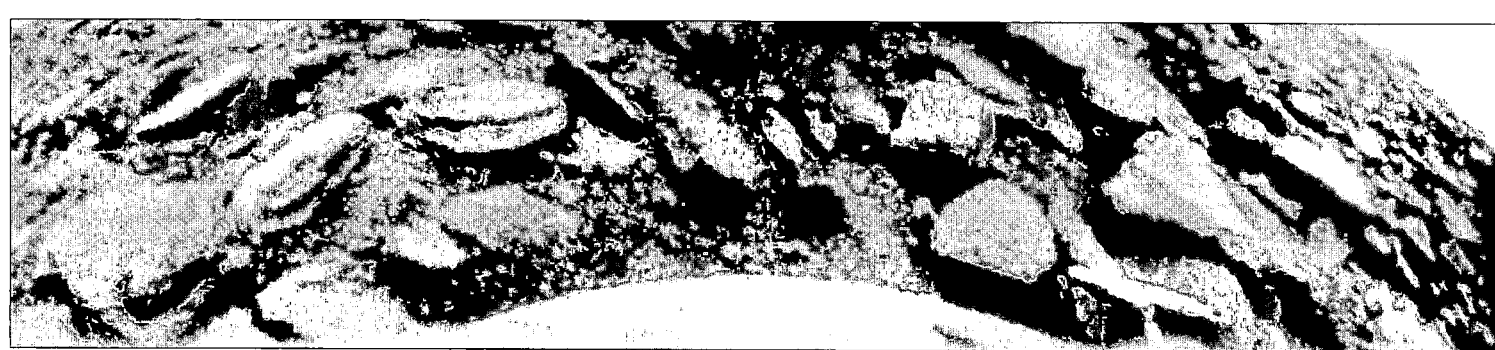

(a)

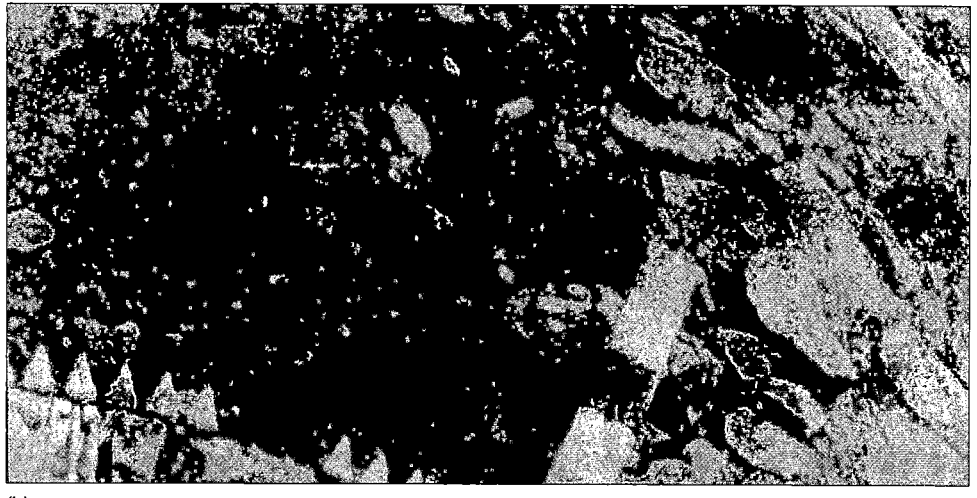

(b)

Figure 9.16 Venus In Situ (a) The first direct view of the surface of Venus, radioed back to Earth from the Soviet *Venera 9* spacecraft, which made a soft landing on the planet in 1975. The amount of sunlight penetrating Venus's cloud cover apparently resembles that on a heavily overcast day on Earth. (b) Another view of Venus, in true color, from *Venera 14*. Flat rocks like those visible in part (a) are present, but there are also many smaller rocks and even fine soil on the surface. This landing site is not far from the *Venera 9* site shown in the previous figure. The peculiar filtering effects of whatever light does penetrate the clouds make Venus's air and ground appear peach-colored—in reality they are most likely gray, like rocks on Earth. *(Russian Space Agency)*

9.5 The Atmosphere of Venus

ATMOSPHERIC STRUCTURE

⚄ Measurements made by the *Venera* and *Pioneer Venus* spacecraft have allowed us to paint a fairly detailed picture of Venus's atmosphere. ∞ (Sec. 6.6) Figure 9.17 shows the run of temperature and pressure with height. Compare this figure with Figure 7.4, which gives similar information for Earth. The atmosphere of Venus is about 90 times more massive than Earth's, and it extends to a much greater height above the surface. On Earth, 90 percent of the atmosphere lies within about 10 km of sea level. On Venus the corresponding (90 percent) level is found at an altitude of 50 km instead. The surface temperature and pressure of Venus's atmosphere are much greater than Earth's. However, the temperature drops more rapidly with altitude, and the upper atmosphere of Venus is actually colder than our own.

Venus's troposphere extends up to an altitude of nearly 100 km. The reflective clouds that block our view of the surface lie between 50 and 70 km above the surface. Data from the *Pioneer Venus* multiprobe indicate that the clouds may actually be separated into three distinct layers within that altitude range. Below the clouds, extending down to an altitude of some 30 km, is a layer of haze. Below 30 km, the air is clear. Above the clouds, a high-speed "jet stream" blows from west to east at about 300–400 km/h, fastest at the equator and slowest at the poles. This high-altitude flow is responsible for the rapidly moving cloud patterns seen in ultraviolet light. Figure 9.18 shows a sequence of three ultraviolet images of Venus in which the variations in the cloud patterns can be seen. Note the characteristic V-shaped appearance of the clouds—a consequence of the fact that, despite their slightly lower speeds, the winds near the poles have a shorter distance to travel in circling the planet and so are always forging ahead of winds at the equator. Near the surface, the dense atmosphere moves more sluggishly—indeed, the fluid flow bears more resemblance to Earth's oceans than to its air. Surface wind speeds are typically less than 2 m/s (roughly 4 mph).

ATMOSPHERIC COMPOSITION

Carbon dioxide is the dominant component of the atmosphere, accounting for 96.5 percent of it by volume. Almost all of the remaining 3.5 percent is nitrogen. Trace amounts of other gases, such as water vapor, carbon monoxide, sulfur dioxide, and argon, are also present. This composition is clearly radically different from Earth's atmosphere. The absence of oxygen is perhaps not surprising, given the absence of life (recall our discussion of Earth's atmosphere in Chapter 7). ∞ (Sec. 7.2) However, there is no sign of the water vapor that we might expect to find if a volume of water equivalent to Earth's oceans had evaporated. If Venus started off with Earthlike composition, something has happened to its water—it is now a very dry planet.

For a long time, the chemical makeup of the reflective cloud layer surrounding Venus was unknown. At first scientists assumed the clouds were water vapor or ice, as on Earth, but the reflectivity of the clouds at different wavelengths didn't match that of water ice. Later infrared observations carried out in the 1970s showed that the clouds (or at least the top layer of clouds) are actually composed of sulfuric acid, created by reactions between water and sulfur dioxide. Sulfur dioxide is an excellent absorber of ultraviolet radiation and could be responsible for many of the cloud patterns seen in ultraviolet light. Spacecraft observations confirmed the presence of these compounds in the atmosphere. They also indicated that there may be particles of sulfur suspended in and near the cloud layers, which may account for Venus's characteristic yellowish hue.

THE GREENHOUSE EFFECT ON VENUS

⑤ Given the distance of Venus from the Sun, the planet was not expected to be such a pressure cooker. As mentioned earlier, calculations based on Venus's orbit and reflectivity indicated a temperature not much different from Earth's, and early measurements of the cloud temperatures seemed to concur. Certainly, scientists reasoned, Venus could be no hotter than the sunward side of Mercury, and it should probably be much cooler. This reasoning was obviously seriously in error.

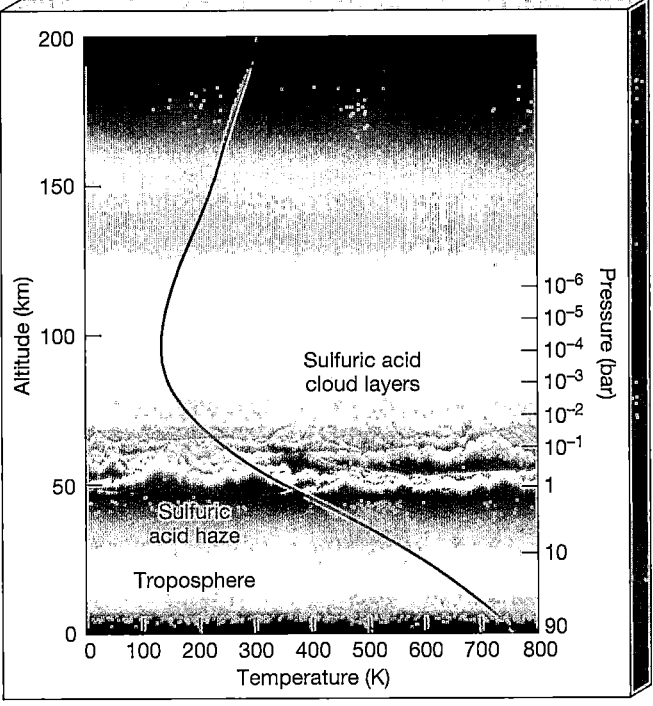

Figure 9.17 Venus's Atmosphere The structure of the atmosphere of Venus, as determined by U.S. and Soviet probes. (One bar is the atmospheric pressure at sea level on Earth.)

Figure 9.18 Atmospheric Circulation Three ultraviolet views of Venus, taken by the *Pioneer Venus* orbiter, showing the changing cloud patterns in the planet's upper atmosphere. The wind flow is in the direction opposite the "V" in the clouds. Notice the motion of the dark region marked by the arrow. Venus's retrograde rotation means that north is at the bottom of these images, and west is to the right. The time difference between the left and right photographs is about 20 hours. *(NASA)*

Why is Venus's atmosphere so hot? And if, as we believe, Venus started off like Earth, why is it now so different? The answer to the first question is fairly easy: Given the present composition of its atmosphere, Venus is hot because of the greenhouse effect. Recall from our discussion in Chapter 7 that "greenhouse gases" in Earth's atmosphere, particularly water vapor and carbon dioxide, serve to trap heat from the Sun. ☞ (Sec. 7.2) By inhibiting the escape of infrared radiation reradiated from Earth's surface, these gases serve to increase the planet's equilibrium temperature, in much the same way as an extra blanket keeps you warm on a cold night. Continuing the analogy a little further, the more blankets you place on the bed, the warmer you will become. Similarly, the more greenhouse gases there are in the atmosphere, the hotter the surface will be.

The same effect naturally occurs on Venus, whose dense atmosphere is made up almost entirely of a primary greenhouse gas, carbon dioxide. As illustrated schematically in Figure 9.19, the thick carbon dioxide blanket absorbs nearly 99 percent of all the infrared radiation released from the surface of Venus, and it is the immediate cause of the planet's sweltering 730 K surface temperature. Furthermore, the temperature is nearly as high at the poles as at the equator, and there is not much difference between the temperatures on the day and night sides. The circulation of the atmosphere spreads energy very efficiently around the planet, making it impossible to escape the blazing heat, even during the planet's two-month-long night.

THE RUNAWAY GREENHOUSE EFFECT

But *why* is Venus's atmosphere so different from Earth's? Why is there so much carbon dioxide in the atmosphere of Venus, and why is the atmosphere so dense? To address these questions, we must consider the processes that created the atmospheres of the terrestrial planets and then determined their evolution. In fact, we can turn the question around and ask instead, "Why is there so *little* carbon dioxide in Earth's atmosphere compared with that of Venus?"

We believe that Earth's atmosphere has evolved greatly since it first appeared. Our planet's secondary atmosphere was outgassed from the interior by volcanic activity 4 billion years ago. ☞ (Sec. 7.2) Since then it has been reprocessed, in part by living organisms, into its present form. On Venus, the initial stages probably took place in more or less the same way, so that at some time in the past, Venus might well have had an atmosphere similar to the primitive secondary atmosphere on Earth, containing water, carbon dioxide, sulfur dioxide, and nitrogen-rich compounds. What happened on Venus to cause such a major divergence from subsequent events on our own planet?

On Earth, nitrogen was released into the air by the action of sunlight on the chemical compounds containing it. Meanwhile, the water condensed into oceans, and much of the carbon dioxide and sulfur dioxide eventually became dissolved in them. Most of the remaining carbon dioxide combined with surface rocks. Thus, much of the secondary outgassed atmosphere quickly became part of the surface of the planet. If all the dissolved or chemically combined carbon dioxide were released back into Earth's present-day atmosphere, its new composition would be 98 percent carbon dioxide and 2 percent nitrogen, and it would have a pressure about 70 times its current value. In other words, apart from the presence of oxygen (which appeared on Earth only after the development of life) and water (the absence of which on Venus will be explained shortly), Earth's atmosphere would be a lot like that of Venus! The real difference between Earth and Venus, then, is that Venus's greenhouse gases never left the atmosphere the way they did on Earth.

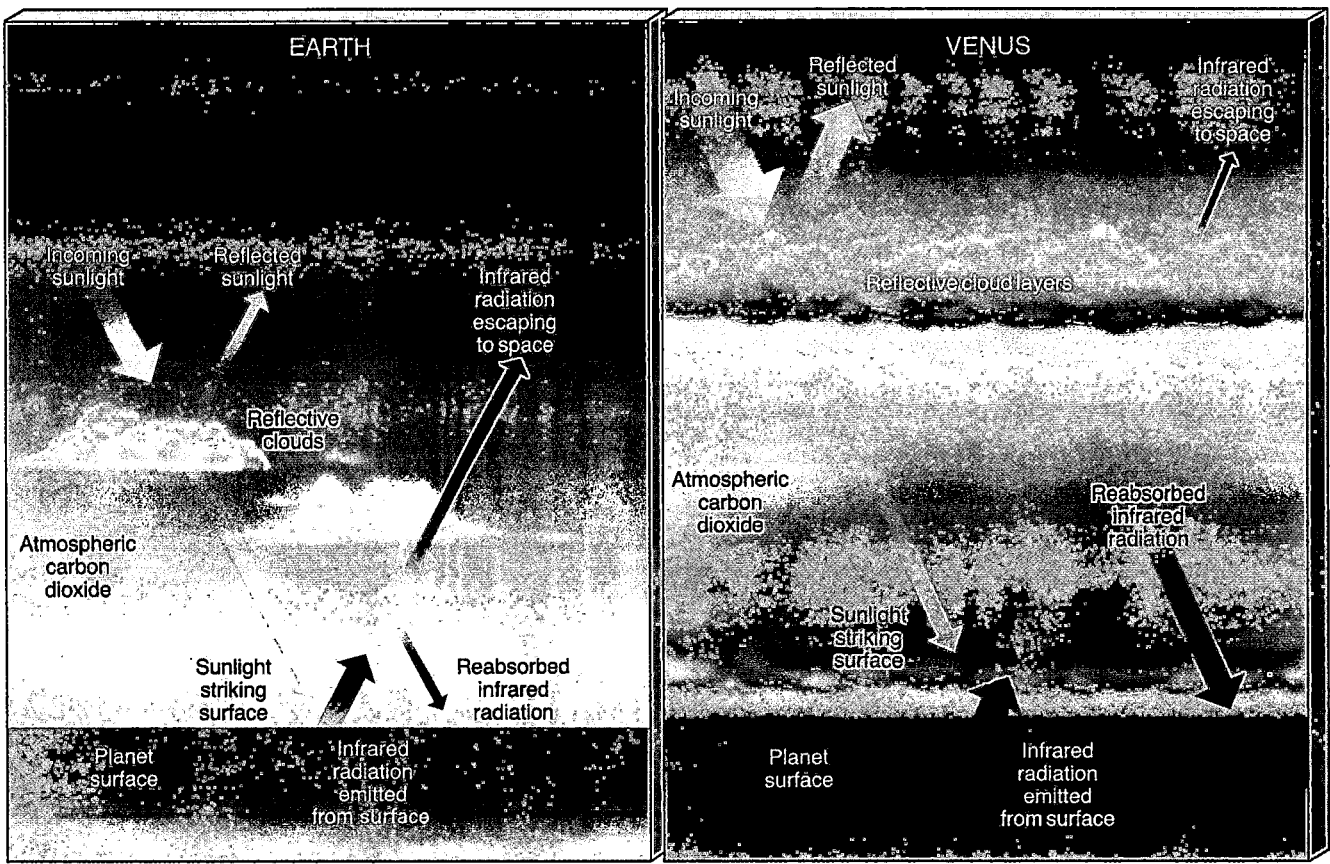

Figure 9.19 Greenhouse Effect on Earth and Venus Because Venus's atmosphere is much thicker and denser than Earth's, a much smaller fraction of the infrared radiation leaving the planet's surface actually escapes into space. The result is a much stronger greenhouse effect than on Earth and a correspondingly hotter planet. The outgoing infrared radiation is not absorbed at a single point in the atmosphere; instead, absorption occurs at all atmospheric levels. (The arrows are meant to indicate only that absorption occurs, not that it occurs at one specific level.)

When Venus's secondary atmosphere appeared, the temperature was higher than on Earth, simply because Venus is closer to the Sun. However, the Sun was probably somewhat dimmer then (see Chapter 22)—perhaps only half its present brightness—so there is some uncertainty as to exactly how much hotter than Earth Venus actually was. If the temperature was already so high that no oceans condensed, the outgassed water vapor and carbon dioxide would have remained in the atmosphere, and the full greenhouse effect would have gone into operation immediately. If oceans did form and most of the greenhouse gases left the atmosphere,* the temperature must still have been sufficiently high that a process known as the **runaway greenhouse effect** came into play.

*In fact, careful study of the Magellan images reveals no sign of ancient seashores or ocean basins, nor evidence of erosion by rivers. However, it is unclear whether such features would have survived the heavy volcanism known to have occurred in the planet's more recent past.

To understand the runaway greenhouse effect, imagine that we took Earth from its present orbit and placed it in Venus's orbit, some 30 percent closer to the Sun. At that distance from the Sun, the amount of sunlight striking Earth's surface would be about twice its present level, so the planet would warm up. More water would evaporate from the oceans, leading to an increase in atmospheric water vapor. At the same time, the ability of both the oceans and surface rocks to hold carbon dioxide would diminish, allowing more carbon dioxide to enter the atmosphere. As a result, the greenhouse heating would increase, and the planet would warm still further, leading to a further increase in atmospheric greenhouse gases, and so on. Once started, the process would "run away," eventually leading to the complete evaporation of the oceans, restoring all the original greenhouse gases to the atmosphere. Although the details are quite complex, basically the same thing would have happened on Venus long ago, ultimately leading to the planetary inferno we see today.

The greenhouse effect on Venus was even more extreme in the past, when the atmosphere also contained water vapor, another greenhouse gas. By intensifying the blanketing effect of the carbon dioxide, the water vapor helped the surface of Venus reach temperatures perhaps twice as hot as at present. At those high temperatures, the water vapor was able to rise high into the planet's upper atmosphere—so high that it was broken up by solar ultraviolet radiation into its components, hydrogen and oxygen. The light hydrogen rapidly escaped, the reactive oxygen quickly combined with other atmospheric gases, and all water on Venus was lost forever.

Although it is highly unlikely that global warming will ever send Earth down the path taken by Venus, this episode highlights the relative fragility of the planetary environment. No one knows how close to the Sun Earth could have formed before a runaway greenhouse effect would have occurred. But in comparing our planet with Venus, we have come to understand that there is an orbital limit, presumably between 0.7 and 1.0 A.U., inside of which Earth would have suffered a similar catastrophic runaway. We must consider this "greenhouse limit" when assessing the likelihood that planets harboring life formed elsewhere in our Galaxy.

☑ Concept Check

■ If Venus had formed at Earth's distance from the Sun, what might its climate be like today?

9.6 Venus's Magnetic Field and Internal Structure

� In 1962, *Mariner 2* flew by Venus, carrying, among other instruments, magnetometers to measure the strength of the planet's magnetic field. None was detected, and subsequent Soviet and U.S. missions, carrying more sensitive detectors, have confirmed this finding. Venus, with an average density similar to Earth's, probably has a similar overall composition and a partially molten iron-rich core. The lack of any detectable magnetic field on Venus, then, is almost surely the result of the planet's extremely slow rotation and consequent lack of dynamo action. ∞ (Sec. 7.5) Having no magnetosphere, Venus has no protection from the solar wind. Its upper atmosphere is continually bombarded by high-energy particles from the Sun, keeping the topmost layers permanently ionized. However, the great thickness of the atmosphere prevents any of these particles from reaching the surface.

None of the *Venera* landers carried seismic equipment, so no direct measurements of the planet's interior have been made, and theoretical models of the interior have little hard data to constrain them. However, to many geologists the surface of Venus resembles the young Earth, at an age of perhaps a billion years. At that time, volcanic activity had begun, but the crust was still relatively thin and the convective processes in the mantle that drive plate tectonic motion were not yet fully established. Measurements of the planet's gravitational field suggest that the planet lacks an asthenosphere, the semisolid part of the upper mantle over which Earth's lithosphere slides. ∞ (Sec. 7.4)

Why has Venus remained in that immature state and not developed plate tectonics like Earth? That question remains to be answered. Some planetary geologists have speculated that the high surface temperature on Venus has inhibited the planet's evolution by slowing the planet's cooling. Possibly the high surface temperature has made the crust too soft for Earth-style plates to develop. Or perhaps the high temperature and soft crust have led to more volcanism, tapping the energy that might otherwise go into convective motion. It may also be that the presence of water plays an important role in lubricating mantle convection and plate motion, so that arid Venus could not evolve along the same path as Earth.

☑ Concept Check

■ If the interior of Venus is quite Earth-like, and Venus has a molten iron core, why doesn't it have a magnetic field like Earth?

Chapter Review

SUMMARY

The interior orbit of Venus means that it never strays far from the Sun in the sky. Because of its highly reflective cloud cover, Venus is brighter than any star in the sky, as seen from Earth. It is so bright that it can be seen even in the daytime.

The extremely thick atmosphere of Venus is nearly opaque to visible radiation, making the planet's surface invisible from the outside. Spectroscopic examination of sunlight reflected from the planet's cloud tops shows the presence of large amounts of carbon dioxide. Venus's atmosphere is nearly 100 times denser than Earth's. The temperature of the upper atmosphere is much like that of Earth's upper atmosphere, but the surface temperature is 730 K. The planet's rotation is slow and retrograde, most likely because of a collision between Venus and some other solar system body during the late stages of the planet's formation.

Venus's surface has been thoroughly mapped by radar from Earth-based radio telescopes and orbiting satellites. The most recent and most thorough survey has been carried out by the U.S. *Magellan* satellite. The planet's surface is mostly smooth, resembling rolling plains with modest highlands and lowlands. Two elevated continent-sized regions are called Ishtar Terra and Aphrodite Terra. There is no evidence for plate tectonic activity as on Earth.

Many **lava domes** (p. 236) and **shield volcanoes** (p. 236) have been found by *Magellan* on Venus's surface, but none of the volcanoes has yet been proven to be currently active. The planet's surface shows no sign of plate tectonics. Features called **coronae** (p. 237) are thought to have been caused by an upwelling of mantle material that, for unknown reasons, never developed into full convective motion. The surface of the planet appears to be relatively young, resurfaced by volcanism every few hundred million years. Some craters on Venus are due to meteoritic impact, but the majority are volcanic in origin. The evidence for currently active volcanoes on Venus includes surface features resembling those produced in Earthly volcanism, fluctuating levels of sulfur dioxide in Venus's atmosphere, and bursts of radio energy similar to those produced by lightning discharges that often occur in the plumes of erupting volcanoes on Earth. However, no actual eruptions have been seen.

Soviet spacecraft that landed on Venus photographed surface rocks with sharp edges and a slablike character. Some rocks on Venus appear predominantly basaltic in nature, implying a volcanic past. Other rocks resemble terrestrial granite and are probably part of the planet's ancient crust.

Venus is comparable in both mass and radius to Earth, suggesting that the two planets started off with fairly similar surface conditions. However, the atmospheres of Earth and Venus are now very different. The total mass of Venus's atmosphere is about 90 times greater than Earth's. The greenhouse effect stemming from the large amount of carbon dioxide in Venus's atmosphere is the basic cause of the planet's current high temperatures. Almost all the water vapor and carbon dioxide initially present in Earth's early atmosphere quickly became part of the oceans or surface rocks. Because Venus orbits closer to the Sun than does Earth, surface temperatures were initially higher, and the planet's greenhouse gases never left the atmosphere. On Venus, the **runaway greenhouse effect** (p. 242) caused all the planet's greenhouse gases—carbon dioxide and water vapor—to end up in the atmosphere, leading to the extreme conditions we observe today.

Venus has no detectable magnetic field, almost certainly because the planet's rotation is too slow for any appreciable dynamo effect to occur. To some planetary geologists, Venus's interior structure suggests that of the young Earth, before convection became established in the mantle.

SELF-TEST: TRUE OR FALSE?

_____ **1.** Venus appears brighter in the sky than any star.

_____ **2.** Venus has a retrograde orbit around the Sun.

_____ **3.** Venus is brightest when it is in its full phase.

_____ **4.** Although Venus is slightly smaller than Earth, its atmosphere is more extensive.

_____ **5.** The average surface temperature of Venus is about 260 K.

_____ **6.** A few large surface features on Venus can be seen in Earth-based images made in the ultraviolet part of the spectrum.

_____ **7.** Water vapor is abundant in the atmosphere of Venus.

_____ **8.** Other than lacking oxygen, Venus's atmosphere is quite similar to that of Earth.

_____ **9.** Venus has roughly the same temperature at its equator as at its poles.

_____ **10.** The surface of Venus is relatively rough, compared with the surface of Earth, with higher highs and lower lows.

_____ **11.** Images from *Magellan* show no sign of tectonic activity on Venus.

_____ **12.** Lava flows are common on the surface of Venus.

_____ **13.** There is strong circumstantial evidence that active volcanism continues on Venus.

_____ **14.** Venus has a magnetic field similar to that of Earth.

_____ **15.** Soviet spacecraft are still sending back data from the surface of Venus.

SELF-TEST: FILL IN THE BLANK

1. Venus's mass has now been well determined through the use of _____ orbiting it.
2. Because Venus has a mass and _____ only slightly lower than Earth's, we might expect that its internal structure and evolution should be Earth-like.
3. Venus's rotation is unusual because it is _____.
4. The most abundant gas in the atmosphere of Venus is _____.
5. The top cloud layers of Venus are composed of _____ droplets.
6. The process that makes Venus so hot is known as the _____.
7. The surface of Venus has been mapped using _____.
8. Ishtar Terra and Aphrodite Terra are two _____ on the surface of Venus.
9. Most craters on the surface of Venus are the result of _____.
10. _____ are huge, roughly circular volcanic regions on the surface of Venus.
11. A meteoroid less than about 1 km in diameter will most likely _____ as it passes through Venus's atmosphere.
12. The scarcity of large impact craters on the surface of Venus indicates that its surface is quite _____.
13. The surface of Venus appears to have been resurfaced by _____ a few hundred million years ago.
14. The main difficulties in using landers to study Venus's surface are the planet's extremely high _____ and _____.
15. This greenhouse effect in the early atmosphere of Venus was most likely intensified by the presence of _____.

REVIEW AND DISCUSSION

1. Why does Venus appear so bright to the eye? Upon what factors does the planet's brightness depend?
2. Explain why Venus is always found in the same general part of the sky as the Sun.
3. Why do astronomers think that the "near resonance" described in the text (between Venus's rotation and revolution, as seen from Earth) is not a true resonance?
4. Describe one observational problem associated with this "near-resonance," real or not.
5. What is our current best explanation of Venus's slow, retrograde spin?
6. If you were standing on Venus, how would Earth look?
7. How did radio observations of Venus made in the 1950s change our conception of the planet?
8. What did ultraviolet images returned by *Pioneer Venus* show about the planet's high-level clouds?
9. Name three ways in which the atmosphere of Venus differs from that of Earth.
10. What are the main constituents of Venus's atmosphere? What are clouds in the upper atmosphere made of?
11. What component of Venus's atmosphere causes Venus to be so hot? Explain why there is so much of this gas in the atmosphere of Venus, compared with Earth's. What happened to all the water that Venus must have had when formed?
12. Earth and Venus are nearly alike in size and density. What primary fact caused one planet to evolve as an oasis for life, while the other became a dry and inhospitable inferno?
13. If Venus had formed at Earth's distance from the Sun, what do you imagine its climate would be like today? Why do you think this?
14. How do the "continents" of Venus differ from Earth continents?
15. How are the impact craters of Venus different from those found on other bodies?
16. What evidence exists that volcanism of various types has changed the surface of Venus?
17. What is the evidence for active volcanoes on Venus?
18. Given that Venus, like Earth, probably has a partially molten iron-rich core, why doesn't it also have a magnetic field?
19. Do you think there might be life on Venus? Explain your answer.
20. Do you think that Earth is in any danger of experiencing a runaway greenhouse effect like Venus?

PROBLEMS  *Algorithmic versions of these questions are available in the Practice Problems module of the Companion Website.*

The number of squares preceding each problem indicates its approximate level of difficulty.

1. ■■ Using the data given in the text, calculate Venus's angular diameter, as seen by an observer on Earth, when the planet is (a) at its brightest, (b) at greatest elongation, and (c) at the most distant point in its orbit.
2. ■■ Seen from Earth, through how many degrees per night (relative to the stars) does Venus move around the time of inferior conjunction (closest approach to Earth)?
3. ■ How long does a radar signal take to travel from Earth to Venus and back when Venus is brightest? Compare this with the round-trip time when Venus is at its closest point to Earth.
4. ■■ What would be the length of a solar day on Venus if the planet's rotation were prograde rather than retrograde?
5. ■■■ Calculate the mean angular orbital speeds of Venus and Mercury, in degrees per day (for simplicity, imagine that the planets have circular orbits). Based on these speeds, how long does it take for Mercury to "lap" Venus—that is, to complete exactly one extra revolution around the Sun? This is the synodic period of Mercury, as seen from Venus.
6. ■ Draw a diagram showing the positions of Earth and Venus on their orbits over the course of one Venus

synodic year (584 Earth days), starting at the point of closest approach of the planets. Mark the locations of the planets at intervals of 73 days, and indicate with an arrow (as in Figure 9.4) the orientation of some point on Venus's surface at each instant.

7. ■■ Compare the magnitude of the tidal gravitational acceleration on Venus (between center and surface) due to Earth at closest approach with Venus's surface gravitational acceleration. Assume circular orbits for both planets. Repeat the question for the tidal acceleration on Venus due to the Sun. What do you conclude about the possibility that Venus's near-resonance with Earth is the result of Earth's tidal influence?

8. ■ What is the size of the smallest feature that could be distinguished on the surface of Venus (at closest approach) by the Arecibo radio telescope at an angular resolution of 1'?

9. ■ Could an infrared telescope with an angular resolution of 0.1" distinguish impact craters on the surface of Venus?

10. ■ When used as a radar instrument, the Arecibo installation can distinguish echoes received as little as 10^{-5} s apart. To what distance does this correspond? This is the effective resolution of Arecibo when making radar observations of Venus.

11. ■ *Pioneer Venus* observed high-level clouds moving around Venus's equator in four days. What was their speed in km/h? In mph?

12. ■ Approximating Venus's atmosphere as a layer of gas 50 km thick, with uniform density 21 kg/m^3, calculate its total mass. Compare your answer with the mass of Earth's atmosphere (Chapter 7, Problem 3), and with the mass of Venus.

13. ■ According to Stefan's law (see Section 3.4), how much more radiation—per square meter, say—is emitted by Venus's surface at 730 K than is emitted by Earth's surface at 300 K?

14. ■■■ In the absence of any greenhouse effect, Venus's average surface temperature, like Earth's, would be about 250 K. In fact, it is about 730 K. Use this information and Stefan's law to estimate the fraction of infrared radiation leaving Venus's surface that is absorbed by carbon dioxide in the planet's atmosphere.

15. ■■ Calculate the orbital period of the *Magellan* spacecraft, moving around Venus on an elliptical orbit with a minimum altitude of 294 km and a maximum altitude of 8543 km above the planet's surface. In 1993 the spacecraft's orbit was changed to have minimum and maximum altitudes of 180 km and 541 km, respectively. What was the new period?

COLLABORATIVE EXERCISES

1. Interplanetary Travel. Traveling at the spacecraft's speed of your group's choosing, calculate the minimum length of time it would take to travel from Earth to Venus.

2. Terminating a Spacecraft. In 1994, NASA elected to "turn-off" the *Magellan Orbiter*, even though it was working perfectly and still sending back valuable data. Does your group agree or disagree with NASA's decision? Explain your reasoning.

3. Venus Surface Features. It is now conventional to name surface features on Venus after famous females. As a group, decide what to name the next five surface features on Venus and justify your selection.

RESEARCHING ON THE WEB

To complete the following exercises, go to the online Destinations module for Chapter 9 on the Companion Website for Astronomy Today *4/e.*

1. Access the "Venus Fact Sheet" page from NASA and determine the range of wind speeds on the Venutian surface.

2. Access the "Venus Nomenclature" page and list the names and diameters for three craters of your choosing.

3. Access the "*Pioneer Venus* Missions" page and describe the five principle components of the *Pioneer Venus* spacecraft.

PROJECTS

1. Is Venus in the morning or evening sky right now? Look for it every few days, over the course of several weeks. Draw a picture of the planet with respect to foreground trees or buildings. If you always observe at the same time every day, you may begin to notice that the planet is getting higher or lower in the sky.

2. Consult an almanac to determine the next time Venus will pass between Earth and Sun. How many days before and after this event can you glimpse the planet with the eye alone?

3. Consult the almanac again to find out the next time Venus will pass on the far side of the Sun from Earth. How many days before and after this event can you see the planet with the naked eye?

4. When Venus ornaments the predawn sky, try keeping track of the planet with your eye alone until it appears in a blue sky, after sunrise. As always, be careful not to look at the Sun!

5. Using a powerful pair of binoculars or a small telescope, examine Venus as it goes through its phases. Note the phase and the relative size of it. (You can compare its size to the field of view in a telescope; always use the same eyepiece for this.) Look at it every few days or once a week. Make a table of the shape of the phase, the size, and the relative brightness to the naked eye. After you have observed it through a significant change in phase, can you see the correlations between these three properties first recognized by Galileo?

SKYCHART III PROJECTS  *The SkyChart III Student Version planetarium program on which these exercises are based is included as a separately executable program on the CD in the back of this text.*

1. ■ Center the display on Venus and zoom in to a 1/20° field of view. Select *DRAW/Sky Background/Color* with *DRAW/ Horizon Mask* deselected. With color selected for the sky, it will be obvious when the sky is too bright to view Venus, and with *Horizon Mask* turned off, Venus will remain visible even when it should be below the horizon. Set time to February 1, 2002. If you cannot juggle the following, then skip forward to the ALTERNATIVE PROCEDURE:

Reduce the size of the window you have just opened so that it takes no more than one fourth of your viewing screen. This is accomplished by clicking on the center of the two boxes (the icon is a box) in the upper right-hand corner of your screen. Be careful: If you inadvertently click on the outermost icon, the *X*, the screen will close and it will be necessary to start again. Then either click on the SkyChart III icon on the desktop, or use *START/Programs* to locate the software, and open a second window running SkyChart III. In similar fashion, reduce the size of this window so that both images are available simultaneously on your screen.

In what follows, attempt to synchronize the running of the two animations. In the second window, center on the Sun and use a field of view of 180°. Under *DRAW*, deselect everything except *Planets, Chart Legend, Scroll Bars, Automatic Limits,* and *Use Color.* Select *DRAW/Sky Background/Black.* You should have a black screen with only the Sun, planets, Moon, and asteroids presented. If you are unfamiliar with the symbols for the planets, click on each one to open the *Object Info* box. You will wish to direct your attention to only the Sun and Venus. Set a trail for Venus with *ANIMATION/Trail For/Venus.* Change the date to also simulate conditions for February 1, 2002, and again, center on the Sun. Reduce screen size to share the screen with the other open window. Set *ANIMATION/1 Week* in each window. Using *CONTROL +A*, start the animation in one window, then immediately click on the second window. When the two dates are synchronized, start the animation in the second window with *CONTROL +A.* You will be able to see the phases of Venus in the first window and its position in orbit in the second window. For more careful study, stop the animations, set time and date the same in both windows, and use F5 and F6 to step through the time changes.

Alternative Procedure Set SkyChart III for February 1, 2002 and center display on Venus, using a 180° field of view

and animation steps of one week. Turn off grid lines, horizon mask, deep space objects and stars. Select *DRAW/Sky Background/Black.* Use *Page Up* on the keyboard to zoom in until the disc of Venus is easily discernible and take note of the phase of the planet. Zoom out to a 180° field of view and step forward in time with F6. Every couple of weeks, zoom in again to observe the change of phase. Correlate the change of phase with position of the Sun. Continue this observation until at least November 1, 2002.

2. ■■ The phases of Venus combined with its brightness were instrumental in arguing for a solar-centered universe. Plot the magnitude of Venus and its size as a function of time. This is most easily done by extracting the ephemeris data (by selecting *FILE/Export/Ephemeris as Text/Select* and entering "Moon", then *Save As "moon.txt"*). Compare with a similar graph for our Moon. Notice that the Moon is orbiting around Earth. Using both graphs, argue that Galileo's explanation makes the most sense.

3. ■■ To determine the size of the solar system, that is, the distance between the planets and the Sun, one needs to measure the astronomical unit (A.U.). This is a task now readily done by bouncing radar off Venus and determining the time it takes for the signal to return to Earth. With the help of Kepler's third law, one can then determine the A.U. (see Chapter 1). The first accurate determination was based on the parallax of Venus during its solar transits of 1761 and 1769. Determine the dates and approximate transit times of Venus. Print out charts that show the location of Venus just after it is between you and the Sun from two extreme positions: the North Pole and the South Pole. Notice the small change of the position of Venus with respect to the Sun. In principle, this allows you to determine the parallax of Venus. Combined with the known distance between the North and South Poles, one can determine the distance to Venus.

4. ■■ It turns out that calculating the distance is not straightforward using Venus's parallax. Instead, measure the parallax of the Sun with respect to the background stars on March 21 from two extreme positions: the North Pole and the South Pole. On this date, the Sun is right above the equator. The baseline is the distance between the poles: 13,000 km. The formula for the distance is: d = baseline/tan Θ, where Θ is the angular change in position with respect to the background stars.

 *In addition to the Practice Problems and Destinations modules, the Companion Website at http://www.prenhall.com/chaisson provides for each chapter an additional true-false, multiple choice, and labeling quiz, as well as additional annotated images, animations, and links to related Websites.*

10 MARS

A Near Miss for Life?

LEARNING GOALS

Studying this chapter will enable you to:

1 Summarize the general orbital and physical properties of Mars.

2 Describe the observational evidence for seasonal changes on Mars.

3 Compare the surface features and geology of Mars with those of the Moon and Earth, and account for these characteristics in terms of Martian history.

4 Discuss the evidence that Mars once had a much denser atmosphere and running water on its surface.

5 Compare the atmosphere of Mars with those of Earth and Venus, and explain why the evolutionary histories of these three worlds diverged so sharply.

6 Describe the characteristics of Mars's moons and explain their probable origin.

 Visit http://www.prenhall.com/chaisson for additional images, animations, and links to related sites for this chapter.

In late 2000, the *Mars Global Surveyor* spacecraft took this remarkable image of what seems to be extensive layering of sand and rock near the Mariner Valley. The view measures only 1.5 km on a side; individual layers are about 10 m thick. At issue is the cause of all this layering, given that the Martian surface is now as dry as any desert on Earth. Were the layers caused by water flowing billions of years ago, when the Martian climate was perhaps warmer, causing sediments to accumulate in ancient lakes? Or was it wind action that carved them over eons of time by transporting massive amounts of dust to form sedimentary strata without the need for water at all? (NASA)

The Big Picture: The search for life on Mars continues unabated; it is one of the oldest, unresolved scientific mysteries. One key ingredient for life anywhere in the universe is probably water, therefore much of the current controversy centers around whether or not Mars was ever much wetter than it is today. Perhaps the issue will be settled only with the arrival of sophisticated robots, or human astronauts, capable of testing the soil directly.

Named by the ancient Romans for their bloody god of war, Mars is for many people the most intriguing of all celestial objects. Over the years it has inspired speculation that life—perhaps intelligent, and possibly hostile—may exist there. With the dawn of the Space Age, those notions had to be abandoned. Visits by robot spacecraft have revealed no signs of life of any sort, even at the microbial level, on Mars. Even so, the planet's properties are close enough to those of Earth that Mars is still widely regarded as the next most hospitable environment for the appearance of life in the solar system, after Earth itself. At about the same time as Earth's "twin," Venus, was evolving into a searing inferno, the Mars of long ago may have had running water and blue skies. If life ever arose there, however, it must be long extinct. The Mars of today appears to be a dry, dead world.

10.1 Orbital Properties

⬛ Mars is the fourth planet from the Sun and the outermost of the four terrestrial worlds in the solar system. It lies outside Earth's orbit, as illustrated in Figure 10.1(a), which shows the orbits of both planets drawn to scale. Because of its exterior orbit, Mars ranges in our sky from a position close to the Sun (for example, when Earth and Mars are at the points marked A in the figure) to one far from the Sun (with the planets at points B). Contrast this

orbit with the nighttime appearances of Mercury and Venus, whose interior orbits ensure that we never see them far from the Sun. From our earthly viewpoint, Mars appears to traverse a great circle in the sky, keeping close to the ecliptic and occasionally executing retrograde loops. ⬿ (Sec. 2.2) The Mars Data box on p. 257 lists some detailed orbital and physical data on the planet.

Mars's orbital eccentricity is 0.093, much larger than that of most other planets—only the innermost and the outermost planets, Mercury and Pluto, have more elongated orbits. Because of this, Mars's perihelion distance from

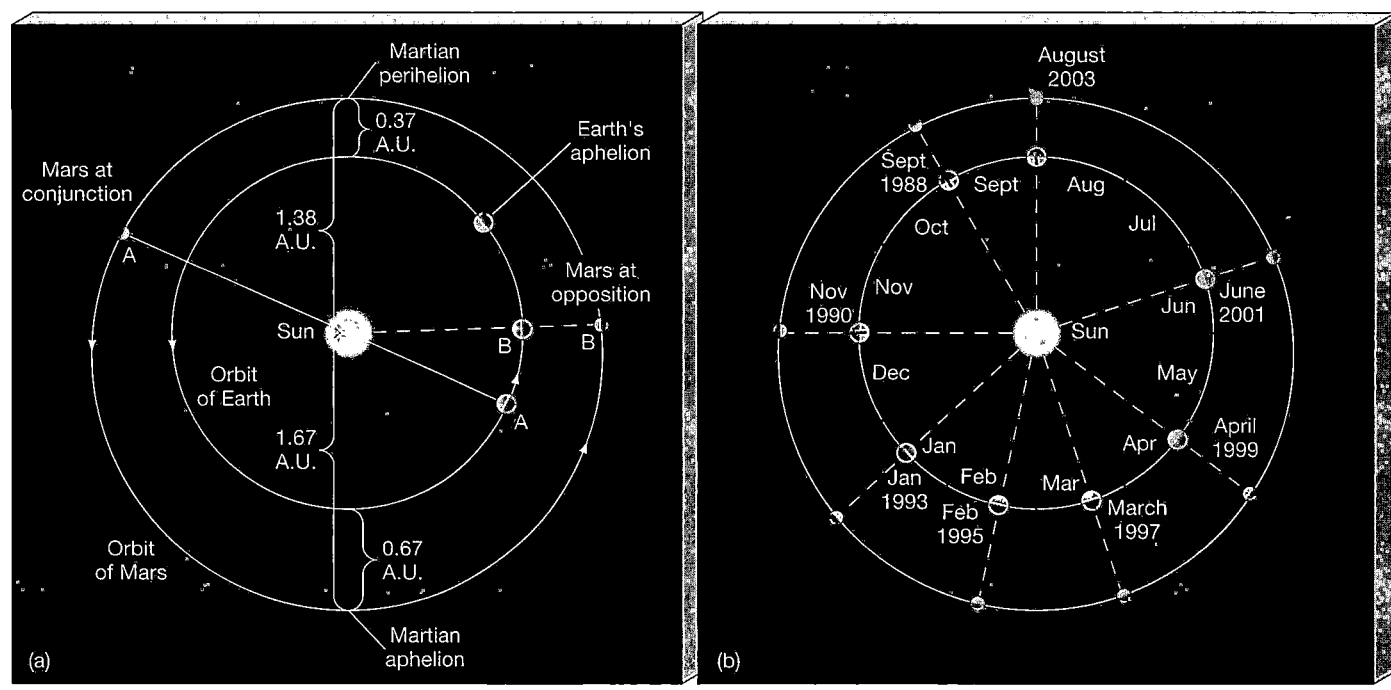

Figure 10.1 Mars Orbit (a) The orbit of Mars compared with that of Earth. Notice that Mars's orbit is noticeably elliptical, unlike Earth's, whose eccentricity is barely perceptible here. When the planets are on opposite sides of the Sun, as at the points marked A, Mars is said to be at conjunction. The planets are at their closest at opposition, when Earth and Mars are aligned and on the same side of the Sun, as at the points marked B. (Note that to get from point A to point B, Earth must travel for nearly 13 months—all the way around its orbit and then some.) (b) Several oppositions of Mars, including the particularly favorable (close) configurations of September 1988 and August 2003 and the unfavorable oppositions of February 1995 and March 1997.

the Sun—1.38 A.U. (207 million km)—is substantially smaller than its aphelion distance—1.67 A.U. (249 million km)—resulting in a large variation in the amount of sunlight striking the planet over the course of its year. In fact, the intensity of sunlight on the Martian surface is almost 45 percent greater when the planet is at perihelion than when it is at aphelion. As we will see, this has a substantial effect on the Martian climate.

Mars is at its largest and brightest in the night sky when it is at *opposition*—that is, when Earth lies between Mars and the Sun (location B on Figure 10.1). If this happens to occur at Martian perihelion, the two planets can come as close as 0.37 A.U. (56 million km). The separation is less than 0.38 A.U. because Earth's orbit is slightly eccentric, and our planet actually lies about 1.01 A.U. from the Sun when such an opposition occurs (in early September; Earth's aphelion occurs in early July). The angular size of Mars under those circumstances is about 25″. Ground-based observations of the planet at those times can distinguish surface features as small as 100 km across—about the same resolution as the unaided human eye can achieve when viewing the Moon.

Such a coincidence of opposition with Martian perihelion is relatively rare, however. It last occurred on September 29, 1988, when the two planets came within 0.39 A.U. of one another. It will next occur on August 30, 2003, very close to Martian perihelion, and Mars will be only 0.37 A.U. distant from Earth. The dates and configurations of some oppositions of Mars are illustrated in Figure 10.1(b), which shows the locations of Earth and Mars at eight oppositions between September 1988 and August 2003. Oppositions occur at roughly 780-day intervals—Mars's synodic period, with corrections for the fact that, in accordance with Kepler's second law, the planets do not move at constant speeds around their orbits. ∞ (Sec. 2.5)

Although Mars is quite bright and easily seen at opposition, the planet is still considerably fainter than Venus. This faintness results from a combination of three factors. First, Mars is more than twice as far from the Sun as is Venus, so each square meter on the Martian surface receives less than one-quarter the amount of sunlight that strikes each square meter on Venus. Second, the surface area of Mars is only about 30 percent that of Venus, so there are fewer square meters to intercept the sunlight. Finally, Mars is much less reflective than Venus—only about 15 percent of the sunlight striking the planet is reflected back into space, compared with nearly 70 percent in the case of Venus. Still, at its brightest, Mars is brighter than any star. Its characteristic red color, visible even to the naked eye, makes it easily identifiable in the night sky.

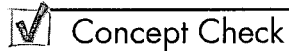 **Concept Check**

■ Why do the closest views of Mars from Earth occur roughly only once every 15 years?

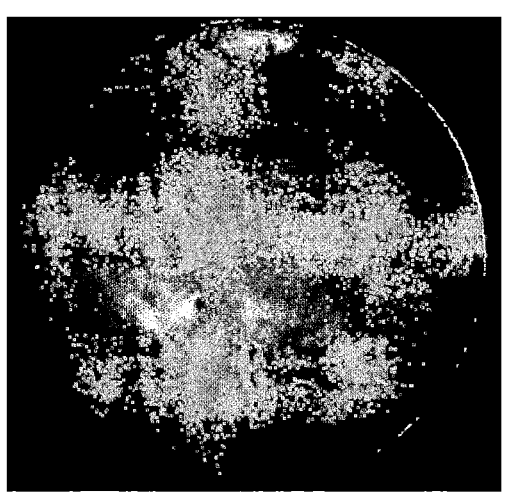

Orbital semimajor axis	1.52 A.U. 227.9 million km
Orbital eccentricity	0.093
Perihelion	1.38 A.U. 206.6 million km
Aphelion	1.67 A.U. 249.2 million km
Mean orbital speed	24.1 km/s
Sidereal orbital period	686.9 solar days 1.881 tropical years
Synodic orbital period	779.9 solar days
Orbital inclination to the ecliptic	1.85°
Greatest angular diameter, as seen from Earth	24.5″
Mass	6.42×10^{23} kg 0.11 (Earth = 1)
Equatorial radius	3394 km 0.53 (Earth = 1)
Mean density	3930 kg/m³ 0.71 (Earth = 1)
Surface gravity	3.72 m/s² 0.38 (Earth = 1)
Escape speed	5.0 km/s
Sidereal rotation period	1.026 solar days $24^h\ 37^m$
Axial tilt	23.98°
Surface magnetic field	approximately 1/800 (Earth = 1)
Magnetic axis tilt relative to rotation axis	—
Mean surface temperature	210 K (range 150–310 K)
Number of moons	2

10.2 Physical Properties

As with Mercury and Venus, we can determine the radius of Mars by means of simple geometry. From the data given earlier for the planet's size and distance, we obtain a radius of about 3400 km. More accurate measurements give a result of 3394 km, or 0.53 Earth radii.

Unlike Mercury and Venus, Mars has two small moons in orbit around it; they are visible (through telescopes) from Earth. Named Phobos (Fear) and Deimos (Panic) for the sons of Ares (the Greek name for the war god known to the Romans as Mars) and Aphrodite (the Greek name for Venus, goddess of love), these moons are little more than large rocks trapped by the planet's gravity. We will return to their individual properties at the end of this chapter. The larger of the two, Phobos, orbits at a distance of just 9378 km from the center of the planet once every 459 minutes. Applying the modified version of Kepler's third law (which states that the square of a moon's orbital period is proportional to the cube of its orbital semimajor axis divided by the mass of the planet it orbits), we find that the mass of Mars is 6.4×10^{23} kg, or 0.11 times that of Earth. ∞ (Sec. 2.7) Naturally, the orbit of Deimos yields the same result.

From the mass and radius, we find that the average density of Mars is 3900 kg/m³, only slightly greater than that of the Moon. If we assume that the Martian surface rocks are similar to those on the other terrestrial planets, this average density suggests the existence of a substantial core of higher than average density within the planet. Planetary scientists now believe that this core is composed largely of iron sulfide (a compound about twice as dense as surface rock) and has a diameter of about 2500 km.

Surface markings easily seen on Mars allow astronomers to track the planet's rotation. Mars rotates once on its axis every 24.6 hours. One Martian day is thus very similar in length to one Earth day. The planet's equator is inclined to the orbit plane at an angle of 24.0°, again very similar to Earth's inclination of 23.5°. Thus, as Mars orbits the Sun, we find both daily and seasonal cycles, just as on Earth. In the case of Mars, however, the seasons are complicated somewhat by variations in solar heating due to the planet's eccentric orbit—southern summer occurs around the time of Martian perihelion and so is significantly warmer than summer in the north.

10.3 Long-Distance Observations of Mars

At opposition, when Mars is closest to us and most easily observed, we see it as full, so the Sun's light strikes the surface almost vertically, casting few shadows and preventing us from seeing any topographic detail, such as craters or mountains. Even through a large telescope Mars appears only as a reddish disk, with some light and dark patches and prominent polar caps. These surface features undergo slow seasonal changes over the course of a Martian year. We saw in Chapter 1 how the inclination of Earth's axis produces similar seasonal changes. ∞ (Sec. 1.3) Figure 10.2 shows some of the best images of Mars ever made from Earth (or

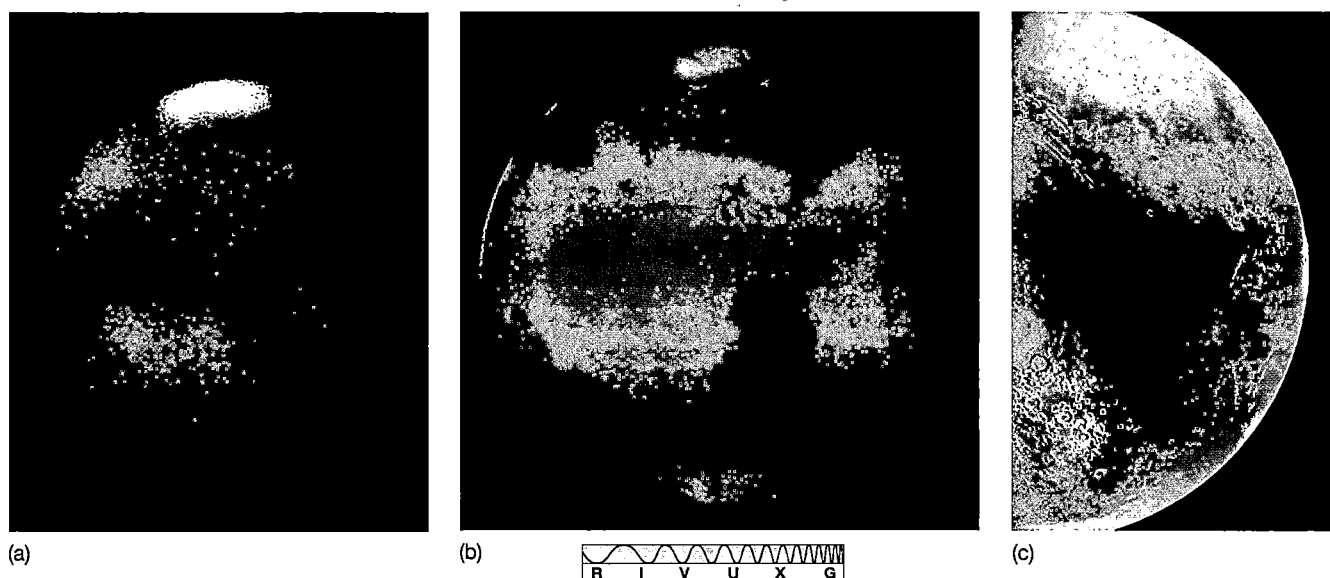

(a) (b) (c)

R I V U X G

Figure 10.2 Mars (a) A deep red (800 nm) image of Mars, taken in 1991 at Pic-du-Midi, an exceptionally clear site in the French Alps. One of the planet's polar caps appears at the top. (b) A visible-light *Hubble Space Telescope* image of Mars, taken while the planet was near opposition in 1997. (c) A view of Mars taken from a *Viking* spacecraft during its approach in 1976. The planet's surface features can be seen clearly at a level of detail completely invisible from Earth. *(Pic-du-Midi Observatory; NASA)*

Earth orbit), along with a photograph taken by one of the U.S. *Viking* spacecraft en route to the planet.

Viewed from Earth, the most obvious Martian surface features are the bright polar caps (see Figure 10.2a). They grow or diminish according to the seasons, almost disappearing at the time of Martian summer. The dark surface features on Mars also change from season to season, although their variability probably has little to do with the melting of the polar ice caps. To the more fanciful observers around the start of the twentieth century, these changes suggested the seasonal growth of vegetation on the planet. It was but a small step from seeing polar ice caps and speculating about teeming vegetation to imagining a planet harboring intelligent life, perhaps not unlike us.

But those speculations and imaginings were not to be confirmed. The caps are mostly frozen carbon dioxide (that is, dry ice), not water ice, as at Earth's North and South poles. The polar caps do contain water, but it remains permanently frozen, and the dark markings seen in Figures 10.2, once thought (by some) to be part of a network of "canals" dug by Martians for irrigation purposes (see *Discovery 10-1*), are actually highly cratered and eroded areas around which surface dust occasionally blows. Repeated covering and uncovering of these landmarks gives the impression (from a distance) of surface variability, but it's only the thin dust cover that changes.

The powdery Martian surface dust is borne aloft by strong winds that often reach hurricane proportions (hundreds of kilometers per hour). In fact, when the U.S. *Mariner 9* spacecraft went into orbit around Mars in 1971, a planetwide dust storm obscured the entire landscape. Had the craft been on a flyby mission (for a quick look) instead of an orbiting mission (for a longer view), its visit would have been a failure. Fortunately, the storm subsided, enabling the craft to radio home detailed information about the surface.

 Concept Check

■ Does Mars have seasons like those on Earth?

10.4 The Surface of Mars

3 Maps of the surface of Mars returned by orbiting spacecraft show a wide range of geological features. Mars has huge volcanoes, deep canyons, vast dune fields, and many other geological wonders. Orbiters have performed large-scale surveys of much of the planet's surface; lander data have complemented these planetwide studies with detailed information on (so far) three specific sites. ∞ (Sec. 6.6)

LARGE-SCALE TOPOGRAPHY

Figure 10.3 shows a planetwide mosaic of thousands of images taken in the 1970s by the *Viking* orbiters, showing some of the planet's topographic features in true color.

More recently, *Mars Global Surveyor* has mapped out the Martian surface to an accuracy of a few meters using an instrument called a laser altimeter. Figure 10.4 shows some results of those measurements, projected onto a Martian globe. Figure 10.5 flattens them out into a more conventional map, and marks some prominent surface features.

A striking feature of the terrain of Mars is the marked difference between the northern and southern hemispheres. The northern hemisphere is made up largely of rolling volcanic plains, not unlike the lunar maria. These extensive lava plains—much larger than those found on Earth or the Moon—were formed by eruptions involving enormous volumes of lava. They are strewn with blocks of volcanic rock as well as with boulders blasted out of impact areas by infalling meteoroids (the Martian atmosphere is too thin to offer much resistance to incoming debris). The southern hemisphere consists of heavily cratered highlands lying some 5 kilometers above the level of the lowland north. Most of the dark regions visible from Earth are mountainous regions in the south. Figure 10.6 contrasts typical terrains in the two hemispheres.

The northern plains are much less cratered than the southern highlands. On the basis of arguments presented in Chapter 8, this smoother surface suggests that the northern surface is younger. ∞ (Sec. 8.5) Its age is perhaps 3 billion years, compared with 4 billion in the south. In places, the boundary between the southern highlands and the northern plains is quite sharp. The surface level can drop by as much as 4 km in a distance of 100 km or so. Most scientists assume that the southern terrain is the original crust of the planet. How most of the northern hemisphere could have been lowered in elevation and subsequently flooded with lava remains a mystery.

The major geological feature on the planet is the Tharsis bulge, visible at the left of Figure 10.3 and marked in Figure 10.5. Roughly the size of North America, Tharsis lies on the Martian equator, and rises some 10 km higher than the rest of the Martian surface. To its east lies Chryse Planitia (the "Plains of Gold"); to the west a region called Isidis Planitia (the "Plains of Isis," an Egyptian goddess). These features are wide depressions, hundreds of kilometers across and up to 3 km deep. If we wished to extend the idea of "continents" from Earth and Venus to Mars, we would conclude that Tharsis is the only continent on the Martian surface. However, as on Venus, there is no sign of plate tectonics—the continent of Tharsis is not drifting like its Earthly counterparts. ∞ (Sec. 7.4) Tharsis appears to be even less heavily cratered than the northern plains, making it the youngest region on the planet. It is estimated to be between 2 and 3 billion years old.

Almost diametrically opposite Tharsis, in the southern highlands, lies the Hellas Basin, which paradoxically contains the lowest point on Mars. (Hellas is clearly visible in Figure 10.4, and is labeled in Figure 10.5.) Some 3000 km across, the floor of the basin lies nearly 9 km below its rim, and over 6 km below the average level of the planet's

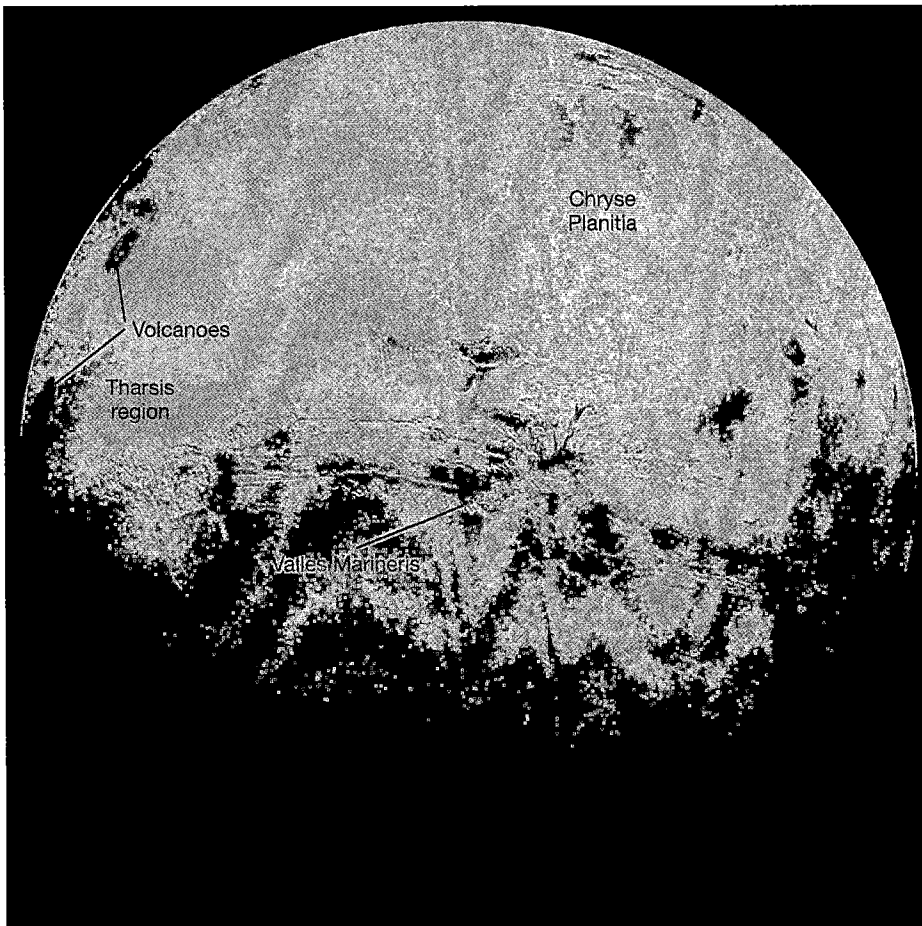

Figure 10.3 Tharsis Some 5000 km across, the Tharsis region bulges out from the planet's equatorial zone, rising to a height of about 10 km. The large volcanoes on the left mark the approximate peak of the bulge. One of the plains flanking the Tharsis bulge, Chryse Planitia, is toward the right. Dominating the center of the field of view is a vast "canyon" known as Valles Marineris. (NASA)

surface. Its shape and structure identify it as an impact feature, similar in many ways to the south pole–Aitken Basin on the far side of Earth's Moon. ⊂⊃ (*Discovery 8-2*) Its formation must have caused a major redistribution of the young Martian crust—maybe even enough to account for a substantial portion of the highlands, according to some researchers. Its heavily cratered floor indicates that the impact occurred very early on in Martian history—perhaps 4 billion years ago.

VOLCANISM

Mars contains the largest known volcanoes in the solar system. Three very large volcanoes are found on the Tharsis bulge, two of them visible on the left-hand side of Figure 10.3. The largest volcano of all is Olympus Mons (Figure 10.7), northwest of Tharsis, lying just over the left (western) horizon of Figure 10.3. It measures some 700 km in diameter at its base—only slightly smaller than the state of Texas—and rises to a height of 25 km above the surrounding plains. The caldera, or crater, at its summit, measures 80 km across. The other three large volcanoes are a little smaller—a mere 18 km high—and lie near the top of the bulge.

Like Maxwell Mons on Venus, these volcanoes are not associated with plate motion—there is none on Mars—but instead are shield volcanoes, sitting atop a hot spot in the underlying Martian mantle (see Section 9.4). All four show distinctive lava channels and other flow features very similar to those found on shield volcanoes on Earth. *Viking* and *Mars Global Surveyor* images of the Martian surface reveal many hundreds of volcanoes. Most of the largest are associated with the Tharsis bulge, but many smaller volcanoes are also found in the northern plains.

The great height of Martian volcanoes is a direct consequence of the planet's low surface gravity. As lava flows and spreads to form a shield volcano, its eventual height depends on the new mountain's ability to support its own weight. The lower the gravity, the less the weight and the higher the mountain. It is no accident that Maxwell Mons on Venus and the Hawaiian shield volcanoes on Earth rise to roughly the same height (about 10 km) above their respective bases—Earth and Venus have similar surface gravity. Mars's surface gravity is only 40 percent that of Earth, so volcanoes rise roughly 2.5 times as high.

Are these volcanoes still active? Scientists have found no direct evidence for recent or ongoing eruptions. However, if these volcanoes were around since the Tharsis uplift (as the formation of the Tharsis bulge is known) and were active as recently as 100 million years ago (an age estimate based on the extent of impact cratering on their slopes),

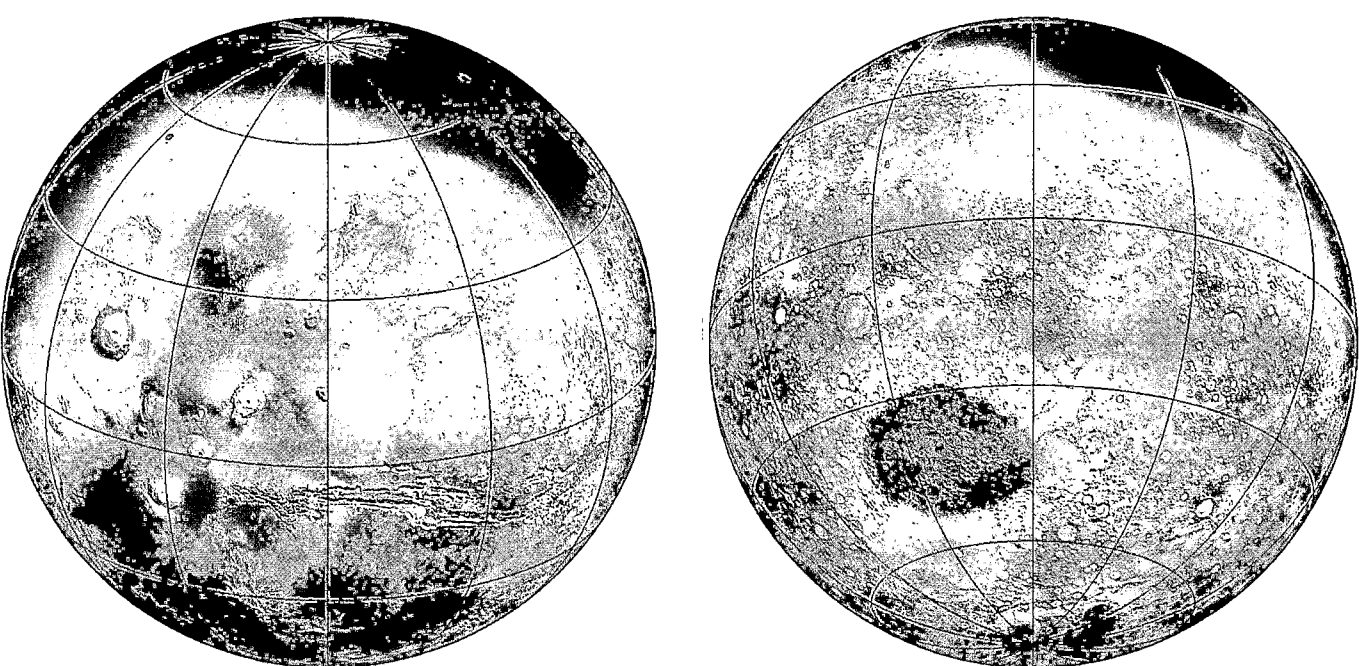

Figure 10.4 Mars Globes Two computer-generated globes of planet Mars, based on detailed measurements made by *Mars Global Surveyor*. Color represents height above (or below) the mean planetary radius, ranging from dark blue (– 8 km), through green, yellow and red (+ 8 km), to white (over 8 km altitude) as shown in the scale above. Frame (a) shows roughly the same hemisphere as Figure 10.3, containing the Tharsis region of Mars. Frame (b) shows the planet's other hemisphere, dominated by the giant Hellas impact basin. *(NASA)*

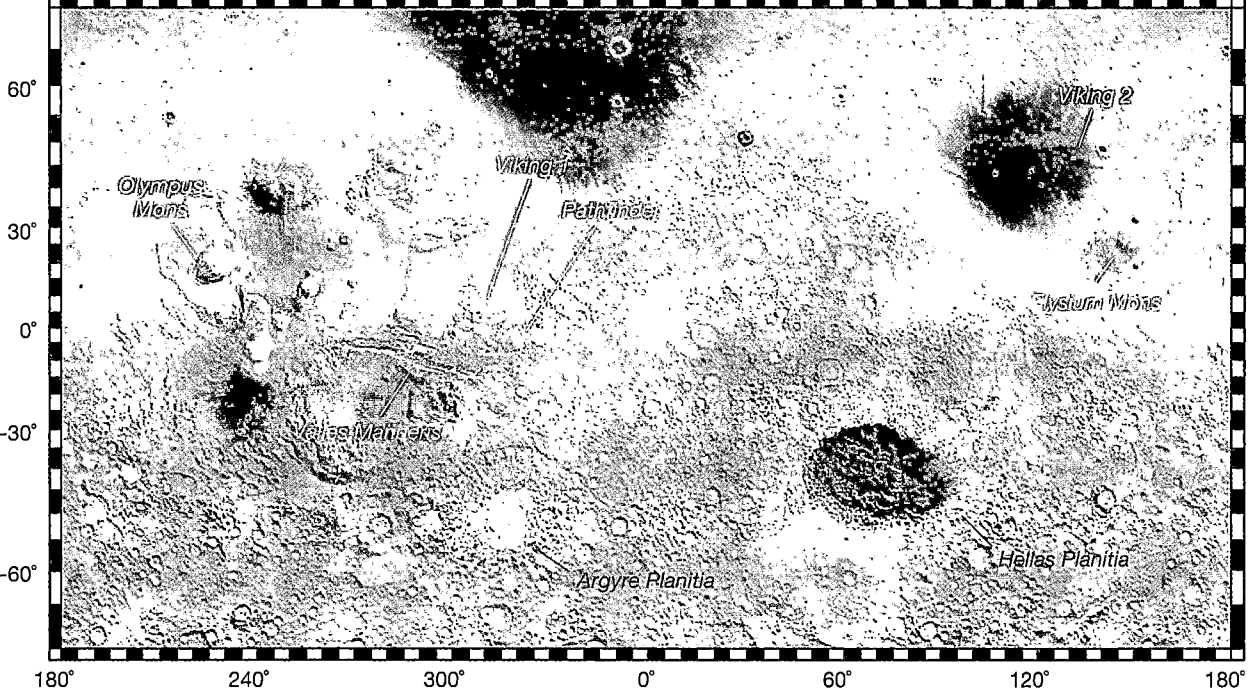

Figure 10.5 Mars Map The *Mars Global Surveyor* data of Figure 10.4, now displayed as a flat map, with some surface features labeled. The *Viking* and *Pathfinder* landing sites are also marked. *(NASA)*

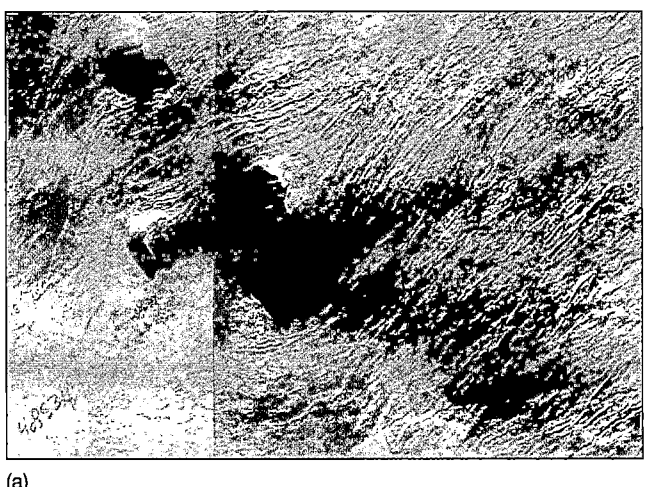

(a)

(b)

R I V U X G

Figure 10.6 Mars Hemispheres (a) The northern hemisphere of Mars consists of rolling, volcanic plains. (b) The southern Martian highlands are heavily cratered (true color). *(NASA)*

some of them may still be at least intermittently active. Millions of years, though, may pass between eruptions.

IMPACT CRATERING

The *Mariner* spacecraft found that the surfaces of Mars and its two moons are pitted with impact craters formed by meteoroids falling in from space. As on our Moon, the smaller craters are often filled with surface matter—mostly dust—confirming that Mars is a dry, desert world. However, Martian craters are filled in considerably faster than their lunar counterparts. On the Moon, ancient craters less than 100 m across (corresponding to depths of about 20 m) have been obliterated, primarily by meteoritic erosion. ∞ (Sec. 8.5) On Mars, there is also a lack of

small craters, but extending to craters 5 km in diameter. The Martian atmosphere is an efficient erosive agent, transporting dust from place to place and erasing surface features much faster than meteoritic impacts alone can obliterate them.

As on the Moon, the extent of large impact cratering (that is, craters too big to have been filled in by erosion since they formed) serves as an age indicator for the Martian surface. The ages quoted earlier, ranging from 4 billion years for the southern highlands to a few hundred million years in the youngest volcanic areas, were obtained in this way.

The detailed appearance of Martian impact craters provides an important piece of information about conditions just below the planet's surface. The ejecta blankets surrounding many Martian craters look quite different

R I V U X G

Figure 10.7 Olympus Mons The largest volcano known on Mars or anywhere else in the solar system. Nearly three times taller than Mount Everest on Earth, this Martian mountain measures about 700 km across the base and 25 km high at the peak. It seems currently inactive and may have been extinct for at least several hundred million years. By comparison, the largest volcano on Earth, Hawaii's Mauna Loa, measures a mere 120 km across and peaks just 9 km above the Pacific Ocean floor. *(NASA)*

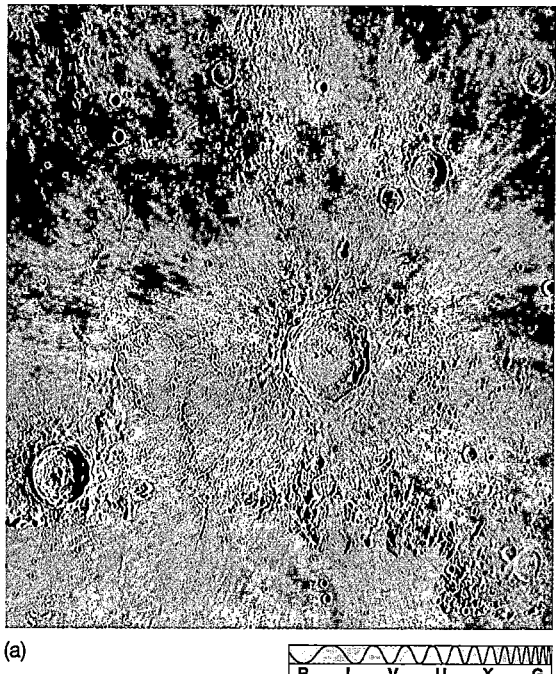

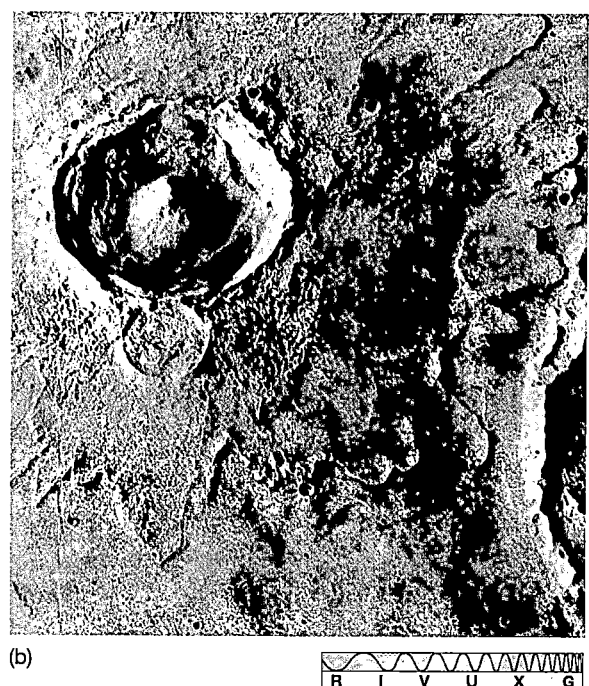

(a)

(b)

R   I   V   U   X   G

Figure 10.8 Crater Comparison (a) The large lunar impact crater Copernicus is quite typical of those found on Earth's Moon. Its ejecta blanket appears to be composed of dry, powdery material. (b) The ejecta from Mars's crater Yuty (18 km in diameter) evidently was liquid in nature. This type of crater is sometimes called a "splosh" crater. (NASA)

from their lunar counterparts. Figure 10.8 compares the Copernicus crater on the Moon with the (fairly typical) crater Yuty on Mars. The material surrounding the lunar crater is just what one would expect from an explosion ejecting a large volume of dust, soil, and boulders. However, the ejecta blanket on Mars gives the distinct impression of a liquid that has splashed or flowed out of the crater. Geologists believe that this *fluidized ejecta* crater indicates that a layer of **permafrost**, or water ice, lies just under the surface. The explosive impact heated and liquefied the ice, resulting in the fluid appearance of the ejecta.

Prior to the arrival of *Mars Global Surveyor*, astronomers believed that all the water below the Martian surface existed in the form of ice. However, in 2000, *Surveyor* mission scientists reported the discovery of numerous small-scale "gullies" in Martian cliffs and crater walls that apparently were carved by running water in the relatively recent past. These features are too small to have been resolved by *Viking* cameras. Figure 10.9 shows one such gully, found in the inner rim of a Martian impact crater in the southern highlands. Its structure has many similarities to the channels carved by flash floods on Earth. The ages of these intriguing features are uncer-

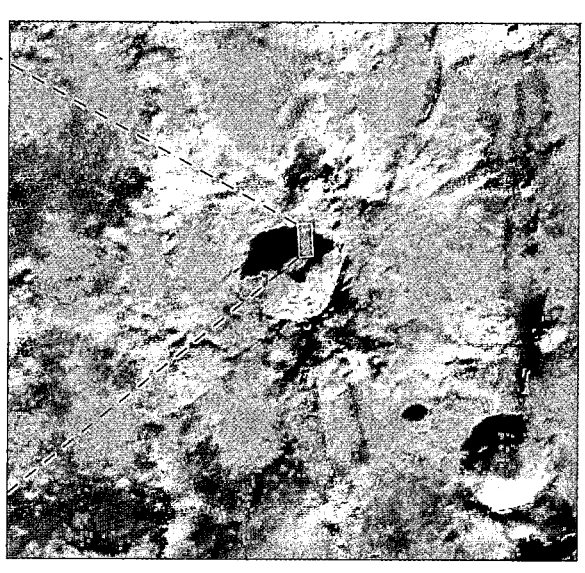

Figure 10.9 Running Water on Mars? This high-resolution *Mars Global Surveyor* view (left) of a crater wall (right) shows evidence of "gullies" apparently formed by running water in the relatively recent past. If this interpretation is correct, it suggests that liquid water may still reside below the Martian surface. (NASA)

R   I   V   U   X   G

tain, and might be as great as a million years in some cases, but the *Surveyor* team speculate that some of them may still be active today, implying that liquid water could exist in some regions of Mars at depths of less than 500 meters.

✓ Concept Check

■ How do we know that the northern Martian lowlands are younger than the southern highlands?

THE MARTIAN "GRAND CANYON"

Yet another feature associated with the Tharsis bulge is a great "canyon" known as Valles Marineris (the Mariner Valley). Shown in its entirety in Figure 10.3 and in more detail in Figure 10.10, it is not really a canyon in the terrestrial sense, because running water played no part in its formation. Planetary astronomers believe that it was formed by the same crustal forces that caused the entire Tharsis region to bulge outward, making the surface split and crack. These cracks, called *tectonic fractures*, are found all around the Tharsis bulge. The Valles Marineris is the largest of them. Cratering studies suggest that the cracks are at least 2 billion years old. Similar (but much smaller) cracks, with similar causes, have been found in the Aphrodite Terra region of Venus. ⚭ (Sec. 9.4) Valles Marineris runs for almost 4000 km along the Mar-

tian equator, about one-fifth of the way around the planet. At its widest, it is some 120 km across, and it is as deep as 7 km in places. Like many Martian surface features, it simply dwarfs Earthly competition. The Grand Canyon in Arizona would easily fit into one of its side "tributary" cracks. Valles Marineris is so large that it can even be seen from Earth—in fact, it was one of the few "canals" observed by nineteenth-century astronomers (see *Discovery 10-1*) that actually corresponded to a real feature on the planet's surface (it was known as the Coprates canal). We must reemphasize, however, that this Martian feature was not constructed by intelligent beings, nor was it carved by a river, nor is it a result of Martian plate tectonics. For some reason, the crustal forces that formed it never developed into full-fledged plate motion as on Earth.

EVIDENCE FOR RUNNING WATER

Although the great surface cracks in the Tharsis region are not really canyons and were not formed by running water, photographic evidence reveals that liquid water once existed in great quantity on the surface of Mars. Two types of flow feature are seen: the *runoff channels* and the *outflow channels*.

The **runoff channels** (one of which is shown in Figure 10.11a) are found in the southern highlands. They are extensive systems—sometimes hundreds of kilometers

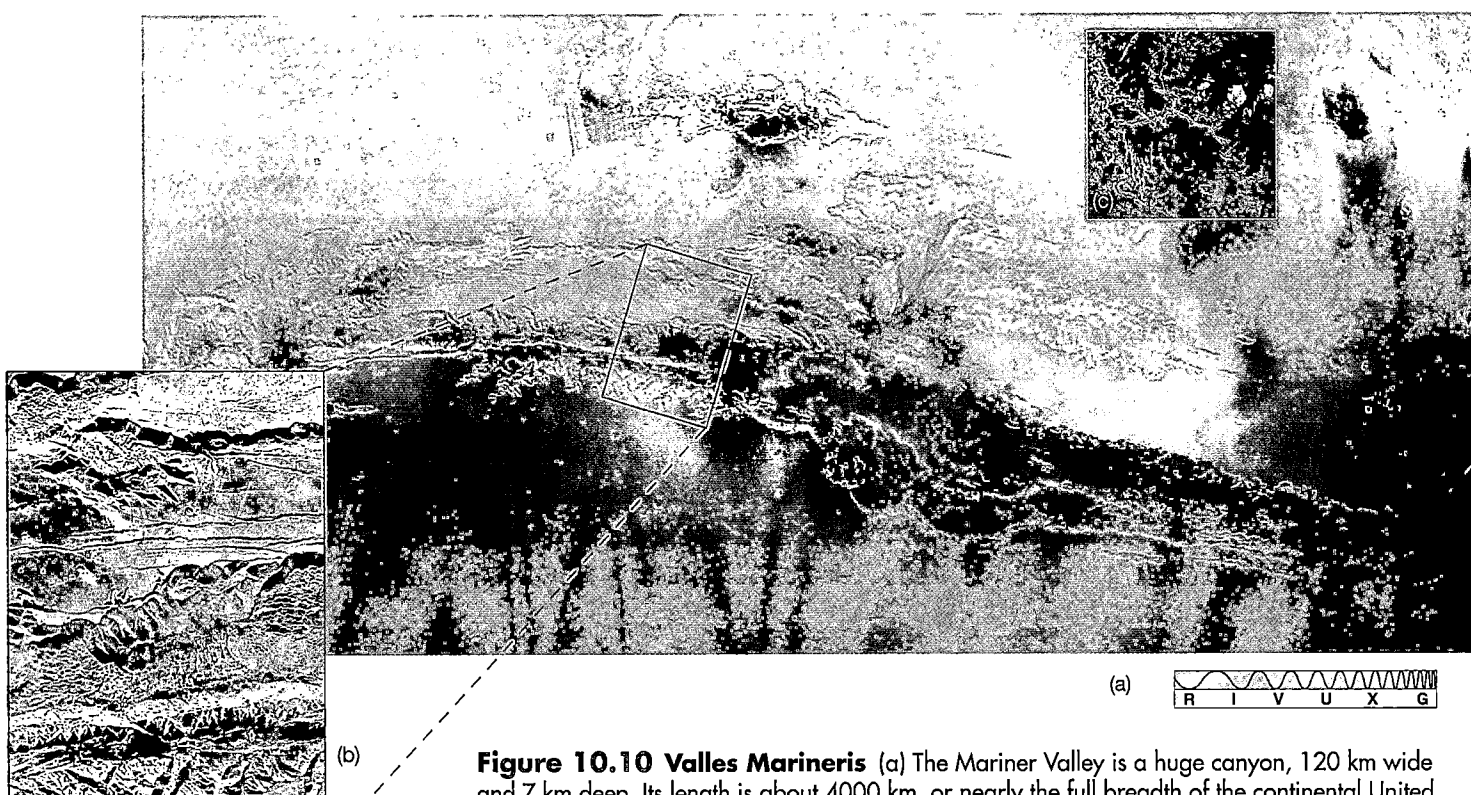

(a)

| R | I | V | U | X | G |

Figure 10.10 Valles Marineris (a) The Mariner Valley is a huge canyon, 120 km wide and 7 km deep. Its length is about 4000 km, or nearly the full breadth of the continental United States. (b) A close-up view shows the complexity of the valley walls and dry tributaries. (c) A comparison, to scale, with Earth's Grand Canyon, which is a mere 20 km wide and 2 km deep, suggests just how big the Mariner Valley is. *(NASA)*

DISCOVERY 10-1

Martian Canals?

The year 1877 was an important one in the human study of the planet Mars. The Red Planet came unusually close to Earth, affording astronomers an especially good view. Of particular note was the discovery, by U.S. Naval Observatory astronomer Asaph Hall, of the two moons circling Mars (see Section 10.7). But most exciting was the report of the Italian astronomer Giovanni Schiaparelli on his observation of a network of linear markings that he termed *canali*. In Italian, *canali* can simply mean "grooves" or "channels," but it can also mean "canals." As far as we know, Schiaparelli did not intend to imply that the canali were anything other than natural, but the word was translated into English as "canals," suggesting that the grooves had been constructed by intelligent beings. Observations of these features became sensationalized in the world's press (especially in the United States), and some astronomers began drawing elaborate maps of Mars, showing oases and lakes where canals met in desert areas.

Percival Lowell (see photo), a successful Boston businessman (and brother of the poet Amy Lowell and Harvard president Abbott Lawrence Lowell), became fascinated by these reports. He abandoned his business and purchased a clear-sky site at Flagstaff, Arizona, where he built a major observatory. He devoted his life to achieving a better understanding of the Martian "canals." In doing so, he championed the idea that Mars was drying out and that an intelligent society had constructed the canals to transport water from the wet poles to the arid equatorial deserts.

Alas, the Martian valleys and channels photographed by robot spacecraft during the 1970s are far too small to be the "canali" that Schiaparelli, Lowell, and others thought they saw on Mars. The entire episode represents a classic case in the history of science—a case in which well-intentioned observers, perhaps obsessed with the notion of life on other worlds, let their personal opinions and prejudices seriously affect their interpretations of reasonable data. The accompanying figures of Mars show how surface features (which were probably genuinely observed by astronomers at the turn of the century) might have been imagined to be connected. The figure on the left is a photograph of how Mars actually looked in a telescope at the end of the nineteenth century. The sketch at right is an interpretation (done at the height of the canal hoopla) of the pictured view. The human eye, under physiological stress, tends to connect dimly observed yet distinctly separated features. Humans saw patterns and canals where none in fact existed.

The chronicle of the Martian canals illustrates how the scientific method requires scientists to acquire new data to sort out sense from nonsense, fact from fiction. Rather than simply believing the claims about the Martian canals, other scientists demanded further observations to test Lowell's hypothesis. Eventually, improved observations, climaxing in the *Mariner* and *Viking* exploratory missions to the Red Planet nearly a century after all the fuss began, totally disproved the existence of canals. It often takes time, but the scientific method does eventually lead to progress in understanding reality.

(Lowell Observatory)

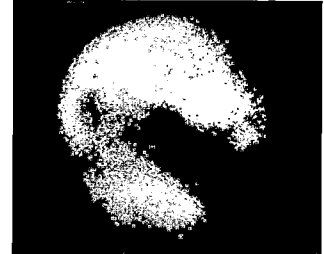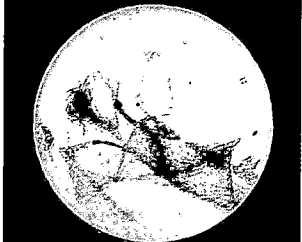

(NASA)

in total length—of interconnecting, twisting channels that seem to merge into larger, wider channels. They bear a strong resemblance to river systems on Earth, and it is believed by geologists that this is just what they are—the dried-up beds of long-gone rivers that once carried rainfall on Mars from the mountains down into the valleys. These runoff channels speak of a time 4 billion years ago (the age of the Martian highlands), when the atmosphere was thicker, the surface warmer, and liquid water widespread.

The **outflow channels** (Figure 10.12a) are probably relics of catastrophic flooding on Mars long ago. They appear only in equatorial regions and generally do not form the extensive interconnected networks that characterize the runoff channels. Instead, they are probably the paths taken by huge volumes of water draining from the southern highlands into the northern plains. The onrushing water arising from these flash floods probably also formed the odd teardrop-shaped "islands"

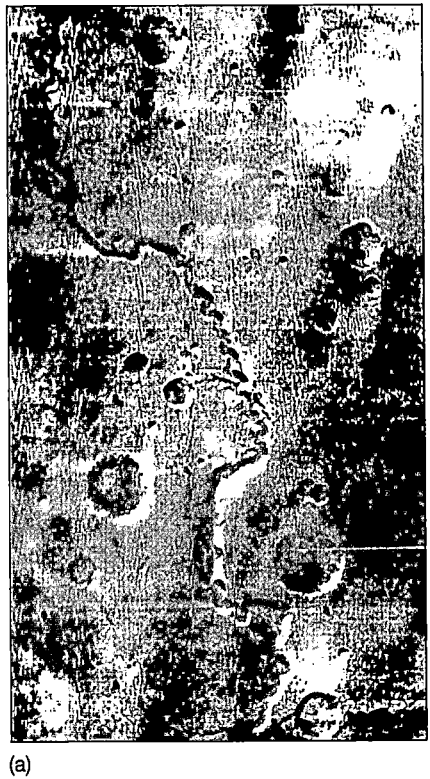

(a)

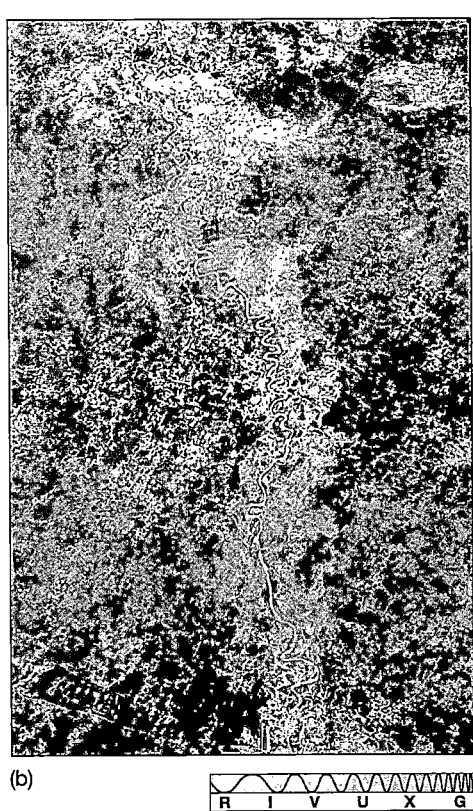

(b)

R I V U X G

Figure 10.11 Martian Channel
(a) This runoff channel on Mars measures about 400 km long and 5 km wide. Here, we compare it with a photograph of the Red River (b) running from Shreveport, Louisiana, to the Mississippi River. The two differ mainly in that there is currently no liquid water in this, or any other, Martian valley. *(NASA)*

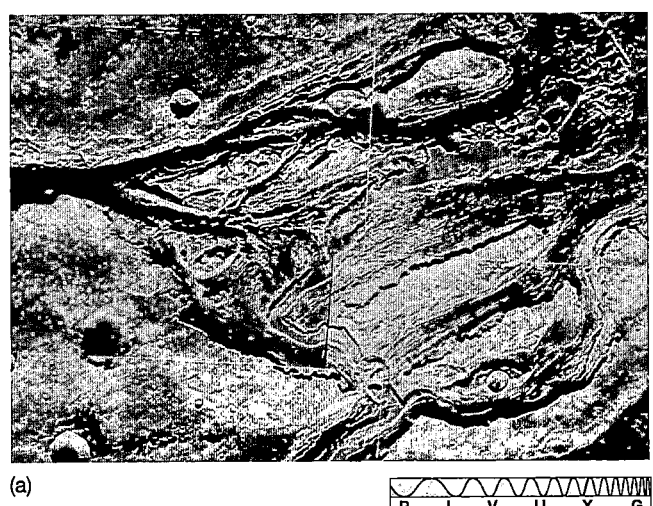

(a)

R I V U X G

Figure 10.12 Martian Outflow (a) An outflow channel near the Martian equator bears witness to a catastrophic flood that occurred about 3 billion years ago. (b) The onrushing water that carved out the outflow channels was responsible for forming these oddly shaped "islands" as the flow encountered obstacles—impact craters—in its path. Each "island" is about 40 km long. *(NASA)*

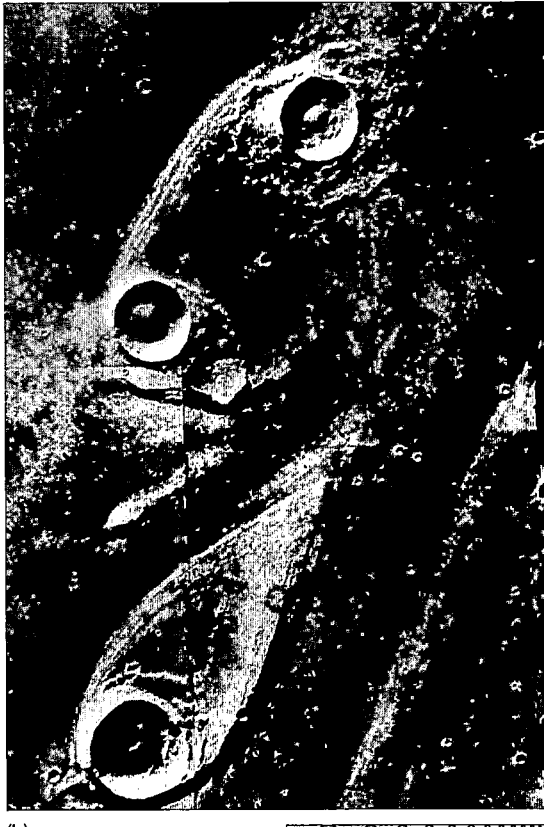

(b)

R I V U X G

(resembling the miniature versions seen in the wet sand of our beaches at low tide) that have been found on the plains close to the ends of the outflow channels (Figure 10.12b). Judging from the width and depth of the channels, the flow rates must have been truly enormous—perhaps as much as a hundred times greater than the 10^5 tons per second carried by the Amazon river, the largest river system on Earth. Flooding shaped the outflow channels about 3 billion years ago, about the same time as the northern volcanic plains formed.

Recent discoveries by *Mars Global Surveyor* seem to support the idea that Mars may have enjoyed an extended early period during which rivers, lakes and even oceans adorned its surface. As shown in Figure 10.13, there is clear evidence of erosion of Martian canyons by running water, and of the formation of terraced "beaches" at the

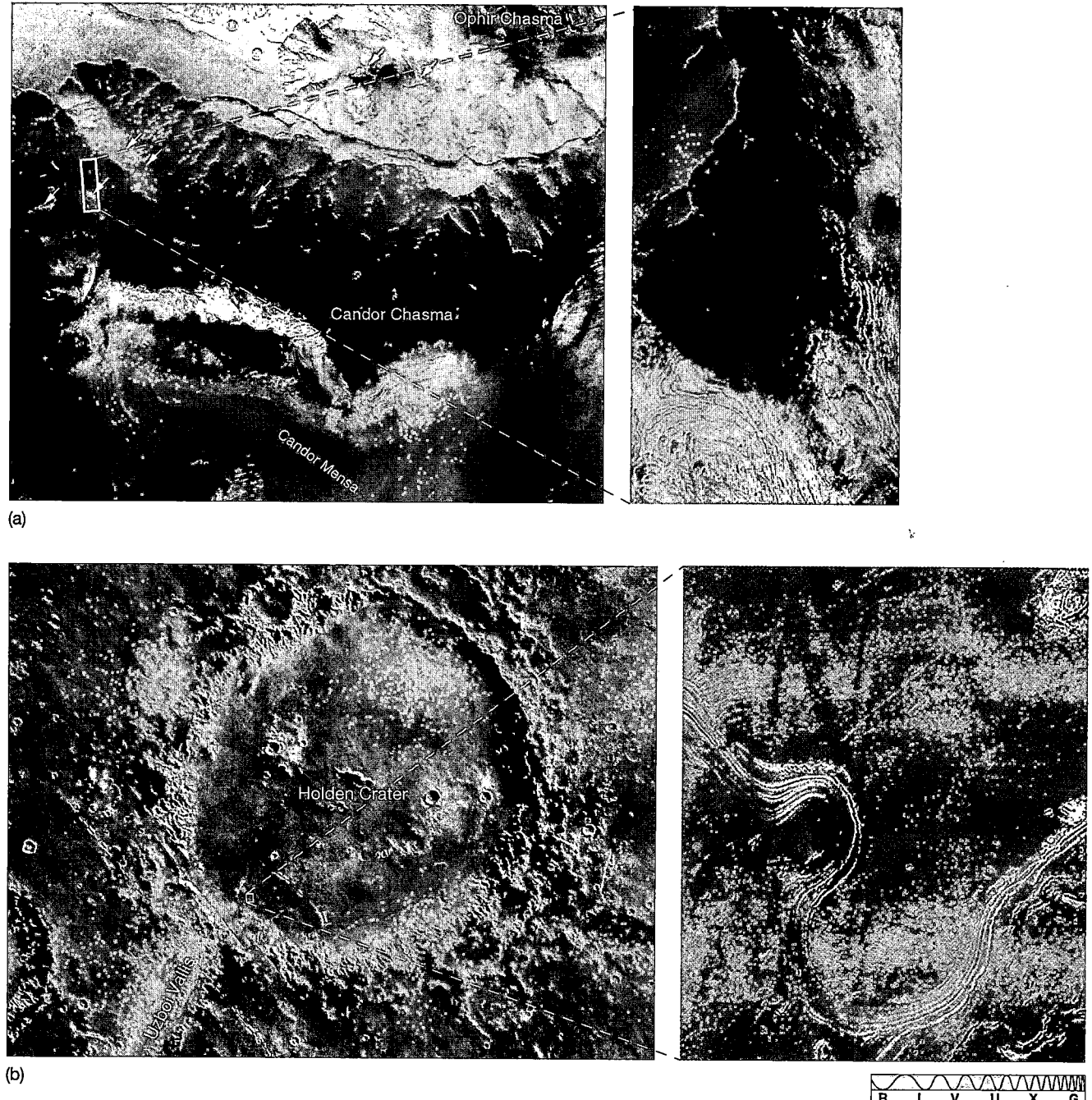

Figure 10.13 Erosion by Water These images of Mars, taken with the *Mars Global Surveyor* camera, show erosion possibly caused by water. In (a), a close-up of a 50-km-long canyon wall in the Mariner Valley (see Figure 10.10) suggests erosion by running water. In (b), we see tentative evidence for standing water erosion in Holden Crater, about 140 km across. *(NASA)*

shores of larger bodies of water. Figure 10.14 is a computer-generated view of the Martian north polar region showing the extent of a possible ancient ocean covering the entire northern lowlands. The Hellas basin (Figure 10.5) is another likely candidate for an ancient Martian sea.

Aside from the gullies mentioned earlier, which are highly suggestive but still not conclusive, astronomers have no direct evidence for liquid water anywhere on the surface of Mars today, and the amount of water vapor in the Martian atmosphere is tiny. Yet the extent of the outflow channels indicates that a huge total volume of water existed on Mars in the past. Where did all that water come from? And where did it all go? The answer may be that virtually all the water on Mars is now locked in the (mostly) permafrost layer under the surface, with perhaps a little more contained in the polar caps. Four billion years ago, as climatic conditions changed, the running water that formed the runoff channels began to freeze, forming the permafrost and drying out the river beds. Mars remained frozen for about a billion years, until volcanic (or some other) activity heated large regions of the surface, melting the subsurface ice and causing the flash floods that created the outflow channels. Subsequently, volcanic activity subsided, the water refroze, and Mars once again became a dry world.

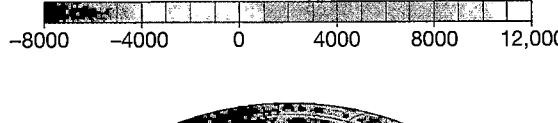

| -8000 | -4000 | 0 | 4000 | 8000 | 12,000 |

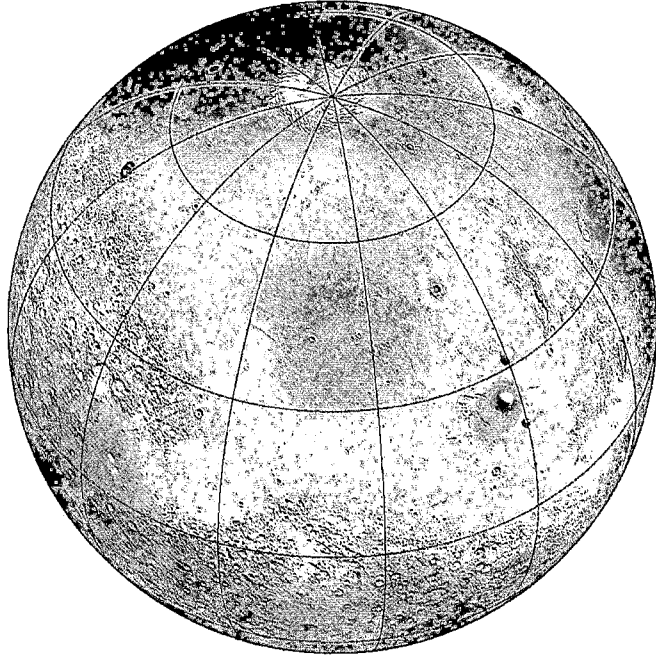

Figure 10.14 Ancient Ocean A possible ancient Martian ocean that spanned the polar regions. The blue regions in this computer-generated map actually indicate elevation, but they also outline quite accurately the extent of the ocean inferred from detailed study of *Mars Global Surveyor* data. (NASA)

POLAR CAPS

We have already noted that the Martian polar caps are composed predominantly of carbon dioxide frost—dry ice—and show seasonal variations. Each cap in fact consists of two distinct parts—the **seasonal cap**, which grows and shrinks each year, and the **residual cap**, which remains permanently frozen. At maximum size, in southern midwinter, the southern seasonal cap is some 4000 km across. Half a Martian year later, the northern cap is at its largest, reaching a diameter of roughly 3000 km. The two seasonal polar caps do not have the same maximum size because of the eccentricity of Mars's orbit around the Sun. During southern winter, Mars is considerably farther from the Sun than half a year later, in northern winter. The southern winter season is longer and colder than that of the north, and the polar cap grows correspondingly larger.

The seasonal caps are composed entirely of carbon dioxide. Their temperatures are never greater than about 150 K ($-120°C$), the point at which dry ice can form. During the Martian summer, when sunlight striking a cap is most intense, carbon dioxide evaporates into the atmosphere, and the cap shrinks. In the winter, atmospheric carbon dioxide refreezes, and the cap reforms. As the caps grow and shrink, they cause substantial variations (up to 30 percent) in the Martian atmospheric pressure—a large fraction of the planet's atmosphere freezes out and evaporates again each year. From studies of these atmospheric fluctuations, scientists can estimate the amount of carbon dioxide in the seasonal polar caps. The maximum thickness of the seasonal caps is thought to be about 1 m.

The residual caps, shown in Figure 10.15, are smaller and brighter than the seasonal caps and show an even more marked north–south asymmetry. The southern residual cap is about 350 km across and, like the seasonal caps, is probably made mostly of carbon dioxide, although it may contain some water ice. Its temperature remains below 150 K at all times. The northern residual cap is much larger—about 1000 km across—and warmer, with a temperature that can exceed 200 K in northern summertime. Planetary scientists believe that the northern cap is made mostly of water ice, an opinion strengthened by the observed increase in the concentration of water vapor above the north pole in northern summer as some small fraction of its water ice evaporates in the Sun's heat. It is quite possible that the northern residual polar cap is a major storehouse for water on Mars.

Why is there such a temperature difference (at least 50 K) between the two residual polar caps, and why is the northern cap warmer, despite the fact that the planet's northern hemisphere is generally cooler than the south (see Section 10.2)? The reason is not fully understood, but it seems to be related to the giant dust storms that envelop the planet during southern summer. These storms, which

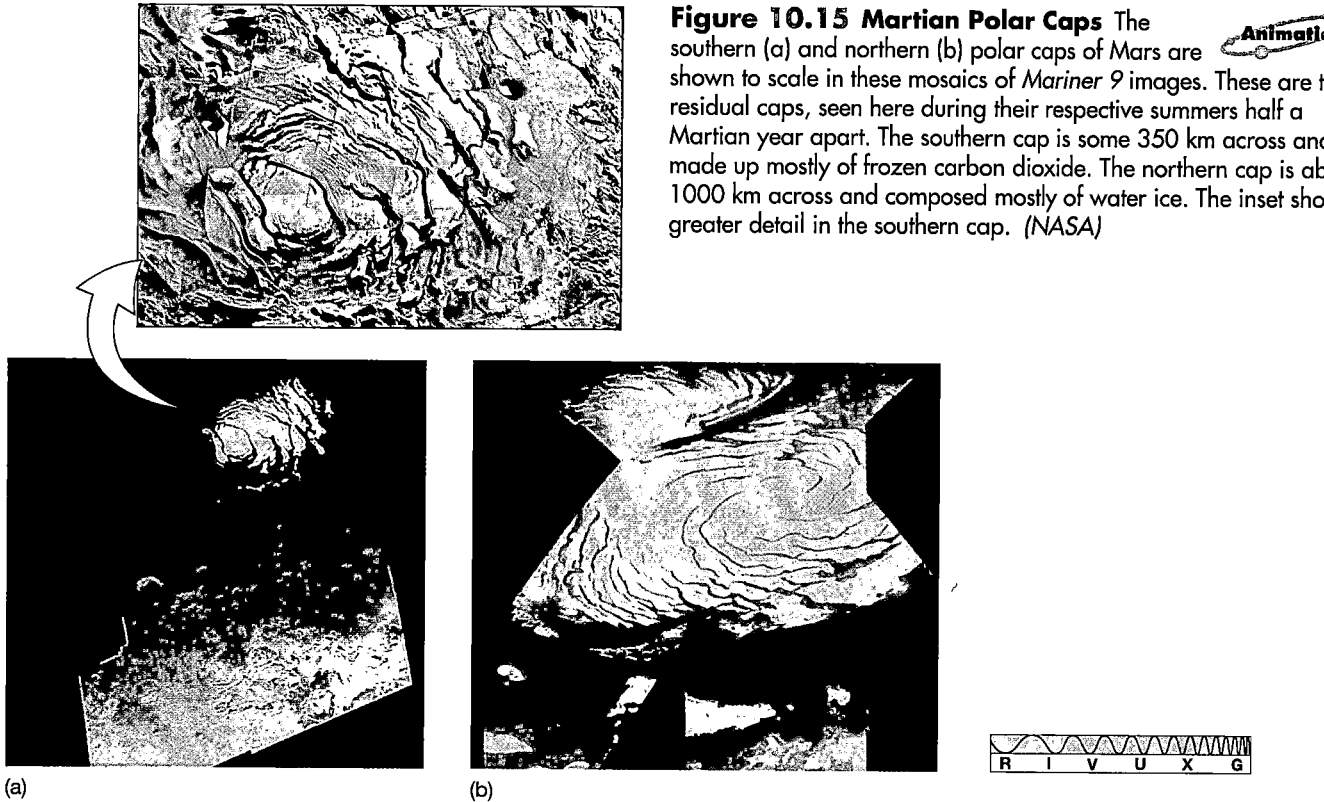

Figure 10.15 Martian Polar Caps The southern (a) and northern (b) polar caps of Mars are shown to scale in these mosaics of *Mariner 9* images. These are the residual caps, seen here during their respective summers half a Martian year apart. The southern cap is some 350 km across and is made up mostly of frozen carbon dioxide. The northern cap is about 1000 km across and composed mostly of water ice. The inset shows greater detail in the southern cap. *(NASA)*

(a)   (b)

last for a quarter of a Martian year (about six Earth months), tend to blow the dust from the warmer south into the cooler northern hemisphere. The northern ice cap becomes dusty and less reflective. As a result, it absorbs more sunlight and warms up.

THE VIEW FROM THE MARTIAN LANDERS

Viking 1 landed in Chryse Planitia, a broad depression to the east of Tharsis. The view that greeted its cameras (Figure 10.16) was a windswept, gently rolling, rather desolate plain, littered with rocks of all sizes, not unlike a high desert on Earth. The surface rocks visible in Figure 10.16 are probably part of the ejecta blanket of a nearby impact crater. *Viking 2* landed somewhat farther north, in a region of Mars called Utopia, chosen in part because mission planners anticipated greater seasonal climatic variations there. The plain on which *Viking 2* landed was flat and featureless. From space, the landing site appeared smooth and dusty. In fact, the surface turned out to be very rocky, even rockier than the Chryse site, and without the dust layer the mission directors had expected (Figure 10.17). The views that the two landers recorded may turn out to be quite typical of the low-latitude northern plains.

The *Viking* landers performed numerous chemical analyses of the Martian regolith. One important finding of these studies was the high iron content of the planet's surface. Chemical reactions between the iron-rich surface soil

Figure 10.16 Viking 1 This is the view from the *Viking 1* spacecraft parked on the surface of Mars. The fine-grained soil and the rock-strewn terrain stretching toward the horizon are reddish. Containing substantial amounts of iron ore, the surface of Mars is literally rusting away. The sky is a pale yellow-pink color, the result of airborne dust. *(NASA)*

Figure 10.17 Viking 2 Another view of the Martian surface, this one rock-strewn and flat, as seen through the camera aboard the Viking 2 robot that soft-landed on the northern Utopian plains. The discarded canister is about 20 cm long. The 0.5-m scars in the dirt were made by the robot's shovel. *(NASA)*

and free oxygen in the atmosphere is responsible for the iron oxide ("rust") that gives Mars its characteristic color. Although the surface layers are rich in iron relative to Earth's surface, the overall abundance is similar to Earth's average iron content. On Earth, much of the iron has differentiated to the center. Chemical differentiation does not appear to have been nearly so complete on Mars.

The most recent visitor to the Martian surface was *Mars Pathfinder*, along with its robot rover *Sojourner*. ∞ (Sec. 6.6) During the unexpectedly long lifetime of its mission (which lasted almost three months instead of the anticipated one month), the lander performed measurements of the Martian atmosphere and atmospheric dust while *Sojourner* carried out chemical analyses of the soil and rocks within about 50 m of the parent craft. In addition, over 16,000 images of the region were returned to Earth.

The *Pathfinder* landing site (Figure 10.18) had been carefully chosen to lie near the mouth of an outflow channel, and the size distribution and composition of the many rocks and boulders surrounding the lander were consistent with their having been deposited there by floodwaters. In addition, the presence of numerous rounded pebbles strongly suggested the erosive action of running water at some time in the past. The soil chemistry in the vicinity of the landing site was similar to that found by the two *Viking* landers. However, analyses of the nearby rocks revealed a chemical makeup different from that of the Martian meteorites found on Earth (see *Discovery 10-2*).

✓ Concept Check

■ Where has all the Martian water gone?

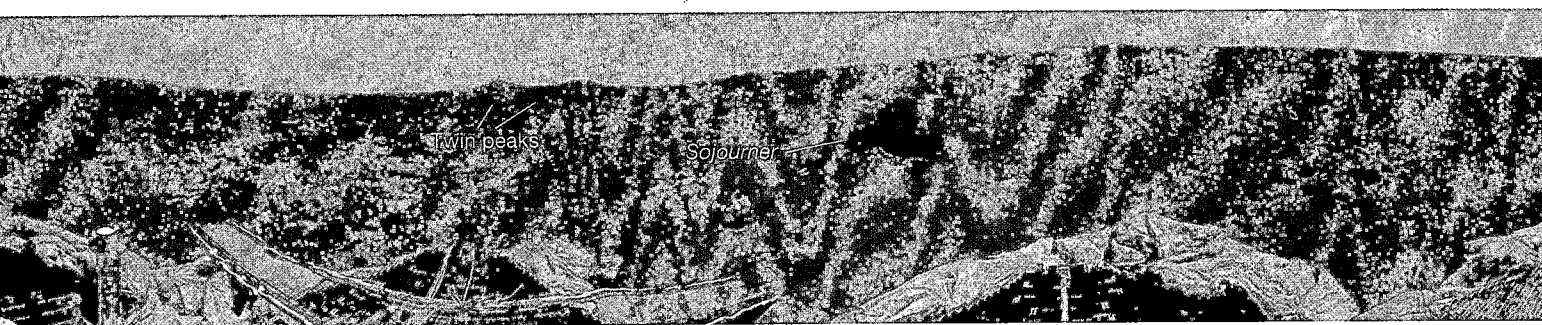

Figure 10.18 Pathfinder This 360° panorama of Mars was taken by the *Pathfinder* robot lander in 1997. Its small roving vehicle, *Sojourner*, can be seen up against a rock at center (see also Fig. 6.4). The "Twin Peaks" on the left center horizon are 1–2 km away. *(NASA)*

10.5 The Martian Atmosphere

COMPOSITION

🜨 Long before the arrival of the *Mariner* and *Viking* spacecrafts, astronomers knew from Earth-based spectroscopy that the Martian atmosphere was quite thin and composed primarily of carbon dioxide. In 1964, *Mariner 4* confirmed these results, finding that the atmospheric pressure is only about 1/150 the pressure of Earth's atmosphere at sea level and that carbon dioxide makes up at least 95 percent of the total atmosphere. With the arrival of *Viking*, more detailed measurements of the Martian atmosphere could be made. Its composition is now known to be 95.3 percent carbon dioxide, 2.7 percent nitrogen, 1.6 percent argon, 0.13 percent oxygen, 0.07 percent carbon monoxide, and about 0.03 percent water vapor. The level of water vapor is quite variable. Weather conditions encountered by *Mars Pathfinder* were quite similar to those found by *Viking 1*.

As the *Viking* landers descended to the surface, they made measurements of the temperature and pressure at various heights. The results are shown in Figure 10.19. The Martian atmosphere contains a troposphere (the lowest-lying atmospheric zone, where convection and "weather" occur), which varies both from place to place and from season to season. ⧀ (Sec. 7.2) The variability of the troposphere arises from the variability of the Martian surface temperature. At noon in the summertime, surface temperatures may reach 300 K. Atmospheric convection is strong, and the top of the troposphere can reach an altitude of 30 km. At night, the atmosphere retains little heat, and the temperature can drop by as much as 100 K. Convection then ceases and the troposphere vanishes.

On average, surface temperatures on Mars are about 50 K cooler than on Earth. The low early-morning temperatures often produce water-ice "fog" in the Martian canyons (Figure 10.20). Higher in the atmosphere, in the stratosphere, temperatures are low enough for carbon dioxide to solidify, giving rise to a high-level layer of carbon dioxide clouds and haze.

For most of the year, there is little day-to-day variation in the Martian weather: The Sun rises, the surface warms up, and light winds blow until sunset, when the temperature drops again. Only in the southern summer does the daily routine change. Strong surface winds (without rain or snow) sweep up the dry dust, carry it high into the stratosphere, and eventually deposit it elsewhere on the planet. At its greatest fury, a Martian storm floods the atmosphere with dust, making the worst storm we could imagine on Earth's Sahara Desert seem inconsequential by comparison. The dust can remain airborne for months at a time. The blown dust forms systems of sand dunes similar in appearance to those found on Earth.

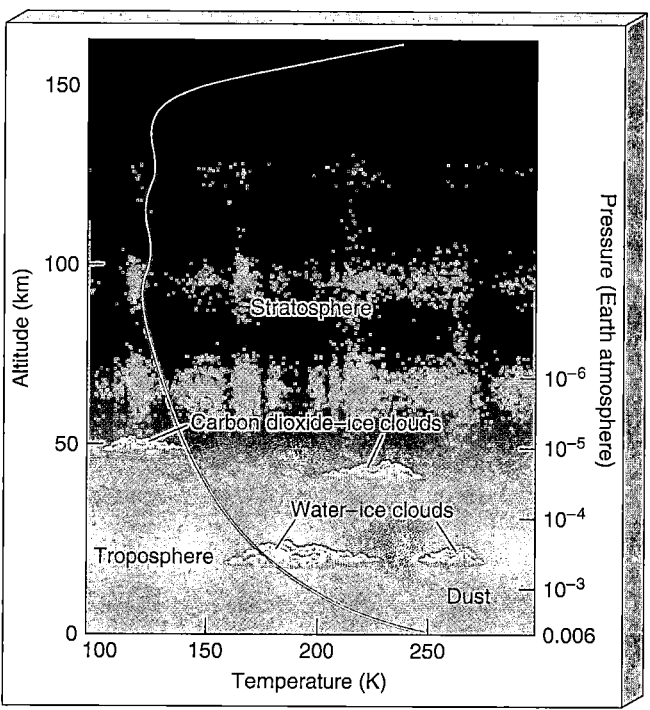

Figure 10.19 Martian Atmosphere Structure of the Martian atmosphere, as determined by *Viking* and *Mars Global Surveyor*. The troposphere rises to an altitude of about 30 km in the daytime. It occasionally contains clouds of water ice or, more frequently, dust during the planetwide dust storms that occur each year. Above the troposphere lies the stratosphere. Note the absence of a higher-temperature zone in the stratosphere, indicating the absence of an ozone layer.

EVOLUTION

🜨 Although there is some superficial similarity in composition between the atmospheres of Mars and Venus, the two planets obviously have quite different atmospheric histories—Mars's "air" is over 10,000 times thinner than that on Venus. As with the other planets we have studied, we can ask *why* the Martian atmosphere is as it is.

Presumably, Mars acquired a secondary outgassed atmosphere quite early in its history, just as the other terrestrial worlds did. Around 4 billion years ago, as indicated by the runoff channels in the highlands, Mars may have had a fairly dense atmosphere, complete with blue skies and rain. Despite Mars's distance from the Sun, the greenhouse effect would have kept conditions fairly comfortable, and a surface temperature of over 0°C seems quite possible. Sometime during the next billion years, most of the Martian atmosphere disappeared. Possibly some of it was lost because of impacts with other large (Moon-sized) bodies in the early solar system. More likely, the Martian atmosphere became unstable, in a kind of reverse runaway greenhouse effect.

On Venus, as we have seen, the familiar greenhouse effect ran away to high temperatures and pressures.

Even before the *Viking* missions reached Mars in 1976, most astronomers had abandoned hope of finding life there. Scientists knew there were no large-scale canal systems, no surface water, almost no oxygen in the atmosphere, and no seasonal vegetation changes. The present lack of liquid water on Mars especially dims the chances for life there now. However, running water and possibly a dense atmosphere in the past may have created conditions suitable for the emergence of life long ago. In the hope that some form of microbial life might have survived to the present day, the *Viking* landers carried out experiments designed to detect biological activity. The accompanying pair of photographs show the robot arm of one of the landers before and after digging a shallow trench to scoop up soil samples.

All three *Viking* biological experiments assumed some basic similarity between hypothetical Martian bacteria and those found on Earth. A *gas-exchange experiment* offered a nutrient broth to any residents of a sample of Martian soil and looked for gases that would signal metabolic activity. A *labeled-release experiment* added compounds containing radioactive carbon to the soil, then waited for results signaling that Martian organisms had either eaten or inhaled this carbon. Finally, a *pyrolitic-release experiment* added radioactively tagged carbon dioxide to a sample of Martian soil and atmosphere, waited a while, then removed the gas and tested the soil (by heating it) for signs that something had absorbed the tagged gas.

Initially, all three experiments appeared to be giving positive signals! However, subsequent careful studies showed that all the results could be explained by inorganic (that is, nonliving) chemical reactions. Thus, we have no

clear evidence for even microbial life on the Martian surface. The *Viking* robots detected peculiar reactions that mimic in some ways the basic chemistry of living organisms, but they did not detect life itself.

A criticism of the *Viking* experiments is that they searched only for life now living. Today, Mars seems locked in an ice age—the kind of numbing cold that would prohibit sustained life as we know it. If bacterial life did arise on an Earth-like early Mars, however, then we might be able to find its fossilized remains preserved on or near the Martian surface. Surprisingly, one place to look for life on Mars is right here on Earth. Scientists think that some meteorites found on Earth's surface come from the Moon and from Mars. These meteorites were apparently blasted off these bodies long ago during an impact of some sort, thrown into space, and eventually trapped by Earth's gravity, ultimately to fall to the ground. The most fascinating of these rocks are surely those from the Red Planet—for one of them may harbor fossil evidence for past life on Mars!

The next figure shows ALH84001, a blackened 2-kg meteorite about 17 cm across, found in 1984 in Antarctica. Based on estimates of the cosmic-ray exposure it received before reaching Earth, the rock is thought to have been blasted off Mars about 16 million years ago. Looking at this specimen through a microscope (inset) scientists can see rounded orange-brown "globules" of carbonate minerals on the rock's shiny crust. Because carbonates form only in the presence of water, the presence of these globules suggests that carbon dioxide gas and liquid water existed near ground level at some point in Mars's history, a conclusion that planetary scientists had earlier drawn from studies of *Viking*'s orbital images of valleys apparently carved by water when the Martian climate was wetter and warmer.

In a widely viewed press conference in Washington, D.C., in the summer of 1996, a group of scientists argued, based on all the data accumulated from studies of ALH84001, that they had discovered fossilized evidence for life on Mars. The key pieces of evidence they presented for primitive Martian life were as follows: (1) Bacteria on Earth can produce structures similar to the globules shown in the inset. (2) The meteorite contains traces of *polycyclic aromatic hydrocarbons*—a tongue-twisting name for a class of complex organic molecules (usually abbreviated PAHs) that, although not directly involved in known biological cycles on Earth, occur among the decay products of plants and other organisms. (3) High-powered electron microscopes show that ALH84001 contains tiny, teardrop-shaped crystals of magnetite and iron sulfide embedded in places where the carbonate has dissolved. On Earth, bacteria are known to manufacture similar chemical crystals. (4) On very small scales, elongated and egg-shaped structures are seen within the carbonate globules. The researchers interpret these minute structures as fossils of primitive organisms.

The photomicrograph at the right shows this fourth, and most controversial, piece of evidence—curved, rodlike structures that resemble bacteria on Earth. Scale is crucial

(NASA)

(Before)

| R | I | V | U | X | G |

(After)

(NASA)

here, however. The structures are only about 0.5 μm across, 30 times smaller than ancient bacterial cells found fossilized on Earth. Furthermore, several key tests have not yet been done, such as cutting through the suspected fossilized tubes to search for evidence of cell walls or semipermeable membranes, or of any internal cavities where body fluids would have resided. Nor has anyone yet found in ALH84001 any amino acids, the basic building blocks of life as we know it (see Chapter 28).

These results remain very controversial. Many experts do not agree that life has been found on Mars—not even fossilized life. Skeptics maintain that all the evidence could be the result of chemical reactions not requiring any kind of biology. Carbonate compounds are common in all areas of chemistry; PAHs are found in many lifeless places (glacial ice, asteroid-belt meteorites, interstellar clouds, and even the exhaust fumes of automobiles); bacteria are not needed to produce crystals; and it remains unclear whether the tiny tubular structures shown are animal, vegetable, or merely mineral. In addition, there is the huge problem of contamination—after all, ALH84001 was found on Earth and apparently sat in the Antarctic ice fields for 13,000 years before being picked up by meteorite hunters.

During 1999, the team released a new analysis of a second meteorite, named Nakhla (shown at right), discovered in the Sahara Desert in 1911 and also believed to have come from Mars. They again report evidence for microbial life, in the form of clusters of minute spheres and ovals found within tiny clay-filled cracks deep inside the meteorite, having similarities in size, shape and arrangement to known nanobacteria on Earth. Since Nakhla is a volcanic basalt rock that solidified about 1.3 billion years ago (compared to 4 billion years for ALH 84001), the new work suggests that life might have spanned the entire history of

Mars. If so, then life might still be present there today. But the opponents remain largely unconvinced.

As things now stand, it's a matter of interpretation—at the frontiers of science, issues are usually not as clear-cut as we would hope. Only additional analysis and new data—perhaps in the form of samples returned directly from Martian surface—will tell for sure if primitive Martian life existed long ago, although most workers in the field seem to have concluded that, taken as a whole, the results do not support the claim of ancient life on Mars. Still, even some skeptics concede that as much as 20 percent of the organic material in ALH84001 could have originated on the Martian surface—although that is a far cry from proving the existence of life there. Should the claim of life on Mars hold up against the weight of healthy skepticism in the scientific community, these findings may go down in history as one of the greatest scientific discoveries of all time. We are—or at least were—not alone in the universe! Maybe.

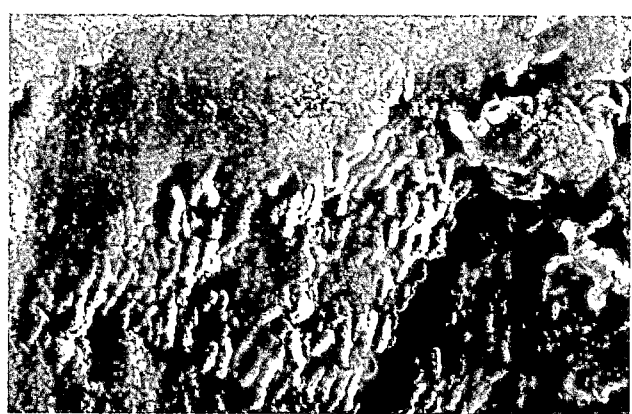

(NASA)

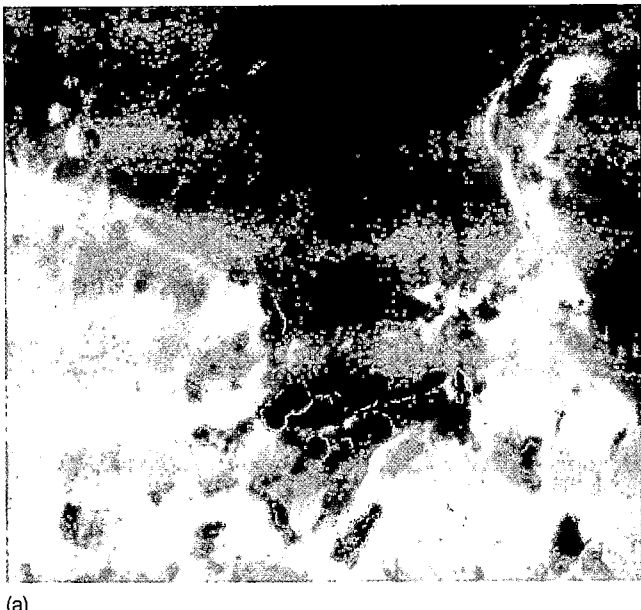

(a)

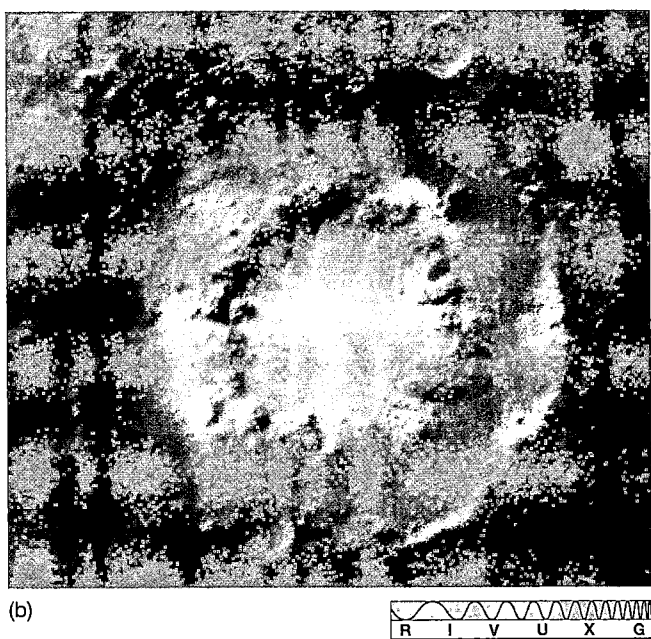

(b)

R   I   V   U   X   G

Figure 10.20 Fog in the Canyons (a) As the Sun's light reaches and heats the canyon floor, it drives water vapor from the surface. When this vapor comes in contact with the colder air above the surface, it condenses again, and a temporary water-ice "fog" results. (b) This is the 200-km-wide Lowell Crater, as seen by *Mars Global Surveyor* in autumn of 2000. *(NASA)*

∞ (Sec. 9.4) Planetary scientists theorize that the early Martian atmosphere also ran away, but in the opposite direction. The presence of liquid water would have caused much of the atmospheric carbon dioxide to dissolve in Martian rivers and lakes (and oceans, if any), ultimately to combine with Martian surface rocks. Recall that on Earth most carbon dioxide is found in surface rocks. Calculations show that much of the Martian atmospheric carbon dioxide could have been depleted in this way in a relatively short period of time, in perhaps as little as a few hundred million years, although some of it might have been replenished by volcanic activity, possibly extending the "comfortable" lifetime of the planet to a half-billion years or so.

As the level of carbon dioxide declined and the greenhouse-heating effect diminished, the planet cooled. The water froze out of the atmosphere, lowering still further the level of atmospheric greenhouse gases, and accelerating the cooling. (Recall from Section 9.4 that water vapor also contributes to the greenhouse effect.)

Since those early times the overall density of the Martian atmosphere has continued to decline through the steady loss of carbon dioxide, nitrogen, and water vapor as solar ultraviolet radiation in the upper atmosphere splits the molecules of these gases into their component atoms, providing some with enough energy to escape. The present level of water vapor in the Martian atmosphere is the maximum possible, given the atmosphere's present density and temperature. Estimates of the total amount of water stored as permafrost or in the polar caps are quite uncertain, but it is likely that if all the water on Mars were to become liquid, it would cover the surface to a depth of several meters.

☑ Concept Check

■ Where did the Martian atmosphere go?

10.6 Martian Internal Structure

3 The *Viking* landers carried seismometers to probe the internal structure of the planet. However, one failed to work, and the other was unable to clearly distinguish seismic activity from the buffeting of the Martian wind. As a result, no seismic studies of the Martian interior have yet been carried out. On the basis of studies of the stresses that occurred during the Tharsis uplift, astronomers estimate the thickness of the crust to be about 100 km.

During its visit to Mars in 1965, *Mariner 4* detected no planetary magnetic field, and for many years the most that could be said about the Martian magnetic field was that its strength was no more than a few thousandths the strength of Earth's field (the level of sensitivity of *Mariner*'s instruments). The *Viking* spacecraft were not designed to make magnetic measurements. In September 1997, *Mars Global Surveyor* detected a very weak Martian field, about 800 times weaker than Earth's. However, this is probably a local anomaly, akin to the magnetic fluctuations detected by *Lunar Prospector* at certain locations on the surface of Earth's Moon, not a global field. ∞ (Sec. 8.7) Because Mars rotates rapidly, the absence of a global magnetic field is taken to mean that the planet's core is nonmetallic, or nonliquid, or both. ∞ (Sec. 7.5) The small size of Mars means that any radioactive (or other internal) heating of its interior would have been less effective at heating and melting the planet than similar heating on Earth. The heat was able to reach the surface and escape more easily than in a larger planet such as Earth or Venus. The evidence we noted earlier for ancient surface activity, especially volcanism, suggests that at least parts of the planet's interior must have melted and possibly differentiated at some time in the past. But the lack of current activity, the absence of any significant magnetic field, the relatively low density (3900 kg/m³), and an abnormally high abundance of iron at the surface all suggest that Mars never melted as extensively as did Earth.

The history of Mars appears to be that of a planet where large-scale tectonic activity almost started but was stifled by the planet's rapidly cooling outer layers. The large upwelling of material that formed the Tharsis bulge might have developed on a larger, warmer planet into full-fledged plate tectonic motion, but the Martian mantle became too rigid and the crust too thick for that to occur. Instead, the upwelling continued to fire volcanic activity, almost up to the present day, but geologically much of the planet apparently died 2 billion years ago.

✔️ Concept Check

■ What is the principal reason for the lack of geological activity on Mars today?

10.7 The Moons of Mars

⑤ Unlike Earth's moon, Mars's moons are tiny compared with their parent planet and orbit very close to it, relative to the planet's radius. Discovered by American astronomer Asaph Hall in 1877, the two Martian moons—Phobos and Deimos—are only a few kilometers across. Their composition is quite unlike that of the planet. Astronomers generally believe that Phobos and Deimos did not form along with Mars but instead are asteroids that were slowed and captured by the outer fringes of the early Martian atmosphere (which, as we have just seen, was probably much denser than the atmosphere today). It is even possible that they are remnants of a single object that broke up during the capture process. They are quite difficult to study from Earth because their proximity to Mars makes it hard to distinguish them from their much brighter parent. The *Mariner* and *Viking* orbiters, however, studied them in great detail.

Both moons, shown in Figure 10.21, are quite irregularly shaped and heavily cratered. The larger of the two is Phobos (Figure 10.21a), which is about 28 km long and 20 km wide and dominated by an enormous 10-km-wide crater named Stickney (after Angelina Stickney, Asaph Hall's wife, who encouraged him to persevere in his observations). The smaller Deimos (Figure 10.21b) is only 16 km long by 10 km wide. Its largest crater is 2.3 km in diameter. The fact that both moons have quite dark surfaces, reflecting no

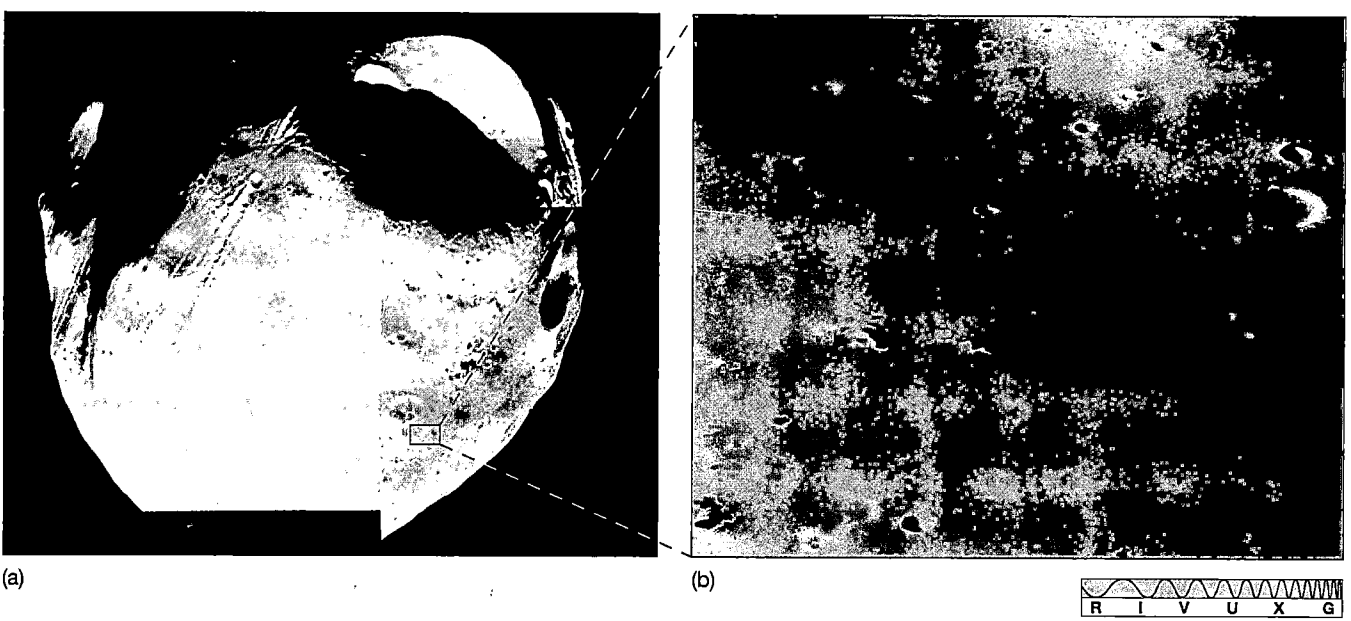

(a)   (b)

Figure 10.21 Martian Moons Mars is accompanied in its trek around the Sun by two tiny moons. Phobos ("fear") and Deimos ("panic") measure about 28 and 16 km long, respectively. Both are shaped like potatoes, with lots of cratering. (a) This is a *Mariner 9* photograph of the irregularly shaped Phobos, not much larger than Manhattan Island. (b) Like Phobos, the smaller moon, Deimos, has a composition unlike that of Mars. Both moons are probably captured asteroids. This photograph was taken by the *Viking* orbiter. The field of view is only 2 km across and most of the boulders shown are about the size of a house. (NASA)

more than 6 percent of the light falling on them, contributes to the difficulty in observing them from Earth.

Phobos and Deimos move in circular, equatorial orbits, and they rotate synchronously (that is, they each keep the same face permanently turned toward the planet). All these characteristics are direct consequences of the tidal influence of Mars. Both moons orbit Mars in the prograde sense—that is, in the same sense (counterclockwise, as seen from above the north celestial pole) as the planet orbits the Sun and rotates on its axis.

Phobos lies only 9378 km (less than three planetary radii) from the center of Mars and, as we saw earlier, has an orbital period of 7 hours and 39 minutes. This orbit period is much less than a Martian day, so an observer standing on the Martian surface would see Phobos move "backward" across the Martian sky—that is, in a direction opposite the apparent daily motion of the Sun. Because the moon moves faster than the observer, it overtakes the planet's rotation, rising in the west and setting in the east, crossing the sky from horizon to horizon in about 5.5 hours. Deimos lies somewhat farther out, at 23,459 km, or slightly less than seven planetary radii, and orbits in 30 hours and 18 min-

utes. Because it completes its orbit in more than a Martian day, it moves "normally," as seen from the ground (that is, from east to west), taking almost 3 days to traverse the sky.

On the basis of measurements of their gravitational effect on the *Viking* orbiters, astronomers have estimated the masses of the two moons. Their densities are around 2000 kg/m^3, far less than that of any world we have yet encountered in our outward journey through the solar system. This is an important reason why astronomers do not believe these moons formed along with Mars. If they are indeed captured asteroids, Phobos and Deimos represent material left over from the earliest stages of the solar system. Astronomers study them not to gain insight into Martian evolution but rather because the moons contain information about the very early solar system, before the major planets had formed.

☑ Concept Check

■ In what ways do Phobos and Deimos differ from Earth's Moon?

Chapter Review

SUMMARY

Mars lies outside Earth's orbit, so it traverses the entire ecliptic plane, as seen from Earth. Its orbit is more elliptical than Earth's, so its distance from the Sun varies more. Mars rotates at almost the same rate as Earth, and its rotation axis is inclined to the ecliptic at almost the same angle as Earth's axis. Because of its axial tilt, Mars has daily and seasonal cycles much like those on our own planet, but they are more complex than those on Earth because of Mars's eccentric orbit. From Earth, the most obvious Martian surface features are the polar caps, which grow and diminish as the seasons change on Mars. The appearance of the planet also changes because of seasonal dust storms that obscure its surface.

Like the atmosphere of Venus, Mars's atmosphere is composed primarily of carbon dioxide. However, unlike Venus, the density of the cool Martian atmosphere is less than 1 percent that of Earth's. Mars may once have had a dense atmosphere, but it was lost, partly to space and partly to surface rocks and subsurface **permafrost** (p. 257) and polar caps. Even today, the thin atmosphere is slowly leaking away. Surface temperatures on Mars average about 50 K cooler than those on Earth. Otherwise, Martian weather is reminiscent of that on Earth, with dust storms, clouds, and fog. The two polar caps on Mars consist of a **seasonal cap** (p. 262), composed of carbon dioxide, which grows and shrinks, and a **residual cap** (p. 262), of water ice, which remains permanently frozen.

In 1971, *Mariner 9* mapped the entire Martian surface, revealing plains, volcanoes, channels, and canyons. *Viking 1* and *Viking 2* reached Mars in 1976 and returned a wealth of data on the planet's surface and atmosphere. Experiments onboard the *Viking* landers detected no evidence for Martian life. There is a marked difference between the two Martian hemispheres. The northern hemisphere consists of rolling volcanic plains and lies several kilometers below the level of the heavily cratered southern hemisphere. The lack of craters in the north suggests that this region is younger. The cause of the north–south asymmetry is not known.

Mars's major surface feature is the Tharsis bulge, located on the planet's equator. It may have been caused by a "plume" of upwelling material in the youthful Martian mantle. Associated with the bulge is Olympus Mons, the largest known volcano in the solar system, and a huge crack, called the Valles Marineris, in the planet's surface. The height of the Martian volcanoes is a direct consequence of Mars's low surface gravity. No evidence for recent or ongoing eruptions has been found. Convection in the Martian interior seems to have been stifled 2 billion years ago by the planet's rapidly cooling and solidifying mantle. On the other side of Mars from Tharsis lies the Hellas basin, the site of a violent meteoritic impact early in the planet's history.

Martian craters differ from those on the Moon by the presence of fluidized ejecta, which provide direct evidence for the permafrost layer beneath the surface, and *Viking* and *Mars Global Surveyor* images indicate that liquid water once existed in great quantity on Mars. Mars may have enjoyed a relatively brief "Earth-like" phase early on in its evolution. The **runoff channels** (p. 258) are the remains of ancient Martian rivers. The **outflow channels** (p. 259) are the paths taken by flash floods that cascaded from the southern highlands into the northern plains. Today, a large amount of that water may be locked up in the polar caps and in the layer of permafrost lying under the Martian surface. *Surveyor* images also suggest that there may still be liquid water below the surface in some places.

Mars has an extremely weak magnetic field. Since the planet rotates rapidly, this implies that its core is nonmetallic, nonliquid, or both. The lack of current volcanism, the absence of any significant magnetic field, the planet's relatively low density, and a high abundance of surface iron all suggest that Mars never melted and differentiated as extensively as did Earth.

The Martian moons Phobos and Deimos are probably asteroids captured by Mars early in its history. Their densities are far less than that of any planet in the inner solar system. They may be representative of conditions in the early solar system.

SELF-TEST: TRUE OR FALSE?

_____ **1.** It is possible, over time, to see Mars at any angular distance from the Sun, at any time of night.

_____ **2.** Seen from Earth, Mars goes through phases, just like Venus and Mercury.

_____ **3.** Mars at times is engulfed by global dust storms.

_____ **4.** Because Mars has such a thin atmosphere, the planet has no significant surface winds.

_____ **5.** Seasonal changes in the appearance of Mars are caused by vegetation on the surface.

_____ **6.** Mars has the largest volcanoes in the solar system.

_____ **7.** The northern hemisphere of Mars is much younger than the southern hemisphere.

_____ **8.** Olympus Mons is the largest impact crater on Mars.

_____ **9.** There are many indications of past plate tectonics on Mars.

_____ **10.** Valles Marineris is similar in size to Earth's Grand Canyon.

_____ **11.** The orange-red color of the Martian surface is primarily due to rust (iron oxide) in its soil.

_____ **12.** Daytime temperatures on Mars can reach 300 K.

_____ **13.** The *Viking* landers are still sending data back to Earth.

_____ **14.** No spacecraft have landed on Mars since the 1970s.

_____ **15.** One of the two moons of Mars is larger than Earth's Moon.

SELF-TEST: FILL IN THE BLANK

1. When Mars is on the opposite side of the Sun from Earth, it is said to be at _____.

2. The seasonal polar ice caps of Mars are composed of _____.

3. The southern hemisphere of Mars consists of heavily _____ highlands.

4. The Tharsis region of Mars is a large equatorial _____.

5. The Hellas region of Mars is a large _____.

6. Several large _____ lie at approximately the center of Tharsis.

7. The great height of Martian volcanoes is a direct result of the planet's low _____.

8. The fluidized ejecta surrounding Martian impact craters is evidence of a layer of _____ just under the surface.

9. Runoff channels carried _____ from the southern mountains into the valleys.

10. Outflow channels are the result of catastrophic _____.

11. Water flowed on the surface of Mars a few _____ years ago.

12. Mars's northern residual polar ice cap consists mostly of _____.

13. Most of the carbon dioxide on Mars is now found in the planet's _____.

14. An extensive _____ may once have covered the northern Martian lowlands.

15. The most recent claims of life on Mars have been based on analyses of _____ found on Earth.

REVIEW AND DISCUSSION

1. Why is opposition the best time to see Mars from Earth?

2. Why are some Martian oppositions better than others for viewing Mars?

3. For a century, there was speculation that intelligent life had constructed irrigation canals on Mars. What did the "canals" turn out to be?

4. Imagine that you will be visiting the southern hemisphere of Mars during its summer. Describe the atmospheric conditions you might face.

5. Describe the two Martian polar caps, their seasonal and permanent composition, and the differences between them.

6. Why is Mars red?

7. Describe the major large-scale surface features of Mars.

8. Why were Martian volcanoes able to grow so large?

9. Why couldn't you breathe on Mars?

10. What is the evidence that water once flowed on Mars?

11. Is there liquid water on Mars today?

12. Is there water on Mars today, in any form?

13. Why do some scientists think Mars once had an extensive ocean? Where was it located?

14. How were the masses of Mars's moons measured, and what did these measurements tell us about their origin?

15. What do measurements of Martian magnetism tell us about the planet's interior?

16. What is the evidence that Mars never melted as extensively as did Earth?

17. How would Earth look from Mars?

18. If humans were sent to Mars to live, what environmental factors would have to be considered? What resources might Mars provide, and which would have to come from Earth?

19. Since Mars has an atmosphere, and it is composed mostly of a greenhouse gas, why isn't there a significant greenhouse effect to warm its surface?

20. Compare and contrast the evolution of the atmospheres of Mars, Venus, and Earth.

PROBLEMS  *Algorithmic versions of these questions are available in the Practice Problems module of the Companion Website.*

The number of squares preceding each problem indicates its approximate level of difficulty.

1. ■▣ By calculating the rate at which Earth overtakes Mars in its orbit (see Chapter 9, Problem 5), verify the value of Mars's synodic period given in Mars Data box on p. 251.

2. ■ Calculate the minimum and maximum angular diameters of the Sun, as seen from Mars.

3. ■■ What is the maximum elongation of Earth, as seen from Mars? (For simplicity, assume circular orbits for both planets.)

4. ■ What will be the minimum size of a Martian surface feature resolvable during the 2003 opposition by an Earth-based telescope with an angular resolution of 0.05"?

5. ■■ Use the reasoning presented in Chapter 1 to calculate the difference in length between the mean Martian solar day and the Martian sidereal day. ∞ (Sec. 1.3)

6. ■ Verify that the surface gravity on Mars is 40 percent that of Earth.

7. ■ What would you weigh on Mars?

8. ■ How long would it take the wind in a Martian dust storm, moving at a speed of 150 km/h, to encircle the planet's equator?

9. ■ The mass of the Martian atmosphere is about 1/150 the mass of Earth's atmosphere and is composed mainly (95 percent) of carbon dioxide. Using the result of Problem 3 in Chapter 7 to determine the mass of Earth's atmosphere, estimate the total mass of carbon dioxide in the atmosphere of Mars. Compare this with the masses of a seasonal polar cap, approximated as a circular sheet of frozen carbon dioxide ("dry ice," having a density of 1600 kg/m^3) of diameter 3000 km and thickness 1 m.

10. ■ Compare the mass of a seasonal polar cap (previous question) to that of a residual cap, of diameter 1000 km, thickness 1 km, and density 1000 kg/m^3.

11. ■ The Hellas impact basin is roughly circular, 3000 km across, and 6 km deep. Taking the Martian crust to have a density of 3000 kg/m^3, estimate how much mass was blasted off the Martian surface when the basin formed. Compare this to the present total mass of the Martian atmosphere (Problem 9).

12. ■■ The outflow channel shown in Figure 10.12 is about 10 km across and 100 m deep. If it carried 10^7 metric tons (10^{10} kg) of water per second, as stated in the text, estimate the speed at which the water must have flowed.

13. ■■ Calculate the total mass of a uniform layer of water covering the entire Martian surface to a depth of 2 m (see Section 10.5). Compare this with the mass of Mars.

14. ■■ Using the data from Section 10.7, compute the time interval between each rising of the moons Phobos and Deimos and the next, as seen from the Martian surface.

15. ■ Using the data given in the text, calculate the maximum angular sizes of Phobos and Deimos, as seen by an observer standing on the Martian surface directly under their orbits. Would a Martian observer ever see a total solar eclipse (see Problem 2)?

COLLABORATIVE EXERCISES

1. Age on Mars. Use the Planetary Data: Mars table to determine how many Martian years old each member of your group would be if they were born on Mars.

2. Evidence for Life. As a group, decide on exactly what evidence you need in order to believe that life, as your group defines it, exists or existed on Mars.

3. Water on Mars. As a group, agree on which image from the text provides the most convincing evidence that water once ran on the surface of Mars. Explain your reasoning.

RESEARCHING ON THE WEB  *To complete the following exercises, go to the online Destinations module for Chapter 10 on the Companion Website for Astronomy Today 4/e.*

1. Access the "Mars Today" page and determine the current wind speed and temperature at the Martian equator.
2. Access the "On the Question of the Mars Meteorite" page and list the three main lines of evidence that suggest the meteorite contains evidence of life on Mars.

3. Access the "Mars Surface Features" page and describe the difference between a *colles* and a *mons*.

PROJECTS

1. Track the motion of the Red Planet in front of the stars for several months following its return to the predawn sky. (Consult an almanac to determine where Mars will be in the sky this year.) You will see that Mars moves rapidly in front of the stars, crossing many constellation boundaries.
2. Several months before opposition, Mars begins retrograde motion. Chart the planet's motion in front of the stars to determine when it stops moving eastward and begins moving toward the west.
3. Notice the increase in Mars's brightness as it approaches opposition. Why is it getting brighter? What other planets

appear in the sky now? How do their brightnesses compare with that of Mars?
4. Look at Mars with as large a telescope as is available to you; binoculars will not be of use. Prepare ahead of time and find out the Martian season that will be occurring at the time of your observation, which hemisphere will be tilted in Earth's direction, and what longitude will be pointing toward Earth at the time of observation. This information can be obtained from many different computer almanacs. Sketch what you see. Look very carefully and take your time. Afterward, try to identify the various features you have seen with known objects on Mars.

SKYCHART III PROJECTS  *The SkyChart III Student Version planetarium program on which these exercises are based is included as a separately executable program on the CD in the back of this text.*

1. ■ Configure SkyChart III with a 90° field of view and center on Mars. Set *ANIMATE/Trail For/Mars* with *Record apparent positions against fixed stars*. Deselect *DRAW/Horizon Mask* and animate with 1-day steps to observe the motion of Mars against the background stars. Beginning on March 1, 2000, run the animation until December 1, 2003. Explain the motion exhibited by Mars between March 1, 2001 and October 1, 2001; and again explain the motion between June 1, 2003 and December 1, 2003. Why do these two motions appear to be different? Better control of the advance can be obtained using keys F5 and F6.

2. ■■ Retrograde motion is an illusion, appearing as Earth overtakes Mars in orbit around the Sun. Create a chart similar to Figure 2.8 illustrating the illusion. Set up SkyChart so that the inner planets can be seen from a position perpendicular to the ecliptic, about 60 AU offset from the Sun (see appendix for setup instructions). Set the *Date & Time* prior to one of the retrograde cycles. *Trail* for both Earth and Mars and adjust the *Time Step* to space out the *Ticks*. Animate forward for several months and print the final result. From the printout, connect the *Ticks* between Earth and Mars in consecutive order, extending the connecting lines to the edge of the page.

 *In addition to the Practice Problems and Destinations modules, the Companion Website at http://www.prenhall.com/chaisson provides for each chapter an additional true-false, multiple choice, and labeling quiz, as well as additional annotated images, animations, and links to related Websites.*

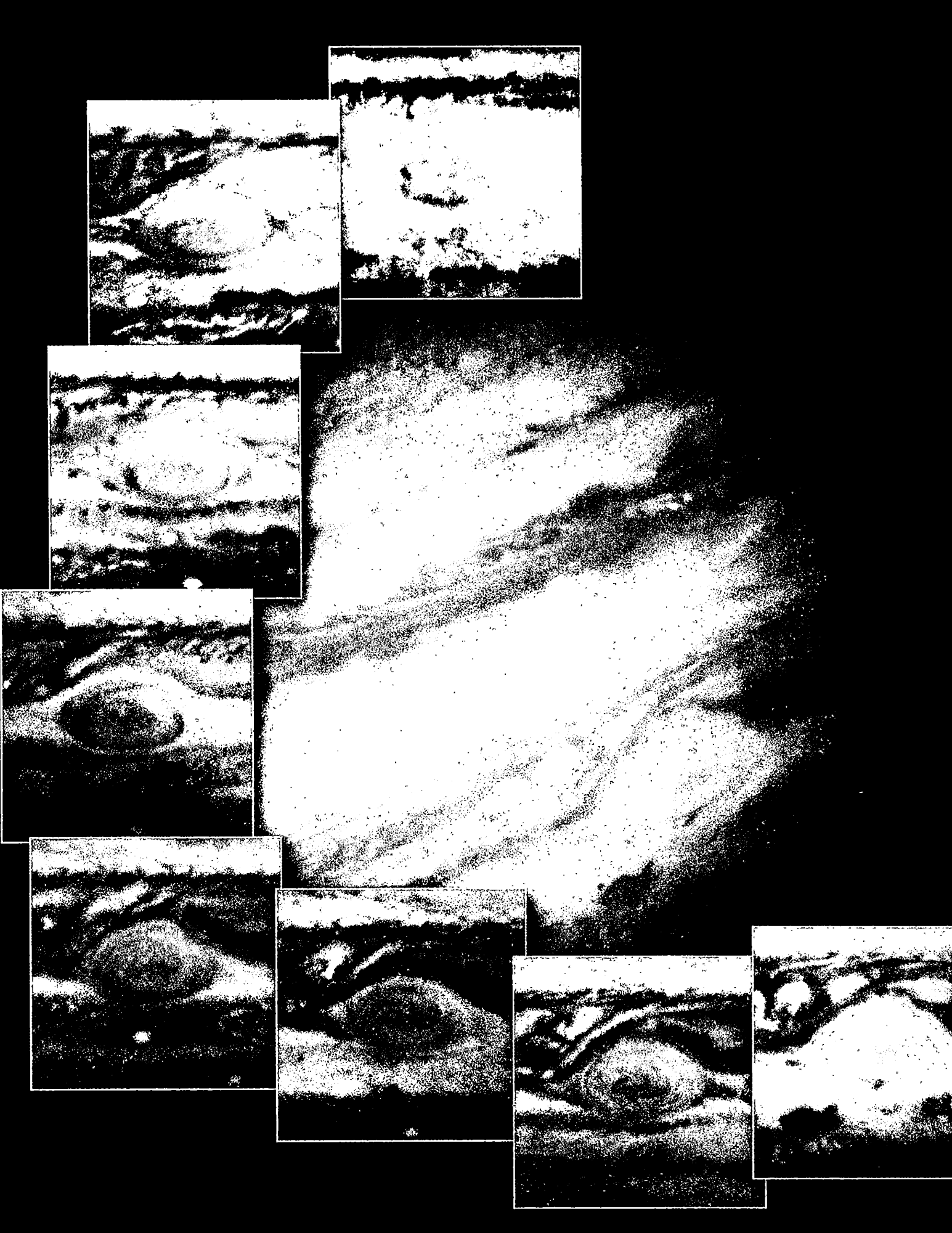

11 JUPITER

Giant of the Solar System

LEARNING GOALS

Studying this chapter will enable you to:

1   Specify the ways in which Jupiter differs from the terrestrial planets in its physical and orbital properties.

2   Discuss the processes responsible for the appearance of Jupiter's atmosphere.

3   Describe Jupiter's internal structure and composition and explain how these are inferred from external measurements.

4   Summarize the characteristics of Jupiter's magnetosphere.

5   Discuss the orbital properties of the Galilean moons of Jupiter and describe the appearance and physical properties of each.

6   Explain how tidal forces can produce enormous internal stresses in a jovian moon, and discuss some results of those stresses.

 Visit http://www.prenhall.com/chaisson for additional annotated images, animations, and links to related sites for this chapter.

Jupiter's Red Spot has been a source of fascination since early seventeenth-century astronomers first turned their telescopes toward this giant planet. More than 300 years later, this vast storm is still there, spinning like a cyclone, the winds within it reaching 400 kilometers per hour. With a diameter of 25,000 km, the Red Spot is the largest known storm in the solar system, in fact it is nearly twice the size of Earth. This mosaic shows a series of images taken by the *Hubble Space Telescope* over a seven-year period during the 1990s. (STScI)

The Big Picture: Astronomers study weather phenomena on other planets partly to gain a better understanding of our own planet, but we also study other planets specifically to learn more about them. Each one seems to display considerably different conditions than those on Earth, and until all of them are thoroughly probed and understood, our inventory of the local part of the universe will be incomplete.

Beyond the orbit of Mars, the solar system is very different from our own backyard. The outer solar system presents us with a totally unfamiliar environment—huge gas balls, peculiar moons, complex ring systems, and a wide variety of physical and chemical phenomena, many of which are still only poorly understood. Although the jovian planets—Jupiter, Saturn, Uranus, and Neptune—differ from one another in many ways, we will find that they have much in common, too. As with the terrestrial planets, we will learn from their differences as well as from their similarities. Our study of these alien places begins with the jovian planet closest to Earth—Jupiter, the largest planet in the solar system and a model for the other jovian worlds.

11.1 Orbital and Physical Properties

▌ Named after the most powerful god of the Roman pantheon, Jupiter is by far the largest planet in the solar system. Ancient astronomers could not have known the planet's true size, but their choice of names was very apt. The Jupiter Data box on p. 289 presents some orbital and physical data on the planet.

THE VIEW FROM EARTH

Jupiter is the third-brightest object in the night sky (after the Moon and Venus), making it very easy to locate and study. As in the case of Mars, Jupiter is brightest when it is near opposition. When this happens to occur near perihelion, the planet can be up to 50″ across, and a lot of detail can be discerned through even a small telescope.

Figure 11.1(a) is a *Hubble Space Telescope* image of Jupiter taken during the opposition of December 1990. Notice the alternating light and dark bands that cross the planet parallel to its equator, and also the large oval at lower right. These atmospheric features are quite unlike anything found on the inner planets. Also in contrast to the terrestrial worlds, Jupiter has many moons that vary greatly in size and other properties (Figure 11.1b). The four largest are visible from Earth with even a small telescope (and, for a few people, with the naked eye). They are known as the **Galilean moons** after Galileo Galilei, who discovered them in 1610. ∞ (Sec. 2.4)

MASS AND RADIUS

Since astronomers have been able to study the motion of the Galilean moons for quite some time, Jupiter's mass has long been known. It is 1.9×10^{27} kg, or 318 Earth masses. ∞ (*More Precisely 6-1*) Jupiter has more than twice the mass of all the other planets combined. Jupiter is such a large planet that many celestial mechanicians—those researchers concerned with the motions of interacting cosmic objects—regard our solar system as containing only two important objects: the Sun and Jupiter. To be sure, in this age of sophisticated and precise spacecraft navigation, the gravitational influence of all the planets must be con-

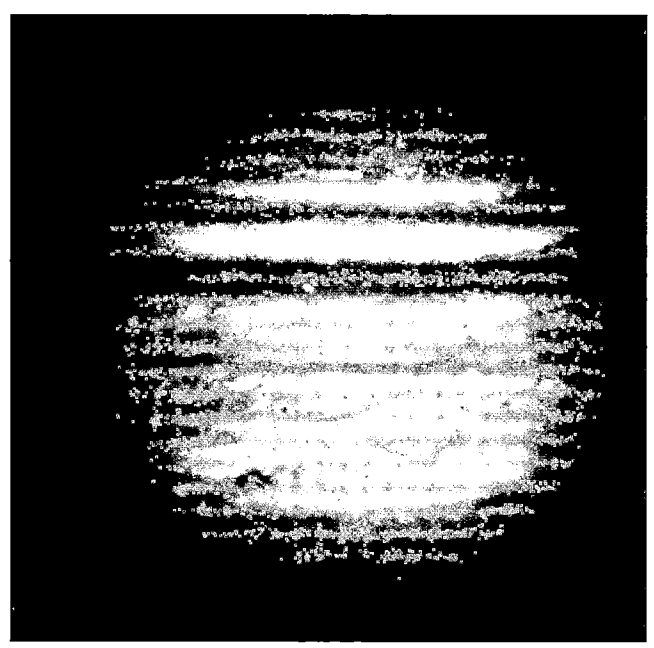

(a)

(b)

Figure 11.1 Jupiter (a) A *Hubble Space Telescope* image of Jupiter, in true color. Features as small as a few hundred kilometers across are resolved. (b) Photograph of Jupiter made with a ground-based telescope, showing several of the planet's moons. *(AURA; NASA)*

sidered, but in the broadest sense, our solar system is a two-object system with a lot of debris. As massive as Jupiter is, though, it is still only a thousandth the mass of the Sun. This makes studies of Jupiter all the more important, for here we have an object intermediate in size between the Sun and the terrestrial planets.

Knowing Jupiter's distance and angular size, we can easily determine its radius. It is 71,500 km, or 11.2 Earth radii. More dramatically stated, more than 1400 Earths would be needed to equal the volume of Jupiter. From the size and mass, we derive an average density of 1300 kg/m³ for the planet. Here (as if we needed it) is yet another indicator that Jupiter is radically different from the terrestrial worlds. It is clear that, whatever Jupiter's composition, it is not made up of the same material as the inner planets (recall from Chapter 7 that Earth's average density is 5500 kg/m³). ∞ (Sec. 7.1) In fact, theoretical studies of the planet's internal structure indicate that Jupiter must be composed primarily of hydrogen and helium. The enormous pressure in the planet's interior greatly compresses these light gases (whose densities on Earth at room temperature and sea level are 0.08 and 0.16 kg/m³, respectively), producing the relatively high average density we observe.

ROTATION RATE

As with other planets, we can attempt to determine Jupiter's rotation rate simply by timing a surface feature as it moves around the planet. However, in the case of Jupiter (and, indeed, all the gaseous outer planets), there is a catch—Jupiter has no solid surface. All we see are cloud features in the planet's upper atmosphere. With no solid surface to "tie them down," different parts of Jupiter's atmosphere move independently of one another. Visual observations and Doppler-

shifted spectral lines indicate that the equatorial zones rotate a little faster ($9^h 50^m$ period) than the higher latitudes ($9^h 55^m$ period). Jupiter exhibits **differential rotation**—the rotation rate is not constant from one location to another. Differential rotation is not possible in solid objects like the terrestrial planets, but it is normal for fluid bodies such as Jupiter.

Observations of Jupiter's magnetosphere provide a more meaningful measurement of the rotation period. The planet's magnetic field is strong and emits radiation at radio wavelengths as charged particles accelerate in response to the planet's magnetic field. Careful studies show a periodicity of $9^h 55^m$ in this radio emission. We assume that this measurement matches the rotation of the planet's interior, where the magnetic field arises. ∞ (Sec. 7.5) Thus, Jupiter's interior rotates at the same rate as the clouds at its poles. The equatorial zones rotate more rapidly.

A rotation period of $9^h 55^m$ is fast for such a large object. In fact, Jupiter has the fastest rotation rate of any planet in the solar system, and this rapid spin has altered Jupiter's shape. As illustrated in Figure 11.2, a spinning object tends to develop a bulge around its midsection. The more loosely the object's matter is bound together, or the faster it spins, the larger the bulge becomes. In objects like Jupiter, which are made up of gas or loosely packed matter, high spin rates can produce a quite pronounced bulge. Jupiter's equatorial radius (71,500 km) exceeds its polar radius (66,900 km) by about 6.5 percent.*

---

*Earth also bulges slightly at the equator because of rotation. However, our planet is much more rigid than Jupiter, and the effect is much smaller—the equatorial diameter is only about 40 km larger than the distance from pole to pole, a tiny difference compared with Earth's full diameter of nearly 13,000 km. Relative to its overall dimensions, Earth is smoother and more spherical than a billiard ball.

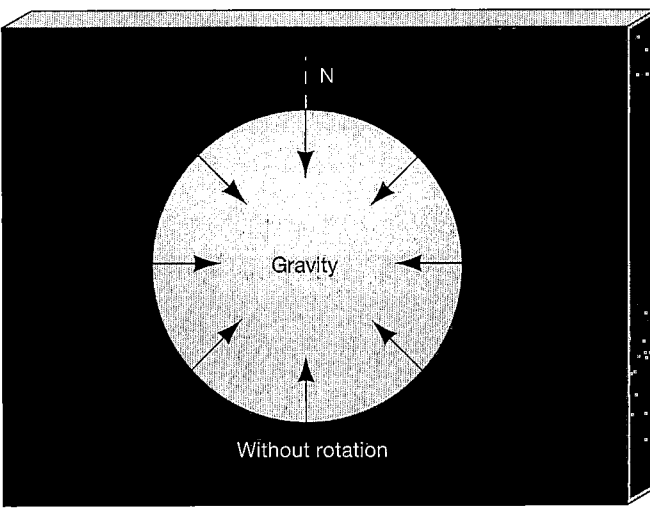

(a)

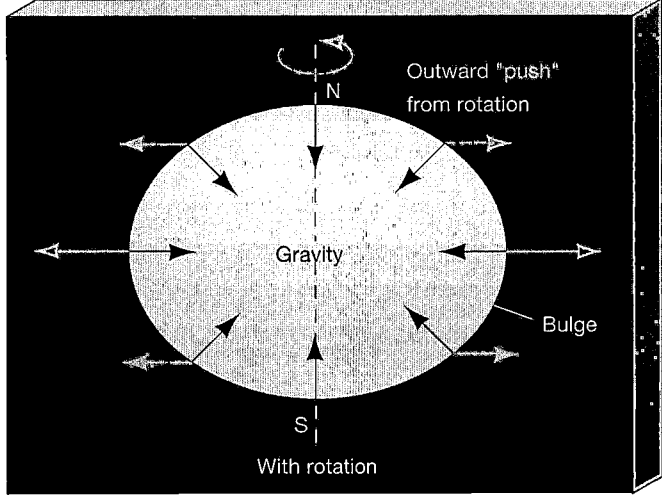

(b)

Figure 11.2 Rotational Flattening All spinning objects tend to develop an equatorial bulge because rotation causes matter to push outward against the inward-pulling gravity. The size of the bulge depends on the mechanical strength of the matter and the rate of rotation. The inward-pointing arrows denote gravity, the outward arrows the "push" due to rotation.

But there is more to the story of Jupiter's shape. Jupiter's observed equatorial bulge also tells us something very important about the planet's deep interior. Careful calculations indicate that Jupiter would be *more* flattened than it actually is if its core were composed of hydrogen and helium alone. To account for the planet's observed shape, we must assume that Jupiter has a dense, compact core, probably of rocky composition, about 10 times the mass of Earth. This is one of the few pieces of data we have on Jupiter's internal structure.

☑ Concept Check

■ How do observations of a planet's magnetosphere allow astronomers to measure the rotation rate of the interior?

11.2 The Atmosphere of Jupiter

② Jupiter is visually dominated by two features: one is a series of ever-changing atmospheric bands arranged parallel to the equator; the other is an oval atmospheric blob called the **Great Red Spot**, or often just the "Red Spot." The cloud bands, clearly visible in Figure 11.1, display many colors—pale yellows, light blues, deep browns, drab tans, and vivid reds among others. Shown in more detail in Figure 11.3, a close-up photograph taken as *Voyager 1* sped past in 1979, the Red Spot is the largest of many features associated with Jupiter's weather. It seems to be an Earth-sized hurricane that has persisted for hundreds of years.

ATMOSPHERIC COMPOSITION

Spectroscopic studies of sunlight reflected from Jupiter gave astronomers their first look at the planet's atmospheric composition. Radio, infrared, and ultraviolet observations later provided more details. The most abundant gas is molecular hydrogen (86.1 percent by number), followed by helium (13.8 percent). Together these two gases make up over 99 percent of Jupiter's atmosphere. Small amounts of atmospheric methane, ammonia, and water vapor are also found. Researchers believe that hydrogen and helium in these same proportions also make up the bulk of the planet's interior.

The abundance of hydrogen and helium on Jupiter is a direct consequence of Jupiter's strong gravity. Unlike the gravitational pull of the terrestrial planets, the gravity of the much more massive jovian planets is powerful enough to have retained even hydrogen. ∞ (*More Precisely 8-1*) Little, if any, of Jupiter's original atmosphere has escaped since the planet formed 4.6 billion years ago.

ATMOSPHERIC BANDS

Astronomers generally describe Jupiter's banded appearance—and, to a lesser extent, the appearance of the other jovian worlds as well—as a series of bright *zones* and dark *belts* crossing the planet. The zones and belts vary in both latitude and intensity during the year, but the general pattern remains. These variations are not seasonal in nature—Jupiter has no seasons—but instead appear to be the result of dynamic motion in the planet's atmosphere. The light-colored zones lie above upward-moving convective currents in Jupiter's atmosphere. The dark belts are regions

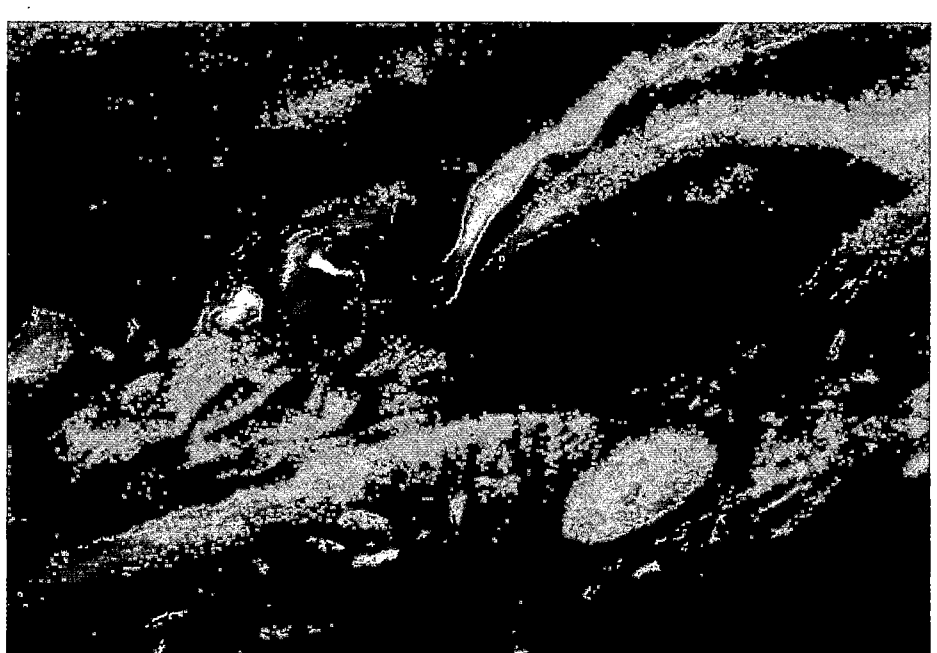

Figure 11.3 Jupiter's Red Spot *Voyager 1* took this photograph of Jupiter's Red Spot (upper right) from a distance of about 100,000 km. The resolution is about 100 km. Note the complex turbulence to the left of both the Red Spot and the smaller white oval vortex below it. (For scale, planet Earth is about the size of the white oval.) *(NASA)*

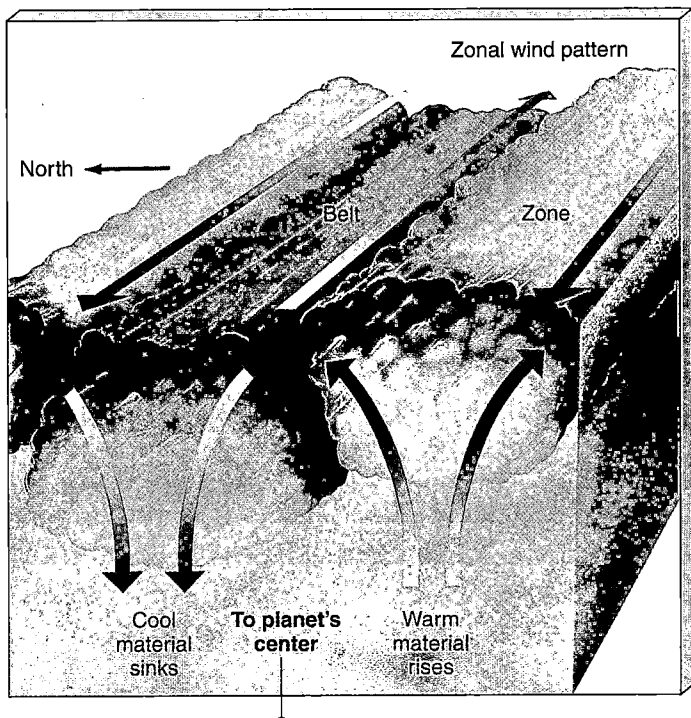

Figure 11.4 Jupiter's Convection Colored bands in Jupiter's atmosphere are associated with vertical convective motion. Upwelling warm gas results in the lighter-colored zones; the darker bands overlie lower-pressure regions where cooler gas sinks back down into the atmosphere. As on Earth, surface winds tend to blow from high- to low-pressure regions. Jupiter's rotation channels these winds into an east–west flow pattern.

representing the other part of the convection cycle, where material is generally sinking downward, as illustrated schematically in Figure 11.4.

Because of the upwelling material below them, the zones are regions of high pressure; the belts, conversely, are low-pressure regions. Thus, belts and zones are Jupiter's equivalents of the familiar high- and low-pressure systems that cause our weather on Earth. However, a major difference between Jupiter and Earth is that Jupiter's rapid rotation has caused these systems to wrap all the way around the planet, instead of forming localized circulating storms, as on our own world. Because of the pressure difference between the two, the zones lie slightly higher in the atmosphere than the belts. The associated temperature differences (recall that temperature increases with depth) and the resulting differences in chemical reactions are the basic reasons for their different colors.

Underlying the bands is an apparently very stable pattern of eastward and westward wind flow, known as Jupiter's **zonal flow**. It is evident in Figure 11.5, which shows the wind speed at different planetary latitudes measured relative to the rotation of the planet's interior (determined from studies of Jupiter's magnetic field). As mentioned earlier, the equatorial regions of the atmosphere rotate faster than the planet, with an average flow speed of some 85 m/s, or about 300 km/h, in the easterly direction. The speed of this equatorial flow is quite similar to that of the jet stream on Earth. At higher latitudes, there are alternating regions of westward and eastward flow, roughly symmetric about the equator, with the flow speed generally diminishing toward the poles.

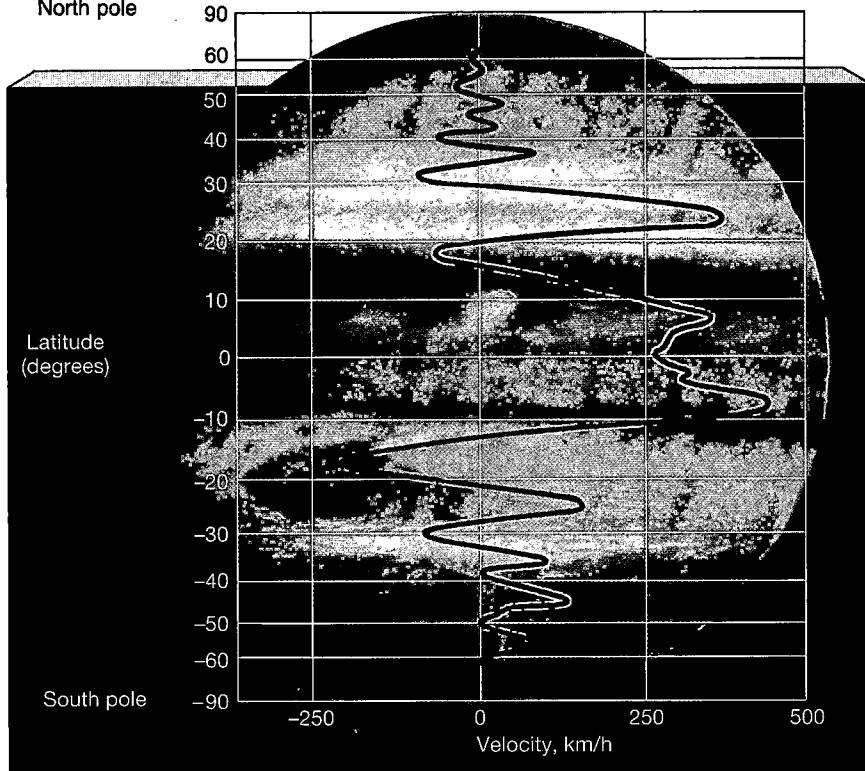

Figure 11.5 Zonal Flow The wind speed in Jupiter's atmosphere, measured relative to the planet's internal rotation rate. Alternations in wind direction are associated with the atmospheric band structure.

As Figure 11.5 shows, the belts and zones are closely related to Jupiter's zonal flow pattern. However, closer inspection shows that the simplified picture presented in Figure 11.4, with wind direction alternating between adjacent bands as Jupiter's rotation deflects surface winds into eastward or westward streams, is really too crude to describe the actual flow. Scientists now believe that the interaction between convective motion in Jupiter's atmosphere and the planet's rapid rotation channels the largest eddies into the observed zonal pattern, but that smaller eddies tend to cause irregularities in the flow. Near the poles, where the zonal flow disappears, the band structure vanishes also.

ATMOSPHERIC STRUCTURE AND COLOR

None of the atmospheric gases listed earlier can, by itself, account for Jupiter's observed coloration. For example, frozen ammonia and water vapor would simply produce white clouds, not the many colors actually seen. Scientists

believe that the cloud colors are the result of complex chemical processes occurring in the planet's turbulent upper atmosphere, although the details are still not fully understood. When we observe Jupiter's colors, we are actually looking down to many different depths in the planet's atmosphere.

Based on the best available data and mathematical models, Figure 11.6 is a cross-sectional diagram of Jupiter's atmosphere. Since the planet lacks a solid surface to use as a reference level for measuring altitude, the top of the troposphere is conventionally taken to lie at 0 km. As on all planets, weather on Jupiter is the result of convection in the troposphere, so the clouds, which are associated with planetary weather systems, all lie at negative altitudes in the diagram. Just above the troposphere lies a thin, faint layer of haze created by photochemical reactions (that is, reactions involving sunlight) similar to those that cause smog on Earth. The temperature at this level is about 110 K; it increases with altitude as the atmosphere absorbs solar ultraviolet radiation.

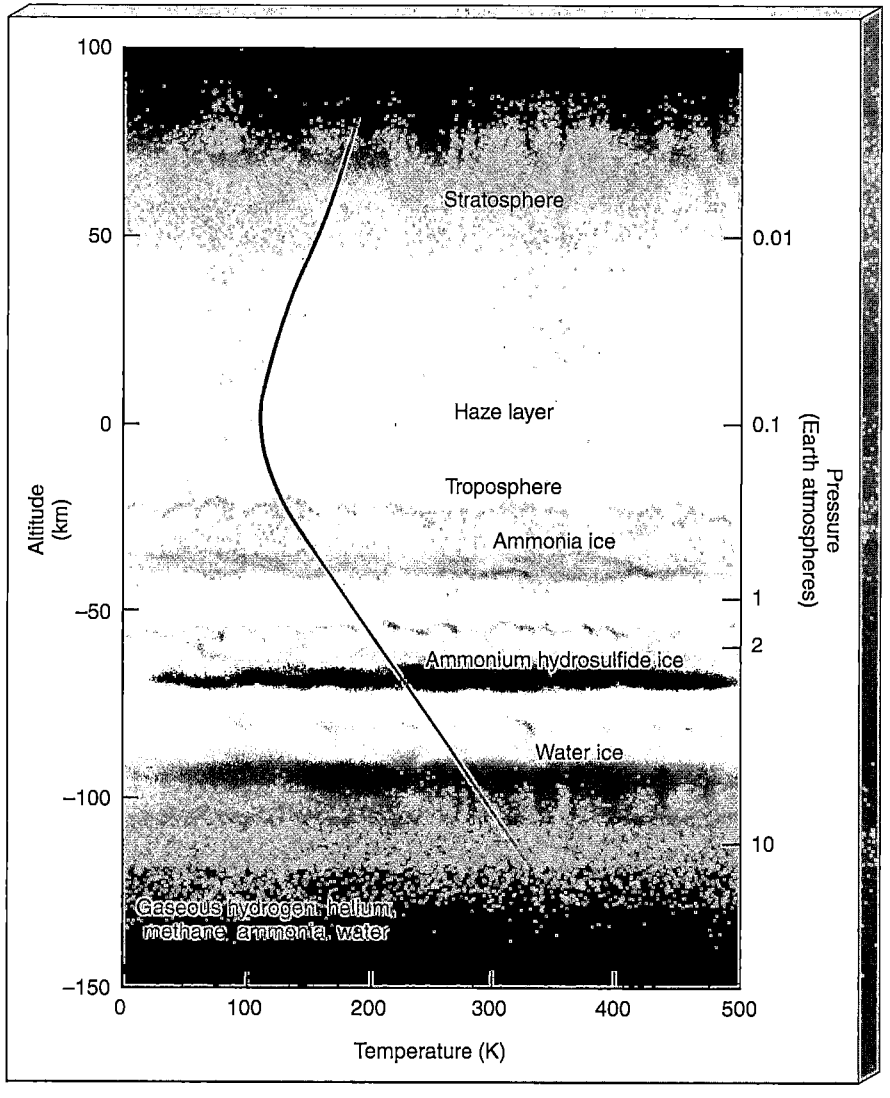

Figure 11.6 Jupiter's Atmosphere
The vertical structure of Jupiter's atmosphere. Jupiter's clouds are arranged in three main layers, each with quite different colors and chemistry. The colors we see in photographs of the planet depend on the cloud cover. The white regions are the tops of the upper ammonia clouds. The yellows, reds, and browns are associated with the second cloud layer, which is composed of ammonium hydrosulfide ice. The lowest cloud layer is water ice and bluish in color. However, the overlying layers are sufficiently thick that this level is not seen in visible light.

Orbital semimajor axis	5.20 A.U.
	778.4 million km
Orbital eccentricity	0.048
Perihelion	4.95 A.U.
	740.7 million km
Aphelion	5.46 A.U.
	816.1 million km
Mean orbital speed	13.1 km/s
Sidereal orbital period	11.86 tropical years
Synodic orbital period	398.88 solar days
Orbital inclination to the ecliptic	1.31°
Greatest angular diameter, as seen from Earth	50″
Mass	1.90×10^{27} kg
	317.8 (Earth = 1)
Equatorial radius	71,492 km
	11.21 (Earth = 1)
Mean density	1330 kg/m³
	0.241 (Earth = 1)
Surface gravity (at cloud tops)	24.8 m/s²
	2.53 (Earth = 1)
Escape speed	59.5 km/s
Sidereal rotation period	0.41 solar days
Axial tilt	3.08°
Surface magnetic field	13.89 (Earth = 1)
Magnetic axis tilt relative to rotation axis	9.6°
Surface temperature	124 K (at cloudtops)
Number of moons	16 (named), 28 (total)

Jupiter's clouds are arranged in three main layers. Below the haze, at a depth of about 40 km (shown as −40 km in Figure 11.6), lies a layer of white wispy clouds made up of ammonia ice. The temperature here is approximately 125–150 K; it increases quite rapidly with increasing depth. A few tens of kilometers below the ammonia clouds, the temperature is a little warmer—over 200 K—and the clouds are probably made up mostly of droplets or crystals of ammonium hydrosulfide, produced by reactions between ammonia and hydrogen sulfide in the planet's atmosphere. At deeper levels in the atmosphere, the ammonium hydrosulfide clouds give way to clouds of water ice or water vapor. This lowest cloud layer, which is not seen in visible-light images of Jupiter, lies some 80 km below the top of the troposphere.

Instead of being white (the color of ammonium hydrosulfide on Earth), Jupiter's middle cloud layer is tawny in color. This is the level at which atmospheric chemistry begins to play a role in determining Jupiter's appearance. Many planetary scientists believe that molecules containing the element sulfur, and perhaps even sulfur itself, are important in influencing the cloud colors—particularly the reds, browns, and yellows, all colors associated with sulfur or sulfur compounds. It is also possible that compounds containing the element phosphorus contribute to the coloration.

Deciphering the detailed causes of Jupiter's distinctive colors is a difficult task, however. The cloud chemistry is complex and very sensitive to small changes in atmospheric conditions, such as pressure and temperature, as well as to chemical composition. The atmosphere is in incessant, churning motion, causing these conditions to change from place to place and from hour to hour. In addition, the energy that powers the reactions comes in many different forms: the planet's own interior heat, solar ultraviolet radiation, aurorae in the planet's magnetosphere, and lightning discharges within the clouds themselves. All of these factors combine to keep a complete explanation of Jupiter's appearance beyond our present grasp.

This description of Jupiter's atmosphere, based largely on *Voyager* data, was put to the test in December 1995 when the *Galileo* atmospheric probe arrived at the planet. ⊂⊃ (Sec. 6.6) The probe survived for about an hour before being crushed by atmospheric pressure at an altitude of −150 km (that is, right at the bottom of Figure 11.6). Initially, the data from the probe appeared to contradict many of the facts we have just reported! Preliminary analysis of the data indicated that Jupiter was much windier, hotter, and drier than expected, and severely depleted in helium. Furthermore, no clear evidence of the three-layered cloud structure depicted in Figure 11.6 was found.

Most of these discrepancies are now known to have been the result of improperly calibrated instruments; *Galileo*'s revised findings on wind speed, temperature, and composition are in good agreement with the picture just presented. In addition, the probe's entry location was in

Jupiter's equatorial zone and, as luck would have it, coincided with an atypical hole almost devoid of upper-level clouds (see Figure 11.7). The warm temperature readings—425 K at 150 km depth, a little higher than indicated by Figure 11.6—are consistent with the probe's having entered a relative clearing in Jupiter's cloud decks, where convective heat can more readily rise (and thus be detected). The low water content may also be normal for the hot, windy regions near Jupiter's equator.

Most experts were somewhat surprised by the depth to which Jupiter's winds continued. *Galileo*'s probe measured high wind speeds throughout its descent into the clouds, not just at the cloud tops, implying that heat deep within the planet, rather than sunlight, drives Jupiter's weather patterns. Finally, complex organic molecules were sought but not found. Some simple carbon-based molecules, such as ethane (C_2H_6), were detected by one of the onboard spectrometers, but nothing suggesting prebiotic compounds or bacteria floating in the atmosphere. That same instrument also detected traces of phosphine (PH_3), which may be a key coloring agent for Jupiter's clouds.

WEATHER ON JUPITER

In addition to the zonal flow pattern, Jupiter has many "small-scale" weather patterns. The Great Red Spot (Figure 11.3) is a prime example. The Great Red Spot was first reported by the British scientist Robert Hooke in the mid-seventeenth century, and we can be reasonably sure that it has existed continuously, in one form or another, for over 300 years. It may well be much older. *Voyager* observations showed the spot to be a region of swirling, circulating

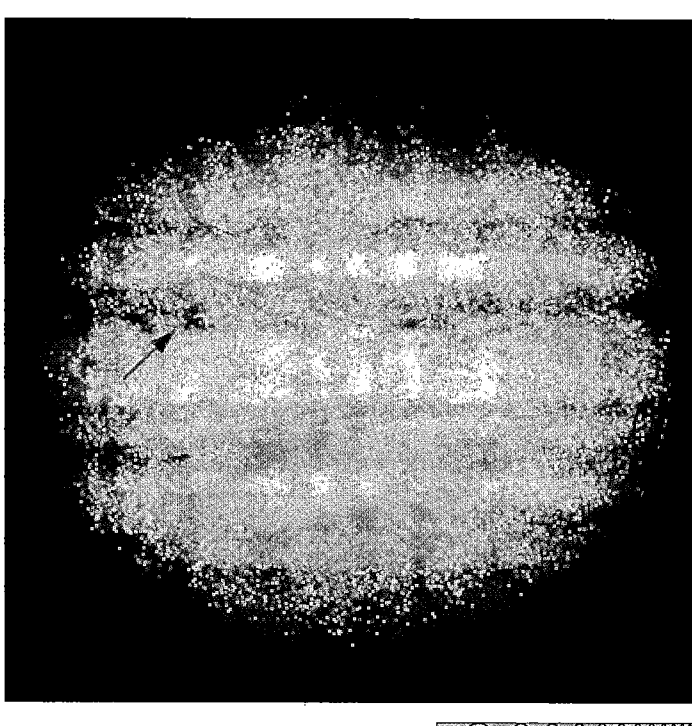

R I V U X G

Figure 11.7 Galileo's Entry Site The arrow on this *Hubble* image shows where the *Galileo* atmospheric probe plunged into Jupiter's cloud deck on December 7, 1995. The entry location was in Jupiter's equatorial zone, and was apparently almost devoid of upper-level clouds. Until its demise, the probe took numerous meteorological measurements, transmitting those signals to the mother ship overhead, which then relayed them to Earth. *(NASA)*

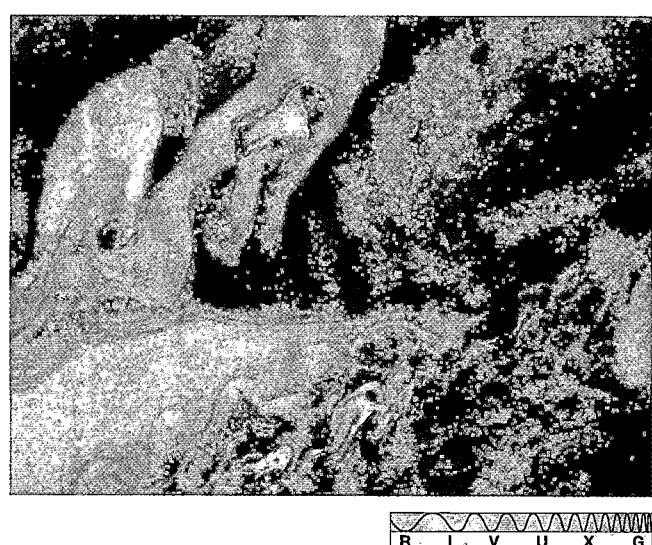

R I V U X G

Figure 11.8 Red Spot Details These *Voyager 2* close-up views of the Great Red Spot, taken four hours apart, show clearly the turbulent flow around its edges. The general direction of motion of the gas north of (above) the Red Spot is westward (to the left), whereas gas south of the Red Spot flows east. The Spot itself rotates counterclockwise, suggesting that it is being "rolled" between the two oppositely directed flows. The colors have been exaggerated somewhat to enhance the contrast. *(NASA)*

winds, rather like a whirlpool or a terrestrial hurricane—a persistent and vast atmospheric storm. The size of the Spot varies, although it averages about twice the diameter of Earth. Its present dimensions are roughly 25,000 km by 15,000 km. It rotates around Jupiter at a rate similar to that of the planet's interior, perhaps suggesting that its roots lie far below the atmosphere.

The origin of the Spot's red color is unknown, as is its source of energy, although it is generally supposed that the Spot is somehow sustained by Jupiter's large-scale atmospheric motion. Repeated observations show that the gas flow around the Spot is counterclockwise, with a period of about six days. Turbulent eddies form and drift away from its edge. The Spot's center, however, remains quite tranquil in appearance, like the eye of a hurricane on Earth. The zonal motion north of the Spot is westward, whereas that to the south is eastward (see Figure 11.8), supporting the idea that the Spot is confined and powered by the zonal flow. However, the details of how this occurs are still a matter of conjecture. Computer simulations of the complex fluid dynamics of Jupiter's atmosphere are only now beginning to hint at answers.

Storms, which as a rule are much smaller than the Red Spot, may be quite common on Jupiter. Spacecraft pho-

tographs of the dark side of the planet reveal both auroral activity and bright flashes resembling lightning. The *Voyager* mission discovered many smaller light- and dark-colored spots that are also apparently circulating storm systems. Note the **white ovals** in Figures 11.3 and 11.8, south of the Red Spot. Like the Red Spot, they rotate counterclockwise. Their high cloud tops give them their color. These particular white ovals are known to be at least 40 years old. Figure 11.9 shows a **brown oval**, a "hole" in the clouds that allows us to look down into the lower atmosphere. For unknown reasons, brown ovals appear only in latitudes around 20°N. Although not as long-lived as the Red Spot, these systems can persist for many years or even decades.

We cannot explain their formation, but we can offer at least a partial explanation for the longevity of storm systems on Jupiter. On Earth a large storm, such as a hurricane, forms over the ocean and may survive for many days, but it dies quickly once it encounters land. Earth's continental landmasses disrupt the flow patterns that sustain the storm. Jupiter has no continents, so once a storm becomes established and reaches a size at which other storm systems cannot destroy it, apparently little affects it. The larger the system, the longer its lifetime.

☑ Concept Check

■ List some similarities and differences between Jupiter's belts, zones, and spots and weather systems on Earth.

Figure 11.9 Brown Oval A close-up view of a brown oval in Jupiter's northern hemisphere. Its color comes from the fact that it is actually a break in the upper cloud layer, allowing us to see deeper into the atmosphere. The oval's length is approximately equal to the diameter of Earth. *(NASA)*

11.3 Internal Structure

⅗ On the basis of Jupiter's distance from the Sun, astronomers had expected to find the temperature of the cloud tops to be around 105 K. At that temperature, they reasoned, Jupiter would radiate back into space exactly the same amount of energy as it received from the Sun. When radio and infrared observations were first made of the planet, however, astronomers found that its blackbody spectrum corresponded to a temperature of 125 K instead. Subsequent measurements, including those made by *Voyager* and *Galileo*, have verified that finding. Although a difference of 20 K may seem small, recall from Chapter 5 that the energy emitted by a planet grows as the *fourth* power of the surface temperature (in Jupiter's case, the temperature of the cloud tops). ∞ (Sec. 3.4) A planet at 125 K therefore radiates $(125/105)^4$, or about twice as much energy as a 105 K planet. Put another way, Jupiter actually emits about twice as much energy as it receives from the Sun. Thus, unlike any of the terrestrial planets, Jupiter must have its own internal heat source.

What is responsible for Jupiter's extra energy? It is not the decay of radioactive elements within the planet—that must be occurring, but not nearly at the rate necessary to produce the temperature we record. Nor is it the process

DISCOVERY 11-1

Almost a Star?

Jupiter has a starlike composition—predominantly hydrogen and helium, with a trace of heavier elements. Did Jupiter ever come close to becoming a star itself? Might the solar system have formed as a double-star system? Probably not. Unlike a star, Jupiter is cold. Its central temperature is far too low to ignite the nuclear fires that power our Sun. Jupiter's mass would have to increase 80-fold before its central temperature would rise to the point where nuclear reactions could begin, converting Jupiter into a small, dim star. Even so, it is interesting to note that although Jupiter's present-day energy output is very small (by solar standards, at least), it must have been much greater in the distant past while the planet was still contracting rapidly toward its present size. For a brief period of time—a few hundred million years—Jupiter might actually have been as bright as a faint star, although its bright-

ness never came within a factor of 100 of the Sun's. Still, seen from Earth at that time, Jupiter would have been about 100 times brighter than the Moon!

What might have happened had our solar system formed as a double-star system? Conceivably, had Jupiter been massive enough, its radiation might have produced severe temperature fluctuations on all the planets, perhaps to the point of making life on Earth impossible. Even if Jupiter's brightness were too low to cause us any problems, its gravitational field (which would be 1/12 that of the Sun if its mass were 80 times greater than its present value) might have made the establishment of stable, roughly circular planetary orbits an improbable event. The size of the "Jupiter," or second largest body, in a newborn planetary system may be a very important factor in determining the likelihood of the appearance of life.

that generates energy in the Sun, nuclear fusion—the temperature in Jupiter's interior, high as it is, is far too low for that (see *Discovery 11-1*). Instead, astronomers theorize that the source of Jupiter's excess energy is the slow escape of gravitational energy released during the planet's formation. As the planet took shape, some of its gravitational energy was converted into heat in the interior. That heat is still slowly leaking out through the planet's heavy atmospheric blanket, resulting in the excess emission we observe. Despite the huge amounts of energy involved—Jupiter's emits about 4×10^{17} watts more energy than it receives from the Sun—the loss is quite slight compared with the planet's total energy. A simple calculation indicates that the average temperature of the interior of Jupiter decreases by only about a millionth of a kelvin per year.

Jupiter's clouds, with their complex chemistry, are probably less than 200 km thick. Below them, the temperature and pressure steadily increase as the atmosphere becomes the "interior" of the planet. Much of our knowledge of Jupiter's interior comes from theoretical modeling. Indeed, apart from data gained following the collision of a comet with Jupiter in 1994 (see Section 14.2), there is very little direct evidence of the planet's internal properties. Planetary scientists use all available bulk data on the planet—mass, radius, composition, rotation, temperature, and so on—to construct a model of the interior that agrees with observations. Our statements about Jupiter's structure are, then, really statements about the model that best fits the facts. However, because the planet consists largely of hydrogen and helium—two simple gases whose physics we think we understand well—we can be fairly confident that Jupiter's internal structure is now understood.

Both the temperature and the density of Jupiter's atmosphere increase with depth below the cloud cover.

However, no "surface" of any kind exists anywhere inside. Instead, Jupiter's atmosphere just becomes denser and denser because of the pressure of the overlying layers. At a depth of a few thousand kilometers, the gas makes a gradual transition into the liquid state (see Figure 11.10). By a depth of about 20,000 km, the pressure is about 3 million

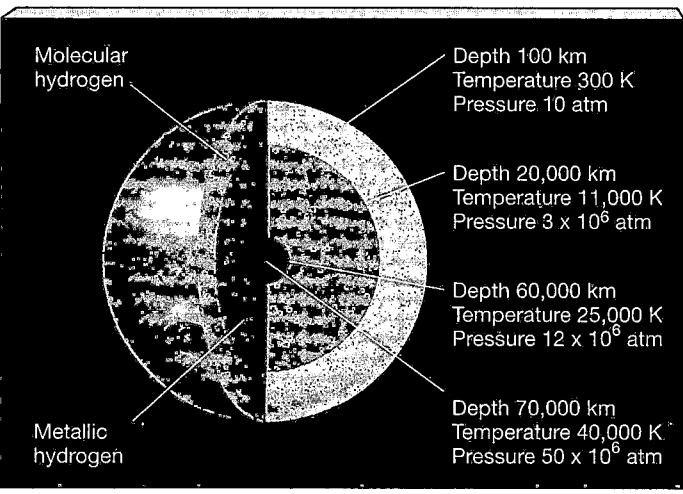

Molecular hydrogen

Metallic hydrogen

Depth 100 km
Temperature 300 K
Pressure 10 atm

Depth 20,000 km
Temperature 11,000 K
Pressure 3×10^6 atm

Depth 60,000 km
Temperature 25,000 K
Pressure 12×10^6 atm

Depth 70,000 km
Temperature 40,000 K
Pressure 50×10^6 atm

Figure 11.10 Jupiter's Interior Jupiter's internal structure, as deduced from *Voyager* measurements and theoretical modeling. The outer radius represents the top of the cloud layers, some 70,000 km from the planet's center. The density and temperature increase with depth, and the atmosphere gradually liquefies at a depth of a few thousand kilometers. Below a depth of 20,000 km, the hydrogen behaves like a liquid metal. At the center of the planet lies a large rocky core, somewhat terrestrial in composition but much larger than any of the inner planets. Although the values are very uncertain, the temperature and pressure at the center are probably about 40,000 K and 50 million (Earth) atmospheres, respectively.

times greater than atmospheric pressure on Earth. Under those conditions, the hot liquid hydrogen is compressed so much that it undergoes another transition, this time to a "metallic" state with properties in many ways similar to those of a liquid metal. Of particular importance for Jupiter's magnetic field (see Section 11.4) is that this metallic hydrogen is an excellent conductor of electricity.

As mentioned earlier, Jupiter's observed flattening requires that there be a relatively small, dense core at its center, of perhaps 10 times Earth's mass. The core's precise composition is unknown, but planetary scientists think that it contains much denser materials than the rest of the planet. Present best estimates indicate that it consists of "rocky" materials, similar to those found in the terrestrial worlds. (Note that the term *rocky* here refers to the chemical composition of the core, not to its physical state. At the high temperatures and pressures found deep in the jovian interiors, the core material bears little resemblance to rocks found on Earth's surface.) In fact, it now appears that all four jovian planets contain similarly large rocky cores, and that the formation of such a large "terrestrial" planetary core is a necessary stage in the process of building up a gas giant.

Because of the enormous pressure at the center of Jupiter—approximately 50 million times that on Earth's surface, or 10 times that at Earth's center—the core must be compressed to a very high density (perhaps twice the core density of Earth). It is probably not much more than 20,000 km in diameter, and the central temperature may be as high as 40,000 K.

☑ Concept Check

■ What does the surface of Jupiter look like?

11.4 Jupiter's Magnetosphere

④ For decades, ground-based radio telescopes monitored radiation leaking from Jupiter's magnetosphere, but only when the *Pioneer* and *Voyager* spacecraft reconnoitered the planet in the mid-1970s did astronomers realize the full extent of the planet's magnetic field. Jupiter, it turns out, is surrounded by a vast sea of energetic charged particles, mostly electrons and protons, somewhat similar to Earth's Van Allen belts but much, much larger. The radio radiation detected on Earth is emitted when these particles are accelerated to very high speeds—close to the speed of light—by Jupiter's powerful magnetic field. This radiation is several thousand times more intense than that produced by Earth's magnetic field. The particles are a potentially serious hazard for manned and unmanned space vehicles alike. Sensitive electronic equipment (not to mention even more sensitive human bodies) requires special protective shielding to operate for long in this hostile environment.

Direct spacecraft measurements show Jupiter's magnetosphere to be almost 30 million km across, roughly a million times more voluminous than Earth's magnetosphere, and far larger than the entire Sun. As with Earth's, the size and shape of Jupiter's magnetosphere is determined by the interaction between the planet's magnetic field and the solar wind. Jupiter's magnetosphere has a long tail extending away from the Sun at least as far as Saturn's orbit (over 4 A.U. farther out from the Sun), as sketched in Figure 11.11. However, on the sunward side, the *magnetopause*—the boundary of Jupiter's magnetic influence on the solar wind—lies only 3 million km from the planet. Near the planet's surface, the planet's magnetic field channels particles from the magnetosphere into the upper atmosphere, forming aurorae vastly larger and more energetic than those observed on Earth (Figure 11.12). ∞ (Sec. 7.5)

The outer magnetosphere appears to be quite unstable, sometimes deflating in response to "gusts" in the solar wind, then reexpanding as the wind subsides. In the inner magnetosphere, Jupiter's rapid rotation has forced most of the charged particles into a flat *current sheet*, lying on the planet's magnetic equator. The portion of the magnetosphere close to Jupiter is sketched in Figure 11.13. Notice that the planet's magnetic axis is not exactly

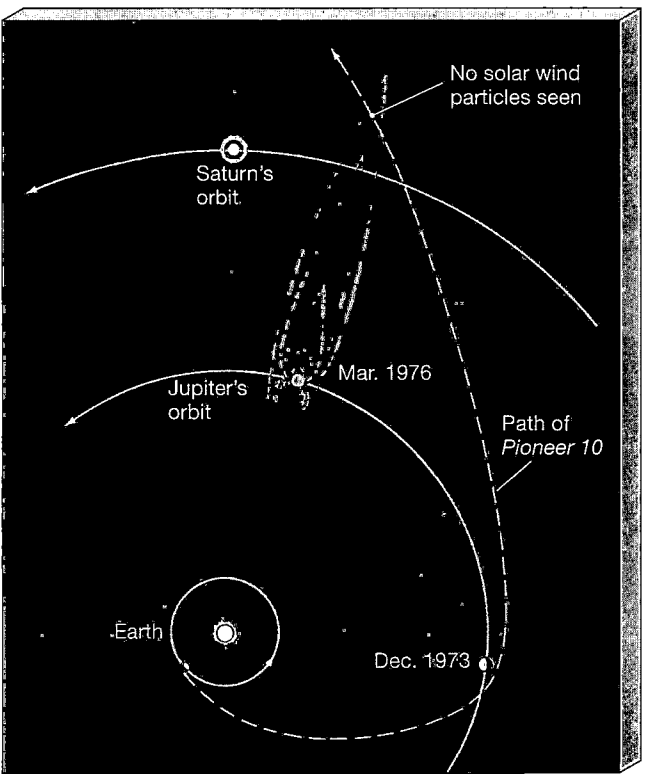

Figure 11.11 *Pioneer 10* Mission The *Pioneer 10* spacecraft did not detect any solar particles while moving behind Jupiter. Accordingly, as sketched here, Jupiter's magnetosphere apparently extends beyond the orbit of Saturn.

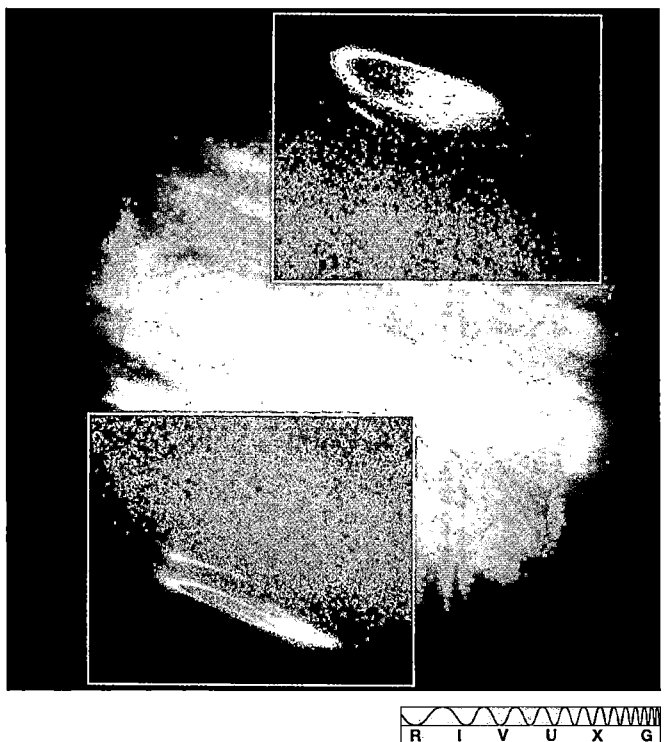

R I V U X G

Figure 11.12 Aurora on Jupiter Aurorae on Jupiter, as seen by the *Hubble Space Telescope*. The main image was taken in visible (true-color) light, but the two insets at the poles were taken in the ultraviolet part of the spectrum. The oval-shaped aurorae, extending hundreds of kilometers above Jupiter's limb, result from charged particles escaping the jovian magnetosphere and colliding with the atmosphere, causing the gas to glow. *(NASA)*

aligned with its rotation axis but is inclined to it at an angle of approximately 10°. The planet's magnetic field happens to be oriented opposite Earth's, with field lines running from north to south, rather than south to north as in the case of our own planet (see Figure 7.21).

Both ground- and space-based observations of the radiation emitted from Jupiter's magnetosphere imply that the *intrinsic* strength of the planet's magnetic field is nearly 20,000 times greater than Earth's. The existence of such a strong field further supports our theoretical model of Jupiter's internal structure. The conducting liquid interior that is thought to make up most of the planet should combine with Jupiter's rapid rotation to produce a large dynamo effect and a strong magnetic field, just as observed. ∞ (Sec. 7.4)

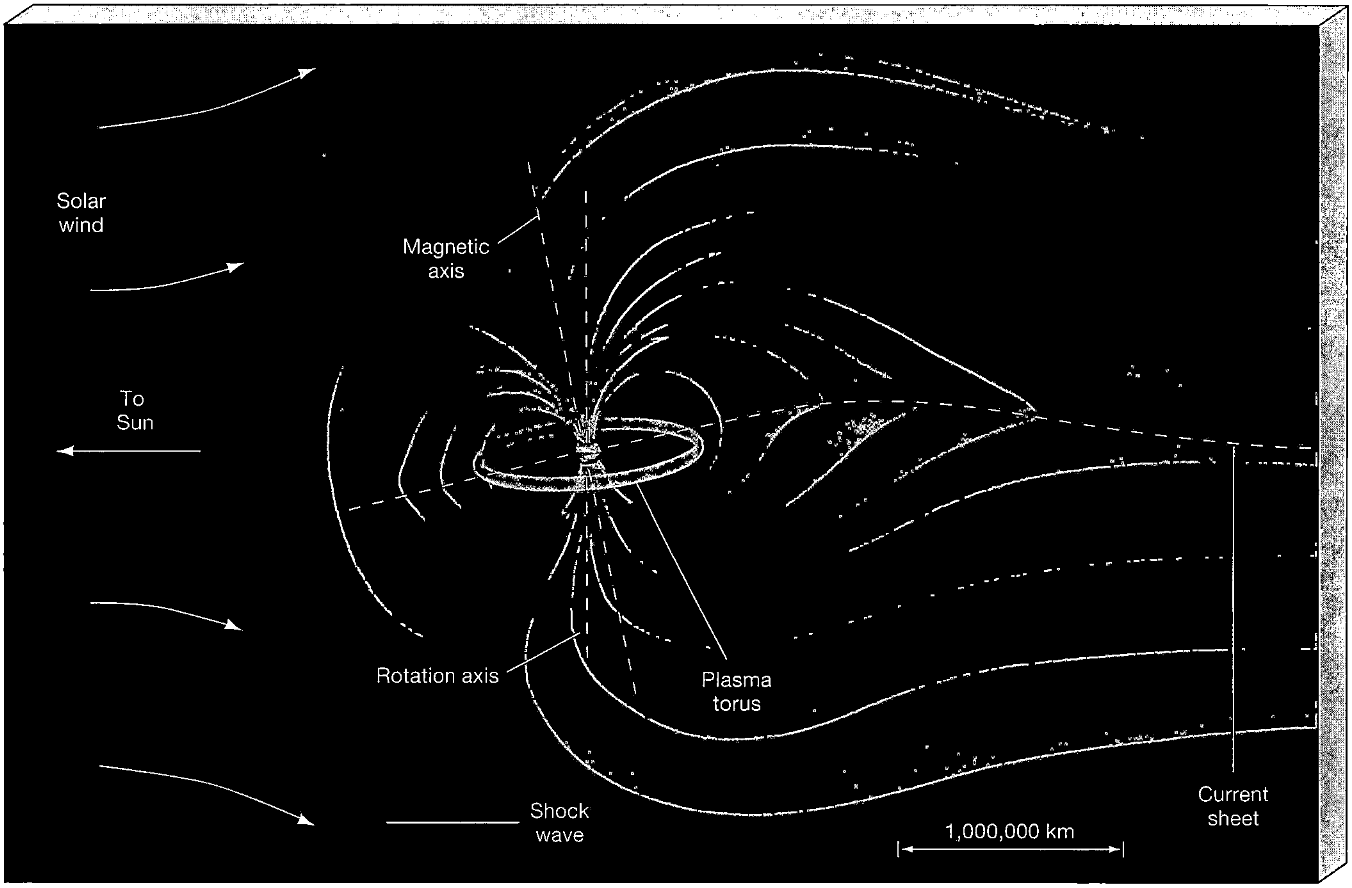

Figure 11.13 Jupiter's Magnetosphere Jupiter's inner magnetosphere is characterized by a flat current sheet, consisting of charged particles squeezed into the magnetic equatorial plane by the planet's rapid rotation. The plasma torus is a ring of charged particles associated with the moon Io; it is discussed in Section 11.5.

Concept Check

□ Why is Jupiter's magnetosphere so much larger than Earth's?

11.5 The Moons of Jupiter

§ At last count, Jupiter has at least 28 satellites. The four largest—the **Galilean moons**—are each comparable in size to Earth's Moon. ∞ (Sec. 2.5) Moving outward from Jupiter, the four are named Io, Europa, Ganymede, and Callisto, after the mythical attendants of the Roman god Jupiter. They move in nearly circular orbits about their parent planet. When the *Voyager 1* spacecraft passed close to the Galilean moons in 1979, it sent some remarkably detailed photographs back to Earth, allowing planetary scientists to discern fine surface features on each. More recently, in the late 1990s, the *Galileo* mission expanded still further our knowledge of these small but complex worlds. We will consider the Galilean satellites in more detail in a moment. Table 11.1 presents some general properties of Jupiter's moon system.

Within the orbit of Io lie four small satellites, all but one discovered by *Voyager* cameras. The largest of the four, Amalthea, is less than 300 km across and is irregularly shaped. E. E. Barnard discovered it in 1892. It orbits at a distance of 181,000 km from Jupiter's center—only 110,000 km above the cloud tops. Its rotation, like that of most of Jupiter's satellites, is synchronous with its orbit because of Jupiter's strong tidal field. Amalthea rotates once per orbit period, every 11.7 hours.

Beyond the Galilean moons lie eight more small satellites, all discovered in the twentieth century but before the *Voyager* missions. They fall into two groups of four moons each. The moons in the inner group move in eccentric, inclined orbits, about 11 million km from the planet. The outer four moons lie about 22 million km from Jupiter. Their orbits too are fairly eccentric, but *retrograde*, moving in a sense opposite all the other moons' orbit (and Jupiter's rotation). It is very likely that each group represents a single body that was captured by Jupiter's strong gravitational field long after the planet and its larger moons originally formed. Both bodies subsequently broke up, either during or after the capture process, resulting in the two families of similar orbits we see today. The masses, and hence the densities, of these small worlds are unknown. However,

TABLE 11.1 The Moons of Jupiter*

NAME	DISTANCE FROM JUPITER (km)	(planet radii)	ORBIT PERIOD (days)	SIZE (longest diameter, km)	MASS** (Earth Moon masses)	DENSITY (kg/m³)	(g/cm³)
Metis	128,000	1.79	0.29	40			
Adrastea	129,000	1.80	0.30	20			
Amalthea	181,000	2.54	0.50	260			
Thebe	222,000	3.10	0.67	100			
Io	422,000	5.90	1.77	3640	1.22	3500	3.5
Europa	671,000	9.38	3.55	3130	0.65	3000	3.0
Ganymede	1,070,000	15.0	7.15	5270	2.02	1900	1.9
Callisto	1,880,000	26.3	16.7	4800	1.46	1900	1.9
Leda	11,100,000	155	239	10			
Himalia	11,500,000	161	251	170			
Lysithea	11,700,000	164	259	24			
Elara	11,700,000	164	260	80			
Ananke	21,200,000	297	−631[†]	20			
Carme	22,600,000	316	−692[†]	30			
Pasiphae	23,500,000	329	−735[†]	36			
Sinope	23,700,000	332	−758[†]	28			

*Does not include the 12 recently discovered small moons described in the text. All are small and move on inclined, eccentric, mainly retrograde orbits some 10–25 million km from the planet.

**Mass of Earth's Moon = 7.4×10^{22} kg = 3.9×10^{-5} Jupiter masses.

[†]Indicates a retrograde orbit.

their appearance and sizes suggest compositions more like asteroids than their larger Galilean companions.

The latest addition to Jupiter's family of satellites is a group of twelve small moons spotted in 2000 by researchers using dedicated large telescopes and specially designed instruments and software to scan large areas of the sky for very faint objects. All the new moons are very small—probably less than 5 or 10 km across—and their masses are unknown. Their orbits have not yet been accurately determined, but are all moderately eccentric, with semimajor axes between 10 and 25 million km, and are inclined at 15–30° to Jupiter's equatorial plane. Most of the orbits are retrograde. In short, the properties of these newcomers seem quite consistent with those of the other eight outlying satellites just described. Very likely they have the same origin—these moons are space "junk" captured by Jupiter, probably long ago.

THE GALILEAN MOONS AS A MODEL OF THE INNER SOLAR SYSTEM

Jupiter's Galilean moons have several interesting parallels with the terrestrial planets. Their orbits are direct (that is, in the same sense as Jupiter's rotation), roughly circular, and lie close to Jupiter's equatorial plane. They range in size from slightly smaller than Earth's Moon (Europa) to slightly larger than Mercury (Ganymede). Figure 11.14 is a *Voyager 1* image of Io and Europa, with Jupiter providing a spectacular backdrop. Figure 11.15 shows the four Galilean moons to scale.

The similarity to the inner solar system continues with the fact that the moons' densities decrease with increasing distance from Jupiter. ∞ (Sec. 6.4) Based largely on detailed measurements made by *Galileo* of the moons' gravitational fields, researchers have built up fairly detailed

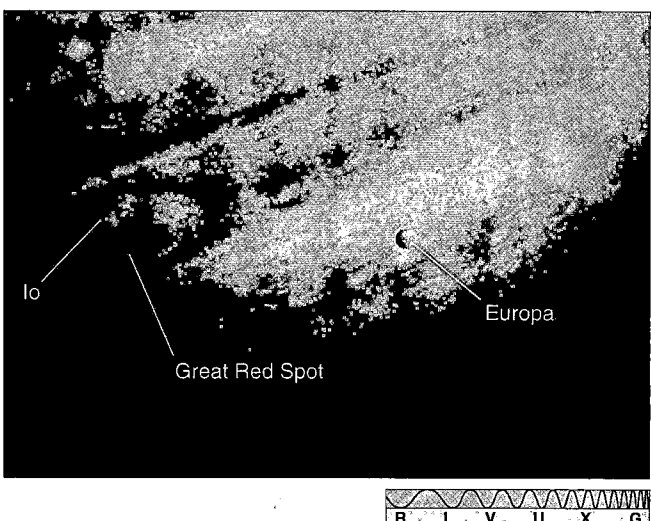

Figure 11.14 Jupiter, Up Close *Voyager 1* took this photo of Jupiter with ruddy Io on the left and pearl-like Europa toward the right. Note the scale of objects here: Both Io and Europa are comparable in size to our Moon, and the Red Spot (seen here to the left bottom) is roughly twice as big as Earth. *(NASA)*

pictures of the moons' compositions and internal structures (Figure 11.16). The innermost two Galilean moons, Io and Europa, have thick rocky mantles, possibly similar to the crusts of the terrestrial planets, surrounding iron/iron sulfide cores. Io's core accounts for about half that moon's total radius. Europa has a water/ice outer shell between 100 and 200 km thick. The two outer moons, Ganymede and Callisto, are clearly deficient in rocky materials. Lighter materials, such as water and ice, may account for as much as half of their total mass. Ganymede appears to have a relatively small metallic core topped by a

Figure 11.15 Galilean Moons The *Galileo* spacecraft photographed each of the four Galilean moons of Jupiter. Shown here to scale, as they would appear from a distance of about 1 million km, they are, from left to right, Io, Europa, Ganymede, and Callisto. *(NASA)*

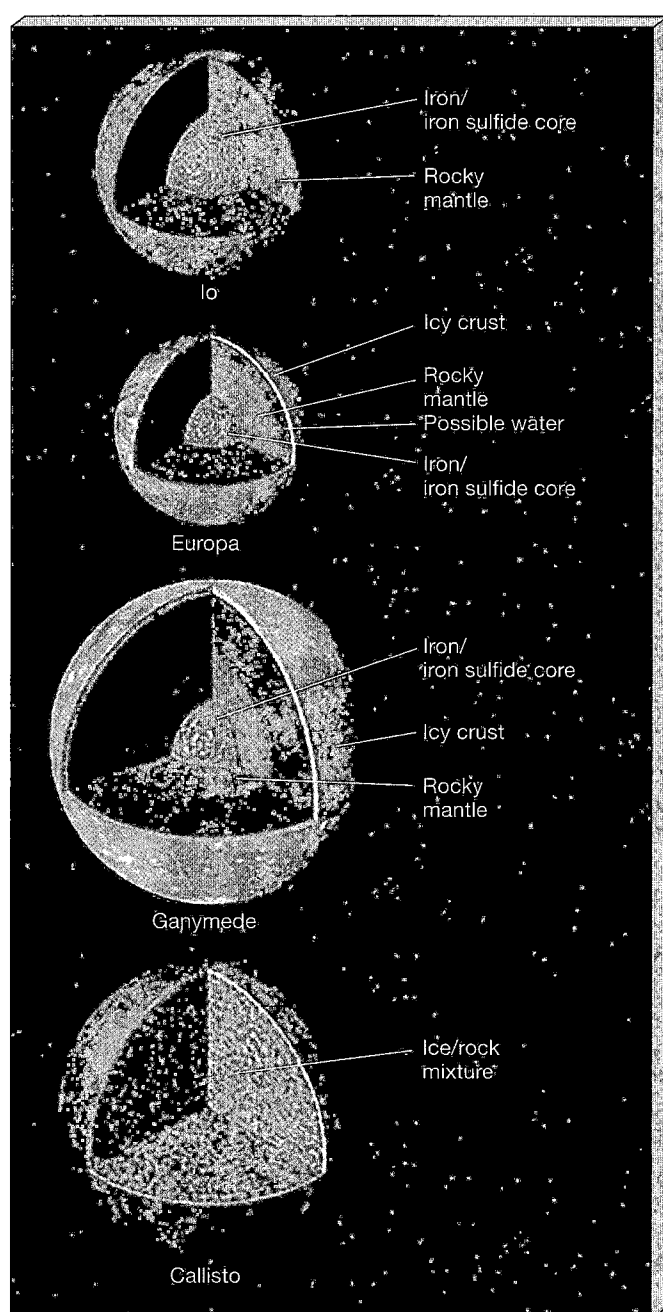

Figure 11.16 Galilean Moon Interiors Cutaway diagrams showing the interior structure of the four Galilean satellites. Moving outward from Io to Callisto, the moons' densities steadily decrease as the composition shifts from rocky mantles and metallic cores in Io and Europa, to a thick icy crust and smaller core in Ganymede, to an almost uniform rock and ice mix in Callisto. Both Ganymede and Europa are thought to have layers of liquid water beneath their icy surfaces.

sion has so far lasted well into 2001.) The Galilean moons were studied at resolutions as fine as a few meters during extremely close passages by the spacecraft.

Not all the properties of the Galilean moons find analogs in the inner solar system, however. For example, all four Galilean satellites are locked into states of synchronous rotation by Jupiter's strong tidal field, so they all keep one face permanently pointing toward their parent planet. By contrast, of the terrestrial planets, only Mercury is strongly influenced by the Sun's tidal force, and even its orbit is not synchronous. ∞ (Sec. 8.4) Finally, inspection of Table 11.1 shows a remarkable coincidence in the orbit periods of the three inner Galilean moons. Their periods are almost exactly in the ratio 1:2:4—a kind of "Bode's law" for Jupiter. ∞ (*Discovery 6-1*) This may be the result of a complex, but poorly understood, three-body resonance in the Galilean moon system, something not found among the terrestrial worlds.

IO: THE MOST ACTIVE MOON

Io, the densest of the Galilean moons, is the most geologically active object in the entire solar system. Its mass and radius are fairly similar to those of Earth's Moon, but there the resemblance ends. Shown in Figure 11.17, Io's surface is a collage of reds, yellows, and blackish browns—resembling a giant pizza in the minds of some startled *Voyager* scientists. As the spacecraft glided past Io, an outstanding discovery was made: Io has active volcanoes! *Voyager 1* photographed eight erupting volcanoes, and six were still erupting when *Voyager 2* passed by four months later.

By the time *Galileo* arrived in 1995, several of the volcanoes observed by *Voyager* had subsided. However, many new ones were seen—in fact, *Galileo* found that Io's surface features can change significantly in as little as a few weeks. In Figure 11.18, one volcano is seen ejecting matter to an altitude of over 100 km. The gases are spewed forth at speeds up to 2 km/s, quite unlike the (relatively) sluggish ooze that emanates from Earth's insides. According to *Galileo's* instruments, lava temperatures generally range from 650 to 900 K, the higher end of the range implying that at least some of the volcanism is similar to that found on Earth. However, temperatures as high as 2000 K—far hotter than any earthly volcano—have been measured at some locations.

The orange color immediately surrounding the volcanoes most likely results from sulfur compounds in the ejected material. In stark contrast to the other Galilean

rocky mantle and a thick icy outer shell. Callisto seems to be a largely undifferentiated mixture of rock and ice.

Many astronomers think that the formation of Jupiter and the Galilean satellites may in fact have mimicked on a small scale the formation of the Sun and the inner planets. For that reason, studies of the Galilean moon system may provide us with valuable insight into the processes that created our own world. We will return to this parallel in Chapter 15. So interested were mission planners in learning more about the Galilean moon system that the already highly successful *Galileo* mission was extended for two more years to allow for even more detailed study, particularly of Europa. (In fact, the renamed *Galileo Europa Mis-*

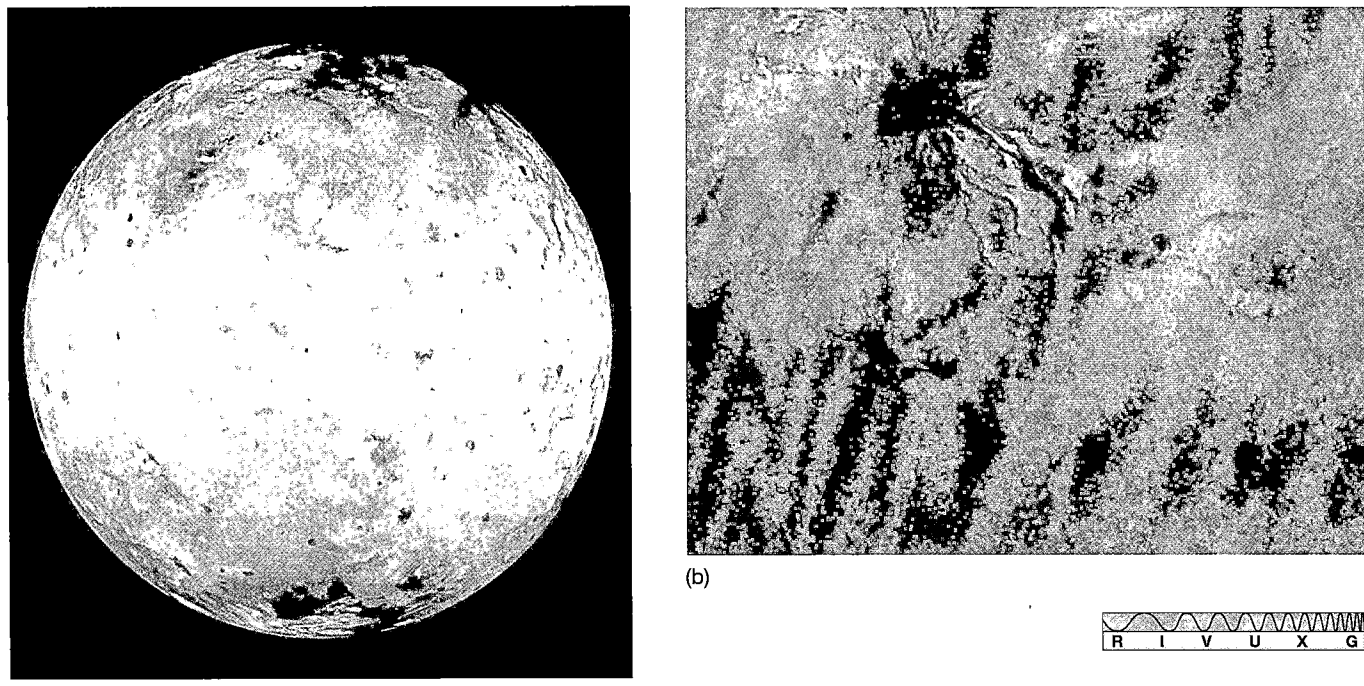

(a)

(b)

R I V U X G

Figure 11.17 Io Jupiter's innermost moon, Io, is quite different in character from the other three Galilean satellites. Its surface is kept smooth and brightly colored by the moon's constant volcanism. The resolution of the *Galileo* photograph in (a) is about 7 km. In the more detailed *Voyager* image (b), features as small as 2 km across can be seen. *(NASA)*

moons, Io's surface is neither cratered nor streaked. (The circular features visible in Figures 11.17 and 11.18 are volcanoes.) Its surface is exceptionally smooth, mostly varying in altitude by less than about 1 km, although some volcanoes are several kilometers high. The smoothness is apparently the result of molten matter that constantly fills in any "dents and cracks." This remarkable moon has the youngest surface of any known object in the solar system. Io also has a thin, temporary atmosphere made up primarily of sulfur dioxide, presumably the result of gases ejected by volcanic activity.

Io's volcanism has a major effect on Jupiter's magnetosphere. All the Galilean moons orbit within the magnetosphere and play some part in modifying its properties, but Io's influence is particularly marked. Although many of the charged particles in Jupiter's magnetosphere come from the solar wind, there is strong evidence that Io's volcanism is the primary source of heavy ions in the inner regions. Jupiter's magnetic field continually sweeps past Io, gathering up the particles its volcanoes spew into space and accelerating them to high speed. The result is the *Io plasma torus* (Figure 11.19; see also Figure 11.13), a doughnut-shaped region of energetic heavy ions that follows Io's orbital track, completely encircling Jupiter. (A plasma is a gas that has been heated to such high temperatures that all its atoms are ionized. In fact, neutral atoms have also been observed in the Io plasma torus.) The plasma torus is quite easily detectable from Earth, but before *Voyager* its origin

was unclear. *Galileo* made detailed studies of the plasma's dynamic and rapidly varying magnetic field. Spectroscopic analysis shows that sulfur is indeed one of the torus's major constituents, strongly implicating Io's volcanoes as its source. As a hazard to spacecraft—manned or unmanned—the plasma torus is formidable. The radiation levels there are lethal.

What causes such astounding volcanic activity on Io? That moon is far too small to have geological activity like Earth. Io should be long dead, like our own Moon. At one time, some scientists suggested that Jupiter's magnetosphere might be the culprit—perhaps the (then-unknown) processes creating the plasma torus were somehow also stressing the moon. We now know that this is not the case. The real source of Io's energy is gravity—Jupiter's gravity. Io orbits very close to Jupiter—only 422,000 km, or 5.9 Jupiter radii, from the center of the planet. As a result, Jupiter's huge gravitational field exerts strong tidal forces on the moon. If Io were the only satellite in the Jupiter system, it would long ago have come into a state of synchronous rotation with the planet, just like our own Moon, for the reasons discussed in Chapter 8. ∞ (Sec. 8.4) In that case, Io would move in a perfectly circular orbit, with one face permanently turned toward Jupiter. The tidal bulge would be stationary with respect to the moon.

But Io is not alone. As it orbits it is constantly tugged by the gravity of its nearest large neighbor, Europa. These tugs are small and not enough to cause any great tidal

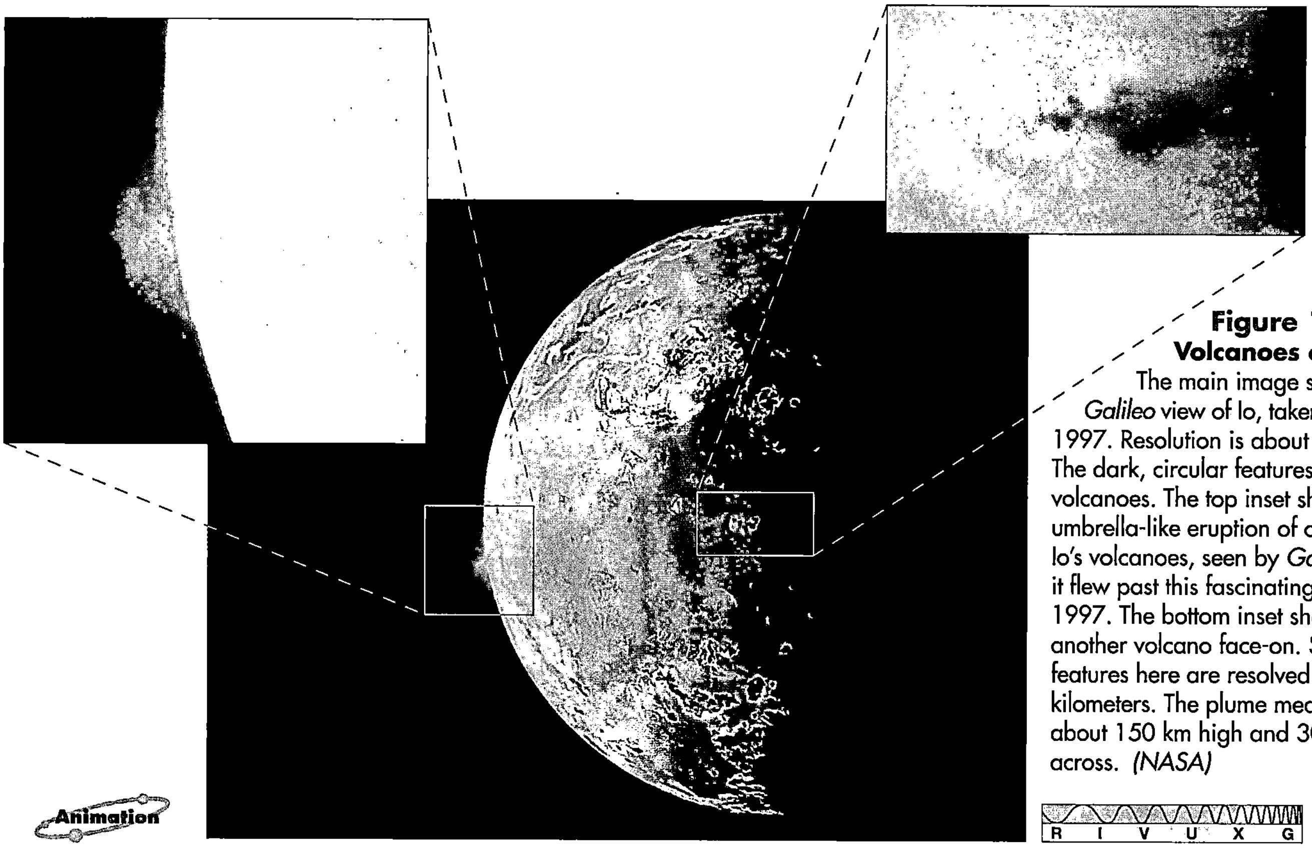

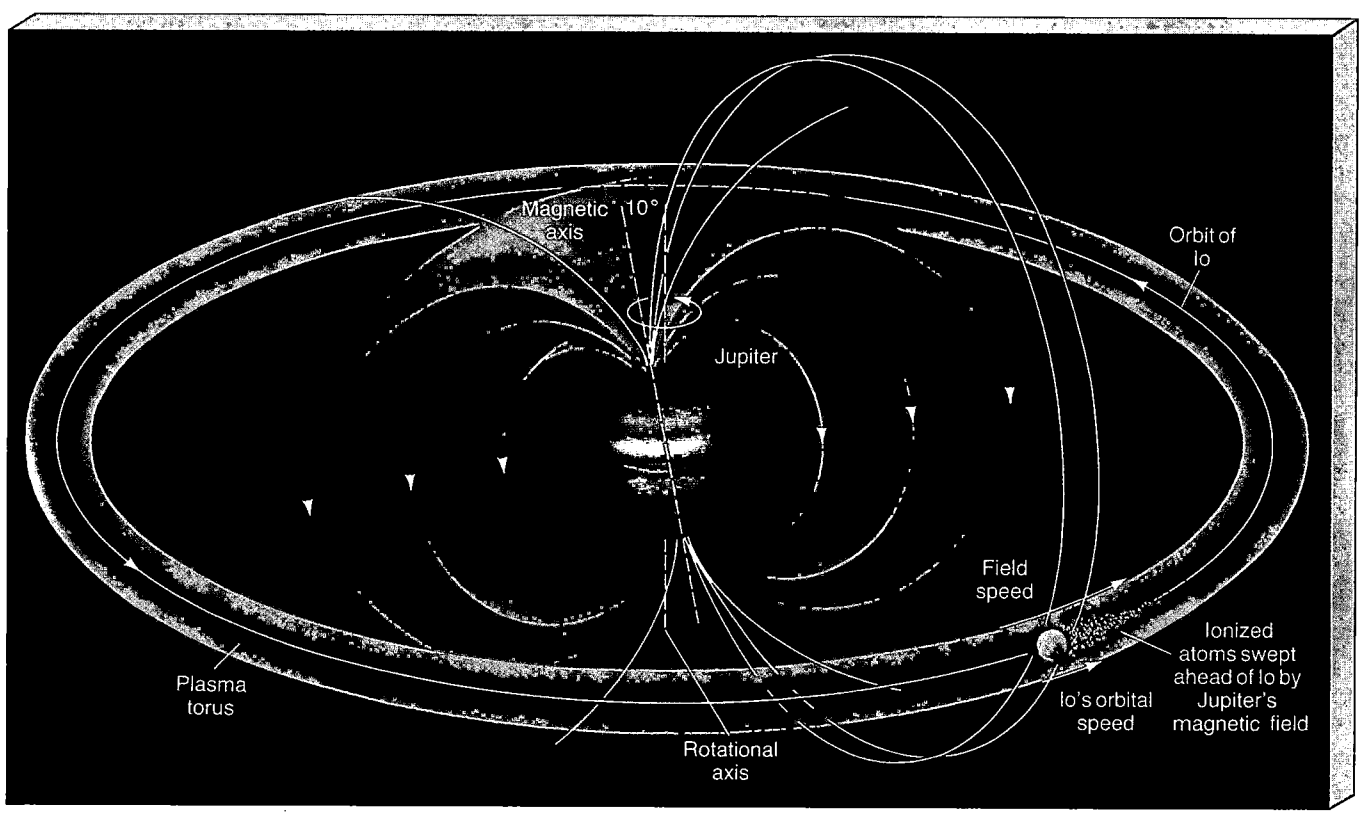

Figure 11.18 Volcanoes on Io The main image shows a *Galileo* view of Io, taken in 1997. Resolution is about 6 km. The dark, circular features are volcanoes. The top inset shows an umbrella-like eruption of one of Io's volcanoes, seen by *Galileo* as it flew past this fascinating moon in 1997. The bottom inset shows another volcano face-on. Surface features here are resolved to a few kilometers. The plume measures about 150 km high and 300 km across. *(NASA)*

R   I   V   U   X   G

Figure 11.19 Io Plasma Torus The torus is the result of material being ejected from Io's volcanoes and swept up by Jupiter's rapidly rotating magnetic field. Spectroscopic analysis indicates that the torus is composed primarily of sodium and sulfur atoms and ions.

effect, but they are sufficient to make Io's orbit slightly noncircular, preventing the moon from settling into a precisely synchronous state. The reason for this effect is exactly the same as in the case of Mercury, as discussed in Chapter 8. ∞ (Sec. 8.4) In a noncircular orbit, the moon's speed varies from place to place as it revolves around its planet, but its rate of rotation on its axis remains constant. Thus it cannot keep one face always turned toward Jupiter. Instead, as seen from Jupiter, Io rocks or "wobbles" slightly from side to side as it moves. The large (100 m) tidal bulge, however, always points directly toward Jupiter, so it moves back and forth across Io's surface as the moon wobbles. These conflicting forces result in enormous tidal stresses that continually flex and squeeze Io's interior.

Just as repeated back-and-forth bending of a piece of wire can produce heat through friction, the ever-changing distortion of Io's interior constantly energizes the moon. This generation of large amounts of heat within Io ultimately causes huge jets of gas and molten rock to squirt out of the surface. *Galileo's* sensors indicated extremely high temperatures in the outflowing material. It is likely that much of Io's interior is soft or molten, with only a relatively thin solid crust overlying it. Researchers estimate that the total amount of heat generated within Io as a result of tidal flexing is about 100 million megawatts. This phenomenon makes Io one of the most fascinating objects in our solar system.

EUROPA: LIQUID WATER LOCKED IN ICE

Europa (Figure 11.20) is a very different world from Io. Lying outside Io's orbit, 671,000 km (9.4 Jupiter radii) from Jupiter, Europa showed relatively few craters on its surface in images taken by *Voyager*, suggesting geologic youth—perhaps just a few million years. Recent activity has erased any scars of ancient meteoritic impacts. The dark areas are rocky deposits that may have come from the moon's interior, or may have been swept up by Europa as it moved in its orbit. Europa's surface also displays a vast

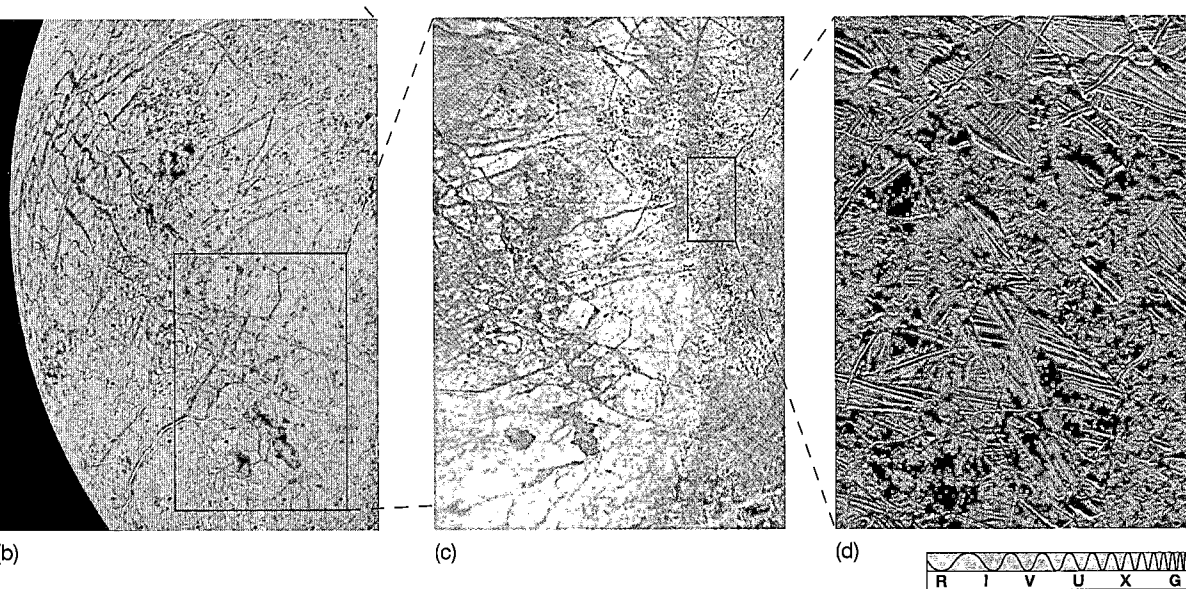

Figure 11.20 Europa The second Galilean moon is Europa. Its icy surface is only lightly cratered, indicating that some ongoing process must be obliterating impact craters soon after they form. The cracks crisscrossing the surface are most likely caused by the tidal effect of Jupiter. The resolution of the *Voyager 2* mosaic in (a) is about 5 km. The two images below it (b and c) display even finer detail. (d) At 50-m resolution this image from the *Galileo* spacecraft shows a smooth yet tangled surface resembling the huge ice floes that cover Earth's polar regions. This region is called Conamara Chaos. *(NASA)*

(a)

(b)   (c)   (d)

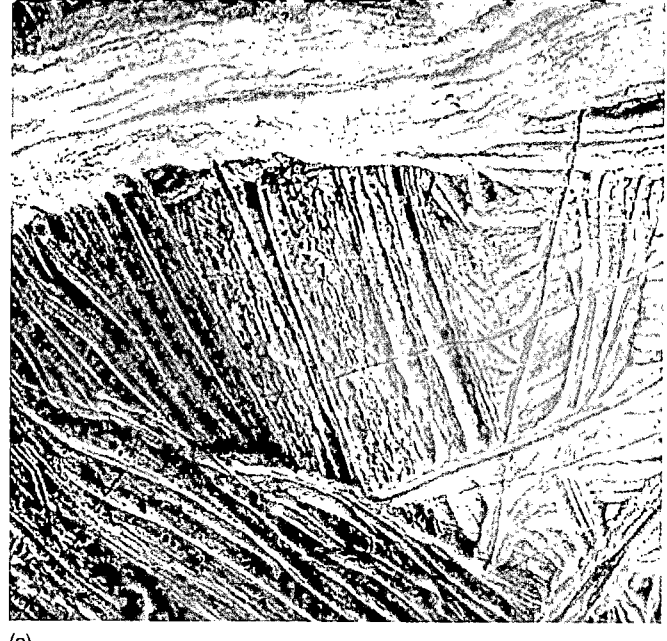

(a)

(b)

RIVUXG

Figure 11.21 Europa Surface Detail Detailed *Galileo* image(s) of Europa, showing (a) "pulled apart" terrain that suggests upwelling material filling in the gaps between separating surface ice sheets, and (b) a region known as Conamara Chaos, where liquid water appears to have flooded a portion of the surface. *(NASA)*

network of lines crisscrossing bright, clear fields of water ice. Some of these linear "bands," or fractures, extend halfway around the satellite and resemble in some ways the pressure ridges that develop in ice floes on Earth's polar oceans.

Before *Galileo*'s arrival, some researchers had theorized that Europa might be completely covered by an ocean of liquid water whose top is frozen at the low temperatures prevailing so far from the Sun. In this view, the cracks in the surface are attributed to the tidal influence of Jupiter and the gravitational pulls of the other Galilean satellites, although these forces are considerably weaker than those powering Io's violent volcanic activity. However, other planetary scientists had contended that Europa's fractured surface was instead related to some form of tectonic activity, one involving ice rather than rock. High-resolution *Galileo* observations now appear to strongly support the former idea. Figure 11.20(d) is a *Galileo* image of this weird moon, showing what look like "icebergs"—flat chunks of ice that have been broken apart, moved several kilometers, and reassembled, perhaps by the action of water currents below. Mission scientists estimate that Europa's surface ice may be several kilometers thick and that there may be a 100-km-deep liquid ocean below it.

Other detailed images of the surface support this hypothesis. Figure 11.21(a) shows a region where Europa's icy crust appears to have been pulled apart and new material has filled in the gaps between the separating ice sheets. Elsewhere on the surface, *Galileo* found what appeared to be the icy equivalent of lava flows on Earth—regions where water apparently "erupted" through the surface and flowed for many kilometers before solidifying. The "puddle" shown in Figure 11.21(b) strongly suggests local flooding of the terrain. The scarcity of impact craters on Europa implies that the processes responsible for these features did not stop long ago. Rather, they must be ongoing.

Further evidence comes from studies of Europa's magnetic field. Magnetic measurements made by *Galileo* on repeated flybys of the moon revealed that Europa has a weak magnetic field that constantly changes strength and direction. This finding is entirely consistent with the idea that the field is generated by the action of Jupiter's magnetism on a shell of electrically conducting fluid about 100 km below Europa's surface—in other words, the salty liquid water layer suggested by the surface observations. These results convinced quite a few skeptical scientists of the reality of Europa's ocean.

The likelihood that Europa has an extensive layer of liquid water below its surface ice opens up many interesting avenues of speculation about the possible development of life there. In the rest of the solar system, only Earth has liquid water on or near its surface, and most scientists agree that water played a key role in the appearance of life here (see Chapter 28). Europa may well contain more liquid water than exists on our entire planet! Of course, the existence of water does not *necessarily* imply the emergence of life. Europa, even in its liquid ocean, is still a hostile environment compared with Earth. Nevertheless, the possibility, even a remote one, of life on Europa was an important motivating factor in the decision to extend the *Galileo* mission for two (now three) more years. Astronomers now eagerly await the launch of NASA's proposed *Europa Orbiter* mission in 2003.

GANYMEDE AND CALLISTO: FRATERNAL TWINS

The two outermost Galilean moons are Ganymede (at 1.1 million km, or 15 planetary radii, from the center of Jupiter) and Callisto (at 1.9 million km, or 26 Jupiter radii). The density of each is only about 2000 kg/m³, suggesting that they harbor substantial amounts of ice throughout and are not just covered by thin icy or snowy surfaces. Ganymede, shown in Figure 11.22, is the largest moon in the solar system, exceeding not only Earth's Moon but also the planets Mercury and Pluto in size. It has many impact craters on its surface and patterns of dark and light markings that are reminiscent of the highlands and maria on Earth's own Moon. In fact, Ganymede's

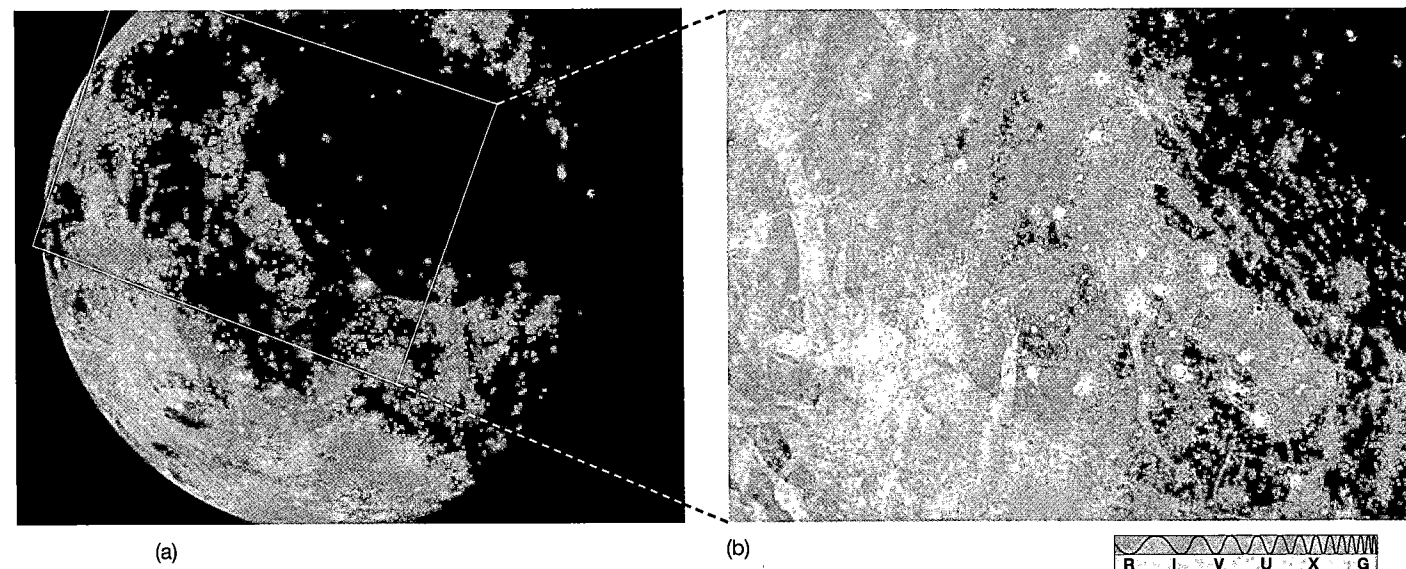

Figure 11.22 Ganymede Jupiter's largest moon, Ganymede, is also the largest satellite in the solar system. The dark regions on the surface are the oldest and probably represent the original icy crust of the moon. The largest dark region visible in the *Voyager 2* image in (a) is called Galileo Regio. It spans some 320 km. The lighter, younger regions are the result of flooding and freezing that occurred within a billion years or so of Ganymede's formation. The light-colored spots are recent impact craters. The resolution of the detailed image in (b) is about 3 km. *(NASA)*

history has many parallels with that of the Moon (with water ice replacing lunar rock). The large, dark region clearly visible in Figure 11.22 is called Galileo Regio.

As with the inner planets, we can estimate ages on Ganymede by counting craters. We learn that the darker regions, like Galileo Regio, are the oldest parts of Ganymede's surface. These regions are the original icy surface of the moon, just as the ancient highlands on our own Moon are its original crust. The surface darkens with age as micrometeorite dust slowly covers it. The light-colored parts of Ganymede are much less heavily cratered, so they must be younger. They are Ganymede's "maria" and probably formed in a manner similar to the way that maria on the Moon were created. Intense meteoritic bombardment caused liquid water—Ganymede's counterpart to our own Moon's molten lava—to upwell from the interior and flood the impacting regions before solidifying.

Not all of Ganymede's surface features follow the lunar analogy. Ganymede has a system of grooves and ridges (shown in Figure 11.23) that may have resulted from crustal tectonic motion, much as Earth's surface undergoes mountain building and faulting at plate boundaries. Ganymede's large size indicates that its original radioactivity probably helped heat and differentiate its interior, after which the moon cooled and the crust cracked. Ganymede seems to have had some early plate tectonic activity, but the process stopped about 3 billion years ago when the cooling crust became too thick. The *Galileo* data

suggest that the surface of Ganymede may be older than was previously thought. With the improved resolution of that spacecraft's images (Figure 11.23c), some regions believed to have been smooth, and hence young, are now seen to be heavily splintered by fractures and thus probably very old.

In 1996, *Galileo* detected a weak magnetosphere surrounding Ganymede, making it the first moon in the solar system on which a magnetic field had been observed, and implying that Ganymede has a modest iron-rich core. The moon's magnetic field is about one percent that of Earth. In December 2000, the magnetometer team reported fluctuations in the field strength similar to those near Europa, suggesting that Ganymede too may have liquid or "slushy" water under its surface. Recent observations of surface formations similar to those attributed to flowing water "lava" on Europa appear to support this view.

Callisto, shown in Figure 11.24, is in many ways similar in appearance to Ganymede, although it has more craters and fewer fault lines. Its most obvious feature is a huge series of concentric ridges surrounding each of two large basins. The larger of the two, on Callisto's Jupiter-facing side, is named Valhalla and measures some 3000 km across. It is clearly visible in Figure 11.24. The ridges resemble the ripples made as a stone hits water, but on Callisto they probably resulted from a cataclysmic impact with an asteroid or comet. The upthrust ice was partially melted, but it resolidified quickly, before the ripples had a

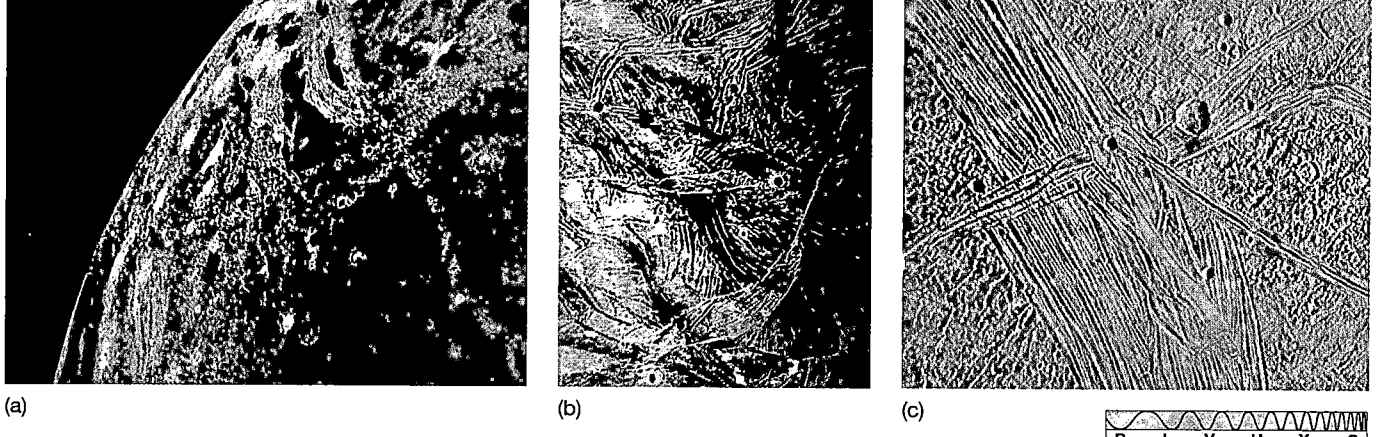

(a)   (b)   (c)

R   I   V   U   X   G

Figure 11.23 Ganymede Surface Detail (a and b) Grooved terrain on Ganymede may have been caused by a process similar to plate tectonics on Earth. This image was captured by *Galileo* in 1997. The area shown here is about 50 km across and reveals a multitude of ever-smaller ridges, valleys, and craters, right down to the resolution limit of *Galileo*'s camera (about 300 m, or about three times the length of a football field). Part (c) shows the grooves at even higher resolution, suggesting erosion of some sort, possibly even caused by water. *(NASA)*

chance to subside. Today, both the ridges and the rest of the crust are frigid ice and show no obvious signs of geological activity (such as the grooved terrain on Ganymede). Apparently, Callisto froze before plate tectonic or other activity could start. The density of impact craters on the Valhalla basin indicates that it formed long ago, perhaps 4 billion years in the past. Yet even on this frozen world there are hints—as yet unconfirmed—from *Galileo*'s magnetometers that there might be a thin layer of water, or more likely slush, deep below the surface.

Ganymede's internal differentiation indicates that the moon was largely molten at some time in the past, yet Callisto is undifferentiated, and hence apparently never melted. Researchers are uncertain why two such similar bodies should have evolved so differently. Complicating things further is Ganymede's magnetic field and possible subsurface liquid water, which suggest that the moon's interior may still be relatively warm. If that is so, then Ganymede's heating and differentiation must have happened relatively recently—less than a billion years ago,

based on recent estimates of how rapidly the moon's heat escapes into space. Scientists have no clear explanation for how this could have occurred. Heating by meteoritic bombardment ended too early, and radioactivity probably could not have provided enough energy at this late time. ∞ (Sec. 7.3) Some astronomers speculate that interactions among the inner moons, possibly related to the 1:2:4 near-resonance mentioned earlier, may have been responsible. These interactions might have caused Ganymede's orbit to change significantly about one billion years ago, and prior tidal heating by Jupiter could have helped melt the moon's interior.

✓ Concept Checks

■ What is the ultimate source of all the activity observed on Jupiter's Galilean satellites?

■ Why are scientists so interested in the existence of liquid water on Europa and Ganymede?

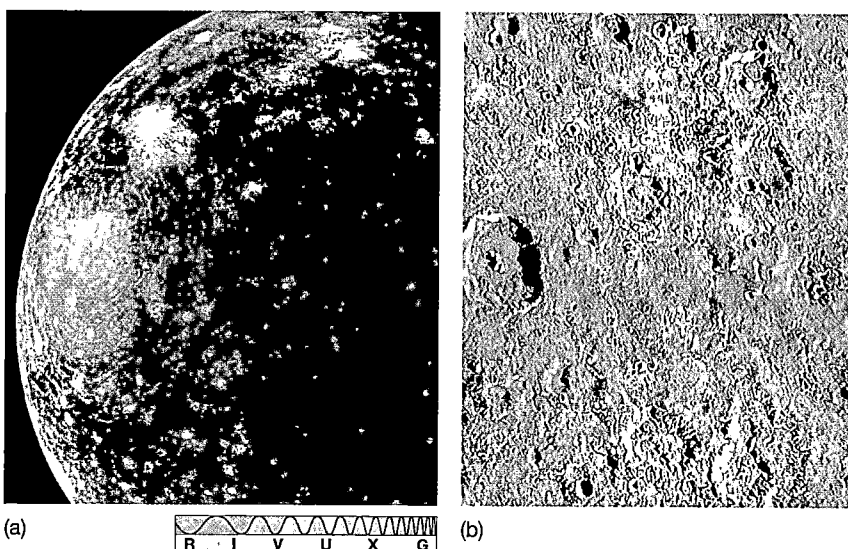

(a)   (b)

R   I   V   U   X   G

Figure 11.24 Callisto The outermost Galilean moon of Jupiter, Callisto, is similar to Ganymede in composition but is more heavily cratered. (a) The large series of concentric ridges visible on the left of the image is known as Valhalla. Extending nearly 1500 km from the basin center, they formed when "ripples" from a large meteoritic impact refroze before they could disperse completely. The resolution in this *Voyager 2* image is around 10 km. (b) The higher-resolution *Galileo* image of Callisto's equatorial region, about 300 × 200 km in area, displays more clearly its heavy cratering. *(NASA)*

11.6 Jupiter's Ring

Yet another remarkable finding of the 1979 *Voyager* missions was the discovery of a faint ring of matter encircling Jupiter in the plane of the planet's equator (see Figure 11.25). This ring lies roughly 50,000 km above the top cloud layer of the planet, inside the orbit of the innermost moon. A thin sheet of material may extend all the way down to Jupiter's cloud tops, but most of the ring is confined within a region only a few thousand kilometers across. The outer edge of the ring is quite sharply defined. In the direction perpendicular to the equatorial plane, the ring is only a few tens of kilometers thick. The small, dark particles that make up the ring may be fragments chipped off by meteorite impacts from two small moons—Metis and Adrastea, discovered by *Voyager*—that lie very close to the ring itself.

Despite differences in appearance and structure, Jupiter's ring can perhaps be best understood by studying the most famous planetary ring system—that of Saturn—so we will postpone further discussion of ring properties until the next chapter.

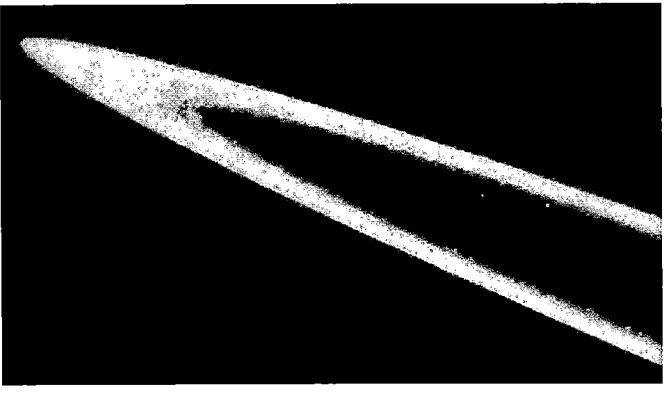

Figure 11.25 Jupiter's Ring Jupiter's faint ring, as photographed (nearly edge-on) by *Voyager 2*. The ring, made up of dark fragments of rock and dust, possibly chipped off the innermost moons by meteorites, was unknown before the two *Voyager* spacecraft arrived at the planet. It lies in Jupiter's equatorial plane, only 50,000 km above the cloud tops. *(NASA)*

Chapter Review

SUMMARY

Jupiter is the largest planet in the solar system. Its mass is more than twice the mass of all the other planets combined, although it is still about 1/1000 the mass of the Sun. It is composed primarily of hydrogen and helium. Jupiter rotates very rapidly, producing a pronounced equatorial bulge. The planet's flattened shape allows astronomers to infer the presence of a large rocky core in its interior. Jupiter displays **differential rotation** (p. 277)—because the planet has no solid surface, the rotation rate varies from place to place in the atmosphere. Measurements of radio emission from Jupiter's magnetosphere provide a measure of the planet's interior rotation rate.

Jupiter's atmosphere consists of three main cloud layers. The colors we see are the result of chemical reactions, fueled by the planet's interior heat, solar ultraviolet radiation, auroral phenomena, and lightning, at varying depths below the cloud tops, seen through "holes" in the overlying clouds. The cloud layers on Jupiter, as on all the jovian worlds, are arranged into bands of bright zones and darker belts crossing the planets parallel to the equator. The bands are the result of convection in the planet's interior and the planet's rapid rotation. The lighter zones are the tops of upwelling, warm currents, and the darker bands are cooler regions where gas is sinking. Underlying them is a stable pattern of eastward or westward wind flow

called a **zonal flow** (p. 279). The wind direction alternates as we move north or south away from the equator. The main weather pattern on Jupiter is the **Great Red Spot** (p. 278), an Earth-sized hurricane that has been raging for at least three centuries. Other, smaller, weather systems—**white ovals** (p. 283) and **brown ovals** (p. 283)—are also observed. They can persist for decades.

Jupiter's atmosphere becomes hotter and denser with depth, eventually becoming liquid. Interior pressures are so high that the hydrogen is "metallic" in nature near the center. The planet has a large "terrestrial" core about 10 times the mass of Earth. Jupiter radiates about twice as much energy into space as it receives from the Sun. The source of this energy is most likely heat released into the planet's interior when Jupiter formed 4.6 billion years ago, now slowly leaking from the surface.

The magnetosphere of Jupiter is about a million times more voluminous than Earth's magnetosphere, and the planet has a long magnetic "tail" extending away from the Sun to at least the distance of Saturn's orbit. Energetic particles spiral around magnetic field lines, accelerated by Jupiter's rotating magnetic field, producing intense radio radiation.

Jupiter and its system of moons resemble a small solar system. Twenty-eight moons have been discovered so far. The outermost eight moons resemble asteroids and have retrograde orbits, suggesting that they may have been captured by Jupiter's gravity long after the planets and largest moons formed. Jupiter's four major moons are called the **Galilean moons** (p. 276), after their discoverer, Galileo Galilei. Their densities decrease with increasing distance from the planet.

The innermost Galilean moon, Io, has active volcanoes powered by the constant flexing of the moon by Jupiter's tidal forces. As Io orbits Jupiter, the moon "wobbles" because of the gravitational pull of Europa. The ever-changing distortion of its interior energizes Io, and geyserlike volcanoes keep its surface smooth with constant eruptions. The material ejected by these volcanoes forms the Io plasma torus in Jupiter's inner magneto-

sphere. Europa has a cracked, icy surface that probably conceals an ocean of liquid water. Its fields of ice are nearly devoid of craters but have extensive large-scale fractures, most likely due to the tidal influence of Jupiter and the gravitational effects of the other Galilean satellites, and small-scale "chaos" caused by the action of the underlying ocean. Ganymede and Callisto have ancient, heavily cratered surfaces. Ganymede, the largest moon in the solar system, shows evidence of past geological activity, but now appears to be unmoving rock and ice, although recent evidence suggests that it too may have subsurface liquid water. Callisto apparently froze before tectonic activity could start there.

Jupiter has a faint, dark ring extending down to the planet's cloud tops. It was discovered in 1979 by *Voyager 1*.

SELF TEST: TRUE OR FALSE?

_____ **1.** Jupiter has over 300 times the mass of Earth, and twice the mass of all the other planets combined.

_____ **2.** The solid surface of Jupiter lies just below the cloud layers visible from Earth.

_____ **3.** There is no evidence to suggest that Jupiter has a dense core of rocky composition.

_____ **4.** Jupiter has only one large (Earth-sized) storm system.

_____ **5.** In general, a storm system in Jupiter's atmosphere is much longer-lived than storms in Earth's atmosphere.

_____ **6.** The element helium plays an important role in producing the colors in Jupiter's atmosphere.

_____ **7.** The magnetosphere of Jupiter is similar in strength to Earth's magnetosphere.

_____ **8.** Most of Jupiter's moons rotate synchronously with their orbits.

_____ **9.** The densities of the Galilean moons increase with increasing distance from Jupiter.

_____ **10.** Io has a noticeable lack of impact craters on its surface.

_____ **11.** The surface of Europa is completely covered by water ice.

_____ **12.** Ganymede shows evidence of ancient plate tectonics.

_____ **13.** Jupiter's ring is made up of icy particles a few meters across.

_____ **14.** Most of the small moons of Jupiter have diameters of several hundred kilometers.

_____ **15.** The thickness of Jupiter's cloud layer is less than one percent of the planet's radius.

SELF TEST: FILL IN THE BLANK

1. The _____ of Jupiter indicates that its overall composition differs greatly from that of the terrestrial planets.

2. The main constituents of Jupiter are _____ and _____.

3. Jupiter's rapid _____ produces a significant equatorial bulge.

4. Jupiter's Great Red Spot has similarities to _____ on Earth.

5. The diameter of the Great Red Spot is about _____ that of Earth.

6. Jupiter's clouds consist of a series of bright _____ and dark _____.

7. Jupiter emits about _____ more radiation than it receives from the Sun.

8. Although often referred to as a gaseous planet, Jupiter is mostly _____ in its interior.

9. Jupiter's magnetic field is generated by its rapid rotation and the element _____, which becomes metallic at the high pressure found in the planet's deep interior.

10. Because of Jupiter's strong tidal field, most of Jupiter's satellites rotate _____ with their orbits around the planet.

11. The Galilean moons make up _____ of the _____ moons of Jupiter. (Give the numbers.)

12. The Galilean moon _____ is larger than the planet Mercury.

13. In contrast with the inner Galilean moons, the outer two Galilean moons have compositions that include significant amounts of _____.

14. Io is the only moon in the solar system with active _____.

15. Scientists speculate that _____ may have liquid water below its frozen surface.

REVIEW AND DISCUSSION

1. In what sense does our solar system consist of only two important objects?

2. What is differential rotation, and how is it observed on Jupiter?

3. What does Jupiter's degree of flattening tell us about its interior?

4. Describe some of the ways in which the *Voyager* mission changed our perception of Jupiter.

5. Describe some of the ways in which the *Galileo* mission changed our perception of Jupiter.

6. What is the Great Red Spot? What is known about the source of its energy?

7. What is the cause of the colors in Jupiter's atmosphere?

8. Why has Jupiter retained most of its original atmosphere?

9. Explain the theory that accounts for Jupiter's internal heat source.

10. What is Jupiter thought to be like beneath its clouds? Why do we think this?

11. What is responsible for Jupiter's enormous magnetic field?

12. In what sense are Jupiter and its moons like a miniature solar system?

13. How does the density of the Galilean moons vary with increasing distance from Jupiter? Is there a trend to this variation? If so, why?

14. What is the cause of Io's volcanic activity?

15. What evidence do we have for liquid water below Europa's surface?

16. Why do scientists think that Ganymede's interior may have been heated as recently as one billion years ago?

17. How does the amount of cratering vary among the Galilean moons? Does it depend on their location? If so, why?

18. Why is there speculation that the Galilean moon Europa might be an abode for life?

19. What might be the consequences of the discovery of life on Europa?

20. Water is relatively uncommon among the terrestrial planets. Is it common among the moons of Jupiter?

PROBLEMS  *Algorithmic versions of these questions are available in the Practice Problems module of the Companion Website.*

The number of squares preceding each problem indicates its approximate level of difficulty.

1. ■ How does the force of gravity at Jupiter's cloud tops compare with the force of gravity at Earth's surface?

2. ■ What are the angular diameters of the orbits of Jupiter's four Galilean satellites, as seen from Earth at closest approach (assuming for definiteness that opposition occurs near perihelion)?

3. ■ Using the figures given in the text, calculate how long it takes Jupiter's equatorial winds to circle the planet, relative to the interior.

4. ■■ Calculate the rotational speed (in km/s) of a point on Jupiter's equator, at the level of the cloud tops. Compare it with the orbital speed just above the cloud tops.

5. ■ Given Jupiter's age and current atmospheric temperature, what is the smallest possible mass the planet could have and still have retained its hydrogen atmosphere? ∞ (*More Precisely 8-1*)

6. ■ If Jupiter had been just massive enough to fuse hydrogen (see *Discovery 11-1*), calculate what the planet's gravitational force on Earth would have been at closest approach, relative to the gravitational pull of the Sun. Assume circular orbits. Also estimate what the magnitude of the planet's tidal effect on our planet would have been, again relative to that of the Sun.

7. ■ Calculate the ratio of Jupiter's mass to the total mass of the Galilean moons. Compare this with the ratio of Earth's mass to that of the Moon.

8. ■■■ At what distance would a satellite orbit Jupiter in the time taken for Jupiter to rotate exactly once, so that the satellite would appear "stationary" above the planet? ∞ (*More Precisely 2-3*)

9. ■ Illustrate the 1:2:4 "resonance" mentioned in the text by drawing a diagram showing the locations of Io, Europa, and Ganymede at various times over the course of one Ganymede orbit. Show the moons' locations at intervals of $\frac{1}{4}$ Io's orbital period.

10. ■■ Estimate the strength of Jupiter's gravitational tidal acceleration on Io, and compare it to the moon's own surface gravity. ∞ (*More Precisely 7-3*) Compare this with the strength of Earth's gravitational tidal force on the Moon, relative to the Moon's own surface gravity.

11. ■■ Estimate the strength of Jupiter's gravitational tidal force on Europa, relative to the moon's own surface gravity.

12. ■■ Estimate the strength of Jupiter's gravitational tidal force on Ganymede, relative to the moon's own surface gravity.

13. ■■ Calculate the strength of Europa's gravitational pull on Io at closest approach, relative to Jupiter's gravitational attraction on Io.

14. ■■ What are the surface gravity and escape speed of Europa?

15. ■■ Compare the apparent sizes of the Galilean moons, as seen from Jupiter's cloud tops, with the angular diameter of the Sun as seen from Jupiter. Would you expect ever to see a total solar eclipse from Jupiter's cloud tops?

COLLABORATIVE EXERCISES

1. Galilean Moons. As a group, agree to which of the Galilean Moons you would send a robotic land-rover mission, and justify your choice.

RESEARCHING ON THE WEB

To complete the following exercises, go to the online Destinations module for Chapter 11 on the Companion Website for Astronomy Today 4/e.

1. Access the "Jupiter Fact Sheet" from NASA and determine the current number of known moons of Jupiter.
2. Access the "Jupiter Nomenclature" page. For what type of mythological god are Io's eruptive centers?

3. Access the "Jupiter Pages in Planetary Collections (JPL)" and determine which spacecraft had the first indications that there was a faint ring around Jupiter.

PROJECTS

1. Are there any stars in the night sky that look as bright as Jupiter? What other differences do you notice between Jupiter and the stars?
2. Use binoculars to look at Jupiter. Can you see any of Jupiter's four largest moons? If you come back the following evening, the moons' relative positions will have changed. Have some changed more than others?

3. Through a telescope, you should be able to see the red-and-tan cloud bands of Jupiter, and you can clearly see some moons. Do the moons orbit in the equatorial plane? Before observing, look up the positions of the Galilean moons in a current magazine such as *Astronomy* or *Sky & Telescope*. Identify each of the moons. Watch Io over a period of an hour. Can you see its motion? Do the same for Europa.

SKYCHART III PROJECTS

The SkyChart III Student Version planetarium program on which these exercises are based is included as a separately executable program on the CD in the back of this text.

1. ■ Re-create the view Galileo saw through his telescope when he first turned his attention to Jupiter, January 7, 1610. Set the time to 7:00 P.M. on this date and set location to Rome. Center on Jupiter. Zoom to 1/2° field of view and set animation for 10 minutes. If your computer is sufficiently fast, you may wish to use one-minute time steps. This re-creates the orientation of Jupiter and its four Galilean moons the way Galileo must have seen them. Run the animation forward and observe how Io and Europa move so fast that in one night of observation their motions can be observed against the background stars and with respect to Jupiter itself. Follow the animation for a few days until you are clear on the motion of the four moons.

2. ■■ Looking to the sky for an accurate clock was a necessity before the invention of mechanical, portable clocks. One such clock can be found in the regular eclipses of Jupiter's Galilean moons. *Center* on Jupiter and set the *Coordinates* to *Ecliptic*. In the *Date & Time* information enter "2000/11/25 00:00:00" and change the *Time Zone* to "0.0" (this data represents an alignment between Jupiter and Earth). Also go to the *COMPUTATION* menu and select *Precision*. In the information box check *Correct for Light Time*. Next, turn the *Object Labels* on and animate the moons using a *Time Step* of *1 Minute*. Watch for Jupiter to begin eclipsing Io (don't confuse an eclipse with a transit). Record the exact time at which the eclipse starts. Continue the animation until Io starts into another eclipse and record the exact time. How long between the beginning of eclipses? Record times of a few more eclipses and show that they occur at regular intervals. (Note: for simplicity ignore Jupiter's umbra—only count an eclipse when Io is actually moving behind Jupiter.)

3. ■■ For the Galilean moons to be useful as clocks, charts must be constructed telling the exact times at which the eclipses start. Construct such a chart predicting future eclipses for the next six months. Use the data gathered from the previous exercise and a spreadsheet program to calculate the dates and times of the eclipses.

4. ■■ Adjust the *Date & Time* and directly check the eclipses predicted from the chart constructed in the previous exercise. You should find that the predictions are close for a month or two but are off by large amounts after six months. In 1675, Ole Roemer discovered similar discrepancies and determined that they had to do with the time it takes light to travel from Jupiter to Earth. From these discrepancies, Roemer calculated the speed of light to be approximately 200,000 km/s. Explain why the predictions do not work and how one might adjust the charts to accurately predict eclipses.

5. ■■ Whenever two bodies orbit each other, we can use Kepler's third law to determine the sum of their masses. Use the modified version of Kepler's third law (Section 2.7) to determine the mass of Jupiter. Use spreadsheet software to construct a graph of angular separation (from Jupiter) versus time for Jupiter's moon Io using *Mouse Coordinates*. From this graph, extract the semimajor axis and period of Io. To scale your measurements to real numbers you will also need to measure Jupiter's angular width and compare this with Jupiter's actual width, 9.5×10^{-4} A.U. How much does the Jupiter-Io system weigh? How does your calculated mass compare with the figure quoted in the text?

6. ■■ Repeat the preceding exercise using the other three Galilean moons, Ganymede, Callisto and Europa. From your calculations, what is the average mass of Jupiter? How far off are you from the actual mass? What are the major factors affecting your measurements?

 In addition to the Practice Problems and Destinations modules, the Companion Website at http://www.prenhall.com/chaisson provides for each chapter an additional true-false, multiple choice, and labeling quiz, as well as additional annotated images, animations, and links to related Websites.

12 SATURN

Spectacular Rings and Mysterious Moons

LEARNING GOALS

Studying this chapter will enable you to:

1   Summarize the orbital and physical properties of Saturn, and compare them with those of Jupiter.

2   Describe the composition and structure of Saturn's atmosphere and interior.

3   Explain why Saturn's internal heat source and magnetosphere differ from those of Jupiter.

4   Describe the structure and composition of Saturn's rings.

5   Define the Roche limit and explain its relevance to the origin of Saturn's rings.

6   Summarize the general characteristics of Titan and discuss the chemical processes in its atmosphere.

7   Discuss some of the orbital and geological properties of Saturn's smaller moons.

 Visit http://www.prenhall.com/chaisson for additional annotated images, animations, and links to related sites for this chapter.

Saturn is often portrayed in garish colors, as in a few figures in this chapter, mainly to highlight certain features of its complex cloud deck. However, this planet actually has very little color; the pastels shown here are the closest approximation of Saturn's natural colors. That's because our view of Saturn is dominated by its high-altitude clouds known to be rich in colorless ammonia ice. (STScI)

The Big Picture: Hardly more than a few decades ago, Saturn was the only planet known to have rings about it. Now, astronomers realize that all the big jovian planets have ring systems, though not as spectacular as Saturn's. We also know that Saturn has more small moons (several dozen) than any other planet, although how one distinguishes a small, genuine moon from the tens of thousands of water-ice debris in the rings is a debatable issue.

Saturn is one of the most beautiful and enchanting of all astronomical objects. Its rings are a breathtaking sight when viewed through even a small telescope, and they are probably the planet's best-known feature. Aside from its famous rings, however, Saturn presents us with another good example of a giant gaseous planet. Saturn is in many ways similar to its larger neighbor, Jupiter, in terms of composition, size, and structure. Yet when we study the two planets in detail, we find that there are important differences as well. A comparison between Saturn and Jupiter provides us with valuable insight into the structure and evolution of all the jovian worlds.

12.1 Orbital and Physical Properties

OVERALL PROPERTIES

① Saturn was the outermost planet known to ancient astronomers. Named after the father of Jupiter in Roman mythology, Saturn orbits the Sun at almost twice the distance of Jupiter. The planet's sidereal orbital period of 29.4 Earth years was the longest natural unit of time known to the ancient world. At opposition, when most of the Earth- (or Earth-orbit) based photographs of Saturn were taken, the planet is at its brightest and can lie within 8 A.U. of Earth. However, its great distance from the Sun still makes it considerably fainter than either Jupiter or Mars. Saturn ranks behind Jupiter, the inner planets, and several of the brightest stars in the sky in terms of apparent brightness.

The Saturn Data box on p. 308 presents orbital and physical data for the planet. Less than one-third the mass of Jupiter, Saturn is still an enormous body, at least by terrestrial standards. As with Jupiter, Saturn's many moons allowed an accurate determination of the planet's mass long before the arrival of the *Pioneer* and *Voyager* missions. Saturn's mass is 5.7×10^{26} kg, or 95 times the mass of Earth. From Saturn's distance and angular size, the planet's radius—and hence the average density—quickly follow. Saturn's equatorial radius is 60,000 km, or 9.5 Earth radii. The average density is 700 kg/m³—less than the density of water (which is 1000 kg/m³). Here we have a planet that would float in the ocean—if Earth had one big enough! Saturn's low average density indicates that, like Jupiter, it is composed primarily of hydrogen and helium. Saturn's lower mass, however, results in lower interior pressure, so these gases are less compressed than in Jupiter's case.

ROTATION RATE

Saturn, like Jupiter, rotates very rapidly and differentially. The rotation period of the interior (as measured from magnetospheric outbursts, which trace the rotation of the planet's core) and at high planetary latitudes (determined by tracking weather features observed in Saturn's atmosphere) is $10^h 40^m$. The rotation period at the equator is $10^h 14^m$, or about 26 minutes shorter. Because of Saturn's lower density, this rapid rotation makes Saturn even more

flattened than Jupiter. In fact, Saturn is the "flattest" planet in the solar system—its polar radius is just 54,000 km, about 10 percent less than the equatorial radius. Careful calculations show that this degree of flattening, large as it is, is less than would be expected for a planet composed of hydrogen and helium alone. Astronomers believe that Saturn also has a rocky core, perhaps 15 times the mass of Earth, or 1.5 times the mass of Jupiter's core.

RINGS

Saturn's best-known feature is its spectacular *ring system*. Because the rings lie in the equatorial plane, their appearance (as seen from Earth) changes seasonally, as shown in Figure 12.1. Saturn's rotation axis is significantly tilted with respect to the planet's orbit plane—the axial tilt is 27°, similar to that of both Earth and Mars. Consequently, as Saturn orbits the Sun, the angles at which the rings are illuminated and at which we view them vary. When the planet's north or south pole is tipped toward the Sun, during Saturn's summer or winter, the highly reflective rings are at their brightest. During Saturn's spring and fall, the rings are close to being edge-on, both to the Sun and to us, so they seem to disappear altogether. The last "ring crossing" occurred in 1995. One important deduction that we can make from this simple observation is that the rings are very thin. In fact, we now know that their thickness is less than a few hundred meters, even though they are over 200,000 km in diameter.

✓ **Concept Check**
- Why do most of the Earth-based images of Saturn in this chapter show the rings seen from above, while Figure 12.3 shows them from below?

12.2 Saturn's Atmosphere

② Saturn is much less colorful than Jupiter. Figure 12.2 shows yellowish and tan cloud belts that parallel the equator, but these regions display less atmospheric structure than do the belts on Jupiter. No obvious large, long-lived "spots" or "ovals" adorn Saturn's cloud decks. Bands and storms do exist (in fact, a large storm is visible in Figure

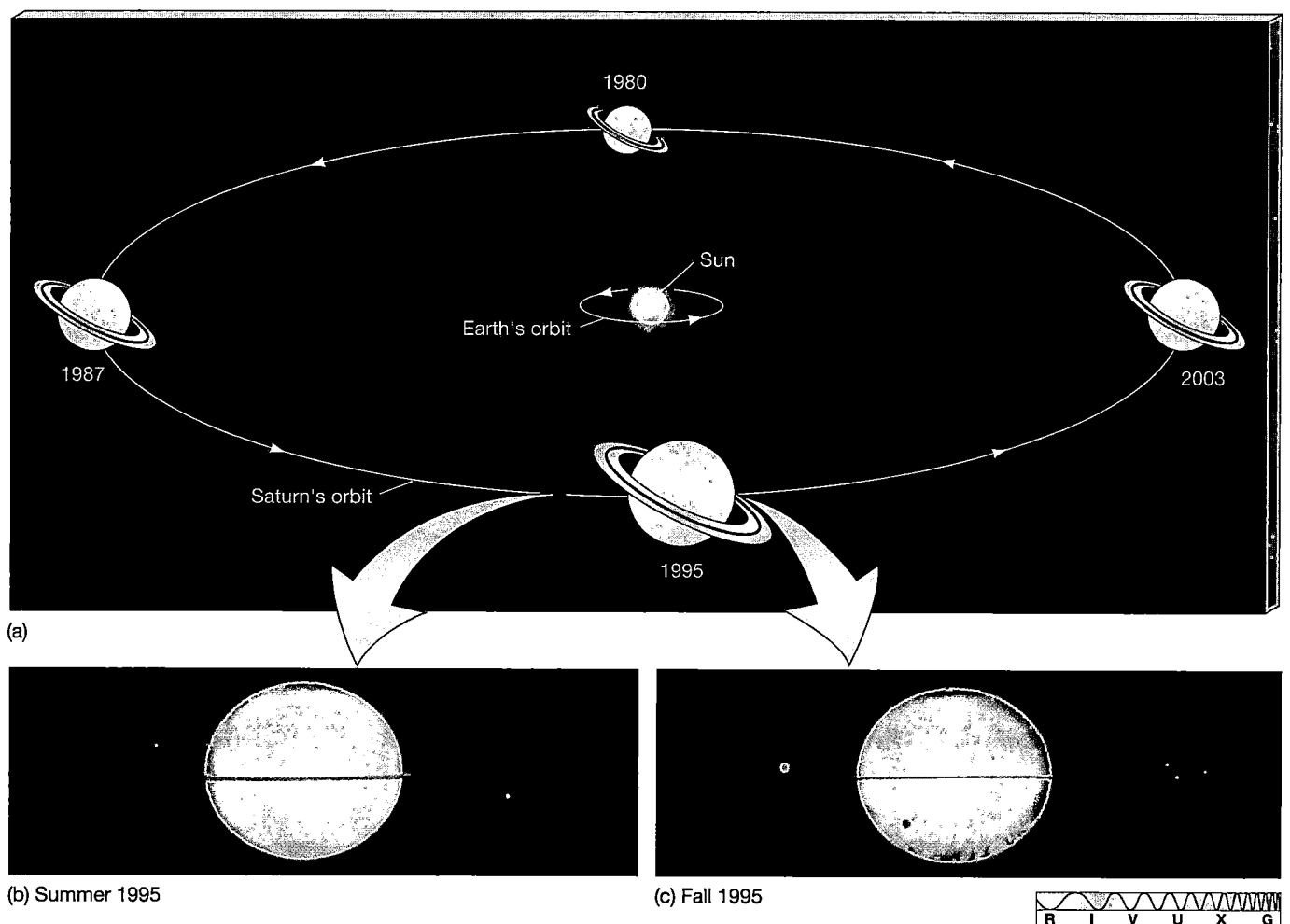

(a)

(b) Summer 1995

(c) Fall 1995

R   I   V   U   X   G

Figure 12.1 Ring Orientation (a) Over time, Saturn's rings change their appearance to terrestrial observers as the tilted ring plane orbits the Sun. At some times during Saturn's 29.5-year orbital period the rings seem to disappear altogether as Earth passes through their plane and we view them edge-on. Numbers along Saturn's orbit indicate the year. The two images are separated by a few months, having been taken (in true color) by the *Hubble Space Telescope* in the summer and fall of 1995. (b) In mid-1995, the rings are seen almost edge-on. (c) Later that year, the ring plane is precisely edge-on as viewed from Earth. Several moons are visible, including Titan's shadow (black dot). See also Figures 12.2 (1994: rings seen from above) and 12.3 (1998: rings seen from below). *(NASA)*

12.2), but the color changes that distinguish them on Jupiter are largely absent on Saturn. The banded structure is more clearly visible in Figure 12.3, which shows a false-color infrared image of the planet.

COMPOSITION AND COLORATION

Astronomers first observed methane in the spectrum of sunlight reflected from Saturn in the 1930s, about the same time that it was discovered on Jupiter. However, it was not until the early 1960s, when more sensitive observations became possible, that ammonia was finally detected. In Saturn's cold upper atmosphere, most ammonia is in the solid or liquid form, with relatively little of it present as a gas to absorb sunlight and create spectral lines. Astronomers finally made the first accurate determinations of the hydrogen and helium content in the late 1960s. These Earth-based measurements were later confirmed with the arrival of the *Pioneer* and *Voyager* spacecraft in the 1970s.

Saturn's atmosphere consists of molecular hydrogen (92.4 percent), helium (7.4 percent), methane (0.2 percent), and ammonia (0.02 percent). As on Jupiter, hydrogen and helium dominate—these most abundant elements never escaped from Saturn's atmosphere because of the planet's large mass and low temperature (see *More Precisely 8-1*). However, the fraction of helium on Saturn is far less than is observed on Jupiter (where, as we saw, helium accounts for nearly 14 percent of the atmosphere) or in the Sun. It is extremely unlikely that the processes that created the outer planets preferentially stripped Saturn of nearly half its helium or that the missing helium somehow escaped from the planet while the lighter hydrogen remained behind. Instead, astronomers believe that at some time in Saturn's past the heavier helium began to sink toward the center of the planet, reducing its abundance in the outer layers and leaving them relatively hydrogen-rich. We will return to the reasons for this differentiation and its consequences in a moment.

303

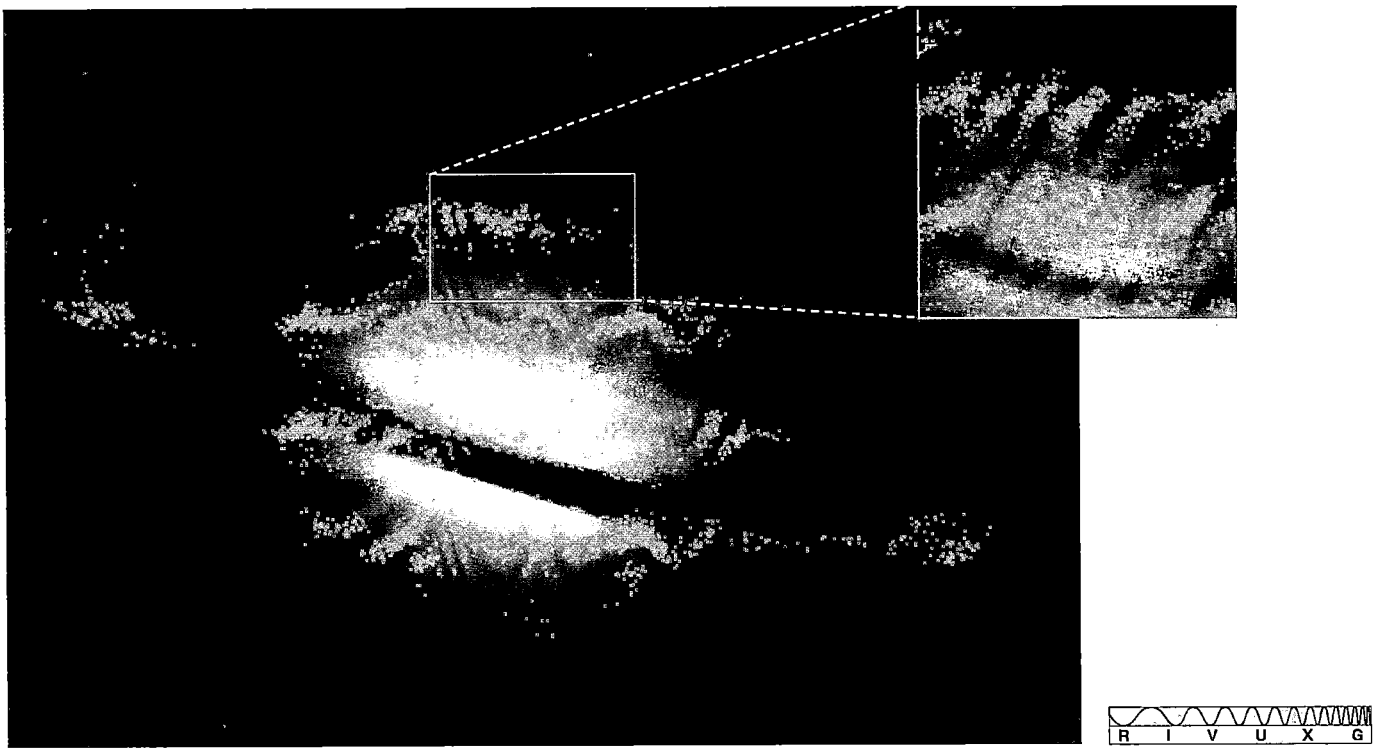

Figure 12.2 Saturn Saturn as seen by the *Hubble Space Telescope* in December 1994. At the time, a rare storm was visible near the planet's equator. The bland colors are approximately true—that is, as the human eye sees things. The insert shows the northern polar region at an earlier time (note the ring segment in the background). *(NASA)*

Figure 12.4 illustrates Saturn's atmospheric structure (compare with the corresponding diagram for Jupiter, Figure 11.6). In many respects, Saturn's atmosphere is quite similar to Jupiter's, except that the temperature is a little lower because of its greater distance from the Sun and because its clouds are somewhat thicker. Since Saturn, like Jupiter, lacks a solid surface, we take the top of the troposphere as our reference level and set it to 0 km. The top of

the visible clouds lies about 50 km below this level. As on Jupiter, the clouds are arranged in three distinct layers, composed (in order of increasing depth) of ammonia, ammonium hydrosulfide, and water ice. Above the clouds lies a layer of haze formed by the action of sunlight on Saturn's upper atmosphere.

The total thickness of the three cloud layers in Saturn's atmosphere is roughly 200 km, compared with about

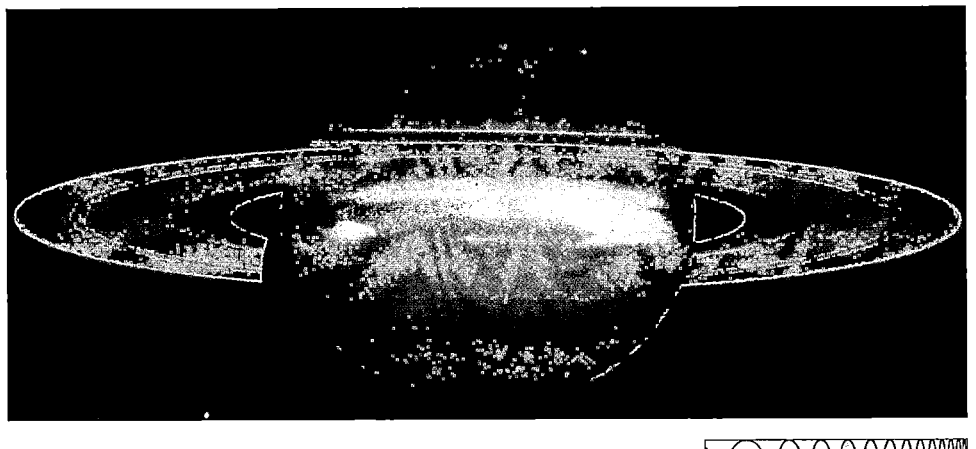

Figure 12.3 Saturn, False Color This image was obtained using an infrared camera on board the *Hubble Space Telescope* in 1998. The false colors were obtained by combining observations made at three different wavelengths: 1.0 μm (blue), 1.8 μm (green), and 2.1 μm (red). Blue coloration indicates regions where the atmosphere is relatively free of haze (see Figure 12.4); green and yellow indicate increasing haze levels; red and orange indicate high-level clouds. Two small storm systems near the equator appear white. Two of Saturn's moons are also visible in the image: Dione (lower left) and Tethys (upper right, just on the "edge" of the planet). *(NASA)*

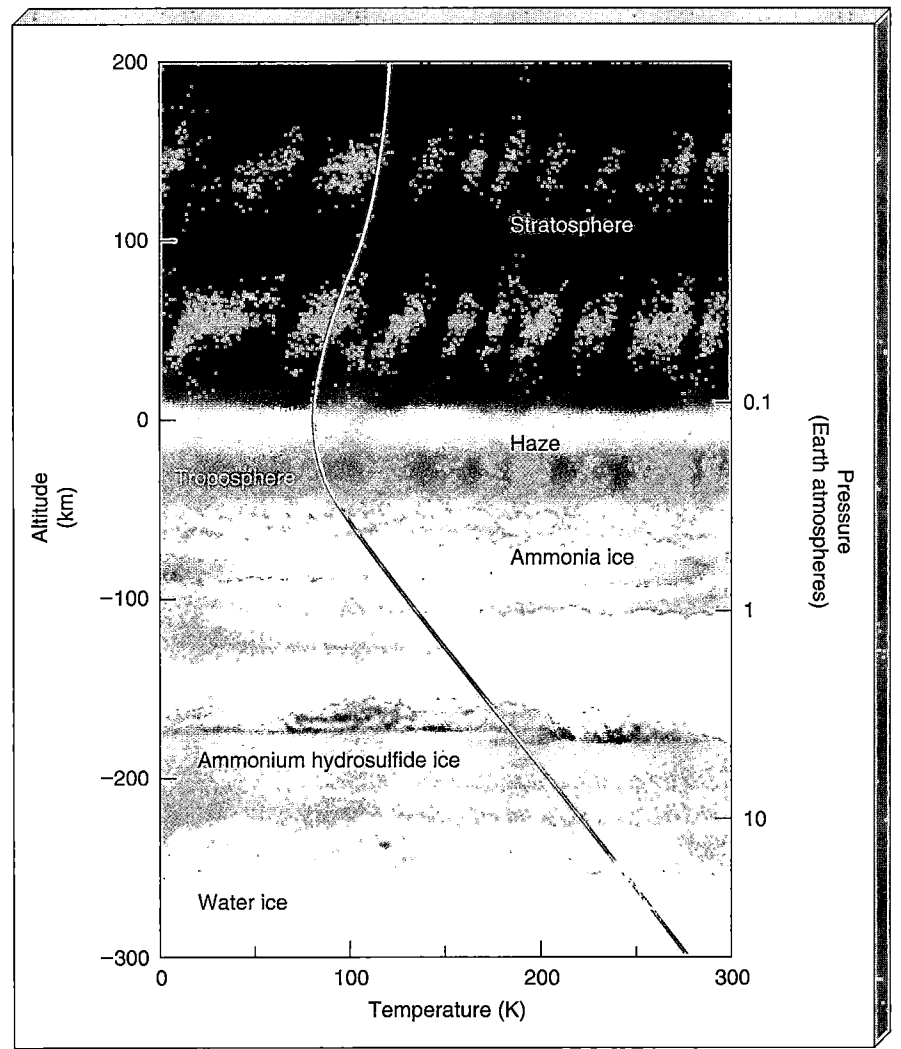

Figure 12.4 Saturn's Atmosphere
The vertical structure of Saturn's atmosphere. As with Jupiter, there are several cloud layers, but Saturn's weaker gravity results in thicker clouds and a more uniform appearance. Colors are intended to represent Saturn's visible-light appearance (note that they are largely unrelated to the infrared "colors" in Figure 12.3).

80 km on Jupiter, and each layer is itself somewhat thicker than its counterpart on Jupiter. The reason for this difference is Saturn's weaker gravity. At the haze level, Jupiter's gravitational field is nearly two and a half times stronger than Saturn's, so Jupiter's atmosphere is pulled much more powerfully toward the center of the planet. Thus Jupiter's atmosphere is compressed more than Saturn's, and the clouds are squeezed more closely together. The colors of Saturn's cloud layers, as well as the planet's overall butterscotch hue, are due to the same basic cloud chemistry as on Jupiter. However, because Saturn's clouds are thicker, there are few holes and gaps in the top layer, so we rarely glimpse the more colorful levels below. Instead, we see only different levels in the topmost layer, which accounts for Saturn's rather uniform appearance.

WEATHER

Saturn has atmospheric wind patterns that are in many ways reminiscent of those on Jupiter. There is an overall east–west zonal flow, which is apparently quite stable. Computer-enhanced images of the planet that bring out

more cloud contrast (see Figure 12.5) clearly show the existence of bands, oval storm systems, and turbulent flow patterns looking very much like those seen on Jupiter. Scientists believe that Saturn's bands and storms have essentially the same cause as does Jupiter's weather. Ultimately, the large-scale flows and small-scale storm systems are powered by convective motion in Saturn's interior and the planet's rapid rotation.

The zonal flow on Saturn is considerably faster than on Jupiter and shows fewer east–west alternations, as can be seen from Figure 12.6 (compare with Figure 11.5). The equatorial eastward jet stream, which reaches a speed of about 400 km/h on Jupiter, moves at a brisk 1500 km/h on Saturn, and extends to much higher latitudes. Not until latitudes 40° north and south of the equator are the first westward flows found. Latitude 40° north also marks the strongest bands on Saturn and the most obvious ovals and turbulent eddies. Astronomers still do not fully understand the reasons for the differences between Jupiter's and Saturn's flow patterns.

In September 1990, amateur astronomers detected a large white spot in Saturn's southern hemisphere, just below the equator. In November of that year, when the

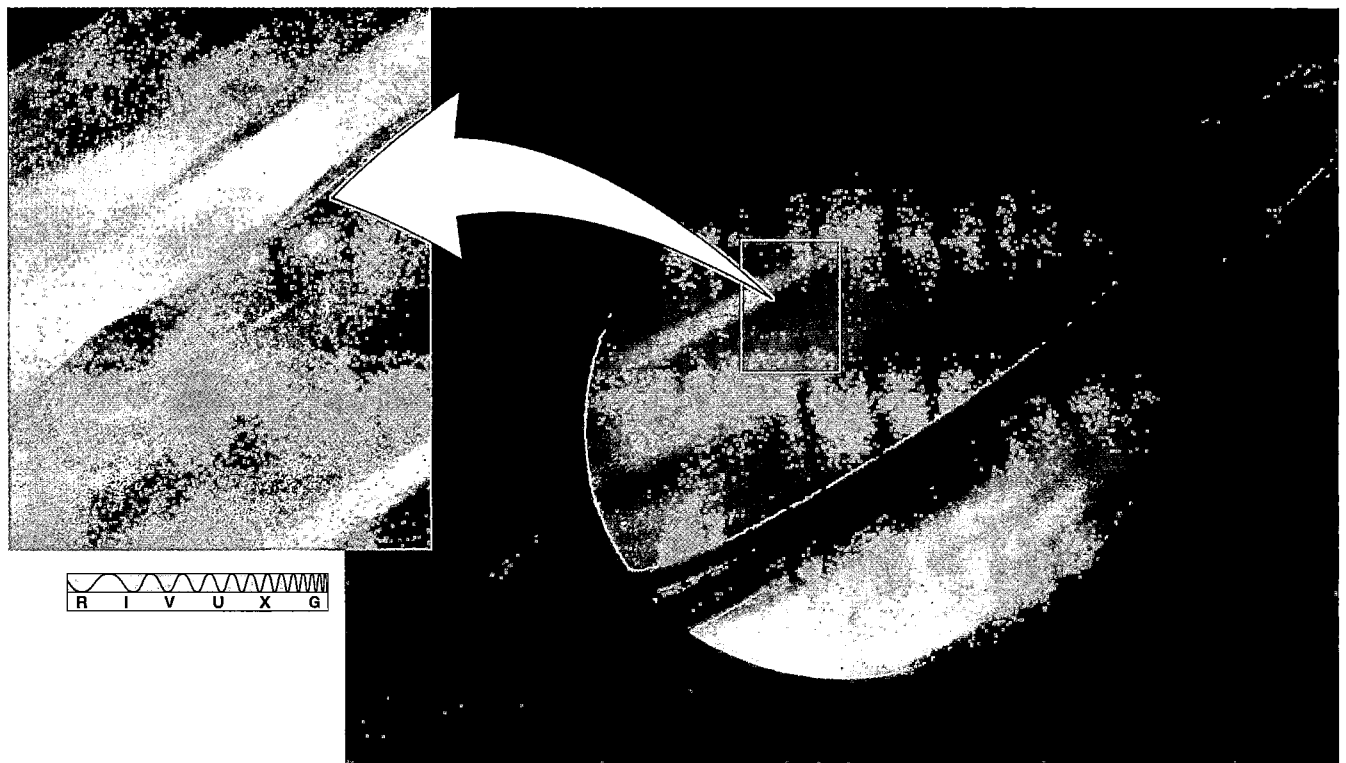

Figure 12.5 Saturn's Cloud Structure We see more structure in Saturn's cloud cover when computer processing and artificial color are used to enhance the image contrast, as in these 1980 *Voyager 1* images of the entire gas ball and of a smaller, magnified piece of it. *(NASA)*

Hubble Space Telescope imaged the phenomenon in more detail, the spot had developed into a band of clouds completely encircling the planet's equator. Some of these images are shown in Figure 12.7. Astronomers believe that the white coloration arose from crystals of ammonia ice formed when an upwelling plume of warm gas penetrated the cool upper cloud layers. Because the crystals were freshly formed, they had not yet been affected by the chemical reactions that color the planet's other clouds.

Such spots are relatively rare on Saturn. The previous one visible from Earth appeared in 1933, but it was much smaller than the 1990 system and much shorter-lived, lasting for only a few weeks. Another large storm system is visible in Figure 12.2, taken in 1994. Figure 12.3 (from 1998) shows two somewhat smaller systems. The turbulent flow patterns seen around the 1990 white spot have many similarities to the flow around Jupiter's Great Red Spot. Scientists speculate that these white spots represent long-lived weather systems on Saturn and hope that routine observations of such temporary atmospheric phenomena on the outer worlds will enable them to gain greater insight into the dynamics of planetary atmospheres.

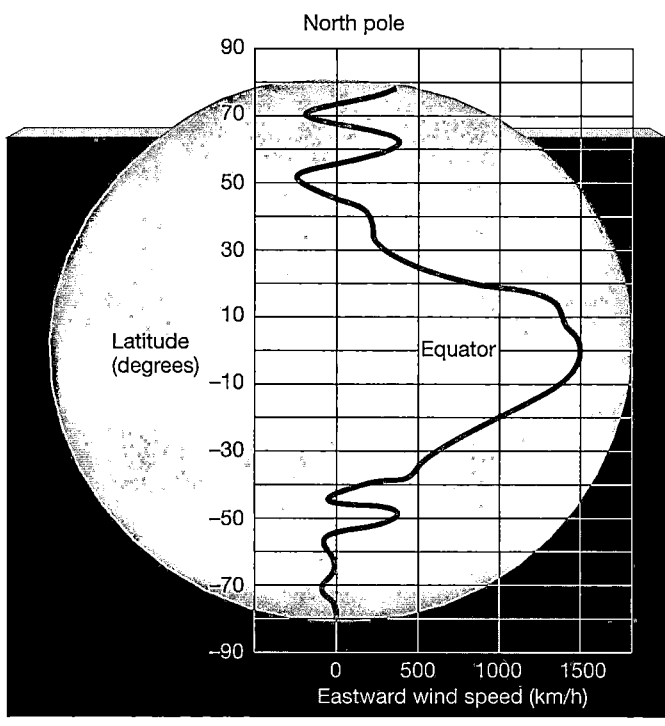

Figure 12.6 Saturn's Zonal Flow Winds on Saturn reach speeds even greater than those on Jupiter. As on Jupiter, the visible bands appear to be associated with variations in wind speed.

✓ **Concept Check**

■ Why are atmospheric features on Saturn generally less vivid than those on Jupiter?

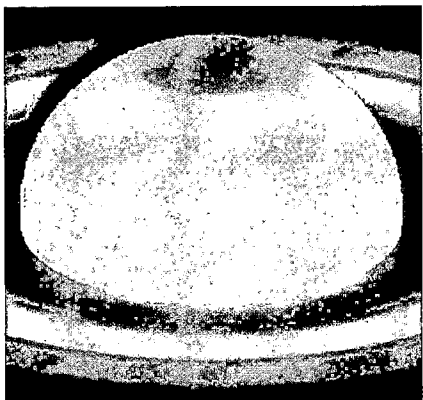

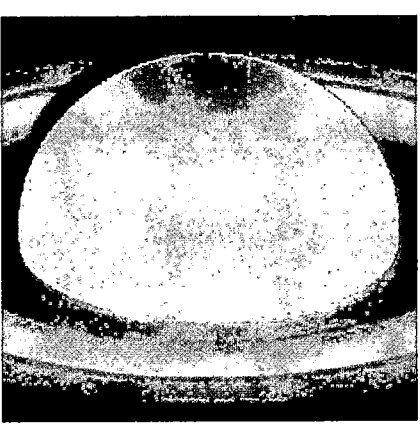

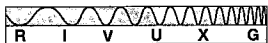

Figure 12.7 Storm on Saturn Circulating and evolving cloud systems on Saturn, imaged by the *Hubble Space Telescope* in 1990 at approximately two-hour intervals. This is the true color of Saturn, the blandest of all the jovian planets. *(NASA)*

12.3 Saturn's Interior and Magnetosphere

INTERIOR STRUCTURE AND INTERNAL HEATING

Figure 12.8 depicts Saturn's internal structure (compare with Figure 11.10 for Jupiter). This picture has been pieced together by planetary scientists using the same tools—*Voyager* observations and theoretical modeling—that they used to infer Jupiter's inner workings. Saturn has the same basic internal parts as Jupiter, but their relative proportions are somewhat different: Saturn's metallic hydrogen layer is thinner, and its core is larger. Because of its lower mass, Saturn has a less extreme core temperature, density, and pressure than Jupiter. The central pressure is around a tenth of Jupiter's—not too different from the pressure at the center of Earth.

Infrared measurements indicate that Saturn's surface (that is, cloud-top) temperature is 97 K, substantially higher than the temperature at which Saturn would reradiate all the energy it receives from the Sun. In fact, Saturn radiates almost three times more energy than it absorbs. Thus Saturn, like Jupiter, has an internal energy source. ∞ (Sec. 11.3) But the explanation behind Jupiter's excess energy—that the planet has a large reservoir of heat left over from its formation—doesn't work for Saturn. Saturn is smaller than Jupiter and so must have cooled more rapidly—rapidly enough that its original supply of energy was used up long ago. What, then, is happening inside Saturn to produce this extra heat?

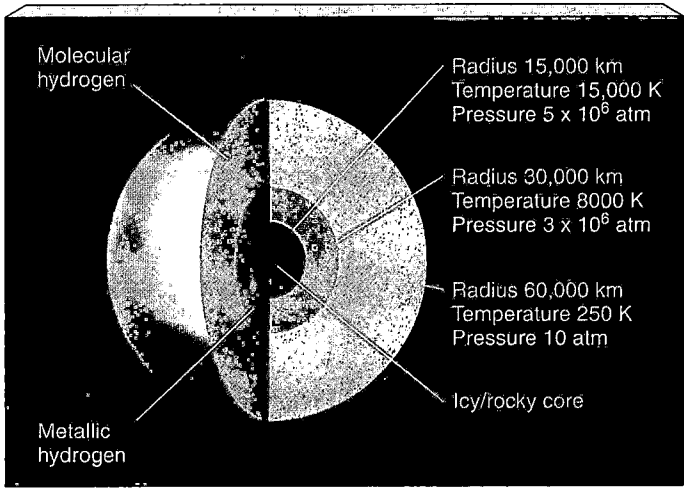

Figure 12.8 Saturn's Interior Saturn's internal structure, as deduced from *Voyager 1* and *2* observations and computer modeling.

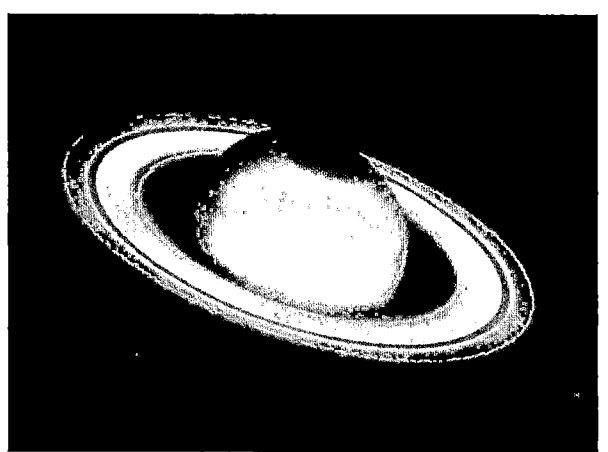

Orbital semimajor axis	9.54 A.U.
	1427 million km
Orbital eccentricity	0.054
Perihelion	9.02 A.U.
	1349 million km
Aphelion	10.05 A.U.
	1504 million km
Mean orbital speed	9.65 km/s
Sidereal orbital period	29.42 tropical years
Synodic orbital period	378.09 solar days
Orbital inclination to the ecliptic	2.49°
Greatest angular diameter, as seen from Earth	21″
Mass	5.68×10^{26} kg
	95.16 (Earth = 1)
Equatorial radius	60,268 km
	9.45 (Earth = 1)
Mean density	687 kg/m³
	0.125 (Earth = 1)
Surface gravity (at cloud tops)	10.4 m/s²
	1.07 (Earth = 1)
Escape speed	35.5 km/s
Sidereal rotation period	0.44 solar days
Axial tilt	26.73°
Surface magnetic field	0.67 (Earth = 1)
Magnetic axis tilt relative to rotation axis	0.8°
Surface temperature	97 K (at cloud tops)
Number of moons	30

The explanation for this strange phenomenon also explains the mystery of Saturn's apparent helium deficit. At the temperatures and high pressures found in Jupiter's interior, liquid helium *dissolves* in liquid hydrogen. In Saturn, where the internal temperature is lower, the helium doesn't dissolve so easily and tends to form droplets instead. The phenomenon is familiar to cooks, who know that it is generally much easier to dissolve ingredients in hot liquids than in cold ones. Saturn probably started out with a fairly uniform solution of helium dissolved in hydrogen, but the helium tended to condense out of the surrounding hydrogen, much as water vapor condenses out of Earth's atmosphere to form a mist. The amount of helium condensation was greatest in the planet's cool outer layers, where the mist turned to rain about 2 billion years ago. A light shower of liquid helium has been falling through Saturn's interior ever since. This **helium precipitation** is responsible for depleting the outer layers of their helium content.

So we can account for the unusually low abundance of helium in Saturn's atmosphere—much of it has rained down to lower levels. But what about the excess heating? The answer is simple: As the helium sinks toward the center, the planet's gravitational field compresses it and heats it up. The gravitational energy thus released is the source of Saturn's internal heat. In the distant future the helium rain will stop, and Saturn will cool until its outermost layers radiate only as much energy as they receive from the Sun. When that happens, the temperature at Saturn's cloud tops will be 74 K. As Jupiter cools, it too may someday experience helium precipitation in its interior, which will cause its surface temperature to rise once again.

MAGNETOSPHERIC ACTIVITY

Saturn's electrically conducting interior and rapid rotation produce a strong magnetic field and an extensive magnetosphere. Probably because of the considerably smaller mass of Saturn's metallic hydrogen zone, the planet's basic magnetic field strength is only about 1/20 that of Jupiter, or about 1000 times greater than that of Earth. The magnetic field at Saturn's cloud tops (roughly 10 Earth radii from the planet's center) is approximately the same as at Earth's surface. *Voyager* measurements indicate that, unlike Jupiter's and Earth's magnetic axes, which are slightly tilted, Saturn's magnetic field is not inclined with respect to its rotation axis. Saturn's magnetic field, like Jupiter's, is oriented opposite that of Earth—that is, an Earth compass needle would point toward Saturn's south pole rather than its north. Figure 12.9 shows an aurora on Saturn, imaged in 1998 by the *Hubble Space Telescope*.

Saturn's magnetosphere extends about 1 million km toward the Sun and is large enough to contain the planet's ring system and the innermost 16 small moons. Saturn's largest moon, Titan, orbits about 1.2 million km from the planet, so it is sometimes found just inside the outer magnetosphere and sometimes just outside, depending on the intensity of

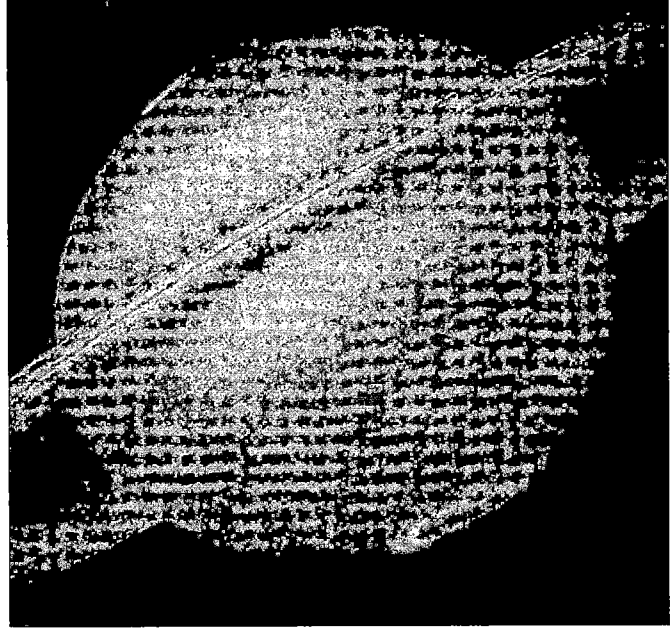

Figure 12.9 Aurora on Saturn An ultraviolet camera aboard the *Hubble Space Telescope* recorded this image of an remarkably symmetrical aurora on Saturn during a solar storm in 1998. *(NASA)*

the solar wind (which tends to push the sunward side of the magnetosphere closer to the planet). Because no major moons lie deep within Saturn's magnetosphere, the details of its structure are different from those of Jupiter's magnetosphere. For example, there is no equivalent of the Io plasma torus. ∞ (Sec. 11.5) Like Jupiter, Saturn emits radio waves, but as luck would have it, they are reflected from Earth's ionosphere (they lie in the AM band) and were not detected until the *Voyager* craft approached the planet.

☑ Concept Check

■ Where did Saturn's atmospheric helium go?

12.4 Saturn's Spectacular Ring System

THE VIEW FROM EARTH

The most obvious aspect of Saturn's appearance is, of course, its *planetary ring system*. Astronomers now know that all the jovian planets have rings, but Saturn's are by far the brightest, the most extensive, and the most beautiful. Galileo saw them first in 1610, but he did not recognize what he saw as a planet with a ring. At the resolution of his small telescope, the rings looked like bumps on the planet, or perhaps components of a triple system of some sort. In 1659 the Dutch astronomer Christian Huygens realized what the "bump" was—a thin, flat ring, completely encircling the planet.

In 1675 the French-Italian astronomer Giovanni Domenico Cassini discovered the first ring feature, a dark band about two-thirds of the way out from the inner edge. From Earth, the band looks like a gap in the ring (an observation that is not too far from the truth, although we now know that there is actually some ring material within it). This "gap" is named the **Cassini Division**, in honor of its discoverer. Careful observations from Earth show that the inner "ring" is in reality also composed of two rings. From the outside in, the three rings are known somewhat prosaically as the **A, B,** and **C rings**. The Cassini Division lies between the A and B rings. The much narrower **Encke gap**, some 300 km wide, is found in the outer part of the A ring. These ring features are marked on Figure 12.10. No finer ring details are visible from our Earthly vantage point. Of the three main rings, the B ring is brightest, followed by the somewhat fainter A ring, and then by the almost translucent C ring. A more complete list of ring

Encke gap
A ring
Cassini Division
B ring
C ring

Figure 12.10 Saturn, Up Close Much fine structure, especially in the rings, appears in this image of Saturn taken while the *Voyager 2* spacecraft approached the planet in 1981. (One of Saturn's moons appears at bottom, and a second casts a black shadow on the cloud tops.) The banded structure of Saturn's atmosphere is also more evident in this photograph. Note the absence of the vivid colors that characterize Jupiter's atmospheric cloudlayers. *(NASA)*

properties appears in Table 12.1. (The D, E, F, and G rings listed in the table are discussed later in this section.)

WHAT ARE SATURN'S RINGS?

A fairly obvious question—and one that perplexed the best scientists and mathematicians on Earth for almost two centuries—is: What are Saturn's rings made of? By the middle of the nineteenth century, various dynamical and thermodynamic arguments had conclusively proved that the rings could not be solid, liquid, or gas! What is left? In 1857 Scottish physicist James Clerk Maxwell, after showing that a solid ring would become unstable and break up, suggested that the rings are composed of a great number of small particles, all independently orbiting Saturn, like so many tiny moons. That inspired speculation was verified in 1895, when Lick Observatory astronomers measured the Doppler shift of sunlight reflected from the rings and showed that the velocities thus determined were exactly what would be expected from separate particles moving in circular orbits in accordance with Newton's law of gravity.

What sort of particles make up the rings? The fact that they reflect most (over 80 percent) of the sunlight striking them had long suggested to astronomers that they were made of ice, and infrared observations in the 1970s confirmed that water ice is indeed a prime ring constituent. Radar observations and later *Voyager* studies of scattered sunlight showed that the diameters of the particles range from fractions of a millimeter to tens of meters, with most particles being about the size (and composition) of a large snowball on Earth.

We now know that the rings are truly thin—perhaps only a few tens of meters thick in places. Stars can occasionally be seen through them, like automobile headlights penetrating a snowstorm. Why are the rings so thin? The answer seems to be that collisions between ring particles tend to keep them all moving in circular orbits in a single plane. Any particle that tries to stray away from this orderly motion finds itself in an orbit that soon runs into other ring particles. Over long periods of time the ensuing jostling serves to keep all particles moving in circular, planar orbits. The asymmetric gravitational field of Saturn (the result of its flattened shape) sees to it that the rings lie in the planet's equatorial plane.

THE ROCHE LIMIT

5 But why a ring of particles at all? What process produced the rings in the first place? To answer these questions, consider the fate of a small moon orbiting close to a massive planet such as Saturn. The moon is held together by internal forces—its own gravity, for example. As we bring our hypothetical moon closer to the planet, the tidal force on it increases. Recall from Chapter 7 that the effect of such a tidal force is to stretch the moon along the direction to the planet—that is, to create a tidal bulge. Recall also that the tidal force increases rapidly with decreasing distance from the planet. ⊂⊃ (Sec. 7.6) As the moon is brought closer to the planet it reaches a point where the tidal force tending to stretch it out becomes *greater* than the internal forces holding it together. At that point the moon is torn apart by the planet's gravity, as shown in Figure 12.11. The pieces of the satellite then pursue their own individual orbits around that planet, eventually spreading all the way around it in the form of a ring.

For any given planet and any given moon, the critical distance, inside of which the moon is destroyed, is known as the *tidal stability limit*, or the **Roche limit**, after the nineteenth-century French mathematician Edouard Roche, who first calculated it. As a handy rule of thumb, if our hypothetical moon is held together by its own gravity and its average density is comparable to that of the parent planet (both reasonably good approximations for Saturn's larger moons), then the Roche limit is roughly 2.4 times the radius of the planet. Thus, for Saturn, no moon can survive within a distance of 144,000 km of the planet's center, about 7000 km beyond the outer edge of the A ring. The main (A, B, C, D, and F) rings of Saturn occupy the region inside Saturn's Roche limit.

These considerations apply equally well to the other jovian worlds. Figure 12.12 shows the location of the ring system of each jovian planet relative to the planet's Roche limit. Given the approximations in our assumptions, we can conclude that all the major planetary rings are found within the Roche limit of their parent planet. Notice that strictly speaking the calculation of this limit applies only to low-density moons massive enough for their own gravity to be the dominant force binding them together. Sufficiently small moons can survive even within the Roche limit because they are held together mostly by interatomic (electromagnetic) forces, not by gravity.

TABLE 12.1 The Rings of Saturn

RING	INNER RADIUS (km)	(planet radii)	OUTER RADIUS (km)	(planet radii)	WIDTH (km)
D	67,000	1.11	74,700	1.24	7700
C	74,700	1.24	92,000	1.53	17,300
B	92,000	1.53	117,500	1.95	25,500
Cassini Division	117,500	1.95	122,300	2.03	4800
A	122,300	2.03	136,800	2.27	14,500
Encke gap*	133,400	2.22	133,700	2.22	300
F	140,300	2.33	140,400	2.33	100
G	165,800	2.75	173,800	2.89	8000
E	180,000	3.00	480,000	8.00	300,000

The Encke gap lies within the A ring.

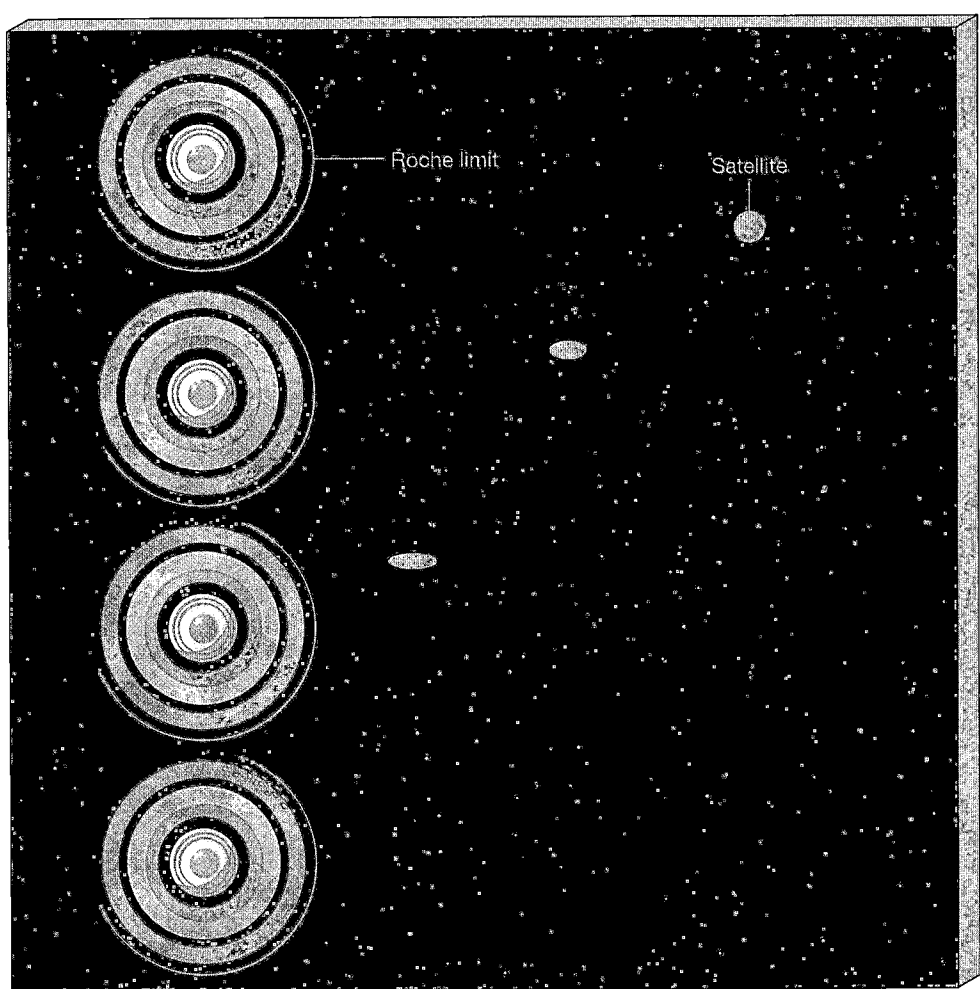

Figure 12.11 Roche Limit The increasing tidal field of a planet first distorts, and then destroys, a moon that strays too close. (The distortion is exaggerated in the second and third panels.)

THE VIEW FROM *VOYAGER*

Thus it was, as *Voyagers 1* and *2* approached Saturn, that scientists on Earth were fairly confident that they understood the nature of the rings. However, there were many surprises in store. The *Voyager* missions changed forever our view of this spectacular region in our cosmic backyard, revealing the rings to be vastly more complex than astronomers had imagined.

As the *Voyager* probes approached Saturn it became obvious that the main rings are actually composed of tens of thousands of narrow **ringlets** (shown in Figure 12.13). Although *Voyager* cameras did find several new gaps in the rings, the ringlets are generally not separated from one another by empty space. Instead, the rings contain concentric regions of alternating high and low concentrations of ring particles—the ringlets are just the high-density peaks. Although the process is not fully understood, it seems that the mutual gravitational attraction of the ring particles (as well as the effects of Saturn's inner moons) enables waves of matter to form and move in the plane of the rings, rather like ripples on the surface of a pond. The wave crests typically wrap around the rings, forming tightly wound spiral patterns called *spiral density waves* that resemble grooves in a huge celestial phonograph record.

Although the ringlets are probably the result of spiral waves in the rings, the true gaps are not. The narrower gaps—about 20 of them—are most likely swept clean by the action of small moonlets embedded in them. These moonlets are larger (perhaps 10 or 20 km in diameter) than the largest true ring particles, and they simply "sweep up" ring material through collisions as they go. Despite many careful searches of *Voyager* images, only one of these moonlets has so far been found—in 1991, after five years of exhaustive study, NASA scientists confirmed the discovery of the eighteenth moon of Saturn (now named Pan) in the Encke gap. Astronomers have found indirect evidence for embedded moonlets, in the form of "wakes" that they leave behind them in the rings, but no other direct sightings have occurred. Despite their elusiveness, however, moonlets are still regarded as the best explanation for the small gaps.

Voyager 2 found a series of faint rings, now known collectively as the **D ring**, inside the inner edge of the C ring, stretching down almost to Saturn's cloud tops. The D ring contains relatively few particles and is so dark that it is

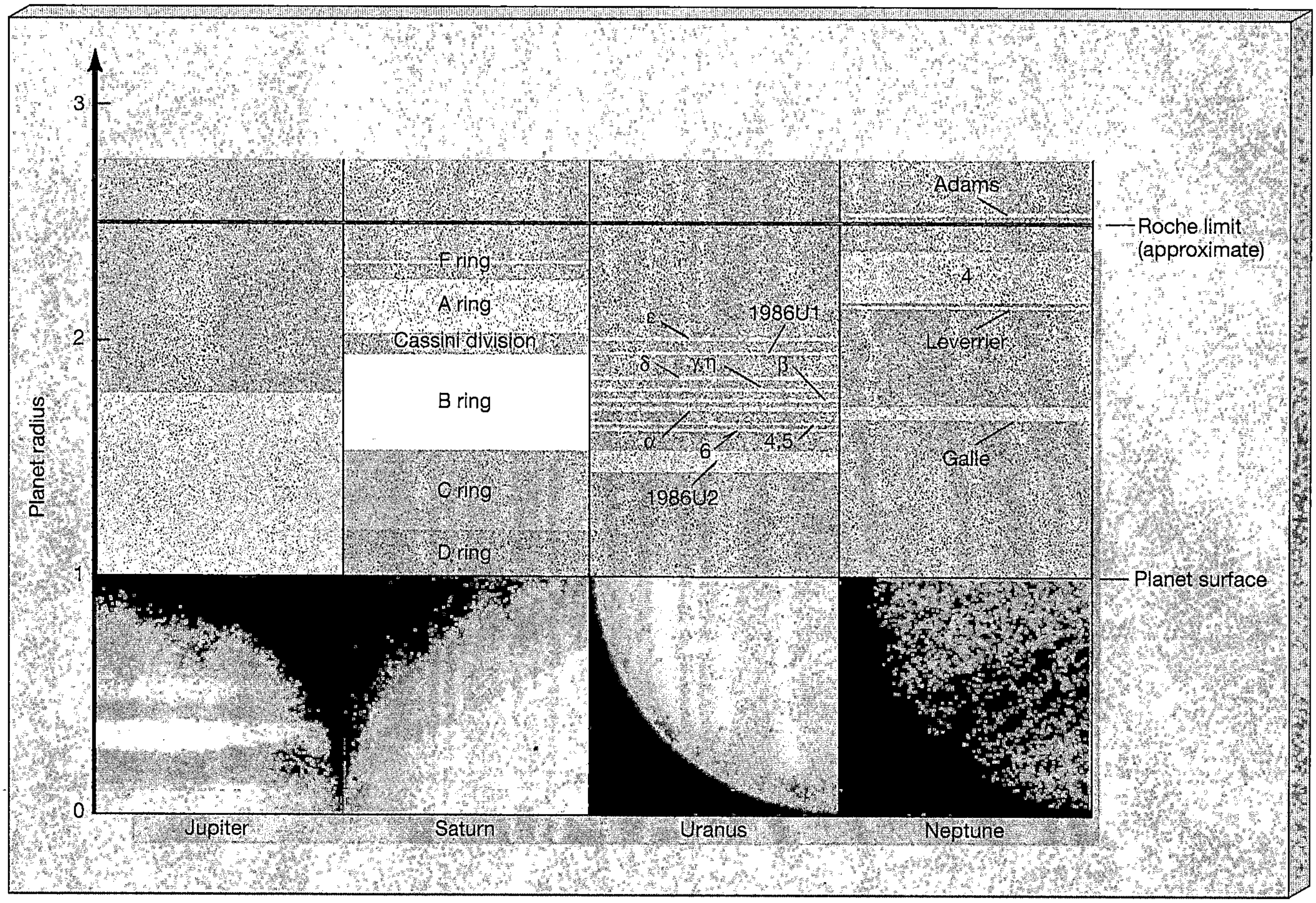

Planet radius

3

2

1

0

E ring
A ring
Cassini division
B ring
C ring
D ring

Adams

Roche limit (approximate)

1986U1

Leverrier

Galle

Planet surface

1986U2

Jupiter   Saturn   Uranus   Neptune

R   I   V   U   X   G

Figure 12.12 Jovian Ring Systems The rings of Jupiter, Saturn, Uranus, and Neptune. All distances are expressed in planetary radii. The red line represents the Roche limit. In all cases the rings lie within the Roche limit of the parent planet.

R   I   V   U   X   G

Figure 12.13 Saturn's Rings, Up Close *Voyager 2* took this close-up of the ring structure just before plunging through the tenuous outer rings of Saturn. The ringlets in the B ring, spread over several thousand kilometers, are resolved here to about 10 km. As *Voyager* approached Saturn, more and more of these tiny ringlets became noticeable in the main rings. The (false) color variations probably indicate different sizes and compositions of the particles making up the thousands of rings. Earth is superposed, to proper scale, for a size comparison. *(NASA)*

completely invisible from Earth. Another faint ring, also a *Voyager 2* discovery, lies well outside the main ring structure. Known as the **E ring**, it appears to be associated with volcanism on the moon Enceladus.

The *Voyager 2* cameras revealed one other completely unexpected feature. A series of dark radial "spokes" formed on the B ring, moved around the planet for about one ring orbit period, and then disappeared (Figure 12.14). Careful scrutiny of these peculiar drifters showed that they were composed of very fine (micron-sized) dust hovering a few tens of meters *above* the plane of the rings. Scientists believe that this dust was held in place by electrostatic forces generated in the ring plane, perhaps resulting from particle collisions there. The electrical fields slowly dispersed, and the spokes faded as the ring revolved. We expect that the creation and dissolution of such spokes is a regular occurrence in the Saturn ring system.

ORBITAL RESONANCES AND SHEPHERD SATELLITES

Voyager images show that the largest gap in the rings, the Cassini Division, is not completely empty of matter. In fact, as shown in Figure 12.15, the Division contains a series of faint ringlets and gaps (and, presumably, embedded moonlets too). The overall concentration of ring particles in the division as a whole is, however, much lower than in the A and B rings. Although its small internal gaps probably result from embedded satellites, the Division itself does not. It owes its existence to another solar system *resonance*, this time involving particles orbiting in the Division

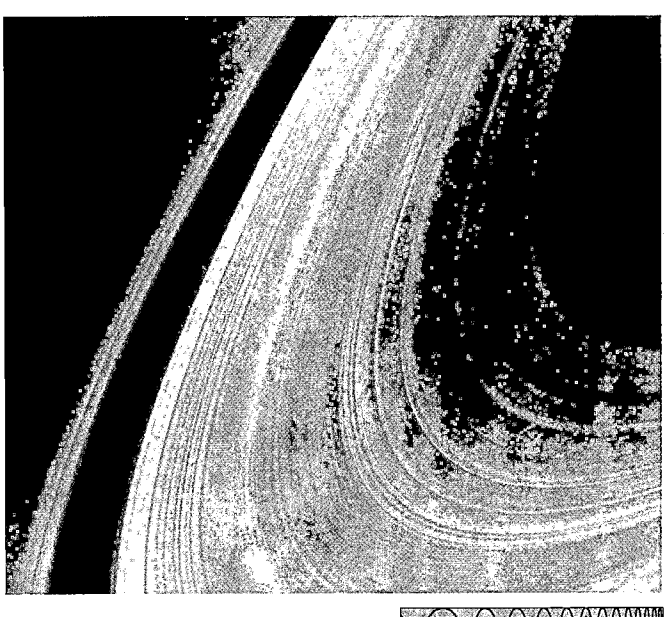

Figure 12.14 Spokes in the Rings Saturn's B ring showed a series of dark temporary "spokes" as *Voyager 2* flew by at a distance of about 4 million km. The spokes were caused by small particles suspended just above the ring plane. *(NASA)*

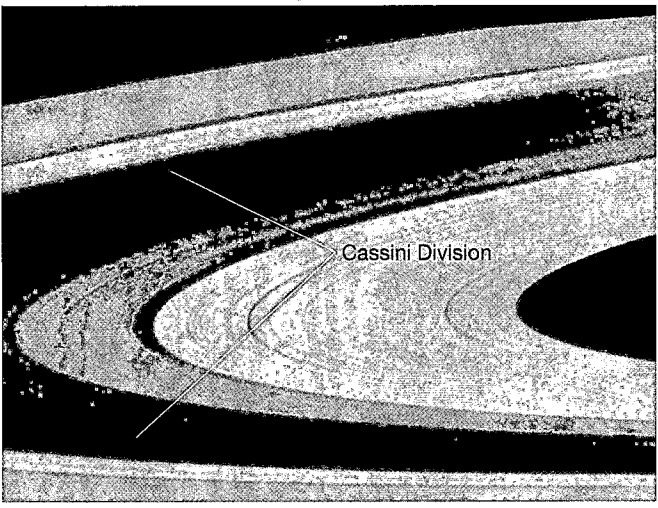

Figure 12.15 Cassini Division Close inspection by *Voyager 2* revealed that the Cassini Division (shown here as the darker color) is not completely empty. It contains a series of faint ringlets and gaps, assumed to be caused by unseen embedded satellites. The density of material in the division is very low, accounting for its dark appearance from Earth. *(NASA)*

and Saturn's innermost major moon, Mimas. ∞ (Sec. 8.4) A ring particle moving in an orbit within the Cassini Division has an orbital period exactly half that of Mimas. Particles in the Division thus complete exactly two orbits around Saturn in the time taken for Mimas to orbit once— a configuration known as a 2:1 resonance. Applying Kepler's third law (recast to refer not to the planets but to Saturn's moons), we can show that this 2:1 resonance with Mimas corresponds to a radius of 117,000 km, the inner edge of the Division. ∞ (*More Precisely 2-3*) The effect of this resonance is that particles in the Division feel a gravitational tug from Mimas at exactly the same location in their orbit every other time around. Successive tugs reinforce one another, and the initially circular trajectories of the ring particles soon get stretched out into ellipses. In their new orbits these particles collide with other particles and eventually find their way into new circular orbits at other radii. The net effect is that the number of ring particles in the Cassini Division is greatly reduced.

Particles in "nonresonant" orbits (that is, at radii where the orbital period is not simply related to the period of Mimas) also experience Mimas's gravitational pull. But the times when the force is greatest are spread uniformly around the orbit, and the tugs cancel out. It's a little like pushing a child on a swing—pushing at the same point in the swing's motion each time produces much better results than do random shoves. Thus, Mimas (or any other moon) has a large effect on the ring at those radii where a resonance exists and little or no effect elsewhere.

We now know that resonances between ring particles and moons play a very important role in shaping the fine structure of Saturn's rings. For example, the sharp outer edge of the A ring is thought to be governed by a 3:2 resonance with Mimas (three ring orbits in two Mimas orbital periods). Most theories of planetary rings predict that the ring system should spread out with time, basically because of collisions among ring particles. Instead, the A ring's outer edge is "patrolled" by a small satellite named Atlas, held in place by the gravity of Mimas, which prevents ring particles from diffusing away. Compare Tables 12.1 and 12.2 and see if you can identify other resonant connections between Saturn's moons, or between the moons and the rings. (You should be able to find quite a few—Saturn is a very complex place!)

Outside the A ring lies the strangest ring of all. The faint, narrow **F ring** (shown in Figure 12.16) was discovered by *Pioneer 11* in 1979, but its full complexity became evident only when *Voyager 1* took a closer look. Unlike the inner major rings, the F ring is narrow, less than a hundred kilometers wide. It lies just inside Saturn's Roche limit, separated from the A ring by about 3500 km. Its narrowness by itself is unusual, as is its slightly eccentric shape, but its oddest feature is that it looks as though it is made up of several separate strands braided together! This remarkable discovery sent dynamicists scrambling in search of an explanation. It now seems as though the ring's intricate structure, as well as its thinness, arise from the influence of two small moons, known as **shepherd satellites**, that orbit on either side of it (Figure 12.17).

These two small, dark satellites, each little more than 100 km in diameter, are called Prometheus and Pandora. They orbit about 1000 km on either side of the F ring, and their gravitational influence on the F-ring particles keeps

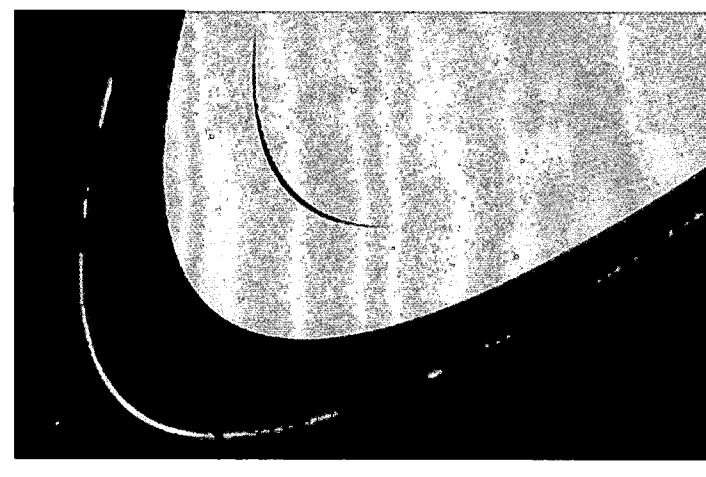

R I V U X G

Figure 12.17 Shepherd Moon The F ring's thinness, and possibly its other peculiarities too, can be explained by the effects of two shepherd satellites that orbit a few hundred kilometers inside and outside the ring. This photo shows one of the shepherding satellites, roughly 100 km in length, at right.

the ring tightly confined in its narrow orbit. As shown in Figure 12.18, any particle straying too far out of the F ring is gently guided back into the fold by one or the other of the moons. (The moon Atlas confines the A ring in a somewhat similar way.) However, the details of how Prometheus and Pandora produce the braids in the F ring and why the two moons are there at all, in such similar orbits, remain unclear. There is some evidence that other eccentric rings found in the gaps in the A, B, and C rings may also result from the effects of shepherding moonlets.

Outside the F ring lies another faint, narrow ring—the **G ring**. Discovered by *Pioneer 11* and imaged in more detail by *Voyager 2*, it apparently lacks ringlets or the peculiar internal structure found in the F ring. Its narrowness and sharp edges suggest the presence of shepherd satellites, but so far none has been found.

THE ORIGIN OF THE RINGS

Two possible origins have been suggested for Saturn's rings. Astronomers estimate that the total mass of ring material is no more than 10^{15} tons—enough to make a satellite about 250 km in diameter. If such a satellite strayed inside Saturn's Roche limit or was destroyed (perhaps by a collision) near that radius, a ring could have resulted. An alternative view is that the rings represent material left over from Saturn's formation stage 4.6 billion years ago. In this scenario, Saturn's tidal field prevented any moon from forming inside the Roche limit, and so the material has remained a ring ever since. Which view is correct?

All the dynamic activity observed in Saturn's rings suggests to many researchers that the rings must be quite

R I V U X G

Figure 12.16 F Ring Saturn's narrow F ring appears to contain kinks and braids, making it unlike any of Saturn's other rings. *(NASA)*

young—perhaps no more than 50 million years old, or 100 times younger than the solar system. There is just too much going on for the ring to have remained stable for billions of years, so they probably aren't left over from the planet's formative stages. If this is so, then either the rings are continuously replenished, perhaps by fragments of Saturn's moons chipped off by meteorites, or they are the result of a relatively recent, possibly catastrophic, event in the planet's system—perhaps a small moon that was hit by a large comet, or even by another moon.

Astronomers prefer not to invoke catastrophic events to explain specific phenomena, but the more we learn of the universe, the more we realize that catastrophe probably plays an important role. For now, the details of the formation of Saturn's ring system simply aren't known.

Concept Check

■ What do the Roche limit and orbital resonances have to do with planetary rings?

12.5 The Moons of Saturn

GENERAL FEATURES

Saturn has the most extensive, and in many ways the most complex, system of natural satellites of all the planets. The planet's 18 named moons are listed in Table 12.2. Observations of sunlight reflected from them suggests that most are covered with snow and ice. Many of them are probably made almost entirely of water ice. Even so, they are a curious and varied lot.

The moons fall into three fairly natural groups. First, there are the many "small" moons—irregularly shaped chunks of ice, all less than 300 km across—that exhibit a bewildering variety of complex and fascinating motion. Second, there are six "medium-sized" moons—spherical bodies with diameters ranging from about 400 to 1500 km—that offer clues to the past and present state of the environment of Saturn while presenting many puzzles regarding their own appearance and history. Finally, there is Saturn's single "large" moon—Titan—which, at 5150 km in diameter, is the second-largest satellite in the solar system (Jupiter's Ganymede is a little bigger). It has an atmosphere denser than Earth's and (some scientists think) surface conditions possibly conducive to life. Notice, incidentally, that Jupiter has no "medium" moons, as just defined. The Galilean satellites are large, like Titan, and all of Jupiter's other satellites are small—no more than 200 km in diameter.

In 1995, researchers using the *Hubble Space Telescope* reported the sighting of two more moons (see *Discovery 12-1*). However, these observations now seem to have been in error—easy to do when dealing with faint objects at the limits of detectability. More recently, in late 2000, astronomers working at observatories on Mauna Kea and at the VLT announced the discovery of 12 more moons orbiting Saturn, bringing the planet's total (for now) to 30. As with the newly discovered satellites of Jupiter, these moons are all very faint (and hence small), and orbit quite far from the planet on rather inclined orbits, much like Saturn's other "small" moons. ∞ (Sec. 11.5) These new moons are

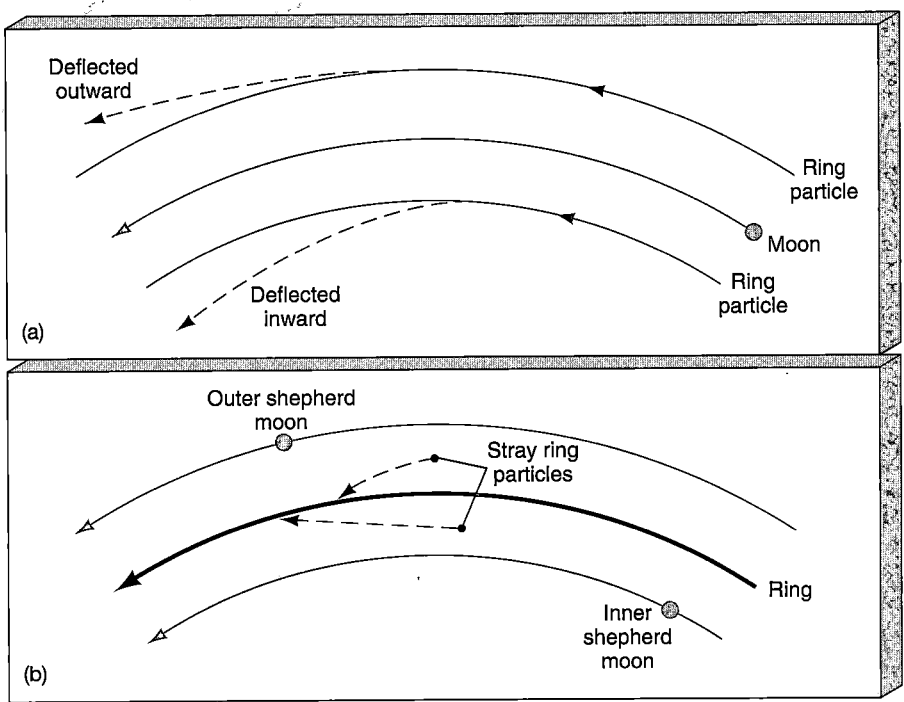

(a)

(b)

Figure 12.18 Moon–Ring Interaction (a) Strange as it may seem, the net effect of the interactions between a moon and ring particles is that the moon tends to push those particles away from it. (b) The F ring shepherd satellites operate by forcing errant F ring particles back into the main ring. Each moon operates as in part (a) so that the ring is confined between the two moons. As a consequence of Newton's third law of motion, the satellites themselves slowly drift away from the ring (unless they in turn are held in place by interactions with other, larger moons).

TABLE 12.2 The Major Moons of Saturn*

NAME	DISTANCE FROM SATURN (km)	(planet radii)	ORBIT PERIOD (days)	SIZE (longest diameter, km)	MASS** (Earth Moon masses)	DENSITY (kg/m³)	(g/cm³)
Pan	133,600	2.22	0.57	20	4×10^{-8}		
Atlas	138,000	2.28	0.60	37			
Prometheus	139,000	2.31	0.61	148	1.9×10^{-6}		
Pandora	142,000	2.35	0.63	110	1.8×10^{-6}		
Epimetheus	151,000	2.51	0.69	138	7.5×10^{-6}		
Janus	151,000	2.51	0.69	199	2.7×10^{-5}		
Mimas	186,000	3.08	0.94	398	0.00051	1100	1.1
Enceladus	238,000	3.95	1.37	498	0.00099	1100	1.1
Calypso	295,000	4.89	1.89	30			
Telesto	295,000	4.89	1.89	30			
Tethys	295,000	4.89	1.89	1060	0.0085	1000	1.0
Dione	377,000	6.26	2.74	1120	0.014	1400	1.4
Helene	377,000	6.26	2.74	32			
Rhea	527,000	8.74	4.52	1530	0.032	1200	1.2
Titan	1,220,000	20.3	16.0	5150	1.83	1900	1.9
Hyperion	1,480,000	24.6	21.3	370			
Iapetus	3,560,000	59.1	79.3	1440	0.022	1000	1.0
Phoebe	13,000,000	215	−550†	230			

*Does not include the 12 recently discovered small moons described in the text. All are small and orbit far from the planet. Their orbits are as yet undetermined.

**Mass of Earth's Moon = 7.4×10^{22} kg = 1.3×10^{-4} Saturn masses.

†Indicates a retrograde orbit.

as yet unnamed, and their orbits are still unknown. If confirmed, they are most likely chunks of debris captured from interplanetary space after close encounters with Saturn.

TITAN

⑥ Perhaps the most intriguing of all Saturn's moons, Titan was discovered by Christian Huygens in 1655. Even through a large Earth-based telescope, this moon is visible only as a barely resolved reddish disk. However, long before the *Voyager* missions, astronomers already knew (from spectroscopic observations) that the moon's reddish coloration is caused by something quite special—an atmosphere. So eager were mission planners to obtain a closer look that they programmed *Voyager 1* to pass very close to Titan, even though that meant the spacecraft could not then use Saturn's gravity to continue on to Uranus and Neptune. (Instead, *Voyager 1* left the Saturn system on a path taking the craft out of the solar system well above the ecliptic plane.)

Scientists believe that Titan's internal composition and structure must have similarities to those of Ganymede and Callisto, because these three moons have quite similar masses and radii, and hence average densities (Titan's den-

sity is 1900 kg/m³). ∞ (Sec.11.5) Titan probably contains a rocky core surrounded by a thick mantle of water ice. However, in view of *Galileo*'s discovery of significant differences between the internal structures of Ganymede and Callisto, the degree of differentiation within Titan is presently unknown. In addition to probing Titan's atmosphere and surface, another important objective of the *Cassini* mission on reaching Saturn in 2004 will be to make detailed measurements of the moon's interior. ∞ (Sec.6.6)

A *Voyager 1* image of Titan is shown in Figure 12.19. Unfortunately, despite the spacecraft's close pass, the moon's surface remains a mystery. A thick, uniform haze layer, similar to the photochemical smog found over many cities on Earth, that envelops the moon completely obscured the spacecraft's view. *Voyager 1* was able to provide mission specialists with detailed atmospheric data, however.

Titan's atmosphere is thicker and denser even than Earth's, and it is certainly far more substantial than that of any other moon. Prior to *Voyager 1*'s arrival in 1980, only methane and a few other simple hydrocarbons had been conclusively detected on Titan. (Hydrocarbons are molecules consisting solely of hydrogen and carbon atoms; methane, CH_4, is the simplest). Radio and infrared observa-

DISCOVERY 12-1

New Moons or Ring Debris at Saturn?

The last five years of the twentieth century saw an un-precedented number of reports of new moons orbiting the outer planets. At least 30 previously unknown satellites were observed, more than 20 of them in just one year—2000. These observations involve searching images of regions of space near the planet for very faint specks of light, usually very close to the telescope's limit of sensitivity. The recent slew of discoveries has come about in large part because of better detectors and improved search techniques.

Such undertakings are not without their problems, however. In 1995, researchers using the *Hubble Space Telescope* announced the discovery of several new moons orbiting Saturn. The findings were reported with great fanfare in the press at the time, for these would have been the first such moons spotted anywhere by *Hubble*. Since then, however, it has become unclear if these sightings are new moons, or even real objects at all.

The original images were processed by removing stray light from near the edge of Saturn's ring, which was minimal, since the observations were made during the ring crossing in 1995, when the rings virtually disappeared from view (see Figure 12.1). After accounting for the positions of all of Saturn's known moons, the field of view still displayed four blobs of light, in addition to the body of Saturn itself. The accompanying four-picture sequence spans 30 minutes of observing time and shows how astronomers might have been fooled into thinking they had made a discovery. Saturn itself appears as a bright white disk at the far right, heavily overexposed in the search for faint new moons; its edge-on rings extend diagonally to the upper left and are barely discernible. What remains is one known moon—Epimetheus, the brightest small blob at left—and several other fainter blobs. The blobs in these images formed the basis for the claims of new moons in the Saturn system.

Subsequent observations by another team of astronomers failed to confirm some of the faint blobs of light in these images but reported some new blobs. What is going on here? It seems that if the blobs are indeed real, then they are probably just clouds of ice and dust orbiting near the planet's rings. Conceivably, the strong tidal pull of Saturn on some of its small moons broke off a piece large enough to be spotted for a time as a transient piece of debris before ultimately dispersing. Perhaps this is even how Saturn's rings are maintained. As for the blobs that cannot be confirmed, they may be only instrumental artifacts, caused by small-scale imperfections in *Hubble*'s mirror, by "hot pixels" within the onboard cameras, or by the computer methods used by astronomers to massage their data. In any case, most astronomers are now of the opinion that these faint blobs, even if real, do not represent new moons around Saturn.

Are the new observations more reliable? Perhaps. They entailed a search of a portion of the sky far from Saturn, much less affected by the planet's light or possible ring debris, and in at least some cases have already been reproduced. Still, we can be sure that these new bodies will be carefully scrutinized before being declared "official."

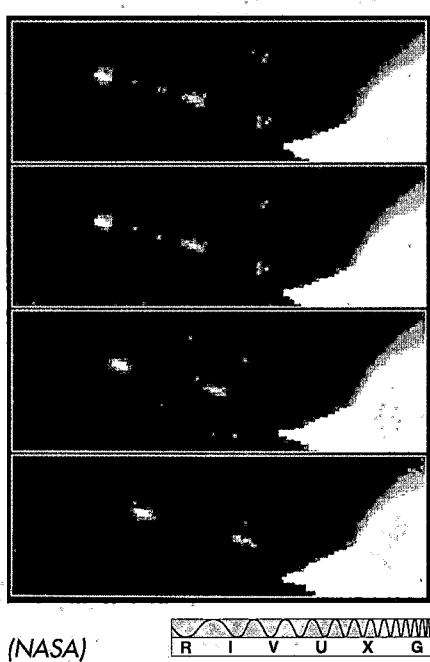

(NASA)

tions from *Voyager 1* showed that the atmosphere is actually made up mostly of nitrogen (roughly 90 percent) and argon (at most 10 percent), with a small percentage of methane. In addition, complex chemistry in Titan's atmosphere maintains steady (but trace) levels of hydrogen gas, the hydrocarbons ethane and propane, and carbon monoxide.

Titan's atmosphere seems to act like a gigantic chemical factory. Powered by the energy of sunlight, it is undergoing a complex series of chemical reactions that ultimately result in the observed smog and trace chemical composition. The upper atmosphere is thick with aerosol haze, and the unseen surface may be covered with organic sediment that has settled down from the clouds. Speculation runs the gamut from oceans of liquid ethane to icy valleys laden with hydrocarbon sludge. Future spacecraft exploration of Titan may present scientists with an opportunity to study the kind of chemistry thought to have occurred billions of years ago on Earth—the prebiotic chemical reactions that eventually led to life on our own planet (see Chapter 28).

Based largely on *Voyager* measurements, Figure 12.20 shows the probable structure of Titan's atmosphere. Despite the moon's low mass (a little less than twice that of

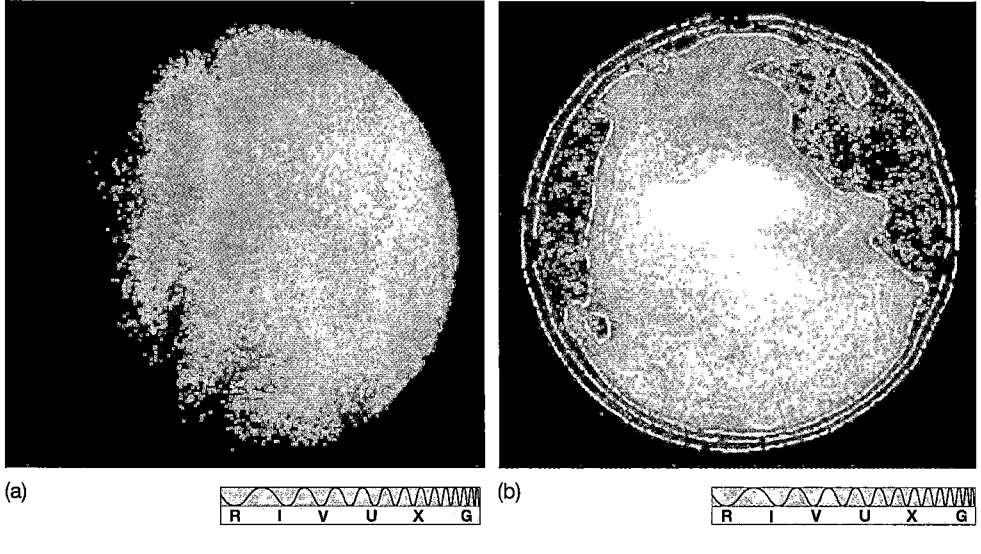

(a)
R I V U X G
(b)
R I V U X G

Figure 12.19 Titan
(a) Larger than the planet Mercury and roughly half the size of Earth, Saturn's largest moon, Titan, was photographed in visible light from only 4000 km away as *Voyager 1* passed by in 1980. All we can see here is Titan's upper cloud deck. For unknown reasons, the northern hemisphere appears slightly brighter than the southern.
(b) However, in infrared, as captured with the adaptive optics system on the Canada-France-Hawaii telescope on Mauna Kea, Titan displays large-scale surface features. The bright regions are thought to be highlands, possibly covered with frozen methane. The most prominent bright area shown here is nearly 4000 km across, about the size of Australia.
(NASA; CFHT)

Earth's Moon) and hence its low surface gravity (one-seventh of Earth's), the atmospheric pressure at ground level is 60 percent greater than on Earth. Titan's atmosphere contains about 10 times more gas than Earth's atmosphere. Because of Titan's weaker gravitational pull, the atmosphere extends some 10 times farther into space than does our own. The top of the main haze layer lies about 200 km above the surface, although there are additional layers, seen primarily through their absorption of ultraviolet radi-

ation, at 300 km and 400 km (Figure 12.21). Below the haze the atmosphere is reasonably clear, although rather gloomy, because so little sunlight gets through.

Titan's surface temperature is a frigid 94 K, roughly what we would expect on the basis of its distance from the Sun. At the temperatures typical of the lower atmosphere, methane and ethane may behave rather like water on Earth, raising the possibility of methane rain, snow, and fog, and even ethane oceans! At higher levels in the atmos-

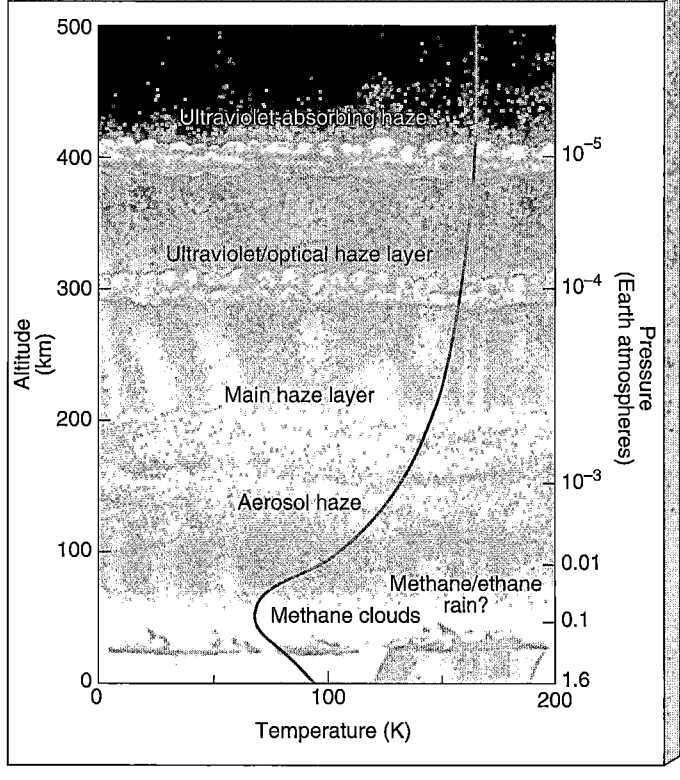

Figure 12.20 Titan's Atmosphere The structure of Titan's atmosphere, as deduced from *Voyager 1* observations.

Figure 12.21 Haze Layers on Titan The haze layers (blue) of Titan's upper atmosphere are visible in these false-color *Voyager 1* images. *(NASA)*

R I V U X G

phere the temperature rises, the result of photochemical absorption of solar radiation.

Why does Titan have such a thick atmosphere, when similar moons of Jupiter such as Ganymede and Callisto have none? The answer seems to be a direct result of Titan's greater distance from the Sun. The moons of Saturn formed at considerably lower temperature than did those of Jupiter. Those conditions would have enhanced the ability of the water ice that makes up the bulk of Titan's interior to absorb methane and ammonia, both of which were present in abundance at those early times. As a result, Titan was initially laden with much more methane and ammonia gas than either Ganymede or Callisto. As Titan's internal radioactivity warmed the moon, the ice released the trapped gases, forming a thick methane-ammonia atmosphere. Sunlight split the ammonia into hydrogen, which escaped into space, and nitrogen, which remained in the atmosphere. The methane, which was less easily broken apart, survived intact. Together with argon outgassed from Titan's interior, these gases form the basis of the atmosphere we see today.

☑ Concept Check

▣ What is unusual about Titan?

SATURN'S MEDIUM-SIZED MOONS

7 Saturn's complement of midsized moons consists (in order of increasing distance from the planet) of Mimas (at 3.1 planetary radii), Enceladus (4.0), Tethys (4.9), Dione (6.3), Rhea (8.7), and Iapetus (59.1). They are shown, to proper scale, in Figure 12.22. All six were known from Earth-based observations long before the space age. The inner five move on circular trajectories, and all are tidally locked by Saturn's gravity into synchronous rotation (so that one side always faces the planet). They therefore all have permanently "leading" and "trailing" faces as they

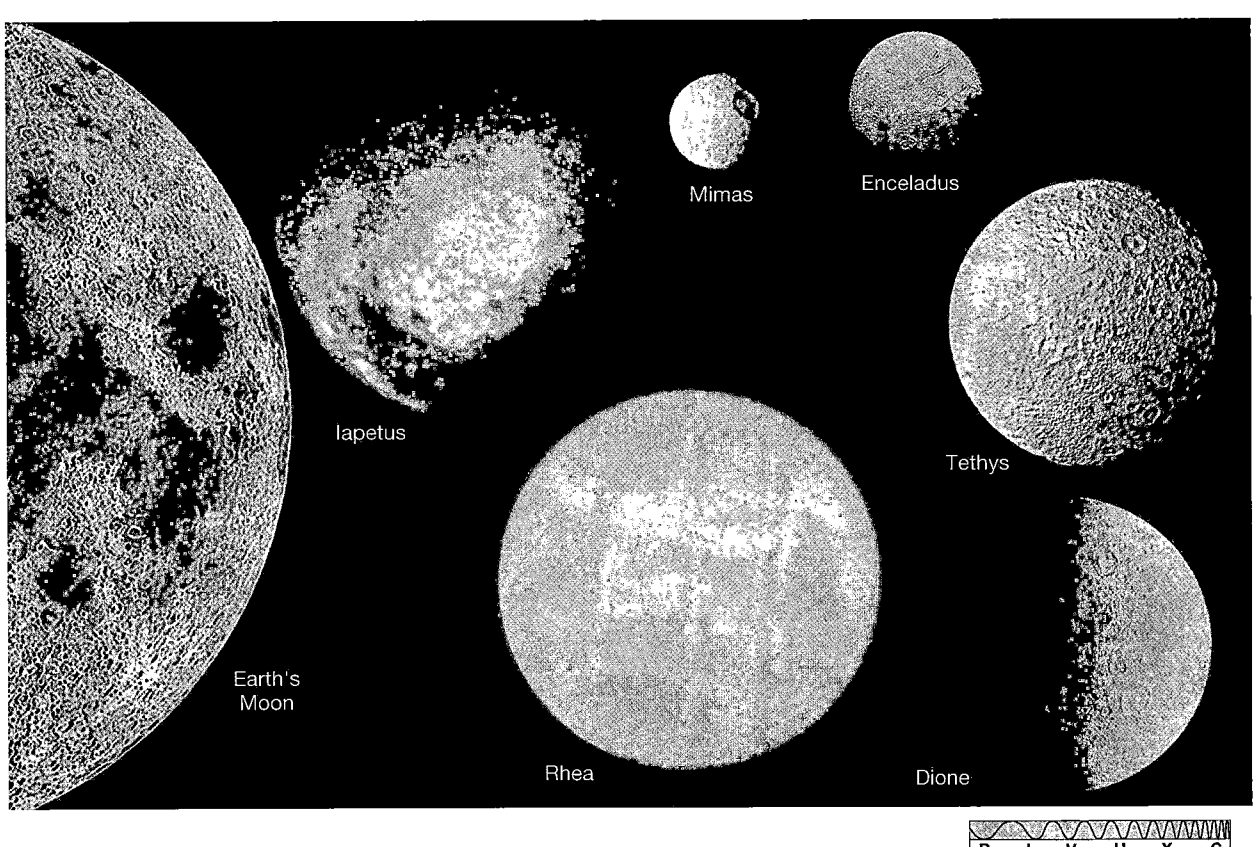

Figure 12.22 Moons of Saturn Saturn's six medium-sized satellites, to scale. All are heavily cratered. Iapetus shows a clear contrast between its light-colored (icy) trailing surface, at the top and center in this image, and its dark leading hemisphere, at the bottom. The light-colored "wisps" covering essentially the whole central portion of this image of Rhea are thought to be water that was released from the moon's interior during some long-ago period of activity and then froze on the surface. Dione and Tethys show evidence of ancient geological activity of some sort. Enceladus appears to still be volcanically active, but the cause of the volcanism is unexplained. Mimas's main surface feature is the large crater Herschel, plainly visible in this *Voyager 2* image; the impact that caused this crater must have come very close to shattering the moon. *(NASA)*

move in their orbits, a fact that is important in understanding their often asymmetrical surface markings.

Unlike the densities of the Galilean satellites of Jupiter, the densities of these six moons do not show any correlation with distance from Saturn. Their densities are all between 1000 and 1400 kg/m^3, implying that nearness to the central planetary heat source was a less important influence during their formation than it was in the Jupiter system. Scientists believe that the midsized moons are composed largely of rock and water ice, as is Titan. Their densities are lower than Titan's, primarily because their lower masses produce less compression of their interiors.

The largest of the six, Rhea, has a mass only 1/30 that of Earth's Moon, and its icy surface is very reflective and heavily cratered (Figure 12.23). At the low temperatures found on its surface, water ice is very hard and behaves rather like rock on the inner planets. For that reason, Rhea's surface craters look very much like craters on the Moon or Mercury. The crater density is similar to that in the lunar highlands, indicating that the surface is old, and there is no evidence of extensive geological activity. Rhea's only real riddle is the presence of so-called *wispy terrain*— prominent light-colored streaks—on its trailing side (see Figure 12.22). The leading face, by contrast, shows no such markings, only craters. Astronomers believe that the wisps were caused by some event in the distant past during which water was released from the interior and condensed on the surface. Any similar markings on the leading side have presumably been obliterated by cratering, which should be more frequent on the satellite's forward-facing surface.

Figure 12.23 Rhea's Polar Cap The north polar region of Rhea, seen here at a superb resolution of only 1 km. The heavily cratered surface resembles that of the Moon, except that we see here craters in bright ice, rather than dark lunar rock. *(NASA)*

Inside Rhea's orbit lie the orbits of Tethys and Dione. These two moons are comparable in size and have masses somewhat less than half the mass of Rhea. Like Rhea, they have reflective surfaces that are heavily cratered, but each shows signs of surface activity, too. Dione's trailing face has prominent bright streaks which are probably similar to Rhea's wispy terrain. Dione also has "maria" of sorts, where flooding appears to have obliterated the older craters. The cracks on Tethys may have been caused by cooling and shrinking of the surface layers or, more probably, by meteoritic bombardment.

The innermost, and smallest, medium-sized moon is Mimas. Despite its low mass—only 1 percent the mass of Rhea—its closeness to the rings causes resonant interactions with the ring particles, resulting most notably in the Cassini Division, as we have already seen. Possibly because of its proximity to the rings, Mimas is heavily cratered. The moon's chief surface feature is an enormous crater, known as Herschel, on the leading face. Its diameter is almost one-third that of the moon itself. The impact that formed Herschel must have come very close to destroying Mimas completely. It is quite possible that the debris produced by such impacts is responsible for creating or maintaining the spectacular rings we see.

Enceladus orbits just outside Mimas. Its size, mass, composition, and orbit are so similar to those of Mimas that one might guess that the two moons would also be very similar to each other in appearance and history. This is not so. Enceladus is so bright and shiny—it reflects virtually 100 percent of the sunlight falling on it—that astronomers believe its surface must be completely coated with fine crystals of pure ice, which may be the icy "ash" of water "volcanoes" on Enceladus. The moon bears visible evidence of large-scale volcanic activity. Much of its surface is devoid of impact craters, which seem to have been erased by what look like lava flows, except that the "lava" is water, temporarily liquefied during recent internal upheavals and now frozen again.

Although no geysers or volcanoes have actually been observed on Enceladus, there seems to be strong circumstantial evidence of volcanism on the satellite. In addition, the nearby thin cloud of small, reflective particles that makes up Saturn's E ring is known to be densest near Enceladus. Calculations indicate that the E ring is unstable because of the disruptive effects of the solar wind, supporting the view that volcanism on Enceladus continually supplies new particles to maintain the ring.

Why is there so much activity on a moon so small? No one knows. Attempts have been made to explain Enceladus's water volcanism in terms of tidal stresses. (Recall the role that Jupiter's tidal stresses play in creating volcanism on Io.) ∞ (Sec. 11.5) However, Saturn's tidal force on Enceladus is only one-quarter the force exerted by Jupiter on Io, and there are no nearby large satellites to force Enceladus away from a circular trajectory. Thus, the ingredients that power Io's volcanoes may not be present on

Enceladus. For now, the mystery of Enceladus's internal activity remains unresolved.

The outermost midsized moon is Iapetus (Figure 12.22). It orbits Saturn on a somewhat eccentric, inclined orbit with a semimajor axis of 3.6 million km. Its mass is about three-quarters that of Rhea. Iapetus is a two-faced moon. The dark, leading face reflects only about 3 percent of the sunlight reaching it, whereas the icy trailing side reflects 50 percent. Spectroscopic studies of the dark regions seem to indicate that the material originates on Iapetus, in which case the moon is not simply sweeping up dark material as it orbits. Similar dark deposits seen elsewhere in the solar system are thought to be organic (carbon-containing) in nature; they can be produced by the action of solar radiation on hydrocarbon (for example, methane) ice. But how the dark markings can adorn only one side of Iapetus in that case is still unknown.

THE SMALL SATELLITES

Finally we come to Saturn's dozen or so small moons. Their masses are poorly known (they are inferred mainly from their gravitational effects on the rings), but they are thought to be similar in composition to the small moons of Jupiter. The outermost small moons, Hyperion and Phoebe, were discovered in the nineteenth century, in 1848 and 1898, respectively. The others were first detected in the second half of the twentieth century. Only the moons in or near the rings themselves were actually discovered by the *Voyager* spacecraft.

Just 10,000 km beyond the F ring lie the so-called *co-orbital satellites* Janus and Epimetheus. As the name implies, these two satellites "share" an orbit, but in a very strange way. At any given instant, both moons are in circular orbits about Saturn, but one of them has a slightly smaller orbital radius than the other. Each satellite obeys Kepler's laws, so the inner satellite orbits slightly faster than the outer one and slowly catches up to it. The inner moon takes about four Earth years to "lap" the outer one. As the inner satellite gains ground on the outer one a strange thing happens. As illustrated in Figure 12.24, when the two get close enough to begin to feel each other's weak gravity, they switch orbits—the new inner moon (which used to be the outer one) begins to pull away from its companion, and the whole process begins again! No one knows how the co-orbital satellites came to be engaged in this curious dance. Possibly they are portions of a single moon that broke up, perhaps after a meteoritic impact, leaving the two pieces in almost the same orbit.

In fact, several of the other small moons also share orbits, this time with larger moons. Telesto and Calypso have orbits that are synchronized with the orbit of Tethys, always remaining fixed relative to the larger moon, lying precisely 60° ahead of and 60° behind it as it travels around Saturn (see Figure 12.25). The moon Helene is similarly tied to Dione. These 60° points are known as **Lagrangian points**, after the French mathematician Joseph Louis Lagrange, who first studied them. Later we will see further examples of this special 1:1 orbital resonance in the motion of some asteroids about the Sun, trapped in the Lagrangian points of Jupiter's orbit.

The strangest motion of all is that of the moon Hyperion, which orbits between Titan and Iapetus, at a distance of 1.5 million km from the planet. Unlike most of Saturn's moons, its rotation is not synchronous with its orbital motion. Because of the gravitational effect of Titan, Hyperion's orbit is not circular, so synchronous rotation cannot occur. In response to the competing gravitational influences of Titan and Saturn, this irregularly shaped satellite constantly changes both its rotation speed and its rotation axis, in a

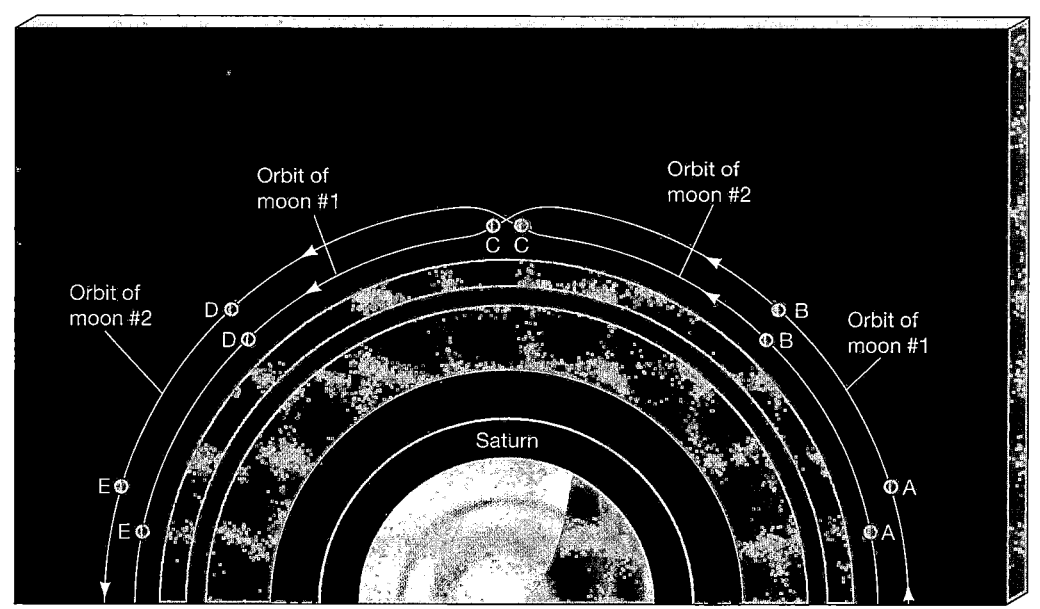

Figure 12.24 Orbit-sharing Satellites The peculiar motion of Saturn's co-orbital satellites, Janus and Epimetheus, which play a never-ending game of tag as they move on their orbits around the planet. The labeled points represent the locations of the two moons at a few successive times. From A to C, satellite 2 gains on satellite 1. However, before it can overtake it, the two moons swap orbits, and satellite 1 starts to pull ahead of satellite 2 again, through points D and E. The whole process then repeats, apparently forever.

Figure 12.25 Synchronous Orbits The orbits of the moons Telesto and Calypso are tied to the motion of the moon Tethys. The combined gravitational pulls of Saturn and Tethys keep the small moons exactly 60° ahead and behind the larger moon at all times, so all three moons share an orbit and never change their relative positions.

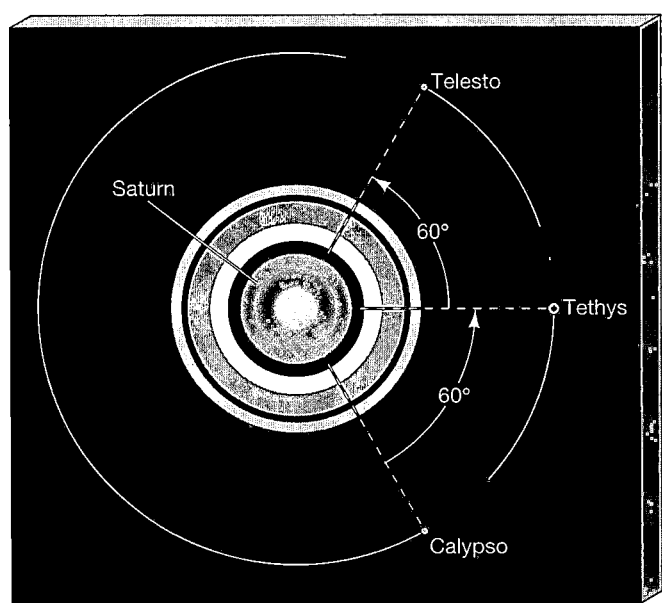

condition known as *chaotic rotation*. As Hyperion orbits Saturn, it tumbles apparently at random, never stopping and never repeating itself, in a completely unpredictable way. Since the 1970s the study of chaos on Earth has revealed new classes of unexpected behavior in even very simple systems. Hyperion is one of the few other places in the universe where this behavior has been unambiguously observed.

 Concept Check

■ Why do Saturn's midsized moons show asymmetric surface markings?

Chapter Review

SUMMARY

Saturn was the outermost planet known to ancient astronomers. Its rings and moons were not discovered until after the invention of the telescope. Saturn is smaller than Jupiter, but still much larger than any of the terrestrial worlds. Like Jupiter, Saturn rotates rapidly, producing a pronounced flattening, and displays differential rotation. Strong radio emission from the planet's magnetosphere allows the rotation rate of the interior to be determined.

Weather systems are seen on Saturn, as on Jupiter, although they are less distinct. Short-lived storms are occasionally seen. Saturn has weaker gravity and a more extended atmosphere than Jupiter. The planet's overall butterscotch hue is due to cloud chemistry similar to that occurring in Jupiter's atmosphere. Saturn, like Jupiter, has bands, ovals, and turbulent flow patterns powered by convective motion in the interior.

Like Jupiter, Saturn emits far more radiation into space than it receives from the Sun. Unlike Jupiter's, Saturn's excess energy emission is the result of **helium precipitation** (p. 308) in the planet's interior, where helium liquefies and forms droplets which then fall toward the center of the planet. This process is also responsible for Saturn's observed helium deficit. Saturn's interior is theoretically similar to that of Jupiter, but with a thinner layer of metallic hydrogen and a larger core. Its lower mass gives Saturn a less extreme core temperature, density, and pressure than Jupiter's. Saturn's conducting interior and rapid rotation produce a strong magnetic field and an extensive magnetosphere that contains the planet's ring system and the innermost 16 moons.

Saturn's rings lie in the planet's equatorial plane, so their appearance from Earth changes as Saturn orbits the Sun. From Earth, the main visible features of the rings are the **A**, **B**, and **C**

rings (p. 309), the **Cassini Division** (p. 309), and the **Encke gap** (p. 309). The Cassini Division is a dark region between the A and B rings. The Encke gap lies near the outer edge of the A ring. The rings are made up of trillions of icy particles ranging in size from dust grains to boulders, all orbiting Saturn like so many tiny moons. Their total mass is comparable to that of a small moon. Both divisions are dark because they are almost empty of ring particles. When the *Pioneer* and *Voyager* probes reached Saturn, they found that the rings are actually made up of tens of thousands of narrow **ringlets** (p. 311). Interactions between the ring particles and the planet's inner moons are responsible for much of the fine structure observed in the main rings.

The **Roche limit** (p. 310) of a planet is the distance within which the planet's tidal field would overwhelm the internal gravity of a moon, tearing it apart and forming a ring. All known planetary ring systems lie inside their parent planets' Roche limits. Saturn's narrow **F ring** (p. 314), discovered by *Pioneer 11*, lies just outside the A ring. It has a kinked, braided structure, apparently caused by two small **shepherd satellites** (p. 314) that orbit close to the ring and prevent it from breaking up. Beyond the F ring is the faint, narrow **G ring** (p. 314), also discovered by *Pioneer 11*. *Voyager 2* discovered the faint **D ring** (p. 311), lying between the C ring and Saturn's cloud layer, and the **E ring** (p. 313), associated with the moon Enceladus. Planetary rings may have lifetimes of only a few tens of millions of years. If so, the fact that we see rings around all four jovian planets means that they must constantly be re-formed or replenished, perhaps by material chipped off moons by meteoritic impact or by the tidal destruction of entire moons.

Saturn's single large moon Titan is the second-largest moon in the solar system. Its thick atmosphere obscures the moon's surface and may be the site of complex cloud and surface chemistry. The existence of Titan's atmosphere is a direct consequence of the cold conditions that prevailed at the time of the moon's formation.

The medium-sized moons of Saturn are made up predominantly of rock and water ice. They show a wide variety of surface terrains and are also heavily cratered. They are all tidally locked by the planet's gravity into synchronous orbits. The innermost midsized moon Mimas exerts influence over the structure of the rings. The Cassini Division, now known to contain faint ringlets and gaps, is the result of resonance between its particles and Mimas. The moon Iapetus has a marked contrast between its leading and trailing faces, while Enceladus has a highly reflective appearance, possibly the result of water "volcanoes" on its surface.

Saturn's small moons exhibit a wide variety of complex motion. Several moons "share" orbits, in some cases lying at the **Lagrangian points** (p. 321) 60° ahead of and behind the orbit of a larger moon. The moon Hyperion tumbles in an unpredictable way as it orbits the planet.

SELF-TEST: TRUE OR FALSE?

____ **1.** Saturn's orbit is almost twice the size of Jupiter's orbit.

____ **2.** The rotation of Saturn is unlike that of Jupiter; it is slow and shows little differential rotation.

____ **3.** Saturn's large axial tilt produces strong seasonal variations in the planet's atmosphere.

____ **4.** Saturn probably does not have a rocky core.

____ **5.** Saturn emits nearly three times more energy in the form of infrared radiation than it receives from the Sun.

____ **6.** No storm systems have ever been seen on Saturn.

____ **7.** The magnetic field of Saturn is much less than Jupiter's and, at the cloud tops, is about as strong as Earth's magnetic field.

____ **8.** Saturn is the only planet with a ring system.

____ **9.** A typical ring particle is 100 m in diameter.

____ **10.** The composition of the ring particles of Saturn is predominantly water ice.

____ **11.** Although Saturn's ring system is tens of thousands of kilometers wide, it is only a few tens of meters thick.

____ **12.** Saturn has small and midsized moons, but no large moons.

____ **13.** Water ice predominates in Saturn's moons.

____ **14.** Titan's atmosphere is 10 times denser than Earth's.

____ **15.** Titan's surface is obscured by thick clouds of water ice.

SELF-TEST: FILL IN THE BLANK

1. Saturn is best known for its spectacular _____ system.

2. Low-lying features in the atmosphere of Saturn, as on Jupiter, are mostly hidden by the upper layer of frozen _____ clouds.

3. Saturn's cloud layers are thicker than those of Jupiter because of Saturn's weaker _____.

4. Relative to Jupiter's atmosphere, Saturn's atmosphere is deficient in _____.

5. Saturn's excess energy emission is caused by _____.

6. As viewed from Earth, Saturn's ring system is conventionally divided into _____ broad rings.

7. The Cassini Division lies between the _____ and _____ rings.

8. The _____ ring of Saturn is the brightest.

9. The rings exist because they lie within Saturn's _____.

10. Two small moons, known as _____ satellites, are responsible for the unusually complex form of the F ring.

11. The composition of Titan's atmosphere is 90 percent _____.

12. Titan's surface may be covered with _____ sediment.

13. Except for Titan, most of the moons of Saturn are heavily cratered, even though their surfaces are mostly water ice. The low _____ of the ice makes it hard as rock.

14. Saturn's _____ ring is probably related to volcanism on the icy moon Enceladus.

15. When small moons share an orbit with a large moon, they are found at the _____ points of the large moon's orbit.

REVIEW AND DISCUSSION

1. Seen from Earth, Saturn's rings sometimes appear broad and brilliant but at other times seem to disappear. Why?

2. What is a ring crossing? When will the next one occur?

3. Why does Saturn have a less varied appearance than Jupiter?

4. What does Saturn's shape tell us about its deep interior?

5. Compare and contrast the atmospheres and weather systems of Saturn and Jupiter, and tell how the differences affect each planet's appearance.

6. Compare the thicknesses of Saturn's various layers (clouds, molecular hydrogen, metallic hydrogen, and core) with the equivalent layers in Jupiter. Why do the thicknesses differ?

7. What mechanism is responsible for the relative lack of helium in Saturn's atmosphere, compared to Jupiter?

8. Is Saturn as a whole deficient in helium relative to Jupiter?

9. When were Saturn's rings discovered? When did astronomers realize what they were?

10. What would happen to a satellite if it came too close to Saturn?

11. What evidence supports the idea that a relatively recent catastrophic event was responsible for Saturn's rings?

12. What effect does Mimas have on the rings?

13. What are shepherd satellites?

14. When *Voyager 1* passed Saturn in 1980, why didn't it see the surface of Saturn's largest moon, Titan?

15. Compare and contrast Titan with Jupiter's Galilean moons.

16. Why does Titan have a dense atmosphere when other large moons in the solar system don't?

17. What is the evidence for geological activity on Enceladus?

18. What mystery is associated with Iapetus?

19. Describe the behavior of Saturn's co-orbital satellites.

20. Imagine what the sky would look like from Saturn's moon Hyperion. Would the Sun rise and set in the same way it does on Earth? How do you imagine Saturn might look?

PROBLEMS  *Algorithmic versions of these questions are available in the Practice Problems module of the Companion Website.*

The number of squares preceding each problem indicates its approximate level of difficulty.

1. ■ What is the angular diameter of Saturn's A ring, as seen from Earth at closest approach?

2. ■ What is the size of the smallest feature visible in Saturn's rings, as seen from Earth at closest approach with a resolution of 0.05″?

3. ■■ What would be the mass of Saturn if it were composed entirely of hydrogen at a density of 0.08 kg/m³, the density of hydrogen at sea level on Earth? Assume for simplicity that Saturn is spherical. Compare your answer with Saturn's actual mass and with the mass of Earth.

4. ■ How long does it take for Saturn's equatorial flow, moving at 1500 km/h, to encircle the planet? Compare this with the wind-circulation time on Jupiter.

5. ■■ If Saturn's surface temperature is 97 K and the planet radiates three times more energy than it receives from the Sun, use Stefan's law to calculate what the surface temperature would be in the absence of any internal heat source.

6. ■ Based on the data given in Sections 12.1 and 12.3 (Figure 12.8), estimate the average density of Saturn's core.

7. ■■ The text states that the total mass of material in Saturn's rings is about 10¹⁵ tons (10¹⁸ kg). Suppose the average ring particle is 6 cm in radius (a large snowball) and has a density of 1000 kg/m³. How many ring particles are there?

8. ■ What is the orbital speed of ring particles at the inner edge of the B ring, in km/s? Compare this with the speed of a satellite in low Earth orbit (500 km altitude, say). Why are these speeds so different?

9. ■ Show that Titan's surface gravity is about one-seventh of Earth's, as stated in the text. What is Titan's escape speed?

10. ■■ Assuming a spherical shape and a uniform density of 2000 kg/m³, calculate how small an icy moon would have to

be before a fastball pitched at 40 m/s (about 90 mph) could escape.

11. ■■■ Calculate the orbital radii of particles having the following properties: (a) a 3:1 orbital resonance with Tethys, that is, orbiting Saturn three times for every orbit of Tethys; (b) a 2:1 resonance with Mimas (two orbits for every orbit of Mimas); (c) a 3:2 resonance with Mimas (three orbits for every two of Mimas); (d) a 2:1 resonance with Dione.

12. ■■ Compare Saturn's tidal gravitational effect on Mimas with Mimas's own surface gravity.

13. ■■■ Compare Saturn's tidal gravitational effect on Titan with Titan's own surface gravity. Based on these numbers and the corresponding numbers for Jupiter's Galilean moons, would you expect significant internal heating in Titan?

14. ■■■ Sunlight reflected back to Earth from a particle in Saturn's rings is Doppler-shifted twice—first because of the relative motion of the source of the radiation (the Sun) and the ring particle, and then again by the relative motion of the particle and the observer on Earth (see Section 3.5). As a result, if Earth, Saturn, and the Sun are roughly aligned (that is, Saturn is near opposition), the observed Doppler shift corresponds to *twice* the particle's orbital speed. A certain solar spectral line, of wavelength 656.112 nm, is reflected from the rings and observed on Earth. If the rings happen to be seen almost edge-on, what is the line's observed wavelength in light reflected from (a) the approaching inner edge of the B ring? (b) the receding inner edge of the B ring? (c) the approaching outer edge of the A ring? (d) the receding outer edge of the A ring?

15. ■■■ Based on the data given in the text, estimate the difference in orbital radii between the two co-orbital satellites.

COLLABORATIVE EXERCISES

1. **Saturn's Rings Scale Model.** Using your tallest group member's outstretched arms, create a complete scale model of all Saturn's rings as listed on Table 12.1 using self-stick notes or tape labels. Measure the maximum distance from nose to fingertip and use this as the scale factor for the rings'

radii. For example, if the distance from nose to fingertip was 40 cm and the outer radius of the E ring is 480,000 km, then the inner radius of the D ring is 67,000 km × (40 cm/480,000 km) = 5.6 cm out from the nose.

RESEARCHING ON THE WEB  *To complete the following exercises, go to the online Destinations module for Chapter 12 on the Companion Website for Astronomy Today 4/e.*

1. Access the "Saturn Fact Sheet" from NASA and determine the current number of known natural satellites (moons) around Saturn.
2. Access the "Saturn Nomenclature" page and determine the size and name of the largest crater on Enceladus.

3. Access the "Saturn Pages in Planetary Collections" at The Nine Planets and determine which space probe was the first to arrive at Saturn and what the next space probe to arrive there is named and when it will arrive.

PROJECTS

1. Saturn moves more slowly among the stars than any other visible planet. How many degrees per year does it move? Look in an almanac to see where the planet is now. What constellation is it in now? Where will it be in one year?
2. Binoculars may not reveal the rings of Saturn, but most small telescopes will. Use a telescope to look at Saturn. Does Saturn appear flattened? Examine the rings. How are they tilted? Can you see a dark line in the rings? This is the Cassini

Division. It once was thought to be a gap in the rings, but the *Voyager* spacecraft discovered that it is filled with tiny ringlets. Can you see the shadow of the rings on Saturn?
3. While looking at Saturn through a telescope can you see any of its moons? They line up with the rings; Titan is often the farthest out, and always the brightest. How many moons can you see? Use an almanac to identify each one you find.

SKYCHART III PROJECTS  *The SkyChart III Student Version planetarium program on which these exercises are based is included as a separately executable program on the CD in the back of this text.*

1. ■ Center on Saturn, and zoom in to a 1/6° field of view. Saturn is much farther than Jupiter, so a greater magnification is required to create the same angular size image of the planet. Deselect horizon mask; turn off stars, constellations, and grid lines. Set animation for 10-minute steps. It may be necessary to lower the fainter limits of displayed objects to bring up some of the moons. To bring up the fainter objects, use F3 or *DRAW/Fainter Limits*. Animate the image to observe the motion of the moons. Notice that they orbit in the same sense as Saturn spins—counterclockwise as seen from above the North Pole.
2. ■ Center on Saturn and set parameters the same as for observing the motion of the moons. Set field of view to 1/20° and animate with one-week time steps. Observe the tilt of Saturn and its ring system. Explain why the planet and its ring should change their tilt so markedly. When will the ring system appear edge-on, and when after that will it next appear edge-on? What is this time interval? How is this time interval related to the period of Saturn's orbit?
3. ■ Galileo trained his telescope on Saturn and initially saw something that he interpreted as two companion planets. He was alarmed when he saw them diminish and disappear,

and wondered whether Saturn had devoured its own children. From Rome in Italy in 1612, follow Saturn over a period of three years. Describe the views that Galileo might have had of Saturn.
4. ■ The first person to propose the correct explanation for "Saturn's children" or the appendages observed by Galileo, was Christian Huygens. Aided by a better telescope, he was able to surmise the nature of these appendages. Observe Saturn in the four years leading up to 1659. What happens? The feat of Huygens was to propose a ring: the first heavenly object that had unmistakably a shape that was not spherical.
5. ■■ Observing from a point directly above the ecliptic plane, determine the period between successive superior conjunctions of Saturn. This determines the synodic orbital period. How long does it take for Saturn to complete one revolution around the Sun? This is the sidereal orbital period. What is the relationship between the synodic and the sidereal periods? It might be helpful to find an analogy in a watch, with the little hand representing Saturn and the large hand Earth.

 *In addition to the Practice Problems and Destinations modules, the Companion Website at http://www.prenhall.com/chaisson provides for each chapter an additional true-false quiz, multiple choice, and labeling quiz, as well as additional annotated images, animations, and links to related Websites.*

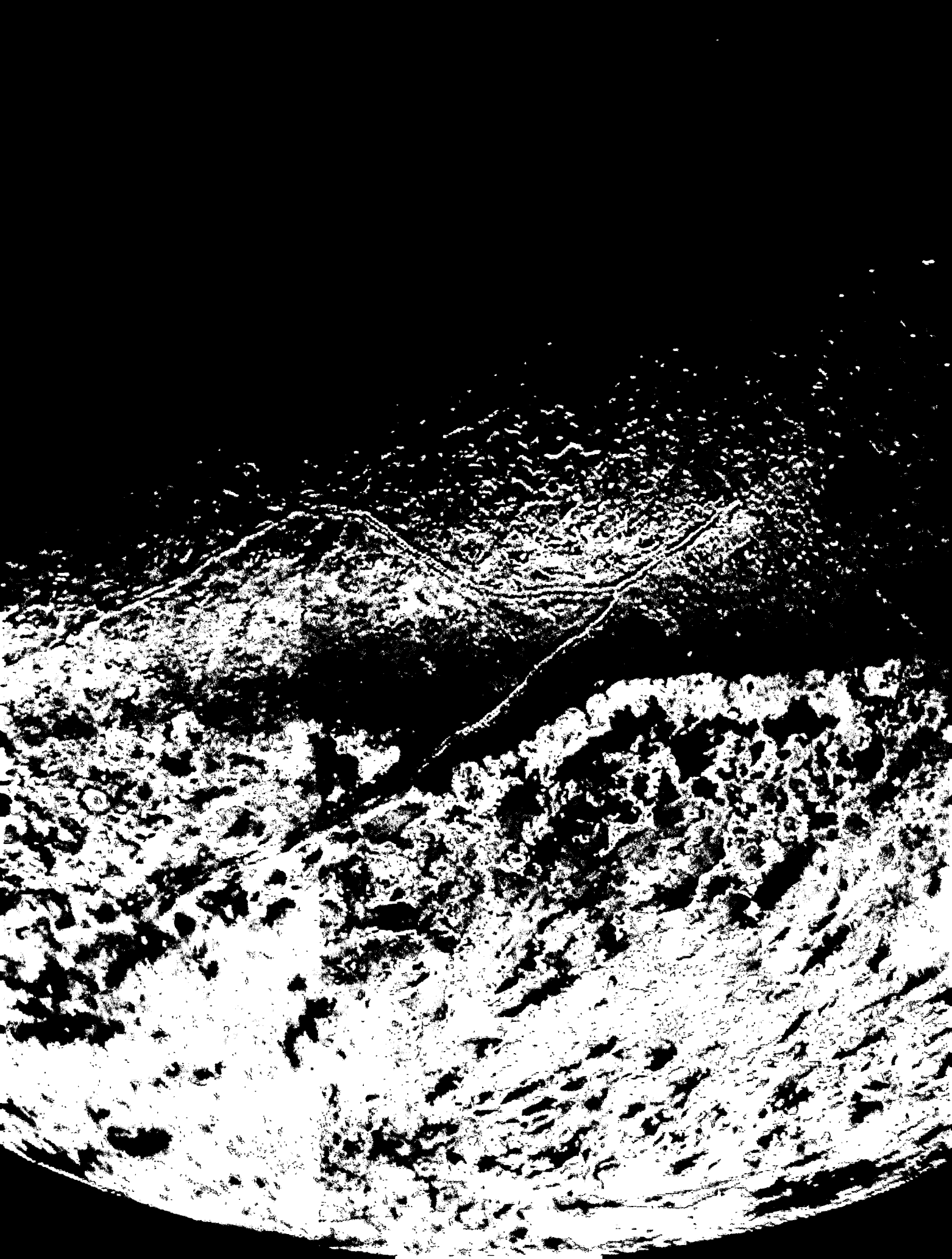

13 URANUS, NEPTUNE, AND PLUTO

The Outer Worlds of the Solar System

LEARNING GOALS

Studying this chapter will enable you to:

1.   Describe how both calculation and chance played major roles in the discoveries of the outer planets.

2.   Summarize the similarities and differences between Uranus and Neptune and compare these planets with the other two jovian worlds.

3.   Explain what the moons of the outer planets tell us about their past.

4.   Contrast the rings of Uranus and Neptune with those of Jupiter and Saturn.

5.   Summarize the orbital and physical properties of Pluto and explain how the Pluto–Charon system differs fundamentally from all the other planets.

 Visit http://www.prenhall.com/chaisson for additional annotated images, animations, and links to related sites for this chapter.

This close-up image of Triton was taken by *Voyager 2*, perhaps the most successful spacecraft ever to tour the solar system. Triton is the largest of the known moons of Neptune, yet has only $\frac{1}{3}$ the mass of the Moon. Pictorial data like these give evidence for fascinating terrain (probably sculpted in ice), a thin (but real) atmosphere, and even evidence for ice volcanoes (the prominent dark streaks are perhaps the result of geyser-like eruptions). *(JPL)*

The Big Picture: The number of known moons in the solar system rapidly increased in the late 1990s. Better telescopes enabled astronomers to take a more complete inventory of all the material objects in our cosmic neighborhood. However, the more closely we examine these alien worlds, the more complex they seem to be. Rather than converging toward well-defined categories of moons, we seem to be finding greater diversity among them.

The three outermost planets were unknown to the ancients—all were discovered by telescopic observations: Uranus in 1781, Neptune in 1846, and Pluto in 1930. Uranus and Neptune have very similar bulk properties, so it is natural to consider them together; they are part of the jovian family of planets. Pluto, by contrast, is not a jovian world. It is very much smaller than even the terrestrial planets and generally seems much more moonlike than planetlike. Indeed, at one time, astronomers even speculated that Pluto was a one-time moon that had somehow escaped from one of the outer planets, most likely Neptune. However, it now seems more probable that Pluto is really the best-known representative of a newly recognized class of objects residing in the outer solar system. Whatever its origin, because of Pluto's location and its similarity to the jovian moons, we study it here along with its larger jovian neighbors.

13.1 The Discovery of Uranus

¶ The planet Uranus was discovered by British astronomer William Herschel in 1781. Herschel was engaged in charting the faint stars in the sky when he came across an odd-looking object that he described as "a curious either nebulous star or perhaps a comet." Repeated observations showed that it was neither. The object appeared as a disk in Herschel's 6-inch telescope and moved relative to the stars, but it traveled too slowly to be a comet. Herschel soon realized that he had found the seventh planet in the solar system.

Since this was the first new planet discovered in well over 2000 years, the event caused quite a stir at the time. The story goes that Herschel's first instinct was to name the new planet "Sidus Georgium" (Latin for "George's star") after his king, George III of England. The world was saved from a planet named George by the wise advice of another astronomer, Johann Bode (of "Titius-Bode law" fame—see *Discovery 6-1*). He suggested instead that the tradition of using names from Greco-Roman mythology be continued and that the planet be named after Uranus, the father of Saturn.

Uranus is in fact just barely visible to the naked eye, if you know exactly where to look. At opposition, it has a maximum angular diameter of 4.1" and shines just above the unaided eye's threshold of visibility. It looks like a faint, undistinguished star. No wonder it went unnoticed by the ancients. Even today, few astronomers have seen it without a telescope.

Through a large Earth-based optical telescope (Figure 13.1), Uranus appears hardly more than a tiny pale greenish disk. With the flyby of *Voyager 2* in 1986, our

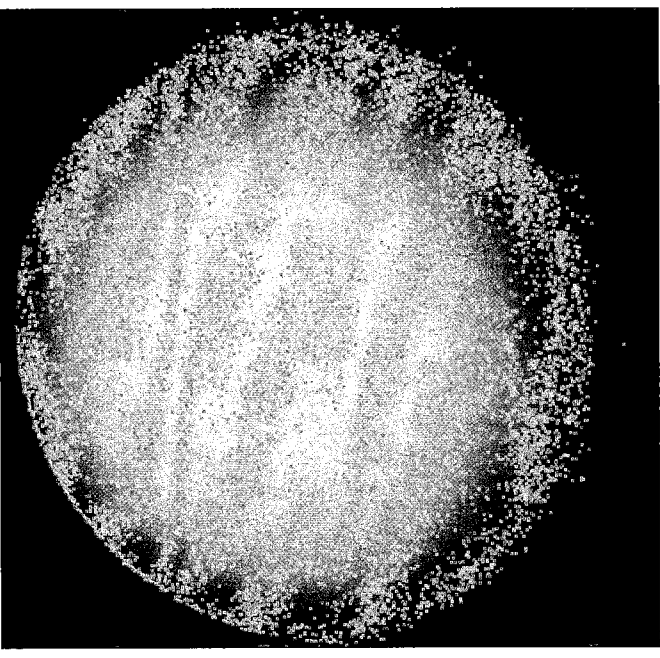

Figure 13.2 Uranus, Close Up This image of Uranus, taken from a distance of about 1 million km, was sent back to Earth by the *Voyager 2* spacecraft as it whizzed past this giant planet at 10 times the speed of a rifle bullet. It approximates the planet's true color, but shows virtually no detail in the largely featureless upper atmosphere, except for a few wispy clouds in the northern hemisphere. *(NASA)*

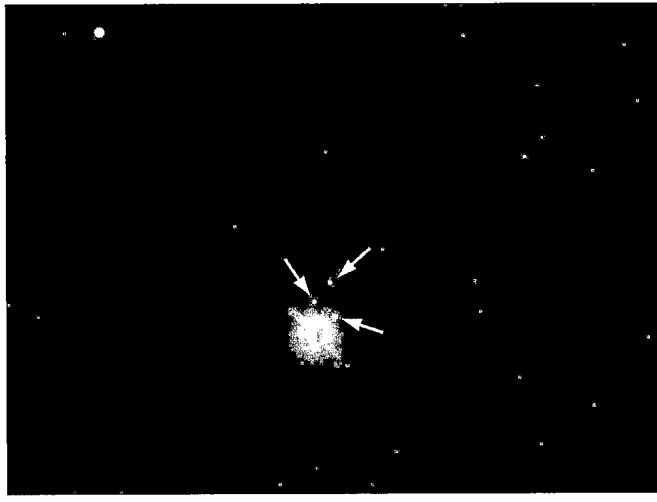

Figure 13.1 Uranus from Earth Details are barely visible on photographs of Uranus made with large Earth-based telescopes. (Arrows point to three of its moons.) *(UC/Lick Observatory)*

knowledge of Uranus increased dramatically, although close-up images of the planet still showed virtually no surface detail (Figure 13.2). The apparently featureless atmosphere of Uranus contrasts sharply with the bands and spots visible on all the other jovian worlds. Some orbital and physical properties of Uranus are presented in the Uranus Data box on p. 332.

13.2 The Discovery of Neptune

¶ Once Uranus was discovered, astronomers set about charting its orbit, and quickly discovered a small discrepancy between the planet's predicted position and where they actually observed it. Try as they might, astronomers could not find an elliptical orbit that fit the planet's trajectory to within the accuracy of their measurements. Half a century after Uranus's discovery, the discrepancy had grown to a quarter of an arc minute, far too big to be explained away as observational error.

The logical conclusion was that an unknown body must be exerting a gravitational force on Uranus—much weaker than that of the Sun, but still measurable. But what body could this be? Astronomers realized that there had to be *another* planet in the solar system perturbing Uranus's motion.

In the 1840s two mathematicians independently solved the difficult problem of determining this new planet's mass and orbit. A British astronomer, John Adams, reached the solution in September 1845; in June of the following year, the French mathematician Urbain Leverrier came up with essentially the same answer. British astronomers seeking the new planet found nothing during the summer of 1846. In September a German astronomer named Johann Galle began his own search from the Berlin Observatory, using a newly completed set of more accurate sky charts. He found the new planet within one or two degrees of the predicted position—on his first attempt. After some wrangling over names and credits, the new planet was named Neptune, and Adams and Leverrier (but not Galle!) are now jointly credited with its discovery.

Neptune's orbital and physical properties are listed in the Neptune Data box on p. 341. With an orbital semimajor axis of 30.1 A.U. and an orbital period of 163.7 years, Neptune has not yet completed one revolution since its discovery. Unlike Uranus, distant Neptune cannot be seen with the naked eye, although it can be seen with a small telescope—in fact, according to his notes, Galileo might actually have seen Neptune, although he had no idea what it really was at the time. Through a large telescope, Neptune appears as a bluish disk, with a maximum angular diameter of 2.4″ at opposition.

Figure 13.3 shows a long Earth-based exposure of Neptune and its largest moon, Triton. Neptune is so distant that surface features are virtually impossible to dis-

Figure 13.3 Neptune from Earth Neptune and two of its moons, Triton (large arrow) and Nereid (small arrow), imaged with a large Earth-based telescope. *(UC/Lick Observatory)*

cern. Even under the best observing conditions, only a few markings can be seen. These are suggestive of multicolored cloud bands—light bluish hues seem to dominate. With *Voyager 2*'s arrival, much more detail emerged, as shown in Figure 13.4. Superficially, at least, Neptune resembles a blue-tinted Jupiter, with atmospheric bands and spots clearly evident.

✓ Concept Check

▪ How did observations of the orbit of Uranus lead to the discovery of Neptune?

13.3 Physical Properties of Uranus and Neptune

② Figure 13.5 shows Uranus and Neptune to scale, along with Earth for comparison. The two giant planets are quite similar in their bulk properties. The radius of Uranus is 4.0 times that of Earth; that of Neptune, 3.9 Earth radii. Their masses (first determined from terrestrial observations of their larger moons and later refined by *Voyager 2*) are 14.5 Earth masses for Uranus and 17.1 Earth masses for Neptune. Uranus's average density is 1300 kg/m³, and Neptune's is 1600 kg/m³. These densities imply that large rocky cores constitute a greater fraction of the planets' masses than do the cores of either Jupiter or Saturn. The cores themselves are probably comparable in size, mass, and composition to those of the two larger giants.

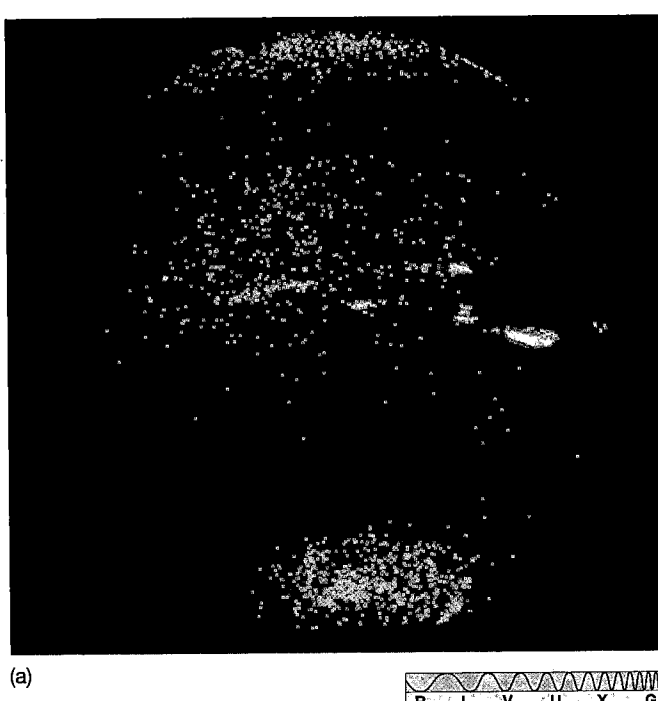

(a)

(b)

Figure 13.4 Neptune, Close Up (a) Neptune as seen by *Voyager 2*, from a distance of roughly 1 million km. A closer view (b), resolved to about 10 km, shows cloud streaks ranging in width from 50 km to 200 km. *(NASA)*

Like the other jovian planets, Uranus has a short rotation period. Earth-based observations of the Doppler shifts in spectral lines first indicated that Uranus's "day" was between 10 and 20 hours long. The precise value of the planet's rotation period—accurately determined when *Voyager 2* timed radio signals associated with its magnetosphere—is now known to be 17.2 hours. Again as with Jupiter and Saturn, the planet's atmosphere rotates differentially. However, Uranus's atmosphere actually rotates faster at the poles (where the period is 14.2 hours) than near the equator (where the period is 16.5 hours).

Each planet in our solar system seems to have some outstanding peculiarity, and Uranus is no exception. Unlike all the other planets, whose spin axes are roughly per-

Jupiter

Earth

Saturn

Uranus

Neptune

Figure 13.5 Jovian Planets Jupiter, Saturn, Uranus, Neptune, and Earth, drawn to scale. Uranus and Neptune are quite similar in their bulk properties. Each probably contains a core about 10 times more massive than Earth. Jupiter and Saturn are each substantially larger, but their rocky cores are probably comparable in mass to those of Uranus and Neptune. *(NASA)*

pendicular to the ecliptic plane, Uranus's rotation axis lies almost within that plane—98° from the perpendicular, to be precise. (Because the north pole lies below the ecliptic plane, the rotation of Uranus, like that of Venus, is classified as retrograde.) Relative to the other planets, we might say that Uranus lies tipped over on its side. As a result, the "north" (spin) pole of Uranus, at some time in its orbit, points almost directly toward the Sun.* Half a "year" later, its "south" pole faces the Sun, as illustrated in Figure 13.6. When *Voyager 2* encountered the planet in 1986, the north pole happened to be pointing nearly at the Sun, so it was midsummer in the northern hemisphere.

The strange orientation of Uranus's rotation axis produces some extreme seasonal effects. Starting at the height of northern summer, when the north pole points closest to the Sun, an observer near that pole would see the Sun move in gradually increasing circles in the sky, completing one circuit (counterclockwise) every 17 hours and dipping slightly lower in the sky each day. Eventually the Sun would begin to set and rise again in a daily cycle, and the nights would grow progressively longer with each passing day. Twenty-one years after the summer solstice, the autumnal equinox would occur, with day and night each 8.5 hours long.

As in Chapter 9, we adopt the convention that a planet's rotation is always counterclockwise as seen from above the north pole (that is, planets always rotate from west to east). ∞ *(Sec. 9.2)*

The days would continue to shorten until one day the Sun would fail to rise at all. The ensuing period of total darkness would be equal in length to the earlier period of constant daylight, plunging the northern hemisphere into the depths of winter. Eventually, the Sun would rise again; the days would lengthen through the vernal equinox and beyond, and in time the observer would again experience a long summer of uninterrupted (though dim) sunshine.

From the point of view of an observer on the equator, by contrast, summer and winter would be almost equally cold seasons, with the Sun never rising far above the horizon. Spring and fall would be the warmest times of year, with the Sun passing almost overhead each day.

No one knows why Uranus is tilted in this way. Some scientists have speculated that a catastrophic event, such as a grazing collision between the planet and another planet-sized body, might have altered the planet's spin axis. There is no direct evidence for such an occurrence, however, and no theory to tell us how we should seek to confirm it.

Neptune's clouds show more variety and contrast than do those of Uranus, and Earth-based astronomers studying them determined a rotation rate for Neptune even before *Voyager 2*'s flyby in 1989. The average rotation period of Neptune's atmosphere is 17.3 hours (quite similar to that of Uranus). Measurements of Neptune's radio emission by *Voyager 2* showed that the magnetic field of the planet, and presumably also its interior, rotates once every

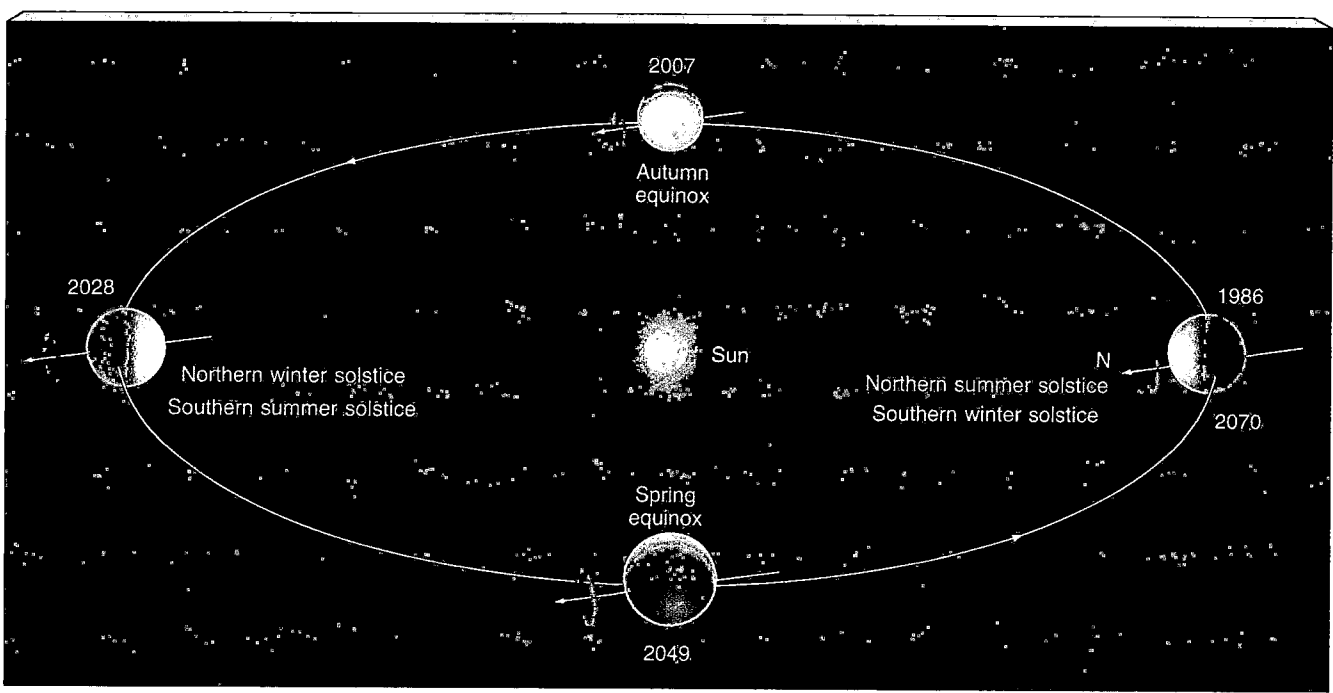

Figure 13.6 Seasons on Uranus The 98° axial tilt of Uranus places its equator almost perpendicular to the ecliptic. As a result, the planet experiences the most extreme seasons known in the solar system. The equatorial regions experience two warm seasons (around the two equinoxes) and two cold seasons (at the solstices) each year; the poles are alternately plunged into darkness for 42 years at a time.

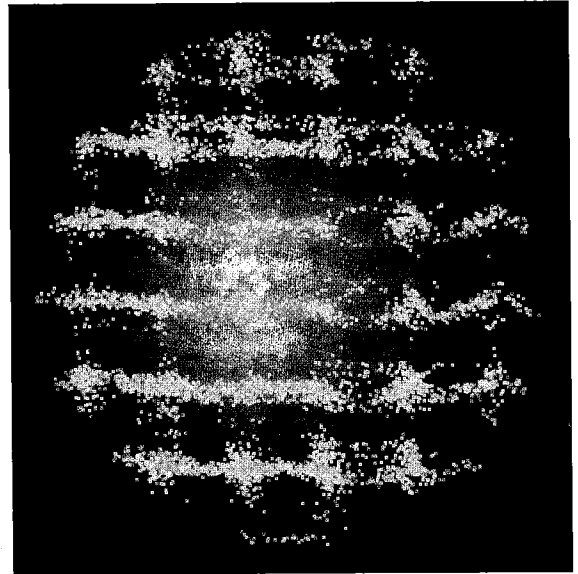

Orbital semimajor axis	19.19 A.U.
	2871 million km
Orbital eccentricity	0.047
Perihelion	18.29 A.U.
	2736 million km
Aphelion	20.10 A.U.
	3006 million km
Mean orbital speed	6.80 km/s
Sidereal orbital period	83.75 tropical years
Synodic orbital period	369.66 solar days
Orbital inclination to the ecliptic	0.77°
Greatest angular diameter, as seen from Earth	4.1″
Mass	8.68×10^{25} kg
	14.54 (Earth = 1)
Equatorial radius	25,559 km
	4.01 (Earth = 1)
Mean density	1271 kg/m^3
	0.230 (Earth = 1)
Surface gravity (at cloud tops)	8.87 m/s^2
	0.91 (Earth = 1)
Escape speed	21.3 km/s
Sidereal rotation period	−0.72 solar days (retrograde)
Axial tilt	97.92°
Surface magnetic field	0.74 (Earth = 1)
Magnetic axis tilt relative to rotation axis	58.6°
Mean surface temperature	58 K
Number of moons	21

16.1 hours. Thus, Neptune is unique among the jovian worlds in that its atmosphere rotates *more slowly* than its interior. Neptune's rotation axis is inclined 29.6° to a line perpendicular to its orbital plane, quite similar to the 27° tilt of Saturn.

☑ Concept Check

☐ What is unusual about the rotation of Uranus?

13.4 The Atmospheres of Uranus and Neptune

COMPOSITION

☝ Spectroscopic studies of sunlight reflected from Uranus's and Neptune's dense clouds indicate that the two planets' outer atmospheres (the parts we actually measure spectroscopically) are quite similar to the atmospheres of Jupiter and Saturn. The most abundant element is molecular hydrogen (84 percent), followed by helium (about 14 percent) and methane, which is more abundant on Neptune (about 3 percent) than on Uranus (2 percent). Ammonia, which plays such an important role in the Jupiter and Saturn systems, is not present in any significant quantity in the outermost jovian worlds.

The abundances of gaseous ammonia and methane vary systematically among the jovian planets. Jupiter has much more gaseous ammonia than methane, but moving outward from the Sun, we find that the more distant planets have steadily decreasing amounts of ammonia and relatively greater amounts of methane. The reason for this variation is temperature. Ammonia gas freezes into ammonia ice crystals at about 70 K. This is cooler than the cloud-top temperatures of Jupiter and Saturn but warmer than those of Uranus (58 K) and Neptune (59 K). Thus, the outermost jovian planets have little or no *gaseous* ammonia in their atmospheres, so their spectra (which record atmospheric gases only) show only traces of ammonia.

The increasing amounts of methane are largely responsible for the outer jovian planets' blue coloration. Methane absorbs long-wavelength red light quite efficiently, so sunlight reflected from the planets' atmospheres is deficient in red and yellow photons and appears blue-green or blue. As the concentration of methane increases, the reflected light should appear bluer. This is just the trend observed: Uranus, with less methane, looks blue-green, while Neptune, with more, looks distinctly blue.

WEATHER

Voyager 2 detected just a few cloud features in Uranus's atmosphere (Figure 13.2), and even those became visible only after extensive computer enhancement. Figure 13.7

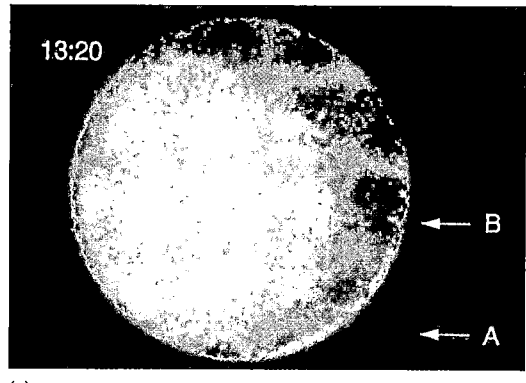

(a)

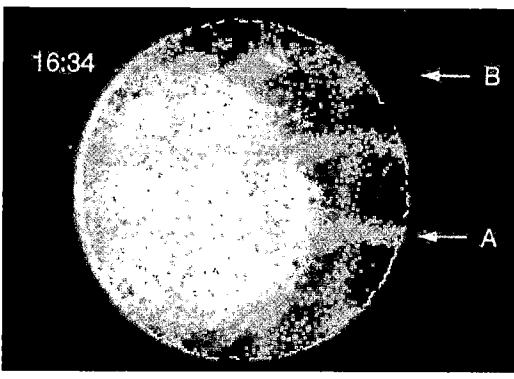

(b)

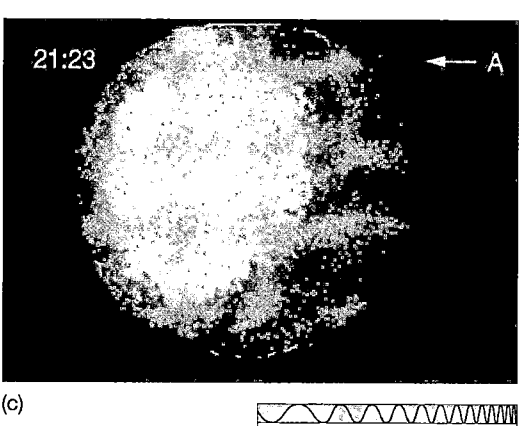

(c)

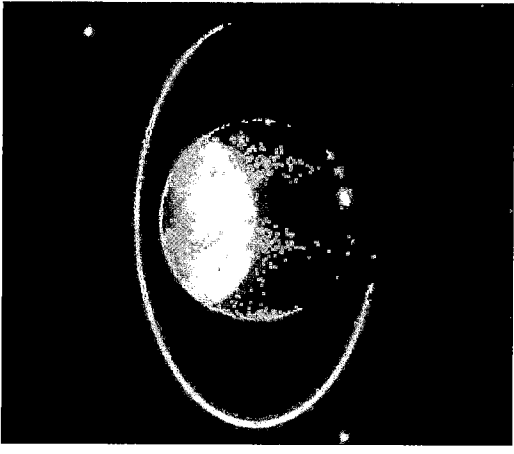

(d)

Figure 13.7 Uranus's Rotation (a), (b), and (c) These computer-enhanced *Hubble Space Telescope* images, taken at roughly four-hour intervals, show the motion of a pair of bright clouds (labeled A and B) in the planet's southern hemisphere. (The numbers at the top give the time of each photo.) (d) A near-infrared image of Uranus, also from *HST*, shows the planet's ring system, as well as a number of clouds (pink and red regions) in the upper atmosphere. The clouds circle the planet at speeds of up to 500 km/h. *(NASA)*

shows some more recent *Hubble Space Telescope* views of the planet. Parts (a) through (c) are heavily processed optical images that show the progress of a pair of bright clouds around the planet. Part (d) shows a false-color, near-infrared rendition of Uranus. The colors in this image generally indicate the depth to which we can see into the atmosphere. Blue-green regions are clear atmospheric regions where astronomers can study conditions down to the lower cloud levels. Yellow-grey colors show sunlight reflecting from higher cloud layers, or from atmospheric haze. Orange-red colors, such as the prominent "spots" on the south (right) edge of this image, indicate very high clouds, much like cirrus clouds on Earth.

Uranus apparently lacks any significant internal heat source, and because of the planet's low surface temperature, its clouds are found only at low-lying, warmer levels in the atmosphere. The absence of high-level clouds means that we must look deep into the planet's atmosphere to see any structure, so the bands and spots that characterize flow patterns on the other jovian worlds are largely "washed out" on Uranus by intervening stratospheric haze.

From computer-processed images such as those shown in Figure 13.7, astronomers have learned that Uranus's atmospheric clouds and flow patterns move around the planet in the same sense as the planet's rotation, with wind speeds ranging from 200 to 500 km/h. In fact, tracking these clouds allowed the measurement of the differential rotation mentioned earlier. Despite the odd angle at which sunlight is currently striking the surface (recall that it is now late summer in the northern hemisphere), the planet's rapid rotation still channels the wind flow into bands reminiscent of those found on Jupiter and Saturn. Wind speeds are faster near the North Pole, possibly because that part of the planet currently receives the greatest amount of solar heating. Even though the predominant wind flow is in the east–west direction, the atmosphere seems to be quite efficient at transporting energy from the heated north to the unheated southern hemisphere. Although much of the south is currently in total darkness, the temperature there is only a few kelvins less than in the north.

Neptune's clouds and band structure are much more easily seen. Although it lies at a greater distance from the Sun, Neptune's upper atmosphere is actually slightly warmer than that of Uranus. Like Jupiter and Saturn, but unlike Uranus, Neptune has an internal energy source—in fact, it radiates 2.7 times more heat than it receives from the Sun. The cause of this heating is still uncertain. Some

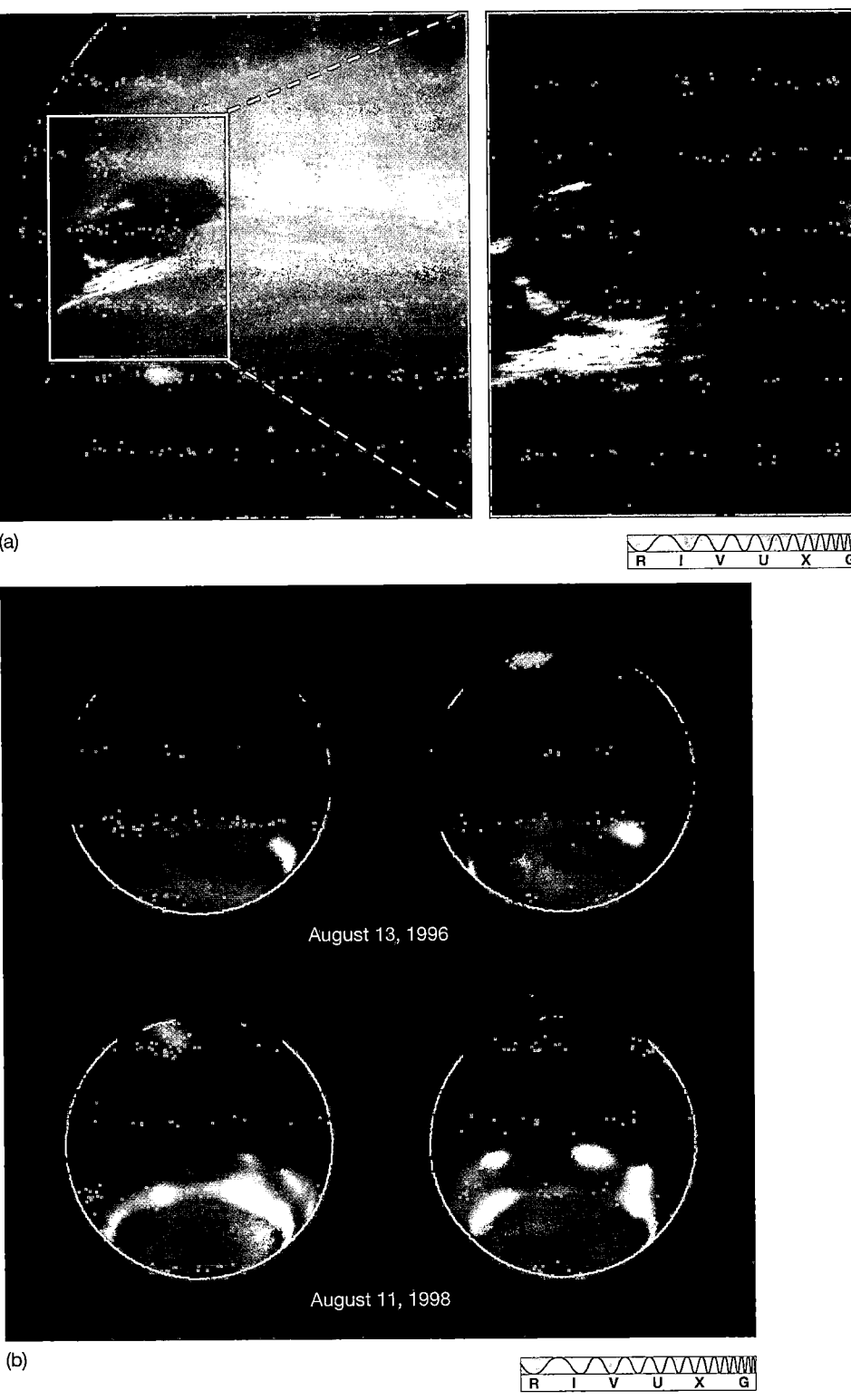

(a)

(b)

August 13, 1996

August 11, 1998

Figure 13.8 Neptune's Dark Spot (a) Close-up views, taken by *Voyager 2* in 1989, of the Great Dark Spot of Neptune, a large storm system in the planet's atmosphere, possibly similar in structure to Jupiter's Great Red Spot. Resolution in the photo on the right is about 50 km; the entire dark spot is roughly the size of planet Earth. (b) These *Hubble Space Telescope* views of Neptune were taken in 1996 and 1998, when the planet was some 4.5 billion km from Earth. The cloud features (mostly methane ice crystals) are tinted pink here because they were imaged in the infrared, but are really white in visible light. Neptune apparently has a remarkably dynamic atmosphere that can change significantly in just a few days. Wind speeds in excess of 1500 km/h have been measured. By the time these images were taken, the Great Dark Spot had disappeared. *(NASA)*

scientists have suggested that Neptune's excess methane has helped "insulate" the planet, tending to maintain its initially high internal temperature. If that is so, then the source of Neptune's internal heat is the same as Jupiter's— it is energy left over from the planet's formation. ∞ (Sec. 11.3) The combination of extra heat and less haze may be responsible for the greater visibility of Neptune's atmospheric features (see Figure 13.8), as its cloud layers lie at higher levels in the atmosphere than do those of Uranus.

Neptune sports several storm systems similar in appearance to those seen on Jupiter (and assumed to be produced and sustained by the same basic processes). The largest such storm, known simply as the **Great Dark Spot**, is shown in Figure 13.8(a). ∞ (Sec. 11.2) Discovered by *Voyager 2* in 1989, the Spot was about the size of Earth, was located near the planet's equator, and exhibited many of the same general characteristics as the Great Red Spot on Jupiter. The flow around it was counterclockwise, as with the Red Spot, and there appeared to be turbulence where the winds associated with the Great Dark Spot interacted with the zonal flow to its north and south. The flow around this and other dark spots may drive updrafts to high altitudes, where methane crystallizes out of the atmosphere to form high-lying cirrus clouds—those visible in Figure 13.8(a) lie some 50 km above the main cloud tops. Astronomers did not have long to study the Dark Spot's properties, however. As shown in Figure 13.8(b), when the *Hubble Space Telescope* viewed Neptune in the 1990s, the Spot had vanished, although several new storms (bright spots) had appeared.

The infrared views in Figure 13.8(b), captured a few hours apart on each of two days separated by two (Earth) years, illustrate Neptune's dynamic weather patterns. Winds blow at speeds in excess of 1500 km/h—almost half the speed of sound in Neptune's upper atmosphere—and with storms the size of Earth more the rule than the exception. The planet's stormy disposition is well established, but very difficult to understand. On Earth, weather systems are driven by the heat of the Sun. However, Neptune lies far from the Sun, in the outer solar system, and the Sun's heating effect is minuscule—nearly a thousand times less than at Earth. How can this planet be so cold, yet so active?

✔ Concept Check

▣ Why are planetary scientists puzzled by the strong winds and rapidly changing storm systems on Neptune?

13.5 Magnetospheres and Internal Structure

Voyager 2 found that both Uranus and Neptune have fairly strong internal magnetic fields—about 100 times stronger than Earth's field and 1/10 as strong as Saturn's. However, because Uranus and Neptune are so much larger than Earth, the magnetic fields at the cloud tops—spread out over far larger volumes than the field on Earth—are actually comparable in strength to Earth's field. Uranus and Neptune each have substantial magnetospheres, populated largely by electrons and protons either captured from the solar wind or created from ionized hydrogen gas escaping from the planets themselves.

When *Voyager 2* arrived at Uranus, it discovered that the planet's magnetic field is tilted at about 60° to the axis

of rotation. On Earth, such a tilt would put the north magnetic pole somewhere in the Caribbean. Furthermore, the magnetic field lines are *not* centered on the planet. It is as though Uranus's field were due to a bar magnet that is tilted with respect to the planet's rotation axis and displaced from the center by about one-third the radius of the planet. Figure 13.9 compares the magnetic field structures of Earth and those of the four jovian planets. The locations and orientations of the bar magnets represent the observed planetary fields, and the sizes of the bars indicate magnetic field strength.

Because dynamo theories generally predict that the magnetic axis should be roughly aligned with the rotation axis—as on Earth, Jupiter, Saturn, and the Sun—the misalignment on Uranus suggested to some researchers that perhaps the planet's field had been caught in the act of reversing. ∞ (Sec. 7.5) Another possibility was that the oddly tilted field was in some way related to the planet's axial tilt—perhaps one catastrophic collision skewed both axes at the same time. Those ideas evaporated in 1989 when *Voyager 2* found that Neptune's field is also inclined to the planet's rotation axis, at an angle of 46° (see Figure 13.9d), and also substantially offset from the center. It now appears that the internal structures of Uranus and Neptune are different from those of Jupiter and Saturn, and this difference changes how their magnetic fields are generated.

Theoretical models indicate that Uranus and Neptune have rocky cores similar to those found in Jupiter and Saturn—about the size of Earth and perhaps 10 times more massive. However, the pressure outside the cores of Uranus and Neptune (unlike the pressure within Jupiter and Saturn) is too low to force hydrogen into the metallic state, so hydrogen stays in its molecular form all the way in to the planets' cores. Astronomers theorize that deep below the cloud layers, Uranus and Neptune may have high-density, "slushy" interiors containing thick layers of water clouds. It is also possible that much of the planets' ammonia is dissolved in the water, accounting for the absence of ammonia at higher levels. Such an ammonia solution would provide a thick, electrically conducting ionic layer that could conceivably explain the planets' misaligned magnetic fields, if the circulating electrical currents that generate the fields occur mainly in regions far from the planets' centers and rotation axes.

At present, we don't know enough about the interiors of Uranus and Neptune to assess the correctness of this picture. Our current state of knowledge is summarized in Figure 13.10, which compares the internal structures of the four jovian worlds.

✔ Concept Check

▣ What is odd about the magnetic fields of Uranus and Neptune?

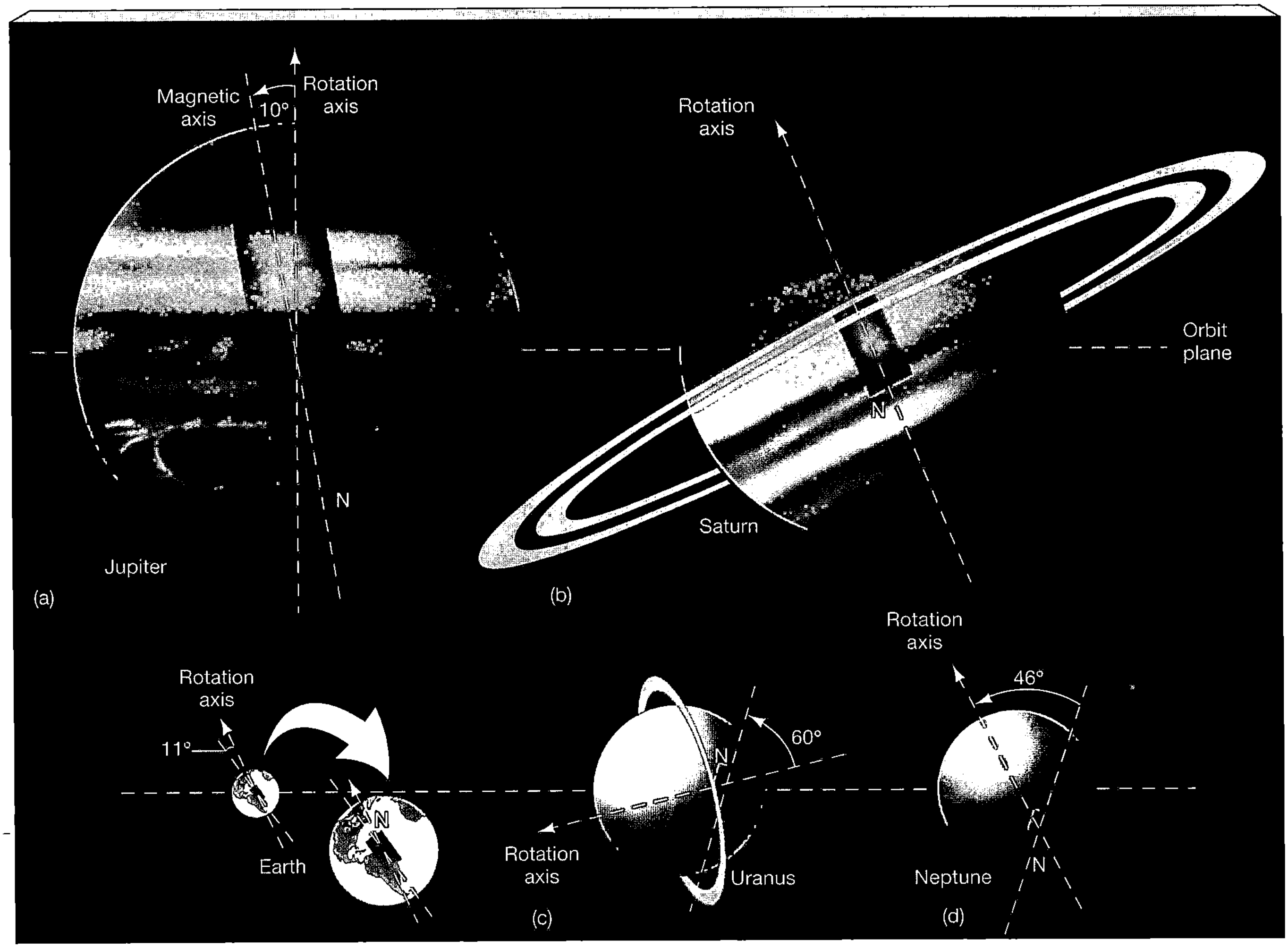

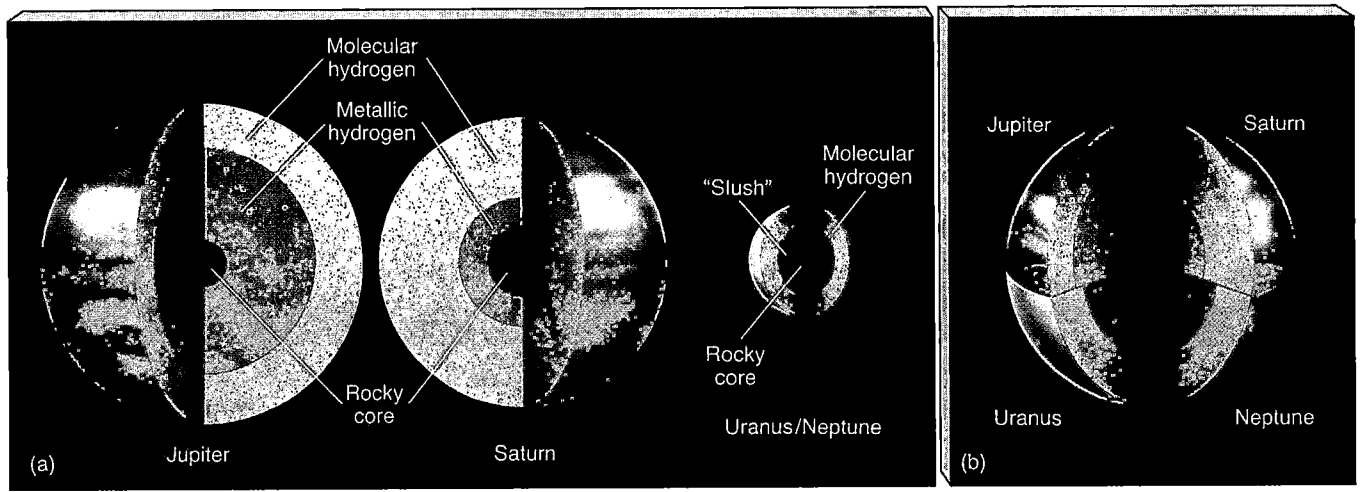

Figure 13.9 Jovian Magnetic Fields A comparison of the magnetic field strengths, orientations, and offsets in the four jovian planets: (a) Jupiter, (b) Saturn, (c) Uranus, (d) Neptune. The planets are drawn to scale, and in each case the magnetic field is represented as though it came from a simple bar magnet. The size and location of each magnet represent the strength and orientation of the planetary field. Notice that the fields of Uranus and Neptune are significantly offset from the center of the planet and are significantly inclined to the planet's rotation axis. Earth's magnetic field is shown for comparison. One end of each magnet is marked N to indicate the polarity of Earth's field.

Figure 13.10 Jovian Interiors A comparison of the interior structures of the four jovian planets. (a) The planets drawn to scale. (b) The relative proportions of the various internal zones.

13.6 The Moon Systems of Uranus and Neptune

3 Like Jupiter and Saturn, both Uranus and Neptune have extensive moon systems, each consisting of a few large moons, long known from ground-based observations, and many smaller moonlets discovered by *Voyager 2* or recently detected from Earth.

URANUS'S MOONS

William Herschel discovered and named the two largest of Uranus's five major moons, Titania and Oberon, in 1789. British astronomer William Lassell found the next largest, Ariel and Umbriel, in 1851. Gerard Kuiper found the smallest, Miranda, in 1948. In order of increasing distance from the planet, they are Miranda (at 5.1 planetary radii), Ariel (7.5), Umbriel (10.4), Titania (17.1), and Oberon (22.8). Ten smaller moons discovered by *Voyager 2* all lie inside the orbit of Miranda. Many of them are intimately related to Uranus's ring system. All these moons revolve in the planet's skewed equatorial plane, almost perpendicular to the ecliptic, in circular, tidally locked orbits, sharing their parent's extreme seasons. The properties of the 17 named moons of Uranus are listed in Table 13.1.

The last two moons listed in the table were discovered in September 1997. Interestingly, they were found not by visiting spacecraft (there have been none since *Voyager 2*, and none are planned), nor by the latest generation of telescopes, such as *Hubble*, Keck, or the VLT. Instead, these newest satellites, now named Caliban and Sycorax, were discovered by astronomers using one of the oldest instruments still in regular use—the 5-m Hale Telescope on Mount Palomar in California. (Sec. 5.1) Their diameters are just 80 km and 160 km, respectively. At the time of their discovery, they were the faintest moons ever imaged with a ground-based telescope.

Since 1997, four even fainter, and still unnamed, moons have been discovered, bringing the total to 21. One, orbiting close to the planet, was found after careful reanalysis of *Voyager 2* images. The others, much farther out (over 6 million km from Uranus), were detected by

TABLE 13.1 The Moons of Uranus*

NAME	DISTANCE FROM URANUS (km)	(planet radii)	ORBIT PERIOD (days)	SIZE (longest diameter, km)	MASS** (Earth Moon masses)	DENSITY (kg/m^3)	(g/cm^3)
Cordelia	49,800	1.95	0.34	26			
Ophelia	53,800	2.10	0.38	32			
Bianca	59,200	2.31	0.43	44			
Cressida	61,800	2.42	0.46	66			
Desdemona	62,700	2.45	0.47	58			
Juliet	64,400	2.52	0.49	84			
Portia	66,100	2.59	0.51	110			
Rosalind	69,900	2.74	0.56	58			
Belinda	75,300	2.94	0.62	68			
Puck	86,000	3.36	0.76	150			
Miranda	130,000	5.08	1.41	480	0.00090	1100	1.1
Ariel	191,000	7.48	2.52	1160	0.018	1600	1.6
Umbriel	266,000	10.4	4.14	1170	0.016	1400	1.4
Titania	436,000	17.1	8.71	1580	0.048	1700	1.7
Oberon	583,000	22.8	13.5	1520	0.041	1600	1.6
Caliban† (S/1997U1)	5,700,000	223	400	80			
Sycorax† (S/1997U2)	5,750,000	225	−415‡	160			

**Does not include the four recently discovered small moons described in the text.*

***Mass of Earth's Moon = 7.4×10^{22} kg = 8.5×10^{-4} Uranus masses.*

†Orbital parameters not well determined.

‡Indicates a retrograde orbit.

techniques similar to those that have proved so successful in identifying new moons of Jupiter and Saturn. ∞ (Sec. 11.5, 12.5) The five outlying moons move on retrograde, highly inclined orbits. Like the outer moons of Jupiter and Saturn, and Phobos and Deimos of Mars, they are most likely debris captured from interplanetary space following a grazing encounter with the planet's atmosphere.

The five largest Uranian moons are similar in many respects to the six midsized moons of Saturn. ∞ (Sec. 12.5) Their densities lie in the range 1100–1700 kg/m³, suggesting a composition of ice and rock, like Saturn's moons, and their diameters range from 1600 km for Titania and Oberon, to 1200 km for Umbriel and Ariel, to 480 km for Miranda. Uranus has no moons comparable to the Galilean satellites of Jupiter, nor to Saturn's single large moon, Titan. Figure 13.11 shows Uranus's five large moons to scale, along with Earth's Moon and Neptune's only midsized moon (named Proteus) for comparison.

The outermost of these moons, Titania and Oberon, are heavily cratered and show little indication of geological activity. Their overall appearance (and quite possibly their history) is comparable to that of Saturn's moon Rhea, except that they lack Rhea's wispy streaks. Also, like all Uranian moons, they are considerably less reflective than Saturn's satellites, suggesting that their icy surfaces are quite dirty.

One possible reason for this may simply be that the planetary environment in the vicinity of Uranus and Neptune contains more small "sooty" particles than does the solar system closer to the Sun. An alternative explanation, now considered more likely by many planetary scientists, cites the effects of radiation and high-energy particles that strike the surfaces of these moons. These impacts tend to break up the molecules on the moons' surfaces, eventually leading to chemical reactions that slowly build up a layer of dark, organic material. This **radiation dark-ening** is thought to contribute to the generally darker coloration of many of the moons and rings in the outer solar system. In either case, the longer a moon has been inactive and untouched by meteoritic impact, the darker its surface should be.

The darkest of the moons of Uranus is Umbriel. It displays little evidence of any past surface activity; its only mark of distinction is a bright spot about 30 km across, of unknown origin, in its northern hemisphere. By contrast, Ariel, similar in size to Umbriel but closer to Uranus, does appear to have experienced some activity in the past. It shows signs of resurfacing in places and exhibits surface cracks a little like those seen on another of Saturn's moons, Tethys. However, unlike Tethys, whose cracks are probably due to meteoritic impact, Ariel's activity probably occurred when internal forces and external tidal stresses (due to the gravitational pull of Uranus) distorted the moon and cracked its surface.

Strangest of all Uranus's icy moons is Miranda, shown in Figure 13.12. Before the *Voyager 2* encounter, astronomers expected that Miranda would resemble Mimas, the moon of Saturn whose size and location it most closely approximates. However, instead of being a relatively uninteresting cratered, geologically inactive world, Miranda displays a wide range of surface terrains, including ridges, valleys, large oval faults, and many other tortuous geological features.

To explain why Miranda seems to combine so many different types of surface features, some researchers have hypothesized that this baffling object has been catastrophically disrupted several times (from within or without), with the pieces falling back together in a chaotic, jumbled way. Certainly, the frequency of large craters on the outer moons suggests that destructive impacts may once have been quite common in the Uranus system. It will be a long time, though, before we can obtain more detailed information to test this theory.

TABLE 13.2 The Moons of Neptune

NAME	DISTANCE FROM NEPTUNE (km)	(planet radii)	ORBIT PERIOD (days)	SIZE (longest diameter km)	MASS* (Earth Moon masses)	DENSITY (kg/m³)	(g/cm³)
Naiad	48,200	1.95	0.29	58			
Thalassa	50,100	2.02	0.31	80			
Despina	52,500	2.12	0.33	150			
Galatea	62,000	2.50	0.43	160			
Larissa	73,500	2.97	0.55	210			
Proteus	118,000	4.75	1.12	440			
Triton	355,000	14.3	−5.88†	2710	0.292	2100	2.1
Nereid	5,510,000	223	360	340	0.0000034	1200	1.2

*Mass of Earth's Moon = 7.4×10^{22} kg = 7.3×10^{-4} Neptune masses.
†Indicates a retrograde orbit.

Figure 13.11 Moons of Uranus and Neptune The five largest moons of Uranus, and Proteus, the sole midsized moon of Neptune, to scale. Earth's moon is shown for comparison. In order of increasing distance from the planet, the Uranian moons are Miranda, Ariel, Umbriel, Titania, and Oberon. The appearance, structure, and history of Titania and Oberon may be quite similar to those of Saturn's moon, Rhea. The smallest details visible on both moons are about 15 km across. Umbriel is one of the darkest bodies in the solar system, although it has a bright white spot on its sunward side. Ariel is similar in size but has a brighter surface, with signs of past geological activity. Resolution is approximately 10 km. *(NASA; Lick Observatory)*

NEPTUNE'S MOONS

From Earth we can see only two moons orbiting Neptune. William Lassell discovered the inner moon, Triton, in 1846. The outer moon, Nereid, was located by Gerard Kuiper in 1949. *Voyager 2* discovered six additional moons, all less than a few hundred kilometers across and all lying within Nereid's orbit. Neptune's known moons are listed in Table 13.2. The planet's only midsized moon (by our previous definition) is shown in Figure 13.11. ⚏ (Sec. 12.5) In its moons we find Neptune's contribution to our

list of solar system peculiarities. Unlike the other jovian worlds, Neptune has no regular moon system—that is, no moons on roughly circular, equatorial, prograde orbits. The largest moon, Triton, is 2700 km in diameter and occupies a circular retrograde orbit 355,000 km (14.3 planetary radii) from the planet, inclined at about 20° to Neptune's equatorial plane. It is the only large moon in our solar system to have a retrograde orbit. The other moon visible from Earth, Nereid, is only 340 km across. It orbits Neptune in the prograde sense, but on an elongated trajectory that brings it as close as 1.4 million km to the

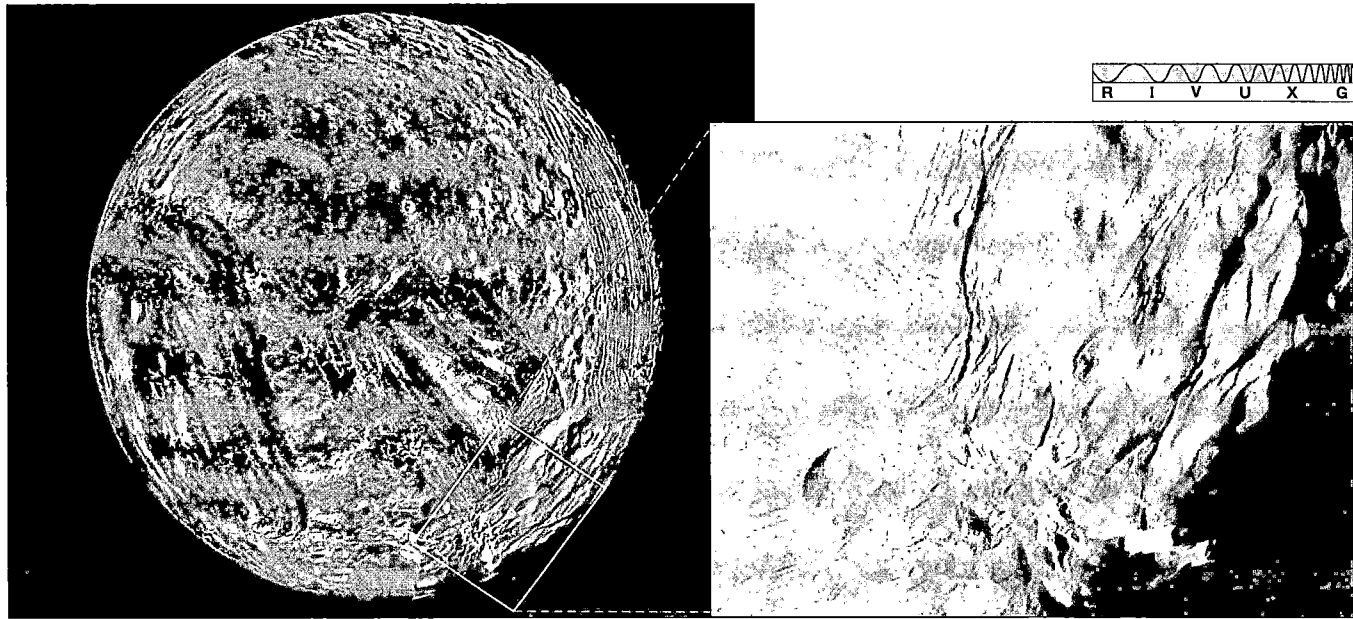

Figure 13.12 Miranda An asteroid-sized moon of Uranus photographed by *Voyager 2*, Miranda has a strange, fractured surface suggestive of a violent past, but the cause of the grooves and cracks is currently unknown. The resolution here is about 1 km. The long "canyon" near the bottom of the inset is about 20 km deep. *(NASA)*

planet and as far away as 9.7 million km. Nereid is probably similar in both size and composition to Neptune's small inner moons.

Voyager 2 approached to within 24,000 km of Triton's surface, providing us with virtually all that we now know about that distant, icy world. Astronomers redetermined the moon's radius (which was corrected downward by about 20 percent) and measured its mass for the first time. Along with Saturn's Titan and the four Galilean moons of Jupiter, Triton is one of the six large moons in the outer solar system. Triton is the smallest of them, with about half the mass of the next smallest, Jupiter's Europa.

Lying 4.5 billion km from the Sun, and with a fairly reflective surface, Triton has a surface temperature of just 37 K. It has a tenuous nitrogen atmosphere, perhaps a hundred thousand times thinner than Earth's, and a surface that most likely consists primarily of water ice. A *Voyager 2* mosaic of Triton's south polar region is shown in Figure 13.13. The moon's low temperatures produce a layer of nitrogen frost that forms and evaporates over the polar caps, a little like the carbon dioxide frost responsible for the seasonal caps on Mars. The frost is visible as the pinkish region on the right of the figure.

Overall, there is a marked lack of cratering on Triton, presumably indicating that surface activity has obliterated the evidence of most impacts. There are many other signs of an active past. Triton's face is scarred by large fissures similar to those seen on Ganymede, and the moon's odd cantaloupe-like terrain may indicate repeated faulting and deformation over the moon's lifetime. In addition, Triton

has numerous frozen "lakes" of water ice (Figure 13.14), which are believed to be volcanic in origin.

Triton's surface activity is not just a thing of the past. As *Voyager 2* passed the moon, its cameras detected two

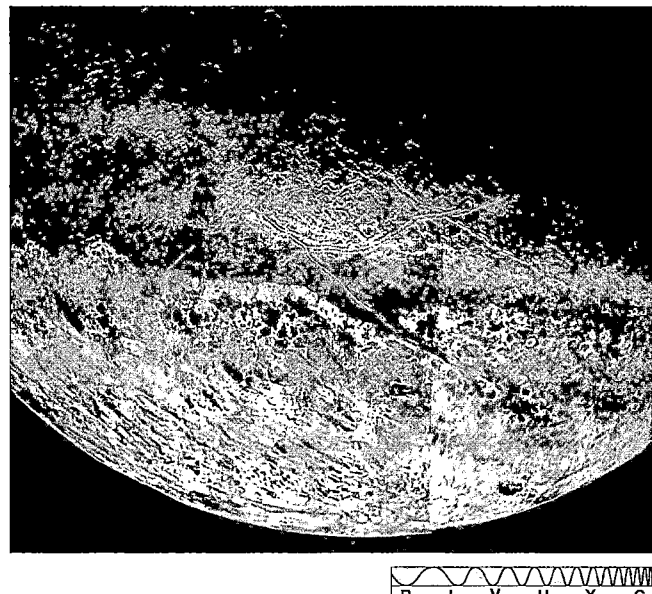

Figure 13.13 Triton The south polar region of Triton, showing a variety of terrains, ranging from deep ridges and gashes to what appear to be frozen water lakes, all indicative of past surface activity. The pinkish region at the right is nitrogen frost, forming the moon's polar cap. Resolution is about 4 km. The long black streaks at bottom left were probably formed by liquid nitrogen geysers on the surface. *(NASA)*

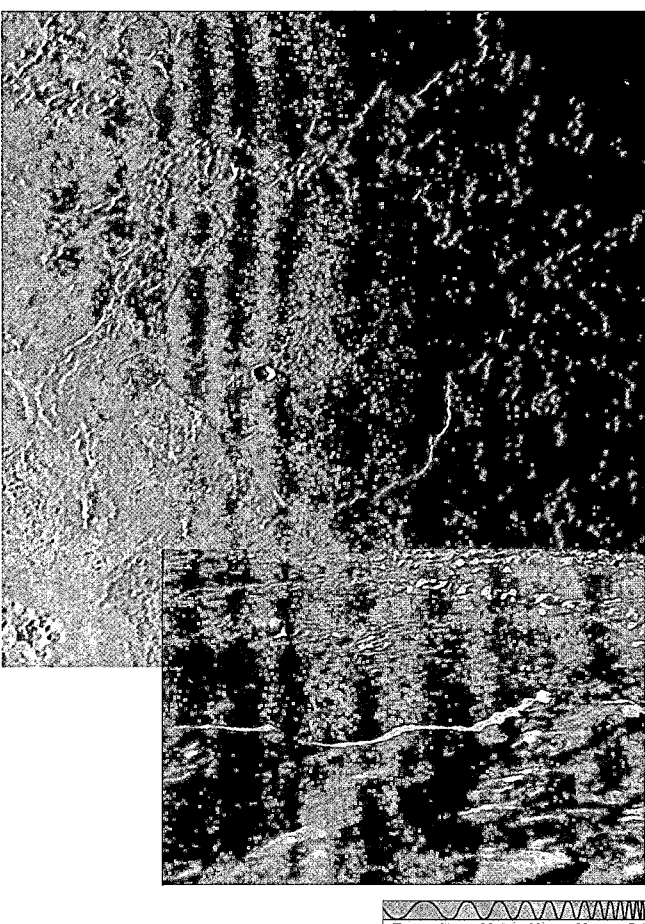

Figure 13.14 Water Ice on Triton Scientists believe that this lakelike feature on Triton may have been caused by the eruption of an ice volcano. The water "lava" has since solidified, leaving a smooth surface. The absence of craters indicates that this eruption was a relatively recent event in Triton's past. The nearly circular feature at the center of this image spans some 200 km in diameter; its details are resolved to a remarkable 1 km. The inset is a computer-generated view illustrating the topographic relief of the same area. *(NASA)*

great jets of nitrogen gas erupting from below the surface and rising several kilometers into the sky. It is thought that these "geysers" result when liquid nitrogen below Triton's surface is heated and vaporized by some internal energy source, or perhaps even by the Sun's feeble light. Vaporization produces high pressure, which forces the gas through cracks and fissures in the crust, creating the displays *Voyager 2* saw. Scientists conjecture that nitrogen geysers may be very common on Triton and are perhaps responsible for much of the moon's thin atmosphere. The long black streaks at the bottom left of Figure 13.13 may have formed when geysers carried dark, carbon-rich material from the moon's interior to the surface. Surface winds in Triton's thin atmosphere may also play a role in spreading the material over the surface.

PLANETARY DATA: Neptune

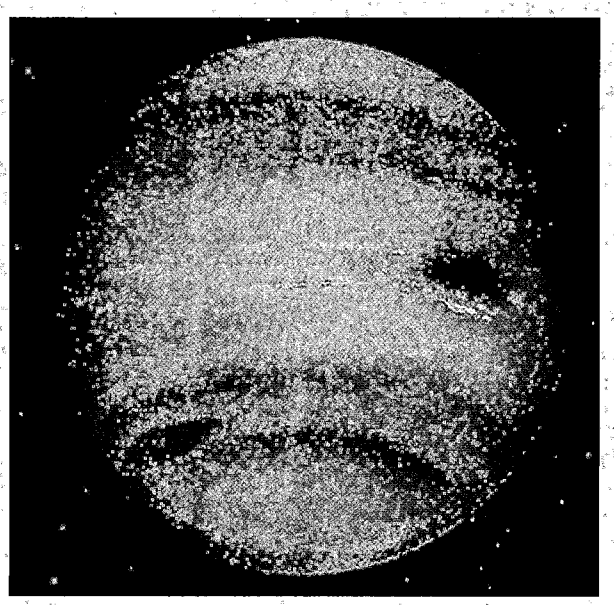

Orbital semimajor axis	30.07 A.U.
	4498 million km
Orbital eccentricity	0.009
Perihelion	29.81 A.U.
	4460 million km
Aphelion	30.33 A.U.
	4537 million km
Mean orbital speed	5.43 km/s
Sidereal orbital period	163.7 tropical years
Synodic orbital period	367.49 solar days
Orbital inclination to the ecliptic	1.77°
Greatest angular diameter, as seen from Earth	2.4″
Mass	1.02×10^{26} kg
	17.15 (Earth = 1)
Equatorial radius	24,766 km
	3.88 (Earth = 1)
Mean density	1638 kg/m^3
	0.297 (Earth = 1)
Surface gravity (at cloud tops)	11.14 m/s^2
	1.14 (Earth = 1)
Escape speed	23.5 km/s
Sidereal rotation period	0.67 solar days
Axial tilt	29.6°
Surface magnetic field	0.43 (Earth = 1)
Magnetic axis tilt relative to rotation axis	46.0°
Mean surface temperature	59 K
Number of Moons	8

The event or events that placed Triton on a retrograde orbit and Nereid on such an eccentric path are unknown, but they are the subject of considerable speculation. Triton's peculiar orbit and surface features suggest to some astronomers that the moon did not form as part of the Neptune system but instead was captured, perhaps not too long ago. Other astronomers, basing their views on Triton's chemical composition, maintain that it formed as a "normal" moon but was later kicked into its abnormal orbit by some catastrophic event, such as an interaction with another similar-sized body. It has even been suggested that the planet Pluto may have played a role in this process (see Section 13.10), although no really convincing demonstration of such an encounter has ever been presented.

The surface deformations on Triton certainly suggest fairly violent and relatively recent events in the moon's past. However, they were most likely caused by the tidal stresses produced in Triton as Neptune's gravity circularized its orbit and synchronized its spin, and they give little indication of the processes responsible for the orbit.

Whatever its past, Triton's future is fairly clear. Because of its retrograde orbit, the tidal bulge Triton raises on Neptune tends to make the moon spiral *toward* the planet rather than away from it (as our Moon moves away from Earth). ∞ (Sec. 7.6) Thus, Triton is doomed to be torn apart by Neptune's tidal gravitational field, probably in no more than 100 million years or so, the time required for the moon's inward spiral to bring it inside Neptune's Roche limit. ∞ (Sec. 12.4) By that time, it is conceivable that Saturn's ring system may have disappeared, so Neptune will then be the planet in the solar system with spectacular rings!

☑ Concept Check

▪ Why is Triton much less heavily cratered than the other moons of Uranus and Neptune?

13.7 The Rings of the Outermost Jovian Planets

THE RINGS OF URANUS

⬛ All the jovian planets have rings. The ring system surrounding Uranus was discovered in 1977, when astronomers observed it passing in front of a bright star, momentarily dimming its light. Such a **stellar occultation** (Figure 13.15) happens a few times per decade and allows astronomers to measure planetary structures that are too small and faint to be detected directly. The 1977 observation was actually aimed at studying the planet's atmosphere by watching how it absorbed starlight. However, 40 minutes before and after Uranus itself occulted the star, the flickering starlight revealed the presence of a set of rings. The discovery was particularly exciting because, at the

time, only Saturn was known to have rings. Jupiter's rings went unseen until *Voyager 1* arrived there in 1979, and those of Neptune were unambiguously detected only in 1989, by *Voyager 2*.

The ground-based observations revealed the presence of a total of nine thin rings. The main rings, in order of increasing radius, are named Alpha, Beta, Gamma, Delta, and Epsilon, and they range from 44,000 to 51,000 km from the planet's center. All lie within the Roche limit of Uranus, which is about 62,000 km from the planet's center. A fainter ring, known as the Eta ring, lies between the Beta and Gamma rings, and three other faint rings, known as 4, 5, and 6, lie between the Alpha ring and the planet itself. In 1986, *Voyager 2* discovered two more even fainter rings, one between Delta and Epsilon and one between ring 6 and Uranus. The main rings are shown in Figure 13.16. More details on the rings are provided in Table 13.3.

The rings of Uranus are quite different from those of Saturn. Whereas Saturn's rings are bright and wide with relatively narrow gaps between them, the rings of Uranus are dark, narrow, and widely spaced. With the exception of the Epsilon ring and the diffuse innermost ring, the rings of Uranus are all less than about 10 km wide, and the spacing between them ranges from a few hundred to about a thousand kilometers. However, like Saturn's rings, all Uranus's rings are less than a few tens of meters thick (that is, measured in the direction perpendicular to the ring plane).

The density of particles within the rings themselves is comparable to that found in Saturn's A and B rings. The particles that make up Saturn's rings range in size from

---

TABLE 13.3 The Rings of Uranus

RING	INNER RADIUS (km)	(planet radii)	OUTER RADIUS* (km)	(planet radii)	WIDTH (km)
1986U2R	37,000	1.45	39,500	1.55	2500
6	41,800	1.64			2
5	42,200	1.65			2
4	42,600	1.67			3
Alpha	44,700	1.75			4–10
Beta	45,700	1.79			5–11
Eta	47,200	1.83			2
Gamma	47,600	1.86			1–4
Delta	48,300	1.90			3–7
1986U1R	50,000	1.96			2
Epsilon	51,200	2.00			20–100

*Most of Uranus's rings are so thin that there is little difference between their inner and outer radii.

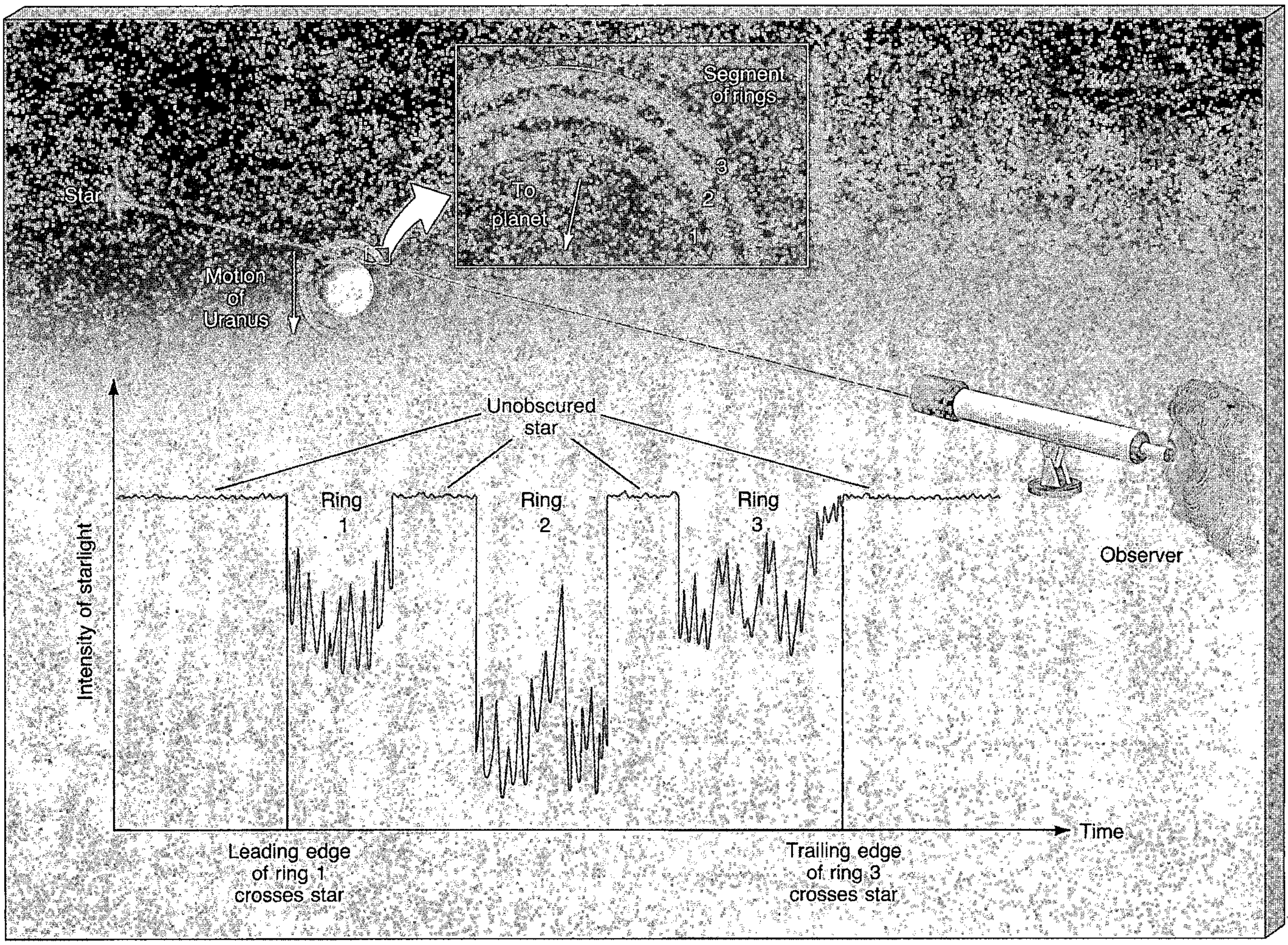

Figure 13.15 **Occultation of Starlight** This technique allows astronomers to detect fine detail on a distant planet. The rings of Uranus were discovered using this system.

dust grains to boulders, but in the case of Uranus, the particles show a much smaller spread—few if any are smaller than a centimeter or so in diameter. The ring particles are also considerably less reflective than Saturn's ring particles, possibly because they are covered with the same dark material as Uranus's moons. The Epsilon ring (shown in detail in Figure 13.17) exhibits properties a little like those of Saturn's F ring. It is slightly eccentric (its eccentricity is 0.008) and of variable width, although no braids were found. It also appears to be composed of ringlets.

Like the F ring of Saturn, Uranus's narrow rings require shepherding satellites to keep them from diffusing away. In fact, the theory of shepherd satellites was first

worked out to explain the rings of Uranus, which had been detected by stellar occultation even before *Voyager 2*'s Saturn encounter. Thus, the existence of the F ring did not come as quite such a surprise as it might have otherwise!

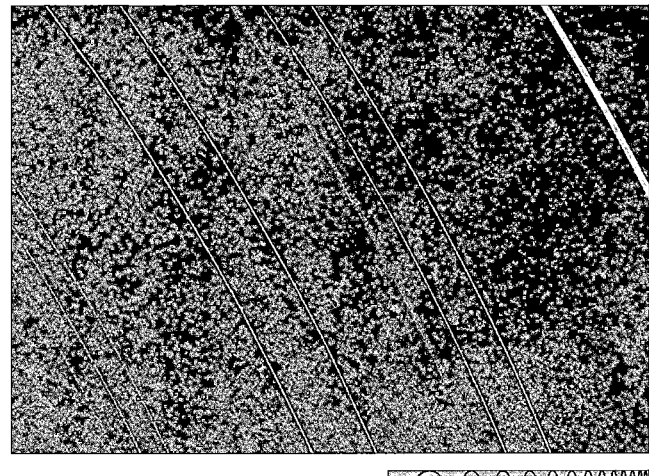

Figure 13.16 **Uranus's Rings** The main rings of Uranus, as imaged by *Voyager 2*. All the rings known before *Voyager 2*'s arrival can be seen in this photo. From the inside out, they are 6, 5, 4, Alpha, Beta, Eta, Gamma, Delta, and Epsilon. Resolution is about 10 km, which is just about the width of most of these rings. The two rings discovered by *Voyager 2* are too faint to be seen here. *(NASA)*

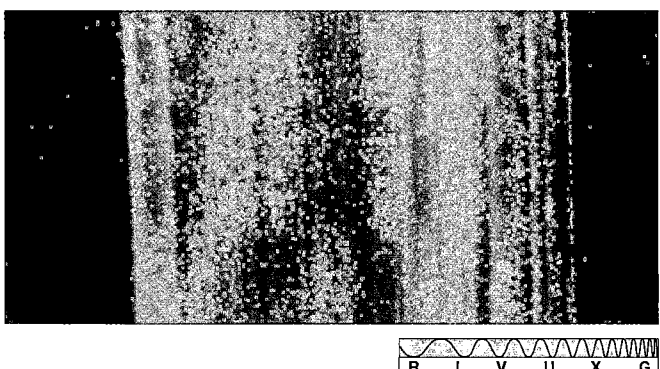

Figure 13.17 Epsilon Ring A close-up of Uranus's Epsilon ring, showing some of its internal structure. The width of the ring averages 30 km; special image processing has magnified the resolution to about 100 m. *(NASA)*

Presumably, many of the small inner satellites of Uranus play some role in governing the appearance of the rings. *Voyager 2* detected the shepherds of the Epsilon ring, Cordelia and Ophelia (see in Figure 13.18). Many other, undetected, shepherd satellites must also exist.

THE RINGS OF NEPTUNE

As shown in Figure 13.19 and presented in more detail in Table 13.4, Neptune is surrounded by five dark rings. Three are quite narrow, like the rings of Uranus; the other two are broad and diffuse, more like Jupiter's ring. The dark coloration probably results from radiation darkening, as discussed earlier in the context of the moons of Uranus. All the rings lie within Neptune's Roche limit. The outermost (Adams) ring is noticeably clumped in places. From Earth we see not a complete ring but only partial arcs—the unseen parts of the ring are simply too thin (unclumped) to

be detected. The connection between the rings and the planet's small inner satellites has not yet been firmly established, but many astronomers believe that the clumping is caused by shepherd satellites.

Although all the jovian worlds have ring systems, the rings themselves differ widely from planet to planet. Is there some "standard" way in which rings form around a planet? And is there a standard manner in which ring systems evolve? Or do the processes of ring formation and evolution depend entirely on the particular planet in question? If, as now appears to be the case, ring systems are relatively short-lived, their formation must be a fairly common event. Otherwise, we would not expect to find rings around all four jovian planets at once. There are also many indications that the individual planetary environment plays an important role in determining a ring system's appearance and longevity. Although many aspects of ring formation and evolution are now understood, it must be admitted that no comprehensive theory yet exists.

✔️ **Concept Check**

▪ What does the Epsilon ring of Uranus have in common with the F ring of Saturn?

TABLE 13.4 The Rings of Neptune

RING	INNER RADIUS (km)	(planet radii)	OUTER RADIUS* (km)	(planet radii)	WIDTH (km)
Galle (1989N3R)	40,900	1.65	42,900	1.73	2000
Leverrier (1989N2R)	53,200	2.15			100
Lassell (1989N4R)†	53,200	2.15	57,200	2.31	4000
Arago (1989N4R)†	57,200	2.31			100
Adams (1989N1R)	62,900	2.54			50

*Three of Neptune's rings are so thin that there is little difference between their inner and outer radii.
†Lassell and Arago were originally identified as a single ring.

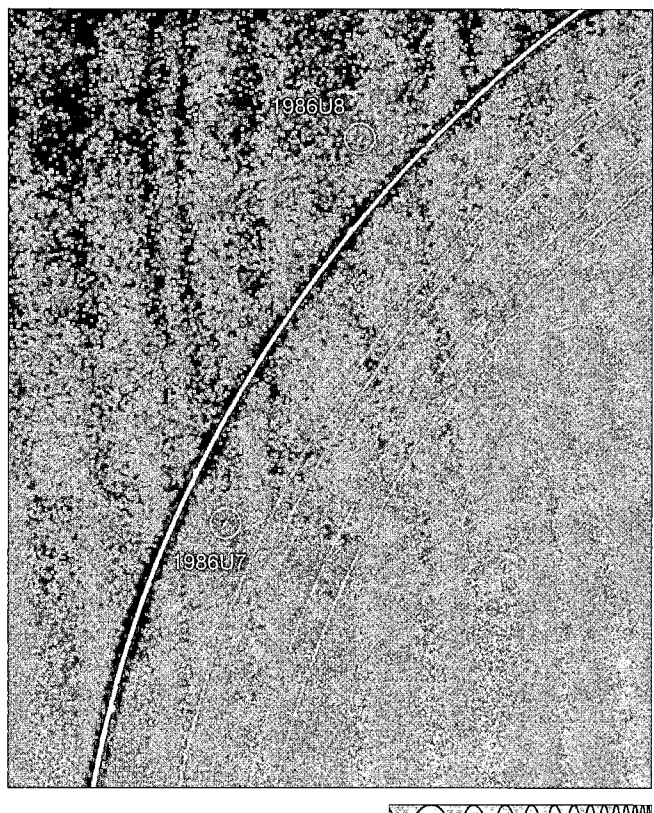

Figure 13.18 Uranian Shepherd Moons These two small moons, discovered by *Voyager 2* in 1986 and now named Cordelia (U7) and Ophelia (U8), tend to "shepherd" the Epsilon ring, keeping it from diffusing away into space. *(NASA)*

Figure 13.19 Neptune's Faint Rings In this long-exposure image, Neptune (center) is heavily overexposed and has been artificially blotted out to make the rings easier to see. One of the two faint rings lies between the inner bright ring (Leverrier) and the planet. The others lie between the Leverrier ring and the outer bright ring (known as the Adams ring). (NASA)

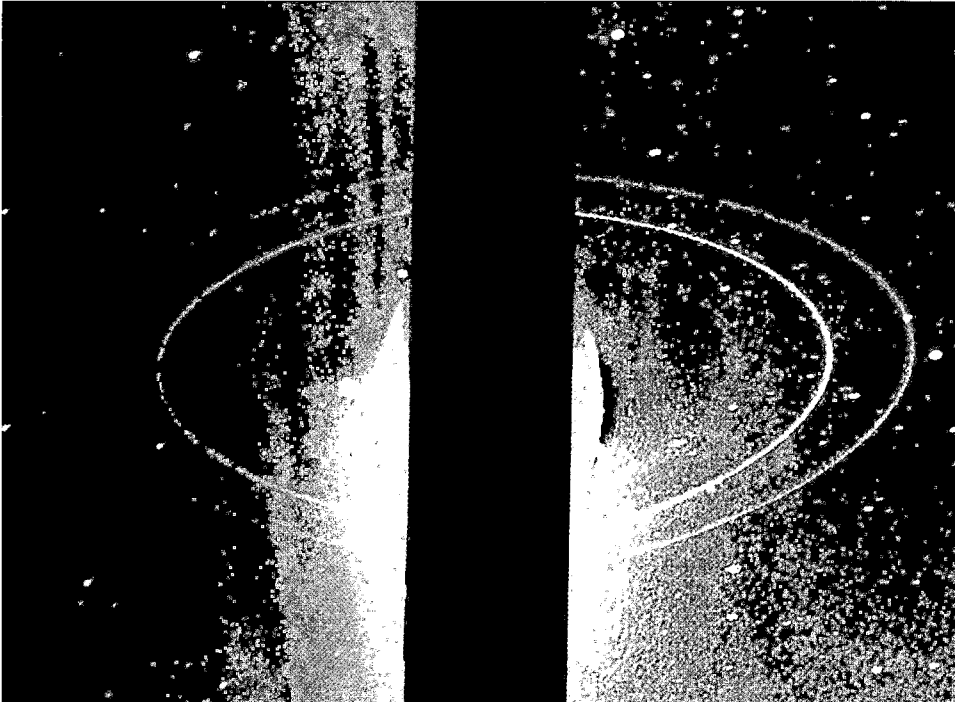

13.8 The Discovery of Pluto

By the end of the nineteenth century, observations of the orbits of Uranus and Neptune suggested that Neptune's influence was not sufficient to account for all the irregularities in Uranus's motion. Furthermore, it seemed that Neptune itself might be affected by some other unknown body. Following their success in the discovery of Neptune, astronomers hoped to pinpoint the location of this new planet using similar techniques. One of the most ardent searchers was Percival Lowell, a capable, persistent observer and one of the best-known astronomers of his day. (Recall that he was also the leading proponent of the theory that the "canals" on Mars were constructed by an intelligent race of Martians—see *Discovery 10-1.*)

Basing his investigation primarily on the motion of Uranus (Neptune's orbit was still relatively poorly determined at the time), and using techniques similar to those developed earlier for Neptune by Adams and Leverrier, Lowell set about calculating where the supposed ninth planet should be. He searched for it, without success, during the decade preceding his death in 1916. Not until 14 years later did American astronomer Clyde Tombaugh, working with improved equipment and photographic techniques at the Lowell Observatory, finally succeed in finding Lowell's ninth planet, only 6° away from Lowell's predicted position. The new planet was named Pluto for the Roman god of the dead who presided over eternal darkness (and also because its first two letters and its astrological symbol ℙ are Lowell's initials). Its discovery was announced on March 13, 1930, Percival Lowell's birthday (and also the anniversary of Herschel's discovery of Uranus).

On the face of it, the discovery of Pluto looked like another spectacular success for celestial mechanics. Unfortunately, it now appears that the supposed irregularities in the motions of Uranus and Neptune did not exist and that the mass of Pluto, not measured accurately until the 1980s, is far too small to have caused them anyway. The discovery of Pluto owed much more to simple luck than to elegant mathematics!

Some orbital and physical data for Pluto are presented in the Pluto Data box on p. 347. Unlike the paths of the other outer planets, Pluto's orbit is quite elongated, with an eccentricity of 0.25. It is also inclined at 17.2° to the plane of the ecliptic. Here already we have some indication that Pluto is unlike its jovian neighbors. Because of its substantial orbital eccentricity, Pluto's distance from the Sun varies considerably. At perihelion, it lies 29.7 A.U. (4.4 billion km) from the Sun, inside the orbit of Neptune. At aphelion, the distance is 49.3 A.U. (7.4 billion km), well outside Neptune's orbit. Pluto last passed perihelion in 1989, and remained inside Neptune's orbit until February, 1999. Its sidereal period is 248.0 years, so the next perihelion passage will not occur until the middle of the 23rd century.

Pluto's orbital period is apparently exactly 1.5 times that of Neptune—the two planets are locked into a 3:2 resonance (two orbits of Pluto for every three of Neptune) as they orbit the Sun. As a result, even though their orbits appear to cross, Pluto and Neptune are in no danger of colliding with each other. Because of the orbital resonance and Pluto's tilted orbit plane, the distance between the two planets at closest

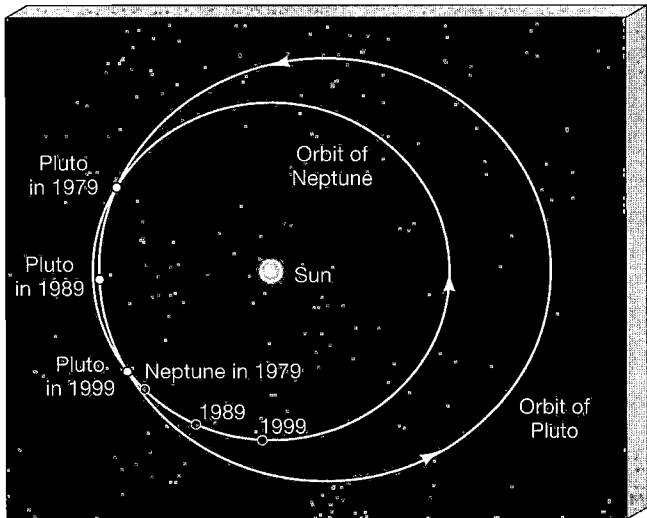

Figure 13.20 Neptune and Pluto The orbits of Neptune and Pluto cross, although Pluto's orbital inclination and a 3:2 resonance prevent the planets from actually coming close to each other. Between 1979 and 1999, Pluto was inside Neptune's orbit, making Neptune the most distant planet from the Sun.

approach is actually about 17 A.U. (compare with Pluto's closest approach to Uranus of just 11 A.U.). The orbits of Neptune and Pluto are sketched in Figure 13.20.

At nearly 40 A.U. from the Sun, Pluto is often hard to distinguish from the background stars. As the two photographs of in Figure 13.21 indicate, the planet is actually considerably fainter than many stars in the sky. Like Neptune, it is never visible to the naked eye.

13.9 Physical Properties of Pluto

Pluto is so far away that little is known of its physical nature. Until the late 1970s, studies of sunlight reflected from its surface suggested a rotation period of nearly a week, but measurements of its mass and radius were very uncertain. All this changed in July 1978, when astronomers at the U.S. Naval Observatory discovered that Pluto has a satellite. It is now named Charon, after the mythical boatman who ferried the dead across the river Styx into Hades, Pluto's domain. The discovery photograph of Charon is shown in Figure 13.22(a). Charon is the small bump near the top of the image. Knowing the moon's orbital period of 6.4 days, astronomers could determine the mass of Pluto to much greater accuracy than had previously been possible. It is 0.0021 Earth masses $(1.3 \times 10^{22}$ kg), far smaller than any earlier estimate—more like the mass of a moon than of a planet. In 1990, the *Hubble Space Telescope* imaged the Pluto–Charon system (Figure 13.22b). The improved resolution of that instrument clearly resolved the two bodies and allowed even more accurate measurements of their properties.

The discovery of Charon also allowed astronomers to measure Pluto's radius very precisely. Pluto's angular size is much less than 1″, so its true diameter is blurred by the effects of Earth's turbulent atmosphere. ∞ (Sec. 5.3) However, Charon's orbit has given astronomers new insight into the system. By pure chance, Charon's orbit over the 6-year period from 1985 to 1991 (less than 10 years after the moon was discovered) happened to be oriented in such a way that viewers on Earth saw a series of eclipses. Pluto and Charon repeatedly passed in front of each other, as

Figure 13.21 Motion of Pluto These two photographs, taken one night apart, show motion of the planet Pluto (arrow) against a field of much more distant stars. Most of Pluto's apparent motion in these two frames is actually due to the orbital motion of Earth rather than that of Pluto. *(UC/Lick Observatory)*

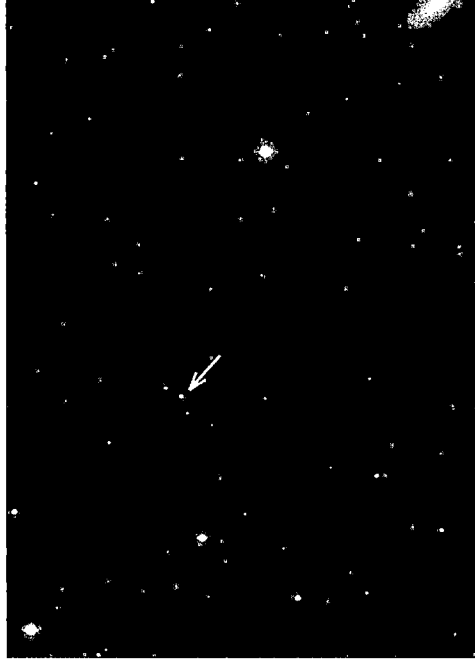

R I V U X G

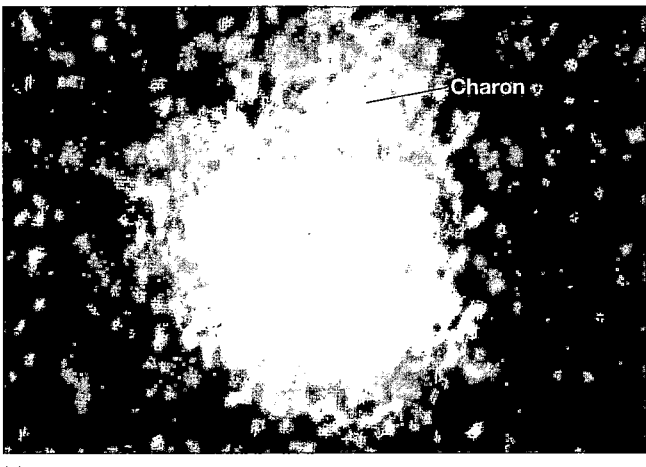

(a)

(b)

Figure 13.22 Pluto and Charon (a) The Animation discovery photograph of Pluto's moon, Charon. The moon is the small bump on the top right portion of the image. (b) The Pluto–Charon system, to the same scale, as seen by the *Hubble Space Telescope*. The angular separation of the planet and its moon is about 0.9″. *(US Naval Observatory; NASA)*

seen from our vantage point. Figure 13.23 sketches this orbital configuration. With more good fortune, these eclipses took place while Pluto was closest to the Sun, making for the best possible Earth-based observations.

Basing their calculations on the variations in reflected light as Pluto and Charon periodically hid each other, astronomers computed their masses and radii and determined their orbit plane. Additional studies of Pluto's surface brightness indicate that the two are tidally locked as they orbit each other. Pluto's diameter is 2270 km, about one-fifth the size of Earth. Charon is about 1300 km across and orbits at a distance of 19,700 km from

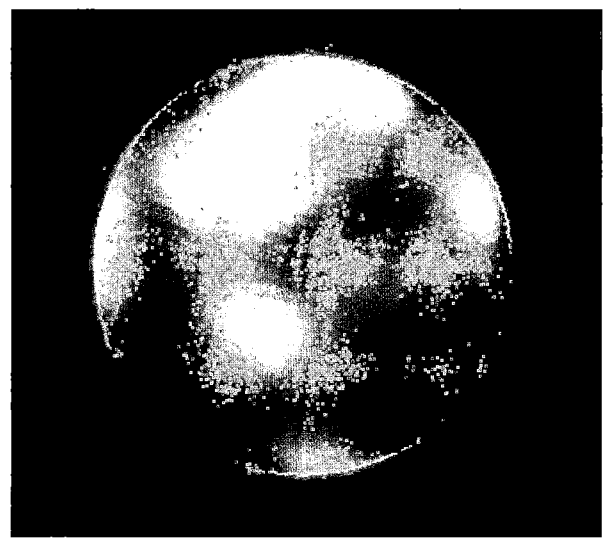

Orbital semimajor axis	39.48 A.U. 5906 million km
Orbital eccentricity	0.249
Perihelion	29.66 A.U. 4437 million km
Aphelion	49.31 A.U. 7376 million km
Mean orbital speed	4.74 km/s
Sidereal orbital period	248.0 tropical years
Synodic orbital period	366.72 solar days
Orbital inclination to the ecliptic	17.15°
Greatest angular diameter, as seen from Earth	0.11″
Mass	1.27×10^{22} kg 0.0021 (Earth = 1)
Equatorial radius	1137 km 0.18 (Earth = 1)
Mean density	2060 kg/m³ 0.374 (Earth = 1)
Surface gravity	0.66 m/s² 0.067 (Earth = 1)
Escape speed	1.2 km/s
Sidereal rotation period	−6.387 solar days (retrograde)
Axial tilt	118°
Surface magnetic field	unknown
Magnetic axis tilt relative to rotation axis	—
Surface temperature	40 to 60 K
Number of moons	1

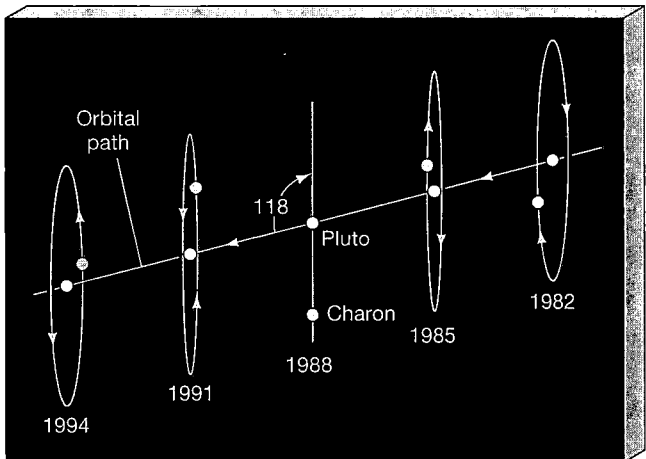

Figure 13.23 Pluto–Charon Eclipses The orbital orientation of Charon produced a series of eclipses between 1985 and 1991. Observations of eclipses of Charon by Pluto and of Pluto by Charon have provided detailed information about the sizes and orbits of both bodies.

Pluto. If planet and moon have the same composition (probably a reasonable assumption), Charon's mass must be about one-sixth that of Pluto, giving the Pluto–Charon system by far the largest satellite-to-planet mass ratio in the solar system.

As shown in Figure 13.24, Charon's orbit is inclined at an angle of 118° to the plane of Pluto's orbit around the Sun. Since the spins of both planet and moon are perpendicular to the plane of Charon's orbit around Pluto, the geographic "north" poles of both bodies lie below the plane of Pluto's orbit. Thus, Pluto is the third planet in the solar system (along with Venus and Uranus) found to have retrograde rotation.

The known mass and radius of Pluto allow us to determine its average density, which is 2100 kg/m^3—too low for a terrestrial planet, but far too high for a mixture of hydrogen and helium of that mass. Instead, the mass, radius, and density of Pluto are just what we would expect for one of the icy moons of a jovian planet. In fact, Pluto is quite similar in both mass and radius to Neptune's large moon, Triton. The planet is almost certainly made up mostly of water ice.

Spectroscopy reveals the presence of *frozen* methane as a major surface constituent. Pluto is the only planet in the solar system on which methane exists in the solid state, implying that the surface temperature on Pluto is no more than 50 K. Pluto may also have a thin methane atmosphere, associated with the methane ice on its surface. Recent computer-generated maps have begun to hint at surface features on Pluto (see *Discovery 13-1*). Similar studies indicate that Charon may have bright polar caps, but their composition and nature are as yet unknown.

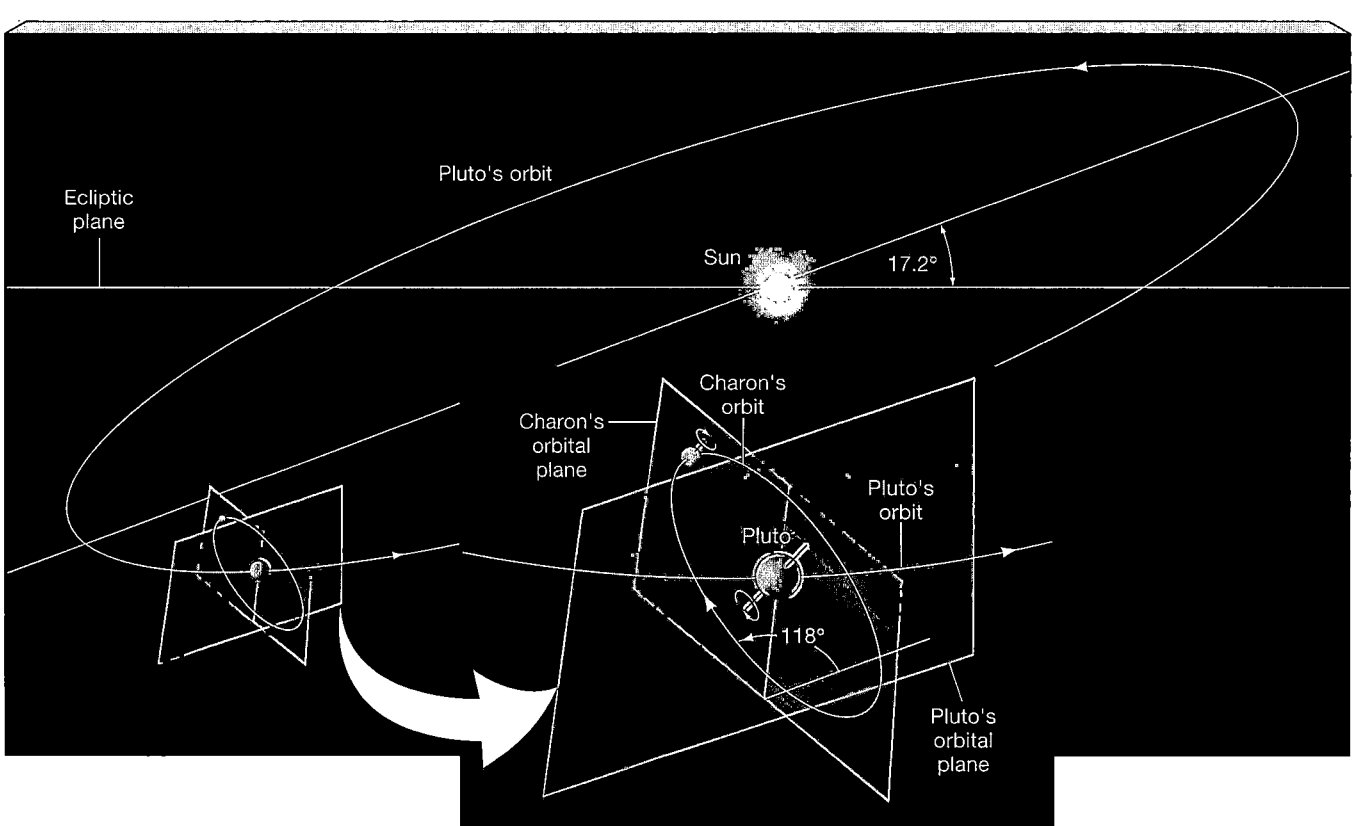

Figure 13.24 Pluto–Charon Orbit Charon's path around Pluto is circular, synchronous, and inclined at 118° to the orbit plane of the Pluto–Charon system about the Sun, which is itself inclined at 17° to the plane of the ecliptic.

13.10 The Origin of Pluto

Because Pluto is neither terrestrial nor jovian in its make-up, and because of its similarity to the ice moons of the outer planets, some researchers suspect that Pluto is not a "true" planet at all. This view is bolstered by Pluto's eccentric, inclined orbit, which is quite unlike the orbits of the other known planets.

Soon after Pluto was discovered, its orbital association with Neptune suggested to some theorists that a catastrophic encounter of some sort might have ejected Pluto from its original orbit around Neptune and perhaps even simultaneously knocked Triton onto its present retrograde path. Although this was an attractive theory at the time, a

plausible sequence of events that could account in detail for the present orbits of those bodies proved elusive. Furthermore, in 1978 the picture was greatly complicated by the discovery of Charon. It was much easier to suppose that Pluto was an escaped moon before we learned that it had a moon of its own. As a result, few astronomers still regard this theory as a likely explanation of Pluto's origin.

Conceivably, Pluto may be just what it seems—a planet that formed in its current orbit, possibly even with its own moon right from the outset. Because we know so little about the environment in the outer solar system, we cannot rule out the possibility that planets beyond Neptune should simply look like Pluto. However, there is mounting evidence that this is not the case either. Astronomers now know of numerous large chunks of ice

DISCOVERY 13-1

Surface Detail on Planet Pluto

Pluto is the only planet for which we have little detailed surface information, because Pluto is the only planet to date that has not been visited by a robot spacecraft. Now new models of Pluto's surface have been constructed, based mainly on images taken by the *Hubble Space Telescope*. The figure at right shows one sample of a dozen snapshots of Pluto, taken at visible and ultraviolet wavelengths throughout the planet's 6.4-day rotation period. This is not an image of Pluto, but a modeled view, that is, a computer-generated sketch of the planet's surface based on a compilation of all available data.

The figure at left is a complete surface map of Pluto, created by carefully collating and analyzing all the data. It is good confirmation of rougher maps made over the previous decade using ground-based observations of eclipses of Pluto by its moon, Charon. Clearly, we have now progressed beyond the stage of seeing Pluto merely as a fuzzy, distant dot of light. With maps like these, astronomers can

monitor its surface detail about as well as you can study nearby Mars with a small backyard telescope.

These models reveal quite a lot of large-scale contrast—surprisingly, for a body composed almost entirely of ice there is more contrast than on any other planet except Earth. About a dozen "provinces" can be identified, including icy-bright polar cap regions, another bright spot seen rotating with the planet itself, a cluster of dark spots, and a few peculiar linear markings. Some of these spots could turn out to be craters or impact basins, such as on Earth's Moon, but much of Pluto's contrast is probably caused by frosts and snows that migrate across the planet as it goes through its seasonal cycles. The reflectivities of the lighter areas resemble that of fresh snow, whereas the darker areas are more akin to the subdued brightness of dirty snow. Let's just hope that people don't start talking about the odd linear markings as being reminiscent of canals!

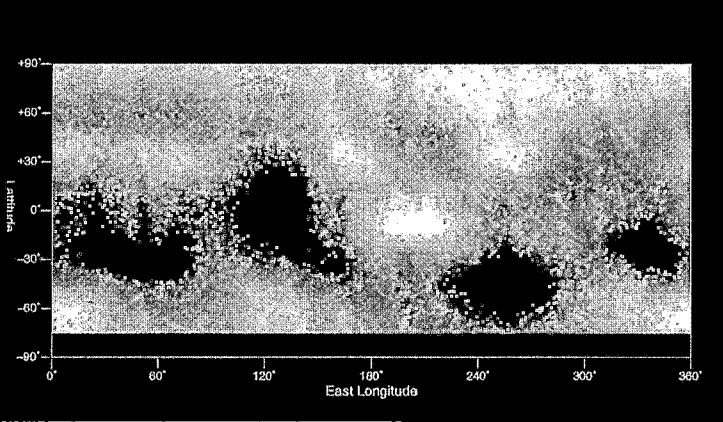
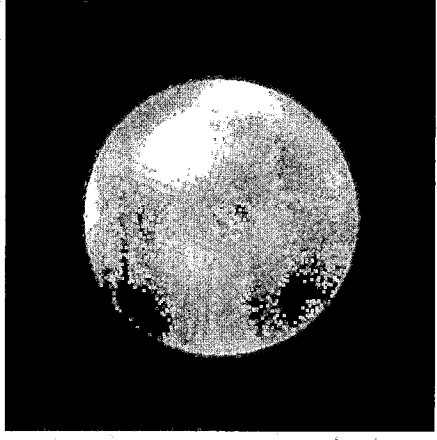

(NASA)

circulating in interplanetary space beyond the orbits of the jovian planets in a region called the *Kuiper belt* (see Chapter 14), and many researchers believe that there may have been thousands of Pluto-sized objects initially present in the outer solar system. Indeed, as we will see in Chapter 15, such objects are the natural building blocks of planets, according to current theories of solar system formation.

Almost all these Pluto-sized objects were "kicked" out to large distances from the Sun following gravitational interactions with Uranus and Neptune. The capture of a few of them by the giant planets could explain some of the strange moons of the outer worlds, especially Triton, and if there were enough moon-sized chunks originally orbiting beyond Neptune, it is quite plausible that Pluto could have captured Charon following a collision (or near-miss) be-

tween the two. Pluto remained behind after most other bodies were captured or ejected, perhaps because its orbital resonance with Neptune prevented it from ever experiencing a close encounter with that (or any other) planet.

At present, our scant knowledge of the compositions of the two bodies does not allow us to confirm or disprove either the "coformation" or the "capture" theory of the Pluto–Charon system. Unfortunately, this uncertainty may persist for some time—there is no present or proposed space mission that might suddenly and radically improve our understanding of these distant worlds.

✔️ Concept Check

◾ What do Pluto and Triton have in common?

Chapter Review

SUMMARY

The outer planets Uranus, Neptune, and Pluto were unknown to ancient astronomers. Uranus was discovered in the eighteenth century, by chance. Neptune was discovered after mathematical calculations of Uranus's slightly non-Keplerian orbit revealed the presence of an eighth planet. At opposition, Uranus is barely visible to the unaided eye. It appears as a pale green disk through a telescope. Neptune cannot be seen with the naked eye, but a telescope shows it as a tiny bluish disk. Today we know the giant planets Uranus and Neptune mainly through data taken by *Voyager 2*. Small, remote Pluto has not been visited by a spacecraft, and our knowledge of it stems from painstaking observations from Earth and the *Hubble Space Telescope*.

The masses of the outer planets are determined from measurements of their orbiting moons. The radii of Uranus and Neptune were relatively poorly known until the *Voyager 2* flybys in the 1980s. Uranus and Neptune have similar bulk properties; their densities imply large, rocky cores making up a greater fraction of the planets' masses than in either Jupiter or Saturn. For unknown reasons, Uranus's spin axis lies nearly in the ecliptic plane, leading to extreme seasonal variations in solar heating on the planet as it orbits the Sun. Surface features are barely discernible on Uranus, but computer-enhanced images from *Voyager 2* revealed atmospheric clouds and flow patterns moving beneath Uranus's haze. Neptune, although farther away from us, has atmospheric features that are clearer because of warmer temperatures and less haze. The **Great Dark Spot** (p. 335) on Neptune had many similarities to Jupiter's Red Spot. It disappeared in 1994.

Unlike the other jovian planets, Uranus has no excess heat emission. The source of Neptune's excess energy, like that of Jupiter's, is most likely heat left over from the planet's formation. Both Uranus and Neptune have substantial magnetospheres. *Voyager 2* discovered that the magnetic fields of both planets are tilted at large angles to the planets' rotation axes. The reason for this is not known.

All but two of Uranus's moons revolve in the planet's equatorial plane, almost perpendicular to the ecliptic, in circular synchronous orbits. Like the moons of Saturn, the medium-sized moons of Uranus are made up predominantly of rock and water ice. Many of them are heavily cratered and in some cases must have come close to being destroyed by the meteoritic impacts whose craters we now see. The strange moon Miranda has geological features that suggest repeated violent impacts in the past. Neptune's moon Triton has a fractured surface of water ice and a thin atmosphere of nitrogen, probably produced by nitrogen "geysers" on its surface. Triton is the only large moon in the solar system to have a retrograde orbit around its parent planet. This orbit is unstable and will eventually cause Triton to be torn apart by Neptune's gravity.

Uranus has a series of dark, narrow rings, first detected from Earth by **stellar occultation** (p. 342)—their obscuration of the light received from background stars. Shepherd satellites are responsible for the rings' thinness. Neptune has three narrow rings like Uranus's and one broad ring, like Jupiter's. They were discovered by *Voyager 2*. The dark coloration of both the rings and the moons of the outer giant planets may be due to **radiation darkening** (p. 338), whereby exposure to solar high-energy radiation slowly causes a dark hydrocarbon layer to build up on a body's icy surface.

Pluto was discovered in the twentieth century after a laborious search for a planet that was supposedly affecting Uranus's orbital motion. We now know that Pluto is far too small to have any detectable influence on Uranus's path. Pluto has a moon, Charon, whose mass is about one-sixth that of Pluto itself. Studies of Charon's orbit around Pluto have allowed the masses and radii of both bodies to be accurately determined. Pluto is too small for a terrestrial planet. Its properties are far more moonlike than planetlike. Most astronomers believe that it is probably the largest (or nearest) member of a class of icy asteroids found in the outer solar system.

SELF-TEST: TRUE OR FALSE?

____ **1.** Uranus was discovered by Galileo.

____ **2.** After the discovery of Uranus, astronomers started looking for other planets and discovered Neptune 10 years later.

____ **3.** Since its discovery Uranus has completed just two-and a half orbits of the Sun.

____ **4.** Uranus and Neptune are comparable in size and are larger than the other two jovian planets.

____ **5.** Uranus and Neptune have very similar rotation rates.

____ **6.** During the northern summer of Uranus, an observer near the north pole would observe the Sun high and almost stationary in the sky.

____ **7.** Both Uranus and Neptune have layers of metallic hydrogen surrounding their central cores.

____ **8.** Relative to the size of the planet, the rocky cores of Uranus and Neptune are larger than the cores of Jupiter and Saturn.

____ **9.** Most of the moons of Uranus have orbits that share the tilt of the planet.

____ **10.** Uranus has no large moons.

____ **11.** The surfaces of the largest Uranian moons are darker than similar moons of Saturn.

____ **12.** Triton's surface is heavily cratered, implying no significant surface activity.

____ **13.** Pluto's moon, Charon, is more than half the diameter of Pluto.

____ **14.** Pluto is larger than Earth's Moon.

____ **15.** Astronomers have firm evidence that Pluto is an escaped moon of Neptune.

SELF-TEST: FILL IN THE BLANK

1. Uranus is about _____ times more distant from the Sun than is Saturn.

2. Uranus's rotation axis is almost _____ to the ecliptic plane.

3. Whereas the abundance of hydrogen and helium for the jovian planets remains more or less constant among the four planets, the relative abundance of _____ increases with increasing distance from the Sun.

4. The planet _____ is dark blue in color, with white cirrus clouds and visible storm systems.

5. The planet _____ is blue-green in color and virtually featureless.

6. The _____ of both Uranus and Neptune are highly tilted relative to their rotation axes and significantly offset from the planets' centers.

7. The Uranian moon that shows the greatest amount of geological activity and disruption over time is _____.

8. All but two of the known moons of Neptune were discovered by _____.

9. Triton's orbit is unusual because it is _____.

10. Nereid's orbit is unusual because it has a high orbital _____.

11. Triton has a thin nitrogen _____.

12. *Voyager 2* observed two nitrogen _____ on Triton's surface.

13. The orbit of Pluto is locked into a 3:2 resonance with _____.

14. Overall, Pluto is most similar to which object in the solar system? _____.

15. The radii of Pluto and Charon were determined accurately by observing a series of _____ during the late 1980s.

REVIEW AND DISCUSSION

1. How was Uranus discovered?

2. Why did astronomers suspect an eighth planet beyond Uranus?

3. How did Uranus come to be spinning "on its side"?

4. What is responsible for the colors of Uranus and Neptune?

5. How are the interiors of Uranus and Neptune thought to differ from those of Jupiter and Saturn?

6. How do the magnetic fields of Uranus and Neptune compare with that of Earth?

7. Describe a day on Titania.

8. What is unique about Miranda? Give a possible explanation.

9. Why are the icy moons of the outer planets so dark?

10. How does Neptune's moon system differ from those of the other jovian worlds? What do these differences suggest about the origin of the moon system?

11. What causes Triton's geysers?

12. What is the predicted fate of Triton?

13. The rings of Uranus are dark, narrow, and widely spaced. Which of these properties makes them different from the rings of Saturn?

14. Why are the Uranian rings so narrow and sharply defined?

15. How do the rings of Neptune differ from those of Uranus and Saturn?

16. Will Pluto and Neptune ever collide?

17. How were the mass and radius of Pluto determined?

18. In what respect is Pluto more like a moon than a jovian or terrestrial planet?

19. Why was the discovery of Uranus in 1781 so surprising? Might there be similar surprises in store for today's astronomers?

20. In what sense were astronomers fortunate to discover Pluto?

PROBLEMS  *Algorithmic versions of these questions are available in the Practice Problems module of the Companion Website.*

The number of squares preceding each problem indicates its approximate level of difficulty.

1. ▨▨ Calculate the time between successive closest approaches of (a) Neptune and Uranus (b) Pluto and Neptune. For simplicity, assume circular orbits and calculate the time taken for the inner plate to "lap" the other one, as in Problem 5, Chapter 9. Compare your answers with Neptune's sidereal orbital period.

2. ▨ What is the gravitational force exerted on Uranus by Neptune, at closest approach? Compare your answer with the Sun's gravitational force on Uranus.

3. ▨▨ What is the angular diameter of the Sun, as seen from Uranus? Compare it with the angular diameter of Titania, seen from the planet's cloud tops. Would you expect solar eclipses to occur on Uranus?

4. ▨▨ If the core of Uranus has a radius twice that of planet Earth and an average density of 8000 kg/m³, calculate the mass of Uranus outside the core. What fraction of the planet's total mass is core?

5. ▨ Estimate the speed of cloud A in Figure 13.7, assuming that it lies near the equator. Is your estimate consistent with the rotation speed of the planet?

6. ▨ Estimate the strength of Neptune's gravitational tidal acceleration on Triton, relative to the moon's own surface gravity. Compare with the corresponding ratio for Jupiter and Io. (Chapter 11, Problem 10).

7. ▨ Add up the masses of all the moons of Uranus, Neptune, and Pluto. (Neglect the masses of the small moons—they contribute little to the result.) How does this sum compare with the mass of Earth's Moon and with the mass of Pluto?

8. ▨▨▨ Astronomers on Earth are observing the occultation of a star by Uranus and its rings (see Figure 13.18). It so happens that the event is occurring when Uranus is at opposition, and the center of the planet appears to pass directly across the star. Assuming circular planetary orbits and, for simplicity, taking the rings to be face-on, calculate (a) how long the 90-km-wide Epsilon ring takes to cross the line of sight, and (b) the time interval between the passage of the Alpha ring and that of the Epsilon ring. (Hint: The apparent motion of Uranus is due to a combination of both the planet's own motion and Earth's motion in their respective orbits around the Sun.)

9. ▨▨ Based on the earlier discussion of planetary atmospheres, would you expect Triton to have retained a nitrogen atmosphere? ⊂⊃ (*More Precisely 8-1*)

10. ▨▨ How long does it take for the inner shepherd moon in Figure 13.18 to "lap" the outer one?

11. ▨▨▨ If Pluto's apparent motion relative to the stars is indeed mainly due to Earth's orbital motion, estimate the angle through which Pluto has moved between the two frames in Figure 13.21.

12. ▨ What would be your weight on Pluto? On Charon?

13. ▨ How close is Charon to Pluto's Roche limit?

14. ▨ From Wien's law, at what wavelength does Pluto's thermal emission peak? In what part of the electromagnetic spectrum does this lie? ⊂⊃ (*More Precisely 3-2*)

15. ▨ What is the round-trip travel time of light from Earth to Pluto (at a distance of 40 A.U.)? How far would a spacecraft orbiting Pluto at a speed of 0.5 km/s travel during that time?

COLLABORATIVE EXERCISES

1. **Scaled Distance to Pluto.** If the distance in meters between your group and the room's exit represents the average distance between the Sun and Earth, determine how far away the planets Uranus, Neptune, and Pluto would be located from where your group sits and ask three group members to stand at the appropriate distances.

RESEARCHING ON THE WEB  *To complete the following exercises, go to the online Destinations module for Chapter 13 on the Companion Website for Astronomy Today 4/e.*

1. Access the "Uranus Fact Sheet" from NASA and determine if its maximum apparent visual magnitude is brighter than the North Star, Polaris.

2. Access the "Neptune Nomenclature" page and describe the appearance of a *sulcus* geologic feature.

3. Access the "Pluto Fact Sheet" from NASA and determine when it will again be closer to the Sun than Neptune.

4. Access the "Plutinos" page and describe the Pluto-Kuiper Express and its current status.

PROJECTS

1. The major astronomy magazines *Sky & Telescope* and *Astronomy* print charts showing the whereabouts of the planets in their January issues. Consult one of these charts and locate Neptune and Uranus in the sky. Uranus may be visible to the naked eye, but binoculars make the search much easier. (Hint: Uranus shines more steadily than the background stars.) With the eye alone, can you detect a color to Uranus? Through binoculars?

2. The search for Neptune requires a much more determined effort! A telescope is best, but high-powered binoculars mounted on a steady support will do. If you can see both planets through a telescope—and they will remain close together on the sky for the rest of this century—contrast their colors. Which planet appears bluer? Through a telescope, does Uranus show a disk? Can you see that Neptune shows a disk, or does it look more like a point of light?

SKYCHART III PROJECTS

The SkyChart III Student Version planetarium program on which these exercises are based is included as a separately executable program on the CD in the back of this text.

1. ▣ Center the display on Uranus and set the time to midnight on September 7, 2003. Zoom to 180° field of view and use left and right arrow keys on keyboard to rotate the image until north is up in your display. Zoom in to a 1/60° field of view. As you zoom in, note the orientation of the distribution of moons and the orientation of the spin axis of Uranus itself. The inclined spin axis should be obvious.

2. ▣ Uranus is a naked-eye object, but only in absence of light pollution. If you have a location with good skies, use SkyChart III to determine when and where to look for it and then see if you can find it. If seeing is not good, use a modest telescope to locate Uranus. Change the date until Uranus is sufficiently high above the horizon that visibility will be good. Note which constellation it will be in at that time, and make an accurate sketch to take into the field. Include in your sketch precise positions of visible stars in the immediate vicinity of Uranus. For instance, in 2003 Uranus will be in Aquarius. Follow the position of Uranus as a few months pass and observe its very slow movement against the background stars, a consequence of its relatively large orbital radius. Deselect horizon mask and animate the scene with time steps of one month. Repeatedly press F6 for better control of time, and observe the relative motion of Uranus with respect to background stars as time advances from sometime in the year 2000 to the year 2005. Explain why there appear to be short periods of retrograde motion.

3. ▣ Leverrier and Adams's predictions of the existence of Neptune and its subsequent discovery were a victory for the scientific method and Newton's theory of gravity. Nevertheless, Neptune was sighted before by Galileo. In January 1613, he was following the moons of Jupiter when he noticed a star that seemed to move. He apparently did not follow up on this. From the location of Rome, center your view on Jupiter, as Galileo would have done. For that particular month, how much does Neptune move relative to Jupiter? How much does it move relative to a nearby star? How faint was Neptune compared to the Galilean moons?

4. ▣▣ Conjunctions occur when the planets appear grouped together in the night sky. Such conjunctions are subject to a great deal of myth and folklore, including catastrophic predictions of Armageddon. Many such conjunctions are of the five "naked eye" planets Mercury, Mars, Venus, Jupiter, and Saturn and usually include the Sun and Moon. Reproduce two past conjunctions on February 5, 1962, May 5, 2000, and a future one on September 8, 2040. What is the angular span of each conjunction (the diameter of an imaginary circle containing the conjunction)?

5. ▣▣ View each of the conjunctions listed in the foregoing exercise from a position perpendicular to the ecliptic, about 60 A.U. offset from the Sun (see Appendix for setup instructions). How does Earth fit into these conjunctions? Is there a true alignment of the planets? How are the outer planets positioned? Explain why these conjunctions have no physical effect on Earth.

 *In addition to the Practice Problems and Destinations modules, the Companion Website at http://www.prenhall.com/chaisson provides for each chapter an additional true-false, multiple choice, and labeling quiz, as well as additional annotated images, animations, and links to related Websites.*

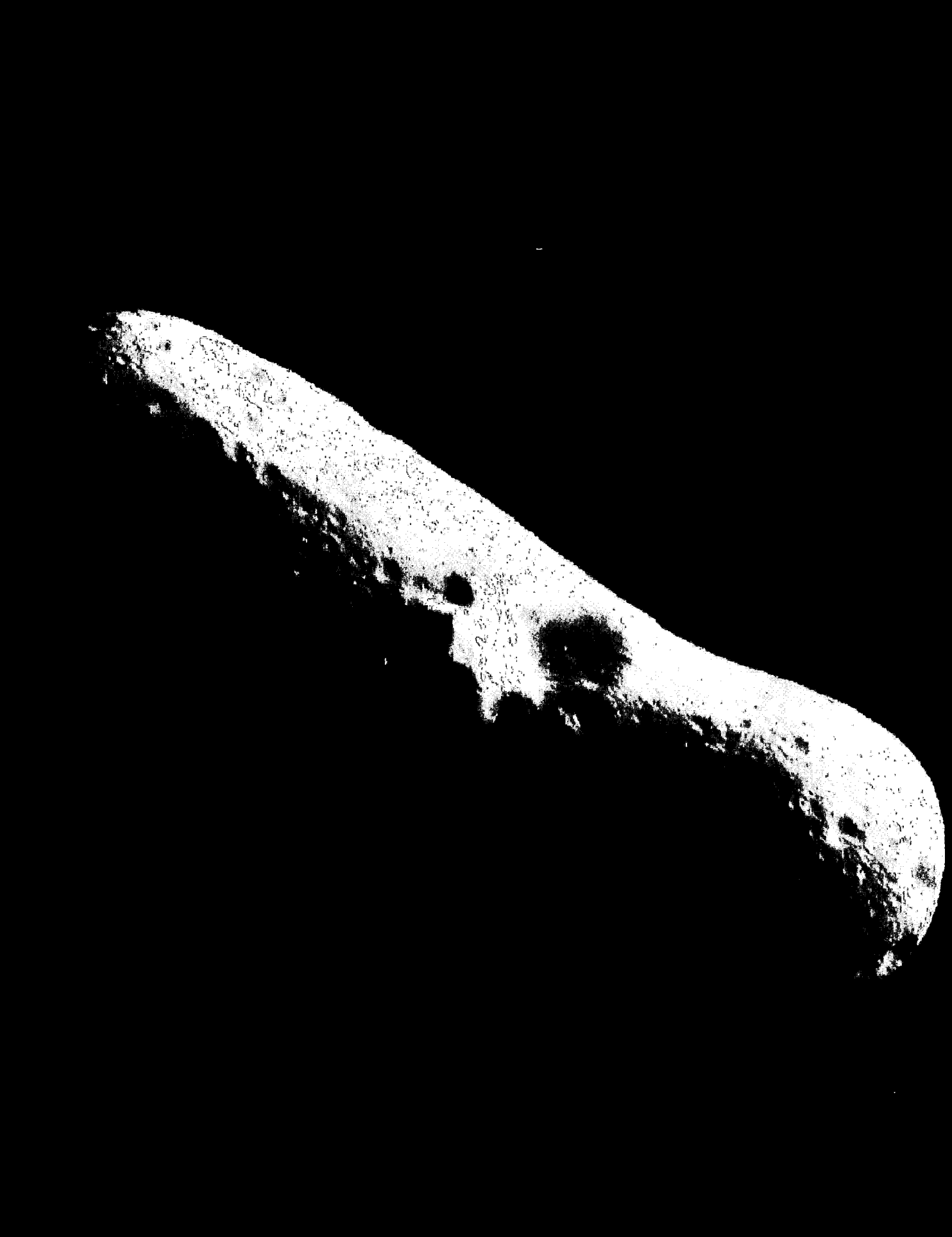

14 SOLAR SYSTEM DEBRIS

Keys to Our Origin

LEARNING GOALS

Studying this chapter will enable you to:

1   Describe the orbital properties of the major groups of asteroids.

2   Summarize the composition and physical properties of a typical asteroid.

3   Explain the effect of orbital resonances on the structure of the asteroid belt.

4   Detail the composition and structure of a typical comet, and explain the formation and appearance of its tail.

5   Discuss the characteristics of cometary orbits and what they tell us about the probable origin of comets.

6   Distinguish among the terms meteor, meteoroid, and meteorite.

7   Summarize the orbital and physical properties of meteoroids, and explain what these suggest about their probable origin.

 Visit http://www.prenhall.com/chaisson for additional annotated images, animations, and links to related sites for this chapter.

The asteroid Eros was imaged here as part of the NEAR (Near Earth Asteroid Rendezvous) space mission in the year 2000. In a remarkable feat of interplanetary navigation, this small robot spacecraft was sent to the asteroid belt about 2.1 A.U. from the Sun, found Eros amid the other debris there, orbited about it for nearly a year, and then landed on it. This image shows the potato-shaped rock; at 34 × 11 × 11 kilometers it is one of the larger known asteroids. (JHU/APL)

The Big Picture: Only within the past couple of decades have scientists taken seriously the idea that life on Earth has been seriously disrupted over the course of billions of years by asteroid and comet impacts. Indeed, the real "motor of evolution" might well be collisions with extraterrestrial bodies. We now have a much better appreciation for Earth's presence in an occasionally hostile environment, and for Earth's fragility in the face of external cosmic dangers.

$\mathbb{A}$ccording to standard definitions, there are only nine planets in the solar system. But several thousand other celestial bodies are also known to revolve around the Sun. These minor bodies—the asteroids and comets—are small and of negligible mass compared with the planets and their major moons. Yet each is a separate world, with its own story to tell about the early solar system. On the basis of statistical deductions, astronomers estimate that there are more than a billion such objects still to be discovered. They may seem to be only rocky and icy "debris," but more than the planets themselves, they hold a record of the formative stages of our planetary system. Many are nearly pristine, unevolved bodies with much to teach us about our local origins.

14.1 Asteroids

Asteroids are relatively small, rocky objects that revolve around the Sun. Their name literally means "starlike bodies," but asteroids are definitely not stars. They are too small even to be classified as planets. Astronomers often refer to them as "minor planets," or sometimes "planetoids."

Asteroids differ from planets in both their orbits and their size. They generally move on quite eccentric trajectories between Mars and Jupiter, unlike the almost circular paths of the major planets. Few are larger than 300 km in diameter, and most are far smaller—as small as a tenth of a kilometer across. The largest known asteroid, Ceres, is just 1/10,000 the mass of Earth and measures only 940 km across. Taken together, the known asteroids amount to less than $\frac{1}{10}$ the mass of the Moon, so they do not contribute significantly to the total mass of the solar system.

ORBITAL PROPERTIES

1 European astronomers discovered the first asteroids early in the nineteenth century as they searched the sky for an additional planet orbiting between Mars and Jupiter, where the Titius-Bode "law" suggested one might be found. ∞ (*Discovery 6-1*) Italian astronomer Giuseppe Piazzi was the first to discover an asteroid. He detected Ceres in 1801 and measured its orbital semimajor axis to be 2.8 A.U.—exactly where the "law" predicted. Within a few years, three more asteroids—Pallas (2.8 A.U.), Juno (2.7 A.U.), and Vesta (3.4 A.U.)—were discovered.

By the start of the twentieth century, astronomers had cataloged several hundred asteroids with well-determined orbits. Now, at the start of the twenty-first century, the list has grown to over 20,000. The total number of known asteroids (that is, including those whose orbits are not yet known with sufficient accuracy to make them "official") now exceeds 100,000. The vast majority of these bodies are found in a region of the solar system known as the **asteroid belt**, located between 2.1 and 3.3 A.U. from the Sun—roughly midway between the orbits of Mars (1.5 A.U.) and Jupiter (5.2 A.U.). All but one of the known asteroids revolve about the Sun in prograde orbits, in the same sense as the planets. The overall layout of the asteroid belt is sketched in Figure 14.1.

Such a compact concentration of asteroids in a well-defined belt suggests that they are either the fragments of a planet broken up long ago or primal rocks that never managed to accumulate into a genuine planet. On the basis of the best evidence currently available, researchers favor the latter view. There is far too little mass in the belt to constitute a planet, and the marked chemical differences among individual asteroids strongly suggest that they could not all have originated in a single body. Instead, astronomers believe that the strong gravitational field of Jupiter continuously disturbs the motions of these chunks of primitive matter, nudging and pulling at them, preventing them from aggregating into a planet.

PHYSICAL PROPERTIES

2 With few exceptions, asteroids are too small to be resolved by Earth-based telescopes, so astronomers must rely on indirect methods to find their sizes, shapes, and composition. Consequently, only a few of their physical and chemical properties are accurately known. To the extent that astronomers can determine their compositions, asteroids have been found to differ not only from the nine known planets and their many moons but also among themselves.

Asteroids are classified by their spectroscopic properties. The darkest, or least reflective, asteroids contain a large fraction of carbon in their makeup. They are known as *C-type* (or *carbonaceous*) asteroids. The more reflective *S-type* asteroids contain silicate, or rocky, material. Generally speaking, S-type asteroids predominate in the inner portions of the asteroid belt and the fraction of C-type bodies steadily increases as we move outward. Overall, about 15 percent of all asteroids are S-type, 75 percent are C-type, and 10 percent are other types (mainly the *M-type* asteroids, containing large fractions of nickel and iron). Many planetary scientists believe that the carbonaceous asteroids consist of very primitive material representative of the earliest stages of the solar system that have not experienced significant heating or chemical evolution since they first formed 4.6 billion years ago.

In most cases, astronomers estimate the sizes of asteroids from the amount of sunlight they reflect and the amount of heat they radiate. These observations are difficult, but size measurements have been obtained in this way for more than 1000 asteroids. On rare occasions, as-

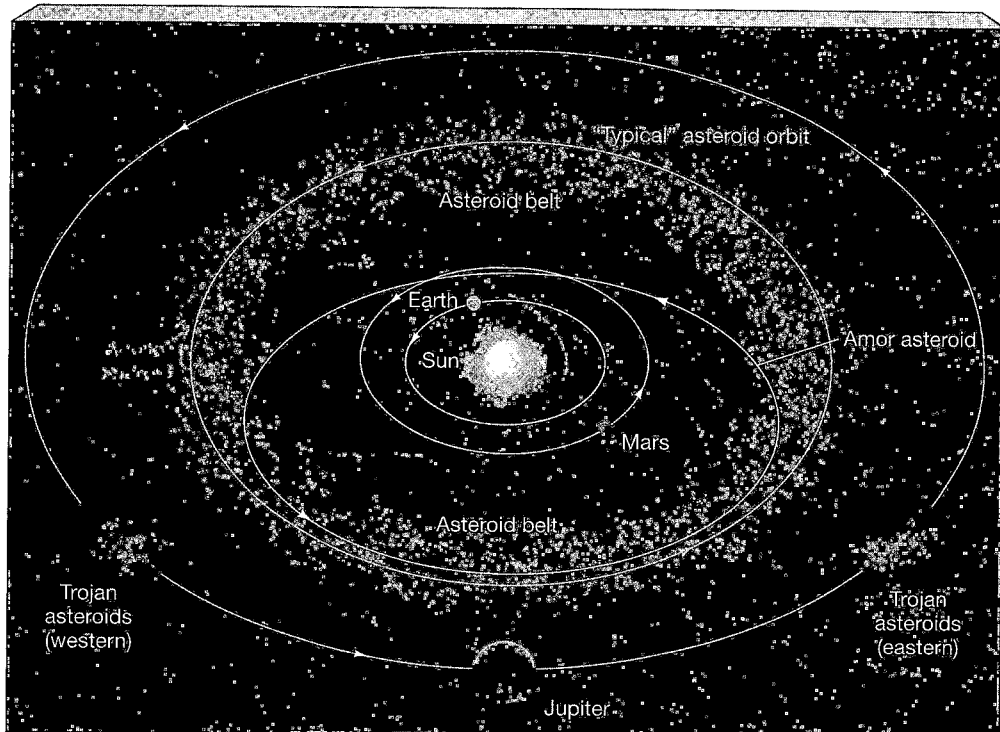

Figure 14.1 Inner Solar System The asteroid belt, along with the orbits of Earth, Mars, and Jupiter (not drawn to scale). The main belt, the Trojan asteroids, and some Apollo (Earth-crossing) and Amor (Mars-crossing) orbits are shown.

tronomers witness an asteroid occulting a star, allowing them to determine its size and shape with great accuracy. The largest asteroids are roughly spherical, but the smaller ones can be highly irregular.

The three largest asteroids, Ceres, Pallas, and Vesta, have diameters of 940 km, 580 km, and 540 km, respectively. Only two dozen or so asteroids are more than 200 km across, and most are much smaller. Almost assuredly, many hundreds of thousands more await discovery. However, observers estimate that they are mostly very small. Probably 99 percent of all asteroids larger than 100 km are known and cataloged, and at least 50 percent of asteroids larger than 10 km are accounted for. Although the vast majority of asteroids are probably less than a few kilometers across, most of the *mass* in the asteroid belt resides in objects greater than a few tens of kilometers in diameter.

Vesta is unique among asteroids in that, despite its small size, it appears to have undergone *volcanism* in its distant past. Numerous meteorites (Section 14.3) found on Earth are believed, based on their orbits and their overall spectral similarities to Vesta, to have been chipped off the asteroid following collisions with other members of the asteroid belt. Remarkably, these meteorites have compositions very similar to terrestrial basalt, indicating ancient volcanic activity. Why Vesta should have experienced such activity, while the larger Ceres and Pallas did not, is not known. Possibly Vesta was once part of a larger body that subsequently broke up. No other fragments of that body have yet been conclusively identified, although the spectrum and surface properties of a tiny 2-km-long asteroid discovered in 1999 suggests that it may once have been part of Vesta.

The first close-up views of asteroids were provided by the Jupiter probe *Galileo* which, on its rather roundabout path to the giant planet, passed twice through the asteroid belt, making close encounters with asteroid Gaspra in October 1991 and asteroid Ida in August 1993 (Figure 14.2). ∞ (Sec. 6.6) Both Gaspra and Ida are S-type asteroids. Technical problems limited the amount of data that could be sent back from the spacecraft during the flybys. Nevertheless, the images produced by *Galileo* showed far more detail than any photographs made from Earth.

Gaspra and Ida are irregularly shaped bodies with maximum diameters of about 20 km and 60 km, respectively. They are pitted with craters ranging in size from a few hundred meters to 2 km across and are covered with a layer of dust of variable thickness. Ida is much more heavily cratered than Gaspra, in part because it resides in a denser part of the asteroid belt. Also, scientists believe that Ida has suffered more from the ravages of time. Ida is about a billion years old, far older than Gaspra, which is estimated to have an age of just 200 million years, based on the extent of cratering. Both asteroids are thought to be fragments of much larger objects that broke up into many smaller pieces following violent collisions long ago.

To the surprise of most mission scientists, closer inspection of the Ida image (Figure 14.2b) revealed the presence of a tiny moon, now named Dactyl, just 1.5 km across, orbiting the asteroid at a distance of about 90 km. A few such *binary asteroids* had previously been observed from Earth. However, the rare binary systems known before the Ida flyby were all much larger than Ida—a moon the size of Dactyl cannot be detected from the ground. Scientists believe that, given

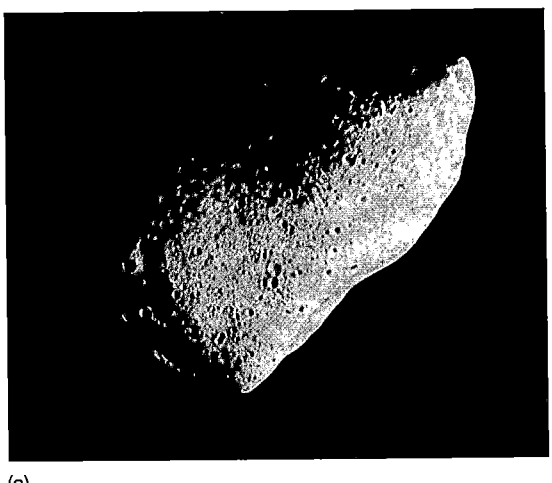
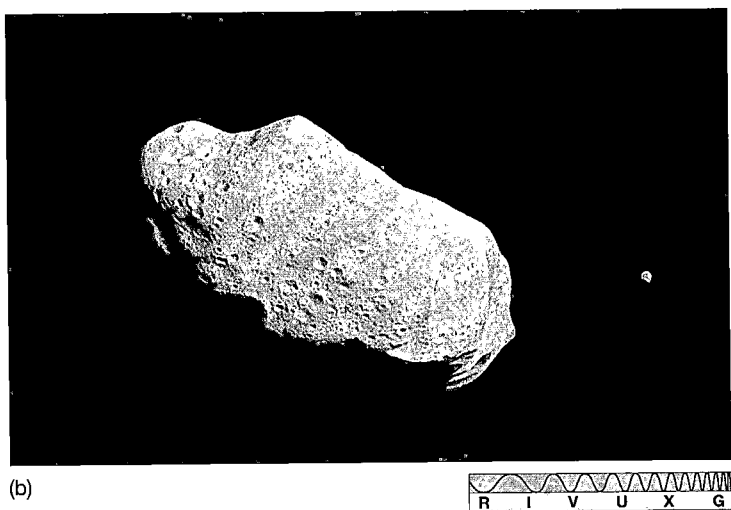

(a)   (b)

Figure 14.2 **Gaspra and Ida** (a) The S-type asteroid Gaspra as seen from a distance of 1600 km by the probe *Galileo* on its way to Jupiter. (b) The S-type asteroid Ida, photographed by *Galileo* from a distance of 3400 km. (Ida's moon, Dactyl, is visible at the right of the photo.) The resolution in these photographs is on the order of 100 m. True-color images showed the surfaces of both bodies to be a fairly uniform shade of gray. Sensors on board the spacecraft indicated that the amount of infrared radiation absorbed by these surfaces varies from place to place, probably as a result of variations in the thickness of the dust layer blanketing them. *(NASA)*

the relative congestion of the asteroid belt, collisions between asteroids may be quite common, providing a source of both interplanetary dust and smaller asteroids and possibly deflecting one or both of the bodies involved onto eccentric, Earth-crossing orbits. The less violent collisions may be responsible for the binary systems we see.

By studying the *Galileo* images, astronomers were able to obtain limited information on Dactyl's orbit around Ida and hence (using Newton's law of gravity—see *More Precisely 2-3*) to estimate Ida's mass at about 5–10 × 10^{16} kg. This information in turn allowed them to measure Ida's density as 2200–2900 kg/m^3, a range consistent with its rocky, S-type classification.

In June 1997 the *Near Earth Asteroid Rendezvous* (*NEAR*) spacecraft visited the C-type asteroid Mathilde on its way to the mission's main target, the S-type asteroid Eros. Shown in Figure 14.3, Mathilde is some 60 km across. By sensing its gravitational pull, *NEAR* measured Mathilde's mass to be about 10^{17} kg, implying a density of just 1400 kg/m^3. To account for this low density, scientists speculate that the asteroid's interior must be quite porous. Indeed, many smaller asteroids seem to be more like loosely bound "rubble piles" than pieces of solid rock. The interior's relatively soft consistency may also explain the unexpectedly large size of many of the craters observed on Mathilde's surface. A solid object would probably have shattered after an impact violent enough to cause such large craters. However, like crumple zones in a car, Mathilde's porous interior could have absorbed and dissipated the impactor's energy, allowing the asteroid to survive the event.

On arrival at Eros on February 14, 2000, *NEAR* (now renamed *NEAR-Shoemaker*) went into orbit around

the asteroid, changing its trajectory several times and coming as close as 5.5 km to the surface (Figure 14.4a). For one year, the spacecraft sent back high-resolution images of Eros (Figure 14.4b), and made detailed measurements of its size, shape, gravitational and magnetic fields, composition, and structure. The craft's various sensors revealed Eros to be a heavily cratered body of

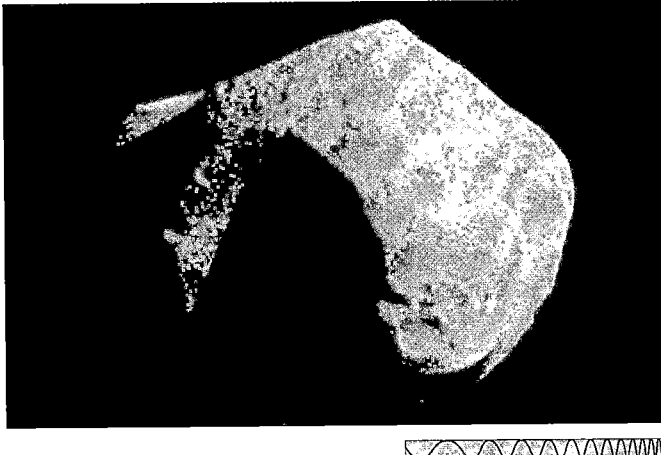

Figure 14.3 **Asteroid Mathilde** The C-type asteroid Mathilde, imaged by the *NEAR* spacecraft en route to the near-earth asteroid Eros. It measures some 60 × 50 km, and rotates every 17.5 days. The largest craters visible in this image are about 20 km across—much larger than the craters seen on either Gaspra or Ida. The reason may be the asteroid's low density (approximately 1400 kg/m^3) and rather soft composition. *(NASA)*

mass 7×10^{15} kg and roughly uniform density around 2700 kg/m^3. The asteroid's interior seems to be solid rock—not rubble, as in the case of Mathilde—although it is extensively fractured due to innumerable impacts in the past. All in all, the measurements are consistent with Eros being a primitive, unevolved sample of material from the early solar system. On February 12, 2001, *NEAR-Shoemaker* landed on Eros, sending back a series

of close-up images as it descended to the surface. Remarkably, despite the lack of landing gear, the spacecraft survived the low-velocity impact. While no further images were obtained, the probe maintained radio contact with Earth for 16 more days before communication finally ceased.

Apart from Ida, Mathilde, and Eros, most asteroid masses are unknown. However, a few of the largest asteroids

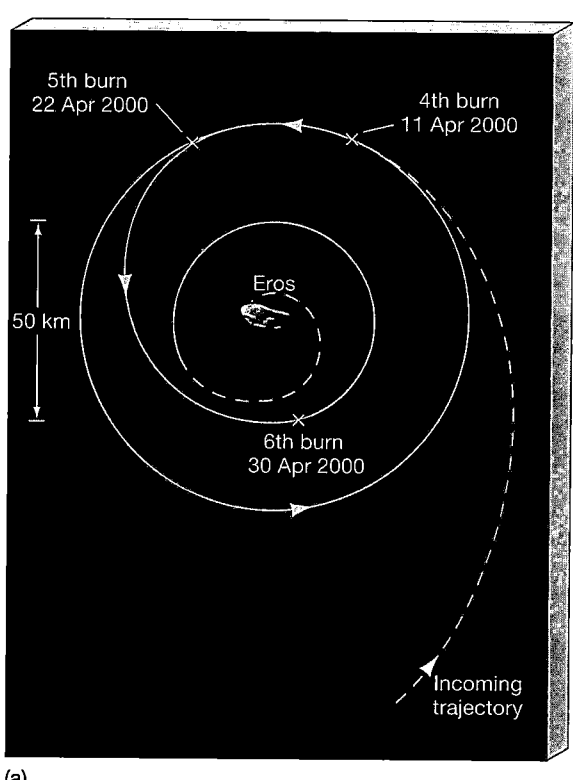

(a)

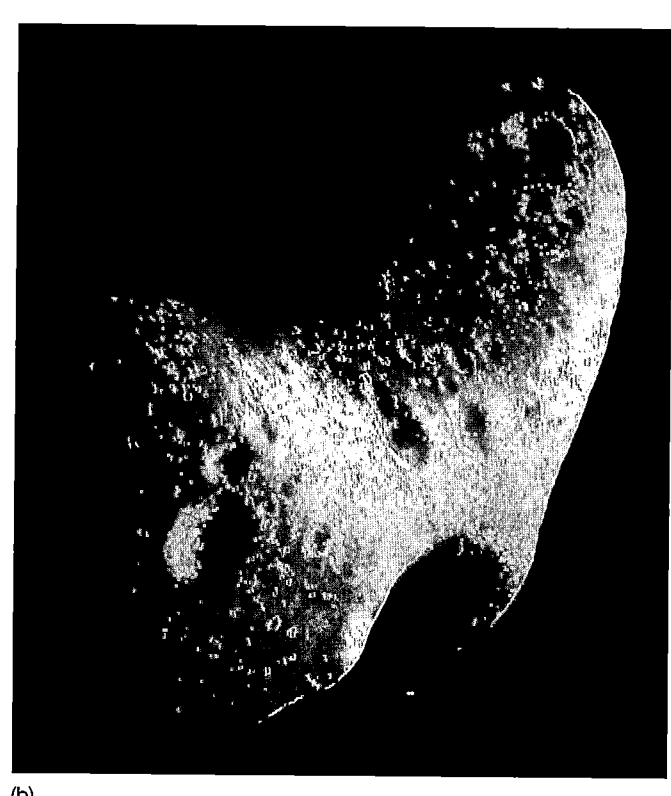

(b)

Figure 14.4 NEAR at Eros (a) The *NEAR-Shoemaker* spacecraft entered orbit around asteroid Eros in February 2000, making a series of orbit corrections during April to bring it closer and closer to the surface before landing on the asteroid on February 12, 2001. (b) A mosaic of detailed images showing the entire asteroid. Craters of all sizes, ranging from 50 m (the resolution of the image) to 5 km, pit the surface. Part (c) shows a close-up image of a "young" section of the surface, where loose material from recent impacts has apparently filled in and erased all trace of older craters. *(JHU/NASA)*

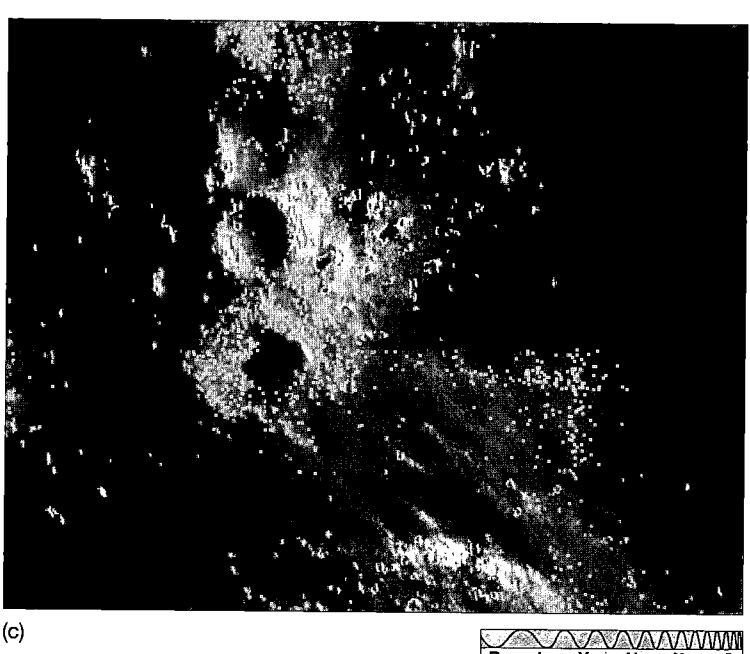

(c)

do have strong enough gravitational fields for their effects on their neighbors to be measured and their masses thereby determined to reasonable accuracy. Their computed densities are generally compatible with the rocky or carbonaceous compositions just described.

☑️ Concept Check

■ Describe some basic similarities and differences between asteroids and the inner planets.

EARTH-CROSSING ASTEROIDS

The orbits of most asteroids have eccentricities lying in the range 0.05–0.3, ensuring that they always remain between the orbits of Mars and Jupiter. Very few asteroids have eccentricities greater than 0.4. Those that do are of particular interest to us, however, as their paths may intersect Earth's orbit, leading to the possibility of a collision with our planet. They are collectively known as **Earth-crossing asteroids**. Those stray asteroids having highly elliptical orbits or orbits well inside the main asteroid belt have probably been influenced by nearby Mars and especially Jupiter. The gravitational fields of those planets can disturb normal asteroid orbits, deflecting them into the inner solar system. Earth-crossing asteroids are termed *Apollo asteroids* (after the first known Earth-crossing asteroid, Apollo) if their orbital semimajor axes exceed 1 A.U., and *Aten asteroids* otherwise. Asteroids whose orbits cross only the orbit of Mars are known as *Amor asteroids* (see Figure 14.1).

As of early 2001, more than 1200 Earth-crossing asteroids are known. Most have been discovered since the late 1990s when, spurred in large part by concerns over possible asteroid or comet impacts, systematic searches for such objects began. Several large, dedicated telescopes now scan the skies for faint objects in our neighborhood. Almost 300 Earth-crossers are officially designated "potentially hazardous," meaning that they are more than about 150 m in diameter (three times the size of the impactor responsible for the Barringer crater shown in Figure 8.19), and move on orbits that could bring them within 0.05 A.U. (7.5 million km) of our planet.

The potential for collision with Earth is real. For example, the perihelion of the Apollo asteroid Icarus (shown in Figure 14.5) is within 0.2 A.U. of the Sun. On its way past Earth in 1968, Icarus missed our planet by "only" 6 million km—a very close call by cosmic standards. More recently, in 1991 an unnamed asteroid (designated 1991 BA) came much closer, passing only 170,000 km from Earth, less than half the distance to the Moon. In December 1994 the Apollo asteroid 1994 XM1 missed us by a mere 105,000 km. All told, between 1990 and 2000, more than 40 asteroids (that we know of!) passed within 10 million km of our planet, and 40 more are expected to pass within that same distance between 2000 and 2010.

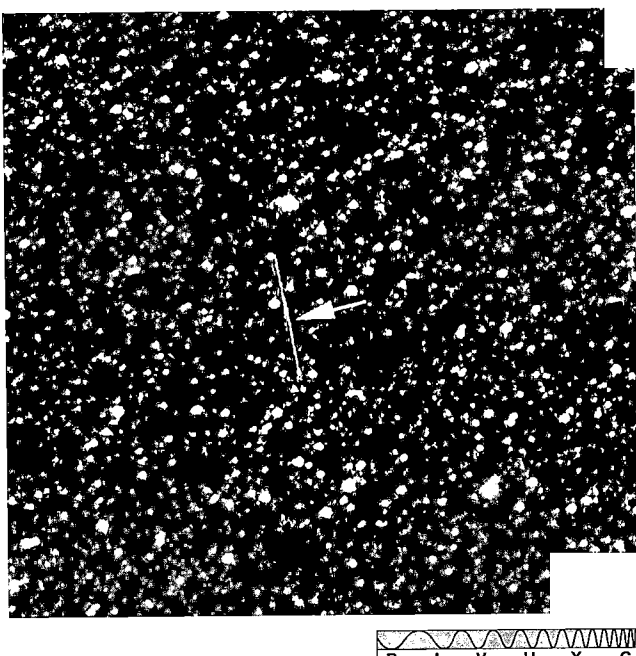

Figure 14.5 Asteroid Icarus The asteroid Icarus has an orbit that passes within 0.2 A.U. of the Sun, well within the orbit of Earth. Icarus occasionally comes close to our own planet, making it one of the better-studied asteroids in the solar system. The asteroid's motion relative to the stars makes it appear as a streak (marked) in this long-exposure photograph, taken in the 1970s. *(Palomar/Caltech)*

Calculations imply that most Earth-crossing asteroids will eventually collide with Earth. On average during any given million-year period our planet is struck by about three asteroids. Because Earth is largely covered with water, on average two of those impacts should occur in the ocean and only one on land. Several dozen large land basins and eroded craters on our planet are suspected to be sites of ancient asteroid collisions (see, for example, Figure 14.19 later in this chapter). The many large impact craters on the Moon, Venus, and Mars are direct evidence of similar events on other worlds.

Most known Earth-crossing asteroids are relatively small—about 1 km in diameter (although one 10 km in diameter has been identified). Even so, a visit of even a kilometer-sized asteroid to Earth could be catastrophic by human standards. Such an object packs enough energy to devastate an area some 100 km in diameter. The explosive power would be equivalent to about a million one-megaton nuclear bombs—a hundred times more than all the nuclear weapons currently in existence on Earth. A fatal blast wave would doubtless affect a much larger area still. Should an asteroid hit our planet hard enough, it might even cause the extinction of entire species—indeed, many scientists think that the extinction of the dinosaurs was the result of just such an impact (see *Discovery 14-1*). Some astronomers take the prospect of an asteroid impact sufficiently seriously that they maintain an "asteroid watch"—an effort to catalog and monitor all Earth-crossing asteroids in order to maximize our warning time of any impending collision.

ORBITAL RESONANCES

3 Although most asteroids orbit in the main belt, between about 2 and 3 A.U. from the Sun, an additional class of asteroids, called the **Trojan asteroids**, orbit at the distance of Jupiter. Several hundred such asteroids are now known. They are locked into a 1:1 orbital resonance with Jupiter by that planet's strong gravity, just as some of the small moons of Saturn share orbits with the medium-sized moons Tethys and Dione, as described in Chapter 12. ∞ (Sec. 12.5) Calculations first performed by the French mathematician Joseph Louis Lagrange in 1772 show that there are exactly five places in the solar system where a small body can orbit the Sun in synchrony with Jupiter (or, for that matter, with any other planet) subject to the combined gravitational influence of both large bodies. These places are known as the *Lagrangian points* of the planet's orbit. As illustrated in Figure 14.6, three of these points (referred to as L_1, L_2, and L_3) lie on the line joining Jupiter and the Sun (or its extension in either direction). The other two—L_4 and L_5—are located on Jupiter's orbit, exactly 60° ahead of and behind the planet. All five Lagrangian points revolve around the Sun at the same rate as Jupiter.

In principle, an asteroid placed at any of the Lagrangian points will circle the Sun in lockstep with Jupiter, always maintaining the same position relative to the planet. However, the three Lagrangian points in line with Jupiter and the Sun are known to be *unstable*—a body displaced,

however slightly, from any of those points will tend to drift slowly away from it, not back toward it. Since matter in the solar system is constantly subjected to small perturbations—by the planets, the asteroids, and even the solar wind—matter does not accumulate in these regions. No asteroids orbit near the L_1, L_2, or L_3 points of Jupiter's orbit.

This is not the case for the other two Lagrangian points, L_4 and L_5. They are both *stable*—matter placed near them tends to remain in the vicinity. Consequently, asteroids tend to accumulate near these points. For unknown reasons, Trojan asteroids tend to be found near Jupiter's leading (L_4) Lagrangian point rather than the trailing (L_5) point. Recently, a few small asteroids have been found similarly trapped in the Lagrangian points of Venus, Earth, and Mars.

The main asteroid belt also has structure—not so obvious as the Trojan orbits or the prominent gaps and ringlets in Saturn's ring system, but nevertheless of great dynamic significance. A graph of the number of asteroids having various orbital semimajor axes (Figure 14.7a) shows that there are several prominently underpopulated regions in the distribution. These "holes" are known as the **Kirkwood gaps**, after their discoverer, the nineteenth-century American astronomer Daniel Kirkwood.

The Trojan asteroids share an orbit with Jupiter—they orbit in 1:1 resonance with that planet. The Kirkwood gaps result from other, more complex, orbital resonances with Jupiter. For example, an asteroid with a semimajor axis of 3.3 A.U. would (by Kepler's third law) orbit the Sun in exactly half the time taken by Jupiter. ∞ (Sec. 2.4) The gap at 3.3 A.U., then, corresponds to a 2:1 resonance. An asteroid at that particular resonance feels a regular, periodic tug from Jupiter at the same point in every other orbit (Figure 14.7b). The cumulative effect of those tugs is to deflect the asteroid into an elongated orbit—one that crosses the orbit of Mars or Earth. Eventually, the asteroid collides with one of those two planets or comes close enough that it is pushed onto an entirely different trajectory. In this way, Jupiter's gravity creates the Kirkwood gaps, and some of the cleared-out asteroids become Apollo or Amor asteroids.

Notice that, although there are many similarities between this mechanism and the resonances that produce the gaps in Saturn's rings (see Chapter 12), there are differences too. ∞ (Sec. 12.4) Unlike Saturn's rings, where eccentric orbits are rapidly circularized by collisions among ring particles, there are no *physical* gaps in the asteroid belt. The in-and-out motion of the belt asteroids as they travel in their eccentric orbits around the Sun means that no part of the belt is actually empty. Only when we look at semimajor axes (or, equivalently, at orbital *energies*) do the gaps become apparent.

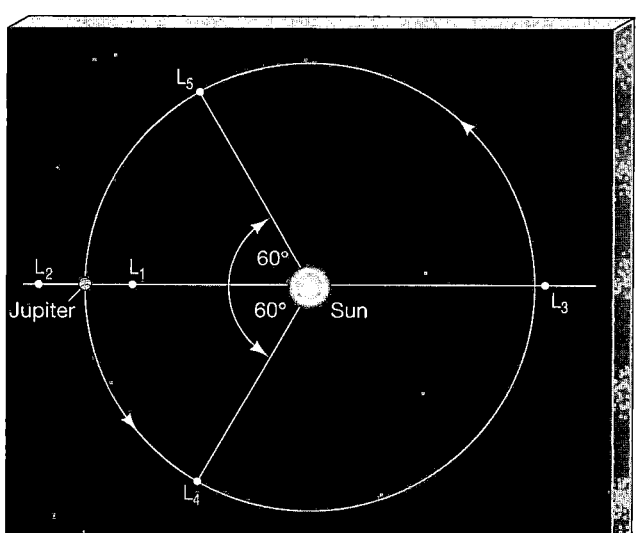

Figure 14.6 Lagrangian Points The Lagrangian points of the Jupiter–Sun system, where a third body could orbit in synchrony with Jupiter on a circular trajectory. Only the L_4 and L_5 points are stable. They are the locations of the Trojan asteroids.

✓ Concept Check

■ Why are astronomers so interested in Earth-crossing asteroids?

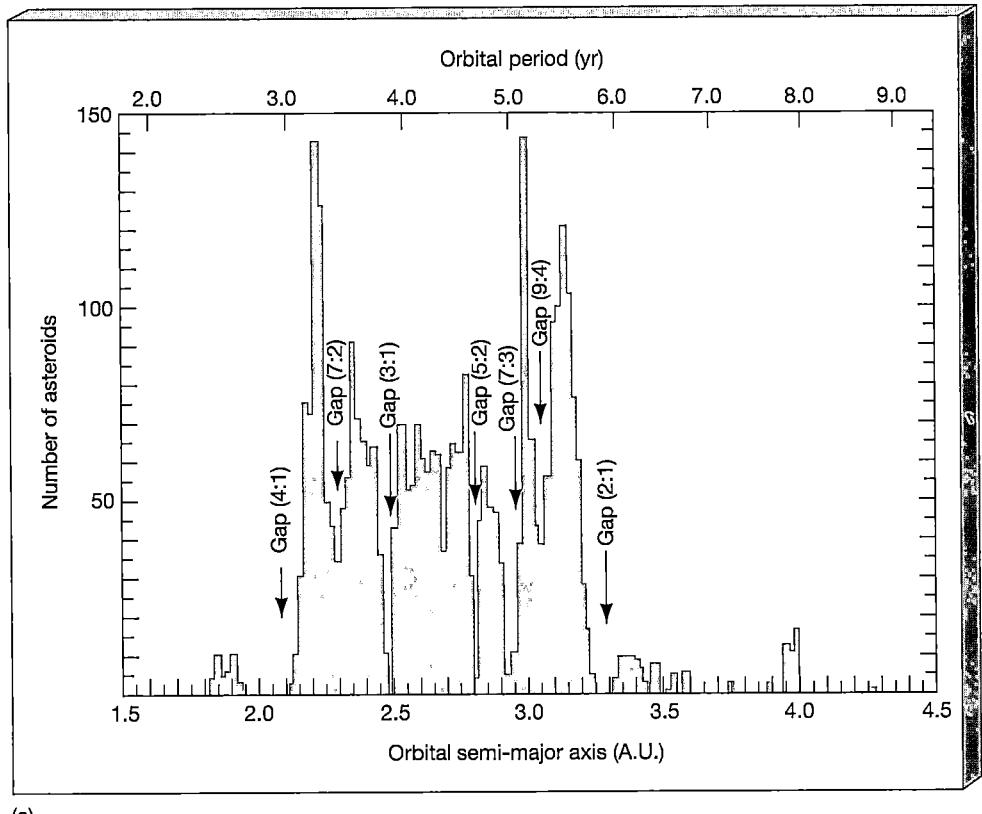

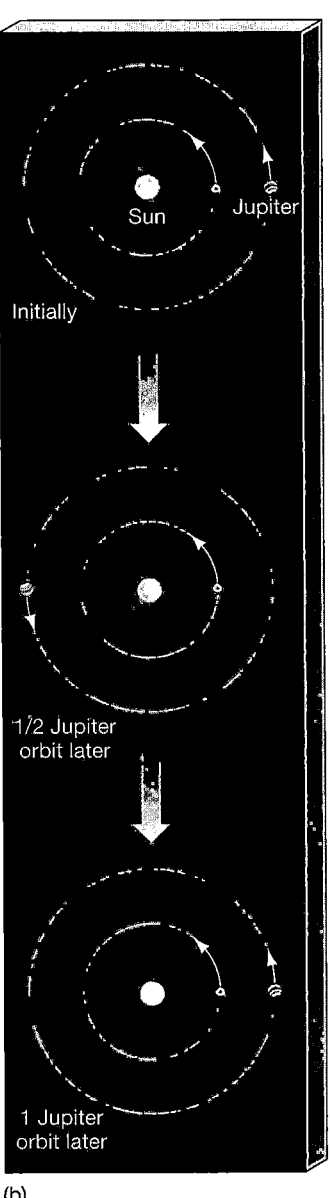

Figure 14.7 Kirkwood Gaps (a) The distribution of asteroid semimajor axes shows some prominent gaps caused by resonances with Jupiter's orbital motion. Note, for example, the prominent gap at 3.3 A.U., which corresponds to the 2:1 resonance—the orbital period is 5.9 years, exactly half that of Jupiter. (b) An asteroid in a 2:1 resonance with Jupiter receives a strong gravitational tug from the planet each time they are closest together (as in panels 1 and 3). Because the asteroid's period is precisely half that of Jupiter, the tugs come at exactly the same point in every other orbit and their effects reinforce each other.

14.2 Comets

Comets are usually discovered as faint, fuzzy patches of light on the sky while still several astronomical units away from the Sun. Traveling in a highly elliptical orbit with the Sun at one focus, a comet brightens and develops an extended **tail** as it nears the Sun. (The name "comet" derives from the Greek word *kome*, meaning "hair.") As the comet departs from the Sun's vicinity, its brightness and its tail diminish until it once again becomes a faint point of light receding into the distance. Like the planets, comets emit no visible light of their own—they shine by reflected (or reemitted) sunlight. Each year a few dozen are detected as they pass through the inner solar system. Many more must pass by unseen.

COMET APPEARANCE AND STRUCTURE

⁴ The various parts of a typical comet are shown in Figure 14.8. Even through a large telescope, the **nucleus**, or main solid body, of a comet is no more than a minute point of light. A typical cometary nucleus is extremely small—only a few kilometers in diameter. During most of the comet's orbit, far from the Sun, only this frozen nucleus exists. When a comet comes within a few astronomical units of the Sun, however, its icy surface becomes too warm to remain stable. Part of it becomes gaseous and expands into space, forming a diffuse **coma** ("halo") of dust and evaporated gas around the nucleus. The coma becomes larger and brighter as the comet nears the Sun. At maximum size, the coma can measure 100,000 km in diameter—almost as large as Saturn or Jupiter.

Engulfing the coma, an invisible **hydrogen envelope**, usually distorted by the solar wind, stretches across millions of kilometers of space. The comet's tail, most pronounced when the comet is closest to the Sun and the rate of sublimation* of material from the nucleus is greatest, is much larger still, sometimes spanning as much as 1 A.U. From Earth, only the coma and tail of a comet are visible to the naked eye. Despite the size of the tail, however, most of the light comes from the coma; most of the comet's mass resides in the nucleus.

Two types of comet tails may be distinguished. The **ion tails** are approximately straight, often made up of glowing, linear streamers like those seen in Figure 14.9(a). Their spectra show emission lines of numerous ionized molecules—molecules that have lost some of their normal complement of electrons—including carbon monoxide, nitrogen, and water, among many others. ∞ (Sec. 4.2) The **dust tails** are usually broad, diffuse, and gently curved (Figure 14.9b). They are rich in microscopic dust particles that reflect sunlight, making the tail visible from afar. (The ion and dust tails are also referred to as *Type I* and *Type II* tails, respectively.)

The tails are in all cases directed *away* from the Sun by the solar wind (the invisible stream of matter and radiation escaping the Sun). Consequently, as depicted in Figure 14.10, the tail always lies outside the comet's orbit and actually leads the comet during the portion of the orbit that is outbound from the Sun.

The ion tails and dust tails differ in shape because of the different responses of gas and dust to the forces acting in interplanetary space. Every tiny particle in space in our solar system—including those in comet tails—follows an orbit determined by gravity and the solar wind. If gravity alone were acting, the particle would follow the same curved path as its parent comet, in accordance with Newton's laws of motion. ∞ (Sec. 2.7) If the solar wind were the only influence, the tail would be swept up by it and would trail radially outward from the Sun. The ion tails are much more strongly influenced by the solar wind than by the Sun's gravity, so those tails always point directly away from the Sun. The heavier dust particles have more of a tendency to follow the comet's orbit, giving rise to the slightly curved dust tails.

COMET ORBITS

5 Comets that survive their close encounter with the Sun—some break up entirely—continue their outward journey to the edge of the solar system. Their highly elliptical orbits take many comets far beyond Pluto, perhaps even as far as 50,000 A.U., where, in accord with Kepler's second law, they move more slowly and so spend most of their time. ∞ (Sec. 2.4)

---

*Sublimation is the process by which a solid changes directly into a gas without passing through the liquid phase. Frozen carbon dioxide—dry ice—is an example of a solid that undergoes sublimation rather than melting and subsequent evaporation. In space, sublimation is the rule, rather than the exception, for the behavior of ice when exposed to heat.

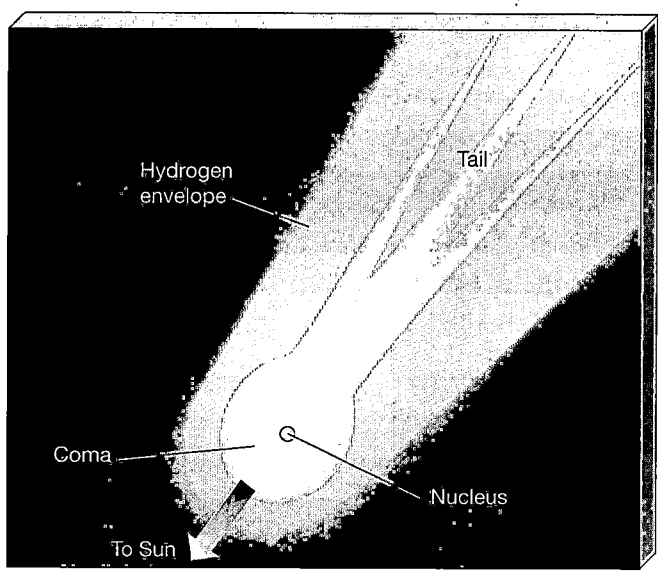

(a)

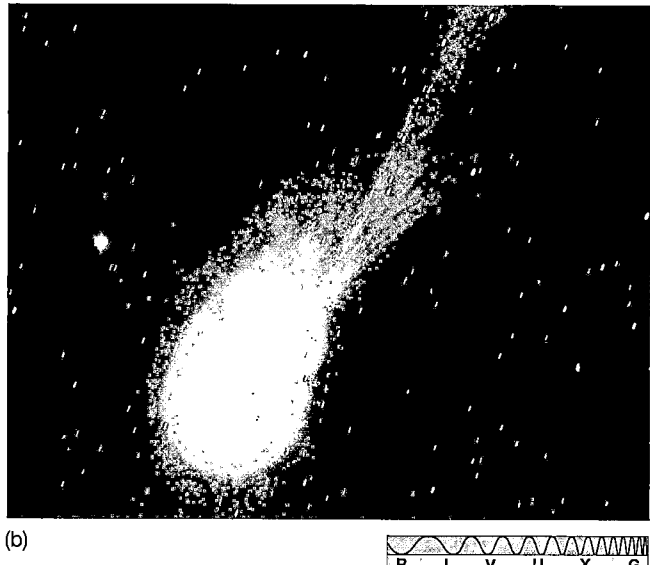

(b)

Figure 14.8 Comet Structure (a) Diagram of a typical comet, showing the nucleus, coma, hydrogen envelope, and tail. The tail is not a sudden streak in time across the sky, as in the case of meteors or fireworks. Instead, it travels through space along with the rest of the comet (so long as the comet is sufficiently close to the Sun for the tail to exist). (b) Halley's comet in 1986, about one month before perihelion. *(NOAO)*

The majority of comets take hundreds of thousands, some even millions, of years to complete a single orbit around the Sun. However, a few *short-period comets* (conventionally defined as those having orbital periods of less than 200 years) return for another encounter within a relatively short time. According to Kepler's third law, the short-period comets do not venture far beyond the distance of Pluto at aphelion.

Unlike the orbits of the other solar system objects we have studied so far, the orbits of comets are not necessarily confined to within a few degrees of the ecliptic plane.

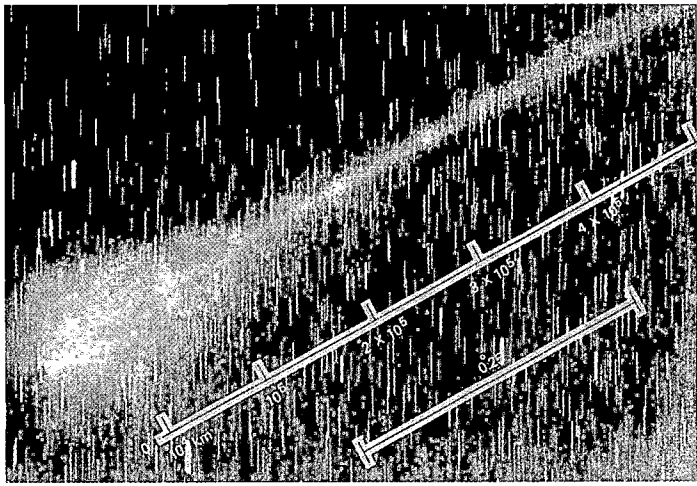

(a)

Figure 14.9 Comet Tails (a) A comet with a primarily ion tail. Called comet Giacobini-Zinner and seen here in 1959, its coma measured 70,000 km across; its tail was well over 500,000 km long. (b) Photograph of a comet having both an ion tail (blue) and a dust tail (white), showing the gentle curvature and inherent fuzziness of the dust (see also *Discovery 14-2*). This is comet Hale-Bopp in 1997. At closest approach to the Sun, its tail stretched nearly 40° across the sky. *(US Naval Observatory; Observers: Herman Mikuz, B. Kambic, Location: Crni Vrh Observatory, Slovenia)*

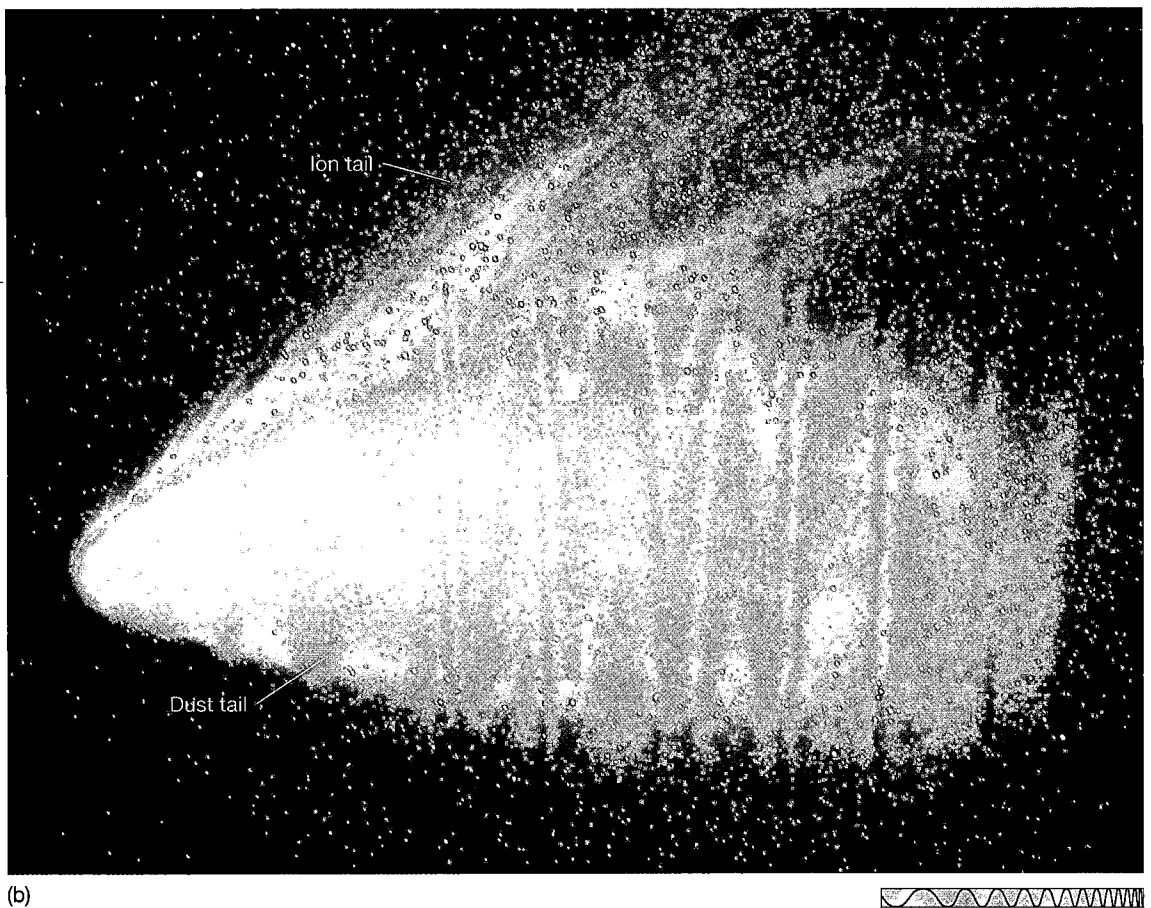

(b)

Short-period comets do tend to have prograde orbits lying close to the ecliptic, but long-period comets exhibit all inclinations and all orientations, both prograde and retrograde, roughly uniformly distributed in all directions from the Sun.

The short-period comets originate beyond the orbit of Neptune, in a region of the solar system called the **Kuiper belt** (named after Gerard Kuiper, a pioneer in infrared and planetary astronomy). A little like the asteroids in the inner solar system, most Kuiper belt comets move in roughly circular orbits between about 30 and 100 A.U.

from the Sun, never venturing inside the orbits of the jovian planets. Occasionally, however, a close encounter between two comets, or (more likely) the cumulative gravitational influence of one of the outer planets, "kicks" a Kuiper belt comet into an eccentric orbit that brings it into the inner solar system, and into our view. The observed orbits of these comets reflect the flattened structure of the Kuiper belt.

What of the long-period comets? How do we account for their apparently random orbital orientations? Only a

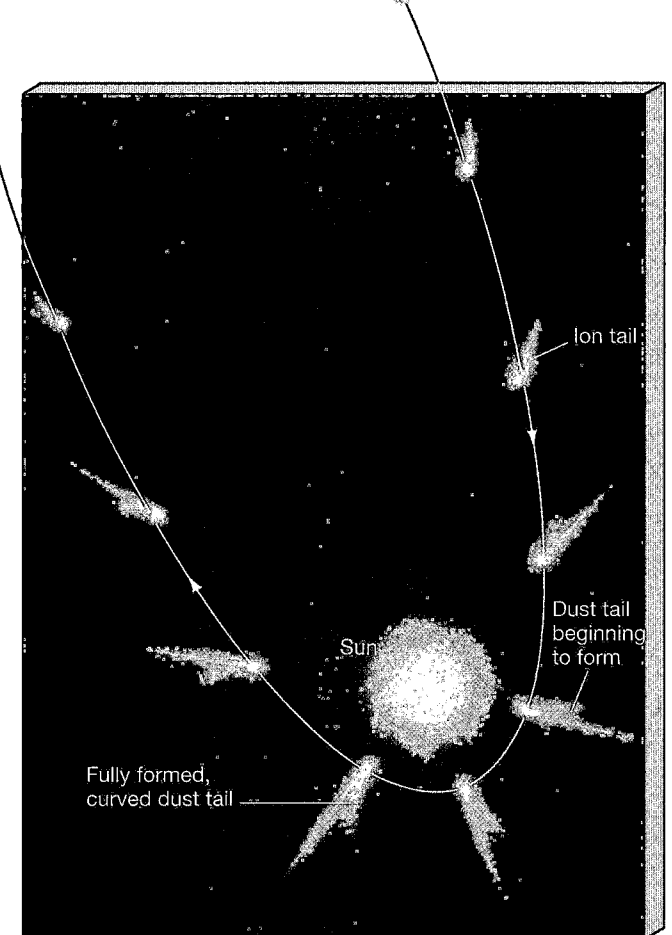

Figure 14.10 Comet Trajectory As it approaches the Sun, a comet develops an ion tail, which is always directed away from the Sun. Closer in, a curved dust tail, also directed generally away from the Sun, may also appear. Notice that although the ion tail always points directly away from the Sun on both the inbound and the outgoing portions of the orbit, the dust tail has a marked asymmetry, always tending to "lag behind" the ion tail.

tiny portion of a typical long-period cometary orbit lies within the inner solar system, so it follows that for every comet we see, there must be many more similar objects at great distances from the Sun. On these general grounds, many astronomers reason that there must be a huge "cloud" of comets far beyond the orbit of Pluto, completely surrounding the Sun. This region, which may contain trillions of comets, of total mass comparable to the mass of the inner planets, is named the **Oort cloud**, after the Dutch astronomer Jan Oort, who first wrote (in the 1950s) of the possibility of such a vast and distant reservoir of inactive, frozen comets. The Kuiper belt and the orbits of some typical Oort cloud comets are sketched in Figure 14.11.

The observed orbital properties of long-period comets have led researchers to conclude that the Oort cloud may be up to 100,000 A.U. in diameter. Like those of the Kuiper belt, however, most of the comets of the Oort cloud never come anywhere near the Sun. Indeed,

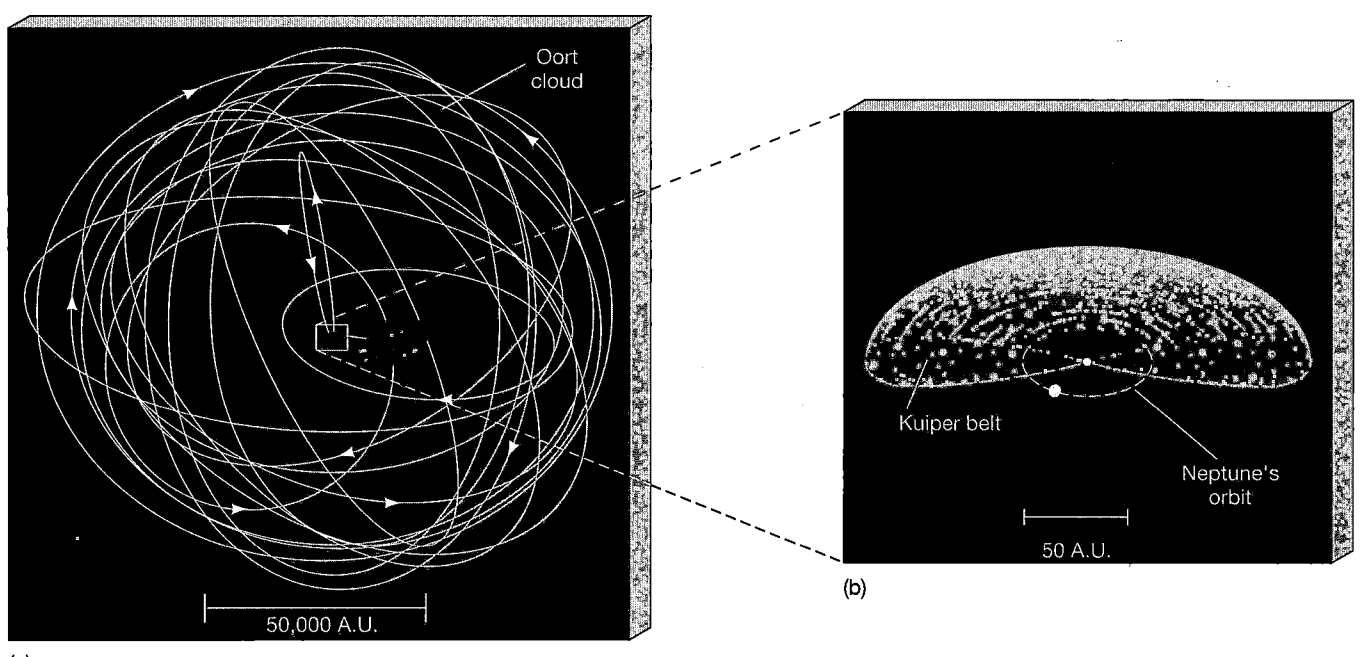

Figure 14.11 Comet Reservoirs (a) Diagram of the Oort cloud, showing a few cometary orbits. Most Oort cloud comets never come close to the Sun. Of all the orbits shown, only the most elongated ellipse represents a comet that will actually enter the solar system (which is smaller than the dot at the center of the figure on this scale) and possibly become visible from Earth. (b) The Kuiper belt, the source of the short-period comets, whose orbits hug the ecliptic plane.

Oort cloud comets rarely approach even the orbit of Pluto, let alone that of Earth. Only when the gravitational field of a passing star happens to deflect a comet into an extremely eccentric orbit that passes through the inner solar system do we actually get to see one of these objects. Because the Oort cloud surrounds the Sun in all directions, instead of being confined near the ecliptic plane like the Kuiper belt, the long-period comets we see can come from any direction in the sky. Despite their great distances and long orbital periods, however, the Oort cloud comets are still gravitationally bound to the Sun. Their orbits are governed by precisely the same laws of motion that control the planets.

☑ Concept Check

■ In what sense are the comets we see very *un*representative of comets in general?

A VISIT TO HALLEY'S COMET

Probably the most famous comet of all is Halley's comet. (Two more recent and widely publicized contenders for that title are described in *Discovery 14-2*.) In 1705 the British astronomer Edmund Halley realized that the 1682 appearance of this comet was not a one-time event. Basing his work on previous sightings of the comet, Halley calculated its path and found that the comet orbited the Sun with a period of 76 years. He predicted its reappearance in 1758. Halley's successful determination of the comet's trajectory and his prediction of its return was an early triumph of Newton's laws of motion and gravity. Although Halley did not live to see his calculations proved correct, the comet was named in his honor.

Once astronomers knew the comet's period, they traced its appearances back in time. Historical records from many ancient cultures show that Halley's comet has been observed at every passage since 240 B.C. A spectacular show, the tail of Halley's comet can reach almost a full astronomical unit in length, stretching many tens of degrees across the sky. Figure 14.12(a) shows Halley's comet as seen from Earth in 1910. Its most recent appearance, in 1986 (Figure 14.12b and also Figure 14.8b), was not ideal for terrestrial viewing, but the comet was closely scrutinized by spacecraft. The comet's orbit is shown in Figure 14.13; its next scheduled visit to the inner solar system is in 2061.

When Halley's comet rounded the Sun in 1986, a small armada of spacecraft launched by the (former) USSR, Japan, and a group of western European countries went to meet it. One of the Soviet craft, *Vega 2*, traveled through the comet's coma and came within 8000 km of the nucleus. Using positional knowledge of the comet gained from the Soviet craft encounter, the European *Giotto* spacecraft (named after the Italian artist who painted an image of Halley's comet not long after its appearance in the year 1301) was navigated to within 600 km of the nucleus. This was a daring trajectory, since at 70 km/s—the speed of the craft relative to the comet—a colliding dust particle becomes a devastating bullet. Debris did in fact damage *Giotto*'s camera, but not before it sent home a wealth of data. Figure 14.14 shows *Giotto*'s view of the comet's nucleus, along with a sketch of its structure.

The results of the Halley encounters were somewhat surprising. Halley's nucleus is an irregular, potato-shaped object, somewhat larger than astronomers had estimated. Spacecraft measurements showed it to be 15 km long by as much as 10 km wide. Also, the nucleus appeared almost jet black—as dark as finely ground charcoal or soot. This solid nucleus was enveloped by a cloud of dust, which scattered light throughout the coma. Partly because of this scattering and partly because of dimming by the dust, none of the visiting spacecraft were able to discern much surface detail on the nucleus.

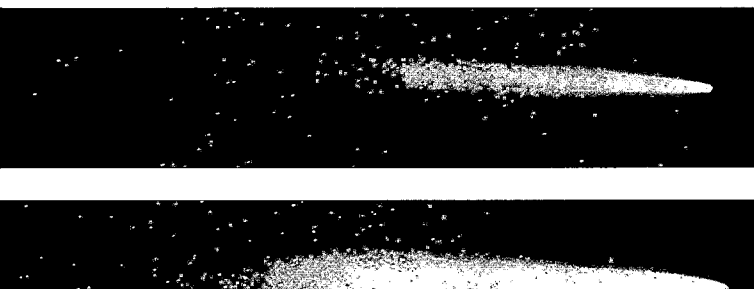

(a)

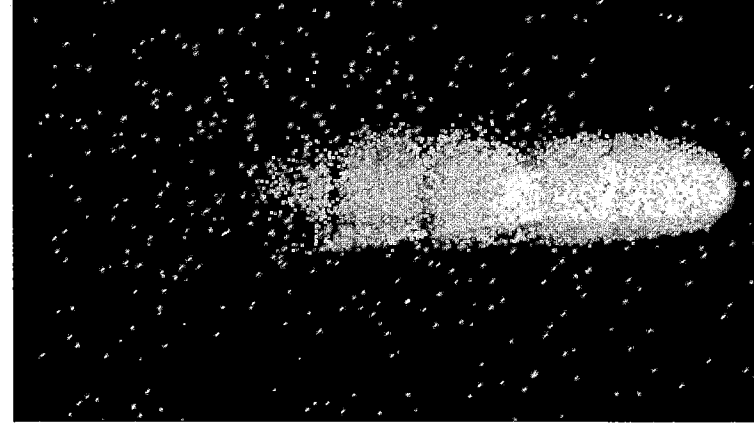

(b)

Figure 14.12 Halley's Comet (a) Halley's comet as it appeared in 1910. Top, on May 10, with a 30° tail, bottom, on May 12, with a 40° tail. (b) Halley, on its return, and photographed with higher resolution, on March 14, 1986. *(Caltech; Mt. Stromlo and Siding Springs Observatories)*

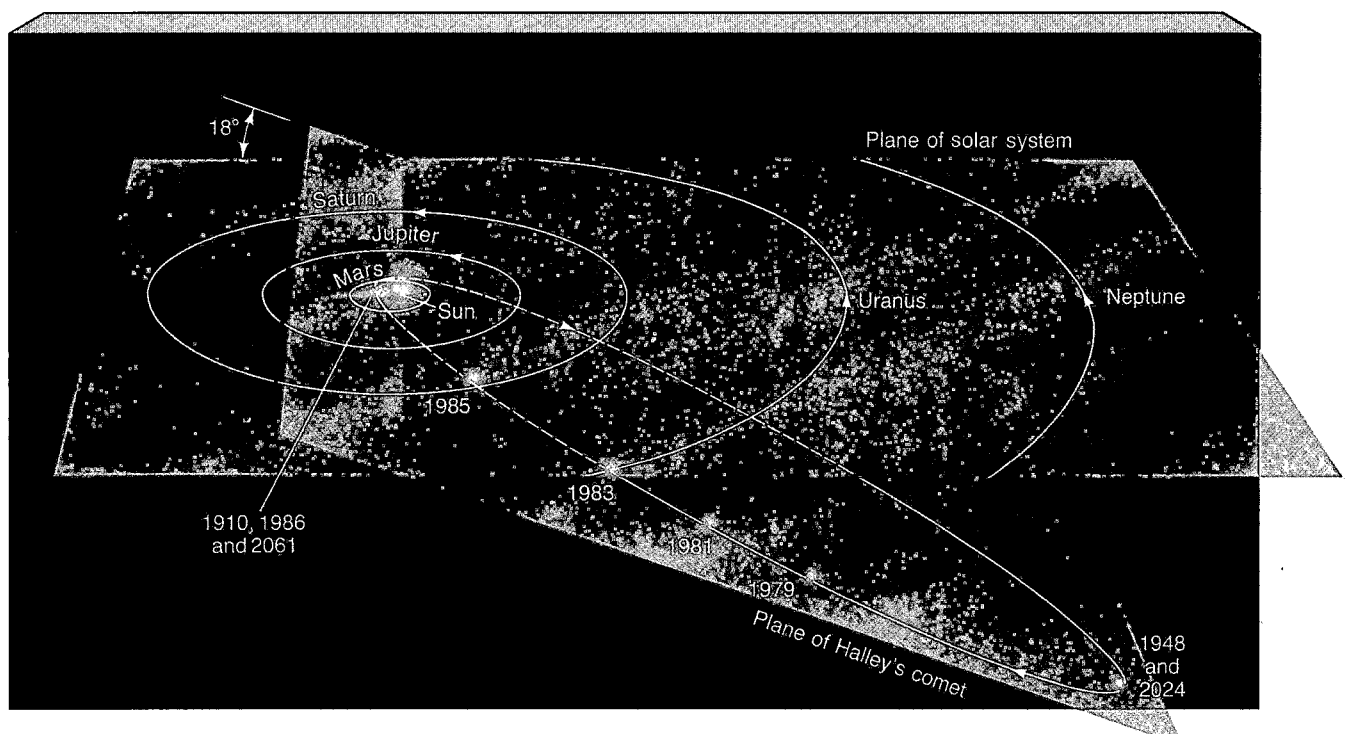

Figure 14.13 Halley's Orbit Halley's comet has a smaller orbital path and a shorter period than most comets, but its orbital orientation is not typical of a short-period comet. Sometime in the past the comet must have encountered a jovian planet (probably Jupiter itself), which threw it into a tighter orbit that extends not to the Oort cloud but merely a little beyond Neptune. Halley applied Newton's law of gravity to predict this comet's return.

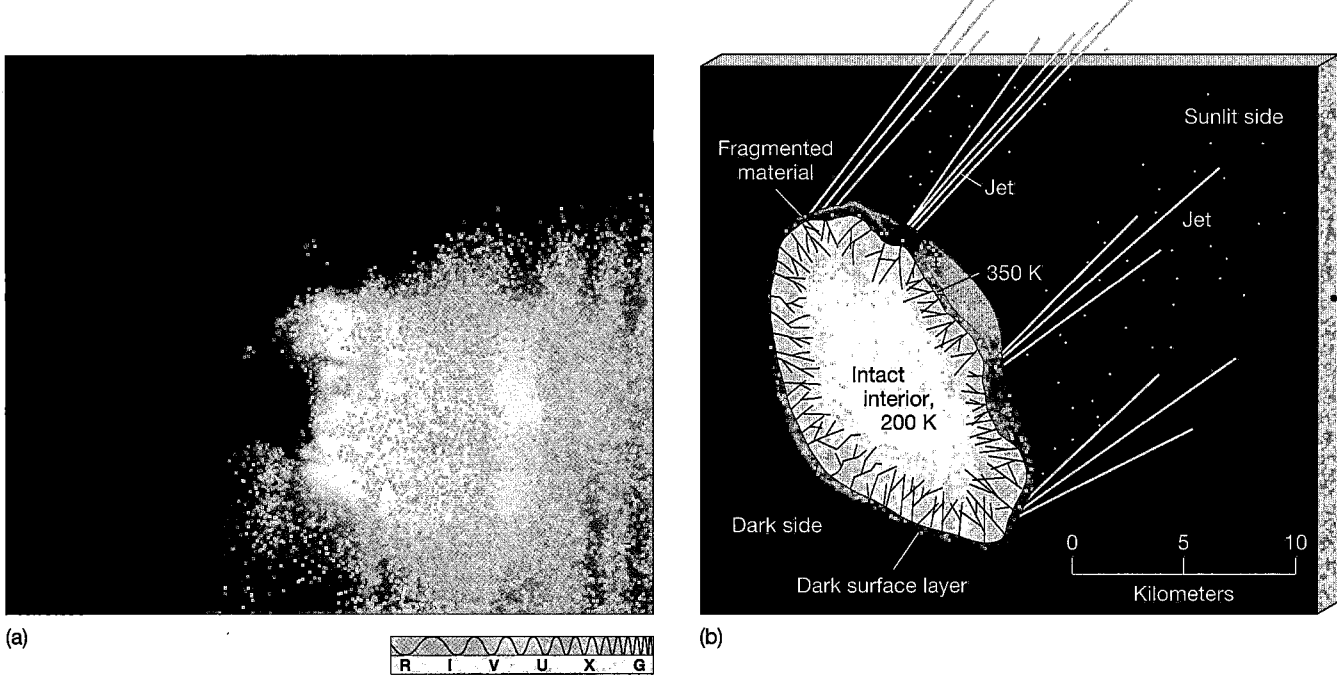

Figure 14.14 Halley, Up Close (a) The *Giotto* spacecraft resolved Halley's comet, showing its nucleus to be very dark, although heavy dust in the area obscured any surface features. Resolution here is about 50 m—half the size of a football field. At the time this image was made, in March 1986, the comet was within days of perihelion, and the Sun was toward the top. The brightest parts of the images are jets of evaporated gas and dust spewing from the comet's nucleus. (b) A diagram of Halley's nucleus, showing its size, shape, jets, and other physical and chemical properties. *(European Space Agency/Max Planck Institut)*

367

DISCOVERY 14-1

What Killed the Dinosaurs?

The name *dinosaur* derives from the Greek words *deinos* (terrible) and *sauros* (lizard). Dinosaurs were no ordinary reptiles. In their prime, roughly 100 million years ago, the dinosaurs were the all-powerful rulers of Earth. Their fossilized remains have been uncovered on all the world's continents. Despite their dominance, according to the fossil record, these creatures vanished from Earth quite suddenly about 65 million years ago. What happened to them?

Until fairly recently, the prevailing view among paleontologists—scientists who study prehistoric life—was that dinosaurs were rather small-brained, cold-blooded creatures. In chilly climates, or even at night, the metabolisms of these huge reptiles would have become sluggish, making it difficult for them to move around and secure food. The suggestion was that they were poorly equipped to adapt to sudden changes in Earth's climate, so that they eventually died out. However, a competing, and still controversial, view of dinosaurs has emerged. Recent fossil evidence suggests that many of these monsters may in fact have been warm-blooded and relatively fast-moving creatures—not at all the dull-witted, slow-moving giants of earlier conception. In any case, no species able to dominate Earth for more than 100 million years could have been too poorly equipped for survival. For comparison, humans have thus far dominated for a little over two million years.

If the dinosaurs didn't die out simply because of stupidity and inflexibility, then what happened to cause their sudden and complete disappearance? Many explanations have been offered for the extinction of the dinosaurs. Devastating plagues, magnetic field reversals, increased tectonic activity, severe climate changes, and supernova explosions have all been proposed. In the 1980s it was suggested that a huge extraterrestrial object collided with Earth 65 million years ago, and this is now (arguably) the leading explanation for the demise of the dinosaurs, although it is by no means universally accepted.

According to this idea, a 10- to 15-km-wide asteroid or comet struck Earth, releasing as much energy as 10 million or more of the largest hydrogen bombs humans have ever constructed and kicking huge quantities of dust (including the pulverized remnants of the asteroid itself) high into the atmosphere (see the figure below). The dust may have shrouded our planet for many years, virtually extinguishing the Sun's rays during this time. On the darkened surface, plants could not survive. The entire food chain was disrupted, and the dinosaurs, at the top of that chain, eventually became extinct.

Although we have no direct astronomical evidence to confirm or refute this idea, we can estimate the chances that a large asteroid or comet will strike Earth today, on the basis of observations of the number of objects presently on Earth-crossing orbits. The second figure shows the likelihood of an impact as a function of the size of the impacting body. The horizontal scale indicates the energy released by the collision, measured in *megatons* of TNT. The mega-

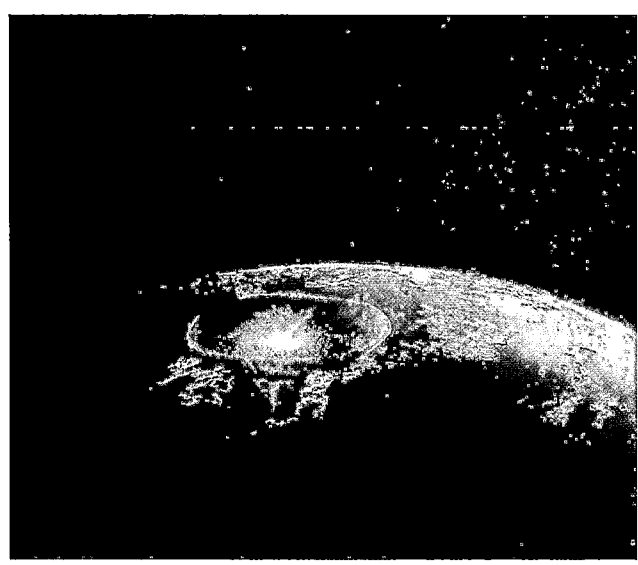

(C. Butler/Astrostock-Sanford)

The spacecraft found direct evidence for several jets of matter streaming from the nucleus. Instead of evaporating uniformly from the whole surface to form the comet's coma and tail, gas and dust apparently vent from small areas on the sunlit side of Halley's nucleus. The force of these jets may be largely responsible for the comet's 53-hour rotation period. Like maneuvering rockets on a spacecraft, such jets can cause a comet to change its rotation rate and even to veer away from a perfectly elliptical orbit. Astronomers had hypothesized the existence of these nongravitational forces on the basis of slight deviations from Kepler's laws observed in some cometary trajectories. However, only during the Halley encounter did astronomers actually see these jets at work.

PHYSICAL PROPERTIES OF COMETS

The mass of a comet can sometimes be estimated by watching how it interacts with other solar system objects or by determining the size of the nucleus and assuming a density characteristic of icy composition. These methods yield typical cometary masses ranging from 10^{12}–10^{16} kg, comparable to the masses of small asteroids. A comet's mass decreases with time because some material is lost each time the comet rounds the Sun as material evaporates from its surface. For comets that travel within an astronomical unit of the Sun, the evaporation rate can reach as high as 10^{30} molecules per second—about 30 tons of cometary material lost for every second the

ton—4.2×10^{16} joules, the explosive yield of a large nuclear warhead—is the only common terrestrial measure of energy adequate to describe the violence of these occurrences.

We see that 100-million-megaton events, like the planetwide catastrophe that supposedly wiped out the dinosaurs, are very rare, occurring only once every 10 million years or so. However, smaller impacts, equivalent to "only" a few tens of kilotons of TNT (roughly equivalent to the bomb that destroyed Hiroshima in 1945), could happen every few years—we may be long overdue for one. The most recent large impact was the Tunguska explosion in Siberia, in 1908, which packed a roughly one-megaton punch (see Figure 14.20).

The main geological evidence supporting this theory is a layer of clay enriched with the element iridium. This layer is found in 65-million-year-old rocky sediments all around our planet. Iridium on Earth's surface is rare because most of it sank into our planet's interior long ago.

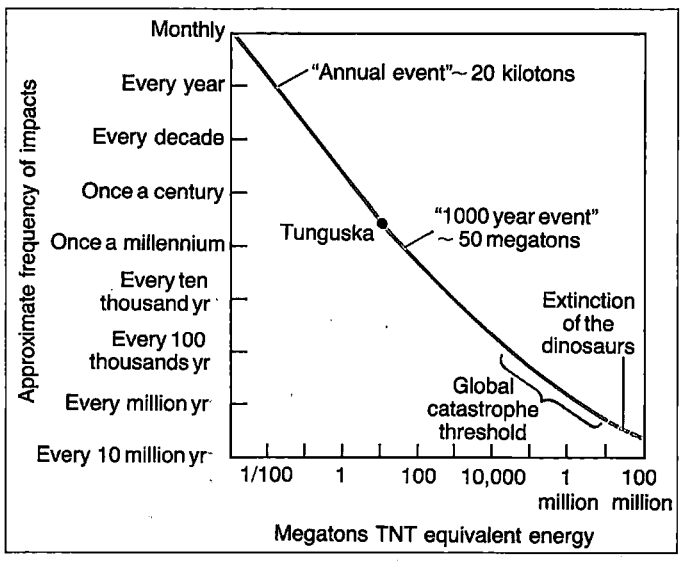

The abundance of iridium in this one layer of clay is about 10 times greater than in other terrestrial rocks, but it matches closely the abundance of iridium found in meteorites (and, we assume, in asteroids and comets too). The site of the catastrophic impact has also been tentatively identified near Chicxulub, in the Yucatan Peninsula in Mexico, where evidence of a heavily eroded, but not completely obliterated, crater of just the right size and age has been found.

The theory is not without its detractors, however. Perhaps predictably, the idea of catastrophic change on Earth being precipitated by events in interplanetary space was rapidly accepted by most astronomers, but it remains controversial among (some) paleontologists and geologists. Opponents argue that the amount of iridium in the clay layer varies greatly from place to place across the globe, and there is no complete explanation of why that should be so. Perhaps, they suggest, the iridium was produced by volcanoes, and had nothing to do with an extraterrestrial impact at all.

Still, in the 20 years since the idea was first suggested, the focus of the debate seems to have shifted. The reality of a major impact 65 million years ago has become widely accepted, and much of the argument now revolves around the question of whether that event actually caused the extinction of the dinosaurs, or merely accelerated a process that was already underway. Either way, the realization that such catastrophic events can and do occur marks an important milestone in our understanding of evolution on our planet. (This realization was doubtless bolstered by the Shoemaker-Levy 9 impact on Jupiter in 1994, as discussed in the text). In addition, there is growing evidence for even larger impacts in the more distant past, with even more sweeping evolutionary consequences.

As a general rule, we can expect that global catastrophes are bad for the dominant species on a planet. As the dominant species on Earth, we are the ones who now stand to lose the most.

comet spends near the Sun (within Earth's orbit, say). Astronomers have estimated that this loss of material will destroy Halley's comet in about 5000 orbits, or 40,000 years. Occasionally, as shown in Figure 14.15, the process can be much more violent, causing a comet suddenly to flare up in brightness, then rapidly fade as its nucleus disintegrates.

In seeking the physical makeup of a cometary body itself, astronomers are guided by the observation that comets have dust that reflects light, as well as gas that emits spectral lines of hydrogen, nitrogen, carbon, and oxygen. Even as the atoms, molecules, and dust particles boil off, creating the coma and tail, the nucleus itself remains a cold mixture of gas and dust, hardly more than a

ball of loosely packed ice with a density of about 100 kg/m^3 and a temperature of only a few tens of kelvins. Experts now consider cometary nuclei to be largely made of dust particles trapped within a mixture of methane, ammonia, carbon dioxide, and ordinary water ice. (These constituents should be fairly familiar to you as the main components of most of the small moons in the outer solar system, discussed in Chapters 12 and 13.) Because of this composition, comets are often described as "dirty snowballs."

In February 1999, NASA launched the *Stardust* mission, with the objective of collecting and returning to Earth the first ever samples of cometary material. On January 15, 2001, the spacecraft used a gravity assist from

Earth to boost it onto a path designed to approach within 150 km of the nucleus of comet P/Wild 2, in January 2004. The comet was chosen because it is a relative newcomer to the inner solar system, having been deflected onto its present orbit by an encounter with Jupiter in 1974. It therefore has not experienced much solar heating or loss of mass by evaporation. If all goes as planned, *Stardust* will return to Earth in January 2006. Mission scientists eagerly await this unprecedented opportunity to study in detail the physical, chemical, and biological properties of a body that most probably has not changed significantly since our solar system formed.

A COMETARY IMPACT

In July 1994, skywatchers were treated to an exceedingly rare event that greatly increased our knowledge of comet composition and structure—the collision of a comet (called Shoemaker-Levy 9, after its discoverers) with the planet Jupiter! When it was discovered in March 1993, Shoemaker-Levy 9 appeared to have a curious, "squashed" appearance. Higher-resolution images (see Figure 14.16a)

revealed that the comet's flattened nucleus was really made up of several pieces, the largest no more than 1 km across. All the pieces were following the same orbit, but they were spread out along the comet's path like a string of pearls a million kilometers long.

What could have caused such an unusual object? Tracing the orbit backward in time, researchers calculated that early in July 1992 the comet had approached to within about 100,000 km of Jupiter, well inside the planet's Roche limit. ∞ (Sec. 12.4) They realized that the objects shown in Figure 14.16a were the fragments produced when a previously "normal" comet was captured by Jupiter and torn apart by its strong gravitational field.

On its next approach to Jupiter, in July 1994, Shoemaker-Levy 9 struck the planet's upper atmosphere, plowing into it at a speed of more than 60 km/s and causing a series of enormous explosions (Figure 14.16b). Every major telescope on Earth, the *Hubble Space Telescope*, *Galileo* (which was only 1.5 A.U. from the planet at the time), and even *Voyager 2* were watching. Each impact created, for a period of a few minutes, a brilliant fireball hundreds of kilometers across and having a temperature of

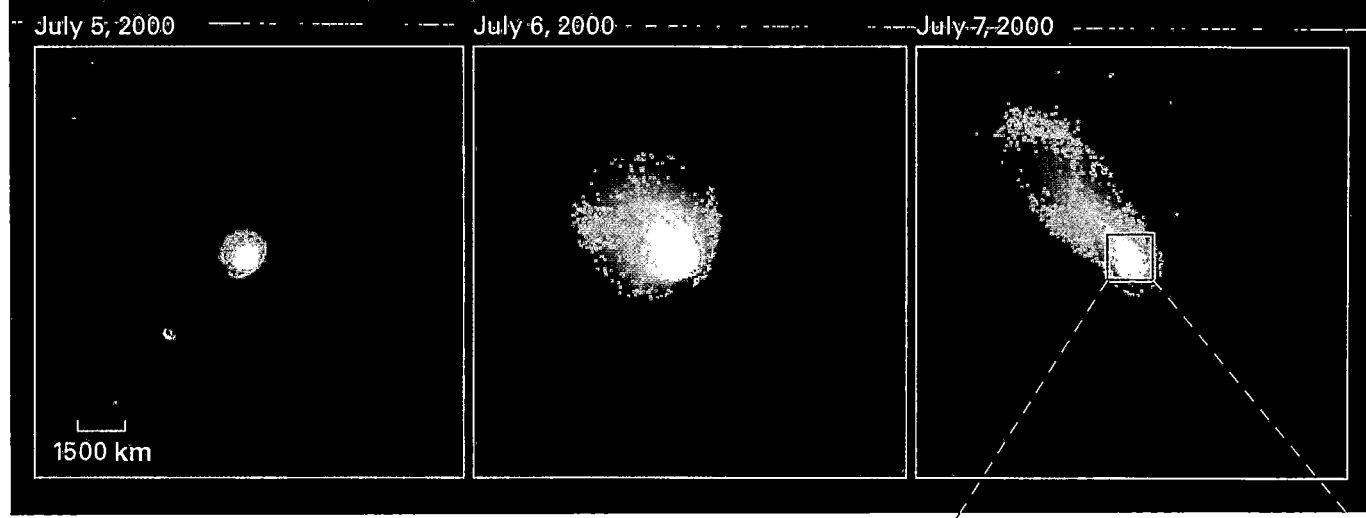

July 5, 2000   July 6, 2000   July 7, 2000

1500 km

Figure 14.15 Comet Flare-up As a comet approaches the heat of the Sun, its icy material vaporizes, causing the spectacular tail we observe. Sometimes, as in this sequence of *Hubble Space Telescope* images captured in July 2000, the vaporization can be explosive, causing a sudden flare-up in brightness. Shortly after these images were taken, the comet's brightness diminished rapidly as its nucleus disintegrated. This was comet LINEAR, named after its "discoverer," the automated Lincoln Lab telescope in New Mexico, conducting a search for Earth-crossing asteroids. The inset shows a dramatic shower of "minicomets" as its nucleus fragmented. *(NASA)*

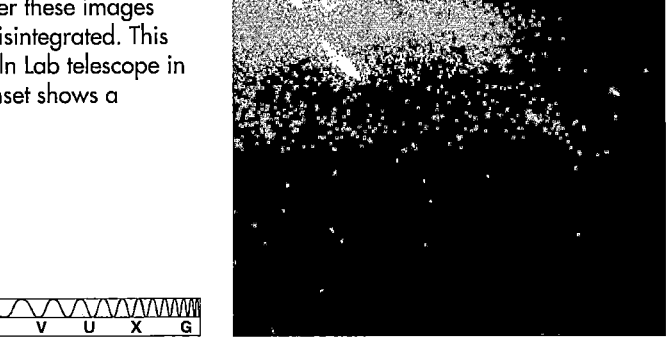

R   I   V   U   X   G

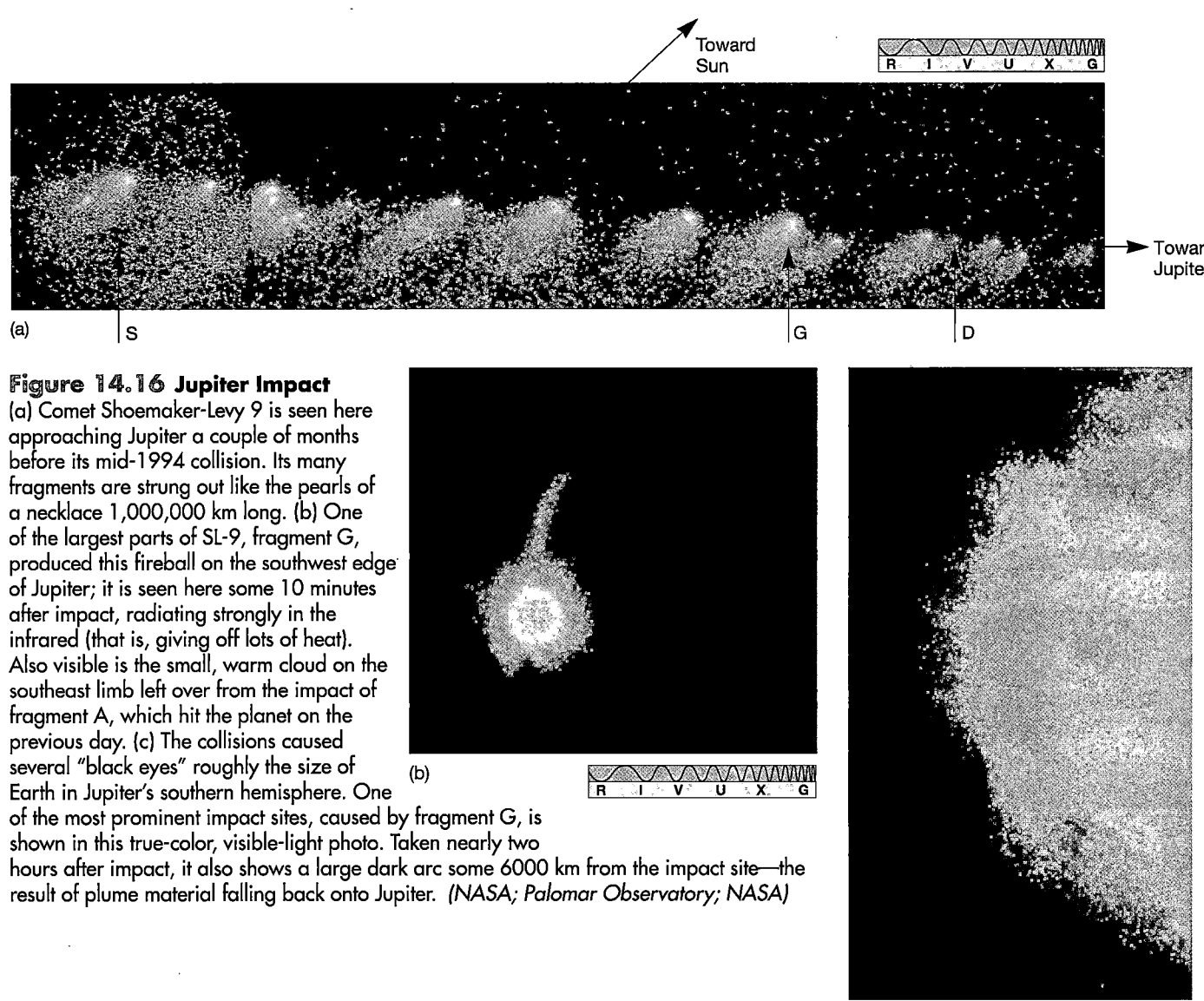

Figure 14.16 Jupiter Impact
(a) Comet Shoemaker-Levy 9 is seen here approaching Jupiter a couple of months before its mid-1994 collision. Its many fragments are strung out like the pearls of a necklace 1,000,000 km long. (b) One of the largest parts of SL-9, fragment G, produced this fireball on the southwest edge of Jupiter; it is seen here some 10 minutes after impact, radiating strongly in the infrared (that is, giving off lots of heat). Also visible is the small, warm cloud on the southeast limb left over from the impact of fragment A, which hit the planet on the previous day. (c) The collisions caused several "black eyes" roughly the size of Earth in Jupiter's southern hemisphere. One of the most prominent impact sites, caused by fragment G, is shown in this true-color, visible-light photo. Taken nearly two hours after impact, it also shows a large dark arc some 6000 km from the impact site—the result of plume material falling back onto Jupiter. *(NASA; Palomar Observatory; NASA)*

many thousands of kelvins. The largest of the fireballs were bigger than planet Earth. The energy released in each explosion was comparable to a billion terrestrial nuclear detonations, rivaling in violence the prehistoric impact suspected of causing the extinction of the dinosaurs on Earth 65 million years ago (see *Discovery 14-1*).

The effects on the planet's atmosphere and the vibrations produced throughout Jupiter's interior were observable for days after the impact. The fallen material from the impacts spread slowly around Jupiter's bands and after five months reached completely around the planet. It took years for all the cometary matter to settle into Jupiter's interior.

As best we can determine, none of the cometary fragments breached the jovian clouds. Only *Galileo* had a direct view of the impacts on the back side of Jupiter, and in every case the explosions seemed to occur high in the atmosphere, above the uppermost cloud layer. Most of the dark material seen in the images is probably pieces of the comet rather than parts of Jupiter. Spectral lines from silicon, magnesium, and iron were detected in the aftermath of the collisions, and the presence of these metals might explain the dark material observed near some of the impact sites (Figure 14.16c). Water vapor was also detected spectroscopically, again apparently from the melted and vaporized comet—which really did resemble a loosely packed snowball.

Concept Check

■ In terms of composition, how do comets differ from asteroids?

14.3 Meteoroids

⑤ On a clear night it is possible to see a few *meteors*—
"shooting stars"—every hour. A **meteor** is a sudden streak
of light in the night sky caused by friction between air
molecules in Earth's atmosphere and an incoming piece of
interplanetary matter—an asteroid, comet, or **meteoroid.**
This friction heats and excites the air molecules, which
then emit light as they return to their ground states, producing the characteristic bright streak shown in Figure
14.17. Note that the brief flash that is a meteor is in no way
similar to the broad, steady swath of light associated with a
comet's tail. A meteor is a fleeting event in Earth's atmosphere, whereas a comet tail exists in deep space and can be
visible in the sky for weeks or even months. Recall from
Section 6.5 that the distinction between an asteroid and a
meteoroid is simply a matter of size. Both are chunks of
rocky interplanetary debris; meteoroids are conventionally
taken to be less than 100 m in diameter.

Before encountering the atmosphere, the piece of debris causing a meteor was almost certainly a meteoroid,
simply because these small interplanetary fragments are far
more common than either asteroids or comets. Any piece
of interplanetary debris that survives its fiery passage
through our atmosphere and finds its way to the ground is
called a **meteorite.**

COMETARY FRAGMENTS

Smaller meteoroids are mainly the rocky remains of bro-
ken-up comets. Each time a comet passes near the Sun,
some cometary fragments dislodge from the main body.
The fragments initially travel in a tightly knit group of
dust or pebble-sized objects, called a **meteoroid swarm,**
moving in nearly the same orbit as the parent comet. Over
the course of time, the swarm gradually disperses along the
orbit, and eventually the **micrometeoroids,** as these small
meteoroids are known, become more or less smoothly
spread all the way around the parent comet's orbit. If
Earth's orbit happens to intersect the orbit of such a young
cluster of meteoroids, a spectacular **meteor shower** can
result. Earth's motion takes it across a given comet's orbit
at most twice a year (depending on the precise orbit of
each body). Intersection occurs at the same time each year
(see Figure 14.18), so the appearance of certain meteor
showers is a regular and (fairly) predictable event.

Table 14.1 lists some prominent meteor showers, the
dates they are visible from Earth, and the comet from

which they are thought to originate. Meteor showers are
usually named for their *radiant,* the constellation from
whose direction they appear to come. For example, the
Perseid shower is seen to emanate from the constellation
Perseus. It can last for several days but reaches maximum
every year on the morning of August 12, when upward of
50 meteors per hour can be observed. Astronomers can use
the speed and direction of a meteor's flight to compute its
interplanetary trajectory. This is how certain meteoroid
swarms have come to be identified with well-known comet
orbits. For example, the Perseid shower shares the same
orbit as comet 1862III, the third comet discovered in the
year 1862 (also known as comet Swift-Tuttle).

STRAY ASTEROIDS

Larger meteoroids—more than a few centimeters in diameter—are usually *not* associated with swarms of cometary
debris. Generally regarded as small bodies that have

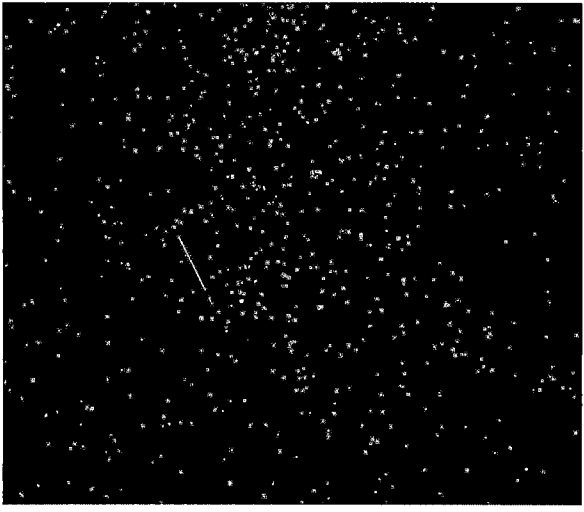

(a)

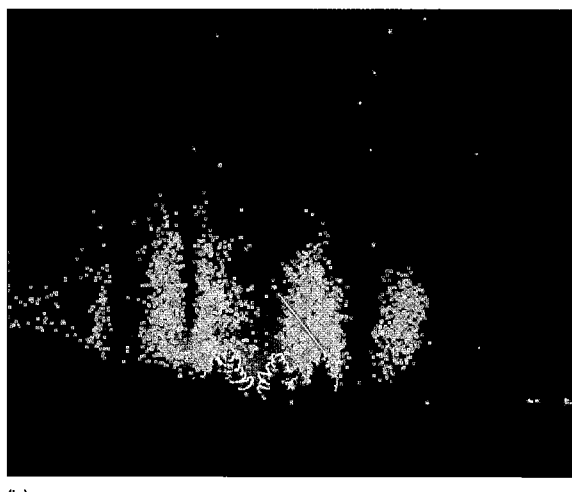

(b)

Figure 14.17 Meteor Trails A bright streak called a meteor is
produced when a fragment of interplanetary debris plunges into the
atmosphere, heating the air to incandescence. (a) A small meteor
photographed against a backdrop of stars. (b) An auroral display
provides the background for a brighter meteor trail. *(P. Parviainen/
Science Photo Library)*

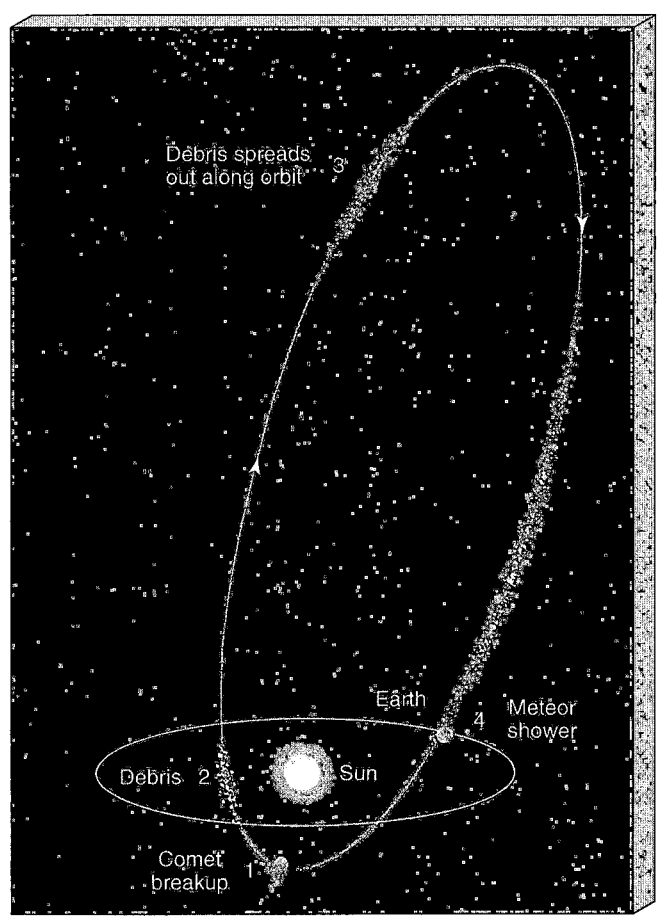

Figure 14.18 Meteor Showers A meteoroid swarm associated with a given comet intersects Earth's orbit at specific locations, giving rise to meteor showers at specific times of the year. We imagine that a portion of the comet breaks up near perihelion, at the point marked 1. The fragments continue along the original comet orbit, gradually spreading out as they go (points 2 and 3). The rate at which the debris disperses around the orbit is actually much slower than depicted here—it takes many orbits for the material to spread out as shown. Eventually, the fragments extend all around the orbit, more or less uniformly. If the orbit happens to intersect Earth's orbit, a meteor shower is seen each time Earth passes through the intersection (point 4).

strayed from the asteroid belt, possibly as the result of asteroid collisions, these objects have produced most of the cratering on the surfaces of the Moon, Mercury, Venus, Mars, and some of the moons of the jovian planets. When these large meteoroids enter Earth's atmosphere with typical velocity of nearly 20 km/s, they produce energetic shock waves, or "sonic booms," as well as bright sky streaks and dusty trails of discarded debris. Such large meteors are sometimes known as *fireballs*. The greater the speed of the incoming object, the hotter its surface becomes and the faster it burns up. A few large meteoroids enter the atmosphere at such high speed (about 75 km/s) that they either fragment or disperse entirely at high altitudes.

The more massive meteoroids (at least a ton in mass and a meter across) do make it to the surface, producing a crater such as the kilometer-wide Barringer Crater shown in Figure 8.19. From the size of this crater, we can estimate that the meteoroid responsible must have had a mass of about 200,000 tons. Only 25 tons of iron

TABLE 14.1 Some Prominent Meteor Showers

MORNING OF MAXIMUM ACTIVITY	SHOWER NAME	ROUGH HOURLY COUNT	PARENT COMET
Jan. 3	Quadrantid	40	—
Apr. 21	Lyrid	10	1861I (Thatcher)
May 4	Eta Aquarid	20	Halley
June 30	Beta Taurid	25	Encke
July 30	Delta Aquarid	20	—
Aug. 11	Perseid	50	1862III (Swift-Tuttle)
Oct. 9	Draconid	up to 500	Giacobini-Zinner
Oct. 20	Orionid	30	Halley
Nov. 7	Taurid	10	Encke
Nov. 16	Leonid	12*	1866I (Tuttle)
Dec. 13	Geminid	50	3200 Phaeton[†]

*Every 33 years, as Earth passes through the densest region of this meteoroid swarm, we see intense showers that can exceed 1000 meteors per minute for brief periods of time. This is next expected to occur in 2032.
[†]Phaeton is actually an asteroid and shows no signs of cometary activity, but its orbit matches the meteoroid paths very well.

DISCOVERY 14-2

Comets Hyakutake and Hale-Bopp

One of the most spectacular comets in recent years was comet Hyakutake 1996. Named after a Japanese amateur astronomer who noticed it as "something odd and out of place" while scanning the skies with a pair of binoculars, Hyakutake grew from a small smudge while still far from the Sun into a splendid display comprising a huge coma nearly the apparent size of the Moon and a tail that eventually stretched a third of the way across the sky. The figure below shows a *Hubble Space Telescope* image of Hyakutake taken in March 1996, when the comet passed closest to Earth—only 15 million km (0.1 A.U.) away. The comet's icy nucleus, the brightest point in the image, is unresolved here (the field of view is about 1000 km across), but radar pulses sent toward Hyakutake did return an echo, indicating that the diameter of the nucleus was 1–3 km. In the image, the Sun is out of the frame at bottom right, and the innermost part of the comet's tail is at upper left (on the side opposite the Sun, as explained in the text).

Other images showed sporadic jets pointing mostly sunward. These jets were gases gushing from the side of

the comet closest to the Sun before wrapping around to become part of the graceful tail. The comet was examined at every conceivable wavelength, but perhaps the most surprising result was the intense X rays emitted from its head. Even the sunward side of the comet was far too cool to emit X rays, which are usually associated with very-high-temperature phenomena. Astronomers speculate that the X rays were produced by shock waves created as the solar wind hit the leading edge of the comet's coma.

Hot on the heels of Hyakutake came another interplanetary vagabond—comet Hale-Bopp 1995. Discovered in 1995 by two American amateur astronomers, it reached its maximum brightness in the spring of 1997. Outshining everything in the night sky except the Moon and the brightest planets and stars, Hale-Bopp was probably the most widely viewed and studied comet in history.

This comet's unusual brightness and long (20°) tail were probably caused by a huge nucleus—about 30–40 km in diameter. That's a very large ball of dirty ice, compared with the average comet core, which measures some 3–5 km

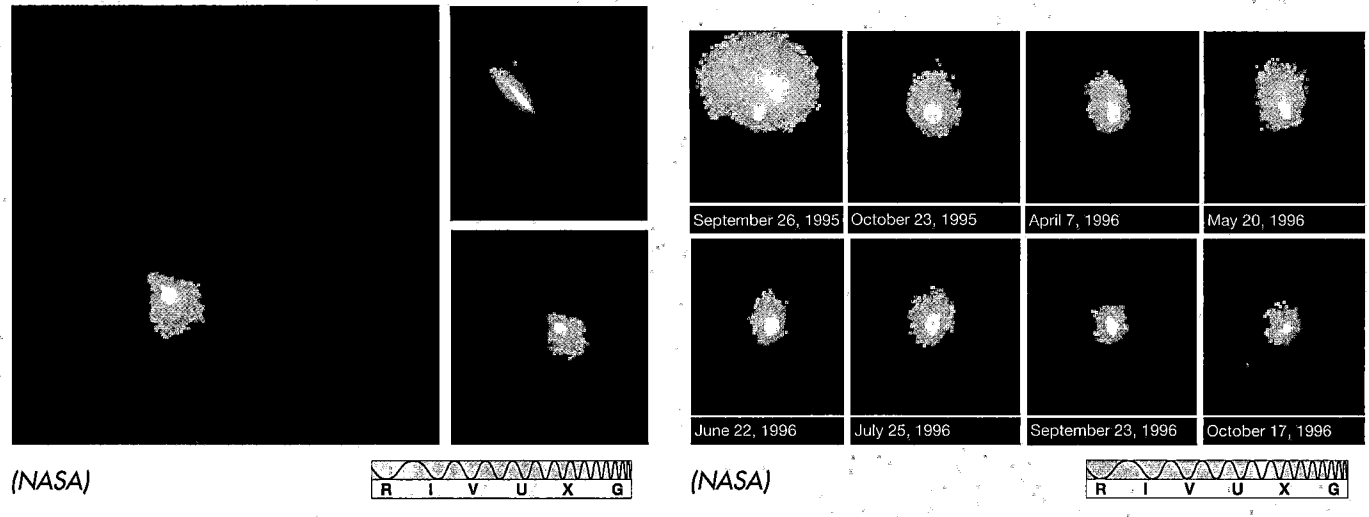

(NASA)

(NASA)

meteorite fragments have been found at the crash site. The remaining mass must have been scattered by the explosion at impact, broken down by subsequent erosion, or buried in the ground.

Currently, Earth is scarred with nearly 100 craters larger than 0.1 km in diameter. Most of these are so heavily eroded by weather and distorted by crustal activity that they can be identified only in satellite photography, as shown in Figure 14.19. Fortunately, such major collisions between Earth and large meteoroids are thought to be rare events now. Researchers believe that, on average, they occur only once every few hundred thousand years (see *Discovery 14-1*).

The orbits of large meteorites that survive their plunge through Earth's atmosphere can be reconstructed in a manner similar to that used to determine the orbits of meteor showers. In most cases, their computed orbits do indeed intersect the asteroid belt, providing the strongest evidence we have that they were once part of the belt before being redirected, probably by a collision with another asteroid, into the Earth-crossing orbit that led to the impact with our planet. Not all meteoroids come from the asteroid belt, however. As we have already seen, some are known to have originated on the surface of Mars. ∞ (*Discovery 10-2*) In addition, detailed composition studies re-

across. The comet or asteroid that struck the Earth 65 million years ago (see *Discovery 14-1*), perhaps causing the extinction of the dinosaurs, is thought to have been about 10–15 km in size. At perihelion, Hale-Bopp's hydrogen envelope was enormous—1 A.U. across.

The second figure at left shows a series of *HST* observations of the inner part of comet Hale-Bopp. The colors are artificial, with white representing the brightest parts of the comet, and red less bright; resolution is nearly 500 km, far larger than the estimated size of the icy nucleus itself. These eight pictures were taken over the course of a year, from September 1995 to October 1996 as the comet neared the Sun.

Astronomers used multiple images like these to chronicle the evolution of the comet's nucleus. Especially interesting is the dust outburst in the first photo at top left—even when Hale-Bopp was still far beyond the orbit of Jupiter. At bottom right, as the comet neared the Sun, multiple jets are seen emanating from the surface. *Hubble* was unable to follow the comet even closer to the Sun (which it rounded on April 1997, at a perihelion distance of 0.9 A.U.), for fear of damaging its delicate optics. The final figure at right shows a spectacular ground-based view of the comet around the time of perihelion.

Apart from its unusual size, Hale-Bopp seems to be quite representative of many long-period comets. Just about all the molecules previously known from other comets were seen in Hale-Bopp's spectrum. New in-

(Aaron Horowitz/Corbis)

frared observing techniques allowed, for the first time, identification of many of the minerals making up the comet's dust. It was found to be very similar to the interplanetary dust found in near-Earth space and collected in Earth's upper atmosphere. What was unique about Hale-Bopp was that, being so large, it could be observed for a very long period of time, even when far from the Sun, allowing observers to keep a continuous and detailed record of its evolution.

Hale-Bopp is now moving out of the inner solar system, its incoming orbit modified by the forces from its jets. Astronomers plan to monitor the comet for several more years as it moves far from the Sun's warmth, and its activity winds down. After a trip well outside the orbit of Pluto, extending some 350 A.U. from the Sun, the comet will return to the inner solar system in about 2400 years.

---

veal that others most likely came from the Moon, blasted off the lunar surface by violent impacts long ago.

Not all meteoroid encounters with Earth result in an impact. One of the most recent meteoritic events occurred in central Siberia on June 30, 1908 (Figure 14.20). The presence of only a shallow depression as well as a complete lack of fragments implies that this Siberian intruder exploded several kilometers above the ground, leaving a blasted depression at ground level but no well-formed crater. Recent calculations suggest that the object in question was a rocky meteoroid about 30 m across. The explosion, estimated to have been equal in energy to a

10-megaton nuclear detonation, was heard hundreds of kilometers away and produced measurable increases in atmospheric dust levels all across the Northern Hemisphere.

METEORITE PROPERTIES

7 One feature that distinguishes the small micrometeoroids, which burn up in Earth's atmosphere, from larger meteoroids, which reach the ground, is composition. The average density of meteoritic fireballs too small to reach the ground (but which can be captured by high-flying aircraft) is about 500–1000 kg/m^3. Such a low density is

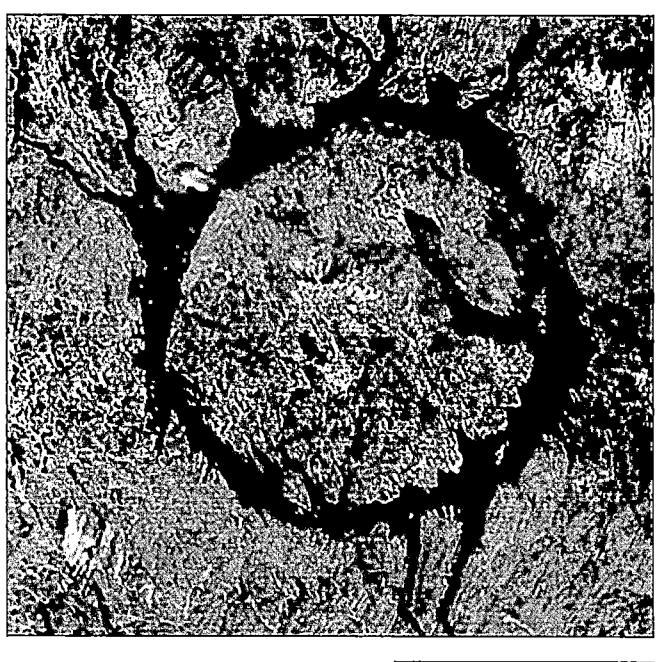

Figure 14.19 Manicouagan Reservoir This photograph, taken from orbit by the U.S. *Skylab* space station, clearly shows the ancient impact basin that forms Quebec's Manicouagan Reservoir. A large meteorite landed there about 200 million years ago. The central floor of the crater rebounded after the impact, forming an elevated central peak. The lake, 70 km in diameter, now fills the resulting ring-shaped depression. *(NASA)*

typical of comets, which are made of loosely packed ice and dust. In contrast, the meteorites that reach Earth's surface are often much denser—up to 5000 kg/m^3—suggesting a composition more like that of the asteroids. Meteorites like those shown in Figure 14.21 have received close scrutiny from planetary scientists—prior to the Space Age, they were the only type of extraterrestrial matter we could touch and examine in terrestrial laboratories.

Most meteorites are rocky in composition (Figure 14.22a), although a few percent are composed mainly of iron and nickel (Figure 14.22b). Their basic composition is much like that of the inner planets and the Moon, except that some of their lighter elements—such as hydrogen and oxygen—appear to have boiled away long ago when the bodies from which the meteorites originated were molten. Some meteorites show clear evidence of strong heating at some time in their past, most likely indicating that they originated on a larger body that either experienced some geological activity or was partially melted during the colli-

sion that liberated the fragments that eventually became the meteorites. Others show no such evidence and probably date from the formation of the solar system.

Most primitive of all are the *carbonaceous* meteorites, so called because of their relatively high carbon content. They are black or dark gray, and they may well be related to the carbon-rich C-type asteroids that populate the outer asteroid belt. (Similarly, the silicate-rich stony meteorites are probably associated with the inner S-type asteroids.) Many carbonaceous meteorites contain significant amounts of ice and other volatile substances, and they are usually rich in organic molecules.

Finally, almost all meteorites are *old*. Direct radioactive dating shows most of them to be between 4.4 and 4.6 billion years old—roughly the age of the oldest lunar rocks. Meteorites, along with some lunar rocks, comets, and perhaps the planet Pluto, provide essential clues to the original state of matter in the solar neighborhood.

Concept Check

- What are meteoroids, and why are they important to planetary scientists?

Figure 14.20 Tunguska Debris The Tunguska event of 1908 leveled trees over a vast area. Although the impact of the blast was tremendous and its sound audible for hundreds of kilometers, the Siberian site was so remote that little was known about the event until scientific expeditions arrived to study it many years later. *(Sovfoto/Eastfoto)*

(a)

(b)

Figure 14.21 Large Meteorites (a) The world's second largest meteorite, the Ahnighito, on display at the American Museum of Natural History in New York serves as a jungle gym for curious children. (b) The Wabar meteorite, discovered in the Arabian desert. Although small fragments of the original meteor had been collected more than a century before, the 2000-kg main body was not found until 1965. *(Corbis-Blair; J. Mandaville/Aramco Magazine)*

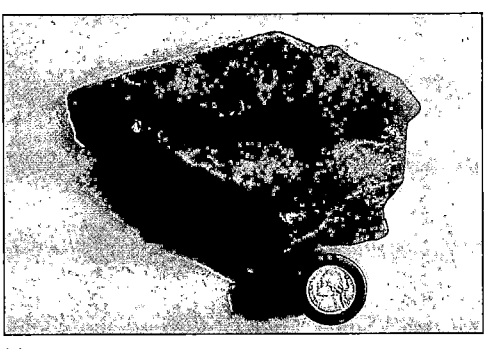

(a)

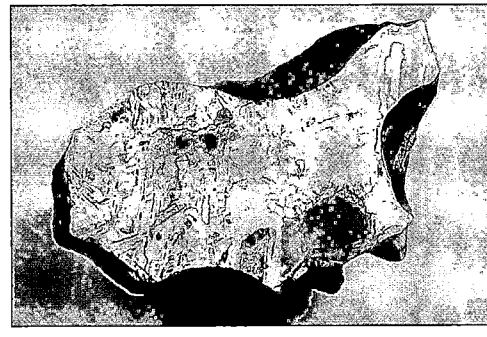

(b)

Figure 14.22 Meteorite Samples (a) A stony meteorite often has a dark fusion crust, created when its surface is melted by the tremendous heat generated during passage through the atmosphere. (b) Iron meteorites, much rarer than stony ones, usually contain some nickel as well. Most such meteorites show characteristic crystalline patterns when their surfaces are cut, polished, and etched with weak acid. *(Science Graphics)*

Chapter Review

SUMMARY

More than 100,000 **asteroids** (p. 356) have been cataloged. Most orbit in a broad band called the **asteroid belt** (p. 356) between the orbits of Mars and Jupiter. They are probably primal rocks that never clumped together to form a planet. The largest asteroids are a few hundred kilometers across. Most are much smaller. The total mass of all asteroids combined is less than $\frac{1}{10}$ the mass of Earth's Moon. Asteroids are classified according to the properties of their reflected light. Brighter S-type (silicate) asteroids dominate the inner asteroid belt, whereas darker C-type (carbonaceous) asteroids are more plentiful in the outer regions. They are believed to have changed little since the solar system formed. Smaller asteroids tend to be irregular in shape, and may have experienced violent collisions in the past.

A few **Earth-crossing asteroids** (p. 360) have orbits that intersect Earth's orbit and will probably collide with our planet one day. The **Trojan asteroids** (p. 361) share Jupiter's orbit, remaining 60° ahead of or behind that planet as it moves around the Sun. The **Kirkwood gaps** (p. 361) in the main asteroid belt have been cleared by Jupiter's gravity.

Comets (p. 362) are fragments of icy material that normally orbit far from the Sun. Unlike the orbits of most other bodies in the solar system, comet orbits are often highly elongated and not confined to the ecliptic plane. Most comets are thought to reside in the **Oort cloud** (p. 365), a vast "reservoir" of cometary material, tens of thousands of astronomical units across, completely surrounding the Sun. A very small fraction of comets happen to have highly elliptical orbits that bring them into the inner solar system. Comets with orbital periods less than about 200 years are thought to originate not in the Oort cloud but in the **Kuiper belt** (p. 364), a broad band lying roughly in the ecliptic plane, beyond the orbit of Neptune.

As a comet approaches the Sun its surface ice begins to vaporize. We see the comet by the sunlight reflected from the dust and vapor released. The **nucleus** (p. 362), or core, of a comet may be only a few kilometers in diameter. It is surrounded by a **coma** (p. 362) of dust and gas. Surrounding this is an extensive invisible **hydrogen envelope** (p. 363). Stretching behind the comet is a long **tail** (p. 362), formed by the interaction between the cometary material and the solar wind. The **ion tail** (p. 363) consists of ionized gas particles and always points directly away from the Sun. The **dust tail** (p. 363) is less affected by the solar wind and has a somewhat curved shape. Spacecraft visited Halley's comet in 1986 and studied its nucleus. All other comet studies have been indirect, usually based on spectroscopic measurements. Comets are icy, dusty bodies, sometimes called "dirty snowballs," that are believed to be leftover material unchanged since the formation of the solar system. Their masses are comparable to the masses of small asteroids.

Meteors (p. 372), or "shooting stars," are bright streaks of light that flash across the sky as a **meteoroid** (p. 372), a piece of interplanetary debris, enters Earth's atmosphere. If any of the meteoroid reaches the ground, it is called a **meteorite** (p. 372). The major difference between meteoroids and asteroids is their size. The dividing line between them is conventionally taken to be 100 m. Comets and stray asteroids are responsible for most of the cratering on the various worlds in the solar system. Earth is still subject to these sorts of collisions. The most recent large impact occurred in 1908, when an asteroid apparently exploded several miles above Siberia. Comet Shoemaker-Levy 9 struck Jupiter in 1994, causing violent explosions in that planet's atmosphere.

Each time a comet rounds the Sun, some cometary material becomes dislodged, forming a **meteoroid swarm** (p. 372)—a group of small **micrometeoroids** (p. 372) following the comet's original orbit. If Earth happens to pass through the comet's orbit, a **meteor shower** (p. 372) occurs. Larger meteoroids are probably pieces of material chipped off asteroids following collisions in the asteroid belt. Meteorite composition is thought to mirror the composition of the asteroids, and the few orbits that have been determined are consistent with an origin in the asteroid belt. Some meteorites show evidence of heating, but the oldest ones do not. Most meteorites are between 4.4 and 4.6 billion years old.

SELF TEST: TRUE OR FALSE?

____ 1. Most asteroids move on almost circular orbits.

____ 2. The Apollo asteroids have perihelion distances of less than 1 A.U.

____ 3. The C-type asteroids are so named because of their heavily cratered surfaces.

____ 4. S-type asteroids are more common in the inner part of the asteroid belt.

____ 5. The least reflective asteroids are C-type.

____ 6. The Kirkwood gaps are two broad zones within the asteroid belt, lying around 2.5 and 3.0 A.U. from the Sun and each about 0.1 A.U. wide, in which no asteroids are found.

____ 7. Some comets travel up to 50,000 A.U. from the Sun.

____ 8. Cometary orbits always lie close to the ecliptic plane.

____ 9. Halley's comet spends most of its time beyond the orbit of Saturn.

____ 10. The Oort cloud is the large cloud of gas surrounding a comet while it is near the Sun.

____ 11. Tails of comets always lie along the path of the orbit.

____ 12. The light known as a meteor is due to the literal burning up of a meteoroid in Earth's atmosphere.

____ 13. Some meteorites found on Earth originally came from the Moon or Mars.

____ 14. Comets are the sources of meteor showers.

____ 15. Astronomers have succeeded in tracing the orbits of some meteorites back into the asteroid belt.

SELF TEST: FILL IN THE BLANK

1. Asteroids are generally _____ in composition.
2. The asteroid belt, where most asteroids are found, lies between the orbits of _____ and _____.
3. The largest asteroids are _____ of kilometers in diameter; the smallest (by definition) are only _____ meters across.
4. The Trojan asteroids share an orbit with _____.
5. The Kirkwood gaps are caused by an orbital _____ with Jupiter.
6. Comets are generally _____ in composition.
7. The process of sublimation is one in which a solid turns into a _____.
8. Comets' orbits are highly _____.
9. The nucleus of a comet is typically _____ kilometers across.
10. Passage of a comet near the Sun may leave a _____ moving in the comet's orbit.
11. The Kuiper belt is the source of the _____ comets.
12. Meteoroids are mostly fragments of _____ and _____.
13. When a meteoroid enters Earth's atmosphere, we see a _____.
14. Meteoroids that impact a planet or moon are called _____.
15. The oldest meteorites are about _____ years old.

REVIEW AND DISCUSSION

1. What are the Trojan, Apollo, and Amor asteroids?
2. How are asteroid masses measured?
3. How have the best photographs of asteroids been obtained?
4. What are the Kirkwood gaps? How did they form?
5. How do the C-type and S-type asteroids differ?
6. Are all asteroids found in the asteroid belt?
7. What are comets like when they are far from the Sun? What happens when they enter the inner solar system?
8. Where are most comets found?
9. Describe the various parts of a comet while it is near the Sun.
10. What are the typical ingredients of a comet nucleus?
11. How do we know what comets are made of?
12. What are some possible fates of comets?
13. Describe two ways in which a comet's orbit may change.
14. In what ways is the Kuiper belt similar to the asteroid belt? In what ways do they differ?
15. Explain the difference between a meteor, a meteoroid, and a meteorite.
16. What causes a meteor shower?
17. What do meteorites reveal about the age of the solar system?
18. Why can comets approach the Sun from any direction, but asteroids generally orbit close to the ecliptic plane?
19. Why do meteorites contain information about the early solar system, yet Earth does not?
20. What might be the consequences if a 10-km-diameter meteorite struck Earth today?

PROBLEMS  *Algorithmic versions of these questions are available in the Practice Problems module of the Companion Website.*

The number of squares preceding each problem indicates its approximate level of difficulty.

1. ■ (a) The asteroid Pallas has an average diameter of 520 km and a mass of 3.2×10^{20} kg. How much would a 100-kg astronaut weigh there? (b) What is the asteroid's escape speed?

2. ■ You are standing on the surface of a spherical asteroid 10 km in diameter, of density 3000 kg/m³. Could you throw a small rock fast enough that it escapes? Give the speed required in km/s and mph.

3. ■■ How large would the asteroid in the previous problem have to be for your weight to be one percent of your weight on Earth?

4. ■ Can you find a simple orbital resonance with Jupiter that can account for the small Kirkwood gap evident in Figure 14.7(a) at semimajor axis around 2.7 A.U.?

5. ■■ The asteroid Icarus (Figure 14.5) has perihelion of 0.19 A.U. and orbital eccentricity of 0.83. Calculate its orbital semimajor axis and aphelion distance from the Sun. Do these figures, by themselves, *necessarily* imply that Icarus will one day collide with Earth?

6. ■■ Calculate the minimum and maximum angular diameter of the Sun, as seen from Icarus (problem 5). How do they compare with the Sun's diameter seen from Earth?

7. ■■ Using the data given in the text, estimate Dactyl's orbital period as it orbits Ida.

8. ■■■ (a) *NEAR*'s initial orbit around Icarus had periapsis (distance of closest approach) of about 100 km from the asteroid's center. If the mass of Eros is 6.7×10^{15} kg and the orbit had an eccentricity of 0.3, calculate the spacecraft's orbital period. (b) Subsequently *NEAR* moved into a closer orbit, with periapsis 14 km and apoapsis (greatest distance) 60 km. What was the new orbital period?

9. ■■ (a) Calculate the orbital period of a comet with a perihelion distance of 0.5 A.U. and aphelion in the Oort cloud, at a distance of 50,000 A.U. from the Sun. (b) A short-period comet has a perihelion distance of 1 A.U. and an orbital period of 125 years. What is its maximum distance from the Sun?

10. ■■■ As comet Hale-Bopp rounded the Sun, nongravitational forces changed its orbital period from 4200 years to 2400 years. By what factor did the comet's semimajor axis change? Given that the perihelion remained unchanged at 0.914 A.U., calculate the old and new orbital eccentricities.

11. ■ Astronomers estimate that comet Hale-Bopp lost mass at an average rate of about 350,000 kg/s during the time it spent close to the Sun—a total of about 100 days. Estimate the total amount of mass lost and compare it with the comet's estimated mass of 5×10^{15} kg.

12. ■■ It has been hypothesized that Earth is under continuous bombardment by house-sized "minicomets" with typical diameters of 10 m, at the rate of some 30,000 per day. Assuming spherical shapes and average densities of 100 kg/ m^3, calculate the total mass of material reaching Earth each year. Compare the total mass received in the past one billion years (assuming all rates were the same in the past) with the mass of Earth's oceans (see Chapter 7, Problem 5).

13. ■ A particular comet has a total mass of 10^{13} kg, 95 percent of which is ice and dust. The remaining five percent is in the form of rocky fragments with an average mass of 100 g.

How many meteoroids would you expect to find in the swarm formed by the breakup of this comet?

14. ■■ It is observed that the number of asteroids or meteoroids of a given diameter is roughly inversely proportional to the square of the diameter. Approximating the actual distribution of asteroids in this way—a single 1000-km body (Ceres), one hundred 100-km bodies, ten thousand 10-km asteroids, and so on—and assuming constant densities of 3000 kg/m^3, calculate the total mass (in units of Ceres's mass) in the form of 1000-km bodies, 100-km bodies, 10-km bodies, 1-km bodies, and 100-m bodies.

15. ■■ A meteoroid was created by a collision in the asteroid belt two billion years ago. Its greatest distance from the Sun is 3 A.U., and its eccentricity is 0.8. How many times has it crossed Earth's orbital radius?

COLLABORATIVE EXERCISES

1. **Comets.** Comets are generally named by the discoverer. Describe what and why each member in your group would name a comet they discovered and how astronomers differentiate between short-period and long-period comets.

2. **Asteroid Excavation.** As a group, decide on what the United States government's policy should be on the mining of minerals from asteroids and justify the policy.

3. **Dinosaur Extinction.** Decide on what evidence would finally settle the "extinction of the dinosaurs" debate once and for all for your group.

RESEARCHING ON THE WEB

 *To complete the following exercises, go to the online Destinations module for Chapter 14 on the Companion Website for* Astronomy Today 4/e.

1. Access the "What is a Meteorite?" page and describe the difference between "find" and "fall" meteorites in terms of how they are discovered and which are most common.

2. Access the "Asteroid Fact Sheet" and list the name, diameter, and rotational period for the four largest asteroids.

3. Access the "Meteor Showers" page list of meteor showers and determine the name and date for the next meteor shower.

4. Access the "Comets Currently Visible" page and list the brightest long-period comet and the brightest short-period comet currently visible with their magnitudes and estimated perihelion dates.

PROJECTS

1. The only way to tell an asteroid from a star is to watch it over several nights. You can detect its movement in front of the star background. The astronomy magazines *Sky & Telescope* and *Astronomy* often publish charts for especially prominent asteroids. Look for the asteroids Ceres, Pallas, or Vesta. They are the brightest asteroids. Use the chart to locate the appropriate star field. Aim binoculars at that location in the sky; you may be able to pick out the asteroid from its location in the chart. If you can't, make a rough drawing of the entire field. Come back a night or two later, and look again. The "star" that has moved is the asteroid.

2. Although a spectacular naked-eye comet comes along only about once a decade, fainter comets can be seen with binoculars and telescopes in the course of every year. *Sky & Telescope* often runs a "Comet Digest" column announcing the whereabouts of comets. A comprehensive list of periodic

comets expected to return in a given year can be found in Guy Ottewell's Astronomical Calendar. This calendar contains a wealth of other sky information as well, including monthly star charts. At this writing, it costs $15 a year and can be purchased from:

> Astronomical Workshop
> Furman University
> Greenville, South Carolina 29613
> (803) 294-2208

3. There are a number of major meteor showers every year, but if you plan to watch one, be sure to notice the phase of the Moon. Bright moonlight or city lights can obliterate a meteor shower. A common misconception about meteor watching is that most meteors are seen in the direction of the shower's radiant point. It's true that if you trace the paths of the meteors backward in the sky, they all can be

seen to come from the radiant. But most meteors don't become visible until they are 20° or 30° from the radiant. Meteors can appear in all parts of the sky! Just relax and let your eyes rove among the stars. You will generally see many more meteors in the hours before dawn than in the

hours after sunset. Why do you suppose meteors have different brightnesses? Can you detect their variety of colors? Watch for meteors that appear to "explode" as they fall, and vapor trails that linger after the meteor itself has disappeared.

SKYCHART III PROJECTS  *The SkyChart III Student Version planetarium program on which these exercises are based is included as a separately executable program on the CD in the back of this text.*

1. ▣ For a better idea of the distribution of asteroids, import into SkyChart III data for asteroids with *FILE/Import* and find the file "5000 Asteroids.txt". Under *DRAW,* deselect *Stars, Deep Sky Objects, Constellations, Grid Lines,* and *Horizon Mask.* Select *VIEW/coordinates/ecliptic* and set *VIEW/5° Field.* Under *COMPUTATION/Location/Select/Sun,* click on *View from Object.* Set *Offset Distance* to 5 A.U. and enter coordinates *Offset R.A.* 18 00 00 and *Offset Dec.* 66 00 00 for the ecliptic North Pole. Center on the Sun and click F3 a few times so that *Magnitudes* on the *Chart Legend* reads at least 9 or *DRAW/Symbol and Grids* and in the *Draw Planets* area set the *Limiting Magnitude* to 9. On the screen you should see two or three plus signs, which indicate asteroids, along with the various planets. Click on the asteroids and you should find that they are Ceres, Pallas, and maybe Vesta. Click on one of the asteroids for name and details. Note the relative positions of the asteroids with respect to the planets. Continue to press F3 and observe the ring of asteroids formed as fainter and fainter objects are brought up.

2. ▣ While difficult to find in the multitude of asteroids, an Apollo asteroid, Toutatis, can be located if you keep pressing F3 until Magnitude 22.5 is indicated on the *Chart Legend.* One of the best ways to find Toutatis in this image is to use *ANIMATION/Trail For/Toutatis.* Run animation with one-week time steps for a short while until a line emerges that distinguishes Toutatis from the rest. Click on the asteroid at the end of the trail, and the details for Toutatis will be presented in the *Object Info* box. What size telescope would be required to see Toutatis, assuming good seeing? Animations run much more efficiently when few objects are presented on the screen because of the many calculations necessary with each change of the screen. Use F4 to reduce the objects presented to only the Sun, planets, and a few asteroids. In addition to setting a trail for Toutatis, set trails for Earth and Ceres. This will reveal the difference in orbits of Apollo asteroids and those with orbits within the asteroid belt.

3. ▣ Typically easier to find are a couple of asteroids, Chiron and Hidalgo, with orbits that extend beyond the orbit of Jupiter. Set up the display as described above and use *ANIMATION/Trail For/Chiron* and *ANIMATION/Trail For/Hidalgo.* Animate with a one-week time step to locate the asteroids. Click on each asteroid to open *Chart Legend* for details, and note the magnitude of each. What size telescope would be required to see these two objects?

4. ▣ With a minimum of objects displayed to only the Sun, planets, and a few asteroids, animate with trails for Earth, Toutatis, Ceres, Chiron, and Hidalgo. Select *VIEW/coordinates/ecliptic.* Observe the differences among the various orbits. This should reveal the variety of orbits possible for asteroids.

5. ▣ The first asteroid was discovered in 1801 and named Ceres. Change the *Date and Time* in SkyChart to the year 1801 and track Ceres's visibility. Return to local time and determine the future visibility of this asteroid from your location.

6. ▣ Comet Halley is one of the most famous comets in existence. Enter "halley" into the *Find Object* search field and track it until it makes its pass around the Sun in 2061. Use a large initial *Time Step,* adjusting it to smaller increments as 2061 approaches. Adjustments to the number of recorded positions may need to be made. When animated in SkyChart, can you explain the strange motion of Halley prior to 2061?

7. ▣ To observe the path followed by Comet Hale-Bopp center SkyChart III on the Sun. Under *COMPUTATION/Location* select *View from Object,* and under *Select* enter in Sun; choose *Exact Match,* enter *Find,* then *Select.* Also under Offset Distance (km) enter 5.00 and click on A.U. For *Offset R.A.* type in 18 00 00 and for *Offset Dec.* use 00 00 00. Under *ANIMATION/Set Trails,* set trails for Hale-Bopp, Earth, Ceres, and Jupiter. Then run the animation in steps of one day from December 1, 1995, until January 1, 1998, to observe how the comet passed through the plane of the ecliptic. Actually, it passed sufficiently close to Jupiter to change its course. If the animation runs too fast, you may wish to step through it with F6 or backwards with F5.

8. ▣▣ SkyChart allows for the addition of asteroids, comets, and satellites in the form of "two-line elements" or "tle" files. Locate the "tle" file for the International Space Station (see Appendix 4 for instructions). Once you have acquired the file choose *FILE/Import/NASA Satellite File* and select the ISS file. The ISS will not be visible in SkyChart. To "see" it, use the *Trail For* option in the *ANIMATION* menu and *Trail* the ISS. Center your animation on Earth and use a *Time Step* of *1 Minute* or less. Under *COMPUTATION/Location,* choose *View from Earth.* Set *Offset Distance* to 0.01 A.U. and enter coordinates *Offset R.A.* 00 00 00 and Offset Dec. 00 00 00. Is the ISS visible from your location?

 *In addition to the Practice Problems and Destinations modules, the Companion Website at http://www.prenhall.com/chaisson provides for each chapter an additional true-false, multiple choice, and labeling quiz, as well as additional annotated images, animations, and links to related Websites.*

15 THE FORMATION OF PLANETARY SYSTEMS

The Birth of Our World

LEARNING GOALS

Studying this chapter will enable you to:

1 Summarize the major features that a theory of solar system origins has to explain.

2 Outline the process by which planets form as natural by-products of star formation.

3 Explain the role played by dust in the currently accepted model of solar system formation.

4 Account for the differences between the terrestrial and the jovian planets.

5 Discuss the role of collisions in determining specific characteristics of the solar system.

6 Outline the properties of known extrasolar planets, and discuss how they fit in with current theories of solar system formation.

 Visit http://www.prenhall.com/chaisson for additional annotated images, animations, and links to related sites for this chapter.

About 1000 light-years from Earth, Herbig-Haro 32 is a young star system in which planets might now be forming. The bright star's high-speed winds and extended jets have recently cleared away much of the surrounding dust and gaseous debris from the central region, thereby exposing the young star to direct view. The jet at the top (green and white) is about 200 A.U. from the star; the bottom jet is fainter, probably because it is still obscured by some of the debris. (STScI)

The Big Picture: The formation of our solar system was a long-ago event, with much of the matter of our primordial Galactic cloud now either wrapped up in the Sun and planets or ejected back out into deep space. Now, some 4.5 billion years later, things are much calmer, and it is not easy to reconstruct what exactly did happen here. Astronomers therefore observe other young star systems, hoping to gain some insight regarding the origins of our own solar system.

Completing these chapters on the planets, you may be struck by the vast range of physical and chemical properties found in the solar system. The planets present a long list of interesting features and bizarre peculiarities, and the list grows even longer when we consider the characteristics of their moons. Every object has its idiosyncrasies, some of them due to particular circumstances, many others the result of planetary evolution. Each time a new discovery is made, we learn a little more about the properties and history of our planetary system. Still, our astronomical neighborhood might seem more like a great junkyard than a smoothly running planetary system. Can we really make any sense of the entire collection of solar system matter? Is there some underlying principle that unifies the knowledge we have gained? The answer, as we will see, is maybe. . . .

15.1 Modeling the Origin of Our Solar System

The origin of the planets and their moons is a complex and as yet incompletely solved puzzle, although the basic outlines of the process involved are becoming understood. Most of our knowledge of the solar system's formative stages has emerged from studies of interstellar gas clouds, fallen meteorites, and Earth's Moon, as well as from the various planets observed with ground-based telescopes and planetary space probes. Ironically, studies of Earth itself do not help much because information about our planet's early stages eroded away long ago. Meteorites and comets provide perhaps the most useful information, for nearly all have preserved within them traces of solid and gaseous matter from the earliest times.

Until the mid-1990s, theories of planetary system formation concentrated almost exclusively on our own solar system, for the very good reason that astronomers had no other examples of planetary systems against which to test their ideas. However, all that has now changed. As of early 2001, we know of literally dozens of **extrasolar planets**—planets orbiting stars other than the Sun—that challenge our theories. And challenge them they do! As we will see, the other planetary systems discovered to date seem to have properties quite different from our own, and may well require us to rethink in some radical ways our concept of how stars and planets form.

Still, we currently have only the barest information on these systems—little more than orbits and mass estimates for the largest planets. Accordingly, we begin our study by outlining the comprehensive theory that accounts, in detail, for most of the observed properties of our own planetary system, the solar system. Later we will return to the observations of extrasolar planets and assess how our theory holds up in the face of these new data.

MODEL REQUIREMENTS

⬚ Any theory of the origin and architecture of our planetary system must adhere to the known facts. We know of nine outstanding properties of our solar system as a whole. They may be summarized as follows:

1. *Each planet is relatively isolated in space.* The planets exist as independent bodies at progressively larger distances from the central Sun; they are not bunched together. In rough terms, each planet tends to be twice as far from the Sun as its next inward neighbor.

2. *The orbits of the planets are nearly circular.* In fact, with the exceptions of Mercury and Pluto, which we will argue are special cases, each planetary orbit closely describes a perfect circle.

3. *The orbits of the planets all lie in nearly the same plane.* The planes swept out by the planets' orbits are accurately aligned to within a few degrees. Again, Mercury and Pluto are slight exceptions.

4. *The direction in which the planets orbit the Sun (counterclockwise as viewed from above Earth's North Pole) is the same as the direction in which the Sun rotates on its axis.* Virtually all the large-scale motions in the solar system (other than comet orbits) are in the same plane and in the same sense. The plane is that of the Sun's equator, and the sense is that of the Sun's rotation.

5. *The direction in which most planets rotate on their axis is roughly the same as the direction in which the Sun rotates on its axis.* This property is less general than the one just described for revolution, as three planets—Venus, Uranus, and Pluto—do not share it.

6. *Most of the known moons revolve about their parent planets in the same direction that the planets rotate on their axes.*

7. *Our planetary system is highly differentiated.* The inner terrestrial planets are characterized by high densities, moderate atmospheres, slow rotation rates, and few or no moons. By contrast, the jovian planets, farther from the Sun, have low densities, thick atmospheres, rapid rotation rates, and many moons.

8. *The asteroids are very old and exhibit a range of properties not characteristic of either the inner or the outer planets or their moons.* The asteroid belt shares, in rough terms, the bulk orbital properties of the planets. However, it appears to be made of primitive, unevolved material, and the meteorites that strike Earth are the oldest rocks known.

9. *The comets are primitive, icy fragments that do not orbit in the ecliptic plane and reside primarily at large distances from the Sun.*

All these observed facts, taken together, strongly suggest a high degree of order within our solar system. The whole system is not a random assortment of objects spinning or orbiting this way or that. Consequently, it hardly seems possible that our solar system could have formed by the slow accumulation of already-made interstellar "planets" casually captured by our Sun over the course of billions of years. The overall architecture of our solar system is too neat, and the ages of its members too uniform, to be the result of random chaotic events. The overall organization points toward a single formation, an ancient but one-time event, 4.6 billion years ago. A convincing theory that explains all the nine features just listed has been a goal of astronomers for centuries.

PLANETARY IRREGULARITIES

It is equally important to recognize what our theory of the solar system does *not* have to explain. There is plenty of scope for planets to evolve after their formation, so circumstances that have developed since the initial state of the solar system was established need not be included in our list. Examples are Mercury's 3:2 spin–orbit coupling, Venus's runaway greenhouse effect, the Moon's synchronous rotation, the emergence of life on Earth and its absence on Mars, the Kirkwood gaps in the asteroid belt, and the rings and atmospheric appearance of the jovian planets. There are many more. Indeed, all the properties of the planets for which we have already provided an *evolutionary* explanation need not be included as items that our theory must account for at the outset.

In addition to its many regularities, our solar system also has many notable *irregularities*, some of which we have already mentioned. Far from threatening our theory, however, these irregularities are important facts for us to consider in shaping our explanations. For example, it is necessary that the explanation for the solar system not insist that *all* planets rotate in the same sense or have *only* prograde moons, because that is not what we observe. Instead, the theory of the solar system should provide strong reasons for the observed planetary characteristics yet be flexible enough to allow for and explain the deviations, too. And, of course, the existence of the asteroids and comets that tell us so much about our past must be an integral part of the picture. That's quite a tall order, yet many researchers now believe that we are close to that level of understanding.

In the following three sections we draw together all the planetary data we have amassed in the last eight chapters and show how both the regularities and the irregularities of our solar system can be largely explained by a single coherent theory.

☑ Concept Check

■ Why is it necessary that a theory of solar system formation make clear statements about how planets arose, yet not be too rigid in its predictions?

15.2 The Condensation Theory

NEBULAR CONTRACTION

2 One of the earliest heliocentric models of solar system formation is termed the **nebular theory**, and may be traced back to the seventeenth-century French philosopher René Descartes. In this model, a large cloud of interstellar gas began to collapse under the influence of its own gravity. As it contracted, it became denser and hotter, eventually forming a star—the Sun—at its center. While all this was going on the outer, cooler, parts of the cloud formed a giant swirling region of matter, creating the planets and their moons as by-products of the star-formation process. This swirling mass destined to become our solar system is usually referred to as the **solar nebula**.

In 1796 the French mathematician-astronomer Pierre Simon de Laplace tried to develop the nebular model in a quantitative way. He was able to show mathematically that the conservation of angular momentum (see *More Precisely 15-1*) demands that an interstellar cloud like the hypothetical solar nebula must spin faster as it contracts. A decrease in the size of a rotating mass must be balanced by an increase in its rotational speed.

The increase in rotation speed, in turn, must have caused the nebula's *shape* to change as it collapsed. In Chapter 11 we saw how a spinning body tends to develop a bulge around its middle. ∞ (Sec. 11.1) The rapidly spinning nebula behaved in exactly this way. As shown in Figure 15.1, the fragment eventually flattened into a pancake-shaped primitive solar system. If we now suppose that planets formed out of this spinning material, we can already begin to understand the origin of some of the large-scale architecture observed in our planetary system today, such as the circularity of the planets' orbits and the fact that they move in nearly the same plane.

Astronomers are fairly confident that the solar nebula formed such a disk because similar disks have been observed (or inferred) around other stars. Figure 15.2(a) shows visible-light images of the region around a star called Beta Pictoris, lying about 50 light-years from the Sun. When the light from Beta Pictoris itself is suppressed and the resulting image enhanced by a computer, a faint disk of warm matter (viewed almost edge-on here) can be seen. This particular disk is roughly 500 A.U. across—about 10 times the diameter of Pluto's orbit. Astronomers believe that Beta Pictoris is a very young star, perhaps only 100 million years old, and that we are witnessing it pass

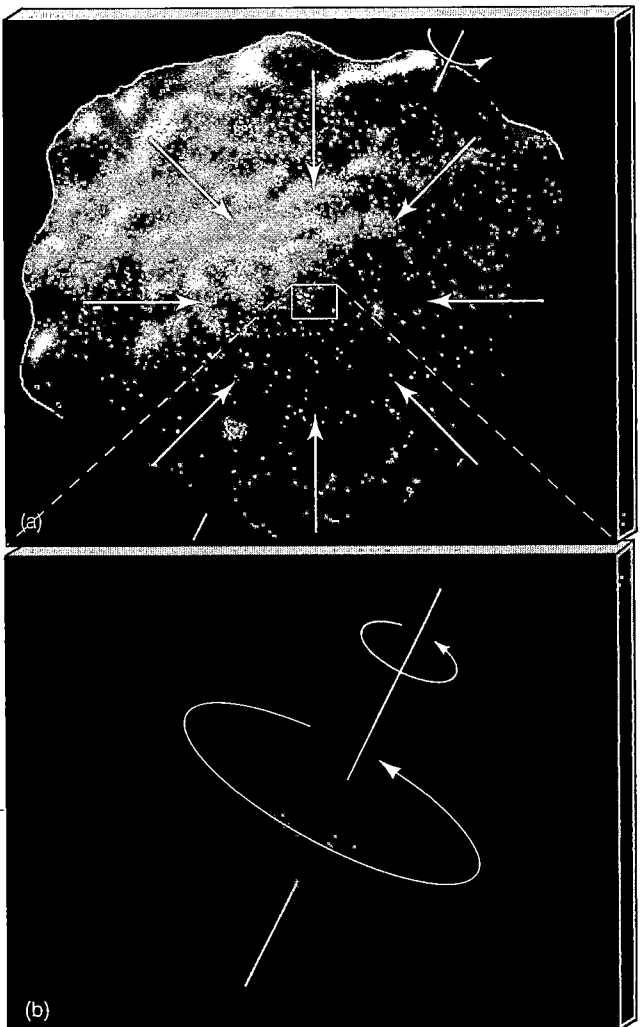

Figure 15.1 Nebular Contraction (a) Conservation of angular momentum demands that a contracting, rotating cloud (a) must spin faster as its size decreases. (b) Eventually, the primitive solar system came to resemble a giant pancake. The large blob at the center would ultimately become the Sun.

through an evolutionary stage similar to the one our own Sun experienced some 4.6 billion years ago. Figure 15.2(b) shows an artist's conception of the disk.

The nebular theory is an example of an *evolutionary* theory, which describes the development of the solar system as a series of gradual and natural steps, understandable in terms of well-established physical principles. Evolutionary theories may be contrasted with *catastrophic* theories that invoke accidental or unlikely celestial events to interpret observations.* Scientists generally try not to

*A good example of such a theory is the collision hypothesis, which imagines that the planets were torn from the Sun by a close encounter with a passing star. This hypothesis enjoyed some measure of popularity during the nineteenth century, in large part due to the inability of other theories to account for the observed properties of the solar system, but no scientist takes it seriously today. Aside from its extreme improbability, it is completely unable to explain the orbits, the rotations, or the composition of the planets and their moons.

invoke catastrophes to explain the universe. However, as we will see, there are instances where pure chance has played a critical role in determining the present state of the solar system.

THE ROLE OF DUST

3 Laplace imagined that as the spinning solar nebula contracted, it left behind a series of concentric rings, each of which would eventually become a planet orbiting a central **protosun**—a hot ball of gas well on its way to becoming the Sun. Each ring then clumped into a **protoplanet**—a forerunner of a genuine planet. The description of the collapse and flattening of the solar nebula is essentially correct, but when modern astronomers began to study the more subtle aspects of the problem, some fatal flaws were found in Laplace's nebular picture.

Calculations show that rings of the sort envisaged in Laplace's theory would probably not form, and even if they did, they would not in most cases condense to form a planet. In fact, computer calculations predict just the opposite: over most of the solar system, the rings would tend to disperse. The protoplanetary matter would be too warm, and no one ring would have enough mass to bind its own matter into a ball. Only in the cool outer regions of the nebula might it be possible for a sufficiently large clump of matter to form and survive. As we will see, such a possibility remains an important part of the modern theory.

The model currently favored by most astronomers is a more sophisticated version of the nebular theory. Known as the **condensation theory**, it combines the good features of the old nebular theory with new information about interstellar chemistry to avoid most of the old theory's problems. The key new ingredient in the modern picture is the presence of *interstellar dust* in the solar nebula. Astronomers now recognize that the space between the stars is strewn with microscopic dust grains, an accumulation of the ejected matter of many long-dead stars (see Chapter 22). These dust particles probably formed in the cool atmospheres of old stars, then grew by accumulating more atoms and molecules from the interstellar gas within the Milky Way Galaxy. The end result is that our entire galaxy is littered with miniature chunks of icy and rocky matter having typical sizes of about 10^{-5} m. Figure 15.3 shows one of many such dusty regions found in the vicinity of the Sun.

Dust grains play an important role in the evolution of any gas. Dust helps to cool warm matter by efficiently radiating its heat away in the form of infrared radiation, reducing the pressure (which is just proportional to the gas temperature) and allowing the gas to collapse more easily under the influence of gravity. Furthermore, the dust grains greatly speed up the process of collecting enough atoms to form a planet. They act as **condensation nuclei**—microscopic platforms to which other atoms can attach, forming larger and larger balls of matter. This is

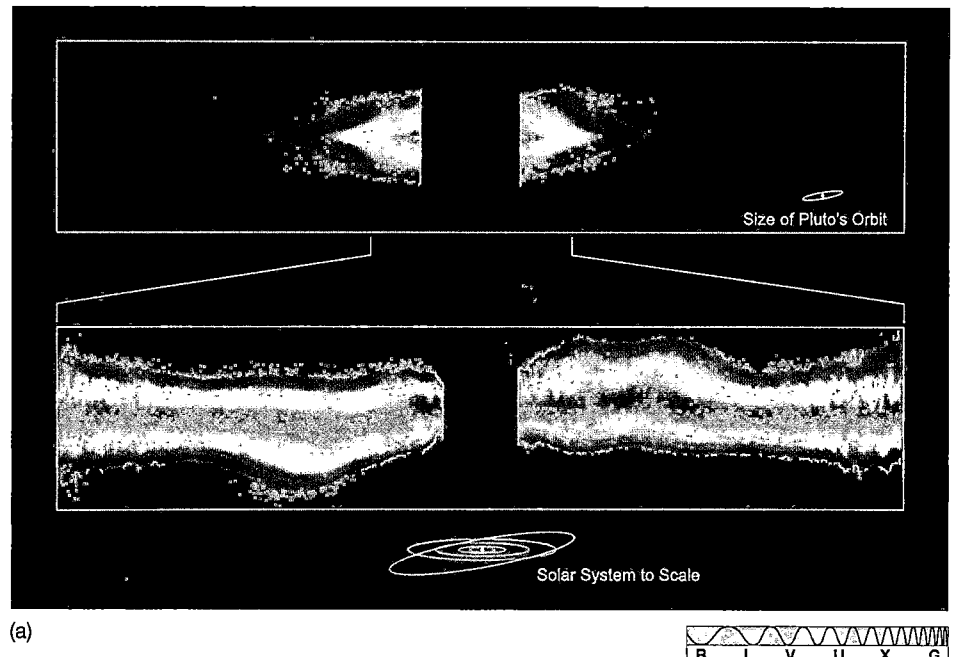

Size of Pluto's Orbit

Solar System to Scale

(a)

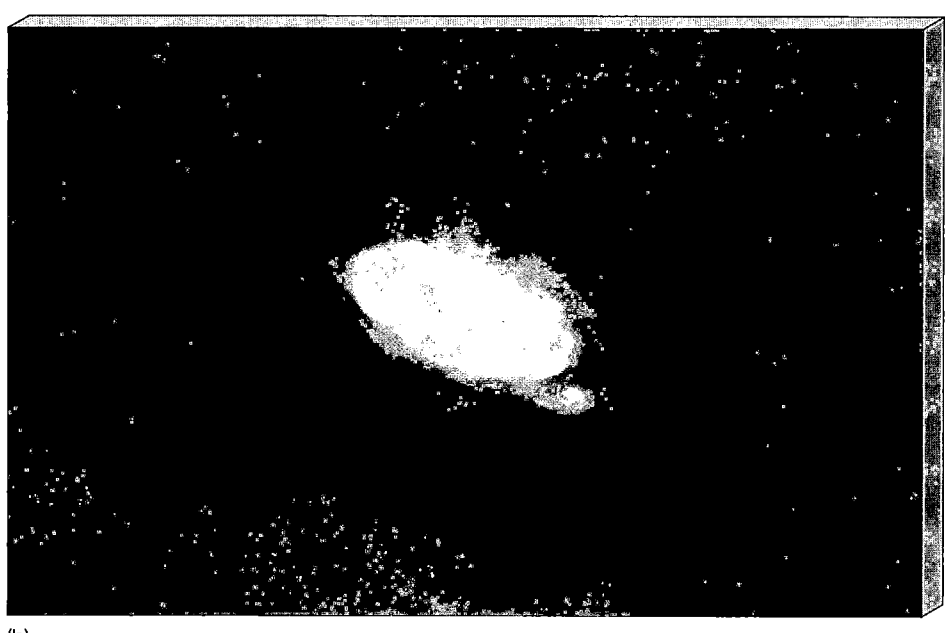

(b)

Figure 15.2 Beta Pictoris
(a) A computer-enhanced view of a disk of warm matter surrounding the star Beta Pictoris. Both images show actual data taken at visible wavelengths, but are presented here in false color to accentuate the details; the bottom image is a close-up of the inner regions of the disk, implying a warp in the disk possibly caused by the gravitational pull of unseen companions. In both images the overwhelmingly bright central star has been removed to let us see the much fainter disk surrounding it. The disk is nearly edge-on to our line of sight. It is made mostly of microscopic dust grains of ices and silicate particles (see Section 15.3) and is illuminated by the reflected light of the central star. For scale, the dimension of Pluto's orbit (78 A.U.) has been drawn adjacent to the images. (b) An artist's conception of the disk of clumped matter, showing the warm disk with a young star at the center and several comet-sized or larger bodies already forming at large radii. The colors are thought to be accurate—generally speaking, if you know the temperature and the density of the nebula, then you can derive its color. At the outer edges of the disk the temperature is low, and the color is a dull red. Progressing inward, the colors brighten and shift to a more yellowish tint as the temperature increases. Mottled dust is seen throughout—such protoplanetary regions are probably very dirty. *(NASA; D. Berry)*

similar to the way that raindrops form in Earth's atmosphere; dust and soot in the air act as condensation nuclei around which water molecules cluster.

ACCRETION AND FRAGMENTATION

Modern models trace the formative stages of our solar system along the following broad lines. Imagine a dusty interstellar cloud fragment measuring about a light-year across. Intermingled with the preponderance of hydrogen and helium atoms in the cloud are some heavy-element gas and dust. Some external influence, such as the passage of another interstellar cloud or perhaps the explosion of a near-

by star, starts the fragment contracting, down to a size of about 100 A.U. As the cloud collapses, it rotates faster and begins to flatten (just as described in the old nebular theory). By the time it has shrunk to 100 A.U., the solar nebula has already formed an extended, rotating disk (Figure 15.4a; see also Figures 15.1b and 15.2b).

According to the condensation theory, the planets formed in three stages. Early on, dust grains in the solar nebula formed condensation nuclei around which matter began to accumulate (Figure 15.4b). This vital step greatly hastened the critical process of forming the first small clumps of matter. Once these clumps formed, they grew rapidly by sticking to other clumps. (Imagine a snowball

Figure 15.3 Dark Cloud Interstellar gas and dark dust lanes mark this region of star formation. The dark cloud known as Barnard 86 (left) flanks a cluster of young blue stars called NGC 6520 (right). Barnard 86 may be part of a larger interstellar cloud that gave rise to these stars. *(D. Malin/Anglo-Australian Telescope)*

thrown through a fierce snowstorm, growing bigger as it encounters more snowflakes.) As the clumps grew larger, their surface areas increased and consequently the rate at which they swept up new material accelerated. They grad-

ually grew into objects of pebble size, baseball size, basketball size, and larger. Figure 15.5 shows an infrared view of a relatively nearby star whose protostellar disk is believed to be in just this state.

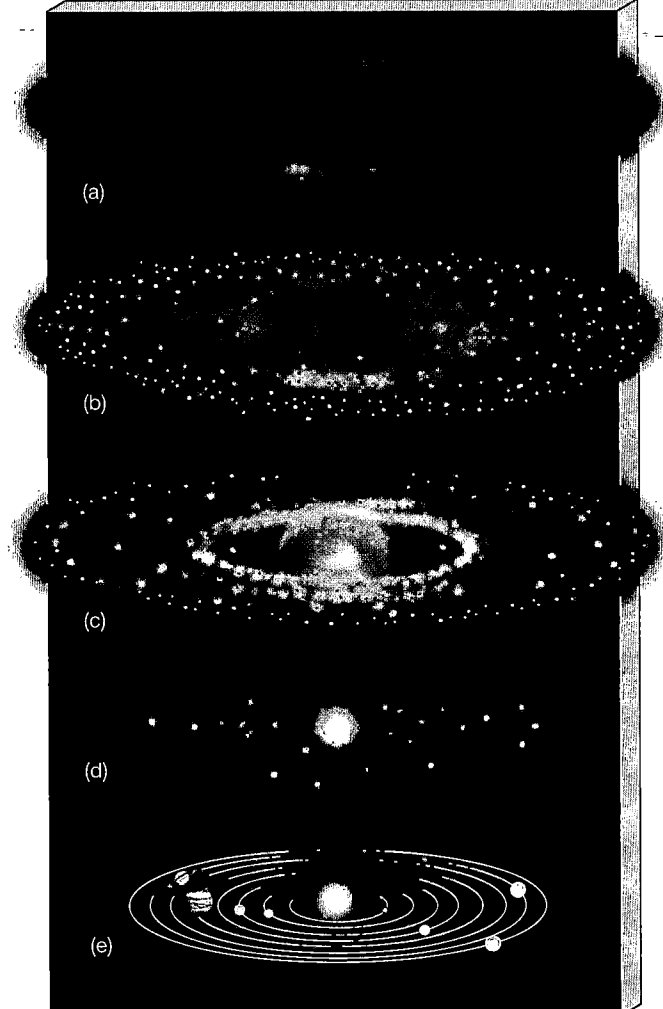

Figure 15.4 Solar System Formation The condensation theory of planet formation (not drawn to scale; Pluto is not shown in part e). (a) The solar nebula after it has contracted and flattened to form a spinning disk (Figure 15.1b). The large blob in the center will become the Sun. Smaller blobs in the outer regions may become jovian planets. (b) Dust grains act as condensation nuclei, forming clumps of matter that collide, stick together, and grow into moon-sized planetesimals. The composition of the grains depends on location within the nebula. (c) Strong winds from the still-forming Sun will soon expel the nebular gas. By this time, some large planetesimals in the outer solar system have already begun to accrete gas from the nebula. (d) With the gas ejected, planetesimals continue to collide and grow. The gas giant planets are already formed. (e) Over the course of a hundred million years or so, planetesimals are accreted or ejected, leaving a few large planets that travel in roughly circular orbits.

Animation

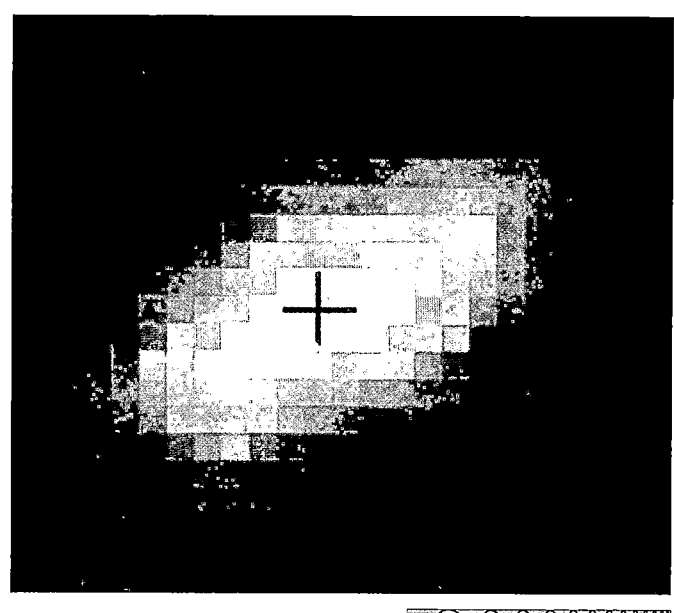

Eventually, this process of **accretion**—the gradual growth of small objects by collision and sticking—created objects a few hundred kilometers across (Figure 15.4c). By that time, their gravity was strong enough to sweep up material that would otherwise not have collided with them, and their rate of growth became faster still. At the end of this first stage, the solar system was made up of hydrogen and helium gas and millions of **planetesimals**—objects the size of small moons, having gravitational fields just strong enough to affect their neighbors.

Figure 15.5 A Disk of Planetesimals? This far-infrared image of a star, HR4796A, more than 200 light-years from Earth is thought to show a circumstellar disk in which the process of planetesimal growth is underway. The star itself is at the position of the cross, and the disk, which is falsely colored to match the dust emission, measures about five times the diameter of our solar system. *(NOAO)*

In the second phase of the accretion process, gravitational forces between the planetesimals caused them to collide and merge, forming larger and larger objects. Because larger objects have stronger gravity, the rich became richer in the early solar system, and eventually almost all the planetesimal material was swept up into a few large protoplanets—the accumulations of matter that would eventually evolve into the planets we know today. Figure 15.6 shows a computer simulation of accretion in the inner solar system. Notice how, as the number of bodies decreases, the orbits of the remainder become more widely spaced and more nearly circular.

As the protoplanets grew, another process became important. The strong gravitational fields produced many high-speed collisions between planetesimals and protoplanets. These collisions led to **fragmentation**, as small objects broke into still smaller chunks, which were then swept up by the protoplanets. Not only did the rich get richer but the poor were mostly driven to destruction! Some of these fragments produced the intense meteoritic bombardment we know occurred during the early evolution of the planets and moons, as we have seen repeatedly in the last few chapters. The fragments that escaped

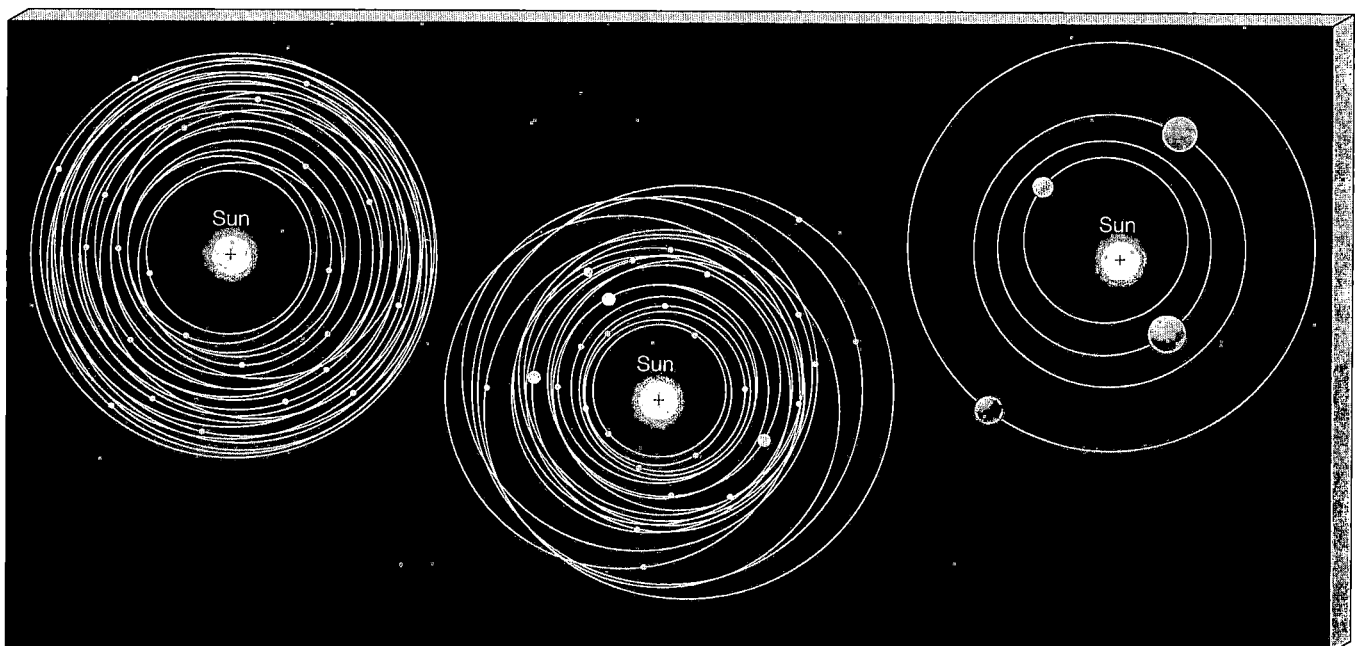

Figure 15.6 Making the Inner Planets Accretion in the inner solar system: Initially, many moon-sized planetesimals orbited the Sun. Over the course of a hundred million years or so, they gradually collided and coalesced, forming a few large planets in roughly circular orbits.

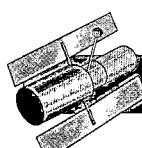

MORE PRECISELY 15-1

The Concept of Angular Momentum

Most celestial objects rotate. Planets, moons, stars, and galaxies all have some *angular momentum*, which we can define as the tendency of a body to keep spinning or moving in a circle. Angular momentum is as important a property of an object as its mass or its energy.

Consider first a simpler motion—*linear momentum*, which is defined as the product of an object's mass and its velocity:

$$\text{linear momentum} = \text{mass} \times \text{velocity.}$$

This is the tendency of an object to keep moving in a straight line in the absence of external forces. Picture a truck and a bicycle rolling equally fast down a street. Each has some linear momentum, but you would obviously find

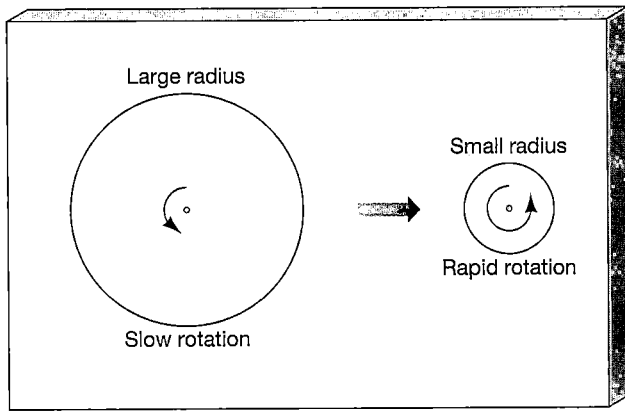

Large radius

Small radius

Rapid rotation

Slow rotation

it easier to stop the less massive bicycle. Although the two vehicles have the same speed, the truck has more momentum. We see that the linear momentum of an object depends on the mass of that object. It also depends on the speed. If two bicycles were rolling down the street at different speeds, the slower one could be stopped more easily.

Angular momentum is an analogous property of objects that are rotating or revolving. It is a measure of the object's tendency to keep spinning, or, equivalently, of how much effort must be expended to stop it. However, in addition to mass and (angular) speed, angular momentum also depends on the way in which an object's mass is distributed. Intuitively, we know that the more massive an object, or the larger it is, or the faster it spins, the harder it is to stop. In fact, angular momentum depends on the object's *mass, rotation rate* (measured in, say, revolutions per second), and *radius*, in a very specific way:

$$\text{angular momentum} \propto \text{mass} \times \text{angular speed} \times \text{radius}^2$$

(the constant of proportionality depends on the details of how the object's mass is distributed).

According to Newton's laws of motion, both types of momentum—linear and angular—must be conserved at all times. In other words, both linear and angular momentum must remain constant before, during, and after a physical change in any object (so long as no external forces act). For example, as illustrated at left, if a spherical object having some spin begins to contract, the relationship above demands that it spin faster, so that the product mass × angular speed

---

capture by a planet or a moon would later become the asteroids and comets.

Mathematical modeling, like the calculation shown in Figure 15.6, indicates that after about 100 million years, the primitive solar system had evolved into nine protoplanets, dozens of protomoons, and the big protosolar mass at the center. Computer simulations generally reproduce the increasing spacing between the planets ("Bode's law"), although the reasons for the regularity seen in the actual planetary spacing remain unclear. ∞ (*Discovery 6-1*) Roughly a billion years more were required to sweep the system clear of interplanetary trash. This was the billion-year period that saw the heaviest meteoritic bombardment, tapering off as the number of planetesimals decreased. ∞ (Sec. 8.5)

MAKING GIANT PLANETS

The accretion picture just described has become the accepted model for the formation of the inner solar system—

the terrestrial planets. However, the origin of the giant jovian worlds is decidedly less clear. Two somewhat different views, with important consequences for our understanding of extrasolar planets, have emerged.

In the first, conventional scenario, the four largest protoplanets became massive enough to enter a third phase of planetary development, their strong gravitational fields sweeping up large amounts of gas directly from the solar nebula. As we will see in a moment, there was more "raw material" available for planet building in the outer solar system, so protoplanets grew fastest there. Ultimately they would become the cores of the jovian worlds. ∞ (Sec. 11.1, 12.1, 13.5) The smaller, inner protoplanets never reached that stage, and as a result their masses remained relatively low. This is the chain of events depicted in Figures 15.4(c) and (d).

In the second scenario, the giant planets formed through instabilities in the cool outer regions of the solar nebula—not so far removed from Laplace's basic

× radius² remains constant. The sphere's mass does not change during the contraction, yet the size of the object clearly decreases. Its rotation speed must therefore increase in order to keep the total angular momentum unchanged.

EXAMPLE 1: Suppose the sphere has radius 1 m and starts off rotating at one revolution per minute. It then contracts to 1/10 its initial size. Conservation of angular momentum says that the sphere's final angular speed A must satisfy

$$\text{mass} \times A \times (0.1\text{m})^2 = \text{mass} \times \\ (1 \text{ rev/min}) \times (1 \text{ m})^2.$$

The mass is the same on either side of the equation and therefore cancels, so we find A = (1 rev/min) × (1 m/0.1 m)² = 100 rev/min, or about 1.7 rev/s.

EXAMPLE 2: Now suppose that the "sphere" is a large interstellar gas cloud about to collapse and form the solar nebula. Initially, let's imagine that it has a diameter of one light-year and rotates very slowly—once every 10 million years. Assuming that the cloud's mass stays constant, the rotation rate must increase to conserve angular momentum as the radius decreases. By the time it has col-

(AP/Wide World Photo)

lapsed to a radius of 100 A.U., the cloud's radius has shrunk by a factor of (0.5 light-year/100 A.U.) ≈ 320. Conservation of angular momentum then implies that its (average) spin rate increases by a factor of 320² ≈ 100,000, roughly one revolution per century, about the orbital period of Uranus. (Of course, this is at best a crude estimate, as the solar nebula did not rotate with a single angular velocity. Instead, it rotated *differentially*, with the inner regions moving faster, in accordance with Kepler's laws.)

The law of conservation of angular momentum also applies to planetary orbits (where now the "radius" is the distance from the planet to the Sun)— in fact, Kepler's second law *is* just conservation of angular momentum, expressed another way. ∞ (Sec. 2.5) Figure skaters also use the principle of angular-momentum conservation too. They spin faster by drawing in their arms (as shown in the figure below) and slow down by extending them. Here the mass of the human body remains the same, but its lateral size changes, causing the body's rotation speed to increase or decrease to keep its angular momentum unchanged.

---

idea—mimicking on small scales the collapse of the initial interstellar cloud. In this view, the jovian protoplanets formed directly, skipping the initial accretion stage and perhaps taking less than a thousand years to acquire much of their mass. These first protoplanets already had gravitational fields strong enough to scoop up more gas and dust from the solar nebula, allowing them to grow into the gas giants we see today. Figure 15.7 illustrates this alternative path to the formation of the jovian worlds.

In either case, once the jovian protoplanets reached the critical size at which they could capture nebular gas, they grew rapidly, their capture rate increasing as their gravitational fields intensified. Their large size today reflects the "head start" they received in the accretion process. If they did not start off with ready-made rock and ice protoplanet cores, these giant worlds had ample time to sweep up rocky and icy planetesimal material after they formed.

Many of the moons of the jovian planets presumably also formed through accretion but on a smaller scale, in the gravitational fields of their parent planets. Once the nebular gas began to accrete onto the large jovian protoplanets, conditions probably resembled a miniature solar nebula, with condensation and accretion continuing to occur. The large moons of the outer planets almost certainly formed in this way. Some of the smaller moons may have been "chipped off" their parent planets during collisions with asteroids; others may be captured asteroids themselves.

Many aspects of the formation of the giant planets remain unresolved. Interactions among the growing planets, and between the planets and their environment, probably played critical roles in determining just how and where the planets formed. One particularly intriguing scenario, believed by some—but certainly not all—planetary scientists, is the possibility that Jupiter, and maybe all four giant planets, formed considerably farther from the Sun than its present orbit, and subsequently "migrated" inward to its current location. This supposed migration is indicated schematically in Figures 15.4 and 15.7 by the locations of the jovian protoplanets.

The idea of planet migration has been around since the mid-1980s, when theorists realized that interactions between massive planets and the nebula in which they moved would have caused just such an inward drift. Observational support came in 1999, when *Galileo* scientists announced that reanalysis of data from the mission's atmospheric probe showed much higher than expected concentrations of the gases nitrogen, argon, krypton, and xenon. ⊂⊃ (Sec. 11.2) These gases, which are thought to have been carried to the planet by captured planetesimals, could not have been retained in the planetesimal ice at temperatures typical of Jupiter's current orbit. Instead, they imply that the planetesimals, and presumably Jupiter too, formed at much lower temperatures. Either the nebula was cooler than previously thought, or Jupiter formed out in what is now the Kuiper belt!

☑ **Concept Check**

■ Why was interstellar dust so important to the formation of our solar system?

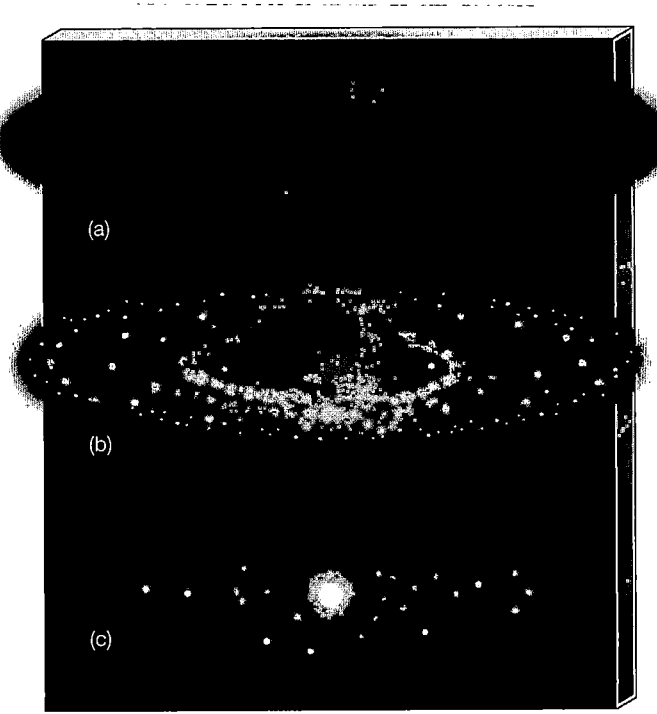

Figure 15.7 Jovian Condensation As an alternative to the growth of massive protoplanetary cores followed by accretion of nebular gas, it is possible that some or all of the giant planets formed directly via instabilities in the cool gas of the outer solar nebula. Part (a) shows the same instant as Figure 15.4(a). (b) Four gas giants have formed, circumventing the accretion process sketched in Figure 15.4. (c) The giant planets have taken their place in the outer solar system.

15.3 The Differentiation of the Solar System

⚅ The condensation theory also allows us to understand the basic differences in content and structure between the terrestrial and the jovian worlds. Indeed, it is in this context that the adjective *condensation* derives its true meaning. To see why a planet's composition depends on its location in the solar system, it is necessary to consider the temperature structure of the solar nebula.

THE ROLE OF HEAT

As the primitive solar system contracted under the influence of gravity, it heated up as it flattened into a disk. The density and temperature were greatest near the central protosun and much lower in the outlying regions. Detailed calculations indicate that the gas temperature near the core of the contracting system was several thousand kelvins. At a distance of 10 A.U., out where Saturn now resides, the temperature was only about 100 K.

In the warmer regions of the cloud, dust grains broke apart into molecules, and they in turn split into excited atoms. Because the extent to which the dust was destroyed depended on the temperature, it also depended on location in the solar nebula. Most of the original dust in the inner solar system disappeared at this stage, but the grains in the outermost parts probably remained largely intact.

The destruction of the dust in the hot inner portion of the solar nebula introduced an important new ingredient into the theoretical mix, one that we omitted from our earlier account of the accretion process. With the passage of time, the gas radiated away its heat and the temperature decreased at all locations, except in the very core, where the Sun was forming. Everywhere beyond the protosun, new dust grains began to condense (or crystallize) from their hotter gas phase to their cooler solid phase, much as raindrops, snowflakes, and hailstones condense from moist, cooling air here on Earth. It may seem strange that although there was plenty of interstellar dust early on, it was mostly destroyed, only to form again later. However, a critical change had occurred. Initially, the nebular gas was uniformly peppered with dust grains. When the dust reformed later, the distribution of grains was very different.

Figure 15.8 plots the temperature in various parts of the primitive solar system just before the onset of the accretion stage. At any given location, the only materials to condense out were those able to survive the temperature there. As marked on the figure, in the innermost regions, around Mercury's present orbit, only metallic grains could form. It was simply too hot for anything else to exist. A little farther out, at about 1 A.U., it was possible for rocky, silicate grains to form, too. Beyond about 3 or 4 A.U., water ice could exist, and so on, with the condensation of

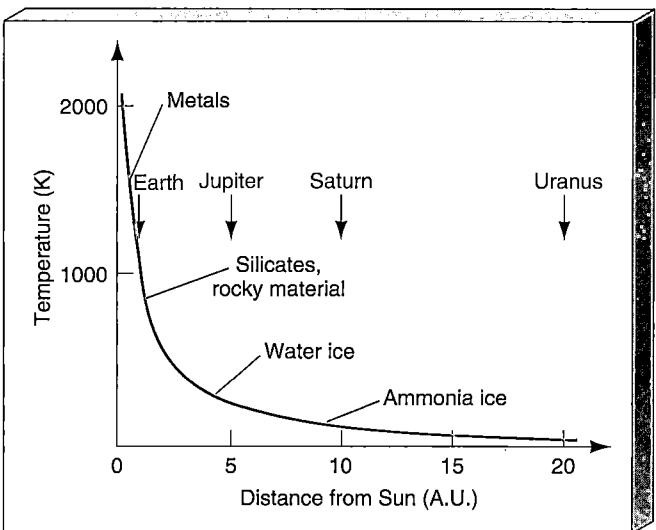

Figure 15.8 Temperature in the Early Solar Nebula
Theoretically computed variation of temperature across the primitive solar nebula. In the hot central regions only metals could condense out of the gaseous state to form grains. At greater distances from the central protosun the temperature was lower, so rocky and icy grains could also form. The labels indicate the minimum radii at which grains of various types could condense out of the nebula.

more and more material possible at greater and greater distances from the Sun. The composition of the material that could condense out at any given radius would ultimately determine the types of planets that formed there.

JOVIAN PLANETS AND COMETS

In the middle and outer regions of the primitive planetary system, beyond about 5 A.U. from the center, the temperature was low enough for the condensation of several abundant gases into solid form. After hydrogen and helium, the most common materials in the solar nebula (as they are today in the universe as a whole) were the elements carbon, nitrogen, and oxygen. The most common chemical compounds were those containing those elements—specifically, water, ammonia, and methane. As we have seen, these compounds are still the primary constituents of jovian atmospheres.

At temperatures of a few hundred kelvins or less, these gases condensed out of the nebula. Consequently, the planetesimals that formed at these distances were formed under cold conditions out of predominantly low-density, icy material. These ancestral fragments were destined to form the cores of the jovian planets. Because more material could condense out of the solar nebula at these radii than in the inner regions near the protosun, accretion began sooner, with more resources to draw on. If they hadn't already formed through instabilities in the cold nebular gas, the outer planets grew rapidly to the

point where they could accrete nebular gas, not just grains, and eventually formed the hydrogen-rich jovian worlds we see today.

With the formation of the four giant jovian planets, the remaining planetesimals were subject to those planets' strong gravitational fields. Over a period of hundreds of millions of years and after repeated "gravity assists" from the giant planets, especially Uranus and Neptune, many of the interplanetary fragments in the outer solar system were flung into orbits taking them far from the Sun (Figure 15.9). Astronomers believe that those fragments now make up the Oort cloud, whose members occasionally visit the inner solar system as comets. ∞ (Sec. 14.2) During this period many icy planetesimals were also deflected into the inner solar system, where they played an important role in the evolution of the inner planets.

A key prediction of this model is that some of the original planetesimals should have remained behind, forming the broad band known as the Kuiper belt, lying beyond the orbit of Neptune. ∞ (Sec. 14.2) In 1993 several such asteroid-sized objects were discovered between 30 and 35 A.U. from the Sun, lending strong support to the condensation theory. Several hundred Kuiper belt objects, having diameters ranging from 50 km to 1000 km, are now known (see *Discovery 15-1*).

Computer simulations indicate that the interactions between the outer planets and the planetesimals also caused significant changes in the orbits of the planets themselves. As shown in Figure 15.9, by the time the outer solar system had been cleared of comets, Jupiter had moved slightly closer to the Sun, its orbital semimajor axis decreasing by a few tenths of an A.U. The other giant planets migrated outward—Saturn by about 1 A.U., Uranus by 3 or 4 A.U., and Neptune by some 7–10 A.U. Note that these migrations occurred long after the putative inward migrations mentioned earlier. Life as a jovian planet is far from simple!

TERRESTRIAL PLANETS AND ASTEROIDS

In the inner regions of the primitive solar system, condensation from gas to solid began when the average temperature was about 1000 K. The environment there was too hot for ices to survive. Many of the abundant heavier elements, such as silicon, iron, magnesium, and aluminum, combined with oxygen to produce a variety of rocky materials. Planetesimals in the inner solar system were therefore rocky in nature, as were the protoplanets and planets they ultimately formed.

These heavier materials condensed into grains in the outer solar system, too, of course. However, they would have been vastly outnumbered by the far more abundant light elements there. Thus the outer solar system is not deficient in heavy elements. The inner solar system is *underrepresented in light material*. The relative scarcity of rocky and metallic elements able to form grains in the hot inner

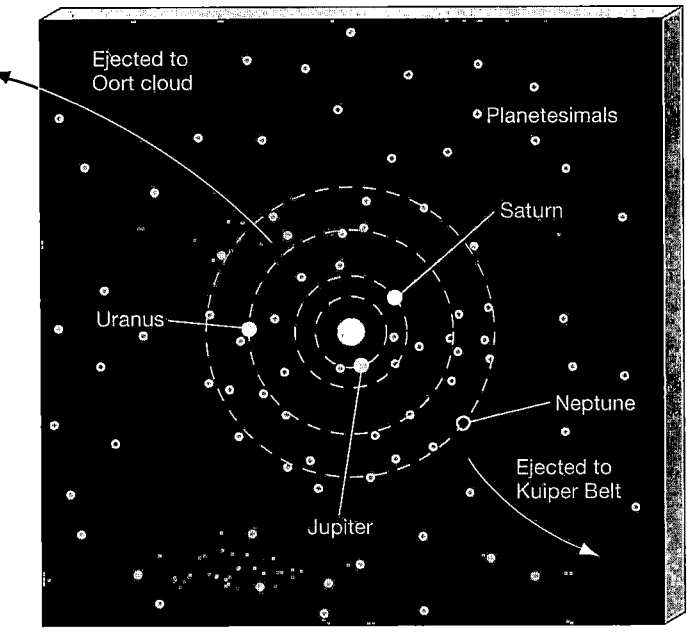

(a)

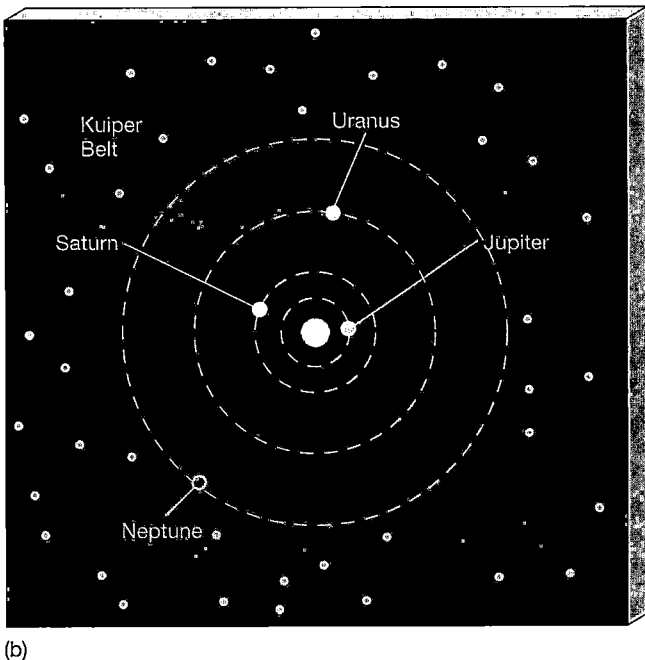

(b)

Figure 15.9 Planetesimal Ejection Ejection of planetesimals to form the Oort Cloud and Kuiper Belt. It is believed that interactions with Jupiter and Saturn "kicked" planetesimals out to very large radii (the Oort cloud). Interactions with Uranus and especially Neptune tended to populate the Kuiper belt but also deflected some planetesimals inward to interact with Jupiter and Saturn. As a result of the inward and outward "traffic," the orbits of all four giant planets were all significantly modified. Neptune was affected most, and may have migrated outward by as much as 10 A.U.

temperatures were far too high, and their gravity too low, to capture or retain those gases. The most likely explanation seems to be that the water and other light gases found on Earth and elsewhere in the inner solar system arrived here in the form of comets from the outer solar system. Deflected onto eccentric orbits as the gravitational fields of the jovian planets cleared the outer solar system of leftover planetesimals, these icy fragments bombarded the newborn terrestrial worlds, supplying them with water *after* their formation.

The myriad rocks of the asteroid belt between Mars and Jupiter failed to accumulate into a planet. Probably, nearby Jupiter's huge gravitational field caused them to collide too destructively to coalesce. Strong Jupiter tides on the planetesimals in the belt would also have hindered the development of a protoplanet. The result is a band of planetesimals, still colliding and occasionally fragmenting, but never coalescing into a larger body—surviving witnesses to the birth of the planets.

CLEANING UP THE DEBRIS

Most of the planetesimals left over after the major planets formed eventually collided with a planet, ended up in the asteroid or Kuiper belts, or were ejected into the Oort cloud. Little solid material remained. But what of the gas that made up most of the original cloud? Why don't we see it today throughout the planetary system? In the outer solar system, some (but not all) of that gas was swept up into planets. But that did not occur in the inner regions, where the terrestrial protoplanets never became massive enough to accrete such light material. Instead, the newly formed Sun took a hand.

All young stars apparently experience a highly active evolutionary stage known as the *T Tauri* phase (Figure 15.10; see also Chapter 19), during which their radiation and stellar winds become very intense. It is thought that much of the nebular gas between the planets was blown away into interstellar space by the solar wind and the pressure exerted by the Sun's radiation when the Sun entered this phase, just before nuclear burning started at its center. The nebular disk was probably only a few million years old when this occurred. Afterward, all that remained were protoplanets, with the gas-rich giant planets

solar nebula is the main reason why the terrestrial planets never became as massive as the jovian worlds. The inner regions of the nebula had to wait for the temperature to drop to the point where grains appeared and accretion could begin, but the outer regions might not have had to wait at all. Accretion in the outer solar system began almost with the formation of the disk itself.

A long-standing puzzle in the condensation theory's account of the formation of the inner planets has been where the water and other volatile gases on Earth and elsewhere originated. At formation, the inner planets' surface

already largely formed, and planetesimal fragments, ready to continue their long evolution into the solar system we know today.

✓ Concept Check

■ Would you expect to find comets and asteroids orbiting other stars?

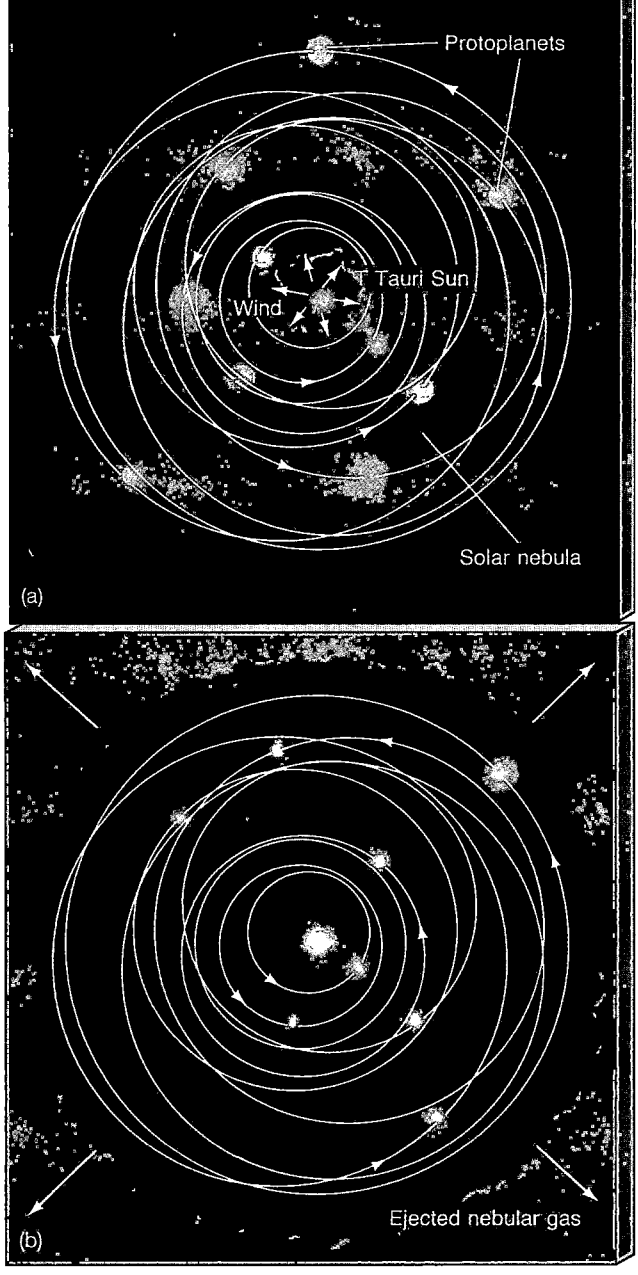

(a)

(b)

Figure 15.10 T Tauri Star (a) Strong stellar winds from newly born stars are responsible for sweeping away any dust and gas left over from the star formation process, (b) leaving only planets and planetesimals behind.

Section 15.4 The Role of Catastrophes **395**

15.4 The Role of Catastrophes

5 The condensation theory accounts for the nine "characteristic" points listed at the start of this chapter. Specifically, the planets' orbits are circular (2), in the same plane (3), and in the same direction as the Sun's rotation on its axis (4) as a direct consequence of the nebula's shape and rotation. The rotation of the planets (5) and the orbits of the moon systems (6) are due to the tendency of the smaller-scale condensations to inherit the nebula's overall sense of rotation. The growth of planetesimals throughout the nebula, with each protoplanet ultimately sweeping up the material near it, accounts for (1), the fact that the planets are widely spaced (even if the theory does not quite explain the regularity of the spacing). The heating of the nebula and the Sun's ignition resulted in the observed differentiation (7), and the debris from the accretion–fragmentation stage naturally accounts for the asteroids (8) and comets (9).

We stressed earlier that an important aspect of any solar system theory is its ability to allow for the possibility of imperfections—deviations from the otherwise well-ordered scheme of things. In the condensation theory, that capacity is provided by the randomness inherent in the encounters that ultimately combined the planetesimals into protoplanets. As the numbers of large bodies decreased and their masses increased, individual collisions acquired greater and greater importance. The effects of these collisions can still be seen today in many parts of the solar system—for example, the large craters on many of the moons we have studied thus far.

Having started with nine regular points to explain, we end our discussion of solar system formation with eight irregular features that still fall within the theory's scope. It is impossible to test any of these assertions directly, but it is reasonable to suppose that some (or even all) of the following "odd" aspects of the solar system can be explained in terms of collisions late in the formative stages of the protoplanetary system. Not all astronomers believe all these explanations. However, most would accept at least some.

1. Mercury's exceptionally large nickel-iron core may be the result of a collision between two partially differentiated protoplanets. The cores may have merged, and much of the mantle material may have been lost. ∞ (Sec. 8.7)

2. Two large bodies could have merged to form Venus, giving it its abnormally low rotation rate. ∞ (Sec. 9.2)

3. The Earth–Moon system may have formed from a collision between the proto-Earth and a Mars-sized object. ∞ (Sec. 8.8)

4. A late collision with a large planetesimal may have caused Mars's curious north–south asymmetry and ejected much of the planet's atmosphere. ∞ (Sec. 10.4)

5. The tilted rotation axis of Uranus may have been caused by a grazing collision with a sufficiently large planetesimal, or by a merger of two smaller planets. ∞ (Sec. 13.3)

6. Uranus's moon Miranda may have been almost destroyed by a planetesimal collision, accounting for its bizarre surface terrain. ∞ (Sec. 13.6)

7. Interactions between the jovian protoplanets and one or more planetesimals may account for the irregular moons of those planets and, in particular, Triton's retrograde motion. ∞ (Sec. 13.6)

8. Pluto may simply be a large representative of the Kuiper belt, and the Pluto–Charon system may be the result of a collision or near-miss between two icy planetesimals before most were ejected by interactions with the jovian planets. ∞ (Sec. 13.9)

☑ Concept Check

■ What is the key "random" element in the condensation theory?

15.5 Planets Beyond the Solar System

⑥ The test of any theory is its applicability and predictive power in situations different from those for which it was originally conceived. ∞ (*Discovery 2-2*) With the discovery in recent years of numerous planets orbiting other stars, astronomers now have the opportunity—indeed, the scientific obligation—to test their theories of solar system formation.

THE DISCOVERY OF EXTRASOLAR PLANETS

The detection of planets orbiting other stars has been a goal of astronomers for decades, if not centuries. Many claims of **extrasolar planets** have been made since the middle of the twentieth century, but before 1995 none had been confirmed, and most have been discredited. Only since the mid-1990s have we seen genuine advances in this fascinating area of astronomy. These advances have come not because of dramatic scientific or technical breakthroughs but rather through steady improvements in telescope and detector technology and computerized data analysis.

It is still not possible to image extrasolar planets. They are just too faint and too close to their parent stars for us to resolve them using current equipment. Instead, the techniques used to find them are *indirect*, based on analysis of the light from the parent star, not from the planet itself. As a planet orbits a star, gravitationally pulling first one way and then the other, the star "wobbles" slightly. The more massive the planet, or the less massive the star, the greater

the star's movement. If the wobble happens to occur along our line of sight to the star (Figure 15.11), then we see small fluctuations in the star's radial velocity, which can be measured using the Doppler effect. ∞ (Sec. 3.5) Those fluctuations allow us to estimate the planet's mass.

Figure 15.12 shows two sets of radial velocity data that betray the presence of planets orbiting other stars. Part (a) shows the line-of-sight velocity of the star 51 Pegasi, a near-twin to our Sun lying some 40 light-years away. These data were acquired in 1994 by Swiss astronomers using the 1.9-m telescope at Haute-Provence Observatory in France, and they were the first substantiated evidence for an extrasolar planet orbiting a Sunlike star.* The regular 50-m/s fluctuations in the star's velocity have since been confirmed by several groups of astronomers and imply that a planet of at least half the mass of Jupiter orbits 51 Pegasi in a circular orbit with a period of just 4.2 days. (For comparison, the corresponding fluctuation in the Sun's velocity due to Jupiter is roughly 12 m/s.) Note that we say "at least half" here because Doppler observations suffer from a fundamental limitation—they cannot distinguish between low-speed orbits seen edge-on and high-speed orbits seen almost face-on (so only a small component of the orbital motion contributes to the line-of-sight Doppler effect). As a result, only lower limits to planetary masses can be obtained.

Figure 15.12(b) shows another set of Doppler data, this time revealing the most complex system of planets discovered to date—a triple-planet system orbiting another nearby Sunlike star named Upsilon Andromedae. The three planets have minimum masses of 0.7, 2.1, and 4.3 times the mass of Jupiter, and orbital semimajor axes of 0.06, 0.83, and 2.6 A.U., respectively. Figure 15.12(c) sketches their orbits, with the orbits of the solar terrestrial planets shown for scale. In all, as of mid-2001, nearly 70 planets have been detected using radial velocity searches.

If the wobble produced in the star's motion is predominantly *perpendicular* to our line of sight, then little or no Doppler effect will be observed, so the radial velocity technique cannot be used to detect a planet. However, in this case, the star's *position* in the sky changes slightly from night to night and, in principle, measuring this transverse motion provides an alternative means of detecting extrasolar planets. Unfortunately, these side-to-side wobbles have proved difficult to measure accurately, as the angles involved are very small and the star in question has to be quite close to the Sun for useful observation to be possible. Several planetary systems have been proposed as candidates, based on observations of this type, but none has yet been placed on the "official" list of confirmed observations.

---

*As we will see in Chapter 22, two other planets having masses comparable to Earth, and one planet comparable to Earth's Moon, had previously been detected orbiting a collapsed star called a pulsar. However, their formation was the result of a chain of events very different from those that formed Earth and the solar system.

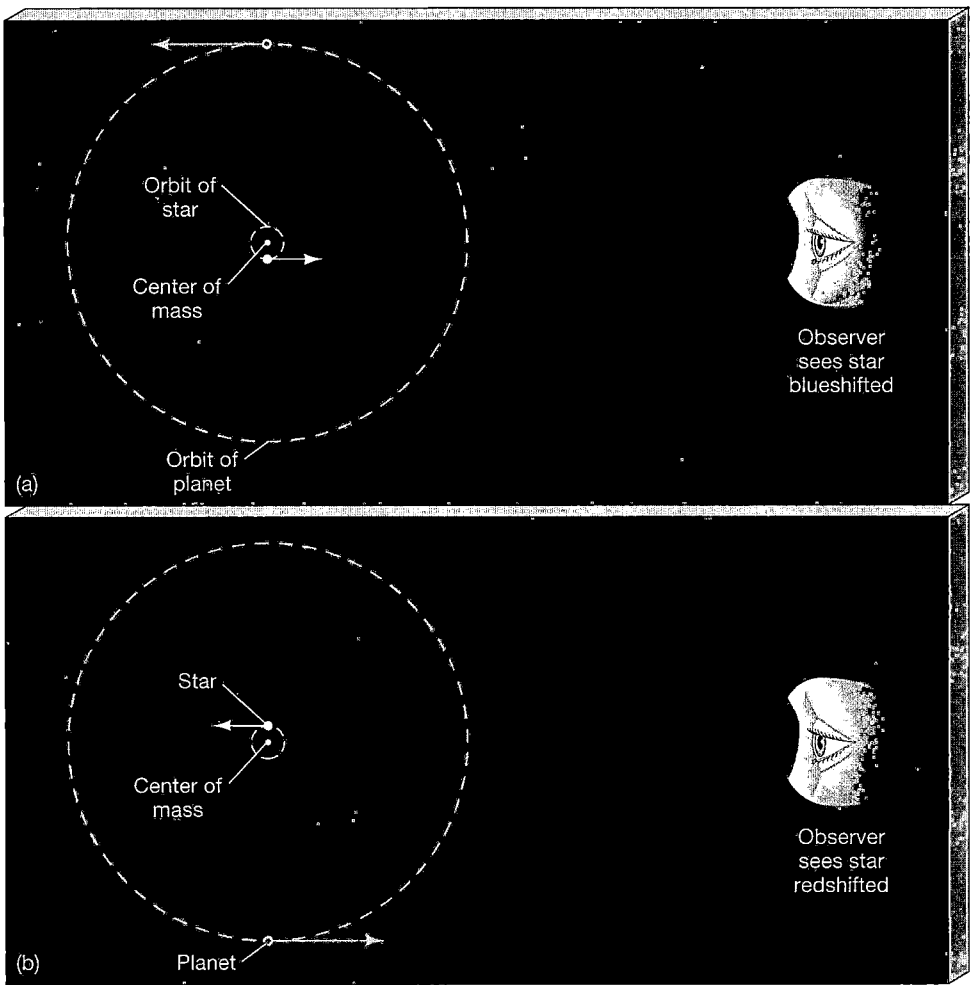

Figure 15.11 Detecting Extrasolar Planets As a planet orbits its parent star, it causes the star to "wobble" back and forth. The greater the mass of the planet, the larger the wobble. The center of mass of the planet–star system stays fixed. If the wobble happens to occur along our line of sight to the star, as shown by the yellow arrow, we can detect it by the Doppler effect.

As just noted, the Doppler technique suffers from the limitation that the angle between the line of sight and the planet's orbital plane cannot be determined. However, in one system originally discovered through Doppler measurements, that is not the case. Observations of a distant solar-type star (known only by its catalog name of HD 209458 and lying some 150 light-years from Earth) reveal a clear drop in brightness each time its 0.6-Jupiter-mass companion, orbiting at a distance of just 7 million km (0.05 A.U.), passes between the star and Earth. The drop in brightness is just 1.7 percent, but it occurs precisely on schedule every 3.5 days, the orbital period inferred from radial velocity measurements. Such planetary *transits* are analogous to the transit of Mercury shown in Figure 2.15. They are rare, as they require us to see the orbit almost exactly edge-on, but when they do occur they allow an unambiguous determination of the planet's mass and radius. The resulting density is just 200 kg/m³, consistent with a high-temperature gas-giant planet orbiting very close to its parent star. Attempts to detect similar brightness fluctuations in other nearby stars due to planetary transits have so far proved unsuccessful.

PLANETARY PROPERTIES

Figure 15.13 shows the orbits of all known extrasolar planets, with the orbits of the terrestrial planets superimposed for comparison. Simply put, *none* of the 60 or so extrasolar planetary systems currently on the "official" list look anything like our own solar system! The observed planets are almost all comparable in mass to Jupiter (although a few have masses more similar to Saturn). However, their orbits are generally much smaller than those of Jupiter and Saturn and usually quite eccentric, often taking them very close (much less than 1 A.U.) to their parent stars. These gas-giant planets orbiting close to their parent stars have come to be known as "hot Jupiters." In addition, although a few two-planet (and one three-planet) systems are known, most of the planets observed appear to be the only massive objects orbiting their star.

The fact that no low-mass planets are observed is not surprising. It is what astronomers call a *selection effect*—lightweight planets simply don't produce large enough velocity fluctuations to be detectable. However, the tight eccentric orbits of the planets detected so far,

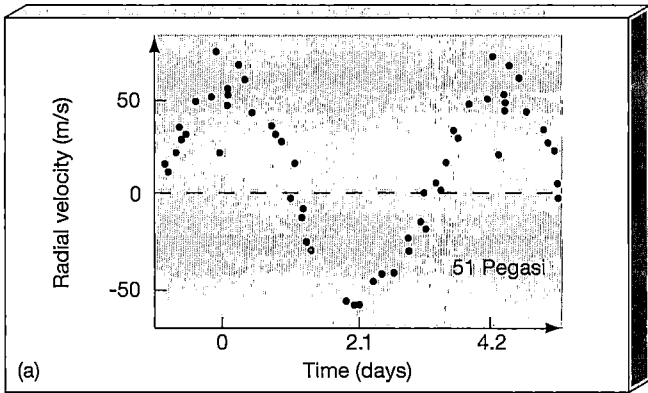

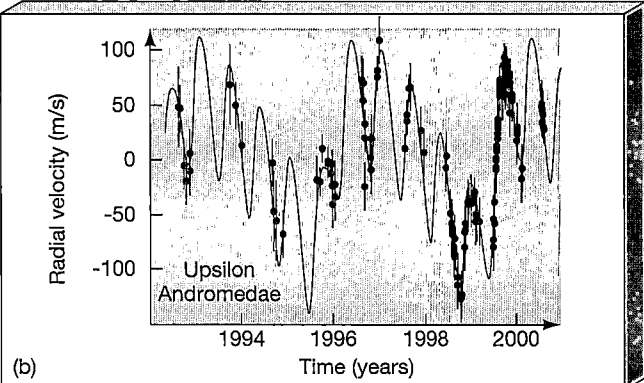

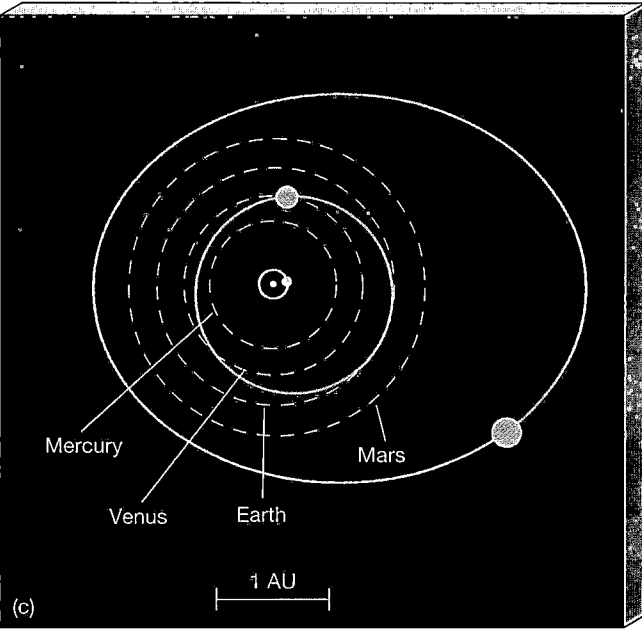

Figure 15.12 Planets Revealed (a) Measurements of the Doppler shift of the star 51 Pegasi reveal a clear periodic signal indicating the presence of a planetary companion of mass at least half the mass of Jupiter. (b) Radial velocity data for Upsilon Andromedae are much more complex, but are well fit (solid line) by a three-planet system orbiting the star. (c) A sketch of the orbits of the three planets inferred from the Upsilon Andromedae system, with the orbits of the terrestrial planets superposed for comparison.

tween genuine Jupiterlike planets and starlike brown dwarfs is unclear, but it is thought to be around fifteen Jupiter masses. According to this view, we just happen to see the orbits almost face-on, greatly reducing the parent star's radial velocity and fooling us into thinking we are observing low-mass planets instead of higher-mass brown dwarfs. There are some problems with this suggestion, however. Since the orientations of the observed orbits are (presumably) random, it is very unlikely that we would just happen to see all of them face-on. And even if we did, there is no observational evidence for the many edge-on (and easier to detect) systems we would then also expect on statistical grounds. Consequently, most astronomers agree that, while there may well be a few brown dwarfs lurking among the list of extrasolar planets, they probably do not constitute a significant fraction of the total.

The orbits and masses of the observed planets spell disaster for any low-mass "terrestrial" planets in these systems. In our solar system, the presence of a massive Jupiter on a nearly circular orbit is known to have a stabilizing influence on the other planetary orbits, tending to preserve the relative tranquility of our planetary environment. In the known extrasolar systems, not only is this stabilization absent, but having a Jupiter-sized planet repeatedly plow through the inner parts of the system means that any terrestrial planets or planetesimals have almost certainly been ejected from the system.

IS OUR SOLAR SYSTEM UNUSUAL?

Not so long ago, many astronomers argued that the condensation scenario described earlier in this chapter was in no way unique to our own system. The same basic processes could have occurred, and perhaps *did* occur, during the formative stages of many of the stars in our Galaxy, so planetary systems like our own should be common. Today we know that planetary systems *are* apparently quite common, but by and large they don't look at all like ours! We can legitimately ask whether our solar system really is as unusual as recent observations seem to imply, and whether these observations invalidate our current theory of solar system formation.

To answer the second question first, current theory in fact provides many ways in which massive planets can end up in short-period or eccentric orbits. Indeed, an important aspect of solar system formation glossed over in the

and the fact that only one planet is usually observed, imply that these systems really do differ from the planets orbiting the Sun.

Eccentric orbits are quite common among double-star systems (see Chapter 17), and some astronomers have suggested that many of the newly found planets are actually *brown dwarfs*—"failed stars" having insufficient mass to become true stars (see *Discovery 19–1*). The dividing line be-

earlier discussion is the fact that many theorists worry about how Jupiter could have remained in a stable orbit after it formed in the protosolar disk! Jupiter-sized planets may be knocked onto eccentric orbits by interactions with other Jupiter-sized planets or by the tidal effects of nearby stars. If they formed by gravitational instability, they would probably have eccentric orbits right from the start (and we then have to explain how those orbits circularized in the case of the solar system). And regardless of how it forms, gravitational interactions between a massive planet and the gas disk in which it moves tend to make it spiral inward, as mentioned earlier, and can easily deposit the planet in an orbit very close to the parent star. Interestingly, it now appears that the presence of Saturn may have helped stabilize Jupiter's orbit against this last effect. Isolated or particularly massive "Jupiters" are precisely the planets one would expect to find on "hot" orbits.

Astronomers would naturally prefer to be more definite, but the simple fact is that the early solar system was a very complicated place, and we currently have no firm observational or theoretical data to tell us which conditions lead to stable planetary orbits like those we enjoy on Earth today, and which lead to the eccentric hot Jupiters observed elsewhere. Both are clearly possible, but the probabilities of each are as yet unknown.

Given that hot Jupiters are indeed possible within existing theory, do the observations imply that they are the norm? The answer is probably no—at least for now. It is clear that the methods employed so far are heavily biased toward finding massive objects orbiting close to their parent stars. These give the strongest signal and are precisely what have been observed. Almost all of the systems detected so far have radial velocities substantially greater than the

12 m/s expected (under the best circumstances) for Jupiter orbiting the Sun. Furthermore, while Jupiter's wobble could be detected with current technology, it is close enough to the instrumental limits that several orbits—that is, several decades' worth of observations—would be needed before a definitive detection could be claimed.

In summary, only a small percentage (less than 10 percent) of the stars surveyed to date have shown evidence for extrasolar planets. The planets observed are precisely those that could have been detected, given today's technology; and their orbits are allowed, if not actually predicted, by current theories of planet formation. The properties of the systems we can see reveal little about those we can't, so for now the planetary properties of the vast majority of stars remain a mystery.

SEARCHING FOR EARTHLIKE PLANETS

Assuming that planetary systems like our own do exist, with Jupiters on circular, or at least nondisruptive, orbits and some stable terrestrial orbits close to their parent star, how might we go about searching for planets like our own? Current estimates suggest that looking for "wobbles" is not likely to be successful. Radial velocity variations will be so small as to be lost within the natural noise of the parent star itself, and side-to-side motions due to Earth-sized objects will be very hard to measure, even from space. Direct imaging of terrestrial planets remains a long-term goal of NASA, but tangible results are probably decades away.

Perhaps surprisingly, the approach given the best chance of success entails searching for planets *transiting* their parent stars. The effect of a transit on a star's brightness is tiny—less

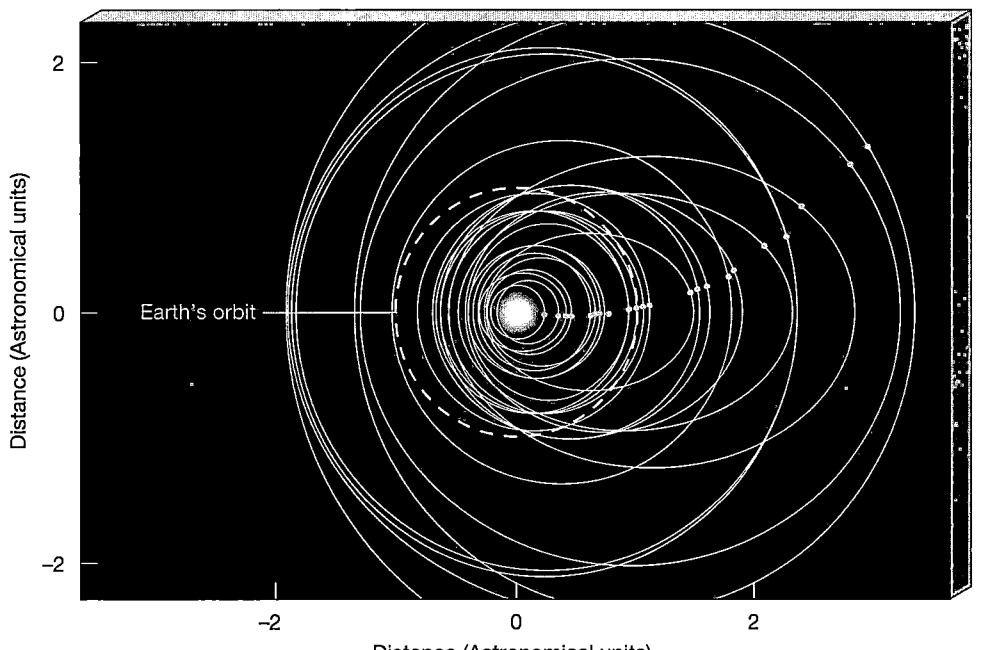

Figure 15.13 Extrasolar Orbits The orbits of all known extrasolar planets residing beyond 0.15 A.U. (as of early 2001), superimposed on a single plot, with Earth's orbit shown for comparison. The extrasolar planets are all comparable in mass to Jupiter— these planetary systems don't look much like ours!

The Kuiper Belt and the Search for a Tenth Planet

No one has ever observed any comets in the faraway Oort cloud—they are just too small and dim for us to see from Earth. But in the 1990s such faint objects began to be inventoried in the relatively nearby Kuiper belt, just beyond Neptune's orbit. ∞ (Sec. 14.2) Ground-based telescopes have led the way in the painstaking work to capture the meager amounts of sunlight reflected from dozens of such dark objects orbiting in the outer solar system. Some of the best available images of a Kuiper belt object are shown below. The fuzzy blob that changes position between one frame and the next is the object in question, which may be almost 1000 km across. In many ways, the Kuiper belt must resemble the asteroid belt, the collection of rocky debris found mainly between Mars and Jupiter. ∞ (Sec. 14.1)

As of early 2001, the current count of trans-Neptunian Kuiper belt objects is 367. They range in diameter from about 50 km to 1000 km, still considerably smaller than Pluto (2300 km), but now extending almost up to Pluto's moon Charon (1100 km) in size. ∞ (Sec. 13.9) Current estimates of the total number of Kuiper-belt objects larger than 100 km exceed 100,000, so the combined mass of all the debris in the Kuiper belt could well be hundreds of times larger than the inner asteroid belt (although still less than the mass of Earth).

While Pluto seems to be the "king of the Kuiper belt," the existence of other Pluto-sized objects orbiting at larger radii has not yet been conclusively ruled out. Systematic faint surveys of a broad swath of the sky including the entire ecliptic plane (as are planned within the next decade) will be needed before such a statement can be made. Even with large ground-based telescopes equipped with sensitive CCD detectors, it is unlikely that objects much smaller than 50 km (which almost certainly are out there, and probably number in the billions) will be found in the near future. (Astronomers using the *Hubble Space Telescope* have reported finding numerous small trans-Neptunian objects, but none has been confirmed after repeated tries.)

A curious feature of the Kuiper belt is that many of its members—almost one-third of all known objects, and perhaps 15 percent of the entire Kuiper belt—orbit in a *3:2 resonance* with Neptune. That is, they orbit the Sun twice for every three Neptune orbits. As noted in Chapter 14, Pluto shares this resonance, and the Kuiper-belt objects orbiting in this manner have accordingly been dubbed "plutinos." ∞ (Sec. 13.8) The presence of one object (Pluto) on such an exceptional orbit might possibly be at-

tributed to chance. It is possible for a planet such as Neptune to "capture" a Pluto-sized object into a resonant orbit, if Pluto's orbit started out close to resonance. However, hundreds of similar orbits require a more coherent explanation.

The leading theory for the existence of the plutinos is that, early in the evolution of the solar system (as described in the text), Neptune's orbit slowly migrated to larger radii as the giant planets interacted with, and ultimately ejected, leftover planetesimals in their vicinity. As Neptune moved outward, the radius corresponding to the 3:2 resonance also swept outward through the surviving planetesimals. Apparently this process was slow enough that many, if not most, of the planetesimals on near-resonant orbits were captured and subsequently carried outward, locked forever in synchrony with Neptune, as the planet's migration continued to its present location. These are the plutinos we see today.

Much of the observational work on the Kuiper belt began as a search for a tenth planet. However, in the process astronomers have come to realize that the Kuiper reservoir is much more than just the source of short-period comets. Pluto, its moon Charon, and the Neptunian moon Triton all reside in or near the Kuiper belt. As we saw in Chapter 13, these three oddballs have very similar physical properties and differ greatly from their neighboring planets and moons. Could this peculiar trio, and now also the large Kuiper-belt object pictured above, be the last survivors of a much larger set of big Kuiper belt objects? Or are more Pluto-sized objects still out there, waiting to be discovered? In either case, Pluto doesn't deserve the status of a genuine planet. It is ironic that work begun as a search to find a tenth planet may yet succeed in reducing the number of "true" solar system planets to eight.

  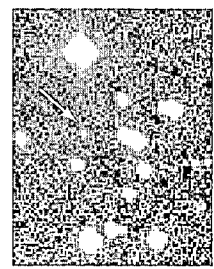

(Lunar and Planetary Lab)

| R | I | V | U | X | G |

than one part in 10^4 for an Earthlike planet crossing the face of the Sun—and also very unlikely, as only nearly edge-on systems will show it. Nevertheless, it is measurable with existing technology. NASA's proposed *Kepler* mission will monitor some 100,000 Sun-like stars for brightness fluctuations caused by Earth-like planets over a four-year period. With optimistic assumptions, some 100 terrestrial transits are expected. If detected, they would imply that planets similar to

our own are common in our Galaxy. NASA will decide on *Kepler*'s fate in 2002.

☑ Concept Check

■ Why is it not too surprising that the extrasolar planetary systems detected thus far have properties quite different from our solar system?

Chapter Review

SUMMARY

Our solar system is an orderly place, making it unlikely that the planets were simply captured by the Sun. The overall organization points toward formation as the product of an ancient, one-time event, 4.6 billion years ago. An ideal theory of the solar system should provide strong reasons for the observed characteristics of the planets yet be flexible enough to allow for deviations.

In the **nebular theory** (p. 385) of the formation of the solar system, a large cloud of dust and gas—the **solar nebula** (p. 385)—began to collapse under its own gravity. As it did so, it began to spin faster, to conserve angular momentum, eventually forming a disk. **Protoplanets** (p. 386) formed in the disk and became planets, and the central **protosun** (p. 386) eventually evolved into the Sun. The **condensation theory** (p. 386) builds on the nebular theory by incorporating the effects of particles of interstellar dust, which helped cool the nebula and acted as **condensation nuclei** (p. 386), allowing the planet-building process to begin. Small clumps of matter grew by **accretion** (p. 389), gradually sticking together and growing into moon-sized **planetesimals** (p. 389), whose gravitational fields were strong enough to accelerate the accretion process. Competing with accretion in the solar nebula was **fragmentation** (p. 389), the breaking up of small bodies following collisions with larger ones. Eventually, only a few planet-sized objects remained. The planets in the outer solar system may have formed directly via instabilities in the disk, or from protoplanet cores which became so large that they could capture the hydrogen and helium gas in the solar nebula.

The condensation theory can explain the basic differences between the jovian and terrestrial planets because the temperature of the solar nebula would be expected to decrease with increasing distance from the Sun. At any given location, the temperature would determine which materials could condense out of the nebula and so control the composition of any planets

forming there. The terrestrial planets are rocky because they formed in the hot inner regions of the solar nebula, near the Sun, where only rocky and metallic materials condensed out. Farther out, the nebula was cooler, and ices of water and ammonia could also form, ultimately leading to the observed differences in composition between the inner and outer solar system.

When the Sun became a star, its strong winds blew away any remaining gas in the solar nebula. Many leftover planetesimals were ejected into the Oort cloud and Kuiper belt by the gravitational fields of the outer planets. They now occasionally revisit our part of the solar system as comets. In the inner solar system, light elements such as hydrogen and helium would have escaped into space. Much, if not all, of Earth's water was carried to our world by comets deflected from the outer solar system. The asteroid belt is a collection of planetesimals that never managed to form a planet, because of Jupiter's gravitational influence. Many "odd" aspects of the solar system may conceivably be explained in terms of collisions late in the formation stages of the protoplanetary system.

Some 70 **extrasolar planets** (p. 384) are now known. All have been discovered by observing their parent star wobble back and forth as the planet orbits, although one has since been observed passing in front of the star, reducing the star's brightness slightly. Most systems found so far contain a single, massive planet comparable in mass to Jupiter on an orbit taking it close to the central star. None look much like our solar system, and the orbits of the observed Jupiter-mass planets would probably have ejected any terrestrial planets long ago. Several theories have been advanced to explain how such "hot Jupiters" might form. Current observations are sensitive only to the sorts of systems seen so far, and it is currently not known whether "hot Jupiters" or our own solar system represent the norm in planetary systems.

SELF-TEST: TRUE OR FALSE?

The following nine questions present properties of the solar system that any model of solar system formation must explain. Which are correctly stated and which are not?

____ **1.** Each planet is relatively isolated in space.

____ **2.** The orbits of the planets are not circular but significantly elliptical.

____ **3.** The orbits of the planets all lie near the ecliptic plane.

____ **4.** The direction of planetary revolution is in the same direction as the Sun's rotation.

____ **5.** Planetary rotation is always in the same direction as the Sun's rotation.

____ **6.** Moons usually revolve in a direction opposite to the rotation of their parent planet.

____ **7.** The planetary system is highly differentiated.

____ **8.** Asteroids were recently formed from the collision and breakup of an object orbiting within the asteroid belt.

____ **9.** Most comets have short periods and orbit close to the ecliptic plane.

True or False?

____ **10.** Water could not have condensed out any closer than 3 or 4 A.U. from the Sun.

____ **11.** The condensation theory offers an evolutionary explanation of the highly tilted rotation axis of Uranus.

____ **12.** According to the condensation theory, the Moon formed from a "mininebula" surrounding the proto-Earth.

____ **13.** Random collisions, inherently a part of the condensation theory, can explain many of the odd properties found among some solar system objects.

____ **14.** Astronomers have no theoretical explanation for the "hot Jupiters" observed orbiting some other stars.

____ **15.** Astronomers know of no extrasolar planetary systems similar to our own.

SELF-TEST: FILL IN THE BLANK

1. The condensation theory, which currently is used to explain the formation of the solar system, is actually just a refined version of the older _____ theory.

2. In the condensation theory, astronomers realized the critical role played by _____ in starting the formation of small clumps of matter.

3. Initially, the accretion of matter into larger bodies occurred through _____ between particles in the solar nebula.

4. By the time planetesimals had formed, the accretion process was accelerated by the effect of _____.

5. In the final stage of accretion, the largest protoplanets were able to attract large quantities of _____ from the solar nebula.

6. The temperature of the inner part (out to 1 or 2 A.U.) of the solar nebula was well above the boiling point of _____.

7. Unlike the terrestrial planets, the planetesimals that formed the jovian planets were made up of _____ material.

8. The large number of leftover planetesimals formed beyond about 5 A.U. were destined to become _____.

9. The water now found on Earth was probably brought here by _____.

10. The reason the planetesimals of the asteroid belt did not form a larger object was probably the gravitational influence of _____.

11. The _____ phase of the early Sun cleared out excess gas not used in planet formation.

12. Angular momentum depends on the mass, angular speed, and _____ of an object.

13. _____ played an important role in determining many of the irregularities in the solar system.

14. Astronomers can detect extrasolar planets by observing the _____ of their parent stars.

15. Most of the extrasolar planets observed so far have masses comparable to the mass of _____ and orbits that take them _____ to their parent stars.

REVIEW AND DISCUSSION

1. List six properties of the solar system that any model of its formation must be able to explain.

2. Explain the difference between evolutionary theories and catastrophic theories of the solar system's origin.

3. Describe the basic features of the nebular theory of solar system formation.

4. Give three examples of how the nebular theory explains some observed features of the present-day solar system.

5. Explain the difference between angular momentum and linear momentum.

6. What is the key ingredient in the modern condensation theory of the solar system's origin that was missing or unknown in the nebular theory?

7. Describe two possible ways in which the jovian planets may have formed.

8. Why are the jovian planets so much larger than the terrestrial planets?

9. What solar system objects, still observable today, resulted from the process of fragmentation?

10. What influence did Earth's location in the solar nebula have on its final composition?

11. How did the temperature structure of the solar nebula determine planetary composition?

12. Why could Earth not have formed out of material containing water? How might Earth's water have gotten here?

13. What are Plutinos, and how did they come to be on their present orbits?

14. What happened in the early solar system when the Sun became a T Tauri star?

15. How did the Kuiper belt and the Oort cloud form?

16. Describe some ways in which random processes played a role in the determination of planetary properties.

17. Describe a possible history of a single comet now visible from Earth, starting with its birth in the solar nebula somewhere near the planet Jupiter.

18. How do astronomers set about looking for extrasolar planets?

19. In what ways do extrasolar planetary systems differ from our own solar system?

20. Do the observed extrasolar planets imply that Earth-like planets are very rare?

PROBLEMS    *Algorithmic versions of these questions are available in the Practice Problems module of the Companion Website.*

The number of squares preceding each problem indicates its approximate level of difficulty.

1. ■■ The orbital angular momentum of a planet in a circular orbit is simply the product of its mass, its orbital speed, and its distance from the Sun. (a) Compare the orbital angular momenta of Jupiter, Saturn, and Earth. (b) Calculate the orbital angular momentum of an Oort cloud comet with mass 10^{13} kg, moving in a circular orbit 50,000 A.U. from the Sun.

2. ■■ An interstellar cloud fragment 0.2 light-years in diameter is rotating at a rate of one revolution per million years. It now begins to collapse. Assuming that the mass remains constant, estimate its rotation period when it has shrunk to (a) the size of the solar nebula, 100 A.U. across, and (b) the size of Earth's orbit, 2 A.U. across.

3. ■■ By what factor would Earth's rotational angular momentum change if the planet's spin rate were to double? By

what factor would Earth's orbital angular momentum change if the planet's distance from the Sun were to double (assuming that the orbit remained circular)?

4. ■■■ We can make a rough model of accretion in the inner solar nebula by imagining a 1-km-diameter body moving at a relative speed of 500 m/s through a collection of similar bodies having a roughly uniform spatial density of 10^{-10} bodies per cubic kilometer. Neglecting any gravitational forces (and hence assuming that the body moves in a straight line until it collides with something), estimate how much time, on average, will be needed for it to experience a collision.

5. ■■ Consider a planet growing by accretion of material from the solar nebula. As it grows, its density remains roughly constant. Does the force of gravity at its surface increase, decrease, or stay the same? Specifically, what would happen to the surface gravity and escape speed as the radius of the planet doubled? Give reasons for your answer.

6. ■ How many 100-km-diameter rocky (3000 kg/m^3) planetesimals would have been needed to form Earth?

7. ■■ Two asteroids, each of mass 10^{18} kg, orbit near the center of the asteroid belt in the plane of the ecliptic on circular paths of radii 2.80 and 2.81 A.U., respectively. Calculate the gravitational force between them at closest approach, and compare it with the tidal force exerted by Jupiter if that planet were also at closest approach at that time.

8. ■■ According to Figure 15.8, the temperature in the early solar nebula at a distance of 1 A.U. from the Sun was about 1100 K. Based on the discussion of surface temperature in Chapter 7, estimate the factor by which the Sun's current energy output would have to increase in order for Earth's present temperature to have this value. ∞ (Sec. 7.2)

9. ■ A typical comet contains some 10^{13} kg of water ice. How many comets would have to strike Earth in order to account for the roughly 2×10^{21} kg of water presently found on our planet? If this amount of water accumulated over a period of 0.5 billion years, how frequently must Earth have been hit by comets during that time?

10. ■ Use the data given in the text to calculate Neptune's orbital period before interactions with planetesimals expanded the orbit to its present size.

11. ■ How many comet-sized planetesimals (10 km diameter, density 100 kg/m^3) would have been needed to form Pluto?

12. ■ The planet orbiting star HD187123 has a semimajor axis of 0.042 A.U. If the star's mass is 1.06 times the mass of the Sun, calculate how many times the planet has orbited its star since the paper announcing it was published on December 1, 1998.

13. ■■ The two planets orbiting the nearby star Gliese 876 are observed to be in a 2:1 resonance. The inner planet has an orbital period of 30 days. If the star's mass is $\frac{1}{3}$ the mass of the Sun, calculate the semimajor axis of the outer planet's orbit.

14. ■■ How close to the Sun would an identical star have to come in order to exert a 0.1 percent tidal perturbation on Jupiter (that is, a tidal force equal to 0.1 percent of the Sun's gravitational attraction on the planet)?

15. ■■■ Not surprisingly, as planet masses decrease, they become harder to detect. Current Doppler techniques can only reliably detect a planet having 0.25 times the mass of Jupiter orbiting a Sun-like star if the planet's *maximum* orbital speed exceeds 40 km/s. Under these conditions, what is the minimum detectable eccentricity for a planet with a five-year orbital period?

COLLABORATIVE EXERCISES

1. Catastrophic Catastrophies. Consider the unusual characteristics of the solar system listed in Section 15.4. Which is the most unlikely and which is the most likely?

2. Pluto: Planet or Kuiper Belt Object. As a group, agree on whether or not Pluto should be classified as a planet. Justify your answer.

RESEARCHING ON THE WEB  *To complete the following exercises, go to the online Destinations module for Chapter 15 on the Companion Website for Astronomy Today 4/e.*

1. Access the "Discovery of a Planetary Orbit Around the Nearby Star 51 Pegasi" page. Why is this an unusual planetary system? Make a sketch to help you find 51 Pegasi in the sky.

2. Access the "Planetary Searches" page and use the current Almanac of Planets to determine how many planetary systems have been confirmed around other stars.

SKYCHART III PROJECTS  *The SkyChart III Student Version planetarium program on which these exercises are based is included as a separately executable program on the CD in the back of this text.*

1. ■ It is impossible to see another planet, but you can locate another solar system. Generate a finder chart for 51 Pegasi, by searching for "51 peg." When does this solar system rise and when does it set? How far away is it?

2. ■ Beta Pictoris is a star around which warm matter has been found. Locate Beta Pictoris (search for "β pic"). What is its declination? When does this system rise and set? Can you ever see this star from where you are? Why or why not?

 *In addition to the Practice Problems and Destinations modules, the Companion Website at http://www.prenhall.com/chaisson provides for each chapter an additional true-false, multiple choice, and labeling quiz, as well as additional annotated images, animations, and links to related Websites.*

16 THE SUN

Our Parent Star

LEARNING GOALS

Studying this chapter will enable you to

1   Summarize the overall properties of the Sun.

2   Explain how energy travels from the solar core, through the interior, and out into space.

3   Name the Sun's outer layers and describe what those layers tell us about the Sun's surface composition and temperature.

4   Discuss the nature of the Sun's magnetic field and its relationship to the various types of solar activity.

5   Outline the process by which energy is produced in the Sun's interior.

6   Explain how observations of the Sun's core challenge our present understanding of fundamental physics.

 Visit http://www.prenhall.com/chaisson for additional annotated images, animations, and links to related sites for this chapter.

This spectacular series of images captures the Sun in the X ray part of the spectrum. They were acquired by *Yohkoh* (meaning "sunbeam" in Japanese), a small satellite launched from Japan in 1991. Designed to study the most active phases of the Sun, this space-based observatory enabled astronomers to watch the Sun on a daily basis as its activity changed—from solar maximum activity at bottom right to minimum activity at top left. *(ISAS/Lockheed Martin)*
The Big Picture: The Sun is our star. It is not merely the center of the solar system or the source of light that distinguishes day from night; it is also the main source of energy that powers weather, climate, and life on Earth. Humans simply wouldn't exist without the Sun. Although we take it for granted each and every day, the Sun is of enormous importance to us in the cosmic scheme of things.

Living in the solar system, we have the chance to study, at close range, perhaps the most common type of cosmic object—a star. Our Sun is a star, and a fairly average star at that, but with one unique feature: It is very close to us—some 300,000 times closer than our next nearest neighbor, Alpha Centauri. Whereas Alpha Centauri is 4.3 light-years distant, the Sun is only eight light-minutes away from us. Consequently, astronomers know far more about the properties of the Sun than about any of the other distant points of light in the universe. A good fraction of all our astronomical knowledge is based on modern studies of the Sun. Just as we studied our parent planet, Earth, to set the stage for our exploration of the solar system, we now study our parent star, the Sun, as the next step in our exploration of the universe.

16.1 Physical Properties of the Sun

1 The Sun is the sole source of light and heat for the maintenance of life on Earth. It is a **star**, a glowing ball of gas held together by its own gravity and powered by nuclear fusion at its center. In its physical and chemical properties, the Sun is very similar to most other stars, regardless of when and where they formed. Indeed, our Sun appears to be a rather typical star, lying right in the middle of the observed ranges of stellar mass, radius, brightness, and composition. Far from detracting from the interest in the Sun, this very mediocrity is one of the main reasons that astronomers study it—they can apply knowledge of solar phenomena to many other stars in the universe.

OVERALL PROPERTIES

The Sun Data box at right lists some basic orbital and physical solar data. The Sun's radius, roughly 700,000 km, is most directly determined by measuring its angular size (0.5°) and then employing elementary geometry. ∞ (Sec. 1.5) The Sun's mass, 2.0×10^{30} kg, follows from application of Newton's laws of motion and gravity to the observed orbits of the planets. ∞ (*More Precisely 2-3*) The average solar density derived from these quantities, approximately 1400 kg/m³, is quite similar to that of the jovian planets and about one-quarter the average density of Earth.

Solar rotation can be measured by timing sunspots ∞ (Sec. 2.4) and other surface features as they traverse the solar disk. These observations indicate that the Sun rotates in about a month, but it does not do so as a solid body. Instead, it spins differentially—faster at the equator and slower at the poles, like Jupiter and Saturn. The equatorial rotation period at the equator is about 25 days. Sunspots are never seen above latitude 60° (north or south), but at that latitude they indicate a 31-day period. Other measurement techniques, such as those discussed in Section 16.2, reveal that the rotation period continues to increase as we approach the poles. The polar rotation period is not known with certainty, but it may be as long as 36 days.

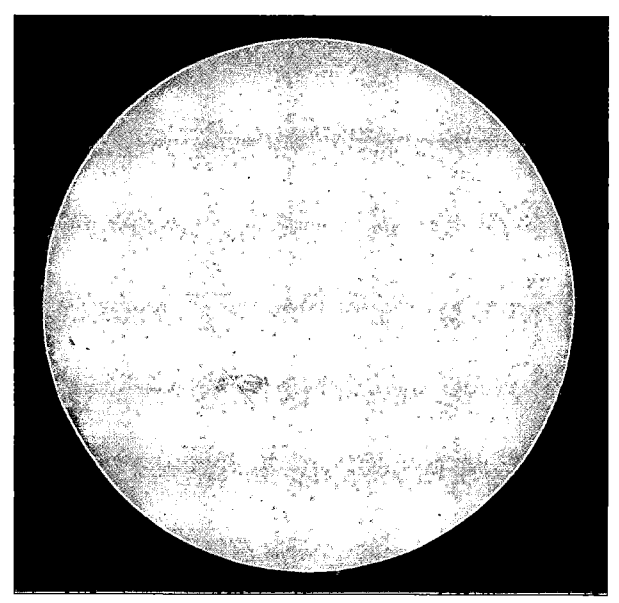

Greatest angular diameter, as seen from Earth	32.5′
Mass	1.99×10^{30} kg 332,000 (Earth = 1)
Equatorial radius	696,000 km 109 (Earth = 1)
Mean density	1410 kg/m³ 0.255 (Earth = 1)
Surface gravity	274 m/s² 28.0 (Earth = 1)
Escape speed	618 km/s
Sidereal rotation period	25.1 solar days (equator) 30.8 solar days (60° latitude) 36 solar days (poles) 26.9 solar days (interior)
Axial tilt	7.25° (relative to ecliptic)
Surface temperature	5780 K (effective temperature)
Luminosity	3.85×10^{26} W

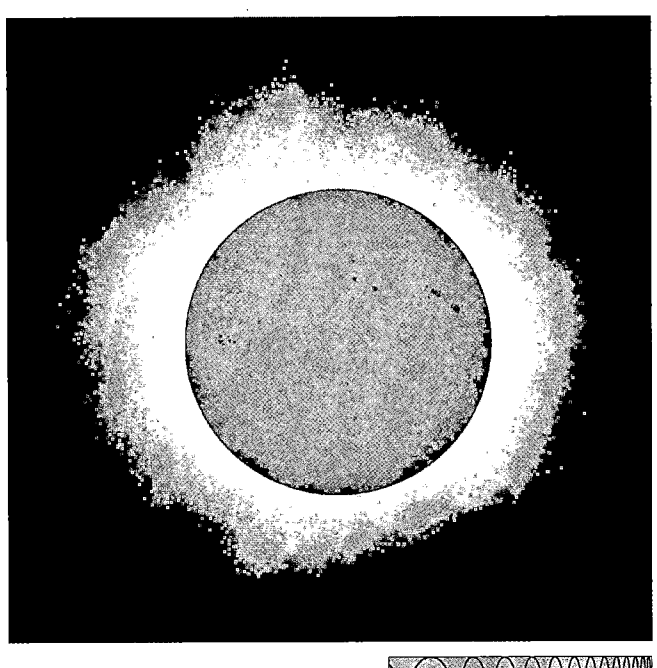

Figure 16.1 The Sun The inner part of this composite image of the Sun shows a sharp solar limb, although our star, like all stars, is made of a gradually thinning gas. The edge appears sharp because the solar photosphere is so thin. The outer portion of the image is the solar corona, normally too faint to be seen, but visible during an eclipse, when the light from the solar disk is blotted out. *(NOAO)*

The Sun's surface temperature is measured by applying the radiation laws to the observed solar spectrum. ∞ (Sec. 3.4) The distribution of solar radiation has the approximate shape of a blackbody curve for an object at about 5800 K. The average solar temperature obtained in this way is known as the Sun's *effective temperature*.

Having a radius of more than 100 Earth radii, a mass of more than 300,000 Earth masses, and a surface temperature well above the melting point of any known material, the Sun is clearly a body very different from any other we have encountered so far.

SOLAR STRUCTURE

The Sun has a surface of sorts—not a solid surface (the Sun contains no solid material) but rather that part of the brilliant gas ball we perceive with our eyes or view through a heavily filtered telescope. This "surface"—the part of the Sun that emits the radiation we see—is called the **photosphere**. Its radius (listed in the Data Box as the equatorial radius of the Sun) is about 700,000 km. However, the thickness of the photosphere is probably no more than 500 km, less than 0.1 percent of the radius, which is why we perceive the Sun as having a well-defined, sharp edge (Figure 16.1).

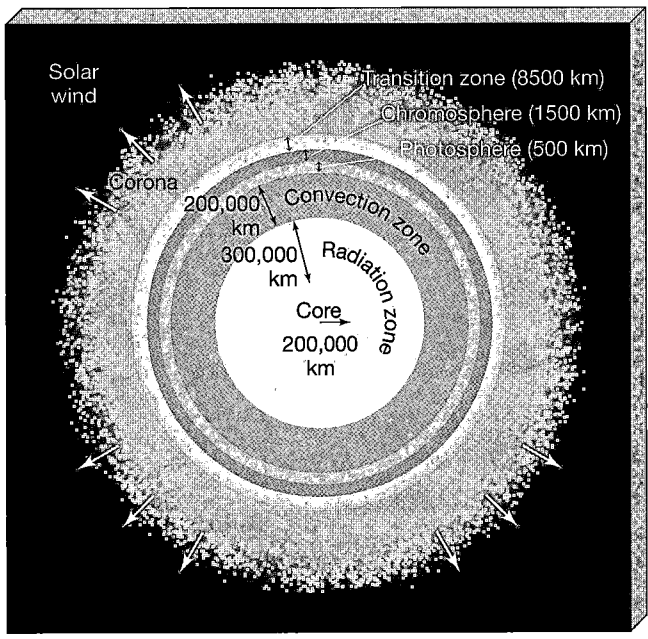

Figure 16.2 Solar Structure The main regions of the Sun, not drawn to scale, with some physical dimensions labeled.

The main regions of the Sun are illustrated in Figure 16.2 and summarized in Table 16.1. Just above the photosphere is the Sun's lower atmosphere, called the **chromosphere**, about 1500 km thick. From 1500 km to 10,000 km above the top of the photosphere lies a region called the **transition zone**, where the temperature rises dramatically. Above 10,000 km, and stretching far beyond, is a tenuous (thin), hot upper atmosphere, the solar **corona**. At still greater distances, the corona turns into the *solar wind*, which flows away from the Sun and permeates the entire solar system. ∞ (Sec. 6.5) Extending down some 200,000 km below the photosphere is the **convection zone**, a region where the material of the Sun is in constant convective motion. Below the convection zone lies the **radiation zone**, where solar energy is transported toward the surface by radiation rather than by convection. The term *solar interior* is often used to mean both the radiation and convection zones. The central **core**, roughly 200,000 km in radius, is the site of powerful nuclear reactions that generate the Sun's enormous energy output.

LUMINOSITY

The properties of size, mass, density, rotation rate, and temperature are familiar from our study of the planets. But the Sun has an additional property, perhaps the most important of all from the point of view of life on Earth—it *radiates* a great deal of energy into space, uniformly (we assume) in all directions. By holding a light-sensitive device—a solar cell, perhaps—perpendicular to the Sun's

TABLE 16.1 The Standard Solar Model

REGION	INNER RADIUS (km)	TEMPERATURE (K)	DENSITY (kg/m³)	DEFINING PROPERTIES
Core	0	15,000,000	150,000	Energy generated by nuclear fusion
Radiation zone	200,000	7,000,000	15,000	Energy transported by electromagnetic radiation
Convection zone	500,000	2,000,000	150	Energy carried by convection
Photosphere	696,000*	5800	2×10^{-4}	Electromagnetic radiation can escape—the part of the Sun we see
Chromosphere	696,500*	4500	5×10^{-6}	Cool lower atmosphere
Transition zone	698,000*	8000	2×10^{-10}	Rapid temperature increase
Corona	706,000	1,000,000	10^{-12}	Hot, low-density upper atmosphere
Solar wind	10,000,000	2,000,000	10^{-23}	Solar material escapes into space and flows outward through the solar system

These radii are based on the accurately determined radius of the photosphere. The other radii quoted are approximate, round numbers.

rays, we can measure how much solar energy is received per square meter of surface area every second. Imagine our detector as having a surface area of one square meter (1 m²) and as being placed at the top of Earth's atmosphere. The amount of solar energy reaching this surface each second is a quantity known as the **solar constant**. Its value is approximately 1400 watts per square meter (W/m²). Most of this energy reaches Earth's surface. Thus, for example, a sunbather's body having a total surface area of about 0.5 m² receives solar energy at a rate of nearly 700 watts, roughly equivalent to the output of a typical electric room heater or about ten 75-W lightbulbs.

Let us now ask about the *total* amount of energy radiated in all directions from the Sun, not just the small fraction intercepted by our detector or by Earth. Imagine a three-dimensional sphere that is centered on the Sun and is just large enough that its surface intersects Earth's center (Figure 16.3). The sphere's radius is 1 A.U., and its surface area is therefore $4\pi \times (1 \text{ A.U.})^2$, or approximately 2.8×10^{23} m². Multiplying the rate at which solar energy falls on each square meter of the sphere (that is, the solar constant) by the total surface area of our imaginary sphere, we can determine the total rate at which energy leaves the Sun's surface. This quantity is known as the **luminosity** of the Sun. It turns out to be just under 4×10^{26} W.

The Sun is an enormously powerful source of energy. *Every second*, it produces an amount of energy equivalent to the detonation of about 100 billion one-megaton nuclear bombs. Six seconds worth of solar energy output, suitably focused, would evaporate all of Earth's oceans. Three minutes would melt our planet's crust. The scale on which the Sun operates simply defies Earthly comparison.

✓ **Concept Check**

■ Why must we assume that the Sun radiates equally in all directions when computing the solar luminosity from the solar constant?

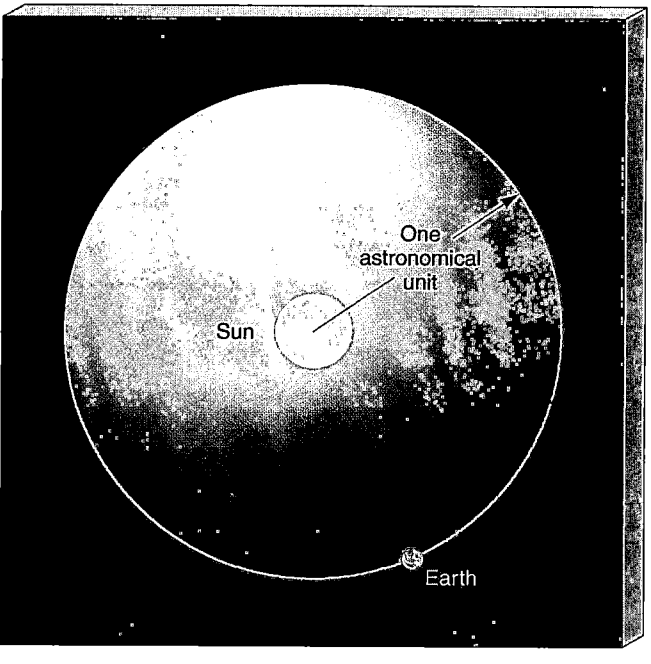

Figure 16.3 Astronomical Unit We can draw an imaginary sphere around the Sun so that the sphere's edge passes through Earth's center. The radius of this imaginary sphere equals one astronomical unit. By multiplying the sphere's surface area by the solar constant, we can measure the Sun's luminosity—the amount of energy it emits each second.

16.2 The Solar Interior

MODELING THE STRUCTURE OF THE SUN

Lacking any direct measurements of the solar interior, astronomers must use more indirect means to probe the inner workings of our parent star. To accomplish this, they construct mathematical models of the Sun, combining all available observations with theoretical insight into solar physics. The result is the **Standard Solar Model**, which has gained widespread acceptance among astronomers.

To test and refine the Standard Solar Model, astronomers are eager to obtain information about the solar interior. However, we have little or no direct information about conditions below the solar photosphere, so we must rely on more indirect techniques. In the 1960s it was discovered that the surface of the Sun vibrates like a complex set of bells. These vibrations, illustrated in Figure 16.4(a), are the result of internal pressure ("sound") waves that reflect off the photosphere and repeatedly cross the solar interior (Figure 16.4b). Because these waves can penetrate deep inside the Sun, analysis of their surface patterns allows scientists to study conditions far below the Sun's surface. This process is similar to the way in which seismologists study the interior of Earth by observing the P- and S-waves produced by earthquakes. ∞ (Sec. 7.3) For this reason, study of solar surface patterns is usually called **helioseismology**, even though solar pressure waves have nothing whatever to do with solar seismic activity—there is no such thing.

The most extensive study of solar oscillations is the ongoing GONG (Global Oscillations Network Group) project. By making continuous observations of the Sun from many clear sites around Earth, solar astronomers can obtain uninterrupted high-quality solar data spanning many days and even weeks—almost as though Earth were not rotating and the Sun never set. The *Solar and Heliospheric Observatory* (*SOHO*), launched by the European Space Agency in 1995 and now permanently stationed between Earth and the Sun some 1.5 million km from our planet (see *Discovery 16-1*), also provides continuous monitoring of the Sun's surface and atmosphere. Analysis of ground- and space-based data provides important additional information about the temperature, density, rotation, and the convective state of the solar interior, allowing detailed comparisons between theory and reality to be made. Direct comparison is possible throughout a large portion of the Sun, and the agreement between model and observations is spectacular—the frequencies and wavelengths of observed solar oscillations are within 0.1 percent of the predictions of the Standard Solar Model.

Some mysteries remain, however. For example, helioseismology indicates that the Sun's rotation speed varies with depth—perhaps not too surprising given the surface

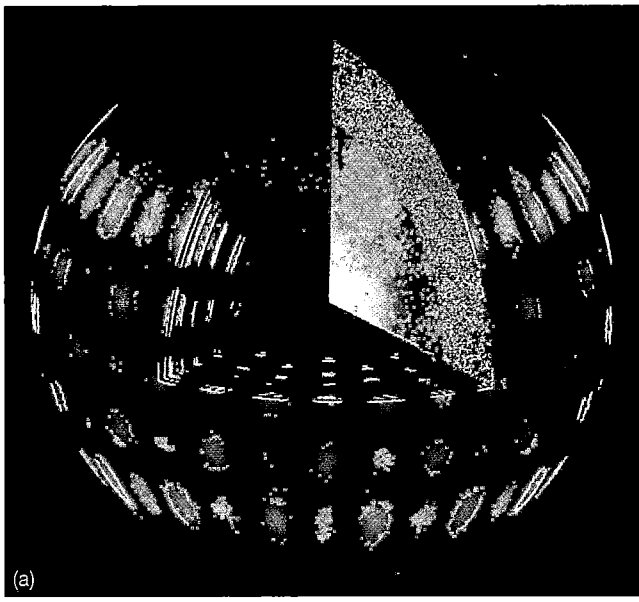

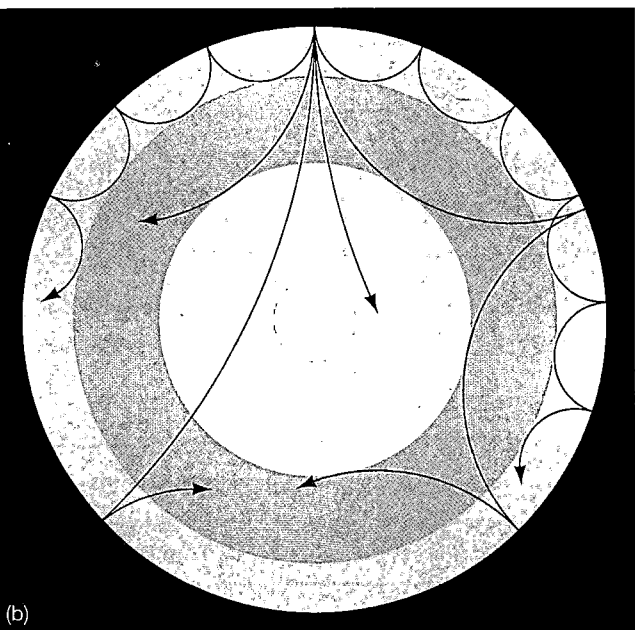

Figure 16.4 Solar Oscillations (a) The Sun has been found to vibrate in a very complex way. By observing the motion of the solar surface, scientists can determine the wavelength and the frequencies of the individual waves and deduce information about the Sun not obtainable by other means. The alternating patches represent gas moving down (red) and up (blue). (See also *Discovery 16-1*.) (b) Depending on their initial directions, the waves contributing to the observed oscillations may travel deep inside the Sun, providing vital information about the solar interior. *(National Solar Observatory)*

differential rotation mentioned earlier and the fact that similar behavior has been noted in the outer planets. What is puzzling, though, is the complexity of the differential motion. The surface layers show a "zonal flow" of sorts, with alternating bands of higher- and lower-than-average

DISCOVERY 16-1

SOHO: Eavesdropping on the Sun

Throughout the few decades of the Space Age, various nations, led by the United States, have sent spacecraft to all but two of the major bodies in the solar system. One of these as-yet-unexplored bodies is Pluto, the most distant planet from the Sun, which has never been visited by a robot orbiter or even a flyby craft—and such a mission is not likely anytime soon. The other unexplored body is the Sun. Currently, the next best thing to a dedicated reconnoitering spacecraft is the *Solar and Heliospheric Observatory* (*SOHO*), which has been radioing back to Earth volumes of new data—and a few new puzzles—about our parent star since its launch in late 1995.

SOHO is a billion-dollar mission operated primarily by the European Space Agency. The two-ton robot is now on-station about 1.5 million km sunward of Earth—about one percent of the distance from Earth to the Sun. This is the so-called L_1 Lagrangian point, where the gravitational pull of the Sun and Earth are precisely equal—a good place to park a monitoring platform. ∞ (Sec. 14.1) There *SOHO* looks unblinkingly at our star 24 hours a day, eavesdropping on the Sun's surface, atmosphere, and interior. This automated vehicle carries a dozen instruments, capable of measuring almost everything from the Sun's corona and magnetic field to its solar wind and internal vibrations. The figure at right shows a false color image of the Sun's lower corona, obtained by combining *SOHO* data obtained at three different extreme ultraviolet wavelengths.

SOHO is positioned just beyond Earth's magnetosphere, so its instruments can study cleanly the charged particles of the solar wind—high-speed matter escaping

from the Sun, flowing outward from the corona. Coordinating these on-site measurements with *SOHO* images of the Sun itself, astronomers now think they can follow solar magnetic field loops expanding and breaking as the Sun prepares itself for mass ejections several days before they actually occur. Given that such coronal storms can play

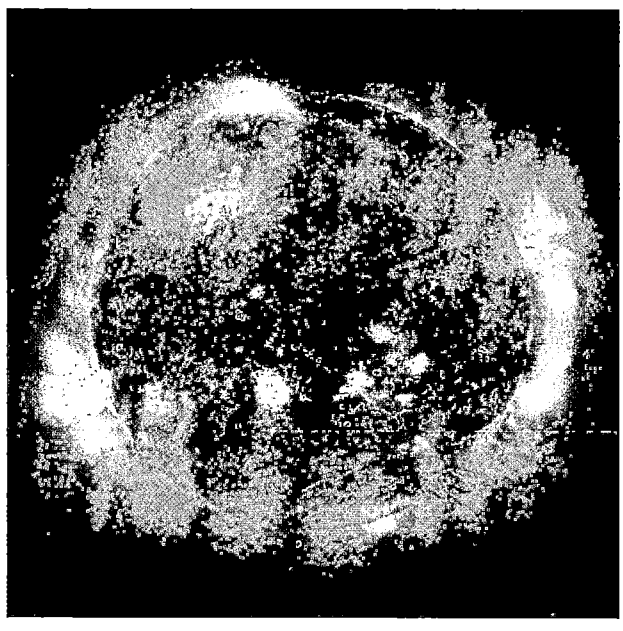

(NASA/ESA)

rotation rates. Just below the surface are wide "rivers" of lower-speed (at the equator) and higher-speed (polar) rotation. The material at the base of the convection zone appears to oscillate in rotation speed, sometimes moving faster (by about 10 percent) than the surface layers, sometimes slower, with a period of about 1.3 years. Deeper still, the radiative interior rotates more or less as a solid body, once every 26.9 days. A full explanation of the Sun's rotation currently eludes theorists.

Figure 16.5 shows the solar density and temperature according to the Standard Solar Model, plotted as functions of distance from Sun's center. Notice how the density drops rather sharply at first, then decreases more slowly near the solar photosphere, some 700,000 km from the center. The variation in density is large, ranging from a core value of about 150,000 kg/m^3, 20 times the density of iron, to an intermediate value (at 350,000 km) of about 1000 kg/m^3, the density of water, to an extremely small photospheric value of 2×10^{-4} kg/m^3, 10,000 times less dense than air at the surface of Earth. Because the density

is so high in the core, roughly 90 percent of the Sun's mass is contained within the inner half of its radius. The solar density continues to decrease out beyond the photosphere, reaching values as low as 10^{-23} kg/m^3 in the far corona—about as thin as the best vacuum physicists can create in laboratories on Earth.

The solar temperature also decreases with increasing radius, but not as rapidly as the density. Computer models indicate a temperature of about 15 million K at the core, consistent with the minimum 10 million K needed to initiate the nuclear reactions known to power most stars, decreasing to the observed value of about 5800 K at the photosphere.

ENERGY TRANSPORT

2 The very hot solar interior ensures violent and frequent collisions among gas particles. Particles move in all directions at high speeds, bumping into one another unceasingly. In and near the core, the extremely high temper-

havoc with communications, power grids, satellite electronics, and other human activity, the prospect of having accurate forecasts of disruptive solar events is a welcome development.

Section 16.2 discussed how astronomers can "take the pulse" of the Sun by measuring its complex rhythmic motions. *SOHO* also has the ability to study the weak sound waves that echo and resonate inside the Sun and can map these vibrations with much higher resolution than was previously possible. It does so not by sensing sound itself but by watching the Sun's surface move up and down ever so slightly. The Sun's "loudest" vibrations are extremely low pitched (0.003 Hz, or one oscillation every five minutes), more like rolling rumbles, just as we might expect from such a huge and massive object.

The second figure is not an image but a radial-velocity map of the Sun's surface (at much higher resolution than the diagram in Figure 16.4), obtained by measuring the Doppler shift of different portions of the solar photosphere. Bright areas are moving toward us, dark areas away. Observations of this sort enable researchers to refine models of the solar interior. For example, they have pinpointed the exact boundary between the convection zone and the radiation zone at 71.3 percent of the radius of the photosphere. The hope is to extend this work to discover how sharply defined the core is and just where its outer boundary lies.

As of mid-2001, *SOHO* is still operating, three years beyond its planned lifetime, having survived assaults from both the Sun and Earth. Twice so far, skillful engineers

have brought the spacecraft back from apparent death after mission controllers mistakenly sent incorrect commands from the ground. It is a good thing that they did, for this remarkable spacecraft has radioed back to Earth a wealth of new scientific insight about our parent star.

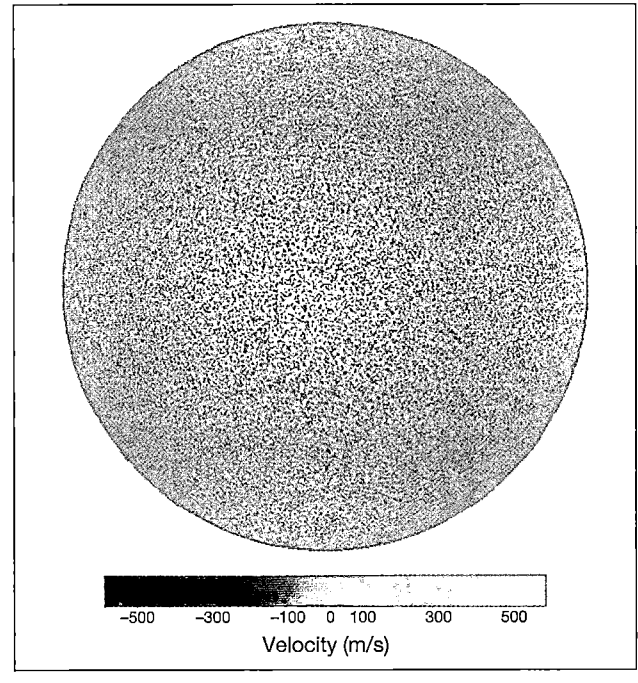

atures guarantee that the gas is completely ionized. Recall from Chapter 4 that under less extreme conditions, atoms absorb photons that can boost their electrons to more excited states. ∞ (Sec. 4.2) With no electrons left on atoms to capture the photons, however, the deep solar interior is relatively transparent to radiation. Only occasionally does a photon encounter and scatter off a free electron or proton. As a result, the energy produced by nuclear reactions in the core travels outward toward the surface in the form of radiation with relative ease.

As we move outward from the core, the temperature falls, atoms collide less frequently and less violently, and more and more electrons manage to remain bound to their parent nuclei. With more and more atoms retaining electrons that can absorb the outgoing radiation, the gas in the interior changes from being relatively transparent to being almost totally opaque. By the outer edge of the radiation zone, 200,000 km below the photosphere, *all* the photons produced in the Sun's core have been absorbed. Not one of them reaches the surface. But what happens to the energy they carry?

The photons' energy must travel beyond the Sun's interior. That we see sunlight—visible energy—proves that energy escapes. That energy reaches the surface by *convection*—the same basic physical process we saw in our study of Earth's atmosphere, although it operates in a very different environment in the Sun. ∞ (Sec. 7.2) Hot solar gas physically moves outward while cooler gas above it sinks, creating a characteristic pattern of convection cells. All through the convection zone, energy is transported to the surface by physical motion of the solar gas. (Remember that there is no physical movement of material when radiation is the energy-transport mechanism; convection and radiation are *fundamentally different* ways in which energy can be transported from one place to another.)

In reality, the zone of convection is much more complex than we have just described. As illustrated in Figure 16.6, there is a hierarchy of convection cells, organized in tiers at different depths. The deepest tier, about 200,000 km below the photosphere, is thought to contain large cells some tens of thousands of kilometers in diameter.

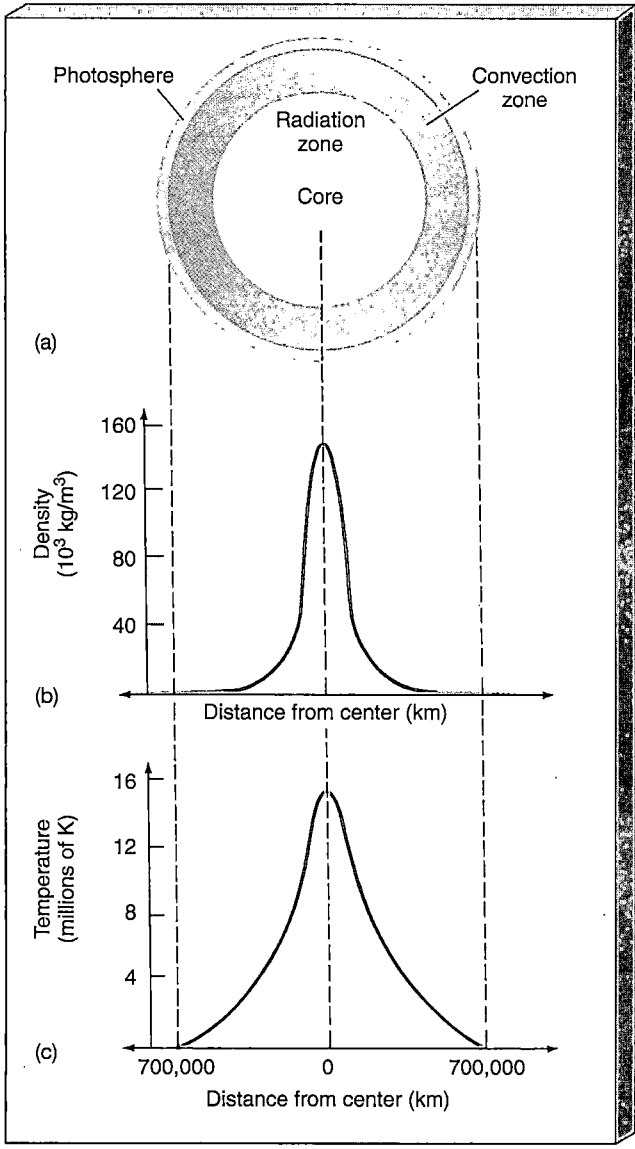

(a)

(b)

(c)

Figure 16.5 Solar Interior Theoretically modeled profiles of density (b) and temperature (c) for the solar interior, presented for perspective in (a). All three parts describe a cross-sectional cut through the center of the Sun.

Heat is then successively carried upward through a series of progressively smaller-sized cells, stacked one on another until, at a depth of about 1000 km, the individual cells are about 1000 km across. The top of this uppermost tier of convection is the visible surface, where astronomers can directly observe the cell sizes. Information about convection below that level is inferred mostly from computer models of the solar interior.

At some distance from the core the solar gas becomes too thin to sustain further upwelling by convection. Theory suggests that this distance roughly coincides with the photospheric surface we see. Convection does not proceed into the solar atmosphere. There is simply not enough gas there. The density is so low that the gas becomes transparent once again, and radiation once again becomes the mechanism of energy transport. Photons reaching the photosphere escape more or less freely into space, and the photosphere emits thermal radiation, like any other hot object. The photosphere is narrow, and the "edge" of the Sun sharp, because this transition from opacity to complete transparency is very rapid. Just below the bottom of the photosphere the gas is still convective, and radiation does not reach us directly. A few hundred kilometers higher, the gas is too thin to emit or absorb any significant amount of radiation.

GRANULATION

Figure 16.7 is a high-resolution photograph of the solar surface. The visible surface is highly mottled, or **granulated**, with regions of bright and dark gas known as *granules*. Each bright granule measures about 1000 km across—comparable in size to a continent on Earth—and has a lifetime of between five and 10 minutes. Together, several million granules constitute the top layer of the convection zone, immediately below the photosphere.

Each granule forms the topmost part of a solar convection cell. Spectroscopic observation within and around the bright regions shows direct evidence for the upward motion of gas as it "boils" up from within. This evidence proves that convection really does occur just below the photosphere. Spectral lines detected from the bright gran-

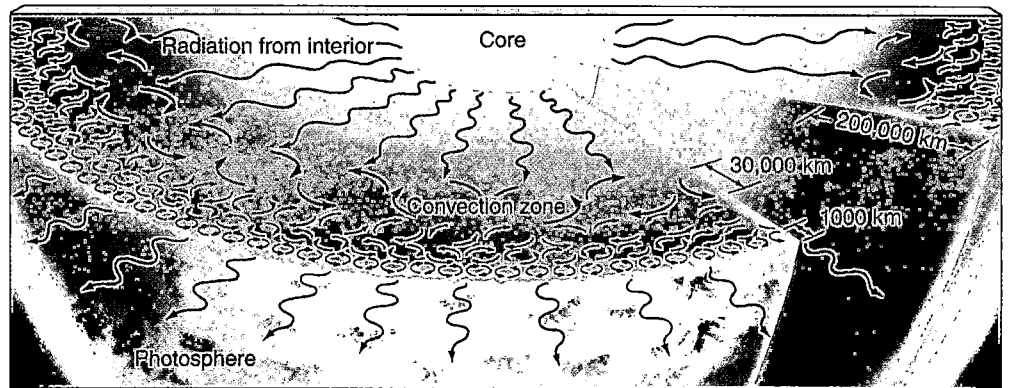

Figure 16.6 Solar Convection Physical transport of energy in the Sun's convection zone. We can visualize the upper interior as a boiling, seething sea of gas. Each convective loop is about 1000 km across. The convective cell sizes become progressively smaller closer to the surface. (This is a highly simplified diagram; there are many different cell sizes, and they are not so neatly arranged.)

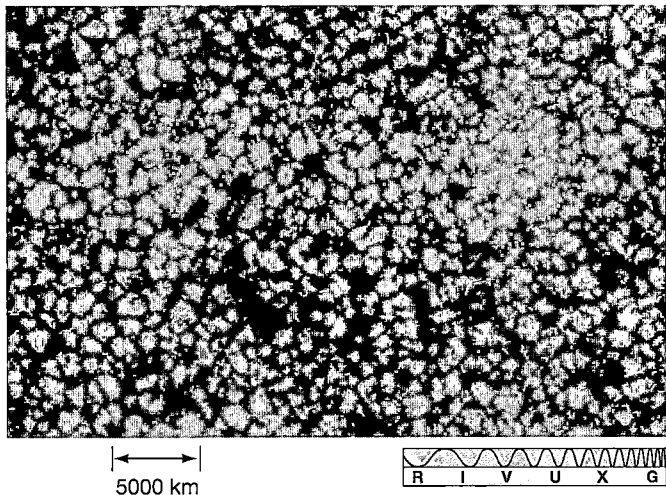

|←——→|

5000 km

R   I   V   U   X   G

Figure 16.7 Solar Granulation Photograph of the granulated solar photosphere. Typical solar granules are comparable in size to Earth's continents. The bright portions of the image are regions where hot material is upwelling from below. The dark regions correspond to cooler gas that is sinking back down into the interior. *(Big Bear Solar Observatory)*

ules appear slightly bluer than normal, indicating Doppler-shifted matter approaching us at about 1 km/s. ∞ (Sec. 3.5) Conversely, spectroscopes focused on the darker portions of the granulated photosphere show the same spectral lines to be redshifted, indicating matter moving away from us.

The brightness variations of the granules result strictly from differences in temperature. The upwelling gas is hotter and therefore emits more radiation than the cooler, downwelling gas. The adjacent bright and dark gases appear to contrast considerably, but in reality their temperature difference is less than about 500 K. Careful measurements also reveal a much larger scale flow on the solar surface. *Supergranulation* is a flow pattern quite similar to granulation except that supergranulation cells measure some 30,000 km across. As with granulation, material upwells at the center of the cells, flows across the surface, then sinks down again at the edges. Scientists believe that supergranules are the imprint on the photosphere of a deeper tier of large convective cells, like those depicted in Figure 16.6.

☑ Concept Check

■ What are the two distinct ways in which energy moves outward from the solar core to the photosphere?

16.3 The Solar Atmosphere

COMPOSITION

③ Astronomers can glean an enormous amount of information about the Sun from an analysis of the absorption lines that arise in the photosphere and lower atmosphere. ∞ (Sec. 4.4) Figure 16.8 (see also Figure 4.4) is a detailed spectrum of the Sun spanning a range of wavelengths from 360 to 690 nm. Notice the intricate dark Fraunhofer absorption lines superposed on the background continuous spectrum.

As discussed in Chapter 4, spectral lines arise when electrons in atoms or ions make transitions between states of well-defined energies, emitting or absorbing photons of specific energies (that is, wavelengths or colors) in the process. ∞ (Sec. 4.2) However, to explain the spectrum of the Sun (and, indeed, the spectra of all stars), we must modify slightly our earlier description of the formation of absorption lines. We explained these lines in terms of cool foreground gas intercepting light from a hot background source. In actuality, both the bright background and the dark absorption lines in Figure 16.8 form in roughly the same location in the Sun—the solar photosphere and lower chromosphere. To understand how line formation occurs, let's reconsider the solar energy–emission process in a little more detail.

Below the photosphere, the solar gas is sufficiently dense, and interactions among photons, electrons, and ions sufficiently common, that radiation cannot escape directly into space. In the solar atmosphere, however, the probability that a photon will escape without further interaction with matter depends on its energy. If that energy happens to correspond to some electronic transition in one of the atoms or ions present in the gas, then the photon may be absorbed again before it can travel very far—the more elements present of the type suitable for absorption, the lower the escape probability. Conversely, if the photon's energy does not coincide with any such transition, then the photon cannot interact further with the gas, and it leaves the Sun headed for interstellar space, or perhaps the detector of an astronomer on Earth.

Thus, as illustrated in Figure 16.9, when we look at the Sun, we are actually peering down into the solar atmosphere to a depth that depends on the wavelength of the light under consideration. Photons with wavelengths well away from any absorption feature tend to come from deep in the photosphere, whereas those at the centers of absorption lines, being much more likely to interact with matter as they travel through the solar gas, mainly escape from higher, cooler levels. The lines are darker than their surroundings because the temperature at the level of the atmosphere where they form is lower than the 5800 K temperature at the base of the photosphere, where most of the continuous emission originates. (Recall that by Stefan's law, the brightness of a radiating object depends on its temperature—the cooler the gas, the less energy it radiates.) ∞ (Sec. 3.4) The existence of the Fraunhofer lines is direct evidence that the temperature in the Sun's atmosphere decreases with height above the photosphere.

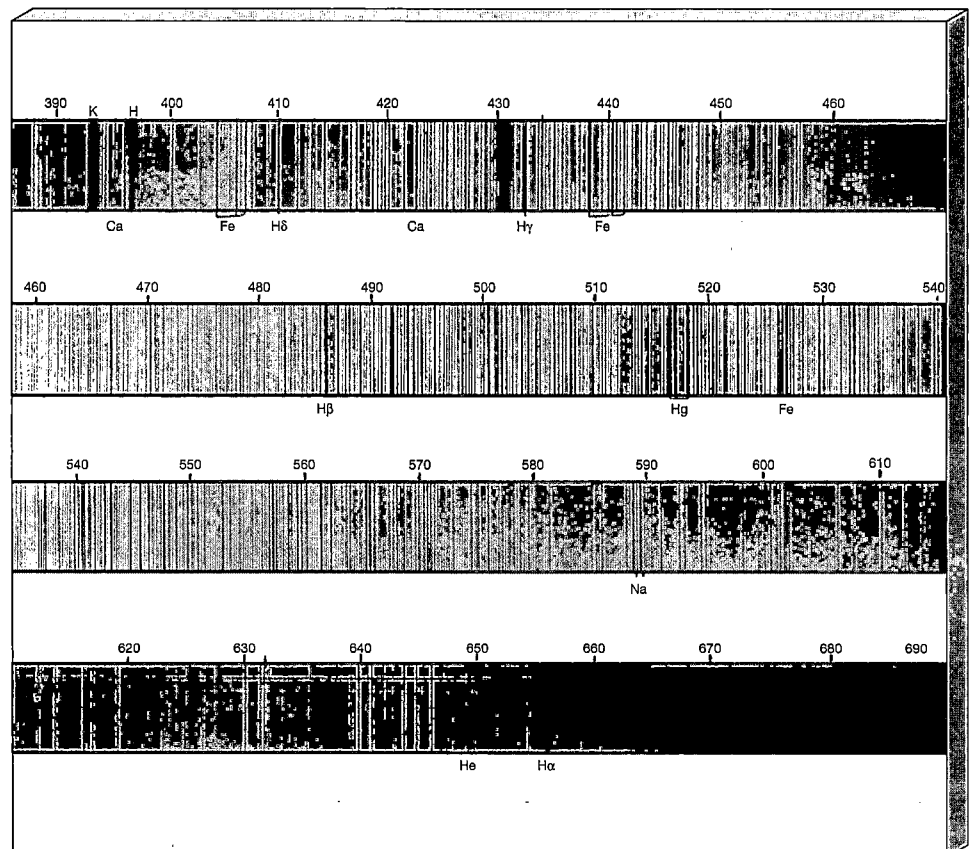

Figure 16.8 Solar Spectrum A detailed spectrum of our Sun shows thousands of Fraunhofer spectral lines which indicate the presence of some 67 different elements in various stages of excitation and ionization in the lower solar atmosphere. The numbers give wavelengths, in nanometers. *(Palomar Observatory/ Caltech)*

Tens of thousands of spectral lines have been observed and cataloged in the solar spectrum. In all, some 67 elements have been identified in the Sun, in various states of ionization and excitation. ∞ (Sec. 4.2) More elements probably exist there, but they are present in such small quantities that our instruments are simply not sensitive enough to detect them. Table 16.2 lists the 10 most common elements in the Sun. Notice that hydrogen is by far the most abundant element, followed by helium. This distribution is just what we saw on the jovian planets, and it is what we will find for the universe as a whole.

Strictly speaking, spectral analysis allows us to draw conclusions only about the part of the Sun where the lines form—the photosphere and chromosphere. However, most astronomers believe that, with the exception of the solar core (where nuclear reactions are steadily changing the composition—see Section 16.5), the data in Table 16.2 are representative of the entire Sun. That belief is strongly supported by the excellent agreement between the Standard Solar Model, which makes precisely this assumption, and helioseismological observations of the solar interior.

THE CHROMOSPHERE

Above the photosphere lies the cooler chromosphere, the inner part of the solar atmosphere. This region emits very little light of its own and cannot be observed visually under normal conditions. The photosphere is just too bright, dominating the chromosphere's radiation. The relative dimness of the chromosphere results from its low density—large numbers of photons simply cannot be emitted by a tenuous gas containing very few atoms per unit volume. Still, although it is not normally seen, astronomers have long been aware of the chromosphere's existence. Figure

TABLE 16.2 The Composition of the Sun

ELEMENT	ABUNDANCE (percentage of total number of atoms)	ABUNDANCE (percentage of total mass)
Hydrogen	91.2	71.0
Helium	8.7	27.1
Oxygen	0.078	0.97
Carbon	0.043	0.40
Nitrogen	0.0088	0.096
Silicon	0.0045	0.099
Magnesium	0.0038	0.076
Neon	0.0035	0.058
Iron	0.0030	0.14
Sulfur	0.0015	0.040

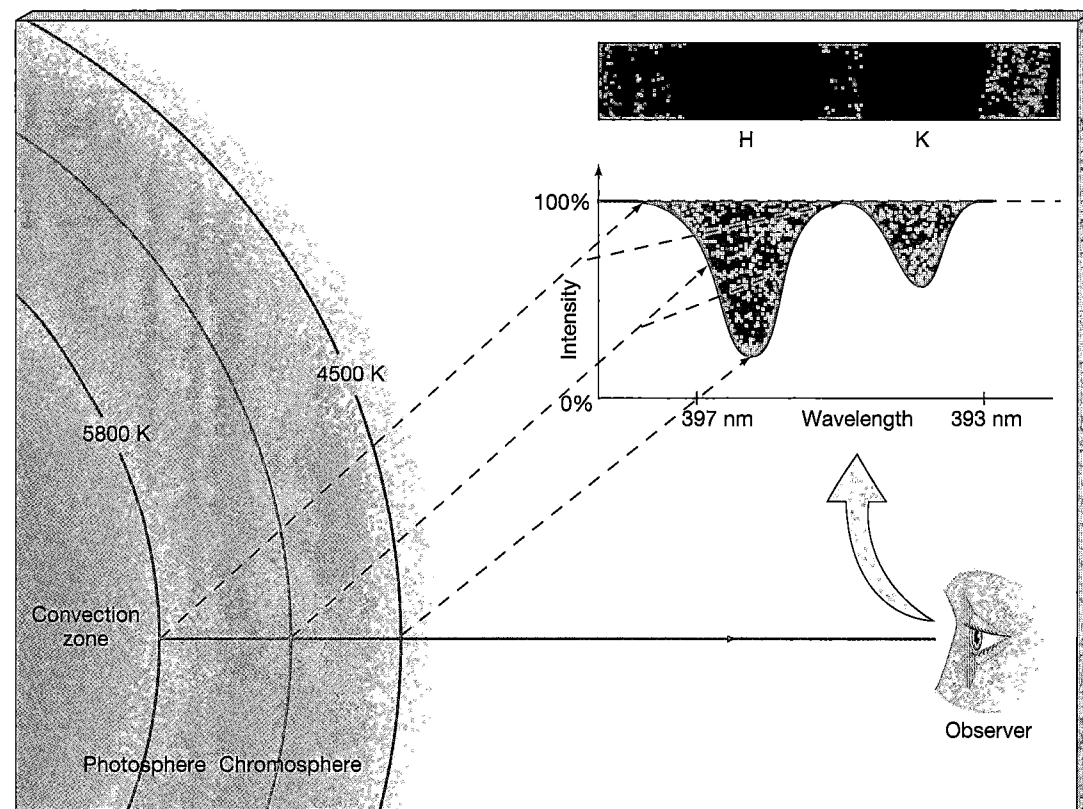

Figure 16.9
Spectral Line Formation Photons with energies well away from any atomic transition can escape from relatively deep in the photosphere, but those with energies close to a transition are more likely to be reabsorbed before escaping, so the ones we see on Earth tend to come from higher, cooler levels in the solar atmosphere. The inset shows a close-up tracing of two of the thousands of solar absorption lines, the "H" and "K" lines of calcium at about 395 nm.

16.10 shows the Sun during an eclipse in which the photosphere—but not the chromosphere—is obscured by the Moon. The chromosphere's characteristic reddish hue is plainly visible. This coloration is due to the red Hα (hydrogen alpha) emission line of hydrogen, which dominates the chromospheric spectrum. ∞ (*More Precisely 4-1*) The chromosphere is far from tranquil. Every few minutes, small solar storms erupt, expelling jets of hot matter known as *spicules* into the Sun's upper atmosphere (Figure 16.11). These long, thin spikes of matter leave the Sun's surface at typical speeds of about 100 km/s, reaching several thousand kilometers above the photosphere. Spicules are not spread evenly across the solar surface. Instead, they cover only about one percent of the total area, tending to accumulate around the edges of supergranules. The Sun's magnetic field is also known to be somewhat stronger than average in those regions. Scientists speculate that the downwelling material there tends.to strengthen the solar magnetic field and spicules are the result of magnetic disturbances in the Sun's churning outer layers.

THE TRANSITION ZONE AND CORONA

During the brief moments of an eclipse, if the Moon's angular size is large enough that both the photosphere and

the chromosphere are blocked, the ghostly solar corona can be seen (Figure 16.12). With the photospheric light removed, the pattern of spectral lines changes dramatically. The intensities of the usual lines alter, suggesting changes in composition or temperature, or both, the spectrum shifts from absorption to emission, and an entirely new set of spectral lines suddenly appears. These new

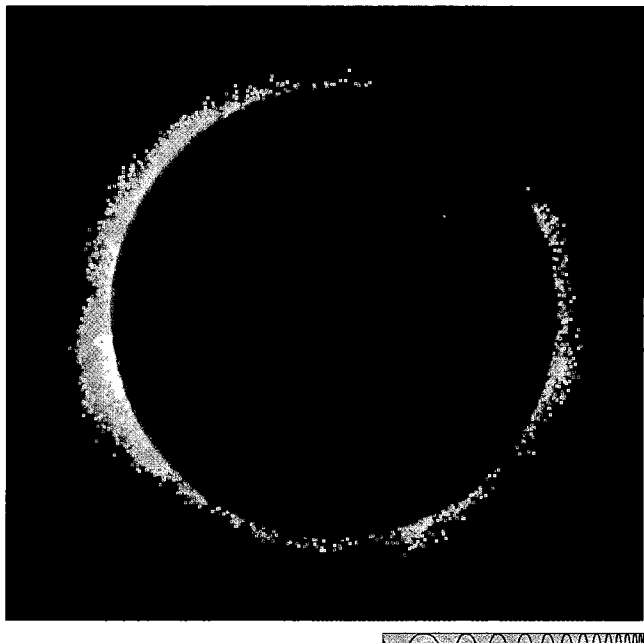

Figure 16.10 Solar Chromosphere This photograph of a total solar eclipse shows the solar chromosphere, a few thousand kilometers above the Sun's surface. *(G. Schneider)*

R   I   V   U   X   G

|←————→|
3000 km

Figure 16.11 Solar Spicules Short-lived narrow jets of gas that typically last mere minutes, spicules can be seen sprouting up from the solar chromosphere in this Hα image of the Sun. The spicules are the thin, dark, spikelike regions. They appear dark against the face of the Sun because they are cooler than the solar photosphere. *(NOAO)*

coronal (and in some cases chromospheric) lines were first observed during eclipses in the 1920s. For years afterward, some researchers (for want of any better explanation) attributed them to a new nonterrestrial element, which they dubbed "coronium."

We now recognize that these new spectral lines do not indicate any new kind of atom. Coronium does not exist. Rather, the new lines arise because atoms in the corona have lost several more electrons than atoms in the photosphere—that is, the coronal atoms are much more highly ionized. Therefore, their internal electronic structures, and hence their spectra, are quite different from the structure and spectra of atoms and ions in the photosphere. For example, astronomers have identified coronal lines corresponding to iron ions with as many as 13 of their normal 26 electrons missing. In the photosphere, most iron atoms have lost only one or two of their electrons.

The cause of this extensive electron stripping is the high coronal temperature. The degree of ionization inferred from

spectra observed during solar eclipses tells us that the gas temperature of the upper chromosphere exceeds that of the photosphere. Furthermore, the temperature of the solar corona, where even more ionization is seen, is higher still.

Figure 16.13 shows how the temperature of the Sun's atmosphere varies with altitude. The temperature decreases to a minimum of about 4500 K some 500 km above the photosphere, after which it rises steadily. About 1500 km above the photosphere, the gas temperature begins to rise rapidly, reaching more than 1 million K at an altitude of 10,000 km. Thereafter, in the corona, the temperature remains roughly constant at around 3 million K, although *SOHO* and other orbiting instruments have detected coronal "hotspots" having temperatures many times higher than this average value.

The cause of this rapid temperature rise is not fully understood. The temperature profile runs contrary to intuition—moving away from a heat source, we would normally expect the heat to diminish, but this is not the case for the lower atmosphere of the Sun. The corona must have another energy source. Astronomers now believe that magnetic disturbances in the solar photosphere are ultimately responsible for heating the corona (Section 16.4).

THE SOLAR WIND

Electromagnetic radiation and fast-moving particles—mostly protons and electrons—escape from the Sun all the time. The radiation moves away from the photosphere at the

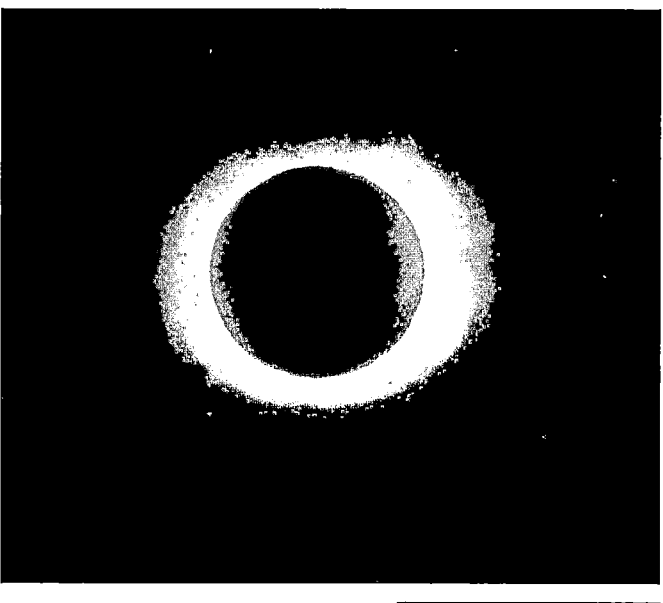

Figure 16.12 Solar Corona When both the photosphere and the chromosphere are obscured by the Moon during a solar eclipse, the faint corona becomes visible. This photograph shows clearly the emission of radiation from the solar corona. *(G. Schneider)*

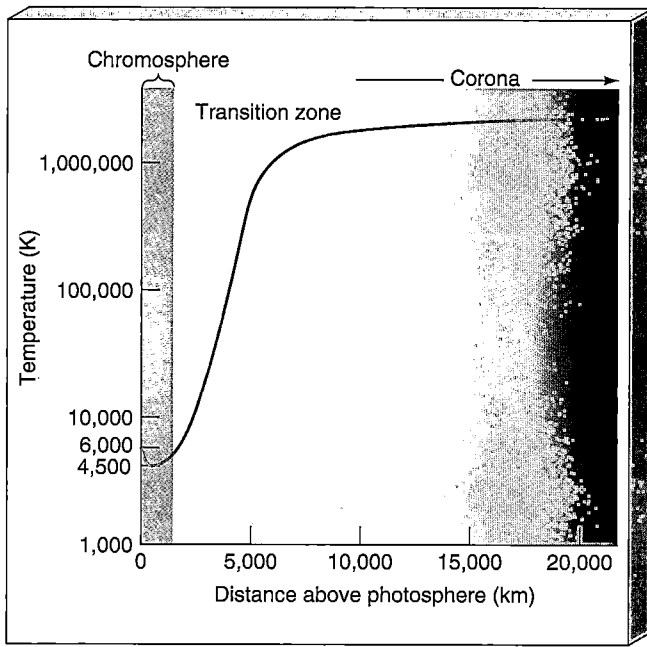

Figure 16.13 Solar Atmospheric Temperature The change of gas temperature in the lower solar atmosphere is dramatic. The minimum temperature occurs in the chromosphere. Beyond that, the temperature rises sharply in the transition zone, finally leveling off at around 3 million K in the corona.

speed of light, taking eight minutes to reach Earth. The particles travel more slowly, although at the still considerable speed of about 500 km/s, reaching Earth in a few days. This constant stream of escaping solar particles is the **solar wind**.

The solar wind results from the high temperature of the corona. About 10 million km above the photosphere, the coronal gas is hot enough to escape the Sun's gravity, and it begins to flow outward into space. At the same time, the solar atmosphere is continuously replenished from below. If that were not the case, the corona would disappear in about a day. The Sun is, in effect, "evaporating"— constantly shedding mass through the solar wind. The wind is an extremely thin medium, however. Although it carries away over a million tons of solar matter each sec-

ond, less than 0.1 percent of the Sun's mass has been lost since the solar system formed 4.6 billion years ago.

THE SUN IN X RAYS

What sort of radiation is emitted by a gas of 3 million K? Unlike the 5800 K photosphere, which emits most strongly in the visible part of the electromagnetic spectrum, the hotter coronal gas radiates at much higher frequencies— primarily in X rays. ∞ (Sec. 3.4) For this reason, X-ray telescopes have become important tools in the study of the solar corona. Figure 16.14 shows several X-ray images of the Sun. The full corona extends well beyond the regions shown, but the density of coronal particles emitting the radiation diminishes rapidly with distance from the Sun. The intensity of X-ray radiation farther out is too dim to be seen here.

In the mid-1970s, instruments aboard NASA's *Skylab* space station revealed that the solar wind escapes mostly through solar "windows" called **coronal holes**. The dark area moving from left to right in Figure 16.14 (which shows more recent data from the Japanese *Yohkoh* X-ray solar observatory) represents a coronal hole. Not really holes, such structures are simply deficient in matter—vast regions of the Sun's atmosphere where the density is about 10 times lower than the already tenuous, normal corona. Note that the underlying solar photosphere looks black in these images because it is far too cool to emit X rays in any significant quantity.

Coronal holes are lacking in matter because the gas there is able to stream freely into space at high speeds, driven by disturbances in the Sun's atmosphere and magnetic field. In coronal holes, the solar magnetic field lines extend from the surface far out into interplanetary space. Charged particles tend to follow the field lines, so they can escape, particularly from the Sun's polar regions, according to recent *SOHO* findings. In other regions of the corona, the solar magnetic field lines stay close to the Sun, keeping charged particles near the surface and inhibiting the outward flow of the solar wind (just as Earth's magnetic field tends to prevent the incoming solar wind from striking Earth), so the density remains relatively high.

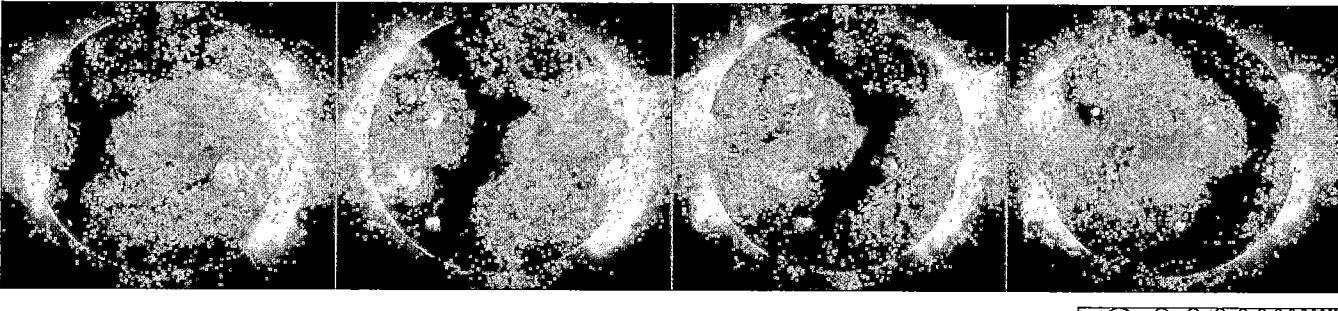

Figure 16.14 Coronal Holes Images of X-ray emission from the Sun observed by the *Yohkoh* satellite. These frames were taken at roughly two-day intervals. Note the dark, V-shaped coronal hole traveling from left to right, where the X-ray observations outline in dramatic detail the abnormally thin regions through which the high-speed solar wind streams forth. *(ISAS/Lockheed Martin)*

The largest coronal holes, like that shown in Figure 16.14, can be hundreds of thousands of kilometers across. Structures of this size are seen only a few times per decade. Smaller holes—perhaps only a few tens of thousand kilometers in size—are much more common, appearing every few hours.

✓ Concept Check

■ Describe two ways in which the spectrum of the solar corona differs from that of the photosphere.

16.4 The Active Sun

🔷 Most of the Sun's luminosity results from continuous emission from the photosphere. However, superimposed on this steady, predictable aspect of our star's energy output is a much more irregular component, characterized by explosive and unpredictable surface activity. Solar activity contributes little to the Sun's total luminosity and probably has no significant bearing on the evolution of the Sun as a star, but it does affect us here on Earth (see *Discovery 16-2*). The size and duration of coronal holes are strongly influenced by the level of solar activity, and hence so is the strength of the solar wind. This in turn directly affects Earth's magnetosphere.

SUNSPOTS

Figure 16.15 is an optical photograph of the entire Sun, showing numerous dark blemishes on the surface. First studied in detail by Galileo, these "spots" provided one of the first clues that the Sun was not a perfect unvarying creation but a place of constant change. ∞ (Sec. 2.4) These dark areas are called **sunspots**. They typically measure about 10,000 km across, about the size of Earth. As shown in the figure, they often occur in groups. At any given time, the Sun may have hundreds of sunspots, or it may have none at all.

Studies of sunspots show an *umbra*, or dark center, surrounded by a grayish *penumbra*. The close-up views in Figure 16.16 show each of these dark areas and the brighter undisturbed photosphere nearby. This gradation in darkness is really a gradual change in photospheric

temperature—sunspots are simply *cooler* regions of the photospheric gas. The temperature of the umbra is about 4500 K, compared with the penumbra's 5500 K. The spots, then, are certainly composed of hot gases. They seem dark only because they appear against an even brighter background (the 5800 K photosphere). If we could magically remove a sunspot from the Sun (or just block out the rest of the Sun's emission), the spot would glow brightly, just like any other hot object having a temperature of roughly 5000 K.

Sunspots are not steady. Most change their size and shape, and all come and go. Figure 16.17 shows a time sequence in which several spots varied—sometimes growing, sometimes dissipating—over a period of several days. Individual spots may last anywhere from 1 to 100 days. A large group of spots typically lasts 50 days.

SOLAR MAGNETISM

What causes a sunspot? Why is it cooler than the surrounding photosphere? The answers to these questions involve the Sun's magnetism. As we saw in Chapter 4, analysis of spectral lines can yield information about the magnetic field where they originate. ∞ (Sec. 4.4) The magnetic field in a typical sunspot is about 1000 times greater than the field in neighboring, undisturbed photospheric regions (which is itself several times stronger than Earth's magnetic field). Scientists believe that sunspots are cooler than their surroundings because these abnormally strong fields tend to block (or redirect) the convective flow of hot gas, which is normally toward the surface of the Sun.

Another indicator of the magnetic nature of sunspots is their grouping. The *polarity* of a sunspot simply indicates which way its magnetic field is directed. We conventionally label spots where field lines emerge from the interior as "S" and those where the lines dive below the photosphere as "N" (so field lines above the surface always run from S to N, as on Earth). Sunspots almost always come in pairs whose members lie at roughly the same latitude and have opposite magnetic polarities. Figure 16.18 illustrates how magnetic field lines emerge from the solar interior through one member (S) of a sunspot pair, loop through the solar atmosphere, then reenter the photosphere through the other member (N).

Despite the irregular appearance of the sunspots themselves, there is a great deal of order in the underly-

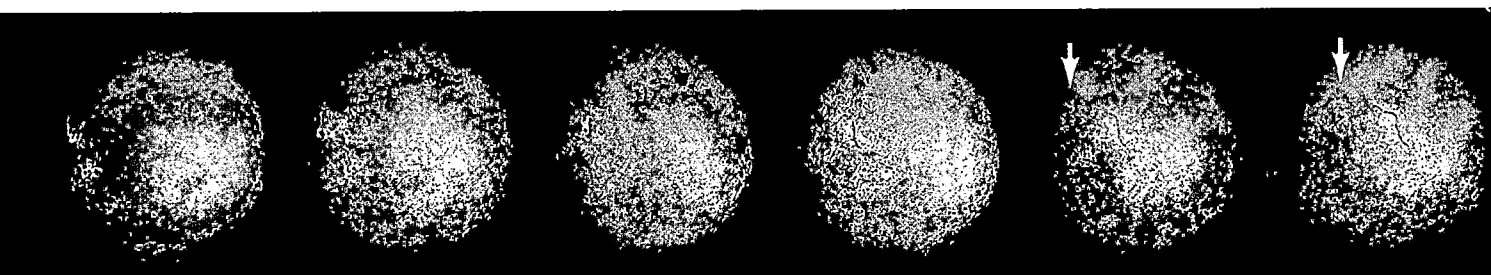

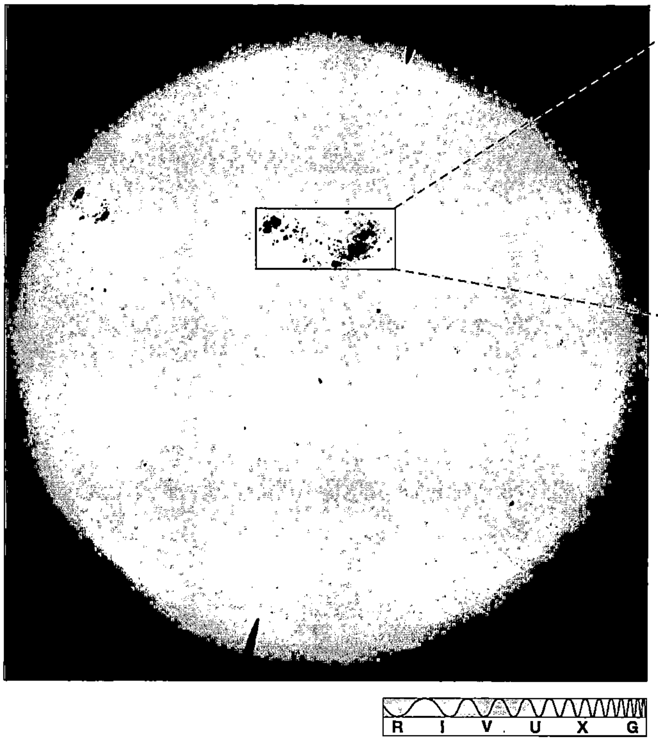

Figure 16.15 Sunspots This photograph of the entire Sun, taken during a period of maximum solar activity, shows several groups of sunspots. The largest spots in this image are over 20,000 km across—twice the diameter of Earth. Typical sunspots are only about half this size. *(Palomar Observatory/ Caltech)*

ing solar field. *All* the sunspot pairs in the same solar hemisphere (north or south) at any instant have the *same* magnetic configuration. That is, if the leading spot (measured in the direction of the Sun's rotation) of one pair has N polarity, as shown in the figure, then all leading spots in that hemisphere have the same polarity. What's more, in the other hemisphere at the same time, all sunspot pairs have the *opposite* magnetic configuration (S polarity leading). To understand these regularities in sunspot polarities, we must look in more detail at the Sun's magnetic field.

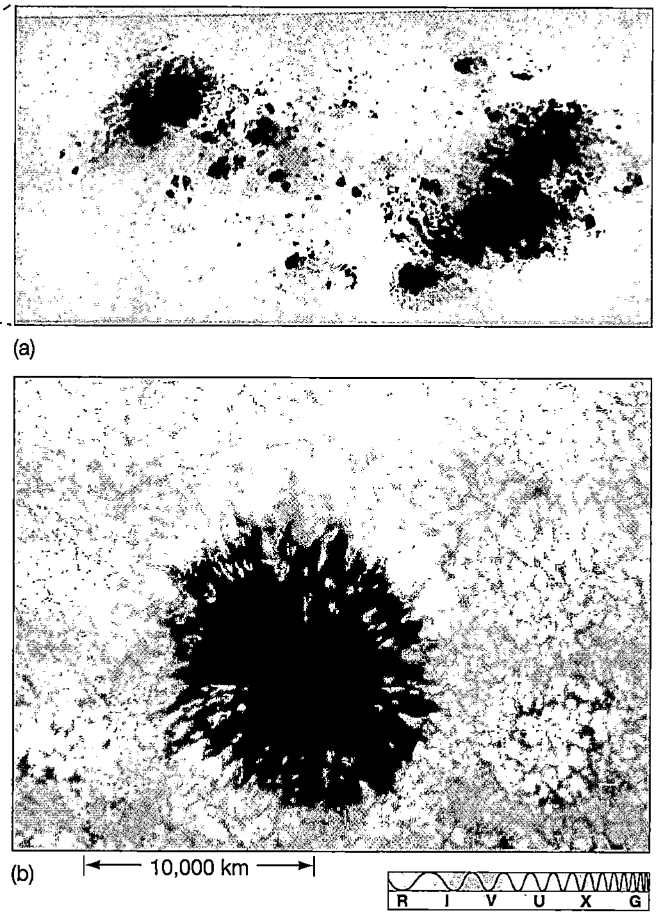

Figure 16.16 Sunspots, Up Close (a) An enlarged photograph of the largest pair of sunspots in Figure 16.15. Each spot consists of a cool, dark inner region called the umbra, surrounded by a warmer, brighter region called the penumbra. The spots appear dark because they are slightly cooler than the surrounding photosphere. (b) A high-resolution, true-color image of a single sunspot shows details of its structure as well as much surface granularity surrounding it. The spot is about the size of Earth. *(Palomar Observatory/Caltech; National Solar Observatory)*

Figure 16.17 Sunspot Rotation The evolution of some sunspots and lower chromospheric activity over a period of 12 days. The sequence runs from left to right. An Hα filter was used to make these photographs, taken from the *Skylab* space station in 1975. An arrow follows one set of sunspots over the course of a week as they are carried around the Sun by its rotation. *(NASA)*

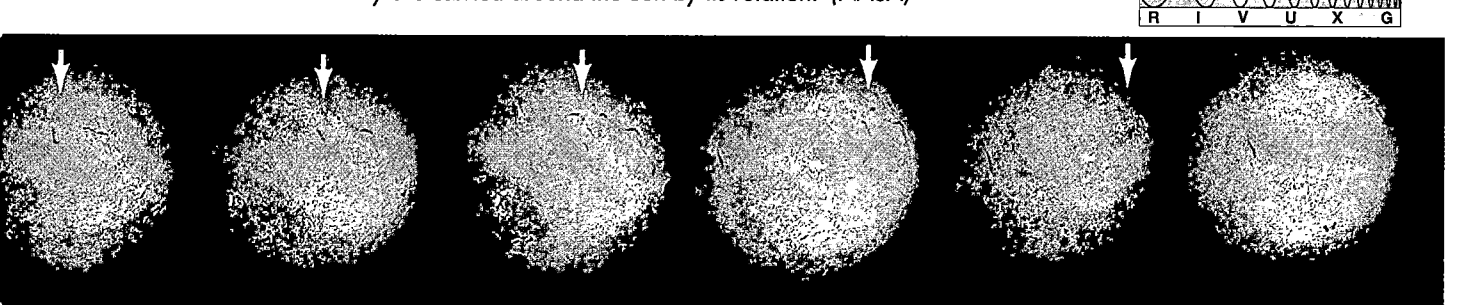

As illustrated in Figure 16.19, the Sun's differential rotation greatly distorts the solar magnetic field, "wrapping" it around the solar equator and eventually causing any originally north–south magnetic field to reorient itself in an east–west direction. At the same time, convection causes the magnetized gas to upwell toward the surface, twisting and tangling the magnetic field pattern. In some places, the field lines become kinked like a twisted garden hose, causing the field strength to increase. Occasionally,

the field becomes so strong that it overwhelms the Sun's gravity, and a "tube" of field lines bursts out of the surface and loops through the lower atmosphere, forming a sunspot pair. The general east–west organization of the underlying solar field accounts for the observed polarities of the resulting sunspot pairs in each hemisphere.

Don't be fooled by the simplicity of the schematic diagram in Figure 16.18(a). Figure 16.18(b) shows a far-ultraviolet image taken by NASA's *Transition Region and Coronal*

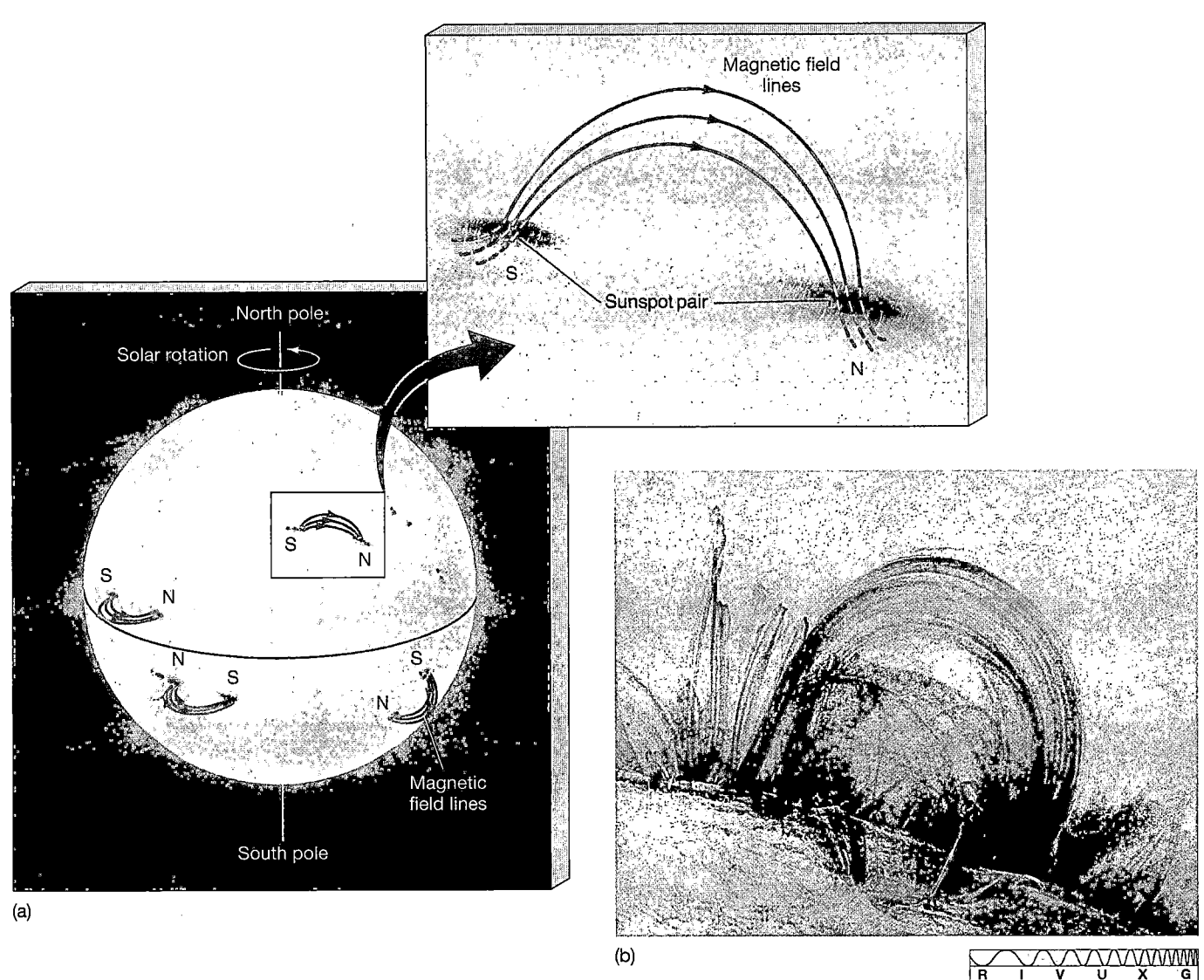

(a)

(b)

| R | I | V | U | X | G |

Figure 16.18 Solar Magnetism (a) Sunspot pairs are linked by magnetic field lines. The Sun's magnetic field lines emerge from the surface through one member of a pair and reenter the Sun through the other member. The leading members of all sunspot pairs in the solar northern hemisphere have the same polarity (labeled N or S, as described in the text). If the magnetic field lines are directed into the Sun in one leading spot, they are inwardly directed in all other leading spots in that hemisphere. The same is true in the southern hemisphere, except that the polarities are always opposite those in the north. The overall direction of the magnetic field reverses itself roughly every 11 years. (b) A far-ultraviolet image taken by NASA's *Transition Region and Coronal Explorer* (*TRACE*) satellite in 1999, showing magnetic field lines arching between two sunspot groups. Note the complex structure of the field lines, which are seen here via the radiation emitted by superheated gas flowing along them. Resolution here is about 700 km. In this negative image (which shows the lines more clearly), the darkest regions have temperatures of about 2 million K. *(NASA)*

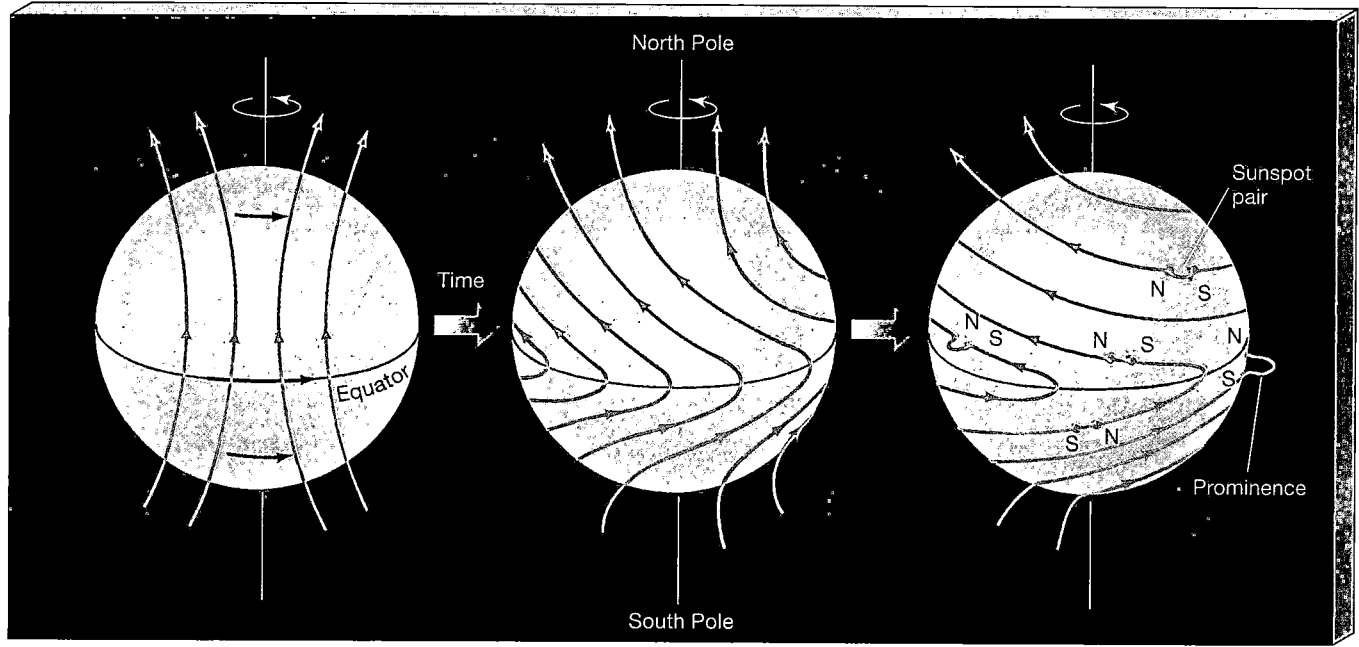

Figure 16.19 Solar Rotation The Sun's differential rotation wraps and distorts the solar magnetic field. Occasionally, the field lines burst out of the surface and loop through the lower atmosphere, thereby creating a sunspot pair. The underlying pattern of the solar field lines explains the observed pattern of sunspot polarities. If the loop happens to occur on the limb of the Sun and is seen against the blackness of space, we see a phenomenon called a prominence, described in Section 16.4.

Explorer (TRACE) satellite in 1999, showing high-temperature gas flowing along a complex network of magnetic field lines connecting two sunspot groups. The gas is thought to be heated by violent events occurring in or near the sunspots themselves.

THE SOLAR CYCLE

Not only do sunspots come and go with time but their numbers and distribution across the face of the Sun also change fairly regularly. Centuries of observations have established a clear **sunspot cycle**. Figure 16.20(a) shows the number of sunspots observed each year during the twentieth century. The average number of spots reaches a maximum every 11 or so years, then falls off almost to zero before the cycle begins afresh.

The latitudes at which sunspots appear vary as the sunspot cycle progresses. Individual sunspots do not move up or down in latitude, but new spots appear closer to the equator as older ones at higher latitudes fade away. Figure 16.20(b) is a plot of observed sunspot latitude as a function of time. At the start of each cycle, at the time of *solar minimum*, only a few spots are seen. They are generally confined to two narrow zones about 25° to 30° north and south of the solar equator. Approximately four years into the cycle, around the time of *solar maximum*, the number of spots has increased markedly. They are found within about 15° to 20° of the equator. Finally, by the end of the cycle, at solar minimum, the total number has fallen again,

and most sunspots lie within about 10° of the solar equator. The beginning of each new cycle appears to overlap the end of the last.

Complicating this picture further, the 11-year sunspot cycle is actually only half of a longer 22-year **solar cycle**. During the first 11 years of the solar cycle, the leading spots of all the pairs in the northern hemisphere have the same polarity, while spots in the southern hemisphere have the opposite polarity (Figure 16.18). These polarities then reverse their signs for the next 11 years, so the full solar cycle takes 22 years to repeat.

Astronomers believe that the Sun's magnetic field is both generated and amplified by the constant stretching, twisting, and folding of magnetic field lines that results from the combined effects of differential rotation and convection, although the details are still not well understood. The theory is similar to the "dynamo" theory that accounts for the magnetic fields of Earth and the jovian planets, except that the solar dynamo operates much faster, and on a much larger scale. ∞ (Sec. 7.4) One prediction of this theory is that the Sun's magnetic field should rise to a maximum, then fall to zero and reverse itself more or less periodically, just as observed. Solar surface activity, such as the sunspot cycle, simply follows the variations in the magnetic field. The changing numbers of sunspots and their migration to lower latitudes are both consequences of the strengthening and eventual decay of the field lines as they become more and more tightly wrapped around the solar equator.

Figure 16.20 Sunspot Cycle (a) Annual number of sunspots throughout the twentieth century, showing the five-year average of the annual data to make long-term trends more evident. The (roughly) 11-year solar cycle is clearly visible. At the time of solar minimum, hardly any sunspots are seen. About four years later, at solar maximum, as many as 100–200 spots are observed per year. The most recent solar maximum occurred in 2000. (b) Sunspots cluster at high latitudes when solar activity is at a minimum. They appear at lower latitudes as the number of sunspots peaks. Finally, they are prominent near the Sun's equator as solar minimum is again approached.

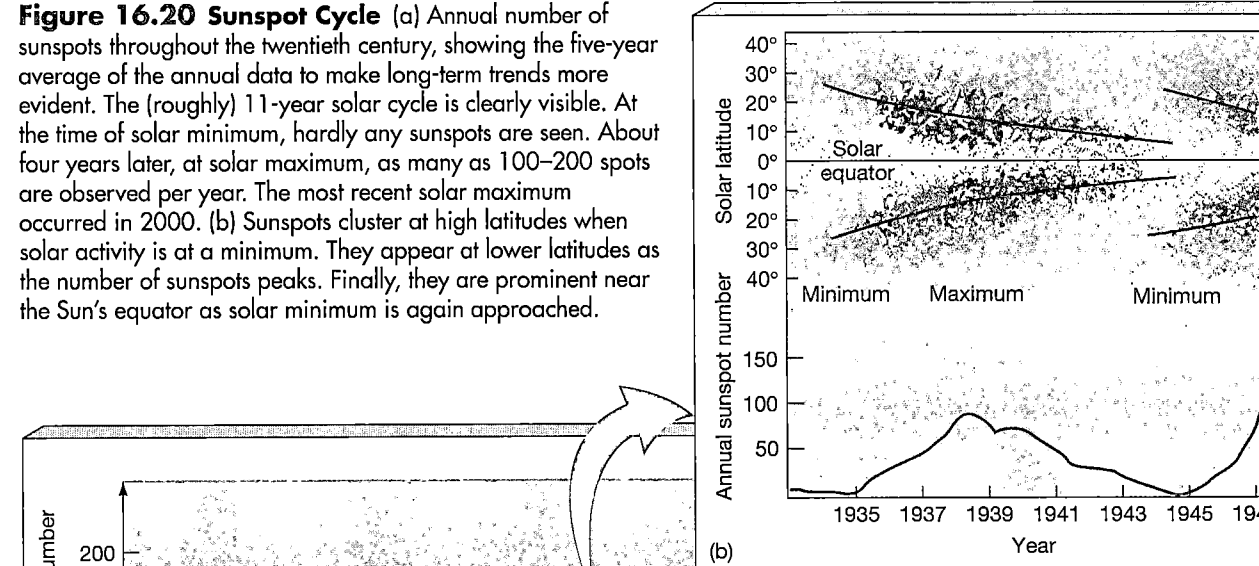

Figure 16.21 plots all sunspot data recorded since the invention of the telescope. As can be seen, the 11-year "periodicity" of the solar sunspot cycle is far from regular. Not only does the period range from 7 to 15 years, but the sunspot cycle has disappeared entirely in the relatively recent past. The lengthy period of solar inactivity that extended from 1645 to 1715 is called the *Maunder minimum*, after the British astronomer who drew attention to these historical records. The corona was apparently also less prominent during total solar eclipses around that time, and Earth aurorae were sparse throughout the late seventeenth century. Lacking a complete understanding of the solar cycle, we cannot easily explain how it could shut down entirely. Most astronomers suspect changes in the Sun's convection zone or its rotation pattern, but the specific cause of the Sun's century-long variations remains a mystery.

ACTIVE REGIONS

The photosphere surrounding a pair or group of sunspots can be a violent place, sometimes erupting explosively, spewing forth into the corona large quantities of energetic particles. The sites of these energetic events are known as **active regions**. Most groups of sunspots have active regions associated with them. Like all other aspects of solar activity, these phenomena tend to follow the solar cycle and are most frequent and violent around the time of solar maximum.

Figure 16.22(a) shows a large solar **prominence**—a loop or sheet of glowing gas ejected from an active region

on the solar surface, moving through the inner parts of the corona under the influence of the Sun's magnetic field. Magnetic instabilities in the strong fields found in and near sunspot groups may cause the prominences, although the details are not fully understood. Figure 16.22(b) shows a filament of hot material soaring high into the solar atmosphere above an active region. The arching magnetic field lines in and around the active region are also easily seen (see also Figure 16.18b). The rapidly changing structure of the field lines and the fact that they can rapidly transport mass and energy from one part of the solar sur-

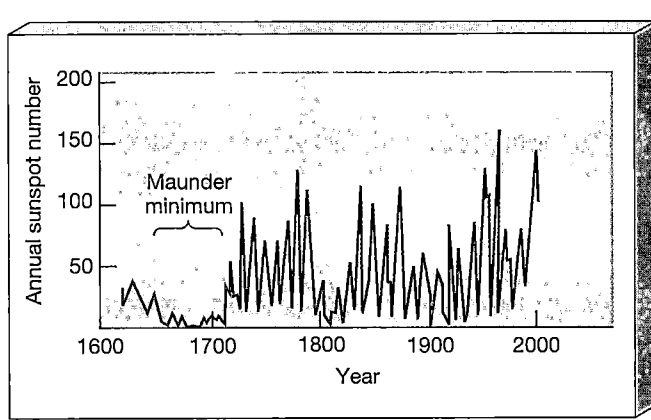

Figure 16.21 Maunder Minimum Number of sunspots occurring each year over the past four centuries. Note the absence of spots during the late seventeenth century.

face to another, possibly tens of thousands of kilometers away, make the theoretical study of active regions an extraordinarily difficult task.

Quiescent prominences persist for days or even weeks, hovering high above the photosphere, suspended by the Sun's magnetic field. *Active prominences* come and go much more erratically, changing their appearance in a matter of hours or surging up from the solar photosphere, then immediately falling back on themselves. A typical solar prominence measures some 100,000 km in extent, nearly 10 times the diameter of planet Earth. Prominences as large as that shown in Figure 16.22(a) (which traversed almost half a million kilometers of the solar surface) are less common and usually appear only at times of greatest solar activity. The largest prominences can release up to 10^{25} joules of energy, counting both particles and radiation—not much compared with the total solar luminosity of 4×10^{26} W but still enormous by terrestrial standards. (All the power plants on Earth would take a billion years to produce this much energy.)

Flares are another type of solar activity observed low in the Sun's atmosphere near active regions. Also the result of magnetic instabilities, flares, like that shown in Figure 16.23(a), are even more violent (and even less well understood) than prominences. They often flash across a region of the Sun in minutes, releasing enormous amounts of energy as they go. Space-based observations indicate that X-ray and ultraviolet emissions are especially intense in the extremely compact hearts of flares, where temperatures can reach 100 million K. So energetic are these cataclysmic explosions that some researchers have likened flares to bombs exploding in the lower regions of the Sun's atmosphere.

A major flare can release as much energy as the largest prominences, but in a matter of minutes or hours rather than days or weeks. Unlike the gas that makes up the characteristic loop of a prominence, the particles produced by a flare are so energetic that the Sun's magnetic field is unable to hold them and shepherd them back to the surface. Instead, the particles are simply blasted into space by the violence of the explosion. Figure 16.23(b) shows the vicinity of a solar active region immediately before a major flare. The bright green region is high temperature gas barely confined by the solar magnetic field. The flare occurred just a few minutes later, ejecting billions of tons of material into interplanetary space. Figure 16.23(c) shows the cooling aftermath of another flare in 2000, where an "arcade" of tightly curved magnetic loops once again confines superhot (but rapidly cooling) gas. An optical image of this region simply shows a sunspot pair, with little indication of the violence that has just occurred.

Figure 16.24 shows a *coronal mass ejection* from the Sun. Sometimes (but not always) associated with flares and prominences, these phenomena are giant magnetic "bubbles" of ionized gas that separate from the rest of the solar atmosphere and escape into interplanetary space. Carrying an enormous amount of energy, they can—if their fields are

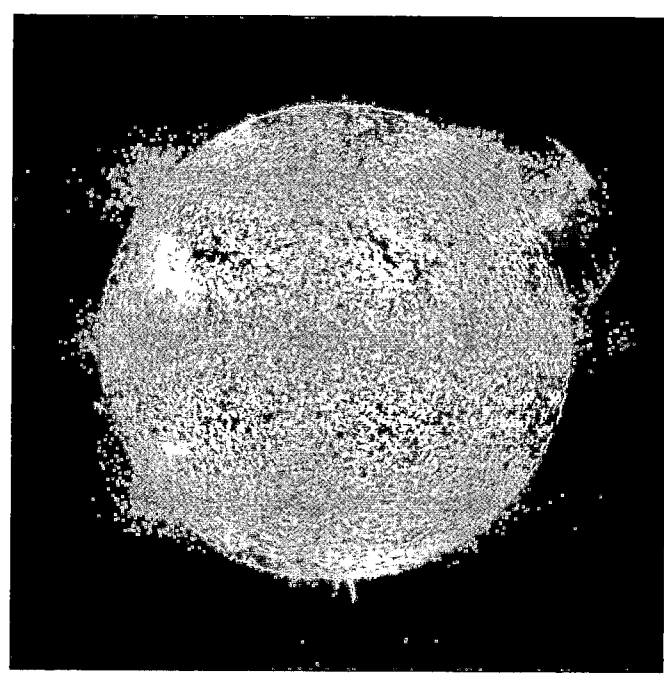

(a)

(b)

Figure 16.22 **Solar Prominences** (a) This particularly large solar prominence was observed by ultraviolet detectors aboard the *SOHO* spacecraft in September, 1999. (b) Like a ' phoenix rising from the solar surface, this filament of hot gas measures more than 100,000 km in length. Earth could easily fit between its outstretched "arms." Dark regions in this *TRACE* image have temperatures less than 20,000 K; the brightest regions are about 1 million K. The ionized gas follows the solar magnetic field lines away from the Sun. Most of it will subsequently cool and fall back to the photosphere. *(NASA)*

properly oriented—connect with Earth's magnetic field, dumping some of their energy into the magnetosphere and potentially causing communications and power disruptions on our planet (see *Discovery 16-2*). Such ejections occur about once per week at times of sunspot minimum, but up to two or three times per day at solar maximum.

(a)

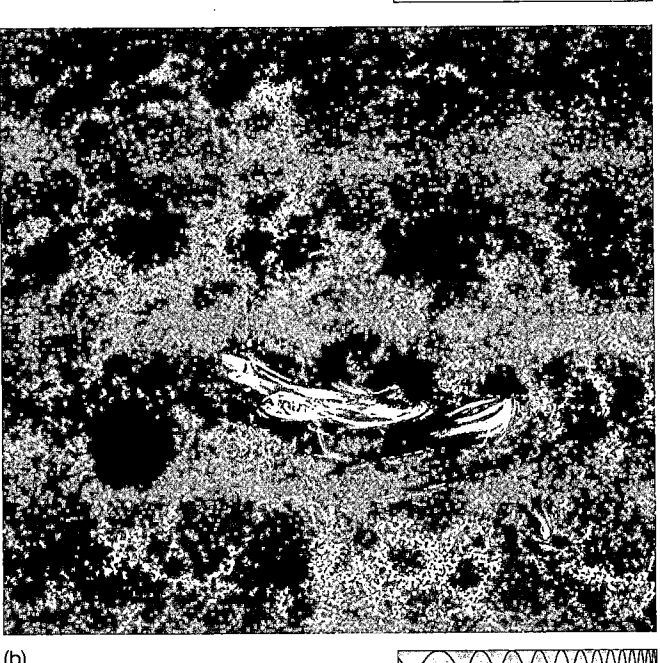

(b)

Figure 16.23 Solar Flares (a) Much more violent than a prominence, a solar flare is an explosion on the Sun's surface that sweeps across an active region in a matter of minutes, accelerating solar material to high speeds and blasting it into space. (b) A flare occurs when hot gas breaks free of the magnetic field confining it and bursts into space. This composite image shows a dark sunspot group (visible light), the surrounding solar photosphere (ultraviolet, shown in red here) and a collection of magnetic loops (extreme ultraviolet, colored green here) confining million-degree gas a few minutes before a major flare in June 2000. (c) This remarkable image, obtained by the *TRACE* satellite, shows an active region shortly after a flare. The green "slinky spring" is actually an arcade of many magnetic field lines arching through the Sun's lower atmosphere roughly from top to bottom, confining hot gas as the Sun's magnetism reestablishes control. *(NASA)*

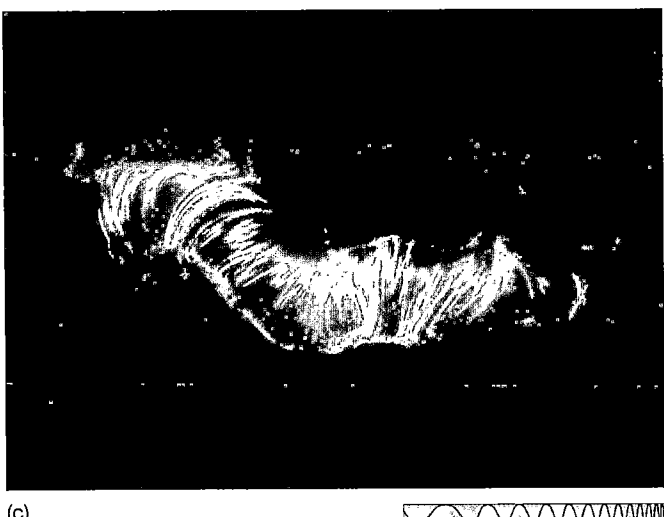

(c)

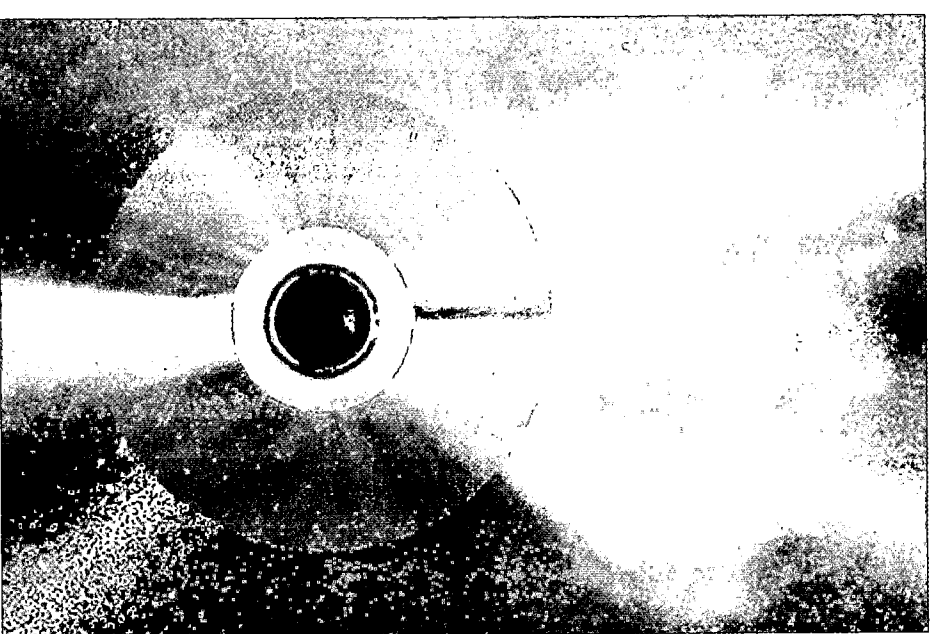

Figure 16.24 Coronal Mass Ejection A few times per week, on average, a giant magnetized "bubble" of solar material detaches itself from the Sun and rapidly escapes into space. Should such a coronal mass ejection encounter Earth, it can severely disrupt our planet's magnetosphere. *(NASA)*

THE CHANGING SOLAR CORONA

The solar corona also varies in step with the sunspot cycle. The photograph of the corona in Figure 16.12 shows the quiet Sun, at sunspot minimum. The corona is fairly regular in appearance and appears to surround the Sun more or less uniformly. Compare this image with Figure 16.25, which was taken in 1991 near a peak in the sunspot cycle. The active corona is much more irregular in appearance and extends farther from the solar surface. The "streamers" of coronal material pointing away from the Sun are characteristic of this phase.

Astronomers think that the corona is heated primarily by solar surface activity, which can inject large amounts of energy into the upper solar atmosphere. *SOHO* observations have implicated a "magnetic carpet" as the source of much of the heating. Using the craft's Doppler imager, astronomers have observed small magnetic loops perpetually sprouting up, then disappearing, all over the solar photosphere—resembling a kind of celestial shag carpet. The whole carpet replenishes itself rapidly, roughly every 40 hours or so, and the loops don't just sink back beneath the surface. Rather, they tend to break open, dumping vast amounts of energy into the lower solar atmosphere. Along with the myriad spicules, these small-scale magnetic loops probably provide most of the energy needed to heat the corona. In addition, more extensive disturbances often move through the corona above an active site in the photosphere, distributing the energy throughout the coronal gas. Given this connection, it is hardly surprising that both the appearance of the corona and the strength of the solar wind are closely correlated with the solar cycle.

☑ Concept Check

■ What do observations of sunspot polarities tell us about the solar magnetic field?

16.5 The Heart of the Sun

🜃 What powers the Sun? What forces are at work in the Sun's core to produce such energy? By what process does the Sun shine, day after day, year after year, eon after eon? Answers to these questions are central to all astronomy. Without them, we can understand neither the physical existence of stars and galaxies in the universe nor the biological existence of life on Earth.

SOLAR ENERGY PRODUCTION

In round numbers, the Sun's luminosity is 4×10^{26} W and its mass is 2×10^{30} kg. We can quantify how efficiently the Sun generates energy by dividing the solar luminosity by the solar mass:

$$\frac{\text{solar luminosity}}{\text{solar mass}} = 2 \times 10^{-4} \text{ W/kg}.$$

This simply means that, on average, every kilogram of solar material yields about 0.2 milliwatts of energy—0.0002 joules (J) of energy every second. This is not much energy—a piece of burning wood generates about a million times more energy per unit mass per unit time than does our Sun. But there is an important difference: The wood will not burn for billions of years.

To appreciate the magnitude of the energy generated by our Sun, we must consider not the ratio of the solar luminosity to the solar mass but instead the total amount of energy generated by each gram of solar matter *over the entire lifetime of the Sun as a star*. This is easy to do. We simply multiply the rate at which the Sun generates energy by the age of the Sun, about five billion years. We obtain a value of 3×10^{13} J/kg. This is the average amount of energy radiated by every kilogram of solar material since the Sun formed. It represents a *minimum* value for the total energy radiated by the Sun, for more energy will be needed for every additional day the Sun shines. Should the Sun endure for another five billion years (as is predicted by theory), we would have to double this value.

This energy-to-mass ratio value is very large. At least 60 trillion joules (on average) of energy must arise from

Figure 16.25 Active Corona Photograph of the solar corona during the July 1991 eclipse, at the peak of the sunspot cycle. At these times, the corona is much less regular and much more extended than at sunspot minimum (compare Figure 16.12). Astronomers believe that coronal heating is caused by surface activity on the Sun. The changing shape and size of the corona are the direct result of variations in prominence and flare activity over the course of the solar cycle. *(National Solar Observatory)*

every kilogram of solar matter to power the Sun throughout its lifetime. But the generation of energy is not explosive, releasing large amounts of energy in a short period of time. Instead, it is slow and steady, providing a uniform and long-lived rate of energy production. Only one known energy-generation mechanism can conceivably power the Sun in this way. That process is **nuclear fusion**—the combining of light nuclei into heavier ones.

NUCLEAR FUSION

We can represent a typical fusion reaction symbolically as

nucleus 1 + nucleus 2 → nucleus 3 + energy.

For powering the Sun, the most important piece of this equation is the energy produced. Where does it come from?

The key point is that during a fusion reaction, the total mass *decreases*—the mass of nucleus 3 is less than the combined masses of nuclei 1 and 2. To understand the consequences of this, we use a very important law of modern physics—the **law of conservation of mass and energy**. Albert Einstein showed at the beginning of the twentieth century that matter and energy are interchangeable. One can be converted into the other, in accordance with Einstein's famous equation:

$$E = mc^2,$$

or

energy = mass × (speed of light)^2.

This equation says that, to determine the amount of energy corresponding to a given mass, simply multiply it by the square of the speed of light (c in the equation). For example, the energy equivalent of 1 kg of matter is $1 \times (3 \times 10^8)^2$, or 9×10^{16} J. The speed of light is so large that even small amounts of mass translate into enormous amounts of energy.

The law of conservation of mass and energy states that the *sum* of mass and energy (properly converted to the same units) must always remain constant in any physical process. There are no known exceptions. According to this law, an object can literally disappear, provided that some energy appears in its place. If magicians really made rabbits disappear, the result would be a flash of energy equaling the product of the rabbit's mass and the square of the speed of light—enough to destroy the magician, everyone in the audience, and probably all of the surrounding state as well! In the case of fusion reactions in the solar core, the energy is produced primarily in the form of electromagnetic radiation. The light we see coming from the Sun means that the Sun's mass must be slowly but steadily decreasing with time.

THE PROTON-PROTON CHAIN

All atomic nuclei are positively charged, so they repel one another. Furthermore, by the inverse-square law, the closer two nuclei come to one another, the greater is the repul-

sive force between them. ∞ (Sec. 3.2) How then do nuclei—two protons, say—ever manage to fuse into anything heavier? The answer is that if they collide at high enough speeds, one proton can momentarily plow deep into the other, eventually coming within the exceedingly short range of the **strong nuclear force**, which binds nuclei together (see *More Precisely 16-1*). At distances of less than about 10^{-15} m, the attraction of the nuclear force overwhelms the electromagnetic repulsion, and fusion occurs. Speeds in excess of a few hundred kilometers per second, corresponding to a gas temperature of 10^7 K or more, are needed to slam protons together fast enough to initiate fusion. Such conditions are found in the core of the Sun and at the centers of all stars.

The fusion of two protons is illustrated schematically in Figure 16.26. In effect, one of the protons turns into a neutron, creating new particles in the process, and combines with the other proton to form a **deuteron**, the nucleus of a special form of hydrogen called *deuterium*. Deuterium (also referred to as "heavy hydrogen") differs from ordinary hydrogen by virtue of an extra neutron in its nucleus. We can represent this reaction as follows:

proton + proton → deuteron + positron + neutrino.

The **positron** particle in this reaction is a positively charged electron. Its properties are identical to those of a normal negatively charged electron, except for its positive charge. Scientists call the electron and the positron a "matter–antimatter pair"—the positron is said to be the *antiparticle* of the electron. These newly created positrons find themselves in the midst of a sea of electrons, with which they interact immediately and violently. The particles and antiparticles annihilate (destroy) one another, producing pure energy in the form of gamma-ray photons.

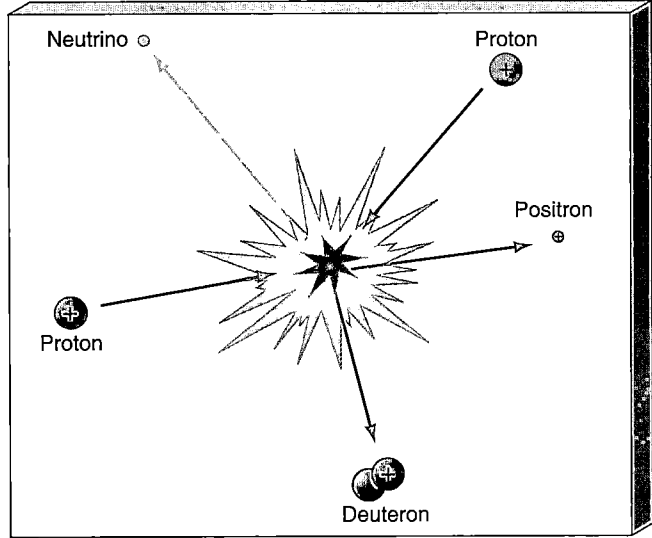

Figure 16.26 Proton Fusion Two protons collide violently, initiating the chain of nuclear fusion that powers the Sun.

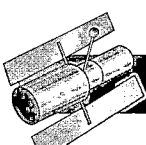

MORE PRECISELY 16-1

Fundamental Forces

Our studies of nuclear reactions have uncovered new ways in which matter can interact with matter at a subatomic level. Let's pause to consider in a slightly more systematic fashion the relationships among the various forces of nature.

As best we can tell, the behavior of all matter in the universe—from elementary particles to clusters of galaxies—is ruled by just four (or fewer) basic forces, which are *fundamental* to everything in the universe. In a sense, the search to understand the nature of the universe is the quest to understand the nature of these forces.

The *gravitational force* is probably the best known. Gravity binds galaxies, stars, and planets together and holds humans on the surface of Earth. As we saw in Chapter 2, its magnitude decreases with distance, according to an inverse-square law. ⇨ (Sec. 2.7) Its strength is also proportional to the masses of each of the two objects involved. Thus, the gravitational field of an atom is extremely weak, but that of a galaxy, consisting of huge numbers of atoms, is very powerful. Gravity is by far the weakest of the forces of nature, but its effect accumulates as we move to larger and larger volumes of space, and nothing can cancel its attractive pull. As a result, gravity is the dominant force in the universe on all scales larger than that of Earth.

The *electromagnetic force* is another of nature's basic agents. Any particle having a net electric charge, such as an electron or a proton in an atom, exerts an electromagnetic force on any other charged particle. The everyday things we see around us are held together by this force. Like gravity, its strength also decreases with distance according to an inverse-square law. ⇨ (Sec. 3.2) However, for subatomic particles, electromagnetism is much stronger than gravity. For example, the electromagnetic force between two protons exceeds their gravitational attraction by a factor of about 10^{36}. Unlike gravity, electromagnetic forces can repel (between like charges) as well as attract (between opposite charges). Positive and negative charges tend to neutralize each other, greatly diminishing their net electromagnetic influence. Above the microscopic level most objects are in fact very close to being electrically neutral. Thus, except in unusual circumstances, the electromagnetic force is relatively unimportant on macroscopic scales.

A third fundamental force of nature is simply termed the *weak nuclear force*. It is much weaker than electromagnetism, and its influence is somewhat more subtle. The weak force governs the emission of radiation from some radioactive atoms; the emission of a neutrino during the first stage of the proton–proton reaction is also the result of a weak interaction. It is now known that the weak force is not really a separate force at all but just a form of the electromagnetic force. Thus, physicists often speak of the "electroweak force." However, when acting in its "weak" mode, the electroweak force does not obey the inverse-square law. Its effective range is less than the size of an atomic nucleus, about 10^{-15} m.

Strongest of all the forces is the *strong nuclear force*. It binds atomic nuclei together and governs the generation of energy in the Sun and all other stars. Like the weak force, and unlike the forces of gravity and electromagnetism, the strong force operates only at very close range. It is unimportant outside a distance of a hundredth of a millionth of a millionth (10^{-14}) of a meter. However, within this range (for example, in atomic nuclei), it binds particles with enormous strength. In fact, it is the range of the strong force that determines the typical sizes of atomic nuclei. Only when two protons are brought within about 10^{-15} m of one another can the attractive strong force overcome their electromagnetic repulsion. Not all particles are subject to all types of force. All particles interact through gravity because all have mass. However, only charged particles interact electromagnetically. Protons and neutrons are affected by the strong force, but electrons are not. Under the right circumstances, the weak force can affect any type of subatomic particle, regardless of its charge.

The final product of the reaction is a particle known as a **neutrino**, a word derived from the Italian for "little neutral one." Neutrinos carry no electrical charge and are of very low mass—at most 1/100,000 the mass of an electron, which itself has only 1/2000 the mass of a proton. (The exact mass of the neutrino is uncertain, although recent evidence suggests that it is not zero.) They move at (or nearly at) the speed of light and interact with hardly anything. They can penetrate, without stopping, several light-years of lead. Their interactions with matter are governed by the **weak nuclear force**, also described in *More Precisely 16-1*. Despite their elusiveness, however, neutrinos can be detected with carefully constructed instruments (see Section 16.6).

Nuclei such as normal hydrogen and deuterium, containing the same number of protons but different numbers of neutrons, represent different forms of the same element—they are known as **isotopes** of that element. Usually, there are about as many neutrons in a nucleus as protons, but the exact number of neutrons can vary, and most elements can exist in a number of isotopic forms. To avoid confusion when talking about isotopes of the same element, nuclear physicists attach a number to the symbol representing the element. This number indicates the total number of particles (protons plus neutrons) in its nucleus. Ordinary hydrogen is denoted by ^{1}H, deuterium by ^{2}H. Normal helium (two protons plus two neutrons) is ^{4}He (also referred to as helium-4), and so on. We will adopt this convention for the rest of this book. We can now rewrite the preceding proton–proton reaction as

$$^1H + {}^1H \rightarrow {}^2H + \text{positron} + \text{neutrino.} \qquad \text{(I)}$$

This equation is labeled (I) because the production of a deuteron by the fusion of two protons is the first step in the fusion process powering most stars. It is the start of the **proton–proton chain**. Gargantuan quantities of protons are fused in this way within the core of the Sun each second. The positrons produced by this reaction quickly encounter electrons and are annihilated, releasing energy in the form of gamma rays.

The next step in solar fusion is the formation of an isotope of helium. A proton interacts with the deuteron particle produced in step (I):

$$^2H + {}^1H \rightarrow {}^3He + \text{energy.} \qquad \text{(II)}$$

This step begins as soon as deuterons appear. The reaction product is an isotope of helium—helium-3 (^3He)—lacking one of the neutrons contained in the normal helium-4 nucleus. Energy is also emitted, again in the form of gamma-ray photons.

The third and final step in the proton–proton chain involves the production of a nucleus of helium-4. Helium-4 (^4He) comes about most often through the fusion of two of the helium-3 nuclei created in step (II):

$$^3He + {}^3He \rightarrow {}^4He + {}^1H + {}^1H + \text{energy.} \qquad \text{(III)}$$

The result is a helium-4 nucleus plus two more protons.

The *net* effect of steps (I) through (III) is this: four hydrogen nuclei (protons; six consumed, two returned) combine to create one helium-4 nucleus, plus some gamma-ray radiation and two neutrinos. The whole process is illustrated in Figure 16.27. Symbolically, we have

$$4(^1H) \rightarrow {}^4He + \text{energy} + 2 \text{ neutrinos.}$$

The gamma rays are slowly degraded in energy as they pass through the solar interior. Photons are repeatedly absorbed by electrons and ions, then reemitted at a wavelength that reflects the temperature of the surrounding gas, in accordance with Wien's law. ∞ (Sec. 3.4) Thus, as the radiation makes its way to the surface through ever-cooler layers, its wavelength decreases. Eventually, it leaves the surface in the form of visible light. The neutrinos escape unhindered into space. The helium stays put in the core.

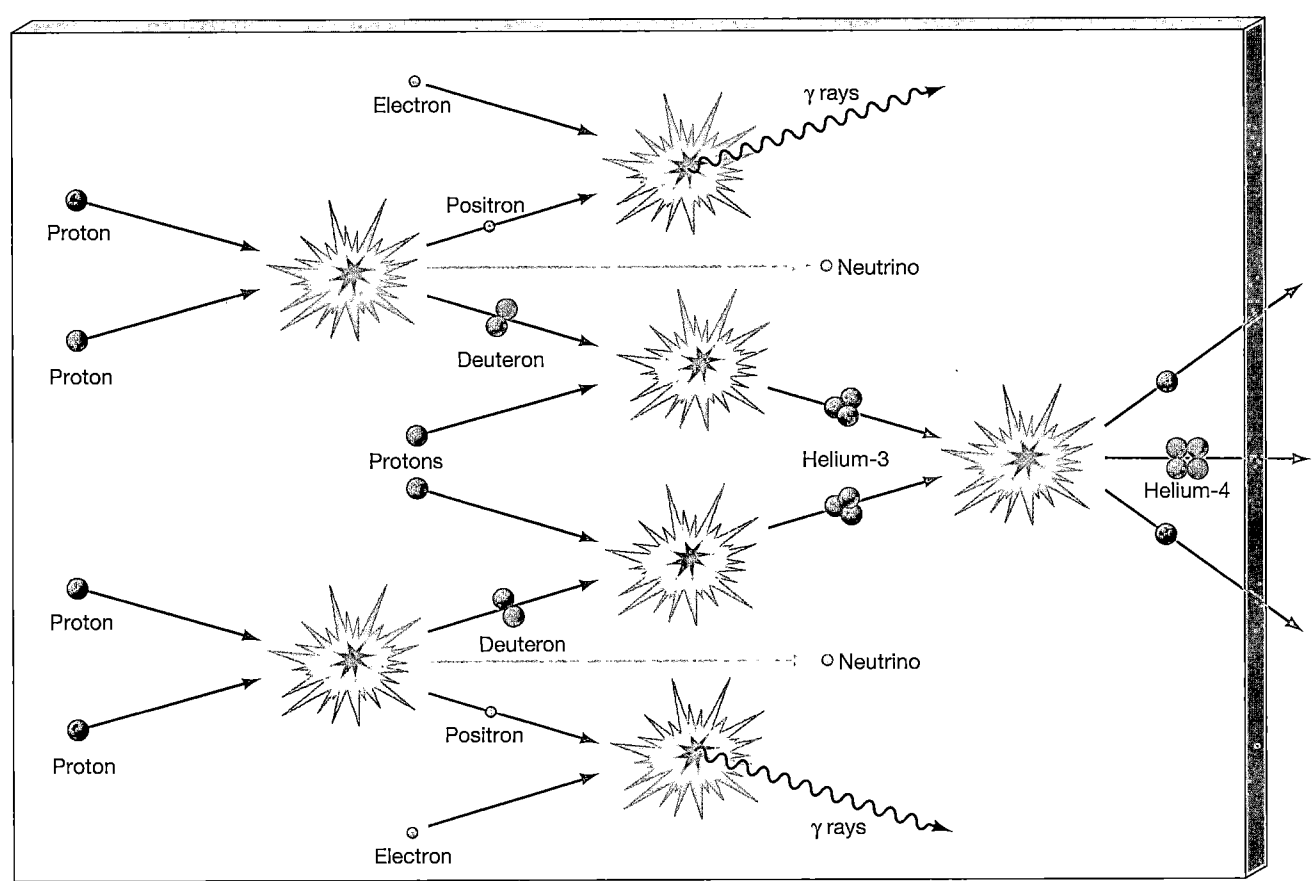

Figure 16.27 Solar Fusion A total of six protons (and two electrons) are converted into two protons, one helium-4 nucleus, and two neutrinos. The two leftover protons are available as fuel for new proton–proton reactions, so the net effect is that four protons are fused to form one helium-4 nucleus. Energy, in the form of gamma rays, is produced in each reaction.

This basic process is responsible for the light emitted by almost all the stars we see. Alternative reactions leading to the same final result exist (see *More Precisely 20-1*), but they are relatively rare in stars like the Sun.

ENERGY GENERATED BY THE PROTON–PROTON CHAIN

Now let's calculate the energy produced in the fusion process and compare it with the 6×10^{13} J/kg needed to account for the Sun's luminosity. Careful laboratory experiments have determined the masses of all the particles involved in the conversion of four protons into a helium-4 nucleus: the total mass of the protons is 6.6943×10^{-27} kg, the mass of the helium-4 nucleus is 6.6466×10^{-27} kg, and the neutrinos are (virtually) massless. The difference between the total mass of the protons and the helium nucleus is 0.048×10^{-27} kg—not much mass, just 0.7 percent of the total, but easily measurable.

Multiplying the vanished mass by the square of the speed of light yields 4.3×10^{-12} J—this is the energy produced by the fusion of 6.7×10^{-27} kg (the rounded-off mass of the four protons) of hydrogen into helium. It follows that fusion of 1 kg of hydrogen would generate 6.4×10^{14} J—more than enough to power the Sun, even for another five billion years. To fuel the Sun's present energy output, hydrogen must be fused into helium in the core at a rate of 600 million ton/s, converting 4.3 million tons of matter into energy every second.

☑ Concept Check

■ Why does the fact that we see sunlight imply that the Sun's mass is slowly decreasing?

16.6 Observations of Solar Neutrinos

Ⓢ Theorists are quite sure that the proton–proton chain operates in the Sun. However, because the gamma-ray energy created in the proton–proton chain is transformed into visible and infrared radiation by the time it emerges from the Sun, astronomers have no direct electromagnetic evidence of the nuclear reactions in the solar core. Instead, the *neutrinos* created in the proton–proton chain are our best bet for learning about conditions in the heart of the Sun. Core neutrinos travel cleanly out of the Sun, interacting with virtually nothing, and escape into space a few seconds after being created. They arrive at Earth's orbit about eight minutes later. Of course, the fact that they can pass through the entire Sun without interacting also makes neutrinos rather difficult to detect here on Earth! Nevertheless, with some knowledge of neutrino physics, it is in fact possible to construct neutrino detectors.

Over the past four decades, four major experiments have sought to detect solar neutrinos reaching Earth's surface. The first, built in the late 1960s, was sited 1.5 km below ground level near the bottom of the Homestake gold mine in South Dakota (to shield it from interference due to cosmic rays and other sources, most of which are unable to penetrate Earth's crust to such a depth). It was designed to measure the changes occurring when a neutrino interacts with a nucleus of chlorine-37, converting it into argon-37. The researchers left a tank containing some 400,000 liters (about 100,000 gallons) of a chlorine-containing chemical—the common cleaning fluid used by dry cleaners—in the mine for months at a time, periodically checking to see if any of the chlorine had been converted into argon, which would signal the absorption of a neutrino.

Given the size of the detector and the physical conditions in the Sun's core implied by the Standard Solar Model, about one solar neutrino of the roughly 10^{16} that streamed through the tank each day should have been detected. The experiment did succeed in detecting solar neutrinos—in itself a remarkable achievement—but the numbers were not as great as predicted. Over the course of the entire experiment, neutrinos were detected about twice per week, on average, not once per day. The neutrino deficit persisted over two decades of almost continuous monitoring until the experiment was terminated in 1993. This clear disagreement between theory and observation has come to be known as the **solar neutrino problem**.

The other three experiments are more recent, and have quite different detector designs. In the 1990s, the Soviet–American Gallium Experiment (SAGE, for short) and the U.S.–European GALLEX collaboration, each using the element gallium to capture solar neutrinos, found a roughly 50 percent shortfall. Figure 16.28(a) shows a large detector located in Kamioka, Japan, constructed in the 1980s (and upgraded in the 1990s) to measure the telltale light emitted when neutrinos stream through it. This device also detects about half the predicted number of neutrinos. The Kamioka detector also has some limited directional capability—Figure 16.28(b) shows a low-resolution image of the Sun produced by this instrument.

Thus, while the four neutrino-detection experiments just described disagree somewhat on the extent of the deficit, each sees significantly fewer than the expected number of solar neutrinos. In fact, the Homestake and Kamioka experiments were sensitive not to neutrinos produced in reaction (I) earlier—the initial step in the proton–proton chain—but to those created by a much less probable sequence of events, occurring only about 0.25 percent of the time, in principle leaving room for theorists to maneuver in their attempts to explain the observations. (But it should be noted that, to many workers, the disagreement between the Homestake and Kamioka results constitutes a second, also unresolved, solar neutrino problem.) However, SAGE and GALLEX could detect neutrinos

(a)

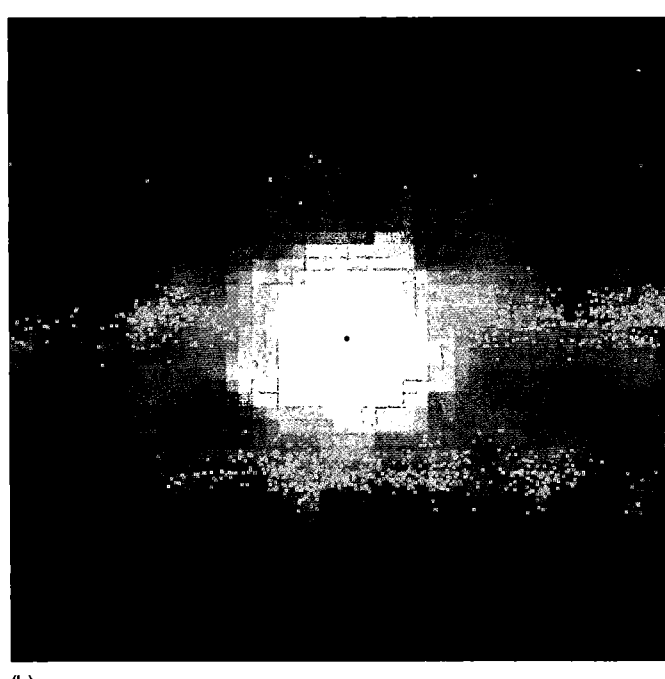

(b)

Figure 16.28 Neutrino Telescope (a) This swimming-pool-sized detector is a "neutrino telescope" of sorts, buried beneath a mountain near Tokyo, Japan. Called Super Kamiokande, it is filled (in operation) with 50,000 tons of purified water, and contains 13,000 light detectors (some shown here being inspected by technicians) to sense the telltale signature of a neutrino passing through the apparatus. (b) An "image" of the Sun made by the Kamiokande instrument. Current neutrino detectors have very low resolution, and the field of view here is about 90° across. The small black dot indicates the actual size of the Sun. *(Institute for Cosmic Ray Research; NASA)*

produced by reaction (I), providing a more direct probe of energy generation in the solar core.

The inescapable conclusion is that, although solar neutrinos are observed (and in fact their measured energies do lie in the range predicted by the Standard Solar Model), there is a real discrepancy between the Sun's theoretical neutrino output and the neutrinos we detect on Earth. How can we explain this contradiction? If, as we think, the detectors are working correctly, there are really only two possibilities. Either neutrinos are not produced as frequently as we think, or not all of them make it to Earth.

Most theorists think it unlikely that the resolution of the solar neutrino problem will be found in the physics of the Sun's interior. The nuclear reactions described earlier are just too well known, and the agreement between the Standard Solar Model and helioseismological observations is too close to permit conditions in the core to deviate much from the model predictions. Instead, the most likely explanation seems to lie in the properties of the neutrinos themselves. If neutrinos have a minute amount of mass, it may be possible for them to change their properties, even to transform into other particles, during their eight-minute flight from the solar core to Earth, through a

process generally known as **neutrino oscillations.** In this view, neutrinos are produced in the Sun at the rate required by the Standard Solar Model, but some turn into something else—actually, other types of neutrinos—on their way to Earth and go undetected. (In the parlance of the field, they are said to "oscillate" into other particles.)

In June 1998, the Kamioka group reported the first experimental evidence of neutrino oscillations (and hence of nonzero neutrino masses), although the observed oscillations did not involve neutrinos of the type produced in the Sun. More recently, in June 2001, measurements made at the Sudbury Neutrino Observatory (SNO) in Ontario, Canada, revealed strong evidence for the "other" neutrinos into which the Sun's neutrinos have been transformed. The total numbers of neutrinos observed are completely consistent with the Standard Solar Model. Apparently the solar neutrino problem has at last been solved.

✓ Concept Check

■ Why do neutrinos give us direct information about conditions in the solar core, while electromagnetic radiation does not?

DISCOVERY 16-2

Solar–Terrestrial Relations

Our Sun has often been worshipped as a god with power over human destinies. Obviously, the steady stream of solar energy arriving at our planet every day is essential to our lives. But over the past century there have also been repeated claims of a correlation between the Sun's activity and Earth's weather. Only recently, however, has the subject become scientifically respectable—that is, more natural than supernatural.

In fact, there do seem to be some correlations between the 22-year solar cycle (two sunspot cycles, with oppositely directed magnetic fields) and periods of climatic dryness here on Earth. For example, near the start of the past eight cycles, there have been droughts in North America—at least within the middle and western plains from South Dakota to New Mexico. The most recent of these droughts, which typically last three to six years, came in the late 1950s. The one expected in the 1980s, however, did not occur as clearly as anticipated.

Other possible Sun–Earth connections include a link between solar activity and increased atmospheric circulation on our planet. As circulation increases, terrestrial storm systems deepen, extend over wider ranges of latitude, and carry more moisture. The relationship is complex, and the subject controversial, because no one has yet shown any physical mechanism (other than the Sun's heat, which does not vary much during the solar cycle) that would allow solar activity to stir our terrestrial atmosphere. Without a better understanding of the physical mechanism involved, none of these effects can be incorporated into our weather-forecasting models.

Solar activity may also influence long-term climate on Earth. For example, the Maunder minimum seems to correspond fairly well to the coldest years of the so-called Little Ice Age that chilled northern Europe during the late 1600s (see Section 16.4). The accompanying "winter" scene actually captured the summer season in 17th-century Holland. How the active Sun, and its abundance of sunspots, may affect Earth's climate is a frontier problem in terrestrial climatology. Measurements of the solar constant made over the past two decades

indicate that the Sun's energy output varies with the solar cycle—paradoxically, the Sun's luminosity is greatest when many dark sunspots cover its surface! Thus, the Maunder minimum does correspond to an extended period of lower-than-average solar emission. However, the observed changes in the Sun's recent luminosity have been small—no more than 0.2 or 0.3 percent. It is not known by how much, if at all, the Sun's output declined during the Maunder minimum, nor how large a change would be needed to account for the climatic changes that occurred.

One correlation that is definitely established, and also better understood, is that between solar activity and geomagnetic disturbances at Earth. The extra radiation and particles thrown off by flares or coronal mass ejections impinge on Earth's environment, overloading the Van Allen belts, causing brilliant auroras in our atmosphere and degrading our communication networks. We are only beginning to understand how the radiation and particles emitted by solar phenomena also interfere with terrestrial radars, power networks, and other technological equipment. Some power outages on Earth are actually caused not by increased customer demand or malfunctioning equipment but by weather on the Sun!

We cannot yet predict just when and where solar flares or coronal mass ejections will occur. However, it would certainly be to our advantage to be able to do so, as this aspect of the active Sun affects our lives. This is a very fertile area of astronomical research and one with clear terrestrial applications.

(Toledo Museum of Art, Toledo, OH)

Chapter Review

SUMMARY

A **star** (p. 406) is a glowing ball of gas held together by its own gravity and powered by nuclear fusion at its center. The main interior regions of the Sun are the **core** (p. 407), where nuclear reactions generate energy, the **radiation zone** (p. 407), where the energy travels outward in the form of electromagnetic radiation, and the **convection zone** (p. 407), where the Sun's matter is in constant convective motion. The sharp solar disk visible from Earth marks the solar **photosphere** (p. 407)—the thin surface layer from which the Sun's light is emitted. Above the photosphere lies the **chromosphere** (p. 407), which is separated from the solar **corona** (p. 407) by a thin **transition zone** (p. 407) in which the temperature increases from a few thousand to over a million kelvins. At about 10–15 solar radii, the gas in the corona is hot enough to escape the Sun's gravity, and the corona begins to flow outward as the **solar wind** (p. 417). Most of the solar wind flows from low-density regions of the corona called **coronal holes** (p. 417).

The Sun's **luminosity** (p. 408) is the total amount of energy radiated from the solar surface per second. It is determined by measuring the **solar constant** (p. 408)—the amount of solar radiation reaching each square meter at Earth's distance from the Sun—and multiplying that amount by the area of an imaginary sphere of radius 1 A.U.

The mathematical model that best fits the observed properties of the Sun is the **Standard Solar Model** (p. 409). Ground and space-based **helioseismology** (p. 409)—the study of oscillations of the solar surface caused by sound waves in the interior—provides considerable insight into the Sun's structure. The effect of the solar convection zone can be seen on the surface in the form of **granulation** (p. 412) of the photosphere. As hotter (and therefore brighter) gas rises and cooler (dimmer) gas sinks, a characteristic "mottled" appearance results. Most of the absorption lines seen in the solar spectrum are produced in the upper photosphere and the chromosphere. Studies of these allow scientists to determine the Sun's composition and the temperature structure of the solar atmosphere.

Solar activity is generally associated with disturbances in the Sun's magnetic field. **Sunspots** (p. 418) are Earth-sized regions on the solar surface that are a little cooler than the surrounding photosphere. They are regions of intense magnetism.

Both the numbers and locations of sunspots vary in an 11-year **sunspot cycle** (p. 421). At solar minimum, only a few spots are typically seen, and they lie far from the solar equator. At solar maximum, the number of spots is much greater, and they generally lie much closer to the equator. The sunspot cycle is quite irregular. Its length varies from 7 to 15 years, but there have been times in the past when no sunspots were seen for long periods. The overall direction of the solar magnetic field reverses from one sunspot cycle to the next. The 22-year cycle that results when the direction of the field is taken into account is called the **solar cycle** (p. 421). Solar activity tends to be concentrated in **active regions** (p. 422) associated with sunspot groups. **Prominences** (p. 422) are looplike or sheetlike structures produced when hot gas ejected by activity on the solar surface interacts with the Sun's magnetic field. The more intense **flares** (p. 423) are violent surface explosions that blast particles and radiation into interplanetary space.

The Sun generates energy by "burning" hydrogen into helium in its core by the process of **nuclear fusion** (p. 426). Nuclei are held together by the **strong nuclear force** (p. 426). When four protons overcome their electromagnetic repulsion and are converted into a helium nucleus in the **proton–proton chain** (p. 428), some mass is lost. The **law of conservation of mass and energy** (p. 426) requires that this mass appear as energy, eventually resulting in the light we see. Very high temperatures are needed for fusion to occur.

Some particles produced during the solar fusion process are the **positron** (p. 426), or anti-electron, which quickly annihilates with electrons in the Sun's core to generate gamma rays, the **deuteron** (p. 426), an **isotope** (p. 427) of hydrogen consisting of a proton and a neutron, and the **neutrino** (p. 427), a near-massless particle that escapes from the Sun without any further interactions once it is created in the core. Neutrinos interact via the **weak nuclear force** (p. 427). Despite their elusiveness, it is possible to detect a small fraction of the neutrinos streaming from the Sun. The observations lead to the **solar neutrino problem** (p. 429)—substantially fewer neutrinos are observed than are predicted by theory. The leading explanation for the deficit is that **neutrino oscillations** (p. 430) convert some neutrinos into other (undetected) particles en route from the Sun to Earth.

SELF-TEST: TRUE OR FALSE?

_____ **1.** The Sun is a rather normal star.

_____ **2.** The Sun's average density is less than that of Earth.

_____ **3.** The Sun's diameter is about 10 times that of Earth.

_____ **4.** Observations of sunspots indicate that the Sun rotates differentially.

_____ **5.** In the solar convection zone, the gas is partly ionized.

_____ **6.** Convection involves cool gas rising toward the solar surface, and hot gas sinking into the interior.

_____ **7.** Most solar absorption lines spectrum are produced in the corona.

_____ **8.** There are as many absorption lines in the solar spectrum as there are elements present in the Sun.

_____ **9.** The faintness of the chromosphere is a direct result of its low temperature.

_____ **10.** The temperature of the solar atmosphere decreases with increasing radius.

_____ **11.** Sunspots are regions of intense magnetic fields.

_____ **12.** Prominences are large flames erupting from the burning surface of the Sun.

_____ **13.** Positrons are the antiparticles of electrons.

_____ **14.** Nuclei are held together by the strong force.

_____ **15.** Neutrinos are hypothetical particles that are believed to exist but have never been detected experimentally.

SELF-TEST: FILL IN THE BLANK

1. The part of the Sun we actually see is called the _____.

2. Traveling outward from the surface, the main regions of the solar atmosphere are the _____, the _____, and the _____.

3. Below the solar surface, in order of increasing depth, lie the _____ zone, the _____ zone, and the _____.

4. _____ seen on the surface of the Sun is evidence of solar convection.

5. The Sun appears to have a well-defined edge because the thickness of the _____ is only 0.1 percent of the solar radius.

6. The most abundant element in the Sun is _____.

7. The second most abundant element in the Sun is _____.

8. The two most abundant elements in the Sun make up about _____ percent of its composition (by number of atoms).

9. The temperature in the solar corona is extremely _____.

10. Sunspots appear dark because they are _____ than the surrounding gas of the photosphere.

11. The sunspot cycle is roughly _____ years long; the solar cycle is _____ years long.

12. The entire solar luminosity is produced in the _____ (give the region) of the Sun.

13. The net result of the proton–proton chain is that _____ protons are fused into a nucleus of _____ , two _____ are emitted, and energy is released in the form of _____.

14. Energy is released in the proton–proton chain because mass is _____ in the process.

15. The solar neutrino problem is the fact that astronomers observe too _____ neutrinos coming from the Sun.

REVIEW AND DISCUSSION

1. Name and briefly describe the main regions of the Sun.

2. How massive is the Sun, compared with Earth?

3. How hot is the solar surface? The solar core?

4. What is luminosity, and how is it measured in the case of the Sun?

5. How do scientists construct models of the Sun?

6. What is helioseismology, and what does it tell us about the Sun?

7. How do observations of the Sun's surface tell us about conditions in the solar interior?

8. Describe how energy generated in the solar core eventually reaches Earth.

9. Why does the Sun appear to have a sharp edge?

10. Give the history of "coronium," and tell how it increased our understanding of the Sun.

11. What is the solar wind?

12. Why do we say that the solar cycle is 22 years long?

13. What is the cause of sunspots, flares, and prominences?

14. Describe how coronal mass ejections may influence life on Earth.

15. What fuels the Sun's enormous energy output?

16. What are the ingredients and the end result of the proton–proton chain in the Sun?

17. Why is energy released in the proton-proton chain?

18. Why are scientists trying so hard to detect solar neutrinos?

19. What is the most likely solution to the solar neutrino problem?

20. What would we observe on Earth if the Sun's internal energy source suddenly shut off? How long do you think it might take—minutes, days, years, millions of years—for the Sun's light to begin to fade? Repeat the question for solar neutrinos.

PROBLEMS _www_ _Algorithmic versions of these questions are available in the Practice Problems module of the Companion Website._

The number of squares preceding each problem indicates its approximate level of difficulty.

1. ■ Use the reasoning presented in Section 16.1 to calculate the value of the "solar constant" (a) on Mercury at perihelion, (b) on Jupiter.

2. ■ Use Wien's law to determine the wavelength corresponding to the peak of the blackbody curve (a) in the core of the Sun, where the temperature is 10^7 K, (b) in the solar convection zone (10^5 K), and (c) just below the solar photosphere (10^4 K). ∞ (Sec. 3.4) What form (visible, infrared, X ray, etc.) does the radiation take in each case?

3. ■■ The largest-amplitude solar pressure waves have periods of about five minutes and move at the speed of sound in the outer layers of the Sun, roughly 10 km/s. (a) How far does the wave move during one wave period? (b) Approximately how many wavelengths are needed to completely encircle the Sun's equator? (c) Compare the wave period with the orbital period of an object moving just above the solar photosphere. ∞ (_More Precisely 2-3_)

4. ■ If convected solar material moves at 1 km/s, how long does it take to flow across the 1000-km expanse of a typical granule? Compare this with the roughly 10-minute lifetimes observed for most solar granules.

5. ■■ Use Stefan's law (flux $\propto T^4$, where T is the temperature in kelvins; see Section 3.4) to calculate how much less energy (as a fraction) is emitted per unit area of a 4500-K sunspot than from the surrounding 5800-K photosphere.

6. ■■ The Sun's differential rotation is responsible for wrapping the solar magnetic field around the Sun (Figure 16.19). Using the data presented in the Sun Data box on p. 406, calculate how long it takes for material at the solar equator to "lap" the material near the poles—that is, to complete one extra trip around the Sun's rotation axis.

7. ■ Use the data presented in this chapter and in *More Precisely 8-1* to estimate the radius at which the speed of protons in the corona first exceeds the solar escape speed.

8. ■ The solar wind carries mass away from the Sun at a rate of about 2 million tons/s (1 ton = 1000 kg). At this rate, how long would it take for all of the Sun's mass to escape?

9. ■ How long does it take for one Earth mass of material to escape the Sun as the solar wind (see Problem 8)?

10. ■ A large coronal mass ejection was observed by *SOHO* as it left the Sun. It reached Earth roughly 36 hours later. What was its average speed? If such ejections occur, on average, three times per week and contain roughly 1 billion tons of matter, calculate the average rate at which they carry off

solar mass. How does this compare to the mass loss rate due to the solar wind?

11. ■■ How long does it take for the Sun to convert one Earth mass of hydrogen into helium?

12. ■■ (a) Assuming constant luminosity, calculate how much equivalent mass (relative to the current mass of the Sun) the Sun has radiated into space in the 4.6 billion years since it formed. How much hydrogen has been consumed? (b) How long would it take the Sun to radiate its own mass into space?

13. ■■ Compare the rate at which the Sun loses mass via the solar wind (Problem 8) with the rate at which the Sun loses mass in the form of radiation.

14. ■■ Assuming that (1) the solar luminosity has been constant since the Sun formed, and (2) the Sun was initially of uniform composition throughout, as described by Table 16.2, estimate how long it would take the Sun to convert all of its original hydrogen into helium.

15. ■■■ The entire reaction sequence shown in Figure 16.27 generates 4.3×10^{-12} J of electromagnetic energy and releases two neutrinos. Assuming that neutrino oscillations transform half of these neutrinos into other particles by the time they travel 1 A.U. from the Sun, estimate the total number of solar neutrinos passing through Earth each second.

COLLABORATIVE EXERCISES

1. Sunspot Counts. Each member of your group should independently count the number of sunspots shown in Figure 16.15. Determine an average number for your group and compare your results to a group sitting next to you.

2. Solar Cycle. Using Figure 16.20, each member of your group should individually determine the number of years between any two sequential sunspot maximums. Average your group's results and compare the average value to the frequently stated 11-year solar cycle.

RESEARCHING ON THE WEB  *To complete the following exercises, go to the online Destinations module for Chapter 16 on the Companion Website for* Astronomy Today *4/e.*

1. Access the "Glossary of Solar-Terrestrial Terms" page and explain how a Wolf Number is calculated.

2. Access the "Big Bear Solar Observatory" page and use the *latest images* hyperlink to make a sketch of where on the Sun sunspots are located today.

3. Access the "Learmonth Solar Observatory Western Australia" page to determine the current solar wind speed as measured by the *ACE* spacecraft.

PROJECTS

The projects given here all require a special solar filter. Such filters are easily purchased from various sources.

NEVER LOOK DIRECTLY AT THE SUN WITHOUT A FILTER!

1. An appropriately filtered telescope will easily show you sunspots. Count the number of sunspots you see on the Sun's surface. Notice that sunspots often come in pairs or groups. Come back and look again a few days later and you'll see that the Sun's rotation has caused spots to move, and the spots themselves have changed. If a sufficiently large sunspot (or, more likely, sunspot group) is seen, continue to watch it as the Sun rotates. It will be out of view for about two weeks. Can you determine the rotation of the Sun from these observations?

2. Solar granulation is not too hard to see. The atmosphere of Earth is most stable in the morning hours. Observe the Sun on a cool morning, one or two hours after it has risen. Use high magnification and look initially at the middle of the Sun's disk. Can you see changes in the granulation pattern? They are there, but they are not always obvious or easy to see.

3. View some solar prominences and flares. Hydrogen-alpha (Hα) filters commercially available for small telescopes are quite expensive, but many science departments will have one. You can often see prominences and flares even during times of sunspot minimum. You are actually viewing the chromosphere rather than the photosphere, so the Sun looks quite different from its normal appearance.

SKYCHART III PROJECTS

The SkyChart III Student Version planetarium program on which these exercises are based is included as a separately executable program on the CD in the back of this text.

1. ■ Because of Earth's 23.5° axial tilt and the slight eccentricity of its orbit about the Sun, the Sun is not directly overhead at noon each day. The actual position of the Sun at the same time each day traces out a pattern in the sky that is called the **analemma**. Set SkyChart III to illustrate this phenomenon. Center on the Sun, set time for 12:00 noon, set field of view to 90°, and set *DRAW/Sky Background/ Black*. Set *ANIMATION/Trail For/Sun* and when *Edit Object Trail* appears, select *Ticks* every one position and *Record* up to 365 positions. Select *Record apparent positions in current coordinate system*, and click on OK. Animate with one-day time steps. Run the animation. The figure painted by the path of the Sun is the analemma. If your computer is slow, animate with time steps of one week.

   Step through the time increments using F5 and F6. Note at which times of the year the Sun is actually overhead at noon, and at which times the sundial would be in most serious disagreement with accurate clocks. Under *DRAW/Symbols & Grids*, turn on *Draw Grid Lines*. Select all grid lines and set *R.A. interval* to 10 minutes. Using this grid spacing, step through the simulation with F5 and F6 until you have an alignment that permits you to estimate how many minutes difference there is between actual noon and noon by a sundial when the difference is the greatest.

2. ◼ Total solar eclipses are certainly one of the most spectacular astronomical events viewable with the naked eye. A total eclipse occured on June 21, 2001, viewable from South Africa. Change the *Location* to Lusaka, Zambia (+ 28 00 00.0 *E*, − 15 00 00.0 *N*, − 1.0 *Time Zone*). *Center* and *Zoom In* on the Sun. You may find it convenient to choose *VIEW/coordinates/ecliptic* for easier viewing. Determine the time and duration of totality.

 *In addition to the Practice Problems and Destinations modules, the Companion Website at http://www.prenhall.com/chaisson provides for each chapter an additional true-false, multiple choice, and labeling quiz, as well as additional annotated images, animations, and links to related Websites.*

17 MEASURING THE STARS

Giants, Dwarfs, and the Main Sequence

LEARNING GOALS

Studying this chapter will enable you to:

1   Explain how stellar distances are determined.

2   Discuss the motions of the stars through space and how these motions are measured from Earth.

3   Distinguish between luminosity and apparent brightness and explain how stellar luminosity is determined.

4   Explain the usefulness of classifying stars according to their colors, surface temperatures, and spectral characteristics.

5   Explain how physical laws are used to estimate stellar sizes.

6   Describe how an H–R diagram is constructed and used to identify stellar properties.

7   Explain how the masses of stars are measured and how mass is related to other stellar properties.

 Visit http://www.prenhall.com/chaisson for additional annotated images, animations, and links to related sites for this chapter.

This rich cluster of perhaps a million stars resides some 150,000 light-years away in the Large Magellanic Cloud, a small galaxy orbiting our own. Known as NGC 1850, it is actually a double cluster—the smaller component can be seen here at roughly the four-o'clock position relative to its larger companion. Both clusters contain many extremely hot, young stars, all much more massive than our Sun. The larger cluster is estimated to be about 40 million years old, the smaller one just 4 million years. The explosive activity of many of the stars in the larger cluster (which is probably responsible for much of the faint, reddish debris surrounding the system) may well have triggered the birth of its smaller companion. *(ESO)*

The Big Picture: The total number of stars, even in our local neighborhood, is virtually beyond our ability to count. Relatively few of them have been studied in detail. Yet astronomers have learned a great deal about stars in general, and their range of properties—their masses, their temperatures, their luminosities, even their ages and (as we shall see in later chapters) their destinies.

437

We have now studied Earth, the Moon, the solar system, and the Sun. To continue our inventory of the contents of the universe, we must move away from our local environment into the depths of space. In this chapter we take a great leap in distance and consider stars in general. Our primary goal is to comprehend the nature of the stars that make up the constellations, as well as the myriad more distant stars we cannot perceive with our unaided eyes. Rather than studying their individual peculiarities, however, we will concentrate on determining the physical and chemical properties they share. There is order in the legions of stars scattered across the sky. Like comparative planetology in the solar system, comparing and cataloging the stars play vital roles in furthering our understanding of the Galaxy and the universe we inhabit.

17.1 The Distances to the Stars

To understand the stars, we must first know how far away they are. We therefore begin our study of stellar astronomy by reviewing the use of simple geometry as a distance-measurement tool.

STELLAR PARALLAX

⫶ Recall from Chapter 1 how surveyors and astronomers use *parallax* to measure distances to terrestrial and solar system objects. Parallax is the apparent shift of a foreground object relative to some distant background as the observer's point of view changes. ∞ (Sec. 1.5) To determine an object's parallax, we observe it from either end of some baseline, and measure the angle through which the line of sight to the object shifts. In astronomical contexts, the angle is usually obtained by comparing photographs made from the two ends of the baseline.

As the distance to the object increases, the parallax becomes smaller and therefore harder to measure. Even the closest stars are so far away that no baseline on Earth is sufficient to allow accurate determination of their distances. Their parallactic shifts, as seen from different points on Earth, are just too small. However, by comparing observations of a star made *at different times of the year*, as shown in Figure 17.1 (compare Figure 1.25), we can extend our baseline to the diameter of Earth's orbit around the Sun, 2 A.U. Only with this enormously longer baseline do some stellar parallaxes become measurable. As indicated in the figure, a star's parallactic angle—or, more commonly, just its "parallax"—is conventionally defined to be *half* its apparent shift relative to the background as we move from one side of Earth's orbit to the other.

Because stellar parallaxes are so small, astronomers generally find it convenient to measure parallax in arc sec-

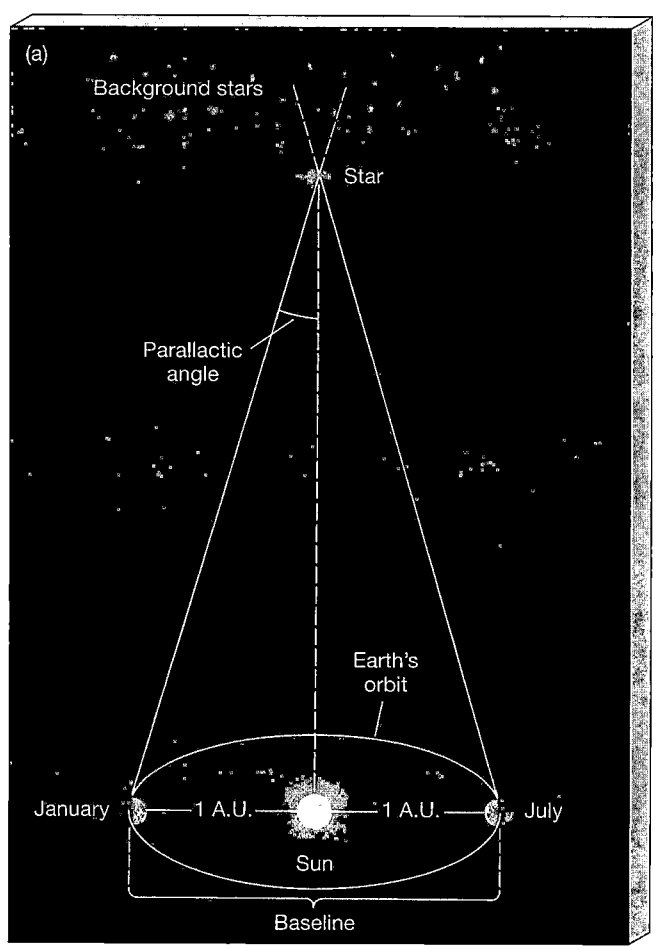

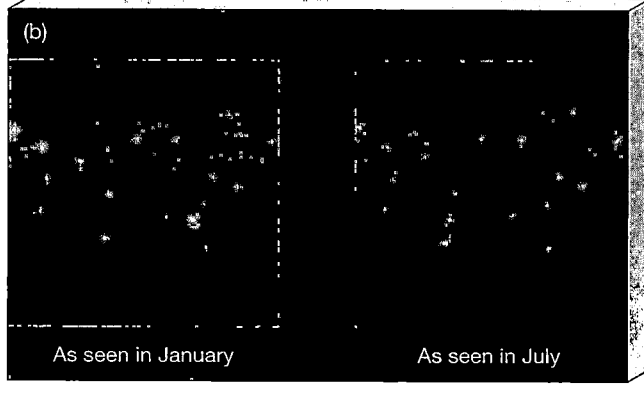

Figure 17.1 Stellar Parallax (a) The geometry of stellar parallax. For observations made six months apart, the baseline is twice the Earth–Sun distance, or 2 A.U. (b) The parallactic angle is usually measured photographically. (The shift is greatly exaggerated in this drawing.)

onds rather than in degrees. If we ask at what distance a star must lie in order for its observed parallax to be exactly 1′, we get an answer of 206,265 A.U., or 3.1 × 10^{16} m. ∞ (*More Precisely 1-4*) Astronomers call this distance 1 **parsec** (1 pc), from "*par*allax in arc *sec*onds." Because parallax decreases as distance increases, we can relate a star's parallax to its distance by the following simple formula:

$$\text{distance (in parsecs)} = \frac{1}{\text{parallax (in arc seconds)}}.$$

Thus, a star with a measured parallax of 1″ lies at a distance of 1 pc from the Sun. The parsec is defined so as to make the conversion between distance and parallactic angle easy. An object with a parallax of 0.5″ lies at a distance of 1/0.5 = 2 pc; an object with a parallax of 0.1″ lies at 1/0.1 = 10 pc, and so on. One parsec is approximately equal to 3.3 light-years.

OUR NEAREST NEIGHBORS

The closest star to Earth (besides the Sun) is called Proxima Centauri. This star is a member of a triple-star system (three separate stars orbiting one another, bound together by gravity) known as the Alpha Centauri complex. Proxima Centauri displays the largest known stellar parallax, 0.76″, which means that it is about 1/0.76 = 1.3 pc away—about 270,000 A.U., or 4.3 light-years. That's the *nearest* star to Earth—at almost 300,000 times the distance from Earth to the Sun! This is a fairly typical interstellar distance in the Milky Way Galaxy.

Vast distances can sometimes be grasped by means of analogies. Imagine Earth as a grain of sand orbiting a marble-sized Sun at a distance of 1 m. The nearest star, also a marble-sized object, is then more than 270 *kilo*meters away. Except for the other planets in our solar system, themselves ranging in size from grains of sand to millimeter-sized pellets and all lying within 50 m of the "Sun," nothing else of consequence exists in the 270 km separating the two stars. Such is the void of interstellar space.

The next nearest neighbor to the Sun beyond the Alpha Centauri system is called Barnard's Star. Its parallax is 0.55″, so it lies at a distance of 1.8 pc, or 6.0 light-years—370 km in our model. Figure 17.2 is a map of our nearest galactic neighbors—the 30 or so stars lying within 4 pc of Earth.

Ground-based images of stars are generally smeared out into a disk of radius 1″ or so by turbulence in Earth's atmosphere. ∞ (Sec. 5.3) However, astronomers have special equipment that can routinely measure stellar parallaxes of 0.03″ or less, corresponding to stars within about 30 pc (100 light-years) of Earth. Several thousand stars lie within this range, most of them of much lower luminosity than the Sun and invisible to the naked eye. Adaptive optics systems allow even more accurate measurements of stellar positions, extending the parallax range to over 100 pc, although such measurements are not yet "routine." As discussed in more detail in *Discovery 17-1*, observations made by the European *Hipparcos* satellite have extended the range of accurately measured parallaxes to over 200 pc, encompassing more than a million stars. Even so, the vast majority of stars in our Galaxy are far more distant.

✓ Concept Check

◻ Why can't astronomers use simultaneous observations from different parts of Earth's surface to determine stellar distances?

17.2 Stellar Motion

2 In addition to the apparent motion caused by parallax, stars have real spatial motion, too. This stellar motion has two components. A star's *radial* velocity—along the line of sight—can be measured using the Doppler effect. ∞ (Sec. 3.5) For many nearby stars, the *transverse* velocity—perpendicular to our line of sight—can also be determined by careful monitoring of the star's position on the sky.

Figure 17.3 compares two photographs of the sky around Barnard's Star. They were made on the same day of the year, but 22 years apart. Note that the star, marked by the arrow, has moved during the 22-year interval shown—if the two images were superimposed, the images of the other stars in the field of view would coincide, but those of Barnard's Star would not. Because Earth was at the same point in its orbit when these photographs were taken, the observed displacement is *not* the result of parallax caused by Earth's motion around the Sun. Instead, it indicates *real* space motion of Barnard's Star relative to the Sun.

The annual movement of a star across the sky, as seen from Earth and corrected for parallax, is called **proper motion**. It describes the transverse component of a star's velocity relative to the Sun. Like parallax, proper motion is measured in terms of angular displacement. Since the angles involved are typically very small, proper motion is usually expressed in arc seconds per year. Barnard's Star moved 227″ in 22 years, so its proper motion is 227″/22 years, or 10.3″/yr.

A star's transverse velocity is easily calculated once its proper motion and its distance are known. At the distance of Barnard's Star (1.8 pc), an angle of 10.3″ corresponds to a physical displacement of 0.00009 pc, or about 2.8 billion km. Barnard's Star takes a year (3.2 × 10^7 s) to travel this distance, so its transverse velocity is 2.8 billion km/3.2 × 10^7 s, or 88 km/s. Even though stars' transverse velocities are often quite large—tens or even hundreds of kilometers per second—their great distances from the Sun mean that it usually takes many years for us to discern their movement across the sky. In fact, Barnard's Star has the largest known proper motion of any star. Only a few hundred stars have proper motions greater than 1″/yr.

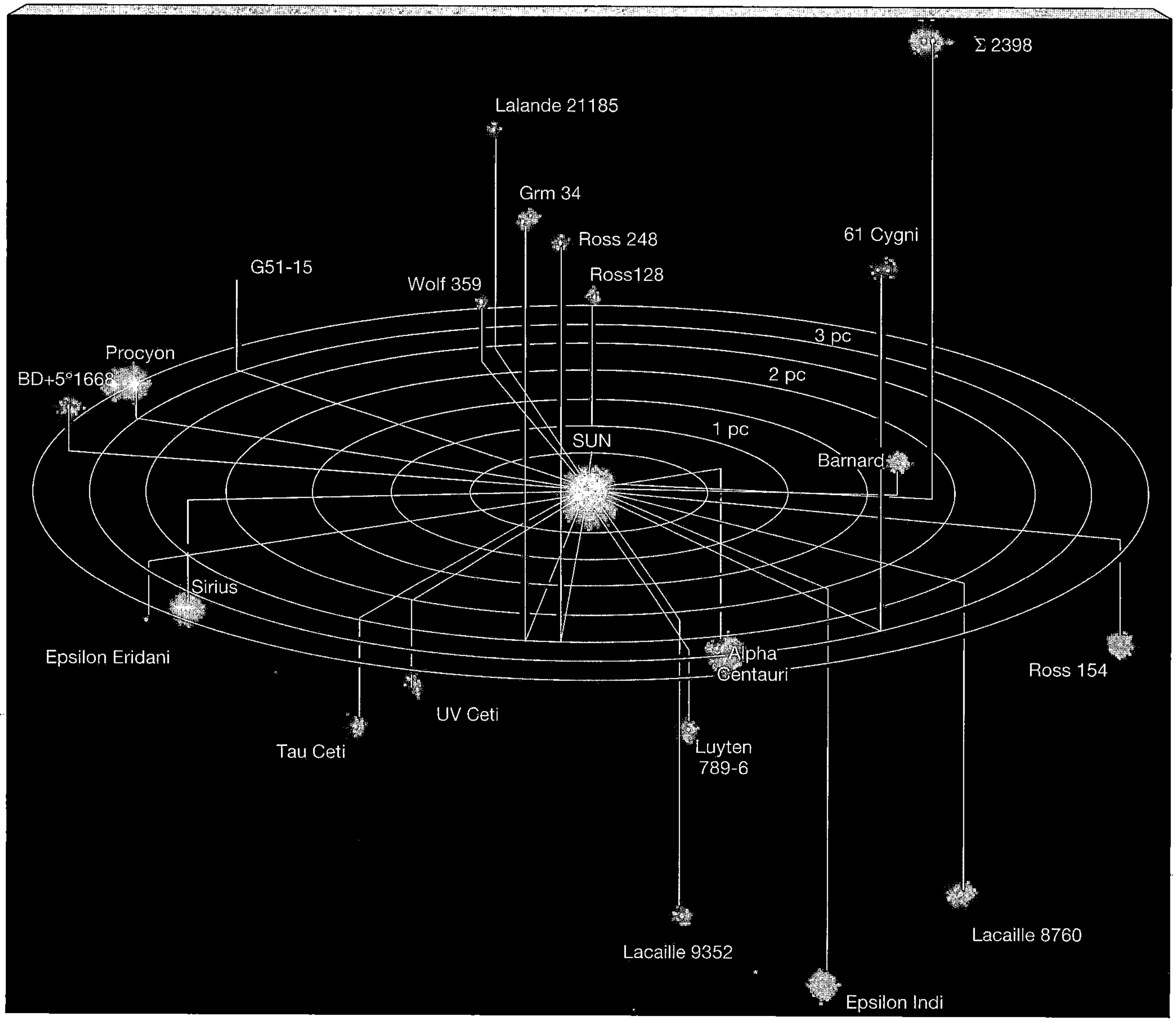

Σ 2398

Lalande 21185

Grm 34

Ross 248

61 Cygni

G51-15

Wolf 359

Ross128

3 pc

Procyon

2 pc

BD+5°1668

1 pc

SUN

Barnard

Sirius

Epsilon Eridani

Alpha
Centauri

Ross 154

UV Ceti

Tau Ceti

Luyten
789-6

Lacaille 9352

Lacaille 8760

Epsilon Indi

Figure 17.2 The Solar Neighborhood A plot of the 30 closest stars to the Sun, projected so as to reveal their three-dimensional relationships. Notice that many are members of double—or multiple—star systems. All lie within 4 pc (about 13 light-years) of Earth.

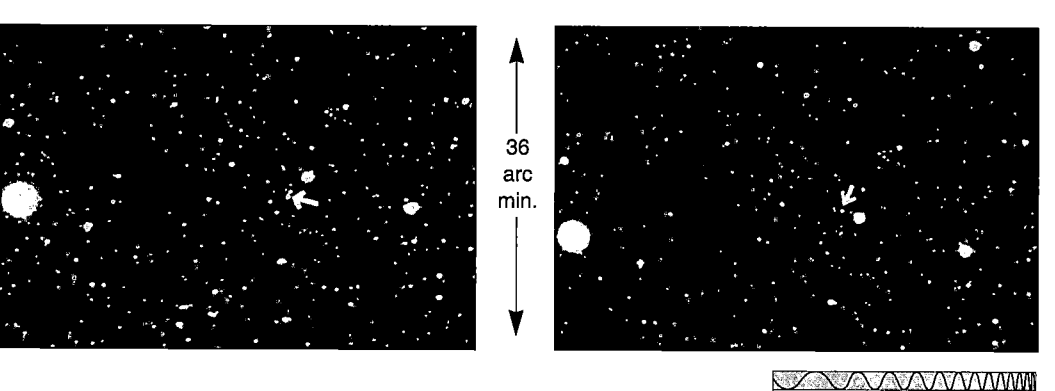

36
arc
min.

Figure 17.3 Proper Motion Comparison of two photographic plates taken 22 years apart shows evidence of real space motion for Barnard's Star (denoted by an arrow). *(Harvard College Observatory)*

R   I   V   .U   X   G

Now consider the three-dimensional motion of our nearest neighbor, the Alpha Centauri system, sketched in Figure 17.4 in relation to our own solar system. Alpha Centauri's proper motion has been measured to be 3.5"/yr. At Alpha Centauri's distance of 1.3 pc, this implies a transverse velocity of 22 km/s. We can determine the other component of motion—the radial velocity—using the Doppler effect. Spectral lines from Alpha Centauri are blueshifted by a tiny amount—about 0.0067 percent—allowing astronomers to measure the star system's radial velocity (relative to the Sun) as 20 km/s toward us. ⊙ (Sec. 3.5) What is the true space motion of Alpha Centauri? Will this alien system collide with our own sometime in the future? The answer is no—Alpha Centauri's transverse velocity will steer it well clear of the Sun. We can combine the transverse and radial velocities according to the Pythagorean theorem. The total velocity is $\sqrt{22^2 + 20^2}$, or about 30 km/s, in the direction shown by the horizontal red arrow in Figure 17.4. As that figure indicates, Alpha Centauri will get no closer to us than about 1 pc, and that won't happen until 280 centuries from now.

✓ Concept Check

▣ Why are the spatial velocities of distant stars generally poorly known?

17.3 Luminosity and Apparent Brightness

⊙ Luminosity is an *intrinsic* property of a star—it does not depend in any way on the location or motion of the observer. It is sometimes referred to as the star's *absolute brightness*. However, when we look at a star, we see not its luminosity but rather its **apparent brightness**—the amount of energy striking unit area of some light-sensitive surface or device (such as a CCD chip or a human eye) per unit time. Apparent brightness is a measure not of a star's luminosity but of the *energy flux* (energy per unit area per unit time) produced by the star, as seen from Earth. It depends on our *distance* from the star. In this section, we discuss in more detail how these important quantities are related to one another.

ANOTHER INVERSE-SQUARE LAW

Figure 17.5 shows light leaving a star and traveling through space. Moving outward, the radiation passes through imaginary spheres of increasing radius surrounding the source. The amount of radiation leaving the star per unit time—the star's luminosity—is constant, so the farther the light travels from the source, the less energy passes through each unit of area. Think of the energy as being spread out over an ever-larger area and therefore

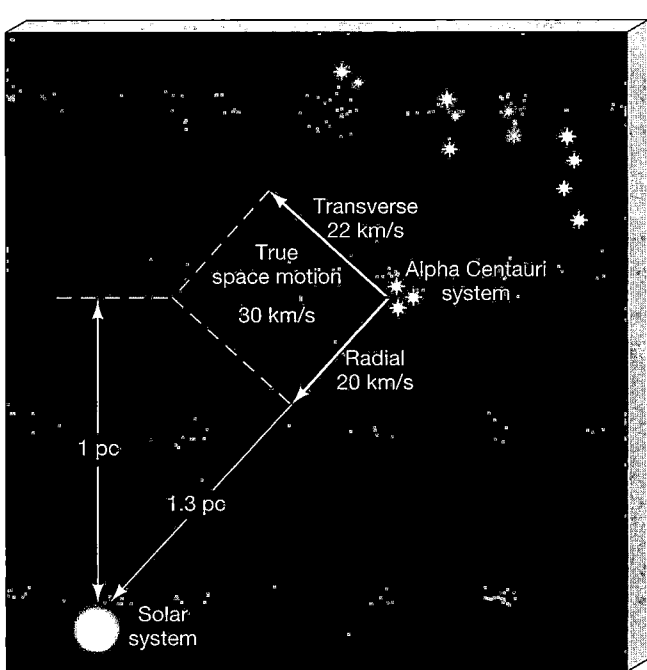

Figure 17.4 Real Space Motion The motion of the Alpha Centauri star system drawn relative to our solar system. The transverse component of the velocity has been determined by observing the system's proper motion. The radial component is measured using the Doppler shift of lines in Alpha Centauri's spectrum. The true space velocity, indicated by the red arrow, results from the combination of the two.

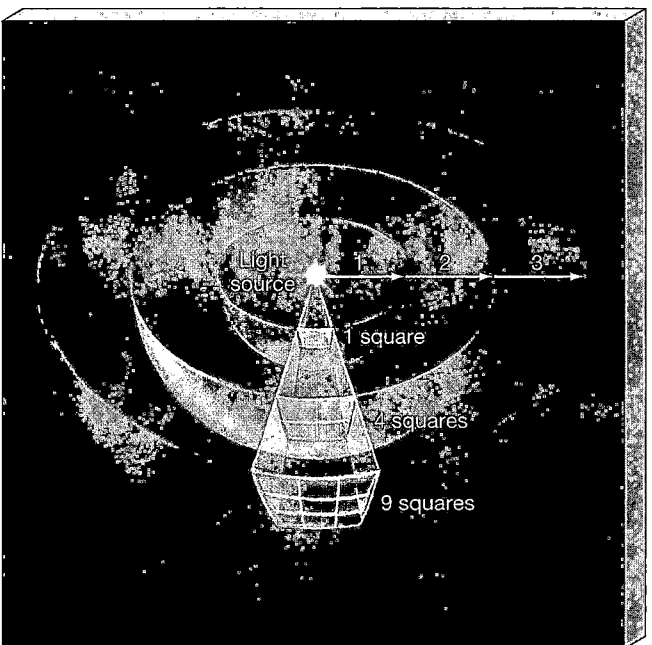

Figure 17.5 Inverse-square Law As it moves away from a source such as a star, radiation is steadily diluted while spreading over progressively larger surface areas (depicted here as sections of spherical shells). Thus, the amount of radiation received by a detector (the source's apparent brightness) varies inversely as the square of its distance from the source.

DISCOVERY 17-1

The *Hipparcos* Mission

In 1989, shortly before the large, complex, and expensive *Hubble Space Telescope* was launched by NASA into Earth orbit, a small, simple, and cheap satellite was also deployed by the European Space Agency (ESA). Called *Hipparcos*—a tortured attempt to honor the ancient Greek astronomer Hipparchus of Nicea, who made the first known star map, by creating an acronym out of *High Precision Parallax Collecting Satellite*—this spacecraft was initially thought to be lost in space. *Hipparcos* went into the wrong orbit, its onboard engine failed to fire, and its solar panels were heavily pelted with particles in Earth's Van Allen belts. However, before the mission ended in 1993, the engineers managed to control the errant spacecraft, and ESA scientists revamped its mission. Nearly a decade after its launch, its massive database was cataloged and released.

The heart of *Hipparcos* (shown here before launch) was a small 29-cm mirror designed to determine the positions of stars very accurately. By measuring those positions from two vantage points in orbit (above Earth's turbulent atmosphere), the satellite was able to determine stellar parallaxes with about 10 times greater accuracy than is possible using telescopes on the ground. Since volume scales as the cube of the size, *Hipparcos*'s prime mission was to measure accurate distances for roughly a million stars out to some 200 pc.

In addition to measuring distances to nearby stars with unprecedented accuracy, some other highlights of the *Hipparcos* mission included the following:

■ In about a million years, the star named Gliese 710 may pose a hazard to planet Earth. Currently this red dwarf star lies 20 pc away in the direction of the Ophiuchus constellation, but its rapid line-of-sight motion will eventually take it within 0.3 pc (less than a light-

year) of the Sun. As it passes perhaps right through the Oort cloud of comets surrounding our solar system, this dim dwarf star will shine as brilliantly as the red giant Betelgeuse does now. It may also perturb the Oort cloud significantly, sending a shower of comets

(ESA)

spread more thinly, or "diluted," as it expands into space. Because the area of a sphere grows as the square of the radius, the energy per unit area—the star's apparent brightness, as seen by our eye or our telescope—is inversely proportional to the square of the distance from the star. Doubling the distance from a star makes it appear 2^2, or 4, times dimmer. Tripling the distance reduces the apparent brightness by a factor of 3^2, or 9, and so on.

Of course, the star's luminosity also affects its apparent brightness. Doubling the luminosity doubles the energy crossing any spherical shell surrounding the star and hence doubles the apparent brightness. We can therefore say that the apparent brightness of a star is directly proportional to the star's luminosity and inversely proportional to the square of its distance:

$$\text{apparent brightness (energy flux)} \propto \frac{\text{luminosity}}{\text{distance}^2}.$$

Thus, two identical stars can have the same apparent brightness if (and only if) they lie at the same distance from Earth. However, as illustrated in Figure 17.6, two nonidentical stars can also have the same apparent brightness if the more luminous one lies farther away. A bright star (that is, one having large apparent brightness) is a powerful emitter of radiation (high luminosity), is near Earth, or both. A faint star (small apparent brightness) is a weak emitter (low luminosity), is far from Earth, or both.

Determining a star's luminosity is a twofold task. First, the astronomer must determine the star's apparent brightness by measuring the amount of energy detected through a telescope in a given amount of time. Second, the star's distance must be measured—by parallax for nearby stars and by other means (to be discussed later) for more distant stars. The luminosity can then be found using the inverse-square law. This is basically the same reasoning we used earlier in our discussion of how astronomers measure the solar lumi-

hurtling in our direction. Astronomers estimate that no star has come this close to the Sun in the past 10 million years or so.

■ Parallax calibrates other "yardsticks" used by astronomers to measure distances in increasingly larger realms and eventually to estimate the size and scale of the entire observable universe. As a result, Hipparcos has affected distances far beyond those it was able to measure directly in our stellar neighborhood. In particular, recalculation of distances to a certain class of variable stars (called Cepheids—see Chapter 23) has put astronomy's entire distance scale on a much firmer foundation.

■ Because of these recalibrations, some distances have changed a little, and others need substantial revision. For example, the Large Magellanic Cloud (Chapter 24) is now thought to be about 10 percent farther away than had previously been estimated. Likewise, the distance to the Andromeda Galaxy has been increased by about a third; its distance is now pegged at 0.8 Mpc (or 2.5 million light-years). ∞ (Sec. 3.1) All distances throughout this text reflect the new values as determined by *Hipparcos*.

■ The new data have also increased the size, and hence the age, of the universe. The best number for that age, based in part on *Hipparcos* measurements, now seems to be about 12–13 billion years (Chapter 24). This is the age used in this book.

■ *Hipparcos* also measured the colors (i.e., temperatures) and apparent brightnesses (and hence luminosities, since the distances are known) of the stars it surveyed. Based on these new data, a major revision

of the Hertzsprung–Russell diagram (see Figures 17.13–17.16) implies that the oldest stars are a little younger than was previously thought. This new analysis of stellar evolutionary tracks seems to reconcile the age of the universe with that of the oldest stars (Chapter 20). Prior to the *Hipparcos* mission, astronomers had faced the embarrassing problem that some stars seemed to be older than the universe itself. No more—this major breakthrough in a heretofore log-jammed subject may well be *Hipparcos*'s greatest contribution to astronomy.

The *Hipparcos* mission was not glamorous. This little satellite sent back no pictures for European citizens (who paid for it) and others to marvel at, and it grabbed few headlines in the world's press. Rather, its catalogs, comprising some 1000 gigabytes of data (or about 1600 CD-ROMs), resemble huge phonebooks—unexciting to look at, but of enormous value for accessing the vital statistics of stars and the entire universe beyond.

Following *Hipparcos*, both NASA and ESA have ambitious plans to expand enormously the scope of stellar measurements. NASA's *Full-Sky Astrometric Mapping Explorer* mission (*FAME*), scheduled for launch in 2004, will obtain accurate parallax and color measurements for roughly 40 million stars within 2500 pc of the Sun. NASA's *Space Interferometry Mission (SIM)* and ESA's *GAIA* project, planned for around 2010, will each have the astonishing range of 25,000 pc—covering our entire Galaxy and encompassing a billion of its brightest stars! In a time span of just three decades the fundamental stellar database upon which almost all of astronomy depends will have increased in size by a factor of a million. The results may be nothing short of revolutionary.

Figure 17.6 Luminosity Two stars A and B of different luminosity can appear equally bright to an observer on Earth if the brighter star B is more distant than the fainter star A.

nosity (in our new terminology, the solar constant is just the apparent brightness of the Sun). ∞ (Sec. 16.1)

THE MAGNITUDE SCALE

Instead of measuring apparent brightness in SI units (for example, watts per square meter, the unit used for the solar constant in Chapter 16), optical (and near-optical) astronomers find it more convenient to work in terms of a construct called the **magnitude scale**. The scale dates from the second century B.C., when the Greek astronomer Hipparchus ranked the naked-eye stars into six groups. The brightest stars were categorized as first magnitude. The next brightest stars were labeled second magnitude, and so on, down to the faintest stars visible to the naked eye, which were classified as sixth magnitude. The range 1 (brightest) through 6 (faintest) spanned all the stars known to the ancients. Notice that a *large* magnitude means a *faint* star.

When astronomers began using telescopes with sophisticated detectors to measure the light received from stars, they quickly discovered two important facts about the magnitude scale. First, the 1–6 magnitude range defined by Hipparchus spans about a factor of 100 in apparent brightness—a first-magnitude star is approximately 100 times brighter than a sixth-magnitude star. Second, the physiological characteristics of the human eye are such that each magnitude change of 1 corresponds to a factor of about 2.5 in apparent brightness. In other words, to the human eye a first-magnitude star is roughly 2.5 times brighter than a second-magnitude star, which is roughly 2.5 times brighter than a third-magnitude star, and so on. (By combining factors of 2.5, you can confirm that a first-magnitude star is indeed $(2.5)^5 \approx 100$ times brighter than a sixth-magnitude star.)

Modern astronomers have modified and extended the magnitude scale in a number of ways. First, we now *define* a change of 5 in the magnitude of an object (going from magnitude 1 to magnitude 6, say, or from magnitude 7 to magnitude 2) to correspond to *exactly* a factor of 100 in apparent brightness. Second, because we are really talking about apparent (rather than absolute) brightnesses, the numbers in Hipparchus's ranking system are called **apparent magnitudes**. Third, the scale is no longer limited to whole numbers—a star of apparent magnitude 4.5 is intermediate in apparent brightness between a star of apparent magnitude 4 and one of apparent magnitude 5. Finally, magnitudes outside the range 1–6 are allowed—very bright objects can have apparent magnitudes much less than 1, and very faint objects can have apparent magnitudes far greater than 6.

Figure 17.7 illustrates the apparent magnitudes of some astronomical objects, ranging from the Sun, at -26.8, to the faintest object detectable by the *Hubble* or *Keck* telescopes, an object having an apparent magnitude of 30—about as faint as a firefly seen from a distance equal to Earth's diameter.

Apparent magnitude measures a star's apparent brightness when seen at the star's actual distance from the Sun. To compare intrinsic, or absolute, properties of stars, however, astronomers imagine looking at all stars from a standard distance of 10 pc. There is no particular reason to use 10 pc—it is simply convenient. A star's **absolute magnitude** is its apparent magnitude when it is placed at a distance of 10 pc from the observer. Because the distance is fixed in this definition, absolute magnitude is a measure of a star's absolute brightness, or luminosity.

When a star farther than 10 pc away from us is moved to a point 10 pc away, its apparent brightness increases and hence its apparent magnitude decreases. Stars more than 10 pc from Earth therefore have apparent magnitudes that are greater than their absolute magnitudes. For example, if a star

at a distance of 100 pc were moved to the standard 10-pc distance, its distance would decrease by a factor of 10, so (by the inverse-square law) its apparent brightness would *increase* by a factor of $10^2 = 100$. Its apparent magnitude (by definition) would therefore decrease by 5. In other words the star's absolute magnitude exceeds its apparent magnitude by 5.

For stars closer than 10 pc, the reverse is true. An extreme example is our Sun. Because of its proximity to Earth, it appears very bright and thus has a large negative apparent magnitude (Figure 17.7). However, the Sun's absolute magnitude is 4.85. If the Sun were moved to a distance of 10 pc from Earth, it would be only slightly brighter than the faintest stars visible in the night sky.

Knowledge of a star's apparent magnitude and distance allows us to compute its absolute magnitude. As discussed further in *More Precisely 17-1*, the numerical difference between a star's absolute and apparent magnitudes is a measure of the distance to the star.

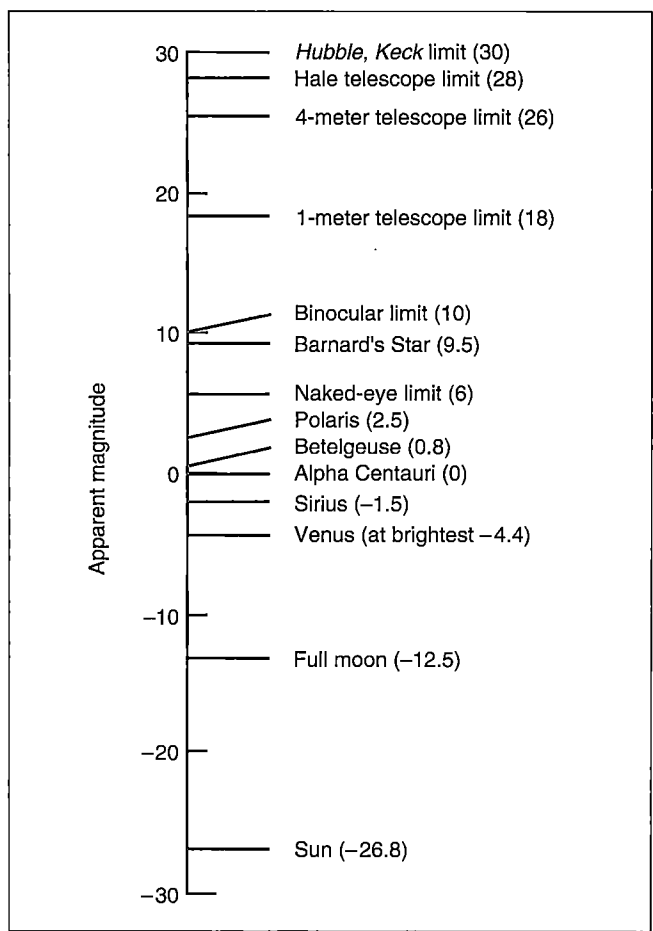

Figure 17.7 Apparent Magnitude This graph illustrates the apparent magnitudes of some astronomical objects. The original magnitude scale was defined so that the brightest stars in the night sky had magnitude 1, and the faintest stars visible to the naked eye had magnitude 6. It has since been extended to cover much brighter and much fainter objects. An increase of 1 in apparent magnitude corresponds to a decrease in apparent brightness by a factor of approximately 2.5.

☑️ Concept Check

■ Two stars are observed to have the same apparent magnitude. Based on this information, what, if anything, can be said about their luminosities?

17.4 Stellar Temperatures

🔆 Looking at the night sky, you can tell at a glance which stars are hot and which are cool. In Figure 17.8, which shows the constellation Orion as it appears through a small telescope, the colors of the cool red star Betelgeuse (α) and the hot blue star Rigel (β) are clearly evident. Note that these colors are intrinsic properties of the stars, and have *nothing* to do with Doppler redshifts or blueshifts. However, to obtain these stars' temperatures (3000 K for Betelgeuse and 15,000 K for Rigel), more detailed observations are required.

COLOR AND THE BLACKBODY CURVE

Astronomers can determine a star's surface temperature by measuring its apparent brightness (radiation intensity) at several frequencies, then matching the observations to the appropriate blackbody curve. ∞ (Sec. 3.4) In the case of the Sun, the theoretical curve that best fits the emission describes a 5800-K emitter. ∞ (Sec. 16.1) The same technique works for any other star, regardless of its distance from Earth.

Because the basic shape of the blackbody curve is so well understood, astronomers can estimate a star's temperature using as few as *two* measurements at selected wavelengths (which is fortunate, as detailed spectra of faint stars are often difficult and time consuming to obtain). This is accomplished through the use of telescope filters that block out all radiation except that within specific wavelength ranges. For example, a B (blue) filter rejects all radiation except for a certain range of violet to blue light. Defined by international agreement to extend from 380 to 480 nm, this range corresponds to wavelengths to which photographic film happens to be most sensitive. Similarly, a V (visual) filter passes only radiation within the 490 to 590 nm range (green to yellow), corresponding to that part of the spectrum to which human eyes are particularly sensitive. Many other filters are also in routine use—a U (ultraviolet) filter covers the near ultraviolet, and infrared filters span longer-wavelength parts of the spectrum.

Figure 17.9 shows how the B and V filters admit different amounts of light for objects of different temperatures. In curve (a), corresponding to a very hot 30,000-K emitter,

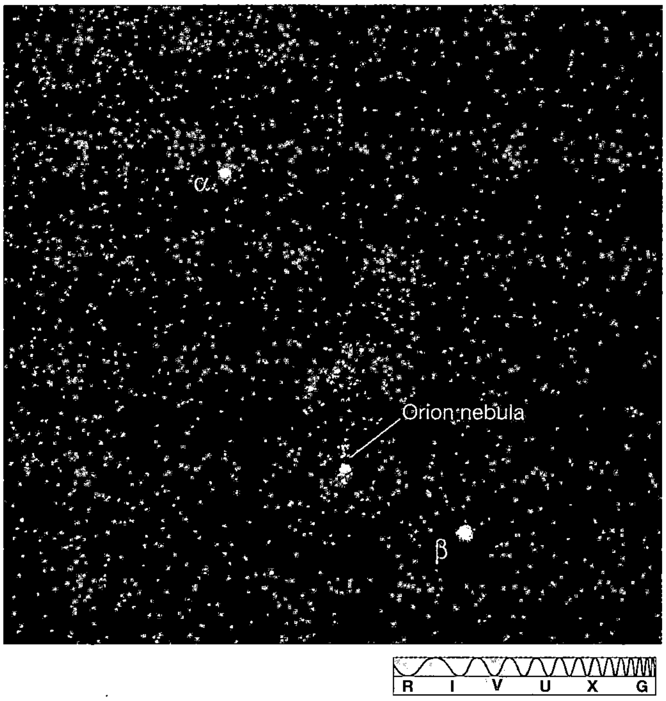

Figure 17.8 Orion The constellation Orion, as it would appear through a small telescope or binoculars. The different colors of the member stars are easily distinguished. The bright red star at the upper left is Betelgeuse (α); the bright blue-white star at the lower right is Rigel (β). (Compare with Figure 1.6.) The scale of the photograph is about 3° across. (*J. Sanford/Astrostock-Sanford*)

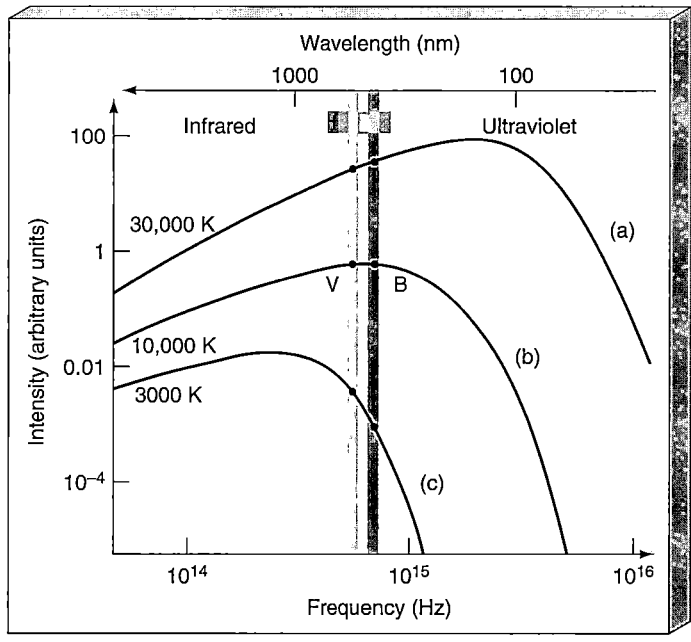

Figure 17.9 Blackbody Curves The locations of the B (blue) and V (visual) filters, with blackbody curves for three different temperatures. Star (a) is very hot—30,000 K—so its B intensity is considerably greater than its V intensity. Star (b) has about the same B and V readings and appears white. Its temperature is about 10,000 K. Star (c) is red; its V intensity greatly exceeds the B value, and its temperature is 3000 K.

MORE PRECISELY 17-1

More on the Magnitude Scale

All the discussion of luminosity, brightness, and energy flux (Section 17.4) can equally well be written in terms of magnitudes. Here we present a little more detail on the "magnitude versions" of two important topics—stellar luminosity and the inverse-square law.

Recall that a star's *absolute magnitude* is the apparent magnitude we would measure if the star were located at a standard distance of 10 pc from us. Absolute magnitude is equivalent to luminosity—an intrinsic property of a star. Given that the Sun's absolute magnitude is 4.85, we can construct a "conversion chart" (shown) relating these two quantities. Since an *increase* in

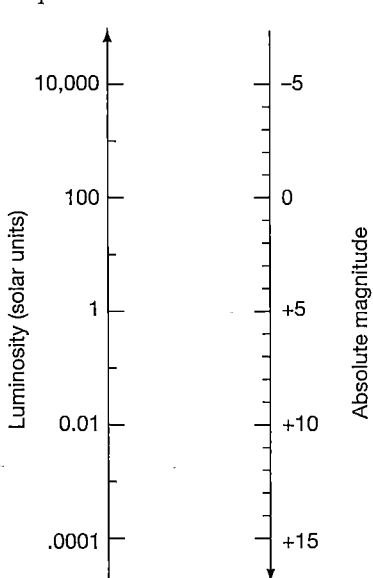

brightness by a factor of 100 corresponds, by definition, to a *reduction* of a star's magnitude by five units, it follows that a star with luminosity 100 times that of the Sun has absolute magnitude $4.85 - 5 = -0.15$, and a 0.01 solar luminosity star has absolute magnitude $4.85 + 5 = 9.85$. We can fill in the gaps by realizing that one magnitude corresponds to a factor of $100^{1/5} \approx 2.512$, two magnitudes to $100^{2/5} \approx 6.310$, and so on. A factor of 10 in brightness corresponds to 2.5 magnitudes. You can use this chart to convert between solar luminosities and absolute magnitudes in many of the figures in this and later chapters.

EXAMPLE: Let's calculate the luminosity (in solar units) of a star having absolute magnitude M (the conventional symbol for absolute magnitude, not to be confused with mass!). The star's absolute magnitude differs from that of the Sun by $(M - 4.85)$ magnitudes so, following the above reasoning, the luminosity L differs from the solar luminosity by a factor of $100^{-(M-4.85)/5}$, or $10^{-2(M-4.85)/5}$. We can therefore write

$$L \text{ (solar units)} = 10^{-0.4(M - 4.85)}.$$

Plugging in some numbers (taken from Appendix 3), we find that the Sun, with $M = 4.85$ of course has $L = 10^0 = $

---

considerably more radiation (about twice as much) is received through the B filter than through the V filter. In curve (b), the temperature is 10,000 K, and the B and V intensities are about the same. In the cool 3000-K curve (c), about five times more energy is received in the V range than in the B range. In each case, it is possible to reconstruct the entire blackbody curve on the basis of only those two measurements because no other blackbody curve can be drawn through both measured points. To the extent that a star's spectrum is well approximated as a blackbody, measurements of the B and V intensities are enough to specify the star's blackbody curve and thus yield its surface temperature.

Operationally, astronomers go about determining a star's temperature by measuring its **color index** (often referred to simply as its "color"), defined as the difference between the star's apparent magnitudes measured through the B and V filters (or, equivalently, the *ratio* of its B to V intensities). Table 17.1 lists a more detailed description of the color index, the surface temperature derived from it, the dominant color perceived in the absence of filters, and some stellar examples. This type of non-spectral-line analysis, in which a star's intensity is measured through each of a set of standard filters, is known as **photometry** (literally meaning "light measurement").

STELLAR SPECTRA

Astronomers often use a more detailed scheme to classify stellar properties, incorporating additional knowledge of stellar physics obtained through spectroscopy. Figure 17.10 compares the spectra of several different stars, arranged in order of decreasing surface temperature (as determined from measurements of their colors). All the spectra extend from 400 to 650 nm, and each shows a series of dark absorption lines superimposed on a background of continuous color, like those of the Sun. ∞ (Sec. 16.3) However, the precise patterns of lines show many differences. Some stars display strong lines in the long-wavelength part of the spectrum (to the left in Figure 17.10). Other stars have their strongest lines at short wavelengths (to the right). Still others show strong absorption lines spread across the whole visible spectrum. What do these differences tell us?

Although spectral lines of many elements are present with widely varying strengths, the differences among the spectra in Figure 17.10 are not due to differences in composition. Detailed spectral analysis indicates that the seven stars shown have very similar elemental abundances—all are more or less solar in makeup. Instead, the differences

1. Sirius A, with $M = 1.4$ has luminosity $10^{1.38} = 24$ solar units, Barnard's Star, with $M = 13.2$ has luminosity $10^{-3.34} = 4.5 \times 10^{-4}$ solar units, Betelgeuse has $M = -5.5$ and luminosity 14,000 Suns, and so on. We can also invert the above relationship and write

$$M = 4.85 - 2.5 \log_{10} L \text{ (solar units)}$$

(where the *logarithm* function is defined by the property that if $a = \log_{10}(b)$, then $b = 10^a$). Thus, a star (Vega) of luminosity 55 times that of the Sun has $M = 4.85 - 2.5 \log_{10} 55 = 4.85 - 2.5(1.74) = 0.5$, one of 0.3 solar luminosities (ε Eridani) has $M = 4.85 - 2.5(-0.52) = 6.1$, and so on.

The *inverse-square law* can also be recast in these terms. Increasing the distance to a star by a factor of 10 decreases its energy flux by a factor of 100 (by the inverse-square law) and hence increases its apparent magnitude by five units. Increasing the distance by a factor of 100 increases the apparent magnitude by 10. Since absolute magnitude is a measure of the energy flux received from a star at a distance of 10 pc, denoting the apparent magnitude by m and the distance by D, we can write:

$$m - M = 5 \log_{10} \left(\frac{D}{10 \text{ pc}} \right)$$

The difference $m - M$ between the apparent and absolute magnitudes of an object is usually referred to as its *distance modulus*. Notice once again that, if D is greater than 10 pc, then m is greater than M, and vice versa. In terms of distance modulus, we can rearrange the preceding equation and rewrite it as:

$$D = 10 \text{ pc} \times 10^{(m-M)/5}.$$

This doesn't look much like the inverse-square law presented in the text! Nevertheless, it contains exactly the same information, and it is in widespread use in this form throughout astronomy.

EXAMPLE: Again using stars listed in Appendix 3, the star Rigel is observed to have apparent magnitude $m = 0.14$ and has a measured distance (using parallax) $D = 240$ pc. From the preceding equation, its absolute magnitude then is $M = 0.14 - 5 \log_{10} 24 = -6.8$, corresponding to a luminosity 46,000 times that of the Sun. Conversely, the star Rigel Kentaurus (Alpha Centauri) has absolute magnitude $M = 4.4$ and is observed to have apparent magnitude $m = 0$. Its distance must therefore be $D = 10 \text{ pc} \times 10^{(0-4.4)/5} = 1.3$ pc (in agreement with the result obtained by parallax!).

are due almost entirely to the stars' *temperatures*. The spectrum at the top of Figure 17.10 is exactly what we would expect from a star with solar composition and a surface temperature of about 30,000 K, the second from a 20,000-K star, and so on, down to the 3000-K star at the bottom.

The spectra of stars with surface temperatures exceeding 25,000 K usually show *strong* absorption lines of singly ionized helium and multiply ionized heavier elements, such

as oxygen, nitrogen, and silicon (these latter lines are not shown in the spectra of Figure 17.10). These strong lines are not seen in the spectra of cooler stars because only very hot stars can excite and ionize these tightly bound atoms. In contrast, the hydrogen absorption lines in the spectra of very hot stars are relatively *weak*. The reason is not a lack of hydrogen—it is by far the most abundant element in all stars. At these high temperatures, however, much of the

TABLE 17.1 Stellar Colors and Temperatures

COLOR INDEX		APPROXIMATE SURFACE TEMPERATURE (K)	COLOR	FAMILIAR EXAMPLES
B magnitude-V magnitude	B intensity/V intensity			
-0.31	1.3	30,000	blue-violet	Mintaka (δ Orionis)
-0.24	1.2	20,000	blue	Rigel
0.00	1.00	10,000	white	Vega, Sirius
0.35	0.72	7000	yellow-white	Canopus
0.65	0.55	6000	yellow	Sun, Alpha Centauri
1.2	0.33	4000	orange	Arcturus, Aldebaran
1.7	0.21	3000	red	Betelgeuse, Barnard's Star

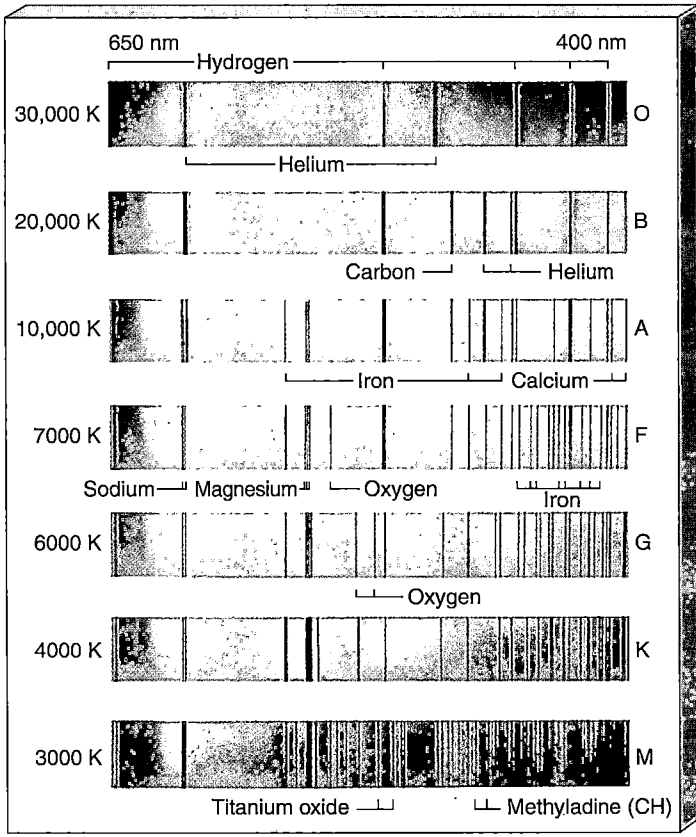

Figure 17.10 Stellar Spectra Comparison of spectra observed for seven different stars having a range of surface temperatures. The hottest stars, at the top, show lines of helium and multiply ionized heavy elements. In the coolest stars, at the bottom, helium lines are not seen, but lines of neutral atoms and molecules are plentiful. At intermediate temperatures, hydrogen lines are strongest. The actual compositions of all seven stars are about the same.

hydrogen is ionized, so there are few intact hydrogen atoms to produce strong spectral lines. What's more, the few hydrogen atoms that have managed to retain their single electron are mostly in such highly excited states that their spectral lines are invisible. ⟨∞ (Sec. 4.4; *More Precisely 4-2*) Hydrogen lines are strongest in stars having intermediate surface temperatures of around 10,000 K. This temperature is just right for electrons to move frequently between hydrogen's second and third orbitals, producing the characteristic red Hα (hydrogen alpha) line at 656.3 nm (see Section 4.2, particularly Figure 4.10). Lines of tightly bound atoms, elements such as helium, oxygen, and nitrogen, which need lots of energy for excitation or ionization, are rarely observed in the spectra of 10,000-K stars, whereas lines from more loosely bound atoms—such as calcium and titanium—are relatively common.

Hydrogen lines are again weak in stars with surface temperatures below about 4000 K. However, unlike the hottest stars, in which most hydrogen atoms are highly excited or ionized, the hydrogen lines in cool stars are weak because most of the electrons are in the ground state. The most intense lines in the spectra of these stars are due to weakly excited heavy atoms; no lines from ionized elements are seen. In fact, the average energy of the photons leaving the surface of the coolest stars is less than that needed to destroy some molecules, and many of the observed absorption lines are produced by molecules rather than by atoms (note the lines of the molecules titanium oxide and methyladine at the bottom of Figure 17.10).

SPECTRAL CLASSIFICATION

Stellar spectra like those shown in Figure 17.10 were obtained for many stars well before the start of the twentieth century as observatories around the world amassed spectra from stars in both hemispheres of the sky. Between 1880 and 1920, researchers correctly identified some of the observed spectral lines on the basis of comparisons between stellar lines and those obtained in the laboratory. Those workers, though, had no firm understanding of how the lines were produced. Modern atomic theory had not yet been developed, so the correct interpretation of the line strengths, as just described, was impossible at the time.

Lacking full understanding of how atoms produce spectra, early workers classified stars primarily according to their hydrogen-line intensities. They adopted an alphabetic A, B, C, D, E. . . scheme in which A stars, with the strongest hydrogen lines, were thought to have more hydrogen than did B stars, and so on. The classification extended as far as the letter P.

In the 1920s, scientists began to understand the intricacies of atomic structure and the causes of spectral lines. Astronomers quickly realized that stars could be more meaningfully classified according to their surface temperature. Instead of adopting an entirely new scheme, however, they chose to shuffle the existing alphabetical categories—those based on the strength of hydrogen lines—into a new sequence based on temperature. In the modern scheme, the hottest stars are designated O, because they have very weak absorption lines of hydrogen and were classified toward the end of the original scheme. In order of decreasing temperature, the original letters now run O, B, A, F, G, K, M. (The other letter classes have been dropped.) These stellar designations are called **spectral classes** (or *spectral types*). Use the time-honored (and politically incorrect) mnemonic "Oh, Be A Fine Girl, Kiss Me" to remember them in the correct order.*

---

*Recently, some astronomers have proposed the addition of two new spectral classes—L and T—for low-mass, low-temperature stars whose odd spectra distinguish them from the M-class stars in the current scheme. For now, at least, the new classification has not been widely adopted. Astronomers are still uncertain whether these new objects are "true" stars, fusing hydrogen into helium in their cores, or if they are "brown dwarfs" (see Chapter 20) that never achieved high enough central temperatures for fusion to begin.

TABLE 17.2 Stellar Spectral Classes

SPECTRAL CLASS	APPROXIMATE SURFACE TEMPERATURE (K)	PROMINENT ABSORPTION LINES	FAMILIAR EXAMPLES
O	30,000	Ionized helium strong; multiply ionized heavy elements; hydrogen faint	Mintaka (O9)
B	20,000	Neutral helium moderate; singly ionized heavy elements; hydrogen moderate	Rigel (B8)
A	10,000	Neutral helium very faint; singly ionized heavy elements; hydrogen strong	Vega (A0), Sirius (A1)
F	7000	Singly ionized heavy elements; neutral metals; hydrogen moderate	Canopus (F0)
G	6000	Singly ionized heavy elements; neutral metals; hydrogen relatively faint	Sun (G2), Alpha Centauri (G2)
K	4000	Singly ionized heavy elements; neutral metals strong; hydrogen faint	Arcturus (K2), Aldebaran (K5)
M	3000	Neutral atoms strong; molecules moderate; hydrogen very faint	Betelgeuse (M2), Barnard's Star (M5)

Astronomers further subdivide each lettered spectral classification into 10 subdivisions, denoted by the numbers 0–9. By convention, the lower the number, the hotter the star. Thus, for example, our Sun is classified as a G2 star (a little cooler than G1 and a little hotter than G3), Vega is a type A0, Barnard's Star is M5, Betelgeuse is M2, and so on. Table 17.2 lists the main properties of each stellar spectral class for the stars presented in Table 17.1.

We should not underestimate the importance of the early work in classifying stellar spectra. Even though the original classification was based on erroneous assumptions, the painstaking accumulation of large quantities of accurate data paved the way for rapid improvements in understanding once a theory came along that explained the observations.

✓ Concept Check

■ Why does a star's spectral classification depend on its temperature?

17.5 Stellar Sizes

DIRECT AND INDIRECT MEASUREMENTS

5 Most stars are unresolved points of light in the sky, even when viewed through the largest telescopes. However, a few are big enough, bright enough, and close enough to allow us to measure their sizes *directly*. One well-known example is the bright red star Betelgeuse (see Figure 17.8), a prominent member of the Orion constellation. In an optical technique known as *speckle interfer-*

ometry, many short-exposure images of a star, each too brief for Earth's turbulent atmosphere to smear it out into a seeing disk, are combined to make a high-resolution map of the star's surface. In some cases, the results are detailed enough to allow a few surface features to be distinguished (Figure 17.11). As adaptive optics techniques continue to improve, it is becoming possible to combine the individual images in real time, again allowing very-high-resolution stellar images to be made. ∞ (Sec. 5.3) Once a star's angular size has been measured, if its distance is also known we can determine its radius by simple geometry. ∞ (Sec. 1.5) With a distance of 130 pc and an angular diameter of 0.045″, Betelgeuse's radius is 630 times that of the Sun. The sizes of a few dozen stars have been measured in this way.

Most stars are too distant or too small for such direct measurements to be possible. Instead, their sizes must be inferred by indirect means, using the radiation laws. ∞ (Sec. 3.4) The radiation emitted by a star is governed by the Stefan-Boltzmann law, which states that the energy emitted *per unit area* per unit time increases as the fourth power of the star's surface temperature. ∞ (*More Precisely 3-2*) To determine the star's luminosity, we must multiply by its surface area—large bodies radiate more energy than do small bodies having the same temperature. Because surface area is proportional to the square of the radius, we have

$$\text{luminosity} \propto \text{radius}^2 \times \text{temperature}^4.$$

This **radius–luminosity–temperature relationship** is important because it demonstrates that knowledge of a star's luminosity and temperature can yield an estimate of its radius—an *indirect* determination of stellar size.

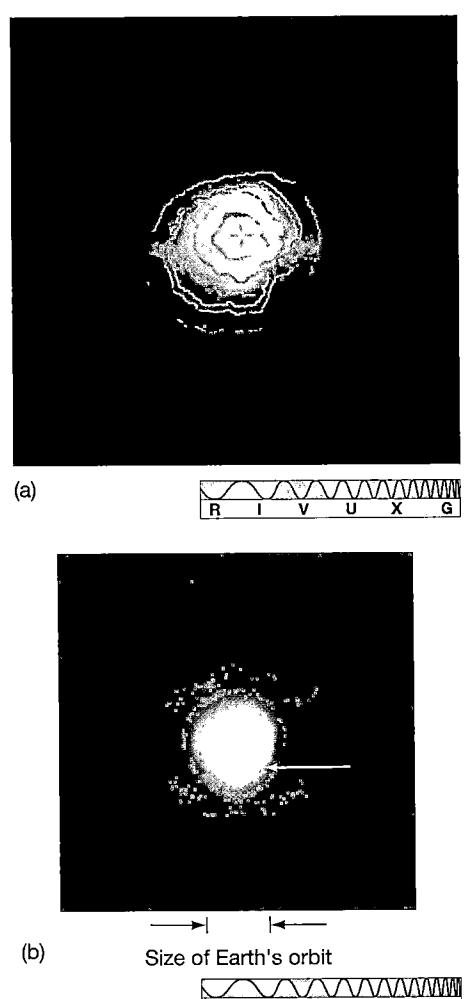

(a)

(b) Size of Earth's orbit

Figure 17.11 Betelgeuse (a) The swollen star Betelgeuse (shown here in false color) is close enough for us to directly resolve its size, along with some surface features thought to be storms similar to those that occur on the Sun. Betelgeuse is such a huge star (some 600 times the size of the Sun) that its photosphere spans roughly the size of Mars's orbit. Most of the surface features discernible here are larger than the entire Sun. (The dark lines are drawn contours, not part of the star itself.) (b) An ultraviolet view of Betelgeuse, in false color, as seen by a European camera on board the *Hubble Space Telescope*. The bright spot at the bottom right of the star (white arrow) is more than 10 times bigger than Earth and at least 2000 K hotter than the surrounding 3000 K photosphere. Most astronomers regard it as a storm of some sort, perhaps similar to those that occur during the active phase of our Sun. *(NOAO; NASA/ESA)*

GIANTS AND DWARFS

Let's consider some examples to clarify these ideas. The star known as Mira (Omicron Ceti) has a surface temperature of about 3000 K and a luminosity of 1.6×10^{29} W. Thus, its surface temperature is half, and its luminosity about 400 times, the corresponding quantities for our Sun.

The radius–luminosity–temperature relationship (see *More Precisely* 17-2) then implies that the star's radius is 80 times that of our Sun. If our Sun were this large, its photosphere would extend as far as the orbit of Mercury. A star as large as Mira is known as a *giant*. More precisely, **giants** are stars having radii between 10 and 100 times that of the Sun. Even larger stars, ranging up to 1000 solar radii in size, are known as **supergiants**. Since any 3000 K object is red in color, Mira is known as a **red giant**.

Now consider Sirius B—a faint companion to Sirius A, the brightest star in the night sky. Sirius B's surface temperature is roughly 24,000 K, four times that of the Sun. Its total luminosity is 10^{25} W, about 0.04 times the solar value. Again using the radius–luminosity–temperature relationship, we obtain a radius of 0.01 solar radii—roughly the size of Earth. Sirius B is much hotter, but smaller and far less luminous than our Sun. Such a star is known as a *dwarf*. In astronomical parlance, the term **dwarf** refers to any star of radius comparable to or smaller than the Sun (including the Sun itself). Because any 24,000 K object glows white, Sirius B is an example of a **white dwarf**.

The radii of the vast majority of stars (mostly measured using the radius–luminosity–temperature relationship) range from less than 0.01 to over 100 times the radius of the Sun. Figure 17.12 illustrates the estimated sizes of a few well-known stars.

✅ Concept Check

■ Can we measure the radius of a star without knowing its distance?

17.6 The Hertzsprung–Russell Diagram

⑥ Astronomers use luminosity and surface temperature to classify stars in much the same way that height and weight serve to classify the bulk properties of human beings. We know that people's height and weight are well correlated—tall people tend to weigh more than short ones. We might naturally wonder if the two basic stellar properties are also related in some way.

Figure 17.13 plots luminosity versus temperature for a few well-known stars. Figures of this sort are called *Hertzsprung–Russell diagrams*, or **H–R diagrams**, after Danish astronomer Ejnar Hertzsprung and U.S. astronomer Henry Norris Russell, who independently pioneered the use of such plots in the second decade of the twentieth century. The vertical scale, expressed in units of the solar luminosity (3.9×10^{26} W), extends over a large range, from 10^{-4} to 10^4; the Sun appears right in the middle of the luminosity range, at a luminosity of 1. Surface temperature is plotted on the horizontal axis, although in

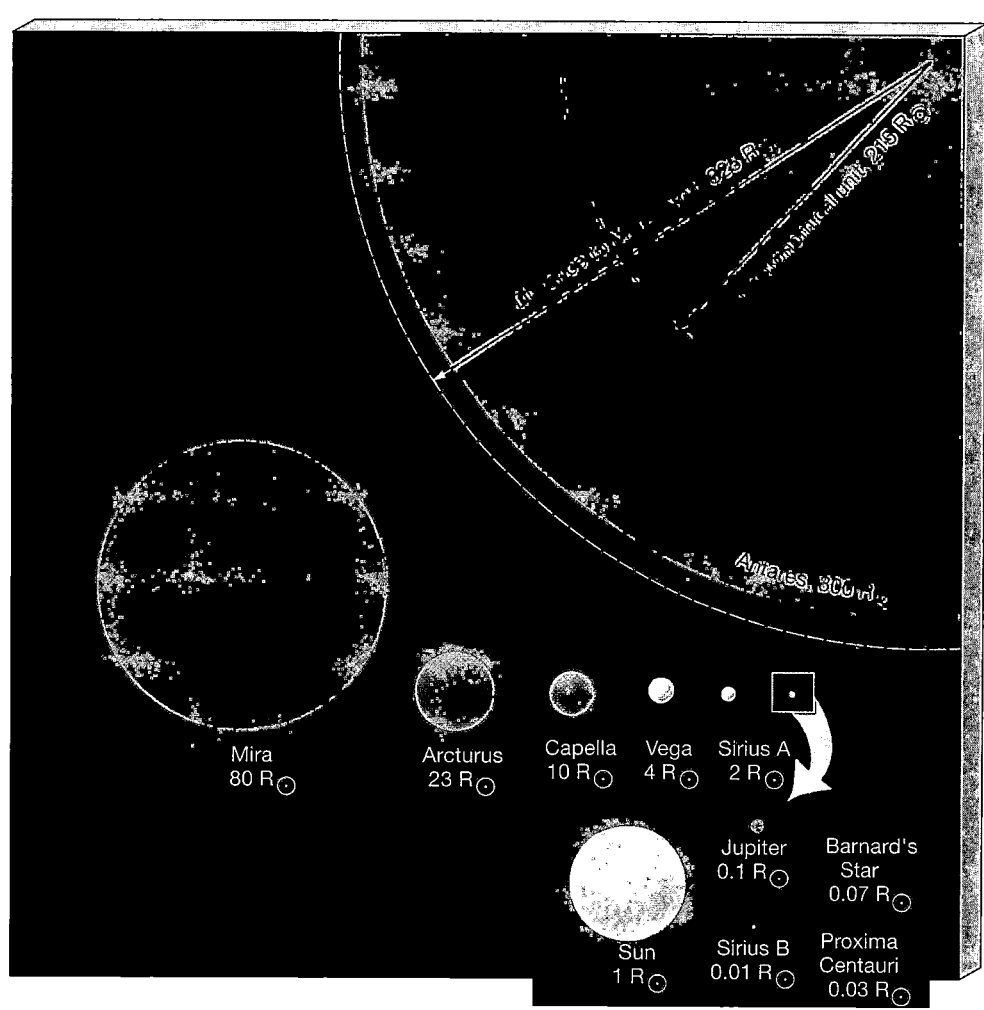

Figure 17.12 Stellar Sizes
Star sizes vary greatly. Shown here are the estimated sizes of several well-known stars, including a few of those discussed in this chapter. Only part of the red-giant star Antares can be shown on this scale, and the supergiant Betelgeuse would fill the entire page.

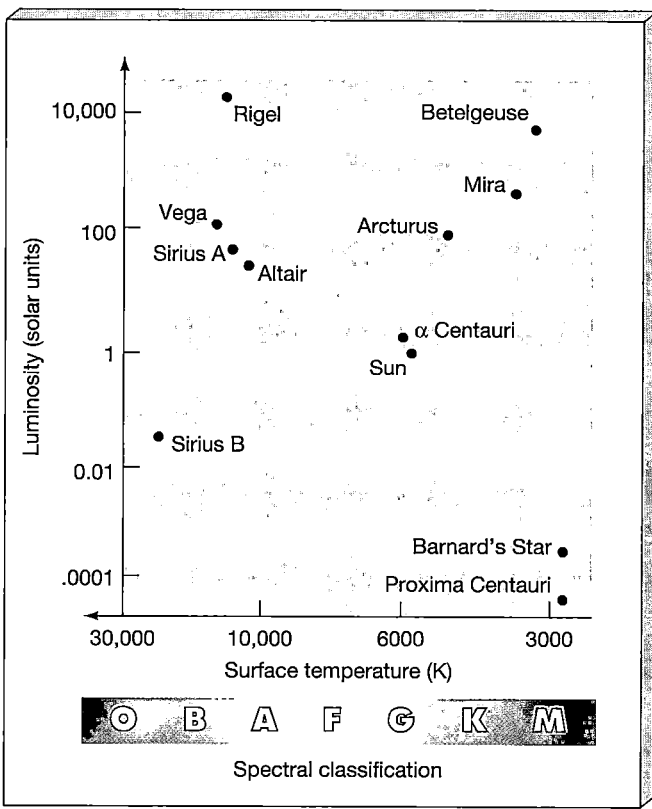

Figure 17.13 H-R Diagram of Prominent Stars
A plot of luminosity against surface temperature (or spectral classification), known as an H–R diagram, is a useful way to compare stars. Plotted here are the data for some stars mentioned earlier in the text. The Sun, of course, has a luminosity of 1 solar unit. Its temperature, read off the bottom scale, is 5800 K—a G-type star. Similarly, the B-type star Rigel, at top left, has a temperature of about 15,000 K and a luminosity more than 10,000 times that of the Sun. The M-type star Proxima Centauri, at bottom right, has a temperature of 3000 K and a luminosity less than 1/10,000 that of the Sun.

the unconventional sense of temperature increasing to the *left* (so that the spectral sequence O, B, A . . . reads left to right). To change the horizontal scale so that temperature would increase conventionally to the right would play havoc with historical precedent.

As we have just seen, astronomers often use a star's *color* to measure its temperature. Indeed, the spectral classes plotted along the horizontal axis of Figure 17.13 are equivalent to the B–V color index. Also, because astronomers commonly express a star's luminosity as an absolute magnitude, stellar *magnitude* instead of stellar luminosity could be plotted vertically (see *More Precisely 17-1*). For these reasons, many astronomers refer to diagrams such as Figure

Estimating Stellar Radii

We can combine the Stefan-Boltzmann law $F = \sigma T^4$ (see *More Precisely 3-2*) with the formula for the area of a sphere, $A = 4\pi R^2$, to obtain the relationship between a star's radius (R), luminosity (L), and temperature (T) described in the text:

$$L = 4\pi\sigma R^2 T^4$$

$$\text{luminosity} \propto \text{radius}^2 \times \text{temperature}^4.$$

If we adopt convenient "solar" units, in which L is measured in solar luminosities (3.9×10^{26} W), R is measured in solar radii (696,000 km), and T in units of the solar temperature (5800 K), we can eliminate the constant $4\pi\sigma$ and write this equation as

L (in solar luminosities)
$$= R^2 \text{ (in solar radii)} \times T^4 \text{ (in units of 5800 K)}.$$

To compute the radius of a star from its luminosity and temperature, we rearrange this equation to read (in the same units)

$$R = \frac{\sqrt{L}}{T^2}.$$

This simple application of the radiation laws is the basis for almost every estimate of stellar size made in this text.

EXAMPLE: Let's compute the radii of the two stars discussed in the text. The star Mira has luminosity $L = 400$ units and temperature $T = 0.5$ units. According to the above equation, its radius therefore is $\sqrt{400}/0.5^2 = 20/0.25 = 80$ solar radii—a giant star. At the opposite extreme, Sirius B has $L = 0.04$, $T = 4$, so its radius is $\sqrt{0.04}/4^2 0.2/16 \approx 0.01$ times the radius of the Sun—a dwarf. The table above lists luminosities, temperatures,

and calculated radii for some of the other stars mentioned in this chapter.

STAR	LUMINOSITY, L	TEMPERATURE, T	RADIUS, R
Sun	1	1	1
Barnard's Star	0.00045	0.5	0.08
Sirius A	24	2.1	1.1
Rigel	66,000	1.9	71
Arcturus	170	0.75	23
Betelgeuse	60,000	0.62	630

Note that some of the luminosities in the table differ from those listed in Appendix 3 because the figures in the Appendix refer to visible light only, whereas the radiation laws refer to *total* luminosities (i.e. at all wavelengths).

Finally, let's follow the chain of reasoning leading from observations of a star to a fairly complete determination of its properties. We use the giant star Canopus, the second brightest star in the (southern) sky, as an example. Canopus is observed to have an apparent magnitude of -0.72. *Hipparcos* (*Discovery 17-1*) has measured its parallax as 0.0104 arc seconds, so its distance is 96 pc. Its absolute magnitude is then -5.6, making its luminosity 17,500 times that of the Sun (*More Precisely 17-1*). The spectral type of Canopus is F0; that information and *Hipparcos* color measurements imply a surface temperature of approximately 7800 K, or 1.3 times the surface temperature of the Sun. The radius–luminosity–temperature relationship then gives a radius of 78 solar radii. The same series of calculations has been performed for each of the more than 100,000 stars whose properties are now known, thanks to *Hipparcos*.

17.13 as **color–magnitude diagrams**. In this book, however, we will cast our discussion strictly in terms of temperature and luminosity measurements.

THE MAIN SEQUENCE

The few stars plotted in Figure 17.13 give little indication of any particular connection between stellar properties. However, as Hertzsprung and Russell plotted more and more stellar temperatures and luminosities, they found that a relationship does in fact exist. Stars are *not* uniformly scattered across the H–R diagram. Instead, most are confined to a fairly well defined band stretching diagonally from top left (high-temperature, high-luminosity) to bottom right (low-temperature, low-luminosity). In other words, cool stars tend to be faint (less luminous)

and hot stars tend to be bright (more luminous). This band of stars spanning the H–R diagram is known as the **main sequence**.

Figure 17.14 shows a more systematic study of stellar properties, covering the 80 or so stars that lie within 5 pc of the Sun. As more points are included in the diagram the main sequence "fills up," and the pattern becomes more evident. The vast majority of stars in the immediate vicinity of the Sun lie on the main sequence.

The surface temperatures of main-sequence stars range from about 3000 K (spectral class M) to over 30,000 K (spectral class O). This relatively small temperature range—only a factor of 10—is determined mainly by the rates at which nuclear reactions occur in stellar cores. In contrast, the observed range in luminosities is very large, covering some eight orders of magnitude (that is, a factor

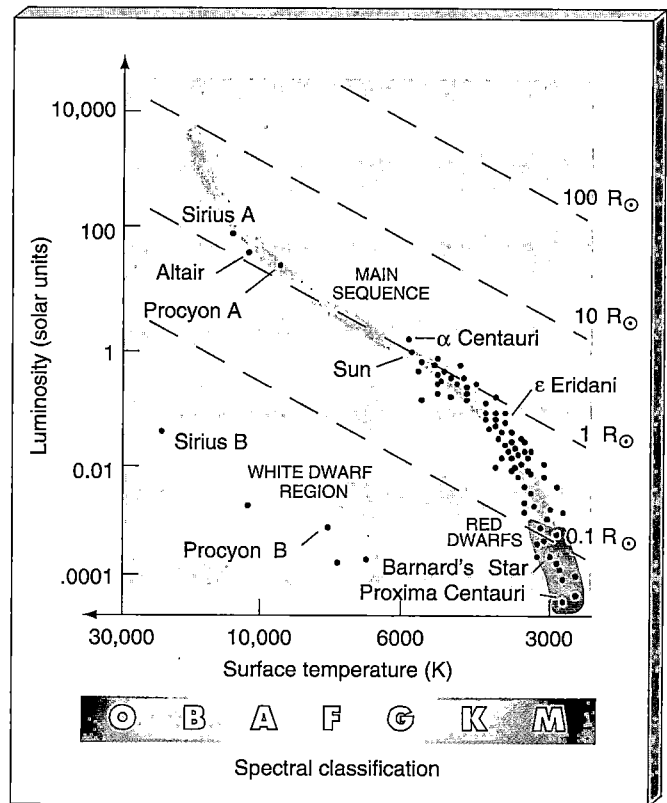

Figure 17.14 H–R Diagram of Nearby Stars Most stars have properties within the shaded region known as the main sequence. The points plotted here are for stars lying within about 5 pc of the Sun. The diagonal lines correspond to constant stellar radius, so that stellar size can be represented on the same diagram as luminosity and temperature. (Here and elsewhere, the symbol ⊙ stands for the Sun.)

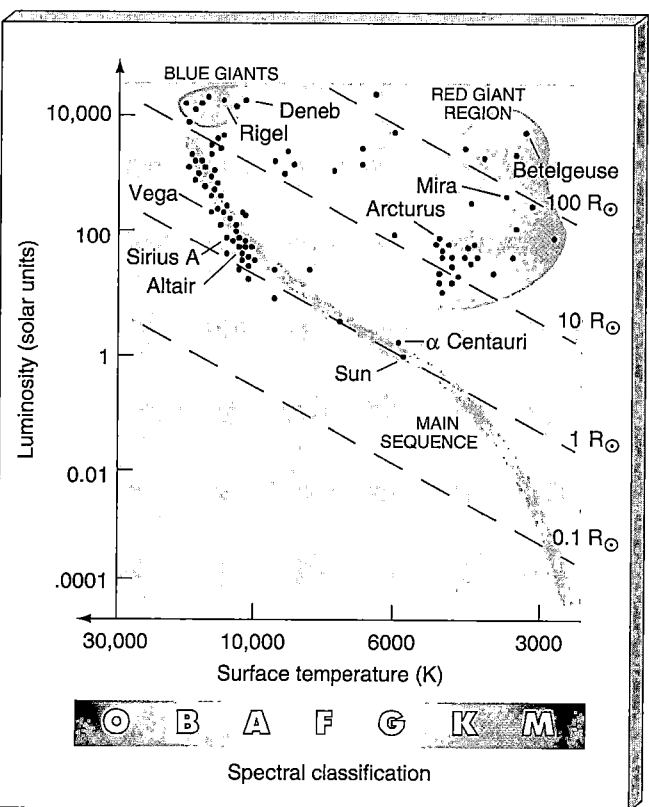

Figure 17.15 H–R Diagram of Bright Stars An H–R diagram for the 100 brightest stars in the sky. Such a plot is biased in favor of the most luminous stars—which appear toward the upper left—because we can see them more easily than we can the faintest stars. (Compare this with Figure 17.14, which shows only the closest stars.)

of 100 million), ranging from 10^{-4} to 10^4 times the luminosity of the Sun.

Using the radius–luminosity–temperature relationship (Section 17.5), astronomers find that stellar radii also vary along the main sequence. The faint, red M-type stars in the bottom right of the H–R diagram are only about 1/10 the size of the Sun, whereas the bright, blue O-type stars in the upper left are about 10 times larger than the Sun. The oblique dashed lines in Figure 17.14 represent constant stellar radius, meaning that any star lying on a given line has the same radius, regardless of its luminosity or temperature. Along a constant-radius line, the radius–luminosity–temperature relationship implies

$$\text{luminosity} \propto \text{temperature}^4.$$

By including such lines on our H–R diagrams, we can indicate stellar temperatures, luminosities, and radii on a single plot.

We see a very clear trend as we traverse the main sequence from top to bottom. At one end, the stars are large, hot, and bright. Because of their size and color, they are referred to as **blue giants**. The very largest are called **blue**

supergiants. At the other end, stars are small, cool, and faint. They are known as **red dwarfs**. Our Sun lies right in the middle.

Figure 17.15 shows an H–R diagram for a different group of stars—the 100 stars of known distance having the greatest apparent brightness, as seen from Earth. Notice the much larger number of very luminous stars at the upper end of the main sequence than at the lower end in Figure 17.15. The reason for this excess of blue giants is simple—we can see very luminous stars a long way off. The stars shown in Figure 17.15 are scattered through a much greater volume of space than those in Figure 17.14, but the sample is heavily biased toward the brightest objects. In fact, of the 20 brightest stars in the sky, only six lie within 10 pc of us; the rest are visible, despite their great distances, because of their high luminosities.

If very luminous blue giants are overrepresented in Figure 17.15, low-luminosity red dwarfs are surely underrepresented. In fact, no dwarfs appear on this diagram. This absence is not surprising, because low-luminosity stars are difficult to observe from Earth. In the 1970s, astronomers began to realize that they had greatly underestimated the number of red dwarfs in our galaxy. As hinted at by the H–R diagram in Figure 17.14, which shows an unbiased sample of stars in the solar neighborhood, red dwarfs are actually the most common type of star in the sky. In fact, they probably account for upward of 80 percent of all stars in the universe (see Figure 17.23). In contrast, O- and B-type supergiants are extremely rare—only about one star in 10,000 falls into this category.

THE WHITE DWARF AND RED GIANT REGIONS

Most stars lie on the main sequence. However, some of the points plotted in Figures 17.13–17.15 clearly do not. One such point in Figure 17.13 represents Sirius B, a white dwarf whose surface temperature (24,000 K) is about four times that of the Sun and whose luminosity is about 0.04 times the solar value. A few more such faint A-type stars can be seen in Figure 17.14 in the bottom left-hand corner of the H–R diagram. This region, known as the **white dwarf region**, is marked on Figure 17.14.

Also shown in Figure 17.13 is Mira (Omicron Ceti), whose surface temperature (3000 K) is about half that of the Sun and whose luminosity is some 400 times greater than the Sun's. Another point represents Betelgeuse (Alpha Orionis), the ninth-brightest star in the sky, a little cooler than Mira but more than 30 times brighter. The

upper right-hand corner of the H–R diagram, where these stars lie (marked on Figure 17.15), is called the **red giant region**. No red giants are found within 5 pc of the Sun (Figure 17.14), but many of the brightest stars seen in the sky are in fact red giants (Figure 17.15). Red giants are relatively rare, but they are so bright that they are visible to very great distances. They form a third distinct class of stars in the H–R diagram, very different in their properties from both main-sequence stars and white dwarfs.

The recent *Hipparcos* mission (*Discovery 17-1*), in addition to determining hundreds of thousands of stellar parallaxes with unprecedented accuracy, also measured the colors and luminosities of more than two million stars. Figure 17.16 shows an H–R diagram based on a tiny portion of the enormous *Hipparcos* dataset. The main sequence and red giant regions are clearly evident. Few white dwarfs appear, however, simply because the telescope was limited to observations of relatively bright objects—brighter than apparent magnitude 12. Almost no white dwarfs lie close enough to Earth that their magnitudes fall below this limit.

About 90 percent of all stars in our solar neighborhood, and probably a similar percentage elsewhere in the universe, are main-sequence stars. About 9 percent of stars are white dwarfs, and 1 percent are red giants.

☑ Concept Check

■ Only a tiny fraction of all stars are giants. Why then do giants account for so many of the brightest stars in the night sky?

17.7 Extending the Cosmic Distance Scale

SPECTROSCOPIC PARALLAX

In Chapter 2 we introduced the first "rung" on a ladder of distance-measurement techniques that will ultimately carry us to the edge of the observable universe. That rung is radar ranging on the inner planets. ⊙⊃ (Sec. 2.6) It establishes the scale of the solar system to great accuracy and, in doing so, defines the astronomical unit. In Section 17.1 we discussed a second rung in the cosmic distance ladder—stellar parallax—which is based on the first. Now, having used the first two rungs to determine the distances and other physical properties of many nearby stars, we can use that knowledge to construct a third rung in the ladder: **spectroscopic parallax.*** As illustrated schematically in Figure 17.17, this new rung expands our cosmic field of view still deeper into space.

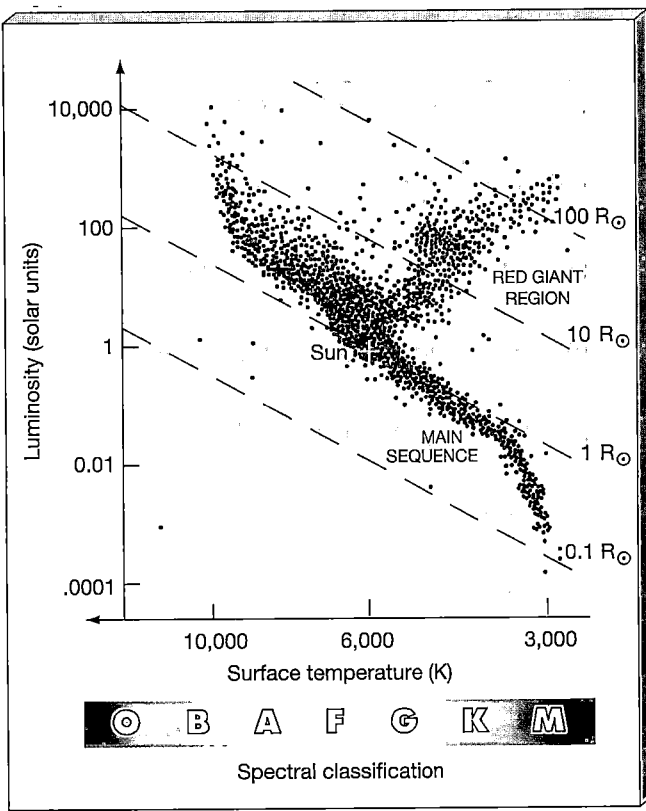

Figure 17.16 *Hipparcos* H–R Diagram This is a simplified version of the most complete H–R diagram ever compiled. It represents more than 20,000 data points, as measured by the European *Hipparcos* spacecraft for stars within about 1000 pc.

This unfortunate name is rather misleading, as the method has nothing in common with stellar (geometric) parallax other than its use as a means of determining stellar distances.

We have already discussed the connections between absolute brightness (luminosity), apparent brightness (energy flux), and distance. Knowledge of a star's apparent brightness and distance allows us to determine its luminosity using the inverse-square law. But we can also turn the problem around. If we somehow knew a star's luminosity and then measured its apparent brightness, the inverse-square law would give us its distance from Earth.

Consider an analogy. Most of us have a rough idea of the approximate brightness and size of a red traffic signal. Suppose we are driving down an unfamiliar street and see a red traffic light in the distance. Our knowledge of the intrinsic luminosity of the light often enables us immediately to make a mental estimate of its distance. A normal traffic light that is relatively dim must be quite distant (assuming it's not just dirty); a bright one must be relatively close. *A measurement of the apparent brightness of a light source, combined with some knowledge of its intrinsic properties, can yield an estimate of the source's distance.* For a star, the trick is to find an independent measure of the luminosity without knowing the distance. The H–R diagram can provide just that.

The main sequence represents a fairly close correlation between temperature and luminosity for most stars, with the exception of a few giants and dwarfs. Thus, the main sequence tells us about the *average* properties of most stars. Let's imagine for a moment that the main sequence is a *line*, rather than a somewhat fuzzy band, in the H–R diagram and that *all* stars lie on the main sequence. From a star's spectrum, we can determine its surface temperature or spectral type. If the star lies on the main sequence, then there is only one possible luminosity corresponding to that temperature. We can read the star's luminosity directly off a graph

such as Figure 17.14, and then determine its distance by measuring the energy flux at Earth and using the inverse-square law. The existence of the main sequence allows us to make a connection between an easily measured quantity (temperature) and the star's luminosity, which would otherwise be unknown. The term *spectroscopic parallax* refers to this whole process of using stellar spectra to infer distances.

Spectroscopic parallax can be used to determine stellar distances out to several thousand parsecs. Beyond that, spectra and colors of individual stars are difficult to obtain. The "standard" main sequence is obtained from H–R diagrams of stars whose distances can be measured by (geometric) parallax, so the method of spectroscopic parallax is calibrated using nearby stars. Note that in using this method, we are assuming (without proof) that distant stars are basically similar to nearby stars, and that *they fall on the same main sequence as nearby stars.* Only by making this assumption can we expand the boundaries of our distance-measurement techniques.

Of course, the main sequence is not really a line in the H–R diagram: it has some thickness. For example, the luminosity of a main-sequence G2-type star (such as the Sun) can range from about 0.5 to 1.5 times the luminosity of the Sun. The main reason for this range is the variation in stellar composition and age from place to place in the Galaxy. As a result, there is an uncertainty in the luminosity obtained by this method and so some uncertainty in the distance. Distances obtained by spectroscopic parallax are probably accurate to no better than 25 percent. Although this may not seem very accurate—a cross-country traveler in the United States would hardly be impressed to be told that the best estimate of the distance between Los Angeles and New York is somewhere between 3000 and 5000 km—it illustrates the point that in astronomy even something as simple as the distance to another star can be very difficult to measure. Still, an estimate with an uncertainty of ±25 percent is far better than no estimate at all. (See also *Discovery 17-1* for future developments that promise radical improvements in this state of affairs.)

LUMINOSITY CLASS

If a star happens to be a red giant or a white dwarf, its distance determined by spectroscopic parallax will be incorrect. We could simply argue that since roughly 90 percent of all stars are on the main sequence, the assumption that a star is a main-sequence star is valid nine out of 10 times, but in fact, astronomers can do much better. Recall from Chapter 4 that the *width* of a spectral line can provide information on the *density* of the gas where the line formed. ∞ (Sec. 4.4) The atmosphere of a red giant is much less dense than that of a main-sequence star, and this in turn is much less dense than the atmosphere of a white dwarf. By studying the width of a star's spectral lines, astronomers can usually tell with a high degree of confidence whether or not it is on the main sequence.

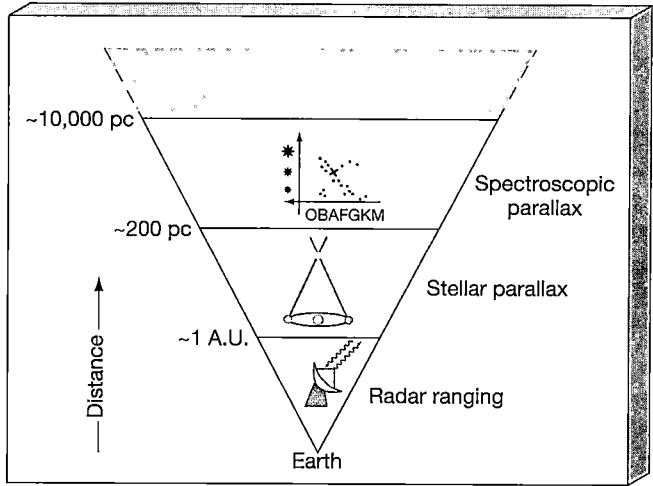

Figure 17.17 Stellar Distances Knowledge of a star's luminosity and apparent brightness can yield an estimate of its distance. Astronomers use this third "rung" in our distance ladder, called spectroscopic parallax, to measure distances as far out as individual stars can be clearly discerned—several thousand parsecs.

TABLE 17.3 Stellar Luminosity Classes

CLASS	DESCRIPTION
Ia	Bright supergiants
Ib	Supergiants
II	Bright giants
III	Giants
IV	Subgiants
V	Main-sequence stars/dwarfs

Astronomers have developed a system for classifying stars according to the width of their spectral lines. Because the line width is particularly sensitive to density in the stellar photosphere, and the atmospheric density in turn is well correlated with luminosity, the class in which a star is categorized has come to be known as its **luminosity class**. This classification provides a means for astronomers to distinguish supergiants from giants, giants from main-sequence stars, and main-sequence stars from white dwarfs by studying a single spectral property—line broadening—of the radiation received.

The standard stellar luminosity classes are given in Table 17.3. Their locations on the H–R diagram are indicated in Figure 17.18. Now we have a way of specifying a star's location in the diagram in terms of properties that are measurable by purely spectroscopic means. *Spectral type and luminosity class define a star just as surely as do temperature and luminosity.* The full specification of a star's spectral properties includes its luminosity class. For example, the Sun, on the main sequence, is of class G2V, Vega is A0V, the red dwarf Barnard's Star is M5V, the red supergiant Betelgeuse is M2Ia, and so on.

Consider, for example, a K2-type star (Table 17.4) with a surface temperature of approximately 5000 K. If the widths of the star's spectral lines tell us that it lies on the main sequence (that is, it is a K2V star), then its luminosity is about 0.3 times the solar value. If its spectral lines are observed to be narrower than lines normally found in main-sequence stars, the star may be recognized as a K2III giant, with a luminosity 100 times that of the Sun (Figure 17.18). If the lines are very narrow, the star may instead be classified as a K2Ib supergiant, brighter by a further factor of 50, at 5000 solar luminosities. In this way the observed

width of the star's spectral lines translates directly into a measure of the star's physical state. Knowledge of luminosity classes allows us to use spectroscopic parallax with some confidence that we are not accidentally counting a red giant or a white dwarf as a main-sequence star and making a huge error in our distance estimate as a result.

☑ Concept Check

◼ Suppose astronomers discover that, due to a calibration error, *all* distances measured by geometric parallax are 10 percent larger than currently thought. What effect would this have on the "standard" main sequence used in spectroscopic parallax?

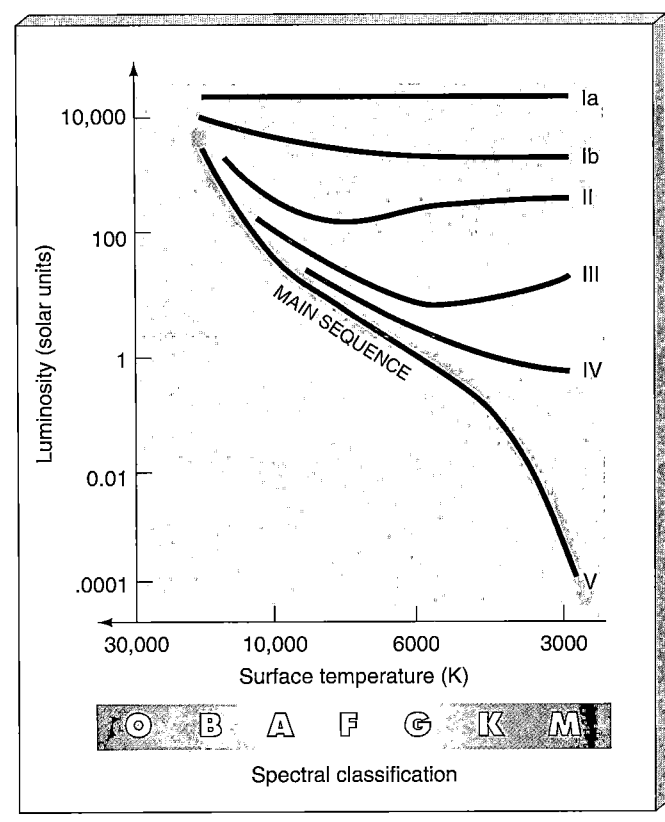

Figure 17.18 Luminosity Classes Stellar luminosity classes in the H–R diagram. A star's location in the diagram can alternatively be specified by its spectral type and luminosity class instead of by its temperature and luminosity.

TABLE 17.4 Variation in Stellar Properties within a Spectral Class

APPROXIMATE SURFACE TEMPERATURE (K)	LUMINOSITY (solar luminosities)	RADIUS (solar radii)	OBJECT	EXAMPLE
5000	0.3	0.8	K2V main-sequence star	ε Eridani
5000	120	23	K2III red giant	Arcturus
5000	5000	100	K2Ib red supergiant	ε Pegasi

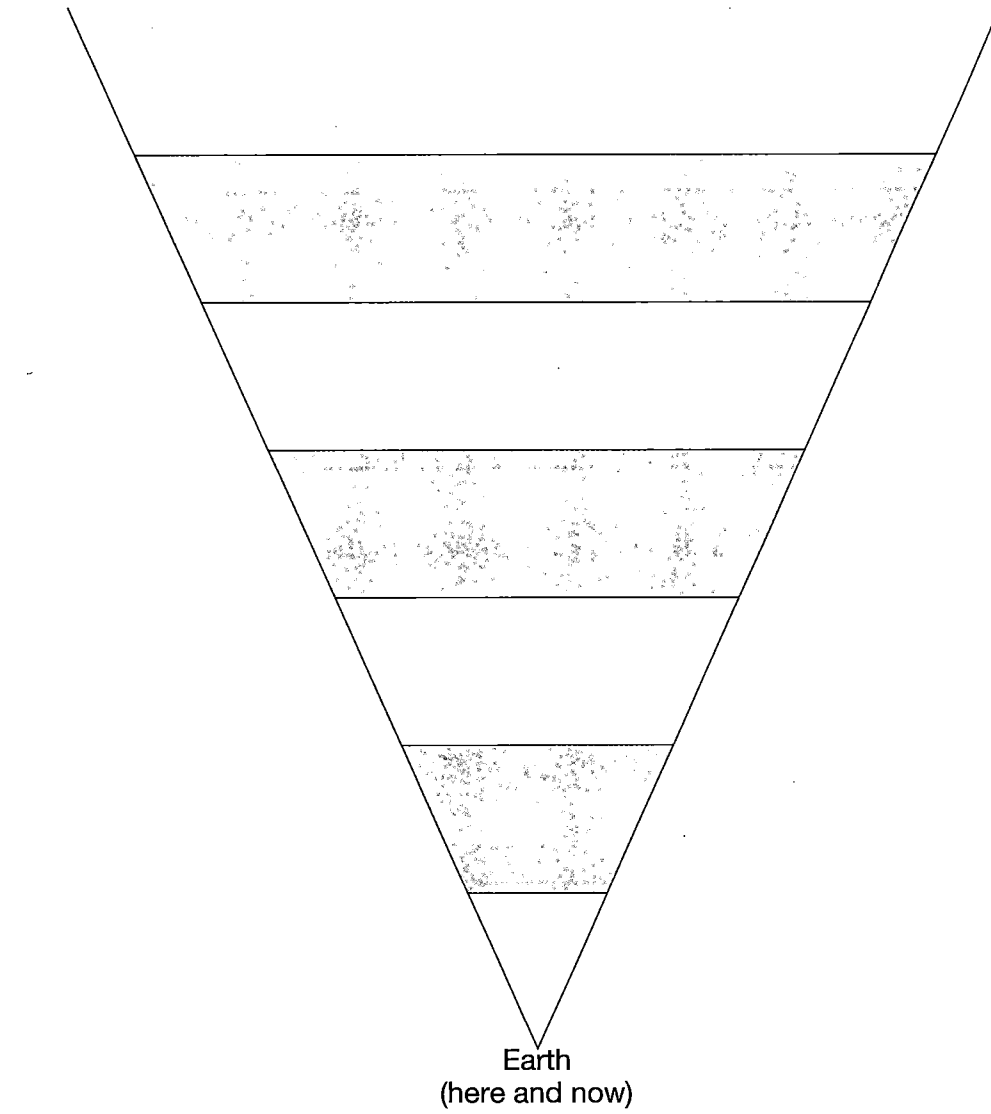

Distance (then and there)

Earth
(here and now)

The distance scale: An inverted pyramid provides a convenient way to
summarize methods used to determine distance in the Universe. We are located
at the bottom apex, looking out toward increasingly large intervals of space, as
well as back toward more ancient intervals of time. Each rung on the ladder
depends on the method below it, thereby creating an intricate set of steps that take
us progressively from close to home to the limits of the observable universe.

17.8 Stellar Masses

⬚ What ultimately determines a star's position on the main sequence? The answer is its *mass* and its *composition*. Mass and composition are fundamental properties of any star. Together they uniquely determine the star's internal structure, its external appearance, and even (as we will see in Chapter 20) its future evolution. The ability to measure these two key stellar properties is of the utmost importance if we are to understand how stars work. We have already seen how spectroscopy is used to determine a star's composition. ⬭ (Sec. 16.3) Now let's turn to the problem of finding a star's mass.

As with all other objects, we measure a star's mass by observing its gravitational influence on some nearby body—another star, perhaps, or a planet. If we know the distance between the two bodies, then we can use Newton's laws to calculate their masses. The extrasolar planetary systems that have recently been detected have not been studied well enough to provide independent stellar mass measurements, and we are a long way from placing our own spacecraft in orbit around other stars. ⬭ (Sec. 15.5) Nevertheless, there are ways of determining stellar masses.

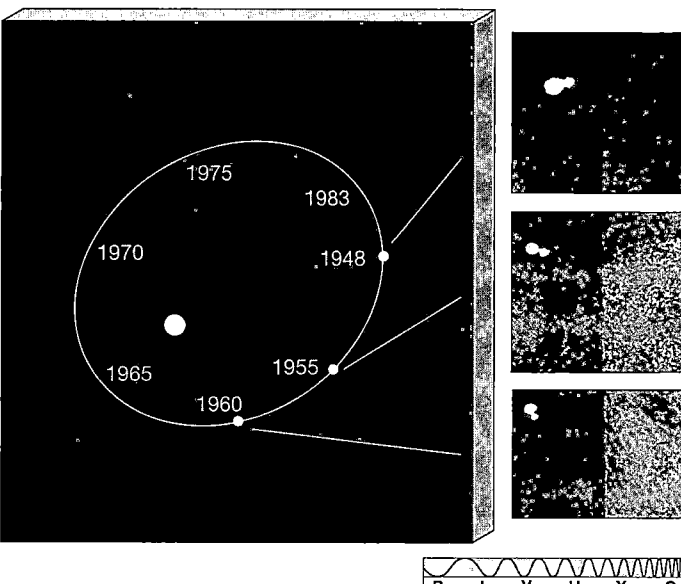

Figure 17.19 Visual Binary The periods and separations of binary stars can be observed directly if each star is clearly seen. This binary star is known as Kruger 60. At left is a schematic diagram of the orbit, at right actual photographs taken in the years indicated. *(Harvard College Observatory)*

BINARY STARS

Most stars are members of *multiple-star systems*—groups of two or more stars in orbit around one another. The majority of stars are found in **binary-star systems**, which consist of two stars in orbit about their common center of mass, held together by their mutual gravitational attraction. ⬭ (Sec. 2.7) Other stars are members of triple, quadruple, or even more complex systems. The Sun is not part of a multiple-star system—if it has anything at all uncommon about it, it may be its lack of stellar companions.

Astronomers classify binary-star systems (or simply *binaries*) according to their appearance from Earth and the ease with which they can be observed. **Visual binaries** have widely separated members that are bright enough to be observed and monitored separately, as shown in Figure 17.19. The more common **spectroscopic binaries** are too distant to be resolved into separate stars, but they can be indirectly perceived by monitoring the back-and-forth Doppler shifts of their spectral lines as the stars orbit each other. Recall that motion toward an observer blueshifts the lines, and motion away from the observer redshifts them. ⬭ (Sec. 3.5)In a *double-line* spectroscopic binary, two distinct sets of spectral lines—one for each component star—shift back and forth as the stars move. Because we see particular lines alternately approaching and receding, we know that the objects emitting the lines are in orbit. In the more common *single-line* systems, such as that shown in Figure 17.20, one star is too faint for its spectrum to be distinguished, so only one set of lines is observed to shift back and forth. This shifting means that the detected star

must be in orbit around another star, even though the companion cannot be observed directly. (If this idea sounds familiar, it should—all of the extrasolar planetary systems discovered to date are extreme examples of single-line spectroscopic binaries.)

In the much rarer **eclipsing binaries**, the orbital plane of the pair of stars is almost edge-on to our line of sight. In this situation, depicted in Figure 17.21, we observe a periodic decrease of starlight as one component passes in front of (transits) the other. By studying the variation of the light from the binary system—the binary's **light curve**—astronomers can derive detailed information not only about the stars' orbits and masses but also about their radii. Thus, eclipsing binaries provide an alternative means of measuring stellar radii that is *independent* of either the direct or the indirect methods described in Section 17.5.

For example, in the sequence shown in Figure 17.21, the maximum brightness (frames 1, 3, and 5) represents the combined brightnesses of the two stars, while the shallower minimum (frame 4) represents the brighter (larger) component only. These two pieces of information allow us to infer the individual brightnesses of the two stars. The deeper minima (frames 2 and 6) occur because the fainter red star partially blocks the light of the much brighter yellow star. The change in brightness tells us what fraction of the brighter star is obscured, and that in turn tells us the ratio of the *areas* of the two stars, and hence (since area is proportional to radius squared) the ratio of their radii. If we also knew the components' orbital *speeds*—from Doppler measurements, say—then the widths of the minima and the

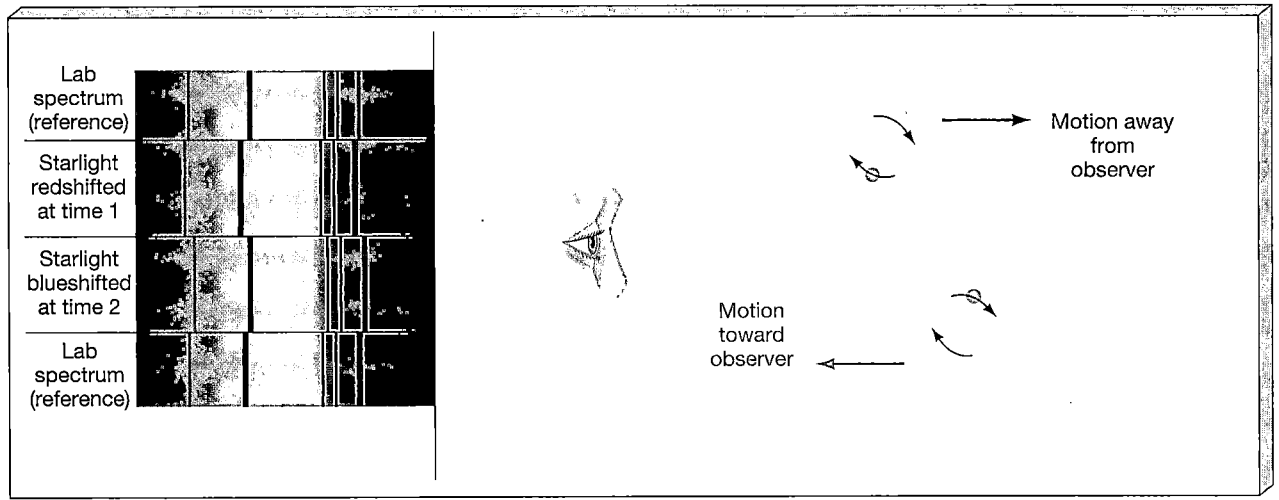

Figure 17.20 Spectroscopic Binary Binary properties can be determined indirectly by measuring the periodic Doppler shift of one star relative to the other as they move in their orbits. The diagram shows a so-called single-line system, in which only one spectrum (from the brighter component) is visible. The observer is situated to the left of the diagram. In the upper frame, the visible component is moving away from the observer and the spectrum is redshifted. In the lower frame, the star is moving toward the observer and the spectrum is blueshifted. Reference lab spectra are shown at top and bottom.

time taken to go from minimum to maximum light would tell us the actual radii of the stars.

These categories of binary-star systems are not mutually exclusive. For example, a single-line spectroscopic binary may also happen to be an eclipsing system. In that case astronomers can use the eclipses to gain extra information about the fainter member of the pair. Occasionally, two unrelated stars just happen to lie close together in the sky, even though they are actually widely separated. These *optical doubles* are just chance superpositions and carry no useful information about stellar properties.

MASS DETERMINATION

By observing the actual orbits of the stars, or the back-and-forth motion of the spectral lines, or the dips in the light

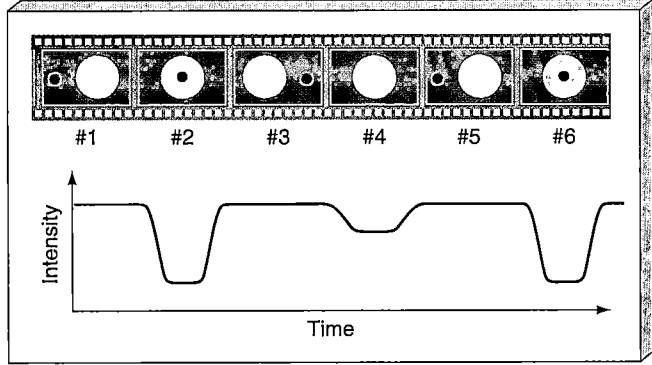

Figure 17.21 Eclipsing Binary If two stars in a binary-star system happen to eclipse one another, information on their radii and masses can be obtained by observing the periodic decrease in starlight as one passes in front of the other.

curve—whatever information is available—astronomers can measure the binary's orbital period. Observed periods span a broad range—from hours to centuries. How much additional information can be extracted depends on the type of binary involved.

If the distance to a *visual* binary is known, the semimajor axis of its orbit can be determined directly, by simple geometry. Knowledge of the binary period and orbital semimajor axis is all we need to determine the combined mass of the component stars, using the modified form of Kepler's third law. ⊂⊃ (Sec. 2.7) Since the orbits of both components can be separately tracked, it is also possible to determine the individual component masses. Recall from Section 2.7 that in any system of orbiting objects, each object orbits the common center of mass. Measuring the distance from each star to the center of mass of a visual binary yields the ratio of the stellar masses. Knowing both the sum of the masses and the ratio of the masses, we can then find the mass of each star.

For *spectroscopic* binaries, it is not possible to determine the semimajor axis directly. Doppler-shift measurements give us information on the orbital velocities of the member stars, but only on their radial components. As a result, we cannot determine the *inclination* of the orbit to our line of sight, and this imposes a limitation on how much information we can obtain—simply put, we cannot distinguish between a slow-moving binary seen edge-on and a fast-moving binary seen almost face-on (so that only a small component of the orbital motion is along the line of sight). We have already encountered this limitation in our study of extrasolar planets. ⊂⊃ (Sec. 15.5)

For a double-line spectroscopic system, individual radial velocities, and hence the ratio of the component masses can be determined, but the uncertainty in the orbital inclination means that only lower limits on the individual masses can be obtained. For single-line systems, even less information is available, and only a fairly complicated relation between the component masses (known as the *mass function*) can be derived. However, if, as is often the case, the mass of the brighter component can be determined by other means (for example, if it is recognized as a main-sequence star of a certain spectral class—see Figure 17.22), a lower limit can then be placed on the mass of the fainter, unseen star.

Finally, if a spectroscopic binary happens also to be an *eclipsing* system, then the uncertainty in the inclination is removed, as the binary is known to be edge-on, or very nearly so. In that case, both masses can be determined for a double-line binary. For a single-line system, the mass function is simplified to the point where the mass of the unseen component is known if the brighter component can be identified.

Despite all these qualifications, individual component masses have been obtained for many nearby binary-star systems. Virtually all we know about the masses of stars is based on such observations. As a simple example, consider the nearby visual binary-star system made up of the bright star Sirius A and its faint companion Sirius B.

Their orbital period is 50 years and their orbital semimajor axis is 20 A.U.—7.5″ at a distance of 2.7 pc—implying that the sum of their masses is 3.2 times the mass of the Sun. Further study of the orbit shows that Sirius A has roughly twice the mass of its companion. It follows that the masses of Sirius A and Sirius B are 2.1 and 1.1 solar masses, respectively.

MASS AND OTHER STELLAR PROPERTIES

Figure 17.22 is a schematic H–R diagram showing how stellar mass varies along the main sequence. There is a clear progression from low-mass red dwarfs to high-mass blue giants. With few exceptions, main-sequence stars range in mass from about 0.1 to 20 times the mass of the Sun. The hot O- and B-type stars are generally about 10 to 20 times more massive than our Sun. The coolest K- and M-type stars contain only a few tenths of a solar mass. The mass of a star at the time of its formation determines its location on the main sequence. Based on observations of stars within a few hundred light-years of the Sun, Figure 17.23 illustrates how the masses of main-sequence stars are distributed. Notice the huge fraction of low-mass stars, as well as the tiny fraction contributed by stars of more than a few solar masses.

Figure 17.24 illustrates how a main-sequence star's radius and luminosity depend on its mass. These two plots, called the *mass–radius* and *mass–luminosity* relations, are based on observations of binary-star systems.

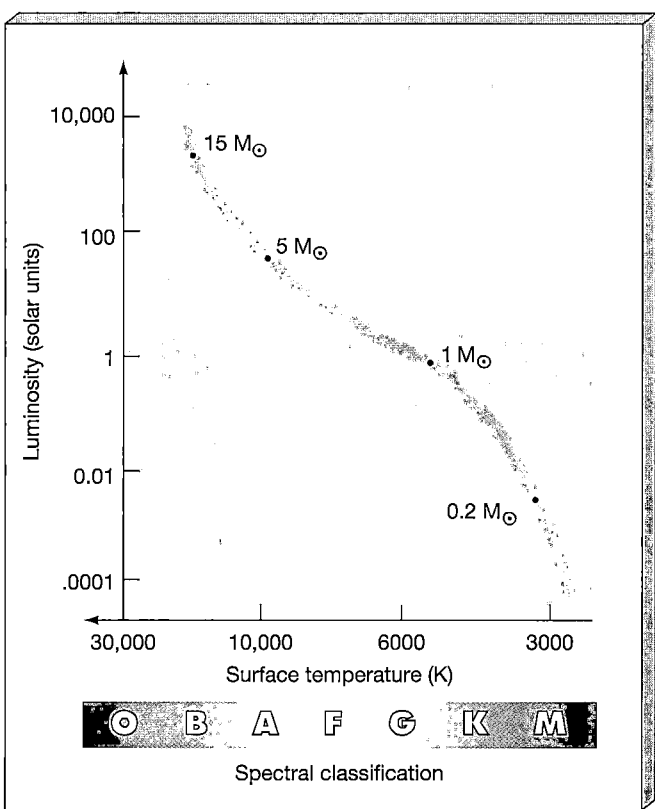

Figure 17.22 Stellar Masses Mass, more than any other stellar property, determines a star's position on the main sequence. Stars that form with low mass will be cool and faint; they lie at the bottom of the main sequence. Very massive stars are hot and bright; they lie at the top of the main sequence.

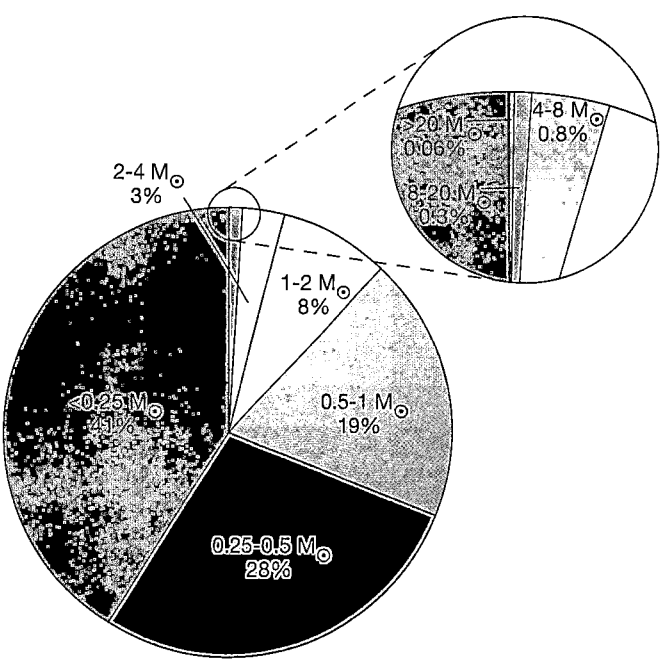

Figure 17.23 Stellar Mass Distribution The distribution of masses of main-sequence stars, as determined from careful measurement of stars in the solar neighborhood.

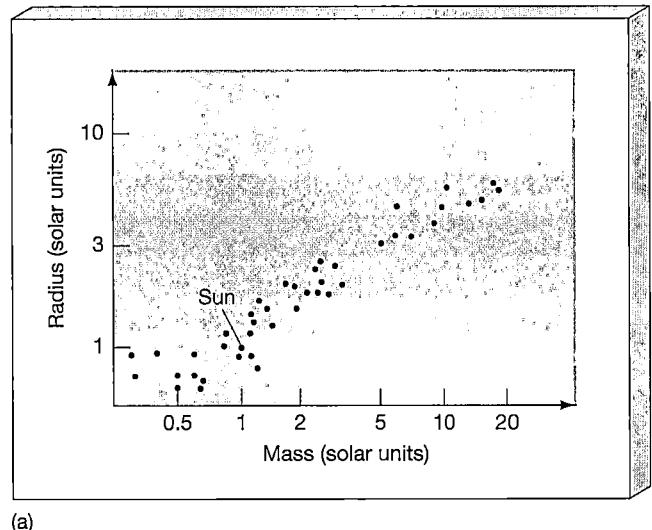

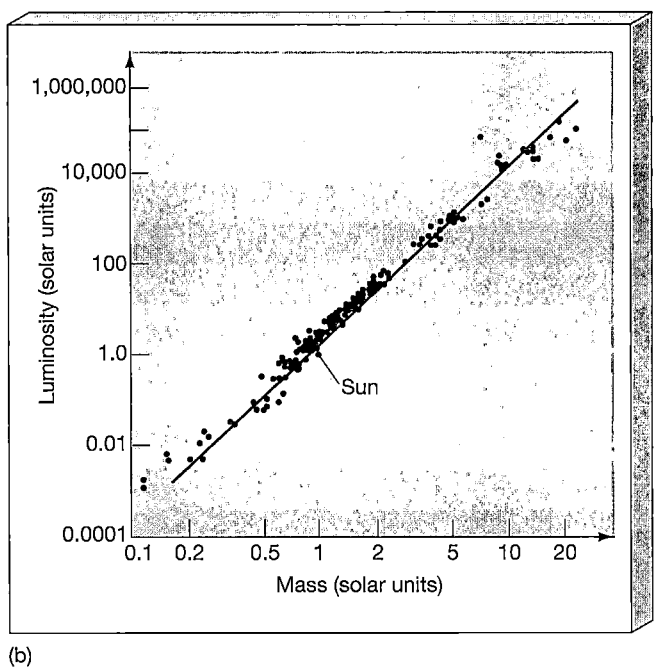

Figure 17.24 Stellar Radii and Luminosities
(a) Dependence of stellar radius on mass for main-sequence stars. The radius increases roughly in proportion to the mass over much of the range. (b) Dependence of luminosity on mass. The luminosity increases much faster than the mass.

Along the main sequence, both radius and luminosity increase with mass. As an approximate rule of thumb, we can say that radius increases proportionally to stellar mass, while luminosity increases much faster—almost as the *fourth power* of the mass (as indicated by the line in Figure 17.24b). Thus, for example, a 2-solar mass main-sequence star has a radius about twice that of the Sun and a luminosity of 16 (2^4) solar luminosities; a 0.2-solar mass main-sequence star has a radius of roughly 0.2 solar radii and a luminosity of around 0.0016 (0.2^4) solar luminosities.

Table 17.5 compares some key properties of several well-known main-sequence stars, arranged in order of decreasing mass. Notice that the central temperature (obtained from mathematical models similar to those discussed in Chapter 16) differs relatively little from one star to another, compared to the large spread in stellar lu-

minosities. ∞ (Sec. 16.2) The rapid rate of nuclear burning deep inside a star releases vast amounts of energy per unit time. How long can the fire continue to burn? We can estimate a main-sequence star's *lifetime* simply by dividing the amount of fuel available (the mass of the star) by the rate at which the fuel is being consumed (the star's luminosity):

$$\text{stellar lifetime} \propto \frac{\text{stellar mass}}{\text{stellar luminosity}}.$$

The mass–luminosity relation tells us that a star's luminosity is roughly proportional to the fourth power of its mass, so we can rewrite this expression to obtain, approximately,

$$\text{stellar lifetime} \propto \frac{1}{(\text{stellar mass})^3}.$$

TABLE 17.5 Key Properties of Some Well-Known Main-Sequence Stars

STAR	SPECTRAL TYPE	MASS, M (solar masses)	CENTRAL TEMPERATURE (10^6 K)	LUMINOSITY, L (solar luminosities)	ESTIMATED LIFETIME (M/L) (10^6 years)
Spica B*	B2V	6.8	25	500	140
Vega	A0V	2.6	21	55	500
Sirius	A1V	2.1	20	23	900
Alpha Centauri	G2V	1.1	17	1.4	8000
Sun	G2V	1.0	15	1.0	10,000
Proxima Centauri	M5V	0.1	0.6	0.00006	16,000,000

*Spica is in fact a binary system comprising a B1III giant primary (Spica A) and a B2V main-sequence secondary (Spica B).

The final column in Table 17.5 lists estimated lifetimes, based on the above proportionality and noting that the lifetime of the Sun (see Chapter 20) is about 10 billion years. For example, the lifetime of a 10-solar mass main-sequence O-type star is roughly $10/10^4 = 1/1000$ of the lifetime of the Sun, or about 10 million years. The nuclear reactions in such a massive star proceed so rapidly that its fuel is quickly depleted, despite its large mass. We can be sure that all the O- and B-type stars we now observe are quite young—less than a few tens of millions of years old. Massive stars older than that have already exhausted their fuel and no longer emit large amounts of energy. They have, in effect, died.

At the opposite end of the main sequence, the cooler K- and M-type stars have less mass than our Sun. With their low core densities and temperatures, their proton–proton reactions churn away rather sluggishly, much more slowly than those in the Sun's core. The small energy release per unit time leads to low luminosities for these stars, so they have very long lifetimes. Many of the K- and M-type stars we now see in the night sky will shine on for at least another trillion years. The evolution of stars—large and small—is the subject of Chapters 20 and 21.

Concept Check

■ How do we know the masses of stars that aren't members of binaries?

Chapter Review

SUMMARY

The distances to the nearest stars can be measured by trigonometric parallax. A star with a parallax of one arc second (1″) is 1 **parsec** (p. 439)—about 3.3 light-years—away. Stars have real motion through space as well as apparent motion as Earth orbits the Sun. A star's **proper motion** (p. 439), its true motion across the sky, is a measure of the star's velocity perpendicular to our line of sight. The star's radial velocity—along the line of sight—is measured by the Doppler shift of its spectral lines.

The **apparent brightness** (p. 441) of a star is the rate at which energy from the star reaches unit area of a detector. Apparent brightness falls off as the inverse square of the distance. Optical astronomers use the **magnitude scale** (p. 443) to express and compare stellar brightnesses. The greater the magnitude, the fainter the star; a difference of five magnitudes corresponds to a factor of 100 in brightness. **Apparent magnitude** (p. 444) is a measure of apparent brightness. The **absolute magnitude** (p. 444) of a star is the apparent magnitude it would have if placed at a standard distance of 10 pc from the viewer. It is a measure of the star's luminosity.

Astronomers often measure the temperatures of stars by measuring their brightnesses through two or more optical filters, then fitting a blackbody curve to the results. The **color index** (p. 446) of a star is the difference in its apparent magnitudes measured through two standard filters. The measurement of the amount of starlight received through each of a set of filters is called **photometry** (p. 446). Astronomers classify stars according to the absorption lines in their spectra. The lines seen in the spectrum of a given star depend mainly on its temperature, and spectroscopic observations of stars provide an accurate means of determining both stellar temperatures and stellar composition. The standard stellar **spectral classes** (p. 448), in order of decreasing temperature, are O, B, A, F, G, K, and M.

Only a few stars are large enough and close enough that their radii can be measured directly. The sizes of most stars are estimated indirectly through the **radius–luminosity–temperature relationship** (p. 449). Stars are categorized as **dwarfs** (p. 450) comparable in size to or smaller than the Sun, **giants** (p. 450) up to 100 times larger than the Sun, and **supergiants** (p. 450) more than 100 times larger than the Sun. In addition to "normal" stars such as the Sun, two other important classes of star are **red giants** (p. 450), which are large, cool, and luminous, and **white dwarfs** (p. 450), which are small, hot, and faint.

A plot of stellar luminosities versus stellar spectral classes (or temperatures) is called an **H–R diagram** (p. 450), or a **color-magnitude diagram** (p. 452). About 90 percent of all stars plotted on an H–R diagram lie on the **main sequence** (p. 452), which stretches from hot, bright **blue supergiants** (p. 453) and **blue giants** (p. 453), through intermediate stars such as the Sun, to cool, faint **red dwarfs** (p. 453). Most main-sequence stars are red dwarfs; blue giants are quite rare. About 9 percent of stars are in the **white dwarf region** (p. 454), and the remaining 1 percent are in the **red giant region** (p. 454).

By careful spectroscopic observations, astronomers can determine a star's **luminosity class** (p. 456), allowing them to distinguish main-sequence stars from red giants or white dwarfs of the same spectral type (or color). Once a star is known to be on the main sequence, measurement of its spectral type allows its luminosity to be estimated and its distance to be measured. This method of distance determination, which is valid for stars up to several thousand parsecs from Earth, is called **spectroscopic parallax** (p. 454).

Most stars are not isolated in space but instead orbit other stars in **binary-star systems** (p. 457). In a **visual binary** (p. 457), both stars can be seen and their orbit charted. In a **spectroscopic binary** (p. 457), the stars cannot be resolved, but their orbital motion can be detected spectroscopically. In an **eclipsing binary** (p. 457), the orbit is oriented in such a way that one star periodically passes in front of the other as seen from Earth and dims the light we receive. The binary's **light curve** (p. 457) is a plot of its apparent brightness as a function of time.

Studies of binary-star systems often allow stellar masses to be measured. The mass of a star determines its size, temperature, and brightness. Fairly well defined mass–radius and mass–luminosity relations exist for main-sequence stars. Hot blue giants are much more massive than the Sun; cool red dwarfs are much less massive. The lifetime of a star can be estimated by dividing its mass by its luminosity. High-mass stars burn their fuel rapidly and have much shorter lifetimes than the Sun. Low-mass stars consume their fuel slowly and may remain on the main sequence for trillions of years.

SELF-TEST: TRUE OR FALSE?

____ **1.** One parsec is a little over 200,000 A.U.

____ **2.** There are no other stars within 1 pc of the Sun.

____ **3.** Parallax can be used to measure stellar distances out to about 1000 pc.

____ **4.** Most stars have radii between 0.5 and 2 times the radius of the Sun.

____ **5.** Star A appears brighter than star B, as seen from Earth. Therefore, star A must be closer to Earth than star B.

____ **6.** Star A and star B have the same absolute brightness (luminosity), but star B is twice as distant as star A. Therefore, star A appears four times brighter than star B.

____ **7.** A star of apparent magnitude 5 looks brighter than a star of apparent magnitude 2.

____ **8.** Differences among stellar spectra are mainly due to differences in composition.

____ **9.** Stars with very weak hydrogen lines in their spectra contain very little hydrogen.

____ **10.** Red giants are very bright because they are extremely hot.

____ **11.** Red dwarfs lie in the lower left part of the H–R diagram.

____ **12.** The brightest stars visible in the night sky are generally all found in the upper part of the H–R diagram.

____ **13.** In a spectroscopic binary, the orbital motion of the component stars appears as variations in the overall apparent brightness of the system.

____ **14.** Astronomers can distinguish between main-sequence and giant stars by purely spectroscopic means.

____ **15.** It is impossible to have a one-billion-year-old O- or B-type main-sequence star.

SELF-TEST: FILL IN THE BLANK

1. Parallax measurements of the distances to the nearest stars use _____ as a baseline.

2. The radial velocity of a star is determined by observing its _____ and using the _____ effect.

3. To determine the transverse velocity of a star, both its _____ and its _____ must be known.

4. The radius of a star can be indirectly determined if the star's _____ and _____ are known.

5. Observations through B and V filters are used to determine stellar _____.

6. The hottest stars show little evidence of hydrogen in their spectra because hydrogen is mostly _____ at these temperatures.

7. The coolest stars show little evidence of hydrogen in their spectra because hydrogen is mostly _____ at these temperatures.

8. The Sun has a spectral type of _____.

9. The H–R diagram is a plot of _____ on the horizontal scale versus _____ on the vertical scale.

10. The band of stars extending from the top left of the H–R diagram to its bottom right is known as the _____.

11. The large, cool stars found at the upper right of the H–R diagram are _____.

12. The small, hot stars found at the lower left of the H–R diagram are _____.

13. _____-star systems are important for providing measurements of stellar masses.

14. Going from spectral class O to M along the main sequence, stellar masses _____.

15. The main-sequence lifetimes of high-mass stars are much _____ than the main-sequence lifetimes of low-mass stars.

REVIEW AND DISCUSSION

1. How is parallax used to measure the distances to stars?

2. What is a parsec? Compare it with the astronomical unit.

3. Explain two ways in which a star's real motion through space translates into motion observable from Earth.

4. How do astronomers go about measuring stellar luminosities?

5. Describe how astronomers measure stellar radii.

6. Describe some characteristics of red giant and white dwarf stars.

7. What is the difference between absolute and apparent brightness?

8. How do astronomers measure stellar temperatures?

9. Briefly describe how stars are classified according to their spectral characteristics.

10. Why do some stars have very few hydrogen lines in their spectra?

11. What information is needed to plot a star on the H–R diagram?

12. What is the main sequence? What basic property of a star determines where it lies on the main sequence?

13. How are distances determined using spectroscopic parallax?

14. Why does the H–R diagram constructed using the brightest stars differ so much from the diagram constructed using the nearest stars?

15. Which stars are most common in the Galaxy? Why don't we see many of them in H–R diagrams?

16. Which stars are least common in the Galaxy?

17. How can stellar masses be determined by observing binary-star systems?

18. High-mass stars start off with much more fuel than low-mass stars. Why don't high-mass stars live longer?

19. In general, is it possible to determine the age of an individual star simply by noting its position on an H–R diagram?

20. Visual binaries and eclipsing binaries are relatively rare compared to spectroscopic binaries. Why is this?

PROBLEMS  *Algorithmic versions of these questions are available in the Practice Problems module of the Companion Website.*

The number of squares preceding each problem indicates its approximate level of difficulty.

1. ■ How far away is the star Spica, whose parallax is 0.013″? What would Spica's parallax be if it were measured from an observatory on Neptune's moon Triton as Neptune orbited the Sun?

2. ■■ A star lying 20 pc from the Sun has proper motion of 0.5″/yr. What is its transverse velocity? If the star's spectral lines are observed to be redshifted by 0.01 percent, calculate the magnitude of its three-dimensional velocity relative to the Sun.

3. ■ What is the luminosity of a star having three times the radius of the Sun and a surface temperature of 10,000 K?

4. ■ A certain star has a temperature twice that of the Sun and a luminosity 64 times greater than the solar value. What is its radius, in solar units?

5. ■■ Two stars—A and B, of luminosities 0.5 and 4.5 times the luminosity of the Sun, respectively—are observed to have the same apparent brightness. Which one is more distant, and how much farther away is it than the other?

6. ■■ Two stars—A and B, of absolute magnitudes 3 and 8, respectively—are observed to have the same apparent magnitude. Which one is more distant, and how much farther away is it than the other?

7. ■■ Calculate the solar energy flux (energy received per unit area per unit time), as seen from a distance of 10 pc from the Sun. Compare it with the solar constant at Earth.

8. ■ Astronomical objects visible to the naked eye range in apparent brightness from faint sixth-magnitude stars to the

Sun with magnitude −27. What range in energy flux corresponds to this magnitude range?

9. ■ A star has apparent magnitude 4.0 and distance 100 pc. What is its absolute magnitude?

10. ■ A star has apparent magnitude 10.0 and absolute magnitude 2.5. How far away is it?

11. ■■ Using the data shown in Figure 17.7, calculate the greatest distance at which a star like the Sun could be seen using (a) binoculars, (b) a typical 1-m telescope, (c) a 4-m telescope, and (d) the *Hubble Space Telescope*.

12. ■■■ In an eclipsing binary, the brightness is observed to drop by 1 percent when the brighter component eclipses the fainter one, and by 10 percent when the fainter component eclipses the brighter one (see Figure 17.21). If the stars lie on the main sequence and their radii follow the law $R \propto M$, what is the ratio of the smaller mass to the larger one?

13. ■■ Two stars in an eclipsing spectroscopic binary are observed to have an orbital period of 25 days. Further observations reveal that the orbit is circular, with a separation of 0.3 A.U., and that one star is 1.5 times the mass of the other. What are the masses of the stars?

14. ■■ Given that the Sun's lifetime is about 10 billion years, estimate the life expectancy of (a) a 0.2-solar mass, 0.01-solar luminosity red dwarf (b) A 3-solar mass, 30-solar luminosity star, (c) A 10-solar mass, 1000-solar luminosity blue giant.

15. ■■■ Assuming the mass–luminosity relation $L \propto M^4$, estimate the mass of the faintest main-sequence star that could be observed at a distance of 50,000 pc by (a) a typical 1-m telescope and (b) the *Hubble Space Telescope*. (See Figure 17.7.)

COLLABORATIVE EXERCISES

1. Inverse-Square Law. Considering where your group is sitting right now, how many times dimmer would an imaginary, super-deluxe, ultrabright flashlight be if it were located at the front door of the group member who lives farthest away as compared to if it were at the front door of the group member who lives closest? Explain your reasoning.

2. Differences in Brightness. As a group, go to Appendix 3: Table 4—The Twenty Brightest Stars, and select two stars.

Refer to the table column providing the apparent visual magnitude of the stars and compare how many times brighter one is as compared to the other.

3. Spectral Class and Temperature. Each group member should select a different star listed in Appendix 3: Table 4—The Twenty Brightest Stars, and estimate its temperature based on its spectral class. Explain your reasoning.

RESEARCHING ON THE WEB  *To complete the following exercises, go to the online Destinations module for Chapter 17 on the Companion Website for* Astronomy Today 4/e.

1. Access the "All-Star Line Up" pages and rank the following stars from largest to smallest in radius: Betelgeuse, Mu Cephei, 40 Eridani, and Aldebaran.

2. Access the "Constellations and Their Stars" pages and determine the names of the three brightest stars in your horoscope birth sign.

3. Access the "Measuring the Brightness of Stars" page and describe the differences in star brightness that can be detected by the human eye.

PROJECTS

1. Every winter, you can find an astronomy lesson in the evening sky. The Winter Circle is an asterism—or pattern of stars—made up of six bright stars in five different constellations: Sirius, Rigel, Betelgeuse, Aldebaran, Capella, and Procyon. These stars span nearly the entire range of colors (and therefore temperatures) possible for normal stars. Rigel is a B-type star, Sirius is an A-type, Procyon is an F-type star, Capella is a G-type star, Aldebaran a K-type star, and Betelgeuse is an M-type star. The color differences of these stars are easy to see. Why do you suppose there is no O-type star in the Winter Circle?

2. In the winter sky, you'll find the red supergiant Betelgeuse in the constellation Orion. It's easy to see because it's one of the brightest stars visible in our night sky. Betelgeuse is a variable star with a period of about 6.5 years. Its brightness changes as it expands and contracts. At maximum size, Betelgeuse fills a volume of space that would extend from the Sun to beyond the orbit of Jupiter. Betelgeuse is thought to be about 10 to 15 times more massive than our Sun, and probably between four and 10 million years old. A similar star can be found shining prominently in midsummer. This is the red supergiant Antares in the constellation Scorpius. Depending on the time of year, can you find one of these stars? Why are they red?

SKYCHART III PROJECTS  *The SkyChart III Student Version planetarium program on which these exercises are based is included as a separately executable program on the CD in the back of this text.*

1. ■ To see examples of the relationship between the B–V color index of a star and its color, locate the following bright stars using SkyChart III: Spica, Rigel, Sirius, Vega, Arcturus, Betelgeuse. The stars are listed from hottest to coolest. Make a table with a column for the B–V color index and another column for the color the star appears to your eyes when it is displayed with SkyChart III. It will be easiest to locate the objects with SkyChart III by setting view for a 5° field, and using *VIEW/Center Object* to place each star in the center of the screen. Click on the star and the *Object Info* box will appear. In that box you will find details, such as color index, on the star. Deselect the horizon mask.

While the color of each star presented on your computer screen may be quite accurate, nothing beats seeing the actual star. Make a third column for the color you perceive it to be when you observe it in the sky, either through a telescope or with the unaided eye. On a clear night observe the stars yourself and relate their color to the color index. A telescope is not necessary for this observation, though a pair of binoculars might help, particularly if you are near city lights. You will find it necessary to work with the date and time settings of your simulation to determine the time of year each star is available to you at a time convenient for observation.

2. ■ Simulate the view through a telescope of the binary pair Albireo and its companion using SkyChart III. Locate Albireo in Cygnus, or more conveniently with *VIEW/Center Object/Albireo*. Choose a field of approximately 1°. Left click on each of the stars of the binary pair to obtain the *Object Info* table. Caution: the Albireo dot is really two close stars (view at 1/120° field to see both). Reclick on another part of the dot if the Albireo statistics do not appear. Note the relationship between color index and the apparent color of each star. When the opportunity presents itself, view this famous binary pair with a telescope and observe the actual colors. When you see these two jewels through a telescope you will appreciate why they are so popular with amateur astronomers.

To simulate the image through a telescope, select *TELESCOPE/Display*. Deselect *Camera field width*. Set *Finder field width* to 3° and *Eyepiece field width* to 1°. Under *TELESCOPE/Connect* select *Emulate Real Telescope Connection*. Close this window and a *Telescope Control* box should open. Click on Albireo and click on *Center* in the *Object Info* box. Then go to the *Telescope Control* and click on *Go To*. The concentric set of rings will slew over to center on Albireo. The center ring presents the image seen through a typical amateur telescope. Zoom out to 5° and you will see the image typically seen through the finder scope depicted by the larger circle.

3. ■■ For this exercise you will determine the number of stars visible to the naked eye. You will take several samples and determine the average number of stars per sample. By imaging the stars as all contained within a celestial sphere, the following formula can be used to determine the total number of stars:

$$\frac{\#\text{ of stars counted}}{0.5-0.5*\cos(R)} = \frac{\text{total }\#\text{ of visible stars}}{1}.$$

Your sample areas only represent a fraction of the total celestial sphere; "0.5 cos(R)", where R is the radius in degrees of your sample area, represents this fraction of the whole. Choose at least 10 sample areas, making a common star like Vega the center of your sample (this makes repeating the exercise much easier). Next you will use the telescope controls to define your sample area. Select *TELESCOPE/ Display* and set the *Finder field width* to 4°, and deselect the *Eyepiece field width* and *Camera field width*. Under *TELESCOPE/Connect* select *Emulate Real Telescope Connection* and click *OK*. The *Telescope Control* box will open. Use the *Find Object* command in the *EDIT* menu to *Select* and *Center* on the star you are looking for. Then go to the *Telescope Control* box and click on *Go To*. After a short time a circle will appear centered around the star you have chosen with a radius of 2°. The last step before counting the stars in the sample area is to adjust the visible limits to roughly match that of a person viewing from a typical suburban area. Select *DRAW/Symbols & grids*. In the upper left corner of the window, change *Faintest magnitude* to 6.0 (check the status bar to make sure 6.0 remains the limit, because if you zoom in or out, the limit may change). About how many visible stars are there?

4. ■■ Repeat the preceding exercise using the exact same sample areas but adjust the visible limits to match that of an average pair of binoculars, a limit of about 10. About how many more stars can be seen with a pair of binoculars than with the naked eye?

5. ■■ Measure parallax. The parallax for all stars is less than 1 arc second or all stars are farther away than 1 pc (see Chapter 1). However, if you were viewing from Pluto, the parallax would increase by roughly a factor of 40. From the planet Pluto, follow Eridani (find "Epsilon Eri") over 250 years by turning on trails and select *COMPUTATION/ Precision*. In the window that appears, select *Correct for Parallax* and set the distance to 1.0. Explain your observations.

 *In addition to the Practice Problems and Destinations modules, the Companion Website at http://www.prenhall.com/chaisson provides for each chapter an additional true-false, multiple choice, and labeling quiz, as well as additional annotated images, animations, and links to related Websites.*

18 THE INTERSTELLAR MEDIUM

Gas and Dust Among the Stars

LEARNING GOALS

Studying this chapter will enable you to:

1   Summarize the composition and physical properties of the interstellar medium.

2   Describe the characteristics of emission nebulae and explain their significance in the life cycle of stars.

3   Discuss the makeup of dark interstellar clouds.

4   Specify the radio techniques used to probe the nature of interstellar matter.

5   Discuss the nature and significance of interstellar molecules.

 Visit http://www.prenhall.com/chaisson for additional annotated images, animations, and links to related sites for this chapter.

Unprecedented detail is seen in this composite image of four exposures taken through six filters with the *Hubble Space Telescope*. This is the so-called Keyhole Nebula, some 8000 light-years away and dominated by a large, approximately circular feature about seven light-years across. The region is notable for its diversity of structure, especially its bright filaments of hot, glowing gas and its dark silhouetted clouds of cold gas and dust, all in rapid, chaotic motion. *(STScI)*

The Big Picture: Interstellar space is the place both where stars are "born" and to which they return at "death." Rich in gas and dust, yet extraordinarily thinly distributed throughout the vast, dark regions among the stars, interstellar matter occasionally glows as nebulae (like that shown at left) or contracts to form new stars (and sometimes planets).

Stars and planets are not the only inhabitants of our Galaxy. The space around us harbors invisible matter throughout the dark voids between the stars. The density of this interstellar matter is extremely low—approximately a trillion trillion times less dense than matter in either stars or planets, far more tenuous than the best vacuum attainable on Earth. Only because the volume of interstellar space is so vast does its mass amount to anything at all. So why bother to study this near-perfect vacuum? We do so for three important reasons. First, there is nearly as much mass in the "voids" among the stars as there is in the stars themselves. Second, interstellar space is the region out of which new stars are born. Third, it is the region into which some old stars explode at death. It is one of the most significant crossroads through which matter passes anywhere in our universe.

18.1 Interstellar Matter

1 Figure 18.1 shows a large region of space, a much greater expanse of universal real estate than anything we have studied thus far. The bright regions are congregations of innumerable stars, some of whose properties we have just studied in Chapter 17. However, the dark areas are not simply "holes" in the stellar distribution. They are regions of space where *interstellar matter* obscures the light from stars beyond. Their very darkness means that they cannot be easily studied by the optical methods used for stellar matter. There is, quite simply, little to see!

From Figure 18.1 (see also Figure 18.5) it is evident that the dark interstellar matter is distributed unevenly throughout space. In some directions the obscuring matter is largely absent, allowing astronomers to study objects literally billions of parsecs from the Sun. In other directions there are small amounts of interstellar matter, so the obscuration is moderate, preventing us from seeing objects more than a few thousand parsecs away, but still allowing us to study nearby stars. Still other regions are so heavily obscured that starlight from even relatively nearby stars is completely absorbed before reaching Earth. (See also *Discovery 18-1*.)

GAS AND DUST

The matter among the stars is called the **interstellar medium**. It is made up of two components—gas and dust—intermixed throughout all space. The gas is made up mainly of individual atoms, of average size 10^{-10} m (0.1 nm) or so, and small molecules, no larger than about 10^{-9} m across. Interstellar dust is more complex. It consists of clumps of atoms and molecules—not unlike chalk dust or the microscopic particles that make up smoke, soot, or fog.

Aside from numerous narrow atomic and molecular absorption lines, the gas alone does not block radiation to any great extent. The obscuration evident in Figure 18.1 is caused by dust. Light from distant stars cannot penetrate

Figure 18.1 Milky Way
A wide-angle photograph of a great swath of space, showing regions of brightness (vast fields of stars) as well as dark regions due to interstellar dust obscuring the light from more distant stars. The field of view is roughly 30° across. A few of the nebulae discussed later in the chapter are labeled. *(Palomar/Caltech)*

the densest accumulations of interstellar dust any more than a car's headlights can illuminate roadside objects in a thick fog. The size of a typical interstellar dust particle—or **dust grain**—is about 10^{-7} m (0.1 μm), comparable in size to the wavelength of visible light.

The ability of a particle to scatter a beam of light depends on both the size of the particle and the wavelength of the radiation involved. ∞ (*More Precisely 7-1*) As a rule of thumb, only particles having diameters comparable to or larger than the wavelength can significantly influence the beam, and the amount of scattering produced by particles of a given size increases with decreasing wavelength. Consequently, dusty regions of interstellar space are transparent to long-wavelength radio and infrared radiation but opaque to shorter-wavelength optical, ultraviolet, and X-ray radiation. This dimming of starlight by interstellar matter is called **extinction**.

Because the interstellar medium is more opaque to short-wavelength radiation than to radiation of longer wavelengths, light from distant stars is preferentially robbed of its higher-frequency ("blue") components. Hence, in addition to being generally diminished in overall brightness, stars also tend to appear redder than they really are, an effect known as **reddening**.

As illustrated in Figure 18.2, extinction and reddening change a star's apparent brightness and color, but absorption lines in the original stellar spectrum are still recognizable in the radiation reaching Earth, so the star's spectral class can be determined. Astronomers can use this fact to study the interstellar medium. From a main-sequence star's spectral and luminosity classes astronomers learn its true luminosity and color. ∞ (Secs. 17.6, 17.7) They then measure the degree to which the starlight has been affected by extinction and reddening en route to Earth, and this, in turn, allows them to estimate both the numbers and the sizes of interstellar dust particles along the line of sight to the star. By repeating these measurements for stars in many different directions and at many different distances from Earth, astronomers have built up a picture of the distribution and overall properties of the interstellar medium in the solar neighborhood.

OVERALL DENSITY

Gas and dust are found everywhere in interstellar space—no part of our Galaxy is truly devoid of matter. However, the density of the interstellar medium is extremely low. Overall, the gas averages roughly 10^6 atoms per cubic meter—just 1 atom per cubic centimeter—although there are large variations from place to place. Densities as great as 10^9 atoms/m^3 and as small as 10^4 atoms/m^3 have been found (*Discovery 18-1*). Matter of such low density is far more tenuous than the best vacuum—about 10^{10} molecules/m^3—ever attained in laboratories on Earth. Interstellar dust is even rarer. On average, there is only one dust particle for every trillion or so atoms—just 10^{-6} dust particles per cubic meter, or 1000 per cubic *kilo*meter. Some parts of interstellar space are so thinly populated that harvesting all the gas and dust in a

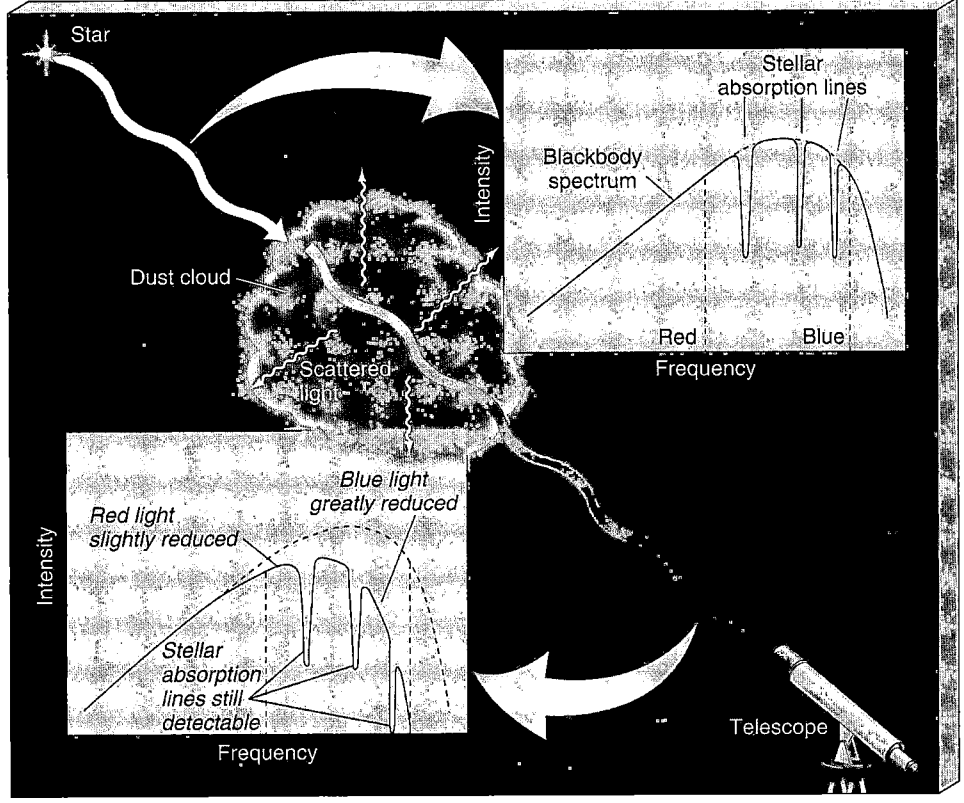

Figure 18.2 Reddening Starlight passing through a dusty region of space is both dimmed and reddened, but spectral lines are still recognizable in the light that reaches Earth. By recognizing stellar spectral features and inferring a star's intrinsic properties, astronomers can estimate the amount of obscuring dust along the line of sight. (Note that this reddening has nothing to do with the Doppler effect—the frequencies of the lines are unchanged, although their intensities may be reduced.)

DISCOVERY 18-1

Ultraviolet Astronomy and the "Local Bubble"

The ultraviolet is that part of the spectrum where we expect to witness events and see objects involving temperatures in the hundreds of thousands, even millions, of kelvins—hot regions like the seething atmospheres or eruptive flares of stars; and violent events, such as massive stars caught in the act of exploding, and active galaxies whose spinning hearts may harbor black holes. Yet, ultraviolet astronomy has also contributed greatly to the study of the interstellar medium by allowing a unique mapping of our local cosmic neighborhood. The line-of-sight technique for analyzing interstellar clouds, described in the text for optical observations, can also be applied to ultraviolet radiation. Astronomers search the ultraviolet spectra of distant, hot stars for narrow absorption lines caused by intervening interstellar clouds.

Short-wavelength ultraviolet radiation is effectively blocked by Earth's atmosphere. To study this region of the electromagnetic spectrum, instruments must be placed above our planet's atmospheric blanket. ∞ (Sec. 5.6) One of the most successful satellites yet launched—placed in geosynchronous orbit in 1978 and finally shut down in late 1996—was the *International Ultraviolet Explorer* (*IUE*). The photograph below shows the *IUE* Science Operations Center at the University of California at Berkeley, where undergraduate students worked alongside scientists and engineers to command the spacecraft and acquire its data. More recently, the *Extreme Ultraviolet Explorer* (*EUVE*), which operated from 1992 until 2000, closed one of the few remaining gaps in the electromagnetic spectrum not well-explored by other means.

Only 25 years ago, astronomers had presumed that the gases filling the spaces among the stars would absorb

virtually all short-wavelength UV radiation before it had a chance to reach Earth. But in 1975, during a historic linkup of the *Apollo* (U.S.) and *Soyuz* (USSR) space capsules, the onboard astronauts and cosmonauts performed a key experiment—they used a small telescope to detect extreme ultraviolet radiation from a few nearby very hot stars. Shortly thereafter, theoretical ideas changed as astronomers began to realize that the interstellar gas is not all uniformly distributed in space. Rather, it is spread very unevenly, in cool dense clumps interspersed by regions of hot, low-density gas shaped like bubbles and tunnels.

Observations made by *IUE* of weak spectral lines from highly excited atoms showed that some regions of interstel-

(UC/Berkeley)

region the size of Earth would yield barely enough matter to make a pair of dice.

How can such fantastically sparse matter diminish light radiation so effectively? The key is size—interstellar space is vast. The typical distance between stars (1 pc or so in the vicinity of the Sun) is much, much greater than the typical size of the stars themselves (around 10^{-7} pc). Stellar and planetary sizes pale in comparison to the vastness of interstellar space. Thus, matter can accumulate, regardless of how thinly spread. For example, an imaginary cylinder 1 m^2 in cross section and extending from Earth to Alpha Centauri would contain more than 10 billion billion dust particles. Over huge distances, dust particles accumulate slowly but surely, to the point at which they can effectively block visible light and other short-wavelength radiation. Even though the density of matter is very low, interstellar space in the vicinity of the Sun contains about as much mass as exists in the form of stars.

Despite their rarity, dust particles make interstellar space a *relatively* dirty place. Earth's atmosphere, by comparison, is about a million times cleaner. Our air is tainted by only one dust particle for about every billion billion (10^{18}) atoms of atmospheric gas. If we could compress a typical parcel of interstellar space to equal the density of air on Earth, this parcel would contain enough dust to make a fog so thick that we would be unable to see our hand held at arm's length in front of us.

COMPOSITION

The composition of interstellar gas is reasonably well understood from spectroscopic studies of absorption lines formed when light from a distant star interacts with gas along the line of sight (see Section 18.3). In most cases, the elemental abundances detected in interstellar gas mirror those found in other astronomical objects, such as the Sun,

lar space are much thinner (5000 atoms/m³) and hotter (500,000 K) than previously expected. Part of the space among the dust clouds and the emission nebulae seems to contain extremely dilute yet seething plasma, probably the result of the concussion and expansion of debris from stars that exploded long ago. Observations by *EUVE* and other space-based instruments have found that these superheated interstellar "bubbles," or *intercloud medium*, may extend far into interstellar space beyond our local neighborhood, and conceivably into the even vaster spaces among the galaxies.

The Sun seems to reside in one such low-density region—a huge cavity called the *Local Bubble*, sketched in the second figure. Only because we live in this vast, low-density bubble can we detect so many stars in the extreme UV; the hot, tenuous interstellar gas is virtually transparent to this radiation. The Local Bubble contains about 200,000 stars and extends for hundreds of trillions of kilometers, or nearly 100 pc. It was probably carved out by multiple supernova explosions that occurred several hundred thousand years ago in the Scorpius-Centaurus association, a rich cluster of young stars. Perhaps ∞ (Sec. 21.3) our hominid ancestors may have seen them—stellar catastrophies as bright as the full Moon.

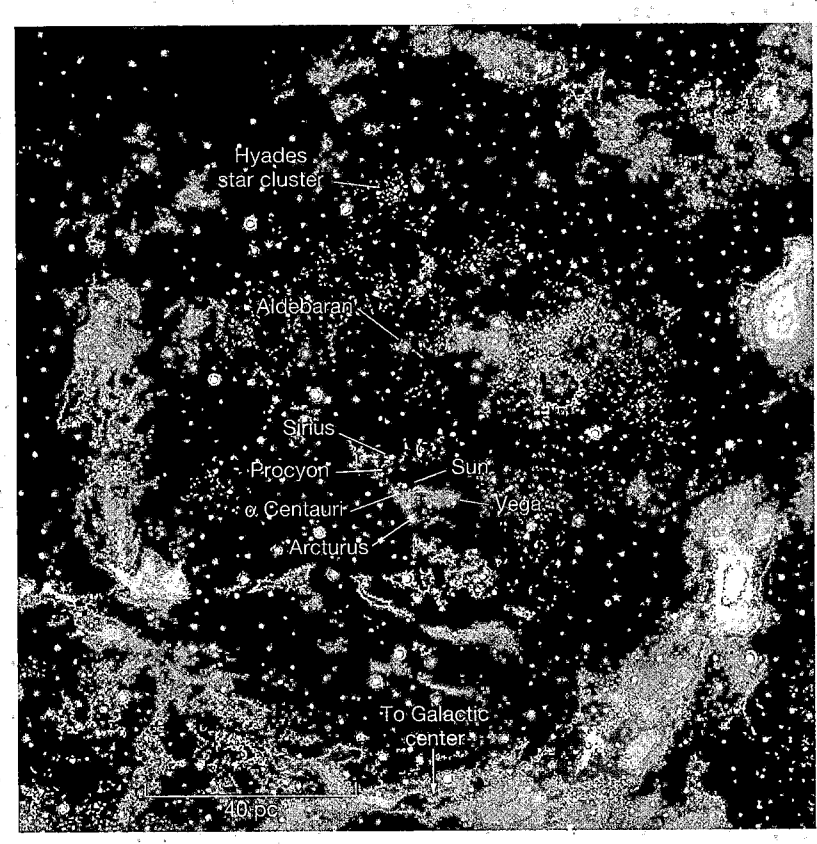

(Rice University)

---

the stars, and the jovian planets. Most of the gas—about 90 percent by number—is atomic or molecular hydrogen; some 9 percent is helium, and the remaining 1 percent consists of heavier elements. The abundances of some heavy elements, such as carbon, oxygen, silicon, magnesium, and iron, are much lower in interstellar gas than in our solar system or in stars. The most likely explanation for this finding is that substantial quantities of these elements have been used to form the interstellar dust, taking them out of the gas and locking them up in a form that is much harder to observe.

In contrast to interstellar gas, the composition of interstellar dust is currently not very well known. We have some infrared evidence for silicates, graphite, and iron—the same elements that are underabundant in the gas—lending support to the theory that interstellar dust forms out of interstellar gas. The dust probably also contains some "dirty ice," a frozen mixture of ordinary water ice contaminated with trace amounts of ammonia, methane, and other chem-

ical compounds. This composition is quite similar to that of cometary nuclei in our own solar system. ∞ (Sec. 14.2)

DUST SHAPE

Curiously, astronomers know the *shapes* of interstellar dust particles better than their composition. Although the minute atoms in the interstellar gas are basically spherical, the dust particles are not. Individual dust grains are apparently elongated or rodlike, as shown in Figure 18.3(a), although recent theoretical studies of how dust particles collide, stick, and break up suggest that their larger-scale structure may be considerably more complex (Figure 18.3b). We can infer this elongated structure because the light emitted by stars is dimmed and partially **polarized**, or aligned, by the dust.

Recall from Chapter 3 that light consists of electromagnetic waves composed of vibrating electric and magnetic

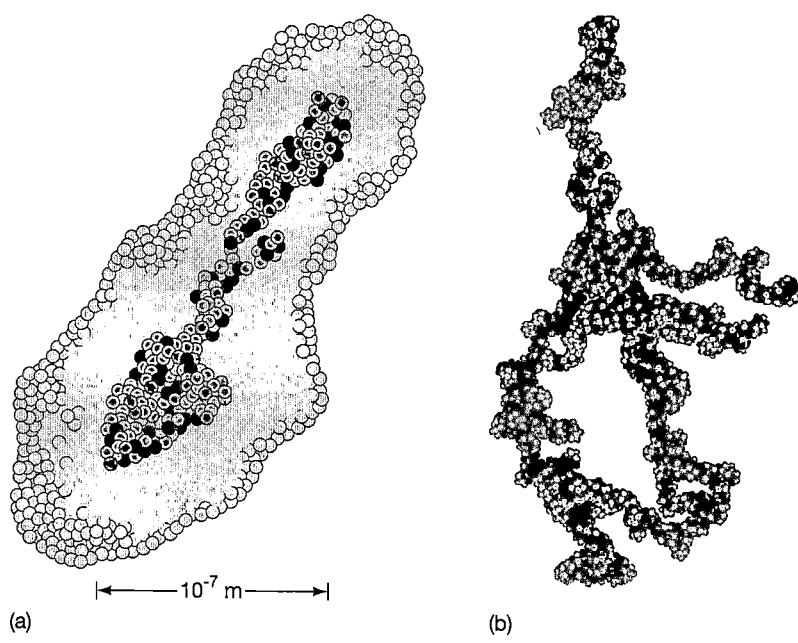

(a)

|←————10^{-7} m————→|

(b)

Figure 18.3 Interstellar Dust (a) A diagram of a typical interstellar dust particle, as inferred from polarization studies. The average size of such particles is only 1/10,000 of a millimeter, yet space contains enough of them to obscure our view in certain directions. Part (b) shows the results of a computer simulation of how grains may grow as dust particles collide, stick, and fragment in interstellar space. The resulting grains are linear, or rodlike, on small scales, but on larger scales may become tangled and twisted in complex ways.

fields. ⊂⊃ (Sec. 3.2) Normally, these waves are randomly oriented, and the radiation is unpolarized. Stars emit unpolarized radiation from their photospheres. Under some circumstances, however, the radiation can become aligned, with all electric fields vibrating in the same plane as the radiation moves through space. ⊂⊃ (*Discovery 3-1*) Polarization of starlight does not occur by chance. If the light detected by our telescope is polarized, it is because some interstellar matter lies between the emitting object and Earth. The polarization of starlight, then, provides another way to study the interstellar medium.

On Earth we can produce polarized light by passing unpolarized light through a Polaroid filter, which has specially aligned elongated molecules that allow the passage of only those waves having electric fields oriented in some specific direction (see Figure 18.4a). Other waves are absorbed and so do not pass through the filter. The alignment of the molecules determines which waves will be transmitted. In interstellar space, dust grains can act like the molecules in the Polaroid filter. If the starlight is polarized, astronomers can conclude that the interstellar dust particles must have an *elongated* shape (like the elongated molecules of the Polaroid filter) and that these molecules are aligned, as shown in Figure 18.4(b). Only then can the dust preferentially absorb certain waves, leaving the remainder (the ones we observe) polarized.

Figure 18.4 Polarization (a) Unpolarized light waves have randomly oriented electric fields. When the light passes through a Polaroid filter, only waves whose electric fields are oriented in a specific direction are transmitted, and the resulting light is polarized. (b) Aligned dust particles in interstellar space polarize radiation in a similar manner. Observations of the degree of polarization allow astronomers to infer the size, shape, and orientation of the particles.

The alignment of the interstellar dust is the subject of ongoing research among astronomers. The current view, accepted by most, is that the dust particles are affected by a weak interstellar magnetic field, perhaps a million times

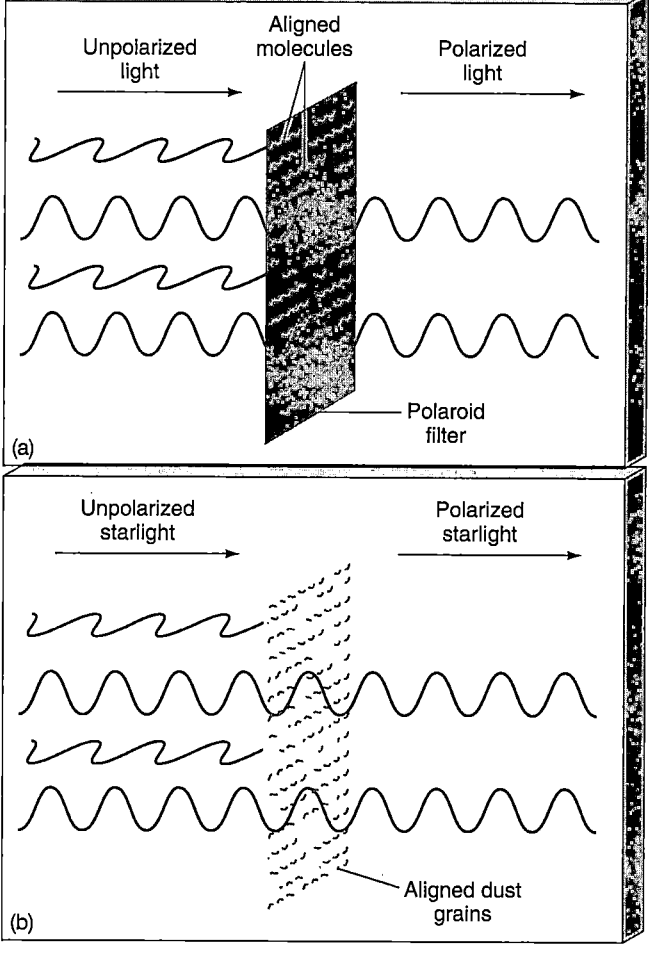

weaker than Earth's field. Each dust particle responds to the field in much the same way that small iron filings are aligned by an ordinary bar magnet. Measurements of the blockage and polarization of starlight thus yield information about the size and shape of interstellar dust particles, as well as about magnetic fields in interstellar space.

Concept Check

■ If space is a near-perfect vacuum, how can there be enough dust in it to block starlight?

18.2 Emission Nebulae

Figure 18.5 is a mosaic of photographs showing a region of the sky even larger than that shown in Figure 18.1. The bright areas are made up of myriad unresolved stars. The dark areas are vast expanses of dust, blocking from our view what would otherwise be a rather smooth distribution of bright starlight. The patchiness of the obscuration is evident. From our vantage point on Earth, this assemblage of stellar and interstellar matter stretches all the way across the sky. On a clear night, it is visible to the naked eye as the Milky Way. In Chapter 23, we will come to recognize this band as the flattened disk, or *plane*, of our own Galaxy.

Figure 18.6 shows a 12°-wide swath of the galactic plane in the general direction of the constellation Sagittarius, as photographed from Earth. The field of view is mottled with stars and interstellar matter. In addition, several large fuzzy patches of light are clearly visible. These fuzzy objects, labeled M8, M16, M17, and M20, correspond to the eighth, sixteenth, seventeenth, and twentieth objects in a catalog compiled by Charles Messier, an eighteenth-century French astronomer.*

Messier was actually more concerned with making a list of celestial objects that might possibly be confused with comets, his main astronomical interest. However, the catalog of 109 "Messier objects" is now regarded as a much more important contribution to astronomy than any comets Messier discovered.

Today they are known as **emission nebulae**—glowing clouds of hot interstellar matter.

OBSERVATIONS OF EMISSION NEBULAE

Historically, astronomers used the term **nebula** to refer to any "fuzzy" patch (bright or dark) on the sky—any region of space that was clearly distinguishable through a telescope but not sharply defined, unlike a star or a planet. We now know that many (though not all) nebulae are clouds of interstellar dust and gas. If they happen to obscure stars lying behind them, we see a dark patch. If something within the cloud—a group of hot young stars, for example—causes it to glow, the nebula appears bright instead. The method of spectroscopic parallax applied to stars visible within the emission nebulae shown in Figure 18.6 indicates that their distances from Earth range from 900 pc (M20) to 1800 pc (M16). ∞ (Sec. 17.8) Thus, all four are near the limit of visibility for any object embedded in the dusty Galactic plane. M16, at the top left, is approximately 1000 pc from M20, near the bottom.

We can gain a better appreciation of these nebulae by examining progressively smaller fields of view. Figure 18.7 is an enlargement of the region near the bottom of Figure 18.6, showing M20 at the top and M8 at the bottom, only a few degrees away. Figure 18.8 is another enlargement of the top of Figure 18.7, presenting a close-up of M20 and its immediate environment. The total area of the close-up view displayed measures some 10 pc across. Emission nebulae are among the most spectacular objects in the entire universe, yet they appear only as small, undistinguished patches of light when viewed in the larger context of the Milky Way, as in Figure 18.5. Perspective is crucial in astronomy.

These nebulae are regions of glowing, ionized gas. At or near the center of each is at least one newly formed hot O- or B-type star producing copious amounts of ultraviolet light. As ultraviolet photons travel outward from the star, they ionize the surrounding gas. As electrons recombine with nuclei, they emit visible radiation, causing the

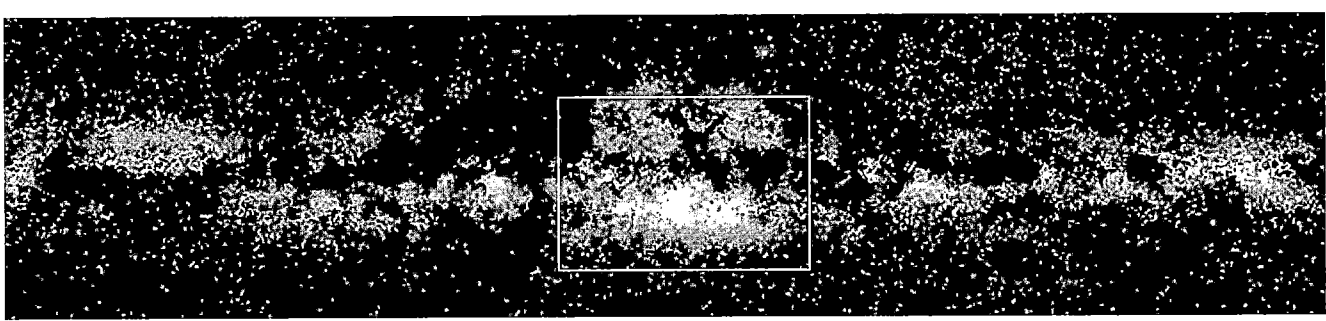

Figure 18.5 Milky Way Mosaic The Milky Way Galaxy, photographed almost from horizon to horizon, thus extending over nearly 180°. This band contains high concentrations of stars as well as interstellar gas and dust. The field of view is several times wider than that of Figure 18.1, whose outline is superimposed on this image. *(Axel Mellinger)*

R I V U X G

gas to fluoresce, or glow. ∞ (Sec. 4.2) Figure 18.9 shows enlargements of two of the other nebulae visible in Figure 18.6. Notice the predominant red coloration of the emitted radiation and the hot bright stars embedded within the glowing nebular gas.

The interaction between stars and gas is particularly striking in Figure 18.9(b). The three dark pillars visible in this spectacular *Hubble Space Telescope* image are part of the interstellar cloud from which the stars formed. The rest of the cloud in the vicinity of the new stars has already been dispersed by their radiation, in a process

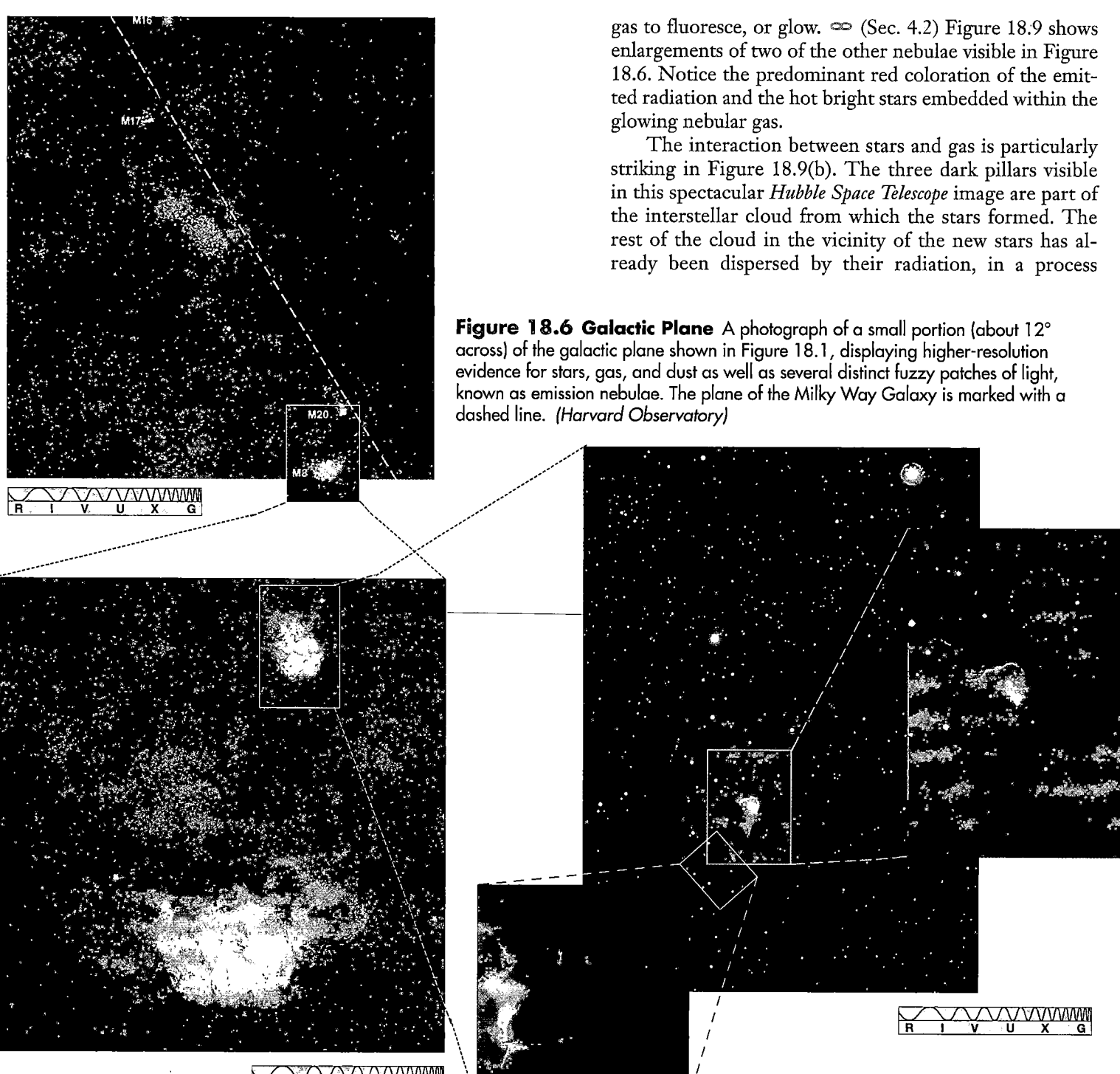

Figure 18.6 Galactic Plane A photograph of a small portion (about 12° across) of the galactic plane shown in Figure 18.1, displaying higher-resolution evidence for stars, gas, and dust as well as several distinct fuzzy patches of light, known as emission nebulae. The plane of the Milky Way Galaxy is marked with a dashed line. *(Harvard Observatory)*

Figure 18.7 M20-M8 Region An enlargement of the bottom of Figure 18.6, showing M20 (top) and M8 (bottom) more clearly. *(Royal Observatory, Edinburgh)*

Figure 18.8 Trifid Nebula Further enlargements of the top of Figure 18.7, showing only M20 and its interstellar environment. The nebula itself (in red) is about 4 pc in diameter. It is often called the Trifid Nebula because of the dust lanes that trisect its midsection (see inset at right). The blue region is unrelated to the red emission nebula and is caused by starlight reflected from intervening dust particles. It is called a *reflection nebula.* The inset at left shows a 0.5-pc-long jet, probably extending from a young star embedded in a pillar of gas and dust. *(AURA; D. Malin; NASA)*

Figure 18.9 Emission Nebulae Enlargements of selected portions of Figure 18.6. (a) M16, the Eagle Nebula. (b) A *Hubble* image of huge columns of cold gas and dust inside M16, delicate sculptures created by the action of stellar ultraviolet radiation on the original cloud. In this image, green represents emission from hydrogen atoms; red, emission from singly ionized sulfur; and blue, emission from doubly ionized oxygen. (c) M8, the Lagoon Nebula. (d) A high-resolution view of the core of M8, a region known as the Hourglass. Notice in all cases the irregular shape of the emitting regions, the characteristic red color of the light, the bright stars within the gas, and the patches of obscuring dust. *(AURA; NASA)*

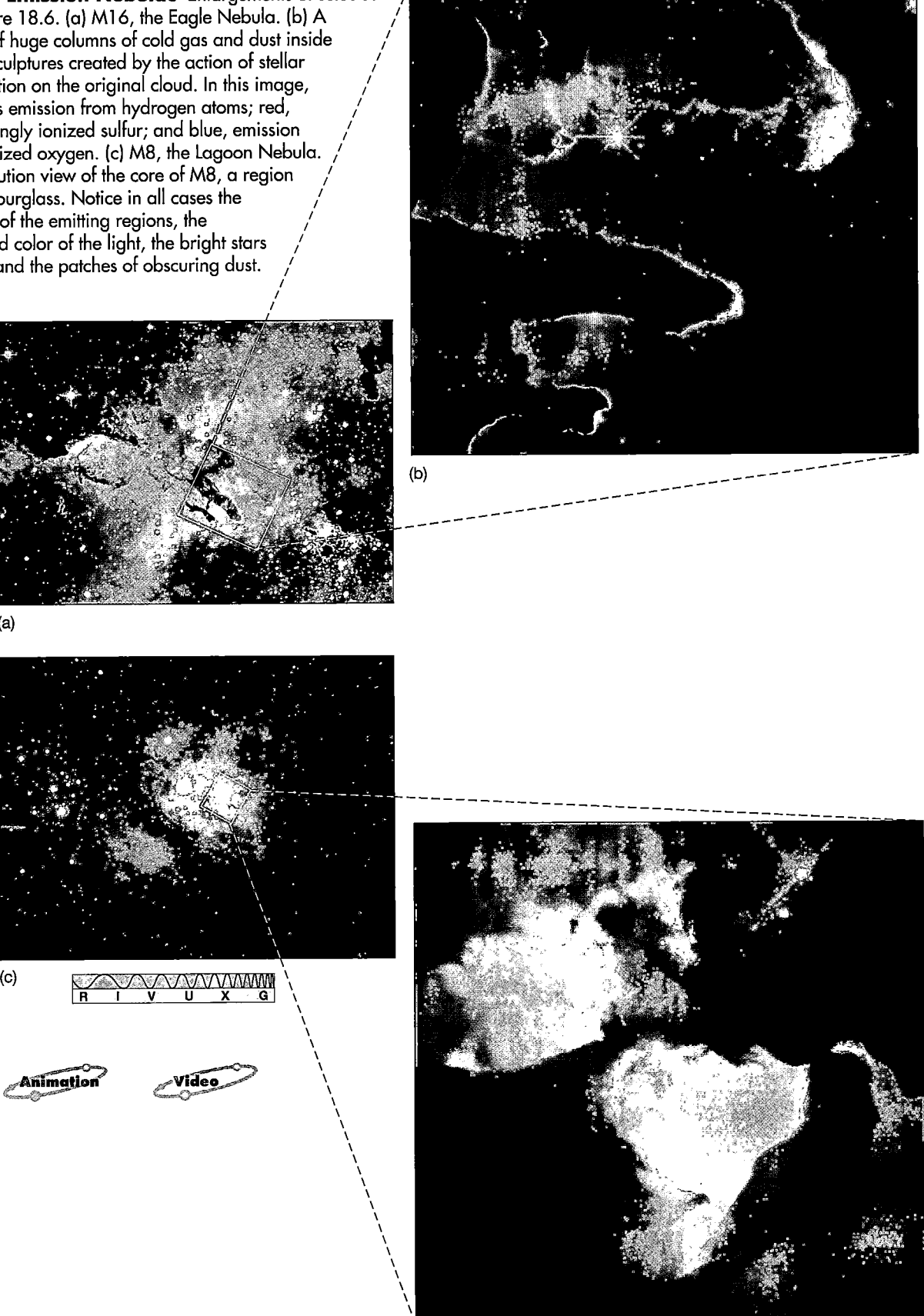

(a)

(b)

(c)

R   I   V   U   X   G

(d)

known as *photoevaporation*. The fuzz around the edges of the pillars, especially at top right and center, is the result of this process. As photoevaporation continues, it eats away the less dense material first, leaving behind delicate sculptures composed of the denser parts of the original cloud, just as wind and water create spectacular structures in Earth's deserts and shores by eroding away the softest rock. The process is a dynamic one—the pillars will eventually be destroyed, but probably not for another hundred thousand or so years.

The reddish hue of these nebulae—and in fact of all emission nebula—results when hydrogen atoms emit light in the red part of the visible spectrum. Specifically, it is caused by the emission of radiation at 656.3 nm—the Hα line discussed in Chapter 4. ∞ (*More Precisely 4-1*) Other elements in the nebula also emit radiation as their electrons recombine, but because hydrogen is so plentiful, its emission usually dominates.

Woven through the glowing nebular gas, and plainly visible in Figures 18.7–18.9, are lanes of dark obscuring dust. Recent studies have demonstrated that these *dust lanes* are part of the nebulae and are not just unrelated dust clouds that happen to lie along our line of sight. This relationship is particularly evident in Figures 18.9(b) and (d), where regions of gas and dust are simultaneously silhouetted against background nebular emission and illuminated by foreground nebular stars.

Spectroscopists often refer to the *ionization state* of an atom by attaching a roman numeral to its chemical symbol—I for the neutral atom, II for a singly ionized atom (missing one electron), III for a doubly ionized atom (missing two electrons), and so on. Because emission nebulae are composed mainly of ionized hydrogen, they are often referred to as *HII regions*. Regions of space containing primarily neutral (atomic) hydrogen are known as *HI regions*.

NEBULAR SPECTRA

Most of the photons emitted by the recombination of electrons with atomic nuclei escape from the nebula. Unlike the ultraviolet photons originally emitted by the embedded stars, they do not carry enough energy to ionize the nebular gas, and they pass through the nebula relatively unhindered. Some eventually reach Earth. Only through these lower-energy photons do we learn anything about emission nebulae.

Nebular spectra tell us a great deal about ionized interstellar gas. Because at least one hot star resides near the center of the nebula, we might think that the combined spectrum of the star and the nebula would be hopelessly confused. In fact, they are not. We can easily distinguish nebular spectra from stellar spectra because the physical conditions in stars and nebulae differ so greatly. In particular, emission nebulae are made of hot thin gas that, as we saw in Chapter 4, yields detectable *emission* lines. ∞ (Sec. 4.2) When our spectroscope is trained on a star, we see a

familiar stellar spectrum, consisting of a blackbody-like continuous spectrum and absorption lines, together with superimposed emission lines from the nebular gas. When no star appears in the field of view, only the emission lines are seen.

Figure 18.10(b) is a typical nebular emission spectrum spanning part of the visible and near-ultraviolet wavelength interval. Numerous emission lines can be seen, and information on the nebula shown in Figure 18.10(a) can be extracted from all of them. The results of analyses of many nebular spectra show abundances close to those derived from observations of the Sun and other stars and elsewhere in the interstellar medium: hydrogen is about 90 percent abundant by number, followed by helium at about 9 percent; the heavier elements together make up the remaining 1 percent.

Unlike stars, nebulae are large enough for their actual sizes to be measurable by simple geometry. Coupling this size information with estimates of the amount of matter along our line of sight (as revealed by the nebula's total emission of light), we can find the nebula's density. Generally, emission nebulae have only a few hundred particles, mostly protons and electrons, in each cubic centimeter—a density some 10^{22} times lower than that of a typical planet. Spectral-line widths imply that the gas atoms and ions have temperatures around 8000 K. ∞ (Sec. 4.4) Table 18.1 lists some vital statistics for each of the nebulae shown in Figure 18.6.

"FORBIDDEN" LINES

When astronomers first studied the spectra of emission nebulae, they found many lines that did not correspond to anything observed in terrestrial laboratories. For example, in addition to the dominant red coloration just discussed, many nebulae also emit light with a characteristic green color (see Figure 18.11). The greenish tint of portions of this nebula puzzled astronomers in the early twentieth century and defied explanation in terms of the spectral lines known at the time, prompting speculation that the nebulae contained elements unknown on Earth. Some scientists even went so far as to invent the term "nebulium" for a new element, much as the name helium came about when that element was first discovered in the Sun (recall "coronium"

TABLE 18.1 Some Nebular Properties

OBJECT	APPROXIMATE DISTANCE (pc)	AVERAGE DIAMETER (pc)	DENSITY (10^6 particles/m³)	MASS (solar masses)	TEMPERATURE (K)
M8	1200	14	80	2600	7500
M16	1800	8	90	600	8000
M17	1500	7	120	500	8700
M20	900	4	100	150	8200

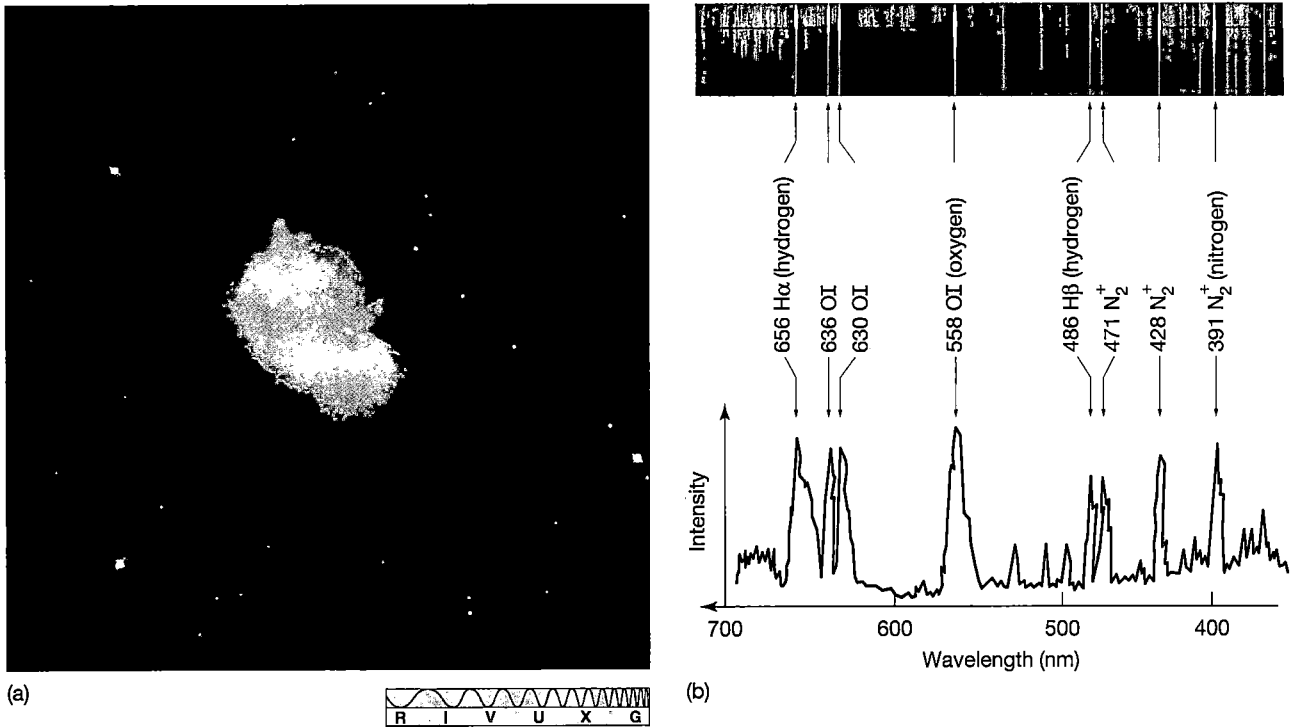

(a)

(b)

Figure 18.10 Emission Nebula Spectrum (a) Nebula NGC 2346 is a glowing patch of gas about 0.2 pc across and residing some 700 pc away. (b) Its emission spectrum shows light intensity over the entire visible portion of the electromagnetic spectrum from red to deep violet. *(NASA; Harvard College Observatory)*

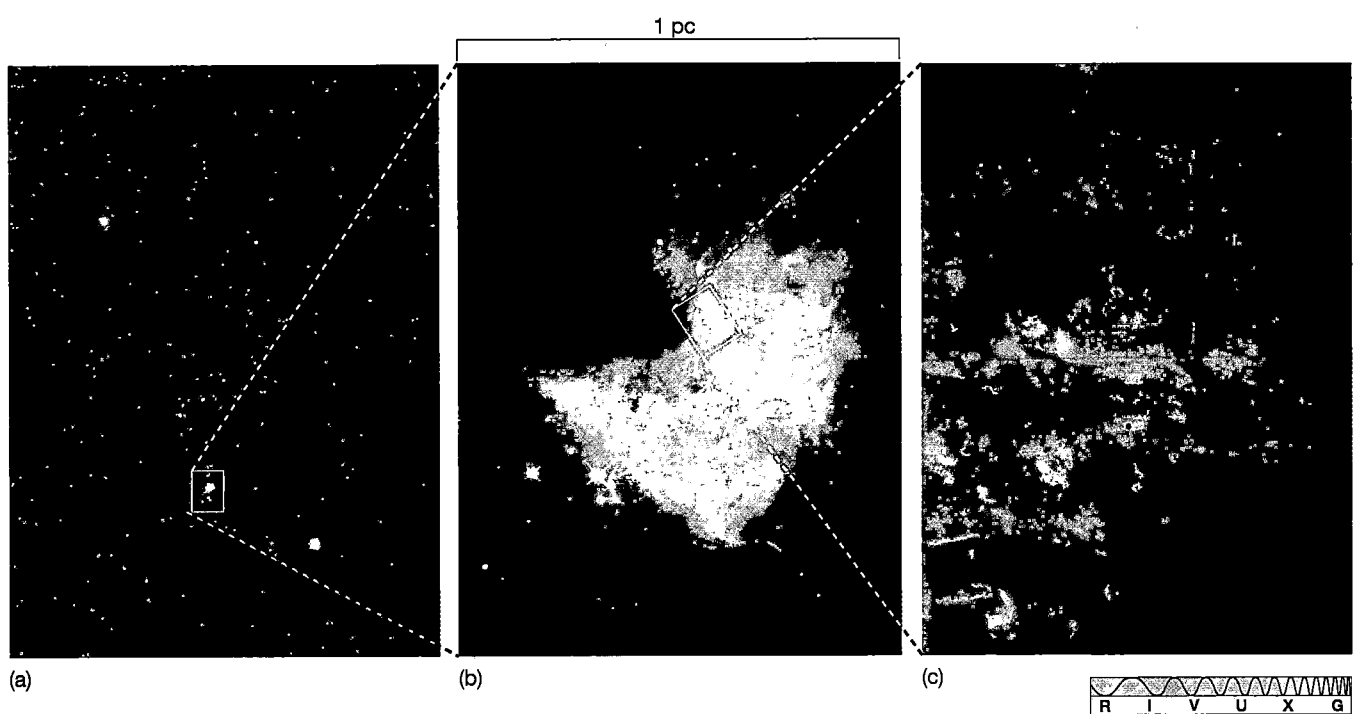

(a)   (b)   (c)

Figure 18.11 Orion Nebula (a) Lying some 450 pc from Earth, the Orion Nebula (M42) is visible to the naked eye as the fuzzy middle "star" of Orion's sword. (b) Like all emission nebulae, it consists of hot, glowing gas powered by a group of bright stars in the center. In addition to the red Hα emission, parts of the nebula show a slight greenish tint, caused by a so-called forbidden transition in ionized oxygen. (c) A high-resolution, approximately true-color image shows rich detail in a region about 0.5 light-year across. Structural details are visible down to a level of 0.1", or six light-*hours*—a scale comparable to our solar system. *(Harvard Observatory; Lick Observatory; NASA)*

from Chapter 16). ⊂∞ (Sec. 16.3) Later, with a fuller understanding of the workings of the atom, astronomers realized that these lines did in fact result from electron transitions within the atoms of familiar elements, but under unfamiliar conditions not reproducible in laboratories. Astronomers now understand that the greenish tint in Figure 18.11(b) and (c) is caused by a particular electron transition in doubly ionized oxygen. However, the structure of oxygen is such that an ion in the higher-energy state for this transition tends to remain there for a very long time—many hours, in fact—before dropping back to the lower state and emitting a photon. Only if the ion is left undisturbed during this time, and not kicked into another energy state, will the transition actually occur and the photon be emitted.

In a terrestrial experiment no atom or ion is left undisturbed for long. Even in a "low-density" laboratory gas, there are many trillions of particles per cubic meter, and each particle experiences millions of collisions every second. The result is that an ion in the particular energy state that produces the peculiar green line in the nebular spectrum never has time to emit its photon in the lab—collisions kick it into some other state long before that occurs. For this reason the line is usually called *forbidden*, even though it violates no laws of physics. It simply occurs on Earth with such low probability that it is never seen.

In an emission nebula, the density is so low that collisions between particles are very rare. There is plenty of time for the excited ion to emit its photon, so the forbidden line is produced. Numerous forbidden lines are known in nebular spectra. They remind us once again that the environment in the interstellar medium is very different from conditions on Earth and warn us of the problems of extending our terrestrial experience to the study of interstellar space.

☑ Concept Check

■ If emission nebulae are powered by ultraviolet radiation from very hot (blue-white) stars, why do they appear red?

18.3 Dark Dust Clouds

③ Emission nebulae are only one small component of interstellar space. Most of space—in fact, more than 90 percent of it—is devoid of nebular regions or superheated "bubbles," and contains no stars. It is simply dark. Look again at Figure 18.5, or just ponder the evening sky. The dark regions are by far the most representative of interstellar space. The average temperature of a typical dark region of interstellar matter is about 100 K. Compare this with 273 K, at which water freezes, and 0 K, at which atomic and molecular motions cease. ⊂∞ (*More Precisely 3-1*) Interstellar space is very cold.

Within these dark voids among the nebulae and the stars lurks another type of astronomical object—the **dark dust cloud**. These clouds are even colder than their surroundings (with temperatures as low as a few tens of kelvins), and thousands or even millions of times denser. Along any given line of sight, their densities can range from 10^7 atoms/m^3 to more than 10^{12} atoms/m^3 (10^6 atoms/cm^3). These latter clouds are generally called *dense* interstellar clouds by researchers, but even these densest interstellar regions are about as tenuous as the best laboratory vacuum. Still, it is because their density is much larger than the average value of 10^6 atoms/m^3 that we can distinguish clouds from the surrounding expanse of interstellar space.

These clouds bear little resemblance to terrestrial clouds. Most are bigger than our solar system, and some are many parsecs across. (Yet even so, they make up no more than a few percent of the entire volume of interstellar space.) Despite their name, these clouds are made up primarily of gas, just like the rest of the interstellar medium. However, their absorption of starlight is due almost entirely to the dust they contain.

OBSCURATION OF VISIBLE LIGHT

Figures 18.12(a) and (b) are optical photographs of a typical interstellar dust cloud. Pockets of intense blackness mark regions where the dust and gas are especially concentrated and the light from background stars is completely obscured. This cloud takes its name from a nearby star, Rho Ophiuchi, and resides relatively nearby—about 300 pc away. Measuring several parsecs across, this cloud is only a tiny part of the grand mosaic shown in Figure 18.5. This cloud clearly is far from spherical. Indeed, most interstellar clouds are very irregularly shaped. Note especially the long "streamers" of (relatively) dense dust and gas in Figure 18.12(b).

The bright patches within the dark region in Figure 18.12(a) are foreground objects—emission nebulae and groups of bright stars. Some of them are part of the cloud itself, where newly formed stars near the surface have created a "hot spot" in the cold, dark gas. Others have no connection to the cloud and just happen to lie along our line of sight. The additional foreground stars in Figure 18.12(b) are too faint to be seen on Figure 18.12(a).

Like all dark dust clouds, the Rho Ophiuchi cloud is too cold to emit any visible light. However, it does radiate strongly at longer wavelengths. Figure 18.12(c) shows an infrared view of the same region, captured by sensitive detectors aboard the *Infrared Astronomy Satellite*. ⊂∞ (Sec. 5.6) These dark and dusty interstellar clouds are sprinkled throughout our Galaxy. We can study them at optical wavelengths only if they happen to block the light emitted by more distant stars or nebulae. The dark outline of Rho Ophiuchi in Figure 18.12(a) and the dust lanes visible in Figures 18.8, 18.9, and 18.12(b)

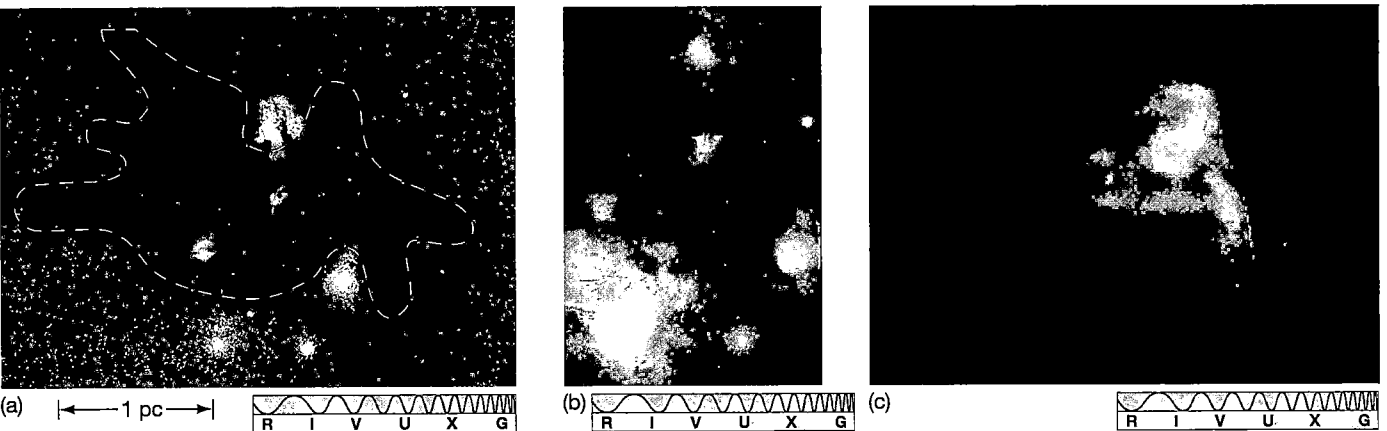

Figure 18.12 Dark Dust Cloud (a) The dark dust cloud Rho Ophiuchi, is "visible" only because it blocks light coming from stars behind it. The dashed line indicates the cloud's approximate outline. (b) Another view of the region, showing fainter foreground objects and more subtle colors. To orient (a) and (b), note the "pentagon" of bright objects clearly visible in each image. The bright star Antares is at the bottom. Up and to its right, near the edge of the image, is a star cluster called M4. Rho Ophiuchi itself is the bright object near the top, surrounded by a blue reflection nebula. Notice the dark dust lanes running across left center. (c) An infrared map of the same region, to roughly the same scale as parts (a) and (b). The very bright source near the top of the cloud is a hot emission nebula, also visible in the optical images. The bright "streamers" at left are the dark dust lanes evident in part (b). (The black diagonal streak at right is an instrumental effect.) *(Harvard Observatory; D. Malin; NASA)*

are good examples of this obscuration. The dust is apparent because it blocks the light coming from behind it. Figure 18.13 shows another striking example of a dark cloud—the Horsehead Nebula in Orion. This curiously shaped finger of gas and dust projects out from the much larger dark cloud in the bottom half of the image and stands out clearly against the red glow of a background emission nebula.

ABSORPTION SPECTRA

Astronomers first became aware of the true extent of dark interstellar clouds in the 1930s as they studied the optical spectra of distant stars. The gas in the cloud absorbs some of the stellar radiation in a manner that depends on the cloud's own temperature, density, and elemental abundance. The absorption lines thus produced contain

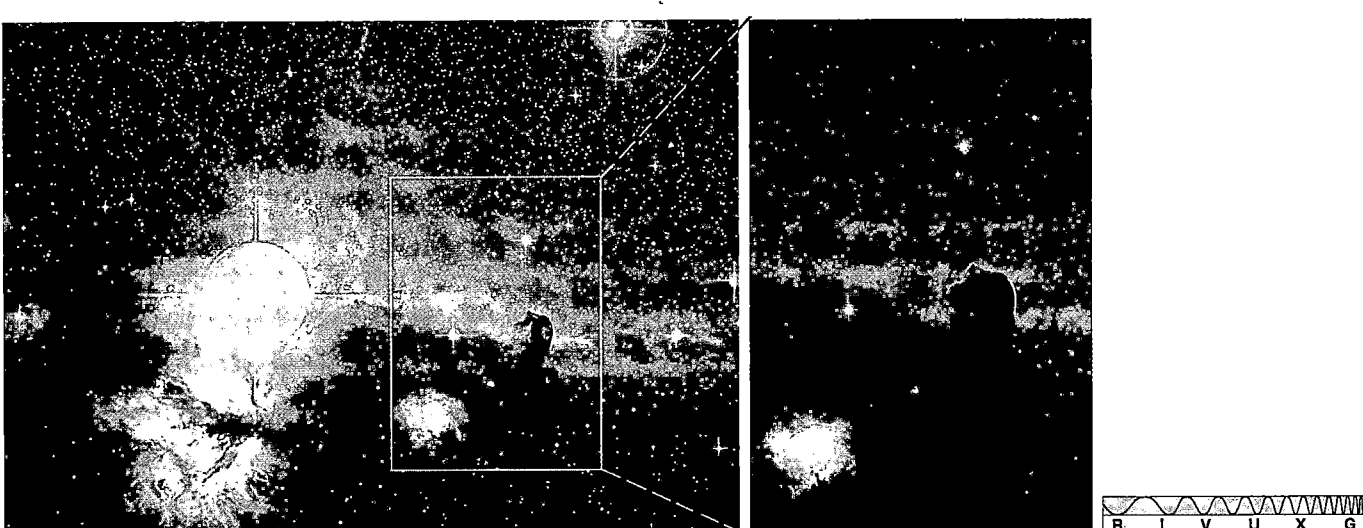

Figure 18.13 Horsehead Nebula Located in the constellation Orion, not far from the Orion Nebula, the Horsehead is a striking example of a dark dust cloud silhouetted against the bright background of an emission nebula. The "neck" of the horse is about 0.25 pc across. The nebular region is roughly 1500 pc from Earth. *(Royal Observatory, Belgium; D. Malin/Anglo-Australian Telescope)*

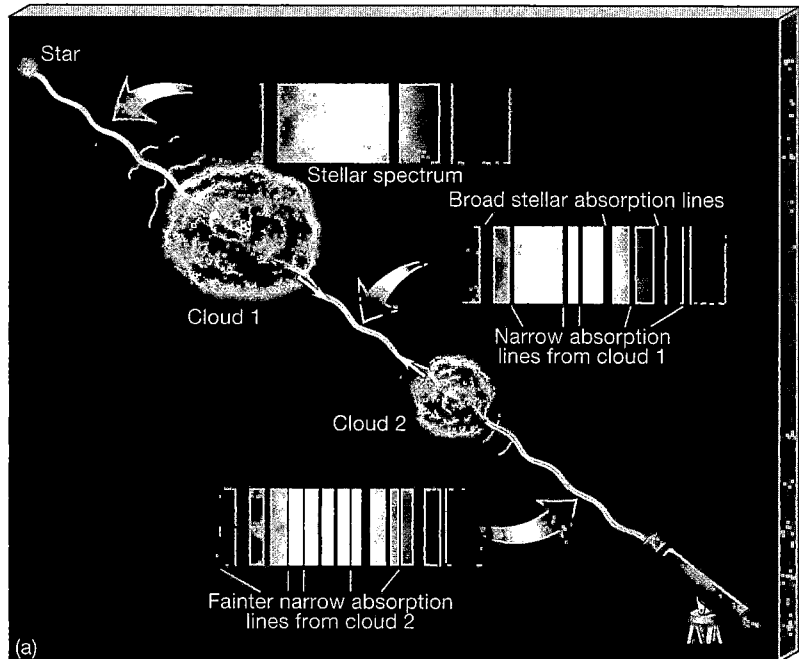

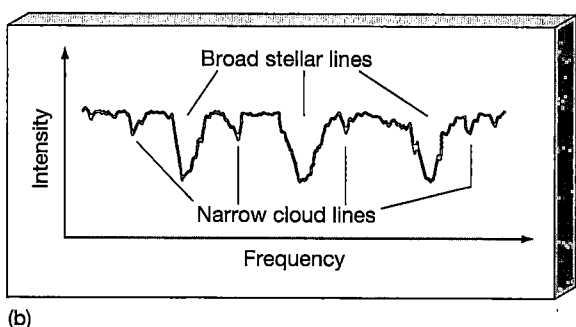

Figure 18.14 Absorption by Interstellar Clouds (a) Simplified diagram of some interstellar clouds between a hot star and Earth. Optical observations might show an absorption spectrum like that traced in (b). The wide, intense lines are formed in the star's hot atmosphere; narrower, weaker lines arise from the cold interstellar clouds. The smaller the cloud, the weaker the lines. The redshifts or blueshifts of the narrow absorption lines provide information on cloud velocities. The widths of all the spectral lines depicted here are greatly exaggerated for the sake of clarity.

information about dark interstellar matter, just as stellar absorption lines reveal the properties of stars. ∞ (Sec. 4.2) Because the interstellar absorption lines are produced by cold, low-density gas, astronomers can easily distinguish them from the much broader absorption lines formed in the star's hot lower atmospheres. ∞ (Sec. 4.4) Figure 18.14(a) illustrates how light from a star may pass through several interstellar clouds on its way to Earth. These clouds need not be close to the star, and indeed they usually are not. Each absorbs some of the stellar radiation in a manner that depends on its own temperature, density, velocity, and elemental abundance. Figure 18.14(b) depicts part of a typical spectrum produced in this way.

Figure 18.15 is a classic example of a dark cloud. Emitting no visible light, the region L977 in the constellation Cygnus is made up of cold gas which is mostly molecular in composition. Invisible to the eye except by the degree to which starlight is dimmed by the cloud, the cloud's molecular emission—in this case in radiation from carbon monoxide (CO) molecules (see Section 18.5)—outlines it clearly at radio wavelengths.

The narrow absorption lines contain information about dark interstellar clouds, just as stellar absorption lines reveal the properties of stars, and nebular emission lines tell us about conditions in hot nebulae. By studying these lines, astronomers can probe the cold depths of interstellar space. In most cases the elemental abundances

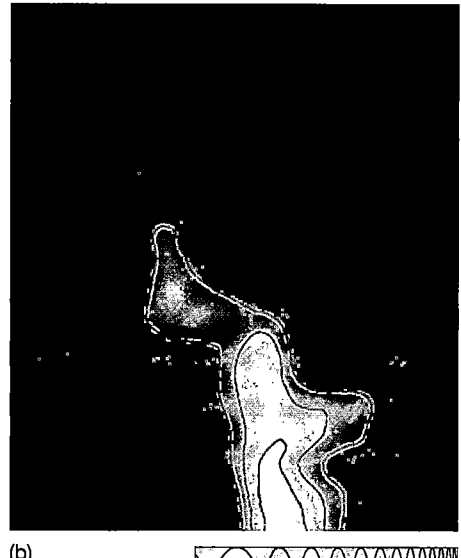

Figure 18.15 Obscuration and Emission (a) At optical wavelengths, this dark dust cloud (known as L977) can be seen only by its obscuration of background stars. (b) At radio wavelengths, it emits strongly in the CO molecular line, with the most intense radiation coming from its densest part. (C. and E. Lada)

detected in interstellar clouds mirror those found in other astronomical objects—perhaps not surprising, since interstellar clouds are the regions that spawn emission nebulae and stars.

☑ Concept Check

■ How do astronomers use optical observations to probe the properties of dark dust clouds?

18.4 21-Centimeter Radiation

⚄ A basic difficulty with the optical technique just described is that we can examine interstellar clouds only along the line of sight to a distant star. To form an absorption line, there has to be a background source of radiation to absorb. The need to see stars through clouds also restricts this approach to relatively local regions, within a few thousand parsecs of Earth. Beyond that distance, stars are completely obscured, and optical observations are impossible. As we have seen, infrared observations provide a means of viewing the emission from some clouds, but they do not completely solve the problem. Only the denser, dustier clouds emit enough infrared radiation for astronomers to study them in that part of the spectrum.

To probe interstellar space more thoroughly, we need a more general, more versatile observational method—one that does not rely on conveniently located stars and nebulae. In short, we need a way to detect cold, neutral interstellar matter anywhere in space through its *own* radiation. This may sound impossible, but such an observational technique does in fact exist. The method relies on low-energy *radio* emissions produced by the interstellar gas itself.

Recall that a hydrogen atom has one electron orbiting a single-proton nucleus. Besides its orbital motion around the central proton, the electron also has some rotational motion—that is, *spin*—about its own axis. The proton also spins. This model is analogous to a planetary system in which, in addition to the orbital motion of a planet about a central star, both the planet (electron) and the star (proton) rotate about their own axes. But bear in mind the crucial difference between planetary and atomic systems: A planet orbiting the Sun is free to move in any orbit and spin at any rate but, within an atom, all physical quantities, such as energy, momentum, and angular momentum (spin), are quantized—they are permitted to take on only specific, distinct values. ∞ (Sec. 4.2)

The laws of physics dictate that there are exactly two possible spin configurations for a hydrogen atom in its ground state. The electron and proton can rotate in the same direction, with their spin axes parallel, or they can rotate with their axes antiparallel (that is, parallel, but oppositely oriented). Figure 18.16 shows these two configu-

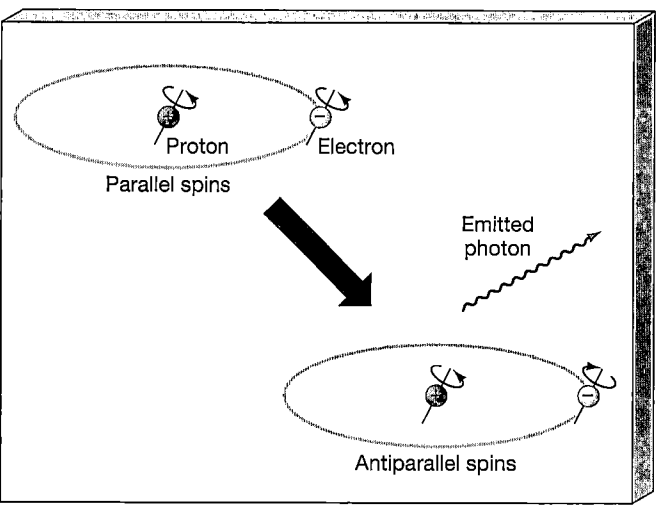

Figure 18.16 Hydrogen 21-cm Emission Diagram of a ground-level hydrogen atom changing from a higher-energy state (electron and proton spins are parallel) to a lower-energy state (spins are antiparallel). The emitted photon carries away an energy equal to the energy difference between the two spin states.

rations. The antiparallel configuration has slightly less energy than the parallel state.

All matter in the universe tends to achieve its lowest possible energy state, and interstellar gas is no exception. A slightly excited hydrogen atom with the electron and proton spinning in the same direction eventually drops down to the less energetic, opposite-spin state as the electron suddenly and spontaneously reverses its spin. As with any other such change, the transition from a high-energy state to a low-energy state releases a photon with energy equal to the energy difference between the two levels.

Because the energy difference between the two states is very small, the energy of the emitted photon is very low. Consequently, the wavelength of the radiation is rather long—in fact, it is 21.1 cm, roughly the width of this book. That wavelength lies in the radio portion of the electromagnetic spectrum. Researchers refer to the spectral line that results from this hydrogen-spin-flip process as **21-centimeter radiation**. It provides a vital probe into any region of the universe containing atomic hydrogen gas. Figure 18.17 shows typical spectral profiles of 21-cm radio signals observed from several different regions of space. These tracings are the characteristic signatures of cold, atomic hydrogen in our Galaxy. Needing no visible starlight to help calibrate their signals, radio astronomers can observe *any* interstellar region that contains enough hydrogen gas to produce a detectable signal. Even the low-density regions between the dark clouds can be studied.

As can be seen in Figure 18.17, actual 21-cm lines are quite jagged and irregular, somewhat like nebular emission lines in appearance. These irregularities arise because there are usually numerous clumps of interstellar gas along any given line of sight. Each has its own density, temperature, radial velocity, and internal motion, so the intensity, width, and Doppler shift of the resultant 21-cm line vary from place to place. All these different lines are superimposed in the signal we eventually receive at Earth, and sophisticated computer analysis is generally required to disentangle them. The "average" figures quoted earlier for the temperatures (100 K) and densities (10^6 atoms/m^3) of the regions between the dark dust clouds are based on 21-cm measurements. Observations of the dark clouds themselves using 21-cm radiation yield densities and temperatures in good agreement with those obtained by optical spectroscopy.

All interstellar atomic hydrogen emits 21-cm radiation. But if all atoms eventually fall into their lowest-energy configuration, why isn't all the hydrogen in the Galaxy in the lower-energy state by now? Why do we see 21-cm radiation today? The answer is that the energy difference between the two states is comparable to the energy of a typical atom at a temperature of 100 K or so. As a result, atomic collisions in the interstellar medium are energetic enough to boost the electron into the higher-energy configuration and so maintain comparable numbers of hydrogen atoms in either state. At any instant, any sample of interstellar hydrogen will contain many atoms in the upper level, and 21-cm radiation will always be emitted.

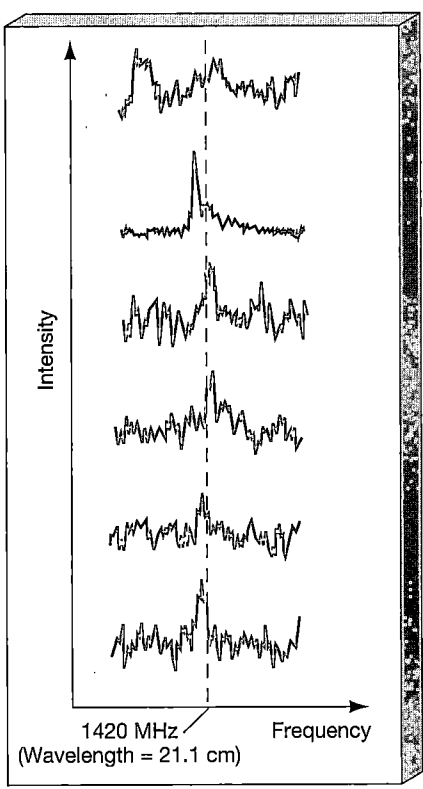

Figure 18.17 21-cm Lines Typical 21.1-cm radio spectral lines observed from several different regions of interstellar space. The peaks do not all occur at a wavelength of exactly 21.1 cm, corresponding to a frequency of 1420 MHz, because the gas in the Galaxy is moving with respect to Earth.

Of great importance, the wavelength of this characteristic radiation is much larger than the typical size of interstellar dust particles. Accordingly, this radio radiation reaches Earth completely unscattered by interstellar debris. The opportunity to observe interstellar space well beyond a few thousand parsecs, and in directions lacking background stars, makes 21-cm observations among the most important and useful in all astronomy. We will see in Chapter 23 how 21-cm maps have played a vital role in mapping out the large-scale structure of our galaxy.

✓ Concept Check

■ Why is 21-cm radiation so useful as a probe of galactic structure?

18.5 Interstellar Molecules

4 5 In certain interstellar regions of cold (typically 20 K) neutral gas, densities can reach as high as 10^{12} particles/m^3. Until the late 1970s, astronomers regarded these regions simply as abnormally dense interstellar clouds, but it is now recognized that they belong to an entirely new class of interstellar matter. The gas particles in these regions are not in atomic form at all; they are molecules. Because of the predominance of molecules in these dense interstellar regions, they are known as **molecular clouds**. They literally dwarf even the largest emission nebulae, which were previously thought to be the most massive residents of interstellar space.

MOLECULAR SPECTRAL LINES

As noted in Chapter 4, molecules can become excited through collisions or by absorbing radiation, much like atoms. ∞ (Sec. 4.3) Furthermore, again like atoms, molecules eventually return to their ground states, emitting radiation in the process. The energy states of molecules are much more complex than those of atoms, however. Molecules, like atoms, can undergo internal electron transitions, but unlike atoms, they can also rotate and vibrate. They do so in specific ways, obeying the laws of quantum physics. Figure 18.18 depicts a simple molecule rotating rapidly—that is, in an excited rotational state. After a length of time that depends on its internal makeup, the molecule relaxes back to a slower rotational rate (a state of lower energy). This change causes a photon to be emitted, carrying an energy equal to the energy difference between the two rotational states involved. The energy differences between these states are generally very small, so the emitted radiation is usually in the radio range.

We are fortunate that molecules emit radio radiation, because they are invariably found in the densest and dusti-

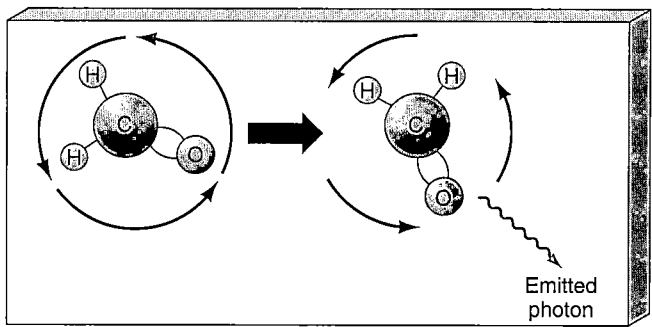

Figure 18.18 Molecular Emission As a molecule changes from a rapid rotation (left) to a slower rotation (right), a photon is emitted that can be detected with a radio telescope. Depicted here is the formaldehyde molecule, H₂CO. The length of the curved arrows is proportional to the spin rate of the molecule.

est parts of interstellar space. These are regions where the absorption of shorter-wavelength radiation is enough to prohibit the use of ultraviolet, optical, and most infrared techniques that might ordinarily detect changes in the energy states of the molecules. Only low-frequency radio radiation can escape.

Why are molecules found only in the densest and darkest of the interstellar clouds? One possible reason is that the dust serves to protect the fragile molecules from the normally harsh interstellar environment—the same absorption that prevents high-frequency radiation from getting out to our detectors also prevents it from getting in to destroy the molecules. Another possibility is that the dust acts as a catalyst that helps form the molecules. The grains provide both a place where atoms can stick and react and a means of dissipating any heat associated with the reaction, which might otherwise destroy the newly formed molecules. Probably the dust plays both roles; the close association between dust grains and molecules in dense interstellar clouds argues strongly in favor of this view, although the details are still being debated.

MOLECULAR TRACERS

In mapping molecular clouds, radio astronomers are faced with a problem. Molecular hydrogen (H₂) is by far the most common constituent of these clouds, but unfortunately, despite its abundance, this molecule does not emit or absorb radio radiation. It emits only short-wavelength ultraviolet radiation, so it cannot easily be used as a probe of cloud structure. Nor are 21-cm observations helpful— they are sensitive only to *atomic* hydrogen, not to the *molecular* form of the gas. Theorists had expected H₂ to abound in these dense, cold pockets of interstellar space, but proof of its existence was hard to obtain. Only when spacecraft measured the ultraviolet spectra of a few stars located near the edges of some dense clouds was the presence of molecular hydrogen confirmed.

With hydrogen effectively ruled out as a probe of molecular clouds, astronomers must use observations of other molecules to study the dark interiors of these dusty regions. Molecules such as carbon monoxide (CO), hydrogen cyanide (HCN), ammonia (NH₃), water (H₂O), methyl alcohol (CH₃OH), formaldehyde (H₂CO), and about 120 others, some quite complex, are now known to exist in interstellar space.* These molecules are found only in very small quantities—they are generally one million to one billion times less abundant than H₂—but they are important as *tracers* of a cloud's structure and physical properties. They are produced by chemical reactions within molecular clouds. When we observe them, we know that the regions under study must also contain high densities of molecular hydrogen, dust, and other important constituents.

The rotational properties of different molecules often make them suitable as probes of regions with different physical properties. Formaldehyde may provide the most useful information on one region, carbon monoxide on another, and water on yet another, depending on the densities and temperatures of the regions involved. These data equip astronomers with a sophisticated spectroscopic "toolbox" for studying the interstellar medium.

For example, Figure 18.19 shows some of the sites where formaldehyde molecules have been detected near M20. At practically every dark area sampled between M16 and M8, this molecule is present in surprisingly large abundance (although it is still far less common than H₂). Analyses of spectral lines at many locations along the 12°-wide swath shown in Figure 18.6 indicate that the temperature and density are much the same in all the molecular clouds studied (50 K and 10¹¹ molecules/m³, on average). Figure 18.20 shows a contour map of the distribution of formaldehyde molecules in the immediate vicinity of the M20 nebula. It was made by observing radio spectral lines of formaldehyde at various locations and then drawing contours connecting regions of similar abundance. Notice that the amount of formaldehyde (and, we assume, the amount of hydrogen) peaks in a dark region well away from the visible nebula.

Radio maps of interstellar gas and infrared maps of interstellar dust reveal that molecular clouds do not exist as distinct and separate objects in space. Rather, they make up huge **molecular cloud complexes**, typically up to 50 pc across and containing enough gas to make a million stars like our Sun. About 1000 such giant complexes are currently known in our Galaxy. Figure 18.21 is a radio map of such a large (330 square degree) region of the sky, made

*Some remarkably complex organic molecules have been found in the densest of the dark interstellar clouds, including formaldehyde (H₂CO), ethyl alcohol (CH₃CH₂OH), methylamine (CH₃NH₂), and formic acid (H₂CO₂). Their presence has fueled speculation about the origins of life, both on Earth and in the interstellar medium—especially since the report by radio astronomers in the mid 1990s (but still unconfirmed) of evidence that glycine (NH₂CH₂COOH), one of the key amino acids that form the large protein molecules in living cells, may also be present in interstellar space.

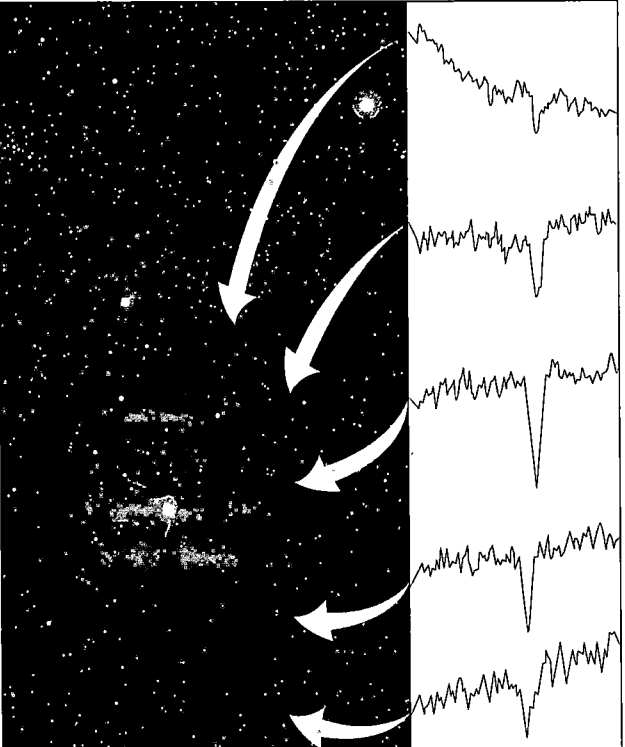

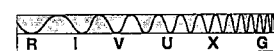

Figure 18.19 Molecular Absorption Near M20 Spectra indicate that formaldehyde molecules exist around M20, as indicated by the arrows. The lines, which are formed by absorption of background radiation, are most intense both in the dark dust lanes trisecting the nebula and in the dark regions beyond the nebula as background radiation passes through. *(background image: AURA)*

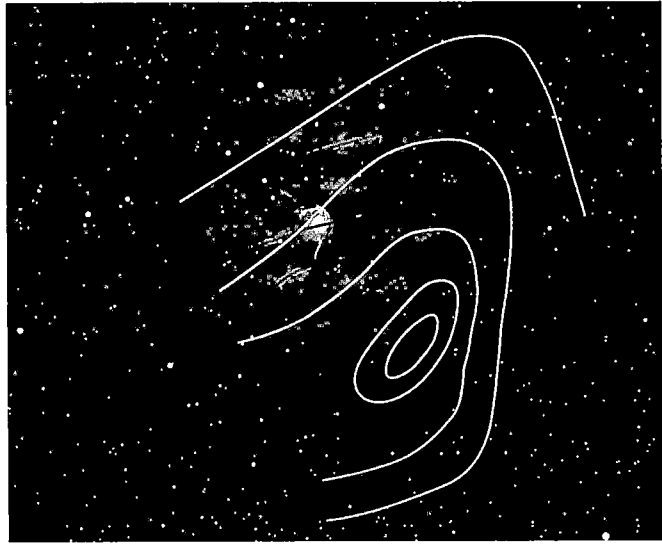

Figure 18.20 M20 Radio Map Contour map of the amount of formaldehyde near the M20 nebula, demonstrating how that gas is especially abundant in the darkest interstellar regions. Other kinds of molecules have been found to be similarly distributed. The contour values increase from the outside to the inside, so the maximum density of formaldehyde lies just to the bottom right of the visible nebula. The green and red contours outline the intensity of the formaldehyde absorption lines at different rotational frequencies. The nebula itself is about 4 pc across. *(background image: AURA)*

using certain spectral lines of the CO molecule, showing numerous molecular cloud complexes stretching across the entire field of view.

The very existence of molecules has forced astronomers to rethink and to reobserve interstellar space. In doing so, they have begun to realize that this active and interesting domain is far from the void suspected by theorists not so long ago. Regions of space recently thought to contain nothing more than galactic "garbage"—the cool, tenuous darkness among the stars—now play a critical role in our understanding of stars and the interstellar medium from which they are born.

☑ Concept Check

■ Why do astronomers use observations of "minority" molecules such as carbon monoxide and formaldehyde when mapping molecular clouds, when these molecules constitute only a tiny fraction of the total?

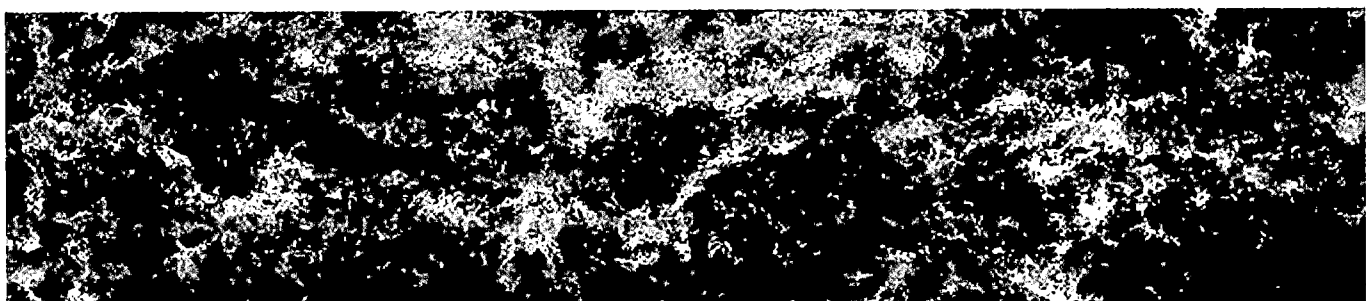

Figure 18.21 Molecular Cloud Complexes This radio map shows the outer portion of the Milky Way (that is, looking in the opposite direction from Figure 18.1) as it appears in CO emission. The bright regions are molecular cloud complexes, dense regions of interstellar space where molecules abound and, apparently, stars are forming. The map extends over a full quadrant of the Galaxy, or 90°, along the galactic plane and was made from 1,696,800 observations of CO spectra. *(Five College Radio Astronomy Observatory)*

Chapter Review

SUMMARY

The **interstellar medium** (p. 468) occupies the space among the stars. It is made up of cold (less than 100 K) gas, mostly atomic or molecular hydrogen and helium, and **dust grains** (p. 469). Interstellar dust is very effective at blocking our view of distant stars, even though the density of the interstellar medium is very low. The spatial distribution of interstellar matter is patchy. The general diminution of starlight by dust is called **extinction** (p. 469). In addition, the dust preferentially absorbs short-wavelength radiation, leading to a distinct **reddening** (p. 469) of light passing interstellar clouds. Interstellar dust is thought to be composed of silicates, graphite, iron, and "dirty ice." Interstellar dust particles are apparently elongated or rodlike. The **polarization** (p. 471) of starlight provides a means of studying them.

A **nebula** (p. 473) is a general term for any fuzzy bright or dark patch on the sky. **Emission nebulae** (p. 473) are extended clouds of hot, glowing interstellar gas. Associated with star formation, they result when hot O- and B-type stars heat and ionize their surroundings. Studies of the emission lines produced by excited nebular atoms allow astronomers to measure the nebula's properties. Some excited atomic states take so long to emit a photon that the spectral lines associated with these transitions are never seen in terrestrial laboratories, where collisions always knock the atom into another energy state before it can emit any radiation. When these lines are seen in nebular spectra, they are called forbidden lines. Nebulae are often crossed by dark dust lanes, part of the larger cloud from which they formed.

Dark dust clouds (p. 478) are cold, irregularly shaped regions in the interstellar medium that diminish or completely obscure the light from background stars. Astronomers can learn about these clouds by studying the absorption lines they produce in starlight that passes through them. Another way to observe cold, dark regions of interstellar space is through **21-centimeter radiation** (p. 481). Such radiation is produced whenever the electron in an atom of hydrogen reverses its spin, changing its energy very slightly in the process. This radio radiation is important because it is emitted by all cool atomic hydrogen gas, even if the gas is undetectable by other means. In addition, 21-cm radiation is not appreciably absorbed by the interstellar medium, so radio astronomers making observations at this wavelength can "see" to great distances.

The interstellar medium also contains many cold, dark **molecular clouds** (p. 482), which are observed mainly through the radio radiation emitted by the molecules they contain. Dust within these clouds probably both protects the molecules and acts as a catalyst to help them form. As with other interstellar clouds, hydrogen is by far the most common constituent, but molecular hydrogen happens to be very hard to observe. Astronomers usually study these clouds through observations of other "tracer" molecules that are less common but much easier to detect. Often, several molecular clouds are found close to one another, forming an enormous **molecular cloud complex** (p. 483) millions of times more massive than the Sun.

SELF-TEST: TRUE OR FALSE?

___ **1.** Interstellar matter is quite evenly distributed throughout the Milky Way Galaxy.

___ **2.** In the vicinity of the Sun, the amount of mass in the form of interstellar matter is comparable to the amount of mass in the form of stars.

___ **3.** There is a lack of heavy elements in interstellar gas because they go into making interstellar dust.

___ **4.** The fact that starlight becomes polarized as it passes through the interstellar medium tells us that interstellar dust particles are spherical in shape.

___ **5.** A typical region of dark interstellar space has a temperature of about 500 K.

___ **6.** An emission nebula is a cloud of dust reflecting the light of a nearby group of stars.

___ **7.** Emission nebulae display spectra almost identical to those of the stars embedded in them.

___ **8.** Forbidden emission lines can occur in emission nebulae because the density of interstellar gas there is extremely low.

___ **9.** Because of the obscuration of visible light by interstellar dust, we can observe stars only within a few thousand parsecs of Earth in any direction.

___ **10.** A typical dark dust cloud is many hundreds of parsecs across.

___ **11.** Because of their low temperatures, dark dust clouds radiate mainly in the far infrared part of the electromagnetic spectrum.

___ **12.** Most interstellar matter exists in the form of molecular clouds.

___ **13.** 21-cm radiation provides astronomers with information on the density, temperature, and internal motions of interstellar gas.

___ **14.** 21-cm radiation can pass unimpeded through the entire Milky Way Galaxy.

___ **15.** Water, formaldehyde, carbon monoxide, and numerous organic molecules have all been found in molecular clouds.

SELF-TEST: FILL IN THE BLANK

1. The interstellar medium is made up of _____ and _____.
2. To scatter a beam of radiation, a particle's size must be _____ the wavelength of the radiation.
3. Extinction is the _____ of starlight by interstellar _____.
4. The density of interstellar matter is very _____.
5. Interstellar gas is composed of 90 percent _____ and 9 percent _____.
6. The temperature of a typical emission nebula is about ____ K.
7. The process in which clouds of cool dense gas (such as in the Eagle Nebula) are eaten away by radiation from nearby hot young stars is called ____.
8. An HII region is another name for _____.

9. Dark dust clouds can have temperatures as low as _____ K.
10. 21-cm radiation is emitted by _____ hydrogen.
11. 21-cm radiation results from a change in the _____ of the electron in a _____.
12. Emissions from molecular clouds are in the _____ part of the electromagnetic spectrum.
13. _____ is thought to play an important role in the formation of molecules in molecular clouds.
14. The most common constituent of molecular clouds is molecular _____.
15. A molecular cloud complex may contain as much as _____ solar masses of gas.

REVIEW AND DISCUSSION

1. Give a brief description of the interstellar medium.
2. What is the composition of interstellar gas? What about interstellar dust?
3. Why is interstellar dust so much more effective than interstellar gas at absorbing starlight?
4. How dense is interstellar matter on average?
5. How is interstellar matter distributed throughout space?
6. What are some methods that astronomers use to study interstellar dust?
7. What is an emission nebula?
8. What is photoevaporation, and how does it change the structure and appearance of an emission nebula?
9. Why are some spectral lines observed in emission nebulae not normally seen in laboratories on Earth?
10. Describe some ways in which we can "see" a dark interstellar cloud.
11. Give a brief description of a dark dust cloud.

12. What is 21-cm radiation? With what element is it associated?
13. Why is 21-cm radiation useful to astronomers?
14. Why can't 21-cm radiation be used to probe the interiors of molecular clouds?
15. How does a molecular cloud differ from other interstellar matter?
16. Why can't astronomers use observations of hydrogen to explore the structure of molecular cloud complexes?
17. How do astronomers explore the structure of molecular cloud complexes?
18. If our Sun were surrounded by a cloud of gas, would this cloud be an emission nebula? Why or why not?
19. Compare the reddening of stars by interstellar dust with the reddening of the setting Sun.
20. Explain what it means for a star's light to be polarized. How does the polarization of starlight provide a means of studying the interstellar medium?

PROBLEMS  *Algorithmic versions of these questions are available in the Practice Problems module of the Companion Website.*

The number of squares preceding each problem indicates its approximate level of difficulty.

1. ■ The average density of interstellar gas within the Local Bubble is much lower than the value mentioned in the text—in fact, it is roughly 10^3 hydrogen atoms/m³. Given that the mass of a hydrogen atom is 1.7×10^{-27} kg, calculate the total mass of interstellar matter contained within a Bubble volume equal in size to planet Earth.

2. ■ Assuming the same average density as in the previous question, calculate the total mass of interstellar hydrogen contained within a cylinder of cross-sectional area 1 m², extending from Earth to Alpha Centauri.

3. ■■ Given the average density of interstellar matter stated in Section 18.1, calculate how large a volume of space would have to be compressed to make a cubic meter of gas equal in density to air on Earth (1.2 kg/m³).

4. ■ Assuming a density of 3000 kg/m³, estimate the mass of the dust particle illustrated in Figure 18.3(a).

5. ■■ A beam of light shining through a dense molecular cloud is diminished in intensity by a factor of two for every 5 pc it travels. By how many magnitudes is the light from a background star dimmed, if the total thickness of the cloud is 60 pc?

6. ■ Interstellar extinction is sometimes measured in magnitudes per kiloparsec (1 kpc = 1000 pc). Light from a star 1500 pc away is observed to be diminished in intensity by a factor of 20 over and above the effect of the inverse-square law. What is the average interstellar extinction along the line of sight, in mag/kpc?

7. ■■ Spectroscopic observations of a certain star reveal it to be a B2II giant, with absolute magnitude –6. ⊂⊃ (Secs. 17.4 and 17.7) The star's apparent magnitude is 14. Neglecting the effects of interstellar extinction, calculate the distance to the star. If the star's distance is known (by other means) to be 5000 pc, calculate the average extinction along the line of sight, in mag/kpc. ⊂⊃ (*More Precisely 17-1*)

8. ■■ A star of apparent magnitude 10 lies 500 pc from Earth. If interstellar absorption results in an average ex-

tinction of 2 mag/kpc, calculate the star's absolute magnitude and luminosity.

9. ▣▪▣ A star of known absolute magnitude −5 has apparent magnitude 10. If interstellar absorption results in an average extinction of 2 mag/kpc, calculate the star's distance. (Note: This problem does not have an algebraic solution. You will have to solve it by numerical means—essentially trial and error on a calculator.)

10. ▣ To carry enough energy to ionize a hydrogen atom, a photon must have a wavelength of less than 9.12×10^{-8} m (91.2 nm). Using Wien's law, calculate the temperature a star must have for the peak wavelength of its blackbody curve to equal this value. ⌐⊃ (Sec. 3.4)

11. ▪▪ Estimate the escape speeds near the edges of the four emission nebulae listed in Table 18.1, and compare them with the average speeds of hydrogen nuclei in those nebulae. ⌐⊃ᵌ *(More Precisely 8-1)* Do you think it is possible that the nebulae are held together by their own gravity?

12. ▪ What would the mass of M8 have to be in order for its escape speed to equal its average molecular speed?

13. ▪▪ If a group of interstellar clouds along the line of sight have radial velocities in the range 75 km/s (receding) to 50 km/s (approaching), calculate the range of frequencies and wavelengths over which the 21.1-cm (1420 MHz) line of hydrogen will be observed. ⌐⊃ (Sec. 3.5)

14. ▪ Calculate the radius of a spherical molecular cloud whose total mass equals the mass of the Sun. Assume a cloud density of 10^{12} hydrogen atoms per cubic meter.

15. ▣▪▪ A cloud of atomic hydrogen has a radius of 1 pc and an average density 10^6 hydrogen atoms per cubic meter. Collisions between atoms ensure that, at any instant, $\frac{3}{4}$ of all atoms are in the upper (parallel spin) state, as discussed in Section 18.4. The transition producing the 21-cm line is very unlikely—the probability that any given atom in the upper state will make the transition during any given second is about 3×10^{-15} (compare with 10^8 for the Hα transition). Use these figures, together with the Planck formula for the energy of the photon emitted in the transition, to estimate the total radio luminosity of the cloud. ⌐⊃ᵌ (Sec. 4.2)

COLLABORATIVE EXERCISES

1. Exploring Density. The interstellar medium has a very low density of about 1000 per cubic kilometer. Estimate the population density of students in the tallest dormitory on campus using units of students per cubic feet and compare to the population density of the classroom. Explain your reasoning.

RESEARCHING ON THE WEB  *To complete the following exercises, go to the online Destinations module for Chapter 18 on the Companion Website for Astronomy Today 4/e.*

1. Access The "Messier Catalog" pages and determine which objects constitute the majority of objects M1 through M10 and which objects for M90 through M100.

2. Access the "Nebulae: Fuzzy Patches In Space" page and define the four types of nebulae.

PROJECTS

1. The constellation Orion the Hunter is prominent in the evening sky of winter. Its most noticeable feature is a short, straight row of three medium-bright stars: the famous belt of Orion. A line of stars extends from the eastmost star of the belt, toward the south. This line represents Orion's sword. Towards the bottom of the sword is the sky's most famous emission nebula, M42, the Orion Nebula. Observe the Orion Nebula with your eye, with binoculars, and with a telescope. What is its color? How can you account for this? With the telescope, try to find the Trapezium, a grouping of four stars in the center of M42. These are hot, young stars; their energy causes the Orion Nebula to glow.

2. Observe the Milky Way on a dark, very clear night. Is it a continuous band of light across the sky or is it mottled? The parts of the Milky Way that appear missing are actually dark dust clouds that are relatively near the Sun. Identify the constellations in which you see these clouds. Make a sketch and compare with a star atlas. Find other small clouds in the atlas and try to find them with your eye or with binoculars.

SKYCHART III PROJECTS  *The SkyChart III Student Version planetarium program on which these exercises are based is included as a separately executable program on the CD in the back of this text.*

1. ▪ One of the most interesting nebula is the Great Orion Nebula, M42. Use SkyChart III to locate Orion, and determine what time of the year it is overhead at a convenient time for viewing. While the nebula is visible to the unaided eye, it is much better with binoculars, and only gets more interesting as you use larger and larger telescopes to observe it.

 *In addition to the Practice Problems and Destinations modules, the Companion Website at http://www.prenhall.com/chaisson provides for each chapter an additional true-false, multiple choice, and labeling quiz, as well as additional annotated images, animations, and links to related Websites.*

19 STAR FORMATION

A Traumatic Birth

LEARNING GOALS

Studying this chapter will enable you to:

1   Discuss the factors that compete against gravity in the process of star formation.

2   Summarize the sequence of events leading to the formation of a star like our Sun.

3   Explain how the process of star formation depends on stellar mass.

4   Describe some of the observational evidence supporting the modern theory of star formation.

5   Explain the nature of interstellar shock waves, and discuss their possible role in the formation of stars.

6   Distinguish between open and globular star clusters, and explain why the study of clusters is important to astronomers.

 Visit http://www.prenhall.com/chaisson for additional annotated images, animations, and links to related sites for this chapter.

The almost featureless emission nebula IC 2944, shown here filling the image in red, provides a backdrop for several small groups of completely dark clouds. Known as "Bok globules," most are totally opaque across their typical light-year dimensions. Such globules are often found associated with nebulae abundant in young hot stars, suggesting that the globules are probably sites of pending star formation. *(AAT)*
The Big Picture: Few issues in astronomy are more basic than knowing how stars form. Stars are the most numerous and obvious residents of the nighttime sky, and astronomers want to know in detail how they originate. The trouble is that stars normally form in the darkness in the hearts of dense interstellar clouds, so optical telescopes can grant only limited insight; telescopes operating in the long-wavelength radio and infrared parts of the spectrum are essential to probe these otherwise invisible regions of space.

We now move from the interstellar medium—the gas and dust among the stars—back to the stars themselves. The next four chapters discuss the formation and evolution of stars. We have already seen that stars must evolve as they consume their fuel supply, and we have extensive observational evidence of stars at many different evolutionary stages. With the help of these observations, astronomers have developed a good understanding of stellar evolution—the complex changes experienced by stars as they form, mature, grow old, and die. We begin by studying the process of star formation, through which interstellar clouds of gas and dust are transformed into the myriad stars we see in the night sky.

19.1 Star-Forming Regions

Our universe is constantly renewing itself. Literally billions of stars have been born, lived out their lives, and died since our Galaxy formed. We do not see this activity when we gaze at the nighttime sky because the time scales on which stars play out this cosmic drama are enormously long by human standards. Even the shortest-lived O-type stars survive for millions of years. ⇔ (Sec. 17.8) Nevertheless, we have plenty of evidence for ongoing stellar evolution throughout the cosmos.

YOUNG STARS IN THE UNIVERSE

Our Sun, and probably most of the stars in our immediate cosmic neighborhood, formed billions of years ago. ⇔ (Sec. 15.2) However, we know that many relatively nearby stars are much younger than this. The magnificent emission nebulae discussed in Chapter 18 and the ultraluminous, short-lived stars that power them are direct proof that star formation is a continuing process. ⇔ (Sec. 18.2) The hottest stars in these regions must have formed less than a few million years ago—the blink of an eye, in cosmic terms—and there is no reason to suppose that galactic star formation has recently and abruptly ceased! Stars are forming all across the Milky Way, even as you read this.

In fact, star-forming regions are observed in all corners of the universe. Figure 19.1 shows one of the largest such regions discovered to date. It lies in a galaxy quite similar to our own, about 1 million parsecs away. Almost 500 pc across, this vast stellar nursery dwarfs any emission nebula known in our Galaxy. Conceivably, the Milky Way may contain similarly large nebulae. If they exist, they are obscured by so much interstellar matter in our galactic plane that we cannot see them. Whatever their size, however—large or small—emission nebulae are the birthplaces of all the stars in our night sky.

How and where do stars form? What factors determine the masses, luminosities, and spatial distribution of stars in our Galaxy and beyond? The association of bright emission nebulae with much larger, dark dust clouds provides the key. ⇔ (Sec. 18.3) Simply put, star formation begins when part of the interstellar medium—one of those cold, dark clouds—starts to collapse under its own weight.

The cloud fragment heats up as it shrinks, and eventually its center becomes hot enough for nuclear fusion to begin. At that point, the contraction stops and a star is born. But what determines which interstellar clouds collapse? For that matter, since all clouds exert a gravitational pull, why didn't they all collapse long ago? To answer these questions and understand the processes leading to the stars we see, we must explore in a little more detail the factors that compete with gravity in determining a cloud's fate.

GRAVITY AND HEAT

1 Consider a small portion of a large cloud of interstellar gas. Concentrate first on just a few atoms, as shown in Figure 19.2. Even though the cloud's temperature is very low, each atom still has some random motion because of the cloud's heat. ⇔ (More Precisely 3-1) Each atom is also influenced by the gravitational attraction of all its neighbors. The gravitational force is not large, however, because the mass of each atom is so small. When a few atoms accidentally cluster for an instant, as shown in Figure 19.2(b), their combined gravity is insufficient to bind them into a lasting, distinct clump of matter. This accidental cluster will disperse as quickly as it forms. The effect of heat—the random motion of the atoms—is much stronger than the effect of gravity.

Now consider a larger group of atoms. Imagine, for example, 50, 100, 1000, even a million atoms, each gravitationally pulling on all the others. With increased mass, the force of gravity is now stronger than before. Will this many atoms exert a combined gravitational attraction strong enough to prevent the clump from dispersing again? The answer—at least under the conditions found in interstellar space—is still no. The gravitational attraction of this mass of atoms is still far too weak to overcome the effect of heat.

We have already seen numerous instances of the competition between heat and gravity. ⇔ (More Precisely 8-1) The temperature of a gas is simply a measure of the average speed of the atoms or molecules in it, so the higher the temperature, the greater the average speed, and hence the higher the pressure of the gas. This is the main reason that the Sun and other stars don't collapse. The outward pressure of their heated gases exactly balances gravity's inward pull.

How many atoms must be accumulated in order for their collective pull of gravity to prevent them from dispersing back into interstellar space? The answer, even for a

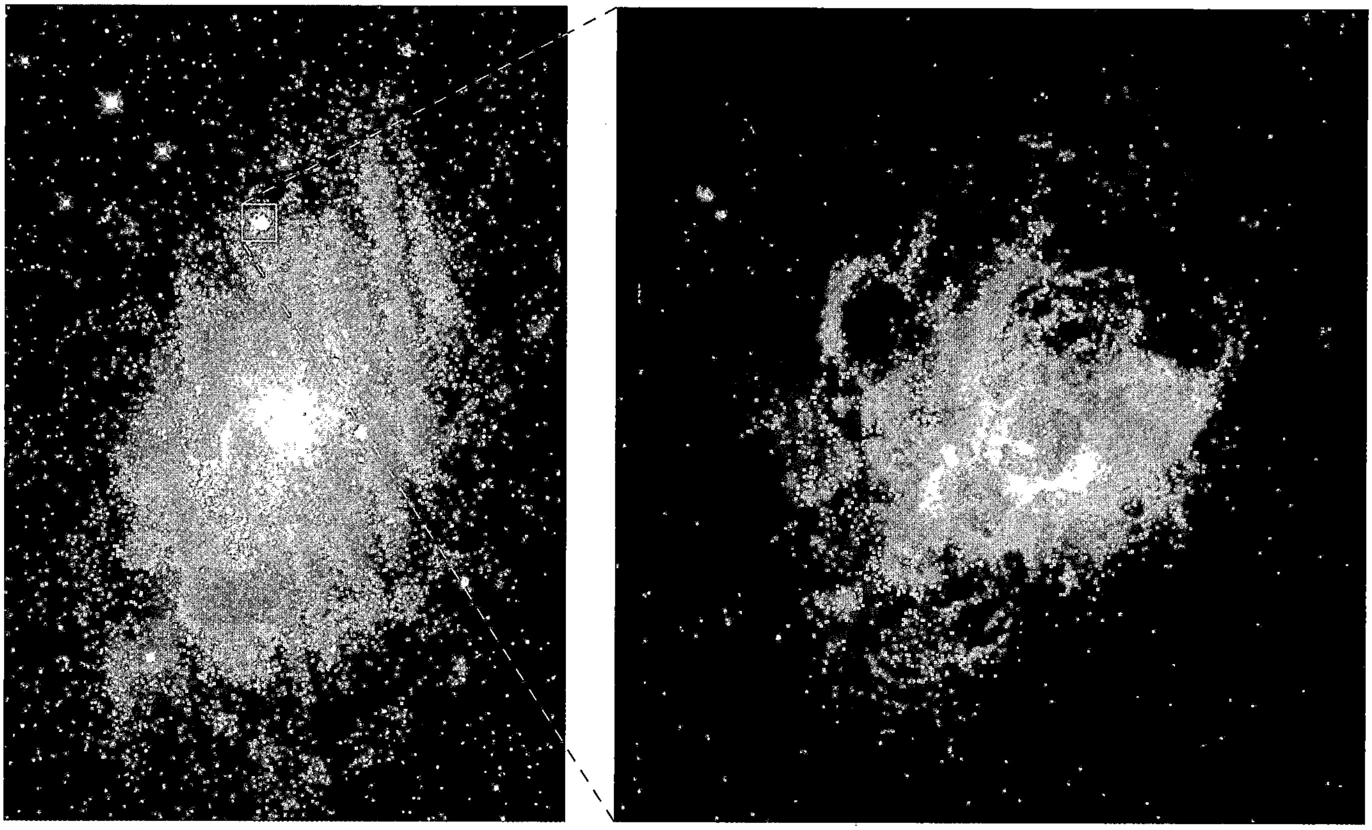

Figure 19.1 Extragalactic Star Formation The giant star-forming region at right, called NGC 604, is roughly 500 pc across. It is found in the nearby galaxy M33, displayed at left on the much larger scale of 40,000 pc across. *(Palomar/Caltech; NASA)*

typical cool (100 K) cloud, is a truly huge number. Nearly 10^{57} atoms are required—much more than the 10^{25} grains of sand on all the beaches of the world, even more than the 10^{51} elementary particles that constitute all the atomic nuclei in our entire planet. There is simply nothing on Earth comparable to a star.

SOME COMPLICATIONS

Heat is not the only factor that tends to oppose gravitational contraction. *Rotation*—that is, spin—can also compete with gravity's inward pull. As we saw in Chapter 15, a contracting cloud having even a small spin tends to develop a bulge around its midsection. ∞ (Sec. 15.2) As the cloud contracts, it must spin faster (to conserve its angular momentum), and the bulge grows—material on the edge tends to fly off into space. (Consider the analogy of mud flung from a rapidly rotating bicycle wheel.) Eventually, as in Figure 15.1, the cloud forms a flattened, rotating disk.

For material to remain part of the cloud and not be spun off into space, a force must be applied—in this case, the force of gravity. The more rapid the rotation, the greater the tendency for the gas to escape, and the greater the gravitational force needed to retain it. It is in this sense that we can regard rotation as opposing the inward pull of gravity. Should the rotation of a contracting gas cloud overpower gravity, the cloud would simply disperse. Thus, more mass is needed for a rapidly rotating interstellar cloud to contract to form a star than is needed for a cloud having no rotation at all.

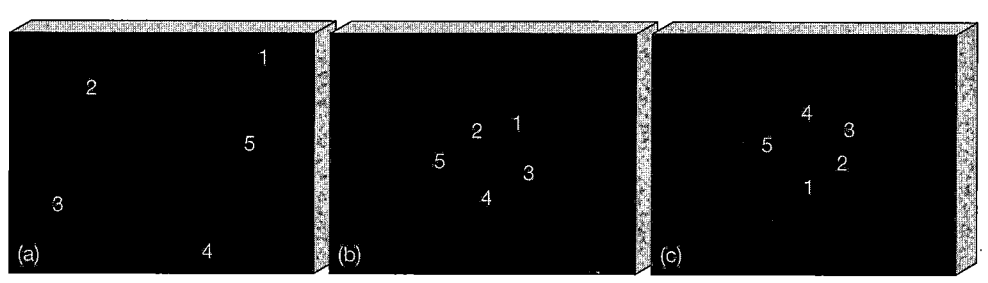

Figure 19.2 Atomic Motion Motions of a few atoms within an interstellar cloud are influenced by gravity so slightly that their paths are hardly changed (a) before, (b) during, and (c) after an accidental, random encounter.

Magnetism can also hinder a cloud's contraction. Just as Earth, the outer planets, and the Sun all have some magnetism, magnetic fields permeate most interstellar clouds. As a cloud contracts, it heats up, and atomic encounters become violent enough to (partly) ionize the gas. As we noted in Chapter 7 when discussing Earth's Van Allen belts, and in Chapter 16 when discussing activity on the Sun, magnetic fields can exert electromagnetic control over charged particles. ∞ (Secs. 7.4, 16.4) In effect, the particles tend to become "tied" to the magnetic field— they are free to move *along* the field lines, but are inhibited from moving *perpendicular* to them.

As a result (Figure 19.3), interstellar clouds may contract in distorted ways. Because the charged particles and the magnetic field are linked, the field itself follows the contraction of a cloud. The charged particles literally pull the magnetic field toward the cloud's center in the direction perpendicular to the field lines. As the field lines are

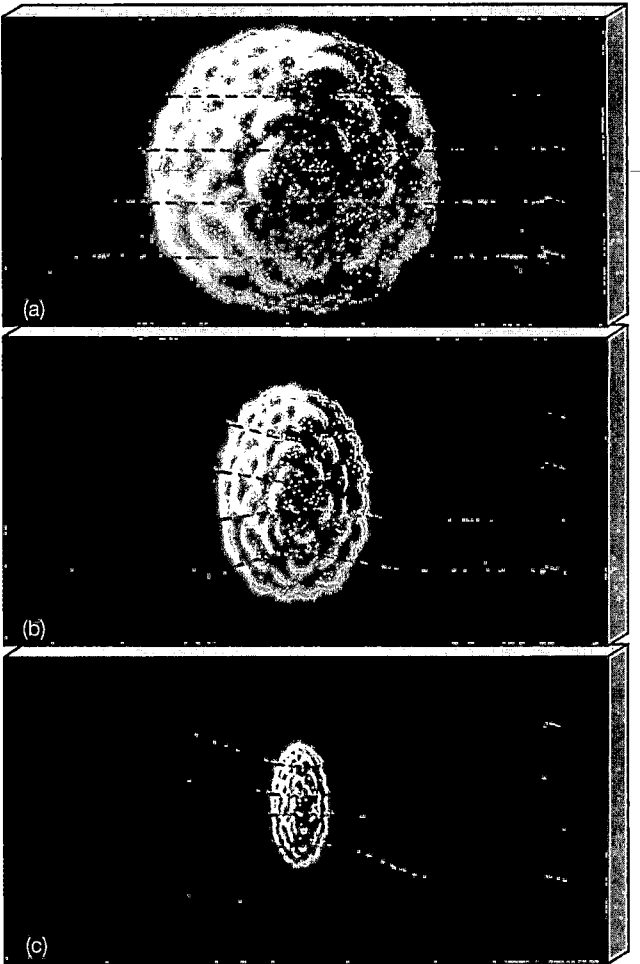

Figure 19.3 Interstellar Magnetic Field Magnetism can hinder the contraction of a gas cloud, especially in directions perpendicular to the magnetic field (solid lines). Frames (a), (b), and (c) trace the evolution of a slowly contracting interstellar cloud having some magnetism.

compressed, the magnetic field strength increases. In this way, the strength of magnetism in a cloud can become much larger than that normally permeating general interstellar space. The primitive solar nebula may have contained a strong magnetic field created in just this way.

Theory suggests that even small quantities of rotation or magnetism can compete quite effectively with gravity and can greatly alter the evolution of a typical gas cloud. Unfortunately, the interplay of these factors is not well understood—both can lead to very complex behavior as a cloud contracts, and the combination of the two is extremely difficult to study theoretically. In this chapter we will try to gain an appreciation for the broad outlines of the star-formation process by neglecting these two complicating factors. Bear in mind, however, that both are probably important in determining the details.

MODELING STAR FORMATION

The next two sections describe the currently accepted theoretical view of star formation, derived in large part from numerical experiments performed on high-speed computers. The results are mathematical predictions of a multifaceted problem incorporating gravity, heat, rotation, magnetism, nuclear reaction rates, elemental abundances, and other physical conditions specifying the state of contracting interstellar clouds.

Scientific theories always develop in response to experimental or observational data, and star formation is no exception. ∞ (*Discovery 2-2*) The theory of star formation has evolved to explain innumerable observations of stars and star-forming regions. However, the phenomenology in this case is so complex and diverse that it is helpful to have a theoretical framework to "connect the dots" between phenomena that might otherwise appear unrelated. Accordingly, we present the theory first, then discuss how and where the observational data fit into and support the theoretical picture.

☑ Concept Check
■ What basic competitive process controls star formation?

19.2 The Formation of Stars Like the Sun

2️⃣ Star formation begins when gravity begins to dominate over heat, causing a cloud to lose its equilibrium and start to contract. Only after the cloud has undergone radical changes in its internal structure is equilibrium finally restored.

Table 19.1 lists seven evolutionary stages that an interstellar cloud goes through in the process of becoming a main-sequence star like the Sun. These stages are charac-

TABLE 19.1 Prestellar Evolution of a Solar-Type Star

STAGE	APPROXIMATE TIME TO NEXT STAGE (yr)	CENTRAL TEMPERATURE (K)	SURFACE TEMPERATURE (K)	CENTRAL DENSITY (particles/m³)	DIAMETER* (km)	OBJECT
1	2×10^6	10	10	10^9	10^{14}	Interstellar cloud
2	3×10^4	100	10	10^{12}	10^{12}	Cloud fragment
3	10^5	10,000	100	10^{18}	10^{10}	Cloud fragment/protostar
4	10^6	1,000,000	3000	10^{24}	10^8	Protostar
5	10^7	5,000,000	4000	10^{28}	10^7	Protostar
6	3×10^7	10,000,000	4500	10^{31}	2×10^6	Star
7	10^{10}	15,000,000	6000	10^{32}	1.5×10^6	Main-sequence star

*For comparison, recall that the diameter of the Sun is 1.4×10^6 km, while that of the solar system is roughly 1.5×10^{10} km.

terized by varying central temperatures, surface temperatures, central densities, and radii of the prestellar object. They trace its progress from a quiescent interstellar cloud to a genuine star. The numbers given in Table 19.1 and the following discussion are valid *only* for stars of approximately the same mass as the Sun. In the next section we will relax this restriction and consider the formation of other stars.

STAGE 1: AN INTERSTELLAR CLOUD

The first stage in the star-formation process is a dense interstellar cloud—the core of a dark dust cloud or perhaps a molecular cloud. These clouds are truly vast, sometimes spanning tens of parsecs ($10^{14}-10^{15}$ km) across. Typical temperatures are about 10 K throughout, with a density of perhaps 10^9 particles/m³. Stage 1 clouds contain thousands of times the mass of the Sun, mainly in the form of cold atomic and molecular gas. (The dust they contain is important for cooling the cloud as it contracts and also plays a crucial role in planet formation, but it constitutes a negligible fraction of the total mass.) ∞ (Sec. 15.2) If such a cloud is to be the birthplace of stars, it must become unstable and eventually break up into smaller pieces. The initial collapse occurs when a pocket of gas becomes gravitationally unstable. Perhaps it is squeezed by some external

event, such as the pressure wave produced when a nearby O- or B-type star forms and ionizes its surroundings, or perhaps its supporting magnetic field leaks away as charged particles slowly drift across the confining field lines. Whatever the cause, theory suggests that once the collapse begins, fragmentation into smaller and smaller clumps of matter naturally follows, as gravitational instabilities continue to operate in the gas. As illustrated in Figure 19.4, a typical cloud can break up into tens, hundreds, even thousands, of fragments, each imitating the shrinking behavior of the parent cloud and contracting ever faster. The whole process, from a single quiescent cloud to many collapsing fragments, takes a few million years.

In this way, depending on the precise conditions under which fragmentation takes place, an interstellar cloud can produce either a few dozen stars, each much larger than our Sun, or a whole cluster of hundreds of stars, each comparable to or smaller than our Sun. There is little evidence of stars born in isolation, one star from one cloud. Most stars—perhaps even all stars—appear to originate as members of multiple systems or large groups of stars. The Sun, which is now found alone and isolated in space, probably escaped from the larger system in which it formed, perhaps after an encounter with another star or some much larger object (such as a molecular cloud).

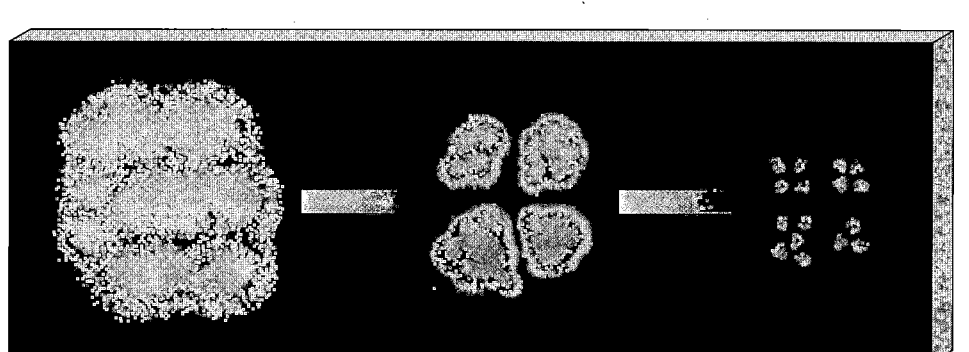

Figure 19.4 Cloud Fragmentation As an interstellar cloud contracts, gravitational instabilities cause it to fragment into smaller pieces. The pieces themselves continue to collapse and fragment, eventually to form many tens or hundreds of separate stars.

STAGE 2: A COLLAPSING CLOUD FRAGMENT

The second stage in our evolutionary scenario represents the physical conditions in just one of the many fragments that develop in a typical interstellar cloud. A fragment destined to form a star like the Sun contains between one and two solar masses of material at this stage. Estimated to span a few hundredths of a parsec across, this fuzzy, gaseous blob is still about 100 times the size of our solar system. Its central density by this time is roughly 10^{12} particles/m^3.

Even though it has shrunk substantially in size, the fragment's average temperature is not much different from that of the original cloud. The reason is that the gas constantly radiates large amounts of energy into space. The material of the fragment is so thin that photons produced within it easily escape without being reabsorbed by the cloud, so virtually all the energy released in the collapse is radiated away and does not cause any significant increase in temperature. Only at the center, where the radiation must traverse the greatest amount of material to escape, is there any appreciable temperature increase. The gas there may be as warm as 100 K by this stage. For the most part, however, the fragment stays cold as it shrinks.

The process of continued fragmentation is eventually stopped by the increasing density within the shrinking cloud. As stage 2 fragments continue to contract, they eventually become so dense that radiation cannot get out easily. The trapped radiation causes the temperature to rise, the pressure to increase, and the fragmentation to cease.

STAGE 3: FRAGMENTATION CEASES

Several tens of thousands of years after it first began contracting, a typical stage 2 fragment has shrunk by the start of stage 3 to roughly the size of our solar system (still 10,000 times the size of our Sun). The inner regions have just become opaque to their own radiation and so have started to heat up considerably, as noted in Table 19.1. The central temperature has reached about 10,000 K—hotter than the hottest steel furnace on Earth. However, the temperature at the fragment's periphery has not increased much. It is still able to radiate its energy into space and so remains cool. The density increases much faster in the core of the fragment than at its periphery, so the outer portions of the cloud are both cooler and thinner than the interior. The central density by this time is approximately 10^{18} particles/m^3 (still only 10^{-9} kg/m^3 or so).

For the first time, our fragment is beginning to resemble a star. The dense, opaque region at the center is called a **protostar**—an embryonic object at the dawn of star birth. Its mass grows as more and more material rains down on it from outside, although its radius continues to shrink because its pressure is still unable to overcome the relentless pull of gravity. After stage 3, we can distinguish a "surface" on the protostar—its *photosphere*. Inside the photosphere, the protostellar material is opaque to the radi-

ation it emits.* From here on, the surface temperatures listed in Table 19.1 refer to the photosphere and not to the "periphery" of the collapsing fragment, whose temperature remains low.

STAGE 4: A PROTOSTAR

As the protostar evolves, it shrinks, its density grows, and its temperature rises, both in the core and at the photosphere. Some 100,000 years after the fragment began to form, it reaches stage 4, where its center seethes at about 1,000,000 K. The electrons and protons ripped from atoms whiz around at hundreds of kilometers per second, yet the temperature is still short of the 10^7 K needed to ignite the proton–proton nuclear reactions that fuse hydrogen into helium. ∞ (Sec. 16.5) Still much larger than the Sun, our gassy heap is now about the size of Mercury's orbit. Heated by the material falling on it from above, its surface temperature has risen to a few thousand kelvins.

Knowing the protostar's radius and surface temperature, we can calculate its luminosity. Surprisingly, it turns out to be several thousand times the luminosity of the Sun. Even though the protostar has a surface temperature only about half that of the Sun, it is hundreds of times larger, making its total luminosity very large indeed—in fact, much greater than the luminosity of most main-sequence stars. Because nuclear reactions have not yet begun, the protostar's luminosity is due entirely to the release of gravitational energy as the protostar continues to shrink, and material from the surrounding fragment (which we called the solar nebula back in Chapter 15) continues to fall onto its surface.

By the time stage 4 is reached, our protostar's physical properties can be plotted on the Hertzsprung–Russell (H–R) diagram, as shown in Figure 19.5. Recall that an H–R diagram is a plot of two key stellar properties—surface temperature (increasing to the left) and luminosity (increasing upward). ∞ (Sec. 17.7) The luminosity scale in Figure 19.5 is expressed in terms of the solar luminosity (4×10^{26} W). Our G2-type Sun is plotted at a temperature of 6000 K and a luminosity of 1 unit. As before, the dashed diagonal lines in the H–R diagrams represent stellar radius, allowing us to follow the changes in a star's size as it evolves. At each phase of a star's evolution, its surface temperature and luminosity can be represented by a point on this diagram. The motion of that point as the star evolves is known as the star's **evolutionary track**. It is a graphical representation of a star's life.

The red track on Figure 19.5 depicts the approximate path followed by our interstellar cloud fragment since it became a protostar at stage 3 (which itself lies off the right-hand edge of the figure). This early evolutionary track is known as the *Kelvin–Helmholtz contraction phase*, after two European physicists (Lord Kelvin and Hermann von

---

*Note that this is the same definition of "surface" that we used for the Sun in Chapter 16. ∞ (Sec. 16.1)

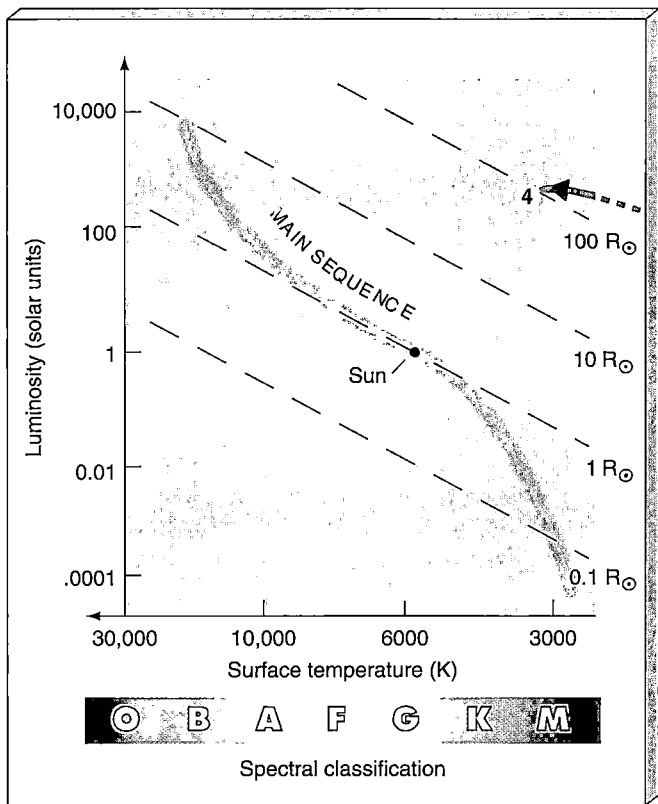

Figure 19.5 Protostar on the H-R Diagram Diagram of the approximate evolutionary track followed by an interstellar cloud fragment before reaching the end of the Kelvin–Helmholtz contraction phase as a stage 4 protostar. The circled numbers on this and subsequent plots refer to the prestellar evolutionary stages listed in Table 19.1 and described in the text.

Helmholtz) who first studied the subject. Figure 19.6 is an artist's sketch of an interstellar gas cloud proceeding along the evolutionary path outlined so far.

Our protostar is still not in equilibrium. Even though its temperature is now so high that outward-directed pressure has become a powerful countervailing influence against gravity's continued inward pull, the balance is not yet perfect. The protostar's internal heat gradually diffuses out from the hot center to the cooler surface, where it is radiated away into space. As a result, the overall contraction slows, but it does not stop completely. From our perspective on Earth, this is quite fortunate: If the heated gas were somehow able to counteract gravity completely before the star reached the temperature and density needed to start nuclear burning in its core, the protostar would simply radiate away its heat and never become a true star. The night sky would be abundant in faint protostars, but completely lacking in the genuine article. Of course, there would be no Sun either, so it is unlikely that we, or any other intelligent life form, would exist to appreciate these astronomical subtleties.

After stage 4, the protostar on the H–R diagram moves down (toward lower luminosity) and slightly to the left (toward higher temperature), as shown in Figure 19.7. Its surface temperature remains almost constant, and it becomes less luminous as it shrinks. This portion of our protostar's evolutionary path running from point 4 to point 6 in Figure 19.7 is often called the *Hayashi track*, after C. Hayashi, a twentieth-century Japanese astrophysicist whose groundbreaking work in the 1960s on the evolution of pre-main-sequence stars still provides the theoretical basis for all studies of star formation.

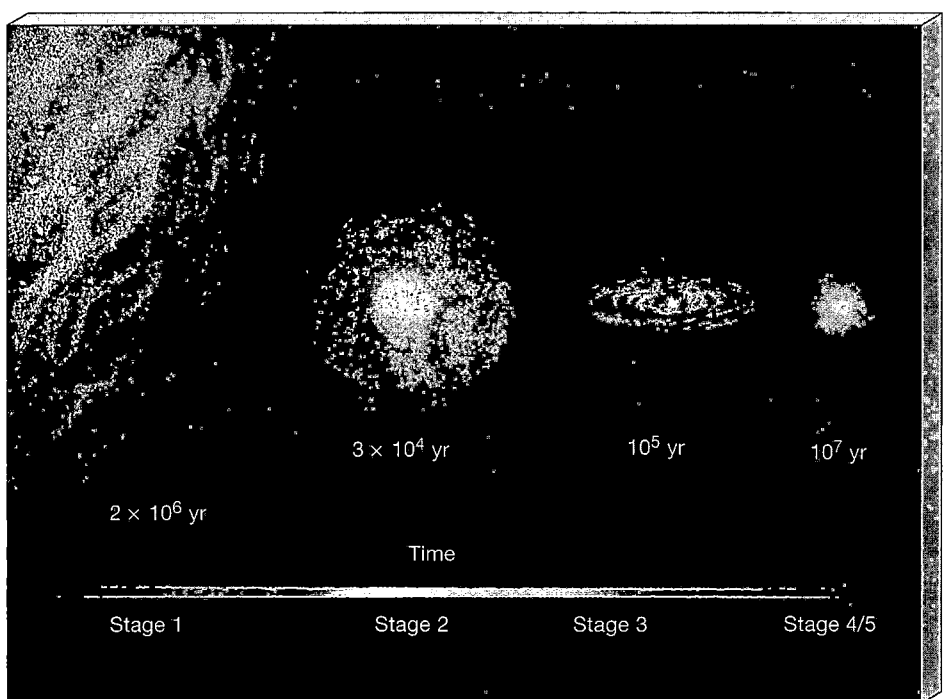

Figure 19.6 Interstellar Cloud Evolution Artist's conception of the changes in an interstellar cloud during the early evolutionary stages outlined in Table 19.1. Shown are a stage 1 interstellar cloud; a stage 2 fragment; a smaller, hotter stage 3 fragment; and a stage 4/stage 5 protostar. (Not drawn to scale.) The duration of each stage, in years, is also indicated.

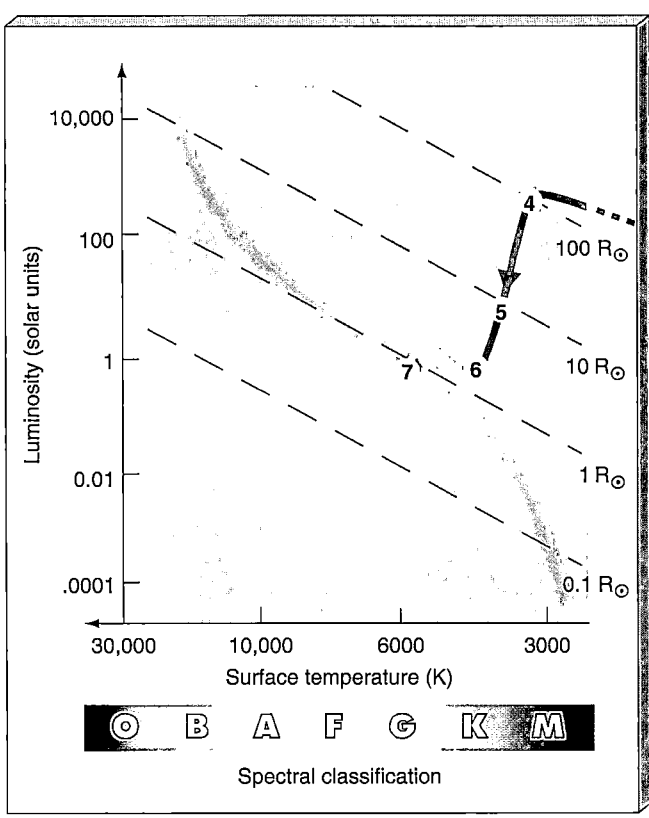

Figure 19.7 Newborn Star on the H-R Diagram
The changes in a protostar's observed properties are shown by the path of decreasing luminosity, from stage 4 to stage 6, often called the Hayashi track. At stage 7, the newborn star has arrived on the main sequence.

Protostars on the Hayashi track often exhibit violent surface activity during this phase of their evolution, resulting in extremely strong protostellar winds, much denser than the solar wind that flows from our own Sun. As mentioned previously, this portion of the evolutionary track is often called the **T Tauri** phase, after T Tauri, the first "star" (actually protostar) to be observed in this stage of prestellar development. ∞ (Sec. 15.3)

STAGE 5: PROTOSTELLAR EVOLUTION

By stage 5 on the Hayashi track, the protostar has shrunk to about 10 times the size of the Sun, its surface temperature is about 4000 K, and its luminosity has fallen to about 10 times the solar value. At this point, the central temperature has reached about 5,000,000 K. The gas is completely ionized by now, but the protons still do not have enough thermal energy to overcome their mutual electromagnetic repulsion and enter the realm of the nuclear binding force. ∞ (Sec. 16.5) The core is still too cool for nuclear fusion to begin.

Events proceed more slowly as the protostar approaches the main sequence. The initial contraction and fragmentation of the interstellar cloud occurred quite rapidly, but by stage 5, as the protostar nears the status of a full-fledged star, its evolution slows. The cause of this slowdown is heat—even gravity must struggle to compress a hot object. The contraction is governed largely by the rate at which the protostar's internal energy can be radiated away into space. The greater this radiation of internal energy—that is, the more rapidly energy moves through the star to escape from its surface—the faster the contraction occurs. As the luminosity decreases, so too does the contraction rate.

STAGE 6: A NEWBORN STAR

Some 10 million years after its first appearance, the protostar finally becomes a true star. By the bottom of the Hayashi track, at stage 6, when our roughly 1-solar-mass object has shrunk to a radius of about 1,000,000 km, the contraction has raised the central temperature to 10,000,000 K, enough to ignite nuclear burning. Protons begin fusing into helium nuclei in the core, and a star is born. As shown in Figure 19.7, the star's surface temperature at this point is about 4500 K, still a little cooler than the Sun. Even though the newly formed star is slightly larger in radius than our Sun, its lower temperature means that its luminosity is somewhat less than (actually, about two-thirds of) the solar value.

STAGE 7: THE MAIN SEQUENCE AT LAST

Over the next 30 million years or so, the stage 6 star contracts a little more. In making this slight adjustment, the central density rises to about 10^{32} particles/m^3 (more conveniently expressed as 10^5 kg/m^3), the central temperature increases to 15,000,000 K, and the surface temperature reaches 6000 K. By stage 7, the star finally reaches the main sequence just about where our Sun now resides. Pressure and gravity are finally balanced, and the rate at which nuclear energy is generated in the core exactly matches the rate at which energy is radiated from the surface.

The evolutionary events just described occur over the course of some 40–50 million years. Although this is a long time by human standards, it is still less than one percent of the Sun's lifetime on the main sequence. Once an object begins fusing hydrogen in its core and establishes a "gravity-in/pressure-out" equilibrium, it is destined to burn steadily for a very long time. The star's location on the H-R diagram—that is, its surface temperature and luminosity—will remain virtually unchanged for the next 10 billion years.

✓ Concept Check
■ What distinguishes a collapsing cloud from a protostar, and a protostar from a star?

19.3 Stars of Other Masses

⑬ The numerical values and the evolutionary track just described are valid only for the case of a 1-solar-mass star. The temperatures, densities, and radii of prestellar objects of other masses exhibit similar trends, but the numbers and the tracks differ, in some cases quite considerably. Perhaps not surprisingly, the most massive fragments within interstellar clouds tend to produce the most massive protostars and eventually the most massive stars. Similarly, low-mass fragments give rise to low-mass stars.

THE ZERO-AGE MAIN SEQUENCE

Figure 19.8 compares the theoretical pre-main-sequence track taken by the Sun with the corresponding evolutionary tracks of a 0.3-solar-mass star and a 3-solar-mass star. All three tracks traverse the H–R diagram in the same general manner, but cloud fragments that eventually form stars more massive than the Sun approach the main sequence along a higher track on the diagram, while those destined to form less massive stars take a lower track. The *time* required for an interstellar cloud to become a main-sequence star also depends strongly on its mass. The most

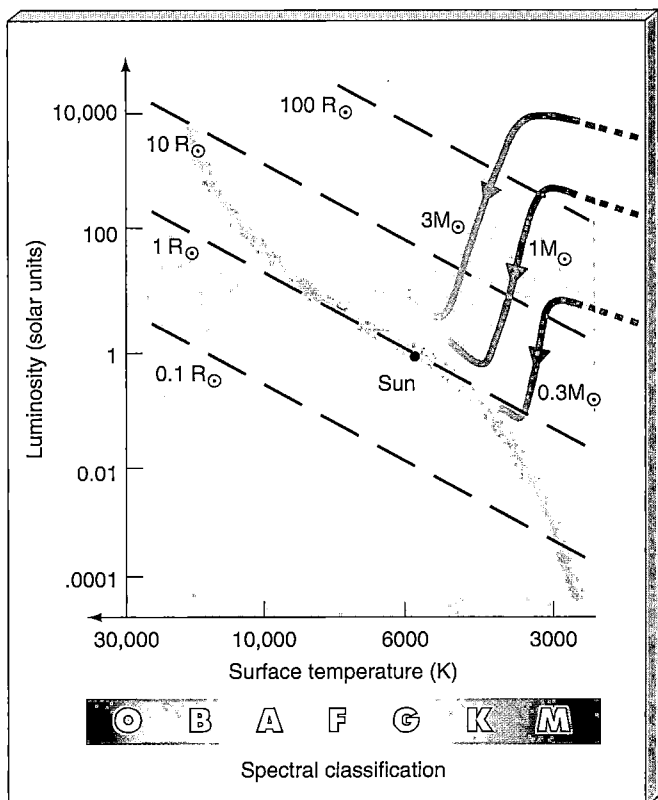

Figure 19.8 Prestellar Evolutionary Tracks Some pre-main-sequence evolutionary paths for stars more massive and less massive than our Sun.

massive cloud fragments heat up to the required 10 million K and become O-type stars in a mere million years, roughly $\frac{1}{50}$ the time taken by the Sun. The opposite is the case for prestellar objects having masses less than that of our Sun. A typical M-type star, for example, requires nearly a billion years to form.

Whatever the mass, the end point of the prestellar evolutionary track is the main sequence. A star is considered to have reached the main sequence when hydrogen burning begins in its core and the star's properties settle down to stable values. The main-sequence line predicted by theory is called the **zero-age main sequence** (ZAMS). It agrees quite well with main sequences observed for stars in the vicinity of the Sun and those observed for stars in more distant star clusters.

If all gas clouds contained precisely the same elements in exactly the same proportions, mass would be the sole determinant of a newborn star's location on the H–R diagram, and the zero-age main sequence would be a well-defined line rather than a broad band. However, the composition of a star affects its internal structure (mainly by changing the opacity of its outer layers), and this in turn affects both its temperature and its luminosity on the main sequence. Stars with more heavy elements tend to be cooler and slightly less luminous than stars that have the same mass but contain fewer heavy elements. As a result, differences in composition between stars "blur" the zero-age main sequence into the broad band we observe.

It is important to realize that *the main sequence is itself not an evolutionary track—stars do not evolve along it*. Rather, it is just a "way station" on the H–R diagram where stars stop and spend most of their lives—low-mass stars at the bottom, high-mass stars at the top. Once on the main sequence, a star stays in essentially the same location in the H–R diagram during its whole time as a stage-7 object. (In other words, a star that arrives on the main sequence as, say, a G-type star can never "work its way up" to become a B- or an O-type main-sequence blue supergiant, or move down to become an M-type red dwarf.) As we will see in Chapter 20, the next stage of stellar evolution occurs when a star moves away from the main sequence. A star leaving the main sequence and entering this next stage has pretty much the same surface temperature and luminosity it had when it arrived on the main sequence millions (or billions) of years earlier.

FAILED STARS

Some cloud fragments are too small ever to become stars. The giant planet Jupiter is a good example. Jupiter contracted under the influence of gravity, and the resultant heat is still detectable, but the planet did not have enough mass for gravity to crush its matter to the point of nuclear ignition. ∞ (Sec. 11.3) It became stabilized by heat and rotation before the central temperature became hot enough to fuse hydrogen. Jupiter never evolved beyond the protostar stage. If Jupiter, or any of the other jovian

DISCOVERY 19-1

Observations of Brown Dwarfs

Cruelly put, brown dwarfs are "failed" stars—objects that formed through the contraction and fragmentation of an interstellar cloud, just like stars, but somehow never reached the critical mass of about 0.08 solar masses (80 times the mass of Jupiter) needed to initiate hydrogen fusion in their cores. We mentioned in the text that interstellar space could contain large numbers of these small, dark, and hard-to-detect objects. Recent observations of star-forming regions have reinforced that idea by providing a plausible way in which the formation process could be stopped before the formation of a true star. Here we examine in a little more detail the observational evidence for these elusive objects.

Until late 1994 this discussion would have been very short indeed, as there was no firm observational evidence for the existence of any brown dwarfs at all! Since then, however, continuing improvements in observational techniques have begun to bear fruit, and several dozen likely brown dwarfs are now known.

Detecting a brown dwarf is no easy task. These objects are so faint that they are difficult—although not impossible—to detect directly. In searching for brown dwarfs in binary-star systems, astronomers have used many of the same techniques employed in the search for extrasolar planets, as discussed in Chapter 15. ⫘ (Sec. 15.5) (Recall from Chapter 17 that most stars are found in

binaries; the same may very well be true of brown dwarfs.) ⫘ (Sec. 17.8) While numerous brown dwarf candidates have been found this way, the observational data seem to show a real deficit of brown dwarfs relative to planets—the same observations that have led to many planet discoveries have turned up only a handful of brown dwarfs. It appears that brown dwarfs are not found as companions to solar-type stars, although recent observations of lower-mass main-sequence stars suggest that the fraction of brown dwarf companions may be higher.

Actually, the dividing line between brown dwarfs and Jupiter-like planets is not completely clear-cut, especially

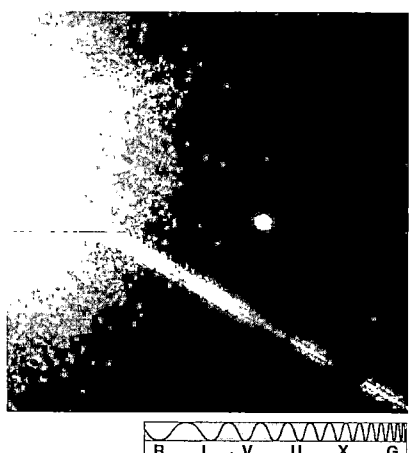

(Palomar; NASA)

planets, had continued to accumulate gas from the solar nebula, it might have become a star (almost certainly to the detriment of life on Earth), but virtually all the matter present during the formative stages of our solar system is gone now, swept away by the solar wind.

Low-mass interstellar gas fragments simply lack the mass needed to initiate nuclear burning. Rather than turning into stars, they will continue to cool, eventually becoming compact, dark "clinkers"—cold fragments of unburned matter—in interstellar space. On the basis of theoretical modeling, astronomers believe that the minimum mass of gas needed to generate core temperatures high enough to begin nuclear fusion is about 0.08 solar masses.

Vast numbers of Jupiter-like objects may well be scattered throughout the universe—fragments frozen in time somewhere along the Kelvin–Helmholtz contraction phase. Small, faint, and cool (and growing ever colder),

they are known collectively as **brown dwarfs**. Our technology currently has difficulty in detecting them, be they planets associated with stars or interstellar cloud fragments far from any star, although recent advances in observational hardware and image-processing techniques have begun to yield impressive results (see *Discovery 19-1*). We can telescopically detect stars and spectroscopically infer atoms and molecules, but astronomical objects of intermediate size outside our solar system remain hard to see. Based on current observations, it seems that up to 100 billion cold, dark Jupiter-sized objects may lurk in the depths of interstellar space—comparable to the total number of "real" stars in our Galaxy.

☑ Concept Check

■ Do stars evolve along the main sequence?

given the properties of the extrasolar planets now known and the possibility that our solar system may not be the only, or even the most likely, type of planetary system to form. Most workers adopt a division at about 12 times the mass of Jupiter. Above this mass (but below 80 Jupiter masses), although core temperatures never become high enough for hydrogen fusion to occur, a contracting fragment will experience a brief phase of *deuterium fusion*, as the core becomes hot enough for deuterium nuclei present in the original cloud to combine. The phase ends once the deuterium is consumed, and the fragment's "nuclear" lifetime is over. Below 12 Jupiter masses, no nuclear fusion of any kind is expected.

The images at left show the binary-star system Gliese 229, first identified as a possible brown dwarf by ground-based infrared observations (left), and subsequently imaged from space (right). The two objects are only 7″ apart; the fainter "star" has a luminosity only a few millionths that of the Sun and an estimated mass about 50 times that of Jupiter. (The bright diagonal streak in the latter image is caused by a hardware problem in the CCD chip used to record it.)

Infrared and spectroscopic studies provide other ways of searching for brown dwarfs, especially those that are not members of binaries. The presence of molecules such as methane, which would be destroyed at the high temperatures found in a star, is a good indicator of a low-temperature object. Infrared observations are particularly effective because brown dwarfs emit most of their radiation in the infrared part of the spectrum, while true stars tend to be

brightest at near-infrared and optical. The second pair of figures below compares visible-light and infrared *Hubble* views of the central part of the Orion nebula (note the four bright stars of the Trapezium), and shows clear evidence for low-mass infrared emitters, many of them probably brown dwarfs.

The presence of lithium, an element that is present in interstellar gas but rapidly depleted in a star once nuclear fusion begins, is another telltale sign of brown dwarfs. Researchers using a spectrometer on the giant Keck telescope in Hawaii have identified several brown dwarf candidates by this means. Recent infrared surveys of the sky have revealed about 100 low-temperature (less than 2000 K) objects, and have so far identified lithium in about a quarter of them.

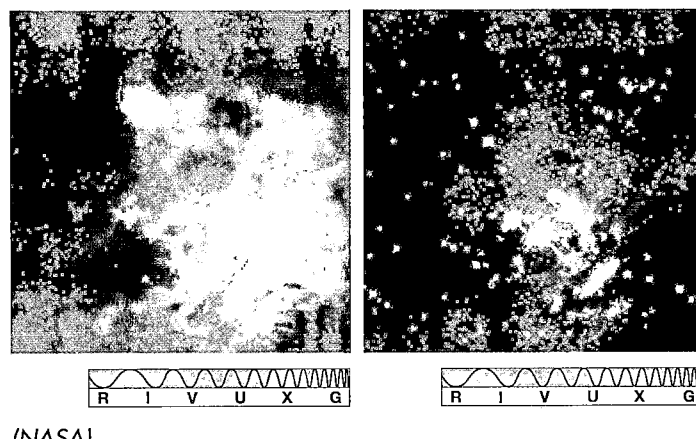

(NASA)

19.4 Observations of Cloud Fragments and Protostars

How can we verify the theoretical picture just outlined? The age of our entire civilization is much shorter than the time needed for a single interstellar cloud to contract and form a star. We can never observe individual objects proceed through the full panorama of star birth. However, we can do the next best thing, by observing many different objects—interstellar clouds, protostars, young stars approaching the main sequence—as they appear today at different stages of their evolutionary paths. Each observation is like part of a jigsaw puzzle. When properly oriented relative to all the others, the pieces can be used to build up a picture of the full life cycle of a star.

EVIDENCE OF CLOUD CONTRACTION

Prestellar objects at stages 1 and 2 are not yet hot enough to emit much infrared radiation, and certainly no optical radiation arises from their dark, cool interiors. The best way to study the early stages of cloud contraction and fragmentation is to observe the radio emission from interstellar molecules within these clouds. Consider again M20, the splendid emission nebula studied in Chapter 18. ∞ (Sec. 18.2) The brilliant region of glowing, ionized gas shown in Figure 18.8 is not our main interest here, however. Instead, the youthful O- and B-type stars that energize the nebula alert us to the general environment where stars are forming. Emission nebulae are signposts of star birth.

The region surrounding M20 contains galactic matter that seems to be contracting. The presence of (optically) invisible gas there was illustrated in Figure 18.20, which showed a contour map of the abundance of the formaldehyde

(H_2CO) molecule. Formaldehyde and many other molecules are widespread in the vicinity of the nebula, especially throughout the dusty regions below and to the right of the emission nebula itself. Further analysis of the observations suggests that this region of greatest molecular abundance is also contracting and fragmenting, well on its way toward forming a star—or, more likely, a star cluster.

The interstellar clouds in and around M20 thus provide tentative evidence of three distinct phases of star formation, as shown in Figure 19.9. The huge, dark molecular cloud surrounding the visible nebula is the stage 1 cloud. Both its density and its temperature are low, about 10^8 particles/m³ and 20 K, respectively. Greater densities and temperatures typify smaller regions within this huge cloud. The totally obscured regions labeled A and B, where the molecular emission of radio energy is strongest, are such denser, warmer fragments. Here, the total gas density is observed to be at least 10^9 particles/m³, and the temperature is about 100 K. The Doppler shifts of the radio lines observed in the vicinity of region B indicate that this portion of M20, labeled "contracting fragment"

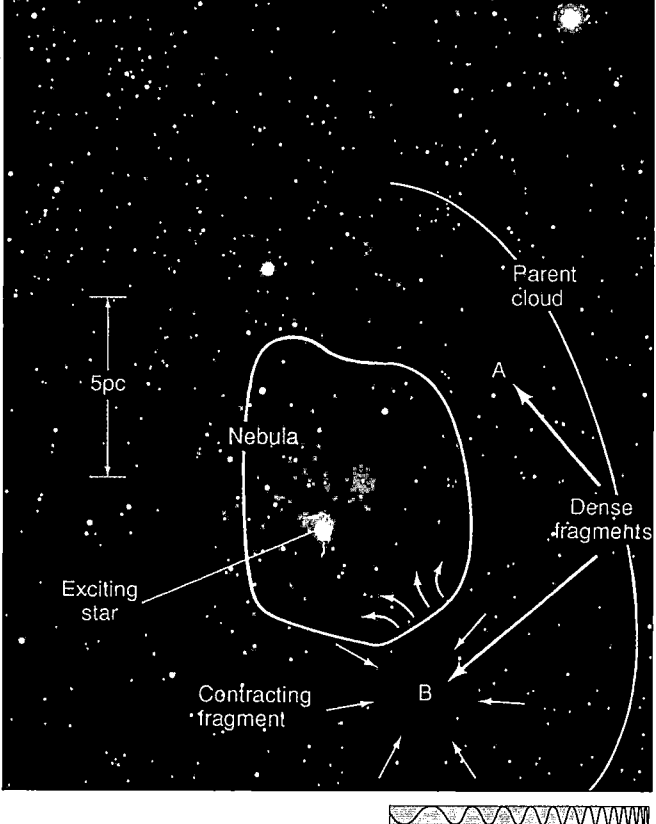

Figure 19.9 Star Formation Phases The M20 region shows observational evidence for three broad phases in the birth of a star: The parent cloud is stage 1 of Table 19.1. The region labeled "contracting fragment" likely lies between stages 1 and 2. Finally, the emission nebula (M20 itself) results from the formation of one or more massive stars (stages 6 and 7). *(AURA)*

in Figure 19.9, is contracting. Less than a light-year across, this region has a total mass over 1000 times the mass of the Sun—considerably more than the mass of M20 itself. It lies somewhere between stages 1 and 2 of Table 19.1.

The third star-formation phase shown in Figure 19.9 is M20 itself. The glowing region of ionized gas results directly from a massive O-type star that formed there within the past million years or so. Because the central star is already fully formed, this final phase corresponds to stage 6 or 7 of our evolutionary scenario.

EVIDENCE OF CLOUD FRAGMENTS

Other parts of our Milky Way Galaxy provide sketchy evidence for prestellar objects in stages 3 through 5. The Orion complex, shown in Figure 19.10, is one such region. Lit from within by several O-type stars, the bright Orion Nebula is partly surrounded by a vast molecular cloud that extends well beyond the roughly 2×3 pc region bounded by the photograph in Figure 19.10(b).

The Orion molecular cloud harbors several smaller sites of intense radiation emitted by molecules deep within the core of the cloud fragment. Their extent, shown in Figures 19.10(c) and (e) measures about 10^{10} km, or 1/1000 of a light-year, about the diameter of our solar system. The gas density of these smaller regions is about 10^{15} particles/m³, much denser than the surrounding cloud. Although their temperature cannot be estimated reliably, many researchers regard these regions as objects well on their way to stage 3. We cannot determine if these regions will eventually form stars like the Sun, but it does seem certain that these intensely emitting regions are on the threshold of becoming protostars.

EVIDENCE OF PROTOSTARS

In the hunt for and study of objects at more advanced stages of star formation, radio techniques become less useful because stages 4, 5, and 6 have increasingly higher temperatures. By Wien's law, their emission shifts toward shorter wavelengths, and so these objects shine most strongly in the infrared. ∞ (Sec. 3.4) One particularly bright infrared emitter, known as the Becklin–Neugebauer object (see Figure 19.23c), was detected in the core of the Orion molecular cloud in the 1970s. Its luminosity is around 1000 times the luminosity of the Sun. Most astronomers agree that this warm, dense blob is a high-mass protostar, probably around stage 4.

Until the *Infrared Astronomy Satellite* (*IRAS*) was launched in the early 1980s, astronomers were aware only of giant stars forming in clouds far away. But *IRAS* showed that stars are forming much closer to home, and some of these protostars have masses comparable to that of our Sun. Figure 19.11 shows two examples of low-mass protostars—both spotted in a rich star-forming region in Orion. Their infrared heat signatures are those expected of an object on the Hayashi track, around stage 5.

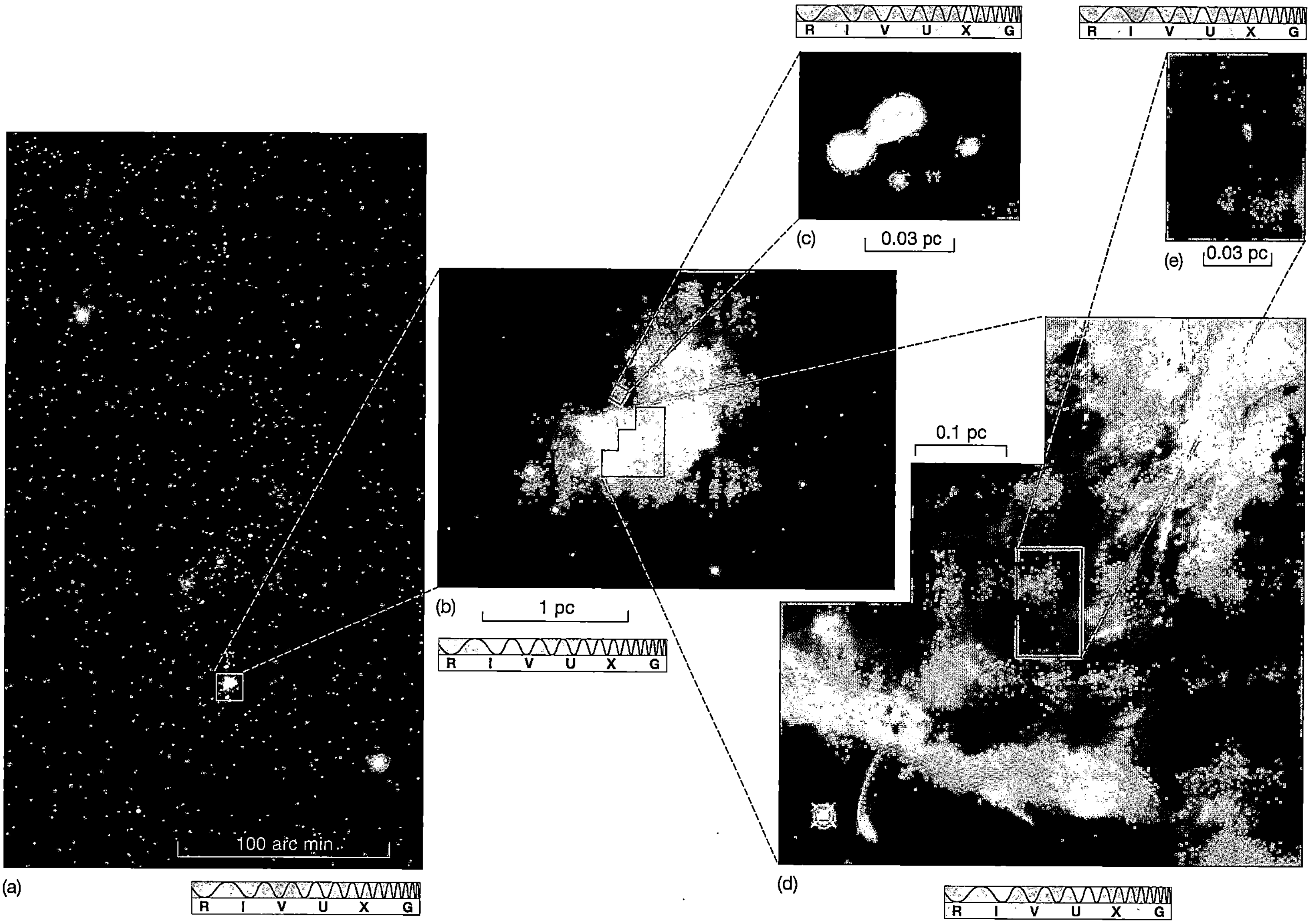

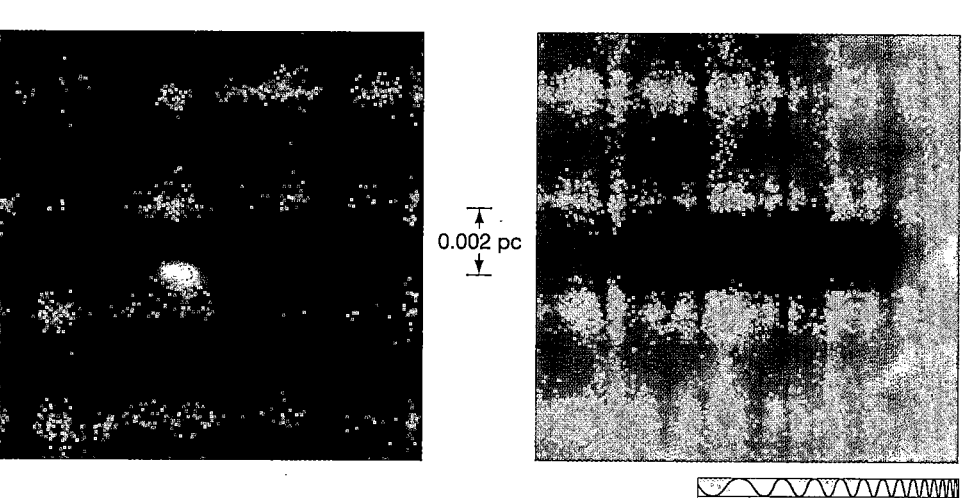

Figure 19.10 Orion Nebula, Up Close (a) The constellation Orion, with the region around its famous emission nebula marked by a rectangle. The Orion Nebula is the middle "star" of Orion's sword. The framed region is enlarged in part (b), suggesting how the nebula is partly surrounded by a vast molecular cloud. Various parts of this cloud are probably fragmenting and contracting, with even smaller sites forming protostars. The three frames at right show some of the evidence for those protostars. (c) A false-color radio image of some intensely emitting molecular sites. (d) A nearly real-color visible image of embedded nebular "knots" thought to harbor protostars. (e) A high-resolution image of several young stars surrounded by disks of gas and dust where planets might ultimately form. *(Astrostock-Sanford; AURA; Harvard-Smithsonian; NASA)*

Figure 19.11 Protostars
Two infrared images of planetary-system-sized dirty disks in the Orion region, showing heat and light emerging from their centers. On the basis of their temperatures and luminosities, these unnamed sources appear to be low-mass protostars on the Hayashi track in the H–R diagram. *(NASA)*

The energy sources for some infrared objects seem to be luminous hot stars that are hidden from optical view by surrounding dark clouds. Apparently, these stars are already so hot that they emit large amounts of ultraviolet radiation, which is mostly absorbed by "cocoons" of dust surrounding them. The absorbed energy is then reemitted by the dust as infrared radiation. These bright infrared sources are known as *cocoon nebulae*. Two considerations support the idea that the hot stars heating the dust have only recently ignited: 1) These dust cocoons are predicted to disperse quite rapidly once their central stars form, and 2) they are invariably found in the dense cores of molecular clouds. The central stars probably lie near stage 6.

PROTOSTELLAR WINDS

Protostars often exhibit strong winds. Radio and infrared observations of hydrogen and carbon monoxide molecules in the Orion molecular cloud have revealed gas expanding

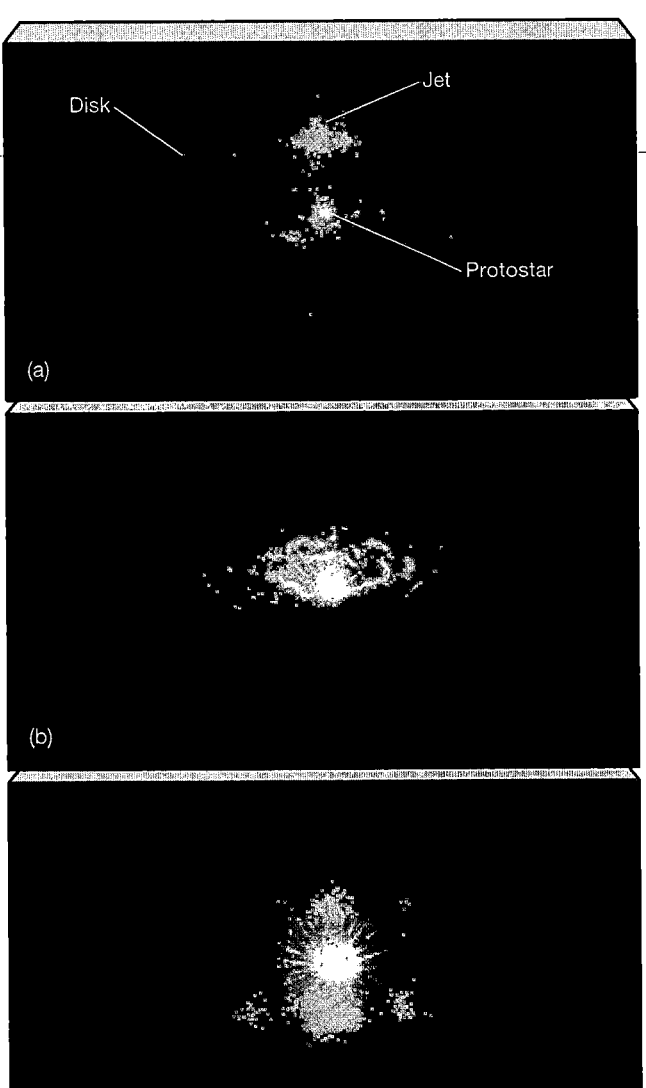

(a)

(b)

(c)

outward at velocities approaching 100 km/s. High-resolution interferometric observations have disclosed expanding knots of water emission within the same star-forming region and have linked the strong winds to the protostars themselves. These winds may be related to the violent surface activity associated with many protostars.

Early in a protostar's life, it may still be embedded in an extensive disk of nebular material in which planets are forming. ∞ (Sec. 15.2) Strong heating within the turbulent disk and a powerful protostellar wind combine to produce a *bipolar flow*, expelling two "jets" of matter in the directions perpendicular to the disk, as illustrated in Figure 19.12. As the protostellar wind gradually destroys the disk, blowing it away into space, the outflow widens until, with the disk gone, the wind flows away from the star equally in all directions. Figure 19.13 shows emission from an especially clear bipolar flow, along with an artist's conception of the system producing it.

These outflows can be very energetic. Figure 19.14 shows a portion of the Orion molecular cloud, south of the Orion Nebula, where a newborn star is seen still surrounded by a bright nebula, its turbulent wind spreading out into the interstellar medium. Below it (enlarged in the inset) are twin jets known as HH1 and HH2. (HH stands for Herbig-Haro, the investigators who first cataloged such objects.) Formed in another (unseen) protostellar disk, these jets have traveled outward for almost half a light-year before colliding with interstellar matter. More Herbig-Haro objects can be seen in the upper right portion of the figure.

✓ Concept Check

■ How can a "snapshot" of the universe today test our theories of the evolution of individual objects?

Figure 19.12 Protostellar Wind (a) The nebular disk around a protostar can be the site of intense heating and strong outflows, forming a bipolar jet perpendicular to the disk. (b) As the disk is blown away by the wind, the jets fan out, eventually (c) merging into a spherical wind. Parts (d) and (e) show two actual circumstellar disks in the Orion region. *(NASA)*

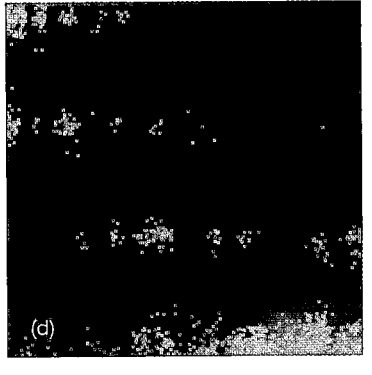

(d)

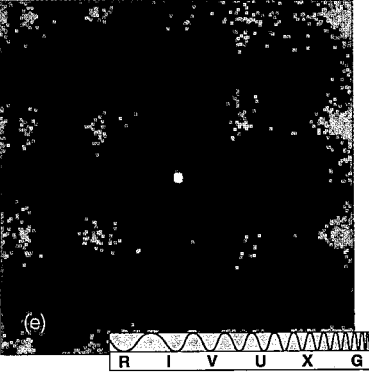

(e)

R I V U X G

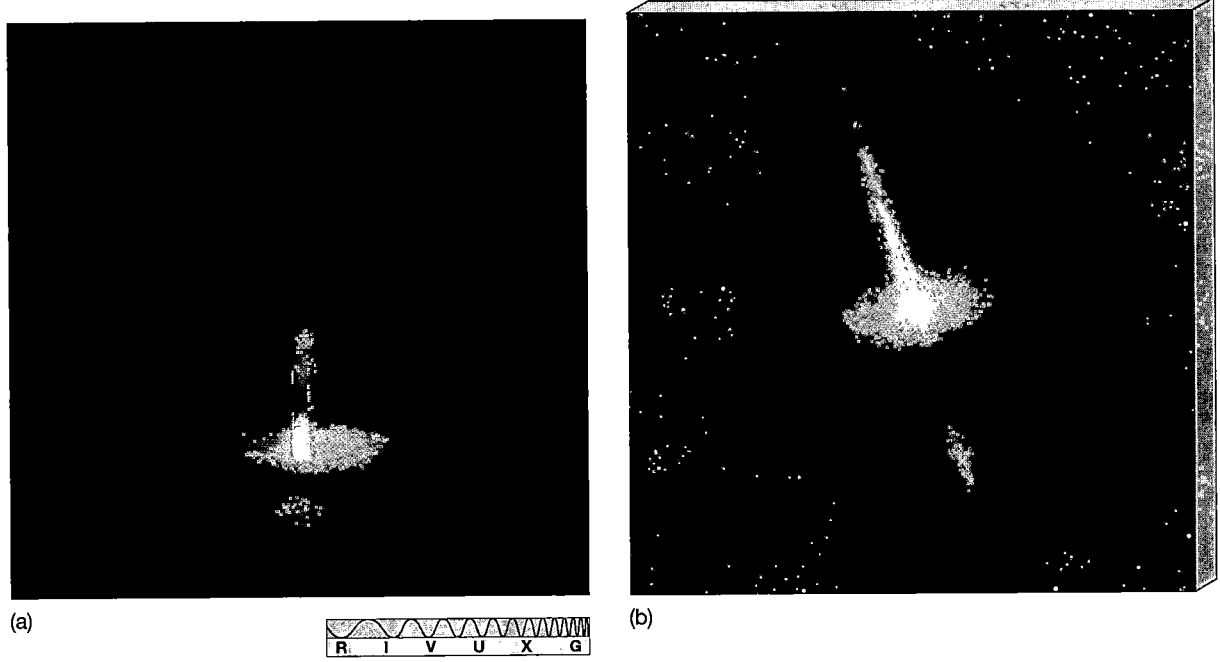

(a)

(b)

R   I   V   U   X   G

Figure 19.13 Bipolar Jets (a) This remarkable image shows two jets emanating from the young *Animation* star system HH30, the result of matter accreting onto an embryonic star near the center. (b) An idealized artist's conception of a young star system, showing two jets flowing perpendicular to the disk of gas and dust rotating around the star. *(NASA; D. Berry)*

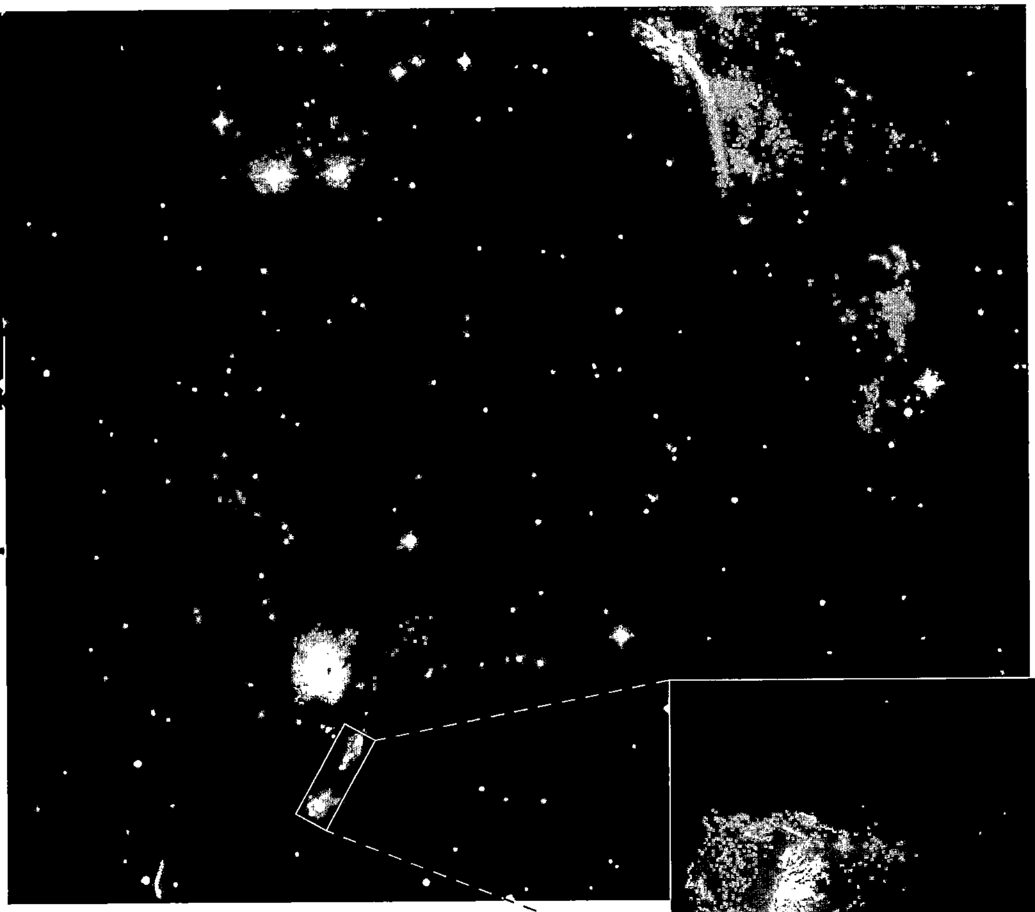

Figure 19.14 Protostellar Outflow and Jets This view of the Orion molecular cloud shows the outflow from a newborn star, still surrounded by nebular gas. The inset shows a pair of jets called HH1/HH2, formed when matter falling onto another protostar (still obscured by the dusty cloud fragment from which it formed) creates a pair of high-speed jets of gas perpendicular to the flattened protostellar disk. The jets are nearly 1 light-year across. Several more Herbig-Haro objects can be seen at top right. One of them, the oddly shaped "waterfall" at top right, may be due to an earlier outflow from the same protostar responsible for HH1 and HH2. *(AURA; NASA)*

R   I   V   U   X   G

19.5 Shock Waves and Star Formation

⑤ The subject of star formation is really much more complicated than the preceding discussion suggests. Interstellar space is populated with many kinds of clouds, fragments, protostars, stars, and nebulae. They interact in a complex fashion, and each type of object can affect the behavior of all the others. For example, the presence of an emission nebula in or near a molecular cloud probably influences the evolution of the entire region. We can easily imagine expanding waves of matter driven outward by the high temperatures and pressures in the nebula. As the waves crash into the surrounding molecular cloud, interstellar gas tends to pile up and become compressed. Such a shell of gas, rushing rapidly through space, is known as a **shock wave**. It can push ordinarily thin matter into dense sheets, just as a plow pushes snow.

Many astronomers regard the passage of a shock wave through interstellar matter as the triggering mechanism needed to initiate star formation in a galaxy. Calculations show that when a shock wave encounters an interstellar cloud, it races around the thinner exterior of the cloud more rapidly than it can penetrate its thicker interior. Thus, shock waves do not blast a cloud from only one direction. They effectively squeeze it from many directions. Atomic bomb tests have experimentally demonstrated this squeezing—shock waves created in the blast tend to surround buildings, causing them to be blown together (imploded) rather than apart (exploded). The "contracting fragment" in Figure 19.9 may well have been triggered by the shock wave from the M20 nebula. Note the correspondence between the shock-compressed region at lower right and the high-density molecular gas revealed by radio studies (Figure 18.20). Once shock waves have begun the compression of an interstellar cloud, natural gravitational instabilities take over, dividing it into the fragments that eventually form stars.

Emission nebulae are by no means the only generators of interstellar shock waves. At least three other driving forces are available—the deaths of old stars (planetary nebulae and supernovae, to be discussed in Chapters 20 and 21), the spiral-arm waves that plow through the Milky Way (Chapter 23), and interactions between galaxies (Chapter 24). Supernovae are by far the most energetic, and probably also the most efficient, way to pile up matter into dense clumps. However, they are relatively few and far between, so the other mechanisms may be more important overall in triggering star formation. Although the evidence is somewhat circumstantial, the presence of young (and thus fast-forming) O- and B-

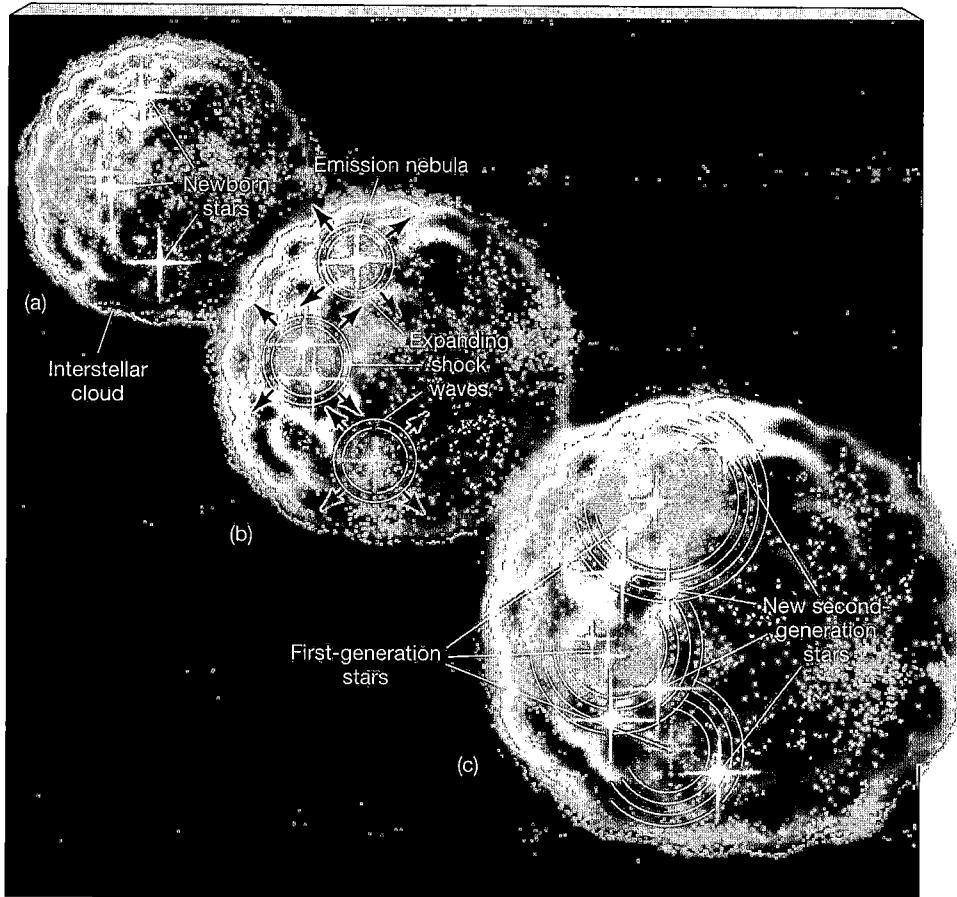

Figure 19.15 Generations of Star Formation (a) Star birth and (b) shock waves lead to (c) more star births and more shock waves in a continuous cycle of star formation in many areas of our Galaxy. Like a chain reaction, old stars trigger the formation of new stars ever deeper into an interstellar cloud.

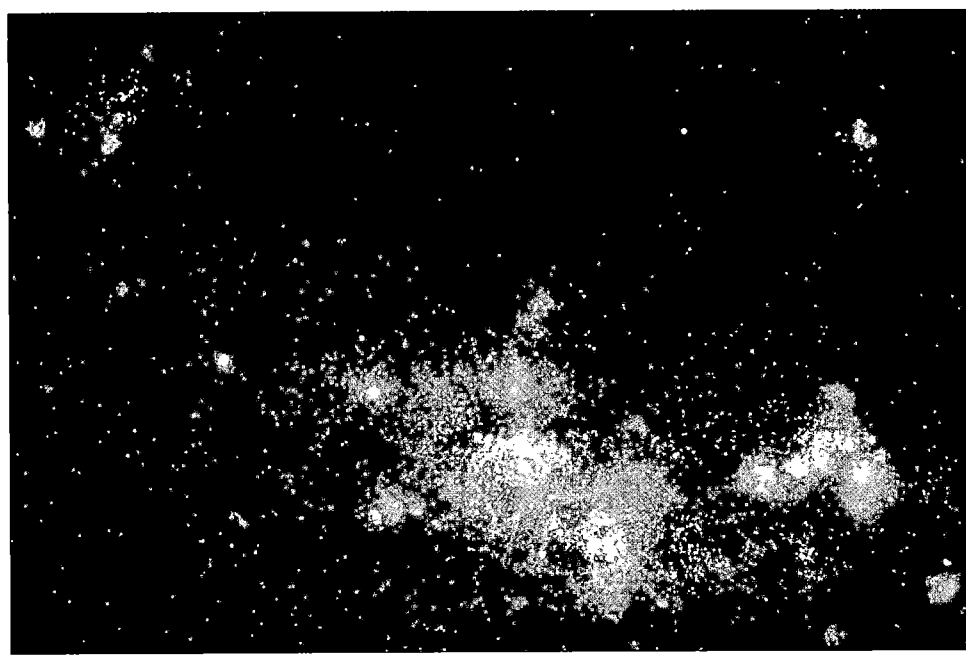

Figure 19.16 A Wave of Star Formation A group of star-forming regions in the galaxy NGC 4214, possibly representing several generations in a chain of star formation. *(NASA)*

type stars in the vicinity of supernova remnants does suggest that the birth of stars is often initiated by the violent, explosive deaths of others.

This picture of shock-induced star formation is complicated by the fact that O- and B-type stars form quickly, live briefly, and die explosively. These massive stars, themselves perhaps born of a passing shock wave, may in turn create new shock waves, either through the expanding nebular gas produced by their births or by their explosive deaths. These new shock waves can produce "second-generation" stars, which in turn will explode and give rise to still more shock waves, and so on. As depicted in Figure 19.15, star formation resembles a chain reaction. Other, lighter stars are also formed in the process, of course, but they are largely "along for the ride." It is the O- and B-type stars that drive the star-formation wave through the cloud.

Observational evidence lends some support to this chain-reaction picture. Groups of stars nearest molecular clouds do indeed appear to be the youngest, whereas those farther away seem to be older. Figure 19.16 shows a spectacular *HST* image of a star-forming region in the galaxy NGC 4214, which lies some 13 million light-years from Earth. A series of bright emission nebulae, powered by hot young stars, can be seen, suggesting that a wave of star formation has recently swept across the region and triggered the sequence seen here.

☑ **Concept Check**

■ Why might we expect multiple episodes of star formation to occur in some locations?

19.6 Star Clusters

⑥ The end result of cloud collapse is a group of stars, all formed from the same parent cloud and lying in the same region of space. Such a collection of stars is called a **star cluster**. Figure 19.17 shows a spectacular view of a newborn star cluster and (part of) the interstellar cloud from which it came. Because all the stars formed at the same time out of the same cloud of interstellar gas, and under the same environmental conditions, clusters are near-ideal "laboratories" for stellar studies—not in the sense that astronomers can perform experiments on them, but because the properties of the stars are very tightly constrained. The only factor distinguishing one star from another in the same cluster is mass, so theoretical models of star formation and evolution can be compared with reality without the complications introduced by broad spreads in age, chemical composition, and place of origin.

CLUSTERS AND ASSOCIATIONS

Figure 19.18(a) shows a small star cluster called the Pleiades, or Seven Sisters, a well-known naked-eye object in the constellation Taurus, lying about 120 pc from Earth. This type of loose, irregular cluster, found mainly in the plane of the Milky Way, is called an **open cluster**. Open clusters typically contain from a few hundred to a few tens of thousands of stars, and are a few parsecs across.

Figure 19.18(b) shows the H–R diagram for stars in the Pleiades. The cluster contains stars in almost all parts of the main sequence—only the very brightest main-sequence stars are missing. (The brightest 6 or 7 stars in the diagram have just left the main sequence, as will be discussed in Chapter 20.) Thus, even though we have no direct evidence of the cluster's birth, we can estimate its age as less than about 100 million years, the lifetime of a main-sequence B-type star. ⊂⊃ (Sec. 17.8) If all the stars in the cluster formed at the same time, then the red stars must be young too. The wisps of leftover gas evident in the photograph are further evidence of the cluster's relative youth. In addition, the system is abundant in heavy elements that (as we will see) could have been created only within the cores of many generations of ancient stars long since perished.

Less massive, but more extended, clusters are known as **associations**. These typically contain no more than a few hundred bright stars, but may span many tens of parsecs. Associations tend to be rich in very young stars. Those containing many pre-main-sequence T Tauri stars are known as *T associations*, whereas those with prominent O- ,and B-type stars, such as the Trapezium in Orion, are called *OB associations*. It is quite likely that the main difference between associations and open clusters is simply the efficiency with which stars formed from the parent cloud.

Figure 19.19(a) shows a very different type of star cluster, called a **globular cluster**. All globular clusters are

roughly spherical (which accounts for their name), are generally found away from the Milky Way plane, and contain hundreds of thousands, and sometimes millions, of stars spread out over about 50 pc. Figure 19.19(b) shows an H–R diagram for this cluster, which is called Omega Centauri. Notice its many differences from Figure 19.18(b)—globular clusters are a very different stellar environment from open clusters like the Pleiades. The distance to this cluster has been determined by a variation on the method of spectroscopic parallax, applied to the entire cluster rather than to individual stars. ⊂⊃ (Sec. 17.7) Omega Centauri lies about 5000 pc from Earth.

The most outstanding feature of globular clusters is their lack of upper-main-sequence stars. Astronomers in the 1920s and 1930s, working with instruments incapable of detecting stars fainter than about 1 solar luminosity at such distances, and having no theory of stellar evolution to guide them, were very puzzled by the H–R diagrams they saw when they looked at globular clusters. Indeed, a comparison of just the top halves of the diagrams (so that the lower main sequences cannot be seen) reveals few similarities between Figures 19.18(b) and 19.19(b).

Most globular clusters contain no main-sequence stars with masses greater than about 0.8 times the mass of the Sun. (The A-type stars in Figure 19.19b are stars at a much

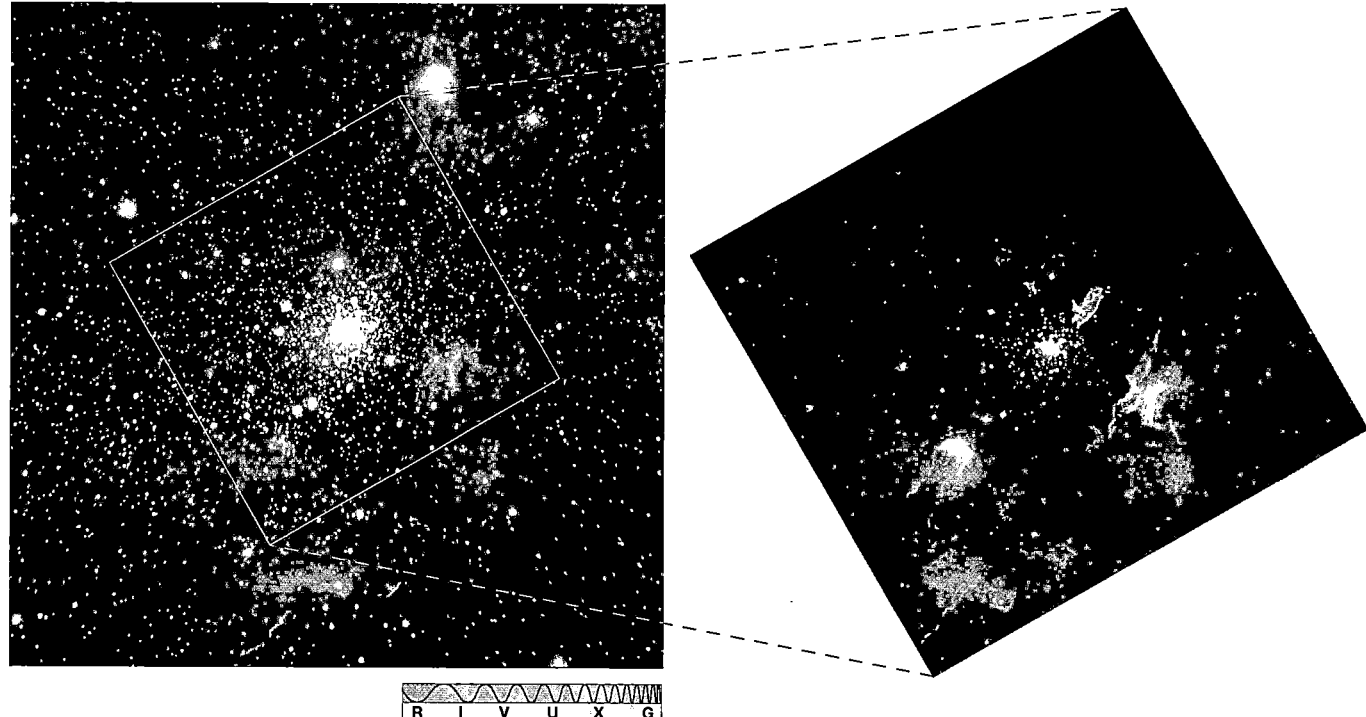

R I V U X G

Figure 19.17 Newborn Cluster The star cluster NGC 3603 and a portion of the larger molecular cloud in which it formed. The cluster contains about 2000 bright stars and lies some 6000 pc from Earth. The field of view shown here spans about 20 light-years. Radiation from the cluster has cleared a cavity in the cloud several light-years across. The inset shows the central area more clearly, including the most massive star in the region (called Sher 25, above and to the left of the cluster) which is already near the end of its lifetime, having ejected part of its outer layers and formed a ring of gas. Many low-mass stars, less massive than the Sun, can also be seen. *(ESO; NASA)*

later stage in their evolution that happen to be passing through the location of the upper main sequence.) Their more massive O- through F-type stars have long since exhausted their nuclear fuel and disappeared from the main sequence (in fact becoming the red giants and other lumi-

nous stars above the main sequence, as we will see in Chapter 20). On the basis of these and other observations, astronomers estimate that most globular clusters are at least 10 billion years old—they contain the oldest known stars in our Galaxy.

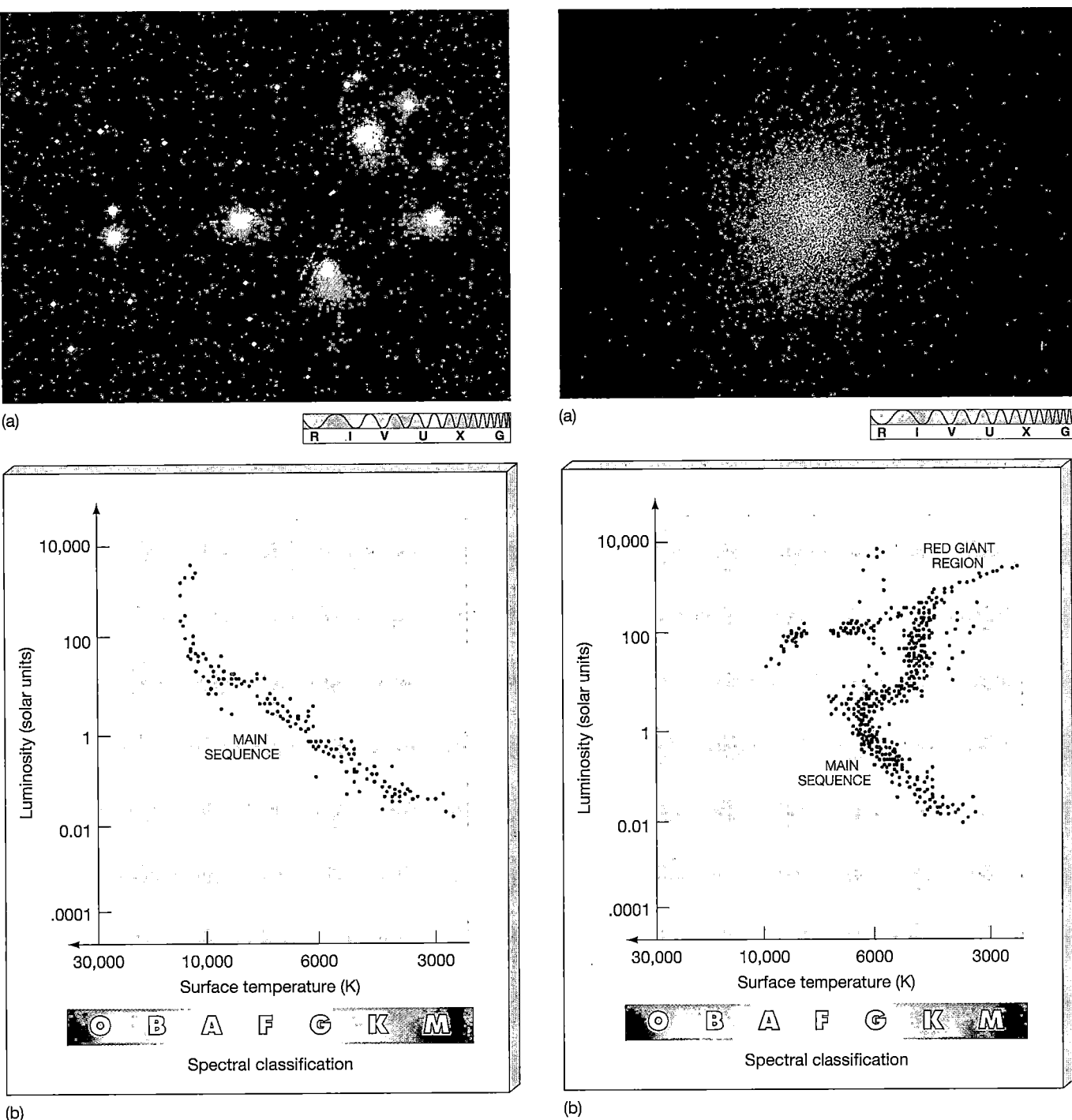

Figure 19.18 Open Cluster (a) The Pleiades cluster (also known as the Seven Sisters, or M45) lies about 120 pc from the Sun. The naked eye can see only six or seven of its brightest stars. (b) The stars of this well-known open cluster yield an H–R diagram. *(AURA)*

Figure 19.19 Globular Cluster (a) The globular cluster Omega Centauri is approximately 5000 pc from Earth and spans some 40 pc in diameter. (b) An H–R diagram for some of its stars. *(AURA)*

Other observations confirm the great ages of these systems. For example, their spectra show few heavy elements, implying that these stars formed in the distant past when heavy elements were much less abundant than they are today (Chapter 21). Astronomers speculate that the 150 or so globular clusters observed today are just the survivors of a much larger population of clusters that formed long ago.

CLUSTERS AND NEBULAE

How many stars form in a cluster, and of what type? How much gas is left over? What does the collapsed cloud look like once star formation has run its course? At present, although the main stages in the formation of individual stars (stages 3–7) are becoming clearer, the answers to these more general questions (involving stages 1 and 2) are still very sketchy. They await a more thorough understanding of the star-formation process.

In general, the more massive the collapsing region, the more stars are likely to form there. In addition, we know from observed H–R diagrams that low-mass stars are much more common than high-mass ones. ⚭ (Sec. 17.8) For every O- or B-type giant, hundreds or even thousands of G-, K-, and M-type dwarfs may form. The precise number of stars of any given mass or spectral type likely depends in a complex (and poorly understood) way upon

conditions within the parent cloud. The same is true of the *efficiency* of star formation—that is, the fraction of the total mass that actually finds its way into stars—which determines the amount of leftover material. However, if, as is usually the case, one or more O- or B-type stars form, their intense radiation and winds will cause the surrounding gas to disperse, leaving behind a young star cluster.

These considerations clearly illustrate the important role played by environment in the star-formation process. The first massive stars to form tend to prevent the formation of additional high-mass stars by disrupting the environment in which stars are growing. This is one reason why low-mass stars are so much more common than high-mass stars. It also helps explain the existence of brown dwarfs, by providing a natural way in which star formation can stop before nuclear fusion begins in the growing stellar core. *Discovery 19-2* describes another system in which the gas-dispersal process may be almost complete.

Until the 1990s, the existence of star clusters within emission nebulae was largely conjecture—the stars cannot be seen optically because they are obscured by dust. However, infrared observations have now clearly demonstrated that stars really are found within star-forming regions! Figure 19.21 compares optical and infrared views of the central regions of the Orion Nebula. The optical image in Figure 19.21(a) shows the Trapezium, the group of four

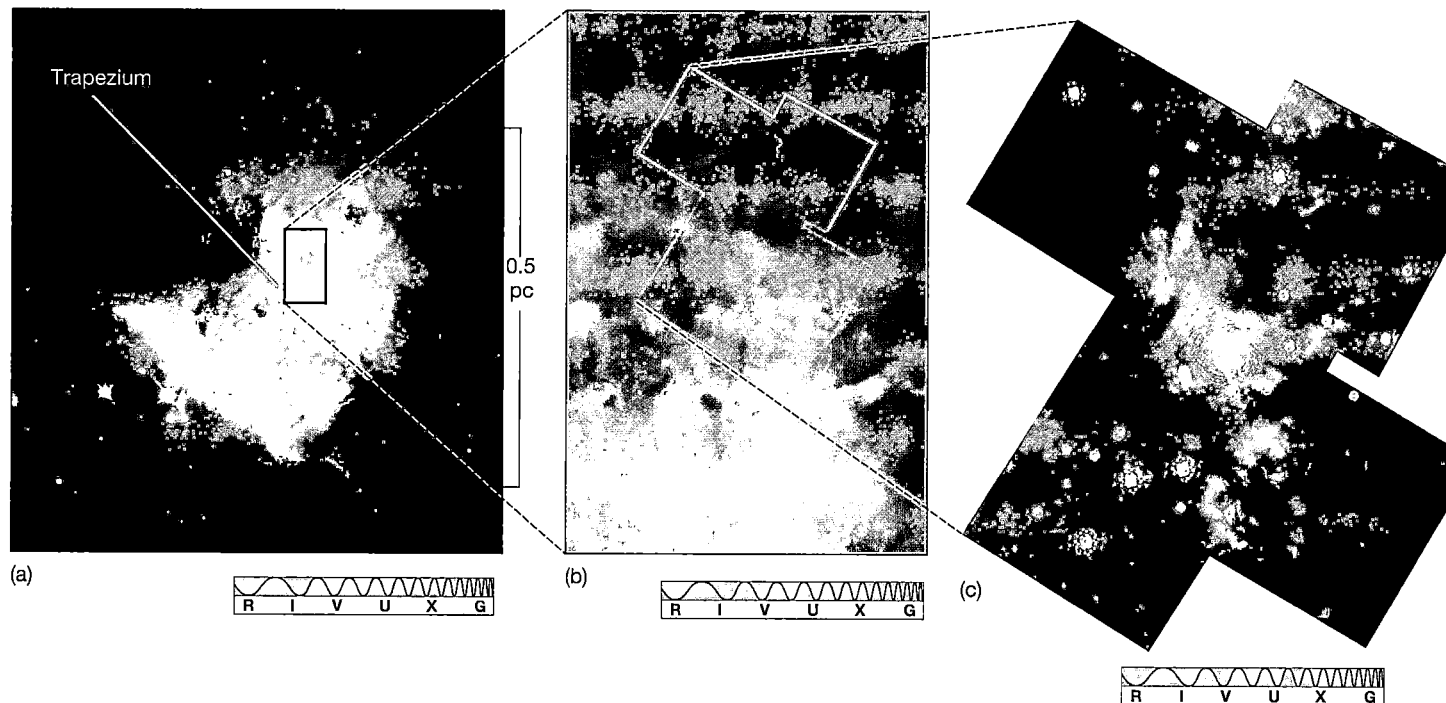

Figure 19.21 Star Formation in Orion Some views of the central regions of the Orion Nebula. (a) A short-exposure visible-light image (using a filter that is transparent only to certain emission lines of oxygen) shows the nebula itself and four bright O-type stars known as the Trapezium. (b) A magnified view of a smaller part of the nebula shows much irregular gas and dust, but few obvious stars which are hidden in the dust. (c) This short-exposure infrared image, acquired by the *Hubble Space Telescope* in 1997, shows several faint red stars emerging from the nebular gas; the brightest star is known as the Becklin–Neugebauer object. *(NASA)*

DISCOVERY 19-2

Eta Carinae

At the heart of the Carina emission nebula (shown in the main figure below) lies a remarkable object called Eta Carinae (right-hand images). With an estimated mass of around 100 times the mass of the Sun and a luminosity of five million times the solar value, Eta Carinae is the most massive star known. Formed probably only a few hundred thousand years ago, the short life of this star has been explosive indeed. In the mid-nineteenth century, Eta Carinae experienced an outburst that made it one of the brightest stars in the southern sky (even though it lies some 3000 pc away from Earth, a very long way compared to most of the bright stars visible in our night sky). The star released as much visible energy as a supernova explosion (see Chapter 21), yet it somehow survived the event. The object shown at bottom right below is the result of that outburst.

The bottom-right image, obtained by *HST* and carefully processed to reveal fine detail, is the highest-resolution view of the explosion obtained to date. Dust lanes, tiny condensations in the outflowing material, and dark radial streaks of unknown origin all appear with exquisite clarity. The star itself is the white dot at the center of the image. The two ends of the

"peanut" (at top right and bottom left) are blobs of ejected material racing away from the star at hundreds of kilometers per second—perhaps enough to expel the surrounding nebular gas and convert the Carina Nebula into the Carina Cluster. Perpendicular to the line joining these two blobs is a thin disk of gas, also moving outward at high speed.

The details of the events leading to this outburst are unclear. Quite possibly such episodes of violent activity are the norm for supermassive stars. However, while a few comparable outbursts have been observed in other galaxies, they are so rare that astronomers still do not know what constitutes "typical" behavior for such exotic objects.

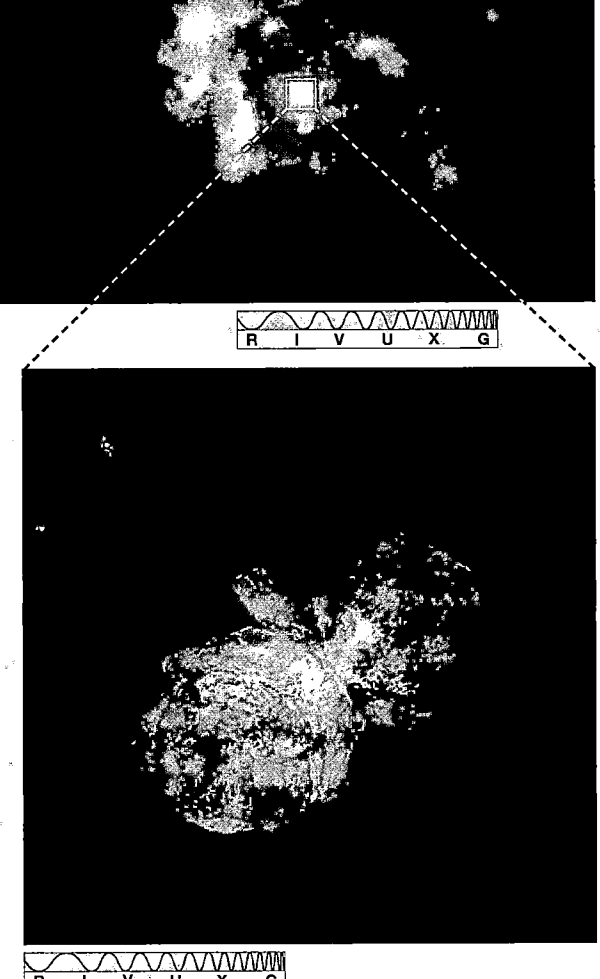

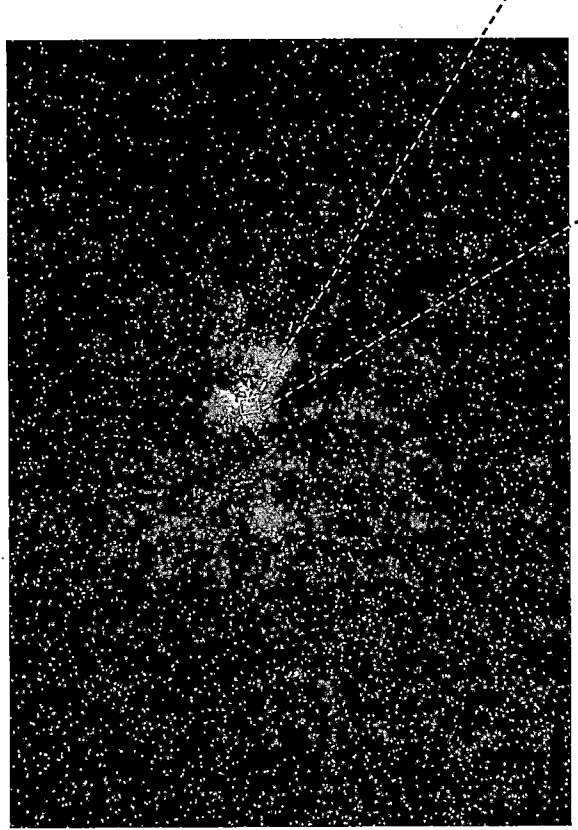

(AURA; NASA)

bright stars responsible for ionizing the nebula; the false-color infrared image in Figure 19.21(c) reveals an extensive cluster of stars within and behind the visible nebula. The Becklin–Neugebauer object (Section 19.4) can be seen as the central yellow spot within this region. It is thought to be a dust-shrouded B-type star just beginning to form its own emission nebula. These remarkable images show many stages of star formation.

CLUSTER LIFETIMES

Eventually, star clusters dissolve into individual stars. In some cases, the ejection of unused gas reduces a cluster's mass so much that it becomes gravitationally unbound and rapidly dissolves. In clusters that survive the early gas-loss phase, stellar encounters tend to eject the lightest stars from the cluster, just as the gravitational slingshot effect can propel spacecraft around the solar system. ∞ (*Discovery 6-2*) At the same time, the tidal gravitational field of the Milky Way Galaxy slowly strips outlying stars from the cluster. Occasional distant encounters with giant molecular clouds also tend to remove cluster stars. A near miss may even disrupt the cluster entirely.

As a result of all these influences, most open clusters break up in a few hundred million years, although the actual lifetime depends on the cluster's mass. Loosely bound associations may survive for only a few tens of millions of years, whereas some very massive open clusters are known from their H–R diagrams to be almost 5 billion years old. In a sense, only when a star's parent cluster has completely dissolved is the star-formation process really complete. The road from a gas cloud to a single, isolated star like the Sun is long and tortuous indeed!

Take another look at the sky one clear, dark evening. Ponder all this cosmic activity while peering upward at the stars. After studying this chapter, you may find you have to modify your view of the night sky. Even the seemingly quiet nighttime darkness is dominated by continual change.

☑ Concept Check

■ If stars in a cluster all form at the same time, how can some influence the formation of others?

Chapter Review

SUMMARY

Stars form when an interstellar cloud collapses under its own gravity and breaks up into pieces comparable in mass to our Sun. Heat, rotation, and magnetism all compete with gravity to influence the cloud's evolution. The evolution of the contracting cloud—the changes in its temperature and luminosity—can be conveniently represented as an **evolutionary track** (p. 494) on the Hertzsprung–Russell diagram. A cold interstellar cloud containing a few thousand solar masses of gas can fragment into tens or hundreds of smaller clumps of matter, from which stars eventually form.

As a collapsing prestellar fragment heats up and becomes denser it eventually becomes a **protostar** (p. 494)—a warm, very luminous object that emits radiation mainly in the infrared portion of the electromagnetic spectrum. At this stage of its evolution, the protostar is also known as a **T Tauri** (p. 496) star, after the first object of this type discovered. Eventually, a protostar's central temperature becomes high enough for hydrogen fusion to begin, and the protostar becomes a star. For a star like the Sun, the whole formation process takes about 50 million years. Protostellar winds encounter less resistance in the directions perpendicular to a protostar's disk. Thus they expel two jets of matter in the directions of the protostar's poles. As the protostellar wind gradually destroys the disk, the jets widen until, with the disk gone, the wind flows away from the star equally in all directions.

More massive stars pass through similar stages, but much more rapidly. Stars less massive than the Sun take much longer to form. The **zero-age main sequence** (p. 497) is the region on the H–R diagram where stars lie when the formation process is over. Mass is the key property for determining a star's characteristics and life span. The most massive stars have the shortest formation times and main-sequence lifetimes. At the other extreme, some low-mass fragments never reach the point of nuclear ignition. The universe may be populated with a vast number of **brown dwarfs** (p. 498)—objects that are not massive enough to fuse hydrogen to helium in their interiors.

Many of the objects predicted by the theory of star formation have been observed in real astronomical objects. The dark interstellar regions near emission nebulae often provide evidence of cloud fragmentation and protostars. Radio telescopes are used for studying the early phases of cloud contraction and fragmentation; infrared observations allow us to see later stages of the process. Many well-known emission nebulae, lit by several O-type stars, are partially engulfed by molecular clouds, parts of which are probably fragmenting and contracting, with smaller sites forming protostars. **Shock waves** (p. 504) can compress other interstellar clouds and trigger star formation. Star birth and the production of shock waves are thought to produce a chain reaction of star formation in molecular cloud complexes.

A single collapsing and fragmenting cloud can give rise to hundreds or thousands of stars—a **star cluster** (p. 505). The formation of the most massive stars may play an important role in suppressing further formation of lower-mass cluster members. **Open clusters** (p. 505), with a few hundred to a few thousand stars, are found mostly in the plane of the Milky Way. They typically contain many bright blue stars, indicating that they formed relatively recently. **Globular clusters** (p. 506) are found mainly away from the Milky Way plane and may contain millions of stars. They include no main-sequence stars much more massive than the Sun, indicating that they formed long ago. Globular clusters are believed to date from the formation of our Galaxy. Loosely bound groups of newborn stars are called stellar **associations** (p. 506). Infrared observations have revealed young star clusters or associations in several emission nebulae. Eventually, clusters break up into individual stars, although the process may take billions of years to complete.

SELF-TEST: TRUE OR FALSE?

____ **1.** Given the typical temperatures found in interstellar space, a cloud containing as few as 1,000,000 atoms has sufficient gravity for it to begin to collapse.

____ **2.** Both rotation and magnetic fields act to accelerate the gravitational collapse of an interstellar cloud.

____ **3.** The time a solar-type star spends in formation is relatively short compared to the time it spends as a main-sequence star.

____ **4.** Most stars form as members of groups or clusters of stars.

____ **5.** A stage 4 protostar may have a luminosity 1000 times that of the Sun.

____ **6.** As it evolves along the Hayashi track from stage 4 to stage 6, a protostar's luminosity stays roughly constant.

____ **7.** The rate of evolution of a stage 5 object is rapid compared with the rates at previous stages.

____ **8.** Brown dwarfs take a long time to form, but will eventually arrive as stars on the lower main sequence.

____ **9.** Stages 1 and 2 of star formation can be observed using optical telescopes.

____ **10.** Shock waves produced by emission nebulae can initiate star formation in nearby molecular clouds.

____ **11.** The formation of the first high-mass stars in a collapsing cloud tends to inhibit further collapse within that cloud.

____ **12.** G-, K-, and M-type stars form more frequently than O- and B-type stars.

____ **13.** The gas in an emission nebula eventually dissipates into space, leaving behind a star cluster.

____ **14.** Star clusters eventually dissipate, leaving behind individual stars like the Sun.

____ **15.** Typical open clusters contain millions of stars.

SELF-TEST: FILL IN THE BLANK

1. Atoms in an interstellar cloud have random motions, with an average velocity determined by the cloud's ____.

2. A(n) ____ plots a star or protostar's changing location on the H–R diagram as the object evolves.

3. In stage 1 of prestellar evolution, a typical interstellar cloud has the following properties: temperature ____ K, size ____ pc, mass ____ solar masses.

4. In stage 2 of prestellar evolution, a contracting interstellar cloud ____ into smaller pieces.

5. During stage 3 of prestellar evolution, as each piece of the original interstellar cloud continues to contract, its central density and temperature ____.

6. At stage 4 of prestellar evolution, a piece of the interstellar cloud becomes a ____.

7. A stage 4 object is plotted in the ____ (upper/lower) ____ (right/left) part of the H–R diagram.

8. At stage 6 the central temperature of the object reaches ____ K.

9. At this temperature, a stage 6 object begins to ____.

10. When hydrogen is fusing stably in the core, the star has reached the ____.

11. The T Tauri phase of a star occurs during stage ____.

12. It takes a star like the Sun a total of about ____ million years to form.

13. More massive stars evolve more ____.

14. Astronomers look for emissions at ____ wavelengths to identify interstellar clouds in stages 1 and 2.

15. At stages 4, 5, and 6, objects emit a great deal of radiation in the ____ part of the electromagnetic spectrum.

REVIEW AND DISCUSSION

1. Briefly describe the basic chain of events leading to the formation of a star like the Sun.

2. What is the role of heat in the process of stellar birth?

3. What is the role of rotation in the process of stellar birth?

4. What is the role of magnetism in the process of stellar birth?

5. What is an evolutionary track?

6. Why do stars tend to form in groups?

7. Why does the evolution of a protostar slow down as it approaches the main sequence?

8. In what ways do the formative stages of high-mass stars differ from those of stars like the Sun?

9. What are brown dwarfs?

10. What are T Tauri stars?

11. Stars live much longer than we do, so how do astronomers test the accuracy of theories of star formation?

12. At what evolutionary stages must astronomers use radio and infrared radiation to study prestellar objects? Why can't they use visible light?

13. Why has it been difficult until recently to demonstrate that stars and protostars actually exist within star-forming regions?

14. What is a shock wave? Of what significance are shock waves in star formation?

15. Explain the usefulness of the H–R diagram in studying the evolution of stars. Why can't evolutionary stages 1–3 be plotted on the diagram?

16. Compare the times necessary for the various stages in the formation of a star like the Sun. Why are some so short and others so long?

17. What do star clusters and associations have to do with star formation?

18. Compare and contrast the observed properties of open star clusters and globular star clusters.

19. How can we tell if a star cluster is young or old?

20. In the formation of a star cluster with a wide range of stellar masses, is it possible for some stars to die out before others have finished forming? Do you think this will have any effect on the cluster's formation?

PROBLEMS  *Algorithmic versions of these questions are available in the Practice Problems module of the Companion Website.*

The number of squares preceding each problem indicates its approximate level of difficulty.

1. ■■ In order for an interstellar gas cloud to contract, the average speed of its constituent particles must be less than half the cloud's escape speed. Will a (spherical) molecular hydrogen cloud of mass 1000 solar masses, radius 10 pc, and temperature of 10 K begin to collapse? Why or why not? (See *More Precisely 8–1*.)

2. ■■ Under the same assumptions as in Problem 1, estimate the minimum mass needed to cause a 1000 K, 1 pc cloud to collapse.

3. ■■ Use the radius–luminosity–temperature relation to explain how a protostar's luminosity changes as it moves from stage 4 (temperature 3000 K, radius 2×10^8 km) to stage 6 (temperature 4500 K, radius 10^6 km). What is the change in absolute magnitude? ⟳ (Sec. 17.5)

4. ■ A protostar on the Hayashi track evolves from a temperature of 3500 K and a luminosity 5000 times that of the Sun to a temperature of 5000 K and a luminosity of 3 solar units. What is its radius (a) at the start, and (b) at the end of the evolution?

5. ■ What is the (approximate) absolute magnitude of a stage-5 protostar? (See Figure 19.7.)

6. ■■ Use the H–R diagrams in this chapter to estimate by what factor a 1000-solar luminosity, 3000 K protostar is larger than a main-sequence star of the same luminosity.

7. ■ By how many magnitudes does a 3-solar-mass star decrease in brightness as it evolves from stage 4 to stage 6? (See Figure 19.8.)

8. ■■ As a simple model of the final stage of star formation, imagine that between stages 6 and 7 a star's surface temperature increases with time at a constant rate while the luminosity remains constant at the stage 7 level. The stage 7 radius is equal to the solar value. Using the temperatures given in Table 19.1, calculate the star's radius at a time exactly halfway between these two stages.

9. ■ What is the luminosity, in solar units, of a brown dwarf whose radius is 0.1 solar radii and whose surface temperature is 600 K (0.1 times that of the Sun)?

10. ■■ What is the maximum distance at which the brown dwarf in the previous problem could be observed by a telescope of limiting apparent magnitude (a) 18, (b) 30?

11. ■ A shock wave from a supernova explosion moves at a speed of about 5000 km/s. How long will such a disturbance take to cross a molecular cloud 20 pc in diameter?

12. ■■ The luminosity of a hypothetical star-forming region is dominated by five bright O-type stars, each of absolute magnitude −8. What is the net absolute magnitude of the region?

13. ■■ If the star-forming region in Problem 12 is in a galaxy 10 Mpc (1 megaparsec = 1,000,000 pc) from Earth, calculate its apparent magnitude.

14. ■■■ An open cluster has a diameter of 5 pc and a mass 1000 times the mass of the Sun. (a) Estimate the typical speed of its component stars. (b) Based on this speed, estimate the number of times a typical star orbits the center of the cluster in the 500 million years it takes for the cluster to dissolve in the galactic tidal field.

15. ■■■ Approximating the gravitational field of our Galaxy as a mass of 10^{11} solar masses at a distance of 8000 pc (see Chapter 23), estimate the "tidal radius" of a 20,000-solar-mass open-star cluster—that is, the distance from the cluster's center, outside of which the galactic tidal force overwhelms the cluster's own gravity.

COLLABORATIVE EXERCISES

1. Prestellar Evolution. As a group, create a large pie chart showing the amount of time spent in each of the stages 1 through 6 listed in Table 19.1 for a solar-type star over a total period of 30 million years. Each group member should label one or two of the stages and clearly describe the characteristics of each.

RESEARCHING ON THE WEB  To complete the following exercises, go to the online Destinations module for Chapter 19 on the Companion Website for Astronomy Today 4/e.

1. Access the "Hertzsprung–Russell Diagram" page and describe the current evolutionary stage of 80 percent of the stars in our galaxy.

2. Access the "All Star Line Up" page and describe how Alpha Centauri relates to our Sun in terms of size, brightness, and spectral class.

3. Access the "What is a Star?" page and describe how stars 10 times more massive and 1/10th the Sun's mass will be different than the Sun in terms of brightness and lifetime.

PROJECT

1. The Trifid Nebula, otherwise known as M20, is a place where new stars are forming. It has been called a "dark night revelation, even in modest apertures." An 8- to 10-inch telescope is needed to see the triple-lobed structure of the nebula. Ordinary binoculars reveal the Trifid as a hazy patch located in the constellation Sagittarius. This nebula is set against the richest part of the Milky Way, the edgewise projection of our own Galaxy around the sky. It is one of many wonders in this region of the heavens. What are the dark lanes in M20? Why are other parts of the nebula bright? There have been reports of large-scale changes occurring in this nebula in the last century and a half. The reports are based on old drawings, which show M20 looking slightly different from how it appears today. Do you think it possible for a cloud in space to undergo a change in appearance on a time scale of years, decades, or centuries?

2. Summer is a good time to search with binoculars for open-star clusters. Open clusters are generally found in the plane of the Galaxy. If you can see the hazy band of the Milky Way arcing across your night sky—in other words, if you are far from city lights and looking at an appropriate time of night and year—you can simply sweep with your binoculars along the Milky Way. Numerous "clumps" of stars will pop into view. Many will turn out to be open-star clusters.

3. Globular star clusters are harder to find. They are intrinsically larger, but they are also much farther away and therefore appear smaller in the sky. The most famous globular cluster visible from the Northern Hemisphere is M13 in the constellation Hercules, visible on spring and summer evenings. This cluster contains half a million or so of the Galaxy's most ancient stars. It may be glimpsed in binoculars as a little ball of light, located about one-third of the way from the star Eta to the star Zeta in the Keystone asterism of the constellation Hercules. Telescopes reveal this cluster as a magnificent, symmetrical grouping of stars.

SKYCHART III PROJECTS  The SkyChart III Student Version planetarium program on which these exercises are based is included as a separately executable program on the CD in the back of this text.

1. ■ Appendix 3 contains a list of the 20 brightest stars. Make your own table of these stars with columns for information on the range of dates each is visible in your area at a time convenient for observation, and another column with information on how to locate each one. Make note of those stars that are not available to you because they are in the Southern Hemisphere. Use SkyChart III and the chart you have made to locate the bright stars available at this time. Along with being able to recognize the constellations, it is rewarding to know the names of the bright stars. A small effort is usually sufficient to develop a satisfying familiarity.

 In addition to the Practice Problems and Destinations modules, the Companion Website at http://www.prenhall.com/chaisson provides for each chapter an additional true-false, multiple choice, and labeling quiz, as well as additional annotated images, animations, and links to related Websites.

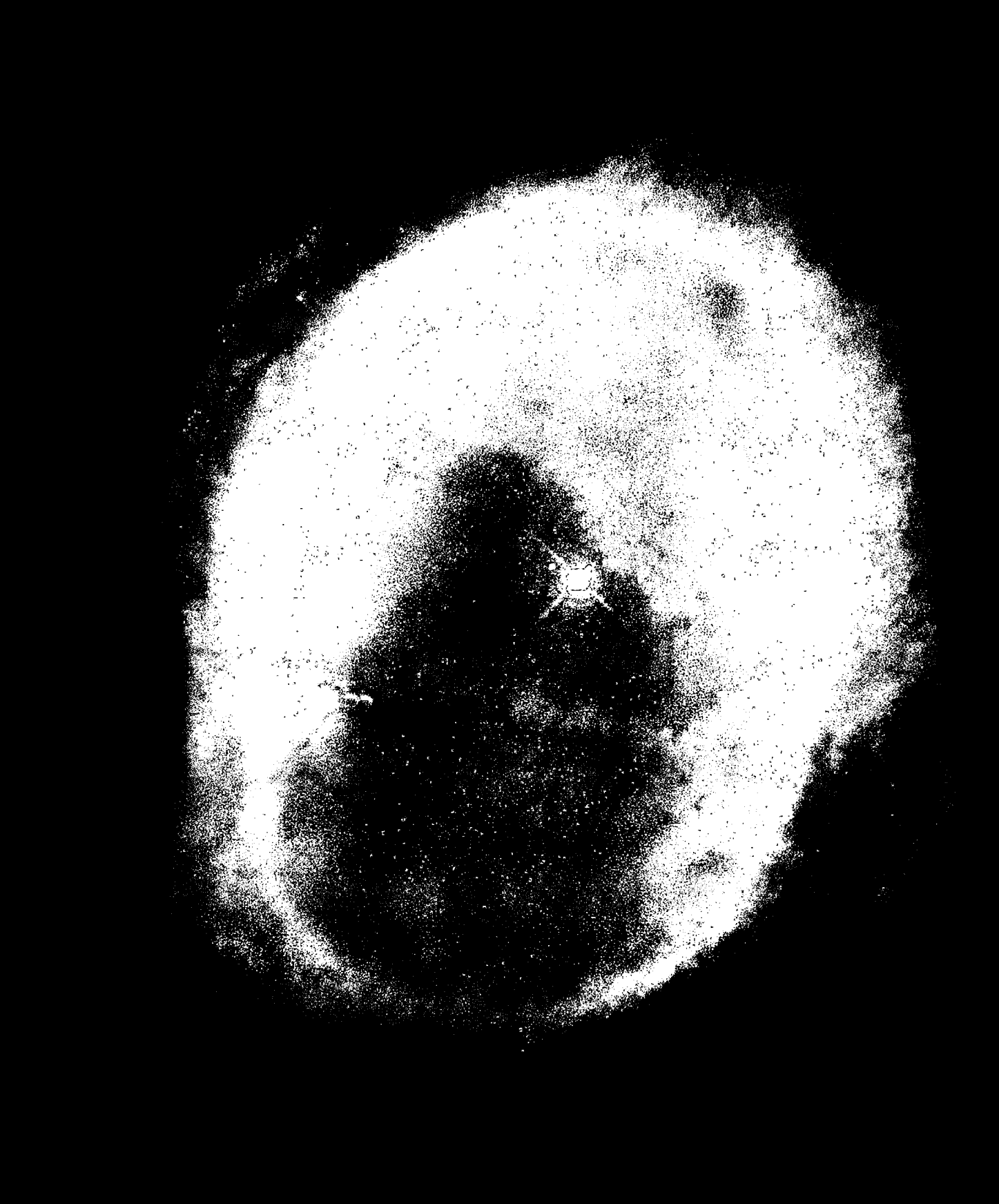

20 STELLAR EVOLUTION

The Life of a Star

LEARNING GOALS

Studying this chapter will enable you to:

1.   Explain why stars evolve off the main sequence.

2.   Outline the events that occur after a Sunlike star exhausts the supply of hydrogen in its core.

3.   Summarize the stages in the death of a typical low-mass star and describe the resulting remnant.

4.   Contrast the evolutionary histories of high-mass and low-mass stars.

5.   Discuss the observations that help verify the theory of stellar evolution.

6.   Explain how the evolution of stars in binary systems may differ from that of isolated stars.

 Visit http://www.prenhall.com/chaisson for additional annotated images, animations, and links to related sites for this chapter.

This striking planetary nebula, named NGC 3132, is an expanding cloud of gas surrounding a dying star some 2000 light-years away. Its diameter is about half a light-year. The glowing material is gently receding from the old (but still white-hot) carbon-rich star seen just to the left of the bright but unrelated star at the center of the ring. (Blue represents the hottest gas, red the coolest.) Our Sun is destined to eject a similar planetary nebula some 5 billion years from now. *(STScI)*

The Big Picture: Stellar evolution is one subject about which astronomers know quite a lot. Surprisingly, given that no one has ever seen even a single star move through all its evolutionary paces, the theory of stellar evolution is rather well understood. Like archaeologists who study bones and artifacts of different ages, astronomers also observe stars of differing ages and then try to assemble the pieces into a consistent picture of how stars evolve over billions of years.

$\mathbf{A}$*fter reaching the main sequence, a newborn star changes little in outward appearance for more than 90 percent of its lifetime. However, at the end of this period, as the star begins to run out of fuel and die, its properties once again change greatly. Aging stars travel along evolutionary tracks that take them far from the main sequence as they end their lives. In this and the next two chapters, we will study the evolution of stars during and after their main-sequence burning stages. We will find that the ultimate fate of a star depends primarily on its mass, although interactions with other stars can also play a decisive role, and that the final states of stars can be strange indeed. By continually comparing theoretical calculations with detailed observations of stars of all types, astronomers have refined the theory of stellar evolution into a precise and powerful tool for understanding the universe.*

20.1 Leaving the Main Sequence

1 Most stars spend most of their lives on the main sequence. A star like the Sun, for example, after spending a few tens of millions of years in formation (stages 1–6 in Chapter 19), resides on or near the main sequence (stage 7) for 10 billion years before evolving into something else. ⬚ (Sec. 19.2) That "something else" is the main topic of this chapter.

Virtually all the low-mass stars that have ever formed still exist as stars. The coolest M-type stars—red dwarfs—consume their nuclear fuel so slowly that not one of them has yet left the main sequence. Some of them will shine steadily for a trillion years or more. ⬚ (Sec. 17.8) Conversely, the most massive O- and B-type stars evolve away from the main sequence after only a few million years. Most of the high-mass stars that have ever existed perished long ago. Between these two extremes, many stars are observed in advanced stages of evolution, their properties quite different from when they formed. By combining these observations with theoretical models, astronomers have built up a comprehensive picture of how stars evolve.

On the main sequence, a star slowly fuses hydrogen into helium in its core. This process is called **core-hydrogen burning**. In Chapter 16 we saw how the proton–proton fusion chain powers the Sun. ⬚ (Sec. 16.5) *More Precisely 20-1* describes another sequence of nuclear reactions, of great importance in stars more massive than the Sun, that accomplishes the same basic result as the proton–proton chain, but in a very different way.

As illustrated schematically in Figure 20.1, a main-sequence star is in a state of *hydrostatic equilibrium*, in which pressure's outward push exactly counteracts gravity's inward pull. This is a stable balance between gravity and pressure, in which a small change in one always results in a small compensating change in the other. For example, a small increase in the star's central temperature leads to an increase in pressure, causing the star to expand; a small temperature decrease leads to a slight contraction. You should keep Figure 20.1 in mind as you study the various stages of stellar evolution described next. Much of a star's complex behavior can be understood in these simple terms.

As the main-sequence star ages, its core temperature slowly increases and both its luminosity and radius in-

crease. These changes are very slow, though—only a factor of three or four in luminosity over the Sun's entire 10-billion-year main-sequence lifetime, for example. As a result, the star's location on the Hertzsprung–Russell diagram remains almost unchanged during this phase (which is why we see the main sequence when we plot an H–R diagram for any reasonably large group of stars). ⬚ (Sec. 17.5) Eventually, as the hydrogen in the core is consumed, the star's internal balance starts to shift and both its internal structure and its outward appearance begin to change more rapidly: the star leaves the main sequence.

Once a star begins to evolve away from the main sequence, its days are numbered. The post-main-sequence stages of stellar evolution—the end of a star's life—depend critically on the star's mass. As a rule of thumb, we can say that low-mass stars die gently, whereas high-mass stars die catastrophically. The dividing line between these two very different outcomes lies around eight times the mass of the Sun, and in this chapter we will refer to stars of more than 8 solar masses as "high-mass" stars. Within both the

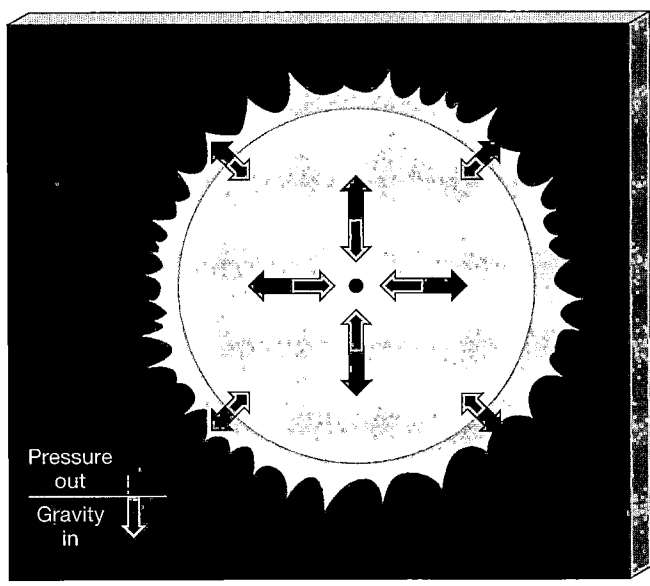

Pressure
out

Gravity
in

Figure 20.1 Hydrostatic Equilibrium In a steadily burning star on the main sequence, the outward pressure of hot gas exactly balances the inward pull of gravity. This is true at every point within the star, guaranteeing its stability.

"high-mass" and the "low-mass" (that is, less than 8 solar masses) categories, there are substantial variations. We will point out some of these variations as we go.

Rather than dwelling on the many details, we will concentrate on a few representative evolutionary sequences. We begin by considering the evolution of a fairly low-mass star like the Sun. The stages described in the next few sections pertain to the Sun as it nears the end of its fusion cycle 5 billion years from now. In fact, most of the qualitative features of the discussion apply to any low-mass star, although the exact numbers vary considerably. Later, we will broaden our discussion to include all stars, large and small.

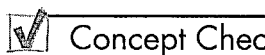 Concept Check

■ In what sense is the Sun stable?

20.2 The Evolution of a Sun-like Star

The surface of a main-sequence star like the Sun occasionally erupts in flares and spots, but for the most part the star does not experience sudden, large-scale changes in its properties. Its average surface temperature remains fairly constant while its luminosity increases very slowly with time. The Sun has roughly the same surface temperature as it had when it formed nearly 5 billion years ago, and is some 30 percent brighter than it was at that time.

THE HELIUM CORE

This stable state of affairs cannot continue indefinitely. Eventually, drastic changes occur in the star's interior structure. After approximately 10 billion years of steady core hydrogen burning, a Sunlike star begins to run out of fuel. The situation is a little like that of an automobile cruising effortlessly along a highway at a constant speed for many hours, only to have the engine suddenly cough and sputter as the gas gauge reaches empty. Unlike automobiles, though, stars are not easy to refuel.

As nuclear fusion proceeds, the composition of the star's interior changes. Figure 20.2 illustrates the increase in helium abundance and the corresponding decrease in hydrogen abundance that take place in the stellar core as the star ages. Three cases are shown: (a) the chemical composition of the original core, (b) the composition after 5 billion years, and (c) the composition after 10 billion years. Case (b) represents approximately the present state of our Sun.

The star's helium content increases fastest at the center, where temperatures are highest and the burning is fastest. The helium content also increases near the edge of the core, but more slowly because the burning rate is less rapid there. The inner, helium-rich region becomes larger

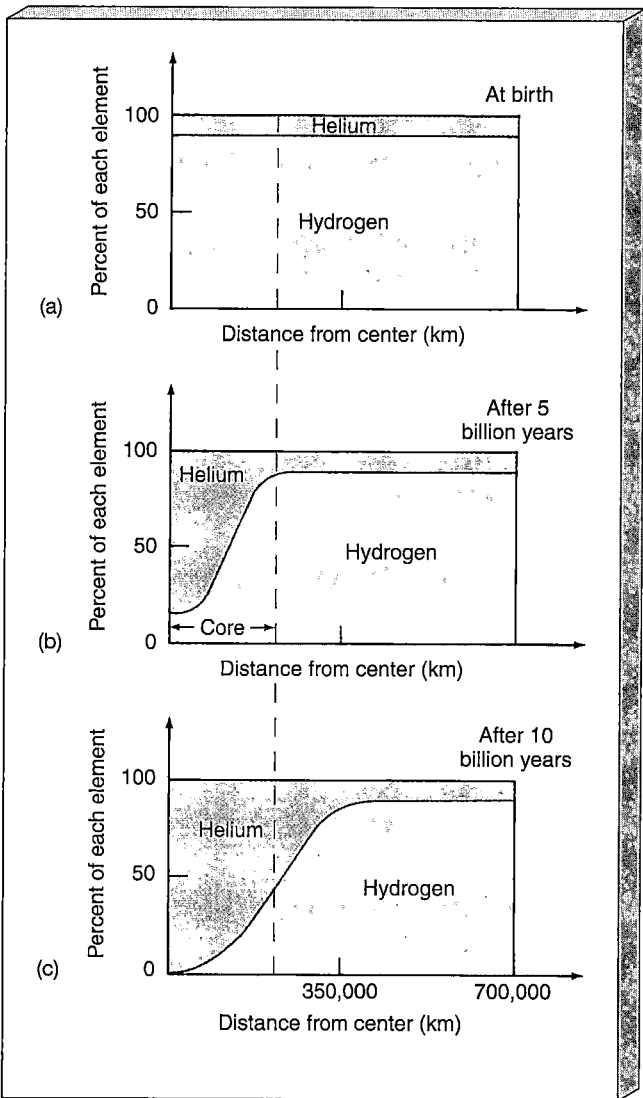

Figure 20.2 Solar Composition Change Theoretical estimates of the changes in a Sunlike star's composition. Hydrogen (yellow) and helium (orange) abundances are shown (a) at birth, on the zero-age main sequence; (b) after 5 billion years; and (c) after 10 billion years. At stage (b) only about five percent of the star's total mass has been converted from hydrogen into helium. This change speeds up as the nuclear burning rate increases with time.

and more hydrogen deficient as the star continues to shine. Eventually, about 10 billion years after the star arrived on the main sequence (Figure 20.2c), hydrogen becomes completely depleted at the center, the nuclear fires there cease, and the location of principal burning moves to higher layers in the core. An inner core of nonburning pure helium starts to grow.

Without nuclear burning to maintain it, the outward-pushing gas pressure weakens in the helium inner core. However, the inward pull of gravity does not. Once the outward push against gravity is relaxed—even a little—

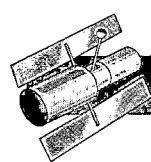

MORE PRECISELY 20-1

The CNO Cycle

The proton–proton chain is not the only nuclear process operating in the Sun and other late-generation stars. Another fusion mechanism capable of converting hydrogen into helium, starting from carbon-12 (^{12}C), proceeds according to the following six steps. Nitrogen (N) and oxygen (O) nuclei are created as intermediate products:

$$^{12}C + {}^{1}H \rightarrow {}^{13}N + \text{energy}$$

$$^{13}N \rightarrow {}^{13}C + \text{positron} + \text{neutrino}$$

$$^{13}C + {}^{1}H \rightarrow {}^{14}N + \text{energy}$$

$$^{14}N + {}^{1}H \rightarrow {}^{15}O + \text{energy}$$

$$^{15}O \rightarrow {}^{15}N + \text{positron} + \text{neutrino}$$

$$^{15}N + {}^{1}H \rightarrow {}^{12}C + {}^{4}He$$

These six steps are termed the *CNO cycle*. Aside from the radiation and neutrinos produced, notice that the sum total of these six reactions is

$$^{12}C + 4({}^{1}H) \rightarrow {}^{12}C + {}^{4}He.$$

In other words, the net result is the fusion of four protons into a single helium-4 nucleus, just as in the proton–proton chain. The carbon-12 acts merely as a *catalyst*, an agent of change that is not itself consumed in the reaction.

The electromagnetic forces of repulsion operating in the CNO cycle are greater than in the proton–proton chain because the charges of the heavy-element nuclei are larger. Accordingly, higher temperatures are required to propel the heavy nuclei into the realm of the strong nuclear force and to ignite fusion. The accompanying figure presents a numerical estimate of the energy released in a

star of solar composition by the proton–proton chain and the CNO cycle, each as a function of gas temperature. The proton–proton chain dominates at lower temperatures, up to about 16 million K. Above this temperature, the CNO cycle is the more important fusion process in stars of solar composition.

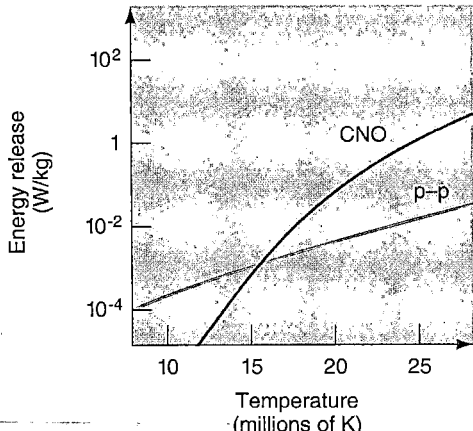

According to our theoretical models of the Sun, the temperature of the solar core is 15 million K, so these curves indicate that the proton–proton cycle is the dominant source of solar energy (notice that each step on the vertical scale corresponds to a *factor* of 100 in energy generation). The CNO cycle contributes no more than 10 percent of the observed solar radiation. However, stars more massive than our Sun often have core temperatures much higher than 20 million K, making the CNO cycle the dominant energy-production mechanism.

---

structural changes in the star become inevitable. As the hydrogen is consumed, the inner core begins to contract. Once all the hydrogen at the center is gone, the process accelerates.

If more heat could be generated, then the core might possibly regain its equilibrium. For example, if helium in the core were to begin fusing into some heavier element, then energy would be created as a by-product of helium burning and the necessary gas pressure would be reestablished. But the helium at the center cannot burn—not yet, anyway. Despite its high temperature, the core is far too cold to fuse helium into anything heavier.

Recall from Chapter 16 that a minimum temperature of about 10^7 K is needed to fuse hydrogen into helium. Only above that temperature do colliding hydrogen nu-

clei (that is, protons) have enough speed to overwhelm the repulsive electromagnetic force between them. ⇌ (Sec. 16.5) Because helium nuclei, with two protons each, carry a greater positive charge, their electromagnetic repulsion is larger, and even higher temperatures are needed to cause them to fuse—at least 10^8 K. A core composed of helium at 10^7 K thus cannot generate energy through fusion.

The shrinkage of the helium core releases gravitational energy, driving up the central temperature and heating the overlying burning layers. The higher temperatures—now well over 10^7 K (but still less than 10^8 K)—cause hydrogen nuclei to fuse even more rapidly than before. Figure 20.3 depicts this situation, in which hydrogen is burning at a furious rate in a shell surrounding the non-

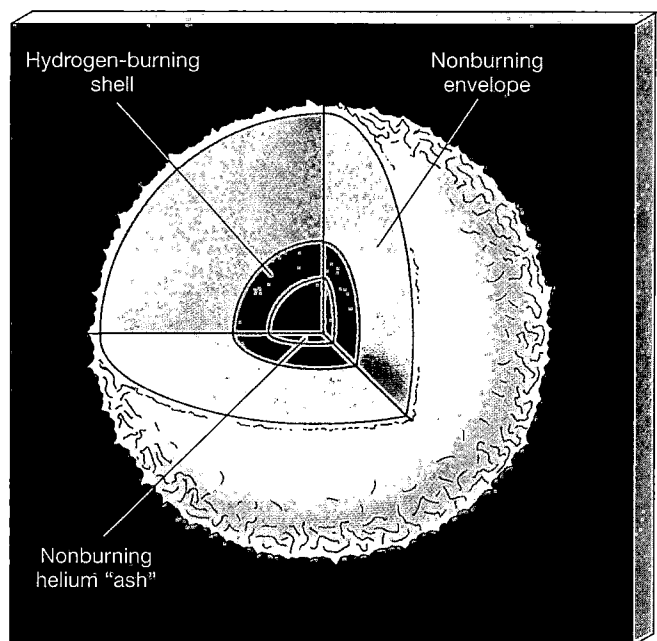

Figure 20.3 Hydrogen-Shell Burning As a star's core loses more and more of its hydrogen, the hydrogen in the shell surrounding the nonburning helium ash burns ever more violently.

burning inner core of helium "ash" in the center. This phase is known as the **hydrogen-shell burning** stage. The hydrogen shell generates energy faster than did the original main-sequence star's hydrogen-burning core, and the shell's energy production continues to increase as the helium core continues to shrink. Strange as it may seem, the star's response to the disappearance of the fire at its center is to get brighter!

THE RED-GIANT BRANCH

Conditions in the aging star have clearly changed from the stable equilibrium that characterized it as a main-sequence object. The helium core is unbalanced and shrinking. The rest of the core is also unbalanced, fusing hydrogen into helium at an ever-increasing rate. The gas pressure produced by this enhanced hydrogen burning causes the star's nonburning outer layers to increase in radius. Not even gravity can stop them. While the core is shrinking and heating up, the overlying layers are expanding and cooling. The star is on its way to becoming a red giant. The change from normal main-sequence star to elderly red giant takes about 100 million years.

Table 20.1 summarizes the key stages through which a solar-mass star evolves. It is a continuation of Table 19.1, except that the density units have been changed from particles per cubic meter to the more convenient kilograms per cubic meter, and we now express sizes as radii rather

than diameters. The numbers in the "Stage" column refer to the evolutionary stages noted in the figures and discussed in the text.

We can trace these large-scale changes on the H–R diagram. Figure 20.4 shows the path away from the main sequence, labeled as stage 7. The star first evolves to the right on the diagram, its surface temperature dropping while its luminosity increases only slightly. The star's roughly horizontal path from its main-sequence location (stage 7) to stage 8 on the figure is called the **subgiant branch**. By stage 8, the star's radius has increased to about three times the radius of the Sun.

The surface temperature at stage 8 has fallen to the point at which much of the interior is opaque to the radiation from within. Beyond this point, convection carries the core's enormous energy output to the surface. One consequence of this fact is that the star's surface temperature remains nearly constant between stages 8 and 9. The almost vertical path followed by the star between stages 8 and 9 is known as the **red-giant branch** of the H–R

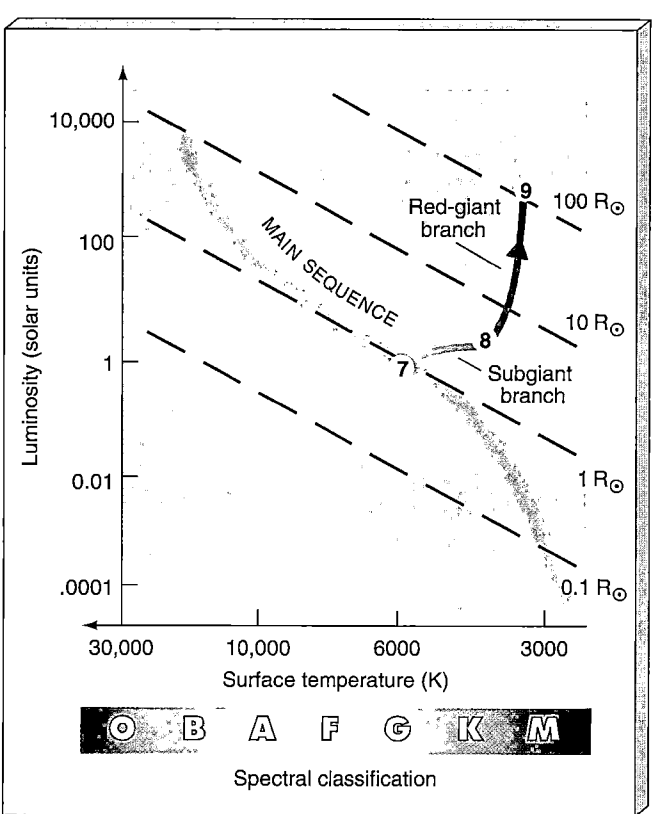

Figure 20.4 Red Giant on the H–R Diagram As the core of helium ash shrinks and the intermediate stellar layers expand, the star leaves the main sequence (stage 7). At stage 8 the star is on its way to becoming a red giant. The star continues to brighten and grow as it ascends the red-giant branch to stage 9, the top of the red-giant branch. As in Chapter 19, the diagonal lines correspond to stars of constant radius, allowing us to gauge the changes in the size of our star.

TABLE 20.1 Evolution of a Sun-like Star

STAGE	APPROXIMATE TIME TO NEXT STAGE (yr)	CENTRAL TEMPERATURE (10^6 K)	SURFACE TEMPERATURE (K)	CENTRAL DENSITY (kg/m^3)	RADIUS (km)	RADIUS (solar radii)	OBJECT
7	10^{10}	15	6000	10^5	7×10^5	1	Main-sequence star
8	10^8	50	4000	10^7	2×10^6	3	Subgiant branch
9	10^5	100	4000	10^8	7×10^7	100	Helium flash
10	5×10^7	200	5000	10^7	7×10^6	10	Horizontal branch
11	10^4	250	4000	10^8	4×10^8	500	Asymptotic giant branch
12	10^5	300	100,000	10^{10}	10^4	0.01	Carbon core
	—		3000	10^{-17}	7×10^8	1000	Planetary nebula*
13	—	100	50,000	10^{10}	10^4	0.01	White dwarf
14	—	Close to 0	Close to 0	10^{10}	10^4	0.01	Black dwarf

*Values refer to the envelope.

diagram. By stage 9, the giant's luminosity is many hundreds of times the solar value. Its radius by this time is around 100 solar radii.

Figure 20.5 compares the relative sizes of a G-type star—like our Sun—and a stage-9 red giant. It also indicates the stages through which the star will evolve. The red giant is huge—about the size of Mercury's orbit. In contrast, its helium core is surprisingly small—only about 1/1000 the size of the entire star, making the core just a few times larger than Earth. The central density is enormous. Continued shrinkage of the red giant's core has compacted its helium gas to approximately 10^8 kg/m^3. Contrast this value with the 10^{-3} kg/m^3 in the giant's outermost layers, with the 5000 kg/m^3 average density of Earth, and with the 150,000 kg/m^3 in the present core of the Sun. About 25 percent of the mass of the entire star is packed into its planet-sized core.

A familiar example of a low-mass star in the red-giant phase is the KIII giant Arcturus (see Figure 17.12), one of the brightest stars in the sky. Its mass is about 1.5 times that of the Sun. Currently in the hydrogen-shell burning stage and ascending the red-giant branch, its radius is some 23 times that of the Sun. It emits almost 200 times more energy than the Sun, much of it in the infrared part of the spectrum.

HELIUM FUSION

2 Should the unbalanced state of a red-giant star continue, the core would eventually collapse, and the rest of the star would slowly drift into space. The forces and pressures at work inside a red giant would literally pull it apart. In fact, for stars less than about one-quarter the mass of the Sun, this is precisely what will eventually happen (in a few hundred billion years—see Section 20.3).

However, for a star like the Sun, this simultaneous shrinking and expanding does not continue indefinitely. A few hundred million years after a solar-mass star leaves the main sequence, something else happens—helium begins to burn in the core. By the time the central density has risen to about 10^8 kg/m^3 (at stage 9), the temperature has reached the 10^8 K needed for helium to fuse into carbon, and the central fires reignite.

The reaction that transforms helium into carbon occurs in two steps. First, two helium nuclei come together to form a nucleus of beryllium-8 (^{8}Be). Beryllium-8 is a very unstable isotope that would normally break up into two helium nuclei in about 10^{-12} s. However, at the high densities found in the core of a red giant, it is possible that the beryllium-8 nucleus will encounter another helium nucleus before this occurs, fusing with it to form carbon-12 (^{12}C). This is the second step of the helium-burning reaction. In part, it is because of the electrostatic repulsion between beryllium-8 (containing four protons) and helium-4 (containing two) that the temperature must reach 10^8 K before this reaction can take place.

Symbolically, we can represent this next stage of stellar fusion as follows:

$$^4\text{He} + {}^4\text{He} \rightarrow {}^8\text{Be} + \text{energy,}$$

$$^8\text{Be} + {}^4\text{He} \rightarrow {}^{12}\text{C} + \text{energy.}$$

Helium-4 nuclei are traditionally known as *alpha particles*. The term dates from the early days of nuclear physics, when the true nature of these particles, emitted by many radioactive materials, was unknown. Because three alpha particles are required to get from helium-4 to carbon-12, the foregoing reaction is usually called the **triple-alpha process**.

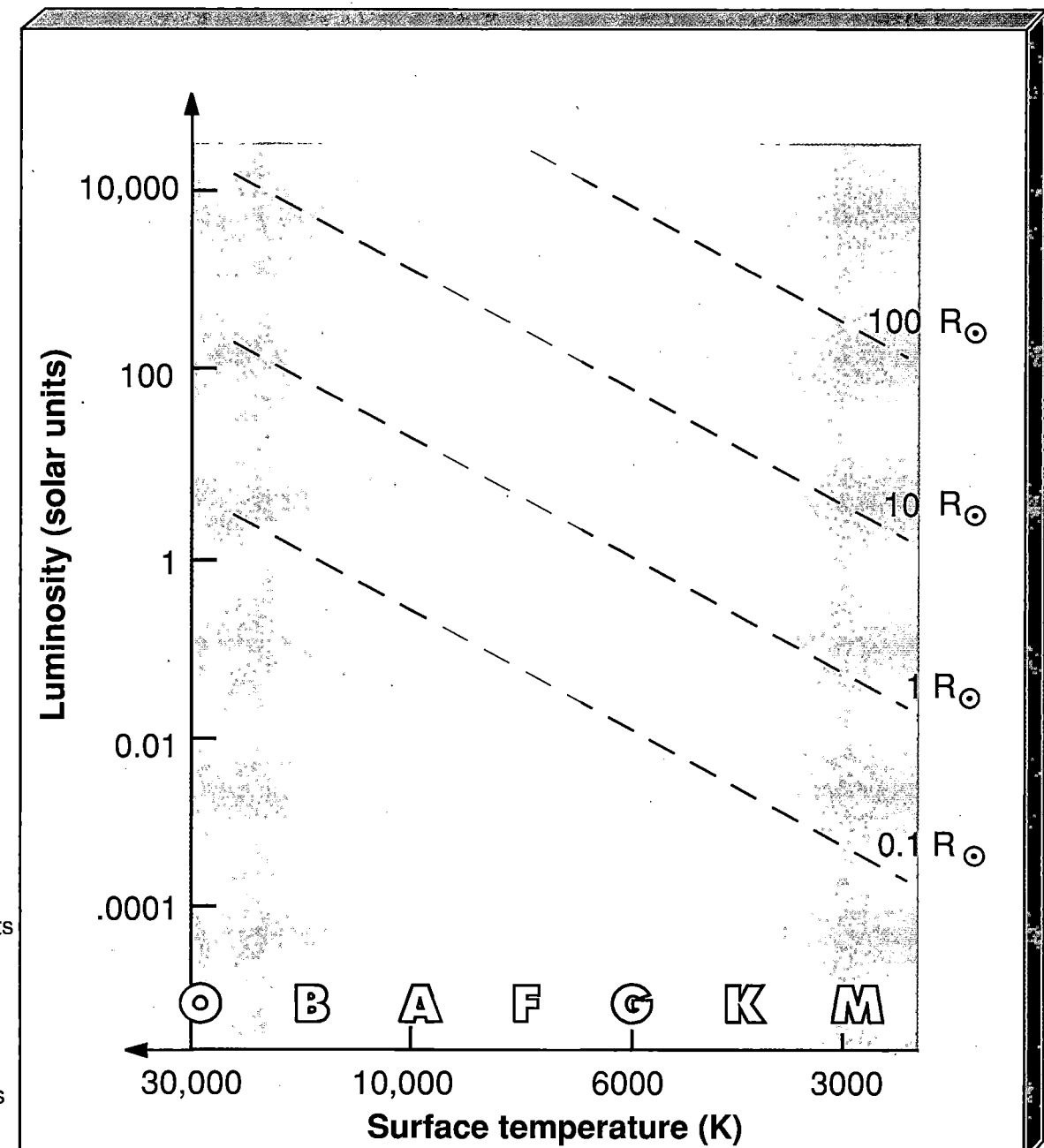

The H-R diagram plots stars by luminosity (vertical axis) and temperature, or spectral class (horizontal axis). The dashed diagonal lines are lines of constant radius.

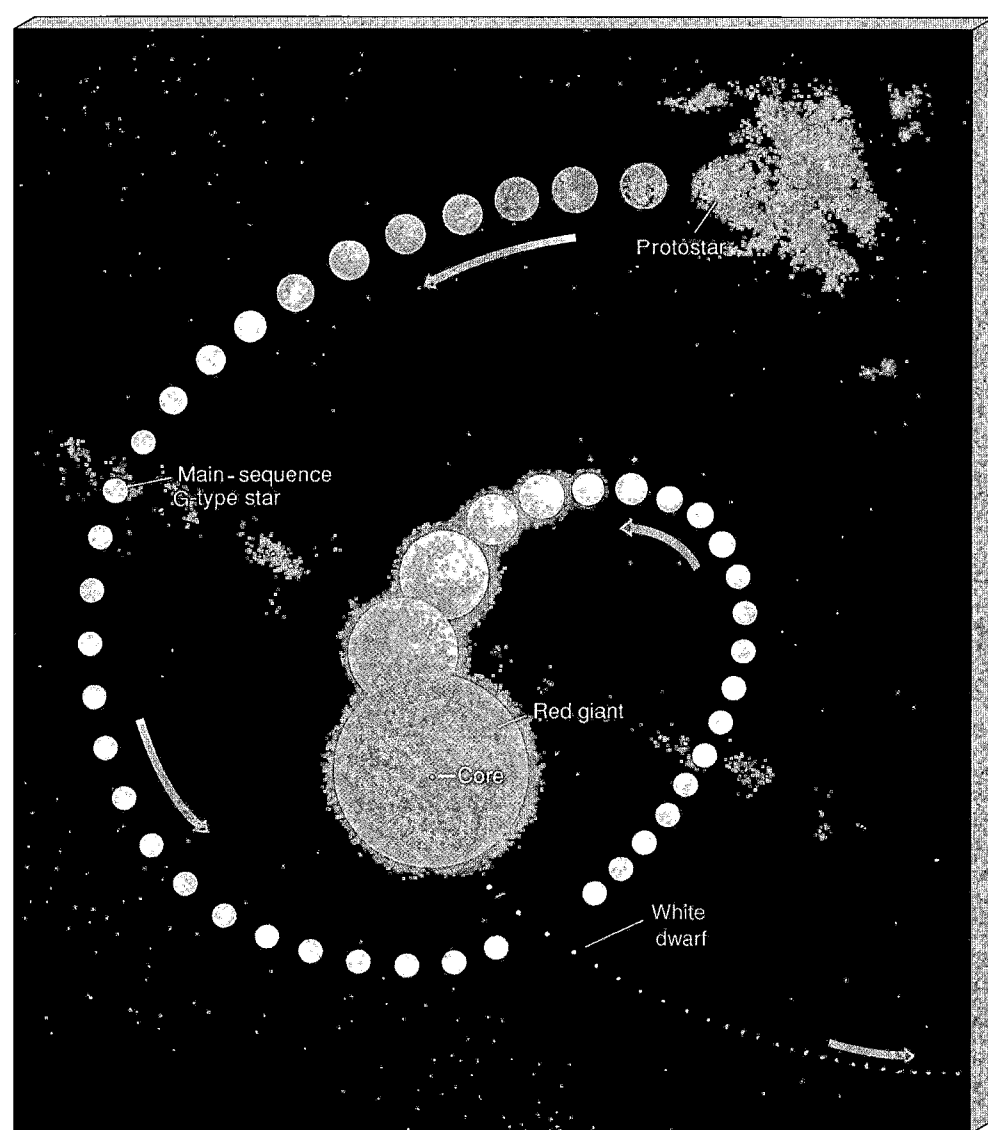

Figure 20.5 G-Type Star Evolution An artist's conception of the relative sizes and colors of a normal G-type star (such as our Sun) in its formative stages, on the main sequence, and while passing through the red-giant and white-dwarf stages. At maximum swelling, the red giant is approximately 70 times the size of its main-sequence parent; the core of the giant is about 1/15 the main-sequence size and would be barely discernible if this figure were drawn to scale. The length of time spent in the various stages— protostar, main-sequence star, red giant, and white dwarf—is roughly proportional to the length of this imaginary trek through space, although the main-sequence lifetime has been deliberately shortened. The star's brief stay on the horizontal branch is not shown here.

THE HELIUM FLASH

For stars comparable in mass to the Sun, there is a complication when helium fusion begins. At the high densities found in the core, the gas has entered a new state of matter whose properties are governed by the laws of quantum mechanics rather than by those of classical physics. Up to now we have been concerned primarily with the nuclei— protons, alpha particles, and so on—that make up virtually all the star's mass and participate in the reactions that generate its energy. However, the star contains another important constituent—a vast sea of electrons stripped from their parent nuclei by the ferocious heat in the stellar interior. At this stage in our story, these electrons play an important role in determining the star's evolution.

Under the conditions found in the stage 9 red-giant core, a rule of quantum mechanics known as the *Pauli exclusion principle* (after Wolfgang Pauli, one of the founding

fathers of quantum physics) prohibits the electrons in the core from being squeezed too close together. In effect, the exclusion principle tells us that we can think of the electrons as tiny rigid spheres that can be squeezed relatively easily up to the point of contact but become virtually incompressible thereafter. This condition is known (for historical reasons) as *electron degeneracy*, and the pressure associated with the contact of the tiny electron spheres is called **electron degeneracy pressure**. It has nothing to do with the thermal pressure (due to the star's heat) that we have been studying up to now. In fact, in our red-giant core, the pressure resisting the force of gravity is supplied almost entirely by degenerate electrons. Hardly any of the core's support results from "normal" thermal pressure, and this has dramatic consequences once the helium begins to burn.

Under normal ("nondegenerate") circumstances, the core could react to and accommodate the onset of helium burning, but in its degenerate state the burning becomes unstable, with explosive consequences. In a star supported by thermal pressure, the increase in temperature produced by the onset of helium fusion would lead to an increase in pressure. The gas would then expand and cool, reducing the burning rate and reestablishing equilibrium, just as discussed earlier. In the electron-supported core of a solar-mass red giant, however, the pressure is largely *independent* of the temperature. When burning starts and the temperature increases, there is no corresponding rise in pressure, no expansion of the gas, no drop in the temperature, and no stabilization of the core. Instead, the core is unable to respond to the rapidly changing conditions within it. The pressure remains more or less unchanged as the nuclear reaction rates increase, and the temperature rises rapidly in a runaway condition called the **helium flash.**

For a few hours, the helium burns ferociously. Eventually, the flood of energy released by this period of runaway fusion heats the core to the point at which normal thermal pressure once again dominates. Finally able to react to the energy dumped into it by helium burning, the core expands, its density drops, and equilibrium is restored as the inward pull of gravity and the outward push of gas pressure come back into balance. The core, now stable, begins to burn helium into carbon at temperatures well above 10^8 K. The helium flash terminates the giant star's ascent of the red-giant branch of the H–R diagram. Yet despite the violent ignition of helium in the core, the flash does *not* increase the star's luminosity. On the contrary, the energy released in the helium flash expands and cools the core and ultimately results in a *reduction* in the energy output. On the H–R diagram, the star jumps from stage 9 to stage 10, a stable state with steady helium burning in the core. As indicated in Figure 20.6, the surface temperature is now higher than it was on the red-giant branch, but the luminosity is considerably less than at the helium flash. This adjustment in the star's properties occurs quite quickly—in about 100,000 years.

At stage 10, our star is now stably burning helium in its core and fusing hydrogen in a shell surrounding it. It resides in a well-defined region of the H–R diagram known as the **horizontal branch**, where core-helium burning stars remain for a time before resuming their journey around the H–R diagram. The star's specific position within this region is determined mostly by its mass—not its original mass, but whatever mass remains after its ascent of the red-giant branch. The two masses differ because during the red-giant stage strong stellar winds eject large amounts of matter from a star's surface (see *Discovery 20-1*). As much as 20–30 percent of the original stellar mass may escape during this period. It so happens that more massive stars have lower surface temperatures at this stage, but all stars have roughly the same luminosity after the helium flash. As a result, stage 10 stars tend to lie along a horizontal line on the H–R diagram, with more massive stars to the right, less massive ones to the left.

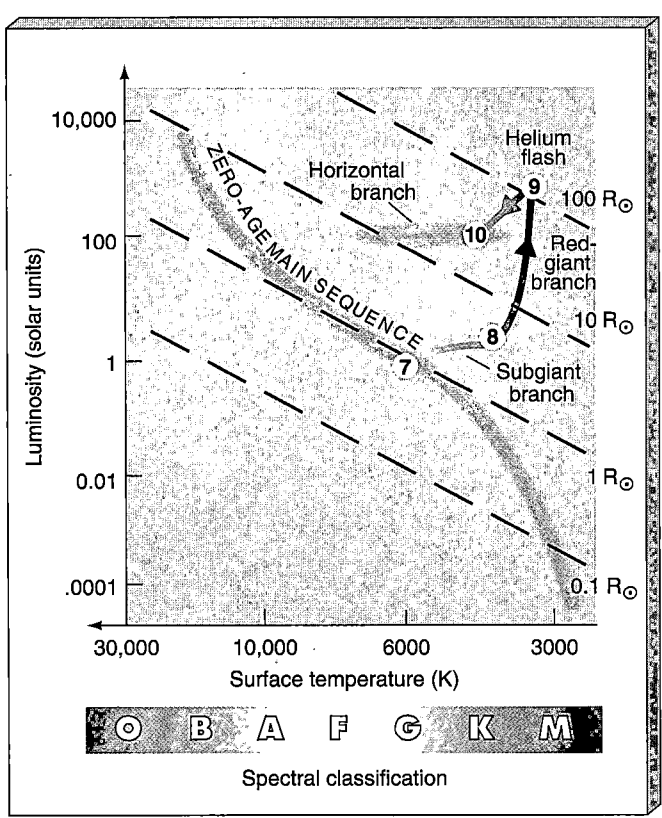

Figure 20.6 Horizontal Branch After its large increase in luminosity while ascending the red-giant branch is terminated by the helium flash, our star settles down into another equilibrium state at stage 10, on the horizontal branch.

THE CARBON CORE

The nuclear reactions in our star's helium core burn on, but not for long. Whatever helium exists in the core is rapidly consumed. The triple-alpha helium-to-carbon fusion reaction—like the proton–proton and CNO-cycle hydrogen-to-helium reactions before it—proceeds at a rate that increases very rapidly with temperature. At the extremely high temperatures found in the horizontal-branch core, the helium fuel doesn't last long—no more than a few tens of millions of years after the initial flash.

As helium fuses to carbon, a new inner core of carbon ash forms, and phenomena similar to the earlier buildup of helium ash begin to occur. Now helium becomes depleted at the very center of the star, and eventually fusion ceases there. In response, the nonburning carbon core shrinks and heats up as gravity pulls it inward, causing the hydrogen- and helium-burning rates in the overlying layers of the core to increase. The star now contains a shrinking carbon core surrounded by a helium-burning shell, which is in turn surrounded by a hydrogen-burning shell. The outer envelope of the star—the nonburning layers surrounding the core—expands, much as it did earlier during the first red-giant stage. By the time it reaches stage 11, the star has become a

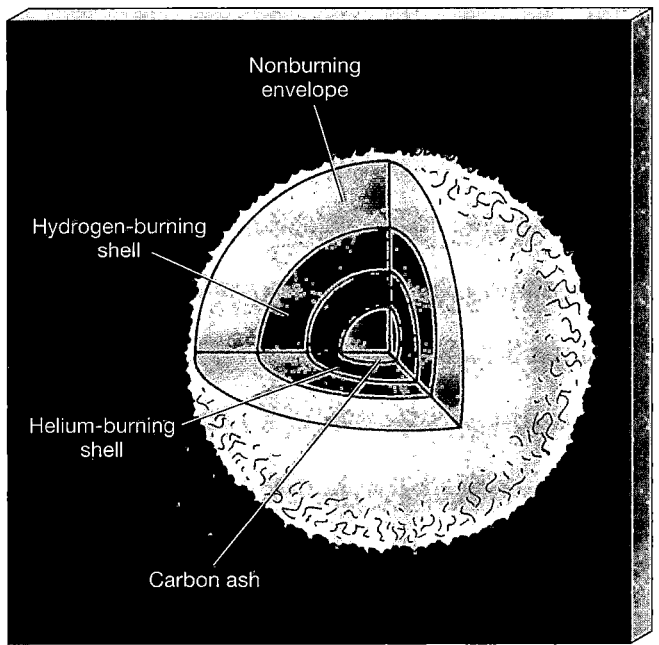

Figure 20.7 Helium-Shell Burning Within a few million years after the onset of helium burning, carbon ash accumulates in the inner core of a star, above which hydrogen and helium are still burning in concentric shells.

swollen red giant for a second time. Figure 20.7 depicts the star's interior structure during this time.

The star's second ascent of the giant branch is shown in Figure 20.8. To distinguish this second track from the first red-giant stage, this phase is sometimes known as the **asymptotic-giant branch**. The burning rates in the shells around the carbon core are much fiercer this time around, and the star's radius and luminosity increase to values even greater than those reached at the helium flash on the first ascent. Our star is now a *red supergiant*. The carbon core grows in mass as more carbon is produced in the helium-burning shell above it but continues to shrink in radius, driving the hydrogen-burning and helium-burning shells to higher and higher temperatures and luminosities.

☑ Concept Check

■ Why does a star get brighter as it runs out of fuel in its core?

20.3 The Death of a Low-Mass Star

THE FIRES GO OUT

🌀 As our red supergiant ascends the asymptotic-giant branch, its envelope swells while its core, too cool for further nuclear burning, continues to contract. If the central temperature could become high enough for carbon fusion to occur, still heavier products could be synthesized, and the newly generated energy might again support the star, restoring for a time the equilibrium between gravity and heat. For solar-mass stars, however, this does not occur. The temperature never reaches the 600 million K needed for a new round of nuclear reactions to occur. The red supergiant is now very close to the end of its nuclear-burning lifetime.

Before the carbon core can attain the incredibly high temperatures needed for carbon ignition, its density reaches a point beyond which it cannot be compressed further. At about 10^{10} kg/m^3, the electrons in the core once again become degenerate, the contraction of the core ceases, and its temperature stops rising. This stage (stage 12 in Table 20.1) represents the maximum compression that the star can achieve—there is simply not enough matter in the overlying layers to bear down any harder.

The core density at this stage is extraordinarily high. A single cubic centimeter of core matter would weigh 1000 kg

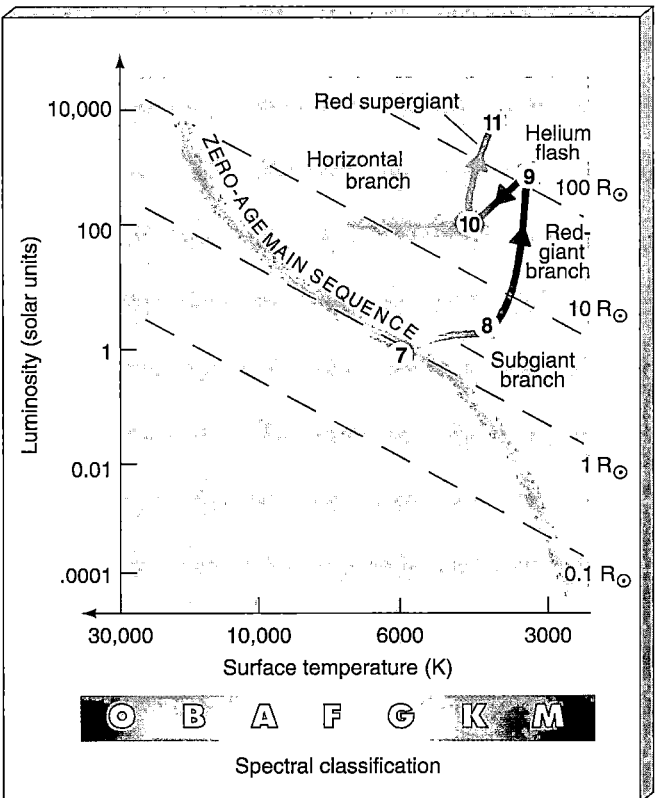

Figure 20.8 Red-Giant Branch Revisited A carbon-core star reascends the giant branch of the H–R diagram—this time on a track called the asymptotic-giant branch—for the same reason it evolved there the first time around: Lack of nuclear burning at the core causes contraction of the core and expansion of the overlying layers.

![DISCOVERY 20-1]

Mass Loss from Giant Stars

Astronomers now know that stars of all spectral types are active and have stellar winds. Consider the highly luminous, hot, blue O- and B-type stars, which have by far the strongest winds. Satellite and rocket observations have shown that their wind speeds may reach 3000 km/s. The result is a yearly mass loss sometimes exceeding 10^{-6} solar masses per year. Over the relatively short span of 1 million years, these stars blow a tenth of their total mass—more than an entire solar mass of material—into space. These powerful stellar winds, driven directly by the pressure of the intense ultraviolet radiation emitted by the stars themselves, hollow out vast cavities in the interstellar gas.

The accompanying figure shows a *Hubble Space Telescope* image of the supergiant star AG Carinae—50 times more massive than the Sun and a million times brighter—shedding its outer atmosphere. It is shown here puffing out vast clouds of gas and dust. (The star, at center, is intentionally obscured to show more clearly the surrounding nebulae; the bright vertical line is also an artifact.)

Observations made with radio, infrared, and optical telescopes have shown that luminous cool stars (for example, K- and M-type red giants) also lose mass at rates comparable to those of the luminous hot stars. Red-giant wind velocities, however, are much lower, averaging merely 30 km/s. They carry roughly as much mass into space as do O-type stellar winds because their densities are generally much greater. Because luminous red stars are inherently cool objects (with surface temperatures of only about 3000 K), they emit virtually no ultraviolet radiation, so the mechanism driving the winds must differ from that in luminous hot stars. We can only surmise that gas turbulence or magnetic fields or both in the atmospheres of the red giants are somehow responsible. The surface conditions in red giants are in some ways similar to those in T Tauri pro-

tostars, which are also known to exhibit strong winds. Possibly the same basic mechanism—violent surface activity—is responsible for both kinds of winds.

Unlike winds from hot stars, winds from these cool stars are rich in dust particles and molecules. Nearly all stars eventually evolve into red giants, so these winds provide a major source of new gas and dust to interstellar space. These stellar winds provide a vital link in the cycle of star formation and the evolution of the interstellar medium.

(NASA)

---

on Earth—a ton of matter compressed into a volume about the size of a grape. Yet despite the extreme compression of the core, the central temperature is "only" about 300 million K. Some oxygen is formed via reactions between carbon and helium at the inner edge of the helium-burning shell:

$$^{12}C + {}^4He \rightarrow {}^{16}O + energy,$$

but collisions among nuclei are neither frequent nor violent enough to create any heavier elements. For all practical purposes, the central fires go out once carbon has formed.

PLANETARY NEBULAE

Our aged, stage-12 star is now in quite a predicament. Its inner carbon core no longer generates energy. The outer-

core shells continue to burn hydrogen and helium, and as more and more of the inner core reaches its final, high-density state, the zone of nuclear burning increases in intensity. Meanwhile, the envelope continues to expand and cool, reaching a maximum radius of about 300 times that of the Sun—big enough to engulf the planet Mars.

Around this time, the burning becomes very unstable. The helium-burning shell is subject to a series of explosive *helium-shell flashes*. These flashes are caused by the enormous pressure there and the extreme sensitivity of the triple-alpha burning rate to small changes in temperature. The flashes produce large fluctuations in the intensity of the radiation reaching the star's outermost layers, causing them to pulsate more and more violently.

Compounding the star's problems, its surface layers are also becoming unstable. As the temperature drops to the point at which electrons can recombine with nuclei to form atoms, each recombination produces additional photons, which tend to push the outer envelope to greater and greater distances from the core. As shown in Figure 20.9, the radius of the star oscillates more and more violently. In less than a few million years, the star's outer envelope is ejected into space at a speed of a few tens of kilometers per second.

In time, a rather unusual looking object results. We say unusual because the "star" now has two distinct parts, both of which constitute stage 12 of Table 20.1. At the center is a small, well-defined core of mostly carbon ash. Hot and dense, only the outermost layers of this core still fuse helium into carbon and oxygen. Well beyond the core lies a spherical shell of cooler, low-density matter—the ejected envelope of the giant—spread over a volume roughly the size of our solar system. Such an object is called a **planetary nebula**. Some well-known examples are shown in Figures 20.10 and 20.11. In all, some 1000 planetary nebulae are known in our Galaxy.

The term *planetary* here is very misleading, for these objects have no association with planets. The name originated in the eighteenth century when, viewed at poor resolution through small telescopes, these shells of gas looked to some astronomers like the circular disks of planets in our solar system. The term *nebula* is also a little confusing, as it suggests kinship with the emission nebulae studied in Chapter 18. ∞ (Sec. 18.2) Although in some ways planetary nebulae do resemble some emission nebulae, and both

undergo similar ionization-recombination processes, these two types of objects are very different. Not only are planetary nebulae much smaller than emission nebulae, they are also associated with much older stars. Emission nebulae are the signposts of recent stellar birth. Planetary nebulae indicate impending stellar death.

The "ring" of a planetary nebula is in reality a three-dimensional shell of warm, glowing gas completely surrounding the core. Its halo-shaped appearance is only an illusion. The shell is the stellar envelope that has been expelled from around the core, but we see it mainly at the edges, where emitting matter accumulates along our line of sight. As illustrated in Figure 20.10(b), the shell is virtually invisible in the direction of the core. Few planetary nebulae are quite as regular as this simple picture might suggest, however the object shown in Figure 20.10(c) perhaps comes closest. Figure 20.11 shows two systems in which the details of the gas-ejection process must have been more complex, evidently playing an important role in determining the planetary nebula's shape and appearance.

The planetary nebula continues to spread out with time, becoming more diffuse and cooler, gradually dispersing into interstellar space. In doing so, it enriches the interstellar medium with atoms of helium, carbon, and oxygen dredged up by convection from the depths of the core into the envelope during the star's final years.

WHITE DWARFS

The carbon core, the stellar remnant at the center of the planetary nebula, continues to evolve. Formerly concealed by the atmosphere of the red-giant star, the core becomes visible as the envelope recedes. Several tens of thousands of years are needed for the core to appear from behind the veil of expanding gas. The core is very small. By the time the envelope is ejected as a planetary nebula, it has shrunk to about the size of Earth (in some cases it may be even smaller than our planet). Its mass is about half the mass of the Sun. Shining only by stored heat, not by nuclear reactions, this small "star" has a white-hot surface when it first becomes visible, although it appears dim because of its small size. The core's temperature and size give rise to its new name—*white dwarf*. This is stage 13 of Table 20.1. The approximate path followed by the star on the H–R diagram as it evolves from stage-11 red supergiant to stage-13 white dwarf is shown in Figure 20.12.

Not all white-dwarf stars are found as the cores of planetary nebulae. Several hundred have been discovered "naked" in our Galaxy, their envelopes expelled to invisibility (or perhaps stripped away by a binary companion—to be discussed shortly) long ago. Figure 20.13 shows an example of a white dwarf, Sirius B, that happens to lie particularly close to Earth; it is the faint binary companion of the much brighter and better-known Sirius A. ∞ (Sec. 17.3) Some properties of Sirius B are listed in Table 20.2. With more than the mass of the Sun packed into a volume

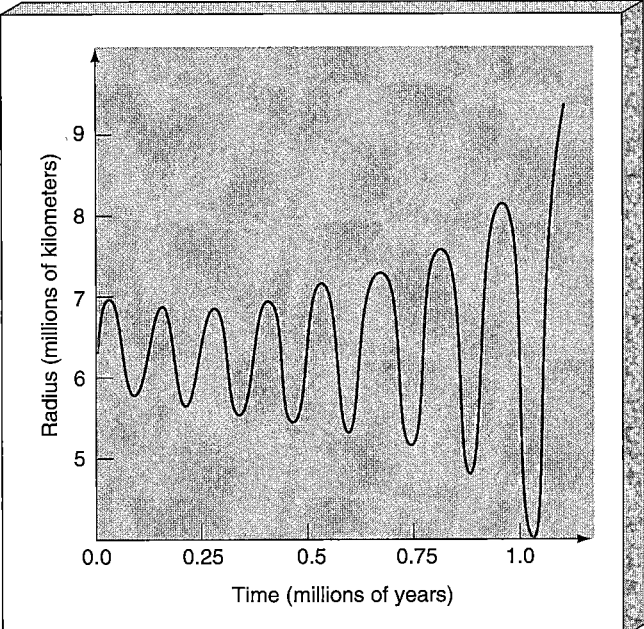

Figure 20.9 Red-Giant Instability Buffeted by helium-shell flashes from within and subject to the destabilizing influence of recombination, the outer layers of a red giant become unstable and enter into a series of growing pulsations. Eventually, the envelope is ejected and forms a planetary nebula.

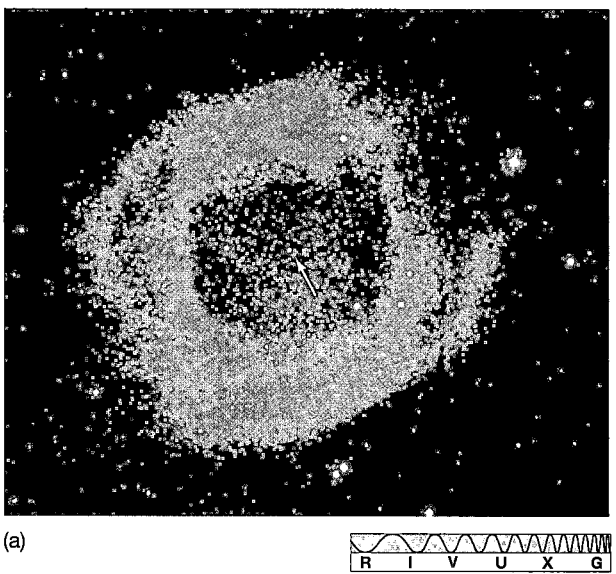

(a)

R   I   V   U   X   G

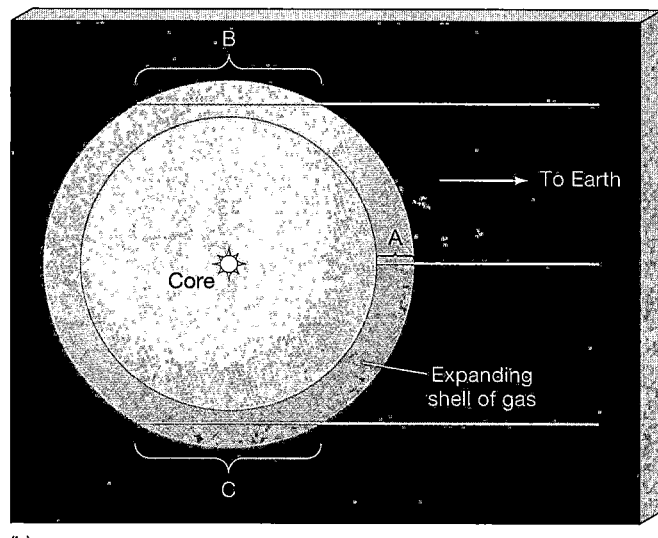

To Earth

Core

Expanding
shell of gas

(b)

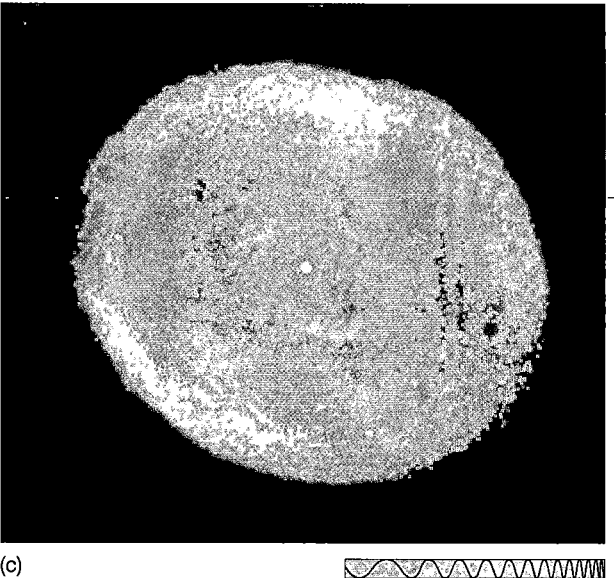

(c)

R   I   V   U   X   G

Figure 20.10 Ejected Envelope A planetary nebula is an object with a small dense core (central blue-white star) surrounded by an extended shell (or partial shells) of glowing matter. (a) The Helix Nebula appears to the eye as a small star (arrow) with a halo around it. About 140 pc from Earth and 0.6 pc across, its apparent size in the sky is roughly half that of the full Moon. (All the other stars visible in the photo are foreground or background objects, unrelated to the planetary nebula.) (b) The appearance of this planetary nebula can be explained once we realize that the shell of glowing gas around the central core is actually quite thin. There is very little gas along the line of sight between the observer and the central star (path A), so that part of the shell is invisible. Near the edge of the shell, however, there is more gas along the line of sight (paths B and C), so the observer sees a glowing ring. (c) This planetary nebula, IC 418, is younger, having expanded from its red-giant phase over only the past few thousand years to its current diameter of 0.03 pc—still more than 10 times the size of our solar system. *(Palomar; NASA)*

smaller than Earth, Sirius B's density is about a million times greater than anything familiar to us in the solar system. In fact, Sirius B has an unusually high mass for a white dwarf—it is believed to be the evolutionary product of a star roughly four times the mass of the Sun. *Discovery*

TABLE 20.2 Sirius B—A Nearby White Dwarf

Mass	1.1 solar masses
Radius	0.008 solar radii (5500 km)
Luminosity (total)	0.04 solar luminosities (1.6×10^{25} W)
Surface temperature	24,000 K
Average density	3×10^9 kg/m^3

20-2 discusses another possible peculiarity of Sirius B's evolution.

Hubble Space Telescope observations of nearby globular clusters have revealed the white-dwarf sequences long predicted by theory but previously too faint to detect at such large distances. Figure 20.14(a) shows a ground-based view of the globular cluster M4, lying 2100 pc from Earth. Part (b) of the figure shows an *HST* closeup of a small portion of the cluster, revealing dozens of white dwarfs (some marked) among the cluster's much brighter main-sequence, red-giant, and horizontal-branch stars. When plotted on an H–R diagram (see Figures 20.15 and 20.20), the white dwarfs fall nicely along the line indicated on Figure 20.12.

Not all white dwarfs are composed of carbon and oxygen. As mentioned earlier, very low-mass stars (less than

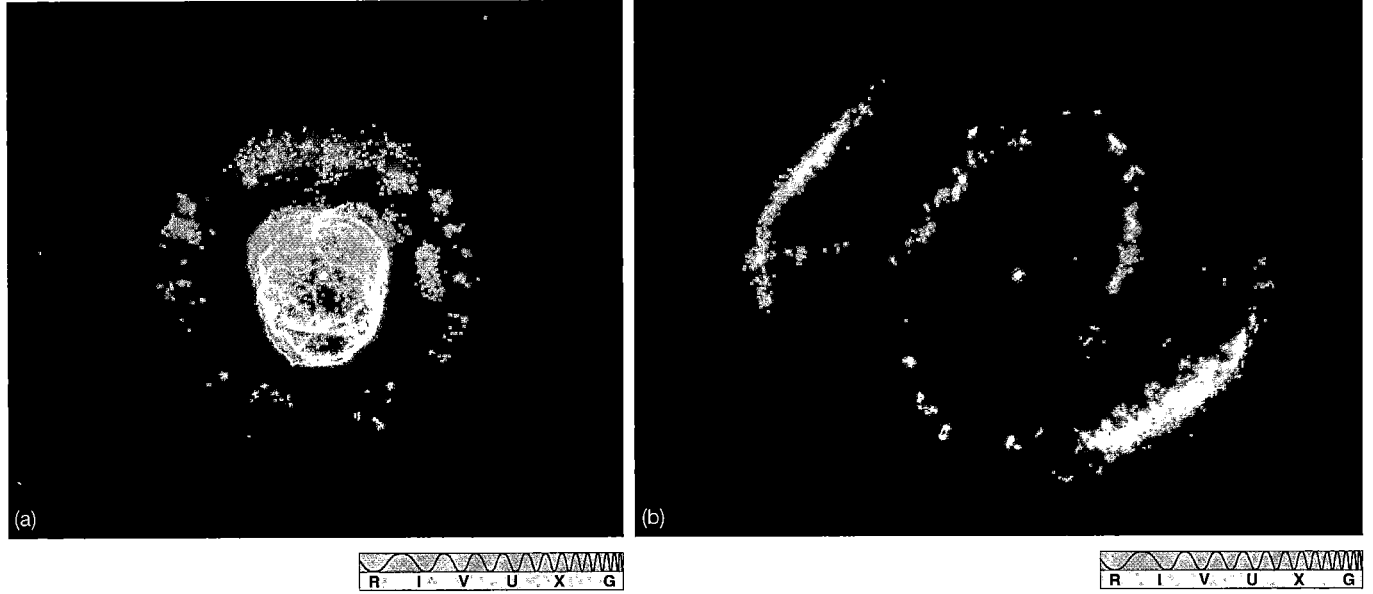

Figure 20.11 Planetary Nebulae (a) The "Eskimo" Nebula more clearly shows several "bubbles" (or shells) of material being blown into space from this planetary nebula. It resides some 1500 pc away in the constellation Gemini. (b) The Cat's Eye Nebula is an example of a much more complex planetary nebula. Intricate structures, including concentric gas shells, jets of high-speed gas, and shock-induced knots of gas are all visible. As usual, red indicates the presence of excited hydrogen. The nebula is about 1000 pc away, in the constellation Draco. It may have been produced by a pair of binary stars (unresolved at the center) that have both shed planetary nebulae. *(NASA)*

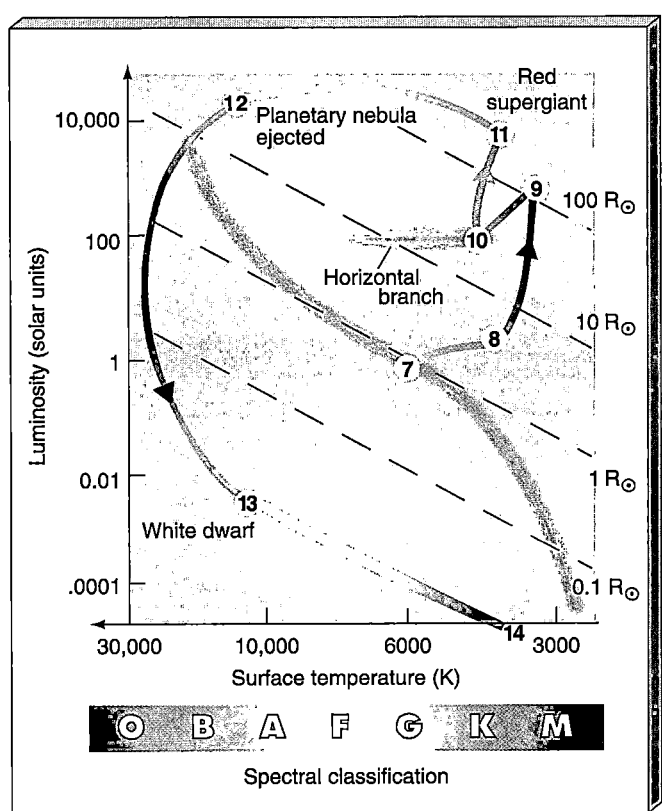

about one-quarter the mass of the Sun) never reach the point of helium fusion. The core of such a star becomes supported by electron degeneracy pressure before its central temperature reaches the 100 million K needed to start the triple-alpha process. Eventually, the star's envelope will

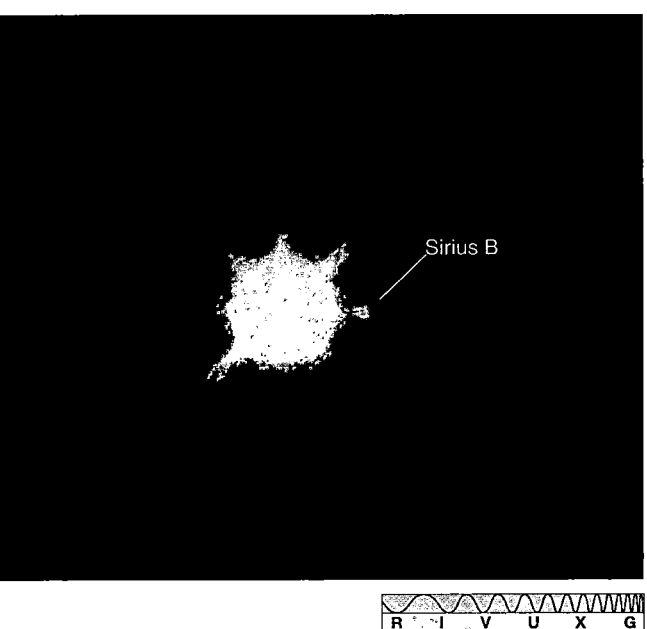

Figure 20.12 White Dwarf on the H–R Diagram A star's passage from the horizontal branch (stage 10) to the white-dwarf stage (stage 13) by way of the asymptotic-giant branch creates an evolutionary path that cuts across the entire H–R diagram.

Figure 20.13 Sirius Binary System Sirius B (the speck of light at right) is a white-dwarf star, a companion to the much larger and brighter star Sirius A. The "spikes" on the image of Sirius A are not real; they are artifacts caused by the support struts of the telescope. *(Palomar Observatory)*

527

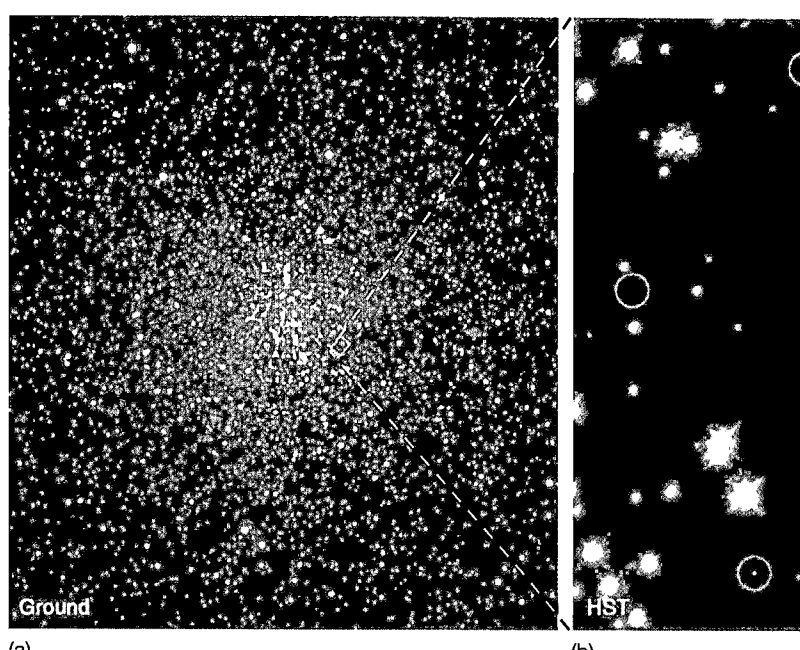

Figure 20.14 Distant White Dwarfs (a) The globular cluster M4, as seen through a large ground-based telescope at Kitt Peak National Observatory in Arizona. M4 is the closest globular cluster to us, at 2100 pc away; it spans some 16 pc. (b) A peek at M4's suburbs by the *Hubble Space Telescope* shows nearly a hundred white dwarfs within a small 0.4-pc area. Some of the brightest ones are circled in blue. *(AURA; NASA)*

be ejected in a manner similar to that of a more massive star, forming a *helium white dwarf*. Of course, the time needed for this to occur is very long—hundreds of billions of years—so no helium white dwarfs have ever actually formed in this way. ∞ (Sec. 17.8) However, if a solar-mass star is a member of a binary system, it is possible for its envelope to be stripped away by a companion (see Section 20.6), exposing the helium core and terminating the star's evolution before helium fusion can start. Several such low-mass white dwarfs have been detected in binary systems. Finally, in stars much more massive than the Sun (close to the 8-solar-mass limit on "low-mass" stars at the time the carbon core forms), temperatures in the core may become high enough that an additional reaction,

$$^{16}C + {}^{4}He \rightarrow {}^{20}Ne + energy,$$

can occur, ultimately leading to the formation of a rare *neon-oxygen white dwarf*.

Once an isolated star becomes a white dwarf, its evolution is over. (As we will see in Chapter 21, white dwarfs in binary systems may have further activity in store.) It continues to cool and dim with time, following the white–yellow–red track near the bottom of the H–R diagram of Figure 20.12, eventually becoming a *black dwarf*—a cold, dense, burned-out ember in space. This is stage 14 of Table 20.1, the graveyard of stars.

The cooling dwarf does not shrink much as it fades away. Even though its heat is leaking away into space, gravity does not compress it further. At the enormously high densities in the star (from the white-dwarf stage on), the resistance of electrons to being squeezed together—the same electron degeneracy that prevailed in the red-

giant core around the time of the helium flash—supports the star, even as its temperature drops almost to absolute zero. As the dwarf cools, it remains about the size of Earth.

COMPARING THEORY WITH REALITY

All the H–R diagrams and evolutionary tracks presented so far are theoretical constructs based largely on computer models of the interior workings of stars. Before continuing our study of stellar evolution, let's take a moment to compare our models with actual observations. Figure 20.15(a) shows the beautiful globular cluster M80, which lies about 8000 pc from Earth. Figure 20.15(b) shows a composite H–R diagram recently constructed using the stars of a number of other globular clusters of roughly the same age and composition as M80. It spans the entire range of stellar luminosities, from supergiants to faint red and white dwarfs. Fitting theoretical models of the main-sequence, giant, and horizontal branches (see Section 20.5) implies an age of about 12 billion years, making these clusters among the oldest-known objects in the Milky Way Galaxy (consistent with the low concentrations of heavy elements observed in their spectra).

The great age of this cluster means that stars more massive than about 0.8 solar masses have already evolved beyond the red-giant stage, mainly becoming white dwarfs. The H–R diagram for this cluster can therefore be compared directly with Figure 20.12, as the giants, supergiants, and horizontal-branch stars are all of roughly 1 solar mass. The similarity between theory and observation is striking—stars in each of the evolutionary stages 7–13 can be seen, in numbers consistent with the theoretical models. Astronomers place great confidence in the theory

(a)

Figure 20.15 Globular Cluster H–R Diagram (a) The globular cluster M80, some 8 kpc away. (b) Combined H–R diagram, based on ground- and space-based observations, for several globular clusters similar in overall composition to M80. The various evolutionary stages predicted by theory and depicted schematically in Figure 20.12 are clearly visible. Note also the blue stragglers—main-sequence stars that appear to have been "left behind" as other stars evolved into giants. They are probably the result of merging binary systems or actual collisions between lower-mass stars in this remarkably dense stellar system. *(NASA)*

(b)

of stellar evolution precisely because its predictions are so often found to be in excellent agreement with plots of real stars. (The points in Figure 20.15(b) are "shifted" a little to the left relative to Figure 20.12 because of composition differences between stars such as the Sun and stars in globular clusters—globular cluster stars tend to be slightly hotter than solar-type stars of the same mass.)

The objects labeled as *blue stragglers* in Figure 20.15(b) appear at first sight to contradict the theory just described. They lie on the main sequence, but in locations that should have evolved into white dwarfs long ago, given the cluster's age of 12 billion years. Blue stragglers are observed in many star clusters. They are main-sequence stars, but they did not form when the cluster did. Instead, they formed much more recently, through *mergers* of lower-mass stars—so recently, in fact, that they have not yet had time to evolve into giants.

How did these stellar mergers occur? In some cases, they are probably the result of stellar evolution in binary systems, as the component stars evolve, grow, and come into contact (see Section 20.6). In others, the mergers are thought to be the result of actual *collisions* between stars. The core of M80 contains a huge number of stars packed into a relatively small volume. For example, a sphere of radius 2 pc centered on the Sun contains exactly four stars, including the Sun itself. ∞ (Sec. 17.1) At the center of M80, the same 2-pc sphere would contain more than 10 *million* stars—our night sky would be ablaze with thousands of objects brighter than Venus! The dense central cores of globular clusters are among the few places in the entire universe where stellar collisions are likely to occur.

☑ Concept Check

■ Why does fusion cease in the core of a low-mass star?

20.4 Evolution of Stars More Massive than the Sun

High-mass stars evolve much faster than their low-mass counterparts. The more massive a star, the more ravenous is its fuel consumption and the shorter its main-sequence lifetime. The Sun will spend a total of some 10 billion years on the main sequence, but a 5-solar-mass B-type star will remain there for only a hundred million years. A 10-solar-mass O-type star will depart in just 20 million years or so. This trend toward much faster evolution for more massive stars continues even after the main sequence. All evolutionary changes happen much more

DISCOVERY 20-2

Learning Astronomy from History

Sirius A, the brighter of the two objects shown in Figure 20.13, appears twice as luminous as any other visible star, excluding the Sun. Its absolute brightness is not very great, but because its distance from us is small (less than 3 pc), its apparent brightness is very large. ⊙ (Sec. 17.3) Sirius has been prominent in the nighttime sky since the beginning of recorded history. Cuneiform texts of the ancient Babylonians refer to the star as far back as 1000 B.C., and historians know that the star strongly influenced the agriculture and religion of the Egyptians of 3000 B.C.

Even though a star's evolution takes such a long time, we might have a chance to detect a slight change in Sirius because the recorded observations of this star go back several thousand years. The chances for success are improved in this case because Sirius A is so bright that even the naked-eye observations of the ancients should be reasonably accurate. Interestingly, recorded history does suggest that Sirius A has changed in appearance, but the observations are confusing. Every piece of information about Sirius recorded between the years 100 B.C. and A.D. 200 claims that this star was *red*. (No earlier records of its color are known.) In contrast, modern observations now show it to be white or bluish white—definitely *not* red.

If these reports are accurate, then Sirius has apparently changed from red to blue-white in the intervening years.

But according to the theory of stellar evolution, no star should be able to change its color in this way in such a short time. A color change such as this should take at least several tens of thousands of years, and perhaps a lot longer. It should also leave some evidence of its occurrence.

Astronomers have offered several explanations for the rather sudden change in Sirius A. These include the suggestions that (1) some ancient observers were wrong and other scribes copied them; (2) a galactic dust cloud passed between Sirius A and Earth some 2000 years ago, reddening the star much as Earth's dusty atmosphere often reddens our Sun at dusk; and (3) the companion to Sirius A, Sirius B, was a red giant and the dominant star of this double-star system 2000 years ago but has since expelled its planetary nebular shell to reveal the white-dwarf star that we now observe.

Each of these explanations presents problems. How could the color of the sky's brightest star have been incorrectly recorded for hundreds of years? Where is the intervening galactic cloud now? Where is the shell of the former red giant? We are left with the uneasy feeling that the sky's brightest star doesn't fit particularly well into the currently accepted scenario of stellar evolution.

---

rapidly for high-mass stars because their larger mass and stronger gravity generate more heat, speeding up all phases of stellar evolution.

Stars leave the main sequence for one basic reason—they run out of hydrogen in their cores. As a result, the early stages of stellar evolution beyond the main sequence are qualitatively the same in all cases: main-sequence hydrogen burning in the core (stage 7) eventually gives way to the formation of a nonburning, collapsing helium core surrounded by a hydrogen-burning shell (stages 8 and 9). A high-mass star leaves the main sequence on its journey toward the red-giant region with an internal structure quite similar to that of its low-mass cousin. Thereafter, their evolutionary tracks diverge.

Figure 20.16 compares the post-main-sequence evolution of three stars having masses 1, 4, and 15 times the mass of the Sun. Note that, whereas stars like the Sun ascend the red-giant branch almost vertically, higher-mass stars move nearly horizontally across the H–R diagram after leaving the upper main sequence. Their luminosities stay roughly constant as their radii increase and their surface temperatures drop.

In stars having more than about 2.5 times the mass of the Sun, helium burning begins smoothly and stably, *not* explosively—there is no helium flash. Calculations indi-

cate that the more massive a star, the lower its core density when the temperature reaches the 10^8 K necessary for helium ignition, and the smaller the contribution to the pressure from degenerate electrons. As a result, above 2.5 solar masses, the unstable core conditions described earlier do not occur. The 4-solar-mass red giant in Figure 20.16 remains a red giant as helium starts to fuse into carbon. There is no sudden jump to the horizontal branch and no subsequent reascent of the giant branch. Instead, the star loops smoothly back and forth near the top of the H–R diagram, as a blue or red supergiant.

A much more important divergence occurs at approximately 8 solar masses—the dividing line between high and low mass mentioned in Section 20.1. A low-mass star never achieves the 600 million K needed to fuse carbon nuclei, and so it ends its life as a carbon-oxygen (or possibly neon-oxygen) white dwarf. A high-mass star, however, can fuse not only hydrogen and helium but also carbon, oxygen, and even heavier elements as its inner core continues to contract and its central temperature continues to rise. The burning rate accelerates as the core evolves. Evolution proceeds so rapidly in the 15-solar-mass star whose evolution is shown in Figure 20.16 that the star doesn't even reach the red-giant region before helium fusion begins. The star achieves a central temperature of 10^8 K

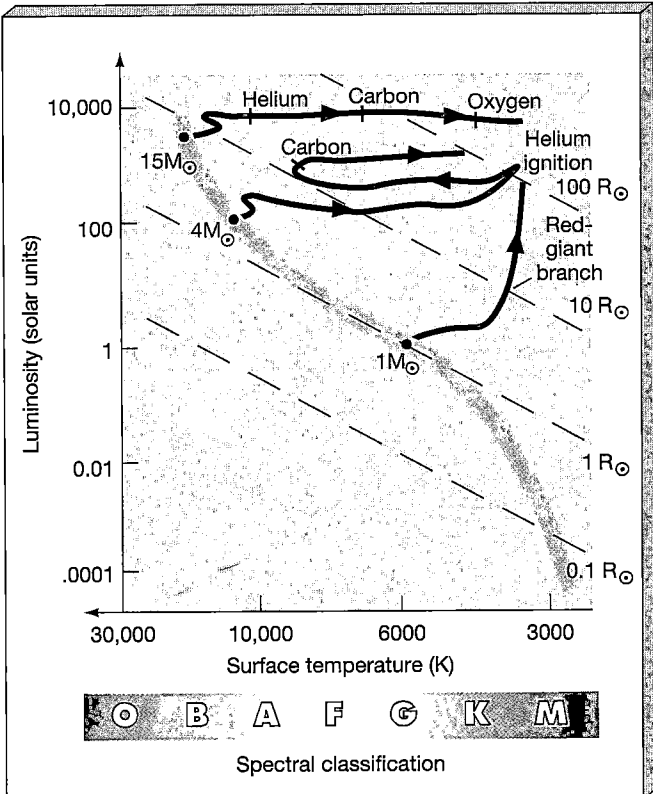

Figure 20.16 High-Mass Evolutionary Tracks
Evolutionary tracks for stars of 1, 4, and 15 solar masses (shown only up to the point of the helium flash in the low-mass cases). Low-mass stars ascend the giant branch almost vertically, whereas high-mass stars move roughly horizontally across the H–R diagram from the main sequence into the red-giant region. The most massive stars experience smooth transitions into each new burning stage. No helium flash occurs for stars more massive than about 2.5 solar masses. Some points are labeled with the element that has just started to fuse in the inner core.

while still quite close to the main sequence, and its evolutionary track continues smoothly across the supergiant region of the H–R diagram, seemingly unaffected by each new phase of burning.

With heavier and heavier elements forming at an ever-increasing rate, the high-mass star shown in Figure 20.16 is very close to the end of its life. We will discuss the evolution and ultimate fate of such a star in more detail in the next chapter, but suffice it to say here that it is destined to die in a violent supernova explosion soon after carbon and oxygen begin to fuse in its core. High-mass stars evolve so rapidly that, for most practical observational purposes, they explode and die shortly after leaving the main sequence.

A good example of a post-main-sequence blue supergiant is the bright star Rigel in the constellation Orion. With a radius some 70 times that of the Sun and a luminosity of around 50,000 solar luminosities, Rigel is thought to have had an original mass about 17 times that

of the Sun, although a strong stellar wind has probably carried away a significant fraction of its mass since it formed. Although still near the main sequence, Rigel is probably already fusing helium into carbon in its core.

Perhaps the best-known red supergiant is Betelgeuse (shown in Figures 17.8 and 17.11), also in Orion and Rigel's rival for the title of brightest star in the constellation. Its luminosity is 10^4 times that of the Sun in visible light, and perhaps five times that in the infrared. Astronomers believe that Betelgeuse is currently fusing helium into carbon and oxygen in its core, but its eventual fate is uncertain. As best we can tell, the star's mass at formation was between 12 and 17 times the mass of the Sun. However, like Rigel and many other supergiants, Betelgeuse has a strong stellar wind, and is known to be surrounded by a huge shell of dust of its own making (*Discovery 20-1*). It also pulsates, varying in radius by about 60 percent. The pulsations and strong wind may be related to the huge spots observed on the star's surface (Figure 17.11). Together, they suggest that Betelgeuse has lost a lot of mass since it formed, but just how much remains uncertain.

Table 20.3 lists some possible outcomes of stellar evolution for stars of different masses. Note that our earlier dividing line of 8 solar masses between "low-mass" and "high-mass" really refers to the mass at the time the carbon core forms. Since very luminous stars often have strong stellar winds, main-sequence stars as massive as 10–12 times the mass of the Sun may still manage to avoid a supernova explosion. Unfortunately, we do not know exactly how much mass either Rigel or Betelgeuse has lost, so we cannot yet tell if they are above or below the threshold for becoming a supernova. Either might explode, or instead become a neon-oxygen white dwarf, but for now we can't say which. We will probably just have to wait and see!

☑ Concept Check

▪ What is the essential evolutionary difference between high-mass and low-mass stars?

TABLE 20.3 End-Points of Evolution for Stars of Different Masses

INITIAL MASS (SOLAR MASSES)	FINAL STATE
less than 0.08	(hydrogen) brown dwarf
0.08–0.25	helium white dwarf
0.25–8	carbon-oxygen white dwarf
8–12 (approx.) *	neon-oxygen white dwarf
greater than 12*	supernova

*Precise numbers depend on the (poorly known) amount of mass lost on and after the main sequence.

20.5 Observing Stellar Evolution in Star Clusters

5 Star clusters provide excellent test sites for the theory of stellar evolution. Every star in a given cluster formed at the same time, from the same interstellar cloud, with virtually the same composition. Only the mass varies from one star to another. This allows us to check the accuracy of our theoretical models in a very straightforward way. Having studied in some detail the evolutionary tracks of individual stars, let's now consider how their collective appearance changes in time.

In Chapter 19 we saw how astronomers estimate the ages of star clusters by determining which of their stars have already left the main sequence. ⊙ (Sec. 19.6) In fact, the main-sequence lifetimes that go into those age measurements represent only a tiny fraction of the data obtained from theoretical models of stellar evolution. Starting from the zero-age main sequence, astronomers can predict exactly how a newborn cluster should look at any subsequent time. Although we cannot see into the interiors of stars to test our models, we can compare stars' outward appearances with theoretical predictions. The agreement—in detail—between theory and observation is remarkably good.

We begin our study shortly after the cluster's formation, with the upper main sequence already fully formed and burning steadily, and lower-mass stars just beginning to arrive on the main sequence, as shown in Figure 20.17(a). The appearance of the cluster at this early stage is dominated by its most massive stars—the bright blue supergiants. Now let's follow the cluster forward in time and see how its H–R diagram evolves.

Figure 20.17(b) shows the appearance of our cluster's H–R diagram after 10 million years. The most massive O-type stars have evolved off the main sequence. Most have already exploded and vanished, as just discussed, but one or two may still be visible as red supergiants. The remaining cluster stars are largely unchanged in appearance—their evolution is slow enough that little happens to them in such a relatively short period of time. The cluster's H–R diagram

Figure 20.17 Cluster Evolution on the H–R Diagram The changing H–R diagram of a hypothetical star cluster. (a) Initially, stars on the upper main sequence are already burning steadily while the lower main sequence is still forming. (b) At 10^7 years, O-type stars have already left the main sequence, and a few red giants are visible. (c) By 10^8 years, stars of spectral type B have evolved off the main sequence. More red giants are visible, and the lower main sequence is almost fully formed. (d) At 10^9 years, the main sequence is cut off at about spectral type A. The subgiant and red-giant branches are just becoming evident, and the formation of the lower main sequence is complete. A few white dwarfs may be present. (e) At 10^{10} years, only stars less massive than the Sun still remain on the main sequence. The cluster's subgiant, red-giant, horizontal, and asymptotic-giant branches are all discernible. Many white dwarfs have now formed.

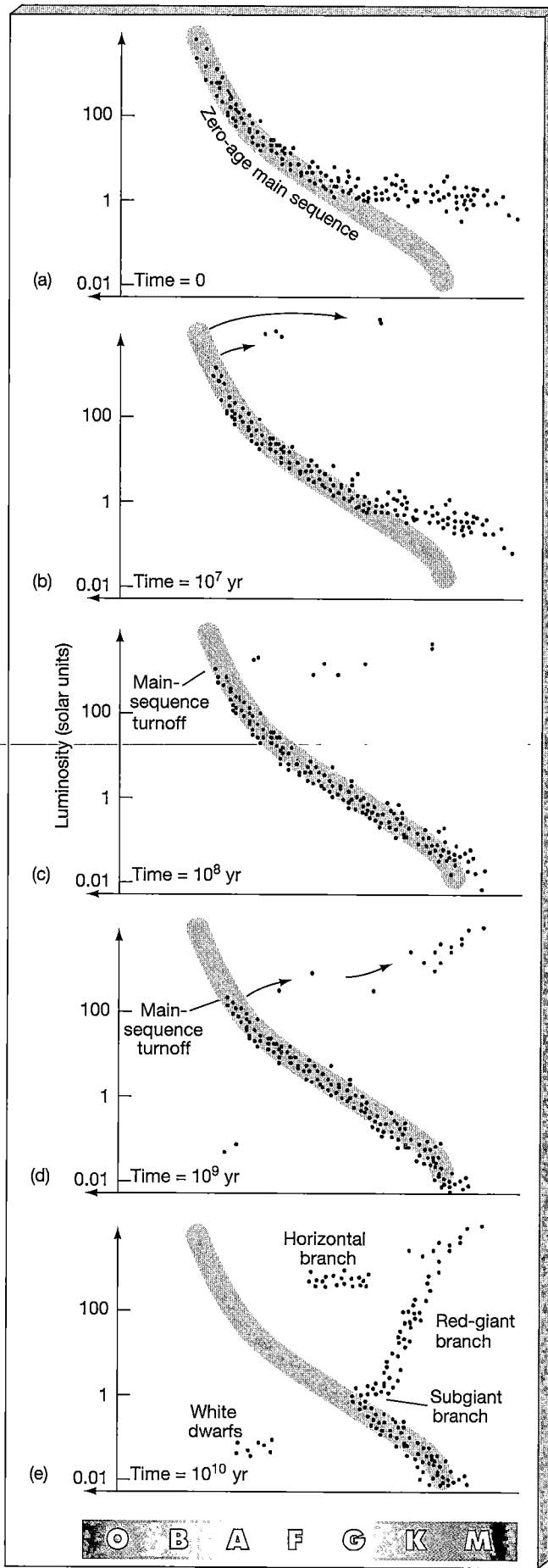

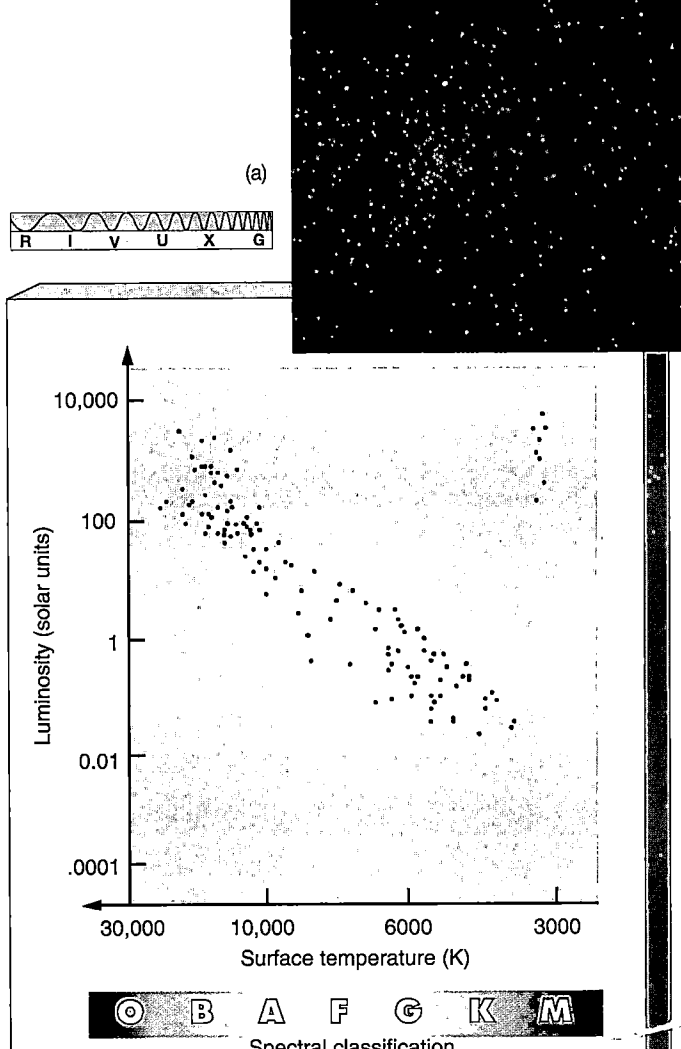

Figure 20.18 Newborn Cluster H-R Diagram (a) The "double cluster" h and chi Persei. (b) The H–R diagram of the pair indicates that the stars are very young—probably only about 10 million years old. *(AURA)*

being "peeled away" from the top down, with fainter and fainter stars turning off and heading for the giant branch as time goes on. Astronomers refer to the high-luminosity end of the observed main sequence as the **main-sequence turnoff**. The mass of the star that is just evolving off the main sequence at any moment is known as the *turnoff mass*.

At 1 billion years, the main-sequence turnoff mass is around 2 solar masses, corresponding roughly to spectral type A2. The subgiant and giant branches associated with the evolution of low-mass stars are just becoming visible, as indicated in Figure 20.17(d). The formation of the lower main sequence is now complete. In addition, the first white dwarfs have just appeared, although they are often too faint to be observed at the distances of most clusters. Figure 20.19 shows the Hyades open cluster and its H–R diagram. The H–R diagram appears to lie between Figures 20.17(c) and 20.17(d), suggesting that the cluster's age is about 600 million years.

At 10 billion years, the turnoff point has reached solar-mass stars of spectral type G2. The subgiant and giant branches are now clearly discernible (see Figure 20.17e), and the horizontal and asymptotic-giant branches appear as distinct regions in the H–R diagram. Many white dwarfs are also present in the cluster. Although stars in all these evolutionary stages are also present in the 1-billion-year-old cluster shown in Figure 20.17(d), they are few in number—typically only a few percent of the total number of stars in the cluster. Also, because they evolve so rapidly, they spend very little time in these regions. Low-mass stars are much more numerous and evolve more slowly, so more of them spend more time in any given region of the H–R diagram, allowing their evolutionary tracks to be more easily discerned.

Figure 20.20 shows the globular cluster 47 Tucanae. By carefully adjusting their theoretical models until the cluster's main sequence, subgiant, red-giant, and horizontal branches are all well matched, astronomers have determined its age to be between 10 and 12 billion years, a little older than our hypothetical cluster in Figure 20.17(e). In fact, globular cluster ages determined in this way show a remarkably small spread—all the globular clusters in our Galaxy appear to have formed between about 10 and 12 billion years ago.

shows the main sequence slightly cut off, along with a rather poorly defined red-giant region. Figure 20.18 shows the twin open clusters h and chi Persei, along with their combined H–R diagram. Comparing Figure 20.18(b) with such diagrams as those in Figure 20.17, astronomers estimate the age of this pair of clusters to be about 10 million years.

After 100 million years (Figure 20.17c) stars brighter than type B5 or so (about 4–5 solar masses) have left the main sequence, and a few more red supergiants are visible. By this time most of the cluster's low-mass stars have finally arrived on the main sequence, although the dimmest M-type stars may still be in their contraction phase. The appearance of the cluster is now dominated by bright B-type main-sequence stars and brighter red supergiants.

At any time during the cluster's evolution the original main sequence is intact up to some well-defined stellar mass, corresponding to the stars that are just leaving the main sequence at that instant. We can imagine the main sequence

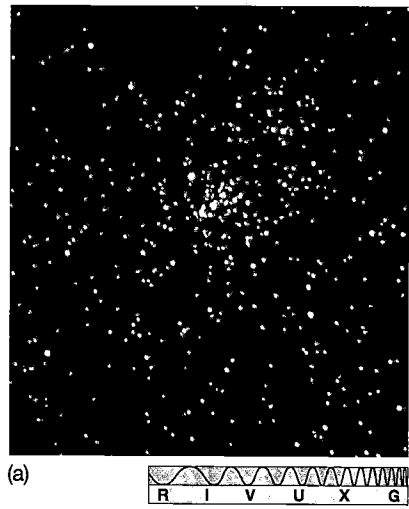

(a)

| R | I | V | U | X | G |

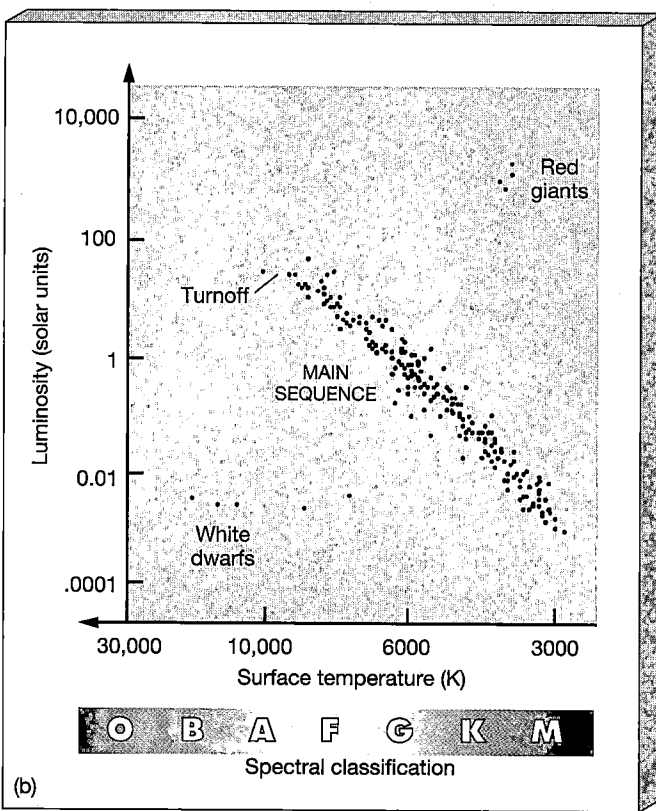

10,000

100

Luminosity (solar units)

1

0.01

.0001

Turnoff

MAIN
SEQUENCE

Red
giants

White
dwarfs

30,000   10,000   6000   3000
Surface temperature (K)

| O | B | A | F | G | K | M |

Spectral classification

(b)

Figure 20.19 Young Cluster H–R Diagram (a) The Hyades cluster, a relatively young group of stars visible to the naked eye. It lies 46 pc away in the constellation Taurus. (b) The H–R diagram for this cluster is cut off at about spectral type A, implying an age of about 600 million years. *(AURA)*

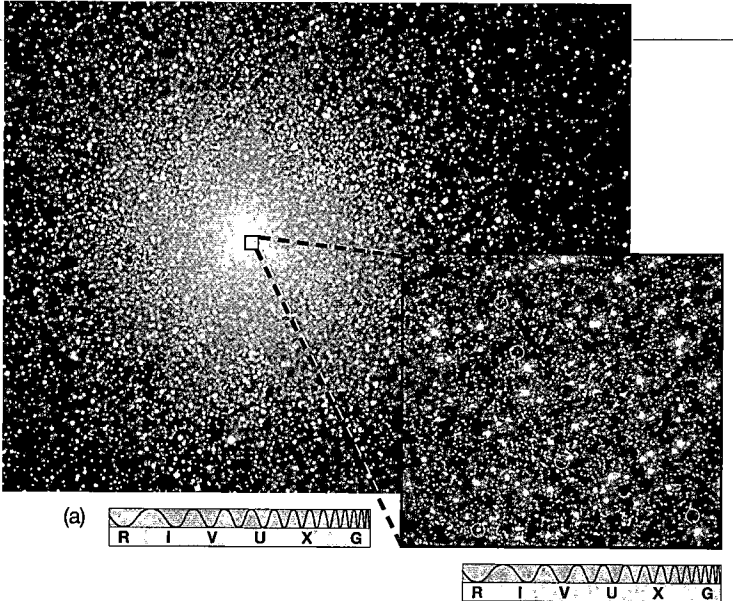

(a)

| R | I | V | U | X | G |

| R | I | V | U | X | G |

10,000

100

Luminosity (solar units)

1

0.01

.0001

Red giants

Blue
stragglers

Turnoff

White
dwarfs

Red
dwarfs

30,000   10,000   6000   3000
Surface temperature (K)

| O | B | A | F | G | K | M |

Spectral classification

(b)

Figure 20.20 Old Cluster H–R Diagram (a) The southern globular cluster 47 Tucanae. (b) Fitting its main-sequence turnoff and its giant and horizontal branches to theoretical models gives 47 Tucanae an age of between 12 and 14 billion years, making it one of the oldest-known objects in the Milky Way Galaxy. The inset is a high-resolution ultraviolet image of 47 Tucanae's core region, taken with the *Hubble Space Telescope* and showing many blue stragglers—massive stars lying on the main sequence above the turnoff point, resulting perhaps from the merging of binary-star systems (see also Figure 20.15). The points representing white dwarfs, some red dwarfs, and blue stragglers have been added to the original data set, based on *Hubble* observations of this and other clusters. The white dwarf data are for the cluster M4 (Figure 20.14). As with M80, the complete H–R diagram for a single cluster is expected to look qualitatively similar. Data on the faintest main-sequence stars shown were obtained using ground-based observations. The thickness of the lower main sequence is due almost entirely to observational limitations, which make it difficult to determine accurately the apparent brightnesses and colors of low-luminosity stars. *(ESO; NASA)*

Stellar evolution is one of the great success stories of astrophysics. Like all good scientific theories, it makes definite testable predictions about the universe while remaining flexible enough to incorporate new discoveries as they occur. Theory and observation have advanced hand in hand. At the start of the twentieth century many scientists despaired of ever knowing even the compositions of the stars, let alone why they shine and how they change. Today, the theory of stellar evolution is a cornerstone of modern astronomy.

☑️ Concept Check

■ Why are observations of star clusters so important to the theory of stellar evolution?

20.6 The Evolution of Binary-Star Systems

⑤ We have noted that most stars in our Galaxy are not isolated objects but are actually members of binary-star systems. However, our discussion of stellar evolution has so far focused exclusively on isolated stars. This prompts us to ask how membership in a binary-star system changes the evolutionary tracks we have just described. Indeed, because nuclear burning occurs deep in the core, does the presence of a stellar companion have any significant effect at all? Perhaps not surprisingly, the answer depends on the distance between the two stars in question.

For a binary system whose component stars are very widely separated—that is, the distance between the stars is greater than perhaps a thousand stellar radii—the two stars evolve more or less independently of one another, each following the track appropriate to an isolated star of its

particular mass. However, if the two stars are closer, then the gravitational pull of one may strongly influence the envelope of the other. In that case the physical properties of both may deviate greatly from those calculated for isolated single stars.

As an example, consider the star Algol (Beta Persei, the second brightest star in the constellation Perseus). By studying its spectrum and the variation in its light intensity, astronomers have determined that Algol is actually a binary (in fact, an eclipsing double-lined spectroscopic binary, as described in Chapter 17), and they have measured its properties very accurately. ∞ (Sec. 17.9) Algol consists of a 3.7-solar-mass main-sequence star of spectral type B8 (a blue giant) with a 0.8-solar-mass red-subgiant companion moving in a nearly circular orbit around it. The stars are 4 million km apart and have an orbital period of about 3 days.

A moment's thought reveals that there is something odd about these findings. On the basis of our earlier discussion, the more massive main-sequence star should have evolved *faster* than the less massive component. If the two stars formed at the same time (as is assumed to be the case), there should be no way that the 0.8-solar-mass star could be approaching the giant stage first. Either our theory of stellar evolution is seriously in error, or something has modified the evolution of the Algol system. Fortunately for theorists, the latter is the case.

As sketched in Figure 20.21, each star in a binary system is surrounded by its own teardrop-shaped "zone of influence," inside of which its gravitational pull dominates the effects of both the other star and the overall rotation of the binary. Any matter within that region "belongs" to the star. It cannot easily flow onto the other component or out of the system. Outside the two regions, it is possible for gas to flow toward either star relatively easily. The two teardrop-shaped regions are called **Roche lobes**, after Edouard Roche, the French mathematician who first studied the binary-system problem in the nineteenth century,

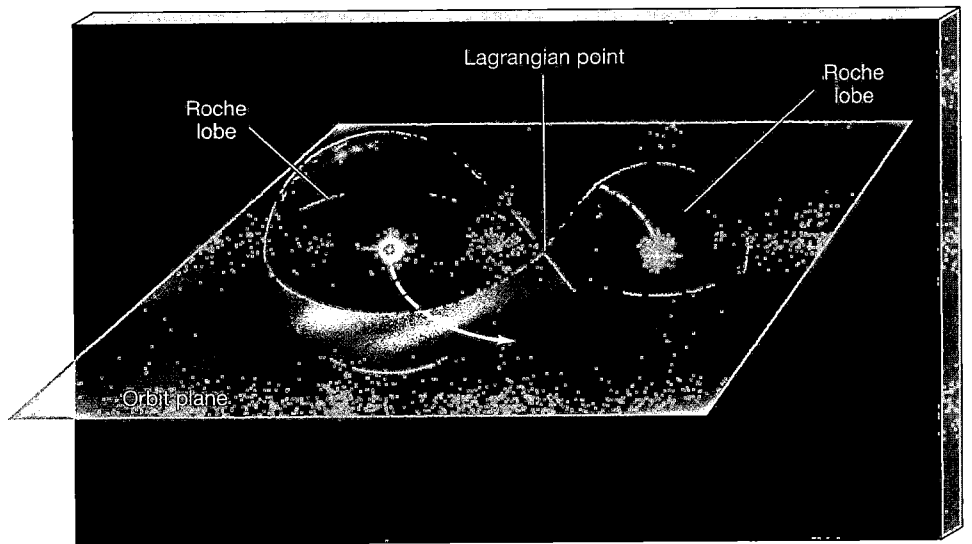

Figure 20.21 Stellar Roche Lobes Each star in a binary system can be pictured as being surrounded by a "zone of influence," or Roche lobe, inside of which matter may be thought of as being "part" of that star. The two teardrop-shaped Roche lobes meet at the Lagrangian point between the two stars. Outside the Roche lobes, matter may flow onto either star with relative ease.

and whose work we have already encountered in the context of planetary rings. ∞ (Sec. 12.4) The Roche lobes of the two stars meet at a point on the line joining them—the inner Lagrangian point (L_1), which we saw in Chapter 14 when discussing asteroid motions in the solar system. ∞ (Sec. 14.1) This Lagrangian point is a place where the gravitational pulls of the two stars exactly balance the rotation of the binary system. The greater the mass of one component, the larger is its Roche lobe and the farther from its center (and the closer to the other star) is the Lagrangian point.

Normally, both stars lie well within their respective Roche lobes, and such a binary system is said to be *detached*, as in Figure 20.22(a). However, as a star evolves off the main sequence and moves toward the giant branch, it is possible for its radius to become so large that it overflows its Roche lobe. Its gas begins to flow onto the companion through the Lagrangian point. The binary in this case is said to be *semidetached* (Figure 20.22b). Because matter is flowing from one star onto the other, semidetached binaries are also known as **mass-transfer binaries.** If, for some reason, the other star also overflows its Roche lobe (either because of stellar evolution or because so much extra material is dumped onto it), the surfaces of the two stars merge. The binary system then consists of two nuclear-burning stellar cores surrounded by a single continuous common envelope—a **contact binary,** shown in Figure 20.22(c).

In a binary system in which the two stars are very close together, neither star has to evolve far off the main sequence before it overflows its Roche lobe and mass transfer begins. In a wide binary, however, both stars may evolve all the way up the giant branch without either surface ever reaching the Lagrangian point, and they evolve just as though they were isolated. Depending on the stars involved and their orbital separations, there are many different possibilities for the eventual outcome of the evolution. Binary evolution is a very complex subject. Let's make these ideas more definite by returning to the question of how the binary star Algol may have reached its present state.

Astronomers believe that Algol started off as a detached binary. For reference, let us label the component that is now the 0.8-solar-mass subgiant as star 1 and the 3.7-solar-mass main-sequence star as star 2. Initially, star 1 was the more massive of the two, having perhaps three times the mass of the Sun. It thus evolved off the main sequence first. Star 2 was originally a less massive star, perhaps comparable in mass to the Sun. As star 1 ascended the

giant branch, it overflowed its Roche lobe and gas began to flow onto star 2. This had the effect of reducing the mass of star 1 and increasing that of star 2, which in turn caused

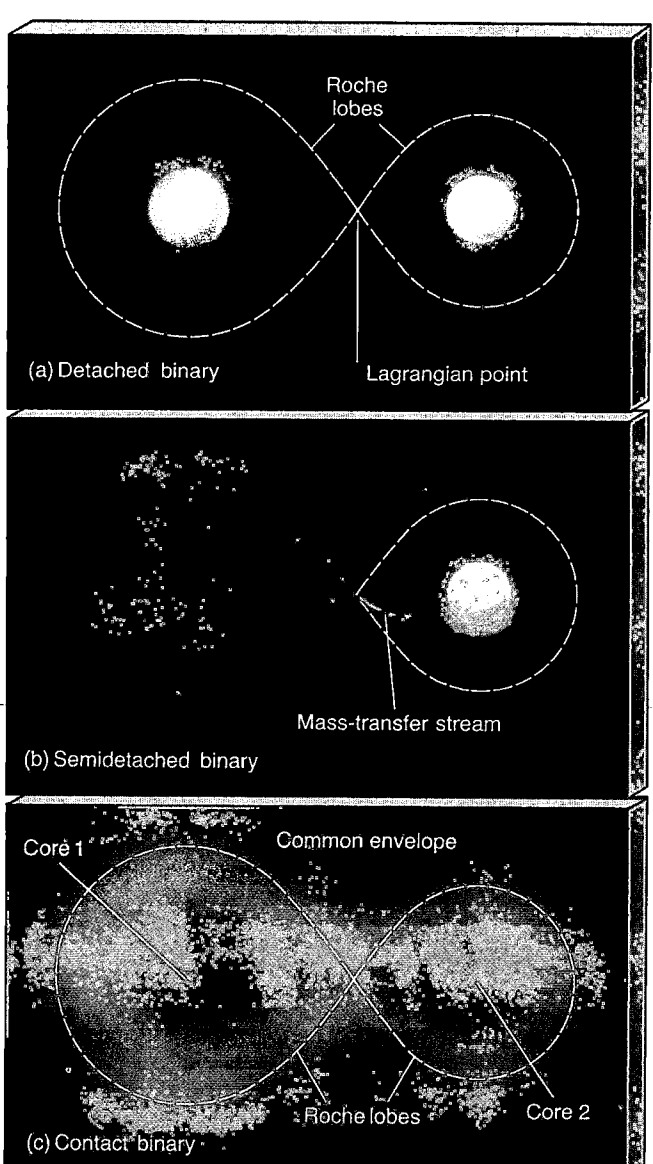

Figure 20.22 Close Binary-Star Systems (a) In a detached binary, each star lies within its respective Roche lobe. (b) In a semidetached binary, one of the stars fills its Roche lobe and transfers matter onto the other, which still lies within its own Roche lobe. (c) In a contact or common-envelope binary, both stars have overflowed their Roche lobes, and a single star with two distinct nuclear-burning cores results.

the Roche lobe of star 1 to shrink as its gravity decreased. As a result, the rate at which star 1 overflowed its Roche lobe increased, and a period of unstable *rapid mass transfer* ensued, transporting most of star 1's envelope onto star 2. Eventually, the mass of star 1 became less than that of star 2. Detailed calculations show that the rate of mass transfer dropped sharply at that point, and the stars entered the relatively stable state we see today. These changes in Algol's components are illustrated in Figure 20.23.

Being part of a binary system has radically altered the evolution of both stars in the Algol system. The original high-mass star 1 is now a low-mass red subgiant, while the roughly solar mass star 2 is now a massive blue-giant main-sequence star. The removal of mass from the envelope of star 1 may prevent it from ever reaching the helium flash. Instead, its naked core may eventually be left behind as a *helium white dwarf.* In a few tens of millions of years, star 2 will itself begin to ascend the giant branch and fill its own Roche lobe. If star 1 is still a subgiant or a giant at that time, a contact binary system will result. If, instead, star 1 has by then become a white dwarf, a new mass-transferring period—with matter streaming from star 2 back onto star 1—will begin. In that case (as we will see), Algol may have a very active and violent future in store.

Just as molecules exhibit few of the physical or chemical properties of their constituent atoms, binaries can display types of behavior that are quite different from either of their component stars. The Algol system is a fairly simple example of binary evolution, yet it gives us an idea of the sorts of complications that can arise when two stars evolve interdependently. A significant fraction of all the binary stars in the Galaxy will pass through some sort of mass-transfer or common-envelope phase. In this chapter we have seen one possible result of mass transfer involving main-sequence stars. We will return to this subject in the next two chapters, when we continue our discussion of stellar evolution and the strange states of matter that may result.

☑ Concept Check

■ Why is it important to understand the evolution of binary stars?

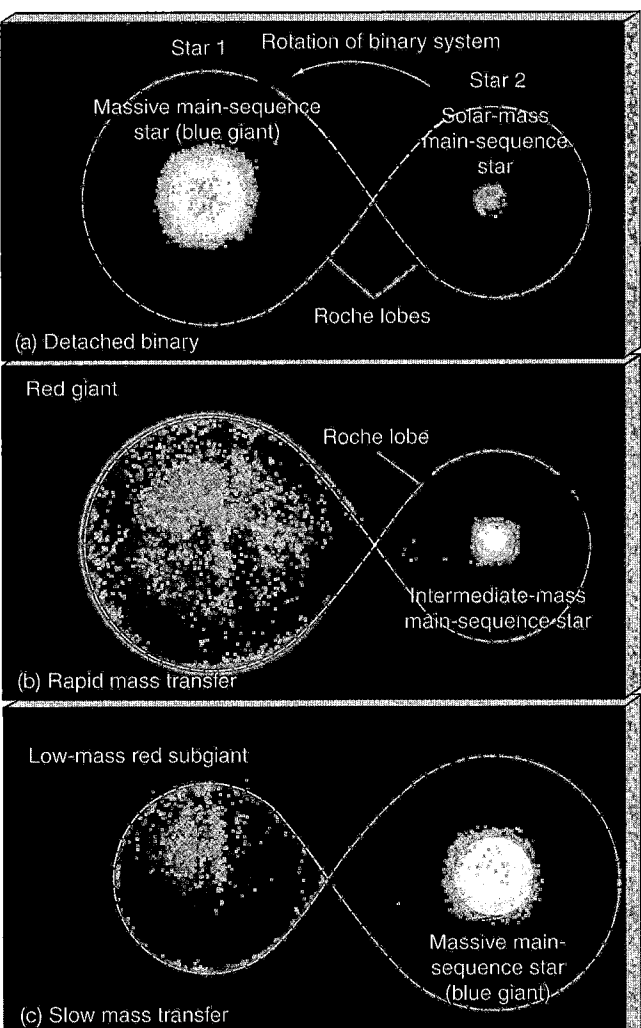

Figure 20.23 Algol Evolution The evolution of the binary star Algol. (a) Initially, Algol was probably a detached binary made up of two main-sequence stars—a relatively massive blue giant and a less massive companion similar to the Sun. (b) As the more massive component (star 1) evolved off the main sequence, it expanded to fill and eventually overflow its Roche lobe, transferring large amounts of matter onto its smaller companion (star 2). (c) Today, star 2 is the more massive of the two, but it is on the main sequence. Star 1 is still in the subgiant phase and fills its Roche lobe, causing a steady stream of matter to pour onto its companion.

Chapter Review

SUMMARY

Stars spend most of their lives on the main sequence, in the **core-hydrogen burning** (p. 516) phase of stellar evolution, stably fusing hydrogen into helium at their centers. Stars leave the main sequence when the hydrogen in their cores is exhausted. For the Sun, which is about halfway through its main-sequence lifetime, this stage will occur about 5 billion years from now. Low-mass stars evolve much more slowly than the Sun, and high-mass stars evolve much faster.

When the central nuclear fires in the interior of a solar-mass star cease, the helium in the star's core is still too cool to fuse into anything heavier. With no internal energy source, the helium core is unable to support itself against its own gravity and begins to shrink. The star at this stage is in the **hydrogen-shell burning** (p. 519) phase, in which the nonburning helium at the center is surrounded by a layer of burning hydrogen. The energy released by the contracting helium core heats the hydrogen-burning shell, greatly increasing the nuclear reaction rates there. As a result, the star becomes much brighter while the envelope expands and cools. A low-mass star like the Sun moves off the main sequence on the H–R diagram first along the **subgiant branch** (p. 519), then almost vertically up the **red-giant branch** (p. 519). As the helium core contracts, it heats up. Eventually, the core of a star more than 0.25 times the mass of the Sun reaches the point at which helium begins to fuse into carbon. The net effect of the fusion reactions is that three helium nuclei (or alpha particles) combine to form a nucleus of carbon in the **triple-alpha process** (p. 520). In a star like the Sun, conditions at the onset of helium burning are such that the electrons in the core have become degenerate—they can be thought of as tiny, hard spheres that once brought into contact, present stiff resistance to being compressed any further. This **electron degeneracy pressure** (p. 521) makes the core unable to "react" to the new energy source, and helium burning begins violently in the **helium flash** (p. 522). The flash expands the core and reduces the star's luminosity, sending it onto the **horizontal branch** (p. 522) of the H–R diagram. The star now has a core of burning helium surrounded by a shell of burning hydrogen.

As helium burns in the core, it forms an inner core of nonburning carbon. The carbon core shrinks and heats the overlying burning layers, and the star once again becomes a red giant. It reenters the red-giant region of the H–R diagram along the **asymptotic-giant branch** (p. 523), becoming an extremely luminous red-supergiant star. The core of a low-mass star never becomes hot enough to fuse carbon. Such a star continues to ascend the asymptotic-giant branch until its envelope is ejected into space as a **planetary nebula** (p. 525). At that point the core becomes visible as a hot, faint, and extremely dense white dwarf. The planetary nebula diffuses into space, carrying helium and some carbon into the interstellar medium. The white dwarf cools and fades, eventually becoming a cold black dwarf. Most white dwarfs are composed of carbon and oxygen, although stars in binary systems may give rise to helium white dwarfs, while more massive stars may become neon-oxygen white dwarfs.

Evolutionary changes happen more rapidly for high-mass stars than for low-mass stars because larger mass results in higher central temperatures. High-mass stars do not experience a helium flash and attain central temperatures high enough to fuse carbon. They form heavier and heavier elements in their cores, at a more and more rapid pace, and eventually die explosively.

The theory of stellar evolution can be tested by observing star clusters, all of whose stars formed at the same time. As time goes by, the most massive stars evolve off the main sequence first, then the intermediate-mass stars, and so on. At any instant, no stars with masses above the cluster's **main-sequence turnoff** (p. 533) mass remain on the main sequence. Stars below this mass have not yet evolved into giants and so still lie on the main sequence. By comparing a particular cluster's main-sequence turnoff mass with theoretical predictions, astronomers can measure the age of the cluster.

Stars in binary systems can evolve quite differently from isolated stars because of interactions with their companions. Each star is surrounded by a teardrop-shaped **Roche lobe** (p. 535), which defines the region of space within which matter "belongs" to the star. As a star in a binary evolves into the giant phase it may overflow its Roche lobe, forming a **mass-transfer binary** (p. 536) as gas flows from the giant onto its companion. If both stars overflow their Roche lobes, a **contact binary** (p. 536) results. Stellar evolution in binaries can produce states not achievable in single stars. In a sufficiently wide binary, both stars evolve as though they were isolated.

SELF-TEST: TRUE OR FALSE?

_____ **1.** "High-mass" stars are conventionally taken to have masses greater than that of the Sun.

_____ **2.** All the red-dwarf stars that ever formed are still on the main sequence today.

_____ **3.** Once on the main sequence, gravity is no longer important in determining a star's internal structure.

_____ **4.** The Sun will get brighter as it begins to run out of fuel in its core.

_____ **5.** As a star evolves away from the main sequence, it gets larger.

_____ **6.** As a star evolves away from the main sequence, it gets hotter.

_____ **7.** As a red giant, the Sun will have a core that is smaller than it was when the Sun was on the main sequence.

_____ **8.** When helium starts to fuse inside a solar-mass red giant, it does so slowly at first, and the rate of fusion increases gradually over many years.

_____ **9.** With the onset of helium fusion, a red giant gets brighter.

____ **10.** A planetary nebula is the disk of matter around a star that will eventually form a planetary system.

____ **11.** For a high-mass star, there is no helium flash.

____ **12.** High-mass stars can fuse carbon and oxygen in their cores.

____ **13.** A star cluster with an age of 500 million years will contain many O-type stars.

____ **14.** In a binary-star system, it is never possible for the lower-mass star to be more evolved than the higher-mass companion.

____ **15.** In a mass-transfer binary, one of the stars has filled its Roche lobe.

SELF-TEST: FILL IN THE BLANK

1. A main-sequence star doesn't collapse because of the outward _____ produced by hot gases in the stellar interior.

2. The Sun will leave the main sequence in about _____ years.

3. While a star is on the main sequence _____ is slowly depleted and _____ builds up in the core.

4. A temperature of at least _____ is needed to fuse helium.

5. At the end of its main-sequence lifetime, a star's core starts to _____.

6. When helium fuses, it produces _____ and releases _____.

7. Just before helium fusion begins in the Sun, the core's outward pressure will be provided mainly by electron _____.

8. As a star ascends the asymptotic-giant branch, its _____ core is shrinking in size and growing in mass.

9. As a red supergiant the Sun will eventually become about _____ times its present size.

10. The various stages of stellar evolution predicted by theory can be tested using observations of stars in _____.

11. By the time the envelope of a red supergiant is ejected, the core has shrunk to a diameter of about _____.

12. A typical white dwarf has the following properties: about half a solar mass, fairly _____ surface temperature, small size, and _____ luminosity.

13. As time goes by, the temperature and the luminosity of a white dwarf both _____.

14. As a star cluster ages, the luminosity of the main-sequence turnoff _____.

15. Whether being a member of a binary-star system will affect the evolution of a star depends largely on the _____ of the two stars in the binary.

REVIEW AND DISCUSSION

1. Why don't stars live forever? Which types of stars live the longest?

2. What is hydrostatic equilibrium?

3. How long can a star like the Sun keep burning hydrogen in its core?

4. Why is the depletion of hydrogen in the core of a star such an important event?

5. What makes an ordinary star become a red giant?

6. Roughly how big (in A.U.) will the Sun become when it enters the red-giant phase?

7. How long does it take for a star like the Sun to evolve from the main sequence to the top of the red-giant branch?

8. Do all stars eventually fuse helium in their cores?

9. What is the helium flash?

10. Describe an important way in which winds from red-giant stars are linked to the interstellar medium.

11. How do the late evolutionary stages of high-mass stars differ from those of low-mass stars?

12. What is the internal structure of a low-mass red supergiant?

13. What is a planetary nebula? Why do many planetary nebulae appear as rings?

14. What are white dwarfs? What is their ultimate fate?

15. Can you think of a way in which a helium white dwarf might exist today?

16. Why are white dwarfs hard to observe?

17. Do many black dwarfs exist in the Galaxy?

18. How can astronomers measure the age of a star cluster?

19. What are the Roche lobes of a binary system?

20. Why is it odd that the binary system Algol consists of a low-mass red giant orbiting a high-mass main-sequence star? How did Algol come to be in this configuration?

PROBLEMS

Algorithmic versions of these questions are available in the Practice Problems module of the Companion Website.

The number of squares preceding each problem indicates its approximate level of difficulty.

1. ■ The Sun will evolve off the main sequence when roughly 10 percent of its hydrogen has been fused into helium. Using the data given in Section 16.5 and Table 16.2, calculate the total amount of mass destroyed (that is, converted into energy) and the total energy released by the fusion of this amount of matter.

2. ■ Use the radius–luminosity–temperature relation to calculate the radius of a red supergiant with temperature 3000 K (half the solar value) and luminosity 10,000 solar luminosities. ∞ (Sec. 17.3) How many planets of our solar system would this star engulf?

3. ■ What would be the luminosity of the Sun if its surface temperature was 3000 K and its radius was (a) 1 A.U., (b) 5 A.U.?

4. ■ Use the radius–luminosity–temperature relation to calculate the radius of a 12,000 K (twice the temperature of the Sun), 0.0004-solar-luminosity white dwarf.

5. ■ Use the graph in *More Precisely 20-1* to estimate the factor by which CNO energy production outstrips proton–proton energy production in a 10-solar-mass star with a central temperature of 25 million K. What do you think the factor would be if the abundances of C, N, and O were just one-tenth the solar value?

6. ■■ A main-sequence star at a distance of 20 pc is barely visible through a certain telescope. The star subsequently ascends the giant branch, during which time its temperature drops by a factor of three and its radius increases 100-fold. What is the new maximum distance at which the star would still be visible using the same telescope?

7. ■ A Sunlike star goes through a rapid luminosity change between stages 8 and 9, when the luminosity increases by about a factor of 100 in 10^5 years. On average, how rapidly does the star's absolute magnitude change, in magnitudes per year? Do you think this change would be noticeable in a distant star within a human lifetime?

8. ■ Calculate the average density of a red-giant core of 0.25 solar mass and radius 15,000 km. Compare this with the average density of the giant's envelope, if it has a 0.5 solar mass and its radius is 0.5 A.U. Compare each with the central density of the Sun. ∞ (Sec. 16.2)

9. ■ How long will it take the Sun's planetary nebula, expanding at a speed of 50 km/s, to reach the orbit of Neptune? How long to reach the nearest star?

10. ■ What are the escape speed (in km/s) and surface gravity (relative to Earth's gravity) of Sirius B? (See Table 20.2.)

11. ■ A 15-solar-mass blue supergiant with a surface temperature of 20,000 K becomes a red supergiant with the same total luminosity and a temperature of 4000 K. By what factor does its radius change?

12. ■■ The radius of Betelgeuse varies by about 60 percent within a period of three years. If the star's surface temperature remains constant, by how much does its absolute magnitude change during this time?

13. ■■ The Sun will reside on the main sequence for 10^{10} years. If the luminosity of a main-sequence star is proportional to the fourth power of the star's mass, what mass star is just now leaving the main sequence in a cluster that formed (a) 400 million years ago, (b) 2 billion years ago?

14. ■■■ In roughly 5 billion years, the Sun will eject its envelope as a planetary nebula. Before then, suppose it loses 20 percent of its mass on the giant branch. (a) If Jupiter stays in a circular orbit while this mass is being lost and its angular momentum stays constant, what will be the planet's orbital radius and period when the Sun's mass has fallen to 0.8 solar masses? (b) When the planetary nebula is formed, the Sun loses a further 0.3 solar masses rapidly enough that we can regard the loss as immediate. Jupiter's instantaneous velocity is unchanged, but the orbit is no longer circular, due to the Sun's smaller mass. The planet's location at that moment becomes the perihelion of the new orbit, and the orbital semimajor axis increases (in this case) by a factor of 2.5. What are the eccentricity and period of the new orbit?

15. ■■■ From the discussions presented in *More Precisely 2-3* and *More Precisely 15-1*, it may be shown that the angular momentum of a circular binary system, of separation r and component masses m_1 and m_2, is proportional to $m_1 \times m_2 \times \sqrt{r}$ (if the total mass is constant). Such a binary has component masses one and two times the mass of the Sun, respectively, and an orbital period of two years. (a) What is its orbital separation r? (b) Mass transfer moves 0.2 solar masses of material from the more massive to the less massive star, keeping the total mass of the system fixed and conserving angular momentum. If the binary remains circular, calculate its new separation and orbital period.

COLLABORATIVE EXERCISES

1. Evolutionary Sequences. As a plot of luminosity vs. temperature, the H–R diagram is useful for describing how stars evolve over time even though "time" is not the label on either axis. As a group, create an imaginary graph of "dollars of financial income" (vertical axis) vs. "weight" (horizontal axis) and use it to describe the past and future life cycle of one of your group members. Clearly label your diagram and provide a figure caption clearly explaining each life phase.

RESEARCHING ON THE WEB  *To complete the following exercises, go to the online Destinations module for Chapter 20 on the Companion Website for* Astronomy Today *4/e.*

1. Access the "Story of a Rarely Seen Stellar Explosion" page and describe what a Sakurai's object is, where it is located in the sky, and how it relates to the evolutionary future of our Sun.

2. Access the "Hertzsprung–Russell Diagram" page and define the four luminosity classes shown on the H–R Diagram.

PROJECTS

1. Can you find the Hyades cluster? It lies about 46 pc away in the constellation Taurus, making up the "face" of the bull. It appears to surround the very bright star Aldebaran, the Bull's eye, which makes it easy to locate in the sky. Aldebaran is a red giant, probably on the asymptotic-giant branch of its evolution. Despite appearances, it is not part of the Hyades cluster. In fact, it lies only about half as far away—some 20 pc from Earth.

2. Now look for the Double Cluster in Perseus, h and chi Persei. These two young clusters probably formed together,

and now move together through space. They lie about 2500 pc away, and are barely visible to the naked eye just east of the "W" of Cassiopeia.

3. Find a library that has the *Astrophysical Journal*. Find an article from the late 1950s and 1960s that gives the photometry of a star cluster like the Pleiades or Hyades. Plot a color–magnitude diagram (V vs. B–V; see Section 17.6). Determine the V magnitude of the main-sequence turnoff, and hence estimate the age of the cluster. Compare your age with that given in the article.

SKYCHART III PROJECTS  *The SkyChart III Student Version planetarium program on which these exercises are based is included as a separately executable program on the CD in the back of this text.*

1. ■■ An H–R diagram in Figure 17.14 relates the surface temperature of a star to its luminosity. In this and the following exercise you will generate an H–R diagram using spreadsheet software. SkyChart does not give us the temperature, so use the color index (see exercise 2 in Chapter 3) as its proxy. You will also need to use the absolute magnitude. SkyChart gives the apparent magnitude of stars and it can be converted to an absolute magnitude if you know the distance. Pick a region in the sky with plenty of stars (a field of about 10° should be sufficient). Select 50 stars and record their magnitude, color index (B–V), and distance in pc. Some information may be missing on a particular star, so just skip it and use another. The first manipulation will be constructing a graph of distance versus apparent magnitude. Looking at your graph, can you identify a general trend? Is there more than one trend at work? Common sense tells us that stars farther away will appear less bright. In fact, the luminosity will decrease as $1/D^2$, where D is the distance to the star, so a star twice as far away will appear one-fourth as bright. Does your graph support this? If it does not, can you give reasons why? What other factors besides distance play a role in a star's apparent magnitude?

2. ■■ Now generate the H–R diagram. To do this, you will have to convert the apparent magnitude, m, to an absolute magni-

tude, M, using the following equation: $M = 5 - 5\log(D) + m$, where D is the distance to the star measured in parsecs, pc. We can easily handle this equation using spreadsheet software (such as Excel). Enter in column A the name of the star, in column B the color index, in column C the distance, and in column D the magnitude. In the cell E2, enter the following expression: $= 5-5*LOG10(C2) + D2$. Select E2 again and select *Copy* from the *Edit* menu. Now highlight E3 through to the end of your data and select *Paste* from the *Edit* menu. You have just figured out the absolute magnitudes for all your data points. The last calculation that needs to be done is to determine the luminosities of the stars from their absolute magnitudes according to: $L = 2.512^{(M_o - M)}$, where L is the luminosity in solar units, M is the absolute magnitude and M_o is the absolute magnitude of the Sun, which is 4.78. In cell F2, enter the equation: $= 2.512^{(4.78 - E2)}$. Now *Copy* the formula from G2 and *Paste* it to cells G3 to the end of the data. Now generate a graph of luminosity versus color index, with the vertical axis on a logarithmic scale. What is the coolest star you recorded? How luminous is it? What is the least luminous star you recorded? Estimate how hot it is. What is the most luminous star you recorded? Estimate how hot it is. How does your H–R diagram compare to the various H–R diagrams in this chapter?

 *In addition to the Practice Problems and Destinations modules, the Companion Website at http://www.prenhall.com/chaisson provides for each chapter an additional true-false, multiple choice, and labeling quiz, as well as additional annotated images, animations, and links to related Websites.*

21 STELLAR EXPLOSIONS

Novae, Supernovae, and the Formation of the Heavy Elements

LEARNING GOALS

Studying this chapter will enable you to:

1 Explain how white dwarfs in binary-star systems can become explosively active.

2 Summarize the sequence of events leading to the violent death of a massive star.

3 Describe the two types of supernovae, and explain how each is produced.

4 Describe the observational evidence that supernovae have occurred in our Galaxy.

5 Explain the origin of elements heavier than helium, and discuss the significance of these elements for the study of stellar evolution.

6 Outline how the universe continually recycles matter through stars and the interstellar medium.

 Visit http://www.prenhall.com/chaisson for additional annotated images, animations, and links to related sites for this chapter.

This spectroscopic X-ray image of the supernova remnant Cassiopeia A was taken by the *Chandra* observatory in Earth orbit. The red, green, and blue regions show where the intensity of low-, medium-, and high-energy X rays, respectively, is greatest. The red matter on the left outer edge is enriched in iron, whereas the bright greenish-white region on the lower left is enriched in silicon and sulfur. In the blue region on the right edge, low- and medium-energy X rays have been filtered out by a cloud of dust and gas within the remnant. *(CXO/SAO)*

The Big Picture: All elements heavier than hydrogen and helium were created in the hearts of stars. Elements heavier than iron, in fact, were synthesized during the explosive deaths of massive stars—in supernovae, which astronomers have observed in many locations across the sky. For some of us, there is something philosophically pleasing about the idea that the deaths of some stars cause the births of others, at the same time creating many of the elements that compose our own world.

What fate awaits a star when it runs out of fuel? For a low-mass star, the white-dwarf stage is not necessarily the end of the road. The potential exists for further violent activity if a binary companion can provide additional fuel. High-mass stars—whether or not they are members of binaries—are also destined to die explosively, releasing vast amounts of energy, creating many heavy elements, and scattering the debris throughout interstellar space. These cataclysmic explosions may trigger the formation of new stars, continuing the cycle of stellar birth and death. In this chapter we will study in more detail the processes responsible for these cataclysmic explosions and the mechanisms that create heavy elements from which we ourselves are made.

21.1 Life after Death for White Dwarfs

¶ Although most stars shine steadily day after day, year after year, some change dramatically in brightness over very short periods of time. One type of star, called a **nova** (plural: *novae*), may increase enormously in brightness—by as much as a factor of 10,000 or more—in a matter of days.

The word *nova* means "new" in Latin, and to early observers these stars did indeed seem new, because they appeared suddenly in the night sky. Astronomers now recognize that a nova is not a new star at all. It is instead a white dwarf—a normally very faint star—undergoing an explosion on its surface that results in a rapid, temporary increase in luminosity. Figure 21.1 illustrates the brightening of a typical nova. Figure 21.2 shows a nova light curve, demonstrating how the luminosity rises dramatically in a matter of days, then fades slowly back to normal over the course of several months. On average, two or three novae

are observed each year. Astronomers also know of many *recurrent novae*—stars that have been observed to "go nova" several times over the course of a few decades.

What could cause such an explosion on a faint, dead star? The energy involved is far too great to be explained by flares or other surface activity, and as we saw in the previous chapter, there is no nuclear activity in the dwarf's interior. ∞ (Sec. 20.3) To understand what happens, we must reconsider the fate of a low-mass star after it enters the white-dwarf phase.

We noted in Chapter 20 that the white-dwarf stage represents the end point of a star's evolution. Subsequently, the star simply cools, eventually becoming a black dwarf—a burned-out ember in interstellar space. This scenario is quite correct for an *isolated* star, such as our Sun. However, should the star be part of a *binary* system, an important new possibility exists. If the distance between the two stars is small enough, then the dwarf's tidal gravitational field can pull matter—primarily hydrogen and helium—away from the surface of its main-sequence or giant companion, as illustrated in Figure 21.3. The system becomes a mass-transferring binary, similar to those discussed in Chapter 20. A stream of gas leaves the companion through the inner (L_1) Lagrangian point and flows onto the dwarf. ∞ (Sec. 20.6)

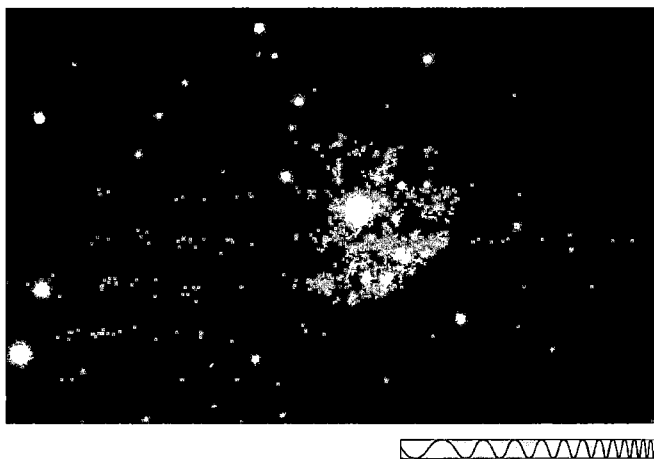

Figure 21.1 Nova A nova is a star that suddenly increases enormously in brightness, then slowly fades back to its original luminosity. Novae are the result of explosions on the surfaces of faint white-dwarf stars, caused by matter falling onto their surfaces from the atmosphere of a larger binary companion. Shown here is Nova Persei some 50 years after having suddenly brightened by a factor of 40,000 in 1901. *(Palomar Observatory/Caltech)*

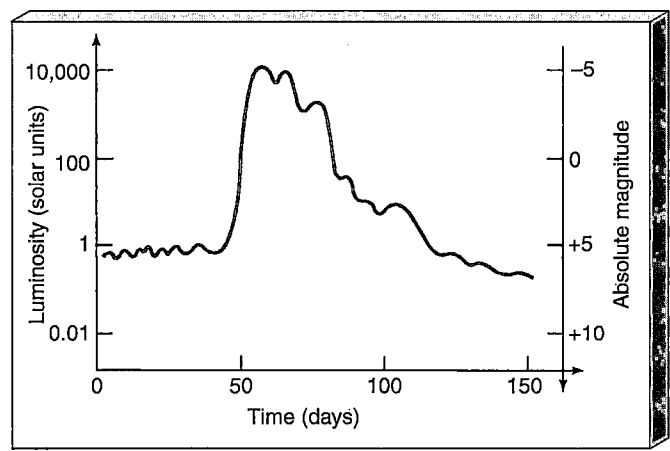

Figure 21.2 Nova Light Curve The light curve of a typical nova. The rapid rise and slow decline in the light received from the star, as well as the maximum brightness attained, are in good agreement with the explanation of the nova as a nuclear flash on a white dwarf's surface.

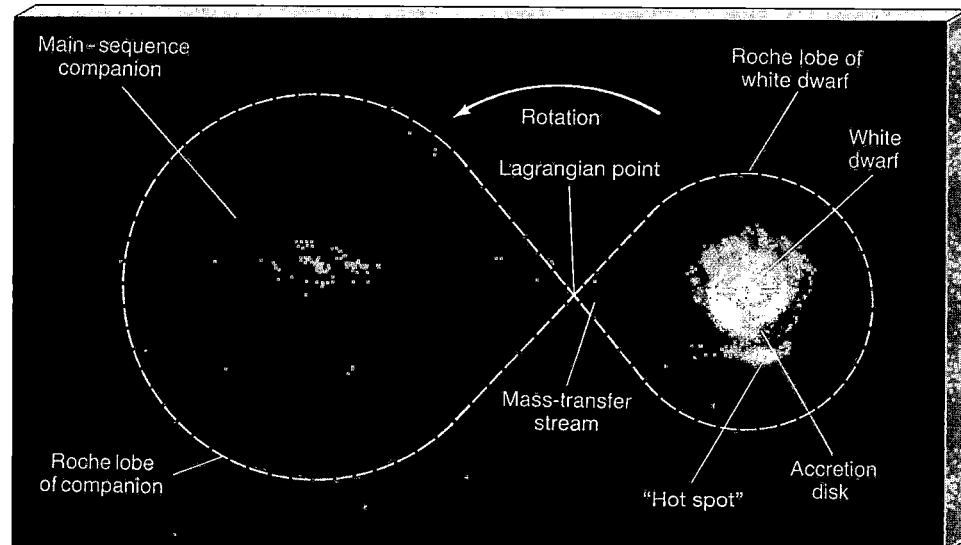

Figure 21.3 Close Binary System A white dwarf in a semidetached binary system may be close enough to its companion (in this case, a main sequence star) that its gravitational field can tear material from the companion's surface. Compare with Figure 20.21. Notice that unlike in the earlier figure, the matter does not fall directly onto the white dwarf's surface. Instead, as discussed a little later in the text, it forms an "accretion disk" of gas spiraling down onto the dwarf.

As it builds up on the white dwarf's surface, the stolen gas becomes hotter and denser. Eventually its temperature exceeds 10^7 K, and the hydrogen ignites, fusing into helium at a furious rate (Figures 21.4a–d). This surface-burning stage is as brief as it is violent. The star suddenly flares up in luminosity then fades away as some of the fuel is exhausted and the remainder is blown off into space. If the event happens to be visible from Earth, we see a nova. Figure 21.4(e) is a photograph of a nova apparently caught in the act of expelling mass from its surface. The nova's decline in brightness results from the expansion and cooling of the dwarf's surface layers as they are blown into space. Studies of the details of these curves provide astronomers with a wealth of information about both the dwarf and its binary companion.

Because of the binary's rotation, material leaving the companion does not fall directly onto the dwarf. Instead, it "misses" the compact star, loops around behind it, and goes into orbit around it, forming a swirling, flattened disk of matter called an **accretion disk** (shown in Figure 21.3). Due to the effects of viscosity (that is, friction) within the gas, the orbiting matter in the disk drifts gradually inward, its temperature increasing steadily as it spirals down onto the dwarf's surface. The inner part of the accretion disk becomes so hot that it radiates strongly in the visible, the ultraviolet, and even the X-ray portions of the electromagnetic spectrum. In many systems the disk outshines the white dwarf itself and is the main source of the light emitted between nova outbursts. X rays from the hot disk are routinely observed in many galactic novae. The point at which the infalling stream of matter strikes the accretion disk often forms a turbulent "hot spot," causing detectable fluctuations in the light emitted by the binary system.

A nova represents one way in which a star in a binary system can extend its "active lifetime" well into the white-dwarf stage. Recurrent novae can, in principle, repeat their violent outbursts many dozens, if not hundreds, of times.

But even more extreme possibilities exist at the end of stellar evolution. Vastly more energetic events may be in store, given the right circumstances.

✔️ Concept Check

■ Will the Sun ever become a nova?

21.2 The End of a High-Mass Star

📐 A low-mass star—one with a mass of less than about 8 solar masses—never becomes hot enough to burn carbon in its core. It ends its life as a carbon-oxygen (or possibly neon-oxygen) white dwarf. ∞ (Sec. 20.3) A high-mass star, however, can fuse not just hydrogen and helium but also carbon, oxygen, and even heavier elements as its inner core continues to contract and its central temperature continues to rise. ∞ (Sec. 20.4) The burning rate accelerates as the core evolves. Can anything stop this runaway process? Is there a stable "white-dwarf-like" state at the end of the evolution of a high-mass star? What is its ultimate fate? To answer these questions, we must look more carefully at fusion in massive stars.

FUSION OF HEAVY ELEMENTS

Figure 21.5 is a cutaway diagram of the interior of a highly evolved star of large mass. Note the numerous layers where various nuclei burn. As the temperature increases with depth, the ash of each burning stage becomes the fuel for the next stage. At the relatively cool periphery of the core, hydrogen fuses into helium. In the intermediate layers, shells of helium, carbon, and oxygen burn to form heavier nuclei. Deeper down reside neon, magnesium, silicon, and other heavy nuclei, all produced by nuclear fusion

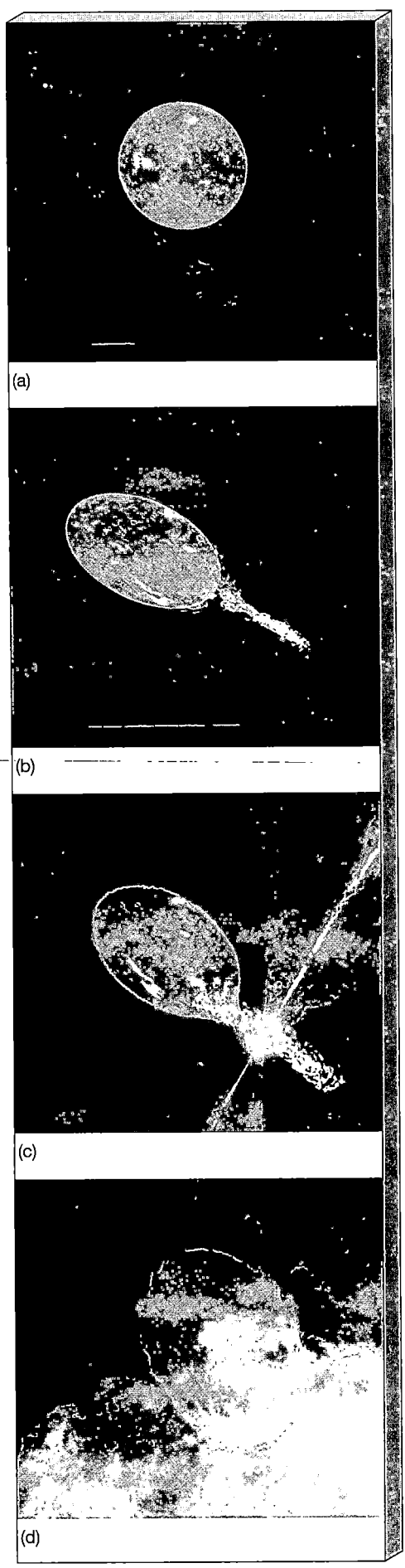

(a)

(b)

(c)

(d)

in the layers overlying the core. The core itself is composed of iron. We will study the key reactions in this burning chain in more detail later in this chapter.

As each element is burned to depletion at the center, the core contracts, heats up, and starts to fuse the ash of the previous burning stage. A new inner core forms, contracts again, heats again, and so on. Through each period of stability and instability, the star's central temperature increases, the nuclear reactions speed up, and the newly released energy supports the star for ever-shorter periods of time. For example, in round numbers, a star 20 times more massive than the Sun burns hydrogen for 10 million years, helium for 1 million years, carbon for 1000 years, oxygen for 1 year, and silicon for a week. Its iron core grows for less than a day.

COLLAPSE OF THE IRON CORE

Once the inner core begins to change into iron, our high-mass star is in trouble. As illustrated in Figure 21.6, iron is the *most stable* element there is. To understand the figure, imagine fusing four protons to form helium-4. According to the figure, the mass per particle of a helium-4 nucleus is less than the mass of a proton, so mass is lost and (in accordance with the law of conservation of mass and energy) energy is released. ∞ (Sec. 16.5) Similarly, combining three helium-4 nuclei to form carbon results in a net mass loss, again releasing energy. In other words, the left side of the figure shows how light elements can fuse to release energy. The right side of the figure shows the opposite process, known as *fission*. Here, combining nuclei will increase the total mass per particle and hence absorb energy, so fusion can't occur. However, splitting a nucleus into lighter nuclei does release energy—this is how nuclear reactors and bombs work.

Iron lies at the dividing line between these two types of behavior, at the lowest point of the curve in Figure 21.6. Iron nuclei are so compact that energy cannot be extracted

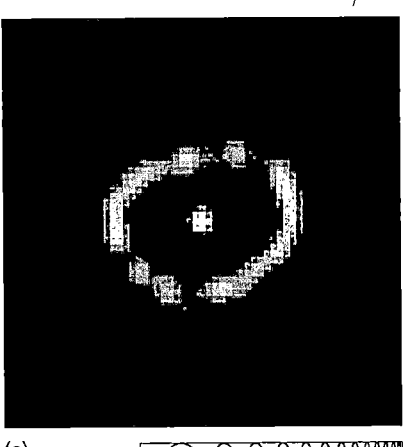

(e)

Figure 21.4 Nova Explosion In this artist's conception, a white-dwarf star (upper left) orbits a cool red giant (a); material accumulates on the white dwarf's surface after being accreted from the companion star (b and c) and then ignites in hydrogen fusion as a nova outburst (d). The real photo in (e) corresponds roughly to the events depicted by the artist at the bottom right of frame (d). This nova is called Nova Cygni (1992) and was imaged here with a European camera on the Hubble Space Telescope. *(D. Berry; NASA/ESA)*

R   I   V   U   X   G

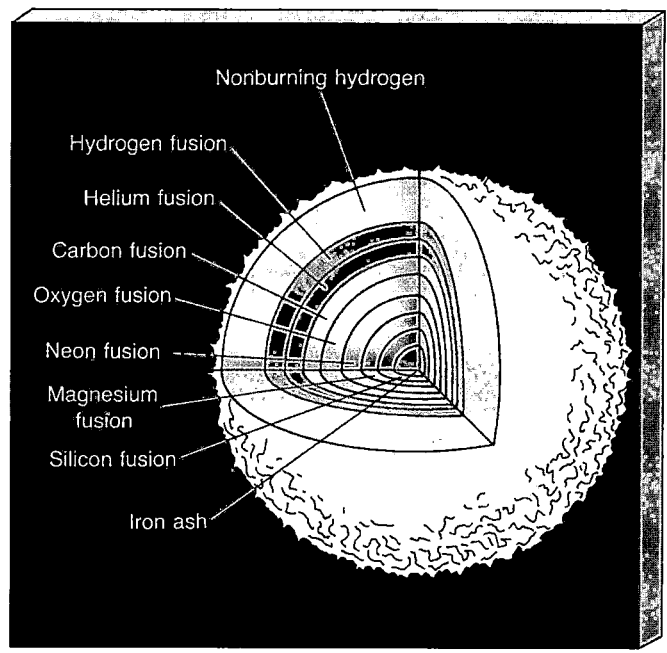

Figure 21.5 Heavy Element Fusion Cutaway diagram of the interior of a highly evolved star of mass greater than 8 solar masses (not to scale). The interior resembles the layers of an onion, with shells of progressively heavier elements burning at smaller and smaller radii and at higher and higher temperatures.

by combining them into heavier elements or by splitting them into lighter ones. In effect, iron plays the role of a fire extinguisher, damping the inferno in the stellar core. With the appearance of substantial quantities of iron, the central fires cease for the last time, and the star's internal support begins to dwindle. The star's foundation is destroyed, and its equilibrium is gone forever. Even though the temperature in the iron core has reached several billion kelvins by this stage, the enormous inward gravitational pull of matter ensures catastrophe in the very near future. Gravity overwhelms the pressure of the hot gas, and the star implodes, falling in on itself.

The core temperature rises to nearly 10 billion K. At these temperatures, individual photons, according to Wien's law, have tremendously high energies, enough to split iron into lighter nuclei and then to break those lighter nuclei apart until only protons and neutrons remain. ∞ (Sec. 3.4) This process is known as *photodisintegration* of the heavy elements in the core. In less than a second, the collapsing core undoes all the effects of nuclear fusion that occurred during the previous 10 million years! But to split iron and lighter nuclei into smaller pieces requires a lot of energy (Figure 21.6, moving from iron to the left). After all, this splitting is just the opposite of the fusion reactions that generated the star's energy during earlier times. Photodisintegration *absorbs* some of the core's thermal energy—in other words, it cools the core and so reduces the

pressure. As nuclei are destroyed, the core of the star becomes even less able to support itself against its own gravity. The collapse accelerates.

Now the core consists entirely of simple elementary particles—electrons, protons, neutrons, and photons—at enormously high densities, and it is still shrinking. As the core density continues to rise, the protons and electrons are crushed together, forming neutrons and neutrinos:

$$p + e \rightarrow n + \text{neutrino}.$$

This process is sometimes called the *neutronization* of the core. Recall from our discussion in Chapter 16 that the neutrino is an extremely elusive particle that hardly interacts at all with matter. ∞ (Sec. 16.5) Even though the central density by this time may have reached 10^{12} kg/m³ or more, most of the neutrinos produced by neutronization pass through the core as if it weren't there. They escape into space, carrying away energy as they go, further reducing the core's pressure support.

The disappearance of the electrons and the escape of the neutrinos make matters even worse for the core's stability. There is now nothing to prevent it from collapsing

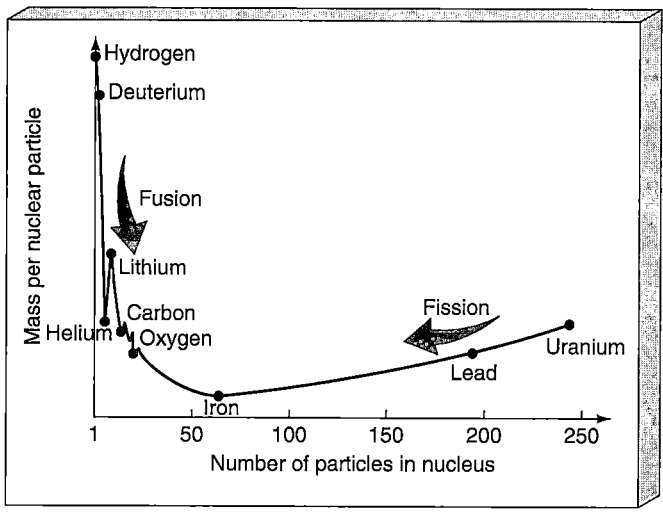

Figure 21.6 Nuclear Masses This graph shows how the masses (per nuclear particle—proton or neutron) of most known nuclei vary with nuclear mass. It contains an enormous amount of information about nuclear structure and stability. When light nuclei fuse (left side of the figure), the mass per particle decreases, and energy is released. ∞ (Sec. 16.5) Similarly, when heavy nuclei split apart (right side) the total mass again decreases, and energy is again released. The nucleus with the smallest mass per nuclear particle is iron. It can neither be fused nor split to release energy. The mass difference between a given nucleus and the value for hydrogen (dashed line) represents the amount of mass lost (or energy released) if free neutrons and protons were combined to form that nucleus—or equivalently, the amount of energy that must be provided to split the nucleus into its component particles. ∞ (*More Precisely 4-1*)

all the way to the point at which the neutrons come into contact with one another, at the incredible density of about 10^{15} kg/m^3. At this point the neutrons in the shrinking core play a role similar in many ways to that of the electrons in a white dwarf. When far apart, they offer little resistance to compression, but when brought into contact, they produce enormous pressures that strongly oppose further gravitational collapse. This *neutron degeneracy pressure*, akin to the electron degeneracy pressure that operates in red giants and white dwarfs, finally begins to slow the collapse. ∞ (Sec. 20.2) By the time the collapse is actually halted, however, the core has overshot its point of equilibrium, and may reach densities as high as 10^{17} or 10^{18} kg/m^3 before turning around and beginning to reexpand. Like a fast-moving ball hitting a brick wall, the core becomes compressed, stops, then rebounds—with a vengeance!

The events just described do not take long. Only about a second elapses from the start of the collapse to the "bounce" at nuclear densities. At that point the core rebounds. An enormously energetic shock wave sweeps through the star at high speed, blasting all the overlying layers—including all the heavy elements just formed outside the iron inner core—into space. Although computer models are still somewhat inconclusive and the details of how the shock reaches the surface and destroys the star remain uncertain, the end result is not. The star explodes, in one of the most energetic events known in the universe (see Figure 21.7). For a period of a few days the exploding star may rival in brightness the entire galaxy in which it resides. This spectacular death rattle of a high-mass star is known as a **core-collapse supernova**.

☑ Concept Check

■ Why does the iron core of a high-mass star collapse?

21.3 Supernova Explosions

NOVAE AND SUPERNOVAE

⑨ Let's compare a supernova with a nova. Like a nova, a **supernova** is a star that suddenly increases dramatically in brightness, then slowly dims again, eventually fading from view. The exploding star is commonly called the supernova's *progenitor*. In some cases, supernovae light curves can appear quite similar to those of novae, and a distant supernova can look a lot like a nearby nova—so much so, in fact, that the difference between the two was not fully appreciated until the 1920s. But novae and supernovae are now known to be very different phenomena. Supernovae are much more energetic events, driven by very different underlying physical processes.*

Well before they understood the causes of either novae or supernovae, astronomers knew of clear observational differences between them. The most important of these is that a supernova is more than a million times brighter than a nova. A supernova produces a burst of light billions of times brighter than the Sun, reaching that brightness within just a few hours after the start of the outburst. The total amount of electromagnetic energy radiated by a supernova during the few months it takes to brighten and fade away is roughly 10^{43} J—nearly as much energy as the Sun will radiate during its *entire* 10^{10}-year lifetime! (Enormous as this energy is, however, it pales in comparison with the energy emitted in the form of neutrinos, which may be 100 times greater.)

---

*Note that, when discussing novae and supernovae, astronomers tend to blur the distinction between the observed event (the sudden appearance and brightening of an object in the sky) and the process responsible for it (a violent explosion in or on a star). The two terms can have either meaning, depending on context.

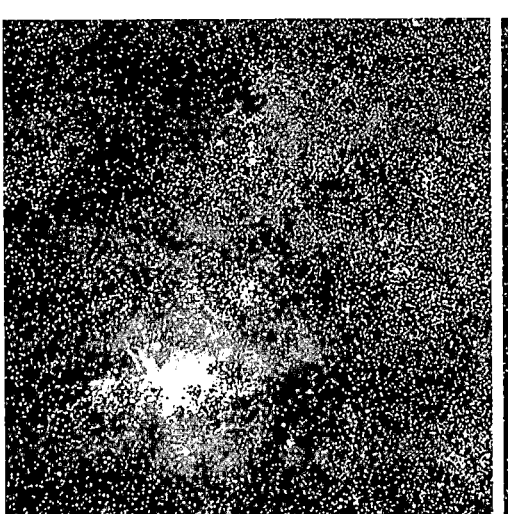 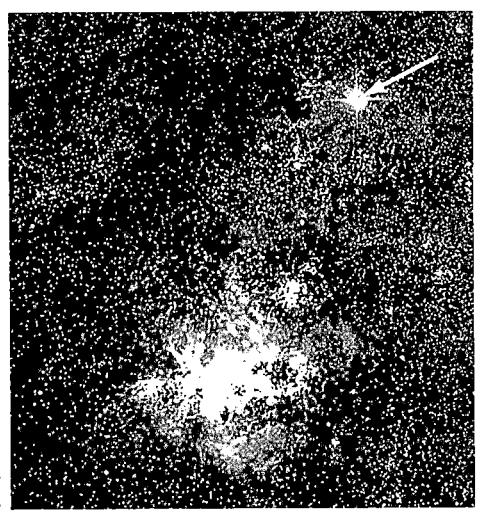

Figure 21.7 Supernova 1987A A supernova called SN 1987A (arrow) was exploding near this nebula (30 Doradus) at the moment the photograph on the right was taken. The photograph on the left is the prior appearance of the star field. (See *Discovery 21-1.*) *(AURA)*

A second important difference is that the same star may become a nova many times, but a star can become a supernova only once. This fact was unexplained before astronomers knew the precise nature of novae and supernovae, but it is easily understood now that we understand how and why these explosions occur. The nova accretion–explosion cycle described earlier can take place over and over again, but a supernova destroys the star involved, with no possibility of a repeat performance.

In addition to the distinction between novae and supernovae, there are also important observational differences *among* supernovae. Some supernovae contain very little hydrogen, according to their spectra, whereas others contain a lot. Also, the light curves of the hydrogen-poor supernovae are qualitatively different from those of the hydrogen-rich ones, as illustrated in Figure 21.8. Based on these observations, astronomers divide supernovae into two classes, known simply as Type I and Type II. **Type I supernovae**, the hydrogen-poor kind, have a light curve somewhat similar in shape to that of typical novae. **Type II supernovae**, whose spectra show lots of hydrogen, usually have a characteristic "plateau" in the light curve a few months after the maximum. Observed supernovae are divided roughly equally between these two categories.

CARBON-DETONATION SUPERNOVAE

What is responsible for these differences among supernovae? Is there more than one way in which a supernova explosion can occur? The answer is yes. To understand the alternative supernova mechanism, we must return to the processes that cause novae and consider the long-term consequences of their accretion–explosion cycle.

Nova explosions eject matter from a white dwarf's surface, but they do not necessarily expel or burn all the material that has accumulated since the last outburst. In other words, there is a tendency for the dwarf's mass to increase slowly with each new nova cycle. As its mass grows and the

internal pressure required to support its weight rises, the white dwarf can enter into a new period of instability—with disastrous consequences.

Recall that a white dwarf is held up not by thermal pressure (heat) but by the degeneracy pressure of electrons that have been squeezed so close together that they have effectively come into contact with one another. ∞ (Sec. 20.3) However, there is a limit to the pressure that these electrons can exert. Consequently, there is a limit to the mass of a white dwarf, above which electrons cannot provide the pressure needed to support the star. Detailed calculations show that the maximum mass of a white dwarf is about 1.4 solar masses, a mass often called the *Chandrasekhar mass*, after the Indian-American astronomer Subramanyan Chandrasekhar, whose work in theoretical astrophysics earned him a Nobel Prize in physics in 1983.

If an accreting white dwarf exceeds the Chandrasekhar mass, the pressure of degenerate electrons in its interior becomes unable to withstand the pull of gravity, and the star immediately starts to collapse. Its internal temperature rapidly rises to the point at which carbon can fuse into heavier elements. Carbon fusion begins everywhere throughout the white dwarf almost simultaneously, and the entire star explodes in another type of supernova—a so-called **carbon-detonation supernova**—comparable in violence to the "implosion" supernova associated with the death of a high-mass star, but very different in cause. In an alternative and (many astronomers think) possibly more common scenario, two white dwarfs in a binary system may collide and merge to form a massive, unstable star. The end result is the same—a carbon-detonation supernova.

We can now understand the differences between Type I and Type II supernovae. The explosion resulting from the detonation of a carbon white dwarf, the descendant of a low-mass star, is a supernova of Type I. Because this conflagration stems from a system containing virtually no hydrogen, we can readily see why the spectrum of a Type I supernova shows little evidence of that element. The appearance of the light curve (as we will soon see) results

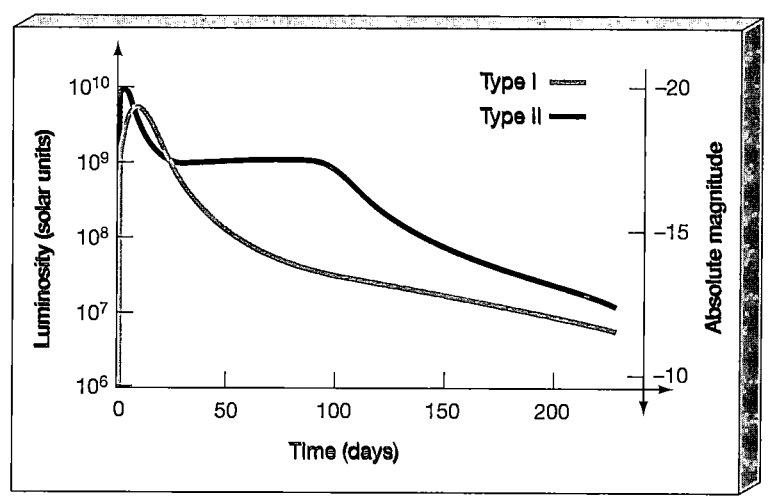

Figure 21.8 Supernova Light Curves
Schematic light curves of typical Type I and Type II supernovae. In both cases the maximum brightness or intensity can reach nearly 10 billion solar luminosities, but there are characteristic differences in the falloff of the luminosity after the initial peak. Type I light curves somewhat resemble those of novae (Figure 21.2). Type II curves have a characteristic bump in the declining phase.

almost entirely from the radioactive decay of unstable heavy elements produced in the explosion itself.

The implosion–explosion of the core of a massive star, described earlier, produces a Type II supernova. Detailed computer models indicate that the characteristic shape of the Type II light curve is just what would be expected from the expansion and cooling of the star's outer envelope as it is blown into space by the shock wave sweeping up from below. The expanding material consists mainly of unburned gas—hydrogen and helium—so it is not surprising that those elements are strongly represented in the supernova's observed spectrum. See *Discovery 21-1* for an account of a well-studied Type II supernova that confirmed many basic theoretical predictions while also forcing astronomers to revise the details of their models.

Figure 21.9 summarizes the processes responsible for the two different types of supernovae. We emphasize that, despite the similarity in the total amounts of energy involved, Type I and Type II supernovae are unrelated to one another. They occur in stars of very different types, under very different circumstances. All high-mass stars become Type II (core-collapse) supernovae, but only a tiny fraction of low-mass stars evolve into white dwarfs that ultimately explode as Type I (carbon-detonation) supernovae. However, there are far more low-mass stars than high-mass

stars, so by a remarkable coincidence the two types of supernova occur at roughly the same rate.

SUPERNOVA REMNANTS

We have plenty of evidence that supernovae have occurred in our Galaxy. Occasionally, the explosions themselves are visible from Earth. In many other cases we can detect their glowing remains, or **supernova remnants**. One of the best-studied supernova remnants is known as the Crab Nebula, shown in Figure 21.10. Its brightness has greatly dimmed now, but the original explosion in the year A.D. 1054 was so brilliant that manuscripts of ancient Chinese and Middle Eastern astronomers claim that its brightness greatly exceeded that of Venus and—according to some (possibly exaggerated) accounts—even rivaled that of the Moon. For nearly a month, this exploded star reportedly could be seen in broad daylight. Native Americans also left engravings of the event in the rocks of what is now the southwestern United States.

The Crab Nebula certainly has the appearance of exploded debris. Even today, the knots and filaments give a strong indication of past violence (and continuing activity— see *Discovery 21-2*). In fact, astronomers have proved that this matter was ejected from some central explosion.

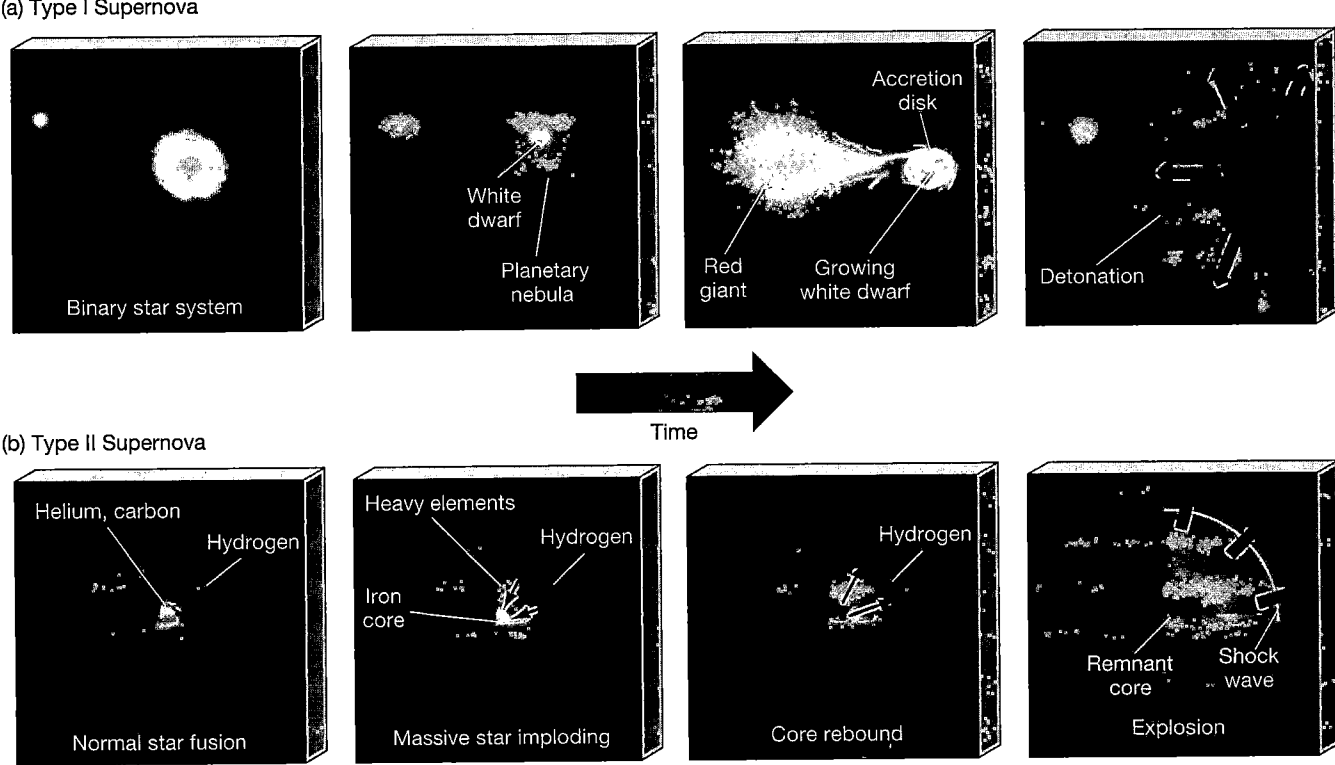

(a) Type I Supernova

Binary star system | White dwarf / Planetary nebula | Accretion disk / Red giant / Growing white dwarf | Detonation

(b) Type II Supernova

Time

Helium, carbon / Hydrogen / Normal star fusion | Heavy elements / Hydrogen / Iron core / Massive star imploding | Hydrogen / Core rebound | Remnant core / Shock wave / Explosion

Figure 21.9 Two Types of Supernova Type I and Type II supernovae have different causes. These sequences depict the evolutionary history of each type. (a) A Type I supernova may result when a carbon-rich white dwarf pulls matter onto itself from a nearby red-giant companion. (b) A Type II supernova occurs when the core of a more massive star collapses, then rebounds in a catastrophic explosion.

Figure 21.10 Crab Supernova Remnant This remnant of an ancient supernova is called the Crab Nebula (or M1 in the Messier catalog). It resides about 1800 pc from Earth and has an angular diameter about one-fifth that of the full Moon. Because its debris is scattered over a region of "only" 2 pc, the Crab is considered to be a young supernova remnant. In A.D. 1054 Chinese astronomers observed the supernova explosion itself. *(ESO; NASA)*

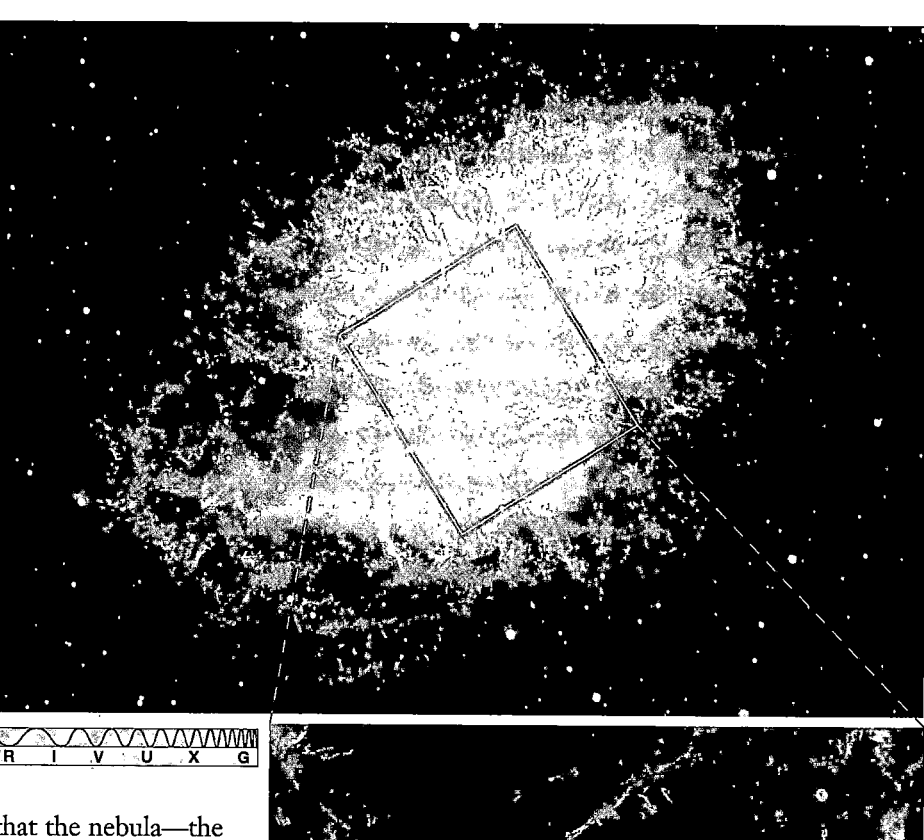

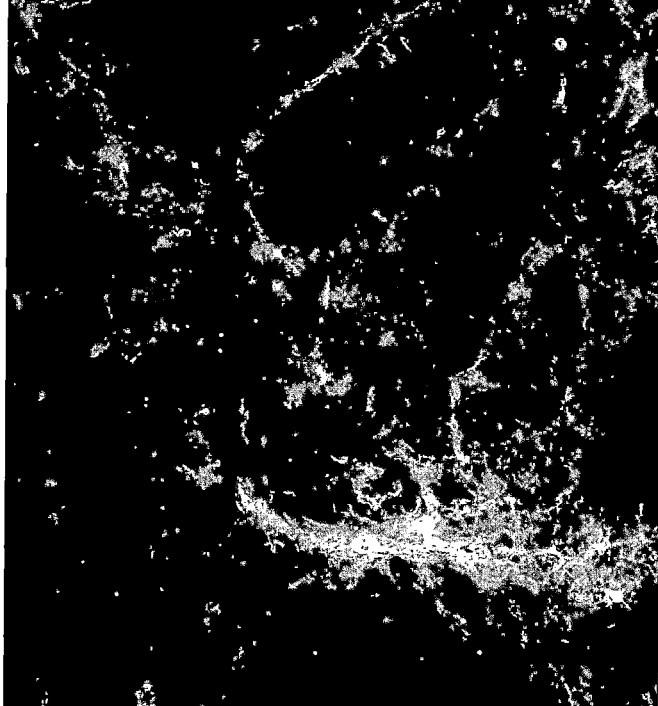

Doppler-shifted spectral lines indicate that the nebula—the envelope of the high-mass star that exploded to create this Type II supernova—is expanding into space at several thousand kilometers per second. A vivid illustration of this fact is provided by Figure 21.11, which was made by superimposing a positive image of the Crab Nebula taken in 1960 and a negative image taken in 1974. If the gas were not in motion, the positive and negative images would overlap perfectly, but they do not. The gas moved outward in the intervening 14 years. Running the motion backward in time, astronomers have found that the explosion must have occurred about nine centuries ago, consistent with the Chinese observations.

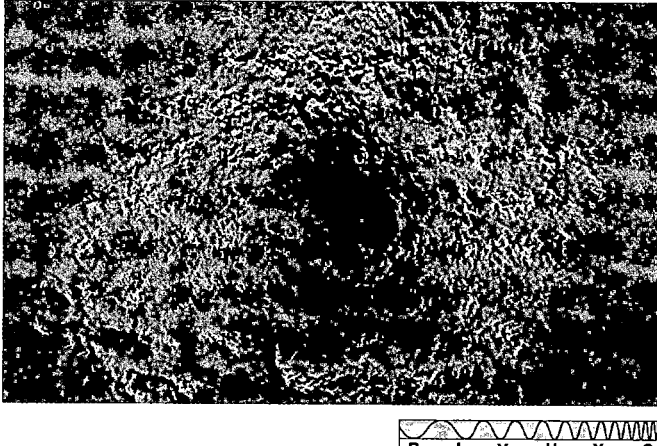

Figure 21.11 The Crab in Motion Positive and negative photographs of the Crab Nebula taken 14 years apart do not superimpose exactly, indicating that the gaseous filaments are still moving away from the site of the explosion. *(Harvard College Observatory)*

The nighttime sky harbors many relics of stars that blew up long ago. Figure 21.12 is another example. It shows the Vela supernova remnant, whose expansion velocities imply that its central star exploded around 9000 B.C. It lies only 500 pc away from Earth. Given its proximity, it may have been as bright as the Moon for several months. We can only speculate what impact such a bright supernova might have had on the myths, religions, and cultures of Stone Age humans when it first appeared in the sky.

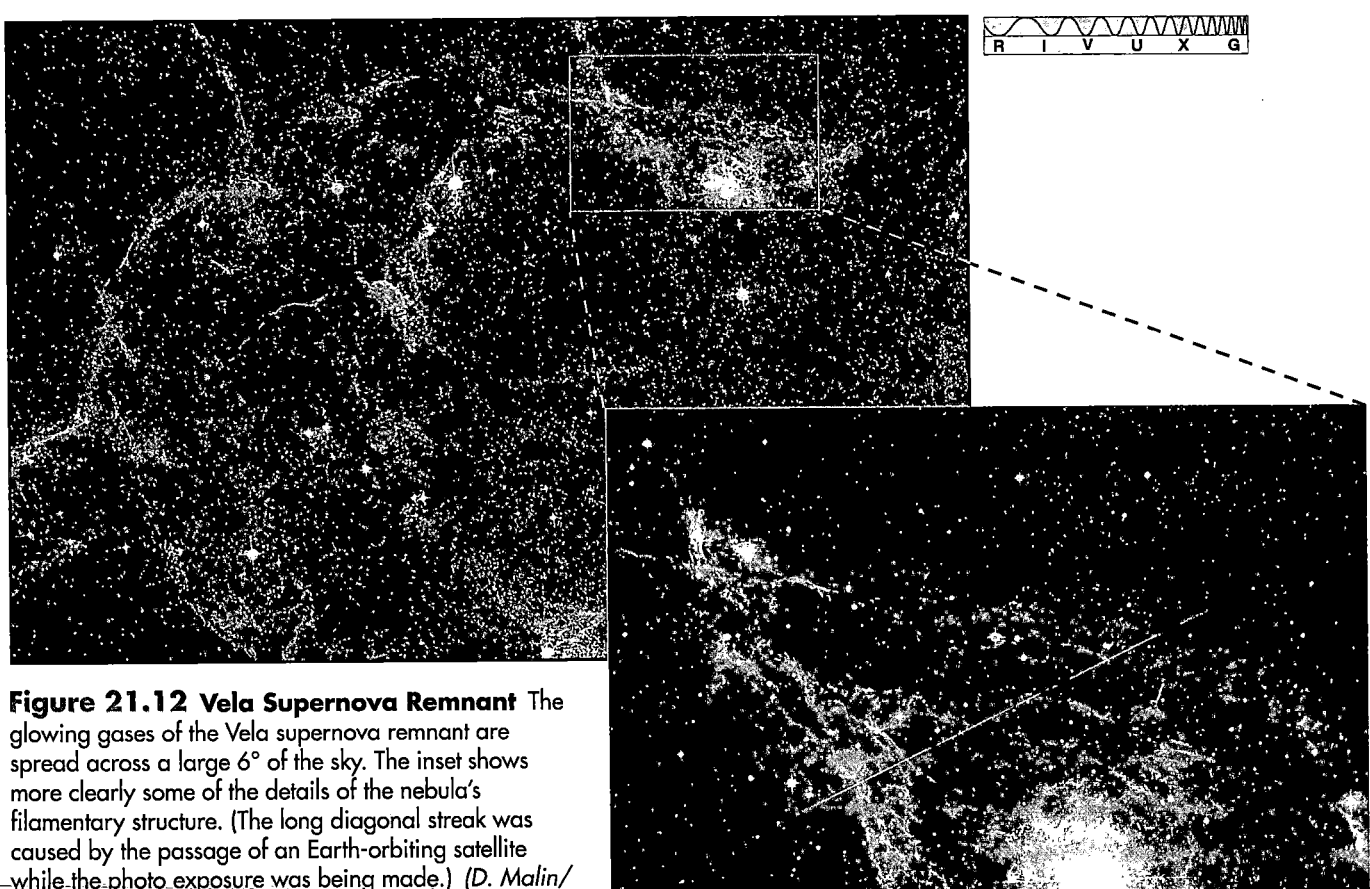

Figure 21.12 Vela Supernova Remnant The glowing gases of the Vela supernova remnant are spread across a large 6° of the sky. The inset shows more clearly some of the details of the nebula's filamentary structure. (The long diagonal streak was caused by the passage of an Earth-orbiting satellite while the photo exposure was being made.) *(D. Malin/Anglo-Australian Observatory)*

Although hundreds of supernovae have been observed in other galaxies during the twentieth century, no one has ever observed with modern equipment a supernova in our own Galaxy. A viewable Milky Way star has not exploded since Galileo first turned his telescope to the heavens almost four centuries ago. Now known as Tycho's supernova, this last supernova observed in our Galaxy caused a worldwide sensation in Renaissance times. The sudden appearance and subsequent fading of this very bright object in the year 1572 helped shatter the Aristotelian idea of an unchanging universe.

Based on stellar evolutionary theory, astronomers calculate that an observable supernova ought to occur in our Galaxy every 100 years or so. Even at a distance of several kiloparsecs a supernova would (temporarily) outshine Venus, the brightest planet in our sky, so it seems unlikely that astronomers could have missed any since the last one nearly four centuries ago. Our part of the Milky Way seems long overdue for a supernova explosion. However, a truly nearby supernova—within a few hundred parsecs, say—would be a very rare event, occurring only every 100,000 years or so. Humanity may be destined to see all supernovae from a distance.

✔️ Concept Check

■ How did astronomers know, even before the mechanisms were understood, that there are at least two distinct physical processes at work in creating supernovae?

21.4 The Formation of the Elements

📓 Up to now we have studied nuclear reactions mainly for their role in stellar energy generation. Now let's consider them again, but this time as the processes responsible for creating much of the world in which we live. The evolution of the elements, combining nuclear physics with astronomy, is a very complex subject, and a very important problem in modern astronomy.

TYPES OF MATTER

We currently know of 115 different elements, ranging from the simplest—hydrogen, containing one proton—to the most complex, called ununoctium and discovered in 1999, with 118 protons and 181 neutrons in its nucleus. (See Appendix 3, Table 2; elements 113, 115, and 117, while believed to exist, have not yet been created experimentally.) All elements exist in different *isotopes*, each having the same number of protons but a different number of neutrons. We often think of the most common or stable isotope as being the "normal" form of an element. Some elements, and many isotopes, are radioactively unstable, meaning that they eventually decay into other, more stable, nuclei.

The 81 stable elements found on Earth make up the overwhelming bulk of matter in the universe. In addition, 10 radioactive elements—including radon and uranium—also occur naturally on our planet. Even though the half-lives (the time required for half the nuclei to decay into something else) of these elements are very long (millions or even billions of years, typically), their steady decay means that they are scarce on Earth, in meteorites, and in lunar samples. ∞ (*More Precisely 7-2*) They are not observed in stars—there is just too little of them to produce detectable spectral lines.

Besides these 10 naturally occurring radioactive elements, 19 more radioactive elements have been artificially produced under special conditions in nuclear laboratories on Earth. The debris collected after nuclear weapons tests also contains traces of some of these elements. Unlike the naturally occurring radioactive elements, these artificial ones decay into other elements quite quickly (in much less than a million years). Consequently, they too are extremely rare in nature. Two other elements round out our list. Promethium is a stable element that is found on our planet only as a by-product of nuclear laboratory experiments. Technetium is an unstable element that is found in stars, but does not exist on Earth—any technetium that existed in our planet at formation decayed long ago.

ABUNDANCE OF MATTER

How and where did all these elements form? Were they always present in the universe, or were they created after the universe formed? Since the 1950s, astronomers have come to realize that the hydrogen and most of the helium in the universe are *primordial*—that is, these elements date from the very earliest times (see Chapter 27). All other elements in our universe result from **stellar nucleosynthesis**—that is, they were formed by nuclear fusion in the hearts of stars.

To test this idea, we must consider not just the list of different kinds of elements and isotopes but also their observed abundances, shown in Figure 21.13. This curve is derived largely from spectroscopic studies of stars, including the Sun. The essence of the figure is summarized in Table 21.1, which combines all the known elements into eight groups based on the total numbers of nuclear particles (protons and neutrons) that they contain. (All isotopes of all elements are included in both Table 21.1 and Figure 21.13, although only a few elements are marked by dots and labeled in the figure.) Any theory proposed for the creation of the elements must reproduce these observed abundances. The most obvious feature is that the heavy elements are much less abundant than most light elements. However, the many peaks and troughs evident in Figure 21.13 also represent important constraints.

HYDROGEN AND HELIUM BURNING

Let's begin by reviewing the reactions leading to heavy-element production at various stages of stellar evolution.

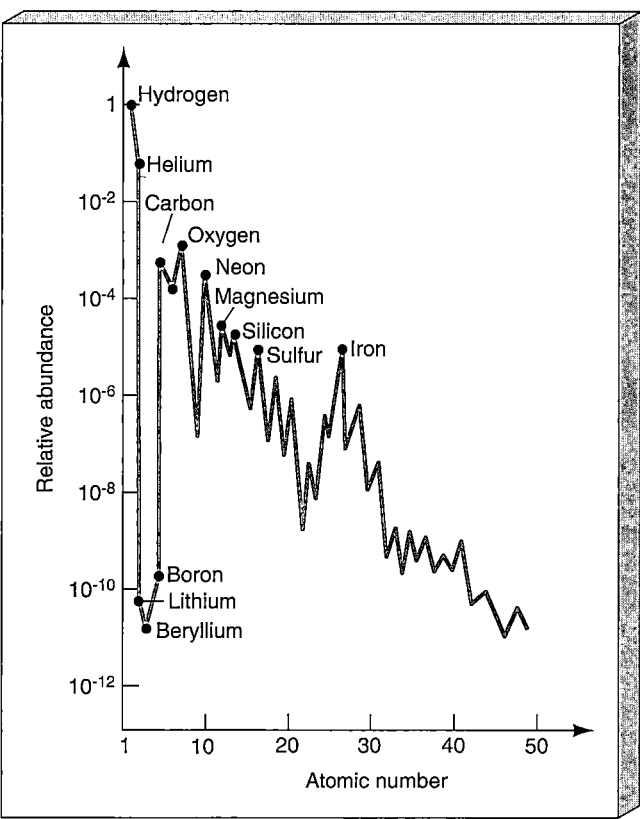

Figure 21.13 Elemental Abundance A summary of the cosmic abundances of the elements and their isotopes, expressed relative to the abundance of hydrogen. The horizontal axis shows atomic number—the number of protons in the nucleus. Notice how many common terrestrial elements are found on "peaks" of the distribution, surrounded by elements that are tens or hundreds of times less abundant. Notice especially the large peak around the element iron. The reasons for these peaks are discussed in the text.

TABLE 21.1 Cosmic Abundances of the Elements

ELEMENTAL GROUP OF PARTICLES	PERCENT ABUNDANCE BY NUMBER*
Hydrogen (1 nuclear particle)	90
Helium (4 nuclear particles)	9
Lithium group (7–11 nuclear particles)	0.000001
Carbon group (12–20 nuclear particles)	0.2
Silicon group (23–48 nuclear particles)	0.01
Iron group (50–62 nuclear particles)	0.01
Middle-weight group (63–100 nuclear particles)	0.00000001
Heaviest-weight group (over 100 nuclear particles)	0.000000001

*The total does not equal 100 percent because of uncertainties in the helium abundance. All isotopes of all elements are included.

Supernova 1987A

In 1987, astronomers were treated to a spectacular supernova in the Large Magellanic Cloud (LMC), a small satellite galaxy orbiting our own. ∞ (Sec. 24.2) Observers in Chile first saw the explosion on February 24, and within a few hours, nearly all Southern Hemisphere telescopes and every available orbiting spacecraft were focused on the object. It was officially named SN 1987A. (The SN stands for "supernova," 1987 gives the year, and A identifies it as the first supernova seen that year.) This was one of the most dramatic changes observed in the universe in nearly 400 years. A 15-solar mass B-type supergiant star with the catalog name SK-69°202 exploded, for a few weeks outshining all the other stars in the LMC combined, as shown in the "before" and "after" images of Figure 21.7.

Because the LMC is relatively close to Earth and because the explosion was detected so soon after it occurred, SN 1987A has provided astronomers with a wealth of detailed information on supernovae, allowing them to make key comparisons between theoretical models and observational reality. By and large, the theory of stellar evolution described in the text has held up very well. Still, SN 1987A did hold some surprises.

According to its hydrogen-rich spectrum, the supernova was a Type II—the core-collapse type—as expected for a high-mass parent star such as SK-69°202. But according to Figure 20.16 (which was computed for stars in our own Galaxy), the parent star should have been a red supergiant at the time of the explosion—not a blue supergiant, as was actually observed. This unexpected finding caused theorists to scramble in search of an explanation. It now seems that, relative to young stars in the Milky Way, the parent star's envelope was deficient in heavy elements. This deficiency had little effect on the evolution of the core and on the supernova explosion, but it did change the star's evolutionary track on the H–R diagram. Unlike a Milky Way star of the same mass, once helium ignited in the core of SK-69°202, the star shrank and looped back toward the main sequence. The star had just begun to return to the right on the H–R diagram following the ignition of carbon, with a surface temperature of around 20,000 K, when the rapid chain of events leading to the supernova occurred.

The light curve of SN 1987A, shown here, also differed somewhat from the "standard" Type II shape (see Figure 21.8). The peak brightness was less than $\frac{1}{10}$ the expected value.

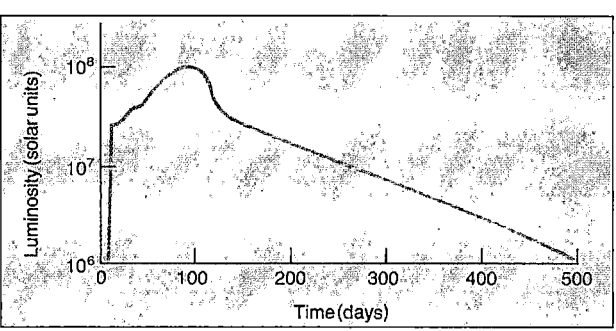

For a few days after its initial detection, the supernova faded as it expanded and cooled rapidly. After about a week the surface temperature had dropped to about 5000 K, at which point electrons and protons near the expanding surface recombined into atomic hydrogen, making the surface layers less opaque and allowing more radiation from the interior to leak out. As a result, the supernova brightened rapidly as it grew. The temperature of the expanding layers reached a peak in late May, by which point the radius of the expanding photosphere was about 2×10^{10} km—a little larger than our solar system. Subsequently, the photosphere cooled as it expanded, and the luminosity dropped as the internal supply of heat from the explosion leaked away into space.

Much of the preceding description would apply equally well to a Type II supernova in our own Galaxy. The differences between the SN 1987A light curve shown here and the Type II light curve in Figure 21.8 are mainly the result of the (relatively) small size of SN 1987A's parent star. The peak luminosity of SN 1987A was less than that of a "normal" Type II supernova because SK-69°202 was small and quite tightly bound by gravity. A lot of the energy emitted in the form of visible radiation (and evident in Figure 21.8) was used up in expanding SN 1987A's stellar envelope, so far less was left over to be radiated into space. Thus, SN 1987A's luminosity during the first few months was lower than expected, and the early peak evident in Figure 21.8 did not occur. The peak in the SN 1987A light curve at about 80–100 days actually corresponds to the "plateau" in the Type II light curve in Figure 21.8.

About 20 hours before the supernova was detected optically, a brief (13-second) burst of neutrinos was simultaneously recorded by underground detectors in Japan and the United States. As discussed in the text, the neutrinos are predicted to arise when electrons and protons in the star's

Look again at Figure 21.6 as we discuss the reactions involved. Stellar nucleosynthesis begins with the proton–proton chain studied in Chapter 16. ∞ (Sec. 16.5) Provided that the temperature is high enough—at least 10 million K—a series of nuclear reactions occur, ultimately forming a nucleus of ordinary helium (^{4}He) from four protons (^{1}H):

$$4(^1\text{H}) \rightarrow {}^4\text{He} + 2 \text{ positrons} + 2 \text{ neutrinos} + \text{energy}.$$

Recall that the positrons immediately interact with nearby free electrons, producing high-energy gamma rays through matter–antimatter annihilation. The neutrinos rapidly escape, carrying energy with them, but playing no direct role in nucleosynthesis. The validity of these reactions has been directly confirmed in nuclear experiments conducted in laboratories around the world during recent decades. In massive stars, the CNO cycle may greatly accelerate the hydrogen-burning process, but the basic

collapsing core merge to form neutrons. The neutrinos preceded the light because they escaped during the collapse, whereas the first light of the explosion was emitted only after the supernova shock had plowed through the body of the star to the surface. In fact, theoretical models consistent with these observations suggest that vastly more energy was emitted in the form of neutrinos than in any other form. The supernova's neutrino luminosity was many tens of thousands of times greater than its optical energy output.

Despite some unresolved details in SN 1987A's behavior, detection of this neutrino pulse is considered to be a brilliant confirmation of theory. This singular event—the detection of neutrinos—may well herald a new age of astronomy. For the first time, astronomers have received information from beyond the solar system by radiation outside the electromagnetic spectrum.

Theory predicts that the expanding remnant of SN 1987A is now on the verge of being resolvable by optical telescopes. The accompanying photographs show the

barely resolved remnant (at left below) surrounded by a much larger shell of glowing gas (in yellow). Scientists reason that the progenitor star expelled this shell during its red-giant phase, some 40,000 years before the explosion. The image we see results from the initial flash of ultraviolet light from the supernova hitting the ring and causing it to glow brightly. As the debris from the explosion itself strikes the ring, it has become a temporary but intense source of X rays. As shown in the *Chandra* image at bottom right, taken in 2000, the fastest-moving ejecta have already reached it, forming the small (1000 A.U. diameter), glowing regions on the left side of the ring.

These images also show core debris moving outward toward the ring. The four insets at bottom left (also the sickle-shaped region in the bottom right image) show material expanding at nearly 3000 km/s. The main image (at left) also revealed, to everyone's surprise, two additional faint rings that might be caused by radiation sweeping across an hourglass-shaped bubble of gas. Why the gas should exhibit this odd structure remains unclear.

Buoyed by the success of stellar-evolution theory and armed with firm theoretical predictions of what should happen next, astronomers eagerly await future developments in the story of this remarkable object.

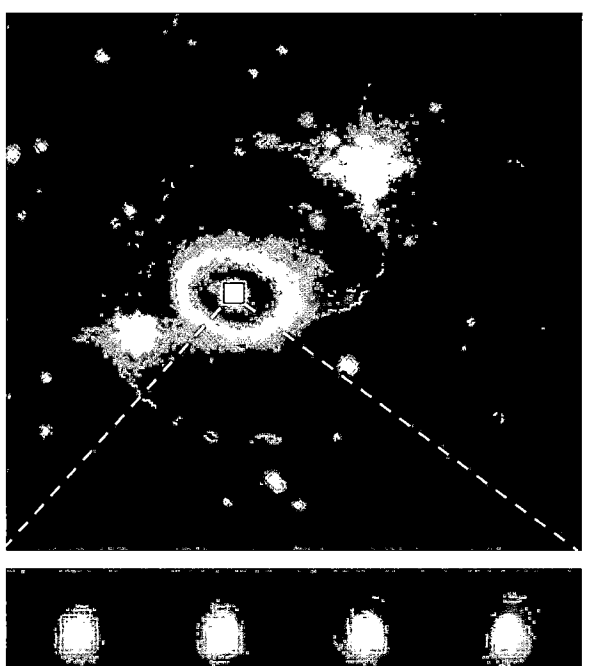

(NASA; NASA)

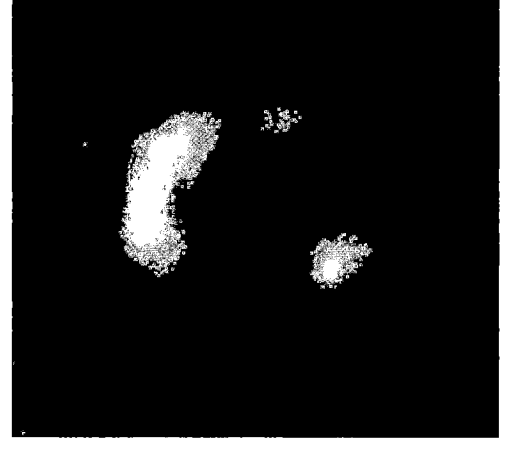

4-protons-to-1-helium-nucleus reaction, illustrated in Figure 21.14, is unchanged. ∞ (*Discovery 20-1*)

As helium builds up in the core of a star, the burning ceases, and the core contracts and heats up. When the temperature exceeds about 100 million K, helium nuclei can overcome their mutual electrical repulsion, leading to the *triple-alpha reaction*, which we discussed in Chapter 20: ∞ (Sec. 20.2)

$$3(^4He) \rightarrow {}^{12}C + energy.$$

The net result of this reaction is that three helium-4 nuclei are combined into one carbon-12 nucleus (Figure 21.15), releasing energy in the process.

CARBON BURNING AND HELIUM CAPTURE

At higher and higher temperatures, heavier and heavier nuclei can gain enough energy to overcome the electrical repulsion between them. At about 600 million K (reached

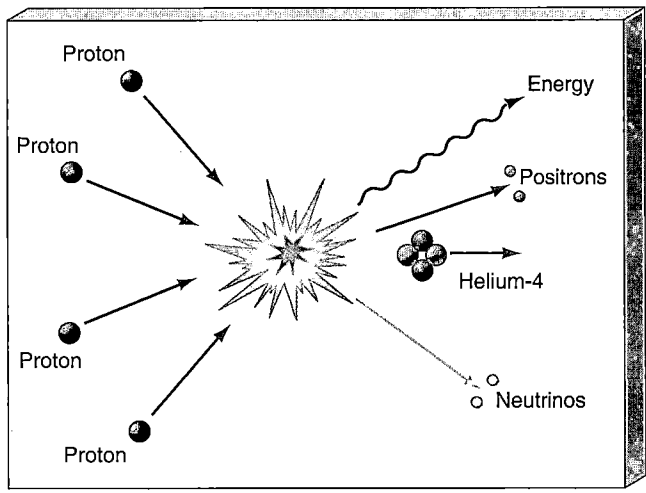

Figure 21.14 Proton Fusion Diagram of the basic proton–proton hydrogen-burning reaction. Four protons combine to form a nucleus of helium-4, releasing energy in the process.

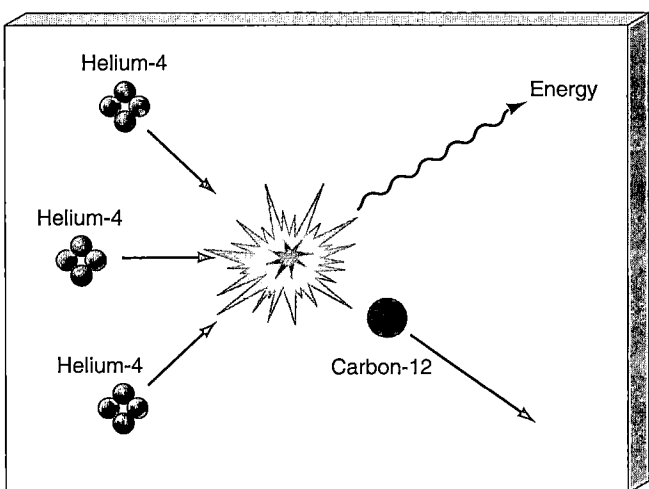

Figure 21.15 Helium Fusion Diagram of the basic triple-alpha helium-burning reaction occurring in post-main-sequence stars. Three helium-4 nuclei combine to form carbon-12.

only in the cores of stars much more massive than the Sun), carbon nuclei can fuse to form magnesium, as depicted in Figure 21.16(a):

$$^{12}C + {}^{12}C \rightarrow {}^{24}Mg + \text{energy.}$$

However, because of the rapidly mounting nuclear charges—that is, the increasing number of protons in the nuclei—fusion reactions between any nuclei larger than carbon require such high temperatures that they

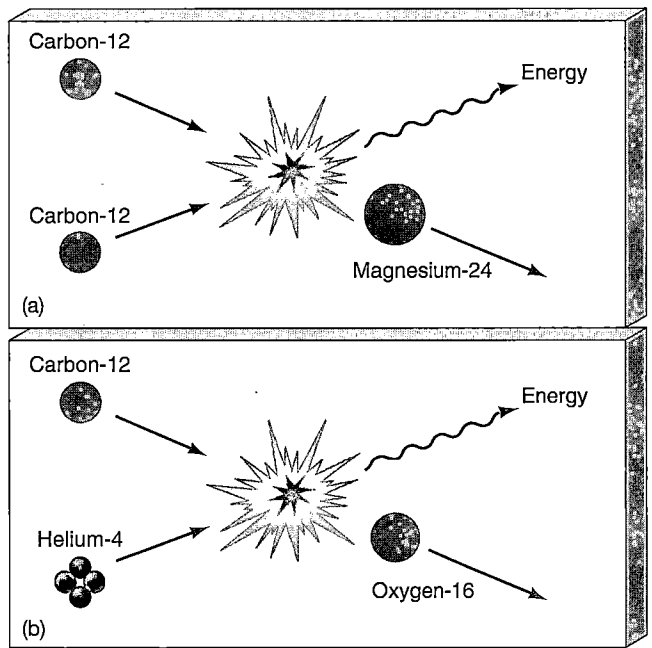

Figure 21.16 Carbon Fusion Carbon can form heavier elements (a) by fusion with other carbon nuclei or, more commonly, (b) by fusion with a helium nucleus.

are actually quite uncommon in stars. The formation of most heavier elements occurs by way of an easier path. For example, the repulsive force between two carbon nuclei is three times greater than the force between a nucleus of carbon and one of helium. Thus, carbon–helium fusion occurs at a lower temperature than that at which carbon–carbon fusion occurs. As we saw in Section 20.3, at temperatures above 200 million K, a carbon-12 nucleus colliding with a helium-4 nucleus can produce oxygen-16:

$$^{12}C + {}^4He \rightarrow {}^{16}O + \text{energy.}$$

If any helium-4 is present, this reaction, shown in Figure 21.16(b), is much more likely to occur than the carbon–carbon reaction.

Similarly, the oxygen-16 thus produced may fuse with other oxygen-16 nuclei at a temperature of about 1 billion K to form sulfur-32,

$$^{16}O + {}^{16}O \rightarrow {}^{32}S + \text{energy,}$$

but it is much more probable that an oxygen-16 nucleus will capture a helium-4 nucleus (if one is available) to form neon-20:

$$^{16}O + {}^4He \rightarrow {}^{20}Ne + \text{energy.}$$

The second reaction is more likely because it occurs at a lower temperature than that necessary for oxygen–oxygen fusion.

Thus, as the star evolves, heavier elements tend to form through **helium capture** rather than by fusion of like nuclei. As a result, elements with nuclear masses of 4 units (that is, helium itself), 12 units (carbon), 16 units (oxygen), 20 units (neon), 24 units (magnesium), and 28 units (silicon) stand out as prominent peaks in Figure 21.13, our chart of cosmic abundances. Each element is built by combining the preceding element and a helium-4 nucleus as the star evolves.

IRON FORMATION

Helium capture is by no means the only type of nuclear reaction occurring in evolved stars. As nuclei of many different kinds accumulate, a great variety of reactions become possible. In some, protons and neutrons are freed from their parent nuclei and are absorbed by others, forming new nuclei with masses intermediate between those formed by helium capture. Laboratory studies confirm that common nuclei, such as fluorine-19, sodium-23, phosphorus-31, and many others, are created in this way. Their abundances, however, are not so great as those produced directly by helium capture, simply because the helium-capture reactions are much more common in stars. For this reason, many of these elements (those with masses not divisible by four, the mass of a helium nucleus) are found in the troughs of Figure 21.13.

Around the time silicon-28 appears in the core of a star, a competitive struggle begins between the continued capture of helium to produce even heavier nuclei and the tendency of more complex nuclei to break down into simpler ones. The cause of this breakdown is heat. By now the star's core temperature has reached the unimaginably large value of 3 billion K, and the gamma rays associated with that temperature have enough energy to break a nucleus apart, as illustrated in Figure 21.17(a). This is the same process of photodisintegration that will ultimately accelerate the star's iron core in its final collapse toward a Type II supernova.

Under the intense heat, some silicon-28 nuclei break apart into seven helium-4 nuclei. Other nearby nuclei that have not yet photodisintegrated may capture some or all of these helium-4 nuclei, leading to the formation of still heavier elements (Figure 21.17b). The process of photodisintegration provides raw material that allows the helium-capture process to proceed to greater masses. The process continues, with some heavy nuclei being destroyed and others increasing in mass. In succession, the star forms sulfur-32, argon-36, calcium-40, titanium-44, chromium-48, iron-52, and nickel-56. The chain of reactions building from silicon-28 up to nickel-56 is

$$^{28}Si + 7(^4He) \rightarrow {}^{56}Ni + energy.$$

This two-step process—photodisintegration followed by the direct capture of some or all of the resulting helium-4 nuclei (or alpha particles)—is often called the *alpha process*.

Nickel-56 is unstable. It decays rapidly, first into cobalt-56, then into a stable iron-56 nucleus. Any unstable nucleus will continue to decay until stability is achieved, and iron-56 is the most stable of all nuclei (Figure 21.6). Thus, the alpha process leads inevitably to the buildup of iron in the stellar core.

Another way of describing Figure 21.6 is to say that iron's 26 protons and 30 neutrons are bound together

Figure 21.17 Alpha Process (a) At high temperatures, heavy nuclei (such as silicon, shown here) can be broken apart into helium nuclei by high-energy photons. (b) Other nuclei can capture the helium nuclei—or alpha particles—thus produced, forming heavier elements by the so-called alpha process. This process continues all the way to the formation of nickel-56 (in the iron group).

Energy

Silicon-28

Helium-4

(a)

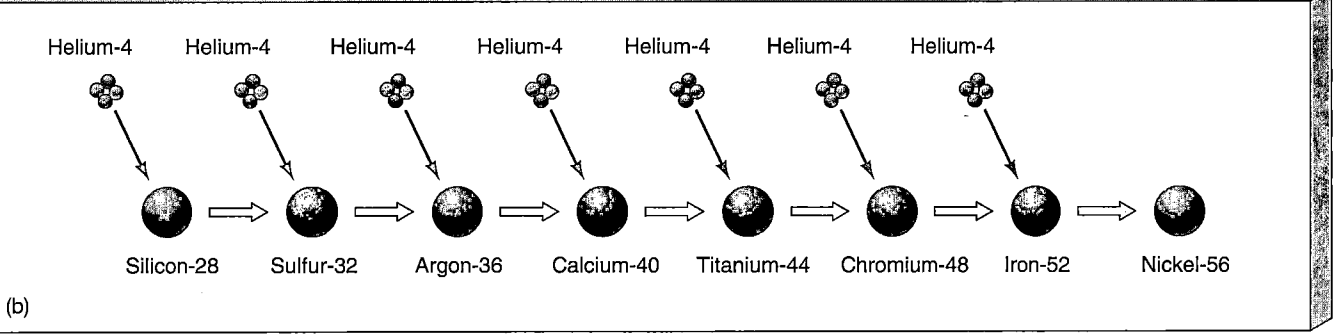

Helium-4   Helium-4   Helium-4   Helium-4   Helium-4   Helium-4   Helium-4

Silicon-28   Sulfur-32   Argon-36   Calcium-40   Titanium-44   Chromium-48   Iron-52   Nickel-56

(b)

DISCOVERY 21-2

The Crab Nebula in Motion

The Crab Nebula is one of the most fascinating and instructive cosmic objects anywhere. Within this remarkable region of space we find a wide variety of physical processes at work, all the result of a supernova explosion that occurred almost a thousand years ago. A famous physics professor at MIT once taught a course exclusively on this object, claiming that if you understand all that transpires within the Crab, then you have mastered much of modern physics. Within this remarkable gas cloud, extending for a couple of parsecs and residing about 1800 pc away, we find applications of particle physics, nuclear physics, electromagnetism, thermodynamics, condensed-matter physics, plasma physics, and gravitational physics, to name but a few.

The accompanying figures present observations suggesting that the Crab Nebula is even more dynamic than had previously been thought. Conditions are changing rapidly among the debris, right before our eyes, giving astronomers a rare chance to study evolutionary changes that actually occur during a single human lifetime.

The first figure consists of two frames. On the left is a true-color, visible-light photo taken with the 5-m Hale telescope on Mount Palomar. ☞ (Sec. 5.1) This image is similar to the main frame of Figure 21.10. The visible light arises in two distinct ways. The outermost filaments seen in red, yellow, and green are literally glowing from the heat and violence of the explosion long ago. However, the faint bluish light toward the center of the nebula arises from a nonthermal process that releases energy as rapidly moving electrons spiral around lines of magnetic field. (This process is not studied in the text until Chapter 25, but you can skip ahead and look at Figure 25.20 to get the gist of it.)

The filamentary structure of the wispy debris leaves little doubt that this is the remnant of an exploded star.

The right frame opposite is a recent image acquired by the *Hubble Space Telescope*, covering the portion of the first image contained within the white box. The image was taken at a wavelength of around 550 nm; the red color was added artificially during computer enhancement.

The central "engine" at the heart of the Crab Nebula is a pulsar—a rapidly spinning, compact starlike object that managed to survive the supernova explosion that created the nebula itself. (We will discuss these strange objects in more detail in Chapter 22.) ☞ (Sec. 22.2) The pulsar can be seen in this image as the left member of the pair of stars near the center of the frame. The pulsar powers the Crab today by accelerating elementary particles into the nebular debris, causing knots and wisps of energetic matter to stream away from the core.

The second set of images allows us to see in action the tiny pulsar powering the Crab Nebula (again, the left star of the pair at upper center). These three images were taken sequentially over the course of a few months. The field of view is an even smaller field than in the right frame of the previous figure—we are now virtually inside the Crab, witnessing the minute changes among the nebular gases near the site of the ancient explosion. Astronomers can now watch the changes in the shock-driven features that brighten and fade, over and over, like ripples in a pond. The light-year-long ripples in the Crab, seen throughout the bottom half of these frames, are moving outward from the pulsar at about half the speed of light. The final frame indicates the direction of the jet from the pulsar's pole (see Chapter 22) that may be the cause of the ripples we see.

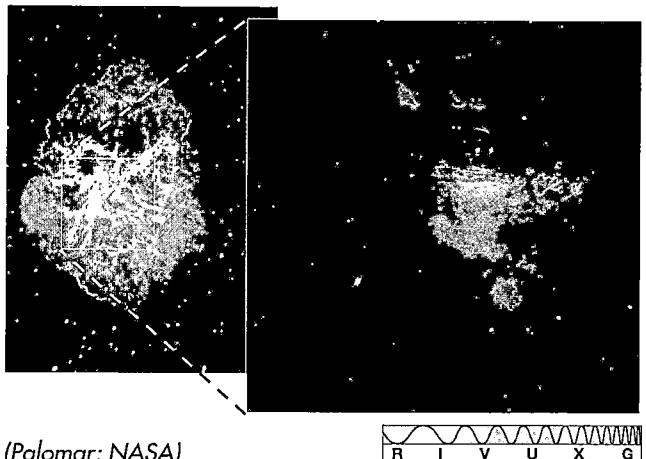

(Palomar; NASA)

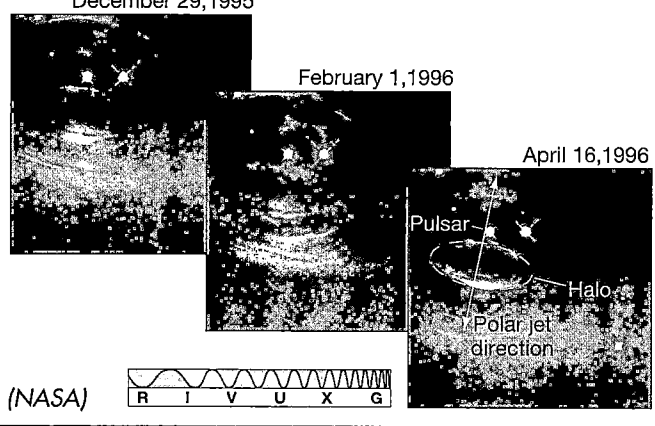

December 29, 1995

February 1, 1996

April 16, 1996

Pulsar
Halo
Polar jet direction

(NASA)

more strongly than the particles in any other nucleus. Iron is said to have the greatest *nuclear binding energy* of any element—more energy per particle is required to break up (unbind) an iron-56 nucleus than the nucleus of any other element. This enhanced stability of iron explains why some of the heavier nuclei in the iron group are more abundant than many lighter nuclei (see Table 21.1 and Figure 21.13)—nuclei tend to "accumulate" near iron as stars evolve.

MAKING ELEMENTS BEYOND IRON

If the alpha process stops at iron, how did heavier elements, such as copper, zinc, and gold, form? To form them, some nuclear process other than helium capture must have been involved. That other process is **neutron capture**—the formation of heavier nuclei by the absorption of neutrons.

Deep in the interiors of highly evolved stars, conditions are ripe for neutron capture to occur. Neutrons are produced as "by-products" of many nuclear reactions, so there are many of them present to interact with iron and other nuclei. Neutrons have no charge, so there is no repulsive barrier for them to overcome in combining with positively charged nuclei. As more and more neutrons join an iron nucleus its mass continues to grow.

Adding neutrons to a nucleus—iron, for example—does not change the element. Rather, a more massive isotope of the same element is produced. Eventually, however, so many neutrons have been added to the nucleus that it becomes unstable and then decays radioactively to form a stable nucleus of some other element. The neutron-capture process then continues. For example, an iron-56 nucleus can capture a single neutron to form a relatively stable isotope, iron-57:

$$^{56}\text{Fe} + \text{n} \rightarrow {}^{57}\text{Fe}.$$

This reaction may be followed by another neutron capture:

$$^{57}\text{Fe} + \text{n} \rightarrow {}^{58}\text{Fe},$$

producing another relatively stable isotope, iron-58. Iron-58 can capture yet another neutron to produce an even heavier isotope of iron:

$$^{58}\text{Fe} + \text{n} \rightarrow {}^{59}\text{Fe}.$$

Iron-59 is known from laboratory experiments to be radioactively unstable. It decays in about a month into cobalt-59, which is stable. The neutron-capture process then resumes: Cobalt-59 captures a neutron to form the unstable cobalt-60, which in turn decays to nickel-60, and so on.

Each successive capture of a neutron by a nucleus typically takes about a year, so most unstable nuclei have plenty of time to decay before the next neutron comes along. Researchers usually refer to this "slow" neutron-capture mechanism as the *s-process*. It is the origin of the copper and silver in the coins in our pockets, the lead in

our car batteries, the gold (and the zirconium) in the rings on our fingers.

MAKING THE HEAVIEST ELEMENTS

The s-process explains the synthesis of stable nuclei up to and including bismuth-209, the heaviest known nonradioactive nucleus, but it cannot account for the heaviest nuclei, such as thorium-232, uranium-238, or plutonium-242. Any attempt to form elements heavier than bismuth-209 by slow neutron capture fails because the new nuclei decay back to bismuth as fast as they form. There must be yet another nuclear mechanism that produces the very heaviest nuclei. This process is called the *r-process* (where r stands for "rapid," in contrast to the "slow" s-process just described). The r-process operates very quickly, occurring (we think) literally during the supernova explosion that signals the death of a massive star.

During the first 15 minutes of the supernova blast, the number of free neutrons increases dramatically as heavy nuclei are broken apart by the violence of the explosion. Unlike the s-process, which stops when it runs out of stable nuclei, the neutron-capture rate during the supernova is so great that even unstable nuclei can capture many neutrons before they have time to decay. Jamming neutrons into light- and middle-weight nuclei, the r-process is responsible for the creation of the heaviest known elements. The heaviest of the heavy elements, then, are actually born *after* their parent stars have died. However, because the time available for synthesizing these heaviest nuclei is so brief, they never become very abundant. Elements heavier than iron (see Table 21.1) are a billion times less common than hydrogen and helium.

OBSERVATIONAL EVIDENCE FOR STELLAR NUCLEOSYNTHESIS

The modern picture of element formation involves many different types of nuclear reactions occurring at many different stages of stellar evolution, from main-sequence stars all the way to supernovae. Light elements—from hydrogen to iron—are built first by fusion, then by alpha capture, with proton and neutron capture filling in the gaps. Elements beyond iron form by neutron capture and radioactive decay. We are reassured of the basic soundness of our theories by three convincing pieces of evidence.

First, the rate at which various nuclei are captured and the rate at which they decay are known from laboratory experiments. When these rates are incorporated into detailed computer models of the nuclear processes occurring in stars and supernovae, the resulting elemental abundances agree extremely well, point by point, with the observational data presented in Figure 21.13 and Table 21.1. The match is remarkably good for elements up through iron and is still fairly close for heavier nuclei. Although the reasoning is indirect, the agreement between theory and observation is so striking that most astronomers regard it as very strong evidence in support of the entire theory of stellar evolution and nucleosynthesis.

Second, the presence of one particular nucleus—technetium-99—provides direct evidence that heavy-element formation really does occur in the cores of stars. Laboratory measurements show that the technetium nucleus has a radioactive half-life of about 200,000 years, a very short time astronomically speaking. No one has ever found even traces of naturally occurring technetium on Earth because it all decayed long ago. The observed presence of technetium in the spectra of many red-giant stars implies that it must have been synthesized in their cores through neutron capture—the only known way in which technetium can form—within the past few hundred thousand years, then transported by convection to the surface. Otherwise, we would not observe it. Many astronomers consider the spectroscopic evidence for technetium as proof that the s-process really does operate in evolved stars.

Third, the study of typical light curves from Type I supernovae indicates that radioactive nuclei form as a result of the explosion. Figure 21.18(a) (see also Figure 21.8) displays the dramatic rise in luminosity at the moment of explosion and the characteristic slower decrease in brightness. Depending on the initial mass of the exploded star, the luminosity takes from several months to many years to decrease to its original value, but the *shape* of the decay curve is nearly the same for all exploded stars. These curves have two distinct features. After the initial peak, the luminosity first declines rapidly, then decreases at a slower rate. This change in the luminosity decay invariably occurs about two months after the explosion, regardless of the intensity of the outburst.

We can explain the two-stage decline of the luminosity curve in Figure 21.18(a) in terms of the radioactive decay of unstable nuclei, notably nickel-56 and its decay product cobalt-56, produced in abundance during the early moments of the supernova explosion. From theoretical models of the explosion we can calculate the amounts of these elements expected to form, and we know their half-lives from laboratory experiments. Because each radioactive decay produces a known amount of energy, we can then determine how the light emitted by these unstable elements should vary in time. The result is in very

good agreement with the observed light curve in Figure 21.18(b)—the luminosity of a Type I supernova is entirely consistent with the decay of about 0.6 solar masses of nickel-56. More direct evidence for the presence of these unstable nuclei was first obtained in the 1970s, when a gamma-ray spectral feature of decaying cobalt-56 was identified in a supernova observed in a distant galaxy.

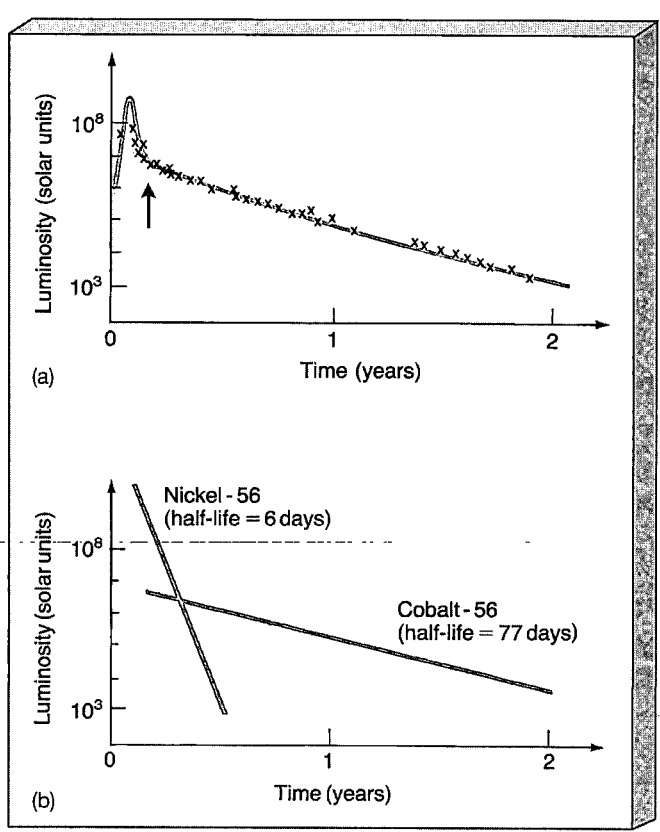

Figure 21.18 Supernova Energy Emission (a) The light curve of a Type I supernova, showing not only the dramatic increase and slow decrease in luminosity but also the characteristic change in the rate of decay about two months after the explosion (after the time indicated by the arrow). This particular supernova occurred in the faraway galaxy IC 4182 in 1938. The crosses are the actual observations of the supernova's light. (b) Theoretical calculations of the light emitted by the radioactive decay of nickel-56 and cobalt-56 produce a light curve very similar to those actually observed in real supernova explosions, lending strong support to the theory of stellar nucleosynthesis.

☑ Concept Check

■ Why are the low-mass elements carbon, oxygen, neon, and magnesium, whose masses are multiples of four, as well as the element iron, so common on Earth?

21.5 The Cycle of Stellar Evolution

⑥ The theory of stellar nucleosynthesis can naturally account for the observed differences in heavy-element abundance between the old globular cluster stars and stars now forming in our Galaxy. Even though an evolved star continuously creates new heavy elements in its interior, the composition changes are largely confined to the core, and the star's spectrum gives little indication of events within. Convection may carry some reaction products (such as the technetium observed in many red giants) from the core into the envelope, but the outer layers largely retain the composition with which the star originally formed. Only at the end of the star's life are its newly created elements released and scattered into space.

Thus, the spectra of the *youngest* stars show the *most* heavy elements, because each new generation of stars increases the concentration of these elements in the interstellar clouds from which the next generation forms. The photosphere of a recently formed star contains a much greater abundance of heavy elements than that of a star that formed long ago. Knowledge of stellar evolution allows astronomers to estimate the ages of stars from spectroscopic studies, even when the stars are isolated and not members of any cluster. ∞ (Sec. 20.5) In the past three chapters we have seen all the ingredients that make up the complete cycle of star formation and evolution in our Galaxy. Let's briefly summarize that process, which is illustrated in Figure 21.19.

1. Stars form when part of an interstellar cloud is compressed beyond the point at which it can support itself against its own gravity. The cloud collapses and fragments, forming a cluster of stars. The hottest stars heat and ionize the surrounding gas, sending shock waves through the surrounding cloud, possibly triggering new rounds of star formation.

2. Within the cluster, stars evolve. The most massive stars evolve fastest, creating heavy elements in their cores and spewing them forth into the interstellar medium in supernova explosions. The lighter stars take longer to evolve, but they too can create heavy elements and may contribute to the "seeding" of interstellar space when they shed their envelopes as planetary nebulae.

3. The creation and explosive dispersal of new heavy elements are accompanied by further shock waves. The passage of these shock waves through the interstellar medium simultaneously enriches the medium and compresses it into further star formation. Each generation of stars increases the concentration of heavy elements in the interstellar clouds from which the next generation forms. As a result, recently formed stars contain a much greater abundance of heavy elements than stars that formed long ago.

In this way, although some material is used up in each cycle—turned into energy or locked up in low-mass stars—the Galaxy continuously recycles its matter. Each new round of formation creates stars with more heavy elements than the preceding generation had. From the old globular clusters, which are observed to be deficient in heavy elements relative to the Sun, to the young open clusters, containing much larger amounts of these elements, we observe this enrichment process in action. Our Sun is the product of many such cycles. We ourselves are another. Without the heavy elements synthesized in the hearts of stars, life on Earth would not exist.

☑ Concept Check

■ Why are supernovae important to life on Earth?

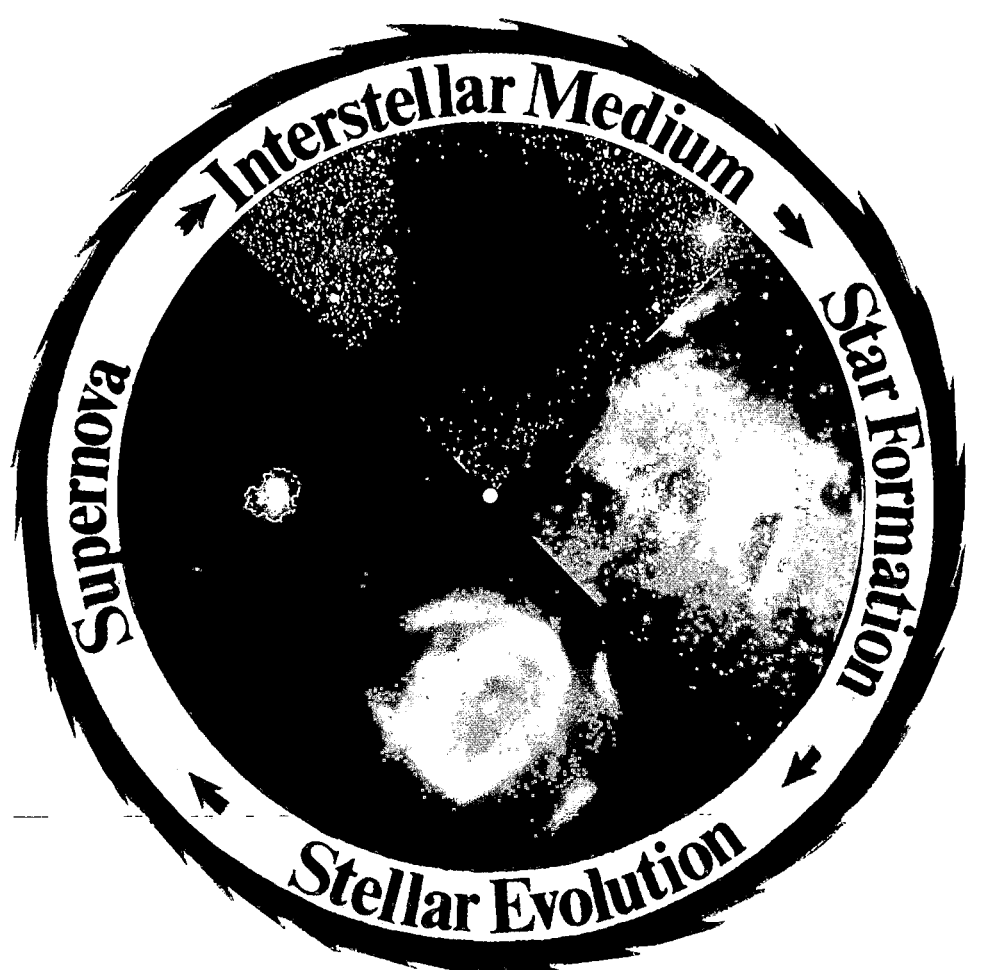

Figure 21.19 Stellar Recycling The cycle of star formation and evolution continuously replenishes the Galaxy with new heavy elements and provides the driving force for the creation of new generations of stars. Clockwise from top, an interstellar cloud (Barnard 68), a star-forming region in our Galaxy (RCW 38), a low-mass planetary nebula (Hen 1357), and a high-mass supernova remnant (G 21.5–0.9).
(L. Chaisson; ESO; NASA)

Chapter Review

SUMMARY

A **nova** (p. 544) is a star that suddenly increases greatly in brightness, then slowly fades back to its normal appearance over a period of months. It results when a white dwarf in a binary system draws hydrogen-rich material from its companion. The gas builds up on the white dwarf's surface, eventually becoming hot and dense enough for the hydrogen to burn explosively, temporarily causing a large increase in the dwarf's luminosity. The matter flowing from the companion star does not fall directly onto the surface of the dwarf. Instead, it goes into orbit around it, forming an **accretion disk** (p. 545). Friction within the disk causes the gas to spiral slowly inward, heating up and glowing brightly as it nears the dwarf's surface.

Stars more massive than about 8 solar masses are able to attain high enough central temperatures to burn carbon and heavier nuclei. As they burn, their cores form a layered structure consisting of burning shells of successively heavier elements. A nonburning core of iron builds up at the center. Iron is special in that its nuclei can neither be fused together nor split apart to produce energy. As a result, stellar nuclear burning stops at iron. As a star's iron core grows in mass it eventually becomes unable to support itself against gravity and begins to collapse. At the enormous densities and temperatures produced during the collapse, iron nuclei are broken down into their constituent particles—protons and neutrons. The protons combine with electrons to form more neutrons. Eventually, when the core has become so dense that the neutrons are effectively brought into physical contact with one another, their resistance to further squeezing stops the collapse and the core rebounds, sending a violent shock wave out through the rest of the star. The star is blown to pieces in a **core-collapse supernova** (p. 548).

Astronomers classify **supernovae** (p. 548) into two broad categories: Type I and Type II. These classes differ by their light curves and by their composition. **Type I supernovae** (p. 549) are hydrogen-poor and have a light curve similar in shape to that of a nova. **Type II supernovae** (p. 549) are hydrogen-rich and have a characteristic bump in the light curve a few months after maximum. A Type II supernova is a core-collapse supernova. A Type I supernova occurs when a carbon-oxygen white dwarf in a binary system exceeds about 1.4 solar masses (the Chandrasekhar mass)—the maximum mass that can be supported against gravity by electron degeneracy pressure. The star collapses and explodes as its carbon ignites. This type of supernova is called a **carbon-detonation supernova** (p. 549).

Theory predicts that a supernova visible from Earth should occur within our Galaxy about once a century, although none has been observed in the last 400 years. We can see evidence of a past supernova in the form of a **supernova remnant** (p. 550), a shell of exploded debris surrounding the site of the explosion and ex-panding into space at thousands of kilometers per second.

All elements heavier than helium formed by **stellar nucleo-synthesis** (p. 553)—the production of new elements by nuclear reactions in the cores of evolved stars. Elements beyond carbon tend to form by **helium capture** (p. 556) rather than by the fusion of two heavy nuclei. Therefore, nuclei whose masses are a multi-ple of the mass of a helium nucleus tend to be more common than others. At high enough core temperatures, photodisintegration breaks apart some heavy nuclei, providing helium-4 nuclei for the synthesis of even more massive elements, leading to a buildup of iron-56 in the core. Elements beyond iron form by **neutron cap-ture** (p. 559) in the cores of evolved stars. With no repulsive elec-tromagnetic barrier to overcome, neutrons can easily combine with nuclei. During a supernova explosion, rapid neutron capture occurs, producing the heaviest nuclei of all. Comparisons between theoretical predictions of element production and observations of element abundances in stars and supernovae provide strong sup-port for the theory of stellar nucleosynthesis.

The processes of star formation, evolution, and explosion form a cycle that constantly enriches the interstellar medium with heavy elements and sows the seeds of new generations of stars. Without the elements produced in supernovae, life on Earth would be impossible.

SELF-TEST: TRUE OR FALSE?

____ **1.** A nova is a sudden outburst of light coming from an old main-sequence star.

____ **2.** Novae occur in binary-star systems.

____ **3.** It takes less and less time to fuse heavier and heavier elements inside a high-mass star.

____ **4.** In a core-collapse supernova, the outer part of the core rebounds from the inner, high-density core, de-stroying the entire outer part of the star.

____ **5.** Most of the energy released during a core-collapse su-pernova is emitted in the form of gamma rays.

____ **6.** A Type I supernova is the same as a nova, but it ap-pears much brighter because it occurs closer to us.

____ **7.** The spectrum of a Type II supernova shows no evi-dence of hydrogen.

____ **8.** It is possible that a recurrent nova will eventually re-sult in a core-collapse supernova.

____ **9.** Once the process gets under way, the core of a massive star collapses in about one second.

____ **10.** Different isotopic forms of an element differ only in the number of protons in the nucleus.

____ **11.** Carbon can fuse with helium more easily than it can fuse with another carbon atom because there is more helium than carbon in stars.

____ **12.** The s-process is one in which heavy elements are formed from silicon.

____ **13.** The r-process occurs only during the first few min-utes of a supernova explosion.

____ **14.** Stellar nucleosynthesis can account for the existence of all elements except hydrogen and helium.

____ **15.** Stellar nucleosynthesis causes the abundance of heavy elements in interstellar gas and newborn stars to in-crease with time.

SELF-TEST: FILL IN THE BLANK

1. In a semidetached binary consisting of a white dwarf and a main-sequence or giant companion, matter leaving the companion forms a(n) _____ around the dwarf.

2. A nova explosion is due to _____ fusion on the _____ of a white dwarf.

3. In the collapsing core of a massive star, photodisintegration is caused by the very high _____ in the core.

4. When a proton and an electron are forced together, they combine to form a _____ and a _____.

5. Core collapse in a massive star is eventually stopped by _____ degeneracy pressure.

6. A Type _____ supernova occurs when the core of a massive star implodes and rebounds.

7. A Type _____ supernova occurs when a white dwarf exceeds the Chandrasekhar mass.

8. The maximum mass for a white dwarf is roughly _____ solar masses.

9. The two types of supernovae can be distinguished observa-tionally by their spectra and by their _____.

10. The first detection of supernova 1987A came in the form of a burst of _____.

11. Supernova 1987A surprised astronomers because its pro-genitor star was a _____ instead of a _____ at the moment of the explosion.

12. Elements differing by four mass units, such as carbon, oxygen, neon, and silicon, are produced in stars by _____.

13. Neutron capture is responsible for the formation of elements heavier than _____.

14. The light curve of a Type I supernova is consistent with energy being released by _____

15. Because of stellar nucleosynthesis, the spectra of old stars show _____ heavy elements than those of young stars.

REVIEW AND DISCUSSION

1. Under what circumstances will a binary star produce a nova?

2. What is an accretion disk, and how does one form?

3. What is a light curve? How can it be used to identify a nova or a supernova?

4. What occurs in a massive star to cause it to explode?

5. How do photodisintegration and neutronization contribute to the demise of a massive star?

6. What is neutron degeneracy pressure?

7. What are the observational differences between Type I and Type II supernovae?

8. What is the Chandrasekhar mass, and what does it have to do with supernovae?

9. How do the mechanisms responsible for Type I and Type II supernovae explain their observed differences?

10. Roughly how often would we expect a supernova to occur in our own Galaxy? How often would we expect to see a galactic supernova?

11. What evidence is there that many supernovae have occurred in our Galaxy?

12. How can astronomers estimate the age of an isolated star?

13. What proof do astronomers have that heavy elements are formed in stars?

14. As a star evolves, why do heavier elements tend to form by helium capture rather than by fusion of like nuclei?

15. Why do the cores of massive stars evolve into iron, not heavier elements?

16. How are nuclei heavier than iron formed?

17. What is the r-process? When and where does it occur?

18. Why was supernova 1987A so important?

19. Why are neutrino detectors important to the study of supernovae?

20. Describe the role played by supernovae in "recycling" galactic matter.

PROBLEMS  *Algorithmic versions of these questions are available in the Practice Problems module of the Companion Website.*

The number of squares preceding each problem indicates its approximate level of difficulty.

1. ■■■ Estimate how close a 0.5-Solar-mass white dwarf must come to the center of a 2-solar-mass subgiant of radius 10 times that of the Sun in order for the white dwarf's tidal field to strip matter from the companion's surface. (Hint: The tidal force approximation given in *More Precisely 7-3* won't work here. Go back to the definition of the tidal force as a difference between two gravitational forces.)

2. ■ Calculate the orbital speed of matter in an accretion disk just above the surface of a 0.6-solar-mass, 15,000-km-diameter white dwarf.

3. ■ A certain telescope could just detect the Sun at a distance of 10,000 pc. What is the apparent magnitude of the Sun at this distance? (For convenience, take the Sun's absolute magnitude to be 5.) What is the maximum distance at which it could detect a nova having a peak luminosity of 10^5 solar luminosities?

4. ■ Repeat the previous calculation for a supernova having a peak luminosity 10^{10} times that of the Sun. What would be the apparent magnitude of the explosion if it occurred at a distance of 10,000 Mpc? Would it be detectable by any existing telescope?

5. ■■ At what distance would a supernova of absolute magnitude −20 look as bright as the Sun? As the Moon? Would you expect a supernova to occur that close to us?

6. ■ A (hypothetical) supernova at a distance of 150 pc has an absolute magnitude of −20. Compare its apparent magnitude with that of (a) the full Moon; and (b) Venus at its brightest. (See Figure 17.7.) Would you expect a supernova to occur this close to us?

7. ■ A supernova's energy is often compared to the total energy output of the Sun over its lifetime. Using the Sun's current energy output, calculate its total energy output, assuming it has a 10^{10} year main-sequence lifetime. How does this compare with the energy released by a supernova?

8. ■■ The *Hubble Space Telescope* is observing a distant Type I supernova of peak apparent magnitude 24. Using the light curve in Figure 21.8, estimate how long after the peak brightness the supernova will become too faint to be seen.

9. ■ The Crab Nebula is now about 1 pc in radius. If it was observed to explode in A.D. 1054, roughly how fast is it expanding? (Assume a constant expansion rate. Is that a reasonable assumption?)

10. ■■ If stars form in our Galaxy at an average rate of 10 per year and all stars greater than 8 solar masses explode as supernovae, use Figure 17.23 to estimate the rate of Type II supernovae in our Galaxy.

11. ■■■ Assuming an interstellar extinction of 2 mag/kpc, calculate the maximum distance at which we could see (with the naked eye, limiting magnitude 6) a Galactic supernova of absolute magnitude −19. (See Chapter 18, problem 9 for more on how to go about solving the equation you obtain here.)

12. ▪▪▪ Repeat the previous question, but for a survey telescope of limiting magnitude 18.

13. ▪▪ As we will see in Chapter 23, the star-forming portion of our Galaxy consists of a highly flattened circular disk about 30 kpc (30,000 pc) in diameter. Interstellar extinction limits our view to within a radius of about 5 kpc of the Sun. If supernovae occur in the Galaxy roughly once every 30 years, on average, and are uniformly spread throughout the disk, calculate how often we should expect to see a supernova.

14. ▪▪ Assuming the data in the previous question, taking all supernovae, for simplicity, to have absolute magnitude −20, and ignoring interstellar extinction for such nearby events, calculate how often we should expect to observe a supernova brighter than the full Moon (apparent magnitude −12.5).

15. ▪▪ Based on the data in Table 21.1, estimate the fraction by mass of "iron-group" elements and the total mass of all elements in the Sun. Compare this with Earth's mass.

COLLABORATIVE EXERCISES

1. Supernova Brightness. Each group member should select a different star listed in Appendix 3: Table 4—The Twenty Brightest Stars—and determine what its new apparent visual magnitude would be if it became a supernova and increased in brightness by a factor of 10,000.

RESEARCHING ON THE WEB  *To complete the following exercises, go to the online Destinations module for Chapter 21 on the Companion Website for Astronomy Today 4/e.*

1. Access the "Supernova Remnants" page and describe the Cygnus loop in terms of what object it is a part of and where it is found in the sky.

PROJECTS

1. In 1758, the French comet hunter Charles Messier discovered the sky's most legendary supernova remnant, now called M1, or the Crab Nebula. It is located northwest of Zeta Tauri, the star that marks the southern tip of the horns of Taurus the Bull. Try to find it—an 8-inch telescope reveals the Crab's oval shape, but it will appear faint; a 10-inch or larger telescope reveals some of its famous filamentary structure.

2. In the *Handbook of Chemistry and Physics*, available in the library reference section, look up the table of isotopes. Pick one or more isotopes and follow their decay into a final stable isotope. For example, choose cobalt-59, formed in the s-process. Note how the isotope decays, what is emitted, and the half-life of the decays. Try this exercise for uranium-235, uranium-238, and plutonium-239.

SKYCHART III PROJECTS  *The SkyChart III Student Version planetarium program on which these exercises are based is included as a separately executable program on the CD in the back of this text.*

1. ▪ The Crab Nebula in Taurus, also known as M1, is a famous diffuse nebula that is not particularly easy to discern under poor seeing conditions. Use SkyChart III to find the location of M1. The easiest way to locate it with the software is with *VIEW/Center Object/Crab Nebula*. Set *VIEW/5° FIELD* so you can see which object is M1. Then zoom out to a field of about 45° so you can see where it is in relation to Pleiades, M45, Orion, and Gemini.

 *In addition to the Practice Problems and Destinations modules, the Companion Website at http://www.prenhall.com/chaisson provides for each chapter an additional true-false, multiple choice, and labeling quiz, as well as additional annotated images, animations, and links to related Websites.*

22 NEUTRON STARS AND BLACK HOLES

Strange States of Matter

LEARNING GOALS

Studying this chapter will enable you to:

1   Describe the properties of neutron stars and explain how these strange objects are formed.

2   Explain the nature and origin of pulsars and account for their characteristic radiation.

3   List and explain some of the observable properties of neutron-star binary systems.

4   Discuss the basic characteristics of gamma-ray bursts, and some theoretical attempts to explain them.

5   Describe how black holes are formed and discuss their effects on matter and radiation in their vicinity.

6   Relate the phenomena that occur near black holes due to the warping of space around them.

7   Discuss the difficulties in observing black holes and explain some of the ways in which the presence of a black hole might be detected.

Visit http://www.prenhall.com/chaisson for additional annotated images, animations, and links to related sites for this chapter.

This peculiar X-ray image of a pulsar shows details heretofore unseen and still not well understood. This is the Vela pulsar (brightest spot at center), located in the midst of the debris field known as the Vela supernova remnant. The double crossbow appearance of the nebular gas around the pulsar is probably caused by the motion (along the arrow toward upper right) of the pulsar through the debris field. Vela is about 800 light-years away; this image spans 0.7 light-year. *(CXO/SAO)*

The Big Picture: Neutron stars, pulsars, and black holes are among the more exotic members of the vast population of stars throughout the universe. They all represent the end states of stellar systems and often exhibit accretion disks and ejected jets, at least for a while after "death." Remarkably, they all fit within our models of stellar evolution; indeed, they were predicted by theory long before they were discovered in space.

Our study of stellar evolution has led us to some very unusual and unexpected objects. Red giants, white dwarfs, and supernova explosions surely represent extreme states of matter completely unfamiliar to us here on Earth. Yet stellar evolution can have even more bizarre consequences. The strangest states of all result from the catastrophic implosion–explosion of stars much more massive than our Sun. The almost unimaginable violence of a supernova explosion may bring into being objects so extreme in their behavior that they require us to reconsider some of our most hallowed laws of physics. They open up a science-fiction writer's dream of fantastic phenomena. They may even one day force scientists to construct a whole new theory of the universe.

22.1 Neutron Stars

⬚ What remains after a supernova explosion? Is the entire progenitor star blown to bits and dispersed throughout interstellar space, or does some portion of it survive? For a Type I (carbon-detonation) supernova, most astronomers regard it as quite unlikely that any central remnant is left after the explosion. The entire star is shattered by the blast. However, for a Type II supernova, involving the implosion and subsequent rebound of a massive star's iron core, theoretical calculations indicate that part of the star may survive. ⬯ (Sec. 21.2) The explosion destroys the parent star, but it may leave a tiny ultracompressed remnant at its center. Even by the high-density standards of a white dwarf, however, the matter within this severely compressed core is in a very strange state, unlike anything we are ever likely to find (or create) on Earth.

Recall from Chapter 21 that during the moment of implosion of a massive star—just prior to the supernova explosion itself—the electrons in the core violently smash into the protons there, forming neutrons and neutrinos. ⬯ (Sec. 21.2) The neutrinos leave the scene at (or nearly at) the speed of light, accelerating the collapse of the neutron core, which continues to contract until its particles come into contact. At that point, neutron degeneracy pressure causes the central portion of the core to rebound, creating a powerful shock wave that races outward through the star, violently expelling matter into space.

The key point here is that the shock wave does not start at the very center of the collapsing core. The innermost part of the core—the region that bounces—remains intact as the shock wave it causes destroys the rest of the star. After the violence of the supernova has subsided, this ball of neutrons is all that is left. Researchers colloquially call this core remnant* a **neutron star**, although it is not a star in any true sense of the word—all its nuclear reactions have ceased forever.

*Astronomers commonly use the term remnant to mean whatever remains of a star's inner core after evolution has ended. Such objects are small and compact—no larger than Earth. They should not be confused with supernova remnants, which are the aftermath of supernova explosions: glowing clouds of debris scattered across many parsecs of interstellar space. ⬯ (Sec. 21.3)

Neutron stars are extremely small and very massive. Composed purely of neutrons packed together in a tight ball about 20 km across, a typical neutron star is not much bigger than a small asteroid or a terrestrial city (see Figure 22.1), yet its mass is greater than that of the Sun. With so much mass squeezed into such a small volume, neutron stars are incredibly dense. Their average density can reach 10^{17} or even 10^{18} kg/m^3, nearly a billion times denser than a white dwarf. A single thimbleful of neutron-star material would weigh 100 million tons—about as much as a good-sized terrestrial mountain. For comparison, the density of a normal atomic nucleus is about 3×10^{17} kg/m^3. In a sense, we can think of a neutron star as a single enormous nucleus, with an atomic mass of around 10^{57}.

Neutron stars are solid objects. Provided that a sufficiently cool one could be found, you might even imagine standing on it. However, this would not be easy, as a neutron star's gravity is extremely powerful. A 70-kg

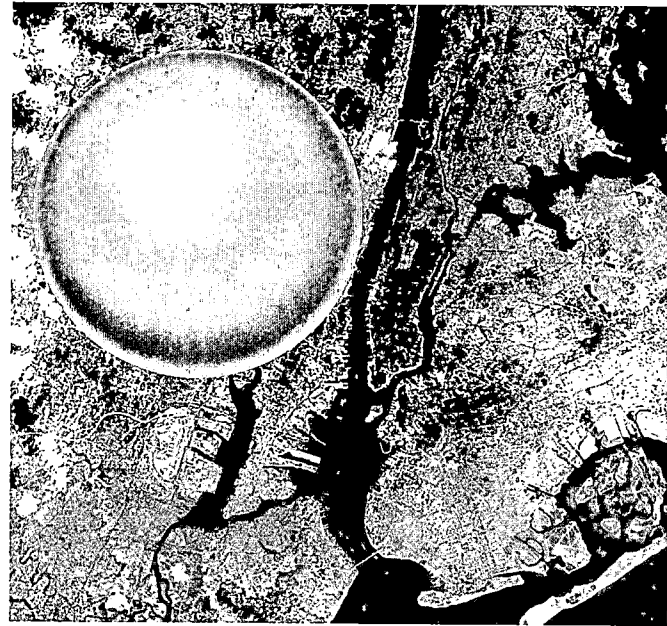

Figure 22.1 Neutron Star Neutron stars are not much larger than many of Earth's major cities. In this fanciful comparison, a typical neutron star sits alongside Manhattan Island. *(NASA)*

(150-pound) human would weigh the Earth-equivalent of about 1 billion kg (1 million tons). The severe pull of a neutron star's gravity would flatten you much thinner than this piece of paper!

In addition to large mass and small size, newly formed neutron stars have two other very important properties. First, they *rotate* extremely rapidly, with periods measured in fractions of a second. This is a direct result of the law of conservation of angular momentum (Chapter 15), which tells us that any rotating body must spin faster as it shrinks. ∞ (*More Precisely 15-1*) Even if the core of the progenitor star were initially rotating quite slowly (once every couple of weeks, say, as is observed in many upper main-sequence stars), it would be spinning a few times per second by the time it had reached a diameter of 20 km. Second, newborn neutron stars have very strong *magnetic fields*. The original field of the progenitor star is amplified by the collapse of the core because the contracting material squeezes the magnetic field lines closer together, creating a magnetic field trillions of times stronger than Earth's field.

In time, theory indicates, our neutron star will spin more and more slowly as it radiates its energy into space, and its magnetic field will diminish. However, for a few million years after its birth, these two properties combine to provide the primary means by which this strange object can be detected and studied.

☑ Concept Check

■ Are all supernovae expected to lead to neutron stars?

22.2 Pulsars

2 Can we be sure that objects as strange as neutron stars really exist? The answer is a confident yes. The first observation of a neutron star occurred in 1967, when Jocelyn Bell, a graduate student at Cambridge University, made a surprising discovery. She observed an astronomical object emitting radio radiation in the form of rapid *pulses*. Each pulse consisted of a 0.01-second (s) burst of radiation, after which there was nothing. Then, 1.34 s later, another pulse would arrive. The time interval between pulses was aston-

ishingly uniform—so accurate, in fact, that the repeated emissions could be used as a very precise clock. Figure 22.2 is a recording of part of the radio radiation from the pulsating object Bell discovered.

Many hundreds of these pulsating objects are now known in the Milky Way Galaxy. They are called **pulsars**. Each has its own characteristic pulse period and duration. The pulse periods of some pulsars are so stable that they are by far the most accurate natural clocks known in the universe—more accurate even than the best atomic clocks on Earth. In some cases the period is predicted to change by only a few seconds in a million years.

When Bell made her discovery in 1967, she did not know what she was looking at. Indeed, no one at the time knew what a pulsar was. The explanation of pulsars as spinning neutron stars won Bell's thesis advisor, Anthony Hewish, a share of the 1974 Nobel Prize in physics. Hewish reasoned that the only physical mechanism consistent with such precisely timed pulsations is a small, rotating source of radiation. Only rotation can cause the high degree of regularity of the observed pulses, and only a small object can account for the sharpness of each pulse. Radiation emitted from different regions of an object larger than a few tens of kilometers across would arrive at Earth at slightly different times, blurring the pulse profile. The best current model describes a pulsar as a compact, spinning neutron star that periodically flashes radiation toward Earth.

Figure 22.3 outlines the important features of this pulsar model. Two "hot spots" on the surface of a neutron star, or in the magnetosphere just above the surface, continuously emit radiation in a narrow "searchlight" pattern. These spots are most likely localized regions near the neutron star's magnetic poles, where charged particles, accelerated to extremely high energies by the star's rotating magnetic field, emit radiation along the star's magnetic axis. The hot spots radiate more or less steadily and the resulting beams sweep through space, like a revolving lighthouse beacon, as the neutron star rotates. Indeed, this pulsar model is often known as the **lighthouse model**. If the neutron star happens to be oriented such that the beam of its pulses sweeps across Earth, we see the pulses. The beams are observed as a series of rapid pulses—each time one of the beams sweeps past Earth, a pulse is seen. The period of the pulses is the star's rotation period.

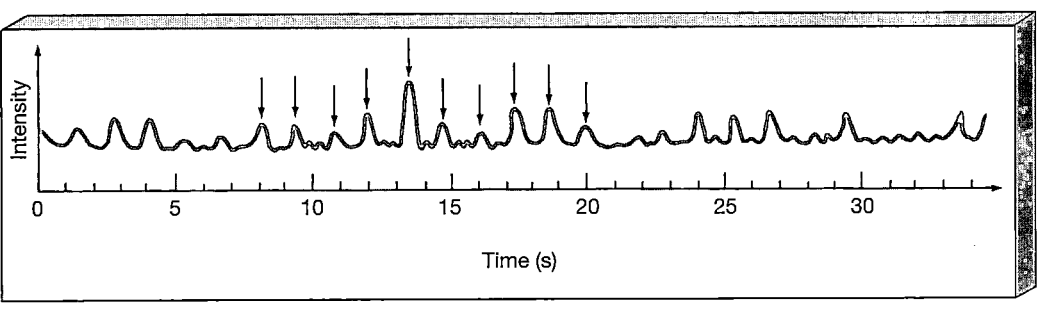

Figure 22.2 Pulsar Radiation Pulsars emit periodic bursts of radiation. This recording shows the regular change in the intensity of the radio radiation emitted by the first such object known. It was discovered in 1967. Some of the pulses are marked by arrows.

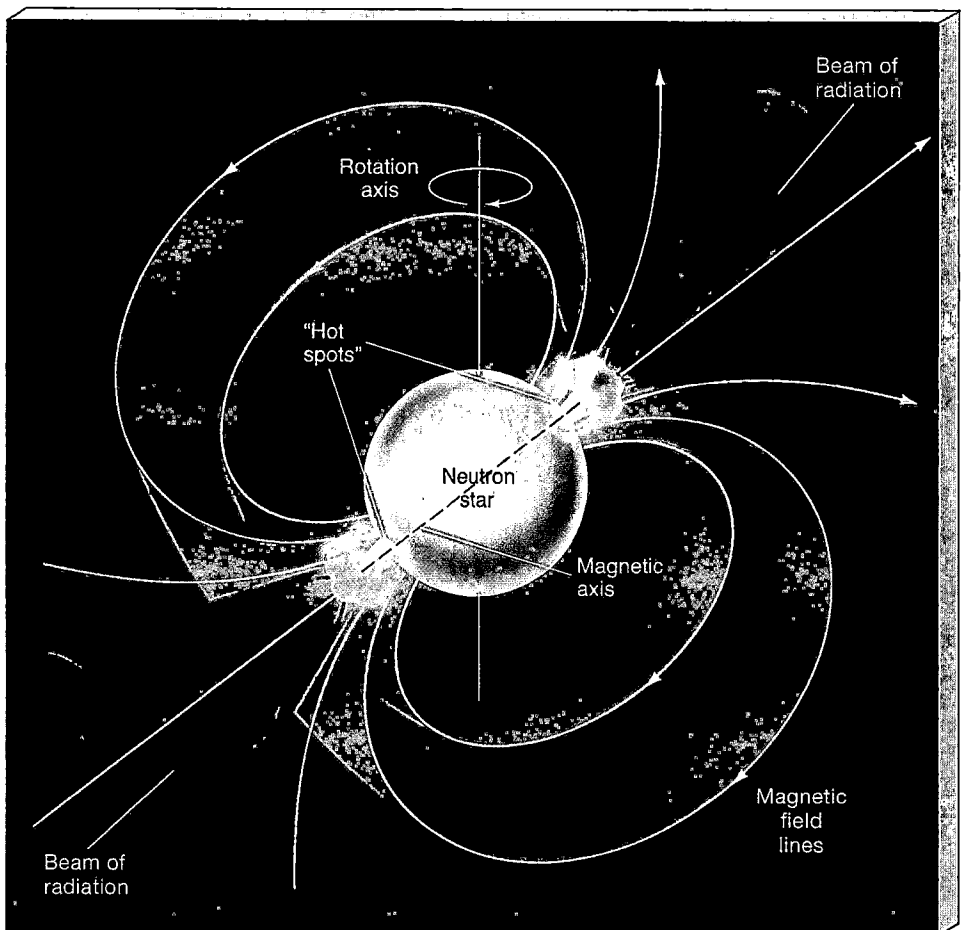

Figure 22.3 Pulsar Model
This diagram of the "lighthouse model" of neutron-star emission accounts for many of the observed properties of pulsars. Charged particles, accelerated by the magnetism of the neutron star, flow along the magnetic field lines, producing radiation that beams outward.

A few pulsars are clearly associated with supernova remnants, although not all such remnants have a detectable pulsar within them. Figure 22.4(a) shows a pair of optical photographs of the Crab pulsar, at the center of the Crab supernova remnant. ∞ (Sec. 21.3) In the left frame, the pulsar is off; in the right frame, it is on. The pulse period is about 33 milliseconds. Figure 22.4(b) shows that the Crab also pulses in X rays. By observing the speed and direction of the Crab's ejected matter, astronomers can work backward to pinpoint the location in space at which the explosion must have occurred and where the supernova core remnant should be located. That is precisely the region of the Crab Nebula from which the pulsating signals arise. The Crab pulsar is evidently all that remains of the once-massive star whose supernova was observed in A.D. 1054. *Discovery 21-2* presents more detail on the complex interaction between the pulsar and the nebula in which it resides.

Most pulsars emit their pulses in the form of radio radiation, but some have been observed to pulse in the visible, X-ray, and gamma-ray parts of the spectrum as well. Figure 22.5 shows the Crab and the nearby Geminga pulsar in gamma rays. Geminga is unusual in that, while it pulsates strongly in gamma rays, it is barely

detectable in visible light and not at all at radio wavelengths. Whatever types of radiation are produced, these electromagnetic flashes at different frequencies all occur at regular, repeated time intervals, as we would expect since they arise from the same object. However, pulses at different frequencies do not necessarily all occur at the same instant in the pulse cycle. The periods of most pulsars are quite short, ranging from about 0.03 s to 0.3 s (flashing between 3 and 30 times per second). The human eye is insensitive to such rapid flashes, making it impossible to observe the flickering of a pulsar by eye, even using a large telescope. Fortunately, instruments can record pulsations of light that the human eye cannot detect.

All pulsars are neutron stars, but not all neutron stars are pulsars, for two reasons. First, the two ingredients that make the neutron star pulse—rapid rotation and strong magnetic field—both diminish with time, so the pulses gradually weaken and become less frequent. Theory indicates that within a few tens of millions of years, the pulsations all but stop. Second, even a young, bright pulsar is not necessarily visible from Earth. The pulsar beam depicted in Figure 22.3 is relatively narrow—perhaps as little as a few degrees across in some

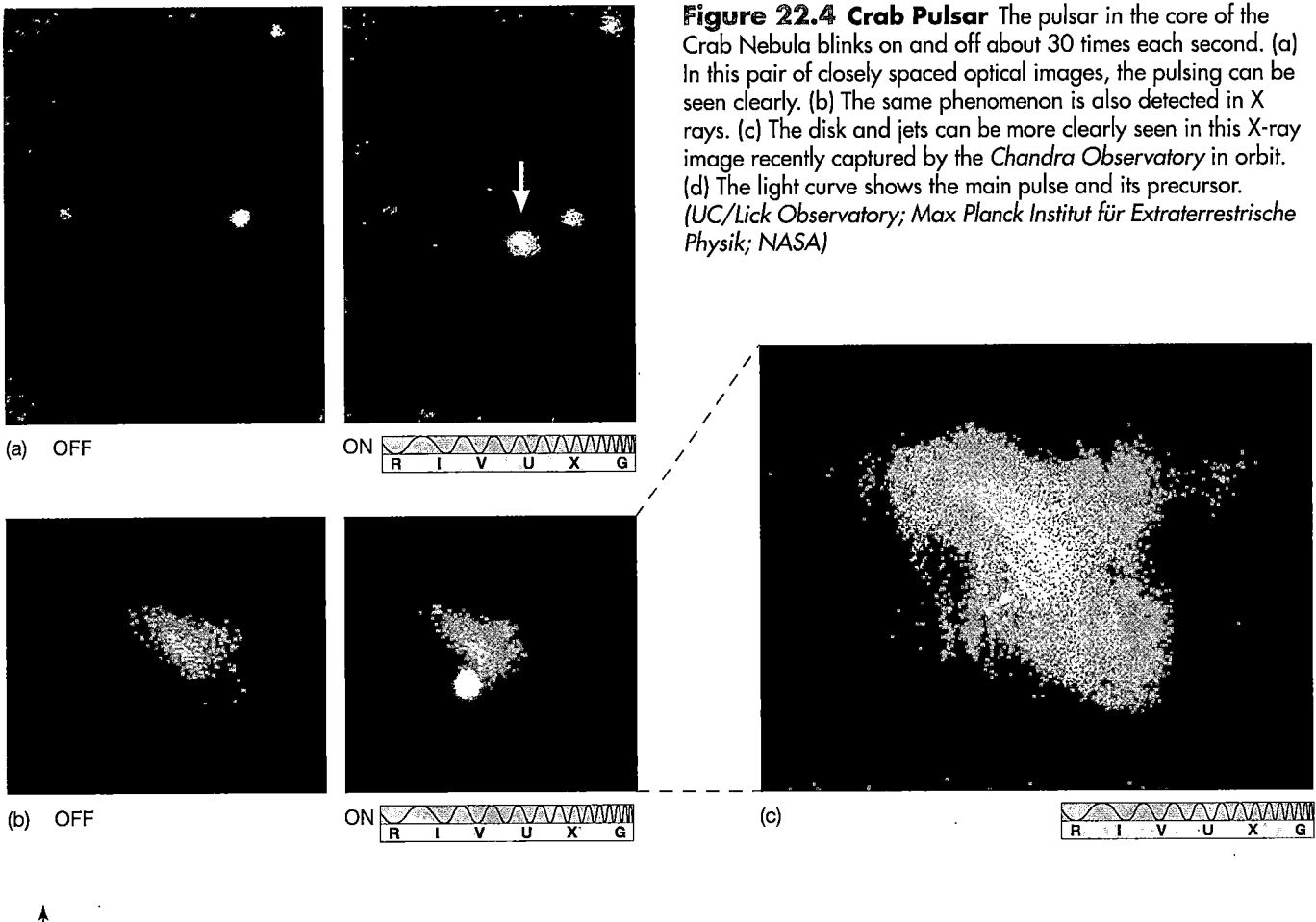

(a) OFF   ON

(b) OFF   ON

(c)

Figure 22.4 Crab Pulsar The pulsar in the core of the Crab Nebula blinks on and off about 30 times each second. (a) In this pair of closely spaced optical images, the pulsing can be seen clearly. (b) The same phenomenon is also detected in X rays. (c) The disk and jets can be more clearly seen in this X-ray image recently captured by the *Chandra Observatory* in orbit. (d) The light curve shows the main pulse and its precursor. *(UC/Lick Observatory; Max Planck Institut für Extraterrestrische Physik; NASA)*

(d)

Intensity (vertical axis)

0   0.01   0.02   0.03

Time (seconds)

cases. Only if the neutron star happens to be oriented in just the right way do we see pulses. When we can see the pulses from Earth, we call the body a pulsar.

Figure 22.6 shows three superposed *Hubble* images of a neutron star that is not a pulsar. First identified by *ROSAT* and *EUVE* observations of its soft X-ray emission, this object is just 30 km across and has a surface temperature of about 700,000 K. Despite this enormous temperature, its small size means that it is very faint—just 25th magnitude in visible light. ∞ (Sec. 17.5). This figure shows the neutron star's proper motion over a period of almost three years. Small wobbles in the motion due to Earth's parallax can also (just) be discerned. The

star is thought to be about 1 million years old, lies about 60 pc away, and is moving at a speed of 110 km/s across our line of sight.

In fact, most known pulsars are observed (usually by Doppler measurements) to have high velocities—much greater than the typical speeds of stars in the Galaxy. The most likely explanation for these anomalously high speeds is that pulsars can receive substantial "kicks" due to asymmetries in the supernovae in which they formed. These asymmetries, which are predicted by theory, are generally not very pronounced, but if the supernova's enormous energy is channeled even slightly in one direction, the newborn neutron star can recoil in the opposite direction with a speed of many tens or even hundreds of kilometers per second.

Given our current knowledge of star formation, stellar evolution, and neutron stars, pulsar observations are consistent with the idea that every high-mass star dies in a Type II supernova, leaving a neutron star behind, and that *all* neutron stars emit beams of radiation, just like the pulsars we actually see.

✓ Concept Check

■ Why don't we see all neutron stars as pulsars?

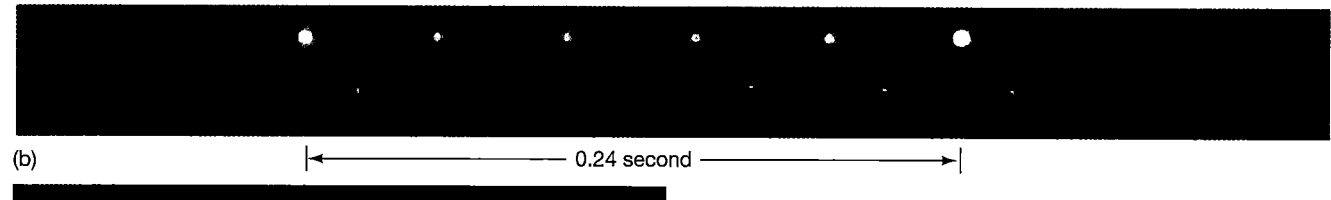

(b) |←——————— 0.24 second ———————→|

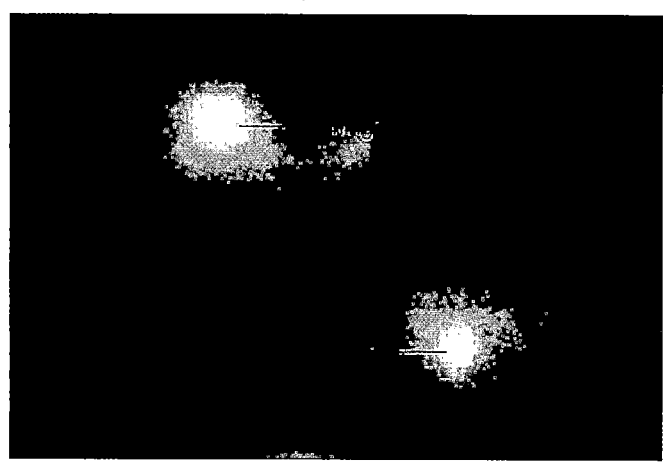

Figure 22.5 Gamma-Ray Pulsars (a) The Crab and Geminga pulsars, which happen to lie fairly close (about 4.5 degrees) to one another in the sky. Unlike the Crab, Geminga is barely visible at optical wavelengths, and undetectable in the radio. (b) Sequence of *Compton Gamma-Ray Observatory* images showing Geminga's 0.24-s pulse period. (The Crab's 33 ms period is too rapid to be resolved by the detector.) *(NASA)*

R I V U X G

(a)

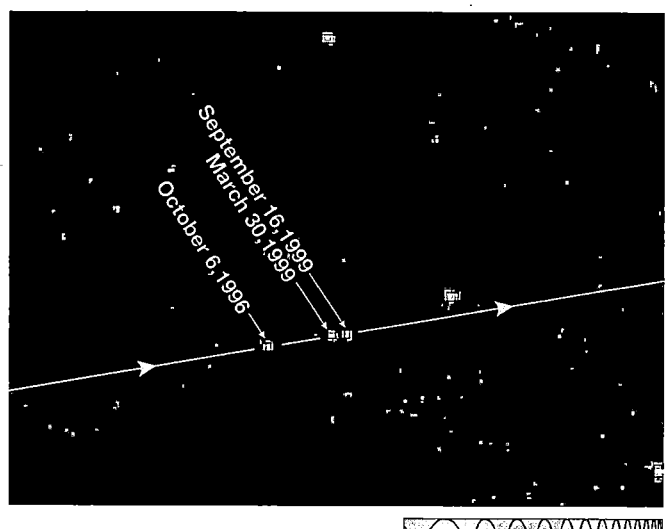

R I V U X G

Figure 22.6 Isolated Neutron Star This lone neutron star was first detected by its X-ray emission, and subsequently imaged by *Hubble*. It lies about 60 pc from Earth. The neutron star is shown here streaking across the sky at more than 100 km/s, in a superposition of three images spanning three years. *(NASA)*

22.3 Neutron-Star Binaries

 We noted in Chapter 17 that most stars are not single but instead are members of binary systems. ∞ (Sec. 17.9) Although many pulsars are known to be isolated (that is, not part of any binary), at least some do have binary companions, and the same is true of neutron stars in general (even the ones not seen as pulsars). One important consequence of this is that some neutron star masses have been

determined very accurately. All the measured masses are fairly close to 1.4 times the mass of the Sun—the Chandrasekhar mass of the stellar core that collapsed to form the neutron-star remnant.

X-RAY SOURCES

The late 1970s saw several important discoveries about neutron stars in binary-star systems. Numerous X-ray sources were discovered near the central regions of our Galaxy and also near the centers of a few rich star clusters. Some of these sources, known as **X-ray bursters**, emit much of their energy in violent eruptions, each thousands of times more luminous than our Sun, but lasting only a few seconds. A typical burst is shown in Figure 22.7.

This X-ray emission is thought to arise on or near neutron stars that are members of binary systems. Matter torn from the surface of the (main-sequence or giant) companion by the neutron star's strong gravitational pull accumulates on the neutron star's surface. As in the case of white-dwarf accretion (see Chapter 21), the material does not fall directly onto the surface. Instead, as illustrated in Figure 22.8(a), it forms an accretion disk. (Compare with Figure 21.2, which depicts the white-dwarf equivalent.) The gas goes into a tight orbit around the neutron star, then slowly spirals inward. The inner portions of the accretion disk become extremely hot, releasing a steady stream of X rays.

As gas builds up on the neutron star's surface its temperature rises due to the pressure of overlying material. Eventually, it becomes hot enough to fuse hydrogen. The result is a sudden period of rapid nuclear burning that releases a huge amount of energy in a brief but intense flash of X rays—an *X-ray burst*. After several hours of renewed accumulation, a fresh layer of matter produces the next

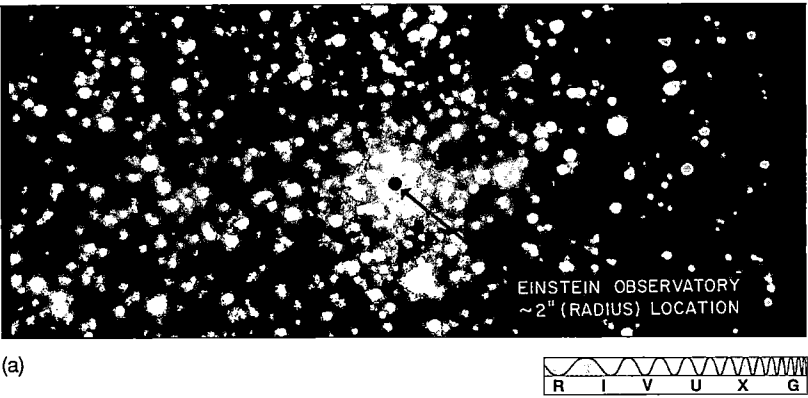

(a)

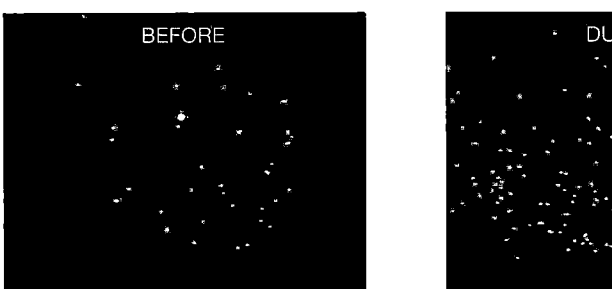

BEFORE

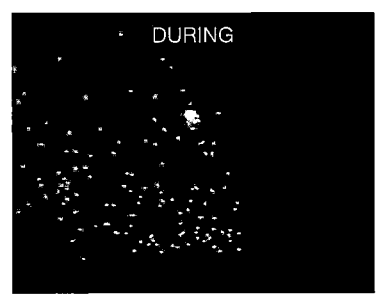

DURING

(b)

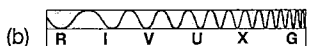

Figure 22.7 X-Ray Burster An X-ray burster produces a sudden, intense flash of X rays, followed by a period of relative inactivity lasting as long as several hours. Then another burst occurs. The bursts are thought to be caused by explosive nuclear burning on the surface of an accreting neutron star, similar to the explosions on a white dwarf that give rise to novae. (a) An optical photograph of the star cluster Terzan 2, showing a 2″ dot at the center where the X-ray bursts originate. (b) X-ray images taken before and during the outburst. The most intense X rays correspond to the position of the dot shown in frame (a). *(SAO)*

burst. Thus, an X-ray burst is much like a nova explosion on a white dwarf, but occurring on a far more violent scale because of the neutron star's much stronger gravity. ⊂⊃ (Sec. 21.1) Not all the infalling gas makes it onto the neutron star surface, however. In at least one case—an object known as SS 433*—we have direct observational evidence that some material is instead shot completely out of the system at enormously high speeds. SS 433 expels more than one Earth mass of material every year in the form of two oppositely directed narrow jets moving roughly perpendicular to the disk. Observations of the Doppler shifts of optical emission lines produced within the jets themselves imply speeds of almost 80,000 km/s—over 25 percent of the speed of light! As the jets interact with the interstellar medium they emit radio radiation, as shown in Figure 22.8(b).

Jets of this sort are apparently quite common in astronomical systems in which an accretion disk surrounds a compact object (such as a neutron star or a black hole). They are believed to be produced by the intense radiation and magnetic fields near the inner edge of the disk, although the details of their formation are still uncertain. Incidentally, they are *not* the "lighthouse" beams of radiation from the neutron star itself (Figure 22.4) that can result in a pulsar. Although SS 433 is the only stellar-mass compact object currently known to produce jets, we will see examples of similar phenomena on much larger scales in later chapters. One of the most important aspects of SS 433 is that we can actually study both the disk and the jets

---

* The name simply identifies it as the 433ʳᵈ entry in a particular catalog of stars with strong optical emission lines.

instead of simply having to assume their existence, as in more distant cosmic objects.

MILLISECOND PULSARS

In the mid-1980s an important new category of pulsars was found—a class of very rapidly rotating objects called **millisecond pulsars**. Several dozen millisecond pulsars are currently known in the Milky Way Galaxy. These objects spin hundreds of times per second (that is, their pulse period is a few milliseconds, 0.001 s). This speed is about as fast as a typical neutron star can spin without flying apart. In some cases, the star's equator is moving at more than 20 percent of the speed of light. This speed suggests a phenomenon bordering on the incredible—a cosmic object of kilometer dimensions, more massive than our Sun, spinning almost at breakup speed, making nearly 1000 complete revolutions *every second*. Yet the observations and their interpretation leave little room for doubt.

The story of these remarkable objects is further complicated because many of them are found in globular clusters. This is odd because globular clusters are known to be very old—10 billion years, at least. Yet Type II supernovae (the kind that create neutron stars) are associated with massive stars that explode within a few tens of *millions* of years after their formation, and no stars have formed in any globular cluster since the cluster itself came into being. Thus, no new neutron star has been produced in a globular cluster in a very long time. But, as mentioned earlier, the pulsar produced in a supernova explosion is expected to slow down in only a few million years, after 10 billion years its rotation should have all but ceased. The rapid rotation of the pulsars found in globular clusters cannot be a relic of their birth. These objects must have been- "spun up"—that is, had their rotation rates increased—by some other, much more recent, mechanism.

The most likely explanation for the high rotation rate of these objects is that the neutron star has been spun up by drawing in matter from a companion star. As matter spirals down onto the star's surface in an accretion disk it provides the "push" needed to make the neutron star spin faster (see Figure 22.9). Theoretical calculations indicate that this process can spin the star up to breakup speed in about a hundred million years. Subsequently, an encounter with another star may eject the neutron star from the binary, or the pulsar's radiation may destroy its companion, so an isolated millisecond pulsar results. This general picture is supported by the finding that, of the 50 or so millisecond pulsars seen in globular clusters, roughly half are known currently to be members of binary systems. These numbers

are quite consistent with the rate at which binaries can be broken up by encounters with other cluster members.

Thus, although a pulsar like the Crab is the direct result of a supernova explosion, millisecond pulsars are the product of a two-stage process. The neutron star was formed in an ancient supernova, billions of years ago. Only relatively recently, through interaction with a binary companion, has it achieved the rapid spin that we observe today. Once again, we see how members of a binary system can evolve in ways quite different from single stars. Notice that the scenario of accretion onto a neutron star from a binary companion is the same scenario that we just used to explain the existence of X-ray bursters. In fact, the two phenomena are very closely linked. Many X-ray bursters may be on their way to becoming millisecond pulsars.

The way in which a neutron star can become a member of a binary system is the subject of active research, because the violence of a supernova explosion would be expected to blow the binary apart in many cases. Only if the supernova progenitor lost a lot of mass before the explosion would the binary system be likely to survive. Alternatively, by interacting with an existing binary and displacing one of its components, a neutron star may become part of a binary system *after* it is formed, as depicted in Figure 22.10. Astronomers are eagerly searching the skies for more millisecond pulsars to test their ideas.

PULSAR PLANETS

Radio astronomers can capitalize on the precision with which pulsar signals repeat themselves to make extremely accurate measurements of pulsar motion. In January 1992, radio astronomers at the Arecibo Observatory

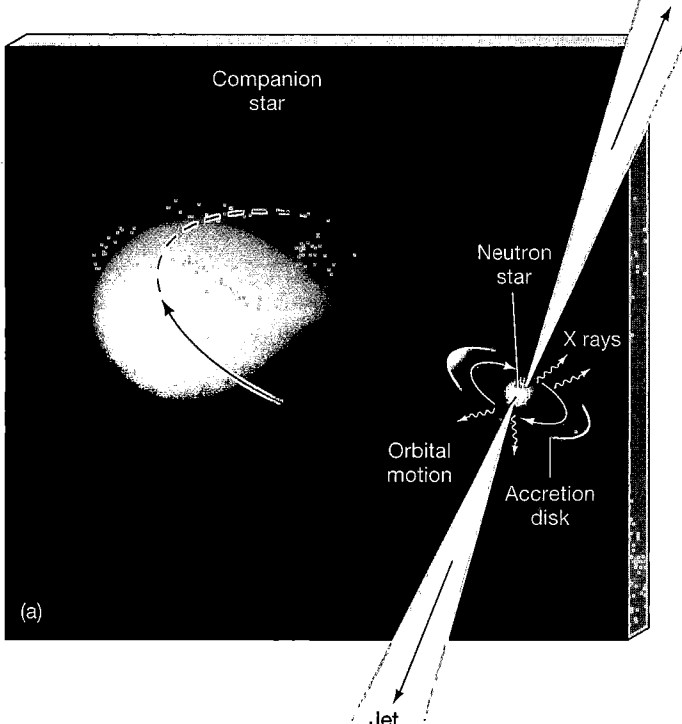

(a)

Figure 22.8 X-Ray Emission (a) Matter flows from a normal star toward a compact neutron-star companion and falls toward the surface in an accretion disk. As the gas spirals inward under the neutron star's intense gravity, it heats up, becoming so hot that it emits X rays. In at least one instance—the peculiar object SS 433—some material may be ejected in the form of two high-speed jets of gas. (b) False-color radiographs of SS 433, made at monthly intervals (left to right), show the jets moving outward and the central source rotating under the gravitational influence of the companion star. *(NRAO)*

(b)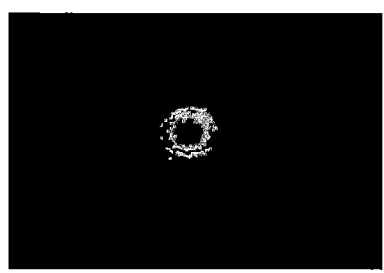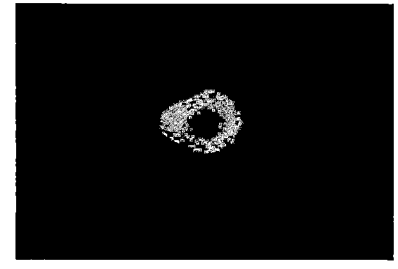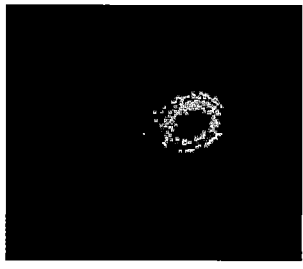

R I V U X G

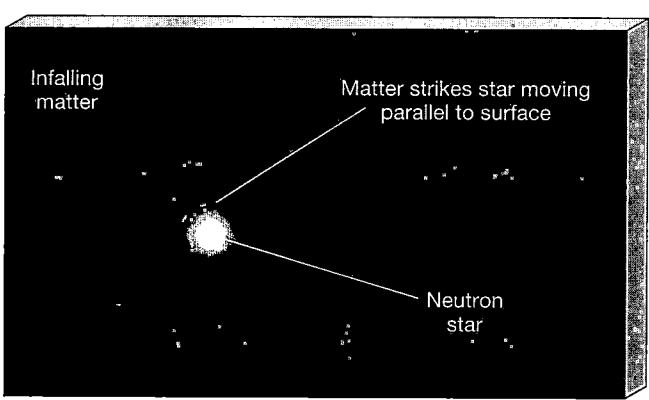

Figure 22.9 Millisecond Pulsar Gas from a companion star spirals down onto the surface of a neutron star. As the infalling matter strikes the star it moves almost parallel to the surface, so it tends to make the star spin faster. Eventually, this process can result in a millisecond pulsar—a neutron star spinning at the incredible rate of hundreds of revolutions per second.

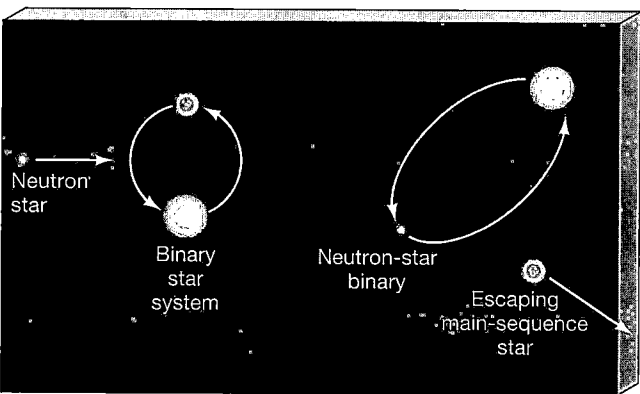

Figure 22.10 Binary Exchange A neutron star can encounter a binary made up of two low-mass stars, ejecting one of them and taking its place. This mechanism provides a means of forming a binary system with a neutron-star component (which may later evolve into a millisecond pulsar) without having to explain how the binary survived the supernova explosion that formed the neutron star.

found that the pulse period of a recently discovered millisecond pulsar lying some 500 pc from Earth varies in an unexpected but quite regular way. Careful analysis of the data has revealed that the period fluctuates on two distinct time scales—one of 67 days, the other of 98 days. The changes in the pulse period are small—less than one part in 10^7—but repeated observations have confirmed their reality.

The leading explanation for these fluctuations holds that they are caused by the Doppler effect as the pulsar wobbles back and forth in space. But what causes the wobble? The Arecibo group believes it is the result of the combined gravitational pulls of not one but *two* planets, each about three times the mass of Earth! One orbits the pulsar at a distance of 0.4 A.U. and the other at a distance of 0.5 A.U. Their orbital periods are 67 and 98 days, respectively, matching the timing of the fluctuations. In April 1994 the group announced further observations that not only confirmed their earlier findings but also revealed the presence of a *third* body, with mass comparable to Earth's Moon, orbiting only 0.2 A.U. from the pulsar.

These remarkable results constituted the first definite evidence of planet-sized bodies outside our solar system. A few other millisecond pulsars have since been found with similar behavior. However, it is unlikely that any of these planets formed in the same way as our own. Any planetary system orbiting the pulsar's progenitor star was almost certainly destroyed in the supernova explosion that created the pulsar. As a result, scientists are still uncertain about how these planets came into being. One possibility involves the binary companion that provided the matter necessary to spin the pulsar up to millisecond speeds. Possibly the pulsar's intense radiation and strong gravity destroyed the companion, then spread its matter out into a disk (a lit-

tle like the solar nebula) in whose cool outer regions the planets might have condensed.

Astronomers have been searching for decades for planets orbiting main-sequence stars like our Sun, on the assumption that planets are a natural by-product of star formation. ⚭ (Sec. 15.2) As we have seen, these searches have now identified many extrasolar planets, although the planetary systems discovered to date don't look much like our own, and nothing comparable in size to Earth has yet been detected. ⚭ (Sec. 15.5) It is ironic that the first and only Earth-sized planets to be found outside the solar system orbit a dead star and have little or nothing in common with our own world.

☑ Concept Check

■ What is the connection between X-ray sources and millisecond pulsars?

22.4 Gamma-Ray Bursts

⚄ Discovered serendipitously in the late 1960s by military satellites looking for violators of the Nuclear Test Ban Treaty, and first made public in the 1970s, **gamma-ray bursts** have developed into one of the deepest mysteries in astronomy today. The bursts consist of bright, irregular flashes of gamma rays typically lasting only a few seconds (Figure 22.11a). Until the 1990s, it was thought that gamma-ray bursts were basically "scaled-up" versions of X-ray bursters, in which matter accreted from the binary companion experienced even more violent nuclear burning, accompanied by the release of the more energetic gamma rays. However, it now appears that this is not the case.

DISTANCES AND LUMINOSITIES

Figure 22.11(b) shows an all-sky plot of the positions of 2704 bursts detected by the *Compton Gamma-Ray Observatory (CGRO)* during its nine-year operational lifetime. ∞ (Sec. 5.6) On average, *CGRO* detected gamma-ray bursts at the rate of about one a day. Note that the bursts are distributed uniformly across the sky (their distribution is said to be "isotropic"), rather than being confined to the relatively narrow band of the Milky Way (compare Figure 5.35). The bursts seemingly never repeat at the same location, show no obvious clustering, and appear unaligned with any known large-scale structure, near or far.

The isotropy of the *CGRO* data has convinced most astronomers that the bursts do not originate within our own Galaxy, as had previously been assumed, but instead are produced far beyond the Milky Way—at so-called cosmological distances. In addition, *Compton* found faint bursts to be less frequent than would be expected from an infinite, unbounded population of such sources. Apparently there is an outer limit of some sort to their distribution in space. However, it remains unclear whether the bursts are associated with the isotropic spread of distant galax-

ies—or whether they are something else yet again, outside of galaxies and somewhere in intergalactic space.

Actual distances to gamma-ray bursts are difficult to determine because astronomers often cannot find any optical counterparts, or any kind of association with known cosmic objects. Gamma rays are far too penetrating to be focused by conventional optics. As a result, *CGRO*'s burst positions were uncertain by several arc degrees, so optical or X-ray telescopes had to scan very large regions of the sky as they searched for the burst. Compounding this problem, the "afterglows" at X-ray or optical wavelengths generally fade quite rapidly, severely limiting the time available to find the counterpart.

The first direct measurement of the distance to a gamma-ray burst was made on May 8, 1997, thanks to a combination of gamma-ray, X-ray, and optical observations of the gamma-ray burst GRB 970508. The Italian-Dutch *BeppoSAX* satellite, launched in 1996, recorded the burst in both the gamma- and X-ray regions of the spectrum. The importance of the X-ray observations is that they allowed a much more accurate determination of the burst's location on the sky—in fact, to within a few arc minutes. That was enough for ground-based optical astronomers to look for and find GRB 970508's optical afterglow. For the first time, astronomers had observed an optical counterpart to a gamma-ray burst.

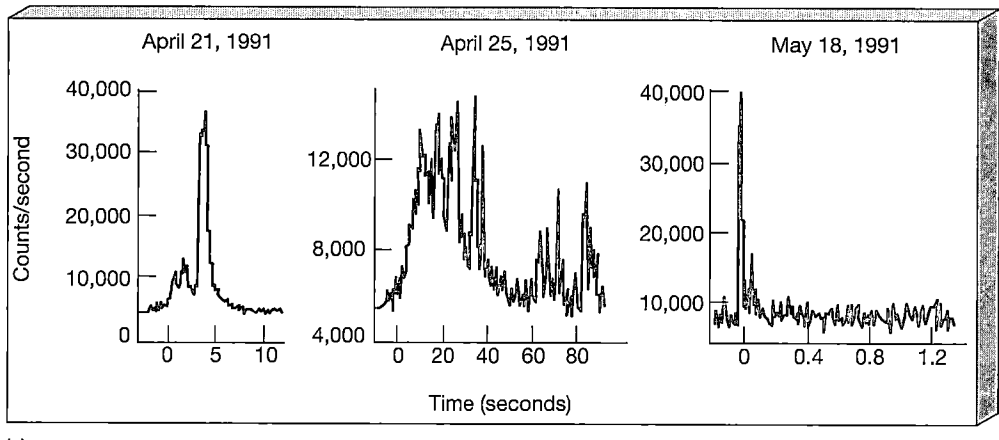

(a)

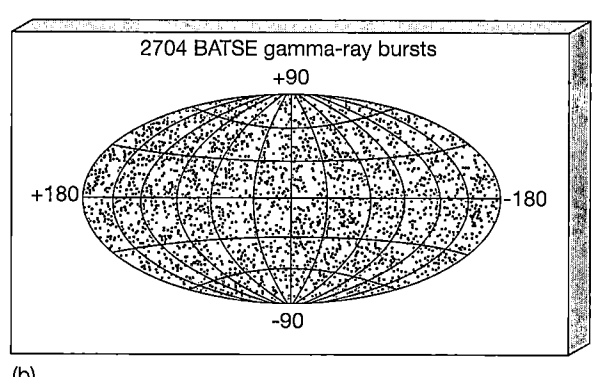

(b)

Figure 22.11 Gamma-Ray Bursts (a) Plots of intensity versus time (in seconds) for some gamma-ray bursts. Note the substantial differences between them. Some bursts are irregular and spiky, whereas others are much more smoothly varying. Whether this wide variation in burst appearance means that more than one physical process is at work is still unknown. (b) Positions on the sky of all the gamma-ray bursts detected by *CGRO* during its nearly nine-year operating lifetime (*BATSE* stands for *Burst and Transient Source Experiment*). The bursts appear to be distributed isotropically (uniformly) across the entire sky. The plane of the Milky Way Galaxy runs horizontally across the center of the map; the direction to the center of our Galaxy is at the center of the map. *(NASA)*

The burst's optical spectrum, obtained using the Keck telescope, revealed a very important piece of information. Several absorption lines of iron and magnesium were identified, but they were redshifted by almost a factor of two in wavelength. Such redshifts, as we will see in Chapter 24, are the result of the expansion of the universe, and they are clear proof—a "smoking gun," if you will—that at least this gamma-ray burst, and presumably all others, really did occur at cosmological distances. The events responsible for GRB 970508 occurred more than 2 *billion* parsecs from Earth.

Figure 22.12(a) and (b) show optical images of GRB 971214 and its host galaxy. According to the redshift of lines observed in the galaxy's spectrum, this burst was almost 5 billion pc away. Figure 22.12(c) and (d) are images of the optical afterglow of GRB 990123, with the host galaxy more readily apparent. In all, some 40 optical or X-ray afterglows of gamma-ray bursts have been detected, and about a dozen distances are known. All are very large, implying that the bursts must be extremely energetic, since otherwise they wouldn't be detectable by our equipment.

If we assume that the gamma rays are emitted equally in all directions (a big assumption!), then we can easily calculate the total energy emitted from the fraction we see, using the inverse-square law. ⚌ (Sec. 17.3) We find that each burst apparently generates more energy—and in some cases hundreds of times more energy—than a typical supernova explosion, all in a matter of seconds!

In October 2000, NASA launched the *High Energy Transient Explorer-2* (*HETE-2*) satellite to continue the work begun by *CGRO*. With accuracy of 10′ in gamma rays and 10″ in X rays, *HETE-2* can instantly relay accurate burst positions to other instruments in space and on the ground, allowing rapid searches for counterparts to be carried out.

WHAT CAUSES THE BURSTS?

What are gamma-ray bursts? Frankly, we don't know! Before the first distance measurements, some theorists had attempted to explain gamma-ray bursts in terms of nearby—and hence much less energetic—events, perhaps

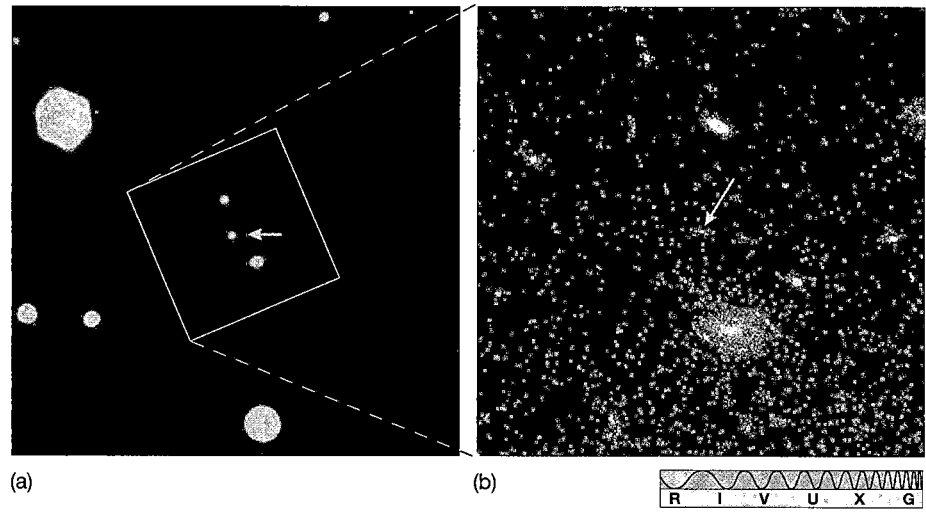

(a)　　　　　　　　(b)

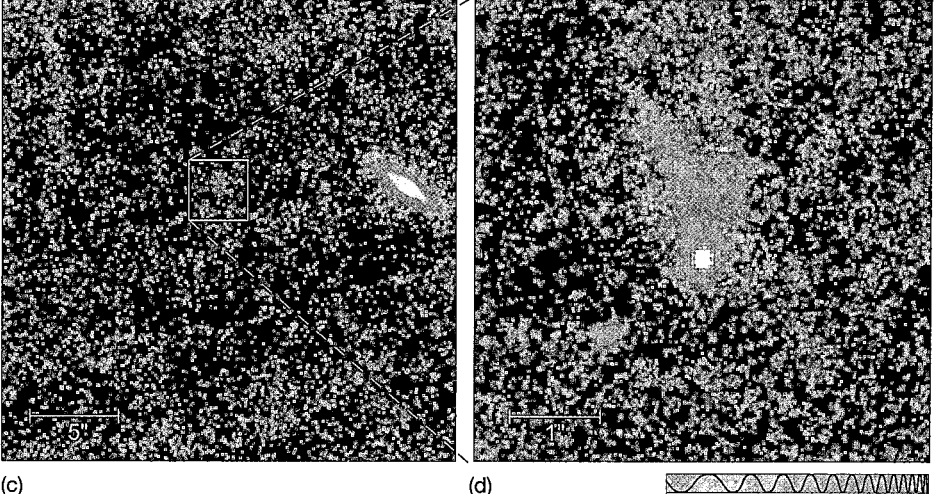

(c)　　　　　　　　(d)

Figure 22.12 Gamma-Ray Burst Counterparts (a and b) Two optical images of the gamma-ray burst GRB 971214. The image at left, taken by the Keck telescope, shows the visible afterglow of the gamma-ray source (arrow) to be quite bright, fully comparable in visual intensity to two other prominent sources in the overlaid box. A Keck spectrum of this source also showed it to be very highly red shifted, placing it near the limits of the observable Universe. By the time the *Hubble* image at right was taken (about two months after the Keck image and four months after the initial burst of gamma rays), the afterglow had faded, but a faint image of a host galaxy remains. (c and d) Similar wide-field and narrow-field optical views of GRB 990123 and its (possible) host galaxy. *(Keck; NASA)*

DISCOVERY 22-1

Gravity Waves

Electromagnetic waves are common, everyday phenomena. Whether they are radio, infrared, visible, ultraviolet, X-ray, or gamma-ray radiation, all electromagnetic waves involve periodic changes in the strengths of electric and magnetic fields. ∞ (Sec. 3.2) They move through space and transport energy. Any accelerating charged particle, such as an electron in a broadcasting antenna or on the surface of a star, generates electromagnetic waves.

The modern theory of gravity—Einstein's theory of relativity (see *More Precisely 22-1*)—also predicts waves that move through space. A *gravity wave* is the gravitational counterpart of an electromagnetic wave. *Gravitational radiation* results from changes in the strength of a gravitational field. In principle, any time an object of any mass accelerates, a gravity wave should be emitted at the speed of light. The passage of a gravity wave should produce small distortions in the space through which it passes. Gravity is an exceedingly weak force compared with electromagnetism, so these distortions are expected to be very small—in fact, much smaller than the diameter of an atomic nucleus for waves that may be produced by any known astrophysical source. Yet many researchers believe that these tiny distortions should be measurable. No one has yet succeeded in detecting gravity waves, but their detection would provide such strong support for the theory of relativity that scientists are eager to search for them.

Theorists are still debating which kinds of astronomical objects should produce gravity waves detectable on Earth. Leading candidates include (1) mergers of binary systems containing black holes, neutron stars, or white dwarfs, (2) the collapse of a star into a black hole, and (3) supernova explosions and neutron-star oscillations. Each of these possibilities involves the acceleration of huge masses, resulting in rapidly changing gravitational fields.

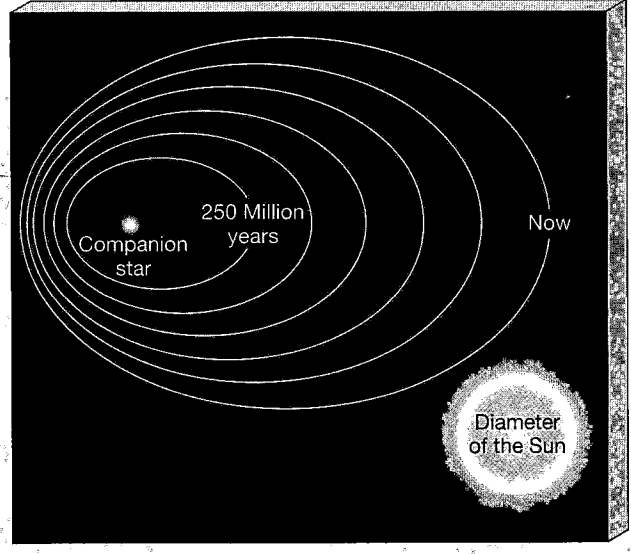

Of the above candidates, the first probably presents the best chance to detect gravity waves, at least for the present. Binary-star systems should emit gravitational radiation as the component stars orbit one another. As energy escapes in the form of gravity waves, the two stars slowly spiral toward one another, orbiting more rapidly and emitting even more gravitational radiation. This runaway situation can lead to the decay and eventual merger of close binary systems in a relatively short time (which, in this case, means tens or hundreds of millions of years, although most of the radiation is emitted during the last few seconds). As we saw in the text (Section 22.3), these mergers could be the origin of gamma-ray bursts, so gravitational

in the halo of our own Galaxy (see Chapter 23) or even in the Oort cloud surrounding the Sun. ∞ (Sec. 14.2) However, it now seems clear that gamma-ray bursts are very distant, extremely energetic events.

What's more, the millisecond flickering in the light curves recorded by *Compton* imply that, whatever their origin, all their energy must come from an extremely small volume—in fact, no larger than a *few hundred kilometers* across. The reasoning is as follows. If the emitting region were, say, 300,000 km—1 light-second—across, even an instantaneous change in intensity at the source would be smeared out over a time interval of 1 s as seen from Earth, because light from the far side of the object would take 1 s longer to reach us than light from the near side. For the gamma-ray variation not to be blurred by the light-travel time, the source cannot be more than 1 light-millisecond, or 300 km, in diameter.

Most theoretical attempts to explain gamma-ray bursts picture the burst as a *relativistic fireball*—an expanding region, possibly a jet, of superhot gas radiating furiously in the gamma-ray part of the spectrum. The complex burst structure and afterglows we see are produced as the fireball expands, cools, and interacts with its surroundings. If the energy is emitted as a jet, then the overall energy of the burst may be reduced to more "manageable" levels, as the energy we see is representative of only a small fraction of the sky. This may help make the huge luminosity of the December 1997 burst somewhat easier to explain. Two leading models for the energy source have emerged, as sketched in Figure 22.13.

The first is the "true" end point of a binary-star system. Suppose that both members of the binary evolve to become neutron stars. As the system continues to evolve,

radiation may provide an alternative means of studying these violent and mysterious phenomena.

Such a slow but steady decay in the orbit of a binary system has in fact been detected. In 1974 radio astronomer Joseph Taylor and his student Russell Hulse at the University of Massachusetts discovered a very unusual binary system. Both components are neutron stars, and one is observable from Earth as a pulsar. This system has become known as the *binary pulsar*. Measurements of the periodic Doppler shift of the pulsar's radiation prove that its orbit is slowly shrinking. Furthermore, the rate at which the orbit is shrinking is exactly what would be predicted by relativity theory if the energy were being carried off by gravity waves. Even though the waves themselves have not yet been detected, the binary pulsar is regarded by most astronomers as a very strong piece of evidence in favor of general relativity (*More Precisely 22-1*). Taylor and Hulse received the 1993 Nobel Prize in physics for their discovery. The figure at left illustrates the scale of and predicted orbital changes in the binary pulsar's orbit.

Gravity waves should contain a great deal of information about the physical events in some of the most exotic regions of space. In 1992, funding was approved for an ambitious gravity-wave observatory called LIGO—short for Laser Interferometric Gravity-Wave Observatory. Twin detectors, one in Hanford, Washington (shown in the figure at right) and, the other in Livingston, Louisiana, will use laser beams to measure the extremely small distortions of space produced by gravitational radiation. The beams will detect the tiny changes (less than one-thousandth the diameter of an atomic nu-

cleus) that will be produced in the lengths of the 4-km-long arms should a gravity wave pass by. The instrument should be capable of detecting gravity waves from many Galactic and extragalactic sources. The detectors were completed in 1999 and will begin taking astronomical data in January 2002. Even more sensitive instruments are planned to come online later in the decade.

If successful, the discovery of gravity waves could herald a new age in astronomy, in much the same way that invisible electromagnetic waves, virtually unknown a century ago, revolutionized classical astronomy and led to the field of modern astrophysics.

(National Science Foundation/LIGO Project)

gravitational radiation (see *Discovery 22-1*) is released and the two ultradense stars will spiral in toward each other. Once they are within a few kilometers of one another, coalescence is inevitable. Such a merger will likely produce a violent explosion comparable in energy to a supernova, and perhaps energetic enough to explain the flashes of gamma rays we observe. The overall rotation of the binary system may channel the energy into a high-speed, high-temperature jet.

The second model, sometimes called a *hypernova*, is a "failed" supernova—but what a failure! In this picture, a very massive star undergoes core collapse just as described earlier for a Type II supernova, but instead of forming a neutron star, the core collapses to a black hole. ∞ (Sec. 21.2) At the same time, the blast wave racing outward through the star stalls. Instead of being blown to pieces,

the star begins to implode onto the black hole, forming an accretion disk and generating a relativistic jet.

The idea of a relativistic fireball is becoming fairly widely accepted among workers in this fast-changing branch of astrophysics. At present, there is no strong evidence for or against either energy-production model, although the possible association of a 1998 supernova in a nearby galaxy with an earlier gamma-ray burst may lend some support to the hypernova scenario. Clearly, the final word on these enigmatic objects has yet to be written.

Concept Check

▣ What are gamma-ray bursts, and why do they pose such a challenge to current theory?

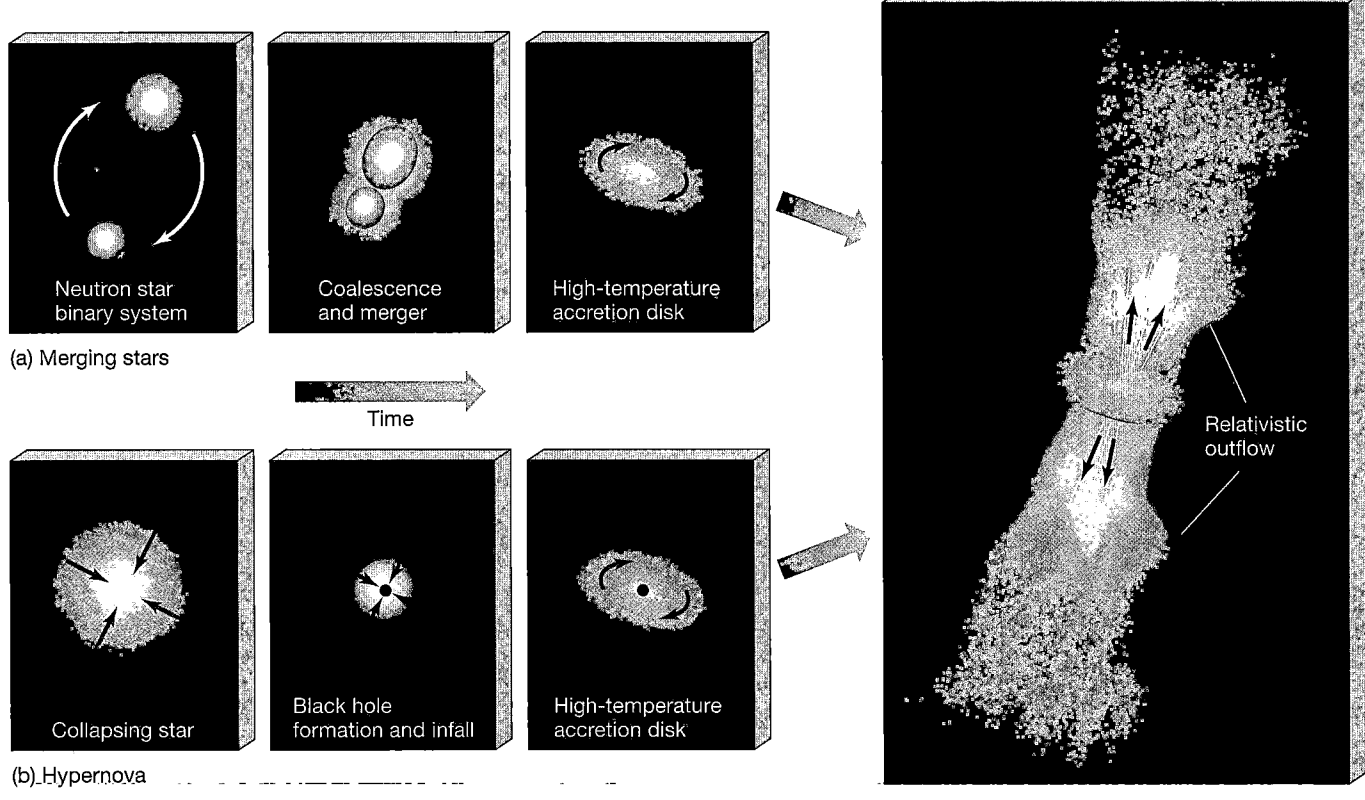

Figure 22.13 Gamma-Ray Burst Models Two models have been proposed to explain gamma-ray bursts. Part (a) depicts the merger of two neutron stars; part (b) shows the collapse of a single star. Both models predict (at right) a relativistic fireball, perhaps in the form of jets, as shown.

22.5 Black Holes

🕭 Neutron stars are supported by the resistance of tightly packed neutrons to further compression. Squeezed together, the neutrons form a hard ball of matter that not even gravity can compress further. Or do they? Is it possible that given enough matter packed into a small enough volume, the collective pull of gravity can eventually crush any opposing pressure? Can gravity continue to compress a massive star into an object the size of a planet, a city, a pinhead—even smaller? The answer, apparently, is yes.

THE FINAL STAGE OF STELLAR EVOLUTION

Although the precise figure is uncertain, mainly because the behavior of matter at very high densities is not well enough understood, most researchers concur that the mass of a neutron star cannot exceed about 3 solar masses. This is the neutron-star equivalent of the white dwarf's Chandrasekhar mass limit discussed in the previous chapter. ∞ (Sec. 21.3) Above this mass, not even tightly packed neutrons can withstand the star's gravitational pull. In fact, we know of *no* force that can counteract grav-

ity beyond the point at which neutron degeneracy pressure is overwhelmed. If enough material is left behind after a supernova explosion that the central core exceeds this limit, gravity wins the battle with pressure once and for all, and the central core collapses forever. Stellar evolution theory indicates that this is the fate of any star whose main-sequence mass exceeds about 25 times the mass of the Sun.

The limit of 3 solar masses is uncertain, in part because it ignores the effects of magnetism and rotation, both of which are surely present in the cores of evolved stars. Because these effects can compete with gravity, they influence stellar evolution. ∞ (Sec. 19.1) In addition, we do not know precisely how the basic laws of physics might change in regions of very dense matter that is both rapidly spinning and strongly magnetized. Generally speaking, theorists expect that the neutron-star mass limit will increase when magnetism and rotation are included, because even larger amounts of mass will then be needed for gravity to compress stellar cores into neutron stars or black holes, but the amount of the increase is not currently known.

As the stellar core shrinks, the gravitational pull in its vicinity eventually becomes so great that even light itself is unable to escape. The resultant object therefore emits no light, no radiation, no information whatsoever. Astronomers call this bizarre end point of stellar evolution, in which a massive core remnant collapses in on itself and vanishes forever, a **black hole.**

ESCAPE SPEED

Newtonian mechanics—up to now our reliable and indispensable tool in understanding the universe—cannot adequately describe conditions in or near black holes. ∞ (*More Precisely 2-2*) To comprehend these collapsed objects, we must turn instead to the modern theory of gravity, Einstein's **general theory of relativity** (see *More Precisely 22-1 and 22-2*). Einstein's description of the universe is equivalent to Newton's for situations encountered in everyday life, but the two theories diverge radically in circumstances where speeds approach the speed of light and in regions of intense gravitational fields.

Yet despite the fact that general relativity is necessary for a proper description of black-hole properties, we can still usefully discuss some aspects of these strange bodies in more or less Newtonian terms. Let's reconsider the familiar Newtonian concept of escape speed—the speed needed for one object to escape from the gravitational pull of another—supplemented by two key facts from relativity: (1) Nothing can travel faster than the speed of light, and (2) all things, *including light*, are attracted by gravity.

Escape speed is proportional to the square root of a body's mass divided by the square root of its radius. ∞ (Sec. 2.7) Earth's radius is 6400 km, and the escape speed from Earth's surface is just over 11 km/s. Now consider a hypothetical experiment in which Earth is squeezed on all sides by a gigantic vise. As our planet shrinks under the pressure its mass remains the same, but the escape speed increases because the radius is decreasing. For example, suppose Earth were compressed to one-fourth its present size. The proportionality just mentioned for escape speed then predicts that the escape speed would double (because $1/\sqrt{\frac{1}{4}} = 2$). Any object escaping from this hypothetically compressed Earth would need a speed of at least 22 km/s.

Imagine compressing Earth some more. Squeeze it by an additional factor of 1000, making its radius hardly more than a kilometer. Now a speed of about 630 km/s would be needed to escape. Compress Earth still further, and the escape speed continues to rise. If our hypothetical vise were to squeeze Earth hard enough to crush its radius to about a centimeter, then the speed needed to escape its surface would reach 300,000 km/s. But this is no ordinary speed—it is the speed of light, the fastest speed allowed by the laws of physics as we currently know them.

Thus, if by some fantastic means the entire planet Earth could be compressed to less than the size of a grape, the escape speed would exceed the speed of light. And because nothing can exceed that speed, the compelling conclusion is that nothing—absolutely nothing—could escape from the surface of such a compressed body. Even radiation—radio waves, visible light, X rays, photons of all wavelengths—would be unable to escape the intense gravity of our reshaped Earth. With no photons leaving, our planet would be invisible and uncommunicative—no signal of any sort could be sent to the universe beyond. The origin of the term *black hole* becomes clear. For all practical purposes, such a supercompact Earth could be said to have disappeared from the universe! Only its gravitational field would remain behind, betraying the presence of its mass, now shrunk to a point.*

THE EVENT HORIZON

Astronomers have a special name for the critical radius at which the escape speed from an object would equal the speed of light and within which the object could no longer be seen. It is the **Schwarzschild radius,** after Karl Schwarzschild, the German scientist who first studied its properties. The Schwarzschild radius of any object is simply proportional to its mass. For Earth, it is 1 cm; for Jupiter, at about 300 Earth masses, it is about 3 m; for the Sun, at 300,000 Earth masses, it is 3 km. For a 3-solar-mass stellar core remnant, the Schwarzschild radius is about 9 km. As a convenient rule of thumb, the Schwarzschild radius of an object is simply 3 km multiplied by the object's mass, measured in solar masses. Every object has a Schwarzschild radius. It is the radius to which the object would have to be compressed for it to become a black hole. Put another way, a black hole is an object that happens to lie within its own Schwarzschild radius.

The surface of an imaginary sphere with radius equal to the Schwarzschild radius and centered on a collapsing star is called the **event horizon.** It defines the region within which no event can ever be seen, heard, or known by anyone outside. Even though there is no matter of any sort associated with it, we can think of the event horizon as the "surface" of a black hole.

A 1.4-solar-mass neutron star has a radius of about 10 km and a Schwarzschild radius of 4.2 km. If we were to keep increasing the star's mass, the star's Schwarzschild radius would grow, although its actual physical radius would not. In fact the radius of a neutron star *decreases* slightly with increasing mass. By the time our neutron star's mass exceeded about 3 solar masses, it would lie just within its own event horizon, and it would collapse of its own accord.

* *In fact, we now know that regardless of the composition or condition of the object that formed the hole, only three physical properties can be measured from the outside—the hole's mass, charge, and angular momentum. All other information is lost once the infalling matter crosses the event horizon. Thus, only three numbers are required to completely describe a black hole's outward appearance. In this chapter we will consider only black holes that formed from nonrotating, electrically neutral matter. Such objects are completely specified once their masses are known.*

MORE PRECISELY 22-1

Einstein's Theories of Relativity

Albert Einstein is probably best known for his two *theories of relativity*, the successors to Newtonian mechanics that form part of the foundation of twentieth-century physics. (The theory of *quantum mechanics*, discussed in Chapter 4 in the context of spectroscopy and atomic structure, forms the other part.)

The *special theory of relativity* (or just *special relativity*), proposed by Einstein in 1905, deals with the preferred status of the speed of light. We have noted that the speed of light c is the maximum speed attainable in the universe. But there is more to it than that. In 1887 a fundamental experiment carried out by two American physicists, A. A. Michelson and E. W. Morley, demonstrated a further important and unique aspect of light—that the measured speed of a beam of light is *independent* of the motion of the observer or the source. No matter what our speed may be relative to the source of the light, we always measure precisely the same value for c—299,792.458 km/s.

A moment's thought leads us to the realization that this is a decidedly nonintuitive statement. For example, if we were traveling in a car moving at 100 km/h and we fired a bullet forward with a speed of 1000 km/h relative to the car, an observer standing at the side of the road would see the bullet pass by at $100 + 1000 = 1100$ km/h, as illustrated in the accompanying figure. However, if we were traveling in a rocket ship at 1/10 the speed of light, $0.1c$, and we shone a searchlight beam ahead of us, the Michelson-Morley experiment tells us that an outside observer would measure the speed of the beam not as $1.1c$, as the preceding example would suggest, but as c. The rules that apply to particles moving at or near the speed of light are different from those we are used to in everyday life.

Special relativity is the mathematical framework that allows us to extend the familiar laws of physics from low speeds (that is, speeds much less than c, which are often referred to as *nonrelativistic*) to very high (or *relat-*

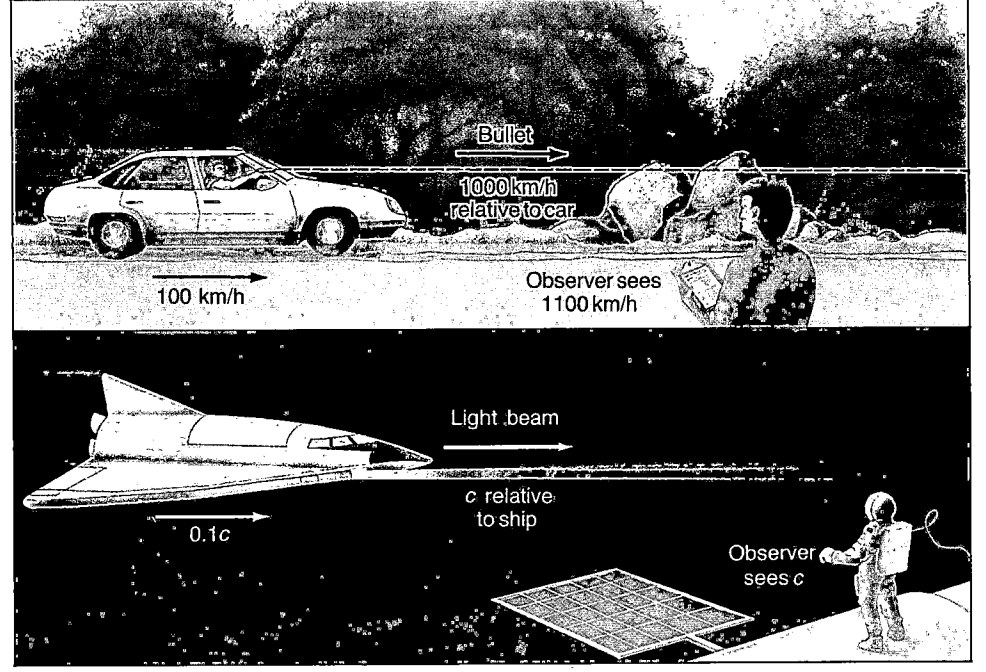

It would not stop shrinking at the Schwarzschild radius—the event horizon is not a physical boundary of any kind, just a communications barrier. The remnant would shrink right past it to ever-diminishing size on its way to being crushed to a point.

Thus, provided that at least 3 solar masses of material remain behind after a supernova explosion, the remnant core will collapse catastrophically, diving below the event horizon in less than a second. The core simply "winks out," disappearing and becoming a small dark region from which nothing can escape—a literal black hole in space. Theory indicates that this is the likely fate of stars having more than about 20–25 times the mass of the Sun.

22.6 Black Holes and Curved Space

⑤ Modern notions about black holes rest squarely on the theory of relativity. Although white dwarfs and (to a lesser extent) neutron stars can be adequately described by the classical Newtonian theory of gravity, only the modern Einsteinian theory of relativity can properly account, for the bizarre physical properties of black holes.

A central concept of general relativity (see *More Precisely 22-1*) is this: Matter—all matter—tends to "warp" or curve space in its vicinity. Objects such as planets and stars

ivistic) speeds, comparable to *c*. Relativity is equivalent to Newtonian mechanics when objects move much more slowly than the speed of light, but it differs greatly in its predictions at relativistic velocities. For example, special relativity predicts that a rapidly moving spacecraft will appear to contract in the direction of its motion, its clocks will appear to run slow, and its mass will appear to increase. Despite their somewhat nonintuitive nature, all the theory's predictions have been repeatedly verified to very high accuracy. Today special relativity is at the heart of all physical science. No scientist seriously doubts its validity.

General relativity is what results when gravity is included in the framework of special relativity. In 1915 Einstein made the connection between special relativity and gravity with the following famous "thought experiment." Imagine that you are enclosed in an elevator with no windows, so that you cannot directly observe the outside world, and the elevator is floating in space. You are weightless. Now suppose that you begin to feel the floor press up against your feet. Weight has apparently returned. There are two possible explanations for this, shown in the accompanying diagram. A large mass could have come nearby, and you are feeling its downward gravitational attraction, *or* the elevator has begun to accelerate upward and the force you feel is that exerted by the elevator as it accelerates you at the same rate. The crux of Einstein's argument is this: There is *no* experiment that you can perform within the elevator, without looking outside, that will let you distinguish between these two possibilities.

Thus, Einstein reasoned, there is no way to tell the difference between a gravitational field and an accelerated frame of reference (such as the rising elevator in the thought experiment). Gravity can therefore be incorporated into special relativity as a general acceleration of all particles. However, another major modification to the theory of special relativity must be made. Central to relativity is

the notion that space and time are not separate quantities but instead must be treated as a single entity—*spacetime*. To incorporate the effects of gravity, the mathematics forces us to the conclusion that spacetime has to be *curved*.

In general relativity, then, gravity is a manifestation of curved space-time. There is no such thing as a "gravitational field," in the Newtonian sense. Instead, objects move as they do because they follow the curvature of space-time, and this curvature of space-time is determined by the amount of matter present. We will explore some of the consequences of this view of gravity in more detail in the text.

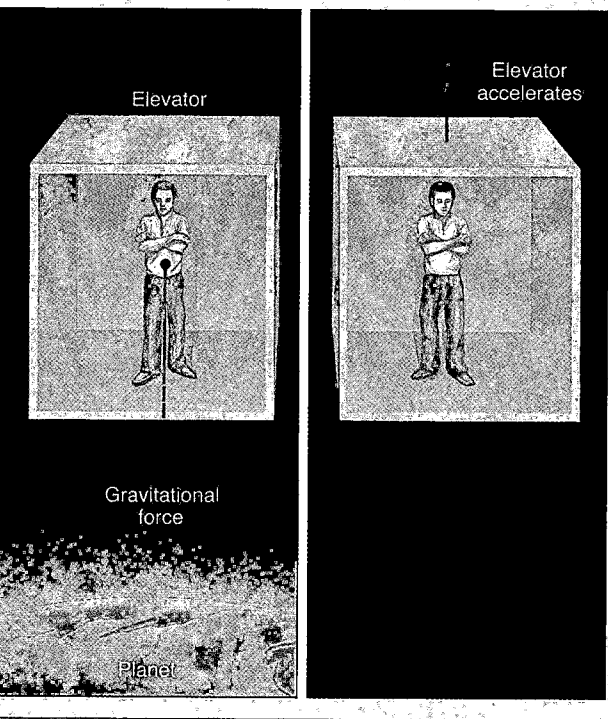

react to this warping by changing their paths. In the Newtonian view of gravity, particles move on curved trajectories because they feel a gravitational force. In Einsteinian relativity, those same particles move on curved trajectories because they are following the curvature of space produced by some nearby massive object. The more the mass, the greater the warping. Close to a black hole, the gravitational field becomes overwhelming and the curvature of space extreme. At the event horizon itself, the curvature is so great that space "folds over" on itself, causing objects within to become trapped and disappear.

Some props may help you visualize the curvature of space near a black hole. Bear in mind, however, that these

props are not real but only tools to help you grasp some exceedingly strange concepts. First, imagine a pool table with the tabletop made of a thin rubber sheet rather than the usual hard felt. As Figure 22.14 suggests, such a rubber sheet becomes distorted when a heavy weight, such as a rock, is placed on it. The otherwise flat sheet sags (becomes distorted), especially near the rock. The heavier the rock, the larger the distortion. Trying to play pool, you would quickly find that balls passing near the rock were deflected by the curvature of the tabletop. In much the same way, both matter *and radiation* are deflected by the curvature of space near a star. For example, Earth's orbital path is governed by the relatively gentle curvature of space

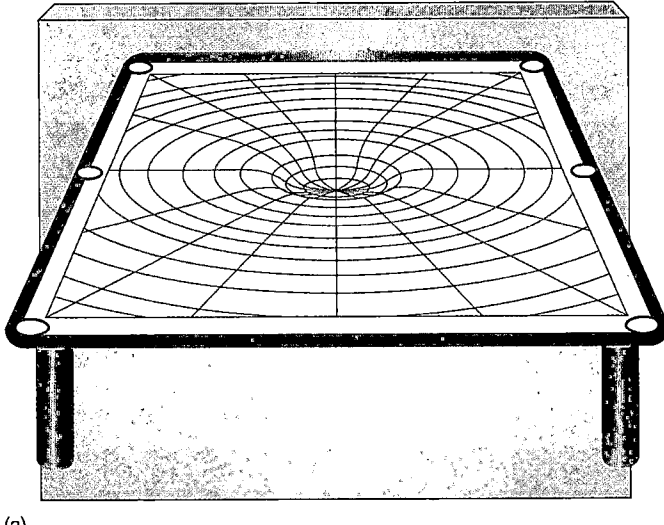

(a)

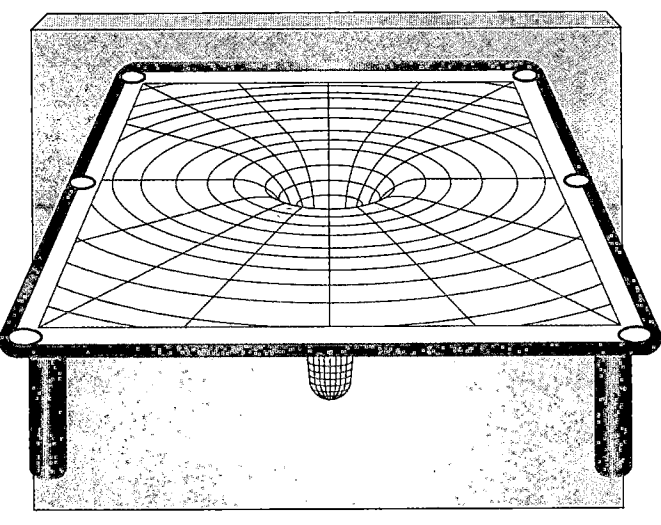

(b)

Figure 22.14 Curved Space (a) A pool table made of a thin rubber sheet sags when a weight is placed on it. Likewise, space is bent, or warped, in the vicinity of any massive object. (b) As the weight increases, so does the warping of space.

created by our Sun. The more massive the object, the more the space surrounding it is curved.

Let's consider another analogy. Imagine a large extended family of people living on a huge rubber sheet—a sort of gigantic trampoline. Deciding to hold a reunion, they converge on a given place at a given time. As shown in Figure 22.15, one person remains behind, not wishing to attend. He keeps in touch with his relatives by means of "message balls" rolled out to him (and back from him) along the surface of the sheet. These message balls are the analog of radiation carrying information through space.

As the people converge, the rubber sheet sags more and more. Their accumulating mass creates an increasing amount of space curvature. The message balls can still reach the lone person far away in nearly flat space, but they arrive less frequently as the sheet becomes more and more warped and stretched—as shown in Figures 22.15(b) and (c)—and the balls have to climb out of a deeper and deeper well. Finally, when enough people have arrived at the appointed spot, the mass becomes too great for the rubber to support them. As illustrated in Figure 22.15(d), the sheet pinches off into a "bubble," compressing the people into oblivion and severing their communications with the lone survivor outside. This final stage represents the formation of an event horizon around the party.

Right up to the end—the pinching off of the bubble—two-way communication is possible. Message balls can reach the outside from within (but at a slower and slower

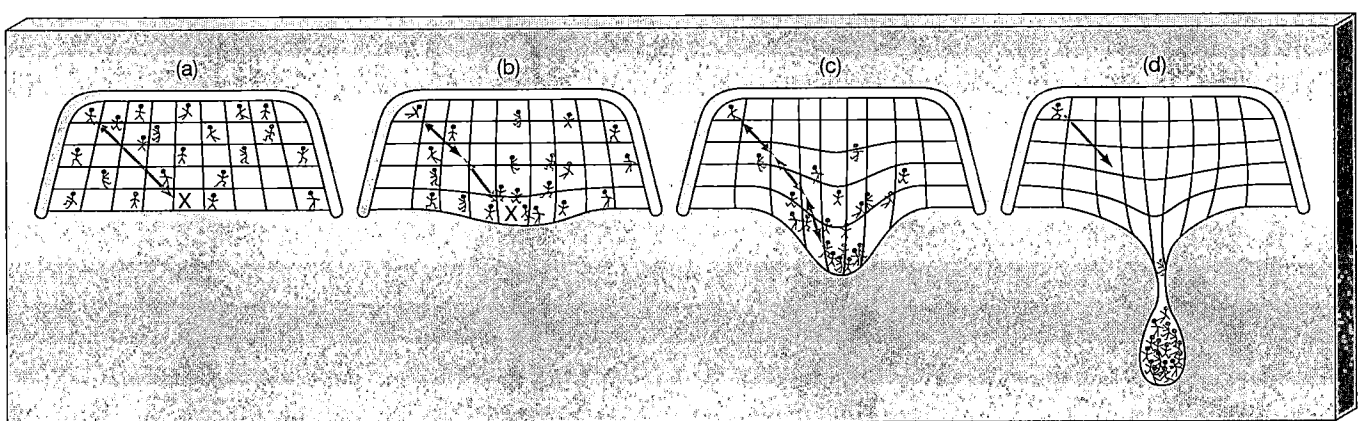

Figure 22.15 Space Warping Any mass causes the rubber sheet (space) to be curved. As people assemble at the appointed spot on the sheet, the curvature grows progressively larger, as shown in frames (a), (b), and (c). The blue arrows represent some directions in which information can be transmitted from place to place. The people are finally sealed inside the bubble (d), forever trapped and cut off from the outside world.

rate as the rubber stretches), and messages from outside can get in without difficulty. However, once the event horizon (the bubble) forms, balls from the outside can still fall in, but they can no longer be sent back out to the person left behind, no matter how fast they are rolled. They cannot make it past the "lip" of the bubble in Figure 22.15(d). This analogy (very) roughly depicts how a black hole warps space completely around on itself, isolating its interior from the rest of the universe. The essential ideas—the slowing down and eventual cessation of outward-going signals and the one-way nature of the event horizon once it forms—all have parallels in the case of stellar black holes.

☑ Concept Check

■ How do Newton's and Einstein's theories differ in their descriptions of gravity?

22.7 Space Travel Near Black Holes

Black holes are *not* cosmic vacuum cleaners. They don't cruise around interstellar space, sucking up everything in sight. The orbit of an object near a black hole is essentially the same as its orbit near a star of the same mass. Only if the object happens to pass within a few Schwarzschild radii (perhaps 50 or 100 km for a typical 5–10-solar-mass black hole formed in a supernova explosion) of the event horizon is there any significant difference between its actual orbit and the one predicted by Newtonian gravity and described by Kepler's laws. Of course, if some matter does happen to fall into one—if its orbit happens to take it too close to the event horizon—it will be unable to get out. Black holes are like turnstiles, permitting matter to flow in only one direction—inward.

Because a black hole will accrete at least a little material from its surroundings, its mass, and hence also the radius of its event horizon, tends to increase over time.

TIDAL FORCES

Matter flowing into a black hole is subject to great tidal stress. An unfortunate person falling feet first into a solar-mass black hole would find herself stretched enormously in height and squeezed unmercifully laterally. She would be torn apart even before she reached the event horizon, for the pull of gravity would be much stronger at her feet (which are closer to the hole) than at her head. The tidal forces at work in and near a black hole are the same basic phenomenon responsible for ocean tides on Earth and the spectacular volcanoes on Io. The only difference is that the tidal forces near a black hole are far stronger than any force we know in the solar system.

As illustrated (with some artistic license) in Figure 22.16, a similar fate awaits any kind of matter falling into a black hole. Whatever falls in—gas, people, space probes—is vertically stretched and horizontally squeezed, in the process being accelerated to high speeds. The net result of all this stretching and squeezing is numerous and violent collisions among the torn-up debris, causing a great deal of frictional heating of the infalling matter. Material is simultaneously torn apart and heated to high temperatures as it plunges into the hole.

So efficient is the heating that, before reaching the hole's event horizon, matter falling into the hole emits radiation of its own accord. For a black hole of mass comparable to the Sun, the energy is expected to be emitted in the form of X rays. In effect, the gravitational energy of matter outside the black hole is converted into heat while that matter falls toward the hole. Thus, contrary to what we might expect from an object whose defining property is that nothing can escape from it, the region surrounding a black hole is expected to be a *source* of energy. Of course, once the hot matter falls below the event horizon, its radiation is no longer detectable—it never leaves the hole.

APPROACHING THE EVENT HORIZON

One safe way to study a black hole would be to go into orbit around it, well beyond the disruptive influence of the hole's strong tidal forces. After all, Earth and the other planets of our solar system all orbit the Sun without falling

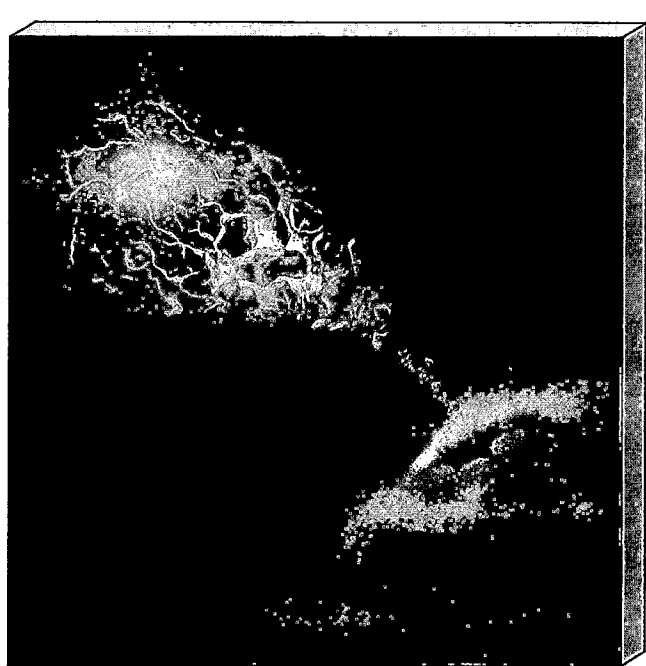

Figure 22.16 Black-Hole Heating Any matter falling into the clutches of a black hole will become severely distorted and heated. This sketch shows an imaginary planet being pulled apart by a black hole's gravitational tides.

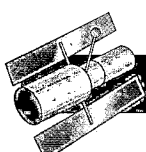
Special relativity is the most thoroughly tested and most accurately verified theory in the history of science. General relativity, however, is on somewhat less-firm experimental ground.

The problem with verifying general relativity is that its effects on Earth and in the solar system—the places where we can most easily perform tests—are very small. Just as special relativity produces major departures from Newtonian mechanics only when velocities approach the speed of light, general relativity predicts large departures from Newtonian gravity only when extremely strong grav-

itational fields are involved—in effect, when orbit speeds and escape velocities become relativistic.

We will encounter other experimental and observational tests of general relativity elsewhere in this chapter. Here we consider just two "classical" tests of the theory. These tests are solar system observations that helped ensure acceptance of Einstein's theory. Later, more accurate measurements confirmed and strengthened these results. Bear in mind, however, that there are no known tests of general relativity in the "strong-field" regime—that part of the theory that predicts black holes, for example—so the full theory has never been experimentally tested.

At the heart of general relativity is the premise that everything, including light, is affected by gravity because of the curvature of space-time. Shortly after he published his theory in 1915, Einstein noted that light from a star should be deflected by a measurable amount as it passes the Sun. The closer to the Sun the light comes, the more it is deflected. Thus, the maximum deflection should occur for a ray that just grazes the solar surface. Ein-

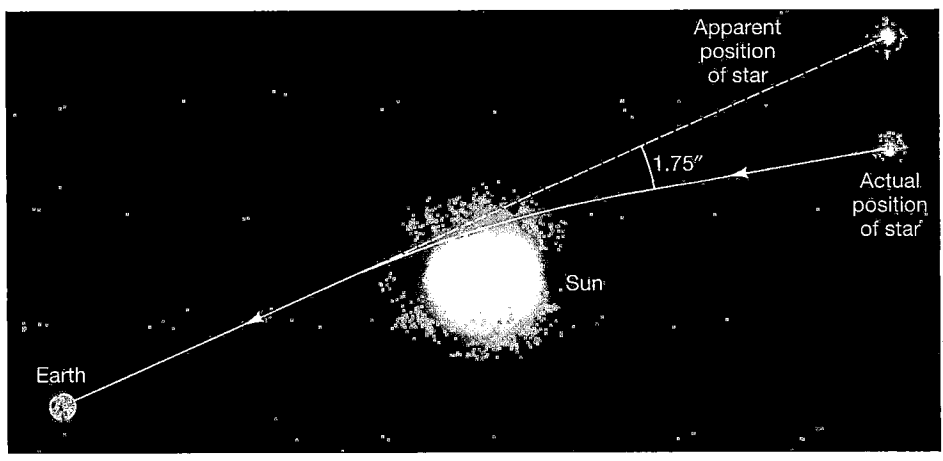

into it and without being torn apart. The gravity field around a black hole is basically no different. However, even from a stable circular orbit, a close investigation of the hole would be unsafe for humans. Human endurance tests conducted on astronauts of the United States and the former Soviet Union indicate that the human body cannot withstand stress greater than about 10 times the pull of gravity on Earth's surface. This breaking point would occur about 3000 km from a 10-solar-mass black hole (which, recall, would have a 30-km event horizon). Closer than that, the tidal effect of the hole would tear a human body apart.

Let's instead send an imaginary indestructible astronaut—a mechanical robot, say—in a probe toward the center of the hole, as illustrated in Figure 22.17. Watching from a safe distance in our orbiting spacecraft, we can then examine the nature of space and time near the hole. Our robot will be a useful explorer of theoretical ideas, at least down to the event horizon. After that boundary is crossed, there is no way for the robot to return any information about its findings.

Suppose, for example, our robot has an accurate clock and a light source of known frequency mounted on it. From our safe vantage point far outside the event horizon, we could use telescopes to read the clock and measure the frequency of the light we receive. What might we discov-

er? We would find that the light from the robot would become more and more redshifted as the robot neared the event horizon. Even if the robot used rocket engines to remain motionless, the redshift would still be detected. The redshift is not caused by motion of the light source. It is not the result of the Doppler effect arising as the robot falls into the hole. Rather, it is a redshift induced by the black hole's gravitational field, clearly predicted by Einstein's general theory of relativity and known as **gravitational redshift**.

We can explain the gravitational redshift as follows. According to general relativity, photons are attracted by gravity. As a result, in order to escape from a source of gravity, photons must expend some energy: they have to work to get out of the gravitational field. They don't slow down at all—photons always move at the speed of light—they just lose energy. Because a photon's energy is proportional to the frequency of its radiation, light that loses energy must have its frequency reduced (or, conversely, its wavelength lengthened). As illustrated in Figure 22.18, radiation coming from the vicinity of a gravitating object will be redshifted by an amount depending on the strength of the gravitational field.

As photons traveled from the robot's light source to the orbiting spacecraft they would become gravitationally redshifted. From our standpoint on the orbiting space-

stein calculated that the deflection angle should be 1.75″—a small, but detectable, amount. Of course, it is normally impossible to see stars close to the Sun. However, during a solar eclipse, when the Moon blocks the Sun's light, the observation becomes possible, as illustrated in the first (highly exaggerated) figure.

In 1919 a team of observers, led by the British astronomer Sir Arthur Eddington, succeeded in measuring the deflection of starlight during an eclipse. The results were in excellent agreement with the prediction of general relativity. Virtually overnight Einstein became world famous. His previous major accomplishments notwithstanding, this single prediction assured him a permanent position as the best-known scientist on Earth! Recently, the high-precision *Hipparcos* satellite ∞ (*Discovery 17-1*) has observed shifts in the apparent positions of many stars, even those whose line of sight is far from the Sun. The shifts are exactly as predicted by Einstein's theory.

Another prediction of general relativity is that planetary orbits should deviate slightly from the perfect ellipses of Kepler's laws. Again, the effect is greatest where gravity is strongest—that is, closest to the Sun. Thus, the largest relativistic effects are found in the orbit of Mercury. Relativity predicts that Mercury's orbit is not a closed ellipse. Instead, its orbit should rotate slowly, as shown in the second (again exaggerated) diagram. The amount of rotation is very small—only 43″ per century—but Mercury's orbit is so well charted that even this tiny effect is measurable.

In fact, the observed rotation rate is 574″ per century, much greater than that predicted by relativity. However, when other (nonrelativistic) gravitational influences, primarily the perturbations due to the other planets, are taken into account, the rotation is in complete agreement with the foregoing prediction.

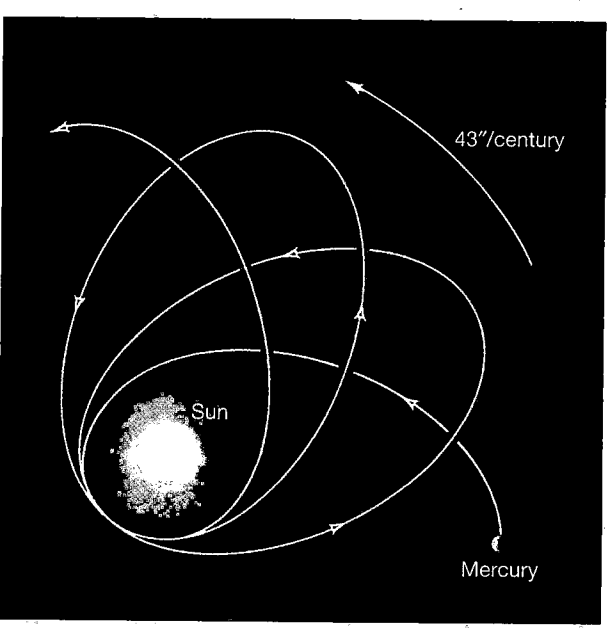

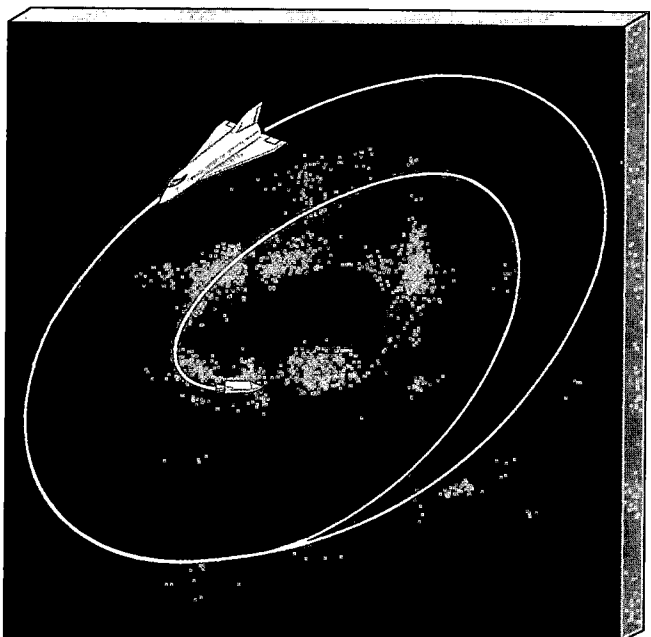

Figure 22.17 Robot Astronaut A hypothetical robot astronaut can travel toward a black hole while performing experiments that humans, farther away, can monitor in order to learn something about the nature of space near the event horizon.

craft, a green light, say, would become yellow and then red as the robot astronaut neared the black hole. From the robot's perspective, the light would remain green. As the robot got closer to the event horizon, the radiation from its light source would become undetectable with optical telescopes. The radiation reaching us in the orbiting spacecraft would by then be lengthened so much that infrared and then radio telescopes would be needed to detect it. Closer still to the event horizon, the radiation emitted as visible light from the robot probe would be shifted to wavelengths even longer than conventional radio waves by the time it reached us.

Light emitted *from the event horizon itself* would be gravitationally redshifted to infinitely long wavelengths. In other words, each photon would use all its energy trying to escape from the edge of the hole. What was once light (on the robot) would have no energy left on arrival at the safely orbiting spacecraft. Theoretically, this radiation would reach us—still moving at the speed of light—but with zero energy. The light radiation originally emitted would be redshifted beyond our perception.

What about the robot's clock? Assuming that we could read it, what time would it tell? Would there be any observable change in the rate at which the clock ticked as it moved deeper into the hole's gravitational field? From the

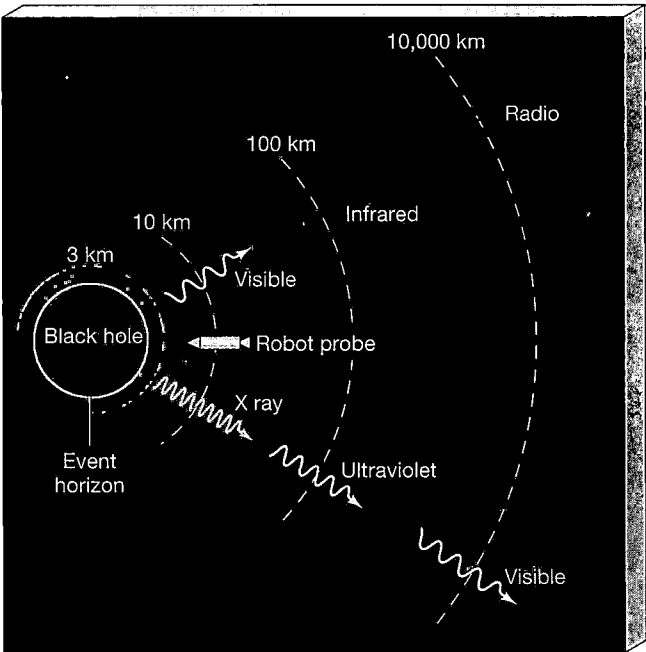

Figure 22.18 Gravitational Redshift As a photon escapes from the strong gravitational field close to a black hole, it must expend energy to overcome the hole's gravity. This energy does not come from a change in the speed at which it travels (this is always 300,000 km/s, even under these extreme conditions). Rather, the photon "gives up" energy by increasing its wavelength. Thus, the photon's color changes. This figure shows the effect on two beams of radiation, one of visible light and one of X rays, emitted from a space probe as it nears the event horizon of a 1-solar-mass black hole. The beams are shifted to longer and longer wavelengths as they move farther from the event horizon.

safely orbiting spacecraft, we would find that any clock close to the hole would appear to tick more *slowly* than an equivalent clock on board the spacecraft. The closer the clock came to the hole, the slower it would appear to run. The clock closest to the hole would operate slowest of all. On reaching the event horizon, the clock would seem to stop altogether. It would be as if the robot astronaut had found immortality! All action would become virtually frozen in time. Consequently, an external observer would never actually witness an infalling astronaut sink below the event horizon. Such a process would appear to take forever.

This apparent slowing down of the robot's clock is known as **time dilation.** It is another clear prediction of general relativity, and in fact it is closely related to the gravitational redshift. To see this connection, imagine that we use our light source as a clock, with the passage of a wave crest (say) constituting a "tick." The clock thus ticks at the frequency of the radiation. As the wave is redshifted, the frequency drops, and fewer wave crests pass the distant observer each second—the clock appears to slow down. This thought experiment demonstrates that the redshift of the radiation and the slowing of the clock are one and the same.

From the point of view of the indestructible robot, however, relativity theory predicts no strange effects at all. To the infalling robot, the light source hasn't reddened, and the clock keeps perfect time. In the robot's frame of reference, everything is normal. Nothing prohibits it from approaching within the Schwarzschild radius of the hole. No law of physics constrains an object from passing through an event horizon. There is no barrier at the event horizon and no sudden lurch as it is crossed; it is only an imaginary boundary in space. Travelers passing through the event horizon of a sufficiently massive hole (such as might lurk in the heart of our own Galaxy, as we will see) might not even know it—at least until they tried to get out!

Most astronomical objects' gravitational fields are far too weak to produce any significant gravitational redshift, although in many cases the effect can still be measured. Delicate laboratory experiments on Earth and in near-Earth orbits have succeeded in detecting the tiny gravitational redshift produced by even our own planet's weak gravity. Sunlight is redshifted by only about a hundredth of an angstrom. A few white-dwarf stars do show some significant reddening of their emitted light, however. Their radii are much smaller than that of our Sun, so their surface gravity is very much stronger than the Sun's. Neutron stars should show a substantial shift in their radiation, although it is currently difficult to disentangle the effects of gravity and magnetism on the observed signals.

DEEP DOWN INSIDE

No doubt you are wondering what lies within the event horizon of a black hole. The answer is simple: No one really knows. However, the question raises some fundamental issues that lie at the forefront of modern physics.

Can an entire star simply shrink to a point and vanish? General relativity predicts that without some agent to compete with gravity, the core remnant of a high-mass star will collapse all the way to a point at which both its density and its gravitational field become infinite—a so-called **singularity.** However, we should not take this prediction of infinite density too literally. Singularities always signal the breakdown of the theory producing them. In other words, the present laws of physics are simply inadequate to describe the final moments of a star's collapse.

As it stands today, the theory of gravity is incomplete because it does not incorporate a proper (that is, a quantum-mechanical) description of matter on very small scales. As our collapsing stellar core shrinks to smaller and smaller radii, we eventually lose our ability even to describe, let alone predict, its behavior. Perhaps matter trapped in a black hole never actually reaches a singularity. Perhaps it just approaches this bizarre state in a manner that we will someday understand as the subject of *quantum gravity*—the merger of general relativity with quantum mechanics—develops.

Having said that, we can at least estimate how small the core can get before current theory fails. It turns out that by the time that stage is reached, the core is already much smaller than any elementary particle. Thus, although a complete description of the end point of stellar collapse may well require a major overhaul of the laws of physics, for all practical purposes the prediction of collapse to a point is valid. Even if a new theory somehow succeeds in doing away with the central singularity, it is very unlikely that the external appearance of the hole, or the existence of its event horizon, will change. Any modifications to general relativity are expected to occur only on submicroscopic scales, not on the macroscopic (kilometer-sized) scale of the Schwarzschild radius.

Singularities are places where the rules break down, and some very strange things may occur near them. Many possibilities have been envisaged—gateways into other universes, time travel, the creation of new states of matter—but none of them has been proved, and certainly none of them has been observed. Because these regions are places where science fails, their presence causes serious problems for many of our cherished laws of physics, from causality (the idea that cause should precede effect, which runs into immediate problems if time travel is possible), to energy conservation (which is violated if material can hop from one universe to another through a black hole). It is unclear whether the removal of the singularity by some future, all-encompassing theory would necessarily also eliminate all of these problematic side effects.

Disturbed by the possibility of such chaos in science, some relativists have even proposed a "principle of cosmic censorship": Nature always hides *any* singularity, such as that found at the center of a black hole, inside an event horizon. In that case, even though physics fails, that breakdown cannot affect us outside, so we are safely insulated from any effects the singularity may have. What would happen if we one day found a so-called *naked singularity* somewhere, a singularity uncloaked by an event horizon? Would relativity theory still hold there? For now, we just don't know.

What sense are we to make of black holes? Do black holes, and all the strange phenomena that occur in and around them, really exist? The basis for understanding these weird phenomena is the relativistic concept that mass warps space—which has already been found to be a good representation of reality, at least for the weak gravitational fields produced by stars and planets (see *More Precisely 22-2*). The larger the mass concentration, the greater the warping, and apparently, the stranger the observational consequences. Although general relativity is not proven, there is presently no reason to disbelieve it, and black holes are one of its most striking predictions. So long as general relativity stands as the correct theory of gravity in the universe, black holes are real.

✓ Concept Check

▫ Why would you never actually witness an infalling object crossing the event horizon of a black hole?

22.8 Observational Evidence for Black Holes

⁊ Theoretical ideas aside, is there any observational evidence for black holes? Can we prove that these strange invisible objects really do exist?

STELLAR TRANSITS?

One way in which we might think of detecting a black hole would be to observe it transiting (passing in front of) a star. Unfortunately, such an event would be extremely hard to see. The 12,000-km-diameter planet Venus is barely noticeable when transiting the Sun, so a 10-km-wide object moving across the image of a faraway star would be completely invisible with either current equipment or any equipment available in the foreseeable future.

Actually, this observation is not even as clear-cut as just suggested. Even if we were close enough to the star to resolve the disk of the transiting black hole, the observable effect would not be a black dot superimposed on a bright background. The background starlight would be deflected as it passed the black hole on its way to Earth, as indicated in Figure 22.19. The effect is the same as the bending of distant starlight around the edge of the Sun, a phenomenon that has been repeatedly measured during solar eclipses throughout the last several decades (see *More Precisely 22-2*). With a black hole, much larger deflections

Figure 22.19 Gravitational Light Deflection The gravitational bending of light around the edges of a small, massive black hole makes it impossible to observe the hole as a black dot superimposed against the bright background of its stellar companion.

would occur. As a result, the image of a black hole in front of a bright companion star would show not a neat, well-defined black dot but rather a fuzzy image virtually impossible to observe, even from nearby.

BLACK HOLES IN BINARY SYSTEMS

A much better way to find black holes is to look for their effects on other objects. Our Galaxy harbors many binary-star systems in which only one object can be seen. Recall from our study of binary-star systems in Section 17.9 that we need to observe the motion of only one star to infer the existence of an unseen companion and measure some of its properties. In the majority of cases, the invisible companion is simply small and dim, nothing more than an M-type star hidden in the glare of an O- or B-type partner, or perhaps shrouded by dust or other debris, making it invisible to even the best available equipment. In either case, the invisible object is not a black hole.

A few close binary systems, however, have peculiarities suggesting that one of their members may be a black hole. Some of the most interesting observations, made during the 1970s and 1980s by Earth-orbiting satellites, revealed binary systems in which the invisible member emits large amounts of X rays. The mass of the emitting object is measured as several solar masses, so we know it is not simply a small, dim star, and radiation pressure from the binary members makes circumstellar debris an unlikely explanation for its invisibility.

One particular binary system drawing much attention lies in the constellation Cygnus. Figure 22.20(a) shows the area of the sky in Cygnus where astronomers have reasonably good evidence for a black hole. The rectangle outlines the celestial system of interest, some 2000 pc from Earth.

The black-hole candidate is an X-ray source called Cygnus X-1, discovered by the *Uhuru* satellite in the early 1970s. The main observational features of this binary system are as follows:

1. The visible companion of the X-ray source—a blue B-type supergiant with the catalog name HDE 226868—was identified a few years after Cyg X-1 was discovered. Assuming that the companion lies on the main sequence, we know that its mass must be around 25 times the mass of the Sun.

2. Spectroscopic observations indicate that the binary system has an orbital period of 5.6 days. Combining this information with further spectroscopic measurements of the visible component's orbital speed, astronomers estimate the total mass of the binary system to be around 35 solar masses, implying that Cygnus X-1 has a mass of about 10 times the mass of the Sun. ∞ (Sec. 17.9)

3. Other spectroscopic studies suggest that hot gas is flowing from the bright star toward an unseen companion.

4. X-ray radiation emitted from the immediate neighborhood of Cygnus X-1 implies the presence of very high temperature gas, perhaps as hot as several million kelvins (see Figure 22.20b).

5. Rapid time variations of this X-ray radiation imply that the size of the X-ray-emitting region of Cygnus X-1 must be less than a few hundred kilometers across. The reasoning is the same as in the discussion of gamma-ray bursts in Section 22.4. X rays from Cygnus X-1 have been observed to vary in intensity on time scales as short as a millisecond. For this variation not to be blurred by the travel time of light across the source, Cygnus X-1 cannot be more than 1 light-millisecond, or 300 km, in diameter.

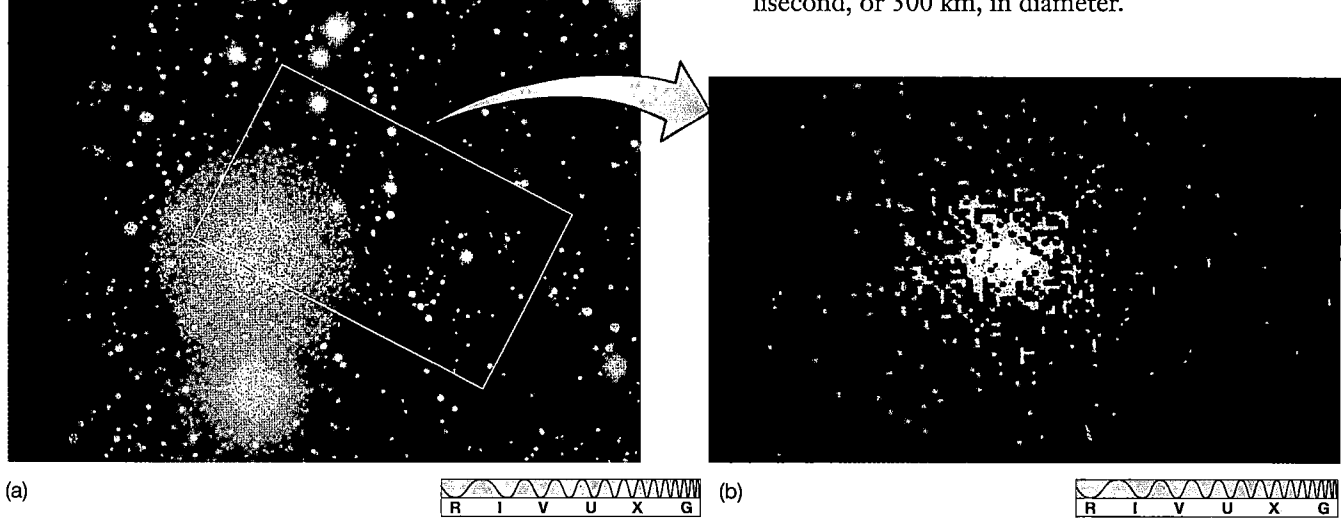

(a)

| R | I | V | U | X | G |

(b)

| R | I | V | U | X | G |

Figure 22.20 Cygnus X-1 (a) The brightest star in this photograph is a member of a binary system whose unseen companion, called Cygnus X-1, is thought to be a good candidate for a black hole. (b) X rays emitted by the Cygnus X-1 source were analyzed by changing them into electronic signals that were then viewed on a video screen, from which this picture was taken. (The field of view here is outlined by the rectangle in part a.) *(Harvard-Smithsonian Center for Astrophysics)*

These properties suggest that the invisible X-ray-emitting companion could be a black hole. The X-ray-emitting region is likely an accretion disk formed as matter drawn from the visible star spirals down onto the unseen component. The rapid variability of the X-ray emission indicates that the unseen component must be compact—a neutron star or a black hole. The mass limit of the dark component argues for the latter, if a neutron star's mass cannot exceed about 3 solar masses. Figure 22.21 is an artist's conception of this intriguing object. As shown, most of the gas drawn from the visible star ends up in a doughnut-shaped accretion disk of matter. As the gas flows toward the black hole, it becomes superheated and emits the X rays we observe just before they are trapped forever below the event horizon.

A few other black-hole candidates are known. For example, the third X-ray source ever discovered in the Large Magellanic Cloud—called LMC X-3—is an invisible object that, like Cygnus X-1, orbits a bright companion star. LMC X-3's visible companion seems to be distorted into the shape of an egg by the unseen object's intense gravitational pull. Reasoning similar to that applied to Cygnus X-1 leads to the conclusion that the compact object LMC X-3 has a mass nearly 10 times that of the Sun, making it too massive to be anything but a black hole. The X-ray binary system A0620-00 has been found to contain an invisible compact object of mass 3.8 times the mass of the Sun. In total, there are perhaps a half-dozen known objects in or near our Galaxy that may turn out to be black holes, although Cygnus X-1, LMC X-3, and A0620-00 probably have the strongest claims.

BLACK HOLES IN THE CENTERS OF GALAXIES

Perhaps the strongest evidence for black holes comes not from binary systems in our own Galaxy but from observations of the centers of many galaxies, including our own.

Using high-resolution observations at wavelengths ranging from radio to ultraviolet, astronomers have found that stars and gas near the centers of many galaxies are moving extremely rapidly, orbiting some very massive, unseen object. Masses inferred from Newton's laws range from millions to billions of times the mass of the Sun. The intense energy emission from the centers of these galaxies, and short-timescale fluctuations in that emission, suggest the presence of massive, compact objects. The leading (and at present, the only) explanation is that these objects are black holes. We will return to these observations, and the question of how such *supermassive black holes* might have formed, in Chapters 23 and 25.

In 2000, X-ray astronomers reported a long-sought but elusive missing link between "stellar-mass" black holes in binaries and the supermassive black holes in the hearts of galaxies. Figure 22.22 shows an unusual looking galaxy called M82, currently the site of an intense and widespread burst of star formation (see Chapter 24). The inset shows a *Chandra* image of the innermost few thousand parsecs of M82, revealing a number of bright X-ray sources close to—but *not* at—the center of the galaxy. Their spectra and X-ray luminosities strongly suggest that they are accreting compact objects with masses ranging from 100 to almost 1000 times the mass of the Sun. If confirmed, they will be the first *intermediate-mass black holes* ever observed.

Too large to be the remnant of a normal star, and too small to warrant the "supermassive" label, these objects present a puzzle to astronomers. Supermassive black holes are thought to have formed long ago, and to have played a role in galaxy formation, but these intermediate-mass black holes appear to have formed recently. Where did they come from? One possible explanation has been suggested by follow-up infrared observations using the Subaru telescope on Mauna Kea, which reveal that more than half of the X-ray sources are apparently associated

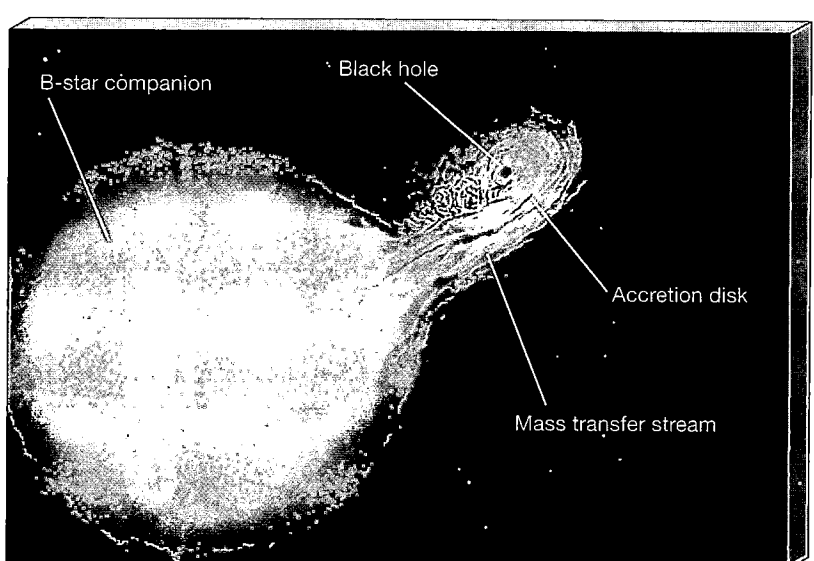

Figure 22.21 Black-Hole Binary Artist's conception of a binary system containing a large, bright, visible star and an invisible, X-ray-emitting black hole. This drawing is based on data obtained from detailed study of Cygnus X-1. *(L. Chaisson)*

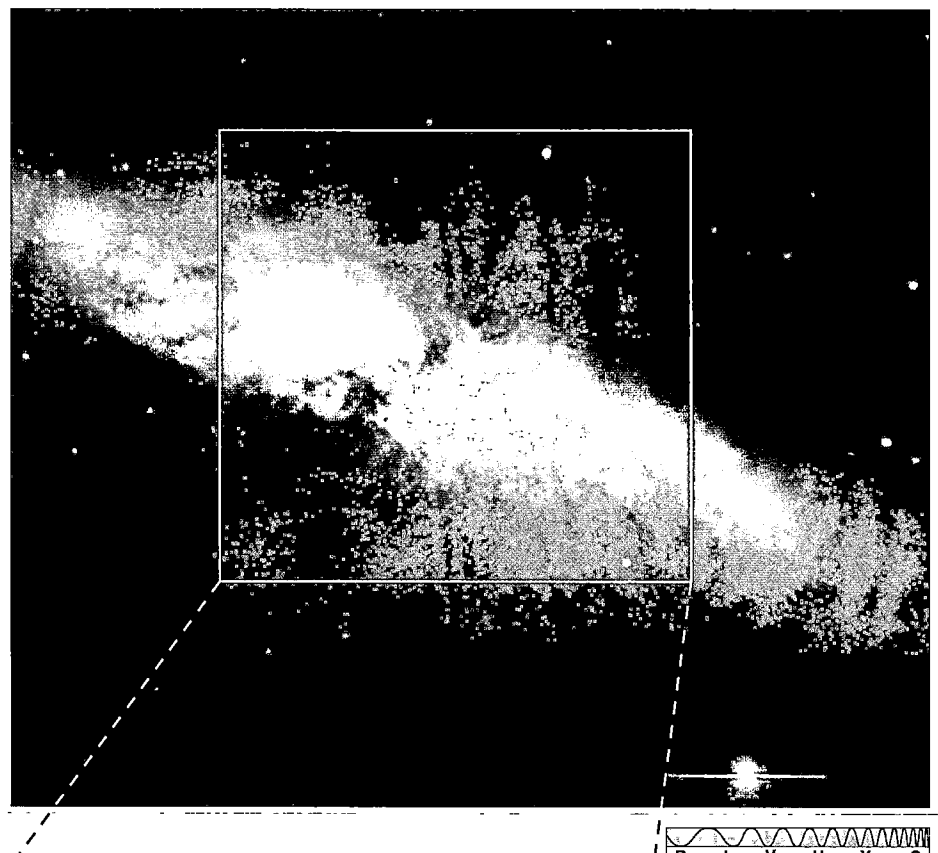

Figure 22.22 Intermediate-Mass Black Holes X-ray observations (inset) of the center of the starburst galaxy M82 reveal a collection of bright sources thought to be the result of matter accreting onto intermediate-mass black holes. The black holes are probably young, have masses between 100 and 1000 times the mass of the Sun, and lie relatively far from the center of M82. *(Subaru Telescope; NASA)*

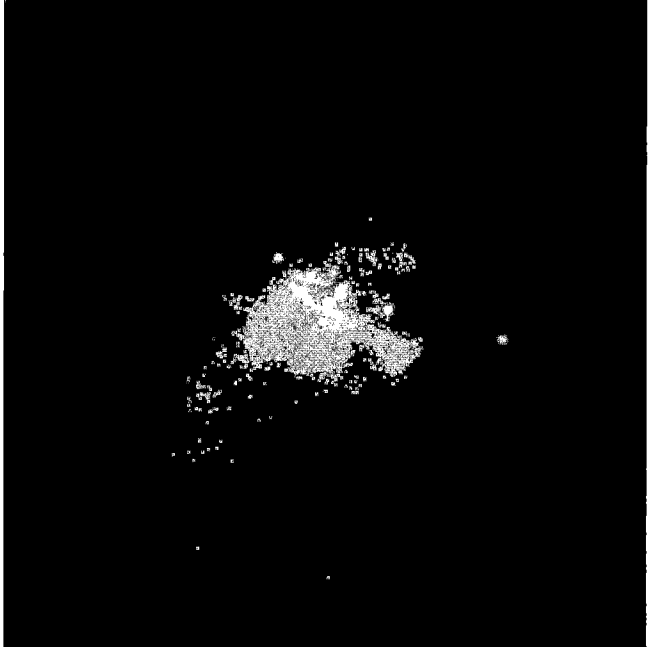

with dense, young star clusters. Theorists speculate that collisions between high-mass main-sequence stars in the congested cores of such a cluster could lead to the runaway growth of an extremely massive and very unstable star, which then collapses to form an intermediate-mass black hole. However, further observations and systematic studies of many other galaxies are clearly needed to fully resolve this issue.

DO BLACK HOLES EXIST?

You may have noticed that the identification of an object as a black hole really proceeds by elimination. Loosely stated, the argument goes: "Object X is compact and very massive. We don't know of anything else that can be that small and that massive. Therefore, object X is a black hole." For the very massive compact objects observed (or inferred) in the centers of galaxies, the lack of viable alternatives means that the black-hole hypothesis has become widely accepted among astronomers. However, Cygnus X-1 and the other suspected stellar-mass black holes in binary systems all have masses relatively close to the dividing line separating neutron stars from black holes. Given the present uncertainties in both observations and theory, might they con-

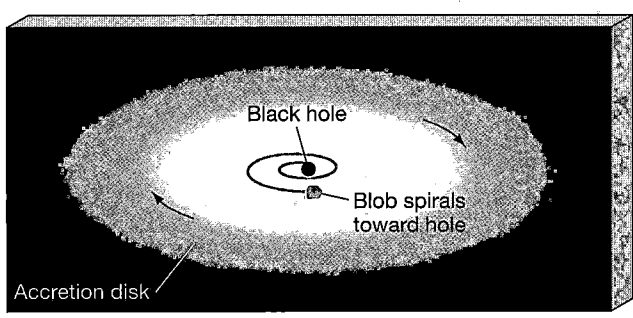

(a)

Figure 22.23 Infalling Matter Observed? (a) As a blob of gas spirals inward from the inner edge of an accretion disk onto a black hole, its luminosity varies in a characteristic way as its decaying orbit loops around the hole. (b) To an outside observer, the result is a series of pulses of radiation, cut off after just a few orbits as the blob crosses the event horizon and is lost forever.

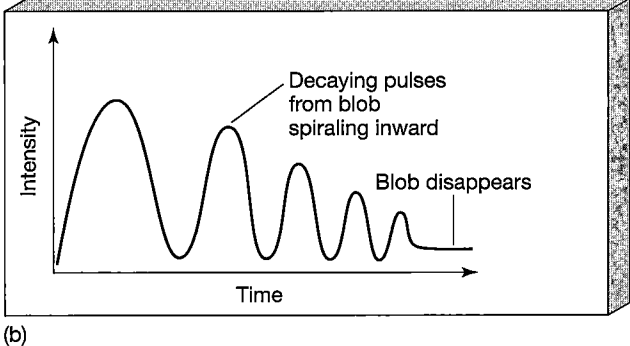

(b)

ceivably be merely dim, dense neutron stars and not black holes at all?

Most astronomers do not regard this as a likely possibility, but it highlights a problem. It is difficult to unambiguously distinguish a 10-solar-mass black hole from, say, a 10-solar-mass neutron star (if one could somehow exist). Both objects would affect a companion star's orbit in the same way; both would tear mass from its surface, and both would form an accretion disk around themselves that would emit intense X rays (although some researchers believe that the accretion disks may differ sufficiently in their detailed properties that the nature of the central object may be identifiable from observations).

In early 2001, however, two groups of researchers independently reported evidence for one of the true defining properties of a black hole—an event horizon. First, using instruments on board *Chandra*, X-ray astronomers found that X-ray novae (mass-transferring binary systems whose X-ray emission suddenly and unpredictably flares up, probably due to instabilities in the inner parts of the accretion disk) suspected of harboring black holes emitted far less radiation between outbursts than those in systems thought to contain neutron stars. Their explanation is that the accreting gas in the latter case glows brightly as it collides violently with the neutron star surface. In the black-hole case, the gas simply crosses the event horizon and vanishes. Second, workers using *Hubble* have observed rapidly fading pulses of ultraviolet light from another

X-ray source, as shown in Figure 22.23. They interpret their observations as clumps of hot gas spiraling down from the inner edge of an accretion disk onto the black hole at the center. The pulses mark the decaying orbit of the blob, and the absence of a flash at the end of the pulse indicates that, at the end of the plunge, the gas simply vanished rather than hitting a solid surface.

So have black holes really been discovered? The answer is probably yes. Skepticism is healthy in science, but only the most stubborn astronomers (and some do exist!) would take serious issue with the reasoning that supports the case for black holes. Can we guarantee that future modifications to the theory of compact objects will not invalidate some or all of our arguments? No, but similar statements could be made in many areas of astronomy—indeed, about any theory in any area of science. We conclude that, strange as they are, black holes have been detected, both in our Galaxy and beyond. Perhaps some day, future generations of space travelers will visit Cygnus X-1 or the center of our Galaxy and (carefully!) test these conclusions firsthand. Until then, we will have to continue to rely on improving theoretical models and observational techniques to guide our discussions of these mysterious objects.

 Concept Check

◘ How do astronomers "see" black holes?

Chapter Review

SUMMARY

A core-collapse supernova may leave behind a remnant, an ultra-compressed ball of material called a **neutron star** (p. 568). The processes that form neutron stars ensure that these stars are rapidly rotating and strongly magnetized at birth. **Pulsars** (p. 569) are objects that appear to emit regular bursts of electromagnetic energy. The accepted explanation for pulsars is the **lighthouse model** (p. 569), in which a rotating neutron star sends a beam of energy into space. If the beam sweeps past Earth, we see a pulsar. The pulse period is the rotation period of the neutron star. Because the pulse energy is beamed into space and because neutron stars slow down as they radiate energy into space, not all neutron stars are seen as pulsars.

A neutron star in a close binary system can draw matter from its companion, forming an accretion disk. The material in the disk heats up even before it reaches the neutron star, and the disk is usually a strong source of X rays. As gas builds up on the star's surface, it eventually becomes hot enough to fuse hydrogen. As with a nova explosion on a white dwarf, when hydrogen burning starts on a neutron star, it does so explosively. An **X-ray burster** (p. 572) results. The rapid rotation of the inner part of the accretion disk causes the neutron star to spin faster as new gas arrives on its surface. The eventual result is a very rapidly rotating neutron star—a **millisecond pulsar** (p. 573). Many millisecond pulsars are found in the hearts of old globular clusters. They cannot have formed recently, so they must have been spun up by interactions with other stars. Careful analysis of the radiation received has shown that some pulsars are orbited by planet-sized objects. The origin of these "pulsar planets" is still uncertain.

Gamma-ray bursts (p. 575) are very energetic flashes of gamma rays that occur about once a day and are distributed uniformly over the entire sky. In some cases, their distances have been measured, placing them at very large distances and implying that they are extremely luminous. The leading theoretical models for these explosions involve the violent merger of neutron stars in a distant binary system, or the recollapse and subsequent violent explosion following a "failed" supernova in a very massive star.

The upper limit on the mass of a neutron star is about 3 solar masses. Beyond that mass, the star can no longer support itself against its own gravity, and it must collapse. No known force can prevent the material from collapsing all the way to a point-like **singularity** (p. 588), a region of extremely high density

where the known laws of physics break down. Surrounding the singularity, at a distance of a few kilometers for a solar-mass object, is a region of space from which even light cannot escape—a **black hole** (p. 581). Astronomers believe that the most massive stars form black holes, rather than neutron stars, after they explode in a supernova.

Conditions in and near black holes cannot be described by Newtonian mechanics. A proper description involves the **theories of relativity** (p. 581) developed by Albert Einstein early in the twentieth century. Even relativity theory fails right at the singularity, however. Relativity theory describes gravity in terms of a warping, or bending, of space by the presence of mass. The more mass, the greater the warping. All particles—including photons—respond to that warping by moving along curved paths. A black hole is a region where the warping is so great that space folds back on itself, cutting off the interior of the hole from the rest of the universe.

The "surface" of a black hole is the **event horizon** (p. 581); its distance from a singularity is called the **Schwarzschild radius** (p. 581). At the event horizon the escape speed equals the speed of light. Within this distance, nothing can escape. Photons passing too close to a black hole are deflected onto paths that cross the event horizon and become trapped. To a distant observer, the clock on a spaceship falling into a black hole would show **time dilation** (p. 588)—it would appear to slow down as the ship approached the event horizon. The observer would never see the ship reach the surface of the hole. At the same time, light leaving the ship would be subject to **gravitational redshift** (p. 586) as it climbed out of the hole's intense gravitational field. Light emitted just at the event horizon would be redshifted to infinite wavelength. Both phenomena are predictions of the theory of relativity. The gravitational redshifts due to both Earth and the Sun are very small but have been detected experimentally.

Once matter falls into a black hole, it can no longer communicate with the outside. However, on its way in, it can form an accretion disk and emit X rays just as in the neutron-star case. The best candidates for black holes are binary systems in which one component is a compact X-ray source. Cygnus X-1, a well-studied X-ray source in the constellation Cygnus, is a long-standing black-hole candidate. There is also substantial evidence for more massive black holes residing in or near the centers of many galaxies, including our own.

SELF-TEST: TRUE OR FALSE?

_____ **1.** The density of a neutron star is comparable to the density of an atomic nucleus.

_____ **2.** As a result of their high masses and small sizes, neutron stars have only weak gravitational pulls at their surfaces.

_____ **3.** Newly formed neutron stars have extremely strong magnetic fields.

_____ **4.** A millisecond pulsar is actually a very old neutron star that has been recently spun up by interaction with a neighbor.

_____ **5.** Planet-sized bodies are not found orbiting pulsars because the supernova that forms a pulsar destroys any planetary system the star might have.

_____ **6.** Gamma-ray bursts mainly occur within the Milky Way Galaxy.

_____ **7.** Nothing can travel faster than the speed of light.

_____ **8.** All things, except light, are attracted by gravity.

_____ **9.** Loosely speaking, a black hole is an object whose escape speed equals or exceeds the speed of light.

____ **10.** Although visible light cannot escape from a black hole, high-energy radiation, like gamma rays, can escape.

____ **11.** It is possible for a beam of light to go into orbit around a black hole.

____ **12.** If you could touch it, the surface of a black hole, the event horizon, would be very hard.

____ **13.** The present laws of physics break down near the center of a black hole.

____ **14.** X rays are emitted by matter accreting onto a stellar-mass black hole.

____ **15.** Thousands of black holes have now been identified in our Galaxy.

SELF-TEST: FILL IN THE BLANK

1. No remnant remains after the explosion of a _____ supernova.

2. A typical neutron star is _____ km in diameter.

3. Neutron stars may be characterized as having _____ rotation rates and _____ magnetic fields.

4. Pulsars were discovered through observations in the _____ part of the electromagnetic spectrum.

5. The pulse period of pulsar radiation tells us the _____ of the neutron star emitting the radiation.

6. All millisecond pulsars are now, or once were, members of _____ systems.

7. X-ray bursters result from accretion of material from a binary companion onto a _____.

8. According to general relativity, space is warped, or curved, by _____.

9. Rapidly moving, very massive objects may emit energy in the form of _____.

10. If the Sun were magically to turn into a black hole of the same mass, Earth's orbit would _____.

11. The radius of the event horizon of a black hole is called the _____.

12. The region of extremely high density at the center of a black hole is called a _____.

13. Photons _____ energy as they escape from a gravitational field.

14. Black holes of stellar origin are believed to have been discovered in several _____ systems.

15. Scientists infer that the energy-emitting region in Cygnus X-1 must be very small because of the rapid _____ of the electromagnetic radiation received from it.

REVIEW AND DISCUSSION

1. How does the way in which a neutron star forms determine some of its most basic properties?

2. What would happen to a person standing on the surface of a neutron star?

3. Why aren't all neutron stars seen as pulsars?

4. Why do many neutron stars move at high speeds relative to their neighbors?

5. What are X-ray bursters?

6. What is the favored explanation for the rapid spin rates of millisecond pulsars?

7. Why do you think astronomers were surprised to find a pulsar with a planetary system?

8. Why do astronomers think that gamma-ray bursts are very distant and very energetic?

9. Describe two leading models for gamma-ray bursts.

10. What does it mean to say that the measured speed of a light beam is independent of the motion of the observer?

11. Use your knowledge of escape speed to explain why black holes are said to be "black."

12. According to special relativity, what is special about the speed of light?

13. Why is it so difficult to test the predictions of general relativity? Describe two tests of the theory.

14. What would happen to someone falling into a black hole?

15. What is an event horizon?

16. What is the principle of cosmic censorship? Do you think it is a sound scientific principle?

17. What makes Cygnus X-1 a good black-hole candidate?

18. What evidence is there for black holes much more massive than the Sun?

19. Imagine that you had the ability to travel at will through the Galaxy. Explain why you would discover many more neutron stars than those known to observers on Earth. Where would you be most likely to find these objects?

20. Do you think that planet-sized objects discovered in orbit around a pulsar should be called planets? Why or why not?

PROBLEMS ~www~ *Algorithmic versions of these questions are available in the Practice Problems module of the Companion Website.*

The number of squares preceding each problem indicates its approximate level of difficulty.

1. ■ The angular momentum of a solid body is proportional to its angular velocity times the square of its radius. ∞ (*More Precisely 15-1*) Using the law of conservation of angular momentum, estimate how fast a collapsed stellar core would spin if its initial spin rate was 1 revolution per day and its radius decreased from 10,000 km to 10 km.

2. ■ What would your mass be if you were composed entirely of neutron-star material, of density 3×10^{17} kg/m^3? (Assume that your average density is 1000 kg/m^3.) Compare this with the mass of a typical 10-km diameter rocky asteroid.

3. ■ Calculate the surface gravitational acceleration and escape speed of a 1.4-solar-mass neutron star with a radius of 10 km. What would be the escape speed from a neutron star of the same mass, and radius 4 km?

4. ■■ Use the radius–luminosity–temperature relation to calculate the luminosity of a 10-km-radius neutron star for temperatures of 10^5 K, 10^7 K, and 10^9 K. At what wavelengths does the star radiate most strongly in each case? What do you conclude about the detectability of neutron stars? Could the brightest of them be plotted on our H–R diagram?

5. ■■ A gamma-ray detector of area 0.5 m² observing a gamma-ray burst records photons having total energy 10^{-8} joules. If the burst occurred 1000 Mpc away, calculate the total amount of energy it released (assuming that the energy was emitted isotropically), and compare it with the Sun's total energy output on the main sequence. How would this figure change if the burst occurred 10,000 pc away instead, in the halo of our Galaxy? What if it occurred within the Oort cloud of our own solar system, at a distance of 50,000 A.U.?

6. ■■ A gamma-ray burst 5000 Mpc away releases 10^{45} joules of energy in the form of gamma rays each of energy 250 keV. ⚭ (*More Precisely 4-1*) They are detected by an instrument in Earth orbit of effective collecting area 0.75 m². How many gamma-ray photons strike the detector?

7. ■ A 10-km-radius neutron star is spinning 1000 times per second. Calculate the speed of a point on its equator, and compare it with the speed of light. (Consider the equator as the circumference of a circle, and recall that circumference $= 2\pi r$.) Also calculate the orbital speed of a particle in a circular orbit just above the surface, if the neutron star's mass is 1.4 times the mass of the Sun.

8. ■ Supermassive black holes are believed to exist in the centers of many galaxies. What would be the Schwarzschild radii of black holes of 1 million and 1 billion solar masses, respectively? How does the first black hole compare in size with the Sun? How does the second compare in size with the solar system?

9. ■■ The angle through which light is deflected as it passes by a gravitating body is proportional to the body's mass and inversely proportional to the distance from the body. Use the data presented in *More Precisely 22-2* to estimate the deflection of a beam of light that just grazes the surface of (a) Earth, (b) Jupiter, (c) Sirius B. (d) A future generation of space astrometry missions will be able to measure accurately angles as small as 10^{-6} arcsec. At what distance from the Sun would this deflection occur?

10. ■■ Gravitational redshift is not confined to the neighborhoods of black holes. It may be shown that the fractional decrease in frequency due to gravitational redshift of a light ray emitted at Earth's surface and detected at height h above the surface is approximately gh/c^2, where $g = 9.80$ m/s² is the gravitational acceleration at Earth's surface and c is the speed of light. Calculate the change in frequency of a 10 GHz (10^{10} Hz) radio signal received by a satellite directly overhead at an altitude of 500 km.

11. ■■ Calculate the tidal acceleration on a 2-m-tall human falling feet-first into a 1-solar-mass black hole—that is, compute the difference in the accelerations (forces per unit mass) on their head and their feet just as the feet cross the event horizon. ⚭ (*More Precisely 7-3*) Repeat the calculation for a 1-million-solar-mass black hole and for a 1-billion-solar-mass black hole (see question 8). Compare these accelerations with the acceleration due to gravity on Earth ($g = 9.8$ m/s²).

12. ■■■ Endurance tests suggest that the human body cannot withstand stress greater than about 10 times the acceleration due to gravity on Earth's surface. At what distance from a 1-solar-mass black hole would the human in the previous question be torn apart?

13. ■■■ Using the data from the previous question, what is the minimum mass of a black hole for which an infalling human could just reach the event horizon intact?

14. ■■ An intermediate-mass black hole in the galaxy M82 has a mass 100 times that of the Sun. What is the radius of its event horizon? Estimate how close to the hole a star like the Sun could approach before being torn apart by tidal forces.

15. ■■ Using the data given in the text, calculate the orbital separation of Cyg X-1. If the companion star's radius is 20 million km, verify (approximately) that the black hole's tidal field is sufficient to draw matter from the companion's surface.

COLLABORATIVE EXERCISES

1. Neutron Star Sizes. The typical neutron star is only about 20 km in diameter. How many neutron stars could fit between the birthplaces of two members in your group? Explain your reasoning.

2. Pulsar Bursts. Using Figure 22.2, each group member should carefully determine the length of time between two different sequential bursts. Estimate the timing to 1/10 second and determine your group's average result. Compare your average value to another group and account for any differences.

RESEARCHING ON THE WEB  *To complete the following exercises, go to the online Destinations module for Chapter 22 on the Companion Website for* Astronomy Today 4/e.

1. Access the "Black Holes FAQ" page and describe what you would observe if you were watching, from a safe distance, a spaceship travel towards the center of a black hole.

2. Access the "Virtual Trips to Black Holes and Neutron Stars" page and view the video, Approaching a Black Hole. Describe what happens to stars when you get close to the black hole.

3. Access the "Movies from the Edge of Spacetime" page and view the video, Schwarzschild Wormhole. What is a Schwarzschild Wormhole?

PROJECTS

1. Many amateur astronomers enjoy turning their telescopes on the magnitude-9 companion to Cygnus X-1, the sky's most famous black-hole candidate. Because none of us can see in X rays, no sign of anything unusual can be seen. Still, it's fun to gaze toward this region of the heavens and contemplate Cygnus X-1's powerful energy emission and strange properties. Even without a telescope, it is easy to locate the region of the heavens where Cygnus X-1 resides. The constellation Cygnus contains a recognizable star pattern, or asterism, in the shape of a large cross. This asterism is called the Northern Cross. The star in the center of the crossbar is called Sadr. The star at the bottom of the cross is called Albireo. Approximately midway along an imaginary line between Sadr and Albireo lies the star Eta Cygni. Cygnus X-1 is located slightly less than 0.5° from this star. With or without a telescope, sketch what you see.

2. Set up a demonstration of the densities of various astronomical objects—an interstellar cloud, a star, a terrestrial planet, a white dwarf, and a neutron star. Select a common object that is easily held in your hand, something that would be familiar to anyone—an apple, for example. For the lowest densities, calculate how large a volume would contain the object's equivalent mass. For high densities, calculate how many of the objects would have to be fit into a standard volume, such as 1 cm^3. This volume is better for this project than 1 m^3 because most people do not appreciate how large a volume 1 m^3 is. Present your demonstration to your class or to some other group of students. Tell them about each astronomical object and how it comes by its density.

SKYCHART III PROJECTS  *The SkyChart III Student Version planetarium program on which these exercises are based is included as a separately executable program on the CD in the back of this text.*

1. ▣ At the center of the Milky Way is an enormous black hole of 3 million solar masses. This is not visible with the naked eye. However, from most places in the United States, you can locate it during the summer months. Find the constellation Sagittarius. A grouping of stars within Sagittarius forms the easily recognizable asterism called the "Teapot." The stars τ and σ Sagittarii (Sgr) form the handle, ζ, ε, δ, π, and λ form the body of the teapot, and the stars ε and γ form the spout. Locate these stars. Draw a line between γ Sgr and X Sgr. About 4/5th from γ to X is the galactic center, containing the black hole.

2. ▣ Change the field of view to 360°. In the *Symbols & Grids* submenu, turn off the stars, constellations and planets. Display the deep sky objects but limit the magnitude for all to

−3 except for the globular clusters, and display the galactic equator and ecliptic. What do you observe? Where are most of the globular clusters found? Now change your location to the Andromeda galaxy. What do you observe? Why does this happen?

3. ▣ Repeat the previous exercise, but now do not display the deep sky objects. Instead, display the stars. Display the planets by symbols. From Earth, what do you observe? From the Andromeda galaxy, what do you observe? This should not make sense. Why? To figure out what goes on, determine the distance to a star that is not very near the Sun. Now change your location to Earth, and figure out the distance to that same star. Did it change?

 *In addition to the Practice Problems and Destinations modules, the Companion Website at http://www.prenhall.com/chaisson provides for each chapter an additional true-false, multiple choice, and labeling quiz, as well as additional annotated images, animations, and links to related Websites.*

23 THE MILKY WAY GALAXY

A Grand Design

LEARNING GOALS

Studying this chapter will enable you to:

1 Describe the overall structure of the Milky Way Galaxy and specify how the various regions differ from one another.

2 Explain the importance of variable stars in determining the size and shape of our Galaxy.

3 Describe the orbital paths of stars in different regions of the Galaxy and explain how these motions are accounted for by our understanding of how the Galaxy formed.

4 Discuss some possible explanations for the existence of the spiral arms observed in our own and many other galaxies.

5 Explain what studies of galactic rotation reveal about the size and mass of our Galaxy and discuss the possible nature of dark matter.

6 Describe some of the phenomena observed at the center of our Galaxy.

 Visit http://www.prenhall.com/chaisson for additional annotated images, animations, and links to related sites for this chapter.

NGC 4603 is a magnificent spiral galaxy that somewhat resembles our own Milky Way. Despite the distance of roughly 100 million light-years, this image affords us a view of bright blue (young) stars in the spiral arms amidst a scattering of giant red (older) stars, a few of them perhaps about to detonate as supernovae. At this distance only the brightest stars can be distinguished as individual objects; the diffuse blue-gray glow arises from myriad fainter stars that cannot be resolved by any telescope yet built. (STScI)
The Big Picture: Stars are not distributed randomly in space; rather, they cluster in gargantuan assemblages. These are the galaxies, of which our Milky Way Galaxy is just one among more than nearly a hundred billion others. We cannot show a photo of our larger home in space because we cannot get outside of it to look back and see it clearly. Our Sun sits in the suburbs of the Milky Way, nearly 30,000 light-years from the central regions, which appear to be a good deal more violent than the part of the Galaxy in which we live.

Looking up on a dark, clear night, we are struck by two aspects of the night sky. The first is a fuzzy band of light—the Milky Way—that stretches across the heavens. From the Northern Hemisphere this band is most easily visible in the summertime, arcing high above the horizon. Its full extent forms a great circle that encompasses the entire celestial sphere. Away from that glowing band, however, our second impression is that the nighttime sky seems more or less the same in all directions. Bunches of stars cluster here and there, but overall, apart from the band of the Milky Way, the evening sky looks pretty uniform. Yet this is only a local impression. Ours is a rather provincial view. When we consider much larger volumes of space, on scales far, far greater than the distances between neighboring stars, a new level of organization becomes apparent as the large-scale structure of the Milky Way Galaxy is revealed.

23.1 Our Parent Galaxy

A **galaxy** is a gargantuan collection of stellar and interstellar matter—stars, gas, dust, neutron stars, black holes—isolated in space and held together by its own gravity. Astronomers are aware of literally millions of galaxies beyond our own. The particular galaxy we happen to inhabit is known as the *Milky Way Galaxy*, or just *the Galaxy*, with a capital *G*.

Our Sun lies in a part of the Galaxy known as the **Galactic disk**—an immense, circular, flattened region containing most of our Galaxy's luminous stars and interstellar matter (and virtually everything we have studied so far in this book). Figure 23.1 illustrates how, viewed from within, the Galactic disk appears as a band of light stretching across our night sky, a band known as the *Milky Way*. As indicated in the figure, if we look in a direction away from the Galactic disk (red arrows), relatively few stars lie in our field of view. However, if our line of sight happens to lie within the disk (white and blue arrows), we see so many stars that their light merges into a continuous blur.

Paradoxically, although we can study individual stars and interstellar clouds near the Sun in great detail, our location within the Galactic disk makes deciphering our Galaxy's large-scale structure from Earth a very difficult task—a little like trying to unravel the layout of paths, bushes, and trees in a city park without being able to leave one particular park bench. In some directions the interpretation of what we see is ambiguous and inconclusive. In others, foreground objects completely obscure our view of what lies beyond, but we cannot move around them to get a better look. As a result, astronomers who study the Milky Way Galaxy are often guided in their efforts by comparisons with more distant, but much more easily observable, systems.

Figures 23.2 and 23.3 show three galaxies thought to resemble our own in overall structure. Figure 23.2 is the Andromeda Galaxy, the nearest major galaxy to the Milky Way Galaxy, lying nearly 800 kpc (roughly 2.5 million light years) away. Andromeda's apparent elongated shape is a consequence of the angle at which we happen to view it. In fact, our Galaxy, like this one, consists of a circular galactic disk of matter that fattens to a **Galactic bulge** at the center. The disk and bulge are embedded in a roughly spherical ball of faint old stars known as the **Galactic halo**. These three basic galactic regions are indicated on the figure (the halo stars are so faint that they cannot be discerned here; see Figure 23.10). Figure 23.3(a) and (b) shows views of two other galaxies—one seen face-on, the other edge-on—that illustrate these points more clearly.

✔ Concept Check

■ Why do we see the Milky Way as a band of light across the sky?

23.2 Measuring the Milky Way

Before the twentieth century, astronomers' conception of the cosmos differed markedly from the modern view. The fact that we live in just one of many enormous "islands" of matter separated by even larger tracts of apparently empty space was completely unknown, and the clear distinction between "our Galaxy" and "the universe" did not exist. The twin ideas that (1) the Sun is not at the center of the Galaxy and (2) the Galaxy is not at the center of the universe required both time and hard observational evidence before they gained widespread acceptance.

STAR COUNTS

In the late eighteenth century, long before the distances to any stars were known, the English astronomer William Herschel tried to estimate the size and shape of our Galaxy simply by counting how many stars he could see in different directions in the sky. Assuming that all stars were of about equal brightness, he concluded that the Galaxy was a somewhat flattened, roughly disk-shaped collection of stars lying in the plane of the Milky Way, with the Sun at its *center* (Figure 23.4). Subsequent refinements to this approach led to much the same picture. Early in the twentieth century, some workers went so far as to estimate the dimensions of this "Galaxy" as about 10 kpc in diameter by 2 kpc thick.

Today the Milky Way Galaxy is known to be several tens of kiloparsecs across, and the Sun lies far from the

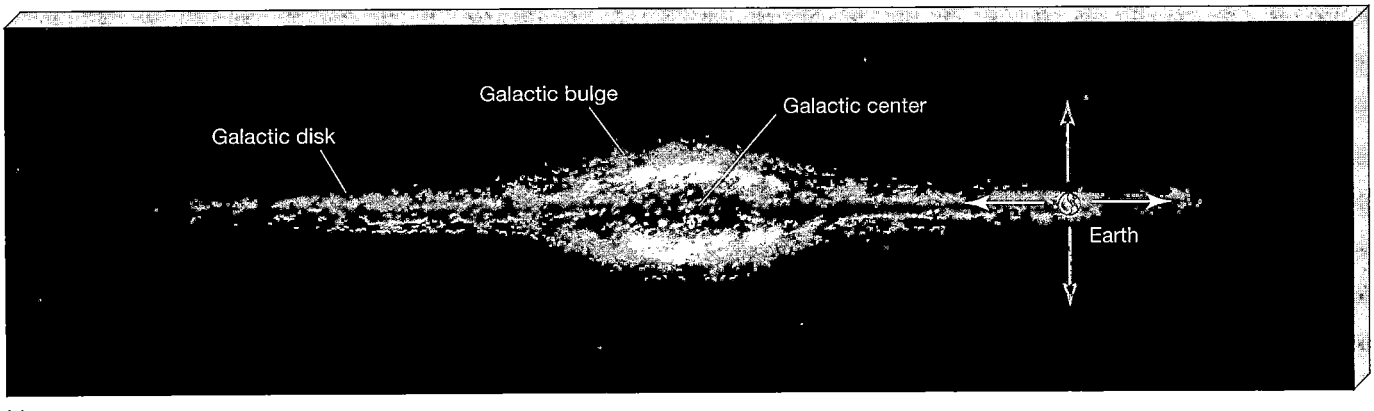

(a)

(b)

| R | I | V | U | X | G |

Figure 23.1 Galactic Plane (a) Gazing from Earth toward the Galactic center (white arrow) in this artist's conception, we see myriad stars stacked up within the thin band of light known as the Milky Way. Looking in the opposite direction (blue arrow), we still see the Milky Way band, but now it is fainter because our position is far from the Galactic center, with the result that we see more stars when looking toward the center than when looking in the opposite direction. Looking perpendicular to the disk (red arrows), we see far fewer stars. (b) This is an enhanced infrared satellite view of the sky all around us; the yellow band is the disk of our Milky Way Galaxy—dimly seen with the naked eye from very dark locations on Earth, but glowing brightly here since the disk is so rich in dust. *(NASA)*

center. How could the older picture have been so flawed? The answer is that the earlier observations were made in the visible part of the electromagnetic spectrum, and astronomers failed to take into account the (then unknown) absorption of visible light by interstellar gas and dust. ∞ (Sec. 18.1) Only in the 1930s did astronomers begin to realize the true extent and importance of the interstellar medium.

Any objects in the Galactic disk more than a few kiloparsecs away from us are hidden from our view by the effects of interstellar dust. The apparent falloff in the density of stars with distance in the plane of the Milky Way is thus not a real thinning of their numbers in space but simply a consequence of the murky environment in the Galactic

disk. The long "fingers" in Herschel's map are directions in which the obscuration happens to be a little less severe than in others. However, because some obscuration occurs in all directions in the disk, the falloff is roughly similar no matter which way we look, so the Sun appears to be more or less at the center. The horizontal extent of Figure 23.4 corresponds approximately to the span of the blue and white arrows in Figure 23.1.

Radiation coming to us from above or below the plane of the Galaxy, where there is less gas and dust along the line of sight, arrives on Earth relatively unscathed. There is still some patchy obscuration, but the Sun happens to lie in a location where the view out of the disk is largely unimpeded by nearby interstellar clouds.

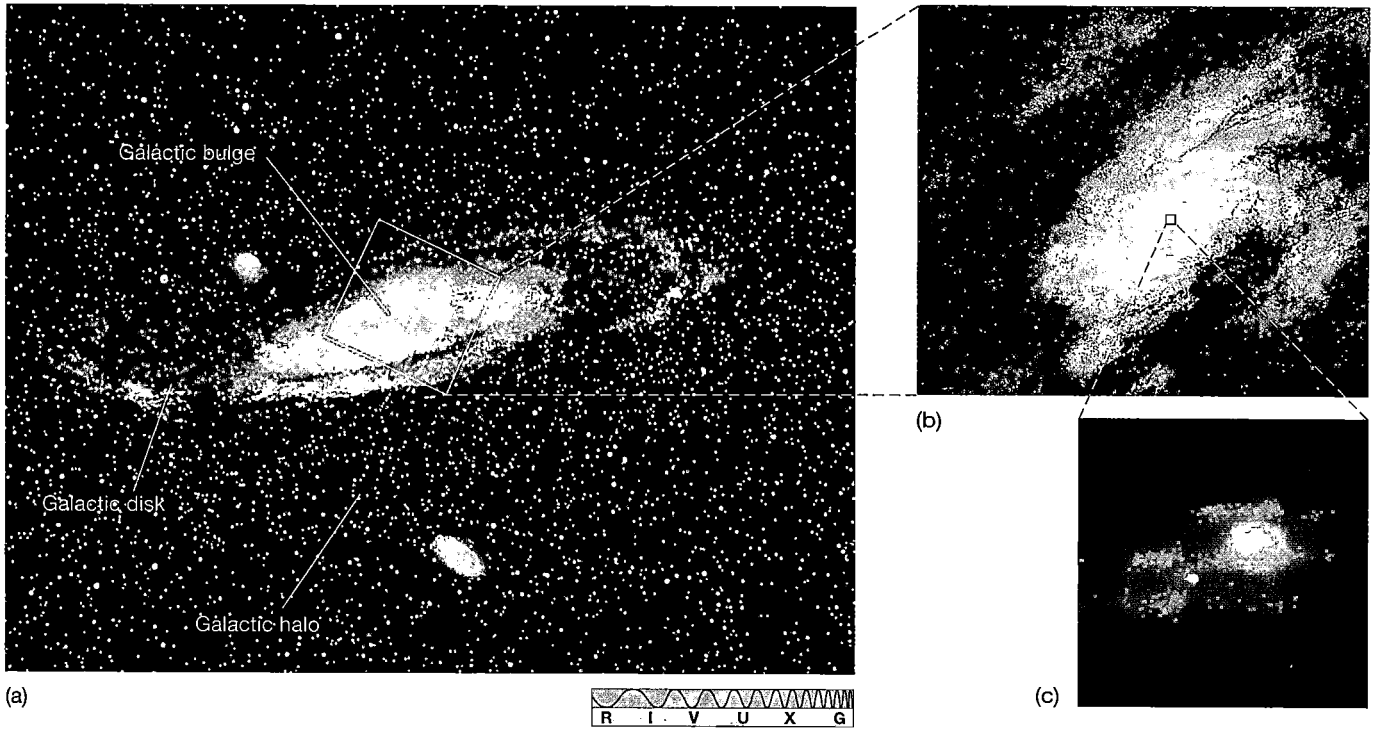

Figure 23.2 Andromeda Structure (a) The Andromeda Galaxy probably resembles fairly closely the overall layout of our own Milky Way Galaxy. The disk and bulge are clearly visible in this image, which is about 30,000 pc across. The faint halo stars cannot be seen. The white stars sprinkled all across this image are not part of Andromeda's halo; they are foreground stars in our own Galaxy, lying in the same part of the sky as Andromeda, but about a thousand times closer. (b) More detail within the inner parts of the galaxy. (c) The galaxy's peculiar—and still unexplained—double core; this inset covers a region only 15 pc across. *(T. Hallas; Palomar/Caltech; NASA)*

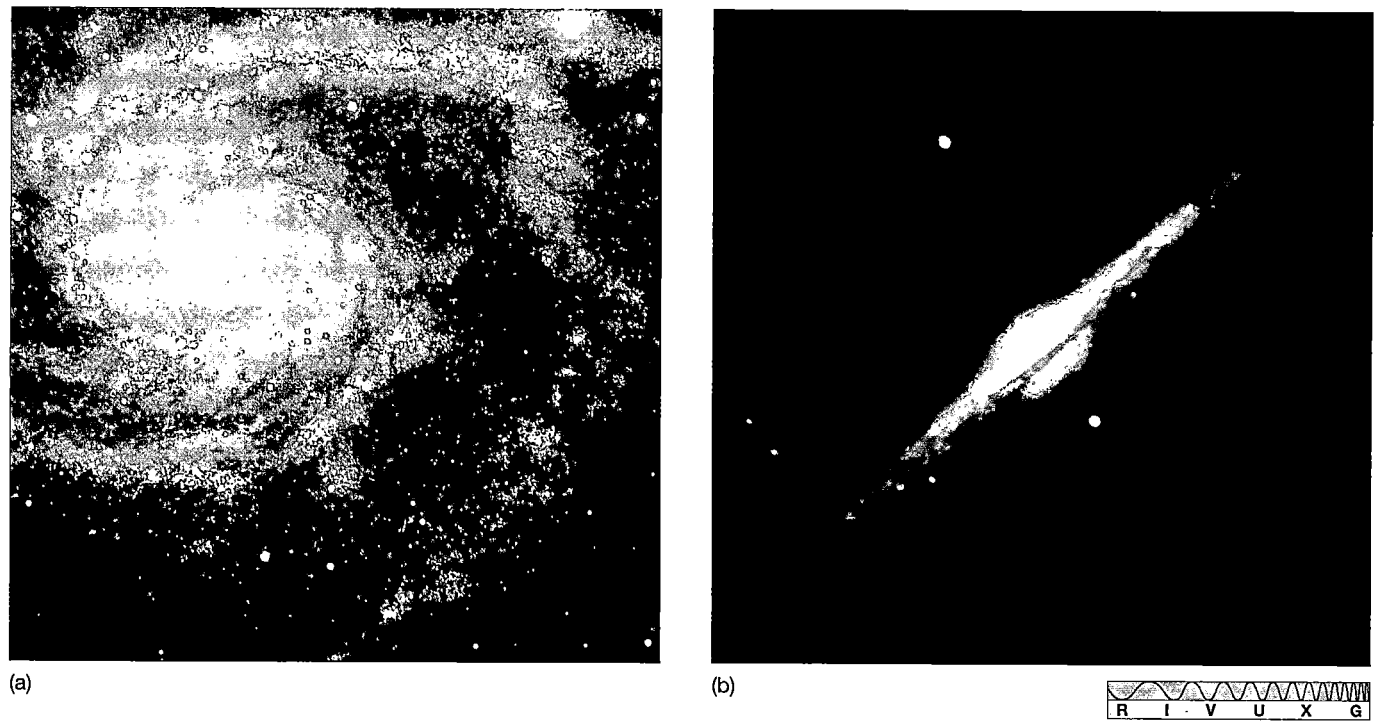

Figure 23.3 Spiral Galaxies (a) This galaxy, catalogued as NGC 2997 and seen nearly face-on, is somewhat similar in its overall structure to our own Milky Way Galaxy and Andromeda. (b) The galaxy NGC 4565 happens to be oriented in such a way that we see it edge-on, allowing us to see clearly its disk and central bulge. *(ESO; SAO)*

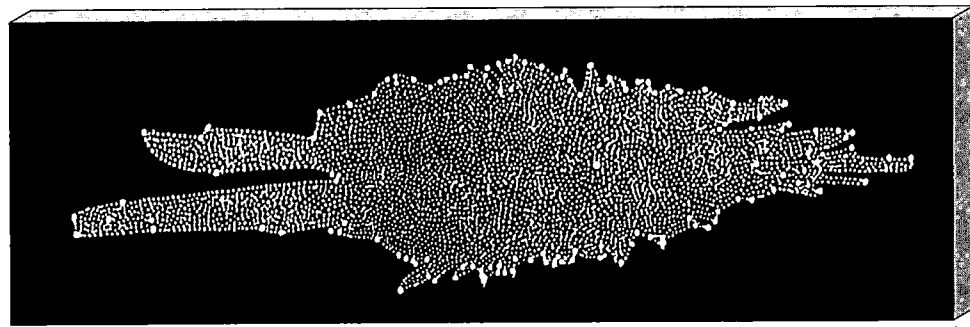

Figure 23.4 Herschel's Galaxy Model Eighteenth-century astronomer William Herschel constructed this "map" of the Galaxy by counting the numbers of stars he saw in different directions in the sky. Our Sun (large yellow dot) appears to lie near the center of the distribution. The long axis of the diagram lies in the plane of the Galactic disk.

SPIRAL NEBULAE AND GLOBULAR CLUSTERS

We have just seen how astronomers' attempts to probe the Galactic disk by optical means are frustrated by the effects of the interstellar medium, whereas looking in other directions, out of the Milky Way plane, we can see to much greater distances. During the first quarter of the twentieth century, studies of the large-scale structure of our Galaxy focused on two particularly important classes of objects, both found mainly away from the Milky Way. The first are *globular clusters*, those tightly bound swarms of old, reddish stars we first met in Chapter 19. ∞ (Sec. 19.6) About 150 are now known. The second were known at the time as *spiral nebulae*. Examples are shown in Figures 23.2(a) and 23.3(a). We know them today as **spiral galaxies**, comparable in size to our own.

Early twentieth-century astronomers had no means of determining the distances to any of these objects. They are too far away to have any observable parallax and, with the technology of the day, main-sequence stars (after the discovery of the main sequence in 1911) could not be clearly identified and measured. For these reasons neither of the techniques discussed in Chapter 17 were applicable. ∞ (Secs. 17.1 and 17.7) As a result, even the most basic properties—size, mass, and stellar and interstellar content—of globular clusters and spiral nebulae were unknown. It was assumed that the globular clusters lay within our own Galaxy, which was thought at the time to be relatively small (using the size estimates just mentioned). The locations of the spiral nebulae were much less clear.

Knowing the distance to an object is vitally important to understanding its true nature. As an example, consider again the Andromeda "nebula" (Figure 23.2). In the late nineteenth century, when improved telescopes and photographic techniques allowed astronomers to obtain images showing detail comparable to Figure 23.2(a), the newly released photographs caused great excitement among astronomers, who thought they were seeing the formation of a star from a swirling gaseous disk! Comparing Figure 23.2(a) with the figures in Chapter 15 (see especially Figure 15.1b), we can perhaps understand how such a mistake could be made—*if* we believed we were looking at a relatively close, star-sized object. Far from demonstrating that Andromeda was distant and large, the new observations seemed to confirm that it was just a small part of our own Galaxy.

Further observations soon made it clear that Andromeda is not a star-forming region. Andromeda's parallax is too small to measure, indicating that it must be at least several hundred parsecs from Earth, and, even at 100 pc—which we now know is vastly less than Andromeda's true distance—an object the size of the solar nebula would be impossible to resolve and simply would not look like Figure 23.2(a). (See Section 22.4 for another, more recent example of how distance measurements directly affect our theoretical understanding of observational data.)

During the first three decades of the twentieth century, both the size of our Galaxy and the distances to the spiral nebulae were hotly debated in astronomical circles. One school of thought maintained that the spiral nebulae were relatively small systems contained within our Galaxy. Other astronomers held that the spirals were much larger objects, lying far outside the Milky Way Galaxy and comparable to it in size. However, with no firm distance information, both arguments were inconclusive. Only with the discovery of a new distance-measurement technique—which we discuss next—was the issue finally settled in favor of the latter view. However, in the process, astronomers' conception of our own Galaxy changed radically and forever.

A NEW YARDSTICK

2 An important by-product of the laborious effort to catalog stars around the turn of the twentieth century was the systematic study of *variable stars*. These are stars whose luminosity changes with time—some quite erratically, others more regularly. Only a small fraction of stars fall into this category, but those that do are of great astronomical significance.

We encountered several examples of variable stars in earlier chapters. Often, the variability is the result of membership in a binary system. Eclipsing binaries, novae, and Type I supernovae are examples. Novae and supernovae collectively are called *cataclysmic variables* because of their sudden, large changes in brightness. In other instances, however, the variability is a basic trait of a star and is not dependent on its being a part of a binary system. We call such a star an *intrinsic variable*.

A particularly important class of intrinsic variables is the class known as **pulsating variable stars**, which vary cyclically in luminosity in characteristic ways. Two types of pulsating variable stars that have played central roles in revealing both the true extent of our Galaxy and the distances to our galactic neighbors are the **RR Lyrae** and **Cepheid** variables. Following long-standing astronomical practice, the names come from the first star of each class to be discovered—in this case the variable star labeled RR in the constellation Lyra, and the variable star Delta Cephei, the fourth brightest star in the constellation Cepheus. (Note, by the way, that these stars have *nothing* whatsoever to do with pulsars!)

RR Lyrae and Cepheid variable stars are recognizable by the characteristic shapes of their light curves. RR Lyrae stars all pulsate similarly (Figure 23.5a), with only small differences in period between one RR Lyrae variable and another. Observed periods range from about 0.5 to 1 day. Cepheid variables also pulsate in distinctive ways (the regular "sawtooth" pattern in Figure 23.5b), but different Cepheids can have very different pulsation periods, ranging from about 1 to 100 days. The period of any given RR Lyrae or Cepheid variable is, to high accuracy, the same from one cycle to the next. The key point is that pulsating variable stars can be recognized and identified *just by observing the variations in the light they emit.*

Why do Cepheids and RR Lyrae variables pulsate? The basic mechanism was first suggested by the British astrophysicist Sir Arthur Eddington in 1941. The structure of any star is determined in large part by how easily radiation can travel from the core to the photosphere—that is, by the *opacity* of the interior, the degree to which the gas hinders the passage of light through it. If the opacity rises, the radiation becomes trapped, the internal pressure increases, and the star "puffs up." If the opacity falls, radiation can escape more easily, and the star shrinks. According to theory, under certain circumstances a star can become unbalanced and enter a state in which the flow of radiation causes the opacity first to rise—making the star expand, cool, and diminish in luminosity—and then to fall, leading to the pulsations we observe.

Thus, pulsating variable stars are normal stars experiencing a brief period of instability as a natural part of stellar evolution. The conditions necessary to cause pulsations are not found in main-sequence stars; they occur in evolved post-main-sequence stars as they pass through a region of the Hertzsprung–Russell diagram known as the *instability strip* (Figure 23.6). When a star's temperature and luminosity place it in this strip, the star becomes internally unstable. Both its temperature and its radius vary in a regular way, causing the pulsations we observe: As the star brightens, its surface becomes hotter and its radius shrinks; as its luminosity decreases, the star expands and cools. As we learned in Chapter 20, high-mass stars evolve across the upper part of the H–R diagram. When their evolutionary tracks take them into the instability strip, they become Cepheid variables. ⊂⊃ (Sec. 20.4) RR Lyrae variables are low-mass horizontal-branch stars that lie within the lower

portion of the instability strip. ⊂⊃ (Sec. 20.2) The importance of these stars to galactic astronomy lies in the fact that once we recognize a star as being of the RR Lyrae or Cepheid type, we can infer its luminosity, and that in turn

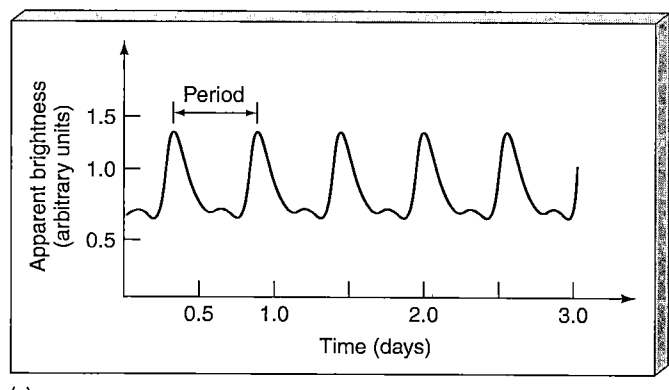

(a)

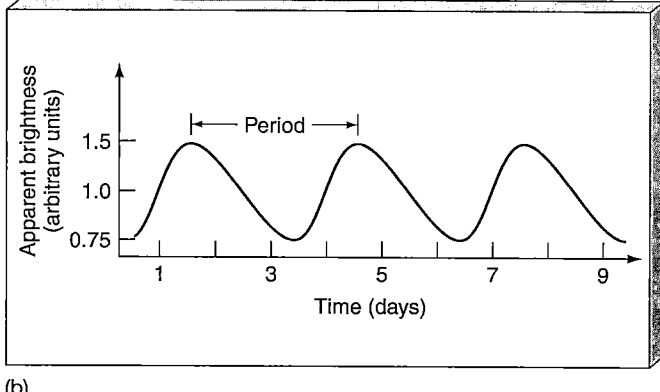

(b)

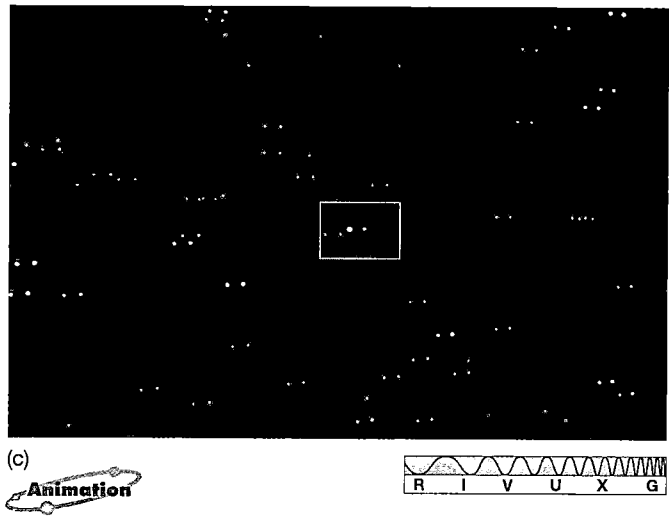

(c)

Figure 23.5 Variable Stars (a) Light curve of the pulsating variable star RR Lyrae. All RR Lyrae-type variables have similar light curves, with periods of less than a day. (b) The light curve of a Cepheid variable star called WW Cygni, having a period of several days. (c) This Cepheid is shown here (boxed) on successive nights, near its maximum and minimum brightness; two photos, one from each night were superimposed and then slightly displaced. *(Harvard College Observatory)*

allows us to measure its distance. The distance calculation is precisely the same as presented in Chapter 17 during our discussion of spectroscopic parallax. ⚌ (Sec. 17.7) Comparing the star's (known) luminosity with its (observed) apparent brightness yields an estimate of its distance, by the inverse-square law: ⚌ (Sec. 17.3)

$$\text{apparent brightness} \propto \frac{\text{luminosity}}{\text{distance}^2}.$$

In this way astronomers can use pulsating variables as a means of determining distances, both within our own Galaxy and far beyond.

How do we infer a variable star's luminosity? For RR Lyrae stars, this is simple. All such stars have basically the same luminosity (averaged over a complete pulsation cycle)—about 100 times that of the Sun—so once a variable star is recognized as being of the RR Lyrae type, its luminosity is immediately known. For Cepheids, we make use of a close correlation between average luminosity and pulsation period, discovered in 1908 by Henrietta Leavitt of Harvard University (see *Discovery 23-1*) and known simply as the **period–luminosity relationship**. Cepheids that vary slowly—that is, have long periods—have large

luminosities; conversely, short-period Cepheids have low luminosities.

Figure 23.7 illustrates the period–luminosity relationship for Cepheids found within a thousand parsecs or so of Earth. Astronomers can plot such a diagram for relatively nearby stars because they can measure the distances using stellar or spectroscopic parallax. Once distances are known, the luminosities of those stars can be calculated. We know of no exceptions to the period–luminosity relationship, and it is consistent with theoretical calculations of pulsations in evolved stars. Consequently, we assume that it holds for all Cepheids, near and far. Thus, a simple measurement of a Cepheid variable's pulsation period immediately tells us its luminosity—we just read it off the plot in Figure 23.7. (The roughly constant luminosities of the RR Lyrae variables are also indicated in the figure.)

This distance-measurement technique works well, provided the variable star can be clearly identified and its pulsation period measured. With Cepheids, this method allows astronomers to estimate distances out to about 25 million parsecs, enough to take us all the way to the nearest galaxies. The less-luminous RR Lyrae stars are not so easily seen as Cepheids, so their useful range is not as great. However, they are much more common so, within their limited range, they are actually more useful than Cepheids.

Figure 23.8 extends our cosmic distance ladder, begun in Chapter 2 with radar ranging in the solar system and expanded in Chapter 17 to include stellar and spectroscopic parallax, by adding variable stars as a fourth method of determining distance. Note that, because the period–luminosity relationship is calibrated using nearby stars, this latest rung inherits any and all uncertainties and errors present in the lower levels. Uncertainties also arise from the "scatter" shown in Figure 23.7. Although the overall connection between period and luminosity is unmistakable, the individual data points do not quite lie on a

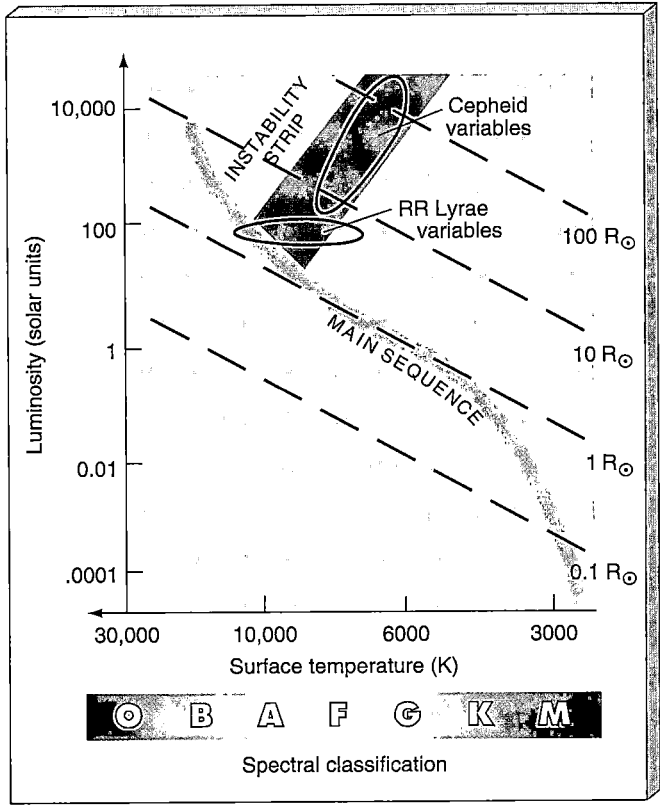

Figure 23.6 Variable Stars on the H–R Diagram
Pulsating variable stars are found in the instability strip of the H–R diagram. As a high-mass star evolves through the strip it becomes a Cepheid variable. Low-mass horizontal-branch stars in the instability strip are RR Lyrae variables.

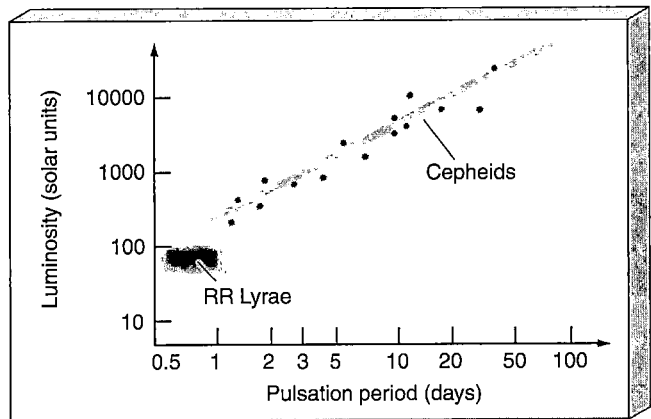

Figure 23.7 Period–Luminosity Plot A plot of pulsation period versus average absolute brightness (that is, luminosity) for a group of Cepheid variable stars. The two properties are quite tightly correlated. The pulsation periods of some RR Lyrae variables are also shown.

DISCOVERY 23-1

Early Computers

A large portion of the early research in observational astronomy focused on monitoring stellar luminosities and analyzing stellar spectra. Much of this pioneering work was done using photographic methods. What is not so well known is that most of the labor was accomplished by women. Around the turn of the twentieth century, a few dozen dedicated women—assistants at the Harvard College Observatory—created an enormous database by observing, sorting, measuring, and cataloging photographic information that helped form the foundation of modern astronomy. Some of them went far beyond their duties in the lab to make several of the basic astronomical discoveries often taken for granted today.

This 1910 photograph shows several of those women carefully examining star images and measuring variations in luminosity or wavelengths of spectral lines. In the cramped quarters of the Harvard Observatory these women inspected image after image to collect a vast body of data on hundreds of thousands of stars. Note the plot of stellar luminosity changes pasted on the wall at the left. The cyclical pattern is so regular that it likely belongs to a Cepheid variable. Known as "computers" (there were no electronic devices then), these women were paid 25 cents an hour.

Beginning in 1880, these workers started a survey of the skies that would be carried on for half a century. Their first major accomplishment was a catalog of the brightnesses and spectra of tens of thousands of stars, published in 1890 under the direction of Williamina Fleming. On the basis of this compilation, several of these women made fundamental contributions to astron-

omy. In 1897 Antonia Maury undertook the most detailed study of stellar spectra to that time, enabling Hertzsprung and Russell independently to develop what is now called the H–R diagram. In 1898 Annie Cannon proposed the spectral classification system (described in Chapter 17) that is now the international standard for categorizing stars. ☞ (Sec. 17.6) In 1908 Henrietta Leavitt discovered the period–luminosity relationship for Cepheid variable stars, which later allowed astronomers to recognize our Sun's true position in our Galaxy, as well as our Galaxy's true place in the universe.

(Harvard Observatory)

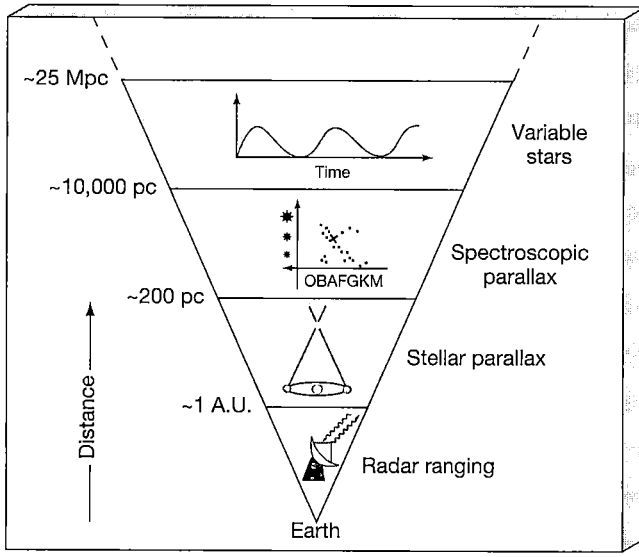

Figure 23.8 Variable Stars on Distance Ladder Application of the period–luminosity relationship for Cepheid variable stars allows us to determine distances out to about 25 Mpc with reasonable accuracy.

straight line. Instead, a range of possible luminosities corresponds to any measured period.

THE SIZE AND SHAPE OF OUR GALAXY

Many RR Lyrae variables are found in globular clusters. Early in the twentieth century the American astronomer Harlow Shapley used observations of RR Lyrae stars to make two very important discoveries about the Galactic globular cluster system. First, he showed that most globular clusters reside at great distances—many thousands of parsecs—from the Sun. Second, by measuring the direction and distance of each cluster, he was able to determine their three-dimensional distribution in space. In this way Shapley demonstrated that the globular clusters map out a truly gigantic, and roughly *spherical*, volume of space, about 30 kpc across.* However, the center of the distribution lies

---

The Galactic globular cluster system and the Galactic halo, of which it is a part, are somewhat flattened in the direction perpendicular to the disk, but the degree of flattening is quite uncertain. The halo is certainly much less flattened than the disk, however.

nowhere near our Sun. It is located nearly 8 kpc away from us, in the direction of the constellation Sagittarius.

In a brilliant intellectual leap, Shapley realized that the distribution of globular clusters maps out the true extent of stars in the Milky Way Galaxy—the region that we now call the Galactic halo. (Since Shapley's time astronomers have identified many individual halo stars that do not belong to any globular cluster.) The hub of this vast collection of matter, 8 kpc from the Sun, is the **Galactic center.** Figure 23.9 shows the distribution, based on modern data, of the 138 globular clusters lying within 20 kpc of the center. We live in the suburbs of this huge ensemble, in the Galactic disk—the thin sheet of young stars, gas, and dust that cuts through the center of the halo.

Shapley's bold interpretation of the globular clusters as defining the overall distribution of stars in our Galaxy was an enormous step forward in human understanding of our place in the universe. Five hundred years ago, Earth was considered the center of all things. Copernicus argued otherwise, demoting our planet to an undistinguished location removed from the center of the solar system. In Shapley's time, as we have just seen, the prevailing view was that our Sun was the center not only of the Galaxy but also of the universe. Shapley showed otherwise. With his observations of globular clusters he simultaneously increased the size of our Galaxy by almost a factor of 10 over earlier estimates and banished our parent Sun to its periphery, virtually overnight!

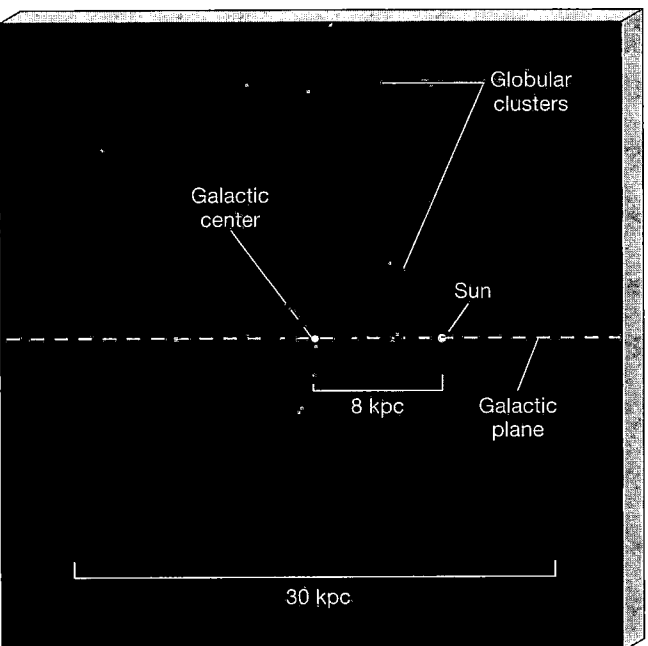

Figure 23.9 Globular Cluster Distribution Our Sun does not coincide with the center of the very large collection of globular clusters. Instead, more globular clusters are found in one direction than in any other. The Sun resides closer to the edge of the collection, which measures roughly 30 kpc across. We now know that the globular clusters outline the true distribution of stars in the Galactic halo.

Curiously, Shapley's dramatic revision of the size of the Milky Way Galaxy and our place in it only strengthened his erroneous opinion that the spiral nebulae were part of our Galaxy and that our Galaxy was essentially the entire universe. He regarded as beyond belief the idea that there could be other structures as large as our Galaxy. Only in the late 1920s was the Copernican principle extended to the Galaxy itself, when American astronomer Edwin Hubble observed Cepheids in the Andromeda Galaxy and finally succeeded in measuring its distance.

✔ Concept Check

■ Can variable stars be used to map out the structure of the Galactic disk?

23.3 The Large-Scale Structure of Our Galaxy

THE SPATIAL DISTRIBUTION OF STARS

⬓ Figure 23.10 illustrates the different spatial distributions of the disk and halo components of the Milky Way Galaxy. As just mentioned, the Sun lies about 8 kpc from the Galactic center. Based on optical, infrared, and radio observations of stars, gas, and dust found within a thousand or so parsecs of the Sun, astronomers estimate that the disk in the vicinity of the Sun is relatively thin—"only" 300 pc thick, or about 1/100 of the galactic diameter. Don't be fooled, though. Even if you could travel at the speed of light, it would take you a thousand years to traverse the thickness of the Galactic disk. The disk may be thin compared with the galactic diameter, but it is huge by human standards.

Actually, the thickness of the Galactic disk depends on the kinds of objects measured. Young stars and interstellar gas are more tightly confined to the plane than are stars like the Sun, and solar-type stars in turn are more tightly confined than are older K- and M-type dwarfs. The reason for this is that stars form in interstellar clouds close to the disk plane but then tend to drift out of the disk over time, mainly due to their interactions with other stars and molecular clouds. Thus, as stars age, their abundance above and below the disk plane slowly increases. Note that these considerations do not apply to the Galactic halo, whose ancient stars and globular clusters extend far above and below the galactic plane. As we will see in a moment, the halo is a remnant of an early stage of our Galaxy's evolution and predates the formation of the disk.

Recently, improved observational techniques have revealed an intermediate category of galactic stars, midway between the old halo and the younger disk, both in age and in spatial distribution. Consisting of stars with estimated ages in the range of 7–10 billion years, this

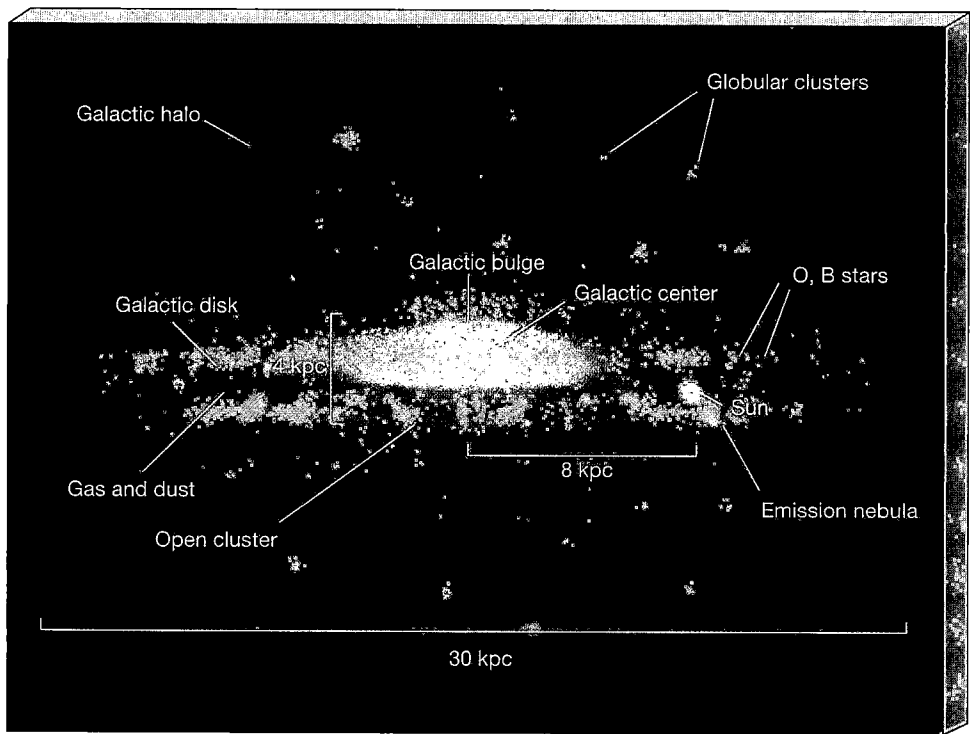

Figure 23.10 Stellar Populations in Our Galaxy Artist's conception of a (nearly) edge-on view of the Milky Way Galaxy, showing the distributions of young blue stars, open clusters, old red stars, and globular clusters.

thick disk component of the Milky Way Galaxy measures some 2–3 kpc from top to bottom. Its thickness is too great to be explained by the slow drift just described. Like the halo, it appears to be a vestige of our Galaxy's distant past.

Also shown in Figure 23.10 is our Galaxy's central bulge, measuring roughly 6 kpc across in the plane of the Galactic disk by 4 kpc perpendicular to the disk plane. Obscuration by interstellar dust makes it difficult to study the detailed structure of the Galactic bulge in optical images of the Milky Way (see, for example, Figure 18.5, which would clearly show a large portion of the bulge were it not for interstellar absorption). However, at longer wavelengths, which are less affected by interstellar matter, a much clearer picture emerges (Figure 23.11; compare with Figure 23.3b). Detailed measurements of the motion of gas and stars in and near the bulge imply that it is football shaped, about half as wide as it is long, with the long axis of the football lying in the galactic plane.

STELLAR POPULATIONS

Aside from their shapes, the three components of the Galaxy—disk, bulge, and halo—have several other properties that distinguish them from one another. First, the halo contains almost *no* gas or dust—just the opposite of the disk and bulge, in which interstellar matter is common. Second, there are clear differences in both *appearance* and *composition* between disk, bulge, and halo stars. Stars in the Galactic bulge and halo are found to be distinctly *redder* than stars found in the disk. Observations of other spiral galaxies also show this trend—the blue-white tint of the

disk and the yellowish coloration of the bulge are evident in Figures 23.2(a) and 23.3(a).

All the bright, blue stars visible in our sky are part of the Galactic disk, as are the young, open star clusters and

| R | I | V | U | X | G |

Figure 23.11 Infrared View of the Milky Way A wide-angle far-infrared image of the disk and bulge of the Milky Way Galaxy, as observed by the *Cosmic Background Explorer* satellite (*COBE*). (*NASA*)

star-forming regions. In contrast, the cooler, redder stars—including those found in the old globular clusters—are more uniformly distributed throughout the disk, bulge, and halo. Galactic disks appear bluish because main-sequence O- and B- type blue supergiants are very much brighter than G-, K-, and M-type dwarfs, even though the dwarfs are present in far greater numbers.

The explanation for the marked difference in stellar content between disk and halo is this: Whereas the gas-rich Galactic disk is the site of ongoing star formation and so contains stars of all ages, all the stars in the Galactic halo are *old*. The absence of dust and gas in the halo means that no new stars are forming there, and star formation apparently ceased long ago—at least 10 billion years in the past, judging from the types of halo stars we now observe. (Recall from Chapter 20 that most globular clusters are thought to be between 10 and 12 billion years old.) ∞ (Sec. 20.5) The gas density is very high in the inner part of the Galactic bulge, making this region the site of vigorous ongoing star formation, and both very old and very young stars mingle there. The bulge's gas-poor outer regions have properties more similar to those of the halo.

Support for this picture comes from studies of the spectra of halo stars, which indicate that these stars are far less abundant in heavy elements (that is, elements heavier than helium) than are nearby stars in the disk. In Chapter 21 we saw how each successive cycle of star formation and evolution enriches the interstellar medium with the products of stellar nucleosynthesis, leading to a steady increase in heavy elements with time. ∞ (Sec. 21.5) Thus, the scarcity of these elements in halo stars is consistent with the view that the halo formed long ago.

Astronomers often refer to young disk stars as *Population I* stars and to old halo stars as *Population II* stars. The idea of two stellar "populations" dates from the 1930s, when the differences between disk and halo stars first became clear. The names are something of an oversimplification, as there is actually a continuous variation in stellar ages throughout the Milky Way Galaxy, not a simple division of stars into two distinct "young" and "old" categories. Nevertheless, the terminology is widely used.

ORBITAL MOTION

3 Now let's turn our attention to the *dynamics* of the Milky Way Galaxy—that is, to the motion of the stars, dust, and gas it contains. Are the internal motions of our Galaxy's members chaotic and random, or are they part of some gigantic "traffic pattern"? The answer depends on our perspective. The motion of stars and clouds we see on small scales (within a few tens of parsecs of the Sun) seems random, but on larger scales (hundreds or thousands of parsecs) the motion is much more orderly.

As we look around the Galactic disk in different directions, a clear pattern of motion emerges (Figure 23.12). Radiation received from stars and interstellar gas clouds in the upper right and the lower left quadrants of Figure

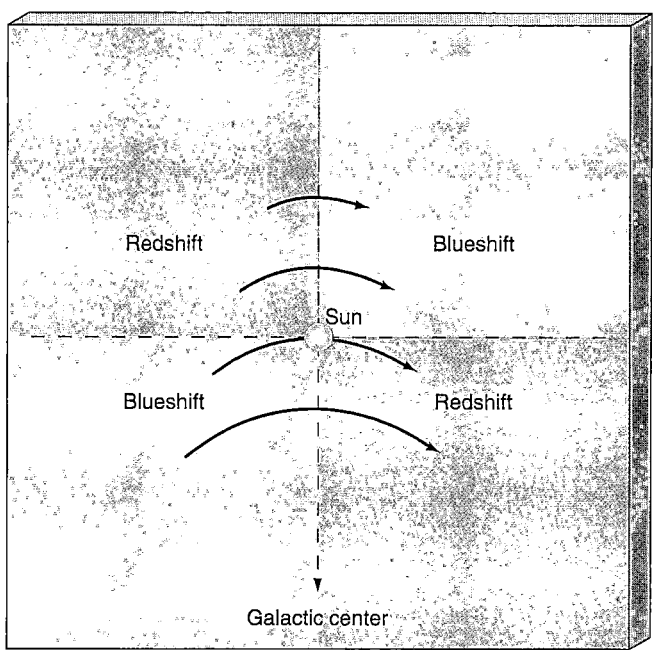

Figure 23.12 Stellar Motion in the Disk Stars and interstellar clouds in the neighborhood of the Sun show systematic Doppler motions. This information tells us that the disk of the Galaxy is spinning in a well-ordered way. These four Galactic quadrants are drawn (as dashed lines) to intersect at the Sun, not at the Galactic center, because we make our observations from the viewpoint of our own planetary system. The longer the arrow, the greater the angular speed of the disk material. Because the Sun orbits faster than stars and gas at larger radii, we are drawing away from material at top left and gaining on that at top right, resulting in the Doppler shifts indicated. Similarly, stars and gas in the bottom left quadrant are gaining on us, while material at bottom right is pulling away.

23.12 is generally *blueshifted*. At the same time, the interstellar regions sampled in the upper left quadrant and the lower right quadrant are *redshifted*. In other words, some regions of the Galaxy (in the blueshifted directions) are approaching the Sun, while others (the redshifted ones) are receding from us.

Careful study of the data leads to the conclusion that the entire Galactic disk is *rotating* about the Galactic center. In the vicinity of the Sun, the orbital speed is about 220 km/s. At the Sun's distance of 8 kpc from the Galactic center, material takes about 225 million years (an interval of time sometimes called 1 *Galactic year*) to complete one circuit. At other distances from the center the rotation period is different—shorter closer to the center, longer at greater distances—that is, the Galactic disk rotates not as a solid object but *differentially*. Similar differential rotation is observed in Andromeda and, in fact, in all other spiral galaxies.

This picture of orderly circular orbital motion about the Galactic center applies only to the Galactic disk. Stars in the Galactic halo and bulge are not so well behaved. The old globular clusters in the halo and the faint, reddish

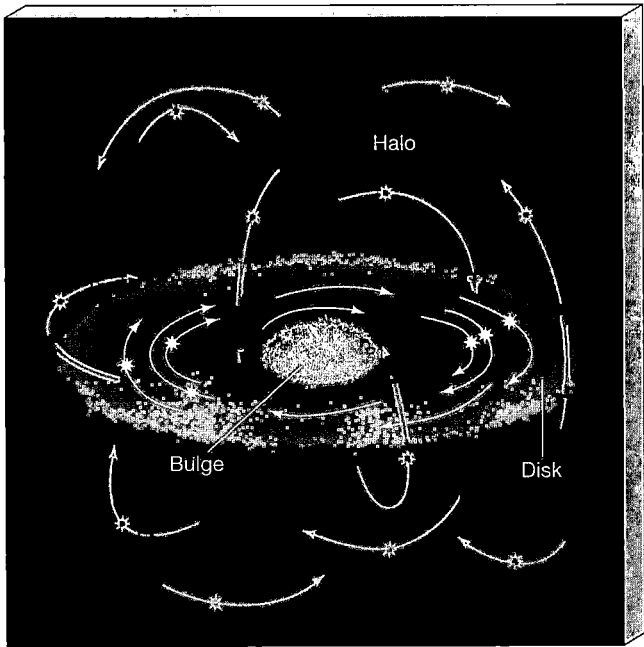

Figure 23.13 Stellar Orbits in Our Galaxy Stars in the Galactic disk move in orderly, circular orbits about the Galactic center. In contrast, halo stars have orbits with largely random orientations and eccentricities. The orbit of a typical halo star takes it high above the Galactic disk, then down through the disk plane, then out the other side and far below the disk. The orbital properties of bulge stars are intermediate between those of disk stars and those of halo stars.

individual stars in both the halo and the bulge do *not* share the disk's well-defined rotation. Instead, their orbital orientations are largely random.* Although they do orbit the Galactic center, they move in all directions, their paths filling an entire three-dimensional volume rather than a nearly two-dimensional disk. At any given distance from the Galactic center, bulge or halo stars move at speeds comparable to the disk's rotation speed at that radius but in *all* directions, not just one. Their orbits carry these stars repeatedly through the disk plane and out the other side. Figure 23.13 contrasts the motion of bulge and halo stars with the much more regular orbits of stars in the Galactic disk. Some well-known stars in the vicinity of the Sun— the bright giant Arcturus, for example—are actually halo stars that are "just passing through" the disk on orbits that take them far above and below the galactic plane.

✔ **Concept Check**

■ Why are there no young halo stars?

---

* Halo stars do, in fact, have some net rotation about the galactic center, but the rotational component of their motion is overwhelmed by the larger random component. the motion of bulge stars also has a rotational component, larger than that of the halo, but still smaller than the random component of stellar motion in the bulge.

23.4 The Formation of the Milky Way

⊛ Table 23.1 compares some key properties of the three basic components of the Galaxy. Is there some evolutionary scenario that can naturally account for the galactic structure we see today? The answer is that there is, and it takes us all the way back to the birth of our Galaxy, more than 10 billion years ago. Not all the details are agreed upon by all astronomers, but the overall picture is now fairly widely accepted. For simplicity we confine our discussion here to the Galactic disk and halo. In many ways the bulge is intermediate in its properties between these two extremes.

Figure 23.14 illustrates the current view of our Galaxy's evolution, starting (not unlike the star-formation scenario outlined in Chapter 19) from a contracting cloud of pregalactic gas. ⊂⊃ (Sec. 19.1) When the first galactic stars and globular clusters formed, the gas in our Galaxy had not yet accumulated into a thin disk. Instead, it was spread out over an irregular, and quite extended, region of space, spanning many tens of kiloparsecs in all directions. When the first stars formed, they were distributed throughout this volume. Their distribution today (the Galactic halo) reflects that fact—it is an imprint of their birth. Many astronomers believe that the very first stars formed even earlier, in smaller systems that later merged to create our Galaxy (Figure 23.14a). Probably many more stars were born during the mergers themselves, as interstellar gas clouds collided and began to collapse. ⊂⊃ (Sec.19.5) Whatever the details, the present-day halo would look much the same in either case.

Since those early times, rotation has flattened the gas in our Galaxy into a relatively thin disk. Physically, this process is similar to the flattening of the solar nebula during the formation of the solar system, as described in Chapter 15, except on a vastly larger scale. ⊂⊃ (Sec. 15.2) Star formation in the halo ceased billions of years ago when the raw materials fell to the galactic plane. Ongoing star formation in the disk gives it its bluish tint, but the halo's short-lived blue stars have long since burned out, leaving only the long-lived red stars that give it its characteristic pinkish glow. The Galactic halo is ancient, whereas the disk is full of youthful activity. The thick disk, with its intermediate-age stars, may represent an intermediate stage of star formation that occurred while the gas was still flattening into the plane.

The chaotic orbits of the halo stars are also explained by this theory. When the halo developed, the irregularly shaped Galaxy was rotating only very slowly, so there was no strongly preferred direction in which matter tended to move. As a result, halo stars were free to travel along nearly any path once they formed (or when their parent systems merged). As the Galactic disk formed, however,

TABLE 23.1 Overall Properties of the Galactic Disk, Halo, and Bulge

GALACTIC DISK	GALACTIC HALO	GALACTIC BULGE
highly flattened	roughly spherical—mildly flattened	somewhat flattened and elongated in the plane of the disk ("football shaped")
contains both young and old stars	contains old stars only	contains both young and old stars; more old stars at greater distances from the center
contains gas and dust	contains no gas and dust	contains gas and dust, especially in the inner regions
site of ongoing star formation	no star formation during the last 10 billion years	ongoing star formation in the inner regions
gas and stars move in circular orbits in the galactic plane	stars have random orbits in three dimensions	stars have largely random orbits but with some net rotation about the Galactic center
spiral arms	no obvious substructure	ring of gas and dust near center; central galactic nucleus
overall white coloration, with blue spiral arms	reddish in color	yellow-white

conservation of angular momentum caused it to spin more rapidly. Stars forming from the gas and dust of the disk inherit its rotational motion and so move on well-defined, circular orbits. Again, the thick disk's orbital properties are consistent with the idea that it formed while gas was still sinking to the Galaxy's midplane.

In principle, the structure of our Galaxy bears witness to the conditions that created it. In practice, however, the

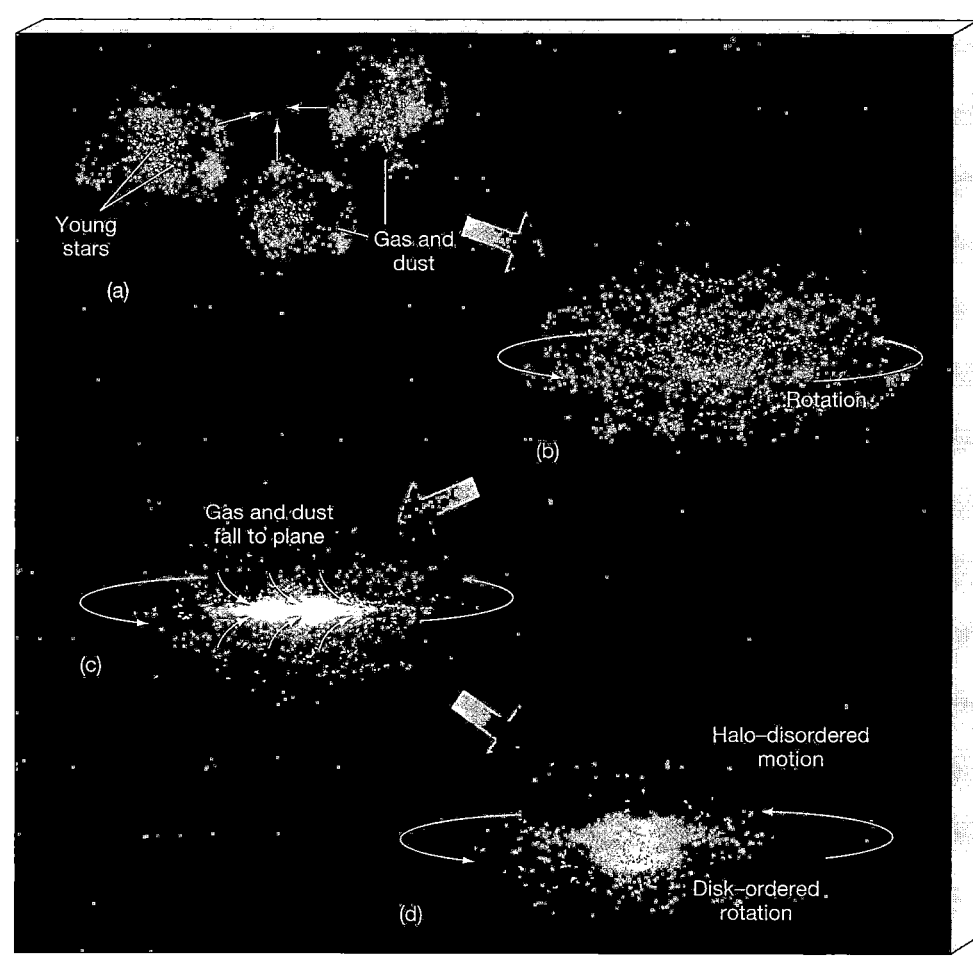

Young stars

(a)

Gas and dust

Rotation

(b)

Gas and dust fall to plane

(c)

Halo–disordered motion

Disk–ordered rotation

(d)

Figure 23.14 Milky Way Formation (a) The Milky Way Galaxy possibly formed through the merger of several smaller systems. (b) Astronomers reason that, early on, our Galaxy was irregularly shaped, with gas distributed throughout its volume. When stars formed during this stage, there was no preferred direction in which they moved and no preferred location in which they were found. (c) In time, rotation caused the gas and dust to fall to the Galactic plane and form a spinning disk. The stars that had already formed were left behind in the halo. (d) New stars forming in the disk inherit its overall rotation and so orbit the Galactic center on ordered, circular orbits.

interpretation of the observations is made difficult by the sheer complexity of the system we inhabit and by the many competing physical processes that have modified its appearance since it formed. As a result, the early stages of the Milky Way are still quite poorly understood. We will return to the subject of galaxy formation in Chapters 24 and 25.

☑ Concept Check

■ How does the current theory of galaxy formation account for the differences between the Galactic halo and disk?

23.5 Galactic Spiral Arms

If we want to look beyond our immediate neighborhood and study the full extent of the Galactic disk, we cannot rely on optical observations, as interstellar absorption severely limits our vision. In the 1950s, astronomers developed a very important tool to explore the distribution of gas in our Galaxy—spectroscopic radio astronomy.

RADIO MAPS OF THE MILKY WAY

The keys to observing galactic interstellar gas are the 21-cm radio-emission line produced by atomic hydrogen and the many radio molecular lines formed in molecular cloud complexes. ∞ (Sec. 18.4) Long-wavelength radio waves are largely unaffected by interstellar dust, so they travel more or less unimpeded through the Galactic disk, allowing us to "see" to great distances. Because hydrogen is by far the most abundant element in interstellar space, the 21-cm signals are strong enough that a large portion of the disk can be observed in this way. As noted in Chapter 18, observations of spectral lines from "tracer" molecules, such as carbon monoxide, allow us to study the distribution of the densest interstellar clouds. ∞ (Sec. 18.5)

Interstellar gas in the Galactic disk exhibits an organized pattern on a grand scale. According to radio studies, the center of the gas distribution coincides roughly with the center of the globular-cluster system, about 8 kpc from the Sun. (In fact, this figure is derived most accurately from radio observations of the distribution of galactic gas, the center of which is normally taken to define the center of our Galaxy.) Near the center, the gas in the disk fattens markedly in the Galactic bulge. Radio-emitting gas has been observed out to at least 50 kpc from the Galactic center. Over much of the inner 20 kpc or so of the disk, the gas is confined within about 100 pc of the galactic plane. Beyond that distance, the gas distribution spreads out somewhat, to a thickness of several kiloparsecs, and shows definite signs of being "warped," possibly because of the gravitational influence of a pair of nearby galaxies (to be discussed in Chapter 24; see also Figure 23.15).

Radio studies provide perhaps the best direct evidence that we live in a spiral galaxy. Figure 23.15 is an artist's conception (based on observational data) of the appearance of our Galaxy as seen from far above the disk, clearly

30 kpc

Figure 23.15 Milky Way Spiral Structure An artist's conception of our Milky Way Galaxy seen face-on. This illustration is based on data accumulated by legions of astronomers during the past few decades, including radio maps of gas density in the Galactic disk. Painted from the perspective of an observer 100 kpc above the Galactic plane, the spiral arms are at their best-determined positions. All the features are drawn to scale (except for the oversized yellow dot near the top, which represents our Sun). The two small blotches to the left are dwarf galaxies, called the Magellanic Clouds. We will study them in Chapter 24. (L. Chaisson)

showing our Galaxy's **spiral arms**, pinwheel-like structures originating close to the Galactic bulge and extending outward throughout much of the Galactic disk. Our Sun lies near the edge of one of these arms, which wraps around a large part of the disk. Notice, incidentally, the scale markers on Figures 23.9, 23.10, and 23.15: The galactic globular-cluster distribution (Figure 23.9), the luminous stellar component of the disk (Figure 23.10), and the known spiral structure (Figure 23.15) all have roughly the *same* diameter—about 30 kpc.

PERSISTENCE OF THE SPIRAL ARMS

The spiral arms in our Galaxy are made up of much more than just interstellar gas and dust. Studies of the Galactic disk within a kiloparsec or so of the Sun indicate that young stellar and prestellar objects—emission nebulae, O- and B-type stars, and recently formed open clusters—are also distributed in a spiral pattern that closely follows the distribution of interstellar clouds. The obvious conclusion is that the spiral arms are the part of the Galactic disk where star formation takes place. The brightness of the young stellar objects just listed is the main reason that the spiral arms of other galaxies are easily seen from afar (Figure 23.3a).

A central problem facing astronomers trying to understand spiral structure is how that structure persists over long periods of time. The basic issue is simple: We know that the inner parts of the Galactic disk rotate more rapidly than do the outer regions; thus stars in the Galactic disk do not move smoothly together, but ceaselessly change their positions relative to one another as they orbit the Galactic center. This differential rotation makes it impos-

sible for any large-scale structure "tied" to the disk material to survive. Figure 23.16 shows how a spiral pattern consisting always of the same group of stars and gas clouds would necessarily "wind up" and disappear within a few hundred million years. Yet spiral arms clearly do exist in our own galaxy, and their prevalence in other disk galaxies suggests that they last for considerably longer than this. Thus, whatever the spiral arms are, they *cannot* simply be dense star-forming regions orbiting along with the rest of the Galactic disk.

Then how do the Galaxy's spiral arms retain their structure over long periods of time in spite of differential rotation? A leading explanation for the existence of spiral arms holds that they are **spiral density waves**—coiled waves of gas compression that move through the Galactic disk, squeezing clouds of interstellar gas and triggering the process of star formation as they go. ⚬ (Sec. 19.5) The spiral arms we observe are defined by the denser-than-normal clouds of gas the density waves create and by the new stars formed as a result of the spiral waves' passage.

This explanation of spiral structure avoids the problem of differential rotation because the wave pattern is not tied to any particular piece of the Galactic disk. The spirals we see are merely patterns moving through the disk, not great masses of matter being transported from place to place. The density wave moves through the collection of stars and gas making up the disk just as a sound wave moves through air or an ocean wave passes through water, compressing different parts of the disk at different times. Even though the rotation rate of the disk material varies with distance from the Galactic center, the wave itself remains intact, defining the Galaxy's spiral arms.

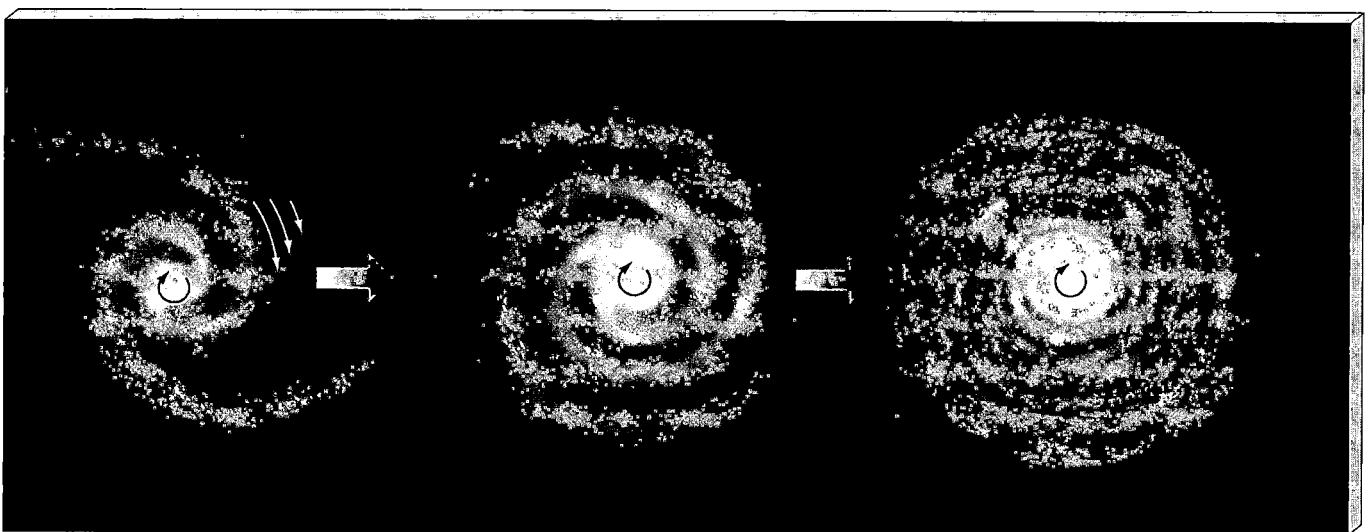

Figure 23.16 Differential Galactic Rotation The disk of our Galaxy rotates differentially—stars close to the center take less time to orbit the Galactic center than those at greater distances. If spiral arms were somehow tied to the material of the Galactic disk, this differential rotation would cause the spiral pattern to wind up and disappear in a few hundred million years. Spiral arms would be too short-lived to be consistent with the numbers of spiral galaxies we observe today.

In the late 1960s American astrophysicists C. C. Lin and Frank Shu proposed a way in which spiral arms in the Galaxy could persist for many Galactic rotations. They argued that the arms themselves contain no "permanent" matter. They should not be viewed as assemblages of stars, gas, and dust moving intact through the disk—those would quickly be destroyed by differential rotation. Instead, a spiral arm should be envisaged as a density wave—a wave of compression and expansion sweeping through the Galaxy.

A wave in water builds up material temporarily in some places (crests) and lets it down in others (troughs). Similarly, as the spiral density wave encounters galactic matter, the gas is compressed to form a region of slightly higher than normal density. Galactic material enters the wave, is temporarily slowed down and compressed as it passes through, then continues on its way. This compression triggers the formation of new stars and nebulae. In this way the spiral arms are formed and re-formed repeatedly, without winding up. Lin and Shu showed that the process can in fact maintain a spiral pattern for very long periods of time.

The accompanying figure illustrates the formation of a density wave in a much more familiar context—a traffic jam on a highway, triggered by the presence of a repair crew moving slowly down the road. As cars approach the crew they slow down temporarily, then speed up again as they pass the worksite and continue on their way. The result, as might be reported by a high-flying traffic helicopter, is a region of high traffic density, concentrated around the location of the work crew and moving with it. An observer on the side of the road, however, sees that the jam never contains the same cars for very long. Cars constantly catch up to the bottleneck, move slowly through it, then speed up again, only to be replaced by more cars arriving from behind.

The traffic jam is analogous to the region of high stellar density in a Galactic spiral arm. Just as the traffic density wave is not tied to any particular group of cars, the spiral arms are not attached to any particular piece of disk material. Stars and gas enter a spiral arm, slow down for a while, then continue on their orbits around the Galactic center. The result is a moving region of high stellar and gas density, involving different parts of the disk at different times. Notice also that, just as in our Galaxy, the wave moves more slowly than, and independently of, the overall traffic flow.

We can extend our traffic analogy a little further. Most drivers are well aware that the effects of a such a tie-up can persist long after the road crew responsible for it has stopped work and gone home for the night. Similarly, spiral density waves can continue to move through the disk even after the disturbance that originally produced them has long since subsided. According to spiral density wave theory, this is precisely what has happened in the Milky Way. Some disturbance in the past produced the wave, which has been moving through the Galactic disk ever since.

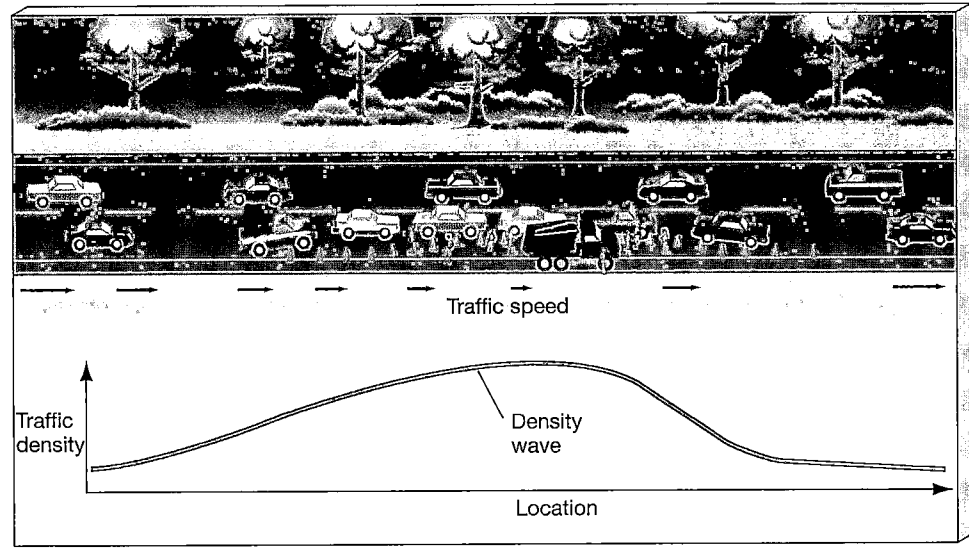

Traffic speed

Traffic density

Density wave

Location

In fact, over much of the visible portion of the Galactic disk (within about 15 kpc of the center), the spiral wave pattern is predicted to rotate *more slowly* than the stars and gas. Thus, galactic material catches up with the wave, is temporarily slowed down and compressed as it passes through, then continues on its way. (For a more down-to-earth example of an analogous process, see *Discovery 23-2*.)

As shown in Figure 23.17, the slowly moving spiral density wave is outrun by the faster-moving disk material. As gas enters the arm from behind, the gas is compressed and forms stars. Dust lanes mark the regions of highest-density gas. The most prominent stars—the bright O- and B-type blue giants—live for only a short time, so young stellar associations, emission nebulae, and open clusters

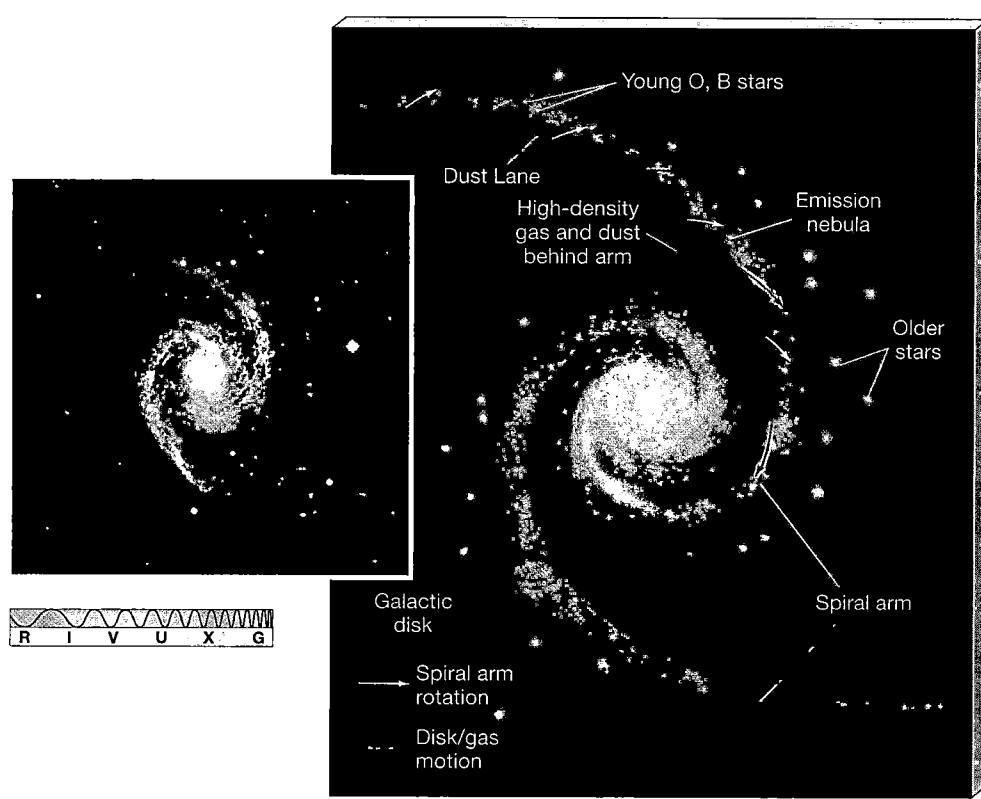

Young O, B stars

Dust Lane

High-density gas and dust behind arm

Emission nebula

Older stars

Galactic disk

Spiral arm

→ Spiral arm rotation

‑‑‑ Disk/gas motion

R I V U X G

Figure 23.17 Density Wave Theory This theory holds that the spiral arms seen in our own and many other galaxies are waves of gas compression and star formation moving through the material of the Galactic disk. In the painting at right, gas motion is indicated by red arrows, and arm motion is indicated by white arrows. Gas enters an arm from behind, is compressed, and forms stars. The spiral pattern is delineated by dust lanes, regions of high gas density, and newly formed O- and B-type stars. The inset shows the spiral galaxy NGC 1566, which displays many of the features described in the text. (D. Malin/AAT)

with long main sequences are found only within the arms, near their birth sites, just ahead of the dust lanes. The brightness of these young systems emphasizes the spiral structure. Farther downstream, ahead of the spiral arms, we see mostly older stars and star clusters. These have had enough time since their formation to outdistance the wave and pull away from it. Over millions of years their random individual motions, superimposed on the overall rotation around the Galactic center, distort and eventually destroy their original spiral configuration, and they become part of the general disk population.

Note, incidentally, that although the spirals shown in Figure 23.17 have two arms each, astronomers are not completely certain how many arms make up the spiral structure in our own Galaxy (see Figure 23.15). The theory makes no strong predictions on this point.

An alternative possibility is that the formation of stars drives the waves, instead of the other way around. Imagine a row of newly formed massive stars somewhere in the disk. As these stars form, the emission nebulae that appear around them send shock waves through the surrounding gas, possibly triggering new star formation. Similarly, when the stars explode in supernovae, more shocks are formed. ⊂⊃ (Sec. 21.2) As illustrated in Figure 23.18(a), the formation of one group of stars thus provides the mechanism for the creation of more stars. Computer simulations suggest that it is possible for the "wave" of star formation thus created to take on the form of a partial spiral and for this pattern to persist for some time. However, this process, sometimes known as *self-propagating star formation*, can produce only pieces of spirals, as are seen in

some galaxies (Figure 23.18b). It apparently cannot produce the galaxywide spiral arms seen in other galaxies and present in our own. It may well be that there is more than one process at work in the spectacular spirals we see.

An important question (but one that unfortunately is not answered by either of the two theories just described) is: Where do these spirals come from? What was responsible for generating the density wave in the first place or for creating the line of newborn stars whose evolution drives the advancing spiral arm? Scientists speculate that (1) the gravitational effects of our satellite galaxies (the Magellanic Clouds, to be discussed in Chapter 24), (2) instabilities in the gas near the Galactic bulge, or (3) the possible barlike asymmetry within the bulge itself may have had a big enough influence on the disk to get the process going.

The first possibility is supported by growing evidence that many other spiral galaxies seem to have experienced gravitational interactions with neighboring systems in the relatively recent past (see Chapter 24). However, many astronomers still regard the other two possibilities as equally likely. For example, they point to *isolated* spirals, whose structure clearly cannot be the result of an external interaction. The fact is that we still don't know for sure how galaxies—including our own—acquire such beautiful spiral arms.

☑ **Concept Check**

■ Why can't spiral arms simply be clouds of gas and young stars orbiting the Galactic center?

Figure 23.18 Self-Propagating Star Formation
(a) In this theory of the formation of spiral arms, the shock waves produced by the formation and later evolution of a group of stars provide the trigger for new rounds of star formation. We have used supernova explosions to illustrate the point here, but the formation of emission nebulae and planetary nebulae is also important. (b) This process may well be responsible for the partial spiral arms seen in some galaxies, such as NGC 3184, shown here in true color. The distinct blue appearance derives from the vast numbers of young stars that pepper its ill-defined spiral arms. (ESO)

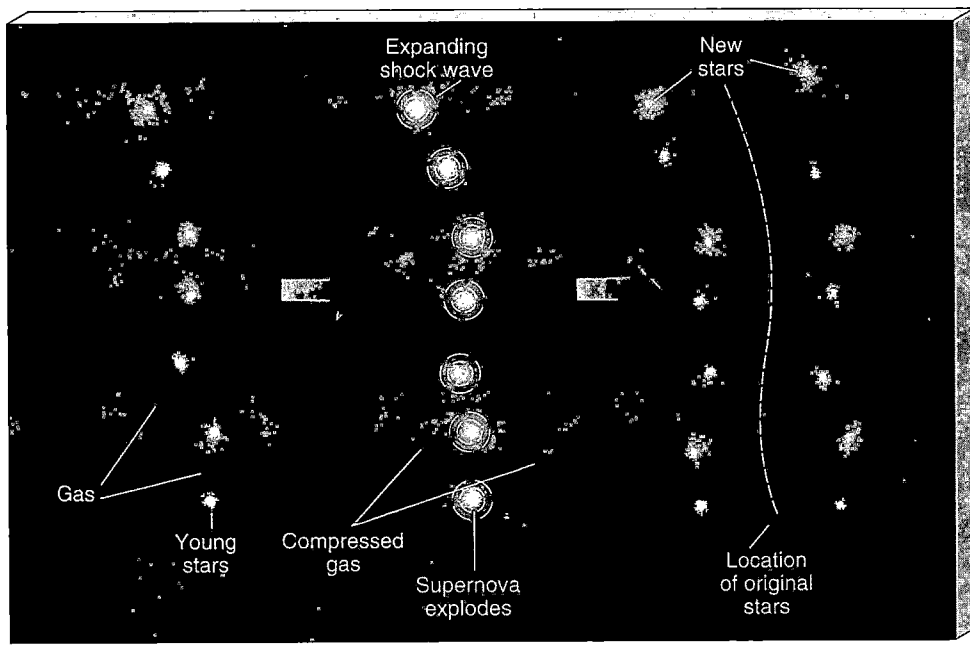

(a)

23.6 The Mass of the Milky Way Galaxy

5 We can measure our Galaxy's mass by studying the motions of gas clouds and stars in the Galactic disk. Recall from Chapter 2 that Kepler's third law (as modified by Newton) connects the orbital period, orbit size, and masses of any two objects in orbit around each other: ⊂⊃ (Sec. 2.7)

$$\text{total mass (solar mass)} = \frac{\text{orbit size (A.U.)}^3}{\text{orbit period (years)}^2}.$$

As we saw earlier, the distance from the Sun to the Galactic center is about 8 kpc, and the Sun's orbital period is 225 million years. Substituting these numbers into the preceding equation, we obtain a mass of almost 10^{11} solar masses—100 *billion* times the mass of our Sun. The Milky Way Galaxy is truly enormous, in mass as well as in size.

But what mass have we just measured? When we performed the analogous calculation in the case of a planet orbiting the Sun, there was no ambiguity: Neglecting the planet's mass, the result of our calculation was the mass of the Sun. ⊂⊃ (Sec. 2.7) However, the Galaxy's matter is not concentrated at the Galactic center (as the Sun's mass is concentrated at the center of the solar system); instead, galactic matter is distributed over a large volume of space. Some of it lies inside the Sun's orbit (that is, within 8 kpc of the Galactic center), and some lies outside, at large distances from both the Sun and the center of the Galaxy. What portion of the Galaxy's mass controls the Sun's orbit? Isaac Newton answered this question three centuries ago: The Sun's or-

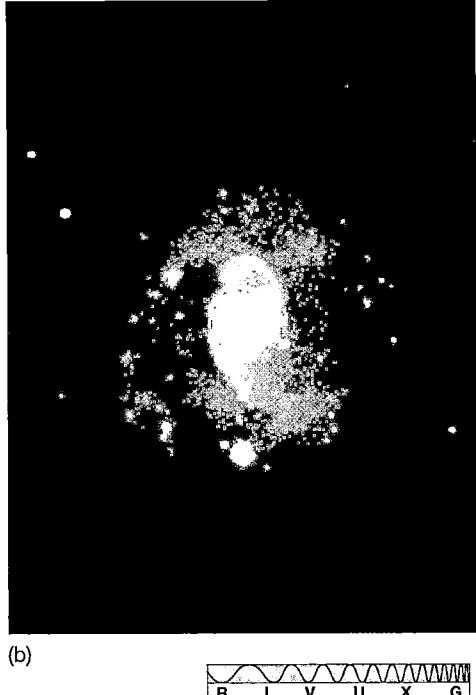

(b)

bital period is determined by the portion of the Galaxy that lies *within the orbit of the Sun*. This is the mass we computed in the equation.

DARK MATTER

To determine the mass of the Galaxy on larger scales—that is, to find how much matter is contained within spheres of progressively larger radii—we must measure the orbital motion of stars and gas farther from the Galactic center than is the Sun. Astronomers have found that the most effective way to do this is to make radio observations of gas in the Galactic disk, because radio waves are relatively unaffected by interstellar absorption and allow us to probe to great distances, far beyond the Sun's orbit. On the basis of these studies, radio astronomers have determined our Galaxy's rotation rate at various distances from the Galactic center. The resultant plot of rotation speed versus distance from the center (Figure 23.19) is called the Galactic **rotation curve**.

Knowing the Galactic rotation curve, we can now repeat our earlier calculation to compute the total mass that lies within any given distance from the Galactic center. We find, for example, that the mass within about 15 kpc from the center—the volume defined by the globular clusters and the known spiral structure—is roughly 2×10^{11} solar masses, about twice the mass contained within the Sun's orbit. Does the distribution of matter in the Galaxy "cut off" at this point, where the luminosity drops off sharply? Surprisingly, it does not.

If all of the mass of the Galaxy were contained within the edge of the visible structure, Newton's laws of motion predict that the orbital speed of stars and gas beyond 15 kpc would decrease with increasing distance from the Galactic center, just as the orbital speeds of the planets diminish as we move outward from the Sun. The dashed line in Figure 23.19 indicates what the rotation curve would look like in that case. However, the true rotation curve is quite different. Far from falling off at larger distances, it *rises* slightly out to the limits of our measurement capabilities. This implies that the amount of mass contained within successively larger radii continues to grow beyond the orbit of the Sun, apparently out to a distance of at least 40 or 50 kpc.

According to the preceding equation, the amount of mass within 40 kpc is approximately 6×10^{11} solar masses. Since 2×10^{11} solar masses lie within 15 kpc of the Galactic center, we have to conclude that at least twice as much mass lies *outside* the luminous part of our Galaxy—the part made up of stars, star clusters, and spiral arms—as lies inside!

Based on these observations of the Galactic rotation curve, astronomers now believe that the luminous portion of the Milky Way Galaxy—the region outlined by the globular clusters and by the spiral arms—is merely the "tip of the Galactic iceberg." Our Galaxy is in reality very much larger. The luminous region is surrounded by an extensive, invisible **dark halo**, which dwarfs the inner halo of stars and globular clusters, and extends well beyond the 15-kpc radius once thought to represent the limit of our Galaxy. But what is the composition of this dark halo? We do not detect enough stars or interstellar matter to account for the mass that our computations tell us must be there. We are inescapably drawn to the conclusion that most of the mass in our Galaxy exists in the form of invisible **dark matter**, which we presently do not understand.

The term *dark* here does not refer just to matter undetectable in visible light. The material has (so far) escaped detection at *all* wavelengths, from radio to gamma rays. Only by its gravitational pull do we know of its existence. Dark matter is not hydrogen gas (atomic or molecular), nor is it made up of ordinary stars. Given the amount of matter that must be accounted for, we would have been able to detect it with present-day equipment if it were in either of those forms. Its nature and its consequences for the evolution of galaxies and the universe are among the most important questions in astronomy today.

Many candidates have been suggested for this dark matter, although none is proven. Stellar-mass black holes may supply some of the unseen mass, although few candidates exist. ⊙ (Sec. 22.8) However, given that black holes are the evolutionary products of (relatively rare) massive stars, it is unlikely that there could be enough of them to hide large amounts of galactic matter. Currently among the strongest "stellar" contenders are brown dwarfs—low-mass prestellar objects that never reached the point of core nuclear burning—white dwarfs, and faint, low-mass red dwarfs. ⊙ (Secs. 19.3 and 20.3) In the jargon of the field, these objects are collectively known as *MA*ssive *C*ompact *H*alo *O*bjects, or MACHOs for short. They could in principle exist in great numbers throughout the Galaxy yet would be exceedingly hard to see.

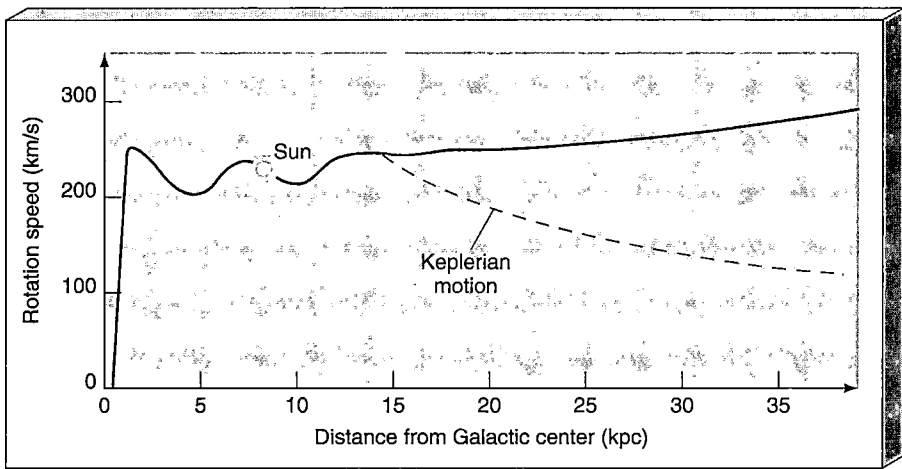

Figure 23.19 Galaxy Rotation Curve The rotation curve for the Milky Way Galaxy plots rotation speed versus distance from the Galactic center. We can use this curve to compute the mass of the Galaxy that lies within any given radius. The dashed curve is the rotation curve expected if the Galaxy "ended" abruptly at a radius of 15 kpc, the limit of most of the known spiral structure and the globular cluster distribution. The fact that the red curve does not follow this dashed line, but instead stays well above it, indicates that there must be additional unseen matter beyond that radius.

Hubble Space Telescope observations of globular clusters seem to argue against at least the last of the above possibilities for MACHOs. Figure 23.20 shows a *Hubble* image of a relatively nearby globular cluster—one close enough that very faint red dwarfs could have been detected if any existed. The *Hubble* data suggest that there is a cutoff at about 0.2 solar masses, below which stars form much less frequently than had previously been supposed. As a result, very-low-mass stars may be unexpectedly rare, at least in the Galactic halo.

A radically different alternative is that the dark matter is made up of exotic *subatomic particles* that pervade the entire universe. In order to account for the properties of dark matter, these particles must have mass (to produce the observed gravitational effects) but otherwise interact hardly at all with "normal" matter (because otherwise we would be able to see them). One class of candidate particles satisfying these requirements have been dubbed *Weakly Interacting Massive Particles*, or WIMPs. Many theoretical astrophysicists believe that such "dark-matter particles" could have been produced in abundance during the very earliest moments of our universe. If they survived to the present day, there might be enough of them to account for all the dark matter we believe must be out there. We will discuss this possibility, and its far-reaching implications, in more detail in Chapter 27. These ideas are hard to test, however, because particles of this nature would necessarily be very difficult to detect. Several detection experiments on Earth have been attempted, so far without success.

THE SEARCH FOR STELLAR DARK MATTER

Recently, researchers have obtained insight into the distribution of stellar dark matter by using a key element of Albert Einstein's theory of general relativity (see Chapter

22)—the prediction that a beam of light can be deflected by a gravitational field, which has already been verified in the case of starlight that passes close to the Sun. ∞ (*More Precisely 22-2*) The effect is small in the case of light grazing the Sun, but it has the potential for making distant and otherwise invisible stellar objects observable from Earth. Here's how:

Imagine looking at a distant star as a faint foreground object (a MACHO, such as a brown or white dwarf) happens to cross your line of sight. As illustrated in Figure 23.21, the intervening object deflects a little more starlight than usual toward you, resulting in a temporary, but quite substantial, *brightening* of the distant star. In some ways, the effect is like the focusing of light by a lens, and the process is known as **gravitational lensing**. The foreground object is referred to as a *gravitational lens*. The amount of brightening and the duration of the effect depend on the mass, distance, and speed of the lensing object. Typically, the apparent brightness of the background star increases by a factor of two to five for a period of several weeks. Thus, even though the foreground object cannot be seen directly, its effect on the light of the background star makes it detectable. (In Chapter 25 we will encounter other instances of gravitational lensing in the universe, but on very much larger scales.)

Of course, stars are very small compared to the distance scale of the Galaxy, and the probability that one star will pass almost directly in front of another, as seen from Earth, is extremely low. But by observing millions of stars every few days over a period of years (using automated telescopes and high-speed computers to reduce the burden of coping with so much data), astronomers have been able to see enough of these events to let them estimate the

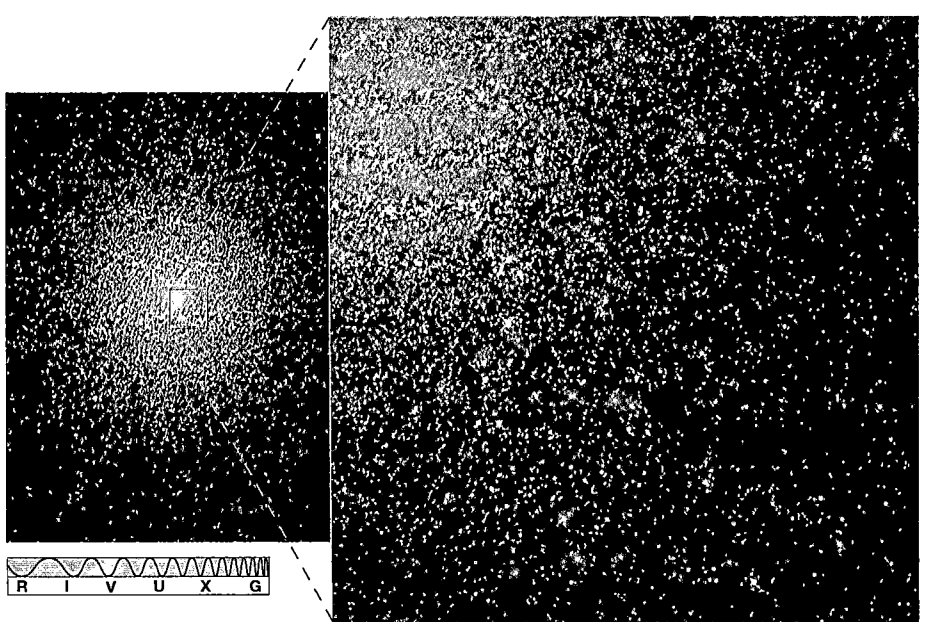

Figure 23.20 Missing Red Dwarfs Very sensitive visible observations with the *Hubble Space Telescope* have apparently ruled out faint red-dwarf stars as candidates for dark matter. The object shown here, the globular cluster 47 Tucanae, is one of many regions searched in the Milky Way. The inset, 0.4 pc on a side, is a high-resolution *Hubble* image of part of the cluster. The red dwarfs that would be expected if they existed in sufficient numbers to account for the dark matter in the Galaxy are not found. (The red stars that are seen are giants.) *(AAT; NASA)*

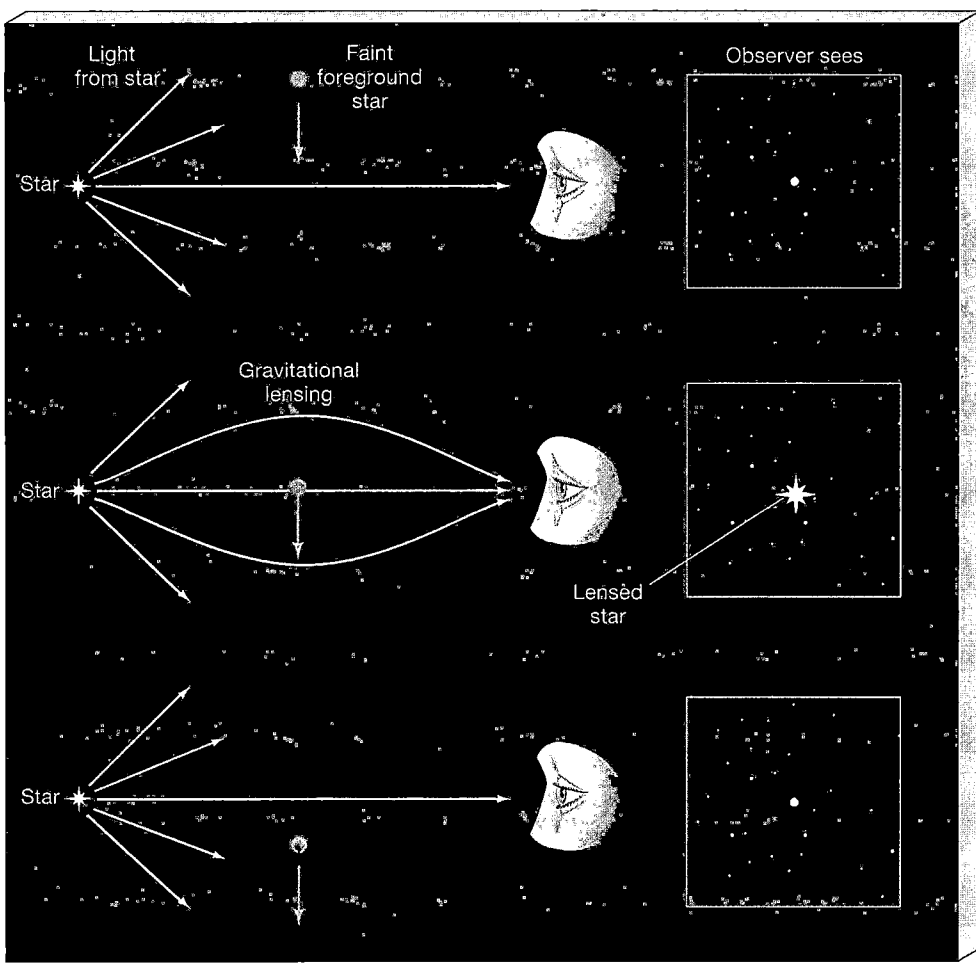

(a)

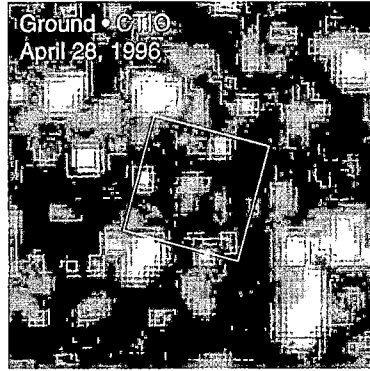

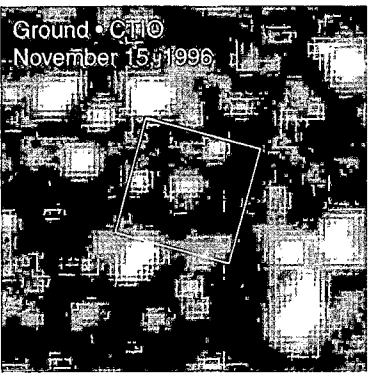

(b)

R I V U X G

Figure 23.21
Gravitational Lensing
(a) Gravitational lensing by a faint foreground object (such as a brown dwarf) can temporarily cause a background star to brighten significantly, providing a means of detecting otherwise invisible stellar dark matter. (b) The brightening of a star during a lensing event, this one implying that a massive, but unseen, object passed in front of the unnamed star at the center of the two boxes imaged six months apart. *(AURA)*

amount of stellar dark matter in the Galactic halo. The technique represents an exciting new means of probing the structure of our Galaxy. The first lensing events were reported in late 1993. Subsequent observations are consistent with lensing by low-mass white dwarfs and suggest that such stars could account for at least half—but, it now seems, apparently not all—of the dark matter inferred from dynamical studies.

Bear in mind, though, that the identity of the dark matter is not necessarily an all-or-nothing proposition. It is perfectly conceivable, and in fact most astronomers think it likely, that more than one type of dark matter exists. For example, it is quite possible that most of the dark matter in the inner (visible) parts of galaxies is in the form of brown dwarfs and very-low-mass stars, while the dark matter farther out may be primarily in the form of exotic particles. We will return to this perplexing problem in later chapters, when we discuss some theories of how galaxies formed and evolve, and how matter in the universe may have come into being.

✓ Concept Check
■ In what sense is dark matter "dark"?

23.7 The Galactic Center

⑤ Theory predicts that the Galactic bulge, and especially the region close to the galactic center, should be densely populated with billions of stars. However, we are unable to see this region of our Galaxy—the interstellar medium in the Galactic disk shrouds what otherwise would be a stunning view. Figure 23.22 shows the (optical) view we do have of the region of the Milky Way to-

ward the Galactic center, in the general direction of the constellation Sagittarius.

With the help of infrared and radio techniques we can peer more deeply into the central regions of our Galaxy than we can by optical means. Infrared observations (Figure 23.23a) indicate that the heart of our Galaxy harbors roughly 50,000 stars per cubic parsec. That's a stellar density about a million times greater than in our solar neighborhood, high enough that stars must experience frequent close encounters and even collisions. Infrared radiation has also been detected from what appear to be huge clouds rich in dust. In addition, radio observations indicate a ring of molecular gas nearly 400 pc across, containing some 30,000 solar masses of material and rotating at about 100 km/s. The origin of this ring is unclear, although researchers suspect that the gravitational influence of our Galaxy's elongated, rotating bulge may well be involved. The ring surrounds a bright radio source that marks the Galactic center.

High-resolution radio observations show more structure on small scales. Figure 23.23(b) shows the bright radio source Sagittarius A, which lies at the center of the boxed region in Figures 23.22 and Figure 23.23(a) and, we think, at the center of our Galaxy. On a scale of about 100 pc, extended filaments can be seen. Their presence suggests to many astronomers that strong magnetic fields operate in the vicinity of the center, creating structures similar in appearance to (but much larger than) those observed on the active Sun. On even smaller scales (Figure 23.23c) the observations indicate a rotating ring or disk of matter only a few parsecs across. And within that, as revealed by the

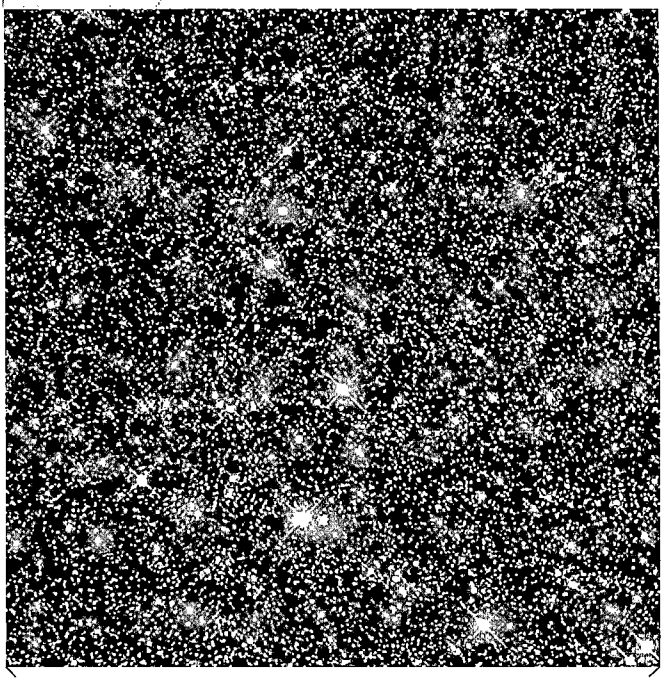

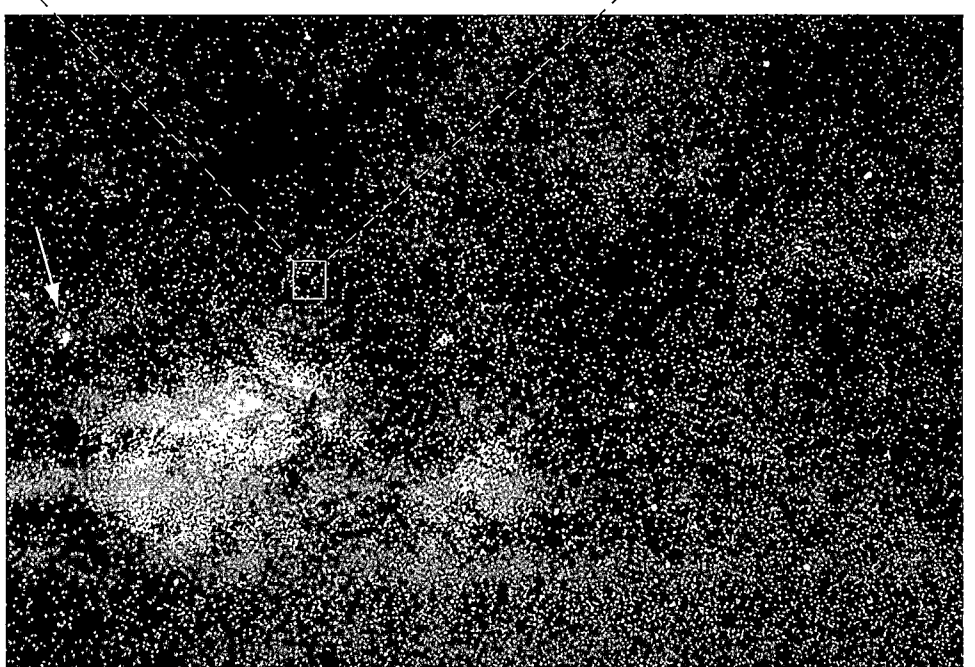

Figure 23.22 Galactic Center in Visible Light A photograph of stellar and interstellar matter in the direction of the Galactic center. Because of heavy obscuration, even the largest optical telescopes can see no farther than one-tenth the distance to the center. The M8 nebula (arrow) can be seen at extreme top center. The field is roughly 20° across, and is a continuation of the bottom part of Figure 18.6; compare also with Figure 18.1. The overlaid box outlines the location of the center of our Galaxy, and the inset shows the best optical view in that direction, but heavily confused by stars along the line of sight. (AURA; NASA)

R I V U X G

Chandra Observatory recently, lies a suspected black hole apparently embedded within a supernova remnant.

What could cause all this activity? An important clue comes from the Doppler broadening of infrared spectral lines emitted from the central swirling whirlpool of gas. The extent of the broadening indicates that the gas is moving very rapidly. In order to keep this gas in orbit, whatever is at the center must be extremely massive—more than a million solar masses. Given the twin requirements of large mass and small size, a leading contender is a black hole. The hole itself is not the source of the energy, of course. Instead, the vast accretion disk of matter being drawn toward the hole by the enormous gravity emits the energy as it falls in, just as we saw (on a much smaller scale) in Chapter 22

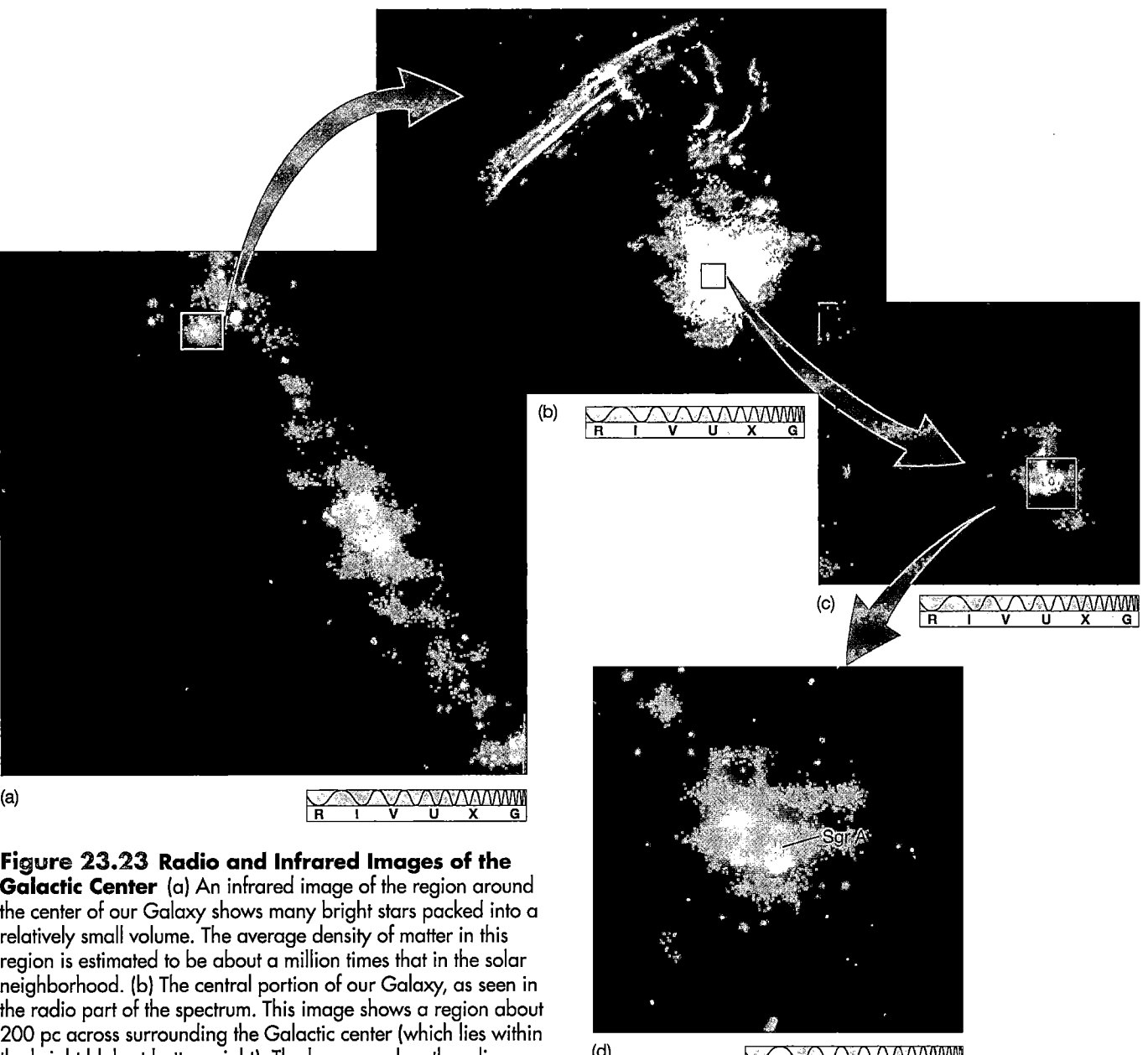

Figure 23.23 Radio and Infrared Images of the Galactic Center (a) An infrared image of the region around the center of our Galaxy shows many bright stars packed into a relatively small volume. The average density of matter in this region is estimated to be about a million times that in the solar neighborhood. (b) The central portion of our Galaxy, as seen in the radio part of the spectrum. This image shows a region about 200 pc across surrounding the Galactic center (which lies within the bright blob at bottom right). The long-wavelength radio emission cuts through the Galaxy's dust, providing an image of matter in the immediate vicinity of the Galaxy's center. (c) The spiral pattern of radio emission arising from Sagittarius A, the center of the Galaxy. The data suggest a rotating ring of matter only 5 pc across. (d) A recent *Chandra* image showing the relation of a hot supernova remnant (red) and Sgr A*, the suspected black hole at the very center. *(UMass/Caltech; NRAO; NASA)*

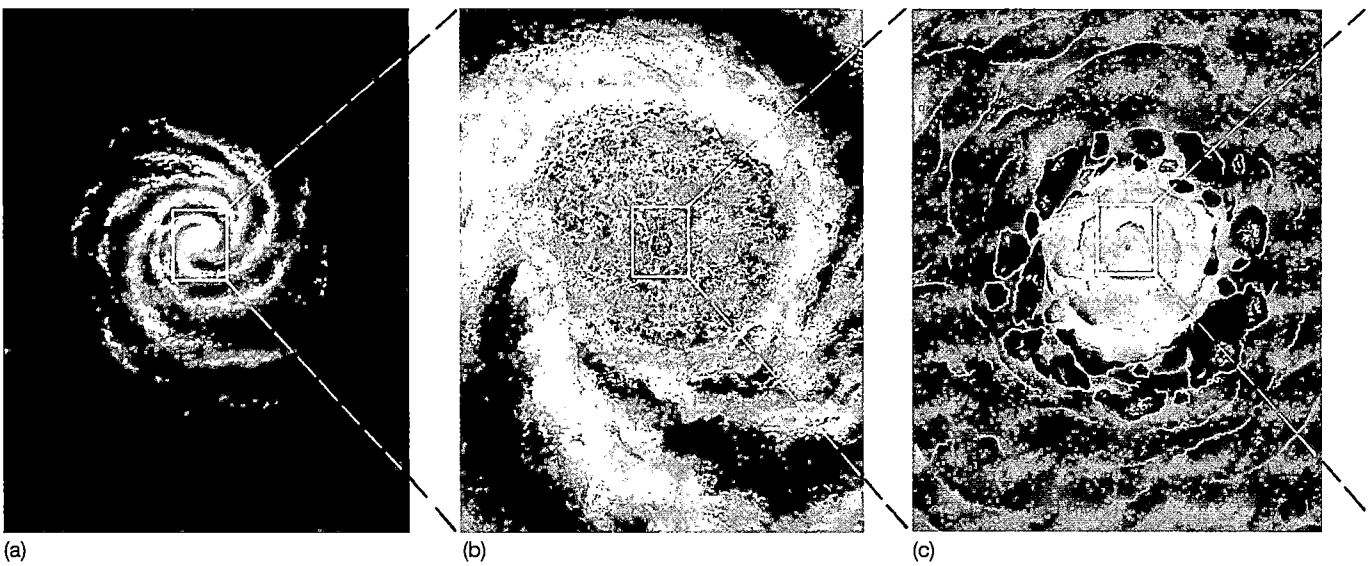

(a)   (b)   (c)

Figure 23.24 Galactic Center Zoom Six artist's conceptions, each centered on the Galactic center and each increasing in resolution by a factor of 10. Frame (a) shows the same scene as Figure 23.15. Frame (f) is a rendition of a vast whirlpool within the innermost parsec of our Galaxy. The data imaged in Figure 23.23 do not closely match these artistic renderings because the Figure 23.23 view is parallel to the Galactic disk, whereas these six paintings portray an idealized version perpendicular to that disk. *(L. Chaisson)*

when we discussed X-ray emission from neutron stars and stellar-mass black holes. ∞ (Secs. 22.3 and 22.8) The strong magnetic fields are thought to be generated within the accretion disk as matter spirals inward, and may act as "particle accelerators," creating the extremely high-energy particles detected on Earth as *cosmic rays* (see *Discovery 23-3*). Astronomers have reason to suspect that similar events are occurring at the centers of many other galaxies.

Figure 23.24 places these findings into a simplified perspective. Each frame is centered on the Galaxy's core, and each increases in resolution by a factor of 10. Frame (a) renders the Galaxy's overall shape, as painted in Figure 23.15. The scale of this frame measures about 100 kpc from top to bottom. Frame (b) spans a distance of 10 kpc from top to bottom and is nearly filled by the great circular sweep of the innermost spiral arm. Moving in to a 1-kpc span, frame (c) depicts the 400-pc ring of matter mentioned earlier. The dark blobs represent giant molecular clouds, the pink patches emission nebulae associated with star formation within those clouds.

In frame (d), at 100 pc, a pinkish region of ionized gas surrounds the reddish heart of the Galaxy. The source of energy producing this vast ionized cloud is assumed to be related to the activity in the Galactic center. Frame (e), spanning 10 pc, depicts the tilted, spinning whirlpool of hot (10^4 K) gas that marks the center of our Galaxy. The innermost part of this gigantic whirlpool is painted in frame (f), in which a swiftly spinning, white-hot disk of gas with temperatures in the millions of kelvins nearly engulfs a massive black hole too small in size to be pictured (even as a minute dot) on this scale.

At the very center of our Galaxy, at the heart of Sagittarius A, is a remarkable object with the odd-sounding name Sgr A* (pronounced "saj · ay · star"). By the standards of the active galaxies to be studied in Chapter 25, this compact *Galactic nucleus* is not particularly energetic. Still, radio observations made during the past two decades, along with more recent X- and gamma-ray observations, suggest that it is nevertheless a pretty violent place. Its total energy output (at all wavelengths) is estimated to be 10^{33} W, which is more than a million times that of the Sun.

VLBI observations using radio telescopes arrayed from Hawaii to Massachusetts imply that Sgr A* cannot be much larger than 10 A.U., and it is probably a good deal smaller than that. This size is consistent with the view that the energy source is a massive black hole, although the case is not airtight. The main alternatives are that Sgr A* could be a multiple supernova of some sort (but we detect no expansion of debris), a rapidly moving neutron star (but Sgr A* seems to be more or less fixed relative to the surrounding Galaxy), a large star cluster (but the energy-emitting region seems far too small), or even an ordinary neutron star embedded within a very unusual radio source (but that seems too ad hoc).

Recent findings support the black hole picture. Figure 23.25 shows a high-resolution infrared image of a 1″ by 1″ (0.04 pc, or 8000 A.U. across) field near the Galactic center, centered on Sgr A*. By combining many short-exposure frames, each too short for atmospheric turbulence to blur the view, U.S. and European researchers have created the first-ever diffraction-limited

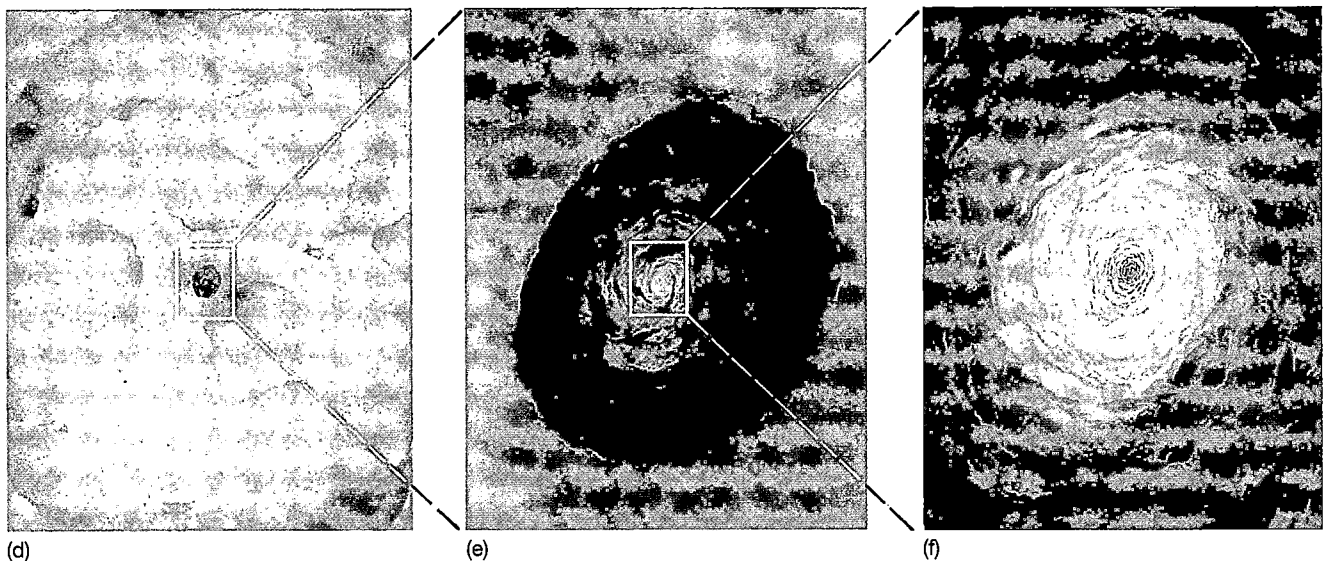

(d) (e) (f)

(0.05" resolution) images of the region. ∞ (Secs. 5.3, 17.5) The background image in Figure 23.25 shows a number of bright stars (the colored blobs) near Sgr A* (indicated by a white star). Overlaid on it are a series of observations of the brightest stars (falsely colored yellow-white here) over a four-year period. The stars' proper motions—their orbits around the Galactic center—can clearly be seen. Their inferred accelerations are consistent with the stars being in orbit around a massive

object at the location of Sgr A*. The two solid lines on the figure show the elliptical orbits that best fit the observations, in accordance with Newton's laws of motion. ∞ (Sec. 2.7) The periods of the orbits shown are about one decade and one century, respectively. These remarkable data imply that Sgr A* contains between 2.3 and 3.3 million solar masses.

Other observations, using adaptive-optics infrared imaging techniques, have revealed a bright source very

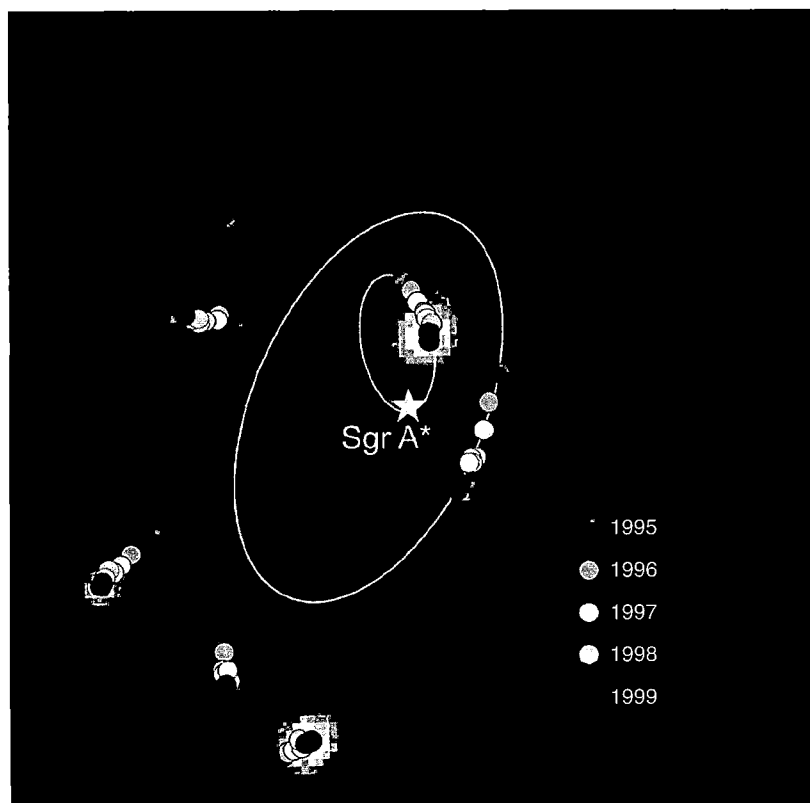

1995
1996
1997
1998
1999

Figure 23.25 Orbits Near the Galactic Center This extremely close-up map of the Galactic center was obtained by infrared speckle imaging—combining many short (0.14-s) exposures to create a high-resolution composite image of the small, 1" by 1", field of view. The fuzzy falsely colored blobs are individual stars orbiting the center. The color-coded circles show the locations of the six brightest stars in the field over a period of four years, from 1995 to 1999. The solid lines show the best-fitting orbits for two of these stars around a black hole of 3 million solar masses, located at Sgr A*. (Adapted from figure by A. Ghez and J. Kormendy)

R I V U X G

DISCOVERY 23-3

Cosmic Rays

In addition to stars, gas, dust, and dark matter, one other type of matter populates our Galaxy, namely, cosmic-ray particles, or cosmic rays for short. Cosmic rays continuously collide with Earth, and they make up our only samples of matter from outside the solar system.

In recent decades high-altitude balloons, rockets, and satellites have enabled astronomers to discover the chemical composition and the energies of cosmic-ray particles. Cosmic rays are actually not "rays" at all. They are subatomic particles. Nearly 90 percent are protons, the nuclei of hydrogen atoms. The nuclei of heavier atoms—helium and a long list of others—make up nine percent. Electrons amount to one percent. The abundances of elements that make up the cosmic rays are close to those we see throughout the Galaxy.

Cosmic-ray particles are very energetic. Some mechanism has accelerated them to extremely high velocities, much higher than any we can achieve in terrestrial laboratories. Virtually all cosmic rays—even the most massive—travel very close to the speed of light. When a Galactic cosmic-ray particle hits our atmosphere, it creates a series of cascades or "showers" of many lower-energy particles. These lower-energy "secondary" particles in turn collide with objects, including humans, on Earth's surface virtually all the time. Your body is being peppered with them right now.

The accompanying photograph shows an actual track of a cosmic-ray particle—in this case, the nucleus of a sulfur atom (shown in red in this false-color image). Entering from the left, the particle collides with the nucleus of an atom in the photographic emulsion, producing a fluorine nucleus (green), other nuclear fragments (blue), and a multitude of subatomic particles (yellow).

On the basis of the numbers of cosmic rays hitting Earth, astronomers estimate the number density of cosmic rays in interstellar space to be on the order of 10^{-3} particle/m^3. That makes them rare, although still more numerous than interstellar dust particles.

What is the source of our Galaxy's cosmic rays? The answer is currently uncertain. The circuitous and complex paths taken by the cosmic-ray particles while traversing the Galaxy, with its tangled magnetic field, prevent us from pinpointing their source by observing their direction of arrival at Earth. Indeed, careful searches have failed to show any preferential direction for cosmic rays. They seem to arrive from all directions with equal likelihood. Candidates for the origin of cosmic rays include (1) violent events at the Galactic center, where strong electromagnetic forces accelerate these particles to high energies; (2) supernova explosions, described in Chapter 21; and (3) the vicinity of neutron stars and black holes, as discussed in Chapter 22.

Most researchers currently favor the second of these explanations, although each of these candidate events probably contributes some particles to the observed hodgepodge of cosmic rays in our Galaxy. Some of the highest-energy cosmic rays may even originate at great distances, in the hearts of galaxies far beyond our own.

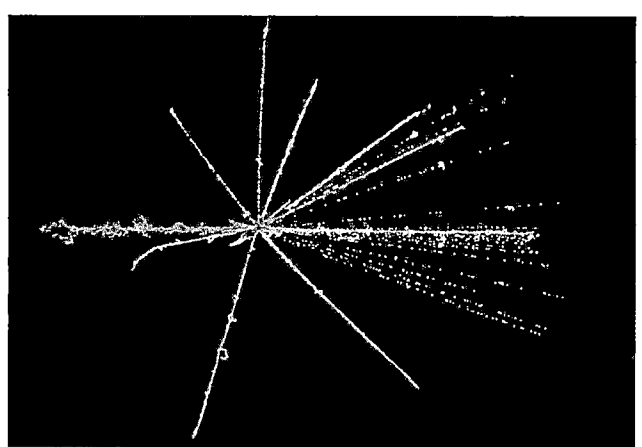

(C. Powell, P. Fowler, and D. Perkins/Science Photo Library/Photo Researchers, Inc.)

---

close to Sgr A* that seems to vary with a 10-minute period. ∞ (Sec. 5.3) This could be a hot spot on the accretion disk that circles the purported hole.

Note that, even with the large mass just mentioned, if Sgr A* is a genuine black hole, the size of its event horizon is still only 0.02 A.U. ∞ (Sec. 22.5) Such a small region, 8 kpc away, is currently unresolvable with any telescope now in existence.

If our knowledge of the Galaxy's center seems sketchy, that's because, despite rapidly improving observations, it *is*

sketchy. Astronomers are still deciphering the clues hidden within its invisible radiation. We are only beginning to appreciate the full magnitude of this strange new realm deep in the heart of the Milky Way.

 Concept Check

■ What is the most likely explanation of the energetic events observed at the Galactic center?

Chapter Review

SUMMARY

A galaxy (p. 600) is a huge collection of stellar and interstellar matter isolated in space and bound together by its own gravity. Because we live within it, the **Galactic disk** (p. 600) of our own Milky Way Galaxy appears as a broad band of light across the sky, a band called the Milky Way. Near the center, the Galactic disk thickens into the **Galactic bulge** (p. 600). The disk is surrounded by a roughly spherical **Galactic halo** (p. 600) of old stars and star clusters. Our Galaxy, like many others visible in the sky, is a **spiral galaxy** (p. 603).

Distant regions of our Galaxy and others can be studied using variable stars, whose luminosity changes with time. **Pulsating variable stars** (p. 604) vary in brightness in a repetitive and predictable way. **RR Lyrae** variables (p. 604) and **Cepheid** variables (p. 604) have characteristic light curves that make them easily recognizable. All RR Lyrae stars have roughly the same luminosity. The **period–luminosity relationship** (p. 605) is a simple correlation between Cepheid period and absolute brightness. The brightest Cepheids can be seen at distances of millions of parsecs. RR Lyrae stars are fainter but much more numerous, making them very useful within the Milky Way. In the early twentieth century, Harlow Shapley used RR Lyrae stars to determine the distances to many of the Galaxy's globular clusters. He found that the clusters have a roughly spherical distribution in space, but the center of the sphere lies far from the Sun. The globular clusters map out the true extent of the luminous portion of the Milky Way Galaxy. The center of their distribution is close to the **Galactic center** (p. 607), which lies about 8 kpc from the Sun. The luminous portion of our Galaxy has a diameter of about 30 kpc.

The Galactic halo lacks gas and dust, so no new stars are forming there. All halo stars are old. The gas-rich disk is the site of current star formation and contains many young stars. Stars in the halo and bulge move on largely random three-dimensional orbits that pass repeatedly through the disk plane but have no preferred orientation. Stars and gas in the disk move on roughly circular orbits around the Galactic center. Halo stars appeared early on, before the Galactic disk took shape, when there was no preferred orientation for their orbits. After the disk formed, stars born there inherited its overall spin and so move on circular orbits in the Galactic plane.

Astronomers use radio observations to explore the Galactic disk because radio waves are largely unaffected by interstellar dust. These observations clearly reveal the extent of our Galaxy's **spiral arms** (p. 613). The spiral arms in spiral galaxies are regions of the densest interstellar gas and are the places where star formation is taking place. The spirals cannot be "tied" to the disk material, as the disk's differential rotation would have wound them up long ago. Instead, they may be **spiral density waves** (p. 615) that move through the disk, triggering star formation as they pass by. Alternatively, the spirals may arise from self-propagating star formation, when shock waves produced by the formation and evolution of one generation of stars triggers the formation of the next.

The Galactic **rotation curve** (p. 617) plots the orbital speed of matter in the disk versus distance from the Galactic center. By applying Newton's laws of motion, astronomers can determine the mass of the Galaxy. They find that the galactic mass continues to increase beyond the radius defined by the globular clusters and the spiral structure we observe. Our Galaxy, like many others, has an invisible **dark halo** (p. 617) containing far more mass than can be accounted for in the form of luminous matter. The **dark matter** (p. 617) making up these dark halos is of unknown composition. Leading candidates include low-mass stars and exotic subatomic particles. Recent attempts to detect stellar dark matter have used the fact that a faint foreground object can occasionally pass in front of a more distant star, deflecting the star's light and causing its apparent brightness to increase temporarily. This deflection is called **gravitational lensing** (p. 618).

Astronomers working at infrared and radio wavelengths have uncovered evidence of energetic activity within a few parsecs of the Galactic center. The leading explanation is that a black hole 2–3 million times more massive than the Sun resides at the heart of our Galaxy.

SELF-TEST: TRUE OR FALSE?

_____ **1.** Globular clusters trace out the large-scale structure of the Galactic disk.

_____ **2.** RR Lyrae stars are a type of cataclysmic variable.

_____ **3.** The Galactic halo contains about as much gas and dust as the Galactic disk.

_____ **4.** The Galactic halo contains only old stars.

_____ **5.** Population I objects are found only in the Galactic halo.

_____ **6.** Up until the 1930s, the main error made in determining the extent of the Galactic disk was due to the neglect of absorption by interstellar dust.

_____ **7.** Astronomers use 21-cm radiation to study galactic molecular clouds.

_____ **8.** Radio techniques are capable of mapping the entire Galaxy.

_____ **9.** In the neighborhood of the Sun, the Galaxy's spiral density wave rotates more slowly than the overall galactic rotation.

_____ **10.** The mass of the Galaxy is determined by counting stars.

_____ **11.** Dark matter is now known to be due to large numbers of black holes.

_____ **12.** Cosmic rays are very energetic photons.

_____ **13.** Most of the mass of our Galaxy exists in the form of dark matter.

_____ **14.** The Galactic center has been extensively studied at radio and infrared wavelengths.

_____ **15.** It is possible to distinguish individual stars near the Galactic center.

SELF-TEST: FILL IN THE BLANK

1. One difficulty in studying our own Galaxy in its entirety is that we live _____.
2. Herschel's attempt to map the Milky Way by counting stars led to an inaccurate estimate of the Galaxy's size because he was unaware of _____.
3. The highly flattened, circular part of the Galaxy is called the Galactic _____.
4. The roughly spherical region of faint old stars and globular clusters in which the rest of the Galaxy is embedded is the Galactic _____.
5. Cepheids and RR Lyrae stars are observed to vary in _____ with periods of days to months.
6. Cepheids and RR Lyrae variables lie in a region of the H–R diagram called the _____.
7. According to the period–luminosity relation, the longer the pulsation period of a Cepheid, the _____ its luminosity.
8. Harlow Shapley determined the distances to the globular clusters using observations of _____.
9. The center of the globular cluster distribution is close to the _____ and about 8000 pc from the _____.
10. Stars and gas in the Galactic disk move in roughly _____ orbits around the _____.
11. At any given location in space, the motions of halo stars are largely _____ in direction.
12. The original cloud of gas from which the Galaxy formed probably had a size and shape similar to the present galactic _____.
13. Rotational velocities in the outer part of the Galaxy are _____ than would be expected on the basis of observed stars and gas, indicating the presence of _____.
14. Observations of the _____ of infrared spectral lines indicate that gas near the Galactic center is orbiting at extremely high speeds.
15. The most likely explanation of the high-speed motion of stars and gas near the Galactic center is that they are orbiting a _____.

REVIEW AND DISCUSSION

1. What evidence do we have that we live in a disk galaxy?
2. Why is it difficult to map out our Galaxy from our vantage point on Earth?
3. What are spiral nebulae? How did they get that name?
4. In what region of the Galaxy are globular clusters found?
5. How are Cepheid variables used in determining distances?
6. Roughly how far out into space (looking out of the Galactic disk) can we use Cepheids to measure distance?
7. What important discoveries were made early in this century using RR Lyrae variables?
8. Of what use is radio astronomy in the study of Galactic structure?
9. Contrast the motions of disk and halo stars.
10. How do we know that the Milky Way Galaxy has spiral arms?
11. Explain why galactic spiral arms are believed to be regions of recent and ongoing star formation.
12. Describe what happens to interstellar gas as it passes through a spiral density wave.
13. What is self-propagating star formation?
14. What do the red stars in the Galactic halo tell us about the history of the Milky Way?
15. What does the rotation curve of our Galaxy tell us about its total mass?
16. What evidence is there for dark matter in the Galaxy?
17. Describe some possible candidates for Galactic dark matter.
18. What is gravitational lensing, and can astronomers use it to search for dark matter?
19. Why can't optical astronomers easily study the center of our Galaxy?
20. Describe some ways in which astronomers can observe the Galactic center.

PROBLEMS  *Algorithmic versions of these questions are available in the Practice Problems module of the Companion Website.*

The number of squares preceding each problem indicates its approximate level of difficulty.

1. ■ Calculate the angular diameter of a prestellar nebula, of radius 100 A.U., lying 100 pc from Earth. Compare this with the roughly 6° diameter of the Andromeda Galaxy (Figure 23.2a).

2. ■■ How close would the nebula in the previous question have to be in order to have the same angular diameter as Andromeda? If the central star had a luminosity 10 times that of the Sun, calculate its apparent magnitude.

3. ■ What is the greatest distance at which an RR Lyrae star of absolute magnitude 0 could be seen by a telescope capable of detecting objects as faint as 20th magnitude?

4. ■ A typical Cepheid variable is 100 times brighter than a typical RR Lyrae star. How much farther away than RR Lyrae stars can Cepheids be used as distance-measuring tools?

5. ■■ The *Hubble Space Telescope* can see a star like the Sun at a distance of 100,000 pc. The brightest Cepheids have luminosities 30,000 times greater than that of the Sun. Taking the Sun's absolute magnitude to be 5, calculate the absolute magnitudes of these bright Cepheids. Neglecting interstellar absorption, how far away can *HST* see them?

6. ■■■ What is the maximum distance at which *Hubble* could see the Cepheid in the previous question if it lay in the Galactic disk, with an average interstellar extinction of 2.5 magnitudes per kiloparsec?

7. ■■ Calculate the proper motion (in arc seconds per year) of a globular cluster with a transverse velocity (relative to the Sun) of 200 km/s and a distance of 3 kpc. Do you think that this motion is measurable?

8. ■■ Calculate the total mass of the Galaxy lying within 20 kpc of the Galactic center if the rotation speed at that radius is 240 km/s.

9. ■■ Using the data presented in Figure 23.19, estimate the distance from the Galactic center at which matter takes (a) 100 million years, (b) 500 million years to complete one orbit.

10. ■■ Consider the motion of the Sun and another star lying 100 pc farther from the Galactic center. Both stars move in circular orbits in the Galactic plane. Initially, the stars lie along the same radial line through the center (so they are 100 pc apart). How far apart will they be after the Sun has completed exactly one orbit?

11. ■■■ Using the data presented in Figure 23.19, calculate how long it takes the Sun to "lap" stars orbiting 15 kpc from the Galactic center. How long does matter at 5 kpc take to lap us?

12. ■ A density wave made up of two spiral arms is moving through the Galactic disk. At the 8-kpc radius of the Sun's orbit around the Galactic center, the wave's speed is 120 km/s, and the Galactic rotation speed is 220 km/s. Calculate how many times the Sun has passed through a spiral arm since forming 4.6 billion years ago.

13. ■ Given the data in the previous question and the fact that O-type stars live at most 10 million years before exploding as supernovae, calculate the maximum distance at which an O-type star (orbiting at the Sun's distance from the Galactic center) can be found from the density wave in which it formed.

14. ■■ Material at an angular distance of 0.2″ from the Galactic center is observed to have an orbital speed of 1200 km/s. If the Sun's distance to the Galactic center is 8 kpc, and the material's orbit is circular and is seen edge-on, calculate the radius of the orbit and the mass of the object around which the material is orbiting.

15. ■■■ Given that the entire field of view in Figure 23.25 is 1 arc second across, assuming a distance of 8 kpc and a black-hole mass of 3 million solar masses, and taking all orbital planes to be perpendicular to the line of sight, estimate the periods of the two orbits shown.

COLLABORATIVE EXERCISES

1. Cepheid Pulsation Periods. For each member of your group, determine the luminosity of stars that have pulsation periods equal to the number of days since each person last ate rice, using Figure 23.8.

PROJECT

1. If you are far from city lights, look for a hazy band of light arching across the sky. This is our edgewise view of the Milky Way Galaxy. The Galactic center is located in the direction of the constellation Sagittarius, highest in the sky during the summer, but visible from spring through fall. Look at the band making up the Milky Way and notice dark regions; these are relatively nearby dust clouds. Sketch what you see. Look for faint fuzzy spots in the Milky Way and note their positions in your sketch. Draw in the major constellations for reference. Compare your sketch with a map of the Milky Way in a star atlas. Did you discover most of the dust clouds? Can you identify the faint fuzzy spots?

 *In addition to the Practice Problems and Destinations modules, the Companion Website at http://www.prenhall.com/chaisson provides for each chapter an additional true-false, multiple choice, and labeling quiz, as well as additional annotated images, animations, and links to related Websites.*

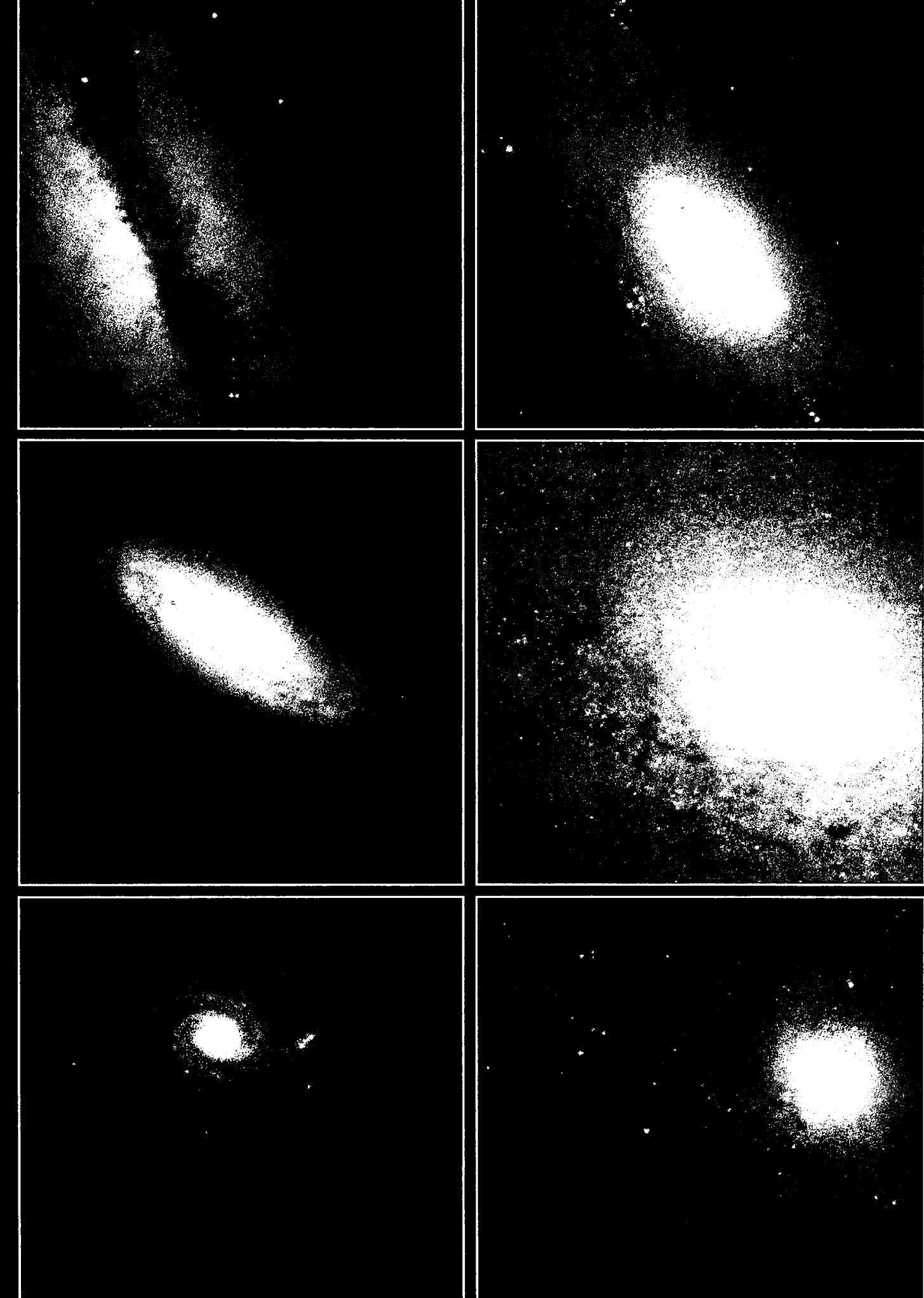

24 NORMAL GALAXIES

The Large-Scale Structure of the Universe

LEARNING GOALS

Studying this chapter will enable you to:

1   Describe the basic properties of the main types of normal galaxies.

2   Discuss the distance-measurement techniques that enable astronomers to map the universe beyond the Milky Way Galaxy.

3   Summarize what is known about the large-scale distribution of galaxies in the universe.

4   Describe some of the methods used to determine the masses of distant galaxies.

5   Explain why astronomers think that most of the matter in the universe is invisible.

6   Discuss some theories of how galaxies form and evolve.

7   State Hubble's law and explain how it is used to derive distances to the most remote objects in the observable universe.

 Visit http://www.prenhall.com/chaisson for additional annotated images, animations, and links to related sites for this chapter.

This collage of distant galaxies was captured in the infrared part of the spectrum. Their names or catalog numbers hardly matter. Most of them are between 20 and 30 million light-years away. Together, they showcase the diversity among spiral galaxies, from edge-on at upper left to face-on at lower left. By observing the infrared radiation they emit, astronomers can better study the dust clouds swirling around their centers. (STScI)

The Big Picture: The farther we probe the distant universe, the more details get lost. We are stuck on (or near) Earth, with no known way ever to travel among the faraway galaxies. When we do look out, we generally see galaxies upon more galaxies; planets become inconsequential when observing truly deep space, stars themselves mere points of hydrogen consumption. Here, we begin our study of distant realms completely unknown to our ancestors of a century ago.

Much of our knowledge of the workings of our own Galaxy is based on observations of other galaxies. We know of literally millions of galaxies beyond our own—many smaller than it, some comparable in size, a few much larger. All are vast, gravitationally bound assemblages of stars, gas, dust, and dark matter; separated from us by almost incomprehensibly large distances. Even a modest-sized galaxy harbors more stars than the number of people who have ever lived on Earth. The light we will receive tonight from the most distant galaxies was emitted long before Earth even existed. By comparing and classifying the properties of galaxies, astronomers have begun to understand their complex dynamics. By mapping out their distribution in space, astronomers trace out the immense realms of the universe. The galaxies remind us that our position in the universe is no more special than that of a boat adrift at sea.

24.1 Hubble's Galaxy Classification

Figure 24.1 shows a vast expanse of space lying about 100 million pc from Earth. Almost every patch or point of light in this figure is a separate galaxy—several hundred can be seen in just this one photograph. Over the years, astronomers have accumulated similar images of many millions of galaxies. We begin our study of these enormous accumulations of matter simply by considering their appearance on the sky.

Seen through even a small telescope, images of galaxies look distinctly nonstellar. They have fuzzy edges, and many are quite elongated—not at all like the sharp, point-like images normally associated with stars. Although it is difficult to tell from the photograph, some of the blobs of light in Figure 24.1 are actually spiral galaxies like the Milky Way Galaxy and Andromeda. Others, however, are definitely not spirals—no disks or spiral arms can be seen.

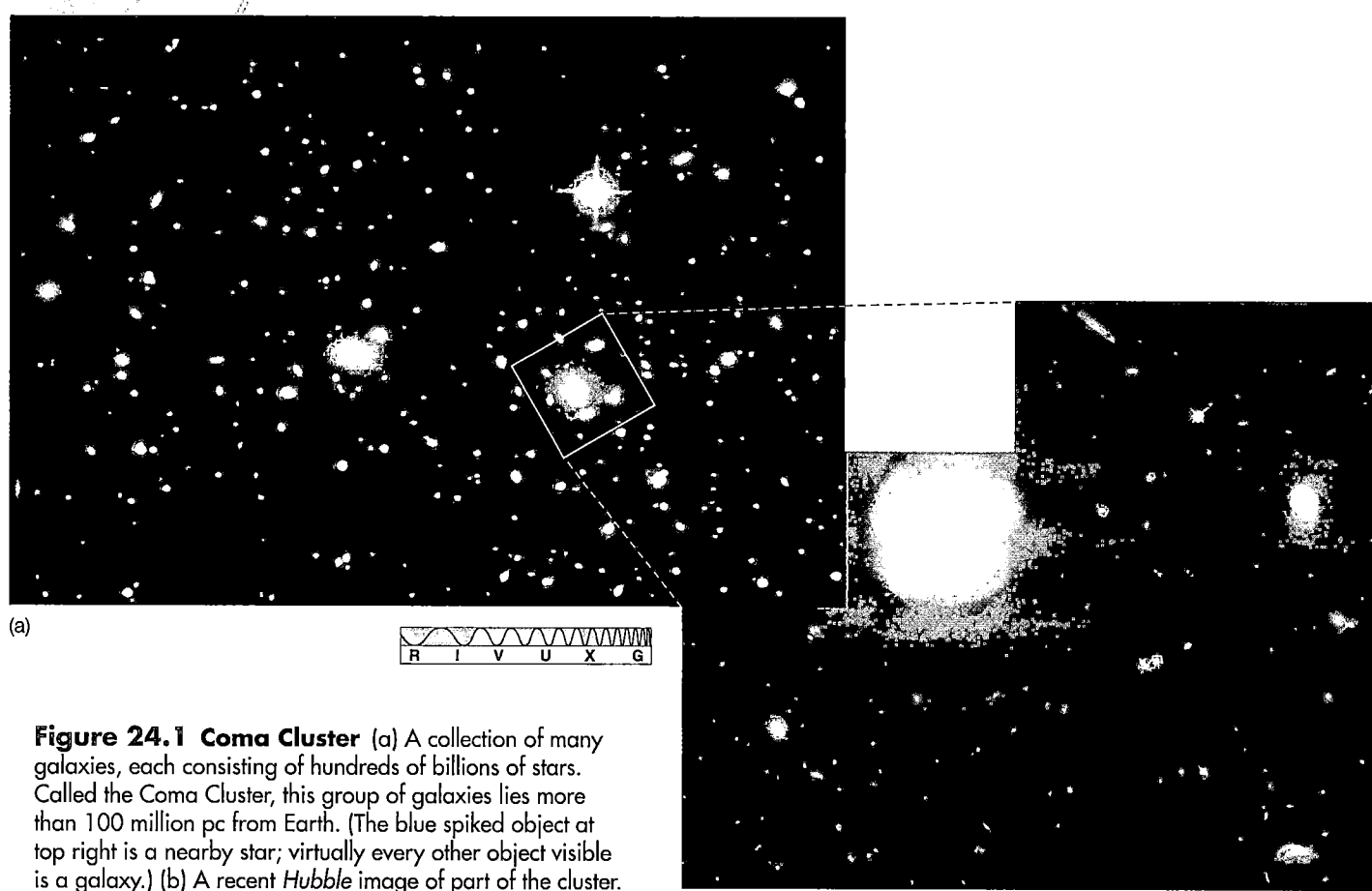

(a)

| R | I | V | U | X | G |

Figure 24.1 Coma Cluster (a) A collection of many galaxies, each consisting of hundreds of billions of stars. Called the Coma Cluster, this group of galaxies lies more than 100 million pc from Earth. (The blue spiked object at top right is a nearby star; virtually every other object visible is a galaxy.) (b) A recent *Hubble* image of part of the cluster. *(AURA; NASA)*

(b)

Even when we take into account their different orientations in space, galaxies do *not* all look the same.

The American astronomer Edwin Hubble was the first to categorize galaxies in a comprehensive way. Working with the then recently completed 2.5-m optical telescope on Mount Wilson, California in 1924, he classified the galaxies he saw into four basic types—*spirals, barred spirals, ellipticals,* and *irregulars*—solely on the basis of appearance. Many modifications and refinements have been incorporated over the years, but the basic **Hubble classification scheme** is still widely used today.

SPIRALS

We saw several examples of **spiral galaxies** in Chapter 23—for example, our own Milky Way Galaxy and our neighbor Andromeda. All galaxies of this type contain a flattened galactic disk in which spiral arms are found, a central galactic bulge, and an extended halo of faint, old stars. ∞ (Sec. 23.3) The stellar density (that is, the number of stars per unit volume) is greatest in the *galactic nucleus*, at the center of the bulge. However, within this general description, spiral galaxies exhibit a wide variety of shapes, as illustrated in Figure 24.2.

In Hubble's scheme, a spiral galaxy is denoted by the letter S and classified as type a, b, or c, according to the size of its central bulge—Type Sa galaxies having the largest bulges, Type Sc the smallest. The size of the bulge is quite well correlated with the tightness of the spiral pattern (although the correspondence is not perfect). Type Sa spiral galaxies, with large central bulges, tend to have tightly wrapped, almost circular, spiral arms. Type Sb galaxies, with smaller bulges, typically have more open spiral arms. Type Sc spirals, with the smallest bulges, often have loose, poorly defined spiral structure. The arms also tend to become more "knotty," or clumped, in appearance as the spiral pattern becomes more open.

Much of our description of the large-scale structure of the Milky Way Galaxy in Chapter 23 applies to spiral galaxies in general. The bulges and halos of spiral galaxies contain large numbers of reddish old stars and globular clusters, similar to those observed in our Galaxy and in Andromeda. Most of the light from spirals, however, comes from A- through G-type stars in the disk, giving these galaxies an overall whitish glow. The spiral arms appear bluish because of the presence of bright blue O- and B-type stars there. We assume that thick disks exist, too, but their faintness makes this assumption hard to confirm—the thick disk in the Milky Way contributes only a percent or so of our Galaxy's total light. ∞ (Sec. 23.3)

Like the disk of the Milky Way, the flat disks of typical spiral galaxies are rich in gas and dust. The 21-cm radio radiation emitted by spirals betrays the presence of interstellar gas, and detailed photographs clearly show obscuring dust in many systems. Stars are forming within the spiral arms, where the interstellar medium is densest, and the arms contain numerous emission nebulae and newly formed O- and B-type stars as discussed in Chapter 23. ∞ (Sec. 23.5) Type Sc galaxies contain the most interstellar gas and dust, Sa galaxies the least. The Sc galaxy NGC 2997 shown in Figure 24.2(c) clearly shows the preponderance of interstellar gas, dust, and young blue stars tracing

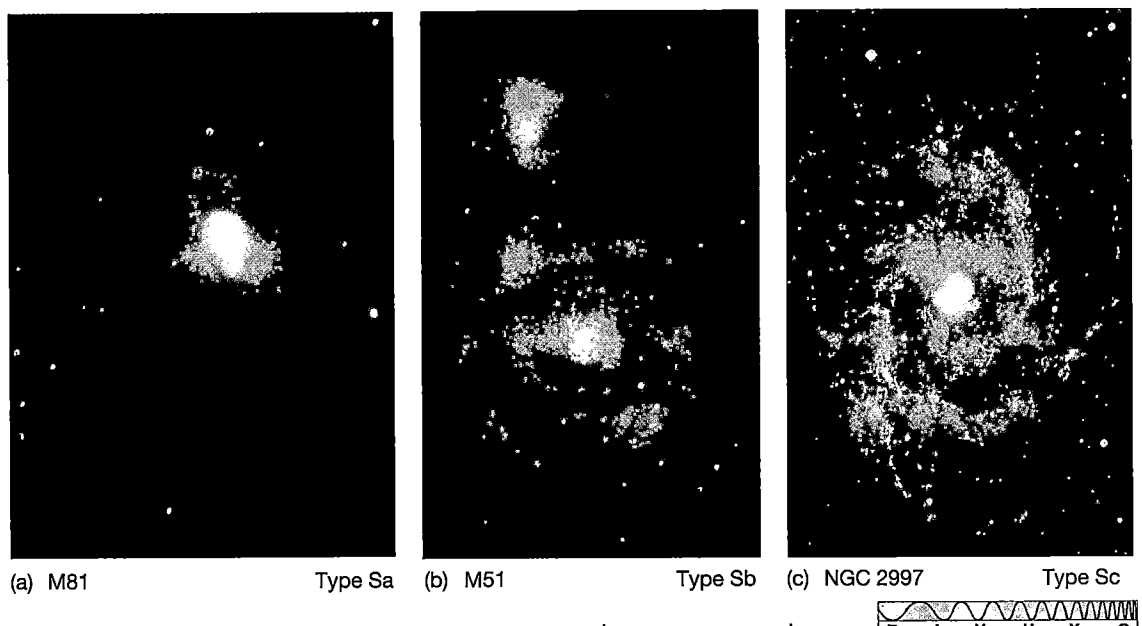

(a) M81   Type Sa   (b) M51   Type Sb   (c) NGC 2997   Type Sc

Figure 24.2 Spiral Galaxy Shapes Variation in shape among spiral galaxies. As we progress from type Sa to Sb to Sc, the bulges become smaller while the spiral arms tend to become less tightly wound. *(W. Keel; NOAO; D. Malin/AAT)*

the spiral pattern. Spirals are not necessarily young galaxies, however. Like our own Galaxy, they are simply rich enough in interstellar gas to provide for continued stellar birth.

Most spirals are not seen face-on as in Figure 24.2. Many are tilted with respect to our line of sight, making their spiral structure hard to discern. However, we do not need to see spiral arms to classify a galaxy as a spiral. The presence of the disk, with its gas, dust, and newborn stars, is sufficient. For example, the galaxy shown in Figure 24.3 is classified as a spiral because of the clear line of obscuring dust seen along its midplane.

A variation of the spiral category in Hubble's classification scheme is the **barred-spiral galaxy**. The barred spirals differ from ordinary spirals mainly by the presence of an elongated "bar" of stellar and interstellar matter passing through the center and extending beyond the bulge, into the disk. The spiral arms project from near the ends of the bar rather than from the bulge (as they do in normal spirals). Barred spirals are designated by the letters SB and are subdivided, like the ordinary spirals, into categories SBa, SBb, and SBc, depending on the size of the bulge. Again like ordinary spirals, the tightness of the spiral pattern is correlated with bulge size. Figure 24.4 shows the variation among barred-spiral galaxies. In the case of the SBc category, it is often hard to tell where the bar ends and the spiral arms begin.

Astronomers often cannot distinguish between spirals and barred spirals, especially when a galaxy happens to be oriented with its galactic plane nearly edge-on toward

Earth, as in Figure 24.3. Because of the physical and chemical similarities of spiral and barred-spiral galaxies, some researchers do not even bother to distinguish between them. Others, however, regard the differences in their structures as very important, arguing that these differences suggest basic dissimilarities in the ways these galaxies formed and evolved. The recent finding that the bulge of our own Galaxy is elongated suggests that the Milky Way is a barred spiral, of type SBb or SBc. ∞ (Sec. 23.3)

ELLIPTICALS

The next major category in the Hubble scheme contains the **elliptical galaxies**. Unlike the spirals, ellipticals have no spiral arms and, in most cases, no flattened galactic disk—in fact, they often exhibit little internal structure of any kind. As with spirals, the stellar density increases sharply in the central nucleus. Ellipticals range in shape from highly elongated to nearly circular in appearance. Denoted by the letter E, these systems are subdivided according to how elliptical they are. The most circular are designated E0, slightly flattened systems are labeled E1, and so on, all the way to the most elongated ellipticals, of type E7 (Figure 24.5).

There is a large range in both the size and the number of stars contained in elliptical galaxies. The largest elliptical galaxies are much larger than our own Milky Way Galaxy. These *giant ellipticals* can range up to a few mega-

Figure 24.3 Sombrero Galaxy
The Sombrero Galaxy, a spiral system seen edge-on. Officially cataloged as M104, this galaxy has a dark band composed of interstellar gas and dust. The large size of this galaxy's central bulge marks it as type Sa, even though its spiral arms cannot be seen. *(ESO)*

R   I   V   U   X   G

(a) NGC 3992　　　　Type SBa　(b) NGC 1365　　　　Type SBb　(c) NGC 6872　　　　Type SBc

Figure 24.4 Barred-Spiral Galaxy Shapes Variation in shape among barred-spiral galaxies. The variation from SBa to SBc is similar to that for the spirals in Figure 24.2, except that now the spiral arms begin at either end of a bar through the Galactic center. In frame (c), the bright star is a foreground object in our own Galaxy, and the object at top center is another galaxy that is probably interacting with NGC 6872. *(NOAO; AAT; ESO)*

parsecs across and contain trillions of stars. At the other extreme, *dwarf ellipticals* may be as small as 1 kpc in diameter and contain only a few million stars. The substantial observational differences between giant and dwarf ellipticals have led many astronomers to conclude that these galaxies are members of separate classes, with quite different formation histories and stellar content. The dwarfs are by far the most common type of ellipticals, outnumbering their brighter counterparts by about 10 to 1. However, most of the *mass* that exists in the form of elliptical galaxies is contained in the larger systems.

Lack of spiral arms is not the only difference between spirals and ellipticals. Most ellipticals contain little or no cool gas and dust. The 21-cm radio emission from neutral hydrogen gas is, with few exceptions, completely absent, and no obscuring dust lanes are seen. In most cases, there is no evidence of young stars or of any ongoing star formation—like the halo of our own Galaxy, ellipticals are made up mostly of rather old, reddish, low-mass stars. Again like the Galactic halo, stellar orbits in ellipticals are disordered, exhibiting little or no overall rotation; stars move in all directions, not in regular, circular paths as in our Galaxy's

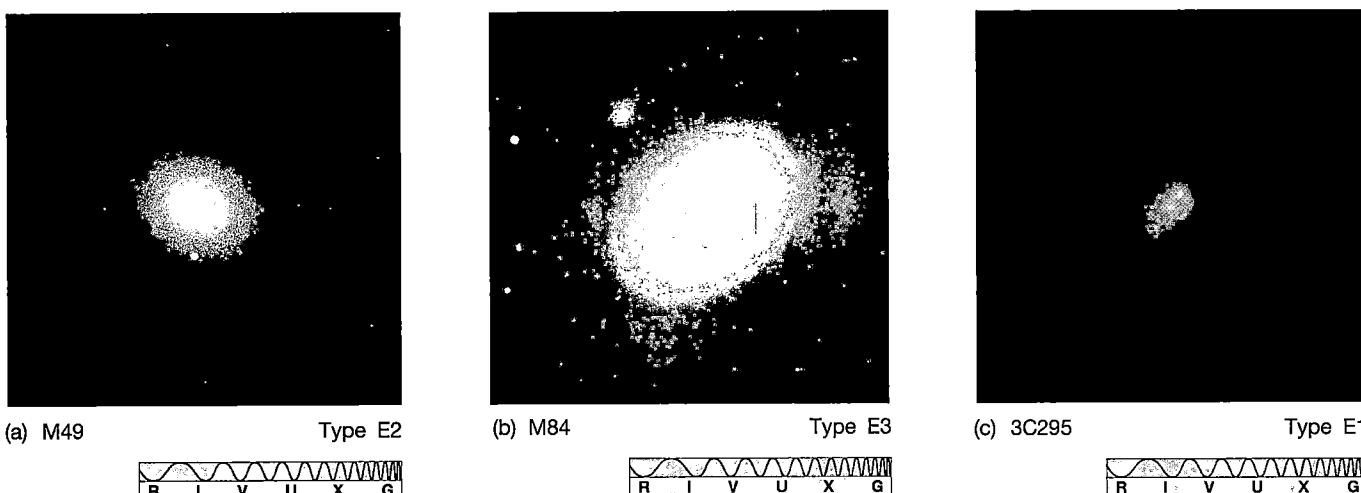

(a) M49　　　　Type E2　(b) M84　　　　Type E3　(c) 3C295　　　　Type E1

Figure 24.5 Elliptical Galaxy Shapes Variation in shape among elliptical galaxies. (a) The E1 galaxy M49 is nearly circular in appearance. (b) M84 is a slightly more elongated elliptical galaxy. It is classified as E3. Both of these galaxies lack spiral structure, and neither shows evidence of cool interstellar dust or gas, although each has an extensive X-ray halo of hot gas that extends far beyond the visible portion of the galaxy. (c) This false-color X-ray image of the giant elliptical galaxy 3C295 displays hot gas (red) well beyond the galaxy itself (white) as well as throughout the cluster of galaxies surrounding 3C295. *(AURA; SAO)*

disk. Ellipticals differ from our Galaxy's halo in at least one important respect, however. X-ray observations reveal large amounts of very *hot* (several million kelvin) interstellar gas distributed throughout their interiors, often extending well beyond the visible portions of the galaxies (Figure 24.5c).

Some giant ellipticals are exceptions to many of the foregoing general statements about elliptical galaxies, as they have been found to contain disks of gas and dust in which stars are forming. Astronomers speculate that these systems may be the results of mergers between gas-rich galaxies. We will return later in this chapter to the important role played by galactic collisions in determining the appearance of the systems we observe today (*Discovery 24-1*).

Intermediate between the E7 ellipticals and the Sa spirals in the Hubble classification is a class of galaxies showing evidence of a thin disk and a flattened bulge, but containing no gas and no spiral arms. Two such objects are shown in Figure 24.6. These galaxies are known as **S0 galaxies** if no bar is evident and **SB0 galaxies** if a bar is present. They look a little like spirals whose dust and gas have been stripped away, leaving behind a stellar disk and bulge. Observations in recent years have shown that many normal elliptical galaxies have faint stellar disks within them, like the S0 galaxies. As with the S0s, the origin of these disks is uncertain, but some researchers suspect that the S0s and ellipticals actually form a continuous sequence, along which the bulge-to-disk ratio varies smoothly.

IRREGULARS

The final galaxy class identified by Hubble is a catch-all category—**irregular galaxies**—so named because their visual appearance does not allow us to place them into any of the other categories just discussed. Irregulars tend to be rich in interstellar matter and young, blue stars, but they lack any regular structure such as well-defined spiral arms or central bulges. They are conventionally divided into two subclasses—Irr I galaxies and Irr II galaxies. The Irr I galaxies often look like misshapen spirals. They tend

to be smaller than spirals but somewhat larger than dwarf ellipticals. They typically contain between 10^8 and 10^{10} stars. The smallest are called *dwarf irregulars*. As with elliptical galaxies, the dwarf type is the most common. Dwarf ellipticals and dwarf irregulars occur in approximately equal numbers and together make up the vast majority of galaxies in the universe. They are often found close to a larger "parent" galaxy.

The much rarer Irr II galaxies, in addition to their irregular shape, have other peculiarities, often exhibiting a distinctly explosive or filamentary appearance. Figure 24.7 shows some examples of these strangely shaped galaxies. Their appearance once led astronomers to suspect that violent events had occurred within them. However, it now seems more likely that in some (but perhaps not all) cases we are seeing the result of a close encounter, or even a collision, between two previously "normal" systems.

Figure 24.8 shows the **Magellanic Clouds**, a famous pair of Irr I galaxies that orbit the Milky Way Galaxy. They are shown to proper scale in Figure 23.15. Studies of Cepheid variables within the Clouds show them to be approximately 50 kpc from the center of our Galaxy. The Large Cloud contains about 6 billion solar masses of material and is a few kiloparsecs across. Both Magellanic Clouds contain lots of gas, dust, and blue stars (and the recent, well-documented supernova discussed in *Discovery 21-2*), indicating youthful activity and ongoing star formation. Both also contain many old stars and several old globular clusters, so we know that star formation has been going on there for a very long time.

Over the years, radio studies have hinted at a possible bridge of hydrogen gas connecting the Milky Way to the Magellanic Clouds, but more observational research is needed to establish this link beyond doubt. It is possible that the tidal force of the Milky Way tore this stream of gas from the Clouds the last time their orbits brought them close to our Galaxy. Of course, gravity works both ways, and many researchers believe that the forces exerted by the Clouds may in turn be responsible for distorting our Galaxy, warping and thickening the outer parts of the Galactic disk. ∞ (Sec. 23.5)

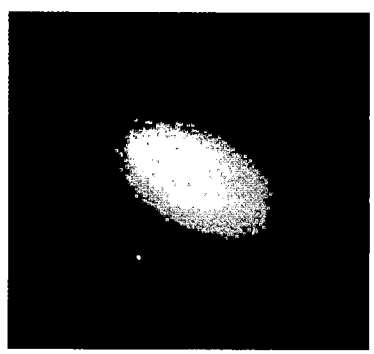

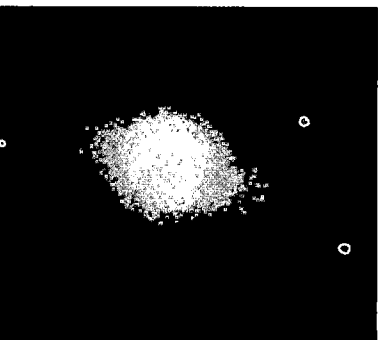

(a) NGC 1201 Type S0   (b) NGC 2859 Type SB0

Figure 24.6 S0 Galaxies (a) S0 galaxies contain a disk and a bulge but no interstellar gas and no spiral arms. They are in many respects intermediate between E7 ellipticals and Sa spirals in their properties. (b) SB0 galaxies are similar to S0 galaxies, except for a bar of stellar material extending beyond the central bulge. *(Palomar/Caltech)*

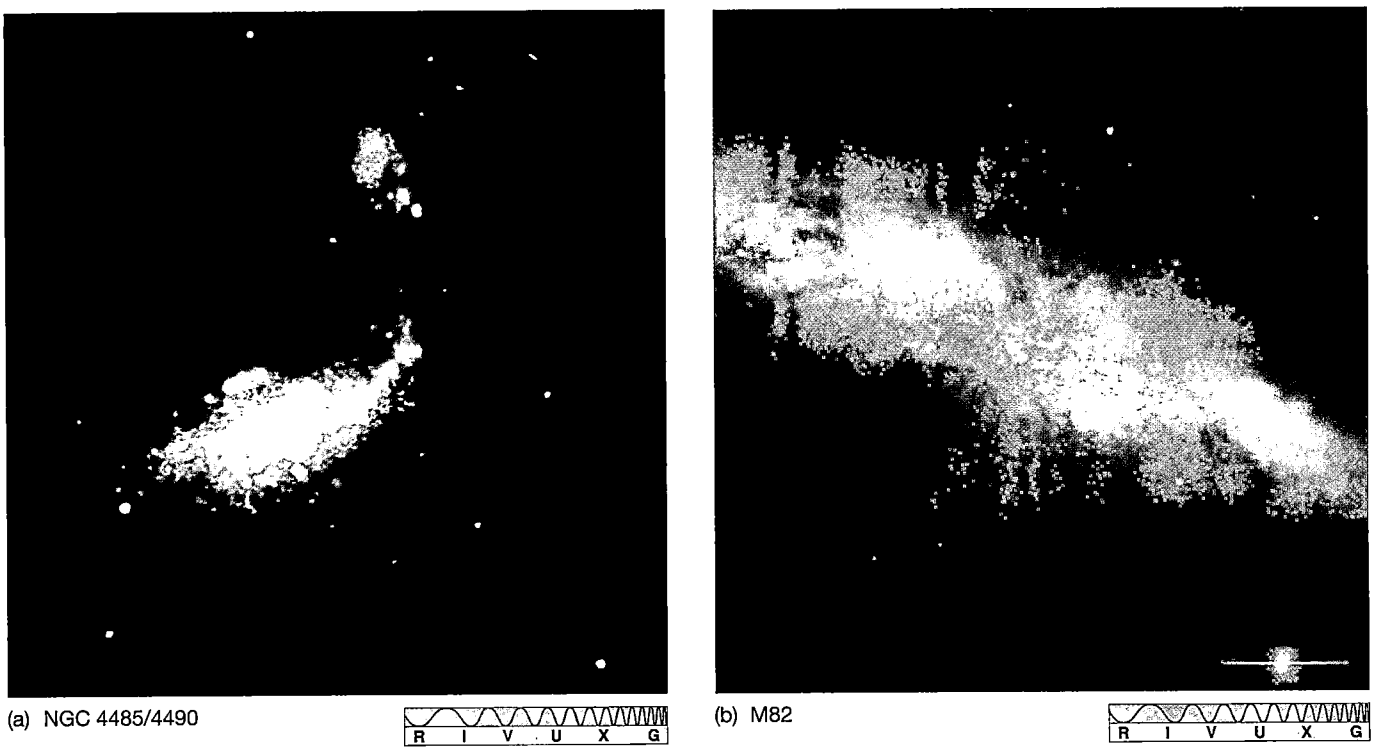

(a) NGC 4485/4490

(b) M82

Figure 24.7 Irregular Galaxy Shapes Some irregular (Irr II) galaxies. (a) The oddly shaped galaxies NGC 4485 and NGC 4490 may be close to each other and interacting gravitationally. (b) The galaxy M82 seems to show an explosive appearance, although interpretations remain uncertain. *(AURA; Subaru)*

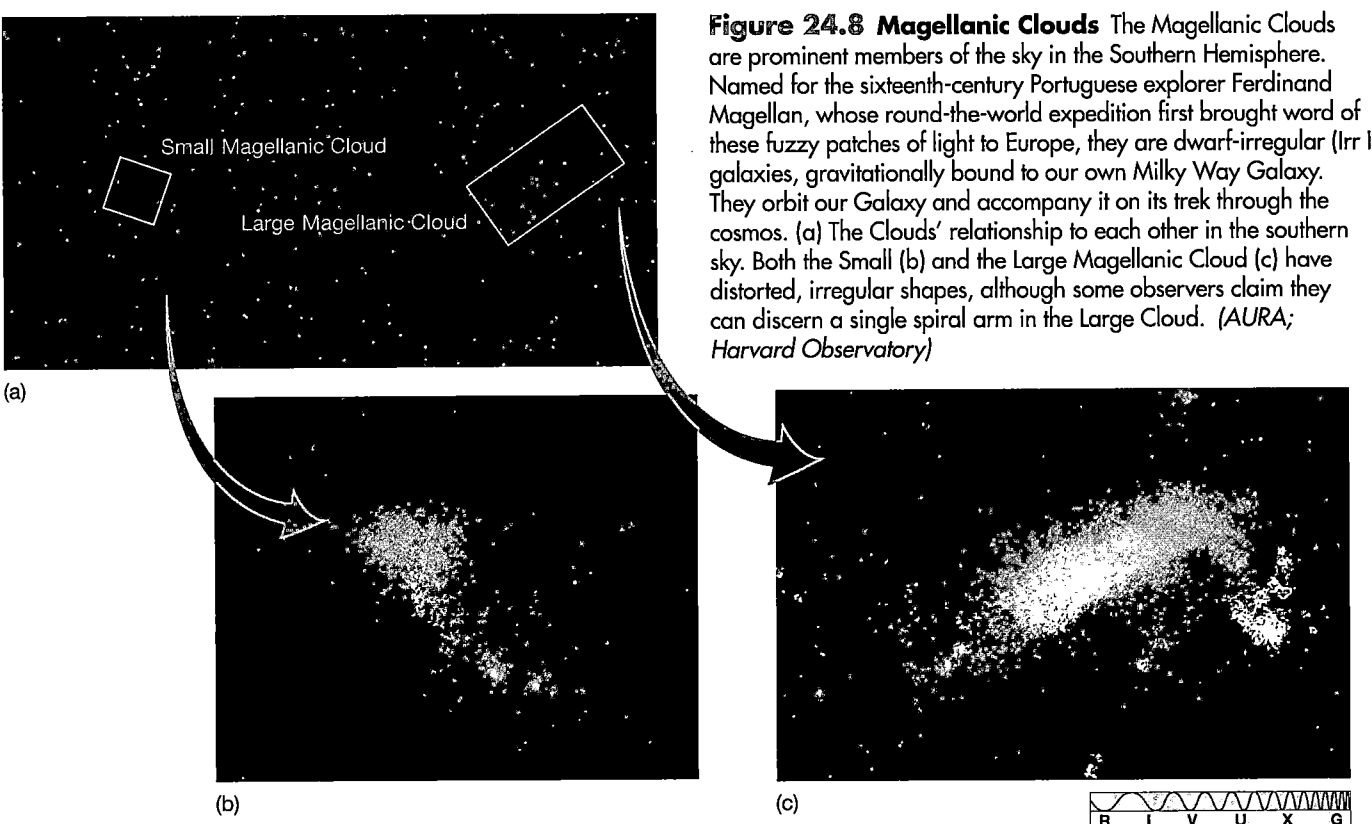

(a)

(b)

(c)

Figure 24.8 Magellanic Clouds The Magellanic Clouds are prominent members of the sky in the Southern Hemisphere. Named for the sixteenth-century Portuguese explorer Ferdinand Magellan, whose round-the-world expedition first brought word of these fuzzy patches of light to Europe, they are dwarf-irregular (Irr I) galaxies, gravitationally bound to our own Milky Way Galaxy. They orbit our Galaxy and accompany it on its trek through the cosmos. (a) The Clouds' relationship to each other in the southern sky. Both the Small (b) and the Large Magellanic Cloud (c) have distorted, irregular shapes, although some observers claim they can discern a single spiral arm in the Large Cloud. *(AURA; Harvard Observatory)*

TABLE 24.1 Basic Galaxy Properties by Type

	SPIRAL/BARRED SPIRAL (S/SB)	ELLIPTICAL (E)	IRREGULAR (Irr)
Shape and structural properties	Highly flattened disk of stars and gas, containing spiral arms and thickening to a central bulge; Sa and SBa galaxies have the largest bulges, and the least obvious spiral structure. SB galaxies have elongated central "bars" of stars and gas. We assume that spirals other than our own also have roughly spherical stellar halos	No disk (but some may contain faint stellar disks). Stars smoothly distributed through an ellipsoidal volume ranging from nearly spherical (E0) to very flattened (E7) in shape. No obvious substructure other than a dense central nucleus	No obvious structure; Irr II galaxies often have an "explosive" appearance
Stellar content	Disks contain both young and old stars; halos consist of old stars only	Contain old stars only	Contain both young and old stars
Gas and dust	Disks contain substantial amounts of gas and dust; halos contain little of either	Contain little or no cool gas and dust, but often have extensive "X-ray halos" of very hot gas	Very abundant in gas and dust
Star formation	Ongoing star formation in spiral arms	No significant recent star formation	Vigorous ongoing star formation
Stellar motion	Gas and stars in disk move in circular orbits around the Galactic center; halo stars have random orbits in three dimensions	Stars have random orbits in three dimensions	Stars and gas have very irregular orbits

AN H–R DIAGRAM FOR GALAXIES?

Table 24.1 summarizes the basic characteristics of the various galaxy types. When he first developed his classification scheme, Hubble arranged the galaxy types into the "tuning fork" diagram shown in Figure 24.9. His aim in doing this was simply to indicate similarities in appearance among galaxies, not to suggest that any connection might exist, but some astronomers have since speculated that perhaps the diagram has deeper significance. Figure 24.9 certainly suggests an evolutionary connection of some sort among galaxy types, but does one really exist? The answer, as best we can tell, is no.

Despite many attempts, no one has ever succeeded in explaining the observed properties of normal galaxies in

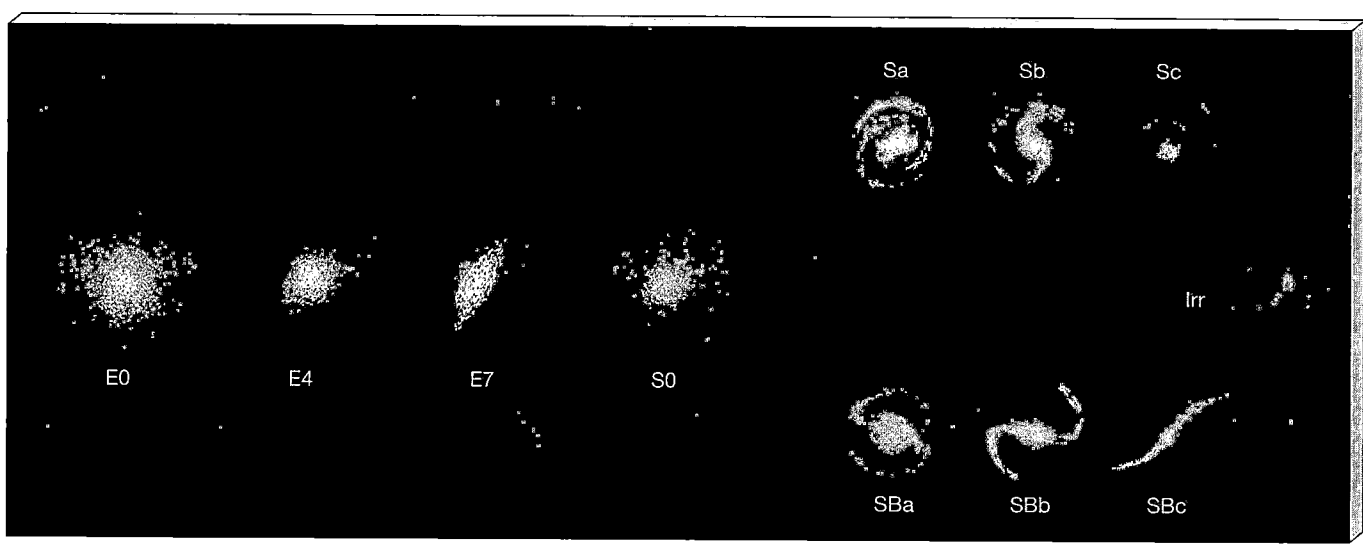

Figure 24.9 Galactic "Tuning Fork" Hubble's tuning fork diagram, showing his basic galaxy classification scheme. The placement of the four basic galaxy types—ellipticals, spirals, barred spirals, and irregulars—in the diagram is suggestive, but the tuning fork has no physical meaning.

evolutionary terms. As far as we can tell, *isolated* normal galaxies do not evolve from one type to another along the Hubble sequence. Spirals are not ellipticals that have grown arms, nor are ellipticals spirals that have somehow expelled their star-forming disks. Some astronomers do suspect that bars may be transient features and that barred-spiral galaxies may therefore evolve into ordinary spirals, but in general, astronomers know of no simple parent-child relationship among normal galaxies. However, the key word above is *isolated*. As we will see in Section 24.4, there is now strong observational evidence indicating that collisions and tidal interactions *between* galaxies may be the main physical processes driving galaxy evolution.

☑️ Concept Check

■ In what sense are large spirals like the Milky Way and Andromeda not representative of galaxies as a whole?

24.2 The Distribution of Galaxies in Space

Now that we have seen some of their basic properties, let us ask how galaxies are spread through the expanse of the universe beyond the Milky Way Galaxy. Are they scattered everywhere throughout intergalactic space all the way out to the very limits of the observable universe, or is there a boundary beyond which galaxies no longer exist? To begin answering this question, we must first know the *distances* to the galaxies.

EXTENDING THE DISTANCE SCALE

② Astronomers estimate that some 40 billion galaxies exist in the observable universe. Some reside close enough for the Cepheid variable technique to work—astronomers have detected and measured the periods of Cepheids in galaxies as far away as 25 Mpc (see Figure 24.10). ∞ (Sec. 23.2) However, most known galaxies lie much farther away. Cepheid variable stars in very distant galaxies simply cannot be observed well enough, even through the world's most sensitive telescopes, to allow us to measure their luminosities and periods. To extend our distance-measurement ladder, therefore, we must find some new object to study. What individual objects are bright enough for us to observe at great distances?

One way in which researchers have tackled this problem is through observations of **standard candles**—easily recognizable astronomical objects whose luminosities are confidently known. The basic idea is very simple. Once an object is identified as a standard candle—by its appearance or by the shape of its light curve, say—its luminosity can be estimated. Comparison of this luminosity with the object's apparent brightness then yields the object's distance and hence the distance to the galaxy in which it resides. Note that, apart from the way in which the luminosity is determined, the Cepheid variable technique relies on identical reasoning. However, the term *standard candle* tends to be applied only to very bright objects.

To be most useful, a standard candle must (1) have a narrowly defined luminosity, so that the uncertainty in estimating its brightness is small, and (2) be bright enough to be seen at large distances. Over the years astronomers have explored the use of many types of objects as standard candles—novae, emission nebulae, planetary nebulae, globular clusters, Type I (carbon-detonation) supernovae, even entire galaxies have

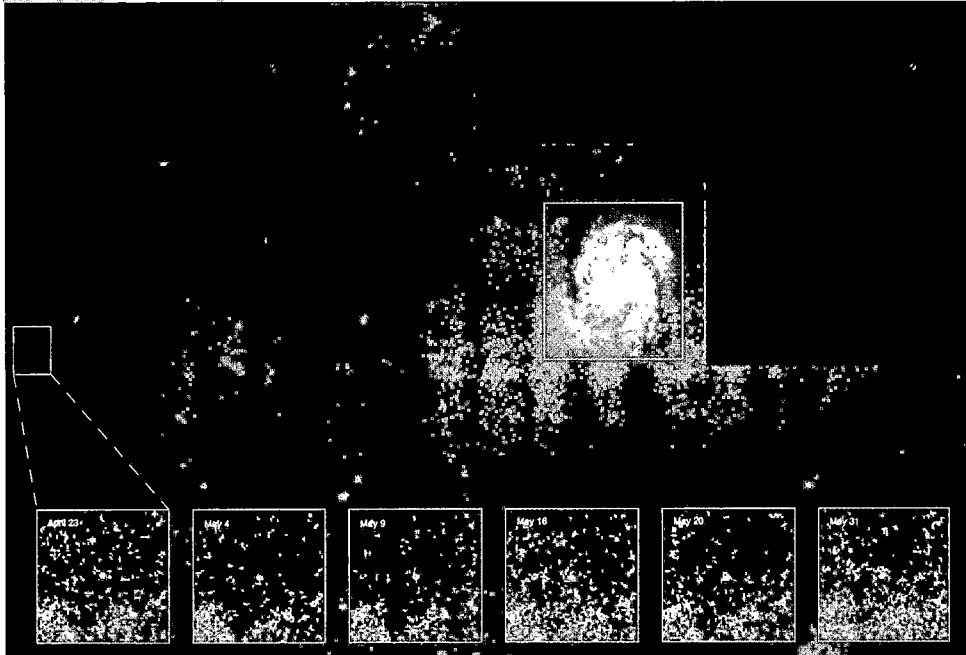

Figure 24.10 Cepheid in Virgo This sequence of six snapshots chronicles the periodic changes in a Cepheid variable star in the spiral galaxy M100, a member of the Virgo Cluster of galaxies. The Cepheid appears at the center of each inset, taken at the different times indicated in 1994. The star looks like a square because of the high magnification by the digital CCD camera—we are seeing individual pixels of the image. The 24th-magnitude star varies by about a factor of two in brightness every seven weeks. (Compare with the image of M100 in *Discovery 5-1.*) *(NASA)*

been employed. Not all have been equally useful. Some have larger intrinsic spreads in their luminosities than others, making them less reliable for measuring distances.

Planetary nebulae and Type I supernovae have proved particularly reliable as standard candles. The latter have remarkably consistent peak luminosities and are bright enough to be seen and identified out to distances of many hundreds of megaparsecs. The small luminosity spread of Type I supernovae is a direct consequence of the circumstances in which these violent events occur. As discussed in Chapter 21, an accreting white dwarf explodes when it reaches the well-defined critical mass at which carbon fusion begins. ⊂⊃ (Sec. 21.3) The magnitude of the explosion is relatively insensitive to the details of how the white dwarf formed, or how it subsequently reached critical mass, with the result that all such supernovae have quite similar properties.*

An important alternative to standard candles was discovered in the 1970s, when astronomers found a close correlation between the rotational speeds and the luminosities of spiral galaxies within a few tens of megaparsecs of the Milky Way Galaxy. Rotation speed is a measure of a spiral galaxy's total mass, so it is perhaps not surprising that this property should be related to luminosity. ⊂⊃ (Sec. 23.6) What *is* surprising, though, is how tight the correlation is. The **Tully-Fisher relation**, as it is now known (after its discoverers), allows us to obtain a remark-

ably accurate estimate of a spiral galaxy's luminosity simply by observing how fast it rotates. As usual, comparing the galaxy's (true) luminosity with its (observed) apparent brightness yields its distance.

To see how the method is used, imagine we are looking edge-on at a distant spiral galaxy and observing one particular emission line, as illustrated in Figure 24.11. Radiation from the side of the galaxy where matter is generally approaching us is blueshifted by the Doppler effect. Radiation from the other side, which is receding from us, is redshifted by a similar amount. The overall effect is that line radiation from the galaxy is "smeared out," or broadened, by the galaxy's rotation. The faster the rotation, the greater the amount of broadening (see also Figure 4.17 for the stellar equivalent). By measuring the amount of broadening, we can therefore determine the galaxy's rotation speed. Once we know that, the Tully-Fisher relation tells us the galaxy's luminosity.

The particular line normally used in these studies actually lies in the radio part of the spectrum. It is the 21-cm line of cold, neutral hydrogen in the galactic disk. ⊂⊃ (Sec. 18.4) It is used in preference to optical lines because (1) optical radiation is strongly absorbed by dust in the disk under study and (2) the 21-cm line is normally very narrow, making the broadening easier to observe. In addition, astronomers often use *infrared*, rather than optical, luminosities to avoid absorption problems caused by dust, both in our own Galaxy and in others.

The Tully-Fisher relation can be used to measure distances to spiral galaxies out to about 200 Mpc, beyond which the line broadening becomes increasingly difficult to measure accurately. A somewhat similar relation exists for elliptical galaxies, linking the broadening of a galaxy's

---

* *Recall from Chapter 21 that a Type II supernova also occurs when a growing stellar core—this time at the center of a massive star—reaches a critical mass. ⊂⊃ (Sec. 21.2) However, the outward appearance of the explosion can be significantly modified by the amount of stellar material through which the blast wave must travel before it reaches the star's surface, resulting in a greater spread in observed luminosities.*

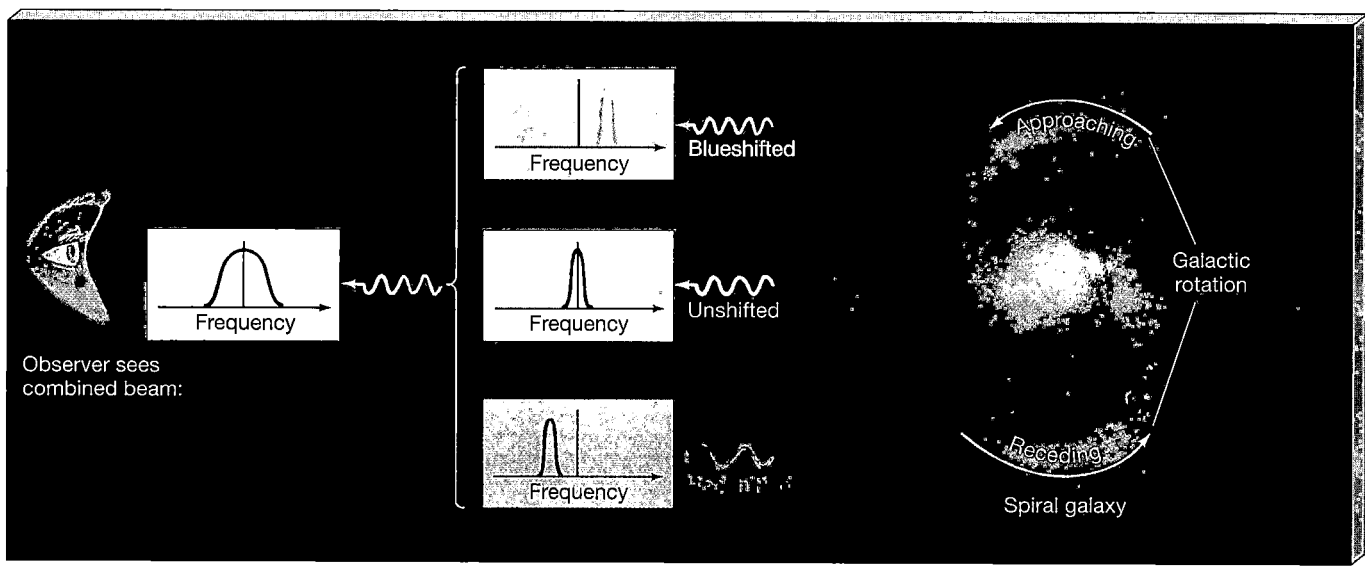

Figure 24.11 Galactic Rotation A galaxy's rotation causes some of the radiation it emits to be blueshifted and some to be redshifted (relative to what the emission would be from a nonrotating source). From a distance, when the radiation from the galaxy is combined into a single beam and analyzed spectroscopically, the redshifted and blueshifted components combine to produce a broadening of the galaxy's spectral lines. The amount of broadening is a direct measure of the rotation speed of the galaxy.

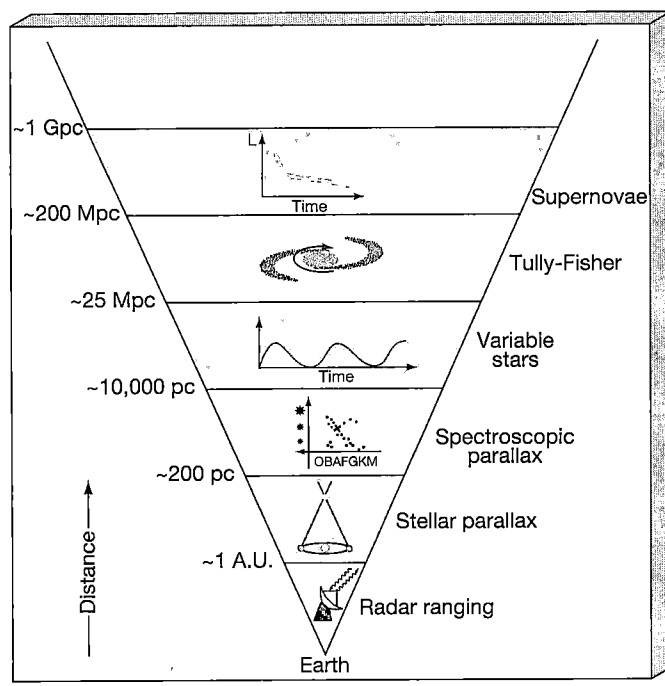

Figure 24.12 **Extragalactic Distance Ladder** An inverted pyramid summarizes the distance techniques used to study different realms of the universe. The techniques shown in the bottom four layers—radar ranging, stellar parallax, spectroscopic parallax, and variable stars—take us as far as the nearest galaxies. To go farther, we must use new techniques, the method of standard candles and the Tully-Fisher relation, each based on distances determined by the four lowest techniques.

Figure 24.12 **Extragalactic Distance Ladder** An inverted pyramid summarizes the distance techniques used to study different realms of the universe. The techniques shown in the bottom four layers—radar ranging, stellar parallax, spectroscopic parallax, and variable stars—take us as far as the nearest galaxies. To go farther, we must use new techniques, the method of standard candles and the Tully-Fisher relation, each based on distances determined by the four lowest techniques.

spectral lines (which, in the case of an elliptical, measures the average random velocity of the stars in the galaxy) and the galaxy's size. By measuring the broadening, astronomers can determine the galaxy's true size, which is then compared with the apparent size to give the distance. ∞ (Sec. 1.5) These methods bypass many of the standard

candles just described and so provide independent means of determining distances to faraway objects.

Standard candles and the Tully-Fisher relation form the fifth and sixth rungs of our cosmic distance ladder (Figure 24.12). In fact, they stand for perhaps a dozen or so related but separate techniques that astronomers have employed in their quest to map out the universe on large scales. Just as with the lower rungs, we calibrate the properties of these new techniques using distances measured by more local means. In this way, the distance-measurement process "bootstraps" itself to greater and greater distances. However, at the same time, the errors and uncertainties in each step accumulate, so the distances to the farthest objects are the least well known.

GALAXY CLUSTERS

3 Figure 24.13 sketches the locations of all the known major astronomical objects within about 1 Mpc of the Milky Way Galaxy. Our Galaxy appears with its dozen or

Figure 24.13 Local Group The Local Group is made up of some 45 galaxies within approximately 1 Mpc of our Milky Way Galaxy. Only a few are spirals; most of the rest are dwarf-elliptical or irregular galaxies, only some of which are shown here. Spirals are colored blue, ellipticals pink, and irregulars white. The inset shows the Milky Way in relation to some of its satellite galaxies.

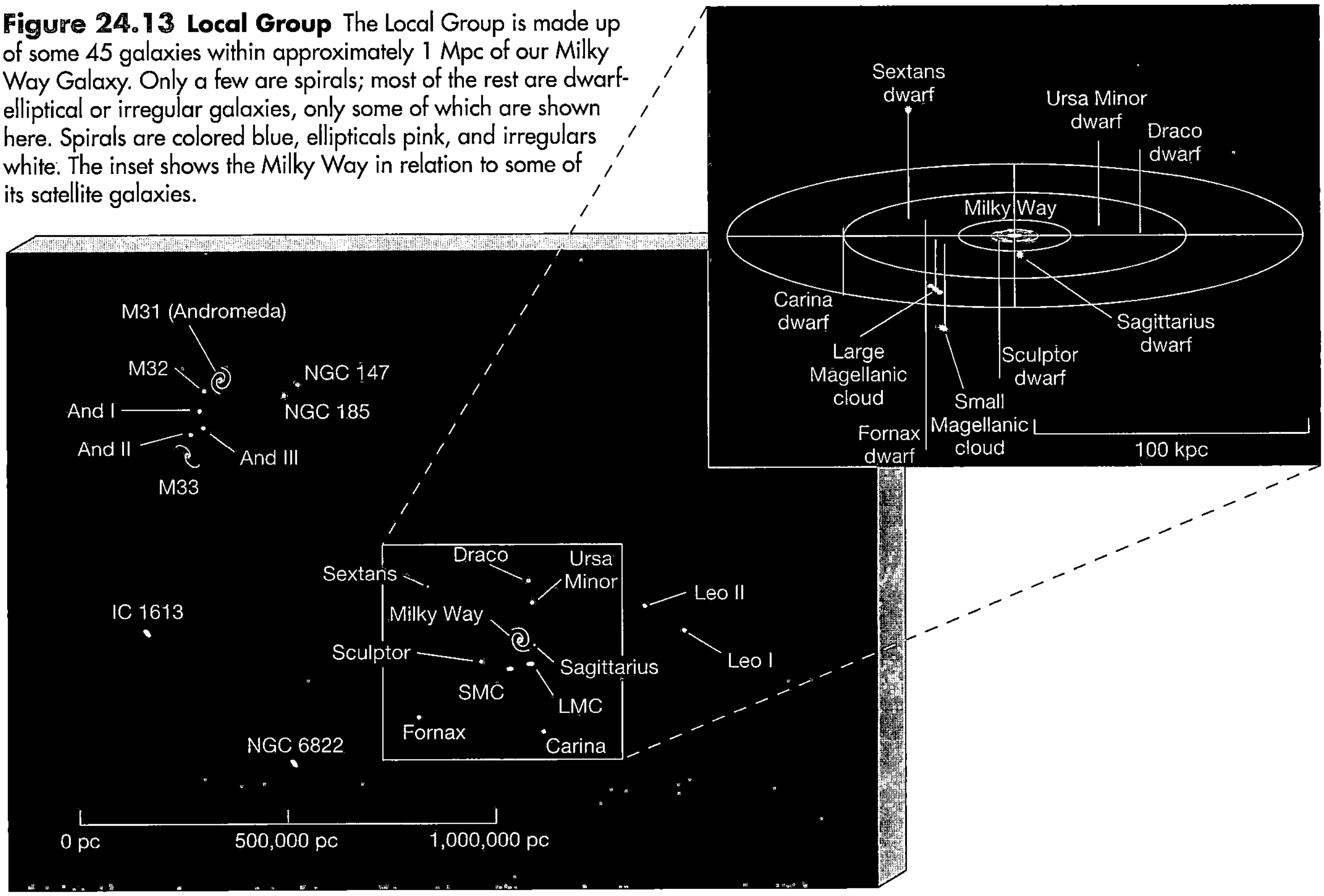

so known satellite galaxies—including the two Magellanic Clouds discussed earlier. The Andromeda Galaxy, lying 800 kpc from us, is also shown, surrounded by 12–15 satellites of its own. Two of Andromeda's galactic neighbors are shown in Figure 24.14; M33 is a spiral, and M32 is a dwarf elliptical, easily seen in Figure 23.2(a) to the bottom right of Andromeda's central bulge.

All told, some 45 galaxies populate our Galaxy's neighborhood. Three of them (the Milky Way, Andromeda, and M33) are spirals; the remainder are dwarf irregulars and dwarf ellipticals. Together, these galaxies form the **Local Group**—a new level of structure in the universe above the scale of our Galaxy. As indicated in Figure 24.13, the Local Group's diameter is roughly 2 Mpc. The Milky Way Galaxy and Andromeda are by far its largest members, and most of the smaller galaxies are gravitationally bound to one or the other of them. The combined gravity of the galaxies in the Local Group binds them together, like stars in a star cluster, but on a millionfold larger scale. More generally, a group of galaxies held together by their mutual gravitational attraction is called a **galaxy cluster**.

Moving beyond the Local Group, we come to the next large concentration of galaxies, the Virgo Cluster (Figure 24.15). It lies about 18 Mpc from the Milky Way Galaxy. Like the Local Group, the Virgo Cluster is held together by the mutual gravitational attraction of its member systems. Unlike the Local Group, however, the Virgo Cluster does not contain a mere 45 galaxies. Instead, it houses approximately 2500 galaxies, each containing 100 billion or

so individual stars. Those galaxies are bound together in a tightly knit group about 3 Mpc across. Figures 24.16(a) and (b) show two views of the locations of Virgo and several other well-defined clusters in our cosmic neighborhood. The region displayed is about 70 Mpc across. Each point in these figures represents an entire galaxy whose distance has been determined by one of the methods described in the text.

Many thousands of galaxy clusters have now been identified and cataloged, and they come in many shapes and sizes. Large, "rich" clusters, like Virgo, contain many thousands of individual galaxies distributed fairly smoothly in space. Small clusters, such as the Local Group, contain only a few galaxies and are quite irregular in shape. A small fraction of galaxies are not members of any cluster. They are apparently isolated systems, moving alone through intercluster space.

CLUSTERS OF CLUSTERS

3 Does the universe have even greater groupings of matter, or do galaxy clusters top the cosmic hierarchy? Most astronomers believe that the galaxy clusters themselves are clustered, forming titanic agglomerations of matter known as **superclusters**.

Together, the galaxies and clusters shown in Figure 24.16 form the *Local Supercluster*, also known as the Virgo Supercluster. Aside from the Virgo Cluster itself, it contains the Local Group and numerous other clusters lying

(a)

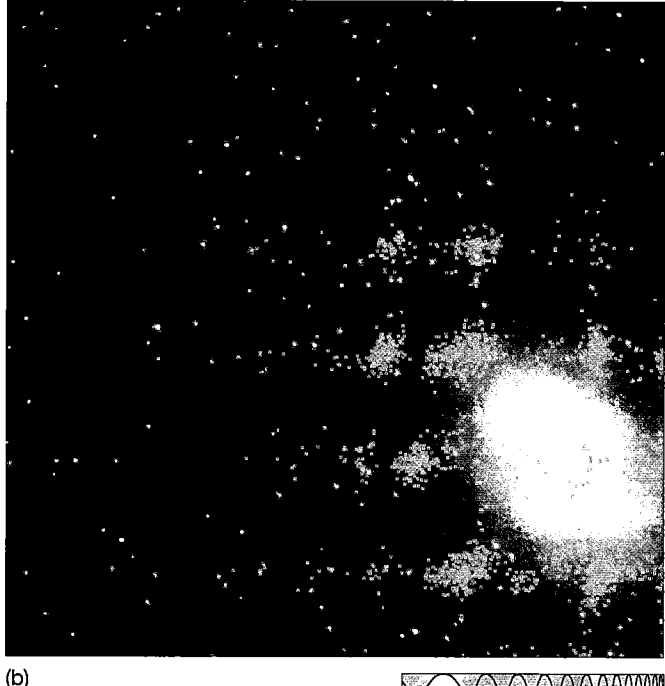

(b)

R I V U X G

Figure 24.14 Satellites of Andromeda Two well-known neighbors of the Andromeda Galaxy (M31): (a) the spiral galaxy M33 and (b) part of the dwarf elliptical galaxy M32 (also visible in Figure 23.2a, a larger-scale view of the Andromeda system). *(M. BenDaniel; NASA)*

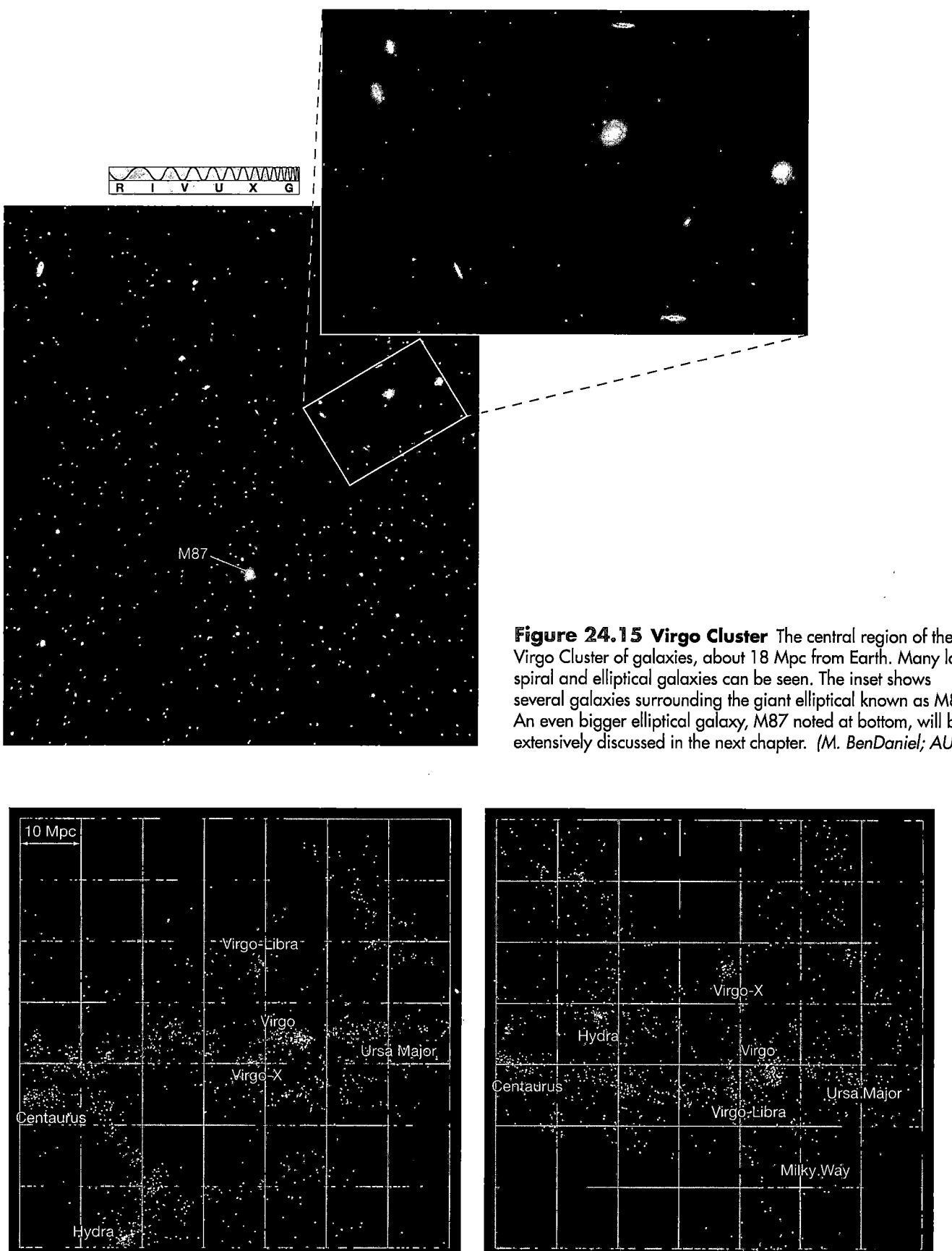

Figure 24.15 Virgo Cluster The central region of the Virgo Cluster of galaxies, about 18 Mpc from Earth. Many large spiral and elliptical galaxies can be seen. The inset shows several galaxies surrounding the giant elliptical known as M86. An even bigger elliptical galaxy, M87 noted at bottom, will be extensively discussed in the next chapter. *(M. BenDaniel; AURA)*

(a)   (b)

Figure 24.16 Local Supercluster The locations of numerous galaxies and galaxy clusters in the vicinity of Virgo. More than 4500 galaxies are represented here, and several prominent galaxy clusters are labeled. Part (a) shows the Virgo Supercluster roughly as it appears from the direction of our own Galaxy, which is located about 20 Mpc (two grid squares) above the page. Part (b) shows a perpendicular view, looking down on the scene in (a) from above and indicating the location of the Milky Way relative to Virgo. Notice the supercluster's irregular, elongated shape. *(Data courtesy B. Tully, U. Hawaii; visualization by S. Levy, NCSA)*

within about 20–30 Mpc of Virgo. Most of the galaxies shown in Figure 24.16 are fairly large spirals and ellipticals; the fainter irregulars and dwarfs are not included. Galaxies are false-colored according to the local galaxy density—white and yellow indicate the most congested regions, green less dense regions, and blue the least dense. The white, yellow and green galaxies trace out approximately the supercluster's extent. All told, the Local Supercluster is about 40–50 Mpc across and contains some 10^{15} solar masses of material. Very irregular in shape and significantly elongated perpendicular to the line joining the Milky Way to Virgo, its center lies near the Virgo Cluster. By now it should perhaps come as no surprise that the Local Group is not found at the heart of the Local Supercluster—we live far off in the periphery, about 18 Mpc from the center.

The Local Supercluster contains a huge number of individual galaxies—several tens of thousands—yet the great majority of known galaxy clusters and superclusters lie far beyond its edge. Figure 24.17 is a long-exposure photograph of one such remote cluster. This rich cluster is far outside the region shown in Figure 24.16. It is only one of many large and distant groups of galaxies scattered throughout the observable universe. The Coma Cluster, shown in Figure 24.1, is another. On and on, the picture is much the same. The farther we peer into deep space, the more galaxies, clusters of galaxies, and superclusters we see.

Is there structure on even larger scales? As we will see in a moment, the answer is still yes. However, before we make our final leap in distance to the limits of the visible universe, let's pause for a moment to take stock of some important properties of galaxies and galaxy clusters and consider a few current ideas on how galaxies form and evolve.

✅ Concept Check

■ Why is it so difficult to measure the distances to far-away galaxies?

24.3 Galaxy Masses

🔍 How can we measure the masses of systems as large as galaxies and galaxy clusters? Surely, we can neither count all their stars nor estimate their interstellar content very well. Galaxies are just too complex to take direct inventory of their material makeup. Instead, we must rely on indirect techniques. As usual, we turn to Newton's law of gravity.

MASS MEASUREMENT

Astronomers can calculate the masses of some spiral galaxies by determining their *rotation curves*, which plot rotation speed obtained by measuring the Doppler shift of various spectral lines versus distance from the galactic center. ∞ (Sec. 23.6) The mass within any given radius then follows directly from Newton's laws. Some rotation curves for nearby spirals are shown in Figure 24.18. They imply masses ranging from about 10^{11} to 5×10^{11} solar masses within about 25 kpc of the center—quite comparable to the results obtained for our own Galaxy using the same technique.

Distant galaxies are generally too far away for such detailed curves to be drawn. Nevertheless, by observing the broadening of spectral lines—as discussed earlier in the context of the Tully-Fisher relation—we can still measure the overall rotation speed. Estimating the galaxy's size then leads to an estimate of its mass. Similar techniques have been applied to ellipticals and irregulars. This approach is useful for measuring the mass lying within about 50 kpc of a galaxy's center—the extent of the electromagnetic emission from stellar and interstellar material.

From studies of the motions of stars and gas within galaxies, astronomers have found that most normal spirals (the Milky Way Galaxy included) and large ellipticals contain between 10^{11} and 10^{12} solar masses of material. Irregular galaxies often contain less mass, about 10^8 to 10^{10} times that of the Sun. Dwarf ellipticals and irregulars can contain as little as 10^6 to 10^7 solar masses of material.

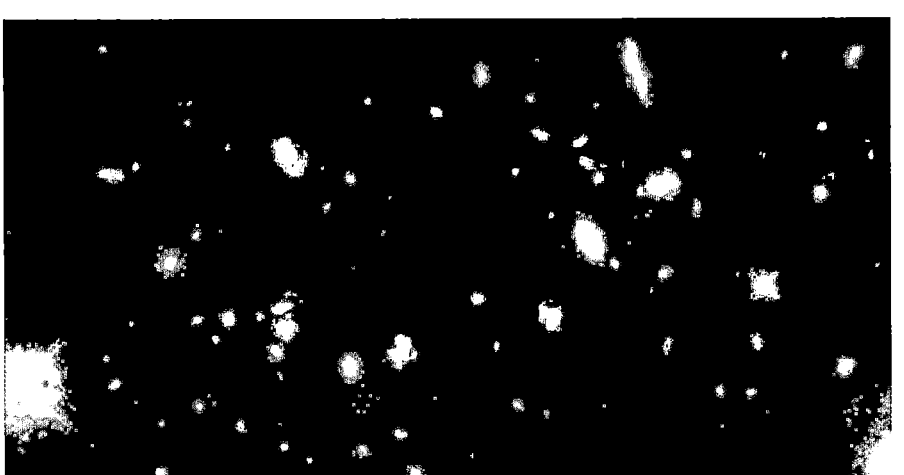

Figure 24.17 Distant Galaxy Cluster The galaxy cluster CL 1358 + 62 contains huge numbers of galaxies and resides roughly a billion parsecs from Earth. Virtually every patch of light in this photograph is a separate galaxy. Thanks to the high resolution of the *Hubble Space Telescope*, we can now discern, even at this great distance, spiral structure in some of the galaxies. We also see many galaxies in collision—some tearing matter from one another, others merging into single systems. *(NASA)*

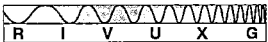

To probe farther from the center, galactic astronomers turn to binary and multiple systems of galaxies, much as stellar astronomers study binary stars to determine stellar masses. ⚭ (Sec. 17.8) Unfortunately, studies of binary galaxies are much less useful than their stellar counterparts, because neither the orbital period nor the orbital orientation can be determined—periods are typically billions of years, so the amount of orbital information that can be obtained is very limited. However, *statistical* estimates of galaxy masses based on the available data are generally consistent with the numbers just quoted.

We can use another statistical technique to derive the combined mass of all the galaxies within a galaxy cluster. As depicted in Figure 24.19, each galaxy within a cluster moves relative to all other cluster members. As with binary galaxies, we cannot watch galaxies move around within a cluster, but we can estimate the cluster's mass by asking how massive it must be in order to bind its galaxies gravitationally. Typical cluster masses obtained in this way lie in the range of 10^{13}–10^{14} solar masses. Notice that this calculation gives us *no* information whatsoever about the masses of individual galaxies. It tells us only about the *total* mass of the entire cluster.

DARK MATTER IN THE UNIVERSE

③ ⑤ Radio observations indicate that the rotation curves of many spiral galaxies, such as those shown in Figure 24.18, remain flat (that is, do not decline and may even rise slightly) far beyond the visible image of the galaxy. We conclude that these spiral galaxies—and perhaps all other spiral galaxies as well—must have invisible dark halos similar to that surrounding the Milky Way Galaxy. ⚭ (Sec. 23.6) Depending on how far out these halos extend (and

observations appear to indicate that they reach at least as far as 50 kpc from the center of a Milky Way–sized galaxy), spiral galaxies may contain 3 to 10 times more mass than can be accounted for in the form of visible matter. Some studies of elliptical galaxies suggest similarly large dark halos surrounding them too, but these observations are still far from conclusive.

Astronomers find an even greater discrepancy when they study galaxy clusters. Calculated cluster masses range from 10 to nearly 100 times the mass suggested by the light emitted by individual cluster galaxies. Stated another way, a lot more mass than we can see is needed to bind galaxy clusters. The problem of dark matter exists, then, not just in our own Galaxy but also in other galaxies and, to an even greater degree, in galaxy clusters as well. It most likely applies to the entire universe. In that case, we must accept the fact that *upward of 90 percent of the universe is composed of dark matter*. As noted in Chapter 23, this matter is not just dark in the visible portion of the spectrum—it is undetected at *any* electromagnetic wavelength.

As discussed in Chapter 23, many possible explanations for the dark matter have been suggested, ranging from stellar remnants of various sorts to exotic subatomic particles. ⚭ (Sec. 23.6) Whatever its nature, however, the dark matter in clusters apparently cannot be simply the accumulation of smaller amounts of dark matter within individual galaxies. Even including the galaxies' dark halos, we still cannot account for all the dark matter in galaxy clusters. As we look on larger and larger scales, we find that a larger and larger fraction of the matter in the universe is dark.

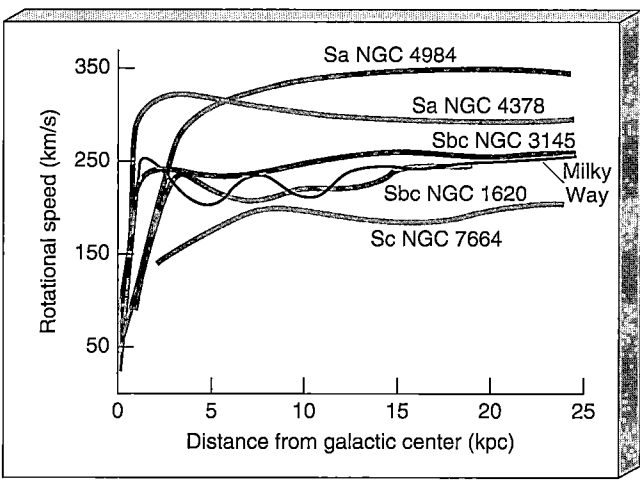

Figure 24.18 Galactic Rotation Curves Rotation curves for some nearby spiral galaxies indicate masses of a few hundred billion times the mass of the Sun. The corresponding curve for our own Galaxy (Figure 23.19) is marked in red for comparison.

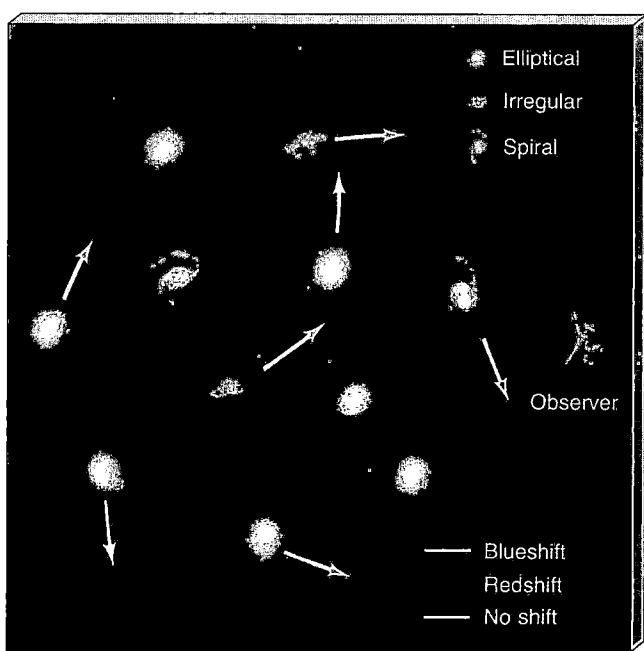

Figure 24.19 Galaxy Cluster Mass The mass of a galaxy cluster may be estimated by observing the motion of many galaxies in the cluster and then calculating how much mass is needed to prevent the cluster from flying apart.

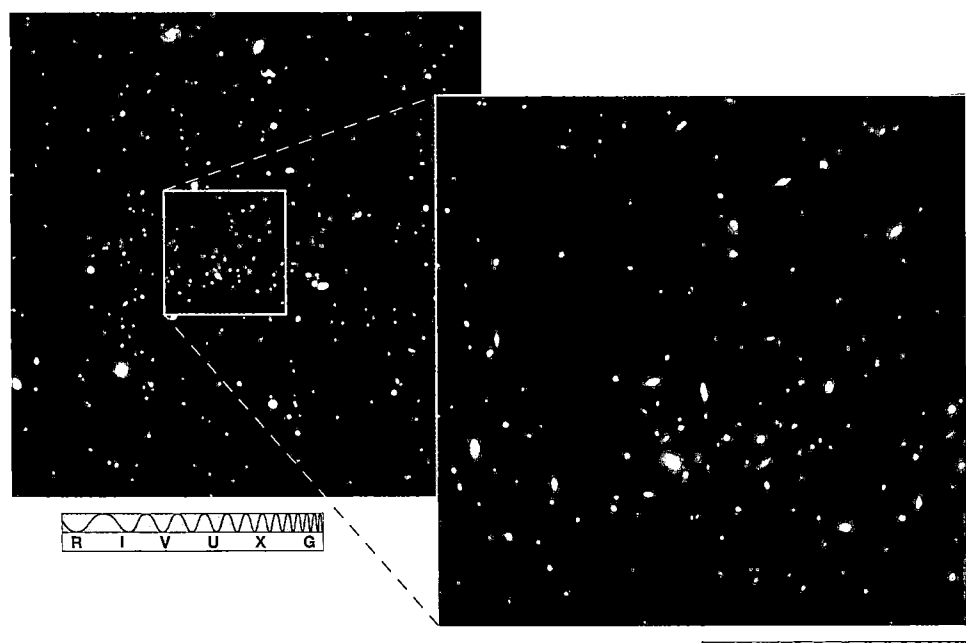

Figure 24.20 Galaxy Cluster X-Ray Emission The image at left is a superposition of infrared and X-ray radiation from a distant galaxy cluster. The X rays are shown as a fuzzy, bluish cloud of hot gas filling the intracluster spaces among the galaxies. At right is a longer infrared exposure of the central region, showing the richness of this cluster, which spans about a million parsecs. *(NASA; ESA)*

INTRACLUSTER GAS

Could the additional dark matter be diffuse intergalactic matter existing among the galaxies within the clusters—that is, intracluster gas? In the late 1970s, satellites orbiting above Earth's atmosphere detected substantial amounts of X-ray radiation in the direction of many galaxy clusters. Figure 24.20 shows a false-color X-ray image of one such cluster. The X-ray-emitting region is centered on, and comparable in size to, the visible image of the cluster. These X-ray observations demonstrated for the first time the existence of large amounts of invisible hot gas—more than 10 million K—within galaxy clusters.

How much matter have the X-ray satellites found? The observations suggest that at least as much matter—and in a few cases substantially more—exists within clusters in the form of hot gas as is visible in the form of stars. This is a lot of material, but it still doesn't solve the dark-matter problem. To account for the total masses of galaxy clusters implied by dynamical studies, we would have to find from 10 to 100 times more mass in gas than in stars. In the mid-1990s, *ROSAT* observations of gas within the Virgo Cluster actually compounded the problem (Figure 24.21). The gas was so hot that the amount of dark matter needed to bind it to the cluster and prevent it from dispersing into intercluster space was far greater than had previously been suspected to exist!

Figure 24.21 Virgo in X Rays This combined optical and X-ray *(ROSAT)* image shows hot (10^7 K) X-ray-emitting gas (falsely colored blue, increasing to yellow and red) within the rich Virgo Cluster of galaxies (see also Figure 24.15). The amount of mass required to bind this gas (especially that surrounding M87, shown in blue) to the cluster greatly exceeds the total mass of the galaxies themselves. *(ESA)*

☑ **Concept Check**

■ Why do astronomers think that most of the matter in the universe is dark?

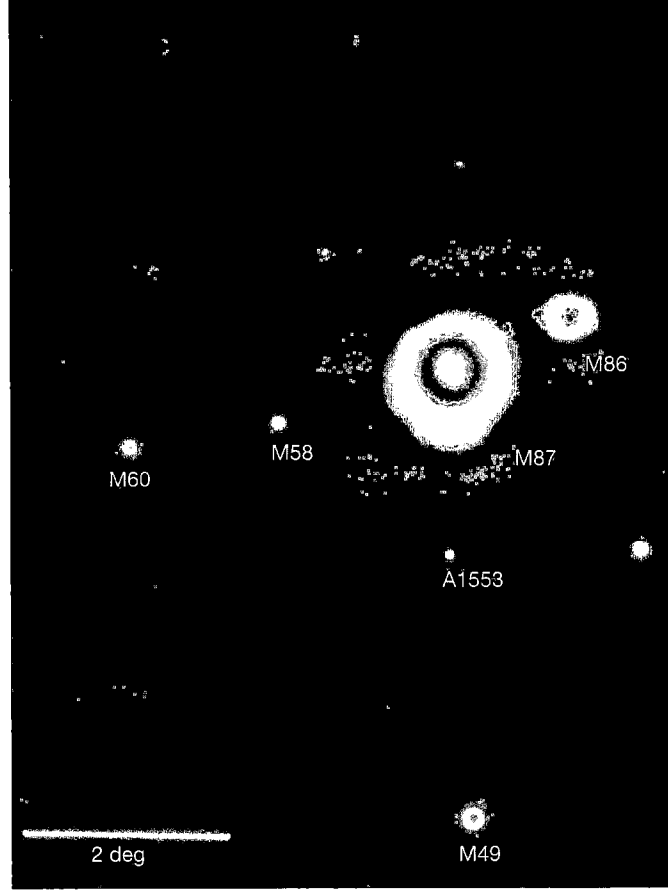

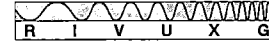

24.4 Galaxy Formation and Evolution

How did the different galaxy types come into being? We know of no simple evolutionary connections among the various categories in the Hubble classification scheme. To address this question, then, we must look more carefully at how galaxies formed.

Unfortunately, compared with the theories of star formation and stellar evolution, the theory of galaxy formation is still in its infancy. Galaxies are much more complex than stars, they are harder to observe, and the observations are harder to interpret. In addition, we have only a partial understanding, and no direct observations, of the conditions in the universe immediately preceding galaxy formation, quite unlike the corresponding situation for stars. ⊂⊃ (Sec. 18.5) Finally, and most important, stars almost never collide with one another, with the result that single stars and binaries evolve in isolation. Galaxies, on the other hand, may suffer many collisions and mergers during their lives (see *Discovery 24-1*), making it much harder to decipher their pasts. Nevertheless, some general ideas have begun to gain widespread acceptance, and we can offer some insights into the processes responsible for the galaxies we see.

MERGERS AND ACQUISITIONS

The seeds of galaxy formation were sown in the very early universe when small density fluctuations in the primordial matter began to grow. For now, let's begin our discussion with these "pregalactic" blobs of gas already formed. The masses of these fragments were quite small—perhaps only a few million solar masses, comparable to the masses of the smallest present-day dwarf galaxies, which may in fact be remnants of this early time. So where did the larger galaxies we find today come from? The key point in our current understanding is the realization that galaxies grow by repeated *merging* of smaller objects, as illustrated in Figure 24.22. Contrast this process with that of star formation, in which a large cloud fragments into smaller pieces that eventually become stars. ⊂⊃ (Sec. 19.2)

Theoretical evidence for this scenario is provided by computer simulations of the early universe, which clearly show merging taking place. Further strong support comes from recent observations that indicate that galaxies at large

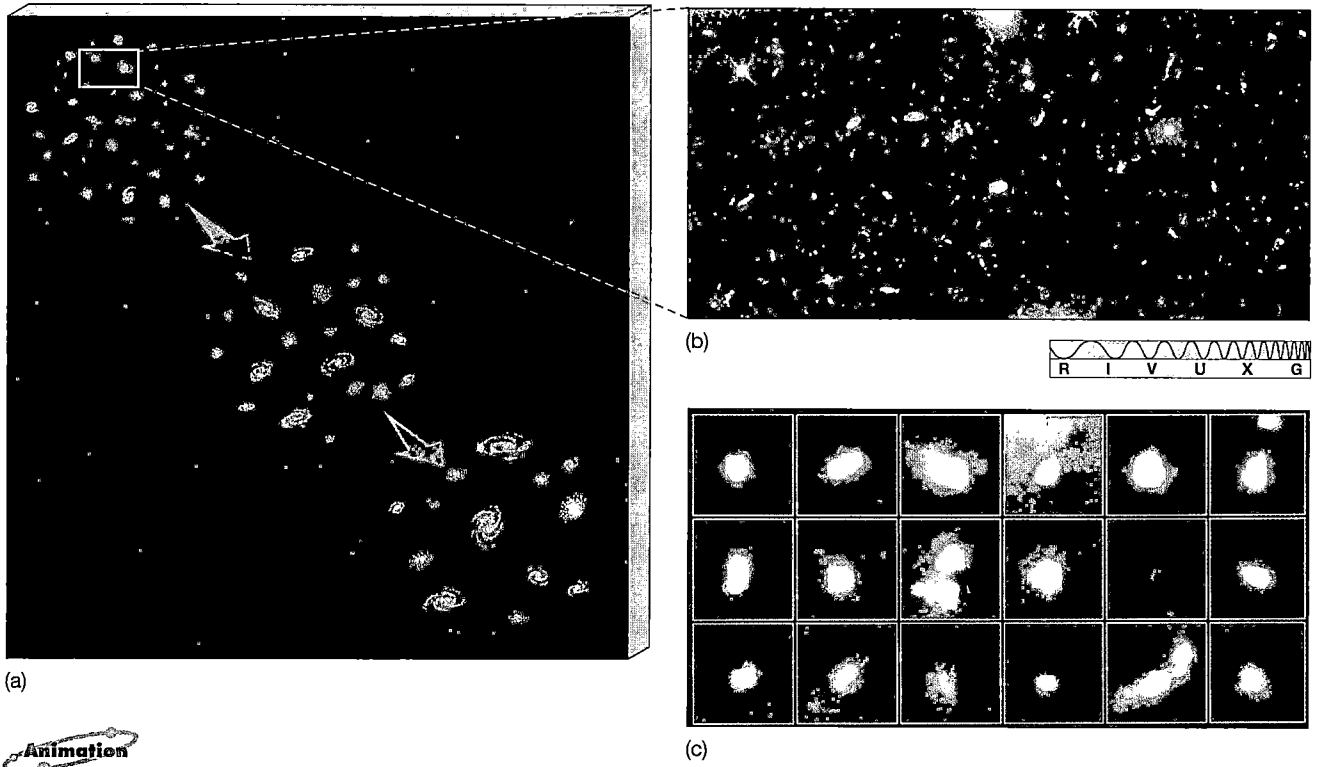

(a)

(b)

(c)

Figure 24.22 Galaxy Formation (a) The present view of galaxy formation holds that large systems were built up from smaller ones through collisions and mergers, as shown schematically in this drawing. (b) This photograph, one of the deepest ever taken of the universe, provides "fossil evidence" for hundreds of galaxy shards and fragments, up to 3000 Mpc distant. (c) Enlargements of selected portions of (b) reveal rich (billion-star) pregalactic fragments all lying within a relatively small volume of space (some 600 kpc across). Their proximity to one another suggests that we may be seeing a group of pregalactic fragments about to merge to form a galaxy. The events pictured took place about 10 billion years ago. *(NASA)*

DISCOVERY 24-1

Colliding Galaxies

Contemplating the congested confines of a rich galaxy cluster (such as Virgo or Coma, with thousands of member galaxies orbiting within a region a few megaparsecs across), we might wonder if collisions among galaxies are common. Gas particles collide in our atmosphere, and hockey players collide in the rink—do galaxies in clusters collide too? The answer is yes.

The first image at right probably shows the aftermath of a bull's-eye collision between a small galaxy (perhaps one of the two at right, although that is by no means certain) and the larger galaxy at left. The result is the Cartwheel Galaxy, about 150 Mpc from Earth, its halo of young stars resembling a vast ripple in a pond. The ripple is most likely a density wave created by the passage of the smaller galaxy through the disk of the larger one (see Section 23.5). The resulting disturbance is now spreading outward from the impact region, creating new stars as it goes.

The image below captures two spiral galaxies apparently passing each other like majestic ships in the night. It is unclear if they are only experiencing a close encounter, or are about to have a head-on collision. The larger, more massive galaxy on the left is NGC 2207; the smaller one on the right is IC 2163. The strong tidal forces of the former are severely distorting parts of the latter, causing stars and gas to be ejected along streamers that stretch nearly 30,000 pc toward the right-hand side of the frame. Analysis of this image suggests that IC 2163 is now swinging past NGC

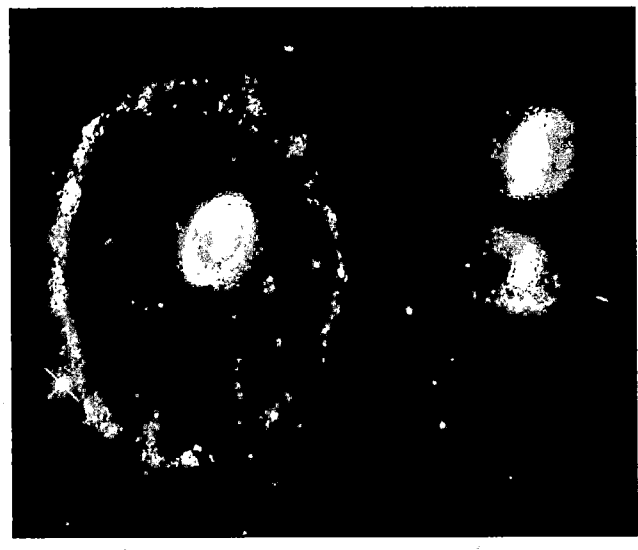

(NASA)

R I V U X G

2207 in a counterclockwise direction, having made a close approach some 40 million years ago. The two galaxies seem destined to undergo further close encounters, as IC 2163 apparently does not have enough energy to escape the gravitational pull of NGC 2207. Each time the two ex-

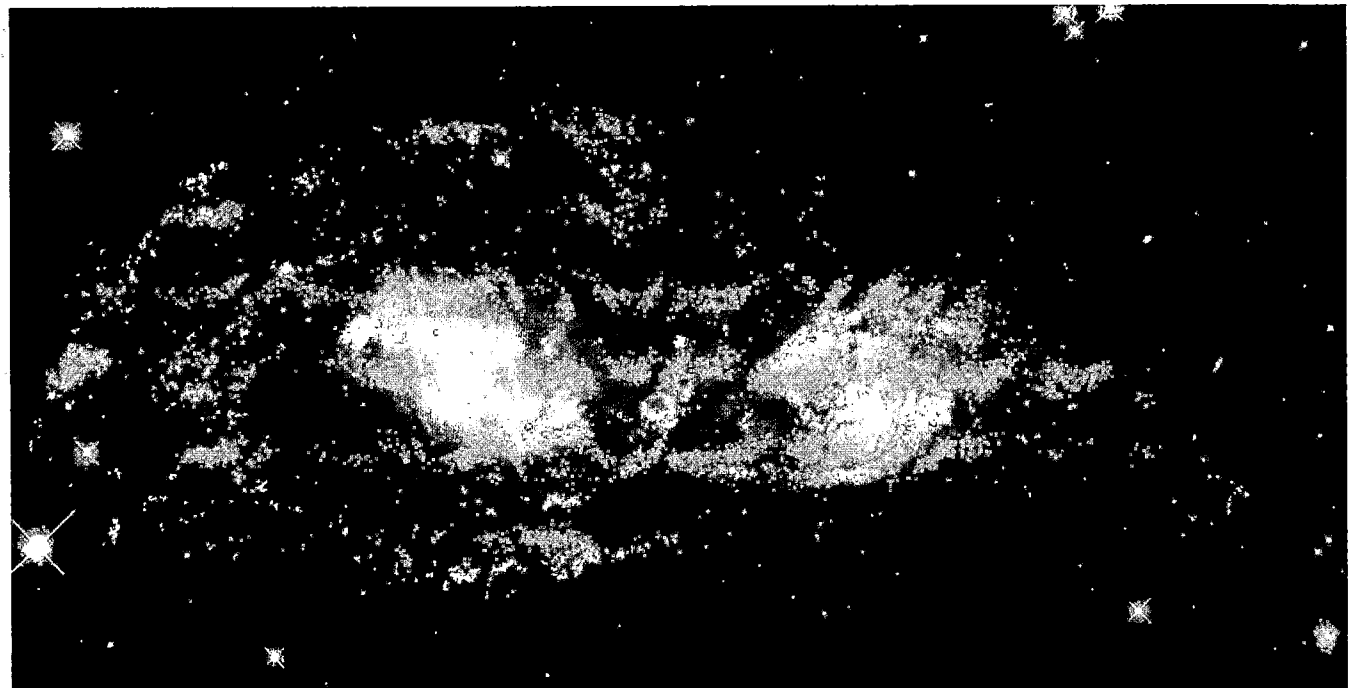

(NASA) *Animation*

R I V U X G

perience a very close brush, bursts of star formation will erupt in both galaxies as the interstellar clouds of gas and dust contained within each are pushed, shoved, and shocked. In roughly a billion years, these two galaxies will probably merge into a single, massive galaxy.

No human will ever witness an entire galaxy collision, for it would last many millions of years. However, computer simulations display formations remarkably similar to observations. The particular simulation shown in the left-hand box of the final pair of images (this page) began with two spiral galaxies, but the details of the original structure have been largely obliterated by the collision. Notice the similarity to the real image of NGC 4038/4039 (the black-and-white image at right), the so-called Antennae galaxies, which show extended tails as well as twin galactic centers only a few hundred parsecs across. Star formation induced by the collision is clearly traced by the blue light from thousands of young, hot stars, evident in the high-resolution *Hubble Space Telescope* image at right. (The *Hubble* field of view is outlined on the black-and-white image.) The simulations suggest that ultimately the two galaxies will merge into one.

Galaxies in clusters apparently collide quite often. In the smaller groups, the galaxies' speeds are low enough that interacting galaxies tend to "stick together," and mergers are a common outcome. In larger groups, galaxies move faster and may pass through one another without sticking. In either case, collisions can have large effects on the galaxies involved. Both "minor" mergers (consumption of a smaller companion by a larger galaxy) and "major" mergers (between galaxies of comparable size) play important roles in governing galaxy evolution. The stellar and interstellar contents of each galaxy are rearranged, and the

merged interstellar matter very likely experiences shock waves that trigger widespread bursts of star formation (see also Figures 24.26 and 24.27).

Curiously, although a collision may wreak havoc on the large-scale structure of the galaxies involved, it has no effect on the individual stars they contain. The stars within each galaxy just glide past one another. Although we have plenty of photographic evidence for galaxy collisions, no one has ever witnessed or photographed a collision between two stars. Stars do sometimes collide in other circumstances—in the dense central cores of galactic nuclei and globular clusters, or as a result of stellar evolution in binary systems—but stellar collisions are a very rare consequence of galaxy interactions.

To understand why individual stars do not collide when galaxies collide, recall that the galaxies within a typical cluster are bunched together fairly tightly. The distance between adjacent galaxies in a given cluster averages a few hundred thousand parsecs, which is only about 10 times greater than the size of a typical galaxy. Galaxies simply do not have that much room to roam around without bumping into one another. By contrast, stars within a galaxy are spread out much more thinly. The average distance between stars within a galaxy is a few parsecs, which is millions of times greater than the size of a typical star. When two galaxies collide, the star population merely doubles for a time, and the stars continue to have so much space that they do not run into one another. The stellar and interstellar contents of each galaxy are certainly rearranged, and the resultant burst of star formation may indeed be spectacular from afar, but for the stars, it's clear sailing.

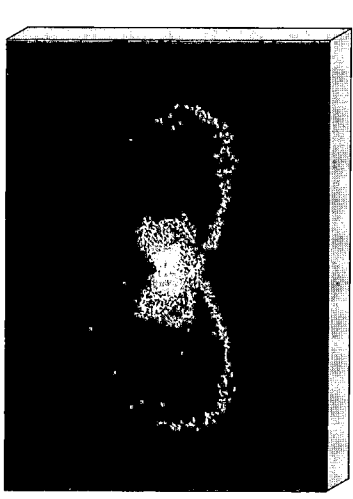

(J. Barnes)

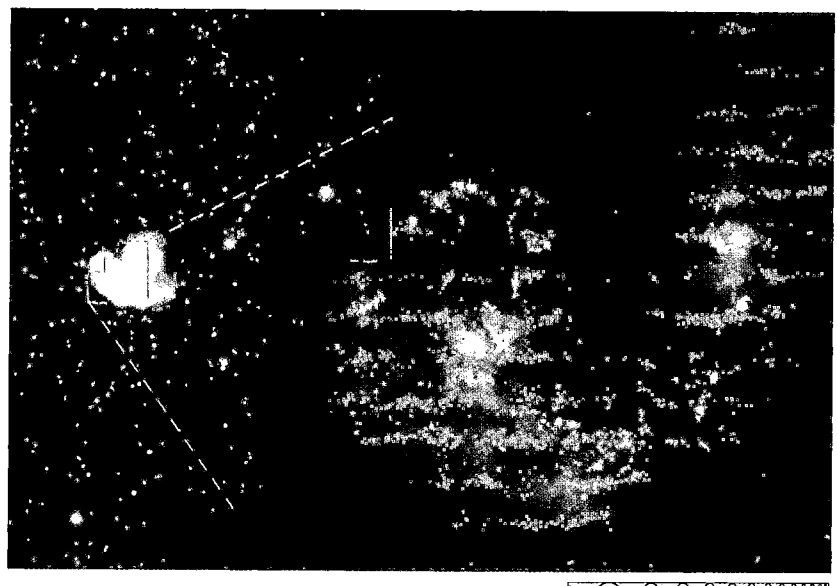

(AURA; NASA)

R I V U X G

distances from us (meaning that the light we see was emitted long ago) appear distinctly smaller and less regular than those found nearby. Figures 24.22(b) and 24.23 show some of these images, which include objects lying up to 5 *billion* parsecs from Earth. The vague bluish patches are separate small galaxies, each containing only a few percent of the mass of the Milky Way Galaxy. Their irregular shape is thought to be the result of galaxy mergers; the bluish coloration comes from young stars that formed during the merger process.

Figure 24.22 (c) shows more detailed views of some of the objects in Figure 24.22(b), all lying in the same region of space, some 600 kpc across and about 3000 Mpc from Earth (see Section 24.5 for more on how such large distances are measured). Each frame spans about one-tenth the size of the Milky Way Galaxy. The 18 objects shown are the small bluish blobs toward the upper left quadrant of Figure 24.22(b). Each blob seems to contain several billion stars spread throughout a distorted spheroid about a kiloparsec across—just about the size and scale expected for a pregalactic fragment. Their decidedly bluish tint suggests that active star formation is already under way. We see them as they were nearly 10 billion years ago, apparently poised to merge into a galaxy-sized object. If this interpretation of the data is correct, then these images lend strong support to the "bottom up" (merger) view of galaxy formation.

When did galaxy formation stop? Astronomers are divided on this issue, which may be more a matter of terminology than of actual astrophysics. Some maintain that there was a fairly well defined time in the past—given by the age of the globular clusters in our own Galaxy, for example—by which most galaxy formation was over. Others point out that many galaxies show evidence of mergers and the accumulation of smaller satellite galaxies over an extended period of time—even up to the present day. These astronomers suggest that the many galaxy interactions we observe today are just part of the same process begun when the first fragments merged. In this view, galaxy formation is still occurring today. In either case, we have ample evidence that galaxies evolve in response to external factors, long after the first pregalactic fragments formed and started to merge.

GALAXY INTERACTIONS

As discussed in *Discovery 24-1*, collisions and interactions between galaxies are relatively commonplace events, basi-

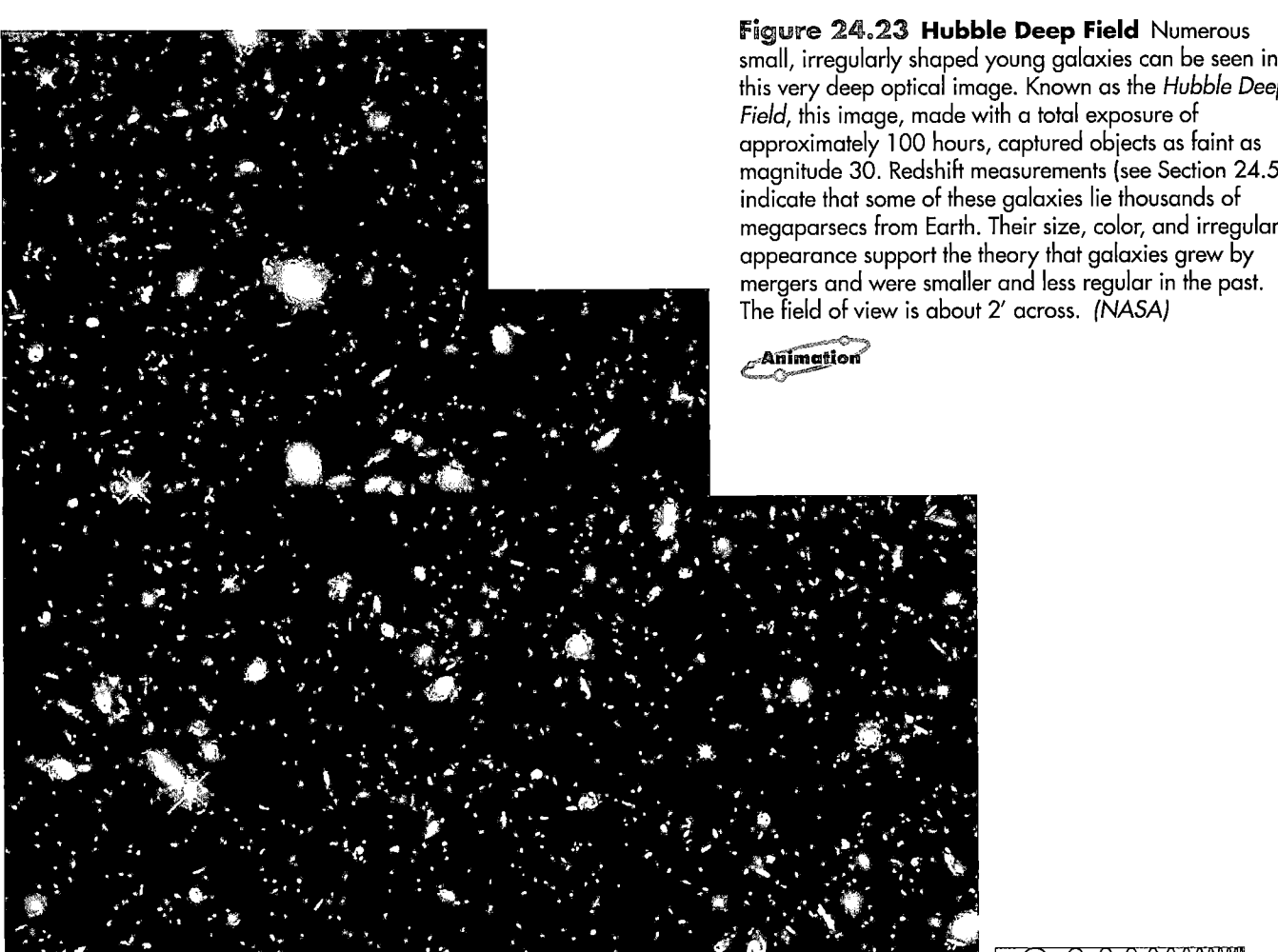

Figure 24.23 Hubble Deep Field Numerous small, irregularly shaped young galaxies can be seen in this very deep optical image. Known as the *Hubble Deep Field*, this image, made with a total exposure of approximately 100 hours, captured objects as faint as magnitude 30. Redshift measurements (see Section 24.5) indicate that some of these galaxies lie thousands of megaparsecs from Earth. Their size, color, and irregular appearance support the theory that galaxies grew by mergers and were smaller and less regular in the past. The field of view is about 2′ across. *(NASA)*

Animation

cally because the distances between galaxies in clusters are not much greater than the sizes of the galaxies themselves. (This is quite unlike the corresponding situation with stars in galaxies—stars are millions of times smaller than typical interstellar distances.) Note that the "sizes" of galaxies in this context include the extensive dark-matter halos known to surround many spirals, including our own, and suspected to surround all galaxies. Computer simulations performed over the past decade have shown that these dark halos are crucial to galaxy interactions.

Consider first two galaxies orbiting each other—a binary galaxy system. As they orbit, the galaxies interact with each other's dark halos, one galaxy stripping halo material from the other by tidal forces. The freed matter is either redistributed within a common envelope or is entirely lost from the binary system. This interaction between the halos changes the orbits of the galaxies, which tend to spiral toward each other, eventually merging. If one galaxy of the pair happens to have a much lower mass than the other, the process is colloquially termed *galactic cannibalism*. Such cannibalism may explain why supermassive galaxies are often found at the cores of rich galaxy clusters. Having dined on their companions, they now lie at the center of the cluster, waiting for more food to arrive. Figure 24.24 is a remarkable combination of images that has apparently captured this process at work.

Now consider two interacting disk galaxies, one a little smaller than the other but each having a mass comparable to that of the Milky Way Galaxy. As shown in the computer-generated frames of Figure 24.25, the smaller galaxy can substantially distort the larger one, causing spiral arms to appear where none existed before. The entire event requires several hundred million years—a span of evolution that supercomputers can model in minutes.

The final frame of Figure 24.25 looks remarkably similar to the double galaxy shown in Figure 24.2(b). Shown there are two galaxies with sizes, shapes, and velocities corresponding very closely to those in the computer simulation. The magnificent spiral galaxy is M51, popularly known as the Whirlpool Galaxy, about 10 Mpc from Earth. Its smaller companion is an irregular galaxy that may have

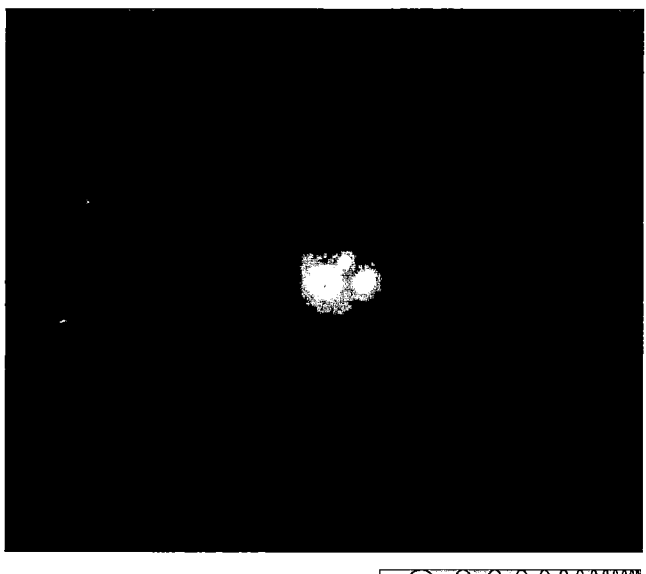

Figure 24.24 Galactic Cannibalism This computer-enhanced, false-color composite optical photograph of the galaxy cluster Abell 2199 is thought to show an example of galactic cannibalism. The large central galaxy of the cluster (itself 120 kpc along its long axis) is displayed with a superimposed "window." (This results from a shorter time exposure, which shows only the brightest objects that fall within the frame.) Within the core of the large galaxy are several smaller galaxies (the three bright yellow images at center) apparently already "eaten" and now being "digested" (that is, being torn apart and becoming part of the larger system). Other small galaxies swarm on the outskirts of the swelling galaxy, almost certainly to be eaten too. *(SAO)*

drifted past M51 millions of years ago. Did this smaller galaxy cause the spiral structure we see in M51? Does the model mirror reality? Perhaps. We need more evidence from other galaxies to confirm the accuracy of these and similar simulations. Still, the computer simulation does demonstrate a plausible way in which two galaxies may have interacted millions of years ago and how spiral arms may be created or enhanced as a result.

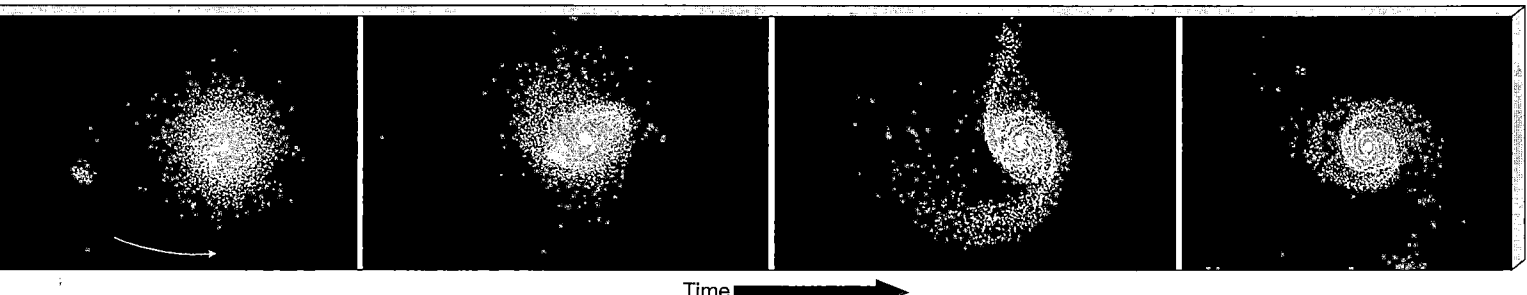

Time ⬛⬛⬛⬛➡

Figure 24.25 Galaxy Interaction Galaxies can change their shapes long after their formation. In this computer-generated sequence, two galaxies closely interact over several hundred million years. The smaller galaxy (red) has gravitationally distorted the larger galaxy (blue), changing it into a spiral galaxy. Compare the result of this supercomputer simulation with Figure 24.2(b), a photograph of M51 and its small companion.

It now seems that many of the most spectacular changes that occur in galaxies result from interactions with other galaxies. Astronomers have cataloged numerous **starburst galaxies**, such as the ones shown in Figure 24.26, where violent events, perhaps a near-collision with a neighbor or an actual merger, appear to have rearranged the galaxy's internal structure and triggered a sudden, intense burst of star formation in the recent past. Several more examples of this phenomenon appear in *Discovery 24-1*.

MAKING THE HUBBLE SEQUENCE

If galaxies form by repeated mergers, can we account for the Hubble sequence, and specifically differences between spirals and ellipticals? The details are still far from certain but, remarkably, the answer now seems to be a qualified yes. Collisions and close encounters between galaxies are random events, and they do not represent a "genuine" evolutionary sequence linking all spirals to all ellipticals and irregulars. However, it is clear that many galaxies and galaxy clusters have evolved greatly since they formed long ago, and computer simulations suggest a plausible way in which the observed Hubble types might have arisen, starting from a universe populated only by irregular, gas-rich galaxy fragments.

The simulations reveal that a "major" merger—a collision between galaxies of comparable size—can destroy a spiral galaxy's disk, processing much of the gas into stars in a galaxy-wide starburst episode. The violence of the merger and the effects of subsequent supernovae eject most of the remaining gas into intergalactic space, creating the hot intracluster gas noted in Section 24.3. Once the burst of star formation has subsided, the resulting object looks very much like an elliptical galaxy. The elliptical's hot X-ray halo is the last vestige of the original spiral's disk. Figure 24.26(c) shows an *HST* image of an Irr II galaxy that may

well be an example of this phenomenon at work. The blue blobs are thought to be young star clusters formed during the starburst, and the explosive appearance suggests that we are witnessing the gas and dust being ejected.

Simulations also indicate that "minor" mergers, in which a small galaxy interacts with and ultimately is absorbed by a larger one, generally leave the larger galaxy intact, with more or less the same Hubble type as it had before the merger. This is the most likely way for large spirals to grow—in particular, our own Galaxy probably formed in this way. Astronomers have identified "streams" of halo stars, all with similar orbits and composition, that may well be the stellar remnants of such mergers in the past. The small Sagittarius dwarf galaxy (Figure 24.13) and the Magellanic clouds (Figure 24.8) will probably be the Milky Way's next victims.

Supporting evidence for this general picture comes from observations that spiral galaxies are relatively rare in regions of high galaxy density, such as the central regions of rich galaxy clusters. This is consistent with the view that their fragile disks are easily destroyed, both by collisions and mergers, which are more common in dense galactic environments. In addition, interactions with the intracluster medium tend to "sweep" the gas out of a galaxy as it moves through the cluster. Spirals also seem to be more common at larger distances (that is, in the past), implying that their numbers are decreasing with time, again presumably as the result of collisions. However, nothing in this area of astronomy is clear-cut. We know of numerous isolated elliptical galaxies in low-density regions of the universe, which are hard to explain as the result of mergers.

In principle, the starbursts associated with these mergers leave their imprint on the star-formation history of the universe, and studies of star formation in distant galaxies have become a very important way of testing and quantifying the details of the merger hypothesis.

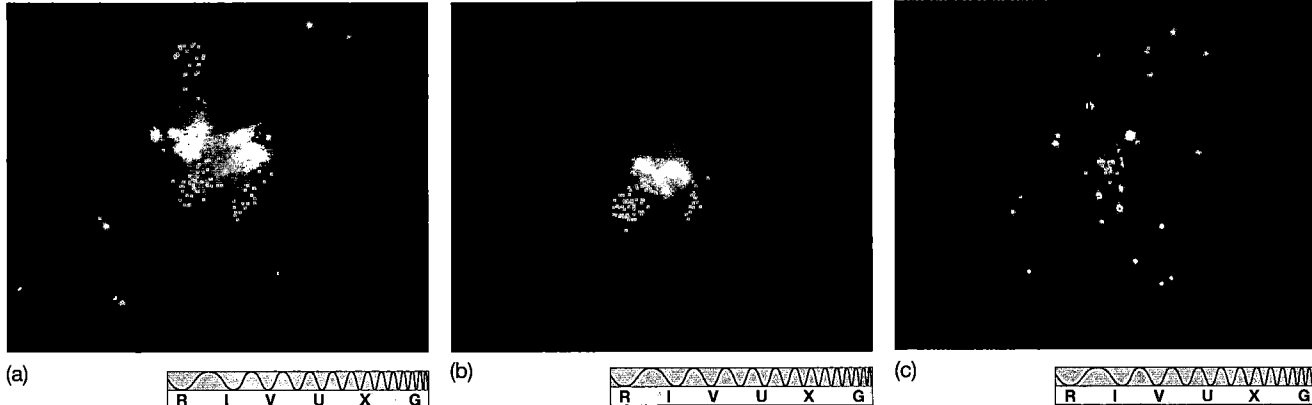

Figure 24.26 Starburst Galaxies (a) This interacting galaxy pair (IC 694 at left, and NGC 3690) shows starbursts now under way in both galaxies—hence the bluish tint. Such intense, short-lived bursts probably last for no more than a few tens of millions of years—a small fraction of a typical galaxy's lifetime. (b) This infrared image of a starburst galaxy (Arp 220) shows a 300-light-year disk, which may be fueling a supermassive black hole within it, and rapid star formation outside it. (c) The peculiar (Irr II) galaxy NGC 1275 contains a system of long filaments that seem to be exploding outward into space. Its blue blobs, as revealed by the *Hubble Space Telescope*, are probably young globular clusters formed by the collision of two galaxies. *(W. Keel; NASA)*

Section 24.5 Hubble's Law **651**

✓ **Concept Check**

■ Other than scale, in what important ways does galaxy evolution differ from stellar evolution?

24.5 Hubble's Law

2 ⁊ Now let's turn our attention to the large-scale *motions* of galaxies and galaxy clusters. Within a galaxy cluster, individual galaxies move more or less randomly. You might expect that, on the largest possible scales, the clusters themselves would also have random, disordered motion—some clusters moving this way, some that, with no overall pattern to the motion. In fact, this is not the case. On the largest scales, galaxies and galaxy clusters alike move in a very *ordered* way.

UNIVERSAL RECESSION

In 1912 the American astronomer Vesto M. Slipher, working under the direction of Percival Lowell, discovered that virtually every spiral galaxy he observed had a redshifted spectrum—it was *receding* from our Galaxy. ∞ (Sec. 3.5) In fact, with the exception of a few nearby systems, *every* known galaxy is part of a general motion away from us in all directions. Individual galaxies that are not part of galaxy clusters are steadily receding. Galaxy clusters too have an overall recessional motion, although their individual member galaxies move randomly with respect to one another. (Consider a jar full of fireflies that has been thrown into the air. The fireflies within the jar, like the galaxies within the cluster, have random motions due to their individual whims, but the jar as a whole, like the galaxy cluster, has some directed motion as well.)

Figure 24.27 shows the optical spectra of several galaxies, arranged in order of increasing distance from the

Radial velocities in km/s	Distance in megaparsecs	Cluster galaxy in
1210	18	Virgo
15,000	230	Ursa Major
21,600	330	Corona Borealis
39,300	600	Bootes
61,200	940	Hydra

Figure 24.27 Galaxy Spectra
Shown on the left are the optical spectra of several galaxies named on the right. Both the extent of the redshift (denoted by the horizontal yellow arrows) and the distance from the Milky Way Galaxy to each galaxy (numbers in center column) increase from top to bottom. The vertical yellow arrow in the top spectrum highlights a particular spectral feature (a pair of dark absorption lines). The horizontal yellow arrows indicate how this feature shifts to longer wavelengths in spectra of more distant galaxies. The lines at top and bottom of each spectrum are laboratory references. (Palomar/Caltech)

R I V U X G

Milky Way Galaxy. These spectra are redshifted, indicating that the galaxies are steadily receding. Furthermore, the extent of the redshift increases progressively from top to bottom in the figure. Because the galaxies' distances from us also increase from top to bottom, we conclude that there is a connection between Doppler shift and distance: The greater the distance, the greater the redshift. This trend holds for nearly all galaxies in the universe. (Two galaxies within our Local Group, including Andromeda, and a few galaxies in the Virgo Cluster display blueshifts and so have some motion toward us, but this results from their random motions within their parent clusters. Recall the fireflies in the jar.)

Figure 24.28(a) shows recessional velocity plotted versus distance for the galaxies of Figure 24.27. Figure 24.28(b) is a similar plot for some more galaxies within about 1 billion pc of Earth. Plots like these were first made by Edwin Hubble in the 1920s and now bear his name— *Hubble diagrams*. The data points generally fall close to a straight line, indicating that a simple relationship connects recessional velocity and distance: The rate at which a galaxy recedes is *directly proportional* to its distance from us. This rule is called **Hubble's law**. We could construct such a diagram for any group of galaxies, provided we could determine their distances and velocities. The universal recession described by the Hubble diagram is sometimes called the *Hubble flow*.

The recessional motions of the galaxies prove that the cosmos is not steady and unchanging on the largest scales. Its contents are in constant relative motion, and the motion is not random. The entire universe is *expanding*. However, let's be clear about what is expanding and what is not. Hubble's law does *not* mean that humans, Earth, the solar system, or even individual galaxies or galaxy clusters, are physically increasing in size. These groups of atoms, rocks, planets, stars, and galaxies are held together by their own internal forces (chemical bonds in the first case, the force of gravity in all the others), and are not themselves getting bigger. Only the largest framework of the universe—the ever-increasing distances separating the galaxies and the galaxy clusters—is expanding.

To distinguish recessional redshift from redshifts caused by motion *within* an object—for example, galaxy orbits within a cluster or explosive events in a galactic nucleus—the redshift resulting from the Hubble flow is called the **cosmological redshift**. Objects that lie so far away that they exhibit a large cosmological redshift are said to be at *cosmological distances*—distances comparable to the scale of the universe.

Hubble's law is an *empirical* discovery—that is, a discovery based strictly on observational results. Its central relationship—a statistical correlation between recessional velocity and distance—is well documented as far as galaxy distances can be reliably determined, but no law of nature demands that all galaxies recede, and no law of physics requires that a link exist between velocity and distance. In that sense, Hubble's "law" is not really a law at all. It is simply a convenient way of noting the observational fact that any galaxy's recessional velocity is directly proportional to its distance from us.

Hubble's law has some fairly dramatic implications. If nearly all galaxies show recessional velocity according to Hubble's law, then does that mean that they all started their journey from a single point? And if we could run time backward, would all the galaxies fly back to this one point, perhaps the site of some explosion in the remote past? The answer is yes—but not in the way you might expect! In Chapters 26 and 27 we explore the ramifications of the Hubble flow for the past and future evolution of our universe. For the remainder of this chapter, however, we set aside its cosmic implications and use Hubble's law simply as a convenient distance-measuring tool.

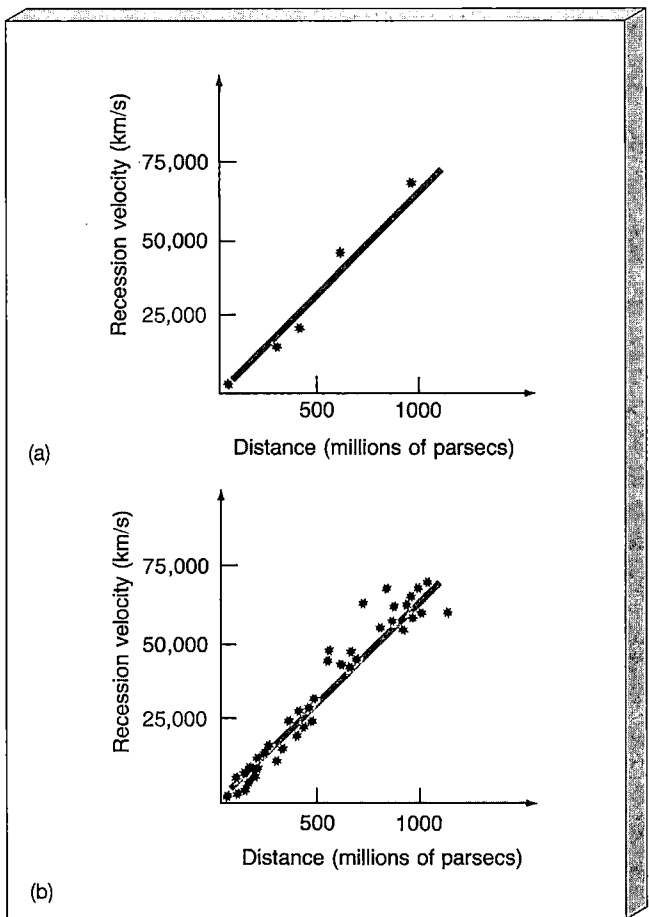

Figure 24.28 Hubble's Law Plots of recessional velocity versus distance (a) for the galaxies shown in Figure 24.28, and (b) for numerous other galaxies within about 1 billion pc of Earth.

HUBBLE'S CONSTANT

The constant of proportionality between recessional velocity and distance in Hubble's law is known as **Hubble's** constant. It is usually denoted by the symbol H_0. The data shown in Figure 24.28 then obey the equation

$$\text{recessional velocity} = H_0 \times \text{distance}.$$

The value of Hubble's constant is the slope of the straight line—recessional velocity divided by distance—in Figure 24.28(b). Reading the numbers off the graph, this value comes to roughly 65,000 km/s divided by 1000 Mpc, or 65 km/s/Mpc (kilometers per second per megaparsec, the most commonly used unit for H_0). Astronomers continually strive to refine the accuracy of the Hubble diagram and the resulting estimate of H_0 because Hubble's constant is one of the most fundamental quantities of nature—it specifies the rate of expansion of the entire cosmos.

The precise value of Hubble's constant is the subject of considerable debate. In the 1970s, astronomers obtained a value of around 50 km/s/Mpc, using a chain of standard candles to extend their observations to large distances. However, in the early 1980s, after the infrared Tully-Fisher technique had become fairly well established, other researchers used it to obtain a measurement of H_0 that was largely independent of methods relying on standard candles. From observations of galaxies within about 150 Mpc, the latter group deduced a value of $H_0 \approx 90$ km/s/Mpc, a result inconsistent with the earlier measurements (even allowing for the estimated uncertainties involved). For some reason, the distances obtained using the Tully-Fisher method were only about half those determined by using standard candles, so the measured value of Hubble's constant nearly doubled using this approach.

As of mid-2001, published measurements of H_0 over the past five years by many different research groups, using different sets of galaxies and a wide variety of distance-measurement techniques, give results mainly between 50 and 80 km/s/Mpc. Most astronomers would be quite surprised if the true value of H_0 turned out to lie outside this range. However, the width of the quoted range is not the result of measurement uncertainties in any one method—there remain real, and as yet unresolved, inconsistencies between the different techniques currently in use. Infrared Tully-Fisher measurements and studies of Cepheid variables, the latter now extended to include the Virgo cluster by researchers using the *Hubble Space Telescope*, generally produce results at the high end of the range, 70–80 km/s/Mpc. However, blue-light Tully-Fisher studies and techniques using standard candles, including Type I supernovae, tend to return lower values, in the range 50–65 km/s/Mpc. Other methods give results scattered between 50 and 80 km/s/Mpc.

For now, astronomers must live with this uncertainty. We will adopt $H_0 = 65$ km/s/Mpc (roughly the median of all recent results) as the best current estimate of Hubble's constant for the remainder of the text. Bear in mind, though, that there is still considerable ambiguity—and dispute—among experts as to the true value of this very important number.

THE COSMIC DISTANCE SCALE

Using Hubble's law, we can derive the distance to a remote object simply by measuring the object's recessional velocity and dividing by Hubble's constant. ⊕ (Sec. 3.5) Using Hubble's law in this way tops our inverted pyramid of distance-measurement techniques (Figure 24.29). This sixth method simply assumes that Hubble's law holds. If this assumption is correct, Hubble's law enables us to measure great distances in the universe—so long as we can obtain an object's spectrum, we can determine how far away the object is. Notice, however, that the current uncertainty in Hubble's constant translates directly into a similar uncertainty in all distances determined by this method.

Many redshifted objects have recessional motions that are a substantial fraction of the speed of light. The most distant objects thus far observed in the universe—some young galaxies and quasars (see Section 25.3)—have redshifts (fractional increase in wavelength) of nearly 6.0, meaning that their radiation has been stretched in wavelength not by just a few percent, as with most of the objects discussed in this chapter, but *sevenfold*. Their ultraviolet spectral lines are Doppler shifted all the way into the infrared part of the spectrum! Such a huge redshift corresponds to a recessional velocity of 96 percent the speed of light. Hubble's law implies that such objects lie more than 8000 Mpc away from us, as close to the limits of the observable universe as astronomers have yet been able to probe. (See *More Precisely 25-2* for a fuller discussion of redshifts and recession velocities comparable to the speed of light.)

The speed of light is finite. It takes time for light or any kind of radiation to travel from one point in space to another. The radiation that we now see from these most distant objects originated long ago. Incredibly, the radiation that astronomers now detect was emitted almost 13 billion years ago, well before our planet, our Sun, and perhaps even our Galaxy came into being. (See again *More Precisely 25-2* for reconciliation of this time with the present distance of 26 billion light-years—8000 Mpc—implied by the previous paragraph.)

LARGE-SCALE STRUCTURE IN THE UNIVERSE

Using Hubble's law, we can complete our census of the large-scale distribution of galaxies. Figure 24.30 shows part of an early survey of the universe performed by astronomers at Harvard University in the 1980s. Using Hubble's law as a distance indicator, the team systematically mapped out the locations of galaxies within about 200 Mpc of the Milky Way in a series of wedge-shaped "slices," each 6° thick, starting in the northern sky. The first slice (shown in the figure) covered a region of the sky containing the Coma Cluster (Figure 24.1), which happens to lie in a direction almost perpendicular to our Galaxy's plane.

The most striking feature of maps such as this is that the distribution of galaxies on very large scales is decidedly

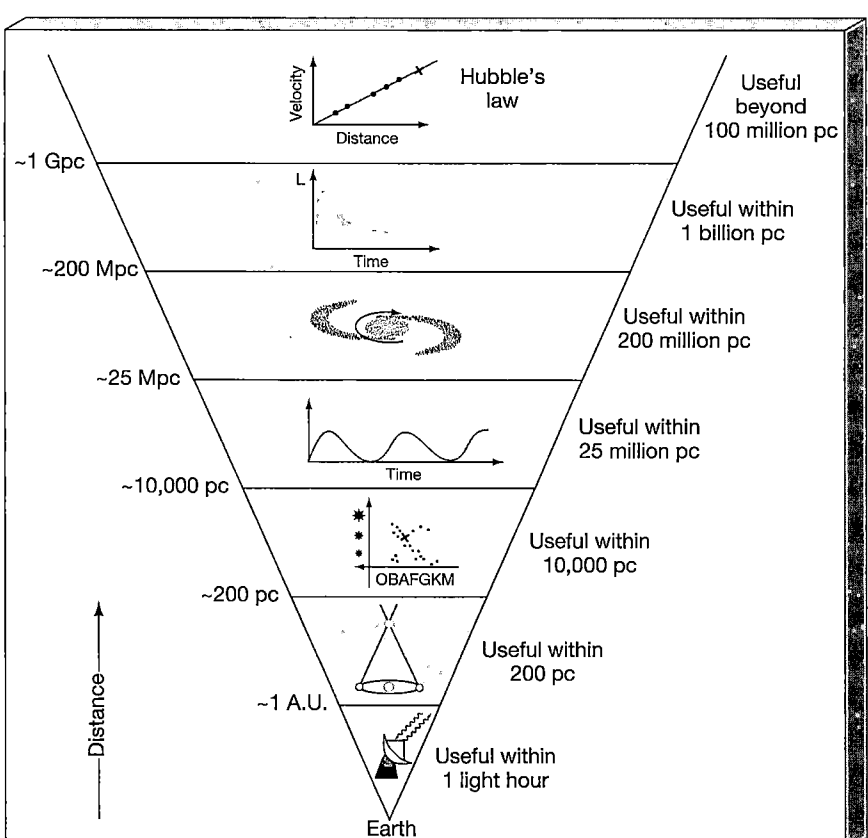

Figure 24.29 Cosmic Distance Ladder Hubble's law tops the hierarchy of distance-measurement techniques. It is used to find the distances of astronomical objects all the way out to the limits of the observable universe.

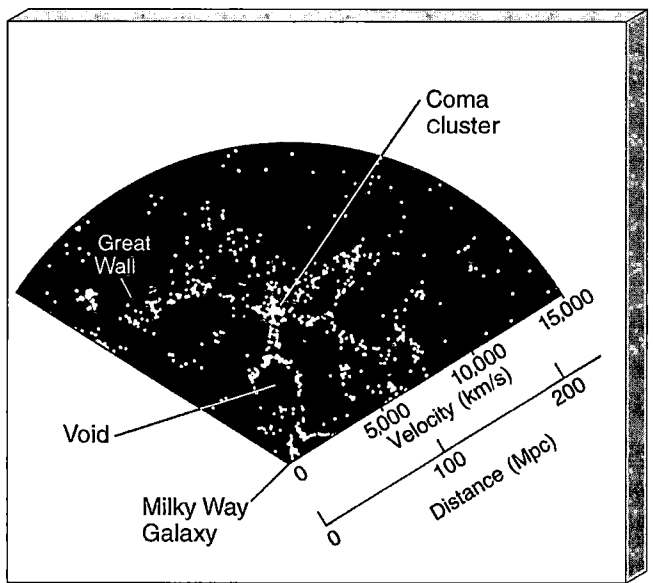

Figure 24.30 Galaxy Survey The first slice of a survey of the universe, covering 1057 galaxies out to an approximate distance of 200 Mpc, clearly shows that galaxies and clusters are not randomly distributed on large scales. Instead, they appear to have a filamentary structure, surrounding vast, nearly empty voids. The distances shown assume $H_0 = 65$ km/s/Mpc. This slice covers 6° of the sky in the direction out of the plane of the paper.

nonrandom. The galaxies appear to be arranged in a network of strings, or filaments, surrounding large, relatively empty regions of space known as **voids**. The biggest voids measure some 100 Mpc across. For a time they were the largest objects in the universe known to astronomers. The best description of the voids and filamentary structure in Figure 24.30 is that the galaxies and galaxy clusters are spread across the surfaces of vast "bubbles" in space. Like suds on soapy water, these gigantic bubbles fill the entire universe. The voids are the interiors of the gigantic bubbles. The densest clusters and superclusters lie in regions where several bubbles meet. The elongated shape of the Virgo Supercluster (see Figure 24.16) is a local example of this same filamentary structure.

Most theorists believe that this "frothy" distribution of galaxies, and in fact all structure on scales larger than a few megaparsecs, traces its origin directly to the conditions found in the very earliest stages of the universe (Chapter 27). As such, studies of large-scale structure are vital to our efforts to understand the origin and nature of the cosmos itself.

The idea that the filaments may be related to even larger structures (the bubble surfaces) was confirmed when the next three slices of the survey, lying above and below the first, were completed. The region of Figure 24.30 indicated by the red outline was found to continue through both the other slices, so we know that it covers *at least* 36°

on the sky perpendicular to the outline in the figure. This extended sheet of galaxies, which has come to be known as the *Great Wall*, measures at least 70 Mpc (out of the plane of the page) by 200 Mpc (across the page). It is one of the largest-known structures in the universe.

Figure 24.31 combines data from several surveys to show the structure of the universe within about 250 Mpc of Earth. Many superclusters and voids can be seen, along with indications of the filamentary structure now thought to exist throughout the cosmos. The Great Wall can be seen arcing across the right center of the figure. Are there still larger structures in the universe? Only with even more extensive surveys can we know for sure. Until the mid-1990s, all large-scale surveys had detected structures comparable in size to the region surveyed—in other words, they have found objects as large as the largest object they could have hoped to find. However, more recent, larger-scale surveys have begun to place strong limits on the extent of the structure. We will discuss these surveys—and their implications for the cosmos—in Chapter 26.

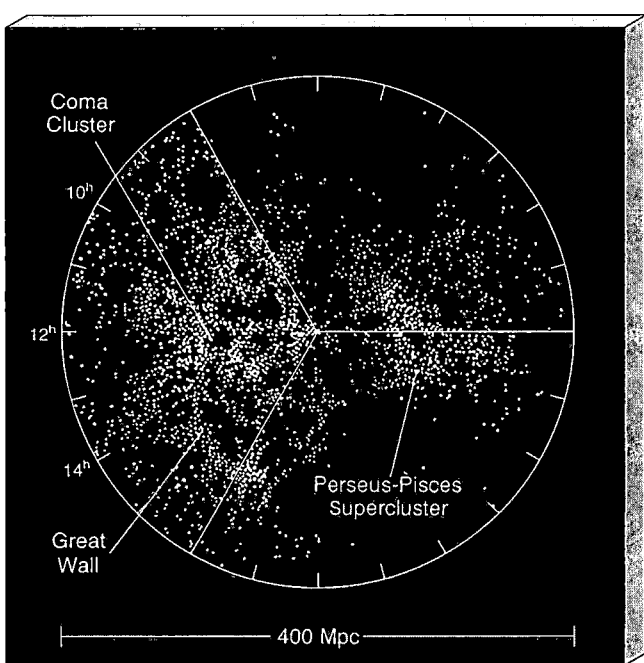

Figure 24.31 Local Universe The large-scale structure of the universe within about 300 Mpc of the Milky Way. The arc on the left is the Great Wall. Note the elongated, filamentary structure of the many superclusters visible here. Many of the distances were obtained using Hubble's law, assuming $H_0 = 65$ km/s/Mpc.

☑ Concept Check

■ How does the expansion of the universe provide a means of measuring distances to galaxies?

Chapter Review

SUMMARY

The **Hubble classification scheme** (p. 631) divides galaxies into several classes, depending on their appearance. **Spiral galaxies** (p. 631) have flattened disks, central bulges, and spiral arms. They are further subdivided on the basis of the size of the bulge and the tightness of the spiral structure. The halos of these galaxies consist of old stars, whereas the gas-rich disks are the sites of ongoing star formation. **Barred-spiral galaxies** (p. 632) contain an extended "bar" of material projecting beyond the central bulge.

Elliptical galaxies (p. 632) have no disk and contain no gas or dust. In most cases, they consist entirely of old stars. They range in size from dwarf ellipticals, which are much less massive than the Milky Way Galaxy, to giant ellipticals, which may contain trillions of stars. **S0** and **SB0 galaxies** (p. 634) have properties intermediate between those of ellipticals and spirals. They have extended halos and stellar disks and bulges (and bars, in the SB0 case), but little or no gas and dust.

Irregular galaxies (p. 634) are galaxies that do not fit into any of the other categories. Some may be the result of galactic collisions or close encounters. Many irregulars are rich in gas and dust and are the sites of vigorous star formation. The **Magellanic Clouds** (p. 634), two small systems that orbit the Milky Way Galaxy, are examples of this type of galaxy.

Astronomers often use **standard candles** (p. 637) as distance-measuring tools. These are objects that are easily recognizable (by their light curves, spectra, or some other directly observable characteristic) and whose luminosities are known to lie in some reasonably well-defined range. Comparing luminosity and apparent brightness, astronomers determine the distance using the inverse-square law. Type I supernovae are particularly useful for measuring distances to faraway galaxies. They are bright and hence easily seen, and have a relatively narrow spread in luminosity. An important alternative is the **Tully-Fisher relation** (p. 638), an empirical correlation between rotational velocity and luminosity in spiral galaxies. By measuring the rotation speed of a spiral and using this relationship, astronomers can determine the galaxy's luminosity and hence its distance.

The Milky Way, Andromeda, and several other smaller galaxies form the **Local Group** (p. 640), a small **galaxy cluster** (p. 640). Galaxy clusters consist of a collection of galaxies orbiting one another, bound together by their own gravity. The nearest large galaxy cluster to the Local Group is known as the Virgo Cluster. Galaxy clusters themselves tend to clump together into **superclusters** (p. 640). The Virgo Cluster, the Local Group, and several other nearby clusters form the Local Supercluster.

The masses of nearby spiral galaxies can be determined by studying their rotation curves. For more distant spirals, masses can be inferred from observations of the broadening of their spectral lines. On larger scales, astronomers use studies of binary galaxies and galaxy clusters to obtain statistical mass estimates of the galaxies involved. As in the Milky Way Galaxy, measurements of the masses of other galaxies and of galaxy clusters reveal the presence of large amounts of dark matter that is presently undetectable at any electromagnetic wavelength. The fraction of dark matter apparently grows as the scale under consideration increases. Large amounts of hot X-ray-emitting gas have been detected among the galaxies in many clusters, but not enough to account for the dark matter inferred from dynamical studies.

Researchers know of no simple evolutionary sequence that links spiral, elliptical, and irregular galaxies. Most astronomers believe that large galaxies formed by the merger of smaller ones

in a process that may be continuing today. Collisions and mergers of galaxies play very important roles in galactic evolution, and interactions between galaxies appear to be very common. A **starburst galaxy** (p. 650) may result when a galaxy experiences a close encounter with a neighbor. The strong tidal distortions caused by the encounter compress galactic gas, resulting in a widespread burst of star formation. Mergers between spirals most likely result in elliptical galaxies.

Distant galaxies are observed to be receding from the Milky Way at rates that increase proportional to their distances from us. This relationship between recessional speed and distance is called **Hubble's law** (p. 652). The constant of proportionality in the law is **Hubble's constant** (p. 653). Its value is believed to lie between 50 and 80 km/s/Mpc; we adopt a value of 65 km/s/Mpc, in the middle of this range. Astronomers use Hubble's law to determine distances to the most remote objects in the universe. The redshift associated with the Hubble expansion is called the **cosmological redshift** (p. 652).

On very large scales, galaxies and galaxy clusters are not spread randomly throughout space. Instead, they are arranged on the surfaces of enormous "bubbles" of matter surrounding vast low-density regions called **voids** (p. 654). The origin of this structure is thought to be closely related to conditions in the very earliest epochs of the universe.

SELF-TEST: TRUE OR FALSE?

____ **1.** Barred-spiral galaxies have the same properties as normal spirals, except for the "bar" feature.

____ **2.** Elliptical galaxies do not contain a flattened gaseous disk.

____ **3.** There is little or no interstellar dust in elliptical galaxies, but there may be substantial amounts of interstellar gas.

____ **4.** Most ellipticals contain only young stars.

____ **5.** Most galaxies are spirals.

____ **6.** Irregular galaxies, although small, have lots of star formation taking place in them.

____ **7.** Elliptical galaxies evolve into spirals.

____ **8.** Type I supernovae can be used to determine distances to galaxies.

____ **9.** Every galaxy is a member of some galaxy cluster.

____ **10.** Galaxy collisions can occur, but they are extremely rare.

____ **11.** Galaxy collisions have little or no effect on the stars and interstellar gas in the galaxies involved.

____ **12.** Distant galaxies appear to be much larger than those nearby.

____ **13.** A typical galaxy cluster contains about a million individual galaxies.

____ **14.** Most galaxies are receding from the Milky Way Galaxy.

____ **15.** Hubble's law can be used to determine distances to the farthest objects in the universe.

SELF-TEST: FILL IN THE BLANK

1. Galaxies are categorized by their _____ classification.

2. Spiral galaxies with tightly wrapped spiral arms tend to have _____ central bulges.

3. Spiral galaxies of type _____ have the least amount of gas; type _____ have the most.

4. Elliptical galaxies generally contain far _____ cool gas and dust than do irregulars.

5. The Milky Way Galaxy, the Andromeda Galaxy, and roughly 40 other galaxies form a small cluster known as the _____.

6. In the Tully-Fisher relation, a galaxy's luminosity is found to be related to the _____ of its 21-cm line.

7. The center of the Local Supercluster lies in _____.

8. When galaxies collide, the star formation rate often _____.

9. Mass determinations from rotation curves, line broadening, and galaxies in binaries and clusters all make use of _____ laws.

10. More than 90 percent of the universe is made up of _____.

11. Intergalactic gas in galaxy clusters emits large amounts of energy in the form of _____.

12. By which process do galaxies form: fragmentation (large objects breaking up into small) or mergers (small objects accumulating into large)? _____.

13. _____ galaxies may be formed by mergers between spiral systems.

14. Hubble's law is a correlation between the redshifts and the _____ of galaxies.

15. On the largest scales, galaxies in the universe appear to be arranged on huge _____ surrounding nearly empty _____.

REVIEW AND DISCUSSION

1. What distinguishes one type of spiral galaxy from another?

2. Describe some similarities and differences between elliptical galaxies and the halo of our own Galaxy.

3. Describe the four rungs in the distance-measurement ladder used to determine the distance to a galaxy lying 5 Mpc away.

4. Describe the contents of the Local Group. How much space does it occupy compared to the volume of the Milky Way?

5. What are standard candles, and why are they important to astronomy?

6. How is the Tully-Fisher relation used to measure distances to galaxies?

7. What is the Virgo Cluster?

8. Describe two techniques for measuring the mass of a galaxy.

9. Why do astronomers believe that galaxy clusters contain more mass than we can see?

10. Why are galaxies at great distances from us generally smaller and bluer than nearby galaxies?

11. What evidence do we have that galaxies collide with one another?

12. Describe the role of collisions in the formation and evolution of galaxies.

13. Give an example of how mergers can transform one type of galaxy into another.

14. Do you think that collisions between galaxies constitute "evolution" in the same sense as the evolution of stars?

15. What are starburst galaxies, and what do they have to do with galaxy evolution?

16. What is Hubble's law?

17. How is Hubble's law used by astronomers to measure distances to galaxies?

18. What is the most likely range of values for Hubble's constant? Why is the exact value uncertain?

19. What are voids?

20. Describe the distribution of galactic matter on very large (more than 100 Mpc) scales.

PROBLEMS  *Algorithmic versions of these questions are available in the Practice Problems module of the Companion Website.*

The number of squares preceding each problem indicates its approximate level of difficulty.

1. ■ A supernova of luminosity one billion times the luminosity of the Sun is used as a standard candle to measure the distance to a faraway galaxy. From Earth the supernova appears as bright as the Sun would appear from a distance of 10 kpc. What is the distance to the galaxy?

2. ■■ A Cepheid variable star in the Virgo cluster has an absolute magnitude of -5 and is observed to have an apparent magnitude of 26.3. Use these figures to calculate the distance to the Virgo cluster.

3. ■ The Andromeda Galaxy is approaching our Galaxy with a radial velocity of 266 km/s. Given the galaxies' present separation of 800 kpc, and neglecting *both* the transverse component of the velocity *and* the effect of gravity in accelerating the motion, estimate when the two galaxies will collide.

4. ■■ Based on the data in Figure 24.18, estimate the mass of the galaxy NGC 4984 inside 20 kpc.

5. ■■ Based on the data in Figure 24.18, estimate the amount of line broadening (maximum minus minimum wavelength) of the 656.3 nm Hα line observed in NGC 4984.

6. ■■ Two galaxies are orbiting each other at a distance of 500 kpc. Their orbital period is estimated to be 30 billion years. Use Kepler's law (as stated in Section 23.6) to find the total mass of the pair.

7. ■■ Use Kepler's third law (Section 23.6) to estimate the mass required to keep a galaxy moving at 750 km/s in a circular orbit of radius 2 Mpc around the center of a galaxy cluster. Given the approximations involved in determining this figure, do you think this is a good estimate of the cluster's true mass?

8. ■■ Assuming that the average speed of the protons in the (ionized) X-ray-emitting gas is the same as the mean orbital speed of the stars, estimate the temperature of the gas in the X-ray halo of an elliptical galaxy of mass 10^{12} solar masses and radius 25 kpc. ∞ (*More Precisely 8-1*)

9. ■■ Calculate the average speed of hydrogen nuclei (protons) in a gas of temperature 20 million K. Compare this with the speed of a galaxy moving in a circular orbit of radius 1 Mpc around a galaxy cluster of mass 10^{14} solar masses.

10. ■ In a galaxy collision, two similar-sized galaxies pass through each other with a combined relative velocity of 1500 km/s. If each galaxy is 100 kpc across, how long will the event last?

11. ■ According to Hubble's law, with $H_0 = 65$ km/s/Mpc, what is the recessional velocity of a galaxy at a distance of 200 Mpc? How far away is a galaxy whose recessional velocity is 4000 km/s? How do these answers change if $H_0 = 50$ km/s/Mpc? If $H_0 = 80$ km/s/Mpc?

12. ■■ According to Hubble's law, with $H_0 = 65$ km/s/Mpc, how long will it take for the distance from the Milky Way Galaxy to the Virgo Cluster to double?

13. ■■■ Assuming Hubble's law with $H_0 = 65$ km/s/Mpc, what would be the angular diameter of an E0 galaxy of radius 40 kpc, if its 656.3 nm Hα line is actually observed at 700 nm?

14. ■■■ A small satellite galaxy is moving in a circular orbit around a much more massive parent, and just happens to be moving exactly parallel to the line of sight as seen from Earth. The recession velocities of the satellite and the parent galaxy are measured to be 6450 km/s and 6500 km/s, respectively, and the two galaxies are separated by an angle of 0.1° on the sky. Assuming $H_0 = 65$ km/s/Mpc, calculate the mass of the parent galaxy.

15. ▪▪▪ Galaxies in a distant galaxy cluster are observed to have recession velocities ranging from 12,500 km/s to 13,500 km/s. The cluster's angular diameter is 55'. Use these data to estimate the cluster's mass. What assumptions do you have to make in order to obtain this estimate? Take $H_0 = 65$ km/s/Mpc.

COLLABORATIVE EXERCISES

1. Sampling the Sky. Figure 24.23 is called the *Hubble Deep Field*. The Picture shows far too many galaxies for one person to easily count. Each group member should count the galaxies in a random area 3 cm × 3 cm and then determine a group average. Since the entire image is approximately 500 cm², multiply your group's average number of galaxies in a 3 cm × 3 cm area by 55.5 to estimate the number of galaxies in the image. How does your value compare to that of another group?

2. Classification Schemes. Edwin Hubble classified galaxies into three broad categories based on their appearance. Classify all the books used this term by the group member who is taking the most classes, using categories such as color, size, thickness, and cost. Use any scheme that adequately describes the collection. Clearly define each category.

RESEARCHING ON THE WEB

To complete the following exercises, go to the online Destinations module for Chapter 24 on the Companion Website for Astronomy Today 4/e.

1. Access the "Galaxy Types" page and make a labeled sketch of the three types of galaxies defined on this page.

2. Access the "Cosmology of the Local Group" page and list the three brightest and the three closest galaxies of the Local Group.

3. Access the "Our Hierarchical Universe" pages and list the top six important questions in cosmology.

PROJECTS

1. Look for a copy of the *Atlas of Peculiar Galaxies* by Halton Arp. It is available in book form or on laser disk. Search for examples of interacting galaxies of various types: (1) tidal interactions, (2) starburst galaxies, (3) collisions between two spirals, and (4) collisions between a spiral and an elliptical. For (1) look for galactic material pulled away from a galaxy by a neighboring galaxy. Is the latter galaxy also tidally distorted? In (2) the surest signs of starburst activity are bright knots of star formation. In what type(s) of galaxies do you find starburst activity? For (3) and (4) how do collisions differ depending on the types of galaxies involved? What typically happens to a spiral galaxy after a near miss or collision? Do ellipticals suffer the same fate?

2. Look for the Virgo Cluster of galaxies. An 8-inch telescope is the perfect size for this project, although a smaller tele-

scope will also work. The constellation Virgo is visible from the United States during much of fall, winter, and spring. To locate the center of the cluster, first find the constellation Leo. The eastern part of Leo is composed of a distinct triangle of stars, Denebola (β), Chort (θ), and Zosma (δ). Go from Chort to Denebola in a straight line east, continue on the same distance as between the two stars and you will be approximately at the center of the Virgo Cluster. Look for the following Messier objects that make up some of the brightest galaxies in the cluster: M49, M58, M59, M60, M84, M86, M87 (a giant elliptical thought to have a massive black hole at its center), M89, and M90. Examine each galaxy for unusual features; some have very bright nuclei.

SKYCHART III PROJECTS  *The SkyChart III Student Version planetarium program on which these exercises are based is included as a separately executable program on the CD in the back of this text.*

1. Use SkyChart III to observe the distribution of visible galaxies. *Deep Sky* and *Grid Lines* should be selected under *DRAW*. Select all grid lines under *DRAW/Symbols & Grids* and set for 180° field of view. Deselect all other parameters under *DRAW*. Set a time near midnight and select a time-step of one month. Repeatedly press the F3 key until the faintest objects are visible. Set *ANIMATION/1 Month* and step the time forward one month at a time by pressing the F6 key. Note the distribution of galaxies, and observe the relationship of the galactic equator to the observed galaxies. Explain the distribution you find.

2. The Large and Small Magellanic Clouds are two irregular galaxies prominent to the unaided eye in the Southern Hemisphere. Set up the software by selecting *VIEW/180° Field*. Select *DRAW/Sky Background/Black*, *DRAW/Constellations* and *DRAW/Deep Sky Objects*; deselect *Stars* and *Horizon Mask*. Change the display with the down arrow key until the Southern Hemisphere is centered in the display. The Large Magellanic Cloud is at about the same hour angle as Orion, and the Small Magellanic Cloud is at about the same hour angle as Cassiopeia. They are present in the display as ovals when *DRAW/Deep Sky Objects* is selected. Select *DRAW/Grid Lines* and observe the relationship between the positions of these irregular galaxies and the Galactic equator.

Alternatively, locate the Magellanic Clouds by using *VIEW/Center Object/Magellanic Cloud*. If this method is used for locating the Magellanic Clouds, then orient the screen with North up and move the display so that you can observe the relative size and position of the clouds to constellations you are familiar with.

3. The Local Group consists of several nearby galaxies orbiting one another in an area about 1 Mpc in diameter. The largest members of this cluster are the Milky Way and Andromeda Galaxies. Print a chart labeling the locations of the nearest members of the Local Group: The Large and small Magellanic Clouds, and the Sculptor (NGC253) and Fornax (NGC1316) galaxies.

4. Locate the Andromeda Galaxy (M31) and print a chart labeling members of the Local Group nearest Andromeda (i.e., M32, M33, NGC 147, NGC 185).

5. Collisions of galaxies are rare and can last many millions of years. Some collisions can result in a change in shape in one or both galaxies, as in the M51 and NGC 5195 galaxies. Others collide more violently, resulting in new star formation, as is the case with galaxies NGC 4038 and NGC 4039, as well as NGC 3690 and IC 694. Find these galaxies in SkyChart III.

 *In addition to the Practice Problems and Destinations modules, the Companion Website at http://www.prenhall.com/chaisson provides for each chapter an additional true-false, multiple choice, and labeling quiz, as well as additional annotated images, animations, and links to related Websites.*

25 ACTIVE GALAXIES AND QUASARS

Limits of the Observable Universe

LEARNING GOALS

Studying this chapter will enable you to:

1   Specify the basic differences between active and normal galaxies.

2   Describe the important features of Seyfert and radio galaxies.

3   Describe the observed properties of quasars and discuss the special properties of the radiation they emit.

4   Explain what drives the central engine thought to power all active galaxies.

5   Discuss the use of quasars as probes of the distant universe.

6   Describe the place of quasars and active galaxies in current theories of galactic evolution.

 Visit http://www.prenhall.com/chaisson for additional annotated images, animations, and links to related sites for this chapter.

3C295 is revealed in this X-ray image to be an active, even explosive, galaxy (white areas at center) enveloped by a vast cloud of loose gas (red) whose temperature reaches some 50 million K. The invisible cloud, which is roughly 5 billion light-years away and 2 million light-years across, contains upward of a hundred galaxies (too cool to be seen in X rays) and enough intergalactic matter to make more than a thousand more galaxies. (CXO/SAO)

The Big Picture: Active galaxies are much more energetic than the normal galaxy—the Milky Way—in which we live. The "central engines" powering them are thought to be supermassive black holes. Many active galaxies are found within huge clusters of galaxies—clusters that contain numerous other galaxies amidst much loose, extremely hot gas. These clusters are among the most massive objects in the universe.

Our journey from the Milky Way to the Great Wall in the past two chapters has widened our cosmic field of view by a factor of 10,000, yet the galaxies that make up the structures we see show remarkable consistency in their properties. The overwhelming majority of galaxies fit neatly into the Hubble Classification Scheme, showing few, if any, unusual characteristics. However, sprinkled through the mix of normal galaxies, even relatively close to the Milky Way Galaxy, are some that are decidedly abnormal in their properties. Although their optical appearance is often quite ordinary, these abnormal galaxies emit huge amounts of energy—far more than a normal galaxy—mostly in the invisible part of the electromagnetic spectrum. Observing such objects at great distances, we may be seeing some of the formative stages of our own galactic home.

25.1 Beyond the Local Realm

Astronomers estimate that some 40 billion galaxies exist in the observable universe. Most lie thousands of megaparsecs from Earth—too far for their Hubble types to be reliably determined with current telescopes—yet, to the extent that their properties can be measured, even very distant galaxies seem basically "normal." Taking into account the steady evolution in galaxy properties outlined in Chapter 24, the luminosities and spectra of distant galaxies are generally consistent with the standard categories in the Hubble classification scheme. ⇨ (Secs. 24.1, 24.4) However, scattered throughout the universe, a few galaxies differ considerably from the norm. They are far more luminous than even the brightest spiral or elliptical galaxies discussed in the previous chapter. Having luminosities sometimes thousands of times greater than that of the Milky Way, they are known collectively as **active galaxies**.

In addition to their greater overall luminosities, active galaxies differ fundamentally from normal galaxies in the *character* of the radiation they emit. Most of a normal galaxy's energy is emitted in or near the visible portion of the electromagnetic spectrum, much like the radiation from stars. Indeed, to a large extent, the light we see from a normal galaxy *is* just the accumulated light of its many component stars. For example, both the Sun and the Milky Way radiate about a million times more energy at optical wavelengths than they emit in the radio part of the spectrum. By contrast, as illustrated schematically in Figure 25.1, the radiation from active galaxies does *not* peak in or near the visible. Most active galaxies do emit substantial amounts of visible radiation, but far more energy is emitted at longer wavelengths. Put another way, the radiation from active galaxies is *inconsistent* with what we would expect if it were the combined radiation of myriad stars. Their radiation is said to be *nonstellar*.

Two important categories of active galaxies are *Seyfert galaxies* and *radio galaxies*. More extreme in their properties are the even more luminous *quasars*. Astronomers conventionally distinguish active galaxies and quasars from normal galaxies based mainly on their appearance and spectra. At visible wavelengths, active galaxies typically *look* like normal galaxies. Indeed, we can think of active galaxies as

being otherwise "normal" systems that happen also to be extremely intense sources of radio and/or infrared radiation. Quasars, on the other hand, are mostly so far away that little internal structure can be discerned. However, the distinction between quasars and active galaxies is not clear-cut, and to some extent dates from the days when the connection between these objects was not understood. Most astronomers now believe that quasars are simply an early stage of galaxy formation, and that the same basic processes power all of these active objects.

Some active galaxies are found locally, sprinkled among the normal galaxies that make up most of our cosmic neighborhood, but as a general rule, most active galaxies lie far from Earth. We might wonder whether this predominance of bright objects at large distances is just an observational selection effect, resulting from our inability to detect faraway, relatively faint normal galaxies (see Figures 17.14 and 17.15 for precisely the same effect for stars

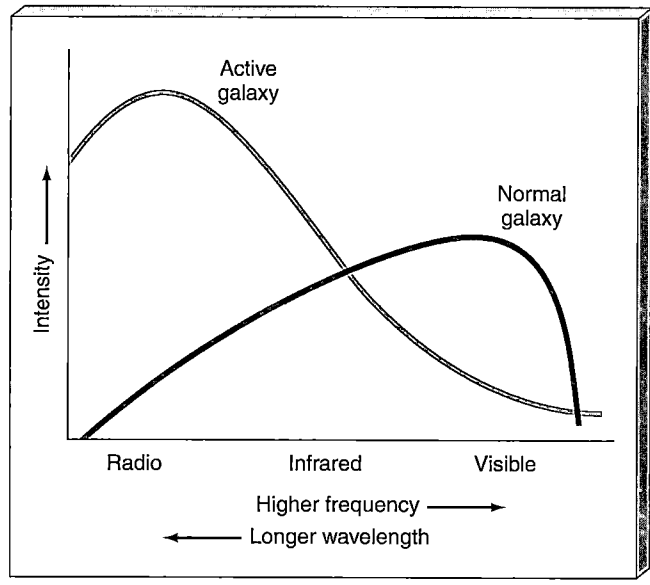

Figure 25.1 Galaxy Energy Spectra The spectrum of the energy emitted from a normal galaxy differs from that of an active galaxy. This plot illustrates the general run of intensity for all galaxies of a particular type and does not represent any single galaxy.

in the vicinity of the Sun). ∞ (Sec. 17.6) Although this observational bias does play a role, it can only partly explain the observations. Bright active galaxies really are more common at great distances, and the most active objects—the quasars—lie farthest away.

Since light travels at a finite speed, looking out to great distances in space is equivalent to looking back in time. The tremendous power and nonstellar radiation of these faraway objects tell astronomers that the universe was once a much more violent place than it is today.

☑ Concept Check

■ What distinguishes active galaxies from normal galaxies?

25.2 Properties of Active Galaxies

SEYFERT GALAXIES

2 In 1943 Carl Seyfert, an American optical astronomer studying spiral galaxies from Mount Wilson Observatory, discovered the type of active galaxy that now bears his name. **Seyfert galaxies** are a class of astronomical objects whose properties lie between those of normal galaxies like the Milky Way and those of the most violent active galaxies known. This fact suggests to many astronomers that Seyferts represent an evolutionary link between these two extremes. The spectral lines of Seyfert galaxies are usually substantially redshifted, telling us that most Seyferts reside at large distances (hundreds of megaparsecs) from us. ∞ (Sec. 24.5) However, a few lie just 20 or 30 Mpc away.

Figure 25.2 shows two optical images of a typical Seyfert galaxy. A casual glance at a long-exposure photograph of a Seyfert reveals nothing strange. Superficially, Seyferts resemble normal spiral galaxies. However, closer study of Seyferts reveals some peculiarities not found in normal spirals.

First, maps of Seyfert energy emission show that nearly all the radiation stems from a small central region known as the **galactic nucleus**. This region lies at the center of the overexposed white patch in Figure 25.2(a); another is shown in Figure 25.2(b). Astronomers suspect that a Seyfert nucleus may be quite similar to the center of a normal galaxy such as the Milky Way or the Andromeda Galaxy, but with one very important difference: The nucleus of a Seyfert is 10,000 times brighter than the center of our Galaxy. Indeed, the brightest Seyfert nuclei are 10 times more energetic than the *entire* Milky Way.

Second, Seyfert galaxies emit their radiation in two broad frequency ranges. The stars in the Seyfert's galactic disk and spiral arms produce about the same amount of *visible* radiation as those of a normal spiral galaxy. However, most of the energy from the Seyfert's nucleus is emitted in the form of *invisible* radio and infrared radiation, which

cannot be explained as coming from stars—it must be non-stellar in origin.

Third, Seyfert spectral lines bear little or no resemblance to those produced by ordinary stars, although they do have many similarities to the spectral lines observed toward the center of our own Galaxy. ∞ (Sec. 23.7) Seyfert spectra contain strong emission lines of highly ionized heavy elements, especially iron. The lines are very broad,

(a)

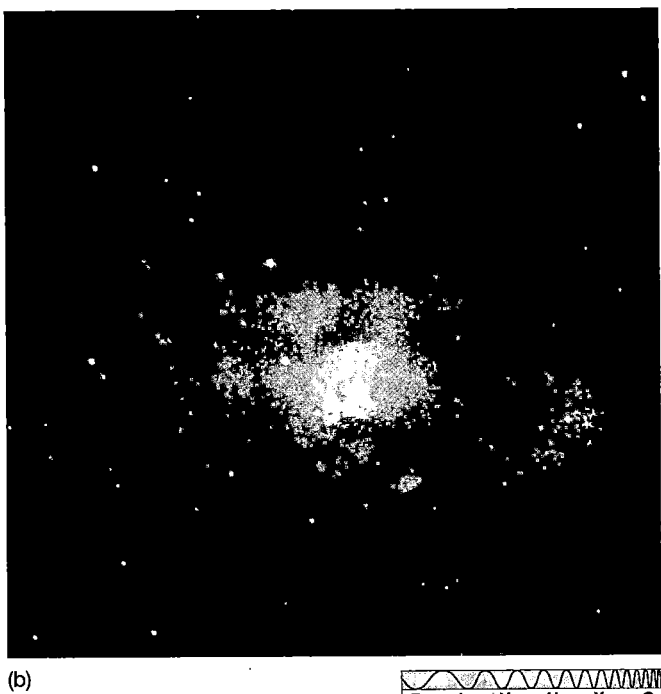

(b)

Figure 25.2 Seyfert Galaxy (a) This image of Seyfert galaxy NGC 7742 resembles a fried egg, with a ring of blue-tinted star-forming regions surrounding a very bright yellow core that spans 1 kpc across. This active spiral galaxy resides about 24 Mpc away. (b) The Circinus galaxy, also a Seyfert with a bright compact core, lies some 4 Mpc away—it is one of the closest active galaxies. *(NASA)*

indicating either that the galaxy's gases are tremendously hot (more than 10^8 K) or that they are rotating very rapidly (at about 1000 km/s) around some central object. ⊂⊃ (Sec. 4.5) The first possibility can be ruled out, since such a high temperature would cause all the gas to be ionized, in which case no spectral lines would be produced. Thus, the broadening indicates rapid internal motion in the nucleus.

Finally, extensive monitoring of Seyfert radiation over long periods of time has shown that the energy emission often varies over time. Figure 25.3 shows an example of luminosity variations for a typical Seyfert. Such radiative changes are unlike anything found in the Milky Way or in any other normal galaxy. A Seyfert's luminosity can double or halve within a fraction of a year.

These rather rapid fluctuations lead us to conclude that the source of energy emissions must be quite compact. As mentioned in Chapter 22 (in the context of luminosity variations in neutron stars and black holes), for astronomers to be able to detect a coherent variation in brightness within a certain time interval, the source of radiation must be smaller in size than the distance traveled by light during that interval. ⊂⊃ (Sec. 22.3) Otherwise, the intensity variations would be blurred, not sharp, as observed. Simply put, an object cannot "flicker" in less time than radiation takes to cross it. Because the rise and fall of a Seyfert's radiation usually occurs within one year, we conclude that the emitting region must be less than 1 light-year across—an extraordinarily small region, considering the huge amount of energy emanating from it. High-resolution interferometric radio maps of Seyfert cores generally confirm this reasoning. ⊂⊃ (Sec. 5.5) Together, the rapid time variability and large radio and infrared luminosities observed in Seyferts imply violent nonstellar activity in their nuclei. This activity may well be similar in *nature* to processes occurring at the center of our own Galaxy, but its *magnitude* is thousands of times greater than the comparatively mild events within our own Galaxy's heart. ⊂⊃ (Sec. 23.7)

RADIO GALAXIES

3 As the name suggests, **radio galaxies** are active galaxies that emit most of their energy in the radio portion of the electromagnetic spectrum. They differ from Seyferts not only in the wavelengths at which they radiate most strongly (Seyferts tend to radiate at shorter radio wavelengths and infrared wavelengths; radio galaxies radiate at longer radio wavelengths) but also in both the appearance and the extent of their emitting regions. Whereas the nonstellar emission from Seyfert galaxies always comes from a small central nucleus, the nonstellar emitting regions of radio galaxies can be much larger—hundreds of kiloparsecs across in some cases.

Figure 25.4 shows the radio galaxy Centaurus A, which lies just 4 Mpc from Earth. Almost none of this galaxy's emission comes from a compact nucleus. Instead, the radio energy is released from two huge extended regions called **radio lobes**—roundish clouds of gas spanning about 1 Mpc, lying well beyond the visible galaxy. Objects such as this are called *lobe-radio galaxies*. (Note that the term "visible galaxy" is commonly used to refer to those components of an active galaxy that emit visible "stellar" radiation, as opposed to the nonstellar and invisible "active" component of the galaxy's emission.)

The radio lobes of lobe-radio galaxies are truly enormous. From end to end, Centaurus A's radio lobes are more than 10 times the size of the Milky Way Galaxy, and are comparable in size to the entire Local Group. The lobes emit no visible light, but their radio luminosities can range from 1/10 to 10 times the energy emitted by our Galaxy at all wavelengths.

Figure 25.5 shows the relationship between the galaxy's visible, radio, and X-ray emission. In visible light, Centaurus A is a rather peculiar-looking object, apparently an E2 galaxy bisected by an irregular band of dust. Numerical simulations suggest that this system is probably the result of a merger between an elliptical galaxy and a smaller spiral galaxy about 500 million years ago. The radio lobes are roughly symmetrically placed, jutting out from the center of the visible galaxy roughly perpendicular to the dust lane. The elliptical galaxy itself is very large—some 500 kpc in diameter.

The lobes of radio galaxies vary in size and shape from galaxy to galaxy, but they are aligned with the center

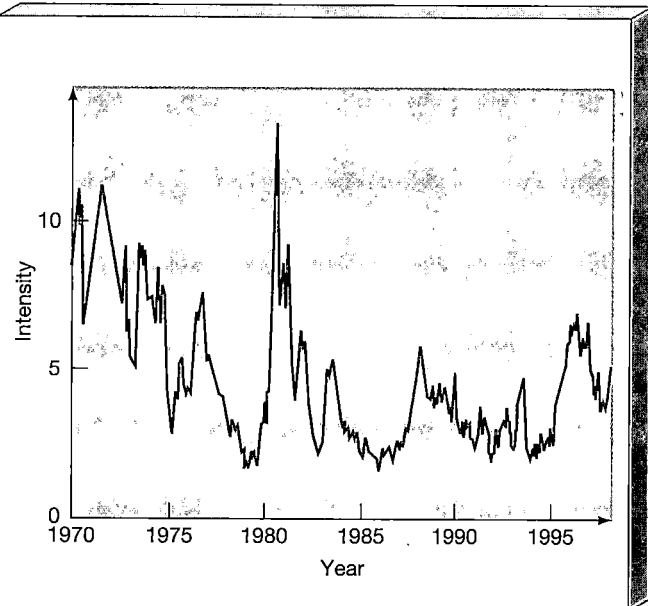

Figure 25.3 Seyfert Time Variability The irregular variations of a particular Seyfert galaxy's luminosity over a period of two decades. Because this Seyfert, called 3C 84, emits most strongly in the radio part of the electromagnetic spectrum, these observations were made with large radio telescopes. The optical and X-ray luminosities vary as well. *(NRAO)*

of the visible galaxy in nearly all cases. This alignment suggests that the lobes consist of material that was somehow ejected in opposite directions by violent events in the galactic nucleus. In the case of Centaurus A, this argument is strengthened by the presence of an additional pair of secondary lobes, smaller than the main lobes (about 50 kpc in length, marked in Figure 25.5) and closer to the visible galaxy. Both pairs of lobes share the same high degree of linear alignment. The inner lobes were most likely expelled from the nucleus by the same basic process as the outer ones, but much more recently, so they have not had time to travel as far. Still higher resolution studies reveal the presence of a roughly 1-kpc-long jet in the center of Centaurus A, again aligned with the larger lobes.

Astronomers believe that Centaurus A's outer lobes were created a few hundred million years ago, quite possibly around the time of the collision thought to be responsible for the galaxy's peculiar optical appearance. Apparently, some violent process at the center of Centaurus A—most probably triggered by the merger—started up around that time and has been intermittently firing jets of matter out into intergalactic space ever since.

Further evidence in favor of the interpretation of radio lobes as material ejected from the center of a galaxy is provided by another object, Cygnus A, shown as an optical image in Figure 25.6(a) and as a high-resolution radio map in Figure 25.6(b). Note the filamentary structure evident in the radio lobes and the thin, radio-emitting line joining the right lobe to the center of the visible galaxy

(the dot at the center of the radio image). These features strongly suggest that we are seeing two oppositely directed, narrow jets of material running into the intracluster gas filling the galaxy cluster of which Cygnus A is a member, then slowing down and spreading out to form the radio lobes. ∞ (Sec. 24.3) In some systems, known as *head–tail* radio galaxies, the lobes seem to form a "tail" behind the main galaxy. For example, the lobes of radio galaxy NGC 1265, shown in Figure 25.7, appear to be "swept back" by some onrushing wind and, indeed, this is the most likely explanation for the galaxy's appearance. If NGC 1265 were at rest, it would be just another double-lobe source, perhaps quite similar to Centaurus A. However, the galaxy is traveling through the intergalactic medium of its parent galaxy cluster (known as the Perseus Cluster), and the outflowing matter forming the lobes tends to be left behind as the galaxy moves.

COMMON FEATURES

Despite their obvious differences, we can establish some important links between Seyferts and radio galaxies. Figure 25.8 shows a *core–halo* radio galaxy, in which most of the energy is emitted from an extremely small central nucleus (which radio astronomers refer to as the *core*) less than 1 pc across, with weaker emission coming from an extended *halo* surrounding the nucleus. The halo typically measures about 50 kpc across, similar in size to the surrounding visible galaxy, which is usually elliptical and often quite faint. The radio luminosity from the nucleus can be

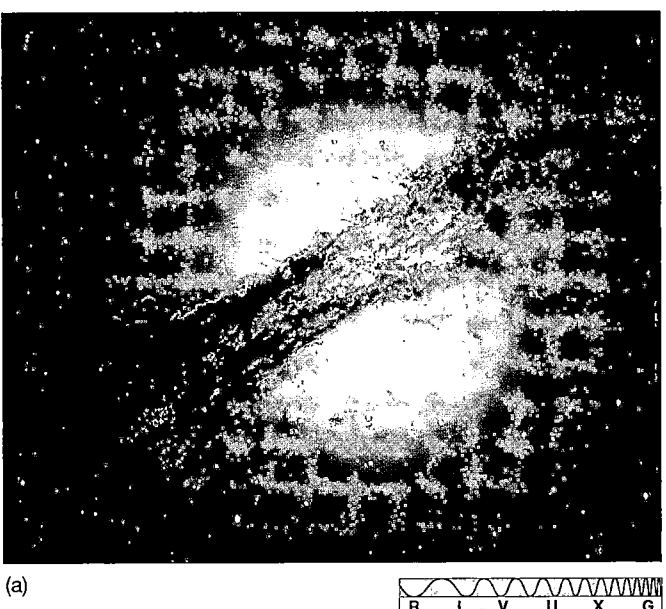

(a)

R   I   V   U   X   G

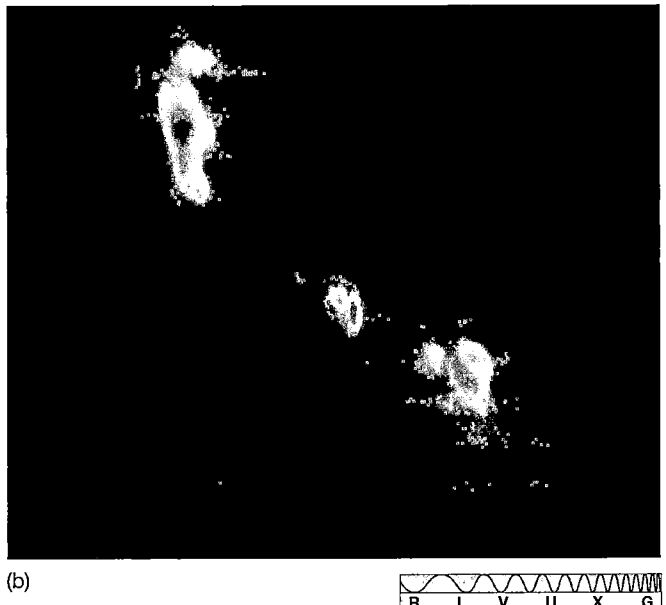

(b)

R   I   V   U   X   G

Figure 25.4 Centaurus A Radio Lobes Lobe-radio galaxies, such as Centaurus A shown here optically in (a), have giant radio-emitting regions (b) extending a million parsecs or more beyond the central galaxy. The lobes cannot be imaged in visible light and are observable only with radio telescopes. The lobes are shown here in false color, with decreasing intensity from red to yellow to green to blue. *(ESO; NRAO)*

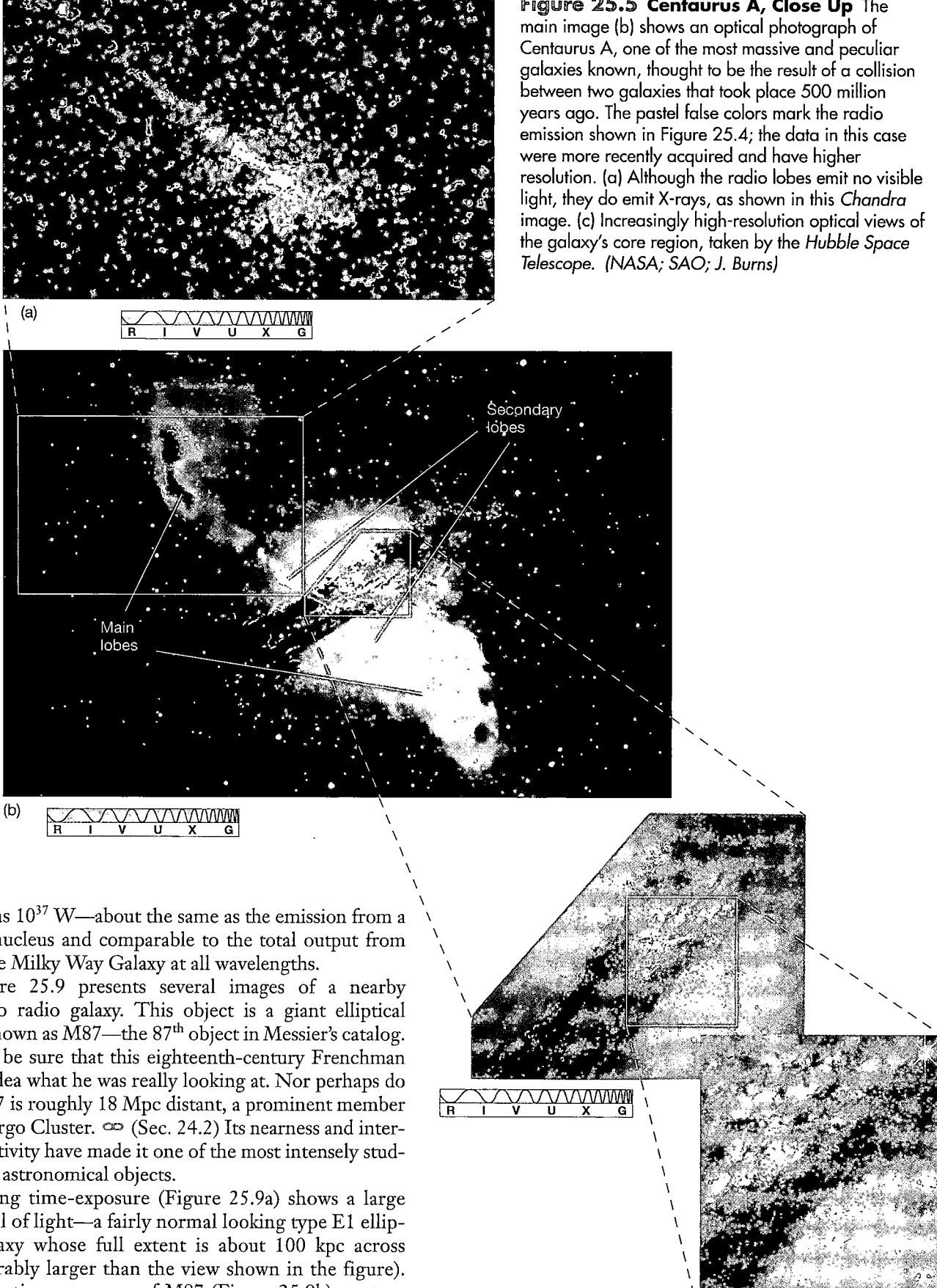

(a)

Figure 25.5 Centaurus A, Close Up The main image (b) shows an optical photograph of Centaurus A, one of the most massive and peculiar galaxies known, thought to be the result of a collision between two galaxies that took place 500 million years ago. The pastel false colors mark the radio emission shown in Figure 25.4; the data in this case were more recently acquired and have higher resolution. (a) Although the radio lobes emit no visible light, they do emit X-rays, as shown in this *Chandra* image. (c) Increasingly high-resolution optical views of the galaxy's core region, taken by the *Hubble Space Telescope*. *(NASA; SAO; J. Burns)*

Secondary lobes

Main lobes

(b)

(c)

as great as 10^{37} W—about the same as the emission from a Seyfert nucleus and comparable to the total output from the entire Milky Way Galaxy at all wavelengths.

Figure 25.9 presents several images of a nearby core–halo radio galaxy. This object is a giant elliptical galaxy known as M87—the 87th object in Messier's catalog. (We can be sure that this eighteenth-century Frenchman had no idea what he was really looking at. Nor perhaps do we!) M87 is roughly 18 Mpc distant, a prominent member of the Virgo Cluster. ∞ (Sec. 24.2) Its nearness and interesting activity have made it one of the most intensely studied of all astronomical objects.

A long time-exposure (Figure 25.9a) shows a large fuzzy ball of light—a fairly normal looking type E1 elliptical galaxy whose full extent is about 100 kpc across (considerably larger than the view shown in the figure). A shorter time-exposure of M87 (Figure 25.9b), capturing only the galaxy's bright inner regions, reveals a long thin *jet* of matter ejected from M87's center. The jet is about 2 kpc long and is traveling outward at very high speed, probably more than half the speed of light. Computer enhancement shows that the jet is made up of a series of distinct "blobs" more or less evenly spaced along its length. This high-speed jet, which itself emits energy at the rate of almost 10^{35} W (250 million times the luminosity of the Sun), has been imaged in the radio and infrared regions of the spectrum (Figures 25.9c and d) as well as in the visible.

Jets such as this are a very common feature of active galaxies. As we will see, they play a vital role in our understanding of these energetic objects. The M87 jet also suggests a simplifying connection between the two types of radio galaxies just discussed. It is likely that the differences between core–halo and lobe-radio galaxies are largely a matter of perspective (Figure 25.10). If we view the jets and lobes from the side, we see a lobe-radio galaxy, but if we view the jet almost head-on—in other words, looking *through* the lobe—we see a core-halo system. *More Precisely 25-1* discusses another curious characteristic of some active galaxies and quasars that also supports this view.

Both Seyferts and radio galaxies emit comparably large amounts of energy, and as we have seen, there is good evidence that the energy source in each is a compact region at the center of an otherwise relatively normal-looking galaxy. In lobe-radio galaxies, that energy is fired out from the nucleus in the form of narrow, high-speed jets of matter that travel into the intergalactic medium and become extended lobes far from the center of the galaxy. As a result, the energy from a lobe-radio galaxy is ultimately emitted (in the form of radio radiation) from a region well outside the visible galaxy. However, in *all* cases studied so far, the central compact nucleus is the place where the energy is actually produced.

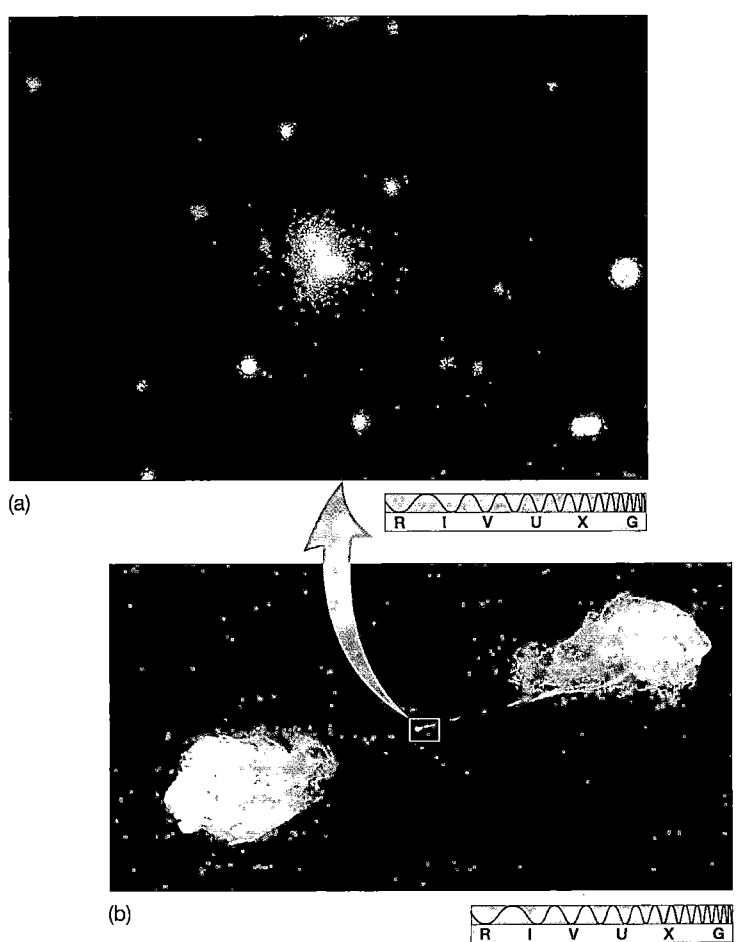

(a)

(b)

Figure 25.6 Cygnus A (a) Cygnus A also appears to be two galaxies in collision. (b) On a much larger scale, it displays radio-emitting lobes on either side of the optical image. The optical galaxy in (a) is about the size of the small dot at the center of (b). Note the thin line of radio-emitting material joining the right lobe to the central galaxy. The distance from one lobe to the other is approximately a million light-years. *(NOAO; NRAO)*

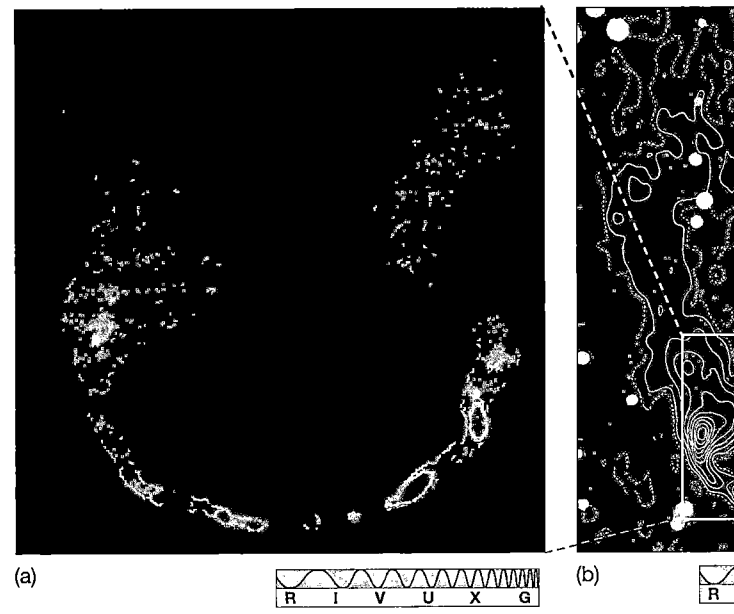

(a)   (b)

Figure 25.7 Head–Tail Radio Galaxy (a) Radiograph, in false color, of the active head–tail galaxy NGC 1265. (b) The same radio data, in contour form, superposed on the optical image of the galaxy. Astronomers reason that this object is moving rapidly through space, trailing a "tail" behind as it goes—a little like a comet, but on a vastly larger scale. *(NRAO; Palomar/Caltech)*

Figure 25.8 Core–Halo Radio Galaxy Radio contour map of a typical core–halo radio galaxy, the one near the center called M86. The radio emission from such a galaxy comes from a bright central nucleus, or core, surrounded by an extended, less intense halo. The radio map is superimposed on an optical image of the galaxy and some of its neighbors, a wider-field version of which was shown previously in Figure 24.15. *(Harvard-Smithsonian Center for Astrophysics)*

Figure 25.9 M87 Jet The giant elliptical galaxy M87 (also called Virgo A) is displayed here at several different wavelengths. (a) A long optical exposure of its halo and embedded central region. (b) A short optical exposure of its core and an intriguing jet of matter, on a smaller scale than (a). (c) A radio image of its jet, on a somewhat expanded scale compared with (b). The red dot at left marks the bright nucleus of the galaxy; the red and yellow blob near the center of the image corresponds to the bright "knot" visible in the jet in (b). (d) A near-infrared image of the jet, at roughly the same scale as (c). *(NOAO; NRAO; NASA)*

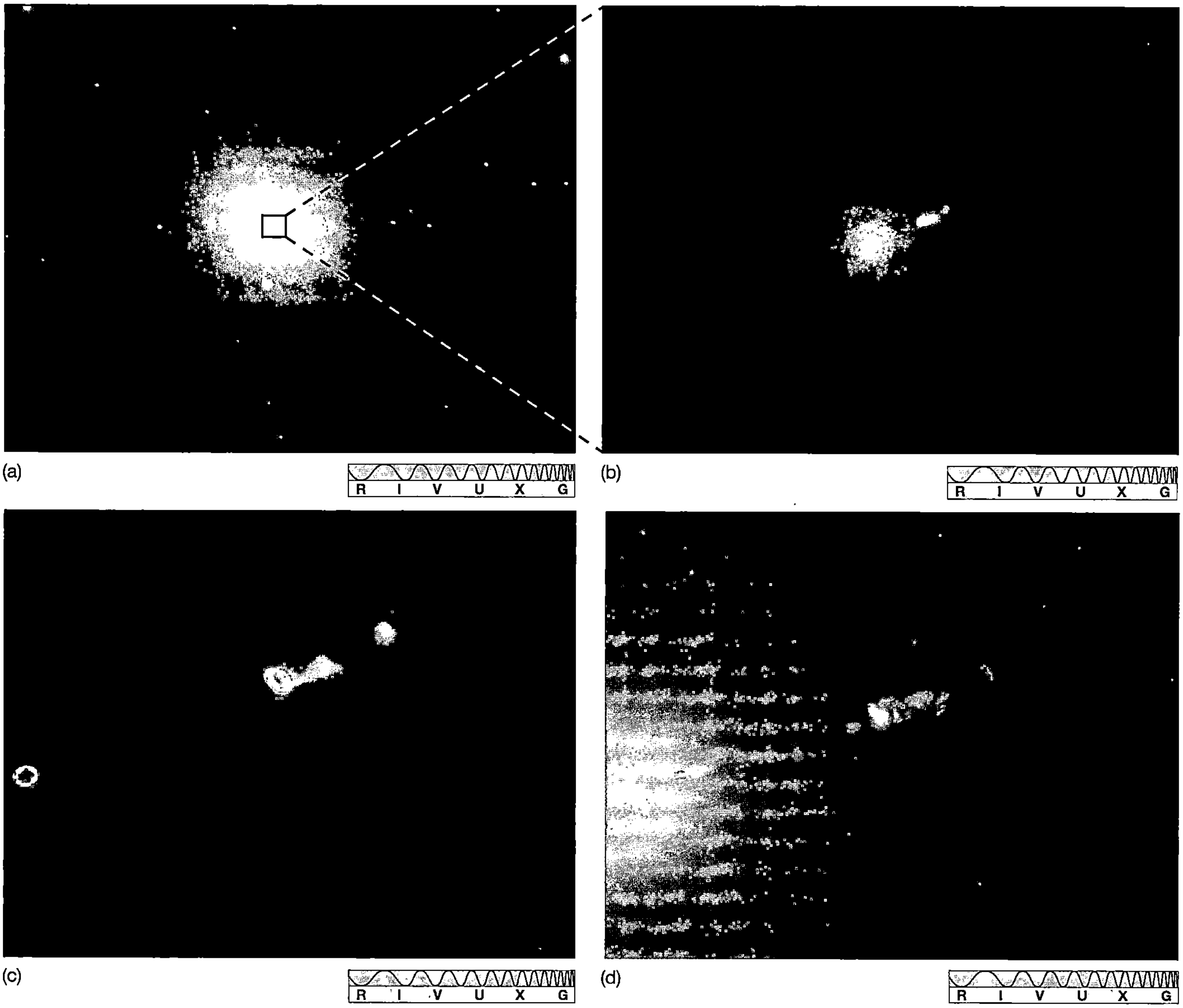

(a)

(b)

(c)

(d)

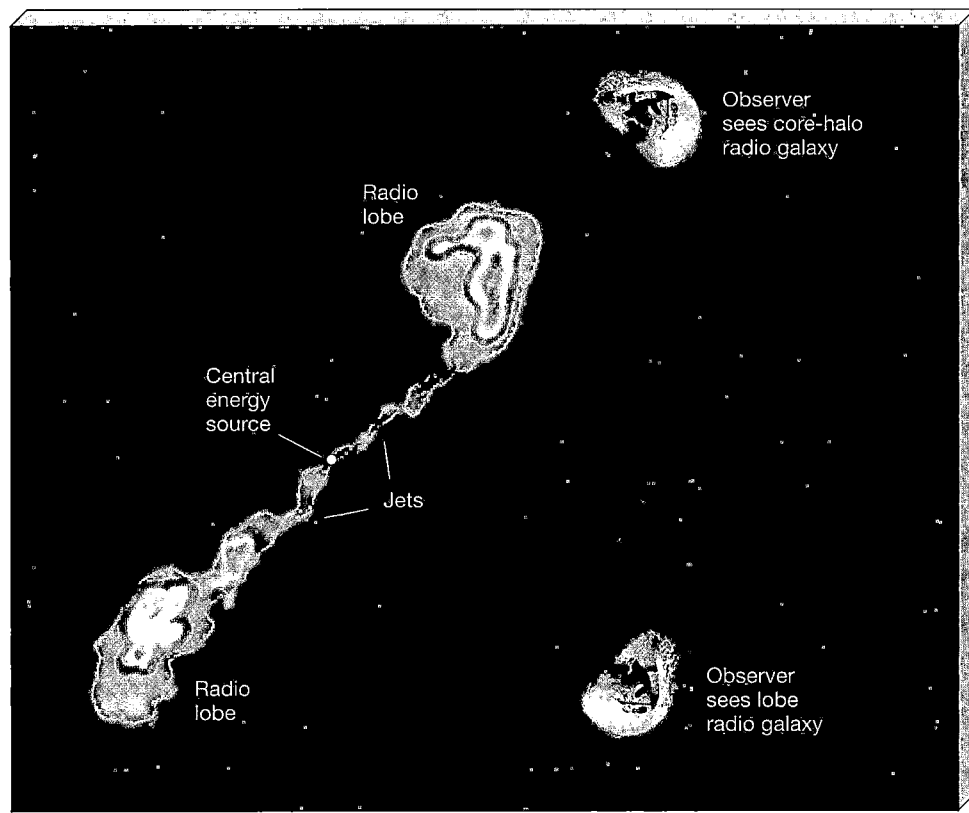

Figure 25.10 Radio Galaxy A central energy source produces high-speed jets of matter that interact with intergalactic gas to form radio lobes. The system may appear to us as either a lobe or a core–halo radio galaxy, depending on our location with respect to the jets and lobes.

Concept Check

■ What evidence do we have that the energy emitted by radio galaxies always originates in the galactic nucleus?

25.3 Quasi-Stellar Objects

In the early days of radio astronomy, many radio sources were detected for which no corresponding visible object was known. By 1960 several hundred such sources were listed in the *Third Cambridge Catalog*, and astronomers were scanning the skies in search of optical counterparts. Their job was made difficult both by the low resolution of the radio observations (which meant that the observers did not know exactly where to look) and by the faintness of these objects at visible wavelengths.

THE DISCOVERY OF QUASARS

In 1960, astronomers detected what appeared to be a faint blue star at the location of the radio source 3C 48 (the 48th object on the *Cambridge* list) and obtained its spectrum. Containing many unknown broad emission lines, the unusual spectrum defied interpretation. 3C 48 remained a unique curiosity until 1962, when another similar-looking, and similarly mysterious, faint blue object with "odd" spectral lines

was discovered and identified with the radio source 3C 273. Several of these peculiar objects are shown in Figure 25.11.

The following year saw a breakthrough when astronomers realized that the strongest unknown lines in 3C 273's spectrum were simply familiar spectral lines of hydrogen redshifted by a very unfamiliar amount—about 16 percent! This large redshift indicated a recessional velocity of about 44,000 km/s (see *More Precisely 25-2*). Figure 25.12 shows the spectrum of 3C 273. Some prominent emission lines, and the extent of their redshift, are marked. Once the reason for the strange spectral lines was known, astronomers quickly found a similar explanation for the spectrum of 3C 48. Its 37 percent redshift implied that it is receding from Earth at almost one-third the speed of light.

These huge speeds mean that neither of the two objects can possibly be members of our Galaxy. Applying Hubble's law (with our adopted value of the Hubble constant $H_0 = 65$ km/s/Mpc), we obtain distances of 700 Mpc for 3C 273 and 1540 Mpc for 3C 48. (*More Precisely 25-2* discusses in more detail how these distances are determined and what they mean.) Clearly not stars (with such enormous redshifts), these objects became known as *quasi-stellar radio sources* (quasi-stellar simply means "starlike"), shortened to **quasars**. Because we now know that not all such highly redshifted, starlike objects are strong radio sources, the term **quasi-stellar object** (or QSO) is more common today. However, the name quasar persists, and we will continue to use it here.

MORE PRECISELY 25-1

Faster-Than-Light Velocities?

The compact sources of emission within radio galaxies and quasars vary not only in intensity but also in structure. High-resolution maps of some active galactic nuclei, made using very-long-baseline interferometry, display dramatic changes in structure, often on time scales as short as months. For example, the accompanying illustration shows five sets of radio images of the core of the well-studied jet in the "nearby" galaxy M87 (see Section 25.2), made over a period of a few years. The images show several large blobs of gas, which move over the course of time (as indicated by the dashed lines). Knowing the distance to M87 (about 18 Mpc) and measuring the angle through which the blobs moved in the course of four years (about 0.08"), astronomers have calculated the blobs' velocities. Astonishingly, the result is nearly 6 times the speed of light!

The notion that the speed of light is the highest attainable velocity is central to modern physics. Scores of predictions made assuming this fact to be true have been verified to high accuracy since Einstein first published his special theory of relativity early in the twentieth century (*More Precisely 22-1*). Astronomers almost universally agree that some reinterpretation of this apparent *superluminal* (that is, faster-than-light) motion is needed.

Several possible explanations have been advanced, but the one most widely accepted among astronomers views superluminal motion as a simple projection effect, produced by blobs of gas moving almost along our line of sight at nearly the speed of light. As illustrated in the accompanying diagram, light emitted by a blob of gas barely outpaces the motion of the blob itself, with the result that light rays emitted at widely separated times may arrive at Earth almost simultaneously, making the transverse motion appear much faster than it really is, possibly even faster than light. This idea is supported by observations of superluminal motion in sources such as M87, where the jet is already known to exist. If we are indeed viewing this core–halo radio galaxy through one of its radio lobes, as astronomers suspect, then the jet feeding that lobe should be aimed almost directly at us, consistent with the superluminal motion we see.

Thus, the most probable explanation for this odd phenomenon is nothing more than basic geometry. Superluminal motion is just a cosmic illusion, made possible by the finite speed of light. The fundamental laws of modern physics have survived another observational assault!

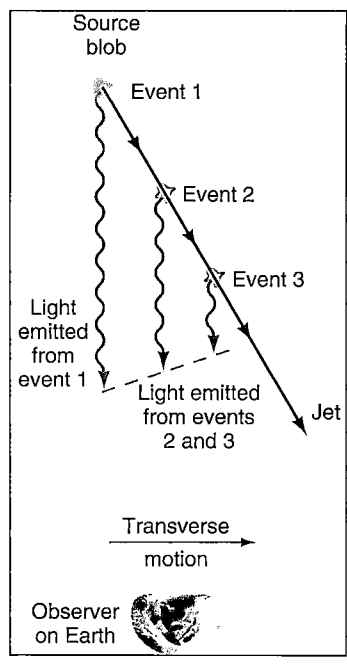

M87 1994 1995 1996 1997 1998

(NASA)

R   I   V   U   X   G

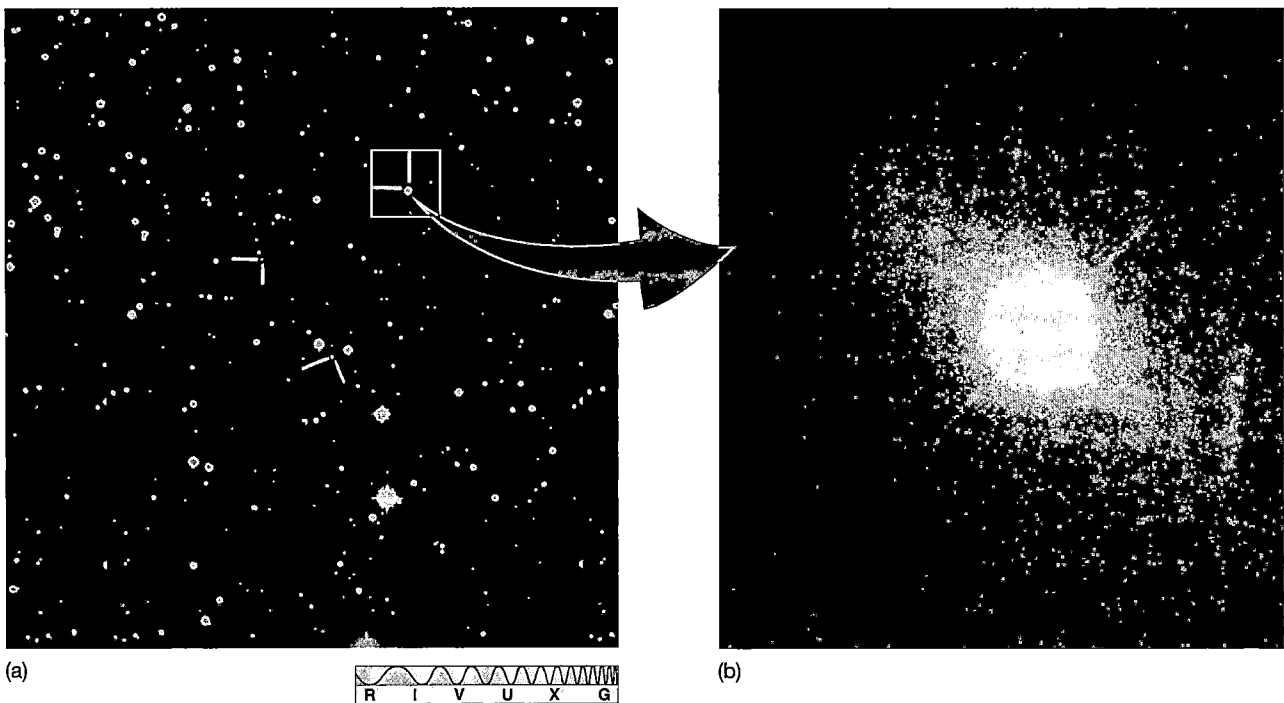

(a)

R I V U X G

(b)

Figure 25.11 Quasars (a) Optical image of a field of quasars (marked), including QSO 1229 + 204, one of the most powerful quasars yet discovered, shown enlarged in (b). Its starlike appearance shows only a hint of structure and gives little outward indication of the object's enormous luminosity. Like 3C 275, its distance from Earth is about 2000 Mpc. *(Palomar Observatory; NASA)*

OBSERVED PROPERTIES OF QUASARS

The most striking characteristic of the several hundred quasars now known is that their spectra all show large redshifts, ranging from 0.06 up to the current maximum of almost 6 (that is, a 600 percent increase in wavelength; see *More Precisely 25-2* for an explanation of the meaning of redshifts greater than 1). Thus, *all* quasars lie at large distances from us—the closest is 270 Mpc away, the farthest

more than 8000 Mpc distant (Table 25.1). The majority of quasars lie more than 1000 Mpc from Earth (with the observed numbers peaking at redshifts of 2–3). We therefore see most quasars as they existed long ago—they represent the universe as it was in the distant past.

Thus, despite their unimpressive optical appearance—see for example, Figure 25.13, which compares a quasar with a spiral galaxy that happens to lie close to it on the

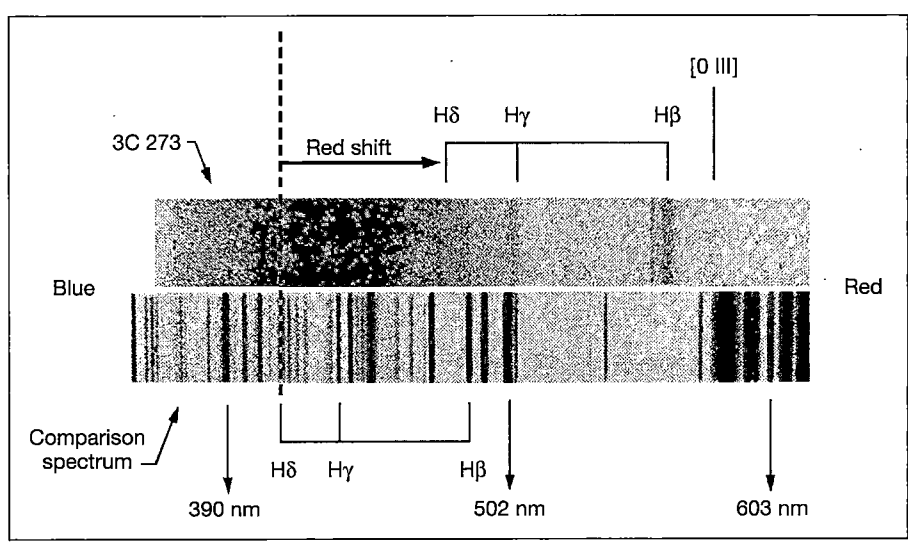

Figure 25.12 Quasar Spectrum
Optical spectrum of the distant quasar 3C 273. Notice both the redward shifts and the widths of the three hydrogen spectral lines marked as Hβ, Hγ, and Hδ. The redshift indicates the quasar's enormous distance. The width of the lines implies rapid internal motion within the quasar itself. (Note that in this figure, red is to the right and blue is to the left.)
(Palomar/Caltech)

sky—the large distances implied by quasar redshifts mean that these faint "stars" are in fact the brightest-known objects in the universe! 3C 273, for example, has a luminosity of about 10^{40} W—20 trillion Suns or 1000 times the luminosity of the Milky Way. In fact, this is a fairly typical value, although observed quasar luminosities range from about 10 to 100,000 times that of the Milky Way. Thus quasars generally outshine the brightest normal and active galaxies by about a factor of 100–1000.

Quasars display many of the same general properties as active galaxies. Their radiation spectrum is nonstellar, and some show evidence of jets and extended emission features (although few quasars are more than a ball of luminous fuzz in visible-light images). Figure 25.14 is an optical photograph of 3C 273. Notice the jet of luminous matter, reminiscent of the jet in M87, extending nearly 3 kpc from the quasar itself. Often, as shown in Figure 25.15, quasar radio radiation arises from regions lying beyond the bright central core, much like the emission from core–halo and lobe-radio galaxies. In other cases the radio emission is confined to the central optical image. Quasars have been observed in the radio, infrared, optical, ultraviolet, and X-ray parts of the electromagnetic spectrum, and some have even been found to emit gamma rays. However, most quasars emit most of their energy in the infrared.

Many quasars have been observed to vary irregularly—and by large factors—in brightness over periods of months, weeks, days, or (in some cases) even hours, in many parts of the electromagnetic spectrum. The same reasoning as we used earlier for active galaxies leads to the conclusion that the region generating the energy must be very *small*—not too much larger than our solar system in some cases.

✓ Concept Check

■ Why did astronomers initially have difficulty recognizing quasars as very luminous, very distant objects?

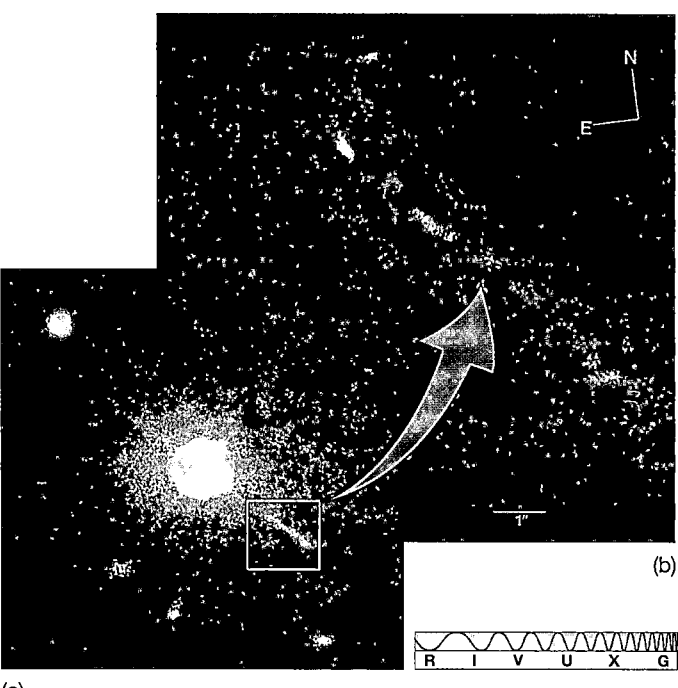

(a)

(b)

Figure 25.14 Quasar 3C 273 (a) The bright quasar 3C 273 displays a luminous jet of matter, but the main body of the quasar is starlike in appearance. (b) The jet extends for about 30 kpc and can be seen better in this high-resolution image. *(AURA)*

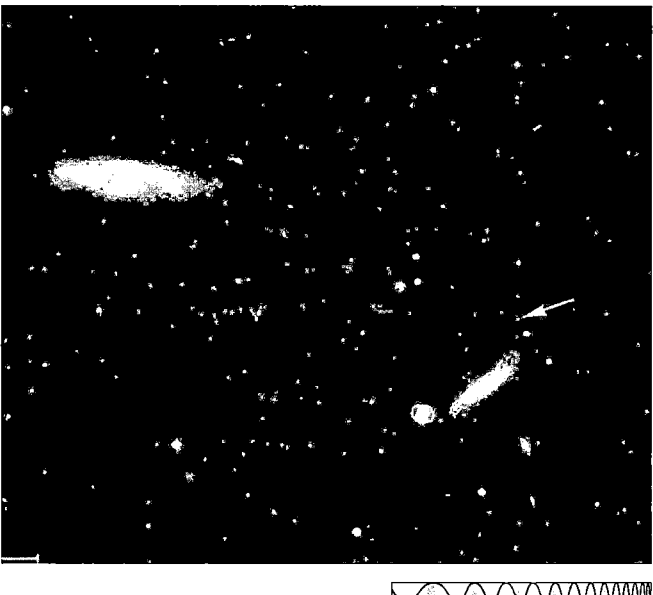

Figure 25.13 Typical Quasar Although quasars are the most luminous objects in the universe, they are often rather unimpressive in appearance. In this optical image, a distant quasar (marked by an arrow) is seen close (on the sky) to a nearby spiral galaxy. The quasar's much greater distance makes it appear much fainter than the galaxy. *(NOAO)*

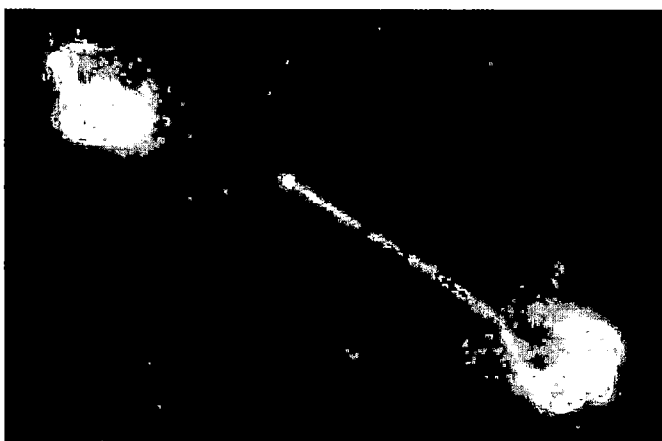

Figure 25.15 Quasar Jets Radio image of the quasar 3C 175, showing radio jets feeding faint radio lobes. The bright (white) central object is the quasar, some 1500 Mpc away. The lobes themselves span approximately a million light-years. *(NRAO)*

25.4 The Central Engine of an Active Galaxy

The behavior of active galaxies and quasars is contrary to that expected from vast collections of stars. Can we explain their enormous nonstellar energy output in terms of known physics? Remarkably, the answer is yes. The present consensus among astronomers is that, despite their great differences in appearance and luminosity, Seyferts, radio galaxies, and quasars almost certainly share a common energy-generation mechanism. The energy can be reprocessed into many different forms before it is finally emitted into intergalactic space, but the basic engine is probably the same in all cases.

As a class, active galaxies and quasars show some or all of the following properties:

1. They have *high luminosities*, generally much greater than the 10^{37} W characteristic of a fairly bright, normal galaxy like the Milky Way.

2. Their energy emission is mostly *nonstellar*—it cannot be explained as the combined radiation of even trillions of stars.

3. Their energy output can be highly *variable*, implying that it is emitted from a small central nucleus much less than a parsec across.

4. They often exhibit *jets* and other signs of explosive activity.

5. Their optical spectra may show broad emission lines, indicative of *rapid internal motion* within the energy-producing region.

The principal questions then are: How can such vast quantities of energy arise from these relatively small regions of space? Why is so much of the energy radiated at low frequencies, especially in the radio and infrared? And what is the origin of the extended radio-emitting lobes and jets? Let us first consider how the energy is produced.

ENERGY PRODUCTION

To develop a feeling for the energetics involved, consider for a moment an object near the "low end" of the range, having a luminosity of 10^{38} W. In and of itself, this energy output is not inconceivably large. The brightest giant ellipticals are comparably powerful. Thus, some 10^{12} stars—a few normal galaxies' worth of material—could *equivalently* power a typical active galaxy. The difficulty arises when we consider that in an active galaxy this energy production is packed into an object much less than a parsec in diameter!

It is difficult to imagine how several Milky Way Galaxies could be compressed into a space no larger than a parsec. Even if we could somehow squeeze that much mass into such

a volume, it would immediately collapse to form a huge black hole, and none of the light it produced could escape. Thus, even neglecting its nonstellar spectrum, the total energy output of an active galaxy simply cannot be explained as the combined energy of many stars. We must think of something else.

The twin requirements of large energy generation and small physical size bring to mind our discussion of X- and gamma-ray sources in Chapter 22. ∞ (Secs. 22.3, 22.8) The presence of the jets in M87 and 3C 273, and the ejection of matter to form radio lobes in Centaurus A and Cygnus A strengthen the connection. Recall that the best current explanation for those "small-scale" phenomena involves the accretion of material onto a compact object—a neutron star or a black hole. Large amounts of energy are produced as matter spirals down onto the central object, and high-speed jets may well be a common by-product of the process (see Figure 22.8). In Chapter 23, we suggested that a similar mechanism, involving a *supermassive black hole*—one with a mass of around a million suns—may also be responsible for the energetic radio and infrared emission observed at the center of our own Galaxy. ∞ (Sec. 23.7)

As illustrated in Figure 25.16, the leading model for the central engine of active galaxies and quasars is a scaled-up version of the same accretion process, only now the black holes involved have masses between a few million and a billion times the mass of the Sun. As with its smaller-scale counterparts, infalling gas forms an accretion disk and spirals down toward the hole. It is heated to high temperatures by friction within the disk and emits large amounts of radiation as a result. In this case, however, the origin of the accreted gas is not a binary companion, as in stellar X-ray sources, but entire stars and clouds of interstellar gas that come too close to the hole and are torn apart by its strong gravity.

Accretion is extremely efficient at converting infalling mass (in the form of gas) into energy (in the form of electromagnetic radiation). Detailed calculations indicate that as much as 10 or 20 percent of the total mass-energy of the infalling matter can be radiated away before it crosses the hole's event horizon and is lost forever. Since the total mass-energy of a star like the Sun—the mass times the speed of light squared—is about 2×10^{47} joules, it follows that the 10^{38}-W luminosity of a bright active galaxy can be accounted for by the consumption of only 1 solar mass of gas per decade by a billion-solar-mass black hole. ∞ (Sec. 22.4) Less luminous active galaxies would require correspondingly less fuel—for example, a 10^{36}-W Seyfert galaxy would devour only one star's worth of material every thousand years.

In this picture the small size of the emitting region is a direct consequence of the compact nature of the central black hole. Even a billion-solar-mass hole has a radius of only 3×10^9 km, or 10^{-4} pc—about 20 A.U.—and theory suggests that the part of the accretion disk responsible for most of the emission is much less than a parsec across.

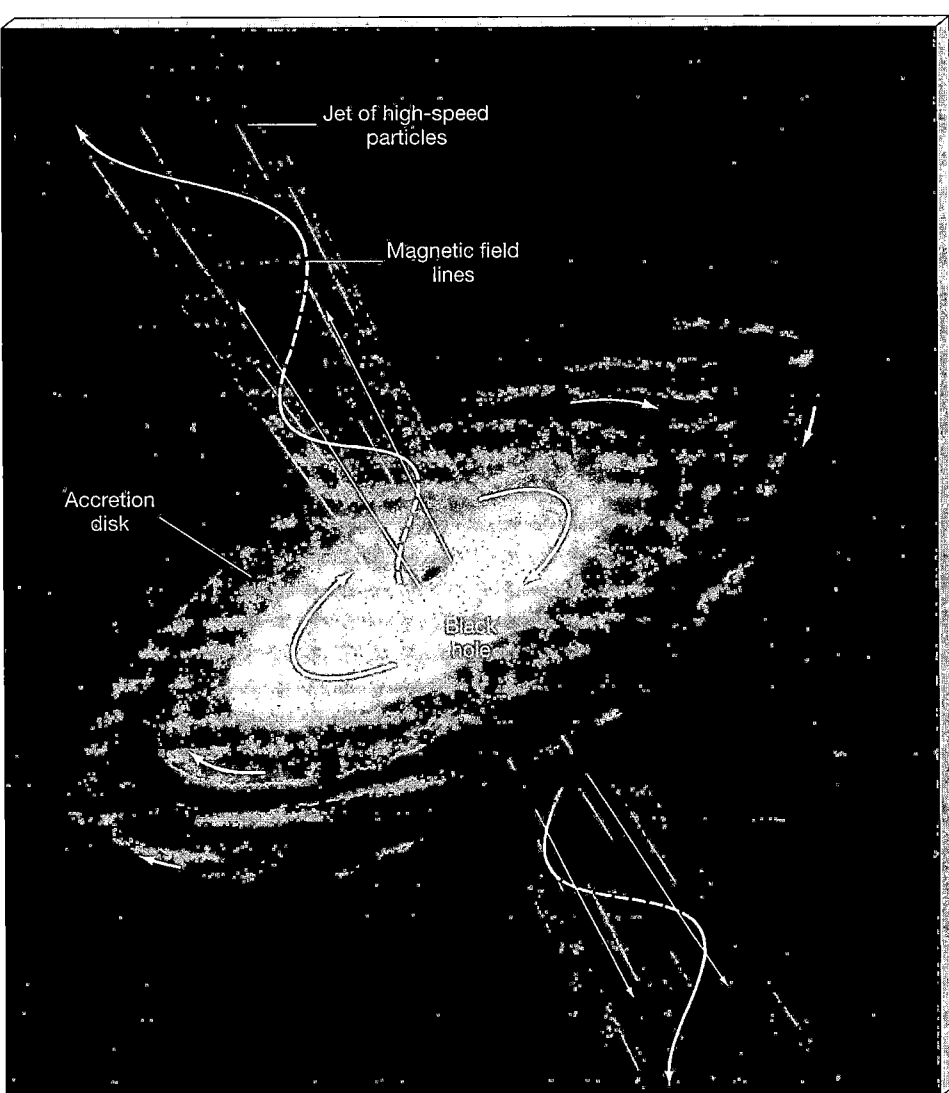

Jet of high-speed particles

Magnetic field lines

Accretion disk

Black hole

Figure 25.16 Active Galaxy Nucleus The leading theory for the energy source in active galactic nuclei (and quasars) holds that these objects are powered by material accreting onto a supermassive black hole. As matter spirals toward the hole, it heats up, producing large amounts of energy. At the same time, high-speed beams of gas may be ejected perpendicular to the accretion disk, forming the jets and lobes seen in many active objects. Magnetic fields generated in the disk are carried by the jets out to the radio lobes, where they play a crucial role in producing the observed radiation.

Instabilities in the accretion disk can cause fluctuations in the energy released, leading to the variability observed in many objects. The broadening of the spectral lines observed in the nuclei of Seyferts and quasars results from the rapid orbital motion of the gas in the hole's intense gravity.

Recent observations of galaxies in the Virgo Cluster by the *Hubble Space Telescope* lend strong support to this general picture. Figure 25.17 shows an image of a disk of gas and dust apparently feeding a possible black hole at the core of a giant elliptical galaxy. As expected from the theory just described, the disk is perpendicular to the huge jets emanating from the center of the active galaxy.

Hubble has also allowed astronomers to probe the fine details of the Virgo Cluster's most prominent object—the huge M87 galaxy (Figure 25.9)—and what they have found is again in excellent agreement with the idea that the energy is produced by accretion onto a large black hole. At M87's distance, *Hubble's* resolution of 0.05" corresponds to

a distance of about 5 pc, so we are still far from seeing the (solar system-sized) central black hole itself, but the improved "circumstantial" evidence has convinced many doubters of the correctness of the theory. Figure 25.18 shows imaging and spectroscopic data that suggest the existence of a rapidly rotating disk of matter orbiting the galaxy's center, perpendicular to the jet. Measurements of the gas velocity on opposite sides of the disk indicate that the mass within a few parsecs of the center is approximately 3×10^9 solar masses—we assume that this is the mass of the central black hole.

Perhaps the most compelling evidence for supermassive black holes in galaxies comes from radio studies. Using the Very Long Baseline Array—a continent-wide network of 10 radio telescopes—a U.S.-Japanese team was able to achieve angular resolution hundreds of times better than that attainable with the *Hubble Space Telescope*. ∞ (Sec. 5.5) VLBA observations of NGC 4258, a spiral galaxy about 6 Mpc away, have uncovered a group of

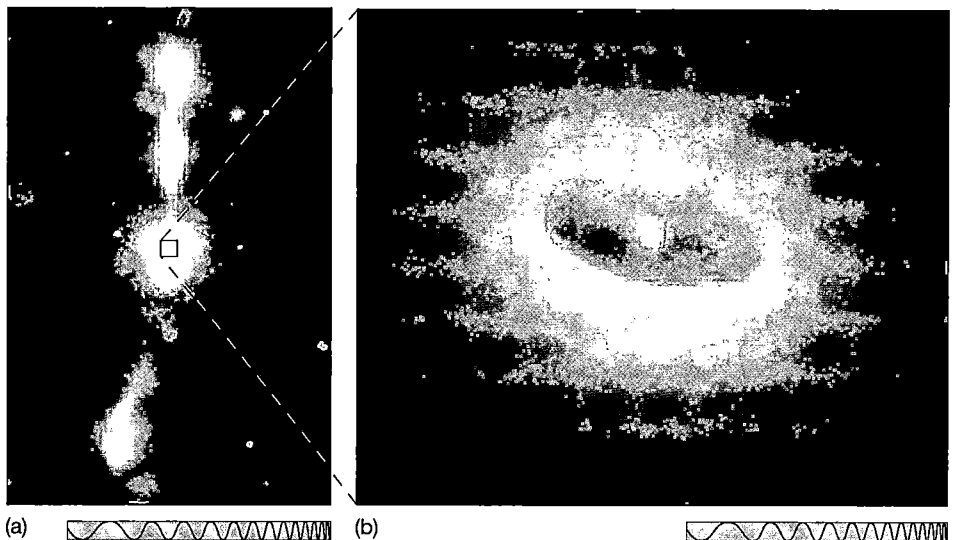

(a)

(b)

Figure 25.17 Giant Elliptical Galaxy (a) A combined optical-radio image of the giant elliptical galaxy NGC 4261, in the Virgo Cluster, shows a white visible galaxy at center from which blue to yellow (false-color) radio lobes extend for about 60 kpc. (b) A close-up photograph of the galaxy's core reveals a 100-pc-diameter disk surrounding a bright hub that presumably harbors a black hole. (NRAO; NASA)

molecular clouds swirling in an organized fashion about the galaxy's core. The astronomers have detected the red- and blueshifts of water-vapor spectral lines in those far-away clouds. As depicted in Figure 25.19, the pattern revealed by the clouds is that of a slightly warped and spinning disk centered precisely on the galaxy's heart. The rotation velocities imply the presence of more than

40 million solar masses all packed within a region less than 0.2 pc across.

ENERGY EMISSION

The model just described is now widely accepted as the correct picture of how active galaxies and quasars generate

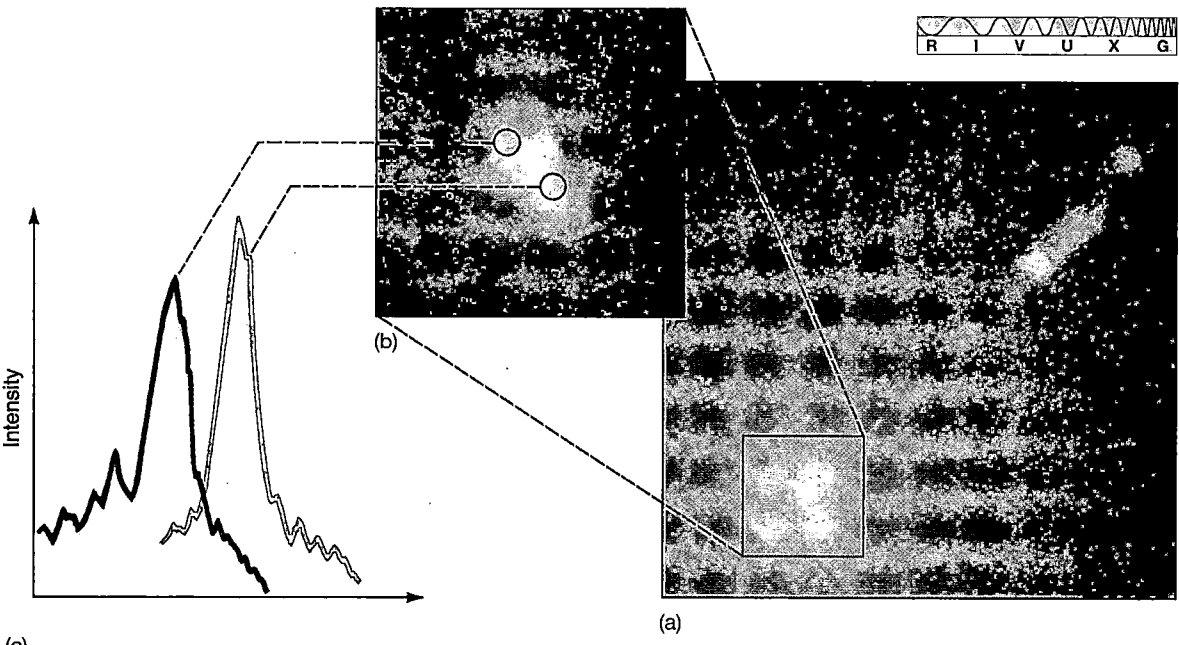

Intensity

(b)

(a)

(c)

Figure 25.18 M87 Disk Imaging and spectroscopic observations of M87 support the idea of a rapidly whirling accretion disk at its core. (a) An image of the central region of M87, similar to that shown in Figure 25.9(d), shows its bright core and jet. The scale is comparable to the scale of Figure 25.9(d). (b) A magnified view of the core suggests a spiral swarm of stars, gas, and dust. (c) Spectral-line features observed on opposite sides of the core show opposite Doppler shifts, implying that the material there is coming toward us on one side and moving away from us on the other. The strong implication is that an accretion disk spins perpendicular to the jet and at its center is a black hole having some 3 billion times the mass of the Sun. (NASA)

their enormous power. It explains the observed facts and has the added advantage of resembling the processes thought to power smaller-scale phenomena, such as stellar X-ray sources and normal galactic nuclei. Having accounted for the source of the energy in active galaxies, let's now turn to the way in which it is eventually emitted into intergalactic space.

In general, the theoretically predicted radiation spectra for matter accreting onto a black hole do not match well with the spectra observed in active galaxies. To account for the details of the spectra of Seyfert galaxies and quasars, it is necessary to modify the mechanism just described by adding the assumption that the energy emitted from the accretion disk is "reprocessed"—that is, absorbed and reemitted at infrared and longer wavelengths—by gas and dust surrounding the nucleus.

•A different reprocessing mechanism is responsible for the spectra of the jets and lobes seen in many active galaxies, especially radio galaxies. The jets consist of material

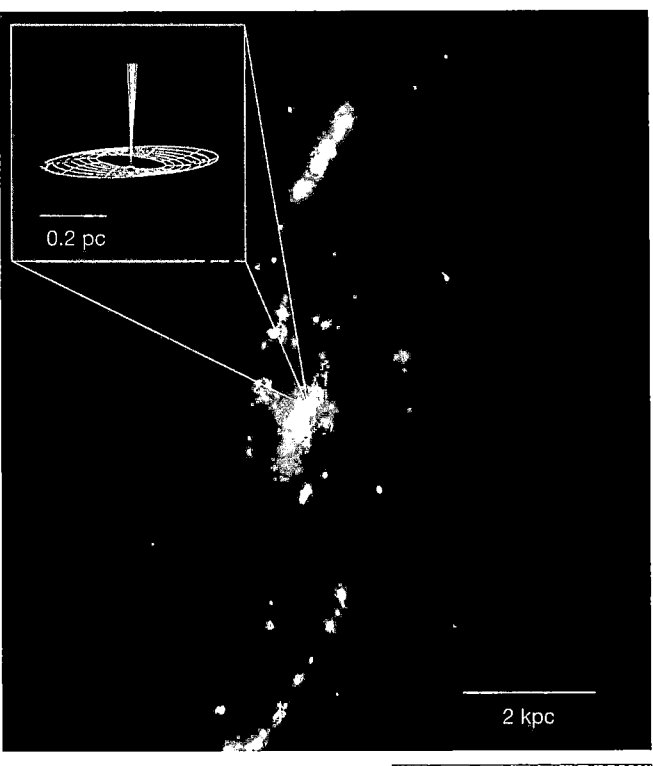

Figure 25.19 Galactic Black Hole A network of radio telescopes has probed the core of the spiral galaxy NGC 4258, shown here in the light of mostly hydrogen emission. Within the innermost 0.2 pc (inset), observations of Doppler-shifted molecular clouds (designated by red, green, and blue dots) show that they obey Kepler's third law perfectly, and have revealed a slightly warped disk of rotating gas (shown here in an artist's conception). At the center of the disk presumably lurks a huge black hole. *(J. Moran)*

(mainly protons and electrons) blasted out into space—and out of the visible portion of the galaxy entirely—from the inner regions of the disk. The details of how jets form remain uncertain, but there is a growing consensus among theorists that jets are a common feature of accretion flows, large and small. The jets also contain strong *magnetic fields* (shown in Figure 25.16), possibly generated by the motion of the gas swirling within the disk; the fields accompany the gas as it leaves the galaxy.

As sketched in Figure 25.20(a), whenever a charged particle (here an electron) encounters a magnetic field, it tends to spiral around the field lines. We have encountered this idea several times previously, in a variety of different contexts (see, for example, the discussion of planetary magnetospheres in Chapters 7 and 11, or solar activity in Chapter 16). As the particles whirl around, they emit electromagnetic radiation, as discussed in Chapter 3. ∞ (Sec. 3.2) The faster the particles move, or the stronger the magnetic field, the greater the amount of energy radiated. In most cases, the fastest-moving particles are the low-mass electrons, so they are responsible for most of the radiation we observe.

The electromagnetic radiation produced in this way—called **synchrotron radiation**, after the type of particle accelerator in which it was first observed—is *nonthermal*: There is no link between the emission and the temperature of the radiating object, so the radiation is not described by a blackbody curve. Instead, its intensity decreases with increasing frequency, as shown in Figure 25.20(b). This is just what is needed to explain the overall spectrum of radiation recorded from active galaxies (compare Figure 25.20b with Figure 25.1). Observations of the radiation received from the jets and radio lobes in active galaxies are completely consistent with this process.

Eventually, the jet is slowed and stopped by the intergalactic medium, the flow becomes turbulent, and the magnetic field grows tangled. The result is a giant radio lobe, like those pictured in Figures 25.6–8, emitting virtually all its energy in the form of synchrotron radiation. Even though the radio emission comes from an enormously extended volume of space that dwarfs the visible galaxy, the *source* of the energy is still the (relatively) tiny accretion disk—a billion billion times smaller in volume than the radio lobe—lying at the galactic center. The jets serve merely as a conduit to transport energy from the nucleus, where it is generated, into the lobes, where it is finally radiated into space.

The existence of the inner lobes of Centaurus A and the blobs in M87's jet imply that jet formation may be an intermittent process. There is also evidence to suggest that much of, if not all, the activity observed in nearby active galaxies could have been sparked by recent interaction with a neighbor. Many nearby active galaxies (Centaurus A, for example) appear to have been "caught in the act" of interacting with another galaxy, suggesting that the fuel supply can sometimes be turned on by a companion. Just

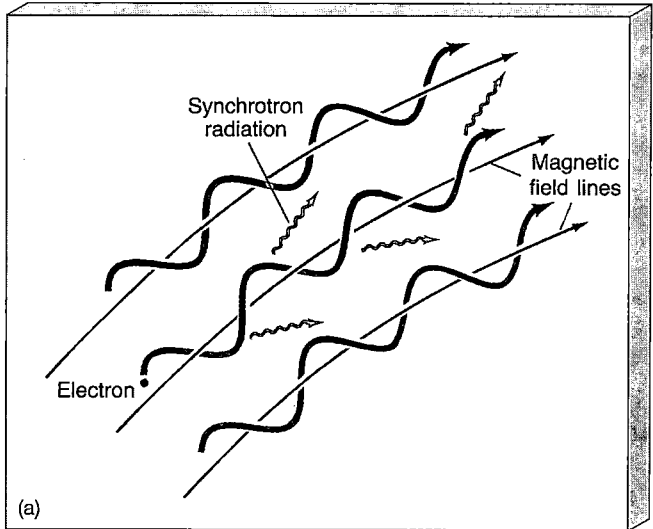

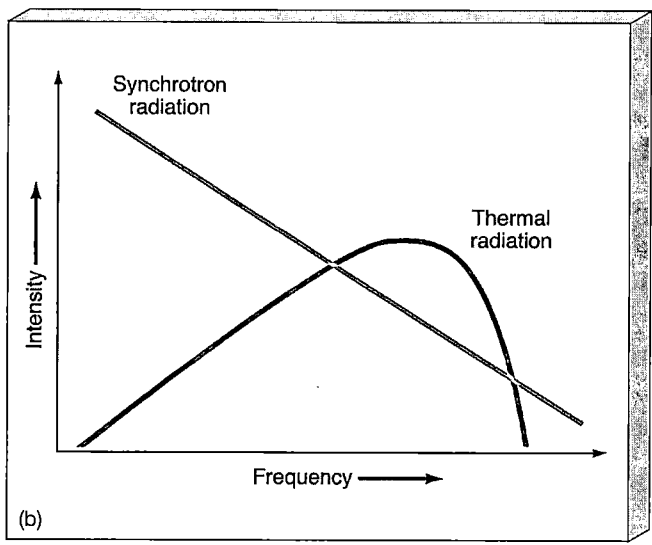

Figure 25.20 Nonthermal Radiation (a) Charged particles, especially fast electrons (red), emit synchrotron radiation (blue) while spiraling along magnetic field lines (black lines). This process is not confined to active galaxies. It occurs, on smaller scales, when charged particles interact with magnetism in Earth's Van Allen belts (see Chapter 7), when charged matter arches above sunspots on the Sun (see Chapter 16), in the vicinity of neutron stars (see Chapter 22), and at the center of our own Galaxy (see Chapter 23). (b) Variation in the intensity of thermal and synchrotron (nonthermal) radiation with frequency. Thermal radiation, described by a blackbody curve, peaks at some frequency that depends on the temperature of the source. Nonthermal synchrotron radiation, by contrast, is most intense at low frequencies. It is independent of the temperature of the emitting object. Compare this figure with Figure 25.1.

as tidal forces can trigger star formation in starburst galaxies (Chapter 24), they may also divert gas and stars into the galactic nucleus, triggering an outburst that may last for many millions of years. ∞ (Sec. 24.4)

QUASAR LIFETIMES

⑤ A 10^8- or 10^9-solar-mass black hole can emit enough energy to power even the brightest (10^{38}-W) radio galaxy by swallowing stars and gas at the relatively modest rate of one star every 10 years. To power a 10^{40}-W quasar, which is 100 times brighter, the hole simply consumes 100 times more fuel—10 stars per year. The reprocessing mechanisms that convert the quasar's power into the radiation we actually detect—namely, the ejection of matter in jets and lobes and the reemission of radiation by surrounding gas and dust—probably operate in much the same manner as in Seyfert and radio galaxies. The most likely explanation for the large luminosities of quasars is simply that there was more fuel available at very early times (Section 25.6). At the distances of most quasars, the galaxies themselves cannot easily be seen. Often, only their intensely bright nuclei are visible from Earth.

In this picture the brightest known quasars devour about 1000 solar masses of material every year. A simple calculation indicates that if they kept up this rate of energy production for the roughly 10-billion-year age of the uni-

verse, a total of 10^{13} stars would have been destroyed. Unless the galaxies housing quasars are much larger than any other galaxy we know of, most of the quasar's parent galaxy would be completely consumed, and the universe should contain many 10^{13}-solar-mass black holes—"burned-out" quasars. We have no evidence of the existence of any such objects.

The most reasonable way around this problem is to suppose that a quasar spends only a fairly short period of time in this highly luminous phase—perhaps a few tens of millions of years. There is theoretical evidence to suggest that black holes tend to eat out "cavities" at the centers of their host galaxies, effectively cutting off their fuel supply through their own greed. A given quasar may have more than one active episode, however. As with nearby active galaxies, quasar outbursts are probably the result of interactions between galaxies in the early universe and may therefore occur many times. The fact that quasars have been observed in some distant galaxy clusters argues in favor of this view, and it is consistent with modern theories of galaxy formation (Section 25.6).

✓ Concept Check

■ How does accretion onto a supermassive black hole power the energy emission from a lobe-radio galaxy?

25.5 Quasars as Cosmic Probes

⑤ Astronomers can take advantage of the great distances, point-like appearance, and large luminosities of quasars to probe the contents and structure of the universe. Since quasars are so far away, the line of sight from Earth to any given quasar has a good chance of intersecting something "interesting." By analyzing quasar images and spectra, it is possible to piece together a partial picture of the intervening space.

This approach is reminiscent of the use of bright stars to probe the interstellar medium near the Sun, and it suffers from the same basic drawback—we can only study regions of the sky where quasars happen to be found. ∞ (Sec. 18.3) However, this problem is expected to diminish in time as ongoing and planned large-scale surveys scan the sky for fainter and fainter objects. For example, the Sloan Digital Sky Survey, an ambitious project currently in progress, expects to construct a map of much of the northern sky by 2005, including several million galaxies and over 100,000 quasars.

QUASAR ABSORPTION LINES

In addition to their own strongly redshifted spectra, many quasars also show additional absorption features that are redshifted by substantially less than the lines from the quasar itself. For example, the quasar known as PHL 938 has an emission-line redshift of 1.955, placing it at a distance of some 5300 Mpc, but it also shows three sets of absorption lines, with redshifts of 1.949, 1.945, and 0.613, respectively. The first two sets may well come from high-speed gas within the quasar itself (the differences in recession velocity are only a few hundred kilometers per second), but the third is interpreted as arising from intervening gas that is much closer to us (only about 2400 Mpc away), so it has a smaller redshift than the quasar itself.

The most likely possibility is that this gas is part of an otherwise invisible galaxy lying along the line of sight.

Quasar spectra, then, afford astronomers a means of probing previously undetected parts of the universe. The absorption lines of atomic hydrogen are of particular interest, because that element makes up so much of all matter in the cosmos. As shown in Figure 25.21, when astronomers observe the spectrum of a high-redshift quasar, they typically see a "forest" of absorption lines starting at the (redshifted) wavelength of the quasar's own Lyman-alpha emission line and extending to shorter wavelengths. ∞ (More Precisely 4-1) These lines are interpreted as Lyman-alpha absorption features produced by gas clouds in foreground structures— galaxies, clusters, and so on—along the line of sight.

The quasar light thus explores an otherwise invisible component of cosmic gas. In principle, every intervening cloud of atomic hydrogen leaves its own characteristic imprint on the quasar's spectrum, in a form that lets us explore the distribution of matter in the universe. By comparing these Lyman-alpha forests with the results of simulations, astronomers hope to refine many key elements of the theories of galaxy formation and the evolution of large-scale structure.

QUASAR "MIRAGES"

In 1979, astronomers were surprised to discover what appeared to be a binary quasar—two quasars with exactly the same redshift and very similar spectra, separated by only a few arc seconds on the sky. Remarkable as the discovery of such a binary would have been, the truth about this pair turned out to be even more amazing. Closer study of the quasars' radio emission revealed that they were not in fact two distinct objects. Instead, they were two separate images of the *same* quasar! Optical views of such a *twin quasar* are shown in Figure 25.22.

What could produce such a "doubling" of a quasar image? The answer is *gravitational lensing*—the deflection

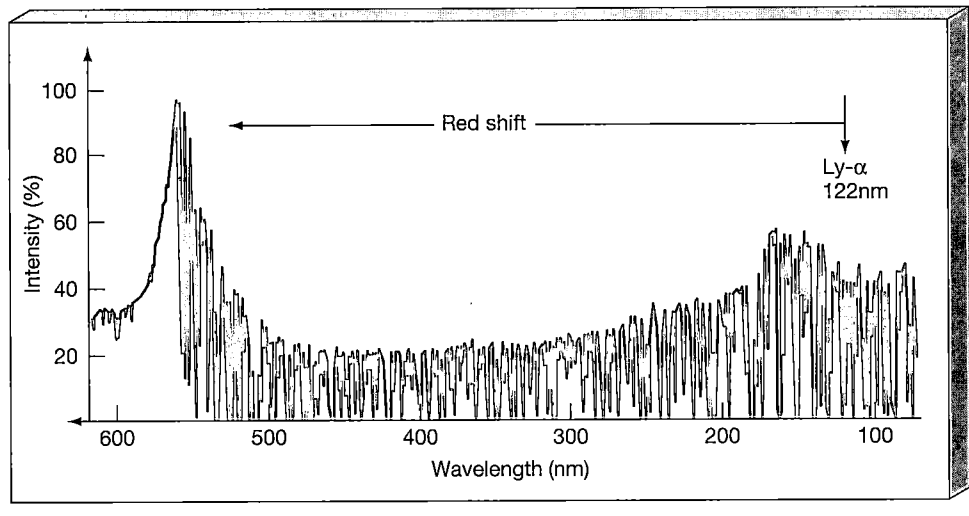

Figure 25.21 Lyman-Alpha Forest When light from a distant quasar passes through a foreground cloud of atomic hydrogen, the absorption lines produced all bear the cloud's redshift. The huge number of absorption lines in the spectrum of this quasar (QSO 1422 + 2309) are the ultraviolet Lyman-alpha lines from hundreds of clouds of foreground hydrogen gas, each redshifted by a slightly different amount (but less than the quasar itself). The peak marks the Lyman-alpha emission line from the quasar, emitted at 122 nm but redshifted here to a wavelength of 564 nm, in the visible range.

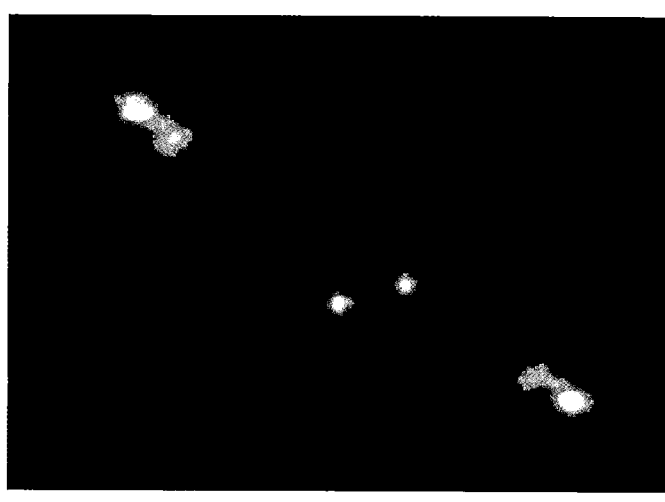

Figure 25.22 Twin Quasar This "twin" quasar (designated AC 114 and located about 2 billion pc away) is not two separate objects at all. Instead, the two large blobs (at upper left and lower right) are images of the same object, created by a gravitational lens; they have striking symmetry. The lensing galaxy itself is probably not visible in this image—the two objects near the center of the frame are thought to be unrelated galaxies in a foreground cluster. *(NASA)*

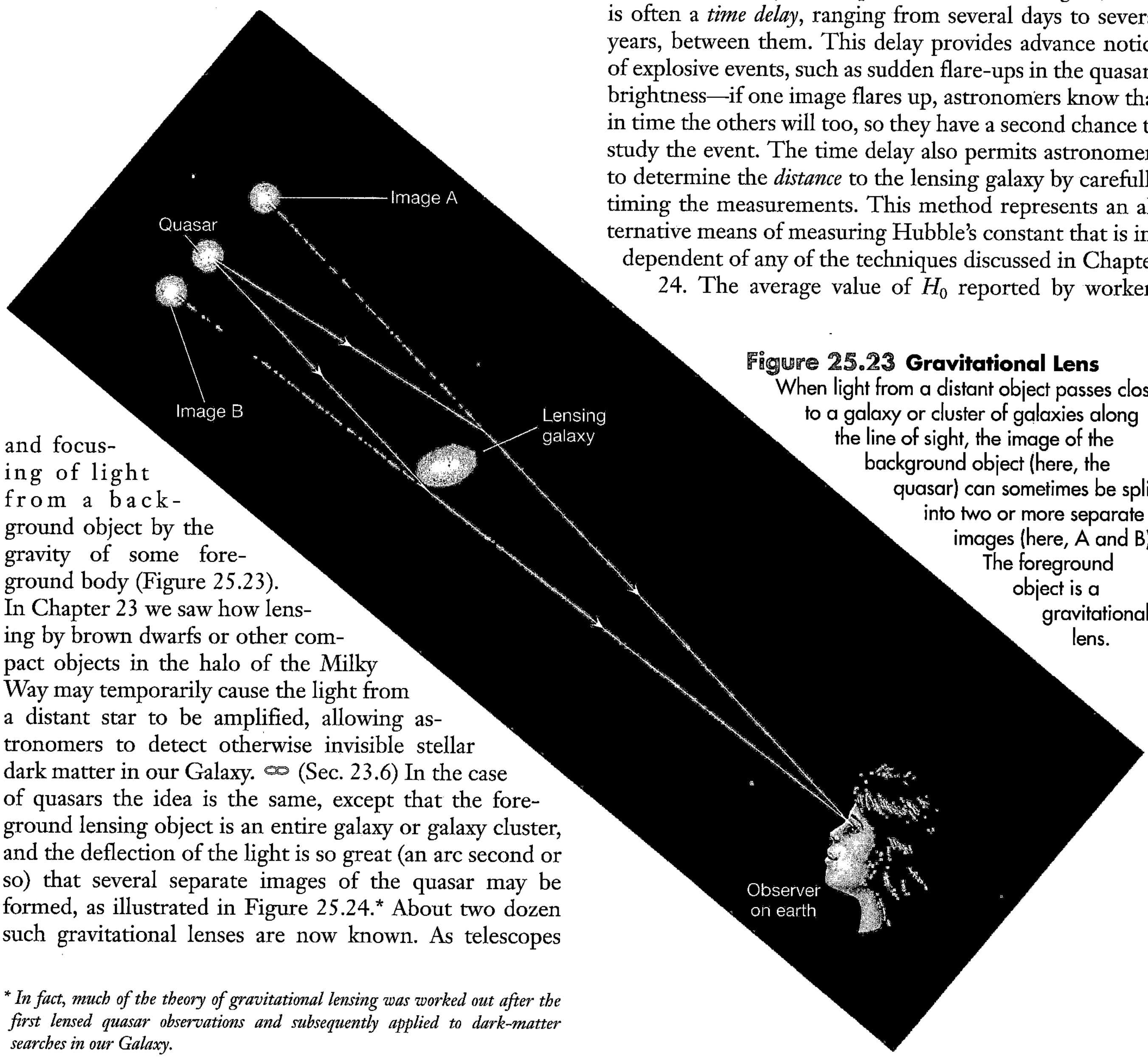

probe the universe with greater and greater sensitivity, astronomers are beginning to realize that gravitational lenses are relatively common features of the cosmos.

The existence of these multiple images provides astronomers with several useful observational tools. First, the lensing tends to amplify the light of the quasar, making it easier to observe. Second, because the light rays forming the images usually follow paths of different lengths, there is often a *time delay*, ranging from several days to several years, between them. This delay provides advance notice of explosive events, such as sudden flare-ups in the quasar's brightness—if one image flares up, astronomers know that in time the others will too, so they have a second chance to study the event. The time delay also permits astronomers to determine the *distance* to the lensing galaxy by carefully timing the measurements. This method represents an alternative means of measuring Hubble's constant that is independent of any of the techniques discussed in Chapter 24. The average value of H_0 reported by workers

and focusing of light from a background object by the gravity of some foreground body (Figure 25.23). In Chapter 23 we saw how lensing by brown dwarfs or other compact objects in the halo of the Milky Way may temporarily cause the light from a distant star to be amplified, allowing astronomers to detect otherwise invisible stellar dark matter in our Galaxy. ∞ (Sec. 23.6) In the case of quasars the idea is the same, except that the foreground lensing object is an entire galaxy or galaxy cluster, and the deflection of the light is so great (an arc second or so) that several separate images of the quasar may be formed, as illustrated in Figure 25.24.* About two dozen such gravitational lenses are now known. As telescopes

Figure 25.23 Gravitational Lens When light from a distant object passes close to a galaxy or cluster of galaxies along the line of sight, the image of the background object (here, the quasar) can sometimes be split into two or more separate images (here, A and B). The foreground object is a gravitational lens.

Image A

Quasar

Image B

Lensing galaxy

Observer on earth

---

* *In fact, much of the theory of gravitational lensing was worked out after the first lensed quasar observations and subsequently applied to dark-matter searches in our Galaxy.*

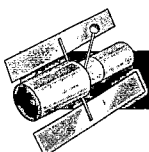

MORE PRECISELY 25-2

Relativistic Redshifts and Look-Back Time

When discussing very distant objects such as quasars, astronomers usually talk about their redshifts rather than their distances. Indeed, it is very common for researchers to speak of an event occurring "at" a certain redshift—meaning that the light received today from that event is redshifted by the specified amount. Of course, because of Hubble's law, redshift and distance are equivalent to each other. However, redshift is the "preferred" quantity, because it is a directly observable property of an object, whereas distance is derived from redshift using Hubble's constant, whose value is not accurately known. (In the next chapter we will see another reason why astronomers favor the use of redshift in studies of the cosmos.)

The redshift of a beam of light is, by definition, the *fractional* increase in its wavelength resulting from the source's motion away from us. ∞ (Sec. 3.5) Thus, a redshift of 1 corresponds to a *doubling* of the wavelength, a redshift of 2 to a tripling, and so on. From the formula for the Doppler shift presented in Chapter 3, the redshift of radiation received from a source moving away from us with speed v is given by

$$\text{redshift} = \frac{\text{observed wavelength} - \text{true wavelength}}{\text{true wavelength}}$$

$$= \frac{\text{recession velocity, } v}{\text{speed of light, } c}.$$

EXAMPLE: Let's illustrate this with two examples, rounding off the speed of light c to 300,000 km/s. A galaxy at a distance of 100 Mpc has a recessional velocity (by Hubble's law) of 65 km/s/Mpc × 100 Mpc = 6500 km/s. Its redshift therefore is 6500 km/s/300,000 km/s = 0.022. Conversely, an object with a redshift of 0.05 has a recessional velocity of 0.05 × 300,000 km/s = 15,000 km/s and hence a distance of 15,000 km/s ÷ 65 km/s/Mpc = 230 Mpc.

Unfortunately, while the foregoing equation is correct for low velocities, it does not take into account the effects of relativity. As we saw in Chapter 22, the rules of everyday physics have to be modified when velocities begin to approach the speed of light, and the formula for the Doppler shift is no exception. ∞ (*More Precisely 22-1*) In particular, although our formula is valid for velocities much less than the speed of light, when $v = c$, the redshift is not 1, as the equation suggests, but is in fact *infinite*. In other words, radiation received from an object moving away from us at nearly the speed of light would be redshifted to almost infinite wavelength.

Thus do not be alarmed to find that many quasars have redshifts greater than 1. This does not mean that they are receding faster than light! It simply means that their recessional velocities are *relativistic*—comparable to the speed of light—and the preceding simple formula is not applicable. Table 25.1 presents a conversion chart relating redshift, recession speed, and present distance. The column headed "v/c" gives recession velocities based on the Doppler effect, taking relativity properly into account (but see Section 26.2 for the correct interpretation of the redshift). All values are based on reasonable assumptions and are usable even for $v \approx c$. We take Hubble's constant to be 65 km/s/Mpc and assume a flat universe (see Section 26.3) in which matter (mostly dark) contributes one-third of the total density. The conversions in the table are used consistently throughout this text. Notice that while the entries in the table agree very well with our earlier simple formulas for small redshifts (less than a few percent, corresponding to objects lying within 100 or 200 Mpc of Earth), they differ greatly for larger distances.

Because the universe is expanding, the "distance" to a galaxy is not very well defined—do we mean the distance when the galaxy emitted the light we see today, or the present distance (as listed in the table), even though we do not see the galaxy as it is today, or is some other measure more appropriate? Largely because of this ambiguity, astronomers prefer to work in terms of a quantity known as the *look-back time* (shown in the last column of Table 25.1), which is simply how long ago an object emitted the radiation we see today. Researchers talk frequently about red-

---

using this approach is, coincidentally, almost exactly 65 km/s/Mpc, as used throughout this text.

Third, *microlensing*—lensing by individual stars in the foreground galaxy—can cause large fluctuations in a quasar's brightness. ∞ (Sec. 23.6) Microlensing allows astronomers to study the stellar content of the lensing galaxy. Finally, by studying the lensing of background quasars and galaxies by foreground galaxy clusters, astronomers can obtain a better understanding of the distribution of dark matter in those clusters, an issue that has great bearing on the large-scale structure of the cosmos.

MAPPING DARK MATTER

Astronomers have extended the ideas first learned from studies of quasars to use lensing of any distant object to probe the universe. Distant faint irregular galaxies—the building blocks of the universe, if current theories are correct—are of particular interest here as they are far more common than quasars, so they provide much better coverage of the sky. ∞ (Secs. 24.2, 24.4)

Figures 25.25(a) and (b) show how the images of faint background galaxies are bent into arcs by the gravity of

shifts and sometimes about look-back times, but hardly ever of the distances to high-redshift objects. Bear in mind, however, that redshift is the only unambiguously measured quantity here. Statements about "derived" quantities, such as distances, look-back times, and recession velocities, all require that we make specific assumptions about how the universe has evolved with time.

For nearby sources, the look-back time is numerically equal to the distance in light-years—the light we receive tonight from a galaxy at a distance of 100 million light-years was emitted 100 million years ago. However, for more distant objects, the look-back time and the present distance in light-years differ because of the expansion of the universe, and the divergence increases dramatically with increasing redshift. For example, a galaxy now located 24 billion light-years from Earth was much closer to us when it emitted the light we now see. Consequently, its light has taken considerably less than 24 billion years—in fact, only about 12.4 billion years—to reach us.

As a simple analogy, imagine an ant crawling across the surface of an expanding balloon at a constant speed of 1 cm/s relative to the balloon's surface. After 10 seconds the ant may think it has traveled a distance of 10 cm, but an outside observer with a tape measure will find that it is actually more than 10 cm from its starting point (measured along the surface of the balloon) because of the balloon's expansion. Furthermore, the difference between the actual distance and 10 cm depends on the details of the balloon's expansion during the journey—the more rapid the expansion, the greater the disparity. In exactly the same way, the present distance to a galaxy with a given redshift depends in detail on how the universe expanded in the past. We will see in Chapter 26 that the details of the past expansion are not well known, so the distance is not well determined. The distances given in the accompanying table are used consistently throughout this book, but realize that they are subject to considerable uncertainty.

Finally, notice that, according to the table, the recessional velocity equals the speed of light—and the redshift

becomes infinite—for objects that emitted their radiation about 14 billion years ago. We will examine the reasons for, and implications of, this fact in the next chapter.

TABLE 25.1 Redshift and Recessional Velocity

REDSHIFT	V/C	PRESENT DISTANCE (Mpc)	(10^6 ly)	LOOK-BACK TIME (millions of years)
0.000	0.000	0	0	0
0.010	0.010	46	150	149
0.025	0.025	115	374	369
0.050	0.049	228	742	724
0.100	0.095	449	1460	1400
0.200	0.180	873	2850	2600
0.250	0.220	1080	3510	3140
0.500	0.385	2000	6520	5340
0.750	0.508	2790	9090	6930
1.000	0.600	3460	11,300	8100
1.500	0.724	4540	14,800	9680
2.000	0.800	5360	17,500	10,700
3.000	0.882	6540	21,300	11,800
4.000	0.923	7350	24,000	12,400
5.000	0.946	7950	25,900	12,700
6.000	0.960	8430	27,500	13,000
10.00	0.984	9620	31,400	13,400
50.00	0.999	12,130	39,600	13,800
100.0	1.000	12,770	41,600	13,900
∞	1.000	14,330	46,700	13,900

nearby galaxy clusters. The degree of bending allows the total mass of the cluster (including the dark matter) to be measured. The (mostly blue) loop- and arc-shaped features visible in Figure 25.25(b) are thought to be multiple images of a single, distant (unseen) object, lensed by the foreground galaxy cluster (the yellow-red blobs in the image). Careful analysis demonstrates the nearly identical shape of each of the blue features, strongly suggesting that they come from a single object, thought to lie nearly twice as far away from us as the intervening cluster. Given the looplike shape and blue color, the distant object may be a normal

spiral or ring-shaped galaxy. Astronomers are now trying to delicately "reassemble" an undistorted image of the distant object by tracing its elongated arcs back to a common point.

It is also possible to reconstruct the foreground mass distribution by carefully analyzing the distortions of background objects observed in images such as Figure 25.25. Systematic surveys of lensed background galaxies thus provide a unique means of tracing out the distribution of (dark) matter on scales far larger than have previously been accessible using conventional techniques. Figure 25.26 shows an example of this process. Part (a) is an optical

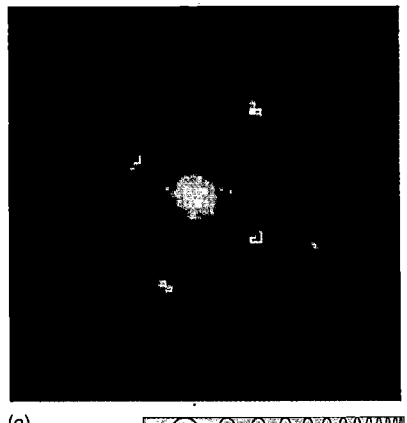

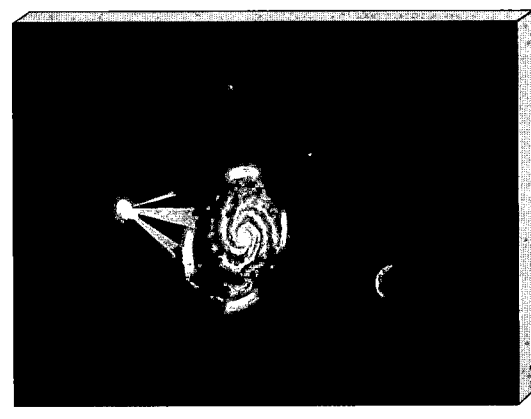

(a)

(b)

Figure 25.24 Einstein Cross
(a) The "Einstein Cross," a multiply-imaged quasar. In this *Hubble* view, spanning only a couple of arc seconds, four separate images of the same quasar have been produced by the galaxy at the center. (b) A simplified artist's conception of what might be occurring here, with Earth at right and the distant quasar at left. *(NASA; D. Berry)*

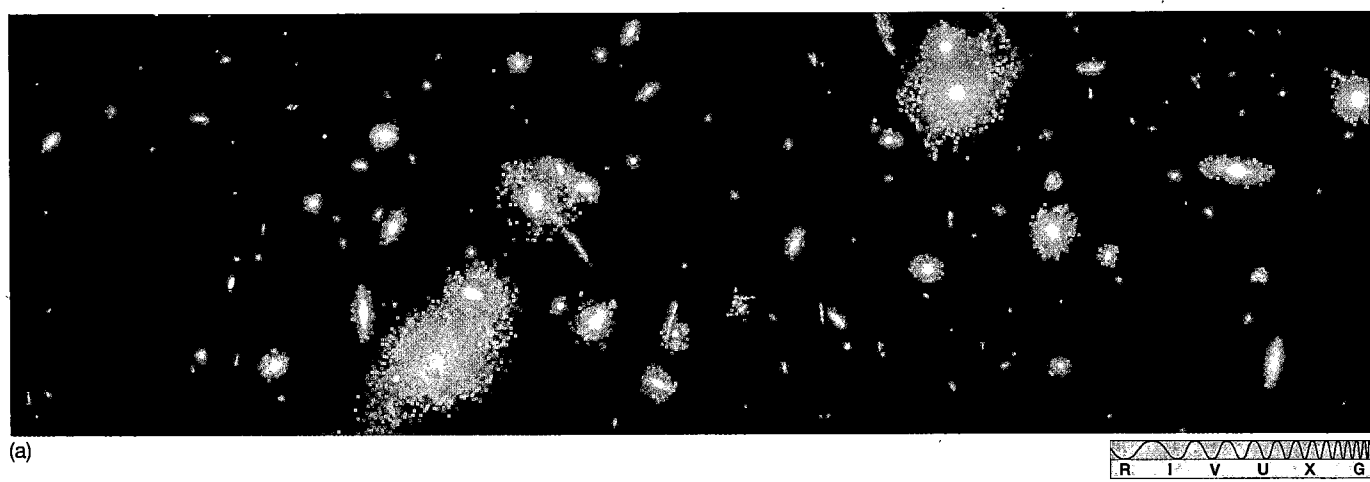

(a)

Figure 25.25 Galaxy Cluster Lensing (a) This spectacular example of gravitational lensing shows more than a hundred faint arcs from very distant galaxies. The wispy pattern spread across the intervening galaxy cluster (A 2218, about a billion parsecs distant) resembles a spider's web, but it is really an illusion caused by the gravitational field of the lensing cluster, which deflects the light from background galaxies and distorts their appearance. By measuring the extent of this distortion, astronomers can estimate the mass of the intervening cluster. (b) An approximately true-color image of the galaxy cluster known only by its catalog name, 0024 + 1654, residing some 1.5 billion pc away. The reddish-yellow blobs are mostly elliptical normal galaxy members of the cluster, concentrated toward the center of the image. The bluish looplike features are thought to be images of a single background galaxy. *(NASA)*

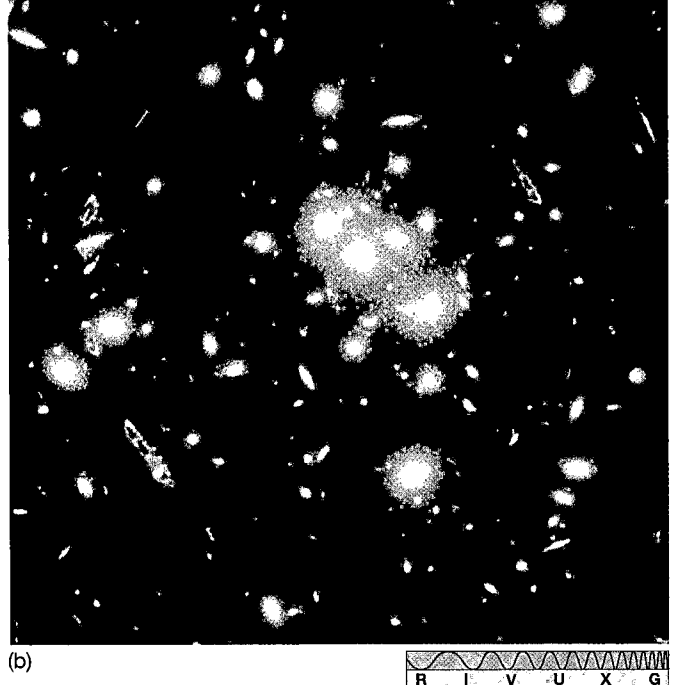

(b)

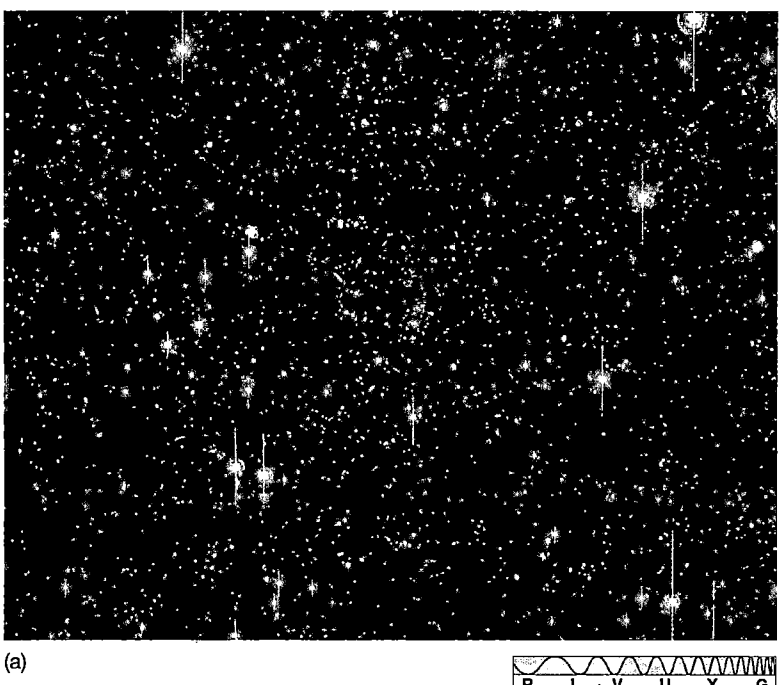

(a)

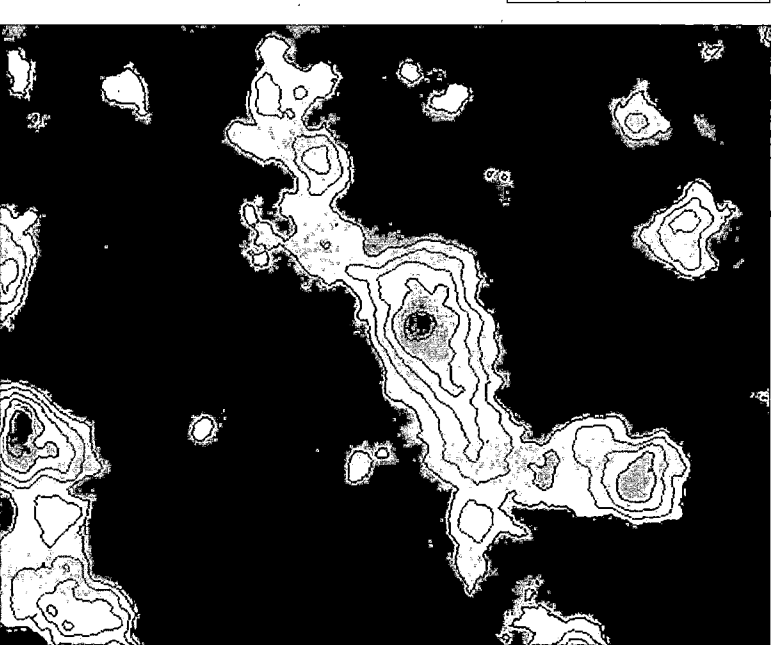

(b)

Figure 25.26 Dark Matter Map By measuring the distortions produced in the images of background objects and modeling the distribution of foreground dark matter required to account for these distortions, astronomers can produce maps of dark matter in the universe, which would otherwise be impossible to detect. (a) An optical view of a region of the sky containing a small galaxy cluster (too diffuse to be easily seen in this image, but easily detectable by its X-ray-emitting intracluster gas). The cluster is the clump of yellowish galaxies near the center of the frame. The fuzzy blue specks scattered across the frame are the much more distant background galaxies whose distortions are used to estimate the cluster's dark-matter content. (b) The distribution of dark matter in and near the visible cluster, obtained by analyzing the images of the background galaxies. Parts (a) and (b) are drawn to the same scale. *(Deep Lens Survey team; data courtesy J. A. Tyson, Bell Labs, Lucent Technologies; NSF/NOAO)*

Animation

image of a (hard to see) foreground galaxy cluster, set against a background of much fainter distant galaxies. Figure 25.26(b) is the reconstructed dark-matter image, revealing the presence of dark mass many megaparsecs from the cluster center. Notice the elongated structure of the dark-matter distribution, reminiscent of the filamentary structure seen in studies of superclusters and large-scale galaxy surveys. ∞ (Secs. 24.2, 24.5)

✓ Concept Check

■ Why are quasars useful as probes of intergalactic space?

25.6 Active Galaxy Evolution

⑥ In Chapters 23 and 24 we addressed the issue of evolutionary change in normal galaxies. ∞ (Secs. 23.4, 24.4) Let's now briefly consider some possible evolutionary links among active galaxies, and between normal and active galaxies. Although the ideas presented in this section are still somewhat speculative, a number of basic themes have begun to gain widespread acceptance among workers in the field.

The fact that most quasars reside at great distances from us demonstrates that these objects were more common in the past than they are today, and most astronomers agree that quasars represent an early stage of galaxy evolution. This opinion is strengthened by the fact that the same black-hole energy-generation mechanism can account for the luminosities of quasars, active galaxies, and the central regions of normal galaxies like our own. Large black holes do not simply vanish, at least in the 15 or so billion years that the universe has existed (see *More Precisely 22-3*). Thus, the presence of supermassive black holes in the centers of many, if not all, normal galaxies is consistent with the idea that galaxies started off as quasars, then eventually "wound down" to become the relatively quiescent normal galaxies we now observe.

As we saw in Section 24.4, the current view of galaxy evolution is that galaxies formed as small irregular clumps of gas, dust, and dark matter, then grew by mergers into the larger systems we see today. Each merger produced an intense burst of star formation, and it is quite possible that a large fraction of the stars formed in the early universe came into being as a result of galaxy collisions. The fact that quasars have been observed with redshifts of almost 6 (the current record, as of mid-2001, is 5.8, but several new high-redshift quasars are discovered every year) means that the process must have started at least 13 billion years ago (see Table 25.1). Let's look a little more closely at how quasars and active galaxies fit into this picture.

To make a quasar, we need a black hole and enough fuel to power it. While fuel was abundant at early times in the form of gas and newly formed stars, black holes were not. They had to form, most probably by the same basic stellar evolutionary processes we saw in Chapters 21 and 22. ∞ (Secs. 21.3, 22.5) Some astronomers believe that

runaway collisions between stars in dense young star clusters, or even the collapse of very massive (hundreds of solar masses) stars, may have accelerated the process. However, regardless of the details, it seems that the building blocks of the supermassive black holes that would ultimately power the quasars were relatively small black holes having masses tens or perhaps a few hundreds of times the mass of the Sun.

These small black holes sank to the center of their still-forming parent galaxy and merged to form a single, much more massive black hole. As galaxies merged, so too did their central black holes, and eventually supermassive (1-million- to 1-billion-solar-mass) black holes existed in the centers of many galaxies. Some supermassive black holes may have formed directly by gravitational collapse of the dense central regions of a protogalactic fragment, or perhaps by a rapid series of mergers in a particularly dense region of the universe. These resulted in the oldest (redshift 6) quasars known, which we see today as they were 13 billion years ago. However, in most cases, all those mergers took time, and the peak quasar redshift of between 2 and 3 corresponds to an epoch 1–2 billion years later. By this time many supermassive black holes had formed, but there was still plenty of merger-driven fuel available to power them.

Young galaxies at this early stage were much fainter than their bright quasar cores. As a result, until quite recently, astronomers were hard-pressed to discern any galactic structure in quasar images. Since the mid-1990s, several groups of astronomers have used the *Hubble Space Telescope* to search for the "host" galaxies of distant quasars. After removing the bright quasar core from the *HST* images and carefully analyzing the remnant light, the researchers have reported that, in every case studied—several

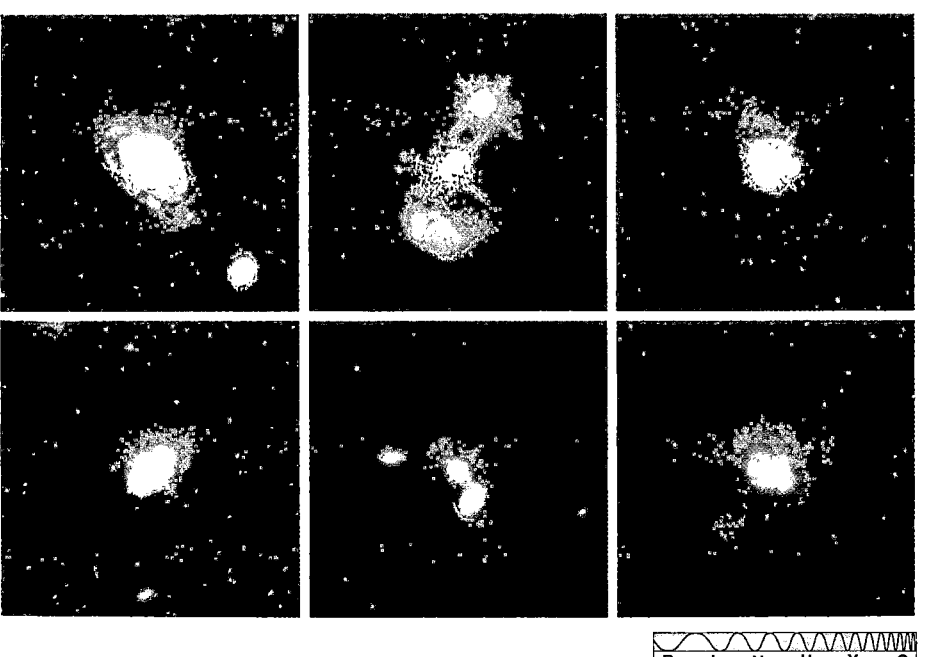

Figure 25.27 Quasar Host Galaxies These long-exposure *Hubble Space Telescope* images of distant quasars clearly show the young host galaxies in which the quasars reside, lending strong support to the view that quasars represent an early, very luminous phase of galactic evolution. The quasar at top left, having the catalog name PG0052+251 and residing roughly 430 Mpc from Earth, is perhaps the best example. Note that several of the quasars appear to be associated with interacting galaxies, consistent with current theories of active galaxy evolution. *(NASA)*

dozen quasars so far—a host galaxy can be seen enveloping the quasar. Figure 25.27 shows some of the longest quasar exposures ever taken. Even without sophisticated computer processing, the hosts are clearly visible.

As we saw in Section 25.4, quasars consume their fuel at a prodigious rate, and a quasar's intense emission is probably quite short lived. Early on, frequent mergers may have replenished the quasar's fuel supply, but as the merger rate declined, these systems spent less and less of their time in the "bright" phase. As the galaxy developed and the luminosity of the central nucleus diminished, the quasar no longer completely overwhelmed the emission from the surrounding stars. The result was an active galaxy—a radio galaxy or a Seyfert—still emitting a lot of energy but now with a definite "stellar" component in its spectrum. This continuous reduction in activity is consistent with observations—weak quasars share some characteristics with some very active galaxies, and the feeblest active galaxies often resemble the most explosive normal galaxies (see also *Discovery 25-1*). Some possible evolutionary connections

among quasars, active, and normal galaxies are illustrated in Figure 25.28.

The galaxy merger scenario also gives us some insight into active galaxies after the "quasar epoch" ended. We know that mergers involving galaxies of similar size process much of the galaxies' gas into stars, and eject most of the rest from the system, eventually resulting in ellipticals. On the other hand, "minor" mergers preserve a spiral's delicate disk, providing a means for spirals to grow. ⊙ (Sec. 24.4) If we assume that the brightest quasars came about through mergers of the largest galaxies, containing the most massive black holes, then this naturally explains why the brightest active galaxies—the radio galaxies—should be associated with ellipticals. Furthermore, the path to spiral galaxies would necessarily have entailed a series of mergers involving smaller galaxies, resulting in the less violent Seyferts.

The central activity continued to decline. Eventually, only the surrounding galaxy remained visible—a normal galaxy, like the majority of those we now see around us. Today the black holes that generated so much youthful

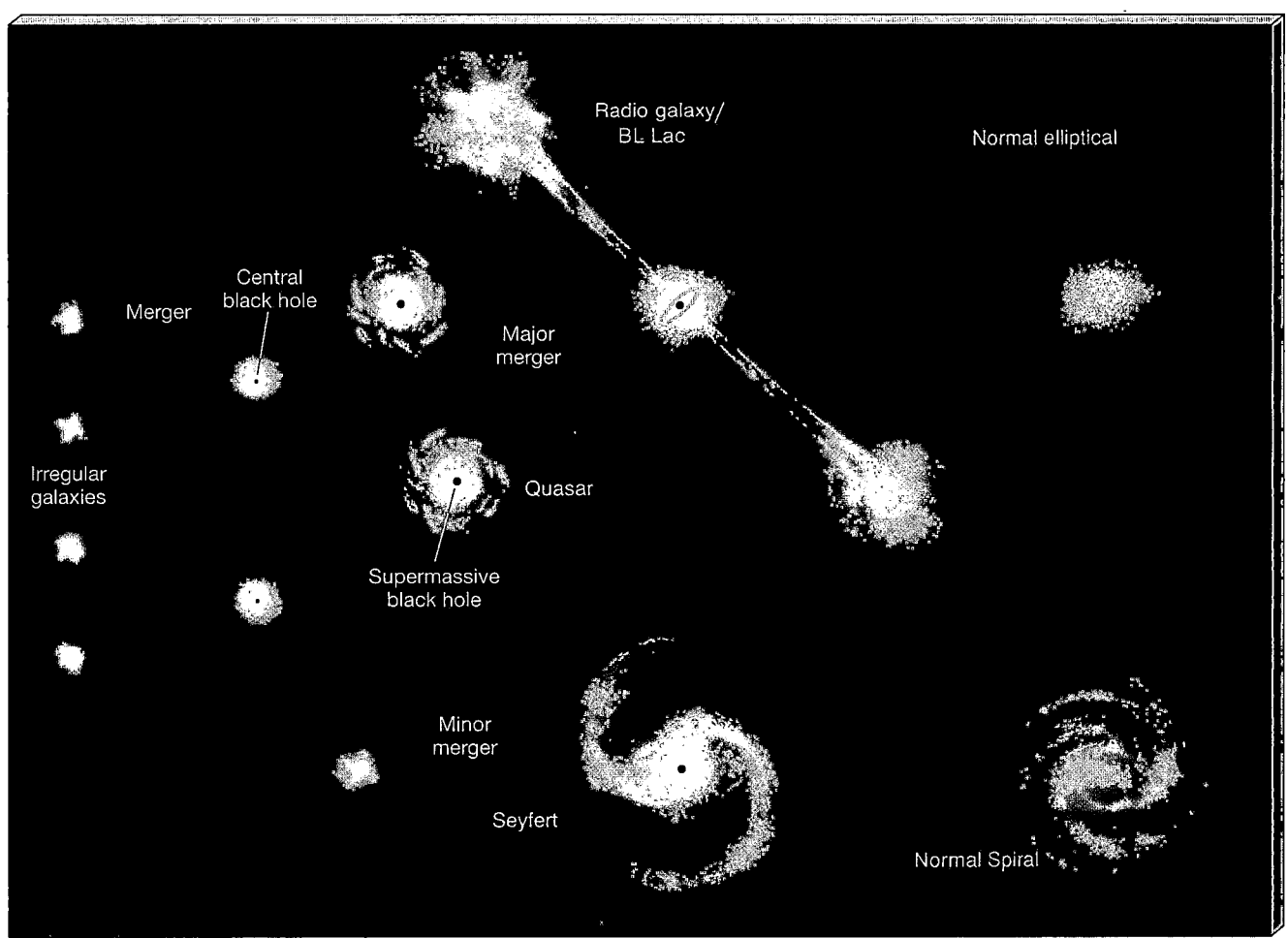

Figure 25.28 Galaxy Evolution A possible evolutionary sequence for galaxies, beginning with galaxy mergers leading to the highly luminous quasars, decreasing in violence through the radio and Seyfert galaxies, and ending with normal ellipticals and spirals. The central black holes that powered the early activity are still there at later times; they simply run out of fuel as time goes on.

energy lie dormant in galactic cores, producing only a relative trickle of radiation. Occasionally, two nearby normal galaxies may interact with each other, causing a flood of new fuel to be directed toward the central black hole of one or both. The engine starts up for a while, giving rise to the nearby active galaxies we observe.

Should this picture be correct—and the confirmation of host galaxies surrounding many quasars lends strong support to this view—then many relatively nearby galaxies (but probably *not* our own Milky Way, whose central black hole is even now only a paltry 3 million solar masses) must once have been brilliant quasars. Perhaps some alien astronomer, thousands of megaparsecs away, is at this very moment observing the progenitor of M87 in the Virgo cluster—seeing it as it was billions of years ago—and is commenting on its enormous luminosity, nonstellar spectrum and high-speed jets, and won-

dering what exotic physical process can possibly account for its violent activity!

When they were first discovered, active galaxies and quasars seemed to present astronomers with insurmountable problems. For a time their dual properties of enormous energy output and small size appeared incompatible with the known laws of physics and threatened to overturn our modern view of the universe. Yet the problems were eventually solved, and the laws of physics remain intact. Far from jeopardizing our knowledge of the cosmos, these violent phenomena have become an integral part of the thread of understanding that binds our own Galaxy to the earliest epochs of the universe in which we live.

☑ Concept Check

■ Do we live in or near a burned-out quasar?

DISCOVERY 25-1

BL Lac Objects

In 1929 an object thought to be a variable star was discovered in the constellation Lacerta. Astronomers gave it the two-letter code BL, so it became known as BL Lacertae, or BL Lac for short. Not until the 1970s, when it became clear that BL Lac was a strong radio source, did anyone question its classification as a star. As more objects like BL Lac were found, astronomers began to realize that they had stumbled across a whole new class of extragalactic sources. *BL Lac objects* seem starlike in a telescope and vary greatly in luminosity over time. They are also powerful compact radio sources, sometimes colloquially referred to as "blazars."

Astronomers originally placed BL Lac objects into a special class because they displayed no spectral lines. Their redshifts could not be determined, and their distances were unknown. However, recent discoveries of extremely faint spectral lines strongly suggest that BL Lac objects are extremely distant, making them nearly as luminous as quasars. Further careful observations have shown that blazars reside at the centers of relatively normal elliptical galaxies. We consider these galaxies active because of the tremendous strength of the blazar radio emission.

The weak spectral lines may be the key needed to unravel the nature of the BL Lac objects. Evidently, the ther-

mal emission from any stars they contain, which would provide spectral lines, is swamped by their nonthermal synchrotron radiation (see Section 25.4), which does not contain spectral features. Other types of active galaxies show a mixture of thermal and nonthermal radiation, so it is more difficult to study the nonthermal radiation alone. BL Lac objects, then, may offer astronomers a chance to study the "bare machine" powering these active galaxies.

In some ways the BL Lac objects seem to represent a "link" or transition phase between radio galaxies and quasars (see Section 25.5). The luminosities of the compact radio sources inside BL Lac objects span the whole range from the relatively weak cores observed in radio galaxies to the stronger ones found in quasars. However, their appearance may be mainly a matter of geometry. Many astronomers think that BL Lacs, like the weaker core–halo radio sources, are lobe-radio galaxies, or perhaps quasars with jets, in which the jet happens to be pointing nearly straight at us. In this view, we are seeing the nucleus and jet *through* the nearer radio lobe; their intense nonthermal radiation overwhelms both the thermal emission of the galaxy and the extended radio emission of the lobe.

Chapter Review

SUMMARY

Active galaxies (p. 662) are much more luminous than normal galaxies and have spectra that are nonstellar in nature, indicating that the energy they emit is not simply the accumulated light of many stars. Most of the energy from active galaxies is emitted in the radio and infrared parts of the electromagnetic spectrum. The fraction of observed galaxies that display activity increases with increasing distance from the Milky Way, indicating that galaxies were generally more active in the past than they are today.

A **Seyfert galaxy** (p. 663) looks like a normal spiral except that it has an extremely bright central **galactic nucleus** (p. 663) whose luminosity can in many cases exceed that of the rest of the galaxy. Spectral lines from Seyfert nuclei are very broad, indicating rapid internal motion. In addition, Seyfert luminosities can vary by large amounts in fractions of a year, implying that the region emitting most of the radiation is much less than one light-year across. **Radio galaxies** (p. 664) are active galaxies that emit most of their energy in the radio part of the spectrum. They are generally comparable to the Seyferts in total energy output. Unlike Seyferts, they are usually associated with elliptical galaxies. In a lobe-radio galaxy, the energy comes from enormous **radio lobes** (p. 664) that dwarf the visible galaxy and lie far outside it. The lobes are usually symmetrically placed with respect to the center of the visible galaxy. In a core–halo radio galaxy, the energy is emitted from a small central nucleus, as in a Seyfert.

Many active galaxies have high-speed jets of matter shooting out from their central nuclei. In lobe-radio galaxies, astronomers believe that the jets transport energy from the nucleus, where it is generated, to the lobes, where it is radiated into space. The jets often appear to be made up of distinct "blobs" of gas, suggesting that the process generating the energy is intermittent.

Quasars (p. 669), or **quasi-stellar objects** (p. 669), were first discovered as starlike radio sources with unknown broad spectral lines. In the early 1960s, astronomers realized that the unfamiliar lines are actually those of familiar elements, but redshifted to wavelengths much longer than normal. Even the closest quasars lie at great distances from us. They are the most luminous objects known.

The generally accepted explanation for the observed properties of active galaxies and quasars is that the energy is generated by accretion of galactic gas onto a supermassive (billion-solar-mass) black hole lying at the center of the nucleus. As material spirals down toward the hole, it heats up and releases enormous amounts of energy. The small size of the accretion disk explains the compact extent of the emitting region, and the high-speed orbits of gas in the hole's intense gravity account for the rapid motion observed. Typical active-galaxy luminosities require the consumption of about 1 solar mass of material every few years. Quasars consume several stars per year. Some of the infalling matter is blasted out into space, producing magnetized jets that create and feed the extended radio lobes. Charged particles spiraling around the magnetic field lines produce **synchrotron radiation** (p. 676) whose spectrum is consistent with the nonstellar radiation observed in radio galaxies and Seyferts. The brightest quasars consume so much fuel that their energy-emitting lifetimes must be relatively short. Quasars probably represent a brief phase of violent activity early in the life of a galaxy.

Quasar spectra can be used as probes of the universe along the line of sight. Some quasars have been observed to have double or multiple images. These result from gravitational lensing, in which the gravitational field of a foreground galaxy or galaxy cluster bends and focuses the light from the more distant quasar. Analysis of this bending provides a means of determining the masses of galaxy clusters—including the dark matter—far beyond the optical images of the galaxies themselves.

Quasars, active galaxies, and normal galaxies may represent an evolutionary sequence. When galaxies began to form and merge, conditions may have been suitable for the formation of large black holes at their centers. A lot of gas was available at those early times, and a highly luminous quasar could have been the result. As the fuel supply diminished, the quasar dimmed, and the galaxy in which it was embedded became visible as an active galaxy. Major mergers may have created elliptical radio galaxies, while minor mergers could have formed the spiral Seyferts. At even later times, the fuel supply declined to the point at which the nucleus became virtually inactive, and a normal galaxy was all that remained.

SELF-TEST: TRUE OR FALSE?

_____ **1.** Active galaxies can emit thousands of times more energy than our own Galaxy.

_____ **2.** The "extra" radiation emitted by active galaxies is due to the tremendous number of stars they contain.

_____ **3.** Active galaxies emit most radiation at visible wavelengths.

_____ **4.** The spectrum of an active galaxy is not well described by a blackbody curve.

_____ **5.** Most core–halo radio galaxies are spirals.

_____ **6.** A billion-solar-mass black hole would be comparable in size to our Galaxy.

_____ **7.** Nearby active galaxies are most likely the result of galaxy interactions.

_____ **8.** A redshift greater than 1 implies a recessional velocity greater than the speed of light.

_____ **9.** All active galaxies are far away.

_____ **10.** Quasar luminosities are so large that astronomers have no idea what process can account for them.

_____ **11.** Other than a small amount of visible light, quasars emit all their radiation at radio wavelengths.

_____ **12.** Many nearby normal galaxies may once have been quasars.

_____ **13.** Quasars emit much more energy than normal galaxies.

_____ **14.** The quasar stage of a galaxy ends because the central black hole swallows up all the matter around it.

_____ **15.** Astronomers have observational evidence that quasars are found in the hearts of young galaxies.

SELF-TEST: FILL IN THE BLANK

1. Active galaxies generally emit more radiation at _____ wavelengths.
2. Seyfert galaxies appear like normal spirals, but with a very bright galactic _____.
3. Lobe-radio galaxies emit radio radiation from regions that are typically much _____ in size than the visible galaxy.
4. Radio lobes are often aligned with the _____ of the visible galaxy.
5. In a core–halo radio galaxy, most of the radio radiation is emitted from the _____.
6. For all types of active galaxies the actual source of the tremendous energy emitted is the galactic _____.
7. The energy source of an active galaxy is unusual in that there is a huge amount of energy emitted from a very _____ region of space.
8. The mass of the black hole responsible for energy production in the active galaxy M87 is thought to be approximately _____ solar masses.
9. The amount of mass that must be consumed by the central engine in order to provide the energy for a typical bright quasar is a few stars per _____ (second/day/year/century).
10. In visible light, quasars have a _____ appearance.
11. Quasar spectra were understood when it was discovered that their radiation is _____ by an unexpectedly large amount.
12. The distance to a quasar in light-years is not simply equal to the time in years since the quasar emitted the light we see, because of the _____ of the universe.
13. Quasars are more common at _____ distances.
14. The fact that a typical quasar would consume an entire galaxy's worth of mass in 10 billion years suggests that quasar lifetimes are relatively _____.
15. The image of a distant quasar can be split into several images because of gravitational lensing by a foreground _____ along the line of sight.

REVIEW AND DISCUSSION

1. Name two basic differences between normal galaxies and active galaxies.
2. Are there any "nearby" active galaxies—within 50 Mpc of Earth, say?
3. Are there any nearby quasars?
4. Describe some of the basic properties of Seyfert galaxies.
5. What distinguishes a core–halo radio galaxy from a lobe-radio galaxy?
6. What is the evidence that the radio lobes of some active galaxies consist of material ejected from the galaxy's center?
7. What conditions result in a head–tail radio galaxy?
8. How do we know that the energy-emitting regions of many active galaxies must be very small?
9. What was it about the spectra of quasars that was so unexpected and surprising?
10. Why do astronomers prefer to speak in terms of redshifts rather than distances to faraway objects?
11. How do we know that quasars are extremely luminous?
12. Briefly describe the leading model for the central engine of an active galaxy.
13. How is the process of synchrotron emission related to observations of active galaxies?
14. What evidence do astronomers have for supermassive black holes in galactic nuclei?
15. How can observations of distant quasars be used to probe the space between us and them?
16. How are BL Lac objects related to other active galaxies?
17. Why do astronomers think that quasars represent an early stage of galaxy evolution?
18. Why do astronomers think that quasars are relatively short-lived phenomena?
19. What happened to the energy source at the center of a quasar?
20. What are the arguments for and against the idea that quasars may in fact be nearby and not at cosmological distances?

PROBLEMS  *Algorithmic versions of these questions are available in the Practice Problems module of the Companion Website.*

The number of squares preceding each problem indicates its approximate level of difficulty.

1. ■ Based on the data presented in the text, calculate the orbital speed of material orbiting at a distance of 0.5 pc from the center of M87.
2. ■■ Spectral lines from a Seyfert galaxy are observed to be redshifted by 0.5 percent and to have broadened emission lines indicating an orbital speed of 250 km/s at an angular distance of 0.1″ from its center. Assuming circular orbits, use Kepler's laws to estimate the mass within this 0.1″ radius. ⏎ (Sec. 23.6)
3. ■■ Calculate the energy flux—that is, the energy received per unit area per unit time—that would be observed at

Earth from a 10^{37}-W Seyfert nucleus located at the Galactic center, 8 kpc away, neglecting the effects of interstellar extinction. Using the data presented in Appendix 3, Table 4, compare this energy flux with that received from Sirius A, the brightest star in the night sky. From what you know about active-galaxy energy emission, is it reasonable to ignore interstellar extinction?

4. ■ Centaurus A—from one radio lobe to the other—spans about 1 Mpc. It lies at a distance of 4 Mpc from Earth. What is the angular size of Centaurus A? Compare this value with the angular diameter of the Moon.
5. ■ Assuming a jet speed of 0.75 c, calculate the time taken for material in Cygnus A's jet to cover the 500 kpc between the galaxy's nucleus and its radio-emitting lobes.

6. ▣ Assuming the upper end of the efficiency range indicated in the text, calculate how much energy an active galaxy would generate if it consumed one Earth mass of material every day. Compare this value with the luminosity of the Sun.

7. ▣ A quasar consumes 1 solar mass of material per year, converting 15 percent of it directly into energy. What is the quasar's luminosity, in solar units?

8. ▣ Assuming an energy-generation efficiency (that is, the ratio of energy released to total mass-energy available) of 10 percent, calculate how much mass a 10^{41}-W quasar would consume if it shone for 10 billion years.

9. ▣▣ A quasar has a luminosity of 10^{40} W and 10^8 solar masses of fuel available. Assuming constant luminosity and 20 percent efficiency, estimate the quasar's lifetime.

10. ▣▣ A certain quasar has a redshift of 0.25 and an apparent magnitude of 13. Using the data from Table 25.1, calculate the quasar's absolute magnitude, and hence its luminosity. Compare the apparent brightness of the quasar, viewed from a distance of 10 pc, with that of the Sun as seen from Earth.

11. ▣ What are the absolute magnitude and luminosity of a quasar with a redshift of 5 and an apparent magnitude of 22?

12. ▣ Based on the data in Table 25.1, estimate the apparent magnitude of a quasar with absolute magnitude -24 and redshift 1.

13. ▣ The spectrum of a quasar with a redshift of 0.20 contains two sets of absorption lines, redshifted by 0.15 and 0.155, respectively. If $H_0 = 65$ km/s/Mpc, estimate the distance between the intervening galaxies responsible for the two sets of lines.

14. ▣▣▣ Light from a distant quasar is deflected through an angle of $3''$ by an intervening lensing galaxy, and is subsequently detected on Earth (Figure 25.23). If Earth, the galaxy, and the quasar are all aligned, the quasar's redshift is 3.0, and the galaxy lies midway between Earth and the quasar, calculate the minimum distance between the light ray and the center of the galaxy.

15. ▣▣ Light from a distant star is deflected by $1.75''$ as it grazes the Sun. ∞ (*More Precisely 22-2*) Given that the deflection angle is proportional to the mass of the gravitating body, and inversely proportional to the minimum distance between the light ray and the body, estimate the mass of the galaxy in the previous question.

RESEARCHING ON THE WEB

To complete the following exercises, go to the online Destinations module for Chapter 25 on the Companion Website for Astronomy Today 4/e.

1. Access the "Active Galaxies" page and follow the hyperlink to "a monster in the center." List the three characteristics an Active Galactic Nucleus must have to be called a blazar.

2. Access the "*Hubble Deep Field*" page and describe how many galaxies are in the picture. Then make a sketch showing in which constellation the image was taken.

PROJECTS

Here are three observational projects that are increasingly challenging.

1. In the previous chapter you were given directions for finding the Virgo Cluster of galaxies. M87, in the central part of this cluster, is the nearest core–halo radio galaxy. M87 has coordinates RA = 12^h 30.8^m, dec = + 12° 24'. At magnitude 8.6 it should not be difficult to find in an 8-inch telescope. Its distance is roughly 18 Mpc. Describe its nucleus; compare what you see with other nearby ellipticals in the Virgo Cluster.

2. NGC 4151 is the brightest Seyfert galaxy. Its coordinates are RA = 12^h 10.5^m, dec = + 39° 24', and it can be found below the Big Dipper in Canes Venatici. At magnitude 10–12 (it is variable), it should be visible in an 8-inch telescope, but it will be challenging to find. Its distance is

13.5 Mpc. As in the case of M87, describe its nucleus and compare with what you have seen for other galaxies.

3. 3C 273 is the nearest and brightest quasar. However, that does not mean it will be easy to find and see! Its coordinates are RA = 12^h 29.2^m, dec = + 2° 03'. It is located in the southern part of the Virgo Cluster but is not associated with it. At magnitude 12–13 (again, it is variable), it may require a 10- or 12-inch telescope to see, but try it first with an 8-inch. It should appear as a very faint star. The significance of seeing this object is that it is 640 Mpc distant. The light you are seeing left this object over 2 billion years ago! 3C 273 is the most distant object observable with a small telescope.

If you can find the three objects listed here, you have started to become an accomplished observer!

SKYCHART III PROJECTS

The SkyChart III Student Version planetarium program on which these exercises are based is included as a separately executable program on the CD in the back of this text.

1. Locate the colliding, lobe-radio galaxies Centaurus A and NGC 1265. Although not much to look at in SkyChart III, Centaurus A is an interesting target for observation.

2. Locate these black-hole candidates: M87, NGC 4261, and NGC 4258.

 *In addition to the Practice Problems and Destinations modules, the Companion Website at http://www.prenhall.com/chaisson provides for each chapter an additional true-false, multiple choice, and labeling quiz, as well as additional annotated images, animations, and links to related Websites.*

26 COSMOLOGY

The Big Bang and the Fate of the Universe

LEARNING GOALS

Studying this chapter will enable you to:

1 State the cosmological principle and explain both its significance and its observational underpinnings.

2 Explain how the age of the universe is determined and discuss the uncertainties involved.

3 Summarize the leading evolutionary models of the universe.

4 Discuss the factors that determine whether the universe will expand forever.

5 Explain what observations of the distant universe reveal about cosmic composition.

6 Explain the relationship between the future of the universe and the overall geometry of space.

7 Describe the cosmic microwave background radiation and explain its importance to our understanding of cosmology.

 Visit http://www.prenhall.com/chaisson for additional annotated images, animations, and links to related sites for this chapter.

This deep field image—a long exposure—was taken with one of the new cameras attached to the Very Large Telescope in Chile. More than a thousand galaxies are crowded into this one image, displaying many different types and shapes; several quasars are also noticeable as bright white points. The blue objects are generally closer, red ones more distant; the green pixelation is mostly detector noise. In all, astronomers estimate that the observable universe contains approximately 40 billion such galaxies. *(ESO)*

The Big Picture: By virtually all accounts, the universe began in a fiery explosion some 10–20 billion years ago. Out of this maelstrom emerged all the energy that would later form galaxies, stars, and planets. The story of the origin and fate of all these systems—and especially of the entire universe—comprises the subject of cosmology.

Our field of view now extends for billions of parsecs into space and billions of years back in time. We have asked and answered many questions about the structure and evolution of planets, stars, and galaxies. At last we are in a position to address the central issues of the biggest puzzle of all: How big is the universe? How long has it been around, and how long will it last? What was its origin, and what will be its fate? Is the universe a one-time event, or does it recur and renew itself, in a grand cycle of birth, death, and rebirth? How and when did matter, atoms, and our Galaxy form? These are basic questions, but they are hard questions. Many cultures have asked them, in one form or another, and have developed their own cosmologies—theories about the nature, origin, and destiny of the universe—to answer them. In this and the next chapter we will see how modern scientific cosmology addresses these important issues and what it has to tell us about the universe we inhabit. After more than 10,000 years of civilization, science may be ready to provide some insight regarding the origin of all things.

26.1 The Universe on the Largest Scales

We saw in Chapter 24 how surveys of the universe have revealed the existence of structures as large as 200 Mpc across. ∞ (Sec. 24.5) The universe shows structure on every scale we have examined so far. Subatomic particles form nuclei and atoms. Atoms form planets and stars. Stars form star clusters and galaxies. Galaxies form galaxy clusters, superclusters, and even larger structures—voids, filaments, and sheets that stretch across the sky. From the quarks in a proton to the galaxies in the Great Wall, we can trace a hierarchy of "clustering" of matter from the very smallest to the very largest scales. It is natural to ask: Does the clustering ever end? Is there some scale on which the universe can be regarded as more or less smooth and featureless? Perhaps surprisingly, given the trend we have just described, most astronomers think the answer is yes.

THE END OF STRUCTURE

Figure 26.1 shows a *redshift survey* of galaxies, similar to that presented in Figure 24.31, except that it includes nearly 24,000 galaxies within about 750 Mpc of the Milky Way, so it covers a much greater volume of space than the earlier figure. The distances to the galaxies are determined assuming Hubble's law, with $H_0 = 65$ km/s/Mpc. ∞ (Sec. 24.5) Numerous voids and "Great Wall-like" filaments can be seen but, apart from the general falloff in numbers of galaxies at large distances—basically because more distant galaxies are harder to see due to the inverse-square law—there is no obvious evidence for any structures on scales larger than about 200 Mpc. Careful statistical analysis confirms this impression.

Plots such as this contain huge amounts of information about the structure and evolution of the universe. Yet, although they cover wide areas of the sky and enormous volumes of space, these studies are still relatively "local," in the sense that they span only about 10 percent of the distance to

the farthest quasars (which lie over 8000 Mpc from Earth). ∞ (Sec. 25.3) The main obstacle to extending these wide-angle surveys to much greater distances is the sheer observational effort of measuring the redshifts of all the galaxies within larger and larger volumes of space. An alternative approach is to narrow the field of view to only a few small patches of the sky and study only very distant galaxies within those patches. The volume surveyed then becomes a long, thin "pencil beam" extending deep into space rather than a wide swath through the local universe. *Discovery 26-1* discusses some recent very deep surveys that have generated a great deal of excitement among astronomers.

The data from both kinds of survey seem to agree that the largest structures in the local universe are only 100–200 Mpc across—no voids or clumps of galaxies much larger than that are seen. Studies of Lyman-alpha forests in quasar spectra lead to generally similar conclusions. ∞ (Sec. 25.5) In short, there is presently *no* evidence for structure in the universe on scales greater than about 200 Mpc.

THE COSMOLOGICAL PRINCIPLE

The results of these large-scale studies strongly suggest that the universe is **homogeneous** (the same everywhere) on scales greater than a few hundred megaparsecs. In other words, if we took a huge cube—300 Mpc on a side, say—and placed it anywhere in the universe, its overall contents would look much the same no matter where it was centered. Some of the galaxies it contained would be clustered and clumped into fairly large structures and some would not, and we would see numerous walls and voids, but the total numbers of these objects would not vary much as the cube was moved from place to place. In this sense, the universe appears *smooth* on the largest scales.

The universe also appears to be **isotropic** (the same in all directions) on these scales. Excluding directions that are obscured by our Galaxy, we count roughly the same number of galaxies per square degree in any patch of the sky we choose to observe, provided we look deep (far) enough that local inhomogeneities don't distort our sample. In other words, any deep pencil-beam survey of the sky

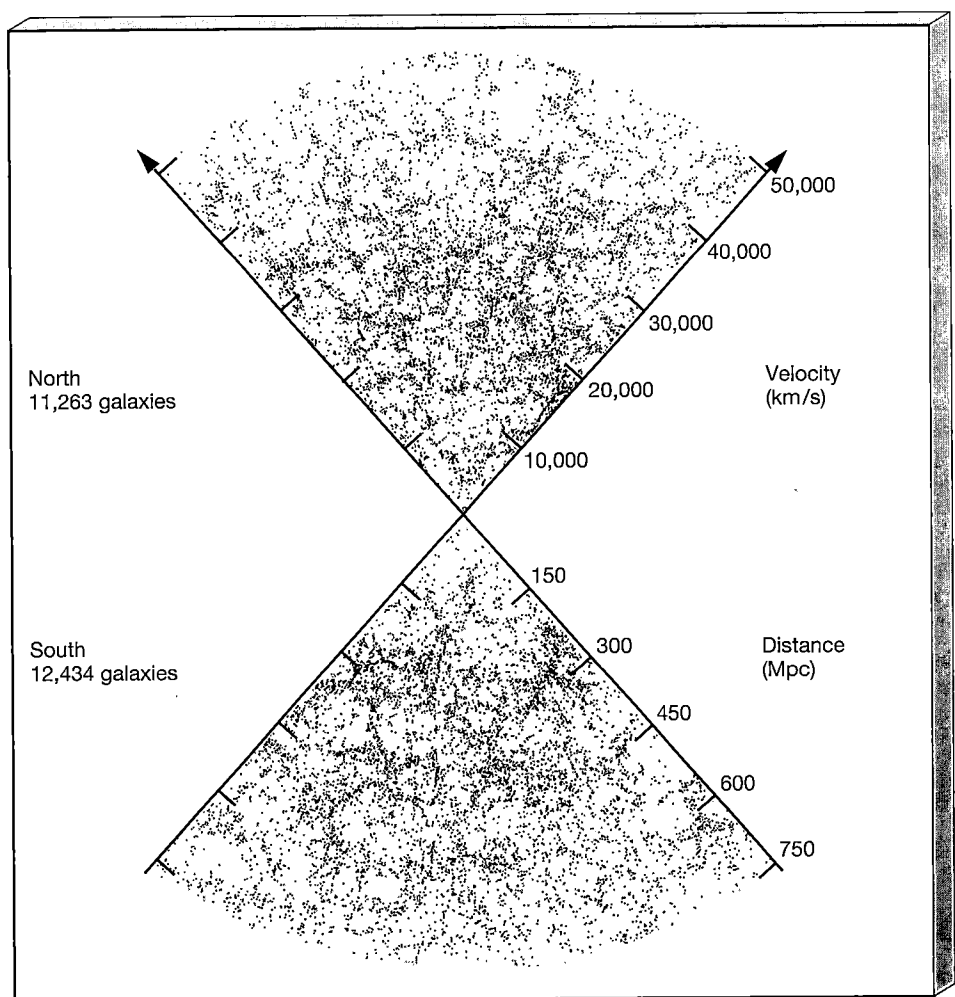

Figure 26.1 Galaxy Survey
This large-scale galaxy survey, carried out at the Las Campanas Observatory in Chile, consists of 23,697 galaxies within about 1000 Mpc, in two 80° × 4.5° wedges of the sky. Many voids and "walls" on scales of up to 100–200 Mpc can be seen, but no larger structures are evident, suggesting that the universe is roughly homogeneous on sufficiently large scales. Preliminary results from even more extensive surveys, now underway, are in general agreement with the findings reported here.

should count about the same number of galaxies, regardless of which patch of the sky is chosen.

In the science of **cosmology**—the study of the structure and evolution of the entire universe—researchers generally assume that the universe is homogeneous and isotropic on sufficiently large scales. No one knows if these assumptions are precisely correct, but we can at least say that they are consistent with current observations, and they provide helpful guidance to our studies of the cosmos. In this chapter we simply assume that they hold. The twin assumptions of homogeneity and isotropy are known as the **cosmological principle**. Note that this principle also includes the important assumption made throughout this book (and indeed throughout astronomy)—that the laws of physics are the same everywhere.

The cosmological principle has far-reaching implications. For example, it implies that there can be no edge to the universe, because that would violate the assumption of homogeneity. Furthermore, it implies that there is no *center*, because that would mean that the universe would not be the same in all directions from any noncentral point, a violation of the assumption of isotropy. This is the familiar Copernican principle expanded to truly cosmic

proportions—not only that are we not central to the universe but that *no one* can be central, because *the universe has no center!* ∞ (Sec. 2.3)

✔ Concept Check

■ In what sense, and on what scale, is the universe homogeneous and isotropic?

26.2 The Expanding Universe

Every time you go outside at night and notice that the sky is dark, you are making a profound cosmological observation. Here's why.

OLBERS'S PARADOX

Let's assume that, in addition to being homogeneous and isotropic, the universe is also infinite in spatial extent and unchanging in time—precisely the view of the universe that prevailed until the early part of the twentieth century. On average, then, the universe is uniformly populated with galaxies filled with stars. In that case, when you look up at the night sky, your line of sight must *eventually* encounter a star, as illustrated in Figure 26.2. The star may lie at an enormous distance in some remote galaxy, but the laws of probability dictate that, sooner or later, any line drawn outward from Earth will run into a bright stellar surface.

Of course, faraway stars appear fainter than those nearby, because of the inverse-square law. ⊂⊃ (Sec. 17.3) However, they are also much more numerous—the number of stars we see at any given distance in fact increases as the *square* of the distance (just consider the area of a sphere of increasing radius). Thus the diminishing brightnesses of distant stars are exactly balanced by their increasing numbers, and stars at all distances contribute equally to the total amount of light received on Earth.

This fact has a dramatic implication. No matter where you look, the sky should be as bright as the surface of a star—the entire night sky should be as brilliant as the surface of the Sun! The obvious difference between this prediction and the actual appearance of the night sky is known as **Olbers's paradox**, after the nineteenth-century German astronomer, Heinrich Olbers, who popularized the idea.

Why is it dark at night? Given that that the universe appears to be homogeneous and isotropic, then one (or both) of the other two assumptions must be false. Either the universe is finite in extent, or it evolves in time. In fact, the answer involves a little of each and is intimately tied to the behavior of the universe on the largest scales.

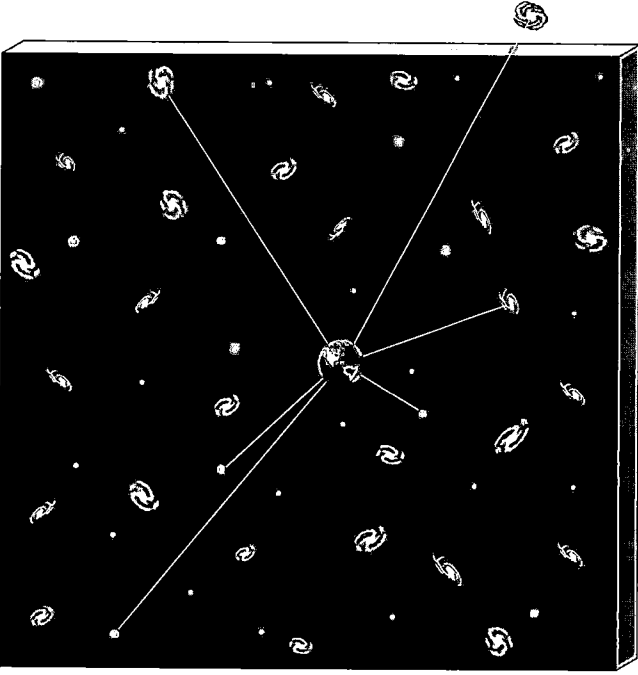

Figure 26.2 Olbers's Paradox If the universe were homogeneous, isotropic, infinite in extent, and unchanging, any line of sight from Earth should eventually run into a star, and the entire night sky should be bright. This obvious contradiction of the facts is known as Olbers's paradox.

THE BIRTH OF THE UNIVERSE

2 We saw in Chapter 24 that all the galaxies in the universe are rushing away from us in a manner described by Hubble's law:

$$\text{recession velocity} = H_0 \times \text{distance},$$

where we take Hubble's constant H_0 to be 65 km/s/Mpc. ⊂⊃ (Sec. 24.5) We have used this relation as a convenient means of determining the distances to galaxies and quasars, but it is much more than that.

Assuming that all velocities have remained constant in time, we can ask a simple question: How long has it taken for any given galaxy to reach its present distance from us? The answer follows from Hubble's law. The time taken is simply the distance traveled divided by the velocity, so

$$\text{time} = \frac{\text{distance}}{\text{velocity}}$$

$$= \frac{\text{distance}}{H_0 \times \text{distance}} \quad \text{(using Hubble's law for velocity)}$$

$$= \frac{1}{H_0}.$$

For $H_0 = 65$ km/s/Mpc, this time is about 15 billion years. Notice that it is *independent* of the distance—galaxies twice as far away are moving twice as fast, so the time they took to cross the intervening distance is the same.

Hubble's law therefore implies that at some time in the past—15 billion years ago, according to the foregoing simple calculation—*all* the galaxies in the universe lay right on top of one another. In fact, astronomers believe that *everything* in the universe—matter and radiation alike—was confined to a single point at that instant. Then the point expanded, flying apart at high speeds. The present locations and velocities of the galaxies are a direct consequence of that primordial blast. This gargantuan explosion, involving everything in the universe, is known as the **Big Bang**. It marked the beginning of the universe.

Thus, by measuring Hubble's constant, we can estimate the age of the universe to be $1/H_0 \approx 15$ billion years. The range of possible error in this age is considerable, both because Hubble's constant is not known precisely and because the assumption that galaxies moved at constant speed in the past is not a good one. We will refine our estimate in a moment, but regardless of the details, the critical fact here is that the age of the universe is *finite*.

The Big Bang provides the resolution of Olbers's paradox. Whether the universe is actually finite or infinite in extent is irrelevant, at least as far as the appearance of the night sky is concerned. We see only a finite part of it— the region lying within roughly 15 billion light-years of us. What lies beyond is unknown—its light has not yet had time to reach us.

Realize that even though it appears to place us at the center of the expansion, Hubble's law does *not* violate the cosmological principle in any way. This can be understood from Figure 26.3, which shows how observers on five separate galaxies might perceive the motion of their neighbors. For simplicity, the galaxies are taken to be equally spaced, 100 Mpc apart, and they are separating in accordance with Hubble's law with $H_0 = 65$ km/s/Mpc. The first pair of numbers beneath each galaxy represents its distance and recessional velocity as measured by the observer on the middle galaxy, number 3.

Now consider how the expansion looks from the point of view of the observer on galaxy 2. Galaxy 4, for example, is moving with velocity 6500 km/s to the right relative to galaxy 3, and galaxy 3 in turn is moving at 6500 km/s to the right as seen by observer 2. Therefore, galaxy 4 is moving at a velocity of 13,000 km/s to the right as seen by the observer on galaxy 2. The distances and velocities that would be measured by observer 2 are noted in the second row. Similarly, the measurements made by an observer on galaxy 1 are noted in the third row. *Each* observer sees an overall expansion described by Hubble's law, and the constant of proportionality—Hubble's constant—is the same in all cases. Far from singling out any one observer as central, Hubble's law is in fact the *only* expansion law possible if the cosmological principle holds.

WHERE WAS THE BIG BANG?

Now we know *when* the Big Bang occurred. Is there any way of telling *where?* We think that the universe is the same everywhere, yet we have just seen that the observed recession of the galaxies described by Hubble's law implies that all the galaxies exploded from a point some time in the past. Wasn't that point, then, different from the rest of the universe, violating the assumption of homogeneity expressed in the cosmological principle? The answer is a definite *no!*

To understand why there is no "center" to the expansion, we must make a great leap in our perception of the universe. If we were to imagine the Big Bang as simply an enormous explosion that spewed matter out into space, ultimately to form the galaxies we see, then the foregoing reasoning would be quite correct—there would be a center and an edge, and the cosmological principle would not apply. But the Big Bang was *not* an explosion in an otherwise featureless, empty universe. The only way that we can have Hubble's law hold *and* retain the cosmological principle is to realize that the Big Bang involved the entire universe—not just the matter and radiation within it, but the universe *itself.* In other words, the galaxies are not flying apart into the rest of the universe. The universe itself is expanding. Like raisins in a loaf of raisin bread that move apart as the bread expands in an oven, the galaxies are just along for the ride.

Let's reconsider some of our earlier statements in light of this new perspective. We now recognize that Hubble's law describes the expansion of the universe itself. Although galaxies have some small-scale, individual random motions, on average they are *not* moving with respect to the fabric of space—any such overall motion would pick out a "special" direction in space and violate the assumption of isotropy. On the contrary, the portion of the galaxies' motion that makes up the Hubble flow is really an expansion

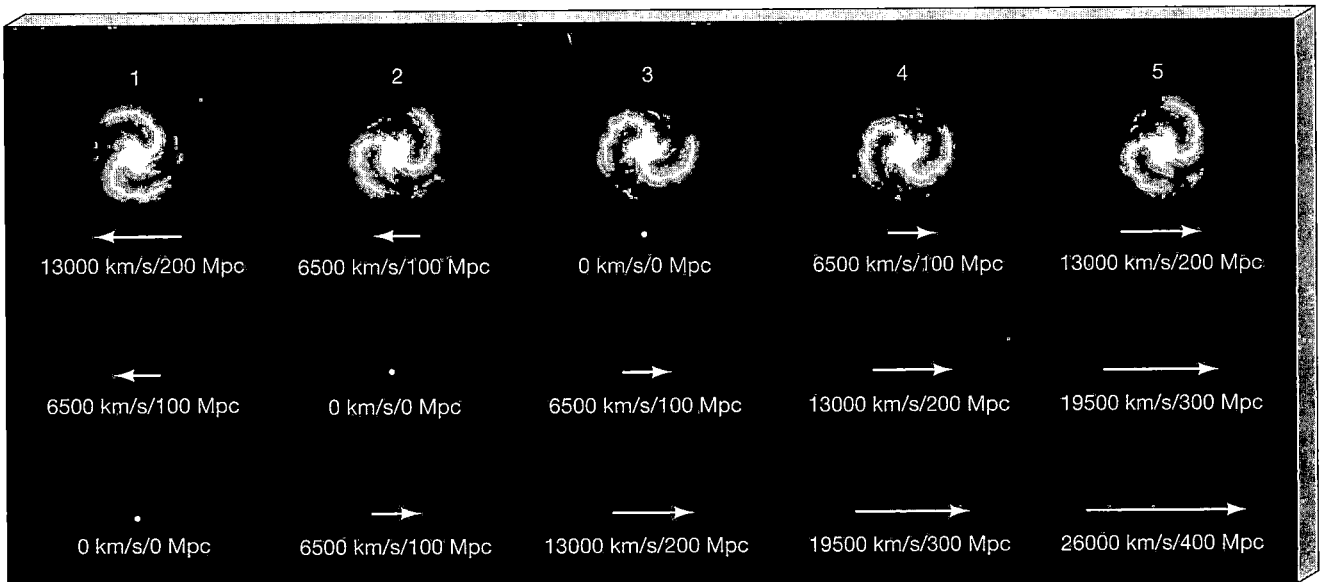

Figure 26.3 Hubble Expansion Hubble's law is the same no matter who makes the measurements. The top numbers are the distances and recessional velocities as seen by an observer on the middle of five galaxies, galaxy 3. The bottom two sets of numbers are from the points of view of observers on galaxies 2 and 1, respectively. In all cases, Hubble's law holds.

DISCOVERY 26-1

Stunning Views of Deep Space

As we probe deep into space, we are looking far back in time. Telescopes are time machines, and astronomers are historians. Now, thanks to NASA's orbiting observatories, we have the deepest, most detailed views of extragalactic space yet obtained. The first image shows the *Hubble Deep Field (HDF)* in approximately true color, that is, approximately what the human eye would see. ⟳ (Sec. 24.4) It was assembled from nearly 300 exposures taken during more than 100 orbits over the course of 10 consecutive days, for a total exposure time of about 100 hours. All the data were combined to make this one spectacular image. (Figure 24.23 shows a smaller view of the same field.) The field is in a region of the sky near the Big Dipper that happens to lie in a direction perpendicular to the congested plane of the Milky Way. The superimposed numbers are redshifts of various galaxies as measured by the Keck Observatory in Hawaii. Note the enormous spread in distances (see Table 25.1) evident in even the small sample of redshifts shown.

Virtually every blob of light in this image is a galaxy. (A few clear exceptions, such as the bright star with the artificial spikes at bottom left of the image, are local stars in our own Milky Way.) In all, approximately 1600 galaxies of all types can be counted in this one image. Hundreds of these galaxies are so faint that they have never been seen before. The image extends to 30th magnitude, making it the most sensitive photograph ever taken. ⟳ (Sec. 17.3) The image is "deep" in the sense that it allows us to see very faint objects. The word *deep* should not be construed to mean that this image goes beyond the realm of ground-based telescopes. *Hubble* does not see significantly farther into the universe than any other large telescope, but it does see more sensitively; hence it is able to pick up light from very dim objects and often to resolve them better than ground-based instruments.

Much good science has been mined from this single, very long exposure, and many new projects have been triggered by it. Indeed, as soon as *HDF* was unveiled, pressure mounted to take another long exposure somewhere else in the sky. The second image is known as the *Hubble Deep Field-South*, or *HDF-S* (the original having been taken in the northern sky). Like *HDF*, it shows a selection of cosmic objects distributed along a narrow corridor of space approximately 3000 Mpc, or 10 billion light-years, in length. It represents an accumulation of data acquired over the course of 10 days of observing during late 1998 near the south celestial pole, in the constellation Tucana.

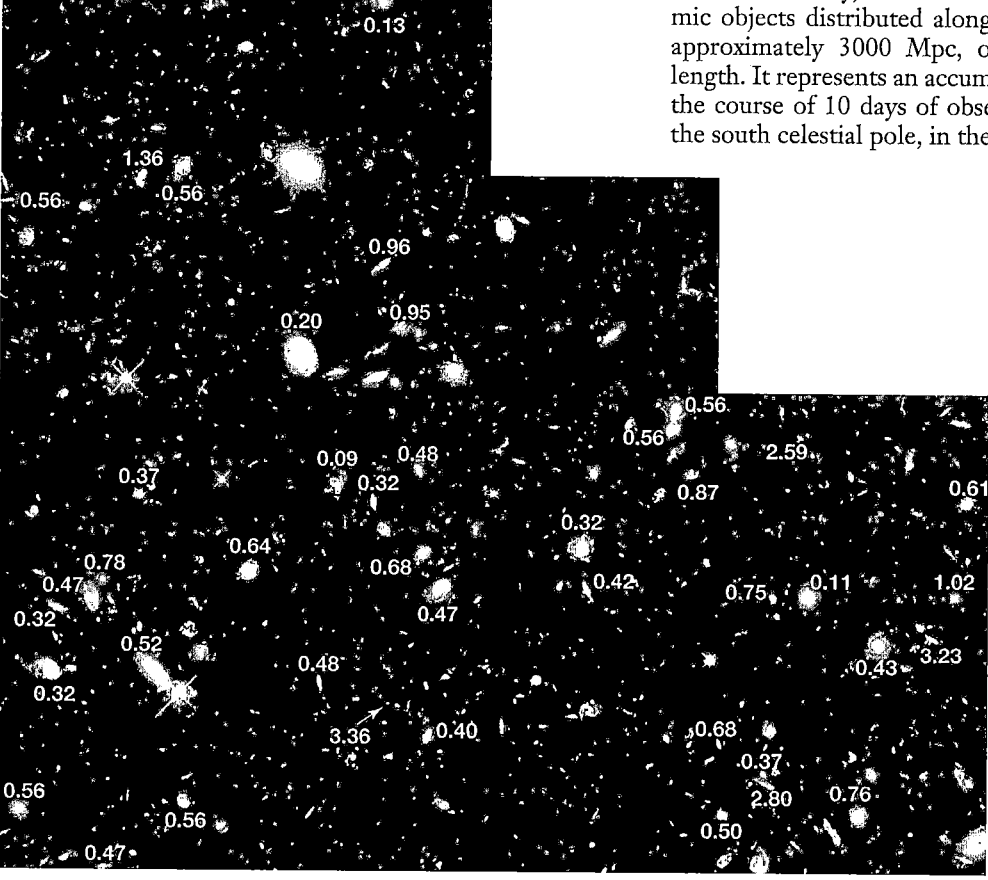

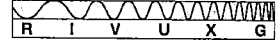

(R. Williams/NASA)

Both images show many galaxies of many ages. Disentangling young, distant galaxies from old, nearby ones is tricky, but as a general rule of thumb (and perhaps a little counterintuitively, given that we have gone to some lengths to stress that distant objects are characterized by large redshifts) the blue objects are probably the farthest away (since we see their most luminous stars in their blazing ultraviolet youth), whereas red objects tend to be closer (we see them by the light of their old, reddened stars). Bright spiral galaxies are most easily seen, while ellipticals appear mostly as reddish blobs. Other galaxies are irregularly shaped, probably the result of collisions or close encounters with neighboring galaxies, seen in this image when the universe was younger and denser.

Of particular interest are the small, faint smudges of blue light in the images, the faraway protogalaxies that gave rise to the galaxies we see today. Some of the faintest galaxies in these images existed when the universe was a mere billion years old, eons before the birth of our solar system. However, only spectroscopic observations can tell if these dim smudges are truly distant. The big ground-based telescopes in Hawaii, Chile, and on Kitt Peak are far better suited than *Hubble* to determine redshifts, and this work is underway.

The *HDF*s cover only a tiny part of the entire sky. The images shown here are just 2' across. To map the entire sky to this depth would take *Hubble* nearly a million years! One way to put these images into a larger perspective is to think of them as core samples, much like those a geologist might take on Earth. By examining the dirt and rocks in their samples, geologists try to reconstruct the history of events that oc-

curred as Earth evolved. Likewise, by studying the galaxies in the *HDF*s, astronomers seek to understand better the history of the universe.

Despite their narrow fields of view, astronomers consider the *HDF*s to be representative of the typical spread of galaxies in space. Statistically, the two images seem virtually identical. This is another way of stating the cosmological principle—the universe on the largest scales looks pretty much the same in all directions. Extrapolating the contents of these images over the entire sky, astronomers then estimate that the total number of galaxies in the observable universe is approximately 40 billion. Contrary to sensationalized news accounts, this is not an astonishingly larger number of galaxies than previously thought. Rather, it is midway between the 10 billion and 100 billion galaxies that astronomers for decades have reasoned should exist in the observable universe.

The final image is known as "Chandra Deep Field-South." It is a one-million-second exposure—the deepest X-ray exposure ever achieved—aimed at the Fornax constellation in the Southern Hemisphere. Most of the objects seen in the image are active galaxies and quasars powered by massive black holes. The image also shows, for the first time, X rays from galaxies, groups of galaxies, and clusters of galaxies that all contribute to a "diffuse X-ray background" radiation. The colors in this false-color image depict the intensities of X rays from lowest to highest as red, yellow, and blue.

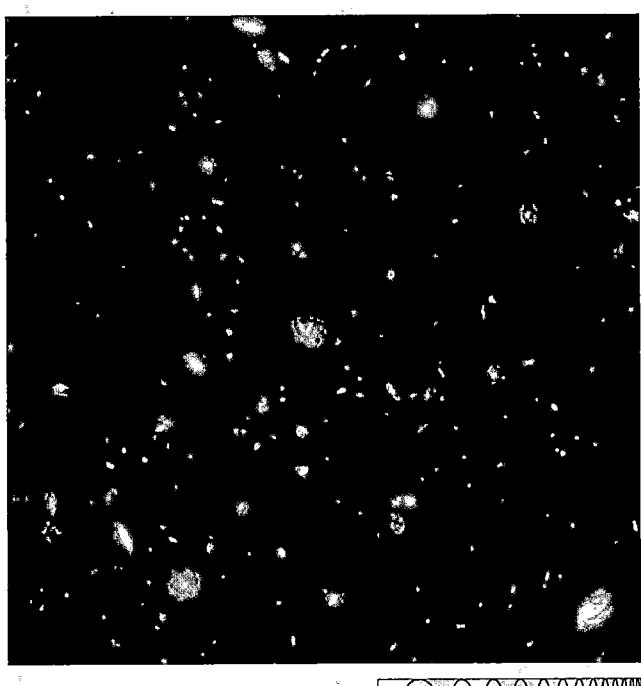

(NASA)

(NASA/JHU/AUI/R. Giacconi et al.)

of space itself. The expanding universe remains homogeneous at all times. There is no "empty space" beyond the galaxies into which they rush. At the time of the Big Bang, the galaxies did not reside at a point located at some well-defined place within the universe. The *entire universe* was a point. That point was in no way different from the rest of the universe; that point *was* the universe. Therefore, there was no one point where the Big Bang "happened"—because the Big Bang involved the entire universe, it happened *everywhere* at once.

To illustrate these ideas, imagine an ordinary balloon with coins taped to its surface, as shown in Figure 26.4. (Better yet, do the experiment yourself.) The coins represent galaxies, and the two-dimensional surface of the balloon represents the "fabric" of our three-dimensional universe. The cosmological principle applies to the balloon because every point on the balloon looks pretty much the same as every other. Imagine yourself as a resident of one of the three coin "galaxies" in the leftmost frame, and note your position relative to your neighbors. As the balloon inflates (that is, as the universe expands), the other galaxies recede from you; more distant galaxies recede more rapidly. (Notice, incidentally, that the coins themselves do *not* expand along with the balloon, any more than people, planets, stars, or galaxies—all of which are held together by their own internal forces—expand along with the universe.) ∞ (Sec. 24.5)

Regardless of which galaxy you chose to consider, you would see all the other galaxies receding from you. Nothing is special or peculiar about the fact that all the galaxies are receding from you. Such is the cosmological principle: No observer anywhere in the universe has a privileged position. There is no center to the expansion and no position that can be identified as the location from which the universal expansion began. Everyone sees an overall expansion described by Hubble's law, with the same value of Hubble's constant in all cases.

Now imagine letting the balloon deflate. This corresponds to running the universe backward from the present time to the Big Bang. *All* the galaxies (coins) would arrive at the same place at the same time—at the instant the balloon reached zero size. But there is no one point on the balloon that could be said to be *the* place where that occurred. The entire balloon expanded from a point, just as the Big Bang encompassed the entire universe and expanded from a point.

This analogy has its shortcomings. The main difficulty here is that we see the balloon, which in our illustration we imagined as two dimensional, expanding into the third dimension of space. This might suggest that the three-dimensional universe is expanding "into" some fourth spatial dimension. It is not, so far as we know. At the very least, if higher spatial dimensions are involved, they are not relevant to our theory of the universe.

THE COSMOLOGICAL REDSHIFT

This view of the expanding universe requires us to reinterpret the cosmological redshift. ∞ (Sec. 24.5) Up to now we have explained the redshift of galaxies as a Doppler shift, a consequence of their motion relative to us. However, we have just argued that the galaxies are *not* in fact moving with respect to the universe, in which case the Doppler interpretation is incorrect. The true explanation is that as a photon moves through space its wavelength is influenced by the expansion of the universe. In a sense we can think of the photon as being attached to the expanding fabric of space, so its wavelength expands along with the universe, as illustrated in Figure 26.5. Although it is common practice in astronomy to refer to the cosmological redshift in terms of recessional velocity, bear in mind that, strictly speaking, this is not the right thing to do. The cosmological redshift is a consequence of the changing size of the universe—it is *not* related to velocity at all.

The redshift of a photon measures the amount by which the universe has expanded since that photon was emitted. For example, when we measure the light from a quasar to have a redshift of five, that means that the light was emitted at a time when the universe was just one-sixth its present size (and that we are observing the quasar as it was at that time). ∞ (*More Precisely* 25-2) In general, the larger a photon's redshift, the smaller the universe was at

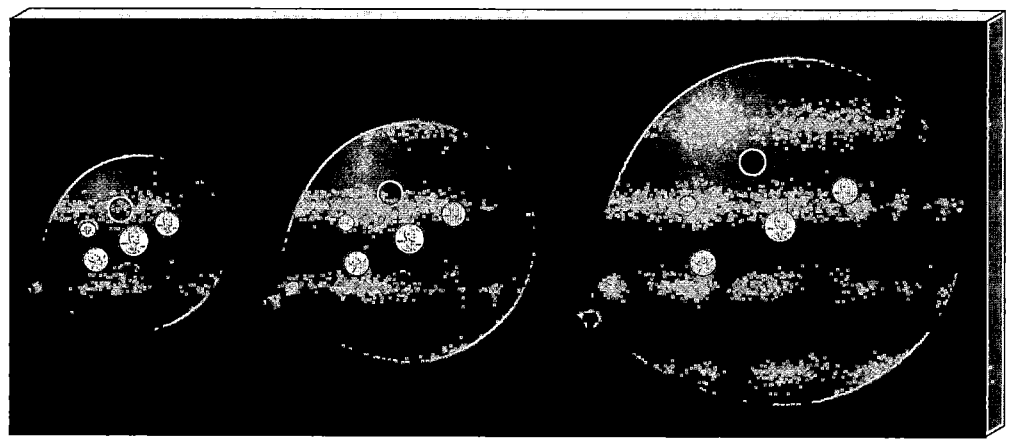

Figure 26.4 Receding Galaxies The coins taped to the surface of a spherical balloon recede from one another as the balloon inflates (left to right). Similarly, galaxies recede from one another as the universe expands. As the coins recede, the distance between any two of them increases, and the rate of increase of this distance is proportional to the distance between them. Thus, the balloon expands according to Hubble's law.

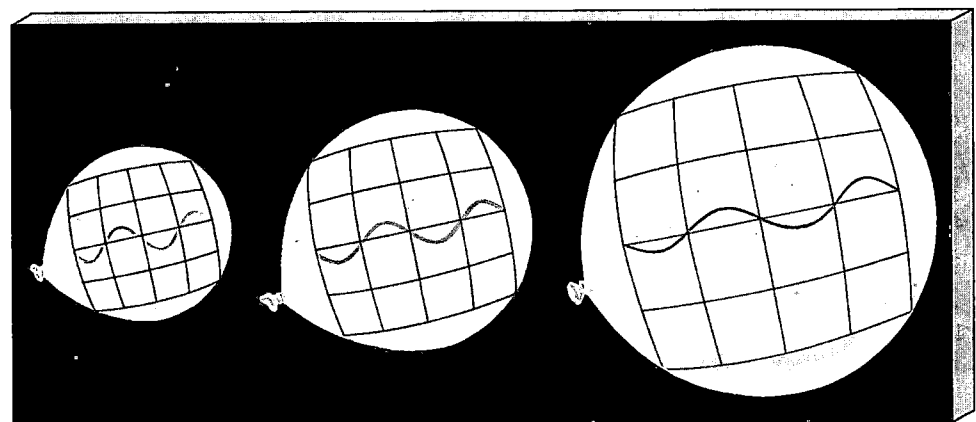

Figure 26.5 Cosmological Redshift As the universe expands, photons of radiation are stretched in wavelength, giving rise to the cosmological redshift.

the time the photon was emitted, and so the longer ago that emission occurred. Because the universe expands with time and redshift is related to that expansion, cosmologists routinely use redshift as a convenient means of expressing time.

RELATIVITY AND THE UNIVERSE

These concepts are difficult to grasp. The notion of the entire universe expanding from a point—with *nothing*, not even space and time, outside—takes some getting used to. Nevertheless, this picture of the universe lies at the heart of modern cosmology. The description of the universe itself (not just its contents) as a dynamic, evolving object is far beyond the capabilities of Newtonian mechanics, which we have used almost everywhere throughout this book. ∞ (Sec. 2.7) Instead, the more powerful techniques of Einstein's *general relativity*, with its built-in notions of warped space and dynamic space-time, are needed.

We encountered general relativity in Chapter 22 (see especially *More Precisely 22-1*) when we discussed the strange properties of black holes. We can loosely summarize its description of the universe by saying that the presence of matter or energy causes a curvature of space (more correctly, *space time*) and that the curved trajectories of freely falling particles within warped space are what Newton thought of as orbits under the influence of gravity. The amount of curvature depends on the amount of matter present, and the orbits of particles in turn depend on the curvature. Put less formally, "space tells matter how to move, and matter tells space how to curve." In a homogeneous universe, the curvature of space must be uniform.

Must we keep this difficult notion of a warped, expanding, homogeneous, universe constantly in the forefront of our thoughts if we are to comprehend the evolution of the cosmos? Perhaps surprisingly, given that relativity is the only theory that properly describes the large-scale behavior of the universe, the answer is no. Although general relativity predicts some curious consequences for the overall *geometry* of space on large scales, as we will see in Section 26.5, the *dynamics* of the universe can

be understood using simple concepts that would have been thoroughly familiar to Newton.

Consider two neighboring points A and B in the expanding universe, as pictured in Figure 26.6. From the perspective of point A, every other point, including B, is rushing away from it, in accordance with Hubble's law. How does the distance between A and B change in time? Does it just keep growing, or does its rate of increase eventually slow down and stop? From a Newtonian viewpoint, we might expect that the overall gravitational pull of the universe would tend to slow the expansion, just as Earth's gravity tends to slow the upward motion of an object projected from the surface. We have just argued that Newtonian physics is inadequate to describe the large-scale dynamics of the universe. However, a surprising prediction of general relativity is that the Newtonian picture gives the right result! The motion of point B relative to point A is *exactly* the same as if A were at the center of a planet and B were a projectile fired vertically upward

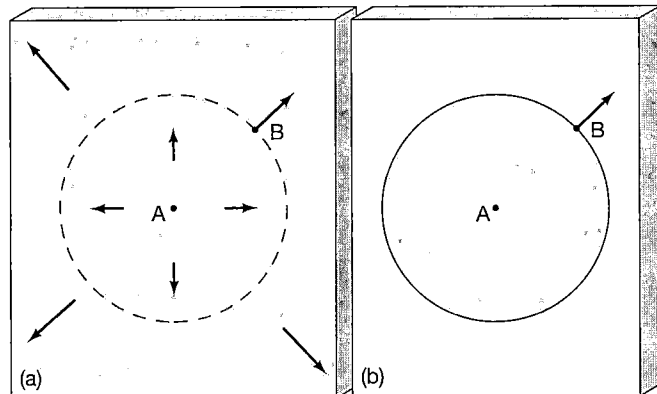

Figure 26.6 Expanding Universe The relative motion of any two points in the expanding universe (a) can be addressed as a problem in Newtonian mechanics, even though Einstein's theory of general relativity is needed to explain why it is correct to do so. If the rest of the universe is ignored—as in part (b)—then the Newtonian calculation of B's motion relative to A gives the same answer as a calculation using general relativity.

from its surface. Thus, we *can* in fact discuss the expansion of the universe in simple Newtonian terms—and we will do so in the next two sections—but we need general relativity to justify our doing so.

☑️ Concept Check

■ Why does Hubble's law imply a Big Bang?

26.3 The Fate of the Universe

③ Will the universe expand forever? This fundamental question about the fate of the universe has been at the heart of cosmology since Hubble's law was first discovered. Until the late 1990s, the prevailing view among cosmologists was that the answer would most likely be found by determining the extent to which gravity would slow, and perhaps ultimately reverse, the current expansion. However, it now appears that the answer is more subtle—and perhaps a lot more profound in its implications—than was hitherto believed.

CRITICAL DENSITY

③ Let's begin with the conventional view, assuming that gravity is the only force that affects the cosmic expansion, and consider as an analogy a rocket ship launched from the surface of a planet. What is the likely outcome of that motion? Now that we know we can apply familiar Newtonian concepts to the problem, we can identify two basic possibilities, depending on the speed of the ship. If the launch speed is high enough, it will exceed the planet's escape speed, and the ship will never return to the surface. ⚬ (Sec. 2.7) The speed will diminish because of the planet's gravitational pull, but it will never reach zero. The spacecraft leaves the planet on an *unbound* trajectory, as illustrated in Figure 26.7(a). Alternatively, if the ship's launch speed is lower than the escape speed, it will reach a maximum distance from the planet, then fall back to the surface. Its *bound* trajectory is shown in Figure 26.7(b).

Similar reasoning applies to the expansion of the universe. Reconsider Figure 26.6, but now imagine that A and B are galaxies at some known distance from each other, with their present relative velocity given by Hubble's law. The same two basic possibilities exist for these galaxies as for our spacecraft—the distance between them can increase forever, or it can increase for a while and then start to decrease. What's more, the cosmological principle says that, whatever the outcome, it must be the same for *any* two galaxies—in other words, the same statement applies to the universe *as a whole*. Thus, as illustrated in Figure 26.8, the universe has only two options: it can continue to expand forever—an *unbound* universe—or the present expansion will someday stop and turn around into a contraction—a *bound* universe.

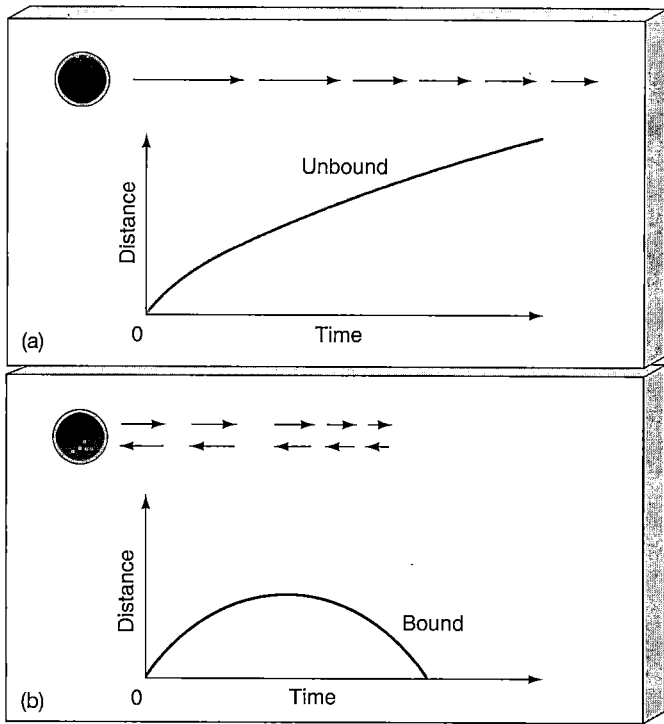

Figure 26.7 Escape Speed (a) A spacecraft (arrow) leaving a planet (blue ball) with a speed greater than the escape speed follows an unbound trajectory. The graph shows the distance between the ship and the planet as a function of time. (b) If the launch speed is less than the escape speed, the ship eventually drops back to the planet. Its distance from the planet first rises, then falls.

The middle curve on Figure 26.8 marks the dividing line between these two possibilities. It shows a *marginally bound* universe that expands forever, but at an ever-decreasing rate, analogous to our rocket ship's leaving the planet with precisely the escape speed. The three curves are drawn so that they all pass through the same point at the present time. All are possible descriptions of the universe given its present size and expansion rate.

What determines which of these possibilities will actually occur? In the case of a rocket ship of fixed launch speed, the *mass* of the planet determines whether or not escape will occur—a more massive planet has a higher escape speed, making it less likely that the rocket can escape. For the universe, the corresponding factor is the *density* of the cosmos. A high-density universe contains enough mass to stop the expansion and eventually cause a recollapse. A low-density universe, conversely, will expand forever. The dividing line between these outcomes, the density corresponding to a marginally bound universe, in which gravity acting alone would be just sufficient to halt the present expansion, is called the **critical density**. For $H_0 = 65$ km/s/Mpc, the critical density is about 8×10^{-27} kg/m^3. That's an extraordinarily low density—just five hydrogen atoms per cubic

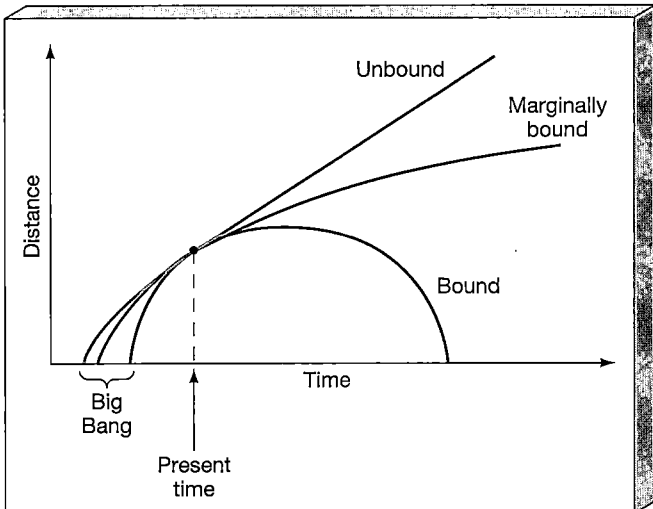

Figure 26.8 Model Universes Distance between two galaxies as a function of time in each of the three possible universes discussed in the text: unbound, bound, and marginally bound. The point where the three curves touch represents the present time.

meter, a volume the size of a typical household closet. In more "cosmological" terms, it corresponds to about 0.1 Milky Way Galaxies (including the dark matter) per cubic megaparsec.

TWO FUTURES

These possible outcomes represent radically different futures for our universe. If the cosmos emerged from the Big

Bang with sufficiently high density, then it contains enough matter to halt its own expansion, and the recession of the galaxies will eventually stop. At some time in the future, astronomers everywhere—on any planet within any galaxy—will announce that the radiation received from nearby galaxies is no longer redshifted. (The light from *distant* galaxies will still be redshifted, however, because we will see them as they were in the past, at a time when the universe was still expanding.) The bulk motion of the universe, and of the galaxies within, will be stilled—at least momentarily.

The expansion may stop, but the pull of gravity will not. The universe will begin to contract. Nearby galaxies will begin to show blueshifts, and both the density and the temperature of the universe will start to rise as matter collapses back onto itself. As illustrated in Figure 26.9(a), the universe will recollapse to a point, requiring just as much time to fall back as it took to rise. First galaxies, then stars will collide with increasing frequency and violence as the available space diminishes and the entire universe shrinks toward a superdense, superhot singularity much like the one from which it originated. The cosmos will ultimately experience a "heat death," in which all matter and life are destined to be incinerated. Some astronomers call the final collapse of this high-density universe the "Big Crunch."

Cosmologists do not know what will happen to the universe if it ever reaches the point of collapse. The laws of physics as we presently understand them are simply inadequate to describe these extreme conditions. However, some theorists have suggested that the universe may not simply end—instead it may "bounce" into a new expansion phase. Figure 26.9(b) depicts a hypothetical—and

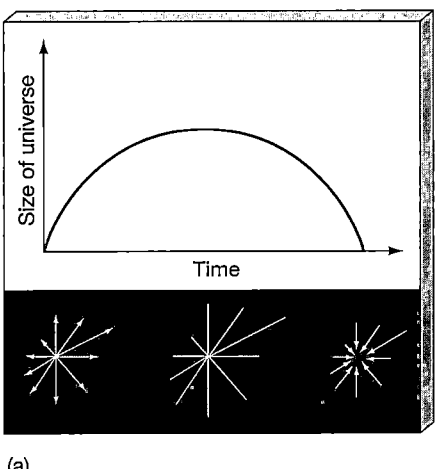

(a)

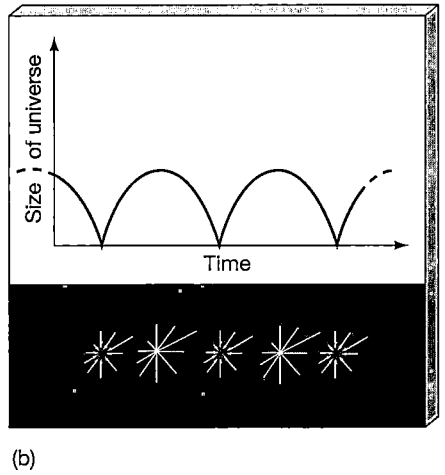

(b)

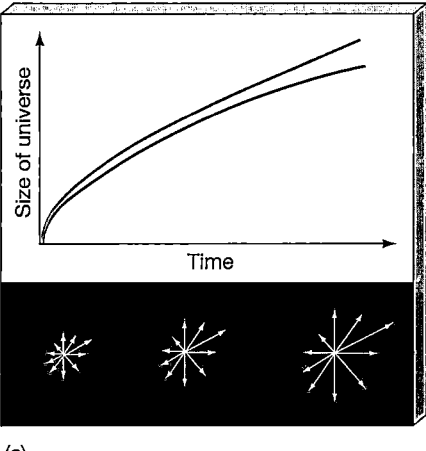

(c)

Figure 26.9 Two Futures (a) A high-density universe has a beginning and an end, and a finite lifetime. The lower frames illustrate its evolution, from explosion, to maximum size, to recollapse.
(b) An oscillating universe has neither beginning nor end. Each expansion–contraction phase ends in a "bounce" that becomes the Big Bang of the next expansion. There is currently no information on whether this can actually occur. (c) A low-density universe expands forever from its explosive beginning. The upper curve represents a universe with density less than the critical value. The lower curve represents a universe with density exactly equal to the critical value.

highly speculative—universe having many such cycles of expansion and contraction.

A quite different fate awaits the universe if its density is below the critical value. In that case its density always has been, and always will be, too small for gravity to cause it to recontract. As illustrated in Figure 26.9(c), such a low-density universe will expand forever. In this scenario the galaxies will continue to recede forever, their radiation weakening with increasing distance. In time an observer on Earth will see no galaxies in the sky beyond the Local Group (which is not itself expanding), even with the most powerful telescope. The rest of the observable universe will appear dark, the distant galaxies too faint to be seen. Eventually, the Milky Way and the Local Group too will peter out as their fuel supply is consumed. This universe will ultimately experience a "cold death." All radiation, matter, and life are eventually destined to freeze.

✓ Concept Check

■ What are the two basic futures for the universe?

26.4 Will the Universe Expand Forever?

Is there any way for us to determine which of these futures just described actually applies to our universe (that is, apart from waiting to find out)? Will the universe end as a small, dense point much like that from which it began? Or will it expand forever? Fortunately, we live at a time when astronomers are subjecting these questions to intensive observational tests.

THE DENSITY OF THE UNIVERSE

🔴 Perhaps the most straightforward way to answer this question is to try to measure directly the average density of the universe. As just noted, for $H_0 = 65$ km/s/Mpc, the critical density separating the two possible futures is 8×10^{-27} kg/m^3. Cosmologists conventionally call the ratio of the universe's actual density to the critical value the *cosmic density parameter* and denote it by the symbol Ω_0 ("omega nought"). In terms of this quantity, then, a critical universe has $\Omega_0 = 1$. A universe with Ω_0 less than 1 has insufficient gravity to halt and reverse the present cosmic expansion, and will expand forever.

How might we determine the density of the universe? On the face of it, it would seem simple—just measure the total mass of the galaxies residing within some large parcel of space, calculate the volume of that space, then divide mass by volume to compute the average density. When astronomers do this, they usually find a little less than 10^{-28} kg/m^3 in the form of luminous matter. Largely independent of whether the chosen region contains many scat-

tered galaxies or only a few rich galaxy clusters, the resulting density is about the same, within a factor of two or three. Galaxy counts thus yield a value of Ω_0 of about 0.01. If that measure were correct, then the universe would expand forever.

But there is a catch. We have noted (Chapters 23 and 24) that most of the matter in the universe is *dark*—it exists in the form of invisible material that has been detected only through its gravitational effect in galaxies and galaxy clusters. ⬤ (Secs. 23.6, 24.3) We currently do not know what the dark matter is, but we *do* know that it is there. Galaxies may contain as much as 10 times more dark matter than luminous material, and the figure for galaxy clusters is even higher—perhaps as much as 95 percent of the total mass in clusters is invisible. Even though we cannot see it, dark matter contributes to the density of the universe and plays its part in opposing the cosmic expansion. Including all the dark matter that is known to exist in galaxies and galaxy clusters increases the value of Ω_0 to 0.2 or 0.3.

Unfortunately, the distribution of dark matter on larger scales is not very well known. We can infer its presence in galaxies and galaxy clusters but we are largely ignorant of its extent in superclusters, voids, or other larger structures. However, there are indications that it accounts for an even greater fraction of the mass on large scales than it does in galaxy clusters. Observations of gravitational lensing by galaxy clusters suggest that dark matter may be considerably more extensive than is indicated by the motions of galaxies within the clusters. ⬤ (Sec. 25.5) One of the few mass estimates for an object much larger than a supercluster comes from optical and infrared observations of the overall motion of galaxies (including the Local Group) within the Local Supercluster. The measured velocities suggest the presence of a nearby huge accumulation of mass known as the *Great Attractor*, with a total mass of about 10^{17} solar masses and a size of 100–150 Mpc. Its average density may be quite close to the critical value.

However, it seems that, when averaged over the entire universe, even objects like the Great Attractor don't raise the overall cosmic density by much. In short, there doesn't seem to be much additional dark matter "tucked away" on very large scales. Most cosmologists agree that the density of matter in the universe is not much greater than about one-third the critical value. Thus, as best we can tell given the current data, the universe is destined to expand forever.

COSMIC ACCELERATION

🔵 🔴 Determining the mass density of the universe is an example of a *local* measurement that provides an estimate of Ω_0. But the result we obtain depends on just how local our measurement is—as we have just seen, there are many uncertainties in the result, especially on large scales. In an attempt to get around this problem, astronomers have devised alternative methods that rely instead on *global* mea-

Figure 26.10 Accelerating Universe? (a) Observations of distant supernovae allow astronomers to measure changes in the expansion rate of the universe. In a decelerating universe (purple and red curves), redshifts of distant objects are greater than would be predicted from Hubble's law (black curve). The reverse is true for an accelerating universe. The points showing observations of some 50 supernovae strongly suggest that the cosmic expansion is accelerating. The vertical scale shows redshift; for small velocities, redshift is just velocity divided by the speed of light. ⇨ (*More Precisely 25-2*) (b) Some of the supernovae used in the study, as seen by the *Hubble Space Telescope* in 1998. The bottom frames show three supernovae (marked by arrows) that exploded in distant galaxies when the Universe was nearly half its current age. The top frames are larger views. These supernovae were originally discovered in 1997 during a ground-based survey with the Canada-France-Hawaii Telescope on Mauna Kea. (*P. Garnavich/Harvard-Smithsonian Center for Astrophysics/NASA*)

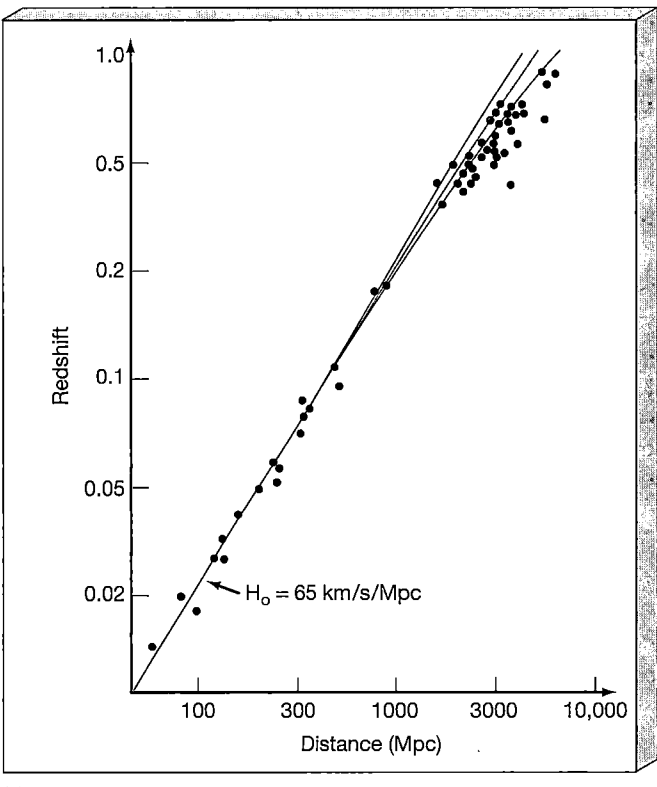

(a)

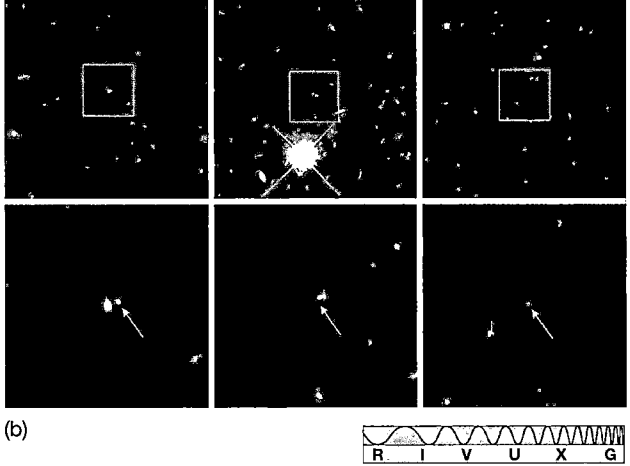

(b)

surements, covering much larger regions of the observable universe. In principle, such global tests should indicate the universe's overall density, not just its value in our cosmic neighborhood.

One such global method is based on observations of Type I (carbon-detonation) supernovae. ⇨ (Sec. 21.3) Recall that these objects are very bright and have a remarkably narrow spread in luminosities, making them particularly useful as standard candles. ⇨ (Sec. 24.2) They can be used as probes of the universe because, by measuring their distances (*without* using Hubble's law) and their redshifts, we can determine the rate of cosmic expansion in the distant past. Here's how the method works.

Suppose the universe is decelerating, as we would expect if gravity were slowing its expansion. Then, because the expansion rate is decreasing, objects at great distances—that is, objects that emitted their radiation long ago—should appear to be receding *faster* than Hubble's law predicts. Figure 26.10(a) illustrates this concept. If the universal expansion were constant in time, recessional velocity and distance would be related by the black line in the figure. (The line is not quite straight because it takes the expansion of the universe properly into account in computing the distance.) ⇨ (*More Precisely 25-2*) In a decelerating universe, the velocities of distant objects should lie *above* the black curve, and the deviation from that curve is greater for a denser universe, in which gravity has been more effective at slowing the expansion.

How does theory compare with reality? In the late 1990s, two groups of astronomers announced the results of independent, systematic surveys of distant supernovae. Some of these supernovae are shown in Figure 26.10(b); the data are marked on Figure 26.10(a). Far from clarifying the picture of cosmic deceleration, however, these findings seem to indicate that the expansion of the universe is not slowing, but actually accelerating! According to the supernova data, galaxies at large distances are receding *less* rapidly than Hubble's law would predict. The deviations from

the decelerating curves appear small in the figure, but they are statistically very significant, and both groups report similar findings. These observations are *inconsistent* with the standard Big Bang model just described and, if confirmed, will necessitate a major revision of our view of the cosmos. The measurements are difficult, and the results depend quite sensitively on just how "standard" the supernovae luminosities really are. Nevertheless, most cosmologists (some reluctantly) have accepted these new results and, for now at least, the acceleration seems to be real.

What could cause an overall acceleration of the universe? Cosmologists do not know, although several possibilities have been suggested. One leading candidate is

Even the greatest minds are fallible. The first scientist to apply general relativity to the universe was, not surprisingly, the theory's inventor, Albert Einstein. When he derived and solved the equations describing the behavior of the universe, Einstein discovered that they predicted a universe that evolved in time. But in 1917 neither he nor anyone else knew about the expansion of the universe, as described by Hubble's law, which would not be discovered for another 10 years. ∞ (Sec. 24.5) At the time, Einstein, like most scientists, believed that the universe was static—that is, unchanging and everlasting. The discovery that there was no static solution to his equations seemed to Einstein to be a near-fatal flaw in his new theory.

To bring his theory into line with his beliefs, Einstein tinkered with his equations, introducing a "fudge factor" describing a hypothetical repulsive force operating on large scales in the universe. This factor is now known as the cosmological constant. As illustrated in the accompanying figure, which shows the effect of introducing a cosmological constant into the equations describing the expansion of a critical-density universe, this factor allows many other solutions to Einstein's equations. One of these solutions describes a "coasting" universe, whose radius does in fact remain constant for an indefinite period of time, Einstein took this to be the static universe he expected.

Instead of predicting an evolving cosmos, which would have been one of general relativity's greatest triumphs, Einstein yielded to a preconceived notion of the way the universe "should be," unsupported by observational evidence. Later, when the expansion of the universe was discovered and Einstein's equations—without the fudge factor—were found to describe it perfectly, he declared that the cosmological constant was the biggest mistake of his scientific career.

For many researchers—Einstein included—the main problem with the cosmological constant was (and still is) the fact that it had no clear physical interpretation. Einstein introduced it to fix what he thought was a problem with his equations, but he discarded it immediately once he realized that no problem actually existed. Scientists are very reluctant to introduce unknown quantities into their equations purely to make the results "come out right." As a result, the cosmological constant fell out of favor among astronomers for many years.

In the 1980s, the concept made something of a comeback with the realization by physicists that the very early universe may have gone through a phase when its evolution was determined by a "cosmological constant" of sorts (see Section 27.4), and this idea is now firmly entrenched in many cosmologists' models of the universe. As discussed in the text, the cosmological constant now seems to have been completely rehabilitated, as it has been identified as one possible candidate for the "dark energy" whose existence has been inferred from studies of the universe on very large scales. The green curves in the figure show how the inclusion of a suitable dark-energy term in Einstein's equations can cause the expansion of the universe to accelerate, instead of slowing down as it would if only gravity were involved.

The figure also shows how the cosmological constant neatly avoids a potential conflict between the age of the universe derived from cosmology and that inferred from studies of star clusters. By adjusting the value of the constant, theorists can construct mathematical models of the universe that did not expand so rapidly in the past, and so are older than the measured value of H_0 suggests. In this way, a Hubble constant of 65 km/s/Mpc in a critical-density universe can be reconciled with a cosmic age of 12 billion years, or even much more.

an additional "vacuum pressure" force associated with empty space and effective only on very large scales. It is known simply as the **cosmological constant** (*Discovery 26-2*). Its influence increases as the universe expands. Thus it was negligible at early times, but today it may be the major factor controlling the cosmic expansion. Models including this force can fit the observational data, but at present astronomers have no clear physical interpretation of what it actually means—it is neither required nor explained by any known law of physics.

Whatever it is, the mysterious cosmic field causing the universe to accelerate is neither matter nor radiation. It is sometimes referred to as **dark energy**. If confirmed, the magnitude of the cosmic acceleration implies that the amount of dark energy in the universe may exceed the total mass-energy of matter (luminous and dark) by a substantial margin. By opposing the attractive force of gravity, the repulsive effect of the dark energy strengthens our earlier conclusion that the universe will expand forever.

COSMIC COMPOSITION

5 In addition to measurements of density and acceleration, astronomers have several other means of estimating the "cosmological parameters" that describe the large-scale properties of our universe.

Theoretical studies of the very early universe (to be discussed in more detail in Chapter 27) strongly suggest that the density of the universe should be exactly critical. These ideas first became widespread in the 1980s, and for many years there seemed to be a major discrepancy between this conclusion and observations that showed a cosmic matter density of 20–30 percent of the critical value, even taking the dark matter into account. The cosmic acceleration just discussed actually resolves that conflict, although not all astronomers are happy at the price of this resolution, which has introduced yet another unknown component into the cosmic mix! Recent detailed measurements of the radiation field now known to fill the

The principal drawback of the cosmological constant is that astronomers have no clear explanation for either its existence or its present value. The leading theories of the structure of matter do in fact predict repulsive forces of this sort, but they generally operate only under extreme conditions, and in any case their "natural" scale is vastly greater (by something like a factor of 10^{120}!) than anything consistent with cosmological observations. An additional problem is that the present value of the repulsive force is roughly comparable to the attractive force of gravity opposing further expansion. Why is that a problem? Simply because, when we calculate the evolution of a universe containing a cosmological constant consistent with current observations (Figure 26.11), we find that this statement was not true in the early universe (when galaxies were forming, say), nor will it be true in 10 or 20 billion years' time. In other words, it seems to suggest that we live at a special time in the history of the universe—a conclusion viewed with great suspicion by astronomers who grew up with the Copernican principle as their guide.

Attempts are underway to construct theories of dark-energy fields that preserve the character of the cosmological constant yet also account for its value in some natural way, perhaps somehow coupling it to the density of matter. Before we make too many sweeping statements about the role of the cosmological constant in cosmology, we should probably remember the experience of its inventor and bear in mind that—at least for now—its physical meaning remains completely unknown.

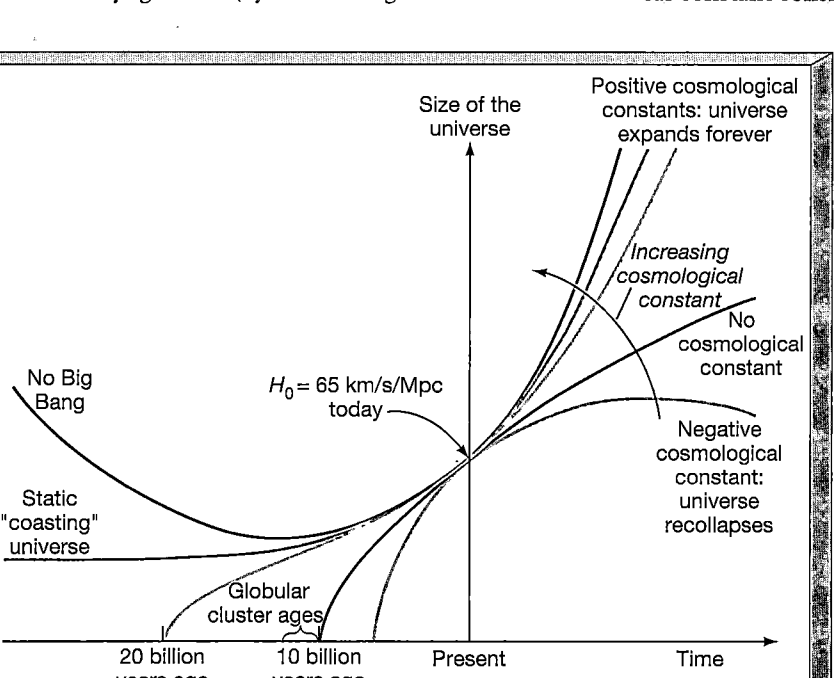

entire cosmos (see Section 26.6 and Chapter 27) support the theoretical prediction of critical density, and are also consistent with the dark energy inferred from the supernova studies.

Further support comes from careful analysis of galaxy surveys like those discussed in Section 26.1, which allow astronomers to measure the growth of large-scale structure in the universe. Simply put, the more mass there is in the universe, the easier it is for clusters, superclusters, walls and voids to grow as gravity gathers matter into larger and larger clumps. Higher density implies more rapid structure formation—or, equivalently, less structure in the past given the structure we see around us today.

Remarkably, all the approaches just described yield consistent results! As of mid-2001, the consensus among cosmologists is that the universe is of precisely critical density, $\Omega_0 = 1$, but that this density is made up of both (mostly dark) matter and dark energy (suitably converted into mass units using the relation $E = mc^2$). ∞ (Sec. 16.5) The

best estimate is that matter accounts for 35 percent of the total, dark energy for the remaining 65 percent. This is the assumption underlying Table 25.1 in *More Precisely 25-2* and used consistently throughout this book.

THE AGE OF THE UNIVERSE

2 In Section 26.2, when we estimated the age of the universe from the value of Hubble's constant, we made the assumption that the expansion speeds of the galaxies were constant in the past. The effects of gravity have tended to slow the universe's expansion, particularly at early times, while the cosmic expansion rate today may actually be increasing. In the absence of a cosmological constant, the universe expanded *faster* in the past than it does today, so the assumption of constant expansion rate leads to an overestimate of the universe's age—such a universe is younger than the 15 billion years we calculated earlier. How much younger depends on how much deceleration has occurred.

Conversely, a repulsive cosmological constant tends to increase the age of the cosmos.

Figure 26.11 illustrates these points. It is similar to Figure 26.8, except that here we have added two extra lines, one corresponding to constant expansion rate at the present value—a completely *empty* 15-billion-year-old universe—the other to the accelerating universe that best fits the supernova data. The age of a critical-density universe with no cosmological constant is about 10 billion years. A low-density, unbound universe is older than 10 billion years but still less than 15 billion years old. The age corresponding to the accelerating universe is about 14 billion years, coincidentally quite close to the value for constant expansion.

Although it is impressive that we can pin down the age of the universe to within less than a factor of two, these numbers highlight a long-standing debate among astronomers. An *independent* estimate of the age of the universe is provided by the ages of the oldest stars and star clusters. Over the years, as stellar and cosmological observations have improved, the corresponding age estimates have also evolved, and there have been periods during which they have given conflicting results—at times it has appeared that the oldest known stars were older than the universe itself!

The situation today is that oldest globular clusters are estimated to be about 12 billion years old, with most cluster ages thought to lie in the 10- to 12-billion-year range. ∞ (Secs. 19.6, 20.5) Without a dark energy component, and given the current preference among theorists for a critical density universe, our adopted value of $H_0 = 65$ km/s/Mpc leads to a cosmological age estimate of 10 billion years—inconsistent with the cluster ages. Thus, unless the universe is of lower density (in fact, $\Omega_0 = 0.3$, the result based just on dark matter, would do very nicely) *or* the value of Hubble's constant is at the low end of the range quoted in Chapter 24 ($H_0 = 50$–55 km/s/Mpc is needed)—and there are well-respected proponents of both those points of view—then we have a clear conflict between these two major areas of astronomy. However, the presently favored cosmology, which includes dark energy, yields an age estimate consistent with cluster ages for all but the very highest values of H_0 now considered "reasonable."

Thus, for $H_0 = 65$ km/s/Mpc, our current best guess of the history of the universe places the Big Bang at 14 billion years ago. The first quasars appeared about 13 billion years ago (at a redshift of 6), the peak quasar epoch (redshifts 2–3) occurred during the next 1 billion years, and the oldest-known stars formed during the 2 billion years after that. These numbers seem to fit together very nicely, but astronomers aren't ready to relax just yet—the history of this subject suggests that there may be a few more unexpected twists and turns in the road before the matter is finally resolved.

☑ Concept Check

■ Why do astronomers think the universe will expand forever?

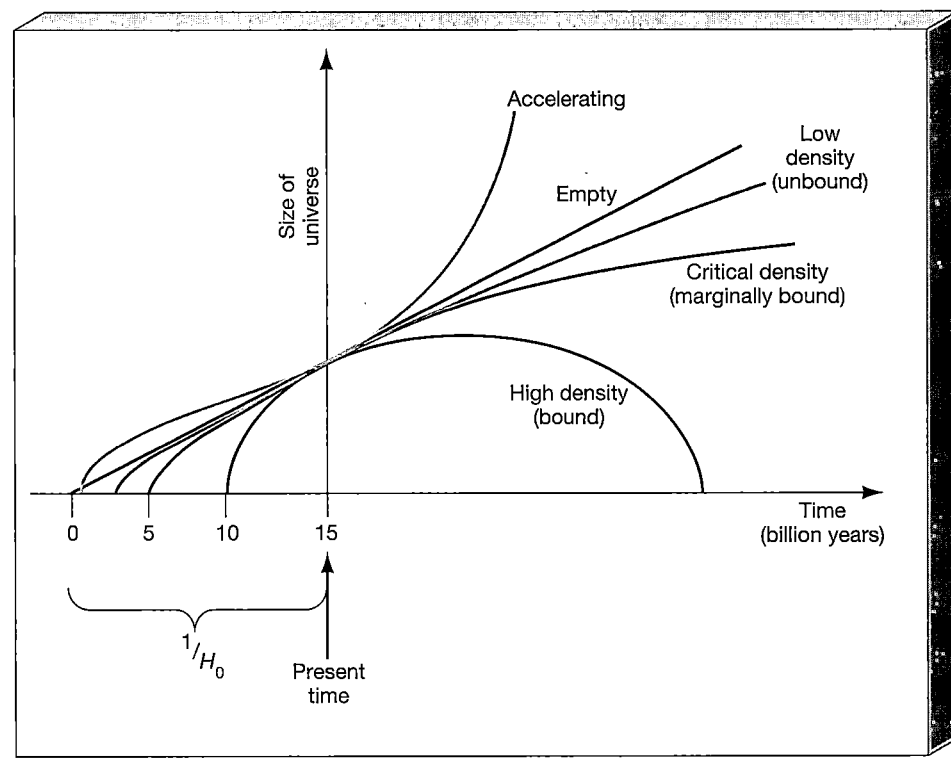

Figure 26.11 Cosmic Age
The age of a universe without a cosmological constant is always less than $1/H_0$, and decreases for larger values of the present-day density. The existence of a repulsive cosmological constant increases the age of the cosmos.

26.5 The Geometry of Space

⑤ We have reverted to the familiar notion of gravity, away from the more correct concept of warped space-time, because speaking in terms of gravity makes our discussion of the evolution of the universe much easier to understand. However, general relativity makes some predictions that do not have a simple description in Newtonian terms. Foremost among these is the fact that space is *curved*, with the degree of curvature determined by the *total density* of the cosmos—including not just matter (luminous and dark), but also radiation and dark energy. In the universe, the curvature must be the same everywhere (assuming homogeneity), so there are really only three possibilities for the large-scale geometry of space. For more information on the different types of geometry involved, see *More Precisely 26-1*.

If the average density of the cosmos is above the critical value, space is curved so much that it bends back on itself and "closes off," making this bound universe *finite* in size. Such a universe is known as a **closed universe**. It is difficult to visualize a three-dimensional volume uniformly arching back on itself in this way, but the two-dimensional version is well known: it is just the surface of a sphere, as with the balloon we discussed earlier. Figure 26.4, then, is the two-dimensional likeness of a three-dimensional closed universe. Like the surface of a sphere, a closed universe has no boundary, yet it is finite in extent.* One remarkable property of a closed universe is illustrated in Figure 26.12. Just as a traveler on the surface of a sphere can keep moving forward in a straight line and eventually return to her starting point, a flashlight beam shone in some direction in space might eventually traverse the entire universe and return from the opposite direction!

The surface of a sphere curves, loosely speaking, "in the same direction" no matter which way we move from a given point. A sphere is said to have *positive curvature*. However, if the average density of the universe is below the critical value, the surface curves like a saddle. It has *negative curvature*. Most people have a good idea of what a saddle looks like—it curves "up" in one direction and "down" in another, but no one has ever seen a uniformly negatively curved surface, for the simple reason that it cannot be constructed in three-dimensional Euclidean space! It is just "too big" to fit. A low-density, unbound, saddle-curved universe is infinite in extent and is usually called an **open universe**.

The intermediate case, in which the density is precisely equal to the critical density, is the easiest to visualize.

Figure 26.12 Einstein's Curve Ball In a closed universe a beam of light launched in one direction may return someday from the opposite direction after circling the universe, just as motion in a "straight line" on Earth's surface will eventually encircle the globe.

This universe, called a **critical universe**, has no curvature. It is said to be "flat," and it is infinite in extent. In this case, and *only* in this case, the geometry of space on large scales is precisely the familiar Euclidean geometry taught in high schools. Apart from its overall expansion, this is basically the universe that Newton knew.

Euclidean geometry—the geometry of flat space—is familiar to most of us because it is a good description of space in the vicinity of Earth. It is the geometry of everyday experience. Does this mean that the universe is flat, which would in turn mean that it has exactly the critical density? The answer is not necessarily. Just as a flat street map is a good representation of a city, even though we know Earth is really a sphere, Euclidean geometry is a good description of space within the solar system, or even the Galaxy, because the curvature of the universe is negligible on scales smaller than about 1000 Mpc. Only on the very largest scales would the geometric effects we have just discussed become evident. However, if the results quoted in the previous section turn out to be correct, then the density is in fact critical and the universe is indeed flat (Euclidean) on the largest scales.

Notice that for the sphere analogy to work, we must imagine ourselves as two-dimensional "flatlanders" who cannot visualize or experience in any way the third dimension perpendicular to the sphere's surface. Flatlanders and their light rays are confined to the sphere's surface, just as we are confined to the three-dimensional volume of our universe

---

☑ Concept Check

■ How is the curvature of space related to the density of the universe?

26.6 The Cosmic Microwave Background

Looking out into space is equivalent to looking back into time. ⟶ (*More Precisely 25-2*) But how far back in time can we probe? Is there any way to study the universe beyond the most distant quasar? How close can we come to perceiving directly the edge of time, the very origin of the universe?

A partial answer to these questions was discovered by accident in 1964, during an experiment designed to improve the U.S. telephone system. As part of a project to identify and eliminate interference in satellite communications, Arno Penzias and Robert Wilson, two scientists at Bell Telephone Laboratories in New Jersey, were studying the Milky Way's emission at microwave (radio) wavelengths, using the horn-shaped antenna shown in Figure 26.13. In their data they noticed a bothersome background "hiss" that just would not go away—a little like the background static on an AM radio station. Regardless of where and when they pointed their antenna, the hiss persisted. Never diminishing or intensifying, the weak signal was detectable at any time of the day, any day of the year, apparently filling all space.

What is the source of this radio noise? And why does it appear to come uniformly from all directions, unchanging in time? Unaware that they had detected a signal of great cosmological significance, Penzias and Wilson sought many different origins for the excess emission, including atmospheric storms, ground interference, equipment short circuits—even pigeon droppings inside the antenna! Eventually, after conversations with colleagues at Bell Labs and theorists at nearby Princeton University, the two experimentalists realized that the origin of the mysterious static was nothing less than the fiery creation of the universe itself. The radio hiss that Penzias and Wilson detected is now known as the **cosmic microwave background**. Their discovery won them the 1978 Nobel Prize in physics.

In fact, researchers had predicted the existence and general properties of the microwave background well before its discovery. As early as the 1940s, physicists had realized that in addition to being extremely dense, the early universe must also have been very hot, and shortly after the Big Bang the universe must have been filled with extremely high-energy thermal radiation—gamma rays of very short wavelength. Researchers at Princeton had extended these ideas, reasoning that the frequency of this primordial radiation would have been redshifted (simply by cosmic expansion) from gamma ray, to X ray, to ultraviolet, eventually all the way into the radio range of the electromagnetic spectrum as the universe expanded and cooled (Figure 26.14). ⟶ (Sec. 3.4) By the present time, they argued, this redshifted "fossil remnant" of the primeval fireball should have a temperature of no more than a few tens of kelvins—peaking in the microwave part of the spectrum. The Princeton group was in the process of constructing a microwave antenna to search for this radiation when Penzias and Wilson announced their discovery.

Figure 26.13 Microwave Background Discoverers This "sugarscoop" antenna, originally built to communicate with Earth-orbiting satellites, was used in discovering the 2.7 K cosmic background radiation. Pictured are Robert Wilson (left) and Arno Penzias, who used the antenna to make the discovery. *(Bell Photographers, Inc.)*

The Princeton researchers confirmed the existence of the microwave background and estimated its temperature at about 3 K. However, because of atmospheric absorption, this part of the electromagnetic spectrum happens to be difficult to observe from the ground, and it was 25 years until astronomers could demonstrate conclusively that the radiation was described by a blackbody curve. In 1989 the *Cosmic Background Explorer (COBE)* satellite measured the intensity of the microwave background at wavelengths straddling the peak of the curve, from a half millimeter up to about 10 cm. The results are shown in Figure 26.15. The solid line is the black-

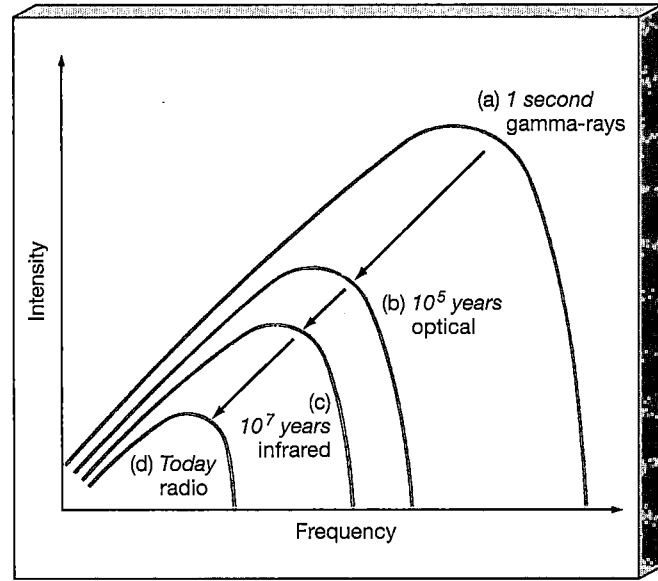

Figure 26.14 Cosmic Blackbody Curves Theoretically derived blackbody curves for the universe (a) one second after the Big Bang, (b) 100,000 years after the Big Bang, (c) 10 million years after the Big Bang, and (d) at present, approximately 10 billion years after the Big Bang.

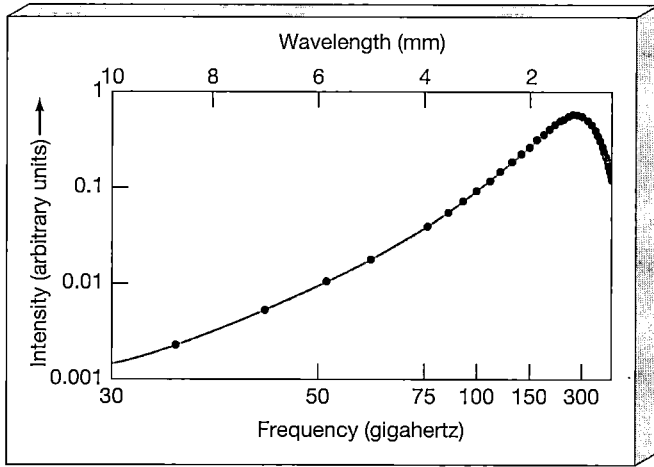

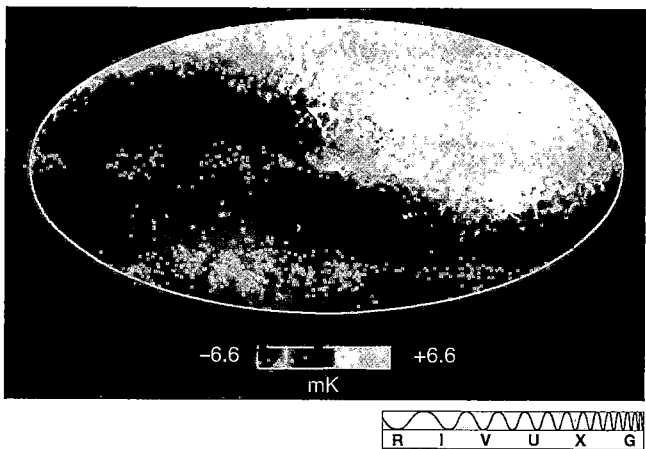

Figure 26.15 Microwave Background Spectrum
The intensity of the cosmic background radiation, as measured by the *COBE* satellite, agrees very well with that expected from theory. The curve is the best fit to the data, corresponding to a temperature of 2.735 K. The experimental errors in this remarkably accurate observation are smaller than the dots representing the data points.

Figure 26.16 Microwave Sky A *COBE* map of the microwave sky reveals that the microwave background appears a little hotter in the direction of the constellation Leo and a little cooler in the opposite direction. The maximum temperature deviation from the average is about 0.0034 K, corresponding to a velocity of 400 km/s in the direction of Leo. *(NASA)*

body curve that best fits the *COBE* data. The near-perfect fit corresponds to a universal temperature of about 2.7 K.

Figure 26.16 shows a *COBE* map of the microwave background temperature over the entire sky. The blue regions are hotter than average, by about 0.0034 K, the red regions cooler by the same amount. This temperature range is not an inherent property of the microwave background, however. Rather, it is a consequence of Earth's *motion* through space. If we were precisely at rest with respect to the universal expansion (like the coin taped to the surface of the expanding balloon in Figure 26.4), then we would see the microwave background as almost perfectly isotropic, as illustrated in Figure 26.17(a). However, if we are moving with respect to that frame of reference, as in Figure 26.17(b), then the radiation from in front of us should be slightly blueshifted by our motion, while that from behind should be redshifted. Thus, to a moving ob-

server, the microwave background should appear a little hotter than average in front and slightly cooler behind.

The data indicate that Earth's velocity is about 380 km/s in the approximate direction of the constellation Leo. Once the effects of this motion are corrected for, the cosmic microwave background is found to be strikingly isotropic. Its intensity is virtually constant (in fact, to about one part in 10^5) from one direction on the sky to another, lending strong support to one of the key assumptions comprising the cosmological principle.

When we observe the microwave background, we are looking almost all the way to the very beginning of the universe. The photons that we receive as these radio waves today have not interacted with matter since the universe was a mere 100,000 years old, when, according to our models, it was less than 1/1000 of its present size. To probe further, back to the Big Bang itself, requires that we enter the world

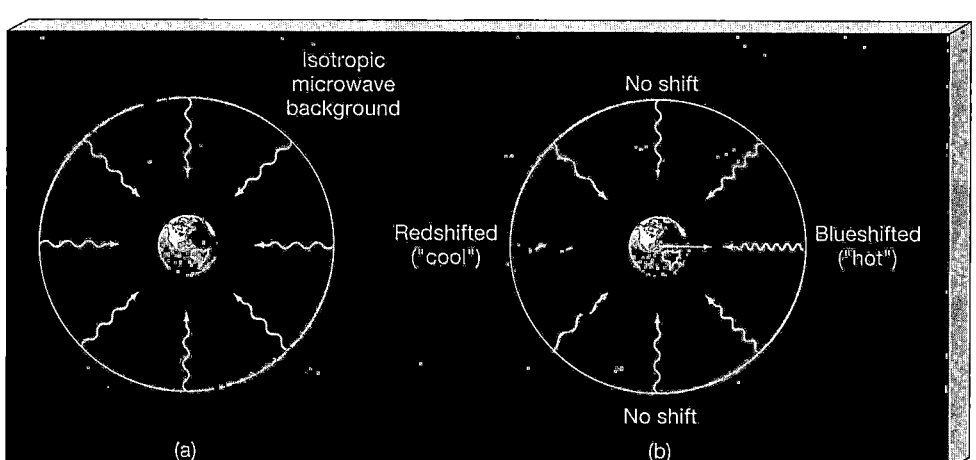

Figure 26.17 Earth's Motion Through the Cosmos
(a) To an observer at rest with respect to the expanding universe, the microwave background appears isotropic. (b) A moving observer measures "hot" blueshifted radiation in one direction (the direction of motion) and "cool" redshifted radiation in the opposite direction.

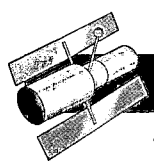
Euclidean geometry is the geometry of flat space—the geometry taught in high schools everywhere. Set forth by one of the most famous of the ancient Greek mathematicians, Euclid, who lived around 300 B.C., it is the geometry of everyday experience. Houses are usually built with flat floors. Writing tablets and blackboards are also flat. We work easily with flat, straight objects, because the straight line is the shortest distance between any two points.

When we construct houses or any other straight-walled buildings on the surface of Earth, the other basic axioms of Euclid's geometry also apply: parallel lines never meet even when extended to infinity; the angles of any triangle always sum to 180°; the circumference of a circle equals π times its diameter. If these rules were not obeyed, walls and roof would never meet to form a house!

In reality, though, the geometry of Earth's surface is not really flat. It is curved. We live on the surface of a sphere, and on that *surface*, Euclidean geometry breaks down. Instead, the rules for the surface of a sphere are those of *Riemannian geometry*, named after the nineteenth-century German mathematician, Georg Friedrich Riemann. There are no parallel "straight" lines on a sphere. The analog of a straight line on a sphere's surface is a "great circle"—the arc where a plane passing through the center of the sphere intersects the surface. Any two such lines must eventually intersect. The sum of a triangle's angles, when drawn on the surface of a sphere, exceeds 180°—in the 90°-90°-90° triangle shown in the accompanying figure, the sum is actually 270°. And the circumference of a circle is less than π times its diameter.

We see that the curved surface of a sphere, governed by the spherical geometry of Riemann, differs greatly from the flat-space geometry of Euclid. The two are approximately the same only if we confine ourselves to a small patch on the surface. If the patch is small enough compared with the sphere's radius, the surface looks "flat" nearby, and Euclidean geometry is approximately valid. This is why we can draw a usable map of our home, our city, even our state, on a flat sheet of paper, but an accurate map of the entire Earth must be drawn on a globe.

When we work with larger parts of Earth, we must abandon Euclidean geometry. World navigators are fully aware of this. Aircraft do not fly along what might appear on most maps as a straight-line path from one point to another. Instead, they follow a great circle on Earth's surface. On the curved surface of a sphere, such a path is always the shortest distance between two points. For example, a flight from Los Angeles to London does not proceed directly across the United States and the Atlantic Ocean, as you might expect

from looking at a flat map. Instead, it goes far to the north, over Canada and Greenland, above the Arctic Circle, finally coming in over Scotland for a landing at London. This is the great circle route—the shortest path between the two cities, as you can easily see if you inspect a globe.

The "positively curved" space of Riemann is not the only possible departure from flat space. Another is the "negatively curved" space first studied by Nikolai Ivanovich Lobachevsky, a nineteenth-century Russian mathematician. In this geometry, there are an *infinite* number of lines through any given point parallel to another line, the sum of a triangle's angles is *less* than 180° (see the accompanying figure), and the circumference of a circle is *greater* than π times its diameter. This type of space is described by the surface of a curved saddle, rather than a flat plane or a curved sphere. It is a hard geometry to visualize!

Most of the local realm of the *three*-dimensional universe (including the solar system, the neighboring stars, and even our Milky Way Galaxy) is correctly described by Euclidean geometry. If the currently favored cosmology described in the text turns out to be correct, then the whole universe is too!

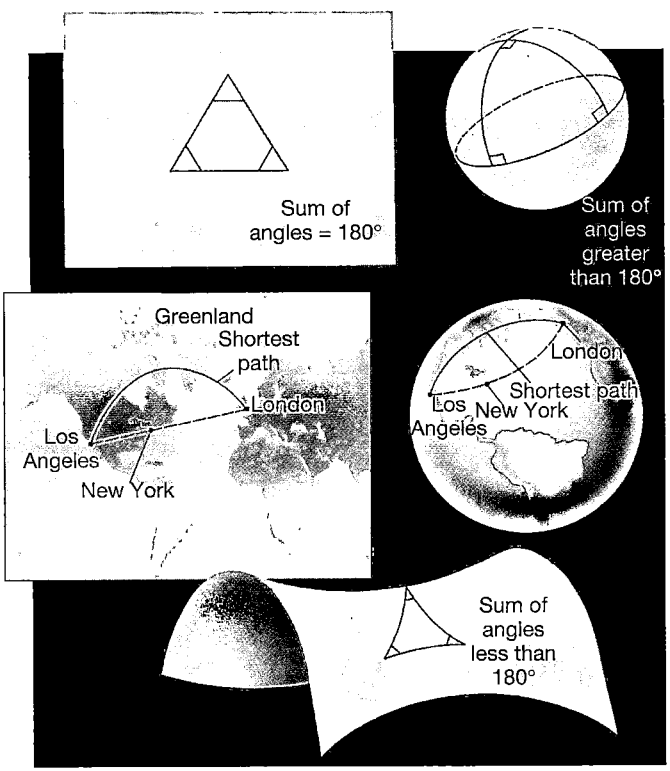

Sum of angles = 180°

Sum of angles greater than 180°

Greenland Shortest path London Los Angeles New York

London Shortest path Los New York Angeles

Sum of angles less than 180°

---

of nuclear and particle physics. The Big Bang was the biggest and the most powerful particle accelerator of all! In the next chapter we will see how studies of conditions in the primeval fireball aid us in understanding the present-day structure and future evolution of the universe in which we live.

✓ Concept Check

■ When was the cosmic microwave background formed?

Chapter Review

SUMMARY

On scales larger than a few hundred megaparsecs, the universe appears roughly **homogeneous** (the same everywhere, p. 692) and **isotropic** (the same in all directions, p. 692). In **cosmology** (p. 693)—the study of the universe as a whole—researchers usually assume that the universe is homogeneous and isotropic. This assumption is known as the **cosmological principle** (p. 693). It implies that the universe cannot have a center or an edge. If the universe were homogeneous, isotropic, infinite, and unchanging, the night sky would be bright because any line of sight would eventually intercept a star. The fact that the night sky is dark is called **Olbers's paradox** (p. 694). Its resolution lies in the fact that, regardless of whether or not the universe is infinite, we see only a finite part of it from Earth—the region from which light has had time to reach us since the universe began.

Tracing the observed motions of galaxies back in time implies that some 15 billion years ago, the universe consisted of a single point that then began to expand rapidly, at the time of the **Big Bang** (p. 694). However, the galaxies are not flying apart into the rest of an otherwise empty universe. Space itself is expanding. The Big Bang did not happen at any particular location in space, because the entire universe was compressed to a point at that instant—the Big Bang happened everywhere at once. The cosmological redshift occurs as a photon's wavelength is "stretched" by cosmic expansion. The extent of the observed redshift is a direct measure of the expansion of the universe since the photon was emitted.

Although Einstein's theory of general relativity is needed for a full description of the cosmos, much of the dynamics of the universe can be understood using simple Newtonian concepts. If the density of the universe is less than the **critical density** (p. 700), there is insufficient cosmic matter to stop the expansion and cause the universe to recollapse. Luminous matter by itself contributes only about one percent of the critical density. When dark matter in galaxies and clusters is taken into account, the figure rises to 20 or 30 percent. The fraction of dark matter on larger scales is uncertain, but it may be even greater than is found in clusters. Most astronomers believe that the present mass density of the universe lies somewhere around one-third of the critical density. Recent observations of distant supernovae suggest that the expansion of the universe may be accelerating, possibly driven by the effects of **dark energy** (p. 704). One candidate for this dark energy is the **cosmological constant** (p. 704), a repulsive force that may exist throughout all space. Its physical nature is unknown.

For $H_0 = 65$ km/s/Mpc, the age of a critical-density universe without dark energy is about 10 billion years. This age estimate conflicts with the 10–12-billion-year ages of globular clusters derived from studies of stellar evolution. If there is no dark energy, then the density of the universe must be significantly less than critical, or Hubble's constant must be less than 65 km/s/Mpc. The inclusion of dark energy increases the age of the universe to 14 billion years, consistent with cluster ages.

General relativity provides a description of the geometry of the universe on the largest scales. The curvature of space-time in a high-density universe is sufficiently large that the universe "bends back" on itself and is finite in extent, somewhat like the surface of a sphere. Such a universe is said to be a **closed universe** (p. 707). A low-density **open universe** (p. 707) is infinite in extent and has a "saddle-shaped" geometry. The intermediate-density **critical universe** (p. 707) will expand forever, but it is spatially flat.

The **cosmic microwave background** (p. 708) is isotropic blackbody radiation that fills the entire universe. Its present temperature is about 3 K. The existence of the microwave background is direct evidence that the universe expanded from a hot, dense state. As the universe has expanded, the initially high-energy radiation has been redshifted to lower and lower temperatures.

SELF-TEST: TRUE OR FALSE?

____ **1.** Cosmic homogeneity can be tested observationally, but cosmic isotropy has no observational test.

____ **2.** Olbers's paradox asks: Why is the night sky dark?

____ **3.** Olbers's paradox is resolved by the cosmological redshift, which reduces the energy of photons received from very distant objects.

____ **4.** Hubble's law implies that the universe will expand forever.

____ **5.** The Big Bang is an expansion only of matter, not of space.

____ **6.** All points in space appear to be at the center of the expanding universe.

____ **7.** The universe appears to be younger than its oldest stars.

____ **8.** The cosmological redshift is a direct measure of the expansion of the universe.

____ **9.** As it travels through space, a photon's wavelength expands at the same rate as the universe is expanding.

____ **10.** According to the standard Big Bang model of the cosmos, the universe has only two possible futures: Either it will continue to expand forever, or it will someday stop expanding and start to contract.

____ **11.** Recent observations of supernovae suggest that the universe is not expanding.

____ **12.** Current estimates suggest that the universe will ultimately end in a "Big Crunch."

____ **13.** A closed universe is geometrically flat.

____ **14.** The spectrum of the cosmic microwave background is that of a blackbody.

____ **15.** The cosmic microwave background is the highly redshifted radiation of the early Big Bang.

SELF-TEST: FILL IN THE BLANK

1. Deep surveys of the universe suggest that the largest structures in space are no larger than about _____ Mpc in size.
2. Homogeneous means "the same _____."
3. Isotropic means "the same in all _____."
4. Together, the assumptions of cosmic homogeneity and isotropy are known as the _____.
5. If the universe had an edge, this would violate the assumption of _____.
6. If the universe had a center, this would violate the assumption of _____.
7. _____ is slowing the expansion of the universe.
8. _____ is accelerating the expansion of the universe.
9. A _____ value for Hubble's constant may yield an age for the universe incompatible with the ages of the oldest globular clusters.

10. Luminous matter makes up about _____ percent of the critical density.
11. Dark matter makes up about _____ percent of the critical density.
12. By observing distant supernovae, astronomers have determined that the expansion of the universe is _____.
13. The surface of a sphere is a two-dimensional example of a _____ curved universe.
14. The _____ background is an isotropic radiation field filling all space.
15. The temperature of the universe, as measured by the background radiation field, is _____ K.

REVIEW AND DISCUSSION

1. What evidence do we have that there is no structure in the universe on very large scales? How large is "very large"?
2. What is the cosmological principle?
3. What is Olbers's paradox? How is it resolved?
4. Explain how an accurate measure of Hubble's constant can lead to an estimate of the age of the universe.
5. We appear to be at the center of the Hubble flow. Why doesn't this violate the cosmological principle?
6. Why isn't it correct to say that the expansion of the universe involves galaxies flying outward into empty space?
7. Where did the Big Bang occur?
8. How does the cosmological redshift relate to the expansion of the universe?
9. What properties of the universe determine whether or not it will expand forever?
10. What will be the ultimate fate of the universe if it does expand forever?
11. Is there enough luminous matter to halt the current cosmic expansion?

12. Is there enough dark matter to halt the current cosmic expansion?
13. What do observations of distant supernovae tell us about the expansion of the universe?
14. What is the cosmological constant, and what does it have to do with the future of the universe?
15. Why are measurements of globular cluster ages important to cosmology?
16. What is the significance of the cosmic microwave background?
17. Why does the temperature of the microwave background fall as the universe expands?
18. How can we measure Earth's motion with respect to the universe?
19. Many cultures throughout history have developed their own cosmologies. Do you think the modern scientific cosmology is more likely to endure than any other? Why or why not?
20. Do you think it constitutes good science to explain the universe mainly in terms of dark matter and dark energy, neither of which is known or understood?

PROBLEMS ⚡**WWW**⚡   *Algorithmic versions of these questions are available in the Practice Problems module of the Companion Website.*

The number of squares preceding each problem indicates its approximate level of difficulty.

1. ■ What is the greatest distance at which a galaxy survey sensitive to objects as faint as 20th magnitude could detect a galaxy as bright as the Milky Way (absolute magnitude -20)?
2. ■■ From Table 25.1 estimate the redshift of the Milky Way at the distance calculated in the previous question.
3. ■ If the entire universe were filled with Milky Way-like galaxies, with an average density of 0.1 per cubic megaparsec, calculate the total number of galaxies observable by the survey in Problem 1, if it covered the entire sky.
4. ■■■ Assuming that the entire universe is uniformly filled with Sunlike stars with a density of five billion stars per cubic megaparsec (corresponding to 50 billion stars per galaxy and critical mass density), calculate how far out into space one would have to look, on average, before the line of sight intersects a star.

5. ■■ Eight galaxies are located at the corners of a cube. The present distance from each galaxy to its nearest neighbor is 10 Mpc, and the entire cube is expanding according to Hubble's law, with $H_0 = 65$ km/s/Mpc. Calculate the recession velocity of one corner of the cube relative to the opposite corner.

6. ■ According to the Big Bang theory described in this chapter, *without a cosmological constant*, what is the maximum possible age of the universe if $H_0 = 50$ km/s/Mpc? 65 km/s/Mpc? 80 km/s/Mpc?

7. ■ For a Hubble constant of 65 km/s/Mpc, the critical density is 8×10^{-27} kg/m^3. (a) How much mass does that correspond to within a volume of one cubic astronomical unit? (b) How large a cube would be required to enclose one Earth mass of material?

8. ■■ The Virgo Cluster is observed to have a recessional velocity of 1200 km/s. Assuming $H_0 = 65$ km/s/Mpc and a critical-density universe, calculate the total mass contained within a sphere centered on Virgo and just enclosing the Milky Way Galaxy. What is the escape speed from the surface of this sphere?

9. ■■ Given the age range for star clusters stated in the text, calculate the *maximum possible* value of Hubble's constant in the cases of (a) constant expansion velocity and (b) critical density without a cosmological constant.

10. ■■ Assuming critical density and using the distances presented in Table 25.1, estimate the total amount of matter in the universe out to a redshift of 6. Express your answer (a) in kilograms, and (b) in solar masses.

11. ■■ The critical density is proportional to the square of Hubble's constant. If the critical density were equal to the

known density of "normal" matter (not dark matter), about 10^{-28} kg/m^3, what would be the corresponding value of Hubble's constant? Is this within the currently accepted range of values?

12. ■ What was the temperature of the cosmic microwave background at the epoch of quasar formation (at a redshift of 6)?

13. ■■ (a) What is the present peak wavelength of the cosmic microwave background? Calculate the size of the universe relative to the present size when the radiation background peaked in (b) the infrared, at 10 μm, (c) in the ultraviolet, at 100 nm, and (d) in the gamma-ray, at 1 nm.

14. ■■ Verify using Wien's law and the Doppler effect that an observer moving at the velocity quoted in the text should indeed see the stated temperature shift in the microwave background.

15. ■■■ Using the information given in Problem 8, calculate the distance from our location to the point that would one day become the center of the Virgo Cluster, at the time when the temperature of the microwave background was equal to the present surface temperature of the Sun.

COLLABORATIVE EXERCISES

1. An Isotropic Universe. Isotropy is the extent to which everything looks the same in every direction. For each group member, imagine where you were on your 10th birthday. Considering buildings, geographic features, and similar objects within a few miles, was the universe isotropic? Fully explain your answer.

2. Hubble's Constant. For each member in your group, determine the maximum age of the universe if the value of Hubble's constant is three times each person's age.

RESEARCHING ON THE WEB      *To complete the following exercises, go to the online Destinations module for Chapter 26 on the Companion Website for Astronomy Today 4/e.*

1. Access the "From the Big Bang to the End of the Universe" page and describe what was occurring three seconds after the Big Bang.

2. Access the "Cosmos in a Computer" page to view the video and describe the nature of 90 percent of the matter in the universe.

3. Access the "Dark Matter" page and describe why dark matter is critical to cosmology.

4. Access the "Introduction to Cosmology" page and describe the matter of which the universe is composed.

PROJECTS

1. Make a model of a two-dimensional universe and examine Hubble's law on it. Find a balloon that will expand into a nice large sphere. Blow it up about halfway and mark dots all over its surface; the dots will represent galaxies. Choose one dot as your home galaxy. Using a measuring tape, measure the distances to various other galaxies, numbering the dots so you will not confuse them later. Now blow the balloon up to full size and measure the distances again, and find the new distances to each dot. Calculate the change in the distances for each galaxy; this is a measure of their velocity (change in position/change in time; the time is the same for all and is arbitrary). Plot their velocities versus their new distances as in Figure 26.10. Do you get a straight-line correlation, i.e., a

"Hubble" law? Does it matter which dot you choose as home? Demonstrate this to your class.

2. Write a paper on the philosophical differences between living in an open, closed, or flat universe. Are there aspects of any of these three possibilities that are hard to accept? It is quite possible that astronomers may determine within your lifetime which is correct. Do you have a preference?

3. Go to your library and read about the *steady-state universe*, which enjoyed some measure of popularity in the 1950s and 1960s. How does it differ from the standard Big Bang model? Why do you think the steady-state model is not widely accepted today?

 *In addition to the Practice Problems and Destinations modules, the Companion Website at http://www.prenhall.com/chaisson provides for each chapter an additional true-false, multiple choice, and labeling quiz, as well as additional annotated images, animations, and links to related Websites.*

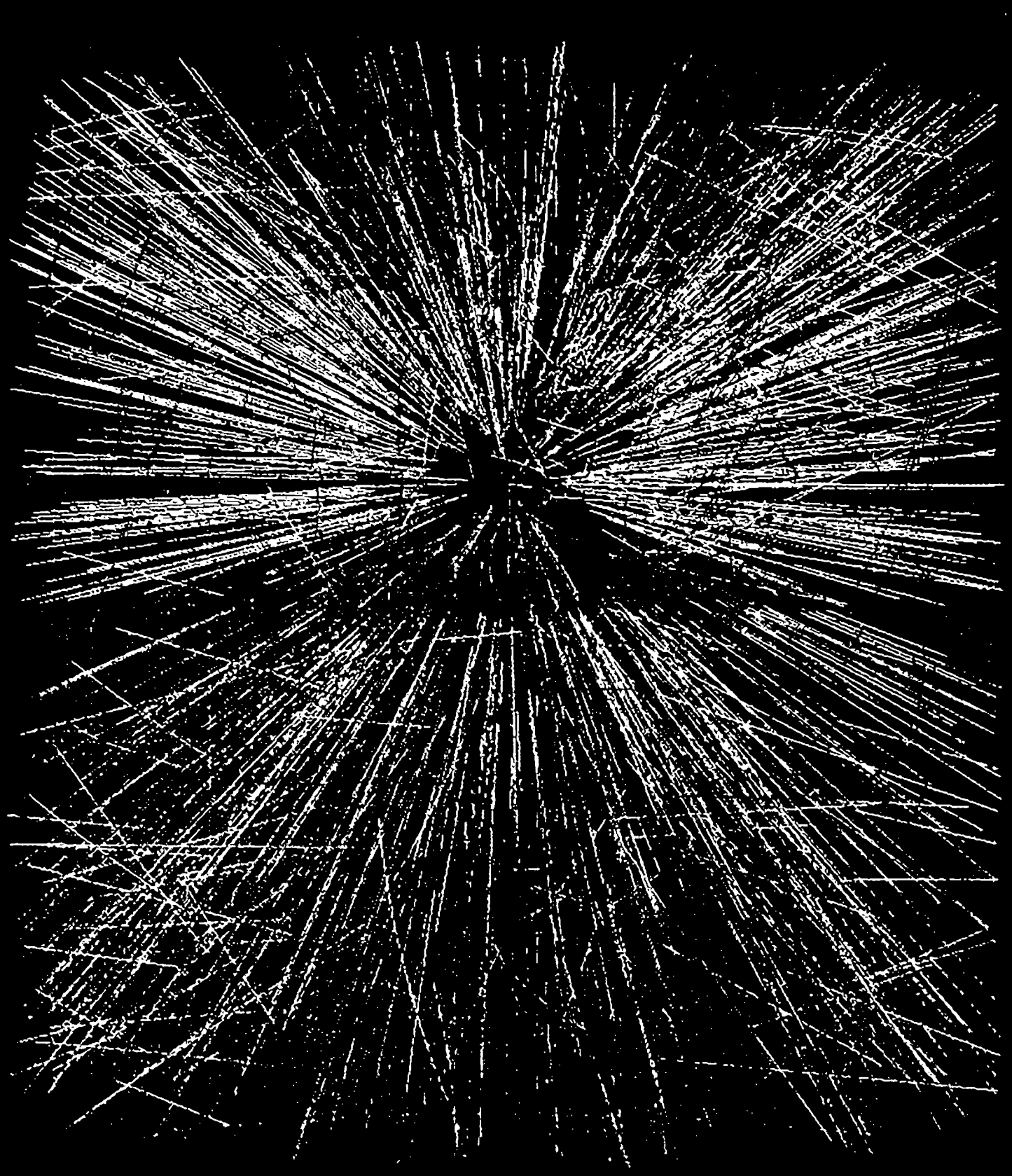

27 THE EARLY UNIVERSE

Toward the Beginning of Time

LEARNING GOALS

Studying this chapter will enable you to:

1. Describe the characteristics of the universe immediately after its birth.

2. Explain how matter emerged from the primeval fireball.

3. Enumerate the epochs in the evolutionary history of the universe and specify the major characteristics of each.

4. Explain how and when the simplest nuclei and atoms formed.

5. Summarize the horizon and flatness problems, and discuss the theory of cosmic inflation as a possible solution to these problems.

6. Explain the formation of large-scale structure in the cosmos, and discuss the observational evidence of our theories of structure formation.

Visit http://www.prenhall.com/chaisson for additional annotated images, animations, and links to related sites for this chapter.

Physicists use accelerators to study matter on the smallest scales and at the highest energies. By smashing together the building blocks of matter and examining the debris from the collisions, scientists infer knowledge about the basic forces and the structure of matter. Here, we see the aftermath of two gold nuclei having been slammed into one another at the Brookhaven National Lab. Scatter diagrams like these show evidence for a "quark soup" similar to conditions that must have existed within the first second of the universe's existence. **The Big Picture:** Direct astronomical observations of the early universe are not possible. The most distant material objects (the quasars) are seen as they were roughly a billion years after the Big Bang, and the most distant radiation (the cosmic microwave background) was launched about 300,000 years after the start of all things. However, particle accelerators on Earth can approximate, for fleeting moments, the incredibly hot and dense conditions that likely prevailed during the earliest epochs of the universe. Such studies comprise the whole new interdisciplinary subject of particle astrophysics. *(RHIC/U.S. Dept. of Energy)*

What were the conditions during the first few seconds of the universe, and how did those conditions change to give rise to the universe we see today? In studying the earliest moments of our universe, we enter a truly alien domain. As we move backward in time toward the Big Bang, our customary landmarks slip away one by one. Atoms vanish, then nuclei, then even the elementary particles themselves. In the beginning the universe consisted of pure energy, at unimaginably high temperatures. As it expanded and cooled, the ancient energy gave rise to the particles that make up everything we see around us today. Modern physics has now arrived at the point where it can reach back almost to the instant of the Big Bang itself, allowing scientists to unravel some of the mysteries of our own beginning.

27.1 Back to the Big Bang

On the very largest scales we can regard the universe as a roughly homogeneous mixture of matter, radiation, and (perhaps) dark energy. ∞ (Sec. 26.4) The overall density of matter is not known with certainty, but it is thought to be at least a few tenths of the critical density of about 8×10^{-27} kg/m^3, below which gravity is insufficient to cause the universe to recollapse. Thus, especially given the repulsive effect of dark energy, the future of the cosmos seems clear—the universe is destined to expand forever. Now let's turn our attention to the past. To begin to understand the early universe, just after the Big Bang, we must look more closely at the respective roles played by matter and radiation in the cosmos.

MATTER AND RADIATION

The matter in the universe consists of the familiar building blocks of atoms—protons, neutrons, and electrons—as well as the mysterious dark matter, whose composition is still hotly debated by astronomers. Most of the radiation in the universe is in the form of the cosmic microwave background, the low-temperature (3 K) radiation field that fills all space. ∞ (Sec. 26.6) Surprisingly, although the microwave background radiation is very weak, it still contains more energy than has been emitted by all the stars and galaxies that have ever existed! The reason for this is that stars and galaxies, though very intense sources of radiation, occupy only a tiny fraction of space. When their energy is averaged out over the volume of the entire universe, it falls short of the energy of the microwave background by at least a factor of 10. For our current purposes, then, we can ignore most of the first 26 chapters of this book and regard the cosmic microwave background as the only significant form of radiation in the universe!

Is matter the dominant component of the universe, or does radiation also play an important role on large scales? In order to compare matter and radiation, we must first convert them to a "common currency"—either mass or energy. Let's choose to compare their masses. We can express the energy in the microwave background as an equivalent density by first calculating the number of photons in any cubic centimeter of space, then converting the total

energy of these photons into a mass using the relation $E = mc^2$. ∞ (Sec. 16.5) When we do this, we arrive at an equivalent density for the microwave background of about 5×10^{-31} kg/m^3. Thus, *at the present moment* the density of matter in the universe far exceeds the density of radiation. In cosmological terminology (which predates the possibility of dark energy in the cosmos), we say that we live in a **matter-dominated** universe.

Was the universe always matter dominated? To answer this question, we must ask how the densities of both matter and radiation change as the universe expands. As shown in Figure 27.1, both decrease, as the expansion dilutes the numbers of atoms and photons alike. But the radiation is also diminished in energy by the cosmological redshift, so its

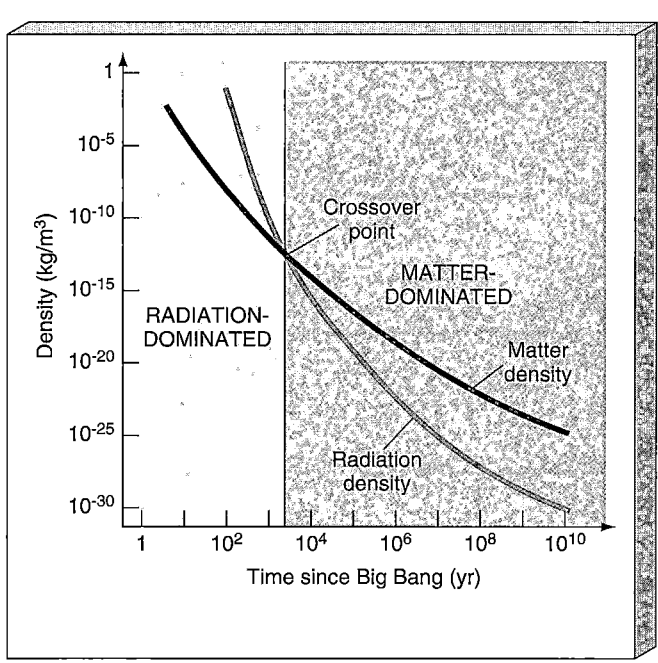

Figure 27.1 Radiation-Matter Dominance As the universe expanded, the number of both matter particles and photons per unit volume decreased. However, the photons were also reduced in energy by the cosmological redshift, reducing their equivalent mass, and hence their density, still further. As a result, the density of radiation fell faster than the density of matter as the universe grew. Tracing the curves back from the densities observed today, we see that radiation must have dominated matter at early times, before the crossover point.

density falls *faster* than that of matter as the universe grows. Conversely, as we look back in time, closer and closer to the Big Bang, the density of the radiation increases faster than that of matter. Accordingly, even though today the radiation density is much less than the matter density, there must have been a time in the past when they were equal. Before that time, radiation was the main constituent of the cosmos. The universe is said to have been **radiation dominated** then. Given our best estimates of the present densities, the crossover point—the time at which the densities of matter and radiation were equal—occurred a few thousand years after the Big Bang, when the universe was about 6000 times smaller than it is today. The temperature of the background radiation at that time was about 15,000 K, so it peaked in the near-ultraviolet portion of the spectrum.

Throughout this book we have been concerned exclusively with the history of the universe long after it became matter dominated—the formation and evolution of galaxies, stars, and planets as the universe thinned and cooled toward the state we see today. In this chapter we consider some important events in the early, hot, radiation-dominated universe, long before any star or galaxy existed, that played no less a role in determining the present condition of the cosmos.

What of dark energy? In Chapter 26 we argued that observations of distant supernovae suggest that this component may currently account for the largest fraction of the total cosmic density. ∞ (Sec. 26.4) However, dark energy is a large-scale phenomenon. It increases in importance as the universe expands. In fact (as usual converting energies and masses into the same units for purposes of comparison), the density associated with the dark energy remains *constant* as the universe expands, while, as we have just seen, the densities of matter and radiation both decrease. We therefore conclude that dark energy, at least as we currently understand it, was probably unimportant at early times.

PARTICLE PRODUCTION IN THE EARLY UNIVERSE

2 The existence of the microwave background implies that the early universe was dominated by an intense radiation field whose temperature fell steadily as the cosmos ex-

panded. The temperatures and densities prevailing at these times were far greater than anything we have encountered thus far, even in the hearts of supernovae. To understand conditions in the universe shortly after the Big Bang, we must delve a little more deeply into the behavior of matter and radiation at very high temperatures.

The key to understanding events at very early times lies in a process called **pair production,** in which two photons give rise to a *particle–antiparticle* pair, as shown in Figure 27.2(a) for the particular case of electrons and positrons. Through pair production, matter is created directly from energy in the form of electromagnetic radiation. The reverse process can also occur—a particle and its antiparticle can *annihilate* each other to produce radiation, as depicted in Figure 27.2(b). In other words, energy in

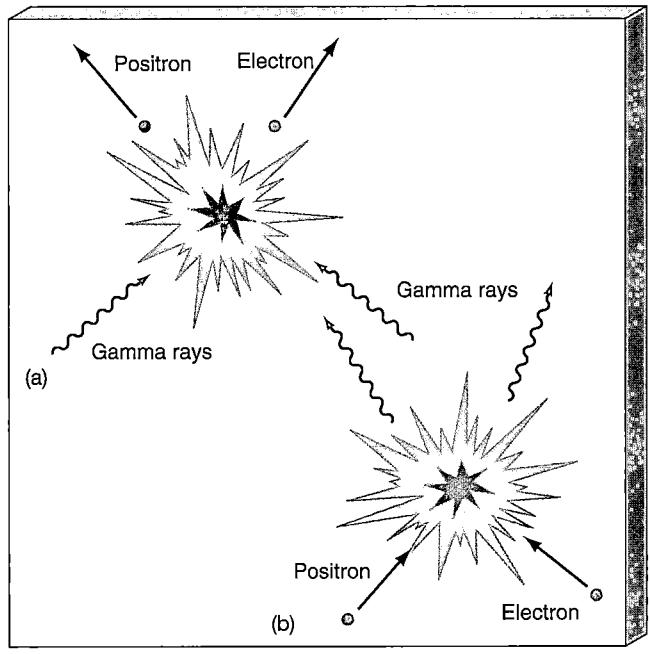

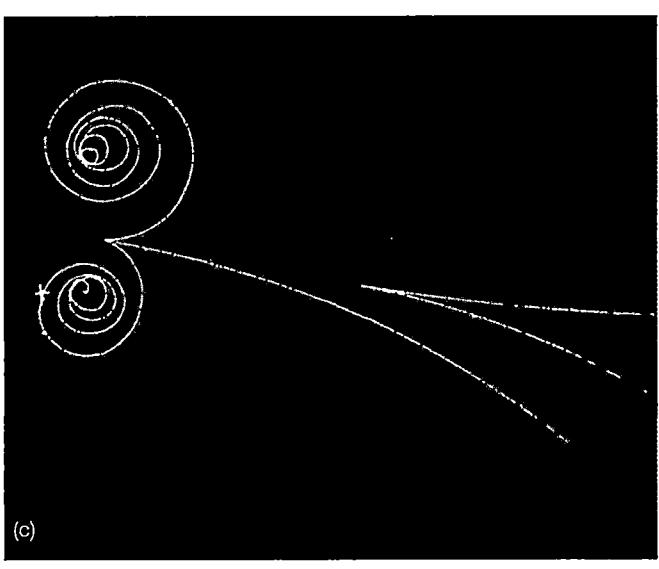

Figure 27.2 Pair Production (a) Two photons can produce a particle–antiparticle pair—in this case an electron and a positron—if their total energy exceeds the mass energy of the particles produced. (b) The reverse process is particle–antiparticle annihilation, in which an electron and positron destroy each other, vanishing in a flash of gamma rays. (c) Tracks in a particle detector allow us to visualize pair creation. Here a gamma ray, whose path is invisible because it is electrically neutral, arrives from the left; it dislodges an atomic electron and sends it flying (the longest path). At the same time it provides the energy to produce an electron–positron pair (the spiral paths, which curve in opposite directions in the detector's magnetic field because of their opposite electric charges). *(Fermi National Laboratory)*

the form of radiation can be freely converted into matter in the form of particles and antiparticles, and particles and antiparticles can be freely converted back into radiation, subject only to the law of conservation of mass and energy.

The higher the temperature of a radiation field, the greater the energy of the typical constituent photons, and the greater the masses of the particles than can be created by pair production. For any given particle, the critical temperature above which pair production is possible, and below which it is not, is called the particle's *threshold temperature*. The threshold temperature increases as the mass of the particle increases. For electrons it is about 6×10^9 K. For protons, which are nearly 2000 times more massive, it is just over 10^{13} K.

As an example of how pair production affected the composition of the early universe, consider the production of electrons and positrons as the universe expanded and cooled. At high temperatures—above about 10^{10} K—most photons had enough energy to form an electron or a positron, and pair production was commonplace. As a result, space seethed with electrons and positrons, constantly created from the radiation field and annihilating one another to form photons again. Particles and radiation are said to have been in *thermal equilibrium*—new particle-antiparticle pairs were created by pair production at the same rate as they annihilated one another. As the universe expanded and the temperature decreased, so did the average photon energy. By the time the temperature had fallen below a billion or so kelvins, photons no longer had enough

energy for pair production to occur, and only radiation remained. Figure 27.3 illustrates how this change took place.

Pair production in the very early universe was directly responsible for all the matter that exists in the universe today. *Everything we see around us was created out of radiation as the cosmos expanded and cooled.* Because we are here to ponder the subject and we ourselves are made of matter, we know that some matter must have survived these early moments. For some reason there was a slight excess of matter over antimatter at early times. A small residue of particles that outnumbered their antiparticles was left behind as the temperature dropped below the threshold for creating them. With no antiparticles left to annihilate them, the number of particles has remained constant ever since. These survivors are said to have *frozen out* of the radiation field as the universe cooled.

The first hundred or so seconds of the universe's existence saw the creation of all of the basic "building blocks" of matter we know today—protons and neutrons froze out when the temperature dropped below 10^{13} K, when the universe was only 0.0001 s old. The lighter electrons froze out somewhat later, about a minute or so after the Big Bang, when the temperature fell below 10^9 K. This "matter-creation" phase of the universe's evolution ended when the electrons—the lightest known elementary particles—appeared out of the cooling primordial fireball. From that point on, matter has continued to evolve, clumping together into more and more complex structures, eventually forming the atoms, planets, stars, galaxies, and large-scale structure we see today, but no new matter has been created since that early time.

☑ Concept Check

■ What does it mean to say that the early universe was radiation dominated?

27.2 The Evolution of the Universe

☉ For the first few thousand years after the Big Bang, the universe was small and dense and dominated by radiation. We will refer to this period as the *Radiation Era*. Some matter existed during this time, but it was a mere contaminant in the blinding gamma-ray light of the primeval Big Bang fireball. Afterwards matter came to dominate, in the *Matter Era*. Atoms, molecules, and galaxies formed as the universe cooled and thinned toward the state we see today.

Let's begin our study of the early universe by summarizing in broad terms the history of the cosmos, starting at the Big Bang. Table 27.1 and Figure 27.4 present the time, density, and temperature spans of eight major epochs in the development of the universe, along with a brief description of the main physical events that dominated the universe during each. The numbers in this

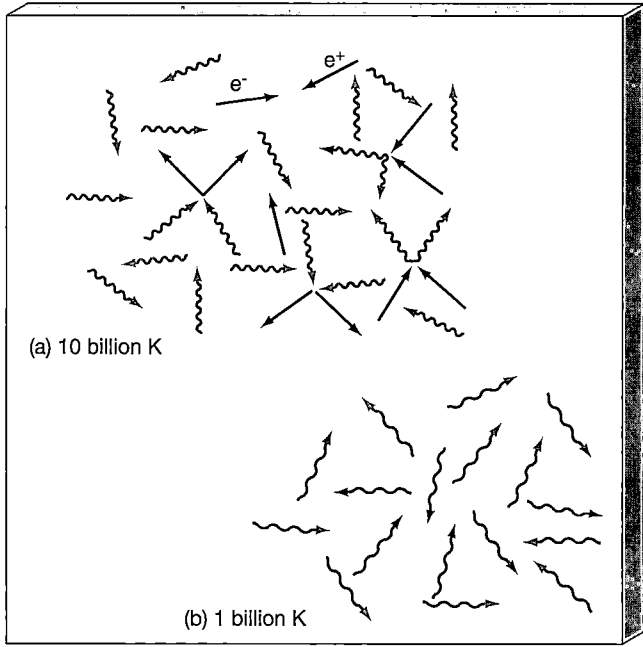

Figure 27.3 Thermal Equilibrium (a) At 10 billion K most photons have enough energy to create particle-antiparticle (electron-positron) pairs, so these particles exist in great numbers in equilibrium with the radiation. (b) Below about 1 billion K, photons have too little energy for pair production to occur.

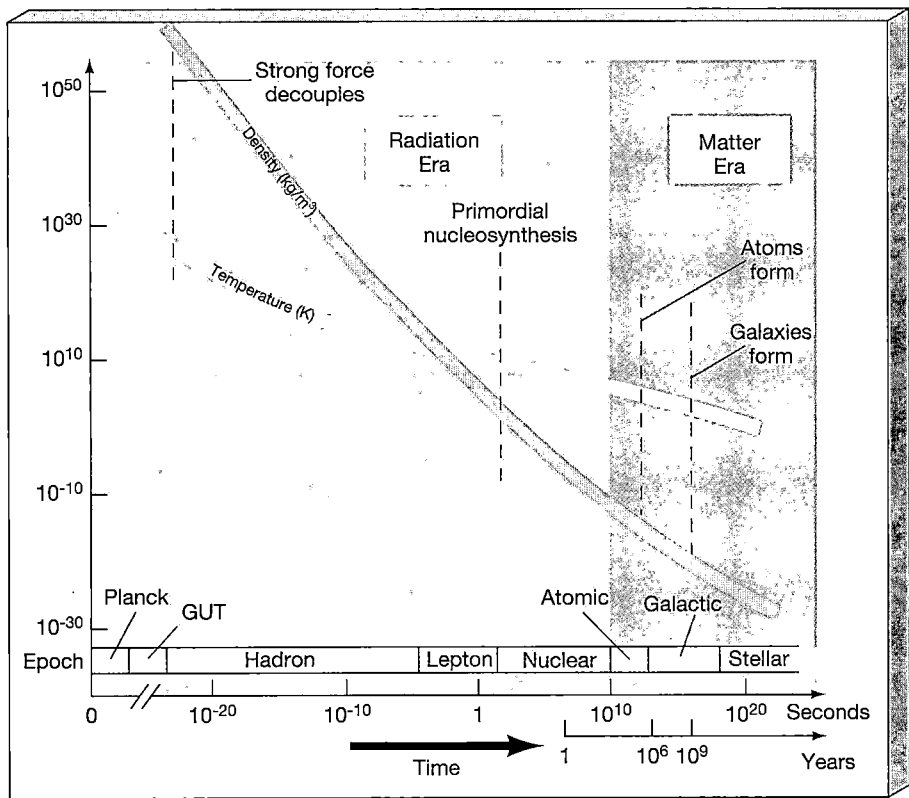

Figure 27.4 Epochs in Cosmic History The average temperature and average density throughout the history of the universe. The epochs listed in Table 27.1 are indicated. (The numerical values for these plots are taken from Table 27.1.) Some key events in the history of the universe are also marked.

table and figure result from pushing the known laws of physics as far back in time as we can. In the next few sections we will expand on some of these epochs in greater detail, but let's not lose sight of the big picture and the place of each epoch in it.

BEFORE THE BIG BANG

The Big Bang was a *singularity* in space and time—an instant when the present laws of physics say the universe had zero size and infinite temperature and density. As we saw in Chapter 22, where we discussed the singularities at the center of black holes, these predictions should not be taken too literally. ∞ (Sec. 22.7) The presence of these singularities signals that, under extreme conditions, the theory—in this case, general relativity—making the predictions has broken down.

At present no theory exists to let us penetrate the singularity at the start of the universe. We have no means of describing these earliest of times, so we have no way of answering the question: What came *before* the Big Bang? Indeed, given the laws of physics as we currently know them, the question itself may be meaningless. The Big Bang represented the beginning of the entire universe—mass, energy, space, *and* time came into being at that instant. Without time, the notion of "before" does not exist. Consequently, some cosmologists maintain that asking what happened before the Big Bang is a little like asking what lies north of the North Pole! Others disagree, however, arguing that when the correct theory of quantum gravity is

constructed, it will remove the singularity and allow us to address the question of what came before.

THE RADIATION ERA

The universe began with an explosion from an incredibly hot and dense state. Precisely what state, we cannot say. And why it exploded, we really don't know. To understand why the universe began expanding, or even more fundamentally, why the universe exists at all, is currently beyond science— there are simply no relevant data. Although ignorant of the moment of creation itself, theorists nevertheless believe that the physical conditions in the universe can be understood in terms of present-day physics back to an extraordinarily short time—a mere 10^{-43} s, in fact—after the Big Bang.

Why can't theorists push our knowledge back to the Big Bang itself? The answer is that we presently have no theory capable of describing the universe at these earliest of times. Under the extreme conditions of density and temperature within 10^{-43} s of the Big Bang, gravity and the other fundamental forces (electromagnetism, the strong force, and the weak force, as described in *More Precisely 27-1*) were indistinguishable from one another—a far cry from the radically different characters we see today. The four forces are said to have been *unified* at this early time— there was, in effect, only one force of nature.

The theory that combines quantum mechanics (the proper description of microscopic phenomena) with general relativity (which describes the universe on the largest

TABLE 27.1 Major Epochs in the History of the Universe

Era	Epoch	Time (after Big Bang)	Density (kg/m³)	Temperature (K)	Main Events
Radiation Era					
		0 s	∞	∞	
	Planck				Unknown physics; quantum gravity
		10^{-43} s	10^{95}	10^{32}	
	GUT				Strong, weak, and electromagnetic forces unified
		10^{-35} s	10^{75}	10^{27}	
	Hadron				Heavy and light particles all in thermal equilibrium
		10^{-4} s	10^{16}	10^{12}	
	Lepton				Only light particles still in thermal equilibrium; neutrinos decouple
		10^{2} s	10^{4}	10^{9}	
	Nuclear				Deuterium and helium formed by fusion of protons and neutrons during first 1000s
		10^{3} yr ($\approx 3 \times 10^{10}$ s)	10^{-13}	6×10^{4}	
Matter Era					
		10^{3} yr ($\approx 3 \times 10^{10}$ s)	10^{-13}	6×10^{4}	
	Atomic				Matter begins to dominate; atoms form; electromagnetic radiation decouples
		10^{6} yr ($\approx 3 \times 10^{13}$ s)	10^{-19}	10^{3}	
	Galactic				Galaxies and larger-scale structure form
		10^{9} yr ($\approx 3 \times 10^{16}$ s)	3×10^{-25}	10	
	Stellar				All galaxies have formed; stars continue to form
		$> 10^{10}$ yr (3×10^{17} s)	10^{-26}	3	

scales) is generically known as quantum gravity. ∞ (Sec. 22.7) The period of time from the beginning to 10^{-43} s is often referred to as the *Planck epoch*, after Max Planck, one of the creators of quantum mechanics. Unfortunately, for now at least, there is no working theory of quantum gravity, so we simply cannot talk meaningfully about the universe during the Planck epoch.

By the end of the Planck epoch the temperature was around 10^{32} K and the universe was filled with radiation and a vast array of subatomic particles created by the mechanism of pair production. At around this time, gravity parted company with the other forces of nature—it became distinguishable from the "quantum" forces, and has remained so ever since. The strong, weak, and electromagnetic forces were still unified. The present-day theories that describe this unification are collectively known as **Grand Unified Theories**, or GUTs for short (see *More Precisely 27-1*). Accordingly, we refer to this period of time as the *GUT epoch*.

Theory indicates that at temperatures below 10^{28} K, the strong nuclear force becomes distinguishable from the electroweak force (the unified weak and electromagnetic force). Once the universe had cooled to this temperature, about 10^{-35} s after the Big Bang, the GUT epoch ended. According to many GUTs, one important legacy of this epoch may have been the appearance and subsequent freeze-out of a veritable "zoo" of very massive (and as yet unobserved) elementary particles that interact only very weakly with normal matter. These "exotic" particles are prime candidates for the dark matter of unknown composition known to exist in abundance both within galaxies and in the unseen depths of intergalactic space. ∞ (Secs. 23.6, 24.3)

Our next major subdivision of the Radiation Era covers the period when all "heavy" elementary particles—that is, all the way down in mass to protons, neutrons, and their constituent quarks—were in thermal equilibrium with the radiation. We refer to this period as the *hadron epoch*, after the general term for particles that interact via the strong force.

The universe continued to expand and cool. At a temperature of about 10^{15} K, 10^{-10} s after the Big Bang, the weak and the electromagnetic components of the electroweak force began to display their separate characters. By about 0.1 milliseconds (10^{-4} s) after the Big Bang, the temperature had dropped well below the 10^{13} K threshold for the creation of protons and neutrons (the lightest hadrons), and the hadron epoch ended. The main constituents of the universe were now lightweight particles—muons, (see *More Precisely 27-1*) electrons, neutrinos, and their antiparticles—all still in thermal equilibrium with the radiation. Compared with the numbers of these lighter particles, only very few protons and neutrons remained at this stage because most had been annihilated.

Electrons, muons, and neutrinos are collectively known as *leptons*, after the Greek word meaning "light" (that is, not heavy). Accordingly, this period in the history of the universe is known as the *lepton epoch*. Quite early on in this epoch, the rapidly thinning universe became transparent to neutrinos. These ghostly particles have been streaming freely through space ever since—most have not interacted with any other particle since the universe was a few seconds old. The lepton epoch ended when the universe was about 100 s old and the temperature fell to about 1 billion K—too low for electron–positron pair production to occur. The density of the universe by this time was about 10 times the density of water.

The final significant event in the Radiation Era occurred when protons and neutrons began to fuse into heavier nuclei. At the start of this period, which we will call the *nuclear epoch*, the temperature was a few hundred million kelvins, and fusion occurred very rapidly, forming deuterium and helium in quick succession before conditions became too cool for further reactions to occur. By the time the universe was about 15 minutes old, much of the helium we observe today had been formed.

THE MATTER ERA

Time passed, the universe continued to expand and cool, and radiation gave way to matter as the dominant constituent of the universe. Our next major epoch extends in time from a few thousand years (the end of the Radiation Era) to about 1 million years after the Big Bang. As the primeval fireball diminished in intensity, a crucial change occurred—perhaps the most important change in the history of the universe. At the end of the nuclear epoch, radiation still overwhelmed matter. As fast as protons and electrons combined, radiation broke them apart again, preventing the formation of even simple atoms or molecules. However, as the universe expanded and cooled, the early dominance of radiation eventually ended. Once formed, atoms remained intact. We will call this period the *atomic epoch*.

The last two epochs together bring us to the current age of the universe. During these late stages, change happened at a much more sedate pace. By the time the universe was about a billion years old, galaxies and large-scale structure had formed. For the first time the visible universe departed from homogeneity on macroscopic scales. The largely uniform universe of the Radiation Era became a universe containing large agglomerations of matter. We call the period from 10^6 to 10^9 years after the Big Bang the *galactic epoch*. At its end, quasars were shining brightly and the first stars were burning and exploding, helping to determine the future shape of their parent galaxies. Since then, stars, planets, and life have appeared in the universe. This final *stellar epoch* has been the subject of the first 25 chapters of this book.

☑ Concept Check

■ Why did lighter and lighter particles "freeze out" of the universe as the cosmos expanded?

27.3 The Formation of Nuclei and Atoms

We now have all the ingredients needed to complete our story of the creation of the elements, begun in Chapter 21 but never quite finished. ∞ (Sec. 21.4) The theory of stellar nucleosynthesis accounts very well for the observed abundances of heavy elements in the universe, but there are discrepancies between theory and observations when it comes to the abundances of the light elements, especially helium. Simply put, the total amount of helium in the universe today—about 25 percent by mass—is far too large to be explained by nuclear-fusion in stars. The accepted explanation is that this base level of helium is *primordial*—that is, it was created during the early, hot epochs of the universe, before any stars had formed. The production of elements heavier than hydrogen by nuclear fusion shortly after the Big Bang is called **primordial nucleosynthesis**.

HELIUM FORMATION IN THE EARLY UNIVERSE

By about 100 s after the Big Bang, the temperature had fallen to about 1 billion K, and apart from "exotic" dark-matter particles, matter in the universe consisted of electrons, protons, and neutrons, with the protons outnumbering the neutrons by about five to one. The stage was set for nuclear fusion to occur. Protons and neutrons combined to produce deuterium nuclei:

$$^1\text{H (proton)} + \text{neutron} \rightarrow {}^2\text{H (deuteron)} + \text{energy.}$$

Although this reaction must have occurred very frequently during the Lepton epoch, the temperature then was still so

MORE PRECISELY 27-1

More on Fundamental Forces

In *More Precisely 16-1* we noted that the behavior of all matter in the universe is ruled by just three *fundamental forces*—gravity, the electroweak force (the unification of the electromagnetic and weak forces), and the strong (nuclear) force. In terrestrial laboratories, these forces display very different properties from one another. Gravity and electromagnetism are long-range, inverse-square forces, whereas the strong and weak forces have very short ranges—10^{-15} and 10^{-17} m, respectively. Furthermore, the forces do not all affect the same particles. Gravity affects everything. The electromagnetic force affects only charged particles. The strong force operates between nuclear particles, such as protons and neutrons, but it does not affect electrons and neutrinos. The weak force shows up in certain nuclear reactions and radioactive decays. The strong force is 137 times stronger than the electromagnetic force, 100,000 times stronger than the weak force, and 10^{39} times stronger than gravity.

In fact, there is more structure below the level of the nucleus. Protons and neutrons are not truly "elementary" in nature but are actually made of subparticles called *quarks*. (The name derives from a meaningless word coined by novelist James Joyce in his book, *Finnegans Wake*.) According to current theory, there are precisely six distinct types of quark in the universe, paired with six distinct types of lepton—the electron and two related "electron-like" particles (called muons and taus), and three types of neutrino. The most massive, and most elusive, of them—the so-called top quark—was discovered at the Fermi National

Laboratory (Fermilab), Illinois, in April 1994. The strong nuclear force is itself just a manifestation of an even more basic force that binds quarks to one another.

On the face of it, one might not imagine that there could be any deep underlying connection between forces as dissimilar as those just described, yet there is growing evidence that they may really be just different aspects of a single basic phenomenon. In the 1960s, theoretical physicists succeeded in explaining the electromagnetic and weak forces in terms of the electroweak force. Shortly thereafter, the first attempts were made at combining the strong and electroweak forces into a single all-encompassing "superforce."

Theories that combine the strong and electroweak forces into one are generically known as *Grand Unified Theories*, or, less formally, GUTs. (Note that the term is plural—no one GUT has yet been proven to be "the" correct description of nature.) One general prediction of GUTs is that the three nongravitational forces are indistinguishable from one another only at enormously high energies, corresponding to temperatures in excess of 10^{28} K. Below that temperature, the superforce splits into two, displaying its separate strong and electroweak aspects. In particle-physics parlance, we say that there is a "symmetry" between the strong and the electroweak forces that becomes broken at temperatures below 10^{28} K, allowing the separate characters of the two forces to become apparent. At "low" temperatures—less than about 10^{15} K, a range that includes almost everything we know on Earth and in the stars—there is a second symmetry breaking, and the

---

high that the deuterium nuclei were broken apart by high-energy gamma rays as soon as they formed. The universe had to wait until it became cool enough for the deuterium to survive. This waiting period is sometimes called the *deuterium bottleneck*.

Only when the temperature of the universe fell below about 900 million K, roughly two minutes after the Big Bang, was deuterium at last able to form and endure. Once that occurred, the deuterium was quickly converted into heavier elements by numerous reactions:

$$^2\text{H} + {}^1\text{H} \rightarrow {}^3\text{He} + \text{energy},$$

$$^3\text{He} + \text{neutron} \rightarrow {}^4\text{He} + \text{energy},$$

$$^2\text{H} + {}^2\text{H} \rightarrow {}^4\text{He} + \text{energy},$$

along with many others. The result was that once the universe passed the deuterium bottleneck, fusion proceeded rapidly and large amounts of helium were formed. In just a few minutes most of the free neutrons were consumed, leaving a universe whose matter content was primarily hydro-

gen and helium. Figure 27.5 illustrates some of the reactions responsible for helium formation. Contrast it with Figure 16.28, which depicts how helium is formed today in the cores of main-sequence stars like the Sun.* ∞ (Sec. 16.5)

We might imagine that fusion could have continued to create heavier and heavier elements, just as in the cores of stars, but this did not occur. In stars, the density and the temperature both *increase* slowly with time, allowing more and more massive nuclei to form, but in the early universe exactly the opposite was true. The temperature and density

---

* *The proton–proton chain that powers the Sun played no significant role in primordial helium formation. The proton–proton reaction that starts the chain is very slow compared with the proton–neutron reaction discussed here and is important in the Sun only because the solar interior contains no free neutrons to make the latter reaction possible.*

Figure 27.5 Helium Formation Some of the reaction sequences that led to the formation of helium in the early universe. Compare this figure with Figure 16.26, which depicts the proton–proton chain in the Sun.

electroweak force splits to reveal its more familiar electromagnetic and weak nature.

The key predictions of the electroweak theory were experimentally verified in the 1970s, winning the theory's originators (Sheldon Glashow, Steven Weinberg, and Abdus Salam) the 1979 Nobel Prize in physics. The GUTs have not yet been experimentally verified (or refuted), in large part because of the extremely high energies that must be reached in order to observe their predictions.

An important idea that has arisen from the realization that the strong and the electroweak forces can be unified is the notion of supersymmetry. It extends the idea of symmetry between fundamental forces to place all particles—those that feel forces (such as protons and electrons), and those that transmit those forces (such as photons and gluons)—on an equal footing. One particularly important prediction of this theory is that all particles should have so-called supersymmetric partners—extra particles that must exist in order for the theory to remain self-consistent. None of these new particles has yet been detected, yet many physicists are convinced of the theory's essential correctness.

These new particles, if they exist, would have been produced in abundance in the Big Bang and should still be around today. They are also expected to be very massive—at least a thousand times heavier than a proton. These so-called supersymmetric relics are among the current leading candidates for the dark matter in the universe (although it must be admitted that recent experimental failures to detect them has dampened some astronomers' early enthusiasm).

Efforts to include gravity within this picture have so far been unsuccessful. Gravitation has not yet been incorporated into a single "SuperGUT," in which all the fundamental forces are united. Some theoretical efforts to merge gravity with the other forces have tried to fit gravity into the quantum world by postulating extra particles—called *gravitons*—that transmit the gravitational force. However, this is a very different view of gravity from the geometric picture embodied in Einstein's general relativity, and combining the two into a consistent theory of quantum gravity has proved very difficult.

A theory currently under very active investigation seeks to interpret all particles and forces in terms of particular modes of vibration of submicroscopic objects known as *strings*. This approach, known as *string theory*, has so far avoided or eliminated many of the thorny problems that have derailed other promising theories. However, it is extremely complex in its mathematical details. For example, it requires that the strings vibrate in no fewer than 11 space-time dimensions—the familiar four, plus seven more "compact" dimensions in which the universe is "rolled up" on very small scales—for the theory to remain consistent. Nevertheless, many theorists feel that it offers the best hope of unifying the forces of nature.

While string theory seems to be our current best bet, at present no theory has succeeded in making any definite statement about conditions in the very early universe. A complete theory of quantum gravity continues to elude researchers.

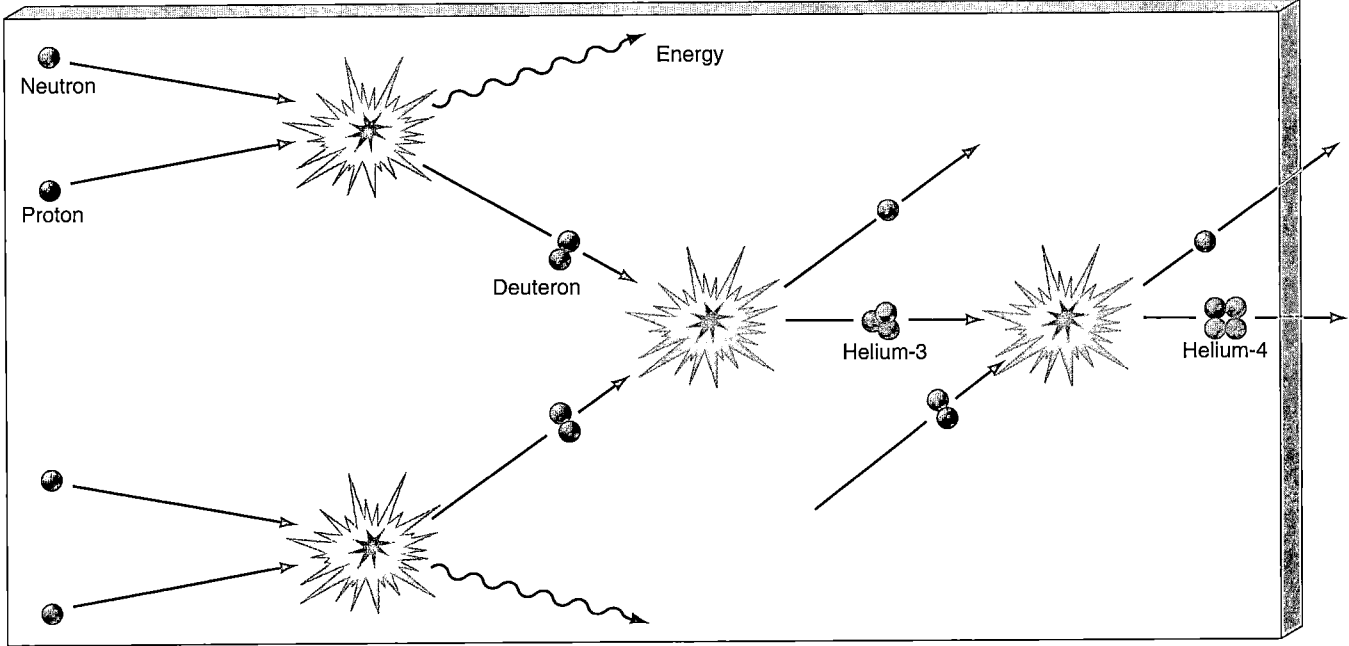

were both *decreasing* rapidly, making conditions less and less favorable for fusion as time went on. Even before the supply of neutrons was completely used up, the nuclear reactions had effectively ceased. Reactions between helium nuclei and protons may also have formed trace amounts of lithium (the next element beyond helium) by this time, but for all practical purposes, the expansion of the universe caused fusion to stop at helium. The brief epoch of primordial nucleosynthesis was over about 15 minutes after it began.

By the end of the period of nucleosynthesis, some 1000 s after the Big Bang, the temperature of the universe was about 300 million K and the cosmic elemental abundances were set. Careful calculations indicate that about one helium nucleus had formed for every 12 protons remaining. Because a helium nucleus is four times more massive than a proton, helium accounted for about one-quarter of the total mass of matter in the universe:

$$\frac{1 \text{ helium nucleus}}{12 \text{ protons} + 1 \text{ helium nucleus}} = \frac{4 \text{ mass units}}{12 \text{ mass units} + 4 \text{ mass units}}$$

$$= \frac{4}{16} = \frac{1}{4}.$$

The remaining 75 percent of the matter in the universe was hydrogen. It would be almost a billion years before nucleosynthesis in stars would change these numbers.

The foregoing calculation implies that all cosmic objects should contain *at least 25 percent* helium by mass. The figure for the Sun, for example, is about 28 percent. However, it is difficult to disentangle the contributions to the present-day helium abundance from primordial nucleosynthesis and later hydrogen burning in stars. Our best hope of determining the amount of primordial helium is to study the oldest stars known, since they formed early on, before stellar nucleosynthesis had had time to change significantly the helium content of the universe. Unfortunately, stars surviving from that early time are of low mass and hence quite cool, making the helium lines in their spectra very weak and hard to measure accurately. ∞ (Secs. 17.6, 17.9) Nevertheless, despite this uncertainty, the observations are generally consistent with the theory just described.

Bear in mind that while all this was going on, matter was just an insignificant "contaminant" in the radiation-dominated universe. Radiation outmassed matter by about a factor of 5000 at the time helium formed. The existence of helium is very important in determining the structure and appearance of stars today, but its creation was completely irrelevant to the evolution of the universe at the time.

DEUTERIUM AND THE DENSITY OF THE COSMOS

During the nuclear epoch, although most deuterium was quickly fused into helium as soon as it formed, a small amount was left over when the primordial nuclear reac-

tions ceased. Observations of deuterium—especially those made by orbiting satellites able to capture deuterium's strongest spectral feature, which happens to be emitted in the ultraviolet part of the spectrum—indicate a present-day abundance of about two deuterium nuclei for every 100,000 protons. However, unlike helium, deuterium is not produced to any significant degree in stars (in fact, deuterium tends to be destroyed in stars), so any deuterium we see today *must* be primordial.

This observation is of great importance to astronomers because it provides them with a sensitive method—and one that is completely independent of the techniques discussed in previous chapters—of probing the present-day density of matter in the universe. According to theory, as illustrated in Figure 27.6, the denser the universe is today, the more particles there were at early times to react with deuterium as it formed, and the less deuterium was left over when nucleosynthesis ended. Comparison of the observed deuterium abundance (marked on the figure) with the theoretical results implies a present-day density of *at most* 3×10^{-28} kg/m³—only a few percent of the critical density.

But before we jump to any far-reaching cosmic conclusions based on this number, we must make a very important qualification. Primordial nucleosynthesis as just

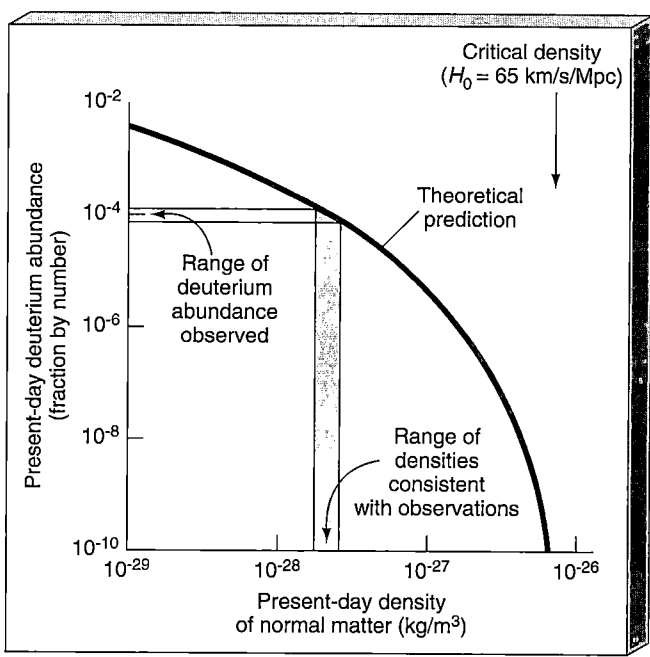

Figure 27.6 Deuterium Abundance The present-day abundance of deuterium depends strongly on the amount of matter present at early times, and this, in turn, determines the present-day density of the universe. Thus, measuring the amount of deuterium in the universe gives us an estimate of the overall density of matter. The best deuterium measurements are marked; they imply that the density of matter in the universe is at most a few percent of the critical value.

described depends *only* on the presence of protons and neutrons in the early universe. Thus, measurements of the abundance of helium and deuterium tell us only about the density of "normal" matter—matter made up of protons and neutrons—in the cosmos. This finding has a momentous implication for the overall composition of the universe. As we saw earlier, astronomers have concluded, on the basis of studies of the motions of galaxies in clusters and superclusters, that the total matter density is about $\frac{1}{3}$ the critical value. ⇨ (Sec. 26.4) If this reasoning is correct, and if the density of normal matter is only a few percent of the critical value, then we are forced to admit that not only is most of the matter in the universe dark but most of the dark matter is *not* composed of protons and neutrons.

Thus, the bulk (about 90 percent) of the matter in the universe apparently exists in the form of elusive subatomic particles (discussed as dark-matter candidates in Chapter 23), whose nature we do not fully understand and whose very existence has yet to be conclusively demonstrated in laboratory experiments. ⇨ (Sec. 23.6) For the sake of brevity, we will adopt from here on the convention that the term "dark matter" refers only to these unknown particles and not to "stellar" dark matter, such as black holes and brown and white dwarfs (also discussed in Chapter 23), which are made of relatively well-understood normal matter.

THE FIRST ATOMS

A few thousand years after the Big Bang, radiation ceased to be the dominant component of the universe. The Matter Era had begun. At the start of the atomic epoch, matter consisted of electrons, protons, helium nuclei (formed by primordial nucleosynthesis), and dark matter. The temperature was several tens of thousands of kelvins—far too hot for atoms of hydrogen to exist, although some helium ions may already have formed. During the next few hundred thousand years, a major change occurred. The universe expanded by another factor of 10, the temperature dropped to a few thousand kelvins, and electrons and nuclei combined to form neutral atoms. By the time the temperature had fallen to 4500 K, the universe consisted of atoms, photons, and dark matter.

The period during which nuclei and electrons combined to form atoms is often called the epoch of **decoupling**, for it was during this period that the radiation background parted company with normal matter. At early times, when matter was ionized, the universe was filled with large numbers of free electrons, which interacted frequently with electromagnetic radiation of all wavelengths. As a result, a photon could not travel far before encountering an electron and scattering off it. In effect, the universe was opaque to radiation. Matter and radiation were strongly "tied," or *coupled*, to one another by these interactions.

After the electrons combined with nuclei to form atoms of hydrogen and helium, only certain wavelengths of radiation—the ones corresponding to the spectral lines of those atoms—could interact with matter. ⇨ (Sec. 4.2) Radiation of other wavelengths could travel virtually forever without being absorbed. The universe became nearly transparent. From that time on, most photons passed generally unhindered through space. As the universe expanded, the radiation simply cooled, eventually becoming the microwave background we see today.

The microwave photons now detected on Earth have been traveling through the universe ever since they decoupled. Their last interaction with matter (at the epoch of decoupling) occurred when the universe was a few hundred thousand years old and roughly 1500 times smaller (and hotter) than it is today—that is, at a redshift of 1500. As illustrated in Figure 27.7, the epoch of atom formation created a kind of "photosphere" in the universe, completely surrounding Earth at a distance of approximately 14,000 Mpc, the distance at which photons last interacted before they decoupled. ⇨ (*More Precisely 25-2*) On our side of the photosphere—that is, since decoupling—the universe is transparent. On the far side—before decoupling—it was opaque. Thus, by observing the microwave background, we are probing conditions in the universe almost all the way back in time to the Big Bang, in much the

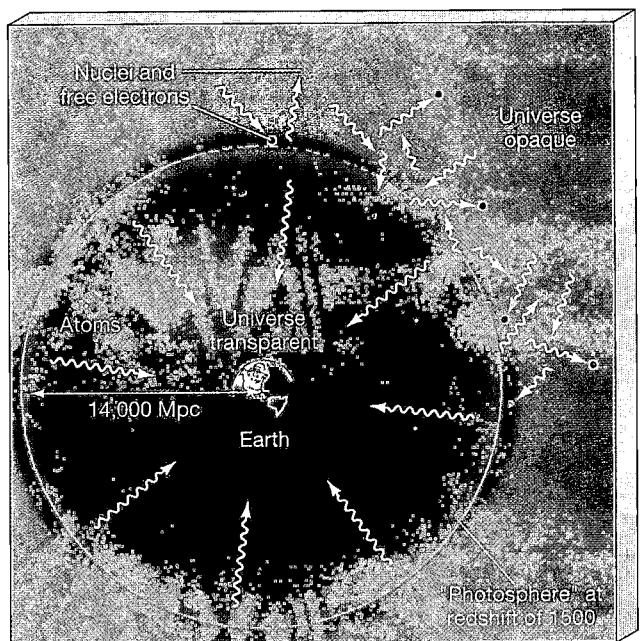

Figure 27.7 Radiation-Matter Decoupling When atoms formed, the universe became virtually transparent to radiation. Thus, observations of the cosmic background radiation allow us to study conditions in the universe around a time at a redshift of 1500, when the temperature dropped below about 4500 K. For an explanation of how we can see a region of space 14,000 Mpc (47 billion light-years) away when the universe is just 15 billion years old, see *More Precisely 25-2*.

same way as studying sunlight tells us about the surface layers of the Sun.

✅ Concept Check

■ How do we know that most of the dark matter in the universe is not of "normal" composition?

27.4 The Inflationary Universe

5 In the late 1970s, cosmologists trying to piece together the evolution of the universe were confronted with two nagging problems that had no easy explanation within the standard Big Bang model. The resolution of these difficulties has caused cosmologists to completely rethink their views of the very early universe.

THE HORIZON AND FLATNESS PROBLEMS

The first problem is known as the **horizon problem,** and it concerns the remarkable isotropy of the cosmic microwave background. ∞ (Sec. 26.6) Recall that the temperature of this radiation is virtually constant, at about 2.7 K, in all directions. Imagine observing the microwave background in two opposite directions on the sky, as illustrated in Figure 27.8. As we have just seen, the radiation

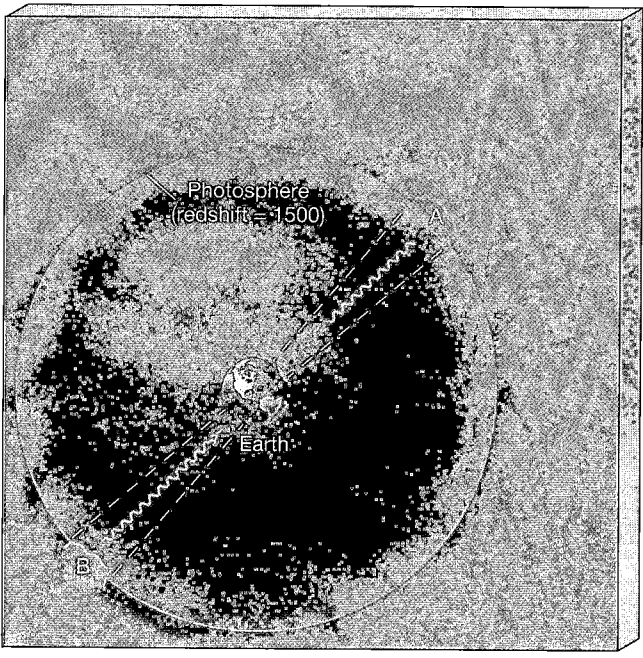

Figure 27.8 Horizon Problem The isotropy of the microwave background indicates that regions A and B in the universe were very similar to each other when the radiation we observe left them, but there has not been enough time since the Big Bang for them ever to have interacted physically with one another. Why then should they look the same?

last interacted with matter in the universe around a redshift of 1500. Thus, in observing these two distant regions of the universe, marked A and B on the figure, we are studying regions that were separated by several million parsecs at the time that they emitted this radiation. The fact that the background radiation is isotropic to high accuracy means that regions A and B had very similar densities and temperatures at the time the radiation we see left them. The problem is, within the Big Bang theory as just described, there is *no* good reason why these regions should be so similar to each other.

To take an everyday example, we all know that heat flows from regions of high temperature to regions of low temperature, but it takes time for this to occur. If we light a fire in one corner of a room, we have to wait a while for the other corners to warm up. Eventually, the room reaches a more or less uniform temperature, but only after the heat from the fire—or, more generally, the *information* that the fire is there—has had time to spread. Similar reasoning applies to regions A and B in Figure 27.8. They are separated by many megaparsecs, and there has not been time for information, which can travel no faster than the speed of light, to travel from one to the other. In cosmological parlance, the two regions are said to be outside each others' *horizon*. But if that is so, then how do they "know" that they are supposed to look the same? With no possibility of communication between them, the only alternative is that regions A and B simply started off looking alike—an assumption that cosmologists are very unwilling to make.

The second problem with the standard Big Bang model is called the **flatness problem.** Whatever the exact value of Ω_0, it appears to be quite close to 1—the total density of the universe is fairly near the critical value. In terms of spacetime curvature, we can say that the universe is remarkably close to being flat. We say "remarkably" here because again there is no good reason *why* the universe should have formed with a density very close to critical. Why not a millionth of or a million times that value? Furthermore, as shown in Figure 27.9, a universe that starts off close to, but not exactly on, the critical curve soon deviates greatly from it, so if the universe is close to critical now, it must have been extremely close to critical in the past. For example, if $\Omega_0 = 0.3$ today (approximately the density of "known" dark matter), then the departure from the critical density at the time of nucleosynthesis would have been only one part in 10^{15} (a thousand trillion).

These observations constitute "problems" because cosmologists want to be able to *explain* the present condition of the universe, not just accept it "as is." They would prefer to resolve the horizon and flatness problems in terms of physical processes that could have taken a universe with no special properties and caused it to evolve into the cosmos we now see. The resolution of both problems takes us back in time even earlier than nucleosynthesis or the formation of any of the elementary particles we know today—back, in fact, almost to the instant of the Big Bang itself.

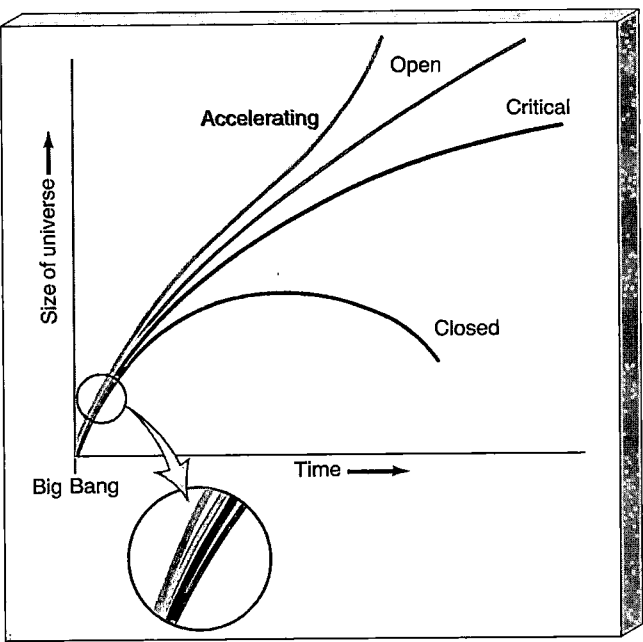

Figure 27.9 Flatness Problem If the universe deviates even slightly from critical density, that deviation grows rapidly in time. For the universe to be as close to critical as it is today, it must have differed from the critical density in the past by only a tiny amount.

FREEZE OUT

Grand Unified Theories predict that three of the four basic forces of nature—electromagnetism and the strong and weak nuclear forces—are in reality aspects of a single, all-encompasing "superforce" (see *More Precisely 27-1*). However, this unification is evident only at enormously high energies, corresponding to temperatures in excess of 10^{28} K. At lower temperatures the superforce reveals its separate electromagnetic, strong, and weak characters.

An essential concept in quantum physics is the idea that forces between elementary particles are exerted, or mediated, by the exchange of another type of particle, generically called a *boson*. We might imagine the two particles as playing a rapid game of catch, using a boson as a ball, as illustrated in Figure 27.10. As the ball is thrown back and forth, the force is transmitted. In ordinary electromagnetism, the boson involved is the photon—a bundle of electromagnetic energy that always travels at the speed of light. The strong force is mediated by particles known as *gluons*. The electroweak theory includes a total of four bosons: the massless photon and three other massive particles, called (for historical reasons) W^+, W^-, and Z^0, all of which have been observed in laboratory experiments. Most of the particles we have encountered so far in this book—electrons, protons, neutrons, neutrinos—play "catch" with at least some of these "balls."

The fundamental forces we know today differ from one another because the particles that mediate these forces

are different. However, in the very early universe, when the basic forces were unified, these particles were indistinguishable from one another. There is said to have been a *symmetry* between the forces of nature at those times. As the universe expanded and cooled, this symmetry was broken, and the separate characters of the forces were revealed.

Section 27.1 discussed how particles "froze out" of the universe as its temperature dropped below the threshold temperature for their creation by pair production. Now that we know that the basic forces of nature are also mediated by particles, we can understand—in general terms, at least—how the fundamental forces froze out too as the universe cooled. The W and Z particles responsible for the electroweak force have masses about 100 times the mass of a proton. The threshold temperature for their production—roughly 10^{15} K—is the temperature at which the weak and electromagnetic forces parted company near the end of the hadron epoch. According to the GUTs, the particle that unifies the strong and electroweak forces is extremely massive—at least 10^{15} times the mass of the proton, and possibly much more. It is because this particle is so massive that the unification of the strong and electroweak forces becomes evident only at extremely high temperatures.

The freezing out of the electroweak force at a temperature of 10^{15} K, some 10^{-10} s after the Big Bang, had little overall effect on the cosmic expansion. By contrast, the freeze out of the strong force, a mere 10^{-35} s after the Big Bang, may have produced one of the strangest events in the history of the cosmos.

COSMIC INFLATION

As the universe expanded and the temperature fell below 10^{28} K, the strong force appeared for the first time as a separate entity—a little like a gas liquefying or water freezing

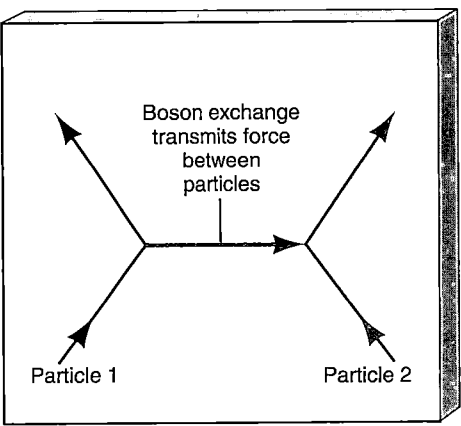

Figure 27.10 Fundamental Forces Forces between elementary particles are transmitted through the exchange of other particles called bosons. As two particles interact, they exchange bosons, a little like playing catch with a submicroscopic ball.

as the temperature drops. Theory suggests that some parts of the universe briefly entered a very odd and unstable state that physicists call the "false vacuum." These regions are of direct interest to us because—if theorists are correct—we live in one!

In essence, due to random fluctuations at the quantum level, these regions found themselves in the "unified" condition a little too long, like water that has been cooled below freezing but has not yet turned to ice. The temporary appearance of the false vacuum within such a region—ours, say—had dramatic consequences. For a short while empty space acquired an enormous *pressure*, which overwhelmed the pull of gravity and caused the region to expand at a greatly accelerated rate. The pressure remained constant as the expansion proceeded, and the acceleration increased with time—in fact, the size of the region *doubled* every 10^{-34} s or so! This period of unchecked expansion, illustrated in Figure 27.11, is known as the **epoch of inflation**.

Eventually, the region reached its normal "true vacuum" state and inflation stopped. The whole episode lasted a mere 10^{-32} s, but during that time the patch of the universe that had become unstable swelled in size by the incredible factor of about 10^{50}. After the inflationary phase, the grand unified force was gone forever. In its place were the more familiar electroweak and strong forces that operate around us in the low-temperature universe of today. The universe once again resumed its (relatively) leisurely expansion, slowly decelerated by the effect of gravity.

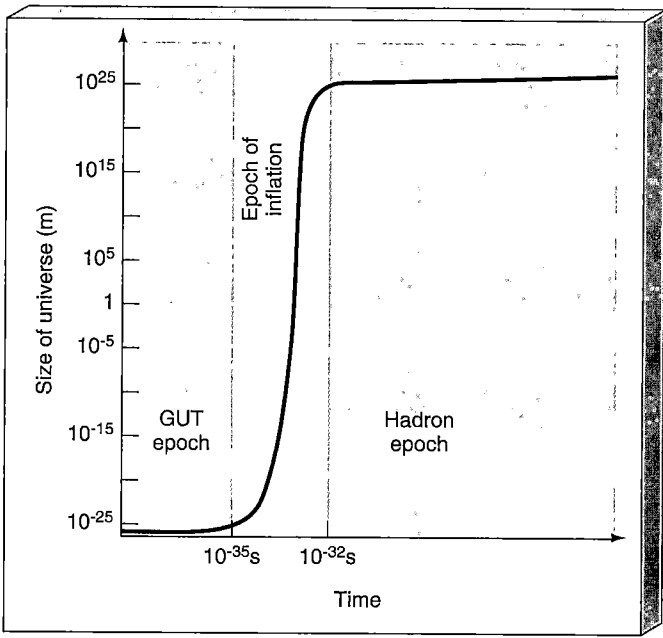

Figure 27.11 Cosmic Inflation During the period of inflation at the end of the GUT epoch, the universe expanded enormously in a very short time. Afterward, it resumed its earlier "normal" expansion rate, except that the size of the cosmos was about 10^{50} times bigger than it was before inflation.

However, a number of important changes had occurred that would have far-reaching ramifications for the evolution of the cosmos.

Note that it is possible—even likely—that not all of the universe experienced inflation. Only some regions became unstable, causing huge inflated "bubbles" to appear in the cosmos. We apparently live in one such bubble; the universe outside is probably unknowable to us. Henceforth we will use the term "universe" to refer to just this bubble and its contents.

The theory just described, and the association of inflation with the end of the GUT epoch, was developed mainly in the 1980s. However, since then researchers have realized that conditions suitable for inflation could have occurred under many different circumstances in the early universe. This generalization actually strengthens inflation as a theory by loosening the restrictions on when it might have occured, although it casts some doubt on exactly when the inflationary epoch leading to "our" universe occurred. Nevertheless, the basic idea of a *quantum fluctuation* expanding to become the universe we know is now quite well established. Some theorists even speculate that we might be living in a sort of "self-creating universe" that erupted into existence spontaneously from inflation in one such random quantum fluctuation! This sort of "statistical" creation of the primal cosmic energy from absolutely nothing has been dubbed "the ultimate free lunch."

Inflation theory makes definite, testable predictions about the large-scale geometry of the universe. In addition, it predicts the existence of residual fluctuations in the cosmic density distribution that are critically important to current theories of structure formation. As we will see in Section 27.5, all these predictions are now being subjected to rigorous scrutiny by astronomers.

IMPLICATIONS FOR THE UNIVERSE

The inflationary epoch provides a natural solution to the horizon and flatness problems described earlier. The horizon problem is solved because inflation took regions of the universe that had already had time to communicate with one another—and so had established similar physical properties—and then dragged them far apart, well out of communications range of one another. Regions A and B in Figure 27.8 have been out of contact since 10^{-32} s after creation, but they were in contact before then. As illustrated in Figure 27.12, their properties are the same today because they were the same long ago, before inflation separated them.

Figure 27.12(a) shows a small piece of the universe just before the onset of inflation. The point that will one day become the site of the Milky Way Galaxy is at the center of the shaded region, which represents the portion of space "visible" to that point at that time—that is, there has been enough time since the Big Bang for light to have traveled

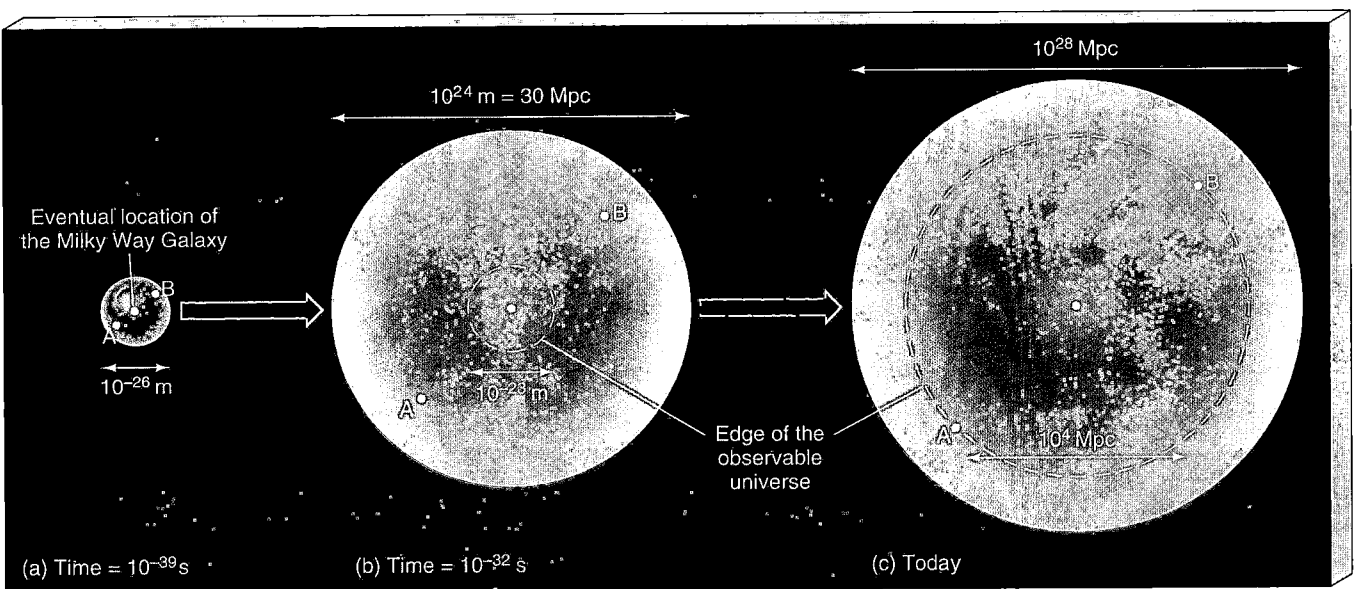

Figure 27.12 Inflation and the Horizon Problem Inflation solves the horizon problem by taking a small region of the very early universe—whose parts had already had time to interact with one another and which had thus already become homogeneous—and expanding it to enormous size. In (a), points A and B are well within the (shaded) homogeneous region of the universe centered on the eventual site of the Milky Way Galaxy. In (b), after inflation A and B are far outside the horizon (indicated by the dashed line), so they are no longer visible from our location. Subsequently, the horizon expands faster than the universe, so that today (c) A and B are just reentering our field of view. They have similar properties now because they had similar properties before the inflationary epoch.

from the edge of this region to its center. That entire region is more or less homogeneous because different parts of it have been able to interact with one another, so any initial differences between them have largely been smoothed out. The points A and B of Figure 27.8 are also marked. They lie within the homogeneous patch, so they have very similar properties. The actual size of the shaded region is about 10^{-26} m—only a trillionth the size of a proton.

Immediately after inflation, as shown in Figure 27.12(b), the homogeneous region has expanded by 50 orders of magnitude, to a diameter of about 10^{24} m, or 30 Mpc—larger than the largest supercluster. By contrast, the visible portion of the universe, indicated by the dashed line, has grown only by a factor of a thousand and is still microscopic in size. In effect, the universe expanded much faster than the speed of light during the inflationary epoch, so what was once well within the horizon now lies far beyond it. In particular, points A and B are no longer visible, either to us or to each other, at this time. (Note that while the theory of relativity restricts matter and energy to speeds less than the speed of light, it imposes no such limit on the universe as a whole.)

Since then, the universe has expanded by a further factor of 10^{27}, so the size of the homogeneous region of space surrounding us is now about 10^{51} m (10^{28} Mpc)—10 trillion trillion times greater than the distance to the most distant quasar. As shown in Figure 27.12(c), the horizon has expanded faster than the universe, so points A and B are just

now becoming visible again. As the portion of the universe now observable from Earth grows in time it remains homogeneous because our cosmic field of view is simply reexpanding into a region of the universe that was within our horizon long ago. We will have to wait a very long time—at least 10^{35} years—before the edge of the homogeneous patch surrounding us comes back into view.

To see how inflation solves the flatness problem, let's return to our earlier balloon analogy. ∞ (Sec. 26.2) Imagine that you are a 1-mm-long ant sitting on the surface of the balloon as it expands, as illustrated in Figure 27.13. When the balloon is just a few centimeters across, you can easily perceive the surface to be curved—its circumference is only a few times your own size. When the balloon expands to, say, a few meters in diameter, the curvature of the surface is less pronounced, but still perceptible. However, by the time the balloon has expanded to a few kilometers across, an "ant-sized" patch of the surface will look quite flat, just as the surface of Earth looks flat to us.

Now imagine that the balloon expands 100 trillion trillion trillion trillion times, as the universe did during the period of inflation. Your local patch of the surface is now completely indistinguishable from a perfectly flat plane, deviating from flatness by no more than one part in 10^{50}. Exactly the same argument applies to the universe. Any curvature the universe may have had before inflation has been expanded so much that space is now virtually flat, at least on the scale of the observable universe (and, in fact,

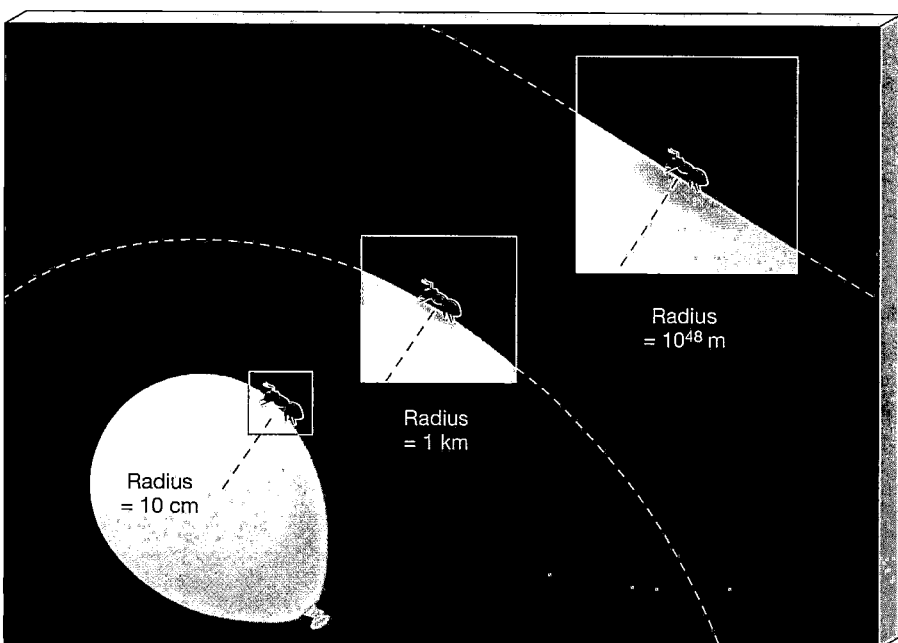

Figure 27.13 Inflation and the Flatness Problem Inflation solves the flatness problem by taking a curved surface, here represented by the surface of the expanding balloon, and expanding it enormously in size. To an ant on the surface, the balloon looks virtually flat when the expansion is over.

Radius = 10^{48} m

Radius = 1 km

Radius = 10 cm

on much larger scales too). In the next section we will see some recent, and very strong, evidence supporting this key prediction of inflation theory.

Notice that this resolution to the flatness problem—the universe appears close to being flat because the universe *is* in fact precisely flat, to very high accuracy—has a very important consequence. Because the universe is flat, the *total* density (including matter, radiation, *and* dark energy) must be exactly the critical value—$\Omega_0 = 1$. ⚬⚬ (Secs. 26.3, 26.5) However, as we have already seen, observations of galaxies and galaxy clusters imply that the cosmic matter density is probably not much greater than about 30 percent of critical, and "normal" matter (like us) comprises at most one-tenth of that. These results, combined with the supernova observations discussed in Chapter 26, leave cosmologists little room to maneuver. ⚬⚬ (Sec. 26.4) We are forced to the conclusion that not only is most matter dark, but *most of the cosmic density isn't made up of matter at all.*

The best match to all available data is a critical-density universe comprising 35 percent (mostly dark) matter and 65 percent dark energy. In a sense, this is the humbling ultimate statement of the Copernican principle. Not only is Earth in no way central to the cosmos, but the stuff of which we are made is unrepresentative of matter in general, and matter itself is the minority constituent of the universe!

☑ **Concept Check**

■ Why does the theory of inflation imply that much of the energy density of the universe may be neither matter nor radiation?

27.5 The Formation of Structure in the Universe

⚬ Just as stars form from inhomogeneities in interstellar clouds, galaxies, galaxy clusters, and larger structures are believed to have grown from small density fluctuations in the matter of the expanding universe. ⚬⚬ (Sec. 19.1) Given the conditions in the universe during the atomic and galactic epochs (Table 27.1), cosmologists calculate that regions of higher-than-average density that contained more than about a million times the mass of the Sun would have begun to contract. There was thus a natural tendency for million-solar-mass "pregalactic" objects to form. In Chapter 24 we saw a little of how these pregalactic fragments might have interacted and merged to form galaxies. ⚬⚬ (Sec. 24.4) Here, we concern ourselves mostly with the formation of structure on much larger scales.

THE GROWTH OF INHOMOGENEITIES

By the early 1980s, cosmologists had come to realize that galaxies could not have formed from the contraction of inhomogeneities involving only *normal* matter. The following lines of reasoning led to this conclusion:

1. Calculations show that before decoupling (which occurred at a redshift of 1500), the intense background radiation would have prevented clumps of normal matter from contracting. Matter and radiation were just too strongly coupled for structure to form. Thus, any such clumps would have had to wait until after decoupling before their densities could start to increase.

2. Because radiation was "tied" to normal matter up until decoupling, any variations in the matter density at that time would have led to temperature variations in the cosmic background radiation—denser regions would have been a little hotter than less dense ones. The high degree of isotropy observed in the microwave background indicates that any density variations from one region of space to another during the epoch of decoupling must have been small—at most a few parts in 10^5. ∞ (Sec. 26.6)

3. Galaxies—or, at least, quasars—are known to have formed by a redshift of 5, and some theorists believe that in order to produce the densest galactic nuclei we see today, the formation process must have already been well established as long ago as a redshift of 10 or 20. ∞ (Sec. 25.6)

4. Because the contracting matter had to "fight" the general expansion of the universe, theory shows contracting pregalactic clumps could have increased in density by a factor of at most 50–100 in the time available (even with fairly optimistic assumptions). As a result, the small inhomogeneities permitted by observations of the microwave background could not have grown into galaxies in the time available—the universe would still have been almost perfectly homogeneous at the time when we know galaxies had already formed.

Put another way, if galaxies had grown from density fluctuations in the normal-matter component of the early universe, then the fluctuations would have had to be so large as to leave a clearly observable imprint on the cosmic microwave background. That imprint is not observed.

DARK MATTER

Fortunately for cosmology (and for life on Earth), much of the universe is made of *dark* matter, which has properties quite different from those of normal matter and provides a natural explanation for the large-scale structure we see today. Whatever the true nature of dark matter, its defining property is that it interacts only very weakly with normal matter and radiation, so its natural tendency to clump and contract under gravity was not hindered by the radiation background. Dark matter started clumping well before decoupling (redshift 1500)—in fact, density inhomogenities in the dark matter component of the universe probably began to grow as soon as matter first began to dominate the universe at a redshift of about 6000. Because the dark matter is not directly tied to the radiation, these inhomogeneities could have been quite large at the time of decoupling without having a correspondingly large effect on the microwave background. In short, dark matter could clump to form large-scale structure in the universe without running into any the problems just described for normal matter.

Thus, as illustrated in Figure 27.14, dark matter determined the overall distribution of mass in the universe and clumped to form the observed large-scale structure without violating any observational constraints on the microwave background. Then, at later times, normal matter was drawn by gravity into the regions of highest density, eventually forming galaxies and galaxy clusters. This picture explains why so much dark matter is found outside the visible galaxies. The luminous material is strongly concentrated near the density peaks and dominates the dark matter there, but the rest of the universe is largely devoid of normal matter. Like foam on the crest of an ocean wave, the universe we can see is only a tiny fraction of the total.

Given that the nature of the dark matter is still unknown, theorists have considerable freedom in choosing its properties when they attempt to simulate the formation of structure in the universe. Cosmologists conventionally distinguish between two basic types of dark matter on the basis of its temperature at the time when galaxies began to form. These types are known as **hot dark matter** and **cold dark matter**, and they lead to quite different kinds of structure in the present-day universe.

Hot dark matter consists of lightweight particles—much less massive than the electron. If neutrinos turn out to have a small mass, as now appears to be the case, they may be leading candidates for hot dark-matter particles. ∞ (Sec. 16.6) Simulations of a universe filled with hot dark matter indicate that large structures, such as superclusters and voids, form fairly naturally, but the computer models cannot account for the existence of structure on smaller scales. Small amounts of hot material tend to disperse, not clump together. Attempts to produce galaxies and clusters by other means after the formation of larger objects have been only partly successful, and most cosmologists have concluded that models based purely on hot dark matter are unable to explain the observed structure of the universe.

Cold dark matter consists of very massive particles, possibly formed during the GUT epoch or even before. Computer simulations modeling the universe with these particles as the dark matter easily produce small-scale structure. With the understanding that galaxies form preferentially in the densest regions, and with a little fine-tuning, these models can also be made to produce large-scale structure comparable to what is actually observed.

Perhaps the best results to date (that is, the results that agree most closely with observations) come from calculations in which a *mixture* of hot and cold dark matter is assumed. Figure 27.15 shows the results of a supercomputer simulation of a mixed dark-matter universe. Compare these images with the real observations of nearby structure shown in Figures 24.16, 24.30, and 25.26. Although calculations like this cannot prove that dark-matter models are the correct description of the universe, the similarities between the models and reality are certainly very striking.

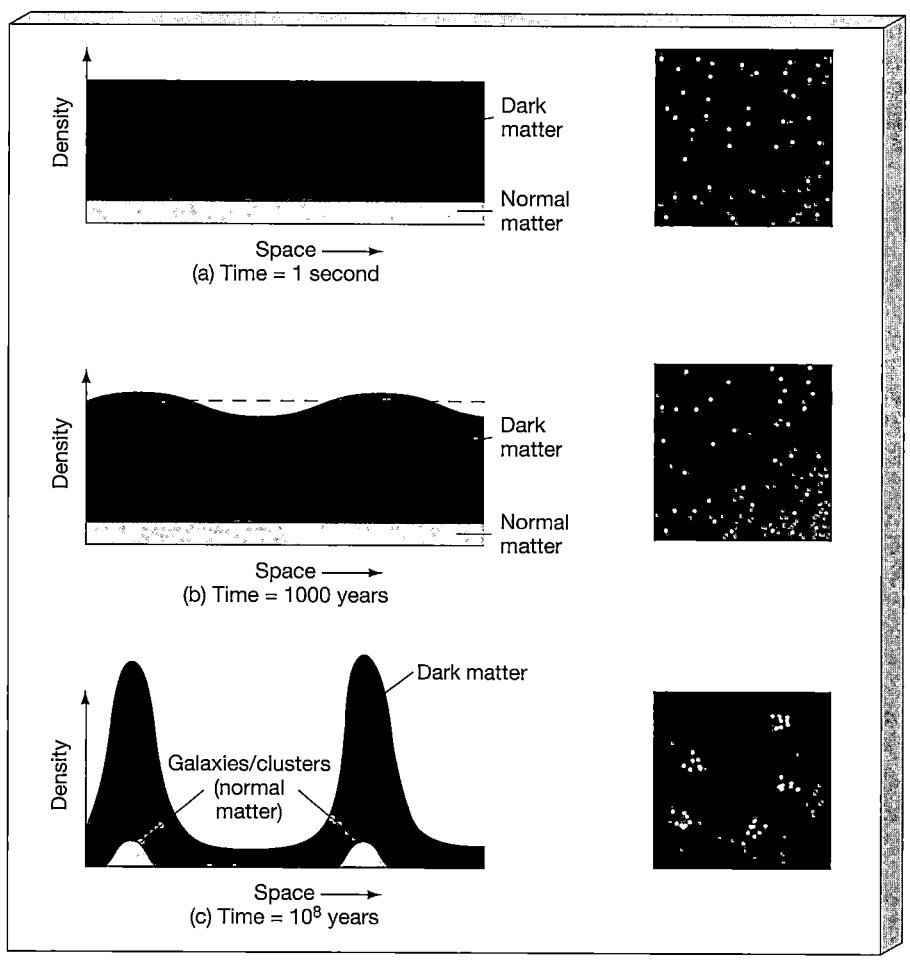

(a) Time = 1 second

(b) Time = 1000 years

(c) Time = 10^8 years

Figure 27.14 Structure Formation The formation of structure in the cosmos depended crucially on the existence of dark matter. (a) The universe started out as a mixture of (mostly) dark and normal matter. (b) A few thousand years after the Big Bang, the dark matter began to clump. (c) Eventually the dark matter formed large structures (represented here by the two high-density peaks) into which normal matter flowed, ultimately to form the galaxies we see today. The three frames at right represent the densities of dark matter (red) and normal matter (yellow) in the graph at left.

THE MICROWAVE BACKGROUND

Because dark matter does not interact directly with photons, its density variations do not cause large (and easily observable) temperature variations in the microwave background. However, the radiation is influenced slightly by the *gravity* of the growing dark clumps, experiencing a slight gravitational redshift that varies from place to place depending on the dark-matter density. As a result, cosmological models predict that there should be tiny "ripples" in the microwave background—temperature variations of only a few parts per million from place to place on the sky.

Until the late 1980s these ripples were too small to be accurately measured, although cosmologists were confident that they would be found. In 1992, after almost two years of careful observation, the *COBE* team announced that the expected ripples had indeed been detected. ∞ (Sec. 26.6) The temperature variations are tiny—only 30–40 millionths of a kelvin from place to place in the sky—but they are there. The *COBE* results are displayed as a temperature map of the microwave sky in Figure 27.16(a). The temperature variation due to Earth's motion (see Figure 26.20) has been subtracted out, as has the radio emission from the Milky Way, and temperature deviations from the average are displayed.

The ripples seen by *COBE*, combined with computer simulations such as that shown in Figure 27.15, predict present-day structure that is consistent with the superclusters, voids, filaments and Great Walls we see around us. While the *COBE* data were limited by relatively low (roughly 7°) resolution, detailed analysis of the ripples also supports the key prediction of inflation theory—that the universe is of exactly critical density and hence spatially flat. For these reasons, the *COBE* observations rank alongside the discovery of the microwave background itself in terms of their importance to the field of cosmology.

The simulation shown in Figure 27.15 implies temperature fluctuations in the microwave background that agree very well with the *COBE* observations, but theorists can (of course) go one stage farther. Figure 27.16(b) presents a prediction, based on the same simulation, of the temperature fluctuations expected from the next generation of (much higher resolution) cosmic microwave background experiments. When the simulation was performed, no such experiment had yet been carried out. However, in late December 1998 and early January 1999, a balloon-borne experiment named Boomerang Balloon Observations of Millimetric Extragalactic Radiation and Geomagnetics) radically improved our view of the microwave background.

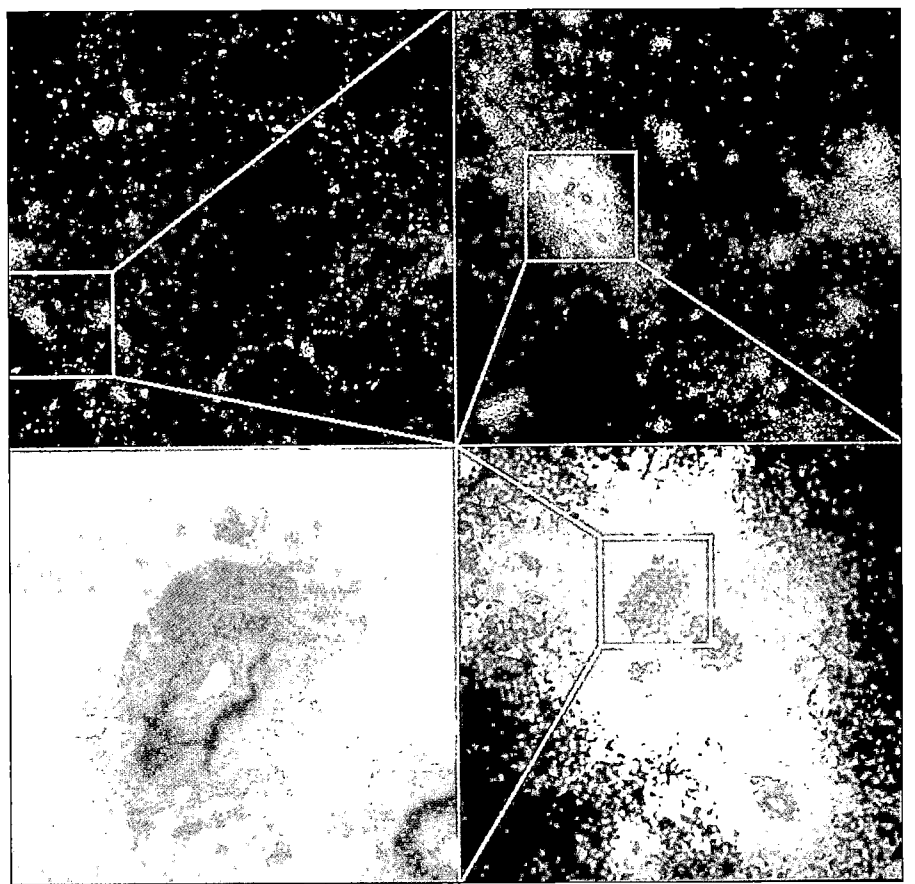

Figure 27.15 Structure Simulated Successively magnified views of a 100 × 100 × 100 Mpc cube in a simulated mixed dark-matter universe with $\Omega_0 = 1$, showing the present-day structure that results from the growth of small density fluctuations in the very early universe. In this particular calculation, 20 percent of the total mass was assumed to be in the form of hot dark-matter particles (actually, neutrinos); the rest was cold dark matter. Colors represent mass density, ranging from the cosmic mean (dark blue), through green, yellow, and red, to 100 times the mean (white). The enlargements zoom in on one particular galaxy in one particular small group of galaxies. The final frame (at bottom left) is roughly 1.5 Mpc across. Notice both the large-scale filamentary structure evident in the top two frames and the extensive dark-matter halos surrounding individual galaxies (the galaxies are roughly the white regions in the bottom two frames). *(E. Bertschinger/ MIT)*

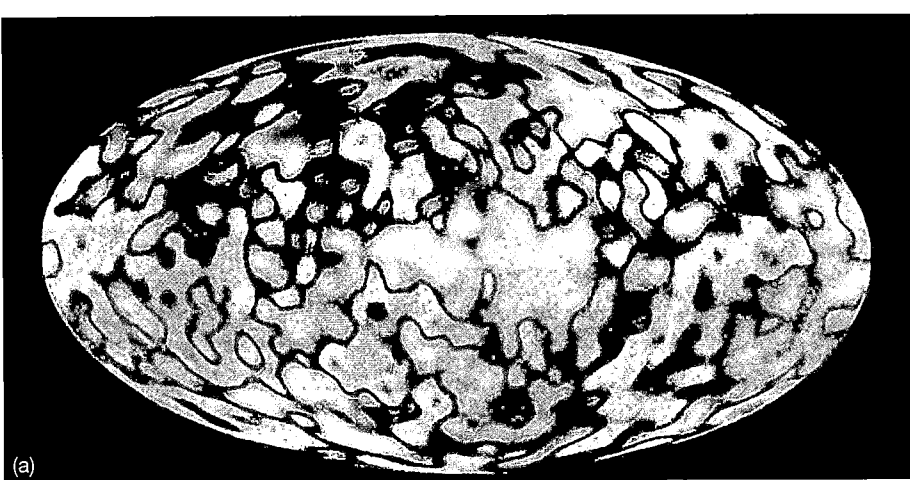

(a)

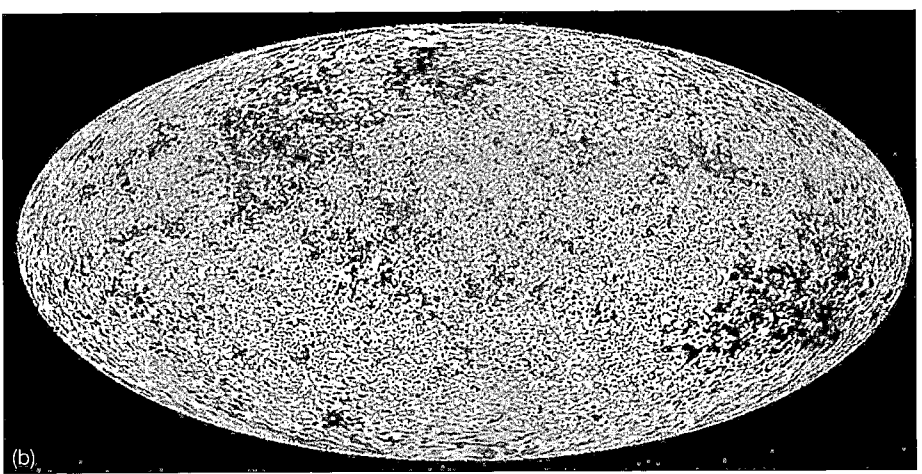

(b)

Figure 27.16 Cosmic Microwave Background Maps (a) *COBE* map of temperature fluctuations in the cosmic microwave background. Hotter-than-average regions are shown in red, cooler-than-average regions in blue. The total temperature range shown is ±200 millionths of a kelvin. (b) Simulated map of microwave background temperature fluctuations corresponding to the simulation shown in Figure 27.15. Dark blue and red represent temperature variations of ±200 millionths of a kelvin from the average, so this map can be compared more or less directly with the *COBE* map in part (a). However, the resolution here is about 0.5 degrees— 20 times sharper than the *COBE* map, and roughly the resolution expected in the next generation of satellite observations. *(NASA; E. Bertschinger)*

R   I   V   U   X   G

Figure 27.17 Boomerang Experiment The
Boomerang instrument (right) and its balloon, on the ground in
Antarctica just before launch in December 1998. The experiment
carried several sensitive microwave detectors above most of
Earth's atmosphere for about 10 days, during which time they
scanned a roughly 900-square-degree patch of the sky at
several frequencies. *(NSF)*

Figure 27.17 shows the Boomerang apparatus on the
ground in Antarctica just before launch. The instrument
consisted of a set of sensitive microwave detectors super-
cooled to a temperature of 0.3 K to minimize their own
thermal emission. The experiment continuously scanned a
roughly $35° \times 25°$ region of the sky away from the galactic
plane at an angular resolution of about 10′—about 40
times sharper than *COBE*, and comparable to the resolution in Figure 27.16(b). The package flew at an altitude of
38 km (keeping it above much of Earth's absorbing atmosphere) and followed the circumpolar winds, returning to

its starting point after a 10-day journey around the South
Pole (explaining the extraordinary effort expended in constructing the experiment's name).

Figure 27.18 shows the patch of the sky covered by the
Boomerang observations. This high-resolution map shows
low- (blue) and high- (red) temperature fluctuations of
about ±300 microkelvins with a characteristic scale of
about one degree. The temperature range is larger than the
fluctuations seen by *COBE* because *COBE*'s low resolution
effectively averaged the data over a large area of the sky.

The most important implications of the data are summarized in Figure 27.19, a complicated-looking plot with a
very simple interpretation. The red dots (with error estimates) show the amount of structure seen in the data on
different angular scales (more technically, they show the
correlation between different data points separated by
those scales), and clearly show a peak at about one degree—
probably not too different from what your eye sees in Figure 27.18. The blue curve shows a theoretical prediction
for a universe with $\Omega_0 = 1$ (actually, with 35 percent matter
and 65 percent dark energy, as quoted in Chapter 26 and
earlier in this chapter). The agreement is striking. In fact,
while the smaller bumps and wiggles depend on other cosmic parameters, the location of the large peak in the theoretical curve is a *direct* measure of the value Ω_0. These data
indicate that Ω_0 is very close to 1, in full agreement with the
inflation prediction, with a very small margin of error.

Ongoing and upcoming space-based missions (NASA's
MAP satellite launched in mid-2001 and ESA's *Planck* mis-

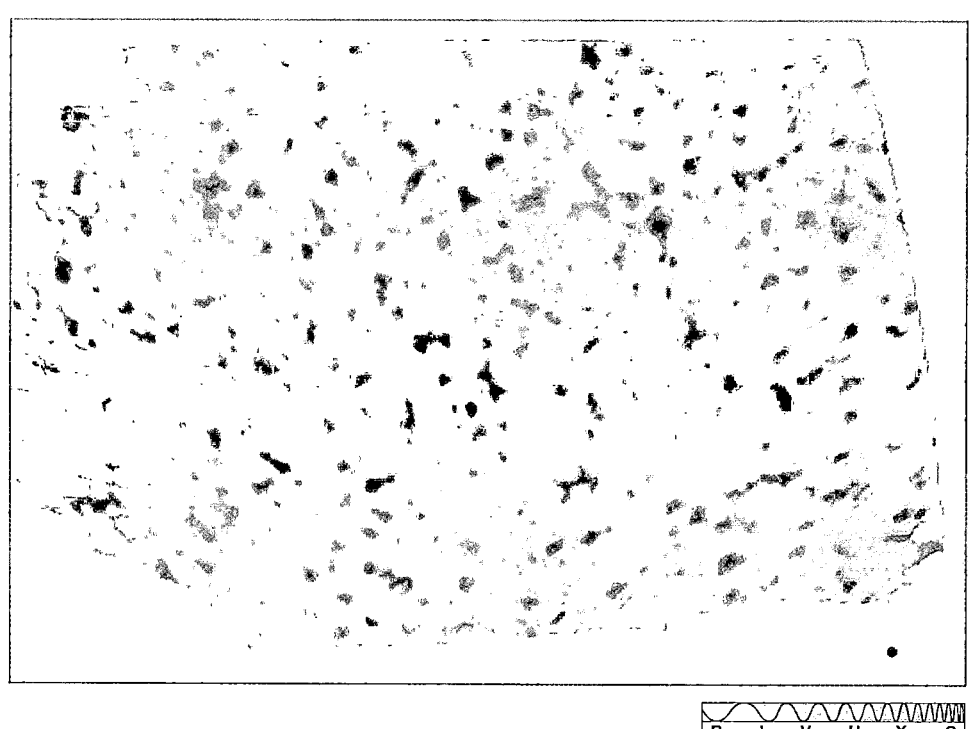

**Figure 27.18 A Flat
Universe** Part of the microwave
sky, as seen by Boomerang at a
frequency of 150 GHz. The region
studied by Boomerang is only a
small part of *COBE*'s full-sky
coverage shown in Figure 27.16(a),
this map covers a 35° by 25°
region located roughly midway
between the Galactic equator and
the south Galactic pole, at
approximately the 5-o'clock
position in Figure 27.16(a).
However, the resolution here,
indicated by the black dot at lower
right, is about 10′ (compared with
7° for *COBE*).

−300 µK ██████████████████ 300 µK
−300   −200   −100   0   100   200   300

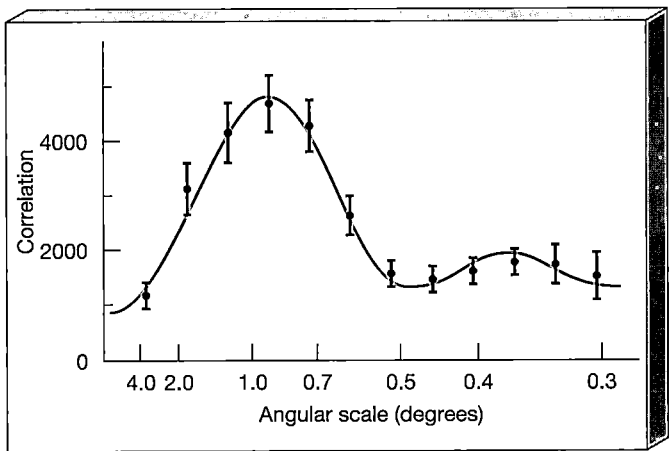

Figure 27.19 Cosmic Structure The most obvious features on the Boomerang map (Figure 27.18) occur on a scale of about 1°. This graph quantifies this by plotting the correlation between different points on the map—in effect measuring the "amount" of structure seen on different angular scales. The red dots are the Boomerang measurements, and the vertical bars are their relative errors. The blue curve is not a fit to the data— rather, it is a theoretical prediction for a flat $(\Omega_0 = 1)$ universe. The excellent agreement in the large peak at 1° (whose position is determined almost entirely by the value of Ω_0) is regarded by most cosmologists as strong evidence that we live in a flat, critical-density universe.

sion in 2005) will further refine these measurements, improving their errors and expanding the coverage to the entire sky. The first decade of the twenty-first century may well see the basic parameters of the universe measured (even if not fully understood) to an accuracy only dreamed of just a few years ago.

☑ Concept Check

■ What do observations of fluctuations in the microwave background tell us about the structure of the universe?

Chapter Review

SUMMARY

At present, the density of matter in the universe greatly exceeds the equivalent mass density of radiation. The universe is **matter dominated** (p. 716). The density of matter was much greater in the past, when the universe was smaller. However, because radiation is redshifted as the universe expands, the density of radiation was greater still. The early universe was **radiation dominated** (p. 717). During the first few minutes after the Big Bang, matter was formed out of the primordial fireball by the process of **pair production** (p. 717). In the early universe, matter and radiation were linked by this process. Particles "froze out" of the radiation background as the temperature fell below the threshold for creating them. The existence of matter today means that there must have been unequal amounts of matter and antimatter early on.

The physical state of the universe can be understood in terms of present-day physics back to about 10^{-43} s after the Big Bang. Before that, the four fundamental forces of nature—gravity, electromagnetism, the strong force, and the weak force—were all indistinguishable. There is presently no theory that can describe these extreme conditions. As the universe expanded and its temperature dropped, the forces became distinct from one another. First gravity, then the strong force, and then the weak and electromagnetic forces separated out.

Only a little of the helium observed in the universe today was formed in stars. Most of it was created by **primordial nucleosynthesis** (p. 721) in the early universe. Some deuterium was also formed at these early times, and it provides a sensitive indicator of the present density of the universe in the form of "nor-

mal" (as opposed to dark) matter. Studies of deuterium indicate that normal matter can account for at most three or four percent of the critical density. The remaining mass inferred from studies of clusters must then be made of dark matter, in the form of unknown particles formed at some very early epoch. When the universe was about 1500 times smaller than it is today, the temperature became low enough for atoms to form. At that time, the (then-optical) radiation background **decoupled** (p. 725) from the matter. The universe became transparent. The photons that now make up the microwave background have been traveling freely through space ever since.

Why should regions of the universe that have not had time to "communicate" with one another look so similar? This is called the **horizon problem** (p. 726). Inflation solves it by taking a small homogeneous patch of the early universe and expanding it enormously in size. The patch is still homogeneous, but it is now much larger than the portion of the universe we can see today. Why should the density of the universe be so near the critical value? This is called the **flatness problem** (p. 726). According to modern **Grand Unified Theories** (p. 720), the three nongravitational forces of nature began to display their separate characters, about 10^{-35} s after the Big Bang. At that time, or possibly earlier, a brief period of rapid cosmic expansion called the **epoch of inflation** (p. 728) occurred, during which the size of the universe increased by a factor of about 10^{50}. Inflation implies that the total cosmic density is in fact exactly critical. Recent observations from the Boomerang mission strongly support this conclusion. In that case, then most of that density is in the form of dark energy, with matter (normal and dark) comprising only about one-third of the total.

The large-scale structure observed in the universe could not have formed out of density fluctuations in normal gaseous matter—there simply has not been enough time, given the twin constraints of the smoothness of the microwave background and the epoch at which the first galaxies and quasars are known to have formed. Instead, dark matter clumped and grew to form the

"skeleton" of the structure now observed. Normal matter then flowed into the densest regions of space, eventually forming the galaxies we now see. Cosmologists distinguish between **hot dark matter** (p. 731) and **cold dark matter** (p. 731), depending on its temperature at the end of the Radiation Era. In order to explain the observed large-scale structure in the universe, much of the dark matter must be cold. In 1992, the *COBE* satellite discovered the expected ripples in the cosmic microwave background associated with the clumping of dark matter at early times.

SELF-TEST: TRUE OR FALSE?

____ **1.** The light emitted from all the stars currently in the universe far outshines the cosmic microwave background.

____ **2.** The time between the Big Bang and 10^{-43} s afterward cannot be described for lack of a theory of quantum gravity.

____ **3.** All matter creation was complete by the time the universe was about one second old.

____ **4.** About 25 percent by mass of normal matter in the universe is primordial helium.

____ **5.** The present-day abundance of deuterium gives us information on the density of dark matter.

____ **6.** Decoupling refers to interactions between matter and antimatter.

____ **7.** The present-day microwave background radiation last interacted with matter around the time of decoupling.

____ **8.** After decoupling, neutral atoms could finally exist.

____ **9.** The horizon problem relates to the isotropy of the microwave background radiation.

____ **10.** The flatness problem is the fact that the observed density of matter is unexpectedly different from the critical density.

____ **11.** The universe grew in size by a factor of about 10,000 during the inflationary period.

____ **12.** If the theory of inflation is correct, then the density of the universe is exactly the critical density.

____ **13.** Physicists have detected cold dark-matter particles in terrestrial laboratories.

____ **14.** Galaxies had already formed by the time the universe was about 1 million years old.

____ **15.** Measurements of the microwave background seem to confirm the predictions of inflation theory.

SELF-TEST: FILL IN THE BLANK

1. Comparing the mass density of radiation and matter, we find that, at the present time, _____ dominates.

2. In the early universe, the total density was dominated by _____.

3. In the process of pair production, two _____ interact to form a particle and an _____.

4. The temperature necessary to form particles _____ as the particles' masses increase.

5. When the universe was a few minutes old, nuclear fusion produced _____ and _____.

6. After about a million years, _____ had formed in the universe.

7. Heavy elements were not formed primordially because, as time passed, the density and temperature of the universe _____.

8. Studies of primordial nucleosynthesis imply that only a tiny fraction of matter in the universe is made up of _____.

9. The cosmic microwave background radiation last interacted with matter when the universe was about _____ times its present size.

10. If the theory of inflation is correct, then the *total* density of the universe must be _____.

11. Combining inflation with measurements of the dark-matter density leads to the conclusion that most of the universe is composed of _____.

12. We know that density fluctuations in the normal-matter component of the early universe must have been very small because otherwise we would see their imprint on the _____.

13. Hot and cold dark matter differ in the masses of their particles. Cold dark matter consists of _____ particles.

14. Theory predicted tiny fluctuations in the _____ of the microwave background; the _____ satellite found them.

15. The Boomerang experiment found that the universe is spatially _____.

REVIEW AND DISCUSSION

1. For how long was the universe dominated by radiation? How hot was the universe when the dominance of radiation ended?

2. What was the role of dark energy in the very early universe?

3. Why is our knowledge of the Planck epoch so limited?

4. When and how did the first atoms form?

5. Describe the universe at the end of the galactic epoch.

6. Why do all stars, regardless of their abundance of heavy elements, seem to contain at least one-quarter helium by mass?

7. Why didn't heavier and heavier elements form in the early universe, as they do in stars?

8. If large amounts of deuterium formed in the early universe, why do we see so little deuterium today?

9. How do measurements of the cosmic deuterium abundance provide a reliable estimate of the density of normal matter?

10. How do we know that most matter in the universe is not "normal"?

11. When did the universe become transparent to radiation?

12. How can we observe the epoch at which the universe became transparent?

13. What is the epoch of inflation, and what happened to the early universe during that time?

14. How does inflation solve the horizon problem?

15. How does inflation solve the flatness problem?

16. What does inflation tell us about the total density of the universe?

17. What is the difference between hot and cold dark matter?

18. What is the connection between dark matter and the formation of large- and small-scale structures?

19. Why were the observations made by the *COBE* satellite so important to cosmology?

20. What key measurement was made by the Boomerang experiment?

PROBLEMS  *Algorithmic versions of these questions are available in the Practice Problems module of the Companion Website.*

The number of squares preceding each problem indicates its approximate level of difficulty.

1. ▣ What was the distance between the points that would someday become the center of the Milky Way Galaxy and the center of the Virgo Cluster at the time of decoupling? (The present separation is 18 Mpc.)

2. ▣▣ What was the equivalent mass density of the cosmic radiation field when the universe was one-thousandth its present size? (Hint: Don't forget the cosmological redshift!)

3. ▣ Assuming critical density today, what were the temperature and density of the universe when the first quasars formed?

4. ▣▣ Of matter and radiation, which dominated the universe, and by what factor in density (assuming critical density today), at the start of (a) decoupling, (b) nucleosynthesis?

5. ▣▣▣ Estimate the temperature needed for electron–positron pair production. The mass of an electron is 9.1×10^{-31} kg. Use $E = mc^2$ to find the energy (Section 16.5), $E = hf$ and $\lambda f = c$ to find the wavelength λ of a photon having that energy (Section 4.2), and finally Wien's law to find the temperature for which a blackbody spectrum peaks at that wavelength (Section 3.4). How does your answer compare with the threshold temperature given in the text?

6. ▣▣ Given that the threshold temperature for the production of electron–positron pairs is about 6×10^9 K and that a proton is 1800 times more massive than an electron, calculate the threshold temperature for proton–antiproton pairs.

7. ▣ At what wavelength did the radiation background peak at the start of the epoch of nucleosynthesis? In what part of the electromagnetic spectrum does this lie?

8. ▣▣ By what factor did the volume of the universe increase during the epoch of primordial nucleosynthesis, from the time when deuterium could first survive until the time at which all nuclear reactions ceased? By what factor did the matter density of the universe decrease during this period?

9. ▣ From Table 25.1, the "photosphere" of the universe corresponding to the epoch of decoupling presently lies some 14,000 Mpc from us (see Figure 27.7). How far away was a point on the photosphere when the background radiation we see today was emitted?

10. ▣▣ Calculate the mass, in kilograms, of the particle that unifies the strong and electroweak forces, if it froze out at a temperature of 10^{28} K. Calculate the mass of a hypothetical particle that might unify gravity with the other forces, given that it froze out at the end of the Planck epoch (Table 27.1).

11. ▣ How many times did the universe double in size during the inflationary period if it expanded by a factor of 10^{50}?

12. ▣▣ What would be the radius of a trillion-solar-mass patch of a homogeneous, critical-density universe (a) today? (b) at decoupling? (c) at the end of the nuclear epoch?

13. ▣▣ The blobs evident in Figure 27.18(a) are about 10° across. If those blobs represent clumps of matter around the time of decoupling, and assuming Euclidean geometry, estimate the size of the clumps (at the time of decoupling).

14. ▣ If dark-matter particles are remnants of the GUT epoch, they must have decoupled from the rest of the matter and radiation in the universe at the end of the GUT epoch. Repeat Problem 9, but for dark-matter particles instead of electromagnetic photons, assuming a present photosphere distance of 14,000 Mpc.

15. ▣▣▣ Assuming critical density today, what size and mass scale correspond to the peak of the Boomerang data (Figure 27.18), at an angular scale of 1° and a redshift of 1500?

RESEARCHING ON THE WEB  *To complete the following exercises, go to the online Destinations module for Chapter 27 on the Companion Website for Astronomy Today 4/e.*

1. Access the "Center for Particle Astrophysics" page, and describe how the Distant Supernova Search Project determines the distance to galaxies.

2. Access the "Frequently Asked Questions About The Cosmic Microwave Background" page, and explain where the cosmic microwave background photons actually originate.

 *In addition to the Practice Problems and Destinations modules, the Companion Website at http://www.prenhall.com/chaisson provides for each chapter an additional true-false, multiple choice, and labeling quiz, as well as additional annotated images, animations, and links to related Websites.*

28 LIFE IN THE UNIVERSE

Are We Alone?

LEARNING GOALS

Studying this chapter will enable you to:

1   Summarize the process of cosmic evolution as it is currently understood.

2   Evaluate the chances of finding life elsewhere in the solar system.

3   Summarize the various probabilities used to estimate the number of advanced civilizations that might exist in the Galaxy.

4   Discuss some of the techniques we might use to search for extraterrestrials and to communicate with them.

 Visit http://www.prenhall.com/chaisson for additional annotated images, animations, and links to related sites for this chapter.

Is there intelligent life elsewhere in the universe? This fanciful painting by noted space artist Dana Berry suggests a plurality of civilizations—some perhaps extinct—on alien worlds well beyond our own solar system. Despite blockbuster movies, science-fiction novels, and a host of claims for extraterrestrial contact, astronomers have found not a shred of evidence for intelligent life anywhere else in the universe. *(Galaxyrise Over Alien Planet courtesy D. Berry)*

The Big Picture: Cosmic evolution is the study of the many varied changes in the assembly and composition of energy, matter, and life in the thinning and cooling universe. This broad, sweeping subject incorporates aspects of astronomy, physics, geology, chemistry, biology, and many of the other natural sciences, thereby providing a powerful world view that helps us evaluate the prospects for life elsewhere in the universe.

739

Are we unique? Is life on our planet the only example of life in the universe? These are difficult questions, for the subject of extraterrestrial life is one on which we have no data, but they are important questions with profound implications for the human species. Earth is the only place in the universe where we know for certain that life exists. In this chapter we take a look at how humans evolved on Earth and then consider whether those evolutionary steps might have happened elsewhere. Having done that, we will assess the likelihood of our having galactic neighbors and consider how we might learn about them if they exist.

28.1 Cosmic Evolution

Figure 28.1 identifies seven major phases in the history of the universe: *particulate, galactic, stellar, planetary, chemical, biological,* and *cultural* evolution. Together, these evolutionary stages make up the grand sweep of **cosmic evolution**—the continuous transformation of matter and energy that has led to the appearance of life and civilization on Earth. The first four phases represent, in reverse order, the contents of this book. We now briefly expand our field of view beyond astronomy to include the other three.

From the Big Bang, to the formation of galaxies, to the birth of the solar system, to the emergence of life, to the evolution of intelligence and culture, the universe has evolved from simplicity to complexity. We are the result of an incredibly complex chain of events that spanned billions of years. Were those events random, making us unique, or are they in some sense *natural,* so that technological civilization is inevitable? Put another way, are we alone in the universe, or are we just one among countless other intelligent life forms in our Galaxy? In this chapter we consider the development of life on Earth and try to assess the likelihood of finding intelligent life elsewhere in the cosmos.

LIFE IN THE UNIVERSE

Before embarking on our study, we need a working definition of *life.* This seemingly simple task is not an easy one—the distinction between the living and the nonliving is not as obvious as we might at first think. Though most physicists would agree on the definitions of matter and energy, biologists have not arrived at a clear-cut definition of life.

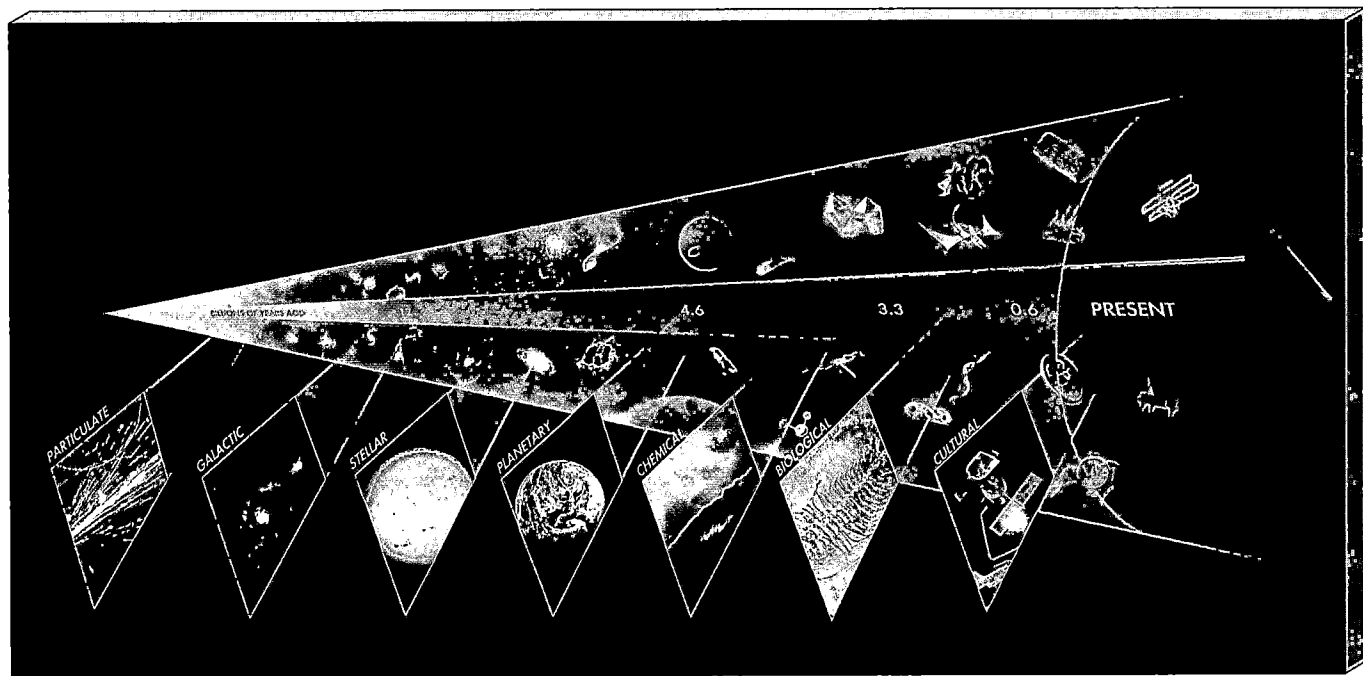

Figure 28.1 Arrow of Time Some highlights of cosmic history are noted along this arrow of time, from the beginning of the universe to the present. Noted along the bottom of the arrow are seven "windows" outlining the major phases of cosmic evolution: evolution of primal energy into elementary particles; evolution of atoms into galaxies and stars; evolution of stars into heavy elements; evolution of elements into solid, rocky planets; evolution of those elements into the molecular building blocks of life; evolution of those molecules into life itself; evolution of advanced life forms into intelligence, culture, and technological civilization. *(D. Berry)*

DISCOVERY 28-1

The Virus

The central idea of chemical evolution is that life evolved from nonlife. But aside from insight based on biochemical knowledge and laboratory simulations of some key events on primordial Earth, do we have any direct evidence that life could have developed from nonliving molecules? The answer is yes. The smallest and simplest entity that sometimes appears to be alive is a virus. We say "sometimes" because viruses seem to have the attributes of both nonliving molecules and living cells. *Virus* is the Latin word for "poison," an appropriate name since viruses are often a cause of disease. Although they come in many sizes and shapes—a typical example is the polio virus, shown here magnified 300,000 times—all viruses are smaller than the size of a typical modern cell. Some are made of only a few thousand atoms. In terms of size, then, viruses seem to bridge the gap between cells that are living and molecules that are not.

Viruses contain some proteins and genetic information (in the form of DNA, or the closely related molecule RNA), but not much else—none of the material by which living organisms normally grow and reproduce. How, then, can a virus be considered alive? When alone, it cannot; a virus is absolutely lifeless when isolated from living organisms. But when it is inside a living system, a virus has all the properties of life. Viruses come alive by transferring their genetic material into living cells. The genes of a virus seize control of a cell and establish themselves as the new master of chemical activity. Viruses grow and reproduce copies of themselves by using the genetic machinery of the invaded cell, often robbing the cell of its usual function.

Some viruses multiply rapidly and wildly, spreading the disease and—if unchecked—eventually killing the invaded organism. In a sense, then, viruses exist within the gray area between the living and the nonliving.

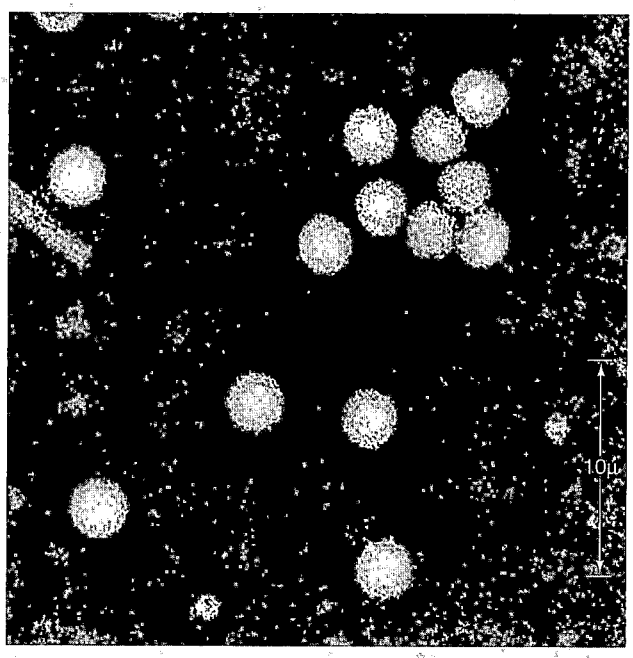

(R. Williams)

Generally speaking, scientists regard the following as characteristics of living organisms: (1) they can *react* to their environment and can often heal themselves when damaged; (2) they can *grow* by taking in nourishment from their surroundings and processing it into energy; (3) they can *reproduce*, passing along some of their own characteristics to their offspring; and (4) they have the capacity for genetic change and can therefore *evolve* from generation to generation and adapt to a changing environment.

These rules are not strict, and there is great leeway in interpreting them. Stars, for example, react to the gravity of their neighbors, grow by accretion, generate energy, and "reproduce" by triggering the formation of new stars, but no one would suggest that they are alive. A virus (see *Discovery 28-1*) is crystalline and inert when isolated from living organisms, but once inside a living system, it exhibits all the properties of life, seizing control of a living cell and using the cell's own genetic machinery to grow and reproduce. Most researchers now believe that the distinction between living and nonliving is more one of structure and complexity than a simple checklist of rules.

The general case in favor of extraterrestrial life is summed up in what are sometimes called the *assumptions of mediocrity*: (1) because life on Earth depends on just a few basic molecules, and (2) because the elements that make up these molecules are (to a greater or lesser extent) common to all stars, and (3) if the laws of science we know apply to the entire universe, as we have supposed throughout this book, then—given sufficient time—life must have originated elsewhere in the cosmos. The opposing view maintains that intelligent life on Earth is the product of a series of extremely fortunate accidents—astronomical, geological, chemical, and biological events unlikely to have occurred anywhere else in the universe. The purpose of this chapter is to examine some of the arguments for each of these viewpoints.

CHEMICAL EVOLUTION

What information do we have about the earliest stages of planet Earth? Unfortunately, not very much. Geological hints about the first billion years or so were largely erased by violent surface activity as volcanoes erupted and meteorites bombarded our planet; subsequent erosion by wind and water has seen to it that little evidence has survived to the present. Scientists believe that the early Earth was barren, with shallow, lifeless seas washing upon grassless, treeless continents. Outgassing from our planet's interior through volcanoes, fissures, and geysers produced an atmosphere rich in hydrogen, nitrogen, and carbon compounds and poor in free oxygen. As Earth cooled, ammonia, methane, carbon dioxide, and water formed. The stage was set for the appearance of life.

The surface of the young Earth was a very violent place. Natural radioactivity, lightning, volcanism, solar ultraviolet radiation, and meteoritic impacts all provided large amounts of energy that eventually shaped the ammonia, methane, carbon dioxide, and water into more complex molecules known as **amino acids** and **nucleotide bases**—organic (carbon-based) molecules that are the building blocks of life as we know it. Amino acids build *proteins*, and proteins control metabolism, the daily utilization of food and energy by means of which organisms stay alive and carry out their vital activities. Sequences of nucleotide bases form *genes*—parts of the DNA molecule—which direct the synthesis of proteins and thus determine the characteristics of the organism. These same genes also carry the organism's hereditary characteristics from one generation to the next in reproduction. In all living creatures on Earth—from bacteria to amoebas to humans—genes mastermind life, and proteins maintain it.

The idea that complex molecules could have evolved naturally from simpler ingredients found on the primitive Earth has been around since the 1920s. The first experimental verification was provided in 1953 when scientists Harold Urey and Stanley Miller, using laboratory equipment somewhat similar to that shown in Figure 28.2, took a mixture of the materials thought to be present on Earth long ago—a "primordial soup" of water, methane, carbon dioxide, and ammonia—and energized it by passing an electrical discharge ("lightning") through the gas. After a few days they analyzed their mixture and found that it contained many of the same amino acids found in all living things on Earth. About a decade later, scientists succeeded in constructing nucleotide bases in a similar manner. These experiments have been repeated in many different forms, with more realistic mixtures of gases and a variety of energy sources, but always with the same basic outcomes.

Although none of these experiments has ever produced a living organism, or even a single strand of DNA, they do demonstrate conclusively that "biological" molecules can be synthesized by strictly *non*biological means, using raw materials available on the early Earth. More advanced experiments, in which amino acids are united under the influence of heat, have fashioned proteinlike blobs that behave to some extent like true biological cells.

Such near-protein material resists dissolution in water (so it would remain intact when it fell from the primitive atmosphere into the ocean) and tends to cluster into small droplets called microspheres—a little like oil globules floating on the surface of water. Figure 28.3 shows some of these proteinlike microspheres. The walls of these laboratory-made droplets permit the inward passage of small molecules, which then combine within the droplet to construct more complex molecules too large to pass back out through the walls. As the droplets "grow," they tend to "reproduce," forming smaller droplets.

Can we consider these proteinlike microspheres to be alive? Almost certainly not. Most biochemists would say that the microspheres are not life itself, but they contain many of the basic ingredients needed to form life. The microspheres lack the hereditary molecule DNA. However, as illustrated in Figure 28.4, they do have similarities to ancient cells found in the fossil record. Thus, while no actual living cells have yet been created "from scratch" in any laboratory, many biochemists feel that the chain of events leading from simple nonbiological molecules almost to the point of life itself has been amply demonstrated.

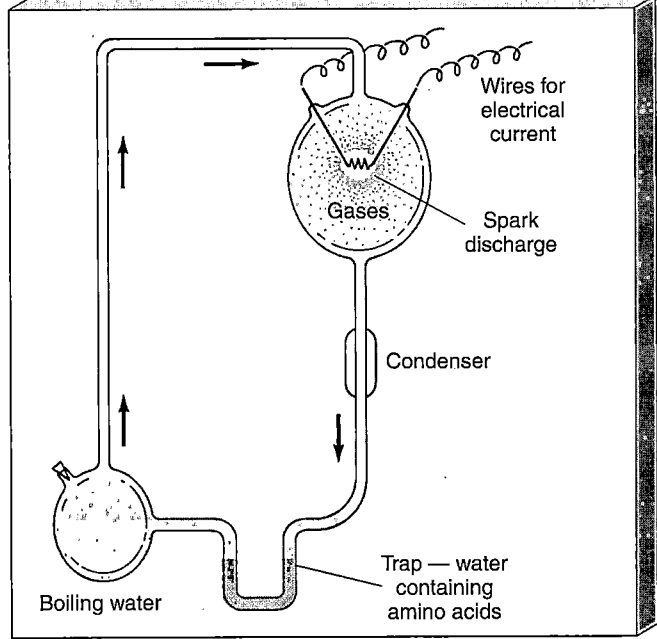

Figure 28.2 Urey-Miller Experiment This chemical apparatus is designed to synthesize complex biochemical molecules by energizing a mixture of simple chemicals. A mixture of gases (ammonia, methane, carbon dioxide, water vapor) is placed in the upper bulb to simulate the primordial Earth atmosphere and then energized by spark-discharge electrodes. After about a week, amino acids and other complex molecules are found in the trap at the bottom, which simulates the primordial oceans into which heavy molecules produced in the overlying atmosphere would have fallen.

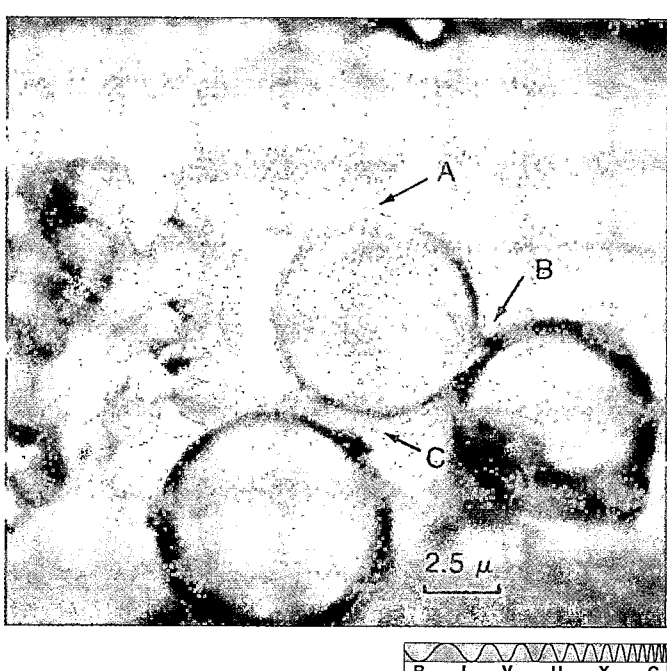

Figure 28.3 Chemical Evolution These carbon-rich, proteinlike droplets display the clustering of as many as a billion amino-acid molecules in a liquid. Droplets can "grow," and parts of droplets can separate from the "parent" to become new individual droplets (as at A, B, C). The scale of 2.5 microns noted here is 1/4000 of a centimeter. *(S. Fox)*

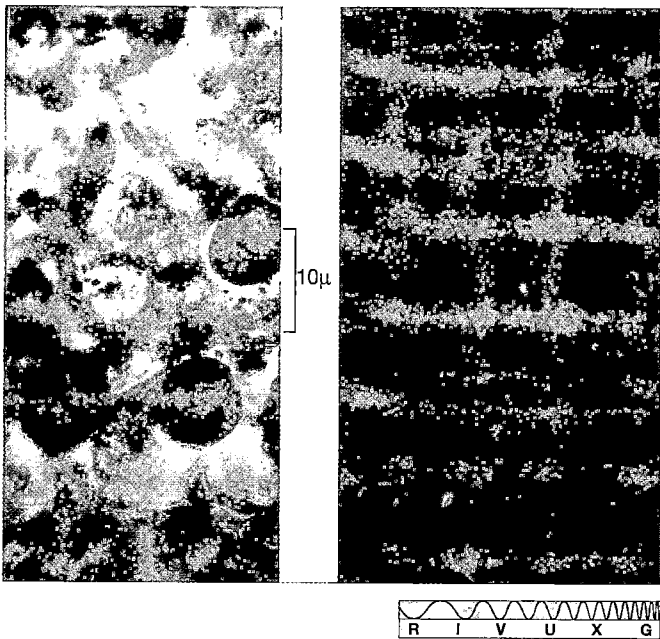

Figure 28.4 Primitive Cells The photograph on the left, taken through a microscope, shows a fossilized organism found in sediments radioactively dated as 2 billion years old. This primitive system possesses concentric spheres or walls connected by smaller spheroids. The roundish fossils here measure about a thousandth of a centimeter. The photograph on the right, also taken through a microscope and on approximately the same scale, displays modern blue-green algae. *(E. Barghoorn)*

AN INTERSTELLAR ORIGIN?

Recently, a dissenting view has emerged. Some scientists have argued that Earth's primitive atmosphere might *not* in fact have been a particularly suitable environment for the production of complex molecules. Instead, they say, there may not have been sufficient energy available to power the chemical reactions, and the early atmosphere may not have contained enough raw material for the reactions to have become important in any case. These researchers suggest that much, if not all, of the organic material that combined to form the first living cells was produced in *interstellar space* and subsequently arrived on Earth in the form of comets, interplanetary dust, and meteors that did not burn up during their descent through the atmosphere.

Several pieces of evidence support this idea. Interstellar molecular clouds are known to contain complex molecules—indeed, there have even been reports (still unconfirmed) of at least one amino acid (glycine) in interstellar space. ∞ (Sec. 18.5) To test this hypothesis, NASA researchers have carried out their own version of the Urey-Miller experiment, in which they exposed an icy mixture of water, methanol, ammonia, and carbon monoxide—representative of many interstellar grains—to ultraviolet radiation to simulate the energy from a nearby newborn star. As shown in Figure 28.5, when they later

placed the irradiated ice in water and examined the results, they found that it had formed droplets, surrounded by membranes, containing complex organic molecules. As with the droplets found in earlier experiments, no amino acids, proteins, or DNA were found in the mix, but the results, repeated numerous times, clearly show that even the harsh, cold vacuum of interstellar space can be a suitable medium in which complex molecules and primitive cellular structure can form.

As described in Chapter 15, these icy interstellar grains are believed to have formed the comets in our own solar system. ∞ (Sec. 15.3) Large amounts of organic material were detected on comet Halley by space probes when Halley last visited the inner solar system, and similarly complex molecules have been observed on many other well-studied comets, such as Hale-Bopp. ∞ (Sec. 14.2, *Discovery 14-2*) Also as discussed in Chapter 15, there is reason to believe that comet impacts were responsible for most of Earth's water, and it is perhaps a small step to imagining that this water already contained the building blocks for life. ∞ (Sec. 15.3)

In addition, a small fraction of the meteorites that survive the plunge to Earth's surface—including perhaps the controversial "Martian Meteorite" discussed in Chapter 10—contain organic compounds. ∞ *(Discovery 10-2)* The

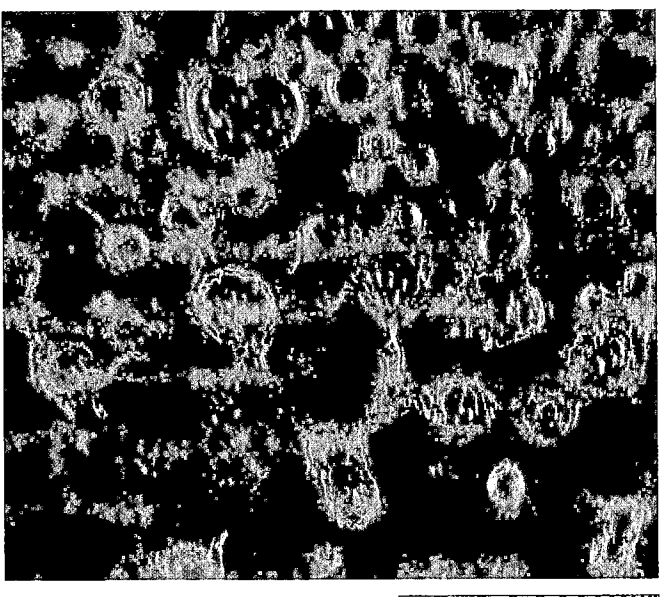

Figure 28.5 Interstellar Globules These oily, hollow droplets rich in organic molecules were made by exposing a freezing mixture of primordial matter to harsh ultraviolet radiation. The larger ones span about 10 microns across and, when immersed in water, show cell-like membrane structure. Although they are not alive, they bolster the idea that life on Earth could have come from space. (NASA)

Murchison meteorite (Figure 28.6), which fell near Murchison, Australia, in 1969, is a particularly well-studied example. Located soon after crashing to the ground, this meteorite has been shown to contain 22 of the amino acids normally found in living cells, although the detailed structures of these molecules indicate potentially important differences between those found in space and those found on Earth. At the very least, though, these discoveries argue that such molecules can form in an interplanetary or interstellar environment, and that they could have reached Earth's surface unscathed after their fiery descent.

Thus, the hypothesis that organic matter is constantly raining down on Earth from space in the form of interplanetary debris is quite plausible. However, whether or not this was the *primary* means by which complex molecules first appeared in Earth's oceans remains unclear.

DIVERSITY AND CULTURE

However the basic materials appeared on Earth, we know that life *did* appear. The fossil record chronicles how life on Earth became widespread and diversified over the course of time. The study of fossil remains shows the initial appearance more than 3.5 billion years ago of simple one-celled organisms such as blue-green algae. These were followed about 2 billion years ago by more complex one-celled creatures, like the amoeba. Multicellular organisms

such as sponges did not appear until about 1 billion years ago, after which there flourished a wide variety of increasingly complex organisms—insects, reptiles, and mammals.

The fossil record leaves no doubt that biological organisms have changed over time—all scientists accept the reality of *biological evolution*. As conditions on Earth shifted and Earth's surface evolved, those organisms that could best take advantage of their new surroundings succeeded and thrived—often at the expense of those organisms that could not make the necessary adjustments and consequently became extinct. What led to these changes? Chance. An organism that happened to have a certain useful genetically determined trait—for example, the ability to run faster, climb higher, or even hide more easily—would find itself with the upper hand in a particular environment. This organism was therefore more likely to reproduce successfully, and its advantageous characteristic would then be more likely to be passed on to the next generation. The evolution of the rich variety of life on our planet, including human beings, occurred as chance mutations—changes in genetic structure—led to changes in organisms over millions of years.

What about the development of intelligence? Many anthropologists believe that, like any other highly advantageous trait, intelligence *is* strongly favored by natural selection. As humans learned about fire, tools, and agriculture, the brain became more and more elaborate. The social cooperation that went with coordinated hunting efforts was another important competitive advantage that developed as brain size increased. Perhaps most important of all was the development of language. Indeed, some anthropologists have gone so far as to suggest that human intelligence

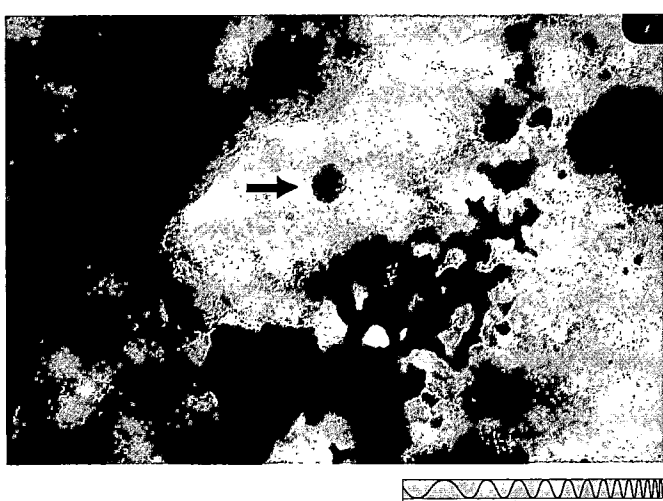

Figure 28.6 Murchison Meteorite The Murchison meteorite contains relatively large amounts of amino acids and other organic material, indicating that chemical evolution of some sort has occurred beyond our own planet. In this magnified view of a meteorite fragment, the arrow points to a microscopic sphere of organic matter. (NASA)

is human language. Through language, individuals could signal one another while hunting food or seeking protection. Even more importantly, now our ancestors could share ideas as well as food and shelter. Experience, stored in the brain as memory, could be passed down from generation to generation. A new kind of evolution had begun, namely, *cultural evolution*, the changes in the ideas and behavior of society. Our more recent ancestors have created, within only the past 10,000 years or so, the entirety of human civilization.

To put all this into historical perspective, let's imagine the entire lifetime of Earth to be 46 years rather than 4.6 billion years. We have no reliable record of the first decade of our planet's existence. Life originated at least 35 years ago, when Earth was about 10 years old. Our planet's middle age is largely a mystery, although we can be sure that life continued to evolve and that generations of mountain chains and oceanic trenches came and went. Not until about six years ago did abundant life flourish throughout Earth's oceans. Life came ashore about four years ago, and plants and animals mastered the land only about two years ago. Dinosaurs reached their peak about one year ago, only to die suddenly about four months later. ⃟ (*Discovery 14-1*) Humanlike apes changed into apelike humans only last week, and the latest ice ages occurred only a few days ago. *Homo sapiens*—our species—did not emerge until about four hours ago. Agriculture was invented within the last hour, and the Renaissance—along with all of modern science—is just three minutes old!

✅ Concept Check

◻ Has chemical evolution been verified in the laboratory?

28.2 Life in the Solar System

2️⃣ Simple one-celled life forms reigned supreme on Earth for most of our planet's history. It took time—a great deal of time—for life to emerge from the oceans, to evolve into simple plants, to continue to evolve into complex animals, and to develop intelligence, culture, and technology. Have those (or similar) events occurred elsewhere in the universe? Let's try to assess what little evidence we have on the subject.

LIFE AS WE KNOW IT

"Life as we know it" is generally taken to mean carbon-based life that originated in a liquid water environment—in other words, life on Earth. Might such life exist elsewhere in our solar system?

The Moon and Mercury lack liquid water, protective atmospheres, and magnetic fields, and so these two bodies are subjected to fierce bombardment by solar ultraviolet

radiation, the solar wind, meteoroids, and cosmic rays. Simple molecules could not possibly survive in such hostile environments. Venus, on the other hand, has far too much protective atmosphere! Its dense, dry, scorchingly hot atmospheric blanket effectively rules it out as a possible abode for life, at least like us.

The jovian planets have no solid surfaces (although some researchers have suggested that life might have evolved in their atmospheres), and Pluto and most of the moons of the outer planets are too cold. However, the possibility of liquid water below Europa's icy surface has refueled speculation about the development of life there, making this moon of Jupiter a prime candidate for future exploration. ⃟ (Sec. 11.5) For now at least, Europa is high on the priority list for missions by both NASA and the European Space Agency. Saturn's moon Titan, with its atmosphere of methane, ammonia, and nitrogen, and possibly with some liquid on its surface, is conceivably a site where life might have arisen, although the results of the 1980 *Voyager 1* flyby suggest that Titan's frigid surface conditions are inhospitable to anything familiar to us. ⃟ (Sec. 12.5)

Still, the planet most likely to harbor life (or to have harbored it in the past) seems to be Mars. This planet seems harsh by Earth standards—liquid water is scarce, the atmosphere is thin, and the lack of magnetism and an ozone layer allows solar high-energy particles and ultraviolet radiation to reach the surface unabated. But the Martian atmosphere was thicker, and the surface warmer and much wetter in the past. ⃟ (Secs. 10.4, 10.5) In the hope that life might once have evolved on Mars as it did on Earth, the *Viking* lander carried a television camera to seek fossilized remnants of large plants or animals. No fossils of any kind were seen. *Mars Pathfinder* also surveyed part of the Martian surface, again without finding any evidence of present or past life on Mars, although the apparent discovery of evidence of recent running water raises some intriguing questions. ⃟ (Sec. 10.4)

The *Viking* landers scooped up Martian soil (Figure 28.7) and tested for life by conducting chemical experiments designed to detect the waste gases and other products of metabolic activity, but no unambiguous evidence of Martian life has emerged. ⃟ (*Discovery 10-2*) Of course, we might argue that the landers touched down on the safest Martian terrain, not in the most interesting regions, such as near the moist polar caps.

Some scientists have suggested that a different type of biology may be operating on the Martian surface. They suggest that Martian microbes capable of eating and digesting oxygen-rich compounds in the Martian soil could also explain the *Viking* results. This speculation would be greatly strengthened if recent announcements of fossilized bacteria in meteorites originating on Mars were confirmed (although it seems that the weight of scientific opinion is currently running against that interpretation of the data). ⃟ (*Discovery 10-2*) The consensus among biologists and chemists today is that Mars does not house any life similar

Figure 28.7 Search for Martian Life Several trenches were dug by the *Viking* robots, such as *Viking 2* seen here on Mars' Utopia Planitia. Soil samples were scooped up and taken inside the robot, where instruments tested them for chemical composition and any signs of life. *(NASA)*

to that on Earth, but a solid verdict regarding life on Mars will likely not be reached until we have thoroughly explored our intriguing neighbor.

When considering the emergence of life under adversity, however, we should perhaps not be too quick to rule out an environment based solely on its extreme properties. Figure 28.8 shows a very hostile environment on Earth, one where life nonetheless thrives under conditions quite unlike anything on our planet's surface. These are sites of undersea volcanic activity where hydrothermal vents spill forth boiling-hot water from vertical tubes a few meters tall. Rich in sulfur and poor in oxygen, the upwelling water feeds "extremeophilic" life by a process known as chemosynthesis—an analog of photosynthesis, yet one that operates in total darkness. Such underground hot springs might conceivably exist on alien worlds, raising the possibility of life forms with much greater diversity over a much wider range of conditions than those known to us on Earth.

ALTERNATIVE BIOCHEMISTRIES

Conceivably, some types of biology might be so different from life on Earth that we do not know how to test for them. What other biologies might exist?

Some scientists have pointed out that the abundant element silicon has chemical properties somewhat similar to those of carbon and have suggested it as a possible alternative to carbon as the basis for living organisms. Ammonia (made of the common elements hydrogen and nitrogen) is sometimes put forward as a possible liquid medium in which life might develop, at least on a planet cold enough

for ammonia to exist in the liquid state. Together or separately, these alternatives would surely give rise to organisms with radically different biochemistries from those we know on Earth. Conceivably, we might have difficulty even recognizing these organisms as alive.

Although the possibility of such alien life forms is a fascinating scientific problem, most biologists would argue that chemistry based on carbon and water is the one most likely to give rise to life. Carbon's flexible chemistry and water's wide liquid temperature range are just what are needed for life to develop and thrive. Silicon and ammonia seem unlikely to fare as well as bases for advanced life forms. Silicon's chemical bonds are weaker than those of carbon and may not be able to form complex molecules—an apparently essential aspect of carbon-based life. Also, the colder the environment, the less energy there is to drive biological processes. The low temperatures necessary for ammonia to remain liquid might inhibit or even completely prevent the chemical reactions leading to the equivalent of amino acids and nucleotide bases.

Still, we must admit that we know next to nothing about noncarbon, nonwater biochemistries, for the very good reason that there are no examples of them to study experimentally. We can speculate about alien life forms and try to make general statements about their characteristics, but we can say little of substance about them.

☑ Concept Check

■ Which solar-system bodies (other than Earth) are the leading candidates in the search for extraterrestrial life?

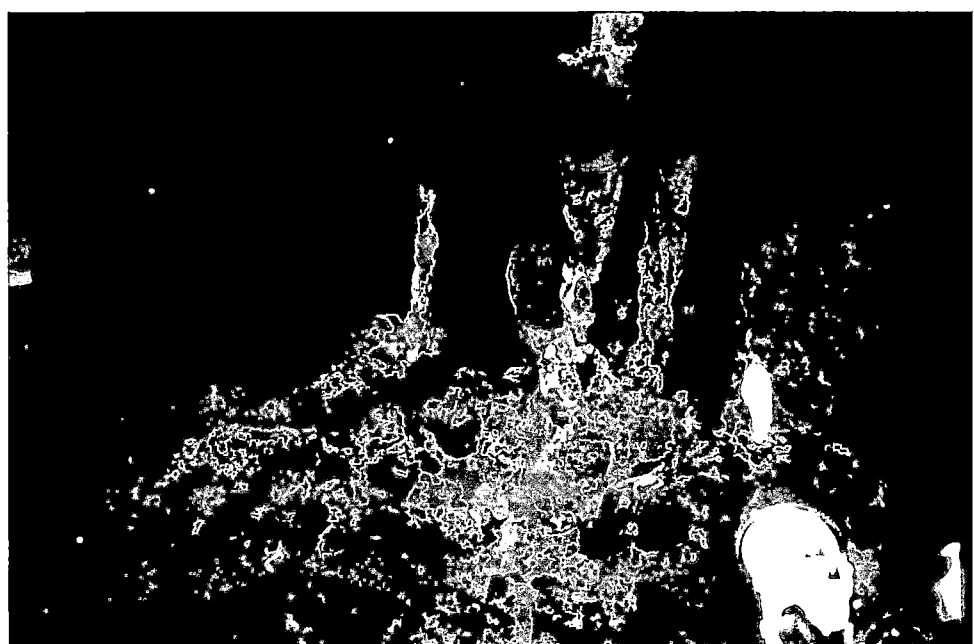

Figure 28.8 Hydrothermal Vents A small, two-person submarine (the *Alvin*, partly seen at bottom right) took this picture of a hot spring, or "black smoker"—one of many along the mid-ocean ridge in the eastern Pacific Ocean. As hot water rich in sulfur pours out of the top of the vent's tube (near center), black clouds billow forth, providing a strange environment for many life forms thriving near the vent. *(WHOI)*

28.3 Intelligent Life in the Galaxy

With humans apparently the only intelligent life in the solar system, we must broaden our search for extraterrestrial intelligence to other stars, perhaps even other galaxies. At such distances, though, we have little hope of actually detecting life with current equipment. Instead, we must ask: How likely is it that life in any form—carbon-based, silicon-based, water-based, ammonia-based, or something we cannot even dream of—exists? Let's look at some numbers to develop estimates of the probability of life elsewhere in the universe.

THE DRAKE EQUATION

An early approach to this statistical problem is known as the **Drake equation**, after the U.S. astronomer who pioneered this analysis (see below).

Several of the terms in this formula are largely a matter of opinion. We do not have nearly enough information to determine—even approximately—every term in the equation, so the Drake equation cannot give us a hard-and-fast answer. Its real value is that it subdivides a large and difficult question into smaller pieces that we can attempt to answer separately. It provides the framework within which the problem can be addressed and parcels out the responsibility for the final solution among many different scientific disciplines. Figure 28.9 illustrates how, as our requirements become more and more stringent, only a small fraction of star systems in the Milky Way are likely to generate the advanced qualities specified by the combination of terms on the right-hand side of the equation.

Let's examine the terms in the equation one by one and make some educated guesses about their values. Bear in mind, though, that if you ask two scientists for their best estimates of any given term, you will likely get two very different answers!

RATE OF STAR FORMATION

We can estimate the average number of stars forming each year in the Galaxy simply by noting that at least 100 billion stars now shine in the Milky Way. Dividing this number by

| number of technological, intelligent civilizations now present | = | rate of star formation, averaged over the lifetime of the Galaxy | × | fraction of stars having planetary systems | × | average number of habitable planets within those planetary systems | × | fraction of those habitable planets on which life arises | × | fraction of those life-bearing planets on which intelligence evolves | × | fraction of those intelligent-life planets that develop technological society | × | average lifetime of a technologically competent civilization. |

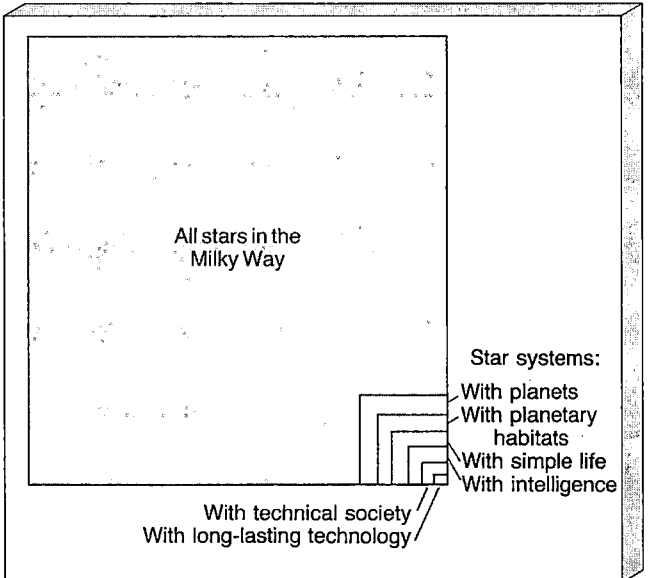

Figure 28.9 Drake Equation Of all the star systems in our Milky Way (represented by the largest box), progressively fewer and fewer have each of the qualities typical of a long-lasting technological society (represented by the smallest box at the lower right corner).

the 10-billion-year lifetime of the Galaxy, we obtain a formation rate of 10 stars per year. This may be an overestimate because we think that fewer stars are forming now than formed at earlier epochs of the Galaxy, when more interstellar gas was available. However, we do know that stars are forming today, and our estimate does not include stars that formed in the past and have since died, so our value of 10 stars per year is probably reasonable when averaged over the lifetime of the Milky Way.

FRACTION OF STARS HAVING PLANETARY SYSTEMS

Many astronomers believe that planet formation is a natural result of the star-formation process. If the condensation theory (Chapter 15) or some variant of it is correct, and if there is nothing special about our Sun, as we have argued throughout this book, then we would expect many stars to have at least one planet. ∞ (Sec. 15.2) Indeed, as we have seen, increasingly sophisticated observations indicate the presence of disks around young stars. Could these disks be protosolar systems? The condensation theory suggests they are, and the short (theoretical) lifetimes of disks imply the existence of many planet-forming systems in the neighborhood of the Sun.

No planets like our own have yet been *seen* orbiting any other star.* The light reflected by an Earthlike planet

* *Recall from Chapter 22 that Earth-mass planets have been observed orbiting some neutron stars. However, as we saw in that chapter, the formation of those planets was very different from that of Earth and the solar system. ∞ (Sec. 22.3) They are not thought to be likely candidates for the emergence of life.*

circling even the closest star would be too faint to detect even with the best equipment. The light would be lost in the glare of the parent star. Large orbiting telescopes may soon be able to detect Jupiter-sized planets orbiting the nearest stars, but even those huge planets will be barely visible. However, as described in Chapter 15, there is now overwhelming evidence for planets orbiting other stars. ∞ (Sec. 15.5) The planets found so far are Jupiter-sized rather than Earth-sized, and generally have rather eccentric orbits, but astronomers are confident that an Earthlike planet (on an Earthlike orbit) will one day be detected, and plans to build the necessary equipment are well underway.

Accepting the condensation theory and its consequences, and without being either too conservative or naïvely optimistic, we assign a value near 1 to this term— that is, we believe that nearly all stars form with planetary systems of some sort.

NUMBER OF HABITABLE PLANETS PER PLANETARY SYSTEM

Temperature, more than any other single factor, determines the feasibility of life on a given planet. The surface temperature of a planet depends on two things: the planet's distance from its parent star and the thickness of its atmosphere. Planets with a nearby parent star (but not too close) and some atmosphere (though not too thick) should be reasonably warm, like Earth or Mars. Planets far from the star and with no atmosphere, like Pluto, will surely be cold by our standards. And planets too close to the star and with a thick atmosphere, like Venus, will be very hot indeed.

Figure 28.10 illustrates how a three-dimensional zone of "comfortable" temperatures—often called a *habitable zone*—surrounds every star. It represents the range of distances within which a planet of mass and composition similar to those of Earth would have a surface temperature

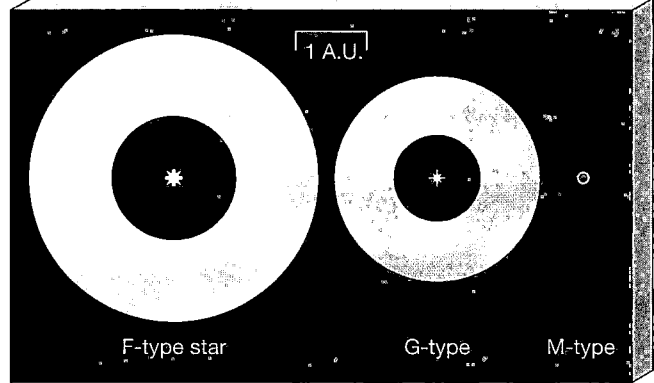

Figure 28.10 Habitable Zones The extent of the habitable zone is much larger around a hot star than around a cool one. For a star like the Sun (a G-type star), the zone extends from about 0.85 A.U. to 2.0 A.U. For an F-type star, the range is 1.2–2.8 A.U. For a faint M-type star only planets orbiting between about 0.02 and 0.06 A.U. would be habitable.

between the freezing and boiling points of water. (Our Earth-based bias is plainly evident here!) The hotter the star, the larger this zone. For example, A- and F-type stars have a rather large habitable zone, but the size of the zone diminishes rapidly as we proceed through G-, K-, and M-type stars. O- and B-type stars are not considered here because they are not expected to last long enough for life to develop, even if they do have planets. M-type stars, despite their large numbers, have such small habitable zones and, in any case, are thought to be prone to such violent surface activity, that they are not generally considered likely hosts to life-bearing planets. ∞ (Sec. 17.8)

Three planets—Venus, Earth, and Mars—reside within the habitable zone surrounding our Sun. However, Venus is too hot because of its exceptionally thick atmosphere and proximity to the Sun. ∞ (Sec. 9.5) Mars is a little too cold, because its atmosphere is much thinner than Earth's and it lies too far from the Sun. But if Venus had Mars's thin atmosphere, and if Mars had Venus's thick atmosphere, both of these nearby planets might conceivably have surface conditions resembling those on Earth.

To estimate the number of habitable planets per planetary system, we first take inventory of how many stars of each type shine in our Galaxy and calculate the sizes of their habitable zones. Then we eliminate binary-star systems because a planet's orbit within the habitable zone of a binary would likely be unstable, as illustrated in Figure 28.11. Given the known properties of binaries in our Galaxy, habitable planetary orbits would probably be unstable in most cases, so there would not be time for life to develop. In addition, we must eliminate almost all of the 10 percent of surveyed stars around which planets have been observed. The large "jovian" planets seen in most cases have eccentric orbits that would destabilize the motion of any inner "terrestrial" world, either ejecting it completely from the system or making conditions so extreme that the chances for the development of life are severely reduced.

However, as pointed out in Chapter 15, we still have insufficient data about most stars to make any definitive statement about their planetary systems. Taking all these factors into account, we assign a value of $\frac{1}{10}$ to this term in our equation. In other words we believe that, on average, there is one potentially habitable planet for every 10 planetary systems that might exist in our Galaxy. Single F-, G-, and K-type stars are the best candidates.

FRACTION OF HABITABLE PLANETS ON WHICH LIFE ARISES

The number of possible combinations of atoms is incredibly large. If the chemical reactions that led to the complex molecules that make up living organisms occurred completely at random, then it is extremely unlikely that those molecules could have formed at all. In that case, life is extraordinarily rare, this term is close to zero, and we are probably alone in the Galaxy, perhaps even in the entire universe.

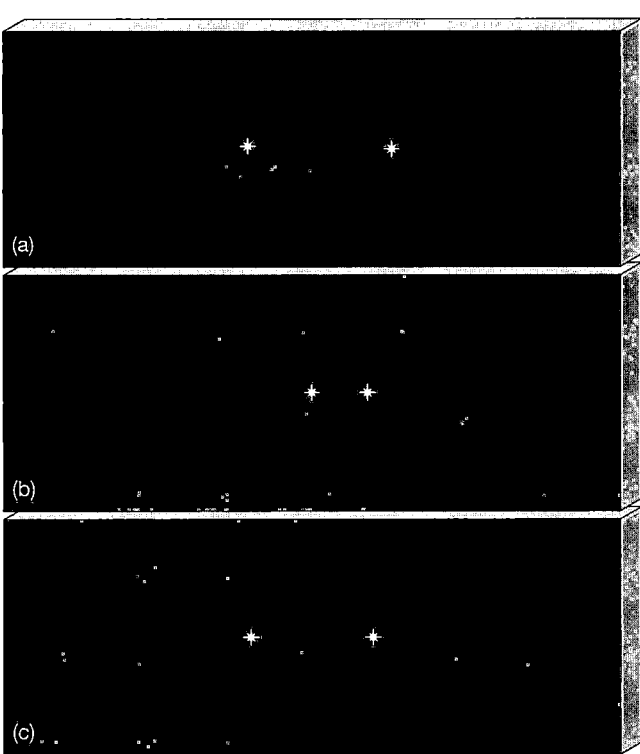

Figure 28.11 Binary-Star Planets In binary-star systems, planets are restricted to only a few kinds of orbits that are gravitationally stable. The orbit in case (a) is stable only if the planet lies very close to its "parent" star, so the gravity of the other star is negligible. Case (b) shows a planet circulating at a great distance about both stars in an elliptical orbit. This orbit is stable only if it lies far from both stars. Another possible, but unstable path, (c) interweaves between the two stars in a figure-eight pattern.

However, laboratory experiments (like the Urey-Miller experiment described earlier) seem to suggest that certain chemical combinations are strongly favored over others—that is, the reactions are not random. Of the billions upon billions of basic organic groupings that could possibly occur on Earth from the random combination of all sorts of simple atoms and molecules, only about 1500 actually do occur. Furthermore, these 1500 organic groups of terrestrial biology are made from only about 50 simple "building blocks" (including the amino acids and nucleotide bases mentioned earlier). This suggests that molecules critical to life may not be assembled by pure chance. Apparently, additional factors are at work at the microscopic level. If a relatively small number of chemical "evolutionary tracks" are likely to exist, then the formation of complex molecules—and hence, we assume, life—becomes much more likely, given sufficient time.

To assign a very low value to this term in the equation is to believe that life arises randomly and rarely. To assign a value close to 1 is to believe that life is inevitable, given the proper ingredients, a suitable environment, and a long enough period of time. No simple experiment can distinguish between these extreme alternatives, and there is little or no middle ground. To many researchers, the discovery of life (past or present) on Mars, Europa, Titan, or some other object in our solar system would convert the appearance of life from an unlikely miracle to a virtual certainty throughout the Galaxy. We will take the optimistic view and adopt a value of 1.

FRACTION OF LIFE-BEARING PLANETS ON WHICH INTELLIGENCE ARISES

As with the evolution of life, the appearance of a well-developed brain is a very unlikely event if only random chance is involved. However, biological evolution through natural selection is a mechanism that generates apparently highly improbable results by singling out and refining useful characteristics. Organisms that profitably use adaptations can develop more complex behavior, and complex behavior provides organisms with the *variety* of choices needed for more advanced development.

One school of thought maintains that, given enough time, intelligence is inevitable. In this view, assuming that natural selection is a universal phenomenon, at least one organism on a planet will always rise to the level of "intelligent life." If this is correct, then the fifth term in the Drake equation equals or nearly equals 1.

Others argue that there is only one known case of intelligence, and that case is life on Earth. For 2.5 billion years—from the start of life about 3.5 billion years ago to the first appearance of multicellular organisms about 1 billion years ago—life did not advance beyond the one-celled stage. Life remained simple and dumb, but it survived. If this latter view is correct, then the fifth term in our equation is very small, and we are faced with the depressing prospect that humans may be the smartest form of life anywhere in the Galaxy. As with the previous term, we will be optimistic and simply adopt a value of 1 here.

FRACTION OF PLANETS ON WHICH INTELLIGENT LIFE DEVELOPS AND USES TECHNOLOGY

To evaluate the sixth term of our equation, we need to estimate the probability that intelligent life eventually develops technological competence. Should the rise of technology be inevitable, this term is close to 1, given long enough periods of time. If it is not inevitable—if intelligent life can somehow "avoid" developing technology—then this term could be much less than 1. The latter possibility envisions a universe possibly teeming with intelligent civilizations, but very few among them ever becoming technologically competent. Perhaps only one managed it—ours.

Again, it is difficult to decide conclusively between these two views. We don't know how many prehistoric Earth cultures failed to develop technology, or rejected its

use. We do know that the roots of our present civilization arose independently at several different places on Earth, including Mesopotamia, India, China, Egypt, Mexico, and Peru. Because so many of these ancient cultures originated at about the same time, it is tempting to conclude that the chances are good that some sort of technological society will inevitably develop, given some basic intelligence and enough time.

If technology is inevitable, then why haven't other life forms on Earth also found it useful? Possibly the competitive edge given by intellectual and technological skills to humans, the first species to develop them, allowed us to dominate so rapidly that other species—gorillas and chimpanzees, for example—simply haven't had time to catch up. The fact that only one technological society exists on Earth does not imply that the sixth term in our Drake equation must be very much less than 1. On the contrary, it is precisely because *some* species will probably always fill the niche of technological intelligence that we will take this term to be close to 1.

AVERAGE LIFETIME OF A TECHNOLOGICAL CIVILIZATION

The reliability of the estimate of each term in the Drake equation declines markedly from left to right. For example, our knowledge of astronomy enables us to make a reasonably good stab at the first term, namely, the rate of star formation in our Galaxy, but it is much harder to evaluate some of the later terms, such as the fraction of life-bearing planets that eventually develop intelligence. The last term on the right-hand side of the equation, the longevity of technological civilizations, is totally unknown. There is only one known example of such a civilization—humans on planet Earth. Our own civilization has survived in its "technological" state for only about 100 years, and how long we will be around before a natural or human-made catastrophe ends it all is impossible to tell. ∞ (*Discovery 14-1*)

One thing is certain: If the correct value for *any one* *term* in the equation is very small, then few technological civilizations now exist in the Galaxy. If the pessimistic view of the development of life or of intelligence is correct, then we are unique, and that is the end of our story. However, if both life and intelligence are inevitable consequences of chemical and biological evolution, as many scientists believe, and if intelligent life always becomes technological, then we can plug the higher, more optimistic values into the Drake equation. In that case, combining our estimates for the other six terms (and noting that $10 \times 1 \times \frac{1}{10} \times 1 \times 1 \times 1 = 1$), we can say:

number of technological, intelligent civilizations now present in the Milky Way Galaxy	=	average lifetime of a technologically competent civilization in years.

Thus, if civilizations typically survive for 1000 years, there should be 1000 of them currently in existence scattered throughout the Galaxy. If they live for a million years, on average, we would expect there to be a million advanced civilizations in the Milky Way, and so on.

Concept Check

■ How does the Drake equation assist astronomers in refining their search for extraterrestrial life?

28.4 The Search for Extraterrestrial Intelligence

Let us continue our optimistic assessment of the prospects for life and assume that civilizations enjoy a long stay on their parent planet once their initial technological "teething problems" are past. In that case, intelligent, technological, and perhaps also communicative cultures are likely to be plentiful in the Galaxy. How might we become aware of their existence?

MEETING OUR NEIGHBORS

For definiteness, let's assume that the average lifetime of a technological civilization is 1 million years—only one percent of the reign of the dinosaurs, but 100 times longer than human civilization has survived thus far. Given the size and shape of our Galaxy and the known distribution of stars in the Galactic disk, we can then estimate the average *distance* between these civilizations to be some 30 pc, or about 100 light-years. Thus, any two-way communication with our neighbors—using signals traveling at or below the speed of light—will take at least 200 years (100 years for the message to reach the planet and another 100 years for the reply to travel back to us).

One obvious way to search for extraterrestrial life would be to develop the capability to travel far outside our solar system. However, this may never be a practical possibility. At a speed of 50 km/s, the speed of the current fastest space probes, the round trip to even the nearest Sunlike star, Alpha Centauri, would take about 50,000 years. The journey to the nearest technological neighbor (assuming a distance of 30 pc) and back would take 600,000 years—almost the entire lifetime of our species! Interstellar travel at these speeds is clearly not feasible. Speeding up our ships to near the speed of light would reduce the travel time, but this is far beyond our present technology.

Actually, our civilization has already launched some interstellar probes, although they have no specific stellar destination. Figure 28.12 is a reproduction of a plaque mounted on board the *Pioneer 10* spacecraft launched in

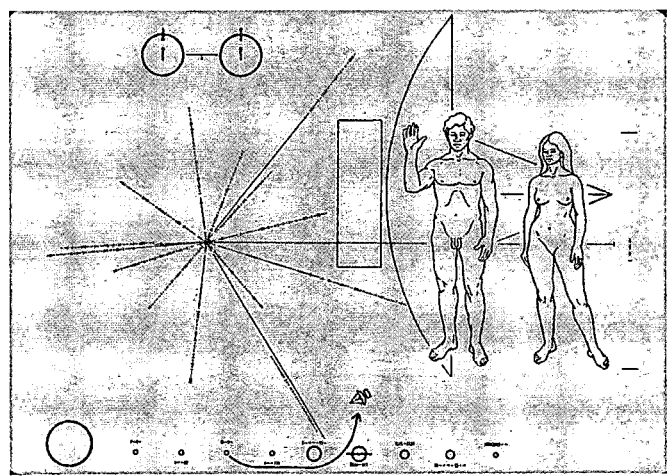

Figure 28.12 Pioneer 10 Plaque A replica of a plaque mounted on board the *Pioneer 10* spacecraft. The important features of the plaque include a scale drawing of the spacecraft, a man, and a woman; a diagram of the hydrogen atom undergoing a change in energy (top left); a starburst pattern representing various pulsars and the frequencies of their radio waves that can be used to estimate when the craft was launched (middle left); and a depiction of the solar system, showing that the spacecraft departed the third planet from the Sun and passed the fifth planet on its way into outer space (bottom). All the drawings have computer-coded (binary) markings from which actual sizes, distances, and times can be derived. *(C. Sagan)*

the mid-1970s and now well beyond the orbit of Pluto, on its way out of the solar system. Similar information was also included aboard the *Voyager* probes launched in 1978. Although these spacecraft would be incapable of reporting back to Earth the news that they had encountered an alien culture, scientists hope that the civilization on the other end would be able to unravel most of its contents using the universal language of mathematics. The caption to Figure 28.12 notes how the aliens might discover from where and when the *Pioneer* and *Voyager* probes were originally launched.

Setting aside the many practical problems of establishing direct contact with extraterrestrials, some scientists have argued that it might not even be a particularly good idea. Our recent emergence as a technological civilization implies that we must be one of the least advanced technological intelligences in the entire Galaxy. Any other civilization that discovers us will almost surely be more advanced than us. Consequently, a healthy degree of caution is warranted. If extraterrestrials behave even remotely like human civilizations on Earth, then the most advanced aliens may naturally try to dominate all others. The behavior of the "advanced" European cultures toward the "primitive" races they encountered on their voyages of discovery in the seventeenth, eighteenth, and nineteenth centuries should serve as a clear warning of the possible undesirable consequences of contact. Of course, the aggressiveness of Earthlings may not apply to extraterrestrials, but given the history of the one intelligent species we know, the cautious approach may be in order.

RADIO COMMUNICATION

A cheaper, and much more practical, alternative is to try to make contact with extraterrestrials using only electromagnetic radiation, the fastest known means of transferring information from one place to another. Because light and other high-frequency radiation are heavily scattered while moving through dusty interstellar space, long-wavelength radio radiation seems to be the natural choice. We would not attempt to broadcast to all nearby candidate stars, however—that would be far too expensive and inefficient. Instead, radio telescopes on Earth would listen *passively* for radio signals emitted by other civilizations. Some preliminary searches of selected nearby stars are now under way, thus far without success.

In what direction should we aim our radio telescopes? The answer to this question at least is fairly easy. On the basis of our earlier reasoning, we should target all F-, G-, and K-type stars in our vicinity. But are extraterrestrials broadcasting radio signals? If they are not, this search technique will obviously fail. If they are, how do we distinguish their artificially generated radio signals from signals naturally emitted by interstellar gas clouds? At what frequency should we tune our receivers? This depends on whether the signals are produced deliberately or are simply "waste radiation" escaping from a planet.

Consider how Earth would look at radio wavelengths to extraterrestrials. Figure 28.13 shows the pattern of radio signals we emit into space. From the viewpoint of a distant observer, the spinning Earth emits a bright flash of radio radiation every few hours. In fact, Earth is now a more intense radio emitter than the Sun. The flashes result from the periodic rising and setting of hundreds of FM radio stations and television transmitters. Each station broad-

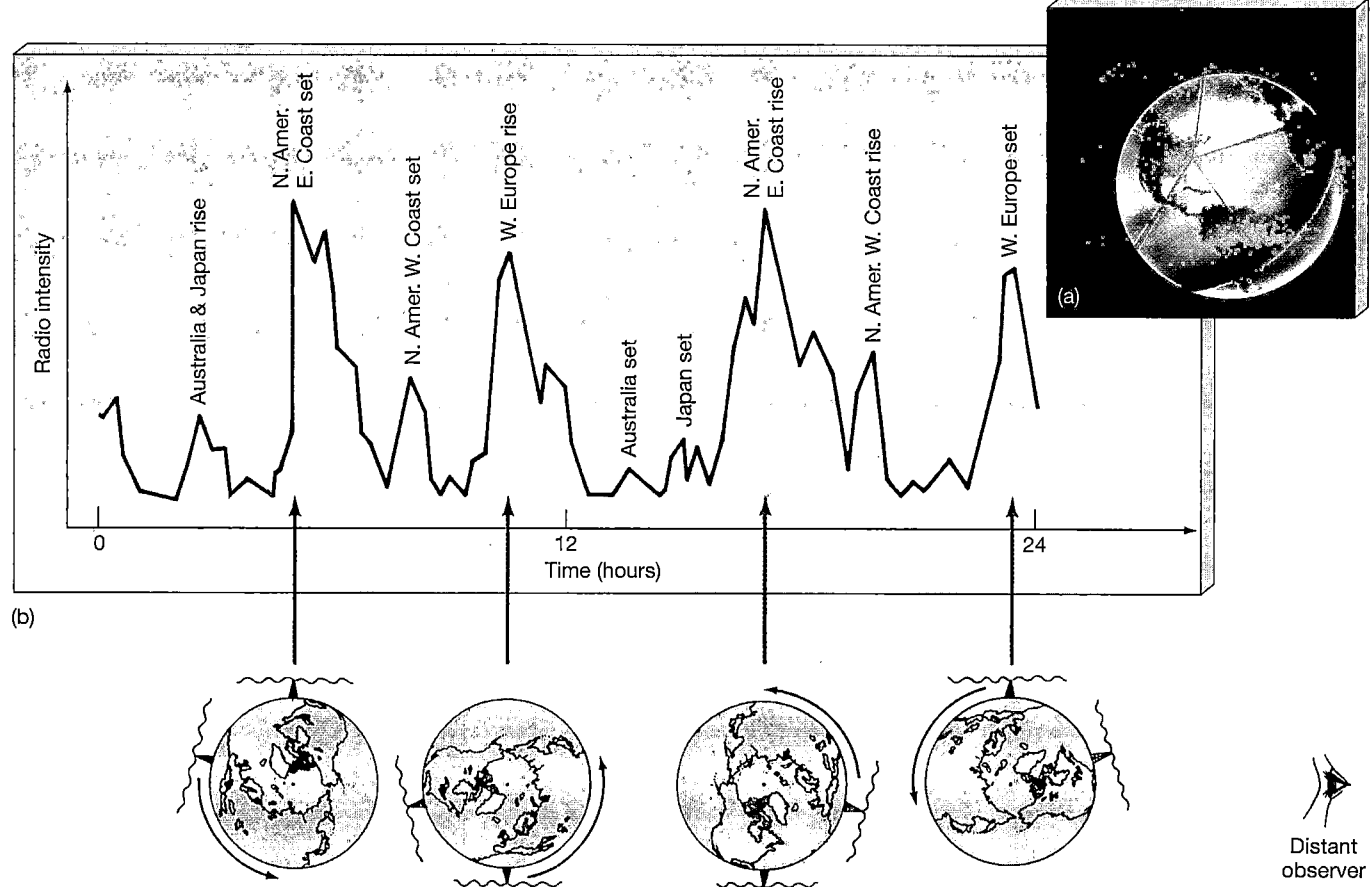

Figure 28.13 Earth's Radio Leakage Radio radiation now leaks from Earth into space because of the daily activities of our technological civilization. (a) FM radio and television transmitters broadcast their energy parallel to Earth's surface, so they send a great "sheet" of electromagnetic radiation into interstellar space, producing the strongest signal in any given direction when they happen to lie on Earth's horizon, as seen from that direction. (b) Because the great majority of transmitters are clustered in the eastern United States and western Europe, a distant observer would detect blasts of radiation from Earth as our planet rotates each day.

casts mostly parallel to Earth's surface, sending a great "sheet" of electromagnetic radiation into interstellar space, as illustrated in Figure 28.13(a). (The more common AM broadcasts are trapped below our ionosphere, so those signals never leave Earth.)

Because the great majority of these transmitters are clustered in the eastern United States and western Europe, a distant observer would detect blasts of radiation from Earth as our planet rotates each day (Figure 28.13b). This radiation races out into space, and has been doing so since the invention of these technologies nearly seven decades ago. Another civilization at least as advanced as ours might have constructed devices capable of detecting this radiation. If any sufficiently advanced (and sufficiently interested) civilization resides on a planet orbiting any of the thousand or so stars within about 65 light-years (20 pc) of Earth, then we have already broadcast our presence to them.

Of course, it may very well be that, having discovered cable, most civilizations' indiscriminate transmissions cease after a few decades. In that case, radio silence becomes the hallmark of intelligence, and we must find an alternate means of locating our neighbors.

THE WATER HOLE

Now let us suppose that a civilization has decided to assist searchers by actively broadcasting its presence to the rest of the Galaxy. At what frequency should we listen for such an extraterrestrial beacon? The electromagnetic spectrum is enormous; the radio domain alone is vast. To hope to detect a signal at some unknown radio frequency is like searching for a needle in a haystack. Are some frequencies more likely than others to carry alien transmissions?

Some basic arguments suggest that civilizations might communicate at a wavelength near 20 cm. As we saw in Chapter 18, the basic building blocks of the universe, namely, hydrogen atoms, naturally radiate at a wavelength of 21 cm. ⟨⟩ (Sec. 18.4) Also, one of the simplest molecules, hydroxyl (OH), radiates near 18 cm. Together, these two substances form water (H_2O). Arguing that water is likely to be the interaction medium for life anywhere, and that radio radiation travels through the disk of our Galaxy with the least absorption by interstellar gas and dust, some researchers have proposed that the interval between 18 and 21 cm is the best wavelength range for civilizations to transmit or monitor. Called the **water hole**, this radio interval might serve as an "oasis" where all advanced galactic civilizations would gather to conduct their electromagnetic business.

This water-hole frequency interval is only a guess, of course, but it is supported by other arguments as well. Figure 28.14 shows the water hole's location in the electromagnetic spectrum and plots the amount of natural emission from our Galaxy and from Earth's atmosphere. The 18- to 21-cm range lies within the quietest part of the spectrum, where the galactic "static" from stars and inter-

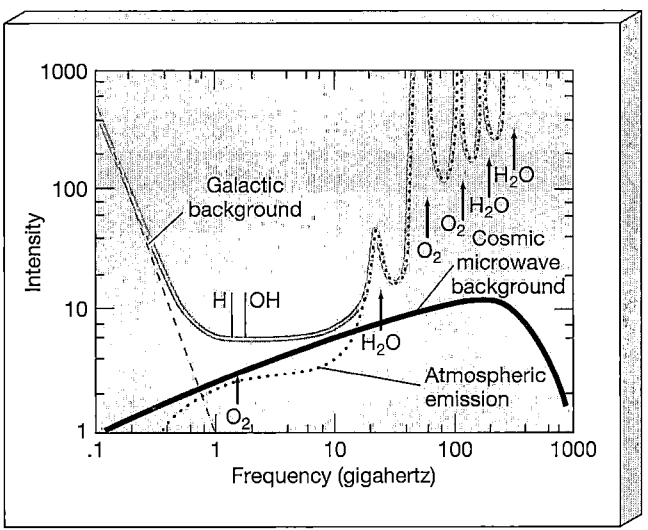

Figure 28.14 Water Hole The "water hole" is bounded by the natural emission frequencies of the hydrogen (H) atom (21-cm wavelength) and the hydroxyl (OH) molecule (18-cm wavelength). The topmost solid (blue) curve sums the natural emissions of our Galaxy (dashed line on left side of diagram) and Earth's atmosphere (dotted line on right side of diagram denoted by various chemical symbols). This sum is minimized near the water-hole frequencies. Perhaps all intelligent civilizations conduct their interstellar communications within this quiet "electromagnetic oasis." *(SETI Institute)*

stellar clouds happens to be minimized. Furthermore, the atmospheres of typical planets are also expected to interfere least at these wavelengths. Thus the water hole seems like a good choice for the frequency of an interstellar beacon, although we cannot be sure of this reasoning until contact is actually achieved.

A few radio searches are now in progress at frequencies in and around the water hole. One of the most sensitive and comprehensive projects in the ongoing search for extraterrestrial intelligence (known to many by its acronym SETI) was Project Phoenix, carried out during the late 1990s. Large radio antennas, such as that in Figure 28.15(a), were used to search millions of channels simultaneously in the 1–3 GHz spectrum. Actually, in these searches, computers do most of the "listening;" humans get involved only if the signals look intriguing. Figure 28.15(b) shows what a typical narrow-band, 1-Hz signal— a potential "signature" of an intelligent transmission— would look like on a computer monitor. However, this observation was merely a test to detect the weak, redshifted radio signal emitted by the *Pioneer 10* robot, now receding into the outer realm of our solar system—a sign of intelligence, but one that we put there. Nothing resembling an extraterrestrial signal has yet been detected.

The space surrounding all of us could be, right now, flooded with radio signals from extraterrestrial civilizations. If only we knew the proper direction and frequency, we might be able to make one of the most startling discoveries of all time. The result would likely provide whole new opportunities to study the cosmic evolution of energy, matter, and life throughout the universe.

☑️ **Concept Check**

■ Why do many researchers regard the water hole as a likely place to search for extraterrestrial signals?

(a)

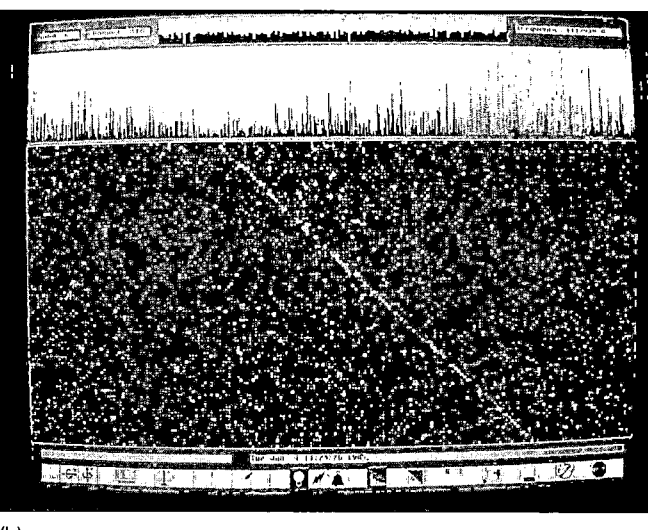
(b)

Figure 28.15 Project Phoenix (a) This large radio telescope at the National Radio Astronomy Observatory in Green Bank, West Virginia, was used by Project Phoenix during the 1990s to search for extraterrestrial intelligent signals. (b) This is a typical recording of such an alien signal—here, as a test, the Doppler-shifted signal detected from the *Pioneer 10* spacecraft, now well beyond the orbit of Pluto. *(NRAO; SETI Institute)*

Chapter Review

SUMMARY

The history of the universe can be divided into seven phases: particulate, galactic, stellar, planetary, chemical, biological, and cultural. **Cosmic evolution** (p. 740) is the continuous process that has led to the appearance of galaxies, stars, planets, and life on Earth.

Living organisms may be characterized by their ability to react to their environment, to grow by taking in nutrition from their surroundings, and to reproduce, passing along some of their own characteristics to their offspring.

Powered by natural energy sources, reactions between simple molecules in the oceans of the primitive Earth are believed to have led to the formation of **amino acids** (p. 742) and **nucleotide bases** (p. 742), the basic molecules of life. Alternatively, some complex molecules may have been formed in interstellar space and then delivered to Earth by meteors or comets.

Organisms that can best take advantage of their new surroundings succeed at the expense of those organisms that cannot make the necessary adjustments. Intelligence is strongly favored by natural selection.

The best hope for life beyond Earth in the solar system is the planet Mars, although no evidence for living organisms has been found. Jupiter's moon Europa and Saturn's Titan may also be possibilities, but conditions there are harsh by terrestrial standards.

The **Drake equation** (p. 747) provides a means of estimating the probability of intelligent life in the Galaxy. The astronomical terms in the equation are the galactic star-formation rate, the likelihood of planets, and the number of habitable planets. Chemical and biological terms are the probability that life appears and the probability that it subsequently develops intelligence. Cultural and political terms are the probability that intelligence leads to technology, and the lifetime of a civilization in the technological state. Taking an optimistic view of the development of life and intelligence leads to the conclusion that the total number of technologically competent civilizations in the Galaxy is approximately equal to the lifetime of a typical civilization expressed in years.

Even with optimistic assumptions, the distance to our nearest intelligent neighbor is likely to be many hundreds of parsecs. Space travel is not now a feasible means of searching for intelligent life. Current programs to discover extraterrestrial intelli-

gence involve scanning the electromagnetic spectrum for signals. So far, no intelligible broadcasts have been received. A technological civilization would probably "announce" itself to the universe by the radio and television signals it emits into space. Observed from afar, our planet would appear as a radio source with a 24-hour period, as different regions of the planet rise and set.

The **water hole** (p. 753) is a region in the radio range of the electromagnetic spectrum, near the 21-cm line of hydrogen and the 18-cm line of hydroxyl, where natural emissions from the Galaxy happen to be minimized. Many researchers regard this as the best part of the spectrum for communications purposes.

SELF-TEST: TRUE OR FALSE?

1. The assumptions of mediocrity argue in favor of the existence of extraterrestrial life.
2. The definition of life requires only that to be considered "alive," you must be able to reproduce.
3. The Urey-Miller experiment produced organic molecules from nonbiological molecules.
4. Organic molecules important for life could have reached Earth's surface via comets.
5. Organic molecules exist only on Earth.
6. Laboratory experiments have created living cells from nonbiological molecules.
7. For most of the history of Earth, life consisted of only single-celled life forms.
8. The *Viking* landers on Mars discovered microscopic evidence of life but found no large fossil evidence.

9. The rate of star formation in the Galaxy is reasonably well known.
10. As yet there is no direct evidence that Earthlike planets orbit other stars.
11. In estimating whether intelligence arises and develops technology, we have only life on Earth as an example.
12. Dinosaurs existed on Earth for 1000 times longer than human civilization has existed to date.
13. Our civilization has launched probes into interstellar space.
14. One disadvantage of interstellar radio communication is that we can do it only with another civilization that has a technology less advanced than our own.
15. We have already broadcast our presence to our interstellar neighbors.

SELF-TEST: FILL IN THE BLANK

1. Amino acids are the building blocks of _____.
2. The naturally occurring molecules present on the young Earth included water, carbon dioxide, _____, and _____.
3. Two sources of energy for chemical reactions on the young Earth were _____ and _____.
4. The fossil record clearly shows evidence of life dating back _____ years.
5. Multicellular organisms did not appear on Earth until about _____ years ago.
6. The Murchison meteorite was discovered to contain relatively large amounts of _____.
7. The Drake equation estimates the number of _____ in the Milky Way Galaxy.
8. Planets in binary star systems are not considered habitable because the planetary orbits are usually _____.

9. The probabilities of the development of life and intelligence on Earth are extremely _____ if random chance is the only evolutionary factor involved.
10. Direct contact between extraterrestrial life forms may be impractical because of the large _____ between civilizations.
11. Radio communication over interstellar distances is practical because the signals travel at the speed of _____.
12. Radio waves can travel throughout the Galaxy because they are not blocked by interstellar _____.
13. Radio waves leaking away from Earth have now traveled a distance of _____ light-years.
14. Radio wavelengths between 18 and 21 cm are referred to as the _____.
15. A two-way communication with another civilization at a distance of 100 light-years will require _____ years.

REVIEW AND DISCUSSION

1. Why is life difficult to define?
2. What is chemical evolution?
3. What is the Urey-Miller experiment? What important organic molecules were produced in this experiment?

4. What other experiments have attempted to produce organic molecules by inorganic means?
5. What are the basic ingredients from which biological molecules formed on Earth?

6. Why do some scientists think life might have originated in space?

7. How do we know anything at all about the early episodes of life on Earth?

8. What is the role of language in cultural evolution?

9. Where else, besides Earth, have organic molecules been found?

10. Where—besides the planet Mars—might we hope to find signs of life in our solar system?

11. Do we know whether Mars ever had life at any time during its past? What argues in favor of the position that it may once have harbored life?

12. What is generally meant by "life as we know it"? What other forms of life might be possible?

13. How many of the terms in the Drake equation are known with any degree of certainty? Which factor is least well known?

14. What factors determine the suitability of a star as the parent of a planet on which life might arise?

15. What is the relationship between the average lifetime of galactic civilizations and the possibility of our someday communicating with them?

16. How would Earth appear at radio wavelengths to extraterrestrial astronomers?

17. Do you think that advanced civilizations would continue to emit large amounts of radio energy as they evolved?

18. What are the advantages in using radio waves for communication over interstellar distances?

19. What is the water hole? What advantages does it offer for interstellar communication?

20. If you were designing a SETI experiment, what parts of the sky would you monitor?

PROBLEMS  *Algorithmic versions of these questions are available in the Practice Problems module of the Companion Website.*

The number of squares preceding each problem indicates its approximate level of difficulty.

1. ▣ If Earth's 4.6-billion-year age were compressed to 46 years, as described in the text, what would be your age, in seconds? How long ago was the end of World War II? The Declaration of Independence? Columbus's discovery of the New World? The extinction of the dinosaurs?

2. ▣▣▣ According to the inverse-square law, a planet receives energy from its parent star at a rate proportional to the star's luminosity and inversely proportional to the square of the planet's distance from the star. ∞ (Sec. 17.3) According to Stefan's law, the rate at which the planet radiates energy into space is proportional to the fourth power of its surface temperature. ∞ (Sec. 3.4) In equilibrium, the two rates are equal. Based on this information and given the fact that (taking into account the greenhouse effect) the Sun's habitable zone extends from 0.6 A.U. to 1.5 A.U., estimate the extent of the habitable zone surrounding a K-type main sequence star of luminosity $\frac{1}{10}$ the luminosity of the Sun.

3. ▣▣ Using the data in the previous problem, how would the inner and outer radii of the Sun's habitable zone change if the solar luminosity increased by a factor of four?

4. ▣▣ The outer edge of the habitable zone corresponds roughly to the freezing point of water (273 K). Using the data in problem 2, how would this radius change if life could survive at temperatures as low as 150 K?

5. ▣▣▣ Using the data in problem 2, what would be the orbital period of a planet orbiting at the outer edge of the habitable zone of an F-type main-sequence star of 1.5 solar masses and luminosity five times that of the Sun?

6. ▪ Based on the numbers presented in the text, and assuming an average lifetime of 5 billion years for suitable stars, estimate the total number of habitable planets in the Galaxy.

7. ▪▪ A planet orbits one component of a binary-star system at a distance of 1 A.U. (see Figure 28.11a). If both stars have the same mass, and their orbit is circular, estimate the minimum distance between the stars for the tidal force due to the companion not to exceed a "safe" 0.01 percent of the gravitational force between the planet and its parent star.

8. ▣ Suppose that each of the "fraction" terms in the Drake equation turns out to have a value of $\frac{1}{10}$, that stars form at an average rate of 20 per year, and that each star has exactly one habitable planet orbiting it. Estimate the present number of technological civilizations in the Milky Way Galaxy if the average lifetime of a civilization is (a) 100 years (b) 10,000 years (c) 1 million years.

9. ▣▣▣ Adopting the estimate from the text that the number of technological civilizations in the Milky Way Galaxy is equal to the average lifetime of a civilization, it follows that the distance to our nearest neighbor decreases as the average lifetime increases. Assuming that civilizations are uniformly spread over a two-dimensional Galactic disk of radius 15 kpc, and all have the same lifetime, calculate the *minimum* lifetime for which two-way radio communication with our nearest neighbor would be possible before our civilization ends. Is the assumption that civilizations are uniformly spread across the disk a reasonable one?

10. ▣▣▣ Repeat the calculation in the previous question for a round-trip personal visit, using current-technology spacecraft that travel at 50 km/s.

11. ▣ How fast would a spacecraft have to travel in order to complete the trip from Earth to Alpha Centauri (a distance of 1.3 pc) and back in less than an average human lifetime (80 years, say)?

12. ▪ Assuming that there are 10,000 FM radio stations on Earth, each transmitting at a power level of 50 kW, calculate the total radio luminosity of Earth in the FM band. Compare this value with the roughly 10^6 W radiated by the Sun in the same frequency range.

13. ▣ Convert the water hole's wavelengths to frequencies. For practical reasons, any search of the water hole must be broken up into channels, much like you find on a television, except these channels are very narrow in radio frequency, about 100 Hz wide. How many channels must astronomers search in the water hole?

14. ■ At what wavelength does the microwave background radiation peak, and what is the corresponding frequency? How does this frequency compare to the waterhole frequency range?

15. ■ There are 20,000 stars within 100 light-years that are to be searched for radio communications. How long will the search take if one hour is spent looking at each star? What if one day is spent per star?

COLLABORATIVE EXERCISES

1. **Definition of Life.** As a group, compose a paragraph everyone agrees with that defines life. It should clearly show that rocks are not alive and that plants are alive. According to your definition, are stars alive? Compare and contrast your group's definition with that from another group.
2. **The Drake Equation.** Independently, each person in your group should estimate the average lifetime of a technologi-

cally competent civilization as described in the Drake equation. Explain the variation in the values your group used.
3. **Extraterrestrial Intelligence.** If your group was appointed to "speak for Earth" upon communication with an extraterrestrial, what would your group say? Write your group's speech and annotate it with explanations of why you chose to say this.

RESEARCHING ON THE WEB

To complete the following exercises, go to the online Destinations module for Chapter 28 on the Companion Website for Astronomy Today 4/e.

1. Access the "All Alone? (Or Not?)" page and describe what evidence the author would like for absolute assurance that life exists elsewhere in the universe.
2. Access the "Search for Extraterrestrial Intelligence" page and describe what happens if the SETI scientists find a signal.

3. Access the "What is Exobiology?" page and define the field of exobiology and list the six biogenic elements in which exobiologists are most interested.
4. Access the "Astrobiology" page and explain the concept of "planetary protection" in the form of a newspaper press release.

PROJECTS

1. Some people suggest that if extraterrestrial life is discovered, it will have a profound effect on people. Interview as many people as you can and ask the following two questions: (1) Do you believe that extraterrestrial life exists? (2) Why? From your results, try to decide whether there will be a profound effect on people if extraterrestrial life is discovered.
2. Conduct another poll, or do it at the same time as the first one. Ask the following question: What one question would you like to ask an extraterrestrial life form in a radio communication? How many responses do you receive that indi-

cate the person is very "Earth-centered" in thinking? How many responses suggest a lack of understanding of how alien an extraterrestrial life form might be? Is your conclusion from the first project different or changed in any way?
3. The Drake equation should be able to "predict" at least one civilization in our Galaxy: us. Try changing the values of various factors so that you end up with at least one. What do these various combinations of factors imply about how life arises and develops? Are there some combinations that just don't make any sense?

SKYCHART III PROJECTS

The SkyChart III Student Version planetarium program on which these exercises are based is included as a separately executable program on the CD in the back of this text.

1. ■ Locate the following planetary candidates: Upsilon Andromedae ("$u" And), which has been shown to have three planets orbiting it, HD 75289, believed to be one of the lowest-mass extrasolar planets, and 16 Cygni AB (HD 186408).

Although not much to look at with SkyChart III, or even the best telescopes currently available, it is exciting to know somebody might be pointing a green finger back at us!

 *In addition to the Practice Problems and Destinations modules, the Companion Website at http://www.prenhall.com/chaisson provides for each chapter an additional true-false, multiple choice, and labeling quiz, as well as additional annotated images, animations, and links to related Websites.*

APPENDIX 1

Scientific Notation

The objects studied by astronomers range in size from the smallest particles to the largest expanse of matter we know—the entire universe. Subatomic particles have sizes of about 0.000000000000001 meter, while galaxies (like that shown in Figure 1.3) typically measure some 1,000,000,000,000,000,000,000 meters across. The most distant known objects in the universe lie on the order of 100,000,000,000,000,000,000,000,000 meters from Earth.

Obviously, writing all those zeros is both cumbersome and inconvenient. More important, it is also very easy to make an error—write down one zero too many or too few and your calculations become hopelessly wrong! To avoid this, scientists always write large numbers using a shorthand notation in which the number of zeros following or preceding the decimal point is denoted by a superscript power, or *exponent*, of 10. The exponent is simply the number of places between the first significant (nonzero) digit in the number (reading from left to right) and the decimal point. Thus, 1 is 10^0, 10 is 10^1, 100 is 10^2, 1000 is 10^3, and so on. For numbers less than 1, with zeros between the decimal point and the first significant digit, the exponent is negative: 0.1 is 10^{-1}, 0.01 is 10^{-2}, 0.001 is 10^{-3}, and so on. Using this notation we can shorten the number describing subatomic particles to 10^{-15} meter, and write the number describing the size of a galaxy as 10^{21} meters.

More complicated numbers are expressed as a combination of a power of 10 and a multiplying factor. This factor is conventionally chosen to be a number between 1 and 10, starting with the first significant digit in the original number. For example, 150,000,000,000 meters (the distance from Earth to the Sun, in round numbers) can be more concisely written at 1.5×10^{11} meters, 0.000000025 meters as 2.5×10^{-8} meter, and so on. The exponent is simply the number of places the decimal point must be moved *to the left* to obtain the multiplying factor.

Some other examples of scientific notation are:

- the approximate distance to the Andromeda Galaxy = 3,000,000 light-years = 3×10^6 light-years
- the size of a hydrogen atom = 0.00000000005 meter = 5×10^{-11} meter
- the diameter of the Sun = 1,392,000 kilometers = 1.392×10^6 kilometers

- the U.S. national debt (as of May 1, 2001) = \$5,668,302,000,000.00 = \$5.688302 trillion = 5.668302×10^{12} dollars.

In addition to providing a simpler way of expressing very large or very small numbers, this notation also makes it easier to do basic arithmetic. The rule for multiplication of numbers expressed in this way is simple: Just multiply the factors and add the exponents. Similarly for division: Divide the factors and subtract the exponents. Thus, 3.5×10^{-2} multiplied by 2.0×10^3 is simply $(3.5 \times 2.0) \times 10^{-2+3} = 7.0 \times 10^1$—that is, 70. Again, 5×10^6 divided by 2×10^4 is just $(5/2) \times 10^{6-4}$, or 2.5×10^2 (= 250). Applying these rules to unit conversions: 200,000 nanometers is $200,000 \times 10^{2-9}$ meter (since 1 nanometer = 10^{-9} meter; see Appendix 2), or $2 \times 10^5 \times 10^{-9}$ meter, or $2 \times 10^{5-9} = 2 \times 10^{-4}$ meter = 0.2 mm. Verify these rules for yourself with a few examples of your own. The advantages of this notation when considering astronomical objects will soon become obvious.

Scientists often use "rounded-off" versions of numbers, both for simplicity and for ease of calculation. For example, we will usually write the diameter of the Sun as 1.4×10^6 kilometers, instead of the more precise number given earlier. Similarly, Earth's diameter is 12,756 kilometers, or 1.2756×10^4 kilometers, but for "ballpark" estimates we really don't need so many digits and the more approximate number 1.3×10^4 kilometers will suffice. Very often, we perform rough calculations using only the first one or two significant digits in a number, and that may be all that is necessary to make a particular point. For example, to support the statement, "The Sun is much larger than Earth," we need only say that the ratio of the two diameters is roughly 1.4×10^6 divided by 1.3×10^4. Since 1.4/1.3 is close to 1, the ratio is approximately $10^6/10^4 = 10^2$, or 100. The essential fact here is that the ratio is much larger than 1; calculating it to greater accuracy (to get 109.13) would give us no additional *useful* information. This technique of stripping away the arithmetic details to get to the essence of a calculation is very common in astronomy, and we use it frequently throughout this text.

Astronomical Measurement

Astronomers use many different kinds of units in their work, simply because no single system of units will do. Rather than the *Système Internationale* (SI), or meter-kilogram-second (MKS), metric system used in most high school and college science classes, many professional astronomers still prefer the older centimeter-gram-second (CGS) system. However, astronomers also commonly introduce new units when convenient. For example, when discussing stars, the mass and radius of the Sun are often used as reference points. The solar mass, written as $M_\odot$, is equal to 2.0×10^{33} g, or 2.0×10^{30} kg (since 1 kg = 1000 g). The solar radius, $R_\odot$, is equal to 700,000 km, or 7.0×10^8 m (1 km = 1000 m). The subscript $\odot$ always stands for Sun. Similarly, the subscript $\oplus$ always stands for Earth. In this book, we try to use the units that astronomers commonly use in any given context, but we also give the "standard" SI equivalents where appropriate.

Of particular importance are the units of length astronomers use. On small scales, the *angstrom* (1 Å = 10^{-10} m = 10^{-8} cm), the *nanometer* (1 nm = 10^{-9} m = 10^{-7} cm), and the *micron* (1 µm = 10^{-6} m = 10^{-4} cm) are used. Distances within the solar system are usually expressed in terms of the *astronomical unit* (A.U.), the mean distance between Earth and the Sun. One A.U. is approximately equal to 150,000,000 km, or 1.5×10^{11} m. On larger scales, the *light-year* (1 ly = 9.5×10^{15} m = 9.5×10^{12} km) and the *parsec* (1 pc = 3.1×10^{16} m = 3.1×10^{13} km = 3.3 ly) are commonly used. Still larger distances use the regular prefixes of the metric system: *kilo* for one thousand and *mega* for one million. Thus 1 kiloparsec (kpc) = 10^3 pc = 3.1×10^{19} m, 10 megaparsecs (Mpc) = 10^7 pc = 3.1×10^{23} m, and so on.

Astronomers use units that make sense within a context, and as contexts change, so do the units. For example, we might measure densities in grams per cubic centimeter (g/cm^3), in atoms per cubic meter (atoms/m^3), or even in solar masses per cubic megaparsec ($M_\odot/Mpc^3$), depending on the circumstances. The important thing to know is that once you understand the units, you can convert freely from one set to another. For example, the radius of the Sun could equally well be written as $R_\odot = 6.96 \times 10^8$ m, or 6.96×10^{10} cm, or 109 $R_\oplus$, or 4.65×10^{-3} A.U., or even 7.36×10^{-8} ly —whichever happens to be most useful. Some of the more common units used in astronomy, and the contexts in which they are most likely to be encountered, are listed below.

LENGTH:		
1 angstrom (Å)	= 10^{-10} m	
1 nanometer (nm)	= 10^{-9} m	atomic physics, spectroscopy
1 micron (µm)	= 10^{-6} m	interstellar dust and gas
1 centimeter (cm)	= 0.01 m	
1 meter (m)	= 100 cm	in widespread use throughout all astronomy
1 kilometer (km)	= 1000 m = 10^5 cm	
Earth radius ($R_\oplus$)	= 6378 km	planetary astronomy
Solar radius ($R_\odot$)	= 6.96×10^8 m	solar system, stellar evolution
1 astronomical unit (A.U.)	= 1.496×10^{11} m	
1 light-year (ly)	= 9.46×10^{15} m = 63,200 A.U.	
1 parsec (pc)	= 3.09×10^{16} m = 206,000 A.U. = 3.26 ly	galactic astronomy, stars and star clusters
1 kiloparsec (kpc)	= 1000 pc	
1 megaparsec (Mpc)	= 1000 kpc	galaxies, galaxy clusters, cosmology

MASS:		
1 gram (g)		
1 kilogram (kg)	= 1000 g	in widespread use in many different areas
Earth mass ($M_\oplus$)	= 5.98×10^{24} kg	planetary astronomy
Solar mass ($M_\odot$)	= 1.99×10^{30} kg	"standard" unit for all mass scales larger than Earth

TIME:		
1 second (s)		in widespread use throughout astronomy
1 hour (h)	= 3600 s	
1 day (d)	= 86,400 s	planetary and stellar scales
1 year (yr)	= 3.16×10^7 s	virtually all processes occurring on scales larger than a star

APPENDIX 3

Tables

TABLE 1 Some Useful Constants and Physical Measurements*

astronomical unit	1 A.U. $= 1.496 \times 10^8$ km (1.5×10^8 km)
light-year	1 ly $= 9.46 \times 10^{12}$ km (10^{13} km; 6 trillion miles)
parsec	1 pc $= 3.09 \times 10^{13}$ km $= 206,000$ A.U. $= 3.3$ ly
speed of light	$c = 299,792.458$ km/s (3×10^5 km/s)
Stefan-Boltzmann constant	$\sigma = 5.67 \times 10^{-8}$ W/m$^2 \cdot$ K^4
Planck's constant	$h = 6.63 \times 10^{-34}$ J s
gravitational constant	$G = 6.67 \times 10^{-11}$ N m^2/kg^2
mass of Earth	$M_\oplus = 5.98 \times 10^{24}$ kg (6×10^{24} kg; about 6000 billion billion tons)
radius of Earth	$R_\oplus = 6378$ km (6500 km)
mass of the Sun	$M_\odot = 1.99 \times 10^{30}$ kg (2×10^{30} kg)
radius of the Sun	$R_\odot = 6.96 \times 10^5$ km (7×10^5 km)
luminosity of the Sun	$L_\odot = 3.90 \times 10^{26}$ W (4×10^{26} W)
effective temperature of the Sun	$T_\odot = 5778$ K (5800 K)
Hubble's constant	$H_0 \approx 65$ km/s/Mpc
mass of an electron	$m_e = 9.11 \times 10^{-31}$ kg
mass of a proton	$m_p = 1.67 \times 10^{-27}$ kg

The rounded-off values used in the text are shown above in parentheses.

Conversions Between Common English and Metric Units

ENGLISH	METRIC
1 inch	= 2.54 centimeters (cm)
1 foot (ft)	= 0.3048 meters (m)
1 mile	= 1.609 kilometers (km)
1 pound (lb)	= 453.6 grams (g) or 0.4536 kilograms (kg) [on Earth]

TABLE 2 Periodic Table of Elements

Legend:
```
 2        ← Atomic number
He        ← Symbol of element
4.003     ← Atomic weight
Helium    ← Name of element
```

Group	1	2	3	4	5	6	7	8	9	10	11	12	13	14	15	16	17	18
Period 1	1 H 1.0080 Hydrogen																	2 He 4.003 Helium
Period 2	3 Li 6.939 Lithium	4 Be 9.012 Beryllium											5 B 10.81 Boron	6 C 12.011 Carbon	7 N 14.007 Nitrogen	8 O 15.9994 Oxygen	9 F 18.998 Fluorine	10 Ne 20.183 Neon
Period 3	11 Na 22.990 Sodium	12 Mg 24.31 Magnesium											13 Al 26.98 Aluminium	14 Si 28.09 Silicon	15 P 30.974 Phosphorus	16 S 32.064 Sulfur	17 Cl 35.453 Chlorine	18 Ar 39.948 Argon
Period 4	19 K 39.102 Potassium	20 Ca 40.08 Calcium	21 Sc 44.96 Scandium	22 Ti 47.90 Titanium	23 V 50.94 Vanadium	24 Cr 52.00 Chromium	25 Mn 53.94 Manganese	26 Fe 55.85 Iron	27 Co 58.93 Cobalt	28 Ni 58.71 Nickel	29 Cu 63.54 Copper	30 Zn 65.37 Zinc	31 Ga 69.72 Gallium	32 Ge 72.59 Germanium	33 As 74.92 Arsenic	34 Se 78.96 Selenium	35 Br 79.909 Bromine	36 Kr 83.80 Krypton
Period 5	37 Rb 85.47 Rubidium	38 Sr 87.62 Strontium	39 Y 88.91 Yttrium	40 Zr 91.22 Zirconium	41 Nb 92.91 Niobium	42 Mo 95.94 Molybdenum	43 Tc (99) Technetium	44 Ru 101.1 Ruthenium	45 Rh 102.90 Rhodium	46 Pd 106.4 Palladium	47 Ag 107.87 Silver	48 Cd 112.40 Cadmium	49 In 114.82 Indium	50 Sn 118.69 Tin	51 Sb 121.75 Antimony	52 Te 127.60 Tellurium	53 I 126.9 Iodine	54 Xe 131.30 Xenon
Period 6	55 Cs 132.91 Cesium	56 Ba 137.34 Barium	71 Lu 174.97 Lutetium *	72 Hf 178.49 Hafnium	73 Ta 180.95 Tantalum	74 W 183.85 Tungsten	75 Re 186.2 Rhenium	76 Os 190.2 Osmium	77 Ir 192.2 Iridium	78 Pt 195.09 Platinum	79 Au 197.0 Gold	80 Hg 200.59 Mercury	81 Tl 204.37 Thallium	82 Pb 207.19 Lead	83 Bi 208.98 Bismuth	84 Po (210) Polonium	85 At (210) Astantine	86 Rn (222) Radon
Period 7	87 Fr (223) Francium	88 Ra 226.05 Radium	103 Lw (257) Lawrencium **	104 Rf (261) Rutherfordium	105 Db (262) Dubnium	106 Sg (263) Seaborgium	107 Bh (262) Bohrium	108 Hs (265) Hassium	109 Mt (266) Meitnerium	110 Uun (269) Ununnilium	111 Uuu (272) Unununium	112 Uub (277) Unununbium	113 Uut (undiscovered) Ununtrium	114 Uuq (285) Ununquadium	115 Uup (undiscovered) Ununpentium	116 Uuh (289) Ununhexium	117 Uus (undiscovered) Ununseptium	118 Uuo (293) Ununoctium

* Lanthanides:

57 La 138.91 Lanthanum	58 Ce 140.12 Cerium	59 Pr 140.91 Praseodymium	60 Nd 144.24 Neodymium	61 Pm (147) Promethium	62 Sm 150.35 Samarium	63 Eu 151.96 Europium	64 Gd 157.25 Gadolinium	65 Tb 158.92 Terbium	66 Dy 162.50 Dysprosium	67 Ho 164.93 Holmium	68 Er 167.26 Erbium	69 Tm 168.93 Thulium	70 Yb 173.04 Ytterbium

** Actinides:

89 Ac (227) Actinium	90 Th 232.04 Thorium	91 Pa (231) Protactinium	92 U 238.03 Uranium	93 Np (237) Neptunium	94 Pu (242) Plutonium	95 Am (243) Americium	96 Cm (247) Curium	97 Bk (249) Berkelium	98 Cf (251) Californium	99 Es (254) Einsteinium	100 Fm (253) Fermium	101 Md (256) Mendelevium	102 No (254) Nobelium

TABLE 3A Planetary Orbital Data

PLANET	SEMI-MAJOR AXIS (A.U.)	SEMI-MAJOR AXIS (10⁶ km)	ECCENTRICITY (e)	PERIHELION (A.U.)	PERIHELION (10⁶ km)	APHELION (A.U.)	APHELION (10⁶ km)
Mercury	0.39	57.9	0.206	0.31	46.0	0.47	69.8
Venus	0.72	108.2	0.007	0.72	107.5	0.73	108.9
Earth	1.00	149.6	0.017	0.98	147.1	1.02	152.1
Mars	1.52	227.9	0.093	1.38	206.6	1.67	249.2
Jupiter	5.20	778.4	0.048	4.95	740.7	5.46	816
Saturn	9.54	1427	0.054	9.02	1349	10.1	1504
Uranus	19.19	2871	0.047	18.3	2736	20.1	3006
Neptune	30.07	4498	0.009	29.8	4460	30.3	4537
Pluto	39.48	5906	0.249	29.7	4437	49.3	7376

PLANET	MEAN ORBITAL SPEED (km/s)	SIDEREAL PERIOD (tropical years)	SYNODIC PERIOD (days)	INCLINATION TO THE ECLIPTIC (degrees)	GREATEST ANGULAR DIAMETER AS SEEN FROM EARTH (arc seconds)
Mercury	47.87	0.24	115.88	7.00	13
Venus	35.02	0.62	583.92	3.39	64
Earth	29.79	1.00	—	0.01	—
Mars	24.13	1.88	779.94	1.85	25
Jupiter	13.06	11.86	398.88	1.31	50
Saturn	9.65	29.42	378.09	2.49	21
Uranus	6.80	83.75	369.66	0.77	4.1
Neptune	5.43	163.7	367.49	1.77	2.4
Pluto	4.74	248.0	366.72	17.2	0.11

TABLE 3B Planetary Physical Data

PLANET	EQUATORIAL RADIUS		MASS		MEAN DENSITY (kg/m³)	SURFACE GRAVITY (Earth = 1)	ESCAPE SPEED (km/s)
	(km)	(Earth = 1)	(kg)	(Earth = 1)			
Mercury	2440	0.38	3.30×10^{23}	0.055	5430	0.38	4.2
Venus	6052	0.95	4.87×10^{24}	0.82	5240	0.91	10.4
Earth	6378	1.00	5.97×10^{24}	1.00	5520	1.00	11.2
Mars	3394	0.53	6.42×10^{23}	0.11	3930	0.38	5.0
Jupiter	71,492	11.21	1.90×10^{27}	317.8	1330	2.53	60
Saturn	60,268	9.45	5.68×10^{26}	95.16	690	1.07	36
Uranus	25,559	4.01	8.68×10^{25}	14.54	1270	0.91	21
Neptune	24,766	3.88	1.02×10^{26}	17.15	1640	1.14	24
Pluto	1137	0.18	1.27×10^{22}	0.0021	2060	0.07	1.2

PLANET	SIDEREAL ROTATION PERIOD (solar days)*	AXIAL TILT (degrees)	SURFACE MAGNETIC FIELD (Earth = 1)	MAGNETIC AXIS TILT (degrees relative to rotation axis)	ALBEDO[†]	SURFACE TEMPERATURE[‡] (K)	NUMBER OF MOONS
Mercury	58.6	0.0	0.011	<10	0.11	100–700	0
Venus	−243.0	177.4	<0.001		0.65	730	0
Earth	0.9973	23.45	1.0	11.5	0.37	290	1
Mars	1.026	23.98	0.001		0.15	180–270	2
Jupiter	0.41	3.08	13.89	9.6	0.52	124	28
Saturn	0.44	26.73	0.67	0.8	0.47	97	30
Uranus	−0.72	97.92	0.74	58.6	0.50	58	21
Neptune	0.67	29.6	0.43	46.0	0.5	59	8
Pluto	−6.39	118	?		0.6	40–60	1

*A negative sign indicates retrograde rotation.

[†]Fraction of sunlight reflected from surface.

[‡]Temperature is effective temperature for jovian planets.

TABLE 4 The Twenty Brightest Stars in Earth's Night Sky

NAME	STAR	SPECTRAL TYPE* A	B	PARALLAX (arc seconds)	DISTANCE (pc)	APPARENT VISUAL MAGNITUDE A	B
Sirius	α CMa	A1V	wd[†]	0.377	2.7	−1.46	+8.7
Canopus	α Car	F0Ib–II		0.010	96	−0.72	
Rigel Kentaurus	α Cen	G2V	K0V	0.770	1.3	−0.01	+1.3
Arcturus	α Boo	K2III		0.091	11	+0.06	
Vega	α Lyr	A0V		0.130	7.8	+0.04	
Capella	α Aur	GIII	M1V	0.077	13	+0.05	+10.2
Rigel	β Ori	B8Ia	B9	0.004	240	+0.14	+6.6
Procyon	α CMi	F5IV–V	wd[†]	0.29	3.5	+0.37	+10.7
Betelgeuse	α Ori	M2Iab		0.008	130	+0.41	
Achernar	α Eri	B5V		0.023	44	+0.51	
Hadar	β Cen	B1III	?	0.006	160	+0.63	+4
Altair	α Aql	A7IV–V		0.190	5.2	+0.77	
Acrux	α Cru	B1IV	B3	0.008	120	+1.39	+1.9
Aldebaran	α Tau	K5III	M2V	0.050	20	+0.86	+13
Spica	α Vir	B1V		0.013	80	+0.91	
Antares	α Sco	M1Ib	B4V	0.005	190	+0.92	+5.1
Pollux	β Gem	K0III		0.10	10	+1.16	
Fomalhaut	α PsA	A3V	?	0.13	7.7	+1.19	+6.5
Deneb	α Cyg	A2Ia		0.001	990	+1.26	
Mimosa	β Cru	B1IV		0.009	110	+1.28	

NAME	VISUAL LUMINOSITY* (SUN = 1) A	B	ABSOLUTE VISUAL MAGNITUDE[†] A	B	PROPER MOTION (arc seconds/yr)	TRANSVERSE VELOCITY (km/s)	RADIAL VELOCITY (km/s)
Sirius	24	0.02	+1.4	+11.5	1.33	16.7	−7.6[‡]
Canopus	1.6×10^4		−5.6		0.02	9.1	20.5
Rigel Kentaurus	1.5	0.4	+4.4	+5.7	3.68	22.7	−24.6
Arcturus	110		−0.3		2.28	119	−5.2
Vega	51		+0.6		0.34	12.6	−13.9
Capella	140	0.01	−0.5	+9.6	0.44	27.1	30.2[‡]
Rigel	4.4×10^4	110	−6.8	−0.3	0.00	1.2	20.7[‡]
Procyon	7.6	0.0006	+2.6	+13.0	1.25	20.7	−3.2[‡]
Betelgeuse	1.0×10^4		−5.2		0.03	18.5	21.0[‡]
Achernar	1100		−2.7		0.10	20.9	19
Hadar	1.2×10^4	560	−5.4	−2.0	0.04	30.3	−12[‡]
Altair	12		+2.2		0.66	16.3	−26.3
Acrux	3500	2200	−4.0	−3.5	0.04	22.8	−11.2
Aldebaran	160	0.002	−0.6	+11.5	0.20	19.0	54.1
Spica	2400		−3.6		0.05	19.0	1.0[‡]
Antares	1.3×10^4	290	−5.5	−1.3	0.03	27.0	−3.2
Pollux	30		+1.2		0.62	29.4	3.3
Formalhaut	17	0.13	+1.8	+7.1	0.37	13.5	6.5
Deneb	2.7×10^5		−8.7		0.003	14.1	−4.6[‡]
Mimosa	3200		−3.9		0.05	26.1	—

*Energy output in the visible part of the spectrum; A and B columns identify individual components of binary-star systems.

[†]"wd" stands for "white dwarf."

[‡]Average value of variable velocity.

TABLE 5 The Twenty Nearest Stars

NAME	SPECTRAL TYPE		PARALLAX (arc seconds)	DISTANCE (pc)	APPARENT VISUAL MAGNITUDE*	
	A	B			A	B
Sun	G2V				−26.72	
Proxima Centauri	M5		0.772	1.30	+11.05	
Alpha Centauri	G2V	K0V	0.750	1.33	−0.01	+1.33
Barnard's Star	M5V		0.545	1.83	+9.54	
Wolf 359	M8V		0.421	2.38	+13.53	
Lalande 21185	M2V		0.397	2.52	+7.50	
UV Ceti	M6V	M6V	0.387	2.58	+12.52	+13.02
Sirius	A1V	wd†	0.377	2.65	−1.46	+8.7
Ross 154	M5V		0.345	2.90	+10.45	
Ross 248	M6V		0.314	3.18	+12.29	
ε Eridani	K2V		0.303	3.30	+3.73	
Ross 128	M5V		0.298	3.36	+11.10	
61 Cygni	K5V	K7V	0.294	3.40	+5.22	+6.03
ε Indi	K5V		0.291	3.44	+4.68	
Grm 34	M1V	M6V	0.290	3.45	+8.08	+11.06
Luyten 789-6	M6V		0.290	3.45	+12.18	
Procyon	F5IV–V	wd†	0.285	3.51	+0.37	+10.7
Σ 2398	M4V	M5V	0.285	3.55	+8.90	+9.69
Lacaille 9352	M2V		0.279	3.58	+7.35	
G51-15	MV		0.278	3.60	+14.81	

NAME	VISUAL LUMINOSITY* (SUN = 1)		ABSOLUTE VISUAL MAGNITUDE*		PROPER MOTION (arc seconds/yr)	TRANSVERSE VELOCITY (km/s)	RADIAL VELOCITY (km/s)
	A	B	A	B			
Sun	1.0		+4.85				
Proxima Centauri	0.00006		+15.5		3.86	23.8	−16
Alpha Centauri	1.6	0.45	+4.4	+5.7	3.68	23.2	−22
Barnard's Star	0.00045		+13.2		10.34	89.7	−108
Wolf 359	0.00002		+16.7		4.70	53.0	+13
Lalande 21185	0.0055		+10.5		4.78	57.1	−84
UV Ceti	0.00006	0.00004	+15.5	+16.0	3.36	41.1	+30
Sirius	24	0.02	+1.4	+11.5	1.33	16.7	−8
Ross 154	0.00048		+13.3		0.72	9.9	−4
Ross 248	0.00011		+14.8		1.58	23.8	−81
ε Eridani	0.30		+6.1		0.98	15.3	+16
Ross 128	0.00036		+13.5		1.37	21.8	−13
61 Cygni	0.082	0.039	+7.6	+8.4	5.22	84.1	−64
ε Indi	0.14		+7.0		4.69	76.5	−40
Grm 34	0.0061	0.00039	+10.4	+13.4	2.89	47.3	+17
Luyten 789-6	0.00014		+14.6		3.26	53.3	−60
Procyon	7.65	0.00055	+2.6	+13.0	1.25	2.8	−3
Σ 2398	0.0030	0.0015	+11.2	+11.9	2.28	38.4	+5
Lacaille 9352	0.013		+9.6		6.90	117	+10
G51-15	0.00001		+17.0		1.26	21.5	—

*A and B columns identify individual components of binary-star systems.
†"wd" stands for "white dwarf."

APPENDIX 4

Using Skychart III

I. GETTING STARTED

Set up SkyChart III and familiarize yourself with its operation. If you have not already done so, use a word processor to display the file *Read Me First!.txt* found on your SkyChart III CD-ROM. Don't be bashful about trying various changes. Explore each of the pull-down menus across the top of the screen. If you make changes that result in a very confusing and undesirable screen simply go to *FILE/Open* and select *Default Setting* to reestablish your original configuration.

One of the most useful features of the software is its ability to run animations. When an animation is described, you will find that sometimes a time step will be recommended. If you find that animations run too fast or too slow, experiment with using smaller or larger time steps. Another way to change the speed is to display more or fewer objects. Be aware that this might mean you will have to adapt other settings to make the exercise work.

II. PERSONALIZING THE SOFTWARE

1. Configure the screen for your local sky by going to *COMPUTATION/Location*. From the list of available locations, select a city near you to synchronize the planetarium view with the view of the sky in your area. For more precision, you can replace longitude and latitude with values for your city, school, or home and change the name accordingly. Click on *Add to List* and *Use as "Here."*

2. Select *COMPUTATION/Date & Time*. Typically, it is most convenient to select *Gregorian* and *Universal Time* and *Automatic daylight savings time correction*.

3. Set *ANIMATION/Real Time* and *ANIMATION/ Animate Forward*.

4. Make any other changes to the screen you wish, such as *VIEW/180°Field* and select the options your prefer under *DRAW/Symbols & Grids*. Sometimes it is helpful to display the *Horizon Line, Ecliptic, Equator*, and *Draw R.A./Dec grid.* The lines will be less intrusive if you set *R.A. Interval (h m):* to 02 00 and *Dec. Interval (° '):* to 15 00.

5. It is common with planetarium software to orient the screen with south, indicated as S on the screen, to the bottom of the screen. This puts East on the left and West on the right. Imagine that you are facing south with the computer positioned in front of you. Now, imagine picking up your computer screen and lifting it over your head. The orientation of the simulated star pattern is the same as the pattern of stars in the sky. The most efficient way to establish this orientation is to select *VIEW/Center Horizon/ South* followed by *VIEW/Center Horizon/Zenith (Up)*.

6. When you have selected all parameter settings to configure the display to your liking, save your screen setting so that you can recreate the settings without going through these configuration steps each time you run the software. It is recommended that you save these settings as your default settings (under *FILE/Save Settings as Default*). Be aware that saving these settings as default replaces the original default settings file.

7. If you edit the position or characteristics of stars, planets, or other bodies and you want to restore the original settings, the easiest way is to reinstall the software.

III. USING THE SOFTWARE

Explore some of the sample configurations supplied with the CD-ROM. A number of very interesting and useful settings files can be accessed from the directory containing the SkyChart III program. Opening *FILES/Open/Settings* should take you to the installed software file where you will find a directory, *Demo Settings Files*. A selection of configuration and animation files has been provided in this directory to demonstrate some of the possibilities for creating simulations of astronomical events. Look at the settings that have been used, and experiment with options to tailor the screen to your liking.

Considerable information has been supplied for most of the displayed objects. If you wish to know the name of an object, or some of the details about a star such as it name, magnitude, or color index, point to the object with the mouse and left click. An information box will appear that can be moved out of the way or closed by clicking on *Close*.

Have fun! Experiment with various other options found in the pull-down menus until you become familiar with all features of the software. Don't worry about putting the screen into some abnormal state. You can always restore your default settings from *FILE/Open/Default Settings*. The confidence you develop by such experimentation will permit you to make maximum use of the software. Note that certain changes, for instance renaming Mars or changing its orbit, are not reversed by restoring the default settings. Reinstallation then becomes necessary to reverse these changes.

The following section is meant to assist with the completion of the SkyChart III exercises and should be used in conjunction with the *HELP* menu in SkyChart III.

Changing Locations

Choose *COMPUTATION/Location* and the *Location* window will appear. To change the location choose from the drop-down menu or enter the site information manually. Use the *Add to List* button to add new sites to the drop-down menu. The *Use as Here* button changes the default location—unless changed, the default location is San Francisco.

Adjusting the Zoom

Select the desired field of view from the *VIEW* menu. The *View/Zoom In* and *View/Zoom Out* commands can be used to change the view at smaller increments. To specify an exact field of view choose *VIEW/Field*.

Zooming and Planet Size

At wide angles of view, the planet symbols are all about the same size—almost as large as the Sun. This makes them easy to find and recognize. As the field width narrows, the size of a planet remains constant until the point at which the planet would actually be visible as a disk. A disk whose diameter matches what you would see through a telescope then replaces the planet symbol. You can magnify the disks of the planets and their moons in *DRAW/Symbols and Grids* submenu. This does not change their orbits around the Sun or one another.

Changing Viewing Directions

Choose *VIEW/Center Horizon* and select a direction, or click and drag the screen (this only works when *DRAW/Mouse Coordinates* is not selected). Note that it is often helpful to change the viewing direction by centering on an object, use *VIEW/Center Object* to do so.

Changing Date and Time

Choose *COMPUTATION/Date & Time*. The date must be entered as *YYYY/MM/DD*. The time is entered using a 24-hour clock to designate a.m. and p.m. and must be entered as *HH:MM:SS*.

Keeping Track of Time

In the lower left corner of the screen is the current time. Use this to measure durations and keep track of the date and time. Note that this is not a clock. To make it function like a clock use the *ANIMATION/Real Time* command. In effect this starts an animation sequence with a time step of one second.

Printing Black & White

In the *DRAW* menu deselect *Use Color*. To change the background color, select *DRAW/Sky Background/White*.

Horizon Mask

Often times the object of interest is not viewable from the current location because it is below the horizon. To remove the horizon and view the entire sky deselect *Horizon Mask* in the *DRAW* menu. Do not confuse this with the

Horizon Line (under *DRAW/Symbols & Grids*). *Horizon Mask* can be drawn without the *Horizon Line* selected and vice versa.

Adding Reference Lines

Select *DRAW/Symbols & Grids*. Choose *Draw Grid Lines* and select which lines to draw; the *Ecliptic* and *Horizon Lines* can be especially useful for determining orientation. Note that the *Horizon Line* in this window is separate from *Horizon Mask* found in the *DRAW* menu. The *Horizon line* can be drawn without *Horizon Mask* selected and vice versa.

Animation

To start an animation choose *ANIMATION/Animate Forward* or *ANIMATION/Animate Backward* to advance or reverse the flow of time. Use *ANIMATION/Animation Off* to stop the sequence. The rate of the animation can be adjusted using the predefined *Time Steps* in the *ANIMATION* menu or by selecting *ANIMATION/Set Time Step*. In the *Set Time Step* window enter a value for the time step in the following format *DD:HH:MM:SS*. Discrete time steps are controlled with the *ANIMATION/Step forward* and *ANIMATION/Step Backward* commands. To "steady" an animation it is often helpful to change the coordinate system to ecliptic coordinates, *VIEW/Coordinates/Ecliptic*.

Changing the Coordinate System

Choose *VIEW/Coordinates* and select a system. One way to make sense of the different coordinate systems is to imagine a Euclidian xyz axis orientated to the current selection. So for *Ecliptic coordinates*, the xy plane would lie in the plane of the ecliptic with z perpendicular to it.

Viewing from Other Objects

Choose *COMPUTATION/Location* and the *Location* dialogue box will appear. Select *View from Object*. Use the *Select* button to *Search for* and select the new viewing object. Once chosen, change the orientation to the new object by adjusting the *Offset Distance*, *R.A.* and *Dec*.

Finding Objects

Use the *EDIT/Find Object* command to *Search for* and *Select* the desired object. Note that the database has limited use of common names. The scientific designation may have to be used in place of a common name.

Trailing Objects

To start a trail for an object select the *Trail On* button from the *Object Info* window or use *ANIMATION/Trail For* to *Search for* and select the desired object. Using either method will bring up the *Edit Object Trail* window. After a trail is turned on and an animation sequence started, a "tick" is drawn on the screen at the end of each time step. In the *Edit Object Trail* window dates can be added to the ticks, the frequency at which the ticks are drawn can be adjusted and, for long animations, the number of recorded positions can be increased. The three radio buttons in the *Edit Object Trail* window have a large impact on the way ticks are drawn during an animation. If an animation yields unexpected results try selecting a different radio button.

Measuring Angular Separation/Distance
Select *DRAW/Mouse Coordinates*. Click on the first object and drag to the second object, the distance between the two is displayed in the upper left corner of the screen in degrees, minutes, and seconds ($x°y'z''$). Note that when *DRAW/Mouse Coordinates* is selected the view can no longer be changed using the "click and drag" method.

Adjusting What You See
Several object types have categories in the *DRAW/Symbols & Grids* submenu—*Stars, Galaxies, Clusters,* and *Nebulae.* These hide or display groups of objects with a single mouse click. When an object is check-marked the objects are shown. When it's not check-marked the objects are hidden.

Viewing Limits
To change the viewing limits, effectively changing the number of stars drawn, use *DRAW/Fainter Limits* and *DRAW/Brighter Limits*. This can be done more precisely using *DRAW/Symbols & Grids*. From the *Symbols & Grids* window set the *Brightest* and *Faintest Magnitudes* for stars, planets, and deep-sky objects. Also set the limits for *Object Labels*. For planets, the disks can be magnified to make them appear larger. Note that changing the disk magnification does not affect the orbit of the planet.

Viewing the Solar System
Select *COMPUTATION/Location/View from Object* and select the Sun. Change the *Offset Distance* to 60 A.U., the *R.A.* to 18 00 00.00, and the *Dec.* to 66 00 00.0. Change the zoom level accordingly.

Exporting Data/Ephemeris
Sometimes it is helpful to export data for a given object and use spreadsheet software to analyze it. To do so, select *File/Export/Ephemeris as Text.* In the *Save Ephemeris* window use the *Select* button to *Search for* and select the desired object. Next specify the *Start* and *End Date & Time* as well as the *Time Step*. Once finished, select the *Save As* button.

Importing TLE Files
Two-Line Elements or "TLEs" are a standard format and are used to insert new satellites, comets, asteroids, etc. into SkyChart III. Many of these can be found on the Internet and a good place to start looking is at NASA's Website. Below is the TLE for the International Space Station:

ISS (ZARYA)
1x25544Ux98067Axxx01029.17112089xx.00037783xx
00000-0xx31449-3x0xx5652
2x25544xx51.5749xx71.2181x0009954xx45.6999xx
59.7877x15.69675324125348

Each x represents a single space. When manually typing this information into a text editor, each x has to be replaced by a single space. The spaces between quantities are important and alterations to the spacing can result in errors when imported into SkyChart. TLE files can be viewed with common text editors like Notepad or Simpletext. Often times these files have .txt extensions. It is recommended that the extension be changed to .tle before importing into SkyChart.

Using Telescope Controls/Telrad Circles
To simulate the view through a telescope select *TELESCOPE/Display*. Set values for the *Finder Field Width* and *Eyepiece Field Width* and deselect *Camera Field Width*. Next choose *TELESCOPE/Connect* and select *Emulate Real Telescope Connection*. Close this window and the *Telescope Control* window will appear. Select the object of interest and click the *Go To* button in the *Telescope Control* window. Two circles will slowly move and center on the chosen object. These are often referred to as "Telrads Circles." To move the circles select another object and again click the *Go To* button in the *Telescope Control* window.

GLOSSARY

A ring One of three Saturnian rings visible from Earth. The A ring is farthest from the planet and is separated from the B ring by the Cassini division. (p. 309)

absolute brightness The apparent brightness a star would have if it were placed at a standard distance of 10 parsecs from Earth. (p. 441)

absolute magnitude The apparent magnitude a star would have if it were placed at a standard distance of 10 parsecs from Earth. (p. 444)

absorption line Dark line in an otherwise continuous bright spectrum, where light within one narrow frequency range has been removed. (p. 88)

acceleration The rate of change of velocity of a moving object. (p. 51)

accretion Gradual growth of bodies, such as planets, by the accumulation of other, smaller bodies. (p. 389)

accretion disk Flat disk of matter spiraling down onto the surface of a neutron star or black hole. Often, the matter originated on the surface of a companion star in a binary-star system. (p. 545)

active galaxies The most energetic galaxies, which can emit hundreds or thousands of times more energy per second than the Milky Way, mostly in the form of long-wavelength nonthermal radiation. (p. 662)

active optics Collection of techniques used to increase the resolution of ground-based telescopes. Minute modifications are made to the overall configuration of an instrument as its temperature and orientation change; used to maintain the best possible focus at all times. (p. 121)

active region Region of the photosphere of the Sun surrounding a sunspot group, which can erupt violently and unpredictably. During sunspot maximum, the number of active regions is also a maximum. (p. 422)

active Sun The unpredictable aspects of the Sun's behavior, such as sudden explosive outbursts of radiation in the form of prominences and flares. (p. 418)

adaptive optics Technique used to increase the resolution of a telescope by deforming the shape of the mirror's surface under computer control while a measurement is being taken; used to undo the effects of atmospheric turbulence. (p. 122)

amino acids Organic molecules which form the basis for building the proteins that direct metabolism in living creatures. (p. 742)

amplitude The maximum deviation of a wave above or below zero point. (p. 64)

angular resolution The ability of a telescope to distinguish between adjacent objects in the sky. (p. 116)

annular eclipse Solar eclipse occurring at a time when the Moon is far enough away from Earth that it fails to cover the disk of the Sun completely, leaving a ring of sunlight visible around its edge. (p. 17)

aphelion The point on the elliptical path of an object in orbit about the Sun that is most distant from the Sun. (p. 46)

Apollo asteroid *See* Earth-crossing asteroid.

apparent brightness The brightness that a star appears to have, as measured by an observer on Earth. (p. 441)

apparent magnitude The apparent brightness of a star, expressed using the magnitude scale. (p. 444)

association Small grouping of (typically 100 or less) bright stars, spanning up to a few tens of parses across, usually rich in very young stars. (p. 506)

asteroid One of thousands of very small members of the solar system orbiting the Sun between the orbits of Mars and Jupiter. Often referred to as "minor planets." (p. 356)

asteroid belt Region of the solar system, between the orbits of Mars and Jupiter, in which most asteroids are found. (p. 356)

asthenosphere Layer of Earth's interior, just below the lithosphere, over which the surface plates slide. (p. 177)

astronomical unit (A.U.) The average distance of Earth from the Sun. Precise radar measurements yield a value for the A.U. of 149,603,500 km. (pp. 47, 177)

astronomy Branch of science dedicated to the study of everything in the universe that lies above Earth's atmosphere. (p. 4)

asymptotic-giant branch Path on the Hertzsprung–Russell diagram corresponding to the changes that a star undergoes after helium burning ceases in the core. At this stage, the carbon core shrinks and drives the expansion of the envelope, and the star becomes a swollen red giant for a second time. (p. 523)

atmosphere Layer of gas confined close to a planet's surface by the force of gravity. (p. 166)

atom Building block of matter, composed of positively charged protons and neutral neutrons in the nucleus surrounded by negatively charged electrons. (p. 91)

aurora Event which occurs when atmospheric molecules are excited by incoming charged particles from the solar wind, then emit energy as they fall back to their ground states. Aurorae generally occur at high latitudes, near the north and south magnetic poles. (p. 185)

autumnal equinox Date on which the Sun crosses the celestial equator moving southward, occurring on or near September 22. (p. 13)

B ring One of three Saturnian rings visible from Earth. The B ring is the brightest of the three, and lies just past the Cassini Division, closer to the planet than the A ring. (p. 309)

barred-spiral galaxy Spiral galaxy in which a bar of material passes through the center of the galaxy, with the spiral arms beginning near the ends of the bar. (p. 632)

baseline The distance between two observing locations used for the purposes of triangulation measurements. The larger the baseline, the better the resolution attainable. (p. 24)

belt Dark, low-pressure region in the atmosphere of a jovian planet, where gas flows downward. (p. 278)

Big Bang Event that cosmologists consider the beginning of the universe, in which all matter and radiation in the entire universe came into being. (p. 694)

binary-star system A system which consists of two stars in orbit about their common center of mass, held together by their mutual gravitational attraction. Most stars are found in binary-star systems. (p. 457)

blackbody curve The characteristic way in which the intensity of radiation emitted by a hot object depends on frequency. The frequency at which the emitted intensity is highest is an indication of the temperature of the radiating object. Also referred to as the Planck curve. (p. 72)

black dwarf The endpoint of the evolution of an isolated, low-mass star. After the white-dwarf stage, the star cools to the point where it is a dark "clinker" in interstellar space. (p. 528)

black hole A region of space where the pull of gravity is so great that nothing—not even light—can escape. A possible outcome of the evolution of a very massive star. (p. 581)

blue giant Large, hot, bright star at the upper-left end of the main sequence on the Hertzsprung–Russell diagram. Its name comes from its color and size. (p. 453)

blueshift Motion-induced changes in the observed wavelength from a source that is moving toward us. Relative approaching motion between the object and the observer causes the wavelength to appear shorter (and hence bluer) than if there were no motion at all. (p. 78)

blue supergiant The very largest of the large, hot, bright stars at the uppermost-left end of the main sequence on the Hertzsprung–Russell diagram. (p. 453)

Bohr model First theory of the hydrogen atom to explain the observed spectral lines. This model rests on three ideas: that there is a state of lowest energy for the electron, that there is a maximum energy beyond which the electron is no longer bound to the nucleus, and that within these two energies the electron can only exist in certain energy levels. (p. 91)

brown dwarf Remnants of fragments of collapsing gas and dust that did not contain enough mass to initiate core nuclear fusion. Such objects are then frozen somewhere along their pre-main-sequence contraction phase, continually cooling into compact dark objects. Because of their small size and low temperature they are extremely difficult to detect observationally. (p. 498)

brown oval Feature of Jupiter's atmosphere that appears only at latitudes near 20 degrees N, this structure is a long-lived hole in the clouds that allows us to look down into Jupiter's lower atmosphere. (p. 283)

C

C ring One of three Saturnian rings visible from Earth. The C ring lies closest to the planet and is relatively thin compared to the A and B rings. (p. 309)

carbon-detonation supernova *See* Type I supernova.

Cassegrain telescope A type of reflecting telescope in which incoming light hits the primary mirror and is then reflected upward toward the prime focus, where a secondary mirror reflects the light back down through a small hole in the main mirror into a detector or eyepiece. (p. 111)

Cassini Division A relatively empty gap in Saturn's ring system, discovered in 1675 by Giovanni Cassini. It is now known to contain a number of thin ringlets. (p. 309)

catastrophic theory A theory that invokes statistically unlikely accidental events to account for observations. (p. 386)

celestial coordinates Pair of quantities—right ascension and declination—similar to longitude and latitude on Earth, used to pinpoint locations of objects on the celestial sphere. (p. 14)

celestial equator The projection of Earth's equator onto the celestial sphere. (p. 8)

celestial sphere Imaginary sphere surrounding Earth to which all objects in the sky were once considered to be attached. (p. 8)

center of mass The "average" position in space of a collection of massive bodies, weighted by their masses. For an isolated sys-

tem this point moves with constant velocity, according to Newtonian mechanics. (p. 54)

Cepheid variable Star whose luminosity varies in a characteristic way, with a rapid rise in brightness followed by a slower decline. The period of a Cepheid variable star is related to its luminosity, so a determination of this period can be used to obtain an estimate of the star's distance. (p. 604)

chaotic rotation Unpredictable tumbling motion that nonspherical bodies in eccentric orbits, such as Saturn's satellite Hyperion, can exhibit. No amount of observation of an object rotating chaotically will ever show a well-defined period. (p. 322)

charge-coupled device (CCD) An electronic device used for data acquisition; composed of many tiny pixels, each of which records a buildup of charge to measure the amount of light striking it. (p. 119)

chromatic aberration The tendency for a lens to focus red and blue light differently, causing images to become blurred. (p. 110)

chromosphere The Sun's lower atmosphere, lying just above the visible atmosphere. (p. 407)

closed universe Geometry that the universe as a whole would have if the density of matter is above the critical value. A closed universe is finite in extent and has no edge, like the surface of a sphere. It has enough mass to stop the present expansion and will eventually collapse. (p. 707)

cold dark matter Class of dark-matter candidates made up of very heavy particles, possibly formed in the very early universe. (p. 731)

collecting area The total area of a telescope capable of capturing incoming radiation. The larger the telescope, the greater its collecting area, and the fainter the objects it can detect. (p. 114)

color index A convenient method of quantifying a star's color by comparing its apparent brightness as measured through different filters. If the star's radiation is well described by a blackbody spectrum, the ratio of its blue intensity (B) to its visual intensity (V) is a measure of the object's surface temperature. (p. 446)

color–magnitude diagram A way of plotting stellar properties, in which absolute magnitude is plotted against color index. (p. 452)

coma An effect occurring during the formation of an off-axis image in a telescope. Stars whose light enters the telescope at a large angle acquire cometlike tails on their images. The brightest part of a comet, often referred to as the "head." (p. 362)

comet A small body, composed mainly of ice and dust, in an elliptical orbit about the Sun. As it comes close to the Sun, some of its material is vaporized to form a gaseous head and extended tail. (p. 362)

comparative planetology The systematic study of the similarities and differences among the planets, with the goal of obtaining deeper insight into how the solar system formed and has evolved in time. (p. 145)

condensation nuclei Dust grains in the interstellar medium which act as seeds around which other material can cluster. The presence of dust was very important in causing matter to clump during the formation of the solar system. (p. 386)

condensation theory Currently favored model of solar system formation which combines features of the old nebular theory with new information about interstellar dust grains, which acted as condensation nuclei. (p. 386)

conservation of mass and energy A fundamental law of modern physics which states that the sum of mass and energy must always remain constant in any physical process. In fusion reactions, the lost mass is converted into energy, primarily in the form of electromagnetic radiation. (p. 426)

constellation A human grouping of stars in the night sky into a recognizable pattern. (p. 6)

contact binary A binary-star system in which both stars have expanded to fill their Roche lobes and the surfaces of the two stars merge. The binary system now consists of two nuclear burning stellar cores surrounded by a continuous common envelope. (p. 536)

continuous spectrum Spectrum in which the radiation is distributed over all frequencies, not just a few specific frequency ranges. A prime example is the blackbody radiation emitted by a hot, dense body. (p. 87)

convection Churning motion resulting from the constant upwelling of warm fluid and the concurrent downward flow of cooler material to take its place. (p. 167)

convection zone Region of the Sun's interior, lying just below the surface, where the material of the Sun is in constant convection motion. This region extends into the solar interior to a depth of about 20,000 km. (p. 409)

Copernican revolution The realization, toward the end of the sixteenth century, that Earth is not at the center of the universe. (p. 39)

core The central region of Earth, surrounded by the mantle. (p. 166); The central region of any planet or star. (p. 407)

core-collapse supernova *See* Type II supernova.

core hydrogen burning The energy burning stage for main-sequence stars, in which the helium is produced by hydrogen fusion in the central region of the star. A typical star spends up to 90 percent of its lifetime in hydrostatic equilibrium brought about by the balance between gravity and the energy generated by core hydrogen burning. (p. 516)

corona One of numerous large, roughly circular regions on the surface of Venus, thought to have been caused by upwelling mantle material causing the planet's crust to bulge outward. (p. 237) The tenuous outer atmosphere of the Sun, which lies just above the chromosphere and, at great distances, turns into the solar wind. (p. 407)

coronal hole Vast regions of the Sun's atmosphere where the density of matter is about 10 times lower than average. The gas there streams freely into space at high speeds, escaping the Sun completely. (p. 417)

cosmic distance scale Collection of indirect distance-measurement techniques that astronomers use to measure distances in the universe. (p. 24)

cosmic evolution The collection of the seven major phases of the history of the universe; namely galactic, stellar, planetary, chemical, biological, cultural, and future evolution. (p. 740)

cosmic microwave background The almost perfectly isotropic radio signal that is the electromagnetic remnant of the Big Bang. (p. 708)

cosmological principle Two assumptions which make up the basis of cosmology, namely that the universe is homogeneous and isotropic on sufficiently large scales. (p. 693)

cosmological redshift The component of the redshift of an object which is due only to the Hubble flow of the universe. (p. 652)

cosmology The study of the structure and evolution of the entire universe. (pp. 36, 693)

crater Bowl-shaped depression on the surface of a planet or moon, resulting from a collision with interplanetary debris. (p. 199)

critical density The cosmic density corresponding to the dividing line between a universe that recollapses and one that expands forever. (p. 700)

critical universe Universe in which the density of matter is exactly equal to the critical density. The universe is infinite in extent and has zero curvature. The expansion will continue forever, but will approach an expansion speed of zero. (p. 707)

crust Layer of Earth which contains the solid continents and the seafloor. (p. 166)

D

D ring Collection of very faint, thin rings, extending from the inner edge of the C ring down nearly to the cloudtops of Saturn. This region contains so few particles that it is completely invisible from Earth. (p. 311)

dark dust cloud A large cloud, often many parsecs across, which contains gas and dust in a ratio of about 1012 gas atoms for every dust particle. Typical densities are a few tens or hundreds of millions of particles per cubic meter. (p. 478)

dark halo Region of a galaxy beyond the visible halo where dark matter is believed to reside. (p. 617)

dark matter Term used to describe the mass in galaxies and clusters whose existence we infer from rotation curves and other techniques, but which has not been confirmed by observations at any electromagnetic wavelength. (p. 617)

declination Celestial coordinate used to measure latitude above or below the celestial equator on the celestial sphere. (p. 14)

decoupling Event in the early universe when atoms first formed, after which photons could propagate freely through space. (p. 725)

deferent A construct of the geocentric model of the solar system which was needed to explain observed planetary motions. A deferent is a large circle encircling Earth, on which an epicycle moves. (p. 37)

degree Unit of angular measure. There are 360 degrees in one complete circle. (p. 11)

density A measure of the compactness of the matter within an object, computed by dividing the mass of the object by its volume. Units are kilograms per cubic meter (kg/m^3), or grams per cubic centimeter (g/cm^3). (p. 147)

deuteron An isotope of hydrogen in which a neutron is bound to the proton in the nucleus. Often called "heavy hydrogen" because of the extra mass of the neutron. (p. 426)

differential rotation The tendency for a gaseous sphere, such as a jovian planet or the Sun, to rotate at a different rate at the equator than at the poles. More generally, a condition where the angular speed varies with location within an object. (p. 277)

differentiation Variation in the density and composition of a body, such as Earth, with low density material on the surface and higher density material in the core. (p. 173)

diffraction The ability of waves to bend around corners. The diffraction of light establishes its wave nature. (p. 66)

Doppler effect Any motion-induced change in the observed wavelength (or frequency) of a wave. (p. 77)

Drake equation Expression that gives an estimate of the probability that intelligence exists elsewhere in the galaxy, based on a number of supposedly necessary conditions for intelligent life to develop. (p. 747)

dust grain An interstellar dust particle, roughly 10^{-7} m in size, comparable to the wavelength of visible light. (p. 469)

dust lane A lane of dark, obscuring interstellar dust in an emission nebula or galaxy. (p. 476)

dust tail The component of a comet's tail that is composed of dust particles. (p. 363)

dwarf Any star with radius comparable to, or smaller than, that of the Sun (including the Sun itself). (p. 450)

dynamo theory Theory that explains planetary and stellar magnetic fields in terms of rotating, conducting material flowing in an object's interior. (p. 186)

E

E ring A faint ring, well outside the main ring system of Saturn, which was discovered by *Voyager* and is believed to be associated with volcanism on the moon Enceladus. (p. 313)

Earth-crossing asteroid An asteroid whose orbit crosses that of Earth. Earth-crossing asteroids are also called Apollo asteroids, after the first asteroid of this type discovered. (p. 360)

earthquake A sudden dislocation of rocky material near Earth's surface. (p. 171)

eccentricity A measure of the flatness of an ellipse, equal to the distance between the two foci divided by the length of the major axis. (p. 44)

eclipse Event during which one body passes in front of another, so that the light from the occulted body is blocked. (p. 15)

eclipse season Times of the year when the Moon lies in the same plane as Earth and Sun, so that eclipses are possible. (p. 18)

eclipsing binary Rare binary-star system that is aligned in such a way that from Earth we observe one star pass in front of the other, eclipsing the other star. (p. 457)

ecliptic The apparent path of the Sun, relative to the stars on the celestial sphere, over the course of a year. (p. 11)

electric field A field extending outward in all directions from a charged particle, such as a proton or an electron. The electric field determines the electric force exerted by the particle on all other charged particles in the universe; the strength of the electric field decreases with increasing distance from the charge according to an inverse-square law. (p. 66)

electromagnetic radiation Another term for light, electromagnetic radiation transfers energy and information from one place to another. (p. 62)

electromagnetic spectrum The complete range of electromagnetic radiation, from radio waves to gamma rays, including the visible spectrum. All types of electromagnetic radiation are basically the same phenomenon, differing only by wavelength, and all move at the speed of light. (p. 70)

electromagnetism The union of electricity and magnetism, which do not exist as independent quantities but are in reality two aspects of a single physical phenomenon. (p. 68)

electron An elementary particle with a negative electric charge; one of the components of the atom. (p. 65)

electron degeneracy pressure The pressure produced by the resistance of electrons to further compression once they are squeezed to the point of contact. (p. 521)

element Matter made up of one particular atom. The number of protons in the nucleus of the atom determines which element it represents. (p. 95)

ellipse Geometric figure resembling an elongated circle. An ellipse is characterized by its degree of flatness, or eccentricity, and the length of its long axis. In general, bound orbits of objects moving under gravity are elliptical. (p. 44)

elliptical galaxy Category of galaxy in which the stars are distributed in an elliptical shape on the sky, ranging from highly elongated to nearly circular in appearance. (p. 632)

emission line Bright line in a specific location of the spectrum of radiating material, corresponding to emission of light at a cer-

tain frequency. A heated gas in a glass container produces emission lines in its spectrum. (p. 87)

emission nebula A glowing cloud of hot interstellar gas. The gas glows as a result of one or more nearby young stars which ionize the gas. Since the gas is mostly hydrogen, the emitted radiation falls predominantly in the red region of the spectrum, because of the hydrogen-alpha emission line. (p. 473)

emission spectrum The pattern of spectral emission lines produced by an element. Each element has its own unique emission spectrum. (p. 88)

Encke gap A small gap in Saturn's A ring. (p. 309)

epicycle A construct of the geocentric model of the solar system which was necessary to explain observed planetary motions. Each planet rides on a small epicycle whose center in turn rides on a larger circle (the deferent). (p. 37)

escape speed The speed necessary for one object to escape the gravitational pull of another. Anything that moves away from a gravitating body with more than the escape speed will never return. (p. 55)

event horizon Imaginary spherical surface surrounding a collapsing star, with radius equal to the Schwarzschild radius, within which no event can be seen, heard, or known about by an outside observer. (p. 581)

evolutionary theory A theory which explains observations in a series of gradual steps, explainable in terms of well-established physical principles. (p. 386)

evolutionary track A graphical representation of a star's life as a path on the Hertzsprung–Russell diagram. (p. 494)

excited state State of an atom when one of its electrons is in a higher energy orbital than the ground state. Atoms can become excited by absorbing a photon of a specific energy, or by colliding with a nearby atom. (p. 92)

extinction The dimming of starlight as it passes through the interstellar medium. (p. 469)

F

F ring Faint narrow outer ring of Saturn, discovered by *Pioneer 11* in 1979. The F ring lies just inside the Roche limit of Saturn, and was found by *Voyager 1* to be made up of several ring strands apparently braided together. (p. 314)

flare Explosive event occurring in or near an active region on the Sun. (p. 423)

flatness problem One of two conceptual problems with the standard Big Bang model, which is that there is no natural way to explain why the density of the universe is so close to the critical density. (p. 726)

fluidized ejecta The ejecta blankets around some Martian craters, which apparently indicate that the ejected material was liquid at the time the crater formed. (p. 257)

focus One of two special points within an ellipse, whose separation from each other indicates the eccentricity. In a bound orbit, planets orbit in ellipses with the Sun at one focus. (p. 44)

forbidden line A spectral line seen in emission nebulae, but not seen in laboratory experiments because, under laboratory conditions, collisions kick the electron in question into some other state before emission can occur. (p. 478)

force Action on an object that causes its momentum to change. The rate at which the momentum changes is numerically equal to the force. (p. 51)

fragmentation The breaking up of a large object into many smaller pieces (for example, as the result of high-speed collisions between planetesimals and protoplanets in the early solar system). (p. 389)

Fraunhofer lines The collection of over 600 absorption lines in the spectrum of the Sun, first categorized by Joseph Fraunhofer in 1812. (p. 88)

frequency The number of wave crests passing any given point in a unit time. (p. 64)

full Moon Phase of the Moon in which it appears as a complete circular disk in the sky. (p. 15)

fusion Mechanism of energy generation in the core of the Sun, in which light nuclei are combined, or fused, into heavier ones, releasing energy in the process. (p. 426)

G

galactic bulge Thick distribution of warm gas and stars around the galactic center. (p. 600)

galactic cannibalism A galaxy merger in which a larger galaxy consumes a smaller one. (p. 649)

galactic center The center of the Milky Way, or any other galaxy. The point about which the disk of a spiral galaxy rotates. (p. 607)

galactic disk Flattened region of gas and dust that bisects the galactic halo in a spiral galaxy. This is the region of active star formation. (p. 600)

galactic halo Region of a galaxy extending far above and below the galactic disk, where globular clusters and other old stars reside. (p. 600)

galactic nucleus Small, central, high-density region of a galaxy. Almost all the radiation from active galaxies is generated within the nucleus. (p. 663)

galaxy Gravitationally bound collection of a large number of stars. The Sun is a star in the Milky Way Galaxy. (p. 600)

galaxy cluster A collection of galaxies held together by their mutual gravitational attraction. (p. 640)

Galilean satellites The four brightest and largest moons of Jupiter (Io, Europa, Ganymede, Callisto), named after Galileo Galilei, the seventeenth-century astronomer who first observed them. (p. 276)

gamma ray Region of the electromagnetic spectrum, far beyond the visible spectrum, corresponding to radiation of very high frequency and very short wavelength. (p. 63)

gamma-ray burst Object that radiates tremendous amounts of energy in the form of gamma rays, possibly due to the collision and merger of two neutron stars initially in orbit around one another. (p. 575)

geocentric model A model of the solar system which holds that Earth is at the center of the universe and all other bodies are in orbit around it. The earliest theories of the solar system were geocentric. (p. 37)

giant A star with a radius between 10 and 100 times that of the Sun. (p. 450)

globular cluster Tightly bound, roughly spherical collection of hundreds of thousands, and sometimes millions, of stars spanning about 50 parsecs. Globular clusters are distributed in the halos around the Milky Way and other galaxies. (p. 506)

Grand Unified Theories Class of theories describing the behavior of the single force that results from unification of the strong, weak, and electromagnetic forces in the early universe. (p. 720)

granulation Mottled appearance of the solar surface, caused by rising (hot) and falling (cool) material in convective cells just below the photosphere. (p. 412)

gravitational field Field created by any object with mass, extending out in all directions, which determines the influence of that object on all others. The strength of the gravitational field decreases as the square of the distance. (p. 53)

gravitational lensing The effect induced on the image of a distant object by a massive foreground object. Light from the distant object is bent into two or more separate images. (p. 618)

gravitational red shift A prediction of Einstein's general theory of relativity. Photons lose energy as they escape the gravitational field of a massive object. Because a photon's energy is proportional to its frequency, a photon that loses energy suffers a decrease in frequency, which corresponds to an increase, or redshift, in wavelength. (p. 586)

gravity The attractive effect that any massive object has on all other massive objects. The greater the mass of the object, the stronger its gravitational pull. (p. 52)

Great Dark Spot Prominent storm system in the atmosphere of Neptune observed by *Voyager 2*, near the equator of the planet. The system was comparable in size to Earth. (p. 335)

Great Red Spot A large, high-pressure, long-lived storm system visible in the atmosphere of Jupiter. The Red Spot is roughly twice the size of Earth. (p. 278)

greenhouse effect The partial trapping of solar radiation by a planetary atmosphere, similar to the trapping of heat in a greenhouse. (p. 170)

ground state The lowest energy state that an electron can have within an atom. (p. 91)

H

heliocentric model A model of the solar system which is centered on the Sun, with Earth in motion about the Sun. (p. 39)

helioseismology The study of conditions far below the Sun's surface through the analysis of internal "sound" waves that repeatedly cross the solar interior. (p. 409)

helium capture The formation of heavy elements by the capture of a helium nucleus. For example, carbon can form heavier elements by fusion with other carbon nuclei, but it is much more likely to occur by helium capture, which requires less energy. (p. 556)

helium flash An explosive event in the post-main-sequence evolution of a low-mass star. When helium fusion begins in a dense stellar core, the burning is explosive in nature. It continues until the energy released is enough to expand the core, at which point the star achieves stable equilibrium again. (p. 522)

helium precipitation Mechanism responsible for the low abundance of helium of Saturn's atmosphere. Helium condenses in the upper layers to form a mist, which rains down toward Saturn's interior, just as water vapor forms into rain in the atmosphere of Earth. (p. 308)

Hertzsprung–Russell diagram A plot of luminosity against temperature (or spectral class) for a group of stars. (p. 450)

high-energy telescope Telescope designed to detect X- and gamma radiation. (p. 133)

highlands Relatively light-colored regions on the surface of the Moon which are elevated several kilometers above the maria. Also called terrae. (p. 199)

homogeneity Assumed property of the universe such that the number of galaxies in an imaginary large cube of the universe is the same no matter where in the universe the cube is placed. (p. 692)

horizon problem One of two conceptual problems with the standard Big Bang model, which is that some regions of the universe which have very similar properties are too far apart to have exchanged information within the age of the universe. (p. 726)

horizontal branch Region of the Hertzsprung–Russell diagram where post-main-sequence stars again reach hydrostatic equilibrium. At this point, the star is burning helium in its core and fusing hydrogen in a shell surrounding the core. (p. 522)

hot dark matter A class of candidates for the dark matter in the universe, composed of lightweight particles such as neutrinos, much less massive than the electron. (p. 731)

Hubble classification scheme Method of classifying galaxies according to their appearance, developed by Edwin Hubble. (p. 631)

Hubble's constant The constant of proportionality which gives the relation between recessional velocity and distance in Hubble's law. (p. 653)

Hubble's law Law that relates the observed velocity of recession of a galaxy to its distance from us. The velocity of recession of a galaxy is directly proportional to its distance away. (p. 652)

hydrogen envelope An invisible sheath of gas engulfing the coma of a comet, usually distorted by the solar wind and extending across millions of kilometers of space. (p. 363)

hydrogen shell burning Fusion of hydrogen in a shell that is driven by contraction and heating of the helium core. Once hydrogen is depleted in the core of a star, hydrogen burning stops and the core contracts due to gravity, causing the temperature to rise, heating the surrounding layers of hydrogen in the star, and increasing the burning rate there. (p. 519)

hydrosphere Layer of Earth which contains the liquid oceans and accounts for roughly 70 percent of Earth's total surface area. (p. 166)

I

image The optical representation of an object produced when light from the object is reflected or refracted by a mirror or lens. (p. 108)

inertia The tendency of an object to continue moving at the same speed and in the same direction, unless acted upon by a force. (p. 51)

inflation Short period of unchecked cosmic expansion early in the history of the universe. During inflation, the universe swelled in size by a factor of about 10^{50}. (p. 728)

infrared Region of the electromagnetic spectrum just outside the visible range, corresponding to light of a slightly longer wavelength than red light. (p. 63)

infrared telescope Telescope designed to detect infrared radiation. Many such telescopes are designed to be lightweight so that they can be carried above (most of) Earth's atmosphere by balloons, airplanes, or satellites. (p. 130)

inner core The central part of Earth's core, believed to be solid, and composed mainly of nickel and iron. (p. 172)

intensity A basic property of electromagnetic radiation that specifies the amount or strength of the radiation. (p. 72)

intercrater plains Regions on the surface of Mercury that do not show extensive cratering but are relatively smooth. (p. 214)

interference The ability of two or more waves to interact in such a way that they either reinforce or cancel each other. (p. 67)

interferometer Collection of two or more telescopes working together as a team, observing the same object at the same time and at the same wavelength. The effective diameter of an interferometer is equal to the distance between its outermost telescopes. (p. 127)

interferometry Technique in widespread use to dramatically improve the resolution of radio and infrared maps. Several telescopes observe the object simultaneously, and a computer analyzes how the signals interfere with each other. (p. 127)

interstellar dust Microscopic dust grains that populate space between the stars, having their origins in the ejected matter of long-dead stars. (p. 386)

interstellar medium The matter between stars, composed of two components, gas and dust, intermixed throughout all of space. (p. 468)

inverse-square law The law that a field follows if its strength decreases with the square of the distance. Fields that follow the inverse-square law decrease rapidly in strength as the distance increases, but never quite reach zero. (p. 52)

Io plasma torus Doughnut-shaped region of energetic ionized particles, emitted by the volcanoes on Jupiter's moon Io and swept up by Jupiter's magnetic field. (p. 290)

ion Atom that has had at least one of its electrons removed. (p. 290)

ion tail Thin stream of ionized gas that is pushed away from the head of a comet by the solar wind. It extends directly away from the Sun. Often referred to as a plasma tail. (p. 363)

ionosphere Layer in Earth's atmosphere above about 100 km where the atmosphere is significantly ionized and conducts electricity. (p. 167)

irregular galaxy A galaxy which does not fit into any of the other major categories in the Hubble classification scheme. (p. 634)

isotopes Nuclei containing the same number of protons but different numbers of neutrons. Most elements can exist in several isotopic forms. A common example of an isotope is deuterium, which differs from normal hydrogen by the presence of an extra neutron in the nucleus. (p. 426)

isotropy Assumed property of the universe such that the universe looks the same in every direction. (p. 692)

J

jovian planet One of the four giant outer planets of the solar system, resembling Jupiter in physical and chemical composition. (p. 150)

K

Kepler's laws of planetary motion Three laws, based on precise observations of the motions of the planets by Tycho Brahe, which summarize the motions of the planets about the Sun. (p. 44)

Kirchhoff's laws Three rules governing the formation of different types of spectra. (p. 89)

Kirkwood gaps Gaps in the spacings of orbital semimajor axes of asteroids in the asteroid belt, produced by dynamical resonances with nearby planets, especially Jupiter. (p. 361)

Kuiper belt A region in the plane of the solar system outside the orbit of Neptune where most short-period comets are thought to originate. (p. 364)

L

Lagrangian point One of five special points in the plane of two massive bodies orbiting one another, where a third body of negligible mass can remain in equilibrium. (p. 321)

lava dome Volcanic formation formed when lava oozes out of fissures in a planet's surface, creating the dome, and then withdraws, causing the crust to crack and subside. (p. 236)

light *See* electromagnetic radiation.

light curve The variation in brightness of a star with time. (p. 457)

lighthouse model The leading explanation for pulsars. A small region of the neutron star, near one of the magnetic poles, emits a steady stream of radiation which sweeps past Earth each time the star rotates. The period of the pulses is the star's rotation period. (p. 569)

light-year The distance that light, moving at a constant speed of 300,000 km/s, travels in one year. One light-year is about 10 trillion kilometers. (p. 4)

line of nodes The line of intersection of the Moon's orbit with the ecliptic plane. (p. 18)

lithosphere Earth's crust and a small portion of the upper mantle that make up Earth's plates. This layer of Earth undergoes tectonic activity. (p. 177)

Local Group The small galaxy cluster that includes the Milky Way Galaxy. (p. 640)

luminosity One of the basic properties used to characterize stars, luminosity is defined as the total energy radiated by star each second, at all wavelengths. (p. 408)

luminosity class A classification scheme which groups stars according to the width of their spectral lines. For a group of stars with the same temperature, luminosity class differentiates between supergiants, giants, main-sequence stars, and subdwarfs. (pp. 408, 456)

lunar eclipse Celestial event during which the moon passes through the shadow of Earth, temporarily darkening its surface. (p. 15)

lunar phase The appearance of the Moon at different points along its orbit. (p. 15)

M

Magellanic Clouds Two small irregular galaxies that are gravitationally bound to the Milky Way Galaxy. (p. 634)

magnetic field Field which accompanies any changing electric field and governs the influence of magnetized objects on one another. (pp. 67, 634)

magnetosphere A zone of charged particles trapped by a planet's magnetic field, lying above the atmosphere. (p. 167)

magnitude scale A system of ranking stars by apparent brightness, developed by the Greek astronomer Hipparchus. Originally, the brightest stars in the sky were categorized as being of first magnitude, while the faintest stars visible to the naked eye were classified as sixth magnitude. The scheme has since been extended to cover stars and galaxies too faint to be seen by the unaided eye. Increasing magnitude means fainter stars, and a difference of

five magnitudes corresponds to a factor of 100 in apparent brightness. (p. 443)

main sequence Well-defined band on the Hertzsprung–Russell diagram on which most stars are found, running from the top left of the diagram to the bottom right. (p. 452)

main-sequence turnoff Special point on the Hertzsprung–Russell diagram for a cluster, indicative of the cluster's age. If all the stars in the cluster are plotted, the lower mass stars will trace out the main sequence up to the point where stars begin to evolve off the main sequence toward the red giant branch. The point where stars are just beginning to evolve off is the main-sequence turnoff. (p. 533)

mantle Layer of Earth just interior to the crust. (p. 166)

mare Relatively dark-colored and smooth region on the surface of the Moon (plural: *maria*). (p. 199)

mass A measure of the total amount of matter contained within an object. (p. 51)

mass–luminosity relation The dependence of the luminosity of a main-sequence star on its mass. The luminosity increases roughly as the mass raised to the third power. (p. 459)

mass–radius relation The dependence of the radius of a main-sequence star on its mass. The radius rises roughly in proportion to the mass. (p. 459)

mass-transfer binary *See* semi-detached binary. (p. 536)

matter-dominated universe A universe in which the density of matter exceeds the density of radiation. The present-day universe is matter dominated. (p. 716)

mesosphere Region of Earth's atmosphere lying between the stratosphere and the ionosphere, 50–80 km above Earth's surface. (p. 167)

meteor Bright streak in the sky, often referred to as a "shooting star," resulting from a small piece of interplanetary debris entering Earth's atmosphere and heating air molecules, which emit light as they return to their ground states. (p. 372)

meteor shower Event during which many meteors can be seen each hour, caused by the yearly passage of Earth through the debris spread along the orbit of a comet. (p. 372)

meteorite Any part of a meteoroid that survives passage through the atmosphere and lands on the surface of Earth. (p. 372)

meteoroid Chunk of interplanetary debris prior to encountering Earth's atmosphere. (p. 372)

meteoroid swarm Pebble-sized cometary fragments dislodged from the main body, moving in nearly the same orbit as the parent comet. (p. 372)

micrometeoroid Relatively small chunks of interplanetary debris ranging from dust-particle size to pebble-sized fragments. (p. 372)

Milky Way Galaxy The spiral galaxy in which the Sun resides. The disk of our Galaxy is visible in the night sky as the faint band of light known as the Milky Way. (p. 600)

millisecond pulsar A pulsar whose period indicates that the neutron star is rotating nearly 1000 times each second. The most likely explanation for these rapid rotators is that the neutron star has been spun up by drawing in matter from a companion star. (p. 573)

molecular cloud A cold, dense interstellar cloud which contains a high fraction of molecules. It is widely believed that the relatively high density of dust particles in these clouds plays an important role in the formation and preservation of the molecules. (p. 482)

molecular cloud complex Collection of molecular clouds that spans as much as 50 parsecs and may contain enough material to make millions of Sun-size stars. (p. 483)

molecule A tightly bound collection of atoms held together by the atoms' electromagnetic fields. Molecules, like atoms, emit and absorb photons at specific wavelengths. (p. 96)

moon A small body in orbit around a planet. (p. 144)

N

nebula General term used for any "fuzzy" patch on the sky, either light or dark. (p. 473)

nebular theory One of the earliest models of solar system formation, dating back to Descartes, in which a large cloud of gas began to collapse under its own gravity to form the Sun and planets. (p. 385)

neutrino Virtually massless and chargeless particle that is one of the products of fusion reactions in the Sun. Neutrinos move at close to the speed of light, and interact with matter hardly at all. (p. 427)

neutrino oscillations Possible solution to the solar neutrino problem, in which the neutrino has a very tiny mass. In this case, the correct number of neutrinos can be produced in the solar core, but on their way to Earth some can "oscillate," or become transformed into other particles, and thus go undetected. (p. 430)

neutron An elementary particle with roughly the same mass as a proton, but which is electrically neutral. Along with protons, neutrons form the nuclei of atoms. (p. 95)

neutron capture The primary mechanism by which very massive nuclei are formed in the violent aftermath of a supernova. Instead of fusion of like nuclei, heavy elements are created by the addition of more and more neutrons to existing nuclei. (p. 559)

neutron star A dense ball of neutrons that remains at the core of a star after a supernova explosion has destroyed the rest of the star. Typical neutron stars are about 20 km across, and contain more mass than the Sun. (p. 568)

new Moon Phase of the moon during which none of the lunar disk is visible. (p. 15)

Newtonian mechanics The basic laws of motion, postulated by Newton, which are sufficient to explain and quantify virtually all of the complex dynamical behavior found on Earth and elsewhere in the universe. (p. 51)

Newtonian telescope A reflecting telescope in which incoming light is intercepted before it reaches the prime focus and is deflected into an eyepiece at the side of the instrument. (p. 111)

north celestial pole Point on the celestial sphere directly above Earth's North Pole. (p. 8)

nova A star that suddenly increases in brightness, often by a factor of as much as 10,000, then slowly fades back to its original luminosity. A nova is the result of an explosion on the surface of a white-dwarf star, caused by matter falling onto its surface from the atmosphere of a binary companion. (p. 544)

nuclear fusion *See* fusion.

nucleotide base An organic molecule, the building block of genes that pass on hereditary characteristics from one generation of living creatures to the next. (p. 742)

nucleus Dense, central region of an atom, containing both protons and neutrons, and orbited by one or more electrons. (p. 91) The solid region of ice and dust that composes the central region of the head of a comet. (p. 362)

O

Olbers's paradox A thought experiment suggesting that if the universe were homogeneous, infinite, and unchanging, the entire night sky would be as bright as the surface of the Sun. (p. 694)

Oort cloud Spherical halo of material surrounding the solar system out to a distance of about 50,000 A.U., where most comets reside. (p. 365)

opacity A quantity that measures a material's ability to block electromagnetic radiation. Opacity is the opposite of transparency. (p. 70)

open cluster Loosely bound collection of tens to hundreds of stars, a few parsecs across, generally found in the plane of the Milky Way. (p. 505)

open universe Geometry that the universe would have if the density of matter were less than the critical value. In an open universe there is not enough matter to halt the expansion of the universe. An open universe is infinite in extent. (p. 707)

outer core The outermost part of Earth's core, believed to be liquid and composed mainly of nickel and iron. (p. 172)

outflow channel Surface feature on Mars, evidence that liquid water once existed there in great quantity; believed to be the relics of catastrophic flooding about 3 billion years ago. Found only in the equatorial regions of the planet. (p. 259)

ozone layer Layer of Earth's atmosphere at an altitude of 20–50 km where incoming ultraviolet solar radiation is absorbed by oxygen, ozone, and nitrogen in the atmosphere. (p. 168)

P

pair production Process in which two photons of electromagnetic radiation give rise to a particle-antiparticle pair. (p. 717)

parallax The apparent motion of a relatively close object with respect to a more distant background as the location of the observer changes. (p. 25)

parsec The distance at which a star must lie in order for its measured parallax to be exactly 1 arc second; 1 parsec equals 206,000 A.U. (p. 439)

partial eclipse Celestial event during which only a part of the occulted body is blocked from view. (p. 15)

penumbra Portion of the shadow cast by an eclipsing object in which the eclipse is seen as partial. (p. 17)

penumbra The outer region of a sunspot, surrounding the umbra, which is not as dark and not as cool as the central region. (p. 418)

perihelion The closest approach to the Sun of any object in orbit about it. (p. 46)

period The time needed for an orbiting body to complete one revolution about another body. (p. 46)

period–luminosity relation A relation between the pulsation period of a Cepheid variable and its absolute brightness. Measurement of the pulsation period allows the distance of the star to be determined. (p. 605)

permafrost Layer of permanently frozen water ice believed to lie just under the surface of Mars. (p. 257)

photometer A device that measures the total amount of light received in all or part of the image. (p. 103)

photometry Branch of observational astronomy in which the brightness of a source is measured through each of a set of standard filters. (p. 446)

photon Individual packet of electromagnetic energy that makes up electromagnetic radiation. (p. 92)

photosphere The visible surface of the Sun, lying just above the uppermost layer of the Sun's interior, and just below the chromosphere. (p. 407)

pixel One of many tiny picture elements, organized into an array, making up a digital image. (p. 120)

Planck curve *See* blackbody curve.

planet One of nine major bodies that orbit the Sun, visible to us by reflected sunlight. (p. 144)

planetary nebula The ejected envelope of a red-giant star, spread over a volume roughly the size of our solar system. (p. 525)

planetary ring system Material organized into thin, flat rings encircling a giant planet, such as Saturn. (p. 309)

planetesimal Term given to objects in the early solar system that had reached the size of small moons, at which point their gravitational fields were strong enough to begin influencing their neighbors. (p. 389)

plate tectonics The motions of regions of Earth's lithosphere, which drift with respect to one another. Also known as continental drift. (p. 176)

polarization The alignment of the electric fields of emitted photons, which are generally emitted with random orientations. (p. 471)

positron Atomic particle with properties identical to those a negatively charged electron, except for its positive charge. This positron is the antiparticle of the electron. Positrons and electrons annihilate one another when they meet, producing pure energy in the form of gamma rays. (p. 426)

precession The slow change in the direction of the rotation axis of a spinning object, caused by some external gravitational influence. (p. 13)

primary atmosphere The chemical components that would have formed Earth's atmosphere. (p. 171)

prime focus The point in a reflecting telescope where the mirror focuses incoming light to a point. (p. 108)

primordial nucleosynthesis The production of elements heavier than hydrogen by nuclear fusion in the high temperatures and densities which existed in the early universe. (p. 721)

principle of cosmic censorship A proposition to separate the unexplained physics near a singularity from the rest of the well-behaved universe. The principle states that nature always hides any singularity, such as a black hole, inside an event horizon, which insulates the rest of the universe from seeing it. (p. 589)

prominence Loop or sheet of glowing gas ejected from an active region on the solar surface, which then moves through the inner parts of the corona under the influence of the Sun's magnetic field. (p. 422)

proper motion The angular movement of a star across the sky, as seen from Earth, measured in seconds of arc per year. This movement is a result of the star's actual motion through space. (p. 439)

proton An elementary particle carrying a positive electric charge, a component of all atomic nuclei. The number of protons in the nucleus of an atom dictates what type of atom it is. (p. 65)

proton–proton chain The chain of fusion reactions, leading from hydrogen to helium, that powers main-sequence stars. (p. 428)

protoplanet Clump of material, formed in the early stages of solar system formation, that was the forerunner of the planets we see today. (p. 386)

protostar Stage in star formation when the interior of a collapsing fragment of gas is sufficiently hot and dense that it becomes opaque to its own radiation. The protostar is the dense region at the center of the fragment. (p. 494)

protosun The central accumulation of material in the early stages of solar system formations, the forerunner of the present-day Sun. (p. 386)

Ptolemaic model Geocentric solar system model, developed by the second century astronomer Claudius Ptolemy. It predicted with great accuracy the positions of the then known planets. (p. 38)

pulsar Object that emits radiation in the form of rapid pulses with a characteristic pulse period and duration. Charged particles, accelerated by the magnetic field of a rapidly rotating neutron star, flow along the magnetic field lines, producing radiation that beams outward as the star spins on its axis. (p. 569)

pulsating variable star A star whose luminosity varies in a predictable, periodic way. (p. 604)

Q

quantization The fact that light and matter on small scales behave in a discontinuous manner, and manifest themselves in the form of tiny "packets" of energy, called quanta. (p. 91)

quarter Moon Lunar phase in which the Moon appears as a half disk. (p. 15)

quasar Starlike radio source with an observed redshift that indicates an extremely large distance from Earth. The brightest nucleus of a distant active galaxy. (p. 669)

quasi-stellar object (QSO) *See* quasar.

quiet Sun The underlying predictable elements of the Sun's behavior, such as its average photospheric temperature, which do not change in time. (p. 425)

R

radar Acronym for RAdio Detection And Ranging. Radio waves are bounced off an object, and the time taken for the echo to return indicates its distance. (p. 48)

radial motion Motion along a particular line of sight, which induces apparent changes in the wavelength (or frequency) of radiation received. (p. 78)

radiation A way in which energy is transferred from place to place in the form of a wave. Light is a form of electromagnetic radiation. (p. 62)

radiation darkening The effect of chemical reactions that result when high-energy particles strike the icy surfaces of objects in the outer solar system. The reactions lead to a buildup of a dark layer of material. (p. 338)

radiation-dominated universe Early epoch in the universe, when the equivalent density of radiation in the cosmos exceeded the density of matter. (p. 717)

radiation zone Region of the Sun's interior where extremely high temperatures guarantee that the gas is completely ionized. Photons only occasionally interact with electrons, and travel through this region with relative ease. (p. 407)

radio Region of the electromagnetic spectrum corresponding to radiation of the longest wavelengths. (p. 63)

radio galaxy Type of active galaxy that emits most of its energy in the form of long-wavelength radiation. (p. 664)

radio lobe Roundish extended region of radio-emitting gas, lying well beyond the center of a radio galaxy. (p. 664)

radio telescope Large instrument designed to detect radiation from space at radio wavelengths. (p. 124)

radioactivity The release of energy by rare, heavy elements when their nuclei decay into lighter nuclei. (p. 175)

radius–luminosity–temperature relationship A mathematical proportionality, arising from Stefan's Law, which allows astronomers to indirectly determine the radius of a star once its luminosity and temperature are known. (p. 449)

red dwarf Small, cool faint star at the lower-right end of the main sequence on the Hertzsprung–Russell diagram. (p. 453)

red giant A giant star whose surface temperature is relatively low so that it glows red. (p. 450)

red-giant branch The section of the evolutionary track of a star corresponding to intense hydrogen shell burning, which drives a steady expansion and cooling of the outer envelope of the star. As the star gets larger in radius and its surface temperature cools, it becomes a red giant. (p. 519)

red-giant region The upper-right corner of the Hertzsprung–Russell diagram, where red-giant stars are found. (p. 454)

redshift Motion-induced change in the wavelength of light emitted from a source moving away from us. The relative recessional motion causes the wave to have an observed wavelength longer (and hence redder) than it would if it were not moving. (p. 79)

red supergiant An extremely luminous red star. Often found on the asymptotic-giant branch of the Hertzsprung–Russell diagram. (p. 523)

reddening Dimming of starlight by interstellar matter, which tends to scatter higher-frequency (blue) components of the radiation more efficiently than the lower-frequency (red) components. (p. 469)

reflecting telescope A telescope which uses a mirror to gather and focus light from a distant object. (p. 108)

refracting telescope A telescope which uses a lens to gather and focus light from a distant object. (p. 108)

refraction The tendency of a wave to bend as it passes from one transparent medium to another. (p. 108)

residual cap Portion of Martian polar ice caps that remains permanently frozen, undergoing no seasonal variations. (p. 262)

retrograde motion Backward, westward loop traced out by a planet with respect to the fixed stars. (p. 36)

revolution Orbital motion of one body about another, such as Earth about the Sun. (p. 9)

right ascension Celestial coordinate used to measure longitude on the celestial sphere. The zero point is the position of the Sun at the vernal equinox. (p. 14)

rille A ditch on the surface of the Moon where molten lava flowed in the past. (p. 212)

ringlet Narrow region in Saturn's planetary ring system where the density of ring particles is high. *Voyager* discovered that the rings visible from Earth are actually composed of tens of thousands of ringlets. (p. 311)

Roche limit Often called the tidal stability limit, the Roche limit gives the distance from a planet at which the tidal force (due to the planet) between adjacent objects exceeds their mutual attraction. Objects within this limit are unlikely to accumulate into larger objects. The rings of Saturn occupy the region within Saturn's Roche limit. (p. 310)

Roche lobe An imaginary surface around a star. Each star in a binary system can be pictured as being surrounded by a tear-drop shaped zone of gravitational influence, the Roche lobe. Any material within the Roche lobe of a star can be considered to be part of that star. During evolution, one member of the binary system can expand so that it overflows its own Roche lobe and begins to transfer matter onto the other star. (pp. 310, 535)

rotation Spinning motion of a body about an axis. (p. 8)

rotation curve Plot of the orbital speed of disk material in a galaxy against its distance from the galactic center. Analysis of rotation curves of spiral galaxies indicates the existence of dark matter. (p. 617)

RR Lyrae star Variable star whose luminosity changes in a characteristic way. All RR Lyrae stars have more or less the same average luminosity. (p. 604)

runaway greenhouse effect A process in which the heating of a planet leads to an increase in its atmosphere's ability to retain heat and thus to further heating, causing extreme changes in the temperature of the surface and the composition of the atmosphere. (p. 242)

runoff channel Riverlike surface feature on Mars, evidence that liquid water once existed there in great quantities. They are found in the southern highlands, and are thought to have been formed by water that flowed nearly 4 billion years ago. (p. 258)

S

S0 galaxy Galaxy which shows evidence of a thin disk and a bulge, but which has no spiral arms and contains little or no gas. (p. 634)

SB0 galaxy An S0-type galaxy whose disk shows evidence of a bar. (p. 634)

scarp Surface feature on Mercury believed to be the result of cooling and shrinking of the crust forming a wrinkle on the face of the planet. (p. 214)

Schwarzschild radius The distance from the center of an object such that, if all the mass were compressed within that region, the escape speed would equal the speed of light. Once a stellar remnant collapses within this radius, light cannot escape and the object is no longer visible. (p. 581)

scientific method The set of rules used to guide science, based on the idea that scientific "laws" be continually tested, and modified or replaced if found inadequate. (p. 44)

seasonal cap Portion of Martian polar ice caps that is subject to seasonal variations, growing and shrinking once each Martian year. (p. 262)

seasons Changes in average temperature and length of day that result from the tilt of Earth's (or any planet's) axis with respect to the plane of its orbit. (p. 13)

secondary atmosphere The chemicals that composed Earth's atmosphere after the planet's formation, once volcanic activity outgassed chemicals from the interior. (p. 171)

seeing A term used to describe the ease with which good telescopic observations can be made from Earth's surface, given the blurring effects of atmospheric turbulence. (p. 118)

seeing disk Roughly circular region on a detector over which a star's pointlike images is spread, due to atmospheric turbulence. (p. 118)

seismic wave A wave that travels outward from the site of an earthquake through Earth. (p. 171)

semimajor axis One-half of the major axis of an ellipse. The semimajor axis is the way in which the size of an ellipse is usually quantified. (p. 44)

Seyfert galaxy Type of active galaxy whose emission comes from a very small region within the nucleus of an otherwise normal-looking spiral system. (p. 663)

shepherd satellite Satellite whose gravitational effect on a ring helps preserve the ring's shape. Examples are two satellites of Saturn, Prometheus and Pandora, whose orbits lie on either side of the F ring. (p. 314)

shield volcano A volcano produced by repeated nonexplosive eruptions of lava, creating a gradually sloping, shield-shaped low dome. Often contains a caldera at its summit. (p. 236)

shock wave Wave of matter, which may be generated by a newborn star or supernova, which pushes material outward into the surrounding molecular cloud. The material tends to pile up, forming a rapidly moving shell of dense gas. (p. 504)

sidereal day The time needed between successive risings of a given star. (p. 9)

sidereal month Time required for the Moon to complete one trip around the celestial sphere. (p. 15)

sidereal year The time required for the constellations to complete one cycle around the sky and return to their starting points, as seen from a given point on Earth. Earth's orbital period around the Sun is one sidereal year (p. 13)

singularity A point in the universe where the density of matter and the gravitational field are infinite, such as at the center of a black hole. (p. 588)

solar constant The amount of solar energy reaching Earth per unit area per unit time, approximately 1400 W/m². (p. 408)

solar core The region at the center of the Sun, with a radius of nearly 200,000 km, where powerful nuclear reactions generate the Sun's energy output. (p. 407)

solar cycle The 22-year period that is needed for both the average number of spots and the Sun's magnetic polarity to repeat themselves. The Sun's polarity reverses on each new 11-year sunspot cycle. (p. 421)

solar day The period of time between the instant when the Sun is directly overhead (i.e., noon) to the next time it is directly overhead. (p. 9)

solar eclipse Celestial event during which the new Moon passes directly between Earth and the Sun, temporarily blocking the Sun's light. (p. 15)

solar interior The region of the Sun between the solar core and the photosphere. (p. 409)

solar maximum Point of the sunspot cycle during which many spots are seen. They are generally confined to regions in each hemisphere, between about 15 and 20 degrees latitude. (p. 421)

solar minimum Point of the sunspot cycle during which only a few spots are seen. They are generally confined to narrow regions in each hemisphere at about 25–30 degrees latitude. (p. 421)

solar nebula The swirling gas surrounding the early Sun during the epoch of solar system formation, also referred to as the primitive solar system. (p. 385)

solar neutrino problem The discrepancy between the theoretically predicted flux of neutrinos streaming from the Sun as a result of fusion reactions in the core, and the flux which is actually observed. The observed number of neutrinos is only about half the predicted number. (p. 429)

solar system The Sun and all the bodies that orbit it—Mercury, Venus, Earth, Mars, Jupiter, Saturn, Uranus, Neptune, Pluto, their moons, the asteroids, and the comets. (p. 145)

solar wind An outward flow of fast-moving charged particles from the Sun. (p. 417)

south celestial pole Point on the celestial sphere directly above Earth's South Pole. (p. 8)

spectral class Classification scheme, based on the strength of stellar spectral lines, which is an indication of the temperature of a star. (p. 448)

spectrograph Instrument used to produce detailed spectra of stars. Usually, a spectrograph records a spectrum on a CCD detector, for computer analysis. (p. 86)

spectrometer Instrument used to produce detailed spectra of stars. Usually, a spectrograph records a spectrum on a photographic plate, or more recently, in electronic form on a computer. (p. 113)

spectroscope Instrument used to view a light source so that it is split into its component colors. (p. 86)

spectroscopic binary A binary-star system which appears as a single star from Earth, but whose spectral lines show back-and-forth Doppler shifts as two stars orbit one another. (p. 457)

spectroscopic parallax Method of determining the distance to a star by measuring its temperature and then determining its absolute brightness by comparing with a standard Hertzsprung–Russell diagram. The absolute and apparent brightness of the star give the star's distance from Earth. (p. 454)

spectroscopy The study of the way in which atoms absorb and emit electromagnetic radiation. Spectroscopy allows astronomers to determine the chemical composition of stars. (p. 89)

speed of light The fastest possible speed, according to the currently known laws of physics. Electromagnetic radiation exists in the form of waves or photons moving at the speed of light. (p. 70)

spin–orbit resonance State that a body is said to be in if its rotation period and its orbital period are related in some simple way. (p. 204)

spiral arm Distribution of material in a galaxy forming a pinwheel-shaped design, beginning near the galactic center. (p. 613)

spiral density wave A wave of matter formed in the plane of planetary rings, similar to ripples on the surface of a pond, which wrap around the rings forming spiral patterns similar to grooves in a record disk. Spiral density waves can lead to the appearance of ringlets. (p. 311)

spiral density wave A proposed explanation for the existence of galactic spiral arms, in which coiled waves of gas compression move through the galactic disk, triggering star formation. (p. 615)

spiral galaxy Galaxy composed of a flattened, star-forming disk component which may have spiral arms and a large central galactic bulge. (pp. 603, 631)

standard candle Any object with an easily recognizable appearance and known luminosity, which can be used in estimating distances. Supernovae, which all have the same peak luminosity (depending on type) are good examples of standard candles and are used to determine distances to other galaxies. (p. 637)

Standard Solar Model A self-consistent picture of the Sun, developed by incorporating the important physical processes that are believed to be important in determining the Sun's internal structure into a computer program. The results of the program are then compared with observations of the Sun and modifications are made to the model. The Standard Solar Model, which enjoys widespread acceptance, is the result of this process. (p. 409)

star A glowing ball of gas held together by its own gravity and powered by nuclear fusion in its core. (p. 406)

star cluster A grouping of anywhere from a dozen to a million stars which formed at the same time from the same cloud of interstellar gas. Stars in clusters are useful to aid our understanding of stellar evolution because, within a given cluster, stars are all roughly the same age and chemical composition and lie at roughly the same distance from Earth. (p. 505)

starburst galaxy Galaxy in which a violent event, such as near-collision, has caused a sudden, intense burst of star formation in the recent past. (p. 650)

Stefan's law Relation that gives the total energy emitted per square centimeter of its surface per second by an object of a given temperature. Stefan's law shows that the energy emitted increases rapidly with an increase in temperature, proportional to the temperature raised to the fourth power. (p. 75)

stellar nucleosynthesis The formation of heavy elements by the fusion of lighter nuclei in the hearts of stars. Except for hydrogen and helium, all other elements in our universe resulted from stellar nucleosynthesis. (p. 553)

stellar occultation The dimming of starlight produced when a solar system object such as a planet, moon, or ring passes directly in front of a star. (p. 342)

stratosphere The portion of Earth's atmosphere lying above the troposphere, extending up to an altitude of 40–50 km. (p. 167)

strong nuclear force Short-range force responsible for binding atomic nuclei together. The strongest of the four fundamental forces of nature. (p. 426)

subgiant branch The section of the evolutionary track of a star that corresponds to changes that occur just after hydrogen is depleted in the core, and core hydrogen burning ceases. Shell hydrogen burning heats the outer layers of the star, which causes a general expansion of the stellar envelope. (p. 519)

summer solstice Point on the ecliptic where the Sun is at its northernmost point above the celestial equator, occurring on or near June 21. (p. 11)

sunspot An Earth-sized dark blemish found on the surface of the Sun. The dark color of the sunspot indicates that it is a region of lower temperature than its surroundings. (p. 418)

sunspot cycle The fairly regular pattern that the number and distribution of sunspots follows, in which the average number of spots reaches a maximum every 11 or so years, then fall off to almost zero. (p. 421)

supercluster Grouping of several clusters of galaxies into a larger, but not necessarily gravitationally bound, unit. (p. 640)

supergiant A star with a radius between 100 and 1000 times that of the Sun. (p. 450)

supergranulation Large-scale flow pattern on the surface of the Sun, consisting of cells measuring up to 30,000 km across, believed to be the imprint of large convective cells deep in the solar interior. (p. 413)

supernova Explosive death of a star, caused by the sudden onset of nuclear burning (Type I), or an enormously energetic shock wave (Type II). One of the most energetic events of the universe, a supernova may temporarily outshine the rest of the galaxy in which it resides. (p. 548)

supernova remnant The scattered glowing remains from a supernova that occurred in the past. The Crab Nebula is one of the best-studied supernova remnants. (p. 550)

synchrotron radiation Type of nonthermal radiation produced by high-speed charged particles, such as electrons, as they are accelerated in a strong magnetic field. (p. 676)

synchronous orbit State of an object when its period of rotations is exactly equal to its average orbital period. The Moon is in

a synchronous orbit, and so presents the same face toward Earth at all times. (p. 202)

synodic month Time required for the Moon to complete a full cycle of phases. (p. 15)

synodic period Time required for a body to return to the same apparent position relative to the Sun, taking Earth's own motion into account; (for a planet) the time between one closest approach to Earth and the next. (p. 215)

T

T Tauri star Protostar in the late stages of formation, often exhibiting violent surface activity. T Tauri stars have been observed to brighten noticeably in a short period of time, consistent with the idea of rapid evolution during this final phase of stellar formation. (p. 496)

tail Component of a comet that consists of material streaming away from the main body, sometimes spanning hundreds of millions of kilometers. May be composed of dust or ionized gases. (p. 362)

telescope Instrument used to capture as many photons as possible from a given region of the sky and concentrate them into a focused beam for analysis. (p. 108)

temperature A measure of the amount of heat in an object, and an indication of the speed of the particles that comprise it. (p. 72)

terrae *See* highlands.

terrestrial planet One of the four innermost planets of the solar system, resembling Earth in general physical and chemical properties. (p. 150)

theories of relativity Einstein's theories, on which much of modern physics rests. Two essential facts of the theory are that nothing can travel faster than the speed of light, and that everything, including light, is affected by gravity. (p. 581)

thick disk Region of a spiral galaxy where an intermediate population of stars resides, younger than the halo stars but older than stars in the disk. (p. 608)

tidal bulge Elongation of Earth caused by the difference between the gravitational force on the side nearest the Moon and the force on the side farthest from the Moon. The long axis of the tidal bulge points toward the Moon. More generally, the deformation of any body produced by the tidal effect of a nearby gravitating object. (p. 187)

tidal force The variation in one body's gravitational force from place to place across another body—for example, the variation of the Moon's gravity across Earth. (p. 187)

tides Rising and falling motion of terrestrial bodies of water, exhibiting daily, monthly and yearly cycles. Ocean tides on Earth are caused by the competing gravitational pull of the Moon and Sun on different parts of Earth. (p. 187)

time dilation A prediction of the theory of relativity, closely related to the gravitational reshift. To an outside observer, a clock lowered into a strong gravitational field will appear to run slow. (p. 588)

total eclipse Celestial event during which one body is completely blocked from view by another. (p. 15)

transition zone The region of rapid temperature increases that separates the Sun's chromosphere from the corona. (p. 407)

transverse motion Motion perpendicular to a particular line of sight, which does not result in Doppler shift in radiation received. (p. 78)

triangulation Method of determining distance based on the principles of geometry. A distant object is sighted from two well-separated locations. The distance between the two locations and the angle between the line joining them and the line to the distant object are all that are necessary to ascertain the object's distance. (p. 24)

triple-alpha process The creation of carbon-12 by the fusion of three helium-4 nuclei (alpha particles). Helium-burning stars occupy a region of the Hertzsprung–Russell diagram known as the horizontal branch. (p. 520)

Trojan asteroid One of two groups of asteroids which orbit at the same distance from the Sun as Jupiter, 60 degrees ahead of and behind the planet. (p. 361)

tropical year The time interval between one vernal equinox and the next. (p. 13)

troposphere The portion of Earth's atmosphere from the surface to about 15 km. (p. 167)

Tully-Fisher relation A relation used to determine the absolute luminosity of a spiral galaxy. The rotational velocity, measured from the broadening of spectral lines, is related to the total mass, and hence the total luminosity. (p. 638)

21-centimeter radiation Radio radiation emitted when an electron in the ground state of a hydrogen atom flips its spin to become parallel to the spin of the proton in the nucleus. (p. 481)

Type I supernova One possible explosive death of a star. A white dwarf in a binary-star system can accrete enough mass that it cannot support its own weight. The star collapses and temperatures become high enough for carbon fusion to occur. Fusion begins throughout the white dwarf almost simultaneously and an explosion results. (p. 549)

Type II supernova One possible explosive death of a star, in which the highly evolved stellar core rapidly implodes and then explodes, destroying the surrounding star. (p. 549)

U

ultraviolet Region of the electromagnetic spectrum, just beyond the visible range, corresponding to wavelengths slightly shorter than blue light. (p. 63)

ultraviolet telescope A telescope that is designed to collect radiation in the ultraviolet part of the spectrum. Earth's atmosphere is partially opaque to these wavelengths, so ultraviolet telescopes are put on rockets, balloons, and satellites to get high above most or all of the atmosphere. (p. 132)

umbra Central region of the shadow cast by an eclipsing body. (p. 17)

umbra The central region of a sunspot, which is its darkest and coolest part. (p. 418)

unbound An orbit which does not stay in a specific region of space, but where an object escapes the gravitational field of another. Typical unbound orbits are hyperbolic in shape. (p. 55)

universe The totality of all space, time, matter, and energy. (p. 4)

V

Van Allen belts At least two doughnut-shaped regions of magnetically trapped, charged particles high above Earth's atmosphere. (p. 183)

variable star A star whose luminosity changes with time. (p. 603)

vernal equinox Date on which the Sun crosses the celestial equator moving northward, occurring on or near March 21. (p. 13)

visible light The small range of the electromagnetic spectrum that human eyes perceive as light. The visible spectrum ranges from about 400–700 nm, corresponding to blue through red light. (p. 62)

visible spectrum The small range of the electromagnetic spectrum that human eyes perceive as light. The visible spectrum ranges from about 4000–7000 angstroms, corresponding to blue through red light. (p. 64)

visual binary A binary-star system in which both members are resolvable from Earth. (p. 457)

void Large, relatively empty region of the universe around which superclusters and "walls" of galaxies are organized. (p. 654)

volcano Upwelling of hot lava from below Earth's crust to the planet's surface. (p. 173)

W

water hole The radio interval between 18 cm and 21 cm, the respective wavelengths at which hydroxyl (OH) and hydrogen (H) radiate, in which intelligent civilizations might conceivably send their communication signals. (p. 753)

wave A pattern that repeats itself cyclically in both time and space. Waves are characterized by the speed at which they move, their frequency, and their wavelength. (p. 63)

wave period The amount of time required for a wave to repeat itself at a specific point in space. (p. 64)

wavelength The distance from one wave crest to the next, at a given instant in time. (p. 64)

weak nuclear force Short-range force, weaker than both electromagnetism and the strong force, but much stronger than gravity; responsible for certain nuclear reactions and radioactive decays. (p. 427)

weird terrain A region on the surface of Mercury with oddly rippled features. This feature is thought to be the result of a strong impact which occurred on the other side of the planet, and sent seismic waves traveling around the planet, converging in the weird region. (p. 215)

white dwarf A dwarf star with sufficiently high surface temperature that it glows white. (p. 450)

white-dwarf region The bottom-left corner of the Hertzsprung–Russell diagram, where white-dwarf stars are found. (p. 454)

white oval Light-colored region near the Great Red Spot in Jupiter's atmosphere. Like the red spot, such regions are apparently rotating storm systems. (p. 283)

Wien's law Relation between the wavelength at which a black-body curve peaks and the temperature of the emitter. The peak wavelength is inversely proportional to the temperature, so the hotter the object, the bluer its radiation. (p. 74)

winter solstice Point on the ecliptic where the Sun is at its southernmost point below the celestial equator, occurring on or near December 21. (p. 11)

X

X ray Region of the electromagnetic spectrum corresponding to radiation of high frequency and short wavelength, far beyond the visible spectrum. (p. 63)

X-ray burster X-ray source that radiates thousands of times more energy than our Sun in short bursts lasting only a few seconds. A neutron star in a binary system accretes matter onto its surface until temperatures reach the level needed for hydrogen fusion to occur. The result is a sudden period of rapid nuclear burning and release of energy. (p. 572)

Z

zero-age main sequence The region on the Hertzsprung–Russell diagram, as predicted by theoretical models, where stars are located at the onset of nuclear burning in their cores. (p. 497)

zodiac The twelve constellations on the celestial sphere through which the Sun appears to pass during the course of a year. (p. 11)

zonal flow Alternating regions of westward and eastward flow, roughly symmetrical about the equator of Jupiter, associated with the belts and zones in the planet's atmosphere. (p. 279)

zone Bright, high-pressure region in the atmosphere of a jovian planet, where gas flows upward. (p. 278)

Chapter 1

1.1 (p. 9) (1) Because the celestial sphere provides a natural means of specifying the locations of stars on the sky. Celestial coordinates are directly related to Earth's orientation in space, but are independent of Earth's rotation. (2) Distance information is lost. 1.2 (p. 13) (1) In the Northern Hemisphere, summer occurs when the Sun is near its highest (northernmost) point on the celestial sphere, or, equivalently, when Earth's North Pole is "tipped" toward the Sun, the days are longest and the Sun is highest in the sky. Winter occurs when the Sun is lowest in the sky (near its southernmost point on the celestial sphere) and the days are shortest. (2) We see different constellations because Earth has moved halfway around its orbit between these seasons and the darkened hemisphere faces an entirely different group of stars. 1.3 (p. 21) (1) The angular size of the Moon would remain the same, but that of the Sun would be halved, making it easier for the Moon to eclipse the Sun. We would expect to see total or partial eclipses, but no annular ones. (2) If the distance is halved, the angular size of the Sun would double and total eclipses would never be seen, only partial or annular ones. 1.4 (p. 27) Because astronomical objects are too distant for direct ("measuring tape") measurements, so we must rely on indirect means and mathematical reasoning.

Chapter 2

2.1 (p. 39) In the geocentric view, retrograde motion is the real backward motion of a planet as it moves on its epicycle. In the heliocentric view, the backward motion is only apparent, caused by Earth "overtaking" the planet in its orbit. 2.2 (p. 42) The discovery of the phases of Venus could not be reconciled with the geocentric model, and observations of Jupiter's moons proved that some objects in the universe did not orbit Earth. 2.3 (p. 47) Yes to both questions (replacing "Sun" by "Jupiter" in the statement of the laws in the latter case), Kepler's laws apply to any body whose orbit around another is governed by the force of gravity. 2.4 (p. 49) Because Kepler determined the overall geometry of the solar system by triangulation using Earth's orbit as a baseline, so all distances were known only relative to the scale of Earth's orbit—the astronomical unit. 2.5 (p. 55) In the absence of any force, a planet would move in a straight line with constant velocity (Newton's first law) and therefore tends to move along the tangent to its orbital path. The Sun's gravity causes the planet to accelerate toward the Sun (Newton's second law), bending its trajectory into the orbit we observe.

Chapter 3

3.1 (p. 70) Light is an electromagnetic wave produced by accelerating charged particles. All waves on the list are characterized by their wave periods, frequencies, and wavelengths, and all carry energy and information from one location to another. Unlike waves in water or air, however, light waves require no physical medium in which to propagate. 3.2 (p. 72) All are electromagnetic radiation and travel at the speed of light. In physical terms, they differ only in frequency (or wavelength), although their effects on our bodies (or our detectors) differ greatly. 3.3 (p. 77) As the switch is turned and the temperature of the filament rises, the bulb's brightness increases rapidly, by Stefan's law, and its color shifts, by Wien's law, from invisible infrared to red to yellow to white. 3.4 (p. 79) According to *More Precisely 2-3*, measuring masses in astronomy usually entails measuring the orbital speed of one object—a companion star, or a planet, perhaps—around another. In most cases, the Doppler effect is an astronomer's only way of making such measurements.

Chapter 4

4.1 (p. 91) They are characteristic frequencies (wavelengths) at which matter absorbs or emits photons of electromagnetic radiation. They are unique to each atom or molecule, and thus provide a means of identifying the gas producing them. 4.2 (p. 92) Electron orbits can occur only at certain specific energies and there is a ground state which has the lowest possible energy. Planetary orbits can have any energy. Planets can stay in any orbit indefinitely; in an atom, the electron must eventually fall to the ground state, emitting electromagnetic radiation in the process. Planets may reasonably be thought of as having specific trajectories around the Sun—there is no ambiguity as to "where" a planet is. Electrons in an atom are smeared out into an electron cloud, and we can only talk of the electron's location in probabilistic terms. 4.3 (p. 96) Spectral lines correspond to transitions between specific orbitals within an atom. The structure of an atom determines the energies of these orbitals, hence the possible transitions, and hence the energies (colors) of the photons involved. 4.4 (p. 97) In addition to changes involving electron energies, changes involving a molecule's vibration or rotation can also result in emission or absorption of radiation. 4.5 (p. 102) Because with few exceptions, spectral analysis is the only way we have of determining the physical conditions—composition, temperature, density, velocity, etc.,—in a distant object. Without spectral analysis, astronomers would know next to nothing about the properties of stars and galaxies.

Chapter 5

5.1 (p. 113) Because reflecting telescopes are easier to design, build, and maintain than refracting instruments. 5.2 (p. 117) The need to gather as much light as possible, and the need to achieve the highest possible angular resolution. 5.3 (p. 123) To reduce or overcome the effects of atmospheric absorption, instruments are placed on high mountains or in space. To compensate for atmospheric turbulence, adaptive optics techniques probe the air above the observing site and adjust the mirror surface accordingly to try to recover the undistorted image. 5.4 (p. 127) Radio observations allow us to see objects whose visible light is obscured by intervening matter, or which simply do not emit most of their energy in the visible portion of the spectrum. 5.5 (p. 129) The long wavelength of radio radiation. Astronomers use the largest radio telescopes possible and interferometry, which combines the signals from two or more separate telescopes to create the effect of a single instrument of much larger diameter. 5.6 (p. 135) Benefits: they are above the atmosphere, so they are unaffected by seeing or absorption; they can also make round-the-clock observations. Drawbacks: cost, smaller size, inaccessibility, vulnerability to damage by radiation and cosmic rays.

Chapter 6

6.1 (p. 146) They provide new tests of our theories of solar-system formation, and new examples of the sorts of planetary systems that are possible. 6.2 (p. 148) By applying the laws of geometry and Newtonian mechanics to observations made from Earth and (more recently) by visiting spacecraft. 6.3 (p. 150) The planets orbit in very nearly the same plane—the ecliptic. Viewed from outside the system, only the orbits of Mercury and Pluto would deviate noticeably from this plane. 6.4 (p. 151) Because the two classes of planet differ in almost every physical property—orbit, mass, radius, composition, existence of rings, and number of moons. 6.5 (p. 153) Because it is thought to be much less evolved than material now found in planets, and hence a better indicator of conditions in the early solar system.

Chapter 7

7.1 (p. 171) It raises Earth's average surface temperature above the freezing point of water, which was critical for the development of life on our planet. Should the greenhouse effect continue to strengthen, however, it may conceivably cause catastrophic climate changes on Earth. 7.2 (p. 175) We would have far less detailed direct (from volcanoes) or indirect (from seismic studies following earthquakes) information on our planet's interior. 7.3 (p. 183) Convection currents in the upper mantle cause portions of Earth's crust—plates—to slide around on the surface. As the plates move and interact, they are responsible for volcanism, earthquakes, the formation of mountain ranges and ocean trenches, and the creation and destruction of oceans and continents. 7.4 (p. 187) It tells us that the planet has a conducting, liquid core in which the magnetic field is continuously generated. 7.5 (p. 190) A tidal force is the *variation* in one body's gravitational force from place to place across another. Tidal forces tend to deform a body, rather than causing an overall acceleration, and they decrease proportional to the inverse cube, rather than the inverse square, of the distance.

Chapter 8

8.1 (p. 199) Both bodies have substantially lower escape speeds than Earth, and any atmosphere they may have once had has escaped into space long ago. 8.2 (p. 201) The maria are younger, denser, and much less heavily cratered than the highlands. 8.3 (p. 206) In the case of the Moon, Earth's tidal force has slowed the spin to the point where the rotation rate is now exactly synchronized with the Moon's orbital period around Earth. For Mercury, the Sun's tidal force has caused the rotation period to become exactly 2/3 of the orbital period, and the rotation axis to be exactly perpendicular to the planet's orbit plane. A synchronous orbit is not possible because of Mercury's eccentric orbit around the Sun. 8.4 (p. 212) Heavy bombardment long ago created the basins which later filled with lava to form the maria. Subsequent impacts created virtually all the lunar craters, large and small, we see today. Meteoritic bombardment is the main agent of lunar erosion, although the rate is much smaller than the erosion rate on Earth. 8.5 (p. 215) They are thought to have formed when the planet's core cooled and contracted, causing the crust to crumple. Faults on Earth are the result of tectonic activity. 8.6 (p. 217) Generation of a magnetic field is thought to require a rapidly rotating body with a conducting liquid core. Neither the Moon nor Mercury rotates rapidly and, while Mercury's core may still be partly liquid, the Moon's probably is not. 8.7 (p. 218) The impact theory holds that a collision created the Moon essentially from Earth's mantle, accounting for the composition similarities. If the collision occurred after Earth had already differentiated and the dense material had formed a core, relatively little of this material would have found its way into the newborn Moon.

Chapter 9

9.1 (p. 230) It is very slow and retrograde. The reason is unknown, but may simply be a matter of chance. 9.2 (p. 232) Because they were observing in the optical and could not see the surface. Their measurements pertained to the upper atmosphere, above the planet's reflective cloud layers. 9.3 (p. 239) No—they are mostly shield volcanoes, where lava upwells through a "hot spot" in the crust. There appears to be no plate tectonic activity on Venus. 9.4 (p. 243) The dynamo model of planetary magnetism implies that both a conducting liquid core and rapid rotation are needed to generate a magnetic field. Venus's rotation is the slowest of any planet in the solar system.

Chapter 10

10.1 (p. 251) Because they occur when Martian opposition happens to coincide approximately with Martian perihelion. Such an alignment happens every 7 Martian synodic years. 10.2 (p. 253) Mars does have seasons, since its rotation axis is inclined to its orbit plane in much the same way as Earth's is. However, the Martian seasons are affected by the planet's eccentric orbit. The appearance of the planet changes seasonally, although the changes have nothing to do with growing cycles, as was once thought. 10.3 (p. 258) The lowlands are much less heavily cratered, implying that they have been resurfaced by volcanism (or smoothed by erosion) since the highlands formed. 10.4 (p. 264) Some of it may have escaped into space. Of the rest, some exists in the form of permafrost (or perhaps liquid water) below the surface. The rest is contained in the Martian polar caps. 10.5 (p. 268) Some of it was lost due to the planet's weak gravity, aided by violent meteoritic impacts. The rest became part of the planet's surface rocks, the polar caps, or the subsurface permafrost. 10.6 (p. 269) The planet's small size, which allowed the planet's internal heat to escape, effectively shutting down the processes that drive mantle convection, volcanism, and plate tectonics. 10.7 (p. 270) They are much smaller, and orbit much closer to the parent planet. In addition, they seem to have a different formation history—their composition differences from Mars suggest that they were captured by the planet long after they (and Mars) formed; for Earth's Moon, this scenario does not seem to work.

Chapter 11

11.1 (p. 278) The magnetic field is generated by the motion of electrically conducting liquid in the deep interior, and therefore presumably shares the rotation of that region of the planet. 11.2 (p. 283) Like weather systems on Earth, the belts and zones are regions of high and low pressure and are associated with convective motion. However, unlike storms on Earth, they wrap all the way around the planet because of Jupiter's rapid rotation. In addition, the clouds are arranged in three distinct layers and the bright colors are the result of cloud chemistry unlike anything operating in Earth's atmosphere. Jupiter's spots are somewhat similar to hurricanes on Earth, but they are far larger and longer-lived. 11.3 (p. 285) Jupiter has no surface. The atmosphere becomes denser and hotter with increasing depth below the cloud layers, and eventually becomes liquid. 11.4 (p. 287) Because Jupiter rotates more rapidly than Earth and because the volume of conducting fluid responsible for the field is much greater. 11.5 (p. 295) (1) Jupiter's gravitational field, via its tidal effect on the moons. (2) Because liquid water is thought to have played a critical role in the appearance of life on Earth, and is presumed to be similarly important elsewhere in the solar system.

Chapter 12

12.1 (p. 302) Because most of the images were taken before 1995, when Earth crossed Saturn's ring plane. Figure 12.3 was obtained in 1998. 12.2 (p. 306) Saturn's cloud and haze layers are thicker than those on Jupiter because of Saturn's lower gravity, so we usually see only the upper level of the atmosphere even though the same basic features and cloud layers are there. 12.3 (p. 309) Roughly half of it has precipitated into the planet's interior, reducing the helium fraction in the outer layers (and releasing gravitational energy as it fell). 12.4 (p. 315) The Roche limit is the radius inside of which a moon will be torn apart by tidal forces. Planetary rings are found inside the Roche limit, (most) moons outside. Orbital resonances between ring particles and moons are responsible for

much of the fine structure in the rings. They may also maintain the orbits of shepherd satellites that keep some ring features sharp. 12.5 (p. 319) It has a thick atmosphere (denser than Earth's), unlike any other moon in the solar system. 12.6 (p. 322) Because they are tidally locked by Saturn's gravity into synchronous orbits, and therefore all have permanently leading and trailing faces that interact differently with the environment around the planet.

Chapter 13

13.1 (p. 329) Uranus's orbit was observed to deviate from a perfect ellipse, leading astronomers to try to compute the mass and location of the body responsible for those discrepancies. That body was Neptune. 13.2 (p. 332) For unknown reasons, Uranus rotates "on its side"—with its rotation axis almost in the plane of the ecliptic. 13.3 (p. 335) Neptune lies far from the Sun and is very cold, so the source of the energy for these atmospheric phenomena is not known. Uranus, closer to the Sun but with a similar atmospheric temperature, shows much less activity. 13.4 (p. 335) In each case, the field is significantly offset from the planet's center and inclined to the rotation axis. 13.5 (p. 342) Triton shows evidence for surface activity—nitrogen geysers and water volcanoes—that seems to have erased most of its impact craters. 13.6 (p. 344) Both are thin rings that require shepherd satellites to prevent them from spreading out and dispersing. 13.7 (p. 350) They have similar masses, radii, composition, and perhaps also similar origins in the Kuiper belt.

Chapter 14

14.1 (p. 360) Similarities: all orbit in the inner solar system, and are solid bodies of generally "terrestrial" composition. Differences: asteroids are much smaller than the terrestrial planets, and their orbits are much less regular. 14.2 (p. 361) Because most Earth-crossing asteroids will eventually come very close to or even collide with our planet, with potentially catastrophic results. 14.3 (p. 366) Most comets never come close enough to the Sun for us to see them. 14.4 (p. 371) Asteroids are generally rocky. Comets are predominantly made of ice, with some dust and other debris mixed in. 14.5 (p. 376) They are fragments of comets or asteroids orbiting the Sun in interplanetary space. They are important to planetary scientists because they generally consist of ancient material, and contain vital information on conditions in the early solar system.

Chapter 15

15.1 (p. 385) Because it must explain certain general features of solar system architecture, while accommodating the fact that there are exceptions to many of them. 15.2 (p. 392) Because dust grains formed condensation nuclei that began the process leading to the formation of planetesimals and eventually planets. 15.3 (p. 395) Yes—if a star formed with a disk of matter around it then the basic processes of condensation and accretion would probably have occurred there too, even if it doesn't have planets like Earth. 15.4 (p. 396) The fact that larger objects were built up through collisions between smaller objects. The general trends are predictable, but the details, at the level of individual collisions, are not. 15.5 (p. 400) Because the search techniques are most sensitive to massive planets orbiting close to their parent stars, which are exactly what have been observed.

Chapter 16

16.1 (p. 408) When we simply multiply the solar constant by the total area to obtain the solar luminosity, we are implicitly assuming that the same amount of energy reaches every square meter of the large sphere in Figure 16.3. 16.2 (p. 413) The energy may be carried in the form of (1) radiation, where energy travels in the form of light, and (2) convection, where

energy is carried by physical motion of upwelling solar gas. 16.3 (p. 418) The spectrum shows (1) emission lines of (2) highly ionized elements, implying high temperature. 16.4 (p. 425) There is a strong field, with a well-defined east–west organization, just below the surface. The field direction in the southern hemisphere is opposite that in the north. However, the details of the fields are very complex. 16.5 (p. 429) Because the Sun shines by nuclear fusion, which converts mass into energy as hydrogen turns into helium. 16.6 (p. 430) Almost all of the neutrinos produced in the solar core reach Earth, with very little chance of interaction with intervening matter. Electromagnetic energy, produced in the form of gamma radiation, interacts frequently with solar matter, taking possibly thousands of years to reach the surface.

Chapter 17

17.1 (p. 439) Because stars are so far away that their parallaxes relative to any baseline on Earth are too small to measure accurately. 17.2 (p. 441) Because the transverse component must be determined by measuring the star's proper motion, which decreases as the star's distance increases, and is too small to measure for most distant stars. 17.3 (p. 445) Nothing—we need to know their distances before the luminosities (or absolute magnitudes) can be determined. 17.4 (p. 449) Because temperature controls which excited states the star's atoms and ions are in, and hence which atomic transitions are possible. 17.5 (p. 450) Yes, using the radius–luminosity–temperature relationship, but only if we can find a method of determining the luminosity that doesn't depend on the inverse-square law (Sec. 17.4). 17.6 (p. 454) Because giants are intrinsically very luminous, and can be seen to much greater distances than the more common main-sequence stars or white dwarfs. 17.7 (p. 456) All stars would be further away, but their measured spectral types and apparent brightnesses would be unchanged, so their luminosities would be greater than previously thought. The main sequence would therefore move vertically upward in the H–R diagram. (We would then use larger luminosities in the method of spectroscopic parallax, so distances inferred by that method would also increase.) 17.8 (p. 461) We don't—we assume that their masses are the same as similar stars found in binaries.

Chapter 18

18.1 (p. 473) Because the scale of interstellar space is so large that even very low densities can add up to a large amount of obscuring matter along the line of sight to a distant star. 18.2 (p. 478) Because the UV light is absorbed by hydrogen gas in the surrounding cloud, ionizing it to form the emission nebulae. The red light is Hα radiation; part of the visible hydrogen spectrum emitted as electrons and protons recombine to form hydrogen atoms. 18.3 (p. 481) By studying absorption lines caused by atoms and molecules in the clouds, and the general extinction due to dust, it is possible to map out a cloud's properties—so long as enough stars are conveniently located behind it. 18.4 (p. 482) Because most of the interstellar matter in the Galactic disk is made up of atomic hydrogen, and 21-centimeter radiation provides a probe of that gas largely unaffected by interstellar absorption. 18.5 (p. 484) Because the main constituent, hydrogen, is very hard to observe, so astronomers must use other molecules as tracers of the cloud's properties.

Chapter 19

19.1 (p. 492) The competing effects of gravity, which tends to make an interstellar cloud collapse, and heat (pressure), which opposes collapse. 19.2 (p. 496) (1) The existence of a photosphere, meaning that the inner part of the cloud becomes opaque to its own radiation, signaling the slowing of the collapse phase. (2) Nuclear fusion in the core and equilibrium between pressure and gravity. 19.3 (p. 498) No—different parts of the main sequence correspond to stars of different masses. A typical star stays at roughly the same location on the main sequence of most of its lifetime. 19.4 (p. 502) We assume that we observe objects at many different evolutionary stages, and that the snapshot therefore provides a representative sample of the evolutionary stages that stars go through. 19.5 (p. 505) Star formation may be triggered by some external event, which might cause several interstellar clouds to start contracting at once. Alternatively, the shock wave produced when an emission nebula forms may be sufficient to send another nearby part of the same cloud into collapse. 19.6 (p. 510) Because stars form at different rates—high-mass stars reach the main sequence and start disrupting the parent cloud long before lower-mass stars have finished forming.

Chapter 20
20.1 (p. 517) Hydrostatic equilibrium means that, if some property of the Sun changes a little, the star's structure adjusts to compensate. Small changes in the Sun's internal temperature or pressure will not lead to large changes in its radius or luminosity. 20.2 (p. 523) Because the nonburning inner core, unsupported by fusion, begins to shrink, releasing gravitational energy, heating the overlying layers and causing them to burn more vigorously, thus increasing the luminosity. 20.3 (p. 529) Because the core's contraction is halted by the pressure of degenerate (tightly packed) electrons before it reaches a temperature high enough for the next stage of fusion to begin. In fact, this statement is true whether the "next round" is hydrogen fusion (brown dwarf), helium fusion (helium white dwarf), or carbon fusion (carbon-oxygen white dwarf). 20.4 (p. 531) Fusion in high-mass stars is not halted by electron degeneracy pressure. Temperatures are always high enough that each new burning stage can start before degeneracy becomes important. Such stars continue to fuse more and more massive nuclei, faster and faster, eventually exploding in a supernova. 20.5 (p. 535) Because a star cluster gives us a "snapshot" of stars of many different masses, but of the same age and initial composition, allowing us to directly test the predictions of the theory. 20.6 (p. 537) Because many, if not most, stars are found in binaries, and stars in binaries can follow evolutionary paths quite different from those they would follow if single.

Chapter 21
21.1 (p. 545) No, because it is not a member of a binary-star system. 21.2 (p. 548) Because iron cannot fuse to produce energy. As a result, no further nuclear reactions are possible, and the core's equilibrium cannot be restored. 21.3 (p. 552) Because the two types of supernova differ in their spectra and their light curves, making it impossible to explain them in terms of a single phenomenon. 21.4 (p. 561) Because they are responsible for creating and dispersing all the heavy elements out of which we are made. In addition, they (or their progenitor high-mass star) may also have played a role in triggering the collapse of an interstellar cloud to form our solar system.

Chapter 22
22.1 (p. 569) No—only Type II supernova. According to theory, the rebounding central core of the original star becomes a neutron star. 22.2 (p. 571) Because (1) not all supernovae form neutron stars, (2) the pulses are beamed, so not all pulsing neutron stars are visible from Earth, and (3) pulsars spin down and become too faint to observe after a few tens of millions of years. 22.3 (p. 575) Some X-ray sources are binaries containing accreting neutron stars, which may be in the process of being spun up to form millisecond pulsars. 22.4 (p. 579) They are energetic bursts of

gamma rays, roughly isotropically distributed on the sky, occurring about once per day. They pose a challenge because they are very distant, and hence extremely luminous, but their energy originates in a region less than a few hundred kilometers across. 22.5 (p. 585) Newton's theory describes gravity as a force produced by a massive object that influences all other massive objects. Einstein's relativity describes gravity as a curvature of space-time produced by a massive object; that curvature then determines the trajectories of all particles—matter or radiation—in the universe. 22.6 (p. 589) Because the object would appear to take infinitely long to reach the event horizon, and its light would be infinitely redshifted by the time it got there. 22.7 (p. 593) By observing their gravitational effects on other objects, and from the X-rays emitted when matter falls into them.

Chapter 23
23.1 (p. 600) The Milky Way is the thin plane of the Galactic disk, seen from within. When our line of sight lies in the plane of the Galaxy, we see many stars blurring into a continuous band. In other directions, we see darkness. 23.2 (p. 607) No, because even the brightest Cepheids are unobservable at distances of more than a kiloparsec or so through the obscuration of interstellar dust. 23.3 (p. 610) Because all the gas and dust in the halo fell to the Galactic disk billions of years ago, so halo star formation has long since ceased. 23.4 (p. 612) The halo formed early in our Galaxy's history, before gas and dust had formed a spinning, flattened disk. Disk stars are still forming today. Consequently, halo stars are old and move in more or less random three-dimensional orbits, while the disk contains stars of all ages, all moving in roughly circular orbits around the Galactic center. 23.5 (p. 615) Because differential rotation would destroy the spiral structure within a few hundred million years. 23.6 (p. 619) Its emission has not been observed at any electromagnetic wavelength. Its presence is inferred from its gravitational effect on stars and gas orbiting the Galactic center. 23.7 (p. 624) Observations of rapidly moving stars gas, and the variability of the radiation emitted suggest the presence of a 2–3 million-solar-mass black hole.

Chapter 24
24.1 (p. 637) Most galaxies are not large spirals—the most common galaxy types are dwarf ellipticals and dwarf irregulars. 24.2 (p. 642) Because distance-measurement techniques ultimately rely upon the existence of very bright objects whose luminosities can be inferred by other means. Such objects become increasingly hard to find and calibrate the farther we look out into intergalactic space. 24.3 (p. 644) Because the amount of luminous matter seen in galaxies and galaxy clusters, even including the hot intracluster gas, falls short of the mass inferred from studies of galaxy rotation curves and galaxy orbits in clusters by at least a factor of 10. 24.4 (p. 651) Stars form by collapse and fragmentation of a large interstellar cloud and subsequently evolve largely in isolation. Galaxies form by mergers of smaller objects, and interactions with other galaxies play a major role in their evolution. 24.5 (p. 655) Hubble's law, which describes the expansion, provides a direct connection between redshift (recession velocity) and distance. Measuring a galaxy's redshift, and assuming that Hubble's law holds, we can immediately determine the galaxy's distance.

Chapter 25
25.1 (p. 663) They are more luminous, emit a large fraction of their energy at radio and infrared wavelengths and, in the case of quasars, all lie at great distances from Earth. 25.2 (p. 669) Many radio galaxies have high-speed jets of material originating in the nucleus and aligned with the radio lobes. 25.3 (p. 672)

Because quasars generally appear starlike, as seen from Earth, and their redshifts were unexpectedly large for objects thought to be members of our own Galaxy. 25.4 (p. 677) The energy is generated in an accretion disk in the galactic nucleus and then transported into the lobes by jets, where it is eventually emitted by the synchrotron process in the form of radio waves. 25.5 (p. 683) Because they are very distant and very bright. Their spectra contain information about galaxies and gas clouds along the line of sight, while quasar lensing provides a probe of dark matter on scales much larger than galaxy clusters. 25.6 (p. 686) Perhaps. Our Galaxy has a massive black hole at its center, which might possibly have supported a high luminosity in the past. However, our central black hole has a mass only a few million times that of the Sun, so our Galaxy probably wasn't extremely luminous in its youth; but many nearby galaxies may well have been very bright quasars in the distant past.

Chapter 26
26.1 (p. 693) On very large scales—more than 300 Mpc—the distribution of galaxies seems to be roughly the same everywhere and in all directions. 26.2 (p. 700) Because, tracing the motion backwards in time, it implies that all galaxies, and in fact everything in the entire universe, were located at a single point at the same instant in the past. 26.3 (p. 702) The universe can expand forever, in which case we die a cold death in which all activity gradually fades away, or the expansion can stop and the universe will recollapse to a fiery Big Crunch. 26.4 (p. 706) There doesn't seem to be enough matter to halt the collapse and, in addition, the observed cosmic acceleration suggests the existence of a large-scale repulsive force in the cosmos that also opposes recollapse. 26.5 (p. 707) A low-density universe has negative curvature, a critical density universe is spatially flat (Euclidean), and a high-density universe has positive curvature (and is finite in extent). 26.6 (p. 710) At the time of the Big Bang. It is the electromagnetic remnant of the primeval fireball.

Chapter 27
27.1 (p. 718) It means that the total mass-energy density of the universe, which today is made up almost entirely of matter and dark energy, was once comprised almost entirely of radiation. We know this because, going back in time toward the Big Bang, the dark energy density stays constant as the universe contracts, the matter density increases because the volume shrinks, but the radiation density increases even faster because of the cosmological redshift. Thus, at sufficiently early times, radiation was the dominant component of the cosmos. 27.2 (p. 721) Once the temperature of the expanding, cooling universe dropped below the point where particle–antiparticle pairs could no longer be created from the radiation background, the particles separated out of the radiation field. Particles and antiparticles annihilated one another, and any leftover "frozen out" matter has survived to the present day. 27.3 (p. 726) Because the amount of deuterium observed in the universe today implies that the present density of normal matter is at most a few percent of the critical value—much less than the density of dark matter inferred from dynamical studies. 27.4 (p. 730) Inflation implies that the universe is flat, and hence that the total cosmic density equals the critical value. However, the matter density seems to be only about one third of the critical value, and the density of electromagnetic radiation (the microwave background) is a tiny fraction of the critical density. The remaining density may be in the form of the "dark energy" thought to be powering the accelerating cosmic expansion (Section 27.3). 27.5 (p. 735) They allow us to measure the value of Ω_0—and in fact imply that it is very close to 1.

Chapter 28

28.1 (p. 745) The formation of complex molecules from simple ingredients by nonbiological processes has been repeatedly demonstrated, but no living cell or self-replicating molecule has ever been created. 28.2 (p. 746) Mars remains the most likely site, although Europa and Titan also have properties that might have been conducive to the emergence of living organisms.   28.3 (p. 751) It breaks a complex problem up into simpler "astronomical," "biochemical," "anthropological," and "cultural" pieces, which can be analyzed separately. It also identifies the types of stars where a search might be most fruitful.   28.4 (p. 754) It is in the radio part of the spectrum, where Galactic absorption is least, at a region where natural Galactic background "static" is minimized, and in a portion of the spectrum characterized by lines of hydrogen and hydroxyl, both of which would likely have significance to a technological civilization.

Chapter 1

True or False? 1.1 T 1.2 T 1.3 F 1.4 F 1.5 F 1.6 T 1.7 T 1.8 F 1.9 T 1.10 F 1.11 T 1.12 F 1.13 F 1.14 F 1.15 T

Fill in the Blanks 1.1 star 1.2 galaxy 1.3 axis 1.4 celestial sphere 1.5 stars 1.6 ecliptic 1.7 winter solstice; southernmost 1.8 celestial equator 1.9 1/60 1.10 quarter 1.11 lunar 1.12 angular diameter 1.13 parallax 1.14 angular diameter 1.15 Earth

Odd-Numbered Problems 1.1 (c) the Moon (384,000 km) 1.3 It would decrease by roughly eight minutes. 1.5 (a) 7.14 solar days; (b) infinite 1.7 1.02 km/s 1.9 (a) 57,300 km; (b) 3.44×10^6 km; (c) 2.06×10^8 km 1.11 6.5×10^{-5} arc sec 1.13 391 1.15 0°

Chapter 2

True or False? 2.1 F 2.2 F 2.3 T 2.4 F 2.5 F 2.6 F 2.7 T 2.8 T 2.9 T 2.10 F 2.11 F 2.12 T 2.13 F 2.14 T 2.15 T

Fill in the Blank 2.1 calendar 2.2 Chinese 2.3 Islamic 2.4 retrograde 2.5 scientific method 2.6 Copernicus 2.7 Earth's 2.8 Brahe 2.9 ellipse; circle 2.10 square; cube 2.11 moons; phases; sunspots 2.12 radar 2.13 force 2.14 product; square 2.15 masses

Odd-Numbered Problems
2.1 (a) 110 km; (b) 44,000 km; (c) 370,000 km 2.3 Pluto perihelion = 29.65 A.U.; Neptune perihelion = 29.80 A.U. 2.5 143 days 2.7 8.1″, if Mercury is at aphelion (0.47 A.U.) and Earth is at perihelion (0.98 A.U.) at the point of closest approach 2.9 9.42 $\times 10^{-4}$ solar = 1.88×10^{27} kg 2.11 10,000 A.U., 16 million years 2.13 7.8 km/s, 7.3 km/s, 4.9 km/s; equal in all cases 2.15 1.7 km/s; 2.4 km/s

Chapter 3

True or False? 3.1 T 3.2 F 3.3 T 3.4 F 3.5 F 3.6 T 3.7 F 3.8 F 3.9 F 3.10 T 3.11 F 3.12 T 3.13 F 3.14 T 3.15 T

Fill in the Blank 3.1 300,000 3.2 wavelength 3.3 frequency 3.4 diffraction 3.5 electric, magnetic 3.6 400 nm, 700 nm 3.7 red 3.8 radio, (some) infrared, visible 3.9 temperature 3.10 0 3.11 273 3.12 6,000 3.13 1,200 3.14 high 3.15 shorter

Odd-Numbered Problems 3.1 1480 m/s 3.3 23 Hz; radio 3.5 0.3 m for a 1 GHz clock; radio 3.7 310; 9.4 microns; infrared 3.9 2.9 microns 3.11 6.4×10^7 W/m^2; 3.9×10^{26} W 3.13 300 km/s away 3.15 1.94×10^{27} kg

Chapter 4

True or False? 4.1 T 4.2 F 4.3 T 4.4 T 4.5 F 4.6 F 4.7 F 4.8 T 4.9 T 4.10 F 4.11 F 4.12 T 4.13 F 4.14 F 4.15 F

Fill in the Blank 4.1 prism 4.2 continuous 4.3 solar spectrum 4.4 dense 4.5 cool, low-density 4.6 particle 4.7 photoelectric effect 4.8 positive, negative 4.9 ionized 4.10 absorbs 4.11 emits 4.12 difference 4.13 infrared 4.14 radio 4.15 broaden

Odd-Numbered Problems 4.1 2.8, 6.2 4.3 620 nm, 12,400 nm, 0.25 nm 4.5 4.1×10^9 4.7 39 microns, 7.7×10^{12} Hz, infrared; 4.5 cm, 6.7 GHz, radio; 46 m, 6.6 MHz, radio 4.9 the first six Balmer lines, ranging in wavelength from 656 nm (Hα) to 410 nm (Hζ) 4.11 137 km/s, approaching 4.13 0.05 nm, to one significant figure 4.15 0.3 rev/day, to one significant figure

Chapter 5

True or False? 5.1 F 5.2 F 5.3 F 5.4 T 5.5 F 5.6 F 5.7 T 5.8 T 5.9 T 5.10 T 5.11 T 5.12 F 5.13 F 5.14 F 5.15 F

Fill in the Blank 5.1 refracting 5.2 reflecting 5.3 reflecting 5.4 area 5.5 diameter, wavelength 5.6 atmosphere 5.7 0.1 5.8 digital 5.9 resolution 5.10 reflecting 5.11 interferometer 5.12 ultraviolet, X-ray, gamma-ray 5.13 infrared 5.14 absorbed 5.15 images

Odd-Numbered Problems 5.1 0.3 arc seconds; 6.8 pixels 5.3 6.7 minutes; 1.7 minutes 5.5 (a) 0.022″, (b) 0.062′ 5.7 160 light-years, using the formula in the text to compute the resolution 5.9 3.3 minutes 5.11 roughly 8 times 5.13 (a) 0.003 arc seconds, (b) 0.005 arc seconds 5.15 No, it is not. The wavelength of a 1-keV photon is 1.2 nm, which would give an angular resolution of about 0.000026″ for the quoted diameter. The effective resolution is complicated by the X-ray mirror arrangement, and is actually determined by other design factors.

Chapter 6

True or False? 6.1 F 6.2 T 6.3 T 6.4 F 6.5 F 6.6 F 6.7 F 6.8 T 6.9 F 6.10 F 6.11 F 6.12 T 6.13 F 6.14 F 6.15 F

Fill in the Blank 6.1 planets 6.2 asteroids 6.3 Mercury, Pluto 6.4 terrestrial 6.5 jovian 6.6 Jupiter 6.7 rocky, icy 6.8 terrestrial 6.9 icy, jovian 6.10 U.S., Mercury 6.11 radar 6.12 Mars 6.13 *Pioneer 10* 6.14 *Voyager 2* 6.15 gravity

Odd-Numbered Problems 6.1 — 6.3 Mercury: 0.31 A.U., 0.47 A.U., Mars: 1.38 A.U., 1.67 A.U., Pluto: 29.7 A.U., 49.3 A.U. 6.5 19″, 16 days 6.7 7×10^{20} kg, 0.01 percent of Earth's mass 6.9 68 A.U., beyond Pluto 6.11 0.77 A.U., taking perihelion as 0.47 A.U. 6.13 0.21, 1.26 A.U, 0.71 years 6.15 2.4 hours, 168,000 km; yes, but only if mission control can predict events half an orbit ahead of time

Chapter 7

True or False? 7.1 F 7.2 F 7.3 F 7.4 T 7.5 T 7.6 T 7.7 F 7.8 T 7.9 F 7.10 T 7.11 F 7.12 F 7.13 F 7.14 F 7.15 T

Fill in the Blank 7.1 less 7.2 crust 7.3 nitrogen, oxygen 7.4 convection 7.5 infrared 7.6 carbon dioxide, water vapor 7.7 volcanoes 7.8 life 7.9 solid, liquid 7.10 granite, basalt 7.11 molten 7.12 plate tectonics 7.13 aurora 7.14 liquid water 7.15 difference

Odd-Numbered Problems 7.1 — 7.3 5.0×10^{18} kg = 8.4×10^{-7} times Earth's mass 7.5 21 m 7.7 (a) 8.4 $\times 10^{-3}$, (b) 0.16, (c) 0.84, (d) 7.0×10^{-3} 7.9 1.9 billion years 7.11 3.1×10^{-6}, 3.1×10^{-7} of surface gravity 7.13 1.6×10^{-7} 7.15 Smaller by a factor of 6.0×10^{-6}. No!

Chapter 8

True or False? 8.1 T 8.2 T 8.3 F 8.4 F 8.5 T 8.6 F 8.7 F 8.8 F 8.9 T 8.10 F 8.11 F 8.12 F 8.13 T 8.14 F 8.15 T

Fill in the Blank 8.1 laser ranging 8.2 sunrise, sunset 8.3 1/4, 2/5 8.4 iron/dense material 8.5 larger 8.6 closer to the Sun 8.7 maria 8.8 meteoritic impact 8.9 radar 8.10 poles 8.11 10 8.12 thicker 8.13 mantle 8.14 magnetic field 8.15 Mars-sized object/planetesimal

Odd-Numbered Problems 8.1 8.8 minutes 8.3 57 kg equivalent 8.5 1.74°, 1.15° 8.7 1.81 hours 8.9 (a) 60.4 km/s, or 6.5°/day, (b) 39.7 km/s, or 2.8°/day 8.11 4.2×10^{23} kg, or roughly 26 percent greater than it is now 8.13 need at least 4.8×10^5 such craters, requiring at least 4.8 trillion years; rate would have to increase by a factor of about 1000 8.15 (a) 4 million years for a 2 cm deep bootprint, (b) 4 million years, accepting literally the factor of 10,000 given in the text (c) 1.6 trillion years, assuming Reinhold is 8 km deep

Chapter 9

True or False? 9.1 T 9.2 F 9.3 F 9.4 T 9.5 F 9.6 F 9.7 F 9.8 F 9.9 T 9.10 F 9.11 T 9.12 T 9.13 T 9.14 F 9.15 F

Fill in the Blank 9.1 spacecraft 9.2 radius 9.3 retrograde 9.4 carbon dioxide 9.5 sulfuric acid 9.6 (runaway) greenhouse effect 9.7 radar 9.8 continents 9.9 volcanism 9.10 cornoae 9.11 burn up 9.12 young 9.13 volcanism 9.14 atmospheric pressure, temperature 9.15 water vapor

Odd-Numbered Problems 9.1 (a) 35.4″,(b) 23.3″,(c) 9.7 9.3 470 seconds, 280 seconds (for a separation of 0.28 A.U.) 9.5 4.1°/day; 1.6°/day; 145 (Earth) days 9.7 tidal acceleration due to Earth is 7.3×10^{-12} times Venus's surface gravity; tidal acceleration due to the Sun is 1.4×10^{-8} times Venus's surface gravity; unlikely that Earth's tidal influence would cause the resonance, as it is much smaller than the tidal force due to the Sun 9.9 yes—the smallest detectable feature would be about 20 km across, much smaller than the largest impact features 9.11 400 km/h, 250 mph 9.13 35 times 9.15 197 minutes, 94 minutes

Chapter 10

True or False? 10.1 T 10.2 F 10.3 T 10.4 F 10.5 F 10.6 T 10.7 T 10.8 F 10.9 F 10.10 F 10.11 T 10.12 T 10.13 F 10.14 F 10.15 F

Fill in the Blank 10.1 conjunction 10.2 carbon dioxide 10.3 cratered 10.4 bulge 10.5 impact crater/basin 10.6 volcanoes 10.7 gravity 10.8 permafrost 10.9 water 10.10 flooding 10.11 billion 10.12 water ice 10.13 surface rocks 10.14 ocean 10.15 meteorites

Odd-Numbered Problems 10.1 780 Earth days 10.3 41° 10.5 2.2 minutes longer 10.7 28 kg equivalent, assuming you weigh 70 kg on Earth 10.9 3.2×10^{16} kg, taking Earth's atmospheric mass to be 5.0×10^{18} kg; 2.8 times the seasonal polar cap mass of 1.1×10^{16} kg 10.11 5×10^{20} kg; 16,000 times the current atmospheric mass 10.13 2.9×10^{17} kg; 4.5×10^{-7} times the mass of Mars 10.15 16.1′, 2.7′; no

Chapter 11

True or False? 11.1 T 11.2 F 11.3 F 11.4 F 11.5 T 11.6 F 11.7 F 11.8 T 11.9 F 11.10 T 11.11 T 11.12 T 11.13 F 11.14 F 11.15 T

Fill in the Blank 11.1 average density 11.2 hydrogen, helium 11.3 rotation 11.4 hurricanes 11.5 twice 11.6 zones, belts 11.7 two times 11.8 liquid 11.9 hydrogen 11.10 synchronously 11.11 4, 28 11.12 Ganymede 11.13 water ice 11.14 volcanoes 11.15 Europa

Odd-Numbered Problems 11.1 2.5 times greater 11.3 61 days 11.5 1.5 percent of the actual mass 11.7 4800 times greater; ratio for Earth–Moon is 81 11.9 — 11.11 3.8×10^{-4}, or 0.038 percent 11.13 3.5 days (retrograde); 7.3×10^{-5} 11.15 Io: 36′, Europa: 18′, Ganymede: 18, Callisto: 9′, Sun: 6′. Yes.

Chapter 12

True or False? 12.1 T 12.2 F 12.3 F 12.4 F 12.5 T 12.6 F 12.7 T 12.8 F 12.9 F 12.10 T 12.11 T 12.12 F 12.13 T 12.14 F 12.15 F

Fill in the Blank 12.1 ring 12.2 ammonia 12.3 gravity 12.4 helium 12.5 helium precipitation 12.6 three 12.7 A, B 12.8 B 12.9 Roche limit 12.10 shepherd 12.11 nitrogen 12.12 hydrocarbon 12.13 temperature 12.14 E 12.15 Lagrangian

Odd-Numbered Problems 12.1 47′ 12.3 7.3×10^{22} kg; 1.3×10^{-4} times the actual mass, 1.2 percent of Earth's mass 12.5 74 K 12.7 1.1×10^{18} 12.9 1.36 m/s^2 (Earth: 9./80 m/s^2); 2.6 km/s 12.11 142,000 km; 117,000 km; 142,000 km; 237,000 km 12.13 Saturn's tidal acceleration is 7.9×10^{-5} times Titan's surface gravity; no—the tidal acceleration is small, comparable to Jupiter's tidal force on Callisto and, regardless of the tidal effect, there are no nearby large moons to perturb Titan's orbit, so the conditions leading to the heating of the inner Galilean moons do not arise 12.15 approximately 50 km

Chapter 13
True or False? 13.1 F 13.2 F 13.3 T 13.4 F 13.5 T 13.6 T 13.7 F 13.8 T 13.9 T 13.10 T 13.11 T 13.12 F 13.13 T 13.14 F 13.15 F
Fill in the Blank 13.1 two 13.2 parallel 13.3 methane 13.4 Neptune 13.5 Uranus 13.6 magnetic fields 13.7 Miranda 13.8 *Voyager 2* 13.9 retrograde (and inclined) 13.10 eccentricity 13.11 atmosphere 13.12 geysers 13.13 Neptune 13.14 Triton 13.15 eclipses
Odd-Numbered Problems 13.1(a) 171 years (1.05 Neptune years), (b) 482 years (2.94 Neptune years) 13.3 1.7', 13.2'; yes, but the next one is not expected until 2007 (see Figure 13.6)! 13.5 10,000 km/h—it moves halfway around the planet in just over eight hours; from the Data Box, rotation speed at the equator is about 10,000 km/h. 13.7 0.44 Earth Moon masses; 3.4 Pluto masses 13.9 Yes, barely—at 37 K, an escape speed of 1.3 km/s would be needed; Triton's escape speed is 1.5 km/s 13.11 2.1', assuming that Pluto is close to perihelion 13.13 7.2 times farther out 13.15 11.1 hours, 20,000 km

Chapter 14
True or False? 14.1 F 14.2 F 14.3 F 14.4 T 14.5 T 14.6 F 14.7 T 14.8 F 14.9 T 14.10 F 14.11 F 14.12 T 14.13 T 14.14 T 14.15 T
Fill in the Blank 14.1 rocky 14.2 Mars, Jupiter 14.3 hundreds, 100 14.4 Jupiter 14.5 resonance 14.6 icy 14.7 gas 14.8 eccentric 14.9 a few; 1–10 14.10 meteor shower 14.11 short-period 14.12 asteroids, comets 14.13 meteor 14.14 meteorites 14.15 4.6 billion
Odd-Numbered Problems 14.1 (a) 3.2 kg equivalent, (b) 0.41 km/s 14.3 230 km diameter 14.5 1.1 A.U., 2.0 A.U.; no—the orbital inclination differs from Earth's 14.7 21 hours, taking the middle of the stated mass range and assuming a circular orbit 14.9 (a) 4.0 million years, (b) 49 A.U. 14.11 3×10^{12} kg; 0.06 percent 14.13 5 trillion 14.15 1.9 billion times (2 crossings per orbit)

Chapter 15
True or False? 15.1 T 15.2 F 15.3 T 15.4 T 15.5 F 15.6 F 15.7 F 15.8 F 15.9 F 15.10 T 15.11 F 15.12 F 15.13 T 15.14 F 15.15 T
Fill in the Blank 15.1 nebular 15.2 dust 15.3 collisions 15.4 gravity 15.5 gas 15.6 water 15.7 icy 15.8 comets 15.9 comets 15.10 Jupiter 15.11 T Tauri 15.12 radius 15.13 Chance/Planetesimal collisions/Random events 15.14 wobbles/changing radial velocities 15.15 Jupiter, close
Odd-Numbered Problems 15.1 (a) 1.9×10^{43}, 7.8×10^{42}, and 2.7×10^{40} kg·m²/s, (b) 1.0×10^{31} kg·m²/s 15.3 2, $\sqrt{2} \approx 1.414$ 15.5 increases proportional to the radius; surface gravity doubles, escape speed doubles 15.7 3.0×10^7 newtons; tidal force = 8.1×10^9 newtons, 270 times larger 15.9 2×10^8; once every 2.5 years 15.11 240 million 15.13 0.21 A.U. 15.15 0.56

Chapter 16
True or False? 16.1 T 16.2 T 16.3 F 16.4 T 16.5 T 16.6 F 16.7 F 16.8 F 16.9 F 16.10 F 16.11 T 16.12 F 16.13 T 16.14 T 16.15 F
Fill in the Blank 16.1 photosphere 16.2 chromosphere, transition zone, corona 16.3 convection, radiation, core 16.4 granulation 16.5 photosphere 16.6 hydrogen 16.7 helium 16.8 99.9 16.9 high 16.10 cooler 16.11 11, 22 16.12 core 16.13 4, helium, neutrinos, gamma rays 16.14 destroyed 16.15 few

Odd-Numbered Problems 16.1 14,600 W/m²; 52 W/m² 16.3 (a) 3000 km, (b) 1500, (c) 1/33 of the 167-minute orbital period 16.5 36 percent (64 percent less) 16.7 5.2 solar radii, assuming a coronal temperature of 3 million K 16.9 95 million years 16.11 310,000 years 16.13 radiation mass loss = 4.3 million tons/s $\approx$ twice the mass loss due to the solar wind 16.15 4×10^{28}

Chapter 17
True or False? 17.1 T 17.2 T 17.3 F 17.4 F 17.5 F 17.6 T 17.7 F 17.8 F 17.9 F 17.10 F 17.11 T 17.12 T 17.13 F 17.14 T 17.15 T
Fill in the Blank 17.1 Earth's orbit 17.2 spectrum, Doppler 17.3 distance, proper motion 17.4 temperature, luminosity 17.5 temperatures 17.6 ionized 17.7 in the ground state 17.8 G2 17.9 temperature/color, luminosity/absolute magnitude 17.10 main sequence 17.11 red giants 17.12 white dwarfs 17.13 Binary 17.14 decrease 17.15 shorter
Odd-Numbered Problems 17.1 77 pc; 0.39^{11} 17.3 80 solar luminosities 17.5 B is three times farther away 17.7 3.3×10^{-10} W/m²; 2.3×10^{-13} times the solar constant 17.9 −1.0 17.11 (a) 100 pc, (b) 4000 pc, (c) 160,000 pc, (d) 1,000,000 pc 17.13 3.5, 2.3 solar masses 17.15 (a) 3.4 solar masses, (b) 0.22 solar masses

Chapter 18
True or False? 18.1 F 18.2 T 18.3 T 18.4 F 18.5 F 18.6 F 18.7 F 18.8 T 18.9 F 18.10 F 18.11 T 18.12 F 18.13 T 18.14 T 18.15 T
Fill in the Blank 18.1 dust, gas 18.2 comparable to 18.3 absorption, dust 18.4 low 18.5 hydrogen, helium 18.6 10,000 18.7 photoevaporation 18.8 an emission nebula 18.9 100 18.10 neutral/atomic 18.11 spin, hydrogen atom 18.12 radio 18.13 Dust 18.14 hydrogen 18.15 a million
Odd-Numbered Problems 18.1 1.9 grams 18.3 7.1×10^{20} m³, or a cube of side 8900 km 18.5 9 18.7 100,000 pc, 2.6 18.9 1.8 kpc 18.11 escape speeds (km/s): 1.8, 1.1, 1.1, 0.80; average molecular speeds: 13.6, 14.0, 14.6, 14.2; No 18.13 frequency: 1419.65–1420.24 MHz, wavelength: 21.1053–21.0965 cm 18.15 photon energy = 9.4×10^{-25} J; 2.6×10^{17} W

Chapter 19
True or False? 19.1 F 19.2 F 19.3 T 19.4 T 19.5 T 19.6 F 19.7 F 19.8 F 19.9 F 19.10 T 19.11 T 19.12 T 19.13 T 19.14 T 19.15 F
Fill in the Blank 19.1 temperature 19.2 evolutionary track 19.3 10, 10, thousands (of) 19.4 fragments 19.5 increase 19.6 protostar 19.7 upper right 19.8 10 million 19.9 fuse hydrogen 19.10 main sequence 19.11 4 19.12 50 19.13 rapidly 19.14 radio 19.15 infrared
Odd-Numbered Problems 19.1 yes, barely: the escape speed is 0.93 km/s, molecular speed is 0.35 km/s 19.3 luminosity decreases by a factor of 7900; absolute magnitude increases by 9.7 19.5 3.2 19.7 8.4 19.9 10^{-6} solar luminosities 19.11 4000 years 19.13 20.25 19.15 37 pc

Chapter 20
True or False? 20.1 F 20.2 T 20.3 F 20.4 T 20.5 T 20.6 F 20.7 T 20.8 F 20.9 F 20.10 F 20.11 T 20.12 T 20.13 F 20.14 F 20.15 T
Fill in the Blank 20.1 pressure 20.2 5 billion 20.3 hydrogen, helium 20.4 100 million K 20.5 contract 20.6 carbon, energy 20.7 degeneracy pressure 20.8 carbon 20.9 500 20.10 star clusters 20.11 10,000 km 20.12 high, low 20.13 decrease 20.14 decreases 20.15 separation
Odd-Numbered Problems 20.1 9.9×10^{26} kg destroyed, 8.9×10^{43} J emitted 20.3 (a) 2900, (b) 73,000 solar luminosities 20.5 roughly a factor of 80; factor of 10 smaller 20.7 5×10^{-5}; not noticeable to

the naked eye, but (just) measurable with a telescope and photometer 20.9 2.9 years, 26,000 years 20.11 increases by a factor of 25 20.13 (a) 2.9 solar masses, (b) 1.7 solar masses 20.15 (a) 2.29 A.U. (b) 1.96 A.U., 1.59 yr

Chapter 21
True or False? 21.1 F 21.2 T 21.3 T 21.4 T 21.5 F 21.6 F 21.7 F 21.8 T 21.9 T 21.10 F 21.11 F 21.12 F 21.13 T 21.14 T 21.15 T
Fill in the Blank 21.1 accretion disk 21.2 hydrogen, surface 21.3 temperature 21.4 neutron, neutrino 21.5 neutron 21.6 II 21.7 I 21.8 1.4 21.9 light curves 21.10 neutrinos 21.11 blue supergiant, red supergiant 21.12 helium capture 21.13 iron 21.14 radioactive decay (of nickel-56 and cobalt-56) 21.15 fewer
Odd-Numbered Problems 21.1 14.7 solar radii 21.3 20, 3.2 Mpc 21.5 0.46 pc; no—there are no O or B stars (in fact no stars at all) within that distance of us. For the Moon: 320 pc; yes, but rarely—see problem 14 21.7 1.2×10^{44} J; about 12 times greater than a supernova's electromagnetic output, and about 1/8 the output in the form of neutrinos 21.9 roughly 1000 km/s; not too bad an assumption—the nebula is moving too fast to be affected much by gravity, although it is probably slowing down as it runs into the interstellar medium 21.11 5.6 kpc 21.13 one per 270 years 21.15 0.43 percent, 8.6×10^{27} kg, or 1400 Earth masses

Chapter 22
True or False? 22.1 T 22.2 F 22.3 T 22.4 T 22.5 F 22.6 F 22.7 T 22.8 F 22.9 T 22.10 F 22.11 T 22.12 F 22.13 T 22.14 T 22.15 F
Fill in the Blank 22.1 Type I (carbon-detonation) 22.2 20 22.3 high, strong 22.4 radio 22.5 rotation period 22.6 binary 22.7 neutron star 22.8 gravity 22.9 gravitational radiation 22.10 stay the same 22.11 Schwarzchild radius 22.12 singularity 22.13 lose 22.14 binary 22.15 fluctuations
Odd-Numbered Problems 22.1 1 million revolutions per day, or 11.6 revolutions per second 22.3 1.9×10^{12} m/s², or 190 billion Earth gravities; 190,000 km/s, or 64 percent of the speed of light; 306,000 km/s 22.5 2.4×10^{44} J, roughly twice the Sun's total energy output; 2.4×10^{34} J; 1.4×10^{25} J 22.7 63,000 km/s = 21 percent of the speed of light; 137,000 km/s 22.9 (a) 0.57 milliarc sec, (b) 0.016 arc sec, (c) 4.1 arc min, (d) 8100 A.U. = 0.039 pc 22.11 2×10^{10} m/s² = 2×10^9 g; 2×10^{-3} g; 2×10^{-9} g 22.13 14,000 solar masses 22.15 3×10^7 km = 0.2 A.U., tidal force due to the hole is 1.4 times greater than the star's gravity at a radius of 20 million km

Chapter 23
True or False? 23.1 F 23.2 F 23.3 F 23.4 T 23.5 F 23.6 T 23.7 F 23.8 T 23.9 T 23.10 F 23.11 F 23.12 F 23.13 T 23.14 T 23.15 T
Fill in the Blank 23.1 within the disk 23.2 light absorption by interstellar dust 23.3 disk 23.4 halo 23.5 brightness 23.6 instability strip 23.7 higher 23.8 RR Lyrae stars 23.9 Galactic center; Sun 23.10 circular, Galactic center 23.11 random 23.12 halo 23.13 larger, dark matter 23.14 Doppler shifts 23.15 massive black hole
Odd-Numbered Problems 23.1 2.0″, much less than the angular diameter of Andromeda 23.3 100 kpc 23.5 −6.2; 17 Mpc 23.7 0.014″/yr; proper motion has in fact been measured for several globular clusters 23.9 (a) 3.5 kpc, (b) 19 kpc 23.11 570 million years (assuming a rotation speed of 240 km/s at 15 kpc); 490 million years (assuming a rotation speed of 200 km/s at 5 kpc) 23.13 1 kpc 23.15 inner: 10 years, outer: 140 years, (approx.)

Chapter 24
True or False? 24.1 T 24.2 T 24.3 T 24.4 F 24.5 F 24.6 T 24.7 F 24.8 T 24.9 F 24.10 F 24.11 F 24.12 F 24.13 F 24.14 T 24.15 T

Fill in the Blank 24.1 Hubble 24.2 large 24.3 Sa, Sc 24.4 less 24.5 Local Group 24.6 width 24.7 the Virgo cluster 24.8 increases 24.9 Newton's 24.10 dark matter 24.11 X-ray radiation 24.12 mergers 24.13 Elliptical 24.14 distances 24.15 walls/sheets/bubbles, voids
Odd-Numbered Problems 24.1 320 Mps 24.3 3 billion years 24.5 1.5 nm, taking the rotation speed to be 350 km/s 24.7 2.6×10^{14} solar masses; reasonable to within a factor of perhaps 2–3 24.9 700 km/s; 660 km/s—comparable 24.11 13,000 km/s, 62 Mpc; 10,000 km/s, 80 Mpc; 16,000 km/s, 50 Mpc 24.13 54 arc seconds 24.15 9×10^{13} solar masses, assuming that the spread in velocities is twice the circular orbit speed around the cluster center at the observed cluster radius, and that the cluster recession speed is 13,000 km/s

Chapter 25
True or False? 25.1 T 25.2 T 25.3 F 25.4 T 25.5 F 25.6 F 25.7 T 25.8 F 25.9 F 25.10 F 25.11 F 25.12 T 25.13 T 25.14 T 25.15 T
Fill in the Blank 25.1 long/radio/infrared 25.2 nucleus 25.3 larger 25.4 center 25.5 core 25.6 nucleus 25.7 small 25.8 1 billion 25.9 year 25.10 starlike 25.11 redshifted 25.12 expansion 25.13 large 25.14 short 25.15 galaxy/galaxy cluster
Odd-Numbered Problems 25.1 5100 km/s 25.3 1.3×10^{-5} W/m²; 1.0×10^{-7} W/m²; yes, since much of the emission from the nucleus is in the infrared 25.5 2.2 million years 25.7 2.2×10^{12} 25.9 11 million years 25.11 −22.5; 8.7×10^{10} solar luminosities 25.13 23 Mpc 25.15 3.6×10^{12} solar masses

Chapter 26
True or False? 26.1 F 26.2 T 26.3 F 26.4 F 26.5 F 26.6 T 26.7 F 26.8 T 26.9 T 26.10 T 26.11 F 26.12 F 26.13 F 26.14 T 26.15 T
Fill in the Blank 26.1 200 26.2 everywhere 26.3 directions 26.4 cosmological principle 26.5 homogeneity 26.6 isotropy 26.7 Gravity 26.8 Dark energy/Cosmological constant 26.9 high 26.10 1 26.11 35 26.12 accelerating 26.13 positively 26.14 cosmic microwave 26.15 2.7
Odd-Numbered Problems 26.1 1000 Mpc 26.3 4×10^{8} 26.5 110 km/s 26.7 (a) 110,000 tons, (b) 1.8 pc 26.9 (a) 82, (b) 55 km/s/Mpc 26.11 7 km/s/Mpc; no 26.13 (a) 1.1 mm, (b) 0.0093, (c) 9.3×10^{-5}, (d) 9.3×10^{-7} 26.15 8.6 kpc

Chapter 27
True or False? 27.1 F 27.2 T 27.3 F 27.4 T 27.5 F 27.6 F 27.7 T 27.8 T 27.9 T 27.10 F 27.11 F 27.12 T 27.13 F 27.14 F 27.15 T
Fill in the Blank 27.1 matter 27.2 radiation 27.3 photons, antiparticle 27.4 increases 27.5 deuterium, helium 27.6 atoms 27.7 fell 27.8 "normal" matter 27.9 1/1500 27.10 the criti-cal density 27.11 dark energy 27.12 cosmic microwave background 27.13 massive 27.14 temperature; COBE 27.15 flat
Odd-Numbered Problems 27.1 12 kpc 27.3 20 K; 3×10^{-24} kg/m³ 27.5 1.2×10^{9} K; 1/5 the temperature in the text 27.7 3×10^{-12} m; hard X ray/gamma ray 27.9 9.3 Mpc 27.11 166 27.13 1.6 Mpc 27.15 160 kpc, 7×10^{18} solar masses

Chapter 28
True or False? 28.1 T 28.2 F 28.3 T 28.4 T 28.5 F 28.6 F 28.7 T 28.8 F 28.9 T 28.10 T 28.11 T 28.12 T 28.13 T 28.14 F 28.15 T
Fill in the Blank 28.1 life 28.2 methane, ammonia 28.3 radioactivity, lightning, volcanism, meteoritic bombardment, solar radiation 28.5 1 billion 28.6 amino acids 28.7 intelligent civilizations 28.8 unstable 28.9 low 28.10 distances 28.11 light 28.12 dust 28.13 70 28.14 water hole 28.15 200
Odd-Numbered Problems 28.1 6.3 seconds (for a 20 year old reader); 16 seconds (in 2001); 71 seconds; 161 seconds; 5.2 days 28.3 They would both increase by a factor of two 28.5 five years 28.7 27 A.U. 28.9 5×10^{8} W (500 times the emission of the Sun) 28.11 32,000 km/s 28.13 1.43×10^{9} to 1.67×10^{9} Hz; 2.4×10^{6} channels 28.15 2.3 years; 55 years

INDEX

A 0620-00 (X-ray binary
 system), 591
A2218 galaxy cluster, 682
Abell 2199, 649
Aberration
 chromatic, 110, 111
 spherical, 122
 of starlight, 42
Absolute brightness (luminosity),
 441–43
Absolute magnitude, 444, 446
Absolute temperature, 74
Absolute zero, 73
Absorption lines, 88–89, 90, 94, 95
 quasar, 678
Absorption spectra, 479–81
Acceleration, 51
 centripetal, 50
 cosmic, 702–4
 due to gravity, 50
Accretion, 387–90
Accretion disk, 545, 567, 572, 574
Active galaxies, 661, 662–69
 central engine of, 673–77
  energy emission, 675–77
  energy production, 673–75
 common features of, 665–67
 evolution of, 683–86
 quasars vs., 662
 radiation emitted by, 662
 radio galaxies, 662, 664–69
 Seyfert galaxies, 662, 663–64,
  667
Active optics, 121
Active prominences, 423
Active regions, 422–24
Active Sun, 418–25
 active regions, 422–24
 changing solar corona, 425
 solar cycle, 421–22, 431
 solar magnetism, 418–19, 421
 sunspots of, 418, 419, 420
Adams, John, 329
Adaptive optics, 122, 124
Adastrea, 296
*Advanced X-Ray Astrophysics Facility
 (AXAF)*, 133, 134
AG Carinae, 524
Ahnighito meteorite, 377
Aine corona, 237, 238
Aitken Basin, 220
Algol, 535–37
ALH84001 meteorite, 266–67
Almagest (Ptolemy), 38
Alpha Centauri, 406, 439, 441, 460
Alpha particles, 520
Alpha-particle spectrometer, 220
Alpha process, 557
ALSEP, 204–5
Altair star, 10
Amalthea, 287
Amino acids, 742
Ammonia, 746
Amor asteroids, 360
Amplitude, 64
Andromeda Galaxy, 62, 70, 114,
 115, 116, 117, 443, 600,
 602, 603, 630, 631, 640

Ångstrom, 64–65
Ångstrom, Anders, 65
Angular diameter, 26
Angular measurement, 8–9, 11
Angular momentum, conservation
 of, 385, 390–91, 569
Angular resolution, 116, 117,
 124–25, 127
Annular eclipse, 17, 18, 19
Antennae galaxies, 647
Antiparticle, 426
Aphelion, 46
Aphrodite Terra, 232, 233, 235,
 258
Apollo asteroids, 360
Apollo Lunar Surface Experiments
 Package (ALSEP), 204–5
Apollo program, 185, 204–5, 207,
 208*n*
Apparent brightness, 441–43, 605
Apparent magnitude, 444
Arab astronomers, 36
Archimedes (crater), 200
Arc minutes, 11, 14
Arc seconds, 11, 14
Arcturus, 520, 610
Arecibo radio telescope, 125, 126,
 203
Ariel (moon), 337, 338, 339
A ring, 309, 310
Aristarchus of Samos, 38
Aristotle, 37, 38, 45, 51
Armstrong, Neil, 204, 205
Associations, 506, 510
Asteroid belt, 144, 149, 356, 357,
 384, 394
Asteroids, 144, 152, 356–62
 Earth-crossing, 360–61
 orbital resonances, 361–62
 properties of, 356–60
 stray, 372–75
 types of, 356, 376
Asthenosphere, 177
Astrologers, 7–8
Astronomical unit (A.U.), 47, 49,
 408
Astronomy
 ancient, 34–36
 birth of modern, 41–42
 defined, 4
 ultraviolet, 470–71
Asymptotic-giant branch, 523
Aten asteroids, 360
Atlas (satellite), 314
Atmosphere. *See also specific planets,
 stars, and other* bodies
 blue sky, reasons for, 169
 composition of, 209
 gravity and, 198, 208–9
 heat and, 199, 208
 primary, 171
 secondary, 171
 solar, 413–18
Atmospheric blurring, 118–19
Atmospheric circulation on Earth,
 431
Atom(s)
 atomic nucleus, 91, 180

formation of, 725–26
 ionization state of, 476
 ionized, 91, 95, 476
 thermal motions of, 100–101
Atomic epoch, 720, 721
Atomic spectra, 94, 95–96
Atomic structure, 91–92
Aurora
 on Earth, 185, 186
 on Jupiter, 285, 286
 on Saturn, 308, 309
Autumnal equinox, 12, 13

Background noise, 120
Ballistic Missile Defense
 Organization, 220
Balmer series, 98–99, 100
Barnard, E.E., 287
Barnard 68 cloud, 562
Barnard 86 cloud, 388
Barnard's Star, 439, 449, 456
Barred-spiral galaxy, 632, 633, 636
Barringer Meteor Crater, 210,
 212, 360, 373
Basalt, 173
Baseline of interferometer, 128
Baseline triangle, 24–25
Becklin-Neugebauer object, 500,
 508, 510
Bell, Jocelyn, 569
Belts, in atmospheric bans, 278–80
BeppoSAX satellite, 576
Beryllium-8, 520
Beta Pictoris star, 385–86, 387
Betelgeuse, 14, 132, 442, 445, 449,
 450, 454, 456, 531
Big Bang, 694, 695–98, 706, 708,
 709–10, 716–18
"Big Crunch," 701
Big Horn Medicine Wheel, 35
Binary asteroids, 357
Binary galaxy system, 649
Binary pulsar, 579
Binary-star systems, 457–58, 459,
 498–99, 544, 749
 black holes in, 590–91
 gravitational radiation from,
  578–79
 neutron-star, 572–75
 stellar evolution in, 535–37
Biochemistries, alternative, 746
Biological evolution, 744
Bipolar flow, 502, 503
Blackbody (Planck) curve, 72, 73,
 74, 75–77, 87, 445–46
Black dwarf, 528
Black hole(s), 567, 579, 580–93,
 617
 active galaxy energy and,
  673–75
 in binary-star systems, 590–91
 in centers of galaxies, 591,
  621–22
 curved space and, 582–85
 defined, 581
 escape speed of, 581
 event horizon of, 581–82,
  585–88, 593

as final stage of stellar
  evolution, 580–81
 intermediate-mass, 591, 592
 observational evidence for,
  589–93
 properties of, 582–85
 quasar energy generation and,
  684
 space travel near, 585–89
 stellar transits of, 589–90
 supermassive, 591, 673–75, 684
 tidal forces of, 585
Blazars, 686
BL Lac objects, 686
Blue giants, 453
Blueshifted radiation, 78–79
Blue stragglers, 529, 534
Blue supergiants, 453, 531
Bode, Johann, 149, 328
Bode's law, 149
Bohr, Niels, 91, 98
Bohr model, 91, 92
"Bok globules," 489
Boomerang Balloon Observations
  of Millimetric Extragalactic
  Radiation and
  Geomagnetics, 732, 734,
  735
Boson, 727
Brackett series, 99
Brahe, Tycho, 43–44
Brightness
 absolute (luminosity), 441–43
 apparent, 441–43, 605
B ring, 309, 310, 312, 313
Brown dwarfs, 398, 498–99, 508,
  617
Brown oval, 283
Bruno, Giordano, 41
Bunsen, Robert, 91

C. Bach (crater), 202
Caldera, volcanic, 236, 237
Calendar, 23
Caliban (moon), 337
Callisto, 288, 289, 293–95, 316, 319
Caloris Basin, 214–15
Calypso, 321, 322
Cambridge Optical Aperture
  Synthesis Telescope
  (COAST), 129
"Canals," Martian, 253, 261
Canis Major (Great Dog), 14
Cannon, Annie, 606
Canopus, 452
Capella binary star, 129
Capricornus constellation, 10
Capture theory, 217
Caracol temple, Mexico, 35
Carbon-12, 520, 555, 556
Carbon-14, 180
Carbonaceous meteorites, 376
Carbon atom, 95
Carbon core of low-mass star,
  522–23, 525
Carbon-detonation (Type I)
  supernovae, 549–50, 568,
  703

Carbon dioxide, 170
Carina Cluster, 509
Carina emission nebula, 509
Cartwheel Galaxy, 646
Cassegrain telescope, 111–12
Cassini, Giovanni Domenico, 309
Cassini Division, 309, 310, 313, 320
Cassini-Huygens mission, 143
Cassini mission, 159
Cassiopeia A, 543
Castor double star, 122, 124
Cataclysmic variables, 603. *See also* Nova/novae; Supernova/supernovae
Catastrophic theories, 386, 395–96
Cat's Eye Nebula, 527
Celestial coordinates, 14
Celestial equator, 8, 12, 14
Celestial mechanics, 154
Celestial poles, 8
Celestial sphere, 8–9
Centaurus A galaxy, 665, 666, 676
Center of mass, 54
Centripetal acceleration, 50
Cepheid variable stars, 443, 604–6, 634, 637
Ceres asteroid, 144, 146, 356, 357
Chandra Deep Field-South, 697
Chandrasekhar, Subramanyan, 549
Chandrasekhar mass, 549, 580
Chandra X-Ray Observatory (CXO), 133, 134
Chaotic rotation, 322
Charge-coupled devices (CCDs), 119–20
Charon (moon), 346–48, 396, 400
Chemical bonds, 96
Chemical evolution, 741–43
Chinese astronomy, ancient, 35–36
Chromatic Aberration, 110, 111
Chromosphere, 407, 408, 413–14
Chryse Planitia, 253, 254, 262
Circinus galaxy, 663
Clementine satellite, 165, 205, 211, 220, 221
Cleopatra crater, 233, 235, 237
Climate on Earth, solar activity and, 431
Closed universe, 707
Cloud(s)
    dark dust clouds, 478–81
    interstellar, 490–92, 493–95, 499–500
    on Jupiter, 281, 284
    Oort, 365–66, 393, 394, 400
    on Saturn, 304–5, 306, 307
Clusters
    galaxy, 5, 639–42, 643, 644, 682
    of galaxy clusters, 640–42
    star
        associations, 506, 510
        blue stragglers in, 529, 534
        emission nebulae and, 508–10
        globular, 506–8, 526, 528–29, 533, 534, 573–74, 603, 606–7, 706
        lifetimes of, 510
        observing stellar evolution in, 532–35

CNO cycle, 518, 554–55
COAST, 129
Cobalt-56, 560
Cocoon nebulae, 502
Coformation (sister) theory, 217
Cold dark matter, 731
Collecting area, 114
Collisional broadening, 102
Collision hypothesis, 386n
Collisions among galaxies, 646–47, 650
Color index, 446
Color-magnitude diagrams, 452
Color of stars, 445–46
Coma cluster, 630, 642
Coma effect, 112
Coma ("halo"), 362
Comet(s), 144, 152–53, 362–71, 374–75, 393. *See also specific comets*
    appearance and structure, 362–63
    collision of, 370–71
    fragments of, 370, 371, 372
    interstellar origin of life and, 743
    nucleus of, 362
    orbits of, 363–66
    physical properties of, 368–70
    solar system knowledge from, 384, 385
    tail of, 363, 364
Comparative planetology, 145
Compton Gamma-Ray Observatory (GRO), 135, 136, 576, 578
Condensation nuclei, 386–87
Condensation theory, 385–92
    catastrophes' role in, 395–96
    on differentiation of solar systems, 392–95
    dust's role in, 386–87, 388
    nebular contraction theory and, 385–86
    of planet formation, 387–92
Conservation of angular momentum, law of, 385, 390–91, 569
Conservation of mass and energy, law of, 426
Constant(s)
    cosmological, 704–5
    gravitational, 52
    Hubble's, 652–53, 679, 694, 705
    Newton's, 52
    Planck's, 92
    solar, 408
    Stefan-Boltzmann, 74
Constellations, 6–8, 10–11, 14
Constructive interference, 67
Contact binary, 536
Continental drift, 176–83
Continuous spectra, 87, 89, 90
Convection, 167–68, 180–81, 411
Convection cells, 168, 411, 412
Convection zone, 407, 408, 411–12
Co-orbital satellites, 321
Copernican principle, 42

Copernican revolution, 33–59
    birth of modern astronomy, 41–42
    defined, 39
    dimensions of solar system, 48–49
    heliocentric model of solar system, 39–40
    laws of planetary motion, 43–47
    Newton's laws and, 49–56
Copernicus, Nicholas, 39, 49, 607
Copernicus crater, 200, 210, 257
Cordelia (U7), 344
Cordillera mountains, 210
Core, solar, 407, 408
Core-collapse supernova, 548
Core-halo radio galaxies, 665–67, 668
Core-hydrogen burning, 516
Core hydrogen burning, 517
Core of earth, 166, 172, 173, 174
Corona, solar, 15, 237, 238, 407, 408, 415–16, 425
Coronal holes, 417–18
Coronal mass ejection, 423, 424
Corpuscular (particle) theory of light, 66
Cosmic Background Explorer (COBE), 708–9, 732–34
Cosmic censorship, principle of, 589
Cosmic density parameter, 702
Cosmic distance scale, 24, 653
Cosmic evolution, 740–45
Cosmic microwave background, 708–10, 726, 732–35
Cosmic rays, 622, 624
Cosmological constant, 704–5
Cosmological distances, 652
Cosmological redshift, 652
Cosmology, 36, 691–713
    Big Bang, 694, 695–98, 708, 709–10, 716–18
    birth of universe, 694–95
    cosmic microwave background, 708–10, 726, 732–35
    cosmological principle, 692–93
    cosmological redshift, 698–99
    expansion of universe, 702–6
    geometry of space, 707
    Olbers's paradox, 693–94
    relativity and, 699–700
Cosmos, density of, 724–25
Coudé focus, 112
Coudé room, 111–12
Crab Nebula, 550–51, 558, 570, 571, 572
Crater-chain pattern, 212, 213
Craters
    lunar, 199, 200, 201, 206–9, 257
    on Mars, 256–58
    on Mercury, 214
    on Venus, 236–39
Crescent Moon, 15, 16
C ring, 309, 310
Critical density, 700–701
Critical universe, 707
Crust of Earth, 166, 173
C-type asteroids, 356, 358, 376

Cultural evolution, 744–45
Current sheet, 285
Curved space, 582–85, 698, 707, 710
Cygnus A galaxies, 665, 667
Cygnus constellation, 590–91
Cygnus X-1, 590, 591, 592

Dactyl moon, 357–58
Dark dust clouds, 478–81
Dark energy, 704, 705, 717
Dark halo, 617
Dark matter, 616–18, 702, 731, 732
    mapping, 680–83
    search for stellar, 618–19
    in universe, 643
Daughter (fission) theory, 217–18
Daughter nuclei, 180
Davy (crater), 213
Death of low-mass star, 523–29
Deceleration, cosmic, 703
Declination, 14
Decoupling, 725
Deep-Space Program Science Experiment, 220
Deferent, 37, 38
Degrees, arc, 11
Deimos, 252, 269, 270
Deneb star, 10
Dense interstellar clouds, 478
Density, 146, 147–48. *See also specific bodies*
    of cosmos, 724–25
    critical, 700–701
    of Earth's interior, 173
    infinite, 588
    of interstellar medium, 469–70
    uncompressed, 151
    of universe, 702, 705
Density waves, spiral, 311, 613–15
Descartes, René, 385
Destructive interference, 67
Detached binary system, 536, 537
Deuterium, 426, 724–25
Deuterium bottleneck, 722
Deuterium fusion, 499
Deuteron, 426, 428
Dialogue Concerning the Two Chief World Systems (Galileo), 41
Diameter, angular, 26
Differential force (tidal force), 187–88, 206
    of black hole, 585
    exerted by Sun, 189
Differential rotation, 277
Differentiation, 173–75
    of solar system, 392–95
Diffraction, 116–17
Diffraction grating, 86–87
Diffraction-limited resolution, 116
Diffraction property of light, 66–67
Dinosaurs, 368–69
Dione, 319, 320
Direct (prograde) motion, 36
Distance modulus, 447
Distances
    cosmological, 652
    measurement of, 24–27
Distance scale, cosmic, 24, 653

Disturbance, 63, 67
Diurnal motion, 9
Dog Star (Sirius), 10, 14, 459, 460
Doppler, Christian, 77
Doppler effect, 77–79, 100–102
    detecting extrasolar planets
        with, 396–97, 398
Doppler gravity experiment,
        220–21
Double-line spectroscopic
        binaries, 457, 458–59
Drake equation, 747, 748, 750
D ring, 311–13
Dust
    dark dust clouds, 478–81
    interstellar, 386–87, 388,
        468–73
    lunar (regolith), 209–11
Dust grain, 469
Dust lanes, 476, 509, 614–15
Dust tails, 363, 364
Dwarf(s), 450
    black, 528
    brown, 398, 498–99, 508, 617
    red, 453–54, 516
    white, 450, 454, 525–28, 534,
        544–45, 549, 580, 588, 638
    helium, 528, 537
Dwarf ellipticals, 633, 634
Dwarf irregulars, 634
Dynamo theory, 174, 186, 420

Eagle Nebula, 475
Early universe, 715–37
    Big Bang, 694, 695–98, 708,
        709–10, 716–18
    evolution of, 718–21
    inflation of, 726–30
    nuclei and atom formation,
        721–25
    particle production in, 717–18
    structure in, 730–35
Earth, 4, 5, 165–93, 330
    atmosphere of, 70–72, 166,
        167–71, 240, 242
    in celestial sphere, 8–9
    core of, 172, 173, 174
    dimensions of, 47
    interior of, 171–75
    magnetic field of, 68, 167, 174
        reversals in, 178–80, 182
    magnetosphere of, 166, 167,
        178–80, 183–87
    nearest stars to, 439, 440
    orbits of, 49
    properties of, 146, 166–67,
        170
    rotation of, 14, 189–90, 230
    surface activity, 175–83
    surface heating of, 168–70
Earth-crossing asteroids, 360–61
Earthquake, 171–72
Eccentricity of ellipse, 44–45
Eclipse(s), 15–21
    annular, 17, 18, 19
    lunar, 15
    solar, 15–20
Eclipse seasons, 18–21
Eclipse year, 21
Eclipsing binaries, 457, 458, 459
Ecliptic, 11, 12

Eddington, Arthur, 587, 604
Eddington crater, 210
Effective temperature of Sun, 407
Einstein, Albert, 78n, 92, 581,
        582–83, 586, 587, 618, 704
Einstein Cross, 682
Einstein Observatory, 133
Ejecta blankets, 207, 257
Electric field, 66
Electromagnetic force, 427
Electromagnetic radiation. *See*
        Radiation
Electromagnetic waves, 67–70
Electromagnetism, 68
Electron(s), 65–67
    production of, 717–18
    spin of, 481
Electron cloud, 91, 93
Electron degeneracy pressure, 521,
        527
Electron transitions, 96, 97
Electron volt (eV), 98
Electroweak force, 720, 722
Elements, 95
    formation of, 552–60
        abundance of matter, 553
        carbon burning and helium
            capture, 555–57
        heaviest elements, 559
        hydrogen and helium
            burning, 553–55
        iron formation, 557–59
        neutron capture, 559
        stellar nucleosynthesis,
            553–60
        types of matter, 552–53
Ellipse, 44–46
Elliptical galaxies, 632–34, 636
Emission lines, 87–88, 89, 90, 94,
        95
Emission nebulae, 93, 95–96,
        473–78, 489
    "forbidden" lines of, 476–78
    interstellar shock waves
        generated by, 504
    nebular spectra, 476, 477
    observations of, 473–76
    star clusters and, 508–10
Emission spectrum, 88
Enceladus, 319, 320–21
Encke gap, 309, 310
Energy
    black hole and production of,
        585
    dark, 704, 705, 717
    generated by proton-proton
        chain, 429
    law of conservation of mass
        and, 426
    nuclear binding, 559
    solar energy production, 408,
        423, 425–26
Energy flux, 74, 441–43
Energy levels, atomic, 98–99
Energy transport in solar interior,
        410–12
Epicycles, 37, 39
Epimetheus, 321
Epsilon ring (Uranus), 342, 343,
        344
Equator, celestial, 8, 12, 14

Equilibrium
    hydrostatic, of main-sequence
        star, 516
    thermal, 718
Equinoxes, 12, 13, 14, 22
Equivalence principle, 50
Eratosthenes (crater), 200
Eratosthenes (philosopher), 27
E ring, 313, 320
Eros asteroid, 355, 358–59
Escape speed, 55, 198, 208, 209,
        581, 700
"Eskimo" Nebula, 527
Eta carinae, 509
Euclid, 26
Euclidean geometry, 707, 710
Europa, 288, 289, 292–93, 745
European Southern Observatory,
        116, 119
European Space Agency, 409, 410,
        442
Event horizon, 581–82, 585–88,
        593
Evolution, cosmic, 740–45
Evolutionary theory, 386
Evolutionary track, 494–96, 531
    prestellar, 497
Excited state, 92, 93, 94
Extinction, 469
Extrasolar planets, 384, 390–92,
        396–400
    discovery of, 396–97
    planetary properties, 397–98
Extraterrestrial intelligence, 750,
        751–54
*Extreme Ultraviolet Explorer
        (EUVE)*, 132, 470, 471
Eyepiece, 109

**Fahrenheit temperature scale,
        73**
Failed stars, 498–99
False color technique, 131
FAME mission, 443
Far Ultraviolet Spectrographic
        Explorer (FUSE), 132
51 Pegasi star, 396, 398
Fireballs, 373, 375
Fission, 546
Fission (daughter) theory, 217–18
Flares, solar, 423, 424, 431
Flatness problem, 726, 727,
        729–30
Fleming, Williamina, 606
Fluidized ejecta, 257
Fluorescence, 94
Focal length, 108
Focus of ellipse, 44
Focus of telescope, 108
"Forbidden" lines of emission
        nebulae, 476–78
Force, 51, 427
Formaldehyde, 483, 484, 499–500
47 Tucanae, 533, 534
Fossil record, 744
Fragmentation, 389–90
Fraunhofer, Joseph von, 88
Fraunhofer lines, 88, 91, 413
Freeze-out, 718, 720, 727
Frequency, 64, 65, 70
    photon energy and, 92

F ring, 314, 315, 343
Full Moon, 15, 16
*Full-Sky Astrometric Mapping
        Explorer (FAME)* mission,
        443
Fundamental forces, 722
FUSE, 132
Fusion
    deuterium, 499
    of heavy elements in high-mass
        star, 545–46, 547
    helium, 520
    nuclear, 426–29

GAIA project, 443
Galactic bulge, 600, 602, 608, 609,
        610, 611, 620
Galactic cannibalism, 649
Galactic center, 126, 607, 608,
        609, 620–24
Galactic disk, 600, 602, 607, 608,
        609, 610–11
    interstellar gas in, 612
    overall properties of, 611
    spiral structure and, 613–15
Galactic epoch, 720, 721
Galactic gas, 612
Galactic halo, 600, 602, 607,
        609–10, 611
Galactic halos, 606n, 607, 608
Galactic nucleus, 622, 631, 663
Galactic rotation, 638
Galactic rotation curve, 617
Galactic year, 609
Galaxies, 4, 5, 629–59. *See also*
        Active galaxies; Milky Way
        Galaxy
    colliding, 646–47, 650
    defined, 600
    distribution in space, 637–42
    formation and evolution of,
        645–50
    galaxy clusters, 5, 639–42, 643,
        644, 682
    H–R diagram for, 636–37
    Hubble classification scheme,
        630–37
        elliptical galaxies, 632–34,
            636
        irregular galaxies, 634–35,
            636, 649
        spiral galaxies, 630, 631–32,
            633, 636, 643, 649, 650
    Hubble's law of, 651–55
    interaction of, 648–50
    large-scale distribution of,
        653–55
    life in, 747–51
    masses of, 642–44
    redshift survey of, 692, 693
    visible, 664
Galaxy cluster, 5, 639–42, 643,
        644, 682
Galaxy spectra, 651–52
Galilean moons, 276, 287, 288–95
    Callisto, 288, 289, 293–95,
        316, 319
    Europa, 288, 289, 292–93,
        745
    Ganymede, 288–89, 293–95,
        315, 316, 319

Ganymede (*continued*)
Io, 147, 287, 288, 289–92, 320–21
Galileo Galilei, 39, 41, 45, 49, 144, 309, 418
Galileo missions, 155, 158, 159, 292, 293, 294, 357
Galle, Johann, 329
Gamma-ray bursts, 575–80
Gamma-ray pulsars, 570, 572
Gamma rays, 63, 65, 92–93, 133–36, 137
Gamma-ray spectrometer, 220
Ganymede, 288–89, 293–95, 315, 316, 319
Gas
greenhouse, 170
interstellar, 468–73, 508–10, 612
intracluster, 644
Gas-exchange experiment, 266
Gaspra asteroid, 357, 358
Gas turbulence, 102
Geminga, 570, 572
Gemini constellation, 10
Gemini North telescope, 114–16, 123
Gemini South telescope, 116, 123
General theory of relativity, 581, 582–83, 588, 589, 699–700
tests of, 586–87
Genes, 742
Geocentric universe, 36–39
Geomagnetic activity at Earth, 431
Geometry
Euclidean, 707, 710
Riemannian, 710
Geysers, nitrogen, 341
Giacobini-Zinner comet, 363
Giant ellipticals, 632–33, 634
Giant planets, 390–92, 393
Giants, 450
blue, 453
red, 450, 454
Gibbous phase of Moon, 15, 16
Giotto spacecraft, 366, 367
Glashow, Sheldon, 723
Gliese 229, 499
Gliese 710, 442
Globular clusters, 506–8, 526, 528–29, 533, 534, 573–74, 603, 606–7, 706
Gluons, 727
Glycine, 483*n*
GONG (Global Oscillations Network Group) project, 409
Grand Canyon, Martian, 258
Grand Unified Theories (GUTs), 720, 722–23, 727
Granite, 173
Granulation, solar, 412–13
Gravitational competition, 490–92
Gravitational constant, 52
Gravitational field, 53
Gravitational force, 427
Gravitational lensing, 618–19, 678–80, 682
Gravitational radiation, 578–79
Gravitational redshift, 586–88
Gravitational slingshot (gravity assist), 154–55

Gravitons, 723
Gravity, 720, 722
atmosphere and, 198, 208–9
Doppler gravity experiment, 220–21
in general relativity, 583
heat and, 490–91
magnetism and, 492
Newton's law of, 49, 50, 52–53, 586
quantum, 588, 688, 719, 720
rotation and, 491
tides and, 187–89
Gravity wave, 578–79
Great Attractor, 702
Great Dark Spot, 334, 335
Great Dog (Canis Major), 14
Great Red Spot, 278, 282–83
Great Wall, 655
Greek astronomers, 36
Greenhouse effect, 170, 171, 240–43
Greenhouse gases, 170
Greenwich mean time, 23
Gregorian calendar, 23
G ring, 314
Ground state, 91
Gula Mons, 236, 237
GUTs, 720, 722–23, 727

HI regions, 476
HII regions. See Emission nebulae
Habitable zone, 748–49
Hadley, Mount, 213
Hadley Rille, 213
Hadron epoch, 720, 721
Hale-Bopp comet, 146, 152, 364, 374–75, 743
Hale telescope, 112, 113, 118, 337
Half-life, 180
Hall, Asaph, 261, 269
Halley, Edmund, 49–51, 366
Halley's comet, 363, 366–68, 369, 743
Haute-Provence Observatory, 396
Hayashi, C., 495
Hayashi track, 495–96
Haystack radio telescope, 125, 127
Head-tail radio galaxies, 665, 667
Heat (thermal energy), 73, 490–91
atmosphere and, 199, 208
differentiation of solar system and, 392–93
Helene (moon), 321
Heliocentric model, 39–40
Helioseismology, 409
Helium, 88, 91, 95, 721–24
Helium burning, 553–55
Helium capture, 556–57
Helium core of low-mass star, 517–19
Helium flash, 521–22
Helium fusion, 520
Helium precipitation, 308
Helium-shell flashes, 524, 525
Helium white dwarf, 528, 537
Helix Nebula, 526
Hellas Basin, 253–54, 255, 260
Helmholtz, Hermann von, 495–96
Hen 1357 nebula, 562
Herbig-Haro 32 system, 383

Herschel, William, 328, 337, 603
Herschel crater, 319, 320
Hertz, Heinrich, 64
Hertz (Hz), 64
Hertzsprung, Ejnar, 450
Hertzsprung–Russell (H-R)
diagram, 443, 450–54, 455
cluster evolution on, 532–33, 534
development of, 606
for galaxies, 636–37
of globular cluster, 528–29
horizontal branch of, 522
instability strip on, 604
of protostar's physical properties, 494, 495, 496
red-giant branch of, 519–20
of stars, 443, 450–54, 455
of white dwarf, 527
Hewish, Anthony, 569
HH1/HH2 (pair of jets), 503
High-Energy Astronomy Observatory (Einstein Observatory), 133
High-energy telescopes, 133–36, 137
High Energy Transient Explorer-2 (HETE-2), 577
Highlands, lunar, 199, 218
High Precision Parallax Collecting Satellite, 442
High-resolution telescopes, 118–24
Himalayas, 178
Hipparchus, 442, 443, 444
Hipparcos satellite, 439, 442–43, 454, 587
History, learning astronomy from, 530
Há line (red line), 94, 96, 99
Holden Crater, 262
Homogeneity, cosmic, 692
Homo sapiens, 745
Hooke, Robert, 282
Horizon problem, 726, 728–29
Horizontal branch, 522
Horsehead Nebula, 479
Hot dark matter, 731
Hot longitudes, 206
Hourglass, 475
Hours, 14
HR4796A, 389
H-R diagrams. See Hertzsprung–Russell (H-R) diagram
Hubble, Edwin, 607, 631, 652
Hubble classification scheme, 630–37
elliptical galaxies, 632–34, 636
irregular galaxies, 634–35, 636, 649
spiral galaxies, 603, 612–15, 630, 631–32, 633, 636, 643, 649, 650
Hubble Deep Field, 648, 696–97
Hubble diagrams, 652
Hubble flow, 652, 695
Hubble's constant, 652–53, 679, 694, 705
Hubble's law, 651–55, 694, 695

Hubble Space Telescope (HST), 107, 119, 120, 122–23, 131, 132, 212, 276, 308, 309, 315, 317, 696
Hulse, Russell, 579
Huygens, Christian, 66, 309, 316
Hyades cluster, 533, 534
Hyakutake comet, 374
Hydrogen, 91, 553–55
energy levels of, 98–99
spectrum of, 87, 88, 93, 94, 97
Hydrogen-burning reaction, proton-proton, 554–55, 556
Hydrogen envelope, 363
Hydrogen-shell burning stage, 519
Hydrogen 21-cm emission, 481–82
Hydrosphere, 166
Hydrostatic equilibrium of main-sequence star, 516
Hydrothermal vents, 746, 747
Hyperion, 321–22
Hypernova, 579
Hypothesis, 45

Iapetus, 319, 321
IC 418 (nebula), 526
IC 2163 (galaxy), 646
IC 2944 (emission nebula), 489
Icarus asteroid, 360
Ice, lunar, 211–12
Ida asteroid, 357, 358
Image processing, 119–20
Images, 108–10
Impact theory, 218
Inclination, 11
Index of Prohibited Books, 39
Indian plate, 177
Inertia, 51
Inferior conjunction, 228
Infinite density, 588
Inflation, cosmic, 727–30
Infrared Astronomy Satellite (IRAS), 131, 132, 500
Infrared radiation, 63, 65
Infrared Space Observatory (ISO), 131
Infrared telescopes, 130–32, 137
Inhomogeneities, growth of, 730–31
Inner core, 172, 174
Intelligence, development of, 744–45, 750
Intensity, 72
Intercloud medium, 471
Intercrater plains, 214
Interference property of light, 67
Interferometry, 127–29
Intermediate-mass black holes, 591, 592
International Astronomical Union, 151
International Ultraviolet Explorer (IUE), 132, 470–71
Interstellar cloud, 490–92, 493–95, 499–500
Interstellar medium, 467–87, 528
composition of, 470–71
dark dust clouds, 478–81
defined, 468

density of, 469–70
emission nebulae, 473–78
gas and dust, 386–87, 388, 468–73, 508–10, 612
interstellar molecules, 482–84
21-centimeter radiation, 481–82
Interstellar space, first living cells produced in, 743
Intracluster gas, 644
Intrinsic variable, 603
Inverse-square law, 52, 441–43, 447
Io, 147, 287, 288, 289–92, 320–21
Ion, 91
Ionization state of atom, 476
Ionized atoms, 91, 95, 476
Ionosphere, 71, 167, 168
Ion tails, 363, 364
Io plasma torus, 290, 291
Iridium, 369
Iron core of high-mass star, collapse of, 546–48
Iron formation, 557–59
Irregular galaxies, 634–35, 636, 649
Ishtar Terra, 232–33, 235
Isidis Planitia, 253
Islamic astronomy, 36
Isotopes, 180, 427, 428, 552
Isotropy
cosmic, 692–93
of cosmic microwave background, 709, 726

Jansky, Karl, 124
Janus, 321
Joule (J), 74
Jovian planets, 150, 151, 390–92, 393. See also Jupiter; Neptune; Saturn; Uranus
Julian calendar, 23
Jumbled (weird) terrain, 216
Juno (asteroid), 356
Jupiter, 196, 275–99, 330, 497–98
atmosphere of, 278–83
bulk properties of, 146, 147
dimensions of, 47
formation of, 391–92
Galileo missions to, 155
internal structure of, 277, 278, 283–85
magnetosphere of, 277, 285–86, 290, 336
mass and radius of, 276–77
moons of, 276, 287–95, 315, 321, 745
orbit of, 361, 362, 399
rings of, 296, 312
rotation rate of, 277–78
Shoemaker-Levy 9 comet collision with, 369, 370–71
spacecraft exploration of, 157–59
starlike composition of, 284
Trojan asteroids orbiting, 361

Keck telescope, 114, 123
Kelvin, Lord, 73, 495
Kelvin-Helmholtz contraction phase, 494–95

Kelvin temperature scale, 73
Kennedy, John F., 204
Kepler, Johannes, 43–47, 49
Kepler mission, 400
Kepler's laws, 44–47, 48, 49, 54
Keyhole Nebula, 467
Kirchhoff, Gustav, 89
Kirchhoff's laws, 89–90, 94–95
Kirkwood, Daniel, 361
Kirkwood gaps, 361, 362
Kuiper, Gerard, 337, 339, 364
Kuiper belt, 350, 364, 365, 393, 394, 396, 400

Labeled-release experiment, 266
Lada Terra, 235
Lagoon Nebula, 475
Lagrange, Joseph Louis, 321, 361
Lagrangian point (L₁), 321, 361, 410, 536
Lakshmi Planum, 233, 236
Language, development of, 744–45
Laplace, Pierre Simon de, 385, 386
Large Magellanic Cloud (LMC), 443, 554, 591, 634, 635
Lassell, William, 337, 339
Lava domes, 236
Lead-206, 180
Lead-207, 181
Leap year, 23
Leavitt, Henrietta, 605, 606
Length measurements, 11
Lensing, gravitational, 618–19, 678–80, 682
Leo constellation, 10
Lepton epoch, 720, 721
Leverrier, Urbain, 329
Lick Observatory, 310
Life, 739–57
in galaxy, 747–51
interstellar origin, evidence of, 483n, 743–44
on Mars, 266–67
search for extraterrestrial intelligence, 750, 751–54
in solar system, 745–46
working definition of, 740–41
Light. See also Radiation
diffraction property of, 66–67
interference property of, 67
obscuration by dark dust clouds of visible, 478–79, 480
particle and wave property of, 94
particle (corpuscular) theory of, 66
polarization of, 67
radiation and, 62–63
speed of, 70, 582, 670
visible, 62–63, 64–65
Light curve, 457, 544, 560
Lighthouse model of neutron-star emission, 569, 570
Light-years, 4, 62
LIGO (Laser Interferometric Gravity-wave Observatory), 579

Lin, C.C., 614
Linear momentum, 390
Line of nodes, 18
Lithosphere, 177
Little Bear (Ursa Minor), 14
Little Ice Age, 431
LMC X-3, 591
Lobachevsky, Nikolai Ivanovich, 710
Lobe radio galaxies, 664–65, 667, 676
Local Bubble, 470–71
Local Group, 639, 640, 642
Local Supercluster, 640–42
Logarithmic scale, 70
Look-back time, 680–81
Lowell, Percival, 261, 345, 651
Lowell Crater, 268
Luminosity
in Hertzsprung–Russell diagrams, 450–54
of nova, 544
period-luminosity relationship, 605–6
of protostar, 494, 495
of stars, 441–44, 446–47, 449–50
of Sun, 407–8
of variable star, 605
Luminosity classes, 455–56
Luna missions, 204
Lunar eclipse, 15
Lunar Orbiter spacecraft, 204
Lunar phases, 15
Lunar Prospector, 205, 211–12, 214, 215, 216, 220–21, 269
Lunar rover, 204
Lyman-alpha forests, 678, 692
Lyman series, 98, 99

M17 (Omega) nebula, 96
M51 galaxy, 129
M83 galaxy, 4
M87 galaxy, 666–67, 668, 674, 675
M100 galaxy, 123
MACHOs (MAssive Compact Halo Objects), 617–18
Magellan, Ferdinand, 635
Magellanic Clouds, 443, 554, 591, 634, 635, 640
Magellan spacecraft, 155–56, 232
Magnetic field, 67–68, 69. See also specific planets
of active galaxy jets, 676
of Earth, 68, 167, 174
reversals in, 178–80, 182
Martian, 268–69
of Mercury, 216–17
of newborn neutron stars, 569
spectral-line broadening and, 102
of Venus, 243
Magnetic poles, 183
Magnetism
gravity and, 492
solar, 418–21
Magnetometers, 220
Magnetopause, 185, 285
Magnetosphere, 166, 167, 178–80. See also under specific planets
Magnitude scale, 443–44, 446–47

Main sequence, 452–54
defined, 452
leaving, 516–17, 529–30
in star formation, 496
zero-age (ZAMS), 497
Main-sequence turnoff, 532, 533, 534
Major axis, 44
Manicouagan Reservoir, 376
Mantle of Earth, 166, 173
MAP satellite, 734
Maria, lunar, 199, 200, 201, 219
Mariner missions, 153–54, 156, 198–99, 201, 202, 243, 253, 256, 261, 263, 264, 268
Mars, 196, 249–70
atmosphere of, 264–68
bulk properties of, 146
dimensions of, 47
Earth-based observations of, 252–53
internal structure of, 268–69
life on, 266–67
moons of, 252, 269–70
orbital properties of, 250–51
planetary data, 251, 252
rotation of, 230
search for life on, 745–46
spacecraft exploration of, 156–57
surface of, 253–64
"canals," 253, 261
"Grand Canyon," 258
"gullies" in cliffs and crater walls, 257–58
impact cratering, 256–58
polar caps, 253, 260–61, 263
topography of, 253–54
viewed by Martian landers, 262–64, 265
volcanism, 254–56
water, 257–60, 262
Mars Climate Orbiter, 157
Mars Express mission, 157
Mars Global Surveyor, 156–57, 253, 254, 255, 262, 269
Mars Observer, 156
Mars Odyssey, 157
Mars Pathfinder mission, 157, 263–64, 265
Mars Polar Lander, 157
Mass(es), 51
center of, 54
Chandrasekhar, 549, 580
of galaxies, 642–44
law of conservation of mass and energy, 426
of Milky Way Galaxy, 616–19
of planets, 146
of stars, 457–61, 497–98
turnoff, 533
Mass function, 459
MAssive Compact Halo Objects (MACHOs), 617–18
Mass-transfer binaries (semidetached binaries), 536, 537, 544
Mathilde (asteroid), 152, 358
Matter, 552–53
dark. See Dark matter

Matter-dominated universe, 716–17
Matter Era, 718, 720, 721
Mauna Kea, Hawaii, 114, 115, 119, 130
Maunder minimum, 422, 423, 431
Maury, Antonia, 606
Maxwell, James Clerk, 233, 310
Maxwell Montes, 233, 235
Mayans, 35
Mead crater, 238
Mean solar day, 22–23
Measurement
    of distance, 24–27
    time, 22–23
Mediocrity, assumptions of, 741
Mercury, 195–225
    atmosphere of, 198–99
    bulk properties of, 146
    dimensions of, 47
    evolutionary history of, 219
    interior of, 216–17, 395
    Mariner 10 flybys of, 153–54
    orbital properties of, 196–97, 587
    properties of, 198, 215
    rotation of, 203–6, 230
    solar transit of, 48
    spacecraft exploration of, 153–54
    surface of, 201–2, 214–15
    tidal forces on, 206
Mercury (element), emission spectrum of, 88
Mergers
    galaxy, 645–48, 650
    of lower-mass stars, 529
Meridian, 22, 23
Mesosphere, 167
Messier, Charles, 473
Meteor, defined, 372
Meteorites, 372, 384
    interstellar origin of life and, 743–44
    properties of, 375–76
Meteoroids, 145, 152, 206–7, 372–77
Meteoroid swarm, 372
Meteors, 144, 206–7
Meteor shower, 372, 373
Methane, on Pluto, 348
Metis, 296
Michelson, A.A., 582
Microcraters, 210
Microlensing, 680
Micrometeoroids, 207, 372
Mid-Atlantic Ridge, 178, 179, 181
Midocean ridges, 178
Migration, planet, 392, 393
Milky Way Galaxy, 4, 136, 137, 468, 599–627, 630, 631, 639–40
    dust particles in, 386
    formation of, 610–12
    Galactic center, 126, 607, 608, 609, 620–24
    mass of, 616–19
    measuring, 600–607
    orbital motion of, 609–10
    overall structure of, 600, 602
    plane of, 473, 474

size and shape of, 606–7
spatial distribution of stars, 607–8
spiral structure of, 612–15
    star counts for, 600–601
    stellar populations, 608–9
Miller, Stanley, 742
Millisecond pulsars, 573–74, 575
Mimas (moon), 313–14, 319, 320
Minutes, 11, 14
Mira, 450, 452, 454
Miranda (moon), 337, 338, 339, 340, 396
Modeling, 24
Molecular clouds/cloud complexes, 482, 483–84
Molecular spectra, 96–97
Molecular spectral lines, 482–83
Molecular speed, 208–9
Molecular tracers, 483–84
Molecules, 96–97
    interstellar, 482–84
Momentum, 385, 390–91, 569
Month, 23
    sidereal, 15, 17
    synodic, 15
Moon(s). *See also specific names of moons*
    of jovian planets, 391
    of Jupiter, 276, 287–95, 315, 321, 745
    of Mars, 252, 269–70
    of Neptune, 338, 339–42, 400
    Newton's law of gravity and, 50
    of Pluto, 346–48, 396, 400
    of Saturn, 308–9, 313–22, 745
    of Uranus, 337–39, 396
Moon, the, 195–225
    angular size of, 11
    atmosphere of, 198–99
    bulk properties of, 146
    craters on, 199, 200, 201, 206–9
    evolutionary history of, 218–19
    explorations of, 204–5, 220–21
    interior of, 215–16
    motion of, 15–21
    nodes of orbit, 18
    orbital properties of, 196
    origin of, 217–18
    properties of, 198
    rotation of, 201, 202–3
    surface composition of, 211–13
    surface features of, 199–201
    tidal bulge in, 202–3, 206
    tides and, 187–90
    volcanism on, 212, 213, 219
Morley, E.W., 582
Motion, Newton's laws of, 49–52
M-type asteroids, 356
Multiple-star systems, 457. *See also* Binary-star systems
Murchison meteorites, 744

N81 emission nebula, 93
Naked singularity, 589
Nakhla meteorite, 267
Nanometer (nm), 64
National Astronomical Observatory of Japan, 114

National Radio Astronomy Observatory, 124, 125
Nazca plate, 177
Neap tides, 189
Near Earth Asteroid Rendezvous (NEAR), 152, 358–59
Nebula, nebulae, 467, 473
    cocoon, 502
    emission. *See* Emission nebulae
    planetary, 524–25, 526, 638
    solar, 385, 388
    spiral, 603
Nebular contraction theory, 385–86
Nebular spectra, 476, 477
Negative curvature, 707
Neon, emission spectrum of, 88
Neon-20, formation of, 556
Neon-oxygen white dwarf, 528
Neptune, 144, 329–44
    atmosphere of, 332–35
    bulk properties of, 146, 329–32
    dimensions of, 47
    discovery of, 329
    Great Dark Spot of, 334, 335
    internal structure of, 335, 336
    magnetosphere of, 335, 336
    mass and radius of, 329
    moons of, 338, 339–42, 400
    planetary data, 341
    rings of, 312, 344–45
    rotation rates of, 331–32
Nereid (moon), 339–40, 342
Neutrino oscillations, 430
Neutrinos, 427, 429–30, 547, 554–55, 568
Neutron capture, 559
Neutron degeneracy pressure, 548, 568, 580
Neutronization, 547
Neutrons, 95
Neutron spectrometer, 220
Neutron-star binaries, 572–75
Neutron stars, 568–80
    distinguishing black hole from, 592
    pulsars, 558, 569–71, 573–74, 575
    rotation of, 569
New Moon, 15
New Technology Telescope (NTT), 121
Newton, Isaac, 49–56, 64, 581, 616
Newtonian mechanics, 51
Newtonian telescope, 111, 112
Newton's constant, 52
Newton's laws, 49–56
NGC 1265 galaxy, 665
NGC 1850 star cluster, 437
NGC 2207 galaxy, 646
NGC 2997 galaxy, 602, 631–32
NGC 3132 nebula, 515
NGC 3603 star cluster, 506
NGC 4214 galaxy, 505
NGC 4258 galaxy, 674–75, 676
NGC 4261 galaxy, 675
NGC4414 galaxy, 4, 5
NGC 4565 galaxy, 602
NGC 4603 galaxy, 599
NGC 604 star-forming region, 491
NGC 6520 stars, 388

NGC 7742 galaxy, 663
Nickel-56, 557, 560
Nitrogen geysers, 341
Noise, background, 120
Nonthermal spectrum of radiation, 75
North celestial pole, 8
Northern Lights (*aurora borealis*), 185
Nova Cygni, 546
Nova/novae, 544–45, 548–49
    X-ray, 593
Nuclear binding energy, 559
Nuclear epoch, 720, 721
Nuclear force, 426, 427, 722
Nuclear fusion, 426–29
Nucleosynthesis
    primordial, 721, 724–25
    stellar, 553–60
Nucleotide bases, 742
Nucleus/nuclei
    atomic, 91, 180
    of comet, 362
    condensation, 386–87
    formation, 721–25
    galactic, 622, 631, 663
Nutation, 21

OB associations, 506
Oberon (moon), 337, 338, 339
Occultation, stellar, 342, 343
Olbers, Heinrich, 694
Olbers's paradox, 693–94
Olympus Mons, 254, 256
Omega Centauri, 76, 77, 507
Omega (M17) nebula, 96
Oort, Jan, 365
Oort cloud, 365–66, 393, 394, 400
Opacity, 70–72
Open star cluster, 505, 507, 510
Open universe, 707
Ophelia (U8), 344
Optical doubles, 458
Optical interferometry, 128–29
Optical telescopes, 108–14, 137, 144–45
Optics, 121, 122, 124
Orbit(s). *See also specific planets*
    planetary, 148–50
    retrograde, 287, 288
    synchronous, 202
Orbital motion of Milky Way Galaxy, 609–10
Orbital period, 146
Orbitals, electron, 91
Orientale Basin, 208, 210
Orion constellation, 6–7, 8, 10, 445, 501
Orion Nebula, 126, 127, 130, 131, 477, 500, 501, 508–10
Outer core, 172, 173
Outflow channels, Martian, 259, 260
Ovda Regio, 233, 235
Oxygen-16, formation of, 556
Ozone layer, 71, 168

Pair production, 717–18
Pallas asteroid, 356, 357
Palomar Observatory Sky Survey, 112, 113

Pandora (moon), 314
Pangaea, 182–83
Pan (moon), 311
Parallax, 25–27
  spectroscopic, 454–55
  stellar, 438–39
Paranal Observatory in Atacama, Chile, 116
Parsec, 439
Partial lunar eclipse, 15
Partial solar eclipse, 15, 18, 19
Particle accelerators, 715
Particle-antiparticle annihilation, 717–18
Particle-antiparticle pair, 717, 718
Particle astrophysics, 715
Particle (corpuscular) theory of light, 66
Particle nature of radiation, 92–94
Particles, charged, 65–67, 68
Paschen series, 99
Pauli, Wolfgang, 521
Pauli exclusion principle, 521
Penumbra, 17, 18, 418
Penzias, Arno, 708
Perihelion, 46
Period, 46
Period-luminosity relationship, 605–6
Permafrost, 257
Persei, 533
Perseid shower, 372
Perseus Cluster, 665
Pfund series, 99
Philosophiae Naturalis Principia Mathematica (The Mathematical Principles of Natural Philosophy) (Newton), 51
PHL 938 quasar, 678
Phobos (moon), 252, 269, 270
Phoebe (moon), 321
Photodisintegration, 547, 557
Photoevaporation, 476
Photometer, 112–13
Photometry, 446
Photons, 92
  particle-antiparticle pair produced by, 717–18
Photosphere, 407, 408, 412, 494
Piazzi, Giuseppe, 356
Pioneer missions, 157–58, 314
  Pioneer 10 spacecraft, plaque mounted on, 751, 753
Pioneer Venus, 155, 231, 232, 237, 240
Pixels, 120
Plains Indians, 35
Planck, Max, 72, 92, 720
Planck (blackbody) curve, 72, 73, 74, 75–77, 87, 445–46
Planck epoch, 720
Planck mission, 734–35
Planck's constant, 92
Planet(s), 4. *See also specific planets*
  brightness of, 37, 46
  condensation theory of formation of, 387–92
  extrasolar, 384, 390–92, 396–400
  giant, 390–92, 393

habitable, 748–50
jovian, 150, 151, 390–92, 393. *See also* Jupiter; Neptune; Saturn; Uranus
  mass of, 146
  orbits of, 148–50
  pulsar, 574–75
  searching for Earthlike, 399–400
  spacecraft exploration of, 153–59
  terrestrial, 150, 151, 393–94
Planetary motion, 36–37
  laws of, 43–47
  Newton's laws and, 53
  properties of orbits, 46
  retrograde, 36
Planetary nebulae, 524–25, 526, 638
Planetary systems, stars with, 748
Planetesimals, 388, 389–90, 393, 394
Planet migration, 392, 393
Planetoids. *See* Asteroids
Planetology, comparative, 145
Plaskett crater, 165
Plasma torus, Io, 290, 291
Plate tectonics, 176–83
  on Ganymede, 294
  on Mars, 269
  on Venus, 237
Plato, 37
Pleiades (Seven Sisters), 6, 505–6, 507
Plutinos, 400
Pluto, 345–50, 410
  bulk properties of, 146, 346–48
  debate over, 151
  dimensions of, 47
  discovery of, 345–46
  moons of, 346–48, 396, 400
  origin of, 349–50
  surface detail, 349
Plutonium 241, 180, 181
Polar caps, on Mars, 253, 260–61, 263
Polaris, 7, 8
Polarity of sunspot, 418
Polarization
  of light, 67
  of starlight by dust, 471–73
Pole Star, 14
Polycyclic aromatic hydrocarbons, 266–67
Population I and II stars, 609
Positive curvature, 707
Positrons, 426, 428, 717–18
Precession, 14, 15
Pressure waves, 171
Primary atmosphere, 171
Primary mirror, 108
Primary waves (*P*-waves), 171, 172, 174
Prime focus, 108–9, 112
Primordial elements, 553
Primordial nucleosynthesis, 721, 724–25
Progenitor of supernova, 548
Prograde (direct) motion, 36
Project Phoenix, 753, 754
Prometheus (moon), 314

Promethium, 553
Prominences, solar, 422, 423
Proper motion, 439, 440
Proteins, 742
Proton-proton chain, 426–29, 518, 554
Protons, 65–67, 481
Protoplanet, 386, 390–92
Protostar, 494–96, 500–502
Protostellar winds, 502, 503
Protosun, 386
Proxima Centauri, 439, 460
Ptolemaic model, 38
Ptolemy, 37–38
Pulsar planets, 574–75
Pulsars, 396*n*, 558, 569–71
  binary, 579
  gamma-ray, 570, 572
  millisecond, 573–74, 575
  Vela, 567
Pulsating variable stars, 603–4
P-waves (primary waves), 171, 172, 174
P/Wild 2 (comet), 370
Pyrolitic-release experiment, 266

"Quantum" forces, 720
Quantum gravity, 588, 688, 719, 720
Quantum mechanics, 91
Quarks, 722
Quarter Moon, 15
Quasar redshifts, 680–81, 684
Quasars, 176, 678–83
  absorption lines, 678
  discovery of, 669, 671
  energy production by, 673, 677, 684
  galaxy evolution and, 684–86
  lifetimes of, 677
  "mirages" of, 678–79
  properties of, 671–72
  superluminal motion of, 670
  twin, 678, 679
Quasi-stellar objects. *See* Quasars
Quiescent prominences, 423

R136 star cluster, 121
Radar, 48
Radial motion, 78
Radial velocity, 100
  of star, 439, 441
Radian, 26
Radiation, 61–83
  defined, 62
  diffraction property of, 66–67
  distribution of, 72–77
  Doppler effect, 77–79
  emitted by active galaxies, 662
  interference property of, 67
  laws of, 72–75
  light and, 62–63
  from Seyfert galaxies, 663–64
  spectrum. *See* Spectrum/spectra
  of Sun, 168–70
  synchrotron, 676, 677
  types of, 62–63, 65
  wave motion of, 63–64, 66–67, 68, 69
Radiation darkening, 338

Radiation-dominated universe, 717
Radiation Era, 718, 719–21
Radiation zone, 407, 408
Radioactive dating, 180–81
Radioactivity, 175, 180
Radio communication, search for extraterrestrial intelligence using, 752–53
Radio galaxies, 662, 664–69
Radio lobes, 664–65
Radio maps of Milky Way, 612–13
Radio radiation, 63, 65
Radio telescopes, 124–27, 137
  interferometer, 127–29
Radius-luminosity-temperature relationship, 449–50, 452, 453
Radius of planet, determining, 146. *See also specific planets*
Radome, 125
Ranger series, 204
Rayleigh, Lord, 169
Rayleigh scattering, 169
RCW 38, 562
Recession velocity, 77
Recurrent novae, 544, 545
Reddening, 469
Red dwarfs, 453–54, 516
Red-giant branch, 519–20
Red-giant instability, 525
Red giant region, 454
Red giants, 450, 454, 521, 524, 560
Red-giant wind, 524
Red line (Há line), 94, 96, 99
Redshifted radiation, 78–79
Redshifts, 577
  cosmological, 652, 698–99
  gravitational, 586–88
  quasar, 680–81, 684
  relativistic, 680–81
Redshift survey of galaxies, 692, 693
Red Spot, 278, 282–83
Red supergiant, 523, 531
Reflecting telescope, 108, 110–12
Reflection nebula, 474
Refracting telescope, 108–10
Refraction, 108
Regolith, 209–11
Regression of the line of nodes, 21
Reinhold crater, 210
Relativistic fireball, gamma-ray bursts as, 578, 579
Relativity
  general, 581, 582–83, 588, 589, 699–700
  tests of, 586–87
  special, 582–83, 586
Relativity theory, 78*n*
Residual cap, 260–61, 263
Resolution, angular, 116, 117, 124–25, 127
Resonance, 400
  orbital, Trojan asteroids and, 361
  of ring particles of Saturn, 313–14
  spin-orbit, 204, 206
Retrograde motion, 36

Retrograde orbits, 287, 288
Retrograde rotation, 229
Rhea, 319, 320
Rho Ophiuchi, 76, 77, 478–79
Riemann, Georg Friedrich, 710
Riemannian geometry, 710
Rigel, 14, 132, 445, 531
Right ascension, 14
Rille, 212, 213
Ringlets in B ring, 311, 312
Rings
    of Jupiter, 296, 312
    of Neptune, 312, 344–45
    of Saturn, 144, 302, 303,
        309–15
    of Uranus, 312, 342–44
Roche, Edouard, 310, 535
Roche limit, 310–11
Roche lobes, 535–36
Rocks, lunar, 211–12
ROSAT X-ray satellite, 133, 135
Rotation, 8
    chaotic, 322
    differential, 277
    gravity and, 491
    molecular, 96–97
    solar, 421
    spectral-line broadening and,
        101–2
Rotation curve
    galactic, 617
    mass measurement of spiral
        galaxies and, 642–43, 644
Rotation period
    of planets, 146, 147
    solar, 406
r-process, 559
RR Lyrae variable stars, 604–6
Runaway greenhouse effect, 241–43
Runoff channels, Martian, 259
Russell, Henry Norris, 450

**SAGE (Soviet-American Gallium
    Experiment), 429–30**
Sagittarius constellation, 10
Salam, Abdus, 723
San Andreas Fault, 177, 179
Saros cycle, 21
Satellite, trajectory of, 55
Saturn, 301–25, 330
    atmosphere of, 302–7
    bulk properties of, 146
    dimension of, 47
    interior structure and heating,
        307–8
    magnetospheres of, 308–9, 336
    moons of, 308–9, 313–22, 745
    physical properties of, 302
    planetary data, 308
    ring debris, 317
    rings of, 144, 302, 303, 309–15
    rotation of, 302
    spacecraft exploration of, 159
SB0 galaxies, 634
Scarp, 214
Schiaparelli, Giovanni, 203, 261
Schmidt, Bernhard, 112
Schmidt telescope, 112, 114
Schmitt, Harrison, 204
Schwarzschild, Karl, 581
Schwarzschild radius, 581, 585

Scientific method, 44–45
Scientific notation, 4
Scorpius-Centaurus association,
    471
Seafloor spreading, 179
Seasonal cap, 260
Seasons and seasonal changes,
    10–13
Secondary atmosphere, 171
Seconds, 11, 14, 22
Seeing, 118–19
Seeing disk, 118
Seismic waves, 171–72, 174
Seismograph, 171, 172
Selection effect, 397
Self-propagating star formation,
    615, 616
Semidetached binaries, 536, 537,
    544
Semi-major axis, 44, 45
Seven Sisters, 6, 505–6, 507
Seyfert, Carl, 663
Seyfert galaxies, 662, 663–64, 667
Sgr A*, 621, 622–24
Shapley, Harlow, 606–7
Shear waves (S-waves), 171–72
Shepherd satellites, 314, 343–44
Shield volcanoes, 236, 237
Shock waves, star formation and,
    504–5, 561
Shoemaker-Levy 9 comet, 369,
    370–71
Shooting stars. *See* Meteors
Short-period comets, 363
Shu, Frank, 614
Sidereal day, 9, 10, 124
Sidereal month, 15, 17
Sidereal orbital period, 47
Sidereal year, 13
Sidereus Nuncius (The Starry
    Messenger) (Galileo), 41
Sif Mons, 236, 237
Silicon-28, 557
SIM (Space Interferometry
    Mission), 129, 443
Single-line spectroscopic binaries,
    457–58
Singularity, 588–89, 719
Sirius A, 530
Sirius B, 450, 454, 525–26, 527,
    530
Sirius (Dog Star), 10, 14, 459, 460
Sister (coformation) theory, 217
Slingshot, gravitational, 154–55
Slipher, Vesto M., 651
Sloan Digital Sky Survey, 678
SN1987A, 553–54
Sodium, emission spectrum of,
    88, 89
SO galaxies, 634
Sojourner, 145, 157, 263
*Solar and Heliospheric Observatory
    (SOHO)*, 409, 410–11, 416,
    417, 425
Solar constant, 408
Solar cycle, 421–22, 431
Solar day, 9, 10, 22
    mean, 22–23
Solar eclipse, 15–20
Solar energy production, 408, 423,
    425–26

Solar magnetism, 418–21
Solar maximum, 421
Solar minimum, 421
Solar nebula, 385, 388
Solar neutrino problem, 429
Solar neutrinos, 429–30
Solar pressure waves, 409
Solar rotation, 421
Solar system, 143–63. *See also* Sun;
        specific planets
    condensation theory of,
        385–92
    debris. *See* Asteroids; Comet(s);
        Meteorites; Meteoroids;
        Meteors
    differentiation of, 392–95
    geocentric model of, 36–39
    heliocentric model of, 39–40
    interplanetary debris, 151–53
    inventory of, 144–46
    irregularities of, 385
    life in, 745–46
    modeling formation of,
        384–85. *See also*
        Condensation theory
    overall layout of, 148–50
    planetary properties, 146–48
    planets beyond, 384, 390–92,
        396–400
    spacecraft exploration of,
        153–59
    terrestrial and jovian planets,
        150–51, 390–94
    as unusual, 398–99
Solar wind, 151, 185, 186, 407,
    408, 410, 416–17
Solstices, 22
Sombrero Galaxy, 632
South celestial pole, 8
Southern lights (*aurora australis*),
    185
*Soviet-American Gallium
    Experiment (SAGE)*, 429–30
Space, curvature of, 698, 707, 710
*Space Infrared Telescope Facility
    (SIRTF)*, 131
Space Interferometry Mission
    (SIM), 129, 443
Spacetime, 583, 699
Special relativity, 582–83, 586
Speckle interferometry, 449
Spectral classes, 448
Spectral-line analysis, 97–102
Spectral lines, molecular, 482–83
Spectrographs (spectrometers), 86
Spectrometers, 113
Spectroscope, 86
Spectroscopic binaries, 457–58
Spectroscopic parallax, 454–55
Spectroscopic radio astronomy, 612
Spectroscopic thermometer,
    98–100
Spectroscopy, 89
Spectrum/spectra, 64, 70–72, 75,
    86–96
    absorption, 479–81
    absorption lines, 88–89, 90, 94,
        95
    astronomical applications of,
        90–91
    atomic, 94, 95–96

blackbody (Planck) curve, 72,
    73, 74, 75–77, 87
    continuous, 87, 89, 90
    emission lines, 87–88, 89, 90,
        94, 95
    formation of, 91–96
    galaxy, 651–52
    hydrogen, 87, 88, 93, 94, 97
    Kirchhoff's laws, 89–90, 94–95
    molecular, 96–97
    of more complex elements,
        95–96
    nebular, 476, 477
    of quasars, 678
    of stars, 446–48
    of Sun, 75, 100, 413–14, 415
    of 3C273, 669, 671
Speed
    escape, 55, 198, 208, 209, 581,
        700
    of light, 70, 582, 670
    molecular, 208–9
Spherical aberration, 122
Spica B, 460
Spicules, solar, 415, 416
Spin-orbit resonance, 204, 206
Spiral arms, 613–15
Spiral density waves, 311, 613–15
Spiral galaxies, 603, 612–15, 630,
        631–32, 633, 636, 643, 649,
        650. *See also* Milky Way
        Galaxy
Spiral nebulae, 603
"Splash" crater, 257
Spring tides, 189
s-process, 559
Sputnik 1, 204
SS 433, 573, 574
Standard candles, 637–38, 639
Standard Solar Model, 408, 409
Standard time, 23
Star(s), 4, 437–65. *See also* Sun
    apparent brightness of, 441–43,
        605
    binary. *See* Binary-star systems
    classification of, 448–49
    clusters of. *See* Star cluster
    defined, 406
    distances to, 438–39
    Hertzsprung–Russell (H-R)
        diagram, 443, 450–54, 455
    lifetimes of, 460–61
    luminosity classes of, 455–56
    luminosity of, 441–44, 446–47,
        449–50
    mass of, 457–61, 497–98
    of Milky Way, spatial
        distribution of, 607–8
    motion of, 9–14
    neutron. *See* Neutron stars
    with planetary systems, 748
    spectroscopic parallax, 454–55
    stellar motion, 439–41
    stellar sizes, 449–50, 451
    in summer sky, 10, 13
    temperature and color of,
        445–49
    variable, 603–6, 634, 637
Starburst galaxies, 650
Star cluster, 437
    associations, 506, 510

blue stragglers in, 529, 534
emission nebulae and, 508–10
globular, 506–8, 526, 528–29, 533, 534, 573–74, 603, 606–7, 706
lifetimes of, 510
observing stellar evolution in, 532–35
Stardust mission, 369–70
Starfire Optical Range, 124
Star formation, 489–513, 561. *See also* Stellar evolution; Stellar explosions
emission nebulae and star clusters, 508–10
evolution of solar-type star, 492–96
failed stars, 498–99
gravitational competition and, 490–92
modeling, 493
observational evidence of, 499–503
rate of, 747–48
self-propagating, 615, 616
shock waves and, 504–5, 561
stars of different masses, 497–98
zero-age main sequence (ZAMS), 497
Starlight, aberration of, 42
Star Wars program (Strategic Defense Initiative), 220
Stefan, Josef, 74
Stefan-Boltzmann constant, 74
Stefan-Boltzmann law, 449
Stefan's law, 74, 75, 168, 413
Stellar epoch, 720, 721
Stellar evolution, 515–41
of binary-star systems, 535–37
cycle of, 561–62
of high-mass stars, 529–31
leaving main sequence, 516–17, 529–30
of low-mass star, 517–29, 531
carbon core, 522–23, 525
death of, 523–29
helium core, 517–19
helium flash, 521–22
helium fusion, 520
planetary nebulae, 524–25, 526
red-giant branch, 519–20
white dwarfs, 525–28, 534
in star clusters, observing, 532–35
Stellar explosions, 543–65. *See also* Elements
end of high-mass star, 545–48
novae, 544–45, 548–49
supernova explosions, 548–52, 554–55, 568
Stellar nucleosynthesis, 553–60
observational evidence for, 559–60
Stellar occultation, 342, 343
Stellar parallax, 438–39
Stellar populations, 609
Stellar transits of black hole, 589–90
Stellar winds, 524

Stickney crater, 269
Stonehenge, 34–35
Strategic Defense Initiative (Star Wars program), 220
Stratosphere, 167, 168
Stray asteroids, 372–75
Strings, 723
Strong nuclear force, 426, 427, 722
Structure in early universe, 730–35
S-type asteroids, 356, 357, 358, 376
Subaru telescope, 114, 123
Subatomic particles, 618, 624
Subduction zones, 177
Subgiant branch, 519
Sublimation, 363n
Sulfur-32, formation of, 556
Summer solstice, 11, 12
Sun, 4, 5, 405–35, 460
absorption lines of, 88–89
active, 418–25
bulk properties of, 146, 406–8
CNO cycle in, 518
composition of, 413–14
evolution and death of star like, 517–29
formation of stars like, 492–96
gravitational pull of, 53
luminosity of, 407–8
motion of, 9–14
nuclear fusion in, 426–29
radiation of, 168–70
solar atmosphere, 413–18
solar energy production, 408, 425–26
solar interior, 409–13
solar neutrinos, observations of, 429–30
solar-terrestrial relations, 431
spectrum of, 75, 100, 413–14, 415
structure of, 407
temperature of, 407, 410, 416, 417
tidal forces exerted by, 189
weight of, 56
in x-rays, 405, 417–18
Sunspot cycle, 421, 422
Sunspots, 41, 418, 419
Superclusters, 640–42
Supergiants, 450, 453, 454, 523, 524, 531
Supergranulation, 413
SuperGUT, 723
Superior conjunction, 228
Super Kamiokande telescope, 430
Superluminal speed, 670
Supermassive black holes, 591, 673–75, 684
Supernova explosions, 548–52, 554–55, 568
Supernova remnant, 134, 550–52, 555, 562, 570
Vela, 551, 552, 567
Supernova/supernovae, 543, 548–50
ancient Chinese data on, 35–36
core-collapse, 548
interstellar shock waves generated by, 504–5

Type I (carbon-detonation), 549–50, 560, 568, 638, 703
Type II, 549, 550, 551, 554, 568, 638n
Supersymmetry, 723
Surveyor mission, 204, 257
S-waves (shear waves), 171–72
Swift-Tuttle (1862III) comet, 372
Sycorax (moon), 337
Synchronous orbit, 202
Synchrotron radiation, 676, 677
Synodic month, 15
Syntaxis (Ptolemy), 38

Tail of comets, 363, 364
T associations, 506
Taylor, Joseph, 579
Technetium, 553
Technetium-99, 560
Tectonic fractures, 258
Telescopes, 41, 107–41
defined, 108
high-energy, 133–36, 137
high-resolution, 118–24
infrared, 130–32, 137
interferometry, 127–29
invention of, 144
light-gathering power of, 114–16
optical, 108–14, 137, 144–45
radio, 124–27, 137
interferometer, 127–29
resolving power of, 116–17
size of, 114–17
ultraviolet, 132, 133, 137
Telesto, 321, 322
Temperature, 72
absolute, 74
of Earth's interior, 173
in Hertzsprung–Russell diagrams, 450–54
line intensity and, 99, 100
on Mars, 264–65
across primitive solar nebula, 392–93
of stars, 445–49
of Sun, 407, 410, 416, 417
threshold, of particle, 718
Terminator line on Moon, 199, 200
Terrestrial planets, 150, 151, 393–94. *See also* Earth; Mars; Mercury; Venus
Tethys (moon), 319, 320, 321, 338
Tharsis bulge, 253, 254, 258
Theory, 44
Thermal energy (heat), 73, 490–91
atmosphere and, 199, 208
differentiation of solar system and, 392–93
Thermal equilibrium, 718
Thermal motions of atoms, 100–101
Thermometer, spectroscopic, 98–100
Thick disk component of Milky Way, 608
39 (heliocentric model), 385
Thorium 232, 180, 181
3C 48 quasar, 669
3C 84 galaxy, 664

3C 175 quasar, 672
3C 273 quasar, 669, 671, 672
3C 275 quasar, 671
Threshold temperature of particle, 718
Tidal bulge, 187, 188, 189, 190
in Moon, 202–3, 206
Tidal force (differential force), 187–88, 206
of black hole, 585
exerted by Sun, 189
Tidal stability limit (Roche limit), 310–11
Tides, 187–90
Time dilation, 588
Time measurement, 22–23
Time zones, 23
Titania (moon), 337, 338, 339
Titan (moon), 308–9, 315, 316–19, 745
Titius, Johann, 149
Titius-Bode law, 149
Tombaugh, Clyde, 345
Total lunar eclipse, 15
Total solar eclipse, 15–17, 18, 19
Tracers, molecular, 483–84
Transition Region and Coronal Explorer (TRACE) satellite, 420–21
Transition zone, 407, 408, 415–16
Transits
planetary, 397, 399–400
solar, 48
Transverse Doppler shift, 78n
Transverse motion, 78
Transverse velocity of star, 439, 441
Trapezium, 508–10
Triangulation, 24, 48
Trifid Nebula, 474
Triple-alpha process, 520, 522
Triple-alpha reaction, 555
Triton (moon), 339, 340–42, 400
Trojan asteroids, 361
Tropical year, 13, 23
Troposphere, 167, 168, 265
T Tauri phase, 394–95
T Tauri star, 496
Tully-Fisher relation, 638–39
Tunguska event of 1908, 369, 376
Turbulence, gas, 102
Turnoff mass, 533
21-centimeter radiation, 481–82
Twin quasar, 678, 679
Tycho's supernova, 552
Type I (carbon-detonation) supernova, 549–50, 560, 568, 638, 703
Type II supernova, 549, 550, 551, 554, 568, 638n

Ultraviolet astronomy, 470–71
Ultraviolet radiation, 63, 65
Ultraviolet telescopes, 132, 133, 137
Umbra, 17, 18, 418
Umbriel (moon), 337, 338, 339
Unbound motion, 55
U.S.-European GALLEX collaboration, 429–30
U.S. National Observatory, 119
Universal recession, 651–52

Universal time (Greenwich mean time), 23
Universe, 4. *See also* Early universe; Life
  age of, 705–6
  birth of, 694–95
  closed, 707
  critical, 707
  density of, 702, 705
  expansion of, 577, 695–98, 702–6
  fate of, 700–702
  geocentric, 36–39
  large-scale structure in, 653–55, 692–93
  matter-dominated, 716–17
  open, 707
  radiation-dominated, 717
  relativity and, 699–700
Upsilon Andromedae, 396
Uraniborg observatory, 43
Uranium-235, 180, 181
Uranium-238, 180–81
Uranus, 144, 328–44
  atmosphere of, 332–35
  bulk properties of, 146, 329–32
  dimensions of, 47
  discovery of, 328–29
  internal structure of, 335, 336
  magnetospheres of, 335, 336
  mass and radius of, 329
  moons of, 337–39, 396
  planetary data, 332
  rings of, 312, 342–44
  rotation rates of, 330–31, 333
  seasons on, 331
Urey, Harold, 742
Ursa Minor (Little Bear), 14
Utopia region (Mars), 262–63

Valhalla basin, 294–95
Valles Marineris (Mariner Valley), 254, 258
Van Allen belts, 183–85, 677
Variable stars, 603–6, 634, 637
Vega 2, 365

Vega star, 10, 449, 456, 460
Vela pulsar, 567
Vela series of satellites, 135
Vela supernova remnant, 551, 552, 567
Velocity
  radial, 100, 439, 441
  recession, 77
  wave, 64
Venera missions, 154–55, 232, 237, 239, 243
Venus, 196, 227–47, 395
  atmosphere of, 238, 240–43
  bulk properties of, 146
  density of, 228–29
  dimensions of, 47
  long-distance observations of, 230–31
  magnetic field and internal structure of, 243
  mass of, 228–29
  orbits of, 49, 228
  phases of, 42
  properties of, 232
  radius of, 228–29
  rotation rate of, 229–30
  spacecraft exploration of, 154–56
  surface of, 232–39
Vernal equinox, 12, 13, 14
Very Large Telescope (VLT), 116, 123
Very Long Baseline Array, 674
Vesta asteroid, 356, 357
Vibration, molecular, 96–97
Viking-Mars Observer, 156
Viking missions, 156, 157, 253, 254, 261, 262–63, 264, 266, 268
Virgo A, 668
Virgo Cluster, 640, 641, 642, 644, 666, 674
Viruses, 741
Visible galaxy, 664
Visible light, 62–63, 64–65
Visible spectrum, 64, 65

Visual binaries, 457, 458
VLA interferometer, 128
VLT (Very Large Telescope), 107
Voids, 654, 692
Volcanoes/volcanism, 173, 175
  on Enceladus, 320–21
  on Io, 289–90, 291
  lunar, 212, 213, 219
  on Mars, 254–56
  shield volcanoes, 236, 237
  on Venus, 236–39
  on Vesta, 357
Voyager missions, 154–55, 158, 310, 311–13, 314, 316–17

Wabar meteorite, 377
Waning Moon, 15, 16
Warm longitudes, 206
Water hole, 753–54
Water on Mars, 257–60, 262
Watt, 74
Wavelength, 64, 65, 70, 74, 77
Wave motion, 63–64, 66–67, 68, 69
Wave period, 64
Wave velocity, 64
Waxing Moon, 15, 16
Weakly Interacting Massive Particles (WIMPs), 618
Weak nuclear force, 427
Weather
  on Jupiter, 282–83
  on Mars, 265
  on Saturn, 305–7
  on Uranus and Neptune, 332–35
Weinberg, Steven, 723
Weird (jumbled) terrain, 215
Whirlpool Galaxy, 649
White dwarf, 450, 454, 525–28, 534, 544–45, 549, 580, 588, 638. *See also* Nova/novae
  helium, 528, 537
White dwarf region, 454
White ovals, 283
Wien's law, 74, 75–76, 131, 168

Wilson, Robert, 708
WIMPs (Weakly Interacting Massive Particles), 618
Wind(s)
  on Jupiter, 279, 283
  on Neptune, 335
  protostellar, 502, 503
  on Saturn, 305, 306
  solar, 151, 185, 186, 407, 408, 410, 416–17
  stellar, 524
Winter solstice, 11, 12
Wispy terrain, 320
Wollaston, William, 88

XM1 asteroid, 360
X-ray bursters, 572–73, 574
X-Ray Multi-Mirror satellite (XMM-Newton), 133
X-ray novae, 593
X-ray radiation, 63
X-rays
  detection methods for, 133–36, 137
  Sun in, 405, 417–18

Year
  eclipse, 21
  leap, 23
  sidereal, 13
  tropical, 13, 23
Yerkes observatory, 111
Yohkoh X-ray solar observatory, 405, 417
Yuty crater, 257

Zeeman effect, 102
Zero-age main sequence (ZAMS), 497
Zodiac, 11, 12
Zonal flow
  on Jupiter, 279
  on Saturn, 305, 306
Zones, in atmospheric bands, 278–80

These star maps show the brighter stars and the prominent constellations as they appear on the dates and at the times indicated. To use these maps, face the south and hold the book overhead with top of the map toward the north and the right-hand edge toward the west. The brightest stars are indicated by the star symbol (☆) and the names are indicated. (Star maps courtesy of Robert Dixon, *Dynamic Astronomy,* 6th ed., Prentice Hall, 1992.)

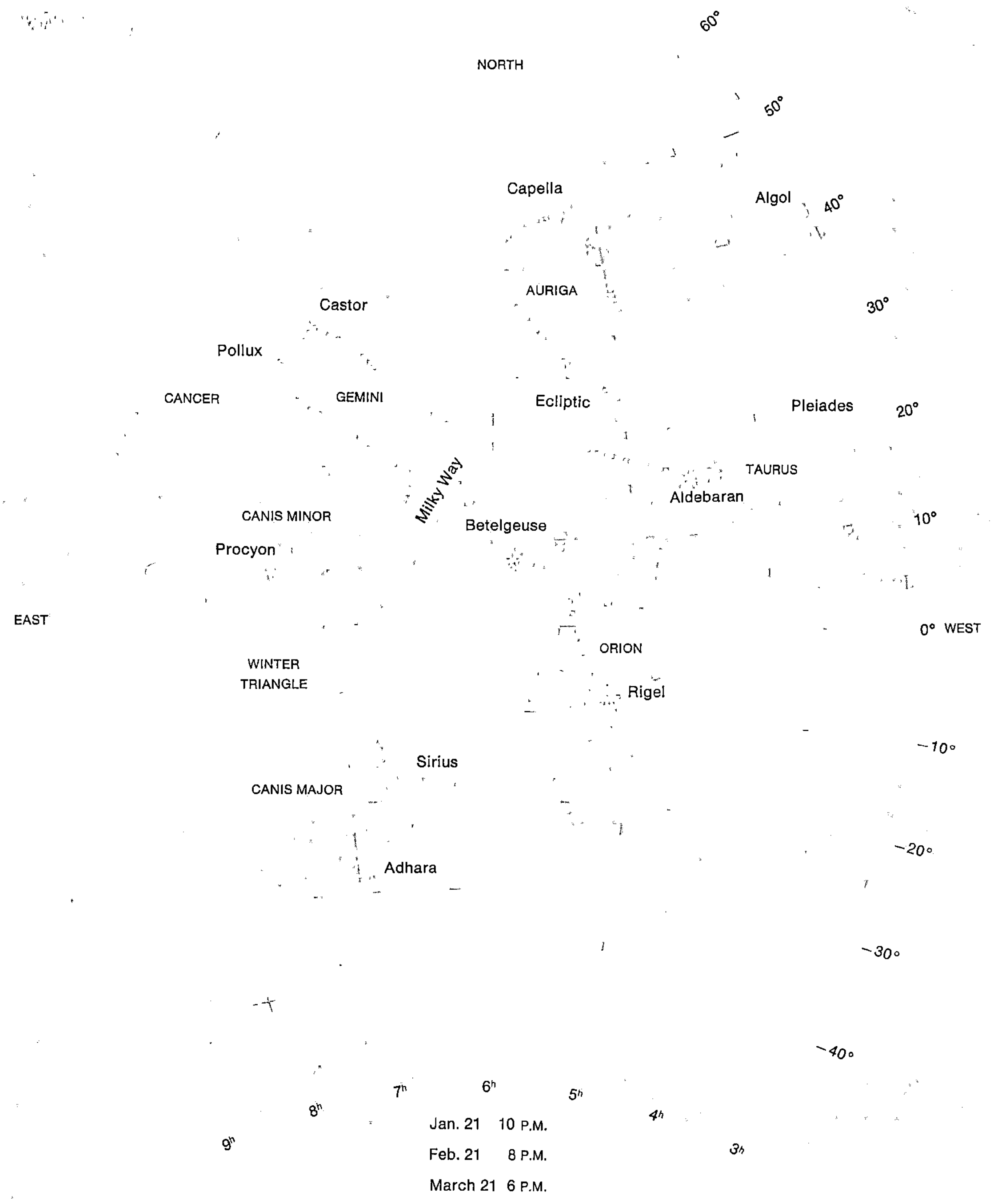

60°

NORTH

50°

Capella

Algol 40°

AURIGA

Castor

30°

Pollux

CANCER   GEMINI   Ecliptic   Pleiades 20°

TAURUS

Milky Way

Aldebaran

CANIS MINOR

Betelgeuse 10°

Procyon

EAST

0° WEST

ORION

WINTER
TRIANGLE

Rigel

−10°

Sirius

CANIS MAJOR

−20°

Adhara

−30°

−40°

7ʰ   6ʰ   5ʰ

8ʰ   Jan. 21   10 P.M.   4ʰ

9ʰ   Feb. 21   8 P.M.   3ʰ

March 21 6 P.M.

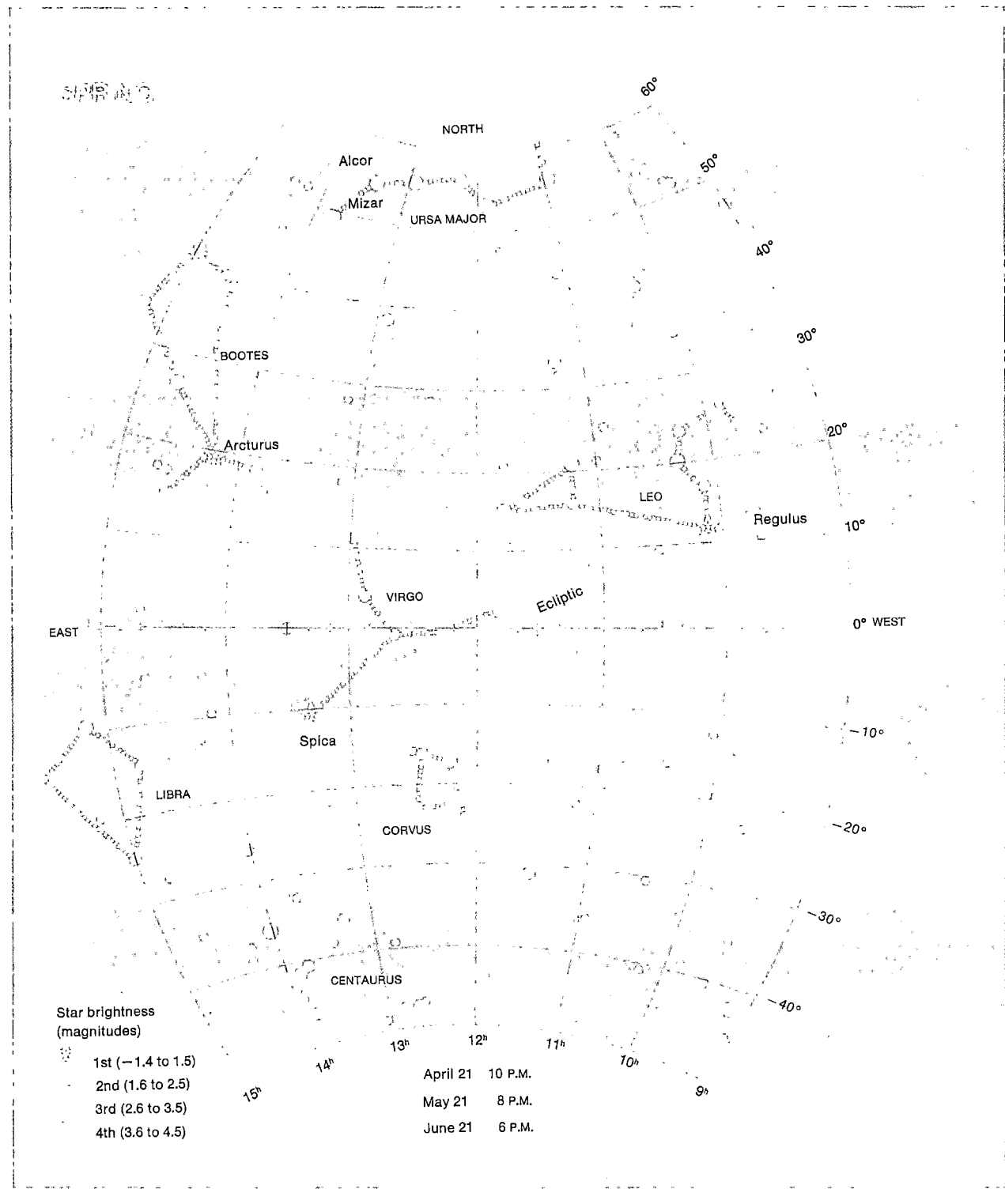

SPRING

NORTH

Alcor

Mizar

URSA MAJOR

60°

50°

40°

30°

20°

BOOTES

Arcturus

LEO

Regulus   10°

VIRGO

Ecliptic

EAST

0° WEST

Spica

−10°

LIBRA

−20°

CORVUS

−30°

CENTAURUS

−40°

Star brightness
(magnitudes)

1st (−1.4 to 1.5)
2nd (1.6 to 2.5)
3rd (2.6 to 3.5)
4th (3.6 to 4.5)

15ʰ   14ʰ   13ʰ   12ʰ   11ʰ   10ʰ   9ʰ

April 21   10 P.M.
May 21   8 P.M.
June 21   6 P.M.

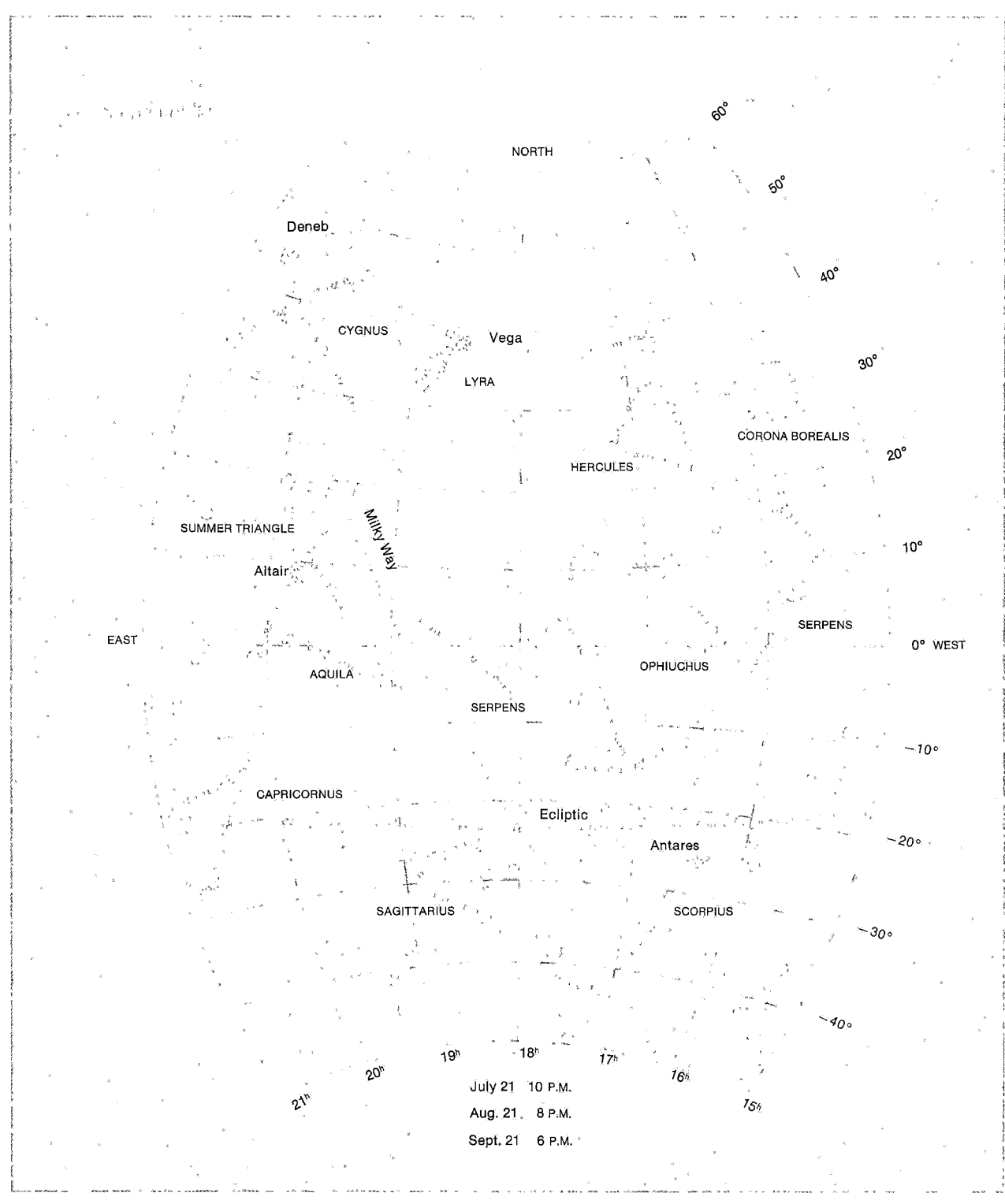

60°

NORTH

50°

Deneb

40°

CYGNUS   Vega

30°

LYRA

CORONA BOREALIS

20°

HERCULES

SUMMER TRIANGLE   Milky Way

10°

Altair

SERPENS

EAST   0° WEST

AQUILA   OPHIUCHUS

SERPENS

−10°

CAPRICORNUS

Ecliptic

−20°

Antares

SAGITTARIUS   SCORPIUS

−30°

−40°

19ʰ   18ʰ   17ʰ

20ʰ   16ʰ

21ʰ   July 21   10 P.M.

15ʰ

Aug. 21   8 P.M.

Sept. 21   6 P.M.

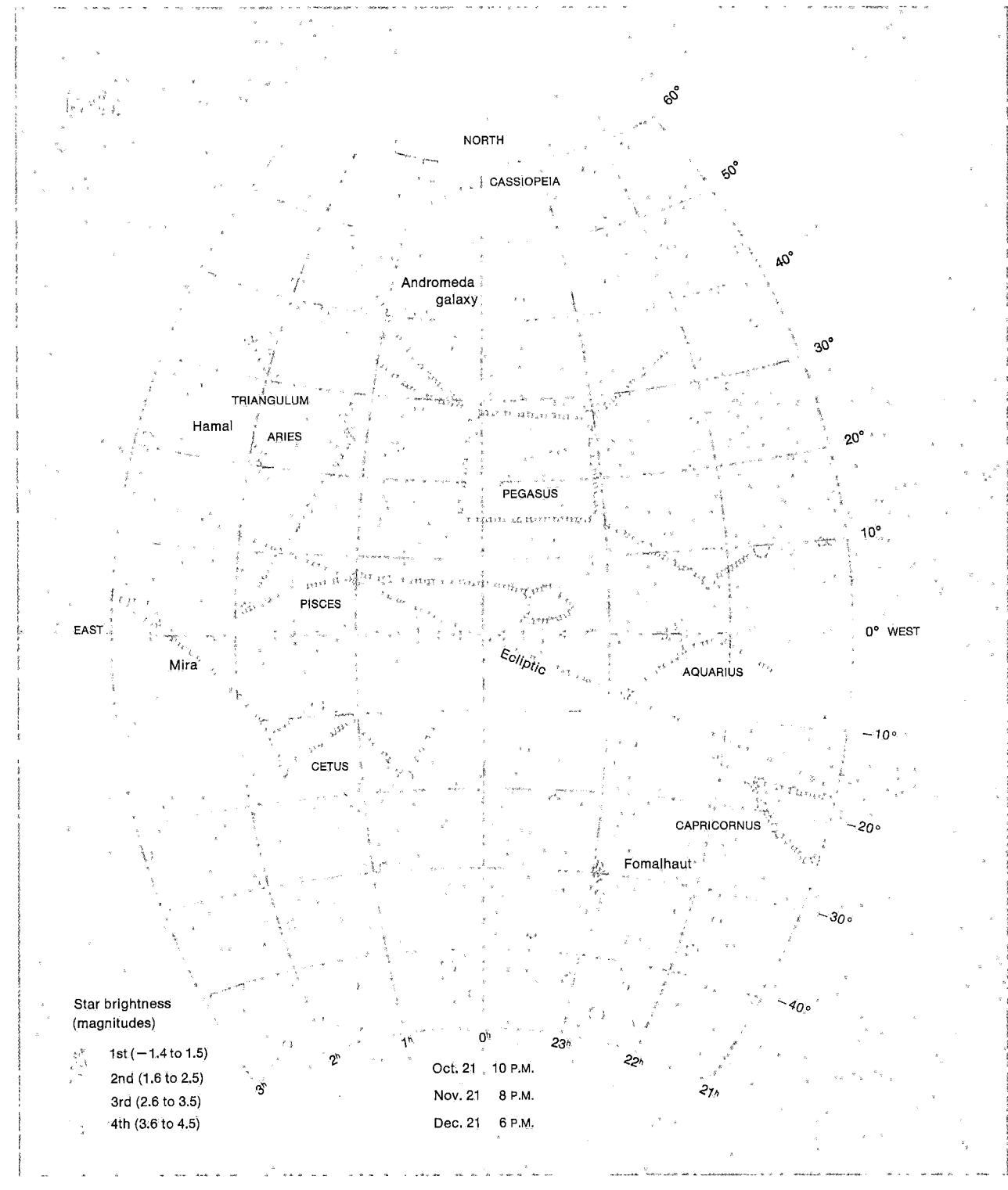

NORTH

CASSIOPEIA

60°

50°

40°

30°

Andromeda galaxy

20°

TRIANGULUM

Hamal

ARIES

10°

PEGASUS

PISCES

EAST

0° WEST

Mira

Ecliptic

AQUARIUS

−10°

CETUS

CAPRICORNUS

−20°

Fomalhaut

−30°

−40°

Star brightness
(magnitudes)

1st (−1.4 to 1.5)
2nd (1.6 to 2.5)
3rd (2.6 to 3.5)
4th (3.6 to 4.5)

1ʰ   0ʰ   23ʰ
2ʰ
3ʰ   22ʰ
                                          21ʰ

Oct. 21   10 P.M.
Nov. 21   8 P.M.
Dec. 21   6 P.M.

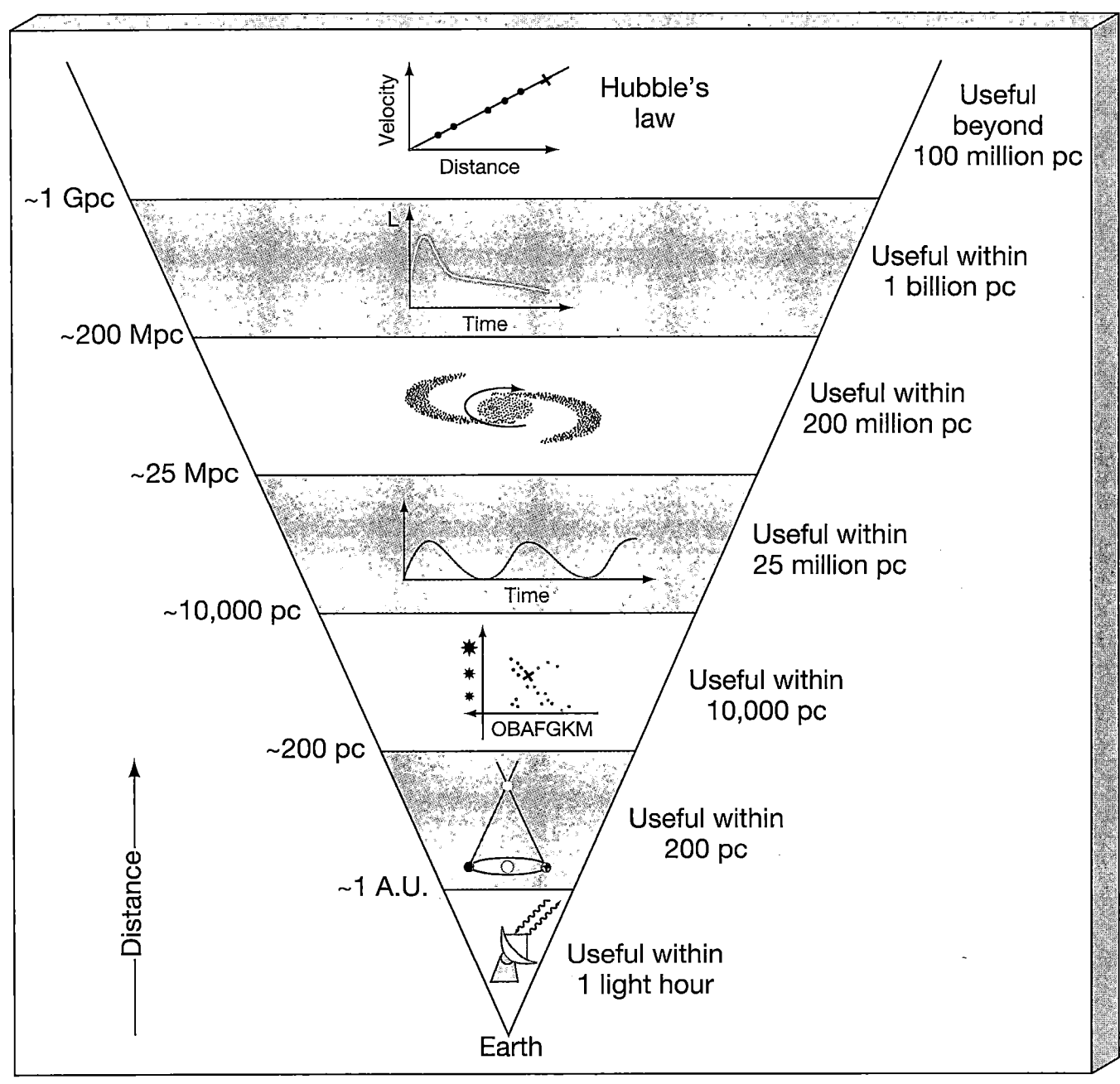

Main-Sequence Stellar Properties by Spectral Class

Spectral Class	Typical Surface Temperature	Color	Mass*	Luminosity*	Lifetime*	Familiar Examples
	(K)		$(M_\odot)$	$(L_\odot)$	$(10^6$ yr $)$	
O	>30,000	Electric Blue	>20	>100,000	<2	
B	20,000	Blue	8	3000	30	Spica (B1)
A	10,000	White	3	75	400	Vega (A0) Sirius (A1)
F	7,000	Yellow-white	1.5	4	4000	Procyon (F5)
G	6,000	Yellow	1.0	1.5	9000	Sun (G2) Alpha Centauri (G2)
K	4,000	Orange	0.5	0.1	60,000	Epsilon Eridani (K2)
M	3,000	Red	0.1	0.005	200,000	Proxima Centauri (M5) Barnard's Star (M5)

*Approximate values for stars of solar composition